PRACTICAL ORGANIC CHEMISTRY

Vogel, Arthur Israel

VOGEL's

TEXTBOOK OF
PRACTICAL ORGANIC CHEMISTRY

FIFTH EDITION
Revised by former and current members of
The School of Chemistry,
Thames Polytechnic, London

Brian S. Furniss

Antony J. Hannaford

Peter W. G. Smith

Austin R. Tatchell

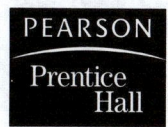
PEARSON
Prentice
Hall

Harlow, England • London • New York • Boston • San Francisco • Toronto • Sydney • Singapore • Hong Kong
Tokyo • Seoul • Taipei • New Delhi • Cape Town • Madrid • Mexico City • Amsterdam • Munich • Paris • Milan

Pearson Education Limited
Edinburgh Gate
Harlow
Essex CM20 2JE
England

and Associated Companies throughout the world

Visit us on the World Wide Web at:
http://www.pearsoned.co.uk

© Longman Group UK Limited 1989

First published 1948

Second Edition 1951

New Impression with addition of Chapter XII on Semimicro Technique 1954

Third Edition 1956

Fourth Edition 1978

Fifth Edition 1989

British Library Cataloguing in Publication Data

Vogel, Arthur Israel
    Vogel's textbook of practical organic chemistry –
    5th ed
    1. Organic chemistry. Laboratory techniques
    I. Title I. Furniss, B. S. (Brian Stanley), 1941 –
    547'0028

ISBN 978-0-582-46236-6

Library of Congress Cataloging-in-Publication Data

Vogel, Arthur Israel.
    Vogel's Textbook of practical organic chemistry – 5th ed. / rev.
    by Brian S. Furniss . . . [et al.]
      p. cm.
    Fourth ed. published in 1978 under title: Vogel's Textbook of
    practical organic chemistry, including qualitative organic analysis.
    Includes bibliographies and indexes.
    ISBN 0-470-21414-7 USA only)
    1. Chemistry, Organic—Laboratory manuals. 2. Chemistry,
    Analytic–Qualitative. I. Furniss, Brain S. (Brian Stanley), 1941–
    II. Vogel, Arthur Israel. Vogel's Textbook of practical organic
    chemistry, including qualitative organic analysis. III. Title.
    QD261.V63   1989
    547—cc19
                                    88-36786
                                     CIP

16
14   13   12

Set in 10/11 pt. Lasercomp Times New Roman
Printed in China
EPC/16

CONTENTS

CHAPTER 9 INVESTIGATION AND CHARACTERISATION OF ORGANIC COMPOUNDS   1196

CHAPTER 10   PHYSICAL CONSTANTS OF ORGANIC COMPOUNDS

APPENDICES

xvi

PREFACE TO FIFTH EDITION

This is the second time that we have prepared a new edition of Vogel's *Textbook of Organic Chemistry* and it is important to reaffirm the aims set out by the late Dr A. I. Vogel in the preface to the first edition. Thus in this new edition every endeavour is made to retain the comprehensive character of the book, and to ensure that it continues to be a one-volume reference text which is of value to practising organic chemists throughout their undergraduate, postgraduate and subsequent careers.

During the preparation of the previous (fourth) edition considerable reorganisation and rewriting of the text was necessary. This arose from the many changes which had taken place in the practice and theory of organic chemistry during the preceding twenty years. Among these changes were the ready availability of a much wider range of substrates and reagents; the development of a whole host of new synthetic reagents; a greater awareness of the hazards associated with handling of organic chemicals; the routine use of chromatographic and spectroscopic techniques; and the use of mechanistic concepts to rationalise and predict the outcome of organic reactions. That revision included a modified chapter on experimental techniques, which was arranged under the headings 'Apparatus and reaction procedures', 'Isolation and purification processes', and 'Determination of physical constants'. New sections were introduced on safe working in organic laboratories, chromatography, and spectroscopic methods. The interpretation of spectroscopic data was discussed in the reorganised chapter entitled Qualitative Organic Analysis. The preparative chapters were reorganised on the basis of aliphatic compounds, aromatic compounds, alicyclic compounds, and heterocyclic compounds, and within these chapters a selection of new synthetic methods was introduced. Following the practice adopted in previous editions these new preparations were checked in the laboratories of the School of Chemistry, Thames Polytechnic.

In this new (fifth) edition, in order to ensure that the book retains its relevance to current teaching of organic chemistry, we have adopted the same structure of the book since we are of the opinion that it provides a bridge between the treatment of organic chemistry theory provided by current standard undergraduate textbooks, and the wider specialist fields of the research literature. Furthermore, in addition to the introduction of new reagents and techniques, the last decade has seen the development of a philosophy of organic synthesis (the strategy of synthesis) which was just beginning to emerge during the writing of the previous edition. It is our intention to reflect this development, since we consider that the teaching of organic chemistry must closely integrate mechanistic theory with the strategy and methodology of synthesis.

The book now commences with Chapter 1, Organic Synthesis, which reviews the important concepts which need to be borne in mind when considering the problem: 'how may compound X be synthesised?' The chapter discusses the structural and stereoisomeric features of molecules, the importance of control, selectivity and protection in organic reaction sequences, and the increasing importance of asymmetric synthesis. These topics are developed further in the subsequent text. The use of computers in organic synthesis is reviewed, and this topic leads to a short summary of the most significant contribution to the teaching of organic synthesis of the recent decade, namely the disconnection (or synthon) approach developed by Dr S. Warren and based upon the original concept of Professor E. J. Corey. This approach has been integrated with the theoretical discussions on preparative procedures which precede the preparative examples in the aliphatic and aromatic chapters. In particular the retrosynthetic analysis of target molecules has provided a framework around which the later alicyclic and heterocyclic chapters have been rewritten. By using this approach we hope that undergraduates and others, who become acquainted with this philosophy in lectures and tutorials, may find this text a useful further source of information.

The first three sections of Chapter 2 deal with Codes of Practice and responsibility, and a summary of hazards which may be encountered in the organic chemistry laboratory. These are intended to acquaint the student with the essential features of safe working practice, and to advise strongly on the importance of consulting with senior members of the laboratory or the appointed safety officer, and of consulting comprehensive specialist texts which should be readily accessible to laboratory users. In particular the advice of safety officers should be sought, since by virtue of their office they have available the latest information. Hazards associated with particular chemicals and procedures are noted in the text as far as information is available. The remainder of Chapter 2, Experimental Techniques, has been up-dated on the availability of new equipment and the applicability of new techniques. An important section on reactions involving air-sensitive compounds has been introduced. The chromatography section now includes the important preparative techniques of flash chromatography, dry column flash chromatography and the 'Chromatotron', together with a fuller discussion on high-performance liquid chromatography.

Spectroscopic methods and the interpretation of spectra are now treated together in Chapter 3. The most significant addition to this chapter is a more detailed coverage, with examples of spectra, of ^{13}C-n.m.r. spectroscopy which is considered alongside ^{1}H-n.m.r. spectroscopy. The inclusion of detailed spectroscopic data in the preparative chapters has vastly increased the opportunities for practise by the reader in problems relating to elucidation of structure. Thus each functional group section in the aliphatic and aromatic chapters includes a summarising statement of the important spectroscopic features, and many of the preparations conclude with a description of key data. These spectroscopic data are only quoted if they provide information for the student from which a definitive structural assignment may be deduced. The information is quoted in several styles. If the spectrum (i.r., n.m.r., m.s., or u.v.–visible) is simple and straightforward the style is of the type 'record and interpret (assign)'. If the spectrum has interesting interpretative features, the spectroscopic details are quoted and some guidance is provided to assist in the elucidation of structure. If, and this is particularly appropriate with aromatic compounds, regioisomers may be compared, these features are noted.

Chapter 4, Solvents and Reagents, has been extended to include important new reagents including alkyllithiums, boranes and new oxidants and reductants, which are used in subsequent chapters.

Approximately 100 new experiments have been introduced into the preparative chapters [Aliphatic (Ch. 5), Aromatic (Ch. 6), Alicyclic (Ch. 7), and Heterocyclic (Ch. 8)]. Of particular interest is the introduction of some illustrative asymmetric syntheses, important protection methods, the use of air- and moisture-sensitive reagents, the wider use of phase transfer reactions, silicon, phosphorus, boron, and titanium reagents, as well as examples of important named reactions. The procedures for these new reactions have been quoted directly from the literature but not checked in the laboratory. The examples have been selected on the basis of their interest, generality and importance. The style of presentation of these new experiments is sometimes somewhat briefer than that of those experiments retained from previous editions, but we believe that they provide a useful introduction for the advanced student to research methods. Many further examples of reagents and techniques have been noted in the theoretical discussions. They have been included since they provide possible project ideas for further investigation by advanced students. It is hoped that all these experiments will provide suitable material for the design of a full range of practical courses. Finally, as noted above, the theoretical discussions have been rewritten to provide an integrated and balanced coverage of mechanisms, methodology and strategy in synthesis.

Chapter 9, Investigation and Characterisation of Organic Compounds, contains much of the chemical information and preparative methods from the chapter previously entitled Qualitative Analysis. The emphasis of this chapter is now on achieving an understanding of chemical behaviour in association with spectroscopic features, and correlating this information to provide a definitive structural elucidation. In this way we hope that the material fulfils the requirements of a range of courses which deal in this topic. The tables of physical constants (Ch. 10) remain unchanged, and the appendices have been up-dated.

We wish to thank Professor B. R. Currell, Ph.D., C.Chem., F.R.S.C., and Dr J. R. Parsonage, C.Chem., F.R.S.C., former and current Heads of School of Chemistry at Thames Polytechnic, for their interest in this project and for granting permission for the use of various facilities. We are indebted to Dr E. Vidgeon for recording the majority of the ^{13}C-n.m.r. spectra, and to Mr V. Kyte and Mr J. Williams for providing information on the availability and usage of laboratory equipment. The assistance of the company representatives of very many manufacturers of chemicals and of laboratory, spectroscopic and chromatographic equipment has been invaluable; general and specific help is noted in the acknowledgements. We are indebted to the referees who made valuable comments on the manuscript which undoubtedly helped to improve the final text.

B.S.F.   A.J.H.   P.W.G.S   A.R.T.
April 1988

xix

PREFACE TO FIRST EDITION

The present volume is an attempt to give to students of practical organic chemistry the benefit of some twenty years' experience in research and teaching of the subject. The real foundations of the author's knowledge of the subject were laid in 1925–1929 when, as a research student at the Imperial College under the late Professor J. F. Thorpe, F.R.S., he was introduced to the methods and experimental technique employed in a large and flourishing school of research in organic chemistry. Since that period the author and his students have been engaged *inter alia* in researches on *Physical Properties and Chemical Constitution* (published in the Journal of the Chemical Society) and this has involved the preparation of over a thousand pure compounds of very varied type. Many of the new procedures and much of the specialised technique developed and employed in these researches are incorporated in this book. Furthermore, new experiments for the elementary student have emanated from these researches; these have been tried out with large classes of undergraduate students over several sessions with gratifying success and have now been included in the present textbook.

In compiling this book, the author has drawn freely from all sources of information available to him – research notes, original memoirs in scientific journals, reference works on organic chemistry, the numerous textbooks on practical organic chemistry, and pamphlets of manufacturers of specialised apparatus. Whilst individual acknowledgement cannot obviously be made – in many cases the original source has been lost track of – it is a duty and a pleasure to place on record the debt the writer owes to all these sources. Mention must, however, be made of *Organic Syntheses*, to which the reader is referred for further details of many of the preparations described in the text.

The book opens with a chapter on the theory underlying the technique of the chief operations of practical organic chemistry: it is considered that a proper understanding of these operations cannot be achieved without a knowledge of the appropriate theoretical principles. Chapter II is devoted to a detailed discussion of experimental technique; the inclusion of this subject in one chapter leads to economy of space, particularly in the description of advanced preparations. It is not expected that the student will employ even the major proportion of the operations described, but a knowledge of their existence is thought desirable for the advanced student so that he may apply them when occasion demands.

Chapters III and IV are confined to the preparation and properties of Aliphatic Compounds and Aromatic Compounds respectively. This division, although perhaps artificial, falls into line with the treatment in many of the existing theoretical textbooks and also with the author's own lecture courses. A short theoretical introduction precedes the detailed preparations of the various classes of organic compounds: it is recommended that these be read concurrently with the student's lecture course and, it is hoped, that with such reading the subject will become alive and possess real meaning. The partition of the chapters in this manner provides the opportunity of introducing the reactions and the methods of characterisation of the various classes of organic compounds; the foundations of qualitative organic analysis are thus laid gradually, but many teachers may prefer to postpone the study of this subject until a representative number of elementary preparations has been carried out by the student. The division into sections will facilitate the introduction of any scheme of instruction which the teacher considers desirable.

Chapters V–X deal respectively with Heterocyclic and Alicyclic Compounds; Miscellaneous Reactions; Organic Reagents in Inorganic and Organic Chemistry; Dyestuffs, Indicators and Related Compounds; Some Physiologically-Active Compounds; and Synthetic Polymers. Many of these preparations are of course intended for advanced students, but a mere perusal of the experimental details of selected preparations by those whose time for experimental work is limited may assist to impress them on the memory. Attention is particularly directed to the chapter on Organic Reagents in Inorganic and Organic Chemistry. It is always a good plan to set advanced students or adequately-trained laboratory assistants on the preparation of those compounds which are required in the laboratory for organic and inorganic analysis; the resulting cost is comparatively low (for o-phenanthroline, for example, it is less than one-tenth of the commercial price) and will serve to promote the use of these, otherwise relatively expensive, organic reagents in the laboratory.

Chapter XI is devoted to Qualitative Organic Analysis. The subject is dicussed in moderate detail and this, coupled with the various Sections and Tables of Physical Constants of Organic Compounds and their Derivatives in Chapters III and IV, will provide a satisfactory course of study in this important branch of chemistry. No attempt that has been made to deal with Quantitative Organic Analysis in this volume.

The textbook is intended to meet the requirements of the student of chemistry throughout the whole of his training. Considerable detail is given in those sections of particular interest to the elementary student; in the author's opinion it is the duty of a writer of a practical textbook to lay a secure foundation of sound experimental technique for the beginner. The subject matter of the book is sufficiently comprehensive to permit the teacher to cover any reasonable course of instruction. It will be observed that the scale of the preparations varies considerably; the instructor can easily adapt the preparation to a smaller scale when such a step is necessary from considerations of cost and time or for other reasons. Quantities of liquid reagents are generally expressed as weights and volumes: the latter refer to a temperature of 20°. The book will be suitable for students preparing for the Pass and Honours (General and Special) B.Sc. of the Universities, the A.R.I.C. and the F.R.I.C. (Organic Chemistry). It will also provide an introduction to research methods in organic chemistry and, it is hoped, may serve as an intermediate reference book for practising organic chemists.

Attention is directed to the numerous references, particularly in Chapter II on Experimental Technique, to firms supplying specialised apparatus. The author has usually had first-hand experience with this apparatus and he feels that some readers may wish to know the present source of supply and also from whom to obtain additional information. It must be mentioned that most of the specialised apparatus has been introduced to the market for the first time by the respective firms after much development research and exhaustive tests in their laboratories. A reference to such a firm is, in the writer's opinion, equivalent to an original literature reference or to a book. During the last decade or two much development work has been carried out in the laboratories of the manufacturers of chemical apparatus (and also of industrial chemicals) and some acknowledgement of the great help rendered to practical organic chemists by these industrial organisations is long overdue; it is certainly no exaggeration to state that they have materially assisted the advancement of the science. A short list of the various firms is given on the next page.*

ARTHUR I. VOGEL
Woolwich Polytechnic, London, SE18
December 1946

---

* In this, the 5th Edition, the list of manufacturers and suppliers is given in Appendix 7.

ACKNOWLEDGEMENTS

We gratefully acknowledge permission to use information contained in the catalogues and brochures of many manufacturers of chemicals and laboratory equipment. The names of these companies are noted at appropriate points in the text, and the full addresses are quoted in Appendix 7. In particular we are grateful to the following who supplied photographs: C. V. Cook and Sons Ltd [hydrogenation apparatus, Fig. 2.63(c)], Electrothermal Engineering Ltd [heating mantles, Fig. 2.47(a), (c) and (d)], Gallenkamp [shakers, Fig. 2.53(a) and (b)], Jencons (Scientific) Ltd (rotary evaporator, Fig. 2.112), and Lab-Marc [heating mantle, Fig. 2.47(b)]. Permission to reproduce from the catalogues' new line diagrams is also acknowledged as follows: Aldrich Chemical Co. Ltd [Figs 2.23, 2.34, 2.71, 2.74(a) and (b), 2.75, 2.115, 2.120 and 2.142], Beckman–RIIC (Figs 3.6 and 3.7), J. Bibby Science Products Ltd [Figs 2.8(d), 2.29, 2.31(a) and (b), 2.33, 2.44 and 2.89] and T. C. Research (Fig. 2.138).

Some of the new spectroscopic data are taken from the collections of the School of Chemistry, Thames Polytechnic, and we wish to acknowledge particularly the assistance of Dr E. Vidgeon for specifically recording the majority of the ^{13}C-n.m.r. spectra. We acknowledge the inclusion of other spectroscopic data from the Sadtler Collections, Standard Infrared Spectra, Standard Ultraviolet Spectra and Standard Nuclear Magnetic Resonance Spectra, Sadtler Research Laboratories, Philadelphia, USA, and from *Compilation of Mass Spectral Data*, A. Cornu and R. Massot (1966–1971), Heyden and Son Ltd in collaboration with Presses Universitaires de France.

We are grateful to the following for permission to reproduce copyright material:

American Chemical Society for two tables giving 'Schematic Representation of Summation Bands', by C. W. Young, R. B. Duval and N. Wright in *Analytical Chemistry* (1951), Vol. 23, and 'Graphical Display of Chemical Shifts for Classes of Paraffins', by L. P. Lindemann and J. Q. Adams in *Analytical Chemistry* (1971), Vol. 32 (10); Butterworths for figures giving 'Decoupled Spectrum of Crotonaldehyde', by W. MacFarlane and R. F. M. White, from *Techniques of High Resolution Nuclear Magnetic Resonance Spectroscopy* (1972); Heyden & Son Ltd, for Figs. 1 and 2 in *Organic Magnetic Resonance* (1972), Vol. 4; Pergamon Press Ltd, for figures from *Organic Chemistry, Applications of NMR* (1969), by L. M. Jackman and S. Sternhell (2nd edn); Sadtler Research Laboratories for ^{13}C nuclear magnetic resonance spectra of anisole, benzyl alcohol and hex-l-ene; the authors for tables taken from *Principles of Organic Mass Spectroscopy*, by D. H. Williams and I. Howe; Verlag Chemie GMBH for figures by E. Breitmaier,

G. Jung and W. Voelter in *Angewandte Chemie* (1971), Vol. 10; John Wiley & Son Inc for a table by R. M. Silverstein, C. G. Bassler and T. C. Merrill from *Spectroscopic Identification of Organic Compounds* (1974), 3rd edn; the authors for tables taken from *Carbon-13 Nuclear Magnetic Resonance for Organic Chemists* (1972) by G. C. Levy and G. L. Nelson; Springer-Verlag for 'Notes for Users of Beilstein' in *Handbook of Organic Chemistry*.

Acta Chemica Scandinavica and the Author, Dr Rolf Carlson for an extract from article by Rolf Carlson, Asa Nilsson & Mats Stromqvist, pp.7–13 *Acta Chem. Scand. B* **37** (1983). Aldrich Chemical Co. Inc. for an extract from '"Dry Column" Flash Chromatography' by L.M. Harwood, p.25 *Aldrichimica Actca* **18** (1985). American Chemical Society for extracts from articles by H.C. Brown p.1816 *J. Am. Chem Soc.* **93** (1971); J.C. Sheenan, p.1156 *J. Am. Chem. Soc.* **80** (1958); modified extracts from articles by H.C. Brown, p.1245 *J. Am. Chem. Soc.* **83** (1961), p.2070 *J. Am. Chem. Soc.* **85** (1963), p.7768 *J. Am. Chem. Soc.* **96** (1974), p.1088 *J. Am. Chem. Soc.* **86** (1964); S. Krishnamurthy & H.C. Brown, p.3384 *J. Am. Chem. Soc.* **98** (1976); A.I. Meyers *et al.*, pp.573–4 *J. Am. Chem. Soc.* **98** (1976); H. Kruizinga *et al.*, p.5188 *J. Am. Chem. Soc.* **103** (1981); J. Klein, p.3613 *J. Am. Chem. Soc.* **81** (1959); W.S. Wadsworth *et al.*, p.1737 *J. Am. Chem. Soc.* **83** (1961); W.H. Pirkle, p.3965 *J. Am. Chem. Soc.* **103** (1981); L.T. Scott, p.6313 *J. Am. Chem. Soc.* **102** (1980); R. Lohrmann, p.832 *J. Am. Chem. Soc.* **88** (1966); P. Gassmann *et al.*, pp. 7606–7 *J. Am. Chem. Soc.* **100** (1978); M.E. Kuehnr *et al.*, pp.4283, 4285 *J. Am. Chem. Soc.* **81** (1959); A.L. Wilds, p.5363 *J. Am. Chem. Soc.* **75** (1953); G. Posner, p.5106 *J. Am. Chem. Soc.* **94** (1972); T. Mukaiyama, p.7508 *J. Am. Chem. Soc.* **96** (1974). Copyright American Chemical Society: extract from article by D.B. Dess *et al.*, p.4155 *J. Org. Chem.* **48** (1983); modified extracts from articles by R. Ratcliffe *et al.*, p.4001 *J. Org. Chem.* **35** (170); B.H Lipshutz, p.3934 *J. Org. Chem.* **49** (1984); E. Juaristi, p.2605 *J. Org. Chem.* **48** (1983); D.E. Bergbreiter, p.219 *J. Org. Chem.* **46** (1981); A. Wissner, p.3972 *J. Org. Chem.* **43** (1978); H.C. Brown, p.5068 *J. Org. Chem.* **47** (1982); C.A. Brown & V. Ahuja, pp.2229–30 *J. Org. Chem.* **38** (1973); S. Krishnamurthy, p.3064 *J. Org. Chem.* **41** (1976); R.O. Hutchins, p.2301 *J. Org. Chem.* **43** (1978); J.C. Stowell, p.561 *J. Org. Chem.* **41** (1976); W.E. Gore *et al.*, p.1708 *J. Org. Chem* **40** (1975); T.H. Chan *et al.*, p.3266 *J. Org. Chem.* **39** (1974); R.P. Thummel, p.304 *J. Org. Chem.* **42** (1977); S. Kim, Y. Moon & K.H. Ahn, p.3315 *J. Org. Chem.* **47** (1982); H.C. Brown *et al.*, pp.5067, 5068 *J. Org. Chem.* **47** (1982); M. Miyashita *et al.*, pp. 3773, 3774 *J. Org. Chem.* **42** (1977); G.A. Olah, pp.1638, 1639 *J. Org. Chem.* **45** (1980); D. Horton *et al.*, p.731 *J. Org. Chem.* **33** (1968); N. Kornblum *et al.*, p.4536 *J. Org. Chem.* **47** (1982); J. McMurry *et al.*, pp.4371, 4373 *J. Org.Chem.* **38** (1973); G.E. Wilson *et al.*, p.2134 *J. Org. Chem.* **33** (1968); J.S. Nimitz, p.212–13 *J. Org. Chem.* **46** (1981); W. Wierenga *et al.*, p.311 *J. Org. Chem.* **44** (1979); N. Kornblum, p.455 *J. Org. Chem.* **22** (1957); K.E. Gilbert, p.660 *J. Org. Chem.* **44** (1979); F. Rolla, p. 4328 *J. Org. Chem.* **47** (1982); S.R. Wann *et al.*, p.2579 *J. Org. Chem.* **46** (1981); H.O. House *et al.*, p.3106 *J. Org. Chem.* **39** (1974); D.R. Maudling *et al.*, pp.2938–9 *J. Org. Chem.* **48** (1983); R.A. Sherrer *et al.*, p. 2129 *J. Org. Chem.* **45** (1980); G. Kabalka *et al.*, p.2387 *J. Org. Chem.* **51** (1986); T. Cohen, p.2056 *J. Org. Chem.* **42** (1977); M.P. Doyle *et al.*, pp.2429, 3497 *J. Org. Chem.* **42** (1977); V. Balogh, p.1341 *J. Org. Chem.* **36** (1971); Z.G. Hajos & D. Parrish, pp.1614, 1619, 1620 *J. Org. Chem.* **39** (1974); E.C. Friedrich, p.4641 *J. Org. Chem.* **50** (1985); J. Wolinsky *et al.*, p.42 *J. Org. Chem.*

30 (1965); S.J. Brois, pp. 3533–4 *J. Org. Chem.* **27** (1962); M.O. Brimeyer *et al.*, p.4255 *J. Org. Chem.* **45** (1980); T. Durst, pp.2043–4 *J. Org. Chem.* **35** (1970); W.T. Brady, p.2847 *J. Org. Chem.* **34** (1969); W.J. Houlihan, p.4512 *J. Org. Chem.* **46** (1981); J.B. Hendrickson *et al.*, p.3345 *J. Org. Chem.* **48** (1983); P.T. Izzo, p.702 *J. Org. Chem.* **24** (1959); G. Olah, p.1247 *J. Org. Chem.* **44** (1979). Copyright American Chemical Society.   American Chemical Society and the authors for modified extracts from articles by Prof. D.N. Harpp, p.324 *J. Org. Chem.* **40** (1975); Prof. A.I. Meyers, p.895–7 *J. Org. Chem.* **43** (1978); Prof. K. Sharpless, p.1924 *J. Org. Chem.* **51** (1986); W. Clark Still, pp.2923–4 *J. Org. Chem.* **43** (1978); Prof. H.H. Wasserman, p.3010 *J. Org. Chem.* **46** (1981). Copyright American Chemical Society.   American Chemical Society, Division of Chemical Education for modified extract from article by E.F. Silversmith, p.645 *J. Chem. Ed.* **63** (1986). Copyright American Chemical Society.   The Chemical Society of Japan for modified extracts from articles by T. Sugita *et al.*, pp. 1481–2 *Chemistry Letters* (1982); T. Yamada *et al.*, p.2403 *Bull. Chem. Soc. Japan* **51** (1978); S. Shiraishi *et al.*, p.1641 *Bull. Chem. Soc. Japan* **57** (1984).   Council of Scientific & Industrial Research, New Delhi for a modified extract from an article by O.P. Vig *et al.*, p.1076 *Indian Journal of Chemistry* **15B** (1977).   Marcel Dekker Inc. for modified extracts from articles by G. Kabalka, p.278 *Synthetic Communications* **9** (1979); A. Rahm, p.485 *Synthetic Communications* **12** (1982).   Pergamon Press Plc for extracts from articles by A. Hulkenberg *et al.*, p.1506 *Tetrahedron Letters* (1982); N. Umino, p.765 *Tetrahedron Letters* (1976); N.R. Ayyanger, p.1099 *Tetrahedron Letters* (1982); O. Repic, p.2731 *Tetrahedron Letters* (1982); G.W. Fleet, p.1439 *Tetrahedron Letters* (1978); modified extracts from articles by E.J. Corey, p.2649 *Tetrahedron Letters* (1975); S.H. Korzeniowski, p.1872 *Tetrahedron Letters* (1977); T.N. Sorrell, p.2474 *Tetrahedron Letters* (1978); J. Tsujf *et al.*, p.3282 *Tetrahedron* **39** (1983); S.S. Jew *et al.*, p.2340 *Tetrahedron* **35** (1979); A.R. Katritzky, p.29 (supplement 1) *Tetrahedron* **37** (1981); N.J. Cusak *et al.*, p.2160 *Tetrahedron* **32** (1976); L. Ghosez, pp.627, 628, 629 *Tetrahedron* **27** (1971); A. McKillop, p.1381 *Tetrahedron* **30** (1974).   Pharmaceutical Society of Japan for modified extracts from articles by S. Yamada, p.995 *Chem. Pharm. Bull.* **13** (1965); p.1142–3 *Chem. Pharm. Bull.* **11** (1963).   Royal Society of Chemistry for modified extracts from articles by A.W. Johnson, p.1015 *J. Chem. Soc* (1946); A. Buss & S. Warren, pp.2311–12, 2319–20 Perkin Trans. 1 *J. Chem. Soc.* (1985); H.R. Hudson, p.106–7 Perkin Trans. 1 *J. Chem. Soc.* (1976); D.N. Gupta *et al.*, p.2972 Perkin Trans. 1 *J. Chem. Soc.* (1981); C.B. Reese *et al.*, p.1490 Perkin Trans. 1 *J. Chem. Soc.* (1980); G. Barger *et al.*, p.1820 *J. Chem. Soc.* (1973); M.C. Whiting, p.1370 Perkin Trans. 1 *J. Chem. Soc.* (1983); R.C. Cookson, p.2448 Perkin Trans. 1 *J. Chem. Soc.* (1979); S.R. Landor, p.2498 *J. Chem. Soc.* (1967); C.D. Maycock *et al.*, p.1855 *J. Chem. Soc.* (1979); M.V. Bombala *et al.*, p.3014 Perkin Trans. 1 *J. Chem. Soc.* (1979); J. Kametani *et al.*, p.2230 Perkin Trans. 1 *J. Chem. Soc.* (1981); P.H. Bentley *et al.*, p.2224 Perkin Trans. 1 *J. Chem. Soc.* (1980); G. Casiraghi *et al.*, p.1865 Perkin Trans. 1 *J. Chem. Soc.* (1981).   Societe Français Caise de Chemie for a translated and modified extract from an article by R. Lombard, p.1462 *Bull Soc. Chem. Fr.* (1958).   Springer-Verlag Heidelberg for modified extracts from 'Notes on the use of the Beilstein Handbook of Organic Chemistry' in Beilstein *Handbook of Organic Chemistry*, adapted extract by M. & A. Bodanszky, p.20 *The Practise of Peptide Synthesis* (1984).   George Thieme Verlag, Stuttgart–New York and the Authors for extracts from articles by F. Bickelhaupt, p.721 *Synthesis* (1983); S. Chandra-sekaran (V. Bhushan *et al.*), p.431 *Synthesis* (1984); Prof. Y.S. Cheng, p.224 *Synthesis* (1980); Prof. J. Villieras, p.302 *Synthesis* (1983); modified extracts from articles by Prof. H.C. Brown (for N.R. De Lue), p.115 *Synthesis* (1976); Prof. H.C. Brown, p.704 *Synthesis* (1979); Roberto Ballini, p.1024 *Synthesis* (1986); Alexander J. Fatiadi, p.357 *Synthesis* (1973); Prof. Tsuneo Imamoto, p.143 *Synthesis* (1983); Dr C.F. Lane, p.136 *Synthesis* (1975); Donald G. Lee, p.462 *Synthesis* (1978); Shunsaku Ohta, p. 833 *Synthesis* (1982); Prof. C. Piechucki, p.870 *Synthesis* 1974); John C. Stowell, p.279 *Synthesis* (1984); Prof. R.S. Tewari, p.331 *Synthesis* (1983); Dr M.C. Venuti, p.62 *Synthesis* (1982).   VCH Verlagsgesell-schaft mbH for translated and modified extracts from articles by E. Bamberger, p.1118 *Chem. Ber.* **52** (1919); W. Huckel, pp. 342–3 *Chem. Ber.* **88** (1955); E. Salmi, p.1805 *Chem. Ber.* **71** (1938).   John Wiley & Sons Inc. for a modified extract by N. Kornblum, p.133 *Organic Reactions* 12 (1962).

Permission for the publication herein of Sadtler Standard Spectra® (Figs. 3.51, 3.53 & 3.54) has been granted, and all rights reserved, by Sadtler Research Laboratories, Division of Bio-Rad Laboratories Inc.

CHAPTER 1
ORGANIC SYNTHESIS

1.1 INTRODUCTION

This book aims to reflect comprehensively the many and varied aspects of practical organic chemistry with which the student will need to become familiar. The synthesis of organic compounds is traditionally an important part of the training of an organic chemist. By undertaking the preparation of a varied range of compounds, and using a representative selection of reaction processes and techniques, the prospective organic chemist becomes familiar with the chemical and physical properties of organic substances and begins to understand more clearly the factors which govern their reactivity. The discussion sections which are sited before the relevant experimental procedures provide a bridge between laboratory work and lecture material. Thus while the synthesis of quite simple compounds is of considerable educational value, particularly if the reactions involved are of a general nature, wider reading of a selection of the excellent review articles that are cited at the ends of chapters, should enable the advanced student of organic chemistry to appreciate the wider implications of the reactions under consideration.

The student will become increasingly aware, during the early years of training, of the very great range of concepts and techniques with which the practising organic chemist has to be familiar. Some of these may be summarised as follows:

1. The preparative, isolation and purification techniques (including the use of a varied range of chromatographic procedures), which are applicable to quantities of reactants and products from milligram to kilogram amounts.
2. The chemical procedures which are used for the structural characterisation of pure compounds, together with the routine application of the various spectroscopic methods.
3. The ever-increasing range of new reagents and reaction procedures which may replace older methods by reason of their greater specificity or increased safety in use.
4. The enormous body of theoretical ideas which enable the manner in which compounds interact to be understood in terms of the reaction mechanism and, if appropriate, stereochemical factors.
5. The applications of computer techniques, initially in the area of literature searching, educational aids, and interfacing with many instrumental techniques, but more recently as an invaluable aid to the design of synthetic routes to complex organic molecules.

1

The attainment of a comprehensive appreciation in each of these overlapping and interrelated areas is a formidable objective for the new student. The integration of the theory of organic chemistry which has been acquired in lectures and tutorials with the practical experience gained in the preparative, chromatographic and spectroscopic laboratories is a challenging prospect.

The key question, 'how may compound X be synthesised?', has been at some time or another applied to the countless numbers of compounds prepared in research and industrial laboratories over a period of the last 150 years. It is therefore an interesting reflection that it is only in the last twenty years or so that consideration has been given to formalising a philosophy and logic of design in organic synthesis. The many brilliant and novel synthetic schemes, often of compounds of great structural complexity, which have been accomplished in the past are attributable largely to the intuitive skills of the research chemist. These skills derived from: (a) a sound understanding of an increasing volume of structural and stereochemical information; and (b) the ability to apply developing mechanistic theory to the transformation of functional groups and to the formation of new carbon–carbon skeletal bonds, to select and devise new reagents to effect these transformations, and to visualise the 'backward' routes from the required target molecule and 'forward' routes from possible starting materials. Rarely, however, in the publication of the final synthesis was space devoted to a description of how the scheme emerged. A notable recent publication[1] has now done much to rectify this omission, by collecting together a series of excellent articles, written by research chemists involved in the work, on the strategy and tactics of the synthesis of some important and complex structures.

Today's synthetic organic chemist now has the opportunity to devise syntheses of graded complexity in a more formalised sense, and to recognise the potential problems associated with such schemes using developing experience in theoretical knowledge and practical expertise. The following sections of this chapter provide an overview of these facets of organic synthesis. The references to this chapter list some of the valuable texts in general organic chemistry,[2] organic reaction mechanism,[3] stereoisomerism,[4] and design (methodology and strategy) in organic synthesis,[5] together with references to original papers which provide some further coverage of the field.

1.2 STRUCTURE OF THE TARGET MOLECULE

Following the postulation of the question 'how may compound X be synthesised?', the first step is to examine the *structure* of the target molecule. Clearly the effectiveness of this examination will depend on the level of structural understanding to which the student has progressed. Compounds (1) to (17) represent a selection of structures of varying (and broadly increasing) complexity. First-year undergraduate students should be able to recognise readily and to name the functional groups in *all* structures, noting for example that both (1) and (14) have a carbon–carbon double bond and a hydroxyl group, and further that the hydroxyl group in (1) is primary, whereas in (14) it is secondary; that in (12) the functional groups that are present are three carbon–carbon double bonds, a keto group, a hydroxyl group and an ester group, etc. The recognition of skeletal structures in compounds (1) to (17) by first-year undergraduates may begin to falter after (10); thus no difficulty should be experienced with (1) to (4) (open carbon chain, or aliphatic), (5) and (6) (aromatic), (7) and (8) (carbocyclic, or alicyclic), (9) and (10) (heteroaromatic). However, structures (11) to (17) are more complex cyclic and polycyclic systems which are not so simply described, although the more advanced student should be able to recognise these latter structures as representative of the important groups of natural products. These are: the terpenes (11); the prostanoids (12); the carbohydrates (13); the steroids (14); the penicillins (15) and cephalosporins; the alkaloids (16); and macrocyclic antibiotics (17).

(1) (2) (3) (4)

(5) (6) (7) (8)

(9) (10) (11)

(12) (13)

(14) (15)

(16)   (17)

The recognition of functional and skeletal features leads to a modification of the question, 'how may compound *X* be synthesised?', to a more specific question, say for (9), 'how may the acetyl derivative of 4-aminopyridine be synthesised?' or for (13), 'what synthetic methods are applied in carbohydrate chemistry to effect acetal formation starting from the readily available parent monosaccharides?' The examination of structure is not yet complete however. The stereoisomeric (optical and geometric) features and the possibility of conformational equilibration, arising from rotation of groups and atoms around single bonds, must also be taken into account. Before considering these features, it is worth pointing out the advisability of model construction using the Dreiding and related molecular models, and in certain instances space-filling models. These are invaluable, from the earliest stages of study, for acquiring an appreciation of the shapes of molecules, their stereoisomeric features, their flexibility, the 'interaction' of groups within the molecule, and the feasibility and the manner of the interaction of the specified compound with a reagent. Computer simulations of molecular structures are also increasingly useful tools for this purpose.

OPTICAL ISOMERISM
A molecule may be defined as *achiral* or *chiral*. Whereas an achiral molecule is one which is identical with and can be superimposed upon its mirror image, a chiral molecule and its mirror image are not superimposable. Each of the pair of non-superimposable mirror image isomers (termed *enantiomers*) rotates the plane of polarised light in equal and opposite (dextrorotatory and laevorotatory) directions. The enantiomers are thus optically active, and a 50:50 mixture is termed a *racemate*. The mirror image related structures (2a and 2b) of the alkyl halide (2) illustrate the usual way of representation of the directional (spatial) arrangement of the four σ-bonds of the saturated (sp^3-hybridised) carbon atom, i.e. two bonds in the plane of the paper, and one pointing forward (wedge shaped) and one pointing back (dotted).

(2a)   (2b)

The term *diastereoisomeric* refers to those molecules having the same structure (functional groups and skeletal arrangement) but which are not mirror image

4

related. Thus tartaric acid (4) exists as two optically active non-superimposable mirror image structures (4a and 4b), each of which is diastereoisomeric with the optically inactive *meso*-tartaric acid (4c).

(4a) (4b) (4c)

Whereas enantiomers (e.g. 4a and 4b) have indentical chemical and physical properties (except their effect on plane polarised light), diastereoisomers (e.g. 4a, or 4b and 4c) frequently differ in their chemical properties, and have different physical properties.

One simple practical method of assessing the possibility of the existence of non-superimposable mirror images, particularly with complex structures, is to construct models of the two molecules. The property of chirality may alternatively be described in terms of the symmetry elements of the molecule. If there is a lack of all elements of symmetry (i.e. a simple axis, a centre, a plane, or an *n*-fold alternating axis) the chiral molecule is *asymmetric*, and will possess two non-superimposable mirror image structures (e.g. 2a and 2b). If, however, the molecule possesses a simple axis of symmetry (usually a C_2 axis) but no other symmetry elements, the chiral molecule is *dissymmetric*. Thus 4a and 4b are dissymmetric and the simple C_2 axis of symmetry, of for example 4a, is shown below. If the molecule possesses a centre of symmetry (C_i) or a plane of symmetry (σ), or an *n*-fold alternating axis of symmetry (S_n), the mirror images of the molecule are superimposable and the molecule is optically inactive. These latter three symmetry elements are illustrated in the case of the molecule 4c.

(4a) (4c)

(4c)

Optical activity was first observed with organic compounds having one or more chiral carbon atoms (or centres) (i.e. a carbon substituted with four different groups). In the structures (1) to (17) the chiral carbons are specified with an asterisk. Subsequently compounds having chiral centres at suitably substituted heteroatoms (e.g. silicon, germanium, nitrogen, phosphorus, arsenic, sulphur, etc.) were also synthesised. Molecular dissymmetry, and hence chirality, also

5

arises in molecules which have a chiral axis [e.g. substituted biphenyls such as (18), or substituted allenes such as (19)], or a chiral plane [e.g. hexahelicenes such as (20), or substituted paracyclophanes such as (21)].

(18)

(19)

(20)

(21)

One of the early major problems in the development of an understanding of the molecular geometry of optical isomers was the formal representation of the individual isomers. It was necessary to show unambiguously the relative spatial arrangements of the groups and of the bonding system (the *configuration*), and to relate these representations to the isolated, purified and characterised compounds (the *configurational assignment*). One important and long-serving convention for specifying the configuration of a chiral compound is attributed to M. A. Rosanoff (1906).[6] In this notation, dextrorotatory glyceraldehyde (3) was arbitrarily specified as having the space arrangement (22). This configuration was denoted by the prefix D (i.e. D-(+)-glyceraldehyde) thereby specifying that the hydroxyl group in this space arrangement lay on the right. The enantiomer was designated L-(−)-glyceraldehyde.

(22)

(23)

It was then possible, by unambiguous chemical interconversion procedures or by physical property comparisons, to relate many compounds having one or more chiral carbons to either of the enantiomers of glyceraldehyde. This enabled a number of chiral compounds to be designated as belonging to the D- or L-series; such assignments were thus made on the basis of *relative configurations*. The X-ray analysis of the sodium rubidium salt of optically pure (+)-tartaric acid in 1951 established its *absolute configuration* as (23).[7] Since this configuration had been previously related to that of D-(+)-glyceraldehyde, all relative configurations of chiral compounds thus assigned were correct in the absolute sense. X-ray analysis is now established as the *only* method that gives directly the absolute configuration of a chiral compound (i.e. the configurational assignment does not depend upon any reference standard).

The D/L convention, though invaluable in the first half of this century, had considerable drawbacks since it was inapplicable to compounds having hetero-atoms as chiral centres, and could not be used for compounds having chiral axes or planes [e.g. structures (18) to (21)]; furthermore, it was difficult to apply to all centres in molecules having numerous chiral carbons. Such difficulties were overcome by the Cahn–Ingold–Prelog [(R/S)] notation.[8] Here, a designated set of rules for assigning priorities to groups attached to chiral centres, chiral axes and chiral planes was proposed, together with a specified way of looking at such centres, axes and planes. Finally, if the circular sequence in moving from groups of high priority to lower priority was clockwise the notation was (R) (rectus, right), or if anticlockwise it was (S) (sinister, left). This is illustrated for (+)- and (−)-glyceraldehyde, which are (R) and (S) respectively, where the priority sequence of groups is $OH > CHO > CH_2OH > H$.

(22a)	(+)-Glyceraldehyde	(22b)	(−)-Glyceraldehyde

The convention has over the last three decades proved to be adaptable, versatile and universal.[9] It should be pointed out, however, that with amino acids (Section 5.14.4, p. 746), and hence in peptide and protein chemistry, and with carbohydrates (Section 5.10), the D/L convention is still the more convenient, mainly because it is used specifically to designate generic relationships between an enormous number of compounds of closely related structure.

Compounds isolated from natural sources are frequently optically pure. Thus camphor (11), cholesterol (14), morphine (16), for example, are isolated in the optically pure state. The parent molecule of (13) is D-glucose, and like camphor and cholesterol is readily available in very large quantities. These, and comparable compounds, form what is now described as a *chiral pool*, i.e. low-cost, readily available, chiral compounds which provide starting materials for conversion into other compounds, of simplified skeletal and functional structure, in which some or all of the chiral features have been retained.

Compounds synthesised in the laboratory *without the use of chiral reagents* (see asymmetric synthesis p. 15) are always obtained as the racemate. In order to separate the individual enantiomers, a *resolution* process needs to be adopted. This aspect is considered in more detail in Section 5.19.

GEOMETRIC ISOMERISM

The target molecule also has to be examined for structural features broadly known as *cis/trans* or geometric isomerism. These isomers differ in all their physical properties and in some, but not all, of their chemical properties (they are in fact diastereoisomers). The structural features leading to this isomerism are the presence of a suitably substituted double bond (carbon–carbon, carbon–nitrogen, nitrogen–nitrogen), or the presence of a suitably substituted cyclic structure [e.g. substituted cyclopropanes such as (8)]. The *cis/trans* notation continues to be used for designating the configuration in carbocyclic structures.

For example, (7) is *trans*-2-methylcyclohexanol; in addition the chirality of C-1 and C-2 should be designated by the (*R/S*) notation, and the correct systematic name is therefore (1*R*,2*R*)-2-methylcyclohexanol (7*a*), since the mirror image molecule, (1*S*,2*S*)-2-methylcyclohexanol (7*b*) is also the *trans* isomer. The *cis* isomer is also chiral, and diastereoisomeric with the *trans* isomer; the two enantiomers would be (1*S*,2*R*)-2-methylcyclohexanol (24*a*) and (1*R*,2*S*)-2-methylcyclohexanol (24*b*).

(7*a*)   (7*b*)   (24*a*)   (24*b*)

In the case of ring-fused systems the *cis/trans* notation is used to designate the geometric nature of the ring junction. Structures (25) and (26) represent partial structures of the ring-fused compounds (13) and (14) respectively, which emphasise that the ring junctions utilise *cis*- or *trans*-orientated bonds.

cis-fused   *trans*-fused

(25)   (26)

In the case of compounds having double bond systems, the configuration is best specified by the (*E/Z*) notation. Here the isomer is designated as (*E*) (entgegen, opposite) if the highest priority groups (as defined by the Cahn–Ingold–Prelog rules) attached to each sp^2-hybridised carbon lie on opposite sides of the double bond; the (*Z*)-isomer (zusammen, together) is that in which the groups of highest priority lie on the same side of the double bond. Thus compound (1) is (*E*)-but-2-en-1-ol; similarly the configurations of the double bond in the side chains of (12) are (*E*) and (*Z*); that in (14) is (*Z*) owing to the constraints of the ring system.

The interconversion of isomers in the case of optical or geometric pairs, if structurally feasible, may only take place by the breaking of σ- or π-bonds. However, there is a further area of stereoisomerism wherein the isomers are interconvertible by *rotation* about a single (σ) bond, and in general the pure stereoisomers are not isolable. These isomers are termed *conformational isomers or conformers*.

CONFORMATIONAL ISOMERISM

The term conformation is used to denote any one of the infinite number of momentary arrangements of the atoms in space which result from the rotation about single bonds. Each of these arrangements will differ in the degree of *intramolecular* interactions and will thus have differing free energies. The term conformer is applied to those discrete spatial arrangements which are at energy minima in the continuous potential energy curve resulting from such rotations in the molecule. Where several minima may be expected, the *preferred conformer* is that with the lowest free energy (i.e. the most stable). The ratio of preferred to other conformers, in an equilibrium aggregate of molecules, may be calculated from their free energy differences. Molecules are often referred to as conformationally *fixed*, *biased*, or *mobile* according to the proportion of the preferred conformer and the energy barrier which needs to be surmounted for mutual interconversion.

The concept of conformational preference is widely applied to open-chain systems and to substituent groups attached to aromatic rings, to explain differences in physical properties (e.g. acidity, basicity, spectroscopic features, etc.), and differences in chemical reactivity (both in terms of reaction rates, and of reaction course leading to alternative products). Its application to alicyclic systems was firstly to cyclohexanes, and to *cis* and *trans* decalins, but it is now applied to all ring systems (except aromatic and heteroaromatic) regardless of structure, size, fusion and bridging. The appreciation of conformational preferences marked the beginning of an important area of study in recent decades, known as *conformational analysis*.[10] Here chemical equilibria, and the rates of chemical reactions, are considered in relation to the possible conformations of reactants and reagents (where appropriate), their mode of approach and the favourable mechanistic pathways to transition states and intermediates, and finally to the conformational and other stereoisomeric features of the products.

To elaborate a little on these general principles, four of the structures from the (1) to (17) group are re-examined. Thus in (5), *N,N*-dimethyl-*p*-nitroaniline, rotation of the substituted amino and nitro groups around the bonds linking them to the benzene ring could be considered. The energy minimum, and hence the most stable arrangement, is that in which the π- and *p*-electrons of the substituent groups overlap with those of the aromatic ring, so that mesomeric interaction is at a maximum (27). The lower basicity of this compound (owing to the electron withdrawing effect of the nitro group) in comparison to aniline may thus be rationalised. If, however, very bulky groups were sited in the *ortho* positions [e.g. a t-butyl group *ortho* to the dimethylamino group (28)], which impose conformational restriction and thereby reduced mesomeric interaction, an increase in basicity results.

(27)   (28)

With compound (7) the chair–chair conformational equilibrium may be represented as (29) ⇌ (30).

(29)   (30)   (31)

9

Although this is a conformationally mobile system, conformation (29) is favoured (more stable) over (30) owing to the 1,3-diaxial interactions arising from both the methyl and hydroxyl groups. Note that the change of conformations does not affect the configurations at the chiral sites since no bonds are cleaved. In the case of *cis*-2-t-butylcyclohexanol (31), the large t-butyl group imposes a favourable bias on the conformation shown even though the hydroxyl group is in an axial position.

Camphor (11), and cholesterol (14), are conformationally rigid, in the former case because of the bridged-ring structure, and in the latter because of the all-*trans* ring fusions.

(11)   (14)

Hence the reactivity of the carbonyl group and of the *endo*- and *exo*-hydrogens in (11), and the stereochemical consequences of such reactions (i.e. the stereo-isomeric nature of the product) are crucially dependent upon the conformationally fixed environment. Likewise the reactivity of the β-orientated hydroxyl group and of the olefinic bond in cholesterol (14) is determined by the fixed conformational environment.

It is perhaps only too obvious that the reactivity of one compound offers the means of the synthesis of another. In essence, a *scheme of synthesis* evolves from a consideration of structural features, aided by an understanding of reaction mechanism, which are applied in the strategy and the methodology of synthesis. The *strategy* of synthesis requires the development of a sequence of ideas as to what needs to be done in functional group modification and skeletal structure assembly or rearrangement, whereas *methodology* of synthesis refers to the procedures, reagents and conditions which are required to bring about efficient reactions in all the intermediate stages of the synthesis. Although current practice is to decide on a strategy and then proceed to a consideration of the methodology to be adopted, the development of ideas in these two areas has historically been the reverse, since methodology has contributed to, and been vastly extended by, the early development of organic reaction mechanisms. In the following discussion the role of methodology is considered first, followed by the important contribution of ideas on the strategy of synthesis.

1.3 REACTION MECHANISM AND THE METHODOLOGY OF SYNTHESIS

Organic reaction mechanisms may be classified into three main groups depending on the manner in which the covalent bonds are cleaved, namely: (a) ionic mechanisms; (b) radical mechanisms; and (c) pericyclic mechanisms.

In an ionic mechanism, the bond breaking process (i) in the organic reactant is in an overall sense *heterolytic*.

$$A—B \longrightarrow A^{\oplus} + :B^{\ominus}$$ (i)

If the *rate determining step* of the reaction is that in which the reactant donates an electron pair to a reagent, the reagent is an electrophile and the overall reaction is *electrophilic*. Similarly if the reactant, in the *rate determining step* of the reaction, accepts an electron pair from the reagent, the reagent is a nucleophile and the reaction *nucleophilic*.

In a radical mechanism, the bond breaking process (ii) in the organic reactant is in an overall sense *homolytic*.

$$A—B \longrightarrow A\cdot + \cdot B$$ (ii)

The reagents in this case are themselves radicals and the subsequent reaction sequence frequently involves a *chain mechanism*.

In the third group, *pericyclic reactions*, there are no intermediate ions or radicals, but the reactant and reagent orientate themselves so as to form a cyclic transition state (usually, but not necessarily, six-membered in type) around which there is an 'electron flow' leading to the formation of the new bonding arrangement [e.g. (iii)].

(iii)

These mechanistic possibilities may be sub-classified into the following reaction types, and since all the reactions described in the text may be assigned a mechanism, the forward cross-references are to typical illustrative examples only.

1. Substitution: nucleophilic (e.g. Section 5.5.1, p. 555 and Section 6.8.2, p. 959), electrophilic (e.g. Section 6.2.1, p. 851), or radical (e.g. Section 6.3.1, p. 861).
2. Addition to multiple bonds: nucleophilic (e.g. Section 5.4.2, p. 532), electrophilic (e.g. Section 5.5.7, p. 574), radical (e.g. Section 5.5.7, p. 575) and pericyclic (e.g. Section 7.6).
3. Elimination: usually nucleophilic (Section 5.2.1, p. 488), but also pericyclic.
4. Rearrangement: this may be either *intra- or inter-molecular*, commonly involving the migration of a carbon species to a carbon (e.g. Section 5.2.1, p. 487), nitrogen (e.g. Section 6.12.6, p. 1047), or oxygen (Section 5.4.3, p. 543) centre. The migrating species may be a nucleophile or a radical, or the reaction may proceed in a pericyclic fashion. The rearrangement of hydrogen (as a proton in prototropy, or in a pericyclic sense) is a widely recognised special case.
5. Oxidation and reduction. These reactions are mechanistically complementary to each other; oxidising reagents (e.g. Section 5.7.1, p. 587) are electrophilic and reducing reagents (e.g. Section 5.4.1, p. 519) are nucleophilic.

Over the past fifty years or more, archetypal cases within each of these reaction types have been studied in great detail. These extensive studies have enabled ideas on the nature and stabilisation of transition states and on the timing of bond-breaking and bond-forming processes to be formulated. Many of the results have led to modifications in techniques, and to the discovery and design of new reactions and reagents, i.e. to the development of the methodology of synthesis.

11

In this developing framework involving mechanisms and methodology, *control* and *selectivity* were recognised as playing a crucial and integral role in reaction pathways.

CONTROL AND SELECTIVITY IN ORGANIC REACTIONS

Most of the reactions described in the synthetic sections of this book give rise to only one major product (ignoring the all too frequent presence of by-products arising from a variety of spurious reactions; see Section 2.2, calculation of chemical yield). However, there are many instances where a compound, under a specified set of reaction conditions, gives rise to two or more different products. In such cases the reaction may proceed under either *kinetic* or *thermodynamic* control.

In a reaction under kinetic control, the product composition via transition states T.S.1 and T.S.2 is determined by the relative rates of the alternative reactions, which are of course governed by the relative free energies of activation (ΔG^*) of the rate determining step of each reaction (Fig. 1.1). Analysis of product composition over the whole time of the reaction will show a constant ratio.

In some reactions, however, product composition does not remain constant, and the ratio of products is significantly different in the early (incomplete) stages of the reaction, from that which is found when all the starting material has reacted. In this case the reaction is said to be under thermodynamic control, where the final ratio of products is determined by their relative thermodynamic stabilities (i.e. their free energy differences, ΔG). Here the products are in equilibrium with either the starting material, or with a common intermediate (as illustrated in Fig. 1.2), so that although the initial product ratio is determined by the relative rates of the reactions, the final ratio reflects the relative stabilities of the products. Sometimes of course the kinetically controlled product is also the more stable, but clearly if this is not the case, attention to detail in specifying reaction conditions may enable the preparative isolation of either of the different products (see for example Section 5.5.7, p. 574, Section 5.10.1, p. 643 and Section 6.4.1, p. 874).

Fig. 1.1

Aspects of selectivity in reaction pathways may be considered under the following topic areas: *chemoselectivity, regioselectivity, protection* and, finally, *stereoselectivity.*

In a bifunctional compound, if a reagent reacts with one functional group preferentially, even though the other is apparently susceptible to the reaction conditions, the reaction is said to be chemoselective. Two illustrative examples are: the reduction of a carbonyl group in the presence of a cyano, nitro or alkoxycarbonyl group (Section 5.4.1, p. 519; see also *Metal hydrides*, Section 4.2.49, p. 445); and the acylation of an aromatic amino group in the presence of a phenolic group (Section 6.9.3, p. 984).

Regioselectivity in a reaction, proceeding without skeletal rearrangements, is observed when a molecule possesses two or more sites of reactivity arising from the presence of *one* functional group, each of which the reagent may attack, resulting in the formation of constitutional isomers. Preferential formation of one of these isomers shows that selection has taken place. In many cases the reactivity of the alternative sites may be enhanced by modification of the experimental conditions. The classical examples are the addition reactions of alkenes (Section 5.4.3, p. 542, Section 5.4.4, p. 545 and Section 5.5.7, p. 574), the addition reactions of α,β-unsaturated carbonyl compounds (e.g. direct carbonyl addition versus conjugate addition), and the aromatic electrophilic substitution reactions leading to either mainly *meta* products, or *ortho/para* products (see Section 6.2.1, p. 852); it should be noted in the latter case that the *ortho/para* ratio may also be subject to regioselective control.

Protection could be regarded as a special instance of a combined chemo- and regio-selectivity, since it embraces aspects of both.[5g] It is implicated when it is necessary to carry out a reaction selectively at one functional group, in the presence of other functional groups, but where the principles of chemoselectivity are not applicable. The accompanying functional groups must therefore be rendered inert to the reaction conditions, i.e. they must be protected. The selective introduction of the protective group must be achieved in good yield, preferably under

Fig. 1.2

13

mild conditions, the protected compound must be readily purified, and inert to the subsequent reaction conditions, and finally the protecting group must be easily removed to regenerate the original functionality. The early chemistry of carbohydrates (Section 5.10) and amino acids (Section 5.14.4, p. 746) provided a sound and substantial foundation from which the application of the principles of protection has permeated into all areas of synthesis.

The term stereoselective is often confused with the term stereospecific, and the literature abounds with views as to the most satisfactory definition. To offer some clarification, it is perhaps timely to recall a frequently used term, introduced a decade or so ago, namely the *stereoelectronic requirements* of a reaction. All concerted reactions (i.e. those taking place in a synchronised process of bond breaking and bond forming) are considered to have precise spatial requirements with regard to the orientation of the reactant and reagent. Common examples are S_N2 displacement reactions (e.g. Section 5.10.4, p. 659), E2 (*anti*) elimination reactions of alkyl halides (e.g. Section 5.2.1, p. 488), *syn* (pyrolytic) elimination reactions (Section 5.2.1, p. 489), *trans* and *cis* additions to alkenes (e.g. Section 5.4.5, p. 547), and many rearrangement reactions. In the case of chiral or geometric reactants, the stereoisomeric nature of the product is entirely dependent on the unique stereoelectronic requirement of the reaction; such reactions are stereospecific.

A stereoselective reaction on the other hand is one in which the stereoelectronic requirement of the reaction mechanism is such that two equally valid alternative pathways are available for the same mechanistic interaction between reactant and reagent. However, either the free energies of activation of the alternative reactions or the thermodynamic stabilities of the products differ, so that one isomer is formed in preference to the other; selection has occurred. An example is provided by the reduction of cholestan-3-one (32). Equatorial attack (i) or axial attack (ii) of the hydride ion is mechanistically equally feasible and stereoelectronically defined. However, steric interactions between the hydride ion source and the conformationally fixed steroid molecule, together with considerations as to whether the reaction was under kinetic or thermodynamic control, would determine that the reaction is proceeding in a stereoselective manner.

A recognition of the concept of stereoselectivity has led the way to a better understanding of the selection of reactants and reagents in order to introduce chirality into a symmetrical molecule, i.e. an *asymmetric synthesis*.

ASYMMETRIC SYNTHESIS

This area of synthetic organic chemistry has seen some of the most interesting developments over the last decade. The progress may be appreciated from the Nobel Symposium 60 on *Asymmetric Organic Synthesis*[11] which highlights the (then) current state of the art. Other texts survey the field in general and specific areas.[4a, b, c, 12]

Before 1940 optically active compounds could only be obtained in stereo-isomerically pure form by isolation from natural sources, by resolution of racemic mixtures, or by a few laboratory controlled enzymic reactions. Many of the chemical reactions described in this book lead to products which contain chiral centres, axes, or planes, but in which the isolated material is the optically inactive (racemic) form. This is a direct consequence of the fact that the reactants, reagents, or solvents are achiral or are themselves racemic. The following selection of reactions drawn from the text illustrate this statement; they may be cross-referenced to the relevant discussion sections, namely: (*a*) Section 5.4.1, p. 519, (*b*) Section 5.4.3, p. 542, (*c*) Section 5.11.7, p. 687, (*d*) Section 8.1.3, p. 1133, (*e*) Section 5.2.4, p. 504 and (*f*) Section 5.4.2, p. 531.

(*a*) $R^1 \cdot CO \cdot R^2 \xrightarrow{[H]} (R)\text{-} + (S)\text{-}R^1 \cdot CHOH \cdot R^2$

(*b*) $R^1 \cdot CH = CH \cdot R^2 \xrightarrow[\text{[O]}]{BH_3} (R)\text{-} + (S)\text{-}R^1 \cdot CH_2 \cdot CHOH \cdot R^2$

(*c*) $R^1 \cdot CH_2 \cdot CO_2H \xrightarrow{\alpha\text{-alkylation}} (R)\text{-} + (S)\text{-}R^1R^2CH \cdot CO_2H$

(*d*) $(Z)\text{-Me} \cdot CH = CH \cdot Me \xrightarrow{R \cdot CO_3H} (R, S)\text{-} + (S, R)\text{-Me} \cdot \overset{\displaystyle O}{\overset{\diagdown}{CH}} \cdot CH \cdot Me$

(*e*) $(R)\text{-} + (S)\text{-}R^1R^2C(OH) \cdot C \equiv CH \xrightarrow{CuBr/HBr} (R)\text{-} + (S)\text{-}R^1R^2C = C = CHBr$

(*f*) $(R)\text{-} + (S)\text{-}R^1R^2CH \cdot CHO \xrightarrow{R^3MgX} (R, R), (S, S), (R, S),$
$(S, R)\text{-}R^1R^2CH \cdot CHOH \cdot R^3$

In order to effect the preferential formation of one stereoisomer (either enantiomer or diastereoisomer) over the other, either the reactant, or the reagent, or the solvent must be the pure enantiomeric form. Three principal categories may be recognised and they may be related to the illustrative examples noted above.

In the first category the reactant is achiral, but either the reagent or the solvent is optically pure; the reaction is then said to be *enantioselective* since one enantiomer is formed in preference to the other [e.g. in the reactions (*a*), (*b*) and (*d*)].

In the second category, a functional site adjacent to that at which an asymmetric reaction is to be effected is reacted with an optically pure reagent (the *chiral auxiliary* or *chiral adjuvant*) to give an optically pure modified reactant. In the subsequent reaction to form the new chiral site, two diastereoisomers would be formed in unequal proportions (the reaction is then said to be *diastereoselective*). When the chiral auxiliary is then subsequently removed, one of the enantiomers would be present in a greater proportion [e.g. (*c*), where the optically pure auxiliary reacts with the carboxyl group, and the subsequent reaction is controlled by the chirality of the auxiliary]. A further point to note is that frequently the mixture of diastereoisomers may be separated readily by one of the latest chromatographic techniques (Section 2.31), in which case removal of the auxiliary leads to the isolation of the pure enantiomers.

15

In the third category the reactant is itself optically pure, and the site of chirality is adjacent to the site of the reaction. It is the chirality of this site which influences the stereoisomeric course of the reaction (diastereoselection) which then gives rise to two diastereoisomers in unequal proportions. For example in (f), if the aldehyde was (R)-Ph(Me)CH·CHO and the Grignard reagent MeMgX, the diastereoisomeric product pair, in unequal proportions, would be (2R,3R)-3-phenylbutan-2-ol and (2S,3R)-3-phenylbutan-2-ol as illustrated below. If the (S)-aldehyde had been used the diastereoisomeric pair would be (2S,3S) and (2R,3S).

Case (e) is interesting because the chirality of the site, coupled with the stereospecific nature of the reaction (Section 5.2.4, p. 504) determines the chirality of the allene which has of course a chiral axis (i.e. the chirality of the chiral site has determined the chirality of the chiral axis).

In all these cases, provided that the reaction is under kinetic control, the proportion of predominant isomer found in the product (whether enantiomeric or diastereoisomeric) is determined by the difference in the free energies of activation of the irreversible steps leading to the alternative diastereoisomeric transition states. In each of the cases noted above, the text gives some amplification of the factors which determine these differences in free energy of activation and, it is hoped, should provide an introduction to the philosophy of asymmetric synthesis. However, a few further general comments may be helpful.

1. It is often helpful to construct models of the reactant and the reagent as a pointer to the stability of the different conformations resulting from intramolecular non-bonded interactions. In particular any internal chelation effects existing in the reagent or reactant, or which might develop during the course of the reaction need to be identified.

2. As detailed a knowledge as possible should be acquired of the mechanism of the reaction, particularly in relation to the *direction* of approach of the reagent to the reactant. For example, in an addition reaction to a carbonyl group, the most (energetically) favourable line of attack of the nucleophilic species (Nu) is at right angles to the *plane* of the σ-bond structure, and at angle of approximately 109° to the *axis* of the C—O σ-bond as illustrated below; the enantiotopic faces (i.e. the re-face or the si-face in $R^1 \cdot CO \cdot R^2$) or the diastereotopic faces (i.e. the re-face or the si-face in the carbonyl group of $R^1 R^2 \cdot CH \cdot CO \cdot R^3$) should be recognised.

In the case of a methylene group, in for example the ketone $Ph \cdot CO \cdot CH_2 \cdot Me$, the enantiotopic hydrogens [i.e. pro-(R) and pro-(S)] should be specified.

$$\text{pro-}R\ \text{H}\cdots\overset{\displaystyle\text{Ph}}{\underset{\displaystyle\text{pro-}S\ \text{H}}{\diagup}}\text{CO·Me} \longrightarrow \text{H}\cdots\overset{\displaystyle\text{Ph}}{\underset{\displaystyle\text{Me}}{\diagup}}\text{CO·Me} + \text{Me}\cdots\overset{\displaystyle\text{Ph}}{\underset{\displaystyle\text{H}}{\diagup}}\text{CO·Me}$$

$$(S) \qquad\qquad (R)$$

3. All the mechanistically acceptable alternative modes of approach of the reactant (or some reactive intermediate derived from it by a preliminary reaction) and the reagent should be considered, with a *qualitative* assessment of the interactions that would develop. In the case of steric interactions, it is convenient to remember that in the case of a reactant having a large (L) and a small (S) group, coming into close proximity to a reagent which also has a large (L') and a small (S') group, the more effective 'packing', i.e. more stable transition state, is (L to S') + (S to L') rather than (L to L') + (S to S'). (An analogy which has been used here is the instance of the packing of a mixture of large and small balls into a box; more effective packing results from alternate size layers.) Another factor worth noting is whether, in one diastereoisomeric transition state, there is a substituent group in either reactant or reagent which fits into a vacant space in the other, whereas in the alternative diastereoisomeric transition state, severe interactions are present.

4. In some cases, and this is particularly noticeable in reactions involving sites in cyclic systems, the stereoisomeric result appears not to be controlled by the steric interactions discussed under (3) (i.e. steric-approach control), but by the stability of the product (product-development control). In these cases the ratio of products enables an assessment to be made of the timing of the bond-breaking and bond-forming processes. Thus in steric-approach control, the transition state (i.e. the energy maximum in the energy profile of the reaction) is reactant-like, and bond-breaking and bond-forming processes are only in their initial stages. In product-development control the transition state is now at an advanced state of bond breaking and bond forming (i.e. the transition state is product-like).

There is now an extensive methodology of asymmetric synthesis intended to achieve the ultimate aim, so effectively realised in nature, of 100 per cent enantiomeric or diastereoisomeric excess. Indeed any new method or new reagent which does not achieve at least 50 per cent excess is now rarely reported. It might be confidently anticipated that many more elegant reactions and reagents will emerge in the coming decade, illustrating the manner in which mechanism and methodology are intimately interwoven.

1.4 REACTION MECHANISM AND THE STRATEGY OF SYNTHESIS

It has been pointed out already that the early design of synthetic routes for the vast number of simple and complex molecules was a largely intuitive operation. Indeed even the most eminent of synthetic chemists rarely recorded in the literature the thought processes which led to the realisation of the successful synthesis of a complex structure. One notable exception was that of the Robinson tropinone synthesis.[13] In this case the molecule was submitted to an 'imaginary hydrolysis' at the points indicated by the dotted line and 'resolved' into succindialdehyde, methylamine and acetone. These reagents were then mixed together under conditions which could reasonably mimic those pertaining to a bio-chemical synthesis, resulting in a low yield of product; the yield was improved to an acceptable level by using acetone dicarboxylate in place of acetone. This was probably the first example of a *target molecule* (TM) being *transformed* into synthetic precursors by a *retrosynthetic* strategy (designated by the symbol $\Rightarrow$), although not described in these terms.

The possibility of using computers in the derivation of synthetic routes was first explored by E. J. Corey[14] following his review of the general philosophy and the methods which were then currently adopted in the synthesis of both simple and complex molecules.[15] Since that time, major computer programs have been devised to assist organic chemists in analysing the strategy of synthesis and selecting the most viable synthetic route to a target molecule.

A review on recent developments classifies these programs into two main categories, the 'passive' programs and the 'active' programs.[16]

The passive programs are in effect computerised libraries, one group of which can locate all compounds that contain a specified sub-structure or a stereo-isomeric arrangement in a given database.[17] A second group relates to reaction-type retrieval programs from databases compiled from *Organic Syntheses*, from *Theilheimer's Synthetic Methods of Organic Chemistry*, or from extensive chemical literature sources.[18]

The major active programs have some similarity between themselves in that most are retrosynthetic and each breaks the target molecule down into synthetic precursor units. These in turn are similarly 'degraded', via retrosynthetic steps, to readily available starting materials. In achieving these aims two different approaches have been adopted, and useful summaries have been published.[19] A third approach has a rather different philosophy since the program operates in a synthetic rather than a retrosynthetic mode.

In the first approach[20] the programs are highly interactive with the user, and intercommunication is achieved via graphical input (stylus and magnetic tablet) and graphical display. In this way the user draws the target molecule and the computer then identifies the structural features, for example the functional groups, the presence of isolated saturated or aromatic rings, fused- and bridged-ring systems, the nature of stereoisomeric and conformational features, etc. A menu is then displayed of the three major strategies, namely: (*a*) group-oriented, (*b*) bond-oriented, or (*c*) long-range strategies, from which a selection is made by the user.

(*a*) The group-oriented strategy involves functional group interconversion (FGI), functional group addition (FGA), functional group removal (FGR), and the unmasking of (latent) functional groups by deprotection or other conversions.

(*b*) The bond-oriented strategy centres attention on the disconnection of bonds in ring systems (bridged or fused), and bonds joining ring atoms to functional groups or other residues.

(c) The long-range strategies encompass those retrosynthetic reactions which result in significant simplification of structure, for example, the Robinson annelation reaction, the Birch reduction, the Diels–Alder reaction, etc.

The selection from this menu is satisfying to the user as it uses intuitive chemical knowledge, but with the option of rejection should the first level of simplification in the retrosynthetic sequence suggested prove to be unpromising. Having selected the first level precursors, each precursor is then treated as a 'target molecule' and processed appropriately. The program then evaluates the overall route selected and recommends reagents, reaction conditions, and suggests possible reaction mechanisms.

The second approach[21] is different in that the programs are not interactive, and are based more upon theoretical considerations. Here the skeletal structure of the target molecule is cleaved into the smallest number of skeletal fragments which can then be related to the c. 5000 readily available starting materials. The program then generates functionality in the skeletal fragments which can lead to consecutive connective constructions from such starting materials based on broad mechanistic principles. Although the programs are capable of suggesting new chemistry and previously unrealised synthetic routes, they do tend to provide a great number of potential precursors and being non-interactive, they require the use of self-limiting guides to ensure chemical viability of the selected synthetic routes.

In the third approach,[22] the programs operate in the synthetic, rather than the retrosynthetic, mode from a database of reaction mechanisms. As such they usefully complement the programs of the previous two groups, offering an answer to the hypothetical question 'if compound Y is subjected to the reaction conditions Z would X, or some other alternative structure, be formed?'.

All of the major chemical companies involved in organic chemical synthesis have access to these programs, and many academic organic research groups utilise these facilities. For undergraduates, the value of the use of computers in organic synthesis is that, for the first time, it is possible to acquire experience in the logic of organic synthetic design. Thus the principles of a retrosynthetic analysis of a target molecule, and the selection (based on sound mechanistic ideas) of the most appropriate synthetic route generated, may be applied without the use of a computer. Hence when the graduate becomes exposed to the computer synthetic programs, familiarity with the concept of logic in synthesis, coupled with a better understanding of the factors which are important in deciding between alternative routes, will enable a transition into new and unfamiliar synthetic problems to be possible.

The principal exponent of the non-computerised approach to retrosynthetic analysis in organic synthesis, and which finds its origin in the early work of E. J. Corey, is S. Warren.[5c, d] A useful shorter account is to be found in the writings of J. Fuhrhop and G. Penzlin.[5e] The non-computerised retrosynthetic analysis has also been termed the *synthon* approach, a term which was first introduced and defined by E. J. Corey.[15]

A synthon may be defined as a structural unit which becomes an idealised fragment as a result of disconnection of a carbon–carbon or carbon–heteroatom bond in a retrosynthetic step (transform). Thus it may be envisaged in general terms, that an open-chain structure in a single disconnection step would give rise to two synthons; two synthons would also arise in a similar disconnection of a

19

bond joining a group to a cyclic structure. The disconnection of a bond within a monocyclic system would be a retrosynthetic ring-opening process; the disconnection of a bond in a bridged structure would give rise to a mono- or disubstituted monocyclic structure. Simultaneous two-bond disconnections may also be possible.

Synthons resulting from single bond disconnections may be ions (cationic or anionic) or radicals according to whether the bond cleavage is heterolytic or homolytic. Usually they are not in themselves reagents, but have to be related to suitable reactants which under appropriate conditions will interact to effect the reverse, synthetic, step. Synthons which are neutral molecules can result from two simultaneous single bond disconnections occurring in a pericyclic manner. The examples which follow are a few illustrative carbon–carbon and carbon–heteroatom disconnections which produce either charged or radical species, or neutral molecules. For a more extensive range of examples the reader's attention is directed to the summaries of retrosynthetic strategies included in the introduction to most of the aliphatic and aromatic functional compounds.

C—C Disconnections

(i)

(ii)

(iii)

(iv)

C—X Disconnections

(i)

(ii)

20

$$\text{(iii)}$$

$$\text{(iv)}$$

The reagents corresponding to the charged species may be deduced by consulting Tables A6.1 and A6.2 in Appendix 6. These tables give lists of commonly encountered synthons generated by heterolytic fission and their most important *reagent equivalents*, together with a cross-reference to those sections of the text which discuss the process in a little more detail.

It should be noted that the nature of the charge on the various synthons shown above corresponds to what would have been expected on the basis of electronegativity, or of inductive or mesomeric stabilisation effects. Not infrequently, however, a disconnection generates a synthon in which the polarity is not what would have been expected on the above grounds; some examples are $R^{\ominus}$, $R\cdot\overset{\ominus}{C}{=}\overset{\oplus}{O}$, $\overset{\oplus}{C}H_2\cdot CO_2R$. These synthons have 'unnatural', or 'reversed' polarity (originally called 'umpolung').[23] However, they are perfectly valid though their reagent equivalents are sometimes not immediately obvious. Some illustrative examples are noted below.

1. The alkyl anion ($R^{\ominus}$) is the simplest example, and it has long been known as the reactive species generated from a Grignard reagent ($\overset{\delta-}{R}{-}\overset{\delta+}{MgX}$), and other related organometallic compounds (see Section 5.4.2, p. 531).

2. The acyl anion ($R\cdot\overset{\ominus}{C}{=}O$) is not stable as such, but when an aldehyde is converted into a 1,3-dithiane by reaction with propane-1,3-dithiol and then treated with base, it forms an *acyl anion equivalent*, and hence is susceptible to attack by electrophilic reagents (see Section 5.9). Two extensive compilations of formyl and acyl anion synthons together with references to their reactions with electrophiles have been published.[24]

3. The cation ($\overset{\oplus}{C}H_2\cdot CO\cdot R$) would arise from an α-halo ketone owing to the good leaving properties of the halogen. In a similar way the α-halo ester, $BrCH_2\cdot CO_2Et$, is a reagent equivalent for the cation, $\overset{\oplus}{C}H_2\cdot CO_2Et$, but in the presence of zinc dust it forms the organometallic reagent, $BrZnCH_2\cdot CO_2Et$, which is the reagent equivalent for the anion, $\overset{\ominus}{C}H_2\cdot CO_2Et$ (see Section 5.14.2, p. 727).

4. The carbon–carbon double bond is attacked by electrophilic reagents; however after conversion to the epoxide the carbons are susceptible to nucleophilic attack, i.e. the donor qualities of the carbons are changed to acceptor qualities (Section 5.4.5, p. 547 and Section 5.18.1, p. 795).

Having considered the broad principles of bond disconnection, and the charged nature of the derived synthons, attention must now be turned to a consideration of the selection of the bond to be cleaved. The forward references noted below enable appropriate examples to be found in the text.

If the target molecule is monofunctional, the disconnection process is classified as a *one-group disconnection*. The bond initially considered for cleavage would be, if present, the α-carbon–heteroatom single bond (i.e. the C—O, C—X, C—N, C—S bonds) as would be found in, for example, alcohols (Section 5.4), alkyl halides (Section 5.5), ethers (Section 5.6), nitroalkanes (Section 5.15), amines (Section 5.16), thiols and thioethers (Section 5.17).

If the functional group is a carbon species [i.e. —C≡C—, —C≡N, —CHO, —CO·R, —CO₂H(R)], then a possible disconnection point would be the bond uniting the α-carbon to the functional group carbon, as is found with alkynes (Section 5.3), aldehydes (Section 5.7), ketones (Section 5.8), carboxylic acids (Section 5.11) or their derived esters (Section 5.12.3, p. 695). Alternative disconnection points which would be worth considering are the α, β- and β, γ-carbon bonds, in for example aldehydes or ketones.

If the target molecule is an acyclic or an alicyclic bifunctional compound in which the functional groups are in a 1,2-, 1,3-, 1,4-, 1,5-, etc., relationship, then the disconnection is termed a *two-group disconnection* and specifically refers to a cleavage of any of the carbon–carbon single bonds which lie between the two functional groups. The examples given in dicarbonyl compounds (Section 5.9), dicarboxylic acids (Section 5.11), keto acids and esters (Section 5.14.3, p. 735), some of the alicyclic (Ch. 7) and heterocyclic (Ch. 8) compounds, all illustrate the value of retrosynthetic analysis, and the derivation of reagent equivalents. The problem of protection of functional groups in either or both of the synthon entities may well need to be considered in the strategy of synthesis of polyfunctional compounds.

If the target molecule is an aromatic or heteroaromatic compound, in which the ring system is found in readily available starting materials (e.g. benzene, pyridine, etc.), disconnection at the bond uniting the ring with the substituent is the most obvious point of cleavage [see for example aromatic nitro compounds (Section 6.2), or aromatic aldehydes (Section 6.10)]. Here of course, with polyfunctional compounds, it would be the *order* in which the substituents were introduced that would be crucial to the synthetic strategy, owing to the directive effects of groups (i.e. the regioselectivity of the individual synthetic steps).

Target molecules which contain a carbon–carbon double bond, in the presence of other functionality or not, may be treated as in a one- or two-group disconnection strategy, as noted above. However, disconnection *at* the double bond may be a valuable retrosynthetic transform, since the synthons may then be related to reagent equivalents for a Wittig-type (Section 5.2.3, p. 495 and Section 5.18) or aldol-type (Section 5.18) synthesis.

It should be pointed out that S. Warren[5c,d] has introduced a 'nomenclature' system for one- and two-group disconnections and for fission at a carbon–carbon double bond. These are valuable when communicating the essence of a retrosynthetic strategy in the analysis of complex target structures.

Pericyclic disconnections and sigmatropic and other *rearrangements* give rise to synthons which are themselves reagents. Such disconnections greatly simplify the target molecule (e.g. the *retro*-Diels–Alder reaction). These disconnections are most commonly applied in alicyclic and heterocyclic systems.

Finally, the target molecule should be inspected for the possibility of effecting a *reconnection* transform. A common example is where a target molecule has two carbonyl groups (aldehydic or ketonic) sited in a 1,5- (or higher) relationship. In such a case the retrosynthetic step is the formation of a carbon–carbon double bond between the carbonyl carbons; the synthetic step would be the cleavage of the double bond by ozonolysis or related methods (e.g. Section 5.7.3, p. 593).

For the undergraduate student, an introduction to the synthon approach is most conveniently explored with acyclic, simple alicyclic and heterocyclic, and aromatic compounds. In each of these groups the presence of polyfunctionality provides increasing experience in the recognition of the way in which a target molecule may be transformed into synthons. Such explorations provide greater knowledge of functional group manipulation, a realisation of the methods of protection which may be necessary, the ordering of operations in the forward synthetic direction, the problems associated with chemoselectivity, regioselectivity and stereoselectivity, and the mechanistic principles upon which the integration of the methodology and strategy of synthesis depends.

1.5 CONCLUSION

This chapter commenced with a review of the numerous theoretical and practical aspects with which the prospective organic chemist would need to become familiar. The major preparative sections of the book, in amplifying the topics which have been covered briefly in this chapter, will serve to stimulate further thought and provide useful, reliable, and interesting syntheses in which experience in preparative techniques, purification procedures, and spectroscopic characterisation may be acquired. The question which was posed at an early stage in this chapter was 'how may compound X be synthesised?'. It is hoped that undergraduates will apply this question to most of the compounds that they encounter, for the solution of these separate riddles, which with many compounds may be checked in the literature, should provide confidence in the approach to the more 'difficult' syntheses encountered in later career.

REFERENCES

1. *Strategies and Tactics in Organic Synthesis* (1984). Ed. T. Lindberg. Orlando: Academic Press.
2. (a) F. A. Carey and R. J. Sundberg. Pt A (1984) *Structure and Mechanism*, 2nd edn; Pt B (1977) *Reactions and Synthesis*. New York: Plenum.
   (b) R. T. Morrison and R. N. Boyd (1987). *Organic Chemistry*, 5th edn. Boston: Allyn and Bacon; T. W. Solomons (1986). *Fundamentals of Organic Chemistry*, 2nd edn. New York: Wiley; A. Streitwiesser, Jr, and C. Heathrock (1985). *Introduction to Organic Chemistry*, 3rd edn. New York: Macmillan.
   (c) J. March (1985). *Advanced Organic Chemistry*, 3rd edn. New York: Wiley-Interscience.
3. (a) P. Sykes (1986). *Guidebook to Mechanism in Organic Chemistry*, 6th edn. Harlow: Longman.
   (b) P. Sykes (1972). *The Search for Organic Reaction Pathways*. Harlow: Longman.
   (c) T. L. Gilchrist and R. C. Storr (1972). *Orbital Reactions and Orbital Symmetry*. London: Cambridge University Press.
   (d) I. Fleming (1976). *Frontier Orbitals and Organic Chemical Reactions*. London: Wiley.

4. (a) E. L. Eliel (1962). *Stereochemistry of Carbon Compounds*. New York: McGraw-Hill.

   (b) J. D. Morrison and H. S. Mosher (1971). *Asymmetric Organic Reactions*. American Chemical Society publication.

   (c) *Topics in Stereochemistry*. Eds E. L. Eliel, S. W. Witen and N. L. Allinger. Interscience, 1967–87, Vols 1–17.

5. (a) S. Turner (1976). *The Design of Organic Synthesis*. Amsterdam: Elsevier.

   (b) R. K. Mackie and D. M. Smith (1982). *Guidebook to Organic Synthesis*. Harlow: Longman.

   (c) S. Warren (1982). *Organic Synthesis: The Disconnection Approach*. Chichester: Wiley.

   (d) S. Warren (1982). *Workbook for Organic Synthesis*. Chichester: Wiley.

   (e) J. Fuhrhop and G. Penzlin (1984). *Organic Synthesis*. Weinheim: Verlag Chemie.

   (f) H. O. House (1972). *Modern Synthetic Reactions*, 2nd edn. California: Benjamin.

   (g) T. W. Greene (1981). *Protective Groups in Organic Synthesis*. New York: Wiley.

   (h) W. Carruthers (1986). *Some Modern Methods of Organic Synthesis*, 3rd edn. Cambridge University Press.

   (i) C. A. Buehler and D. E. Pearson. *Survey of Organic Synthesis*, Vol. 1 (1970) and Vol. 2 (1977). New York: Wiley-Interscience.

6. M. A. Rosanoff (1906). *J. Am. Chem. Soc.*, **28**, 114.

7. J. M. Bijvoet, A. F. Peerdeman and A. J. van Bommel (1951). *Nature*, **168**, 271.

8. R. S. Cahn and C. K. Ingold (1951). *J. Chem. Soc.*, 612; R. S. Cahn, C. K. Ingold and V. Prelog (1956). *Experientia*, **12**, 81.

9. K. R. Hanson (1966). *J. Am. Chem. Soc.*, **88**, 2731; H. Hirschmann and K. R. Hanson (1971). *J. Org. Chem.*, **36**, 3293; IUPAC (1970). 'Tentative rules for the nomenclature of organic chemistry; Section E Fundamental stereochemistry'. *J. Org. Chem.*, **35**, 2849; J. F. Stoddart (1973). 'Stereochemistry', in *MPT International Review of Science*. Ed. W. D. Ollis. London: Butterworths, Vol. 1; J. F. Stoddart (1979). 'Stereochemistry', in *Comprehensive Organic Chemistry*. Eds D. H. R. Barton and W. D. Ollis. Oxford: Pergamon, Vol. 1.

10. (a) D. H. R. Barton (1950). *Experientia*, **6**, 316.

    (b) E. L. Eliel, N. L. Allinger, S. J. Angyal and G. A. Morrison (1966). *Conformational Analysis*. New York: Interscience-Wiley.

    (c) M. Hanack (1965). 'Conformational Theory', in *Organic Chemistry: a series of monographs*. Ed. A. T. Blomquist. New York: Academic Press, Vol. 3.

11. The Nobel Symposium 60, Sweden. *Asymmetric Organic Synthesis* (1984). Ed. S. Gronowitz. New York: Cambridge University Press.

12. M. Nogradi (1986). *Stereoselective Syntheses*. Weinheim: VCH.

13. R. Robinson (1917). *J. Chem. Soc.*, **111**, 762.

14. E. J. Corey and W. T. Wipke (1969). *Science*, **166**, 178; E. J. Corey (1971). *Quarterly Reviews*, **25**, 555.

15. E. J. Corey (1967). *Pure Appl. Chem.*, **14**, 19.

16. N. J. Hrib (1986). *Ann. Rep. in Med. Chem.*, **21**, 303.

17. CAS Online, available from Chemical Abstract Service, Columbus, Ohio. MACCS and MACCS-3D, available from Molecular Design Ltd, Hayward, California.

18. RECCAS, available from Molecular Design Ltd, Hayward, California. P. A. Zurer (1985). *Chem. Eng. News*, **63**, 21; SYNLIB will be available from Distributed Graphics Ltd, J. Haggin (1983). *Chem. Eng. News*, **61**, 7.

19. See Reference 16; A. K. Long, S. D. Rubenstein and L. J. Joncas (1983). *Chem. Eng. News*, **61**, 22.

20. OCCS (Organic Chemical Synthesis Simulation) but now replaced by LHASA (Logistics and Heuristics Applied to Synthetic Analysis), see reference 14; E. J. Corey, A. K. Long, T. W. Greene and J. W. Miller (1985). *J. Org. Chem.*, **50**, 1920; E. J. Corey, A. K. Long, S. D. Rubenstein (1985), *Science*, **228**, 408. For a student approach see R. D. Stolow and L. J. Joncas (1980). *J. Chem. Ed.*, **57**, 868; SECS (Simulation and

Evaluation in Chemical Synthesis) and CASP, P. Gund, E. J. J. Grabowski, D. R. Hoff, G. M. Smith, J. D. Andose, J. B. Rhodes and W. T. Wipke (1980). *J. Chem. Info. Comput. Sci.*, **20**, 88; W. Sieber, Abstracts of 191st ACS National Meeting, New York, 14 April 1986, ORGN 88.

21. IGOR (Interactive Generation of Organic Reactions), J. Bauer, R. Herges, E. Fountain and I. Ugi (1985). *Chimia*, **39**, 43; SYNGEN (Synthesis Generation), J. B. Hendrickson, E. Braun-Keller and G. A. Toczko (1981). *Tetrahedron*, **37**, Suppl. 1, 359; J. B. Hendrickson, D. L. Grier and A. G. Toczko (1985). *J. Am. Chem. Soc.*, **107**, 5228; SYNCHEM-2, H. L. Gelernter, A. F. Saunders, D. L. Larsen, K. K. Agarwal, R. H. Boivie, G. A. Spritzer and J. E. Searleman (1977). *Science*, **197**, 1041; H. L. Gelernter, S. S. Bhagwat, D. L. Larsen and G. A. Miller (1983). *Anal. Chem. Symp. Ser.*, **15**, 35.

22. CAMEO (Computer-Assisted Mechanistic Evaluation of Organic Reactions), T. D. Salatin and W. L. Jorgensen (1980). *J. Org. Chem*, **45**, 2043.

23. D. Seebach (1979). *Agnew. Chem., Int. Edn Engl.*, **18**, 239; O. W. Lever Jr (1976). *Tetrahedron*, **32**, 1943.

24. T. A. Hase and J. K. Koskimies (1981). *Aldrichemica Acta*, **14**, 73; T. A. Hase and J. K. Koskimies (1982). Ibid., **15**, 35.

CHAPTER 2
EXPERIMENTAL TECHNIQUES

2.1 GENERAL INSTRUCTIONS FOR SAFE WORKING IN ORGANIC CHEMICAL LABORATORIES

INTRODUCTION

Chemistry laboratories need not be dangerous places in which to work, despite the many potential hazards associated with them, provided that certain elementary precautions are taken and that all workers conduct themselves with common sense and alertness.

There will almost invariably be a senior person assigned to be in charge of a chemical laboratory, irrespective of the nature of the work to be done there. However, it must be emphasised that the exercise of care and the adoption of safe working procedures is the responsibility of each and every person in that laboratory. If there is any doubt as to the safety of a proposed experiment, advice should be sought from an experienced person rather than just hoping for the best.

All workers must adopt a responsible attitude to their work and avoid any thoughtless, ignorant or hurried behaviour which may lead to an accident and possible harm to themselves or to others. They should always pay attention to what is going on around them and be aware of the possible dangers arising from the work of others as well as from their own experiments.

Laboratory accidents are often caused by attempts to obtain results in too great a hurry. Laboratory workers must therefore adopt a thoughtful, careful and methodical approach to what they are doing. In particular, concentration on the job in hand and not allowing the attention to be distracted must be particularly commended. Similarly the unnecessary distraction of others is to be deplored. In routine experiments and operations, it is important to remember the truth of the old adage 'familiarity breeds contempt' and to be on one's guard against the feeling that 'it cannot happen to me'.

PERSONAL PROTECTION

No worker should be allowed in a laboratory without a full-length protective coat, preferably white, since spillages and stains are then more readily detected. *Furthermore, all personnel, including visitors, must wear safety spectacles or goggles at all times.* There are several good lightweight goggles and spectacles available from laboratory suppliers which provide, in routine use, good coverage of the eyes and upper face; many designs are suitable for use over prescription glasses. Prescription safety glasses are of course readily available through specialised sources, and although more expensive, would be appropriate to the full-time laboratory worker. Although contact lenses give some protection against mechanical damage to the eye, the wearing of protective spectacles is still essential. It should be noted however that in the event of an accident, these conventional safety spectacles provide varying degrees of protection against flying fragments, but often very little protection against the splashing or spraying of hot, corrosive or toxic liquids and gases. In the event of chemical splashes, first-aid should be confined to thorough irrigation of the eyes with clean water. Contact lenses may restrict effective irrigation and their removal should be undertaken only by qualified medical staff. Close-fitting safety goggles or, preferably, a visor covering the whole face afford a much greater measure of protection in these circumstances. Whenever experiments or operations that are suspected as being hazardous are being carried out, the additional protection of safety screens is strongly advised. In any case all experiments involving toxic reagents or solvents must be conducted in an efficient fume cupboard, and disposable plastic gloves should be carefully inspected to ensure that they are free from pinholes.

CONDUCT IN THE LABORATORY
Except in an emergency, running, or any over-hurried activity, should be forbidden in and around the laboratories, as should be practical jokes or other irresponsible behaviour. Eating, drinking and smoking in the laboratory should also be forbidden; these constitute a further, avoidable, risk of the ingestion of toxic substances, and in the case of smoking an obvious fire hazard.

TIDINESS AND CLEANLINESS
Coupled with a general consideration for the safety of others, tidiness is a major factor in laboratory safety; the laboratory must be kept clean and tidy at all times. Passageways between and around the benches and near exits must not be blocked with equipment or furniture. Floors must be kept in good condition to prevent slipping or tripping, i.e. they must be kept free from oil or water, and from any protrusion. Any spillage on a floor or bench should be cleaned up immediately.

Some indication of a chemist's practical ability is apparent from the appearance of the working bench. This should always be kept clean and dry; this is easily done if suitable wet and dry rags are kept at hand. Apparatus not immediately required should be kept as far as possible in a cupboard beneath the bench; if it must be placed on the bench, it should be arranged in a neat and orderly manner. Dirty apparatus can be placed in a plastic bowl away from the working area until it can be cleaned and put away. *Solid waste and filter papers must not be thrown in the sink.*

All glassware should be scrupulously clean and, for most purposes, dry before being employed in preparative work in the laboratory. The use of rigorously dried apparatus for reactions involving highly moisture-sensitive compounds is described in Section 2.17.8, p. 126. It is advisable to develop the habit of cleaning all glass apparatus immediately after use as the nature of the contaminant will, in general, be known. Furthermore, the cleaning process becomes more difficult if the dirty apparatus is allowed to stand for a considerable period, particularly if volatile solvents have evaporated in the meantime.

It must be emphasised that there is no universal cleaning mixture. The chemist must take into account the nature of the substance to be removed and the amount of deposit, and act accordingly. Thus if the residue in the flask is known to be basic in character, dilute hydrochloric or sulphuric acid may dissolve it completely; similarly, dilute sodium hydroxide solution may be employed for acidic residues. In these instances the acidic or basic aqueous solution may be washed down the drain with liberal quantities of water. If the residue is known to dissolve in an inexpensive organic solvent, this should be employed; in such a case the solution must be poured into the appropriate 'residues' bottle, *not down the sink*. Solvent suitable specifically for cleaning use may be recovered by distilling accumulated residues. Remaining heavily contaminated material should be suitably disposed of (Section 2.1, p. 30).

The simplest method for gross deposits, when access by a test-tube brush is possible, is to employ a commercial household washing powder containing an abrasive which does not scratch glass (e.g. 'Vim', 'Ajax', etc.). The washing powder is either introduced directly into the apparatus and moistened with a little water or else it may be applied to the dirty surface with a test-tube brush which has been dipped into the powder; the glass surface is then scrubbed until the dirt has been removed. The operation should be repeated if necessary. Finally, the apparatus is thoroughly rinsed with distilled water. If scrubbing with the water–washing powder mixture is not entirely satisfactory, the powder may be moistened with an organic solvent, such as acetone.

Three alternative cleaning solutions are worth a trial if the above methods fail:

1. A warm 15 per cent trisodium phosphate solution to which a little abrasive powder, such as pumice, has been added; this reagent is not suitable for the removal of tarry residues.
2. The highly effective surface active agent, Decon 90, which is claimed to be suitable for virtually all laboratory cleaning applications. It is totally rinsable, phosphate free, biodegradable and non-toxic. It is particularly suitable for silicone oils, greases, polymeric residues and tars.
3. The 'chromic acid' cleaning mixture. This is essentially a mixture of chromium trioxide (for precautions in its use see Section 2.3.2, p. 37) and concentrated sulphuric acid, and possesses powerful oxidising and solvent properties. A convenient method of preparation is as follows. Five grams of sodium dichromate are dissolved in 5 ml of water in a 250 ml beaker; 100 ml of concentrated sulphuric acid are then added slowly with constant stirring. The temperature will rise to 70–80 °C. The mixture is allowed to cool to about 40 °C and then transferred to a dry, glass-stoppered, clearly labelled, bottle. Before using this mixture for cleaning purposes, the vessel to be cleaned should be rinsed with water to remove water-soluble organic matter, and particularly reducing agents, as far as possible. After draining away as much of the water as is practicable, a quantity of the cleaning mixture is introduced into the vessel, the soiled surface thoroughly wetted with the mixture, and the main quantity of the cleaning mixture returned to the stock bottle. After standing for a short time with occasional rotation of the vessel to spread the liquid over the surface, the vessel is thoroughly rinsed successively with tap and distilled water.

The use of any of the above methods, as detailed, may be regarded as safe and satisfactory for the removal of gross residues; *no other chemical treatment should be attempted owing to the possibility of explosion hazard.* The final removal of trace residues is most effectively accomplished by the ultrasonic bath. The tanks for laboratory use have capacities from 2.7 to 85 litres; the tank fluid is usually Decon 90. Vessels containing gross residues should not be cleaned in these baths as the fluid will rapidly become greatly contaminated and its efficiency will be diminished. A further advantage of these baths is their ability to free ground-glass joints when these have become 'fused' by degraded chemical contaminants.

After cleaning and rinsing with distilled water, small glass apparatus may be dried by leaving it in an electrically heated oven maintained at 100–120 °C for about 1 hour. However, much organic apparatus is too bulky for oven drying and, moreover, is generally required soon after washing; other methods of drying are therefore used. If the apparatus is wet with water, the latter is drained as completely as possible, then rinsed with a little industrial spirit or acetone. For reasons of economy, the wet industrial spirit or acetone should be collected in suitably labelled Winchester bottles for future recovery by distillation and re-use. After rinsing with the organic solvent, the subsequent drying is more conveniently done by means of a warm air blower (available, for example, from Gallenkamp). It consists of a power driven blower which draws air through a filter, passes it through a heater, and forces it through upward pointing tubes which support the apparatus, and which are specially constructed to accommodate flasks and cylinders having narrow necks which make other means of drying difficult; each apparatus support tube has a number of holes at its end to ensure good distribution of heated air. Cold air may be circulated if required.

ACCIDENT PROCEDURE
Every person working in a laboratory should ensure that he or she knows where the exits and fire escapes are situated and that there is free access to them. All workers should also note the positions of fire extinguishers, fire blankets and drench showers, and make sure that they know how to use them. The checking of such equipment should be carried out by the proper authorities at regular intervals. The worker should also be familiar with the location of the first-aid equipment provided for emergency use, and the position of the nearest telephone; the numbers of the appropriate medical teams, hospitals and fire brigades should be displayed in a prominent position. In addition to the above general common-sense approach to accidents, individual industrial, research, and teaching establishments publish accident procedure regulations to meet their own specific requirements; every worker must be fully conversant with such regulations.

AFTER-HOURS WORKING
No person should ever work alone in a laboratory. Experiments which must be left running overnight are best sited in a specially constructed overnight room (see Section 2.16), but if this is not possible, the precautions recommended in Section 2.16 should be adopted and the apparatus labelled clearly as to the nature of the reaction and the likely hazards. Clear instructions must be left so that an unqualified person can terminate the experiment in an emergency. 'Please leave on' notices should be left alongside any service which is to be left running (water, electricity). Here again individual establishments publish specific regulations relating to after-hours working and overnight experiments, and these must be strictly adhered to.

STORAGE OF CHEMICALS IN LABORATORY AREAS

Chemicals should never be allowed to accumulate on benches or in fume cupboards, but should always be returned to their proper places on robust storage shelves; incompatible chemicals should be separated from one another. Heavy containers and bottles of dangerous chemicals should be returned as soon as possible to the main chemical store where specific regulations for safe storage apply. Strict regulations also apply to the quantity of solvents that may be stored in a specific laboratory; furthermore such solvents must be stored in a fire-proof steel cabinet fitted with a vapour-seal door, and an area must be assigned and properly equipped for the safe dispensing of flammable and toxic solvents. Any Winchester bottle, whether containing dangerous or innocuous chemicals, that needs to be carried a short distance, should be supported at the base *and* at the neck – *never* at only one of these points. For longer distances the special carriers which are available should be used.

Fume cupboards must be kept free from surplus chemicals and discarded apparatus. If stocks of noxious chemicals which are in frequent use have to be stored in a fume cupboard they should all be assigned to one which is set aside for this purpose and is properly fitted with shelving. Chemicals which are carcinogenic, but the use of which is permitted, are always retained in a main store in specially provided sealed cabinets; definitive authorisation is required for their use.

All containers (bottles, ampoules, vials, etc.) of chemicals purchased commercially bear a clear label indicating the nature of the contents, and a hazard symbol, together with risk and safety summaries specifying the possible danger associated with the contents. Appendix 5 gives a list of the more important regulatory European Economic Community (EEC) hazard symbols. Charts summarising the meaning of these symbols should be placed prominently in every laboratory. When chemicals are transferred to another container the same hazard code must be placed on the new container. In the case of vessels which have lost their labels, the contents should be positively identified and the container relabelled; should there be any doubt, the material should be disposed of safely. Since gummed labels readily dry out and drop off, it is a good idea to seal them to the vessel with transparent adhesive tape. Since many chemicals deteriorate with age, it is also a good idea to write the date on the label when the bottle arrives from the distributor.

GLASSWARE

Glass apparatus should be carefully examined before use and any which is cracked, chipped, flawed or dirty should be rejected. Minute cracks in glassware for use in evacuated systems are particularly dangerous.

Many apparently simple manipulations such as the cutting of glass tubing or rod, the insertion of glass tubing or thermometers into rubber bungs or bark corks, or the removal of tight stoppers from bottles, can lead to serious cuts. Care should be taken to adopt the correct procedures (Sections 2.9 and 2.10). All apparatus and clean glassware not in use should be stored away and not allowed to accumulate on benches.

WASTE DISPOSAL

This is one of the most important but difficult aspects of laboratory use and management. Waste material must *never* be allowed to accumulate in the laboratory; it should be removed regularly from the laboratory area for storage in suitable containers so that it can be disposed of appropriately. There should be separate bins with properly fitting lids for broken glassware and for flammable materials such as paper or cloths which may have been used to mop up flammable liquids. Innocuous waste solids should be placed in bins provided, toxic solids should be sealed in a plastic bag and placed in a separate bin; both bins should be clearly labelled. Waste solvents should be placed in suitable containers and appropriately labelled, but indiscriminate mixing of solvents *must* be avoided. Halogenated solvents in particular should be kept apart from other solvents.

Most large laboratory complexes will have arrangements by which the accumulated waste material is disposed of appropriately; they may for example have facilities for the combustion of quantities of flammable organic material. Smaller establishments however may have to rely on the services of specialised contractors. The problems associated with the disposal of small quantities of toxic or hazardous unwanted chemicals can be lessened by the individual laboratory worker taking intelligent action. Guidelines for suitable disposal methods have been comprehensively dealt with in, for example, the *Aldrich Catalogue of Fine Chemicals*. Local regulations relating to the disposal of chemicals down the main drains are stringent: *under no circumstances* should untreated wastes and water-insoluble organic solvents be thrown down the sink.

2.2 PLANNING OF EXPERIMENTS AND RECORDING OF RESULTS

LEVELS OF RESPONSIBILITY IN EXPERIMENT PLANNING

Before commencing work in the laboratory, sound planning of the experiment to be undertaken is essential. The detail of such planning, and the level of personal responsibility involved by the laboratory worker, depends on the degree of expertise that has been previously acquired. Three levels of experience may be recognised for the user of this book, although it should be realised that the transition between them is a continuous process.

Junior technical staff and students in sub-degree and first-year undergraduate courses need considerable guidance, for their own safety, and for that of other users of the laboratory. Experiments are selected in a sequence by supervisory staff to ensure increasing experimental competence and confidence. Glassware and other apparatus for the specific experiment are usually provided from a central source, the authenticity of reagents and solvents is checked by technical staff, hazards (which should be minimal at this stage) are clearly delineated, and the instruction sheets or experimental details by means of a book page reference should be clear, detailed and unambiguous. Frequently a short list of questions relating to the experiment is provided to reinforce the understanding of the practical and theoretical aspects of the experiment. In these circumstances experiment planning requires that the worker should carefully study the complete details and advice provided, and form a clear idea of what is to be done and precisely how it is proposed to do it.

At a second stage the student or technician accepts more responsibility in the experiment planning. Thus although instruction sheets (or book references) are suitably detailed, a level of technical competence in routine operations acquired from earlier work is now expected. Furthermore, the assembly of apparatus, the

31

verification of starting material authenticity [by spectroscopic or chromato-graphic methods, or by physical means (m.p., $[\alpha]_{\lambda}^{t}$, etc.)], the assessment of hazards from the instruction sheets and from other sources, the safe disposal of reaction residues, the decision as to which techniques are required for the verifi-cation of product purity, and finally the timing of the operations within the period(s) available (including if necessary the planning of overnight operations), all require decision by the worker. Usually senior staff, in discussion with the junior worker, assess the soundness of the proposed plan before it is put into operation. The student should regard the verification of starting material auth-enticity as most valuable experience; time and effort will not then be wasted in the not unknown event of mistaken identity, and also it provides an opportunity to acquire a personal 'library' of spectroscopic, chromatographic and other reference data.

The third (but not of course the last) stage of increasing expertise, usually re-quires the planning of experiments which involve the use of advanced techniques or equipment, and the planning of 'open-ended' experiments and undergraduate projects leading to postgraduate-type work. In addition to the responsibilities outlined at the two earlier stages, it now becomes necessary for the worker to check the availability and cost of reagents and solvents, to relate this knowledge to alternative synthetic procedures, to ensure that adequate advice is available for the operation of advanced equipment and techniques, and to accept respons-ibility for the assessment of hazards to themselves, to other laboratory users, and particularly to junior personal assistants.

All major chemical companies, and most teaching establishments, now have a computerised stock-control system which enables information about 'on-site' availability and cost of chemicals to be rapidly obtained. Many companies are now linked to a central database (e.g. 'Chemquest', Pergabase Inc.), which pro-vides information (via text or graphics) on the world-wide commercial availabil-ity (from over fifty chemical catalogues) and on relative costs, etc., of fine chemicals. The hazards in use and disposal of all chemicals (see Section 2.3) may be readily checked from the catalogues of major chemical suppliers and other important sources.

RECORDING OF RESULTS
This is an important part of any chemical experiment, since careful observation, allied to accurate reporting, is the very essence of any scientific exercise.

The guiding principle in writing up an experiment is to record all the details which would enable another person to understand what was done and to repeat the entire experiment exactly without prior knowledge. Thus, in addition to a written account of the work done, including notes on any special apparatus used, details of all volumes, weights, temperatures, times, chromatographic pro-cedures (e.g. t.l.c., g.l.c.) and conditions and results, etc., must all be recorded. The writing up of all laboratory work must be done at the time of the work, in a stiff-covered notebook of adequate size; a loose-leaved notebook is *not* suitable. It is important that numerical results such as yields, titration volumes, melting points and boiling points, etc., are entered directly into the notebook and not on scraps of paper. The latter are liable to be lost and their use encourages untidy practical habits.

The recommended format is to use a fresh double-sided page for each new ex-periment or part of an experiment. The right-hand page should be used for a descriptive account of what was done and what was observed at the time – the page must be dated. This account may be continued, if necesary, overleaf on the next right-hand (dated) page. The left-hand page should be reserved for equations, calculations of yields, melting points, reaction mechanisms, etc., and possibly a later commentary. A properly written experimental account should be generously spaced out so that the different sections are discernible at a glance. It is not a good idea to stick spectra and g.l.c. traces into the notebook, since it will rapidly become very bulky and the binding will be damaged. Rather they should be kept in a separate folder and cross-referenced with the numbered and dated pages of the notebook.

During a reaction, all unexpected happenings and anything not understood should be carefully recorded at the time. Experiments sometimes go wrong, even with well-known procedures. (Note that there are often mistakes in the instructions published in the chemical literature, e.g. decimal points in the wrong places, leading to the use of incorrect weights or volumes.) In such a case, the worker should always try to unravel the reason for failure of the reaction and to make suitable changes in the procedure, rather than just hurrying to repeat the experiment without modification. Indeed a properly conducted 'post-mortem' of an experimental result, leading, for example, to explanations of failure, planning to improve yields, or planning for further experimental development, can only be effective if the full details are available so that decisions can be made on the basis of reliable observation.

It should be borne in mind in the case of project and postgraduate students, when embarking on their programme of work, that their experimental record may be the source material for subsequent publication in the chemical literature, the basis upon which the project report or thesis will be written, and the framework from which further studies may develop.

CALCULATION OF YIELDS AND THE MONITORING OF REACTIONS

The *theoretical yield* in an organic reaction is the weight of product which would be obtained if the reaction had proceeded to completion according to the chemical equation. The *yield* is the weight of pure product which is isolated from the reaction. The *percentage yield* may be expressed thus:

$$\text{Yield } (\%) = \frac{\text{weight of product}}{\text{theoretical yield}} \times 100$$

All the reactions in this book, and most that are published in the chemical literature, give not only the quantities of reactants and reagents (weights in grams or milligrams, volumes/densities), but also their molar proportion (in moles, or millimoles, or the molar concentration of a reagent in solution). This allows a ready means for the calculation of the theoretical yield which is expected on the basis of the chemical reaction, and also an assessment as to which reagents are being used in excess. If molar proportions are not quoted, these must always be calculated. Note of course that the calculation of the theoretical yield is based upon the molar quantity of the reactant or reagent in least amount.

The success of a reagent is judged from the percentage yield, being excellent ($>90\%$), very good ($>80\%$), good ($>70\%$), fair ($>50\%$) and poor ($<40\%$). The aim of the organic chemist when working on a reaction is to improve the yield, and this requires judgement as to which of the following factors are in operation which may lead to a reduction in the amount of product formed. These include: (a) the fact that the reaction may be reversible and the position of equilibrium unfavourable to the product; (b) the incursion of side reactions leading to the formation of by-products; (c) the premature work-up of the reaction before its completion; (d) the volatilisation of products during reaction or during work-up; (e) the loss of product due to incomplete extraction, inefficient recrystallisation or distillation operations, or other work-up procedures; and (f) the presence of contaminants in the reactants or reagents leading to a less efficient reaction.

In order to assess the contribution of each of these factors, and hence to optimise yields (i.e. thereby to specify the quantities, reaction conditions and isolation sequence to achieve the highest feasible yield), all stages in the reaction and isolation operations need to be monitored. The *qualitative* monitoring of a reaction, usually by t.l.c. or g.l.c. analytical procedures (Section 2.31), and less frequently by n.m.r. spectroscopic analysis (see, for example, Expt 5.165), is a valuable means of establishing the complete disappearance of starting material, the number and significance of by-products, and the effectiveness of purification procedures. Frequently, however, it is desired to carry out the same reaction under a graded series of conditions (i.e. using different times, temperatures, solvent mixtures, molar ratios of reactants, etc.), or to survey the effectiveness of a reagent with a range of substrates, or of a range of reagents with the same substrate. To simplify the workload of such surveys, and to avoid numerous work-up sequences which may introduce mechanical variables, *quantitative* g.l.c. analysis is used if appropriate to give an accurate yield of product (g.l.c. % yield). (see Section 2.31). It should be pointed out that g.l.c. yields when quoted in the literature are always greater than isolated yields, often by a very significant amount. Therefore, while g.l.c. yields provide valuable information on the optimisation of conditions, the skill of the organic chemist lies in the technical ability to isolate pure products in high yields.

Finally, when working in the field of asymmetric synthesis, the organic chemist needs to quote both the chemical yield and the *optical yield*. The percentage optical yield or optical purity [enantiomeric excess (ee) %], is calculated thus:

$$ee(\%) = \frac{[\alpha]_\lambda^t \text{ of product}}{[\alpha]_\lambda^t \text{ of pure enantiomer}} \times 100$$

It is essential that the specific rotation of the product and the pure enantiomer be measured in the same solvent, at the same wavelength and temperature, and at a similar concentration; if possible both measurements should be made at the same time. Optical yields in enzymic reactions carried out under laboratory conditions approach 100 per cent; an asymmetric chemical synthesis may be regarded as promising if the optical yield ranges upwards from 20 per cent.

2.3   HAZARDS IN ORGANIC CHEMISTRY LABORATORIES

2.3.1   INTRODUCTION

It is emphasised in Sections 2.1 and 2.2 that all workers in a laboratory must act responsibly in the interests of their own safety and that of their colleagues. These sections include general guidelines on safe practice for common laboratory operations, and should be seen to supplement the individual safety information booklets which should be provided to all new members of a laboratory by the Safety Committee of the appropriate organisation.* In addition, members of a laboratory should have ready access to appropriate books dealing with commonly encountered hazards. Two important publications are *Hazards in the Chemical Laboratory*, and *Guide to Safe Practices in Chemical Laboratories*, both of which are published by The Royal Society of Chemistry.[1,2] Other important texts which should be consulted are noted in the list of references.[3-5]

In teaching laboratories, and in laboratories concerned with routine work, the practical courses and operations will have been assessed for inherent hazards, appropriate cautionary advice provided, and periodic checks on the effectiveness of such advice made. In research and other laboratories, workers should always consult senior staff and the Safety Officer responsible for the laboratory or laboratory complex in those cases involving chemicals or operations that represent a departure from previous experience.

Each laboratory must be equipped with a first-aid box of a size and content appropriate to the number of members of the laboratory and the nature of the work being carried out. The HSE publishes guidelines on the contents of such boxes.[6] A wall chart of emergency treatment should be located adjacent to the first-aid box and all members of the laboratory should be familiar with this information. Furthermore, the members of the laboratory should be aware of the name or names of the nominated trained first-aiders and where they may be contacted. A trained first-aider should be summoned in all cases of personal injury.

2.3.2 EXPLOSION AND FIRE HAZARDS

GENERAL ASPECTS

Explosive and highly flammable substances or mixtures of substances quite commonly have to be used in organic chemistry laboratories. Ignorance of the hazards which are likely to be encountered all too frequently leads to explosions and fires, but these may usually be avoided and the experiment conducted with a reasonable measure of safety if, in addition to the general rules for laboratory practice mentioned under Sections 2.1 and 2.2, the following guidelines are followed.

1. The use of a substance known to be explosive should be avoided if a safer alternative can be used.
2. If an explosive or dangerously reactive substance has to be used, then it should be used in the smallest possible quantity and with all the appropriate precautions which are indicated below.
3. Workers should try to foresee and avoid the situation where a dangerously reactive chemical is likely to come into contact with combustible material, or where an explosive substance is likely to be subjected to the stimulus of shock or excess heat.
4. Reactions known or likely to involve explosion or fire hazards should always be tried out on a small scale first, and only then carefully scaled up in stages if

* Regulations for the Control of Substances Hazardous to Health (COSHH) come into force in the UK in October 1989; those with responsibility for laboratory management should be familiar with these regulations.

no warning signs of danger are apparent (e.g. no undue rise in temperature or evolution of gas, etc.). Since for a reaction vessel the surface area per unit volume decreases with increasing volume, scaled-up reactions may exhibit unexpectedly large and possibly dangerous temperature rises. If a small-scale reaction procedure is known to be safe, it is better to repeat it several times to acquire the required stock of product, rather than to attempt to scale-up the process to achieve this in one step.

5. For notably exothermic reactions involving dangerously active reagents, the safest procedure is to add the reagent dropwise, with rapid stirring, at the same rate as it is used up. *Overcooling* must be avoided since this may inhibit the reaction sufficiently to allow a dangerous accumulation of the reagents; if the temperature is then allowed to rise, a violent reaction may occur. It may actually be safer to heat such a reaction to ensure complete consumption of each drop of reagent as it is being added.

EXPLOSIVE COMPOUNDS
The following compounds or groups of compounds are likely to be dangerously explosive in their own right. They may explode under the stimulus of heat, impact or friction, or apparently spontaneously.

1. Acetylene gas and the acetylide salts of heavy metals; silver and copper acetylides are extremely shock-sensitive. Polyacetylenes and some halogenated acetylenes.
2. Hydrazoic acid and all azides, both organic and inorganic (only sodium azide is safe); aryl azides and silver azide may be inadvertently formed during some reactions (see below, p. 37).
3. Diazonium salts (when solid) and diazo compounds.
4. Inorganic nitrates, especially ammonium nitrate. The nitrate esters of polyhydric alcohols.
5. Polynitro compounds, e.g. picric acid (and heavy metal picrates), trinitrobenzene (TNB), trinitrotoluene (TNT); all these substances are safe when damp with water.
6. Metal salts of nitrophenols.
7. Peroxides; these are a common cause of explosions due to their formation in ether solvents (see below, p. 404). Concentrated aqueous hydrogen peroxide solution, see Section 4.2.41, p. 439.
8. Nitrogen tribromide, trichloride and triiodide; these are all highly sensitive and violently explosive, and should never be prepared or used unless absolutely necessary.

POTENTIALLY DANGEROUS MIXTURES
Powerful oxidants are particularly dangerous when mixed with easily oxidised organic substances such as simple alcohols, polyhydric alcohols, carbohydrates and cellulose-containing materials such as paper, cloth or wood. They are also dangerous when mixed with elements such as sulphur and phosphorus, and with finely divided metals such as magnesium powder. The following are common examples:

1. Perchloric acid, chlorates and perchlorates.
2. Chromium trioxide ('chromic anhydride'), chromates and dichromates. Concentrated nitric acid and nitrates.

3. Permanganates.
4. Concentrated hydrogen peroxide.
5. Liquid oxygen and liquid air.

SOME SPECIFIC DANGERS OF EXPLOSION

Peroxides in ether solvents. This is one of the commonest causes of explosions in organic chemistry laboratories. Simple dialkyl ethers such as diethyl ether and di-isopropyl ether, and cyclic ethers such as 1,4-dioxane and tetrahydrofuran, form less volatile peroxides on exposure to air and light. If therefore one of these solvents is purified by distillation, the peroxide content in the residue is progressively increased and eventually a violent explosion may occur. In view of this: (i) such solvents should not be stored for long periods or in half empty bottles; containers should be of dark glass; (ii) before the solvents are distilled a peroxide test should be carried out, and, if positive, the peroxide must be removed (Section 4.1.*11*, p. 402 and Section 4.1.*15*, p. 404); and (iii) since purified ethers in contact with air rapidly peroxidise again (10 minutes in the case of tetrahydrofuran) they should be retested for peroxides and purified if necessary immediately before use.

Solid sodamide and potassium metal. Both of these substances undergo surface oxidation to give oxide films which may initiate explosions when the samples are handled. In the case of potassium, surface oxidation occurs even when the metal is stored under oil, and the act of paring off the oxide film with a knife may initiate an explosion. Samples of potassium which are heavily encrusted with oxide should not be used but should be carefully destroyed by adding the lumps to a large excess of propan-2-ol. Similarly, old or obviously encrusted (yellow) lumps of sodamide (Section 4.2.*67*) should not be ground in a pestle and mortar, but should be destroyed by mixing with solid ammonium chloride.

Alkali metals with chlorinated solvents. The alkali metals sodium, potassium and lithium (and also other metals, e.g. aluminium and magnesium, especially when finely divided), are all violently reactive towards halogenated organic compounds, notably the common chlorinated solvents such as carbon tetrachloride. Lumps or chips of these metals should *never* be washed with halogenated solvents – a violent explosion can result.

Perchloric acid. This can react violently with organic material such as cork, cloth, rubber or wood. In addition the fumes which are readily evolved from the liquid acid are easily absorbed by these substances which are thus rendered violently flammable or explosive. For this reason, perchloric acid should not be stored in a wood-framed fume cupboard or near to any organic material.

Chromic acid and nitric acid as cleaning agents. Violent explosions have ensued when attempts have been made to remove tarry residues from reaction flasks by adding chromic acid mixtures, or concentrated nitric acid, and heating. If such residues are not removed by chromic acid mixtures in the cold (even after prolonged treatment with several changes of acid) followed by scrubbing with scouring powder, then the only safe course is to throw the flask away.

Azides. Explosive aryl azides may be formed inadvertently during the Sandmeyer and other diazonium reactions.

Explosive silver azide forms when solutions of ammoniacal silver nitrate

(Tollen's reagent) are allowed to stand before use. This is extremely dangerous; Tollen's reagent should always be freshly prepared following the procedure given on p. 1219, taking care not to exceed the recommended concentrations. Unused reagent should be destroyed by the addition of aqueous sodium chloride.

Liquid nitrogen. Liquid nitrogen (b.p. $-196\,°C$) contains some liquid oxygen (b.p. $-183\,°C$) as an impurity and therefore evaporation leads to an increasing proportion of liquid oxygen, so that before complete evaporation occurs the residual liquid may contain *up to 80 per cent of liquid oxygen*. Contact of this residue with organic or combustible material of almost any sort is likely to cause an explosion. If Dewar flasks containing liquid nitrogen have been used as cooling baths, great care must be taken to ensure that all liquid nitrogen and oxygen has evaporated *completely* before the Dewar flask is used for another purpose (e.g. as an acetone–solid carbon dioxide cooling bath).

Glass vacuum assemblies. Before using any glass apparatus for vacuum distillation or sublimation it should be examined to ensure that, (i) it is of the correct thickness and type (thin-walled glassware and conical shaped flasks are not suitable), and (ii) that it is free from cracks and flaws. Vacuum desiccators should always be used in the smallest suitable size and should be encased in wire safety cages. Dewar flasks can cause considerable damage since they may collapse violently ('implode') if they are maltreated. All Dewar flasks should therefore be bound, over their entire length, with adhesive tape to contain flying fragments of glass in the event of an implosion.

Opening glass ampoules. Ampoules of volatile chemicals must be thoroughly cooled before opening. Cooling must be effected with care, particularly if the contents are highly reactive (e.g. boron trichloride). If cooling is mishandled the glass may crack and the release of the contents into the cooling bath may lead to a violent explosion. Ampoules should not be cooled to a low temperature *too quickly*. Cooling in ice–water initially, followed by ice–salt, will usually be satisfactory; cooling to solid carbon dioxide temperatures is not necessary.

If the contents have a tendency to decompose, considerable pressure may develop in the sealed ampoule on storage, and great care should be taken during opening. The cooled ampoule should be removed from the cooling bath and wrapped in strong cloth behind a safety screen. A clean scratch should be made in the neck of the ampoule with a sharp file or glass knife and the neck cracked off by touching the scratch with the molten end of a thin red-hot glass rod.

Ampoules should be well cooled before resealing. Resealing should be avoided if possible however; it is best to obtain a smaller size of ampoule and use the whole of the contents for one experiment.

Compressed gas cylinders. Under certain circumstances, cylinders of compressed gas may constitute major explosion and fire hazards and, despite their apparent robust construction, they should always be handled with care.[7]

In view of the high pressures involved, any possibility of slow leakage from a cylinder of a flammable or toxic gas should be carefully guarded against. Thus, gas should never be drawn from a cylinder unless the appropriate reduction valve has been correctly fitted. The main cylinder valve should never be opened more than is necessary to provide the required gas flow (two full turns of the spindle at the most *); when the cylinder is not in use the gas should be shut off at the main valve and not at the regulator, which should then be bled of surplus pressure and closed. A suspected leak may be tested for by brushing with the approved leak detection solution, usually 1 per cent aqueous Teepol.

The valves and screw threads of cylinders and regulators should *never* be greased since this may lead to an explosion. If a cylinder has a very stiff spindle valve or if the screw threads are damaged, it should be returned to the suppliers for replacement. Similarly, defective regulators and pressure gauges should never be used.

The possibility of the sudden release of the entire contents of a cylinder must be guarded against. Apart from obvious dangers in the release of flammable or toxic gases, the sudden release of any gas can transform a cylinder into a lethal jet-propelled missile. Thus any weakening of a cylinder by damage, particularly to the valve, must be prevented. Cylinders must never be allowed to stand free in an upright position where they might be knocked over. They should be supported either by strapping to a bench or wall, or kept in one of the special mobile trolleys available for the purpose. Cylinders should only be moved by the use of these trolleys.

Gas cylinders should be stored upright in specially designated areas of moderate temperature which are adequately ventilated and entirely weather-proof. These areas should not be used to store any other chemicals and should be kept free from water, oils, or any corrosive liquids or vapours. The appropriate regulations should be consulted for the approved segregation of cylinders of different gases. Large notices labelled FULL and EMPTY should be prominently displayed on cylinders to prevent confusion and mistakes. A cylinder colour code data sheet should be permanently fixed to the wall of the cylinder store. All cylinders now carry a label which is designed to give appropriate safety information.

As a general point, the regulations strongly recommend that no cylinder should be used in a laboratory, although small nitrogen cylinders for vacuum distillations, or small cylinders (lecture bottle size) of laboratory gases (e.g. chlorine, ethylene, etc.) may be acceptable exceptions providing appropriate precautions are taken in their use (see for example the BDH catalogue). Compressed gas supplies, for example for g.l.c. equipment, should be piped into the laboratory from cylinders located in a specially constructed area outside the laboratory. In the case of ammonia, when used in organic synthesis, the recommendations of Section 2.17.7, p. 116 should be noted. Acetylene cylinders should also be located in a specially constructed area adjacent to the dangerous operations laboratory while in use, but then immediately returned to the main cylinder store. Academic and research institutions have their own specific regulations which should be complied with.

FIRE HAZARDS
Fire hazards in organic chemistry laboratories are often considerable due to the quantities of volatile and flammable chemicals, particularly solvents, which are commonly used. Specific methods for dealing with the more notable hazards are given below.

---

* Spindles should never be fully unscrewed since some are not captive and will be blown out by the full gas pressure if they are unscrewed completely.

Flammable solvents. Particular care should be taken when handling flammable solvents (and other chemicals) which are also *highly volatile*. The vapour may drift to a distant ignition source and burn back to ignite the main bulk of the liquid. An important rule is never to allow any vapour of a volatile chemical to escape into the open laboratory (in addition to fire hazards many vapours are toxic; see Section 2.3.4, p. 44). Strict regulations apply to the total quantity of solvents which may be stored in a laboratory, and with the exception of small bench reagent bottles, storage must be in approved cabinets. Bulk dispensing should take place in a designated area. Large quantities of solvent should not be allowed to accumulate on the work bench but returned to the storage area. If spillage of solvent or accidental release of flammable vapour occurs, the whole laboratory should be ventilated as soon as possible. The design and location of the main solvent store is subject to strict regulation.

A measure of the flammability of a compound is given by the *flash point* (the temperature at which the liquid gives rise to ignitable vapour). Any liquid with a flash point of less than 15 °C should be regarded as dangerously flammable and treated accordingly. If a solvent also has a low *autoignition temperature* (the temperature at which the vapour will spontaneously ignite in air), it should be treated with particular care.

Some highly flammable common solvents are given in Table 2.1 in order of increasing flash point. This does not include other dangerously flammable substances not commonly used as solvents.

Table 2.1 Flash points of common solvents

	(°C)		(°C)
Pentane and light petroleum	−49	Butan-2-one	−7
(b.p. 40–60 °C)		Ethyl acetate	−4.4
Diethyl ether	−45	Heptane	−4
Cyclopentane	−37	Methylcyclohexane	−4
Carbon disulphide	−30*	Toluene	4.4
Di-isopropyl ether	−28	1,2-Dimethoxyethane	4.5
Hexane and light petroleum	−23	Acetonitrile	6
(b.p. 60–80 °C)		Pentan-2-one	7
Cyclohexane	−20	Methanol	10
Acetone	−18	1,4-Dioxane	12
Tetrahydrofuran	−17	Propan-2-ol	12
Benzene	−11	Ethanol	12
Methyl acetate	−9	Ethylbenzene	15

* Carbon disulphide has the very low autoignition temperature of 100 °C. The vapour may therefore ignite on contact with steam pipes or with boiling water baths.

Ignition sources. Naked flames should rarely be used in organic chemistry laboratories. The heating of reaction mixtures is much more safely accomplished by means of a steam bath, an electric heating mantle or an oil bath heated by means of a small electric immersion heater (or, less safely, by a hot plate). If Bunsen flames have to be used, they should be lit only after a careful survey of neighbouring apparatus and chemicals has revealed no fire hazard. The flame should be turned out whenever it is not actually in use; a gas–air Bunsen flame may be invisible in bright sunlight and thus the cause of a fire or burning accident.

If flammable vapour is allowed to accumulate in the vicinity of electrical devices such as thermostats, stirrer motors, vacuum pumps, drying ovens, etc., it may be ignited by sparking from electrical contacts; this may be minimised by good laboratory ventilation and the prevention of the local build-up of solvent vapours. It should be noted that stirrers are available which are 'spark-free'; air-driven stirrers are also manufactured.

Sparking of contacts has caused serious explosions and fires when domestic-type refrigerators have been used to store volatile substances, even in small quantities. There is now available an excellent range of specially designed laboratory refrigerators. Since a volatile compound can have an appreciable vapour pressure at $0\,^{\circ}$C a dangerous concentration of vapour (which may be toxic) can accumulate within the cabinet of the refrigerator; volatile compounds should therefore be stored in clearly labelled glass containers, with well-fitting stoppers.

Leaking oxygen cylinders. If a cylinder containing oxygen is allowed to leak over a period when normal ventilation is turned off, the concentration of oxygen in the air may become great enough to cause a very fierce fire in the event of an ignition source being present and of there being flammable materials in the vicinity. All compressed oxygen cylinders should be tested for leaks by brushing the valve joints with the approved leak detection solution. Leaking cylinders should be suitably labelled and returned to the suppliers.

Sodium residues. Bottles containing sodium wire previously used for solvent drying constitute a fire and explosion hazard. The sodium, sometimes heavily coated with hydroxide or oxide film, should be covered with propan-2-ol and set aside with occasional swirling until all the sodium particles are destroyed (at least 2 hours). The contents of the bottle should then be poured into a large excess of water (water should *not* be added to the bottle) and the bottle washed out several times with industrial spirit. Only then can the bottle be safely rinsed with water.

Fires involving sodium metal are very hot and localised and are best dealt with by smothering with sand or by using a dry powder extinguisher, *not* a carbon tetrachloride or carbon dioxide extinguisher.

Metal hydrides. Lithium hydride, sodium hydride, potassium hydride and lithium aluminium hydride all react violently with water liberating hydrogen; the heat of reaction may cause explosive ignition. Excess metal hydride from a reaction must be destroyed by the careful addition of ethyl acetate or acetone.

Partially or fully alkylated metal hydrides (e.g. diethylaluminium hydride), are also pyrophoric and precautions in their use are noted under specific compounds in Section 4.2; handling techniques are discussed in Section 2.17.8, p. 120.

THE DANGEROUS OPERATIONS LABORATORY

It is strongly recommended that all reactions involving any possible hazard from explosive, flammable, dangerously reactive or highly toxic substances should be carried out in a special laboratory solely designed for the purpose. Such a laboratory should not be used for any routine teaching or research purposes, or for the storage of chemicals or apparatus apart from those required for specific hazardous reactions (e.g. autoclaves, furnaces, etc.). Ideally the laboratory should be purpose-built and should incorporate the following safety features:

1. Water-proof and vapour-proof electric lamps, switches and power points.

41

2. Fume cupboards fitted with powerful extractor fans capable of rapidly changing all the air in the laboratory.
3. Fire-resistant doors and walls.
4. An adequate supply of protective clothing including safety visors and goggles, protective gloves, rubber aprons and boots.
5. Good quality safety shields and screens for guarding potentially violent reactions.
6. An automatic carbon dioxide fire-extinguishing system.
7. The following should also be provided immediately outside the laboratory: storage facilities for gas masks and self-contained breathing apparatus; fire blankets, and buckets of sand for spilled liquids and for smothering fires; large carbon dioxide and dry powder extinguishers if no automatic fire extinguishing system is installed; a telephone with a clear notice beside it listing procedures and numbers to be dialled in an emergency.

If no laboratory or other room is available for conversion to a dangerous operations laboratory, then a semi-permanent structure of adequate design sited at a safe distance from regularly occupied laboratories and offices may be constructed from lightweight and fire-proof building materials (cf. unattended operations, Section 2.16).

THE CONDUCT OF EXPLOSIVE OR VIOLENT REACTIONS

There is a common tendency to regard fume cupboards as the proper sites for potentially explosive or violently reactive processes. This is not to be recommended since the glass windows of fume cupboards may not be of sufficient quality and thickness to withstand an explosion, and the confinement of gaseous reaction products by the sides and top of the fume cupboard increases the severity of the blast. All potentially violent reactions should therefore be conducted on an open bench, with the apparatus surrounded by safety shielding on all sides but open at the top. It has been shown that even flimsy protection at the top dramatically reduces the efficiency of such side screens to contain an explosion. The best design of safety shield is a flat plate of polycarbonate (minimum thickness 3 mm) suspended in a vertical plane from above, and heavily weighted along the bottom edge. The performance of conventional curved free-standing shields may be considerably improved by heavily weighting the bottom edge to prevent the whole shield being blown over in an explosion.

2.3.3　REACTIVE INORGANIC REAGENTS

Many inorganic reagents used in organic chemistry laboratories are highly reactive (and hence have 'corrosive' properties), causing immediate and severe damage if they are splashed or spilled on to the skin, or when they are inhaled as vapours, dusts or mists. In addition, their high reactivity may cause a rapid evolution of heat when they are mixed with other chemicals, including water, resulting in a corrosive and possibly toxic mixture being sprayed and splashed about; sometimes fires or explosions follow. When using such chemicals suitable protective clothing including gloves should be worn. Adequate protection of the eyes is absolutely essential and safety spectacles, or preferably goggles or a visor, must always be worn. When there is any possibility of inhalation of reactive vapours or dusts, all operations should normally be conducted in a fume cupboard. Additonal protection may be provided by a gas mask or well-fitting dust mask.

If any corrosive liquid or solid is spilled on to the skin it should be immediately washed off with copious quantities of water; in cases of splashes in the eyes, every second counts. Any spillages should be cleaned up without delay, preferably with the aid of sand. Flooding a spillage on a floor or bench with water is not always advisable if this is likely to spread the corrosive material and cause it to lodge in crevices and between floorboards. In cleaning up extensive spillage where noxious fumes are involved, full protective clothing including respirators should be used.

Some highly reactive chemicals and their dangerous properties are listed below. Those which give off highly corrosive irritant and/or toxic vapours or, if solids, are similarly hazardous in the form of dusts, are marked with an asterisk (*) and should only be used in fume cupboards. More details concerning the properties of many of these are given in Section 4.2. Specific information on the hazardous properties of individual chemicals is collected in several comprehensive works.[8-11]

STRONG ACIDS

All of the following react violently with bases and most give off very harmful vapours.

* Hydrobromic acid and hydrogen bromide.
* Hydrochloric acid and hydrogen chloride.
* Hydrofluoric acid and hydrogen fluoride – both react readily with glass and quickly destroy organic tissue. New thick rubber or plastic gloves should be worn after carefuly checking that no holes are present. Skin burns must receive immediate and specialised medical attention.
* Nitric acid (concentrated and fuming).
* Perchloric acid (explosion danger, see Section 2.3.2, p. 37.

Sulphuric acid (concentrated and 'oleum') – should always be mixed with water very carefully, by pouring into cold water as a thin stream to prevent acid splashes or spray. 'Chromic acid' cleaning mixtures have the corrosive properties of concentrated sulphuric acid as well as the dangerous oxidising properties of the chromic acid.
* Chlorosulphonic acid – this is a highly corrosive liquid which reacts violently with water.

STRONG BASES

Calcium oxide, potassium hydroxide and sodium hydroxide – these react violently with acids, generate heat on contact with water, and have a powerful corrosive action on the skin, particularly the corneal tissue of the eye.
* Ammonia (gas and concentrated aqueous solution, d 0.880). Concentrated hydrazine solutions (and hydrazine salts); * hydrazine vapour is harmful.
* Sodamide – usually obtained in a granular form which reacts violently with water; it is irritant and corrosive in a finely divided form. Old and highly coated samples should not be crushed for use but should be destroyed (see Section 2.3.2, p. 37).

---

* Highly corrosive, irritant and/or toxic vapours or dusts.

HALOGENS

All are toxic and corrosive. Great care should be exercised when working with fluorine, which is violently reactive towards a wide range of substances. The interhalogen compounds are also powerfully reactive.

REACTIVE HALIDES

All of the following are highly reactive, particularly towards water; ampoules òf liquids should be opened in a fume cupboard after cooling, observing the precautions detailed in Section 2.3.2, p. 38.

* Boron trichloride; * phosphorus tribromide; * trichloride and * pentachloride; * silicon tetrachloride. * Aluminium chloride and * titanium(IV) chloride are rather less reactive.

CHROMIUM TRIOXIDE, CHROMATES AND DICHROMATES

All these form corrosive dusts; those from water-soluble chromates are particularly dangerous since they dissolve in nasal fluid and in perspiration. Long-term exposure can lead to ulceration and cancer. Chromium trioxide may cause sensitisation by skin contact.

2.3.4 HAZARDS DUE TO TOXIC CHEMICALS

A very large number of compounds encountered in organic chemistry laboratories are poisonous, i.e. 'toxic'. Indeed, nearly all substances are toxic to some extent and the adoption of safe and careful working procedures which prevent the entry of foreign substances into the body is therefore of paramount importance, and should become second nature to all laboratory workers. Toxic substances can enter the body by the following routes:

Ingestion (through the mouth). This is fortunately not common in laboratories, but can occur through the accidental contamination of food, drink or tobacco, and by misuse of mouth pipettes. It is strongly recommended that no one should ever eat, drink or smoke in a laboratory. The practice of storing bottles of milk or beer in laboratory refrigerators is to be strongly condemned.

Workers should always wash their hands thoroughly on leaving a laboratory and before eating. All pipetting by mouth should be avoided since there are excellent rubber bulb and piston-type pipette fillers available commercially.

In addition to the ingestion hazard associated with smoking, the vapours of many volatile compounds yield toxic products on pyrolysis when drawn through a lighted cigarette or pipe (e.g. carbon tetrachloride yields phosgene).

Inhalation (into the lungs). This is a more common pathway for the absorption of toxic chemicals; these may be in the form of gases, vapours, dusts or mists. All toxic powders, volatile liquids and gases should only be handled in efficient fume cupboards. The practice of sniffing the vapours of unknown compounds for identification purposes should be conducted with caution.

Direct absorption (through the skin into the bloodstream). This is also a common route for the absorption of a toxic substance whether liquid, solid or gaseous. The danger may be reduced by wearing rubber or plastic gloves, in addition to the usual laboratory white coat. However, clean and careful working procedures are still necessary despite these precautions. Protective gloves are often permeable to organic solvents and are easily punctured; they should therefore be frequently inspected and replaced when necessary. If a toxic substance is accidentally spilled on the skin, it should be washed off with copious quantities of cold water with the aid of a little soap where necessary. The use of solvents for washing spilled chemicals off the skin is best avoided since this may hasten the process of absorption through the skin.

Repeated contact of solvents and many other chemicals with the skin may lead to dermatitis, an unsightly and irritating skin disease which is often very hard to cure. In addition, sensitisation to further contact or exposure may occur.

The toxic effects of chemical compounds can be classified as either 'acute' (short term) or 'chronic' (long term). Acute effects, as exemplified by powerful and well-known poisons such as hydrogen cyanide and chlorine, are immediately obvious, well appreciated by most laboratory workers, and are therefore fairly easily avoided. However, many chemicals exhibit chronic toxic effects which may only come to light after long-term exposure to *small quantities*. This type of insidious poisoning is harder to detect (and therefore prevent) since the results may only manifest themselves after months or even years of exposure (or even long after exposure has ceased). Chronic poisoning may also cause symptoms which are not easily recognisable as such, e.g. sleeplessness, irritability, memory lapses and minor personality changes. It must be stressed, however, that the final results of chronic poisoning may be very serious and can lead to premature death. Every effort should be made by the laboratory worker to guard against these possibilities by adopting a rigorous approach to the avoidance of breathing all vapours and dusts, and of any contact between the skin and liquids or powders.

There are available several designs of atmosphere-monitoring equipment. These employ an extremely wide range of individual detector tubes, which are specifically sensitive to commonly encountered gas and vapour contaminants that may be present in laboratory and industrial sites. The equipment is appropriately designed for short-term (instantaneous) and long-term monitoring; personal monitoring equipment for solvent vapour is also available (further information from Drager Ltd, or from Vinten Instruments Ltd).

The guiding principle for all workers should be to treat *all* chemicals as potentially harmful. The following discussion (pp. 46–48) centres on compounds with acute toxic properties which are likely in organic laboratories; those which give rise to particularly severe chronic effects are noted on pp. 48–51. Substances marked **C** are also known to be carcinogenic, but the discussion of their availability or control is noted.

An indication of the hazard associated with the use of a toxic material in the form of a vapour or dispersed dust is given by a limit value. The *threshold limit value* (TLV, expressed as p.p.m. or mg m^{-3}) represents a level under which it is believed nearly all workers may be repeatedly exposed to on a day-to-day basis without adverse effect. These values are up-dated annually and recommended by the American Conference of Governmental Industrial Hygienists (ACGIH).[12] Since 1984, in the UK, the Health and Safety Executive (HSE) has adopted two types of limits, but only for those compounds which are available and used in the UK.[13] These are the *recommended limit* (RL, as p.p.m. or mg m^{-3}) which represents good practice and realistic levels for the degree of exposure, and the *control limit* (CL, as p.p.m. or mg m^{-3}) which is applied to the relatively smaller number of substances having unusually serious toxic effects.

The control limits should not be exceeded and wherever possible reduced, and in any case these substances are subject to specific Regulations or Codes of Practice. The RL values are annually reviewed, but the CL values may be altered at shorter time intervals. The TLV value is used in the UK if a RL or CL value is not available. With some substances, *long-term* or *short-term exposure limit values* (LTEL or STEL) are published. Finally it should be noted that the Council of Europe is proposing to introduce a harmonised list of exposure standards for adoption by the EEC countries.

For detailed information on the toxicological properties of individual substances reference should be made to the specialist monographs on the subject,[14, 15] and to the comprehensive works already noted.

HIGHLY TOXIC SOLIDS
Even small quantities of these substances are likely to rapidly cause serious illness or even death. Particular care should be taken to avoid inhalation of dusts and absorption through the skin as well as the more obvious hazards of direct ingestion.

	RL ($mg\,m^{-3}$)
Arsenic compounds	0.2 (as As)
Inorganic cyanides	5 (as CN)
Mercury compounds, particularly alkyl mercurials*	0.01
Osmium tetroxide (hazardous vapour)	0.002
Oxalic acid (and its salts)	1
Selenium and its compounds	0.2 (as Se)
Thallium salts	0.1 (as Tl)
Vanadium pentoxide	0.5

* Dangerously chronic toxic effects (see p. 50).

DANGEROUSLY TOXIC GASES
All operations involving the use or liberation of these substances must be carried out in an efficient fume cupboard. In most cases contact with the skin must be prevented.

	RL ($mg\,m^{-3}$)	RL (p.p.m.)
Boron trifluoride	3	1
Carbon monoxide	55	50
Chlorine	3	1
Cyanogen	20	10
Diazomethane, C	0.4	0.2
Fluorine	2	1
Hydrogen cyanide	10	10
Hydrogen fluoride	2	3
Hydrogen sulphide	14	10
Nitrogen dioxide (nitrous fumes) and nitrosyl chloride	5	3
Ozone	0.2	0.1
Phosgene	0.4	0.1
Phosphine	0.4	0.3

DANGEROUSLY TOXIC LIQUIDS AND SEVERE IRRITANTS
These substances have dangerously toxic vapours and are also harmful through skin absorption. Prolonged exposure to small amounts is likely to give rise to

chronic effects. The vapours of many are powerful irritants particularly to the respiratory system and to the eyes.

		RL (mg m^{-3})	RL (p.p.m.)
Acetyl chloride		—	—
Acrylaldehyde (acrolein)		0.25	0.1
Alkyl (and aryl) nitriles		—	—
Allyl alcohol		5	2
Allyl chloride		3	1
Benzene*, C		30	10
Benzyl bromide (and chloride)		5	1
Boron tribromide (and trichloride)		10	1
Bromine		0.7	0.1
Bromomethane (methyl bromide)		60	5
Carbon disulphide	(CL)	30	10
2-Chloroethanol (ethylene chlorohydrin)		3	1
3-Chloropropanoyl chloride		—	—
Crotonaldehyde		6	2
Diketen		—	—
Dimethyl sulphate, C (and diethyl sulphate)		0.5	0.1
Fluoroboric acid		—	—
Hydrofluoric acid		2.5	3
Isocyanatomethane (methyl isocyanate)*	(CL)	0.02	—
Nickel carbonyl, C		0.35	0.05
Oxalyl chloride		—	—
Pentachloroethane		—	—
Tetrabromoethane	(TLV)	15	1
Tetrachloroethane	(TLV)	35	5
Trimethylchlorosilane		—	—

* Dangerous chronic toxic effects (see p. 50).

OTHER HARMFUL SUBSTANCES
The following compounds and groups of compounds have generally harmful effects when inhaled as vapours or dusts, or absorbed through the skin, or both; some are also corrosive. All should be regarded as potentially harmful by long-term exposure to small quantities.

Alkyl bromides and chlorides. Many simple alkyl bromides and chlorides, and poly-halogenated methanes and ethanes fall into this category, including some common solvents. All should be treated as potentially harmful, but in addition to those already listed above the following may be regarded as some of the more dangerous.

		RL (mg m^{-3})	RL (p.p.m.)
Bromoethane (ethyl bromide)		890	200
Bromoform		5	0.5
3-Bromopropyne (propargyl bromide)		—	—
Carbon tetrachloride		65	10
Chloroform, C		50	10
Dichloromethane	(CL)	350	200
1,2-Dibromoethane (ethylene dibromide), C		145	20
1,2-Dichloroethane (ethylene dichloride)		40	10
Iodomethane, C (methyl iodide)		28	5

Aromatic and aliphatic amines Simple aliphatic primary, secondary and tertiary amines have toxic vapours, e.g. diisopropylamine (RL 5 p.p.m., 20 mg m^{-3}),

47

dimethylamine (RL 10 p.p.m., 18 mg m^{-3}), ethylamine (RL 10 p.p.m., 18 mg m^{-3}) and triethylamine (RL 10 p.p.m., 40 mg m^{-3}).

Likewise many aromatic amines are extremely harmful as vapours and by skin absorption. The following list includes some representative examples, but *all* aromatic amines, including alkoxy-, halogeno- and nitro-amines should be treated as potentially harmful. In addition, many aromatic amines are known to be powerful cancer-causing agents (carcinogens; see below) and the use of some of these is legally controlled.

	RL (mg m^{-3})	RL (p.p.m.)
Aniline	10	2
Anisidines (aminoanisoles)	0.5	0.1
Chloroanilines	—	—
Chloronitroanilines	—	—
N,N-Diethylaniline	—	—
N,N-Dimethylaniline	25	5
N-Ethylaniline	—	—
N-Methylaniline (TLV)	2	0.5
p-Nitroaniline (and isomers)	6	1
p-Phenylenediamine (and isomers)	0.1	—
o-Toluidine (and isomers)	9	2
Xylidines	10	2

Phenols and aromatic nitro compounds. As with aromatic amines, very many phenolic compounds and aromatic nitro compounds exhibit toxic properties. They give off harmful vapours, are readily absorbed through the skin and, particularly the phenols, have corrosive properties. All phenols and aromatic nitro compounds should therefore be handled with care and assumed to have the properties listed above.

	RL (mg m^{-3})	RL (p.p.m.)
Phenol	19	5
Cresols	22	5
Catechol and resorcinol	20	5
Chlorophenols and dichlorophenols	—	—
Nitrobenzene	5	1
p-Nitrotoluene (and isomers)	30	5
m-Dinitrotoluene (and isomers)	1	0.15
2,4-Dinitrotoluene (and isomers)	1.5	—
p-Chloronitrobenzene (and isomers)	1	—
Dichloronitrobenzenes	—	—
Nitrophenols	—	—
Dinitrophenols and dinitrocresols	0.2	—
Picric acid	0.1	—

CARCINOGENIC SUBSTANCES

Many organic compounds have been shown to cause cancerous tumours in man, although the disease may not be detected for several years. The manufacture and use of some of these substances is forbidden in factories in Great Britain (according to *The Carcogenic Substances Regulations* 1967)[16] and in the USA. It is essential that Safety Officers be fully aware of current regulations with regard to the use of these compounds in academic and research laboratories.

When handling known or suspected carcinogens, every effort should be made to avoid inhalation of their vapours and contamination of the skin. They must only be handled in fume cupboards using protective gloves. It is essential that bottles or vials containing the compounds should be properly labelled with suitable warnings. Supplies of carcinogenic compounds should be kept in a locked container, preferably in a fume cupboard.

The ACGIH (USA) has developed a triple classification of carcinogenicity of substances, and this classification has been adopted by the HSE (UK). These are 'human carcinogens', 'suspected carcinogens' and 'experimental carcinogens'. In the UK, substances classified in the first category are subject to legal prohibition of manufacture and/or use; substances in the second category are subject to control limits; substances in the third category are those in which less stringent precautions may be permissible. It is essential that the Safety Officer be conversant with the regulations which pertain to the laboratory within his or her responsibility.[16-19]

The most dangerous known carcinogens which are most likely to be encountered in laboratory work are noted below.

Aromatic amines, substituted hydrazines and their derivatives. These should *all* be treated as potentially carcinogenic, and as a group probably constitute the greatest danger to the organic chemist since even a slight exposure may initiate the formation of tumours. The following list includes some of the most hazardous substances.

2-Acetylaminofluorene
2-Aminoazotoluene
4-Aminobiphenyl* (and 4-nitrobiphenyl)*
4-Aminostilbene
3-Amino-1,2,4-triazole
Auramine and Magenta
4,4'-Diaminobiphenyl (benzidine)*
4,4'-Diamino-3,3'-dichlorobiphenyl†
4,4'-Diamino-3,3'-dimethylbiphenyl†
4,4'-Diamino-3,3'-dimethoxybiphenyl† (dianisidines)†
Dimethylaminoazobenzene (Butter Yellow)
1,1-Dimethylhydrazine
Hydrazine
Methylhydrazine
N-Phenyl-2-naphthylamine
2-Naphthylamine*
1-Naphthylamine†

N-Nitroso compounds. *All* nitrosamines [R'·N(NO)·R] and nitrosamides [R'·N(NO)CO·R] should be regarded as potentially powerful carcinogens, since most compounds of these types have been shown to possess high activity in experimental animals. The following are some of the more likely to be encountered in the laboratory.

N-Methyl-N-nitrosoaniline
N-Methyl-N-nitrosourea

---

* Importation and use in manufacture of these compounds is prohibited in the UK except if present at less than 1 per cent in another material.
† The use of these compounds is controlled by legislation in the UK.

N-Methyl-N-nitrosourethane
N-Nitrosodimethylamine†
N-Nitrosopiperidine

Alkylating agents
Aziridine (and some of its derivatives)
Bis(chloromethyl) ether
Chloromethyl methyl ether*
Diazomethane
Dimethyl sulphate
Epichlorhydrin
Iodomethane (methyl iodide)
Nitrogen mustards (i.e. $R \cdot N(CH_2 \cdot CH_2Cl)_2$)
β-Propiolactone

Monocyclic and polycyclic aromatic hydrocarbons
Benzene
Benz[a]pyrene
Dibenz[a,h]anthracene
Dibenz[c,g]carbazole
7,12-Dimethylbenz[a]anthracene

Halogenated hydrocarbons
Carbon tetrachloride
Chloroform
1,2-Dibromoethane
1,4-Dichlorobutene
Hexachlorobutadiene
Vinyl chloride

Phosphorus- and sulphur-containing compounds
Hexamethylphosphoramide
1,3-Propanesultone
Thioacetamide
Thiourea

Asbestos Inhalation of asbestos dust and fibres can cause 'asbestosis', a crippling and eventually fatal lung disease which often becomes lung cancer in its later stages. The industrial use of asbestos is strictly controlled in the UK by the Asbestos Regulations 1969.[20]

SUBSTANCES WITH VERY HARMFUL CUMULATIVE EFFECTS
The substances noted here present chronic health hazards which are usually manifest after continued exposure over a relatively long period. In the case of workers handling this group of compounds, biological checks under the supervision of a medical practitioner, to determine whether unacceptable absorption into the body is taking place, may be prudent.

Benzene (RL 10 p.p.m.). Inhalation of benzene vapour has a chronic cumulative effect leading to acute anaemia and may lead to leukaemia. Very few people can smell benzene in vapour concentrations of less than 75 p.p.m. (i.e. seven times the

---

*† For footnotes, see p. 49.

RL). If therefore one can smell benzene, it is being inhaled in harmful quantities. For general solvent use, benzene can in nearly all cases be replaced by the less volatile and less toxic toluene (RL 100 p.p.m.).

Lead compounds. These are powerful cumulative poisons and ingestion of even small amounts must be guarded against. Organic lead compounds (e.g. lead tetra-ethyl) are volatile and inhalation of their vapours must be avoided; they are also dangerous by skin absorption.

Mercury and mercury compounds. These vary greatly in toxicity. Generally, mercury(II) salts are more toxic than mercury(I) salts. Liquid organic mercury compounds are highly poisonous and dangerous by inhalation and absorption through the skin, whereas solid organomercurials are less toxic. However, all mercury compounds should be treated with caution and any long-term exposure avoided.

Elemental mercury readily evolves the vapour which constitutes a severe cumulative and chronic hazard.* No mercury surface should ever be exposed to the atmosphere but should be covered with water. All manipulations involving mercury should be carried out in a fume cupboard and over a tray to collect possible spillage. Spilt mercury is best collected using a glass nozzle attached to a water suction pump via a bottle trap; the contaminated areas should be spread with a paste of sulphur and lime.

A severe mercury vapour hazard may occur through misuse of mercury-containing vacuum gauges (e.g. the 'vacustat', Section 2.30) attached to oil vacuum pumps. If the gauge is turned about its axis too quickly, mercury may be sucked into the pump and circulated with the hot oil to release large quantities of mercury vapour into the atmosphere. If there is any possibility of this having happened, the pump must not be used and should be stripped down and cleaned as soon as possible.

Isocyanates. These are highly toxic compounds, particularly toluene diisocyanate, which cause permanent lung damage on repeated exposure. Furthermore, over a period of time, the worker may become sensitised to isocyanates, such that subsequent undetectable concentrations produce severe symptoms.

2.3.5 ELECTRICAL SAFETY

Concern with the hazards associated with the use of flammable and toxic chemicals in the laboratory often causes the dangers from electrical equipment to be overlooked. However, many accidents are caused by the malfunctioning of electric appliances and by thoughtless handling.

New equipment should be carefully inspected to check that the plug has been correctly fitted, otherwise a 'live' chassis will result. International standards for Great Britain and Europe stipulate the following colours for electric cables:

Live, Brown; *Neutral*, Blue; *Earth*, Green/yellow.

In the USA (and for equipment imported from the USA) the colours are:

Live, White; *Neutral*, Black; *Earth*, Green.

---

* The normal vapour pressure of mercury at room temperature is many times the CL value of 0.05 mg m^{-3}.

Before any electric appliance is used, it should be inspected to ensure that: (*a*) it is in good condition with no loose wires or connections; (*b*) it is properly earthed; (*c*) it is connected to the correct type of plug by good quality cable with sound insulation; and (*d*) that it is protected by a fuse of the correct rating. Loose or trailing electric cables should be avoided and if the appliance has to be sited some way from the power source, the cable should run neatly along the side of a bench and preferably be secured with adhesive tape. Cable hanging across the aisle between two benches should never be permitted. Any items of equipment (e.g. stirrer motors or heating mantles) which have had any chemicals spilled on them should not be used until they have been thoroughly cleaned and dried.

In the handling and setting up of electrical equipment, the operator must ensure that the apparatus is set up on a dry bench. It is essential to assemble the apparatus first, and only then to plug into the mains and switch on. The apparatus should be switched off before any attempts are made to move or adjust it.

High voltage equipment (e.g. for use in electrophoresis, or in the generation of ozone) requires special precautions. Ideally, such apparatus should be isolated within an enclosure equipped with an interlocking device so that access is possible only when the current is switched off.

2.3.6   ULTRAVIOLET RADIATION

Ultraviolet (u.v.) lamps, arcs and other high intensity light sources which emit u.v. radiation should never be viewed directly or eye damage will result. Special close-fitting goggles which are opaque to u.v. radiation should be worn, and protective screens placed around the apparatus assembly (e.g. in a photochemical reaction) which incorporates the u.v. source; the need to avoid the inadvertent viewing of reflected u.v. light should also be borne in mind and the viewing of chromatographic columns or plates may be hazardous. Exposure of the skin to intense u.v. radiation gives rise to burns (cf. sunburn) and prolonged exposure may give rise to more extensive tissue damage. Protective gloves should therefore be worn during work involving such exposure risks. Adequate ventilation must also be provided to prevent possible build-up of the highly irritant and toxic ozone which is produced when oxygen is irradiated with u.v. light in the 185 nm region.

APPARATUS AND REACTION PROCEDURES

2.4   INTERCHANGEABLE GROUND GLASS JOINTS

The commercial development of glass manipulation, coupled with the use of borosilicate glass having low expansion, resistance to heat, corrosion and thermal shock, has made available truly interchangeable ground glass joints at moderate cost. These, fitted to apparatus for standard and specialised operations, have made laboratory work in practical organic chemistry, and indeed in many other branches of chemistry, easily accomplished. The advantages of the use of ground glass joints include:

1. No corks or rubber stoppers are, in general, required and the selection, boring and fitting of corks is largely eliminated, thus resulting in a considerable sav-

ing of time. Furthermore, contamination of chemicals as the result of contact with corks or bungs is therefore avoided.
2. Corrosive liquids and solids (concentrated acids, acid chlorides, bromine, phosphorous pentachloride, etc.) are easily manipulated, and no impurities are introduced into the product from the apparatus.
3. As all joints are made to accurate standards they should all fit well; this is particularly valuable for systems operating under reduced pressure.
4. By employing a few comparatively simple units, most of the common operations of organic chemistry may be carried out.
5. Wider passages are provided for vapours, thus diminishing the danger in violent reactions and reducing the dangers of flooding from condensing vapours in distillations, especially under reduced pressure.

As illustrated (Figs 2.1 and 2.6), the types of ground glass joints which are manufactured to precise specifications are either *conical joints* or *spherical joints* respectively.

The interchangeability of conical joints (cone and socket joints) is ensured by the use of a standard taper of 1 in 10 on the diameter in accordance with the recommendations of the International Organisation for Standardisation and of the various national standardising authorities. The brand name 'Quickfit' is a registered trademark (J. Bibby Science Products Ltd) and used to describe

Table 2.2   Dimensions of British Standard interchangeable ground glass conical joints (supplied by J. Bibby Science Products Ltd)

Size designation	Nominal diameter of wide end (mm)	Nominal diameter of narrow end (mm)	Nominal length of engagement (mm)
5/13	5.0	3.7	13
7/16	7.5	5.9	16
10/19	10.0	8.1	19
12/21	12.5	10.4	21
14.23	14.5	12.2	23
19/26	18.8	16.2	26
24/29	24.0	21.1	29
29/32	29.2	26.0	32
34/35	34.5	31.0	35
40/38	40.0	36.2	38
45/40	45.0	41.0	40
50/42	50.0	45.8	42
55/44*	55.0	50.6	44
60.46	60.0	55.4	46
7/11	7.5	6.4	11
10/13	10.0	8.7	13
14/15	14.5	13.0	15
19/17	18.8	17.1	17
24/20	24.0	22.0	20
55/29*	55.0	52.1	29
24/10	24.0	23.0	10
40/13	40.0	38.7	13
50/14	50.0	48.6	14

* Supplement to the ISO range.

Table 2.3 Dimensions of USA standard interchangeable ground glass joints

Size designation	Approximate diameter at small end (mm)	Computed diameter at large end of ground zone (mm)	Approximate length of ground zone (mm)
Full-length joints			
7/25	5	7.5	25
10/30	7	10.0	30
12/30	9.5	12.5	30
14/35	11	14.5	35
19/38	15	18.8	38
24/40	20	24.0	40
29/42	25	29.2	42
34/45	30	34.5	45
40/50	35	40.0	50
45/50	40	45.0	50
50/50	45	50.0	50
55/50	50	55.0	50
60/50	55	60.0	50
71/60	65	71.0	60
103/60	97.0	103.0	60
Medium-length joints			
5/12	3.8	5.0	12
7/15	6.0	7.5	15
10/18	8.2	10.0	18
12/18	10.7	12.5	18
14/20	12.5	14.5	20
19/22	16.6	18.8	22
24/25	21.5	24.0	25
29/26	26.6	29.2	26
34/28	31.7	34.5	28
40.35	36.5	40.0	35

Table 2.4 Dimensions of spherical joints (supplied by J. Bibby Science Products Ltd)

Size designation	Nominal diameter (mm)	Minimum diameter of wide end (mm)	Maximum diameter of narrow end (mm)
S 13	12.700	12.5	7.0
S 19	19.050	18.7	12.5
S 29	28.575	28.0	19.0
S 35	34.925	34.3	27.5
S 41	41.275	40.5	30.0
S 51	50.800	50.0	36.0

Table 2.5 Dimensions of some semi-ball connections (these dimensions taken from USA sources)

Semi-ball size	Ball diameter (mm)	Tube bore (mm)	Semi-ball size	Ball diameter (mm)	Tube bore (mm)
12/2	12	2	28/15	28	15
12/3	12	3	35/20	35	20
12/5	12	5	35/25	35	25
18/7	18	7	40/25	40	25
18/9	18	9	50/30	50	30
28/12	28	12	65/40	65	40

ground glass joints manufactured to the British Standards Institution (BS 572 and 2761). In the USA interchangeable ground glass joints conforming to the specifications of the National Bureau of Standards (Commercial Standard CS 21–58) are designated by the symbol $\mathbf{\check{S}}$.* It should be noted that the term 'cone' is used for the part which is inserted and the term 'socket' for the part into which the cone is inserted.

The dimensions of conical joints are indicated by a numerical code which incorporates the nominal diameter of the wide end and the length of the ground zone, e.g. in British usage 14/23 indicates a wide end diameter of nominally 14.5 mm, and a length of engagement of 23 mm; the USA code is similarly assigned. However, the length of engagement of the joints differs somewhat in the British and American specifications; consequently UK and European joints are *not* compatible with those manufactured in the USA, a point of particular note when ordering new, and perhaps specialised, apparatus. The dimensions of British and American interchangeable conical joints are listed in Tables 2.2 and 2.3, respectively.

The sizes of standard spherical joints (semi-ball or ball and socket joints) are designated by a code which indicates approximately the nominal diameter of the ground hemisphere, e.g. in British usage S35 designates a spherical joint of nominal diameter 34.925 mm. Sizes of spherical joints are collected in Tables 2.4 and 2.5.

2.5 TYPES OF GROUND GLASS JOINTS

All ground glass joints are usually constructed of high resistance borosilicate glass such as Pyrex. The most common form of conical joint is shown in Fig. 2.1 and is the type largely encountered in practice. That shown in Fig. 2.2 is similar but has reinforcing glass bands about the female joint which greatly add to the mechanical strength of the walls. Figure 2.3 depicts a ground glass joint with glass hooks, to which light springs may be attached. Figure 2.4 illustrates drip cones for condensers and the like. Figure 2.5 illustrates a cone joint with stem for use, for example, as a gas or steam inlet.

---

* The specifications are obtainable from the British Standards Institution, 2 Park Street, London, W1A 2BS; and from the Superintendent of Documents, US Government Printing Office, Washington 25, DC.

Fig. 2.1   **Fig. 2.2**   **Fig. 2.3**   **Fig. 2.4**   **Fig. 2.5**

Fig. 2.6   **Fig. 2.7**

The spherical or semi-ball joint is shown in Fig. 2.6 which includes one type of special clamp for holding the two halves of the joint together. This connection cannot freeze or stick (as conical joints sometimes do) and it introduces a degree of flexibility into the apparatus in which it is used. The area of contact between the ground surfaces is relatively small so the joints are not intended to provide for considerable angular deflection. The main application is in conjunction with conical joints rather than as a substitute for them. The conical–spherical adapters shown in Fig. 2.7 provide a means of inserting a spherical joint while retaining the conical joint principle.

Precision made *stopcocks*, with interchangeable ground glass keys, are usually fitted with key retainers to prevent accidental loss or breakage during storage and use.

2.6 CARE AND MAINTENANCE OF GROUND GLASS JOINTS

Great care must be taken to keep all ground glass surfaces free from grit or dust; during storage a strip of paper should be inserted into joints and stopcocks to prevent jamming. For work at atmospheric pressure, no lubricant should be required; it is advisable, however, in order to reduce the danger of sticking to apply a *slight* smear around the upper part of each ground joint of a rubber grease, Apiezon grease L or M, or Silicone stopcock grease.

When salt solutions or alkaline substances may come into contact with ground glass surfaces, light lubrication of the surfaces is essential. When greasing stopcocks, only the outer parts of the plug should be lightly smeared with lubricant; in those cases where the lubricant is not harmful, the whole of the plug may be given a very thin smear of the grease but particular care must be taken to avoid the entrance of the lubricant into the bore of the plug. If necessary the bore of a stopcock can be cleaned conveniently with a pipe-cleaner.

Lubrication of all ground glass surfaces is essential for distillations under reduced pressure. Suitable lubricants are Apiezon grease L, M or N and Silicone stopcock grease. The use of synthetic joint linings particularly in distillation assemblies is a very useful alternative to lubrication with grease and completely removes any possibility of lubricant contaminating the contents of the apparatus. Joint linings (or sleeves) are available in polytetrafluoroethylene (PTFE), a polymer extremely resistant to chemical attack and to heat (up to 250 °C); a PTFE film may also be applied to the joint surface using an aerosol spray.

SEIZING OF GROUND GLASS JOINTS

Provided adequate care is exercised to use only joints that fit well, and the ground surfaces are suitably lubricated and parted after use while still warm, sticking will rarely occur. If, however, a ground joint should seize up or freeze the following suggestions may be found useful:

1. Set the joint in a vertical position and apply a layer of glycerine or penetrating oil to the upper surface. The glycerine will slowly penetrate into the joint, thus permitting the separation of the ground surfaces.
2. If procedure (1) is unsuccessful, direct a stream of hot air from a blower on to the outer surface of the joint for a few seconds and gently draw the members apart with a twisting action; gentle tapping on the edge of a wooden bench is sometimes helpful.
3. Introduce the joint into a small luminous Bunsen flame for a few seconds, and then gently draw the ground surfaces apart. If the glass is of Pyrex (or of any other heat-resisting variety), there is very little danger of a crack resulting from this process. The object of the heating is to cause the glass of the socket to expand before any appreciable change has occurred in the inner cone. More even, gentle and local heating of the joint, which is particularly suitable for stoppered flasks containing volatile, flammable or corrosive liquids, may be achieved by the following procedure. Wrap the flask in several layers of cloth with the neck protruding and hold securely within a fume cupboard; wrap, with one turn, a piece of fibrous string around the outside of the ground glass joint, and with a to-and-fro pulling action on the two string ends (gently at first until the rhythm of motion is acquired), allow the circle of string

around the joint to move smoothly along its length. Two operators are required, the one holding the flask steady should ensure that the stopper is directed away from both in case of accidents; both should be wearing safety spectacles. After a few minutes the joint will have heated sufficiently by friction and to a temperature leading to a smaller risk of decomposition of substances held between the joint surfaces than by the flame method; frequently the stopper may be removed by pulling with a twisting motion.

4. Frozen stopcock joints, or adapters or still-heads in *empty* flasks, may frequently be freed by immersion in an ultrasonic cleaning bath (Section 2.1, p. 29).

2.7 APPARATUS WITH INTERCHANGEABLE GROUND GLASS JOINTS SUITABLE FOR GENERAL USE IN PREPARATIVE ORGANIC CHEMISTRY

In considering the following typical standard units of equipment fitted with ground glass joints, it must be borne in mind that while a particular piece of glass equipment of certain capacity or dimensions may be fitted with alternative joint sizes, the range is usually restricted in relation to their relative proportions. When equipping a laboratory, it is usually convenient to limit the range of socket sizes thus permitting interchangeability with the minimum number of adapters. For example, with Quickfit, 14/23, 19/26, 24/29 and 34/35 joints are suitable for macro scale experiments, and 10/19 and 14/23 for semimicro scale experiments; a similar selection would be appropriate from the USA coded sizes.

In Fig. 2.8(a)–(d), the various designs of *flasks* are collected. Type (a) is a pearshaped flask, the capacity range being usually 5 ml to 100 ml, the joint sizes are in the range 10/19 to 24/29. Type (b) is a round-bottomed flask (short-necked), the capacity range being 5 ml to 10 litres, joint sizes being in proportion; medium- and long-necked designs are also available. Type (c) illustrates a range of wide-necked reaction flasks which are useful in semimicro and in pilot scale experiments and which are fitted with large diameter flat-flange joints, the capacities range from 250 ml to 20 litres, the flange bore being 75 mm to 100 mm respectively; the multi-socket lids are illustrated in Fig. 2.18(a) and (b). The advantages of this type of reaction vessel are that (i) the lids are easily detachable, (ii) large stirrers are readily accommodated, (iii) the vessels are cleaned readily and (iv) the removal or addition of solids and viscous fluids is facilitated; the ground flange joints are fully interchangeable. Special clamps are available for the support of such flasks. Type (d) of Fig. 2.8 is a jacketed flange flask which

(a)   (b)   (c)   (d)

Fig. 2.8

Fig. 2.9

is ideally suited for reactions requiring accurate temperature control. The example illustrated is available from J. Bibby Science Products Ltd. Other designs are available from, for example, Wheaton Scientific.

Various types of *multi-necked round-bottomed flasks* are illustrated in Fig. 2.9(a)–(d); designs with pear-shaped flasks are available. The centre socket is usually the larger and the side sockets are generally smaller; type (d) shows the side socket being employed for the insertion of a capillary tube necessary in a vacuum distillation assembly (see Section 2.27).

Ground glass *stoppers* of all standard sizes are available and may be of the design shown in Fig. 2.10; the flat head is preferred since the stopper may be stood on end when not in use, thus avoiding contamination of the ground surface; an additional refinement is the provision of a finger grip.

Often in the assembly of apparatus, joint adapters are required if the joint sizes of the various parts are not compatible. A *reduction adapter* is illustrated in Fig. 2.11 and an *expansion adapter* in Fig. 2.12; numerous combinations are of course possible, but it must be emphasised, however, that in a well-designed assembly of apparatus the number of adapters should be reduced to a minimum and, best of all, completely eliminated.

Distillation heads (or *still-heads*) are shown in Fig. 2.13 (a)–(c). Type (a) is a bend ('knee-tube') which is frequently employed for those distillations which merely require the removal of solvent. Type (b) is a simple distillation head; when fitted into a flask with a ground glass socket, the assembly is virtually a distillation flask. For some purposes, a thermometer may be fitted into a one-hole rubber stopper of correct taper and then inserted into the socket; the area of rubber which is exposed to the organic vapour is relatively so small that the amount of contamination thus introduced is negligible. If, however, all rubber stoppers must be absent because of the highly corrosive nature of the vapour, a thermometer fitted with an appropriate size cone is employed. Alternatively the

Fig. 2.10   **Fig. 2.11**   **Fig. 2.12**   **Fig. 2.13**

socket of a distillation head may be fitted with a *screw-capped adapter* (see Fig. 2.32) through which a thermometer may be inserted. Type (*c*) is a Claisen distillation head; the left-hand socket accommodates the capillary tube for use in distillations under vacuum (see Section 2.27) and the right-hand socket a suitable thermometer.

Frequently for semimicro and micro work it is more convenient to use the pear-shaped flask designs which incorporate the distillation heads (e.g. Fig. 2.14(*a*) and (*b*)).

Multiple adapters provide for additional entries into a single-necked flask when a multi-necked flask is not available. Either double-necked or triple-necked adapters (Fig. 2.15 and Fig. 2.16(*a*) and (*b*) are commonly used having a range of socket and cone sizes. The 'swan-neck adapter' of Fig. 2.17 is useful for vacuum distillations as it permits the insertion of a capillary tube through the screw thread joint. This joint may also be used for insertion of a thermometer or a gas inlet in the narrow neck and a reflux condenser into the ground joint; this device virtually converts a three-necked flask into a four-necked flask.

Multiple socket lids for fitment to the flange flasks (illustrated in Fig. 2.8(*c*)) are shown in Fig. 2.18(*a*) and (*b*). These allow for the introduction of a great variety of standard equipment for stirring, temperature measurement, the inlet of gas, etc. The sockets may be vertical or angled at 5, 10 or 15 degrees from the vertical axis.

Several types of *condensers* are widely used (Figs 2.19–2.23). An improved form of Liebig's condenser, sometimes termed a West condenser, is shown in Fig. 2.19; it has an inner tube with very thin walls and the space between it and the heavy-walled outer tube is small, consequently there is a rapid heat transfer to the fast-flowing cooling water leading to greater efficiency. The length of the jacket is usually 6 to 50 cm and the design is available in a range of joint sizes.

(*a*)　　　　(*b*)

Fig. 2.14　　　　　　　**Fig. 2.15**

(*a*)　　　　(*b*)

Fig. 2.16　　　　　**Fig. 2.17**

(a)   (b)

Fig. 2.18

Figure 2.20 (Davies type) and Fig. 2.21 (double coil type) are examples of efficient double surface condensers. Figure 2.22 depicts a 'screw' type of condenser (Friedrich pattern); this highly efficient condenser is employed for both reflux and downward distillation. The ice or dry ice–acetone condenser (Fig. 2.23) is useful for volatile liquids.

The water inlet and outlet side-arms on the condensers illustrated are of the standard olive all-glass type. Breakage can easily occur, often resulting in serious hand injury, when attaching or detaching rubber or plastic water hoses. A recent design (Bibby Science Products) incorporates a screw-thread at side-arm ends on to which a plastic hose connector may be screwed to give a watertight seal. Not only is this safer in the hands of less experienced workers, but it allows for more rapid apparatus assembly.

Various forms of *receiver adapters* or *connectors* for attachment to the end of condensers when used in a distillation assembly are shown in Figs 2.24–2.26.

Fig. 2.19   **Fig. 2.20**   **Fig. 2.21**   **Fig. 2.22**   **Fig. 2.23**

61

Fig. 2.24   Fig. 2.25   Fig. 2.26   Fig. 2.27   Fig. 2.28

The simplest form (Fig. 2.24) carries glass hooks for securing it to the condenser by means of a rubber band from the side tube to the hook; an improved form, incorporating two glass joints, is shown in Fig. 2.25. A useful adapter is illustrated in Fig. 2.26; when employed at atmospheric pressure, a drying tube may be attached to the side tube, if desired; in a distillation under reduced pressure, the side tube is connected to a vacuum pump.

Cone/rubber tubing adapters ('*take-off*' adapters), shown in Fig. 2.27(*a*) and (*b*), fulfil a number of useful purposes in preparative organic operations, for example where very small volumes of solvents need to be rapidly removed.

A *calcium chloride guard-tube* is illustrated in Fig. 2.28 which is widely used for protecting apparatus assemblies from the ingress of moisture.

For many operations the globular form of *dropping, addition* or *separatory funnel* having a suitable cone joint fitted to the stem is convenient, but when required on either a multiple-necked flask or with a multiple adapter, the cylindrical design (Fig. 2.29) is preferred; this is similarly provided with a cone on the stem and a ground socket. Figure 2.30 illustrates a cylindrical funnel with pressure-equalising tube; this is invaluable for reactions which are conducted in an atmosphere of inert gas. Either funnel may be fitted with an all-glass or a Rotaflo stopcock; the latter gives excellent liquid flow control. Jacketed dropping funnels for use with ice–water or dry ice–acetone slurry coolants are available and are useful when reagents to be added to a reaction mixture need to be kept at low temperatures. Dropping funnels are also available with a design of stopcock which allows infinite control of the rate of addition.

(*a*)   (*b*)

Fig. 2.29   Fig. 2.30   Fig. 2.31

The two designs of the Dean and Stark apparatus (Fig. 2.31(*a*) and (*b*); available from Bibby Science Products) carry a flask on the lower cone and a reflux condenser on the upper socket. They are used for the automatic separation of two immiscible components in a distillate and the subsequent return of the upper layer (*a*) or the lower layer (*b*) to the reaction flask.

2.8   OTHER TYPES OF INTERCHANGEABLE JOINTS AND STOPCOCKS

In some reactions the presence of a ground glass surface may initiate a rapid decomposition of reaction products and the danger of an explosion. One such reaction is the formation of diazomethane (Section 4.2.*25*, p. 430). In these cases the glass 'Clear-seal' joint (Wheaton Scientific) is essential. This type of joint needs no lubrication or plastic seals, does not allow seepage of solutions into the joint, and being transparent allows for ready observation of thermometer scales that cannot be seen with the ground glass type.

The useful screw-thread connector is becoming widely available. The simplest is the screw-capped adapter (Fig. 2.32) where the screw-capped joint is associated with a ground glass cone. The figure shows the silicone rubber ring and PTFE washer, and the adapter is useful for the insertion of gas tubes, thermometers or stirrer shafts. The screw-thread connection principle has been applied to the connection of water hoses to condensers and vacuum lines to Buchner flasks, to securing simple sealing septa to flasks, etc.

Stopcocks of the standard design are available with interchangeable Teflon plugs (or keys) fitted to the glass barrels. The Rotaflo design (Fig. 2.33) is now a commonly used alternative. It consists of a fully interchangeable PTFE key, incorporating a unique locking device to avoid total accidental unscrewing. The general purpose (GP) design is used in dropping funnels, burettes, etc., and provides excellent liquid flow control. A high performance (HP) design, which operates at reduced pressures down to 10 mmHg, is invaluable for incorporation into high vacuum systems.

The '*O-ring joint*' (manufactured by, for example, J. Young and Co. Ltd) (Fig. 2.34) is particularly suitable for incorporation into vacuum line assemblies. The joint consists of a slightly tapered cone, with a terminal annular indentation carrying a replaceable PTFE ring seal, which is inserted into a suitably-sized socket. The PTFE ring provides a seal under vacuum and also allows for a degree of flexibility in the vacuum line which facilitates assembly. The O-ring seal principle has also been incorporated into the design of, for example, stop-

Fig. 2.32   Fig. 2.33   Fig. 2.34   Fig. 2.35

cocks, screw-thread connectors, and dry, lubricant-free seals in vacuum desiccators.

Rubber (red or silicone) septa (Fig. 2.35), are widely used for fitting to sockets of flasks to allow injection or removal of liquids or gases via syringes or cannulae (Section 2.17.8, p. 122). The annular serrations depress against the glass wall providing an excellent seal, and the turnover flange is moulded to grip the outside of the container neck.

2.9 THE USE OF CORKS AND RUBBER STOPPERS

Although these have been largely replaced by ground glass joints, corks and rubber stoppers still find occasional use in the laboratory.

Two points must be borne in mind when selecting a cork stopper. In the first place, the cork should be examined for flaws; unless corks of the highest quality are employed, they are liable to have deep holes, which render them useless. In the second place the cork should fit into the socket to only about one-quarter of its length. It should then be softened by rolling in a cork press or by wrapping it in paper and rolling under the foot.

To bore a cork, a borer should be selected which gives a hole only very slightly smaller than that desired. The cork borer is moistened with water or alcohol or better still with glycerol. The borer is held in the right hand and the cork in the left hand. The hole is started at the narrow end with a continuous rotary motion. Beginners should bear in mind that the borer is a cutting instrument and not a punch, and on no account should it be allowed to burst its way through the cork because the borer, upon emerging, will almost invariably tear the surface of the cork. It is a good plan to examine the borer from time to time as it advances through the cork to see that it is cutting a straight hole. Boring should be stopped when it is half through the cork and the tool removed from the hole. The cork plug is pushed out with the aid of the solid metal rod supplied with the set of borers, and the remainder of the hole is bored from the other end. If the holes are carefully aligned, a clean cut hole is obtained. Experienced laboratory workers frequently complete the whole boring operation from one side, but beginners usually tear the edges of the cork by this method. A well-fitting cork should slide over the tube (side-arm of distilling flask, thermometer, lower end of condenser, etc.) which is to pass through it with only very moderate pressure. The bored cork should be tested for size; if it is too small, the hole should be enlarged to the desired diameter with a small round file. When the correct size is obtained, the tube is held near the end and inserted into the cork. The tube is then grasped near the cork and cautiously worked in by gentle twisting. Under no circumstances should the tube be held too far from the cork nor should one attempt to force a tube through too small an opening in a cork; neglect of these apparently obvious precautions may result in a severe cut in the hand from the breaking of the glass tube. The sharp edges of freshly cut glass tubing must be smoothed by fire polishing (Section 2.10).

For consistently successful results in cork boring, a sharp cork borer must be used. The sharpening operation will be obvious from Fig. 2.36. The borer is pressed gently against the metal cone, while slight pressure is applied with the cutter A at B; upon slowly rotating the borer a good cutting edge will be obtained. If too great pressure is applied either to the borer or to the 'cutter', the

Fig. 2.36

result will be unsatisfactory and the cutting circle of the borer may be damaged. To maintain a cork borer in good condition, it should be sharpened every second or third time it is used.

To bore a rubber stopper, it is essential to employ a very sharp cork borer of the same size as the tube to be inserted into the hole. The borer is lubricated with a little glycerol or alcohol and steadily rotated under only very slight pressure. The operation requires a good deal of patience and time and frequent lubrication may be necessary; if too much pressure is exerted on the borer, a hole of irregular shape and diminishing size will result.

The insertion of a glass tube into the rubber stopper or into rubber tubing is greatly facilitated by moistening the rubber with a little alcohol. Some grades of synthetic polymer tubing are only semi-flexible and are best fitted on to a glass tube after softening by immersion of the ends in a boiling water bath.

After some use rubber may stick to glass and great care must be taken not to break the glass tube when removing it. Frequently the exertion of gentle pressure on the rubber stopper by means of the two thumbs while the end of the tube rests vertically on the bench will loosen the stopper; this operation must, however, be conducted with great care. Another method is to slip the smallest possible cork borer, lubricated with a little glycerol, over the tube, and gradually to rotate the borer so that it passes between the stopper and the glass tube without starting a new cut. In cases of difficulty it is *always* safer, and in the end more economic, to cut the rubber stopper away from the glass tube.

2.10 CUTTING AND BENDING OF GLASS TUBING

Many students tend to forget the practical details learnt in elementary courses of chemistry; they are therefore repeated here. *To cut a piece of glass tubing*, a clean scratch is first made with a triangular file, sharp glass knife or diamond pencil. The tubing is held in both hands with the thumbs on either side of the scratch, but on the side opposite to it. The tubing is then pulled gently as though one wanted to stretch the tube and also open the scratch. A break with a clean edge will result. The cut edge must then be rounded or smoothed by *fire polishing*. With soda-glass the end of the tube is heated in the Bunsen flame until the edges melt and become quite smooth; the tube is steadily rotated all the time so as to ensure even heating. Overheating should be avoided as the tube will then partially collapse. Manipulations with Pyrex glass tubing, either fire polishing, bending, or drawing into a capillary leak for vacuum distillations (Section 2.27), have to be carried out in the flame of an oxygen-gas blowpipe.

A 'batswing' or 'fish-tail' burner is generally used for the *bending of soda glass tubing*. Both hands are used to hold the tube in the length of the flame (Fig.

Good bend   Bad bend
(a)   (b)   (c)

Fig. 2.37

2.37(a)) so that 5–8 cm are heated: the tube must be slowly rotated about its axis so as to heat all sides equally. As soon as the glass is felt to be soft, it is bent to the required shape. This is best done by removing it from the flame and allowing one end to fall gradually under its own weight, while being guided so that it is in the same plane as the rest of the tube. The glass must never be forced, otherwise a bad bend with a kink will be obtained as in Fig. 2.37(c).

2.11 GENERAL LABORATORY APPARATUS

Apart from the ground glass apparatus discussed in the previous sections, the student should also be aware of the range of other equipment available from general laboratory suppliers which is used in the course of preparative and analytical work. It should be remembered however that often the practical worker has to design, from the available equipment, pieces of apparatus to carry out a specific operation. This is particularly necessary when handling semimicro and micro quantities of material. Instances are cited in the later sections on isolation and purification procedures where designs are suggested.

Various types of *flasks* are shown in Fig. 2.38(a)–(e). Types (a) and (b) are flat-bottomed flasks (the Florence flask) with or without a wide neck, of capacities between 50 ml and 20 litres, the larger sizes having a tooled ring neck (b) to increase mechanical strength; type (c) is the round-bottomed flask having capacities of between 50 ml and 20 litres; type (d) is the short-necked boiling flask (the so-called bolt-head flask) with a tooled ring neck, of capacities between 50 ml and 10 litres. Type (e) is the familiar Erlenmeyer or conical flask obtainable in narrow and wide mouth designs, with and without graduations, in sizes which range from 5 ml to 6 litres.

The usual Griffin form of *beaker* with spout, Fig. 2.39(a), is widely used. Sizes between 5 ml and 6 litres are available. The conical beaker, Fig. 2.39(b), occasionally finds use in preparative work. Some designs have a Teflon (PTFE) rim

(a)   (b)   (c)   (d)   (e)

Fig. 2.38

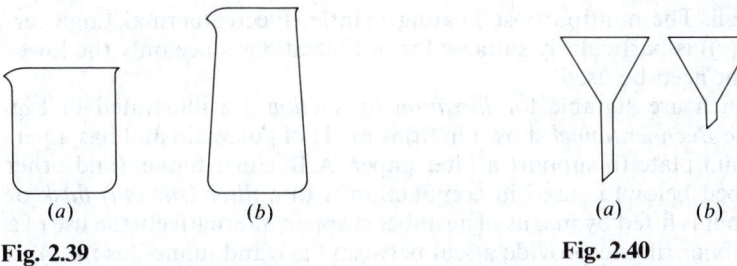

(a)   (b)   (a)   (b)

Fig. 2.39   **Fig. 2.40**

to enable aqueous solutions to be poured safely, drop by drop, if necessary. Polypropylene and polyethylene beakers are available in various sizes but have more limited use being unable to withstand temperatures above 120 °C and being unsuitable for use with many organic solvents. PTFE beakers are very much more expensive but are able to withstand temperatures up to 300 °C and are inert to most chemicals.

Various kinds of *funnels* are depicted in Figs 2.40(a) and (b), and 2.41(a)–(c). Figure 2.40(a) and (b) are long and short stem (0.5–1 cm) designs of ordinary filtration funnels respectively having a 60° angle; a wide-stemmed design of (b) is useful when transferring powders. The funnels in Fig. 2.41 are known as *separatory funnels* (obtainable in capacities between 25 ml and 5 litres); type (a) is the pear-shaped form and is the most widely used; types (b) and (c) are the conical and cylindrical separatory funnels respectively; the latter being available in a graduated form. The use of separatory funnels fitted with Teflon or with Rotaflo stopcocks reduces the problems associated with tap seizure and also avoids the use of lubricants which may contaminate the products.

When it is required to filter hot solutions, as for example in recrystallisations employing decolourising charcoal (Section 2.20), a heated jacket around the filtration funnel is necessary. A simple, inexpensive form is shown in Fig. 2.42; it consists of a coil made of copper of about 10 mm diameter forming a 60° cone which may be readily constructed in the laboratory workshop. Hot water or steam is passed through the coil, hence it is very suitable for the filtration of flammable liquids. The coil may also be used in 'cold filtration' and in some forms of sublimation apparatus (Section 2.21) by circulating ice-cooled water

(a)   (b)   (c)

Fig. 2.41

Fig. 2.42

through the coil. The multipurpose heating mantle (Electrothermal Engineering) (Fig. 2.47(c)) is particularly suitable for hot filtrations since only the lower heating element need be used.

Funnels which are suitable for *filtration by suction* are illustrated in Fig. 2.43(a)–(f). The *Buchner funnel* shown in (a) is made of porcelain and has a perforated porcelain plate to support a filter paper. A Buchner funnel (and other funnels described below) is used in conjunction with a *filter (suction) flask* or *tube* into which it is fitted by means of a rubber stopper; alternatively the use of a flat annular rubber ring to provide a seal between flask and funnel (as in (c)) is often more convenient. The side-arm of the flask or tube is attached by means of thick-walled rubber tubing ('pressure tubing') via a suitable trap to a water pump (Section 2.19). The *Hirsch funnel* shown in (b) has sloping sides and is designed to deal with a smaller amount of precipitate than is the Buchner funnel. The smallest size will accommodate filter papers 3–4 mm in diameter. The '*slit sieve' funnel* (d) is constructed entirely of glass (Jena or Pyrex) and therefore possesses obvious advantages over the opaque (porcelain) Buchner or Hirsch funnel. Similar advantages are apparent with the sintered glass funnel (e), which is available in a number of porosities (coarse, medium and fine).

The Buchner and Hirsch funnels, and the filter flasks and tubes, are available with interchangeable ground glass joints (for example, Fig. 2.43(f)). The Buchner funnel is also available with a screw-thread side-arm for easy attachment via a plastic hose connector to the vacuum line (Bibby Science Products).

| (a) | (b) | (c) |
| (d) | (e) | (f) |

Fig. 2.43

Spatulas in stainless steel, nickel, flexible horn, or moulded polythene are available commercially. One such design of palette knife has a flexible stainless steel blade provided with a wooden handle. Nickel spatulas may have turned-up ends, or spoon- or blade-shaped ends, and are available in sizes from 100 mm × 5 mm (for semimicro work) to 200 mm × 10 mm (for macro work).

Solids which are moist with either water or organic solvents are routinely dried in a *vacuum desiccator* at room temperature. Several high dome safety desiccators are available commercially in sizes which range from 76 mm (micro) to 305 mm in diameter. Those in which a glass or ebonite side entry tap socket of the interchangeable type (usually 34/35 joint size) is fitted to the lid, are to be preferred. Figure 2.44 shows one such desiccator with a Rotaflo stopcock assembly with plastic safety connector (Bibby Science Products); in this case the air inlet to the desiccator terminates in a hooked extension which serves to ensure that the air flow when the vacuum is released is directed in an even upward spread to prevent dispersal of the sample. The joint between the lid and the base may be an interchangeable ground flange and this joint needs lubrication (e.g. with Apiezon grease) before the desiccator is evacuated. In this type the lid is removed by side pressure after the vacuum has been released. A 'Dry-Seal' joint (developed and marketed by Jencons) is that in which a groove in the top flange of the desiccator base accommodates a removable elastomer sealing ring. This ring becomes flattened by the lid when the vacuum is applied, the design being such that over-compression is avoided. No lubricants are required, and when the vacuum is released the lid is simply lifted off – the removal of desiccator lids in the ground flange type can cause considerable difficulty. *In use, all vacuum desiccators must be sited in an appropriately sized and totally enclosed wire-mesh desiccator cage; desiccator implosion may occur at any time when it is under vacuum, and represents a serious hazard.*

Ordinary (i.e. atmospheric pressure) desiccators are available in the Dry-Seal or ground flange range and have limited use for storage of samples in a dry atmosphere.

The nature of the charge in a desiccator, which is placed in the lower compartment below the metal gauze plate, is dependent on whether water or organic solvents are to be removed and whether acidic or basic vapours are likely to be evolved during the drying process. Suitable charges are discussed in Section 2.20.

Fig. 2.44

2.12 COOLING OF REACTION MIXTURES

It is often necessary to obtain temperatures below that of the laboratory. Finely crushed ice is used for maintaining the temperature at 0–5 °C; it is usually best to use a slush of crushed ice with sufficient water to provide contact with the vessel to be cooled and to stir frequently. It is of course essential to insert a thermometer into the reaction mixture to ensure that the desired temperature is attained. For temperatues below 0 °C, the commonest freezing mixture is an intimate mixture of common salt and crushed ice: a mixture of one part of common salt and three parts of ice will produce a temperature of about −5 to −18 °C. Greater cooling may be obtained by the use of crystalline calcium chloride; temperatures of −40 to −50 °C may be reached with five parts of $CaCl_2.6H_2O$ and 3.5–4 parts of crushed ice.

If ice is temporarily not available, advantage may be taken of the cooling effect attending the solution of certain salts or salt mixtures in water. Thus a mixture produced by dissolving 1 part of NH_4Cl and 1 part of $NaNO_3$ in 1–2 parts of water causes a reduction in temperature from 10 to −15 °C to −20 °C; 3 parts of NH_4Cl in 10 parts of water from 13 to −15 °C; 11 parts of $Na_2S_2O_3.5H_2O$ in 10 parts of water from 11 to −8 °C; and 3 parts of NH_4NO_3 in 5 parts of water from 13 to −13 °C.

Solid carbon dioxide (Dry Ice, Drikold, Cardice) is employed when very low temperatures are required. The commercially available blocks are stored in specially insulated containers. Since frostbite may result from handling solid carbon dioxide, it is advisable to either wear gloves or to cover the hands with a thick cloth. Conveniently small-sized lumps may be obtained by hammering, with a wooden or polyethylene mallet, a suitable large piece wrapped in a cloth or contained within a stout canvas bag. The small pieces are carefully added to either ethanol or acetone in a plastic bowl until the lumps of solid carbon dioxide no longer evaporate vigorously. The temperatures attained are in the region of −50 to −70 °C according to the efficiency of the lagging around the freezing bath. In order to keep the freezing mixture for hours or overnight, it should be prepared in a Dewar flask.

The use of cooling baths employing other solvents with solid carbon dioxide enables other temperatures to be attained. An extensive list has been published[21] from which the following have been selected: ethylene glycol/Cardice, −15 °C; acetonitrile/Cardice, −42 °C; diethyl ether/Cardice, −100 °C. A steady state temperature cooling bath may also be obtained by adding solid carbon dioxide to o-xylene : m-xylene mixtures[22]; the volume fraction of o-xylene determines the temperature of the bath. For example, m-xylene/Cardice, −72 °C; o-xylene (0.4) : m-xylene (0.6), −58 °C; o-xylene (0.8) : m-xylene (0.2), −32 °C.

The attainment of temperatures lower than −100 °C requires the use of baths employing liquid nitrogen,[23] either alone, or admixed with other solvents. The hazards of using liquid nitrogen are pointed out in Section 2.3.2, p. 38, and the use of such cooling baths is *not* advised except in the hands of experienced workers.

2.13 HEATING OF REACTION MIXTURES

Heating of aqueous solutions is most conveniently carried out using a Bunsen burner with the glass vessel suitably supported on a tripod and ceramic-centred gauze; it is essential to use a heat resistant bench mat, and *under no circumstances* should such apparatus be left unattended. It is also imperative that no other worker using flammable solvents is in the vicinity.

In the case of solutions of flammable liquids having a boiling point below 100 °C, the stainless steel electrically-heated *water bath* or *steam bath* provided with a constant-level device must be used. The individual circular type is provided with a series of concentric rings in order to accommodate flasks and beakers of various sizes. A rectangular type, suitable for use in student classes, has several holes each fitted with a series of concentric rings. In both cases the water bath is fitted with an immersion heating element controlled by a suitable regulator.

For temperatures above 100 °C, *oil baths* are generally used. Medicinal paraffin may be employed for temperatures up to about 220 °C. Glycerol and dibutyl phthalate are satisfactory up to 140–150 °C; above these temperatures fuming is usually excessive and the odour of the vapours is unpleasant. For temperatures up to about 250 °C, 'hard hydrogenated' cottonseed oil, m.p. 40–60 °C, is recommended: it is clear, not sticky and solidifies on cooling; its advantages are therefore obvious. Slight discoloration of the 'hard' oil at high temperature does not affect its value for use as a bath liquid. The Silicone fluids, e.g. Dow Corning 550, are probably the best liquids for oil baths but are somewhat expensive for general use. This Silicone fluid may be heated to 250 °C without appreciable loss or discoloration. Oil baths should be set up in the fume cupboard wherever possible. A thermometer should always be placed in the bath to avoid excesive heating. Flasks, when removed from an oil bath, should be allowed to drain for several minutes and then wiped with a rag. Oil baths may be heated by a gas burner but the use of an electric immersion heater is safer and is to be preferred. A satisfactory bath suitable for temperatures up to about 250 °C may be prepared by mixing four parts by weight of 85 per cent ortho-phosphoric acid and one part by weight of meta-phosphoric acid; the mixed components should first be heated slowly to 260 °C and held at this temperature until evolution of steam and vapours has ceased. This bath is liquid at room temperatures. For temperatures up to 340 °C, a mixture of two parts of 85 per cent ortho-phosphoric acid and one part of meta-phosphoric acid may be used: this is solid (or very viscous) at about 20 °C.

High temperatures may be obtained also with the aid of *baths of fusible metal alloys*, e.g. *Woods metal* – 4 parts of Bi, 2 parts of Pb, 1 part of Sn and 1 part of Cu – melts at 71 °C; *Rose's metal* – 2 of Bi, 1 of Pb and 1 of Sn – has a melting point of 94 °C; a eutectic mixture of lead and tin, composed of 37 parts of Pb and 63 parts of Sn, melts at 183 °C. Metal baths should not be used at temperatures much in excess of 350 °C owing to the rapid oxidation of the alloy. They have the advantage that they do not smoke or catch fire; they are, however, solid at ordinary temperature and are usually too expensive for general use. It must be remembered that flasks or thermometers immersed in the molten metal must be removed before the metal is allowed to solidify.

One of the disadvantages of oil and metal baths is that the reaction mixture cannot be observed easily; also for really constant temperatures, frequent adjustment of the source of heat is necessary. These difficulties are overcome when comparatively small quantities of reactants are involved, in the apparatus shown in Fig. 2.45 (not drawn to scale).

A liquid of the desired boiling point is placed in the flask A which is heated with an electric mantle (see below). The liquid in A is boiled gently so that its vapour jackets the reaction tube BC; it is condensed by the reflux condenser at D and returns to the flask through the siphon E. Regular ebullition in the flask is ensured by the bubbler F. The reaction mixture in C may be stirred mechanically. It is convenient to have a number of flasks, each charged with a different liquid; changing the temperature inside C is then a simple operation. A useful assembly consists of a 50 ml flask A with a 19/26 joint, a vapour jacket about 15 cm long, a 34/35 joint at B and a 19/26 or 24/29 joint at D.

The following liquids may be used (boiling points are given in parentheses): pentane (35 °C); acetone (56 °C); methanol (65 °C); carbon tetrachloride (77 °C); trichloroethylene (86 °C); toluene (110 °C); chlorobenzene (132–133 °C); bromo-benzene (155 °C); p-cymene (176 °C); o-dichlorobenzene (180 °C); methyl benzoate (200 °C); tetralin (207 °C); ethyl benzoate (212 °C); 1,2,4-trichlorobenzene (213 °C); isopropyl benzoate (218 °C); methyl salicylate (223 °C); propyl benzoate (231 °C); diethyleneglycol (244 °C); butyl benzoate (250 °C); diphenyl ether (259 °C); dimethyl phthalate (282 °C); diethyl phthalate (296 °C); benzophenone (305 °C); benzyl benzoate (316 °C).

An air bath is a very cheap and convenient method of effecting even heating of small distillation flasks (say, 25 ml or 50 ml), where the use of a micro Bunsen burner, results in fluctuations in the level of heating due to air draughts. It may be readily constructed from two commercial tin cans (not aluminium) (those from tinned fruit or food are quite suitable), of such sizes that one fits into the other to provide a small concentric gap as an air insulator. The cleaned large can is cut to the same height as the small can, and the base is then removed. The

Fig. 2.45   **Fig. 2.46**

cleaned smaller can has a number of holes punched in the base. The edges of both cans must be smoothed and any ragged pieces of metal removed. A circular piece of ceramic paper (1 mm thickness) of the same diameter of the smaller can is inserted over the holes. A piece of reinforced calcium silicate matrix (6 mm thickness) of diameter slightly greater than the larger can is then obtained and a hole of suitable diameter made in its centre; the sheet is than cut diametrically. The two halves which constitute the cover of the air bath, will have the shape shown in Fig. 2.46(*b*). The diameter of the hole in the lid should be approximately equal to the diameter of the neck of the largest flask that the air bath will accommodate. The air bath, supported on a tripod and wire gauze, is heated by means of a Bunsen burner; the position of the flask, which should be clamped, is shown in Fig. 2.46(*a*). The flask should not, as a rule, rest on the bottom of the bath. The advantages of the above air bath are: (*a*) simplicity and cheapness of construction; (*b*) ease of temperature control; (*c*) rapidity of cooling of contents of the flask either by removing the covers or by completely removing the air bath; and (*d*) the contents of the flask may be inspected by removing the covers.

Heating mantles provide one of the most convenient means of controlled heating of reaction vessels. They consist of a heating element enclosed within a knitted glass-fibre fabric which is usually protected with a safety earth screen (Fig. 2.47(*a*), Electrothermal Engineering). The heating unit is enclosed within an outer rigid housing (often of polypropylene or aluminium) which is appropriately insulated so that the mantle may be handled at a low outer case temperature. Heating control is by in-built or external energy regulators. Fixed sizes for round-bottomed flasks having capacities from 50 ml to 5 litres are standard (e.g. Fig. 2.47(*b*), Isopad). In addition a multipurpose unit is now available which will accept a variety of different sized flasks of round-bottomed or pear-shaped design (Fig. 2.47(*c*), Electrothermal Engineering); this unit has a bottom outlet to accept 60° angle funnels to be heated in hot filtrations, in which case only the lower section of the heating element need be activated. Further designs of mantle (Electrothermal Engineering) are the fully enclosed flexible heating mantle with elastic neck entry which is often convenient when the apparatus assembly does not allow the satisfactory support of the encased type, and the heating mantle with in-built stirrer (Fig. 2.47(*d*)). Other manufacturers are Glas-Col and Thermolyne Corp, and the units are available through Aldrich Chemical Co. Ltd.

Electric hot plates may also be employed in the case of flat-bottomed vessels, and are provided with suitable energy regulators. Various sizes are available for individual use or for groups of students. The heating surface may be either cast-iron aluminium-sprayed, or a glass–ceramic surface. In the former case it is often advisable to interpose a sheet of ceramic paper between the metal top and the vessel to be heated, particularly if the contents of the latter are liable to 'bump'. *Electic hot plates should not be used with low boiling, flammable liquids (e.g. ether, light petroleum, etc.) contained in open beakers since ignition can frequently occur when the heavier vapour spills on to the heated surface.*

2.14 MECHANICAL AGITATION

Mechanical stirring is not necessary in work with homogeneous solutions except when it is desired to add a substance portion wise or dropwise and to bring it as

(a)

(b)

Fig. 2.47(a)–(d)

(c)

(d)

rapidly as possible into intimate contact with the main bulk of the solution. This applies particularly in those cases where a precipitate is formed and adsorption may occur, or where heat is generated locally which may decompose a sensitive preparation. In such cases the solution must be continuously agitated by manual shaking or, preferably, by mechanical stirring. When large quantities of material are to be dealt with, it is much easier and much more efficient to employ mechanical stirring. The importance of mechanical agitation cannot be over-estimated where heterogeneous mixtures are involved. In many preparations the time required for completion of the reaction is shortened, temperatures are more readily controlled, and the yields are improved when mechanical agitation is employed. No apology is therefore needed for discussing this subject in some detail.

Stirring in open vessels, such as beakers or flasks, can be effected with the aid of a stirrer attached directly to a small electric motor by means of a chuck or a short length of 'pressure' tubing, although in these circumstances some form of stirrer guide is advisable; for example, a sleeve of glass tubing, lubricated with glycerol, having a diameter slightly greater than that of the stirrer shaft and supported in a clamp.

Excellent stirring units are available commercially, and include 'explosion-proof' and 'air-driven' designs (Gallenkamp, Jencons, etc.). The geared-drive model (the Citenco stirrer, available from, for example, Jencons) gives good control of stirring performance. These stirrers may often be fitted with a flexible drive shaft so that the motor may be placed some distance from the stirrer head and reaction vessel, thus enabling the assembly to be used for flammable, corrosive or fuming liquids without damage to the motor.

Stirrers are usually made of glass, but those of Monel metal, stainless steel or Teflon also find application in the laboratory. An important advantage of a stirrer with a Teflon blade is that it is comparatively soft and merely bends if it hits the glass even at high speed; furthermore, it can be shaped to fit the bottom of the vessel, thus rendering the stirring of small volumes of liquid in a large flask possible. A few typical stirrers are shown in Fig. 2.48 (*a*)–(*e*); *types* (*a*) and (*b*) may be easily constructed from a glass rod. Types (*c*) and (*d*) are the Teflon and the glass-link stirrer respectively; because of the flexible end they possess the advantage that they may be inserted through a narrow neck; in (*c*) the half-moon shape allows it to be employed for stirring liquids in round-bottomed or flat-bottomed vessels (the latter by turning the blade over). Type (*e*) is a stainless steel propeller blade stirrer.

A useful stirrer – sometimes termed a *Hershberg stirrer* – for efficient agitation in round-bottomed vessels, even of pasty mixtures, is presented in Fig. 2.49. It consists of a glass rod to which a glass ring is sealed; the glass ring is threaded with chromel or nichrome or tantalum wire (about 1 mm diameter). By sealing another glass ring at right angles to the first and threading this with wire, even better results will be obtained. The stirrer is easily introduced through a narrow opening, and in operation follows the contour of the flask; it is therefore particularly valuable when it is desired to stir a solid which clings obstinately to the bottom of a round-bottomed flask.

Some form of suitable *stirrer seal* must be provided in any of the following operations: (*a*) simultaneous stirring and refluxing of a reaction mixture; (*b*) stirring the contents of a closed vessel; (*c*) agitation with the prevention of the escape of gas or vapour; and (*d*) stirring in an inert atmosphere, such as nitrogen.

A simple rubber sleeve gland (the *Kyrides seal*) is illustrated in Fig. 2.50; the short length of rubber tubing attached to the ground glass cone projects to form a tight seal around the stirrer shaft (5–6 mm). Glycerol (or Silicone grease) is applied at the point of contact of the glass and rubber to act as a lubricant and sealing medium. This seal may be used under reduced pressures down to about 10–12 mmHg; it is not dependable for stirring operations lasting several hours since the rubber tubing may stick to the shaft and may also be attacked by

(*a*)   (*b*)   (*c*)   (*d*)   (*e*)

Fig. 2.48   **Fig. 2.49**

(a)   (b)   (c)

Fig. 2.50   Fig. 2.51

organic vapours, causing it to swell and allow the escape of vapours. In these circumstances the screwcap gland, Fig. 2.32, may be used when the Silicone rubber ring should be similarly lubricated; this type of seal is not suitable for use at high speeds or under vacuum.

Precision ground stirrer/stirrer guide units (Jencons) eliminate the need for any form of additional seal and can be used very satisfactorily under vacuum, in which case a trace of Silicone grease as lubricant is recommended. The long length of the ground bearing substantially reduces the problem of 'whipping' and vibration. Figure 2.51(*a*) shows the general purpose type with a cone joint; a design in which the water-cooling jacket surrounds the precision ground tubes is also available. A useful variant is illustrated in Fig. 2.51(*b*) in which a vapour 'take off' outlet is provided; it is also of use for reactions performed under an inert atmosphere.

Ground sleeve glands have largely replaced the conventional *mercury-sealed stirrer*; one design of the latter which is suitable for general laboratory use is illustrated in Fig. 2.51(*c*). It is not as versatile as the ground sleeve type; high speeds introduce the hazard arising from spattering mercury (in which case glycerol is a safer alternative) and the seal is not suitable for use under vacuum.

A type of stirrer, known as a *Vibro-mixer* and of particular value for closed systems, is illustrated in Fig. 2.52, fitted into the central neck of a flask. The enclosed motor, operating on alternating current, vibrates the stirrer shaft at the same frequency as the a.c. mains, moving up and down in short, powerful strokes. A control knob at the top of the stirrer housing is provided for adjusting the stroke length from gentle stokes (0.2 mm) to powerful strokes (*c.* 2 mm in thrust). As it is a non-rotating stirrer, a hermetic seal with the reaction flask can be made easily. Several type of stirrer blades are available; two 'plate' stirrers are shown in Fig. 2.52(A and B). In A the holes taper upward; the liquid will (on the downstroke) flow up through the wide lower orifices to be violently expelled through the narrower orifices at the top. The principle involved is similar to what happens when water flowing through a pipe suddenly enters a narrower pipe; the speed of flow is greatly increased. In B the holes taper downward; excellent mixing is thus obtained for solids, etc., at the bottom of the vessel. The base

77

Fig. 2.52

of the stand upon which the stirrer is mounted rests upon anti-vibration mats in order to reduce vibration to a minimum. No guide for the stirrer shaft is necessary; the stirring is very efficient and particularly suitable for PTC experiments.

Several forms of *mechanical shaking machines* are employed for automatic mixing of heterogeneous systems and find many applications in the organic chemistry laboratory. Two of these are illustrated in Fig. 2.53(*a*) and (*b*). Type (*a*) is the bottle/flask shaker (Gallenkamp) in which the carrier will accommodate six 2-litre Winchester pattern bottles held horizontally; an interchangeable platform enables the shaker to be converted to accept a number of conical flasks, up to 77 × 250 ml or 20 × 2 litres. Type (*b*) is a 'wrist-shaker' which will accommodate up to 8 × 50 ml flasks; although not all positions need to be occupied, the load must be balanced.

Magnetic stirring is now widely used. A rotating field of magnetic force is employed to induce variable speed stirring action within either closed or open vessels. The stirring is accomplished with the aid of small permanent magnets ('followers') sealed in Pyrex glass, polypropylene or Teflon.* The principle of magnetic stirring will be evident from Fig. 2.53(*c*). A permanent bar magnet, mounted horizontally, is attached to the shaft of an electric motor; the whole is mounted in a cylindrical housing with flat metal top and heavy cast metal base. An energy regulator (which is frequently incorporated in the housing) is provided to control the rate of stirring. To use the apparatus, the regulator, which is

---

* The 'followers' are available in different lengths and of varying designs to fulfil a range of stirring requirements, such as would be necessary with differently shaped vessels or with fluids of different viscosities.

(a)

(b)

Magnetic
follower

Bar magnet

(c)

Fig. 2.53(a)–(c)

79

initially in the off-position, is slowly rotated (this increases the motor speed) until the required rate of stirring is attained. When the experiment is complete, the regulator is returned to the zero position, the 'stirrer' allowed to come to rest and removed with the aid of a pair of forceps.†

Magnetic stirring has many obvious applications, but the most important are probably to stirring in closed systems, e.g.: (*a*) where gas volume changes must be observed as in catalytic hydrogenations; (*b*) where exclusion of air is desirable to prevent oxidation; (*c*) where reactions are to be carried out in an anhydrous environment; and (*d*) where small containers are used and the introduction of a propeller shaft is inconvenient.

Many forms and sizes of magnetic stirring apparatus are available commercially. These include those fitted with an electric hot plate attached to the flat top, the temperature of which is controlled by an energy regulator. Electric heating mantles are also available in which magnetic stirring is incorporated so that both heating and stirring is possible in an apparatus assembly using round-bottomed flasks [e.g. solid-state Stirrer Mantle (EMA) from Electrothermal Engineering Ltd, Fig. 2.47(*d*)]. This design gives control over a range of volumes and viscosities of fluids in the flask, and additionally bi-directional or reverse stirring at 20 second intervals is a further facility. Magnetic stirring may not be effective if the medium is excessively viscous, or if substantial amounts of solids are present. In such cases mechanical stirrers must be used.

2.15 TYPICAL GROUND GLASS JOINT ASSEMBLIES FOR STANDARD REACTION PROCEDURES

It is hoped that the account of the interchangeable ground glass joint apparatus already given will serve as an introduction to the subject. For the numerous applications of such apparatus, the reader is referred to the catalogues of the manufacturers listed in Appendix 7. Most of the simpler operations in practical organic chemistry may be carried out with a set of apparatus which can be purchased for a comparatively modest sum. Indeed, such a set is the first in a progressive series of 'Quickfit' sets which are specifically designed to cover the experimental requirements of practical chemistry in many different fields and in scales ranging from gram to kilogram quantities. For student work, in either introductory or advanced multi-stage experiments, the glassware is supplied in boxes fitted with integral plastic trays so that each item fits into a moulded well marked with the appropriate catalogue number. Checking equipment and its storage is therefore greatly simplified.

Some typical assemblies are collated in the following diagrams, which represent most of the basic reaction procedures which are adopted in the later experimental sections.

An assembly for *heating a reaction mixture under reflux* is illustrated in Fig. 2.54. The precise design of condenser depends upon the volatility of the reaction liquid, low boiling liquids ($< 60\,°C$) require the use of a double surface condenser. Additionally a calcium chloride tube may be inserted at the end of the condenser if the reaction mixture contains moisture-sensitive components. It is

---

† A useful recovery device for followers is made by forcing a small bar magnet into the end of stiff polymer tubing.

Fig. 2.54   **Fig. 2.55**   **Fig. 2.56**

important that 'boiling chips' (carborundum, or broken porcelain) are added *before* heating (whether by gas or by electrical means, or whether using an oil or a water bath) is commenced.

For the purpose of *heating a reaction mixture under reflux with addition of liquid* the assembly shown in Fig. 2.55 is suitable. The two-way adapter and flask shown here may, of course, be replaced by a two-necked flask. Manual agitation may be required at intervals to ensure good mixing, alternatively the vigour of boiling may automatically aid the mixing process. The selection of condenser design, the presence of calcium chloride protecting tubes (on condenser and separatory funnel) and the mode of heating depend upon the nature of the reactants.

It should be emphasised that the apparatus assemblies described above are for those cases where the reaction mixture is a homogeneous liquid. With reaction mixtures having suspended solid components or with reaction mixtures involving immiscible liquids stirring is essential. An assembly for *heating under reflux with the addition of liquid and with stirring* is shown in Fig. 2.56. The stirrer shown may of course be omitted and agitation effected with a magnetic stirrer incorporated in an electric hot plate or heating mantle (Section 2.14).

In an experiment where the *addition of a moisture-sensitive solid* (e.g. anhydrous aluminium chloride) in small portions to a reaction mixture is required, but where no requirement of heating under reflux is necessary, the simplest assembly is that shown in Fig. 2.57. Here the side socket is fitted with a length of wide, thin-walled rubber tubing and a 100 or 250 ml conical flask containing the reagent is inserted into the other end of the tubing. The solid is readily added in portions by raising the flask; the latter can be cut off from the reaction mixture by 'kinking' the rubber tube. The diagram also shows that the central cone has been fitted with a suitable stirrer (see Section 2.14) and the third socket may be fitted either with a calcium chloride tube or with an appropriate nitrogen inlet system (see Fig. 2.58) if an inert atmosphere is to be preserved. The stirring

Fig. 2.57

is of course essential and is far more efficient and convenient than manual agitation.

An alternative design for the addition of solid is shown in Fig. 2.58 which also illustrates a nitrogen inlet system. It may be constructed, if desired, from a small pear-shaped flask, and a broken pipette; the connections are bored rubber of a size appropriate to fit into the sockets. The solid is charged into the conical reservoir; by raising the plunger to the appropriate height, any desired amount of solid may be made to flow into the reaction vessel, and the flow can be completely stopped by merely twisting down the plunger until the rubber ring seals the opening. The rubber ring should be, say, 3 mm thick and 5 mm wide; if solvents which attack rubber are present a neoprene gasket may be used. The rubber tubing at the top is lubricated with glycerol (Kyrides seal) to make an

N₂ inlet

To escape valve

Fig. 2.58

air-tight joint which will allow free movement of the plunger. The hopper may be recharged during the reaction without breaking the seal by lifting the stopper while holding down the pipette.

Where the *addition of solid* is required to a reaction mixture, which is being *heated under reflux with stirring*, either of these devices may be attached to the upper socket of the reflux condenser, which must, of course, be of the Liebig type. In these circumstances a second reflux condenser must be fitted with a guard-tube or with a nitrogen inlet system (cf. Fig. 2.60) to allow for equilibration of pressure within the apparatus.

The *addition of a gas to a reaction mixture* (commonly the hydrogen halides, fluorine, chlorine, phosgene, boron trifluoride, carbon dioxide, ammonia, gaseous unsaturated hydrocarbons, ethylene oxide) requires the provision of safety precautions which may not be immediately apparent. Some of these gases may be generated *in situ* (e.g. diborane in hydroboration reactions), some may be commercially available in cylinders, and some may be generated by chemical or other means (e.g. carbon dioxide, ozone). An individual description of the convenient sources of these gases will be found under Section 4.2.

For those gases which are generated *in situ*, no further points need be noted here as the apparatus is described in the appropriate section.

For those gases which are available in cylinders, the manufacturer's notes about precautions against any hazards in their use should be carefully followed and the appropriate antidotes should be available for immediate use. It is important to realise that cylinders of all gases are potentially very dangerous owing to the possibility of the valve being broken off should the cylinder be knocked over; cylinders must therefore always be securely strapped, or supported, in frames whether in storage or in use. Cylinders of toxic gases (e.g. chlorine, sulphur dioxide) are often of such dimensions that they may be easily accommodated and suitably supported in an efficient fume cupboard containing the reaction apparatus. With the larger cylinders of toxic or hazardous gases (e.g. ammonia, acetylene) it is essential that these should be located on the laboratory wall outside the building and piped as appropriate through holes in the wall directly into the fume cupboard.* Cylinders so positioned should have suitable protection from the weather and should be easily accessible. Large cylinders of inert gases (e.g. nitrogen) should be supported in suitable stands.

For gases which are generated by chemical or other means, it is essential that the flow of gas be maintained at as steady a rate as possible, by suitable control of the generating apparatus. *Such apparatus should not be left unattended, even for a comparatively short space of time.*

The gases provided by the above sources should be led to the reaction vessel via a train of Drechsel bottles, suitably charged to effect prior drying or purification, and to provide a trap if 'sucking back' of the reaction mixture occurs (e.g. in the rapid uptake of hydrogen chloride). Should 'suck-back' occur the experiment is not then ruined, nor does the reaction mixture either cause a potential hazard with the chemicals in the generating apparatus, or cause irreparable damage to the cylinder valve. A suitable apparatus set-up is shown in Fig. 2.59, where it should be observed that the gas is released just under the surface of the reaction liquid by means of a glass tube fitted with a wide pore size glass frit; this improves gas–liquid contact and aids the absorption of the gaseous reactant, but

---

* See Section 2.3.2 for recommendations on the provision of a Dangerous Operations Laboratory.

Fig. 2.59

Fig. 2.60

Fig. 2.61

should not be used if a solid reaction product that may block the pores is likely to be formed.

It is sometimes necessary (e.g. in reactions involving organolithium compounds or in certain Grignard preparations) to carry out a *reaction in an atmosphere of an inert gas*, such as nitrogen (see also Section 2.17.8, p. 120). A suitable set-up is shown in Fig. 2.60. Dry nitrogen is introduced at the top of the condenser and initially can be allowed to sweep through the apparatus and escape at the mouth of the dropping funnel; it will be noted that the latter has a pressure-equalising side tube. After a few minutes the flow of inert gas may be reduced. The level of mercury (alternatively mineral oil or a high-boiling ester, e.g. dibutyl phthalate) in the escape valve should be such that a slight pressure of gas within the apparatus is maintained when the funnel is closed. This arrangement* is economical in nitrogen, obviates the evaporation of solvent and is to be preferred to the use of a continuous stream of inert gas.

In some reactions a gas is evolved which may be of an irritant or corrosive nature (e.g. hydrogen chloride in Friedel–Crafts reactions, sulphur dioxide–hydrogen chloride in acid chloride preparations) and it is advisable to employ a suitable *gas absorption trap*. Either of the gas traps depicted in Fig. 2.61(*a*) and (*b*) are used when limited quantities of water-soluble gases are to be absorbed. For larger volumes of gas, or where the gas is rapidly evolved, the gas traps shown in Fig. 2.61(*c*) and (*d*) are very satisfactory. In (*c*) the gas is passed into a wide tube through which a stream of water (usually from a reflux condenser) flows into a large filter flask and overflows at constant level, which is above the lower end of the wide tube; a water seal is thus provided which prevents the escape of gas into the atmosphere, and the heat of solution of the gas is dissipated. A convenient size for (*d*) is a tube 80–100 cm long and 25 mm diameter.

A highly efficient gas-absorption apparatus† is depicted in Fig. 2.62. The overall length is about 40 cm; two inlets for obnoxious gases are provided, but one can be readily closed if not required. The waste water from a water condenser may be employed. The water enters in the middle of the apparatus and

---

* This nitrogen inlet may also be used with other apparatus assemblies where an inert atmosphere is to be preserved in the reaction vessel, for example as illustrated in Fig. 2.58. 'Bubblers' are also commercially available.
† Designed in the Research Laboratories of May and Baker Ltd, Dagenham.

Fig. 2.62

passes up the outer annulus, spraying out at the top of the tower on to 9 mm Raschig or similar rings. It then passes down the column and through the water trap at the bottom of the apparatus to waste through a side tube fitted with a siphon-breaking device. The contaminated gas enters at either side of the two inlet connections and is absorbed by the water passing down the column.

2.16 PRECAUTIONS FOR UNATTENDED REACTIONS

It is in the nature of organic reactions that quite frequently prolonged reaction times are necessary for their successful completion. Obviously a considerable saving in working time can be achieved if such a reaction can be left running unattended, and in particular overnight. Other prolonged operations include, for example, constant extraction with solvents (Section 2.22) and the elution of chromatographic columns (Section 2.31). For all these, essential services such as water (for cooling purposes) and electricity (for heating, stirring or for the operation of instruments) may have to be left on, and certain elementary but essential precautions are necessary to minimise the danger of damage due to fire or flood. In general it is desirable that the addition of all reactants should be completed before the reaction is left unattended overnight; once any obviously exothermic process has moderated the reaction can usually be safely left if necessary.

The safest arrangement for dealing with the problem of overnight reactions, particularly those carried out on a fairly large scale, is to have available some form of simple shelter constructed of angle iron and heat-resistant sheeting fitted with power points and a water supply; this may be sited on any convenient flat area away from the laboratory building. Alternatively a small room fitted with efficient extractor fans and with an automatic fire extinguishing system, or even a large tiled fume cupboard kept free of extraneous apparatus, could be set aside for such reactions. Adequate provision must be made for drainage should flooding occur; in the event, for example, of condenser leads splitting and becoming unattached.*

For reactions which merely require stirring without heating it is necessary to ensure that the stirrer shaft is rotating freely in a rigidly clamped stirrer guide to avoid the possibility of the stirrer breaking the reaction vessel. If a stirrer seal of the Kyrides type (Section 2.14) has to be used it must be well lubricated to avoid seizure. An electromagnetic stirrer can of course be used if the stirring it provides is adequate. Although modern shakers are designed so as to prevent the possibility of their 'travelling' over a bench or floor surface, nevertheless it is advisable to take steps to prevent movement and to ensure that the moving parts cannot accidentally touch other apparatus in the vicinity, or move against a wall surface.

The commonest operation which needs to be carried out during an overnight period is probably a reaction which requires heating under reflux, with or without stirring. Any form of gas heating is highly hazardous and electric heating mantles must be used; it is essential that the controls should be sited well away from the reaction assembly. Rubber tubing used for connecting reflux condensers, etc., to the water supply should be inspected for latent defects and should be securely wired on to the taps and glass apparatus inlets and outlets. Waste water should be led away through a rubber tube which has fitted at its end a glass tube, which projects well into the main drain. Before the reaction is finally left, one should check that the cooling water is flowing freely but not violently through all condensers.

2.17 APPARATUS FOR SPECIAL REACTION TECHNIQUES

2.17.1 CATALYTIC HYDROGENATION

Many classes of organic compounds may be efficiently reduced by molecular hydrogen in the presence of a suitable catalyst (*catalytic hydrogenation*).[24] Depending upon the nature of the functional group which it is required to reduce, the experimental conditions necessary for hydrogenation may vary widely; for example, hydrogenation may be carried out either at room temperature or at temperatures up to about 300 °C, with the use of hydrogen at atmospheric pressure or at pressures up to about 350 atmospheres. The successful hydrogenation of a particular functional group depends also upon the correct choice of a suitable catalyst, and a wide range of formulations is available for the preparation of catalysts in a suitably active form. The main commonly used catalyst preparations together with the functional groups which they most effectively reduce are discussed briefly below; their preparation is dealt with

---

* A clear notice should be displayed near the apparatus indicating the type of reaction being carried out, the nature of the solvent (if any) being used, and whether the water and/or electricity supply is required to be left on. In certain cases it may be desirable to indicate what steps should be taken if any of the essential services fail.

under Section 4.2. The nature of the solvent used may also influence the course of the hydrogenation. Acceptable solvents must not of course themselves be reduced under the hydrogenation conditions; cyclohexane, ethanol, acetic acid or ethyl acetate are widely used. Small quantities of acids or bases added to neutral solvents may have a significant effect on the course of some hydrogenations.

HYDROGENATION CATALYSTS

Platinum metal group. These are powerfully active catalysts which are used at normal or slightly elevated temperatures and pressures.

Platinum in a finely divided form is obtained by the *in situ* reduction of hydrated platinum dioxide (*Adams catalyst*); finely divided platinum may also be used supported on an inert carrier such as decolourising carbon. Finely divided *palladium* prepared by reduction of the chloride is usually referred to as *palladium black*. More active catalysts are obtained however when the palladium is deposited on decolourising carbon, barium or calcium carbonate, or barium sulphate. Finely divided *ruthenium* and *rhodium*, usually supported on decolourising carbon or alumina, may with advantage be used in place of platinum or palladium for some hydrogenation reactions.

The platinum metal group of catalysts readily reduce most olefinic and acetylenic multiple bonds at normal temperatures and pressures. The selective semi-hydrogenation of an alkyne to an alkene is efficiently carried out using deactivated palladium catalysts (e.g. *Lindlar's catalyst*). The reduction of aliphatic aldehydes and ketones to alcohols is a little more difficult to achieve; palladium is virtually ineffective but platinum which has been promoted by the addition of a little iron(II) sulphate works well. On the other hand, aryl aldehydes and ketones are readily reduced over palladium, but the products are the corresponding hydrocarbons rather than alcohols (i.e. *hydrogenolysis* of the intermediate alcohols, which are benzylic in nature, occurs). Such alcohols are best prepared using ruthenium, which does not promote hydrogenolysis. The platinum metal group are not the catalysts of choice for the reduction of aromatic ring systems, although ruthenium and rhodium are very effective, and Adams' catalyst in the presence of acetic acid may be used. This latter catalyst system is also useful for the reduction of pyridine rings; somewhat elevated pressures of hydrogen are used. An important use of deactivated palladium catalysts in the selective hydrogenolysis of acid chlorides to aldehydes (the *Rosenmund* procedure).

Nickel. Nickel catalysts for hydrogenation can be prepared in a range of activities (*Raney nickel catalysts*; see Section 4.2.*50*, p. 450). The most active grades are comparable with platinum and palladium in many of the reductions mentioned above. The less active grades are frequently used however and usually require moderate or high temperatures and pressures, depending upon the nature of the group which is to be reduced. These catalysts are particularly useful for the reduction of nitrogen-containing functional groups such as $-NO_2$, $-C\equiv N$, $>C=NOH$ to primary amines; nickel catalysts are not deactivated (poisoned) by amino compounds as are the platinum metal group. Nickel also effects hydrogenation of benzene rings; for catalysts of a moderate degree of activity, temperatures about 100 °C and pressures of about 100 atmospheres are usually adequate. An important application of Raney nickel, which is related to its insensitivity to the action of catalyst poisons, is in the reductive fission of

C—S bonds, a process which is exploited in the desulphurisation of organic sulphur-containing compounds.

Copper-chromium oxide. This is a catalyst of uncertain composition prepared by the ignition of basic copper(II) ammonium chromate. It is an approximately equimolar combination of copper(II) chromite and copper(II) oxide ($CuCr_2O_4 \cdot CuO$) but is evidently not a simple mixture of these two components. The catalytic activity is enhanced by the incorporation of some barium chromite; hydrogenations require however the use of relatively high temperatures and pressures. This catalyst, which may be regarded as complementary to Raney nickel, is generally useful for the reduction of oxygen-containing functions, and is the catalyst of choice for converting esters into primary alcohols. It may be used for the reduction of amides; it does not usually reduce an aromatic ring unless the conditions are exceptionally severe.

APPARATUS FOR CATALYTIC HYDROGENATION

The equipment described below presents potential hazards from the usage of hydrogen gas and of vessels maintained at pressures greater than atmospheric. All equipment should be located therefore in a suitably constructed laboratory (Section 2.3.2, p. 41). Furthermore the apparatus should be tested regularly for leaks (using an inert gas such as nitrogen), and in the case of pressure vessels inspected at appropriate intervals by a qualified engineer.

Hydrogenation at atmospheric pressure: standard procedure. This method for effecting hydrogenation at atmospheric pressure employs the apparatus shown semi-diagrammatically in Fig. 2.63(a); it is supported on a suitable metal rod framework. The essential features are a long-necked hydrogenation flask A fitted to the apparatus with sufficient flexible tube to allow shaking, a series of

Fig. 2.63(a)

water-filled burettes and reservoirs (two are shown, B and C), a manometer D and mercury safety trap E. The size of the burettes will be appropriate to the scale of operation for which the apparatus is designed; a suitable combination of 2 litre, 1 litre, 250 ml or 100 ml sizes may be used. The various parts of the apparatus are connected as far as is possible by ground glass joints lubricated with Silicone grease. Flexible tubing in contact with hydrogen is of polyvinyl chloride (PVC); the hydrogenation flask should be screened by a laminated safety glass shield. *No flames should be allowed in the laboratory while the hydrogenation apparatus is in use.*

The procedure for conducting the hydrogenation is as follows:

1. Disconnect the hydrogenation flask A, open taps H, K and L, and fill the burettes with water by raising the reservoirs. Close taps K and L and lower the reservoirs.
2. Charge the hydrogenation flask with the catalyst, and with the solution to be hydrogenated, taking care that the solution washes down traces of catalyst which might be adhering to the sides of the flask so that finally all of the catalyst is covered by solution. Attach the flask to the apparatus. Connect the apparatus to a water pump via a trap (not shown) and the three-way tap F and attach a hydrogen cylinder to G via a reducing valve.
3. Close tap M and evacuate the apparatus via tap F.
4. Close tap F and fill the apparatus with hydrogen to atmospheric pressure, as indicated by the manometer D, by slowly opening tap M; close tap M.
5. Re-evacuate the apparatus via tap F and then close tap F; repeat steps 4 and 5 once more.
6. Refill the apparatus with hydrogen via tap M, open taps K and L and allow the burettes to fill with hydrogen, if necessary lowering the reservoirs further. Close tap M.
7. With taps K, L and H open, adjust the levels of water in the reservoirs to just above those in the burettes and momentarily open the three-way tap to the atmosphere so that the pressure of hydrogen in the system reaches atmospheric. Record the water levels in the burettes. Close tap L.
8. Shake the flask A to initiate the hydrogenation, and adjust the reservoir of the burette B periodically so that the pressure of hydrogen is slightly above atmospheric. When the hydrogen in B is all used up, close tap K and open tap L to use the hydrogen in burette C.
9. When hydrogen uptake ceases adjust the level of the reservoir for burette C and read the burette. Close tap L.
10. Stop the shaker and swirl the flask manually to wash down below the surface of the solution all traces of catalyst which may be adhering to the sides of the flask and evacuate the apparatus via tap F. Admit air* through F and detach the hydrogenation flask.
11. Correct the total volume of hydrogen used to standard temperature and pressure to determine the uptake in moles.
12. Filter off the spent catalyst on a small Hirsch funnel and wash it with a little of the solvent. The damp used catalyst should be transferred immediately to a residues bottle for subsequent recovery (Sections 4.2.*54* and *61*, pp. 452 and

---

* Provided that the catalyst is covered with solution there is little danger of an explosion occurring when air is admitted to the apparatus; however, it is wise to ensure that appropriate precautions have been taken.

459 respectively); used hydrogenation catalysts should not be allowed to become dry on the filter paper as they are liable to inflame. The filtrate should then be worked up in a manner appropriate to the nature of the product.

The apparatus may be tested and the activity of the catalyst assessed by carrying out a hydrogenation of cinnamic acid (Expt 6.138) or maleic acid. The procedure for the latter is described below.

Hydrogenation of maleic acid. Place 20 mg of Adams' platinum dioxide catalyst (Section 4.2.*61*, p. 459) in a hydrogenation flask, introduce a solution of 0.58 g (0.005 mol) of maleic acid in 15 ml of ethanol and attach the flask to the adapter of the atmospheric hydrogenation apparatus (Fig. 2.63(*a*)). Fill the flask and gas burettes with hydrogen by the procedure discussed above; note the volumes in the gas burettes and then gently agitate the flask contents by means of the shaker. When uptake of hydrogen ceases note the total volume of hydrogen absorbed; this should be in the region of 115 ml. Follow the procedure discussed above for replacing the hydrogen in the apparatus with air; disconnect the hydrogenation flask, filter off the catalyst and wash it with a little ethanol (do not allow the catalyst to become dry, but after the washing operation remove the filter paper and rinse the catalyst into the residues bottle with water). Evaporate the ethanol to leave a residue of succinic acid, 0.58 g, m.p. 184 °C; the m.p. is unaffected after recrystallisation from 2.5 ml hot water.

The Brown[2] hydrogenator. A convenient alternative procedure for carrying out atmospheric pressure hydrogenations involves the use of the Brown hydrogenator.[25] The design of apparatus found in many laboratories is of a standard assembly for the hydrogenation of about 1 to 100 g of material, or in a larger version for the hydrogenation of 100 to 1000 g. The procedure uses the reaction of acetic acid with sodium borohydride to provide a convenient source of pure hydrogen, and thus avoids the problems associated with the usage of hydrogen gas cylinders; the apparatus is designed to allow the automatic generation of the gas. The catalyst (usually a highly active form of platinum) is prepared *in situ* immediately before use by the reduction of the metal salt with sodium borohydride.

The apparatus can be used in two ways:

1. Where hydrogen is generated in one flask and hydrogenation of the substrate in the presence of the catalyst is effected in another ('*external hydrogenation*').
2. Where hydrogen generation and hydrogenation are effected in the same flask ('*internal hydrogenation*').

External hydrogenation. The apparatus for this operation is shown in Fig. 2.63(*b*). It consists basically of three glass vessels, a hydrogen generator A, a hydrogenation flask B and a pressure control bubbler C, which are connected in series by means of air-tight O-ring joints.

The mercury bubbler C acts as a safety vent and controls the pressure in the apparatus. A ball valve near the top of the inlet tube prevents the mercury being sucked into flask B in the event of the automatic control valve D becoming blocked.

The hydrogenation vessel B is an Erlenmeyer flask with a slightly convex base, which is attached by means of wire springs to the inlet adapter E which incorporates a port closed with a serum cap to allow the introduction of appropriate solutions from a syringe.

Fig. 2.63(b)

The hydrogen generator flask A is a similarly shaped Erlenmeyer flask. The inlet adapter to which this flask is attached incorporates a mercury valve D which controls the rate at which the sodium borohydride solution contained in the 250 ml burette G is allowed to flow into the flask A via the syringe needle H (gauge 17 or 19) and the vent holes F in the control valve D.

Efficient stirring in the flasks A and B is provided by means of suitably sized Teflon-covered magnetic followers fitted with half-inch Teflon collars; the magnetic stirrer units serve also to further support the apparatus which should be securely clamped to a rack.

The following stock solutions are required:

Stabilised sodium borohydride solutions. (a) 1.00 M *Aqueous solution:* dissolve 0.8 g of sodium hydroxide in 150 ml of water, add 7.71 g of sodium borohydride (assuming 98% purity) and stir to dissolve. Dilute the solution to 200 ml and filter. (b) 2.50 M *Aqueous solution:* repeat the procedure under (a) exactly but increasing the amount of sodium borohydride to 19.25 g. (c) 1.00 M *Ethanol solution:* dissolve 0.8 g of sodium hydroxide in 10 ml of water, dilute to 200 ml with absolute ethanol and add 7.71 g of sodium borohydride. Stir until solution is effected and filter.

0.2 M *Ethanolic chloroplatinic acid solution:* dissolve 1.00 g of chloroplatinic acid (40% platinum metal) in 10 ml of absolute ethanol.

The procedure for conducting the hydrogenation of 0.5 mol of a compound is described below. (Flask sizes and reagent quantities suitable for the hydrogenation of other molar amounts of substrate are listed in Table 2.6.)

CAUTION: *All the following operations involving catalyst preparation and hydrogenation should be conducted in the fume cupboard; in particular, large quantities of hydrogen are evolved in step 2.*

1. Remove the flask B (500 ml capacity) from the apparatus assembly and add 100 ml of absolute ethanol, 5.0 ml of 0.2 M ethanolic chloroplatinic acid and 5 g of decolourising charcoal; insert a 38 mm Teflon-covered follower bar.

2. Place the flask on a magnetic stirrer unit and stir vigorously while adding 25 ml of 1.0 M ethanolic sodium borohydride as rapidly as possible without allowing the contents of the flask to foam over. After about one minute, add 20 ml of glacial acetic acid or concentrated hydrochloric acid to destroy excess of sodium borohydride.

3. Add 0.5 mol of the compound to be hydrogenated either neat or in ethanolic solution and reconnect flask B to the apparatus but do not commence stirring.

4. Charge the burette G with 1.00 M aqueous sodium borohydride solution.

5. Place 20 ml of glacial acetic acid in flask A (250 ml capacity) equipped with a Teflon-covered follower bar. Reconnect flask A to the apparatus and stir magnetically whilst injecting from a syringe through the side port of the inlet adapter 30 ml of the aqueous sodium borohydride solution. The rate of addition should be such as to effectively flush the apparatus with hydrogen without ejecting mercury from the bubbler C.

6. Open the stopcock of burette G; the depth of mercury in the control valve D is sufficient to support the column of borohydride solution. Now begin vigorous magnetic stirring of the contents of flask B when hydrogenation will begin. As the pressure drops in the system the valve allows sodium borohydride solution to be drawn into flask A via the vent holes F and the syringe needle H (gauge 17). The hydrogenation will then continue automatically until it has been completed. Finally note the volume of sodium borohydride solution which has run in from the burette.

7. Disconnect flask B, remove the catalyst by filtration and isolate the reduction product by suitable work-up procedures.

8. Calculate the uptake of hydrogen from the recorded volume of sodium borohydride solution used.

$$250 \text{ ml } 1.00 \text{ M } NaBH_4 \equiv 1.00 \text{ mol } H_2$$

If the substrate is insoluble in ethanol or if it is sensitive to protic media another solvent must of course be used. Ethyl acetate, tetrahydrofuran or diglyme are suitable alternatives but not dimethylformamide or acetonitrile which poison the catalyst. If an alternative solvent is needed the catalyst is prepared in ethanol as in step 1 and 2, but the procedure thereafter is modified as follows.

(a) Pour the contents of flask B into a sintered glass Buchner funnel, and remove most of the ethanol with gentle suction until the catalyst is left covered by an approximately 3 mm layer of the solvent.

(b) Add 50 ml of ethanol to the catalyst in the funnel, stir with a spatula and remove most of the ethanol by suction as in (a).

(c) Similarly wash the catalyst three times with 50 ml portions of the new solvent. The catalyst must not be allowed to become dry at any time during the above filtration procedure.

(d) Wash the catalyst into the hydrogenation flask with 100 ml of new solvent with the aid of a wash bottle. Continue with step 3 of the standard procedure described above.

Table 2.6 Specifications for representative hydrogenations (external)

	Hydrogenation flask							Hydrogen generator				
Compound (mmol)	Flask (ml)	Absolute ethanol (ml)	Charcoal (g)	H$_2$PtCl$_6$ ml, 0.2 M	Conc. acid (ml)	Length of stirring bar (mm)	1.00 M Ethanolic NaBH$_4$ (ml)	Flask (ml)	Acetic acid (ml)	Aqueous NaBH$_4$ to flush (ml)	Molarity of aqueous NaBH$_4$ for hydrogenation	Needle gauge
2000	2000	400	20	20	80	50.8	100	500	80	40 (2.5 M)	2.5	17
1000	1000	200	10	10	40	50.8	50	500	40	25 (2.5 M)	2.5	17
500	500	100	5	5	20	50.8 or 38.1	25	250	20	30 (1.0 M)	1.0	17
250	250	50	2.5	2.5	10	38.1	12.5	125	10	15 (1.0 M)	1.0	19
100	125	25	1.0	1.0	4	38.1	5.0	125	10	10 (1.0 M)	1.0	19
50	125	25	1.0	1.0	4	25.4	5.0	125	10	10 (1.0 M)	1.0	19

Table 2.7 Specifications for representative hydrogenations (internal)

Compound (mmol)	Flask (ml)	Absolute ethanol (ml)	Charcoal (g)	H$_2$PtCl$_6$ ml, 0.2 M	1.00 M Ethanolic NaBH$_4$ for catalyst prep. (ml)	Conc. acid (ml)	Length of stirring bar (mm)	Needle gauge
1000	2000	200	10.0	10.0	80	50	50.8	17
500	1000	100	5.0	5.0	40	25	50.8 or 38.1	17
250	500	50	5.0	2.5	20	10	38.1	19
100	250	50	1.0	1.0	10	10	38.1	19
50	125	25	1.0	1.0	5	5	25.4	19

Internal hydrogenation. For this mode of operation the Erlenmeyer flask B and the inlet adapter E are omitted from the assembly shown in Fig. 2.63(*b*) and the pressure control bubbler C is connected directly to the inlet adapter fitted to flask A. Catalyst preparation, hydrogen generation and hydrogenation are all carried out in flask A. The procedure for the hydrogenation of 0.5 mol of a compound is described below; this may be modified for other molar quantities of substrate as indicated in Table 2.7.

CAUTION: *The entire operation should be conducted in the fume cupboard.*

1. Place 100 ml of absolute ethanol, 5.0 ml of 0.2 M ethanolic chloroplatinic acid and 5 g of decolourising charcoal in flask A (1000 ml capacity); insert a 38 mm Teflon-covered follower bar.
2. Connect the flask to the inlet adapter and support it on a magnetic stirrer unit.
3. Charge the burette G with 1.00 M ethanolic sodium borohydride and stir the contents of the flask vigorously. To prepare the catalyst rapidly inject 40 ml of 1.00 M ethanolic sodium borohydride solution through the inlet port by means of a syringe and after one minute inject 25 ml of glacial acetic acid to destroy excess sodium borohydride.
   CAUTION: *A large volume of hydrogen is evolved.*
4. Open the stopcock of the burette, inject from a syringe through the inlet port 0.5 mol of the compound to be hydrogenated as a liquid or as an ethanolic solution when hydrogenation will proceed automatically as in the external hydrogenation technique.
5. At the conclusion of the hydrogenation, record the volume of sodium borohydride solution used, remove the catalyst by filtration and isolate the product by suitable work-up procedures.

Hydrogenation under pressure. The following account refers primarily to commercial apparatus suitable for conducting hydrogenations under pressure; the apparatus can of course be employed for other reactions under pressure (Section 2.17.2, p. 97), but some modifications of experimental procedure will then be necessary.

The apparatus shown in the photograph (Fig. 2.63(*c*), Chas. W. Cook & Sons Ltd) is designed for use at temperatures up to 70 °C and at working pressures up to 60 p.s.i. when using a glass reaction bottle (available with a capacity of either 500 ml or 1 litre). Stainless steel reaction bottles can be used at pressures up to 300 p.s.i. and if necessary at temperatures up to 200 °C. The bottle fits into an aluminium carrier fitted with an aluminium alloy cover carrying a sulphur-free rubber sealing ring. A metallic heating unit which surrounds the bottle is provided; the lid is fitted with a thermocouple well which dips into the reaction bottle. The carrier is pivoted in a support frame to allow controlled rocking by a geared motor with an eccentric drive. The reaction bottle is connected by a flexible PTFE tube to a mild steel hydrogen reservoir with a capacity of 4.3 litres and which as normally supplied has a maximum operating pressure of 200 p.s.i. (Reservoirs operating at pressures up to 500 p.s.i. for hydrogenations in stainless steel reaction bottles are also available.) At the forward end a double valve provides for: (*a*) the evacuation of the bottle using a water pump; (*b*) the controlled charging of the bottle with hydrogen with the aid of a pressure gauge marked in pounds; and (*c*) the release of pressure in the bottle without loss of gas from the storage tank. The uptake of hydrogen may be computed from the change in pressure as the result of the hydrogenation.

Fig. 2.63(c)

A similar design of apparatus is available from Parr Instrument Co. Both manufacturers publish detailed instructions on the use of the equipment for carrying out a hydrogenation reaction.

The essential features of an excellent high-pressure autoclave are illustrated in Fig. 2.63(d) (Baskerville and Lindsay). The special feature of this apparatus, constructed almost entirely of stainless steel, is the incorporation of a totally enclosed agitator in the form of a plunger which is operated electro-magnetically; agitation efficiency is at least as high as is achieved with shaking autoclaves and is very affective for hydrogenation purposes. The apparatus is stationary, has no external moving parts, and can be made compact and convenient to use. The reaction vessel B is made of F.M.B. stainless steel machined out of the solid and is provided with a cover fitted respectively with a thermometer or thermocouple pocket T, a central vertical tube, and an outer vessel nut with compression screws for making the pressure joint between the cover and the vessel. Sd is a solenoid operated through the contactor C, Bd is a bursting disc, G is a pressure gauge, V_1 is a control valve, V_2 is an evacuation valve (the last-named is connected through VP to a vacuum pump for complete evacuation of the apparatus). The agitator A consists of a stainless steel rod at the lower end of which is secured a circular stainless steel plate; at the upper end of the rod passing through the centre of the vertical tube is a stainless steel sheathed armature which, in its lowest position, just enters the lower end of the solenoid coil surrounding the central tube. The solenoid Sd through the contactor C operates at a rate between 20 and 90 cycles per minute controlled by an adjustable screw on the contactor, resulting in a vertical reciprocating movement in the agitator rod. The whole autoclave is placed in an electrically heated air bath H. Autoclaves are available in capacities ranging from 20 ml to 2 litres for use with pressures up to 350 atmospheres and temperatures as high as 300 °C; special liners of Pyrex

Fig. 2.63(d)

glass are supplied for use with substances which attack stainless steel or are affected by it.

High-pressure autoclaves are also available from Parr Instrument Co.; stirring is by turbine-type impellers. The reaction vessel is provided with a liquid-sampling valve which enables monitoring procedures. Both manufacturers provide detailed operating and servicing advice.

2.17.2 REACTIONS UNDER PRESSURE

Reactions which require the use of substantially increased pressures are usually carried out in a high-pressure metal autoclave. It should be emphasised again that these autoclaves should be isolated in a purpose-designed laboratory building, and that inspection and testing should be carried out at appropriate intervals by a suitably qualified engineer. The apparatus described under 'Catalytic hydrogenation' (Section 2.17.1) is suitable for many reactions on a moderate scale, i.e. when the total volume of reactants is compatible with the size of the vessels available. Similar specially designed assemblies for small-scale reactions (volumes from 5 to 20 ml) are also available. Reactions involving corrosive materials require vessels provided with a resistant lining such as an acid-resisting enamel or Pyrex glass.

A cheap and effective small-scale pressure vessel designed and constructed in the editors' laboratories is shown in Fig. 2.64(a). The main sections of the apparatus are constructed from a high grade of stainless steel (EN 58J). The main

97

Fig. 2.64

cylindrical vessel A has a raised thread (3 cm of thread of 3 mm pitch, Unified form) on the open end. The top section B fits tightly into A and the joint is sealed by a 'Viton' O ring, C; B is clamped firmly into place by the threaded collar D which is screwed home hand-tight on to A. A hole drilled through B leads to the 'pop'-type safety valve E* and a threaded channel through B carries the adjustable 'Hoke' valve F.

To use the bomb, the vessel A is set in an upright position and charged with reactants and solvent; B is then pushed home into A with the valve F open and the securing collar D is screwed down hand-tight.† The vessel is evacuated with a vacuum pump attached to the nozzle of F, the valve is closed and the vacuum line removed. The bomb is then positioned behind adequate safety shielding, preferably in an isolated position, and heated in an oil bath or preferably by an electrical heating tape wound around the barrel A.

On completion of the reaction, the bomb should be allowed to cool to room temperature and the barrel cooled further to about $-15\,°C$ in an ice–salt or acetone–Cardice bath. The valve F may then be opened to release any pressure (fume cupboard), and the bomb dismantled and the contents removed for work-up.

When one or more of the reactants is highly volatile the barrel A must be cooled thoroughly before the reactants are added. This may be achieved by standing A upright in an ice–salt or, if necessary, an acetone–Cardice bath. During cooling it is desirable to prevent the condensation of atmospheric mois-

---

* This valve may be set to vent at various pressures by changing the strength of the internal spring; in this apparatus, the valve is set to about 300 p.s.i. For insurance purposes, the vessel must be subjected to a hydrostatic pressure test under the supervision of a Chartered Engineer. The vessel described above was subjected to a pressure of 60 bar ($70\,kgf\,cm^{-2}$) at ambient temperature and at the maximum operating temperature ($200\,°C$ using a 'Viton' O-ring seal).

† The securing collar D must not be over-tightened since this will produce an unnecesary additional load on the thread.

ture on the inside surface of A by closing the opening of A with a rubber bung carrying a calcium chloride guard-tube. After cooling the bung is removed, the cold reactants and solvent are added, and the apparatus is quickly assembled and evacuated.

When a reactant is a gas at room temperature (e.g. 1,3-butadiene, Expt 7.23) the following procedure, which should be conducted in the fume cupboard, may be adopted to liquefy and transfer it to the pre-cooled vessel A. The apparatus consists of a purpose built acetone–Cardice condenser Fig. 2.64(b) with a 34/35 upper socket and a 24/29 cone at the lower end on to which is fitted a two-necked round-bottomed flask. The side-arm of the flask carries a screw cap adapter through which is passed a length of glass tubing so that it just protrudes into the flask. The length of the tube should be such that it can be repositioned with the end reaching to the bottom of the flask. The outlet of this tube is connected to a calcium chloride guard-tube via polyethylene tubing.

Add a small quantity of an appropriate drying agent in granular form to the flask and grease the joints lightly and protect from the ingress of moisture by wrapping with absorbent cotton wool and sealing with adhesive tape.

Charge the inner vessel of the condenser with acetone–Cardice, surround the flask with a cooling bath of acetone–Cardice, and allow the gaseous reagent to flow slowly through the condenser inlet from a preparative assembly or from a compressed gas cylinder.

When sufficient reagent has been condensed in the flask, shut off the supply of gas and connect the condenser inlet to a supply of nitrogen. Loosen the screw cap slightly, push the glass tube to the bottom of the flask, and re-tighten the screw cap. Remove the drying tube and apply a slight pressure of nitrogen to the condenser inlet to drive the condensed reagent directly into the precooled pressure vessel A via the plastic tubing. The pressure vessel is then quickly sealed and evacuated.

2.17.3 UNCATALYSED AND CATALYSED VAPOUR PHASE REACTIONS

Thermal decompositions (pyrolyses) and catalysed reactions in the vapour phase are widely used large-scale industrial techniques. These vapour phase reactions often lead to more economic conversions than the smaller batchwise laboratory methods, because relatively inexpensive catalyst preparations (compared to the often expensive reagents required in laboratory procedures) may be used, and because the technique lends itself to automated continuous production. In undergraduate laboratory courses the technique has not achieved widespread use. The discussion below of the various apparatus designs, to meet a range of experimental conditions, may be regarded as an introduction to this topic.

A number of research groups have explored vapour phase pyrolytic reactions using apparatus designs (available from, for example, Aldrich), which allow for controlled vapour flow (and hence heat contact time), under moderate to high vacuum conditions. These techniques enable preparative procedures to be developed which involve the trapping of reaction intermediates (e.g. carbenes, nitrenes, arynes, etc.), and also the study of very short-lived species by the interfacing of the pyrolysis unit with, for example, a mass spectrometer. This specialist area has been comprehensively reviewed.[26]

Examples which are cited in the following experiments are the depolymerisa-

Fig. 2.65(a)

tion of dicyclopentadiene to cyclopentadiene (Expt 7.24), the thermal decomposition (pyrolysis) of acetone to keten (Expt 5.140), the pyrolysis of 1,5-
diacetoxypentane to 1,4-pentadiene (Expt 5.14), and the formation of symmetrical and unsymmetrical ketones by reaction of carboxylic acid vapours with a
manganese(II) oxide catalyst (Expts 5.92 and 5.93). In each case the apparatus
incorporates a reservoir containing the reactant, a heated reaction chamber
which may or may not contain a catalyst and into which reactant vapours are
led, and a collection flask in which the product is trapped. In some cases provision is made for separating the product from unchanged reactant and returning
the latter to the reservoir for recycling.

The simplest apparatus is that shown in Fig. 2.65(a) where decomposition of
the reactant occurs at reflux temperature without the aid of a catalyst, the products being more volatile than the reactant. This apparatus assembly uses the
component parts of the vacuum drying pistol illustrated in Fig. 2.86. The reactant in the reservoir A is heated to gentle reflux by means of a heating mantle or
oil bath. The vapour, which consists of product and reactant, passes into the
chamber B, when the undecomposed reactant condenses on the finger C and
returns to A; the more volatile product passes to the condenser D to collect
in the receiver flask E. The liquid in the finger C must have a boiling point
above that of the product but well below that of the reactant; it will boil and
reflux during the progress of the experiment and hence boiling chips in C will be
required.

When higher temperatures for pyrolysis are required in an uncatalysed
vapour phase reaction, the apparatus illustrated in Fig. 2.65(b) could be used.
This was originally designed for the pyrolysis of acetone vapour which when
passed over a nichrome filament heated at 700–750 °C gives keten in yields
exceeding 90 per cent.

The construction of the filament will be apparent from the enlarged inset.

Fig. 2.65(b)

About 350 cm of 24 gauge Nichrome wire* is formed into a tight spiral by wind-ing the wire round a glass rod 3 mm in diameter and stretching the coil so formed to a length of 70 cm. The filament is held in position on 1.5-cm-long platinum hooks A sealed into the Pyrex glass rod B which supports them. The three platinum hooks at the bottom of the rod are placed 120° apart; two platinum hooks support the filament at a distance of 11 cm above the lower end. The ends of the filament C are connected to tungsten leads by means of nickel or brass connectors D, 10 mm in length and 3.5 mm in internal diameter. The tung-sten leads (24 gauge) are sealed into the glass at the points E; the leads are insulated by means of 6 mm glass tubing F, which are held by a cork stopper G. If desired, the tungsten leads may be soldered immediately above the glass seal to copper wires (24 gauge) which are passed through the glass tubing F. The tungsten or copper wire leads are connected to the 240–250 volt a.c. mains through a variable transformer (Variac).

All the glass in the apparatus is in Pyrex and connections are made by means of standard glass joints of appropriate size. Chamber H is constructed from a 25-cm length of glass tubing of 70 mm internal diameter; the joint J is 55/44. The connecting tube K is in 12–15 mm tubing, the side arm L is of 15 mm tubing; the condensers M and N are efficient double surface condensers, 50 cm and 90 cm long respectively (the sizes are not critical); O is a liquid trap, constructed of 35 mm tubing and is 120 mm long, with side tube of 8 mm diameter; the stopcock is for the removal of liquid from the trap.

The operation of the apparatus for the preparation of keten (1) is as follows. Acetone is placed in the flask which is heated in an electric mantle until the liquid gently refluxes from the condenser M. After a few minutes the U-tube attached to K will fill with acetone and this provides a liquid trap which ensures that all the acetone vapour passes through H. After heating under reflux for a

---

* US: B. and S. gauge 24 Chromal A wire, an alloy of 80 per cent Ni and 20 per cent Cr.

further five minutes to drive air from the chamber H, the filament current is switched on so that the filament C attains a dull red glow (700–750 °C). Keten is formed almost immediately and is allowed to pass directly via the three-way tap shown into the reaction flask (2). The apparatus requires little attention apart from occasionally removing the condensed acetone from the trap O. At the end of the run, the following operations must be carried out rapidly in this order: (i) remove the source of heat from the flask; (ii) turn off the filament current; and (iii) open the stopcock on O.

Notes. (1) *Keten is a poisonous gas having a toxicity comparable with phosgene*; leaks from the apparatus, which must be contained in a fume cupboard, are recognised by a pungent odour resembling acetic anhydride. For this reason it is customary to attach a second receiver flask containing a compound which readily reacts with keten, e.g. aniline, to the other arm of the three-way stopcock. In this way when the reaction in the main flask is complete the keten gas may be diverted to the second receiver flask while the apparatus is switched off and allowed to cool. Escape to the atmosphere of keten is thereby avoided.

(2) The yield of keten may be determined by weighing the acetanilide formed by passing keten through excess aniline for a measured period of time.

For pyrolyses which proceed best in a heated tube, which may with advantage be packed with glass beads or with porcelain chips to increase the heated surface area, and for reactions which occur on the surface of a heated catalyst, the basic apparatus shown in Fig. 2.65(c) is often suitable. A pressure-equalising

Fig. 2.65(c)

Fig. 2.65(d)

funnel A allows the reactant or reactants to be dropped at a constant rate into the combustion tube B, which is about 100 cm long and made of Pyrex tube (23 mm outer diameter and 16 mm inner diameter) and contains either glass beads or a suitable catalyst. Frequently it is necessary to conduct the reaction in the absence of air, and the adapter with T-connection C provides a means of displacing the air in the apparatus with nitrogen and of sweeping the products of the reaction through the combustion tube into the condenser and thence to the collection flask D. C also provides the means of introducing a gaseous reactant into the combustion tube should this be required. The Drechsel bottles E may serve a number of different purposes according to the nature of the experiment; for example: (a) they may monitor the flow of nitrogen gas from the inlet C; (b) they may be used to check the complete displacement of air from the apparatus by using Fieser's solution (Section 4.2.*52*, p. 452); and (c) they may be used to absorb unwanted gaseous products to prevent contamination of the atmosphere. The tube is heated by a furnace (F) (e.g. from Aldrich), or by a custombuilt thermally insulated heated jacket (e.g. from Electrothermal Enginering, or Glas-Col).

GENERAL PROCEDURE FOR OPERATION

The combustion tube B is packed with glass beads or catalyst held in position with plugs of glass wool, and inserted centrally into the furnace. After fitting the remaining apparatus components, the air in the apparatus is displaced with nitrogen, the furnace is allowed to heat to the required temperature, and the combustion tube is allowed to reach temperature equilibrium. In those cases where the catalyst requires heat treatment (as in the case of manganese(II) carbonate on pumice) adequate time must be allowed for the activation process to reach completion. The reactant is then allowed to drop into the combustion tube (the flow of nitrogen must be stopped if the rate of formation of gaseous products is to be observed) at a rate of about one drop every 3–4 seconds. The apparatus subsequently requires little attention and the passage of say 750 ml of reactant requires a period of addition of between 48 and 72 hours.

The isolation and purification of the reaction products which collect in D will of course be determined by their chemical nature, and details are given in the appropriate sections.

In some cases a single pass of reactants through the combustion tube gives only a low conversion into products and hence it is necessary to provide a means of recycling unreacted material while continuously removing product to avoid its decomposition. Some adaption to the above apparatus is then required and one such assembly is shown in Fig. 2.65(*d*). The reactant is heated in the flask A and the vapour passes upwards through the combustion tube B; the reactant and products are swept by a slow nitrogen flow into the Vigreux fractionating column. Unreacted material is collected in the Dean and Stark side tube, and returned as appropriate to the flask A. The Vigreux column is surmounted by a still head fitted with a condenser and collector flask.

2.17.4 OZONOLYSIS

The cleavage of a carbon–carbon double bond by oxidation with ozone (as ozonised oxygen) followed by hydrolysis to yield carbonyl compounds is a reaction sequence of considerable importance. This reaction, for example, can be used for the determination of the structure of an unsaturated compound by identification of the carbonyl fragments, or it may be used in suitable cases for the preparation of aldehydes or ketones which are not readily available by other means. Ozonisation of carbon–carbon multiple bonds, unlike oxidation with excess potassium permanganate or chromic acid which for example will also oxidise primary and secondary alcohols, is a relatively specific process.

When ozonised oxygen is passed through a solution of an ethylenic compound in an inert solvent (e.g. methanol, ethyl acetate, glacial acetic acid, chloroform or hexane) preferably at a low temperature (-20 to $-30\,°C$), ozone adds on readily and quantitatively to the double bond to give an ozonide (1):

$$\underset{(1)}{\overset{R^1\quad R^3}{\underset{R^2\quad R^4}{\diagup\diagdown}}} \xrightarrow{O_3} \underset{(1)}{\overset{R^1\quad O\quad R^3}{\underset{R^2\quad O—O\quad R^4}{}}} \xrightarrow{H_2O} \underset{(2)}{R^1R^2CO} + \underset{(3)}{R^3R^4CO} + H_2O_2$$

Excess ozone should be avoided since further oxidation may occur. Conveniently a wash bottle charged with potassium iodide solution and acetic acid is attached to the outlet of the reaction vessel; the completion of ozonolysis is indicated by the sudden extensive separation of iodine. Alternatively the flow of ozonised oxygen may be interrupted, and a drop of the solution removed and placed upon a white porcelain tile and allowed to mix with a drop of tetranitromethane when the production of a yellow coloration is indicative of the presence of unreacted alkene.

The ozonides are usually not isolated since they are generally viscid oils or glasses, *frequently with violently explosive properties, particularly upon warming.* They can however, be smoothly converted into carbonyl compounds (2) and (3) by hydrolysis, preferably under reducing conditions (e.g. zinc dust and aqueous acetic acid), by hydrogenation over platinium on calcium carbonate or by treatment with dimethyl sulphide or thiourea. (Section 5.7.3, p. 592.) These conditions prevent the further oxidation (by the hydrogen peroxide formed during hydrolysis) of any aldehydic products to the corresponding carboxylic acids; if the acids are in fact the desired products, the decomposition of the ozonides may be carried out oxidatively, e.g. in the presence of hydrogen peroxide or potassium permanganate.

A simple semimicro *laboratory ozoniser* is illustrated in Fig. 2.66; this gives reasonably satisfactory results for small quantities (2–4 g) of organic compounds. It consists of a wash bottle or small bubbler A to indicate the rate of flow of the oxygen from a cylinder fitted with a reducing valve, a Berthelot tube B for the generation of ozone, a vessel C to hold the solution of the compound to be ozonised, and a flask D containing 5 per cent potassium iodide in aqueous acetic acid. Since ozone is markedly *toxic* and is also a *lung irritant* the outlet from D should be led by means of PVC tubing to the extraction vent of the fume cupboard. The Berthelot tube is charged with dilute copper(II) sulphate solution and is connected by a copper or stainless steel wire (2–4 mm in diameter) to the high voltage terminal of a transformer (7500–10 000 volts). The second electrode is the earthed aluminium foil covering most of the exterior of the Berthelot tube and is bound with insulating tape. As a precaution all high-voltage connections are heavily insulated with rubber tape and the lead to the top of the electrode is covered with PVC or equivalent tubing. The main dimensions are shown in the

High
Voltage

Outer tube 35 mm outside dia.
Inner tube 28 mm outside dia.
Annular space 2 mm

Oxygen

18cm

←|3cm|→

C   D

Aluminium
foil

A   B

Fig. 2.66

figure. The ozoniser should be constructed of soft soda-glass tubing (Pyrex glass is unsatisfactory): the glass should be thoroughly cleaned and the annular space through which the oxygen passes should be as uniform as possible. *The complete apparatus should be placed in a fume cupboard behind a shatter-proof screen of laminated safety glass.*

Commercially available ozonisers (e.g. from UVP International) use either air or oxygen and are capable of yielding about 0.005 mol O_3/hour (with air) or 0.5 mol O_3/hour (with oxygen). A commercial ozoniser may be incorporated, by means of PVC tubing, in place of the Berthelot tube on an apparatus assembly similar to that above.

Should it be necessary to estimate the amount of ozone produced by the ozone generator, the vessel C is charged with 50 ml of a 5 per cent solution of potassium iodide in aqueous acetic acid (1:1 v/v) and the ozonised oxygen allowed to pass for a set period, say 1 hour, at a steady and measured flow rate (e.g. bubbles/second counted by means of the bubbler A). The iodine which is liberated is determined by washing the contents of C into a conical flask and titrating the liberated iodine with 0.1 M sodium thiosulphate solution; the yield of ozone may be calculated in mol/hour for the particular flow rate selected.

$$O_3 + 2I^{\ominus} + 2H_3O^{\oplus} \longrightarrow O_2 + I_2 + 3H_2O$$
$$I_2 + 2S_2O_3^{2\ominus} \longrightarrow 2I^{\ominus} + S_4O_6^{2\ominus}$$

On completion of the ozonisation of the olefinic material the method of decomposition of the resulting ozonide and the subsequent work-up procedure will be determined by whether the object of the reaction is preparative in nature, or whether it is required to identify the carbonyl compounds produced as an aid to the determination of the structure of the alkene.

In the latter case the following procedure is recommended. Wash the contents of the reaction vessel into a round-bottomed flask, add zinc dust and acetic acid and fit the flask with a steam distillation assembly (Section 2.25) ensuring that the receiver adapter outlet is just below the surface of an aqueous acidic solution of 2,4-dinitrophenylhydrazine (p. 1218) contained in a conical flask. Steam distil the solution and collect the volatile carbonyl compounds until no further precipitate of 2,4-dinitrophenylhydrazone is observed with fresh portions of reagent. Extract the combined distillate–reagent solutions with dichloromethane, dry and evaporate the extract, dissolve the residue in a little toluene and chromatograph the solution on a column of alumina using toluene as the developing solvent. Evaporate the eluate and characterise the 2,4-dinitrophenylhydrazone derivative. Extract the residual liquors from the steam distillation with ether or dichloromethane, wash, dry, evaporate and convert the residue into a 2,4-dinitrophenylhydrazone derivative for examination in a similar manner.

When ozonolysis of the olefinic material is to be carried out for preparative purposes, the initial ozonisation should be conducted in dry methanol and the ozonide decomposed by hydrogenation over palladium hydroxide on calcium carbonate in the following manner. Rinse the contents of the reaction vessel with methanol into the hydrogenation flask containing palladium hydroxide on calcium carbonate catalyst (see Section 4.2.54, p. 452) and a magnetic stirrer follower (Section 2.14), attach the flask to the hydrogenation apparatus (Fig. 2.63(a)) and immerse the hydrogenation flask in an ice bath placed upon a magnetic stirrer plate. This cooling is essential to avoid an undue rise in temperature of the solution during hydrogenation, which is exothermic, since this may lead to the alternative formation of a carboxylic acid at the expense of aldehyde. Charge the apparatus with hydrogen and hydrogenate the solution as detailed in Section 2.17.1, p. 89. Emphasis should be placed upon the importance of placing the hydrogenation vessel behind appropriate shatter-proof screens. When hydrogenation is complete, filter off the catalyst, remove the solvent on a rotary evaporator, and purify the product by crystallisation or distillation as appropriate.

2.17.5   ORGANIC PHOTOCHEMISTRY

Although it has long been recognised that chemical change can be effected by means of ultraviolet (200–400 nm) and visible light (400–750 nm), studies in this area of chemistry have until quite recently been largely the province of the physical chemist. However, a rapidly increasing number of investigations since 1960 have shown that many novel and synthetically useful reactions including dimerisation, cycloaddition, rearrangement, oxidation, reduction, substitution and elimination may be consequent upon the absorption of light by organic molecules. Many chemical transformations can be effected which would otherwise require a large number of steps by standard chemical procedures. This progress in synthetic organic photochemistry has been aided by the commercial development of suitable light sources, by advances in procedures available for the separation and identification of the components of mixtures and, not least, by the realisation that many photochemical reactions occur quite cleanly to give good yields of the desired product. In addition the photochemical experiments can often be carried out much more simply than many standard chemical reactions.

An understanding of organic photochemistry requires a knowledge of the energy transitions which a molecule may undergo following irradiation with electromagnetic radiation. Some consideration of these energy transitions is given in Chapter 3 in relation to the use of u.v. and i.r. spectroscopy in structural elucidation. The following account is intended to provide sufficient theoretical background to allow some appreciation of photochemical reactions, of which illustrative practical examples are given in Expts 5.36 and 7.24. A detailed treatment of photochemical processes may be found in a number of recent books on photochemistry.[27]

The total energy of a molecule is the sum of its electronic, vibrational, rotational and translational energies. Whereas the translation energy increases continuously with the temperature of the system, the first three energy states are quantised and excitation to higher energy levels requires the absorption of discrete amounts of energy (quanta) which can be supplied by electromagnetic radiation. The amount of energy associated with such radiation depends on its wavelength, the longer the wavelength the smaller the energy (p. 385). Excitation of a molecule to higher rotational and vibrational energy levels can thus occur on absorption of radiation in the far infrared, and in the infrared regions of the spectrum respectively (i.e. the low energy portion of the spectrum), and is associated with relatively small increases in the energy of the molecule ($\sim 0.5 - 42\,kJ\,mol^{-1}$). Absorption of ultraviolet (200–400 nm) and visible (400–700 nm) radiation by a molecule is associated with an increase in energy in the range $600–160\,kJ\,mol^{-1}$ and results in the excitation of its valence electrons to higher energy levels. The energy associated with a photon of radiation in the ultraviolet region is of the same order as the bond energies of many of the bonds present in organic molecules (e.g. C–H, $410\,kJ\,mol^{-1}$). It is thus not surprising that absorption of light in this region can result in chemical reactions and that the reactions of molecules in such electronically excited states are often quite novel.

Excitation of a molecule to a higher energy level involves promotion of an electron from a bonding (σ or π) or a non-bonding (n) orbital to an antibonding (σ^* or π^*) orbital. Four types of transitions are possible and the energy associated with each (which can be represented diagrammatically in Fig. 2.67(a)) decreases in the order $\sigma \rightarrow \sigma^* > n \rightarrow \sigma^* > \pi \rightarrow \pi^* \approx n \rightarrow \pi^*$.

The $\sigma \rightarrow \sigma^*$ and $n \rightarrow \sigma^*$ transitions are of little signficance in organic photochemical synthesis as they occur in the far ultraviolet (< 200 nm), a region which is not readily accessible practically owing to the absorption of radiation in this region by oxygen. The $\pi \rightarrow \pi^*$ and $n \rightarrow \pi^*$ transitions occur in the ultraviolet region and are responsible for the vast majority of useful photochemical reactions; in simple ketones the $n \rightarrow \pi^*$ transition occurs at ~ 270 nm with an associated energy of $443.1\,kJ\,mol^{-1}$, and the $\pi \rightarrow \pi^*$ transition of butadiene occurs at 217 nm with an associated energy of $551.5\,kJ\,mol^{-1}$. While the overall magnitude of the energy required to effect these electron transitions explains why bonds may be broken during irradiation, a more detailed consideration of these electronically excited states is necessary to understand the various possible ways in which energy absorbed may be dissipated. In particular the importance of the concept of singlet and triplet states must be considered.

Most organic molecules have an even number of electrons and these are paired (spins in opposite direction); energy states with paired electrons are called singlet (S) states (no net electronic magnetic moment and hence only *one* possible energy state in a magnetic field). The ground state of a molecule is referred to as S_0 and the higher excited singlet states as S_1, S_2, S_3, etc. Inversion of the spin of one electron results in the formation of a different electronic state having two unpaired electrons (same spin); this is referred to as a triplet (T) state

Fig. 2.67(a)–(c)

(a net electronic magnetic moment and hence *three* possible energy states in a magnetic field). For each possible excited singlet state (S_x) there is a corresponding lower energy triplet state (T_x). According to quantum mechanical theory, transitions between states of the same multiplicity are *allowed* whereas transitions between states of different multiplicity are formally *forbidden*. The ground state and lowest singlet and triplet states are represented schematically in Fig. 2.67(b).

Following the absorption of radiation and the promotion of an electron (π or n) to the vibrationally excited single state S_1† (which occurs very rapidly in $\sim 10^{-15}$ s, the dissipation of this energy may take place in a variety of ways itemised below, some of which may be represented diagrammatically by means of a Jablonski diagram (Fig. 2.67(c)).

1. Initially excess vibrational energy is rapidly lost by radiationless processes, such as collision with solvent molecules, to give the thermally equilibrated excited singlet molecules S_1. This has a short lifetime ($\sim 10^{-8}$ s) and may then lose its energy by any of the processes 2–5 below.

---

† Absorption of sufficient energy can also cause excitation to higher singlet states such as S_2; however, these generally decay very rapidly by radiationless transitions to the highest vibrationally excited state S_1. This is possible since the S_0 and S_1 states are separated by the greatest energy gap, whereas the higher energy states have progressively smaller energy differences and overlapping potential energy surfaces which allow radiationless loss of energy. In certain cases the energy input may of course by sufficient to cause immediate bond dissociation.

2. Emission of light from the excited molecule may occur which then returns to the ground state, i.e. fluorescence is observed.
3. Thermal dissipation of the energy to surrounding molecules may occur; this is a radiationless process (*internal conversion*).
4. Chemical reaction can occur.
5. Conversion to the lower energy triplet (T_1) by spin inversion (*intersystem crossing*); although formally forbidden this can occur with very high efficiency when the energy difference between the two states is small. It is most notable in carbonyl and aromatic compounds (e.g. intersystem crossing occurs with 100 per cent efficiency in the case of benzophenone).

An excited molecule in the triplet state also has a number of ways in which its energy may be dissipated.

1'. Emission of light from the excited molecule may occur with return to the ground state, but at longer wavelength than fluorescence, i.e. phosphorescence is observed. Although this transition is formally forbidden, as spin inversion is involved, it does eventually occur with the important consequence that the T_1 state has a very much longer lifetime (10^{-6} s → several seconds) than the S_1 state.
2'. The species may decay by internal conversion.
3'. Chemical reaction may occur – the longer lifetime of the triplet state compared to the S_1 state means that chemical reaction is a much more important feature, and is of prime importance in synthetic photochemistry.
4'. Energy transfer to a neighbouring (different) molecule may occur so that the acceptor molecule is promoted to a triplet state of either equal or lower energy than the donor triplet species, which itself undergoes spin inversion and returns to the ground state S_0. Such a transfer will occur only if the acceptor molecule has an available lower energy excited level.

There are two types of photochemical processes which lead to these various transitions and thence to a realisation of the synthetic possibilities of the processes 4 and 3' above.

(a) *Direct photolysis* – where the incident radiation is directly absorbed by a substrate X, which is thus promoted to the excited singlet state X* which then loses its energy by the processes outlined above.

$$X \xrightarrow{\ h\nu\ } X^* \longrightarrow \text{product(s), etc.}$$

(b) *Indirect or sensitised photolysis* – where a photo-excited donor molecule (D*)in the singlet or triplet state, referred to as a *sensitiser* and produced by absorption of the incident radiation, transfers its energy to the substrate X which is thereby promoted to an excited state (e.g. see 4' above). In this process the sensitiser returns to the ground state, is chemically unchanged, and may be further excited by incident radiation.

$$D \xrightarrow{\ h\nu\ } D^*$$
$$D^* + X \longrightarrow D + X^*$$
$$X^* \longrightarrow \text{product(s), etc.}$$

Many compounds (e.g. alkenes) do not undergo intersystem crossing from the singlet state to the synthetically more useful triplet state as the energy differ-

Fig. 2.67(d)

ence between the two states is large. However, provided that the energy of the triplet state of the sensitiser molecule is about 20.9 kJ mol^{-1} greater than that of the triplet state of the substrate, energy may be transferred to provide excited molecules in the triplet state which may then undergo chemical reaction. The procedure is also useful for populating triplet states of a compound whose singlet state is in an inaccessible part of the ultraviolet spectrum (i.e. < 200 nm).* This sensitising process is represented schematically in Fig. 2.67(d).

PHOTOCHEMICAL APPARATUS AND EXPERIMENTAL PROCEDURES

Prior to a study of the possible photochemical reaction of a compound, its spectrum in the ultraviolet or visible region must be determined in order that a light source emitting the appropriate wavelength of radiant energy may be selected. In the case of a sensitised photochemical reaction the spectrum of the sensitiser should be determined.

Light sources. In early photochemical work sunlight was the source of radiant energy, and it still remains a useful and cheap source in favourable climates for reactions requiring irradiation at wavelengths down to 320 nm. Ordinary high wattage tungsten lamps may also be used for reactions proceeding under the influence of visible light. However, photochemical reactions on a preparative scale are most often effected with radiant energy of wavelength 220–380 nm, and for these purposes mercury arc lamps are used almost exclusively. There are essentially three types available.

1. *Low-pressure mercury arc lamps,* which operate at a mercury vapour pressure of about 10^{-3} mmHg and emit mainly at 254 nm and 184 nm (about 80–95 per cent of this radiation is produced at 254 nm). Low-pressure mercury arc lamps with phosphor coatings on the interior walls are available which give maximum light emission at longer wavelengths over broad selected regions, e.g. centring at 300 nm or 350 nm.
2. *Medium-pressure mercury arc lamps,* which operate at internal pressures of from 1 to 10 atmospheres and emit radiation over the region 200–1400 nm, with particularly intense emission at 313 nm, 366 nm, 435.8 nm and 546.1 nm.

---

* Triplet and singlet excited state energy values for a large number of substances have recently been collected.[28]

3. *High-pressure mercury arc lamps,* which operate at internal pressures of from 100 to several hundred atmospheres and give almost continuous emission over the whole spectrum from about 200–1400 nm. The radiant energy is particularly rich in visible light.

Low-pressure mercury arc lamps operate at near room temperature. Much of the energy input of medium and especially high-pressure lamps however is converted into heat so that these lamps must be cooled. The medium-pressure lamps have been used most extensively for synthetic work on account of their high light output, ease of handling and broad spectrum emission. The full arc spectrum of these lamps is often employed in preparative photochemistry, but if necessary removal of unwanted regions of the spectrum can be effected by surrounding the lamp with chemical or glass (e.g. Corex, Vycor) filters; unwanted light below 300 nm can of course be removed by irradiation through Pyrex glass. For a comprehensive account of light filters and commercially available lamps, the reader should consult the monograph of Schönberg, Schenck and Neumüller.[29] Details of the energy output at the emitted wavelengths can usually be obtained from the lamp suppliers but it should be appreciated that these values change on ageing.

The vast majority of photochemical syntheses have been conducted in the liquid phase, hence the apparatus assembly for a photochemical reaction must take into account the light transmission characteristics of the material from which the reaction vessels are made. Pyrex glass transmits most of the incident light above about 300 nm and may be used in the construction of apparatus for reactions which require light above this wavelength. Quartz vessels, transparent down to 200 nm, must be used for reactions which require light below 300 nm. Certain types of quartz allow transmission below 200 nm.

There are basically two assemblies for carrying out preparative photochemical syntheses. The light source may either be placed outside the vessel containing the solution of substrate (*external irradiation*), or it may be placed inside the vessel containing the solution (*internal irradiation*); in this latter case the solution is subjected to the full output of the lamp and therefore this is to be preferred. Both assemblies are available commercially from Hanovia Lamps Ltd for internal irradiation in reactor sizes of 1 and 10 litres; this manufacturer also supplies the 'Reading' photochemical reactor which is designed for external irradiation. Apparatus designed for preparative photochemistry is also available from Applied Photophysics Ltd, and the Southern New England Ultraviolet Co. A convenient apparatus for internal irradiation which has been used in these laboratories will be described and is shown diagrammatically in Fig. 2.67(*e*). The apparatus consists of a three-necked Pyrex reaction flask of approximately 1 litre capacity with a central 45/50 socket to which is fitted the light source unit. The two side-necks of the flask allow the flushing of the reaction mixture with nitrogen and the attachment of a reflux condenser and/or a protective drying tube; the reactants are stirred magnetically. The light source unit has an outer quartz jacket A, an inner quartz jacket B and a mercury arc lamp C. Both jackets are made of fused quartz and each has a 75-mm-long zone made from pure synthetic quartz sited opposite the light-emitting region of the arc lamp; these zones allow about 80 per cent light transmission for every 1 mm thickness of quartz. Cooling water or air may be passed through the annular space between the two jackets via the inlet tube D and outlet tube E which are built

111

Fig. 2.67(e)

Fig. 2.67(f)

into the inner jacket. Similar jackets are available in Pyrex for reactions which proceed on irradiation above 300 nm.

The arc lamp C is attached by metal clips to the hollow support tube F which also allows flushing of the lamp area by nitrogen via inlet G and outlet H. The support tube F, the electrical leads to the arc lamp (not shown) and terminals for the cable from the power control unit (also not shown) are all attached to the terminal block J which is fitted with an insulating cover. The terminal block is attached to the inner quartz jacket by means of a rubber sleeve K and two sleeve clips, so that all the live parts are protected. The lamp and inner quartz jacket B can be withdrawn from the 40/38 socket attached to the outer quartz jacket A shown, and the entire unit can be withdrawn from the central socket of the reaction flask.

Two mercury arc lamps with associated power units are available: (a) a 2 watt low-pressure U-shaped lamp which emits mainly at 186 nm and 254 nm, and (b) a 100 watt medium-pressure straight-tube lamp emitting predominantly at 254 nm, 265 nm, 297 nm, 313 nm and 366 nm with intense emission also in the visible region; both lamps have synthetic quartz envelopes.

Water cooling via D and E is essential with the medium-pressure lamp whereas with the low-pressure lamp gas cooling is usually sufficient. The latter may be conveniently effected by drawing filtered air through the annular space between the jackets by connecting H to a water pump, which should be situated in the fume cupboard in order to vent any ozone formed. Light-filter solutions may replace the cooling water if it is required to remove any particular regions of light emission, and these must of course be circulated and cooled in an arrangement external to the apparatus.

Both lamps generate ozone and oxides of nitrogen in air, hence the inner lamp area should be flushed slowly with nitrogen via G and H as described above. It should be noted however that low-pressure lamps give maximum light output at a wall temperature of 40 °C so excessive cooling in the lamp region is to be avoided; this is particularly important when flushing with gas from a cylinder as the gas is likely to be cold due to expansion. In addition, it is of course essential that only dry gas be used because of the live connections in this region.

Reactor vessels may readily be constructed to cater for smaller volumes of reactants. Figure 2.67(f) shows such a vessel having a capacity of about 110 ml. The outer quartz jacket A of the light source unit (Fig. 2.67(e)) fits into the 45/40 centre socket, a reflux condenser and/or drying tube may be attached to one of the side 14/23 sockets while the other allows nitrogen flushing via the Teflon tube M attached to the drawn-out cone. Stirring is by means of the magnetic follower bar N. Even smaller volumes may be accommodated by taping suitably sized tubes to the side of the outer jacket A of the light source unit and surrounding the whole with aluminium foil.

A number of important aspects should be borne in mind when planning and executing a photochemical synthesis and these are outlined below.

1. *Safety*. Ultraviolet light is extremely dangerous to the eyes and also harmful to the skin so that proper precautions must be taken when conducting a photochemical experiment (see Section 2.3.6, p. 52). Ideally the apparatus assembly should be situated in the fume cupboard and aluminium foil wrapped around the reaction vessel (which also additionally serves as a light reflector) when using an internal irradiation arrangement; the whole should be surrounded by a light shield made from board. Rubber tubes for cooling water should be wired on and a suitable cut-out device for the lamp incorporated into the circuit. This latter precaution is essential in case the water supply should fail since the heat generated by a medium-pressure lamp could lead to fracture of the apparatus, loss of material and possibly fire when flammable solvents are used. Suitable precautions should also be observed with regard to the electrical equipment.

2. *Degassing*. Dissolved oxygen should normally be removed by passing nitrogen or other inert gas through the reaction solution for about 0.5 hour prior to irradiation and a nitrogen atmosphere should be maintained throughout the experiment.

3. *Stirring*. Relatively concentrated solutions are often used in preparative photochemistry and in consequence most of the light is absorbed by a very thin layer of solution adjacent to the lamp. Some appreciation of this fact may be obtained from the following. Consider a 0.1 cm layer of a 0.01 M solution of a compound having a molar absorptivity (ε) of $1000 \, l \, \text{mol}^{-1} \, \text{cm}^{-1}$. From the Beer–Lambert Law, p. 385, the light energy (I) transmitted through this layer is given by the expression:

$$I = \frac{I_0}{10^{\varepsilon cl}} = \frac{I_0}{10^{1000 \times 0.01 \times 0.1}} = \frac{I_0}{10}$$

where I_0 is the intensity of the incident radiation, c is the molar concentration of the compound and l the length of the absorbing solution in centimetres. Thus it can be seen that 90 per cent of the light is absorbed in this 0.1 cm layer of solution; vigorous stirring is therefore essential to change this layer continually. For volumes up to 1 litre this can usually be effected adequately by magnetic stirring. Stirring may also be effected by means of nitrogen introduced through a medium porosity fritted glass plate sealed to the bottom of the gas inlet tube; this

113

is particularly useful with small volumes of reactants. It is important, using internal irradiation with equipment such as described above, that the solution level should be above the light-emitting region of the lamp.

4. *Time of reaction.* A measure of the efficiency of a photochemical synthesis is given by the quantum yield (ϕ) for the product, which is defined as:

$$\phi = \frac{\text{number of molecules of product formed}}{\text{number of quanta absorbed}}$$

Quantum yields for reactions which proceed by way of a free radical chain mechanism may be as high as many thousands, whereas reactions which do not occur through such a mechanism have quantum yields in the region 0–1. Thus a knowledge of ϕ, and the number of useful quanta emitted per second by the lamp, could allow calculation of the approximate time of photolysis. However, since these values are often not known it is usual to monitor the progress of the reaction, e.g. by using a suitable chromatographic procedure (Section 2.31), or by noting the disappearance of a characteristic absorption band in the u.v. spectrum of the starting material.

Depending on the concentration of the substrate(s) the reaction time required may be of the order of days rather than hours. This can often of course be shortened by irradiating with a higher-intensity lamp; 500-watt medium-pressure lamps are commonly used.

To ensure as rapid a reaction as possible it is essential that all light-transmitting surfaces are kept absolutely clean, and handling of the quartz envelope of the mercury lamp should be avoided as finger marks will gradually 'burn' into the surface thereby reducing transmission. A thin film of polymer may occasionally be deposited on the outer surface of A adjacent to the lamp and this should be removed, otherwise light input to the reaction medium will be severely reduced. An ingenious apparatus has been described[30] using fluorescent tubes as light source, which allows the radiation to fall directly on to a moving thin film of the reaction solution thus avoiding this difficulty and also obviating the need for expensive quartz apparatus.

Irradiation of a solid substrate often gives a single product, whereas in solution a number of isomers may be formed (e.g. in dimerisation reactions); this is presumably due to the ordered arrangement of the substrate molecules in the solid phase. Such reactions may be carried out in a variety of ways. When irradiation by a mercury arc lamp is necessary, the material may be deposited as a thin film on the inside wall of a container (such as a large glass gas jar) by evaporating its concentrated solution in a volatile solvent; the lamp unit may then be inserted into the jar. Alternatively the finely powdered material may be placed in petri dishes under an arc lamp and stirred occasionally to provide a fresh surface for irradiation. When sunlight is a suitable source of radiant energy, exposure of the powdered material contained in large petri dishes may be employed, or a round-bottomed flask may be coated on the inside with a thin layer of material by the evaporation procedure and exposed to the sun, occasionally rotating the flask to ensure even exposure.

Details of a very large number of photochemical reactions have been published.[29, 31]

2.17.6 ELECTROLYTIC (ANODIC) SYNTHESES (THE KOLBÉ REACTION)

Electro-organic chemistry is the study of the oxidation and reduction of organic molecules and ions, dissolved in a suitable solvent, at an anode and cathode respectively in an electrolysis cell, and the subsequent reactions of the species so formed. The first experiment of this type was reported in 1849 by Kolbé, who described the electrolysis of an aqueous solution of a carboxylate salt and the isolation of a hydrocarbon. The initial step involves an anodic oxidation of the carboxylate anion to a radical which then dimerises to the alkane.

$$R \cdot CO_2^{\ominus} \xrightarrow{-e} R^{\bullet} + CO_2$$
$$2R^{\bullet} \longrightarrow R - R$$

Following the study of the simple coupling of radicals derived from the salt of a single carboxylic acid, it was found that the electrolysis of a mixture of carboxylate anions or of the salts of half esters of dicarboxylic acids increased the synthetic value of the method. This arises from the possibility of the formation of symmetrical and unsymmetrical coupled products of the derived radicals. These anodic syntheses are illustrated in the synthesis of hexacosane (Expt 5.11), sebacic acid (decanedioic acid), octadecanedioic acid and myristic acid (tetradecanoic acid), in Expt 5.131.

The electrolysis cell used for these conversions may be readily constructed in the laboratory (Fig. 2.68) and provides a simple introduction to the technique of electro-organic chemistry which is of increasing importance in organic synthesis. The cell consists of a cylindrical Pyrex glass vessel (16 cm × 7 cm) fitted with an internal cooling coil so that the temperature of the electrolyte may be controlled; a working temperature range of 30 to 35 °C is usually satisfactory, but excessive cooling may cause some of the product to crystallise. With a smaller electrolytic cell constructed from a large boiling tube the internal coil may be omitted and cooling effected by external means only. In either case no stirring is required since the evolution of carbon dioxide during electrolysis provides adequate agitation of the electrolyte. The electrodes consist of two platinum plates (4 cm × 2.5 cm × 0.3 mm) set about 2 mm apart to each of which is attached a platinum wire sealed into a glass tube containing mercury by which contact with the d.c. electrical circuit is effected. The carboxylic acid is dissolved in methanol (con-

Fig. 2.68

taining sufficient sodium methoxide to convert about 2 per cent of the added acids into the carboxylate salts), and the solution electrolysed with a current of between 1 and 2 amperes until the electrolyte becomes slightly alkaline. The length of time required to complete the electrolysis may be roughly estimated for carboxylic acids and half acid methyl esters to be between 20 and 50 per cent in excess of the theoretical value calculated from the number of Faradays required in relation to the amount of acid employed, i.e. from the equation $It/96\,500 = $ mol $R \cdot CO_2^{\ominus}$, where I is the current in amperes and t the time in seconds. It is often advantageous occasionally to reverse the direction of the current to dislodge insoluble deposits on one or other of the electrodes; if this is not done the current will be observed to drop prematurely, leading to a slowing down of the electrolysis.

Some general considerations which require variations in the simple electrolysis cell construction described above to meet the requirements for electrolytic oxidations and reductions of a wide range of organic compounds may be briefly summarised, but attention is drawn to the very extensive surveys which are available.[32]

The first general comment relates to the solvent system. In those cases where the electrolysis substrate does not exist in an aqueous–ethanolic or methanolic solution in a suitable ionic form, it is necessary to provide a solvent system of low electrical resistance which will dissolve the substrate, and also a supporting electrolyte whose function is to carry the current between the electrodes. Examples of such solvents are dioxane, glyme, acetonitrile, dimethylformamide and dimethyl sulphoxide; supporting electrolytes include the alkali metal halides and perchlorates, and the alkylammonium salts (e.g. perchlorates, tetrafluoroborates, toluene-p-sulphonates). With these electrolysis substrates, mass transfer to the electrode surface is effected by efficient stirring.

Although not relevant to the Kolbé reaction, a second comment relates to the necessity of ensuring that products formed at the working electrode (either anode or cathode) do not migrate to, and react at, the counter electrode or indeed react with compounds formed at the counter electrode. Recent cell designs therefore incorporate anode and cathode compartments separated by a rigid porous membrane. The difficulties of so selecting a glass frit of porosity sufficient to ensure transport of current, but not of electrolysis substrate or products, has in part been overcome by the use of a frit which supports a gel membrane.[33] This gel is formed by impregnating the frit with a hot solution of methylcellulose in dimethylformamide and allowing to cool. These membranes appear to be stable for the length of time required for an electrolysis when acetonitrile and 1,2-dimethoxyethane are used as solvent systems.

The electrode material frequently has crucial consequences on the course of electrolytic oxidation and reduction processes. Although platinum is the commonest electrode material, carbon, mercury and copper have all been used in numerous specific conversions. Selection of electrode material should therefore be based upon previously established characteristics when new conversions are to be studied.

2.17.7 LIQUID AMMONIA TECHNIQUES

Many important synthetic organic reactions are carried out in liquid ammonia (b.p. $-33\,°C$); this is a good solvent for many organic compounds having a range of polarities, and also for the metals lithium, potassium, sodium and calcium.

Solutions of these metals in liquid ammonia effect (i) the reduction of a range of functional groups such as carbonyl and acetylenic and also conjugated and aromatic systems, and (ii) cleavage of benzyl and allyl ethers and thioethers. These reactions are usually carried out by the general procedure of adding the metal to a solution of the substrate in liquid ammonia to which dry methanol or ethanol or t-butanol has been added to provide a ready proton source (alcohols are more acidic than ammonia).[34]

A second principal use of liquid ammonia involves forming a suspension of an alkali metal amide ($LiNH_2$, KNH_2 or $NaNH_2$) by adding the appropriate metal to liquid ammonia containing a trace of iron(III) ions (added as iron(III) nitrate) as a catalyst.

$$2Na + 2NH_3 \longrightarrow 2NaNH_2 + H_2$$

The amide ions are powerful bases and may be used (i) to dehydrohalogenate halo-compounds to alkenes and alkynes, and (ii) to generate reactive anions from terminal acetylenes, and compounds having reactive α-hydrogens (e.g. carbonyl compounds, nitriles, 2-alkylpyridines, etc.); these anions may then be used in a variety of synthetic procedures, e.g. alkylations, reactions with carbonyl components, etc. A further use of the metal amides in liquid ammonia is the formation of other important bases such as sodium triphenylmethide (from sodamide and triphenylmethane).

Although these amides are frequently used as a suspension in liquid ammonia, an inert co-solvent (such as ether or tetrahydrofuran) may be added should the organic substrate not be readily soluble in liquid ammonia. Alternatively after amide formation the liquid ammonia may be allowed to evaporate completely during simultaneous addition of the inert solvent; subsequently the organic substrate may be added in the same solvent to the alkali amide.

Liquid ammonia is supplied in cylinders (Section 4.2.5, p. 417) which incorporate a simple tap valve with a screw-thread wide-bore outlet. Although a special gas-reducing valve may be obtained from the suppliers, for most purposes it is adequate to screw on to the outlet a wide-bore metal-tube adapter (Fig. 2.69(a)) to which may be fitted wide-bore rubber or stout polyethylene tubing, secured by copper wire. When the cylinder is upright only ammonia gas

(a)   (b)

Fig. 2.69

will be released when the valve is opened. To obtain liquid ammonia, the cylinder needs to be supported valve downwards at an angle of about 60° from the vertical position, with the outlet valve above the level of the vessel into which the liquid ammonia is to be discharged. The cylinder should be securely supported in a purpose-designed scaffolding, an example of which is illustrated in Fig. 2.69(*b*). Current safety regulations recommend that no cylinder should be located within a laboratory area. It may be found convenient to locate the cylinder along the outside wall adjoining the fume cupboard of a dangerous operations laboratory (Section 2.3.2, p. 41); the inlet tube may then be led through a hole in the wall directly into the fume cupboard. Although this outside area for the cylinder should be suitably protected against the weather, the cylinder should nevertheless be returned to the main cylinder store after use.

The valve tap on the cylinder is very tightly closed; it is best released by attaching the valve lever and gently tapping the lever end with a hammer in short sharp blows with gradually increasing force until the ammonia starts to escape. This method is easier and is to be preferred to continuously applied hand pressure.

The rubber or plastic outlet tube should be depression-free and lead via an adapter into the receiver flask (see below) which must be sited in the fume cupboard, and it is advised that this delivery tube be additionally supported by means of a retort stand and clamp. This latter precaution is necessary since the initial force of ammonia release may cause it to flex with considerable thrust which may lead to apparatus damage.

Until the cylinder valve, valve outlet and adapter, and rubber or plastic tubing have cooled to $-33\,°C$, only ammonia gas emerges. Eventually however liquid ammonia will flow into the flask and when this has cooled to $-33\,°C$ it may be filled with the appropriate quantity; this may be gauged by pre-marking the vessel to the volume of liquid required. The flask selected should be of such size that it is only half full. This cooling effect leads to the final liquid ammonia sample containing traces of moisture (between 0.1 and 0.5 g/l), but this impurity is not harmful in the subsequent reactions. Such traces of moisture may be detected and removed by the addition of a little sodium metal with stirring; a rapid disappearance of the initial blue colour indicates some water present while a persistent blue colour indicates its absence. The laborious and often wasteful procedure of redistilling liquid ammonia using an acetone–Cardice charged condenser is not usually worth while (see, however, Section 7.5, p. 1114).

The simplest apparatus assembly is shown in Fig. 2.69(*c*). It consists of a three-necked round-bottomed flask of appropriate size equipped with a mechanical stirrer unit fitted with a Hershberg wire stirrer or a Teflon or a glass stirrer sited in the central neck. A pressure-equalising dropping funnel fitted with a soda-lime-filled guard-tube (not calcium chloride) is placed in one side-neck. The second side-neck accommodates the inlet adapter for the liquid ammonia supply, and this may subsequently be closed with a glass stopper; the addition of metals, other solid reagents, or gaseous reagents such as acetylene is also made through this side-neck. The flask is surrounded by a box containing cork chips, vermiculite or other insulating material; the outside of the flask rapidly acquires a coating of ice when being charged with liquid ammonia and this provides additional insulation. The flask contents may be viewed by the removal from time to time of a section of the external ice coating by pouring a few drops of acetone or ethanol on to the outside of the flask from a wash bottle.

Fig. 2.69

Alternatively the flask may be surrounded by an acetone–Cardice bath if reaction temperatures in the region of $-78\,°C$ are required, or where there may be some danger of the reaction product being of such a volatility as to be swept away in the stream of ammonia vapour.

This basic apparatus assembly may be modified to meet a range of alternative circumstances. For example, if no pressure-equalising dropping funnel is available, an adapter with T-connection may be inserted between the flask and a normal dropping funnel; the outlet of the T-connection is attached to a soda-lime guard-tube. Alternatively a dropping funnel may be used directly, in which case the second side arm of the flask should be closed with a soda-lime guard-tube; if stirring of the reaction mixture is vigorous it may be necessary to interpose a short air condenser between this guard-tube and the flask.

A further modification may be required if the reagent to be added from the dropping funnel is sensitive to the action of ammonia vapour. In such a case not only should the atmosphere around the reagent be of some inert gas (e.g. nitrogen) but also the reagent should be added below the surface of the liquid ammonia. The modification shown in Fig. 2.69(d) may be employed in this case. The adapter illustrated is a stirrer guide with gas inlet connection; the dropping funnel is of the pressure-equalising type; a normal dropping funnel requires that the second side-neck of the flask permits the escape of gas to the atmosphere via a guard-tube.

The individual sequence of operations subsequent to half-filling the flask with liquid ammonia depends on the nature of the reactions involved. Specific details are given in Expts 5.23, 5.26 and 5.43. A general comment however is worth noting that during the course of the reaction ammonia gas is continuously escaping, and in all but prolonged reaction sequences it is advisable to ensure an initial adequate supply of liquid ammonia in the reaction flask to avoid the necessity of topping up the level. In some reactions it has been noted that considerable foaming occurs during the addition of reagents. This may be controlled to some extent by the addition of a little ether, or by raising the stirrer blade so that it agitates the surface of the liquid and hence assists in the rapid breakdown of foam. If these measures are unsuccessful, reagent addition should be stopped until foaming abates, and continued addition should then take place more slowly.

The simplest work-up procedure for non-volatile reaction products and products which are obtained in the form of alkali metal salts is to allow the ammonia to evaporate overnight through the guard-tube into a vented fume cupboard. Usually in these cases an inert solvent such as ether or tetrahydro-furan is added to the flask before evaporation commences. If time is important the flask may be placed on a warm water bath (c. 45–50 °C) and the flask contents stirred; this should also be carried out in a fume cupboard. In either case a stream of nitrogen may be introduced into the flask if the product is sensitive to atmospheric oxidation. The subsequent treatment of the residue will depend upon the particular reaction and individual experiments should be consulted for typical isolation procedures.

With volatile products which may arise in reactions which do not lead to the initial formation of metal salts (i.e. dehydrohalogenations, alkylations leading to terminal acetylenes) the following procedure is recommended.[35]

The three-necked reaction flask is fitted with a stopcock in one side-neck, a stopper in the second side-neck and a wide-bore glass tube (4–5 mm internal diameter) fitted by means of a rubber bung or screw-capped adapter to the central neck. This glass tube, which reaches to within a few millimetres from the flask bottom, is fitted with a plastic tube which terminates below the surface of an ample quantity of crushed ice contained in a large conical flask. Some extraction solvent which is to be used in the work-up is added to the crushed ice. When the stopcock is turned off the pressure build-up forces the liquid ammonia solution into the ice–solvent mixture, the flow being controlled by periodically opening the stopcock. Additional ice is added to the conical flask as required until the transfer of liquid ammonia solution is complete. The reaction flask is rinsed with extraction solvent and this is added to the main bulk. If the reaction mixture contains large amonts of salts (e.g. sodium halides) it is necessary to swirl the flask in order to keep the salts in suspension and so prevent the tube becoming blocked – if this should happen the stopcock must be opened immediately. An alternative procedure is to allow the salts to settle to the bottom of the reaction flask and gradually lower the glass tube through the liquid as transfer is effected; when all the supernatant liquid has been transferred crushed ice and extraction solvent is added to the residue which is then combined with the main bulk.

2.17.8 REACTIONS INVOLVING AIR-SENSITIVE COMPOUNDS

It is often necessary to carry out reactions in an inert atmosphere or to handle starting materials, intermediates or products which are air-sensitive. Such reactions can be carried out successfully without specialised apparatus, such as Schlenk tubes and purpose-designed vacuum lines,[36] provided attention is given to the detail of the procedure. The procedures described in this section should be routinely within the capacity of advanced students and of the practising organic chemist. Suppliers of air-sensitive materials provide guidance on handling techniques which should always be followed carefully. Particularly useful is the pamphlet published by Aldrich Chemical Company.[37] Reactions involving organolithium reagents (e.g. Expts 5.10; 6.158; 7.3), organoboranes[38] (e.g. Expt 5.44) and Wittig reagents (Section 5.17) are among those which require these standard procedures. The yield from Grignard reactions (e.g. Expts 5.39–5.42) may well be improved if the reactions are carried out in this manner.

Even if reactions of this sort are carried out only relatively infrequently, it is advisable to set up a simple nitrogen (or argon) line which will facilitate many of the operations described in this section. The line leading from the gas cylinder is shown diagrammatically in Fig. 2.70.

Fig. 2.70

It consists of a pressure reducing valve, a T-piece to allow flushing of syringes and cannulae, T-pieces to allow flow of inert gas to the apparatus and to allow evacuation of the apparatus, and a bubbler to indicate the flow of the inert gas and to allow a slight positive pressure of the inert gas to be maintained in the apparatus. Nitrogen can be led to the apparatus directly by rubber or preferably polythene tubing. It is important that this tubing is dry and flushed with nitrogen before attaching it to the apparatus. It is advisable to keep a selection of tubing specifically for this purpose and not to mix it with tubing used for water condensers and other purposes. An alternative and convenient method of leading nitrogen to the apparatus is to attach a syringe needle to the nitrogen line by means of a wired-in Luer lock fitting. Nitrogen can thus be fed easily into a reaction set-up via a rubber septum.

To carry out reactions successfully in an inert atmosphere it is necessary to ensure each of the following:

1. That the 'inert atmosphere' is in fact inert and does not react with any of the reagents or intermediates.
2. That before the reaction is started the apparatus is thoroughly dry and filled with the inert gas.
3. That all reagents and solvents are dry and oxygen free.
4. That solvents and reagents are transferred to the reaction vessel in such a way that they do not come into contact with air or moisture.
5. That the air-sensitive materials are handled in an inert atmosphere throughout.

How to achieve each of these requirements is described in the following sections.

INERT GAS
Nitrogen is commonly used as the inert gas and the commercially available material is satisfactory for most purposes without pretreatment. If there is any doubt about the purity of the nitrogen gas it may be purified as described in Section 4.2.52. An alternative inert gas to nitrogen may be required in cases where one or more of the reagents (e.g. lithium metal) reacts with nitrogen. In such cases argon may be used although it is considerably more expensive than nitrogen.

DRYING THE APPARATUS AND FILLING WITH INERT GAS
One of the following procedures should be used to ensure that the apparatus to be used is dry and filled with inert gas.

Method 1. Dry the apparatus in the oven (120 °C) for several hours, and preferably overnight. Allow the apparatus to cool in a desiccator and then assemble with a stream of nitrogen flowing through it.

Method 2. After over-drying, assemble the apparatus hot and allow it to cool while nitrogen flows through it.

Method 3. Assemble the apparatus cold and flame it thoroughly with a Bunsen flame to drive out moisture while purging with nitrogen and allow to cool with continuing nitrogen flow.

Method 4. Evacuate the apparatus using a vacuum pump and refill it with inert gas. Repeat the evacuation and refilling process twice more. This procedure is described in more detail below.

DRYING SOLVENTS AND REAGENTS
Procedures for drying reagents and solvents are described in Chapter 4. Dry solvents which have not been redistilled under nitrogen should be de-oxygenated by bubbling a stream of nitrogen through them using the gas delivery tube shown in Fig. 2.59. This procedure can be carried out conveniently in a two-necked flask equipped with the gas inlet tube and a rubber septum into which a syringe needle is inserted to allow nitrogen to escape after bubbling through the solvent. Solvent may be removed using a syringe or cannula inserted into the septum as described below. An alternative method of ensuring that relatively small volumes of solvent are oxygen free is to use the repeated freeze/thaw cycle. In this procedure the liquid is placed in the reaction set-up and frozen with liquid nitrogen. The apparatus is then evacuated, isolated from the vacuum line and allowed to warm up. Gas absorbed in the liquid is released. The liquid is again frozen and the apparatus is evacuated to pump off the released gases. This process is repeated once more and finally the reaction apparatus is filled with nitrogen. Many dried and redistilled solvents absorb air and moisture very readily and it is therefore advisable to transfer them by the procedure described below using a syringe or cannula so that they are not exposed to the atmosphere.

TRANSFER OF SOLVENTS AND REAGENTS
Liquids. Routine transfer of air-sensitive liquids is carried out with a syringe or with a long flexible wide-bore, double-tipped needle (cannula), which is inserted into reaction vessels and storage containers via a rubber septum. A septum is simply a flexible rubber stopper with a sleeve which folds down over the neck of the container. The septum can be pierced by a syringe needle and reseals itself when the needle is removed. It is thus an essential piece of equipment for reactions in an inert atmosphere. They are available in a range of sizes to fit standard ground glass joints, stopcock arms and containers. Septa gradually deteriorate when exposed to organic solvents for a long period. Exposure to solvents should therefore be limited as far as possible, for example by separating them from refluxing solvent by attaching them to a flask via an interposed stopcock rather than directly into the neck of the flask. After repeated insertion of needles into the septum it will eventually cease to be gas-tight and will need to be discarded.

Glass syringes are available in a wide range of sizes and with a variety of fittings. Syringes for the transfer of small volumes of liquids are individually ground and calibrated and do not have interchangeable barrels and plungers. It is important that the barrels and plungers are used only as a matched pair, otherwise accuracy will be lost and there will be a danger of the plunger sticking in the barrel. Larger volume syringes have interchangeable plungers, although in this case the accuracy with which the volume of liquid can be measured is reduced. Plastic syringes are also available and although they are suitable for transferring many aqueous solutions they are not recommended for general use in the organic laboratory.

A degree of care is necessary when handling syringes if expensive damage is to be avoided and accurate delivery of volumes of liquid ensured. Care is also needed to ensure that the needle is firmly fitted to the syringe. If it should become detached with the syringe full of air-sensitive or pyrophoric compound the consequences could be disastrous. Two further commonly encountered problems are the attachment of the needle to the barrel of the syringe, and plungers sticking in the barrel of the syringe. With the simple syringe design a tight seal between the syringe and the needle is best obtained by attaching the needle to the barrel when both are hot after drying in the oven, and allowing them to cool together. The greater contraction of the metal needle ensures a good fit. In order to remove the needle subsequently it may be necessary to warm its neck gently in the Bunsen flame. This should not be done until all organic material has been removed by flushing with a suitable low-boiling solvent and 'pumping' to evaporate the remaining traces of liquid. Syringes with Luer lock fittings should not present any difficulty in attaching the needle to the syringe. In order to ensure that the plunger moves freely in the barrel and to minimise the dangers of sticking, the plunger on interchangeable syringes can be lubricated with a small amount of silicone grease. Normally syringe needles have bevelled tips in order to penetrate the rubber septum easily. In some cases it is desirable to have a needle with a flat tip, for example in the flask-to-flask transfer of liquids described below when it is desired to remove the last drops of liquid from a container. These are available commercially, but it is necessary to puncture the septum initially with a normal bevel-tipped needle before attempting to pass the flat-tipped needle through.

A typical procedure for transferring a liquid (5–100 ml) from one septum-capped vessel to another with a syringe is as follows. Select a syringe with a capacity of the same order as the volume of liquid to be transferred. The needle should be long enough to reach below the surface of the liquid to be transferred and also sufficiently long and flexible to allow the syringe to be inverted while still inserted in the septum. Ensure that the syringe and needle are clean. Dry them in an oven at 120 °C for 3–4 hours and allow them to cool in a desiccator. The needle can be attached to the barrel of the syringe while it is hot so that as it cools a tight fit is assured. Flush the syringe with nitrogen as follows. Push the plunger to the bottom of the barrel of the syringe and insert the needle through the septum on the nitrogen line (see Fig. 2.70). Slowly withdraw the plunger to the maximum graduation on the barrel to fill it with nitrogen and then withdraw the needle from the septum. Expel the nitrogen from the syringe and repeat the process of filling with nitrogen and emptying twice more. The syringe is now ready for use in transferring air-sensitive liquids. Ensure that the vessel from which the liquid is to be removed is connected to the nitrogen line and that there is a slight positive pressure of nitrogen. This can be achieved most easily by inserting a syringe needle attached to the nitrogen line through the rubber septum. Now hold the needle of the prepared syringe close to its tip and insert it through the septum into the vessel containing the liquid to be transferred. It should be inserted sufficiently far for the tip of the needle to be below the surface of the liquid in the vessel. Slowly withdraw the plunger until slightly more than

123

the required volume has been taken up into the syringe. It is quite likely that in this process bubbles of gas will be generated in the liquid and accumulate at the surface of the liquid in the syringe. This gas can be expelled, assuming a sufficiently long and flexible needle has been attached to the syringe initially, by raising the tip of the needle above the surface of the liquid in the flask, carefully inverting the syringe and then expelling the gas into the vessel (Fig. 2.71). The syringe should now contain only liquid and can be returned to the vertical position. The required volume of liquid can be obtained by depressing the plunger appropriately. The needle can now be withdrawn from the septum and the liquid transferred via a septum to the reaction vessel or dropping funnel. The plunger should be depressed gradually to the bottom of the barrel.

If the transfer of an accurately measured volume is required the liquid remaining in the needle should not be expelled by pumping the plunger of the syringe. Syringes are calibrated to deliver a given volume when the plunger is fully depressed. In any event, the volume of liquid in the needle will obviously vary significantly depending on the length and diameter of the needle. Care should be taken in disposing of any reactive material remaining in the syringe needle by carefully drawing into the syringe suitable clean, dry solvent and then expelling the syringe contents into a container for treatment or disposal. The accuracy with which a volume of liquid can be measured with a syringe is limited. With a syringe up to 1 ml an accuracy of 1–2% may be achievable with practice and this should be satisfactory, for example, for the removal of aliquot portions in the estimation of solutions of alkyllithiums (Section 4.2.47, p. 442). With larger volumes the accuracy of measurement decreases. If a precisely measured volume of liquid is required for addition to a reaction it is advisable to pre-calibrate a dropping funnel specifically for this purpose or to transfer an accurately measured volume or weighed amount, rather than rely on the accuracy of a syringe or a graduated dropping funnel.

An alternative to the syringe technique for the transfer of liquids uses the double-ended needle and is suitable when it is not necessary to know precisely the volume of liquid being transferred, or when the volume is being measured in some other way, e.g. by transfer from a septum-capped measuring cylinder. This

N₂ inlet via syringe needle

Septum

Septum

Fig. 2.71

Fig. 2.72

method is particularly useful for the transfer of an air-sensitive intermediate from the flask in which it has been prepared into a second reaction set-up, or for the transfer of commercially available air-sensitive material supplied in an air-tight container. The procedure is illustrated in Fig. 2.72. The cannula (double-ended needle) is inserted just through the septum of flask A which contains the liquid to be transferred, through which nitrogen is flowing, and purged by allowing nitrogen to flow through it for a minute or two. The other end of the needle is then inserted through the septum of flask B, previously flushed with nitrogen. The tip of the needle in the delivery flask A is then lowered below the level of the liquid. The rate of transfer of liquid can then be controlled by adjusting the flow of nitrogen into the delivery flask with the three-way stopcock. If it is desired to transfer the total contents of the delivery flask a cannula with a flat end rather than the normal bevelled end should be used. The liquid could be transferred if required via a septum to the dropping funnel of the second reaction set-up rather than directly into the flask. On occasions it is necessary to transfer intermediates which have been prepared at low temperature and which would deteriorate if allowed to warm up. This problem may be solved in one of two ways depending on the temperature sensitivity of the intermediate. The first method is to transfer the solution in small batches to the dropping funnel (which could if necessary be jacketed) with the majority of the solution remaining in the cooled delivery flask. Alternatively the cannula itself can be looped and passed through a cooling bath held at the required temperature.

Solids. Some solids can be added to the reaction flask at the start of the reaction before the apparatus is filled with inert gas and therefore present little problem. Examples include magnesium for Grignard reactions and metal hydrides. In the latter case it may be necessary to remove oil in which the hydride is suspended (Section 4.2.49, p. 445) and this can be achieved using the procedures described below for separating solids and liquids in an inert atmosphere. If it is necessary to add a solid to the reaction mixture during the course of a reaction, a modification of the apparatus shown in Fig. 2.57 can be used in which a flask with a ground glass neck is linked to the reaction flask by an adapter with two male ground glass joints. Addition of the solid is achieved by rotation of the adapter in the neck of the flask. If the solid is to be added under reflux, the apparatus shown in Fig. 2.58 can be used.

Gases. Gases can be added to reactions in an inert atmosphere by an extension of the standard procedures noted in Section 2.15. If the quantity of gas does not

N₂ inlet via
syringe needle

N₂

Reaction flask

Condensed gas or
liquid reagent

Fig. 2.73

have to be measured (i.e. it is simply required to produce a saturated solution), the gas can be bubbled into the liquid in the normal way as shown in Fig. 2.59, with the appropriate precautions being taken to ensure that the inert atmosphere is maintained. Alternatively a balloon inflated with the required gas and attached to the reaction flask can be used. If it is necessary to add a known quantity of gas the method to be used will depend on the boiling point of the gas. For gases which can be condensed relatively easily, e.g. with a dry-ice condenser, a given volume of gas can be condensed into a measuring cylinder and then allowed to boil off gradually via a cannula into the reaction flask, Fig. 2.73.

For precise determination of the amount of the condensed gas, the container should be weighed before and after the gas is condensed. For gases which are not easily condensed, the volume can be measured using a calibrated gas burette of the type shown in Fig. 2.63(a) and the gas then allowed to pass gradually into the flask after suitable drying or other purification.

MAINTAINING AN INERT ATMOSPHERE IN THE APPARATUS

Normally it is not necessary to have a continuous flow of the inert gas through the apparatus. If the reaction lasts for several hours this can be very expensive in the amount of gas used. It is sufficient to ensure that no air is allowed to enter the apparatus. This is usually achieved by maintaining a slight positive pressure of inert gas in the apparatus by means of a bubbler. It is important that all the ground glass joints are properly secured with spring clips, elastic bands or by being wired on, in order to ensure that they do not leak under the small positive pressure of the inert gas or, more importantly, come apart during the course of

Fig. 2.74

the reaction. An alternative and simpler way of maintaining the inert atmosphere is to use the balloon technique in which a rubber balloon is inflated with the inert gas and wired to the reaction flask.* A number of different types of bubblers is available. The simplest type is shown in Fig. 2.74(a) and has been noted in Fig. 2.60.

The positive pressure of inert gas is determined by the height of the liquid (mineral oil or mercury) in the tube. There is a danger in this type of bubbler that a sudden reduction in the pressure in the reaction flask will cause a suck-back of the liquid from the bubbler into the reaction flask. This can be avoided by using a bubbler with an internal reservoir (Fig. 2.74(b)). Although this type of bubbler will prevent suck-back of the liquid in the bubbler into the reaction vessel, it will not prevent the ingress of air. This can be prevented by using a bubbler with a valve which closes when the pressure differential is reversed.

Apparatus for reactions in an inert atmosphere. The experimental set-up will be determined by the nature of the reaction, but in all cases it will be necessary to meet the requirements set out above for the conduct of any reaction in an inert atmosphere. For the majority of purposes commercially available glassware with ground glass joints is adequate. A number of adapters are particularly useful for the addition of reagents via a septum. Several of these are illustrated in Fig. 2.75(a)–(e).The use of some of these adaptors is illustrated in the experimental set-ups which are described below. Either (a) or (b) can be fitted directly to the reaction flask to allow direct addition of reagents to the reaction vessel thus avoiding any undesired contamination which might arise from the use of a dropping funnel. Adapters (a) and (b) also allow removal and transfer of air-sensitive liquids. Adapter (c) can be used to accommodate a thermometer and to allow direct addition to the reaction flask via the septum side-arm. The range of adapters available allows considerable flexibility in arranging experimental set-

---

* The balloon technique is often recommended in suitable reactions to maintain an inert and dry atmosphere within the reaction flask. It should be noted however that air and moisture infuse rather rapidly and that this atmosphere may only be maintained for half-an-hour or so.

(a) (b) (c) (d) (e)

Fig. 2.75

ups for reactions in an inert atmosphere. The three-way stopcock which can be equipped with a rubber septum on one of its arms is particularly useful.

Stirring. The most convenient method of stirring, particularly in small-scale reactions, is to use a Teflon-coated magnetic stirring bar (Section 2.14). However, care must be taken to avoid over-rapid stirring which may cause the stirrer bar to fly free from the guiding magnet and penetrate the wall of the flask, with potentially disastrous results. A mechanical stirrer (such as that shown in Fig. 2.51(a) is to be preferred in any of the following circumstances: (i) large volumes of liquids are used; (ii) the reaction mixture is likely to become viscous; (iii) there are large amounts of suspended solids; (iv) very rapid stirring is required.

A very simple apparatus for reaction in an inert atmosphere is a one-necked flask equipped with a magnetic stirrer and three-way stopcock (Fig. 2.76). The vertical arm of the stopcock is fitted with a rubber septum and the horizontal arm leads to a second three-way stopcock giving access to the nitrogen supply and to the vacuum line. This apparatus is suitable for reactions which take place at or below room temperature, do not require addition of solid reagent once the

Fig. 2.76

reaction is under way and usually where no excessive gaseous by-products are formed, and where stirring can be accomplished satisfactorily with a magnetic stirrer. The use of this set-up is relatively limited, but the procedures described for filling the apparatus with inert gas and for adding liquids via the septum are of general applicability.

A typical procedure for the use of the apparatus is as follows. Dry the flask in an oven at 120 °C for four hours or overnight and allow it to cool in a desiccator. The stopcocks should be dry and the barrels lightly greased. Charge the flask with any dry solid reagent required at the start of the reaction and with a Teflon-coated magnetic stirring bar. Insert into the neck of the flask the three-way stop-cock A fitted with a rubber septum on the vertical arm. Attach the horizontal arm of the stopcock to a second three-way stopcock B, one arm of which is attached to the nitrogen line and one to the vacuum line. Care must be taken in adjusting the stopcocks to ensure that the vacuum is never applied directly to the nitrogen line. Turn stopcock B so that the apparatus is connected to the vacuum line and gradually open stopcock A so that the flask is evacuated. The small volume of air between the barrel of stopcock A and the rubber septum should also be evacuated and replaced by nitrogen by appropriate manipulation of the stopcock. Carefully adjust stopcock B so that the apparatus is connected to the nitrogen line and gradually fill the flask with nitrogen. **CAUTION:** *The nitrogen flow through the bubbler should be increased and the stopcock turned very slowly in order to avoid any suck-back through the bubbler.*

Once the flask has been filled with nitrogen turn the stopcock B so that the apparatus is connected to the vacuum line and evacuated once more. The process of evacuation and filling with nitrogen should be carried out three times. Adjust stopcock B so that the apparatus is connected to the nitrogen line and reduce the flow of nitrogen to a level sufficient to maintain a slight positive pressure in the system with the occasional bubble of nitrogen passing through the bubbler. Turn stopcock A so that the flask is open to the septum arm and to the nitrogen line, and introduce solvents or liquid reagents into the flask via the septum with a syringe or cannula which passes through the orifice of the stopcock and into the flask. It is advisable to isolate the septum arm from the reaction flask by appropriate adjustment of stopcock A, except when additions are being made, in order to avoid any possibility of leakage through the septum. The reaction flask should however always be connected to the nitrogen line. If the reaction flask should require cooling during the course of the reaction care should be taken to increase the flow of nitrogen through the bubbler to avoid the possibility of suck-back. If any gases are evolved during the course of the reaction they will escape from the exit tube of the bubbler. If these gases present any hazard the exit tube should be passed to a gas absorption trap (Section 2.15) or passed into a fume cupboard extraction vent.

An alternative to the use of the one-necked flask described above is to use the apparatus shown in Fig. 2.77. In this case a two-necked flask is used, with the septum which is used for the addition or removal of liquids fitted to one arm of the flask and a three-way stopcock to the other. With this arrangement it is not possible to isolate the septum from the reaction flask. Should this be required the septum is fitted to a stopcock in the second arm of the flask. Flasks with an integral stopcock are available commercially.

An experiment set-up which is more generally useful is shown in Fig. 2.78. It incorporates a reflux condenser, a pressure-equalising dropping funnel, a ther-

Fig. 2.77

mometer and a magnetic stirrer. If a mechanical stirrer is required the thermometer and reflux condenser can be accommodated in one side-arm using an adapter (Fig. 2.15). This set-up can be flushed with nitrogen at the start of the reaction procedure by opening the stopcock on the dropping funnel and inserting a syringe needle in the septum to allow outflow of nitrogen. On occasions it may be found convenient to use a flask with four necks. Variations in these set-ups can be achieved using the adapters and the reflux condensers described in Section 2.7. Thus, for example, a Dewar condenser (Fig. 2.23) is used for reactions involving liquids with low boiling points.

Fig. 2.78

SEPARATION OF LIQUIDS AND SOLIDS

On occasions it is necessary to separate a liquid and a solid in an inert atmosphere. If the liquid and solid separate easily and distinctly with the solid settling to the bottom of the flask, the process of separation can be accomplished by careful use of a syringe or cannula to remove the supernatant liquid. The liquid may be discarded, stored or transferred to another reaction set-up for further reaction as appropriate. The solid remaining in the flask can be washed if necessary by the addition and subsequent removal of further quantities of dry, oxygen-free solvent through the septum. If the solid is desired solvent-free, residual solvent can be removed by connecting the flask to the vacuum line and condensing the solvent in a suitable trap. If the solid and liquid do not separate easily and the solid remains in suspension, filtration will be necessary. The simplest way of achieving this is with the gas dispersion or filter tube with a sintered glass end (Fig. 2.59). The technique is simlar to that described in Section 2.20 (Fig. 2.81) for recrystallisation at low temperature. The filter tube can be fitted to the flask at the start of the reaction or subsequently inserted (after thorough drying) with a rapid stream of nitrogen flowing through the neck of the flask into which the tube is to be inserted. If the solid is required it remains in the flask after further washing. If the liquid is required for further reaction it can be led directly to a second reaction flask.

ISOLATION AND PURIFICATION TECHNIQUES

2.18 GENERAL CONSIDERATIONS

At the conclusion of a reaction the pure product must be isolated from the reaction mixture by a sequence of operations collectively termed the 'work-up'. As well as the required product the reaction mixture may contain, for example, solvent which has been used as the reaction medium, excess reactants or reagents, unwanted reaction products (by-products) arising from alternative reaction pathways and so on. The planning of the isolation operations and application to such complex mixtures is therefore an exacting test of the expertise of the chemist. Frequently a student fails to bring a successful reaction to a fruitful conclusion by using an ill-considered work-up procedure, which results in loss of the required product either by decomposition during attempted isolation, or from a premature discard of product because of lack of appreciation of its physical or chemical properties. It should be emphasised that even when a detailed published procedure is being followed it is unwise to discard any liquid or solid fractions separated during work-up until the final product has been isolated and adequately characterised.

Because of the length of time that a complete isolation process often takes, it is wise practice, particularly with new syntheses carried out for the first time, to monitor the progress of the reaction. Thus the disappearance from a reaction mixture of one of the reactants or the build-up of the reaction product, measured on small aliquot portions removed at convenient time intervals from the bulk reaction mixture, can yield valuable information on the progress of a reaction. Usually the former is to be preferred since the physical properties (e.g. spectroscopic information, Chapter 3), chemical reactivity (e.g. characteristic tests of functional groups, Section 9.5) and chromatographic behaviour (Section 2.31) of the reactant, and the influence of solvents or other reactants on the reliability of the chosen monitoring processes may be readily checked before the reaction is commenced.

The adoption of a particular isolation procedure will depend to a large extent upon the physical and chemical properties of the product. Some guidelines for useful general approaches may however be given with regard to the physical state at ambient temperature of the crude mixture resulting from the reaction, i.e. whether it is a *one-phase* (either *solid* or *liquid*) or a *two-phase* (*solid/liquid* or *liquid/liquid*) system.

In the case of the *one-phase solid system* if the organic product is neutral and insoluble in water, washing with water may be used to remove soluble impurities such as inorganic salts. Alternatively the crude solid may be extracted with a suitable organic solvent (Section 2.22), filtered, and the extract washed with water. Further washing successively with dilute aqueous acid and dilute aqueous alkali removes basic and acidic impurities. Removal of solvent after drying (Sections 2.23 and 2.24) leads to the recovery of the purified solid for recrystallisation from a suitable solvent (Section 2.20). Continuous extraction of the solid (e.g. in a Soxhlet apparatus) may be necessary if the required product is only sparingly soluble in convenient organic solvents.

If the crude solid product contains the required product in the form of a salt (e.g. the alkali metal salt of a phenol) and is therefore water soluble, acidification of the aqueous solution (or basification in the case, for example, of amine salts) liberates the free acidic compound (or base) which may be recovered by filtration or solvent extraction as appropriate.

The *one-phase liquid system* is more frequently encountered since many organic reactions are carried out in solution. Direct fractional distillation may separate the product, if it is a liquid, from the solvent and other liquid reagents, or concentration or cooling may lead to direct crystallisation of the product if this is a solid. However, it is often more appropriate, whether the required product is a liquid or solid, to subject the solution to the acid/base extraction procedure outlined above and considered in detail on p. 162. This acid/base extraction procedure can be done directly if the product is in solution in a water-immiscible solvent. A knowledge of the acid-base nature of the product and of its water solubility is necessary to ensure that the appropriate fraction is retained for product recovery. In those cases where the reaction solvent is water miscible (e.g. methanol, ethanol, dimethylsulphoxide, etc.) it is necessary to remove all or most of the solvent by distillation and to dissolve the residue in an excess of a water-immiscible solvent before commencing the extraction procedure. The removal of solvent from fractions obtained by these extraction procedures is these days readily effected by the use of a rotary evaporator (p. 185) and this obviates the tedium of removal of large volumes of solvent by conventional distillation.

A crude reaction mixture consisting of *two phases* is very common. In the case of a *solid/liquid* system, it will of course be necessary to make certain in which phase the required product resides. A simple example is where the product may have crystallised out from the reaction solvent; the mixture therefore only requires to be cooled and filtered for the bulk of the product to be isolated. The filtrate should then routinely be subjected to suitable concentration or extraction procedures to obtain the maximum yield of product.

Direct filtration would also be employed when the solid consists of unwanted reaction products, in which case the filtrate would be treated as the single-phase liquid system above. Where it is evident that the product has crystallised out admixed with contaminating solid material a separation might be effected if the mixture is reheated and filtered hot (p. 139).

Liquid/liquid two-phase systems are often encountered; for example, they result from the frequent practice of quenching a reaction carried out in an organic solvent by pouring it on to ice or into dilute acid. A further instance of a liquid/liquid system arises from the use of steam distillation (Section 2.25) as a preliminary isolation procedure. This is particulalry suitable for the separation of relatively high-boiling liquids and steam volatile solids from inorganic contaminants, involatile tars, etc. The subsequent work-up procedure normally presents no additional problems since the phases are usually readily separable and can be treated in a manner appropriate to the chemical or physical properties of the required product by procedures already outlined.

All these preliminary procedures give solid or liquid products which are rarely of high purity; the degree of purity may be checked by chromatographic and spectroscopic methods. Purification may often be successfully accomplished by recrystallisation or sublimation for solids (Sections 2.20 and 2.21); fractional distillation under atmospheric or reduced pressure for liquids or low melting solids (Sections 2.26 and 2.27); molecular distillation for high-boiling liquids (Section 2.28). In those cases where the use of these traditional methods does not yield product of adequate purity, resort must be made to preparative chromatographic procedures (p. 199). Here a knowledge of the chromatographic behaviour obtained from small-scale trial experiments will be particularly valuable.

The final assessment of the purity of a known product is made on the basis of its physical constants (Sections 2.33 to 2.37 and Chapter 3) in comparison with those cited in the literature. In the case of a new compound the purity should be assessed and the structural identity established by appropriate chromatographic and spectroscopic methods.

2.19 FILTRATION TECHNIQUES

Filtration of a mixture after completion of a reaction will often be necessary either to isolate a solid product which has separated out or to remove insoluble impurities or reactants, in which case the desired product remains in solution. In this section the filtration of cold solutions is described; the filtration of hot solutions is considered in Section 2.20.

When substantial quantities of a solid are to be filtered from suspension in a liquid, a Buchner funnel of convenient size is employed. The ordinary Buchner funnel (Fig. 2.43(*a*)) consists of a cylindrical porcelain funnel carrying a fixed, flat, perforated porcelain plate. It is fitted by means of a rubber stopper, rubber cone or flat rubber ring into the neck of a thick-walled filtering flask (also termed a filter flask, Buchner flask or suction flask) (Fig. 2.43(*c*)), which is connected by means of thick-walled rubber tubing (rubber 'pressure' tubing) to a similar flask or safety bottle, and the latter is attached by rubber 'pressure' tubing to a filter pump; the safety bottle or trap is essential since a sudden fall in water pressure may result in the water being sucked back and contaminating the filtrate. The use of suction renders rapid filtration possible and also results in a more complete removal of the mother-liquor than filtration under atmospheric

133

pressure. A filter paper is selected (and trimmed, if necessary) of such size that it covers the entire perforated plate, but its diameter should be slightly less than the inside diameter of the funnel; the filter paper should never be folded up against the sides of the funnel. The filter paper is moistened with a few drops of clear supernatant liquor and the suction of the pump applied, when the filter paper should adhere firmly to, and completely cover, the perforated plate of the funnel and thus prevent any solid matter from passing under the edge of the paper into the flask below. As much of the supernatant liquor as possible is now poured into the funnel and filtered before the bulk of the residual slurry is transferred to the filter funnel; this procedure is often quicker than initially bringing all the solid into suspension and pouring it directly on to the filter. Furthermore initial gentle suction often leads to more effective filtration than powerful suction since in the latter case the finer particles of solid may reduce the rate of filtration by being drawn into the pores of the filter paper. Any solid remaining in the reaction flask is easily transferred by rinsing with a little of the filtrate* and well stirring to remove solids which may be adhering to the sides of the reaction flask. This operation may be repeated until all solid material has been transferred to the filter. The suction is continued until most of the liquid has passed through and this is facilitated by pressing the solid down with a wide glass stopper to leave a uniformly flat, pressed surface. The filter cake is then washed with an appropriate solvent and again sucked dry. If the filter cake is the required product, then this must be subjected to purification using suitable recrystallisation procedures (Section 2.20). If the filtrate contains the reaction product, further suitable isolation procedures would then of course be adopted (e.g. Sections 2.22, 2.25 and 2.31).

Some modification to the above general technique of isolation by filtration may be necessary in the light of the chemical nature of the reaction mixture, of the particle size of the solid, or of the ratio of the amount of solid to liquid material to be filtered.

For example, strongly alkaline or strongly acidic reaction mixtures weaken cellulose filter papers. Acid-hardened grades which are more chemically resistant are commercially available (e.g. Whatman filter papers) but for maximum resistance to chemical attack, glass-fibre paper (e.g. Whatman) or a glass funnel fitted with a fixed sintered glass plate (Section 2.11) may be used.

The filtration of very finely divided suspended material is often very tedious as a result of the filter paper pores becoming clogged. In such a case the addition of a suitable filter aid (e.g. a high grade diatomaceous earth such as Celite 545, or Whatman filter aids) to the suspension overcomes the problem; alternatively the suspension may be filtered through a bed of filter aid prepared by pouring a slurry of it in a suitable solvent into the filter funnel fitted with the required size of filter paper. The initial application of gentle suction in the filtration is in this case vital. A glass-fibre filter paper, supported on a conventional filter paper in a Buchner or Hirsch funnel, is useful for the rapid removal of finely divided solid impurities from a solution.

The selection of a funnel appropriate to the amount of solid rather than the total volume of liquor to be filtered is important. When the volume of liquid is large relative to the amount of solid, the apparatus shown in Fig. 2.79 may be

---

* The filter flask must be disconnected from the pump before the latter is turned off otherwise suck-back will occur which may ruin the reaction products should these be present in the filtrate.

→ To pump

Fig. 2.79

used; here the funnel should be of a size appropriate to the amount of solid to be finally collected. When the receiver is about two-thirds full, atmospheric pressure is restored by suitably rotating the 'three-way' stopcock; the filtrate may then be removed by opening the tap at the lower end. The apparatus is again exhausted and the filtration continued.

For the suction filtration of small quantities of solid (less than 5 g) contained in a small volume of liquid, a small conical Buchner funnel, known as a Hirsch funnel, is employed (see Fig. 2.43(b)); the filtrate is collected either in a small filter flask or in a test tube with side-arm, the arrangement being illustrated in Fig. 2.43(f). A small sintered funnel, or a slit-sieve funnel may also be employed. The procedure for filtration is similar to that already given for the Buchner funnel.

Small volumes of solution (up to 2 ml in one operation) may conveniently be filtered through a dropping (Pasteur) pipette into the constriction of which has been rammed a small piece of paper tissue (about 3 cm square). The pipette is supported vertically and the solution is added from a second Pasteur pipette. Pressure to accelerate the filtration process may then be applied from a rubber bulb attached to the top of the pipette.

This method is particularly useful for the preparation of solutions of samples for spectroscopic examination where it is important to remove all insoluble impurities. The method may be adapted for the decolourisation of small samples by incorporating a short column of decolourising carbon above the paper plug (cf. Section 2.20).

2.20 RECRYSTALLISATION TECHNIQUES

Solid organic compounds when isolated from organic reactions are seldom pure; they are usually contaminated with small amounts of other compounds (impurities) which are produced along with the desired product. The purification of impure crystalline compounds is usually effected by crystallisation from a suitable solvent or mixture of solvents.

The purification of solids by crystallisation is based upon differences in their solubility in a given solvent or mixture of solvents. In its simplest form, the crystallisation process consists of: (i) dissolving the impure substance in some suitable solvent at or near the boiling point; (ii) filtering the hot solution from particles of insoluble material and dust; (iii) allowing the hot solution to cool

135

thus causing the dissolved substance to crystallise out; and (iv) separating the crystals from the supernatant solution (or mother-liquor). The resulting solid, after drying, is tested for purity (usually by a melting point determination, Section 2.33, but also by spectroscopic methods, Chapter 3, or by thin-layer chromatography, Section 2.31), and if found impure is again recrystallised from fresh solvent. The process is repeated until the pure compound is obtained; this often means until the melting point is unchanged, but confirmation by the other methods specified above is desirable.

The theory underlying the removal of impurities by crystallisation may be understood from the following considerations. It is assumed that the impurities are present in comparatively small proportion – usually less than 5 per cent of the whole. Let the pure substance be denoted by A and the impurities by B, and let the proportion of the latter be assumed to be 5 per cent. In most instances the solubilities of A (S_A) and of B (S_B) are different in a particular solvent; the influence of each compound upon the solubility of the other will be neglected. Two cases will arise for any particular solvent: (i) the impurity is more soluble than the compound which is being purified ($S_B > S_A$) and, (ii) the impurity is less soluble than the compound ($S_B < S_A$). It is evident that in case (i) several recrystallisations will give pure sample of A, and B will remain in the mother-liquors. Case (ii) can be more clearly illustrated by a specific example. Let us assume that the solubilities of A and B in a given solvent at the temperature of the laboratory (15 °C) are 10 g and 3 g per 100 ml of solvent respectively. If 50 g of the crude material (containing 47.5 g of A and 2.5 g of B) are dissolved in 100 ml of the hot solvent and the solution allowed to cool to 15 °C, the mother-liquor will contain 10 g of A and 2.5 g (i.e. the whole) of B; 37.5 g of pure crystals of A will be obtained.

The most desirable characteristics of a solvent for recrystallisation are as follows:

1. A high solvent power for the substance to be purified at elevated temperatures and a comparatively low solvent power at the laboratory temperature or below.
2. It should dissolve the impurities readily or to only a very small extent.
3. It should yield well-formed crystals of the purified compound.
4. It must be capable of easy removal from the crystals of the purified compound, i.e. possess a relatively low boiling point.

It is assumed, of course, that the solvent does not react chemically with the substance to be purified. If two or more solvents appear to be equally suitable for recrystallisation, the final selection will depend upon such factors as ease of manipulation, toxicity, flammability and cost.

Some common solvents available for the recrystallisation are collected in Table 2.8, broadly in the order of decreasing polarity. Their purification is included in Section 4.1.

The use of ether as a solvent for recrystallisation should be avoided wherever possible, partly owing to its great flammability and partly owing to its tendency to creep up walls of the containing vessel, thus depositing solid matter by complete evaporation instead of preferential crystallisation. Carbon disulphide, b.p. 46 °C, should *never* be used if an alternative solvent can be found; it has a dangerously low flash point and forms very explosive mixtures with air.

Other recrystallisation solvents include tetrahydrofuran (THF), b.p. 65–66 °C;

Table 2.8   Common solvents for recrystallisation

Solvent	b.p. (°C)	
Water (distilled)	100	To be used whenever suitable
Methanol*	64.5	Flammable; toxic
Ethanol	78	Flammable
Industrial spirit	77–82	Flammable
Rectified spirit	78	Flammable
Acetone	56	Flammable
Ethyl acetate	78	Flammable
Acetic acid (glacial)	118	Not very flammable, pungent vapours
Dichloromethane (methylene chloride)*	41	Non-flammable; toxic
Chloroform*	61	Non-flammable; vapour toxic
Diethyl ether	35	Flammable, avoid whenever possible
Benzene*†	80	Flammable, vapour highly toxic
Dioxane*	101	Flammable, vapour toxic
Carbon tetrachloride*	77	Non-flammable, vapour toxic
Light petroleum	40–60	Flammable‡
Cyclohexane	81	Flammable

* **CAUTION:** *The vapours of these solvents are toxic and therefore recrystallisations involving their use must be conducted in an efficient fume cupboard; excessive inhalation of any vapour should be avoided.* For notes on cumulative toxic effects refer to Section 2.3.
† Toluene is much less toxic than benzene and should be used in place of the latter whenever possible.
‡ Other fractions available have b.p. 60–80, 80–100 and 100–200 °C; when the boiling point exceeds 120 °C the fraction is usually called 'ligroin'. Pentane, b.p. 36 °C, and heptane, b.p. 98 °C, are also frequently used recrystallisation solvents.

butan-2-one (ethyl methyl ketone), b.p. 80 °C; 1,2-dichloroethane* (ethylene chloride), b.p. 84 °C; acetonitrile* (methyl cyanide), b.p. 80 °C; toluene*†, b.p. 110 °C; pyridine*, b.p. 115.5 °C; chlorobenzene*, b.p. 132 °C; cellosolve* (2-ethoxyethanol), b.p. 134.5 °C; dibutyl ether, b.p. 141 °C; 1,1,2,2-tetrachloroethane*, b.p. 147 °C; dimethylformamide* (DMF; formdimethylamide), b.p. 153 °C; dimethyl sulphoxide, b.p. 189 °C (d); nitrobenzene*, b.p. 209.5 °C; and ethyl benzoate, b.p. 212 °C.

The following rough generalisations may assist the student in the selection of a solvent for recrystallisation, but it must be clearly understood that numerous exceptions are known (for a further discussion see Section 9.2):

1. A substance is likely to be most soluble in a solvent to which it is most closely related in chemical and physical characteristics.
2. In ascending a homologous series, the solubilities of the members tend to become more and more like that of the hydrocarbon from which they may be regarded as being derived.
3. A polar substance is more soluble in polar solvents and less soluble in non-polar solvents. The solvents in Table 2.8 have been listed broadly in order of decreasing polar character.

* **CAUTION:** *The vapours of these solvents are toxic and therefore recrystallisations involving their use must be conducted in an efficient fume cupboard; excessive inhalation of any vapour should be avoided.* For notes on cumulative toxic effects refer to Section 2.3.
† Toluene is much less toxic than benzene and should be used in place of the latter whenever possible.

In practice the choice of a solvent for recrystallisation must be determined experimentally if no information is already available. About 0.1 g of the powdered substance* is placed in a small test tube (75 × 11 mm or 110 × 12 mm) and the solvent is added a drop at a time with continuous shaking of the test tube. After about 1 ml of the solvent has been added, the mixture is heated to boiling, due precautions being taken if the solvent is flammable. If the sample dissolves easily in 1 ml of cold solvent or upon gentle warming, the solvent is unsuitable. If all the solid does not dissolve, more solvent is added in 0.5 ml portions, and again heated to boiling after each addition. If 3 ml of solvent is added and the substance does not dissolve on heating, the substance is regarded as sparingly soluble in that solvent, and another solvent should be sought. If the compound dissolves (or almost completely dissolves†) in the hot solvent, the tube is cooled to determine whether crystallisation occurs. If crystallisation does not take place rapidly, this may be due to the absence of suitable nuclei for crystal growth. The tube should be scratched below the surface of the solution with a glass rod; the fine scratches on the walls (and the minute fragments of glass produced) may serve as excellent nuclei for crystal growth. If crystals do not separate, even after scratching for several minutes and cooling in an ice–salt mixture, the solvent is rejected. If crystals separate, the amount of these should be noted. The process may be repeated with other possible solvents, using a fresh test tube for each experiment, until the best solvent is found; the approximate proportions of the solute and solvent giving the most satisfactory results should be recorded.

If the substance is found to be far too soluble in one solvent and much too insoluble in another solvent to allow of recrystallisation, *mixed solvents* or 'solvent pairs' may frequently be used with excellent results. The two solvents must, of course, be completely miscible.‡ Recrystallisation from mixed solvents is carried out near the boiling point of the mixture. The compound is dissolved in the solvent in which it is very soluble, and the hot solvent, in which the substance is only sparingly soluble, is added cautiously until a slight turbidity is produced. The turbidity is then just cleared by the addition of a small quantity of the first solvent and the mixture is allowed to cool to room temperature; crystals will separate. Pairs of liquids which may be used include: alcohols and water; alcohols and toluene; toluene and light petroleum; acetone and light petroleum; diethyl ether and pentane; glacial acetic acid and water; dimethylformamide with either water or toluene.

When the best solvent or solvent mixture and the appropriate proportions of solute and solvent have been determined by these preliminary tests or have been obtained from reference books containing solubility data,[39,40] the solid substance is placed in a round-bottomed flask of suitable size fitted with a reflux condenser (Fig. 2.54) and slightly less than the required quantity of solvent is added together with a few pieces of porous porcelain to prevent 'bumping' (see

---

* With practice the student should be able readily to perform trial recrystallisations with much smaller quantities of material (e.g. 5 mg) using a small ignition or centrifuge tube and correspondingly smaller quantities of solvents.

† If the crude substance contains an insoluble impurity, difficulty may be experienced at a later stage in estimating how much solute has crystallised from the cold solution. The hot solution should therefore be filtered into another tube through a very small fluted filter paper contained in a small short-stemmed funnel. The solution must always be clear before cooling is attempted.

‡ Solvent pairs selected from the extremes of the list Table 2.8 are not usually sufficiently miscible to be satisfactory, e.g. methanol and light petroleum.

Section 2.24). The mixture is heated to boiling on a water bath (if the solvent boils below 80 °C) or with an electric heating mantle, and more solvent is added down the condenser until a clear solution, apart from insoluble impurities*, is produced. If the solvent is not flammable, toxic or expensive, recrystallisation may be carried out in a conical flask, into the neck of which a funnel with a short stem is inserted, which is heated on an electric plate.

FILTRATION OF THE HOT SOLUTION
The boiling or hot solution must be rapidly filtered before undue cooling has occurred. (If a flammable solvent has been used, all flames in the vicinity must be extinguished.) This is usually done through a fluted filter paper (see below) supported in a relatively large funnel with a short wide stem; separation of crystals in and clogging of the stem is thus reduced to a minimum. The funnel should be warmed in an electric or steam oven before filtration is started, when it should be supported in a conical flask of sufficient size to hold all the solution; the conical flask is stood on an electric hotplate or steam bath and the filtrate is kept boiling gently so that the warm solvent vapours maintain the temperature of the solution undergoing filtration, and thus prevent premature deposition of crystals on the filter or in the neck of the funnel. If solid does separate out on the filter it must be scraped back into the first flask, redissolved and refiltered. The filtered solution is covered with a watch- or clock-glass, and then set aside to cool undisturbed. If large crystals are desired, any solid which may have separated from the filtered solution should be redissolved by warming (a reflux condenser must be used for a flammable solvent), the flask wrapped in a towel or cloth, and allowed to cool slowly. If small crystals are required, the hot saturated solution should be stirred vigorously and cooled rapidly in a bath of cold water or of ice. It should be noted that large crystals are not necessarily purer than small ones; generally very impure substances are best purified by slow recrystallisation to give large crystals, followed by several rapid recrystallisations to give small crystals.

If large quantities of hot solution are to be filtered, the funnel (and fluted filter paper) should be warmed externally during the filtration. The heating mantle illustrated in Fig. 2.47(c) is particularly suitable, using the lower heating element; no flames should be present while flammable solvents are being filtered through this funnel. When dealing with considerable volumes of aqueous or other solutions which do not deposit crystals rapidly on cooling, a Buchner funnel preheated in an oven may be used for filtration (see Section 2.19). The filter paper should be of close-grained texture and should be wetted with solvent before suction is applied; the solution may then be poured on to the filter.

PREPARATION OF A FLUTED FILTER PAPER
The filter paper is first folded in half and again in quarters, and opened up as shown in Fig. 2.80(a). The edge 2,1 is then folded on to 2,4 and edge 2,3 on to 2,4, producing, when the paper is opened, new folds at 2,5 and 2,6. The folding is continued, 2,1 to 2,6 and 2,3 to 2,5, thus producing folds at 2,7 and 2,8 respectively (Fig. 2.80(b)); further 2,3 to 2,6 giving 2,9, and 2,1 to 2,5 giving 2,10 (Fig. 2.80(c)). The final operation consists in making a fold in each of the eight segments – between 2,3 and 2,9, between 2,9 and 2,6, etc. – in a direction *opposite* to

---

* The undissolved material will be readily recognised if preliminary solubility tests have been correctly interpreted.

Fig. 2.80

the first series of folds, i.e. the folds are made outwards instead of inwards as at first. The result is a fan arrangement (Fig. 2.80(d)), and upon opening, the fluted paper (Fig. 2.80(e)) is obtained.

USE OF DECOLOURISING CARBON

The crude product of an organic reaction may contain a coloured impurity. Upon recrystallisation, this impurity may dissolve in the boiling solvent and be partly adsorbed by the crystals as they separate upon cooling, yielding a coloured product. Sometimes the solution is slightly turbid owing to the presence of a little resinous matter or a very fine suspension of an insoluble impurity, which cannot always be removed by simple filtration. These impurities can be removed by boiling the substance in solution with a little decolourising charcoal for 5–10 minutes, and then filtering the solution while hot as described above. The decolourising charcoal adsorbs the coloured impurity and holds back resinous, finely divided matter, and the filtrate is usually free from extraneous colour, and therefore deposits pure crystals. The decolourisation takes place most readily in aqueous solution but can be performed in almost any organic solvent; the process is least effective in hydrocarbon solvents. It must be pointed out that boiling in a solvent with decolourising carbon is not always the most effective method of removing the colour; if this is only partially effective, it is often worth while to pass the cold solution of the substance (preferably in an organic solvent such as ethanol) through a small amount of decolourising carbon supported on a wad of cotton wool in the stem of a funnel – this is effectively a chromatographic procedure.

An excessive quantity of decolourising agent must be avoided, since it may also adsorb some of the compound which is being purified. The exact quantity to be added will depend upon the amount of impurities present; for most purposes 1–2 per cent by weight of the crude solid will be found satisfactory. If this quantity is insufficient, the operation should be repeated with a further 1–2 per cent of fresh decolourising charcoal. Sometimes a little charcoal passes through the close-grained filter paper; the addition, before filtration of a filter aid (filter-paper pulp or Celite), will give a clear filtrate. Attention is directed to the fact that the decolourising charcoal should not be added to a superheated solution as the latter may foam excessively and boil over. Excellent decolourising carbons are marketed under the trade names 'Norit' (from birch wood), 'Darco' and 'Nuchar', and are widely available.

140

DIFFICULTIES ENCOUNTERED IN RECRYSTALLISATION

The separation of a second liquid phase, commonly known as an 'oil', instead of the expected crystalline solid, sometimes occurs during recrystallisation. The oil often solidifies on standing, although at times a considerable period may elapse before crystallisation occurs. The resulting crystals will probably occlude some of the mother-liquor, and the purity will therefore not be high. The separation of the oil may be avoided by diluting the solution considerably, but this will lead to large losses. It is probably best to re-heat the mixture until a clear solution is obtained, and allow it to cool spontaneously; immediately the oil commences to separate, the mixture is vigorously stirred so that the oil is well dispersed in the solution. Eventually, crystals will separate and these will grow in the bulk of the solution and not in a pool of oil, so that occlusion of the mother-liquor is considerably reduced. When all the oil has disappeared, stirring may be stopped and the crystals allowed to accumulate. Sometimes the addition of a minute quantity of the crude compound in order to 'seed' the solution may facilitate the initial crystallisation.

Occasionally substances form supersaturated solutions from which the first crystals separate with difficulty; this is sometimes caused by the presence of a little tar or viscous substance acting as a protective colloid. The following methods should be tried in order to induce crystallisation:

1. By scratching the inside of the vessel with a glass rod. The effect is attributed to the breaking off of small particles of glass which may act as crystal nuclei, or to the roughening of the surface, which facilitates more rapid orientation of the crystals on the surface.

2. By inoculating (*seeding*) the solution with some of the solid material or with isomorphous crystals, crystallisation frequently commences and continues until equilibrium is reached. The 'seed crystals' may be obtained by cooling a very thin film of liquid to a low temperature. Several drops of the solution are placed in a test tube or beaker and spread into a thin film by rotating the container, the latter is then cooled in a mixture of ice and salt or in some other suitable freezing mixture. A better procedure, which avoids the necessity of subsequently scraping the surface to remove the 'seed crystals' and the attendant melting if the compound is impure or of low melting point, is to moisten a small glass bead with the supersaturated solution, place it in a test tube, cool the latter in a freezing mixture and thus form crystals on the surface of the bead. The glass bead can then be rolled out of the tube into the vessel containing the main bulk of the solution. Seed crystals may sometimes be formed when a few drops of the solution are placed on a watch glass and the solvent is gradually allowed to evaporate while at the same time the film is rubbed with a glass rod.

3. By cooling the solution in a freezing mixture (ice and salt, ice and calcium chloride, or solid carbon dioxide and acetone). It must be borne in mind that the rate of crystal formation is inversely proportional to the temperature; cooling to very low temperatures may render the mass very viscous and thus considerably hinder crystallisation. In such a case, the mixture should be allowed to warm slowly so that it may be given the opportunity to form crystals if it passes through an optimum temperature region for crystal formation. Once minute crystals have been formed, it is very probable that their size will be increased by keeping the mixture at a somewhat higher temperature.

4. By adding a few lumps of solid carbon dioxide; this produces a number of cold spots here and there, and assists the formation of crystals.
5. If all the above methods fail, the solution should be left in an ice chest (or a refrigerator) for a prolonged period. The exercise of considerable patience is sometimes necessary so as to give the solute every opportunity to crystallise.

The product of a chemical reaction which has been isolated by solvent extraction and subsequent removal of solvent (see Sections 2.22 and 2.24), and which normally should be crystalline, is sometimes an oil due to the presence of impurities. It is usually advisable to attempt to induce the oil to crystallise before purifying it by recrystallisation. Methods 1 and 2 (previous paragraph) may be applied; method 2 cannot always be used because of the difficulty of securing the necessary seed crystals, but should these be available, successful results will usually be obtained. Another procedure is to add a small quantity of an organic solvent in which the compound is sparingly soluble or insoluble, and then to rub with a stirring rod or grind in a mortar until crystals appear; it may be necessary to continue the rubbing for an hour before signs of solidification are apparent. Another useful expedient is to leave the oil in a vacuum desiccator over silica gel or some other drying agent. If all the above methods fail to induce crystallisation, direct recrystallisation may be attempted: the solution should be boiled with decolourising carbon as this may remove some of the impurities responsible for the difficulty of crystal formation. Occasionally, conversion into a simple crystalline derivative is applicable; subsequent regeneration of the original compound will usually yield a pure, crystalline solid. Instances will occur however when assessment by thin-layer chromatography (Section 2.31) of the number of probable impurities in the isolated reaction mixture and of their relative amounts is advisable. It may then be judged whether some prior purification by suitable preparative chromatography (Section 2.31) or by solvent extraction (Section 2.22), should be performed before crystallisation is attempted.

The technique for the removal of solids by filtration with suction has already been described (Section 2.19). The same technique will of course be applied to the collection of recrystallised compounds. Additionally, however, it should be noted that the mother-liquor from a recrystallisation is often of value for the recovery of further quantities of product, and should be transferred to another vessel after the crystals have been drained and washed with solvent. The mother-liquor may be then subsequently concentrated (Section 2.24; suitable precautions being taken, of course, if the solvent is flammable), and a further crop of crystals obtained. Occasionally yet another crop may be produced. The crops thus isolated are generally less pure than the first crystals which separate, and they should be combined and recrystallised from fresh solvent; the purity is checked by a melting point determination.

After the main filtrate has been removed for such treatment, the crystals on the filter pad should be washed to remove remaining traces of mother-liquor which, on drying, would contaminate the crystals. The wash liquid will normally be the same solvent or solvent mixture used for recrystallisation and must be used in the smallest amount compatible with efficient washing, in order to prevent appreciable loss of the solid. With the suction discontinued the crystals are treated with a small volume of the chilled solvent and cautiously stirred with a spatula or with a flattened glass rod (without loosening the filter paper) so that the solvent wets all the crystals. The suction is then applied again, and the crystals are pressed down with a wide glass stopper as before. The washing is repeated, if necessary, after connection to the filter pump has again been broken.

If the solvent constituting the crystallisation medium has a comparatively high boiling point, it is advisable to wash the solid with a solvent of low boiling point in order that the ultimate crystalline product may be easily dried; it need hardly be added that the crystals should be insoluble or only very sparingly soluble in the volatile solvent. The new solvent must be completely miscible with the first, and should not be applied until crystals have been washed at least once with the original solvent.

RECRYSTALLISATION AT VERY LOW TEMPERATURES

This technique is necessary either when the solubility of the compound in the requisite solvent is too high at ordinarily obtained temperatures (refrigerator to room temperatures) for recovery to be economic, or when handling compounds which are liquid at room temperature but which may be recrystallised from a solvent maintained at much lower temperatures (say -10 to $-40\,°C$). In this latter case, after several successive low temperature recrystallisations, the compound will revert to a liquid on storage at room temperature, but the purification process by recrystallisation will have been achieved.

The following crude, but none the less relatively effective, procedure may be adopted in those cases where the compound is not moisture sensitive, and where the amount of product is relatively large (say 5 to 50 g). A round-bottomed or conical flask protected with a calcium chloride tube and containing the solvent in which the compound has been previously dissolved is placed in a suitable cooling mixture (see Section 2.12) until crystallisation is complete. A second flask, also protected by a calcium chloride tube and containing the washing solvent, is also chilled in the same cooling bath. A Buchner funnel is fitted with a suitable filter paper, attached to the filter flask and filled with powdered solid carbon dioxide. Immediately prior to filtration the solid carbon dioxide is tipped out (for very low temperature filtrations some proportion of the solid carbon dioxide may be retained in the funnel since this helps to maintain the low temperature of the filtering mixture) and the solution is filtered as rapidly as possible using the previously chilled solvent for rinsing and washing. One must work as rapidly as possible, returning the flasks to the cooling bath at every opportunity and ensuring that before pouring from either flask the outside is wiped with a cloth, otherwise some drops of the cooling-bath mixture may drain on to the filter cake.

It will be clear from the above account that recrystallisation and filtration at low temperatures are attended by two inherent difficulties: (1) moisture is rapidly deposited on the chilled compound, the solvent and the Buchner funnel, and (2) it is difficult to maintain the apparatus, product and solvent at the required temperature throughout the filtration process.

To overcome these difficulties some ingenuity in apparatus design is necessary, and Fig. 2.81 illustrates one possible assembly.

Here a three-necked, pear-shaped flask A is fitted with a condenser and calcium chloride tube B, a filter stick* C (this being a glass tube having a sintered

---

* Filter sticks of a range of dimensions and porosities are available from J. Bibby Science Products Ltd. The choice of size of flasks and filter sticks will be governed by the scale of the recrystallisation. Broadly speaking pear-shaped flasks are more suitable for the range 100 mg to 5 g, but may be replaced by round-bottomed flasks for operations on a larger scale.

Fig. 2.81

glass frit at its end), a stirrer guide D with flexible rubber or polythene tubing to allow vertical and circular movements of the filter stick, a two-necked flask E fitted with a nitrogen gas inlet system, and a delivery tube (with stopcock) F. The thoroughly dried assembled apparatus, containing the compound to be recrystallised in flask A and with the filter stick drawn into the upper part of the flask, is flushed through with nitrogen. The nitrogen inlet tube is removed, solvent is introduced into flask E, the nitrogen inlet tube is replaced, and by suitable control of the stopcock F (1) the required amount of solvent is allowed to flow into flask A (2). Flask A is now heated by suitable means until solution of the compound is complete, and the liquors are allowed to cool to near room temperature before both flasks A and E are immersed in the appropriate cooling bath. When recrystallisation is complete, the filter stick C is lowered to the level of the mother-liquors and suction applied slowly (3). The horizontal and circular movement allowed by the flexible connection D enables the crystalline solid to be pressed down into a filter cake and the mother-liquor to be efficiently removed. *Without disconnecting the suction*, washing solvent is allowed to flow into the flask A; by adjustment of the position of the delivery tube it is possible to direct solvent flow on to the outside of the filter-stick tube in order to remove contaminating mother-liquor. These initial washings also serve the purpose of rinsing the inside of the filter stick so that subsequent drainage which may occur when suction is discontinued will not cause contamination of the crystalline material. If further washing is required the suction is discontinued and the solvent allowed to flow on to the crystals which are stirred with the filter stick; finally suction is reapplied. Further recrystallisations may of course be carried out without removal of material from flask A. The entire assembly should be allowed finally to reach room temperature with the dry atmosphere maintained within, so that when the apparatus is disconnected the purified product will not be contaminated with condensing water vapour from the laboratory atmosphere.

Notes. (1) The nitrogen inlet system should be of the type suggested in Fig. 2.60 with a sufficiently great enough head of mercury or mineral oil in the escape valve to force the solvent from flask E to A. Control of solvent flow should be by stopcock F, and the outlet of the solvent delivery tube should be above the final level of solution.

(2) The amount of solvent used will have been estimated from the trial recrystallisations carried out in ignition or test tubes using a cooling bath to effect crystallisation.

(3) The filter stick should be attached to a suction pump via a filter trap so that the mother-liquor may be collected, and, if need be, concentrated for further crystal crops; furthermore it is essential that the suction be carefully controlled (by the use of additional stopcocks which are not shown) so that the filter stick may be lowered at the same rate as the level of mother-liquor falls. Undue immersion of the filter stick leads to some troublesome difficulties in removing contaminating mother-liquor from the outside of the filter stick.

RECRYSTALLISATION IN AN INERT ATMOSPHERE

Substances which decompose, or otherwise undergo structural modification, on contact with air must be recrystallised in an indifferent atmosphere, which is usually nitrogen but may on occasions be carbon dioxide, or rarely, hydrogen (see Section 2.17.8, for hazards). The apparatus assembly shown in Fig. 2.81 is suitable with the modification that the calcium chloride tube is replaced by a second nitrogen inlet system. The apparatus is flushed with nitrogen, and the solid material is quickly transferred to flask A; the apparatus is flushed with nitrogen before solvent transfer from flask E to A is carried out. Subsequent operations are as described above using such cooling methods as are appropriate to the recrystallisation process. Finally the material is thoroughly dried under nitrogen in flask A by allowing the gas to pass through the system before the apparatus is disconnected. Short periods of exposure of the crystalline material to the atmosphere is not always harmful and hence it may usually be transferred rapidly to a suitable container for storage under nitrogen (e.g., a nitrogen-filled desiccator).

In those cases where even a short exposure to the atmosphere is harmful, the recrystallisation and filtration processes may be carried out in a nitrogen-filled manipulator glove box (available, for example, from Gallenkamp, Miller-Howe) which has been adapted to accommodate the services required for a normal recrystallisation procedure. The size of the glove box itself and the dimensions of the outlet panels will naturally limit the scale on which recrystallisation can be carried out in this manner.

TECHNIQUE OF SEMIMICRO AND MICRO RECRYSTALLISATIONS

The student in the later stages of his training will certainly be required to recrystallise quantities of solid material within the range of 1 g to fractions of a milligram. These small quantities could arise from: (i) small-scale preparations involving very expensive materials; (ii) preparations of derivatives of small amounts of natural products; (iii) by-products isolated from a reaction process; (iv) chromatographic separation procedures (column and thin-layer techniques), etc. For convenience the experimental procedure to be adopted for recrystallisation of small quantities may be described under three groups:

1. A scale in the range of 1 g to 20 mg.
2. A scale in the range of 20 mg to 1 mg.
3. A scale below 1 mg.

The scale of the recrystallisation envisaged in group (1) means that the operations are carried out in the conventional manner but in apparatus of reduced size. Thus small conical flasks (5 to 20 ml), pear-shaped flasks (5 to 20 ml), semimicro test tubes (75×10 mm or 100×12 mm) or centrifuge tubes (1 to 5 ml) are employed; it is best that the solutions in these receptacles be heated on a water or oil bath rather than directly with a semimicro burner since the heating process can thus be better controlled. Operations involving hot flammable solvents should be performed under reflux using semimicro interchangeable ground glass joint apparatus (7/11 and 10/19). The crystals which separate on cooling are removed by filtration using a small Hirsch funnel, or a small conical glass funnel fitted with a perforated or sintered glass filtration plate; it is often advantageous to place a small filter paper upon the sintered glass plate since complete removal of the crystalline material is thus facilitated. Typical filtration assemblies are shown in Figs 2.82 and 2.43(f). In Fig. 2.82 a rubber cone is used with the Hirsch funnel and boiling tube (150 × 25 mm) having a side-arm for attachment to the suction pump; the filtrate is collected in a centrifuge tube or in a semimicro test tube resting upon a wad of cotton wool. In Fig. 2.43(f) the assembly incorporates ground glass joints and is suitable for the 20–100 mg scale.

An apparatus for the filtration of quantities of crystals within the scale of both *groups* (1) and (2) incorporates the so-called *Willstätter 'filtration nail'*. The latter consists of a thin glass rod flattened at one end. It is readily constructed by heating the end of a short glass rod in the blowpipe flame and pressing vertically upon a heat resistant board. The 'nail' is fitted into a small glass funnel (Fig. 2.83(a)) which is attached to the filter tube by a rubber ring or cone, or the funnel may incorporate a ground glass joint. The nail head may be covered with a circle of filter paper cut with the aid of a cork borer of appropriate size; alternatively, with a good fit between the nail-head edge and the funnel surface and with well-formed crystals, an initial layer of crystals is held at the join and provides a filter medium. The latter technique is valuable for the final stages in the preparation of dust- and fibre-free samples for subsequent elemental analysis. The dimensions given in Fig. 2.83(a) are a guide to the construction of a 'filtration nail' and funnel capable of handling up to 1 g of solid; a smaller size will handle correspondingly smaller quantities (20 mg) just as efficiently. It is often convenient to make several 'nails' with different sized heads which will fit a single funnel so that the 'nail' and filter-paper diameter appropriate to the amount of solid to be collected may be employed. For the larger sizes of 'nail' which accommodate filter papers of 15–25 mm diameter, it is advisable to corrugate the head of the 'nail' as shown (somewhat exaggerated) in Fig. 2.83(b) in

Fig. 2.82   **Fig. 2.83**

order to permit drainage of filtrate over the entire paper; these corrugations are easily produced by pressing the hot glass on the surface of an old single-cut file of coarse grade.

Should it be necessary, before crystallisation takes place, to filter the hot or boiling solutions to remove dust, fibre, etc., the Pyrex micro filter (8 ml capacity) shown in Fig. 2.84 is suitable since the long cylindrical tube reduces evaporation. The filter would be used in an assembly similar to that shown in Fig. 2.82.

It is clear that when the scale of the operation falls within group (2) (20 mg to 1 mg), losses arising from transference of material from flask to filter become more serious and contamination by dirt and filter-paper fibre more likely. For these reasons one of the following variations in technique should be employed.

The solution and crystallisation processes may be conducted in a centrifuge tube; when crystallisation is complete the tube and its contents should be centrifuged (with a suitable hand-operated or electric centrifuge) to pack the crystalline mass. Filtration is now performed by introducing the end of a dropper pipette (previously drawn out to a capillary of approximately 0.1 mm diameter), down the inside surface of the centrifuge tube until it reaches the bottom; in this way there is less chance of breakage of the capillary because of the support provided by the side of the tube. The mother-liquor is now drawn into the pipette, the capillary end serving as the filter. Drops of washing solvent from another dropper pipette are now directed, first to the inside of the tube contaminated with mother-liquor, and then on to the surface of the crystalline mass. The solvent percolates through the crystalline mass and is drawn up into the capillary – the filled capillary pipette is withdrawn and returned empty as frequently as is necessary. Further recrystallisation may be performed without removing the solid from the centrifuge tube, and finally the centrifuge tube containing the purified compound is placed in a vacuum desiccator or drying pistol. When completely solvent-free it will be found that the product may be cleanly removed from the centrifuge tube.

When the volume of solvent required for the recrystallisation is large, compared to the quantity of solid material, initial removal of mother-liquor from the centrifuge tube by this capillary pipette method may be tedious. In such a case, centrifugation should be more prolonged to get more effective packing of the

Fig. 2.84   Fig. 2.85

(a)   (b)

solid, and the bulk of mother-liquor may be drained without disturbance of the solid by inverting the tube over another receptacle in a smooth jerk-free motion. While the tube is in the inverted position the lip is now rinsed with solvent from a dropper pipette; finally wash solvent is introduced into the tube and on to the solid. Subsequent filtration may now be by the capillary-ended pipette as described above.

An alternative means of filtration of quantities of material in the 1 to 5 mg region is to effect solution and crystallisation in a narrow glass tube (6 mm) sealed at one end and with a slight constriction sited at such a position to give the required capacity. After crystallisation is complete, a glass rod flattened at one end to give a type of Willstätter 'nail' (Fig. 2.85(a)) is introduced, the tube is inverted into a centrifuge tube (Fig. 2.85(b)) and the whole is centrifuged. Washing may be effected by removing the 'nail' from the crystallisation tube, introducing wash solvent with a capillary pipette, reintroducing the 'nail' and filtering by centrifuging in the inverted position as before.

Where a recrystallisation has been conducted in a centrifuge or small diameter tube and too much solvent has been added initially, concentration of the solution may be effected by the following technique. A small carborundum chip is introduced and the tube is heated very carefully over a water or oil bath of suitable temperature to give gentle boiling. A flow of nitrogen is directed to the surface of the boiling liquid by means of a capillary-end pipette; the process is continued until the required concentration is effected. The technique of concentration may also be used in those cases where further crystalline crops are required from the mother-liquor.

Where it is clear that the solid to be purified contains dust, or insoluble impurities, then it is necessary to use initially a greater volume of solvent (to prevent prior crystallisation) and to filter the solution in the micro-filter shown in Fig. 2.84. The filtered solution is concentrated using the technique described above until the required smaller volume is reached using first, if necessary, larger test tubes and transferring in stages the concentrated solutions by dropper pipettes to finally the centrifuge or small diameter tube of an appropriate size.

It should be emphasised that in all these operations the laboratory worker should have readily to hand racks, retort stands, clamps and bosses to hold and support the centrifuge tubes and dropping pipettes, and watch glasses to cover tube ends and prevent the entry of dust.

Recrystallisation of quantities of materials less than 1 mg (group (3)) is carried out in melting point or capillary tubes (e.g. 1.5 mm diameter), and is a much simpler operation than might be imagined. All manipulations should be carried out on a bench top covered with a piece of white glazed card or opaque glass. This ensures that all dust and dirt may be immediately observed and removed, and that the glass apparatus does not accidentally pick up dirt, bench polish, etc., small quantities of which could seriously contaminate the product.

The solid is introduced into one of the two open ends of the capillary tube (c 1 mm i.d.) by pushing the end into the solid contained in a specimen tube or on a watch glass. The *solvent* is introduced by holding the same open end of the capillary tube against a drop of solvent suspended from the end of a capillary dropping pipette. The amount of solvent taken up depends on the balance between surface tension and capillary forces, and the force of gravity, which in turn is determined by the angle at which the capillary tube is held and by the portion of the suspended drop which is touched by the capillary end. In general a horizontally-held tube rapidly takes up a column of several centimetres of liquid, while a tube held vertically over the top of a suspended drop takes up only a few millimetres of liquid. Practice with empty tubes results in the technique of solvent introduction being easily acquired. The solute/solvent-free end is now sealed in the flame of a micro-burner, allowed to cool, and the tube contents transferred to the newly sealed end by placing the capillary tube in a centrifuge tube and centrifuging. It is advisable to cover the centrifuge tube with a soft rubber cap through which some pin holes have been pierced; the capillary tubes inserted through these holes are then supported adequately to prevent breakage. The open end is now washed clean by the introduction of more solvent. This is accomplished by passing the midsection of the tube through a micro-burner flame and then introducing the end into a drop of liquid; as the tube cools liquid is drawn into the tube. This additional liquid may be centrifuged down, or if not required to augment the volume of solvent already present it may be removed from the tube by wrapping the open end in a filter or paper tissue and passing the midsection of the tube again through the micro-burner flame. The cleaned end is now sealed, and solution of the solute in the solvent effected by immersing the end of the tube in an appropriate heating bath. The tube is now cooled and when crystallisation is complete [a cooling bath (1) may be necessary to induce crystallisation], the tube is centrifuged to compact the crystalline mass. Removal of mother-liquor may be achieved by cutting off the top of the capillary tube (2) and using a very fine capillary pipette; washing solvent is introduced by another capillary pipette. Finally the end of the tube containing the purified solid may be severed, dried in a vacuum desiccator, and the contents removed by holding the glass section in a pair of tweezers and using a short piece of platinum wire sealed into a glass rod as a spatula.

Notes. (1) For the support of these tubes in heating and cooling baths it is recommended that the top of the bath be covered with a wire mesh large enough to allow insertion of the tube; the upper end of the tube should be pushed into a small rubber collar which then allows the tube to be suspended satisfactorily.

(2) Scratching of tube of this size is accomplished by a carborundum glass cutter, which is easily constructed by fusing a carborundum chip in the end of a glass rod. Alternatively a diamond pencil may be used.

DRYING OF RECRYSTALLISED MATERIAL

The conditions for drying recrystallised material depend upon the quantity of product, the nature of the solvent to be removed and the sensitivity of the product to heat and to the atmosphere.

With large-scale preparations of stable compounds, moist with non-toxic solvents which are volatile at room temperature (e.g. water, ethanol, ethyl acetate, acetone), the Buchner funnel is inverted over two or three thicknesses of drying paper (i.e. coarse-grained, smooth-surfaced filter paper) resting upon a pad of newspaper, and the crystalline cake removed with the aid of a clean spatula; several sheets of drying paper are placed on top and the crystals are pressed firmly. If the sheets become too damp with solvent, the crystals should be transferred to fresh paper. The crystals are then covered by a piece of filter paper perforated with a number of holes or with a large clock glass or sheet of glass supported upon corks. The air drying is continued until only traces of solvent remain (usually detected by smell or appearance) and final drying is accomplished by placing the solid in an electric oven controlled at a suitable tempera-

149

ture.* The disadvantage of this method of drying is that the crystallised product is liable to become contaminated with filter-paper fibre.

With smaller amounts (e.g. 1–20 g) of more valuable recrystallised material the filter cake is transferred to a tared watch glass, broken down into small fragments without damaging the crystalline form, and air dried under another suitably supported watch glass before being placed into a temperature-controlled oven.

With low melting solids, the best method of drying is to place the crystals on a watch glass in a desiccator (Section 2.11) charged with an appropriate substance to absorb the solvent. For general purposes, water vapour is absorbed by a charge of granular calcium chloride, concentrated sulphuric acid† or silica gel.‡ Methanol and ethanol vapours are absorbed by granular calcium chloride or silica gel. Vapours from diethyl ether, chloroform, carbon tetrachloride, benzene, toluene, light petroleum and similar solvents are absorbed by a charge of freshly cut shavings of paraffin wax; since the sample may contain traces of moisture, it is advisable to insert also a dish containing a suitable desiccant. If the compound is moist with glacial acetic acid (e.g. from a recrystallisation of some 2,4-dinitrophenylhydrazones), or with concentrated hydrochloric acid (e.g. from a recrystallisation of an amine hydrochloride), a dual charge of silica gel or concentrated sulphuric acid together with a separate receptacle containing flake sodium hydroxide is necessary to absorb the water and acid vapours respectively. Samples which are to be used subsequently in reactions requiring anhydrous conditions are best dried in a desiccator charged with phosphoric oxide.§

Drying is more rapid in a *vacuum desiccator* of the design shown in Section 2.11. When exhausting a desiccator a filter flask trap should always be inserted between the desiccator and the pump. The vacuum should be applied gradually and the precaution taken of surrounding the desiccator by a cage of fine-mesh steel wire (Desiguard¶); the collapse of the desiccator will then do no harm.

When using a vacuum desiccator, the vessel containing the substance (clock glass, etc.) should be covered with an inverted clock glass. This will protect the finer crystals from being swept away should the air, accidentally, be rapidly admitted to the desiccator. In actual practice the tube inside the desiccator leading from the stopcock is bent so that the open end points in the direction of the lid, hence if the tap is only slightly opened and air allowed to enter slowly, there is little danger of the solid being blown from the clock glass or other receptacle.

---

* Many students place carefully recrystallised samples into a heated oven maintained at a temperature higher than the melting point of their solid with inevitable results; this leads to undue waste of effort and chemicals. Even if the melting point is known, it is always advisable to make a trial with a small quantity on a watch glass. In fact, a temperature of about 50 °C over a period of 1–2 hours is usually adequate for the removal of the common organic solvents mentioned in Table 2.8. If the material can be left overnight at this temperature complete removal of water will occur.

† If a solution of 18 g of barium sulphate in 1 litre of concentrated sulphuric acid is employed, a precipitate of barium sulphate will form when sufficient water has been absorbed to render it unfit for drying; recharging will then, of course, be necessary.

‡ It is usual to employ blue self-indicating silica gel crystals which turn a pale pink colour when regeneration by heating in an electric oven is required.

§ Phosphoric oxide coated on an inert carrier is available from B.D.H. and other suppliers, and is an efficient agent for desiccators. An indicator is incorporated which turns from colourless to blue as water absorption increases.

¶ Desiccator cages manufactured by Jencons (Scientific) Ltd.

Fig. 2.86

Frequently the water or other solvent is so firmly held that it cannot be completely removed in a vacuum desiccator at ordinary temperatures. Large quantities of material (100 g upwards) must therefore be dried in a vacuum oven at higher temperatures, using one of the commercial designs which are available. For smaller amounts of recrystallised material a convenient laboratory form of vacuum oven is the so-called 'drying pistol'. An interchangeable glass joint assembly* is shown in Fig. 2.86 where vapour from a boiling liquid in the flask A rises through the jacket surrounding the drying chamber B (holding the substance) and returns to the flask from the condenser; the drying chamber B is connected to the vessel C containing the drying agent; C is attached to a suction pump. The liquid in A is selected according to the temperature required, e.g. chloroform (61 °C), trichloroethylene (86 °C), water (100 °C), perchloroethylene (120 °C), 1,1,2,2-tetrachloroethane (147 °C), etc. The charge in C consists of phosphoric oxide distributed on glass wool† (to prevent 'caking') when water is to be removed, potassium hydroxide for removal of acid vapours, paraffin wax for removal of organic solvents such as chloroform, carbon tetrachloride, benzene, etc.

EXERCISES IN RECRYSTALLISATION
To gain experience in recrystallisation technique the student should carry out the following experiments.

Choice of solvent for recrystallisation Obtain small samples (about 0.5 g) of the following compounds from the storeroom: (i) salicylic acid; (ii) acetanilide; (iii) m-dinitrobenzene; (iv) naphthalene; and (v) toluene-p-sulphonamide. Use the following solvents: distilled water, industrial spirit, rectified spirit, acetone, toluene, glacial acetic acid and hexane.

Place 0.1 g of the substance in a semimicro test tube (75 × 10 mm or 100 × 12 mm) and proceed systematically with the various solvents as detailed on

---

* Available from J. Bibby Science Products Ltd. This apparatus is also referred to as the *Abderhalden vacuum drying apparatus.*
† Alternatively the granular desiccant of phosphoric oxide coated on a mineral carrier may be used.

p. 138. Finally, summarise your results, and indicate the most suitable solvent or solvents for the recrystallisation of each of the above compounds.

Acetanilide from water. Weigh out 4.0 g of commercial acetanilide into a 250 ml conical flask. Add 80 ml of water and heat nearly to the boiling point on an electric hotplate. The acetanilide will appear to melt and form an 'oil' in the solution. Add small portions of hot water, while stirring the mixture and boiling gently, until the solid has dissolved (or almost completely dissolved). If the solution is not colourless, allow to cool slightly, add about 0.1 g of decolourising carbon and continue the boiling for a few minutes in order to remove the coloured impurities. Filter the boiling solution through a fluted filter paper (for preparation, see p. 139) supported in a short-necked funnel; if the solution cannot be filtered in a single operation, keep the unfiltered portion hot by returning the conical flask to the hotplate. Alternatively, the solution may be filtered through a hot-water funnel (Fig. 2.42 or 2.47(*c*)). Collect the filtrate in a 250 ml conical flask. When all the solution has been filtered, cover the flask containing the hot filtrate with a clock glass and cool rapidly with swirling. Allow to stand for about 30 minutes to complete the separation of the solid. Filter with suction through a small Buchner funnel (Section 2.19), wash the crystals twice with 5 ml portions of cold water (to remove the adhering mother-liquor) and press them in the funnel with a spatula or the back of a flat glass stopper. Remove the funnel from the filter flask, invert it on two thicknesses of filter or absorbent paper resting upon a pad of newspaper and if necessary dislodge the pad of crystals by tapping the funnel; allow the crystals to dry in the air. It is advisable in air-drying to cover the crystals with a large clock glass resting upon corks, or the crystals may be covered with a large filter paper perforated with a number of holes in order to allow the solvent to evaporate. For more rapid drying, the crystals may be placed on a clock glass or in an evaporating basin in an oven held at a temperature of about 80 °C. Weigh the yield of recrystallised material and determine the melting point. If the recrystallised product is not sufficiently pure (melting point low or melting over a range of several degrees), repeat the recrystallisation. Pure acetanilide has m.p. 114 °C.

If an m.p. determination is required soon after recrystallisation, a small quantity may be rapidly dried by pressing it several times upon a pad of several thicknesses of filter or absorbent paper and placing it upon a watch glass in a warm place. A piece of unglazed porous plate may also be used.

Optional or alternative experiments are the recrystallisation of 3.0 g of crude benzoic or salicylic acid from water.

Naphthalene from alcohol (crystallisation from a flammable solvent). Weigh out 5.0 g of commercial naphthalene into a 100 ml round-bottomed flask. Add 25 ml of rectified spirit (or industrial spirit), 2–3 fragments of porous porcelain, and fit a reflux condenser (compare Fig. 2.54; a guard-tube is not required here). Heat the mixture on a water or steam bath or in an electric heating mantle until the solvent boils. Add successive small volumes (each 2–3 ml) of the solvent, and boil gently after each addition, until the naphthalene has dissolved (apart from insoluble impurities). If the solution is coloured, remove it from the heat source, and when it has cooled somewhat add 0.2–0.3 g of decolourising charcoal and mix thoroughly. Boil the mixture for several minutes. Filter the hot solution through a fluted filter paper or through a hot water funnel (**CAUTION:** *All flames in the vicinity must be extinguished*), and collect the filtrate in a conical flask. Cover the receiver with a watch glass and cool it in cold water. Stir or shake the solution as cooling proceeds. After 30 minutes, filter off the crystals through a small Buchner funnel at the water pump; wash all the crystals into the funnel by rinsing the flask with some of the filtrate. Discontinue the suction and wash the crystals with two 5-ml portions of chilled rectified or industrial spirit. Continue the suction and press the crystals down firmly with a spatula or a flat glass stopper. Dry the crystals on filter paper as in *Acetanilide from water*, above. When dry, determine the weight and also the m.p. of the purified naphthalene. Pure naphthalene has m.p. 80 °C.

Alternative experiments: (a) Recrystallisation of crude benzoic acid (5.0 g) from methanol (30 ml); the wash liquid should be 50 per cent acqueous methanol. (b) Recrystallisation of acetanilide (5 g) from toluene (100 ml); filter through a preheated funnel.

Sulphanilic acid from water. Use 5.0 g of crude (grey) sulphanilic acid and proceed as in *Acetanilide from water*, above. Add 0.2 g of decolourising carbon to the solution at 70–80 °C, and continue the boiling for several minutes. If the filtered solution is not colourless it must be boiled with a further 0.2 g of decolourising carbon. Filter the cold solution at the water pump, wash with a little cold water, dry and weigh the yield of recrystallised product.

2.21 SUBLIMATION TECHNIQUES – FREEZE DRYING

Purification of some organic compounds may frequently be achieved by the technique of sublimation as an alternative, or in addition, to recrystallisation. The success of the method depends upon the compound having a high enough vapour pressure at a temperature below the melting point, so that the rate of vaporisation from the solid will be rapid and the vapour may be condensed back to the solid upon a cooled surface. Impurities should have materially different vapour pressures to the compound undergoing purification so that they may be either removed with the initial sublimate or allowed to remain in the residue. The yield of sublimate will be greatly improved if the sublimation is carried out under reduced pressure, and further under these conditions the lower temperature employed reduces the possiblity of thermal degradation. Substances having low vapour pressures at their melting points can only be sublimed under greatly diminished pressures (10^{-3} to 10^{-6} mmHg).

The theory of the sublimation process has been discussed in detail elsewhere[41]; the following describes the practical aspects of this technique which is applicable down to a few milligrams of material.

The simplest form of apparatus for the sublimation at atmospheric pressure of quantities of material in the region of 10 to 25 g consists of a porcelain dish covered with a filter paper which has been perforated with a number of small holes; a watch glass of the same size, convex side uppermost, is placed upon the filter paper. The substance is placed inside the dish, and the latter heated with a minute flame on a wire gauze or sand bath in a fume cupboard. The sublimate collects on the watch glass, and the filter paper below prevents the sublimate from falling into the residue. The watch glass may be kept cool by covering it with several pieces of damp filter paper and moistening these from time to time. A modification, for use with larger quantities of material, employs an inverted

153

Fig. 2.87   **Fig. 2.88**

glass funnel with a plug of glass wool in the stem in place of the watch glass and supported on the porcelain dish by a narrow ring of heat resistant board fitted near the rim. Upon heating the dish gently the vapour of the pure compound passes through the holes in the filter paper and condenses on the inside walls of the funnel; care must be taken that the heat supply is adjusted so that the funnel does not become more than lukewarm. An inverted water jacket (cf. Fig. 2.42), filled with cold water, gives excellent results.

For sublimation under reduced pressure (*vacuum sublimation*), several designs of apparatus are available to suit a range of differently scaled operations. The sublimation of quantities of materials in the region of 5 g is usually conducted in an apparatus having the basic design shown in Fig. 2.87.* The impure substance is placed in the bottom of the wider tube, the cold finger is inserted and connected to the water supply; the T-connection is attached to an oil vacuum pump. Evacuation of the system should be carried out slowly since the sudden removal of traces of moisture, solvent or air from the crystalline mass could cause spattering of the solid on to the cold finger, leading to contamination of the final sublimate. The outer tube is then heated by immersion in a bath of liquid paraffin or silicone oil, not by a direct flame, which would be difficult to control and which may lead to decomposition due to too high a temperature being attained. The temperature should be allowed to rise slowly and held at that level at which sublimation is seen to occur. Usually a 'misting' of the cold finger – provided that the sample was completely dry – is the first indication of sublimation. The process should not be hurried, either by raising the temperature too rapidly in the initial stages or by using finally too high a temperature. As a general guide the temperature should be in the region of 30 °C below the melting point of the solid or lower if sublimation takes place reasonably smoothly.

With care a mixture may be fractioned by sublimation; when the amount of sublimate formed at a particular temperature no longer seems to increase, the sublimation process should be stopped and the sublimate removed. The cleaned cold finger is then reintroduced and the sublimation is continued at a higher

---

* The apparatus has been constructed to the editor's specification by R. B. Radley & Co Ltd.

temperature when further less volatile fractions may be obtained. The temperature and pressure of sublimation should be recorded, together with the melting point of the impure substance and of the sublimate, for reference purposes.

Another design is illustrated in Fig. 2.88 and may be purchased from appropriate suppliers or made to specification; the sizes of the pot and of the cold finger are appropriate to the quantity of material to be sublimed, which may be as low as 20 mg. Frequently these assemblies, sometimes with slight modification, may be used for the high vacuum micro-distillation of viscous liquids (Section 2.28).

For the sublimation of quantities of materials in the region of a few milligrams, the vacuum sublimation block supplied by Reichert-Jung is particularly suitable. This block is located on the hot-stage microscope (see p. 240), the sublimation chamber carefully evacuated, the temperature raised slowly, and the process of sublimation observed through the microscope.

FREEZE DRYING

This process, frequently called lyophilisation, is necessary when water is to be removed from solutions containing heat-labile materials so that conventional distillation, even under reduced pressure, would cause extensive losses by decomposition. Examples are to be found in the removal of water from aqueous solutions of enzymes, polysaccharides, peptides, etc. In principle the aqueous solution is frozen in a suitable solid carbon dioxide freezing mixture (see Section 2.12), and the ice is sublimed off to leave a dry residue.[42]

Figure 2.89 illustrates a commercially available Quickfit lyophiliser (Bibby Science Products) accommodating a single flask of such a size that the volume of aqueous solution to be treated is one-quarter its total capacity. The charged flask is rotated in a dry ice–acetone bath so that an even layer of frozen solution is obtained over the inside. The flask is immediately attached to the refrigerant chambers which are filled with a Cardice-acetone mixture. An oil vacuum pump is connected to the refrigerant chamber via the supplementary trap, which if possible should be immersed in a Dewar flask filled with liquid nitrogen; such a cooled trap provides maximum protection for the vacuum pump. Vacuum is

Fig. 2.89

applied to the apparatus and sublimation of ice takes place over a period of several hours (best carried out overnight). Air should be readmitted to the apparatus very slowly as the dried material is frequently a light 'fluffy' powder and is liable to become dispersed. The flask is removed and the ice allowed to melt and drain off through the stopcock.

Other designs of apparatus, which are variations on this pattern, are available for freeze-drying smaller quantities of solution using test tubes having ground glass joints or round-bottomed flasks of up to 50 ml capacity.

2.22 SOLVENT EXTRACTION

As was pointed out in Section 2.18, the crude products of most organic reactions are multicomponent mixtures, and a convenient initial isolation procedure, for the first stages of both the separation of such mixtures and of the purification of the components, may involve solvent extraction processes. The general cases which are discussed below to illustrate the technique of solvent extraction are selected to cover many of the commonly met systems. The student is recommended to refer to the comments in Section 2.18 on the necessity of assessing the chemical and physical nature of the components of a particular reaction mixture with regard to their solubilities in solvents, and to their acidic, basic or neutral characteristics.

EXTRACTION OF LIQUIDS
Batch-extraction processes. Perhaps one of the most frequent cases that is encountered is the separation of a neutral organic compound (or compounds) from a solution or suspension (as either a solid or liquid) in an aqueous medium, by shaking with an organic solvent in which the compound is soluble and which is immiscible (or nearly immiscible) with water.

The solvents generally employed for extraction are diethyl ether or di-isopropyl ether, toluene, dichloromethane and light petroleum. The solvent selected will depend upon the solubility of the substance to be extracted in that solvent and upon the ease with which the solvent can be separated from the solute. Diethyl ether, owing to its powerful solvent properties and its low boiling point (35 °C) thus rendering its removal extremely facile, is very widely used; its chief disadvantage lies in the great fire hazard attending its use, but this may be reduced to a minimum by adopting the general precautions given in Section 2.3.2, p. 39. The fire hazard is reduced also by employing di-isopropyl ether (b.p. 67.5 °C), but this solvent is much more expensive than diethyl ether.

If prior information is not available, solvent selection should be based on some small-scale trials. A few millimetres of suspension or solution to be extracted are placed in a small test tube and shaken with an equal volume of diethyl ether, when dissolution of suspended material clearly indicates that the solvent would be satisfactory. If the solution to be extracted is homogeneous initially, then the ether solution is removed with a dropper pipette on to a watch glass, and the ether is allowed to evaporate to determine whether material has been extracted. A little experience soon enables the student to differentiate between organic liquids so extracted and traces of water simultaneously removed during the extraction process. If extraction with diethyl ether proves unsatisfactory the experiment is repeated with a fresh sample of reaction mixture using dichloromethane (b.p. 40 °C) as the extraction solvent. If necessary, the other solvents are tried similarly until a suitable solvent has been selected.

By way of illustration, the technique of the bulk batch-extraction of an aqueous solution with diethyl ether* is as follows. A separatory funnel (conical or pear-shaped which short stem and fitted with a ground glass interchangeable stopper) is selected of about twice the volume as that to be extracted and mounted in a ring on a stand with a firm base. The barrel and plug of the stop-cock are dried with a linen cloth and in the case of glass stopcocks lightly treated with a suitable lubricant (Apiezon or silicon grease). The solution and the extraction solvent (usually about one-third of the volume of the solution, but see theory of extraction, below) are introduced into the funnel, and the latter stoppered. All naked flames in the immediate vicinity should be extinguished. The funnel with the stopper firmly held in place is then shaken gently (so that the excess vapour pressure† will be developed slowly), inverted and the stopcock opened in order to relieve the excess pressure. The stopcock is again closed, the funnel again shaken and the internal pressure released. When the atmosphere inside the funnel is saturated with ether vapour, further shaking develops little or no additional pressure. At this stage, the funnel is vigorously shaken for 2–3 minutes to ensure the maximum possible transfer of the organic substance to the ether layer, and then returned to the stand in order to allow the mixture to settle. When two sharply defined layers have formed, the lower aqueous layer is run off and separated as completely as possible. The residual ethereal layer is then poured out through the upper neck of the funnel; contamination with any drops of the aqueous solution still remaining in the stem of the funnel is thus avoided. The aqueous solution may now be returned to the funnel and the extraction repeated, using fresh ether on each occasion until the extraction is complete. Not more than three extractions are usually required, but the exact number of extractions will naturally depend upon the partition coefficient of the substance between water and ether. The completeness of the extraction can always be determined by evaporating a portion of the last extract on the water bath and noting the amount of residue. The combined ethereal solutions are dried with an appropriate reagent (Section 2.23), and the ether removed on a water bath (Sections 2.24 and 2.27). The residual organic material is now further purified, depending upon its properties and the organic impurities removed in the extraction, by chromatography, by recrystallisation or by distillation. It is also important to retain the aqueous solution until the final purified product is isolated so that incorrect observations on the solubility characteristics of the required product do not lead to premature discarding of the product.

Occasionally *emulsions* are formed in the extraction of aqueous solution by organic solvents, thus rendering a clean separation impossible. Emulsion formation is particularly liable to occur when the aqueous solution is alkaline, and when dichloromethane is the extracting solvent. The emulsion may be broken by any of the following devices, but in general its occurrence may be minimised

---

* The diethyl ether (frequently abbreviated to ether) should be reasonably free from 'peroxides', see Section 4.1.*15*, p. 404.

† When ether is poured into a funnel containing an aqueous solution, a two-liquid phase system is formed. If the funnel is stoppered and the mixture shaken the vapour pressure of the ether (300–500 mm according to the temperature) is ultimately added to the pressure of the air (about 760 mm) plus water vapour, thus producing excess of pressure inside the funnel. Hence the necessity for shaking gently and releasing the pressure from time to time until the air has been expelled.

by using a very careful swirling action in the shaking of the separatory funnel during the initial extraction; only in the final extraction is a more vigorous action adopted.

1. Mechanical means, such as agitating the end of a glass rod at the interface of the emulsion with the growing liquid phase, or alternatively gentle rocking of the funnel or gentle swirling, may be successful. Slow filtration through a compacted pad of glass wool in a Hirsch or Buchner funnel is often satisfactory. The emulsion may be transferred to a conical flask and the flask placed in an ultrasonic cleaning bath.
2. An increase in concentration of ionic species may be helpful as the result of the addition of sodium chloride, sodium sulphate or potassium carbonate, for example. With extractions involving alkaline solutions the addition of dilute sulphuric acid may be helpful, providing that complete neutralisation or acidification does not take place since this may result in a change in the chemical nature of some of the components (see below).
3. Emulsions may sometimes be broken by the addition of a few drops of alcohol or other suitable solvent from a dropper pipette, the outlet of which is sited at the emulsion–liquid interface.
4. A satisfactory separation is frequently obtained if the mixture is simply allowed to stand for some time.

   In the isolation of organic compounds from aqueous solutions, use is frequently made of the fact that the solubility of many organic substances in water is considerably decreased by the presence of dissolved inorganic salts (sodium chloride, calcium chloride, ammonium sulphate, etc.). This is the so-called *salting-out effect*. A further advantage is that the solubility of partially miscible organic solvents, such as ether, is considerably less in the salt solution, thus reducing the loss of solvent in extractions.

   The process of extraction is concerned with the *distribution law* or *partition law* which states that if to a system of two liquid layers, made up of two immiscible or slightly miscible components, is added a quantity of a third substance soluble in both layers, then the substance distributes itself between the two layers so that the ratio of the concentration in one solvent to the concentration in the second solvent remains constant at constant temperature. It is assumed that the molecular state of the substance is the same in both solvents.* If c_A and c_B are concentrations in the layers A and B, then, at constant temperature:

$$c_A/c_B = \text{constant} = K$$

The constant K is termed the *distribution* or *partition coefficient*. As a very rough approximation the distribution coefficient may be assumed equal to the ratio of the solubilities in the two solvents. Organic compounds are usually relatively more soluble in organic solvents than in water, hence they may be extracted from aqueous solutions. If electrolytes, e.g. sodium chloride, are added to the aqueous solution, the solubility of the organic substance is lowered, i.e. it will be *salted out*: this will assist the extraction of the organic compound.

   The problem that arises in extraction is the following. Given a limited quant-

---

* For a theoretical treatment involving association or dissociation in one solvent, suitable texts should be consulted.[43,44]

ity of the solvent, should this be used in one operation or divided into several portions for repeated extractions in order to secure the best result? A general solution may be derived as follows. Let the volume v ml of the aqueous solution containing w_0 grams of the dissolved substance be repeatedly extracted with fresh portions of s ml of the organic solvent, which is immiscible with water. If w_1 grams is the weight of the solute remaining in the aqueous phase after the first extraction, then the concentrations are w_1/v g per ml in the aqueous phase and $(w_0 - w_1)/s$ g per ml in the organic solvent layer. The partition coefficient K is given by:

$$\frac{w_1/v}{(w_0 - w_1)/s} = K$$

or

$$w_1 = w_0 \frac{Kv}{Kv + s}$$

Let w_2 grams remain in the aqueous layer after the second extraction, then:

$$\frac{w_2/v}{(w_1 - w_2)/s} = K$$

or

$$w_2 = w_1 \frac{Kv}{Kv + s}$$

$$= w_0 \left(\frac{Kv}{Kv + s} \right)^2 \tag{i}$$

Similarly if w_n grams remain in the aqueous layer after the nth extraction:

$$w_n = w_0 \left(\frac{Kv}{Kv + s} \right)^n$$

We desire to make w_n as small as possible for a given weight of solvent, i.e. the product of n and s is constant, hence n should be large and s small; in other words, the best results are obtained by dividing the extraction solvent into several portions rather than by making a single extraction with the whole quantity. It must be emphasised that the expression deduced above applies strictly to a solvent which may be regarded as completely immisicible with water, such as toluene, dichloromethane or carbon tetrachloride; if the solvent is slightly miscible, e.g. ether, the equation (i) is only approximate, but is nevertheless useful for indicating the qualitative nature of the results to be expected.

   Let us consider a specific example, viz. the extraction of a solution of 4.0 g of butanoic acid in 100 ml of water at 15 °C with 100 ml of benzene at 15 °C. The partition coefficient of the acid between benzene and water may be taken as 3 (or $\frac{1}{3}$ between water and benzene) at 15 °C. For a single extraction with benzene, we have:

$$w_n = 4 \left(\frac{\frac{1}{3} \times 100}{\frac{100}{3} + 100} \right) = 1.0 \, \text{g}$$

159

For three extractions with 33.3 ml portions of fresh benzene:

$$w_n = 4\left(\frac{\frac{1}{3} \times 100}{\frac{100}{3} + 33.3}\right)^3 = 0.5\,\text{g}$$

Hence one extraction with 100 ml of benzene removes 3.0 g (or 75%) of the butanoic acid, while three extractions remove 3.5 g (or 87%) of the total acid. This clearly shows the greater efficiency of extraction obtainable with several extractions when the total volume of solvent is the same. Moreover, the smaller the distribution coefficient between the organic solvent and the water, the larger the number of extractions that will be necessary.

   The above considerations apply also to the removal of a soluble impurity by extraction (or washing) with an immiscible solvent. Several washings with portions of the solvent give better results than a single washing with the same total volume of the solvent.

Continuous extractions of liquids. When the organic compound is more soluble in water than in the organic solvent (i.e. the distribution coefficient between the organic solvent and water is small), very large quantities of organic solvent must be employed to obtain even a moderately efficient extraction. This may be avoided by the use of an apparatus for continuous extraction where only relatively small volumes of solvent are required.

   Two types of apparatus are available according to whether the aqueous solution is to be extracted with an organic solvent which is heavier or lighter than water. The precise design varies according to the manufacturer (e.g. Bibby Science Products; Aldrich, etc.); the following description and diagrams may be regarded as typical.

Fig. 2.90   Fig. 2.91   Fig. 2.92

Figures 2.90, 2.91 and 2.92 illustrate the apparatus employed for the extraction of an aqueous solution by solvents lighter than water, such as ether or toluene (*liquid–liquid extraction by upward displacement*). That shown in Fig. 2.90 is suitable for small amounts of aqueous solution (6–8 ml); the solvent distils from a flask (attached to the lower end) and condenses in a reflux condenser (attached to the upper end), passes through the funnel down a narrow tube partially open at the lower end into the aqueous solution held in a glass thimble, then rises to the surface and returns to the flask via the small hole at the thimble top, having during its passage extracted some portion of the dissolved material from it. The apparatus shown in Fig. 2.91 is useful for volumes of aqueous solution in the range 10 to 500 ml; its mode of operation is similar to the previous case, but this apparatus has additionally baffle discs to assist the even distribution of rising solvent droplets and hence increase the efficiency of the extraction. The liquid–liquid extraction apparatus shown in Fig. 2.92 is available in capacities up to 2.5 litres. This design incorporates a sintered glass disc at the end of the condensed solvent delivery tube, which serves the purpose of dispersing the solvent into small droplets. It is important that solvent is poured continuously down the delivery tube to escape through the sinter while the disc is lowered through the aqueous solution during assembly.

In the use of all these extractors, it should be borne in mind that the extraction process takes place over several hours and due precautions should be taken as the apparatus will be unattended (see Section 2.16). Furthermore as the extraction solvent is more than likely to be flammable, due attention to fire hazards should be taken. The flasks of solvent should be heated with a heating mantle and the level of the solvent in the flask should be above the ring of contact between the mantle element and the outer glass surface, otherwise there is danger of prolonged and possibly harmful overheating of material left as a ring above the solvent level. A further point to note with the use of all these types of apparatus is that the level of aqueous solution should be substantially below the thimble holes, or the extractor side-arms, even when the baffle discs, etc, are in position. This is necessary since the volumes of aqueous solution may increase, either owing to the small solubility in water of the extracting solvent (and this is particularly noticeable with diethyl ether) or to a small rise in the temperature of the aqueous solution during extraction leading to expansion; this could lead to some of the aqueous solution being carried over into the solvent flask.

Figures 2.93, 2.94 and 2.95 illustrate apparatus employed for the extraction of aqueous solutions by solvent heavier than water such as dichloromethane or carbon tetrachloride (*liquid–liquid extraction by downward displacement*).

That shown in Fig. 2.93, when fitted with flask and condenser, is suitable for extractions of about 10 ml of solution. The condensed solvent drops through the funnel and thence down through the solution and escapes via the side-arm sealed into the bottom of the extractor thimble. When assembling the apparatus it is advisable to pour a few millimetres of the extracting solvent into the thimble before pouring in the liquid to be extracted; in this way contamination of the solvent in the flask by a carry-over of solution to be extracted is minimised. The apparatus shown in Fig. 2.94, when fitted with a flask and condenser, is suitable for the extraction of about 50 ml of aqueous solution; the baffle discs improve the dispersion of solvent into droplets. Some of the solvent should be placed in the extraction vessel first, then the baffle plates and finally the aqueous solution; further addition of solvent to prevent the passage of aqueous solution down the

161

Fig. 2.93   **Fig. 2.94**   **Fig. 2.95**

solvent return tube may be necessary. The apparatus shown in Fig. 2.95 is suitable for the extraction of up to 1 litre of aqueous solution. Here the solvent vapours pass through the holes at the top of the solvent delivery tube to condense in the reflux condenser, and pass through the delivery tube and thence through the sintered disc which should be located just under the surface of the solution to be extracted. The fine droplets collect at the base of the extractor which should have been previously loaded with sufficient solvent so that with the weight of aqueous solution, the solvent level in the side-arm coincides with the horizontal portion of the solvent return tube.

Extraction by chemically active solvents. Not infrequently the crude organic product from a reaction may contain a mixture of acidic (phenols and carboxylic acids), basic and neutral components in various combinations. Some of these components may of course be impurities, but none the less, whether as a preliminary purification stage or as a means of separating the mixture, a carefully planned solvent extraction procedure may be adopted using acidic and basic reagents which react chemically with the basic and acidic components of the mixture respectively. The following full account of a typical procedure may be abbreviated in practice according to the complexity of the mixture to be handled.

The multicomponent organic mixture is dissolved in a suitable solvent; this should be diethyl ether if at all possible for the reasons stated above, but any low boiling, water-immiscible solvent (light petroleum, dichloromethane, etc.) may be used. This solution is now shaken in a separatory funnel (see above) with several successive portions of dilute hydrochloric acid (1M) or dilute sulphuric acid (1M; (1)). Basic components are thus extracted into the aqueous acidic extract, and the combined extracts are washed once with the clean organic solvent to remove traces of the original organic phase which may have been carried over into the aqueous extract (this is called a *'back-extraction' process*, (2)). The basic components are recovered (unless they only represent known impurities which are not required), by cooling the aqueous extract in an ice bath, basifying it carefully by adding an aqueous solution of sodium hydroxide (5M; see (3)) dropwise and with stirring, extracting the precipitated oil or solid with organic solvent, and drying (Section 2.23) and evaporating (Section 2.27) the extract. The original organic solution from which the basic components have been removed is now extracted with several successive portions of dilute aqueous sodium hydroxide or sodium carbonate solution (1M). Acidic components will be extracted into the aqueous alkaline layer. After 'back-extraction' with fresh organic solvent, the acidic components may be recovered (if necessary) by cooling the alkaline extract, acidifying by the careful dropwise addition of hydrochloric acid (5M; (4)), extracting the precipitated solid or liquid with an organic solvent, and drying and evaporating the extract (5). The original organic solution now only contains neutral components; these may be recovered by washing the solution first with a little dilute aqueous hydrochloric acid to remove traces of alkali, then with distilled water until the washings are neutral. The organic solution is finally dried and the solvent is evaporated.

Notes. (1) The volumes of organic and aqueous solvents to be used depend, of course, on the quantities of material to be handled. As a guide, a 5 g mixture may be dissolved in 30 ml of organic solvent and extracted with three successive portions of 10 ml of aqueous acid. The student should always check the completeness of the extraction by removing a little of the final extract into a test tube and adding a little concentrated sodium hydroxide solution to make the solution alkaline; a cloudiness suggests that further extractions of the original solution with aqueous acid are necessary. Because of the conversion of the base component into its water-soluble salt, almost complete removal of the base from the mixture is achieved in relatively few extractions.

(2) The solvent washings are best returned to the original solvent solution which in any case may require 'topping-up' in a prolonged extraction process.

(3) The reason for using a concentrated solution of alkali is to keep the final total volume of aqueous solution to a minimum to facilitate the subsequent recovery of the basic components. If no precipitate is visible but the aroma of an amine is noticeable, this implies some degree of water solubility; recovery is then best attempted using one of the continuous extraction techniques.

(4) If aqueous sodium carbonate has been used then considerable effervescence will accompany this acidification process. It is advisable that a flask which is large compared to the volume of solution to be treated should be employed and that the solution be shaken vigorously during the addition of acid.

(5) Extraction of the original solution with sodium hydroxide will have removed phenols, enols and carboxylic acids. Separation of these may be readily accomplished by redissolving the acidic components in diethyl ether (or other suitable solvent). Extraction with saturated aqueous sodium hydrogen carbonate will remove the carboxylic acids, enabling the phenolic (or enolic) components to be recovered by evaporating the dried organic phase. Acidification of the aqueous extract will liberate the carboxylic acid components which may then be isolated by extraction in the usual way.

While the above details provide a general procedure for handling mixtures of acidic, basic and neutral components, other selective extraction reagents may be utilised in certain special instances. For example, cold concentrated sulphuric acid will remove unsaturated hydrocarbons (alkenes and alkynes) present in saturated hydrocarbons, or alcohols and ethers present in alkyl halides. In the former case soluble sulphonated products are formed while in the latter case alkyl hydrogen sulphates or addition complexes that are soluble in the concentrated acid are produced. Another example is provided by the removal of contaminating benzaldehyde from the benzyl alcohol obtained by the Cannizzaro reaction (Expt 6.133).

EXTRACTION OF SOLIDS

The process is generally applied to the removal of natural products from dried tissue originating from plants, fungi, seaweed, mammals, etc. The steam-volatile natural products (e.g. those occurring in the essential oils) such as the alcohols, esters and carbonyl compounds of the aliphatic (both acyclic and alicyclic) and the simpler aromatic systems, are removed by steam distillation (Section 2.25). The non-steam-volatile compounds may be removed by solvent extraction using a batch or continuous process. Not infrequently a comprehensive study of the range of organic substances in a particular tissue requires extraction with a succession of solvents starting with light petroleum (b.p. $\approx 40\,°C$) for the removal of the least polar components (e.g. higher homologues of terpenes, steroids, etc.), progressing through to more polar solvents such as diethyl ether, acetone, ethanol and finally water for the sequential removal of the more polar compounds (e.g. amino acids, carbohydrates, etc.).

The *batch process*, which tends to be less efficient than the continuous extraction process, involves macerating the tissue with the appropriate solvent in a Waring Blender, soaking for a short time (1), filtering in a suitable size of Buchner funnel and then returning the residue to fresh solvent for further extraction. The combined solvent extracts are then evaporated, usually under reduced pressure and the residue submitted to appropriate fractionation procedures (2).

Notes. (1) Warming the suspended solid in the solvent may be necessary by removing the 'porridge' to a suitable flask and heating under reflux. Care must be taken if such is the case to supervise this operation carefully as there may be considerable tendency towards 'bumping'. It should also be borne in mind that this batch-extraction process uses open-type vessels and usually large volumes of solvents; precautions must therefore be taken in relation to the possible fire and toxic hazards involved in the use of a particular solvent. (2) As a first step, this procedure would involve solvent extraction procedures to divide the multicomponent mixture into acidic, basic and neutral fractions (see above). Subsequently chromatography, fractional crystallisation, etc., would be employed as appropriate.

For the *continuous extraction of a solid* by a hot solvent, it is better to use a Soxhlet extraction apparatus such as that shown in Fig. 2.96. The solid substance is placed in the porous thimble A (made of tough filter paper) and the latter is placed in the inner tube of the Soxhlet apparatus. The apparatus is then fitted to a round-bottomed flask C of appropriate size containing the solvent and boiling chips, and to a reflux condenser D (preferably of the double surface type). The solvent is boiled gently; the vapour passes up through the tube E, is condensed by the condenser D, and the condensed solvent falls into the thimble A and slowly fills the body of the Soxhlet.* When the solvent reaches the top of

---

* For solids of low density, the top of the porous thimble A should be above the siphon tube F, otherwise the solid may tend to float out of the thimble and pass down the siphon tube; a plug of glass wool may also be placed within the top of the thimble.

Fig. 2.96   **Fig. 2.97**

the tube F, it siphons over into the flask C, and thus removes that portion of the substance which it has extracted in A. The process is repeated automatically until complete extraction is effected. The extracted compound may be isolated from its solution in C by any of the usual methods. One minor disadvantage of this apparatus is that the temperature of the liquid in A differs considerably from the boiling point of the solvent; extraction is thus effected by the lukewarm liquid and is therefore relatively slow, particularly if the solubility of the substance increases markedly with temperature. This disadvantage is absent in the modifications illustrated in Fig. 2.97 in which the part of the apparatus housing the extraction thimble is surrounded by the vapour of the solvent: extraction is accordingly effected by the hot solvent. The capacity of the Soxhlet extractor is quoted in terms of the siphoning volume; sizes ranging from 6 ml to 5 litre are available.

2.23 DRYING OF LIQUIDS OR OF SOLUTIONS OF ORGANIC COMPOUNDS IN ORGANIC SOLVENTS

Organic liquids, or solutions of organic substances in organic solvents such as would be obtained from solvent extraction procedures described in Section 2.22, are usually dried by direct contact with a solid drying agent. The selection of the desiccant will be governed by the following considerations: (i) it must not combine chemically with the organic compound; (ii) it should have a rapid action and an effective drying capacity; (iii) it should not dissolve appreciably in the liquid; (iv) it should be as economical as possible; and (v) it should have no catalytic effect in promoting chemical reactions of the organic compound, such as polymerisation, condensation reactions and auto-oxidation. The various common drying agents are discussed in detail below; the drying of solvents, including the use of molecular sieves, together with other special techniques of solvent purification is discussed in Section 4.1.

165

It is generally best to shake the liquid with small amounts of the drying agent until no further action appears to take place; too large an excess is to be avoided in order to keep adsorption losses down to a minimum. If sufficient water is present to cause the separation of a small aqueous phase (e.g. with calcium chloride), this must be removed* and the liquid treated with a fresh portion of the desiccant. If time permits, the liquid, when apparently dry, should be filtered and left overnight in contact with fresh drying agent. The desiccant should, in general, be separated by filtration (best through a fluted filter paper) before the distillation of the liquid. This is particularly necessary with many reagents whose drying action depends upon the formation of hydrates (e.g. sodium sulphate, magnesium sulphate and calcium chloride): at higher temperatures the vapour pressures above the salts become appreciable and unless the salts are removed, much, if not all, of the water may be returned to the distillate. However, with some desiccating agents (e.g. calcium oxide, phosphoric oxide), the reaction products with water are quite stable and filtration is not essential.

A *list of the common drying agents* with their practical limitations and their important applications follows.

ANHYDROUS CALCIUM CHLORIDE
This reagent is widely employed because of its high drying capacity and its cheapness. It has a high water-absorption capacity (since it forms $CaCl_2.6H_2O$ below $30\,°C$) but is not very rapid in its action; ample time must therefore be given for desiccation. The slowness of the action is attributed to the blanketing of the particles of calcium chloride with a thin layer of the solution formed by the extraction of the water present; on standing, the water combines forming a solid lower hydrate, which is also a desiccating agent.

The industrial process for preparing the reagent usually permits a little hydrolysis to occur, and the product may contain some free calcium hydroxide or basic chloride. It cannot therefore be employed for drying acids or acidic liquids. Calcium chloride combines with alcohols, phenols, amines, amino acids, amides, ketones and some aldehydes and esters, and thus cannot be used with these classes of compounds.

MAGNESIUM SULPHATE
The most effective commercially available form of this desiccant is the monohydrate; a cheaper grade contains from 30 to 40 per cent of water but this retains useful desiccating action (the fully hydrated form is the heptahydrate). It is an excellent neutral desiccant, rapid in its action, chemically inert and fairly efficient, and can be employed for most compounds including those (e.g. esters, aldehydes, ketones, nitriles, amides) to which calcium chloride is not applicable.

ANHYDROUS SODIUM SULPHATE
This is a neutral drying agent, is inexpensive, and has a high water-absorption capacity (forming $Na_2SO_4.10H_2O$, below $32.4\,°C$). It can be used on almost all occasions, but the drying action is slow and not thorough. The desiccant is valuable for the preliminary removal of large quantities of water. Sodium sulphate is

---

* The aqueous phase may be removed rapidly and conveniently by filtration through a Whatman Phase Separating Paper and washing with a small quantity of the dry solvent. The water-repellent paper retains the aqueous layer.

an inefficient drying agent for solvents such as benzene and toluene and is use-less as a desiccant above 32.4 °C, at which temperature the decahydrate begins to lose water of crystallisation.

ANHYDROUS CALCIUM SULPHATE

When the dihydrate $CaSO_4.2H_2O$ or the hemihydrate $2CaSO_4.H_2O$ is heated in an oven at 230–240 °C for about three hours, anhydrous calcium sulphate is obtained. It is sold commercially under the name of '*Drierite*' (not to be confused with '*Dehydrite*', which is anhydrous magnesium perchlorate). The reagent is ex-tremely rapid and efficient in its action, is chemically inert, and is insoluble in organic solvents; it may therefore be used with most organic compounds. The only disadvantage is its limited capacity for absorption of water since it passes into the hemihydrate $2CaSO_4.H_2O$, and should theoretically absorb only 6.6 per cent of its weight of water to retain its maximum efficiency; where extreme desiccation is not essential, the porous commercial product may absorb up to about 10 per cent of its weight of water. It is recommended that the solution or liquid be subjected to a preliminary drying with magnesium or sodium sulphate, before using anhydrous calcium sulphate.

ANHYDROUS POTASSIUM CARBONATE

This drying agent possesses a moderate efficiency and drying capacity (the dihydrate is formed). It is applied to the drying of nitriles, ketones, esters and some alcohols, but cannot be employed for acids, phenols and other acidic sub-stances. It also sometimes replaces sodium hydroxide or potassium hydroxide for amines, when a strongly alkaline reagent is to be avoided. Potassium carbon-ate frequently finds application in the salting-out of water-soluble alcohols, amines and ketones, and as a preliminary drying agent. In many cases it may be replaced by the desiccant magnesium sulphate.

SODIUM AND POTASSIUM HYDROXIDES

The use of these efficient reagents should usually be confined to the drying of amines (soda lime, barium oxide or calcium oxide may also be employed), potassium hydroxide is somewhat superior to the sodium compound. These bases react with many organic compounds (e.g. acids, phenols, esters and amides) in the presence of water, and with some common solvents (e.g. chloro-form) so that their use as desiccants is very limited.

CALCIUM OXIDE

This reagent is commonly used for the drying of alcohols of low molecular weight; its action is improved by preheating to 700–900 °C in an electric furnace. Both calcium oxide and calcium hydroxide are insoluble in the solvents, stable to heat, and practically non-volatile, hence the reagent need not be removed before distillation.* Owing to its high alkalinity, it cannot be used for acidic compounds or for esters; the latter would undergo hydrolysis.

---

* Some finely divided particles of solid may be carried over during the distillation from calcium oxide. It is recommended that the head of the ground glass distillation assembly leading to the con-denser be filled with purified glass wool in order to retain the finely-divided solid. The purified glass wool is prepared by boiling commerical glass wool with concentrated nitric acid for about 15 minutes, washing thoroughly with distilled water, and drying at 120 °C. Alternatively it may be more convenient to use a splash head (Fig. 2.102).

167

PHOSPHORIC OXIDE

This is an extremely efficient reagent and is rapid in its action. Phosphoric oxide is difficult to handle, channels badly, is expensive and tends to form a protective syrupy coating on its surface. A preliminary drying with anhydrous magnesium sulphate, etc., should precede its use. Phosphoric oxide is only employed when extreme desiccation is required. It may be used for hydrocarbons, ethers, alkyl and aryl halides and nitriles, but not for alcohols, acids, amines and ketones.

Table 2.9 Common drying agents for organic compounds*

Alcohols	Anhydrous potassium carbonate; anhydrous calcium sulphate or magnesium sulphate; calcium oxide.
Alkyl halides **Aryl halides**	Anhydrous calcium chloride; anhydrous calcium sulphate or magnesium sulphate; phosphoric oxide.
Saturated and aromatic hydrocarbons Ethers	Anhydrous calcium chloride; anhydrous calcium sulphate; phosphoric oxide.
Aldehydes	Anhydrous calcium sulphate; magnesium sulphate or anhydrous sodium sulphate.
Ketones	Anhydrous calcium sulphate; magnesium sulphate or anhydrous sodium sulphate; anhydrous potassium carbonate.
Organic bases (amines)	Solid potassium or sodium hydroxides; calcium oxide or barium oxide.
Organic acids	Anhydrous calcium sulphate; magnesium sulphate or anhydrous sodium sulphate.

* For a discussion on the use of molecular sieves, see Section 4.1.

DRYING BY DISTILLATION

In most cases the distillation of organic preparations before drying is regarded as bad technique, but in a number of instances of solvents or liquids, which are practically insoluble in water, the process of distillation itself effects the drying. In short, advantage is taken of the formation of binary and ternary mixtures of minimum boiling point. Thus if moist benzene is distilled, the first fraction consists of a mixture of benzene and water (the constant boiling point mixture, b.p. 69 °C, contains 9 per cent of water); after the water has been removed, dry benzene distils. Other solvents which may be dried in this manner include carbon tetrachloride, toluene, xylene, hexane, heptane, light petroleum, 1,4-dioxane and ethylene dichloride. The dry solvent should not be collected until after about 10 per cent of the main bulk has passed over, since it is necessary to eliminate also the moisture absorbed by the walls of the flask and the condenser. If moist aniline (b.p. 184 °C) or moist nitrobenzene (b.p. 210 °C) is distilled, the moisture is rapidly removed in the first portion of the distillate and the remainder of the liquid passes over dry. Sometimes a moist liquid preparation, which is sparingly soluble in water, is dried by admixture with a solvent (formerly benzene† now

---

† Benzene should be replaced by the much less toxic toluene if at all possible; this forms a binary azeotrope with water, b.p. 85 °C, containing 20 per cent of water.

usually toluene) immiscible with water, and the resulting mixture is distilled. Thus when a mixture of pentanoic acid, water and benzene is distilled, the mixture of benzene and water passes over first (b.p. 69.3 °C), this is followed by dry benzene (b.p. 80 °C), and finally by dry pentanoic acid (b.p. 186 °C). This method has been used for the drying of commercial preparations of 3-methylbutanoic acid and higher aliphatic carboxylic acids by distillation with about 40 per cent of the weight of benzene until the temperature of the vapours reaches 100 °C. The dehydration of crystallised oxalic acid by distillation with carbon tetrachloride is sometimes regarded as another example of the use of a binary mixture for the removal of water (see also pinacol from pinacol hydrate, Expt 5.35).

The following is an example of the use of a ternary mixture in the drying of a solid. D-Fructose (laevulose) is dissolved in warm absolute ethanol, benzene is added, and the mixture is fractionated. A ternary mixture, alcohol–benzene–water, b.p. 64 °C, distils first, and then the binary mixture, benzene–alcohol, b.p. 68.3 °C. The residual, dry alcoholic solution is partially distilled and the concentrated solution is allowed to crystallise: the anhydrous sugar separates.

2.24 DISTILLATION AT ATMOSPHERIC PRESSURE[45]

A typical assembly for the purification of liquids by simple distillation at atmosphere pressure is shown in Fig. 2.98. The flask may be of any appropriate size, although small quantities of liquid (between 5 and 25 ml) are best distilled in pear-shaped flasks; the flask when charged with liquid should be one-half to two-thirds full. The screw-cap adapter on the still-head allows the bulb of the thermometer to be located slightly below the level of the side tube. If the boiling point of the liquid is likely to be above 150 °C the water-cooled condenser shown is replaced by an unjacketed tube fitted with ground glass joints at each end to act as an air-cooled condenser. A drying tube attached to the side-arm adapter may be filled with anhydrous calcium chloride held in position by loose plugs of cotton wool if it is desired to protect the distillate from moisture in the

Fig. 2.98

atmosphere. If the liquid is flammable the side-arm should be fitted with a piece of rubber tubing, the outlet of which leads to an open window, fume cupboard or drain. When the distillation flask has been charged with liquid, a few fragments of unglazed porous porcelain (porous pot) to promote regular ebullition in the subsequent heating are added; they should *never* be added to the hot liquid.* The flask may be heated on a ceramic-centred wire gauze or preferably in a bath (Section 2.13) appropriate to the boiling point of the liquid undergoing distillation. For small quantities of liquid contained in pear-shaped flasks careful heating with a semi-luminous Bunsen flame may be used. Heating may be rather rapid until boiling commences; the rate of heating must then be reduced and the source of heat adjusted so that the distillate is collected at the rate of one or two drops per second. It must be borne in mind that at the commencement of the distillation it takes an appreciable time for the vapour to heat the upper part of the flask and the thermometer. The distillation should not be conducted too slowly, for the thermometer may momentarily cool from lack of a constant supply of fresh vapour on the bulb, and an irregular thermometer reading will result.

It will be found that the temperature will first rise rapidly until it is near the boiling point of the liquid, then slowly, and finally will remain practically constant. At this point a clean, weighed receiver should be connected to the apparatus and the distillate collected until only a small volume of liquid remains in the flask; the temperature should be noted at regular intervals. If the liquid being distilled is not grossly impure most of it will pass over within a narrow temperature range (within 2–3 degrees).

Should the temperature rise steadily, instead of remaining virtually constant, it is then clear that this simple distillation procedure is unsuitable for the purification of the sample and some form of fractional distillation (Section 2.26) will have to be used.

For the distillation of quantities of liquid in the range 0.5 to 5 ml several designs of apparatus are suitable. The pear-shaped flask with fixed side-arm condenser (Fig. 2.99) having capacities of 2 ml or 5 ml (joint size 7/11) is still available in many research laboratories, and in any case may be easily constructed by a competent glass blower. This design has the advantage of minimising losses due to retention of the distillate as a film on the glass surface. The side-arm has a tip on the drip end to facilitate drainage and collection of distillate. A design is also available (Aldrich) for quick micro-distillation in which the liquid condensing on a cold finger collects in a small cup (Fig. 2.100); a thermometer may be inserted in the side joint, although it is essential that liquid condensing on the mercury bulb drips into the collecting cup. The apparatus is equally effective under atmospheric or reduced pressure. Controlled heating is best achieved with both apparatus designs by means of a suitably-sized air bath or an oil bath. Finally the Kugelrohr bulb-to-bulb distillation unit (Aldrich) (see Section 2.28) may be operated at atmospheric pressure.

The assembly shown in Fig. 2.101 is useful for distilling off solvent from solutions, as would be obtained for example from solvent extraction procedures. The

---

* Other aids to regular boiling include the addition of the following: fragments of pumice stone or of carborundum; small strips of Teflon tape 19 mm wide or of shredded Teflon (these may be washed with an organic solvent, dried and reused); small pieces of platinum wire (use is made of the well-known property of platinum in absorbing large quantities of gases).

Fig. 2.99   **Fig. 2.100**

Fig. 2.101

solution is placed in the separatory funnel and is allowed to drop into the flask, initially about half-full with the solution, at approximately the same rate as that at which the solvent distils into the receiver. The use of a large flask for distilling the solution is thus rendered unnecessary. The distilling flask (alternatively, a Claisen flask with fractionating side-arm – see Fig. 2.108 – may be used, particularly if the residue is to be ultimately distilled under diminished pressure) should have a capacity of about twice the estimated volume of the residue after the removal of the solvent. The removal of the solvent in this manner is sometimes termed *flash distillation*.

Relatively large volumes of solvents are conveniently removed by 'stripping' under reduced pressure using a rotary evaporator (Section 2.27, Fig. 2.112).

2.25 STEAM DISTILLATION

Steam distillation[46,47] is a means of separating and purifying organic compounds. Essentially the operation consists of volatilising a substance by passing steam into a mixture of the compound and water. Provided the organic compound has an appreciable vapour pressure (at least 5–10 mm at 100 °C), it will distil with the steam. Steam distillation takes place at a temperature below the boiling point of water and hence, in numerous cases, well below the boiling point of the organic substance. This renders possible the purification of many substances of high boiling point by low-temperature distillation, and is particularly valuable when the substances undergo decomposition when distilled alone at atmospheric pressure. It is also of importance in the separation of the desired organic compound:

(a) from non-volatile tarry substances which are formed as by-products in many reactions;

(b) from aqueous mixtures containing dissolved inorganic salts;

(c) in those cases where other means of separation might lead to difficulties (e.g. the direct ether extraction of aniline, produced by the reduction of nitrobenzene by tin, etc., leads to troublesome emulsion formation owing to the alkali and the tin compounds present: Expt 6.48);

(d) from compounds which are not appreciably volatile in steam (e.g. o-nitrophenol from p-nitrophenol: Expt 6.102); and

(e) from certain by-products which are steam volatile (e.g. biphenyl and excess of unreacted starting materials from the less volatile triphenylcarbinol: Expt 5.42).

A simple apparatus for steam distillation is shown in Fig. 2.102. Flask A contains the liquid to be steam distilled; it is fitted with the 'splash-head' B which prevents the carry-over of the contents of the flask A into the receiver. To carry out a steam distillation, the solution (or mixture of the solid with a little water) is placed in the flask A, and the apparatus is completely assembled. Steam is passed into flask A, which is itself heated by means of a flame to prevent too rapid an accumulation of water. If the substance crystallises in the condenser and tends to choke it, the water should be run out of the condenser for a few minutes until the solid material has been melted and carried by the steam into the receiver; the water should then be cautiously readmitted to the hot condenser. It is best to use a condenser of the double surface type if the rate at which the steam distillation is carried out is rapid; if necessary two such condensers connected in series may be used since in most steam distillations best results are obtained when the process of distillation is carried out rapidly. The passage of steam is continued until no appreciable amount of water-insoluble material is

Fig. 2.102

Fig. 2.103

detected in the distillate (1). To discontinue the distillation the supply of steam is disconnected from the splash-head and the source of heat removed from flask A. The method of isolation of the organic compound from the distillate will depend upon the physical state and upon its water solubility. For example, a solid compound which is virtually insoluble in water would be removed by filtration; liquids and water-soluble solids would be isolated by batch or continuous solvent extraction procedures as described in Section 2.22. For the steam distillation of small quantities of material a suitable design of apparatus is shown in Fig. 2.103.

Note. (1) With water-insoluble materials distillation will usually be continued until the distillate is quite clear. For water-soluble materials a suitable chemical procedure for detection must be used, e.g. for aldehydic or ketonic compounds, portions of the distillate would be tested with an *aqueous* acidic solution of 2,4-dinitrophenylhydrazine.

2.26 FRACTIONAL DISTILLATION AT ATMOSPHERIC PRESSURE

Unless the boiling points of the components of a mixture are widely different it is usual to employ a fractionating column to attempt the separation of liquid mixtures by distillation. Apparatus for precision fractionation, which can successfully separate mixtures in which the components have boiling points which differ by only a few degrees, is available, although careful operation and an appreciation of the factors which influence the efficiency of the fractionating column chosen are needed.

A fractionating column consists essentially of a long vertical tube through which the vapour passes upward and is partially condensed; the condensate flows down the column and is returned eventually to the flask. Inside the column the returning liquid is brought into intimate contact with the ascending vapour and a heat interchange occurs whereby the vapour is enriched with the more volatile component at the expense of the liquid, in an attempt to reach equilibrium within the liquid–vapour system. The conditions necessary for a good separation are:

(a) comparatively large amounts of liquid continually returning through the column;
(b) thorough mixing of liquid and vapour;
(c) a large active surface of contact between liquid and vapour.

Excessive cooling should be avoided; this difficulty is particularly apparent with liquids of high boiling point and may be overcome by suitably insulating or lagging the outer surface of the column or, if possible, by surrounding it with a vacuum jacket or an electrically heated jacket.

The assembly shown in Fig. 2.104 illustrates a set-up for simple fractionation using a *Vigreux column* which has moderate fractionating efficiency and is probably one of the most widely used columns. The column consists of a glass tube with a series of indentations such that alternate sets of indentations point downwards at an angle of 45° in order to promote the redistribution of liquid from the walls to the centre of the column. The mixture to be fractionated is placed in a flask of convenient size (it should be one-third to one-half full), a few fragments of porous porcelain added and a water condenser attached to the side-arm. The distillate is collected in small flasks or in test tubes. The bulb of the thermometer

Fig. 2.104

should be just below the level of the side-arm. It is advisable to lag the column to minimise the effect of draughts in producing excessive cooling. Lagging of the column is essential if the boiling point of any of the components exceeds 100 °C. The flask is then heated in an air bath, an oil bath, or a heating mantle (Section 2.13), to ensure a uniform heating. The initial heating must not be hurried, as owing to the considerable extra condensation which occurs while the column is warming up, the latter may easily choke with liquid. Once distillation has commenced, the rate of heating is adjusted so that the liquid passes over at the rate of one drop every two or three seconds. Under these conditions fairly efficient fractionation should be obtained. When the low boiling point fraction has passed over, distillation should cease. The heating is then slowly increased, and a sharp rise in boiling point should occur as the second fraction commences to distil; it is assumed, of course, that the fractionating system is capable of effecting a sharp fractionation of the components of the mixture. If the set-up is inefficient, a relatively large intermediate fraction may be obtained. It is desired to emphasise the fact that the distillation must be conducted slowly; no time is usually saved by distilling rapidly since a second fractionation will then be necessary.

Other designs of fractionating columns commonly used are illustrated in Fig. 2.105(a)–(c). The *all-glass Dufton column* (Fig. 2.105(a)) is a satisfactory fractionating column for general use. The glass spiral must be carefully ground to fit the outer tube in order to prevent appreciable leakage of vapour past the spiral. The length of the spiral is usually 15 or 30 cm, the internal diameter of the tube is 15–20 mm and the distance between the turns of the spiral is 9–13 mm; the cone and socket are 19/26 or 24/29. This type of column has the advantage of a small hold-up (i.e. a low volume of liquid is retained within the column compared to the flask charge), but it is of relatively low efficiency.

The *pear-bulb column* (Fig. 2.105(b)) is a precision-bore tube with accurately fitting removable bulbs which fit closely to form a liquid seal between the bulbs and the walls of the column. Additional mixing between the ascending vapour

(a)   (b)   (c)

Fig. 2.105

and the descending liquid is provided by inserting small glass 'bubbles' between the separate pear-bulbs.

The *Hempel column* (Fig. 2.105(c)) is a single glass tube, 25 to 75 cm long and 15 to 25 mm diameter, fitted with either 24/29, 29/32 or 34/35 ground glass joints, which may be filled to within 5 cm of the top with a suitable packing. This packing is supported by a small glass spiral of appropriate size. A number of excellent column packings are available commercially. The simplest and cheapest, yet very efficient, packing consists of hollow glass rings (*Raschig rings*), of 6 or 9 mm length and 6 or 9 mm diameter; similar hollow porcelain rings are almost equally effective.

Single-turn glass helices (*Fenske rings*) are an alternative but somewhat more expensive packing for Hempel fractionating columns. A convenient size for single-turn helices is 4.0 mm external diameter and 0.50 mm rod thickness: one kilo of these occupies almost 2 litres. These helices form a closely-spaced packing providing maximum contact between the continuously moving liquid and vapour streams and at the same time allowing for good 'through-put' of descending liquid and ascending vapour.

A detailed discussion of the theory of fractional distillation is to be found in other texts,[48,49] but a brief description of the terms used in discussing fractionating columns and the chief desiderata for efficient columns will be given. The *capacity* of a column is a measure of the quantity of vapour and liquid which can be passed counter-current to each other in a column without causing it to choke or flood. The *efficiency* of a column is the separating power of a definite length of the column; it is measured by comparing the performance of the column with that calculated for a theoretically perfect plate column under similar conditions. A *theoretical plate* is defined as the length of distilling column such that the vapour leaving the plate has the same composition as the vapour which would be in stationary equilibrium with liquid at that temperature, as obtained from the vapour–liquid phase diagram. Since the efficiency of the column depends on the establishment of equilibrium conditions between ascending vapour and descending liquid by thorough and intimate mixing, it is clear that the removal of the more volatile component from the top of the column should be as slow as possible. The number of theoretical plates cannot be determined from the dimensions of the fractionating column; it is computed from the separation effected by distillation of a liquid mixture (e.g. benzene and toluene; benzene and carbon tetrachloride; benzene and dichloroethane; heptane and methylcyclohexane), the vapour and liquid compositions of which are accurately known. An ordinary 1-cm tube 1 metre long might be equivalent to only one theoretical plate, while the same tube filled with a suitable packing can give the equivalent of twenty of more theoretical plates. A column with twelve theoretical plates is satisfactory for the practical separation of a mixture of benzene and toluene (Δb.p. 30 °C); where the two components of a mixture differ in b.p. by only about 3 °C, a column with approximately 100 theoretical plates would be required. The effectiveness of a column depends upon the height as well as upon the packing or internal construction, hence the efficiency is frequently expressed in terms of the *height equivalent per theoretical plate* (*HETP*). It is obtained by dividing the height by the number of theoretical plates, and is usually stated in centimetres. For the comparison of the relative efficiencies of fractionating columns, the operating procedure should be standardised.

The ideal fractionation yields a series of sharply defined fractions, each distilling at a definite temperature. After each fraction has distilled, the temperature rises rapidly, no liquid being distilled as an intermediate fraction. If the temperature is plotted against the volume of the distillate in such an ideal fractionation, the graph obtained is a series of alternate horizontal and vertical lines resembling a staircase. A more or less sloping break reveals the presence of an intermediate fraction and the amount of such fraction can be used as a qualitative criterion of the performance of different columns. The ultimate aim in the design of efficient fractionating columns is to reduce the proportion of the intermediate fractions to a minimum. The most important factors which influence the separation of mixtures into sharp fractions are the following:

1. *Time of distillation.* For any column there is always an optimum time of distillation below which accuracy is sacrificed and above which the slightly improved separation does not justify the extra time taken. For most laboratory columns this will vary between 1 hour and 8–10 hours.
2. *Hold-up of column.* The hold-up of liquid should be reduced to a minimum compatible with scrubbing effectiveness and an adequate column capacity. The ratio of charge of the still to the hold-up of the column should be as large as possible; in general, the still charge should be at least twenty times the hold-up.
3. *Thermal insulation.* Even slight heat losses considerably disturb the delicate equilibrium of an efficient column, and almost perfect thermal insulation is required for the separation of compounds with boiling points only a few degrees apart. Theoretically, the greatest efficiency is obtained under adiabatic conditions. If the components boil below 100 °C, a silvered vacuum jacket is satisfactory; the efficiency of such a jacket will depend upon the care with which it is cleaned, silvered and exhausted. In general, the most satisfac-

tory insulation is provided by the application of heat to balance the heat loss. An electrically-heated insulating tape is fitted round the column; the temperature of the tape, which should be controlled by means of an external resistance or a variable voltage transformer (Variac), should be adjusted within 5 °C of the temperature of the vapour condensing at the upper end of the column.

4. *Reflux ratio.* This is defined as the ratio between the number of moles of vapour returned as refluxed liquid to the fractionating column and the number of moles of final product (collected as distillate), both per unit time. The reflux ratio should be varied according to the difficulty of fractionation, rather than be maintained constant; a high efficiency of separation requires a high reflux ratio.*

Otherwise expressed, the number of theoretical plates required for a given separation increases when the reflux ratio is decreased, i.e. when the amount of condensed vapour returned to the column is decreased and the amount distilled off becomes greater. The variation in the reflux ratio is achieved by the use of a suitable take-off head (or still-head), usually of the *total condensation variable take-off* type. In use, all the vapour is condensed and the bulk of the condensate is returned to the fractionating column, small fractions of the condensate being allowed to collect in a suitable receiver. The design may be appreciated from the line diagram shown in Fig. 2.107 in which the controlled collection of distillate is by the socket-cone screw-operated valve sited just below the condenser drip end.

Figure 2.106 shows a generally useful fractional distillation unit employing a packing of glass helices (Gallenkamp). The column is provided with an electrically-heated jacket the temperature of which may be adjusted with an energy regulator. The still-head is of the total condensation variable take-off type; all vapour at the top of the column is condensed, a portion of the condensate is returned to the column by means of the special stopcock (which permits fine adjustment of the reflux ratio) and the remainder is collected in the receiver. The advantages of the still-head are that true equilibrium conditions can be established before any distillate is collected; this is particularly important when the jacket temperature must be controlled. Furthermore changing from a lower to a higher boiling point fraction is comparatively easy. The stopcock is closed and the liquid is allowed to reflux until the thermometer records the lowest temperature possible; at this point the column is effecting its maximum degree of separation and an equilibrium condition is reached. The tap is then partially opened and the distillate is collected in the receiver until the temperature begins to rise. The stopcock is then closed and equilibrium conditions again established, and a further fraction is removed. In this way sharper separations may be obtained. Further improvement results from the use of a capillary tube to drain the condensate into the receiver. The reflux ratio may be measured approximately by counting the number of drops of liquid which fall back into the col-

---

* The more difficult the fractionation, the greater the reflux ratio to be employed. Thus for compounds differing only slightly in boiling point, this may be as high as 50 to 1; for liquids of wider boiling point range, thus permitting of fairly easy separation, a reflux ratio of 5 or 10 to 1 may be used.

Beyond certain limits increase of the reflux ratio does not appreciably increase the separating power or efficiency of the column. As a rough guide, if the column has an efficiency of n plates at total reflux, the reflux ratio should be between $2n/3$ and $3n/2$.

Fig. 2.106

umn as compared with the number of drops which fall into the receiver flask (the liquid drops falling off the slanting ends of the drip tubes are readily observable).

The vacuum distillation adapter also shown in Fig. 2.106 allows the collection of fractions when distilling under reduced pressure. Its operation is similar to the 'Perkin triangle' or equivalent device ('intermediate receiver adapter') described in Section 2.27. The general technique of conducting a fractional distillation is as follows:

1. Charge the flask with the mixture, and attach to the column. Set the still-head for total reflux and heat the flask until the material begins to reflux into the column. Then heat the column very slowly until the refluxing liquid reaches the top of the column and the boiling point registers on the thermometer. Adjust the temperature in (or near the top of) the jacket, as recorded on the thermometer (just visible in diagram), adjacent to the column until it is just below (i.e. within 5 °C) the boiling point recorded in the vapour. With random packings, such as Fenske or Raschig rings, the column should first be flooded in order to coat the packing completely with liquid; it is then oper-

ated under total reflux until the equilibrium is attained (about 1 hour per ten theoretical plates).

2. When the column has reached equilibrium, adjust the head to give the desired reflux ratio, change the receiver and collect the lowest boiling point component over an appropriate distillation range, say 1–2 °C. During the distillation, maintain as high a rate of reflux as possible consistent with prevention of flooding the column; under these conditions the reflux ratio is controlled by the rate of take-off. As the lowest boiling component is removed, the proportion of it in the distillation flask gradually decreases and eventually a mixture of two components reaches the top of the column, and this will be indicated by a slight rise in boiling point. When this occurs, gradually increase the reflux ratio, i.e. decrease the rate of take-off: this will make it possible to collect the lowest boiling point fraction over a narrow range; eventually a point will be reached when even with a high reflux ratio the boiling point rises. At this stage, change the receiver and commence the collection of the intermediate fraction.

3. During the distillation of the intermediate fraction, keep the rate of take-off very slow. The boiling point will rise and eventually either remain constant or increase very slowly. At this point, change the receiver, adjust the temperature of the heating jacket again and collect the second fraction over a narrow distillation range – rapidly so long as the temperature remains essentially constant, then more slowly until finally the second intermediate fraction is reached again while distilling very slowly. Change the receiver, collect the intermediate fraction and proceed as before for the third component, etc.

The following general comments upon situations which may arise during fractionation may be helpful:

(a) The sharper the fractionation, the smaller, of course, is the intermediate fraction. If the difference in boiling points of the components being separated is considerable, the separation will be so facile that practically all the lower boiling point component will be removed whilst the boiling point remains essentially constant. Eventually the upper part of the column will begin to run dry, distillation will slow up and finally stop, while the reflux at the bottom of the column will be heavy. The vapour temperature may begin to fall until it is below the temperature at the top of the heating jacket. Mere increase of the bath temperature may result in the flooding of the column: the power input to the heating jacket must be gradually increased until reflux again reaches the top of the column, the boiling point begins to rise and eventually becomes constant; the temperature in the jacket is maintained just below the boiling point of the vapour.

(b) As the rate of take-off is reduced near the end of a fraction, a slight lowering of the bath temperature may be necessary to avoid flooding of the column. Also as the boiling point rises during the collection of the intermediate fraction, the power input to the jacket must be increased in order to hold its temperature just below the boiling point.

(c) If the column is flooding near the top and there is little reflux at the bottom, the jacket temperature is too high. If there is normal heavy reflux at the bottom of the column and there is flooding at the top, the bath temperature is probably too high. If the column is flooding near the bottom and there is little reflux near the top, the jacket temperature is too low.

(d) If it is desired to collect the liquid remaining in the column at the end of the fractionation (constituting the 'hold-up'), the column may be stripped by the addition of a '*chaser*' at the beginning of the fractional distillation in a quantity somewhat greater than the estimated 'hold-up'. The boiling point of the 'chaser' should be at least 20 °C higher than the final boiling point of the material being fractionated. For this operation the bath temperature is kept sufficiently high to distil the end component, and the jacket temperature is carefully and slowly raised above the boiling point of the component. 'Chasers' should be chemically inert, inexpensive, and should not form azeotropic mixtures with the components of the mixture undergoing fractionation; examples are: toluene, b.p. 110 °C; *p*-cymene, b.p. 175 °C; tetralin, b.p. 207 °C; diphenyl ether, b.p. 259 °C.

When it is required to separate by fractional distillation components of a mixture which differ in their boiling points by only a few degrees, the *spinning-band fractionating column* offers the best chance of success. The fractionating column consists of a vertical glass tube into which is accurately fitted throughout its length a spiral of Teflon or metal gauze (platinum, stainless steel or Monel) which is fixed to a central Teflon or metal rod and which has a diameter very slightly less than the internal bore diameter of the tube. The spiral extends through the reflux condenser of the specially designed, integral, total-condensation variable take-off still-head, and may be spun by means of either direct or magnetic couplings to an electric motor (Fig. 2.107). The central rod of the spinning band extends into the distillation flask and terminates in a Teflon

Motor

Magnet
Glass cap

Fig. 2.107

stirrer to ensure smooth boiling. The column may be jacketed with a Nichrome heating element or alternatively the column and still-head may be vacuum jacketed and silvered. The advantage of the spinning band (which rotates at selected speeds between 600 and 3000 r.p.m.) is that it increases vapour–liquid contact upon which the column efficiency depends, by causing the vapour to be thrown on to the column walls, and into contact with the liquid descending as a thin film. Furthermore these columns have little tendency to flood and have a low hold-up which therefore allows their high efficiency to be realised to the full. When deliberately flooded they rapidly clear and liquid–vapour equilibrium is re-attained. The spinning band has the further advantage of assisting the passage of vapours through the column. This reduces the pressure difference which always exists between the top (lower pressure) and bottom (higher pressure) regions of a fractionating column. This pressure difference depends upon the dimensions of the column, the nature of the column packing and the rate of distillation. A large pressure drop is undesirable since it leads to a higher heat input at the distillation flask being needed to sweep the vapours to the still-head. With a spinning-band column this difference may be as low as 0.23 mmHg. This small pressure drop is a feature which makes this column design particularly suitable for fractional distillations under reduced pressure (see Section 2.27).

2.27 DISTILLATION UNDER DIMINISHED PRESSURE ('VACUUM' DISTILLATION)[50]

Many organic substances cannot be distilled satisfactorily under atmospheric pressure because they undergo partial or complete decomposition before the normal boiling point is reached. By reducing the external pressure to 0.1–30 mmHg, the boiling point is considerably reduced and the distillation may usually be conducted without danger of decomposition.

In a *vacuum distillation* apparatus certain features should be present to facilitate the ease of operation and these have been incorporated into the assembly illustrated in Fig. 2.108. A is a pear-shaped Claisen–Vigreux flask, the left-hand neck of which carries a screw-cap adapter through which is inserted a glass tube B of appropriate diameter drawn out to a capillary C, at its lower end (1). The tube B carries at its upper end a short piece of pressure tubing and a screw clip D.* The condenser carries a three-limbed multiple receiver and adapter E, frequently called a 'pig', the outlet being connected via a suitable trap and manometer (Section 2.30) to either a water or an oil pump (Section 2.29). The pig adapter permits the collection of three individual fractions without breaking the vacuum and interrupting the progress of the distillation. The flask is heated either by means of an air bath or by means of a water or oil bath as appropriate†

---

* After some experience it will be found that a drawn-out capillary tube of the correct size may be prepared; the rubber tubing and the screw clip D are then omitted.

If pressure tubing is used, it is advisable to insert a short length of thin metal wire (e.g. copper wire, 22 gauge) to prevent the tubing being closed completely by the screw clip.

† Experienced laboratory workers sometimes employ a large free flame for liquids which tend to froth considerably; by directing the flame for the most part at approximately the level of the surface of the liquid and heating the circumference evenly with a 'rotating' flame, the frothing may be reduced and the distillation carried out with comparative safety. Boiling points which are slightly high may be obtained by the use of a free frame unless the liquid is distilled slowly.

Fig. 2.108

(Section 2.13); in the latter case the bulb of the flask is immersed at least two-thirds into the bath, which should contain a thermometer.

To carry out a distillation (under the reduced pressures obtainable with a water pump), the liquid is poured into the Claisen flask so that it is about one-half full (the apparatus is completely assembled as in Fig. 2.108) and the water supply to the condenser is turned on. The water pump is then allowed to reach its maximum capacity with the screw clip D almost fully closed (2). The latter is then adjusted so that a fine stream of air bubbles passes through the liquid in order to minimise 'bumping' in the subsequent distillation.* (The introduction of a gas (air) tends to prevent a delay in the appearance of vapour and thus to prevent superheating; the volume of air introduced in the form of minute bubbles is small so that the effect of the partial pressure upon the boiling point will usually be negligible.) When the mercury level in the manometer (Section 2.30) is steady, the pressure in the system is noted. If the pressure is unsatisfactory, the apparatus must be carefully tested for leaks and these eliminated before the distillation can be commenced; special attention should be paid to ensure that all the glass joints are firmly in position and not contaminated by grit, and that the rubber pressure tubing fits tightly over the glass connections. When a satisfactory vacuum has been achieved the flask is heated. With a water or oil bath, the temperature of the bath should be 20–25 °C above the boiling point of the liquid at the recorded pressures. If an air bath is employed, the temperature is slowly raised until the liquid commences to distil, and the heating is maintained at this intensity so that the liquid distils at the rate of 1–2 drops per second. (For high boiling point liquids, it is advantageous to lag the neck of the flask below the outlet tube.) The readings on the thermometer and manometer are taken frequently during the course of distillation. If the initial distillate boils

---

* For air-sensitive compounds the capillary leak should be connected to a suitable nitrogen gas supply.

at a lower temperature than that expected, the heating is continued until the thermometer records a temperature near that anticipated, and the receiver is then changed by rotation of the pig to bring a clean flask under the condenser outlet. For a pure compound the boiling point will not rise more than a degree or two during the whole of the distillation, even when the bath temperature has to be raised considerably towards the end to drive off the last of the liquid. At the conclusion of the distillation the heating bath is removed, the 'vacuum' is gradually released and the screw clip on D is fully opened (this will prevent any liquid entering the capillary).

If the pressure during distillation is not exactly that given in the recorded boiling point, it may be estimated *very approximately* for the working pressures of a water pump (10–25 mm) by assuming that a difference of 1 mm in pressure corresponds to one degree difference in the boiling point. Table 2.10 may be found useful as a guide to the *approximate* boiling point under diminished pressure when the boiling point under atmospheric pressure is known; it will enable the student to select the thermometer employed in the distillation.

Table 2.10 Approximate boiling points ($^\circ$C) at reduced pressures

Pressure (mmHg)	Water	Chlorobenzene	Benzaldehyde	Ethyl salicylate	Glycerol	Anthracene
760	100	132	179	234	290	354
50	38	54	95	139	204	225
30	30	43	84	127	192	207
25	26	39	79	124	188	201
20	22	34.5	75	119	182	194
15	17.5	29	69	113	175	186
10	11	22	62	105	167	175
5	1	10	50	95	156	159

Notes. (1) The capillary when drawn out should be sufficiently robust so that it is not broken during the vigorous boiling but should have a degree of flexibility to permit some movement of the capillary during distillation; this is particularly advantageous when round-bottomed flasks are used. Furthermore the bore should be such that only a fine stream of bubbles is admitted to the flask when the vacuum is initially applied. The successful construction of a suitable capillary requires some practice and it is usually helpful to perform the operation in two stages. Initially a length of Pyrex tubing (c. 15 cm × 5 mm) is rotated in the flame of an oxygen-gas burner so that about 2 cm in the middle of the tube is heated to dull redness. The softened glass is allowed to thicken gently before it is removed from the flame and extended by a few centimetres (Fig. 2.109(*a*)). The second stage is to reheat with a needle flame a narrow section of the thickened portion which when pliable is extended by a steady pulling action (Fig. 2.109(*b*)). Experience will determine how the speed and length of extension affects the dimensions of the final capillary. The capillary is then cut to the length required so that when it is inserted into the flask the end comes within 1–2 mm of the bottom.

(2) If the material in the flask contains traces of volatile solvents, it is advisable to allow the passage of a comparatively large volume of air through the liquid while warming the

(*a*) (*b*)

Fig. 2.109

flask slightly; this drives off the last traces of volatile solvents, which are carried down the water pump. If this is not done, the pressure obtained when testing out the apparatus will be above the real capacity of the pump, and the student will erroneously assume either that the pump is not functioning efficiently or that leaks are present in the apparatus. When all traces of volatile solvents have been removed, the screw clip D is almost completely closed or otherwise adjusted.

When it is necessary to use an oil vacuum pump to attain lower pressures, it is essential to prevent large volumes of solvent vapour from passing into the pumping system. The oil pump should therefore be guarded with a suitable trap; furthermore distillation at water pump pressures should first be used to remove the bulk of low boiling solvent before the oil pump is brought into operation.

For the *fractional* distillation of mixtures under diminished pressure, when a more efficient fractionating column is necessary, the pig type of adapter should be replaced by the more versatile *Perkin-type receiver adapter* to enable the more numerous fractions to be collected conveniently. The complete apparatus for vacuum distillation is depicted in Fig. 2.110. The two-necked round-bottomed flask A is fitted with a Hempel column (packed with Fenske rings) connected via a still-head with thermometer to a water-cooled condenser which terminates in the Perkin receiver adapter. The stopcocks may be of the ground glass type as shown or of the Rotaflo (HP) design. The capillary leak shown illustrates a simple alternative arrangement to that described above. Here the take-off adapter B (see also Fig. 2.27(a)) carries a short length of pressure tubing C, fitted with a screw clip D. A convenient length of fine capillary tube, which must be flexible yet reasonably robust, is threaded through the tubing and located in position by tightening the clip D.

To carry out the distillation the flask is charged and the apparatus assembled. Before evacuating the system the tap F is closed to isolate the receiver E from the flask J, but the taps H and G are turned so as to connect the pump to the receiver E (via H) and to the flask J (via G). The apparatus is now evacuated

Fig. 2.110

until a steady pressure reading is obtained on the manometer (checking for leaks as previously) and the flask is heated by an air or oil bath until distillation commences. The initial distillate is allowed to run directly into the flask J by turning the tap F; as soon as observation of the thermometer shows that a steady boiling fraction is distilling, the tap F is closed and the required fraction is allowed to collect in E, noting the thermometer and manometer readings. While distillation is taking place, the three-way tap G is turned to admit air to the flask J, which is then removed and replaced by a clean receiver. The vacuum in J is now restored, by first isolating the distillation unit from the pumping system by closing the tap H, and then connecting J to the pump via the tap G. When a steady pressure is once more attained, the tap H is opened and the contents of E allowed to flow into J by opening F. When the boiling point of the distillate indicates that a new fraction is beginning to distil, this is isolated in E by means of the tap F and the procedure for changing the receiver J is repeated.

For vacuum fractional distillation of liquids having close boiling points, which necessitates the use of a total condensation variable take-off head, the receiver adapter modification noted in Fig. 2.106 is employed. Its operation is similar to that of the Perkin triangle; the column is operated under total reflux while the receivers are being changed.

The high efficiency and small pressure drop of the spinning-band columns (Fig. 2.107) makes them very suitable for precision vacuum fractional distillation. The still-head is provided with a type of Perkin triangle assembly which allows the receivers to be changed without disturbing the column equilibrium.

Vacuum distillation on the semimicro scale (1–8 ml) is conveniently carried out using the apparatus illustrated in Fig. 2.111. Although this design is not now available commercially, many research laboratories still have this useful piece of

Fig. 2.111

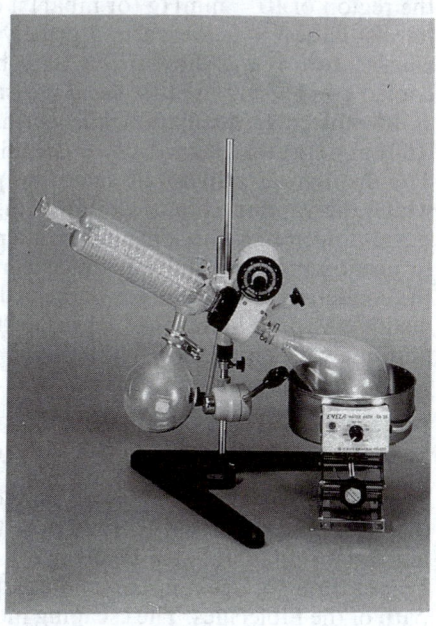

Fig. 2.112

equipment, and in any case, should the need arise it may be readily constructed by a competent glass blower. It should be pointed out also that quantities of liquid within this range, containing components of close boiling point, may be more easily separated by preparative g.l.c. (Section 2.31), and this may often be the method of choice.

The rapid removal of a large quantity of volatile solvent from a solution of an organic compound (i.e. from a solvent extraction process) is conveniently effected by using the *Rotary Film Evaporator* (Fig. 2.112, Jencons). Evaporation is conducted under reduced pressure (a water pump is the most convenient) and therefore at relatively low temperatures. The evaporator flask is heated to the appropriate temperature on a water bath, and is rotated during evaporation; this ensures thorough mixing, prevents bumping and also exposes a relatively large liquid film. The solvent distils from the evaporating surface, is condensed by the spiral condenser and runs off into the receiving flask. Further quantities of the solution may be added through the feed-tube controlled by the stopcock. A quick-action jack with built-in tension spring is sometimes fitted; the operator may lift or lower the assembly and secure it in any desired position within pre-set upper and lower limits. Various sizes of specially shaped flasks are available or, alternatively, round-bottomed flasks with appropriate ground glass joints may be used satisfactorily.

2.28 HIGH VACUUM DISTILLATION – MOLECULAR DISTILLATION

The apparatus designs which have been described in the previous section for distillation under reduced pressure are virtually useless for the distillation of compounds having very high boiling points, which need to be distilled at pressures in the region of 10^{-5} mmHg (or mbar) if decomposition is to be avoided. Successful distillation is achieved in a greatly simplified distillation unit in which the chief feature is the short direct path between a heated liquid surface and the cooled condensing area (*molecular distillation, short path distillation*).[51, 52]

In molecular distillation, the permanent gas pressure is so low (less than 0.001 mmHg) that it has very little influence upon the speed of the distillation. The distillation velocity at such low pressures is determined by the speed at which the vapour from the liquid being distilled can flow through the enclosed space connecting the still and condenser under the driving force of its own saturation pressure. If the distance from the surface of the evaporating liquid to the condenser is less than (or of the order of) the mean free path of a molecule of distillate vapour in the residual gas at the same density and pressure, most of the molecules which leave the surface will not return. The mean free path of air at various pressures is as follows:

Pressure (mmHg)	1.0	0.1	0.01	0.001
Mean free path (cm)	0.0056	0.0562	0.562	5.62

The mean free path of large organic molecules is shorter; it is evident, therefore, that the condenser must be quite close to the evaporating surface. Strictly speaking, a molecular still may be defined as a still in which the distance between the evaporating surface and the cold condensing surface is less than the mean free path of the molecules. The escaping molecules will, for the most part, proceed in a straight path to the condenser; by maintaining the temperature of the latter comparatively low, the amount of reflection of molecules from the condensing surface is reduced. The great advantage of distillation under a high vacuum is that the 'boiling point' is considerably reduced – in some cases by as much as 200–300 °C – thus rendering possible the distillation of substances which decompose at higher temperatures, of substances which are very sensitive to heat, and also of compounds of very high boiling point and large molecular weight.

When the evaporating liquid is a single substance, the rate of evaporation will be $\rho c/s$ grams per square cm per second, where ρ is the density of the saturated vapour at the given temperature, c is the mean molecular velocity and s the mean free path of a distillate molecule. If the liquid is a mixture, the rate of evaporation of the rth component will be $\rho_r c_r/s$ grams per square cm per second. The separation obtained in a molecular distillation thus depends upon the quantity $\rho_r c_r$, unlike the separation obtained in ordinary distillation, where the vapour is in equilibrium with the liquid, which depends upon ρ_r. Since c_r is inversely proportional to the square root of the molecular weight, and the magnitude of ρ_r is in general greatest for the components of least molecular weight, $\rho_r c_r$ is greatest for constituents of least molecular weight. Molecular distillation (sometimes termed *evaporative distillation*) is the only method by which substances of high molecular weight can be distilled without decomposition. According to Langmuir (1917) the theoretical rate of distillation can be written in the form:

$$w = \rho \sqrt{\frac{1}{2\pi MRT}}$$

where w is the weight of substance evaporating per square cm of liquid surface per second, M is the molecular weight of the liquid, R the gas constant and T the absolute temperature. In practice, lower values are obtained because of the reflection of molecules from the condensing surface.

The vacuum sublimation apparatus (Fig. 2.88) is particularly suitable when only small quantities (10–50 mg) of fairly viscous high boiling liquids need to be distilled. The design offers the least hindrance to the flow of vapour from the evaporating to the condensing surface. The rate of distillation is determined by the rate at which the liquid surface is able to produce vapour. Since a liquid sample may almost certainly contain dissolved gases, or solvents which have been used to aid its transference to the distillation chamber, even greater care must be taken in applying the vacuum than is the case with the sublimation of solids. Initially a stopper should be used in place of the cold finger. To avoid excessive frothing and splashing the vacuum must be reduced *very gradually* and the temperature increased in careful stages. Initially the vacuum attainable with a water pump is employed and the temperature increased slowly by immersion of the distillation unit in a water bath at a suitably controlled temperature; gentle agitation of the unit during the heating will aid the removal of solvent and keep frothing to a minimum. When ebullition has ceased the water pump is replaced by an oil pump and the vacuum slowly reapplied and the gentle heating continued. Only when it is clear that no further volatile material is being removed (and often this may take up to an hour or so) is the stopper replaced by the cold finger, the apparatus connected to the source of high vacuum and the molecular distillation commenced. The vacuum required is that provided by a suitable vapour diffusion pump, and the complete assembly required for the distillation is illustrated schematically in Fig. 2.113. The individual components are

187

Fig. 2.113

the distillation unit A, the vapour traps B, a simple uncalibrated mercury manometer C, the vapour diffusion pump D (Section 2.29), a phosphoric oxide moisture trap E, an oil immersion rotatory 'backing' pump F (Section 2.29) and the
McLeod gauge or Pirani/Penning gauge G (Section 2.30).* These components
are connected with wide-bore glass tubing having the minimum number of
bends and fitted with ground glass joints sealed with Apiezon wax W, or better
with joints of the O-ring seal type (Section 2.8) as appropriate. The wide-bore
vacuum taps H–M are sited at the points indicated to enable sections of the
assembly to be isolated and to facilitate vacuum control and measurement. The
entire apparatus, without of course the unit A, is usually permanently assembled
on laboratory scaffolding with the section between taps K and M, together with
the McLeod gauge, kept permanently under vacuum. A manostat (Section 2.30)
may be attached to the outlet from tap J if a pressure higher than that produced
from the pumping system is required.

In use, with all the taps in the closed position, the distillation unit A is
attached to the ground glass joint which is fitted to tap H; the Dewar flasks surrounding the vapour traps are filled with suitable coolant (Cardice–acetone, or
liquid nitrogen), and the condenser water to the vapour diffusion pump and to
the cold finger (or condenser) of A turned on. The backing pump is switched on
and the taps M, L, K and H are opened in sequence so that the system is evacuated to the pressure attainable with this pump. The pressure in the system is
indicated by the auxiliary manometer C, which can also be used to check for
leaks in the apparatus by closing M and noting any fluctuations in its mercury
level. When the backing pump has been reconnected to the system by turning
tap M, the heat supply to the vapour diffusion pump is brought into operation
and the system allowed to reach the minimum pressure as indicated by the
McLeod gauge (see Section 2.30 for the operation of this gauge). It is at this
stage that the manostat is operated, if required, by connecting it to the system
via tap J. The distillation unit A is now heated slowly in an oil bath until misting
of the cold finger is observed when the temperature of the oil bath should be
noted and maintained at this level. The reading on the McLeod or Pirani/
Penning gauge should be checked periodically during the progress of the distillation. At the conclusion of the distillation the heat supply to the vapour diffusion pump is disconnected and the unit A isolated by closing tap H. After
allowing several minutes to elapse to allow the temperature of the diffusion
pump fluid† and the distillation unit A to drop substantially, tap K is closed and

---

* It is highly desirable that a portable safety screen suitably located should be provided between the
operator and the distillation unit and vacuum system illustrated.
† This is particularly important in those cases where the fluid in the vapour diffusion pump is
Apiezon oil and is heated by electrical means. The capacity of the heating element is such that overheating of the oil occurs leading to 'cracking' with the formation of lower boiling components which
diminish the efficiency of the vapour diffusion pump in any subsequent operation.

Fig. 2.114

the backing pump switched off after connecting it to the atmosphere via the three-way tap M. The pin-hole three-way tap I is opened carefully to admit air to the traps B which are then dismantled for cleaning, drying and reassembly. Tap H is opened and the unit A dismantled when it has reached room temperature. The water supply to the condensers is finally turned off.

In cases where the quantity of material to be distilled is such that there is a danger of drainage of droplets of condensed liquid from the cold finger, a small collection cup may be attached to the cold finger by means of platinum wire suitably fused to the two glass surfaces (Fig. 2.114). Alternatively the Wheaton–Hickman flask (Aldrich) (Fig. 2.115) may be used; note in this case the screw thread joint.

An apparatus for the high vacuum distillation of larger quantities is the Hickman vacuum still shown in Fig. 2.116; it is about 600 mm in diameter, 45 mm high and will hold about 40 ml of liquid. The roof of the still is filled with ice-water or any appropriate freezing mixture. A modification which permits continuous flow of cooling liquid over the roof of the still is shown in Fig. 2.117.

Small quantities (0.1–2 g) of material may be distilled using the distillation unit shown in Fig. 2.118 which is readily constructed from 9-mm-diameter Pyrex tubing, the bulbs being made to a size appropriate to the size of sample. The material to be distilled is diluted with a little solvent so that it can be introduced

Fig. 2.115   **Fig. 2.116**   **Fig. 2.117**

Fig. 2.118

in stages directly into the distillation bulb, without contaminating the sides of the tube, by means of a capillary pipette. After each addition the solvent is removed in the usual careful manner and the residue prepared for the final vacuum distillation as in the former cases. The unit is supported by a retort stand and clamp so that the receiver tube slopes slightly downward and the distillation bulb is encased in an air bath or immersed in an oil bath. The vacuum is allowed to reach a steady value before heating of the air or the oil bath by means of a controlled flame is commenced. Liquid distils into the first indentation and the temperature and pressure is noted. The indentations further along the tube prevent further distilled material flowing to waste.

A design of apparatus* which has found particular use in the editors' laboratories for the distillation of viscous high boiling monosaccharide derivatives under high vacuum (10^{-5} mmHg) in relatively large amounts (up to 100 g) is shown in Fig. 2.119. Preliminary removal of volatile solvents from the material is carried out in a round-bottomed flask on a rotary evaporator under water-pump pressure and then by means of an oil vacuum pump which is fitted with a series of suitable cooled solvent traps. The hot solvent-free material, while still in an adequately fluid state, is then poured into the distillation retort using a pre-warmed, long-necked, wide-tube funnel, the end of which reaches into the distillation bulb. In this way contamination of the inside of the unit is avoided, and removal of last traces of solvent, which would be tiresome on this scale, obviated. A pine splint is inserted into the flask to prevent 'bumping' during the distillation, which is effected by evacuating the flask with the aid of a suitable high vacuum source and heating it in an oil bath.

These latter pieces of equipment may be called *bulb-to-bulb distillation units*. A commercial form is the Kugelrohr apparatus (Aldrich). The material to be distilled is placed in a suitably sized round-bottomed flask (one-third full) and attached to the receiver flask train (Fig. 2.120). This arrangement is connected to a horizontal drive shaft which enables the bulbs to be gently rocked by a safe air- or vacuum-operated oscillating motor, which speeds distillation and prevents bumping. The distillation flask may be heated up to 225 °C in an i.r.-radiant heater. The hollow drive shaft may be connected to a vacuum pumping system and the apparatus operates down to 1×10^{-5} mmHg.

Fig. 2.119   **Fig. 2.120**

---

*   This apparatus has been made to specification by R. B. Radley & Co. Ltd; the size of the distillation bulb is either 200 ml, 50 ml or 10 ml.

2.29 VACUUM PUMPS

The two principal operations in an organic preparative laboratory which require the use of a vacuum pump are those of filtration and distillation under reduced pressure. The effectiveness of the vacuum attained by a pumping system may be quoted as centimetres/millimetres of mercury, or as a torr value, although recently the use of the millibar (mbar), has become widespread. The interrelationship of these units is:

760 mmHg (0 °C) = 1013.2 mbar, and 1 torr = 1 mmHg (0 °C)

WATER PUMPS.

The high-pressure water supply is employed for the operation of the ordinary 'filter pump', which finds so many applications in the laboratory. Several types of water-jet pumps of glass, plastic or metal construction are available from most laboratory suppliers. These are often fitted with a suitable non-return valve to prevent the apparatus being flooded as a result of fluctuating water pressure. Connection to the water tap in the case of the metal pump is by a direct screw-threaded joint; with the glass or plastic models high-pressure tubing of suitable bore is wired to the tap and to the pump.

It is routinely desirable to interpose a large pressure bottle A (Fig. 2.121) fitted with a rubber bung between the pump and the apparatus to act as a trap in the event of failure of the non-return valve and to serve as a pressure equalising reservoir. Connection to the apparatus and to a manometer (see also Fig. 2.124) is via a three-way tap B which allows for the release of the vacuum as required; the two-way tap C permits the manometer to be isolated from the system when necessary.

Theoretically, an efficient water pump should reduce the pressure in the system to a value equal to the vapour pressure of the water at the temperature of the water supply mains. In practice this pressure is rarely attained (it is usually 4–10 mm, or 5.3–13.3 mbar, higher) because of the leakage of air into the apparatus and the higher temperature of the laboratory. The vapour pressure at 5, 10, 15, 20 and 25 °C is 6.5, 9.2, 12.8, 17.5 and 23.8 mm (or 8.2, 12.3, 17.1, 23.3 and

Fig. 2.121

31.7 mbar) respectively. It is evident that the 'vacuum' obtained with a water pump will vary considerably with the temperature of the water and therefore with the season of the year. The water pump vacuum is routinely used for filtration, for removal of solvent using a rotary evaporator, and for many distillations under reduced pressure.

OIL IMMERSION PUMPS.

These now find extensive use in an organic laboratory, either as individual units or as a large-capacity unit connected to numerous points in the laboratory (satisfactory installation units are supplied by Edwards High Vacuum Ltd). Commercially available single-stage pumps may evacuate down to 0.1 mmHg (0.133 mbar); somewhat higher pressures are satisfactory for many laboratory purposes. To take advantage of the low pressure produced by a good oil pump, narrow-bore connections in the apparatus assembly should be avoided by using ground glass joints, or 'O'-ring joints if appropriate, wherever possible. Rubber tubing connections should be as short as possible.

For convenience in laboratory use, the pump is mounted on a suitable trolley (e.g. that from Gallenkamp), which also houses a vacustat (Section 2.30) and a pair of glass vapour traps (Fig. 2.122); these are essential since they protect the pump against the intake of moisture or chemical vapours, either of which would be harmful to the pumping efficiency if allowed to contaminate the oil of the pump. Before the pump is brought into operation therefore, the trapping vessels are filled with a suitable coolant (dry ice–acetone). After use the traps *must* be cleaned and dried. If there is a possibility that the vapours are corrosive then it is essential that a more elaborate trapping system should be employed.

VAPOUR DIFFUSION PUMPS.

To attain pressures lower than that produced by the oil immersion pump, a vapour diffusion pump is employed which gives pressures down to 5 ×

Fig. 2.122

Fig. 2.123

10^{-6} mmHg (6.6×10^{-6} mbar). The principles of the operation of a vapour diffusion pump are illustrated by reference to the schematic diagram shown in Fig. 2.123. Vapour molecules ascending from the boiler emerge in a downward direction through the various orifices sited under baffle plates attached to the central tube. Gas molecules diffuse into this descending stream and are thereby propelled downwards and removed by a subsidiary backing pump of the oil immersion rotatory type, after the vaporising fluid has been condensed on the cooled jacket which surrounds the unit. The condensed fluid drains into the boiler to be revaporised.

Mercury-charged vapour diffusion pumps, in which the boiler unit is constructed of quartz or Pyrex and designed to be heated with a gas flame or by an electric element, are available commercially (e.g. Jencons); the remainder of the unit is constructed in Pyrex glass. More robust and highly efficient vapour diffusion pumps of all-metal construction designed for use with either mercury or suitable grades of Apiezon or Silicone oil and electrically heated are also commercially available. Diffusion pumps, whether of mercury or of oil, when first brought into operation do not usually reach their lowest pressure; this will only be achieved after continuous operation for 24 hours, during which time dissolved gases, etc., are removed from the fluid, together with occluded gases from the glass surfaces.

2.30 MANOMETERS, VACUSTATS, VACUUM GAUGES AND MANOSTATS

A frequently used simple *mercury manometer* which is employed for the measurement of pressure in the range 0.5–17 cm (6.6 to 226 mbar) is the U-tube design illustrated in Fig. 2.124 (Anschutz manometer, Gallenkamp). It consists of a U-tube charged with mercury and mounted in a wooden stand. The scale B, graduated in millimetres and sometimes made of mirror glass in order to eliminate errors due to parallax, is sited between the two arms and is movable. This

Fig. 2.124   **Fig. 2.125**

enables adjustment of the scale so that one of the mercury levels coincides with a convenient point on the scale, and facilitates the reading of the difference in the height of the mercury in the two arms which gives the pressure directly. A tap A is usually provided so that the manometer may be isolated from the distillation apparatus. Great care should be exercised when using this manometer; if air is allowed to enter the exhausted apparatus rapidly when the tap A is open, the mercury may rise to the top of the closed end with sufficient velocity to break it. It is advisable, therefore, to open tap A only when the pressure needs to be measured during an experiment, and at the conclusion of the distillation to open the tap *very slowly* after the pressure in the apparatus has been restored to atmospheric. A model is also available (Aldrich) in which the U-tube is totally enclosed in a glass chamber to contain the mercury in the case of U-tube breakage; in this case the scale is graduated in torr and mbar units.

The *vacustat* (Edwards High Vacuum) (Fig. 2.125) is a very useful pressure gauge which is usually employed in conjunction with an oil pump; two models which cover the ranges 10 to 10^{-2} mbar and 1 to 10^{-3} mbar are available. It is direct reading, compact, and is charged with only about 150 g of triple distilled mercury. The gauge must be rotated carefully to the vertical position (as shown) when reading the pressure; it is then returned to the horizontal position equally carefully, otherwise there is a tendency for some of the mercury to spill over into the tubing which connects the vacustat to the apparatus via an 'O'-ring glass-to-metal adapter (insert A), and hence into the pump (p. 51). The gauge does not automatically record a variable pressure.

The *MacLeod gauge* (Manostat Corporation), illustrated in Fig. 2.126, is widely found in established research laboratories, and is used for the measurement of pressures down to about 5×10^{-6} mmHg (6.5×10^{-6} mbar). The gauge, mounted on a suitable stand, is connected to the vacuum system between the diffusion pump and the vapour traps (see Fig. 2.113) by means of a ground glass joint permanently sealed with Apiezon wax W, or preferably by means of an O-ring joint (Section 2.8). The gauge may be isolated from the vacuum system by the tap A, Fig. 2.126. The side-arm of the mercury reservoir is connected via a three-way stopcock B to a suitable auxiliary vacuum system (usually the vacuum achieved with a water-jet pump is adequate).

To take a reading after the gauge has been newly installed, the three-way tap B is closed and the gauge is connected to the vacuum system by opening the tap A and allowing the mercury to be partially drawn upward into the bulb. The three-way tap B connection to the auxiliary vacuum supply is then opened to allow the mercury to be drawn down from the bulb. When the gauge and the mercury reservoir have been completely evacuated by several such successive operations, the gauge is isolated by closing taps A and B. The pressure in the system may now be recorded by carrying out the following sequence. The tap A is opened carefully, and after a pause of a minute or so to allow the gauge to be finally evacuated to the pressure in the remainder of the system, tap B is cautiously opened to admit air into the reservoir which allows mercury to rise into the bulb. As the mercury approaches the bottom end of the closed capillary the three-way tap should be adjusted so that mercury rises at a very slow rate until the level in the reference capillary is coincident with the zero on the scale, at which point tap B is closed. The pressure reading is now recorded by the level of the mercury in the closed capillary. The gauge is then isolated once more by closing tap A and the mercury reservoir is evacuated via the three-way tap B to

Fig. 2.126

withdraw mercury well below the bulb in the gauge. While the bulb-trap at the top of the gauge prevents overflow of mercury into the vacuum system it is advisable to carry out all the operations needed to record the pressure in the system carefully and methodically, since violent 'bumping' in the gauge due to the incautious inlet of air could lead to breakage and the hazard associated with spilling mercury. Finally it is customary to keep the gauge and reservoir permanently under vacuum.

The major disadvantage of the MacLeod gauge is the very large quantity of mercury that is required (over 3.6 kg). Not only is this expensive but, if the apparatus breaks in use or storage, there would be a considerable toxic hazard. For these reasons the use of a *Pirani* or a *Penning* gauge (Edwards High Vacuum) is to be recommended. The former is suitable for pressures in the region of 5 to 10^{-3} mbar, and the latter for pressures in the region of 10^{-2} to 10^{-7} mbar. These gauges operate on the principle that when gas molecules in the region of a detector are struck by electrons from a heated filament (the ion source), they become electrically charged, and are thus attracted towards a detector site where they cause an electric current. This current is amplified and calibrated to indicate a pressure which is displayed on a meter or by means of a digital readout. These gauges may be inserted directly into the vacuum line, as

195

discussed for the MacLeod gauge, via ground glass joint adapters or 'O'-ring seals. It is essential that these gauges can be isolated from the vacuum line via wide-bore stopcocks, which should not be opened to the vacuum line until the pressure in the system has been reduced. A stopcock should then only be opened when a pressure reading is required otherwise the filament may become unnecessarily contaminated.

The maintenance of a constant pressure in a system during distillation under reduced pressure is of great practical importance if steady distillation at constant temperature is to be achieved, and a trustworthy boiling point recorded. Devices which maintain a constant pressure in a system that is higher than the minimum pressure that the pump will give are termed *manostats*. One commercially available form of instrument is the *Cartesian manostat, model* 7 (Manostat Corporation) (Fig. 2.128). Its operation may be explained with the aid of the Fig. 2.127. Mercury is introduced into the container until the disc of the float just makes contact with the orifice, when the pressure is equalised inside and outside the float. The device is connected to the pump and to the system by way of a large reservoir and a manometer. With the stopcock open, the pressure is reduced by way of a by-pass between the pump and the system until the desired value as read on the manometer is reached, then both the stopcock and by-pass are closed; the device will automatically maintain the desired pressure. If the system is vacuum-tight, the pressure will maintain itself; a slight leak, which may be introduced intentionally, will cause the pressure to rise slightly. This will produce a displacement of the mercury level downward outside the float and a corresponding displacement upward inside the float; the buoyant force on the float is consequently diminished and when this reduction in buoyancy becomes sufficient to overcome the suction force at the orifice due to the pressure differential, the disc will break away from the orifice and permit the pump to evacuate sufficient gas from the system to restore the original pressure. When the original pressure is restored, the disc will return to its former position and seal off the orifice The cycle is repeated indefinitely if the size of the leak in the system does not exceed the capacity of the gas flow that is possible through the orifice and the pump is of sufficient rating to carry the load.

Needle valves (Edwards High Vacuum) are available which provide a fine control of a *gas bleed* into the vacuum line, and are suitable for gas admissions

Fig. 2.127

Fig. 2.128

down to 10^{-5} mbar. They may be inserted into the vacuum line; it is further recommended that a filter be attached to the inlet of the valve to prevent damage from the uptake of dust or grit.

2.31 CHROMATOGRAPHY

Chromatography is a separation process which depends on the differential distributions of the components of a mixture between a mobile bulk phase and an essentially thin film stationary phase.[53-56] The stationary phase may be either in the form of a packed column (*column chromatography*) through which a mobile phase is allowed to flow, or in the form of a thin layer adhering to a suitable form of backing material (*thin-layer chromatography*) over which the mobile phase is allowed to ascend by capillary action.

The thin film stationary phase may be either a liquid or a solid, and the mobile phase a liquid or a gas. Possible combinations of these phases then give rise to the principal chromatographic techniques in general use.

In *partition chromatography* the stationary phase is a thin liquid film adsorbed on the surface of an essentially inert support. The mobile phase may be either a liquid (*liquid–liquid partition chromatography*) or a gas (*gas–liquid partition chromatography* or *gas chromatography*). In either system the separation depends largely upon partition between the two phases although the separation process may be complicated by the incursion of adsorption effects involving the inert support and the compounds undergoing chromatographic separation. *Paper chromatography* is an example of partition chromatography in which filter paper serves as a support for the immobile liquid phase.

In *adsorption chromatography* the mobile phase is usually a liquid and the stationary phase is a finely-divided solid adsorbent (*liquid–solid chromatography*). Separation here depends on the selective adsorption of the components of a mixture on the surface of the solid. Separations based on gas–solid chromatographic processes are of limited application to organic mixtures. The use of *ion-exchange resins* as the solid phase constitutes a special example of liquid–solid chromatography in which electrostatic forces augment the relatively weak adsorption forces.

Apart from partition and adsorption processes, chromatographic separations may also be based upon differences in molecular size (*gel permeation chromatography*, or *gel filtration*). In this technique gel-like material, which is commercially available in a range of porosities, serves as the stationary phase, and separation is achieved through differential diffusion into the pores of the matrix, of molecules which are not large enough to be completely excluded.

The chromatographic techniques which are principally of use to the synthetic organic chemist are described in the following sections. These are:

1. Thin-layer chromatography (t.l.c.).
2. Liquid–solid column chromatography.
3. Gas–liquid chromatography (g.l.c.).
4. High performance liquid chromatography (h.p.l.c.).

These techniques may be variously used as analytical tools to establish the complexity of mixtures and the purity of samples, and as preparative tools for the separation of mixtures into individual components.

197

The selection of a particular technique which might be expected to be the most appropriate for a given situation is to some extent a matter of experience, and the following considerations may be used as a guide to assist in the choice of method.

In general preparative work, the chromatographic techniques cited above may be used: (a) to establish the purity and authenticity of starting materials and (if appropriate) reagents; (b) to monitor the reaction, particularly in the case of new reactions, or in the optimisation of experimental conditions to achieve the highest possible yield of product; (c) to check the isolation and purification procedures; (d) to achieve the separation of product mixtures should this not be possible by means of distillation, recrystallisation, or sublimation procedures; (e) to provide a further check on the authenticity of the final product in addition to that provided by the comparison of physical constants (e.g. m.p., b.p., n_D^t, $[\alpha]_D^t$, etc.) and spectroscopic data with those quoted in the literature.

It should be noted of course that with well authenticated preparations, the application of chromatographic procedures at the stages noted above would not usually be necessary; however, when for some unknown reason the expected product is not obtained, or when new reactions are being studied, chromatography is invaluable.

The most convenient and economic techniques of choice for the rapid analysis of starting materials and for the assessment of purity of a crude reaction product are t.l.c. and g.l.c. These techniques may also be used to monitor the progress of a reaction for which optimum conditions are uncertain, as may be the case when an established published procedure is used as the basis for carrying out other preparations of a similar nature. In these cases the reaction is *monitored* by the periodic removal from the reaction mixture of test portions for suitable chromatographic study. Clearly the chromatographic behaviour of starting materials and, if possible, expected products, needs to be established prior to the commencement of the reaction. For t.l.c. this would include solvent and thin layer selection, a detection method, and an appraisal of sensitivity of detection with respect to the concentration of components in the reaction medium. For g.l.c. preliminary experiments would be required to select a suitable column and the appropriate operating conditions.

In the case of solid products and starting materials it is often convenient and sufficient to load a sequence of appropriately sized samples of the homogeneous reaction mixture, taken at various times during the progress of the reaction, directly on to a thin-layer chromatographic plate. However the work-up procedure of the reaction mixture should be carefully studied to determine whether sample pretreatment is required. For example, the product may be present in the reaction mixture as an acid or base salt, or as a chelated complex, and the isolation procedure would then include a step to ensure decomposition to the required product. In such a case, sample pretreatment is essentially the operation which would be followed in the final work-up procedure (carried out on the small scale in test tubes and with Pasteur pipettes) to give a final solution containing reactants and products for loading on to the t.l.c. plate.

With liquid reactants and products some simple form of sample pretreatment is often essential. This arises from the fact that a g.l.c. column may rapidly become contaminated with involatile and possibly highly acidic or basic material which would render it useless within a short period of time.

By way of illustration the following simple example of a suitable procedure may be cited. In the preparation of an alkyl halide from the alcohol (Expt 5.54) using constant boiling hydrobromic acid and sulphuric acid, an aliquot portion of the reaction mixture is removed by means of a capillary pipette after a suitable time interval from the start of the reaction (say 30 minutes) and transferred to a micro test tube. A few drops of ether or dichloromethane are added and the contents of the tube shaken. The lower organic layer is removed to another micro test tube with a capillary pipette and washed successively by shaking it with dilute aqueous sodium hydrogen carbonate and water, the aqueous layers being removed with the aid of a pipette. The organic layer is dried by the addition of a little magnesium sulphate desiccant, the tube centrifuged if necessary, and a sample of the clear upper layer submitted to gas chromatographic investigation.

When the success or outcome of a reaction is uncertain or unknown, chromatographic methods are invaluable for assessing the success of the purification process. They may well reveal that the procedures adopted are unsuited to the required isolation and purification of the reaction product. In these cases the chromatographic behaviour of the components in the system which has been revealed by these preliminary small-scale studies provides a basis upon which purification by preparative chromatographic methods may be achieved. Thin-layer chromatographic behaviour may be reasonably closely duplicated by employing a similar stationary phase and a similar mobile phase in either a 'wet' or a 'dry' column technique, which can then be readily scaled up to accommodate the bulk of the reaction product. Preparative gas chromatography has the advantage that usually little further investigational work is required to accomplish the separation of larger quantities, but as will be seen later a preparative scale separation may require the use of an automated apparatus using a multiple cycling procedure.

In recent years h.p.l.c. has become a valuable chromatographic tool for analytical and preparative scale work. In this latter area the separation of isomers (structural, diastereoisomeric, and enantiomeric) has been possible by the selection of appropriate column packing material and solvent systems. However, the equipment, operating costs, and column packing materials are more expensive than those in t.l.c., g.l.c. and conventional liquid–solid column chromatography.

THIN-LAYER CHROMATOGRAPHY (t.l.c.)[57, 58]
In this technique it is usual to employ glass plates coated with layers of the solid stationary phase, which adhere to the plates, generally by virtue of a binding agent, such as calcium sulphate, which is incorporated. The prepared thin layer on glass is often called a *chromaplate*.

The most commonly used stationary phases, which are available in grades specially prepared for t.l.c. use, include silica gel, alumina, kieselguhr and cellulose powder; many of these are available with a fluorescent compound (e.g. zinc sulphide) incorporated in order to facilitate the detection of the resolved components of the mixture which is then achieved by viewing the plates under ultraviolet light. Other materials suitable for special applications are polyamides, modified celluloses with ion exchange properties and the various forms of organic gel having molecular sieving properties (e.g. Sephadex, Bio-Gel P). Stationary phases for t.l.c. may be obtained from the following suppliers: Aldrich Chemical Co., BDH, Eastman Kodak, Fluka AG., Johns-Manville, E. Merck, Phase Separations, Sigma Chemical Co., M. Woelm, etc.

199

Preparation of plates. Before glass plates are coated with adsorbent they must be carefully cleaned with laboratory detergent, using a test-tube brush to remove adhering particles, rinsed thoroughly with distilled water, placed in a suitable metal rack and dried in an oven. Subsequent to the treatment with detergent solution the plates should only be handled by the edges or by the under-surface which is not to be coated with adsorbent. Failure to observe this precaution may result in the formation of a mechanically unstable layer which is liable to flaking due to grease spots on the glass surface. In severe cases of grease contamination it may be necessary to use a chromic acid cleaning mixture (Section 2.2).

Small plates suitable for preliminary exploration of the chromatographic process with regard to the selection of a suitable stationary phase or the selection of a solvent system are conveniently prepared from microscope slides using a dipping technique; *this operation is conducted in a fume cupboard.*

A slurry is prepared by the slow addition with shaking of 30 g of adsorbent (most usually silica gel or alumina) to 100 ml of dry dichloromethane contained in a wide-necked capped bottle. A pair of microscope slides is held together and dipped into the slurry, slowly withdrawn and allowed to drain momentarily while held over the bottle. The slides are parted carefully and placed horizontally in a rack sited in a fume cupboard to dry for approximately 10 minutes. The surplus adsorbent is then removed by means of a razor blade drawn down the glass edges. It may be desirable to activate the adsorbent further by heating it at 110 °C; since the activity of the adsorbent varies with the heat treatment and the subsequent storage conditions of the prepared plate this further treatment should be carefully standardised. For the attainment of a high degree of reproducibility it is usually best to activate the adsorbent and allow the plate to cool in a desiccator cabinet immediately before use.

Larger single-glass plates (i.e. 20 × 5 cm) may be coated conveniently using the easily assembled apparatus shown in Fig. 2.129. It consists of a sheet of plate glass (20 × 30 cm) at the upper and lower ends of which two glass plates (20 × 5 cm) are secured by means of a cement for glass: the plate to be coated is placed in the central depression and is held in position by two uncoated plates one on either side. The thickness of the layer can be adjusted to the thickness required (say 0.25–0.3 mm) by wrapping both ends of a glass rod 14 cm long and 7.5 mm diameter with equal lengths (12.5 cm) of 2.5 cm Sellotape: this is the 'spreader'.

The exact composition of a suitable slurry for spreading depends on the nature of the adsorbent; this should therefore generally be prepared according to the procedure recommended by the supplier. The composition of the slurry may

Fig. 2.129

have to be varied, as for example in the preparation of buffered silica gel or kieselguhr plates, or of plates for reversed phase thin-layer chromatography which require coatings of hydrophobic layers. A slurry from about 2 g of dry adsorbent is sufficient to coat one plate of the size stated here. The slurry is poured on to the upper end of the central glass plate (Fig. 2.129) and spread evenly over the plate with the special applicator or spreader: this should be completed in 15 seconds. The layer is allowed to stand for 5 minutes in order to set; cellulose and polyamide plates are allowed to dry at room temperature and are then stored in a dust-free cabinet – they are not normally heated. Inorganic adsorbents are activated and stored under standard conditions (see above).

Figure 2.130 illustrates diagrammatically one form of commercially available spreader (Shandon) which consists of a flat frame capable of holding rigidly five 20 × 20 cm glass plates or the equivalent number of 20 × 10 cm or 20 × 5 cm plates. The construction of the frame is such that when the plates have been positioned edge to edge the upper surfaces are all aligned in the same plane. This flat surface, free from ridges at plate joints, ensures a uniform thickness of adsorbent layer over all the plates and provides a smooth path for the metal hopper (Fig. 2.131) when drawn across the surface. It is usually best to place at either end of the line of plates to be coated a 20 × 5 cm glass plate to provide an area upon which the coating process may be started and finished, since frequently the layer thickness at the immediate start and finishing points is not uniform. The prepared slurry is poured evenly into the rectangular well of the hopper located on the end plate and this is then drawn steadily over the glass plate surface so that the slurry flows evenly through the gap provided on the following edge. The thickness of the layer may be preselected by loosening the nuts and adjusting the accurately machined metal gate by means of a 'feeler' gauge, the hopper resting on a flat glass surface. This enables layers of thickness 200–2000 μm to be selected; in fact a thickness of 250 μm is usually the most suitable for routine use. When the full capacity of the spreader is to be utilised and when thick layers are to be spread, necessitating a fairly large volume of slurry, it is sometimes helpful to place a glass rod of suitable length immediately along the gap between the glass surface and the metal gate to prevent excessive flow of slurry while the hopper is being charged prior to the spreading operation.

Loading of plates. In order to load the prepared chromoplate (say 20 × 5 cm size) with the sample to be investigated the following procedure should be followed; this may be modified appropriately when the larger plates and the micro plates are used. Wipe any excess adsorbent from the back and edges of the plate.

Fig. 2.130

Fig. 2.131

Form a sharp boundary by scoring with a metal scriber parallel with and 5 mm from the shorter edge (in the case of 5- and 10-cm-wide plates) of the plate and carefully remove surplus adsorbent with the aid of the flat side of a spatula and blowing carefully. Align the lower edge of the commercially available template (Fig. 2.132(A)) along this edge of adsorbent layer; draw the metal scriber through the adsorbent layer using the upper edge (B) of the template as a guide. Carefully blow surplus adsorbent from this cut which then provides a finishing line for the subsequent solvent front; if not clear of adsorbent the solvent will flow unevenly across the line and lead to an unsatisfactory evaluation of the chromatogram. With the template still in position, lightly mark a series of starting points on the adsorbent surface with the metal scriber through the application holes in the template; these are usually located 15 cm from the finishing line. (These starting points are 1 cm apart and the design of the template is such that four may be symmetrically accommodated on a 5-cm-width plate.)

In a plastic-stoppered glass sample tube prepare a solution of the mixture to be investigated having a concentration in the range 0.5–3.0 per cent by dissolving 3 mg in from 0.1–0.6 ml of solvent depending on its solubility. The selected solvent should be reasonably volatile (e.g. chloroform or light petroleum, b.p. 40–60 °C); in the subsequent application of a drop of this solution to the adsorbent rapid evaporation of the solvent is desirable, since this leads to the formation of a small-diameter spot which results in a better separation of the components in the subsequent chromatographic development process. Similar volumes of aqueous solutions give larger spots which result in a more diffuse chromatogram – if aqueous solutions need to be used a technique of multiple application (see below) may be necessary. Solutions of pure compounds thought to be present in the mixture (e.g. starting materials, possible reaction products) are similarly prepared for application to the adsorbent on the same plate and alongside the mixture spot – these act as reference compounds to permit more ready interpretation of the chromatogram.

The solutions are applied individually to the marked points on the adsorbent layer by means of a sample applicator. This is prepared by drawing out a melting point capillary tube in a micro-Bunsen flame and snapping the drawn-out portion in two after scratching with the edge of a fragment of unglazed porcelain to ensure a clean break. The applicator is charged by dipping the capillary end into the solution and after withdrawing, touching the end on a piece of filter paper until the volume is reduced to about 0.5 μl. Using the template as a hand rest, the solution is transferred to the plate by touching the tip of the capillary on to the adsorbent layer, taking care not to disturb the surface unduly. If the 0.5 μl volume has been estimated with reasonable accuracy the size of the spot

Fig. 2.132

should be about 3 mm diameter. Alternatively the use of commercially available disposable micro-pipettes which contain standard volumes of from 1 to 5 μl and which can easily be partly filled to contain the required amount of solution, or the use of calibrated syringes delivering from 0.1 to 5 μl, allows the volume of solution to be applied to the plate to be judged more accurately. With more dilute solutions or where a heavier loading of material is required, larger volumes may be applied by allowing the solvent to evaporate during the intermittent addition of the solution to the plate (the use of a commercial electric warm-air blower is recommended) so that the diameter of the spot never exceeds 3 mm. A suitable cipher is inscribed on the adsorbent layer beyond the finishing line and opposite the point of application to identify the nature of the solution applied. After use the capillary may be cleaned by dipping it into pure solvent, draining by touching the tip on to a filter-paper and repeating the operation two or three times.

Selection of the solvent system. If the chromatographic behaviour of the substance under investigation is unknown, the most satisfactory developing solvent must be ascertained by preliminary trial runs using micro-plates in 4-oz wide-mouthed screw-topped bottles. It is convenient to set up a series of such bottles containing solvent systems of increasing polarity. For example, hexane, toluene, carbon tetrachloride, dichloromethane, diethyl ether, ethyl acetate, acetone, methanol. Identically loaded micro-plates are developed separately using the chosen solvents, dried and sprayed with the appropriate reagent (see below), and the chromatographic mobility of the individual components noted. Solvents which cause all the components to remain near to the spot origin or to move near to the solvent front are clearly unsatisfactory. If it is seen that no single solvent gives a satisfactory chromatogram, with well-spaced compact spots, it is necessary to examine the effect of using mixtures of solvents to provide systems having a range of intermediate polarity. For example, mixtures of toluene and methanol, or hexane and ethyl acetate, are often suitable when the pure solvents are unsatisfactory.

Development of plates. Individual 20 × 5 cm plates are conveniently developed in a cylindrical glass jar (Fig. 2.133) (Shandon). Larger plates, 20 × 10 cm and 20 × 20 cm, require a rectangular glass tank of suitable dimensions such as that shown in Fig. 2.134; such a tank can also be used to allow the simultaneous de-

Fig. 2.133

Fig. 2.134

velopment of several of the smaller-sized plates. Micro-plates are easily accommodated individually in 4-oz wide-mouthed screw-capped glass bottles.

Line the inside of the jar with filter paper, leaving a gap for viewing the chromaplate. Saturate the filter paper with the selected developing solvent, close the jar and allow to stand for about 10 minutes so that the atmosphere in the jar becomes saturated with solvent vapour. Insert the plate with the origins spot towards the bottom of the jar, tilted as shown, so that the uncoated face is uppermost. Carefully pour down the side of the jar more of the developing solvent so that the bottom of the adsorbent layer is well immersed; the solvent level should not however reach as far as the spots. Recap the jar and allow the solvent to ascend by capillary action to the finishing line which has been scored across the plate. The time required to complete this development varies greatly with the composition of the solvent and the nature of the adsorbent. If the system is inconveniently slow-running the development process may be terminated before the solvent reaches the finishing line, provided that the position of the solvent front is marked on the adsorbent layer *immediately* the plate is removed from the development tank.

After removal, the plate is dried suitably depending upon the volatility and toxicity of the solvent system; for example, dry the plate in the fume cupboard (if necessary) with a warm-air blower or dry in a temperature-controlled oven, etc.

Location of spots. The positions of coloured components can of course usually be seen without any difficulty providing that the concentration in the initial spot is sufficiently high and that excessive spreading of the component during development has not occurred. Viewing the plate under an ultraviolet lamp will reveal u.v. fluorescent compounds the positions of which must of course be marked with the scriber on the surface of the adsorbent. Non-fluorescent compounds can be detected by virtue of their fluorescent quenching effect when they are chromatographed on adsorbents into which a fluorescent indicator has been incorporated (e.g. Silica gel GF_{254}). Routine inspection of plates under ultraviolet light is to be recommended before any further detection processes are applied.

A useful general, but unspecific, detecting agent for most organic compounds is iodine vapour. The dried plate is allowed to stand in a closed tank containing a good supply of iodine crystals scattered over the tank bottom; usually the spots are revealed as brown stains. Their positions should be marked as soon as the plate has been removed from the iodine tank since standing in air for a short while causes the iodine to evaporate and the stains to disappear.

Another general locating procedure applicable in the main only to plates coated with inorganic adsorbents and for the detection of organic material is to spray the plate with concentrated sulphuric acid or with a solution of concentrated sulphuric acid (4 ml) in methanol (100 ml), and then to heat the plate in an oven to about 200 °C until the organic materials are revealed as dark charred spots. Other general spray reagents are potassium permanganate (2%) in an aqueous solution of sodium hydrogen carbonate (4%), or, phosphomolybdic acid (10%) in ethanol.

Spray guns (of glass or plastic), operated by compressed air or a rubber bulb-type hand blower (e.g. Fig. 2.135) are available from, for example, Bibby Science, or Shandon. This spraying operation must of course be carried out with considerable care and it is advisable to place the plate at the bottom of a large rectangular glass tank, placed on its side and located in a fume cupboard before spraying is attempted. It is also good practice to wear a suitable protective face mask. It is essential to cover the plate evenly with spray but without so saturating the adsorbent layer that the liquid visibly flows over the surface, since this will cause distortion of the zones.

Chemical methods for the detection of colourless compounds by the use of a suitable chromogenic spray reagent are widely used. The most commonly required spray reagents are available commercially with spray guns operated by an aerosol propellent (e.g. Sigma Chemical Co.). Many of these are selective for a particular functional group or groups and may be extremely sensitive; e.g. the ninhydrin reagent for the detection of amino acids. Other such spray reagents are more general in their application; e.g. indicators may be used in sprays for the detection of acids and bases. Such chemical locating agents may be usually applied with advantage after successively viewing the plate under ultraviolet light, exposing the plate to iodine vapour and allowing the iodine to evaporate. This extended treatment gives a much more comprehensive picture of the composition of the mixture of components on the chromatogram than does a single non-selective method.

The selectivity and sensitivity of a wide range of spray formulations may be found in the many specialist monographs on thin-layer chromatography.

Provided that the experimental conditions are reproducible the movement of any substance relative to the solvent front in a given chromatographic system is constant and characteristic of the substance. The constant is the R_F value and is defined as:

$$R_F = \frac{\text{distance moved by substance}}{\text{distance moved by the solvent front}}$$

Figure 2.136 indicates the method of measurement of R_F values of each of the components of a typical chromatogram:

$$R_F = \frac{a}{b}; \quad R_F' = \frac{a'}{b}; \quad R_F'' = \frac{a''}{b}$$

True reproducibility in R_F values is however rarely achieved in practice due to minor changes in a number of variables such as:

(a) the particle size of different batches of adsorbent;
(b) the solvent composition and the degree of saturation of the tank atmosphere with solvent vapour;
(c) prior activation and storage conditions of the plates;
(d) the thickness of adsorbent layer, etc.

Fig. 2.135

Fig. 2.136

It is therefore *not* desirable to use an R_F value in isolation as a criterion of identity; marker spots, when reference compounds are available, should always be run on the same plate as the mixture to substantiate the identification of the components.

Precoated plates. Glass plates precoated with the most commonly used adsorbents, with or without fluorescent indicators, can be purchased from the specialist suppliers noted above. Having been prepared under carefully standardised conditions these have a much higher degree of reproducibility in the subsequent analysis than hand-coated plates, since some of the factors noted above which cause variation in R_F values, e.g. particle size, layer thickness, etc., have been eliminated. In particular mention should be made of the high performance plates (h.p.t.l.c.); these are of layers of silica gel having a smaller particle size with a narrower particle size distribution, namely in the range 5–10 μm. The chromatographic resolution and compactness of spots is far superior with these h.p.t.l.c. plates.

Precoated plates are also available for reversed-phase liquid–liquid partition thin-layer chromatography. Here the silica gel has been treated with an octadecyl silylating reagent thus coating the particles with a non-polar chemically-bonded thin film. The solvent employed is more polar than the film and chromatographic development results from partition between these two phases.

Reference must also be made to the use of layers precoated on to flexible sheets which are also available commercially. The backing material may be either of aluminium foil or more usually a solvent-resistant polyester sheet. These sheets can be cut to the desired size with a pair of scissors and activated if necessary. Several millimetres'-width of adsorbent are then scraped off the sheet, on the sides which will be parallel to the direction of solvent flow, using a spatula guided by the edge of a steel rule. This step is necessary to prevent a solvent film travelling along the edge and in the space created between the adsorbent layer and the backing sheet by the scissor cutting. The prepared sheet is then used as one would a normal thin-layer glass plate, but it has the added advantage that the adsorbent layer is less likely to be accidentally damaged. A further advantage with these flexible plates is that the developed chromatogram may be stored in a notebook, etc., after use.

Two-dimensional chromatography. When complex mixtures are to be studied, the R_F values of the individual components may be so close that a clear-cut separation of the components is not achieved. In such a case a two-dimensional thin-layer chromatographic separation can be used with advantage.

A single spot of the mixture is applied near to one of the corners of a 20 × 20 cm plate and the chromatogram developed in one direction as usual. The plate is then removed and dried and the chromatogram re-developed in a second solvent system so that the direction of solvent flow is at right angles with respect to the first (Fig. 2.137(a) and (b)). These illustrations point the need for correctly placing the origin spot with respect to the edge of the chromatogram so that the solvent levels in both development processes do not cover the applied spot or the individual components which separate in the first development.

It is usual to employ two solvent systems in both of which the individual components have an adequately wide range of R_F values since in this way a good separation of components over the whole plate is observed. The spot location is achieved be means previously described; each component will be characterised

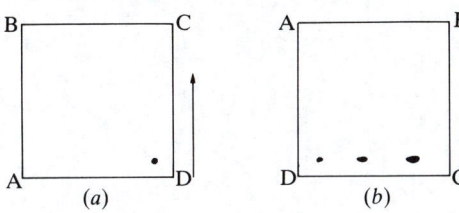

Fig. 2.137

by two R_F values. The incorporation of an individually placed marker spot cannot of course be applied in the two-dimensional chromatogram. If the chromatographic behaviour of reference compounds under the operating conditions needs to be determined, these must be run on a second two-dimensional chromatogram which is developed along with the chromatogram of the unknown mixture.

Preparative thin-layer chromatographic techniques. The simplest form of preparative thin-layer chromatography is to use coated plates in which the adsorbent thickness is in the range 1–2 mm, although layer thickness up to 4 mm may be employed. The sample solution is loaded on to the plate as a continuous streak by means of a syringe; there are commercially available applicators which enable uniform loading to be achieved. After development in the selected solvent system, the dried plate is viewed under u.v. light and the position of the separated components (seen as bands across the plate) carefully marked with the scriber. The adsorbent layers corresponding to the located bands are then separately removed and eluted with a suitable solvent to recover the individual components. A convenient method is to use a glass tube carrying a sintered glass disc within its length, one end of which is attached to tubing leading to a water- or oil-vacuum pump, and the other end narrowed and shaped to act as a 'vacuum cleaner' nozzle. With this arrangement the adsorbent collected then constitutes a small chromatographic column which facilitates solvent elution.

The '*Chromatotron*' is a novel and highly convenient piece of equipment for preparative thin-layer separations. The equipment was designed by I. and S. Harrison (authors of *Compendium of Organic Synthetic Methods*) and is manufactured and marketed by TC Research. Basically (Fig. 2.138) it consists of a slanted circular glass plate which is spun about a central shaft by means of an electric motor. The apparatus casing enables the plate to be enclosed in a nitrogen atmosphere, although it may be viewed through the transparent Teflon lid. The glass plate carries the adsorbent layer of 1, 2 or 4 mm thickness, and sample loading and solvent delivery are via an off-centre inlet. Since the plate is spun during loading and solvent development, the sample and separated constituents are as radial bands. A continuous, regulated, solvent flow results in eluting solvent being spun off the edge of the plate and collected via a radial channel to drain into an outlet tube. Direct observation with coloured compounds, or observation under u.v. light, enables the separated fractions to be easily collected. The particular advantages of the equipment are that up to 2 g of a mixture may be separated in 5–20 minutes, the nitrogen atmosphere prevents oxidation of sensitive compounds, and the adsorbent layers are easy to prepare and in any case may be regenerated *in situ* and reused.

Fig. 2.138

Exercises in thin-layer chromatography. *Separation of amino acids.* Prepare solutions of DL-alanine, L-leucine and L-lysine hydrochloride by dissolving 5 mg of each separately in 0.33 ml of distilled water, measured with a graduated 1 ml pipette (leucine may require warming to effect solution). Mix one drop of each solution to provide a mixture of the three amino acids and dilute the remainder of each solution to 1 ml to give solutions of the respective amino acids. The latter will contain about 5 μg of each amino acid per μl. Apply approximately 0.5 μl of each of the solutions to a Silica Gel G plate and allow to dry in the air (i.e. until the spots are no longer visible).

Prepare the developing solvent by mixing 70 ml of propan-1-ol with 30 ml of concentrated aqueous ammonia (d 0.88). Line the inside of the jar with filter paper reaching to within 3 cm of the bottom and moisten with the developing solvent. Insert the prepared plate into the jar and carefully introduce by means of a pipette sufficient of the developing solvent so that the lower edge of the adsorbent layer is immersed in the solvent; put the cover in position in the mouth of the jar, and allow the chromatogram to develop.

Remove the chromatogram, dry it at 100 °C for 10 minutes and spray with ninhydrin reagent [0.2% solution in butan-1-ol, (1)]; heat at 110 °C for 5–10 minutes in order to develop the colour. Mark the centre of each spot with the metal scriber and evaluate and record the R_F values.

Note. (1) Ninhydrin (p. 630) is the 2-hydrate of indane-1,2,3-trione. It reacts with α-amino acids to yield highly coloured products. Contact with the skin should be avoided since it produces a rather long-lasting purple discoloration.

Separation of 2,4-dinitrophenylhydrazones. The solutions are prepared by dissolving 10 mg of each of the 2,4-dinitrophenylhydrazones of acetone, butan-2-one and hexan-3-one (or hexan-2-one) in 0.5 ml of ethyl acetate. Prepare a flexible silica gel sheet of dimensions 20 × 5 cm in the manner already described and apply c. 0.5 μl of each of the three solutions to give the marker spots of a diameter of between 2 and 3 mm. A mixed spot is conveniently obtained by loading sequentially to the same area further 0.5 μl aliquot portions of each of the solutions and allowing the solvent to evaporate completely between each addition.

Charge the paper-lined jar with the developing solvent (toluene: light petroleum b.p. 40–60 °C, 3 : 1), insert the loaded flexible sheet and allow the development to proceed. Air dry the developed chromatogram and record directly the R_F values of the components.

LIQUID–SOLID COLUMN CHROMATOGRAPHY

Separations on a preparative scale employing this technique are generally accomplished by loading the substance on to a cylindrical column of the solid stationary phase, and developing the chromatogram by allowing the liquid mobile phase contained in a suitable reservoir to flow through the column under gravity, or under pressure applied to the top of the solvent reservoir.

In the conventional technique, continuous passage of a single eluting solvent through the column may eventually result in the emergence from the bottom of the column of the individual components of the mixture so that they can be individually collected and recovered. A refinement is progressively to increase the polarity of the mobile phase to assist in the displacement of the individual fractions from the stationary phase and hence to speed up the overall time of operation. If, however, actual elution from the column of the components of the mixture by such means is impracticable (e.g. because the time required for elution is too prolonged), it may be necessary to drain off surplus solvent so that the column packing material may be extruded in one piece on to a glass sheet and portions of the column cut off and separately extracted and examined. In the so-called 'dry-column' technique described below (p. 216 – which differs from the conventional technique in details of chromatographic development) this column cutting and extraction is the method employed for recovery of separated components.

If the desired compound or compounds are coloured (or strongly fluorescent under ultraviolet light), their location on the column or in selected eluent fractions presents no problems. Hence suitable fractions are combined and concentrated to recover the purified material.

Colourless compounds in eluate fractions are usually detected by one of the following generally applicable procedures:

1. Provided that the mobile phase is a relatively volatile organic solvent, the simplest procedure for assessing the progress of the chromatographic separation is to collect the eluate as a series of fractions of equal volume* and to evaporate to dryness each fraction in a rotary evaporator and to weigh the residues obtained. A graphical plot of weight versus fraction number then gives a profile of the chromatographic separation which has been achieved; the total weight eluted from the column at any one time should always be compared to the amount of mixture loaded on to the column to provide a guide to recovery of loaded material. The homogeneity of the residues should be further examined by t.l.c. to decide whether any further fractionation is required or whether any of the residues contain essentially one and the same component and may therefore be combined.
2. Each of the individual fractions collected could of course be examined directly by t.l.c. (using one of the non-selective detecting agents, e.g. iodine vapour),

---

* Large numbers of fractions having preselected volumes of between less than 1 ml and 50 or 100 ml are most easily collected with the aid of one of the many designs of automatic collectors.

but it may be somewhat difficult to estimate the possible concentration of material in the eluate, and hence to determine the loading required on the chromaplate. Multiple application of some of the individual fractions to the same area of the plate may be necessary to detect compounds in relatively high dilution. These t.l.c. results will then determine in what manner the fractions may be combined and if necessary further treated.

Either of these methods could be expected to cover the majority of cases encountered in qualitative or rough quantitative studies applicable to the development of preparatively useful laboratory synthetic procedures. Nevertheless in this context the monitoring of a column chromatographic separation by ultraviolet spectroscopic methods may frequently be the less tedious method provided that certain conditions are met. Firstly the components of the mixture should adsorb in the ultraviolet, and the wavelengths used for screening the fractions should be selected so that all of them are detected. Where measurement of the absorption of fractions at a single wavelength is inadequate determination of the adsorption at two or more wavelengths should be made. Secondly the method is only conveniently used with a single solvent development, and is only applicable when solvents which are transparent at the wavelength of ultraviolet light selected have been employed for chromatogram development. The reference cell will contain the pure solvent and in preparative work aliquot portions of the fractions containing components will need to be diluted with pure solvent in order to obtain an on-scale absorbance reading. The optical activity of suitable organic compounds may also be exploited for their detection in chromatographic fractions, particularly in the case of the separation of mixtures of natural products.

In the detection of colourless and non-fluorescent compounds on extruded column packing material, the technique of applying a thin streak of a suitable test reagent, if appropriate, lengthwise down the extruded column may be used. The colours which appear on the surface of the column at the place touched by the reagent indicate the positions of the zones; that part of the column packing containing colour test reagent can be readily shaved off so that it may be discarded before the bulk of the material is separated and the component recovered by extraction. Alternatively a strip of Sellotape may be placed momentarily lengthwise down the extruded column surface; the thin layer so removed is then sprayed with the chromogenic reagent and the revealed bands located on the main column by suitable alignment. When no such colour test is available the extruded column must simply be divided into arbitrary segments, each of which must be extracted, the solvent removed by evaporation and the residue examined further (i.e. weight, t.l.c.).

The technique of conventional column chromatography. The essential part of the apparatus consists of a long narrow glass tube (10–90 cm long and 1–4.8 cm diameter); these dimensions give columns which hold between from 25 to 400 g of column packing material. Figure 2.139 depicts an assembly with one of the smaller columns (10–40 cm long and 1–1.8 cm diameter) fitted with ground glass joints at its ends to allow for the attachment of a separatory funnel (to act as a solvent reservoir) and a Buchner flask via an adapter with tap for the collection of the eluate fractions. This design incorporates a sintered glass disc (porosity 0) to retain the column packing. Figure 2.140 illustrates the larger chromatography column which has ground glass joints fitted to both ends; in use the

Fig. 2.139   **Fig. 2.140**

bottom of the column is closed with a ground glass joint having a sealed sintered glass disc incorporated (see insert). This design allows for the column packing to be easily extruded.

When substances which are oxidised by air need to be handled, the separatory funnel is additionally fitted with a screw-capped adapter with delivery tube attached to a suitably controlled supply of an inert gas (e.g. nitrogen – see Fig. 2.60); additionally the receiver should be a three-necked flask of suitable size, the two side joints enabling a flow of inert gas to be passed through the receiver.

It is not usually desirable in this conventional system to attach a vacuum supply to the receiver since this results in tight packing of the column material which in turn slows the rate of solvent flow. Application of slight pressure to the top of the separatory funnel by means of a cone adapter attached to a suitably controlled supply of compressed air is the more satisfactory means of increasing the rate of solvent flow.

Column packing materials. The selection of suitable column packing materials (from the suppliers noted above) is made on the basis of the chromatographic process which needs to be employed for a particular separation, e.g. adsorption or ion-exchange processes or gel filtration. The most widely used and generally applied process in preparative chemistry, where the specific need is frequently the purification of starting materials (when only impure technical products are available) and the complete or partial resolution of components of reaction mixtures, is adsorption chromatography. Chromatographic separation on ion-exchange columns is a useful analytical and preparative technique for the resolution of mixtures of acids or bases (e.g. amino acids, amino-phenols, etc.) and for

211

the isolation of neutral organic material from aqueous solutions containing cationic and anionic substances (the technique of 'de-salting', see Expt 5.183 for the removal of chloride ion by a resin in the isolation of amino acids). The procedures for packing columns in both these groups differ little in principle. Gel filtration is an invaluable chromatographic process for the quantitative analysis of mixtures of naturally occurring compounds of high molecular weight (e.g. proteins, peptides, enzymes, hormones, nucleic acids, etc.) available in small-sized samples; its use with these groups of compounds on a preparative scale is becoming increasingly important but a discussion of the specialised apparatus used and examples of specific applications are outside the terms of reference of this volume.

The most widely used column packing for adsorption chromatography is aluminium oxide (alumina). The particle size of the commercially available grade is in the range 50–200 μm (70–290 mesh) which allows for relatively even packing of adsorbent during column packing, for reasonable solvent flow under the force of gravity, and for the rapid attainment of equilibrium distribution of the adsorbate between the surface of the adsorbent and the mobile liquid phase. Alumina may be obtained in basic (pH 10), neutral (pH 7) and acidic (pH 4) forms (1), and it is important to ensure that the correct type is employed because of catalytically induced reactions which each may cause with particular functional compounds. For example, basic alumina may lead to hydrolysis of esters, acidic alumina may lead to dehydration of alcohols (particularly tertiary alcohols) or may cause isomerisation of carbon–carbon double bonds; in these circumstances neutral alumina is to be recommended. The activity of all three forms of alumina, which is broadly regarded as relating both to the magnitude of the attractive forces between the surface groups on the adsorbent and the molecules being adsorbed, and to the number of sites at which such attraction takes place, is classified into five grades (*the Brockmann scale*).[59] Grade I is the most active (i.e. it retains polar compounds most strongly), and is obtained by heating the alumina at about 300–400 °C for several hours. Successively less active grades, II–V, are then obtained by the addition of appropriate amounts of water (II, 3–4%; III, 5–7%; IV, 9–11%; V, 15–19%) (2). The activity grade is assessed by determining the chromatographic behaviour of specified dyes, loaded in pairs on to an alumina column (5 cm long and 1–5 cm diameter) under carefully standardised conditions and developing the chromatogram with benzene-light petroleum (b.p. 60–80 °C) (3). By comparing the results with those given in Table 2.11 the grade of alumina may be assigned.

Silica gel (pH 7) may also be graded according to the amount of water added to the most active grade, obtained by heating for several hours at temperatures not exceeding 300 °C; these are II (5%), III (15%), IV (25%), V (38%). These activity gradings are assigned using the same dye pairs as for the grading of alumina.[60] Addition of larger amounts of water leads to the formation of a substantial film of surface water so that a column prepared from such material may be used to effect separations by partition rather than adsorption.

Less frequently employed adsorbents include magnesium silicate, magnesium oxide, magnesium carbonate, calcium carbonate, barium carbonate, calcium hydroxide, calcium sulphate, lactose, starch, cellulose and Fuller's earth. Activation by drying obviously requires careful control in many of these cases and clearly some would be quite unsuitable for separations involving compounds possessing certain functional groups. The inorganic adsorbents of this group are

Table 2.11 Grading of activated alumina

	Grade of activity		
Dye position	I	II	III
Column, top 1 cm	p-methoxy-azobenzene		Sudan Red*
Column, bottom 1 cm	azobenzene	p-methoxy-azobenzene	Sudan Yellow†	Sudan Yellow
Eluate	—	azobenzene	—	p-methoxy-azobenzene

Dye position	IV		V
Column, top 1 cm		p-amino-azobenzene	p-hydroxy-azobenzene
Column, bottom 1 cm	Sudan Red	Sudan Red	p-amino-azobenzene
Eluate	Sudan Yellow	—	—

* Sudan Red (Sudan III) has the structure

† Sudan Yellow is 1-(phenylazo)-2-naphthol (Expt 6.82).

Table 2.12 Graded series of adsorbents

1. Sucrose, starch
2. Inulin
3. Talc
4. Sodium carbonate
5. Calcium carbonate
6. Calcium phosphate
7. Magnesium carbonate
8. Magnesium hydroxide
9. Calcium hydroxide
10. Silica gel
11. Magnesium silicate (Florisil)
12. Alumina
13. Fuller's earth

usually obtained as very fine powders and the solvent flow through columns prepared from such material is extremely slow; the column performance in this respect may be improved by mixing the adsorbent before column preparation with diatomaceous earth filter-aids (trade names: Filter-Cel, Super-Cel, Clara-Cel, etc.), which have only low adsorbent activity.

Table 2.12, due largely to Strain,[61] gives a list of adsorbents in increasing order of adsorptivity.

Notes. (1) Neutral alumina may be obtained from the basic form by stirring it with excess water and heating to 80 °C; dilute hydrochloric acid is added dropwise with stirring until slightly acid (pH 6.5) and the heating maintained for 1 hour. The supernatant liquor is decanted and the alumina stirred with aqueous ammonia (2%) at 70–80 °C for 30 minutes. The alumina is recovered by filtration, washed with distilled water until the filtrate is free of chloride ion and dried at 120 °C.

(2) The deactivation of an adsorbent by the addition of water is achieved by simply mixing the appropriate quantities in a stoppered flask and shaking in a mechanical shaker for about 1 hour to ensure equilibration.

(3) Prepare two glass columns of approximately 10 cm length from 1.5-cm-diameter glass tubing as follows: select a 22 cm length of tubing and heat the mid-section in a broad blow-pipe flame (see Section 2.10) until it has softened and thickened. Draw the ends apart by about 5 cm to give a constricted portion of about 0.5 cm diameter with walls sufficiently thick so that when cut into two portions these constricted ends are not fragile; anneal both ends of each glass column. Plug the constricted end of one tube with a little glass wool, clamp vertically in a retort and add alumina, while at the same time tapping the glass tube gently with a wooden rod, to give finally a 5 cm length of adsorbent. Place a small conical beaker under the column outlet. Dissolve 5 mg of each of the pair of dyes selected in 5 ml of benzene (**CAUTION**) (warming on a water bath may be necessary) and when solution is complete add 20 ml of light petroleum (b.p. 60–80 °C); fill a 10 ml pipette, fitted with a rubber suction bulb, with the dye solution, and with the pipette top touching the inner part of the glass column about 1 cm from the adsorbent surface, allow the liquid to discharge from the pipette so as not to disturb the adsorbent surface. (It is frequently desirable to place a piece of filter paper on to the top of the adsorbent surface to prevent it being disturbed by the flow of solution – in this case the pipette may be allowed to discharge directly on to the filter-paper cover.) While the solution soaks into the column fill a clean 10 ml pipette with eluting solvent (benzene : light petroleum, 1 : 4) and as the column liquid level just falls to the level of the adsorbent, carefully rinse the inside of the glass tube with a few ml of the solvent. When these washings have drained into the adsorbent surface add the remainder of the 10 ml portion of developing solvent. The chromatogram is assessed when the solvent portion has drained. The operation is repeated in precisely the same manner with a fresh alumina column and with another dye pair if necessary until the activity grading of the alumina has been established.

Selection of solvents for adsorption chromatography. The choice of solvent for transferring the mixture to be chromatographed to the column will naturally depend upon the solubility characteristics of the mixture. If it is already in solution, for example as an extract, this is usually evaporated to dryness under reduced pressure and the residue dissolved in the minimum volume of the most non-polar solvent suitable. As concentrated a solution as possible is desirable to achieve a compact band at the top of the column of adsorbent, so that during subsequent development the separation will hopefully proceed with formation of discrete bands.

Generally adsorption on to the adsorbent takes place most readily from non-polar solvents, such as light petroleum or toluene, and least from more highly polar solvents such as esters and alcohols. Frequently the most non-polar solvent for introducing the mixture on to the column and the initial solvent for chromatogram development are the same. Initial adsorption therefore takes place rapidly and development may if necessary be accelerated by progressively increasing the polarity of the eluting solvent using the 'eluotropic' series given below as a guide to sequential solvent selection; all these solvents have sufficiently low boiling points to permit ready recovery of eluted material:

Hexane, cyclohexane, carbon tetrachloride, trichloroethylene, toluene, dichloromethane, chloroform, diethyl ether, ethyl acetate, acetone, propanol, ethanol, methanol.

Rather than effecting a sharp change in solvent composition it is usual to introduce on to the column gradually increasing concentrations of the more polar solvent until a complete change has been effected. This may be carried out in practice by using successively mixtures of a non-polar and a polar solvent in which the proportions of the components are in a ratio of, say, 90:10, 70:30, 50:50, 30:70, 10:90, 0:100, or by continuously dripping the more polar solvent into the reservoir containing the non-polar solvent, which should be fitted with a stirring device so that the composition changes gradually.

The order in which components of a mixture are eluted from a column is related to their relative polarty. Thus with a mixture of two components of differing polarity, e.g. a hydrocarbon and a ketone, separation is achieved because the more polar ketone is adsorbed more strongly on the adsorbent and hence the hydrocarbon may be eluted with a relatively non-polar solvent; the ketone is then eluted by changing to a more polar solvent. The ease of elution of the adsorbate may be broadly in the following order:

Saturated hydrocarbons > alkenes, alkynes, aromatic hydrocarbons > esters, aldehydes and ketones > amines, alcohols, thiols > phenols, carboxylic acids.

In a comprehensive study of a mixture having unknown chromatographic characteristics it is frequently desirable to be initially guided in the selection of adsorbents and solvents from information obtained by t.l.c. analysis using alumina or silica gel on microscope slides. Only if these prove unsatisfactory would recourse be made to the other adsorbents.

It should be noted that the resolution obtained on a t.l.c. plate is rather better than would be obtained on a conventional adsorption column (see however other column techniques below) and hence further trials should be made with the various activity grades and with controlled solvent composition changes before the bulk of material is submitted to this type of separation. It is in these trials that careful attention to the chromatographic profile obtained from suitable analysis of the eluate fractions, and to the total recovery of material from the column, is so important.

Adsorption column preparation and loading. In order to obtain satisfactory results, the tube must be uniformly packed with the adsorbent; uneven distribution may lead to the formation of cracks and channels and to considerable distortion of adsorption band shapes. If there is any doubt concerning the uniformity of particle size of the adsorbent powder it should be sifted before use to remove the larger particles; fines are removed from the adsorbent using a sedimentation procedure immediately prior to column packing. In this the alumina or silica gel adsorbent is stirred into between five to ten times its volume of the selected solvent or solvent system, allowed to settle for five minutes and the supernatant liquor decanted off; the procedure is repeated until the supernatant liquid is clear.

As a rough guide the amount of adsorbent used should normally be 25–50 times the weight of the material to be separated. A slurry of the adsorbent in the solvent (approximately 1:10) is poured through a funnel into a clean dry column clamped vertically, in a position away from draughts or warm air currents from a radiator or electric oven, etc. The adsorbent will settle evenly and free of air bubbles if assisted by gentle tapping of the tube with a wooden rod. For packing large columns the slurry is best contained in a separatory funnel and stirred with a link-type stirrer while it is allowed to flow into the column. Solvent is removed via the tap fitted to the adapter at the column end and more slurry is added until the required length of column is obtained. Some workers recommend that a second column is fitted to the top of the first so that all the slurry can be added in one portion thereby yielding a more perfectly uniform column of adsorbent on settling. Fresh solvent is allowed to flow through the column under the hydrostatic pressure that is envisaged for subsequent chromatographic development, until no further settling is apparent. At no time during the column preparation nor in subsequent use should the level of liquid fall below the level of adsorbent.

The top of the column is frequently covered with a circle of filter paper or a layer of clean sand to prevent disturbance of the surface during subsequent loading. A suitably concentrated solution of the mixture is added from a pipette, the liquid is allowed to drain just to the surface of the adsorbent and the inside of the tube is rinsed with a small quantity of the solvent which is again allowed to drain just on to the column. Finally the column space above the adsorbent is filled with solvent and a dropping funnel filled with solvent is attached.

The subsequent chromatographic development, analysis of fractions and recovery of separated components is as described above.

The conventional technique just described has been widely used for many decades for the successful separation of functional isomers, and a number of examples of its use in preparative work are included in the following text. However, its drawbacks became very apparent following the observation that the resolution of mixtures on t.l.c. plates is far superior using the same adsorbents and solvent systems. Furthermore the period of time necessary to complete a column chromatogram may be lengthy, and band tailing may reduce the effectiveness of the separation. Three techniques have been developed to overcome these difficulties, namely *dry-column chromatography*, *flash chromatography*, and *dry-column flash chromatography*.

Dry-column chromatography. This technique has been developed by Loev and Goodman.[62] In this method alumina (100–200 μm; activity II or III, with incorporated fluorescent indicator) or silica gel (100–250 μm; activity III, with incorporated fluorescent indicator) (1) is packed dry (using an approximate ratio of 1 g adsorbate : 300 g of adsorbent) into a glass column (2 to 5 cm diameter) fitted with an adapter incorporating a sintered disc (Fig. 2.140) or into nylon tubing (2 to 5 cm diameter) which has been welded or sealed at the lower end and into which a piece of glass wool is then placed. The packing process in the case of a glass column is assisted by the use of an ultra-vibrator moved alongside the tubing; to fill the nylon tube the bag is filled to about one-third its length and then allowed to fall vertically from a distance of a few inches on to the bench surface to assist good packing; an ultra-vibrator may be used but it not essential and the column obtained is sufficiently rigid to be handled as a filled glass tube. During packing and until the chromatogram development is complete the tap in the glass assembly is kept open; with the nylon tubing the bottom end is perforated by a needle since the packing is retained by the glass wool plug.

The column may be loaded by the technique of dissolving the liquid or solid in the minimum volume of solvent, and by means of a capillary pipette distributing this solution evenly on to the top of the column and allowing the solvent to completely drain. An alternative, and indeed preferable, method is to dissolve the material to be chromatographed in a suitable volatile solvent (ether, light petroleum, dichloromethane) adding column adsorbent equal to about five times its weight, and evaporating the solvent. The loaded adsorbent is then added to the column top and packed into an even layer. With both techniques the column surface is covered with clean, acid-washed sand to a depth of between 0.5 and 1 cm.

The column is developed by gravity flow with a solvent head of between 3 to 5 cm. This is conveniently achieved by placing the solvent (2) in a dropping funnel, closing with a stopper, and with the outlet tube touching the surface layer of sand, carefully opening the stopcock. Solvent will slowly escape until the drop in pressure in the funnel prevents further flow. The funnel outlet is allowed to remain under the level of liquid in the column, since as the level falls air bubbles will rise into the funnel and allow further portions of solvent to escape. Development is complete when the solvent front reaches the bottom of the column (between 15 and 30 minutes).

Progress of the development is readily observable by viewing the nylon columns under ultraviolet light which penetrates the nylon covering; the results are less satisfactory with glass columns. The column is extruded from the glass assembly; with the nylon columns the tube is laid horizontally on a sheet of glass and the nylon covering slit lengthways with a razor blade. Detection of the zones and extraction of the separated components is as described previously (p. 217).

Notes. (1) The success of the dry-column technique for the resolution of mixtures is completely dependent on the use of the correct activity grade of adsorbent. It should not be assumed that the grade purchased is activity I, to which the appropriate amount of water could be added (see earlier). It is therefore essential to determine the activity by the use of the method previously described, or by using the elegant micro-method described by Loev and Goodman in their definitive paper.
(2) The solvent selected will be that which has proved successful with the trial examination on t.l.c. analysis using microscope slides. Preferably this should be a single solvent system. If a mixed solvent system is necessary Loev and Goodman suggest that the deactivated adsorbent should be mixed with about 10 per cent by weight of the solvent system and equilibrated by shaking before being used to prepare the dry column.

Flash chromatography. This technique has been described by W. C. Still *et al.*[63] and allows for the rapid separation (10 to 15 minutes) of mixtures of components having ΔR_F values greater or equal to 0.15 and with a sample loading of 0.01–10.0 g, although the latter is not regarded as being a limiting value. The essential features are the use of Silica gel 60, 40–63 μm (40–230 mesh), in a column through which the solvent is passed by the application of positive air pressure applied to the column head; resolution is sensitive to eluant flow rate, and with the solvent mixture ethyl acetate/light petroleum (b.p. 30–60 °C), this is 2.0 $\pm$ 0.1 in/minute (50.8 $\pm$ 2.5 mm). This solvent system is recommended as a good general purpose one, although for extremely polar compounds, acetone/light petroleum or acetone/dichloromethane are useful alternatives.

The apparatus consists of a chromatography column and a flow controller (Fig. 2.141). This figure is reproduced from the original literature report, to illustrate the description of the technique below, but the equipment is commercially available (e.g. Aldrich Chemical Co., May and Baker Ltd). The flow controller is a simple variable bleed device for precise regulation of the elution rate and is constructed from a glass/Teflon needle valve. Eluate fractions are collected in

Fig. 2.141

test tubes (20 × 150 mm), and separated components detected by t.l.c. in a manner similar to that described above for conventional column chromatography.

The *general procedure* is as follows.[63] First a low viscosity solvent system (e.g. ethyl acetate/light petroleum b.p. 30–60 °C) (1) is found which separates the mixture and moves the desired components on analytical t.l.c. to an R_F of 0.35 (2). If several compounds are to be separated which run very close on t.l.c., adjust the solvent to put the midpoint between the components at R_F 0.35. If the compounds are widely separated, adjust the R_F of the less mobile component to 0.35.

Having chosen the solvent, a column of appropriate diameter (see Table 2.13) is selected and a small plug of glass wool is placed in the tube connecting the

Table 2.13

Column diam. (mm)	Vol. of eluant (ml)	Sample typical loading (mg) $\Delta R_F \geqslant 0.2$	$R_F \geqslant 0.1$	Typical fraction size (ml)
10	100	100	40	5
20	200	400	160	10
30	400	900	360	20
40	600	1600	600	30
50	1000	2500	1000	50

stopcock to the column body (A). Two telescoping lengths of glass tubing (6 and 8 mm o.d.) make placement of the glass wool plug easy. Next a smooth 3-mm layer of 50–100 mesh sand is added to cover the bottom of the column and dry 40–63 μm silica gel is poured into the column in a single portion to give a depth of c. 140 mm. With the stopcock open, the column is gently tapped vertically on the bench top to pack the gel. Next a 3-mm layer of sand is carefully placed on the flat top of the dry silica gel bed and the column is clamped for pressure packing and elution. The solvent chosen as above is then poured carefully over the sand to fill the column completely. The needle valve (B) of the flow controller is opened all the way and the flow controller is fitted tightly to the top of the column and secured with strong rubber bands. The main air-line valve leading to the flow controller is opened slightly and a finger is placed fairly tightly over the bleed-port (C). This will cause the pressure above the adsorbent bed to climb rapidly and compress the silica gel as solvent is rapidly forced through the column. It is important to maintain the pressure until *all* the air is expelled and the lower part of the column is cool; otherwise the column will fragment and should be repacked unless the separation is a trivial one. Particular care is necessary with large diameter columns. The pressure is then released and excess eluant is forced out of the column above the adsorbent bed by partially blocking the bleed port (C). The top of the silica gel column should not be allowed to run dry. Next the sample is applied by pipette as a 20–25 per cent solution in the eluant to the top of the adsorbent bed and the flow controller is briefly placed on top of the column to push all of the sample into the silica gel (3). The solvent used to pack the column is ordinarily reused to elute the column. The walls of the column are washed down with a few millilitres of fresh eluant, the washings are pushed into the gel as before, and the column is carefully filled with eluant so as not to disturb the adsorbent bed. The flow controller is finally secured to the column and adjusted to cause the surface of the solvent in the column to fall 51 mm/minute. This seems to be the optimum value of the flow rate for most low viscosity solvents for any diameter with the 40–63 μm silica gel. Fractions are collected until all the solvent has been used (see Table 2.13 to estimate the amount of solvent and sample size). It is best not to let the column run dry since further elution is occasionally necessary. Purified components are identified as described above by t.l.c. If the foregoing instructions are followed exactly, there is little opportunity for the separation to fail.

Although fresh columns are used for each separation, the expense of large-scale separations makes it advantageous to reuse large diameter columns. Column recycling is effected by first flushing (rate = 51 mm/minute) the column with approximately 130 mm of the more polar component in the eluant (generally ethyl acetate or acetone) and then with 130 mm of the desired eluant. If the eluant is relatively non-polar (e.g. ≤ 10 per cent ethyl acetate/light petroleum), it may be more advantageous to use a flushing solvent (e.g. 20–50 per cent ethyl acetate/light petroleum) which is somewhat less polar than the pure high polarity component.

Notes. (1) Solvents were distilled prior to use. Thin-layer chromatograms were run on glass supported silica gel 60 plates (0.25-mm layer) (E. Mark No. 5765).
(2) If this R_F is given by a solvent having < 2 per cent of the polar component, a slightly less polar eluant is desirable. Thus if 1 per cent ethyl acetate/light petroleum gives a compound an R_F of 0.35 on t.l.c., the column is run with 0.5 per cent ethyl acetate.
(3) If the sample is only partially soluble in the eluant, just enough of the more polar com-

ponent is added to give complete dissolution. Large quantities of very polar impurities are best removed prior to chromatography so that excessive quantities of solvent or large increases in solvent polarity will be unnecessary for sample application.

Dry-column flash chromatography.   This technique has been developed from flash chromatography by L. M. Harwood.[64] The principal feature is that suction is applied to the column packing, and eluting solvents are added in predetermined volumes with the column being allowed to run dry before the next fraction is added. Furthermore the apparatus is both simple and of easy operation.

The *general procedure* is as follows.[64] The apparatus (Fig. 2.142) is set up for filtration using a porosity 3 cylindrical sinter. Table 2.14 gives guidelines for the choice of sinter size and amounts of silica, sample and solvent. Columns longer than 55 mm are neither practical nor necessary, since reduction in efficiency may be observed on large-scale set-ups.

To water aspirator

Fig. 2.142

Table 2.14

Sinter (size diam./ length, mm)	Silica (wt, g)	Sample	Solvent fraction (ml)
30/45	15	15–500 mg	10–15
40/50	30	0.5–3 g	15–30
70/55	100	2–15 g	20–50

Fill the sinter to the lip with Silica gel 60 (400–230 mesh), and apply suction pressing the silica with extra care at the circumference. Still pressing, level the surface, tapping the sides of the sinter firmly to obtain a totally level surface and a head space for the addition of solvents.

Under vacuum, pre-elute the column with the least polar combination of the required solvents in which the product mixture is readily soluble. If possible, use a single least-polar component (e.g. use pentane for pentane/ether gradient elution). If the silica has been packed correctly, the solvent front will be seen descending in a horizontal line. If channelling occurs, suck the column dry and repeat the packing procedure. Keep the surface of the silica covered with solvent during the pre-elution until solvent is seen passing into the receiver. Then allow the silica to dry under suction.

Load the sample mixture (dissolved in the minimum amount of the pre-elution system) evenly on to the surface of the silica and elute the products by adding successive portions of increasing-polarity solvent mixtures, allowing the column to be sucked dry after each addition (the silica surface is only slightly disturbed on addition of solvent). Generally a solvent gradient whereby the more polar component is increased by 5–10 per cent is the most convenient. Under these conditions, the product is usually eluted by that solvent mixture in which it would have R_F 0.5 on t.l.c. For quantities greater than 100 mg, elution is often accompanied by frothing on the underside of the sinter. With a little experience, separations of the same efficiency as t.l.c. are easily possible. There is minimal material loss and the technique is economical in time and solvents.

During prolonged chromatography with volatile solvents, atmospheric moisture may condense on the apparatus; although this does not affect the efficiency of the separation, it may be diminished by substituting less volatile solvents (e.g. hexane for pentane). No solvent is particularly disfavoured for this technique but combinations of pentane (hexane), ether, ethyl acetate and methanol are adequate for most separations. Low diffusion of the product bands during chromatography usually means that each component is eluted in one or two fractions, resulting in minimal loss of product in cross-contaminated fractions.

Exercises in conventional column adsorption chromatography. *Purification of anthracene.* Dissolve, with warming if necessary, 50 mg of crude anthracene (usually yellowish in colour) in 50 ml of hexane. Prepare a 20 cm column of activated alumina using a slurry containing 50 g of activated alumina (Grade II) in hexane in a glass chromatographic column of 1.8 cm diameter and 40 cm length fitted with a sintered disc, adapter with tap and separatory funnel. Add the anthracene solution to the top of the column from a separatory funnel, rinse the funnel with a little hexane and then develop the chromatogram with 200 ml of hexane. Examine the column from time to time in the light of an ultraviolet lamp (Section 2.3.6, p. 52). A narrow, deep blue fluorescent zone (due to carbazole) will be observed near the top of the column; the next zone down the column is a yellow non-fluorescent zone due to naphthacene, the intensity of which will depend upon the purity of the sample of anthracene used. The anthracene forms a broad, blue-violet fluorescent zone in the lower part of the column. Continue to develop the chromatogram until the anthracene begins to emerge from the column, reject the first runnings since these contain the less strongly adsorbed paraffin-like impurities and change to a clean receiver. Continue to elute the column with hexane until the removal of the anthracene is complete (1); the yellow zone should not reach to the bottom of the column. Concentrate the eluate fraction containing the anthracene under reduced pressure on a rotary evaporator to about 2 ml, cool the flask in an ice–salt bath and by means of ice-cold hexane quantitatively transfer the crystals and solution to a filter funnel. Wash the pure anthracene crystals with chilled hexane; the product (30 mg), which is fluorescent in daylight, has m.p. 215–216 °C.

Note. (1) When the anthracene band begins to emerge from the bottom of the column its elution may be accelerated by changing to a mixture of hexane:benzene (1:1) **(CAUTION)**. In this case the fractions containing the anthracene are evaporated to dryness under reduced pressure and then redissolved in hexane and concentrated to low bulk

221

as described in the main text. This additional operation is necessary owing to the solubility of anthracene in benzene.

*Separation of cholestenone from cholesterol.**

Place a mixture of 1.0 g of purified cholesterol and 0.2 g of copper(II) oxide in a test tube clamped securely at the top, add a fragment of Cardice in order to displace the air by carbon dioxide and insert a plug of cotton wool in the mouth of the tube. Heat in a metal bath at 300–315 °C for 15 minutes and allow to cool; rotate the test tube occasionally in order to spread the melt on the sides. Warm the cold residue with a few ml of benzene and pour the black suspension directly into the top of a previously prepared chromatographic column (1); rinse the test tube with a little more benzene and pour the rinsings into the column. With the aid of slight pressure (c. 3–4 cm of mercury), allow the solution to drain into the alumina column; stir the top 0.5 cm or so with a stout copper wire at frequent intervals to prevent blockage by the finely divided copper compounds. When all the black liquid has run in, there should be free flow without the necessity of further stirring. Continue the development with benzene until a distinctly yellowish diffuse zone approaches the bottom of the column; some 150 ml of liquid will have been collected. Now collect 5 ml fractions until the yellow band is completely removed. Evaporate each of these fractions separately; the earlier ones yield oils (giving a yellow 2:4-dinitrophenylhydrazone) and the later ones will crystallise upon rubbing (cholestenone). Continue the elution with a further 400 ml of benzene; the latter upon evaporation yields most of the cholestenone. Isolate the remaining cholestenone by continuing the elution with benzene containing 0.5 per cent of absolute ethanol until a dark brown band approaches the bottom of the column. Collect all the crystalline residues with the aid of a little light petroleum, b.p. 40–60 °C, into a small flask and remove the solvent. Dissolve the residue in 40–50 ml of hot methanol, add 0.2 g of decolourising carbon, filter through a small bed of alumina (6 mm × 6 mm), concentrate to about 20 ml and leave to crystallise overnight. The yield of cholestenone, m.p. 82 °C, is 0.5 g.

Note. (1) Prepare the column for chromatography by mixing 90 g of chromatographic alumina (Spence) with sufficient benzene to form a thin slurry when stirred. Pour this, stirring briskly, into a tube (40 cm long and 18 mm internal diameter) having a sintered glass disc, and rinse with a little more benzene. An even packed column, about 35 cm long, should result. Allow to drain until the supernatant benzene is within 1 cm of the alumina before adding the solution to be chromatographed. Under no circumstances should the column be allowed to drain so that the liquid level falls below that of the alumina.

---

* The experimental details were kindly supplied by Professor D. H. R. Barton, F.R.S. and Dr W. Rigby; because of the toxicity of benzene this chromatographic column should be set up in the fume cupboard.

Separation of (Z)-azobenzene from the (E)-isomer.

$$\underset{(E)}{\overset{\displaystyle Ph}{\underset{\displaystyle Ph}{\ddot{N}=\ddot{N}}}} \quad \rightleftharpoons \quad \underset{(Z)}{\overset{\displaystyle Ph \qquad Ph}{\ddot{N}=\ddot{N}}}$$

Dissolve 1.0 g of azobenzene [Expt 6.90; this is the (E)-form] in 50 ml of light petroleum, b.p. 40–60 °C, in a 200-ml beaker. Irradiate the solution for 30 minutes with ultraviolet light; this is conveniently carried out by supporting a Hanovia fluorescent lamp, model 16, about 13 cm above the surface of the liquid in the beaker. Meanwhile prepare a chromatographic column from 50 g of activated acid alumina (Grade I) as a slurry in light petroleum, b.p. 40–60 °C, to give a column of approximate dimensions 20 cm × 1.8 cm. After the column has been formed in this way, place a well-fitting filter paper at the top of the column and pour the solution, immediately after it has been irradiated, slowly down a glass rod on to the filter paper until the column is filled with liquid; take great care not to disturb the upper portion of the column. Develop the chromatogram with 100 ml of light petroleum, b.p. 40–60 °C. A sharp coloured band [(Z)-form], c. 2 cm in length, makes its appearance at the top of the column while a diffuse coloured region [containing the (E)-form] moves down the column. The upper portion of the column should be screened from light by wrapping it with black paper, held in position by a rubber band, during the development process; this will largely prevent the reconversion of the (Z)- into the (E)-form. Extrude the column and remove the coloured 2 cm band from the column and shake it with 150 ml of light petroleum, b.p. 40–60 °C, containing 1.5 ml of absolute methanol (1); filter off the alumina, with suction, and wash the filtrate with two 15 ml portions of water to remove the methanol present. Dry the light petroleum extract by shaking it with about 1 g of anhydrous sodium sulphate for 10 minutes, filter and evaporate the solvent under reduced pressure. The residual coloured solid, m.p. 71.5 °C, is practically pure (Z)-azobenzene. Its individuality and its purity may be confirmed by recording the ultraviolet absorption spectrum in ethanol solution as soon as possible after its isolation; (Z)-azobenzene has λ_{max} 281 nm, ε 5260, (E)-azobenzene has λ_{max} 320 nm, ε 21 300 in ethanol solution.

Note. (1) Alternatively the (Z)-isomer may be removed from the column by changing the eluting solvent to one of light petroleum, b.p. 40–60 °C, containing 1 per cent of methanol. The eluate fraction is then washed with water to remove the methanol in the manner described in the main text.

GAS–LIQUID CHROMATOGRAPHY (g.l.c.)
Providing the components are adequately volatile gas–liquid chromatography (g.l.c.) is perhaps the most powerful technique for the rapid and convenient analysis of the composition of mixtures of organic compounds.[65, 66] It is based upon the partition of components between a mobile gas phase and a stationary liquid phase retained as a surface layer on a suitable solid supporting medium.

The basic design of instrument is schematically illustrated in Fig. 2.143 and consists of a metal or glass column approximately 2–3 m in length and 2–4 mm internal diameter, which is in the form of a circular spiral of three or four turns, and packed with supporting medium impregnated with a stationary phase; the column is located in a temperature controlled oven. The mobile gas phase

Fig. 2.143

(usually but not necessarily nitrogen) enters the column at one end, which also incorporates an injection port for the introduction of the sample. The components of the sample are carried down the column, separation being dependent upon their individual partition coefficients between the stationary and the mobile phase, and emerge in turn into a detector system which is attached to the column end. The signal from the detector is suitably amplified and fed into a pen recorder. The qualitative (and if necessary the quantitative) composition of the mixture is assessed from an examination of the graphical traces so produced.

A simple, but typical chromatogram (Fig. 2.144) for a homologous mixture of ketones illustrates the powerful nature of this analytical tool. The chromatographic behaviour of each component of a mixture under a given set of experimental conditions is usually recorded in terms of its retention time (t_R). This represents the time required for the component to emerge from the column after injection. On the older-type instruments this is evaluated by measuring the distance on the pen recorder chart paper from the point of injection to the centre of the peak and dividing this by the chart speed. A point to note from this trace is that an increase in retention times leads in general to a broadening of the peaks. From a practical point of view this fact is of value in the interpretation of chromatograms of mixtures having an unknown number of components. If insufficient time has been allowed for the slowest running component to emerge and a second injection is made, the appearance of a peak which is anomalously broad in relation to its retention time strongly suggests that it represents a slow running component from a previous injection.

Provided that the detector is equally sensitive to each of the components of the mixture, and that therefore the ratio of peak areas is equal to the ratio of weights of components in the mixture (a reasonable assumption when com-

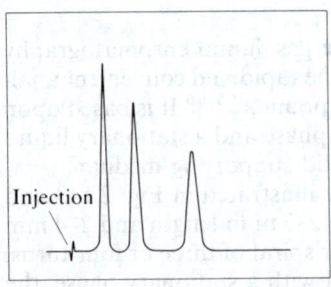

Fig. 2.144

pounds of the same functional type are being studied), and provided that all the components are volatile and have been eluted, the percentage of each component can be determined by measuring the area under each peak and expressing this as a proportion of the total area.

Thus, for a two-component mixture (A + B):

$$\text{Per cent A} = \frac{\text{area of peak A}}{\text{area of [peak A + peak B]}} \times 100, \text{ etc.}$$

If standard mixtures of known weight ratios of components do not give the same ratio of areas, the detector is not responding equivalently to each component. In this case a quantitative analysis of the mixture requires preliminary experimentation. One method is the use of an internal standard (for a more detailed discussion of other procedures, specialist texts should be consulted[65,66]). Thus for a two-component mixture (A + B), a third component C is selected as the internal standard. Mixtures of A + C and B + C are prepared in which the weights of each component in each mixture are known. The relevant chromatogram is recorded and the detector response factors calculated from the following relationships:

$$\frac{\text{area of A}}{\text{area of C}} = \frac{\text{wt of A}}{\text{wt of C}}; \therefore \text{ wt of A} = \frac{\text{area of A} \times \text{wt of C}}{\text{area of C}}$$

or

wt of A = response factor (A/C) × wt of C;

similary for B.

Known weights of C with the mixture (A + B) are mixed, the chromatogram recorded, the areas of all three peaks measured, and the areas of A and B converted into weights by use of the response factors, from which the percentages of A and B may be calculated.

The success of the analysis is dependent on the selection of a suitable internal standard. Some important criteria for such selection are as follows: (a) that it is not chemically reactive towards any of the components of the mixture; (b) that it has a retention time which gives a base line separation from the other components, including any impurities; (c) that the retention time is comparable with that of the components of the mixture; (d) that the peak profile is symmetrical and therefore does not exhibit either 'fronting' or 'tailing'; and (e) that the detector response to the internal standard is such that neither an excessive, nor a minute, weight of standard compared to the weight of mixture needs to be used.

Mention must be made of the use of an internal standard to monitor a reaction by g.l.c. analysis, and also to calculate the g.l.c. yield. Here, a known weight of the standard, inert to the reaction conditions and conforming to the other criteria of selection noted above, is added initially to the reaction mixture. In the case where samples of the mixture can be removed and loaded directly on to the column, the subsequent analysis presents no problem and may be deduced from the discussion above. In the case of samples which require evaporation of solvent prior to chromatographic examination, it is only necessary to ensure that the standard, and indeed the components to be analysed, do not volatilise under the conditions of concentration. If the samples require more involved solvent extraction procedures, then further experimentation is required to establish that the internal standard and the components are being equally and efficiently recovered.

The present generation of gas–liquid chromatographic instruments have many important design features, and the manufacturers' literature should be consulted for precise details (e.g. from Perkin-Elmer, Philips Analytical, Varian Associates, Waters Chromatography Division). However, some key features may be summarised briefly: (a) an oven design for subambient, isothermal, or temperature programming operations, which may be selected via a software key user interaction; (b) a versatile injection system for liquid or gas samples, suitable for packed columns or for capillary columns – these latter require what is termed a split/splitless injector with automatic vent control to allow only a minute sample to be loaded on to this type of column[67]; (c) a wide range of interchangeable detectors including flame ionisation, thermal conductivity, electron capture, and specific detectors for nitrogen, phosphorus and sulphur-containing compounds; (d) a video display unit recording all the parameters of the chromatographic run, and data handling software for specifying retention times and for the calculation of peak areas and detector responses; and (e) finally, interfacing with, for example, a Fourier Transform infrared spectrophotometer[68] or a mass spectrometer – in these cases identification of the components of the mixture may be by visual comparison with standard spectra, or by automatic searching with an appropriate database.

The examples of the use of g.l.c. in the preparative sections of this text employ the conventional packed columns, and provide the undergraduate with the opportunity of becoming familiar with the technique. The most usual *solid support material* is kieselguhr which is available from commercial suppliers in various standard particle sizes. Pretreated grades are available which either have been acid washed to remove acid-soluble materials from the support surface, or have been both acid and base washed to remove additionally any acidic organic contaminants. The polar character of the support material, which is due to the presence of surface hydroxyl groups, may be modified by treatment with dimethyldichlorosilane. Examples of two other types of support materials, namely GC Porasil and Poropak, are available from Waters, although other manufacturers market their own proprietary products. The former material is a spherical silica packing of controlled pore-volume and surface area. This packing material may be used either with or without a stationary phase. The other is a polymeric ethylvinyl–divinylbenzene which is available in a range of modified structures and may be used as column packing directly without the need for a stationary phase. Capillary columns are of fused silica, usually in lengths of 15, 25 or 50 m and of 0.20 or 0.32 mm internal diameter. The stationary phase is either adsorbed on to the inside surface, giving a layer thickness of 0.20 μm (e.g. Flexica columns, Pierce Chemical Co.), or chemically bonded on to the inside surface (e.g. Flexibond columns, Pierce Chemical Co.). The latter have a longer life, since they may be rinsed and reconditioned at appropriate intervals. Stationary phases of different polarities are available and this, coupled with the high efficiency (resolution) of the capillary system, makes these columns of very great value in routine analytical laboratories.

The choice of *stationary phase* will be influenced by the polar character of the components of the mixture. In general, mixtures with components of high polarity separate better on chromatography when the more polar stationary phases are used. The chromatographic separation of a mixture is judged to be successful if the peaks are well separated (i.e. good resolution) and the peak shape is symmetrical and lacking in extensive trailing. The oven temperature has to be related to the overall volatility of the components and this in turn also influences the choice of the stationary phase; for example, it would be useless to endeavour to separate a mixture of high boiling compounds on a column with a stationary phase of low thermal stability.

The range of stationary phases which is now available commercially is very extensive and offers an adequate choice of polar character and thermal stability. Table 2.15 lists some of the more generally useful stationary phases, their maximum operating temperatures and the classes of compounds for which satisfactory separations have most usually been found.

It is frequently desirable when assessing the purity of new compounds or evaluating the complexity of liquid mixtures to observe the chromatographic behaviour on more than one type of column-packing material, since rarely will two compounds have identical retention times on two substantially different types of stationary phases.

Table 2.15 A selection of commonly used stationary phases for gas–liquid chromatography columns

Stationary phase	Temperature limit (°C)	Solvent code*	Applications
Squalane	150	A	General; hydrocarbons, halogenated compounds
Benzylbiphenyl	100	A	
Apiezons	150–300	A	
Silicone GE SE-30	350	A	High temperature general use; when on silanised
Silicone GE SE-52	300		support material used for trimethylsilyl ethers of polyhydroxy compounds, polyamines, etc.
Carbowax, grades 550–20 M	100–200	B	Alcohols, aldehydes, ketones, ethers
Diglycerol	150	C	Alcohols, carbonyl compounds, etc.
Dinonyl phthalate	150	D	Alcohols, carbonyl compounds, etc.
Diethylene glycol succinate	200	B	Esters; general use with polar compounds

* Solvent code: these solvents have been found to be the most suitable for the preparation of solutions of the stationary phase prior to the addition of support material during the coating process. A = toluene; B = dichloromethane; C = methanol; D = acetone.

While the *resolution* obtainable for a given mixture is largely determined by the nature of the stationary phase, the length of the column, and the efficiency of the column packing process, some modifications to the appearance of the final chromatogram to enable a better evaluation of retention times and peak areas (if this is being done by the operator rather than by the software of the instrument) may be achieved by certain simple operations. The best resolution that a particular column is capable of achieving may only be realised with the lowest sample loading, e.g. of the order of 0.1 μl with conventional packed columns; it is for this reason that capillary columns which operate with a fraction of this amount are so efficient. Lowering the temperature of the column increases the retention times of all components and may make marginal improvements to the resolution of the peaks; this delayed emergence of components leads to broader peaks and this in turn may simplify the calculation of peak areas if these are being measured by the operator. Reduction of the flow rate of gas also causes an increase in retention time. The present generation of instruments have associated software as a means to attenuate peaks and make base-line corrections. They also incorporate temperature programming which enables the more volatile components of a mixture to be eluted with good resolution at an appropriate low initial column temperature; the column temperature is then raised at a predetermined rate to a higher level to allow the less volatile components to be eluted more quickly. The chromatogram profile obtained in this way is superior to that obtained at constant temperature (i.e. isothermally).

A provisional *identification* of the components of the mixture may be made from a comparison of the retention times with those obtained for the pure components, if available (e.g. a solvent or reactant used in the original preparation, etc.). Identification must however be confirmed by careful co-chromatography of the mixture with each of the suspected components, added in turn. An enhancement of the appropriate peak will confirm the presence in the mixture of the added component. It must be emphasised that the amount of reference compound added should be related to the amount suspected to be present in the mixture, and that several separate additions of say 20, 50 and 100 per cent should be planned. If care is not taken in this way the peak corresponding to a trace component whose identity is required may be swamped by that of an added reference compound and its identification made unreliable. Figure 2.145(a)–(e) illustrates a sequence of results which might be expected to arise from adding increasing amounts of component B to the mixture, consisting of a major component A with a trace impurity, the original chromatogram of which is shown in (a); (b) and (c) are chromatograms which could be expected on the addition of compound B, the suspected trace impurity, in 50 and 100 per cent of the amount of trace component, and the fact that the peak enhancement is proportional to the added component confirms that the trace component is in fact B; had B not been identical with the trace component, addition may have given a chromatogram such as (d) revealing the separation of the trace component from the added component B. The effect of adding too great a proportion of B is illustrated in (e), where it is clear that it cannot be established that the coincidence of the peaks of the trace impurity and the added reference compound B is exact.

Compounds which are not sufficiently volatile to be analysed directly by the g.l.c. technique, may often be converted into a volatile compound by some suitable *derivatisation* process. Although the conversion of polar functional groups (e.g. OH, SH, NH_2, NH, CO_2H) into the corresponding ethers, esters, or

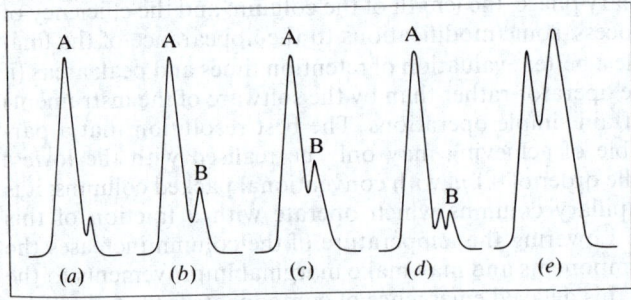

Fig. 2.145

N-alkyl and N-acyl derivatives has been widely used, the method of choice is that of trimethylsilylation. The procedure is widely applicable to natural products (e.g. carbohydrates, amino acids, steriods, alkaloids, polyhydroxyphenols, fatty acids, etc.), and has found great use in analytical and preparative laboratories which handle such compounds.

Among the more important trimethylsilylation reagents are: trimethylchlorosilane (TMCS); hexamethyldisilazane [$(Me_3Si)_2NH$, (HMDS)]; N-trimethylsilylimidazole (TMSI) (1); N,O-bis(trimethylsilyl)acetamide (BSA) (2); N,O-bis(trimethylsilyl)trifluoroacetamide (BSTFA) (3); and N-methyl-N-(t-butyldimethylsilyl)trifluoroacetamide (MTBSTFA) (4).

These reagents in the presence of a solvent (such as pyridine, acetonitrile, dimethylformamide) rapidly effect trimethylsilylation, often at room temperature, of hydroxyl, amino, imino, amido, thiol and carboxyl groups whether in mono- or polyfunctional compounds. Frequently a sample of the reaction mixture may be loaded directly on to the g.l.c. column, although in some cases removal of the small amount of solvent which is present may improve subsequent analysis of the chromatographic trace. Proprietary reagent formulations are available from Pierce Chemical Co., pioneers in this field and publishers of detailed procedures suited to particular groups of compounds.

The exercise described below, which has been used in the editors' laboratory, is illustrative of the analysis of monosaccharides as their trimethylsilyl derivatives formed in pyridine with a mixture of TMCS and HMDS. It should be noted with this method that most aldoses give two peaks corresponding to a mixture of anomeric derivatives, and ketoses give additional peaks which are thought to correspond to derivatives having different ring sizes. This problem may be largely overcome by the use of TMSI in pyridine as the silylating reagent. Comprehensive reviews have been published on the silylation of organic compounds and on other suitable derivatisation procedures for use in g.l.c. analysis.[69, 70]

Preparative gas chromatography. The purified fractions from a gas chromatographic column can in principle be collected by interposing between the column end and the detector a splitting device which diverts most of the effluent through suitably cooled traps in which the components are individually condensed. The microlitre sample injection employed in a conventional analytical chromatogram does not of course yield useful amounts of purified components. However, with the aid of longer columns (2, 4.5 or 9 m) of a somewhat larger diameter (c. 1 cm), larger sample loads (20–100 μl) and an automatic repeating cycle of sample injections, a truly preparative scale separation of components may be achieved.

Exercises in gas chromatography. In most laboratories which routinely use the gas chromatographic technique, a range of columns packed with a selection of stationary phases on appropriate support material will of course be available. Furthermore the range and sophistication of the available instruments is now very wide, each having particular merits in relation to the nature of the work being carried out. The following two exercises may therefore be regarded as only an introduction to some of the principal practical features of the technique including column preparation, qualitative and quantative analysis, and an example of derivatisation, as applied to monosaccharides.

Analysis of a mixture of ketones. (a) *Qualitative analysis.* The following instructions for the packing of a standard 1.5-m length, 4-mm diameter glass column with a 10 per cent dinonyl phthalate on Chromosorb W support can be taken as indicative of the general technique appropriate to the packing of a column.

Dissolve 1.5 g of dinonyl phthalate in 40 ml of acetone in a 250 ml round-bottomed flask and add 13.5 g of Chromosorb W, mesh 60/80, slowly and with swirling. Allow the mixture to stand for 2 hours with occasional gentle agitation, remove the solvent on a rotary evaporator under reduced pressure and finally heat the residue at 100 °C on a water bath for 1 hour under water-pump vacuum. Insert a plug of glass wool into the end of the shorter glass limb of the column and fit the metal connector. Attach the metal connector to a vacuum line and apply suction to compress the plug. Support the column in a suitable clamp and connect a glass funnel with a piece of rubber tubing to the longer glass limb which is held vertically. With the vacuum applied, pour successive small portions of the column-packing material into the column via the funnel and tap the glass spiral with a wooden rod to assist packing. When the column is full to within 5.5 cm from the open end, remove the funnel and insert a plug of glass wool to retain the packing. Condition the column for use by heating it for 24 hours in an oven held at 110 °C while a steady stream of nitrogen is passed through (1).

Fit two similarly prepared columns into the instrument according to the manufacturer's instructions and adjust the controls so that an oven temperature of 80 °C and a flow rate of 40 ml/min of the carrier gas is maintained. With the attenuator control of the instrument set at a relatively high value (say, 20×10^4, which represents a fairly low sensitivity) and with the recorder switched on, prepare to load the front column with a 1 μl sample of acetone in the following way. Insert the needle of a 1 μl micro-syringe into a sample of acetone, and withdraw the plunger to beyond the 1 μl calibrated mark. Remove the syringe and adjust the plunger lead to the mark, wipe the outside of the needle with paper tissue and insert the full needle length into the column via the injection port – some resistance will be felt during passage through the rubber septum. Simultaneously with the injection, activate the pen recorder by operating the jet polarity switch on the amplifier to indicate the injection time. After a few seconds' pause remove the syringe needle and evaporate residual acetone by operating the plunger a few times. Both during insertion and withdrawal of the needle some support should be given to it with the finger to prevent damage by bending.

Observe the peak obtained and estimate any necessary adjustments to the sensitivity controls to modify its size so that its maximum is between one-half and three-quarters of the chart paper width. Check this adjustment by injecting a further 1 μl sample and then repeat until a reproducible peak is obtained.

Finally increase the sensitivity setting by a factor of 10 and reduce the injection volume to 0.1 μl – similarly check the reproducibility of this injection by repeating several times (2).

Notes. (1) With columns having other stationary phases the temperature of the column conditioning process is usually 10–15 °C above that of the temperature at which the column will be subsequently operated; it should not exceed the temperature limit value quoted in Table 2.15 (p. 227).
(2) If difficulty is experienced in achieving the required reproducibility it is possible that the injection septum is leaking, thus allowing some of the injected sample to escape – a variation in retention times may also be noticed as a result of leakage of carrier gas. The septum should be replaced. Unreproducible results may also be due to leakage around the syringe plunger.

Inject 0.1 μl samples of each of the following straight chain aliphatic ketones; butan-2-one, pentan-3-one, pentan-2-one and heptan-3-one. After each injection clean the syringe by filling it with acetone and expelling the latter several times and finally evaporating the residual acetone as before.

Plot a graph of $\log_{10} t_R$ against the molecular weight or alternatively the carbon number (i.e. the number of carbon atoms in the molecule) of the ketone. Estimate the retention time of a ketone containing six carbon atoms and check your result by a suitable injection. Finally examine the chromatographic behaviour of 3-methylbutan-2-one, 4-methylpentan-2-one and 5-methylhexan-2-one.

(*b*) *Quantitative analysis.* Prepare a standard series of mixtures of pentan-2-one and heptan-3-one having the approximate composition 20:80, 50:50 and 75:25, w/w, by weighing accurately appropriate quantities into a semimicro tube.

Inject successively and in duplicate 0.1 μl samples of each of the mixtures into the column and record the chromatographic traces. Determine the areas under the peaks by both of the following procedures and compare the results with those given by the instrument software, if fitted.

1. Measure the heights of each peak from the extrapolated base line, and the width of the peak at half peak height; the approximate area is the product of these two values.
2. Trace each peak carefully on to another sheet of paper, cut out and weigh the area enclosed between the trace and the base line.

Using the relationships noted previously (p. 224) establish that there is an equal response by the detector to each component.

Inject a 0.1 μl sample of a given pentan-2-one and heptan-3-one mixture, measure the area of each peak and calculate the percentage composition as noted above.

Analysis of a monosaccharide. Prepare a column packing using acid-washed, silanised Chromosorb W as the solid support and a stationary phase (5%) of either Silicone GE-SE52 or OV17. These have a temperature tolerance of up to 300 °C, and are therefore also suitable for oligosaccharide analyses. A stationary phase of ethylene glycol succinate ester (EGS) may be used if the maximum temperature is 170 °C. Columns having a length of 1.5 m and a carrier gas (nitrogen) flow rate of 40 ml/min are satisfactory.

Place 10 mg of the monosaccharide in a small sample tube and by means of a graduated syringe add 0.25 ml of pyridine followed by 0.05 ml of HMDS and 0.02 ml of TMCS. (**CAUTION:** *This operation should be performed in a fume cupboard. The syringe should be carefully rinsed with dichloromethane and dried between use with each reagent*.) Stopper the sample tube with a plastic cap, shake the tube and allow to stand for 20 minutes. Centrifuge the tube at low speeds to compact the solid deposit and chromatograph a sample (0.1 μl) of the supernatant liquor. Several monosaccharides should be derivatised and chromatographed in this manner; an unknown sample (or mixture) should be supplied and its identity established with the aid of peak enhancement experiments.

If excessive tailing of the pyridine peak in the chromatographic trace causes difficulty in interpretation (e.g. with pentoses) proceed as follows. Transfer the supernatant liquid of the centrifuged solution by means of a dry dropper pipette to a small test tube and attach a suitable adapter which is fitted to a cooled vacuum trap and pump. Remove the pyridine under reduced pressure; continuous agitation of the tube in a water bath held at about 50 °C is advisable. Dilute the viscous residue with 0.5 ml of dichloromethane and re-chromatograph.

HIGH PERFORMANCE LIQUID CHROMATOGRAPHY (h.p.l.c.)

This technique has become, within the last decade, the prime analytical method for those compounds which are involatile or thermally unstable so that they are not amenable to g.l.c. analysis.[71-73] They include, to name a few groups only, the natural products (carbohydrates, steroids, alkaloids, peptides and amino acids, antibiotics, nucleosides, etc.), and the synthetic and naturally occurring compounds arising from research in the pharmaceutical, agricultural and food industries. Prior to the development of h.p.l.c., analysis of compounds within these groups was, for example, either by derivatisation followed by g.l.c. analysis, by quantitative t.l.c. or paper chromatographic anaylsis, or by tedious and time-consuming conventional liquid–solid or liquid–liquid (partition) chromatography.

In principle, h.p.l.c. arose from conventional liquid column chromatography, following the development of g.l.c. and realisation that it was a rapid and accurate analytical method. This led to a reappraisal of the liquid column chromatographic system, which in turn resulted in research developments in instrument design and in the manufacture of column-packing materials. These now have precise specifications to make them suitable for adsorption, normal and reversed phase partition, ion exchange, gel permeation, and more recently affinity chromatography.

The essential features of the equipment may be briefly summarised, but the catalogues of the manufacturers (e.g. Beckmann, Macherey-Nagel through Field Analytical, Philips Analytical, Perkin-Elmer, Varian, Waters, etc.) and the comprehensive textbooks and monographs should be consulted for detailed discussions of equipment design and theoretical aspects of this technique.

The column consists of a stainless steel tube which for analytical work may be up to 20 cm in length and in the region of 5 mm internal diameter, although the tendency in recent times is for even narrower columns. For preparative operations the column length may be up to 50 cm and the internal diameter up to 20 mm. The column-packing material is of a very uniform particle size, and may be, for example, 10, 7, 5 or 4 μm; the smaller particle sized material is preferred for analytical applications, and the larger particle size for preparative work. The uniformity of grading of these fine particles is one particular feature upon which the resolving power of the system depends. Furthermore, because of the expense of these column-packing materials, it is usually advisable to fit a guard column, which is interposed between the injector system and the analytical column; its function is to remove particulate materials and components (impurities) which would be strongly retained on the packing material of the analytical (or preparative) column and hence ultimately decrease its lifetime. The guard column is much shorter in length (say 2.5 cm for analytical columns) and is usually packed with the same column-packing material as the main column, but of larger particle size.

A consequence of the column dimensions, and the physical nature of the packing material, is the need for a pulse-free pressurised liquid flow through the column. The pressures are maintained by a suitable pumping system up to a value of 6000 p.s.i. The liquid flow through the column may be from 1 μl/min to 10 μl/min according to the conditions required for the analysis; 2 μl/min appears the normal level for many analyses. Preparative columns are often operated at very much higher flow rates. The software provided with the instrument allows for convenient selection and alteration of flow rates. The solvent delivered to the top of the column (or guard column) by the pump, must be of high purity (available for example from BDH, Pierce, Romil, etc.) and usually degassed; the latter is required to prevent air-locks being generated during the course of the chromatographic run. Most instruments are equipped with a solvent delivery system which allows for gradient elution; with some instruments the facility for handling up to four different solvent sources is available.

The injector system is often of the loop type. Here the main solvent delivery tube to the column top is by-passed in a loop, which may be isolated and depressurised, and injected with sample via a septum. After injection the liquid in the loop is released into the main solvent flow. The loop volume is of comparable capacity to the injection volume. Most instruments are designed for auto-sampling in the case of multiple analyses, the operation being controlled by the instrument software.

The column eluant flows into the detector, usually via a splitter device. The most commonly used dual detector system is a u.v. detector in association with a refractive index detector. The latter is required for those compounds which do not absorb in the u.v., although the versatility of the former is extended by employing a variable wavelength detector which will operate over the entire range, e.g. 190–700 nm. At the present time it has been suggested that over 70 per cent of analyses are carried out with these two detectors. Other detectors are available for more specialised purposes, and include a fluorescence detector and a conductivity detector. The signal from the detector is, as with the g.l.c. instruments, fed to a video display unit; the software of the instrument allows all the details of the chromatographic run to be recorded, the peaks attenuated without the need of repeating the analysis, the results calculated, etc. In the case of preparative procedures the column eluant may be collected in appropriate fractions, and the separated compounds isolated in the usual manner.

Interfacing of h.p.l.c. with other instruments, for the purpose of identifying the structure of unknown compounds in a mixture, is a major development which might be expected over the next few years.

Given that the above instrumental features are appropriate, the success of an analysis or preparative separation depends on the selection of the column-packing material. The manufacturers (e.g. Macherey-Nagel, Merck, Phase Separation, Pierce, Waters, etc.) supply their own trademarked products, and the following is a broad general account of the types of packing available. The most widely used packings are based on silica, where the particle shape may be spherical or irregular. The importance of particle size distribution is noted above. The silica surface may be unmodified, for use in adsorption chromatography, or modified with chemically bonded phases, for use in normal phase and reversed phase chromatography. Polar bonded phases such as cyano-propyl, or amino-propyl, with a less polar solvent system, would be an example of the former; non-polar bonded phases such as a C_{18} alkyl chain, or an aromatic residue, with a more polar solvent (e.g. acetonitrile/water) would be an example of the latter. Some of these polar bonded phases may also be used in ion exchange chromatography. An important development in chemically bonded phases is the attachment to the silica surface of an optically pure (chiral) moiety, which results in the facility for analysis (and preparative separation) of enantiomers (see Section 5.19). Glass beads with a porous silica surface, alumina, cellulose acetate and polyamide packing materials are also available from the manufacturers, but are not as widely used as the aforementioned.

The packing of h.p.l.c. columns is a far more critical operation than is the case with g.l.c. columns, and a considerable degree of expertise and experience is required. No attempt is made here to describe the detailed operation. In principle, however, the column-packing material is made into a stable suspension in a suitable solvent (e.g. mixtures of toluene:dioxane and cyclohexane). Such suspensions must not agglomerate or sediment during packing, and this is achieved by adjusting the viscosity of the solvent mixture, and by ultrasonic treatment. The suspension is transferred to a reservoir, to one end of which is attached a pulse-free pumping system, and to the other the empty column. Careful control of the pressure during the subsequent packing operation and during the column conditioning process by solvent elution is essential in order to obtain columns of excellent resolving characteristics.

2.32 STORAGE OF SAMPLES

After preparation, all pure products should be stored in suitable clearly and permanently labelled containers. The bulk samples of stable solids and liquids may be stored in screw-capped and ground glass stoppered bottles respectively. Hygroscopic samples, or those liable to decomposition by contact with atmospheric moisture, should either be stored in a desiccator or in a bottle which is sealed by painting over the closure with molten paraffin wax. Where there is a possibility of photochemical decomposition of chemicals it is general good practice to keep them out of direct sunlight and to store them in brown bottles.

Small specimens of all products, including reaction intermediates isolated from reaction sequences, and particularly samples of fractions isolated as the result of lengthy chromatographic or other purification procedures, should invariably be retained for reference purposes. The commercially available straight-sided specimen tubes with polyethylene plug seals, which are available in a range of sizes, are suitable in the case of solid samples. It is usually advantageous to label them with the name and a code reference to enable physical data (elemental analysis, spectroscopic information, etc.) to be located in the laboratory note-books.

Liquid samples may be sealed in specially prepared glass ampoules for pro-longed storage or when specimens need to be sent away, for example, for elemental analysis. Ready-made ampoules may be obtained commercially in a range of sizes or can be made by the following method. A short length of moderately thick-walled tubing of internal diameter suited to the sample size is cleaned by immersion in a narrow cylinder containing chromic acid cleaning mixture (Section 2.2), thoroughly washed with distilled water, followed by a little acetone and is then dried by passing a current of warm air through it. One end is then sealed off in the blowpipe flame as in Fig. 2.146(a). The constriction is then made by carefully rotating the tube in a small blowpipe flame; it is important that the wall of the tubing remains uniformly thick at this point. The sample is most con-veniently introduced into the ampoule from a capillary pipette or micro-syringe needle passing through a protective tube previously inserted into the ampoule via the constriction (Fig. 2.146(b)). It is advantageous for the end of this protect-ive tube, which should not reach below the final level of the liquid to be intro-duced, to be slightly rounded off. This prevents any of the liquid sample which adheres to the syringe or capillary pipette from contaminating the outer side of the protecting tube. This may be safely withdrawn without danger of introduc-ing sample on to the inside surface of the constriction. The ampoule may then be sealed off in the usual way using a small flame.

If the sample needs to be sealed in an atmosphere of nitrogen, the inlet of the protecting tube is connected to a low-pressure supply of the gas when the syringe or pipette has been withdrawn. When all the air has been displaced the tube may be slowly withdrawn and the ampoule sealed. Filled ampoules con-taining volatile samples should be thoroughly chilled in a suitable cooling bath

(a)   (b)

Fig. 2.146

before such air displacement, and before sealing. Care must be exercised when sealed ampoules, particularly those which have been stored for some time, are reopened, and the precautions outlined in Section 2.3.2, p. 38 should be noted.

DETERMINATION OF PHYSICAL CONSTANTS

2.33 DETERMINATION OF MELTING POINT — MIXED MELTING POINTS

A pure crystalline organic compound has, in general, a definite and sharp melting point; that is, the melting point range (the difference between the temperature at which the collapse of the crystals is first observed and the temperature at which the sample becomes completely liquid) does not exceed about 0.5 °C. The presence of small quantities of miscible, or partially miscible, impurities will usually produce a marked increase in the melting point range and cause the commencement of melting to occur at a temperature lower than the melting point of the pure substance. The melting point is therefore a valuable criterion of purity for an organic compound.

A sharp melting point is usually indicative of the high purity of a substance. There are, however, some exceptions. Thus a eutectic mixture of two or more compounds may have a sharp melting point, but this melting point may be changed by fractional crystallisation from a suitable solvent or mixture of solvents. The number of exceptions encountered in practice is surprisingly small, hence it is reasonable to regard a compound as pure when it melts over a range of about 0.5 °C (or less) and the melting point is unaffected by repeated fractional crystallisation.

The experimental method in most common use is to heat a small amount (about 1 mg) of the substance in a capillary tube inserted into a suitable melting point apparatus and to determine the temperature at which melting occurs. The *capillary melting point tubes are prepared* either from soft glass test tubes or from wide glass tubing (c. 12 mm diameter).* A short length of glass tubing or glass rod is firmly fused to the closed end of the test tube. The test tube (or wide glass tubing) must first be thoroughly washed with distilled water to remove dust, alkali and products of devitrification which remain on the surface of the glass, and then dried. The closed end of the test tube is first heated while being slowly rotated in a small blowpipe flame; the glass rod or tube is simultaneously heated in the same manner. When the extremities of both pieces of glass are red hot, they are firmly fused togeher, twisting of the joint being avoided, and then removed momentarily from the flame until the seal is just rigid enough that no bending occurs. The test tube is then immediately introduced into a large 'brush' flame, so that a length of about 5 cm is heated, and the tube is rotated uniformly in the flame. When the heated portion has become soft and slightly thickened as the result of the heating, the tube is removed from the flame and, after a second or two, drawn, slowly at first and then more rapidly, as far apart as the arms will permit (or until the external diameter of the tube has been reduced to 1–2 mm).

---

* Pyrex glass is preferable, but this requires an oxygen-gas blowpipe for manipulations. Suitable melting point tubes may be purchased from dealers in scientific apparatus or chemicals. It is, however, excellent practice for the student to learn to prepare his own capillary tubes.

If the operation has been successfully performed, a long capillary of regular bore throughout most of its length will be obtained. The long thin tube is then cut into lengths of about 8 cm by touching it lightly with a file and then tapping gently with the flat portion of the file; after a little practice, no difficulty should be experienced in dividing the long capillary into suitable lengths without crushing the fragile tubing. It will be found that a short length of tubing ('glass spindle'), sufficiently rigid to act as a holder, will remain attached to the test tube after the long capillary has been cut off. The operation may then be repeated. When the test tube becomes too short to be handled at the open end, a piece of glass tubing or rod may be fused on, in the manner previously described, to act as a convenient handle. In this way a large number of capillary tubes may be prepared from one test tube. One end of each of the capillary tubes should be sealed by inserting it horizontally into the extreme edge of a small Bunsen flame for a few seconds, and the capillary tube rotated meanwhile; the formation of a glass bead at the end of the tube should be avoided. The prepared capillary tubes should be stored either in a large specimen tube or in a test tube closed with a cork.

The capillary tube is then filled as follows. About 25 mg of the dry substance is placed on a glass slide or upon a fragment of clean porous porcelain plate and finely powdered with a clean metal or glass spatula, and then formed into a small mound. The open end of the capillary tube is pushed into the powder, 'backing' the latter, if necessary, with a spatula. The solid is then shaken down the tube by tapping the closed end on the bench or by gently drawing the flat side of a triangular file (a pocket nail file is quite effective) along the upper end of the tube. The procedure is repeated until the length of lightly-packed material is 3–5 mm, and the outside of the tube is finally wiped clean.

The two principal types of *melting point apparatus* in common use are those in which heating of the capillary tube is by means of a heated liquid bath and those in which heating is carried out in, or on, an electrically-heated metal block.

A convenient form of bath is the *Thiele apparatus* (Fig. 2.147) in which the liquid (1) is contained in a tube with a closed bent side-arm. The thermometer (2) is located as shown, being inserted through a cork, a section of which having

Fig. 2.147   **Fig. 2.148**

been cut away (see inset) so that the thermometer scale is visible, and also to allow for expansion of the air in the apparatus.

Notes. (1) The safest and most satisfacory bath liquids are the highly stable and heat-resistant Silicone oils. A cheaper alternative is medicinal paraffin; it has a low specific heat, is non-flammable and is non-corrosive, but it can only be safely heated to about 220 °C; above this temperature it begins to decompose and becomes discoloured. Concentrated sulphuric acid has been suggested for use as a melting point bath fluid but is *not* recommended.

(2) For melting point (and also for boiling point) determinations, it is convenient to use thermometers which have been calibrated by partial immersion to a distance marked on the stem (30 mm is suitable for melting points, 80 mm for boiling points). If a thermometer calibrated by total immersion is used, an error is introduced resulting from the cooling of the mercury thread which is not heated in the apparatus but is exposed to the cooler laboratory atmosphere. The necessary *stem correction* to be added to the observed melting point to give the 'corrected' value is given by the expression $0.000\,16\,N(t_1 - t_2)$, where N is the length of the exposed mercury thread in degrees, t_2 is the mean temperature of the exposed mercury thread determined on an auxiliary thermometer placed alongside with its bulb at the middle of the exposed thread, and t_1 is the observed temperature on the thermometer scale.

The filled melting point tube is attached to the lower end of the thermometer in such a way that the substance is at the level of the middle of the mercury bulb (which has previously been wetted with the bath liquid); the moistened capillary is then slid into position. Providing that the length of capillary tube above the level of the bath liquid exceeds the length immersed, advantage is taken of the surface tension of the bath liquid to hold the melting-point tube in position by capillary attraction. A thin rubber band prepared by cutting narrow rubber tubing may be used to attach the capillary tube near its open end to the thermometer; alternatively the tube may be held in position securely with the aid of fine wire. The thermometer, with capillary attached, is inserted into the centre of the main tube of the Thiele apparatus.

On heating the bent side-arm, the heated liquid circulates and raises the temperature of the sample in such a way that no stirring of the bath liquid is required.

The melting point apparatus is heated comparatively rapidly with a small flame until the temperature of the bath is within 15 °C of the melting point of the substance, and then slowly and regularly at the rate of about 2 °C per minute until the compound melts completely. The temperature at whch the substance commences to liquefy and the temperature at which the solid has disappeared, i.e. the melting point range, is observed. For a pure compound, the melting point range should not exceed 0.5–1 °C; it is usually less. Any sintering or softening below the melting point should be noted as well as any evolution of gas or any other signs of decomposition.* If the approximate melting point is not known, it is advisable to fill two capillaries with the substance. The temperature of the bath may then be raised fairly rapidly using one capillary tube in order to determine the melting point approximately; the bath is then allowed to cool about 30 °C, the second capillary substituted for the first and an accurate determination made.

---

* A substance which commences to soften and pull away from the sides of the capillary tube at (say) 120 °C, with the first appearance of liquid at 121 °C, and complete liquefaction at 122 °C with bubbling, would be recorded as m.p. 121–122 °C (decomp.), softens at 120 °C.

It should be noted that a second determination of the melting point should not be made as the bath liquid cools by observing the temperature at which the molten material in the capillary tube solidifies, or by reheating the bath after the solidification has occurred. This is because, in many cases, the substance may partially decompose, and in some instances it may undergo a change into another crystalline form possessing a different melting point. A freshly-filled capillary tube should always be employed for each subsequent determination. Substances which sublime readily are sometimes heated in melting point capillaries sealed at both ends. For compounds which melt with decomposition, difficulties sometimes arise in the melting point determination; it is best to insert the capillary tube into the bath when the temperature is only a few degrees below the melting and decomposition point of the material. This avoids decomposition, with consequent lowering of the melting point, during the time that the temperature of the bath liquid is being raised.

A liquid heating bath may be dispensed with by the use of an apparatus employing *electrical heating*. An electrically heated aluminium or copper block is convenient for this purpose. The essential features of the apparatus are shown in Fig. 2.148. The large hole in the centre is for a thermometer and the three smaller holes are for the melting-point capillaries which can be observed simultaneously through a suitable magnifying eye-piece; the melting point tubes are suitably illuminated by one, or two (as shown), side lights. Although the conventional mercury thermometer is still used in many laboratories, the current design of apparatus (Gallenkamp, Jencons, etc.) employs a digital display of temperature from a platinum resistance thermometer which responds more readily to temperature changes. Electrothermal Engineering have designed an electric melting-point apparatus with a digital display, incorporating a hold-switch which enables the operator to 'freeze' the display at the moment of melting and hence achieve precision measurement while viewing the sample. The rate of heating is controlled from the front of the apparatus which allows for both rapid heating and variable control. A water-cooled plug is available for lowering the temperature of the heating block between tests should this be necessary.

The *Kofler hot bench* (C. Reichert Optische Werke AG, available through Cambridge Instruments Ltd), illustrated in Fig. 2.149, consists of a metal alloy band with a corrosion-free steel surface, 36 cm long and 4 cm wide, heated electrically at one end, the other end remaining unheated to give a moderate almost constant temperature gradient. Fluctuations in the mains voltage are compensated for by a built-in stabiliser. The graduations cover the range 50 to 260 °C in 2 °C. Provision is made for variations in room temperature by adjustment on the

Fig. 2.149

reading device moving over the scale. The current must be switched on for about 40 minutes before the apparatus is required and the latter should be surrounded by a screen to protect it from draughts.

The hot bench should be calibrated before use with the aid of several of the substances supplied by the manufacturers. The test substances include: azobenzene, m.p. 68 °C; benzil, m.p. 95 °C; acetanilide, m.p. 114.5 °C; phenacetin, m.p. 134.5 °C; benzanilide, m.p. 163 °C; *p*-acetamidophenyl salicylate (salophene), m.p. 191 °C; and saccharin, m.p. 228 °C. The melting point is readily determined by sprinkling a few small crystals of the substance on the hot bench; these may be moved along the bench by the brass lancet attached to the reading device. Usually a sharp division occurs between the solid and liquid, and the temperature corresponding to the line of demarcation is read off on the scale. For maximum accuracy, the apparatus should be recalibrated with two test substances with melting points close to that of the unknown. The procedure is clearly rapid and is very useful for substances which tend to decompose upon gradual heating. It is important not to allow any of the molten substance to remain on the corrosion-resistant steel surface for long periods; it should be wiped away with paper tissues immediately after the experiment.

The *microscope hot stage type* of melting point apparatus (C. Reichert Optische Werke AG), which is essentially an electrically-heated block on a microscope stage, is of particular value when the melting point of a very small amount (e.g. of a single crystal) has to be determined. Further advantages include the possibility of observation of any change in crystalline form of the crystals before melting. The main features of the apparatus are shown in Fig. 2.150. The apparatus also incorporates a polariser which facilitates the observation of melting. The rate of heating is controlled by means of a rheostat; the temperature may be measured with a mercury thermometer (two are usually provided covering the ranges 20 to 230 °C and 120 to 350 °C), or with a platinum resistance thermometer and a digital display unit. A microscope *cold stage* is of similar construction and enables melting points in the range −50 to +80 °C to be determined; cooling below 0 °C is provided by a supply of liquid carbon dioxide.

MIXED MELTING POINTS

In the majority of cases the presence of a 'foreign substance' will lower the melting point of a pure organic compound. This fact is utilised in the so-called mixed melting point test for the identification of organic compounds. Let us suppose that an organic compound X having a melting point of 140 °C is suspected to be *o*-chlorobenzoic acid. Its identity may be established by performing a melting point determination on a mixture containing approximately equal weights of X

Fig. 2.150

and of an authentic specimen of o-chlorobenzoic acid (A). If the melting point of the mixture is 140 °C, then X is o-chlorobenzoic acid, but if the melting point is is depressed by several degrees A and X cannot be identical. It is recommended that at least three mixtures containing, say, 20 per cent X + 80 per cent A: 50 per cent X + 50 per cent A; and 80 per cent X + 20 per cent A be prepared, and the melting points be determined.

Cases may arise in which the melting points of certain mixtures are higher than the individual components, e.g. if an addition compound of higher melting point is formed or if the two compounds are completely soluble in the solid state forming solid solutions. The mixed melting point test therefore, although a great practical value, is not infallible and should accordingly be used with reasonable regard to these possibilities.

EXERCISES IN THE DETERMINATION OF MELTING POINTS AND MIXED MELTING POINTS
The melting points of pure samples of the following compounds are determined using a melting point bath or an electrically heated apparatus, in the manner detailed above. The correct melting points of the pure substances are given in parentheses.

1. p-Nitrotoluene (54 °C) or azobenzene (68 °C).
2. 1-Naphthol (96 °C) or catechol (104 °C) or benzil (95 °C).
3. Benzoic acid (122 °C) or 2-naphthol (123 °C) or urea (133 °C).
4. Salicylic acid (159 °C) or phenylurea (mono) (148 °C).
5. Succinic acid (185 °C) or p-tolylurea (mono) (180 °C).
6. p-Nitrobenzoic acid (239 °C) or s-diphenylurea (242 °C).

By working in the above order, it will not be necessary to wait for the apparatus to cool between consecutive determinations.

In order to gain experience in the determination of mixed melting points the following simple experiment should be carried out.

Determine the melting point of pure cinnamic acid (133 °C) and pure urea (133 °C). Approximately equal weights (c. 50 mg) of the two compounds are placed on a clean porous porcelain tile. These are now ground together and intimately mixed with the aid of the flat side of a micro-spatula. The melting point tube filled with this mixture is placed in the melting-point apparatus alongside melting-point tubes filled with each of the two components. In this way careful observation of the melting behaviour of the mixture and of the pure components will clearly show the considerable depression of melting point.

Similar experiments may be carried out on a mixture of benzoic acid (122 °C) and 2-naphthol (123 °C), or a mixture of acetanilide (114 °C) and antipyrin (113 °C).

2.34 DETERMINATION OF BOILING POINT

When reasonable amounts of liquid components are available (> 5 ml) the boiling point is readily determined by slowly distilling the material from a pear-shaped flask in an apparatus assembly shown in Fig. 2.98, and recording the temperature at which the bulk of the compound distils. Due attention should be paid to the experimental procedure which was discussed in detail in Section 2.24.

For smaller quantities of liquid compounds (0.5–3.0 ml) the material should be distilled in the apparatus assembly shown in Fig. 2.99.

When only minute quantities of liquid are available, either of the two *micromethods for the determination of the boiling point* may be used.

Method 1. (*Siwoloboff's method*, 1886). Two tubes, closed at one end, are required; one, an ordinary melting point capillary, 90–110 mm long and 1 mm in diameter, and the other, 80–100 mm long and 4–5 mm in diameter. The latter may be prepared from 4–5 mm glass tubing and, if desired, a small thin bulb, not exceeding 6 mm in diameter, may be blown at one end. A small quantity of the liquid, 0.25–0.5 ml (depending upon the boiling point), is placed in the wider tube, and the capillary tube, with sealed end uppermost, is introduced into the liquid. The tube is then attached to the thermometer by a rubber band (Fig. 2.151) and the thermometer is immersed in the bath of a melting-point apparatus. As the bath is gradually heated there will be a slow escape of bubbles from the end of the capillary tube, but when the boiling point of the liquid is attained a rapid and continuous escape of bubbles will be observed. The reading of the thermometer when a rapid and continuous stream of bubbles first emerges from the capillary tube is the boiling point of the liquid. Unless the temperature is raised very slowly in the vicinity of the boiling point of the liquid, the first determination may be slightly in error. A more accurate result is obtained by removing the source of heat when the rapid stream of bubbles rises from the end of the capillary tube; the speed at which bubbles are given off will slacken and finally, when the last bubble makes its appearance and exhibits a tendency to suck back, the thermometer is read immediately. This is the boiling point of the liquid because it is the point at which the vapour pressure of the liquid is equal to that of the atmosphere. As an additional check on the latter value, the bath is allowed to cool a few degrees and the temperature slowly raised; the thermometer is read when the first continuous series of bubbles is observed. The two thermometer readings should not differ by more than 1 °C. It should, however, be remembered that the Siwoloboff method gives trustworthy results only for comparatively pure liquids; small amounts of volatile impurities such as ether or water may lead to boiling points being recorded which approximate to those of the volatile component.

Method 2. (*Emrich's method*). A capillary tube about 10 cm long and of about 1 mm bore is used. One end is drawn out by means of a micro flame into a capillary with a very fine point and about 2 cm long as in Fig. 2.152(*a*). Such a capillary pipette may also be constructed by suitably drawing out soft glass tubing of 6–7 mm diameter. The tube (*a*) is then dipped into the liquid of which the boiling point is to be determined; the liquid will rise slowly by capillary attraction, and the tube is removed when the liquid has filled the narrow conical portion. The capillary end is then sealed by merely touching with a minute flame. A small air bubble is formed in the point of the capillary; it should be examined with a lens to make sure that it is not too large. A convenient size is 1–3 mm long (Fig. 2.152(*b*)). The prepared capillary tube is then attached to a thermometer as in a melting point determination (Section 2.33) and slowly heated in a Thiele apparatus. The capillary is best observed with a lens. When the bubble enlarges (as in Fig. 2.152(*c*)) and begins to exhibit signs of upward motion, the flame is removed or considerably lowered. The temperature at which the bubble reaches the surface of the bath liquid is the boiling point of the liquid. The bath is allowed to

Fig. 2.151   Fig. 2.152

cool about 10 °C below the first observed boiling point, then slowly heated again, and a second determination of the boilng point is made.

EXERCISES IN BOILING POINT DETERMINATION
The following pure liquids offer a convenient selection of compounds having a range of boiling points: (a) carbon tetrachloride (77 °C); (b) ethylene dibromide (132 °C) or chlorobenzene (132 °C); (c) aniline (184.5 °C); and (d) nitrobenzene (211 °C). The boiling points may be determined using conventional distillation apparatus assemblies. The compounds could also be employed to give practice in the determination of boiling point by the Siwoloboff or Emrich methods.

2.35 DETERMINATION OF MOLECULAR WEIGHT

Mass spectrometry offers the most refined method for the evaluation of the molecular weight of those compounds having vapour pressures higher than 0.1 mmHg at 350 °C. With instruments of high resolving power, the molecular weight is obtained to an accuracy of ± 5 p.p.m. These accurate molecular weights may be used to deduce possible molecular formulae with the aid of Mass and Abundance Tables[74] which list the accurate mass (up to 500) of all likely combinations of C, H, O and N (see also Section 3.3).

Occasionally the characterisation of an organic compound by means of an approximate molecular weight determination may be useful. Methods based upon ebullioscopic or cryoscopic procedures are often too time consuming for routine use. However, the high freezing point depression of camphor permits molecular weights to be determined rapidly and with reasonable accuracy (between 1% and 5%) using an ordinary melting-point apparatus (*Rast's camphor method*).

Support a small clean test tube (e.g. 75 × 10 mm) in a hole bored in a cork so that it will stand conveniently on the pan of a balance. Weigh the tube. Introduce about 50 mg of the compound of which the molecular weight is to be determined and weigh again. Then add 500–600 mg of pure, resublimed camphor (e.g. the micro-analytical reagent) and weigh again. Stopper the test tube loosely and melt the contents by placing it in an oil bath previously heated to about 180 °C;

stir the liquid with a platinum wire, but do not heat the liquid for more than 1 minute or camphor will sublime from the solution. Allow to cool, transfer the solid to a clean watch glass and powder the solid. Introduce some of the powder into a thin capillary tube of which the closed end is carefully rounded; press the solid down into the closed end with the aid of a platinum wire or with a closed capillary tube of smaller diameter. The height of the solid should not exceed 2 mm. Determine the melting point of the mixture in a liquid melting point bath using, preferably, a 100–200 °C thermometer graduated in 0.1 or 0.2 °C, or in an electrically-heated apparatus (Section 2.33). Good illumination and very careful control of the rate of heating is essential. The melting point is taken as that temperature at which the last fragment of solid disappears. To make sure that the mixture is homogeneous repeat the melting-point determination with a second sample; if the two differ appreciably, prepare a new mixture. Then determine the melting point of the original camphor. The difference in melting points gives the depression of the melting point of camphor caused by the addition of the compound. The molecular weight M can then be calculated from the formula:

$$M = \frac{K \times w \times 1000}{\Delta T \times W}$$

where K is the molecular depression constant of camphor (39.7), w is the weight of the compound, W is the weight of the camphor and ΔT is the depression of the melting point.

Note. The solute concentration should be above 0.2 M; in dilute solution K increases from 39.7 to about 50.

The Rast camphor method, although very simple, is nevertheless liable to some limitations. One serious difficulty is that the melting point of camphor is itself rather high and this may lead to decomposition of the compound whose molecular weight is to be determined. Another difficulty is the limited solubility of many classes of compound in liquid camphor, and this severely restricts its general applicability. Some useful alternative solvents, having high molar freezing point depression constants are given in Table 2.16.

Table 2.16 Solvents for molecular weight determination by depression of freezing point

Compound	Melting point (°C)	Molar depression constant
Cyclohexanol	24.7	42.5
Camphene	49	31
Cyclopentadecanone	65.6	21.3
Bornylamine	164	40.6
Borneol	202	35.8
cis-4-Aminocyclohexane-1-carboxylic acid lactam	196	40

2.36 DETERMINATION OF OPTICAL ROTARY POWER

Compounds which rotate the plane of polarised light around its axis, whether they are in the gaseous, liquid or molten state, or in solution, are said to be *optically active*. This property arises from the lack of certain elements of symmetry in the molecule (i.e. a centre, a plane or an n-fold alternating axis of symmetry) with the result that the molecule and its mirror image are non-superimposable. Although first observed with compounds having one or more chiral carbon atoms (i.e. a carbon substituted with four different groups), optically active compounds having chiral centres including atoms of silicon, germanium, nitrogen, phosphorus, arsenic, sulphur, etc., have also been prepared. Molecular dissymmetry, and hence optical activity, also arises in molecules, such as certain substituted biphenyls, allenes, etc., which have chiral axes or chiral planes rather than chiral atoms as such. The study of optically active coordination complexes is a more recent, important and expanding field of study.

Some of these topics have been elaborated in Section 1.2, where the reader is referred to the many excellent monographs and articles which are available, and which fully explore the current aspects of the stereochemistry of molecules and the importance of stereochemical considerations of appropriate reaction processes. This section is devoted to the experimental determination of optical rotatory power.

When a beam of monochromatic light is passed through a crystal of Iceland spar, two beams are transmitted, each vibrating in one plane which is perpendicular to the other. A *Nicol prism* is composed of two sections of Iceland spar so cut, and again sealed with Canada balsam, that one of these rays is refracted to the side (this is absorbed by the black surroundings of the prism) so that the light which finally passes through the prism is vibrating in one plane only. This light is said to be *plane polarised*. This polarised light is allowed to pass through another Nicol prism similarly orientated and the light viewed from a point remote from, but in line with, the light source. It will be found now that on rotating the second prism the field of view appears alternately light and dark and the minimum of brightness follows the maximum as the prism is rotated through an angle of 90°; the field of view will appear dark when the axes of the two prisms are at right angles to one another. The prism by which the light is polarised is termed the *polariser*, and the second prism, by which the light is examined, is called the *analyser*.

If, when the field of view appears dark, a tube containing a solution of an optically active compound is placed between the two prisms, the field lights up; one of the prisms must then be turned through a certain angle α before the original dark field is restored. Since the plane of vibration of polarised light may have to be rotated either clockwise or anti-clockwise, it is necessary to observe a convention to designate the direction of rotation. When, in order to obtain darkness, the analyser has to be turned clockwise (i.e. to the right), the optically active substance is said to be *dextrorotatory*, or ($+$); it is *laevorotatory*, or ($-$), when the analyser must be rotated anti-clockwise (i.e. to the left).

The obvious disadvantage of the above simple instrument (*polarimeter*) is the difficulty of determining the precise 'end-point' or the point of maximum darkness. The human eye is a poor judge of absolute intensities, but is capable of matching the intensities of two simultaneously viewed fields with great accuracy. For this reason all precision polarimeters (such as those obtainable from Bellingham and Stanley Ltd) are equipped with an optical device that divides the field into two or three adjacent parts (half-shadow or triple-shadow polarimeter) such that when the 'end-point' is reached the sections of the field become of the same intensity. A very slight rotation of the analyser will cause one part to become lighter and the other darker. The increase in sensitivity so attained is

245

Fig. 2.153

illustrated by the fact that an accuracy of at least 0.01° is easily obtained with the use of an 'end-point' device, whereas with the unaided eye the settings are no more accurate than 4–5°.

A half-shadow polarimeter (*Lippich type*) is illustrated diagrammatically in Fig. 2.153. Here two polarised rays are produced by means of the main Nicol prism P and a small Nicol prism P′; the latter covers half the field of the larger polariser P and its plane of polarisation is slightly inclined to that of P. The angles between the planes of polarisation may be altered by a slight rotation of the polariser P. Upon rotating the analyser A, a position will be found at which one beam will be completely, the other only partially, extinguished; the one half of the field of view will therefore appear dark, while the other will still remain light when viewed with the eyepiece E as in Fig. 2.154(*a*). Upon rotating the analyser A still further, a second position will be found at which only the second beam will be extinguished and the field will have the appearance shown in (*c*). When, however, the analyser occupies an intermediate position, the field of view will appear of uniform brightness (as in *b*), and this is the position to which the analyser must be set, and the reading from the circular scale S which also incorporates a vernier carefully noted.

In Fig. 2.153, B is a collimator, T the trough (shown without cover) which houses the polarimeter tube C, E the eyepiece and D the heavy support stand for the apparatus. N is a device for moving P and thus altering the 'half-shadow angle'; this has the advantage of increasing the intensity of light which is transmitted by the polarising prism and this may be essential when the optical activity of coloured solutions is to be determined. On the other hand the reproducibility with which the accurate position of uniform brightness of field may be ascertained diminishes.

Two forms of polarimeter tube are shown in Fig. 2.155. The common type (*a*) consists of tube of thick glass with accurately ground ends: the tube is closed by

(*a*)   (*b*)   (*c*)

(*a*)

(*b*)

Fig. 2.154   **Fig. 2.155**

means of circular plates of glass with parallel sides, which are pressed against the ends of the tube by means of screw caps. The caps must not be screwed so tightly as to cause strain in the glass end plates as this would cause a rotation; the glass plates at the end must be clear and free from finger marks and the exposed surface must be dry. In a modification, the tube is surrounded by a jacket to permit the circulation of water at constant temperature by means of a pump. Tube (b) has a cup opening in the centre of the glass tube; the glass plate ends are in this case sealed on to the tube with a cement which is inert to most solvents.

The unit of length in polarimetry is 1 dm, hence the tubes are generally made in lengths which are fractions or multiples of this quantity, e.g. 0.5, 1, 2 or 4 dm. The tube bore is usually 8 mm and hence the capacity of the 1 dm tube is 5.02 ml. Frequently 'neat' liquids or solutions having volumes in the range 1–2 ml need to be examined for rotatory power and in these cases polarimeter tubes (usually 0.5 dm) having a tube bore of 3 or 5 mm are employed. Polarimeter tubes having smaller bores (1, 2 or 4 mm) are available, together with other sizes of tubes, from Bellingham and Stanley Ltd. Jacketed tubes are available from Optical Activity Ltd.

The filling of these polarimeter tubes requires some attention to detail. Tubes which have the wider bore can conveniently be of the centre-filling type (Fig. 2.155(b)) and it is merely necessary to carefully pour the solution or liquid, which must be completely free of suspended particles or droplets of immiscible solvents, into the central opening, carefully rocking the tube to disperse air-locks; both the tube and the liquid should be at the temperature of the laboratory before filling and the final level of liquid should be within the central cup. The end-filling tubes (Fig. 2.155(a)), no matter which size of bore, require a little practise to avoid the presence of air-bubbles in the finally sealed tube which would hamper the field of view. Having ensured that the tube and the circular glass plates (which should be handled by the edges) are clean and dry, one of the ends of the tube is securely capped. The tube is then placed vertically on a bench surface, sealed end downwards, and the tube is filled nearly to the top of the open end with liquid. In the case of the narrow-bore tube a capillary pipette is used. In both cases, to avoid air-locks, the end of the pipette is carefully lowered to the bottom of the tube (without touching the glass end plate surface) and raised as the liquid is allowed to flow into the tube. The final stage of filling requires that the liquid is allowed to flow slowly from the pipette or capillary pipette, until the liquid surface is just 'proud' of the glass tube; the circular glass plate is then slid horizontally into position, as the pipette is withdrawn, sweeping surplus liquid away and providing a seal free of air bubbles. The cap is carefully screwed into position and the filled tube placed horizontally in the polarimeter trough T.

For accurate work it is essential to determine the 'zero' position of the instrument with the empty polarimeter tube; its position in the polarimeter and the exact position of the end plates should be registered by suitable markings. The readings on the circular scale (using the vernier) should be noted on about ten successive determinations in which the analyser prism is returned to the position of uniform brightness both from a clockwise and an anti-clockwise direction and the results averaged. The solution in the filled tube, exactly orientated as previously, is then examined and the average of ten successive readings of the position of uniform brightness from clockwise and anti-clockwise direction determined. Subtraction gives the optical rotation of the liquid or solution.

The magnitude of the optical rotation depends upon: (i) the nature of the substance; (ii) the length of the column of liquid through which the light passes; (iii) the wavelength of the light employed; (iv) the temperature; and (v) the concentration of the optically active substance, if a solute. In order to obtain a measure of the rotatory power of a substance, these factors must be taken into account. As a rule the wavelength employed is either that for the sodium D line, 5893 Å (obtained with a sodium vapour lamp) or the mercury green line, 5461 Å (produced with a mercury vapour lamp provided with a suitable filter). The temperature selected is 20 °C or that of the laboratory t °C. The *specific rotation for a neat active liquid* at a temperature t for the sodium line is given by:

$$[\alpha]_D^t = \frac{\alpha}{ld}$$

where α is the angular rotation, l is the length of the column of liquid in decimetres and d is the density at a temperature t. The *specific rotation for a solution of an optically active substance* is likewise given by:

$$[\alpha]_D^t = \frac{100\alpha}{lc} = \frac{100\alpha}{lpd}$$

where l is the length of the column of liquid in decimetres, c is the number of grams of the substance dissolved in 100 ml of the solution, p is the number of grams of the substance dissolved in 100 g of the solution and d is the density of the solution at the temperature t. In expressing the specific rotation of a substance in solution, the concentration and the solvent used (the nature of which has an influence on the rotation) must be clearly stated.

The *molecular rotation* is given by the expression:

$$[M]_D^t = [\phi]_D^t = \frac{[\alpha]_D^t \times M}{100} = \frac{\alpha}{lc \, (\text{mol}/100\,\text{ml})}$$

where M is the molecular weight. For example, natural camphor has $[\alpha]_D^{20} - 44.3°$ (c 3.6 in EtOH) and $[M]_D^t - 67.3°$.

Measurement of optical rotation, particularly if many samples have to be examined, is often more convenient using one of the various types of photoelectric polarimeters now available commercially. These detect the balance point electronically, and the optical rotation at a single wavelength may be read on a micrometer scale or may be displayed on a digital readout. Some instruments display a warning light if the solution being examined does not allow sufficient transmission of light for a rotation to be measured (i.e. in the case of dark coloured solutions). These instruments (e.g. from Bellingham and Stanley, Optical Activity Ltd, or from Japan Spectroscopic Co. Ltd) can measure the optical rotation of a few milligrams of sample with high accuracy. Most usually the light source is a soldium lamp, but the option of using any of the major mercury emission wavelengths from a mercury lamp is available.

Automatic spectropolarimeters are available for the measurement of optical rotation as a function of wavelength (in the region 180–700 nm), enabling optical rotatory dispersion (ORD) curves to be recorded. Models are also available (e.g. Japan Spectroscopic Co. Ltd) for the measurement of circular dichroism (CD) curves in the wavelength region of 180–1000 nm, and 700–2000 nm. Authoritative accounts of the value of ORD and CD data in studies on the structure of organic molecules should be consulted for more information about these techniques.[75]

2.37 DETERMINATION OF REFRACTIVE INDEX

The refractive index of a liquid is a frequently quoted physical constant, which together with the boiling point, and chromatographic and spectroscopic features, provide the means which aid the characterisation of organic liquids.

The refractive index of a liquid is conveniently determined with an Abbé refractometer (available from, for example, Bellingham and Stanley Ltd). This refractometer has the advantage that it requires only a drop or so of the sample and the refractive index (usually from 1.3000 to 1.7000) may be read directly to 0.001 and by estimation to 0.0001.

The principle of a refractometer is the observation of the 'critical angle' for total reflection between glass of high refractive index (e.g. flint glass, n_D 1.75) and the substance to be examined. The glass is in the form of a right-angled prism upon the hypotenuse face AB of which the compound to be investigated is placed as a thin film (about 0.15 mm thick) and then covered with a second similar prism (Fig. 2.157). The face AC of the prism plays a part in the refraction of the light, and it is the angle of emergence (α) from this face which is measured, the scale of the instrument being, however, divided to read the refractive index directly. The ray shown in Fig. 2.156 and in Fig. 2.157 is that which enters the face AB at grazing incidence and corresponds to the edge of the dark part of the field of view observed in the telescope of the instrument (Fig. 2.158). The direction of the ray after entering the face AB depends upon its wavelength, and thus the scale of the refractive index will vary with the light employed. That selected is for the sodium emission line 589.3 nm (the D line), and in high-precision instruments a monochromatic light source is employed. Routine laboratory instruments incorporate an optical compensating system which enables a white light source to be used.

To determine the refractive index of a liquid at 20 °C,* water from a thermostatic bath is circulated through the jacket which surrounds the two prisms, until the temperature on the thermometer of the instrument has remained steady for at least 10 minutes. The hinged prism is unclamped and both prism surfaces are wiped carefully with paper tissue suitable for optical surfaces. An appropriate amount of liquid (usually only a drop) is applied to one prism surface and the prisms reclamped. The field of view through the telescope is observed and the control knob adjusted so that coincidence of the borderline between the light

Fig. 2.156

Fig. 2.157

---

*Bellingham and Stanley provide detailed instructions on the operation of the Abbé refractometer; the above account is a general outline of the procedure.

249

Fig. 2.158

and dark parts is located on the cross-wires. The refractive index is read directly from a scale which is observed through a second telescope. Immediately the determination has been completed, the prisms are again separated and carefully wiped clean using a tissue moistened with a suitable solvent. The accuracy of the instrument may be checked by measuring the refractive index of distilled water ($n_D^{10°}$ 1.3337, $n_D^{20°}$ 1.3330, $n_D^{30°}$ 1.3320, $n_D^{40°}$ 1.3307).

REFERENCES

1. *Hazards in the Chemical Laboratory* (1986). 4th edn. Ed. L. Bretherick. London: The Royal Society of Chemistry.
2. *Guide to Safe Practices in Chemical Laboratories* (1987). London: The Royal Society of Chemistry.
3. *Safety Measures in Chemical Laboratories* (1981). 4th edn. London: HMSO.
4. *Laboratory Hazards Bulletin* (1981). Ed. S. Templer. Nottingham: The Royal Society of Chemistry (published monthly).
5. *Safety Representatives and Safety Committees* (1977). London: Health and Safety Executive.
6. *First Aid at Work* (1981). Health and Safety Series Booklet (HS(R)11. HMSO.
7. *Safe Under Pressure* (1987). London: The British Oxygen Co.
8. L. Bretherick (1985). *Handbook of Reactive Chemical Hazards*, 3rd edn. London: Butterworths.
9. N. I. Sax (1984). *Dangerous Properties of Industrial Materials*, 6th edn. New York: Reinhold.
10. *The Sigma-Aldrich Library of Chemical Safety Data* (1985). Ed. R. E. Lenga. Wisconsin: Sigma-Aldrich Corp.
11. *Compendium of Safety Data Sheets for Research and Industrial Chemicals* (1985). Eds L. H. Keith and D. B. Walters. Weinheim: VCH.
12. *Threshold Limit Values for Chemical Substances and Physical Agents in the Workroom Environment* (1985). Cincinnati: ACGIH (updated annually).
13. *Occupational Exposure Limits* (1985). HSE Guidance Note EH 40/85. London: HMSO.
14. Ethel Browning (1965). *Toxicity and Metabolism of Industrial Solvents*. Amsterdam: Elsevier.

250

15. *Handbook of Toxicology* (1955–58). Vols I–IV. Philadelphia: Saunders.
16. *The Carcinogenic Substances Regulations* (1967). No. 879. London: HMSO.
17. N. I. Sax (1981). *Cancer-Causing Chemicals.* New York: Reinhold.
18. *Guide to Recognising and Handling Carcinogens in a Veterinary Laboratory* (1975). Weybridge MAFF Central Laboratory.
19. *Safety in Biological Laboratories* (1977). Eds E. Hartree and V. Booth. London: The Biochemical Society.
20. *The Asbestos Regulations* (1969). No. 690. London: HMSO.
21. A. J. Gordon and R. A. Ford (1972). *The Chemists Companion.* New York: Wiley-Interscience, p.451.
22. A. M. Phipps and D. N. Hume (1968). *J. Chem. Ed.,* **45**, 664.
23. R. E. Randeau (1966). *J. Chem. Eng. Data,* **11**, 124.
24. R. L. Augustine (1965). *Catalytic Hydrogenation: Techniques and Application in Organic Synthesis.* London: Arnold; M. Hudlicky (1984). *Reduction in Organic Chemistry.* Chichester: Horwood; P. N. Rylander (1985). 'Hydrogenation methods', in *Best Synthetic Methods.* Eds A. R. Katritzky, O. Meth-Cohn and C. V. Rees. London: Academic Press.
25. C. A. Brown and H. C. Brown (1966). *J. Org. Chem.,* **31**, 3989.
26. R. F. C. Brown (1980). 'Pyrolytic methods in organic chemistry', in *Organic Chemistry: A Series of Monographs.* Ed. H. H. Wasserman, Vol. 41. New York: Academic Press.
27. J. M. Coxon and B. Halton (1986). *Organic Photochemistry.* London: Cambridge University Press; N. J. Turro (1981). *Modern Molecular Photochemistry.* New York: Benjamin-Cummings; A. Padwa (1979–87). *Organic Photochemistry,* Vols 1–9. New York: Dekker; *Photochemistry in Organic Synthesis.* Ed. J. D. Coyle. Royal Society of Chemistry. Special publication No. 57 (1986).
28. See reference 21, p. 351.
29. A. Schönberg, G. O. Schenck and O. A. Neumüller (1968). *Preparative Organic Photochemistry.* Berlin: Springer-Verlag.
30. S. D. Cohen, M. V. Mijovic, G. A. Newmann and E. Pitts (1967). *Chem. Ind. (London),* 1079.
31. *Organic Photochemical Syntheses* (1971). Ed. R. Srinivasan. New York: Wiley-Interscience, Vol. 1; P. Margaretha (1982). 'Preparative organic photochemistry', in *Topics in Current Chemistry.* Ed. J-M Lehn, Vol. 103; *Synthetic Organic Photochemistry* (1984). Ed. W. M. Horspool. New York: Plenum.
32. A. J. Fry (1972). *Synthetic Organic Electrochemistry.* New York: Harper and Row; D. K. Kyriacou (1981). *Basics of Electroorganic Synthesis.* New York: Wiley-Interscience; *Recent Advances in Electroorganic Synthesis* (1987). Ed. S. Torii. Studies in Organic Chemistry 30. Amsterdam: Elsevier.
33. G. Dryhurst and P. J. Elving (1967). *Anal. Chem.,* **39**, 607.
34. H. O. House (1972). *Modern Synthetic Reactions,* 2nd edn. California: Benjamin, p. 145.
35. L. Brandsma (1971). *Preparative Acetylenic Chemistry.* Amsterdam: Elsevier.
36. D. F. Shriver and M. A. Drezdzon (1986). *The Manipulation of Air-Sensitive Compounds,* 2nd edn. New York: Wiley.
37. Aldrich Chemical Company (1983). *Handling Air-Sensitive Reagents.* Technical Information Bulletin AL-134.
38. G. W. Kramer, A. B. Levy and M. M. Midland (1975) in *Organic Syntheses via Boranes.* Ed. H. C. Brown. New York: Wiley.
39. *Dictionary of Organic Compounds* (1982). 5th edn. Ed. J. Buckingham. Supplements 1–4, 1983–86. New York: Chapman and Hall.
40. *Handbook of Chemistry and Physics* (1986–87). 67th edn. Ed. R. C. Weast. Florida: CRC Press Inc.
41. (a) R. S. Tipson (1965). 'Sublimation', in *Technique of Organic Chemistry.* Ed. A. Weissberger. Interscience, Vol. IV, p. 661,

(b) N. D. Cheronis (1954). 'Distillation, sublimation and extraction', in *Technique of Organic Chemistry*. Ed. A. Weissberger. New York: Interscience, Vol. VI, p. 84.

42. G. Broughton (1956). 'Freeze drying', in *Technique of Organic Chemistry*, 2nd edn. Ed. A. Weissberger. New York: Interscience, Vol. III, p. 831.

43. S. Glasstone and D. Lewis (1963). *Elements of Physical Chemistry*, revised 2nd edn. London: Macmillan, pp. 379–82.

44. L. C. Craig and D. Craig, 'Laboratory extraction and countercurrent distribution'. Reference 42, p. 149.

45. E. Knell (1963). *Handbook of Laboratory Distillation*, 2nd edn. Ed. E. C. Lumb. Amsterdam: Elsevier.

46. C. S. Carlson and J. Stewart (1965). 'Extractive and azeotropic distillation', in *Technique of Organic Chemistry*, 2nd edn. Ed. A. Weissberger. New York: Wiley-Interscience, Vol. IV, p. 49.

47. L. H. Horsley (1973). 'Azeotropic Data-III', in *Advances in Chemistry Series* No. 116. American Chemical Society Applied Publication.

48. F. E. Williams (1965). 'Ordinary fractional distillation', in *Technique of Organic Chemistry*, 2nd edn. Ed. A. Weissberger. New York: Interscience, Vol. IV, p. 299.

49. F. L. J. Sixma and H. Wynberg (1964). *A Manual of Physical Methods in Organic Chemistry*. New York: Wiley, p. 107.

50. R. S. Tipson, *Distillation under Moderate Vacuum*. Reference 48, p. 511.

51. E. S. Perry, *Distillation under High Vacuum*. Reference 48, p. 535.

52. P. Ridgeway Watt (1963). *Molecular Stills*. London: Chapman Hall.

53. H. G. Cassidy (1963). 'Fundamentals of chromatography', in *Technique of Organic Chemistry*. Ed. A. Weissberger. New York: Interscience, Vol. X.

54. *Chromatography* (1974). 3rd edn. Ed. E. Heftmann. New York: Reinhold.

55. B. L. Karger, L. R. Snyder and C. Horvath (1973). *An Introduction to Separation Science*. New York: Wiley-Interscience.

56. A. Braithwaite and F. J. Smith (1985). *Chromatographic Methods*, 4th edn. London: Chapman Hall.

57. J. C. Touchstone and M. F. Dobbins (1983). *Practice of Thin-Layer Chromatography*, 2nd edn. New York: Wiley.

58. B. Fried and S. H. Sherma (1986). *Thin-Layer Chromatography*, 2nd edn. Chromatographic Science Series, Vol. 35. New York: Dekker.

59. H. Brockmann and H. Schodder (1941). *Chem. Ber.*, **74B**, 73.

60. R. Hernandez, R. Hernandez and L. R. Axelrod (1961). *Anal. Chem.*, **33**, 370.

61. H. H. Strain (1945). *Chromatographic Adsorption Analysis*, 2nd edn. New York: Interscience.

62. B. Loev and M. M. Goodman (1967). *Chem. Ind. (London)*, 2026.

63. W. C. Still, M. Kahn and A. Mitra (1978). *J. Org. Chem.*, **43**, 2923; for details of a similar technique see D. F. Taber (1982). *J. Org. Chem.*, **47**, 1351.

64. L. M. Harwood (1985). *Aldrichimica Acta*, **18**, 25.

65. J. A. Perry (1981). *Introduction to Analytical Gas Chromatography*. Chromatographic Science Series, Vol. 14. Ed. G. Cazes. New York: Dekker.

66. *Modern Practice of Gas Chromatography* (1985). 2nd edn. Ed. R. L. Grob. New York: Wiley-Interscience.

67. (a) M. L. Lie, F. J. Yang and K. D. Bartle (1984). *Open Tubular Column Gas Chromatography*. New York: Wiley.
(b) K. Grob (1986). *Classical Split and Splitless Injection in Capillary Gas Chromatography*. *Chromatographic Methods*, Eds W. Bertsch, W. G. Jennings and R. E. Kaiser, Heidelberg: Verlag.

68. W. Herres (1987). *Capillary Gas Chromatography – Fourier Transform Infrared Spectroscopy*. *Chromatographic Methods*, Eds. W. Bertsch, W. G. Jennings and R. E. Kaiser. Heidelberg: Verlag.

69. J. Drozd (1981). 'Chemical derivatisation in gas chromatography' in *Journal of Chromatography Library*, Vol. 19. Amsterdam: Elsevier.

252

70. A. E. Pierce (1979). *Silylation of Organic Compounds*. Pierce Chemical Company; see also the 1987 Pierce Handbook and General Catalogue.
71. R. J. Hamilton and P. A. Sewell (1982). *Introduction to High Performance Liquid Chromatography*, 2nd edn. London: Chapman Hall.
72. *Techniques of Liquid Chromatography* (1982). Ed. C. F. Simpson. Chichester: Wiley.
73. *Practice of High Performance Liquid Chromatography* (1986). Ed. H. Engelhardt. Berlin: Springer-Verlag.
74. (a) J. H. Benyon and A. E. Williams (1963). *Mass and Abundance Tables for Use in Mass Spectrometry*. New York: Elsevier.
   (b) J. Lederberg (1964). *Computation of Molecular Formulae for Mass Spectrometry*. San Francisco: Holden Day.
75. For example: W. Klyne (1960). 'Optical rotatory dispersion and the study of organic structures', in *Advances in Organic Chemistry*. Eds R. A. Raphael, E. C. Taylor, H. Wynberg. New York: Interscience, Vol. 1, p. 239; *Optical Rotatory Dispersion and Circular Dichroism in Organic Chemistry* (1967). Ed. G. Snatzke. London: Heyden and Son; P. Crabbe and A. C. Parker (1972). 'Optical rotatory dispersion and circular dichroism', in *Techniques of Chemistry*. Eds A. Weissberger and B. W. Rossiter. New York: Wiley, Vol. 1, Part IIIc, p. 183.

CHAPTER 3
SPECTROSCOPIC METHODS AND THE INTERPRETATION OF SPECTRA

INTRODUCTION

Spectroscopic methods are used routinely by the organic chemist to provide information about the nature of chemical substances. Early in their undergraduate course students will be expected to record infrared and ultraviolet spectra of compounds they prepare in the laboratory and will be provided with proton and carbon-13 nuclear magnetic resonance spectra and mass spectra of these and other compounds. The use of these techniques is an essential adjunct to current practice of organic chemistry.[1] This chapter provides an introduction to the use of these techniques in organic chemistry; in addition, spectroscopic data are provided in Chapters 5 to 8 for many of the compounds whose preparations are described. Students should thus develop the habit of recording such spectra routinely and checking the key features against the correlation tables given in this chapter and in the appendices. The use of spectroscopic information alongside that derived from chemical investigation in order to determine the structure of unknown organic compounds is described in Chapter 9.

The scope and power of spectroscopic techniques has been increased substantially in recent years by major advances in instrumentation.* The more significant of these advances have been the development of Fourier Transform methods in both nuclear magnetic resonance and infrared spectroscopy, the use of computers as an integral part of instruments to store and analyse data, and the linking of chromatographic and spectroscopic instruments, the so-called hyphenated techniques such as gas chromatography–mass spectrometry (gc/ms), gas chromatography–Fourier Transform infrared (gc/FT-IR). Each of these developments has provided the chemist with still more powerful techniques to investigate the nature and behaviour of organic compounds.

The techniques considered in this chapter are *infrared spectroscopy* (or *vibrational spectroscopy*), *nuclear magnetic resonance spectroscopy*, *ultraviolet–visible spectroscopy* (or *electronic spectroscopy*) and *mass spectrometry*. Absorption of infrared radiation is associated with the energy differences between vibrational states of molecules; nuclear magnetic resonance absorption is associated with changes in the orientation of atomic nuclei in an applied magnetic field; absorption of ultraviolet and visible radiation is associated with changes in the energy states of the valence electrons of molecules; and mass spectrometry is concerned

---

* The instruments are available from, for example, Beckmann-RIIC Ltd, Gow-Mac Instrument Co. Ltd, Jasco International Co. Ltd, Jeol and Co. Ltd, Mattson Instruments Ltd, Perkin-Elmer Ltd, Phillips Analytical and Varian Associates.

with the analysis of the ionic fragments produced when molecules are caused to disintegrate, usually under electron bombardment.

The availability of low-cost spectrophotometers for the measurement of absorption of electromagnetic radiation in the ultraviolet–visible and in the infrared regions has resulted in the wide use in recent years of these instruments in most laboratories. Intermediates in synthetic sequences and purified reaction products are examined routinely to characterise them by comparison of their spectra with those of authentic material, or by careful interpretation of the significance of the main absorption frequencies with reference to correlation tables, which list the characteristic frequencies associated with the most commonly encountered bonding systems. The recognition of structural features by these spectroscopic methods supplements the information derived from chemical tests and from derivative preparations; the use of these techniques extends further to include monitoring the progress of chemical reactions, quantitative analysis of mixtures of chemical compounds and studies on the effects of structural modifications on the characteristic group absorption frequencies (e.g. solvent interactions, etc.).

Nuclear magnetic resonance spectroscopy developed later than ultraviolet–visible and infrared spectroscopy but is now an important tool for the organic chemist. Although the instruments are still comparatively expensive, all laboratories require access to the facility. This is often provided by a central service which records the spectra of compounds provided by the organic chemist. However, simplification of the operation of routine instruments makes it feasible for the organic chemist to have direct access to them. The more sophisticated instruments, such as those operating at high magnetic fields and those linked to gas chromatographs require specialised technical support. All organic chemists, however, need to be able to interpret nuclear magnetic resonance spectra.

In a similar way the routine interpretation of mass spectrometric data by the laboratory worker is to be expected, even though the measurement of the spectra requires the facilities of a central expert service.

The electromagnetic spectrum – units. The wavelengths of electromagnetic radiation of interest vary from metres for the radiofrequency range to about 10^{-10} m for X-rays. A wave has associated with it both wavelength, λ, and frequency, v, which are related by the equation:

$$v\lambda = c$$

where c is the velocity of the electromagnetic radiation (3×10^8 m s^{-1}). Hence the wavelength is the distance between adjacent crests while the frequency is the number of crests which pass a fixed point in a given time.

The specific regions and the phenomena they produce are correlated with the wavelength and the frequency in Table 3.1.

Because of the great difference in wavelength of the various regions, it is inconvenient to use the same units throughout to specify a particular position in the spectrum. In the ultraviolet–visible regions the wavelengths are expressed in nanometres (nm, 10^{-9} m; formerly this wavelength unit was called a millimicron, mμ). In the infrared region the wavelengths are expressed in micrometres (μm, 10^{-6} m; formerly this wavelength unit was called a micron, μ), or as the reciprocal wavelength in centimetres, $1/\lambda$, termed the wavenumber, $\bar{v}$. In the radiofrequency region absolute frequencies are used rather than the wave-

Table 3.1 Regions of the electromagnetic spectrum

Spectral region	Wavelength	Frequency in wavenumbers (cm^{-1})	Special phenomena
Gamma rays	0.0001–0.01 nm		Nuclear reactions
X-rays	0.01–2 nm		Inner electron transitions
Vacuum ultraviolet	2–200 nm	5 000 000–50 000	Ionisation of atoms and molecules
Ultraviolet	200–400 nm	50 000–25 000 ⎫	Outer electron transitions
Visible	400–750 nm	25 000–13 333 ⎭	
Infrared	0.75–25 μm	13 333–400	⎧ (Stretching) Molecular vibrations
Far infrared	25 μm–1 mm	400–10	⎩ (Bending)
Microwave	1 mm–30 cm	10–0.033	⎰ Molecular rotation ⎱ Electron spin resonance
Radio ⎰ Short wave	10–50 m		Nuclear magnetic resonance
Medium wave	190–555 m		Nuclear quadrupole
⎱ Long wave	1000–2000 m		resonance

The limitations on the extent of the various regions given above are, of course, arbitrary.

numbers. For example, a wavelength of 5 metres corresponds to a frequency of c/λ or 6×10^7 Hz (hertz, defined as cycles per second) which may be written as 60 MHz.

3.1 INFRARED SPECTROSCOPY[2]

The region of infrared spectrum which is of greatest importance to the organic chemist is that which lies between 4000 and 660 cm^{-1}. Absorption bands in the spectrum result from energy changes arising as a consequence of molecular vibrations of the bond *stretching* and *bending* (*deformation*) type. The positions of atoms in molecules may be regarded as mean equilibrium positions, and the bonds between atoms may be considered as analogous to springs, subject to stretching and bending. Each atom or group of atoms in a molecule oscillates about a point at which attraction of nuclei for electrons balances the repulsion of nuclei by nuclei, and electrons by electrons. These oscillations have natural periods which depend upon the masses of the atoms and the strengths of the bonds involved. The amplitude of the oscillations, but not the frequency, can be increased by supplying energy by means of electromagnetic radiation. Nuclei and electrons bear electric charges, the force required can be supplied by the oscillating electric vector of an electromagnetic wave of frequency and phase which *match* those of a particular molecular vibration. Transfer of energy in this way is possible if a change in the amplitude of that vibration results in a change of molecular dipole moment (the dipole moment may be regarded as analogous to the coupling mechanism of a spring); radiant energy is then absorbed and the intensity of radiation at this particular wavelength is decreased on passing through the compound. The intensity of absorption bands depends upon the magnitude of the change in oscillating dipole moment of the bonds during the transition, and also is directly proportional to the number of bonds in the mole- cule responsible for that particular absorption. Thus hydrogen or carbon bonded to oxygen or nitrogen gives rise to strong infrared absorption because of the polarity of these particular bonds. In contrast, no absorption results from stretching vibrations in a homonuclear double bond or triple bond which is symmetrically substituted; such vibrations are termed *infrared inactive*. The recognition of such bonds is, however, made possible by an examination of the Raman spectra of such molecules (i.e. the vibrations are *Raman active*).

There are two main types of molecular vibrations: stretching and bending. A stretching vibration is a vibration along a bond axis such that the distance between the two atoms is decreased or increased. A bending vibration involves a change in bond angles.

For a diatomic molecule A—B, the only vibration that can occur is a periodic stretching along the A—B bond. The masses of the two atoms and their connecting bond may be treated, to a first approximation, as two masses joined by a spring and Hooke's law may be applied. This leads to the expression for the frequency of vibration $\bar{v}$ in wavenumbers (cm^{-1}):

$$\bar{v} = \frac{1}{2\pi c}\left(\frac{f}{m_A m_B/m_A + m_B}\right)^{\frac{1}{2}}$$

where c is the velocity of light ($m\,s^{-1}$), f is the force constant of the bonds ($N\,m^{-1}$) and m_A and m_B the masses (in g) of the atoms A and B respectively. The value of f is c. 500 $N\,m^{-1}$ for single bonds and about two or three times this value for double and triple bonds respectively: it is a measure of the resistance of the bond to stretching and is roughly proportional to the energy of the bond. Application of this equation to the case of the stretching of a C—H bond, and using 19.9×10^{-24} g and 1.67×10^{-24} g as the mass values for carbon and hydrogen respectively, together with the accepted values for c and f, gives a frequency of 3020 cm^{-1}. The stretching of a carbon–hydrogen bond in a methyl or a methylene group is actually observed in the regions about 2975 and 2860 cm^{-1} respectively; the slight deviation from the calculated value is a reflection of the fact that modifications to the frequency of vibration arise from the strengths and polarities of the bonds associated with the carbon atom, and these have been ignored in this calculation.

With polyatomic molecules many more fundamental vibrational modes are possible. A qualitative illustration of the stretching and bending modes for the methylene group is shown in Fig. 3.1. Arrows indicate periodic oscillations in the directions shown; the $\oplus$ and $\ominus$ signs represent, respectively, relative move-

| Symmetric stretching | Asymmetric stretching | Scissoring (in-plane bending) |

| Rocking (in-plane bending) | Wagging (out-of-plane bending) | Twisting (out-of-plane bending) |

Fig. 3.1 Stretching and bending (deformation) vibrational modes for the methylene group, which is typical of an XY_2 system.

ment at right angles to the surface of the page. A *symmetrical stretching mode*, where the hydrogens are vibrating in phase towards and away from the carbon nucleus, requires less energy than the corresponding *asymmetric stretching mode* and therefore absorbs at a slightly lower wavenumber. Bending vibrations, which are descriptively termed *scissoring, rocking, twisting* or *waging* modes, absorb at considerably lower wavenumbers since the energy associated with these deformations is much less.

The infrared spectrum therefore consists of a number of absorption bands arising from infrared active fundamental vibrations; however, even a cursory inspection of an i.r. spectrum reveals a greater number of absorptions than can be accounted for on this basis. This is because of the presence of *combination bands, overtone bands* and *difference bands*. The first arises when absorption by a molecule results in the excitation of two vibrations simultaneously, say v_1, and v_2, and the combination band appears at a frequency of $v_1 + v_2$; an overtone band corresponds to a multiple ($2v$, $3v$, etc.) of the frequency of a particular absorption band. A difference band arises when absorption of radiation converts a first excited state into a second excited state. These bands are frequently of lower intensity than the fundamental absorption bands but their presence, particularly the overtone bands, can be of diagnostic value for confirming the presence of a particular bonding system.

INSTRUMENTAL FEATURES OF INFRARED SPECTROPHOTOMETERS

Until quite recently the commonest forms of infrared spectrophotometers were grating or prism instruments in which the sample was exposed only to narrow bands of radiation at any one time and the amount of absorption at that wavelength was determined by comparison with a reference beam. The wavelength of radiation to which the sample was exposed was varied to cover the desired range over a period of time. These instruments are still in common use and are described below. Increasingly the prism and grating instruments are being superseded by Fourier Transform infrared spectrophotometers. One report suggested that by 1986 there was only one remaining manufacturer of grating instruments in the United States as compared with ten manufacturers of Fourier Transform instruments. The advantages of FT-IR instruments are speed and sensitivity, enabling coupling with g.l.c. equipment to be increasingly used.

The essential features of the prism or grating spectrophotometer, both for infrared and u.v.–visible regions of the spectrum, are a *source of radiant energy* covering the entire region to be measured; a *monochromator* and *slit system* to isolate monochromatic or narrow wavelength bands of radiant energy derived from the source; a compartment to hold both a *sample cell* and a *reference cell* (frequently in u.v.–visible spectroscopy the sample is dissolved in a suitable solvent and the reference cell therefore contains the neat solvent; in infrared spectroscopy there is often no reference cell, comparison being made between the sample and air); a *detector* to differentiate between the intensity of the reference beam and the beam passing through the sample; an *amplifier* for increasing the magnitude of the resultant signal; and a *recorder*. A hard copy of the spectrum is usually provided by a pen recorder in which the movement of the chart paper is geared to the monochromator or its associated mirror system so that the position of the pen on the printed chart paper corresponds to the wavelength of radiant energy passing through the sample and reference cells. Common features of modern spectrophotometers are the presentation of spectra on a *visual display unit, microprocessor control* of instrument conditions and the storage and analysis of the recorded spectrum using a *computer.*

The source of radiant energy in infrared spectrophotometers is a glowing ceramic rod maintained at approximately 1700 °C (Nernst filament). Although many spectrophotometers which are used for the routine examination of samples employ a sodium chloride prism,* the best results for high resolution spectra are achieved by the use of two diffraction gratings to cover the ranges 4000–1300 cm^{-1} (2.5–7.7 μm) and 2000–650 cm^{-1} (5.0–15.4 μm) which may be selected either at the control panel of the instrument or by an automatic interchange at 2000 cm^{-1}. All modern grating or prism instruments are double beam recording spectrophotometers and the energy difference of the beams emerging from the reference cell and the sample cell is measured by the optical null method. In this method the resultant signal from the detector is amplified and used to drive mechanically a comb device (known as an attenuator) into the reference beam to reduce its intensity to that of the sample beam; at this point the detector emits no signal and movement of the attenuator stops. The detector is a device capable of measuring small differences in temperature of the two beams and may be either a thermistor, a thermocouple or a Golay cell. The degree of compensation to the reference beam to balance it with the sample beam is of course a measure of the absorption by the sample. This method is in contrast to the ratio-recording method in which the detector compares directly the intensity of the two emergent beams. The cell windows for infrared spectroscopy are made of compressed sodium chloride (most usually), or of potassium bromide, silver chloride or caesium bromide.

DETERMINATION OF INFRARED SPECTRA

Infrared radiation refers broadly to the wavelength region 0.5–1000 μm. The limited portion of infrared radiation between 2 and 15 μm (5000–660 cm^{-1}) is of greatest practical use to the organic chemist.

As noted above, for interaction to be possible between the electromagnetic radiation and the bonding system of a molecule, leading to uptake of energy and therefore to an increase in the amplitude of the appropriate stretching or bending vibration, two conditions have to be met:

1. There must be a change in the charge distribution within the bond undergoing stretching or bending, i.e. the dipole moment of the bond must vary during vibration so that interaction with the alternating electric field of the radiation is possible.
2. The frequency of the incident radiation must exactly correspond to the frequency of the particular vibrational mode.

As the frequency range is scanned the various infrared active vibrations (i.e. those involving a dipole moment change) will sequentially absorb radiation as the energy equivalence of the radiation and the particular vibrational mode is met, giving rise to a series of absorptions.

A single vibrational energy change is accompanied by a number of rotational

---

* Lithium fluoride, calcium fluoride and potassium bromide prisms are used to study with high resolution the absorption characteristics of compounds in specified regions (usually in conjunction with diffraction gratings), e.g. 4000–1700, 4200–1300, 1100–385 cm^{-1} respectively.

energy changes and consequently vibrational spectra occur as bands rather than as lines. Absorption band maxima are presented either in wavelengths (micrometres, μm) or in wavenumbers ($\bar{\nu}$, expressed as reciprocal centimetres, cm^{-1}). Band intensities are expressed either as the transmittance T (the ratio of the radiant power transmitted by a sample to the radiant power incident on the sample I/I_0) or the absorbance A ($\log_{10} I_0/I$). The intensities of the absorption bands in an infrared spectrum are usually indicated qualitatively, e.g. as very strong (vs), strong (s), medium (m) or weak (w), etc.

SAMPLE PREPARATION

The infrared spectrum of a *liquid* may conveniently be recorded as a thin film of the substance held in the infrared beam between two infrared-transparent discs without the need for a diluting solvent. It is customary to use polished plates of sodium chloride as the support material; this material is adequately transparent in the region 2–15 μm. Spectra in the longer wavelength region (12–25 μm) can be recorded using potassium bromide plates. Sealed cells (p. 267) should be used for volatile liquids.

Great care must be exercised in the handling and use of these plates since in particular traces of moisture will cause polished surfaces to become 'fogged', thus causing undesirable scattering of the transmitted radiation. Plates are stored in air-tight containers containing small bags of a suitable desiccant, e.g. silica gel; they should only be handled by the edges and as far as possible under a radiant heater. After use the plates should be first wiped with a paper tissue, rinsed with a jet of dichloromethane, wiped again with a tissue and finally allowed to dry under a radiant heater. Washings should be collected in a suitable container for subsequent recovery. Plates which have become fogged through misuse need to be carefully repolished using one of the commercially available polishing kits.

In order to determine the spectrum of a neat liquid sample, a capillary film of the pure dry material is formed between a pair of plates by carefully placing three small drops on the polished surface of one plate, covering them with the second plate and exerting gentle pressure with a slight rotatory motion to ensure that the film contains no air bubbles. The prepared plates are then placed in the demountable cell holder, ensuring that the gaskets are properly located, and the quick release nuts are firmly screwed down (but not too tightly, otherwise the liquid will exude from between the plates). The whole assembly is then located in the sample beam path of the infrared spectrophotometer.

With most instruments, before the spectrum is actually recorded, it can readily be ascertained whether the thickness of the film is adequate to provide a satisfactory spectrum by scanning the spectrum rapidly, noting the movement of the pen recorder without allowing it to make a permanent trace. This may sometimes be done by moving the chart paper manually. If it is judged that the band intensities are too high, the cell must be dismantled and a thinner film prepared. When the spectrum has been recorded, calibration of the wavelength scale may be checked by superimposing on the recorded spectrum characteristic peaks from the spectrum of a polystyrene film which is inserted into the instrument in place of the sample cell. Suitable intense bands of the spectrum occur at 3027, 2851, 1602, 1028 and 907 cm^{-1}.

Solids are generally examined as a *mull* or in a pressed *alkali halide disc* (usually potassium bromide). For mulls, Nujol (a high boiling fraction of petroleum) is most commonly used, although when it is desired to study frequency ranges in which Nujol absorption bands appear, Fluorolube (perfluoro-kerosene, a mixture of fluorinated hydrocarbons), or hexachlorobutadiene is employed. The infrared spectra of the mulling agents should be recorded and kept available for reference purposes.

Mulls are prepared by grinding about 2.5 mg of the solid sample with one or two drops of Nujol in a small agate pestle and mortar. The mixture must be thoroughly ground for at least five minutes to ensure that a fine particle size is obtained, so that the light scattering is reduced to a minimum and damage to the halide plates by scratching is avoided. The paste is spread on one plate of the demountable cell, covered with the other and the sample thickness varied by rotating and squeezing the plates to force out excess material. The plates are inserted into the cell holder and placed in the sample beam path of the spectro-photometer as described above. Rapid scanning of the spectrum in conjunction with a knowledge of the spectrum of the relevant mulling agent will give an indication as to whether the concentration of the sample in the mull is sufficient to provide a satisfactory spectrum. If necessary the concentration should be adjusted either by adding more of the sample followed by regrinding or by diluting with more of the mulling agent.

In the pressed disc technique a known weight of sample is intimately ground with pure, dry potassium bromide and the mixture inserted into a special die and subjected to pressure under vacuum. The concentration of sample in the disc is usually in the region of 1.0 per cent. The disc so produced may be mounted directly in the sample beam path of the spectrophotometer and the spectrum recorded. This method has the advantage that the spectrum so produced is entirely due to the sample since pure dry potassium bromide is infrared transparent in the 2–25 μm region. To eliminate the possibility of impurities in the potassium bromide, however, a blank disc (no sample) can be made and mounted in the reference beam path of the spectrophotometer. Care should be taken to ensure that both discs are of equal thickness otherwise inverse peaks may occur if the potassium bromide is damp or impure, and this will be particularly noticeable if the reference disc is thicker than the sample disc.

The halide used must be of AnalaR grade and should be pre-powdered to a particle size which will pass through a 70-mesh sieve; sieving is not absolutely necessary provided that each batch of powder is tested to show that it does subsequently produce good discs. Pre-powdering may be carried out in a mechanical grinding machine or by hand in an agate pestle and mortar. Drying of the powder is best done by leaving it in a shallow dish in an oven at 120 °C for at least 24 hours. It may then be transferred to a loosely stoppered bottle which should be kept in a desiccator.

As a general procedure, 500 mg of pre-ground potassium bromide is weighed and mixed with the appropriate quantity of the sample (i.e. 5 mg for a 1% disc) whose spectrum is to be recorded. The mixture is further intimately ground in a vibration mill*; the time required for the grinding depends on the degree of pre-powdering but is usually in the region of 1–2 minutes.

Pressing of the disc is usually carried out in a commercially available stainless steel disc die assembly† under pressure of the order of 8–9 tons/square inch. The

---

* The Vibromill supplied by Beckman-RIIC Ltd.
† The Evacuable KBr Die supplied by Beckman-RIIC Ltd.

die is normally kept in a vacuum desiccator and placed under a radiant heater for at least 30 minutes before use. Dies must be scrupulously cleaned after use to remove all traces of alkali halide which may corrode the stainless steel. To prepare the disc proceed as follows:

1. Push the body of the die (A) on to the base (B) (Fig. 3.2).
2. Insert one of the stainless steel pellets (C) into the barrel of the die, polished face *upwards*.
3. Introduce a measured quantity of the ground sample (150–250 mg) into the bore of (A).
4. Distribute the powder evenly over the polished surface of the pellet (C) by slowly introducing the plunger (D) into the barrel with careful rotation, and then rotating it a few times while simultaneously exerting gentle pressure with the fingers.
5. Slowly withdraw the plunger taking care not to disturb the powder which should show a surface which is perfectly smooth and free of pits or cracks.
6. Insert the second pellet into the barrel of the die, polished face *downwards*, and bring into contact with the powder by pressing down lightly with the plunger; complete the assembly by pushing the O-ring seal on the plunger into contact with the die surface (Fig. 3.3).
7. Connect the die to the vacuum pump and evacuate for a period of at least 3 minutes; place the die (while still attached to the vacuum pump) under the hydraulic press* and apply a pressure not exceeding 8–9 tons for 1 minute.
8. Release the pressure and vacuum after this time, invert the die, and remove the base (B) (the plunger is kept in position with the fingers).
9. Place the perspex cylinder on the top of the assembly, which is then returned to the hydraulic press, and apply sufficient pressure to lift the lower pellet and the disc clear of the barrel (Fig. 3.4).
10. Finally release the press and remove the die body (A) and the upper steel pellet from the face of the disc and then remove the disc itself with tweezers (the disc should never be handled with the fingers), and mount it in the specially designed holder. These latter operations are best conducted under a radiant heater. The disc holder is located in position in the sample beam path of the spectrophotometer; if required a blank potassium bromide disc, similarly prepared, is introduced into the reference beam path.

Providing that care has been taken in the disc preparation the final disc should be slightly opaque due to the presence of the sample (the blank disc should be transparent). Should the disc show a number of white spots, it is probable that the mixture has been unevenly ground. If the disc shows a tendency to flake, then excessive grinding of the powder is indicated. If after being removed from the die the disc becomes cloudy this is indicative of the uptake of water; to avoid this difficulty it is necessary to ensure that the die is evacuated for a sufficiently long period and that the removal of the disc from the die is carried out under a radiant heater.

Several commercially available small-scale, manually-operated presses are available for use when the number of halide discs which require to be prepared does not warrant the purchase of the more elaborate hydraulic press. They operate on the principle of achieving the necessary pressure on the powdered sample

---

* Beckman-RIIC Ltd.

Fig. 3.2   Fig. 3.3   Fig. 3.4

(about 50–100 mg) by spreading it evenly between the optically polished faces of two bolts which are then screwed into opposite ends of a cylinder and tightened relative to each other. After removal of the bolts, the barrel containing the disc is mounted in the specially designed cell slide in the sample beam path of the spectrophotometer. A suitably adjusted metal comb attenuator must be introduced into the reference beam path to compensate for the fact that the barrel structure of the press which is held in the sample beam restricts the amount of radiation which can pass through the halide disc, otherwise even at a wavelength in which no absorption is taking place the reference and sample beams will be out of balance. The barrel and bolts are cleaned by a stream of tap water, rinsed with ethanol, then with dichloromethane and finally dried with paper tissue and placed under a radiant heater. Care should be taken not to damage the polished surfaces of the bolts by scraping off adhering particles of halide with a spatula or by allowing the faces of the bolts to come into contact within the barrel.

Solutions of either solids or liquids are normally handled in cells of 0.025 mm to 1 mm thickness using concentrations of 20 per cent to 0.5 per cent respectively in cases of compounds having a molecular weight of about 150. Compounds of higher molecular weight are examined at correspondingly higher concentrations.

Pure dry chloroform, carbon tetrachloride or carbon disulphide (Section 4.1 6,7,32, pp. 399 and 411) are the solvents most commonly employed. Their selection is based upon the fact that they exhibit relatively few intense absorption bands in the region 5000–650 cm^{-1} (Fig. 3.5(a), (b) and (c)).

In the region of these intense absorption bands it is not possible to record the absorption due to the solute, even with the compensating effect of an identical path length of the pure solvent in the reference beam, since virtually no radiation reaches the detector system and the pen recorder is not activated.

In the region of the less intense solvent absorption bands, the use of pure solvent and solution in matched fixed path length cells enables the absorption due to the solute to be recorded satisfactorily, providing a sufficiently concentrated solution requiring a short path length cell is used. (Clearly a more dilute solution necessitating longer path length cells causes the less intense solvent absorption bands to become more prominent, thus reducing the overall transmittance in this region.)

For routine use liquid cells are of two types: (a) the demountable cells in which the path length may be varied by utilising spacers of lead or Teflon of appropriate thickness; these cells have the advantage that they may be easily dismantled after use for cleaning and if necessary for repolishing of the cell win-

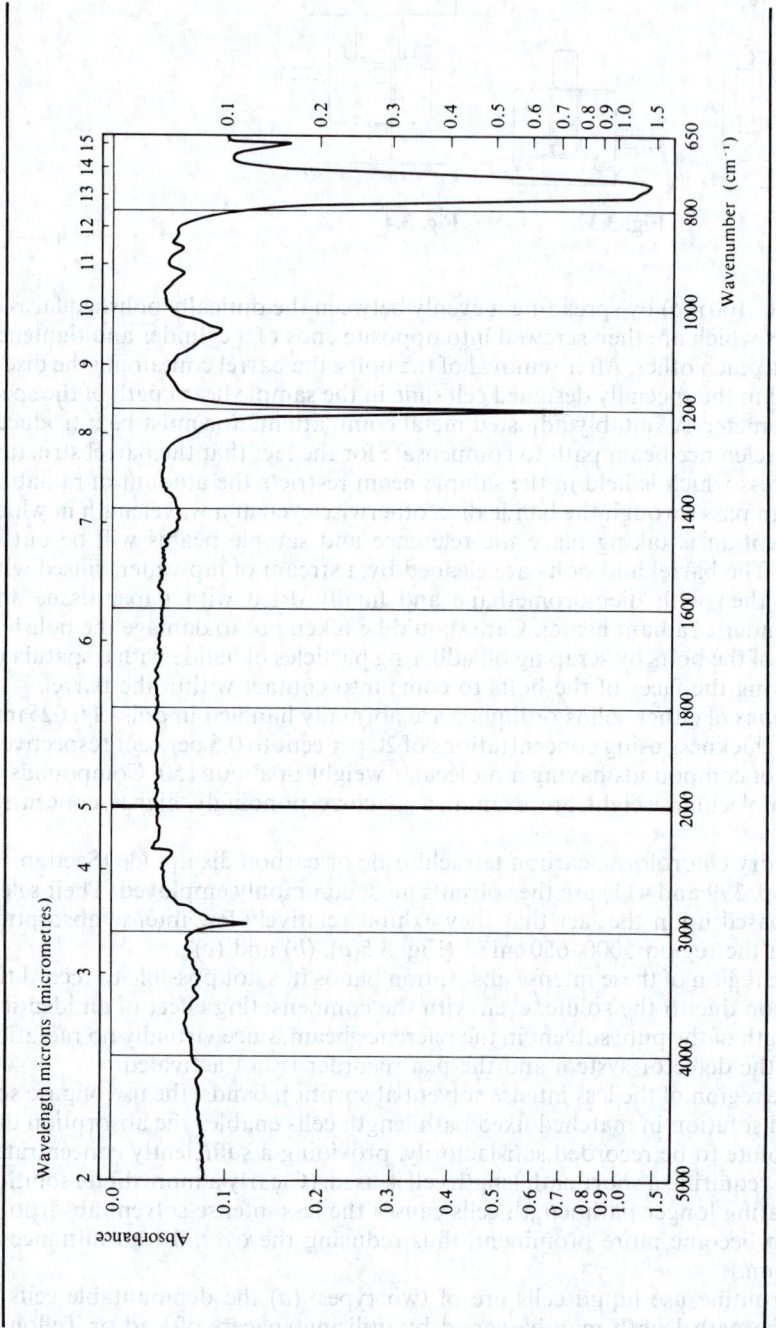

Fig. 3.5(a) Chloroform.

Fig. 3.5(a)–(c) Spectra of solvents for infrared spectroscopy.

Fig. 3.5(b)   Carbon tetrachloride.

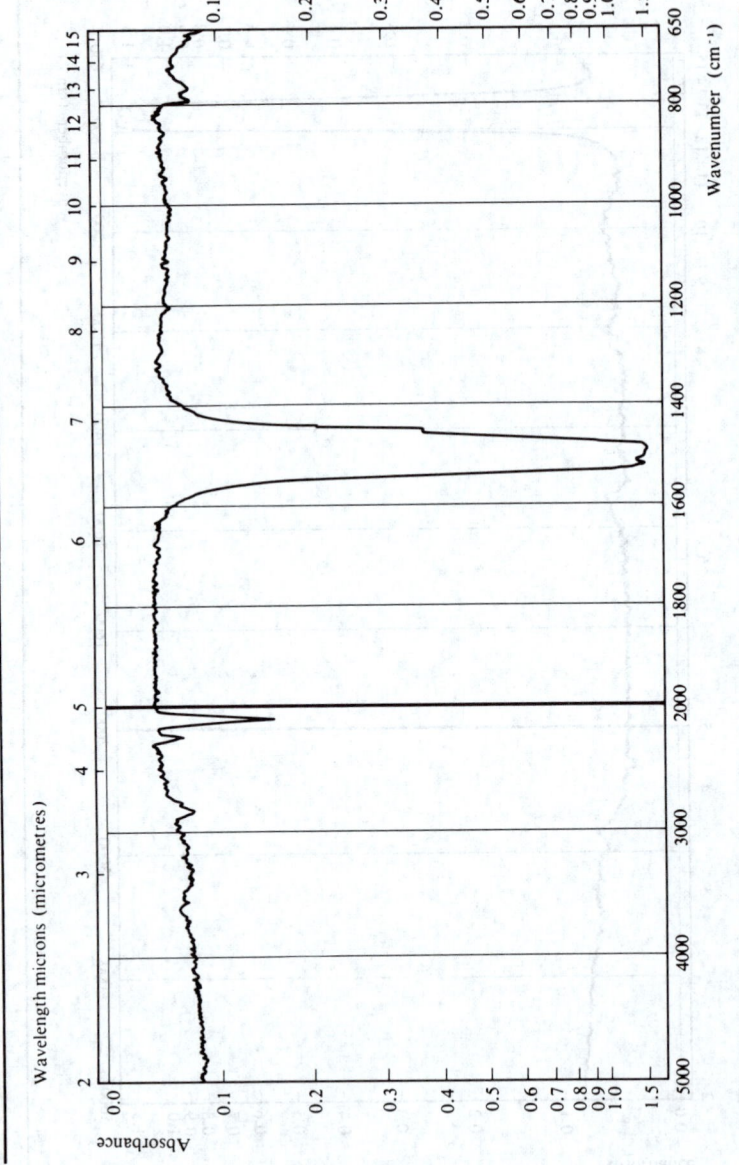

Fig. 3.5(c)   Carbon disulphide (recorded in a variable path length cell, set at 0.015 mm path length).

dows; (b) the fully assembled sealed cells of fixed path length (available as matched pairs) which use an amalgamated lead spacer forming a permanent leak-proof seal; these cells are particularly useful for volatile samples but they require much greater care in use since deterioration of the inside plate surface necessitates an expensive overhaul. The overall design of these two types is similar and one such example is illustrated in Fig. 3.6. The cell design incorporates inlet and outlet ports by means of which the liquid may be introduced. For this operation the cell is placed horizontally on the bench with the ports uppermost, a syringe needle is inserted into one of the ports, and the solution is injected. The passage of fluid across the cavity is easily observed and care must be taken to avoid the presence of trapped air bubbles. Teflon stoppers are inserted into the ports to keep the liquid within the cell which is then mounted in the sample beam path of the spectrophotometer. The neat solvent is similarly introduced into the matched cell and placed in the reference beam path.

The cells are emptied by attachment of one of the ports to a vacuum line incorporating a suitable trapping system, cleaned by several rinses with neat solvent and then with dichloromethane. Solvent is removed by a short period of suction and the cell is finally dried under a radiant heater and stored in a desiccator. Prolonged passage of air through the cell must be avoided otherwise fogging of the cell windows from atmospheric moisture may occur.

The most accurate way of compensating for solvent absorption is to use the more expensive variable path length cells (Fig. 3.7) in which one of the plates which constitute the liquid cell can be moved with the aid of a micrometer device to allow adjustment to any required path length. This allows the accurate matching in the spectrophotometer of two such cells filled with the appropriate solvent. One cell may then be emptied, cleaned and refilled with solution so that the spectrum of the solute may be recorded.

On occasions it is desirable to investigate the spectroscopic properties of solutions at high dilution (i.e. 0.005 M), as for example in the study of inter- and intra-molecular hydrogen bonding of compounds containing hydroxyl groups.

Fig. 3.6   Fig. 3.7   Fig. 3.8

Such concentrations require cell path lengths in the region of 1.5 cm. The construction of a cell of these dimensions is relatively simple and a suitable design which has found use in these laboratories is shown in Fig. 3.8.

FEATURES OF AN INFRARED SPECTRUM

A typical i.r. spectrum is that of acetophenone shown in Fig. 3.9. Some general features illustrative of the philosophy relating to the interpretation of spectra and the correlation of absorption bands with the presence of particular groupings should be noted.

It will be immediately apparent that this spectrum may be divided into two parts, the first between 4000 and 1600 cm^{-1} and the second from 1600 to 660 cm^{-1}. In the former there are relatively few absorption bands, but in the latter a great number of absorptions is observed. Indeed, although as noted below aromatic and aliphatic compounds may be recognised from the general spectral profile, all organic compounds exhibit this apparent segregation of bands into these two main regions. The second region is frequently referred to as the 'fingerprint region' since complete superimposability of two spectra in this region provides confirmation of identity.

The former region could be termed the 'functional group region' since, as will be noted below, the fundamental vibrational modes of most of the principal functional groups absorb in this region. Thus all compounds containing a carbonyl group (whether it be an aldehyde, ketone, carboxylic acid, acid chloride, amide, ester, etc.) will exhibit strong absorption in the 1700 cm^{-1} region.

Aromatic compounds, of which the spectrum in Fig. 3.9 is typical, always exhibit sharp and often numerous bands in the fingerprint region. Aliphatic compounds on the other hand give rise to far fewer, broad, bands in this region. These differences in the profile of the spectra of aromatic and aliphatic compounds provide a valuable first step in spectral interpretation.

Correlation charts and tables. Central to the philosophy of i.r. spectral interpretation is the fact that many stretching and bending modes in a molecule are virtually independent of changes of structure at more remote sites. Structural modifications closer to the absorbing centre do of course affect the energy associated with the absorption, and lead to a shift of the absorption band to higher or lower frequencies. These frequency shifts, however, have been found to lie within defined limits and the numerical value often provides valuable information on the structural environment of the associated group. For example, the absorption maximum for the carbon–hydrogen stretching frequency lies in the general region around 3000 cm^{-1}; if the carbon is sp^3-hybridised the maximum is just below 3000 cm^{-1}; if the carbon is sp^2-hybridised the position is just above 3000 cm^{-1} and if the carbon is sp-hybridised the position is at about 3250 cm^{-1}. The remarkable constancy of these absorption positions for the carbon–hydrogen stretching mode in all organic compounds examined enables the reverse deduction to be made, i.e. absorption bands exhibited by an unknown compound in the region of 2800–2900 cm^{-1} and in the region of 3040 cm^{-1} would indicate the presence of both saturated carbon–hydrogen bonds and carbon–hydrogen bonds in an alkene or an aromatic system.

The band positions for all the major structural bonding types have been determined and correlation charts and tables are available which give the ranges within which particular bonding types have been observed to absorb. A simpli-

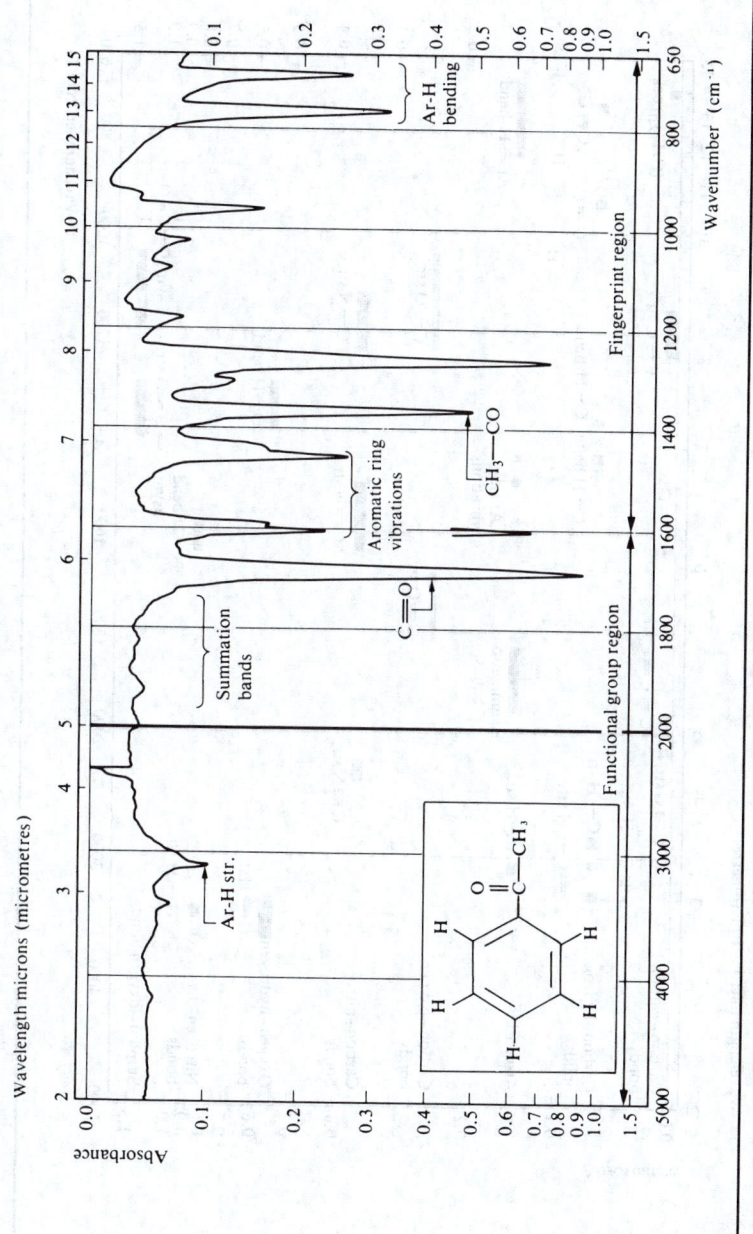

Fig. 3.9 Infrared spectrum of acetophenone recorded as a thin film.

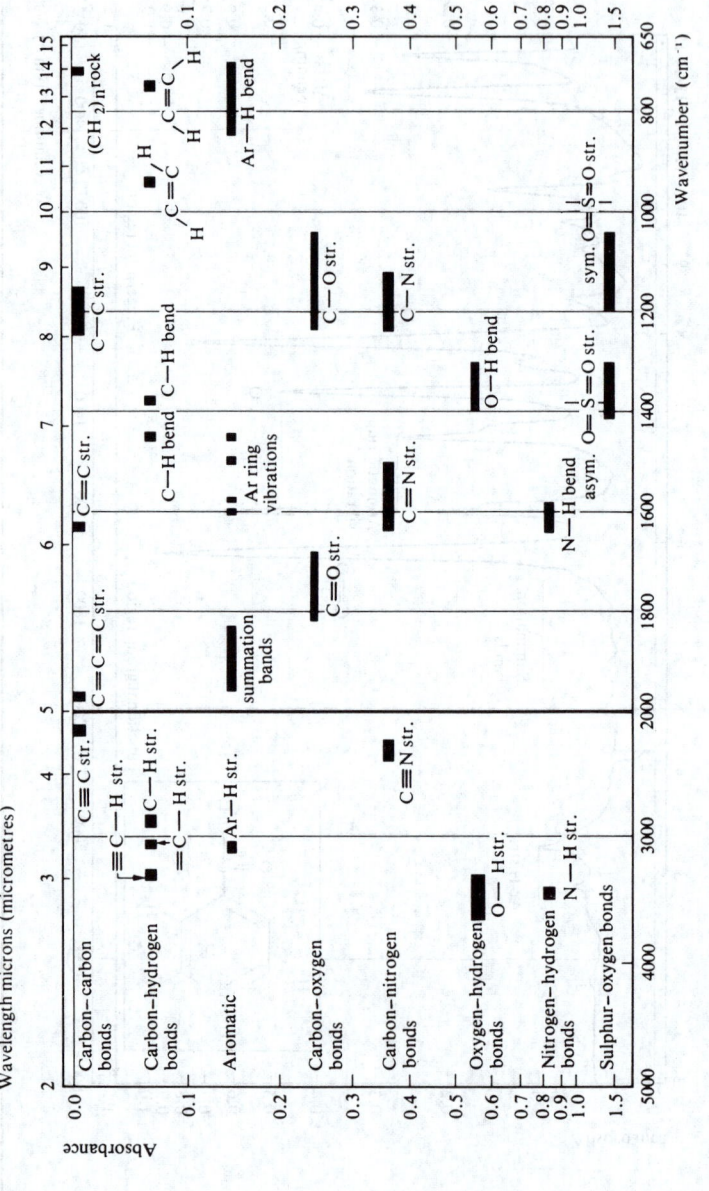

Fig. 3.10 Simplified correlation chart of absorption positions of important bonding types.

fied correlation chart is provided by Fig. 3.10 which indicates the ranges within which the stretching and bending absorptions have been observed. This chart has been prepared on a typical spectral grid since rapid recognition of significant absorption bands is usually achieved by such visual familiarity with wavelength regions. The alternative and more accurate and informative way of presenting correlation information is by means of tables. These have been collected in Appendix 2, Tables A2.1–A2.13. It is to these tables that reference should be made when endeavouring to elucidate the structure of a compound from the infrared spectrum.

INTERPRETATION OF AN INFRARED SPECTRUM

The spectrum of acetophenone (Fig. 3.9) provides the opportunity for illustrating one possible method which may be adopted to correlate the absorption bands in a spectrum with the bonding types from which they arise. It is important that as much information as possible should be extracted from the functional group region first, then further information sought in the fingerprint region as a result of these conclusions. It is usually unwise to haphazardly relate intense peaks to specific structural features and the following represents a more logical approach.* The correlation chart above may be used in this simple illustration.

Consideration of the general profile provides circumstantial evidence that the compound may be aromatic. By looking first at the absorption band in the $3000\,\mathrm{cm}^{-1}$ region it is apparent that there are present sp^2-hybridised carbon–hydrogen bonds (absorption just above $3000\,\mathrm{cm}^{-1}$) as well as sp^3-hybridised carbon–hydrogen bonds (absorption just below $3000\,\mathrm{cm}^{-1}$). The only other significant absorption in the region above $1600\,\mathrm{cm}^{-1}$ is the band at $1680\,\mathrm{cm}^{-1}$. This is clearly due to the presence of a carbonyl group. Thus the compound may be either an olefinic or an aromatic compound containing a carbonyl system.

The differentiation is easily made by noting the characteristic absorption bands corresponding to the aromatic ring system around $1600–1450\,\mathrm{cm}^{-1}$ and $850–660\,\mathrm{cm}^{-1}$, and the characteristic pattern in the overtone region around $2000–1800\,\mathrm{cm}^{-1}$. The origin of these absorptions is discussed in detail below; their presence in this spectrum clearly shows that the structure is a monosubstituted aromatic compound.

The precise position of the carbonyl band, which is at the lower end of the frequency range, may now be rationalised. It implies that the carbonyl function is in conjugation with the aromatic π-electron system leading to reduced double bond character, a weaker carbon–oxygen bond and hence a lower absorption frequency. This conjugation is confirmed by observing the increased intensity of the $1580\,\mathrm{cm}^{-1}$ band associated with the aromatic ring vibration.

It is now necessary to deduce the nature of the carbonyl function (i.e. whether it is an aldehyde, ketone, ester, etc.). Each of these functional groups (with the exception of ketones) exhibits further characteristic and identifiable absorption bands due to the attachment of atoms other than carbon to the carbonyl carbon atom. Thus an aldehyde should exhibit a double band in the region of $2830–2700\,\mathrm{cm}^{-1}$, due to stretching of the C—H bond in the aldehydic group. A car-

---

* This approach, which is developed in more detail below, is based on the method used by the late Dr L. J. Bellamy, CBE, in his lectures and tutorials on i.r. spectroscopy to postgraduate students in the School of Chemistry and with which the editors were privileged to be associated.

boxylic ester would exhibit a pair of intense absorptions near 1300 and 1100 cm^{-1} due to C—O stretching modes. Logical, sequential and careful searching in the two regions of the spectrum for the presence or absence of diagnostic bands as indicated above would finally lead to the conclusion that the compound was a monosubstituted aromatic ketone. It is in this situation that negative information is as important as positive information, i.e. the absence of the double band at about 2800 cm^{-1} eliminates from consideration the possibility of an aldehyde. That the compound is likely to be acetophenone could be deduced from the relatively weak absorption in the region below 3000 cm^{-1} and the presence of absorption at 1370 cm^{-1}, the latter being characteristic of one of the deformation modes of a methyl group. Positive identification would be most readily achieved by consideration of the p.m.r. spectrum.

The following summary provides a recommended approach to the interpretation of an unknown spectrum which may be adopted until experience has developed an intuitive appreciation of the characteristics of infrared spectra. It should be used in association with the more detailed notes which follow, describing the way in which characteristic group frequencies arise and the variations in frequency position which accompany environmental changes.

1. Obtain a satisfactory spectrum of the unknown dry compound using the techniques which have been described above. If the spectrum has been supplied, make a careful note of the conditions under which the spectrum was recorded, any solvents used, etc.
2. Qualitatively assess from the spectrum profile whether the compound is likely to be aliphatic or aromatic in type.
3. Mark any absorptions apparent in the spectrum which are known to arise from the solvent or mulling agent used.
4. Inspect the C—H stretching region and identify the bands as either of aliphatic or aromatic/olefinic origin.
5. Evaluate the degree of carbon chain branching by approximately assessing the methyl:methylene ratio from the relative intensity of the absorption bands in the saturated C—H region below 3000 cm^{-1}. This may necessitate re-recording the spectrum on a grating instrument to obtain better resolution.
6. Search the high frequency end of the spectrum, i.e. the region 4000–3000 cm^{-1}, for the presence of bands arising from the presence of —O—H, —N—H and ≡C—H bonds.
7. Extract from the spectrum information provided by the presence of relatively intense absorption bands in the region 2500–1600 cm^{-1}. This should provide evidence for the presence or absence of C≡C, C≡N, C=O, C=C.
8. As a result of the conclusions deduced from 4–7, attempt to classify the compound; on the basis of this classification search the fingerprint region for specific evidence to support the postulated structure. Examples are: (a) if an aromatic compound is suspected because of =C—H absorption, confirm by examination of the region 1600, 1580, 1500 and 1450 cm^{-1} and then endeavour to establish the substitution pattern by looking specifically in the 850–650 cm^{-1} region, and then in the overtone region, 2000–1800 cm^{-1}; (b) if an alkene is suspected, search for evidence of its substitution type; (c) if a carbonyl group is present, deduce its nature by searching for evidence of the presence of associated groups, etc.

9. If no absorption bands are present in the functional group region, with the exception of those arising from carbon–hydrogen stretching modes, consider the possibilities of ethers, alkyl halides, sulphur compounds, tertiary amines and nitro compounds as detailed in the sections below.

10. Relate the structural information deduced in this way with that obtained by other spectral methods or by appropriate chemical tests.

CHARACTERISTIC GROUP FREQUENCIES

Alkanes, cycloalkanes and alkyl groups. The diagnostically important bands in these compounds arise from C—H stretching and bending vibrations, although some bands due to C—C skeletal vibrations are also of value.

Alkanes. Methyl and methylene groups both have asymmetric and symmetric C—H stretching vibration modes, giving rise to four absorption bands just below $3000\,\mathrm{cm}^{-1}$; the CH_3 vibration modes are shown in Fig. 3.11 and the CH_2 vibrations are those depicted in Fig. 3.1. The absorption bands are not normally resolved by prism spectrophotometers, and in the spectrum of decane (Fig. 3.12) recorded on such an instrument the C—H vibrations are revealed as two overlapping bands just below $3000\,\mathrm{cm}^{-1}$. With grating instruments the absorption bands are resolved so that the CH_3 asymmetric and symmetric vibrations which occur near 2962 and $2872\,\mathrm{cm}^{-1}$ respectively, and the CH_2 asymmetric and symmetric vibrations which occur near 2926 and $2853\,\mathrm{cm}^{-1}$ respectively, are clearly visible. These absorption positions do not vary much in the case of unsubstituted alkanes. However, very useful qualitative information can be obtained, regarding the relative number of CH_3 and CH_2 groups in an alkane, by inspection of the relative intensities of these bands, since these are dependent on the number of such groups present in a compound. This is illustrated by the part spectra recorded on a high resolution (grating) spectrophotometer shown in Fig. 3.13(a)–(c). Spectrum (a) is that of hexane which has a similar number of methyl and methylene hydrogens so that the corresponding bands are of approximately equal intensity. In decane (spectrum (b)), on the other hand, the two CH_2 bands are much more intense than the CH_3 bands; in cyclohexane

Symmetric stretching

Asymmetric stretching

Symmetric deformation

Asymmetric deformation

Fig. 3.11 Stretching and bending vibrational modes for a methyl group.

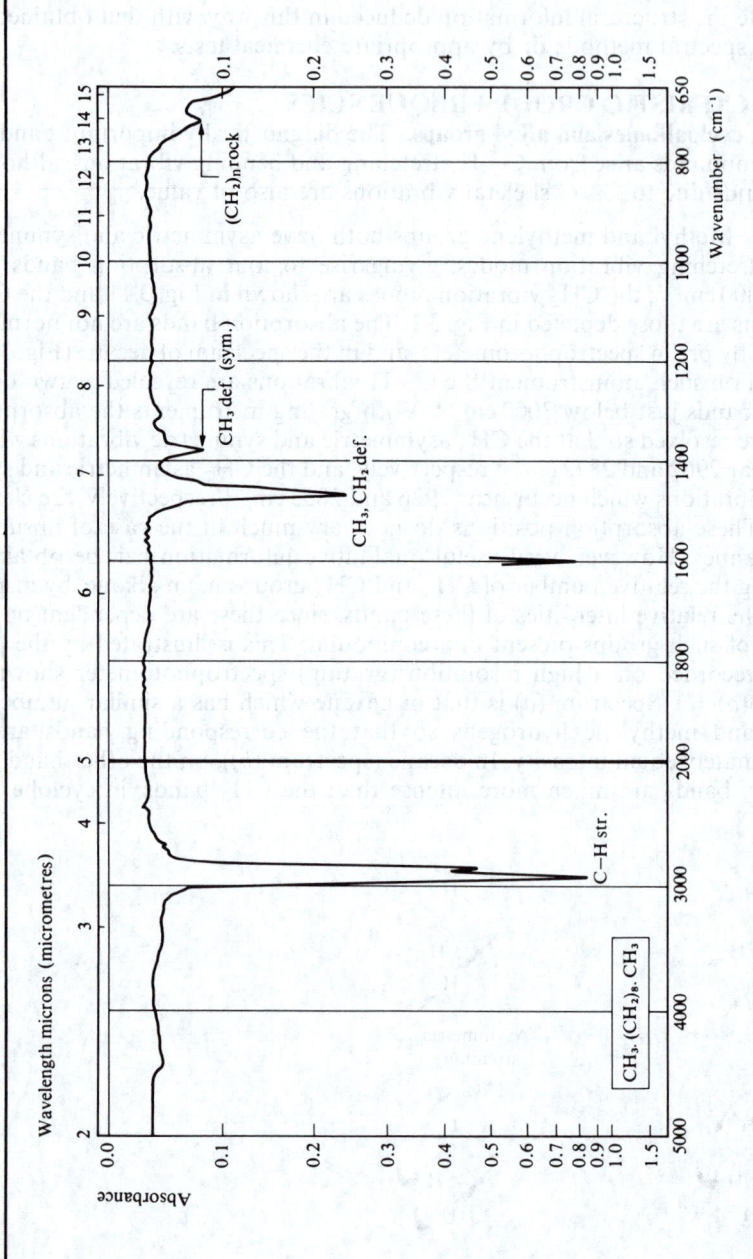

Fig. 3.12   Infrared spectrum of decane recorded as a liquid film on a prism spectrophotometer.

Fig. 3.13(a)   Part i.r. spectrum of hexane recorded as a liquid film.

(spectrum (c)) only the two CH_2 absorption bands are observed. A highly branched alkane will thus show strong CH_3 and weak CH_2 bands. Tertiary C—H stretching vibrations produce a weak band near $2890\,\text{cm}^{-1}$ which is often masked by the other C—H bands.

The CH_3 group has two C—H deformation vibrational modes which are shown in Fig. 3.11; the asymmetric vibration gives a band near $1450\,\text{cm}^{-1}$ and the symmetric 'umbrella-like' vibration a band near $1375\,\text{cm}^{-1}$. The four possible bending modes of the CH_2 group are those shown in Fig. 3.1; the scissoring vibration gives a band near $1465\,\text{cm}^{-1}$ which overlaps with the asymmetric band near $1450\,\text{cm}^{-1}$. The position of the CH_3 band ($1375\,\text{cm}^{-1}$) is remarkably constant when attached to carbon and this allows ready recognition of the C—CH_3 group in a molecule. When a second methyl group is attached to the same carbon atom as in the isopropyl group, splitting of this band occurs to give two bands of approximately equal intensity. The $1375\,\text{cm}^{-1}$ band arising from a t-butyl group is also split, but in this case the intensities of the two bands are in the ratio of approximately 2:1 with the less intense band at higher frequencies.

Compounds containing at least four adjacent methylene groups, i.e. —$(CH_2)_n$—, $n \geqslant 4$, show a weak band near $725\,\text{cm}^{-1}$ due to the four groups rocking in phase; this band increases in intensity with increasing length of the chain.

Fig. 3.13(b)   Part i.r. spectrum of decane recorded as a liquid film.

Cycloalkanes. The C—H stretching vibrations of unstrained ring systems give rise to bands in the same region of the spectrum as acyclic compounds; as the size of the ring decreases however there is a shift to higher frequency and cyclopropanes give a band in the 3060–3040 cm^{-1} region of the spectrum. The absence of CH$_3$ stretching bands in the spectrum of cyclohexane has been commented on above, and there will not of course be any bands in the methyl C—H bending region; the absence of the C—H bending bands may also be noted in the spectrum of cyclohexanecarboxaldehyde, Fig. 3.29.

Alkyl groups. The symmetric stretching vibration of a methyl group when attached to nitrogen or oxygen results in absorption at a lower frequency than when attached to carbon; additionally, the symmetric deformation vibration in compounds containing NCH$_3$ and OCH$_3$ groups leads to absorption at a higher frequency. These frequency shifts are of some diagnostic value for the identification of such groups (e.g. the spectrum of anisole, Fig. 3.25, and the spectrum of *N*-methylaniline, Fig. 3.27), although reliable confirmation would be obtained from a p.m.r. spectrum.

   The attachment of a methyl or methylene group to a carbonyl group results in the C—H symmetric bending deformations becoming more intense and the bands appear at slightly lower frequency than normally.

Fig. 3.13(c)   Part i.r. spectrum of cyclohexane recorded as a liquid film.

The correlation tables for alkanes, cycloalkanes and alkyl groups are in Appendix 2, Table A2.1.

Alkenes. The presence of unsaturation in a molecule can usually be readily recognised by the presence of a small sharp band just above $3000\,\text{cm}^{-1}$ due to the unsaturated $=$C—H stretching vibration. Aromatic C—H bonds also show weak absorption in this region, but there are, however, other distinguishing features in the i.r. of aromatic compounds which readily enable them to be differentiated from alkenes. The precise position of the band depends on the nature of the alkene. For example, the terminal methylene group in vinyl (—CH$=$CH$_2$) and *gem*-disubstituted ($>$C$=$CH$_2$) alkenes absorb in the 3095–$3075\,\text{cm}^{-1}$ range, and the $=$C—H bond in *cis*-, *trans*- and trisubstituted alkenes absorbs at 3040–$3010\,\text{cm}^{-1}$, a region which may be masked by strong absorption due to saturated C—H stretching vibration bands when alkyl groups are present in the compound.

The C—H out-of-plane bending or wagging vibrations of hydrogens attached to unsaturated carbons give rise to important absorption bands in the 1000–$800\,\text{cm}^{-1}$ region. These are frequently strongest in the spectra of simple alkenes, and they readily allow one to ascertain the substitution pattern of the alkene; thus the vinyl group shows two strong bands near $990\,\text{cm}^{-1}$ and

277

$910 \, \text{cm}^{-1}$, whereas a *trans*-disubstituted alkene shows only one strong band near $965 \, \text{cm}^{-1}$.

In-plane bending of the unsaturated C—H bond gives rise to absorption in the $1420–1290 \, \text{cm}^{-1}$ region which is frequently of weak intensity. As this absorption occurs in the region of the spectrum associated with C—C stretching and saturated C—H bending vibrations, it is of little diagnostic value, but can be of use in confirming the presence of a double bond, e.g. the in-plane bending vibration (or scissoring) of the $=C\begin{smallmatrix}H \\ H\end{smallmatrix}$ group produces a band near $1415 \, \text{cm}^{-1}$.

The C=C stretching vibration gives rise to an absorption band in the $1680–1620 \, \text{cm}^{-1}$ region in simple alkenes. The band is of variable intensity, but is much less intense than that from the C=O stretching vibration which also leads to absorption in this region, a consequence of the very much less polar character of the olefinic bond. In general, the more highly substituted the double bond, the higher is the frequency at which it absorbs, and the lower is its intensity; thus the vinyl group gives a relatively strong band near $1640 \, \text{cm}^{-1}$ and can be readily seen in the spectrum of oct-1-ene (Fig. 3.14), which is typical for a simple vinyl alkene.

Attachment of a polar group normally lowers the C=C stretching frequency, so that, for example, vinyl chloride absorbs at $1610 \, \text{cm}^{-1}$ and vinyl bromide at $1593 \, \text{cm}^{-1}$; slight frequency shifts outside the above range can also occur when the double bond is exocyclic to a ring system; thus a methylene group attached to six-, five-, four- and three-membered ring systems absorbs respectively at $1651 \, \text{cm}^{-1}$, $1657 \, \text{cm}^{-1}$, $1678 \, \text{cm}^{-1}$ and $1736 \, \text{cm}^{-1}$, a shift which is associated with increasing ring strain. It should be noted that there will be no C=C stretching absorption band in the spectrum of symmetrically substituted *trans*-olefinic compounds such as *trans*-1,2-dichloroethylene and fumaric acid. Despite the fact that there are highly polar bonds in each of these compounds, because of the symmetry of the molecules, stretching of the C=C bond does not result in any change in the oscillating dipole moment, so this vibration is infrared inactive. The vibrations of these bonds may however be readily observed in the Raman spectrum.

$$\begin{array}{cc} H & Cl \\ \diagdown \diagup \\ C{=}C \\ \diagup \diagdown \\ Cl & H \end{array} \qquad \begin{array}{cc} H & CO_2H \\ \diagdown \diagup \\ C{=}C \\ \diagup \diagdown \\ HO_2C & H \end{array}$$

When a C=C bond is conjugated with a carbonyl group, another double bond or an aromatic ring, the bond has less double-bond character, i.e. is weaker, and the absorption shifts to longer wavelength (lower wavenumber), while the increased polarity of the double bond results in a considerable enhancement in intensity. With conjugated aliphatic systems, the number of absorption bands observed is the same as the number of conjugated double bonds; thus dienes, trienes and tetraenes show two, three and four bands respectively in the $1650–1600 \, \text{cm}^{-1}$ region.

The correlation table for alkenes is in Appendix 2, Table A2.2.

Aromatic compounds. A characteristic feature of the i.r. spectra of aromatic compounds is the presence of a relatively large number of sharp bands, and particularly diagnostic are those near $3030 \, \text{cm}^{-1}$ due to =C—H stretching vibra-

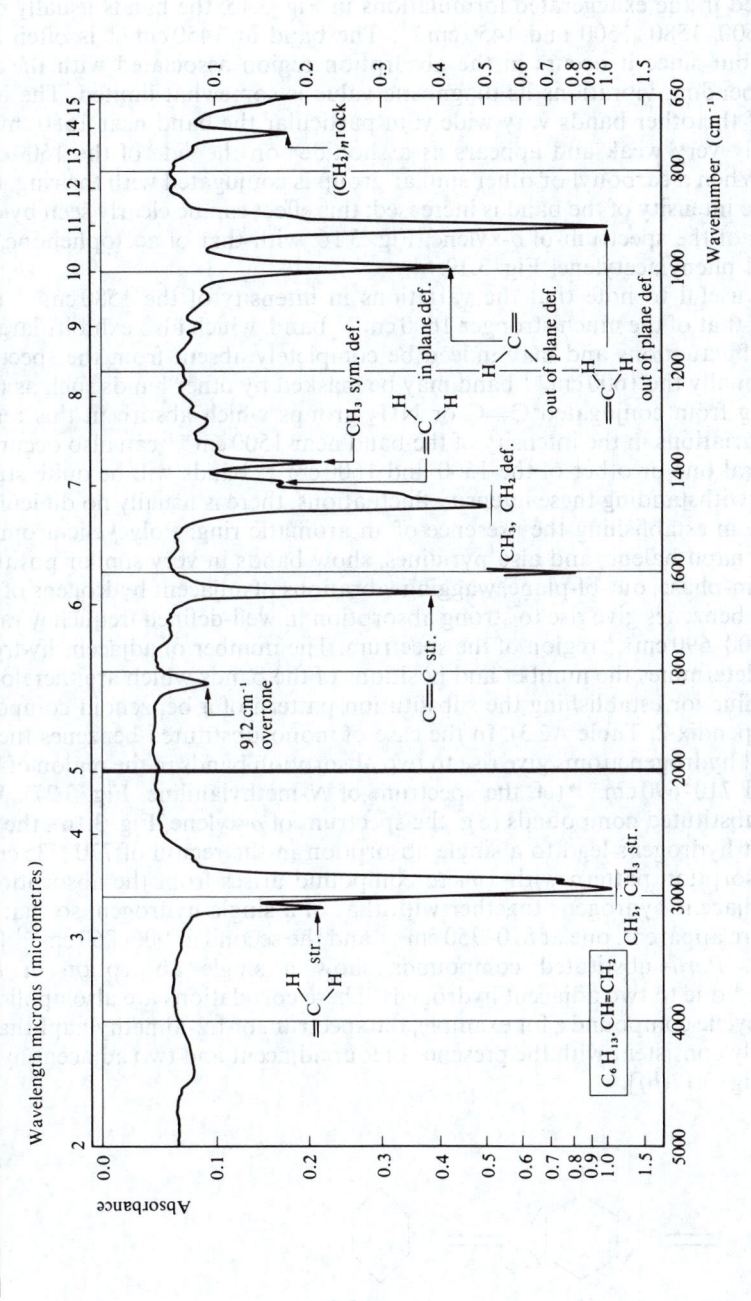

Fig. 3.14 Infrared spectrum of oct-1-ene recorded as a liquid film.

tions and those in the 1600–1450 cm^{-1} region which result from the in-plane skeletal vibrations of the aromatic ring. These latter vibrations involve expansion and contraction of the carbon–carbon bonds within the ring of the type indicated in the exaggerated formulations in Fig. 3.15; the bands usually occur near 1600, 1580, 1500 and 1450 cm^{-1}. The band at 1450 cm^{-1} is often quite strong but since it occurs in the absorption region associated with the alkyl C—H bending vibrations its diagnostic value is somewhat limited. The intensities of the other bands vary widely; in particular the band near 1580 cm^{-1} is normally very weak and appears as a shoulder on the side of the 1600 cm^{-1} band. When a carbonyl or other similar group is conjugated with the ring, however, the intensity of the band is increased; this effect can be clearly seen by comparison of the spectrum of o-xylene, Fig. 3.16, with that of acetophenone, Fig. 3.9, and phenylacetylene, Fig. 3.19.

It is useful to note that the variations in intensity of the 1580 cm^{-1} band parallel that of the much stronger 1600 cm^{-1} band, which also exhibits large intensity fluctuations and may indeed be completely absent from the spectrum. Occasionally the 1600 cm^{-1} band may be masked by other bands such as those resulting from conjugated C=C or NH$_2$ groups which absorb in this region. Wide variations in the intensity of the band near 1500 cm^{-1} can also occur, but in general one or other of the 1500 and 1600 cm^{-1} bands will be quite strong, and notwithstanding these intensity fluctuations, there is usually no difficulty in practice in establishing the presence of an aromatic ring. Polycyclic aromatics such as naphthalene, and also pyridines, show bands in very similar positions.

The in-phase, out-of-plane, wagging vibrations of adjacent hydrogens of substituted benzenes give rise to strong absorption in well-defined frequency ranges in the 900–690 cm^{-1} region of the spectrum. The number of adjacent hydrogen atoms determines the number and positions of the bands which are therefore of great value for establishing the substitution pattern of a benzenoid compound (see Appendix 2, Table A2.3). In the case of monosubstituted benzenes the five adjacent hydrogen atoms give rise to two absorption bands in the region of 770–730 and 710–690 cm^{-1} (cf. the spectrum of N-methylaniline, Fig. 3.27). With ortho-substituted compounds (e.g. the spectrum of o-xylene, Fig. 3.16), the four adjacent hydrogens lead to a single absorption in the region of 770–735 cm^{-1}. The absorption pattern with a meta compound arises from the absorption of three adjacent hydrogens together with that of a single hydrogen, so that two bands are apparent, one at 810–750 cm^{-1} and the second at 900–860 cm^{-1} [Fig. 3.17(a)]. Para-substituted compounds show a single absorption at 860–800 cm^{-1} due to two adjacent hydrogens. These correlations are also applicable to polycyclic compounds; for example, the spectrum of 1,2-dimethylnaphthalene is entirely consistent with the presence of four adjacent and two adjacent hydrogens [Fig. 3.17(b)].

Fig. 3.15   Some infrared active aromatic ring vibrations (exaggerated for illustration).

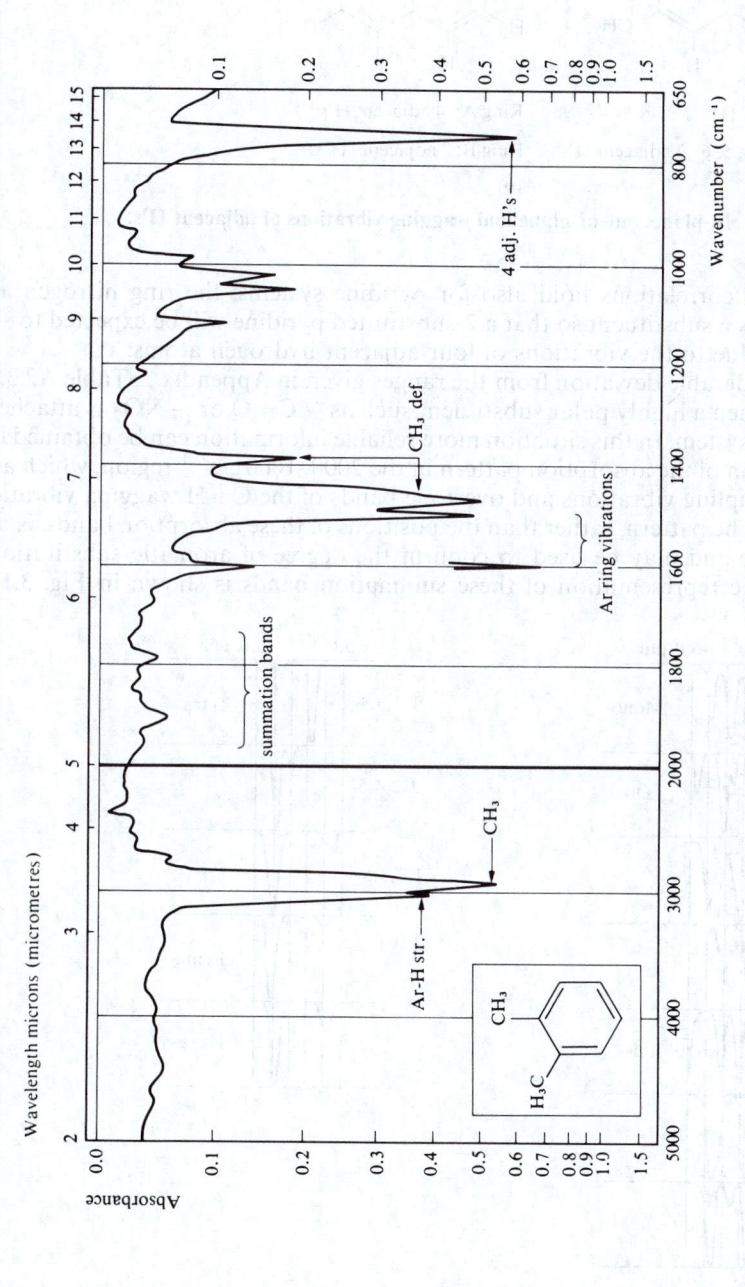

Fig. 3.16   Infrared spectrum of *o*-xylene recorded as a liquid film.

Position 2: 1H   Ring A: 4 adjacent H's

Positions 4, 5, 6: 3 adjacent H's   Ring B: 2 adjacent H's

(a)   (b)

Fig. 3.17 In-plane, out-of-plane and wagging vibrations of adjacent H's.

These correlations hold also for pyridine systems; the ring nitrogen atom counts as a substituent so that a 2-substituted pyridine will be expected to show a band due to the vibrations of four adjacent hydrogen atoms.

Considerable deviation from the ranges given in Appendix 2, Table A2.3, can occur when a highly polar substituent such as —C=O or —NO$_2$ is attached to the ring system. In this situation more reliable information can be obtained from inspection of the absorption pattern in the 2000–1600 cm^{-1} region, which arises from coupling vibrations and overtone bands of the C—H wagging vibrational modes. The pattern, rather than the positions of these absorption bands, is characteristic and may be used to confirm the degree of aromatic substitution. A schematic representation of these summation bands is shown in Fig. 3.18. It

Fig. 3.18 Schematic representation of summation bands. Data reproduced from C. W. Young, R. B. DuVall and N. Wright (1951). *Analyt. Chem.,* 23, 709.

should be noted that the absorption is extremely weak and the bands are frequently barely visible when the spectrum is recorded under normal conditions. Hence to enable the absorption patterns to be recognised, neat liquids or concentrated solutions of solids ($\sim 10\%$) should be examined in a 1.0-mm cell.

The correlation table for aromatic compounds is in Appendix 2, Table A2,3.

Alkynes and allenes. Absorptions arising from multiple bond stretching vibrations are important features in the i.r. spectra of both these types of compounds.

Alkynes. Monosubstituted alkynes are characterised by a strong sharp absorption band near $3320\,cm^{-1}$ arising from the $\equiv C-H$ stretching vibration. Bonded N—H and O—H bands also appear in this region but they are, in contrast, quite broad and cannot be confused with the $\equiv C-H$ band. The $C\equiv C$ stretching vibration gives rise to a weak absorption in the $2260-2100\,cm^{-1}$ region of the spectrum, and a frequency difference of about $100\,cm^{-1}$ between mono- and disubstituted alkynes allows them to be differentiated. The intensity of the $C\equiv C$ band is variable, and while it is readily observed in the spectrum of a monosubstituted alkyne, it may be very weak or absent from the spectrum of a disubstituted alkyne, depending on the nature of the substituents. Symmetrically disubstituted alkynes, such as acetylenedicarboxylic acid, do not exhibit absorption while, in contrast, if the two substituents are sufficiently different in character that the $C\equiv C$ bond is made more polar, a relatively strong band may be observed. The C—H bending absorption of monosubstituted acetylenes occurs in the range $680-610\,cm^{-1}$ and is usually quite strong; aromatic acetylenic compounds show two bands in this region. A broad band in the $1300-1200\,cm^{-1}$ range is believed to be an overtone or combination band derived from the C—H bending vibration. The spectrum of phenylacetylene, Fig. 3.19, is an instructive example; in addition to the strong $\equiv C-H$ band at $3310\,cm^{-1}$ and $C\equiv C$ stretching band at $2100\,cm^{-1}$, other noteworthy and clearly visible features include the sharp aromatic C—H stretching band at $3070\,cm^{-1}$, the fairly strong 'ring breathing' vibration bands in the $1600-1420\,cm^{-1}$ region and the strong C—H in-phrase wagging bands at $770\,cm^{-1}$ and $690\,cm^{-1}$; these latter give rise to weak overtone and combination bands in the $2000-1650\,cm^{-1}$ region which are characteristic of a monosubstituted benzene.

Allenes. Allenes show a moderately intense band (sometimes as a double peak) at $2000-1900\,cm^{-1}$ due to the asymmetric $C=C=C$ stretching vibration, which can be seen (near $1970\,cm^{-1}$) in the spectrum of 1-bromo-3-methylbuta-1,2-diene (Fig. 3.20). A terminal $=CH_2$ group gives rise to a strong band near $850\,cm^{-1}$, with an overtone near $1700\,cm^{-1}$, and is the result of the out-of-plane CH_2 wagging vibration analogous to the CH_2 wagging vibration in vinyl compounds.

The correlation tables for alkynes and allenes are in Appendix 2, Table A2.4.

Alcohols and phenols. Both these classes of compounds are characterised by the strong absorption resulting from the O—H stretching modes; the position and shape of the bands are sensitive to the electronic and steric features of the compound and also to the physical state of the sample. Absorption bands arising from C—O stretching and O—H bending vibrations are also of diagnostic value.

Examination of dilute solutions of simple alcohols or phenols in a non-polar solvent such as carbon tetrachloride reveals the free O—H stretching band in

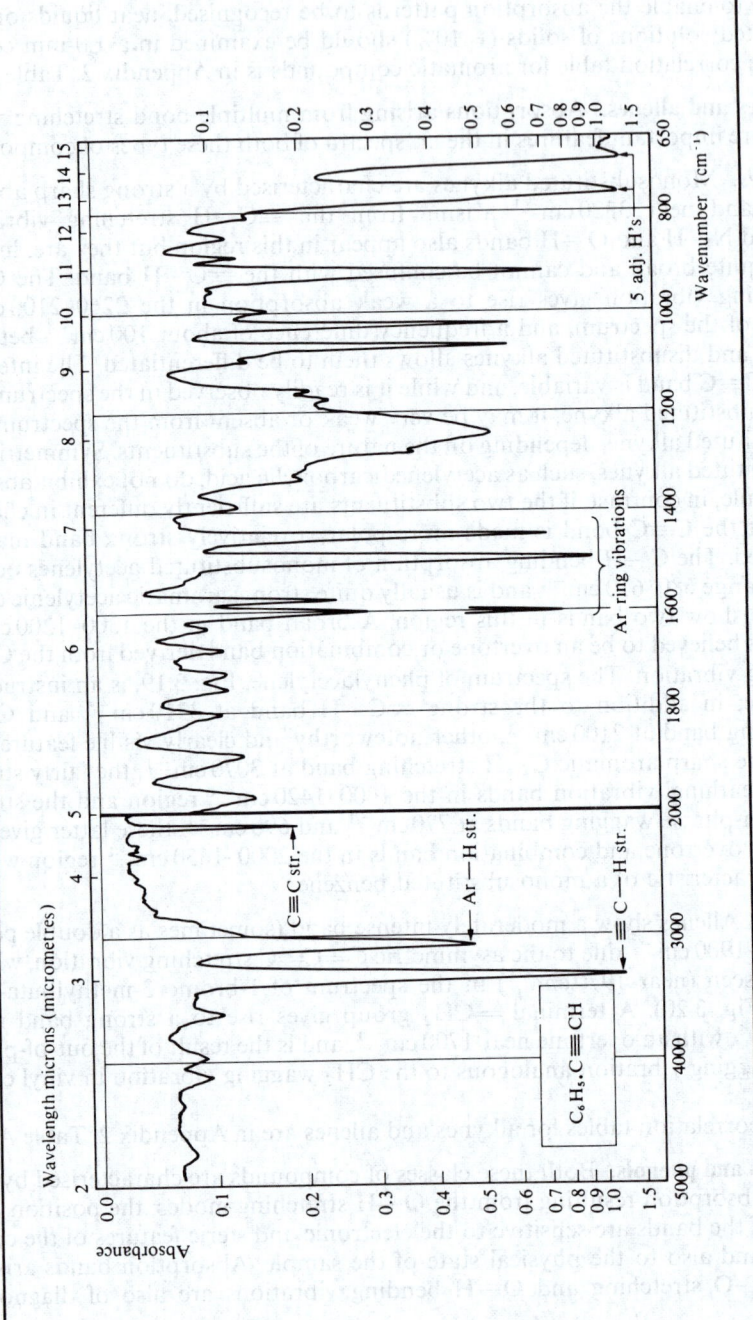

Fig. 3.19 Infrared spectrum of phenylacetylene recorded as a liquid film.

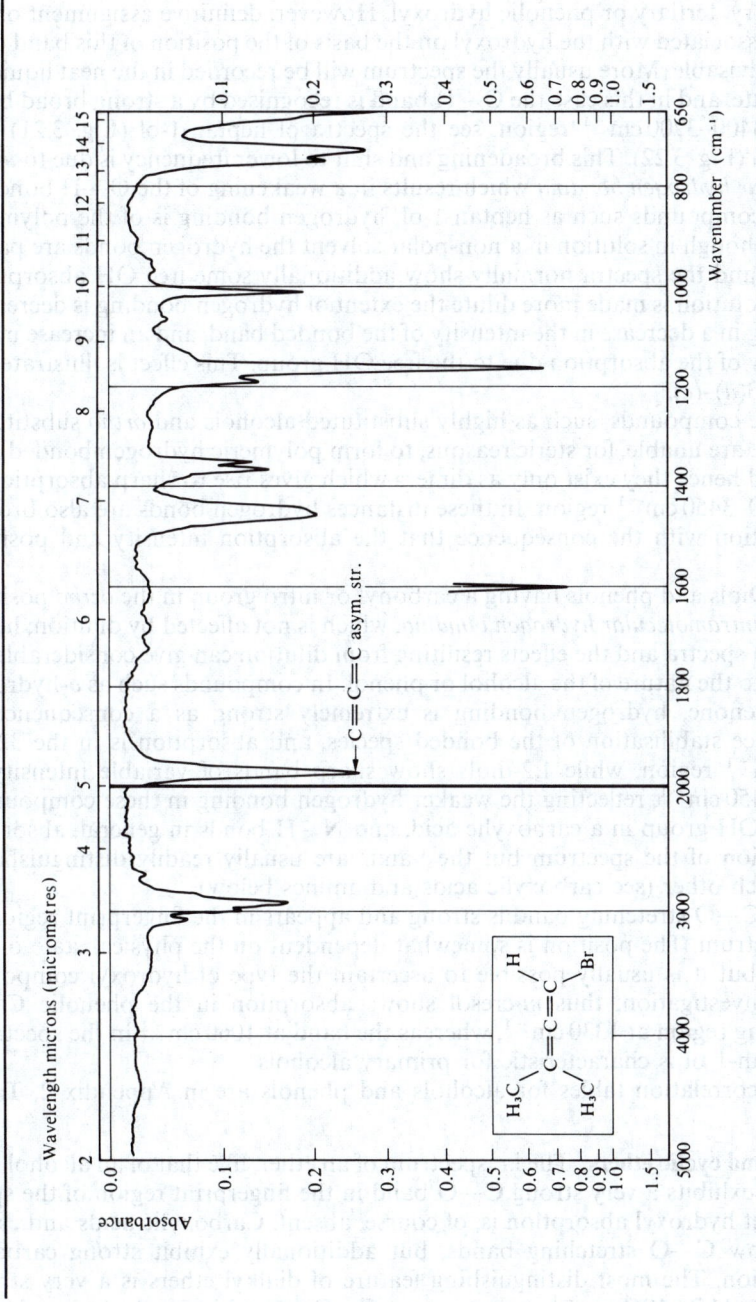

Fig. 3.20   Infrared spectrum of 1-bromo-3-methylbuta-1,2-diene recorded as a liquid film.

the 3650–3590 cm^{-1} region; the precise position of the band has been correlated with the nature of the carbon atom to which the hydroxyl group is attached. Thus the absorption frequency shifts to lower values in the order primary, secondary, tertiary or phenolic hydroxyl. However, definitive assignment of the group associated with the hydroxyl on the basis of the position of this band only is not advisable. More usually the spectrum will be recorded in the neat liquid or solid state, and in this case the O—H band is recognised by a strong broad band in the 3400–3200 cm^{-1} region; see the spectra of heptan-1-ol (Fig. 3.21) and *m*-cresol (Fig. 3.22). This broadening and shift to lower frequency is due to *inter-molecular hydrogen bonding* which results in a weakening of the O—H bond. In simple compounds such as heptan-1-ol, hydrogen bonding is of the polymeric type, although in solution in a non-polar solvent the hydrogen bonds are partly broken and the spectra normally show additionally some free OH absorption. As the solution is made more dilute the extent of hydrogen bonding is decreased resulting in a decrease in the intensity of the bonded band, and an increase in the intensity of the absorption due to the free OH group. This effect is illustrated in Fig. 3.23(*a*)–(*c*).

Some compounds, such as highly substituted alcohols and *ortho* substituted phenols, are unable, for steric reasons, to form polymeric hydrogen bonded species, and hence they exist only as dimers which gives rise to sharp absorption in the 3550–3450 cm^{-1} region. In these instances hydrogen bonds are also broken on dilution with the consequence that the absorption intensity and position change.

1,2-Diols and phenols having a carbonyl or nitro group in the *ortho* position exhibit *intramolecular hydrogen bonding*, which is not affected by dilution; hence solution spectra and the effects resulting from dilution can give considerable insight into the nature of the alcohol or phenol. In compounds such as *o*-hydroxy-acetophenone, hydrogen bonding is extremely strong as a consequence of resonance stabilisation of the bonded species, and absorption is in the 3200–2500 cm^{-1} region, while 1,2-diols show sharp bands of variable intensity at 3570–3450 cm^{-1}, reflecting the weaker hydrogen bonding in these compounds.

The OH group in a carboxylic acid, and N—H bonds in general, absorb in this region of the spectrum but the bands are usually readily distinguishable from each other (see carboxylic acids and amines below).

The C—O stretching band is strong and appears in the fingerprint region of the spectrum. The position is somewhat dependent on the physical state of the sample but it is usually possible to ascertain the type of hydroxyl compound under investigation; thus *m*-cresol shows absorption in the phenolic C—O stretching region at 1330 cm^{-1}, whereas the band at 1060 cm^{-1} in the spectrum of heptan-1-ol is characteristic for primary alcohols.

The correlation tables for alcohols and phenols are in Appendix 2, Table A2.5.

Ethers and cyclic ethers. The i.r. spectrum of an ether, like that of an alcohol or a phenol, exhibits a very strong C—O band in the fingerprint region of the spectrum but hydroxyl absorption is, of course, absent. Carboxylic acids and esters also show C—O stretching bands, but additionally exhibit strong carbonyl absorption. The most distinguishing feature of dialkyl ethers is a very strong band at 1150–1060 cm^{-1} (asymmetric C—O stretching) which is seen at 1120 cm^{-1} in the spectrum of dibutyl ether (Fig. 3.24). Like the C—O stretching

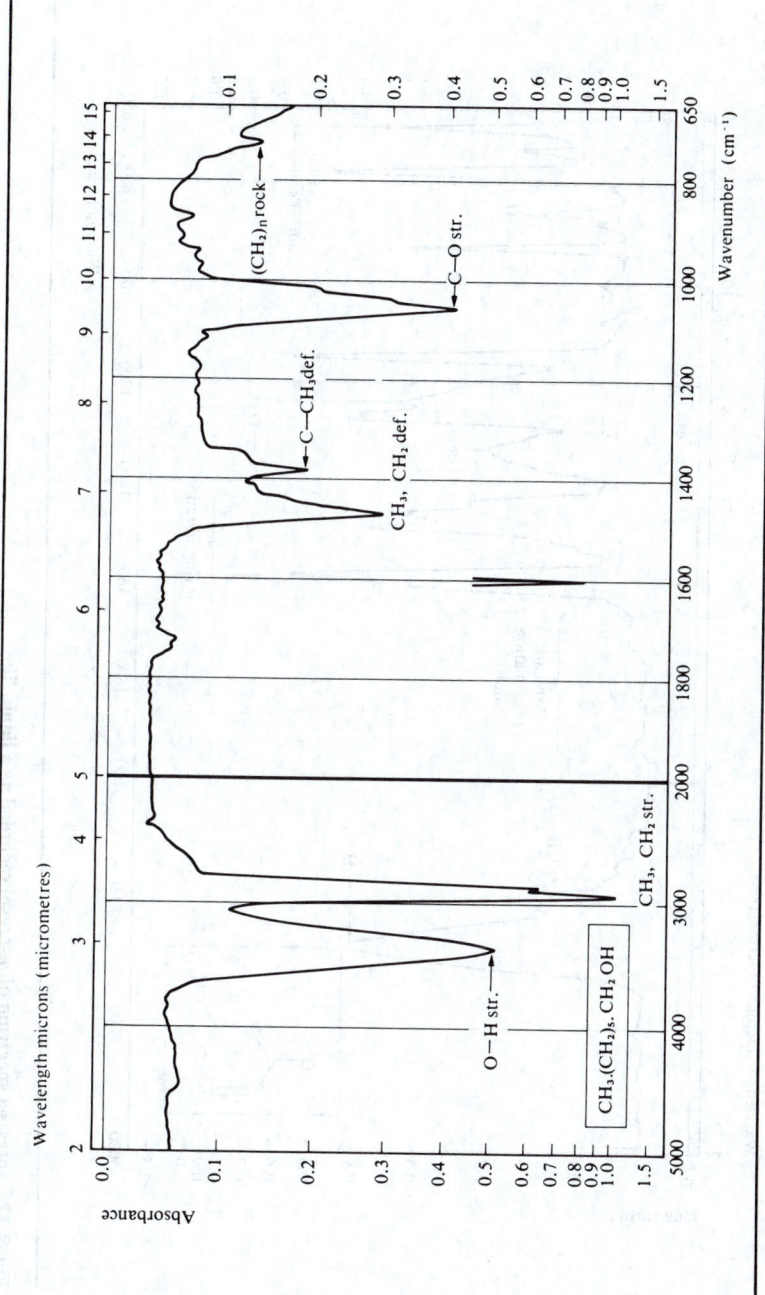

Fig. 3.21 Infrared spectrum of heptan-1-ol recorded as a liquid film.

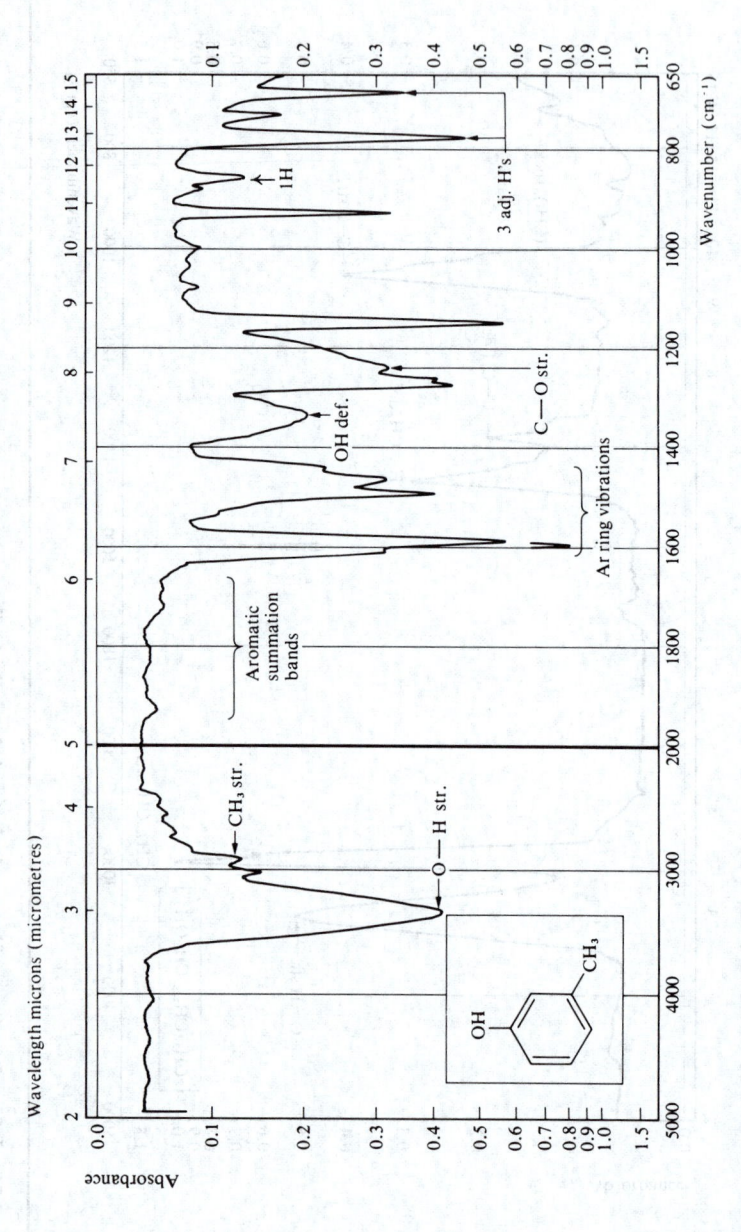

Fig. 3.22   Infrared spectrum of *m*-cresol recorded as a liquid film.

Fig. 3.23(b)

Fig. 3.23(a)

Fig. 3.23(a–c) Effect of dilution with carbon tetrachloride on the solution spectrum of heptan-1-ol, (a) 2.5 per cent w/v; (b) 1 per cent w/v; (c) 0.5 per cent w/v.

Fig. 3.23(c)

bands in alcohols, the ether C—O—C bands are complex in origin and involve strong coupling with other vibrations in the molecule. In dialkyl ethers the symmetric vibration is usually very weak and only the asymmetric stretching band is observed. The spectra of aralkyl ethers exhibit a band at 1270–1230 cm^{-1}, attributed to the C—O—C asymmetric stretching vibration and also a band at 1075–1020 cm^{-1} arising from the symmetric C—O—C vibration. The high frequency band may be considered to be the aromatic C—O stretching band since it is the only one observed (at 1240 cm^{-1}) in diphenyl ether. Strong absorption in this high frequency region is also observed with vinyl ethers. One of the reasons for the higher absorption frequency of the aryl–oxygen bond is the strengthening of the bond by resonance. The two bands can be identified in the spectrum of anisole (Fig. 3.25) at 1240 cm^{-1} and 1040 cm^{-1}.

The correlation tables for ethers and cyclic ethers are in Appendix 2, Table A2.6.

Amines. In dilute solution primary amines show two absorption bands, one near 3500 cm^{-1} and the other near 3400 cm^{-1}, arising from the asymmetric and symmetric stretching vibrations of the two NH bonds (cf. the vibrations of a methylene group, Fig. 3.1). Secondary amines show just one band near 3300 cm^{-1} due to the single N—H stretching vibration, while tertiary amines do not absorb in this region. These characteristic absorptions allow one to dis-

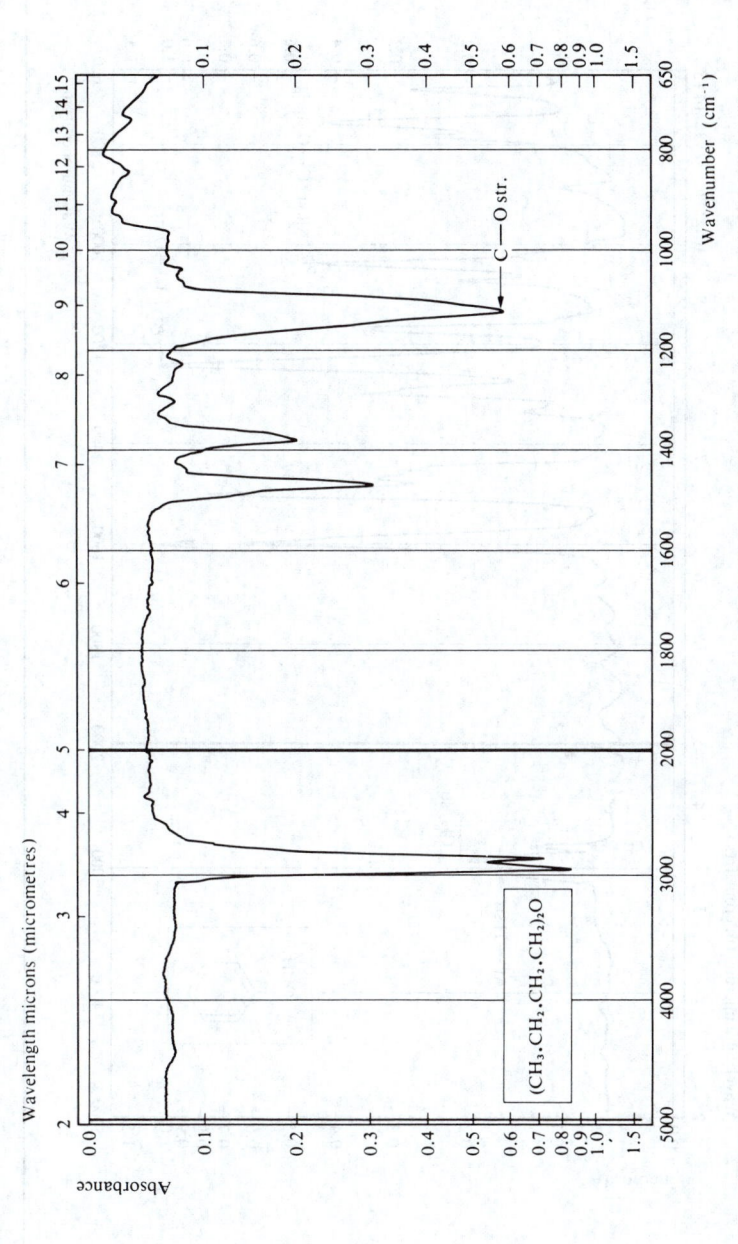

Fig. 3.24 Infrared spectrum of dibutyl ether recorded as a liquid film.

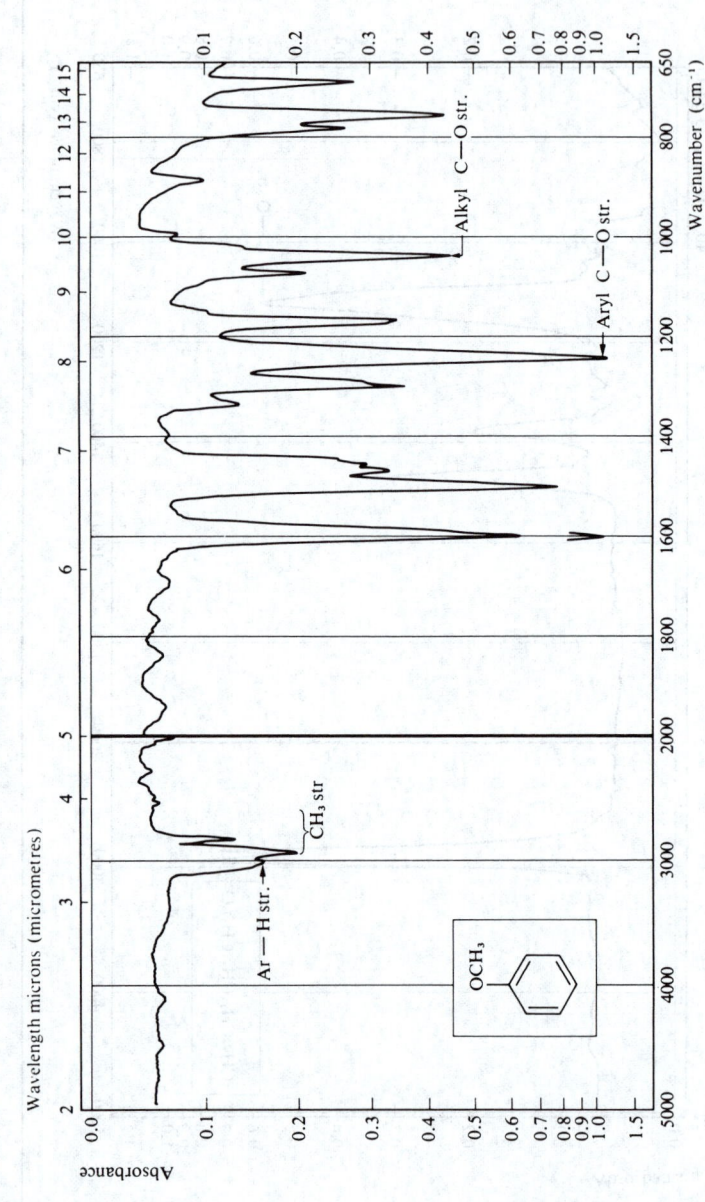

Fig. 3.25   Infrared spectrum of anisole recorded as a liquid film.

tinguish readily between the three classes of amines. Imines also show a single band in the 3500–$3200\,\text{cm}^{-1}$ region. The bands are shifted to lower frequencies in the condensed phase as a result of hydrogen bonding which is, however, much weaker than in hydroxyl compounds because of the lower electronegativity of nitrogen. Because of the weaker hydrogen bonding this frequency shift is not so great and the absorption bands tend to be appreciably narrower than the corresponding bonded OH absorption bands. Aromatic primary and secondary amines absorb at slightly higher frequencies than the corresponding aliphatic amines, and the separation between the symmetric and asymmetric stretching bands in the aromatic primary amines is greater; these spectral differences have been rationalised on the basis of interaction of the lone electron pair on the nitrogen atom with the aromatic ring in arylamines resulting in a shorter, stronger, bond.

A medium to strong broad band in the 1650–$1590\,\text{cm}^{-1}$ region is characteristic of aliphatic primary amines and arises from the NH_2 scissoring vibration; additionally these amines show a medium–strong, broad, multiple absorption band at 850–$750\,\text{cm}^{-1}$ arising from NH_2 twisting and wagging deformations. These absorptions are clearly seen in the spectrum of butylamine (Fig. 3.26), which also shows the characteristic rather broad hydrogen bonded NH_2 stretching band as a closely spaced doublet at $3350\,\text{cm}^{-1}$.

In aliphatic secondary amines the N—H bending band is usually absent from the spectrum or else it is very weak and appears at 1650–$1550\,\text{cm}^{-1}$. In aromatic secondary amines the band is of medium intensity, and appears in the same region, but the assignment is complicated by the presence of aromatic ring vibrations which also occur here, and which are sometimes intensified when a nitrogen atom is attached to the ring.

Carbon–nitrogen stretching vibrations in aliphatic amines occur in the 1190–$1020\,\text{cm}^{-1}$ region of the spectrum, and the exact position of the band has been correlated with the nature of the amino group and the degree of branching at the α-carbon atom. In butylamine, Fig. 3.26, this vibration is observed as a band of medium intensity at $1080\,\text{cm}^{-1}$, the region expected for an aliphatic amino compound with a primary α-carbon atom. The C—N stretching band is also present in aromatic amines and appears as a medium–strong absorption at somewhat higher frequency, i.e. 1360–$1250\,\text{cm}^{-1}$, as a result of the increased strength of the bond in aromatic amines; the position of the absorption is dependent on the nature of the amine. Secondary aromatic amines for example exhibit a strong band in the 1350–$1280\,\text{cm}^{-1}$ region, as in the spectrum of N-methylaniline, Fig. 3.27. Unfortunately although these aromatic C—N bands appear within fairly constant frequency ranges for the primary, secondary and tertiary amines these ranges overlap and unambiguous identification is not possible. Nevertheless, as in the case of N-methylaniline, for example, the appearance of the band in the appropriate C—N stretching region in association with the sharp absorption near $3400\,\text{cm}^{-1}$ can be taken as definitive evidence for the presence of a secondary amino group.

Amine salts and also amino acids are characterised by strong absorptions between 3200 and $2800\,\text{cm}^{-1}$ due to the N—H stretching bands of the ions $-\overset{\oplus}{N}H_3$, $>\overset{\oplus}{N}H_2$, etc. (cf. the methyl, methylene and methine stretching bands in this region), as well as by multiple combination bands in the 2800–$2000\,\text{cm}^{-1}$ region and N—H bending vibrations in the 1600–$1400\,\text{cm}^{-1}$ region.

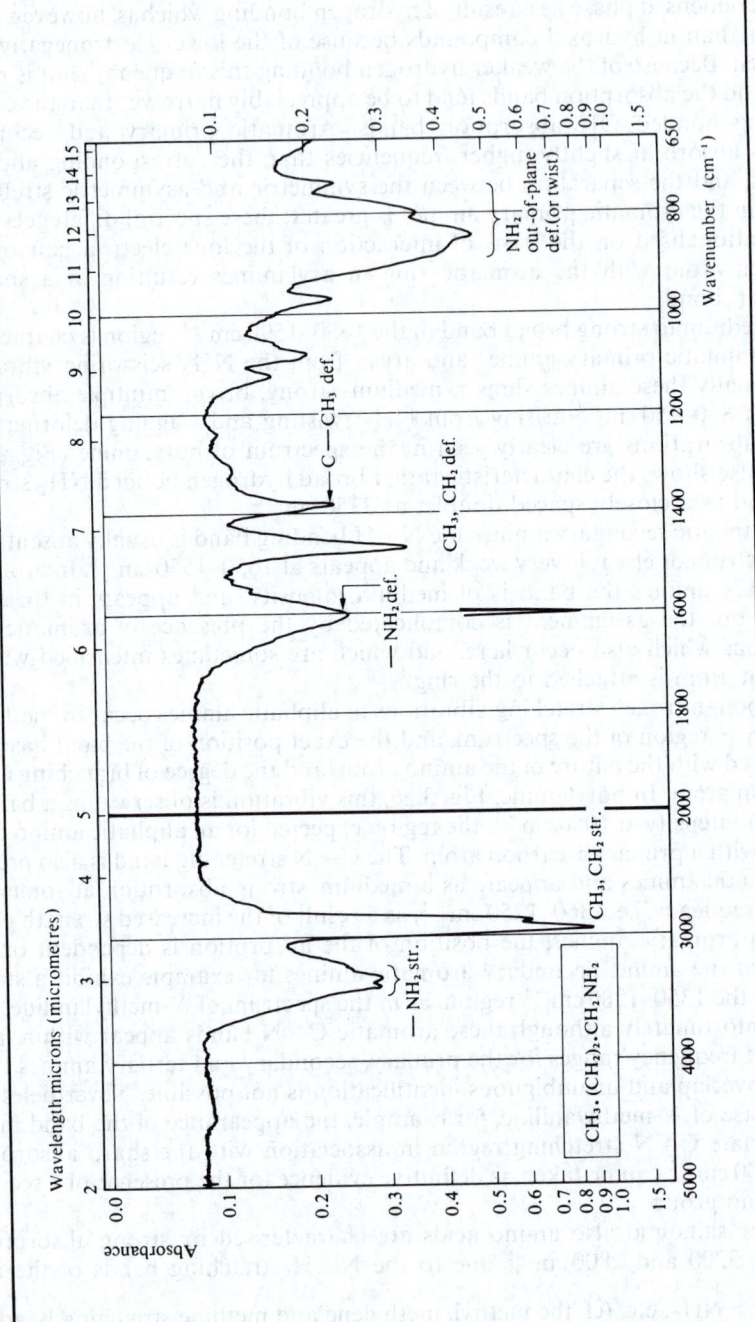

Fig. 3.26 Infrared spectrum of butylamine recorded as a liquid film.

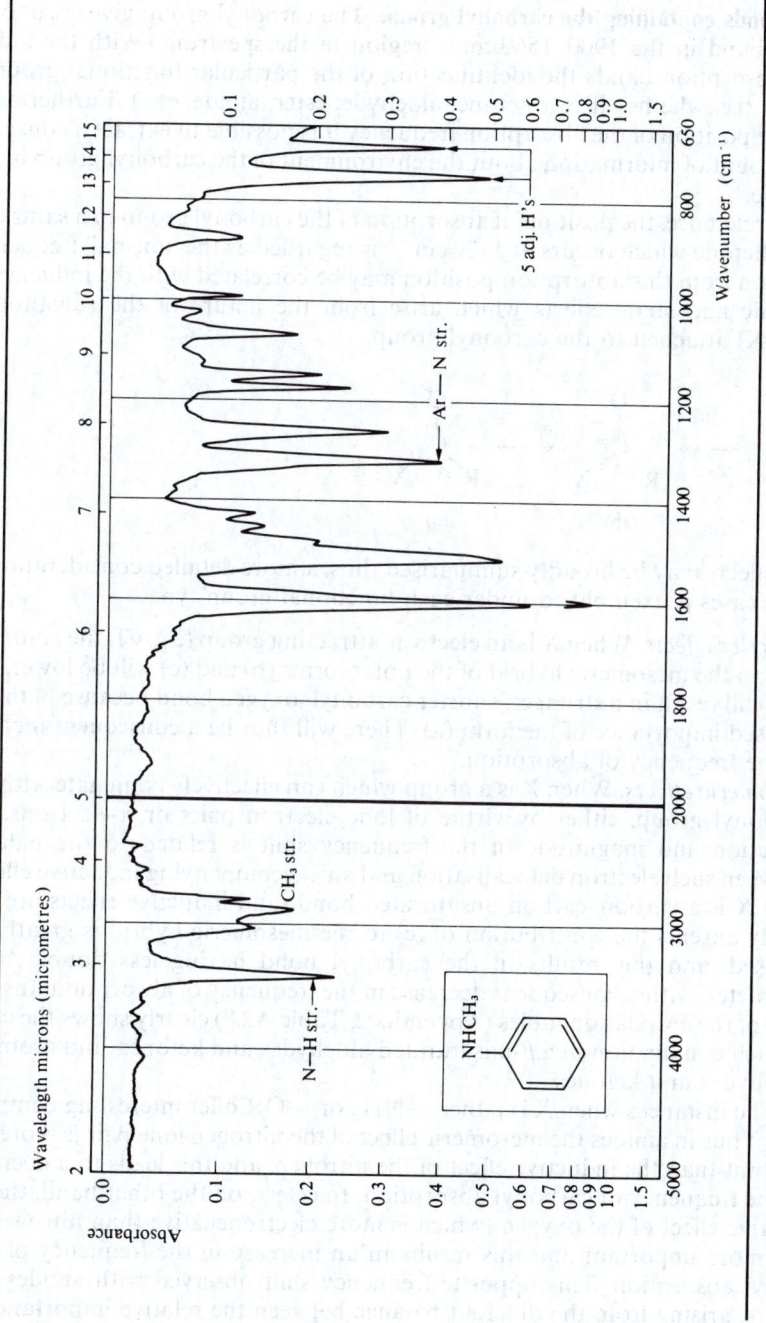

Fig. 3.27 Infrared spectrum of *N*-methylaniline recorded as a liquid film.

The correlation tables for amines and amine salts are in Appendix 2, Table A2.7.

Compounds containing the carbonyl group. The carbonyl group gives rise to an intense band in the $1900–1560\,\text{cm}^{-1}$ region of the spectrum. With the aid of other absorption bands the identification of the particular functional group is possible (i.e. whether it is a ketone, aldehyde, ester, amide, etc.). Furthermore, from the position of the absorption frequency it is possible to extract a considerable amount of information about the environment of the carbonyl group in the molecule.

As a reference, the position of absorption of the carbonyl group in a saturated acyclic ketone which occurs at $1720\,\text{cm}^{-1}$ is regarded as the 'normal' frequency. Deviation from this absorption position may be correlated with the influence of electronic and steric effects which arise from the nature of the substituents (R and X) attached to the carbonyl group.

These effects may be broadly summarised thus; a more detailed consideration of specific cases is exemplified under each functional group type.

1. *Inductive effects.* When X is an electron-attracting group (e.g. Cl) the contribution to the mesomeric hybrid of the polar forms (b) and (c) will be lower, and this will result in a stronger, shorter carbonyl–oxygen bond because of the increased importance of the form (a). There will thus be a consequent increase in the frequency of absorption.
2. *Mesomeric effects.* When X is a group which can effectively conjugate with the carbonyl group, either by virtue of lone electron pairs or π-electrons, the direction and magnitude of the frequency shift is related to the balance between such electron delocalisation and any accompanying inductive effects.

    If X is a carbon–carbon unsaturated bond (and inductive effects are virtually absent) the contribution of (c) to the mesomeric hybrid is greatly increased, and this results in the carbonyl bond having less double bond character, with a consequent decrease in the frequency of absorption. Inspection of the correlation tables (Appendix 2, Table A2.8) clearly shows the effect of such conjugation in α,β-unsaturated aldehydes and ketones, and aromatic aldehydes and ketones.

    The instances when X is either —NH$_2$ or —OR offer interesting comparison. Thus in amides the mesomeric effect of the nitrogen lone pair is more important than the inductive effect of the nitrogen, and this leads to a decrease in the frequency of carbonyl absorption. In esters, on the other hand, the inductive effect of the oxygen (which is more electronegative than nitrogen) is the more important and this results in an increase in the frequency of carbonyl absorption. This opposite frequency shift observed with amides and esters, arising from the different balance between the relative importance of mesomeric and inductive effects is consistent with the relative chemical reactivity which these two groups exhibit.

3. *Bond angle effects.* When the carbonyl-carbon is part of a ring system containing 3, 4 or 5 carbon atoms, the decrease in the bond angle of the two sp^2-hybridised orbitals results in steric strain effects which cause the frequency of absorption of the carbonyl group to be shifted to higher values.

The correlation tables for compounds containing the carbonyl group are in Appendix 2, Table A2.8.

Ketones. Normal acyclic ketones can be recognised by a strong band at $1720\,\text{cm}^{-1}$ (e.g. see the spectrum of 4-methylpentan-2-one, Fig. 3.28). Branching at the α-carbon atoms results in an increase in the C—ĈO—C bond angle and this results in a decrease in frequency of absorption from the normal value of $1720\,\text{cm}^{-1}$ to, for example, $1697\,\text{cm}^{-1}$ as in di-t-butylketone. Conversely as the C—ĈO—C bond angle is decreased the absorption frequency rises, thus cyclopentanone and cyclobutanone absorb at $1750\,\text{cm}^{-1}$ and $1775\,\text{cm}^{-1}$ respectively.

Conjugation of a carbonyl group with a C=C linkage results in a lowering of the frequency of absorption as a result of the decreased double bond character of the carbonyl group. An aliphatic C=C bond in conjugation with a carbonyl group reduces its frequency by about $40\,\text{cm}^{-1}$ to a value of $1680\,\text{cm}^{-1}$; conjugation with acetylenic bonds produces comparable shifts. Most α,β-unsaturated acyclic ketones can exist in two conformations, the s-*cis* and the s-*trans* forms, and two carbonyl absorption bands are shown in compounds of this type. Thus methyl vinyl ketone absorbs at 1716 and $1686\,\text{cm}^{-1}$, and it is assumed that the lower frequency band is due to the *trans* form in which electron delocalisation is expected to be more effective. The effects of conjugation and ring size are additive and may be used to predict the positions of absorption in more complex compounds.

s-*trans*   s-*cis*

Conjugation with an aryl group shifts the frequency to $1715–1695\,\text{cm}^{-1}$; acetophenone itself absorbs at $1692\,\text{cm}^{-1}$ (Fig. 3.9). The band position is dependent also on the nature and position of any ring substituents.

Introduction of a halogen on the α-carbon atom of a ketone leads to a shift to higher frequency provided that the halogen can rotate to eclipse the carbonyl group. The frequency of carbonyl absorption is unaffected by the presence of further halogens on the α-carbon but it can, however, be increased by further substitution on the α'-carbon atom. The magnitude of these frequency shifts increases in the order Br, Cl, F and there is no doubt that the shifts arise from a field effect. The *cis–trans* isomers of 4-t-butylchlorocyclohexanone can be distinguished in this way. In the preferred conformation of the *cis* isomer the carbonyl group and the equatorial halogen are eclipsed, resulting in an increase of the frequency of absorption of the carbonyl group. On the other hand, in the preferred conformation of the *trans* isomer the halogen is axial and the absorption frequency of the carbonyl group in this case is similar to 4-t-butylcyclohexanone.

297

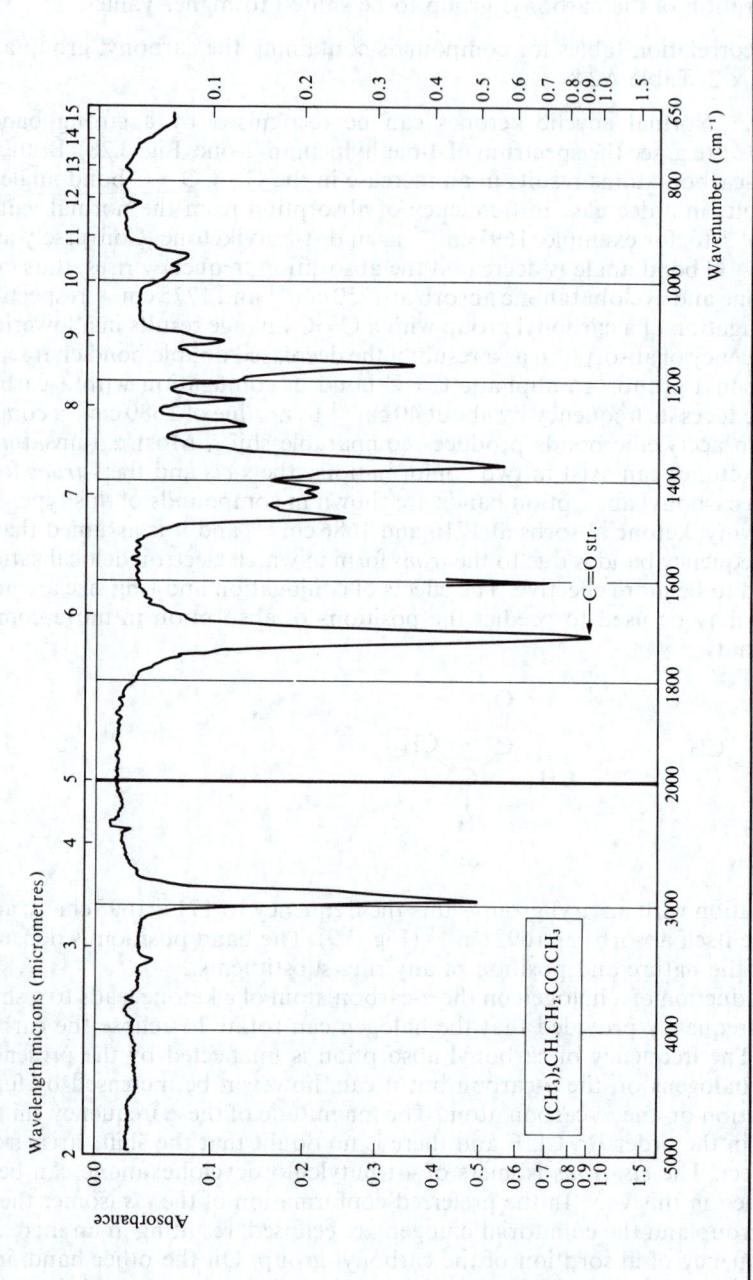

Fig. 3.28 Infrared spectrum of 4-methylpentan-2-one.

$(CH_3)_2CH.CH_2.CO.CH_3$

C=O str.

cis-isomer   trans-isomer

Aldehydes. The carbonyl group in saturated aldehydes absorbs at slightly higher frequencies, 1740–1730 cm^{-1}, than are observed in saturated ketones. As expected, α,β-unsaturation, or attachment of the carbonyl carbon to an aromatic ring, causes a shift of absorption to lower frequencies; aromatic aldehydes generally absorb near 1715–1695 cm^{-1}. Special structural features can, however, cause a large frequency shift of the carbonyl absorption, as in salicylaldehyde, where internal (chelated) hydrogen bonding results in absorption at 1666 cm^{-1}.

Aldehydes can readily be distinguished from ketones by means of aldehydic C—H stretching absorption, which results in two weak bands near 2820 cm^{-1} and 2720 cm^{-1}. The appearance of two bands, rather than one, is due to *Fermi resonance* between the fundamental aldehyde C—H stretching vibration, and the first overtone of the aldehyde C—H deformation band (at 975–780 cm^{-1}). The higher frequency band is frequently not observed in the spectrum of aliphatic aldehydes because of masking by strong saturated C—H stretching vibrations in this region. The 2720 cm^{-1} band can normally be seen, however, as in the spectrum of cyclohexanecarboxaldehyde (Fig. 3.29). Both the bands are visible in the spectrum of anisaldehyde (Fig. 3.30), which has only weak C—H absorption arising from the OCH$_3$ group in this region; note also the lower frequency (1690 cm^{-1}) of the C=O band in this case compared with cyclohexanecarboxaldehyde.

Carboxylic acids. Even in quite dilute solution in non-polar solvents, acids exist essentially as dimeric species, which may be readily explained on the basis of the electronic structure of the carboxyl group. Powerful hydrogen bonding between the molecules, and the strength of these bonds, has been accounted for on the basis of a large contribution of an ionic resonance structure.

As a consequence, the spectrum of a carboxylic acid in the condensed phase (KBr disc, a mull or liquid film), or in concentrated solution, exhibits absorption due to dimeric species and even in very dilute solution only a small proportion of the monomer is present. Under these sample conditions the C=O stretching band appears at 1725–1700 cm^{-1} for saturated acyclic carboxylic acids, but is

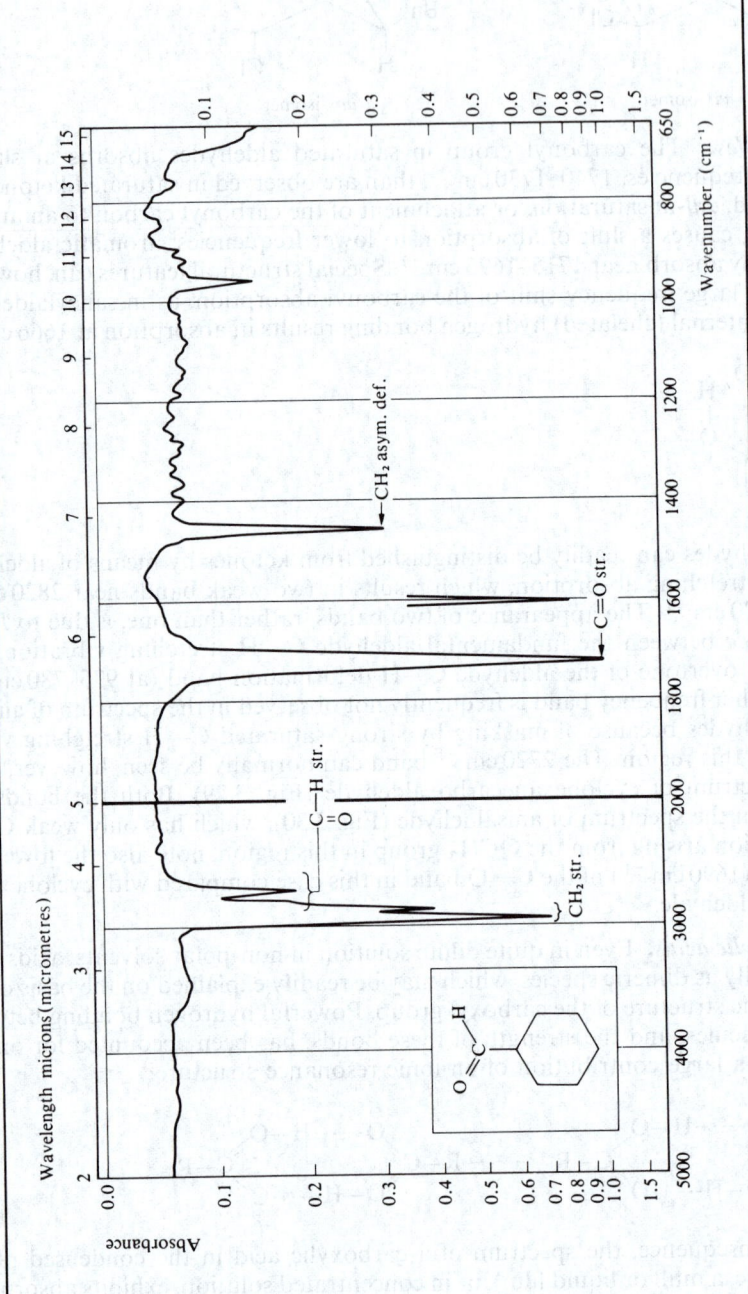

Fig. 3.29 Infrared spectrum of cyclohexanecarboxaldehyde recorded as a liquid film.

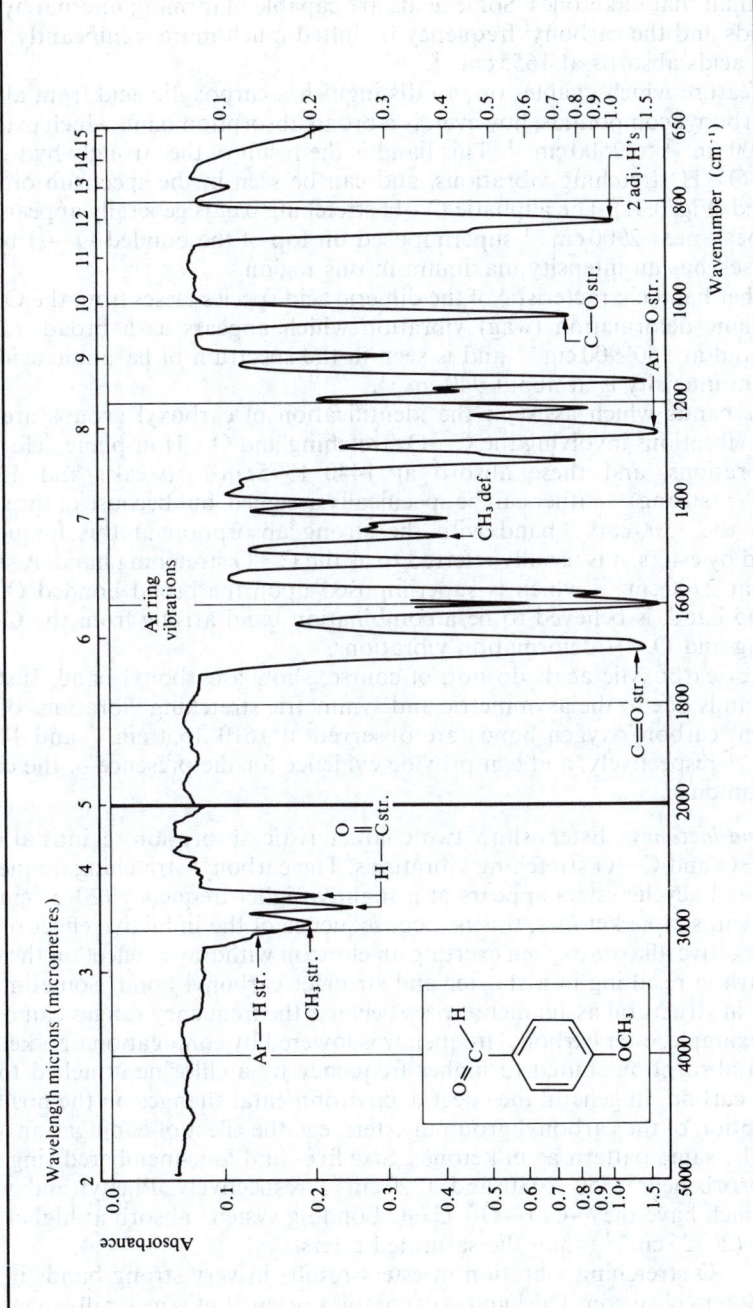

Fig. 3.30 Infrared spectrum of anisaldehyde recorded as a liquid film.

shifted to lower frequencies (by about 20–30 cm^{-1}) when conjugated with a double bond or an aromatic ring. The intensity of the absorption is generally greater than that of ketones. Some acids are capable of forming internal hydrogen bonds and the carbonyl frequency is shifted much more significantly; thus salicylic acids absorbs at 1655 cm^{-1}.

The feature which enables one to distinguish a carboxylic acid from all the other carbonyl compounds, however, is a broad absorption band which extends from 3300 cm^{-1} to 2500 cm^{-1}. This band is the result of the strongly hydrogen bonded O—H stretching vibrations, and can be seen in the spectrum of hexanoic acid (Fig. 3.31). The aliphatic C—H stretching bands generally appear as a jagged peak near 2900 cm^{-1} superimposed on top of the bonded O—H band, which itself has an intensity maximum in this region.

Another band characteristic of the dimeric acid species arises from the O—H out-of-plane deformation (wag) vibration which appears as a broad, rather weak, band at 950–900 cm^{-1} and is seen in the spectrum of hexanoic acid; its maximum intensity is at about 940 cm^{-1}.

Other bands which assist in the identification of carboxyl groups are the coupled vibrations involving the C—O stretching and O—H in-plane deformation vibrations, and these absorb at 1440–1395 cm^{-1} (weak) and 1320–1211 cm^{-1} (strong). Neither can be specifically assigned, but because of the similarity of the 1300 cm^{-1} band with the strong absorption at this frequency exhibited by esters, it is usually referred to as the C—O stretching band. A small band near 2700 cm^{-1}, which is superimposed upon the broad bonded O—H stretching band, is believed to be a combination band arising from the C—O stretching and O—H deformation vibrations.

Salts of carboxylic acids do not, of course, show a carbonyl band. Instead strong bands due to the asymmetric and symmetric stretching vibrations of the equivalent carbon–oxygen bonds are observed at 1610–1550 cm^{-1} and 1420–1300 cm^{-1} respectively, and can provide evidence for the presence of the carboxylate anion.

Esters and lactones. Esters show two characteristic absorption regions arising from C=O and C—O stretching vibrations. The carbonyl stretching frequency in saturated acyclic esters appears at a slightly higher frequency (20–30 cm^{-1}) than that in simple ketones; this is a consequence of the inductive effect of the electronegative alkoxy-oxygen exerting an electron withdrawal effect on the carbonyl carbon resulting in a shorter and stronger carbonyl bond. Sometimes a problem in structural assignment arises because the frequency ranges can overlap; for example, ester carbonyl frequency is lowered by conjugation or a ketone carbonyl absorption shifted to higher frequency by a chlorine attached to an adjacent carbon. In general the effect of environmental changes on the position of absorption of the carbonyl group in esters, e.g. the effect of conjugation, etc., follows the same pattern as in ketones. Six- five- and four-membered ring lactones absorb near 1750, 1780 and 1820 cm^{-1} respectively. Phenyl and vinyl esters, which have the —CO—Ö—C=C bonding system, absorb at higher frequencies (20–25 cm^{-1}) than the saturated esters.

The C—O stretching vibration in esters results in very strong bands in the 1300–1100 cm^{-1} region. The band is of complex origin, but is generally regarded as arising mainly from the acyl-oxygen bond. Unfortunately its diagnostic use is rather limited since other strong bands, e.g. ether C—O stretching, also appear

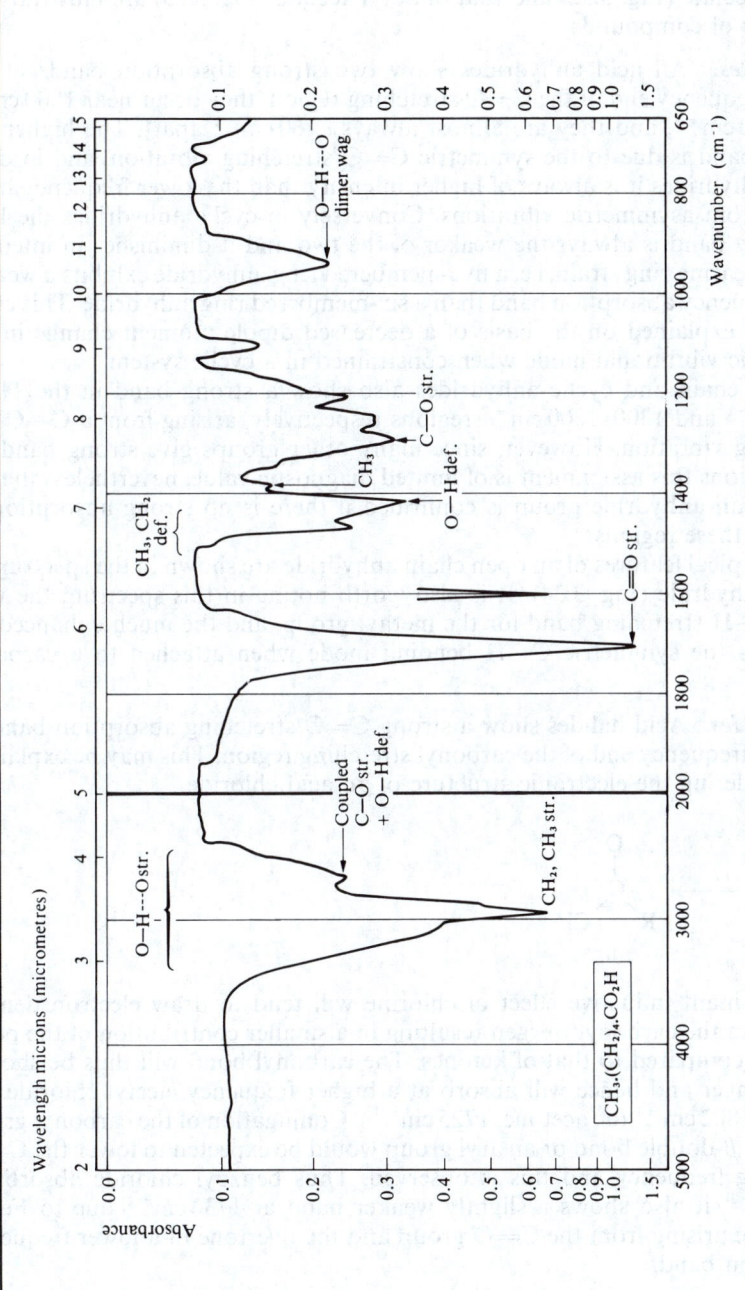

Fig. 3.31 Infrared spectrum of hexanoic acid recorded as a liquid film.

in this region. However, it has been shown that simple esters such as formates, acetates and butyrates all show a strong band near $1200\,cm^{-1}$. The spectrum of phenyl acetate (Fig. 3.32) and that of butyl acetate (Fig. 3.33) are illustrative of this class of compounds.

Anhydrides. All acid anhydrides show two strong absorption bands at the higher frequency end of the C=O stretching region; they occur near $1800\,cm^{-1}$ and $1750\,cm^{-1}$, and they are almost always *c.* $60\,cm^{-1}$ apart. The higher frequency band is due to the symmetric C=O stretching vibration, and in open chain anhydrides it is always of higher intensity than the lower frequency band arising from asymmetric vibrations. Conversely in cyclic anhydrides the high frequency band is always the weaker of the two and it diminishes in intensity with increasing ring strain, i.e. a five-membered ring anhydride exhibits a weaker high frequency absorption band than a six-membered ring anhydride. This effect has been explained on the basis of a decreased dipole moment change in the symmetric vibrational mode when constrained in a cyclic system.

Open chain and cyclic anhydrides also show a strong band in the 1170–$1050\,cm^{-1}$ and 1300–$1200\,cm^{-1}$ regions respectively, arising from a C—O—C stretching vibration. However, since many other groups give strong bands in these regions this assignment is of limited diagnostic value; nevertheless the absence of an anhydride group is confirmed if there is no strong absorption in either of these regions.

The typical features of an open chain anhydride are shown in the spectrum of acetic anhydride (Fig. 3.34). It is also worth noting in this spectrum the very weak C—H stretching band for the methyl group, and the much enhanced intensity of the symmetric C—H bending mode when attached to a carbonyl group.

Acid halides. Acid halides show a strong C=O stretching absorption band at the high frequency end of the carbonyl stretching region. This may be explained by considering the electronic structure of an acid chloride.

The dominant inductive effect of chlorine will tend to draw electron density away from the carbonyl oxygen resulting in a smaller contribution of the polar form (b) compared to that of ketones. The carbonyl bond will thus be shorter and stronger and hence will absorb at a higher frequency; acetyl chloride absorbs at $1802\,cm^{-1}$ (cf. acetone, $1725\,cm^{-1}$). Conjugation of the carbonyl group with an α,β-double bond or an aryl group would be expected to lower the C=O stretching frequency and this is observed. Thus benzoyl chloride absorbs at $1773\,cm^{-1}$; it also shows a slightly weaker band at $1736\,cm^{-1}$ due to Fermi resonance arising from the C=O group and the overtone of a lower frequency absorption band.

Amides. All amides are characterised by a strong carbonyl absorption band, referred to as the 'amide I' band. Primary and secondary amides additionally show bands arising from N—H stretching and bending vibrations. The N—H

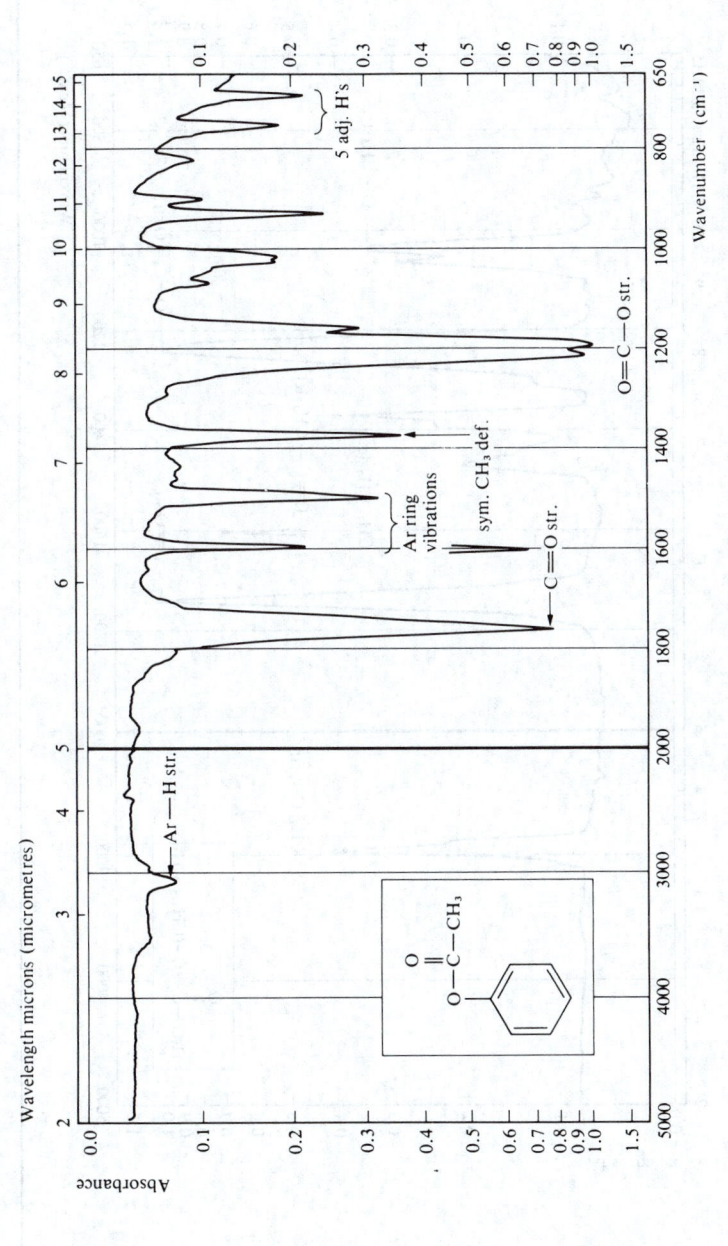

Fig. 3.32   Infrared spectrum of phenyl acetate recorded as a liquid film.

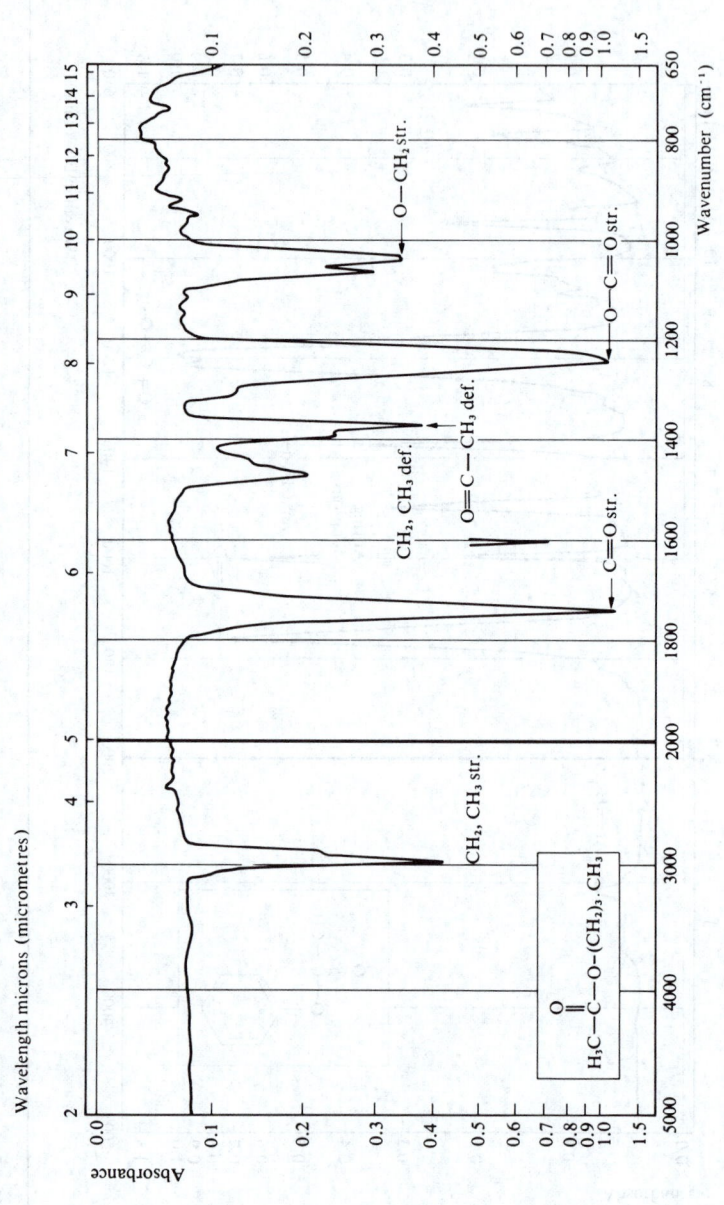

Fig. 3.33 Infrared spectrum of butyl acetate recorded as a liquid film.

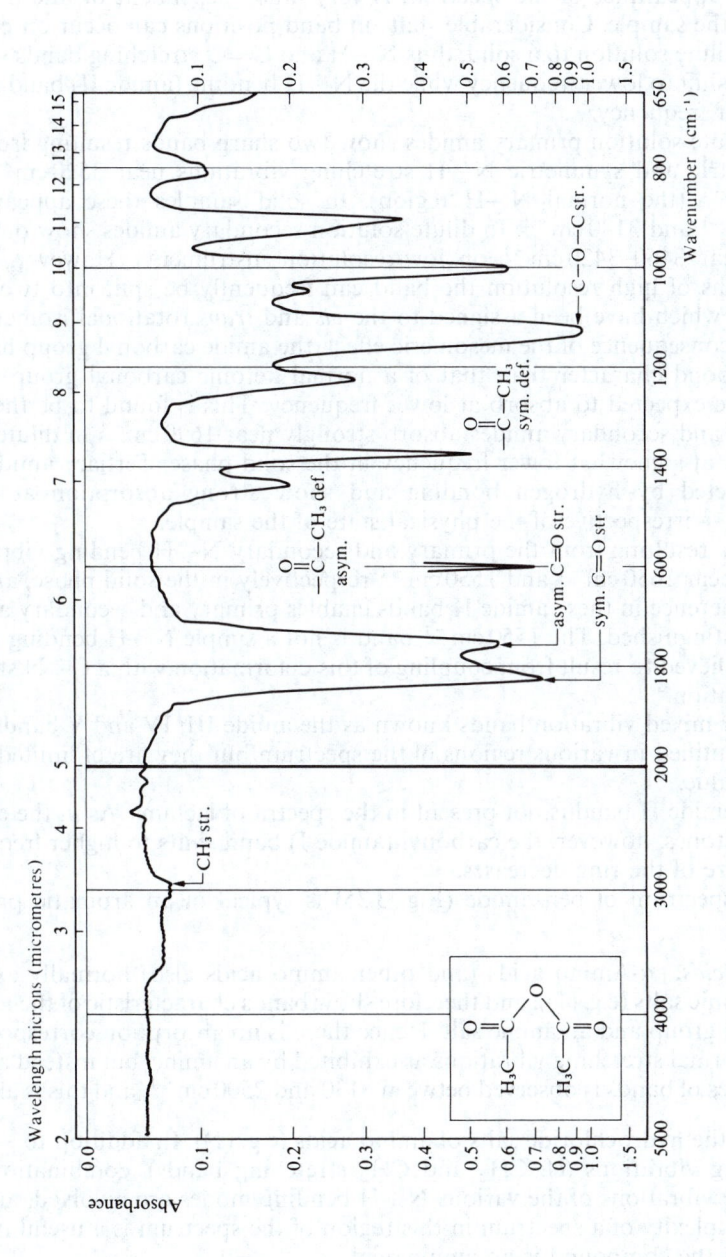

Fig. 3.34 Infrared spectrum of acetic anhydride recorded as a liquid film.

bending absorption is generally at slightly lower frequency than the carbonyl absorption and is referred to as the 'amide II' band.

Amides have a very strong tendency to self-associate by hydrogen bonding, and the appearance of the spectrum is very much dependent on the physical state of the sample. Considerable shifts in band positions can occur on passing from a dilute solution to a solid, thus N—H and C=O stretching bands show a marked shift to lower frequency while the N—H bending (amide II) band moves to higher frequency.

In dilute solution primary amides show two sharp bands resulting from the asymmetric and symmetric N—H stretching vibrations near $3520\,cm^{-1}$ and $3400\,cm^{-1}$ (the normal N—H region). In solid samples these appear near $3350\,cm^{-1}$ and $3180\,cm^{-1}$. In dilute solution secondary amides show only one band near $3460-3420\,cm^{-1}$ on low-resolution instruments. However, under conditions of high resolution the band can frequently be split into two components which have been assigned to the *cis* and *trans* rotational isomers.

As a consequence of the mesomeric effect, the amide carbonyl group has less double bond character than that of a normal ketonic carbonyl group and it would be expected to absorb at lower frequency. This is found to be the case; primary and secondary amides absorb strongly near $1690\,cm^{-1}$ in dilute solution and at somewhat lower frequency in the solid phase. Tertiary amides are not affected by hydrogen bonding and show strong absorption at $1670-1630\,cm^{-1}$ irrespective of the physical state of the sample.

Bands resulting from the primary and secondary N—H bending vibrations appear near $1650\,cm^{-1}$ and $1550\,cm^{-1}$ respectively in the solid phase, and the large difference in these amide II bands enables primary and secondary amides to be distinguished. The $1550\,cm^{-1}$ band is not a simple N—H bending mode, but is believed to result from coupling of this deformation with a C—N stretching vibration.

Other mixed vibration bands known as the amide III, IV and V bands have been identified in various regions of the spectrum but they are of limited diagnostic value.

The amide II band is not present in the spectra of lactams. As in the case of cyclic ketones, however, the carbonyl (amide I) band shifts to higher frequency as the size of the ring decreases.

The spectrum of benzamide (Fig. 3.35) is typical of an aromatic primary amide.

Amino acids. α-Amino acids (and other amino acids also) normally exist as zwitterionic salts (e.g. (1)), and therefore show bands characteristic of the ionised carboxyl group and an amine salt. Hence there is no absorption corresponding to the normal stretching vibrations as exhibited by an amine, but instead a complex series of bands is observed between 3130 and $2500\,cm^{-1}$, and this is also the case for the hydrochloride salts of amino acids (e.g. (2)). In addition to $-\overset{\oplus}{N}H_3$ stretching vibrations (cf. CH_3 and CH_2 stretching bands), combination and overtone vibrations of the various N—H bending modes are involved, and the very complexity of a spectrum in this region of the spectrum is a useful indication that the compound is an amino acid.

For *N*-substituted compounds, such as proline, only the $-\overset{\oplus}{N}H_2$ stretching vibrations are involved and these appear at lower frequency. A relatively promin-

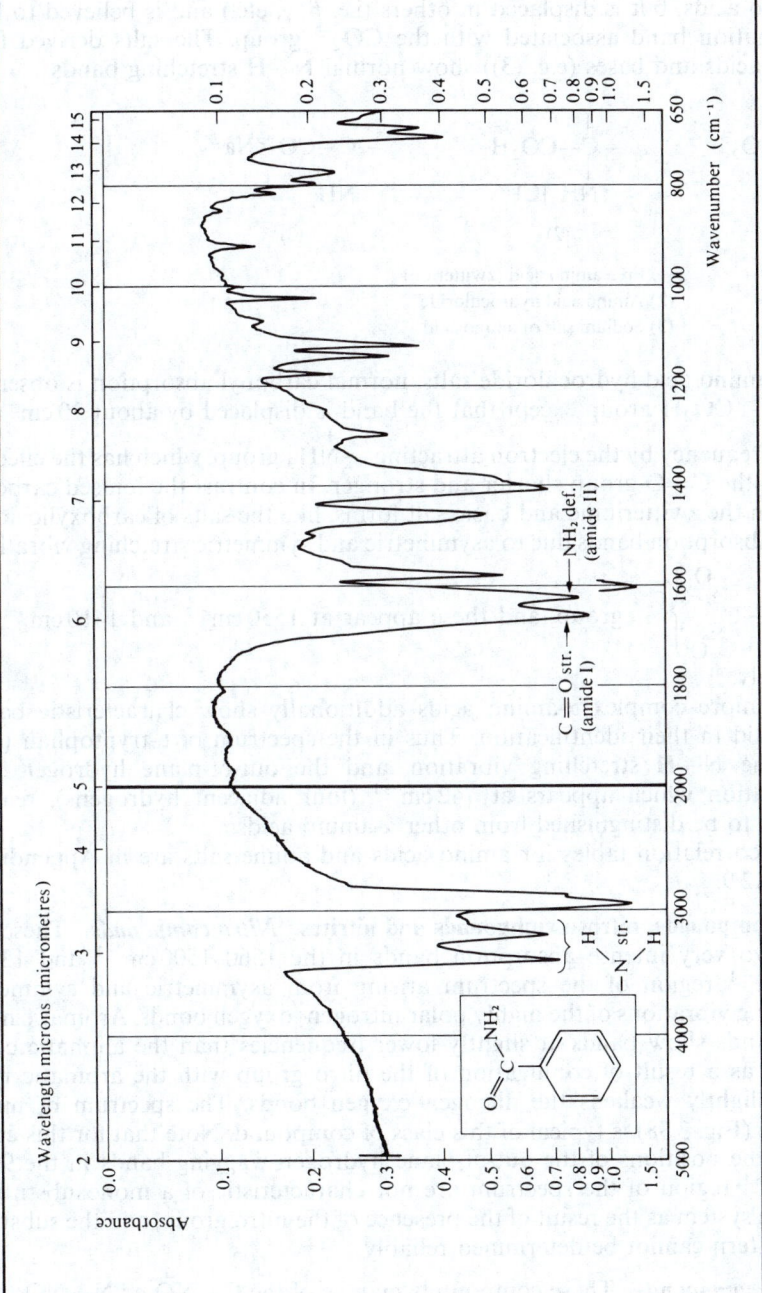

Fig. 3.35   Infrared spectrum of benzamide recorded as a Nujol mull.

ent band between 2200 and 2000 cm^{-1} is found in the spectrum of most amino acids and their hydrochloride salts and can be clearly seen in the spectrum of ($\pm$)-valine (Fig. 3.36) at 2130 cm^{-1}. The band appears at this position in all α-amino acids, but is displaced in others (i.e. β, γ, etc.) and is believed to be a combination band associated with the $CO_2^{\ominus}$ group. The salts derived from amino acids and bases (e.g. (3)) show normal N—H stretching bands.

$$
\begin{array}{ccc}
-\overset{|}{\underset{\underset{\oplus}{\overset{|}{NH_3}}}{C}}-CO_2^{\ominus} & -\overset{|}{\underset{\underset{\oplus}{\overset{|}{NH_3}\}Cl^{\ominus}}}{C}}-CO_2H & -\overset{|}{\underset{\overset{|}{NH_2}}{C}}-CO_2^{\ominus}Na^{\oplus} \\
(1) & (2) & (3)
\end{array}
$$

(1) Free amino acid (zwitterion)
(2) Amino acid hydrochloride
(3) Sodium salt of amino acid

In the amino acid hydrochloride salts, normal carbonyl absorption is observed for the —CO$_2$H group except that the band is displaced by about 20 cm^{-1} to higher frequency by the electron attracting —$\overset{\oplus}{NH_3}$ group, which has the effect of making the C=O group shorter and stronger. In contrast the ionised carboxyl group in the zwitterionic and basic salt forms, like the salts of carboxylic acids, shows absorption bands due to asymmetric and symmetric stretching vibrations

of the $-C\begin{smallmatrix} \nearrow O \\ \\ \searrow O \end{smallmatrix}\Big\}^{\ominus}$ group, and these appear at 1550 cm^{-1} and 1410 cm^{-1} respectively.

The more complex α-amino acids additionally show characteristic bands which aid in their identification. Thus in the spectrum of L-tryptophan (Fig. 3.37) the N—H stretching vibration, and the out-of-plane hydrogen wag deformation which appears at 742 cm^{-1} (four adjacent hydrogens), readily allow it to be distinguished from other α-amino acids.

The correlation tables for amino acids and amine salts are in Appendix 2, Table A2.9.

Nitro compounds, nitroso compounds and nitrites. *Nitro compounds.* These exhibit two very intense absorption bands in the 1560–1500 cm^{-1} and 1350–1300 cm^{-1} region of the spectrum arising from asymmetric and symmetric stretching vibrations of the highly polar nitrogen–oxygen bonds. Aromatic nitro compounds show bands at slightly lower frequencies than the aliphatic compounds as a result of conjugation of the nitro group with the aromatic ring, which slightly weakens the nitrogen–oxygen bonds. The spectrum of nitrobenzene (Fig. 3.38) is typical of this class of compound. Note that for this compound the positions of the out-of-plane hydrogen wagging bands in the 900–700 cm^{-1} region of the spectrum are not characteristic of a monosubstituted benzene system as the result of the presence of the nitro group, and the substitution pattern cannot be determined reliably.

Nitroso compounds. These compounds may be of the C—NO or N—NO type. Tertiary C-nitroso compounds tend to dimerise, and secondary and primary C-nitroso compounds readily rearrange to oximes. In the monomeric state they

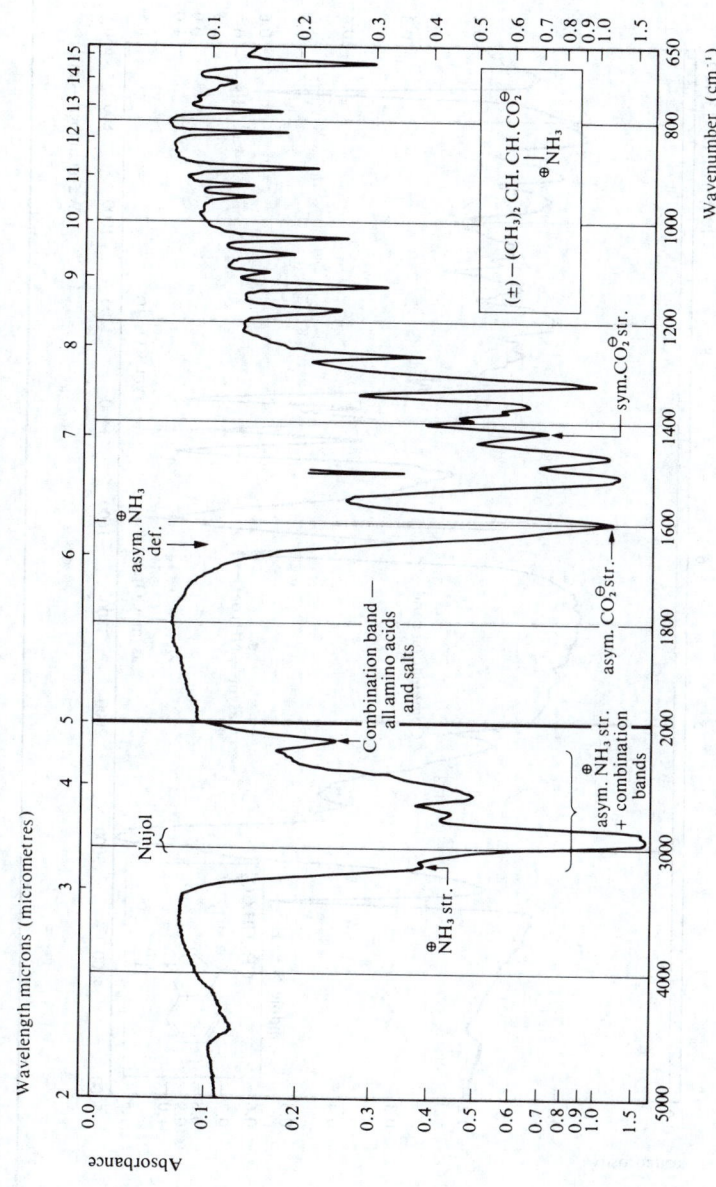

Fig. 3.36   Infrared spectrum of (±)-valine recorded as a Nujol mull.

Fig. 3.37   Infrared spectrum of L-tryptophan recorded as a Nujol mull.

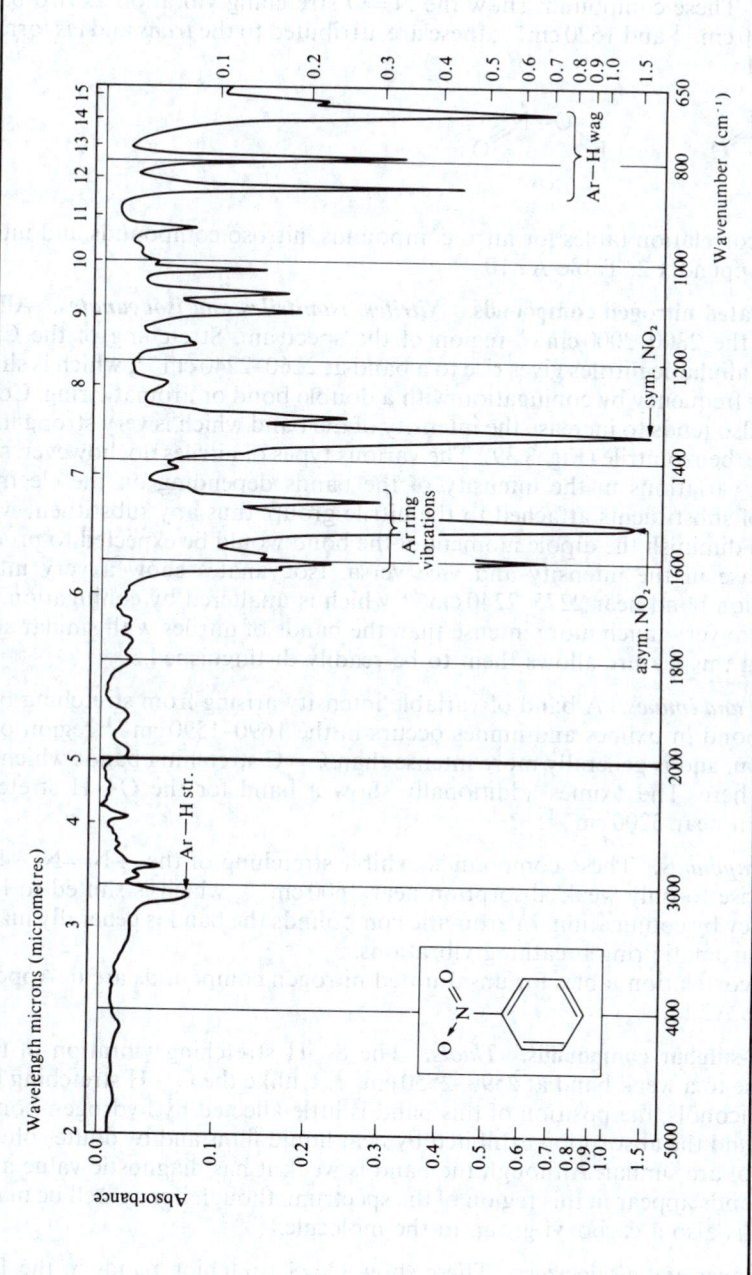

Fig. 3.38 Infrared spectrum of nitrobenzene recorded as a liquid film.

absorb in the 1600–1500 cm^{-1} region, but in solution they exist preferentially as dimers and then absorb near 1290 cm^{-1} (*cis*) or 1400 cm^{-1} (*trans*). *N*-Nitroso compounds show a band near 1450 cm^{-1} in solution in carbon tetrachloride.

Nitrites. These compounds show the N=O stretching vibration as two bands near 1660 cm^{-1} and 1620 cm^{-1}; these are attributed to the *trans* and *cis* forms of the nitrite.

The correlation tables for nitro compounds, nitroso compounds and nitrites are in Appendix 2, Table A2.10.

Unsaturated nitrogen compounds. *Nitriles, isonitriles and isocyanates.* All absorb in the 2300–2000 cm^{-1} region of the spectrum. Stretching of the C≡N bond in aliphatic nitriles gives rise to a band at 2260–2240 cm^{-1}, which is shifted to lower frequency by conjugation with a double bond or aromatic ring. Conjugation also tends to increase the intensity of the band which is very strong in, for example, benzonitrile (Fig. 3.39). The various types of nitriles do, however, show marked variations in the intensity of the bands depending on the electronic effects of substituents attached to the nitrile group; thus any substituent which tends to diminish the dipole moment of the bond would be expected to produce a decrease in the intensity and vice versa. Isocyanates show a very intense absorption band near 2275–2240 cm^{-1} which is unaltered by conjugation. The bands are very much more intense than the bands of nitriles with similar structure and this feature allows them to be readily distinguished.

Oximes and imines. A band of variable intensity arising from stretching of the C=N bond in oximes and imines occurs in the 1690–1590 cm^{-1} region of the spectrum, and is generally more intense than C=C stretching bands which also appear here. The oximes additionally show a band for the O—H stretching vibration near 3200 cm^{-1}.

Azo compounds. These compounds exhibit stretching of the —N=N— bond giving rise to only weak absorption near 1600 cm^{-1}, which is shifted to lower frequency by conjugation. In aromatic compounds the band is generally masked by the aromatic ring breathing vibrations.

The correlation tables for unsaturated nitrogen compounds are in Appendix 2, Table A2.11.

Organo-sulphur compounds. *Thiols.* The S—H stretching vibration of thiols gives rise to a weak band at 2590–2550 cm^{-1}. Unlike the O—H stretching band in the alcohols, the position of this band is little affected by hydrogen-bonding effects, and the absorption exhibited by neat liquid films and by dilute solutions of a thiol are similar. Although the band is weak it has diagnostic value as few other bands appear in this region of the spectrum, though it may well be masked if there is also a carboxyl group in the molecule.

Thioketones and dithioesters. These show C=S stretching bands in the 1270–1190 cm^{-1} region. Since the C=S bond is not as polar or as strong as the C=O bond the absorption band is not very intense and appears in the low frequency

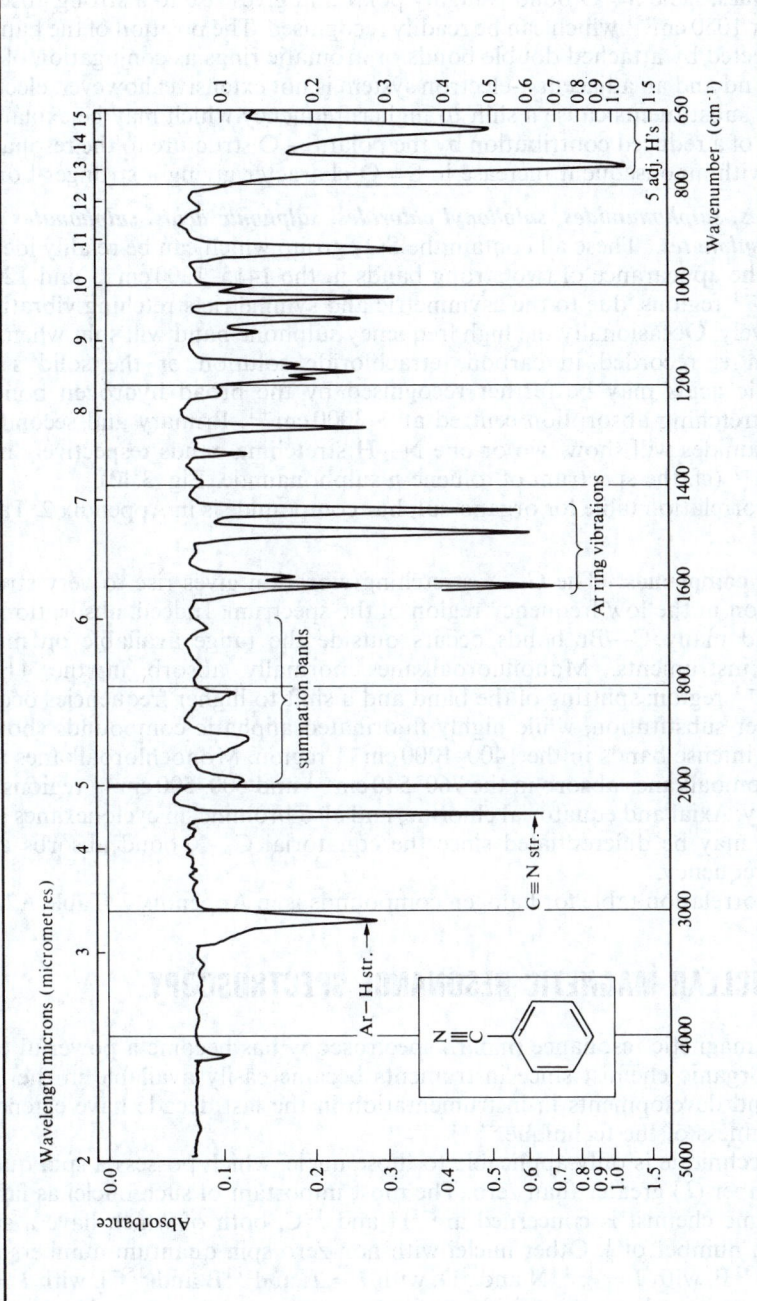

Fig. 3.39   Infrared spectrum of benzonitrile recorded as a liquid film.

315

region of the spectrum; coupling with other bands in this region can make identification difficult.

Sulphoxides. The S=O bond is highly polar and gives rise to a strong absorption near $1050\,cm^{-1}$ which can be readily recognised. The position of the band is little affected by attached double bonds or aromatic rings as conjugation of the S=O bond and an adjacent π-electron system is not extensive; however, electronegative substituents cause a shift to higher frequency which may be explained in terms of a reduced contribution by the polar S—O structure to the resonance hybrid with a consequent increase in S=O character giving a stronger bond.

Sulphones, sulphonamides, sulphonyl chlorides, sulphonic acids, sulphonates and organic sulphates. These all contain the SO_2 group which can be readily identified by the appearance of two strong bands in the 1415–$1300\,cm^{-1}$ and 1200–$1120\,cm^{-1}$ regions, due to the asymmetric and symmetric stretching vibrations respectively. Occasionally the high frequency sulphone band will split when the spectrum is recorded in carbon tetrachloride solution or the solid state. Sulphonic acids may be further recognised by the broad hydrogen bonded O—H stretching absorption centred at $\sim 3000\,cm^{-1}$. Primary and secondary sulphonamides will show two or one N—H stretching bands respectively near $3300\,cm^{-1}$ (cf. the spectrum of toluene-*p*-sulphonamide, Fig. 3.40).

The correlation table for organo-sulphur compounds is in Appendix 2, Table A2.12.

Halogen compounds. The C—X stretching vibration gives rise to very strong absorption in the low frequency region of the spectrum. Indeed, absorption by C—I and many C—Br bonds occurs outside the range available on many routine instruments. Monofluoroalkanes normally absorb in the 1100–$1000\,cm^{-1}$ region; splitting of the band and a shift to higher frequencies occurs on further substitution, while highly fluorinated aliphatic compounds show a series of intense bands in the 1400–$1000\,cm^{-1}$ region. Monochloroalkanes and monobromoalkanes absorb in the 760–$540\,cm^{-1}$ and 600–$500\,cm^{-1}$ regions respectively. Axial and equatorial chlorine, and also bromine, in cyclohexanes and steroids, may be differentiated since the equatorial C—X bond absorbs at a higher frequency.

The correlation table for halogen compounds is in Appendix 2, Table A2.13.

3.2 NUCLEAR MAGNETIC RESONANCE SPECTROSCOPY

Nuclear magnetic resonance (n.m.r.) spectroscopy has become a powerful tool for the organic chemist since instruments became easily available in the late 1950s, and developments in instrumentation in the last decade have extended the usefulness of the technique.[3a–g]

The technique is only applicable to those nuclei which possess a spin quantum number (I) greater than zero. The most important of such nuclei as far as the organic chemist is concerned are ^1H and ^{13}C, both of which have a spin quantum number of $\frac{1}{2}$. Other nuclei with non-zero spin quantum numbers are ^{19}F and ^{31}P, with $I = \frac{1}{2}$; ^{14}N and ^2D, with $I = 1$; and ^{11}B and ^{35}Cl, with $I = \frac{3}{2}$. All of these have been extensively studied by n.m.r. spectroscopy. Nuclei with non-zero spin quantum numbers can be thought of as tiny spinning bar

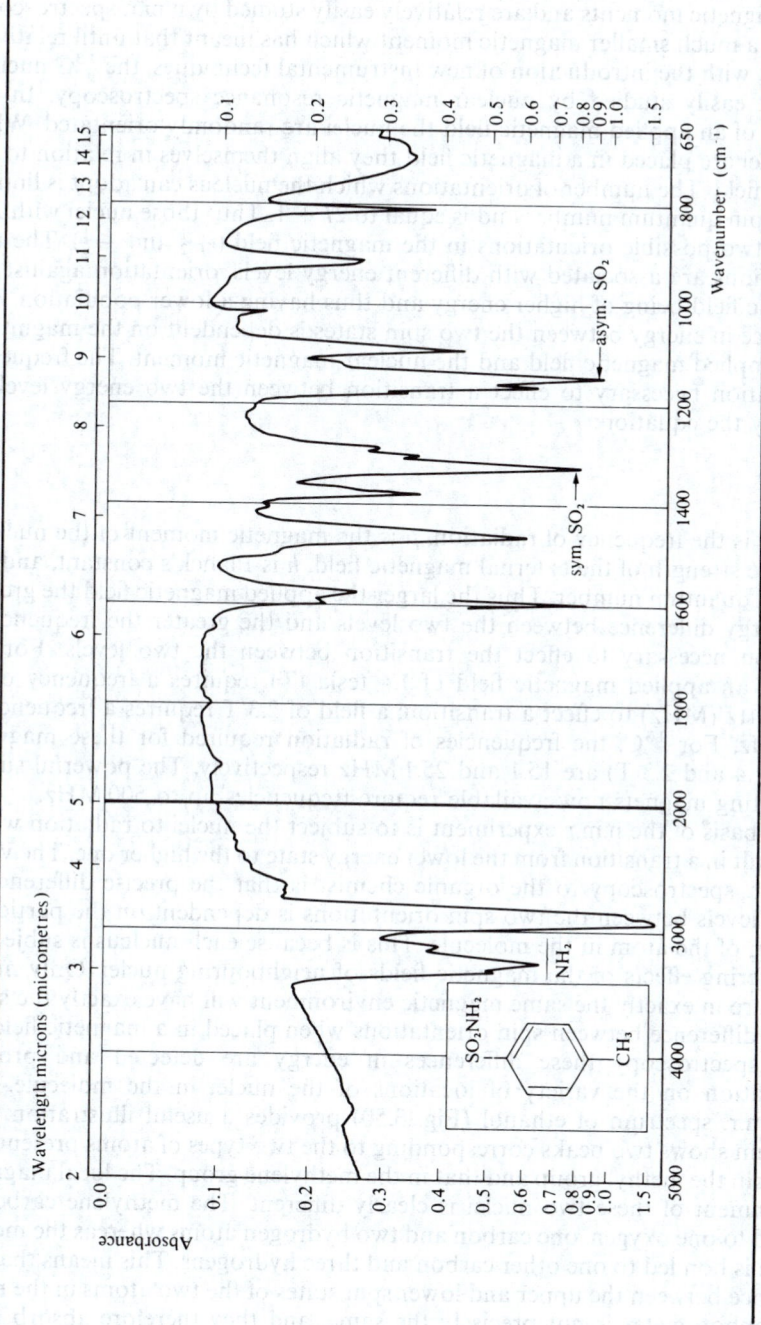

Fig. 3.40 Infrared spectrum of toluene-*p*-sulphonamide recorded as a Nujol mull.

magnets. This spinning magnetic field creates a magnetic dipole, the magnitude of which is given by the nuclear magnetic moment μ. ^1H and ^{19}F have relatively large magnetic moments and are relatively easily studied by n.m.r. spectroscopy. ^{13}C has a much smaller magnetic moment which has meant that until relatively recently, with the introduction of new instrumental techniques, the ^{13}C nucleus was not easily studied by nuclear magnetic resonance spectroscopy. In the absence of an applied magnetic field the nuclei are randomly orientated. When the nuclei are placed in a magnetic field they align themselves in relation to the applied field. The number of orientations which the nucleus can adopt is limited by the spin quantum number and is equal to $2I + 1$. Thus those nuclei with $I = \frac{1}{2}$ have two possible orientations in the magnetic field ($+\frac{1}{2}$ and $-\frac{1}{2}$). The two orientations are associated with different energy levels, orientation against the magnetic field being of higher energy and thus having a lower population. The difference in energy between the two spin states is dependent on the magnitude of the applied magnetic field and the nuclear magnetic moment. The frequency of radiation necessary to effect a transition between the two energy levels is given by the equation:

$$v = \frac{\mu B_0}{hI}$$

where v is the frequency of radiation, μ is the magnetic moment of the nucleus, B_0 is the strength of the external magnetic field, h is Planck's constant, and I is the spin quantum number. Thus the larger the applied magnetic field the greater the energy difference between the two levels and the greater the frequency of radiation necessary to effect the transition between the two levels. For the proton, an applied magnetic field of 1.4 tesla (T) requires a frequency of 60 megahertz (MHz) to effect a transition; a field of 2.3 T requires a frequency of 100 MHz. For ^{13}C, the frequencies of radiation required for these magnetic fields (1.4 and 2.3 T) are 15.1 and 25.1 MHz respectively. The powerful superconducting magnets now available require frequencies up to 500 MHz.

The basis of the n.m.r experiment is to subject the nuclei to radiation which will result in a transition from the lower energy state to the higher one. The value of n.m.r. spectroscopy to the organic chemist is that the precise difference in energy levels between the two spin orientations is dependent on the particular location of the atom in the molecule. This is because each nucleus is subject to the differing effects of the magnetic fields of neighbouring nuclei. Only nuclei which are in exactly the same magnetic environment will have exactly the same energy difference between spin orientations when placed in a magnetic field. In n.m.r. spectroscopy these differences in energy are detected and provide information on the variety of locations of the nuclei in the molecule. The ^{13}C-n.m.r. spectrum of ethanol (Fig. 3.50) provides a useful illustration. The spectrum shows two peaks corresponding to the two types of atoms present, the carbon in the methyl group and that in the methylene group. The local magnetic environment of these two nuclei is clearly different. The methylene carbon is bonded to one oxygen, one carbon and two hydrogen atoms whereas the methyl carbon is bonded to one other carbon and three hydrogens. This means that the difference between the upper and lower spin states of the two atoms in the n.m.r spectrophotometer is not precisely the same, and they therefore absorb radiation at slightly different frequencies.

The n.m.r experiment is carried out by placing the compound in solution in a glass tube (typically 0.5 mm diameter) between the poles of a magnet. The tube is spun rapidly to minimise any magnetic anisotropy. The sample is subject to irradiation at radiofrequency and the absorption of radiation is detected. There are two methods by which this process is carried out. Until relatively recently, the most common method was to modify the magnetic field very slightly over time while holding the radiofrequency constant and to detect absorption of radiation as the field varied (*field sweep*). This is easier to achieve than to hold the field constant and to vary the radiofrequency (*frequency sweep*). Since the field is directly proportional to the radiofrequency required for transition, the result could be represented as a spectrum of absorption against frequency. This is *Continuous Wave nuclear magnetic resonance spectroscopy* (CW-n.m.r.) and many instruments operating on this principle are still in use. The Continuous Wave method has proved satisfactory for many routine uses but it has a number of limitations. One is the time taken to run the spectrum and the consequent limitation on accumulating large numbers of spectra on the same sample. In the case of ^{13}C where the isotope constitutes only 1.1 per cent of the carbon nuclei, many accumulated scans are required in order to produce a spectrum in which the absorption peaks can be readily distinguished from background noise.

An alternative technique which has been developed more recently is to irradiate the sample with all radiofrequencies in the desired range (i.e. covering all ^1H absorption frequencies or all ^{13}C frequencies) at once, while holding the magnetic field constant. This causes all the protons or all the carbon nuclei to absorb at their individual frequencies. The interaction of all the frequencies results in an interferogram which is of no direct value for interpretation. It is converted by Fourier Transform to a spectrum showing absorption against frequency. The technique is known as *pulsed-Fourier Transform nuclear magnetic resonance spectroscopy*, often abbreviated to FT-n.m.r. The pulses of radiation are very short lived, usually about 20 microseconds (μs). The excited nuclei then emit radiation on falling back to the ground state. The pulse cannot be repeated until there is a sufficient population in the ground state once again, otherwise no radiation will be absorbed. When there is no longer an excess of nuclei in the ground state the compound is said to be saturated, and the phenomenon is known as *saturation*. (This phenomenon can occur in the Continuous Wave method when very slow scan speeds are used.) The process of returning to the ground state is known as *relaxation*, which can occur by a number of mechanisms. The rate of return to the ground state is characterised by the *relaxation time*. In pulsed FT-n.m.r. pulses are repeated frequently (0.1–1 s, the acquisition time) depending on the relaxation time of the nucleus, which allows a large number of spectra to be recorded and stored in the instrument's computer in the space of a few minutes. Thus it is possible by this technique to obtain good spectra of nuclei such as ^{13}C which are present naturally in low abundance and which have relatively small nuclear magnetic moments. It also allows n.m.r. spectra to be obtained with very much smaller samples (down to about 5 μg) than is possible with Continuous Wave instruments.

FEATURES OF A NUCLEAR MAGNETIC RESONANCE SPECTRUM
The two types of spectra most frequently encountered by the organic chemist are ^{13}C and ^1H spectra. Considerably greater sensitivity is required for ^{13}C than for ^1H due to the low natural abundance of ^{13}C and the lower magnetic moment compared to that of the proton. However, greater resolution is possible with

319

^{13}C. This is illustrated by the ^{1}H and ^{13}C spectra of 3-methylheptane shown in Fig. 3.41. The proton spectrum only distinguishes between the methyl and methine protons, whereas the ^{13}C spectrum shows seven distinct peaks. However, the two types of spectra have many features in common which may be illustrated by an examination of the ^{1}H and ^{13}C spectra of toluene, Fig. 3.42 and Fig. 3.43 respectively.

By convention, frequency, and therefore magnetic field strength, increase from left to right in the n.m.r. spectrum. Thus moving from left to right is *upfield* and moving from right to left is *downfield*. Upfield absorptions are said to be more *shielded*, downfield absorptions are *deshielded*. The position of an absorption in the n.m.r. spectrum may be represented either on a frequency scale (Hz), or on a scale of magnetic field (tesla), and again by convention the frequency scale is used. In typical n.m.r. spectra the differences in the positions of absorption are usually small compared to the absolute value of the frequency which is around 10^7 Hz. In the case of ^{1}H spectra the differences are of the order of a few hundred Hz and for ^{13}C they are a few thousand Hz. The position of an absorption in the n.m.r. spectrum is therefore denoted, not by the absolute frequency value, but by its relationship to the absorption frequency of a reference compound, i.e. Hz upfield or downfield from the reference. The normal reference compound for both ^{13}C and ^{1}H spectra is tetramethylsilane [$(CH_3)_4Si$, TMS]. All the protons in TMS are equivalent, and the proton nuclei in the vast majority of organic compounds absorb downfield from the single TMS signal. Similarly the vast majority of ^{13}C nuclei in organic compounds absorb downfield from the carbon absorption in TMS. Since tetramethylsilane is not soluble in

Fig. 3.41 Proton magentic resonance spectrum (top) and ^{13}C nuclear magnetic resonance spectrum (bottom) of 3-methylheptane. Data reproduced from G. C. Levy and C. L. Nelson (1972). *Carbon-13 Nuclear Magnetic Resonance for Organic Chemists*. New York; Wiley-Interscience, p. 39.

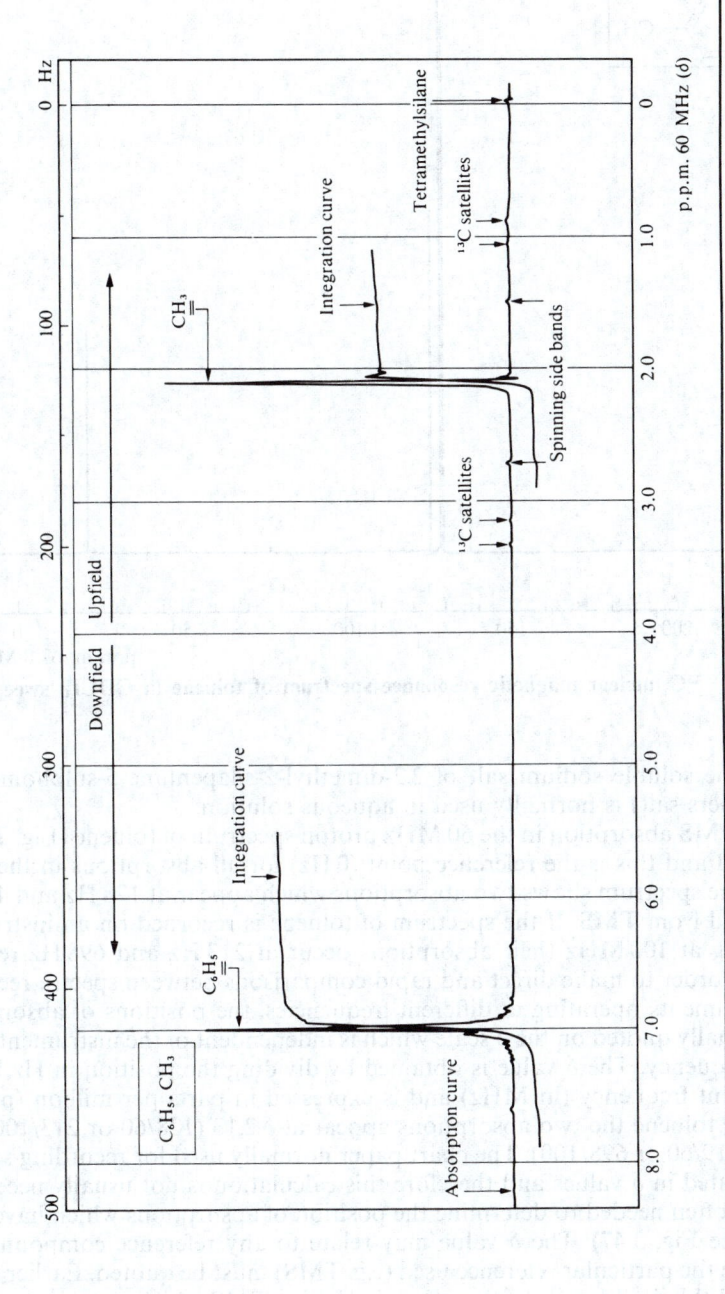

Fig. 3.42 Proton magnetic resonance spectrum of neat toluene; sweep width 500 Hz.

Fig. 3.43 **¹³C nuclear magnetic resonance spectrum of toluene in CDCl₃; sweep width 250 p.p.m.**

water, the soluble sodium salt of 2,2-dimethyl-2-silapentane-5-sulphonic acid (DSS, Tiers salt) is normally used in aqueous solution.

The TMS absorption in the 60 MHz proton spectrum of toluene (Fig. 3.42) is indicated and this is the reference point (0 Hz) for all absorptions in the spectrum. The spectrum shows two absorptions which appear at 128 Hz and 419 Hz downfield from TMS. If the spectrum of toluene is recorded on an instrument operating at 100 MHz their absorptions occur at 213 Hz and 698 Hz respectively. In order to make direct and rapid comparisons between spectra recorded on instruments operating at different frequencies, the positions of absorptions are normally quoted on the δ scale which is independent of the instrument operating frequency. The δ value is obtained by dividing the position in Hz by the instrument frequency (in MHz) and is expressed in parts per million (p.p.m.). Thus for toluene the two absorptions appear at δ 2.13 (128/60 or 213/100) and δ 6.98 (419/60 or 698/100). The chart paper normally used for recording spectra is calibrated in δ values and therefore this calculation is not usually necessary, but it is often needed to determine the position of absorptions which have been offset (see Fig. 3.47). The δ value may relate to any reference compound and therefore the particular reference used (e.g. TMS) must be quoted. Earlier literature used the similar τ scale; this related solely to TMS which was given a value of 10. The two scales can be readily interconverted since $\tau = 10 - \delta$.

The 67.8 MHz ¹³C spectrum of toluene (Fig. 3.43) is also recorded on chart paper calibrated in δ values. In this case the reference point (δ 0) is the carbon absorption of TMS. ^{13}C nuclei in most organic compounds appear within the range of 250 p.p.m. downfield from TMS. The spectrum of toluene shows five absorptions at 137.5, 128.9, 128.1, 125.2 and 21.1. The spectrum was recorded as a solution in $CDCl_3$. The three peaks centred at about δ 77 are due to the carbon atom in the solvent coupling with the adjacent deuterium.

It is usual for the p.m.r. absorption spectrum to be accompanied by an integration curve. This normally appears above the absorption curve on the chart and consists of a series of steps, each step being related to a particular absorption. The height of each step is a measure of the area under each absorption peak, and this in turn is a measure of the relative number of proton nuclei giving rise to that absorption. Thus the ratios of the heights of the steps on the integration curve gives the ratio of the different types of nuclei in the compound. Some care is needed in using the values obtained from integration measurements, since frequently the steps are not well defined and it is not possible to measure their heights accurately. The values for the ratio of nuclei obtained may then be difficult to relate to whole numbers of nuclei and several possibilities must therefore be considered. It cannot be emphasised too strongly that all the likely ratios must be considered since an unjustified adherence to one set of values may lead to completely incorrect deductions being made subsequently. If the measurements do not lead to a realistic value for the proton ratio, the step heights in the integration curve should be carefully reconsidered.

The heights of the steps in the integration curve in the toluene spectrum are in the ratio of 3:5 for peaks at δ 2.13 and δ 6.98 respectively. It is thus possible to assign the former to the methyl protons and the latter to the protons on the benzene ring.

In contrast to ^{1}H spectra, it is not possible to determine the relative ratio of carbon atoms in a compound by integration of the peak areas in the ^{13}C FT-n.m.r. spectrum. There are two reasons for this. The first results from the different relaxation times of carbon atoms in different environments. This means that some atoms with long relaxation times may still be partly saturated when the next pulse of radiation is received, and the resulting absorption peak areas will not be proportional to the number of different carbon atoms. Carbon atoms without hydrogen attached have longer relaxation times and are therefore likely to give rise to peaks of lower intensity in the spectrum. The second reason is due to the *Nuclear Overhauser Effect* (NOE). This is the enhancement of some signals in the ^{13}C spectrum as a result of the spin-decoupling process (see below) which is used to produce the normal, noise-decoupled spectrum by removing the interaction between carbon and hydrogen nuclei. The NOE is not the same for all nuclei. The maximum effect is for carbon atoms with hydrogen attached. The consequence is that carbon atoms without hydrogen attached appear without any NOE enhancement. As a result of these two effects it is often possible to identify by inspection, as a result of their low intensity, those peaks in the ^{13}C spectrum which result from carbon atoms not attached to hydrogen, including those in aromatic rings which carry a substituent.

A number of other features apparent in the toluene proton spectrum are worthy of note at this stage. Each absorption is accompanied by a number of small satellite peaks equally spaced on either side of the main absorptions. These may be *spinning side-bands* or ^{13}C *satellites* (p. 342). The spinning side-bands are caused by inhomogeneities in the magnetic field and in the sample tube. They

can normally be identified easily since they are of much weaker intensity than
the main signal, and furthermore they appear in pairs, equally spaced on either
side of the main absorption band. The identification of these satellite peaks as
spinning side-bands can be confirmed by re-recording the spectrum using a
faster or slower rate of spinning of the sample tube; the spinning side-bands will
then move respectively further from or nearer to the main absorption.

The n.m.r. spectrum of an organic compound recorded as a solution (usually
5–10%) sometimes exhibits small peaks arising from the solvent. The solvents
normally chosen do not contain hydrogen; those commonly used include CCl_4,
$CDCl_3$, $(CD_3)_2SO$, $(CD_3)_2CO$, C_6D_6, D_2O, C_5D_5N. Usually, however, the
deuterated solvents contain a small proportion of isotopically isomeric molecules containing hydrogen instead of deuterium, and this will give rise to an
additional peak or peaks in the proton spectrum. A list of the position of these
absorptions for some of the commoner solvents is given in Appendix 3, Table
A3.10. Care should be taken in comparing chemical shifts from spectra obtained
using different solvents since variations may result, particularly when values
obtained with aromatic solvents, such as benzene or pyridine, are compared
with those obtained using saturated solvents.

THE CHEMICAL SHIFT

The position of an absorption peak relative to that of the reference compound is
known as the *chemical shift*. Each nucleus in a different environment experiences
a slightly different local magnetic field due to the circulation of electrons in
neighbouring bonds and to through-space effects. A slightly different applied
magnetic field is therefore required for resonance and absorption occurs in different regions of the spectrum. Modification of the local magnetic environment
at the nucleus, for example by the introduction of a substituent group, will alter
the position of the absorption band. Effects which cause shifts to lower fields
(downfield) are termed *deshielding*; the opposite effect (upfield shift) is termed
shielding. Figure 3.44 shows approximate ranges of proton chemical shifts and
Fig. 3.45 for ^{13}C chemical shifts.

Measurement of the chemical shift. When a nucleus (or set of equivalent nuclei,
see below) gives rise to a single absorption peak in the spectrum it is a simple
matter to determine the chemical shift from a measurement of its separation
from the reference peak. In ^1H-spectra when coupling of the nucleus results in a
first order multiplet (see below) measurement of the separation from the reference peak must be made to the mid-point of the multiplet. In more complex
spin–spin interactions it is not possible to determine directly the chemical shift
by measurement in this way, and resort must be made to the application of
mathematical methods of analysis.

^1H **chemical shifts.** More detailed chemical shift data for a wide range of proton
environments is given in Appendix 3, Tables A3.1, A3.3 and A3.4. In particular
the chemical shift values quoted in Table A3.1 show that an electronegative substituent in aliphatic systems causes a downfield shift; the greater the electronegativity the more substantial the shift. When two substituents are attached to
the same carbon atom there is a greater downfield shift, but not as great as the
sum of the two substituents separately. The approximate position of absorption
in such cases can be predicted on the basis of the empirical parameters shown in
Appendix 3, Table A3.2.

324

Fig. 3.44 Approximate ranges of proton chemical shifts (R = H or alkyl; Y = SR, −NR₂; X = −OR, −NHCO·R, −O·CO·R, halogen). Data reproduced from L. M. Jackman and S. Sternhell (1969). *Applications of Nuclear Magnetic Resonance in Organic Chemistry.* **2nd edn. London: Pergamon Press, p. 161.**

A comparison of the proton spectrum of anisole (Fig. 3.46) with that of toluene (Fig. 3.42) provides an illustration of the way in which substituents may influence the values of the chemical shift. The spectrum of toluene reveals the typical chemical shifts of aromatic protons and of methyl protons attached to an aromatic system. The five aromatic protons appear as a sharp single band; this is typical of those cases where the substituent is neither strongly shielding or de-shielding (e.g. alkyl or substituted alkyl groups). The spectrum of anisole (C_6H_5·O·CH_3) shows the deshielding effect of the electronegative oxygen causing the methyl absorption to be shifted downfield by a further 1.3δ as compared to toluene. In contrast the absorptions due to the aromatic protons are shifted upfield and split into two groups, the intensities being in the ratio 2:3. The group shifted furthest upfield is due to the *ortho* and *para* protons, and the group only slightly shifted is due to the two *meta* protons. Substituents which have this shielding effect are those which are *ortho/para* directing in electrophilic substitution reactions, e.g. hydroxy, alkoxy, amino and substituted amino groups (see Appendix 3, Table A3.5).

The proton spectrum of phenylacetic acid (C_6H_5·CH_2·CO_2H; Fig. 3.47) exhibits three absorptions in the ratio 1:2:5 due to the carboxylic acid, methylene and phenyl protons respectively. The carboxylic acid proton has been offset by

Fig. 3.45 ¹³C chemical shifts in organic compounds. Data reproduced from E. Breitmaier, G. Jung and W. Voelter (1971). *Angew. Chem. Int. Edn. Engl.*, 10, 679.

Fig. 3.46   Proton magnetic resonance spectrum of neat anisole; sweep width 500 Hz.

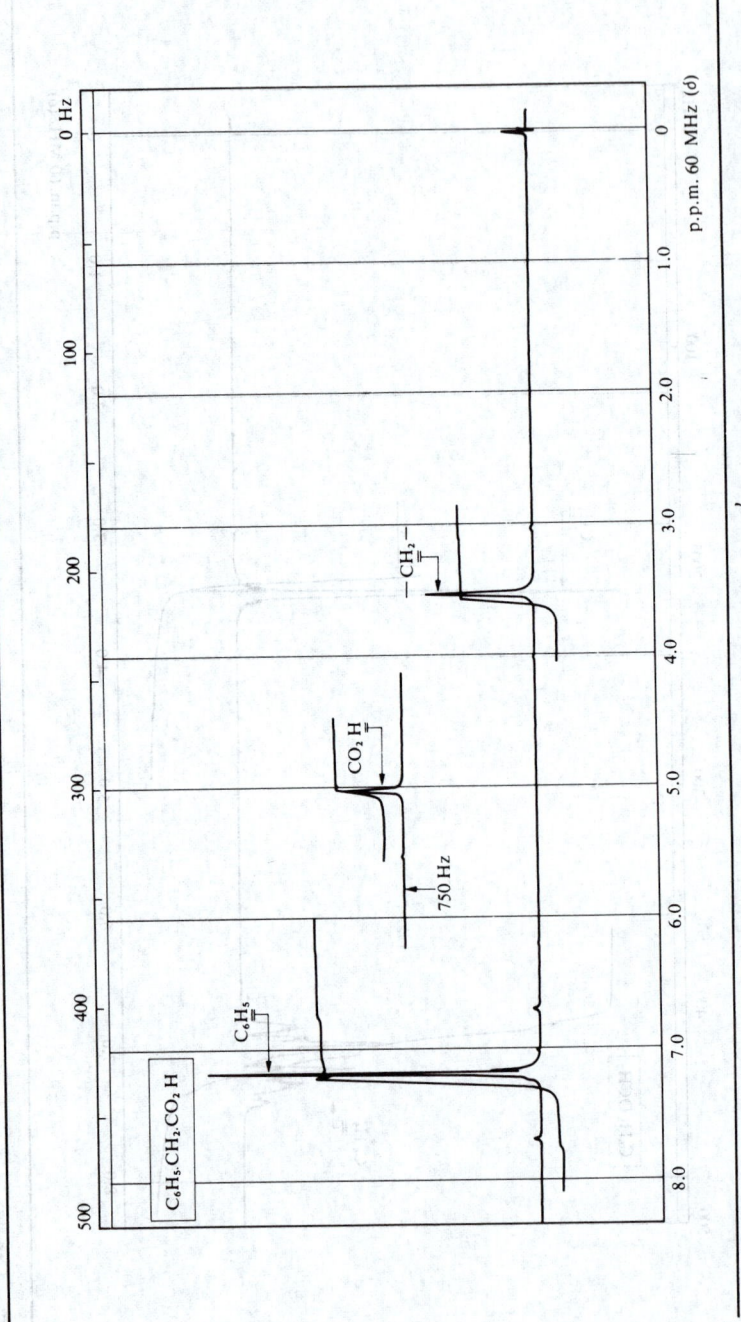

Fig. 3.47 Proton magnetic resonance spectrum of phenylacetic acid in CDCl$_3$ solution; sweep width 500 Hz. Inset: sweep offset 400 Hz.

328

400 Hz so that it can be recorded on the chart which specifies a sweep width of 500 Hz. The actual absorption position of this proton, $\delta\,11.67$, is calculated by adding the amount by which the absorption has been offset (400 Hz) to the position of absorption recorded on the chart (300 Hz), i.e. 700 Hz, and dividing by the operating frequency (60 MHz).

In the proton spectrum of acetophenone ($C_6H_5{\cdot}CO{\cdot}CH_3$; Fig. 3.48) the methyl absorption occurs at $\delta\,2.40$, which should be compared with the corresponding absorption in toluene, since this illustrates the greater deshielding effect of the carbonyl group relative to that of the aromatic ring. The protons on the aromatic ring in this case are shifted downfield as compared with those in toluene, and appear in two groups centred at $\delta\,7.8$ and $\delta\,7.3$, the intensities being in the ratio 2:3. The low-field group is due to the two protons *ortho* to the carbonyl and the high-field group is due to the *meta* and *para* protons. Substituents which exhibit this strongly deshielding effect are those which are *meta* directing on electrophilic substitution reactions, e.g. acyl, carboxyl, alkoxycarbonyl, nitro, etc. (see Appendix 3, Table A3.5).

^{13}C chemical shifts. A considerable amount of data is available which correlates the position of absorptions in the ^{13}C n.m.r. spectrum with the structure of an organic molecule, and it is these empirical correlations which provide the main basis for the use of the technique in structure determination. Figure 3.45 shows the general relationships between structure and chemical shift. The values for the chemical shift are normally related to the tetramethylsilane carbon absorption, with positive values increasing to lower field (corresponding to the δ scale in p.m.r. spectroscopy). The vast majority of absorptions fall in a range of 200 p.p.m. between the carbonyl absorptions at low field and the methyl absorptions at high field. The position of absorption of some of the commonly used solvents is also included in Fig. 3.45.

Hybridisation of the carbon atom has a significant effect on the chemical shift: sp^3-hybridised carbon absorbs at high field (0–60 p.p.m. downfield from TMS), sp^2-carbon at low field (80–200 p.p.m.) and sp-carbon at intermediate values. The precise position of absorption of a particular atom is largely determined by the electronic effects of any substituents, and the fact that these are approximately additive enables fairly accurate predictions of chemical shifts to be made, provided that similar compounds of known structure are available for reference purposes.

Saturated compounds. The position of absorptions of methyl, methylene, methine and quaternary carbon atoms in the alkanes is shown in Fig. 3.49. Within each group the exact position of absorption is determined by the number and nature of substituents on the β and γ carbons. Replacement of a proton by CH_3 results in a downfield shift of c. 8 p.p.m. at C-1, and c. 10 p.p.m. at C-2, and an upfield shift at C-3 of c. 2 p.p.m. Polar substituents result in a downfield shift in the position of absorption; Table A3.12 in Appendix 3 shows the effect on ^{13}C chemical shifts of replacing a methyl group by various polar substituents.

Alkenes and aromatics. The resonances for these classes of compounds appear in the same region (80–140 p.p.m. downfield from TMS) since in both cases the carbon atoms are sp^2-hybridised. Empirical rules for calculating the position of absorption in acyclic alkenes have been developed; the appropriate substituent parameter is added to the value for carbon in ethylene (123.3 p.p.m.).

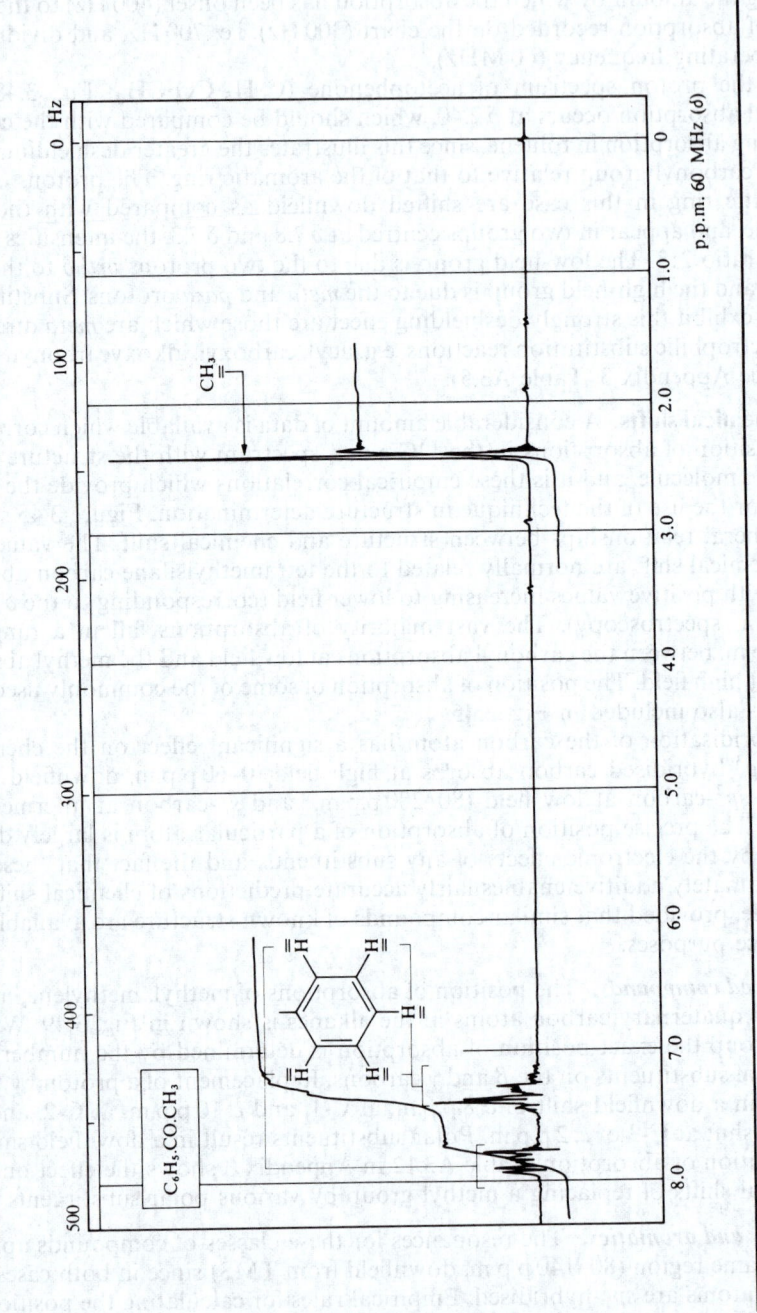

Fig. 3.48 Proton magnetic resonance spectrum of neat acetophenone.

Fig. 3.49 Graphical display of chemicals shifts for classes of paraffins. Data reproduced from L. P. Lindemann and J. Q. Adams (1971). *Analyt. Chem.*, **43**(10), 1251.

Alkene substituent parameters

$$\overset{\gamma'}{C}-\overset{\beta'}{C}-\overset{\alpha'}{C}-\overset{*}{C}=\overset{\alpha}{C}-\overset{\beta}{C}-\overset{\gamma}{C}-C$$

α 10.6;	β 7.2;	γ −1.5;
α′ −7.9;	β′ −1.8;	γ′ 1.5

Additional correction factors for *cis* double bond, —1.1.

Substituents attached to the olefinic carbon atoms exert a substantial effect on the chemical shift of both of these carbon atoms. These effects are exemplified by the chemical shift values for monosubstituted alkenes shown in Appendix 3, Table A3.13.

^{13}C chemical shifts in aromatic compounds are dependent on the polarity of the substituent. Appendix 3, Table A3.14 shows the substituent effects for a range of substituted benzenes. The ^{13}C spectra of substituted benzenes can often be interpreted on the basis of these substituent parameters in association with data from off-resonance decoupled spectra.

Some data for chemical shifts in heteroaromatic compounds are shown in Appendix 3, Table A3.15.

Organic functional groups. The general chemical shift range for carbonyl and other functional groups is shown in Fig. 3.45. Although there is considerable overlap, distinct regions of absorption can be identified. It is interesting to note that there is not a linear relationship between ^{13}C chemical shift and the carbonyl stretching frequency in the infrared. The ^{13}C absorptions are shifted upfield by up to 10 p.p.m. by the introduction of an α-halogen or α,β-unsaturation.

The following examples of ^{13}C spectra illustrate some of the general features discussed above.

The ^{13}C spectrum of ethanol (CH$_3$·CH$_2$·OH; Fig. 3.50) illustrates the effect of an electronegative substituent on the chemical shift of saturated carbon atoms. The effect of the substituent is felt beyond the atom to which the substituent is directly bonded. This can be seen by comparing the chemical shift of the methyl carbon in ethanol (δ 17.0) with that of the methyl carbon in hex-1-ene (δ 14.0), Fig. 3.51. The methylene carbon in ethanol appears at δ 56.5, a downfield shift of about 35 p.p.m. compared to the penultimate methylene carbon in hex-1-ene.

The ^{13}C spectrum of hex-1-ene (CH$_2$=CH·CH$_2$·CH$_2$·CH$_2$·CH$_3$; Fig. 3.51) shows six distinct absorptions associated with each of the six carbon atoms. The

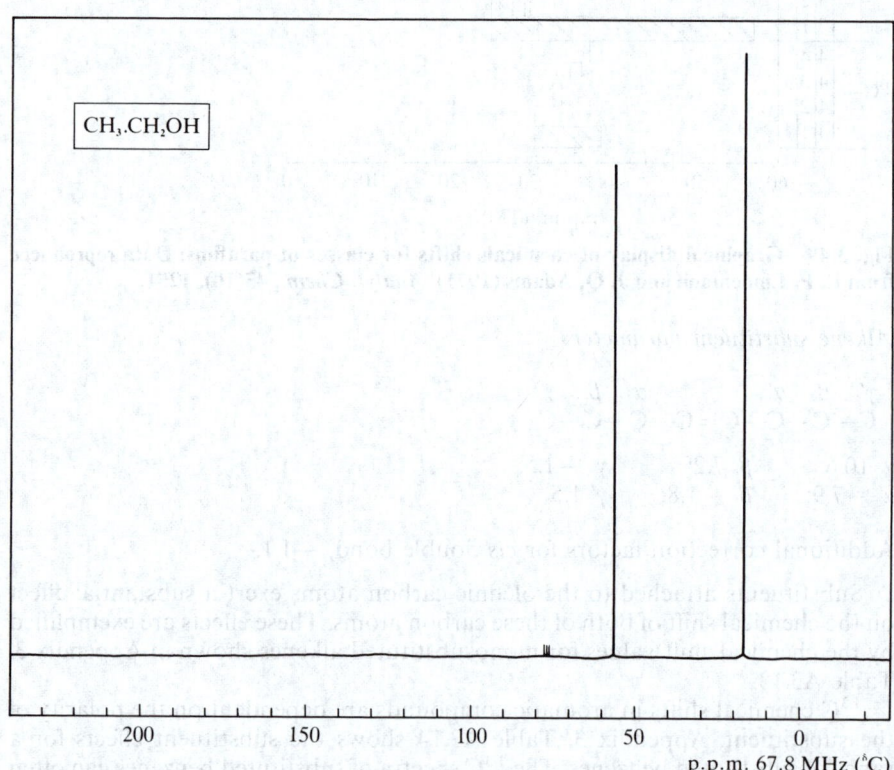

CH$_3$.CH$_2$OH

p.p.m. 67.8 MHz (^{13}C)

Fig. 3.50 ^{13}C nuclear magnetic resonance spectrum of ethanol in CDCI$_3$; sweep width 250 p.p.m.

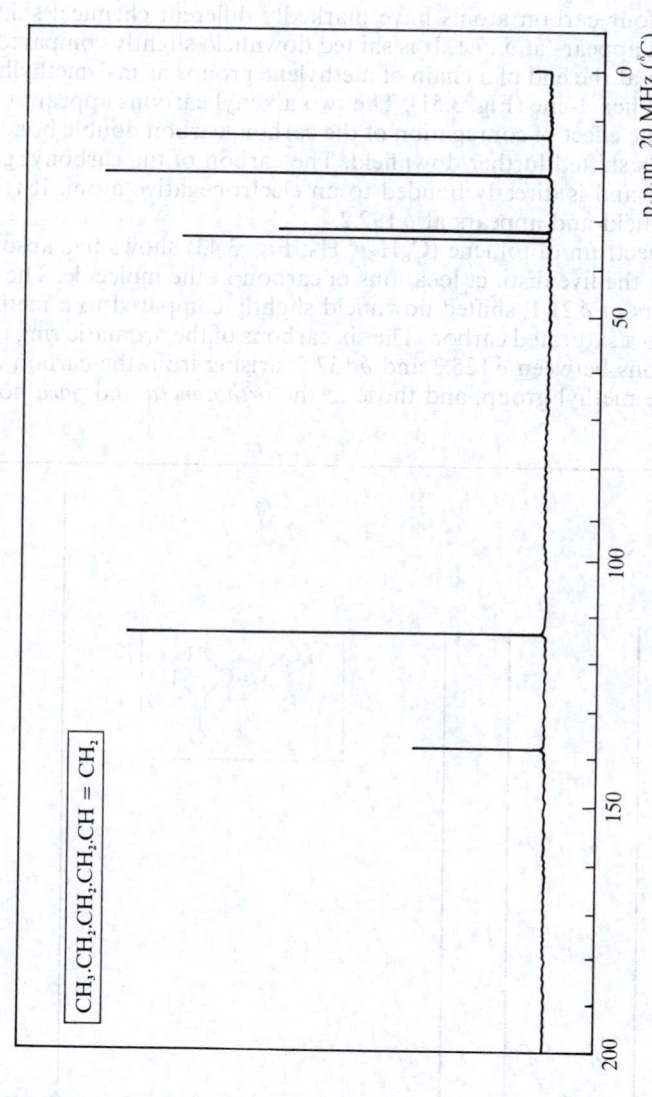

Fig. 3.51 ^{13}C nuclear magnetic resonance spectrum at hex-1-ene in CDCl$_3$; sweep width 200 p.p.m. Data reproduced from the Standard Carbon-13NMR Spectra Collection, © Sadtler Research Laboratories, Division of Bio-Rad Laboratories, Inc.

four sp^3-hybridised carbons appear between δ 14.0 and δ 33.8 and the two sp^2-hybridised carbons appear downfield at δ 114.3 for the terminal carbon and δ 139.2 for the substituted alkenyl carbon.

The ^{13}C spectrum of crotonaldehyde (CH$_3$·CH=CH·CHO; Fig. 3.52) provides a good example of the way in which the ^{13}C chemical shift is determined both by the state of hybridisation of the carbon atom and the nature of the substituent. The four carbon atoms have markedly different chemical shifts. The methyl carbon appears at δ 17.1. It is shifted downfield slightly compared to the methyl carbon at the end of a chain of methylene groups as in 3-methylheptane (Fig. 3.41) and hex-1-ene (Fig. 3.51). The two alkenyl carbons appear at δ 133.4 and δ 152.9. The effect of conjugation of the carbon–carbon double bond is that the β-carbon is shifted further downfield. The carbon of the carbonyl group is sp^2-hybridised and is directly bonded to an electronegative atom. It is shifted furthest downfield and appears at δ 192.2.

The ^{13}C spectrum of toluene (C$_6$H$_5$·CH$_3$; Fig. 3.43) shows five absorptions resulting from the five distinct locations of carbon in the molecule. The methyl carbon appears at δ 21.1, shifted downfield slightly compared to a methyl carbon bonded to a saturated carbon. The six carbons of the aromatic ring result in four absorptions between δ 125.2 and δ 137.5, arising from the carbon directly bonded to the methyl group, and those in the *ortho*, *meta* and *para* positions.

Fig. 3.52 ^{13}C nuclear magnetic resonance spectrum of crotonaldehyde in CDCl$_3$; sweep width 250 p.p.m.

334

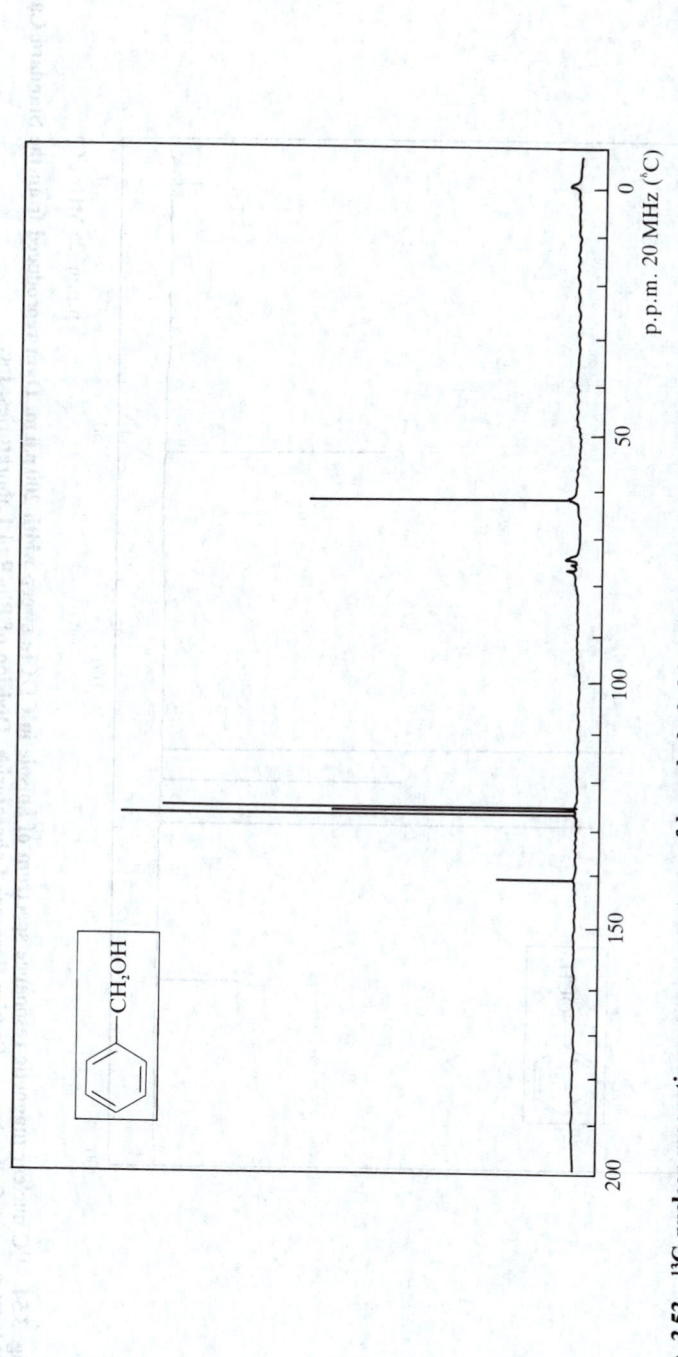

Fig. 3.53 ^{13}C nuclear magnetic resonance spectrum of benzyl alcohol in CDCl$_3$; sweep width 200 p.p.m. Data reproduced from the Standard Carbon-13NMR Spectra Collection, © Sadtler Research Laboratories, Division of Bio-Rad Laboratories, Inc.

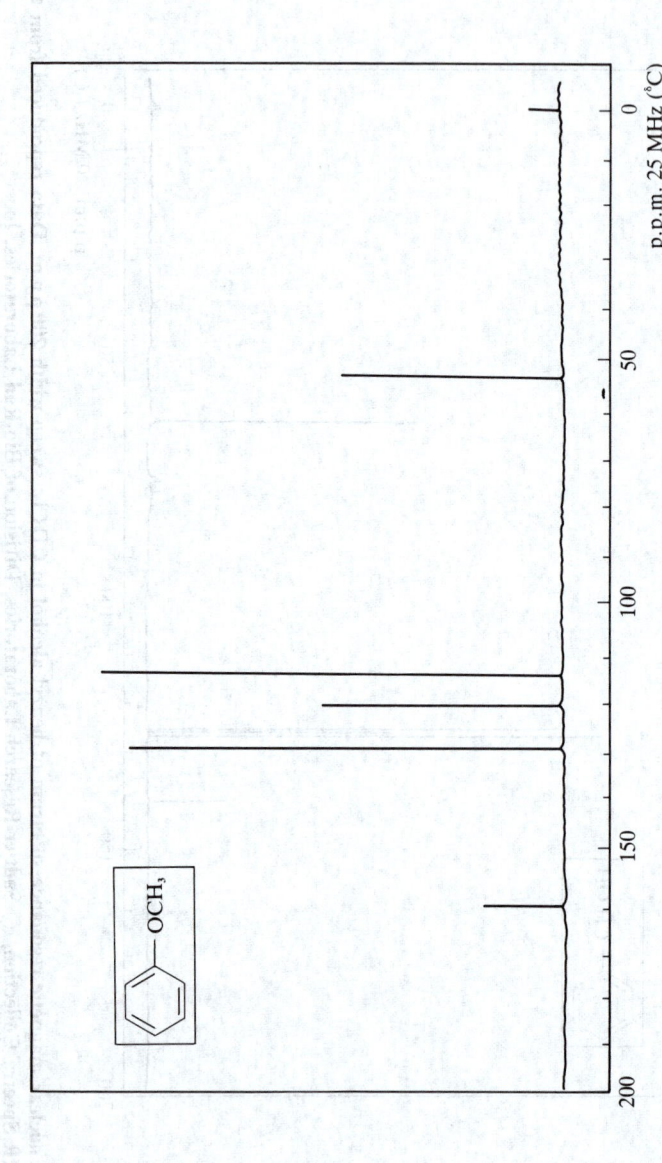

p.p.m. 25 MHz (°C)

Fig. 3.54. [13]C nuclear magnetic resonance spectrum of anisole in CDCl₃; sweep width 200 p.p.m. Data reproduced from the Standard Carbon-13NMR Spectra Collection, © Sadtler Research Laboratories, Division of Bio-Rad Laboratories, Inc.

Note the weak absorption of the aromatic carbon atom which carries the substituent and therefore has no hydrogen atoms attached.

The aromatic carbons in the ^{13}C spectrum of benzyl alcohol ($C_6H_5 \cdot CH_2 \cdot OH$; Fig. 3.53) result in a similar absorption pattern to that of toluene. The aliphatic carbon appears at δ 64.3, a downfield shift of more than 40 p.p.m. as a result of the replacement of one of the hydrogens of the methyl group by a hydroxyl group. The six carbons of the aromatic ring give a group of four absorptions between δ 126.9 and δ 140.9.

The ^{13}C spectrum of anisole ($C_6H_5 \cdot OCH_3$; Fig. 3.54) further illustrates the deshielding effect which results from an electronegative atom such as oxygen. The methyl carbon of anisole appears at δ 54.8 a downfield shift of more than 30 p.p.m. compared with the methyl carbon in toluene. The aromatic carbons have all been shifted downfield; the carbon directly bonded to the oxygen is most affected appearing at δ 159.9, a downfield shift of about 22 p.p.m. compared with the equivalent carbon in toluene. The *ortho*, *meta* and *para* carbon atoms appear at δ 114.1, 129.5 and 120.7 respectively.

The aromatic carbon atoms in phenylacetic acid ($C_6H_5 \cdot CH_2 \cdot CO_2H$; Fig. 3.55) result in a similar absorption pattern to that described above for benzyl alcohol. The methylene carbon appears at δ 41.1 compared with δ 64.2 in benzyl alcohol. It is instructive to compare the chemical shift of the carbonyl carbon in phenylacetic acid (δ 178.3), crotonaldehyde (δ 192.2) (Fig. 3.52) and 4-methyl-

Fig. 3.55 ^{13}C nuclear magnetic resonance spectrum of phenylacetic acid in CDCl₃; sweep width 250 p.p.m.

337

(a)

(b)

Fig. 3.56 (a) The 5000 Hz FT c.m.r. spectrum of 0.2 M cholesterol; pulse interval 0.4 seconds (total time, 3.4 hours). (b) The 1800 Hz plot expansion of the data in (a). All resonances assigned. Data reproduced from G. C. Levy and G. L. Nelson (1972). *Carbon-13 Nuclear Resonance for Organic Chemistry*. New York; Wiley-Interscience, pp. 164 and 165.

pentan-2-one $(CH_3)_2CH\cdot CH_2\cdot CO\cdot CH_3$ (δ 195.9). This variation in the position of the carbonyl carbon in aldehydes, ketones and carboxylic acids is represented in general terms in the schematic diagram of ^{13}C chemical shifts, Fig. 3.45.

The spectrum of cholesterol reproduced in Fig. 3.56 shows the individually resolved lines of the 27 carbon atoms in the molecule, and provides an impressive example of the extent of the structural information which can be obtained about complex organic molecules with the aid of ^{13}C-n.m.r.[3m, 3n]

SPIN–SPIN SPLITTING

The ^{13}C spectra discussed above consist of sets of single peaks each one due to a carbon atom in a particular magnetic environment. The majority of proton spectra normally encountered by the organic chemist are much less simple in appearance. Proton absorptions frequently appear, not as sharp single peaks as in the case of the ^{13}C spectra, but as multiplets of varying complexity. The multiplets result from interaction, or *coupling* between neighbouring nuclei. These multiplets can provide valuable information about the structure of the molecule. It should be noted that carbon atoms also couple with neighbouring nuclei but the results of these couplings are not observed in the ^{13}C spectra normally used by the organic chemist. The coupling between carbon and hydrogen has been removed by the process of noise decoupling in order to simplify the spectrum. The coupling between two ^{13}C nuclei is not significant because of the low natural abundance of the ^{13}C isotope. The phenomenon of spin–spin splitting is therefore discussed in terms of proton spectra.

Equivalent protons. All hydrogens which are in identical environments have the same chemical shift and therefore absorb at the same frequency; they are said to be *chemically equivalent*. This can arise in two ways. Firstly, the protons are equivalent if they are bonded to the same carbon atom which is also free to rotate. For example, the three protons in a methyl group are equivalent and appear as a singlet (see the spectra of toluene, anisole or acetophenone above), and the two protons of a methylene group, provided that it can rotate freely, are identical and appear as a singlet (see the spectrum of phenylacetic acid above); frequently this is not the case with methylene groups in cyclic systems where rotation is restricted.

Secondly, hydrogens on different carbon atoms will have the same chemical shift if they are structurally indistinguishable. Thus the spectrum of *p*-xylene exhibits two signals of relative intensity 3:2. There are six methyl protons in identical environments which appear at $\delta\,2.3$ and the four aromatic protons are identical and appear at $\delta\,7.0$.

Methylene protons adjacent to a chiral centre will be non-equivalent, despite the fact that there is free rotation about the carbon–carbon bond. Such protons are described as *diastereotopic*, since replacement of either of the two hydrogens in turn by a group X produces a pair of diastereoisomers. Such is the case of the two methylene protons (H_a and H_b) in 1-phenylpropan-2-ol, which are non-equivalent and therefore have different chemical shifts in the p.m.r. spectrum.

Nuclei may be chemically equivalent but magnetically non-equivalent. To be *magnetically equivalent* nuclei must couple in exactly the same way to all other nuclei in the system. Thus the two protons in 1,1-difluoroethylene are magnetically *non-equivalent* since the coupling of F_a (and also of F_b) to each of H_a and H_b is different.

$$F_a\diagdown\diagup H_a$$
$$C{=}C$$
$$F_b\diagup\diagdown H_b$$

First-order spin–spin interactions. Protons bonded to adjacent carbon atoms or to carbon atoms connected by a conjugated system interact with each other so that the resulting signal appears as a multiplet rather than a singlet. If the difference between the chemical shift values of the coupled protons is reasonably large in comparison with the value of the coupling constant, a first-order spin–spin coupling pattern results. For two groups of interacting nuclei the coupling constant J is given by the separation between any two adjacent peaks in a first-order multiplet; the value, which is quoted in Hz, is independent of the operating frequency of the instrument. The magnitude of the coupling constant is dependent on the relative positions of the two coupled nuclei in the molecule and frequently gives valuable information about the structure of the compound.

It should be noted that if the two groups of protons are coupled to cause spin–spin splitting, the J value can be measured from the multiplet arising from either of the sets of protons. Thus in the spectrum of ethanol, the sample of which contains a trace of acid (Fig. 3.68) the coupling between the methyl and methylene protons can be determined by measurement of the separation between any two adjacent lines in the triplet resulting from the methyl group, or any two adjacent lines in the quartet resulting from the methylene group. Measurement of all coupling constants can frequently show which groups of protons are coupled with each other, and hence give valuable structural information. A simple example of this procedure is seen in the spectrum of pure ethanol (Fig. 3.67) in which the methylene protons couple with both the methyl protons and with the hydroxyl protons, but with different coupling constants, 7 Hz and 5 Hz respectively. Typical values of coupling constants are given in the correlation tables, Appendix 3, Tables A3.7, A3.8 and A3.9.

The first-order multiplets arising from spin–spin coupling can be analysed on the basis of two simple rules which give information on the number of peaks in the multiplet and on the relative intensity of the peaks in the multiplet.

1. *The number of peaks.* If a proton is coupled with N other *equivalent* protons the number of peaks in the multiplet is N + 1.

2. *Relative intensities in each multiplet.* These can be deduced using Pascal's Triangle shown below. The value assigned to each position in the triangle is derived by adding together the values of adjacent positions in the preceding level; the outside position at each level is always unity.

```
                    1
                 1     1
              1     2     1
           1     3     3     1
        1     4     6     4     1
     1     5    10    10     5     1
```

Spin–spin splitting patterns arising from typical groups of protons are shown in Fig. 3.57.

When a proton is coupled to two *non-equivalent* sets of neighbouring protons more complex multiplets result; this is illustrated by the spectrum of pure ethanol (Fig. 3.67). Thus the methylene protons are coupled to the three methyl protons, giving rise to a quartet, and are further coupled to the hydroxyl proton which therefore causes each of the peaks of the quartet to appear as a doublet. The multiplet therefore consists of eight peaks due to the two overlapping quartets i.e. $(N + 1)(M + 1)$. On occasions there may be difficulty in recognising the components of these more complex multiplets, as some peaks may be superimposed.

Fig. 3.57   Typical spin-spin splitting patterns.

Long-range coupling, i.e. coupling beyond three bonds, may be observed in some circumstances, especially in conjugated systems. The coupling constants are then usually small (0–3 Hz) in comparison with geminal or vicinal coupling constants (cf. Appendix 3, Tables A3.7 and A3.8). Commonly encountered systems which exhibit long-range coupling are the allylic system

(H—C=C—CH) and the corresponding acetylenic system (H—C≡C—CH),

and aromatic and heteroaromatic rings. Conjugated acetylenes are capable of spin–spin interaction over as many as nine bonds. Long-range coupling is often observed between nuclei which are linked by a conjugated system arranged in a zig-zag manner thus:

The conjugated system is often part of a cyclic or polycyclic compound, e.g.

Coupling between protons and other nuclei is often useful in interpreting spectra of appropriate compounds. Weak satellites, arising from ^{13}C—^1H coupling, may sometimes be observed near strong signals in the proton spectrum. They can be recognised since they appear as doublets 50 to 100 Hz on either side of the main signal and are unaffected by the rate of spinning of the sample (unlike spinning side-bands). ^{13}C satellites are visible in the spectrum of toluene (Fig. 3.42).

Values for coupling constants between protons and other atoms are given in Appendix 3, Table A3.8.

Distortion of multiplets. The two rules for analysing spin–spin splitting patterns can only be applied exactly when the difference between the chemical shifts of the coupling nuclei is substantially greater (about 10 times) than the value of the coupling constants. As the chemical shift difference decreases the simple first-order analysis starts to break down. For groups of protons which are still well separated this results in a distortion of the multiplets. This distortion results in the inner peaks (i.e. those nearer the other multiplet) of the two multiplets increasing in intensity while the outer peaks decrease in intensity. This is often a useful guide for establishing which groups of multiplets are related by spin–spin coupling. This distortion is diagrammatically illustrated in the calculated spectra shown in Fig. 3.58 where $\Delta v/J$ is 14 and 5.7.

More complex spin–spin interactions. As the chemical shift difference becomes similar to the value of the coupling constant, the first-order analysis breaks down completely. It is then not possible to measure either the chemical shift or the coupling constants directly from the spectrum and resort must be made to more rigorous analytical procedures which are beyond the scope of this book. However, there are available a variety of computer programs specifically designed to analyse complex n.m.r. spectra.[3h]

Fig. 3.58   Calculated AB spectra. J_{AB} **was set at 7 Hz and** ΔvJ **was varied to the ratios shown. Data reproduced from L. M. Jackman and S. Sternhell (1969).** *Applications of Nuclear Magnetic Resonance Spectroscopy in Organic Chemistry.* **2nd edn. London; Pergamon Press, p. 130.**

Frequently, however, it is necessary only to recognise complex patterns arising from common groupings and not to analyse them completely. The nomenclature which is adopted for naming spin–spin systems follows from the number of interacting nuclei and the magnitude of the chemical shift differences as compared to the coupling constants. Thus nuclei are represented by letters of the alphabet (A, B, M, X, Y, etc.), the first letters of the alphabet being used for nuclei at lowest field. Nuclei which are chemically identical but magnetically non-equivalent are differentiated by primes (A,A', B,B', etc.). Nuclei which have a large chemical shift difference as compared with the coupling constants are represented by letters well separated in the alphabet giving rise to systems such as AX, AMX, etc. If the chemical shift differences are of the same order of magnitude as the coupling constant, letters adjacent in the alphabet are used, so that AX becomes AB, the AMX system becomes ABC, etc. Some examples are given in Appendix 3, Table A3.11. A useful compilation for the recognition of spin systems has been published,[3i] but the following systems are those commonly encountered and they are illustrated with appropriate spectra.

Two-spin system (**AB**). In all cases (except when $\Delta v/J = 0$ and hence the spectrum is a single absorption line, A_2) the splitting pattern shows two doublets (a doublet of doublets) distorted to a greater or lesser extent; this is often referred to misleadingly as the 'AB quartet'. The coupling constant J_{AB} is the separation between the lines of either doublet. The change in the splitting pattern as the chemical shift difference (Δv) approaches the value of the coupling constant J is shown in Fig. 3.58.

Three-spin systems. (*a*) AB_2 and AX_2. The AX_2 system results in a doublet and

343

Fig. 3.59 Calculated AB_2 spectra. Data reproduced from L. M. Jackman and S. Sternhell (1969). *Applications of Nuclear Magnetic Resonance Spectroscopy in Organic Chemistry.* 2nd edn. London; Pergamon Press, p. 131.

triplet as described above. The change in the appearance of the spectrum as Δv approaches J is shown in Fig. 3.59. Note that the number of peaks in the spectrum increases. The spectrum of pyrogallol (Fig. 3.60) illustrates a typical AB_2 spectrum; the inset is the scale-expanded spectrum of the aromatic region.

(*b*) *AMX, ABX, ABC*. The spectrum of pyrrole-2-carboxylic acid (Fig. 3.61) illustrates a first-order AMX system. The spectrum contains three multiplets in the region $\delta\,6\text{–}7$ due to the C—H protons of the heteroaromatic ring. The expanded spectrum (50 Hz sweep width) shows that each of these multiplets consists of a doublet of doublets. Two coupling constants can be extracted from each multiplet by measurement of the separation between the first and second peaks, and between the first and third peaks of each multiplet; this is shown in the inset. The vinyl group often gives rise to an ABX spectrum as shown in the spectrum of styrene (Fig. 3.62). If accurate values for the coupling constants and chemical shifts are required from such a spectrum, resort must be made to more rigorous methods of analysis. However, a qualitative approach is frequently all that is required and on this basis the spectrum can be analysed easily. The X proton couples with protons A and B to give a distorted doublet of doublets, which form the four lines in the X portion of the spectrum. Protons A and B each give rise to doublets due to coupling with the X proton, and each of these four peaks is further split by the small coupling between the A and B protons. The ABC spectrum is very complex and does not readily yield useful information.

Aromatic compounds. The magnitude of the coupling constants due to coupling between protons attached to the aromatic ring varies in the order $o > m \gg p$; *para* coupling constants are in fact frequently not discernible in the spectrum.

Fig. 3.60   Proton magnetic resonance spectrum of pyrogallol (AB₂) in D₂O solution; reference DSS, and sweep width 500 Hz. Inset: sweep offset 380 Hz, sweep width 50 Hz. The signal at δ 5 is due to HOD and arises from exchange of the three phenolic hydrogens with the solvent.

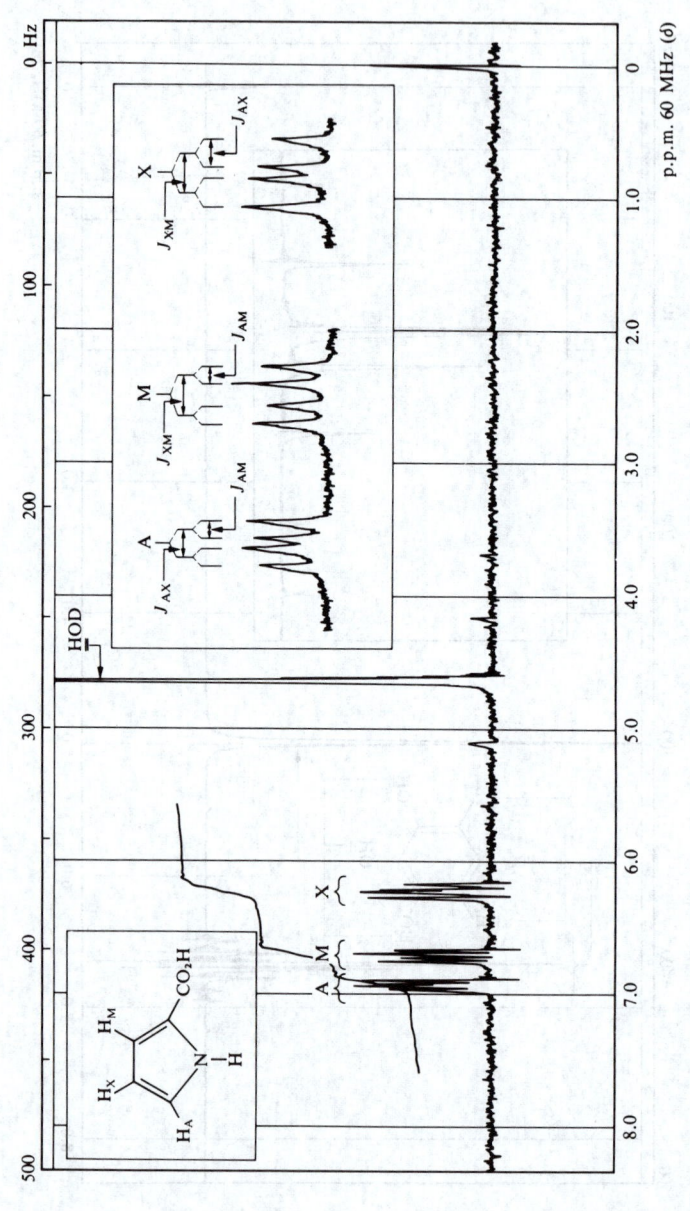

Fig. 3.61 Proton magnetic resonance spectrum of pyrrole-2-carboxylic acid in D_2O/NaOD solution; reference DSS, and sweep width 500 Hz. Inset: expansion of aromatic proton absorptions, sweep width 50 Hz.

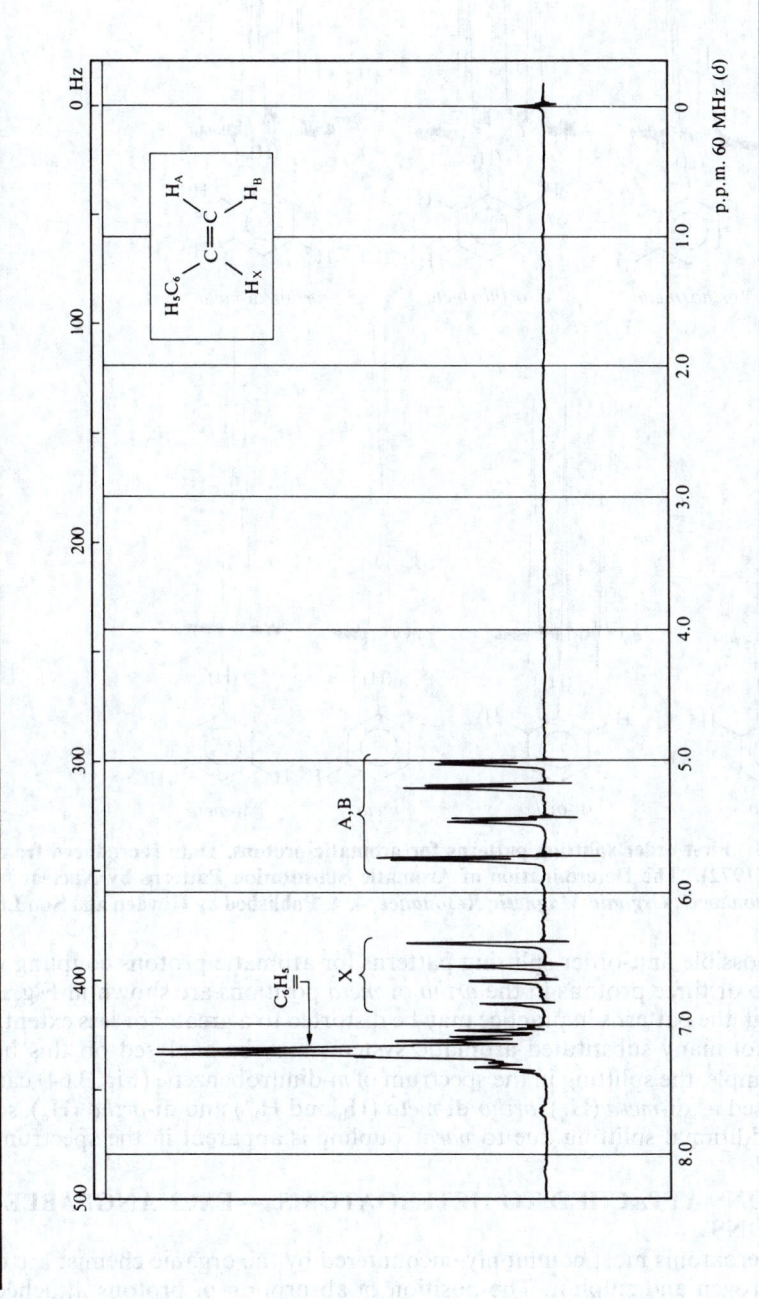

Fig. 3.62 Proton magnetic resonance spectrum of styrene in CDCl₃ solution; sweep width 500 Hz.

347

Fig. 3.63 First order splitting patterns for aromatic protons. Data reproduced from M. Zanger (1972). 'The Determination of Aromatic Substitution Patterns by Nuclear Magnetic Resonance', *Organic Magnetic Resonance,* **4, 4. Published by Heyden and Son Ltd.**

Seven possible first-order splitting patterns for aromatic protons coupling with one, two or three protons in the *ortho* or *meta* positions are shown in Fig. 3.63. Although the patterns in practice may be distorted to a greater or less extent, the spectra of many substituted aromatic systems may be analysed on this basis. For example, the splitting in the spectrum of *m*-dinitrobenzene (Fig. 3.64) can be recognised as di-*meta* (H$_a$), *ortho*/di-*meta* (H$_b$ and H$_b'$) and di-*ortho* (H$_c$); some small additional splitting due to *para* coupling is apparent in the spectrum.

PROTONS ATTACHED TO HETEROATOMS — EXCHANGEABLE PROTONS

The heteroatoms most commonly encountered by the organic chemist are oxygen, nitrogen and sulphur. The position of absorption of protons attached to these atoms is not normally sufficiently reliable for interpretative purposes, although there are exceptions to this general rule (e.g. carboxylic acids, enols,

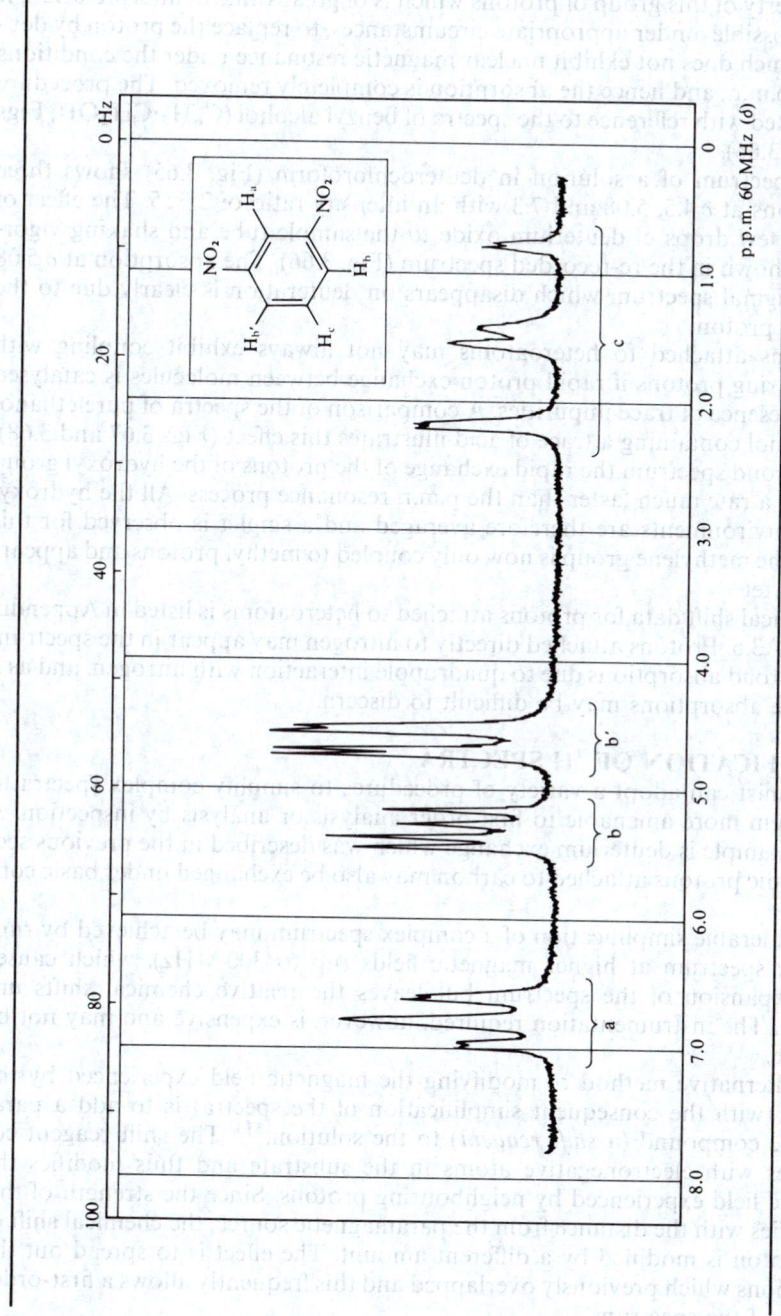

Fig. 3.64 Proton magnetic resonance of *m*-dinitrobenzene in CDCl₃ solution; sweep offset 460 Hz, sweep width 100 Hz. The δ scale is not applicable in this case; actual δ values may be obtained by dividing the chemical shift values in Hz by 60 (cf. Fig. 3.47) and relevant discussion p. 325).

etc., see Appendix 3, Table A3.6). The position of the absorption is often highly dependent on the nature of the solvent and the concentration. However, there is one property of this group of protons which is of great value in interpretation; it is often possible, under appropriate circumstances, to replace the proton by deuterium which does not exhibit nuclear magnetic resonance under the conditions used for p.m.r., and hence the absorption is completely removed. The procedure is illustrated with reference to the spectra of benzyl alcohol ($C_6H_5\cdot CH_2OH$; Figs 3.65 and 3.66).

The spectrum of a solution in deuterochloroform (Fig. 3.65) shows three absorptions at δ 4.5, 5.08 and 7.3 with an intensity ratio of 2:1:5. The effect of adding a few drops of deuterium oxide to the sample tube and shaking vigorously is shown in the re-recorded spectrum (Fig. 3.66). The absorption at δ 5.08 in the original spectrum which disappears on deuteration is clearly due to the hydroxyl proton.

Protons attached to heteroatoms may not always exhibit coupling with neighbouring protons if rapid proton exchange between molecules is catalysed by the presence of trace impurities. A comparison of the spectra of pure ethanol and ethanol containing a trace of acid illustrates this effect (Figs 3.67 and 3.68). In the second spectrum the rapid exchange of the protons of the hydroxyl group occurs at a rate much faster than the p.m.r. resonance process. All the hydroxyl proton environments are therefore averaged and a singlet is observed for this proton; the methylene group is now only coupled to methyl protons and appears as a quartet.

Chemical shift data for protons attached to heteroatoms is listed in Appendix 3, Table A3.6. Protons attached directly to nitrogen may appear in the spectrum as very broad absorptions due to quadrupole interaction with nitrogen, and as a result the absorptions may be difficult to discern.

SIMPLIFICATION OF ¹H SPECTRA

The chemist can adopt a variety of procedures to simplify complex spectra to make them more amenable to first-order analysis or analysis by inspection. A simple example is deuterium exchange which was described in the previous section. Acidic protons attached to carbon may also be exchanged under basic conditions.

Considerable simplification of a complex spectrum may be achieved by running the spectrum at higher magnetic fields (up to 300 MHz), which causes linear expansion of the spectrum but leaves the relative chemical shifts unchanged. The instrumentation required, however, is expensive and may not be available.

An alternative method of modifying the magnetic field experienced by the protons (with the consequent simplification of the spectra) is to add a paramagnetic compound (a *shift reagent*) to the solution.[3j,k] The shift reagent coordinates with electronegative atoms in the substrate and thus modifies the magnetic field experienced by neighbouring protons. Since the strength of this field varies with the distance from the paramagnetic source, the chemical shift of each proton is modified by a different amount. The effect is to spread out the absorptions which previously overlapped and this frequently allows a first-order analysis of the spectrum.

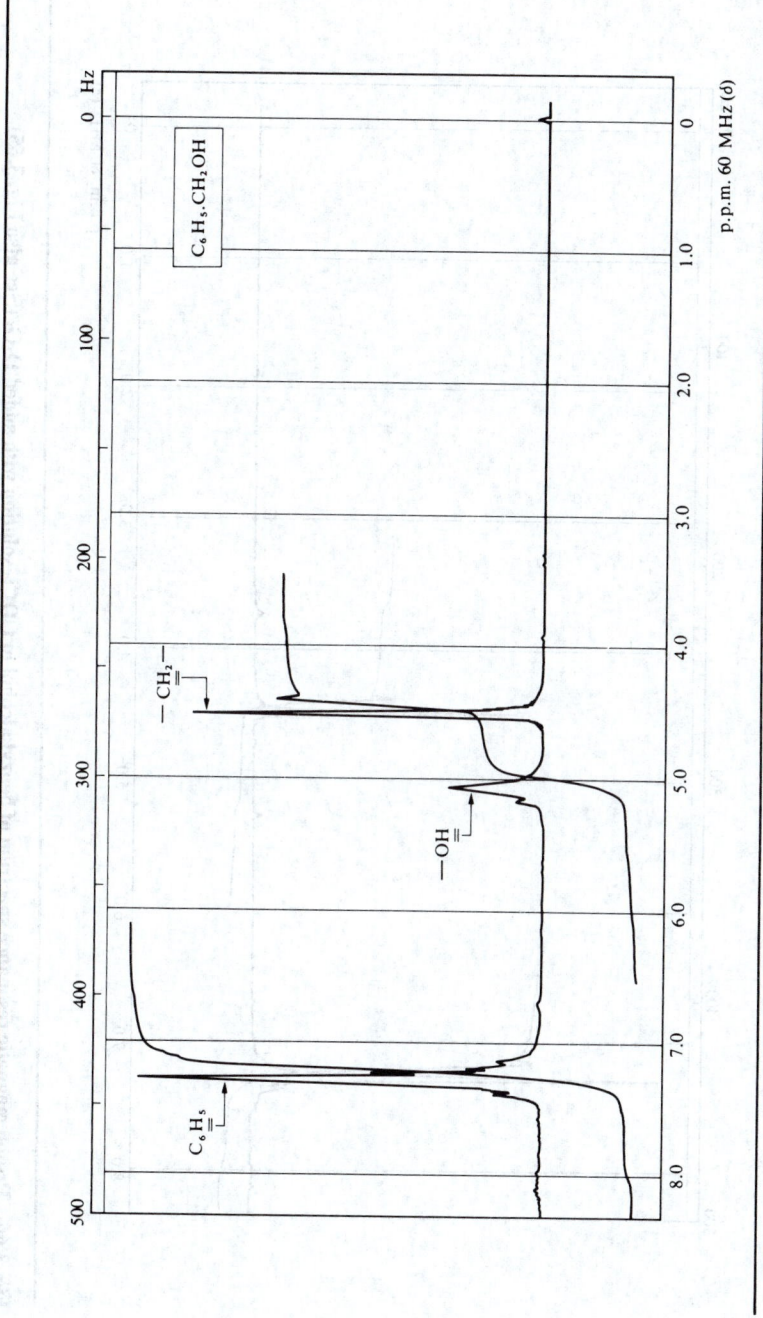

Fig. 3.65 Proton magnetic resonance spectrum of benzyl alcohol in CDCl₃ solution; sweep width 500 Hz.

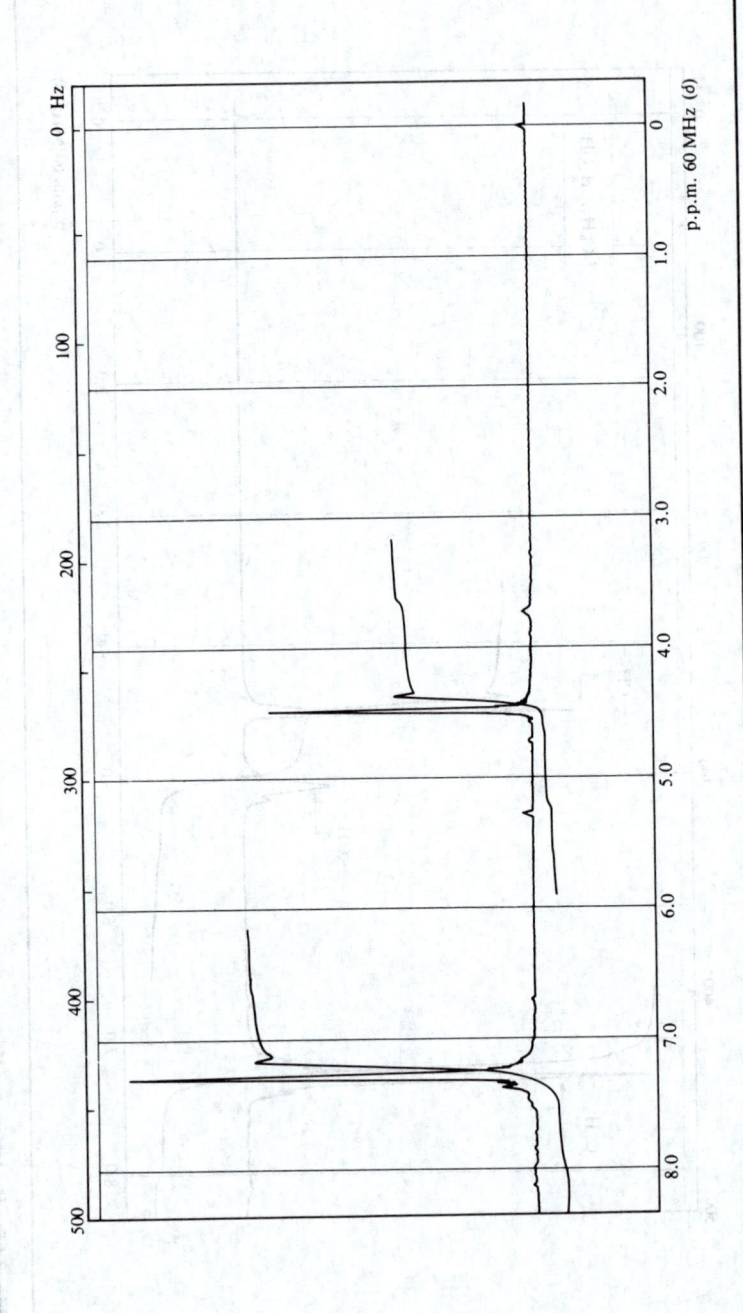

Fig. 3.66   Proton magnetic resonance spectrum of benzyl alcohol in CDCl₃ solution with added D₂O. (See also Fig. 3.65).

Fig. 3.67 Proton magnetic resonance spectrum of pure ethanol; sweep width 500 Hz.

Fig. 3.68   Proton magnetic resonance spectrum of ethanol with a trace of acid; sweep width 500 Hz.

The most commonly used shift reagents are tris-chelates of lanthanide ions with the β-diketones, 2,2,6,6-tetramethylheptane-3,5-dione (dipivaloylmethane, (1)) and 1,1,1,2,2,3,3-heptafluoro-7,7-dimethyloctane-4,6-dione (2). Typical reagents are tris-(dipivaloylmethanato) europium and tris-1,1,1,2,2,3,3-heptafluoro-7,7-dimethyloctane-4,6-dionato europium, the names of which are normally abbreviated to Eu(dpm)$_3$ and Eu(fod)$_3$.

$$(CH_3)_3C\diagdown$$
$$C{=}O$$
$$H_2C\diagup$$
$$C{=}O$$
$$(CH_3)_3C\diagup$$
(1)

$$CF_3{\cdot}CF_2{\cdot}CF_2\diagdown$$
$$C{=}O$$
$$H_2C\diagup$$
$$C{=}O$$
$$(CH_3)_3C\diagup$$
(2)

A wide range of lanthanide shift reagents is now commercially available including some derived from diketones in which all the protons have been replaced by deuterium, thus preventing any interference in the p.m.r. spectrum.

The extent of the lanthanide-induced shift is dependent on the basicity of the functional group and on the nature, purity and concentration of the shift reagent. Alcohols and amines generally exhibit the largest shift, but many other compounds such as ethers, carbonyl compounds, nitriles, sulphoxides, oximes, etc., exhibit useful shifts. Of the commonly used reagents Eu(fod)$_3$ normally causes the greatest shifts as it is a stronger Lewis acid. It also has the advantage of a much higher solubility. The magnitude of the lanthanide-induced shift is considerably decreased by the presence of a small amount of water, and since many of the reagents are hygroscopic, care should be taken in their handling and storage.

The two major applications of lanthanide shift reagents are firstly the simplification of the spectrum, and secondly the confirmation of the assignment of signals by relating the extent of the shift to the concentration of the shift reagent.

The effect on the spectrum of 4-methylpentan-2-one caused by the addition of increasing amounts of a shift reagent is shown in Fig. 3.69. The shift reagent used in this case is the europium chelate of 1,1,1,2,2-pentafluoro-6,6-dimethyl-3,5-heptanedione [Eupfd, $CF_3{\cdot}CF_2{\cdot}CO{\cdot}CH_2{\cdot}CO{\cdot}C(CH_3)_3$]. In the absence of a shift reagent the absorption of three of the groups of protons virtually overlaps; addition of the shift reagent spreads out these absorptions to allow a ready analysis of the spectrum as indicated. Those protons closest to the donor atom (in this case the carbonyl oxygen) are shifted by a larger extent for a given amount of reagent added. This relationship is illustrated in Fig. 3.70, which shows that the methyl and methylene groups bonded to the carbonyl are shifted the most, followed by the methine proton, and the methyl groups furthest from the carbonyl are shifted least.

The assignment of absorptions in the n.m.r. spectrum is greatly assisted by the use of shift reagents. Figure 3.71 shows the proton shifts of the tricyclic dilactam (3) as a function of the concentration of Eu(fod)$_3$. Linear plots with varying slopes are obtained, the largest slope resulting from the four protons at C-5 and C-11 which are closest to the lanthanide ion. The protons at C-2 and C-8 which were initially at lower field are shifted least and hence the lines intersect.

Fig. 3.69 60 MHz proton magnetic resonance spectra of methyl isobutyl ketone (10.6 mg, 1.1×10^{-4} mol) in CCl_4 (0.5 ml) containing various amounts of Eu(pfd)$_3$; A, 0.0 mg; B, 13.5 mg; C, 21.1 mg; D, 29.0 mg. Data reproduced from H. E. Francis and W. F. Wagner (1972). 'Induced Chemical Shifts in Organic Molecules; Intermediate Shift Reagents', *Organic Magnetic Resonance*, 4, 190 (Fig. 1), Heyden and Son Ltd.

Fig. 3.70 Induced shifts of the proton resonances of methyl isobutyl ketone as a function of added Eu(pfd)$_3$. Data reproduced from H. E. Francis and W. F. Wagner (1972), 'Induced Chemical Shifts in Organic Molecules; Intermediate Shift Reagents', *Organic Magnetic Resonance*, 4, 190 (Fig. 2), Heyden and Son Ltd.

Fig. 3.71 **Effect of addition of Eu(fod)$_3$ on the chemical shift values for the protons in** *N,N'*-dimethyl-*cis,trans,cis*-3,9-diaza-tricyclo[6.4.0^{2,7}]dodecane-4,10-dione (3).

Optically active lanthanide shift reagents such as tris-(3-trifluoromethyl-hydroxymethylene-(+)-camphorato) europium (4) are commercially available. They can be used for the direct determination of optical purity and for the measurement of enantiomeric composition. The differences in the lanthanide-induced shift between enantiomers can be as high as 1.8 p.p.m. depending on the geometry of the molecule.

Spin decoupling.[31, m, n] Spin decoupling is a technique for determining which nuclei are coupled together by observing the effect in the spectrum when the coupling is removed. The decoupling is achieved by irradiating the substrate with a strong radiofrequency signal corresponding to the resonance frequency of one of the nuclei; the spectrum resulting from the remaining nuclei is scanned to observe any simplification which results. Successful decoupling is not achieved if the separation of the coupled multiplets is less than about 20 Hz for a 100-MHz instrument.

357

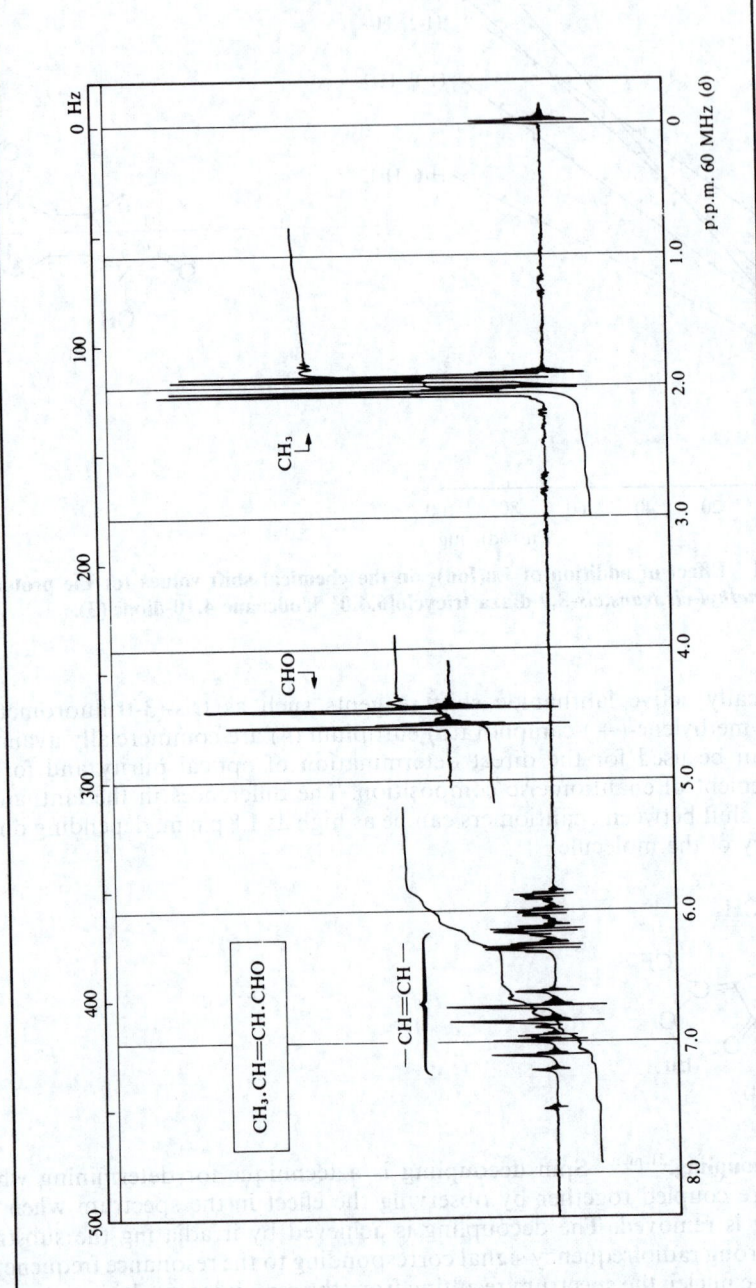

Fig. 3.72   Proton magnetic resonance spectrum of crotonaldehyde in CDCl₃ solution; sweep offset 300 Hz, sweep width 500 Hz.

Fig. 3.73 Decoupled spectrum of crotonaldehyde. Data reproduced from W. McFarlane and R. F. M. White (1972). *Techniques of High Resolution Nuclear Magnetic Resonance Spectroscopy*. London; Butterworths, p. 28.

Figures 3.72 and 3.73(*a*) and (*b*) illustrate the application of spin decoupling to the simplification of the spectrum of crotonaldehyde [(5); spectrum 3.72].

(5)

Figure 3.73(*a*) shows the spin–spin splitting in the olefinic region. The multiplet at low field is two slightly overlapping quartets which arise from the coupling of the proton H_a with the methyl protons (H_c) to give a quartet which is then split into a pair of quartets by coupling with the proton H_b. The multiplet at higher field is more complex and arises from the proton H_b. Coupling of H_b to H_a gives a doublet, each signal of which is split into a pair of doublets by coupling with the aldehydic proton H_d. Each of these signals is then split by the methyl protons to give the observed four closely-spaced quartets. Irradiation at the methyl protons causes all the quartets to collapse to single peaks, and the re-run, simplified spectrum, Fig. 3.73(*b*), now shows the low-field doublet corresponding to H_a and the pair of doublets for H_b, where coupling to H_a and H_d only is observed.

INTERPRETATION OF THE P.M.R. SPECTRUM

It is not possible to prescribe a set of rules which is applicable on all occasions. The amount of additional information available will most probably determine the amount of information it is necessary to obtain from the p.m.r. spectrum. However, the following general procedure will form a useful initial approach to the interpretation of most spectra.

1. Make a table of the chemical shifts of all the groups of absorptions in the spectrum. In some cases it will not be possible to decide whether a particular

group of absorptions arises from separate sets of nuclei, or form a part of one complex multiplet. In such cases it is probably best initially to include them under one group and to note the spread of chemical shift values.

2. Measure and record the heights of the integration steps corresponding to each group of absorptions. With overlapping groups of protons it may not be possible to measure these exactly, in which case a range should be noted. Work out possible proton ratios for the range of heights measured, by dividing by the lowest height and multiplying as appropriate to give integral values. If the accurate measurement of integration steps is not possible a range of proton ratios should be calculated, and noted down.

3. Note any obvious splitting of the absorptions in the table (e.g. doublet, triplet, quartet, etc.). For spectra which appear to show first-order splitting, the coupling constants of each multiplet should be determined by measuring the separation between adjacent peaks in the multiplet. Any other recognisable patterns which are not first order should be noted.

4. Note any additional information such as the effect of shaking with D_2O, use of shift reagents, etc.

5. Attempt a preliminary assignment of the nature (e.g. alkyl, alkenyl, aryl) of each of the groups of absorptions on the basis of their chemical shift.

6. By considering both the relative intensities and the multiplicities of the absorptions attempt to determine which groups of protons are coupled together. The magnitude of the coupling constant may give an indication of the nature of the protons involved.

7. Relate the information thus obtained to any other information available on the compound under consideration.

8. Having arrived at a possible structure or partial structure consult the correlation tables and work out the position and nature of the absorptions expected from the postulated structure.

9. Decide whether additional information, e.g. spin decoupling or the use of shift reagents, is required.

10. Repeat steps 6, 7, 8 until a self-consistent set of results is obtained.

FURTHER INFORMATION FROM ^{13}C SPECTRA

Only one of the three items of information normally available from p.m.r. spectra (i.e. chemical shift, coupling constant and relative numbers of absorbing nuclei) is routinely available from the ^{13}C spectrum, and that is the chemical shift. Quantitative coupling constants are not normally obtained, and relative numbers of nuclei cannot usually be derived from measurement of peak areas. The large ^{13}C—^1H coupling constant (125–200 Hz for directly bonded protons) results in multiplets which overlap to a considerable extent, and in the absence of decoupling makes the spectrum difficult to analyse. Spectra are therefore normally spin-decoupled and each absorption appears as a sharp singlet; this technique is known as wide-band or noise decoupling. Although the sensitivity is thus increased, all the information normally available from spin–spin splitting patterns is lost. An alternative method of decoupling (off-resonance decoupling) does however allow coupling of directly bonded carbon and hydrogen to be observed, although the separation of the peaks of the multiplets produced by this method is not equal to the true ^{13}C—^1H coupling constant. It is thus possible to identify carbon atoms associated with methyl, methylene and methine groups since the absorptions appear as quartets, triplets and doublets respectively, provided that the bonded hydrogens are equivalent. The use of the off-resonance procedure is illustrated in Figs 3.74(*a*) and 3.74(*b*), which shows the noise decoupled and off-resonance decoupled spectra of butane-1,3-diol and 2,2′-bipyridyl respectively. The methyl carbon of the diol appears as a quartet at high field; the two methylene carbons appear as triplets, the one bonded to oxygen being at lower field, and the low-field doublet is due to the methine carbon. In the case of 2,2′-bipyridyl each carbon bonded to one hydrogen appears as a doublet; the two carbons bonded to nitrogen appear at lower field.

3.3 MASS SPECTROMETRY

It is unlikely that the laboratory organic chemist will be required to record mass spectra of compounds produced in the laboratory as they will normally be obtained through a centralised service. This section therefore concentrates on the interpretation of spectra rather than on the techniques for obtaining the spectra. For further information on this aspect of mass spectrometry the reader should consult the sources listed in the references at the end of this chapter.[4]

Probably the most common use of mass spectrometry by the organic chemist is for the accurate determination of molecular weight. A second important use is to provide information about the structure of compounds by an examination of the fragmentation pattern.

Fig. 3.74(*a*) ^{13}C nuclear magnetic resonance spectrum of butane-1,3-diol in CDCl$_3$; signal assignment by off-resonance decoupling.

361

Fig. 3.74(b) ¹³C signal assignment by off-resonance decoupling for 2,2′-bipyridyl, 25.2 MHz. The numbers by the signals indicate the numbering of the carbon atoms. Values relative to TMS = 0. Data reproduced from E. Breitmaier, G. Jung and W. Voelter (1971). *Angew. Chem. Internat. Edn.*, **10**, 667.

THE MASS SPECTRUM

In a typical mass spectrometer, an organic compound under high vacuum is bombarded with electrons (of about 70 eV energy). Loss of an electron from the molecule followed by various fission processes gives rise to ions and neutral fragments. The positive ions are expelled from the ionisation chamber and resolved by means of a magnetic or an electric field.

The mass spectrum is a record of the current produced by these ions as they arrive at a detector. The intensity of a peak in the spectrum is thus an indication of the relative number of ions; the larger the peak the more abundant the ion producing it. Many mass spectrometers produce up to five traces simultaneously of differing sensitivity to allow weaker peaks to be studied, while also allowing intense peaks to be recorded on the chart. Figure 3.75 shows part of a

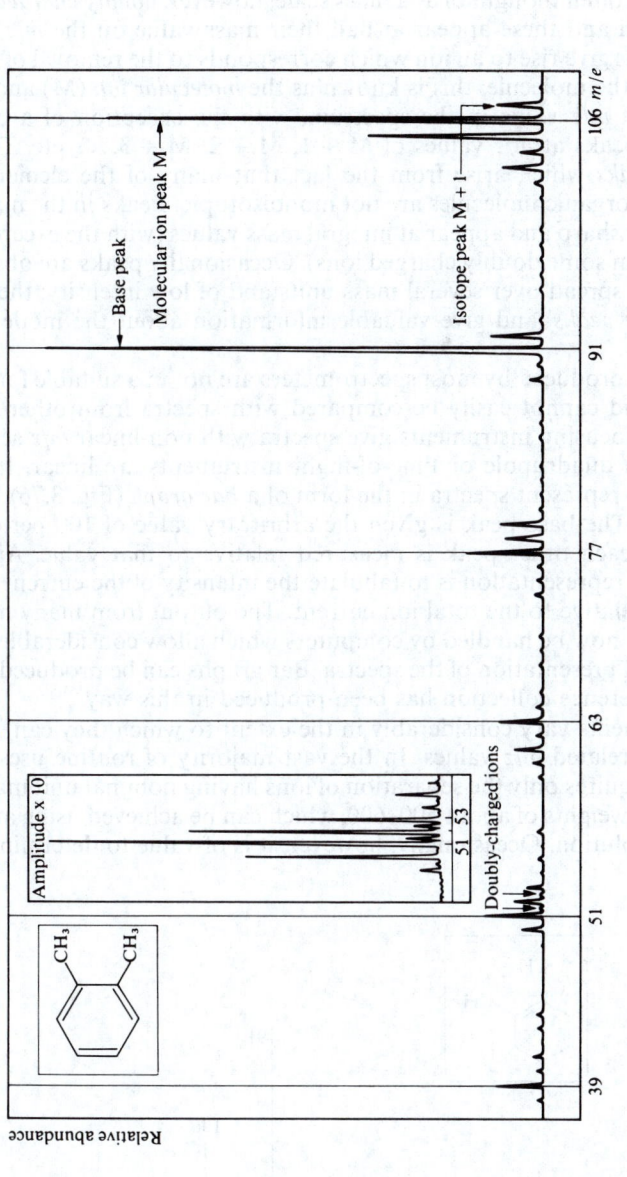

Fig. 3.75   Part of a low resolution mass spectrum of *o*-xylene.

363

low resolution spectrum and a number of features should be noted. The most intense peak in the spectrum is known as the *base peak*. Ions produced in the fragmentation of the organic compound are separated according to their mass: charge ratio (m/z) (formerly m/e). Since the majority of ions are singly charged the scale is often thought of as a mass scale; however, *doubly charged ions* are not uncommon and these appear at half their mass value on the m/z scale. Many compounds give rise to an ion which corresponds to the removal of a single electron from the molecule; this is known as the *molecular ion* (M) and usually has the highest m/z value in the spectrum, with the exception of a characteristic group of peaks at m/z values of M + 1, M + 2, M + 3,..., etc. The latter are *isotope peaks* which arise from the fact that many of the elements normally present in organic molecules are not monoisotopic. Peaks in the mass spectrum are usually sharp and appear at integral mass values (with the exception of those arising from some doubly charged ions). Occasionally peaks are observed which are broad, spread over several mass units and of low intensity; these are called 'metastable peaks' and give valuable information about the mode of fragmentation.

Spectra produced by most spectrometers are not in a suitable form for reproduction and cannot easily be compared with spectra from other instruments. Magnetic focusing instruments give spectra with non-linear m/z scales whereas those from quadrupole or time-of-flight instruments are linear. It is common practice to represent spectra in the form of a *bar graph* (Fig. 3.76) with a linear m/z scale. The base peak is given the arbitrary value of 100 per cent and the height of each other peak is measured relative to that value. An alternative method of representation is to tabulate the intensity of the current arising from each ion relative to the total ion current. The output from many mass spectrometers can now be handled by computers which allow considerable flexibility in the form of presentation of the spectra. Bar graphs can be produced directly and a large reference collection has been produced in this way.[6h]

Instruments vary considerably in the extent to which they can separate ions of closely related m/z values. In the vast majority of routine uses the organic chemist requires only the separation of ions having nominal unit masses of up to molecular weights of about 500–600, which can be achieved using an instrument of low resolution. Occasionally, however, it is of value to determine the precise

Fig. 3.76   Mass spectrum of *o*-xylene as a bar graph.

mass of particular ions accurately (up to six places of decimals) and for this purpose an instrument of high resolution is required.

Molecular formulae. Probably the most useful single piece of information for the organic chemist which can be derived from the mass spectrum is the molecular formula. Provided that the molecular ion can be identified, there are in principle two methods for deriving the molecular formula, using either high or low resolution. The most reliable method, although requiring the more sophisticated high-resolution instrumentation, is the accurate mass measurement of the molecular ion. Since atomic masses are not exact integers (see Appendix 4, Table A4.1) each combination of atoms will have a unique non-integral value. For example, CH_2O and C_2H_6 both have an integral mass of 30 but the accurate masses are 30.010 565 and 30.046 950 respectively. Accurate mass measurement will therefore distinguish between these two molecules. Tables are available which reduce the problem of relating accurate masses to possible molecular formulae.[6i] Accurate mass determination is most useful to the organic chemist in confirming the identity of a specific molecular ion rather than in suggesting possible formulae for completely unknown molecules.

An alternative method of determination of the molecular formula which utilises low-resolution spectra is based on the measurement of the *intensities of the isotope peaks*. The natural abundance of the stable isotopes of some common elements is shown in Appendix 4, Table A4.2. The data are presented in two ways, firstly as a percentage of the total isotopes present, and secondly as a percentage of the most abundant isotope. Each combination of atoms will thus give rise to a group of isotope peaks of predictable intensities. Taking methane as an example, the ratio $^{12}CH_4 : ^{13}CH_4 = 100 : 1.08$. Thus the intensity of the M + 1 peak will be 1.08 per cent of the intensity of the molecular peak, although there will also be a very small contribution from $^{12}C^1H_3{}^2H$. Table 3.2 lists some of the intensities of the M + 1 and M + 2 peaks for various combinations of C, H, N, O having a nominal mass of 120.

Table 3.2 Intensities of isotope peaks for the combinations of C, H, N, O of mass 120

Formulae	M + 1	M + 2
$C_2H_4N_2O_4$	3.15	0.84
$C_2H_6N_3O_3$	3.52	0.65
$C_3H_{12}N_4O$	5.00	0.31
$C_4H_{12}N_2O_2$	5.36	0.52
$C_6H_6N_3$	7.72	0.26
C_9H_{12}	9.92	0.44

Extensive compilations of such data are available: they can easily be modified to include elements other than C, H, O and N.

One limitation on the use of isotope peak intensities to determine the molecular formula is that the molecular ion must be relatively intense, otherwise the isotope peaks will be too weak to be measured with the necessary accuracy. Difficulty may also arise from spurious contributions to the isotope peak intensities from the protonated molecular ion, from weak background peaks or from impurities in the sample. In any event the method is only reliable for molecules having molecular weights up to about 250–300.

Deductions from isotope abundances. Assuming that the molecular ion has been identified correctly and intensities of the isotope peaks measured, the next stage in the analysis is to work out all possible molecular formulae which are consistent with this information. The common elements can be divided into three groups according to their isotopic composition. Firstly, those elements with a single natural isotope, e.g. hydrogen, fluorine, phosphorus, iodine. Hydrogen is placed in this group since the contribution from ^2H is extremely small. Secondly, those elements with a second isotope of one mass unit higher than the most abundant isotope, e.g. carbon and nitrogen. Thirdly, those with an isotope two mass units higher than the most abundant, e.g. chlorine, bromine, sulphur, silicon and oxygen.

Members of the last group, and especially chlorine and bromine, are most easily recognised from the characteristic patterns of the peaks, spaced at intervals of two mass units, which they produce in the spectrum. Typical patterns for combinations of bromine and chlorine atoms are shown in Figs 3.77 and 3.78. It may be difficult to estimate the number of oxygen atoms due to the low natural abundance (0.20%) of ^{18}O.

The intensity of the M + 1 peak allows an estimate to be made of the number of carbon and nitrogen atoms (however, if Cl, Br, S, or Si is present, loss of a proton from the M + 2 may enhance the intensity of the M + 1 peak). For a molecule not containing nitrogen, the maximum number of carbon atoms can be deduced by dividing the relative intensity of the M + 1 peak by 1.1 Thus a molecule having twelve C atoms will give an M + 1 peak of 13.2 per cent. If nitrogen is present its contribution to the M + 1 peak will amount to 0.36 ×

Number of bromine atoms

Number of chlorine atoms

Fig. 3.77 Isotope abundances for combinations of chlorine and bromine atoms.

Number of chlorine atoms

5 6 7

Number of chlorine atoms

8 9 10

Fig. 3.78   Isotope abundances for combinations of chlorine atoms.

number of nitrogen atoms; this figure must be subtracted from the measured relative intensity of the M + 1 peak before calculating the number of carbon atoms. An indication of the number of nitrogen atoms present may be deduced with the aid of the 'nitrogen rule' (see below).

Once the numbers of these two groups of elements have been estimated, the remainder of the mass of the ion must be due to the monoisotopic elements, the numbers of which can then usually be deduced.

A study of isotope abundance can give information about the elemental composition of other ions in the spectrum as well as the molecular ion. However, care must be taken that the intensities being measured arise solely from the isotopic contribution and not from other ions of different elemental composition.

Recognition of the molecular ion. Since the molecular formula is normally the most important piece of information to be derived from the mass spectrum it is necessary to be as certain as possible that the molecular ion within the molecular cluster (M, M + 1, M + 2, etc.) has been correctly identified. A number of tests can be applied which will show if an ion is not the molecular ion.

The ion must be an odd-electron ion since the molecular ion is produced by loss of one electron from the neutral molecule: the converse is not true since there may well be odd-electron ions other than the molecular ion in the spectrum, arising from rearrangement reactions. If the elemental composition of the ion can be determined, the index of hydrogen deficiency (the sum of multiple bonds and ring systems) can be used to determine whether the ion is an odd-electron ion. The *index of hydrogen deficiency* is the number of pairs of hydrogen atoms which must be removed from the saturated open-chain formula to give the observed molecular formula. For a molecule $I_yII_nIII_zIV_x$:

the index of hydrogen deficiency = $x - y/2 + z/2 + 1$

where   I = any monovalent atom
       II = O, S or any other divalent atom
       III = N, P or any other trivalent atom
       IV = C, Si or any other tetravalent atom

For example, thiophene, C_4H_4S, has an index of hydrogen deficiency

367

of $(4 - \frac{4}{2} + 1) = 3$. The index of hydrogen deficiency must be a whole number for an odd-electron ion. For an even-electron ion the value will be non-integral.

A second test which can be applied is the *nitrogen rule*. If a molecule (or ion) contains an odd number of nitrogen atoms it will have an odd numerical value for the molecular weight, whereas if it contains zero or an even number of nitrogen atoms it will have an even-numbered molecular weight. The rule applies to all compounds containing C, H, O, N, S, halogens, P, B, Si. Thus for a species with zero or even number of nitrogens, odd-electron ions will have an even mass number and even-electron ions will have an odd mass number.

A third indication that an ion is indeed the molecular ion may be obtained from an examination of the fragment ion peaks in the vicinity of the ion. Mass losses of between 3 and 15 and between 20 and 26 are highly unlikely, and if they are observed would suggest that the putative molecular ion is in fact a fragment ion.

Alteration of instrumental conditions may also provide evidence to confirm the recognition of the molecular ion. The use of maximum sensitivity may show up a very weak molecular ion. Alternatively, if the energy of the electron beam is decreased the intensity of the fragment ions will decrease relative to the molecular ion; this also applies to fragment ions arising from impurities. Alternative methods of ionisation such as chemical ionisation and field ionisation are very much more likely to produce a molecular ion cluster than the electron ionisation method, and should be used if they are available.

Intensity of the molecular ion. The lower the energy required for ionisation of the molecule, and the more stable the molecular ion, the more intense will be the peak in the mass spectrum. Structural features within the molecule have characteristic values of ionisation energy and hence determine the amount of energy required to form the molecular ion. Table 3.3 gives a general indication of the intensity of the molecular ion for various types of compounds. It must be borne in mind that if the molecule contains a readily cleaved bond the molecular ion peak will be much less intense. In general the intensity of the molecular ion increases with unsaturation and with the number of rings, but decreases with chain branching. The presence of heteroatoms with easily ionised outer-shell electrons increases the intensity of the molecular ion.

Table 3.3   Intensity of the molecular ion in the mass spectrum

Strong	Medium	Weak or non-existent
Aromatic hydrocarbons	Aromatic bromides and iodides	Aliphatic alcohols, amines and nitriles
Aromatic fluorides, chlorides, nitriles and amines	Conjugated alkenes	Branched chain compounds
Saturated cyclic compounds	Benzyl and benzoyl compounds	Nitro compounds
	Straight chain ketones and aldehydes, acids, esters, amides	
	Ethers	
	Alkyl halides	

Fragmentation. Although it may be of very low abundance, the molecular ion provides vital information about the identity of the molecule. Further information must be derived from the fragmentation pattern, i.e. the pattern of ions produced by decomposition of the molecular ion. Not all ions are of equal importance and some guidelines and rationalisations are needed to enable the organic chemist to derive the information required from the mass spectrum. Firstly, as discussed above, the molecular ion is the most important in the spectrum. Secondly, odd-electron ions are generally of more significance than even-electron ions of similar mass or abundance, since they are generally formed via a rearrangement reaction which may be characteristic of a particular class of compound. Thirdly, ions of high mass are likely to give more useful information than those at lower mass, since they are likely to have been formed as the result of a simple rational fragmentation. Fourthly, metastable ions (see later) may give useful information on the nature of the fragmentation processes.

There are two important factors which determine the intensities of fragment ions in the mass spectrum: the stability of the ion, and the energy relationships of the bonds broken and formed in the reactions leading to the ion. Although the conditions in the mass spectrometer (very low pressure, unimolecular reactions) differ substantially from those normally encountered in organic chemistry, the fundamental ideas of physical organic chemistry, and in particular those concerned with carbocation stability, can be used effectively in the rationalisation of the appearance of the mass spectrum. Thus the following common fragmentations all give rise to typically stable carbocations.

$$\left[\begin{array}{c} CH_3 \\ | \\ R{-}C{-}CH_3 \\ | \\ CH_3 \end{array} \right]^{\oplus\cdot} \longrightarrow \overset{\cdot}{R} + \begin{array}{c} CH_3 \\ | \\ CH_3{-}C^{\oplus} \\ | \\ CH_3 \end{array}$$

A stable tertiary carbocation is formed. The order of stability of saturated carbocations decreases in the order: tertiary > secondary > primary > methyl.

$$R{-}CH_2{-}CH\overset{\cdot}{{-}}CH_2 \longrightarrow \overset{\cdot}{R} + \overset{\oplus}{C}H_2{-}CH{=}CH_2 \longleftrightarrow CH_2{=}CH{-}\overset{\oplus}{C}H_2$$

Formation of a resonance-stabilised allylic carbocation.

$$C_6H_5{\cdot}CH_2{-}R]^{\oplus\cdot} \longrightarrow \overset{\cdot}{R} + \begin{array}{c} \oplus \end{array}$$

The aromatic seven-membered cyclic tropylium ion $C_7H_7^{\oplus}$ is formed.

$$^1R{-}CH_2{-}\overset{\oplus}{X}{-}{^2}R \longrightarrow {^1}\overset{\cdot}{R} + CH_2{=}\overset{\oplus}{X}{-}{^2}R \longleftrightarrow \overset{\oplus}{C}H_2{-}\overset{\cdot\cdot}{X}{-}{^2}R$$

The carbocation is stabilised by delocalisation of the lone-pair electrons on the adjacent heteroatom.

$$\begin{array}{c} ^1R \\ \diagdown \\ \diagup \\ ^2R \end{array} C{=}\overset{\oplus}{\ddot{O}}{:} \longrightarrow {^1}\overset{\cdot}{R} + {^2}R{-}C{\equiv}\overset{\oplus}{O}{:} \longleftrightarrow {^2}R{-}\overset{\oplus}{C}{=}\ddot{O}{:}$$

The resonance-stabilised acylium ion is formed in this case.

The molecular ion is formed by removal from the molecule of the electron of lowest ionisation potential. The energy required to remove an electron varies in the order

lone-pair < conjugated π < non-conjugated π < σ

A radical ion is thus formed which can fragment in a variety of ways. Simple bond cleavage may occur to give a neutral and an ionic fragment. Alternatively, a number of rearrangement processes may take place which are then followed by bond cleavage reactions. The important types of fragmentations and rearrangements are summarised and exemplified below.

Fragmentation by movement of one electron. Bonds are broken by movement of one electron, represented by a fish-hook arrow ($\frown$).

(a) *σ-cleavage* (sigma cleavage)

$$^1\dot{R}—^2R]^{\oplus\bullet} \longrightarrow {}^1\dot{R} + {}^2R^{\oplus}$$

Ionisation results in removal of a σ-electron and the σ-bond then breaks preferentially to give a stable carbocation with the ejection of the largest possible group as the radical.

(b) *α-cleavage* (alpha cleavage)

$$^1R—CH_2—\overset{\oplus}{\underset{\bullet\bullet}{O}}—^2R \longrightarrow {}^1\dot{R} + CH_2=\overset{\oplus}{\underset{\bullet\bullet}{O}}—^2R$$

$$\begin{array}{c}^1R\\ \\ {}^2R\end{array}\!\!\!\!CH—\overset{\oplus}{NH_2} \longrightarrow {}^1\dot{R} + {}^2R\cdot CH=\overset{\oplus}{NH_2}$$

$$\begin{array}{c}^1R\\ \\ {}^2R\end{array}\!\!\!\!C=\overset{\oplus}{\underset{\bullet\bullet}{O}} \longrightarrow {}^1\dot{R} + {}^2R\cdot C\equiv\overset{\oplus}{\underset{\bullet\bullet}{O}}$$

$$^1R—CH_2—CH—\overset{\oplus}{CH_2} \longrightarrow {}^1\dot{R} + CH_2=CH—\overset{\oplus}{CH_2}$$

In all of these processes an uncharged alkyl radical is lost enabling the residual electron to pair with that associated with the original radical ion, to form an even-electron ionic species.

Fragmentation by movement of an electron pair. Bonds are broken by movement of two electrons towards the positive charge, and represented by a normal 'curly arrow' ($\curvearrowright$).

$$^1R—\overset{\oplus}{\underset{\bullet\bullet}{O}}—^2R \longrightarrow {}^1\overset{\oplus}{R} + :\overset{}{\underset{\bullet\bullet}{O}}—^2R$$

$$CH_2=\overset{\oplus}{\underset{\bullet\bullet}{O}}—R \longrightarrow CH_2=\overset{}{\underset{\bullet\bullet}{O}}: + \overset{\oplus}{R}$$
(formed from α-cleavage
in an ether)

$$R—C\equiv\overset{\oplus}{O}: \longrightarrow \overset{\oplus}{R} + :C=\overset{\bullet\bullet}{\underset{\bullet\bullet}{O}}:$$
(formed from α-cleavage of
an aldehyde or ketone)

$$R—\overset{\oplus}{OH_2} \longrightarrow \overset{\oplus}{R} + H_2O$$

An electron pair is donated to the charge site. The electron pair may come from the bond adjacent to the charge site.

Rearrangements. These may yield odd-electron ions which are normally easily identified in the spectrum (cf. the nitrogen rule above). They are thus useful aids in the interpretation of the spectrum. Owing to the large excess of energy normally available in the ion source, molecular rearrangements are extremely common (see below). They may be random rearrangements (scrambling) which result in the general redistribution of certain atoms in the molecule, or more specific rearrangements, frequently involving a transfer of a hydrogen atom, which are characteristic of a certain type of molecular structure and give rise to easily recognisable ions in the mass spectrometer. It is the latter type which are particularly useful in the elucidation of molecular structure.

The most frequently encountered example is the *McLafferty rearrangement* which involves the transfer of a γ-hydrogen atom in an unsaturated system via a low-energy six-membered transition state:

The ionic fragment may be either the alkene or the fragment containing the heteroatom; this is determined by the relative ionisation potentials of the two groups. The rearrangement is general for this type of functional grouping and also occurs with oximes, hydrazones, ketimines, carbonates, phosphates, sulphites, alkenes and phenylalkanes. A similar rearrangement may occur in saturated systems; in this case a small cyclic transition state is permitted since it does not have to accommodate the double bond.

371

Peaks which arise from *metastable ion decomposition* are normally broad and of low intensity. They arise from the fragmentation of ions which have already been accelerated out of the ion source but have not yet reached the magnetic field. They are thus displaced from the position in the spectrum which would correspond to their true mass. The position of the metastable peak (m^*) is related to the mass of the precursor ion (m_1) and the mass of the product ion (m_2) by the equation.

$$m^* = \frac{(m_2)^2}{m_1}$$

The existence of a metastable ion and its relationship to m_1 and m_2 thus confirm that the ion m_2 was in fact formed directly from m_1. There are in theory many possible solutions to the equation, but the actual solution is normally obtained by inspection of the spectrum using major peaks, usually of similar intensity, as possible values for m_1 and m_2. For spectrometers which have an exponential mass scan this is a simple operation since the distances between m^* and m_2, and m_2 and m_1 will be identical.

Although they may be of low abundance, ions at the high mass end of the spectrum are of major significance in providing information about molecular structure. They result from the loss of small neutral fragments and are least likely to be the result of random rearrangements. Thus M − 1, M − 15, M − 18 peaks normally arise from loss of H, CH_3 and H_2O respectively. A list of some common neutral fragments is given in Appendix 4, Table A4.3.

APPEARANCE OF THE MASS SPECTRUM

The number of abundant ions in the mass spectrum and their distribution is indicative of the type of molecule. As discussed above, the mass and the relative abundance of the molecular ion gives an indication of the size and general stability of the molecule. An abundant molecular ion is expected, for example, from aromatic and saturated polycyclic molecules, provided that no easily cleaved group is present. A spectrum consisting of a few prominent ions suggests there are only a few favoured decomposition pathways indicating a small number of labile bonds or stable products.

The presence of particular *series of ions* in the spectrum is often indicative of certain types of molecules. Compounds with large saturated hydrocarbon groups give series of ions separated by fourteen mass units, corresponding to CH_2, since all the carbon–carbon and carbon–hydrogen bonds are of similar energy. The abundance of ions at the lower end of the spectrum steadily increases for straight-chain alkyl groups as the result of secondary reactions (see the spectrum of decane. Fig. 3.79(a)). The sequence in straight-chain alkanes appears at $\overset{\oplus}{C_nH_{2n+1}}$ (29, 43, 57, 71,...), but for compounds containing functional groups the positions are shifted due to the presence of heteroatoms. Some of the common series are shown in Table 3.4. Unfortunately the series for aldehydes and ketones overlaps the alkyl series since CO and C_2H_4 are both of mass 28. Complex molecules may show more than one series.

Characteristic ion series are also produced by aromatic compounds, the exact positions being dependent on the nature of the substituent (Table 3.5).

Certain types of compounds give characteristic ions in the mass spectrum which are often readily picked out and are useful indicators of possible struc-

Table 3.4 Ion series: aliphatic compounds

Compound type	General formula	Ion series
Alkyl	C_nH_{2n+1}	29, 43, 57, 71 …
Alkylamines	$C_nH_{2n+2}N$	30, 44, 58, 72 …
Aliphatic alcohols and ethers	$C_nH_{2n+1}O$	31, 45, 59, 73 …
Aliphatic aldehydes and ketones	$C_nH_{2n+1}CO$	43, 57, 71, 85 …
Aliphatic acids and esters	$C_nH_{2n-1}O_2$	59, 73, 87, 101 …
Alkyl chlorides	$C_nH_{2n}{}^{35}Cl$	49, 63, 77 …
	$C_nH_{2n}{}^{37}Cl$	51, 65, 79 …

Table 3.5 Ion series: aromatic compounds

Electron withdrawing substituent:	38, 39	50, 51	63, 64	75, 76
Electron donating substituent and heterocyclic compounds	39, 40	51, 52	65, 66	77, 78, 79

tures. These include m/z 30 (amines), 31 (primary alcohol), 74 (methyl alkanoates), 91 (benzyl), 149 (phthalic acid and esters). The possible compositions of some common fragment ions are listed in Appendix 4, Table A4.4. Some caution must be adopted in the use of these tables.

INTERPRETATION OF THE MASS SPECTRUM

The following scheme is suggested as a general approach to the interpretation of the mass spectrum. Each spectrum presents its own challenge and therefore too rigid adherence to any scheme is unwise. Reference should be made to the appropriate paragraph of this section for fuller details of each step.

1. Identify the molecular ion.
2. Determine the elemental composition and the index of hydrogen deficiency, i.e. the number of double bonds and rings.
3. Make any deductions which are possible from the general appearance of the spectrum; identify any ion series and characteristic ions.
4. Note possible structures of neutral fragments from the presence of high mass ions.
5. Identify any odd-electron ions and consider possible rearrangements (see *Rearrangements*, p. 371).
6. Suggest a feasible structure on the basis of the mass spectral and any other evidence. Predict the mass spectrum of the postulated compound and compare with the unknown spectrum. Make any modification to the proposed structure which appears necessary. Check the mass spectral behaviour of compounds of similar structures by consulting appropriate reference collections.

MASS SPECTRA OF CLASSES OF ORGANIC COMPOUNDS

Many types of organic compounds exhibit characteristic mass spectral behaviour, a knowledge of which is useful in the interpretation of their spectra. The following section provides an introduction to the interpretation of mass spectra of simple organic compounds but readers should consult the texts listed in the references.[4] Some caution is needed in the application of this information since the incorporation of additional substituents or functional groups into a molecule may well prevent 'characteristic' fragmentation.

Hydrocarbons. *Saturated hydrocarbons.* A high energy for ionisation is required for alkanes and the ions thus formed undergo random rearrangements. The molecular ion is normally present although it may be weak. The spectra normally consist of clusters of peaks separated by fourteen mass units corresponding to a difference of a CH_2 group. The $M - CH_3$ ion is frequently missing and for unbranched alkanes the intensity of other ions increases steadily to reach a maximum at m/z 43 ($C_3H_7^{\oplus}$) or m/z 57 ($C_4H_9^{\oplus}$); these peaks are mainly due to the highly branched ions resulting from molecular rearrangements; the spectrum of decane, Fig. 3.79(*a*), is typical. Branched chain hydrocarbons show intense peaks corresponding to preferential cleavage at a tertiary or quaternary carbon atom; thus the spectrum of 2,6-dimethyloctane, Fig. 3.79(*b*), shows an intense peak at m/z 113 due to loss of an ethyl group and formation of the secondary carbocation. Alicyclic hydrocarbons generally show a more abundant molecular ion, but the spectra are more difficult to interpret due to random rearrangement.

Alkenes. These give spectra in which the molecular ion peak is usually distinct and there is an increased abundance of the $C_nH_{2n-1}^{\oplus}$ ion series as compared with alkanes, as illustrated by the spectrum of hex-1-ene (Fig. 3.80). The location of

(*a*) Decane

(*b*) 2,6-Dimethyloctane

Fig. 3.79 Comparison of the mass spectra of straight chain and branched chain saturated hydrocarbons (*a*) decane; (*b*) 2,6-dimethyloctane.

Fig. 3.80 **Mass spectrum of hex-1-ene.**

the double bond in alkenes is often difficult to determine due to the occurrence of facile rearrangements. Cyclic alkenes undergo a characteristic *retro*-Diels–Alder fragmentation.

Aromatic hydrocarbons. Generally these hydrocarbons give rise to a prominent molecular ion as the result of the stabilising effect of the ring; doubly charged ions are also often apparent as low-intensity peaks at half integral mass values. Alkyl-substituted benzenoid compounds (for example *o*-xylene, Fig. 3.76) usually give rise to a base peak at m/z 91 due to the tropylium ion, $C_7H_7^{\oplus}$. This may eliminate a neutral acetylene molecule to give a peak at m/z 65.

$$\longrightarrow \quad C_5H_5^{\oplus} + HC{\equiv}CH$$
$$m/z\ 65$$

Aromatic compounds with alkyl groups having a chain of at least three carbon atoms can undergo a shift of a γ-hydrogen probably via a type of McLafferty rearrangement, giving rise to a prominent peak at m/z 92.

The characteristic aromatic cluster of ions in alkylbenzenes occurs at m/z 77, 78 and 79 (cf. the spectrum of *o*-xylene, Fig. 3.76).

Alcohols, phenols, ethers. The molecular ion of *alcohols* is weak or undetectable. Characteristic ions result from alpha-cleavage giving rise to resonance-stabilised carbocations ions; the loss of the largest alkyl group is the preferred pathway although ions resulting from losses of the other groups may also be observed.

375

Fig. 3.81   Mass spectrum of propan-1-ol.

Primary alcohols in particular give an M − 18 peak due to loss of water from the molecular ion although this peak may partly arise from thermal decomposition of the alcohol in the ion source. Initial migration of a hydrogen on the alkyl chain is followed by cleavage of the carbon–oxygen bond, see, for example, the spectrum of propan-1-ol, Fig. 3.81, which shows strong peaks at m/z 59, 42, 31 due to the loss of H, H_2O, and formation of $CH_2 {=} \overset{\oplus}{O}H$ respectively.

$$R {\cdot} \overset{H}{\underset{(H_2C)_n - CH_2}{CH}} \quad \overset{\oplus}{O}H$$

$$\downarrow$$

$$R {\cdot} \overset{\cdot}{\underset{(H_2C)_n - CH_2}{CH}} \quad \overset{\oplus}{O}H_2 \quad \xrightarrow{-H_2O} \quad R {\cdot} \overset{\cdot}{\underset{(H_2C)_n - \overset{\oplus}{CH_2}}{CH}}$$

$$\swarrow \qquad\qquad\qquad\qquad\qquad \downarrow$$

$$R {\cdot} \overset{\oplus}{\underset{(CH_2)_n}{CH - CH_2}} \qquad\qquad R {\cdot} \overset{\cdot}{CH} {\cdot} (CH_2)_{n-2} {\cdot} \overset{\oplus}{CH_2} + CH_2 {=} CH_2$$

Phenols usually give a strong molecular ion. Typical peaks in the spectrum arise from M − 28 (CO), which is a useful odd-electron ion, and M − 29 (CHO).

The molecular ion peak of *ethers* is weak or negligible. There are two fragmentation processes which are typical of ethers. A characteristic fragmentation is the cleavage of the carbon–oxygen bond. This often gives rise to the most abundant ion in the spectrum of aliphatic ethers.

$$^1R {\overset{\curvearrowleft}{\underset{}{{-}}} \overset{\oplus}{\overset{..}{O}} {-} {^2}R} \longrightarrow {^1}\overset{\oplus}{R} + {:}\overset{..}{O} {-} {^2}R$$

Alternatively cleavage of the α,β-bond (α-cleavage) may occur.

$$^1R \diagdown \atop ^2R \diagup CH\overset{\oplus}{-}\overset{\displaystyle ..}{O}-CH_2 \cdot CH_2 \cdot {}^3R \longrightarrow {}^1R \cdot CH=\overset{\oplus}{O}-CH_2 \cdot CH_2 \cdot {}^3R + {}^2\dot{R}$$

$$m/z\ 45,\ 59,\ 73,\ \text{etc.}$$

This type of ion may then break down further:

$$^1R \cdot CH=\overset{\oplus}{O}-CH_2 \atop \underset{H-CH \cdot {}^3R}{|} \longrightarrow {}^1R \cdot CH=\overset{\oplus}{O}H + CH_2=CH \cdot {}^3R$$

See for example the spectrum of diethyl ether, Fig. 3.82, which shows strong peaks at m/z 59, 45 and 31 due to $CH_2=\overset{\oplus}{O}\cdot CH_2 \cdot CH_3$, $CH_3 \cdot CH=\overset{\oplus}{O}H$ and $CH_2=\overset{\oplus}{O}H$ respectively.

Suitably substituted aromatic ethers will undergo a McLafferty rearrangement in the same way as alkylbenzenes.

Thiols and thioethers. The molecular ion is normally much more abundant in the case of sulphur compounds than with the corresponding oxygen compounds due to the lower ionisation energy of the non-bonding sulphur electrons. The presence and number of sulphur atoms is usually indicated by the contribution of ^{34}S to the M + 2 peak, and in addition homologous series of fragments containing sulphur are present having four mass units higher than those of the hydrocarbon series.

Thiols show similar fragmentations to those of alcohols, typical ions arising from α-cleavage and from loss of hydrogen sulphide (M − 34).

$$^1R \diagdown \atop ^2R \diagup CH\overset{\oplus}{-}\overset{\displaystyle ..}{S}H \longrightarrow {}^2R \cdot CH=\overset{\oplus}{S}H + {}^1\dot{R}$$

$$m/z\ 47,\ 61,\ 75$$

Aldehydes and ketones. Normally the molecular ion is observable for these compounds. Characteristic peaks in the spectra of ketones arise from cleavage α to

Fig. 3.82   Mass spectrum of diethyl ether.

the carbonyl group which gives two possible acylium ions, followed by loss of carbon monoxide giving the corresponding carbocations.

The more abundant acylium ion is normally produced by loss of the largest alkyl group.

In aromatic aldehydes and ketones the base peak usually arises from $Ar \cdot \overset{\oplus}{C}{\equiv}O$; compare the spectra of benzaldehyde and acetophenone (Fig. 3.83(a) and (b)) which both show intense peaks due to $C_6H_5 \cdot \overset{\oplus}{C}{\equiv}O$, and further fragmentations typical of aromatic compounds. The α-cleavage reaction is normally less significant for aldehydes than for ketones, although a prominent peak at m/z 29 (CHO) is sometimes observed.

McLafferty rearrangements are common for aliphatic aldehydes and ketones, providing that an alkyl group of at least three carbons long is attached to the carbonyl group. Odd-electron ions are formed which are useful in the analysis of the spectrum.

(a) Benzaldehyde

(b) Acetophenone

Fig. 3.83 Comparison of the mass spectra of (a) benzaldehyde and (b) acetophenone.

m/z 58, 72, 86

Thus the spectrum of 4-methylpentan-2-one (Fig. 3.84) shows a strong peak at *m/z* due to the odd-electron ion $(CH_3C(OH)=CH_2)^{\ddagger}$ resulting from the McLafferty rearrangement.

Fig. 3.84 Mass spectrum of 4-methylpentan-2-one.

Carboxylic acids. Monocarboxylic acids normally show the molecular ion in the spectrum. Cleavage of bonds adjacent to the carbonyl group (α-cleavage) results in formation of fragments of mass M − 17 (OH) and M − 45 (CO_2H). Characteristic peaks arise from the McLafferty rearrangement.

m/z 60, 74 . . .

379

Carboxylic esters. The molecular ion of the ester $^1R \cdot CO_2 \, ^2R$ is usually observed in those cases where the alkyl group, ^2R, is smaller than C_4. The characteristic ions in the spectrum arise from McLafferty rearrangements, which can occur with either the acyl– or alkoxy–alkyl group, providing they are at least three or two carbon atoms long respectively.

A characteristic ion formed from esters of long-chain alcohols results from re-arrangement of two hydrogen atoms ('McLafferty + 1' rearrangement).

The peaks at m/z 56 and 61 in the mass spectrum of butyl acetate (Fig. 3.85) can be explained by the above rearrangements. The mass spectrum of ethyl butanoate, Fig. 3.86, shows two important peaks due to odd-electron ions at m/z 88 and 60, resulting from two successive McLafferty rearrangements.

Fig. 3.85   Mass spectrum of butyl acetate.

Fig. 3.86   Mass spectrum of ethyl butanoate.

Cleavage of the alkoxyl group gives rise to the abundant ion m/z 71 $(M - OC_2H_5; R{\cdot}C{\equiv}\overset{\oplus}{O})$ which is a good diagnostic ion for esters.

$$^1R \underset{^2R{\cdot}O}{\overset{}{>}}C{=}\overset{\oplus}{\underset{..}{O}}: \longrightarrow \ ^1R{-}C{\equiv}\overset{\oplus}{O}: + \overset{\bullet}{O}{}^2R$$

Amines. The molecular ion is weak or negligible for aliphatic amines, although aromatic amines show an intense molecular ion.

The characteristic cleavage reactions of amines are similar to those of alcohols and ethers.

$$R{\overset{\frown}{-}}CH_2{-}\overset{\oplus}{N}H_2 \longrightarrow CH_2{=}\overset{\oplus}{N}H_2 + \overset{\bullet}{R}$$
$$m/z\ 30$$

In α-substituted primary amines, loss of the largest alkyl group is preferred. Similar ions are formed from secondary and tertiary amines.

$$^1R{-}\underset{^2R}{\overset{\oplus}{\underset{|}{CH}}}{-}\overset{\oplus}{N}H{-}CH_2{\cdot}CH_2{\cdot}{}^3R \longrightarrow {}^1R{-}CH{=}\overset{\oplus}{N}H{-}CH_2 \longrightarrow$$
$$H{-}CH{\cdot}{}^3R$$
$$^1R{-}CH{=}\overset{\oplus}{N}H_2 + {}^3R{\cdot}CH{=}CH_2$$
$$m/z\ 44,\ 58,\ 72\ldots$$

The spectrum of diethylamine, Fig. 3.87, is typical.

Fig. 3.87 **Mass spectrum of diethylamine.**

Amides. Primary amides exhibit behaviour similar to the corresponding acid and methyl esters; substituted amides resemble the higher alkyl esters. There is a common tendency to form M + 1 ions by ion–molecule reactions. Primary amides generally give a strong peak at m/z 44:

$$R—C\overset{\overset{\oplus}{O}:}{\underset{NH_2}{\Big\langle}} \longrightarrow \dot{R} + H_2N—C\equiv\overset{\oplus}{O}: \longleftrightarrow H_2\overset{\oplus}{N}=C=O$$
$$m/z\ 44$$

A characteristic fragmentation is via the McLafferty rearrangement.

$$m/z\ 59,\ 73,\ 87\ldots$$

Nitriles. The molecular ion peak is weak or non-existent in aliphatic nitriles but strong in aromatic compounds. Interpretation of the spectrum is often difficult since skeletal rearrangements are common and the resulting ion series (m/z 41, 55, 69, etc.) overlaps with that arising from hydrocarbons. Thus in the McLafferty rearrangement:

$$\longrightarrow {}^3R\cdot CH=C=\overset{\oplus}{N}H + {}^1R\cdot CH=CH\cdot{}^2R$$

Halogens. The patterns of isotope peaks should indicate the nature and number of halogen atoms in the molecule. This is especially useful for aromatic halogen compounds, but may be less valuable for aliphatic compounds which often exhibit a weaker molecular ion peak.

A typical fragmentation of alkyl chlorides and bromides is the loss of an alkyl group with the formation of a halonium ion. The ion forms the base peak in

382

straight-chain compounds but the intensity is considerably reduced if the chain is branched.

X = Br, Cl

Substituted aromatics – the *ortho* effect. Aromatic compounds bearing substituents with an appropriately placed hydrogen atom will undergo a facile rearrangement involving a second substituent in the *ortho* position, e.g.

Y = CH_2, O, NH
X = OH, OR, NH_2

The ion m/z 149, which is characteristic of phthalic acid esters, arises as the result of the operation of this *ortho* effect.

3.4 ULTRAVIOLET–VISIBLE SPECTROSCOPY[5]

The information concerning the structure of an organic molecule which may be gained from an ultraviolet–visible spectrum is more limited than in the case of i.r., n.m.r. and mass spectra. The principal features which may be detected are multiply bonded systems, conjugated systems and aromatic (and hetero-aromatic) nuclei. The electronic transitions in these systems which give rise to absorption in the 200–700 nm region are $\pi \to \pi^*$ and $n \to \pi^*$ and these are of diagnostic value (see also Section 2.17.5, p. 106). The $n \to \sigma^*$ electronic transition which arises in saturated compounds containing the heteroatoms sulphur, nitrogen, bromine and iodine is of less importance since it leads to absorption just below 200 nm. The corresponding transition in compounds containing oxygen and chlorine leads to absorption at somewhat shorter wavelengths; indeed, the transparency of alcohols and chloroalkanes in the region 200–700 nm makes them ideal solvents for u.v. spectral determinations. The $\sigma \to \sigma^*$ electronic transition in saturated hydrocarbons leads to absorption in the far-ultraviolet. Apart from the fact that special techniques are required to record the absorption, the information gained is in any event of little diagnostic value.

INSTRUMENTAL FEATURES OF ULTRAVIOLET–VISIBLE SPECTROPHOTOMETERS

The basic features of ultraviolet–visible spectrophotometers are similar in many respects to those described in Section 3.1 for i.r. spectrophotometers.

Commercial instruments for u.v. usually also cover the visible region and therefore have two light sources – a deuterium or hydrogen discharge tube for the region 200–370 nm, and a tungsten filament lamp for the region 325–750 nm; with recording instruments there is an automatic interchange at 370 nm. The monochromator incorporates a quartz prism or diffraction grating. In normal circumstances the lower limit of measurement is 190 nm owing to the fact that oxygen absorbs radiation below 190 nm, and quartz becomes less transparent in this region. Below 190 nm, measurements require the use of diffraction gratings and special vacuum techniques.

Modern instruments are double-beam recording spectrophotometers in which the light beam is split into two parallel beams which pass through the sample and reference cells and thence to the detector system. As with i.r. spectro-photometers, modern instruments include such features as the presentation of spectra on a visual display unit, microprocessor control of instrumental conditions and the storage and analysis of the recorded spectrum using a computer. Cells used in ultraviolet–visible spectroscopy are made entirely of quartz and are available in sizes which provide a path length (i.e. the length of sample in the beam of the radiation) varying from 0.5 to 10 cm.

DETERMINATION OF ULTRAVIOLET–VISIBLE SPECTRA

When a molecule absorbs ultraviolet or visible light of frequency v or wavelength λ, an electron undergoes a transition from a lower to a higher energy level in the molecule. The energy difference ΔE is given by the expression:

$$\Delta E = hv = \frac{hc}{\lambda}$$

where h is Planck's constant and c the velocity of the radiation. Multiplication by Avogadro's number, N_A (6.02×10^{23} mol^{-1}), will express the energy absorbed per mole. By inserting the numerical values for h (6.63×10^{-37} kJs), for c (3×10^{10} cm s^{-1}) and using the conversion factor (4.184) to convert kJ into kcal, the expression becomes:

$$\Delta E_{kJ\,mol^{-1}} = \frac{N_A hc}{\lambda} = \frac{6.02 \times 10^{23} \times 6.63 \times 10^{-37} \times 3 \times 10^{10}}{\lambda_{cm}}$$

$$= \frac{119.75 \times 10^3}{\lambda_{nm}}$$

or

$$\Delta E_{kcal\,mol^{-1}} = \frac{119.75 \times 10^3}{\lambda_{nm} \times 4.184}$$

For the region 200–750 nm, therefore, the energy required for electron transitions is in the range 600–160 kJ mol^{-1}; for the ultraviolet region 200–400 nm, the energy is of the same order of magnitude as the bond energies of common covalent bonds (e.g. C—H bond energy is ≈ 410 kJ mol^{-1}). For this reason prolonged exposure of the sample to ultraviolet radiation during measurement should be avoided to minimise possible decomposition of a proportion of the sample.

Energies of these magnitudes are associated with the promotion of an electron from a non-bonding (n) orbital or a π-orbital, to an antibonding π-orbital (π^*) or to an antibonding σ-orbital (σ^*). The most important transitions in organic compounds are:

(a) $\pi \rightarrow \pi^*$ transitions; these are usually associated with the multiple bonds of carbon with carbon, nitrogen, oxygen, sulphur, etc., and they generally give rise to high intensity absorption;

(b) $n \rightarrow \pi^*$ transitions; these are usually associated with groups such as carbonyl, thiocarbonyl, nitroso, etc., and generally the intensity of absorption is very much lower than that arising from (a).

LAWS OF LIGHT ABSORPTION
The Beer–Lambert law states that the proportion of light absorbed by a solute in a transparent solvent is independent of the intensity of the incident light and is proportional to the number of absorbing molecules in the light path:

$$\log_{10}\left(\frac{I_0}{I}\right) = A = \varepsilon c l$$

where
  I_0 = intensity of incident light
  I = intensity of transmitted light
  ε = molar absorptivity or molar extinction coefficient
  c = concentration of solute in moles/litre
  l = cell (path) length (cm)
  A = absorbance

It will therefore be seen that ε is a measure of the absorbance of the solution at a concentration of 1 mole per litre in a 1 cm cell. Beer's law is a limiting law and is strictly valid only at low concentrations.

When the molecular weight (M) of the absorbing substance is unknown, the extinction coefficient of a 1 per cent solution in a 1 cm cell ($A_{1cm}^{1\%}$) is generally used for comparison of absorption intensities:

$$A_{1cm}^{1\%} = \frac{A}{cl}$$

where c is now in grams per 100 ml and l is in cm. It is related to the molar absorptivity by the expression

$$\varepsilon = A_{1cm}^{1\%} \times \frac{M}{10}$$

Both ε and $A_{1cm}^{1\%}$ are independent of concentration or cell length provided the Beer–Lambert law is obeyed; the latter constant does not involve the molecular weight and is therefore used for compounds of unknown or uncertain constitutions.

The u.v. spectrum is a plot of the wavelength or frequency of absorption against the absorbance ($\log_{10} I_0/I$) or the transmittance (I/I_0). Spectral data are also presented in which the absorbance is expressed as the molar absorptivity, ε or $\log \varepsilon$, i.e. as a graphical plot of λ versus ε or $\log \varepsilon$. The intensity of an absorption band in the u.v. spectrum is usually expressed as the molar absorptivity (ε) at maximum absorption (λ_{max}). The smaller the difference between ground and excited states the longer will be the wavelength of absorption. The latter follows from the expression $\lambda = hc/\Delta E$. Thus absorption of light in the visible region, which is responsible for the colour of certain compounds, involves a lower energy transition as compared with the absorption of light in the ultraviolet region.

SOLVENTS FOR ULTRAVIOLET SPECTROSCOPY

Organic compounds generally absorb too strongly for their ultraviolet spectra to be directly determined, and dilute solutions must be prepared with solvents that are transparent to u.v. light over the wavelengths of interest. Fortunately a suitable solution of a compound whose u.v. spectrum is required to be recorded can usually be prepared, since a reasonably wide selection of solvents which are transparent down to about 205 nm is available. These include hexane,* heptane,* cyclohexane,* iso-octane (2,4,4-trimethylpentane),* chloroform,* tetrahydrofuran, 1,4-dioxane,* propan-2-ol,* ethanol,* methanol,* and water.* In the far u.v., suitable solvents are hexane and heptane. Those solvents which are commercially available as 'spectroscopically pure' (from, for example, Romil Chemicals) are marked with an asterisk, although for many purposes analytical grade reagents are satisfactory if the cell length is small. It should be noted that the value of λ_{max} may be dependent upon solvent polarity, and hence the solvent used for the spectral determination should always be specified.

SOLUTION PREPARATION

To obtain accurate absorbance values in the region of maximum sensitivity of the spectrophotometer it is usual to prepare a solution having a concentration which would give an A value at λ_{max} in the region of about 0.5 for the manual instrument, or about 0.9 for the automatic recording instrument. For a compound having an ε value of the order of 15 000, as is found for example in the case of crotonaldehyde (λ_{max} 220 nm, ε 15 000, M 70) and using a path length cell of 1 cm, substitution in the Beer–Lambert equation:

$$\log_{10} I_0/I = \varepsilon cl$$

gives $0.5 = 15\,000 \times c \times l$;

whence $c = \dfrac{0.5}{15\,000} = 3.33 \times 10^{-5}\,\text{mol dm}^{-3}$

or $= \dfrac{70 \times 0.5}{15\,000} = 2.33 \times 10^{-3}\,\text{g dm}^{-3}$

A solution of this concentration is most usually prepared by weighing (say) 23 mg (or 2.3 mg) of the substance and dissolving it in 100 ml (or 10 ml) of solvent using a graduated flask; 1 ml of the solution is then diluted exactly to 100 ml with the same solvent.

OPTICAL CELLS AND THEIR CARE

Although cells constructed of glass are suitable for the determination of spectra in the visible range, this material is not sufficiently transparent in the u.v. region and quartz cells must be used. These commonly have path lengths of from 0.5 to 10 cm; a 1-cm-square cell requires about 3 ml of solution.

Cells may become contaminated as a result of evaporation of solvent from solutions, and also by acquiring a film of grease as a result of careless handling. Such films and dust particles decrease transmission and can also contaminate liquids placed subsequently in the cell. Immediately after use therefore cells must be emptied and rinsed with clean solvent and then cleaned with a suitable detergent solution and stored in distilled water. A brush which might scratch the optical surface should *never* be used. Solid contaminants must always be removed by the following wet cleaning procedure:

(a) stand the cell in cold detergent for 15 minutes;
(b) rinse several times with distilled water;
(c) rinse with ethanol and store in covered containers containing distilled water or dry under a radiant lamp. Do not allow the cell to dry until the cleaning procedure is complete.

The following precautions in handing cells should also be observed:

(a) cells should only be handled by means of the etched surfaces;
(b) when wiping the outside surfaces of the cell, prior to placing in the instrument, paper tissues only should be used;
(c) cells should be removed from storage under water with the aid of suitably protected tongs.

The cells are filled bearing in mind the following points. The clean dry cell is rinsed with the appropriate solvent and then with the prepared solution before being finally filled. For solutions made up with volatile solvents the filled cell should be closed with the fitted lid provided; a reference cell is similarly filled with the neat solvent. If a cell has been stored under water it is first rinsed well with the appropriate solvent and then with a little of the prepared solution before being finally filled. If the solvent used is immiscible with water a preliminary washing with ethanol is necessary.

For accurate work a pair of matched cells should be used and each should be placed in the instrument so that the incident radiation enters via the same optical face every time.

It is not considered appropriate here to give the details of the operation of the spectrophotometer; the manual provided by the manufacturers for a particular instrument available should be consulted.

FEATURES OF AN ULTRAVIOLET–VISIBLE SPECTRUM

In the spectra of simple molecules the absorptions due to $n \rightarrow \pi^*$ transitions lie at longer wavelengths than those arising from the $\pi \rightarrow \pi^*$ excitations which is of course a measure of the lower energy required for electron promotion in the former case. The absorption occurs over a range of wavelengths about discernible maxima leading most frequently in solution to a broad absorption curve. This is because the spacing between the rotational and vibrational transitions of a polyatomic molecule are relatively small (about 0.5–$4.2\,\text{kJ}\,\text{mol}^{-1}$) and electron transitions occur (with corresponding slight differences of energy) from a range of vibrational–rotational levels in the ground state to a range of such levels in the excited state.

The term *chromophore* is used to describe any structural feature which leads to absorption in the ultraviolet–visible region and includes groups in which $\pi \rightarrow \pi^*, n \rightarrow \pi^*$ and $n \rightarrow \sigma^*$ transitions are possible. The term *auxochrome* is used to designate groups possessing non-bonding electron pairs which are conjugated with a π-bond system; an example is a hydroxyl or amino group attached to an aromatic ring system. Such n–π conjugation leads to a shift of absorption to longer wavelengths, referred to as a *bathochromic shift*; conversely a shift to shorter wavelengths is descibed as a *hypsochromic shift*.

Apart from the wavelength of maximum absorption (λ_{max}), the intensity of absorption (ε, the molar extinction coefficient) is of value in elucidating structural features. A high ε value (5000–$10\,000$) reflects a high probability of the occurrence of the relevant electronic transition and is observable in systems in which the relative symmetry of the ground and the excited state is such as to lead to a change in the transition moment. When this symmetry requirement is not met, the transition is regarded as being 'forbidden', and as a consequence the probability is low and the ε value is usually less than 100. For example, the $n \rightarrow \pi^*$ transition in a ketone, such as acetone, leads to absorption at $\lambda_{max}\,280\,\text{nm}$, $\varepsilon\,13$. Changes in the molecular environment adjacent to a chromophore may either increase or decrease the intensity of the absorption and such correlations may be of help in the interpretation of spectra. The magnitude of the ε value is also dependent upon such factors as the polarity of the solvent and the extent to which the ground and excited states are stabilised by solvation.

The carbon–carbon multiple bond. An isolated carbon–carbon double bond absorbs near $180\,\text{nm}$ ($\varepsilon \approx 15\,000$) as a result of a $\pi \rightarrow \pi^*$ electronic transition. Alkyl groups cause small bathochromic shifts so that a tetrasubstituted acyclic alkene absorbs in the region of $200\,\text{nm}$ (see Table 3.6).

An isolated carbon–carbon triple bond similarly absorbs in the $180\,\text{nm}$ region (acetylene, $\lambda_{max}\,173\,\text{nm}$, $\varepsilon \approx 6000$); the absorption maximum is shifted to longer wavelengths by the presence of alkyl groups (Table 3.6).

Conjugation of multiple carbon–carbon double bonds leads to significant changes in λ_{max} and in the ε value. For example, buta-1,3-diene has $\lambda_{max}\,217\,\text{nm}$, ε

Table 3.6 Approximate absorption positions of isolated carbon–carbon multiply bonded systems

Structure	λ_{max} (nm)	Structure	λ_{max} (nm)
$R \cdot CH = CH_2$	177	$R \cdot C \equiv CH$	185
$trans$-$R \cdot CH = CH \cdot R$	180	$R \cdot C \equiv C \cdot R$	196
cis-$R \cdot CH = CH \cdot R$	183		
$R_2 C = C \cdot R_2$	200		

Table 3.7 Approximate absorption positions of conjugated carbon–carbon multiply bonded systems

Structure	λ_{max} (nm)	Structure	λ_{max} (nm)
$CH_3 \cdot (CH = CH)_3 \cdot CH_3$	275	$CH_3 \cdot (C \equiv C)_3 \cdot CH_3$	207
$CH_3 \cdot (CH = CH)_4 \cdot CH_3$	310	$CH_3 \cdot (C \equiv C)_4 \cdot CH_3$	234
$CH_3 \cdot (CH = CH)_5 \cdot CH_3$	342	$CH_3 \cdot (C \equiv C)_5 \cdot CH_3$	261
$CH_3 \cdot (CH = CH)_6 \cdot CH_3$	380	$CH_3 \cdot (C \equiv C)_6 \cdot CH_3$	284

21 000, and there is a regular bathochromic shift with an increase in the number of conjugated double bonds (Table 3.7). Alkyl substituents attached to the multiply bonded carbon atoms have also been found to lead to uniform increments (5 nm) in the wavelength of absorption.

Conjugation of the carbon–carbon triple bond with other triple bonds (polyynes) or with carbon–carbon double bonds (polyenynes) also leads to progressive shifts of λ_{max} to higher wavelengths. In acyclic systems such absorptions are frequently of diagnostic value in deciding the extent of conjugation.

The regularity of wavelength shifts associated with changes in the extent of conjugation, the degree of substitution and the geometrical relationship of the double-bond system has led to the formulation of a set of empirical rules which enable the absorption maxima of substituted conjugated dienes to be predicted. Hence in cases where two isomeric structures are feasible, comparison of the calculated and experimental λ_{max} often enables a structural assignment to be made.

In formulating the rules (Table 3.8) for cyclic dienes, these are classified as either heteroannular or homoannular, e.g.

Heteroannular   Homoannular

The base λ_{max} value for the former is taken as 214 nm and for the latter as 253 nm. To these values are added increments according to the nature of the substituent present and other structural features as shown in the table. In those cases where homoannular and heteroannular chromophores coexist in conjugation, the base value for the homoannular contribution is used, to which is added the appropriate increment for extended conjugation. It should also be noted that homoannular dienes in ring systems other than the six-membered ring do not give a satisfactory correlation between calculated and experimentally determined λ_{max} values. The rules however are adequate for acyclic or heteroannular

Table 3.8 Fieser–Woodward rules for conjugated diene absorption

	λ (nm)
Parent heteroannular open chain diene	214
Parent homoannular diene	253
Increments for:	
Double bond extending conjugation	30
Alkyl substituent or ring residue	5
Exocyclic double bond	5
Polar groupings: O·CO·CH$_3$	0
OR	6
SR	30
Cl, Br	5
—NR$_2$	60
Solvent correction	0
	Total = λ_{max} calc.

dienes providing that there exists in the structure no geometrical constraint which leads to a departure from coplanarity of the σ-bond system and to a consequent reduction in π-orbital overlap.

The carbonyl group. Acetone (in cyclohexane solution) exhibits two absorption bands; one appears at 190 nm (ε 1860) and corresponds to the $\pi \rightarrow \pi^*$ transition, while the second is at 280 nm (ε 13) and corresponds to the $n \rightarrow \pi^*$ transition. The absorption maxima of these bands are solvent-dependent. Ultraviolet spectra of saturated aldehydes, carboxylic acids, esters and lactones exhibit a similar absorption profile, and in general are of little diagnostic value.

Conjugation of the carbonyl group with a carbon–carbon double bond, however, significantly alters the absorption pattern. Thus the absorption corresponding to electron promotion from the carbon–carbon π-system to the antibonding orbital of the carbonyl group (the *electron-transfer* or *E.T. band*) is found in the region of 220 to 250 nm (ε 10 000–15 000) for simple enones; the term E.T. band is reserved for $\pi \rightarrow \pi^*$ transitions in which the conjugated chromophores are dissimilar. A weak band (ε 50–100) is also to be found in the region 310–330 nm and is due to the displaced $n \rightarrow \pi^*$ transition of the carbonyl group.

The position of the E.T. band depends in a predictable manner upon the extent of conjugation, the degree of substitution, etc., and may be calculated following rules which are analogous to those available for the prediction of absorption characteristics of conjugated dienes and which are set out in Table 3.9. The base values selected are 215 nm for an enone in an acyclic or six-membered ring system, or 202 nm for an enone system in a five-membered ring, or 207 nm for an α,β-unsaturated aldehyde.

Aromatic compounds. These compounds exhibit characteristic absorption in the ultraviolet–visible region of the spectrum, and although they are frequently easily recognised from their other spectroscopic properties, examination of their electronic spectra can often lead to the elucidation or confirmation of some of the detailed structural features.

Hydrocarbons. Benzene (in hexane solution) exhibits three absorption bands, λ_1, λ_2 and λ_3, which occur at 184 nm (ε 60 000), 204 nm (ε 7400) and 254 nm

Table 3.9   Woodward–Fieser rules for enone absorption

(1) $\overset{\beta}{>}C=\overset{\alpha}{C}-C=O$

(2) $>\overset{\delta}{C}=\overset{\gamma}{C}-\overset{\beta}{C}=\overset{\alpha}{C}-C=O$

		λ (nm)
Parent enone in an acyclic or six-membered ring		215
Parent enone in a five-membered ring		202
Parent $\alpha\beta$-unsaturated aldehyde		207
Increments for:		
Double bond extending conjugation		30
Alkyl substituent or ring residue	α	10
	β	12
	γ and higher	18
Polar groupings: —OH	α	35
	β	30
	γ	50
—O·CO·CH$_3$	α, β, γ	6
—OCH$_3$	α	35
	β	30
	γ	17
	δ	31
—Cl	α	15
	β	12
—Br	α	25
	β	30
—NR$_2$	β	95
Exo double bond		5
Homodiene component		39
Solvent correction (see Table 3.10)		
		Total = λ_{max} calc.

Table 3.10   Solvent correction values

Solvent	Correction/nm
Ethanol	0
Methanol	0
Dioxane	+5
Chloroform	+1
Ether	+7
Hexane	+11
Cyclohexane	+11
Water	−8

(ε 204) respectively which are due to the various allowed $\pi \rightarrow \pi^*$ transitions. Alkyl substituents cause a bathochromic shift of the λ_2 and λ_3 bands with little change in the ε value (see Table 3.11).

Polycyclic aromatic hydrocarbons with both angular and linear types of ring fusion show absorption curves of a similar profile to that of benzene but with the absorption maxima shifted to longer wavelengths; the greater the number of rings the more pronounced the shift.

Substituted benzenoid systems. As noted above, alkyl substituents cause a small bathochromic shift in the λ_2 and λ_3 bands, and a similar effect is observed in the case of halogen substituents. However, substituents which contain multiply bonded groups (C=O, NO$_2$, C≡N, C=C, C≡C), and to a lesser degree substituents having non-bonding electrons in conjugation with the aromatic π-system, cause very pronounced bathochromic shifts of the two bands; this is frequently coupled with an increase in the ε value of the λ_3 band. In some cases additional bands will appear in the spectrum as a result of electron transitions associated with the substituent group (e.g. acetophenone $n \rightarrow \pi^*$, λ_{max} 320, ε 50). The effects of a selection of these groups on the positions of the λ_1, λ_2 and λ_3 bands may be gathered from the data cited in Table 3.11.

Table 3.11   Absorption characteristics of aromatic systems and their substituted derivatives.*

Compound (solvent †)	λ_1 (nm)		λ_2 (nm)		λ_3 (nm)	
Benzene[a]	184	60 000	204	7 400	254	200
Toluene[b]	—	—	207	7 000	254	160
o-Xylene[b]	—	—	210	8 300	263	300
m-Xylene[b]	—	—	212	7 200	265	300
p-Xylene[b]	193	54 000	212	8 000	274	460
Naphthalene[c]	220	100 000	275	5 700	312	250
Anthracene[c]	253	200 000	375	8 000	obscured by λ_2	
Tetracene[c]	278	200 000	474	13 000	obscured by λ_2	
Pentacene[c]	310	270 000	580	15 000	obscured by λ_2	
Chlorobenzene[d]	—	—	210	7 400	264	190
Bromobenzene[d]	—	—	210	7 900	261	192
Benzaldehyde[d]	—	—	250	11 400	280	1 100
Acetophenone[c]	—	—	246	9 800	280	1 100
Benzophenone[c]	—	—	252	18 000	obscured by λ_2	
Nitrobenzene[d]	—	—	269	7 800	obscured by λ_2	
Benzonitrile[d]	—	—	224	13 000	271	1 000
Styrene[d]	—	—	247	10 000	281	540
Phenylacetylene[c]	—	—	235	10 000	c. 280	300
Phenol[d]	—	—	211	6 200	270	1 450
Anisole[d]	—	—	217	6 400	269	1 480
Aniline[d]	—	—	230	8 600	280	1 430

* Some of this data is abstracted from *Physical Methods in Organic Chemistry*, ed. J. C. P. Schwarz, Oliver and Boyd, Edinburgh, 1964, p. 147.
† The solvents employed are (*a*) hexane, (*b*) methanol, (*c*) ethanol and (*d*) water containing sufficient methanol to ensure miscibility.

The origin of these bathochromic shifts lies in the more extensive mesomerism that exists in these derivatives, which thus reduces the energy difference between the ground and the excited states, and hence shifts the absorption to longer wavelength. Such aspects are dealt with in more detail in the specialist texts.[5]

Heterocyclic systems. Pyridine exhibits an absorption spectrum very similar to that of benzene with an additional absorption band at 270 nm which is assigned to the transition involving the nitrogen lone pair. Similarly quinoline and isoquinoline have spectral profiles closely analogous to naphthalene.

The five-membered heterocycles (furan, thiophene and pyrrole), despite their aromaticity, show distinct spectral differences from benzene. For example, furan has λ_{max} 200, 205 and 211 nm; thiophene has λ_{max} 235 nm, and pyrrole has λ_{max} 210 and 240 nm.

REFERENCES

1. **Interpretation of the spectra of organic compounds:** (a) R. M. Silverstein, G. C. Bassler and T. C. Morrill (1981). 4th edn. *Spectrometric Identification of Organic Compounds.* New York: Wiley.
(b) W. Kemp (1987). *Organic Spectroscopy,* 2nd edn. Basingstoke: MacMillan.
(c) D. H. Williams and I. Flemming (1973). *Spectroscopic Methods in Organic Chemistry,* 4th edn. London: McGraw-Hill.
(d) S. Sternhell and J. R. Kalman (1986). *Organic Structures from Spectra.* Chichester: Wiley.
(e) S. F. Dyke, A. J. Floyd, M. Sainsbury and R. S. Theobold (1978). *Organic Spectroscopy,* 2nd edn. London: Longman.
2. **Infrared spectroscopy:** (a) L. J. Bellamy. *The Infra-red Spectra of Complex Molecules.* Vol. 1, 3rd edn (1975); Vol. 2 (1980), formerly published as *Advances in Infrared Group Frequencies.* London: Chapman and Hall.
(b) A. D. Cross and R. A. Jones (1969). *An Introduction to Practical Infrared Spectroscopy.* London: Butterworths.
(c) G. Socrates (1980). *Infra-red Characteristic Group Frequencies.* New York: Wiley.
(d) W. O. George and P. S. McIntyre (1987). *Infrared Spectroscopy.* Chichester: Wiley.
3. **Nuclear magnetic resonance spectroscopy:** (a) W. Kemp (1986). *NMR in Chemistry.* Basingstoke: MacMillan.
(b) L. M. Jackman and S. Sternhell (1969). *Applications of Nuclear Magnetic Resonance in Organic Chemistry,* 2nd edn. Oxford: Pergamon.
(c) W. McFarlane and R. F. M. White (1972). *Techniques of High Resolution Nuclear Magnetic Resonance.* London: Butterworths.
(d) D. A. R. Williams (1986). *Nuclear Magnetic Resonance Spectroscopy.* Chichester: Wiley.
(e) G. C. Levy, R. L. Lichter and R. L. Nelson (1980). *Carbon-13 NMR,* 2nd edn. New York: Wiley-Interscience.
(f) E. Breitmaier and W. Voelter (1987). *Carbon-13 NMR Spectroscopy,* 3rd edn. Weinheim: VCH.
(g) R. J. Abrahams and P. Loftus (1978). *Proton and Carbon-13 NMR Spectroscopy: an Integrated Approach.* Chichester: Wiley.
(h) C. W. Haigh (1971).'Computer programs in the analysis of NMR spectra', in *Annual Reports on NMR Spectroscopy.* Ed. E. F. Mooney. New York: Academic Press.
(i) K. B. Wilberg and B. J. Nist (1962). *Interpretation of NMR Spectra.* New York: Benjamin.
(j) T. J. Wenzel (1987) *NMR Shift Reagents.* Boca Raton, Florida: CRC Press.
(k) T. C. Morrill (1986). *Lanthanide Shift Reagents in Stereochemical Analysis.* Weinheim: VCH.
(l) M. Oki (1985). *Applications of Dynamic NMR Spectroscopy to Organic Chemistry.* Weinheim: VCH.
(m) J. L. Marshall (1983). *Carbon–Carbon and Carbon–Proton NMR Couplings: Applications to Organic Stereochemistry and Conformational Analysis.* Deerfield Beach, Florida: Verlag Chemie International.
(n) A. E. Derome (1987). *Modern NMR Techniques for Chemistry Research.* Oxford: Pergamon.

4. **Mass spectrometry:** (a) H. C. Hill (1974). *Introduction to Mass Spectrometry*, 2nd edn. London: Heyden.

(b) G. W. A. Milne (1971). *Mass Spectrometry: Techniques and Applications.* New York: Wiley-Interscience.

(c) I. Howe, D. H. Williams and R. D. Bowden (1981). *Mass Spectrometry*, 2nd edn. New York: McGraw-Hill.

(d) J. R. Chapman (1985). *Practical Mass Spectrometry.* New York: Wiley.

(e) H. C. Budzikiewicz, C. Djerassi and D. H. Williams (1967). *Mass Spectrometry of Organic Compounds.* San Francisco: Holden-Day.

(f) F. W. McLafferty (1973). *Interpretation of Mass Spectra*, 2nd edn. London: Benjamin.

(g) R. Davis and M. Frearson (1987). *Mass Spectrometry.* Chichester: Wiley.

(h) A. G. Harrison (1983). *Chemical Ionisation Mass Spectrometry.* Boca Raton, Florida: CRC Press.

5. **Ultraviolet–visible spectroscopy:** (a) R. Friedel and M. Orchin (1958). *Ultraviolet Spectra of Aromatic Compounds.* New York: Wiley.

(b) H. H. Jaffe and M. Orchin (1962). *Theory and Applications of Ultraviolet Spectroscopy.* New York: Wiley.

(c) A. I. Scott (1964). *Interpretation of Ultraviolet Spectra of Natural Products.* Oxford: Pergamon.

(d) R. C. Denney and R. Sinclair (1987). *Visible and Ultraviolet Spectroscopy.* Chichester: Wiley.

6. **Collections of spectroscopic data:** (a) *Standard Spectra Collection.* Sadtler Research Laboratories, Philadelphia. (The 1980 index covers 48 000 grating i.r. spectra, 59 000 prism i.r. spectra, 48 000 ultraviolet/visible spectra, 32 000 60 MHz n.m.r. spectra, and 7000 carbon-13 n.m.r. spectra.)

(b) C. J. Pouchert (1984). *Aldrich Library of Infrared Spectra.* Milwaukee: Aldrich Chemical Co. (12 000 i.r. spectra).

(c) C. J. Pouchert (1985). *The Aldrich Library of FT-IR Spectra.* Milwaukee: Aldrich Chemical Co. (10 600 FT–IR spectra).

(d) E. Breitmaier, G. Haas and W. Voelter (1979). *Atlas of Carbon-13 NMR Data.* London: Heyden.

(e) W. Bremser, L. Ernst, B. Franke, R. Gerhards and A. Hardt (1982). *Carbon-13 NMR Spectra.* Weinheim: Verlag Chemie. (Microfiche collection of 30 000 spectra).

(f) C. J. Pouchert (1983). *Aldrich Library of NMR Spectra.* Milwaukee: Aldrich Chemical Co.

(g) *Eight Peak Index of Mass Spectra*, 3rd edn. London : Mass Spectrometry Data Centre, Royal Society of Chemistry. (Gives the eight principal ions in the mass spectra of 52 332 compounds.)

(h) S. Abrahamson, E. Stenhagen and F. W. McLafferty (1974). *Registry of Mass Spectral Data.* New York: Wiley.

(i) J. H. Beynon and A. E. Williams (1963). *Mass Abundance Tables for Use in Mass Spectrometry.* London: Elsevier.

(j) J. P. Phillips, D. Bates, H. Feuer and B. S. Thyagarajan (1986). *Organic Electronic Spectral Data*, Vol. XXII, 1980. (Annual abstract of published u.v./visible spectroscopic data.)

(k) J. Zupan (1986). *Computer-Supported Spectroscopic Databases.* Chichester: Ellis Horwood.

CHAPTER 4
SOLVENTS AND REAGENTS

4.1 THE PURIFICATION OF COMMON ORGANIC SOLVENTS[1]

Physical data of solvents which are commonly used in the organic chemistry laboratory are tabulated in Appendix 5.

Commercially available grades of organic solvents are of adequate purity for use in many reactions provided that the presence of small quantities of water (the most widespread impurity in all organic solvents) is not harmful to the course of the reaction, and also that the presence of other impurities (e.g. ethanol in diethyl ether) is unlikely to cause undesirable side reactions. The commercially available grades for general use are often accompanied by specifications indicating the amount and nature of any impurities present.

When the levels of impurities, including moisture, are unacceptable for particular reactions, and when large volumes of such solvents are likely to be required, it is frequently more economic to purify the commercial grades than to purchase the more expensive AnalaR grades. Solvents of the appropriate grade of purity should also be used in isolation (extraction) and purification (recrystallisation) processes, particularly in the latter stages immediately prior to spectroscopic and/or elemental analysis.

An account of the more important common drying agents and their use in drying solutions of organic compounds has already been given (Section 2.23). Although the drying efficiency of the individual desiccants cited differs considerably, many of them are of value for the preliminary drying procedures for the vast majority of organic solvents. This preliminary treatment is in fact essential, unless it is certain that the water content is very low, before using the more powerful drying agents (such as a reactive metal, e.g. sodium, or a metal hydride, e.g. calcium hydride, lithium aluminium hydride). These latter drying agents remove the remaining traces of water from solvents which are required for reactions necessitating strictly anhydrous conditions. Attention should be drawn to the considerable fire or explosion hazards of these highly reactive drying agents, particularly at the end of a solvent distillation when residual material has to be disposed of. Recommended methods of safely destroying sodium metal and metal hydrides are given in Section 2.3.

While the drying agent selected for preliminary and final drying must of course have no chemical action on the organic solvent, with some solvents a specific chemical treatment is necessary to remove impurities other than water before drying is attempted. Apart from impurities arising during the manufacturing processes, many organic solvents undergo autoxidation on standing with the formation of dangerously explosive peroxides. Such solvents should always be tested for the presence of peroxides, and if present these should be removed according to the methods detailed below under individual examples, before other purification processes are attempted.

The purified and dried solvent is then distilled; separation from contaminants of similar boiling point may require the use of an efficient fractionating column, and high boiling solvents may need to be distilled under reduced pressure. It must be remembered that many of the common organic solvents are markedly toxic – benzene and the halogenated hydrocarbons are notorious examples, but many others are potentially hazardous to varying degrees (see Section 2.3), and the inhalation of vapour should always be avoided. Almost all organic solvents are also flammable; the more volatile compounds, notably ether and carbon disulphide, are particularly hazardous in consequence. Apart from taking the obvious precautions of avoiding all flames in the vicinity of a solvent distillation, it must be remembered that faulty electrical connections or even contact with hot metal surfaces may ignite the vapour of volatile solvents. The capacity of the condensing system used must be fully adequate to cope with the volume of solvent distilled, and double surface condensers are essential in many cases.

Rigorously dried organic solvents are frequently markedly hygroscopic. The distillation assembly may be protected by a suitable drying tube, but the distillation is better carried out under an atmosphere of nitrogen and the distilled solvent similarly stored under nitrogen. If the solvent is stored in a multi-necked flask fitted with a nitrogen inlet and outlet tube, a measured amount of solvent can be removed by syringe with little or no contamination of the solvent with air or moisture. The reader is referred to Section 2.17.8, p. 120 for a detailed description of these procedures. When quantities of dry solvents are required regularly, it is convenient to store the solvent over a drying agent and under a nitrogen atmosphere. Dry solvent is then distilled off and used immediately as required. An example is described below for dioxane, Section 4.1.*20*, p. 407.

It is often convenient to remove final traces of water with the aid of a molecular sieve and to store the dried solvent in the presence of the sieve. The descriptive term, *molecular sieve*, applies to a group of dehydrated synthetic sodium and calcium aluminosilicate adsorbents (*zeolites*)* which have a crystal lattice structure incorporating uniformly sized holes or pores which are able to accept molecules smaller than a limiting dimension; the larger molecules do not diffuse into the lattice structure. This selectivity, based upon molecular shape and dimension, accounts for the sieving action and is particularly valuable for the removal from gases and liquids of water, which readily diffuses into the pores and is retained by a strong adsorptive attraction.† The pore size is determined by the nature of the manufacturing process and currently four principal types are available in bead, pellet or powder forms, namely Types 3A, 4A, 5A and 13X, representing an effective pore diameter of approximately 0.3, 0.4, 0.5 and 1.0 nm respectively. All these are stable over a pH range 5–11 but interaction with

* The first synthetic zeolites were known as Linde Molecular Sieves but are now marketed as 'Union Carbide' Molecular Sieves; they are available from Union Carbide International Company, USA, or Union Carbide (UK) Ltd directly, or through the usual chemical suppliers.
† Booklets giving detailed information on the structure, action and applications of molecular sieves are available from most suppliers of laboratory chemicals.

strong acids is to be avoided; grades are available which are more resistant to the action of acid (e.g. AW-300, AW-500).

Types 3A, 4A and 5A are those which are most usually employed for drying purposes; 5A is also capable of adsorbing the higher homologues of straight chain alkanes, alkenes and alcohols. Their adsorptive capacity for water is higher than that of silica gel, alumina or activated charcoal. The use of powdered molecular sieves has recently been recommended as an efficient method for drying some of the lower alcohols. After use molecular sieves may be readily regenerated by heating between 150 and 300 °C in a suitable oven or in a stream of dry air and then cooling in a desiccator.

Solvents used for the preparation of solutions for spectroscopic examination (particularly infrared measurements) need to be rigorously purified or spectro-scopic grades must be purchased (see Chapter 3). Even in this latter case, and particularly with the more hygroscopic solvents, the solvent may become contaminated with moisture during usage of the solvent from a previously opened bottle. It is therefore advisable to dry the solvent immediately before use by means of a molecular sieve.

SATURATED ALIPHATIC HYDROCARBONS

1. LIGHT PETROLEUM*
The fractions of refined petroleum which are commonly used have b.p. 40–60, 60–80, 80–100 and 100–120 °C. It is not advisable to employ a fraction with a wider b.p. range than 20 °C, because of possible loss of the more volatile portion during its use in recrystallisation, etc., and consequent different solubility relationships with the higher boiling residue. For some purposes the presence of unsaturated (chiefly aromatic) hydrocarbons in light petroleum is undesirable. Most of the unsaturated hydrocarbons may be removed by shaking two or three times with 10 per cent of the volume of concentrated sulphuric acid (for details, see under *Benzene*); vigorous shaking is then continued with successive portions of a concentrated solution of potassium permanganate in 10 per cent sulphuric acid until the colour of the permanganate remains unchanged. The solvent is then thoroughly washed with sodium carbonate solution and then with water, dried over anhydrous calcium chloride and distilled. If required perfectly dry, it should be allowed to stand over sodium wire (see *15. Diethyl ether* below), or calcium hydride.

More recently a convenient method of purification has been recommended† which is to decant the solvent, previously treated with sulphuric acid, directly on to a basic alumina (Grade I) column using about 50 g of adsorbent for each 100 ml of solvent; the first 5 per cent of eluate is discarded. The column receiver should be suitably protected from the ingress of moisture by the attachment of a calcium chloride tube. This method also removes any peroxides which may be present.

Light petroleum fractions free from aromatic hydrocarbons are marketed, as

---

* Also known as petroleum ether and sometimes termed ligroin when referring to fractions of b.p. above 100 °C.
† Details abstracted from M. Woelm information leaflets. It should be particularly noted that peroxides remain unchanged on alumina, which must therefore not be heated; the used alumina should be thoroughly wetted before throwing away.

are the pure homologues, *pentane, hexane, heptane, octane, etc.* While some of these latter are available in spectroscopically pure grades, their purification for spectroscopic use may be readily achieved by passing through a chromatographic column having silica gel (Grade I) in the lower section and basic alumina (Grade I) in the upper section.

Similar purification procedures apply to *cyclohexane, methylcyclohexane* and the *decalins*.

The purity of all these hydrocarbon solvents may be checked by gas–liquid chromatography (Section 2.31) using an Apiezon, a Silicone oil or an SE-52 silicone rubber gum chromatographic column.

AROMATIC HYDROCARBONS

2. BENZENE

Benzene has been identified as a carcinogen. (**CAUTION**: *All procedures involving benzene must be carried out in a well-ventilated fume cupboard, and protective gloves should be worn.*) The analytical reagent grade benzene is satisfactory for most purposes; if required dry, it is first treated with anhydrous calcium chloride, filtered and then placed over sodium wire (for experimental details, see under *15. Diethyl ether*) or a Type 5A molecular sieve. Phosphorus pentoxide, lithium aluminium hydride or calcium hydride may be used as alternatives to sodium wire.

Commercial benzene may contain thiophene C_4H_4S, b.p. 84 °C, which cannot be separated by distillation or by fractional crystallisation. The presence of thiophene may be detected by shaking 3 ml of benzene with a solution of 10 mg of isatin in 10 ml of concentrated sulphuric acid and allowing the mixture to stand for a short time: a bluish-green coloration is produced if thiophene is present. The thiophene may be removed from benzene by shaking with concentrated sulphuric acid, advantage being taken of the fact that thiophene is more readily sulphonated then benzene. The technical benzene is shaken repeatedly with about 15 per cent of its volume of concentrated sulphuric acid in a stoppered separatory funnel* until the acid layer is colourless or very pale yellow on standing, or until the thiophene test is negative. After each shaking lasting a few minutes, the mixture is allowed to settle and the lower layer is drawn off. The benzene is then shaken twice with water in order to remove most of the acid, once with 10 per cent sodium carbonate solution, again with water and finally dried with anhydrous calcium chloride. After filtration, the benzene is distilled through an efficient column and the fraction, b.p. 80–81 °C, collected. If required perfectly dry the distilled benzene may either be stored over sodium wire or left in the presence of a Type 5A molecular sieve. Pure benzene has b.p. 80 °C/760 mmHg and m.p. 5.5 °C.

3. TOLUENE

Toluene free from sulphur compounds may be purchased. Commercial toluene may contain methyl thiophenes (thiotolenes), b.p. 112–113 °C, which cannot be removed by distillation. It may be purified with concentrated sulphuric acid in a similar manner to the purification of benzene, but care must be taken that the

---

* Alternatively, the mixture may be stirred mechanically for 20–30 minutes. After three such treatments, the acid usually has only a pale colour.

temperature is not allowed to rise unduly ($< 30\,°C$) as toluene is sulphonated more easily then benzene. If required perfectly dry the distilled toluene may be stored over sodium wire, phosphorus pentoxide, or calcium hydride, or left in the presence of a Type 5A grade of molecular sieve. Pure toluene has b.p. $110.5\,°C/760\,mmHg$.

4. XYLENES

For solvent purposes various grades of xylenes (the mixture of isomers and ethylbenzene) are available; purification and drying procedures are similar to those used for benzene and toluene. For chemical purposes the commercially available pure isomeric xylenes are usually available in at least 99 per cent purity.

HALOGENATED HYDROCARBONS

5. DICHLOROMETHANE (*Methylene chloride*)

The commercial grade is purified by washing with portions of concentrated sulphuric acid until the acid layer remains colourless, and then with water, sodium carbonate solution and water again. It is dried initially over calcium chloride and then distilled from calcium hydride before use. The fraction b.p. 40–$41\,°C$ is collected. Dichloromethane should be stored in a brown bottle away from light over Type 3A molecular sieve.

Dichloromethane is a useful substitute for diethyl ether in extraction processes when it is desired to employ a solvent which is more dense than water.

6. CHLOROFORM

(**CAUTION**: *Chloroform is a suspect carcinogen; wherever possible it should be replaced by dichloromethane as an extraction solvent.*) The commercial product contains up to 1 per cent of ethanol which is added as a stabiliser. The ethanol may be removed by any of the following procedures:

(*a*) The chloroform is shaken five or six times with about half its volume of water, then dried over anhydrous calcium chloride for at least 24 hours, and distilled.

(*b*) The chloroform is shaken two or three times with a small volume (say 5%) of concentrated sulphuric acid, thoroughly washed with water, dried over anhydrous calcium chloride or anhydrous potassium carbonate and distilled.

(*c*) The chloroform is passed through a column of basic alumina (Grade I; 10 g per 14 ml of solvent), a procedure which also removes traces of water and acid; the eluate may be used directly.

Alternative drying agents for chloroform are phosphorus pentoxide, calcium sulphate, or powdered Type 4A molecular sieve.

Pure chloroform has b.p. $61\,°C/760\,mmHg$. *It must not be dried by standing with sodium or an explosion may occur. The solvent, when free of alcohol, should be kept in the dark in order to avoid the photochemical formation of dangerous quantities of phosgene.*

7. CARBON TETRACHLORIDE

CAUTION: *Carbon tetrachloride is a suspect carcinogen; avoid breathing vapour and contact with the skin or eyes.*

399

The analytical reagent product is sufficiently pure for most purposes; the carbon disulphide content does not usually exceed 0.005 per cent. The technical product may contain up to 4 per cent of carbon disulphide; this may be removed by the following method. One litre of commercial carbon tetrachloride is treated with potassium hydroxide (1.5 times the quantity required to combine with the carbon disulphide) dissolved in an equal weight of water and 100 ml of rectified spirit, and the mixture is shaken vigorously for 30 minutes at 50–60 °C. After washing with water, the process is repeated with half the quantity of potassium hydroxide. Ethanol is then removed by shaking several times with 500 ml of water, followed by shaking with small portions of concentrated sulphuric acid until there is no further coloration. The carbon tetrachloride is then washed with water, dried over anhydrous calcium chloride and distilled. Further purification may be effected, if necessary, by passing the distilled solvent through a column of alumina and then allowing it to stand in the presence of a Type 5A molecular sieve and finally distilling before use. The pure compound has b.p. 76.5 °C/760 mmHg. *Carbon tetrachloride must not be dried over sodium, as an explosion may result. Fire extinguishers containing this solvent cannot be applied to a fire originating from sodium or similarly reactive metals.*

Carbon tetrachloride is one of the solvents which may be dried relatively efficiently by simple distillation, rejecting the first 10 per cent of distillate, until the distillate is clear (compare Section 2.23).

ALIPHATIC ALCOHOLS

8. METHANOL
The synthetic methanol now available is suitable for most purposes without purification: indeed some manufacturers claim a purity of 99.85 per cent with not more than 0.1 per cent by weight of water and not more than 0.02 per cent by weight of acetone.

Most of the water may be removed from commercial methanol by distillation through an efficient fractionating column (Fig. 2.106); no constant boiling point mixture is formed as is the case with ethanol. Anhydrous methanol can be obtained from the fractionally distilled solvent by standing over a Type 3A molecular sieve or by treatment with magnesium metal using the procedure given for super-dry ethanol described below. It should be stored over Type 3A molecular sieve beads. Pure methanol has b.p. 65 °C/760 mmHg.

If the small proportion of acetone present in synthetic methanol is objectionable it may be removed when present in quantities up to 1 per cent by the following procedure (Morton and Mark, 1934). A mixture of 500 ml of methanol, 25 ml of furfural and 60 ml of 10 per cent sodium hydroxide solution is refluxed in a 2-litre round-bottomed flask, fitted with a double surface condenser, for 6–12 hours. A resin is formed which carries down all the acetone present. The alcohol is then fractionated through an efficient column, the first 5 ml which may contain a trace of formaldehyde being rejected. The recovery of methanol is about 95 per cent.

9. ETHANOL
Ethanol of a high degree of purity is frequently required in preparative organic chemistry. For some purposes ethanol of c. 99.5 per cent purity is satisfactory; this grade may be purchased (the 'absolute alcohol' of commerce), or it may be

conveniently prepared by the dehydration of rectified spirit with calcium oxide. *Rectified spirit* is the constant boiling point mixture which ethanol forms with water, and usually contains 95.6 per cent of ethanol by weight. Whenever the term 'rectified spirit' is used in this book, approximately 95 per cent ethanol is to be understood. Ethanol which has been denatured by the incorporation of certain toxic additives, notably methanol, to render it unfit for consumption, constitutes the *industrial spirit* (*industrial methylated spirit, IMS*) of commerce; it is frequently a suitable solvent for recrystallisations.

Dehydration of rectified spirit by calcium oxide.   Pour the contents of a Winchester bottle of rectified spirit (2–2.25 litres) into a 3-litre round-bottomed flask and add 500 g of calcium oxide which has been freshly ignited in a muffle furnace and allowed to cool in a desiccator. Fit the flask with a double surface condenser carrying a calcium chloride guard-tube, reflux the mixture gently for 6 hours (preferably using a heating mantle) and allow to stand overnight. Reassemble the condenser for downward distillation via a splash head adapter to prevent carry-over of the calcium oxide in the vapour stream. Attach a receiver flask with a side-arm receiver adapter which is protected by means of a calcium chloride guard-tube. Distil the ethanol gently discarding the first 20 ml of distillate. Preserve the absolute ethanol (99.5%) in a bottle with a well fitting stopper.

'Super-dry' ethanol.   The yields in several organic preparations (e.g. malonic ester syntheses, reductions involving sodium and ethanol, etc.) are considerably improved by the use of ethanol of 99.8 per cent purity or higher. This very high grade ethanol may be prepared in several ways from commercial absolute alcohol or from the product of dehydration of rectified spirit with calcium oxide.

The method of Lund and Bjerrum depends upon the reactions:

$$Mg + 2EtOH \longrightarrow H_2 + Mg(OEt)_2 \tag{1}$$

$$Mg(OEt)_2 + 2H_2O \longrightarrow Mg(OH)_2 + 2EtOH \tag{2}$$

Reaction (1) usually proceeds readily provided the magnesium is activated with iodine and the water content does not exceed 1 per cent. Subsequent interaction between the magnesium ethanolate and water gives the highly insoluble magnesium hydroxide; only a slight excess of magnesium is therefore necessary.

Fit a dry 1.5- or 2-litre round-bottomed flask with a double surface condenser and a calcium chloride guard-tube. Place 5 g of clean dry magnesium turnings and 0.5 g of iodine in the flask, followed by 50–75 ml of commercial absolute ethanol. Warm the mixture until the iodine has disappeared: if a lively evolution of hydrogen does not set in, add a further 0.5 g portion of iodine. Continue heating until all the magnesium is converted into ethanolate, then add 900 ml of commercial absolute ethanol and reflux the mixture for 30 minutes. Distil off the ethanol directly into the vessel in which it is to be stored, using an apparatus similar to that described for the dehydration of rectified spirit. The purity of the ethanol exceeds 99.95 per cent provided adequate precautions are taken to protect the distillate from atmospheric moisture. The super-dry ethanol is exceedingly hygroscopic; it may with advantage be stored over a Type 4A molecular sieve.

If the alcohol is required for conductivity or other physico-chemical work and traces of bases are objectionable, these may be removed by redistillation from a

401

little 2,4,6-trinitrobenzoic acid. This acid is selected because it is not esterified by alcohols, consequently no water is introduced into the alcohol.

10. PROPAN-1-OL

The purest available commercial propan-1-ol (*propyl alcohol*) should be dried with anhydrous potassium carbonate or with anhydrous calcium sulphate, and distilled through an efficient fractionating column. The fraction, b.p. 96.5–97.5 °C/760 mmHg, is collected. If the propan-1-ol is required perfectly dry, it may be treated with magnesium activated with iodine by the method described above for ethanol.

11. PROPAN-2-OL

Two technical grades of propan-2-ol (*isopropyl alcohol*) are usually marketed having purities of 91 per cent and 99 per cent respectively. The former has a b.p. of about 80.3 °C and is a constant boiling point mixture with water. Propan-2-ol may contain peroxide, which if present must be removed before dehydration is attempted. Therefore *test for peroxide* by adding 0.5 ml of propan-2-ol to 1 ml of 10 per cent potassium iodide solution acidified with 0.5 ml of dilute (1:5) hydrochloric acid and mixed with a few drops of starch solution just prior to the test: if a blue (or blue-black) coloration appears in one minute, the test is positive. *To remove peroxide* heat under reflux 1 litre of propan-2-ol with 10–15 g of solid tin(II) chloride for half an hour. Test a portion of the cooled solution for peroxide: if iodine is liberated, add further 5 g portions of tin(II) chloride and heat under reflux for half-hour periods until the test is negative. Add about 200 g of calcium oxide and heat under reflux for 4 hours, and then distil, discarding the first portion of distillate. The water content may be further reduced by allowing the distillate to stand over calcium metal or a Type 5A molecular sieve for several days, followed by further fractionation. Anhydrous propan-2-ol has b.p. 82–83 °C/760 mmHg. It should be noted that peroxide generally redevelops during several days.

12. HIGHER ALCOHOLS

Absolutely dry butan-2-ol (s-butyl alcohol) is required for the estimation of alkyllithium reagents (Section 4.2.27, p. 443). It can be dried by standing for 3 to 6 hours over powdered Type 3A molecular sieve. Removal of the last traces of powder may require distillation or filtration, but standing for several days allows the desiccant to settle. 2-Methylpropan-2-ol (t-butyl alcohol) may be dried in a similar manner. These and other higher alcohols may also be purified by drying with anhydrous potassium carbonate or with anhydrous calcium sulphate, and fractionated after filtration from the desiccant in apparatus with ground glass joints. The boiling points of the fractions to be collected are as follows:

Butan-1-ol (*butyl alcohol*), b.p. 116.5–118 °C/760 mmHg.
2-Methylpropan-1-ol (*isobutyl alcohol*), b.p. 106.5–107.5 °C/760 mmHg.
Butan-2-ol (*s-butyl alcohol*), b.p. 99–100 °C/760 mmHg.
2-Methylpropan-2-ol (*t-butyl alcohol*), b.p. 81.5–82.5 °C/760 mmHg, m.p. 25.5 °C.
Pentan-1-ol (*amyl alcohol*), b.p. 136–137.5 °C/760 mmHg.
3-Methylbutan-1-ol (*isoamyl alcohol*), b.p. 130–131 °C/760 mmHg.
Hexan-1-ol (*hexyl alcohol*), b.p. 156.5–157.5 °C/760 mmHg.

If perfectly anhydrous alcohols are required these may in general be obtained by treatment with sodium followed by addition of the corresponding alkyl succinate or phthalate. Sodium alone cannot be used for the complete removal of water in an alcohol owing to the equilibrium between the resulting sodium hydroxide and the alcohol:

$$NaOH + ROH \rightleftharpoons RONa + H_2O$$

The purpose of adding the ester is to remove the sodium hydroxide by the saponification reaction:

$$\begin{array}{c} CH_2 \cdot CO_2R \\ | \\ CH_2 \cdot CO_2R \end{array} + 2NaOH \longrightarrow \begin{array}{c} CH_2 \cdot CO_2Na \\ | \\ CH_2 \cdot CO_2Na \end{array} + 2ROH$$

Typically 7 g of sodium metal are added to 1 litre of butan-2-ol (having no more than 0.5% of water) contained in a two-necked flask fitted with a double surface condenser. When all the metal has reacted (some warming may be necessary to increase the speed of reaction), 33 g of pure 2-butyl succinate or 41 g of pure 2-butyl phthalate are added and the mixture is heated under gentle reflux for 2 hours. Distillation through a Vigreux column affords a distillate containing not more than 0.05 per cent of water.

13. MONO-ALKYL ETHERS OF ETHYLENE GLYCOL, $R \cdot O \cdot CH_2 \cdot CH_2OH$

The monomethyl, ethyl and butyl ethers are inexpensive and are known as *methyl cellosolve, cellosolve* and *butyl cellosolve* respectively. They are completely miscible with water, and are excellent solvents. The commercial products are purified by drying over anhydrous potassium carbonate or anhydrous calcium sulphate, followed by fractionation after the removal of the desiccant. See 'carbitols' below for removal of peroxide. The boiling points of the pure products are:

Ethylene glycol monomethyl ether (or *2-methoxyethanol*), b.p. 124.5 °C/ 760 mmHg
Ethylene glycol monoethyl ether (or *2-ethoxyethanol*), b.p. 135 °C/760 mmHg
Ethylene glycol monobutyl ether (or *2-butoxyethanol*), b.p. 171 °C/ 760 mmHg

14. MONO-ALKYL ETHERS OF DIETHYLENE GLYCOL, $R \cdot O \cdot CH_2 \cdot CH_2 \cdot O \cdot CH_2 \cdot CH_2OH$

The monomethyl, ethyl and butyl ethers are inexpensive commercial products and are known as *methyl carbitol, carbitol* and *butyl carbitol* respectively. They are all completely miscible with water and are purified as already described for the 'cellosolves' above. The boiling points of the pure compounds are:

Diethylene glycol monomethyl ether, b.p. 194 °C/760 mmHg
Diethylene glycol monoethyl ether, b.p. 198.5 °C/760 mmHg
Diethylene glycol monobutyl ether, b.p. 230.5 °C/760 mmHg

The cellosolve and carbitol solvents may contain traces of peroxide. These can be removed either by heating under reflux over anhydrous tin(II) chloride (see *11. Propan-2-ol*) or by filtration under slight pressure through a column of activated basic alumina (Grade I); the used alumina should be saturated with water before being discarded.

ETHERS

15. DIETHYL ETHER (Ether)*

The chief impurities in commercial ether are water and ethanol. Furthermore, when ether is allowed to stand for some time in contact with air and exposed to light, slight oxidation occurs with the formation of the highly explosive diethyl peroxide (Et_2O_3). The danger from this unstable compound becomes apparent at the conclusion of the distillation of impure ether, when the comparatively non-volatile peroxide becomes concentrated in the residue in the distillation flask, and a serious explosion may then result if an attempt is made to evaporate the ether to dryness. It is perhaps worthy of comment in this connection that in the extraction of an organic compound with ether and the subsequent removal of the solvent, the presence of the residual compound seems largely to eliminate the danger due to traces of peroxide, due presumably to its catalytic effect which leads to a more controlled decomposition. Nevertheless ether which has been standing for several months in a partially filled bottle exposed to light and air should be tested for peroxide by the procedure described under *11. Propan-2-ol*. If present, the peroxide may be removed by shaking 1 litre of ether with 10–20 ml of a concentrated solution of an iron(II) salt prepared either by dissolving 60 g of iron(II) sulphate in a mixture of 6 ml of concentrated sulphuric acid and 110 ml of water, or by dissolving 100 g of iron(II) chloride in a mixture of 42 ml of concentrated hydrochloric acid and 85 ml of water.†

Peroxide may also be removed by shaking with an aqueous solution of sodium sulphite or with solid tin(II) chloride (see *11. Propan-2-ol*) or by passage through a column of alumina. *It is worthy of note that all dialkyl ethers have a tendency to form explosive peroxides and they should be routinely tested before further purification leading to a final distillation process is attempted.*

Apart from the dangers inherent in the use of diethyl ether due to the presence of peroxide, attention must be directed to the fact that ether is *highly flammable* and also extremely volatile (b.p. 35 °C), and great care should be taken that there is no naked flame in the vicinity of the liquid (see Section 2.3). *Under no circumstances should ether be distilled over a bare flame*, but always from a steam bath or an electrically-heated water bath and with a highly efficient double surface condenser. Ether vapour has been known to ignite on contact with a hot plate or even a hot tripod upon which a water bath has previously been heated.

Purification of commercial ether. Divide the contents of a Winchester bottle of ether into approximately two equal volumes and shake each in a large separatory funnel with 10–20 ml of the above iron(II) sulphate solution diluted with 100 ml of water. Remove the aqueous solution and combine the two ether portions in a clean dry Winchester bottle and add 100–200 g of anhydrous calcium chloride. Allow this mixture to stand for 24 hours with occasional shaking; the water and ethanol are largely removed during this period. Filter the ether through a large fluted filter paper into another clean dry Winchester

---

* In this book, and in most others, the term *ether* implies reference to diethyl ether; other homologues are referred to by their systematic or trivial names.
† Traces of aldehydes are produced. If ether of high purity is required, it should be further shaken with 0.5 per cent potassium permanganate solution (to convert the aldehyde into acid), then with 5 per cent sodium hydroxide solution, and finally with water.

Fig. 4.1

bottle. (**CAUTION**: *All flames in the vicinity must be extinguished.*) Fine sodium wire (about 7 g) is then introduced into the ether with the aid of a sodium press (Fig. 4.1). The latter consists of a rigid metal framework, which can be attached to the bench by means of a single bolt (as in the figure). An adjustable bottle stand is provided so that bottles up to a capacity of one Winchester quart can be used and their necks brought up to the underside of the mould. The plunger is of stainless steel as is also the one-piece mould and die. (A number of dies of various sizes, thus giving sodium wire of different diameters, are usually available for alternative use.) The die is nearly filled with lumps of clean sodium (Section 4.2.68, p. 462), then placed in position in the press, and the plunger slowly screwed down. As soon as the sodium wire emerges from the die, the Winchester bottle containing the ether is held immediately beneath the die, and the plunger is gradually lowered until all the sodium has been forced as a fine wire into the ether. The Winchester bottle is then closed by a rubber stopper carrying a calcium chloride guard-tube to exclude moisture and to permit the escape of hydrogen; the ether is allowed to stand for about 24 hours. The steel die must be removed from the press after use, any residual sodium destroyed by immersion in industrial spirit, and then thoroughly washed with water and finally dried, preferably in a heated oven; the plunger should also be swabbed with a rag or filter paper soaked in industrial spirit. If, on the following day, no bubbles of hydrogen rise from the sodium in the ether and the latter still possesses a bright surface, the Winchester bottle is closed by its own screw-capped stopper or by a rubber stopper, and preserved in the dark (to check the formation of peroxide as far as possible) in a cool place remote from flames. If, however, the surface of the sodium wire is badly attacked, due to insufficient drying with the calcium chloride, the ether must be filtered through a fluted filter paper into another clean, dry Winchester bottle and the treatment with sodium repeated.* The absolute diethyl ether thus prepared is suitable for use, for example, for Grignard reactions. If a fresh supply of high-grade ether of analytical reagent quality is available, the treatment with ferrous salt solution prior to the final sodium drying procedure may be omitted.

An alternative method of drying, following the removal of peroxide, is to heat the ether under reflux under nitrogen with sodium benzophenone ketyl (use 10 g of benzophenone, 10 g of sodium and 1 litre of ether) until the dark blue or

---

* The sodium residues in the bottle should be destroyed as recommended on p. 41.

purple colour persists. The pure ether can then be distilled immediately before use.

16. DI-ISOPROPYL ETHER

The commercial product usually contains appreciable quantities of peroxide; this should be removed by treatment with an acidified solution of an iron(II) salt or with a solution of sodium sulphite (see under *15. Diethyl ether*). The di-isopropyl ether is then dried over anhydrous calcium chloride and distilled, the fraction b.p. 68.5 °C/760 mmHg being collected. Di-isopropyl ether should be stored in brown bottles away from the light. A small amount of hydroquinone (2×10^{-5} M) may be added as a peroxide inhibitor.

17. DIBUTYL ETHER

Technical dibutyl ether does not usually contain appreciable quantities of peroxide, unless it has been stored for a prolonged period. It should, however, be tested for peroxide, and, if the test is positive, the solvent should be shaken with an acidified solution of an iron(II) salt or with a solution of sodium sulphite (see under *15. Diethyl ether*). The dibutyl ether is dried with anhydrous calcium chloride and distilled through a fractionating column; the portion b.p. 140–141 °C is collected. Alternatively, it may be distilled from calcium hydride or sodium metal. If a fraction of low boiling point is obtained, the presence of butan-1-ol is indicated and may be removed by shaking twice with an equal volume of concentrated hydrochloric acid (see, however, Expt 5.70), followed by washing with water and drying. Pure dibutyl ether has b.p. 142 °C/760 mmHg.

18. DI-ALKYL ETHERS OF MONOETHYLENE AND DIETHYLENE GLYCOL

The dimethyl ether of ethylene glycol (dimethoxyethane, $Me \cdot O \cdot CH_2 \cdot CH_2 \cdot O \cdot Me$, frequently referred to as DME or *dimethyl cellosolve* or *glyme*) has b.p. 85 °C/760 mmHg, is miscible with water, is a good solvent and an excellent inert reaction medium. The diethyl ether of ethylene glycol (*diethyl cellosolve*) is partially miscible with water (21% at 20 °C) and has b.p. 121.5 °C/760 mmHg.

The dimethyl ether of diethylene glycol ($Me \cdot O \cdot (CH_2 \cdot CH_2 \cdot O \cdot)_2 Me$, *diglyme*) has b.p. 62 °C/17 mmHg; the corresponding diethyl ether (*diethyl carbitol*) has b.p. 186 °C/760 mmHg.

All these are excellent solvents for organic compounds; they are purified by initial storage over sodium hydroxide pellets and then heated under reflux with calcium hydride, lithium aluminium hydride, sodium hydride, or sodium, before being fractionally distilled (under reduced pressure if necessary) in an atmosphere of nitrogen.

19. TETRAHYDROFURAN (*THF*)

The commercial grade of this solvent is obtainable in greater than 99.5 per cent purity, in which water and peroxides are the major impurities; an inhibitor for peroxide formation may have been added by the manufacturers. Peroxide, if present, *must* be removed by passage through a column of alumina (see *1. Light petroleum* for footnote on the disposal of used alumina), or by shaking with iron(II) sulphate solution as described under diethyl ether before drying and further purification is attempted. If the latter method is employed the solvent should then be dried initially over calcium sulphate or solid potassium hydroxide,* before being heated under reflux over calcium hydride or lithium aluminium hydride. The solvent is finally fractionally distilled. Pure tetrahydrofuran has b.p. 65–66 °C/760 mmHg; it should be stored over calcium hydride under nitrogen, and labelled with the date of purification. It should not be stored for more than a few days unless an antioxidant such as 2,6-di-t-butyl-4-methylphenol (0.025%) is added.

20. DIOXANE (*1,4-Dioxane*; *diethylene dioxide*)

The commercial grade usually contains small quantities of acetaldehyde and appreciable amounts of glycol acetal (ethylene acetal), Me—⟨ (O, O) ⟩ , together with some water. Upon keeping, the acetal tends to undergo hydrolysis and the liberated acetaldehyde leads to some peroxide formation. Purification may be effected by decomposing the acetal with dilute acid, followed by drying and fractionation. One litre of technical dioxane, 14 ml of concentrated hydrochloric acid and 100 ml of water are heated in a fume cupboard (dioxane vapour is highly toxic) under reflux for 6–12 hours while a slow stream of nitrogen is bubbled through the solution to remove the acetaldehyde formed. The cold solution is treated with excess potassium hydroxide pellets with shaking until some remain undissolved, and the strongly alkaline aqueous layer is run off; most of the residual water is removed by keeping the dioxane over fresh potassium hydroxide pellets for 24 hours. This treatment is followed by heating the decanted solvent under reflux over excess of sodium for 6–12 hours, i.e. until reaction ceases and the sodium remains bright. Finally, the dioxane is distilled from sodium preferably into a receiver encased in black paper; it should be stored out of contact with air and in the dark. The pure compound has b.p. 101.5 °C/760 mmHg and m.p. 12 °C.

Distillation of the purified dioxane from calcium hydride or preferably passage through a column of basic activated alumina (1) before use, ensures the removal of any peroxide which may develop on storage.

Dioxane is a very useful solvent for a variety of organic compounds; its solvating properties are similar to, and often better than, those of diethyl ether. It is miscible with water in all proportions and is extremely hygroscopic.

Note. (1) An alternative to passage through alumina consists of heating the dioxane under reflux over sodium followed by the addition of benzophenone which gives a deepblue purple solution of a disodium benzophenone complex when the dioxane becomes anhydrous. The solvent should be distilled from this mixture under nitrogen, the fractionating column and receiving flask being covered with black polythene film.

KETONES

21. ACETONE

Although major impurities in the commercial grades of acetone are methanol, acetic acid and water, the analytical reagent generally contains less than 0.1 per

---

* It has been reported that a serious explosion may occur when impure THF containing peroxide is treated with solid potassium hydroxide or with a concentrated aqueous solution of potassium hydroxide.[2]

cent of the organic impurities although the water content may be as high as 1 per cent.

Commercial acetone may be purified in several ways:

(a) The acetone is heated under reflux with successive quantities of potassium permanganate until the violet colour persists. It is then dried with anhydrous potassium carbonate or anhydrous calcium sulphate,* filtered from the desiccant and fractionated; precautions are taken to exclude moisture.

(b) To 700 ml of acetone, b.p. 56–57 °C, contained in a litre bottle, a solution of 3 g of silver nitrate in 20 ml of water is added, followed by 20 ml of 1 M sodium hydroxide solution, and the mixture is shaken for about 10 minutes. The mixture is then filtered, dried with anhydrous calcium sulphate and distilled.

(c) When only a relatively small quantity of pure, dry acetone is required, it may be purified through the bisulphite complex: the latter is decomposed with sodium carbonate solution (for details, see under *22. Ethyl methyl ketone*), dried over anhydrous calcium sulphate and distilled. A more convenient procedure is to make use of the addition compound with sodium iodide ($NaI,3C_3H_6O$), which decomposes on gentle heating and is particularly well adapted for the preparation of pure acetone. One hundred grams of finely powdered sodium iodide are dissolved under reflux in 440 g of boiling commercial acetone, and the solution is cooled in a mixture of ice and salt (-8 °C). The crystals are filtered off and quickly transferred to a dry distilling flask, connected to an efficient condenser and to a receiver cooled in ice. Upon gentle warming, the acetone distils rapidly.

Acetone purified by these means, or the analytical reagent grade, may have the water content reduced by storage over a Type 4A molecular sieve. Silica gel or alumina should not be used as an aldol type reaction is initiated with the formation of water as a by-product. Pure acetone has b.p. 56.2 °C/760 mmHg and is highly flammable.

22. ETHYL METHYL KETONE (*Butan-2-one*)

This excellent solvent has properties similar to those of acetone but it has a somewhat higher boiling point and is therefore less flammable. A preliminary purification is effected by drying the commercial product with anhydrous potassium carbonate or anhydrous calcium sulphate, filtering from the desiccant, and fractionating through an efficient column; the fraction, b.p. 79–80 °C, is collected separately, and is quite satisfactory for recrystallisations. This may be further purified either through the bisulphite addition compound or through the sodium iodide addition compound. The ethyl methyl ketone, b.p. 79–80 °C, is shaken with excess of saturated sodium bisulphite solution until reaction ceases, cooled to 0 °C, the bisulphite complex filtered off, the filter cake well drained, washed with a little ether, and then dried in the air. The dry bisulphite complex is decomposed with a slight excess of sodium carbonate solution, and distilled in steam. The ketone is salted out from the distillate with potassium carbonate, separated, dried with anhydrous potassium carbonate (this will also remove traces of sulphur dioxide and carbon dioxide present) and, after filtration,

---

* Anhydrous calcium chloride should not be used as some chemical combination occurs.

allowed to stand for several hours over anhydrous calcium sulphate. It is then distilled. In the sodium iodide method, the ketone is saturated with sodium iodide by boiling under reflux, the solution is filtered through a hot-water funnel, cooled in a freezing mixture and white crystals (which have a m.p. of 73–74 °C) filtered off. Gentle heating of the crystals in a fractional distillation assembly gives pure ethyl methyl ketone of b.p. 79.5 °C/760 mmHg.

ESTERS

23. METHYL ACETATE

An anhydrous product of 99 per cent purity (b.p. 56.5–57 °C) is available commercially; this is comparatively cheap so that purification of inferior products is not worth while. It is appreciably soluble in water (c. 24% at 20 °C). The pure compound has b.p. 57 °C/760 mmHg.

If it is desired to purify an inferior product, 1 litre is heated under reflux for 6 hours with 85 ml of acetic anhydride and then distilled through a fractionating column: the liquid passing over at 56–57 °C is collected. The distillate is shaken with 20 g of anhydrous potassium carbonate for 10 minutes, filtered and redistilled. The resulting methyl acetate has a purity of 99.9 per cent.

24. ETHYL ACETATE

Various grades of ethyl acetate are marketed. The anhydrous compound, b.p. 76–77 °C, is of 99 per cent purity, is inexpensive and is suitable for most purposes. The 95–98 per cent grade usually contains some water, ethanol and acetic acid, and may be purified in the following manner. A mixture of 1 litre of the commercial ethyl acetate, 100 ml of acetic anhydride and 10 drops of concentrated sulphuric acid is heated under reflux for 4 hours and then fractionated. The distillate is shaken with 20–30 g of anhydrous potassium carbonate, filtered and redistilled from calcium hydride. The final product has a purity of about 99.7 per cent and boils at 77 °C/760 mmHg.

NITROGEN-CONTAINING SOLVENTS

25. FORMAMIDE

Formamide is an excellent solvent for many polar organic compounds and for a selection of inorganic salts. It is very hygroscopic and readily hydrolysed by acids or bases. The commercial product frequently contains formic acid, water and ammonium formate. Purification may be effected by passing ammonia gas into the solvent until a slight alkaline reaction is obtained; addition of dry acetone then precipitates the ammonium formate. The filtered solution is dried over magnesium sulphate and fractionally distilled under reduced pressure; distillation at atmospheric pressure causes decomposition. Pure formamide has b.p. 105 °C/11 mmHg.

26. N,N-DIMETHYLFORMAMIDE (DMF)

N,N-Dimethylformamide is a widely used solvent for many recently developed synthetic procedures because of its powerful solvating properties and its chemical stability in the absence of acidic or basic catalysts. However, distillation at atmospheric pressure (b.p. 149–156 °C) or contact with desiccants such as solid sodium or potassium hydroxide or calcium hydride causes varying degrees of decomposition.

Commercial grades of DMF may be purified initially by azeotropic distillation with benzene (**CAUTION**). Distil a mixture of 1 litre of DMF and 100 ml of benzene at atmospheric pressure and collect the water:benzene azeotrope which distils between 70 and 75 °C. Shake the residual solvent with powdered barium oxide or with activated alumina (Grade I), filter and distil under nitrogen at reduced pressure; collect the fraction having b.p. 76 °C/39 mmHg or 40 °C/10 mmHg. The distillate is best stored over a Type 4A molecular sieve.

An alternative method of purification which avoids the hazards associated with the use of benzene is to dry the DMF over calcium sulphate or Type 3A molecular sieve for 72 hours, followed by distillation under reduced pressure. The solvent thus obtained is suitable for most purposes. If azeotropic distillation is required toluene may be substituted for benzene.

27. ACETONITRILE
This is another highly versatile solvent for many synthetic procedures. Although some commercial grades may be available having purities equal to or greater than 99.5 per cent, the usual contaminants in lower purity grades are water, acetamide, ammonium acetate and ammonia. Water may be removed with activated silica gel or Type 4A molecular sieves (not solid potassium hydroxide since this causes decomposition; calcium sulphate and calcium chloride are inefficient in this instance). This partially dried solvent is stirred with calcium hydride (1) which is added portionwise until hydrogen evolution ceases. The solvent is decanted from the solid and fractionally distilled at atmospheric pressure using a high efficiency column (Fig. 2.106); the pure solvent has b.p. 81–82 °C/760 mmHg.

Note. (1) Phosphorus pentoxide has been reported to be a more effective desiccant than calcium hydride.

28. N-METHYLPYRROLIDONE
This versatile solvent has good chemical stability in the absence of acids and bases which catalyse the cleavage of the lactam ring. It is most conveniently dried by initial azeotropic distillation with previously dried benzene or toluene as described for DMF, and the residual liquid is shaken with barium oxide, the desiccant is removed and the solvent is fractionally distilled under reduced pressure (c. 20 mm). The pure solvent has b.p. 94–96 °C/20 mmHg, or 202 °C/760 mmHg.

29. PYRIDINE*
The analytical reagent grade (>99.5% purity) will satisfy most requirements. If required perfectly dry, it should be heated under reflux over calcium hydride, potassium hydroxide or sodium hydroxide pellets or over barium oxide, and then distilled with careful exclusion of moisture. It is hygroscopic and forms a hydrate of b.p. 94.5 °C. Pure pyridine has b.p. 115.3 °C/760 mmHg. It should be stored over calcium hydride, barium oxide or Type 4A molecular sieve.

Pure pyridine may be prepared from technical coal-tar pyridine in the following manner. The technical pyridine is first dried over solid sodium

---

* Pyridine should be handled with extreme **CAUTION**. It should be dispensed in an efficient fume cupboard and disposable gloves should be worn.

hydroxide, distilled through an efficient fractionating column, and the fraction, b.p. 114–116 °C, collected. Four hundred ml of the redistilled pyridine are added to a reagent prepared by dissolving 340 g of anhydrous zinc chloride in a mixture of 210 ml of concentrated hydrochloric acid and 1 litre of absolute ethanol. A crystalline precipitate of an addition compound (probable composition $2C_5H_5N,ZnCl_2,HCl$*) separates and some heat is evolved. When cold, this is collected by suction filtration, and washed with a little absolute ethanol. The yield is about 680 g. It is recrystallised from absolute ethanol to constant m.p. (151.8 °C). The base is liberated by the addition of excess of concentrated sodium hydroxide solution (c. 40%) to the complex followed by steam distillation until the distillate is no longer alkaline to litmus (c. 1000 ml). The steam distillate is treated with 250 g of solid sodium hydroxide, the upper layer separated and the aqueous layer extracted with two 250 ml portions of ether. The combined upper layer and ether extracts are dried with anhydrous potassium carbonate, the ether removed on a water bath and the pyridine distilled through a fractionating column. Further drying is effected as described for the analytical grade reagent. The pure pyridine has b.p. 115.3 °C/760 mmHg.

30. QUINOLINE

Quinoline is little used as a solvent. It may be dried if required with potassium hydroxide pellets and fractionally distilled under reduced pressure; b.p. 114 °C/17 mmHg.

31. NITROBENZENE

Nitrobenzene of analytical reagent quality is satisfactory for most purposes. The technical product may contain dinitrobenzene, the nitrotoluenes and aniline. Most of the impurities are retained in the residue after addition of dilute sulphuric acid and steam distillation: the nitrobenzene in the distillate is separated, dried with calcium chloride and distilled under reduced pressure. The pure solvent has b.p. 210 °C/760 mmHg and m.p. 5.7 °C.

Nitrobenzene is an extremely versatile solvent, and may sometimes be employed for the crystallisation of compounds which do not dissolve appreciably in the common organic solvents. The vapour is *very toxic*, so that recrystallisations must be carried out in the fume cupboard. After the crystals have been collected, they should be washed with a volatile solvent, such as ethanol or ether, to remove the excess of nitrobenzene (compare Section 2.20). A disadvantage of nitrobenzene as a solvent is its pronounced oxidising action at the boiling point.

SULPHUR-CONTAINING SOLVENTS

32. CARBON DISULPHIDE

When working with this solvent, its toxicity (it is a blood and nerve poison) and particularly its high flammability should be borne in mind. Distillation of

---

* There appear to be at least two zinc chloride complexes of pyridine, one of m.p. 207 °C and composition $2C_5H_5N,ZnCl_2$, and the other of m.p. 152 °C and probable composition $2C_5H_5N,ZnCl_2,HCl$. The former is slightly soluble in water and in hot ethanol: the latter passes into the former in aqueous solution, is readily soluble in hot ethanol and can therefore be readily recrystallised from this solvent.

411

appreciable quantities of carbon disulphide should be carried out in a water bath at 55–65 °C; it has been known to ignite as the result of being overheated on a steam bath.

The analytical reagent grade is suitable for most purposes. The commercial substance may be purified by shaking for 3 hours with three portions of potassium permanganate solution (5 g per litre), twice for 6 hours with mercury and finally with a solution of mercury(II) sulphate (2.5 g per litre). It is then dried over anhydrous calcium chloride, and fractionated from a water bath at 55–65 °C. The pure compound boils at 46.5 °C/760 mmHg.

33. DIMETHYL SULPHOXIDE (DMSO)*
Dimethyl sulphoxide has recently come to be recognised as a highly useful water-miscible solvent for many preparative procedures and for spectroscopic work; a study of its chemistry has revealed its potential as an important synthetic reagent. It is hygroscopic and distillation at atmospheric pressures causes some decomposition. The commercial grade may be dried by standing overnight over freshly activated alumina, calcium hydride, barium oxide or calcium sulphate. The filtered solvent is then fractionally distilled over calcium hydride under reduced pressure (c. 12 mmHg) and stored over a Type 4A molecular sieve. An additional stage involving cooling the solvent to about 5 °C, filtering the partially crystallised mass and fractionally distilling the melted crystals has been recommended. The pure solvent has b.p. 75–76 °C/12 mmHg, m.p. 18–19 °C.

34. SULPHOLANE (Tetrahydrothiophene-1,1-dioxide)
This is a further useful water-miscible aprotic solvent of moderately high dielectric constant but with weak acidic and basic characteristics. Although not soluble in saturated hydrocarbon solvents it is a good solvent for most other classes of organic compounds. Purification is effected by passage through a column of activated alumina and distillation under reduced pressure or by repeated vacuum distillation from sodium hydroxide pellets. The pure compound has b.p. 113–117 °C/6 mmHg and m.p. 27 °C; some decomposition occurs at the boiling point (287 °C) at atmospheric pressure. The solvent may be stored over a Type 4A molecular sieve.

PHOSPHORUS-CONTAINING SOLVENTS

35. HEXAMETHYLPHOSPHORIC TRIAMIDE (HMPT)
HMPT is now thought to be a potent carcinogen. *It should be handled with extreme care; inhalation of the vapour and contact with the skin should be avoided.* This solvent is miscible both with water and with many polar and non-polar organic solvents with the exception of saturated aliphatic hydrocarbons. It forms a complex with chlorinated solvents by which means it may be removed from aqueous solutions. The solvent may be dried by shaking with calcium hydride or barium oxide followed by distillation under reduced pressure and

---

* There have been reports of several serious explosions resulting from dimethyl sulphoxide being allowed to come into contact with periodic acid, magnesium perchlorate or perchloric acid. A mixture of sodium hydride (4.5 mol) and dimethyl sulphoxide (18.4 mol) may also explode after about 1 hour.

storage over a Type 4A molecular sieve. The pure solvent has b.p. 127 °C/20 mmHg.

4.2 PREPARATION AND PURIFICATION OF REAGENTS[3]

An account is given in this section of a selection of inorganic and organic reagents, many of which are required for the preparations described later. The preparation of those reagents which can be purchased at reasonable cost will not normally be described, but in those cases where the purified reagents are somewhat expensive, the methods of purification of the technical products will be outlined. Some comments on precautions for handling hazardous reagents are also included, but the relevant passages in Section 2.3 should be also noted. When a reagent is used for the first time, the potential hazards associated with the reagent should be carefully checked with appropriate sources (e.g. manu-facturers' catalogues, i.e. BDH, Aldrich, etc.). Particular attention is drawn to those gaseous reagents which are available commercially in cylinders; the comments in Section 2.3.2, p. 38 regarding the hazards of storage and handling of gas cylinders and of the precautions to be taken in their use should be noted.

1. ALKYL NITRITES
These are readily prepared by the interaction at 0 °C of the alcohol with sodium nitrite in the presence of excess of concentrated sulphuric acid.

$$ROH + HNO_2 \xrightarrow{\;H_2SO_4\;} R{\cdot}ONO + H_2O$$

Alkyl nitrites are valuable nitrosating agents (e.g. Expt 5.100); isopentyl nitrite is the common commercially available example, but several illustrative prepar-ations are given below. For *pentyl nitrite*, equip a 1-litre three-necked flask with a powerful mechanical stirrer, a separatory funnel with stem extending to the bottom of the flask and a thermometer. Cool the flask in a mixture of ice and salt. Place a solution of 95 g (1.38 mol) of 'AnalaR' sodium nitrite in 375 ml of water in the flask and stir. When the temperature has fallen to 0 °C (or slightly below) introduce slowly from the separatory funnel a mixture of 25 ml of water, 62.5 g (34 ml) of concentrated sulphuric acid and 110 g (135 ml, 1.25 mol) of pentan-1-ol, which has previously been cooled to 0 °C. The rate of addition must be controlled so that the temperature is maintained at ± 1 °C; the addition takes 45–60 minutes. Allow the mixture to stand for 1.5 hours and then filter from the precipitated sodium sulphate (1). Separate the upper yellow pentyl nitrite layer, wash it with a solution containing 1 g of sodium hydrogen carbonate and 12.5 g of sodium chloride in 50 ml of water, and dry it with 5–7 g of magnesium sulphate. The resulting crude pentyl nitrite (107 g, 73%) is satisfactory for many purposes (2). Upon distillation, it passes over largely at 104 °C with negligible decomposition. The b.p. under reduced pressure is 29 °C/40 mmHg.

Notes. (1) Care must be exercised in handling pentyl and the other alkyl nitrites; inhalation of the vapour may cause severe headache and heart excitation. The preparation must therefore be conducted in an efficient fume cupboard.
(2) Alkyl nitrites decompose slowly upon standing and should be kept in a cool place. They should preferably be used within a few days or, at most, within two weeks of their preparation. The decomposition products include water, oxides of nitrogen, the alcohol and polymerisation products of the aldehyde.

Hexyl nitrite may be prepared similarly by using 95 g (1.38 mol) of 'AnalaR' sodium nitrite in 375 ml of water; a mixture of 25 ml of water, 62.5 g (34 ml) of concentrated sulphuric acid and 127.5 g (156 ml, 1.25 mol) of hexan-1-ol. The yield of crude product is 124 g (76%). B.p. 129–130.5 °C or 52 °C/44 mmHg.

For *butyl nitrite* use the same quantities as for hexyl nitrite, but with 92.5 g (114 ml, 1.25 mol) of butan-1-ol replacing the hexan-1-ol. The yield of crude product is 110 g (85%). Butyl nitrite boils at 76.5–77.5 °C at atmospheric pressure with slight decomposition, but distils unchanged at 27 °C/43 mmHg.

Ethyl nitrite may be prepared by dissolving 38 g (0.55 mol) of sodium nitrite in 120 ml of water in a 500-ml flask equipped as above. Dilute 23 g (29 ml, 0.5 mol) of ethanol with an equal volume of water, carefully add 25 g (13.5 ml) of concentrated sulphuric acid and dilute to 120 ml with water. Cool both solutions to −10 °C in an ice–salt bath and add the acid–alcohol mixture to the nitrite solution slowly with constant stirring during about 30 minutes. Transfer the reaction mixture to a cooled separating funnel, run off the lower aqueous phase, wash the ethyl nitrite layer rapidly with ice-cold 2 per cent sodium hydrogen carbonate solution and dry over anhydrous sodium sulphate. The product may be kept at 0 °C as a 50 per cent solution in absolute ethanol if required but should be used as soon as possible. The b.p. of pure ethyl nitrite is 17 °C.

2. ALUMINIUM ALKOXIDES

Aluminium t-butoxide is prepared as follows. In a 500-ml round-bottomed flask fitted with a reflux condenser protected by a calcium chloride guard-tube, place 16 g (0.59 mol) of aluminium foil, 50 g (63.5 ml, 0.67 mol) of anhydrous t-butyl alcohol (2-methylpropan-2-ol) and 2 g aluminium isopropoxide (see below) to remove traces of water. Heat the mixture (electric mantle) to boiling, add about 0.1 g of mercury(II) chloride and shake vigorously: the object of the shaking is to distribute the mercury(II) chloride and thus assist an even amalgamation of the aluminium. Continue heating; the colour of the reaction mixture gradually changes from clear to milky to black and hydrogen is evolved. When the mixture is black, allow the reaction to proceed for an hour without further heating, and then add 61 g (77 ml, 0.82 mol) of anhydrous t-butyl alcohol and 50 ml of anhydrous benzene (**CAUTION**). Heat gently to restart the reaction; it will continue vigorously without further heating for about 2 hours: when the reaction subsides heat the mixture under reflux for 12 hours. Remove the benzene and unreacted t-butyl alcohol by distillation under reduced pressure (water pump) using a rotary evaporator, taking care to remove the final traces of solvent as far as possible. Add 250 ml of anhydrous ether; dissolve the solid aluminium t-butoxide by heating under reflux for a short time. After cooling, add 9 ml of undried ether and immediately shake vigorously; the small amount of water thus introduced forms aluminium hydroxide, which assists the precipitation of the black suspended material. Allow to stand for 2 hours, centrifuge the mixture for 30 minutes to remove aluminium hydroxide, unused aluminium and mercury. After centrifugation the solution should be colourless or almost so; if it is still dark in colour, add a further 6 ml of undried ether and centrifuge again. Now remove the solvent under reduced pressure (water pump) using a rotary evaporator. Allow the flask to cool with drying tube attached, crush the product with a spatula and transfer it to a small bottle: seal the latter against moisture. The yield of white or pale grey aluminium t-butoxide is 105 g (72%).

Aluminium isopropoxide is prepared as follows. Place 27 g (1 mol) of clean aluminium foil in a 1-litre round-bottomed flask containing 235 g (300 ml, 3.91 mol) of dried isopropyl alcohol (*Propan-2-ol*, Section 4.1.*11*, p. 402) and 0.5 g of mercury(II) chloride. Attach an efficient double surface reflux condenser carrying a calcium chloride guard-tube. Heat the mixture on a water bath or heating mantle to reflux. Then add 2 ml of carbon tetrachloride (**CAUTION**) (a catalyst for the reaction between aluminium and dry alcohols) through the condenser, and continue the heating. The mixture turns grey and, within a few minutes, a vigorous evolution of hydrogen commences. Discontinue the heating: it may be necessary to moderate the reaction by cooling the flask in ice-water or in running tap water. After the reaction has moderated heat the mixture under reflux until all the metal has reacted (6–12 hours). The mixture becomes dark because of the presence of suspended particles.* Pour the hot solution into a 500-ml flask fitted with a still-head, a condenser and a 250-ml receiving flask. Add a few fragments of porous porcelain and heat the flask in an oil bath at 90 °C under slightly diminished pressure (water pump). When nearly all the isopropyl alcohol has distilled, raise the temperature of the bath to 170 °C and lower the pressure gradually to the full vacuum of the water pump. Immediately the temperature of the distillate rises above 90 °C, stop the distillation and remove the condenser. Attach a 500-ml receiving flask with adapter directly to the still-head, add a few fresh boiling chips and distil; use either an oil bath, at 180–190 °C, or an air bath. The aluminium isopropoxide passes over, as a colourless viscid liquid, at 140–150 °C/12 mmHg; the yield is 190 g (90%). Pour the molten aluminium isopropoxide into a wide-mouthed, glass-stoppered bottle and seal the bottle with paraffin wax (or with sealing tape) to exclude moisture. Generally the alkoxide (m.p. 118 °C) crystallises out, but the substance exhibits a great tendency to supercool and it may be necessary to cool to 0 °C for 1–2 days before solidification occurs.

The reagent is conveniently stored as a solution in isopropyl alcohol. The molten (or solid) alkoxide is weighed out after distillation into a glass-stoppered bottle or flask and is dissolved in sufficient dry isopropyl alcohol to give a 1 M solution. This solution may be kept without appreciable deterioration provided the glass stopper is sealed with paraffin wax or sealing tape. Crystals of aluminium isopropoxide separate on standing, but these may be redissolved by warming the mixture to 65–70 °C.

For many reductions it is not necessary to distil the reagent. Dilute the dark solution, prepared as above to the point marked with an asterisk, to 1 litre with dry isopropyl alcohol; this gives an approximately 1 M solution. Alternatively, prepare the quantity necessary for the reduction, using the appropriate proportions of the reagents.

3. ALUMINIUM AMALGAM

Place 100 g of thin aluminium foil (*c*. 0.05 mm thickness), cut into strips about 15 cm long and 2.5 cm wide and loosely folded, in a 3-litre flask and cover with a 10 per cent solution of sodium hydroxide; warm the flask on a water bath until a vigorous evolution of hydrogen has taken place for several minutes (**CAUTION**). Wash the foil thoroughly with water and with rectified spirit; this

---

* This crude solution may be used directly for some preparations (see later). In general the presence of these particles has no influence on reactions using aluminium isopropoxide.

operation produces an exceptionally clean surface for amalgamation. Add sufficient of a 2 per cent solution of mercury(II) chloride to cover the aluminium completely and allow to react for about 2 minutes; pour off the supernatant solution and wash the amalgam with water, rectified spirit and finally with moist ether. Cover the amalgam with about 1.5 litres of moist ether, when it is then ready for immediate use. If another solvent, e.g. methyl or ethyl acetate, is to be employed in the reduction with moist aluminium amalgam, the ether may of course be replaced by this solvent.

4. ALUMINIUM CHLORIDE (ANHYDROUS)

This reagent may be purchased in the form of small pellets which are easily and quickly crushed in a mortar immediately before addition to the reaction flask (as for example in a Friedel–Crafts reaction, Expt 6.5), and covered with solvent. In these cases the small amount of hydrolysis which occurs during the grinding operation is not unduly harmful to the reaction yield. This pelleted material does unfortunately deteriorate fairly rapidly on reaction with atmospheric moisture on continued opening and closing of the reagent bottle. The material should therefore be carefully inspected before use and if a large amount of powdery white material is present a fresh bottle should be used.

However, in some reactions a very high grade of anhydrous aluminium chloride may be required. This is conveniently prepared by placing the crushed pellets in a suitably sized round-bottomed flask fitted with a simple distillation bend to which is attached a two-necked round-bottomed receiver flask; the second outlet is connected to a water pump via a drying tower similar to that shown in Fig. 4.2 and filled with granular calcium chloride. The distillation flask is heated cautiously with a brush flame and the aluminium chloride sublimes under reduced pressure. It is inadvisable to use an oil immersion rotary pump because of possible corrosion damage even with suitably placed protection traps.

Fig. 4.2

5. AMMONIA

Gaseous ammonia is conveniently obtained from a cylinder of the liquefied gas; the cylinder must be equipped with a reducing valve. The rate of flow of the gas may be determined by passage through a bubble counter containing a small quantity of concentrated potassium hydroxide solution (12 g of KOH in 12 ml of water). A safety bottle should be inserted between the cylinder and the bubble counter and the reaction vessel; the gas may be dried by passage through a column loosely packed with soda lime or calcium oxide lumps (cf. Fig. 4.2). For reactions which require the use of liquid ammonia see Section 2.17.7, p. 116.

Small quantities of ammonia may if necessary be prepared with the aid of the apparatus depicted in Fig. 4.2. Concentrated ammonia solution (d 0.88) is gently heated in the flask surmounted by an efficient reflux condenser. The gas is further dried by passage through a column which is loosely packed with soda lime or calcium oxide lumps,* and is then passed through a Drechsel bottle to act as a safety trap.

6. BENZOYL PEROXIDE

The commercial product is cheap and is usually supplied moistened with approximately 25 per cent of water. Small quantities may be prepared in the laboratory from benzoyl chloride and hydrogen peroxide in the presence of alkali.

$$2Ph{\cdot}COCl + H_2O_2 \xrightarrow{\ominus OH} Ph{\cdot}CO{\cdot}O{\cdot}O{\cdot}CO{\cdot}Ph + 2HCl$$

Immerse a 600-ml beaker containing 50 ml (0.175 mol) of 12 per cent (40 volume) hydrogen peroxide and equipped with a mechanical stirrer, in an ice bath sited in a fume cupboard. Support two dropping funnels, containing respectively 30 ml of 4 M sodium hydroxide solution and 30 g (25 ml, 0.214 mol) of redistilled benzoyl chloride (*lachrymatory*), with their stems inside the beaker. Add the two reagents alternately a few drops at a time, taking care that the temperature does not rise above 5–8 °C and that the solution is maintained faintly alkaline throughout. When all the reagents have been added, stir the solution for a further half an hour; by this time the odour of the benzoyl chloride should have disappeared. Filter off the flocculent precipitate at the pump, wash it with a little cold water and air dry upon filter paper. The yield of benzoyl peroxide is 12 g (46%). It may be purified by dissolving in chloroform at *room temperature* and adding twice the volume of methanol. Benzoyl peroxide should *not* be recrystallised from hot chloroform, as a serious explosion may result. The compound melts at 106 °C with decomposition. Like all organic peroxides, benzoyl peroxide should be handled with care behind shatter-proof screens, and horn or moulded polyethylene (*not* nickel) spatulas should be used. It is very shock sensitive.

To determine the exact peroxide content of benzoyl peroxide (and of other organic peroxides), the following procedure may be employed. Dissolve about 0.5 g, accurately weighed, of benzoyl peroxide in 15 ml of chloroform in a 350-ml conical flask. Cool to −5 °C, and add 25 ml of a 0.1 M solution of sodium methoxide in methanol in one portion with cooling and shaking. After 5 minutes at −5 °C, add 100 ml of iced water, 5 ml of 10 per cent sulphuric acid and 2 g of

---

* The column packing is retained by a plug of glass wool resting on the sintered glass disc fitted at the base of the glass column; a plug of glass wool is also inserted on the top of the column packing.

potassium iodide in 20 ml of 10 per cent sulphuric acid, in the order mentioned, with vigorous stirring. Titrate the liberated iodine with standard 0.10 M sodium thiosulphate solution:

1 ml of 0.10 M $Na_2S_2O_3 \equiv 0.0121$ g of benzoyl peroxide

7. BORANES
The complexes and derivatives of borane (BH_3) are now among the most useful and versatile reagents available to the organic chemist. The reader is referred to key monographs and reviews for a full account of the variety and scope of these reagents.[4] A very wide range of boranes is now available commercially, but it is often convenient and less expensive to prepare them as and when needed. Most reactions with boranes should be carried out under an inert atmosphere. Section 2.17.8, p. 120 should be consulted for details of the experimental procedures for these reactions.

Borane can be prepared *in situ* as described in Expts 5.44 and 5.89. Alternatively, a separate solution of *borane in tetrahydrofuran* can be prepared by adding a solution of sodium borohydride to boron trifluoride-etherate and sweeping the resulting borane into tetrahydrofuran with the aid of a nitrogen gas stream.

$$3NaBH_4 + 4BF_3 \longrightarrow 4BH_3 + 3NaBF_4$$

Fit the central neck of a 500-ml round-bottomed two-necked flask (the generating flask), supported on a magnetic stirrer unit, with a pressure-equalising dropping funnel connected to a nitrogen supply via a suitable mercury safety valve (cf. Fig. 2.66). Connect with Tygon or polyethylene tubing the side-neck of the generating flask to a Drechsel wash bottle, the outlet of which leads into a 500-ml two necked round-bottomed flask via a gas inlet tube terminating in a glass frit (cf. Fig. 2.59). Charge this flask with 400 ml of dry tetrahydrofuran (Section 4.1.*19*, p. 406) and close the second neck with a calcium chloride guard-tube. Dissolve 13.3 g (0.35 mol) of sodium borohydride (Section 4.2.*49*, p. 447) in 350 ml of dry diglyme (Section 4.1.*18*, p. 406) and place 300 ml (0.3 mol) of this solution in the dropping funnel. Use the remainder of the solution to adequately charge the Drechsel wash bottle so that the borane subsequently generated is scrubbed free of boron trifluoride. Place 85 g (0.6 mol) of redistilled boron trifluoride etherate (Section 4.2.*8*, p. 421) in the generating flask, insert a follower bar and allow nitrogen to pass through the assembly for a period of about 15 minutes to displace all the air. Generate the borane by adding the sodium borohydride solution steadily to the stirred boron trifluoride-etherate. When all the solution has been added continue to pass the nitrogen gas for a further period of 15 minutes to ensure that all the borane has been swept into the tetrahydrofuran solution. This will also ensure that little borane gas will escape to the atmosphere when the apparatus is disconnected; borane is highly toxic and may contain impurities which can lead to spontaneous ignition.

The final tetrahydrofuran solution prepared in this way is approximately 1 M with respect to borane; the solution is reasonably stable if stored between 0 and 5 °C.

An alternative technique (which uses essentially the same apparatus assembly as described above) involves passage of borane directly into a solution of the organic substrate in tetrahydrofuran, diglyme or ether to effect immediate reaction. This requires the additional facility of magnetic stirring and of cooling the reaction mixture.

Borane–methyl sulphide (BH$_3$·SMe$_2$) is a complex of borane which is available commercially and should be handled in a fume cupboard. It is a flammable, moisture-sensitive liquid and is supplied in sealed containers under nitrogen. It offers a number of advantages over the tetrahydrofuran complex.[5] The concentration of borane in solution is about ten times that in tetrahydrofuran. The solution is more stable than the tetrahydrofuran complex and can be stored for several months under nitrogen at room temperature without decomposition. The solution is soluble in and does not react with a range of aprotic solvents such as ether, hexane, toluene and dichloromethane thus widening the range of available solvents for hydroboration reactions. Experimental work is easy since the by-products formed by hydrolysis on addition of water are easily removed. Methyl sulphide which is distilled off from the reaction mixture should be destroyed by pouring the distillate slowly into a large excess of 5 per cent sodium hypochlorite. The resulting solution may be safely poured down the drain after about 30 minutes. Examples of the use of borane–methyl sulphide complex in synthesis are described in Expts 5.34 and 5.66.

Borane is also available commercially as a complex with ammonia and with a wide range of organic bases.

The majority of mono- and di-substituted alkenes undergo hydroboration to give trialkylboranes which are then available for a variety of synthetic procedures. With some alkenes, alkylation of borane does not go to completion and the resulting mono- or di-alkyl boranes are useful in synthesis as modified boranes. Thus, 2-methylbut-2-ene gives *bis(3-methyl-2-butyl)borane* (*disiamylborane*).

$$2Me_2C{=}CH{\cdot}Me + BH_3 \longrightarrow [Me_2CH{\cdot}CHMe]_2BH$$

This borane reagent may be prepared as follows.[6] In a dry 500-ml flask equipped with a condenser, a thermometer and a pressure-equalising dropping funnel are placed 80 ml of diglyme, 23.1 g of 2-methylbut-2-ene (0.33 mol) in 20 ml of diglyme and 4.7 g of sodium borohydride (0.125 mol). The flask is immersed in an ice bath and 23.5 g of boron trifluoride-etherate (0.165 mol) is added dropwise over a period of 30 minutes. The semi-solid reaction mixture containing 0.165 mol of disiamylborane is permitted to remain an additional 15 hours at 0–5 °C, and then used for hydroboration. Examples of the use of disiamylborane are described in Sections 5.4.3 p. 542 and 5.7.1, p. 588.

Addition of borane to 2,3-dimethylbut-2-ene gives *2,3-dimethyl-2-butylborane* (*thexylborane*) which is also a synthetically useful substituted borane.

$$Me_2C{=}CMe_2 + BH_3 \longrightarrow Me_2CH{\cdot}C(Me)_2BH_2$$

It can be prepared as a solution in either (a) tetrahydrofuran or (b) diglyme, as follows.[7]

(a) In a 100-ml three-necked flask fitted with a thermometer, a pressure-equalising dropping funnel, a condenser and a side-arm capped with a rubber septum to permit removal of samples is placed 4.2 g (50 mmol) of 2,3-dimethylbut-2-ene in 33 ml of tetrahydrofuran. The flask is cooled to −10 °C, then 27.5 ml of a 0.91 M solution of borane in tetrahydrofuran is added slowly to the olefin. After completion of the borane addition, the flask is maintained at 20–25 °C.

(b) In a 100-ml flask (as above) are placed 18 ml of diglyme, 9.4 ml of a 1 M solution of sodium borohydride in diglyme and 8.3 ml of a 1.5 M solution of 2,3-dimethylbut-2-ene (12.5 mol) in diglyme. The flask is immersed in an ice bath, then 7.0 ml of a 1.83 M solution of boron trifluoride-diglymate is added to the well-stirred reaction mixture over 15 minutes. The reaction is allowed to proceed for an additional hour at 0–3 °C. An example of the use of thexylborane in synthesis is described in Section 5.4.3, p. 452. Kits for the preparation of up to 0.1 mol of disiamylborane and thexylborane are available commercially from the Aldrich Chemical Company. They consist of sealed containers of solutions of borane–tetrahydrofuran complex and a solution of the appropriate alkene.

Reaction of borane with the cyclic diene, cycloocta-1,5-diene, is used to prepare the synthetically useful borane *9-borabicylo[3.3.1]nonane (9-BBN)*

It may be prepared as follows.[8] A 1-litre flask with an injection port is equipped with a magnetic stirring bar and a reflux condenser. To the top of the condenser is fitted a pressure-equalising dropping funnel which is connected to a source of nitrogen. The flask is cooled to −5 °C in an ice–salt bath and charged with 505 ml of 1.98 M borane in tetrahydrofuran (1 mol). The dropping funnel is charged with 108 g of cycloocta-1,5-diene (1 mol) (distilled from lithium aluminium hydride before use, b.p. 45–48 °C/19 mmHg) which is added dropwise over 1 hour to the stirred borane solution. The cooling bath is removed and the mixture heated under reflux for 1 hour. The resulting clear solution is cooled slowly, without stirring, to room temperature. Cooling in an ice–water bath completes precipitation of the product. The supernatant liquid is decanted using a double-ended 15-gauge needle (Section 2.17.8, p. 124). The solid is washed twice with 50 ml portions of ice-cold pentane, decanting the supernatant liquid as before. After drying at 50 °C *in vacuo*, the yield of fine white powder is approximately 75 per cent (85–95 g), m.p. 152–155 °C (in a sealed, evacuated capillary). The thermal stability of 9-BBN is remarkable. Samples can be heated under nitrogen to 200 °C for 24 hours with only minor loss of hydride activity. The stability of 9-BBN towards air oxidation is unique among the dialkylboranes and is comparable to that of lithium aluminium hydride. Even so, some loss of purity attends exposure to the atmosphere. Consequently, for quantitative studies it is recommended that manipulation of solid 9-BBN be carried out as much as possible under a nitrogen atmosphere, in order to maintain the maximum hydride activity and purity. Solutions of 9-BBN on the other hand are highly reactive towards both oxygen and water and should be rigorously protected from these reactants. The reagent can be stored either as the solid or in tetrahydrofuran solution for indefinite periods without any noticeable change in activity provided that an inert atmosphere is maintained.

A further synthetically useful borane, *catecholborane* (1,3,2-benzo-dioxaborole) is obtained by the reaction of borane with 1,2-dihydroxybenzene (catechol).

420

It is prepared as follows.[9] A 2 M solution of borane in tetrahydrofuran (100 ml, 200 mmol), maintained under nitrogen, is placed in a dry 500-ml flask which is vented to a fume cupboard through a mercury bubbler. The flask is immersed in an ice bath and a solution of 1,2-dihydroxybenzene (catechol) (22 g, 200 mmol) in tetrahydrofuran (50 ml) is added over 30 minutes with efficient stirring to the borane solution at 0 °C. After the completion of the addition, the reaction mixture is stirred at 25 °C for an additional 30 minutes. Distillation provides 19.2 g (80%) of catecholborane, b.p. 76–77 °C/100 mmHg. An example of the use of catecholborane in synthesis is described in Expt 6.4.

More recently, boranes such as diisopinocampheylborane and α-iso-pinocampheyl-9-borabicyclo[3.3.1]nonane have been shown to be valuable reagents for the synthesis of chiral compounds.[10] The reaction of borane with α-pinene gives *diisopinocampheylborane*

This reagent may be prepared in a state of high optical purity as follows.[11] A 250-ml flask equipped with a septum inlet, a magnetic stirring bar, and a distillation condenser connected to a receiver cooled in a dry ice–acetone bath is charged with 5.05 ml (50 mmol) of borane–methyl sulphide complex and 15 ml of tetrahydrofuran, and 15.9 ml of α-pinene (100 mmol) (1) is added dropwise at 0 °C. After the contents are stirred at 0 °C for 3 hours, a mixture of dimethyl sulphide and tetrahydrofuran (13 ml) is removed under vacuum (0 °C/30 mmHg), and the flask brought to atmospheric pressure by flushing with nitrogen gas and charged with 18 ml of tetrahydrofuran and 2.4 ml of α-pinene (15 mmol). It is then stored in the cold room at 0 °C. An example of the use of diisopinocampheylborane in synthesis is described in Expt 5.45.

Note. (1) Either (+)- or (−)-α-pinene may be used to give the corresponding optically active Ipc₂BH derivatives. (+)-α-Pinene (Aldrich) was 92 per cent *ee* (see p. 34).

8. BORON TRIFLUORIDE

This gas, b.p. − 101 °C, is available in cylinders and can be bubbled directly into the reaction mixture through a suitable safety trap. It is highly irritant and toxic and should only be handled in a suitable fume cupboard. For many preparative purposes it is convenient to use the commercially available boron trifluoride-etherate [BF₃·Et₂O] which contains 48 per cent w/w of boron trifluoride. This reagent is a colourless liquid which frequently becomes very dark on storage, but is readily purified before use by adding 2 per cent by weight of dry ether, to ensure an excess, and distilling from calcium hydride under reduced pressure (b.p. 46 °C/10 mmHg). Further convenient sources of boron trifluoride are the liquid boron trifluoride–acetic acid complex* which contains about 40 per cent w/w of boron trifluoride, boron trifluoride–methanol† containing c. 51 per cent w/w BF_3, and boron trifluoride–dimethyl sulphide complex† containing 12 per cent w/w BF_3.

9. BROMINE

CAUTION: *Bromine is extremely corrosive and must be handled with great care (preferably using protective gloves) and always in the fume cupboard.* The liquid produces painful burns and the vapour is an extremely powerful irritant. Bromine burns should be treated immediately with a liberal quantity of glycerol. If the vapour is inhaled, relief may be obtained by soaking a handkerchief in ethanol and holding it near the nose. The commercial product may be dried (and partially purified) by shaking with an equal volume of concentrated sulphuric acid, and then separating the acid. Chlorine, if present, may be removed by fractionation in an all-glass apparatus from pure potassium bromide: the b.p. of pure bromine is 59 °C/760 mmHg.

10. N-BROMOSUCCINIMIDE, $CH_2 \cdot CO \cdot NBr \cdot CO \cdot CH_2$

This reagent is available commercially and may be further purified by recrystallisation as rapidly as possible from ten times its weight of hot water or from glacial acetic acid.

It can be prepared from succinimide by dissolving the latter in a slight molar excess of chilled sodium hydroxide solution (of approximately 3 M strength) and adding rapidly with vigorous stirring one molar proportion of bromine dissolved in an equal volume of carbon tetrachloride (**CAUTION**). A finely crystalline white product is obtained which may be collected, washed with ice-cold water, dried and used directly or recrystallised as detailed above.

11. t-BUTYL HYPOCHLORITE

This compound is a useful chlorinating reagent, and although relatively stable, its purification by distillation is not normally attempted. Primary and secondary hypochlorites readily explode when exposed to light, and even in the absence of light rapid decomposition occurs at room temperature. t-Butyl hypochlorite may be prepared by the interaction of t-butyl alcohol, sodium hydroxide and chlorine at a carefully controlled low temperature. A less hazardous synthetic method is described in *Organic Syntheses* Coll. Vol. 5, p. 184.

$$Me_3COH + Cl_2 + NaOH \longrightarrow Me_3COCl + H_2O + NaCl$$

The procedure is as follows. Assemble in a fume cupboard a 1-litre three-necked flask fitted with a gas inlet tube extending almost to the bottom of the flask, a sealed stirrer unit and a gas outlet adapter (cf. Fig. 6.2) incorporating a thermometer extending well into the flask. Place in the flask a solution of 40 g (1 mol) of sodium hydroxide in 250 ml of water, and add 37 g (47 ml, 0.5 mol) of t-butyl alcohol followed by sufficient water (about 250 ml) to give a homogeneous solution. Surround the flask with a bath of iced water, and pass chlorine (from a cylinder of the gas) steadily into the solution, with stirring, as long as the gas is absorbed, and then slowly for about 30 minutes more. **CAUTION**: *the*

---

* Available from BDH Chemicals Ltd.
† Available from Aldrich Chemical Co. Ltd.

temperature of the reaction mixture must not exceed 20°C throughout the addition. Transfer the product to a separating funnel, remove and discard the aqueous phase, and wash the upper layer with 25 ml portions of 10 per cent sodium carbonate solution until the washings are no longer acidic to Congo red indicator. Finally wash the product three times with cold water and dry over anhydrous calcium chloride. The yield of t-butyl hypochlorite, which is usually about 98 per cent pure, is 40–50 g (74–92%). *It should be stored in a securely stoppered dark bottle in the cold;* the vapour is powerfully lachrymatory.

12. CADMIUM CHLORIDE

Cadmium compounds are very toxic and a suitable face mask should be worn to prevent inhalation of dust. Anhydrous cadmium chloride can be obtained from the commercially available hydrated compound in the following way: dry the hydrate to constant weight at 110 °C; grind finely, dry again for 2–3 hours at. 110 °C and then place in a screw-capped bottle and keep in a desiccator over calcium chloride. The use of anhydrous cadmium chloride in synthesis is described in Expts 5.94 and 6.126.

13. CARBON DIOXIDE

This gas is conveniently generated from calcium carbonate chips (marble) and dilute hydrochloric acid (1:1) in a Kipp's apparatus; it should be passed through a Drechsel bottle containing water or sodium hydrogen carbonate solution to remove acid spray and, if required dry, through two further Drechsel bottles charged with concentrated sulphuric acid.

Large quantities of carbon dioxide may be obtained from a cylinder of the liquefied gas; the gas should be dried by passing it through two Drechsel bottles containing concentrated sulphuric acid. A little air is present in the gas.

For some purposes (e.g. in the Grignard reaction) solid carbon dioxide marketed as 'Cardice', 'Dry Ice' or 'Drikold' may be employed. Lumps of solid carbon dioxide should not be picked up with unprotected hands otherwise frostbite may result. If powdered material is required, the larger lumps should be wrapped in cloth and hammered. Solid carbon dioxide affords a convenient regulated supply of the gas when suitably sized lumps are allowed to evaporate in a Buchner flask attached to a sulphuric acid drying train.

14. CARBON MONOXIDE

Carbon monoxide is available in cylinders. *It is very poisonous and all operations involving its generation and use must be carried out in an efficient fume cupboard.* The gas may be prepared in the laboratory by the action of concentrated sulphuric acid upon concentrated formic acid (d 1.2; about 90 per cent w/w $H \cdot CO_2H$) at 70–80 °C. The apparatus shown in Fig. 4.3 employs a 500-ml round-bottomed flask fitted with a socket/cone adapter with 'T' connection and a stoppered pressure-equalising funnel. The gas is dried by passage through two Drechsel bottles containing concentrated sulphuric acid. The round-bottomed flask is charged with 125 g of concentrated sulphuric acid and 85 g of formic acid is slowly added from the dropping funnel; the pressure-equalising side-arm allows the addition process to be carefully controlled and this minimises excessive frothing. Frothing may be further controlled by the addition of a small quantity of liquid paraffin. The steady stream of carbon monoxide which is evolved may contain traces of carbon dioxide and sulphur dioxide; these

To drying train

Fig. 4.3

impurities may be removed, if desired, by passage upwards through a glass column filled with potassium hydroxide pellets (cf. Fig. 4.2).

15. CHLORINE
Chlorine is a very toxic and irritant gas and all operations involving its generation and use must be carried out in an efficient fume cupboard. For relatively large quantities of chlorine, cylinders of the gas fitted with an appropriate reducing valve should be used. The gas should be dried by passage through two Drechsel bottles containing concentrated sulphuric acid and then through a Drechsel bottle filled with glass wool to remove acid spray. Small quantities of chlorine are readily prepared by the action of concentrated hydrochloric acid upon potassium permanganate in the apparatus assembly shown in Fig. 4.3. The calculated quantity of potassium permanganate ($0.367\,g\,KMnO_4 \equiv 0.412\,g\,Cl_2$) is placed in the round-bottomed flask and a slight excess of concentrated hydrochloric acid is placed in the pressure-equalising funnel which should then be stoppered and the stopper secured with an elastic band ($1.000\,g\,KMnO_4$ requires 6.2 ml of concentrated hydrochloric acid). The chlorine is passed through a Drechsel bottle containing water to remove hydrogen chloride and is then dried by means of another Drechsel bottle charged with concentrated sulphuric acid; it is advisable to insert an empty Drechsel bottle to act as a safety trap between the reaction vessel and the dry chlorine generating source. The acid is allowed to drop slowly upon the permanganate crystals; the flask should be shaken from time to time. When about half the acid has been added, the evolution of gas tends to slow down; the flask should then be warmed slightly; after all the acid has been added the mixture is boiled gently to complete the evolution of chlorine.

16. CHLOROSULPHONIC ACID ($Cl \cdot SO_3H$)
This reagent must be handled with great care; it is very corrosive to the skin and clothing, and reacts with water with great violence. If the specimen is impure, or discoloured, it should be distilled in an all-glass apparatus and the fraction, b.p. 148–150 °C/760 mmHg, collected; due precautions should be taken to protect the distillate from moisture.

17. CHLOROSULPHONYL ISOCYANATE (ClSO₂NCO)

Chlorosulphonyl isocyanate is a clear, colourless, mobile lachrymatory and corrosive liquid, m.p. $-43\,°C$, b.p. $107–108\,°C$ ($38\,°C/50\,mmHg$).[12] **CAUTION**: *Contact with the skin for more than a few seconds may result in severe burns. It should always be handled in a fume cupboard.* The reagent may be stored for short periods in glass containers with rubber stoppers or for longer periods in sealed glass ampoules. Ground glass stoppers soon become frozen even when thoroughly lubricated with grease. Chlorosulphonyl isocyanate is thermally stable. It is the most chemically reactive isocyanate known. It fumes in moist air, reacts very violently with water, and is unstable in other protic solvents. It can be used in a number of aprotic solvents including aliphatic, aromatic and chlorinated hydrocarbons, ether (but not tetrahydrofuran) and acetonitrile. An example of its use in synthesis is described in Expt 8.8.

18. CHROMIUM(VI) OXIDISING AGENTS

Chromium(VI) species are widely used as oxidising agents in organic chemistry. *Chromium(VI) oxide* (chromium trioxide, chromic anhydride) is a deliquescent, red crystalline substance. It is very soluble in water and in sulphuric acid, and is a very powerful oxidising agent which must be handled with care (Section 2.3.2, p. 35). It is often used as a solution in acetic acid (Expts 6.123 and 6.130) or acetic anhydride (Expt 6.117). The solution of chromium(VI) oxide in acetic anhydride is prepared by adding chromium trioxide portionwise to the well-cooled acetic anhydride. Addition of the anhydride to the oxide in bulk may lead to explosive decomposition. *Sodium dichromate* in aqueous sulphuric acid (see Expt 5.86 for preparation) is also used for a variety of oxidations including simple primary alcohols to aldehydes (Expt 5.74 and Section 9.6.4; p. 1241) secondary alcohols to ketones (Expts 5.86 and 5.87), and alkylboranes to ketones (Expt 5.89). Acidic dichromate is not a generally useful reagent for the oxidation of primary alcohols since the aldehydes produced are readily oxidised in turn to carboxylic acids. Selective oxidation with chromium(VI) species can however be carried out successfully in aprotic media, using recently developed chromium(VI) species. Two complexes of chromium(VI) oxide with pyridine are useful selective oxidants for primary alcohols and other alcohols containing acid-sensitive groups: pyridinium dichromate, and pyridinium chlorochromate.

One procedure for the oxidation of primary and secondary alcohols with *dipyridine chromium(VI) oxide (pyridinium dichromate, PDC)* is as follows.[13] Chromium trioxide 6.00 g (60 mmol) is added to a magnetically stirred solution of 9.49 g (120 mmol) of dry pyridine in 150 ml of dry dichloromethane. The flask is stoppered with a drying tube containing drierite, and the deep burgundy solution is stirred for 15 minutes at room temperature. At the end of this period a solution of the alcohol (10 mmol) in a small volume of dichloromethane is added in one portion. A tarry black deposit separates immediately; after stirring for an additional 15 minutes at room temperature, the solution is decanted from the residue, which is washed with 200 ml of ether. The combined organic solutions are washed with three 100 ml portions of 5 per cent aqueous sodium hydroxide solution, 100 ml of 5 per cent aqueous hydrochloric acid, 100 ml of 5 per cent aqueous sodium hydrogen carbonate solution, 100 ml of saturated aqueous sodium chloride and dried over magnesium sulphate. Alternatively, the decanted dichloromethane solution is concentrated *in vacuo* and the residue taken up in ether, filtered to remove insoluble chromium salts, washed with dilute aqueous base and saturated brine, and dried over magnesium sulphate. Evaporation of the solvent at reduced pressure affords the crude aldehyde or ketone product. Examples of the use of this procedure are: octan-2-ol (97% yield of crude carbonyl compound); octan-1-ol (90%); benzyl alcohol (89%); borneol (89%); cinnamyl alcohol (96%).

Pyridinium chlorochromate (*PCC*) is a stable complex formed from chromium trioxide and pyridine in the presence of hydrochloric acid. Pyridinium chlorochromate is mildly acidic and therefore not suitable for acid-sensitive substrates or products. It provides an alternative and complementary selective oxidising agent to pyridinium dichromate. It is commercially available, but can easily be prepared as follows.[14] To 184 ml of 6 M hydrochloric acid (1.1 mol) is added 100 g (1 mol) of chromium(VI) oxide rapidly with stirring. After 5 minutes the homogeneous solution is cooled to 0 °C and 79.1 g (1 mol) of pyridine (**CAUTION**) is carefully added over 10 minutes. Re-cooling to 0 °C gives a yellow–orange solid which is collected on a sintered glass funnel and dried for 1 hour *in vacuo* (yield 180.8 g, 84%). The solid is not appreciably hygroscopic and can be stored for extended periods at room temperature without change. An example of the use of pyridinium chlorochromate is the oxidation of heptan-1-ol to heptanal (Expt 5.76).

Pyridinium chlorochromate adsorbed on alumina is convenient for some applications.[15] To a solution of chromium trioxide (6 g) in 6 M hydrochloric acid (11 ml) is added pyridine (4.75 g) within 10 minutes at 40 °C. The mixture is kept at 10 °C until a yellow–orange solid forms. Reheating to 40 °C gives a solution. Alumina (50 g) is then added to the solution with stirring at 40 °C. After evaporation in a rotary evaporator, the orange solid is dried *in vacuo* for 2 hours at room temperature. The reagent can be kept for several weeks under vacuum in the dark without losing its activity. A typical procedure for the use of this reagent is the oxidation of carveol to carvone (Expt 5.88).

19. COPPER

Copper powder is prepared by dissolving 100 grams of recrystallised copper(II) sulphate in 350 ml of hot water in a 1-litre beaker; a magnetic stirrer is provided. After cooling to the laboratory temperature, the stirrer is set in motion and 35 g (or more, if necessary) of purified zinc powder (see 4.2.80, *Zinc*) are gradually added until the solution is decolourised. The precipitated copper is washed by decantation with water. Dilute hydrochloric acid (5%) is added to the precipitate in order to remove the excess of zinc, and stirring is continued until the evolution of hydrogen ceases. The copper powder is filtered, washed with water and kept in a moist condition (as a paste) in a stoppered bottle.

Activated copper bronze is essential for satisfactory yields in the Ullmann reaction, and uniform results can be obtained by the following activation process. One hundred grams of copper bronze are treated with 1 litre of a 2 per cent solution of iodine in acetone for 5–10 minutes. This results in the production of a rather greyish colour due to the formation of copper iodide. The product is filtered off on a Buchner funnel, removed and washed by stirring with 500 ml of 1:1 solution of concentrated hydrochloric acid in acetone. The copper iodide dissolves, and the residual copper bronze is filtered and washed with acetone. It is then dried in a vacuum desiccator. The activated copper bronze should be used immediately after preparation.

20. COPPER–CHROMIUM OXIDE CATALYST ('COPPER CHROMITE' CATALYST)

A reactive form of copper–chromium oxide suitable for hydrogenations (see Section 2.17.1, p. 87) may be obtained by the decomposition of basic copper ammonium chromate; the main reactions may be written as:

$$2Cu(NO_3)_2 + Na_2Cr_2O_7 + 4NH_3 + 3H_2O$$

$$\xrightarrow{55°C} 2CuNH_4(OH)CrO_4 + 2NaNO_3 + 2NH_4NO_3$$

$$\xrightarrow{300°C} CuO·CuCr_2O_4 + N_2 + 5H_2O$$

The most active forms contain barium chromite, which is incorporated by adding barium nitrate to the reaction mixture. The barium in the catalyst gives protection against sulphate poisoning and is said to stabilise the catalyst against reduction.

Dissolve 15.5 g of barium nitrate (AnalaR) and 130 g of copper(II) nitrate trihydrate (AnalaR) in 450 ml of water at 80 °C. Prepare a solution of sodium chromate by dissolving 89 g of recrystallised sodium dichromate dihydrate in 200 ml of water and adding 112.5 ml of concentrated ammonia solution (d 0.880). Add the warm solution (80 °C) of nitrates in a thin stream, with stirring, to the sodium chromate solution (at 25 °C). Collect the orange precipitate by suction filtration, wash it with two 50 ml portions of water, drain well and dry at 75–80 °C for 12 hours; powder finely.

Equip a 500-ml three-necked flask with a funnel for introducing a solid, a wide air condenser and a stainless steel stirrer with crescent blade, 1 cm long and 8 cm wide, so shaped that it conforms to the bottom of the flask. Immerse the flask in a metal bath at 350 °C. Add the powder through the funnel, with rapid stirring, during a period of 15 minutes. Heat with stirring at a bath temperature of 350 °C for 20 minutes after all the solid has been added. Leach the product by stirring for 30 minutes with 300 ml of 10 per cent acetic acid at room temperature (1). Allow to settle, decant the solution and wash the residue with six 50–60 ml portions of water. Filter with suction on a Buchner funnel, dry at 125 °C for 12 hours and grind finely in a mortar. The yield of catalyst (a brownish-black powder) is 85 g. No special precautions are necessary in handling or storing the catalyst since it is unaffected by exposure to air or moisture.

Note. (1) This treatment removes a substantial proportion of the excess of copper(II) oxide and renders the catalyst more active for hydrogenation purposes.

A *copper–chromium oxide on pumice* catalyst has particular value for the dehydrogenation of primary and secondary alcohols to the corresponding carbonyl compounds (see Section 5.6.1, p. 587). Dissolve 10.4 g of barium nitrate (AnalaR) in 280 ml of water at about 80 °C and add to this hot solution 87 g of copper(II) nitrate trihydrate (AnalaR); stir the mixture and heat until a homogeneous solution results. Prepare a solution of 50.4 g of recrystallised ammonium dichromate in a mixture of 200 ml of water and 75 ml of concentrated ammonia solution (d 0.880). To the ammonium chromate solution at 25–30 °C add the hot (80 °C) nitrate solution in a thin stream with stirring. Allow the mixture to cool and filter off the yellowish-brown precipitate with suction; press with a glass stopper and suck as dry as possbile. Transfer the precipitate of copper barium ammonium chromate to a large evaporating dish, add sufficient water to form a moderately thick paste and introduce pumice (4–8 mesh) with stirring until most of the paste has been transferred to the pumice; about 300 g of pumice are required. Heat on an electric hot-plate until the particles of pumice no longer adhere one to another. Remove the impregnated pumice (yellowish-brown) to a small evaporating dish and heat, by means of a Bunsen flame, with stirring until the colour changes through brown to a uniform black.

21. COPPER(I) BROMIDE

Dissolve 45 g (0.18 mol) of copper(II) sulphate pentahydrate and 19 g (0.19 mol) of sodium bromide in 150 ml of water. Add, with stirring, a solution of 11.8 g of sodium metabisulphite in 120 ml of water to the hot solution during 5 minutes (1). If the blue colour is not completely discharged, add a little more sodium metabisulphite. Cool the mixture and decant the supernatant liquor. Wash the precipitate by decantation with water containing a little dissolved sulphur dioxide to prevent oxidation. A solution of copper(I) bromide may be prepared by dissolving the moist solid in 30 ml of constant boiling point hydrobromic acid (48% w/w HBr). If however the solid salt is required collect the precipitate in a Buchner funnel, wash with water containing a little dissolved sulphur dioxide, followed by ethanol and then ether each containing sulphur dioxide and press with a glass stopper. Remove residual solvent and finally dry the solid in a vacuum desiccator over sulphuric acid and potassium hydroxide.

$$4CuSO_4 + 4NaBr + Na_2S_2O_5 + 3H_2O \longrightarrow 4CuBr + 6NaHSO_4$$

Note. (1) Alternatively pass a stream of sulphur dioxide through the heated solution (60 °C) for about 2 hours.

An alternative means of preparation of a solution of copper(I) bromide involves heating under reflux a mixture of 63 g (0.25 mol) of copper(II) sulphate pentahydrate, 20 g (0.314 mol) of copper turnings, 114 g (1.109 mol) of sodium bromide, 30 g (16.3 ml) of concentrated sulphuric acid and 1 litre of water for 3–4 hours. If the colour of the solution has not become yellowish after this period of heating, a few grams of sodium sulphite should be added to complete the reduction.

$$CuSO_4 + Cu + 2NaBr \longrightarrow 2CuBr + Na_2SO_4$$

22. COPPER(I) CHLORIDE

Dissolve 35 g (0.14 mol) of copper(II) sulphate pentahydrate and 9.2 g (0.157 mol) of pure sodium chloride in 125 ml of water; warming may be necessary. Add a solution of 8.4 g (0.044 mol) of sodium metabisulphite (1) in 90 ml of water to the hot solution during about 5 minutes with constant shaking. Cool to room temperature (use an ice bath if neccessary) and decant the supernatant liquor from the colourless copper(I) chloride. Wash the precipitate twice by decantation with water containing a little dissolved sulphur dioxide, the latter to prevent oxidation. For most purposes a solution of copper(I) chloride is required: the moist copper(I) chloride is dissolved in 60 ml of concentrated hydrochloric acid. The solution should be used within 24 hours of its preparation as it tends to oxidise (and therefore darken) on keeping. If it is not to be used immediately, the solution is kept in a tightly stoppered bottle (2). If the dry solid copper(I) chloride is required, the moist solid should be washed several times by decantation with water containing sulphur dioxide, collected on a Buchner funnel, washed several times with small portions of glacial acetic acid and dried in an air oven at 100–120 °C until the odour of acetic acid has disappeared. The copper(I) chloride thus obtained has a pure white colour and should be kept in a tightly stoppered bottle. The yield is almost quantitative.

$$4CuSO_4 + 4NaCl + Na_2S_2O_5 + 3H_2O \longrightarrow 4CuCl + 6NaHSO_4$$

Notes. (1) The reduction may also be effected with an alkaline solution of sodium sulphite prepared from 8.4 g of sodium metabisulphite and 7.0 g of sodium hydroxide pellets in 90 ml of water. The reaction is probably:

$$2CuSO_4 + 4NaCl + Na_2SO_3 + H_2O \longrightarrow 2CuCl + 3Na_2SO_4 + 2HCl$$

(2) An alternative means of preparation of a solution of copper(I) chloride is given under Expt 6.71.

23. COPPER(I) CYANIDE

The preparation is based essentially upon the reaction formulated below. It should be carried out in the fume cupboard and great caution exercised in the handling of the cyanides.

$$4CuSO_4 + 4NaCN + Na_2S_2O_5 + 3H_2O \longrightarrow 4CuCN + 6NaHSO_4$$

Place 500 g (2 mol) of powdered copper(II) sulphate pentahydrate in a 3-litre beaker or round-bottomed flask equipped with a stirrer and dissolve in 1600 ml of water at 40–50 °C. Prepare solutions of 140 g (0.74 mol) of commercial sodium metabisulphite in 400 ml of water (A) and of 140 g (2.15 mol) of potassium cyanide (96–99% purity) (**CAUTION**) in 400 ml of water (B), and filter if necessary from small amounts of insoluble matter. Warm solutions A and B separately to 60 °C. Make the copper(II) sulphate solution faintly acid to Congo red with dilute sulphuric acid and add solution A with mechanical stirring during 1–2 minutes, followed immediately by solution B. There is a slight frothing, a little sulphur dioxide is evolved, but no appreciable amount of cyanogen or hydrogen cyanide. After about 10 minutes, filter the hot solution and wash the product thoroughly with boiling water, and finally with rectified spirit. Dry at 100–110 °C to a fine soft powder (24–36 hours). The yield is 167 g (93%).

The most satisfactory method of preparation of a *copper(I) cyanide solution* is to dissolve the copper(I) cyanide (90 g, 1 mol) in a solution of sodium cyanide (125 g, 2.5 mol) (**CAUTION**) in 600 ml of water. If it is desired to avoid the preparation of solid copper(I) cyanide, the following procedure may be adopted. Copper(I) chloride, prepared from 35 g of copper(II) sulphate pentahydrate as described under 22 above, is suspended in 60 ml of water contained in a 500-ml round-bottomed flask, which is fitted with a mechanical stirrer. A solution of 18.5 g of sodium cyanide (96–98%) in 30 ml of water is added and the mixture is stirred. The copper(I) chloride passes into solution with considerable evolution of heat. As the copper(I) cyanide is usually employed in reactions with solutions of aryl diazonium salts it is usual to cool the resulting copper(I) cyanide solution in ice.

429

24. DIALKYL SULPHATES

Dimethyl sulphate [Me$_2$SO$_4$] is a liquid of b.p. 188.5 °C and is practically without odour. **CAUTION**: *Both vapour and liquid dimethyl sulphate are highly poisonous and the substance should only be used in a fume cupboard with a good draught, and rubber gloves should be worn.* Inhalation of vapour may lead to giddiness or even more serious results. The liquid itself is readily absorbed through the skin, with toxic results. If the liquid is accidentally splashed upon the hands, wash immediately with much concentrated ammonia solution in order to hydrolyse the compound before it can be absorbed through the skin; then rub gently with a wad of cotton wool soaked in ammonia solution.

Commercial dimethyl sulphate may be purified by, (*a*) allowing it to stand over anhydrous potassium carbonate until it is neutral to Congo red paper, or (*b*) by washing, just before use, with an equal volume of ice-water, followed by one-third of its volume of cold, saturated sodium hydrogen carbonate solution, and finally drying over calcium oxide. In both cases the purified dimethyl sulphate is fractionally distilled under reduced pressure from calcium oxide and the fraction having b.p. 72–73 °C/13 mmHg collected.

Diethyl sulphate [Et$_2$SO$_4$] is somewhat less poisonous than dimethyl sulphate, but similar precautions in the use and handling of this reagent should be observed; all operations should be conducted in the fume cupboard and rubber gloves should be worn. If the diethyl sulphate is dark in colour, it should be placed in a separatory funnel of suitable size and washed first with ice-water, then with aqueous sodium hydrogen carbonate solution until free of acid and finally dried over successive portions of calcium oxide. The dried diethyl sulphate is fractionally distilled from calcium oxide and has b.p. 93 °C/13 mmHg.

25. DIAZOMETHANE

CAUTION: *Liquid diazomethane, CH$_2$N$_2$, b.p. −24 °C, is an explosive compound and explosions may also occur in the gaseous state if the substance is dry and undiluted.* The gas may be handled with safety by diluting it with nitrogen. For synthetic work, a dry dilute ethereal solution of the gas is employed and this can be handled with reasonable safety, although such solutions have also been known to explode. *However, due regard must be paid to the highly toxic character of the gas by carrying out all operations in an efficient fume cupboard.*

An ethereal solution of diazomethane is usually prepared immediately before it is required for reaction. A convenient precursor for the generation of diazomethane is *N*-methyl-*N*-nitrosotoluene-*p*-sulphonamide which is prepared by the action of nitrous acid on *N*-methyltoluene-*p*-sulphonamide. This latter is formed from toluene-*p*-sulphonyl chloride and methylamine in alkaline solution. The methylnitrosamide affords diazomethane on reaction with potassium hydroxide solution.

$$p\text{-Me·C}_6\text{H}_4\text{·SO}_2\text{Cl} \xrightarrow[\ominus\text{OH}]{\text{Me·NH}_2} p\text{-Me·C}_6\text{H}_4\text{·SO}_2\text{NHMe} \xrightarrow{\text{HNO}_2}$$

$$p\text{-Me·C}_6\text{H}_4\text{·SO}_2\text{N(NO)Me}$$

$$p\text{-Me·C}_6\text{H}_4\text{·SO}_2\text{N(NO)Me} \xrightarrow{\ominus\text{OH}} p\text{-Me·C}_6\text{H}_4\text{·SO}_3^{\ominus} + \text{CH}_2\text{N}_2 + \text{H}_2\text{O}$$

N-*Methyl*-N-*nitrosotoluene*-p-*sulphonamide*.* Divide 320 g (1.68 mol) of purified toluene-*p*-sulphonyl chloride, m.p. 68–69 °C (Section 4.2.78, p. 466) into three portions of 190, 90 and 40 g. Prepare a solution of 70 g (1.75 mol) of sodium hydroxide in 70 ml of water and cool to room temperature. Place 210 ml (2.25 mol) of 33 per cent aqueous methylamine solution (or 174 ml of the 40% aqueous solution) in a 1-litre round-bottomed flask and add the 190 g of toluene-*p*-sulphonyl chloride in portions with swirling during about 5 minutes. The mixture becomes warm. Allow the temperature to rise to 80–90 °C in order to maintain the *N*-methyltoluene-*p*-sulphonamide (m.p. 78 °C) in a molten condition, otherwise the latter may form a hard cake and reaction may be incomplete; also do not permit the temperature to rise above 90 °C as appreciable loss of methylamine may result. The mixture should be acid to litmus within 5 minutes after the completion of the first addition of the sulphonyl chloride (1). Then add 50 ml of the 50 per cent sodium hydroxide solution carefully with swirling, followed immediately by 90 g of the sulphonyl chloride in portions as before. When the mixture has again become acidic (1), introduce 25 ml of the sodium hydroxide solution, followed by 40 g of toluene-*p*-sulphonyl chloride with vigorous swirling. After the mixture has again become acidic, add the remainder of the sodium hydroxide solution. The liquid phase of the final mixture should be alkaline; if it is acidic, indicating excessive loss of methylamine, add sufficient methylamine solution to render the mixture basic.

Rinse the walls of the flask with a little water and complete the reaction by heating the mixture (which consists of two layers and a precipitate of sodium chloride) on a boiling water bath for 15 minutes with vigorous mechanical stirring. Pour the hot reaction mixture into 1500 ml of glacial acetic acid contained in a 4-litre round-bottomed flask; rinse the flask with 250 ml of acetic acid. Cool the solution in an ice bath to 5 °C (2), stir mechanically and add a solution of 125 g (1.8 mol) of sodium nitrite in 250 ml of water from a dropping funnel during about 45 minutes: maintain the temperature below 10 °C and continue the stirring for 15 minutes after the addition is complete. The nitroso derivative separates as a yellow crystalline solid during the reaction. Add 1 litre of water to the reaction mixture and collect the precipitate by suction filtration; press it on the funnel and wash with about 500 ml of water. Transfer the product to a beaker, stir it well with about 400 ml of water, then filter and wash again on the funnel until the odour of acetic acid is no longer apparent. Dry to constant weight in a vacuum desiccator over concentrated sulphuric acid. The yield of *N*-methyl-*N*-nitrosotoluene-*p*-sulphonamide, m.p. 58–60 °C, is 325 g (90%). This is sufficiently pure for the preparation of diazomethane. It should be kept in a dark bottle. It may be recrystallised by dissolution in boiling ether (1 ml/g), addition of an equal volume of light petroleum, b.p. 40–60 °C, and cooling in a refrigerator. This reagent may be stored for long periods at room temperature in a dark bottle without appreciable decomposition.

Notes. (1) Occasionally the liquid may not become acidic after the first or second addition, even though the sulphonyl chloride has reacted completely (this is due to a smaller loss of methylamine than is expected). If such is the case, no more than 5 minutes should be allowed between successive additions of sulphonyl chloride and alkali. The whole procedure occupies about 30 minutes.

---

* These experimental conditions are reproduced by kind permission of Professor H. J. Backer.

(2) A reaction temperature below 0 °C should be avoided because the total volume of acetic acid is just sufficient to keep the N-methyltoluene-p-sulphonamide in solution above 0 °C.

Preparation of diazomethane. CAUTION: *This preparation should be carried out only in a fume cupboard provided with a powerful exhaust system. The use of a screen of safety glass in addition to wearing safety spectacles is essential.* Since the explosive decomposition of diazomethane is frequently initiated by rough or sharp surfaces the use of apparatus with ground glass joints and of boiling chips in the subsequent distillation is *not* recommended. New glass apparatus with rubber stopper connections carrying fire-polished glass connecting tubes, or glass apparatus with Clearjoints should be used. A further precaution against explosion is to ensure that the apparatus assembly, located in a fume cupboard, is not exposed to direct sunlight or placed near a strong artificial light.

The following procedures may be used for the preparation of ethereal solutions of diazomethane containing ethanol; they differ slightly according to whether large or small quantities are required. The presence of ethanol is not harmful in many uses of diazomethane.

Method 1. Add 50 ml of 96 per cent ethanol to a solution of 10 g (0.18 mol) of potassium hydroxide in 15 ml of water. Place this solution in a 200-ml distillation flask equipped with a dropping funnel and an efficient double surface condenser. Fit by means of rubber bungs and glass tube connectors two receiving flasks (of 500 ml and 100 ml capacity) in series to the end of the condenser; cool both the flasks in an ice–salt mixture and arrange that the inlet tube leading into the second smaller receiver flask dips below the surface of a charge of 40 ml of ether. This will serve to trap any uncondensed diazomethane–ether vapour escaping from the first receiver. Heat the distilling flask in a water bath at 60–65 °C; place a solution of 43 g (0.2 mol) of N-methyl-N-nitrosotoluene-p-sulphonamide in about 250 ml of ether in the dropping funnel and introduce it into the flask over a period of 45 minutes. Adjust the rate of addition so that it is about equal to the rate of distillation. When the dropping funnel is empty, add more ether (c. 30 ml) gradually until the ether distilling over is colourless. The combined ethereal solutions in the receivers contain 5.9–6.1 g (70%) of diazomethane (1).

Method 2. For smaller quantities of diazomethane, the use of a dropping funnel is unnecessary. Dissolve 2.14 g of N-methyl-N-nitrosotoluene-p-sulphonamide in 30 ml of ether, cool in ice and add a solution of 0.4 g of potassium hydroxide in 10 ml of 96 per cent ethanol. If a precipitate forms, add more ethanol until it just dissolves. After 5 minutes, distil the ethereal diazomethane solution from a water bath. The ethereal solution contains 0.32–0.35 g of diazomethane (1).

Method 3. An ethereal solution of diazomethane free from ethanol may be prepared by this method: such a solution is required, for example, in the Arndt–Eistert reaction with acid chlorides (compare Expt 5.130). In a 100-ml distilling flask provided with a dropping funnel and an efficient downward condenser, place a solution of 6 g of potassium hydroxide in 10 ml of water, 35 ml of carbitol (diethyleneglycol monoethyl ether) and 10 ml of water: connect the condenser to two conical flasks in series containing 10 and 35 ml of ether respectively and cooled in an ice–salt bath (see *Method 1*). Heat the mixture on a water bath at 70–75 °C in a beaker of water placed upon a hot-plate incorporating a magnetic stirring unit with a Teflon-coated follower situated in the flask. As soon as the ether commences to distil, add a solution of 21.5 g of N-methyl-N-nitrosotoluene-p-sulphonamide in 125 ml of ether through the dropping funnel during a period of about 15 minutes, stirring electromagnetically. After the addition of the nitrosoamide, add 30–40 ml of ether through the dropping funnel and distil until the distillate is colourless. The ethereal solution in the conical flasks contains about 3.4 g of diazomethane (1).

The diazomethane–ether solutions prepared by all these methods should be dry. If in doubt, it may be dried with potassium hydroxide pellets. The anhydrous ethereal solution may be stored in the fume cupboard in a smooth glass flask or bottle for a day or so at −70 °C; since slow decomposition occurs with liberation of gas, the containing vessel should be protected by a calcium chloride guard-tube.

Note. (1) *To determine the exact diazomethane content*, allow an aliquot portion of the ethereal diazomethane solution to react with an accurately weighted amount (say, about 1 g) of benzoic acid (AnalaR) in 50 ml of anhydrous ether. The solution should be completely decolourised, thus showing that the benzoic acid is present in excess. Dilute the solution with water and titrate the excess of benzoic acid with standard 0.1 M alkali using phenolphthalein as indicator.

General methylation procedure with diazomethane. *The reaction must be carried out in the fume cupboard.* Dissolve 2–3 g of the compound (say, a phenol or a carboxylic acid) in a little anhydrous ether or absolute methanol, cool in ice and add the ethereal solution of diazomethane in small portions until gas evolution ceases and the solution acquires a pale yellow colour. Test the coloured solution for the presence of excess of diazomethane by removing a few drops into a test tube and introducing a glass rod moistened with glacial acetic acid: immediate evolution of gas should occur. Evaporate the solvent, and purify the product by distillation or crystallisation.

26. DIETHYL CARBONATE
Commercial diethyl carbonate may be purified by the following process. Wash 100 ml of diethyl carbonate successively with 20 ml of 10 per cent sodium carbonate solution, 20 ml of saturated calcium chloride solution and 25 ml of water. Allow to stand for 1 hour over anhydrous calcium chloride with occasional shaking, filter into a dry flask containing 5 g of the same desiccant, and allow to stand for a further hour. Distil and collect the fraction boiling at 125–126 °C. Diethyl carbonate combines with anhydrous calcium chloride slowly and prolonged contact should therefore be avoided. Anhydrous calcium sulphate may also be used.

27. N,N-DICYCLOHEXYLCARBODIIMIDE (DCC; $C_6H_{11}\cdot N = C = N\cdot C_6H_{11}$)
The reagent is a potent skin irritant and may cause sensitisation; it must be handled with care. Material of good quality (c. 99% pure) is available commercially as a waxy low-melting solid, m.p. 34–35 °C. It is most convenient to liquefy the reagent by the use of a hot-air blower or by standing the reagent bottle in a little warm water to assist weighing out.

In most of its applications the reagent acts as a dehydrating condensing agent and is often recovered as dicyclohexylurea. Although the reagent is not unduly expensive, it may be desirable to make use of the recovered urea; the latter may be converted into the diimide by treatment in pyridine solution with toluene-*p*-sulphonyl chloride, phosphorus oxychloride or phosphoric oxide.

To prepare DCC from dicyclohexylurea recrystallise the recovered urea from ethanol; m.p. 234 °C. Add dropwise with stirring 17.1 g (47 ml, 0.11 mol) of phosphorus oxychloride to 22.5 g (0.1 mol) of dicyclohexylurea in 50 ml of pyridine at 50 °C, and heat at 60–90 °C for 1.5 hours. Pour the reaction product on to crushed ice, extract with light petroleum (b.p. 60–80 °C) and dry the extract over anhydrous sodium sulphate. Remove the solvent using a rotary evaporator and distil the residual oil under reduced pressure. The yield of diimide, b.p. 157–159 °C/15 mmHg (131 °C/3–4 mmHg), is about 14 g (68%).

28. ETHYLENE OXIDE
The reagent has b.p. 11 °C and is supplied either in 100-ml sealed tubes or in 100-ml cylinders equipped with an appropriate valve. The gas, which is highly flammable, has no very distinctive smell and must be regarded as a hazardous toxic reagent which must not be inhaled or allowed to come into contact with the skin and eyes. Precautions in the use of ethylene oxide are described in Expt 5.39, which may be regarded as typical.

29. FORMALDEHYDE
Commercial formalin is an aqueous solution containing 37–40 per cent w/v of formaldehyde (0.37–0.40 g HCHO per ml), stabilised by the addition of 12 per cent of methanol.

When dry gaseous formaldehyde is required it may be obtained by the depolymerisation of paraformaldehyde at 180–200 °C. For reactions at elevated temperatures dried paraformaldehyde may be used as the *in situ* source of formaldehyde (Expt. 5.39).

30. GIRARD'S REAGENTS 'T' AND 'P'
Girard's reagent 'T' is carbohydrazidomethyltrimethylammonium chloride (1) (trimethylaminoacetohydrazide chloride) and is prepared by the reaction of the quaternary ammonium salt formed from ethyl chloroacetate and trimethylamine with hydrazine hydrate in alcoholic solution.

$$Me_3N + ClCH_2 \cdot CO_2Et \longrightarrow Cl^{\ominus}\{Me_3\overset{\oplus}{N}\cdot CH_2 \cdot CO_2Et \xrightarrow{NH_2 \cdot NH_2}$$

$$Cl^{\ominus}\{Me_3\overset{\oplus}{N}\cdot CH_2 \cdot CO \cdot NH \cdot NH_2 + EtOH$$
$$(1)$$

Girard's reagent 'P' is the corresponding pyridinium compound, prepared by replacing the trimethylamine by pyridine. The reagent 'T', unlike the reagent 'P', is very deliquescent, but is nevertheless widely used for laboratory work because of its greater solubility. The quaternary ammonium grouping imparts water solubility.

The main use of the Girard reagents 'T' and 'P' is for the isolation of small amounts of ketones from admixture with other organic matter contained in, for example, various natural products; the carbonyl derivatives are water-soluble. The ketonic material, dissolved in ethanol containing 10 per cent acetic acid, is heated for 30–60 minutes with the reagent in slight excess, the volume being adjusted to give a 5 to 10 per cent solution of the reagent. The cooled solution is diluted with water containing enough alkali to neutralise 90 per cent of the acid and to give an alcohol content of 10–20 per cent. It is then exhaustively extracted with ether to remove non-ketonic compounds; the water-soluble hydrazone derivatives are decomposed by the addition of mineral acid up to a concentration of 0.5 M and, after about 1 hour at room temperature, the liberated ketonic compound is isolated by extraction with ether.

$$\text{Reagent } T + R^1 \cdot CO \cdot R^2 \underset{HCl}{\overset{EtOH, AcOH}{\rightleftharpoons}} Cl\{Me_3\overset{\oplus}{N} \cdot CH_2 \cdot CO \cdot NH \cdot N{=}CR^1R^2 + H_2O$$

To prepare Girard's reagent 'T', place a solution of 98.5 g (84.5 ml, 0.8 mol) of ethyl chloroacetate and 200 ml of absolute ethanol in a 1-litre, three-necked flask, fitted with a thermometer, stirrer and an ice-cooled condenser (Fig. 2.23). Cool the solution to 0 °C by stirring in an ice–salt bath, stop the stirrer and add 49 g (74 ml, 0.83 mol), measured after precooling to −5 °C, of trimethylamine all at once. Control the exothermic reaction sufficiently by external cooling so that the temperature of the mixture rises to 60 °C during about 1 hour. When there is no further evolution of heat, allow the reaction mixture to stand at room temperature for 20–24 hours. Remove the condenser, replace the thermometer by a dropping funnel and add 40 g (0.8 mol) of 100 per cent hydrazine hydrate (Section 4.2.*31*) (**CAUTION**) with stirring during 10–15 minutes. Stir for a further 45 minutes, cool the solution slightly and, unless crystallisation commences spontaneously, scratch the walls of the vessel with a glass rod to induce crystallisation. The product separates in fine colourless needles. Cool in an ice bath, collect the highly hygroscopic salt rapidly on a Buchner funnel, wash with 150 ml of cold absolute ethanol and press dry. Dry the product in a vacuum desiccator over concentrated sulphuric acid; the yield is 105 g (85%), m.p. 175–180 °C (decomp.). (This material (1) contains a small amount of the symmetrical dihydrazide, but is quite satisfactory as a reagent for the separation of ketones.) A further crop of 12 g may be obtained after distilling off 200–300 ml of solvent from the mother-liquor and washings at the pressure of the water pump.

Note. (1) The deliquescent solid must be stored in a dry, tightly-stoppered container. If exposed to the air it deteriorates rapidly, developing an unpleasant odour. Samples that have been kept for some time are best recrystallised from absolute ethanol before use.

For Girard's reagent 'P' ($C_5H_5\overset{\oplus}{N}\cdot CH_2\cdot CO\cdot NH\cdot NH_2\}\overset{\ominus}{Cl}$) equip a 1-litre, three-necked flask, with a sealed stirrer unit, a dropping funnel and a reflux condenser, and add 200 ml of absolute ethanol, 63 g (64.5 ml, 0.8 mol) of pure anhydrous pyridine (**CAUTION**) and 98.5 g (84.5 ml, 0.8 mol) of ethyl chloroacetate. Heat the mixture under reflux for 2–3 hours until the formation of the quaternary salt is complete; acidify a small test-portion with dilute sulphuric acid; it should dissolve completely and no odour of ethyl chloroacetate should be apparent. Cool the mixture in ice and salt. Run in rapidly from a dropping funnel a solution of 40 g (0.8 mol) of 100 per cent hydrazine hydrate (**CAUTION**) in 50 ml of absolute ethanol all at once. A vigorous exothermic reaction soon develops and is accompanied by vigorous effervescence. The product separates almost immediately. When cold, filter with suction, wash with ice-cold ethanol and dry in the air. The yield of Girard's reagent 'P' is 135 g

(90%); this is satisfactory for the isolation of ketones. A pure product may be obtained by recrystallisation from methanol.

31. HYDRAZINE HYDRATE

CAUTION: *Hydrazine is a suspected carcinogen and therefore should be handled with appropriate precautions.* Hydrazine hydrate containing 60 per cent w/w and 98–100 per cent w/w $NH_2 \cdot NH_2 \cdot H_2O$ is available commercially (1). A hydrazine solution, 60 per cent w/w of hydrazine hydrate, may be concentrated as follows. A mixture of 150 g (144 ml) of the solution and 230 ml of xylene is distilled from a 500-ml round-bottomed flask through a well-lagged Hempel (or other efficient fractionating) column in an atmosphere of nitrogen. All the xylene passes over with about 85 ml of water. Upon distillation of the residue, about 50 g of 90–95 per cent hydrazine hydrate (2) are obtained.

Notes. (1) Solutions of hydrazine hydrate are extremely corrosive; protective gloves and safety spectacles should always be worn when handling this reagent.
(2) Hydrazine hydrate may be titrated with standard acid using methyl orange as indicator or, alternatively, against standard iodine solution with starch as indicator. In the latter case about 0.1 g, accurately weighed, of the hydrazine hydrate solution is diluted with about 100 ml of water, 2–3 drops of starch indicator added, and immediately before titration 5 g of sodium bicarbonate are introduced. Rapid titration with iodine gives a satisfactory end-point.

$$5NH_2 \cdot NH_2 \cdot H_2O + 2I_2 \longrightarrow 4NH_2 \cdot NH_2 \cdot HI + 5H_2O + N_2$$

Anhydrous hydrazine may be obtained by refluxing 100 per cent hydrazine hydrate (or the 95% material which has previously been allowed to stand overnight over 20% w/w of potassium hydroxide and filtered) with an equal weight of sodium hydroxide pellets for 2 hours and then distilling in a slow stream of nitrogen; b.p. 114–116 °C. Distillation in air may lead to an explosion.

32. HYDRIODIC ACID

This is supplied as an azeotrope with water (constant boiling hydriodic acid), b.p. 125.5–126.5 °C/760 mmHg, *d* 1.70, which contains 55–57 per cent w/w HI (0.936 to 0.99 g HI per ml). Additional grades available contain 45 per cent w/w HI and 67 per cent w/w HI; the latter is stabilised by the addition of 0.03 per cent w/w of hypophosphorous acid. Once a bottle is opened the contents tend to deteriorate, becoming discoloured in a few days. This can be avoided by purging the container with nitrogen before resealing.

The constant boiling azeotrope may be conveniently prepared in the laboratory, if required, by the following procedure. In a fume cupboard a 1.5-litre, three-necked flask is charged with a mixture of 480 g of iodine and 600 ml of water. The central socket is fitted with an efficient mechanical stirrer which leads almost to the bottom of the flask, and the smaller sockets respectively with a lead-in tube for hydrogen sulphide extending to well below the surface of the liquid and with an exit tube attached to an inverted funnel just dipping into 5 per cent sodium hydroxide solution. The mixture is vigorously stirred and a stream of hydrogen sulphide (either from a freshly-charged Kipp's apparatus or from a cylinder of the gas) passed in as rapidly as it can be absorbed. After several hours the liquid assumes a yellow colour (sometimes it is almost colourless) and most of the sulphur sticks together in the form of a hard lump (1). The sulphur is removed by filtration through a funnel plugged with glass wool (or through a sintered glass funnel), and the filtrate is boiled until the lead acetate paper test for hydrogen sulphide is negative. The solution is filtered again, if necessary. The hydriodic acid is then distilled from a 500-ml Claisen flask, and the fraction, b.p. 125.5–126.5 °C/760 mmHg, is collected. The yield of the constant boiling acid containing 57 per cent w/w HI is 785 g (90%).

$$H_2S + I_2 \longrightarrow 2HI + S$$

Note. (1) The hard lump of sulphur remaining in the flask is best removed by boiling with concentrated nitric acid in the fume cupboard.

33. HYDROBROMIC ACID
This is supplied as an azeotrope with water (constant boiling hydrobromic acid), b.p. 126 °C, d 1.46–1.49, which contains 47–48 per cent w/w HBr (0.695 to 0.715 g HBr per ml). A grade of hydrobromic acid is available containing 60 per cent w/w HBr (1.007 g HBr per ml).

34. HYDROCHLORIC ACID
This is supplied in grades containing either 32 per cent w/w HCl or 36 per cent w/w HCl (0.371 g HCl per ml and 0.424 g HCl per ml respectively).

35. HYDROFLUORIC ACID
This is supplied as a 40 or 48 per cent solution. It is advisable to wear acid-resisting rubber gloves and protective goggles when handling hydrofluoric acid. The acid in contact with the skin produces extremely painful burns. In case of accident, the burned surface, which becomes white, is held under running water until the natural colour returns. A paste made from magnesium oxide and glycerine should be applied immediately; this is said to be helpful in preventing the burn becoming serious. A calcium-gluconate-containing antidote gel is available commercially as an immediate treatment for HF burns. Medical advice should be sought in the case of such burns.

36. HYDROGEN
Cylinders of compressed hydrogen (99.9998% purity) may be purchased or hired; the impurities comprise traces of nitrogen, oxygen, moisture and hydrocarbons. Less pure grades containing slightly larger amounts of contaminating oxygen are however quite suitable for most purposes. Since hydrogen is chiefly employed for catalytic reductions and oxygen has, in general, no harmful effect upon the reduction, no purification is usually necessary. If, however, oxygen-free hydrogen is required, the oxygen may be removed by passage through Fieser's solution (see under *52. Nitrogen*), and then through a wash bottle containing concentrated sulphuric acid to which some silver sulphate has been added; the latter will detect and remove any hydrogen sulphide that may have formed from the decomposition of Fieser's solution.

37. HYDROGEN BROMIDE
Hydrogen bromide is most conveniently prepared by the action of bromine upon tetrahydronaphthalene (tetralin) if the quantity required does not justify the purchase of a cylinder of the gas.

$$C_{10}H_{12} + 4Br_2 \longrightarrow C_{10}H_8Br_4 + 4HBr$$

Only half of the added bromine is recovered as hydrogen bromide; a yield of about 45 per cent of hydrogen bromide is evolved computed on the weight of bromine taken. It is essential that the tetralin is pure and perfectly dry; it should therefore be allowed to stand over magnesium sulphate or anhydrous calcium sulphate for several hours, filtered and distilled under reduced pressure. The tetralin is placed in a round-bottomed flask fitted with a socket/cone adapter with 'T' connection and a dropping funnel (this may be of the pressure-equalising type (cf. Fig. 4.3)). Bromine is allowed to drop in from the funnel at a regular rate and the contents of the flask are gently swirled from time to time to ensure a steady evolution of hydrogen bromide. The traces of bromine carried over with the gas may be removed by allowing it to bubble through a Drechsel bottle charged with dry tetralin. A safety trap should always be interposed between the generating apparatus and the reaction vessel.

A solution of *hydrogen bromide in glacial acetic acid* (45% w/v) is a convenient commercially available source, which is suitable for many preparations.

38. HYDROGEN CHLORIDE

Method 1 (*from concentrated sulphuric acid and fused ammonium chloride*). The most convenient procedure is to allow concentrated sulphuric acid to react with lumps of fused ammonium chloride in a Kipp's apparatus.* The gas may be dried by passage through a Drechsel bottle containing concentrated sulphuric acid; the latter should be followed by an empty Drechsel bottle as a precaution against 'sucking back' of the contents of the reaction vessel.

Method 2 (*from concentrated sulphuric acid and concentrated hydrochloric acid*). The apparatus shown in Fig. 4.4 is employed. The upper funnel has a capacity of 100 ml and has an appropriate length of capillary tubing fused to the outlet tube; the lower funnel has a capacity of 500 ml. When the capillary tube is filled with concentrated hydrochloric acid, there is sufficient hydrostatic pressure to force the hydrochloric acid into the sulphuric acid. The Drechsel bottle (A) contains concentrated sulphuric acid and the other acts as a safety trap; the whole apparatus must be mounted on a heavy stand.

About 150 ml of concentrated sulphuric acid is placed in the larger funnel and 100 ml of concentrated hydrochloric acid in the smaller separatory funnel. The latter is raised from the adapter until the capillary tube is above the sulphuric acid, the capillary tube is carefully filled with concentrated hydrochloric acid, and the funnel is then lowered into the adapter socket. The rate of evolution of hydrogen chloride is controlled by regulation of the supply of hydrochloric acid: this will continue until a volume of hydrochloric acid equal to that of the concentrated sulphuric acid has been used. The diluted sulphuric acid should then be removed and the apparatus recharged. The yield is 31–33 g of hydrogen chloride per 100 ml of concentrated hydrochloric acid.

39. HYDROGEN CYANIDE
CAUTION: *Great care must be exercised in the preparation of this gas for it is a*

---

* An alternative and quickly assembled apparatus employs a Buchner flask fitted with a ground glass joint to which is attached a dropping funnel. Ammonium chloride is placed in the flask, moistened with concentrated hydrochloric acid, and concentrated sulphuric acid added dropwise from the funnel at such a rate that the evolution of gas may be controlled.

Fig. 4.4

dangerous poison; all operations must be conducted in a fume cupboard provided with an efficient draught. The appropriate antidotes should be available. The apparatus used is similar to, but generally on a smaller scale than, that described for *Hydrogen chloride, Method 2,* in which the larger separating funnel is replaced by a round-bottomed flask of suitable size. A saturated solution of sodium cyanide (prepared by dissolving 50 g of commercial sodium cyanide, about 96% purity, in sufficient water to make 125 ml of solution) is added 1 cm below the surface of 125 ml of aqueous sulphuric acid (50% by volume) contained in the flask. Any residual hydrogen cyanide may be expelled by warming the flask on a water bath. The gas may be used directly, or may be collected in the liquid form (b.p. 26 °C) by passing through a glass coil, 4–5 cm bore and 50 cm long, surrounded by ice; a freezing mixture must not be used as this may result in solidification of the hydrogen cyanide (m.p. − 15 to − 14.5 °C) and consequent clogging of the apparatus. If the hydrogen cyanide is required anhydrous, it should be passed through a train of three large U-tubes filled with anhydrous calcium chloride, the tubes being immersed in a water bath maintained at 30–40 °C.

40. HYDROGEN IODIDE
This gas may be conveniently prepared by allowing a solution of two parts of iodine in one part of hydriodic acid (*d* 1.7) to drop on to excess of red phosphorus; the apparatus assembly employed could be of the general type illustrated in Fig. 4.3. The evolution of hydrogen iodide takes place in the cold; when the evolution of gas slackens considerably, the mixture should be gently warmed.

41. HYDROGEN PEROXIDE
Aqueous solutions of hydrogen peroxide having concentrations of 6, 12, 30, 50 and 70 per cent w/v are available from most chemical suppliers. Higher

439

concentrations of hydrogen peroxide, i.e. 86 per cent w/v, are also available* ('High Test Peroxide'). Care should be taken with all concentrations of hydrogen peroxide since explosions can occur on contact with organic material or with transition metals.

Even the highly concentrated solutions of hydrogen peroxide ($> 50\%$) may be handled safely providing certain basic precautions are observed (see also Expt 5.190, Note (1)). Firstly it is advisable to wear protective spectacles and rubber or plastic gloves. Since these high-strength solutions may ignite textiles, it is also strongly advised that a rubber or plastic apron be worn and that all operations involving transfer of such solutions be conducted in the fume cupboard, with the additional precaution that all the apparatus is sited in a plastic or aluminium tray containing water to catch accidental spillage. Inhalation of vapour arising from use at elevated temperatures from these higher concentrations may cause inflammation of the nose and throat, and exposure of the eyes leads to ulceration of the cornea. Solutions of hydrogen peroxide spilt on the skin should be washed off immediately with water, when only temporary discomfort will then be experienced. Indeed, a ready supply of water should be to hand to wash away splashes and leakages.†

Frequently the strengths of solutions of hydrogen peroxide are quoted according to available oxygen content; thus 30 per cent w/v $\equiv$ 100 volumes, i.e. 1 ml of the H_2O_2 solution when fully decomposed by heat gives 100 ml of oxygen at s.t.p. This solution is approximately 8.82 M and hence 1 mol is contained in 113 ml. The strength of an aqueous solution of hydrogen peroxide is conveniently measured by the volumetric procedure of titration with standard sodium thiosulphate of the iodine released when aqueous hydrogen peroxide is treated with acidified potassium iodide.[16]

42. HYDROGEN SULPHIDE

This very poisonous gas should only be prepared and used in an efficient fume cupboard. It is usually generated from iron(II) sulphide and dilute hydrochloric acid (1:3) in a Kipp's apparatus; it should be washed with water to remove acid spray. The resulting hydrogen sulphide contains hydrogen because of the presence of free iron in commercial iron(II) sulphide.

Pure hydrogen sulphide (99.6%) can be obtained commercially in cylinders.

43. HYDROGEN TETRAFLUOBORATE (*Fluoroboric acid*)

A purified grade of the reagent is available commercially containing 42 per cent w/w HBF_4. It is usually more convenient to prepare the reagent before use; the details are given in Expt 6.77. *Great caution should be exercised in handling this reagent.*

44. IODINE MONOCHLORIDE

In a fume cupboard, pass dry chlorine gas into 127 g (0.5 mol) of iodine in a distilling flask until the weight has increased by 34.5 g (0.49 mol). The chlorine should be fed in via a glass frit located at or below the surface of the iodine while

---

* From Interox Chemicals Ltd, who supply notes on the safe handling of these concentrated solutions, and advise on the use of hydrogen peroxide and peroxyacids (Section 4.2.56, p. 455).
† A freshly prepared solution (5%) of sodium ascorbate should be readily available for use as an eye wash.

the flask is gently shaken. Distil the iodine chloride in an ordinary distillation apparatus; protect the receiver flask from the atmospheric moisture by a calcium chloride guard-tube. Collect the fraction, b.p. 97–105 °C; the yield is 140 g (88%). Preserve the iodine monochloride in a dry, glass-stoppered bottle. *Care should be taken when handling this compound since the liquid is corrosive and gives off a harmful vapour. If it should come into contact with the skin, an effective antidote is dilute hydrochloric acid (1:1).*

45. LEAD TETRA-ACETATE

The reagent as supplied by manufacturers is moistened with glacial acetic acid to prevent hydrolytic decomposition. It may be prepared in the laboratory by warming lead oxide (red lead, Pb_3O_4) with acetic acid in the presence of sufficient acetic anhydride to combine with the water formed.

$$Pb_3O_4 + 8Me\cdot CO_2H \longrightarrow (Me\cdot CO_2)_4Pb + 2(Me\cdot CO_2)_2Pb + 4H_2O$$

The filtrate which contains lead acetate may be treated with chlorine:

$$2(Me\cdot CO_2)_2Pb + Cl_2 \longrightarrow (Me\cdot CO_2)_4Pb + PbCl_2$$

and the resulting lead tetra-acetate separated from the accompanying lead chloride by recrystallisation from glacial acetic acid.

A mixture of 550 g of glacial acetic acid and 185 g of acetic anhydride is placed in a 1-litre, three-necked flask provided with a thermometer and a sealed stirrer unit. The liquid is vigorously stirred, heated to 55–60 °C and 300 g of dry red lead powder are added in portions of 15–20 g. A fresh addition is made only after the colour due to the preceding portion has largely disappeared. The temperature should not be allowed to rise above 65 °C. Towards the end it may be necessary to warm the flask cautiously to about 80 °C in order to complete the reaction. At the end of the reaction, the thick and somewhat dark solution is cooled, and the precipitated lead tetra-acetate is filtered off (the mother-liquor, *A*, is put aside) and washed with glacial acetic acid. The crude product, without being dried, is dissolved in hot glacial acetic acid containing a little acetic anhydride, the solution treated with a little decolourising carbon, filtered through a hot-water funnel and cooled. The colourless crystalline product is filtered off and dried in a vacuum desiccator over potassium hydroxide pellets. The yield is about 150 g (78%).

A further 100 g of lead tetra-acetate may be obtained from the mother-liquor (*A*) by returning it to the original flask, heating to about 75 °C with stirring, and passing through it a stream of dry chlorine. When the reaction is complete a few grams of decolourising carbon are added, the mixture is maintained at 75 °C for a few minutes, and the hot suspension is filtered with suction through a preheated Buchner funnel. The residue, consisting largely of lead chloride, is washed with hot glacial acetic acid. The filtrate, on cooling, deposits lead tetra-acetate in colourless needles; it is collected and dried as described above. Although contaminated with a little (<5%) lead chloride the resulting lead tetra-acetate is satisfactory for most purposes; if required perfectly pure, it should be recrystallised as detailed above.

46. LITHIUM

CAUTION: *See safety comments given for sodium (at 4.2.68) below. Lithium is somewhat less reactive than sodium but similar precautions should be observed.*

Lithium is available as the dry solid packed under argon and in the form of shot or as a dispersion in liquid paraffin. A convenient method of preparing lithium from the rod form is as follows. Place a piece of lithium weighing about 3 g and slightly moist with paraffin oil on a dry surface (slate or tiles) and pound it with a clean hammer or 500-g weight into a thin sheet about 0.5 mm thick. Cut the sheet into thin strips about 2–3 mm wide and transfer it to a beaker containing anhydrous ether. Weigh out the quantity of lithium required under dry ether or paraffin oil. Dry each strip with filter paper, cut it by means of a pair of scissors into small pieces about 1 mm wide and allow the small pieces to fall directly into anhydrous ether in a reaction flask. The lithium thus retains its bright lustre.

The lithium may also be pressed into wire of about 0.5 mm diameter; a rather sturdy press is necessary. The wire may be collected directly in sodium-dried ether.

Lithium shot or dispersion in liquid paraffin can be exposed to air during handling without deterioration. It may be transferred by pouring through a wide-necked funnel. Small quantities of the dispersion may be destroyed by washing with water to allow the lithium metal to react with water. Larger quantities should be suspended in ether and treated in a fume cupboard with dry t-butyl alcohol. Hydrogen is liberated in this reaction.

47. LITHIUM REAGENTS
Alkyllithium reagents are used routinely in organic synthesis. They must be handled with care and protective clothing should be worn. Solutions of alkyllithiums are extremely flammable and may ignite on contact with moist air; they should therefore always be stored and handled in an inert atmosphere (argon or nitrogen). They are measured and transferred most conveniently by means of a hypodermic syringe or cannula (see Section 2.17.8, p. 122). A number of alkyllithiums are available commercially and the manufacturers' instructions for their handling should be followed carefully. The lithium reagent may be stored in the original container by placing a rubber septum on the neck and removing portions of the reagent as required by means of a hypodermic syringe; the volume of reagent removed should then be replaced by an equal volume of nitrogen similarly introduced by means of the syringe. An alternative means of transferring the reagent to a two- or three-necked flask for storage is as follows. Equip the necks with stopcocks for nitrogen inlet and outlet. The outlet should be of such a design that when it is open the needle of a hypodermic syringe can be passed through its centre into the flask. The transfer of the lithium reagent to the flask is then achieved by inserting into the neck of the reagent bottle a tight fitting rubber bung previously fitted with two glass tubes: a short inlet tube for connection to a supply of inert gas, and an exit tube which reaches to the bottom of the reagent bottle. The exit tube is connected by polythene tubing to the flask which has been previously dried and flushed with nitrogen. The reagent is then transferred by means of a slight positive pressure of inert gas. Whenever the reagent is removed a stream of inert gas should be passed through the flask before closing the stopcocks. Butyllithium can be stored for an extended period at room temperature under nitrogen. The concentration of the reagents should be checked periodically by one of the methods described below.

Methyllithium is available commercially as a solution in diethyl ether. For some applications it is desirable to have methyllithium free of halide, and a method has been described for the preparation of such halide-free solutions.[17]

442

Butyllithium is available as a solution in hexane, but can be prepared by a literature method if necessary.[18]

Ethyllithium may be prepared as follows.[19] Lithium dispersion (30% in mineral oil, 1% sodium, 18 g, 0.77 mol) is placed in a 500-ml three-necked round-bottomed flask containing a magnetic stirrer bar and kept under an argon atmosphere (see Section 2.17.8, p. 129). The lithium is washed with dry hexane (4 × 20 ml) and then 20 ml of dry hexane is added to the lithium and the temperature lowered to −5 °C. A dry-ice condenser is attached and ethyl chloride (12 ml, 167 mmol), which had been previously condensed in a separate two-necked pear-shaped flask (see Section 2.17.8, p. 126), is diluted with 20 ml of hexane and added via a cannula over a 3-hour period to the lithium maintained at −5 °C. After the addition is complete, the mixture is allowed to come to room temperature very slowly (1). After stirring for 3 hours at room temperature the slurry takes on a purplish tint. Filtration through a sintered glass funnel is achieved via a cannula (maintaining an inert atmosphere – see Section 2.17.8, p. 131). The flask is rinsed with an additional 10 ml of dry ether. The resulting solution is colourless. The solvent is then removed by using a vacuum pump and a dry-ice trap. Ethyllithium is present as a clean white solid. The solvent is then replaced with 20 ml of dry ether. Titration (using the method described below) indicates the solution to be 3.80 M (79.8 mmol). The ethyllithium is then stored in the freezer and used within a few weeks.

Note. (1) Even after the addition is complete and the mixture is at ambient temperature, the reaction may suddenly become exothermic. Hence, an ice bath should be kept ready at all times.

Propyllithium may be prepared by the following method.[19] Lithium dispersion (30% in mineral oil, 1% sodium, 14 g, 0.6 mol) is placed in a 250-ml two-necked round-bottomed flask containing a magnetic stirring bar, under an argon atmosphere (see Section 2.17.8, p. 129). The lithium is washed with dry hexane (5 × 20 ml) followed by addition of 15 ml of dry hexane. The slurry is cooled to −10 °C and 1-chloropropane, distilled previously from calcium hydride, is placed in a pressure-equalising dropping funnel, dissolved in 30 ml of dry hexane, and added slowly over 1.5 hours while maintaining the temperature in the reaction vessel at −10 °C. After addition is complete, the reaction is warmed to room temperature and stirring is continued for 5 hours. The solution becomes purplish-brown in colour. Filtration via a cannula through a sintered glass funnel (see Section 2.17.8, p. 131) affords a clear solution which is titrated (see below) and stored in the refrigerator.

Estimation of solutions of alkyllithium and Grignard reagents. These compounds may deteriorate with time and it is important for their use in synthesis to know the actual concentration of the reagent.

Method 1. Use of 2,2′-bipyridyl as indicator for the estimation of alkyllithiums.[20] Alkyllithium reagents and Grignard reagents form coloured charge transfer complexes with certain aromatic nitrogen heterocycles such as 2,2′-bipyridyl, 1,10-phenanthroline and 2,2′-biquinolyl. When the organometallic reagent is decomposed by compounds having an acidic hydrogen, in this case butan-2-ol, the colour disappears. The complexes thus form suitable indicators for the titration of the organometallic reagents. The titration is performed under an atmosphere of dry, oxygen-free nitrogen in a 50-ml, three-necked flask fitted with a nitrogen inlet and outlet and containing a glass-covered magnetic stirrer follower bar. The third neck is fitted with a bung (or rubber septum) through the centre of which passes the delivery tip of the burette; the tip should be well clear of the bottom of the bung.

Dry the flask in the oven and allow to cool while passing a stream of nitrogen (fit a stopper in the third neck during this operation). Place about 2 mg of 2,2′-bipyridyl (1) and 10 ml of anhydrous benzene (**CAUTION**) in the flask and flush the flask with nitrogen for about 20 minutes. While maintaining a rapid flow of nitrogen, remove the nitrogen exit briefly and add a 5.00 ml aliquot portion (2) of the organometallic reagent to the solution. Titrate the resulting solution with a 1 M standard solution of butan-2-ol in *p*-xylene (3) until the colour of the charge transfer complex is discharged (4).

1 mol butan-2-ol ≡ 1 mol of organometallic reagent.

Notes. (1) 1,10-Phenanthroline or 2,2′-biquinolyl may be used as alternative indicators. The colour of the complex is dependent upon the complexing agent. Thus butyllithium gives a greenish-yellow colour with 2,2′-biquinolyl, reddish-brown with 1,10-phenanthroline and reddish-purple with 2,2′-bipyridyl.
(2) Measure the aliquot portion of the organometallic reagent with a pipette which has been previously warmed in the oven and allowed to cool with a stream of nitrogen flowing through it. The use of a suction bulb is essential.
(3) Depending on the anticipated concentration of the organometallic reagent, a more or a less concentrated solution may be required. Both the *p*-xylene and butan-2-ol should be freshly dried and distilled (Sections 4.1.4, p. 399 and 4.1.*12*, p. 402 respectively).
(4) The solution remains clear throughout the titration since the lithium but-2-oxide which is formed in the reaction is soluble in the solvent mixture.

Method 2.   Use of 4-biphenylmethanol as indicator for the estimation of alkyllithiums.[21] This reagent gives a sharp colour change (colourless to orange–red) at the titration end-point. It is recrystallised from hexane/chloroform (4:1) and dried *in vacuo* before use. A typical procedure is as follows. In a dry 25 ml conical flask is placed 200–300 mg of 4-biphenylmethanol and a dry bar magnet. The flask is capped with a rubber septum and flushed with nitrogen before the introduction of 10 ml of dry tetrahydrofuran. From a graduated syringe is added dropwise the alkyllithium solution, until the appearance of the orange–red colour. The amount consumed contains 1 equivalent of alkyllithium relative to 4-biphenylmethanol.

Method 3.   Use of N-phenyl-1-naphthylamine for the estimation of organolithium and organomagnesium reagents.[22] This reagent gives better-defined end-points than other reagents when alkylmagnesium halides are titrated. It can be kept in a nitrogen-flushed bottle for at least one year, but a solution in xylene gradually turns yellow–brown. A typical procedure is as follows. To a flame-dried 50 ml round-bottomed flask under nitrogen containing a magnetic stirring bar is added 2 ml of an ethereal solution of methylmagnesium bromide. This flask is then connected to a burette containing a xylene solution of N-phenyl-1-naphthylamine (0.005 M) and butan-2-ol (0.25 M) by means of a hypodermic needle connected to a Luer fitting on the burette. While the flask is vented to a mineral oil bubbler, the titrant is added dropwise to the stirred solution. A yellow–orange colour which initially forms continues to deepen until enough butan-2-ol had been added to react with all the amide and Grignard reagent. At this point the colour disappears and a cloudy white suspension forms.

It is sometimes desirable to detect the presence of an alkyllithium in a reaction mixture. For this purpose a few milligrams of 2,2'-bipyridyl or biphenylmethanol can be added to the reaction mixture as an indicator. A reddish or reddish-purple coloration indicates the presence of an alkyllithium.

48. MANGANESE DIOXIDE

The principal organic use for manganese dioxide is for the selective oxidation of allylic and benzylic primary and secondary alcohols to the corresponding carbonyl compounds. The activity of the reagent varies with its method of preparation; material of good activity may be obtained by oxidising manganese(II) ions with an excess of permanganate under alkaline conditions*:

$$3Mn^{2\oplus} + 2MnO_4^{\ominus} + 4\overset{\ominus}{O}H \longrightarrow 5MnO_2 + 2H_2O$$

Add simultaneously during 1 hour (a) a solution of 223 g (1 mol) of manganese(II) sulphate tetrahydrate in 300 ml of water, and (b) 240 ml (2.5 mol) of 40 per cent aqueous sodium hydroxide solution to a hot stirred solution of 190 g (1.2 mol) of potassium permanganate in 1200 ml of water. Continue stirring for a further 1 hour, isolate the fine brown precipitate of manganese dioxide (preferably by centrifugation) and wash it thoroughly with water until the washings are colourless. Dry the product at 100–120 °C and grind finely. Alternatively remove as much of the wash-water as possible by prolonged suction and activate the damp cake (in 25 g portions) by removing most of the remaining water by azeotropic distillation with 150 ml of benzene[24] (**CAUTION**).

Some commercial samples of precipitated manganese dioxide may be active enough for use directly in an oxidation process. *To assess the activity of a sample of manganese dioxide*, dissolve 0.25 g of pure cinnamyl alcohol in 50 ml of dry light petroleum (b.p. 40–60 °C) and shake the solution at room temperature for 2 hours with 2 g of the sample of manganese dioxide (previously dried over phosphoric oxide). Filter, remove the solvent by evaporation and treat the residue with an excess of 2,4-dinitrophenylhydrazine sulphate in methanol† (Section 9.6.13, p. 1257). Collect the cinnamaldehyde 2,4-dinitrophenyl-hydrazone and crystallise it from ethyl acetate. An active dioxide should give a yield of the derivative, m.p. 255 °C (decomp.), in excess of 0.35 g (60%).

49. METAL HYDRIDES

A large number of metal hydrides are now available to the organic chemist. They vary in the range of their application, ease of handling and cost. They are often the reagent of choice for the reduction of a wide range of organic functional groups.[25] Deuterium analogues of several of the complex hydrides discussed below are also available commercially. They allow the introduction of a deuterium atom into organic compounds at known positions, giving compounds which are of value in the investigation of metabolic pathways and reaction mechanisms.

Lithium aluminium hydride is a powerful reducing agent which will reduce a wide range of functional groups rapidly. A typical example, the reduction of an

---

* The preparation of an active manganese dioxide/charcoal catalyst has been described.[23]
† The determination of the extent of oxidation by measurement of i.r. absorption at 3 μm has been described.[26]

ester to an alcohol, is described in Expt 5.38. This reactivity makes it a very useful reagent for the organic chemist, but limits its use for selective reductions in a molecule containing more than one functional group. The reagent is usually supplied in powdered form contained in individual plastic bags sealed within a metal can. It is also available as a solution in ether, diglyme, tetrahydrofuran, or toluene, in the last case as the THF complex. Lithium aluminium hydride reacts violently with water, liberating hydrogen, and any contact with even traces of moisture must be avoided since the heat of reaction which ensues may cause the material to ignite.* Great caution should therefore be employed in handling the reagent, taking particular care not to inhale the powder during weighing or transfer operations, etc. Residual powdered reagent may be safely destroyed by suspending it in light petroleum (b.p. 60–80 °C) contained in a vessel sited behind a safety screen, carefully adding dropwise ethyl acetate until apparent reaction ceases, and standing the mixture overnight. The procedure should be repeated using first ethanol and then water. Finally, the aqueous layer may be separated and rinsed down the drain and the organic layer placed in the 'recovery' bottle. *All operations involving the reagent should be conducted behind a safety screen in a fume cupboard; plastic gloves should be worn.*

Lithium aluminium hydride reductions are usually carried out in a suitable ethereal solvent, such as rigorously dried diethyl ether (Section 4.1.*15*, p. 404) or tetrahydrofuran (Section 4.1.*19*, p. 406); the solubility of the reagent is 25–30 g and 13 g per 1000 g of solvent respectively. These solutions invariably contain a voluminous insoluble residue arising from impurities which may be formed by the action of moisture on the hydride during handling operations; this however constitutes not more than 1 per cent of the added reagent, and does not normally interfere with the subsequent reduction processes. In those cases which require the use of accurately known quantities of hydride it is necessary to prepare a standardised ethereal solution of the reagent. An ethereal solution (about 0.5 M) may be prepared by heating under reflux for 1 hour in an atmosphere of nitrogen about 20 g of lithium aluminium hydride with 1 litre of rigorously dried ether in a 2-litre, round-bottomed flask fitted with a double surface reflux condenser protected with a calcium chloride tube. The solution is cooled and filtered through a sintered glass Buchner funnel previously rinsed with dried ether and the residue is washed with a small portion of dried ether. The filtrate is transferred to the reservoir of a suitable automatic burette previously flushed with nitrogen from which aliquot portions may be dispensed for standardisation, and for use in subsequent reductions. Care should be exercised in the treatment of the residue in the Buchner funnel; it is safely destroyed by adding it to ethyl acetate contained in a beaker; the funnel is then rinsed with further quantities of ethyl acetate. The other vessels (Buchner flask, round-bottomed flask, etc.) are best rinsed with ethanol.

To standardise the ethereal solution of lithium aluminium hydride, an aliquot portion is added to an excess of a standardised solution of iodine in benzene (**CAUTION**), when the following reaction ensues:

$$2I_2 + LiAlH_4 \longrightarrow LiAlI_4 + 2H_2$$

Unreacted iodine is then titrated with standard sodium thiosulphate solution.

---

* Solid lithium aluminium hydride fires should only be extinguished using dry sand or powdered limestone. Water or carbon dioxide or chemically filled extinguishers must never be used.

Add 5 ml of an approximately 0.5 M ethereal solution of lithium aluminium hydride to 50 ml of a standardised approximately 0.2 M iodine solution in benzene. Shake the solution gently and allow to stand for 5 minutes, add 50 ml of distilled water followed by about 2 ml of glacial acetic acid. Titrate the excess of iodine with a standardised 0.2 M solution of sodium thiosulphate using starch solution in the final stages to determine the end-point.

1 ml M $Na_2S_2O_3 \equiv 0.12692$ g iodine

0.03795 g $LiAlH_4 \equiv 0.50768$ g iodine

The experimental conditions for reductions using lithium aluminium hydride are similar to those for the Grignard reaction. For compounds which are readily soluble in ether, a solution of the compound in dry ether is added to an ethereal solution of lithium aluminium hydride (slight excess) at such a rate that the reaction mixture boils gently. For compounds which are slightly or sparingly soluble in ether, a Soxhlet apparatus (Fig. 2.96) is inserted between the reaction flask and the reflux condenser and the compound is placed in the Soxhlet thimble. When the reduction is complete, the excess of reagent is decomposed by the cautious addition of moist ether, or an ethanol–ether mixture or by the dropwise addition of cold water with vigorous stirring; when water is used, it is desirable to employ a large flask because of the foaming which takes place. On the whole it is best to employ ethyl acetate, as its reduction product (ethanol) does not interfere in the subsequent isolation and no hydrogen is evolved. The reaction mixture is then poured gradually into excess of ice-cold dilute sulphuric acid to decompose the complex aluminium compounds and to dissolve the precipitated aluminium hydroxide; the product is usually in the ethereal layer but, if it is water-soluble, it must be isolated from the aqueous solution. For bases, after extraction of any neutral or acidic products, the solution is rendered alkaline with 10 M sodium hydroxide and the whole (including the precipitated aluminium hydroxide) is extracted with ether. As an alternative decomposition procedure in this latter case, and also for use with products which are acid-sensitive, the excess of lithium aluminium hydride is best decomposed by the careful dropwise addition of just sufficient water to produce a granular precipitate of lithium aluminate ($LiAlO_2$) which is readily filterable. An excess of water produces a gelatinous precipitate of aluminium hydroxide and should be avoided.

Sodium borohydride is a milder reducing agent than lithium aluminium hydride. Among the carbonyl compounds, it normally only reduces aldehydes and ketones readily; see Section 5.4.1, p. 519, for further discussion. Examples of reductions by sodium borohydride which illustrate various aspects of selectivity are described in Expts 5.31 (chloral and *m*-nitrobenzaldehyde), 5.117 (carbohydrates), 6.50 (nitro to amino group) and 6.58 (reduction of a Schiff base). The use of sodium borohydride as a hydrogen source for small-scale hydrogenations is described in Section 2.17.1, p. 87. Sodium borohydride is available as the solid reagent or as a solution in diglyme, triglyme, or aqueous sodium hydroxide. It is also available on alumina or silica gel as solid supports. Unlike lithium aluminium hydride, sodium borohydride is insoluble in ether (but soluble in dioxane) and is normally used as a reducing agent in aqueous or aqueous alcoholic solution. The stability of the aqueous solution is increased by the addition of alkali; indeed, a stabilised 12 per cent solution of sodium borohydride in 43 per cent aqueous sodium hydroxide is available commercially as *Sodium Borohydride—SWS* (Aldrich).

The hydride is rather readily decomposed by methanol or ethanol; isopropyl alcohol and t-butyl alcohol are the preferred alcoholic solvents.

An alkaline borohydride solution may be assayed by addition of excess standard potassium iodate solution leading to its decomposition in accordance with the equation:

$$3\overset{\ominus}{B}H_4 + 4IO_3^{\ominus} \longrightarrow 4I^{\ominus} + 3H_2BO_3^{\ominus} + 3H_2O$$

Unreacted potassium iodate is then determined by addition of potassium iodide followed by acidification when the iodine liberated is estimated by titration with standard sodium thiosulphate solution.

Lithium borohydride is intermediate in activity as a reducing agent between lithium aluminium hydride and sodium borohydride. In addition to the reduction of aldehydes and ketones it will readily reduce esters to alcohols. It can be prepared *in situ* by the addition of an equivalent quantity of lithium chloride to a 1 M solution of sodium borohydride in diglyme. Lithium borohydride should be handled with as much caution as lithium aluminum hydride. It may react rapidly and violently with water; contact with skin and clothing should be avoided.

Lithium triethylborohydride (Super-Hydride) is a much more powerful reducing agent than lithium aluminium hydride. It is useful for the reductive dehalogenation of alkyl halides, but unlike lithium aluminium hydride does not affect aryl halides. It is available as solution in tetrahydrofuran in sealed containers under nitrogen. The solutions are flammable and moisture sensitive and should be handled with the same precautions as are taken with other organometallic reagents (see Section 4.2.*47*, p. 442).

Sodium cyanoborohydride is a very mild reducing agent. The electron withdrawing nature of the nitrile group in the cyanoborohydride anion reduces the tendency to donate hydride ion and therefore sodium cyanoborohydride is a milder reducing agent than sodium borohydride. It also has the particular advantage that it is stable in mildly acid solutions (down to pH 3). It is soluble in tetrahydrofuran, methanol, water and in some dipolar aprotic solvents such as dimethylformamide, but insoluble in hydrocarbons. It is particularly useful for reductive amination and for the reduction of tosylhydrazones used in the conversion of carbonyl to methylene groups. The commercially available material is suitable for most purposes. If necessary, it can be prepared by the following method.[27] To a rapidly stirred slurry of sodium borohydride (80.2 g, 2.09 mol) in tetrahydrofuran (1000 ml) in a 2-litre flask is added a solution of hydrogen cyanide in tetrahydrofuran (294 g containing 58.8 g of hydrogen cyanide) at 25 °C (**CAUTION**: See Section 4.2.*39*, p. 438 for precautions in handling hydrogen cyanide). Evolution of hydrogen occurs slowly during the addition. Following the addition, the reaction mixture is stirred for 1 hour at 25 °C and then heated at reflux until hydrogen evolution has ceased. Filtration followed by vacuum removal of tetrahydrofuran gives white, solid sodium cyanoborohydride; yield 120 g (91%). The product prepared by this method does not require purification for most uses. If further purification of this or the commercial product is required it can be accomplished by dissolving the sodium cyanoborohydride in tetrahydrofuran (20% w/v), filtering and reprecipitating by the addition of a four-fold volume of dichloromethane. Sodium cyanoborohydride is a white amorphous powder, m.p. 240–242 °C (decomp.). Contact with air should be kept to a minimum because the compound is very hygroscopic.

Bis(triphenylphosphine)copper(I) tetrahydroborate $[(Ph_3P)_2CuBH_4]$ has found use as a reagent for the reduction of aliphatic and aromatic acid chlorides to the corresponding aldehyde, and is an alternative to the standard Rosenmund procedure. This is illustrated in Expt 6.120. The reagent may be prepared by either of the following two methods; both preparations should be conducted in an efficient fume cupboard as hydrogen is evolved.

(a)[28] Copper(I) chloride (1 g, 0.01 mol) is dissolved in 500 ml of chloroform (**CAUTION**) containing 12 g of triphenylphosphine. A solution of sodium borohydride (0.76 g, 0.02 mol) in 100 ml of 95 per cent ethanol is added dropwise. After stirring for an hour, the solution is filtered and concentrated using a rotatory evaporator to give a white crystalline precipitate of $(Ph_3P)_2CuBH_4$, which is collected by filtration and washed with ethanol and water. The air-dried product weighs 6 g (100%), m.p. 175 °C (decomp.).

(b)[29] Hydrated copper(II) sulphate (16 g, 64 mmol) is stirred in a solution of triphenylphosphine (54 g, 200 mmol) in ethanol (1400 ml) to give after 1 hour, a pale-green solution. Finely powdered sodium borohydride (12 g, 320 mmol) is added with stirring; hydrogen is evolved and a white precipitate formed. After 1 hour, the crude product is collected by filtration and dissolved in chloroform (400 ml) (**CAUTION**). The reaction mixture is filtered and ethanol (800 ml) added to the filtrate. The precipitate is collected by filtration and washed with ethanol and ether to give bis(triphenylphosphine)copper(I) tetrahydroborate, 32 g, 53 mmol (83% based on copper sulphate) as white needles, m.p. 174–6 °C.

Sodium hydride is a white, crystalline, free-flowing powder, which must be kept in air-tight containers for protection against moisture and oxygen. If exposed to the air unduly, traces of sodium hydroxide formed on the surface render the material hygroscopic; rapid absorption of moisture may then take place, and the heat generated by the reaction with water may suffice to ignite the solid. Great care must be taken therefore in handling sodium hydride and all operations involving the manipulation of the dry solid material should be conducted in a manipulator glove box in an atmosphere of dry nitrogen. Sodium hydride residues are best destroyed by careful treatment with ethyl acetate [Section 2.3.2, p. 41].

For most purposes it is much more convenient to use the commercially available 55–60 per cent dispersion of the hydride in white mineral oil, which being less susceptible to atmospheric moisture is much less hazardous. For reactions which are performed in those organic solvents in which the oil is soluble (e.g. light petroleum, benzene, toluene, etc.) the presence of the mineral oil does not impair the reactivity of the hydride and the dispersion may be used directly. If removal of the oil prior to reaction is necessary, as for example in those cases where its separation from the reaction product may prove difficult, this can be accomplished by washing the dispersion by decantation with a rigorously dried solvent (e.g. light petroleum), the operation being conducted in a manipulator glove box, or in the apparatus described in Section 2.17.8, p. 131.

Potassium hydride is more basic and more reactive than sodium hydride. It is available commercially as a 35 per cent dispersion in mineral oil. The precautions described above for the handling of sodium hydride must also be adopted with potassium hydride, but because of its greater reactivity greater care needs to be taken. In particular care needs to be taken to ensure that any potassium hydride suspended in organic solvent is not allowed to come into contact with water as a fire will result. It is recommended that the organic layers from the washing of potassium hydride, for example in the removal of the mineral oil, are treated with one of the lower alcohols such as butanol to destroy any small particles of potassium hydride. Similar precautions need to be taken with any glassware, spatulas or other equipment which might have come into contact with potassium hydride. The mineral oil is soluble in a range of organic solvents (e.g. ether, hexane, toluene, dibutyl ether) and these can be used as reaction solvents without affecting the activity of the reagent. In the absence of the protective oil potassium hydride should be protected from air and moisture.

Diisobutylaluminium hydride is a useful agent for various homogeneous reductions such as alkynes to alkenes, esters or nitriles to aldehydes. It is available commercially as the neat reagent and as a solution in several different solvents (hexane, dichloromethane, tetrahydrofuran, etc.). The neat reagent is a liquid and is supplied in metal cylinders. It is pyrophoric and highly reactive and therefore must be handled with great care. The molar solutions in organic solvents are more stable and easier to handle provided that the normal precautions are taken for working with air-sensitive compounds (Section 2.17.8, p. 120). They must all be handled in an atmosphere free from oxygen and water. Commonly used solvents for diisobutylaluminium hydride are dichloromethane, diglyme, dioxane and tetrahydrofuran. Even when the substrate is not completely soluble in these solvents successful reductions can be accomplished.

Tri-n-butyltin hydride is a useful reducing agent for the conversion of alkyl halides to hydrocarbons. It is moisture sensitive and has an irritant vapour. It is a colourless liquid which can be kept for some time providing that moisture is excluded. It is available commercially in sealed containers under nitrogen but can be prepared by the following method if required.[30] A solution of lithium aluminium hydride (4.75 g, 0.125 mol) in 400 ml of anhydrous ether is placed in a three-necked round-bottomed flask equipped with a mechanical stirrer, a condenser and a pressure-equalising dropping funnel containing 100 g (0.308 mol) of tributyltin chloride. The system is flushed with nitrogen and then tributyltin chloride is added dropwise at a rate which maintains a gentle reflux of the ether. The resulting mixture is then stirred for 7 hours. *Careful* addition of water to the cooled mixture destroys the excess lithium aluminium hydride. The ether solution is decanted, washed with three 100-ml portions of water and dried over calcium sulphate. Evaporation of the solvent and distillation under vacuum give 68 g (78%) of tributyltin hydride (b.p. 68–69 °C/0.3 mmHg, n_D^{25} 1.4688) which is kept under argon.

50. NICKEL

The introduction by Raney (1927) of a new form of catalyst (the *Raney nickel catalyst*) with enhanced activity for hydrogenation at low pressures and temperatures in comparison with the usual form of nickel catalyst as employed by Sabatier and Senderens opened up a new field of controlled catalytic hydrogenation. A special alloy, prepared essentially by the fusion of approximately equal parts of aluminium and nickel at 1200–1500 °C, is treated with alkali which dissolves the aluminium and leaves the nickel as a finely-divided black suspension. The catalyst is thoroughly washed to free it from alkali, is stored under absolute ethanol in an air-free container and is measured in the form of the suspension; it must be handled under a solvent at all times as it is highly pyrophoric.

The advantages of this catalyst are that it is cheaper and less delicate than platinium, fairly large quantities of organic substrate may be hydrogenated and the process is reasonably rapid. The following method gives a catalyst of moderate activity (*W2 Raney nickel*).

Place a solution of 190 g of sodium hydroxide in 750 ml of water in a 2-litre beaker equipped with an efficient stirrer (1), cool in an ice bath to 10 °C and add 150 g of nickel–aluminium alloy in small portions, with stirring, at such a rate that the temperature does not rise above 25 °C. If excessive foaming is encountered, add 1 ml of octan-1-ol. When all the alloy has been introduced (about 2 hours), stop the stirrer, remove the beaker from the ice bath and allow the contents to attain room temperature. When the evolution of hydrogen becomes slow, heat the reaction mixture gradually (2) on a water bath, until the evolution again becomes slow (about 8–12 hours); add distilled water to restore the original volume, stir the mixture, allow to settle and decant the supernatant liquid. Transfer the nickel to a stoppered graduated cylinder with the aid of distilled water, and decant the water again. Add a solution of 25 g of sodium hydroxide in 250 ml of water, shake to disperse the catalyst thoroughly, allow to settle and decant the alkali solution. Wash the nickel by suspension in distilled water and decantation until the washings are neutral to litmus, then ten times more to remove the alkali completely (25–40 washings are required) (3). Repeat the washing process three times with 100 ml of rectified spirit (95% ethanol) and three times with absolute ethanol. Store the catalyst in bottles which are completely filled with absolute ethanol and tightly stoppered; the product is highly pyrophoric and must be kept under liquid at all times. The Raney nickel contained in this suspension weights about 75 g.

In the practical applications of Raney nickel it is more convenient to measure the catalyst than to weigh it. The product, prepared as above, contains about 0.6 g of the catalyst per millilitre of settled material: a level teaspoonful is about 3 g of nickel.

Notes. (1) The stirrer should be provided with a motor which will not ignite the hydrogen – an induction motor or an air stirrer is suitable. The stirrer itself may be of glass, Monel metal or stainless steel (cf. Fig. 2.48).
(2) The heating should not be too rapid initially or the solution may froth over.
(3) The number of washings may be reduced to about twenty, if time is allowed for diffusion of the alkali from the surface of the catalyst into the surrounding wash-water. Use 750 ml of water in each washing, allow diffusion to proceed for 3–10 minutes, stir again and decant the supernatant liquid as soon as the catalyst settles to the bottom.

51. NITRIC ACID

The commercial concentrated acid, *d* 1.42, is a constant boiling azeotrope with water, b.p. 120.5 °C/760 mmHg, containing about 70 per cent w/w HNO_3 (0.989 g HNO_3 per ml). Colourless concentrated acid may be obtained free of coloured impurities (due to oxides of nitrogen or to nitrous acid) by warming to about 60 °C and passing in a stream of dust-free nitrogen; the addition of a little urea considerably accelerates the process.

$$2HNO_2 + (NH_2)_2CO \longrightarrow CO_2 + 2N_2 + 3H_2O$$

The so-called fuming nitric acid, *d* 1.5, contains about 95 per cent w/w HNO_3

(1.419 g HNO_3 per ml) and is available commercially; it has a yellow colour due to the presence of oxides of nitrogen which may be removed as detailed above. The acid may be prepared by distilling a mixture of equal volumes of concentrated nitric acid and concentrated sulphuric acid in a distillation assembly which incorporates a splash-head fitment to the distillation flask to act as a trap for acid spray. The volume of distillate collected should be slightly less than one-half of concentrated nitric acid originally used.

52. NITROGEN
Cylinders of compressed nitrogen may be purchased or hired. The gas may contain traces of oxygen which may be removed, if necessary, by passage either through an alkaline solution of pyrogallol (15 g of pyrogallol dissolved in 100 ml of 50% sodium hydroxide solution) or through Fieser's solution, which consists of an alkaline solution of sodium dithionite to which sodium anthraquinone-2-sulphonate is added. *Fieser's solution* is prepared by dissolving 20 g of potassium hydroxide in 100 ml of water, and adding 2 g of sodium anthraquinone-2-sulphonate and 15 g of commercial sodium dithionite (*c*. 85%) to the warm solution and stirring until dissolved: the blood-red solution is ready for use when it has cooled to room temperature, and will absorb about 750 ml of oxygen. The exhaustion of this solution is indicated by the change in colour to dull-red or brown, or when a precipitate appears. Oxygen-free nitrogen in cylinders is available commercially, but is, of course, more expensive than the normal commercial compressed gas.

53. OSMIUM TETROXIDE (*Osmic acid*)
This is supplied as the solid in sealed ampoules or as a solution in water or 2-methylpropan-2-ol (t-butyl alcohol). It must be handled in a fume cupboard. *It is extremely irritating and toxic and constitutes a severe eye injury hazard*. The solution in t-butyl alcohol (Expt 5.47) must be prepared and dispensed in an efficient fume cupboard, with the added protection of gloves and goggles. This solution is reasonably stable (e.g. the decomposition after one month is about 20%), provided that no 2-methylprop-1-ene arising from the t-butyl alcohol is present as an impurity. Formation of black coloidal osmium, which can catalyse the decomposition of hydrogen peroxide, for example in the hydroxylation of alkenes, is rapid.

54. PALLADIUM CATALYSTS
To prepare *palladium on charcoal* (5% Pd), heat a solution of 1.7 g of palladium chloride (1) in 1.7 ml of concentrated hydrochloric acid and 20 ml of water on a water bath for 2 hours or until solution is complete, and add this to a solution of 30 g of sodium acetate trihydrate in 200 ml of water contained in a 500-ml hydrogenation flask. Add 20 g of acid-washed activated charcoal (2) and hydrogenate in an atmospheric hydrogenation apparatus (Fig. 2.63(*a*)) until absorption ceases. Collect the catalyst on a Buchner funnel and wash it with five 100 ml portions of water and suck as dry as possbile. Dry the catalyst at room temperature (3) over potassium hydroxide pellets or anhydrous calcium chloride in a vacuum desiccator. Powder the catalyst (about 20 g) and store in a tightly stoppered bottle.

For *palladium on charcoal* (30% Pd) prepare a solution of 8.25 g of palladium chloride (1) in 5 ml of concentrated hydrochloric acid and dilute with 50 ml of distilled water. Cool the solution in an ice–salt bath and add 50 ml of 40 per cent formaldehyde solution and 11 g of acid-washed activated charcoal (2). Stir the mixture mechanically and add a solution of 50 g of potassium hydroxide in 50 ml of water, keeping the temperature below 5 °C. When the addition is complete, raise the temperature to 60 °C for 15 minutes. Wash the catalyst thoroughly by decantation with water and finally with dilute acetic acid, collect on a suction filter and wash with water until free from chloride or alkali. Dry at 100 °C and store in a desiccator.

Palladium black is prepared as follows. Dissolve 5 g of palladium chloride in 30 ml of concentrated hydrochloric acid and dilute with 80 ml of water; cool in an ice–salt bath and add 35 ml of 40 per cent formaldehyde solution. Add a cold solution of 35 g of potassium hydroxide in 35 ml of water dropwise during 30 minutes to the vigorously stirred palladium solution. Warm to 60 °C for 30 minutes and then wash the palladium precipitate six times by decantation with water. Filter on a sintered crucible, wash with 1 litre of water and suck dry and transfer to a desiccator charged with silica gel. The yield is 3.1 g.

Palladium on barium sulphate (5% Pd) is prepared as follows. Dissolve 4.1 g of palladium chloride (1) in 10 ml of concentrated hydrochloric acid and dilute with 25 ml of water. Add all at once 60 ml of 3 M sulphuric acid to a rapidly stirred, hot (80 °C) solution of 63.1 g of barium hydroxide octahydrate in 600 ml of water contained in a 2-litre beaker. Add more 3 M sulphuric acid to render the suspension just acid to litmus. Introduce the palladium chloride solution and 4 ml of 40 per cent formaldehyde solution into the hot mechanically-stirred suspension of barium sulphate. Render the suspension slightly alkaline with 30 per cent sodium hydroxide solution, continue the stirring for 5 minutes longer and allow the catalyst to settle. Decant the clear supernatant liquid, replace it by water and resuspend the catalyst. Wash the catalyst by decantation 8–10 times and then collect it on a medium-porosity sintered glass funnel, wash it with five 25 ml portions of water and suck as dry as possible. Dry the funnel and contents at 80 °C, powder the catalyst (48 g) and store it in a tightly stoppered bottle.

For use in the Rosenmund reduction (Expt 6.120) the catalyst is moderated by the addition of the appropriate quantity of a quinoline–sulphur poison prepared in the following manner. Heat under reflux 1 g of sulphur with 6 g of quinoline for 5 hours and dilute the resulting brown liquid to 70 ml with xylene which has been purified by distillation over anhydrous aluminium chloride. Thiourea (about 20% by weight of the palladium–barium sulphate catalyst) may also be used as a catalyst poison.

The preparation of *Lindlar's catalyst* (palladium on calcium carbonate moderated by treating with lead acetate and quinoline)[31] has been described in detail; it is used for effecting the partial reduction of an acetylenic bond to an olefin.

Where it is advantageous to maintain the neutrality of the hydrogenation mixture, *palladium on barium carbonate* catalyst is recommended. For the preparation of this catalyst the experimental details noted above for the barium sulphate based catalyst are used, but the barium hydroxide and sulphuric acid are replaced by 46.5 g of precipitated barium carbonate and the volume of hydrochloric acid is reduced to 4.1 ml.

For *palladium hydroxide on calcium carbonate* mix hot solutions of 55 g of anhydrous calcium chloride and 53 g of anhydrous sodium carbonate each dissolved in 150 ml of distilled water. Filter the resulting precipitate of calcium carbonate, wash it well with water and suspend it in 200 ml of distilled water. Dissolve 1 g of palladium chloride in 2.4 ml of concentrated hydrochloric acid and dilute with 30 ml of distilled water. Adjust the pH of the solution (4) to 4.0–4.5 by the cautious addition with stirring of 3 M sodium hydroxide solution. No permanent precipitate should be obtained at this stage. Add this solution to the calcium carbonate suspension, warm to 80 °C with stirring until conversion to the insoluble palladium hydroxide is complete, i.e. until the supernatant liquors are colourless. Wash several times with distilled water by decantation, filter with suction and wash sparingly with distilled water until the washings are chloride free. Dry over silica gel in a vacuum desiccator and preserve in a tightly stoppered bottle.

Notes. (1) Alternatively, the equivalent quantity of palladium chloride dihydrate may be used.

(2) Any of the commercial forms of activated carbon ('Norit', 'Darco', etc.) may be employed; the carbon should be heated on a steam bath with 10 per cent nitric acid for 2–3 hours, washed free from acid with water and dried at 100–110 °C before use.

(3) Heating may cause ignition of the carbon.

(4) The pH adjustment is most conveniently followed with the aid of a pH meter.

55. PERIODIC ACID

Periodic acid has a selective oxidising action upon compounds having two hydroxyl groups or a hydroxyl and an amino group attached to adjacent carbon atoms, which is characterised by the cleavage of the carbon–carbon bond (*Malaprade reaction*):

$$R \cdot CH(OH) \cdot CH(OH) \cdot R' + HIO_4 \longrightarrow R \cdot CHO + R' \cdot CHO + HIO_3 + H_2O$$

$$R \cdot CH(OH) \cdot CH(NH_2) \cdot R' + HIO_4 \longrightarrow R \cdot CHO + R' \cdot CHO + HIO_3 + NH_3$$

No oxidation occurs unless the hydroxyl groups or a hydroxyl and an amino group are attached to adjacent carbon atoms, hence the reaction may be employed for testing for the presence of contiguous hydroxyl groups (e.g. 1,2-diols) and hydroxyl and amino groups. Carbonyl compounds in which the carbonyl group is contiguous to a hydroxyl group or a second carbonyl group are also oxidised, e.g., α-hydroxy-aldehydes or -ketones, 1,2-diketones and α-hydroxy acids:

$$R \cdot CH(OH) \cdot CO \cdot R' + HIO_4 \longrightarrow R \cdot CHO + R' \cdot CO_2H + HIO_3$$

$$R \cdot CO \cdot CO \cdot R' + HIO_4 + H_2O \longrightarrow R \cdot CO_2H + R' \cdot CO_2H + HIO_3$$

The oxidation may proceed through the hydrated form of the carbonyl group $> CH(OH)_2$. The rate of oxidation is 1,2-glycols $>$ α-hydroxy aldehydes $>$ α-hydroxy ketones $>$ α-hydroxy acids.

Periodic acid is available in two grades, one contains a minimum of 95 per cent w/w $HIO_4 \cdot 2H_2O$, the other is an aqueous solution of periodic acid containing 50 per cent w/w $HIO_4 \cdot 2H_2O$.

Sodium periodate, and less frequently potassium periodate, are employed in oxidations which are to be carried out within the pH range 3–5. *Sodium metaperiodate* ($NaIO_4$) has a solubility in water of c. 0.07 g per ml; the addition of alkali leads to the precipitation of the far less soluble *sodium paraperiodate* ($Na_2H_3IO_6$, disodium trihydrogen orthoperiodate, water solubility c. 0.20%). Sodium paraperiodate is available commercially and may be converted into the metaperiodate salt by dissolving 100 parts in a mixture of 150 parts of water and 45 parts of concentrated nitric acid, warming to effect solution, filtering through a sintered glass funnel if necessary, and allowing the sodium metaperiodate to crystallise overnight at room temperature. Potassium metaperiodate has a lower solubility in water than sodium metaperiodate.

In cases where the organic substrate is insoluble in water it may be necessary to carry out the oxidation in ethanol, methanol, dioxane or acetic acid which have been diluted with water.

It is essential that only a slight excess of the calculated amount of reagent required to effect the oxidation is used otherwise oxidation of the reaction products may become a significant undesirable side reaction. The course of the oxidation may be followed by removing aliquot portions of the reaction mixture and determining the amount of unused oxidant. Details of an iodimetric procedure are given in Expt. 5.117, Note (1).

The use of *sodium metaperiodate supported on silica gel* as an oxidant for diols that are insoluble in an aqueous medium is of value. The preparation of the reagent is reported as follows.[32] A mixture of sodium metaperiodate (2.00 g), silica gel (20.0 g) (1) and deionised water (50 ml) was stirred at 20 °C for 30 minutes. The water was then removed at 10 mmHg using a rotatory evaporator. Benzene (50 ml, **CAUTION**) was added and was evaporated similarly. The reagent was finally dried for 16 hours *in vacuo* (0.5 mmHg) in an oven at 120 °C. The reagent generally contained, by iodimetry, *c.* 0.36 mmol of sodium metaperiodate per gram.

Note. (1) The silica gel was from BDH, chromatographic grade 60–120 mesh.

56. PEROXYACIDS

CAUTION: *All reactions involving hydrogen peroxide solutions should be carried out behind a shatter-proof safety screen* (see also Section 4.2.41, p. 439).

Performic acid is prepared as required by treating 30 per cent hydrogen peroxide solution ('100-volume') with excess 88–90 per cent formic acid.

For many purposes a reagent containing varying amounts of *peracetic acid* can be prepared *in situ* if convenient by adding 30 per cent hydrogen peroxide (1 part) to glacial acetic acid (3 parts) in the presence of a catalytic amount of sulphuric acid. Alternatively an approximately 40 per cent solution of peracetic acid in acetic acid containing a little sulphuric acid is available commercially.* The peracetic acid content of the solution may be determined by the iodometric procedure described under perbenzoic acid. In use it is sometimes desirable to neutralise the sulphuric acid with a stoichiometric amount of sodium acetate; the neutralised reagent should not be allowed to stand but should be used immediately.

Pertrifluoroacetic acid may be prepared by the reaction of hydrogen peroxide with trifluoroacetic acid. The following procedure gives an anhydrous solution of the reagent. Add trifluoroacetic anhydride (25 ml, 0.18 mol) dropwise to a stirred suspension of 86 per cent hydrogen peroxide (4.1 ml, 0.15 mol) in ice-cold dichloromethane (70 ml). On completion of the addition stir at 0 °C for a further 10 minutes, dry with anhydrous sodium sulphate and use the solution

---

* From Interox Chemicals Ltd, who supply notes on safe handling of these concentrated solutions, and advise on the use of hydrogen peroxide and peroxyacids.

without delay. An oxidation reaction with pertrifluoroacetic acid is described in Expt 5.190.

$$(F_3C \cdot CO \cdot)_2O + H_2O_2 \longrightarrow F_3C \cdot CO_3H + F_3C \cdot CO_2H$$

For *perbenzoic acid* place 5.2 g (0.225 mol) of sodium in a 500 ml dry conical flask provided with a reflux condenser and add 100 ml of absolute methanol; slight cooling may be necessary to moderate the vigour of the reaction. Cool the resulting solution of sodium methoxide to $-5\,°C$ in a freezing mixture of ice and salt: remove the condenser. Add a solution of 50 g (0.206 mol) of freshly recrystallised benzoyl peroxide (Section 4.2.6, p. 417) (1) in 200 ml of chloroform (**CAUTION**), with shaking and cooling, at such a rate that the temperature does not rise above $0\,°C$. Keep the mixture in the ice–salt bath for 5 minutes with continuous shaking; it turns milky but no precipitate appears. Transfer the reaction mixture to a 1-litre separatory funnel and extract the sodium perbenzoate with 500 ml of water containing much crushed ice. It is essential that the separation be carried out as rapidly as possible and the temperature kept as near $0\,°C$ as feasible, especially before the free acid is liberated from the sodium salt. Separate the chloroform layer, and extract the aqueous layer twice with 100 ml portions of cold chloroform to remove the methyl benzoate. Liberate the perbenzoic acid from the aqueous solution by the addition of 225 ml of ice-cold 0.5 M sulphuric acid and extract it from solution with three 100 ml portions of cold chloroform. Dry the moist chloroform solution (about 308 ml) with a little anhydrous sodium sulphate, transfer to a polyethylene container (*not* glass), and keep it in an ice box or a refrigerator until required (2); it contains about 24 g (84%) of perbenzoic acid. *To determine the exact perbenzoic acid content* of the solution, proceed as follows. Dissolve 1.5 g of sodium iodide in 50 ml of water in a 250-ml reagent bottle, and add about 5 ml of glacial acetic acid and 5 ml of chloroform. Introduce a known weight or volume of the chloroform solution of perbenzoic acid and shake vigorously. Titrate the liberated iodine with standard 0.1 M sodium thiosulphate solution in the usual manner.

1 ml of 0.1 M $Na_2S_2O_3 \equiv 0.006\,9$ g of perbenzoic acid

Notes.   (1) It is essential to use freshly recrystallised benzoyl peroxide since the commercial material usually gives poor results. The material may be assayed as described in Section 4.2.6, p. 417.
(2) Perbenzoic acid is used for the conversion of olefinic compounds into epoxides.

The number of olefinic linkages in a given compound can be established with accuracy by quantitative titration with perbenzoic acid. A solution of the substance and excess perbenzoic acid in chloroform is allowed to stand for several hours at a low temperature and the amount of unreacted perbenzoic acid in solution is determined: a blank experiment is run simultaneously.

To obtain *crystalline perbenzoic acid* the following procedure may be adopted, *the operation being conducted behind a shatter-proof screen.* Dry the moist chloroform solution with a little sodium sulphate or magnesium sulphate for an hour, filter and wash the desiccant with a little dry chloroform. Remove the chloroform under reduced pressure at the ordinary temperature while carbon dioxide is introduced through a capillary tube. Dry the white or pale-yellow residue for several hours at 30–35 °C under 10 mmHg pressure. The yield of crystalline perbenzoic acid, m.p. about 42 °C, which is contaminated with a little benzoic acid, is 22 g.

Perbenzoic acid may be recrystallised by dissolving it in a mixture of 3 parts of light petroleum (b.p. 40–60 °C; freed from alkenes, Section 4.1.*1*, p. 397) and 1 part ether using about 4–5 ml per gram, seeding and cooling to −20 °C. (*Use a shatter-proof screen.*) Long white needles, m.p. 41–42 °C, are obtained. It is moderately stable when kept in the dark at low temperatures (−20 °C); it is very soluble in chloroform, ethyl acetate and ether, but only slightly soluble in cold water and in cold light petroleum.

Solutions of *m-chloroperbenzoic acid* (MCPBA) are stable at moderate temperatures for prolonged periods. It is therefore a versatile oxidising agent for organic synthesis. *m*-Chloroperbenzoic acid of 99 per cent assay can be obtained by washing the commercially available technical product (80–85%) with a phosphate buffer of pH 7.5 and drying the residue under reduced pressure. The per-acid can be assayed iodometrically before use.

A convenient general method for the conversion of aliphatic and aromatic carboxylic acids into the corresponding peroxyacids involves reaction with 70 per cent hydrogen peroxide in the presence of methanesulphonic acid.[33]

57. PHOSGENE (*Carbonyl chloride*)

Phosgene may be purchased in cylinders or in the form of a solution (*c.* 12.5% by weight) in toluene in glass ampoules. **CAUTION**: *Owing to the very poisonous character of the gas (b.p. 8 °C), all operations with it must be conducted in a fume cupboard provided with a powerful draught, and all excess of phosgene must be absorbed in 20 per cent sodium hydroxide solution.** The preparation of the gas is rarely undertaken in the laboratory, but small quantities may be prepared by the following procedure. The apparatus (assembled in a fume chamber) consists of a flask with a short reflux condenser carrying in the top cone a socket/cone adapter with 'T' connection fitted with a dropping funnel. The 'T' connection is attached to a train of Drechsel bottles as shown in Fig. 4.5; A and C act as safety traps, the phosgene is absorbed in the toluene contained in B, and the hydrogen chloride and traces of phosgene are absorbed in the 20 per cent aqueous sodium hydroxide solution contained in D. Concentrated sulphuric acid, to which 2 per cent by weight of ignited kieselguhr has been added, is placed in the flask; carbon tetrachloride (**CAUTION**) is introduced into the dropping funnel. The sulphuric acid is heated to 120–130 °C in an oil bath and the carbon tetrachloride is allowed to drop in slowly; the resulting phosgene is absorbed in the toluene (B), while the hydrogen chloride is retained in D.

$$3CCl_4 + 2H_2SO_4 \longrightarrow 3COCl_2 + 4HCl + S_2O_5Cl_2$$

---

* It is usually advisable to suspend in the fume cupboard in which operations involving phosgene are being carried out, several filter-papers dipped in an ethanolic solution containing 5 per cent of *p*-dimethylaminobenzaldehyde and 5 per cent of colourless diphenylamine. A dangerous quantity of phosgene in the atmosphere is indicated by a colour change from yellow to deep orange. An additional precaution which is recommended is that the worker should wear a suitable gas mask; this is essential when the apparatus is disconnected for cleaning.

Fig. 4.5

58. PHOSPHORIC ACID

Commercial syrupy orthophosphoric acid has an approximate composition of 88–90 per cent w/w H_3PO_4 (d 1.75; 1.57 g H_3PO_4 per ml; 65% P_2O_5 equivalent). An approximately 100 per cent w/w H_3PO_4 (anhydrous orthophosphoric acid) is also marketed (72% P_2O_5 equivalent) but may be prepared from the 90 per cent H_3PO_4 by mixing with cooling four parts by weight of 90 per cent H_3PO_4 with one part of phosphoric oxide.

Polyphosphoric acid (tetraphosphoric acid), having an approximate composition $2P_2O_5 \cdot 3H_2O$, has a phosphoric oxide equivalent of 82–84 per cent. As supplied commercially it is a very viscous liquid and difficult to handle at laboratory temperature. Warming on a steam bath produces a mobile liquid which can readily be poured. Alternatively a satisfactory reagent can be prepared in the laboratory by dissolving 1.8 parts by weight of phosphoric oxide in 1 part by weight of 88–90 per cent H_3PO_4; this reagent has a phosphoric oxide equivalent of 87 per cent.

59. PHOSPHORUS (RED)

Commercial red phosphorus is usually contaminated with small quantities of acidic products. It should be boiled for 15 minutes with distilled water, allowed to settle, decanted through a Buchner funnel and then washed two or three times with boiling water by decantation. Finally the phosphorus is completely transferred to the Buchner funnel and washed with hot water until the washings are neutral. It is dried at 100 °C, and kept in a desiccator or in a tightly stoppered bottle.

60. PHOSPHORUS TRIBROMIDE

Into a 500-ml three-necked flask, provided with a sealed mechanical stirrer (1), a dropping funnel and a reflux condenser, are placed 28 g of purified red phosphorus (59. above), and 200 ml of carbon tetrachloride (dried over anhydrous calcium chloride) (**CAUTION**). Dry bromine (198 g; 63.5 ml) is placed in the dropping funnel and added to the vigorously stirred contents of the flask at the rate of about 3 drops per second. A little hydrogen bromide is evolved and the preparation should be carried out in a fume cupboard. After all the bromine has been added, the mixture is refluxed for 15 minutes by immersing the flask in a water bath at 80–90 °C. The clear solution is then decanted through a fluted filter paper, and the carbon tetrachloride is distilled off through a short column (e.g. the all-glass Dufton column, Section 2.26); the residue, upon distillation through the well-lagged column, boils at 166–169 °C (mainly at 168 °C). The yield is 190 g (78%).

Note. (1) A precision-ground glass stirrer unit is most satisfactory.

61. PLATINUM DIOXIDE (*Adams' catalyst*)

Platinum dioxide for use in hydrogenations is available commercially. It may alternatively be prepared by either of the following methods.

Method 1 (*from ammonium chloroplatinate*). Place 3.0 g of ammonium chloroplatinate and 30 g of sodium nitrate (AnalaR) (1) in a Pyrex beaker or porcelain dish and heat gently at first until the rapid evolution of gas slackens and then more strongly until a temperature of about 300 °C is reached. This operation occupies about 15 minutes, and there is no spattering. Maintain the fluid mass at 500–530 °C for 30 minutes, and allow the mixture to cool. Treat the solid mass with 50 ml of water. The brown precipitate of platinum oxide ($PtO_2 \cdot H_2O$) settles to the bottom. Wash it once or twice by decantation, filter through a hardened filter paper and wash on the filter until practically free from nitrates. Stop the washing process immediately the precipitate tends to become colloidal (2): traces of sodium nitrate do not affect the efficiency of the catalyst. Dry the oxide in a desiccator, and weigh out portions of the dried material as required.

Method 2 (*from chloroplatinic acid*). Dissolve 3.5 g of the purest commercial chloroplatinic acid in 10 ml of water contained in a 250-ml Pyrex beaker or porcelain basin, and add 35 g of sodium nitrate (AnalaR) (1). Evaporate the mixture to dryness by heating gently over a Bunsen flame while stirring with a glass rod. Then raise the temperature to 350–370 °C within about 10 minutes: fusion will occur accompanied by the evolution of brown oxides of nitrogen and the gradual separation of a precipitate of brown platinum oxide. If foaming occurs, stir the mixture more vigorously and direct an additional flame at the top of the reaction mixture, if necessary. If the burner beneath the beaker is removed when frothing commences, the top of the fused mass solidifies and material may be carried over the sides of the vessel. After 15 minutes, when the temperature has reached about 400 °C, the evolution of gas decreases considerably. Continue the heating until at the end of 20 minutes the temperature is 500–550 °C; at this stage the evolution of oxides of nitrogen has practically ceased and there is a gentle evolution of gas. Maintain the temperature at this point (best with the full force of a Bunsen burner) for about 30 minutes, by which time fusion is complete. Allow the mass to cool (the Pyrex beaker may crack), add 50 ml of water and proceed as in *Method 1*.

Notes. (1) The use of an equivalent quantity of potassium nitrate (AnalaR) is said to produce a more active catalyst.
(2) It is advisable to test a small portion of the filtrate for platinum by acidifying with hydrochloric acid and adding a few drops of tin(II) chloride solution: a yellow or brown colour develops according to the quantity of platinum present. The yellow colour is soluble in ether, thus rendering the test more sensitive. If platinum is found, treat the filtrate with excess of formaldehyde and sodium hydroxide solution and heat; platinum black separates on standing and may be filtered and worked up with other platinum residues.

Platinum residues from hydrogenation reactions should be carefully preserved and subsequently recovered by conversion into ammonium chloroplatinate by the following method. Dissolve the platinum or platinum residues in *aqua regia*, evaporate just to dryness several times with concentrated hydrochloric acid, dissolve the final residue in a little water and filter. Precipitate ammonium chloroplatinate from the filtrate by addition of excess of a saturated solution of ammonium chloride. Filter and dry the precipitate at 100 °C.

62. POTASSIUM
CAUTION: *Great care must be taken in the handling of potassium* and the following precautions must be rigorously observed. Cut the metal under light petroleum (which has been dried over sodium wire) contained in a mortar: do not use a beaker or a crystallising dish because it is too fragile. Cut off the outer oxide-coated surface and immediately transfer the scraps with tweezers to a second mortar containing dry light petroleum. Weigh the freshly cut potassium by removing it with tweezers to a filter paper, blot it rapidly and introduce it into a tared beaker containing dry light petroleum. Introduce the weighed potassium into the reaction mixture. The scraps of potassium should not be stored; they must be decomposed immediately by transferring the mortar to the rear of an empty fume cupboard and adding t-butyl alcohol (not methanol or ethanol) in small portions from a dropper pipette at such a rate that the reaction does not become vigorous. Keep a square of heat-resistant sheet, large enough to cover the mortar, at hand; if the liquid should catch fire, it may be extinguished easily by covering the mortar with the heat-resistant sheet. Add sufficient t-butyl alcohol to react completely with all the potassium. Any specks of potassium remaining in the first mortar used for the cutting operation or small scraps that adhere to the knife must be disposed of in the fume cupboard by cautious treatment with t-butyl alcohol as described above.

63. POTASSIUM FLUORIDE
Finely powdered, anhydrous potassium fluoride may be prepared as follows (see Expt 5.61). Grind finely pure laboratory grade, anhydrous potassium fluoride, and heat it in an electrically-heated oven at 180–210 °C; store in a desiccator. Before use, dry the powdered salt at 180 °C for 3 hours and grind again in a warm (c. 50 °C) glass mortar.

64. SELENIUM DIOXIDE
The preparation of any quantity of selenium dioxide from selenium is hardly worth while, and it is better to purify the commercially available dioxide. *Extreme care should be taken in this operation, which should be conducted in an efficient fume cupboard, because of the very poisonous properties of selenium*

compounds. The crude dioxide is sublimed in one of the assemblies described in Section 2.21, appropriate to the quantity of material needed. The sublimate of pure selenium dioxide is obtained in colourless, long, needle-like crystals which should be stored in a tightly stoppered bottle.

65. SILVER NITRITE

Warm concentrated solutions of silver nitrate (containing 48 g of AgNO$_3$) and potassium nitrite (containing 30 g of KNO$_2$) are mixed, and the mixture is allowed to cool. The silver nitrite which separates is filtered off and washed with water. It may be recrystallised from water at 70 °C, and is dried either in a vacuum desiccator or in an air oven at about 40 °C; the yield is about 90 per cent. Silver nitrite should be stored in a tightly-stoppered amber bottle.

66. SILYLATING AGENTS

A range of reagents is available to introduce silyl groups into organic molecules for use in synthesis, or in gas liquid chromatography to achieve appropriate volatility.

Chlorotrimethylsilane is a valuable reagent for the protection of the hydroxyl function in organic synthesis (Sections 5.4.6, p. 550 and 5.11.8, p. 690). It is a corrosive, flammable liquid, b.p. 57 °C. It should be distilled from calcium hydride (or tributylamine) under nitrogen before use, and stored and weighed under nitrogen. Use of the reagent without purification has been reported to lead to explosions. Chlorotrimethylsilane may contain dichlorodimethylsilane as an impurity. This may be removed before distillation by very cautious treatment with a small amount of water, which hydrolyses the dichloro compound more rapidly. Excess chlorotrimethylsilane in a reaction mixture may be destroyed by the very careful addition of aqueous sodium hydrogen carbonate solution. Use of the reagent in synthesis is described in Expts 5.56 and 7.10. Chlorotrimethylsilane has been used, in conjunction with other silylating agents, for the preparation of volatile derivatives of a range of biologically active compounds for analysis by gas chromatography, but it has been largely superseded for this purpose by other silylating agents such as *bis-trimethylsilylacetamide* and *bis-trimethylsilyltrifluoroacetamide* (Section 2.31).

Alternative silyl protecting groups can be introduced by the appropriate choice of reagents. *t-Butyldimethylsilyl chloride* is a useful reagent for the introduction of the t-butyldimethylsilyl protecting group. It is a flammable, corrosive solid, m.p. 86–89 °C, b.p. 125 °C. The use of the reagent for the preparation of *t-butyldimethylsilyl esters* is illustrated by the following typical example.[34] To a solution of 13.0 g (100 mmol) of heptanoic acid and 15.82 g (105 mmol) of t-butyldimethylsilyl chloride in 20 ml of dry DMF was added 13.96 g (205 mmol) of imidazole. The solution was stirred overnight, poured into water, and extracted with petroleum ether. The organic solution was washed with a saturated solution of sodium hydrogen carbonate and dried over magnesium sulphate. The solvent was removed and the residue was distilled (95–100 °C/1.5 mmHg) giving 21.0 g (86%) of *t-butyldimethylsilyl heptanoate*; p.m.r. spectrum (CDCl$_3$/TMS) δ 2.30 (t, 2H, CH$_2$CO), 1.30 (m, 8H, —(CH$_2$)$_4$—), 0.97 (s, 12H, SiC(CH$_3$)$_3$, terminal CH$_3$), 0.27 (s, 6H, Si(CH$_3$)$_2$; i.r. spectrum (neat) 1730 cm^{-1}.

67. SODAMIDE

Sodamide is available in a granular form having a purity in the region of 80–90 per cent, but it is rather difficult to pulverise which in any case must be done by grinding in a glass mortar under an inert hydrocarbon solvent (toluene, xylene, etc.). Powdered sodamide of high activity, free from sodium and oxygenated components, is also commercially available (e.g. from May and Baker) in 25-g bottles or 500-g tins. Sodamide should never be stored in a stoppered bottle from which samples are to be removed intermittently, since dangerous mixtures may result when the substance is exposed for 2–3 days to even limited amounts of air at ordinary temperature. As a safe practice, sodamide should be used immediately the container is opened and should not be kept longer than 12–24 hours unless under an inert solvent. In all cases where the sodamide has been seen to become yellowish or brown in colour, due to the formation of oxidation products, *the resulting mixture should not be used*, as it may be highly explosive; it should be destroyed by covering with toluene and slowly adding, with stirring, ethanol diluted with toluene. Small amounts of unused sodamide should be destroyed by the addition of methanol or ethanol; see also p. 37.

In many reactions involving the use of sodamide it is frequently more convenient to prepare the reagent *in situ* by the iron(III) ion catalysed reaction of sodium (see 68 below) with liquid ammonia. In a 500-ml three-necked flask, equipped with a removable glass stopper, sealed stirrer unit of the Hershberg type and a reflux condenser with soda-lime guard-tube, place 300 ml of anhydrous liquid ammonia (see Section 2.17.7, p. 116). Add just sufficient sodium (0.5 g) to the stirred liquid to produce a permanent blue colour, then 0.5 g of powdered iron(III) nitrate (to catalyse the conversion of sodium into sodamide), followed by 13.3 g of clean sodium metal (cut into small pieces) over a period of 30 minutes. When the sodium has been converted into sodamide (as indicated by the change from a blue solution to a grey suspension) allow the ammonia to evaporate (Section 2.17.7, p. 116) adding sufficient anhydrous ether through a dropping funnel to keep the volume of the liquid at about 300 ml. After nearly all the ammonia has evaporated, stir the suspension of sodamide and heat under reflux for 5 minutes, and then cool to room temperature. A suspension of 23.4 g of sodamide in dry ether is thus obtained; the conversion is practically quantitative.

68. SODIUM

CAUTION: *Sodium must be handled with great care and under no circumstances should the metal be allowed to come into contact with water as a dangerous explosion may result.* Sodium is stored under solvent naphtha (petroleum distillates, b.p. 152–204 °C) or xylene; it should *not* be handled with the fingers but with tongs or tweezers. Waste or scrap pieces of sodium should be placed in a bottle provided for the purpose and containing solvent naphtha or xylene; they should never be thrown into the sink or into the waste box. If it is desired to destroy scrap sodium, it should be added in small portions to a rather large quantity of industrial spirit. The preparation of *sodium wire* is described under the technique of drying of ether, Section 4.1.*15*, p. 404. *Granulated sodium* (also termed *molecular sodium*, *powdered sodium* or *sodium sand*) may be conveniently prepared by either of the following methods which require clean sodium. Commercial sodium is invariably covered with a non-metallic crust which is usually shaved off with a knife under a dry inert solvent (e.g. ether, xylene) before

462

use. As this procedure is rather wasteful the alternative method recommended is to be preferred.[35] Here, lumps of sodium metal are immersed in dry xylene contained in a wide-mouthed conical flask and heated carefully on an electric hot-plate with gentle swirling until the sodium just melts and flows away from the contaminating surface oxide. The flask is then removed from the hot-plate and upon cooling the sodium melt solidifies in globules which may then be removed with a pointed spatula to be immediately reimmersed under fresh inert solvent. The residual crust, after decantation of the xylene, is covered with industrial spirit to ensure the safe destruction of the remaining traces of sodium metal, see also p. 41.

Method 1. Twenty-three grams of clean sodium* are introduced into a 750- or 1000-ml, round-bottomed Pyrex flask containing about 200 ml of sodium-dried xylene, or at least sufficient xylene to cover the sodium completely, and the flask is placed on a sand bath. Two or three thicknesses of dry cloth (or a thick towel), sufficient to envelop the whole flask, are placed in a convenient position on the bench. The sand bath is heated cautiously and the ring of condensed vapour of the xylene is carefully watched. When the ring of condensed vapour has risen to within 2.5 cm from the neck of the flask the flame beneath the burner is extinguished. A well-fitting rubber stopper or good quality bark cork is rapidly fitted into the flask neck and the flask is completely wrapped in the previously prepared cloth. The stopper is held firmly in place through the enveloping cloth and the flask is shaken vigorously for 30–60 seconds or until the molten sodium is converted into a fine dispersion. The flask is then placed on a cork ring and the stopper immediately removed. The sodium is thus obtained in the form of small spheres the size of which is controlled by the time and rapidity of the shaking. Any particles of sodium adhering to the sides of the flask are washed under the xylene. When the contents of the flask have cooled to room temperature, the xylene may be decanted and the sodium washed twice with 100 ml of sodium-dried ether to remove traces of xylene; finally the finely-divided sodium is covered with absolute ether. A bucket, half full of dry sand, should be kept at hand in case of breakage of the flask. Not more than 25–30 g of sodium should be powdered at one time by this procedure.

Method 2. It is often convenient to prepare the powdered sodium in the flask in which the subsequent reaction is to be carried out; this is usually a three-necked flask. Into a 1-litre three-necked flask fitted with a reflux condenser (protected by a drying tube containing soda lime), a sealed stirrer and a dropping funnel are placed 23 g of clean sodium and 150–200 ml of sodium-dried xylene. The flask is surrounded by a mantle and heated until the sodium has melted. The stirrer is started and, after the sodium is suitably granulated, the mantle is removed. When the contents of the flask have cooled to the laboratory temperature, the stirrer is stopped. The xylene may then be decanted, and the sodium washed with two 100 ml portions of sodium-dried ether to remove traces of xylene as in *Method 1*. Large quantities of molecular sodium may be prepared by this method.

---

* Clean sodium should be weighed under a dry inert solvent; sodium-dried ether or light petroleum (b.p. 60–80 °C) is usually used.

69. SODIUM ACETATE

The anhydrous salt is prepared from the crystallised sodium acetate, $Me\cdot CO_2Na\cdot 3H_2O$, by heating in a large porcelain evaporating basin over a small free flame. The salt first liquefies, steam is evolved and the mass solidifies as soon as most of the water of crystallisation has been driven off. To remove the residual water, the solid is carefully heated with a larger flame, the burner being constantly moved until the solid just melts. Care must be taken that the solid is not overheated; too strong heating will be recognised by the evolution of combustible gases and charring of the substance. The fused salt is allowed to solidify and is removed from the vessel while still warm with a knife or spatula. It is immediately powdered and stored in a tightly stoppered bottle.

Commercial fused sodium acetate is usually satisfactory, but if necessary it can be melted and maintained in the fused state for several minutes in order to remove the water absorbed during storage.

70. SODIUM AMALGAM

The amalgam which is generally employed for reductions contains from 1 to 3 per cent of sodium. Amalgams with a sodium content greater than 1.2 per cent are solid at ordinary temperature and can be powdered in a mortar; the 1.2 per cent amalgam is semi-solid at room temperature, but is completely fluid at 50 °C. Clean sodium (22.8 g) is placed in a 500-ml round-bottomed flange flask, provided with a dropping funnel (containing 750 g of mercury) in the central socket, and inlet and outlet tubes for dry nitrogen in two side sockets. The air is displaced by nitrogen. About 10 ml of mercury are added and the flask is warmed gently with a free flame until reaction commences. The flame is then removed and the reaction is maintained by the slow addition of the mercury. When about half the mercury has been introduced, the amalgam will commence to solidify; it should be kept molten by heating with occasional shaking. After the addition of all the mercury, the hot molten amalgam is poured on to a uralite board, powdered in a mortar* and stored in a tightly stoppered bottle. Amalgams of 1 and 2 per cent strength may be prepared similarly.

71. SODIUM CYANIDE

If necessary, the powdered commercial material (c. 98% pure) may be dried in a vacuum desiccator over potassium hydroxide pellets. *Sodium cyanide is very poisonous and must be handled with great care.* Residual solutions containing alkali cyanides should be rendered innocuous by the addition of an excess of sodium hypochlorite before being washed down the main drain of the laboratory with a liberal supply of water; they should never be treated with acid. (Expt 5.157).

72. SODIUM ETHOXIDE

This is a flammable, moisture-sensitive solid which is available commercially. It may be prepared conveniently *in situ* as required by the method described in Expt 5.103. A solution of sodium ethoxide in ethanol is required for many

---

* The mortar should be provided with a tightly fitting rubber cover to protect the powdered material from oxidation in air. Alternatively a manipulator glove box previously flushed with nitrogen could be used.

reactions and is readily prepared by the reaction of sodium with ethanol (Expt 5.95). A 21 per cent w/w solution in ethanol is also available commercially.

73. TIN(II) CHLORIDE

The anhydrous reagent is prepared from the hydrate as follows. Crystalline tin(II) chloride, $SnCl_2 \cdot 2H_2O$, is heated for 1 hour in an oil bath at 195–200 °C, the cooled melt is powdered and kept in a desiccator or a tightly stoppered bottle. The resulting product, although satisfactory in many instances, is not entirely dependable. The following procedure (Stephen, 1930) invariably gives an excellent product. In a 400-ml beaker are placed 102 g (89.5 ml; 1 mol) of redistilled acetic anhydride, and 123 g of analytical reagent grade tin(II) chloride dihydrate (0.5 mol) are added whilst the liquid is stirred either manually or mechanically; dehydration is almost instantaneous. The operation must be conducted in the fume cupboard as much heat is evolved and the acetic anhydride may boil. After about 1 hour, the anhydrous tin(II) chloride is filtered off on a Buchner or sintered glass funnel, washed free from acetic acid with two 30 ml portions of anhydrous ether and dried overnight in a vacuum desiccator. Anhydrous tin(II) chloride may be kept for an indefinite period in a desiccator; it may also be stored in a tightly stoppered bottle.

The anhydrous compound is not appreciably hygroscopic, is readily soluble in acetone and pentan-1-ol, and insoluble in benzene, toluene, xylene and chloroform; it is also readily soluble in absolute methanol or ethanol, but a trace of water causes immediate hydrolysis with the formation of an opalescent precipitate.

74. SULPHUR DIOXIDE

The liquefied gas is commercially available* in aluminium canisters (net weight 500 g) which are provided with screw-operated valves.

75. SULPHURIC ACID

Ordinary concentrated acid, d 1.84, is a constant boiling mixture, b.p. 338 °C/760 mmHg, and it contains 98 per cent w/w H_2SO_4 (1.799 g H_2SO_4 per ml). The 100 per cent acid may be obtained by addition of the calculated quantity of oleum[36]; it is also available commercially.

Oleum is marketed in a range of strengths up to $c.$ 65 per cent SO_3. From 0 to 30 per cent free SO_3, it is a liquid; from 30 to 55 per cent free SO_3 it is a solid (maximum m.p. 35 °C at 45% free SO_3); from 60 to 70 per cent free SO_3, it is a liquid. The acid must be kept in ground glass stoppered thick-walled bottles. If it is required to melt the acid, the stopper is removed, a watch glass placed on the mouth of the bottle, and the bottle is placed on a uralite board in a warm temperature-controlled oven at 40 °C. The liquid should be removed from the bottle with the aid of an automatic dispenser fitted into the neck; this procedure is more satisfactory than that of pouring the liquid acid from the bottle.

76. SULPHURYL CHLORIDE

The technical product should be fractionated in an all-glass apparatus; the fraction, b.p. 69–70 °C, is collected. The pure substance has b.p. 69 °C/760 mmHg.

---

* For example, from BDH Ltd.

77. THIONYL CHLORIDE

The technical product frequently contains traces of acids, sulphur chlorides and sulphuryl chloride; it is essential to remove these before using the reagent for the preparation of acid chlorides, etc. Commercial purified thionyl chloride is satisfactory for most purposes. A colourless product of high purity may be obtained by either of the following methods.

Method 1. Commercial thionyl chloride is first fractionated in an all-glass apparatus from quinoline in order to remove acid impurities (50 g of thionyl chloride from 10 ml of quinoline); the receiver is protected from the entrance of moisture by a guard-tube, filled with anhydrous calcium chloride. The distillate is then refractionated as before from boiled linseed oil (50 g of thionyl chloride from 20 ml of linseed oil), the fraction, b.p. 76–78 °C, being collected.

Method 2. This method of purification[37] is more economic. Here technical thionyl chloride is placed in a distilling flask and not more than 5 per cent w/w of dipentene (p-mentha-1,8-diene) added with swirling. The mixture is immediately distilled at atmospheric pressure (not reduced pressure), using a gas burner (not a heating mantle), without a fractionating column, and with a thermometer dipping into the liquid mixture. Distillation is stopped when the temperature of the liquid reaches 84–86 °C; between 80 and 90 per cent of the original quantity of thionyl chloride having b.p. 76–78 °C will be collected. Redistillation of the product, to which 1–2 per cent of linseed oil has been added, through a short fractionating column gives pure colourless thionyl chloride having b.p. 77 °C/760 mmHg. This reagent must be stored in a well-fitting glass-stoppered bottle.

78. TOLUENE-p-SULPHONYL CHLORIDE

Unless the reagent has been recently purchased it may contain substantial amounts of toluene-p-sulphonic acid. The most satisfactory procedure for the purification of the chloride involves dissolving it in the minimum amount of chloroform (about 2.5 ml per g) and diluting with 5 volumes of light petroleum (b.p. 40–60 °C), which precipitates impurities. The filtered solution is treated with decolourising charcoal, filtered and concentrated to small volume when colourless crystals of the pure reagent, m.p. 68 °C, are obtained; these should be washed with chilled light petroleum (b.p. 40–60 °C).

79. TRIPHENYLMETHYL HALIDES (Trityl halides)

Triphenylmethyl chloride is commercially available but may contain a proportion of triphenylmethanol formed by hydrolysis during storage. It may be purified by dissolving in about one-half its weight of hot benzene (**CAUTION**) containing 10–20 per cent of acetyl chloride, diluting with two volumes of light petroleum, b.p. 60–80 °C, and cooling. The product is filtered rapidly, washed with light petroleum and dried in a vacuum desiccator over paraffin wax shavings and silica gel to remove solvent. The pure compound has m.p. 112–113 °C; it should be stored in a well-stoppered bottle sealed against the ingress of moisture.

Alternatively it may be prepared from triphenylmethanol (10 g) by heating under reflux in dry benzene (5 ml) with redistilled acetyl chloride (6.0 ml) for 30 minutes. The mixture is cooled, diluted with light petroleum (10 ml, b.p.

40–60 °C), chilled on an ice bath and the crystals collected and recrystallised and stored as described above.

Triphenylmethyl bromide, m.p. 153–154 °C, may be prepared in a similar manner from triphenylmethanol and acetyl bromide.

80. ZINC

Commercial *zinc powder* is usually about 90 per cent pure and requires acid treatment to remove surface oxide which reduces its activity. The zinc may be activated by stirring, say, 400 g of powder with 150 ml of 10 per cent hydrochloric acid for 2 minutes, filtering and washing with 300 ml of water followed by 100 ml of acetone. The zinc powder should be analysed, if the amount of zinc required in a reaction procedure is critical, by one of the published methods.[38] Zinc is also available in the form of sheet, wire and wool.

One method of preparation of a *zinc–copper couple* is described in Expt 6.155. It may also be prepared by stirring activated zinc dust with 2 per cent aqueous copper sulphate solution and washing successively and thoroughly with water, absolute ethanol and dry ether, Expt 5.21.

Zinc amalgam (for Clemmensen reduction) may be prepared by either of the following two methods.

Method 1. Two hundred grams of zinc wool are placed in a 2-litre, three-necked flask and covered with a 10–15 per cent solution of sodium hydroxide. The flask is gently warmed on a water bath until hydrogen is vigorously evolved (**CAUTION**); the sodium hydroxide solution is then immediately poured off (it may be necessary to dilute with water first in order to moderate the vigour of the reaction), and the zinc is washed repeatedly with distilled water until most, if not all, of the sodium hydroxide has been removed. The zinc is then covered with a 1 per cent solution of mercury(II) chloride and allowed to stand for 30–60 minutes with occasional shaking. The mercury(II) chloride solution is then poured off, and the amalgamated zinc is washed twice with distilled water. The amalgamated zinc is then covered with 500 ml of concentrated hydrochloric acid and 100 ml of water. The compound (about 0.3–0.4 mol) to be reduced is then added, and the reaction is allowed to proceed while a current of hydrogen chloride gas is passed through the liquid.

Method 2.[39] A mixture of 200 g of zinc wool, 15 g of mercury(II) chloride, 10 ml of concentrated hydrochloric acid and 250 ml of water is stirred or shaken for 5 minutes. The aqueous solution is decanted, and the amalgamated zinc is covered with 150 ml of water and 200 ml of concentrated hydrochloric acid. The material (about 0.3–0.4 mol) to be reduced is then added immediately and the reaction is commenced.

81. ZINC CYANIDE

The preparation of zinc cyanide should be carried out in the fume cupboard and great caution exercised in the handling of the cyanides. Prepare solutions containing 100 g of technical sodium cyanide (97–98% NaCN) (**CAUTION**) in 125 ml of water and 150 g of anhydrous zinc chloride in the minimum volume of 50 per cent ethanol (1). Add the sodium cyanide solution rapdily, with agitation, to the zinc chloride solution. Filter off the precipitated zinc cyanide at the pump, drain well, wash with ethanol and then with ether. Dry the product in a desiccator or in an air bath at 50 °C, and preserve in a tightly stoppered bottle.

The yield is almost quantitative and the zinc cyanide has a purity of 95–98 per cent. It has been stated that highly purified zinc cyanide does not react in the Adams' modification of the Gattermann reaction (compare Expt 6.113). The product, prepared by the above method, is, however, highly satisfactory. Commercial zinc cyanide may also be used.

Note. (1) It is important in this preparation to ensure an excess of zinc chloride over sodium cyanide. If the latter is in excess, the zinc cyanide generally precipitates as a sticky mass, which is difficult to filter and unsatisfactory for the preparation of phenolic aldehydes.

REFERENCES

1. Methods for the purification of all solvents of value are collected in J. A. Riddick and W. B. Burger (1970). 'Organic solvents', in *Techniques of Chemistry*, 3rd edn. Ed. A. Weissberger, New York: Wiley-Interscience, Vol. II.
2. *Organic Syntheses* (1963). Wiley, Coll. Vol. 4, pp. 474 and 792.
3. More detailed accounts of a wide range of reagents used in organic synthesis can be found in the following reference texts and monographs:
   (a) L. F. Fieser and M. Fieser (1967–86). *Reagents for Organic Synthesis*. New York: Wiley, Vols 1–12.
   (b) *Organic Syntheses*. New York: Wiley. Annual volumes and periodic collective volumes.
   (c) D. D. Perrin, W. L. F. Armarego and D. R. Perrin (1980). *Purification of Laboratory Chemicals*, 2nd edn. Oxford: Pergamon Press.
   (d) G. Cainelli and G. Cardillo (1984). *Chromium Oxidations in Organic Chemistry*. Berlin: Springer-Verlag.
   (e) E.-I. Negishi (1980). *Organometallics in Organic Synthesis*. New York: Wiley.
   (f) W. P. Weber (1983). *Silicon Reagents for Organic Synthesis*. Berlin: Springer-Verlag.
   (g) P. Hodge and D. C. Sherrington (1981). *Polymer-supported Reactions in Organic Synthesis*. New York: Wiley.
   (h) M. Freifelder (1978). *Catalytic Hydrogenation in Organic Synthesis*. New York: Wiley.
   (i) M. Hudlicky (1984). *Reductions in Organic Synthesis*. Chichester: Ellis Horwood.
4. (a) E.-I. Negishi and M. J. Idacavage (1985). *Organic Reactions*, **33**, 1.
   (b) H. C. Brown, with techniques by G. W. Kramer, A. B. Levy and N. M. Midland (1975). *Organic Synthesis via Boranes*. New York: Wiley-Interscience.
5. (a) G. G. Saucy (ed.) (1985). *Organic Syntheses*, **63**, 136. New York: Wiley.
   (b) H. C. Brown, Y. M. Choi and S. Narasimhan (1982). *J. Org. Chem.*, **47**, 3153.
   (c) C. F. Lane (1975). *Aldrichimica Acta*, **8**, 20.
   (d) C. F. Lane (1974). *J. Org. Chem.*, **39**, 1437.
6. H. C. Brown and G. Zweifel (1961). *J. Am. Chem. Soc.*, **83**, 1241.
7. G. Zweifel and H. C. Brown (1963). *J. Am. Chem. Soc.*, **85**, 2066.
8. H. C. Brown, E. F. Knight and C. G. Scouten (1974). *J. Am. Chem. Soc.*, **96**, 7765.
9. H. C. Brown and S. K. Gupta (1971). *J. Am. Chem. Soc.*, **93**, 1816.
10. (a) H. C. Brown (1985) in *Asymmetric Organic Synthesis. Nobel Symposium No. 60*. London: Cambridge University Press.
    (b) H. C. Brown in R. Sheffield (ed.) (1986). *Modern Synthetic Methods*, **4**, 307.
11. H. C. Brown, M. C. Desai and P. K. Jadhav (1982). *J. Org. Chem.*, **47**, 5065.
12. W. A. Szabo (1977). *Aldrichimica Acta*, **10**, 23.
13. (a) R. Ratcliffe and R. Rodenhurst (1970). *J. Org. Chem.*, **35**, 4000.
    (b) For a review of the use of this reagent see *Synthesis*, **1982**, 245.

14. E. J. Corey and J. W. Suggs (1975). *Tetrahedron Lett.*, 2647.
15. Y.-S. Cheng, W.-L. Liu and S. Chen (1980). *Synthesis*, 223.
16. *Vogel's Textbook of Quantitative Inorganic Analysis* (1978). 4th. edn. Eds J. Bassett, R. C. Denney, G. H. Jeffrey and J. Mendham. London: Longman, p. 381.
17. M. L. Semmelhack (ed.) (1984). *Organic Syntheses*, Vol. 62, p. 101. New York: Wiley.
18. M. L. Semmelhack (ed.) (1984). *Organic Syntheses*, Vol. 62, p. 4. New York: Wiley.
19. B. H. Lipshutz, R. S. Wilhelm, J. A. Kozlowski and D. Parker (1984). *J. Org. Chem.*, **49**, 3928.
20. S. C. Watson and J. F. Eastham (1967). *J. Organomet. Chem.*, **9**, 165.
21. E. Juaristi, A. Martinez-Richa, A. Garcia-Rivera and J. S. Cruz-Sanchez (1983). *J. Org. Chem.*, **48**, 2603.
22. D. E. Bergbreiter and E. Pendergrass (1981). *J. Org. Chem.*, **46**, 219.
23. I. M. Goldman (1969). *J. Org. Chem.*, **34**, 1979.
24. L. A. Carpino (1970). *J. Org. Chem.*, **35**, 3971.
25. H. C. Brown and S. Krishnamurthy (1979). *Tetrahedron*, **35**, 567.
26. R. J. Gritter and T. J. Wallace (1959). *J. Org. Chem.*, **24**, 1051.
27. (a) C. F. Lane (1975). *Synthesis*, 135.
    (b) R. C. Wade, E. A. Sullivan, J. R. Berschied Jr. and K. F. Purcell (1970). *Inorg. Chem.*, **9**, 2146.
28. T. N. Sorrell and R. J. Spillane (1978). *Tetrahedron Lett.*, 2473.
29. G. W. J. Fleet, C. J. Fuller and P. J. C. Harding (1978). *Tetrahedron Lett.*, 1437.
30. L. Ghosez, R. Montaigne, A. Roussel, H. Vanlierde and P. Mollet (1971). *Tetrahedron*, **27**, 615.
31. R. L. Augustine (1965). *Catalytic Hydrogenation: Techniques and Applications in Organic Synthesis*. London: Arnold, p. 69.
32. D. N. Gupta, P. Hodge and J. E. Davies (1981). *J. Chem. Soc., Perkin Trans, I*, 2970.
33. L. S. Silbert, E. Siegel and D. Swern (1962). *J. Org. Chem.*, **27**, 1336.
34. A. Wissner and C. V. Grudzinskas (1978). *J. Org. Chem.*, **43**, 3972.
35. See reference 3(a), Vol. 1, p. 1022.
36. See reference 16, p. 308.
37. W. Rigby (1969). *Chem. and Ind. (London)*, **42**, 1508.
38. See reference 16, p. 488.
39. E. L. Martin (1936). *J. Am. Chem. Soc.*, **58**, 1438.

CHAPTER 5
ALIPHATIC COMPOUNDS

5.1 ALKANES

The acyclic aliphatic hydrocarbons have the general formula C_nH_{2n+2}; cyclic saturated hydrocarbons (alicyclic hydrocarbons) have the general formula C_nH_{2n} (if monocyclic), C_nH_{2n-2} (if bicyclic), etc. The structures of these hydrocarbons may be represented in the chemical literature in various ways; some common conventions which are used freely in texts are illustrated here.

Alkyl groups ($-C_nH_{2n+1}$) up to four carbon atoms are frequently abbreviated thus: $-CH_3 \equiv Me$; $-C_2H_5 \equiv Et$; $-C_3H_7 \equiv Pr$ or Pr^i; $-C_4H_9 \equiv Bu$; or Bu^i, or Bu^t. Illustrative representations of:

butane (1);
2-methylbutane (2);
2,3-dimethylpentane (3);
and 2,2,3-trimethylpentane (4);

are given below. The presence of chiral carbons atoms (marked with an asterisk) may be significant; thus (1) and (2) are achiral molecules, but (3) and (4) are chiral.

$CH_3 \cdot CH_2 \cdot CH_2 \cdot CH_3 \equiv$

(1)

$CH_3 \cdot CH_2 \cdot CH(CH_3)_2 \equiv$

(2)

$CH_3 \cdot CH_2 \cdot CH(CH_3) \cdot CH(CH_3)_2 \equiv$

(3)

$CH_3 \cdot C(CH_3)_2 \cdot CH(CH_3) \cdot CH_2 \cdot CH_3 \equiv$

(4)

470

Particularly with alicyclic compounds the Me notation shown above for a methyl group is often omitted; on the other hand the skeletal structures for the other alkyl groups may be drawn out in full. This is illustrated below for 1-methyl-4-isopropylcyclohexane (5), and 1-butyl-4-ethylcyclohexane (6).

(5)

(6)

The synthesis of alkanes is exemplified by the following typical procedures.

1. The catalytic hydrogenation and the chemical reduction of alkenes (Expts 5.1 and 5.2).
2. The hydrolysis of alkylmagnesium halides (Expt 5.3), and the hydrogenolysis of alkyl halides and of alkyl methane- or toluene-p-sulphonates (Expt 5.4).
3. The reduction of aldehydes and ketones (Expts 5.5 and 5.6).
4. Coupling reactions (a) using organometallic compounds (Expts 5.7 to 5.10), and (b) at the anode (Expt 5.11).

SUMMARY OF RETROSYNTHETIC STRATEGIES
Functional group addition (FGA) (methods 1, 2 and 3)

X = Br, I, or MgX
Me·SO$_2$·O—
p-Me·C$_6$H$_4$·SO$_2$·O—

(TM)

Disconnection (methods 4a and 4b)

X = Br, I

471

SPECTROSCOPIC FEATURES

The i.r. absorption spectra of the open-chain alkanes and of the saturated cyclic hydrocarbons show few, but characteristic, absorption bands. These are found just below $3000\,cm^{-1}$ (C—H str.), at approximately 1450 and $1375\,cm^{-1}$ (C—H deform.) and near $725\,cm^{-1}$ [$(CH_2)_n$, $n \geqslant 4$, rock); no other absorption is present [e.g. Fig. 3.13(b) decane]. The diagnostic value of i.r. absorption is limited to a broad assessment of the degree of branching in the carbon chain from inspection of high resolution spectra of the carbon-hydrogen stretching region (p. 273). The p.m.r. spectra of alkanes are of very limited value for structural assignment; ^{13}C-n.m.r. however has become of particular use in this respect (see Fig. 3.41, p. 320). The m.s. of alkanes may give some structural information on the degree of branching (p. 374); the spectra of decane and 2,6-dimethyl-octane [Fig. 3.79(a) and 3.79(b)] are illustrative. Alkanes do not absorb in the accessible u.v.-visible region.

5.1.1 THE CATALYTIC HYDROGENATION AND CHEMICAL REDUCTION OF ALKENES

Conversion of an alkene (or an alkyne) into an alkane is readily achieved by shaking it under hydrogen at room temperature and at atmospheric pressure in the presence of a platinium or palladium catalyst. With a Raney nickel catalyst somewhat higher temperatures and pressures are employed (see Section 2.17.1, p. 88).

$$R^1 \cdot CH{=}CH \cdot R^2 \xrightarrow[Pt]{H_2} R^1 \cdot CH_2 \cdot CH_2 \cdot R^2 \xleftarrow[Pt]{H_2} R^1 \cdot C{\equiv}C \cdot R^2$$

The method is illustrated by the conversion of 2-methylbut-2-ene (Expt 5.12) into 2-methylbutane (Expt 5.1). The experiment has been incorporated to illustrate the handling of low boiling point liquids, and also to illustrate some aspects of interpretative infrared spectroscopy.

The *rate* of catalytic hydrogenation of a carbon–carbon double bond depends to some extent on the degree of substitution and environmental hindrance. In general the more highly substituted (or highly hindered) alkene will be hydrogenated more slowly. Such differences in rate have been exploited in the hydrogenation of dienes to effect a conversion into a mono-ene. A good selective catalyst is P-2 nickel boride,[1] generated *in situ* from nickel(II) acetate and sodium borohydride, and its use in the large-scale conversion of dicyclo-pentadiene into 5.6-dihydro-*endo*-dicyclopentadiene[2] is described in Expt 5.2. Details are also included (Note 3 to Expt 5.2) of a smaller scale general procedure for use as a preliminary step in the study of a new diene system.[2] Selective chemical reduction of mono- and di-substituted alkenes has also been effected by a complex species produced by the reaction of cobalt(II) chloride and sodium borohydride.[3]

Experiment 5.1 2-METHYLBUTANE

$$Me \cdot CH{=}CMe_2 \xrightarrow[Pt]{H_2} Me \cdot CH_2 \cdot CHMe_2$$

Place 100 mg of Adams' platinum dioxide catalyst (Section 4.2.61, p. 459) and 9.8 g (0.14 mol) of 2-methylbut-2-ene in a 100-ml hydrogenation flask (Section 2.17.1, p. 89). Attach the flask to the adapter of the atmospheric hydrogenation apparatus (Fig. 2.63(*a*)) and cool the lower part in an ice-water bath. Fill the flask and gas burettes with hydrogen by the procedure discussed in Section 2.17.1, p. 89; note the volumes in the gas burettes, remove the cooling bath and gently agitate the flask. When uptake of hydrogen ceases (the catalyst often coagulates and collects at the bottom of the flask at this stage), note the total volume of hydrogen absorbed; this should be in the region of 3 litres. Cool the flask contents and follow the procedure for replacing the hydrogen in the apparatus with air. Disconnect the hydrogenation flask and with a suitably sized dropper pipette transfer the liquid to a small distillation flask leaving the catalyst in the hydrogenation flask (1). Distil the 2-methylbutane (b.p. 30 °C) using a small ice-cooled water condenser with the receiver flask immersed in an ice–salt cooling bath (2). The yield is 7 g (70%). Record the i.r. spectrum and compare it with the spectrum of the starting material (3). Note (*a*) the disappearance upon hydrogenation of the absorption bands at 810 and 1675 cm^{-1} (due to the out-of-plane deformation of the =C—H bond and the stretching of the carbon–carbon double bond respectively) in the alkene, and (*b*) the replacement of the peak at 1380 cm^{-1} in the alkene (—CH$_3$ bending mode) by a doublet at 1385 and 1375 cm^{-1} ((CH$_3$)$_2$C— plus terminal —CH$_3$). The m.s. shows principal fragment ions at m/z 72 (M), 57 (M—CH$_3$), 43 (M—C$_2$H$_5$, base peak) and at 29 (C$_2$H$_5^\oplus$).

Notes. (1) This technique, rather than conventional filtration, is to be preferred in this case owing to the high volatility of the hydrocarbon. For details on the disposal and recovery of the catalyst see Section 2.17.1, p. 89, *Hydrogenation at atmospheric pressure* (12).
(2) An alternative and convenient arrangement is to supply the condenser with cooled water from an ice/water reservoir by means of a peristaltic pump.
(3) In both cases a fixed path length cell (0.025 mm) should be employed.

Experiment 5.2 5,6-DIHYDRO-*ENDO*-DICYCLOPENTADIENE[2]

Dicyclopentadiene is chilled until a major proportion has solidified. The material is filtered and the solid is pressed on a Buchner funnel with a spatula to remove as much liquid as possible. The solid is warmed just to melt and the process is repeated. The resulting solid has m.p. 28–30 °C (lit.[4] m.p. 27–28 °C).

In a 2-litre flask, 37.5 g (150 mmol) of nickel(II) acetate tetrahydrate (1) are dissolved in 200 ml of 95 per cent ethanol. The flask is attached to a Brown[2] hydrogenator and flushed with hydrogen. With rapid stirring 150 ml of 1.0 M sodium borohydride solution in ethanol (2) is added to reduce the nickel acetate to P-2 nickel. The purified *endo*-dicyclopentadiene (407 g, 3.08 mol) is melted with 200 ml of ethanol and the mixture injected into the hydrogenator. With vigorous stirring, the reaction proceeds smoothly; hydrogen uptake ceases when 3.08 mol of hydrogen has been absorbed.

Five grams of activated carbon are added to the reaction mixture to aid in catalyst removal, and the warm (40 °C) mixture is filtered through a thin (*c.* 3 mm) pad of carbon on a Buchner funnel. The filter pad is washed with

2×100 ml of warm acetone. The combined organic layers are distilled to remove solvent. The residue is distilled through a short Vigreux column to give 370 g, 2.76 mol (90%) of 5,6-dihydro-*endo*-dicyclopentadiene, b.p. 178–180 °C, m.p. 48–50 °C. Recrystallisation from methanol gives m.p. 50 °C (lit.[5] 48.5–50 °C); p.m.r. (CCl$_4$, TMS), δ 1.25 (s, 3.8H), 1.45 (s, 2.2H), 2.0–3.2 (m, broad, 6.2H), 5.70 (s, broad, 2.0H). The product is free from starting material by g.l.c. analysis.

Notes. (1) The nickel(II) acetate tetrahydrate was AnalaR grade.
(2) A stabilised solution suitable for catalytic reduction was prepared by dissolving 4.0 g of sodium borohydride powder in a mixture of 95 ml of absolute ethanol and 5 ml of 2 M sodium hydroxide and filtering the resulting solution. This solution is best prepared freshly the day of use for maximum catalyst reproducibility but may be utilised satisfactorily for up to 5 days if kept refrigerated. Formation of small amounts of sediment under refrigeration is not harmful.
(3) Details of a procedure for smaller scale exploratory experiments are given as follows.[2] Nickel(II) acetate tetrahydrate (1.24 g, 5.00 mmol) is dissolved in $(50-n)$ ml (n = volume of substrate to be added) of 95 per cent ethanol in a 125 ml Erlenmeyer flask (modified for high stirring). The flask is attached to a borohydride hydrogenator, which is then flushed with nitrogen. With vigorous stirring, 5.0 ml of 1.0 M sodium borohydride solution in ethanol (see above) is injected over 15 seconds. When gas evolution from the mixture ceases, the catalyst is ready for use. The hydrogenator is purged with hydrogen and the reaction initiated by injecting the substrate. Addition of solid substrates is accomplished with the stirrer stopped, just before purging with hydrogen. Samples for analysis are withdrawn from the reactor with a syringe and stainless steel needle. Gas chromatographic columns which are found to be useful have stationary phases of either adiponitrile, triethylene glycol–silver nitrate, tris-(cyanoethoxy)propane, UCON 50 HB 2000, and squalene.

5.1.2 THE HYDROLYSIS OF ALKYLMAGNESIUM HALIDES, AND THE HYDROGENOLYSIS OF ALKYL HALIDES AND OF ALKYL METHANE- OR TOLUENE-*p*-SULPHONATES

One of the earliest methods for the conversion of alkyl halides and also, indirectly, alcohols into alkanes is the hydrolysis of the corresponding alkylmagnesium halides (Expt 5.3, the preparation of hexane from hexylmagnesium bromide). More recently complex borohydrides (in particular lithium triethylborohydride, 'Superhydride'),[6] and also a reducing system formed from sodium hydride, an alkoxide and a metallic salt [e.g. FeCl$_3$, Ni(OAc)$_2$, ZnCl$_2$ or CdCl$_2$] have been used for the hydrogenolysis of alkyl halides.[7]

$$RX \xrightarrow{\text{Mg}} R \cdot MgX \xrightarrow{\text{H}_2\text{O}} RH \xleftarrow{\text{[H]}} RX$$

Superhydride has also been used for the hydrogenolysis of alkyl toluene-*p*-sulphonates, and an illustrative example is given in Expt 5.4 for the preparation of cyclooctane from cyclooctyl toluene-*p*-sulphonate. A sodium borohydride hydrogenolysis of alkyl halides and of alkyl methane- and toluene-*p*-sulphonates under PTC conditions provides an interesting variant of this reaction.[9]

Selective removal of a halogen or tosyl group in a bifunctional molecule has been effected with lithium aluminium hydride in differing solvent systems.[10]

$$Me \cdot (CH_2)_9 \cdot CH_2OTs \xleftarrow[\text{diglyme}]{\text{LiAlH}_4} BrCH_2 \cdot (CH_2)_9 \cdot CH_2OTs \xrightarrow[\text{ether}]{\text{LiAlH}_4}$$

$$BrCH_2 \cdot (CH_2)_9 \cdot Me$$

Experiment 5.3 HEXANE

$$C_5H_{11} \cdot CH_2Br \xrightarrow{\text{Mg}} C_5H_{11} \cdot CH_2MgBr \xrightarrow{\text{H}_2\text{O}} C_5H_{11} \cdot Me$$

In a suitably equipped 500 ml three-necked flask prepare an ether solution (100 ml) of hexylmagnesium bromide from 12 g (0.5 mol) of magnesium turnings and 82.5 g (70.5 ml, 0.5 mol) of dry 1-bromohexane using the conditions and procedure described in Expt 5.39. When most of the magnesium has disappeared (about 4 hours) add AnalaR ammonium chloride (27 g) and leave the reaction mixture overnight. Cool the flask in ice and add slowly a large excess of dilute hydrochloric acid; the precipitate will dissolve completely. Separate the upper ethereal layer, and wash it successively with dilute hydrochloric acid and water; dry with magnesium sulphate or anhydrous calcium sulphate. Distil the ethereal solution through an efficient fractionating column (e.g. a Hempel column filled with 6-mm glass or porcelain rings, or a 30-cm all-glass Dufton column; see Figs 2.105(c) and 2.105(a)). After the ether has passed over, hexane will distil at 67–70 °C (13–14 g, 30–33%). Record the i.r. spectrum and compare it with Fig. 3.13(a). The m.s. shows principal fragment ions at m/z 86 (M), 57 ($M - C_2H_5$, base peak), 43 ($M - C_3H_7$) and 29 $(C_2H_5^{\oplus})$.

Experiment 5.4 CYCLOOCTANE[8]

An oven-dried 300-ml flask, equipped with a side-arm fitted with a silicone rubber septum, a magnetic stirrer bar, and a reflux condenser connected to a mercury bubbler, is cooled to room temperature under a stream of dry nitrogen. Tetrahydrofuran (20 ml) is introduced, followed by 7.1 g (25 mmol) of cyclooctyl tosylate (1). The mixture is cooled to 0 °C (ice bath). To this stirred solution, lithium triethylborohydride (Section 4.2.49, p. 448) [33.3 ml (50 mmol) of a 1.5 M solution in tetrahydrofuran] is added, and the ice bath removed. The mixture is stirred for 2 hours (c. 25 °C). Excess hydride is decomposed with water. The organoborane is oxidised with 20 ml of 3 M sodium hydroxide solution and 20 ml of 30 per cent hydrogen peroxide [(2) and (3)]. Then the tetrahydrofuran layer is separated. The aqueous layer is extracted with 2 × 20 ml portions of pentane. The combined organic extracts are washed with 4 × 15 ml portions of water to remove ethanol produced in the oxidation. The organic extract is dried (MgSO$_4$) and volatile solvents removed by distillation (2). Distillation of the residue yields 2.27 g (81%) of cyclooctane as a colourless liquid, b.p. 142–146 °C, $n_D^{20} 1.4630$.

Notes. (1) A procedure for the preparation of alkyl toluene-p-sulphonates is described in Expt 6.46.
(2) The editors emphasise the **CAUTIONARY** notes relating to the handling of hydrogen peroxide (p. 439), and to the distillation of extracts following the use of tetrahydrofuran/hydrogen peroxide in oxidation procedures (p. 552).
(3) The editors suggest that the procedure described in Expt 5.44 for the oxidation of the organoborane be followed.

5.1.3   THE REDUCTION OF ALDEHYDES AND KETONES

One procedure for the conversion of a carbonyl group to a methylene group is the *Clemmensen reduction*, and involves the use of zinc amalgam in the presence of concentrated hydrochloric acid.

$$R^1 \cdot CO \cdot R^2 \xrightarrow[HCl]{Zn/Hg} R^1 \cdot CH_2 \cdot R^2$$

This represents a somewhat unusual reduction of a carbonyl group, which might be expected to give an alcohol with the metal–acid reducing system (cf. Section 5.4.1, p. 519). Alcohols are stable to the Clemmensen conditions and are not therefore intermediates in the reduction process, which is thought to proceed by a mechanism involving the formation of organo-zinc intermediates. A typical example is provided by the conversion of heptan-2-one to heptane (Expt 5.5).

The method is often used for the reduction of aromatic carbonyl compounds (see 6.1.1, p. 827) which are reduced in good yield. Here the procedure is modified[11] by the addition of a solvent immiscible with hydrochloric acid, i.e. toluene.

When the Clemmensen method fails, or when strongly acidic conditions are precluded owing to the presence of acid-sensitive functional groups, the *Wolff–Kishner reduction* or the Huang–Minlon modification of it may succeed. The latter method is also discussed in Section 6.1.1, p. 827, and illustrated in Expt 6.4, *Method A*.

An alternative procedure is to convert the carbonyl compound into the toluene-*p*-sulphonylhydrazone,[12] followed by reduction with either sodium borohydride in acetic acid,[13] or with catecholborane, followed by decomposition of the intermediate with sodium acetate or tetrabutylammonium acetate.[14] The former method is illustrated by the conversion of undecan-6-one into undecane (Expt 5.6), and the latter by the conversion of acetophenone into ethylbenzene (Expt 6.4, *Method B*). A feature of these methods is that with α, β-unsaturated ketones, migration of the carbon–carbon multiple bond occurs; thus the tosylhydrazone of isophorone gives 3,3,5-trimethylcyclohex-1-ene, and the tosylhydrazone of oct-3-yn-2-one gives octa-2,3-diene.

Experiment 5.5   HEPTANE

$$(Me \cdot CH_2 \cdot CH_2)_2 CO \longrightarrow Me \cdot (CH_2)_5 \cdot Me$$

Place 100 g (1.53 mol) of zinc wool in a 1-litre three-necked flask and amalgamate it in accordance with *Method 1* in Section 4.2.80, p. 467. Fit the flask with a sealed stirrer unit, an efficient double surface condenser and a lead-in tube dipping almost to the bottom of the flask for the introduction of hydrogen chloride gas (compare Figs 4.4 and 2.59); insert an empty wash bottle between the hydrogen chloride generator and the flask. Introduce through the condenser 250 ml of concentrated hydrochloric acid and 50 ml of water, set the stirrer in motion and then add 40 g (0.35 mol) of heptan-4-one (Expt

5.92). Pass a slow current of hydrogen chloride through the mixture; if the reaction becomes too vigorous, the passage of hydrogen chloride is temporarily stopped. After 2–3 hours most of the amalgamated zinc will have reacted. Leave the reaction mixture overnight, but disconnect the hydrogen chloride gas supply first. Remove the stirrer and the condenser from the flask. Arrange for direct steam distillation from the flask by fitting a stopper into one neck, a knee tube connected to a downward condenser in the central aperture and connect the lead-in tube to a source of steam. Stop the steam distillation when the distillate passes over as a clear liquid. Separate the upper layer, wash it twice with distilled water, dry with magnesium sulphate or anhydrous calcium sulphate and distil through a short fractionating column. Collect the fraction, b.p. 97–99 °C (1). The yield of heptane is 26 g (74%).

Note. (1) The products of most Clemmensen reductions contain small amounts of unsaturated hydrocarbons. These can be removed by repeated shaking with 10 per cent of the volume of concentrated sulphuric acid until the acid is colourless or nearly so; each shaking should be of about 5 minutes duration. The hydrocarbon is washed with water, 10 per cent sodium carbonate solution, water (twice), dried with magnesium sulphate or anhydrous calcium sulphate and finally fractionally distilled over sodium.

Cognate preparation. *Octane.* Use 45 g (0.35 mol) of octan-2-one (Expt 5.86) and 100 g of amalgamated zinc. Collect the fraction, b.p. 124–126 °C; the yield is 23 g (58%).

Experiment 5.6 UNDECANE[13]

To a slurry of undecan-6-one toluene-*p*-sulphonylhydrazone (5.08 g, 15 mmol) (1) in 50 ml of glacial acetic acid is added sodium borohydride pellets (*c.* 5.67 g, 150 mmol, 24 pellets) (2) at such a rate that foaming is not a problem (*c.* 1 hour). The solution is stirred at room temperature for 1 hour and then at 70 °C for 1.5 hours. The solution is then poured into crushed ice, made basic with aqueous sodium hydroxide and extracted with three portions of pentane. The pentane solution is dried and concentrated in a rotary evaporator, and the residue distilled at reduced pressure (Kugelrohr apparatus) to obtain 1.96 g (84%) of undecane. Undecane has b.p. 87 °C/20 mmHg.

Notes. (1) The hydrazones are prepared by the following general procedure.[12] The carbonyl compound and a 10 per cent molar excess of toluene-*p*-sulphonylhydrazine in absolute ethanol (*c.* 2 ml per gram of carbonyl compound) are heated on a steam bath until a clear solution results (15 minutes). Cooling affords crystalline products in good to excellent yields. Recrystallisation is accomplished from ethanol or aqueous acetone. For hindered ketones periods of up to 14 hours of reflux are suggested.
(2) Obtainable from Alfa Inorganics.

5.1.4 COUPLING REACTIONS

Disconnection at the various carbon–carbon bonds in an alkane molecule will give a range of symmetrical and/or unsymmetrical synthons. For example, the simple molecule hexane, if disconnected in either a radical or ionic fashion, will give rise to: (i) a one-carbon synthon and a five-carbon synthon; (ii) a two-carbon synthon and a four-carbon synthon; and (iii) two equivalent three-carbon synthons. Branched chain alkanes offer more numerous possibilities.

Me ~~~~~ Me

(iii) (ii) (i)

The synthetic methods described below illustrate the use of various reagent equivalents to such synthons which give a good yield of the required product.

COUPLING REACTIONS USING ORGANOMETALLIC COMPOUNDS
A long established method for the preparation of alkanes involves heating an alkyl halide with sodium metal (the *Wurtz synthesis*).

$$2RX + 2Na \longrightarrow R-R + 2NaX$$

This type of symmetrical coupling of two molecules of an alkyl halide will obviously give better yields than the alternative unsymmetrical coupling process which must be used, for example, to make a straight chain alkane having an odd number of carbon atoms.

$$R^1X + R^2X + 2Na \longrightarrow R^1-R^2 + R^1-R^1 + R^2-R^2 + 2NaX$$

The symmetrical Wurtz coupling has been found to give particularly good yields in the case of the higher alkane homologues. Selected illustrative examples (hexane, octane, decane and dodecane) are to be found in Expt 5.7. A related reaction is to be found in the synthesis of hexa-1,5-diene (Expt 5.7, cognate preparation), which illustrates the symmetrical coupling of two molecules of allyl iodide by the action of sodium metal.

'Directed' coupling reactions, where no cross-coupling may occur, requires the reagents derived from the appropriate synthons to be ionically different at the reactive site. The carbon–carbon coupling reaction to form hept-1-ene (Expt 5.8), illustrates the importance of selecting the reactive, and readily available, allyl bromide (cationic site) with butylmagnesium bromide (anionic site) (A), rather than the other alternative (B).

A multistage synthesis which well illustrates the use of Grignard reagents in carbon–carbon formation is the synthesis of 7-acetoxyheptanal (7).[15] Retrosynthetic analysis reveals the necessity of aldehyde group protection as a 1.3-dioxane (8) (see p. 624) prior to the reaction process.

478

(9)   (10)

The synthesis is described in Expt 5.9 where (10) is formed from tetrahydrofuran (cf. p. 559), and the Grignard reagent (9), from 2-(2-bromoethyl)-1,3-dioxane and magnesium turnings in tetrahydrofuran.[16]

Unsymmetrical coupling reactions leading to alkanes can also be achieved by the use of copper 'ate' complexes. The organocopper reagents are prepared most conveniently by the reaction of 2 mol of an organolithium reagent with 1 mol of the appropriate copper(I) halide. Reaction of the organocuprate with an alkyl halide results in the displacement of halogen by an S_N2 mechanism.

$$2R^1Li + CuX \longrightarrow R_2^1CuLi + LiX$$

Lithium dialkyl
cuprate

$$R_2^1CuLi + R^2X \longrightarrow R^1-R^2 + R^1Cu + LiX$$

An example of this coupling reaction is provided by the preparation of undecane (Expt 5.10).

Experiment 5.7   OCTANE

$$2C_4H_9Br + 2Na \longrightarrow C_8H_{18} + 2NaBr$$

Weigh out 23 g (1 mol) of clean sodium under sodium-dried ether (Section 4.1.*15*, p. 404), cut it up rapidly into small pieces and introduce the sodium quickly into a dry 750- or 1000-ml round-bottomed flask. Fit a dry 30-cm double surface condenser (e.g. of the Davies type) into the flask and clamp the apparatus so that the flask can be heated on a wire gauze. Weigh out 68.5 g (53 ml, 0.5 mol) of butyl bromide (Expt 5.54) previously dried over anhydrous sodium sulphate. Introduce about 5 ml of the bromide through the condenser into the flask. If no reaction sets in, warm the flask gently with a small luminous flame; remove the flame immediately reaction commences (the sodium will acquire a blue colour). When the reaction subsides, shake the contents of the flask well; this will generally produce further reaction and some of the sodium may melt. Add a further 5 ml of butyl bromide, and shake the flask. When the reaction has slowed down, repeat the above process until all the alkyl bromide has been transferred to the flask (about 1.5 hours). Allow the mixture to stand for 1–2 hours. Then add down the condenser by means of a dropping funnel 50 ml of rectified spirit dropwise over 1.5 hours, followed by 50 ml of 50 per cent aqueous ethanol during 30 minutes, and 50 ml of distilled water over 15 minutes; shake the flask from time to time. Add 2–3 small pieces of porous porcelain and reflux the mixture for 3 hours; any unchanged butyl bromide will be hydrolysed. Add a large excess (500–750 ml) of water, and separate the upper layer of crude octane (17–18 g). Wash it once with an equal volume of water, and dry it with magnesium sulphate. Distil through a short fractioning side-arm (cf. Fig. 2.104) and collect the fraction, b.p. 123–126 °C (15 g, 52%) (1).

479

Note. (1) All hydrocarbons prepared by the Wurtz reaction contain small quantities of unsaturated hydrocarbons. These may be removed by shaking repeatedly with 10 per cent of the volume of concentrated sulphuric acid until the acid is no longer coloured (or is at most extremely pale yellow); each shaking should be of about 5 minutes' duration. The hydrocarbon is washed with water, 10 per cent sodium carbonate solution, water (twice), and dried with magnesium sulphate or anhydrous calcium sulphate. It is then distilled from sodium; two distillations are usually necessary.

Cognate preparations.   *Hexane.*   Use 23 g (1 mol) of sodium and 61.5 g (45.5 ml, 0.5 mol) of propyl bromide. It is advisable to employ two efficient double surface condensers in series. Collect the fraction, b.p. 68–70 °C (10 g, 47%).

Decane.   Use 23 g (1 mol) of sodium and 75.5 g (62 ml, 0.5 mol) of 1-bromopentane (Expt 5.54) or 99 g (65.5 ml, 0.5 mol) of 1-iodopentane (Expt 5.59). Collect the fraction, b.p. 171–174 °C (28 g, 79%). The i.r. and mass spectra are shown in Figs 3.13(b) and 3.79(a) respectively.

Dodecane.   Use 23 g (1 mol) of sodium and 82.5 g (70.5 ml, 0.5 mol) of 1-bromohexane (Expt 5.55). Collect the fraction, b.p. 94 °C/13 mmHg (37 g, 87%).

Hexa-1,5-diene.   Place 55 g (2.4 mol) of clean sodium cut into small pieces (Section 4.2.68, p. 462) in a 500-ml round-bottomed flask fitted with two 25- or 30-cm double surface condensers in series. Weigh out 136 g (72 ml, 0.8 mol) of freshly distilled allyl iodide, b.p. 99–101 °C (Expt 5.56). Introduce about one-quarter of the allyl iodide through the condensers. Warm the flask gently until the sodium *commences* to melt and immediately remove the flame. A vigorous reaction sets in and a liquid refluxes in the condensers. Add the remainder of the allyl iodide in small portions over a period of 2 hours. Allow the mixture to cool during 3 hours and arrange the flask for distillation (compare Fig. 2.98). Distil from an oil bath maintained at 90–100 °C when most of the hydrocarbon will pass over; finally raise the temperature of the bath to 150 °C in order to recover the product as completely as possible. Redistil from a 50-ml flask with fractionating side arm and containing a little sodium and collect the fraction of b.p. 59–50 °C; the yield of hexa-1,5-diene is 26 g (79%).

Experiment 5.8   HEPT-1-ENE

$$\text{BuMgBr} + \text{Br·CH}_2\text{·CH=CH}_2 \longrightarrow \text{Bu·CH}_2\text{·CH=CH}_2 + \text{MgBr}_2$$

In a 1-litre three-necked flask prepare the Grignard reagent, butylmagnesium bromide, from 12.2 g (0.5 mol) of dry magnesium turnings, a small crystal of iodine, 68.5 g (53 ml, 0.5 mol) of butyl bromide and 260 ml of anhydrous ether, following the experimental details given in Expt 5.39. Equip a 500-ml three-necked flask with a sealed stirrer unit, a 100-ml separatory funnel and a double surface condenser. Force the solution of the Grignard reagent with the aid of pure, dry nitrogen and a tube containing a plug of purified glass wool (1) into the 500-ml flask through the top of the double surface condenser. Charge the separatory funnel with a solution of 50 g (35 ml, 0.42 mol) of allyl bromide (Expt 5.54) in 25 ml of anhydrous ether; place calcium chloride drying tubes into the top of the double surface condenser and of the dropping funnel. Immerse the flask containing the Grignard reagent in cold water, stir vigorously, and add the allyl bromide at such a rate that the ether boils gently; cool momentarily in ice if the reaction becomes too vigorous. It is important that the allyl bromide reacts when added, as indicated by gentle boiling of the solution (2). When all the allyl bromide has been introduced, continue stirring for 45 minutes while refluxing gently by immersing the flask in a bath of warm water. Allow to cool (3). Pour the reaction mixture cautiously on to excess of crushed ice contained in a large beaker. Break up the solid magnesium complex and decompose it with ice and dilute sulphuric acid or concentrated ammonium sulphate solution. Separate the ether layer, wash it with ammoniacal ammonium sulphate solution to remove any dissolved magnesium salts and dry over magnesium sulphate. Distil the dry ethereal solution through a fractionating column: after the ether has passed over, collect the hept-1-ene at 93–95 °C. The yield is 29 g (71%).

Notes. (1) Solid magnesium must be absent to avoid the formation of biallyl via allyl magnesium bromide; the insertion of a short plug of glass wool effectively removes any finely divided magnesium or alternatively use a tube terminating in a glass frit. (2) If reaction does not occur when a little allyl bromide is first introduced, further addition must be discontinued until the reaction has commenced. Remove 2–3 ml of the Grignard solution with a dropper pipette, add about 0.5 ml of allyl bromide and warm gently to start the reaction; after this has reacted well, add the solution to the main portion of the Grignard reagent.
(3) A slight excess of Grignard reagent should be present at this stage. Test for the presence of the reagent as follows. Remove 0.5 ml of the clear liquid with a dropper pipette and add 0.5 ml of a 1 per cent solution of Michler's ketone [4,4'-bis(dimethyl-amino)benzophenone] in benzene, followed by 1 ml of water and 3–4 drops of 0.01 M iodine in glacial acetic acid; shake. A greenish-blue colour results if a Grignard reagent is present. In the absence of iodine, the colour fades. A dye of the diphenylmethane type is produced.

$$p\text{-}Me_2N\cdot C_6H_4\cdot CO\cdot C_6H_4\cdot NMe_2\text{-}p \xrightarrow{RMgX}$$
$$p\text{-}Me_2N\cdot C_6H_4\cdot CR(OMgX)\cdot C_6H_4\cdot NMe_2\text{-}p \xrightarrow{H^\oplus}$$
$$p\text{-}Me_2\overset{\oplus}{N}{=}C_6H_4{=}CR\cdot C_6H_4\cdot NMe_2\text{-}p$$
(coloured cation)

Experiment 5.9   7-ACETOXYHEPTANAL[15]

1-Acetoxy-4-iodobutane. A 500-ml flask is fitted with an addition funnel, condenser and magnetic stirrer. Tetrahydrofuran (50 ml, 0.61 mol), acetic anhydride (100 ml, 1.0 mol), zinc dust (0.1 g, 1.5 mmol) are placed in it. Aqueous hydriodic acid (55%, 40 ml, 0.29 mol) is added dropwise at a rate that keeps the exothermic reaction at reflux. After the addition is complete, the mixture is allowed to cool for 1 hour and then poured into 200 ml of saturated aqueous sodium carbonate. The mixture is extracted with dichloromethane, dried with magnesium sulphate and the solvent evaporated. The residual oil is distilled to give the product, yield 56.7 g (80%), b.p. 76–78 °C/0.5 mmHg; p.m.r. (CDCl$_3$, TMS) δ 1.88 (m, 4H), 2.02 (s, 3H), 3.20 (t, 2H, $J = 6$ Hz, —CH$_2$I) 4.05 (t, 2H, $J = 6$ Hz, —CH$_2$·OAc).

2-(6-Acetoxyhexyl)-1,3-dioxane. Copper(I) iodide (0.60 g, 3.1 mmol) and 1-acetoxy-4-iodobutane (12.1 g, 50.0 mmol) are cooled in a dry ice/propan-2-ol bath under a nitrogen atmosphere (1). The Grignard reagent (2), prepared from 2-(2-bromoethyl)-1,3-dioxane[16b] (12.19 g, 62.5 mmol) in 50 ml of tetrahydrofuran, is added dropwise to the cooled solution. This is stirred at −80 °C for 30 minutes, slowly raised to reflux temperature over a 2-hour period, and then heated at reflux for 6 hours. The tetrahydrofuran is removed by rotary evaporation and the residue poured into aqueous ammonia/ammonium chloride solution (800 ml). The product is extracted with ether (3 × 50 ml), dried with magnesium sulphate and distilled. A small amount of 1-acetoxy-4-iodobutane is recovered, and then the product is obtained in a yield of 8.86 g (77%), b.p. 90–100 °C/0.3 mmHg.

7-Acetoxyheptanal. 2-(6-Acetoxyhexyl)-1,3-dioxane (4.60 g, 20.0 mmol), methanol (150 ml), glacial acetic acid (35 ml) and concentrated hydrochloric acid (1 ml) are stirred and left to stand for 3 days. This mixture is poured cautiously into saturated sodium hydrogen carbonate solution (500 ml) and the product extracted with ether (3 × 200 ml). The extract is dried with magnesium sulphate and evaporated to give the crude dimethyl acetal, which is then dissolved in glacial acetic acid (100 ml), water (10 ml) and hydrochloric acid (1 ml) and left to stand for 3 days. The solution is poured cautiously into saturated aqueous sodium hydrogen carbonate (1 litre), made basic with additional sodium hydrogen carbonate, and the product extracted with ether (3 × 50 ml). The extract is dried with magnesium sulphate, evaporated and the residue distilled to give the product, yield 3.10 g (90%), b.p. 93–98 °C/0.7 mmHg.

Notes. (1) The editors suggest that the apparatus consists of a 100-ml two-necked flask fitted with a silicone rubber septum into which is inserted a syringe needle connected to a nitrogen feed (Section 2.17.8, p. 120) and through which the reagents may be syringe-injected; the second arm of the flask is fitted with a reflux condenser the outlet of which is connected to a mercury bubbler.
(2) The Grignard reagent is prepared by the following procedure[16a] with suitable adjustment of the quantities of reagents employed. A 50-ml flask is equipped with a reflux condenser, a nitrogen atmosphere and magnetic stirring. In it are placed magnesium turnings (0.97 g, 40 mmol), dry tetrahydrofuran (25 ml) and 2-(2-bromoethyl)-1,3-dioxane[16b] (5.85, 30 mmol) (or from Aldrich). This is heated to reflux and the heat immediately removed. The exothermic reaction is moderated at reflux by the occasional application of an ice bath. After 10 minutes, heat is applied to maintain refluxing for an additional 10 minutes. After cooling to room temperature the solution is drawn up into a 50-ml syringe leaving excess magnesium behind.

Experiment 5.10 UNDECANE

$$2BuLi + CuI \longrightarrow (Bu)_2CuLi + LiI$$

$$C_6H_{13}\cdot CH_2I \xrightarrow[-78°C]{(Bu)_2CuLi} C_6H_{13}\cdot CH_2\cdot Bu$$

Place 9.53 g (0.05 mol) of copper(I) iodide in a two-necked, 500-ml round-bottomed flask containing a glass-covered magnetic follower bar. Fit a rubber septum to one of the necks of the flask and connect the other neck to one arm of a three-way stopcock. Connect the second arm of the stopcock to a supply of dry, oxygen-free nitrogen (cf. Fig. 2.77) and the remaining arm to a vacuum pump. Evacuate the flask by opening the stopcock to the vacuum pump (1) and then, with a rapid stream of nitrogen flowing, *carefully* open the stopcock to the nitrogen supply to fill the flask with nitrogen. Repeat this process twice; flame the flask gently on the final occasion it is evacuated. Maintain a static atmosphere of nitrogen in the flask throughout the reaction by passing a slow stream of nitrogen through the nitrogen line. Cool the flask in an acetone/Cardice bath and transfer 100 ml of dry tetrahydrofuran (2) to the flask via the septum using a hypodermic syringe (3). Stir the cooled ($-78°C$) suspension and add 52.0 ml of a 1.92 molar solution of butyllithium in hexane (4) from a hypodermic syringe. Stir the flask solution at $-78°C$ for 1 hour, and then add a solution of 3.39 g (0.015 mol) of 1-iodoheptane (Expt 5.60) in 10 ml of dry tetrahydrofuran dropwise from the syringe. Stir the solution at $-78°C$ for 1 hour and then at $0°C$ (ice bath) for a further 2 hours ((5) and (6)). Hydrolyse the reaction mixture by pouring it carefully into 100 ml of 1 M hydrochloric acid. Separate the upper organic layer and extract the aqueous layer with two 50 ml portions of pentane. Wash the combined organic layers with water (50 ml), dry over magnesium sulphate and evaporate the solvents on the rotary evaporator. Distil the residue at atmospheric pressure using a semimicro scale distillation unit fitted with a short fractionating side-arm packed with glass helices (cf. Fig. 2.111). Undecane has b.p. 194–197 °C; the yield is 1.24 g (53%).

Notes. (1) This operation should be carried out behind a safety screen.

(2) Tetrahydrofuran should be freshly distilled from lithium aluminium hydride. It is convenient to store a supply of peroxide-free tetrahydrofuran (Section 4.2.*19*, p. 406) under nitrogen over lithium aluminium hydride and distil appropriate quantities as required.

(3) The syringes and long flexible needles can be obtained from Aldrich Chemical Co.

(4) Solutions of butyllithium, and other lithium reagents, may be purchased from Aldrich. See Section 4.2.*47*, p. 442 for notes on the storage, handling and analysis of alkyllithium reagents.

(5) The progress of the reaction may be conveniently followed by g.l.c. Remove samples (*c*. 2–3 ml) from the reaction mixture by means of a hypodermic syringe and submit them to a small-scale hydrolysis and extraction procedure similar to that described for the main reaction mixture. Analyse the organic layer on a 10 per cent squalane on Chromosorb W column held at 140 °C. Under these conditions 1-iodoheptane has a slightly longer retention time than undecane.

(6) Complete reaction of the alkyllithium can be tested by means of the Gilman test described in Expt 5.8, Note (3).

ANODIC COUPLING REACTIONS

The synthesis of alkanes which involves the electrolysis of salts of carboxylic acids was first reported by Kolbé in 1849. The technique and the apparatus has been described in Section 2.17.6, p. 115, and the illustrative example relevant to this section is that of the preparation of hexacosane (the experimental details were originally supplied by Dr R. P. Linstead, C.B.E., F.R.S.). (See also Section 5.11.5, p. 677).

Experiment 5.11 HEXACOSANE

$$2CH_3 \cdot (CH_2)_{12} \cdot CO_2^{\ominus} \xrightarrow{-2e} C_{26}H_{54} + 2CO_2$$

Dissolve 5.0 g of pure myristic acid (tetradecanoic acid) in 25 ml of absolute methanol to which 0.1 g of sodium has been added. Place the solution in a cylindrical cell (25 cm long, 3 cm diameter) provided with two platinum plate electrodes (2.5 × 2.5 cm) set 1–2 mm apart (cf. Section 2.17.6, p. 115). Electrolyse at about 1 amp until the electrolyte is just alkaline (pH 7.5–8). Cool the cell in an ice bath during the electrolysis. Reverse the current from time to time; this will help to dislodge the coating of insoluble by-products on the electrodes. Neutralise the cell contents by adding a few drops of glacial acetic acid, and evaporate most of the solvent under reduced pressure using a rotary evaporator. Pour the residue into the water and extract the crude product with ether. Wash the ethereal solution with dilute sodium hydroxide solution, dry (magnesium sulphate) and evaporate the solvent. Recrystallise the residue from light petroleum (b.p. 40–60 °C). The yield of hexacosane, m.p. 57–58 °C is 2.4 g (65%).

5.2 ALKENES

The location of a carbon–carbon double bond in the carbon chain of a mono-alkene may be terminal ($R \cdot CH = CH_2$) or non-terminal ($R \cdot CH = CH \cdot R$). With non-terminal alkenes the possibility of $(E)/(Z)$ geometric isomerism is present in disubstituted alkenes (e.g. $R \cdot CH = CH \cdot R$, where the two alkyl substituents may be the same or different), and may be present in trisubstituted alkenes (e.g. $R^1R^2CH = CH \cdot R^1$) and tetrasubstituted alkenes (e.g. $R^1R^2C = CR^1R^2$ or $R^1R^2C = CR^3R^4$) according to the nature of the alkyl group. The location of two or more double bonds (dienes, trienes, etc.) in a carbon chain may be remote (isolated) (e.g. $Me \cdot CH = CH \cdot CH_2 \cdot CH = CH \cdot Me$, hepta-2,5-diene), conjugated (e.g. $Me \cdot CH = CH \cdot CH = CH \cdot Et$, hepta-2,4-diene), or cumulative (e.g. $Me \cdot CH = C = C(Me) \cdot Et$, 4-methylhexa-2,3-diene). With isolated and conjugated dienes various combinations of $(E)/(Z)$ isomerism are possible as shown below:

Hepta-2,5-diene

| (E, E) | (E, Z) | (Z, Z) |

Hepta-2,4-diene

Me $\diagdown\diagup\diagdown\diagup\diagdown$ Et

(E, E)

Me

(Z, Z)

Me $\diagdown\diagup\diagdown\diagup\diagdown$

(E, Z)

Me

(Z, E)

Suitably substituted cumulative dienes (the allenes) have a chiral axis result-ing in non-superimposition of mirror images and hence optical activity, as for example in the instance below:

4-Methylhexa-2,3-diene

Me Et : Et Me

H   Me : Me   H

The construction of models emphasises all these stereoisomeric features; also note that in a homologous sequence of *cumulative* polyenes, the presence of a chiral axis, or of geometric isomerism, alternates [i.e. diene (R/S), triene (E/Z), tetraene (R/S), etc.]

The introduction of a carbon–carbon double bond into a molecule may be effected by the following typical procedures.

1. 1,2-Elimination processes (β-elimination) (Expts 5.12 to 5.15).
2. The partial hydrogenation of alkynes (Expt 5.16).
3. Wittig and related reactions (Expts 5.17 to 5.20).
4. Selected rearrangements of alkynes to allenes (Expts 5.21 and 5.22).

SUMMARY OF RETROSYNTHETIC STRATEGIES

Functional group interconversion (FGI) (methods 1 and 2)

$$R^1 \cdot C \equiv C \cdot R^2 \underset{(2)}{\Longleftarrow} R^1 \cdot CH = CH \cdot R^2 \underset{(1)}{\Longrightarrow} R^1 \cdot CHX \cdot CH_2 \cdot R^2$$

(TM)

$X = OH, Hal, \overset{\oplus}{N}R_3$
$O \cdot CS \cdot SMe,$
$O \cdot CO \cdot Me, N(O)R_2$

[Continued overleaf]

Disconnection (method 3)

$R^3 = R^4 = H$

(TM)

SPECTROSCOPIC FEATURES

In addition to alkyl group absorption, the characteristic *i.r.* absorption bands of monoalkenes are in the region just above $3000 \, cm^{-1}$ and in the region of $1000–800 \, cm^{-1}$, due to the stretching and out-of-plane bending vibrations respectively of the sp^2-carbon–hydrogen bond. Frequently the substitution pattern and geometrical isomerism may be ascertained from inspection of this latter region (p. 277). An absorption band arising from the carbon–carbon double bond stretching vibration occurs in the region of $1680–1620 \, cm^{-1}$. The spectrum of oct-1-ene (Fig. 3.14) illustrates these features. The protons attached to the unsaturated carbons are deshielded and resonate in the *p.m.r.* spectra at low fields ($\delta \, 5.00–6.00$); illustrative spectra are provided for styrene (Fig. 3.62) and crotonaldehyde (Figs 3.72 and 3.73). If these protons have sufficiently different chemical shift values for their coupling constants to be measured, geometric configuration, if appropriate, may be assigned. The ^{13}C-*n.m.r.* spectrum of hex-1-ene is shown in Fig. 3.51. The interpretation of the fragmentation pattern in the *m.s.* of alkenes is discussed on p. 374. Conjugated and cumulative alkenes absorb in the accessible region of the *u.v.–visible* spectrum, and the number of double bonds present may often be deduced from the positions of the absorption maxima (p. 388).

5.2.1   1,2-ELIMINATION PROCESSES (β-ELIMINATION)

Many convenient methods for the introduction of carbon–carbon double bonds into a saturated carbon chain involve the removal of two atoms or groups from adjacent carbon atoms. Usually, but not invariably, one of these groups is hydrogen (i.e. the removal of HX). Two main types of elimination reactions are recognised – heterolytic processes in solution and pyrolytic reactions in the gas phase. A detailed discussion of the mechanisms of these reactions may be found in all standard and advanced textbooks; in each of the reactions discussed below the probable mechanism is noted in relation to the aim of obtaining good yields of regio- or stereoisomerically pure compounds.

THE DEHYDRATION OF ALCOHOLS

This dehydration proceeds under acidic conditions and is a widely used olefin-forming reaction. In the laboratory phosphoric acid is the reagent of choice; sulphuric acid, which is often used, can lead to extensive charring and oxidation and hence to lower yields of alkene. Other reagents include potassium hydrogen sulphate and anhydrous copper(II) sulphate.[17] Passage over heated alumina is also effective.

Tertiary alcohols undergo elimination the more readily, and most certainly by an E1 process.

$$R_2C(OH)\cdot CR_2 \xrightarrow{+H^\oplus} \xrightarrow{-H_2O} \xrightarrow{-H^\oplus} R_2C{=}CR_2$$

In the preparative example of the dehydration of 2-methylbutan-2-ol [$Me_2C(OH)\cdot CH_2\cdot Me$], the loss of a proton from the intermediate carbocation (2) may take place from either of the two adjacent positions giving a mixture of regioisomers, 2-methylbut-2-ene (3) and 2-methylbut-1-ene (1) (Expt 5.12).

$$CH_2{=}C\cdot CH_2\cdot Me \xleftarrow{(i)} CH_2{-}\overset{\oplus}{C}{-}CH\cdot Me \xrightarrow{(ii)} Me\cdot C{=}CH\cdot Me$$
$$\quad\quad\ |\quad\quad\quad\quad\quad\quad\quad |\quad\quad\quad\quad\quad\quad\quad\ |$$
$$\quad\quad Me\quad\quad\quad\quad\quad\quad\ Me\quad\quad\quad\quad\quad\quad Me$$
$$\quad\quad (1)\quad\quad\quad\quad\quad\quad\ (2)\quad\quad\quad\quad\quad\quad (3)$$

The most highly substituted (and hence the thermodynamically more stable) alkene is formed preferentially (*Saytzeff rule*); in this case the ratio of products, (3):(1), is 4:1 (g.l.c. analysis). Application of the reaction sequence to the aliphatic tertiary alcohols prepared in Expts 5.41 and 5.42 will reveal which alcohol gives a mixture of regioisomers and which will give a single alkene. In predicting these product compositions, due attention should be paid to the possibilities of stereoisomers [(E)/(Z)] being formed. Those cases where a mixture of regio- and stereoisomers of similar stability is obtained are clearly of less preparative value.

The dehydration of ditertiary alcohols in the presence of hydrobromic acid may lead to dienes (e.g. pinacol to 2,3-dimethylbuta-1,3-diene, cognate preparation in Expt 5.12), although in this case some concomitant rearrangement to t-butyl methyl ketone (pinacolone, Expt 5.98) occurs under the acidic conditions employed.

$$Me_2C(OH)\cdot C(OH)Me_2 \xrightarrow{HBr} CH_2{=}C(Me)\cdot C(Me){=}CH_2 + Bu^t\cdot CO\cdot Me$$

The dehydration of primary or secondary alcohols, when proceeding via a secondary or primary carbocation respectively (i.e. the E1 pathway), may be preparatively less satisfactory owing to intramolecular rearrangements involving a hydrogen or an alkyl group. The driving force for such rearrangements is of course the relative thermodynamic stability of the carbocations (t > s > p, cf. p. 829). This would give rise to a greater complexity of possible products. With primary alcohols and sulphuric acid or phosphoric acid, the corresponding intermediate esters are formed, from which ethers result due to a competing substitution pathway (Section 5.6.1, p. 581).

One useful example is the dehydration of the secondary alcohol 2-methyl-cyclohexanol, which yields a mixture of two regioisomers in which the more highly substituted alkene predominates; in this case the mixture may be separated by careful fractional distillation (cognate preparation in Expt 5.12).

THE DEHYDROHALOGENATION OF ALKYL HALIDES
The removal of a molecule of a hydrogen halide from an alkyl halide to yield an alkene is effected under strongly basic conditions, e.g. a concentrated alcoholic solution of sodium or potassium hydroxide or alkoxide. This overall reaction has been submitted to most rigorous mechanistic studies. Most of the factors (temperature, nature of base, structure of substrate, solvent, etc.) which control product composition have been evaluated. It thus appears that under the conditions noted above, an E2 process, in which the participating sites adopt an *anti*-periplanar conformation leading to an *anti*-elimination process, is generally favoured.

The complexity of the mixture of regio- and stereoisomers of alkenes which may arise in these dehydrohalogenations may be illustrated by a consideration of the possible elimination routes in the case of 3-bromohexane.

It is clear therefore that this preparative procedure may only be of value for the formation of a restricted range of alkene target molecules. Recent procedures include the use of the non-nucleophilic base, 1,8-diazabicyclo[5.4.0]undec-7-ene (DBU),[18] and the use of PTC methods.[19] One useful example is provided by the ready conversion of 3-bromocyclohexene (Expt 5.68) into cyclohexa-1,3-diene (Expt 5.13) where the base employed is quinoline.

The dehydrohalogenation of 1,2- (and 1,1-) dihalides leads to the introduction of a triple bond into a molecule (Section 5.3.1, p. 509).

THERMAL DECOMPOSITION OF QUATERNARY AMMONIUM HYDROXIDES

An alternative bimolecular elimination process involves the thermal decomposition in an atmosphere of nitrogen of a quaternary ammonium hydroxide (*Hofmann exhaustive methylation procedure*).

$$R{-}NH_2 \xrightarrow{3MeI} R{-}\overset{\oplus}{N}Me_3 \; \overset{\ominus}{I} \xrightarrow{AgOH} R{-}\overset{\oplus}{N}Me_3 \; \overset{\ominus}{OH}$$

$$\underset{HO}{\overset{\ominus}{\;}} \; R{-}\overset{\oplus}{N}Me_3 \;\; \longrightarrow \;\; \underset{H}{\overset{R}{{>}}}{=}CH_2 + NMe_3 + H_2O$$

The reaction has been extensively used for the determination of the structure of naturally occurring bases (e.g. the alkaloids). However it has rather limited preparative value, even though the elimination reaction occurs without any rearrangement of the carbon skeleton, and the regioisomer which predominates in the product is the less highly substituted alkene (*Hofmann rule*; contrast the Saytzeff rule). Such alkenes are now more usually prepared by other procedures noted below.

PYROLYSIS OF ACETATE AND XANTHATE ESTERS

Two examples are provided of these pyrolytic eliminations, both of which proceed by a similar mechanistic pathway which is classified as a *syn*-elimination process.

The advantage of these methods is that although somewhat high temperatures are employed, the product is not usually contaminated by compounds arising from double bond migration or from skeletal rearrangement. For example the thermal decomposition of *O*-[1,1,2-trimethylpropyl] *S*-methyldithiocarbonate (Expt 5.209) gives a satisfactory yield of 3,3-dimethylbut-1-ene (Expt 5.15). This product is homogeneous by g.l.c. analysis and the i.r. spectrum clearly confirms the unrearranged skeletal structure. In contrast, the acid-catalysed dehydration of the corresponding alcohol, proceeding by way of an E1 mechanism and a carbocation intermediate, would inevitably result in rearrangement and formation of 2,3-dimethylbut-2-ene as the main product.

$$Me_3C{\cdot}CH(OH){\cdot}Me \xrightarrow[-H_2O]{+H^{\oplus}} Me_3C{\cdot}\overset{\oplus}{C}H{\cdot}Me \xrightarrow{rearr.} Me_2\overset{\oplus}{C}{\cdot}CHMe_2 \xrightarrow{-H^{\oplus}}$$
$$Me_2C{=}CMe_2$$

Acetate pyrolysis is illustrated by the synthesis of penta-1,4-diene (Expt 5.14) from 1,5-diacetoxypentane. This is prepared from 5-chloropentyl acetate which is conveniently obtained from tetrahydropyran by treatment with acetyl chloride; this ring-opening reaction offers interesting comparison with the methods for the preparation of α, ω-dihalides discussed in Section 5.5.1, p. 555.

The pyrolysis of *N*-oxides to yield alkenes proceeds at lower temperatures than the above reactions.[20]

$$\text{Me}_2\text{N} \underset{\text{H}_2\text{O}_2}{\overset{\text{H}}{\longrightarrow}} \text{Me}_2\overset{\oplus}{\text{N}} \overset{\text{O}^\ominus}{\underset{}{\text{H}}} \underset{-\text{Me}_2\text{NOH}}{\longrightarrow}$$

The tertiary amine, $R \cdot CH_2 \cdot NMe_2$, may be prepared by the reaction of a Grignard or lithium reagent (RMgX or RLi) with Eschenmoser's salt $(\overset{\oplus}{CH_2}\!=\!\overset{\ominus}{NMe_2}\}I)$.[21] A recent convenient preparation of the dimethyl-(methylene)ammonium iodide reagent from trimethylsilyl iodide and N,N,N',N'-tetramethylenediamine[22] could result in greater value in future being placed on this reaction for the preparation of terminal alkenes.

Experiment 5.12   **2-METHYLBUT-2-ENE** (*in admixture with 2-METHYLBUT-1-ENE*)

$$Me_2C(OH)\cdot CH_2\cdot Me \xrightarrow{H_3PO_4} Me_2C\!=\!CH\cdot Me + CH_2\!=\!C(Me)\cdot CH_2\cdot Me$$

Place 25.0 g (31 ml, 0.28 mol) of 2-methylbutan-2-ol and 10 ml of 85 per cent orthophosphoric acid in a 100-ml, round-bottomed flask and swirl to mix thoroughly. Fit the flask with a 20-cm fractionating column filled with glass helices, a Claisen still-head and a condenser leading to a 50-ml receiving flask cooled in a beaker of iced water (Fig. 2.104). Add a few pieces of porous porcelain and heat the reaction mixture gently with a Bunsen burner. Collect the alkene fraction which distils in the range 35–38 °C during a period of 30 minutes. Dry the distillate with 1–2 g of magnesium sulphate. Wash and dry the distillation apparatus, decant the dried distillate into a 50-ml flask and redistil in the reassembled apparatus. Collect the fraction boiling at 37–38 °C; the yield is 12.5 g (64%). Record the infrared spectrum of the product using a fixed path-length cell (0.025 mm). The stretching bands of the terminal $(1645\,cm^{-1})$ and non-terminal $(1670\,cm^{-1})$ carbon–carbon double bonds can both be observed; bands at 890 and $805\,cm^{-1}$ ($=$C—H deformation) also establish the presence of both terminal and non-terminal olefinic systems. Analyse the product by g.l.c. on a Silicone oil column at 30 °C; 2-methylbut-1-ene appears first, closely followed by 2-methylbut-2-ene; the areas under the peaks are in the ratio of 1:4.

Cognate preparations.   *2,3-Dimethylbuta-1,3-diene.*   In a 1-litre round-bottomed flask, surmounted by an efficient fractionating column (e.g. one of the Hempel type, see Section 2.26) place 177 g (1.5 mol) of anhydrous pinacol (Expt 5.35), 5 ml of constant boiling point hydrobromic acid (c. 47–48% w/w HBr) and a few fragments of porous porcelain. Attach a condenser and a receiver to the column. Heat the flask gently in an oil bath; the rate of distillation should be 20–30 drops per minute. Collect the distillate until the temperature at the top of the column is 95 °C (60–70 minutes). Separate the upper non-aqueous layer, wash it twice with 50 ml portions of water, add 0.25 g of hydroquinone as an inhibitor and dry it overnight with 7–8 g of anhydrous calcium chloride. Transfer to a 500-ml flask and distil through the same column. Collect the following fractions: (*a*) 69–70.5 °C (70 g); (*b*) 70.5–105 °C (7 g); and (*c*) 105–106 °C (35 g). Fraction (*a*) is pure dimethylbutadiene (yield 57%), (*b*) is an intermediate fraction and (*c*) is pinacolone.

Dimethylbutadiene may be kept for a limited period in an ice box or in a refrigerator; it is advisable to add about 0.2 g of hydroquinone as an inhibitor.

Cyclohexene. Fit a 500-ml three-necked flask with a fractionating column (e.g. a Hempel column filled with 6-mm glass or porcelain rings) carrying a thermometer at its upper end, and a separatory funnel; close the third neck with a stopper. Attach an efficient double surface condenser to the column; use a filter flask, cooled in ice, as receiver. Place 50 g of 85 per cent ortho-phosphoric acid in the flask and heat it in an oil bath at 160–170 °C. Add, through the funnel, 250 g (2.5 mol) of cyclohexanol over a period of 1.5–2 hours. When all the cyclohexanol has been introduced, raise the temperature of the bath to about 200 °C and maintain it at this temperature for 20–30 minutes. The temperature at the top of the column should not rise above 90 °C. Saturate the distillate with salt, separate the upper layer and dry it with magnesium sulphate. Distil the crude cyclohexene through an efficient col-umn and collect the fraction boiling at 81–83 °C; the residue is largely cyclo-hexanol. The yield of cyclohexene is 165 g (80%).

1-Methylcyclohexene. Use 20 g (0.18 mol) of 2-methylcyclohexanol and 5 ml of phosphoric acid. Proceed as for 2-methylbut-2-ene but use a shorter frac-tionating column (12 cm); it is not necessary to cool the receiving flask. Collect the fraction boiling between 100 and 112 °C and dry and redistil using a simple distillation apparatus. Collect the purified alkene fraction boiling at 103–110 °C; the yield is 14.2 g (76%). Analyse the product by g.l.c. on a Sili-cone oil column at 60 °C (Section 2.31); 3-methylcyclohexene (20%) appears first followed by 1-methylcyclohexene (80%). The major product, 1-methyl-cyclohexene, b.p. 110 °C, can be separated from the 3-methyl isomer, b.p. 103 °C, by careful fractionation through a spinning band column (Fig. 2.107).

Experiment 5.13 CYCLOHEXA-1,3-DIENE

In a 100-ml round-bottomed flask place 16.1 g (0.10 mol) of 3-bromocyclo-hexene (Expt 5.68), 38.7 g of dried, redistilled quinoline and a magnetic stirrer follower. Attach to the flask a Claisen still-head fitted with a thermometer and a condenser set for downward distillation. Protect the apparatus from atmospheric moisture by a calcium chloride guard-tube attached to the side-arm of the receiver. Support the reaction flask in an oil bath which rests on a magnetic stirrer/hotplate, commence stirring rapidly and heat the oil bath to 160–170 °C. The cyclohexadiene steadily distils as colourless liquid, b.p. 80–82 °C, over a period of about 30 minutes, yield 5.4 g (68%). The product is 99 per cent pure by g.l.c. analysis; use either a 2.7 m column of 10 per cent poly-ethyleneglycol adipate on Chromosorb W held at 60 °C with a flow rate of carrier gas of 40 ml/minute, t_R 2.4 minutes, or, a 1.5 m column of 10 per cent Silicone oil on Chromosorb W held at 60 °C with a flow rate of carrier gas of 40 ml/minute, t_R 1.5 minutes.

Experiment 5.14   PENTA-1,4-DIENE (*Divinylmethane*)

1,5-Diacetoxypentane. Equip a 5-litre three-necked flask with an efficient sealed stirrer unit of the Hirschberg type, a double-surface reflux condenser and screw-capped adapter carrying a thermometer. Add 516 g (582 ml, 6.0 mol) of tetrahydropyran, 480 g (434 ml, 6.0 mol) of acetyl chloride and 4 g of powdered fused zinc chloride. Heat the mixture with a heating mantle with vigorous stirring under reflux, and continue to heat until the temperature of the reaction mixture reaches 150 °C (about 3–5 hours). Allow the reaction mixture to cool to near room temperature and add 980 g (10 mol) of solid potassium acetate and 20 g of sodium iodide. Heat the mixture again with stirring to 160 °C and maintain the temperature for 12 hours. Cool to room temperature, add 1 litre of light petroleum (b.p. 40–60 °C), filter under suction and wash the residue thoroughly with more light petroleum. Combine the filtrate and washings and remove the solvent on a rotary evaporator. Fractionally distil the residue under reduced pressure and collect the 1,5-diacetoxypentane at 108–114 °C/3.5–4 mmHg. The yield is about 790 g (70%); the product is sufficiently pure for use in the next stage, but a sample may be purified by redistillation and then has b.p. 102–104 °C/3 mmHg (or 92–94 °C/1.0 mmHg).

Fig. 5.1

Penta-1,4-diene. Assemble the pyrolysis apparatus [see Fig. 2.65(*c*)] with the pyrolysis tube packed with glass beads but with the acetone–Cardice cooling trap arrangement shown in Fig. 5.1 attached to the water condenser. Pass a gentle stream of nitrogen gas through the apparatus and heat the combustion tube to 580 °C when 900 g of 1,5-diacetoxypentane is added dropwise, during about 12 hours (1). Fractionally distil the total pyrolysate carefully and slowly at atmospheric pressure and collect the fraction boiling below 50 °C which should be refractionated to give penta-1,4-diene of b.p. 24–27 °C.

The yield is 170 g (52%). Assign the i.r. absorptions which occur at 3100, 1840, 1650, 995 and 910 cm^{-1}. The ^{13}C-n.m.r. (CDCl$_3$, TMS) reveals signals at δ 38.1 (C$_3$), 115.5 (C$_1$ and C$_5$), 136.6 (C$_2$ and C$_4$).

Note. (1) Solidified acetic acid often blocks the flow of nitrogen in the cooling trap. It can be dislodged by careful warming with a warm-air blower.

Experiment 5.15 3,3-DIMETHYLBUT-1-ENE

$$Me_3C \cdot CHMe \cdot O \cdot CS \cdot SMe \xrightarrow[-MeSH]{-S=C=O} Me_3C \cdot CH = CH_2$$

CAUTION: *This preparation should be conducted in an efficient fume cupboard.*

From 51 g (63 ml, 0.5 mol) of 3.3-dimethylbutan-2-ol prepare *O*-[1,2,2-trimethylpropyl]*S*-methyl dithiocarbonate (the xanthate ester) using the procedure described in Expt 5.209, up to and including the removal of toluene and residual alcohols with a rotary evaporator. Transfer the red residual crude xanthate ester to a 250-ml round-bottomed flask. Fit a vertical air condenser (about 12 inches), and attach a still-head with thermometer leading to an efficient double surface condenser terminating in a receiver adapter and flask. Immerse the receiver flask in an ice–salt bath and arrange for the water supply to the condenser to be cooled in an ice–salt bath. Heat the xanthate ester to boiling using a Bunsen burner; smooth decomposition occurs forming the alkene which slowly distils. On completion of the pyrolysis wash the distillate with three 10 ml portions of ice-cold 20 per cent potassium hydroxide solution, then with ice-cold water and finally dry the organic layer with anhydrous calcium chloride. Distil and collect 3,3-dimethylbut-1-ene as a fraction having b.p. 40–42 °C (cool the receiver flask in ice). The yield is 23.5 g (53%) (1). Further purification may be effected by redistillation from sodium metal. The product shows the characteristic absorption bands in the i.r. at 3050 ($=$C—H), 1820 ($=$CH$_2$ overtone), 1640 (C$=$C), 1385 (m), 1365 (s) ((CH$_3$)$_3$C), 995 and 915 cm^{-1} (CH and CH$_2$ deformation). Assign the ^{13}C-n.m.r. signals which occur at δ 29.4, 33.7, 109.0 and 149.8.

Note. (1) Gas–liquid chromatography analysis using 10 per cent Silicone oil column (1.5 m) at 40 °C with a nitrogen flow rate of 40 ml per minute gives t_R 1.18 minutes.

5.2.2 THE PARTIAL HYDROGENATION OF ALKYNES

Alkynes readily undergo catalytic hydrogenation to yield first the corresponding alkene and then the alkane.

$$R' \cdot C \equiv C \cdot R^2 \longrightarrow R^1 \cdot CH = CH \cdot R^2 \longrightarrow R^1 \cdot CH_2 \cdot CH_2 \cdot R^2$$

With most hydrogenation catalysts of the platinum and nickel groups (Sections 4.2.*61*, p. 459 and 4.2.*50*, p. 450), a mixture of products is obtained even if an attempt is made to stop the reaction at the half-way stage. An alkene may only be obtained in good yield if particular attention is paid to the selection of a deactivated hydrogenation catalyst. An early, highly effective formulation of a deactivated palladium catalyst is Lindlar's catalyst (Section 4.2.*54*, p. 453), but palladium-on-barium sulphate in the presence of quinoline, or palladium-on-calcium carbonate, is also recommended. In these cases it is advisable to

493

monitor the rate of hydrogenation when it will be noted that the reaction slows significantly after the uptake of one molar proportion of hydrogen.

The partial hydrogenation process is stereospecific, syn-addition of hydrogen leading to the (Z)-alkene in a virtually stereochemically pure form. The reaction is illustrated by the conversion of but-2-ynoic acid into (Z)-but-2-enoic acid using the Lindlar catalyst (Expt 5.16); g.l.c. analysis shows the crude product to be 95 per cent stereochemically pure, the remainder being the (E)-isomer. The first cognate preparation in Expt 5.16 illustrates the use of a palladium-on-barium sulphate catalyst for the partial hydrogenation of but-2-yn-1-ol to give the corresponding (Z)-alkenol.[23] In the second cognate preparation, but-2-yne-1,4-diol is hydrogenated with a palladium-on-calcium carbonate catalyst.[24]

Although these catalytic partial hydrogenations of alkynes may well be regarded as the procedure of choice for (Z)-alkenes,[25] other catalytic systems have been explored. These include a sodium hydride–sodium alkoxide–nickel(II) acetate reagent,[26] and a sodium borohydride–palladium chloride–polyethylene glycol system.[27] Diisobutylaluminium hydride (DIBAL) has also been used for the conversion of alkynes into (Z)-alkenes.[28] (E)-Alkenes are formed when the internal triple bond is reduced with sodium in liquid ammonia.[29]

Experiment 5.16 (Z)-BUT-2-ENOIC ACID

$$Me \cdot C \equiv C \cdot CO_2H \xrightarrow[\text{Lindlar catalyst}]{H_2} \underset{H \quad\quad H}{\overset{Me \quad\quad CO_2H}{C=C}}$$

In a 250-ml hydrogenation flask (Section 2.17.1, p. 89) place 100 mg of Lindlar catalyst (1) and a solution of 2 g (0.024 mol) of but-2-ynoic acid (Expt 5.216) in 100 ml of ethanol. Attach the flask to the adapter of the atmospheric hydrogenation apparatus [Fig. 2.63(a)], and fill the flask and gas burettes with hydrogen by the procedure discussed in Section 2.17.1, p. 89; note the volumes in the gas burettes. Agitate the flask and note the burette readings when uptake of hydrogen practically ceases (the volume of hydrogen absorbed should be in the region of 540 ml). Follow the procedure given in Section 2.17.1, p. 89 for replacing the hydrogen in the apparatus with air. Disconnect the hydrogenation flask, and filter the solution to remove the catalyst which should then be transferred to a residue bottle. Evaporate the ethanol on a rotary evaporator; a white solid is obtained if the temperature of the flask is below 15 °C. Gas–liquid chromatography analysis (using a 1.5 m column of 10 per cent diisodecyl phthalate containing 1 per cent phosphoric acid on Chromosorb W held at 90 °C with a nitrogen flow rate of 20 ml per minute) of a solution of the acid in ether shows the crude product to contain 95 per cent of (Z)-but-2-enoic acid (2). The pure acid, m.p. 15 °C, may be obtained by low temperature recrystallisation (Section 2.20) using light petroleum (b.p. 40–60 °C). The p.m.r. spectrum should be recorded and compared with that of crotonic acid (Expt 5.216).

Notes. (1) See Section 4.2.54, p. 453; a suitable preparation is available from, for example, Fluka-A.G. or Aldrich.
(2) Under these conditions isomerisation to the more stable (E) isomer is not apparent. The retention times for (Z)-but-2-enoic acid, (E)-but-2-enoic acid and but-2-ynoic acid are 6.8, 10.0 and 19.8 minutes respectively.

Cognate preparations. (Z)-But-2-en-1-ol.[23] The semihydrogenation of but-2-yn-1-ol (10 g) is performed in a Parr hydrogenation apparatus at 1–7 lb pressure and 27 °C with methanol (260 ml) as the solvent, and 5 per cent palladium-on-barium sulphate (260 mg) poisoned with synthetic quinoline (260 mg) as the catalyst. Distillation of the product gives (Z)-but-2-en-1-ol (7.1 g, 71%), b.p. 56 °C/40 mmHg; p.m.r. (CDCl$_3$, TMS) δ 1.95 (d of d, 3H), 4.22 (m, 2H), 5.74 (m, 2H); g.l.c. analysis of the reduction product on Carbowax 20M at 120 °C gives resolved peaks for the (E) and (Z) isomers and on the basis of peak areas indicates an (E)/(Z) ratio of 2:98.

(Z)-But-2-ene-1,4-diol.[24] But-2-yne-1,4-diol (20 g) is dissolved in methanol (350 ml) and hydrogenated at atmospheric pressure in the presence of 0.5 per cent palladium-on-calcium-carbonate (2.0 g). After 24 hours one equivalent of hydrogen (5650 ml) will be absorbed. The catalyst is then separated and the filtrate distilled, giving (Z)-but-2-ene-1,4-diol as a colourless oil (15.7 g, 77%), b.p. 134–135 °C/15 mmHg, n_D^{25} 1.4716. The product forms a dibenzoate (prisms from aqueous methanol), m.p. 69–70 °C.

5.2.3 THE WITTIG AND RELATED REACTIONS

Quaternisation of triphenylphosphine with an alkyl halide gives a quaternary phosphonium halide (4) which under the influence of a strong base eliminates hydrogen halide to give an alkylidenephosphorane [(5), an ylide]. The latter reacts with an aldehyde or ketone to give first an intermediate betaine (6), which rearranges to the oxaphosphetane (7), which then under the reaction conditions eliminates triphenylphosphine oxide to form an alkene.

$$Ph_3P: \curvearrowright CH_2 - X \longrightarrow \overset{\oplus}{Ph_3P} - CH_2 \cdot R^1 \overset{X^\ominus}{} \xrightarrow{\text{base}} \overset{\oplus}{Ph_3P} - \overset{\ominus}{C}H \cdot R^1 \longleftrightarrow Ph_3P = CH \cdot R^1$$

(4)   (5)

$$Ph_3P = CH \cdot R^1 \longrightarrow \left[\overset{\oplus}{Ph_3P} - CH \cdot R^1 \atop \overset{\ominus}{O} - CH \cdot R^2 \right] \longrightarrow \left[Ph_3P - CH \cdot R^1 \atop O - CH \cdot R^2 \right] \longrightarrow$$

O=CH·R²

(6)   (7)

$$Ph_3PO + R^1 \cdot CH = CH \cdot R^2$$

The overall sequence is the *Wittig reaction*, and its great value lies in the fact that the mild conditions do not normally promote structural isomerisation; for this reason alkenes of unambiguous structure may be synthesised.

Since its discovery in 1953, the reaction has been extensively studied with respect to structural variation in the phosphonium ylide, in the range of functionalised aldehydes and ketones, in the nature of the bases and the polarity of the solvents that may be used, and in the mechanism of the reaction and the factors which influence the (E/Z) ratio. Many thousands of simple and complex syntheses have been effected and the method is widely exploited in research laboratories and in the industrial processes for syntheses in the steroid and carotenoid field.

The preparative examples which are discussed below, and elsewhere in the text, have been selected to illustrate some of the considerations needed in the selection of starting materials, and some of the experimental conditions of the reaction that have been employed.

A retrosynthetic analysis for the alkene $R^1R^2C{=}CH{\cdot}R^3$ shows two possible carbonyl compounds as starting materials.

$$\overset{\oplus}{Ph_3P}{-}\overset{R^1}{\underset{R^2}{\overset{\ominus}{\diagup}}} \quad O{=}\overset{R^3}{\underset{H}{\diagdown}} \quad \overset{A}{\Longleftarrow} \quad \overset{R^1}{\underset{R^2}{\diagdown}}{\overset{R^3}{\diagup}} \quad \overset{B}{\Longrightarrow} \quad \overset{R^1}{\underset{R^2}{\diagdown}}{=}O \quad \overset{\ominus}{\underset{\overset{\oplus}{PPh_3}}{\diagup}}\overset{R^3}{\diagdown}$$

In the synthesis of methylenecyclohexane (Expt 5.17), *route B* using cyclohexanone and methylenetriphenylphosphorane ($R^1{=}H$) is preferred, since the carbonyl compound of *route A*, formaldehyde ($R^3{=}H$), is more difficult to handle, even though no difficulty would be envisaged with the formation of cyclohexylidenetriphenylphosphorane. In other cases the relative availability of the carbonyl compounds and haloalkanes would need to be considered. Since the alkylidenetriphenylphosphoranes have high carbanionic reactivity they rapidly interact with oxygen and water. These reactions are therefore carried out under nitrogen and under anhydrous conditions; strong bases such as the methylsulphinyl carbanion ($Me{\cdot}SO{\cdot}CH_2^{\ominus}$) generated from dimethyl sulphoxide with sodium hydride, or butyllithium or phenyllithium (e.g. cognate preparation) are used.

In the case of phosphonium salts having the structure $Ph_3\overset{\oplus}{P}{\cdot}\overset{\ominus}{C}HR^1R^2X$, where R^1 and R^2 contain electron-withdrawing and mesomerically stabilising groups (e.g. Ar, CO_2R, CO, C=C), the derived carbanion is more stable, more tolerant of water, and may be formed with less strong bases. Hence sodium alkoxides, sodium hydroxide, or even sodium carbonate may be used. However such carbanions have lower carbanionic reactivity, and although reacting readily with aldehydes, ketones are frequently not attacked. An example of the use of sodium hydroxide as the base is provided by the conversion of 9-anthraldehyde into (E)-1-(9-anthryl)-2-phenylethene by reaction with benzyltriphenylphosphonium bromide[30] (Expt 5.18). The reactivity of the ylide towards ketones may be improved upon by preparing a Wittig-type reagent by reacting triethyl phosphite with the relevant halide and treating the resulting alkyl phosphonate with base (*Horner–Emmons or Wadsworth–Emmons reaction*).[31]

$$(EtO)_3P + R^1{\cdot}CH_2X \xrightarrow{-EtX} (EtO)_2P(O){\cdot}CH_2{\cdot}R^1 \xrightarrow{base} (EtO)_2P(O){\cdot}\overset{\ominus}{C}H{\cdot}R^1$$

$$R^2R^3CO + (EtO)_2P(O){\cdot}\overset{\ominus}{C}H{\cdot}R^1 \longrightarrow R^2R^3C{=}CH{\cdot}R^1 + (EtO)_2P(O){\cdot}O^{\ominus}$$

Examples of this reaction are to be found in Sections 5.18.2, p. 798 and 5.18.3, p. 804 for the synthesis of unsaturated carbonyl compounds and unsaturated esters.

In the above discussion the problem of stereoisomeric control has been set aside. Total stereoisomeric purity is in fact difficult to achieve, and extensive mechanistic studies have been carried out to elucidate the factors which influence the (E/Z) ratio. It thus appears that, in the formation of the alkene $R^1{\cdot}CH{=}CH{\cdot}R^2$, thermodynamic control leading to predominance of the (E)-isomer is promoted by elevated temperatures, apolar solvents, stabilising salts (lithium or sodium halides), carbanion stabilisation, an electron-rich phosphorus atom [—$PO(OR)_2$], and excess base. On the other hand, kinetic control leading to a predominance of the (Z)-isomer, is promoted by low temperatures, polar (aprotic) solvents, salt-free solutions, no carbanion stabilisation, and an electrophilic phosphorus atom (PPh_3).[32]

A further structural variation in the phosphorus ylide was developed in the *Wittig–Horner reaction*.[33] Here the anion-stabilising group is the diphenyl-phosphinoyl group (Ph_2PO), and the anion is derived from the alkyl (or aralkyl) diphenylphosphine oxide $(Ph_2P(O)\cdot CH_2R)$. These compounds are formed by the hydrolysis of the corresponding phosphonium halide with aqueous alkali.

$$Ph_3P + R\cdot CH_2X \longrightarrow Ph_3\overset{\oplus}{P}\cdot CH_2\cdot R \xrightarrow{\overset{\ominus}{O}H} Ph_2P(O)\cdot CH_2\cdot R + PhH$$

Taking a simple general example to explain the usefulness of this reaction, treatment with base (usually butyllithium) of the alkyldiphenylphosphine oxide gives the anion which then reacts with an aldehyde to form the corresponding-*erythro*:*threo* diastereoisomeric mixture of alcohols (cf. p. 517) in unequal proportions; only one enantiomer of each diastereoisomeric pair is formulated [(8) and (9)].

erythro (major) *threo* (minor)
   (8)   (9)

These alcohols, unlike the betaine–oxaphosphetane intermediates of the Wittig reaction, may usually be isolated in a stable crystalline form, and the diastereo-isomers are frequently separable by chromatography. The *erythro–threo* proportion is greatly biased in favour of the *erythro*-isomer. Conversion into the alkenes is then effected by reaction with sodium hydride in dimethylformamide which proceeds by a *syn*-elimination of the water-soluble $Ph_2PO_2^{\ominus}$ species, probably via a four-membered cyclic transition state similar to the corresponding Wittig reaction.[34] Thus the *erythro*-alcohol (8) gives the (Z)-alkene and the *threo*-alcohol (9) gives the (E)-alkene; owing to the *erythro*-selectivity in alcohol formation and chromatographic separability of diastereoisomers, the overall reaction sequence provides a preparatively useful route to (Z)-alkenes.

erythro (8)   (Z)   *threo* (9)   (E)

An alternative route to the *erythro*:*threo* diastereoisomeric mixture of alcohols results from the acylation of the alkyldiphenylphosphine oxide with an ester or lactone to yield the β-ketophosphine oxide, followed by reduction with sodium borohydride. This reduction shows *threo*-selectivity, so that, following separation of the diastereoisomers, a preparatively useful route to (E)-alkenes is achieved.

$$\underset{Ph_2P}{\overset{\displaystyle O}{\|}}\overset{R^1}{\diagup} \quad \xrightarrow[\text{(ii) } R^2\cdot CO_2R^3]{\text{(i) BuLi}} \quad \underset{O}{\overset{\displaystyle O}{\underset{\|}{Ph_2P}}}\overset{R^1}{\underset{R^2}{\diagdown}} \quad \xrightarrow{\text{NaBH}_4} \quad \underset{HO}{\overset{\displaystyle O}{\underset{H}{\overset{\|}{Ph_2P}}}}\overset{H}{\underset{R^2}{\diagdown}}R^1$$

threo (major)

The illustrative example (Expt 5.19) of these two complementary routes is the preparation of (Z)- and (E)-isosafrole[35]; the formulation of the specific reaction stages is given in Expt 5.19.

The *Peterson reaction* may be regarded as the silicon-based analogue of the Wittig reaction.[36] In the simplest and earliest examples, trimethylsilylmethyl-magnesium chloride ($Me_3Si\cdot CH_2MgCl$, a source of α-silylcarbanion), was allowed to react with a carbonyl compound to give a β-silylalcohol (10). Subsequently the alcohol was converted into the oxyanion with either sodium hydride or sodium hydroxide which decomposes in refluxing tetrahydrofuran to the alkene by a *syn*-elimination process.

$$Me_3Si\cdot CH_2Cl \xrightarrow{\text{Mg}} Me_3Si\cdot CH_2MgCl$$

$$R_2CO + Me_3Si\cdot CH_2MgCl \longrightarrow R_2C(OMgCl)\cdot CH_2SiMe_3 \xrightarrow[\text{NH}_4Cl]{\text{H}_2O}$$

$$\underset{(10)}{R_2C(OH)\cdot CH_2\cdot SiMe_3} \xrightarrow{\text{NaH}} \underset{O^{\ominus}SiMe_3}{\overset{R_2C-CH_2}{\underset{|}{\overset{|}{\quad}}}} \longrightarrow R_2C{=}CH_2 + {\overset{\ominus}{O}}SiMe_3$$

The final elimination step may also be effected under acidic conditions, but the possibility of rearrangement to other alkene regioisomers makes this a less satisfactory procedure. The scope of this reaction, which was initially of particular value for the preparation of terminal alkenes, has been increased by the development of other methods for the preparation of α-silylcarbanions.[37]

In the preparation of 2,6-dimethylhepta-1,5-diene, Expt 5.20, the final *in situ* elimination step is effected with thionyl chloride and is thought to proceed via a six-membered cyclic transition state.[38]

$$\underset{OMgX}{\overset{R_2C-CH_2\cdot SiMe_3}{\underset{|}{\overset{|}{\quad}}}} \xrightarrow[-\text{MgXCl}]{\text{SOCl}_2} \underset{\underset{Cl}{\overset{|}{S}}}{\overset{CH_2}{R_2C\diagdown \underset{O \diagdown \diagup O}{\quad}SiMe_3}} \longrightarrow R_2C{=}CH_2 + Me_3SiO\cdot SOCl$$

Experiment 5.17   METHYLENECYCLOHEXANE

$$Ph_3P + MeBr \longrightarrow Ph_3\overset{\oplus}{P}\cdot Me\}Br^{\ominus} \xrightarrow[-H^{\oplus}]{\text{Base}} Ph_3\overset{\oplus}{P}{-}\overset{\ominus}{C}H_2$$

$$Ph_3\overset{\oplus}{P}{-}\overset{\ominus}{C}H_2 + \bigcirc{=}O \longrightarrow \bigcirc{=}CH_2 + Ph_3PO$$

Triphenylmethylphosphonium bromide.   Dissolve 68.1 g (0.26 mol) of triphenylphosphine in 60 ml of dry benzene (**CAUTION**). Place the solution in a stout-walled wide-necked glass bottle of about 1-litre capacity and cool in an ice–salt bath. Add 33.4 g (20 ml, 0.35 mol) of methyl bromide (1) and seal the bottle with a rubber bung secured with wire. Remove the bottle from the cooling bath and leave for one day at room temperature. Cool the bottle in an ice–salt bath before opening, filter off the white crystalline solid under suction and wash with hot benzene (fume cupboard). Dry the product in the oven at 100 °C and store in a desiccator over phosphorus pentoxide. The yield is 91.5 g (99%), m.p. 232–233 °C.

Note. (1) Methyl bromide (b.p. 4.5 °C) is supplied in sealed vials. The vial must be thoroughly cooled in an ice–salt bath before any attempt is made to open it. It is advisable to wear a face mask and thick gloves during the operation. Wrap a cloth around the vial and score around the neck with a glass-knife. Place several layers of cloth around the scored neck and break off the end of the neck.

Methylenecyclohexane. Fit a 250-ml three-necked flask containing a magnetic stirring bar with a condenser, nitrogen inlet and a thermometer; surround the flask with a water bath supported on a hot-plate/magnetic stirrer unit. Pass a steady stream of dry nitrogen through the flask and maintain the flow throughout the reaction. Place 125 ml of dry dimethyl sulphoxide (Section 4.1.*33*, p. 412) in the flask and stir vigorously for 30 minutes to remove oxygen gas. Briefly remove the condenser from the flask and add 4.8 g (0.1 mol) of a 50 per cent suspension of sodium hydride in oil by means of a solids funnel. Warm the mixture to 65 °C on the water bath and stir until all the sodium hydride dissolves (about 1.5 hours). Cool the mixture to room temperature in a cold-water bath. Remove the condenser, place a solids funnel in the central neck of the flask and add 35.7 g (0.1 mol) of triphenyl-methylphosphonium bromide in portions over 15 minutes; the reaction is slightly exothermic. Replace the condenser and stir at room temperature for 45 minutes to complete the formation of the methylenephosphorane. Add 9.8 g (0.1 mol) of cyclohexanone in one portion; an exothermic reaction ensues and the temperature rises to about 70 °C. Stir the mixture at 50 °C for 1 hour and then transfer to a single-neck 250-ml flask, add a boiling stick and attach a still-head and condenser arranged for vacuum distillation. A calcium chloride drying tube should be placed in the line between the water pump and receiving flask to prevent water vapour condensing in the flask. Distil and collect the volatile products boiling below 75 °C at 50 mmHg in a receiver cooled in an acetone–Cardice bath; the distillate consists of methylenecyclo-hexane contaminated with a small amount of benzene. Redistil the crude product at atmospheric pressure through a 7-cm fractionating column packed with glass helices. Reject the material (mainly benzene) which boils below 99 °C, and collect the pure methylenecyclohexane as a fraction of b.p. 99–100 °C; the yield is 6.2 g (64%). Confirm its homogeneity by g.l.c. analysis on a Silicone oil column at 60 °C.

Cognate preparation. *3,3,5-Trimethyl-1-methylenecyclohexane.* Prepare an ethereal solution of phenyllithium by the procedure described in Expt 8.32, using 2.7 g (0.39 mol) of lithium shavings, 26 g (17.5 ml, 0.16 mol) of dry redistilled bromobenzene and 85 ml of anhydrous ether. After the conversion to phenyllithium is complete dilute the solution with a 15 ml portion of anhydrous ether and decant the solution from any unreacted lithium [Expt 8.32 Note (2)] into a clean three-necked 250-ml flask equipped with a nitrogen inlet, a solids addition funnel and a reflux condenser. Add at room temperature 53.5 g (0.15 mol) of triphenylmethylphosphonium bromide in portions over 15 minutes; temporarily replace the solids addition funnel with a stopper if the reaction mixture shows a tendency to boil. Stir the suspension at room temperature for 4 hours and then add 21 g (0.15 mol) of 3,3,5-trimethylcyclohexanone; stir vigorously to effect solution. Heat the reaction mixture with stirring overnight and then add 25 ml of water. Remove the precipitate by filtration, wash it with ether and dry the combined ether extracts. Remove the solvent on a rotary evaporator and distil the residue through a fractionating column; collect the fraction having b.p. 155–160 °C. Refractionate and collect the pure product having b.p. 156–158 °C; the yield is 11.3 g (53%). The purity may be checked by g.l.c. analysis using a Carbowax 20 M (10%) column held at 80 °C.

Experiment 5.18 (E)-1-(9-ANTHRYL)-2-PHENYLETHENE[30]

In a 25-ml Erlenmeyer flask is placed a stirring bar, 0.97 g (2.5 mmol) of benzyltriphenylphosphonium chloride, 0.57 g (2.5 mmol) of 9-anthraldehyde (Aldrich), and 3 ml of dichloromethane. The mixture is stirred as vigorously as possible, and 1.3 ml of 50 per cent aqueous sodium hydroxide are added from a dropper at the rate of 1 drop per 7 seconds. Stirring is continued for 30 minutes. The mixture is transferred to a small separating funnel. Dichloromethane and water (10 ml of each) are used to rinse the flask. The funnel is shaken and the organic layer removed. The aqueous layer is extracted with 5 ml of dichloromethane. The dried (1 g of calcium chloride) organic layer is concentrated on a water bath, and the remaining solvent removed in a vacuum desiccator. The resulting yellow semi-solid is recrystallised from 15 ml of propan-1-ol to give 0.52 g (74%) of thin yellow plates, m.p. 131–133 °C; p.m.r. (CDCl$_3$, TMS) δ 7.3–8.5 (m, 14H, aromatic protons), 6.82 (d, 1H, $J = 17$ Hz, vinylic proton), 7.82 (d, 1H, embedded in the aromatic signal). The J value of 17 Hz proves the structure to be (E). The i.r. spectrum shows a strong band at 962 cm^{-1}, indicative of an (E) configuration; the band at 680 cm^{-1} is indicative of a monosubstituted benzene ring.

Experiment 5.19 (Z)- and (E)-ISOSAFROLES[35]

$$Ar =$$

(Z)-Isosafrole. (*1RS,2SR*)-*2-Diphenylphosphinoyl-1-(3,4-methylenedioxy-phenyl)-propan-1-ol.* n-Butyllithium (1.5 M in hexane; 5.8 ml) is added drop-wise from a syringe (1) to a stirred solution of ethyldiphenylphosphine oxide (2.0 g, 8.69 mmol) (2) in dry tetrahydrofuran (30 ml) at 0 °C (3). After 30 minutes the red reaction solution is cooled to − 78 °C (acetone–solid carbon dioxide) and a solution of 3,4-methylenedioxybenzaldehyde (1.3 g, 8.69 mmol) in dry tetrahydrofuran (10 ml) is added dropwise from a syringe. The rate of addition is such that the internal temperature is maintained at − 78 °C. The orange solution is allowed to warm to room temperature over 2 hours and then water (25 ml) is added. The tetrahydrofuran is removed under reduced pressure and the aqueous residue is diluted with brine (15 ml) before extraction with dichloromethane (3 × 50 ml). The combined organic extracts are dried (magnesium sulphate) and evaporated to dryness to give the prod-uct as a crystalline mixture of diastereoisomers which are separated by flash column chromatography (elution with ethyl acetate) (4). The first diastereo-isomer to be eluted from the column is the (1*RS,2SR*)-adduct (*erythro*-isomer), microcrystals (2.5 g, 75.8%), m.p. 137–140 °C [recrystallised from ethyl acetate–light petroleum (b.p. 60–80 °C)], R_F 0.6 (blue fluorescence) (4), i.r. 3400 (OH), 1235 (C—O), and 1150 cm^{-1} (P=O); p.m.r. (CDCl$_3$, TMS) δ 8.1–7.4 (m, 10H, Ph$_2$PO), 6.8 and 6.75 (two s, 3H, aryl Hs), 5.9 (s, 2H, OCH$_2$), 5.2 (dd, 1H, J_{HH} = 1 Hz, J_{HP} = 9 Hz, CHOH), 4.5 (broad s, 1H, OH), 2.55 (ddq, J_{HH} = 1 Hz, J_{HMe} = J_{HP} = 7 Hz, CHMe), and 1.05 (dd, 3H, J_{HMe} = 7 Hz, J_{MeP} = 17 Hz, Me), m.s. $M^{\oplus}$ 380.1157 (C$_{22}$H$_{21}$O$_4$P requires M 380.1178), m/z 380 (9%), 230 (63%, Ph$_2$POEt) and 202 (100%, Ph$_2$POH). The second diastereoisomer to be eluted from the column is the (1*RS,2RS*)-adduct (*threo*-isomer) (277 mg, 8.4%), m.p. 197–199 °C [from ethyl acetate–light petroleum (b.p. 40–60 °C)], R_F 0.45, i.r. (Nujol mull) 3275 (OH), 1440 (P—Ph) and 1165 cm^{-1} (P=O); p.m.r. (CDCl$_3$, TMS) δ 8.0–7.2 (m, 10H, Ph$_2$PO), 6.8 and 6.65 (two s, 3H, aryl Hs), 5.8 (s, 2H, OCH$_2$), 5.55 (broad d, 1H, J_{HOH} c. 3 Hz, OH), 4.7 (broad dd, 1H, J_{HH} = J_{HP} = 9 Hz, CHOH), 2.8 (ddq, 1H, J_{HMe} = 7 Hz, J_{HH} = J_{HP} = 9 Hz, CHMe) and 0.75 (dd, 3H, J_{HMe} = 7 Hz, J_{MeP} = 17 Hz, Me); m.s. $M^{\oplus}$ 380.1177 (C$_{22}$H$_{21}$O$_4$P requires M, 380.1177), m/z 381 (22%, M + 1), 380 (68%), 379 (51%, M − 1), 230 (76%, Ph$_2$POEt), 202 (100%, Ph$_2$POH) and 201 (92%, Ph$_2$PO$^{\oplus}$).

(Z)-Isosafrole. Sodium hydride (80% dispersion in oil; 24 mg, 0.79 mmol) is added in one portion to a stirred solution of the (1*RS,2SR*)-phosphine oxide (*erythro*-isomer) (300 mg, 0.79 mmol) in dry dimethylformamide (30 ml) (3).

The clear reaction solution is warmed to $50\,°C$ for 30 minutes by which time a white solid precipitates from solution. The reaction mixture is cooled and the precipitate dissolved by the addition of water (25 ml). The mixture is diluted with brine (15 ml) and extracted with ether (3 × 40 ml). The combined organic extracts are washed with water (3 × 40 ml), dried (magnesium sulphate), and the solvent removed under reduced pressure. Bulb-to-bulb distillation (Kugelrohr apparatus) gives (Z)-isosafrole (108 mg, 84.4%) as a colourless liquid, R_F 0.75 (fluorescent), i.r. (liquid film) 1500 (aryl-H), 1450, 1260, and 1040 (C—O), 940 and $820\,cm^{-1}$; p.m.r. $(CDCl_3, TMS)$ δ 6.8 (m, 3H, aryl Hs), 6.35 (dq, 1H, $J_{HMe} = 2\,Hz$, $J_{HH} = 11\,Hz$, CH=CHMe), 5.95 (s, 2H, OCH_2), 5.7 (dq, 1H, $J_{HMe} = 7\,Hz$, $J_{HH} = 11\,Hz$, $CHMe$) and 1.9 (dd, 3H, $J_{HMe} = 2$ and $7\,Hz$). Gas–liquid chromatography analysis (15% LAC-2R-446 on Chromosorb W, 2.7 m × 3.1 mm) shows that the product contains c. 4 per cent of the (E)-isomer.

Notes. (1) All reactions in non-aqueous solutions are carried out under a nitrogen atmosphere.
(2) Preparation of alkyldiphenylphosphine oxides. General procedure from phosphonium salts. Triphenyl phosphine is heated under reflux with an excess of alkyl halide. The precipitated phosphonium salt is filtered off, washed well with ether, and then heated with 30 per cent w/w aqueous sodium hydroxide (c. 4 ml/g) until all the benzene has distilled out. The mixture is cooled and extracted with dichloromethane, and the extracts are dried (magnesium sulphate) and evaporated to dryness. In this way ethyldiphenylphosphine oxide is obtained from triphenyl phosphine (65.6 g, 0.25 mol) and iodoethane (42.9 g, 0.275 mol) in dry toluene (250 ml) to give first the phosphonium salt (102.4 g, 97.9%) after 3.5 hours, from which the phosphine oxide is obtained as needles (53.2 g, 92.5%), m.p. 123–124 °C (from ethyl acetate); p.m.r. δ $(CDCl_3,$ TMS) 7.9–7.3 (m, 10H, Ph_2PO), 2.3 (m, 2H, CH_2) and 1.2 (dt, 3H, $J_{HMe} = 7$ Hz, $J_{MeP} = 17\,Hz$, Me).
(3) Dry tetrahydrofuran is freshly distilled from sodium wire using benzophenone radical as an indicator. Toluene and ether are dried by distillation from sodium wire and are stored over sodium. Dimethylformamide is dried by distillation from 4A molecular sieves and stored over 4A molecular sieves.
(4) Thin-layer chromatograms are run on commercially prepared pre-coated plates (Merck, Kieselgel $60F_{254}$) and eluted with ethyl acetate. Flash column chromatography is carried out using a 150 × 51 mm bed of Merck Kieselgel 60 (230–400 mesh) silica. Optimum separation of diastereoisomeric phosphine oxides is achieved by eluting with solvent which gives the midpoint between the isomers as c. R_F 0.45.

(E)-Isosafrole. *3,4-Methylenedioxybenzoic acid.* A solution of potassium permanganate (18 g, 0.114 mol) in water (360 ml) is added over 45 minutes to a mixture of 3,4-methylenedioxybenzaldehyde (12 g, 0.08 mol) and water (300 ml) stirred at 80 °C. The reaction mixture is stirred a further 1 hour at 80 °C, after which 10 per cent (w/w) aqueous potassium hydroxide (25 ml) is added and the mixture then filtered while hot. The filtrate is cooled to room temperature and the product precipitated by acidification (pH 2) with concentrated hydrochloric acid. The precipitate is collected and washed with water to give the acid as needles (10 g, 75.4%), m.p. 229–231 °C (from 95% ethanol), R_F 0.6, i.r. (Nujol mull) 2650–2500 (OH) and $1670\,cm^{-1}$ (C=O): p.m.r. $[(CD_3)_2SO, TMS]$ δ 12.4 (broad s, 1H, OH), 7.5 (dd, 1H, $J_{HH} = 2$ and 8 Hz, aryl H), 7.25 (d, 1H, $J_{HH} = 2\,Hz$, aryl H), 6.8 (d, 1H, $J_{HH} = 8\,Hz$, aryl H) and 6.0 (s, 2H, OCH_2).

Methyl 3,4-methylenedioxybenzoate. A solution of the foregoing acid (1.66 g, 0.01 mol) in methanol (40 ml) containing 5 drops of concentrated sulphuric acid is heated under reflux for 18 hours. The reaction is cooled to room temperature, basified with saturated sodium hydrogen carbonate, and the methanol removed under reduced pressure. The aqueous residues are diluted with water (50 ml), extracted with dichloromethane (3 × 20 ml) and the combined organic extracts dried (magnesium sulphate) and evaporated to dryness to give the ester (1.5 g, 83.3%), m.p. 49–51 °C [from light petroleum (b.p. 60–80 °C)], R_F 0.65; i.r. (Nujol mull) 1710 cm^{-1} (C=O); p.m.r. (CDCl$_3$, TMS) δ 7.6 (dd, 1H, J_{HH} = 2 and 8 Hz, aryl H), 7.4 (d, 1H, J_{HH} = 2 Hz, aryl H), 6.75 (d, 1H, J_{HH} = 8 Hz, aryl H), 5.95 (s, 2H, OCH$_2$) and 3.8 (s, 3H, Me).

2-Diphenylphosphinoyl-1-(3,4-methylenedioxyphenyl)propan-1-one. n-Butyllithium (1.5 M in hexane; 6.7 ml) is added dropwise from a syringe to a stirred solution of ethyldiphenylphosphine oxide (2.3 g, 0.01 mol) in dry tetrahydrofuran (35 ml) at 0 °C [see Notes (1), (2) and (3) above]. After 30 minutes the red reaction solution is cooled to − 78 °C (acetone–solid carbon dioxide) and a solution of methyl 3,4-methylenedioxybenzoate (900 mg, 5 mmol) in dry tetrahydrofuran (15 ml) is added dropwise from a syringe. The pale yellow solution is allowed to warm to room temperature before addition of water (20 ml) and removal of the tetrahydrofuran under reduced pressure. The aqueous residues are diluted with brine (15 ml) and extracted with dichloromethane (3 × 30 ml). The combined organic extracts are dried (magnesium sulphate) and evaporated to dryness to give an oil that crystallises with time. Flash column chromatography (elution with ethyl acetate–acetone, 2:1) gives the ketone as needles (1.6 g, 84.7% based on the ester), m.p. 163–165 °C (from ethyl acetate), R_F 0.35; i.r. (Nujol mull) 1765 (C=O, 1245 (C—O) and 1195 cm^{-1} (P=O); p.m.r. (CDCl$_3$, TMS) δ 8.1–7.2 (m, 12H, Ph$_2$PO, aryl Hs), 6.7 (d, 1H, J_{HH} = 8 Hz, aryl H), 5.95 (s, 2H, OCH$_2$), 4.5 (dq, 1H, J_{HMe} = 7 Hz, J_{HP} = 16 Hz, CHMe) and 1.5 (dd, 3H, J_{HMe} = 7 Hz, J_{MeP} = 16 Hz, Me); m.s. M$^{\oplus}$, 378.1043 (C$_{22}$H$_{19}$O$_4$P requires M, 378.1021), m/z 379 (3%, M + 1), 378 (33%), 201 (45%, Ph$_2$PO$^{\oplus}$) and 149 (100%, M − Ph$_2$POCH$_2$CH$_2$).

Reduction of the α-diphenylphosphinoyl ketone. Sodium borohydride (110 mg, 2.91 mmol) is added in one portion to a stirred solution of the ketone (1.1 g, 2.91 mmol) in ethanol (30 ml). The reaction mixture is heated under reflux for 3 hours, cooled to room temperature, and then saturated aqueous ammonium chloride (15 ml) is added. The ethanol is removed under reduced pressure and several drops of dilute hydrochloric acid are added to the aqueous residues. After dilution with brine (20 ml), the aqueous reaction mixture is extracted with dichloromethane (3 × 50 ml) and the combined organic extracts are dried (magnesium sulphate) and evaporated to dryness to give the product as a mixture of diastereoisomers. Separation by flash column chromatography (elution with ethyl acetate–acetone, 4:1) gives the (1RS,2SR)-phosphine oxide (*erythro*-isomer) (65 mg, 5.9%) and the (1RS,2RS)-phosphine oxide (*threo*-isomer) (1.007 g, 90.7%).

(E)-Isosafrole. The procedure is the same as that described for the preparation of the (Z)-isomer above. The (1RS,2RS)-phosphine oxide (*threo*-isomer) (500 mg, 1.32 mmol) and sodium hydride (80% dispersion in oil; 79 mg, 2.63 mmol) give, after distillation, (E)-isosafrole (184 mg, 86.4%) as a colour-

less liquid, R_F 0.75; i.r. (liquid film) 1500 (aryl-H), 1440, 1250 (C—O), 965 (C—H out-of-plane deform.), 940 and 785 cm^{-1}; p.m.r. (CDCl$_3$, TMS) δ 6.8 and 6.65 (two s, 3H, aryl Hs), 6.3 (d, 1H, J_{HH} = 15 Hz, CH=CHMe), 6.05–5.7 (m overlain by s at 5.8, total 3H, CHMe and OCH$_2$) and 1.8 (d, 3H, J_{HMe} = 5 Hz, Me). The (Z)-isomer is not detected by g.l.c. (column noted above).

Experiment 5.20   2,6-DIMETHYLHEPTA-1,5-DIENE[38]

To a stirred solution of trimethylsilylmagnesium chloride, made from chloro-methyltrimethylsilane (2.5 g, 0.02 mol) and magnesium (0.5 g, 0.02 mol) in sodium-dried ether (25 ml), is added a solution of 6-methylhept-5-en-2-one (2.5 g, 0.02 mol) (Aldrich) in sodium-dried ether (5 ml). The addition is drop-wise at such a rate that a gentle reflux of the mixture is maintained. The reaction mixture is refluxed with continued stirring. After 3 hours, the reaction mixture is cooled in an ice bath and thionyl chloride (1.8 ml, 0.025 mol) is added. The ice bath is removed and stirring continued at room temperature. After 1 hour, the reaction mixture is hydrolysed by the dropwise addition of a saturated solution of ammonium chloride. The coagulated solid is filtered off and washed with ether. Distillation of the combined filtrate-washings after drying gives 1.4 g (57%) of 2,6-dimethylhepta-1,5-diene as a clear colourless liquid, b.p. 135–136 °C; n_D^{23} 1.4385; p.m.r. (neat) δ 1.52 (s, 3H), 1.62 (s. 6H), 2.06 (m, 4H), 4.64 (s, 2H), and 5.06 (m, 1H).

5.2.4   SELECTED REARRANGEMENTS OF ALKYNES TO ALLENES

Unsaturated compounds containing two cumulated double bonds, i.e. $>$ C=C=C $<$, are known as allenes. In addition to methods involving elimination reactions and other reactions analogous to those used in the synthesis of alkenes, compounds of this type may be prepared by the rearrangement of suitably substituted alkynes. A particularly useful method of preparing simple allenic hydrocarbons involves the treatment of secondary or tertiary propargyl halides (e.g. (11)) with a zinc–copper couple in boiling ethanol. The mechanism of the reaction, which involves replacement of a halogen by hydrogen with propargylic bond migration, is not fully understood but a cyclic mechanism involving an organozinc intermediate (13) and the solvent has been postulated.

The process is illustrated by the conversion of 3-chloro-3-methylpent-1-yne into 3-methylpenta-1,2-diene (Expt 5.21). Isolation of the allene is facilitated by the fact that it forms an azeotrope with ethanol, which may be removed by washing with water. The required tertiary chloride is obtained by shaking 3-methylpent-1-yn-3-ol (14) with cold concentrated hydrochloric acid. Some of the 1-chloroallene (16) is additionally formed via the resonance stabilised carbo-cation ion (15). However, 1-chloroallenes are reduced by a zinc–copper couple to the corresponding allenes and hence the crude unsaturated chloride may be used directly for the allene synthesis.

Many secondary and tertiary propargyl alcohols give 1-haloallenes in good yield when treated with concentrated aqueous halo-acid in the presence of the corresponding copper(I) halide. Thus tertiary propargyl alcohols (17) react with 48 per cent w/w hydrobromic acid in the presence of a catalyst consisting of copper(I) bromide, copper powder and ammonium bromide over a temperature range of 0–35 °C to form the corresponding 1-bromoallene (18) in yields up to 80 per cent. A mechanism involving the intermediacy of an acetylene–copper(I) π-complex from which the 1-bromoallene is formed by a stereospecific $S_N i'$ process in which the configuration is retained, has been proposed.

$$HBr + Cu^IBr \rightleftharpoons H^\oplus + Cu^IBr_2^\ominus$$

$$(HO)\overset{\ominus}{C}u^IBr \xrightarrow{H^\oplus} Cu^IBr + H_2O$$

An example of this reaction is provided by the conversion of 2-methylbut-3-yn-2-ol to 1-bromo-3-methylbuta-1,2-diene (Expt 5.22).

Secondary propargyl alcohols (17; $R^2 = H$) require rather stronger acid and a longer reaction time, and the resulting bromoallenes may contain up to 5 per cent of the 3-bromoacetylenes, which may, however, be removed by fractional distillation.

Experiment 5.21 3-METHYLPENTA-1,2-DIENE

$$Et \cdot C(Me)(OH) \cdot C{\equiv}CH \xrightarrow{HCl} Et \cdot C(Me)(Cl) \cdot C{\equiv}CH \xrightarrow{Zn-Cu}$$
$$Et \cdot C(Me){=}C{=}CH_2$$

3-Chloro-3-methylpent-1-yne. Place 73.6 g (0.75 mol) of 3-methylpent-1-yn-3-ol and 750 ml of concentrated hydrochloric acid (d 1.18) in a 2-litre separating funnel and shake the mixture vigorously. Periodically remove portions of the upper organic layer and record the infrared spectrum as a liquid film; continue the shaking until the broad band centred at 3390 cm^{-1}, arising from the O—H stretching vibration of the alcohol, has completely disappeared (about 15 minutes). Discard the dark lower aqueous acid layer and dry the pale brown organic layer over anhydrous potassium carbonate for about 2 hours; a fresh portion of anhydrous potassium carbonate should be added after 30 minutes. Remove the drying agent by filtration. The yield of deep straw-coloured liquid, which consists of 3-chloro-3-methylpent-1-yne (c. 73%) and 1-chloro-3-methylpenta-1,2-diene together with traces of other rearrangement products, is about 70 g; it can be used without purification for the next stage. The chloroacetylene has $t_R \sim 2.2$ minutes on a 1.5-m column 10 per cent Silicone oil on Chromosorb W column held at 70 °C with a nitrogen flow rate of 40 ml/minute.

3-Methylpenta-1,2-diene.   Prepare a zinc–copper couple as follows. Place 65.4 g (1 mol) of zinc dust in a 250-ml conical flask and wash successively with four 50 ml portions of 3 per cent hydrochloric acid, two 50 ml portions of distilled water, two 100 ml portions of 2 per cent aqueous copper(II) sulphate, two 50 ml portions of distilled water, 100 ml of 95 per cent ethanol and 100 ml of absolute ethanol. Each washing operation should last for about 1 minute and the liquid should then be drained as thoroughly as possible by decantation before continuing with the next liquid wash. Finally transfer the zinc–copper couple by washing with 100 ml of absolute ethanol into a 250-ml three-necked, round-bottomed flask fitted with a reflux condenser, separatory funnel and an efficient (e.g. paddle-type) sealed stirrer. Heat the stirred zinc–ethanol suspension to boiling, remove the source of heat and then add the crude chloroacetylene (69.5 g) dropwise from the separatory funnel at such a rate so as to maintain gentle reflux; this should take about 50 minutes. Then heat the stirred reaction mixture under gentle reflux for 45 minutes, allow to cool slightly and rearrange the condenser for downward distillation. Distil the reaction mixture with stirring and collect the allene–ethanol azeotrope which passes over between 64 and 78 °C. Wash the colourless distillate with three portions of water (1 × 150 ml; 2 × 50 ml) and then separate the upper organic layer and dry over magnesium sulphate. Distil the crude product through a fractionating column (15 cm × 2 cm) filled with glass helices. Reject any material distilling below 70 °C, which contains appreciable quantities of acetylenic material, and collect 3-methylpenta-1,2-diene ($\geqslant$95% purity) as a colourless liquid, b.p. 70–72 °C; the yield is about 18.9 g (approximately 31% overall yield from 3-methylpent-1-yn-3-ol). Record the infrared spectrum as a liquid film. The strong absorption bands at 1964 cm^{-1} and 840 cm^{-1} are characteristic of the asymmetric C=C=C stretching and =CH$_2$ out-of-plane wagging vibrations respectively of a terminal allene. The presence of a trace of acetylenic impurity may be observed as a weak absorption at 3305 cm^{-1} (≡C—H stretch). The purity of the 1,2-diene may be checked by g.l.c. on a 1.5-m column of 10 per cent Silicone oil on Chromosorb W held at 61 °C with a nitrogen flow rate of 40 ml/minute; the compound has t_R 1.7 minutes.

Experiment 5.22 1-BROMO-3-METHYLBUTA-1,2-DIENE

$$Me_2C(OH)\cdot C\equiv CH \xrightarrow[\text{HBr/Cu}]{\text{NH}_4\text{Br/CuBr}} Me_2C=C=CHBr$$

In a 250-ml two-necked, round-bottomed flask fitted with a stirrer and a dropping funnel, place a mixture of 14.3 g (0.1 mol) of powdered copper(I) bromide, 9.8 g (0.1 mol) of powdered ammonium bromide, 0.64 g (0.01 mol) of copper powder and 80 ml ($c. 0.66$ mol) of 46–48 per cent w/w concentrated hydrobromic acid (d 1.46–1.49). Stir the dark green mixture vigorously at room temperature and add 25.2 g (0.3 mol) of 2-methylbut-3-yn-2-ol dropwise during 5 minutes; maintain the vigorous stirring for 1.5 hours (1). Filter the reaction mixture through a sintered glass Buchner funnel and rinse the reaction vessel and funnel with 10 ml of light petroleum (b.p. 40–60 °C). Transfer the filtrate to a separatory funnel, discard the lower aqueous layer and remove any unchanged alcohol and copper salts from the almost colourless organic layer by shaking with three 10 ml portions of concentrated hydrobromic acid. In the final washing the lower acid layer should show no trace of violet coloration. Dry the combined organic layers over a mixture of solid sodium hydrogen carbonate and magnesium sulphate, filter through a sintered glass Buchner funnel; rinse the flask and the inorganic solids with 5 ml of light petroleum (b.p. 40–60 °C). Transfer the filtrate to a 50-ml distillation flask fitted with a 9-cm fractionating column filled with glass helices and distil under reduced pressure, using a nitrogen bleed, from a bath held at 80–90 °C. Collect the 1-bromo-3-methylbuta-1,2-diene as a fraction, b.p. 52–54 °C/50 mmHg, in a flask cooled in ice-water (2). The yield is 28.1 g (64%). The purity of the product may be checked by g.l.c. analysis using a 1.5-m column of 10 per cent Silicone oil on Chromosorb W held at 80 °C with a nitrogen flow rate of 40 ml/minute; the material has t_R 3.5 minutes. The infrared spectrum (liquid film) shows the characteristic strong allene (C=C=C) absorption band at 1950 cm^{-1} (see Fig. 3.20). On standing 1-bromo-3-methylbuta-1,2-diene tends to isomerise to 1-bromo-3-methylbuta-1,3-diene and polymerise.

Notes. (1) The progress of the reaction may be monitored by occasionally halting the stirring and removing a portion of the clear upper organic layer for infrared examination. The reaction is complete when a sample does not show hydroxyl and alkynyl hydrogen absorption at 3400 and 3300 cm^{-1} respectively.
(2) In common with the distillation of all unsaturated compounds in general, the residue in the distillation flask should be cooled to room temperature before admitting air into the system.

5.3 ALKYNES

The location of the carbon–carbon triple bond in the carbon chain may be terminal (R·C≡CH) or non-terminal (R^1·C≡C·R^2). Illustrative representations are shown below for:

hex-1-yne (1) $CH_3\cdot CH_2\cdot CH_2\cdot CH_2\cdot C\equiv CH$;
and hex-2-yne (2) $CH_3\cdot CH_2\cdot CH_2\cdot C\equiv C\cdot CH_3$.

Me ⌇⌇⌇ ≡   Me ⌇⌇⌇ ≡

    (1)   (2)   Me

The synthesis of acetylenic compounds is exemplified by the following typical procedures.

1. The dehydrohalogenation of *vic*- and *gem*-dihalides (Expts 5.23 and 5.24).
2. The oxidation of dihydrazones of 1,2-diketones (Expt 5.25).
3. Alkylation of a terminal alkyne (Expt 5.26).
4. Coupling reactions leading to diynes (Expts 5.27 and 5.28).

SUMMARY OF RETROSYNTHETIC STRATEGIES

Functional group interconversion (FGI) (methods 1 and 2)

Disconnection (methods 3 and 4)

SPECTROSCOPIC FEATURES

The sharp characteristic *i.r.* absorption band in the region of $2200\,cm^{-1}$, arising from the stretching of the carbon–carbon triple bond, is of variable intensity (p. 283), but the absorption is less intense than in the case of the nitrile group. Terminal alkynes are easily recognised from the strong sharp absorption band in the region of $3320\,cm^{-2}$ arising from the stretching vibration of the carbon–hydrogen bond (the spectrum of phenylacetylene, Fig. 3.19). The alkynyl protons and the α-protons of alkyl groups attached to the alkynyl carbons are relatively more shielded due to magnetic anisotropic effects, and therefore resonate in the *p.m.r.* spectra at a higher field than the corresponding protons in substituted alkenes. The m.s. of alkynes show a fragmentation behaviour which bears a close resemblance to that of alkenes (p. 374). Conjugated poly-ynes and enynes show absorption in the *u.v.–visible* spectra, and the position of the absorption maxima may provide a valuable guide to the extent of conjugation (p. 388).

Illustrative spectroscopic interpretations are given in some of the preparative sections below.

5.3.1 THE DEHYDROHALOGENATION OF *VIC-* and *GEM*-DIHALIDES

A simple method for introducing a triple bond into an organic compound is to treat an appropriate dihalide with a strong base. Since vicinal dihalides (usually the bromide) are readily formed by reaction of bromine with an alkene, and geminal dihalides from aldehydes or ketones with phosphorus pentachloride, the method is a useful general procedure for the preparation of terminal and non-terminal alkynes from readily available starting materials.

$$R^1 \cdot CH{=}CH \cdot R^2 \longrightarrow R^1 \cdot CHBr \cdot CHBr \cdot R^2 \longrightarrow R^1 \cdot C{\equiv}C \cdot R^2$$
$$R^1 \cdot CO \cdot CH_2 \cdot R^2 \longrightarrow R^1 \cdot CCl_2 \cdot CH_2 \cdot R^2 \longrightarrow R^1 \cdot C{\equiv}C \cdot R^2$$

The products from this reaction are not invariably homogeneous since the first formed alkyne may isomerise, probably by way of an intermediate allene. With alcoholic potassium hydroxide at 170–180 °C, for example, terminal alkynes tend to rearrange to internal alkynes. The rearrangement process may be minimised by using sodamide in liquid ammonia as the reagent.

$$R \cdot CH_2 \cdot C{\equiv}CH \rightleftharpoons R \cdot CH{=}C{=}CH_2 \rightleftharpoons R \cdot C{\equiv}C \cdot CH_3$$

Examples of the use of alcoholic potassium hydroxide are given in Expt 5.23 for the conversion of 10,11-dibromoundecanoic acid (itself prepared from undec-10-enoic acid and bromine), into undec-10-ynoic acid, and of ethyl cinnamate into phenylpropynoic acid.

Recently PTC methods for the dehydrohalogenation process have been developed. One uses tetrabutylammonium hydrogen sulphate with 50 per cent sodium hydroxide and pentane-dichloromethane as the organic solvent[39]; a second uses 18-crown-6 with solid potassium t-butoxide and light petroleum as solvent.[40] A third procedure uses tetraoctylammonium bromide as the phase transfer catalyst with solid potassium hydroxide in a light petroleum medium.[41] In general, with these procedures, 1,2-dihalides, and 1,1-dihalides from aldehydes, give alk-1-ynes; internal geminal dihalides from symmetrical ketones give internal alkynes; 2,2-dihalides from methyl ketones give alk-1-ynes only if C-3 is a tertiary carbon.

The loss of a bromide ion when situated in a β-position to a carboxylate group occurs very readily under mildly basic conditions. Thus in an alternative synthesis of phenylacetylene (Expt 5.24), cinnamic acid dibromide is converted into β-bromostyrene under the influence of hot aqueous sodium carbonate solution. Dehydrobromination to yield the acetylenic compound is then achieved in the usual manner with potassium hydroxide.

509

Experiment 5.23 UNDEC-10-YNOIC ACID

$$CH_2\!=\!CH\cdot(CH_2)_8\cdot CO_2H \xrightarrow{Br_2} CH_2Br\cdot CHBr\cdot(CH_2)_8\cdot CO_2H \xrightarrow{KOH}$$
$$CH\!\equiv\!C\cdot(CH_2)_8\cdot CO_2H$$

Purify commercial undecylenic acid by distillation of, say, 250 g under diminished pressure and collect the fraction, b.p. 152–154 °C/6 mmHg; this has a freezing point of 23 °C. Dissolve 108 g (0.58 mol) of the purified undecylenic acid in 285 ml of dry carbon tetrachloride (**CAUTION**, p. 399) in a 1-litre three-necked flask provided with a sealed stirrer unit, a dropping funnel and a reflux condenser. Cool the flask in a freezing mixture of ice and salt, stir the solution and add 96 g (31 ml, 0.6 mol) of dry bromine (Section 4.2.9, p. 422) during a period of 1 hour: allow the mixture to warm up gradually to the temperature of the laboratory. Remove the carbon tetrachloride with a rotary evaporator and pour the residue into a large evaporating dish (fume cupboard). Upon standing 1–2 days (more rapidly when left in a vacuum desiccator over silica gel), the dibromo acid crystallises completely. The yield is quantitative.

Transfer the solid dibromo acid to a 2-litre round-bottomed flask attached to a reflux condenser, add a solution of 263 g of potassium hydroxide in 158 ml of water, and heat in an oil bath at 150–160 °C for 8 hours. Considerable frothing occurs, but this is reduced by the addition of small quantities (about 0.1 g) of a suitable detergent, e.g. sodium dodecyl benzenesulphonate, from time to time. Allow the mixture to stand overnight, add 1500 ml of water, shake until all the solid dissolves and acidify with dilute sulphuric acid to Congo red. A solid cake of acid separates on the surface of the liquid after standing for several hours. Extract with four 250 ml portions of ether, dry with anhydrous sodium or magnesium sulphate and remove the ether by distillation on a water bath. Transfer the residue to a 250-ml flask fitted with a Claisen still-head and distil cautiously under diminished pressure using a free flame. A little ether and water pass over first and the temperature rises rapidly to 175 °C/15 mmHg. Collect separately the fractions (a) b.p. 177–182 °C/15 mmHg (52 g) and (b) 182–200 °C/15 mmHg (15 g). The flask contains a large residue, which is discarded. Fraction (a) solidifies completely on cooling and has m.p. 37–41 °C; upon recrystallisation from light petroleum, b.p. 60–80 °C, 34 g (32%) of pure undec-10-ynoic acid, m.p. 41–42 °C, are obtained. A further quantity of product is obtained from fraction (b), which solidifies to a slightly sticky solid: upon recrystallisation from light petroleum, b.p. 60–80 °C, a sticky solid separates, which, after spreading upon a porous tile, becomes colourless and has m.p. 41–52 °C (3 g).

Cognate preparations. *Phenylpropynoic acid.* Place a solution of 88 g (84 ml, 0.5 mol) of ethyl cinnamate (Expt 6.137) in 50 ml of carbon tetrachloride in a 500-ml round-bottomed flask. Immerse the flask in ice and add 80 g (25.5 ml, 0.5 mol) of bromine from a separatory funnel slowly with frequent shaking. The halogen will disappear rapidly at first, but more slowly towards the end of the reaction; no hydrogen bromide is evolved and the time of the addition is about 20–25 minutes. Allow the mixture to stand for 1 hour, pour the solution into a large evaporating dish and permit the excess of bromine and the carbon tetrachloride to evaporate spontaneously in the fume cupboard. The crude ethyl 2,3-dibromo-3-phenylpropanoate will remain as a solid cake; this can be dried by pressing between large filter papers. The yield of crude ester, m.p. 66–71 °C, is 140 g (83%) (1).

Dissolve 85 g of potassium hydroxide in 400 ml of rectified spirit by heating in a 1500-ml round-bottomed flask, provided with a reflux condenser, on a water bath. Cool to 40–50 °C, and add 112 g (0.33 mol) of the crude dibromo ester; when the initial exothermic reaction has subsided, heat the mixture on a water bath for 5–6 hours. Pour the contents of the flask into a large beaker and, when cold, add concentrated hydrochloric acid with stirring until neutral to litmus. Cool, filter the precipitated solids at the pump and wash with a little alcohol. Set the solids (A) aside. Transfer the filtrate to the original flask and distil the liquid until the temperature of the vapour reaches 95 °C. Combine the residue in the flask with the precipitated solids (A), dissolve in 270 ml of water, add about 300 g of crushed ice and cool the flask in an ice bath. Stir the mixture mechanically, and add 20 per cent sulphuric acid slowly until the solution is strongly acid to Congo red. Allow to stand for 20 minutes, filter off the dark-coloured crude phenylpropynoic acid at the pump and wash it with three 15 ml portions of 2 per cent sulphuric acid. Dissolve the solid in about 300 ml of 5 per cent sodium carbonate solution, add 6 g of decolourising charcoal and heat on a water bath for 30 minutes with occasional shaking. Filter through a fluted filter paper, cool the filtrate in ice and then add 70 g of crushed ice. Stir the solution mechanically and add 20 per cent sulphuric acid slowly until acid to Congo red. After 20 minutes, filter the precipitated acid by suction, wash with 15 ml of 2 per cent sulphuric acid, then with a little water, and dry in the air. The yield of pure phenylpropynoic acid, m.p. 134–135 °C, is 23 g (47%).

Note. (1) To obtain the pure dibromo ester, recrystallise from light petroleum, b.p. 60–80 °C; the recovery of the pure ester, m.p. 75 °C, is 85 per cent.

Experiment 5.24 PHENYLACETYLENE

$$\text{Ph·CH=CH·CO}_2\text{H} \xrightarrow{\text{Br}_2} \text{Ph·CHBr·CHBr·CO}_2\text{H} \xrightarrow{\text{Na}_2\text{CO}_3}$$
$$\text{Ph·CH=CHBr} \xrightarrow{\text{KOH}} \text{Ph·C≡CH}$$

Dissolve 74 g (0.5 mol) of cinnamic acid in 300 ml of hot dichloromethane in a 500-ml flask and cool the solution in ice-water with shaking. As soon as the solid begins to crystallise out, add a solution of 80 g (26 ml, 0.5 mol) of bromine (*care in handling*) in 50 ml of dichloromethane rapidly in three portions with vigorous shaking and cooling. Allow the flask and contents to stand in an ice-bath for 30 minutes to allow complete crystallisation of the product; collect the latter by filtration. Obtain a pure specimen of 2,3-dibromo-3-phenylpropanoic acid (m.p. 204 °C decomp.) by recrystallising a small sample of the crude product from aqueous ethanol. Boil the bulk of the crude bromoacid under reflux with 750 ml of 10 per cent aqueous sodium carbonate solution, cool and separate the layer of crude β-bromostyrene. Extract the aqueous phase with two 75 ml portions of ether, combine this extract with the organic phase, dry over anhydrous calcium chloride and remove the ether on a rotary evaporator. About 65–70 g of crude β-bromostyrene is obtained.

Place 100 g of potassium hydroxide pellets in a 500-ml flask, moisten the pellets with about 2 ml of water and fit the flask with a still-head carrying a dropping funnel and a condenser set for downward distillation. Heat the flask in an oil bath maintained at 200 °C and add the crude β-bromostyrene drop-wise on to the molten alkali at a rate of about 1 drop per second. Phenyl-acetylene begins to distil over; slowly raise the bath temperature to about 220 °C and keep it at this point until the addition is complete. Then continue to heat at about 230 °C until no more product distils over. Separate the upper layer of the distillate, dry it over potassium hydroxide pellets and redistil. Collect the phenylacetylene at 142–144 °C: the yield is 25 g (49%). The p.m.r. spectrum (CDCl$_4$, TMS) exhibits signals at δ 3.08 (s, 1H) and 7.35 (m, 5H).

5.3.2 THE OXIDATION OF DIHYDRAZONES OF 1,2-DIKETONES

This method has been of particular value for the synthesis of cyclic acetylenes, the diketones having been prepared from the corresponding acyloins (Section 5.9.1, p. 628).

An early procedure used triethyl phosphate directly on the diketone, but better yields are obtained by the oxidation of the corresponding dihydrazone.[42] The oxidant may be mercury(II) oxide, which is rather expensive;[43] alternatively copper(I) chloride in dichloromethane and pyridine is oxidised with oxygen gas, and the derived complex is then used to oxidise the dihydrazone to the acetyl-enic group with the evolution of nitrogen.[44] The reaction is illustrated by the conversion of benzil dihydrazone into diphenylacetylene (Expt 5.25).

Experiment 5.25 DIPHENYLACETYLENE[44]

CAUTION: *Great care should be taken in the handling of hydrazine hydrate since it is highly corrosive and a suspected carcinogen; protective gloves and a face visor should be worn.*

Benzil dihydrazone. A solution of 105.1 g (0.5 mol) of benzil in 325 ml of propan-1-ol is prepared in a round-bottomed flask which is fitted with a reflux condenser. To this solution 76 g (1.30 mol) of 85 per cent hydrazine hydrate is added and the mixture heated under reflux for 60 hours. The solution is cooled in an ice bath and the benzil dihydrazone is separated by suc-tion filtration. The crystals are washed with 200 ml of cold absolute ethanol and dried on the suction filter for 1 hour. The yield of benzil dihydrazone is 99–106 g (83–89%), m.p. 150–151.5 °C.

Diphenylacetylene. Reagent grade copper(I) chloride (0.403 g, 4.0 mmol) is dissolved in a mixture of dry dichloromethane (20 ml) and pyridine (2 ml, **CAUTION**) to give a yellow–green clear solution. Oxygen gas from a gas burette is introduced to the solution with vigorous stirring. Rapid absorption of oxygen (about 20 ml) ceases in 15 minutes, the mixture becoming a dark green suspension. To the mixture is added dropwise a solution of benzil dihydrazone (0.237 g, 1.0 mmol) in dichloromethane (10 ml) over a period of 10 minutes at room temperature with stirring. The slightly exothermic reaction proceeds smoothly with nitrogen evolution. Stirring is continued for 1 hour under an oxygen atmosphere to ensure the completion of the oxidation. The mixture is hydrolysed with 3 M hydrochloric acid (40 ml), the organic layer is separated, and the aqueous layer extracted twice with dichloromethane. The combined organic layer and extracts are washed with brine, dried over magnesium sulphate, and evaporated to leave the crude diphenylacetylene, which is purified by column chromatography (silica gel, hexane), to give 0.173 g (97%) of pure diphenylacetylene, m.p. 59–60 °C.

5.3.3 ALKYLATION OF A TERMINAL ALKYNE

Terminal alkynes by virtue of the presence of an acidic hydrogen atom can be converted into the corresponding alkynylsodium or alkynylmagnesium halide (see Section 5.4.2, p. 531) which may then be alkylated to give a homologous alkyne.

$$R^1 \cdot C{\equiv}CH + NaNH_2 \longrightarrow R^1 \cdot C{\equiv}C^{\ominus}Na^{\oplus} \xrightarrow{R^2Br} R^1 \cdot C{\equiv}C \cdot R^2$$

Acetylene itself can form both the mono- or di-sodium salt; the former is the main product when a large excess of acetylene is used in the reaction with sodamide in liquid ammonia.

In these alkylation reactions primary alkyl halides (the bromide for preference) should be used as the alkylating agents, since secondary and tertiary halides undergo extensive olefin-forming elimination reactions in the presence of the strongly basic acetylide ion. A typical synthesis is that of hex-1-yne (Expt 5.26).

A modified procedure which is suitable for the introduction of a secondary or tertiary alkyl group uses the trialkynylalane as the intermediate. This is formed from aluminium chloride and an alkynyllithium, and then subsequently allowed to react with the tertiary or secondary halide.[45]

$$Bu \cdot C{\equiv}CH \xrightarrow{BuLi} Bu \cdot C{\equiv}C^{\ominus}Li^{\oplus} \xrightarrow{\frac{1}{3}AlCl_3} Bu \cdot C{\equiv}C \cdot Al/3$$

$$Bu \cdot C{\equiv}C \cdot Al/3 + Me_3CCl \longrightarrow Bu \cdot C{\equiv}C \cdot Bu^t$$

Experiment 5.26 HEX-1-YNE (*Butylacetylene*)

$$CH{\equiv}CH + NaNH_2 \xrightarrow{liq.\ NH_3} CH{\equiv}C^{\ominus}Na^{\oplus} + NH_3$$

$$BuBr + {}^{\ominus}C{\equiv}CH \xrightarrow{liq.\ NH_3} Bu \cdot C{\equiv}CH + {}^{\ominus}Br$$

Equip a 5-litre three-necked flask (or a 5-litre flange flask fitted with a multiple head adapter) with a Herschberg or other efficient stirrer, a soda-lime guard-tube and a liquid ammonia inlet tube as described in Section 2.17.7, (p. 116). Observing the precautions noted in this section, run in liquid ammonia until the flask is about two-thirds full (c. 3.5 litres). Prepare a suspension of sodamide from 138 g (6 mol) of clean sodium (Section 4.2.68, p. 462) using 0.5 g of finely-powdered crystallised iron(III) nitrate as the catalyst as described in detail in Section 4.2.67 (p. 462). It may be necessary to add further quantities of liquid ammonia to maintain the volume of 3.5 litres. When the conversion of sodium into sodamide is complete, replace the ammonia-addition tube by a wide tube reaching almost to the bottom of the flask to allow for the passage of acetylene through the suspension of sodamide. The acetylene gas from a cylinder should be freed from acetone by passing through two Drechsel bottles half-filled with concentrated sulphuric acid; when the acid in the second wash bottle becomes discoloured the wash bottles should be recharged with fresh acid. An empty Drechsel bottle (to act as a safety trap) and a mercury escape valve (cf. Fig. 2.60) should be interposed between the reaction flask and the Drechsel bottles charged with sulphuric acid. Surround the reaction flask with an acetone–Cardice cooling bath and pass acetylene rapidly (2–3 litres per minute) into the sodamide suspension until a uniformly black liquid is formed (usually 4–5 hours) (1). Carefully watch the mercury escape valve in case the wide entry tube becomes blocked by deposition of solid. If this should happen, temporarily stop the acetylene flow and clear the tube by inserting a glass rod of appropriate diameter. It may also become necessary to introduce some more liquid ammonia. Replace the soda-lime guard-tube with a pressure-equalising separatory funnel charged with 685 g (538 ml, 5 mol) of butyl bromide which is introduced over a period of 1.5–2 hours with stirring and while a slow stream of acetylene (c. 500 ml per minute) continues to pass through the reaction mixture. The reaction is exothermic so that it will be necessary to maintain the cooling bath at $-50\,°C$ by continued addition of solid carbon dioxide. When all the alkyl bromide has been added, discontinue the supply of acetylene and allow the ammonia to evaporate. Before evaporation is complete add cautiously 60 g of ammonium chloride with stirring to decompose the excess of sodium acetylide (or sodamide) if present. Now introduce 500 g of crushed ice followed by about 1.5 litres of distilled water. Steam distil the reaction mixture, separate the hydrocarbon layer in the distillate, dry over magnesium sulphate and fractionally distil through a Hempel column filled with Fenske rings. Collect the hex-1-yne, b.p. 71–72 °C. The yield is 280 g (68%). The i.r. absorptions occur at 3300 ($\equiv$CH) and 2120 cm^{-1} (C$\equiv$C). The ^{13}C-n.m.r. (CDCl$_3$, TMS) has absorptions at δ 13.7 (C$_6$), 18.3 (C$_5$), 22.1 (C$_4$), 30.9 (C$_3$), 68.4 (C$_2$) and 84.5 (C$_1$).

Note. (1) Occasionally the reaction mixture does not become completely black nor free from suspended solid; here the acetylide is an insoluble (or sparingly soluble) form, but it gives satisfactory results in the preparation of hex-1-yne. The saturated solution of the soluble forms of sodium acetylide in liquid ammonia at $-34\,°C$ is about 4.1 M.

5.3.4 COUPLING REACTIONS LEADING TO DIYNES

Oxidative coupling of a terminal alkyne is a particularly easily performed carbon–carbon σ-bond forming reaction, which results in a good yield of the symmetrical diacetylene. A widely used procedure involves the oxidation of the alkyne with air or oxygen in aqueous ammonium chloride in the presence of a copper(I) chloride catalyst (*Glaser oxidative coupling*).

$$2R \cdot C \equiv CH + [O] \xrightarrow{Cu^{\oplus}} R \cdot C \equiv C \cdot C \equiv C \cdot R + H_2O$$

A modified procedure illustrated by the synthesis of a diynediol (Expt 5.27) from the acetylenic alcohol, 2-methylbut-3-yn-2-ol, uses methanol and pyridine as the solvent, the latter acting also as a complexing agent for the copper(I) ions. An alternative effective coupling system involves the use of copper(II) acetate in pyridine which does not require the use of air or oxygen.

A coupling procedure particularly suited to the synthesis of unsymmetrical diacetylenes involves the reaction of a terminal acetylene with a 1-bromo-acetylene in the presence of a catalyst consisting of a solution of copper(I) chloride in a primary amine to which small quantities of hydroxylamine hydrochloride is added (the *Cadiot–Chodkiewicz coupling*).

$$R^1 \cdot C \equiv CH + Cu^{\oplus} \longrightarrow R^1 \cdot C \equiv CCu + H^{\oplus}$$

$$R^1 \cdot C \equiv CCu + BrC \equiv C \cdot R^2 \longrightarrow R^1 \cdot C \equiv C \cdot C \equiv C \cdot R^2 + Cu^{\oplus} + Br^{\ominus}$$

The organic base serves to remove the liberated protons and to assist solution of the copper(I) catalyst by the formation of a complex. Hydroxylamine hydrochloride helps to maintain an adequate concentration of copper(I) ions, which however is best kept rather low otherwise unwanted self-coupling of the bromoalkyne occurs. The reaction is illustrated in Expt 5.28 by the coupling of ω-bromophenylacetylene with 2-methylbut-3-yn-2-ol.

The coupling of terminal acetylenic compounds with the reactive allylic bromides in the presence of 1,8-diazabicyclo[5.4.0]undec-7-ene (DBU) and copper(I) iodide appears to be a useful route to the non-conjugated enyne system, e.g. the formation of hex-5-en-2-yn-1-ol.[46]

$$\text{HO} \diagdown \text{—C} \equiv \text{CH} + \text{Br} \diagup \diagdown \text{CH}_2 \xrightarrow[\text{CuI}]{\text{DBU}} \text{HO} \diagdown \text{—C} \equiv \text{C} \diagdown \diagup \text{=CH}_2$$

Experiment 5.27 2,7-DIMETHYLOCTA-3,5-DIYNE-2,7-DIOL

$$2Me_2C(OH) \cdot C \equiv CH \xrightarrow[\text{pyridine}]{O_2: CuCl} Me_2C(OH) \cdot C \equiv C \cdot C \equiv C \cdot (OH)CMe_2$$

Place in a 250-ml conical flask 17.4 g of 2-methylbut-3-yn-2-ol (20 ml, 0.207 mol), 20 ml of methanol, 6 ml of dry pyridine (0.074 mol, **CAUTION**), 1 g of copper(I) chloride (0.010 mol) and a magnetic following bar. Then, with fairly vigorous stirring, bubble oxygen through the solution for 2 hours, at a rate of at least 10 litre/hour. The solution becomes warm and rapidly turns dark green. Cool the reaction mixture using an ice-water cooling bath and add 10 ml of concentrated hydrochloric acid to neutralise the pyridine and to keep copper salts in solution. Add 50 ml of a saturated aqueous sodium chloride solution, cool and filter off the precipitated product. Wash the filtered

515

solid with a little ice-cold water to remove any remaining colour and recrystallise it from dry toluene. This also serves to dry the product if the toluene/water azetrope is removed by using a Dean and Stark water-separator [Fig. 2.31(*a*)]. Cool the toluene solution and filter the colourless crystalline product; 12 g (70%) of alkynol, m.p. 132–133 °C, is obtained. Further recrystallisation from toluene raises the m.p. to 133–133.5 °C.

Experiment 5.28 2-METHYL-6-PHENYLHEXA-3,5-DIYN-2-OL

$$Ph \cdot C \equiv CH + HOBr \longrightarrow Ph \cdot C \equiv CBr + H_2O$$

$$Me_2C(OH) \cdot C \equiv CH + BrC \equiv C \cdot Ph \xrightarrow[\substack{(CH_3)_2NH \\ NH_2OH}]{H_2O,\ CuCl}$$

$$Me_2C(OH) \cdot C \equiv C \cdot C \equiv C \cdot Ph + HBr$$

ω-**Bromophenylacetylene.** Prepare a solution of sodium hypobromite as follows. To a 250-ml conical flask fitted with a ground glass joint add 60 g of ice, 30 ml of 10 M aqueous sodium hydroxide (0.30 mol) and 21.8 g (7.0 ml, 0.136 mol) of bromine. Swirl for a few seconds to dissolve the bromine, then add 13.0 g (14.0 ml, 0.127 mol) of phenylacetylene (Expt 5.24) (1). Securely stopper the flask, and shake vigorously for 5 hours on a mechanical shaker. Extract the product from the reaction mixture with three 50 ml portions of ether, wash the combined extracts with water and dry the ethereal layer with anhydrous sodium sulphate. Filter the solution and remove the ether on a rotary evaporator. ω-Bromophenylacetylene (2) sufficiently pure for use in the following reaction is obtained as a light yellow oil (3); the yield is 20.4 g (89%).

2-**Methyl-6-phenylhexa-3,5-diyn-2-ol.** In a 100-ml conical flask place 40 ml of 25–30 per cent aqueous dimethylamine solution (0.22–0.27 mol), 200 mg (0.002 mol) of copper(I) chloride (Section 4.2.22, p. 428) and 650 mg (0.009 mol) of hydroxylamine hydrochloride. Add 12 ml (10.42 g, 0.124 mol) of 2-methylbut-3-yn-2-ol, add a magnetic follower bar and chill in an ice-water bath. Then, whilst stirring magnetically, and with continued cooling, add from a dropping pipette 20.4 g (0.113 mol) of ω-bromophenylacetylene and ensure complete transfer of the bromo compound by rinsing several times the pipette and container with reaction mixture. When the addition has been completed, continue stirring and cooling for a further 10 minutes and then add 25 ml of 5 per cent aqueous potassium cyanide to complex the copper salts. Extract the solution with four 40 ml portions of ether and dry the combined extracts with anhydrous sodium sulphate. Filter and evaporate the ether solution using a rotary evaporator. Purify the solid product by recrystallisation from light petroleum (b.p. 80–100 °C), and wash the filtered product with light petroleum (b.p. 40–60 °C). The yield of pure diynol (4), m.p. 55–58 °C, is 15.3 g (73%).

Notes. (1) *Phenylacetylene has a penetrating and persistent odour and is mildly lachrymatory. It is recommended that it be handled in the fume cupboard and that protective gloves be worn.*
(2) ω-*Bromophenylacetylene has a very penetrating and persistent odour and is mildly lachrymatory. It should be stored in the fume cupboard and care taken during subsequent handling.*

(3) Gas–liquid chromatography analysis using a 1.5 m S.E. 52 chromatographic column, at 120 °C with a nitrogen flow rate of 45 ml/minute, gives retention times for phenylacetylene and ω-bromophenylacetylene of 2.16 and 0.76 minutes respectively.
(4) The purity of this alkynol may be checked by g.l.c. analysis using an S.E. 52 chromatographic column. With a 1.5-m column, at 170 °C and with a nitrogen gas flow rate of 45 ml/minute, the retention time is 3.84 minutes. Similar conditions using an S.E. 30 column give a retention time of 3.16 minutes. The two reactants have very much shorter retention times [see Note (3) above].

5.4 ALIPHATIC ALCOHOLS

The hydroxyl group in saturated alcohols may be located on a primary carbon atom ($R \cdot CH_2OH$), a secondary carbon atom (R_2CHOH) or a tertiary carbon atom (R_3COH). The hydroxyl-carrying carbon (the α-carbon) in secondary and in tertiary alcohols may be a chiral site (*), as shown below for:
(S)-hexan-2-ol (1) $CH_3 \cdot CH_2 \cdot CH_2 \cdot CH_2 \cdot CH(OH) \cdot CH_3$;
and (S)-2,3-dimethylpentan-3-ol (2) $CH_3 \cdot CH(CH_3) \cdot C(CH_3)(OH) \cdot CH_2 \cdot CH_3$.

Vinylic alcohols [$R \cdot CH = CH(OH)$ or $R \cdot CH = C(OH) \cdot R$] are transient tautomeric isomers of the corresponding aldehydes ($R \cdot CH_2 \cdot CHO$) or ketones ($R \cdot CH_2 \cdot CO \cdot R$) respectively. Allylic alcohols (e.g. $R \cdot CH = CH \cdot CH_2OH$) and homoallylic alcohols (e.g. $R \cdot CH = CH \cdot CH_2 \cdot CH_2OH$) are important species and are considered in Section 5.18.1, p. 794.

The stereoisomerism of vicinal diols of the type $R^1 \cdot CH(OH) \cdot CH(OH) \cdot R^2$ is often denoted by the *erythro/threo* notation [e.g. formulae of pentane-2,3-diol, (3) and (4) below], which originates from the configuration of erythrose and threose (p. 639). This notation is applied in other instances (see p. 497).

erythro ($\pm$)-pair
(3)

threo ($\pm$)-pair
(4)

The synthesis of alcohols is exemplified by the following typical procedures.

1. The reduction of aldehydes, ketones and esters (Expts 5.29 to 5.38).
2. The interaction of carbonyl-containing compounds with organometallic reagents (Expts 5.39 to 5.43).
3. The hydroboration–oxidation of alkenes (Expts 5.44 and 5.45).
4. The oxymercuration–demercuration of alkenes (Expt 5.46).
5. The hydroxylation of alkenes (Expts 5.47 and 5.48).

Methods for the protection of the hydroxyl group are considered in Section 5.4.6, p. 550.

SUMMARY OF RETROSYNTHETIC STRATEGIES

Functional group interconversion (FGI) (methods 1, 3, 4 and 5)

Disconnection (method 2)

Reagent equivalents:

$$R^{\ominus} \equiv RMgX$$

$$\underset{R'}{\overset{R}{>}}\!\!\overset{\oplus}{-}OH \equiv H^{\oplus} + \underset{R'}{\overset{R}{>}}\!\!=\!O$$

$$\overset{\oplus}{CH_2}OH \equiv H^{\oplus} + \triangle\!O$$

$$R\!-\!\overset{OH}{\underset{2\oplus}{\diagdown}} \equiv H^{\oplus} + \left[R\!-\!\overset{\oplus}{C}\!=\!O \equiv R\!-\!\overset{O}{\diagdown}_{OEt} \right]$$

$$R\!-\!\overset{OH}{\underset{H}{\overset{\oplus}{<}}} \equiv H^{\oplus} + R\!-\!\overset{O}{\underset{H}{\diagdown}}$$

$$\overset{3\oplus}{C}\!-\!OH \equiv H^{\oplus} + \left[\overset{2\oplus}{C}\!=\!O \equiv \underset{EtO}{\overset{O}{\diagdown}}\!\!\overset{}{\diagdown}_{OEt} \right]$$

R = H or alkyl

SPECTROSCOPIC FEATURES

An alcohol, either neat or in relatively concentrated solution, shows characteristic strong broad absorption in the 3400–3200 cm^{-1} region of the i.r. spectrum, due to the stretching of the intermolecular hydrogen-bonded hydroxyl group. In very dilute solution the non-hydrogen-bonded hydroxyl group shows sharp absorption in the region of 3600 cm^{-1} [e.g. heptan-1-ol, Fig. 3.23(a)–(c)]. The hydroxyl-hydrogen is readily recognised in the p.m.r. spectrum since it undergoes exchange with deuterium oxide under suitable conditions (e.g. ethanol, p. 348); the signals corresponding to the alkyl group (or groups) may sometimes be used for structural characterisation (see p. 341 for spin–spin splitting patterns). The ^{13}C-n.m.r. spectra of ethanol and butane-1,4-diol are shown in Fig. 3.50 and Fig. 3.74 respectively. The molecular ion is not usually observed in the m.s. and the highest fragment ion has a m/z value corresponding to $(M - H_2O)$ [e.g. propan-1-ol (Fig. 3.81)]. Alcohols do not absorb in the accessible region of the u.v.–visible spectrum. Illustrative examples of spectroscopic interpretation are provided in the preparative examples below.

5.4.1 THE REDUCTION OF ALDEHYDES, KETONES AND ESTERS

Primary and secondary alcohols may be synthesised by the reduction of the corresponding carbonyl compounds by a great variety of reagents.

$$R\cdot CHO \xrightarrow{[H]} R\cdot CH_2OH \qquad R^1\cdot CO\cdot R^2 \xrightarrow{[H]} R^1\cdot CH(OH)\cdot R^2$$

Two reductive systems for ketones, which have the merit of being economic for large-scale preparations, are sodium and absolute ethanol (Expt 5.29), and zinc dust and aqueous sodium hydroxide (Expt 5.30).

These dissolving metal reductions have the disadvantage of being relatively unselective. Potassium borohydride and sodium borohydride however each show a considerable degree of selectivity; thus aldehydes and ketones may be reduced to alcohols while halogeno, cyano, nitro, amido and alkoxycarbonyl groups remain unaffected. The reductions of chloral and m-nitrobenzaldehyde (Expt 5.31) to the corresponding substituted alcohols are illustrative of this selectivity. The reagents are used in aqueous ethanolic solution. The essential step in the mechanism of these reactions involves a hydride ion transfer to the carbonyl carbon from the borohydride anion, which is capable of reducing 4 moles of carbonyl compound. Decomposition of the resulting anionic complex with water or dilute acid liberates the required alcohol.

$$R \cdot \overset{\underset{H}{|}}{C}{=}O \quad H{-}\overset{\ominus}{B}H_3 \longrightarrow R \cdot CH_2O^\ominus + BH_3 \longrightarrow R \cdot CH_2O\overset{\ominus}{B}H_3$$

$$R \cdot CH_2O\overset{\ominus}{B}H_3 \xrightarrow{3R \cdot CHO} (R \cdot CH_2O)_4\overset{\ominus}{B} \xrightarrow{H_3O^\oplus} 4R \cdot CH_2OH + B(OH)_3$$

PTC procedures using tetrabutylammonium borohydride have been described.[47]

Aldehydes and ketones can be selectively reduced to the corresponding alcohols by aluminium alkoxides. The most satisfactory alkoxide for general use is aluminium isopropoxide.

$$3R^1 \cdot CO \cdot R^2 + [Me_2CHO]_3Al \rightleftharpoons [R^1R^2CHO]_3Al + 3Me_2CO$$

The carbonyl compound to be reduced is heated with aluminium isopropoxide in excess isopropyl alcohol (propan-2-ol) under a simple fractionating column with provision for slow distillation until no more acetone is detected in the distillate; the alcoholic reduction product is recovered from the reaction mixture after acidification. The process is usually termed the *Meerwein–Ponndorf–Verley reduction*. It is a mild method of reducing carbonyl compounds in good yield, and is particularly valuable since other groups, e.g. a conjugated double bond, a nitro group or a halogen atom, are unaffected. Experimental details for the reduction of crotonaldehyde (Expt 5.32) are given. The above reversible equation indicates that 1 mol of aluminium isopropoxide will reduce directly 3 moles of the carbonyl compound. It is generally desirable to use excess of the reductant except for aromatic aldehydes; for the latter, side reactions (e.g. $2R \cdot CHO \rightarrow R \cdot CO_2 \cdot R$; Tishchenko reaction) tend to occur with excess of the reagent.

The mechanism of the reaction involves the coordination of the carbonyl compound with the aluminium atom in aluminium isopropoxide followed by an intramolecular transfer of a hydride ion:

$$[Pr^iO]_2Al \cdot OCHR^1R^2 \xrightarrow[-Me_2CO]{+2R^1R^2CO} [R^1R^2CHO]_3Al \xrightarrow{H_3O^\oplus} 3R^1R^2CHOH$$

The product is the racemic $[(R)/(S)]$ alcohol since the free energies of the two diastereoisomeric transition states, resulting from hydride attack on the *si*-face of the ketone as shown (order of priorities $O > R^1 > R^2$, p. 16) or the *re*-face, are identical. The use of an aluminium alkoxide, derived from an optically pure secondary alcohol, to effect a stereoselective reaction (albeit in low *ee%*) was one of the first instances of an asymmetric reduction.[48] Here (S)-(+)-butan-2-ol, in the form of the aluminium alkoxide, with 6-methylheptan-2-one was shown to give rise to two diastereoisomeric transition states [(5), (R,S); and (6), (S,S)] which lead to an excess of (S)-6-methylheptan-2-ol [derived from transition state (6)], as expected from a consideration of the relative steric interactions. Transition state (5) has a less favourable Me—Me and Et—Hex interaction and hence a higher free energy of activation; it therefore represents the less favourable reaction pathway (see p. 15).

(5)   (6)

More recently a very great variety of chiral complexed aluminium hydride reagents, derived from lithium aluminium hydride, have been studied in an endeavour to increase the differences in steric interactions which could develop in diastereoisomeric transition states.[49] One of the most recent is the lithium aluminium hydride complex from either (R)-(+)- or (S)-(−)-1,1'-bi-2-naphthol (Section 6.1.3, p. 836).[50] This chiral inducing agent is expensive but has been successfully used for chiral reductions in the syntheses of various pheromones, prostaglandins and related biologically important compounds,[51] since high ee per cent values (often greater than 80%) are obtained.

A useful illustrative example of stereoselective reduction, which has the merit of being economic, and which reduces a range of dialkyl, alkyl alkynyl, or alkyl aryl ketones in good chemical and optical yield is given in Expt 5.33.[52] For the asymmetric reduction of acetophenone, lithium aluminium hydride is allowed to react with 3-O-benzyl-1,2-O-cyclohexylidene-α-D-glucofuranose (Expt 5.115) to give the complex (7). The most favourable transition state for the reduction of the ketone with (7) is that in which a hydride ion (from the less screened H″) is transferred to the re-face of acetophenone to give (S)-(−)-1-phenylethanol (ee 33%). If however the complex (7) is first allowed to react with ethanol to form the trialkoxy-complex (8), reduction of acetophenone results from the transfer of a hydride ion from the more screened H′ to the si-face of acetophenone to give (R)-(+)-1-phenylethanol (ee 53%).

(7)   (8)

Stereoselective reductions based on complexed borohydrides have also proved of value in many instances; in particular they have been of use in the synthesis of epimeric cyclic alcohols. For example, the reduction of 4-t-butylcyclohexanone to the cis-alcohol [99.5%, arising from equatorial hydride ion attack (i)] is effected by L-Selectride (lithium tri-s-butylborohydride, cf. Section 4.2.49, p. 448), or LS-Selectride[53] (lithium trisiamylborohydride, cf. Section 4.2.49, p. 448) but to the trans-alcohol [98%, arising from axial hydride ion attack (ii)] with lithium butylborohydride.[54] The experimental details of these reductions are given in Expt 5.34.

cis-   trans-

1,2-Diols [e.g. tetramethylethylene glycol or pinacol, $Me_2C(OH)\cdot C(OH)Me_2$] may be formed from ketones (e.g. acetone) by reactions with amalgamated magnesium (Expt 5.35). The pinacolic coupling of aldehydes and ketones has also been effected with Ti(III) species and the method has been adapted to effect intramolecular coupling reactions.[55] Retrosynthetically this bimolecular reduction may be regarded as a *reconnection* process and is particularly useful with aromatic ketones, when the reaction may be carried out photolytically, as in the case of the conversion of benzophenone into benzopinacol in the presence of propan-2-ol (Expt 5.36). Either sunlight or a medium-pressure mercury arc lamp (p. 110) may be used as a source of radiation.

$$Ph_2CO \longrightarrow Ph_2CO^*(singlet) \longrightarrow Ph_2CO^*(triplet) \xrightarrow{Me_2CHOH}$$
$$Ph_2\dot{C}OH + Me_2\dot{C}OH$$

$$Me_2\dot{C}OH \xrightarrow[-Me_2CO]{Ph_2CO} Ph_2\dot{C}OH \quad then \quad 2Ph\dot{C}OH \longrightarrow Ph_2C(OH)\cdot C(OH)Ph_2$$

Primary alcohols may be conveniently prepared by the reduction of esters with sodium and absolute ethanol (the *Bouveault–Blanc reduction*, Expt 5.37).

$$R\cdot CO_2Et + 4[H] \xrightarrow{Na/EtOH} R\cdot CH_2OH + EtOH$$

The method has also been applied to the esters of dicarboxylic acids for the preparation of α, ω-diols.

Esters may alternatively be reduced to primary alcohols either using hydrogen under pressure in the presence of a copper chromite catalyst,[56] or with lithium aluminium hydride (Expt 5.38), but not with sodium borohydride which is insufficiently reactive. However it has been found recently that sodium borohydride in mixed solvents (methanol/tetrahydrofuran) reduces β-ketoesters to 1,3-diols, and this method offers a convenient route to this type of compound.[57]

Experiment 5.29 HEPTAN-2-OL

$$Me\cdot(CH_2)_4\cdot CO\cdot Me \xrightarrow[Na/EtOH]{2[H]} Me\cdot(CH_2)_4\cdot CHOH\cdot Me$$

Place a mixture of 114 g (140 ml, 1.0 mol) of heptan-2-one (1), 300 ml of rectified spirit (95% ethanol) and 100 ml of water in a 1.5-litre three-necked flask fitted with an efficient double surface condenser and a thermometer dipping into the reaction mixture. Through the third neck add 65 g (2.8 mol) of clean sodium, preferably in the form of wire (Section 4.2.68, p. 462) although small pieces may be used with somewhat inferior results, gradually and at such a rate that the reaction is under control; cool the flask in running water or in ice during the addition. The temperature should not rise above 30 °C. When the sodium has *completely* reacted, add 1 litre of water and cool the mixture to about 15 °C. Separate the upper layer, wash it with 25 ml of dilute hydrochloric acid (1:1), then with 25 ml of water, and dry with anhydrous potassium carbonate or anhydrous calcium sulphate. Distil through an efficient fractionating column and collect the heptan-2-ol at 156–158 °C. The yield is 75 g (65%).

Note. (1) The ketone may be synthesised as in Expt 5.95; it is also available commercially. The latter should first be dried, redistilled, and the fraction, b.p. 150–152 °C, collected.

Cognate preparations. *Hexan-2-ol.* Dissolve 100 g (123 ml, 1 mol) of hexan-2-one (Expt 5.95) in 750 ml of ether, add 150 ml of water and stir the mixture vigorously. Introduce 69 g (3 mol) of clean sodium in the form of wire (or small pieces) as rapidly as possible; the reaction must be kept under control and, if necessary, the flask must be cooled in ice or in running water. When all the sodium has reacted, separate the ethereal layer, wash it with 25 ml of dilute hydrochloric acid (1:1), then with water, dry with anhydrous potassium carbonate or with anhydrous calcium sulphate and distil through a fractionating column. Collect the fraction of b.p. 136–138 °C. The yield of hexan-2-ol is 97 g (95%).

Cyclopentanol. Use cyclopentanone and proceed as for hexan-2-ol. Collect the cyclopentanol as a fraction of b.p. 139–142 °C.

Experiment 5.30 BENZHYDROL (*Diphenylmethanol*)

$$Ph \cdot CO \cdot Ph \xrightarrow[Zn/^\ominus OH]{2[H]} Ph \cdot CHOH \cdot Ph$$

In a 1-litre three-necked flask, equipped with a reflux condenser, a mechanical stirrer and a thermometer dipping into the reaction mixture, place 50 g (0.275 mol) of benzophenone (Expt 6.121), 500 ml of rectified spirit, 50 g of sodium hydroxide and 50 g (0.76 mol) of zinc powder. Stir the mixture; the temperature slowly rises to about 70 °C. After 3 hours, when the temperature has commenced to fall, filter the reaction mixture with suction and wash the residue twice with 25 ml portions of hot rectified spirit. Do not allow the residual zinc powder to become dry as it is flammable. Pour the filtrate into 2 litres of ice water acidified with 100 ml of concentrated hydrochloric acid. The benzhydrol separates as a white crystalline mass. Filter at the pump and dry in the air. The yield of crude benzhydrol, m.p. 65 °C, is 49 g. Recrystallise from 50 ml of hot ethanol and cool in a freezing mixture of ice and salt. Collect the colourless crystals and dry in the air; 36 g of pure benzhydrol, m.p. 68 °C, are obtained. Dilute the mother-liquor with water to precipitate the residual benzhydrol, and recrystallise this from a small quantity of hot alcohol.

Experiment 5.31 2,2,2-TRICHLOROETHANOL

$$CCl_3 \cdot CH(OH)_2 \xrightarrow[NaBH_4]{2[H]} CCl_3 \cdot CH_2OH + H_2O$$

Dissolve 16.5 g (0.1 mol) of chloral hydrate in 20 ml of water in a 200-ml beaker. Place a solution of 1.3 g (0.03 mol) of sodium borohydride in 20 ml of cold water in a small dropping funnel. Cool the chloral hydrate in an ice-water bath; add the borohydride solution dropwise (while stirring with a thermometer) at such a rate that the temperature of the solution is maintained at 20–30 °C – *the reaction is strongly exothermic*. When the borohydride solution has been added, allow the reaction mixture to stand at room temperature for 10 minutes; stir occasionally. Then add 2 ml of 2.5 M hydrochloric acid dropwise and with stirring to destroy any residual borohydride and finally add a further 5 ml of the acid. Add sufficient ether to form two distinct layers, separate the ether layer, wash it with a little water and dry over magnesium sulphate. Remove the ether on a rotary evaporator and distil the residue from an air bath. Collect the 2,2,2-trichloroethanol at 151–153 °C. The yield is 9.8 g (65%).

Cognate preparation. *m-Nitrobenzyl alcohol*. Clamp a 500-ml three-necked flask, equipped with a mechanical stirrer, a thermometer and a burette, above the bench so that an ice bath can be placed beneath it. Place a solution of 15.1 g (0.1 mol) of *m*-nitrobenzaldehyde (Expt 6.19) in 100 ml of methanol in the flask and, while stirring, add a solution of sodium borohydride (1.4 g, 0.037 mol $NaBH_4$ in 2 ml of 2 M sodium hydroxide diluted with 18 ml of water) at the rate of 0.5 ml per minute, with occasional cooling to keep the reaction at 18–25 °C. When about three-quarters of the solution has been added, there is no further tendency for the temperature to rise, and the addition is stopped. Treat a small portion of the reaction mixture with dilute sulphuric acid: hydrogen should be evolved.

Remove most of the methanol by distillation on a steam bath, and dilute the residue with 100 ml of water. Extract the mixture with ether, wash the upper layer with water and dry it rapidly with a little anhydrous magnesium sulphate. Remove the ether by flash distillation and distil the residual pale yellow oil under diminished pressure. Collect the *m*-nitrobenzyl alcohol at 183–185 °C/17 mmHg; it solidifies to a pale yellow solid, m.p. 30 °C, when cooled in ice. The yield is 13 g (85%).

Experiment 5.32   BUT-2-EN-1-OL (*Crotyl alcohol*)

$$\text{Me} \diagup\!\!\diagdown\!\!\diagup \overset{\text{H}}{\underset{\text{O}}{||}} + \text{Me}_2\text{CHOH} \xrightarrow{\text{Al(OPr}^i)_3} \text{Me} \diagup\!\!\diagdown\!\!\diagup \text{OH} + \text{Me}_2\text{CO}$$

Prepare a solution of aluminium isopropoxide (see Section 4.2.2, p. 415) from 23.5 g (0.83 mol) of aluminium, 0.5 g of mercury(II) chloride and 250 ml of dry isopropyl alcohol (Section 4.1.11, p. 402); add 105 g (1.5 mol) of redistilled crotonaldehyde, b.p. 102–103 °C, and 500 ml of dry isopropyl alcohol. Attach an efficient fractionating column to the flask and arrange for distillation from an oil bath so that the acetone distils as it is formed. Maintain the temperature of the bath at about 110 °C and the temperature at the top of the column at 60–70 °C. When the distillate no longer gives a test for acetone (8–9 hours) (1), reflux for a further 10 minutes and then distil off most of the remaining isopropyl alcohol, preferably under reduced pressure. Cool the residue to 40 °C and add 450 ml of cold 3 M sulphuric acid (from 72.5 ml of concentrated sulphuric acid and 395 ml of water); cooling is necessary. Separate the upper oily layer, wash it once with water and distil at 60–70 °C while lowering the pressure slowly from about 275 mm to 60 mm; then continue the distillation to 110 °C and 20 mm. In this way the crotyl alcohol (A) is separated from the higher boiling polymerisation products. Combine the aqueous layers and distil until the distillate no longer gives a test for unsaturation with a dilute solution of bromine in carbon tetrachloride. Saturate the aqueous distillate with potassium carbonate, separate the oily layer and add it to (A). Dry with 5 g of anhydrous potassium carbonate, decant the oil and distil through an efficient fractionating column. Collect the crotyl alcohol at 119–121 °C. The yield is 55 g (50%). Record the p.m.r. (CDCl$_3$) and compare the spectrum with that of (Z)-but-2-en-1-ol, p. 495. Assign the ^{13}C-n.m.r. absorptions which occur at δ 17.6, 63.0, 127.1 and 130.7.

Note. (1) For the acetone test reagent see p. 1218, *Method 2*.

Experiment 5.33 (S)-(−)- AND (R)-(+)-1-PHENYLETHANOL

(R) *(S)*

All reagent additions should be made with a suitable syringe (Section 2.17.8, p. 120).

(S)-(−)-1-Phenylethanol. Equip an oven-dried three-necked 250-ml round-bottomed flask with a precision-ground stirrer unit, a reflux condenser connected to a mercury bubbler, and a rubber septum with a syringe needle attached to a dry nitrogen supply. Flush the apparatus with nitrogen. Transfer 53.2 ml of an 0.5 M ethereal solution of lithium aluminium hydride (0.026 mol) (1) to the flask. Prepare a solution of 8.8 g (0.025 mol) of 3-O-benzyl-1,2-O-cyclohexylidene-α-D-glucofuranose (Expt 5.115) in 50 ml of ether and add it dropwise to the stirred hydride solution. Heat the mixture under reflux with stirring for 30 minutes, then add a solution of 3.0 g (0.025 mol) of acetophenone in 50 ml of dry ether. Heat under reflux for a further 2 hours, cool the reaction mixture, and decompose excess hydride by the cautious dropwise addition of 5 ml of water. Filter the ethereal solution using a Celite filter pad, dry the filtrate over magnesium sulphate and concentrate on a rotatory evaporator. Transfer the concentrated solution to a small fractional distillation unit (Fig. 2.111), and first remove the remaining solvent before distilling the secondary alcohol under reduced pressure. The yield of (S)-(−)-1-phenylethanol is 2.5 g (83%), b.p. 79 °C/3 mmHg, [α]$_D^{20}$ − 14.4° (neat), *ee* 33.5% (maximum reported specific rotation − 42.86°).

(R)-(+)-1-Phenylethanol. Using a 500-ml three-necked round-bottomed flask equipped as in the foregoing experiment, react 117 ml of an 0.5 M ethereal solution of lithium aluminium hydride (0.058 mol) (1) with 8.8 g (0.025 mol) of the monosaccharide derivative in 50 ml of dry ether, and heat under reflux as above for 90 minutes. Then add dropwise a solution of 4.5 g (0.098 mol) of ethanol and heat under reflux for a further 1 hour before the dropwise addition of a solution of 1.5 g (0.012 5 mol) of acetophenone in 50 ml of dry ether. Heat under reflux for a further 2 hours, cool and add cautiously 15 ml of

water. Work up the reaction mixture as in the foregoing experiment to obtain 1.2 g (80%) of (R)-(+)-1-phenylethanol, $[\alpha]_D^{20}$ + 22.9° (ee 53.4%).

Note. (1) See Section 4.2.49, p. 445, for the standardisation of ethereal solutions of lithium aluminium hydride; adjust the volume of addition to ensure that the recommended molar quantity of reagent is used.

Experiment 5.34   CIS- AND TRANS-4-t-BUTYLCYCLOHEXANOL

| | Method A | 99.5% | 0.5% |
| | Method B | 2% | 98% |

Method A. **cis-4-t-Butylcyclohexanol.**[53]   An oven-dried 250-ml two-necked flask, equipped with a Teflon stopcock, a magnetic stirrer bar, and a reflux condenser connected to a mercury bubbler, is cooled to room temperature under a dry stream of nitrogen. Lithium trisiamylborohydride solution in tetrahydrofuran (LS-Selectride, 67 ml, 28 mmol) is introduced into the reaction flask and cooled to − 78 °C (dry ice–acetone). Then 3.7 g (24 mmol) of 4-t-butylcyclohexanone dissolved in 25 ml of tetrahydrofuran (maintained at 0 °C) is added. The resulting mixture is stirred vigorously for 2 hours at − 78 °C and then allowed to equilibrate to room temperature (1 hour). The reaction mixture is hydrolysed with 4 ml of water and 15 ml of ethanol is added; the organoborane is oxidised with 10 ml of 6 M sodium hydroxide and 15 ml of 30 per cent hydrogen peroxide (**CAUTION**) (1). The aqueous phase is saturated with anhydrous potassium carbonate, the organic phase separated, and the aqueous phase extracted with two 20-ml portions of ether. The combined extracts are dried (MgSO₄). Gas–liquid chromatography analysis indicates the presence of the *cis*-alcohol in >99.5 per cent isomeric purity. The volatile solvents and the isoamyl alcohol were removed under reduced pressure to give 3.65 g (98%) of essentially pure *cis*-4-t-butylcyclohexanol as a snow-white solid, m.p. 80 °C (lit. m.p. 82 °C).

Method B. **trans-4-t-Butylcyclobutylcyclohexanol.**[54]   *Preparation of lithium n-butylborohydride in tetrahydrofuran–hexane.* In a 100-ml flask with a magnetic stirrer bar and a rubber septum under a dry nitrogen atmosphere is placed borane–dimethyl sulphide complex (1.0 ml, 10.0 mmol). Tetrahydrofuran (32.5 ml) is added and the flask immersed in an ice bath. Butyllithium in hexane (1.5 M, 6.6 ml, 9.9 mmol) is slowly added to the flask with vigorous stirring, and the resulting solution is stirred for an additional 30 minutes to give a solution of lithium n-butylborohydride (0.25 M) in tetrahydrofuran–hexane. The hydride concentration is determined by hydrolysing a known aliquot of the solution with a mixture of tetrahydrofuran and 0.5 M sulphuric acid at room temperature and measuring the volume of hydrogen evolved.

In a 25-ml flask with a magnetic stirrer and a rubber septum under a dry nitrogen atmosphere 4-t-butylcyclohexanone (72 mg, 0.47 mmol) is placed, and tetrahydrofuran (2.7 ml) is added. To the resulting solution in a dry-ice–acetone bath under a dry nitrogen atmosphere is added dropwise a solution of lithium n-butylborohydride (1.8 ml of a 0.25 M solution, 0.45 mmol) in tetrahydrofuran–hexane. After 2 hours of being stirred at −78 °C, the reaction mixture is hydrolysed with water (0.5 ml), allowed to warm to room temperature, and oxidised with 10 per cent sodium hydroxide solution (3 ml) and 30 per cent hydrogen peroxide (2 ml, **CAUTION**) (1) by stirring overnight at room temperature. After diethyl ether (10 ml) is introduced, the aqueous layer is separated and extracted three times with ether. The combined organic layers are washed with sodium hydrogen sulphite solution and saturated sodium chloride solution, dried over anhydrous magnesium sulphate, and evaporated to dryness to give the corresponding alcohol (72 mg, 98%). The product is subjected to g.l.c. analysis (2.1 m × 3.7 mm, 10% Carbowax 20 M column at 125 °C) which shows 98 per cent of the *trans* alcohol and 2 per cent of the *cis* alcohol.

Note. (1) See Section 5.4.6, p. 552, for the precautions that must be taken before evaporating extracts arising from reactions involving the use of hydrogen peroxide in the presence of tetrahydrofuran.

Experiment 5.35 2,3-DIMETHYLBUTANE-2,3-DIOL (*Pinacol*)

$$2Me_2CO \xrightarrow{\text{Mg/Hg}} Me_2C\!-\!CMe_2\}Mg^{2\oplus} \xrightarrow{\text{H}_2\text{O}} Me_2C(OH)\cdot C(OH)Me_2\cdot 6H_2O$$
$$\qquad\qquad\qquad\; O^{\ominus}\; O^{\ominus}$$

CAUTION: *This preparation should be conducted in an efficient fume cupboard.*

Pinacol hydrate. Place 20 g (0.83 mol) of dry magnesium turnings and 200 ml of anhydrous benzene (**CAUTION**) in a dry, 1-litre two-necked flask, fitted with a dropping funnel and an efficient double surface condenser (Fig. 2.55) and carrying calcium chloride guard-tubes. Place a solution of 22.5 g of mercury(II) chloride (**POISONOUS**) in 100 g (127 ml, 1.72 mol) of dry AnalaR acetone in the funnel and run in about one-quarter of this solution; if the reaction does not commence in a few minutes, as indicated by a vigorous ebullition, warm the flask on a water bath and be ready to cool the flask in running water to moderate the reaction. Once the reaction has started, no further heating is required. Add the remainder of the solution at such a rate that the reaction is as vigorous as possible and yet under control. When all the mercury(II) chloride solution has been run in and while the mixture is still refluxing, add a mixture of 50 g (63.5 ml, 0.86 mol) of dry AnalaR acetone and 50 ml of dry benzene. When the reaction slows down, warm the flask on a water bath for 1–2 hours. During this period the magnesium pinacolate swells up and nearly fills the flask. Cool slightly, disconnect the flask from the condenser and shake until the solid mass is well broken up: it may be necessary to use a stirrer. Attach the condenser and reflux for about 1 hour, or until the magnesium has disappeared.

Now add 50 ml of water through the dropping funnel and heat again on the water bath for 1 hour with occasional shaking. This converts the magnesium pinacolate into pinacol (soluble in benzene) and a precipitate of magnesium hydroxide. Allow the reaction mixture to cool to 50 °C and filter at the pump. Return the solid to the flask and reflux with a fresh 125 ml portion of benzene for 10 minutes in order to extract any remaining pinacol; filter and combine with the first filtrate. Distil the combined extracts to one-half the original volume in order to remove the acetone: treat the residual benzene solution with 75 ml of water and cool in an ice bath, or to at least 10–15 °C, with good stirring. After 30–60 minutes, filter the pinacol hydrate which has separated at the pump and wash it with benzene to remove small quantities of mercury compound present as impurities. Dry the pinacol hydrate by exposure to air at the laboratory temperature. The yield is 90 g (48%), m.p. 45.5 °C. This product is sufficiently pure for most purposes. The crude pinacol hydrate may be purified by dissolving it in an equal weight of boiling water, treating with a little decolourising charcoal if necessary, filtering the hot solution and cooling in ice; the recovery is over 95 per cent.

Pinacol. Pinacol hydrate may be dehydrated in the following manner (compare Section 2.23, p. 168, *Drying by distillation*). Mix 100 g of pinacol hydrate with 200 ml of benzene (**CAUTION**) and distil; a mixture of water and benzene passes over. Separate the lower layer and return the upper layer of benzene to the distilling flask. Repeat the process until the benzene distillate is clear. Finally distil the anhydrous pinacol and collect the fraction boiling at 169–173 °C (50 g). The pure pinacol has m.p. 43 °C, but on exposure to moist air the m.p. gradually falls to 29–30 °C and then rises to 45–46 °C when hydration to the hexahydrate is complete.

Experiment 5.36 BENZOPINACOL

$$2Ph_2CO \xrightarrow[Me_2CHOH]{hv} Ph_2C(OH)\cdot C(OH)Ph_2 + Me_2CO$$

Method A. **Irradiation with sunlight.** Dissolve 10 g (0.055 mol) of benzophenone in 50 ml of propan-2-ol in a 100-ml round-bottomed Pyrex flask by slight warming, and add one drop of glacial acetic acid. Add further quantities of propan-2-ol, cooling to room temperature, until the solution is about 6 mm below the bottom of the flask joint. Stopper the flask, taking care that none of the solution contaminates the joint, cover the stopper and joint with aluminium foil and place the flask in direct sunlight. Colourless crystals begin to separate within 24 hours. Allow the flask to remain in the sunlight until no further solid appears to separate (about 8 days). Cool the solution in icewater and collect the product by suction filtration, wash it with about 10 ml of ice-cold propan-2-ol and dry. About 9.28 g (92%) of almost pure benzopinacol, m.p. 180–182 °C, is obtained. It may be recrystallised from glacial acetic acid (about 80 ml) from which it separates as colourless needles, m.p. 185–186 °C; the yield is 8.1 g (81%).

Method B. **Irradiation with a mercury arc lamp.** Use the 100 watt mediumpressure mercury arc lamp with the Pyrex outer and inner jackets (Section 2.17.5, p. 111) and a reaction vessel of approximately 110 ml capacity (Fig. 2.67(*f*)). Make up a solution of 10 g (0.055 mol) of benzophenone in about 110 ml of propan-2-ol containing one drop of glacial acetic acid as described above. Place this solution in the reactor vessel, together with a magnetic follower bar, and irradiate the vigorously stirred solution under an atmosphere of nitrogen. Benzopinacol begins to separate within half an hour. As the quantity of product increases it gradually collects on the surface of the Pyrex jacket which thus restricts the amount of light reaching the solution. Therefore, after about 2–3 hours, switch off the lamp, raise the lamp insert from the reaction mixture and carefully scrape off the solid into the reactor vessel. Collect the product (about 4 g) by vacuum filtration, wash it with a few ml of cold propan-2-ol and dry. Return the filtrate to the reaction vessel. Clean the lamp insert with paper tissue moistened with ethanol and dry, and replace it in the reaction vessel. Continue the irradiation as above until a further appreciable quantity of product has separated (about 2 hours) and collect as before. Repeat this procedure until no further solid separates from the reaction mixture. About four crops of material will be obtained during a total irradiation period of 8 hours. The yield of benzopinacol, m.p. 178–182 °C, is about 8.65 g (75%). It may be purified by recrystallisation from glacial acetic acid as described above.

Experiment 5.37 PENTAN-1-OL (*Amyl alcohol*)

$$\text{Bu·CO}_2\text{Et} \xrightarrow[\text{Na/EtOH}]{\text{[H]}} \text{Bu·CH}_2\text{OH} + \text{EtOH}$$

Fit the central neck of a 1-litre two-necked flask with an efficient double surface condenser and close the side-neck with a stopper. Place 52 g (59.5 ml, 0.4 mol) of ethyl pentanoate (Expt 5.152) and 800 ml of super-dry ethanol (1) (Section 4.1.9, p. 401) in the flask. Add 95 g (4.1 mol) of clean sodium (Section 4.2.68, p. 462) in small pieces through the aperture at such a rate that the vigorous refluxing is continuous (20–30 minutes). Reflux the mixture in an oil bath for 1 hour in order to be certain that all the sodium has dissolved. Replace the reflux condenser by an efficient fractionating column (e.g. Hempel or all-glass Dufton column, etc.) and set the condenser for downward distillation. Fractionate the mixture from an oil bath; about 250 ml of absolute ethanol are thus recovered. Treat the residue, consisting of pentanol and sodium ethoxide, with 330 ml of water and continue the distillation (oil bath at 110–120 °C) until the temperature at the top of the column reaches 83 °C, indicating that practically all the ethanol has been removed; about 600 ml of approximately 90 per cent ethanol are recovered. Remove the fractionating column and steam distil the mixture (Fig. 2.102); about 200 ml must be collected before all the pentan-1-ol is removed. Separate the crude pentanol, dry it over anhydrous potassium carbonate or anhydrous calcium sulphate and distil through a short column. Collect the fraction boiling at 137–139 °C. The yield of pentan-1-ol is 25 g (71%).

Note. (1) The ethanol must be absolute; a lower grade gives a poor yield.

Cognate preparations. *2-Phenylethanol.* Prepare a suspension of 42 g (1.83 mol) of sodium in 120 ml of sodium-dried toluene in a 3-litre three-necked flask following the procedure described in *Method 2* under *Sodium*, Section 4.2.68, p. 462. Do not decant the toluene; when the mixture has cooled to about 60 °C, add a solution of 50 g (0.30 mol) of ethyl phenylacetate (Expt 5.152) in 150 g (190 ml) of super-dry ethanol (Section 4.1.9, p. 401) as rapidly as possible without allowing the reaction to get out of control. Then add a further 200 g (253 ml) of super-dry ethanol. When the reaction has subsided, heat the flask in a water bath until the sodium is completely dissolved. Distil off the ethanol and toluene under reduced pressure using a rotary evaporator. Dilute the residue with water and extract the phenylethanol with ether, dry the extract with magnesium sulphate, remove the solvent and distil the residual oil under reduced pressure. Collect the 2-phenylethanol at 116–118 °C/25 mmHg. The yield is 25 g (67%).

The alcohol may be purified by conversion into the calcium chloride addition compound. Treat it with anhydrous calcium chloride; much heat is evolved and the addition compound is formed. After several hours, remove any oil which has not reacted by washing with petroleum ether (b.p. 60–80 °C). Decompose the solid with ice-water, separate the alcohol, dry and distil.

Butane-1,4-diol (*Tetramethylene glycol*). Place 60 g (2.6 mol) of clean sodium in a 3-litre three-necked flask fitted with two efficient double surface condensers and a dropping funnel protected by a calcium chloride tube. Add from the dropping funnel a solution of 35 g (0.2 mol) of diethyl succinate in 700 ml of super-dry ethanol (Section 4.1.9, p. 401) as rapidly as possible consistent with the reaction being under control; it may be necessary to immerse the flask momentarily in a freezing mixture. When the vigorous action has subsided, warm the mixture on a water bath or in an oil bath at 130 °C until all the sodium has reacted (30–60 minutes). Allow to cool and cautiously add 25 ml of water (1); reflux for a further 30 minutes to bring all the solid into solution and to hydrolyse any remaining ester. Add 270 ml of concentrated hydrochloric acid to the cold reaction mixture, cool in ice, filter off the precipitated sodium chloride and treat the filtrate with 300 g of anhydrous potassium carbonate to free it from water and acid. Filter the alcoholic solution through a large sintered glass funnel, and extract the solid twice with boiling ethanol. Distil off the ethanol from the combined solutions; towards the end of the distillation solid salts will separate. Add dry acetone, filter and distil off the acetone. Distil the residue under diminished pressure, and collect the butane-1,4-diol at 133–135 °C/18 mmHg. The yield is 13 g (72%). Record the p.m.r. spectrum (DMSO-d_6) and assign the signals that are observed at δ 1.43 [m(A_2B_2), 4H], 3.39 (d of t, 4H) and 4.34 (t, 2H).

Note. (1) Alternatively, the following procedure for isolating the diol may be used. Dilute the partly cooled mixture with 250 ml of water, transfer to a distilling flask, and distil from an oil bath until the temperature reaches 95 °C. Transfer the hot residue to an apparatus for continuous extraction with ether (e.g. Fig. 2.93). The extraction is a slow process (36–48 hours) as the diol is not very soluble in ether. Distil off the ether and, after removal of the water and ethanol, distil the diol under reduced pressure.

Experiment 5.38 HEXANE-1,6-DIOL

All the apparatus and reagents must be thoroughly dry. Set up a dry bowl to serve later as a cooling bath in a fume cupboard, a 1500-ml three-necked flask

with a sealed stirrer unit, a 250-ml dropping funnel and a double surface condenser (see Fig. 2.56); attach guard-tubes containing calcium chloride to the
open ends of the condenser and the dropping funnel. The mechanical stirrer
should be a powerful one. It must be emphasised that all operations, including weighing, with solid lithium aluminium hydride must be conducted in the
fume cupboard; during weighing, etc., the front of the fume chamber is pulled
down so that there is a narrow opening to allow the hands to enter (see also
Section 4.2.*49*, p. 446, for additional precautions and methods for removal of
traces of reagent).

Remove the dropping funnel from the flask neck and replace it by a funnel
with a very short wide stem and introduce 10.5 g (0.263 mol) of powdered lithium aluminium hydride into the flask through this funnel, and use about
300 ml of sodium-dried ether to transfer the last traces into the flask. Replace
the dropping funnel and guard-tube. Set the stirrer in motion, and place a
solution of 50.5 g (0.25 mol) of freshly distilled diethyl adipate, b.p. 133–
135 °C/14 mmHg (Expt 5.144), in 150 ml of anhydrous ether in the dropping
funnel. After stirring for 10 minutes (some of the lithium aluminium hydride
may remain undissolved), add the diethyl adipate solution so that the ether
refluxes gently; the reaction mixture rapidly becomes viscous and four 50 ml
portions of anhydrous ether must be added during the reduction to facilitate
stirring. Continue the stirring for 10 minutes after the diethyl adipate has
been added. Decompose the excess of lithium aluminium hydride by the
dropwise addition, with sitrring, either of 75 ml of water, or, preferably, by the
more rapid addition of 22 g (24.5 ml) of ethyl acetate (1). Filter the reaction
product from the sludge through a sintered glass funnel; dry the ethereal solution with magnesium sulphate and distil off the ether with a rotary evaporator. The colourless viscous residue (18.5 g) solidifies completely on cooling
and has m.p. 41–42 °C, i.e. is pure hexane-1,6-diol. Dissolve the sludge
remaining in the filter funnel in 20 per cent sulphuric acid, extract the resulting solution with six 100 ml portions of ether, or use a continuous ether
extractor (Fig. 2.92). Remove the ether by means of a rotary evaporator; the
residue (6 g) crystallises completely on cooling, m.p. 41–42 °C. The total yield
of hexane-1,6-diol is 24.5 g (91%).

Note. (1) Before adding water, remove the calcium chloride tubes and fit the reflux
condenser with a long tube extending to the duct at the top of the fume cupboard; this
will carry the escaping hydrogen above the motor of the stirrer. A spark-proof stirring
motor is recommended and should be used, if available. The dropwise addition of
water must be conducted while the mixture is stirred vigorously; foaming may occur
and the reaction may be moderated by filling the bath surrounding the reaction vessel
with cold water.

5.4.2 THE INTERACTION OF CARBONYL-CONTAINING COMPOUNDS WITH ORGANOMETALLIC REAGENTS

Primary, secondary or tertiary alcohols may be prepared by the reaction of a
suitable carbonyl-containing compound with an alkylmagnesium halide (a
Grignard reagent). The reagent is usually obtained quite readily by adding an
alkyl halide (the bromide is frequently preferred) to a suspension of magnesium
turnings in anhydrous ether. Initiation of the reaction may require the addition
of a few crystals of iodine, the purpose of which may be to form a catalytic

amount of magnesium iodide or possibly simply to etch the metal surface. As it is important that the iodine should be concentrated at the metal surface the mixture should not be stirred at this stage. The resulting alkylmagnesium halide is soluble in the ether solvent as a result of coordination of two ether molecules onto the magnesium, and may be represented by formula (9).

$$RX + Mg \xrightarrow{Et_2O} \underset{(9)}{\overset{R}{\underset{X}{\diagdown}}Mg\overset{OEt_2}{\underset{OEt_2}{\diagup}}}$$

Other ethereal solvents (e.g. tetrahydrofuran, dibutyl ether) may also be used; indeed tetrahydrofuran is the required solvent for the formation of vinylmagnesium halides.

The true structural nature of the reactive species in solution is uncertain and for convenience the reagent may be represented as a polarised species, $\overset{\delta-}{R}—\overset{\delta+}{MgX}$ [see p. 21 for a discussion of reversed polarity ('umpolung')]. The reaction with the carbonyl group of an aldehyde, ketone, or ester may then be represented as a nucleophilic addition process in the following way.

The summary of retrosynthetic strategies (p. 518) indicates the molar equivalents of Grignard reagent required for the formation of: (a) primary alcohols from either formaldehyde (obtained by the depolymerisation of paraformaldehyde), or from ethylene oxide; (b) symmetrical or unsymmetrical secondary alcohols from aldehydes, or symmetrical secondary alcohols from ethyl formate; and (c) various symmetrical and unsymmetrical tertiary alcohols from ketones, or from esters other than ethyl formate, or from diethyl carbonate. Full representative equations showing the intermediates for these reactions are given in the preparative accounts, namely Expt 5.39 [for (a)], Expt 5.40 [for (b)], and Expts 5.41 to 5.43 [for (c)]. Table 5.1 summarises the possible disconnections to give the specific reagents required for the various preparations which are described; the choice of route is based on the availability or convenience of handling of the reagents. Table 5.1 also includes examples for the preparation of alkenols and alkynols from vinylic and acetylenic Grignard reagents. An alternative route to alkynols involves the use of an alkynylsodium which is formed by the action of sodamide on a terminal alkyne.

Two principal approaches to the synthesis of an optically pure chiral secondary or tertiary alcohol from the reaction of an organometallic reagent with an aldehyde or ketone respectively are of current interest. In the first approach an alkyllithium or dialkylmagnesium is initially complexed with a chiral reagent which then reacts with the carbonyl compound. In this way two diastereoisomeric transition states are generated, the more stable of which leads to an enantiometric excess of the optically active alcohol. This approach is similar in principle to the asymmetric reductions discussed in Section 5.4.1 (see also p. 15). Two chiral catalysts may be noted as successful examples, (10) derived

Table 5.1

Alcohol	Disconnection (i)	Disconnection (ii) (preferred route)
Primary (Expt 5.39)		
(cyclohexyl)–CH$_2$OH (ii)	—	(cyclohexyl)–MgX H$_2$C=O
Me⌒⌒OH (ii) (i)	C$_5$H$_{11}$MgX H$_2$C=O	BuMgX (epoxide)
Bu⌒⌒OH (ii) (i)	C$_8$H$_{17}$MgX H$_2$C=O	C$_7$H$_{15}$MgX (epoxide)
Secondary (Expt 5.40)		
Me–CH(Me)–CH(OH)–Me (ii)(i)	MeMgX Me$_2$CH·CHO	Pr^{i}MgX (O=CH–Me)
Bu–CH(OH)–Bu (i)(ii)	BuMgX Bu·CHO	2[BuMgX] (H–C(=O)–OEt)
Tertiary (Expts 5.41 to 5.43)		
Bu–C(OH)(Me)(Me) (ii)(i)	2[MeMgX] (Bu–C(=O)–OEt)	BuMgX (Me–C(=O)–Me)
Me⌒–C(OH)(Me)(Me) (ii)(i)	MeMgX (Pr–C(=O)–Me)	PrMgX (Me–C(=O)–Me)
(1,3,3-trimethylcyclohexyl with C≡CH) (ii)	—	CH≡CMgX (trimethylcyclohexanone)
Ph·C≡C–C(OH)(Ph)(Ph) (ii)(i)	PhMgX Ph·C≡C–C(=O)–Ph	Ph·C≡C–MgX (Ph–C(=O)–Ph)
Me⌒–C(OH)(Me)(Me) (ii)(ii)	—	(ii) 2[MeMgX] (Pr–C(=O)–OEt); (ii) PrMgX (Me–C(=O)–Me)
(cyclobutyl)(OH)(CH=CH$_2$) (ii)	—	CH$_2$=CHMgX (cyclobutanone)
Et–C(OH)(Et)(Et) (i)(ii)	EtMgX (Et–C(=O)–Et)	3[EtMgX] (EtO–C(=O)–OEt)

533

from (R,R)-2,2'-bis(bromomethyl)-1,1'-binaphthyl and 1,2-diaminoethane,[58] and (11) derived from (S)-proline[59]; many other ligands have been studied.[60]

(10)   (11)

In the second approach the carbonyl function is incorporated in a chiral adjuvant (or auxiliary) which then stereoselectively directs a preferred attack of the organometallic reagent on the *si*- or *re*-faces of the carbonyl group, as determined by steric and electronic interactions. This results in two diastereoisomeric products in a ratio dependent on the relative free energies of activation. One such auxiliary is (12), derived from the readily available and optically pure (+)-pulegone.

(12)   (13)

In a multistage reaction (12) is converted into the acyloxathiane (13); subsequent reaction with a Grignard reagent (R^2MgX) takes place preferentially at the *re*-face of the carbonyl group. The product of this reaction is decomposed to give the α-hydroxy aldehyde, $R^1R^2C(OH)\cdot CHO$, this then being reduced to the diol, $R^1R^2C(OH)\cdot CH_2OH$, from whence the primary hydroxyl function is elaborated to a methyl group by reduction of the monotosylate. In this way tertiary alcohols of the type $R^1R^2C(OH)\cdot Me$ may be synthesised in high optical yield.[61]

Other chiral auxiliaries have been employed in this strategy for the synthesis of optically active alcohols.[62]

Experiment 5.39   CYCLOHEXYLMETHANOL (*primary alcohol from formaldehyde*)

Equip a 2-litre three-necked flask with a sealed stirrer unit, a 500-ml separating funnel [which is replaced by a solids addition system (Fig. 2.57) in the latter part of the experiment], and an efficient double surface reflux condenser; place calcium chloride guard-tubes on the top of the funnel and on the condenser. All parts of the apparatus must be thoroughly dry. Arrange the flask so that it may be heated on an oil bath. Place 26.7 g (1.1 mol) of magnesium turnings (1) and a crystal of iodine in the flask. Measure out in separate vessels 118.5 g (121 ml, 1.0 mol) of cyclohexyl chloride (Expt 5.51) and 450 ml of dibutyl ether. Introduce about 100 ml of pure, dry, dibutyl ether (Section 4.1.*17*, p. 406) and 15 ml of the chloride into the flask. Warm the flask gently to initiate the reaction and if necessary add a further crystal of iodine. The onset of the reaction is accompanied by the disappearance of the iodine colour, the development of cloudiness and bubbles being released from the metal surface. When the reaction is progressing well, add sufficient dibutyl ether to cover the magnesium and set the stirrer in motion. Add dropwise the remainder of the cyclohexyl chloride dissolved in the remainder of the dibutyl ether at such a rate that the reaction proceeds smoothly. When the solution commences to cool and only a small amount of metal remains, remove the funnel and replace it with the solids addition system containing 100 g (c. 3.3 mol) of dry paraformaldehyde (2). Heat the reaction mixture to 100–110 °C and add the paraformaldehyde in small portions over 2 hours to the well-stirred solution. Continue the heating and stirring for a further half-an-hour. Cool the mixture, add 300 g of finely crushed ice and agitate the mixture until decomposition is complete. Add twice the theoretical quantity of 30 per cent sulphuric acid to dissolve the magnesium hydroxide, and then steam distil the mixture until no more oil passes over (2000–2500 ml). Saturate the distillate with sodium chloride and separate the upper organic layer. Dry with anhydrous potassium carbonate and distil under reduced pressure using a fractionating column; collect the first fraction of dibutyl ether and then a second fraction of cyclohexylmethanol. Add 5 g of freshly dehydrated calcium oxide to the alcohol fraction and heat on a water bath for 30 minutes; this will remove last traces of unreacted halogen compound. Filter, redistil under reduced pressure and collect the fraction of b.p. 88–93 °C/18 mmHg; the yield is 70 g (66%). The boiling point of cyclohexylmethanol at atmospheric pressure is 182 °C.

Note. (1) Commerical magnesium turnings for the Grignard reaction should be washed with sodium-dried ether to remove traces of surface grease which may be present, dried at 100 °C and allowed to cool in a desiccator.
(2) Dry the material in a desiccator over phosphoric oxide for 2 days.

Cognate preparations. *Hexan-1-ol* (*primary alcohol from ethylene oxide*)

$$\text{BuMgBr} + \overset{O}{\triangle} \longrightarrow [\text{Bu·CH}_2\text{·CH}_2\text{·OMgBr}] \xrightarrow{\text{H}_3\text{O}^\oplus} \text{Bu·CH}_2\text{·CH}_2\text{OH}$$

CAUTION: *Owing to the toxicity of ethylene oxide and of benzene all operations must be carried out in an efficient fume cupboard.*

Equip a 2-litre three-necked flask as above; the solids addition system is not required but the gas inlet system described below should be assembled for use in the latter stages of the experiment. An appropriate water bath should be used for heating. Place 37.5 g (1.54 mol) of washed and dried magnesium turnings and 300 ml of sodium dried ether in the flask and add a small crystal of iodine. Prepare a solution of 205.5 g (161 ml, 1.50 mol) of pure dry butyl bromide (Expt 5.54) in 300 ml of dry ether in the separatory funnel and introduce about 25 ml of the solution into the flask. As soon as the reaction commences (disappearance of the iodine colour), set the stirrer in motion and add the remainder of the butyl bromide solution at such a rate that steady refluxing of the reaction mixture is maintained (if the reaction becomes too vigorous it should be moderated by momentarily cooling the flask in an ice bath). When the addition is complete, maintain gentle reflux, with the aid of a hot-water bath if necessary, until most of the magnesium has reacted (about 15–30 minutes). Cool the flask in a freezing mixture of ice and salt. Remove the separatory funnel and replace it by a tube, 4 mm in diameter, the end of which is about 2 cm above the surface of the liquid. Attach this delivery tube to a flask fitted with 'wash bottle' tubes, the long tube being nearer the three-necked flask and the other end being connected to a supply of dry nitrogen. Cool this flask in a mixture of ice and salt and introduce rapidly 90 g (2.02 mol) of ethylene oxide (1) from a 100 g sealed bulb of the reagent (Section 4.2.28, p. 434); the latter must, of course, be cooled in an ice and salt mixture before opening (2). Gradually introduce the ethylene oxide into the reaction flask over a period of 1.5–2 hours; the temperature should not rise above 10 °C. When all has been added, remove the freezing mixture surrounding the three-necked flask. The temperature of the mixture will gradually rise and the reaction mixture will boil gently. When boiling ceases, reflux on a water bath for 30 minutes. Allow to cool, insert a thermometer into a neck of the flask, arrange the condenser for downward distillation and collect 250 ml of ether in a measuring cylinder; do not collect a larger volume of ether as a violent reaction may set in, apparently due to a rearrangement of the initial reaction product, and considerable loss may ensue. Change the receiver, and introduce 250 ml of sodium-dried benzene (**CAUTION**) into the reaction mixture. Continue the distillation with stirring until the temperature of the distilling vapour reaches 65 °C. Then boil the mixture under reflux for 30 minutes; generally by this time the mixture has become so viscous that stirring is no longer very effective. Allow to cool. Decompose the reaction mixture with 500 ml of an ice–water mixture, and dissolve the precipitated magnesium hydroxide with 30 per cent sulphuric acid; add sufficient finely-crushed ice to keep the mixture cold. Steam distil and collect about 2 litres of distillate. Separate the oily layer (*A*), and distil the aqueous layer until free of hexan-1-ol; add the oil so obtained to (*A*). Stir the crude hexan-1-ol on a water bath with 250 ml of 20 per cent sodium hydroxide solution, and steam distil again as before. Dry the oil with a little anhydrous calcium sulphate, distil through an efficient fractionating column and collect the fraction, b.p. 154–157 °C. The yield of hexan-1-ol is 90 g (49%).

Notes. (1) It is advisable to cool and open the ampoule of ethylene oxide behind a safety screen in a fume cupboard, and to wear plastic gloves and goggles.
(2) Instead of adding the liquid ethylene oxide (b.p. 10.5 °C), the latter may be dissolved in 100 ml of ice-cold anhydrous ether; this solution is added directly to the reaction mixture during 15–30 minutes. The yield however is somewhat lower.

Nonan-1-ol (*primary alcohol from ethylene oxide*).   Prepare a Grignard reagent from 24.5 g (1 mol) of magnesium turnings, 179 g (157 ml, 1 mol) of 1-bromoheptane (Expt 5.55) and 300 ml of dibutyl ether as described above. Cool the solution to 0 °C and, with vigorous stirring, add an excess of ethylene oxide. Maintain the temperature to 0 °C for 1 hour after the ethylene oxide has been introduced, then allow the temperature to rise to 40 °C and maintain the mixture at this temperature for 1 hour. Finally heat the mixture on a water bath for 2 hours. Decompose the addition product by pouring the cooled reaction mixture into ice-water, acidify with sulphuric acid to dissolve the precipitated magnesium hydroxide and isolate the reaction products as described in the procedure for hexan-1-ol. Collect the nonan-1-ol at b.p. 95–100 °C/12 mmHg; the yield is 95 g (69%).

Experiment 5.40   3-METHYLBUTAN-2-OL (*secondary alcohol from an aldehyde*)

$$Pr^iMgBr \xrightarrow{Me \cdot CHO} Pr^i \cdot CH(OMgBr) \cdot Me \longrightarrow Pr^i \cdot CH(OH) \cdot Me$$

Prepare a solution of isopropylmagnesium bromide from 37.5 g (1.54 mol) of magnesium turnings and 185 g (142 ml, 1.50 mol) of isopropyl bromide in 200 ml of anhydrous ether; use a 1-litre flask equipped as in Expt 5.39, and follow the broad experimental details given in the cognate preparation of Expt 5.39. Cool the resulting Grignard reagent to − 10 to − 5 °C in a freezing mixture of crushed ice and anhydrous calcium chloride. Remove the separatory funnel and re-attach to the flask via a two-necked adapter to allow the introduction of a thermometer for monitoring the reaction temperature. Add a solution of 67 g (83.5 ml, 1.53 mol) of freshly distilled acetaldehyde (b.p. 20.5–21 °C) in 90 ml of anhydrous ether over a period of 30 minutes. Do not allow the temperature to rise above − 5 °C. When all the acetaldehyde has been added, pour the reaction product upon 700 g of crushed ice; any excess of magnesium should remain in the flask. Dissolve the basic magnesium bromide by the addition of 350 ml of 15 per cent sulphuric acid. Separate the ethereal solution and extract the aqueous layer with four 50 ml portions of ether. Dry the combined ethereal solutions over 8 g of anhydrous potassium carbonate (or the equivalent quantity of anhydrous calcium sulphate), and fractionally distil through an all-glass Dufton (or other efficient fractionating) column. Collect the 3-methylbutan-2-ol at 110–111.5 °C. The yield is 70 g (52%). The ^{13}C-n.m.r. absorption (CDCl$_3$, TMS) occur at δ 18.2 (Me$_2$), 19.9 (C$_1$), 35.2 (C$_3$) 72.5 (C$_2$).

Cognate preparation. *Nonan-5-ol* (*secondary alcohol from ethyl formate*)

$$BuMgBr \xrightarrow{H \cdot CO_2Et} [Bu \cdot CH(OMgBr)OEt] \xrightarrow{-Mg(OEt)Br} Bu \cdot CHO \xrightarrow{BuMgBr}$$

$$[Bu_2CHOMgBr] \longrightarrow Bu_2CHOH$$

Prepare a solution of butylmagnesium bromide from 12.2 g (0.50 mol) of magnesium turnings, 69 g (54 ml, 0.50 mol) of butyl bromide (Expt 5.54) and 250 ml of dry ether in a 1-litre flask. Cool the flask containing the resulting Grignard reagent in an ice bath. Place a solution of 18.5 g (20 ml, 0.25 mol) of pure ethyl formate (Expt 5.143) in 40 ml of anhydrous ether in the separatory funnel. Stir the solution of the Grignard reagent and run in the ethyl formate solution at such a rate that the ether refluxes gently (10–15 minutes). Remove the ice bath and continue the stirring for 10 minutes.

Place 35 ml of water in the separatory funnel and run it into the *vigorously stirred* reaction at such a rate that rapid refluxing occurs. Follow this by a cold solution of 15.5 ml of concentrated sulphuric acid in 135 ml of water. Two practically clear layers will now be present in the flask. Decant as much as possible of the ethereal layer (*A*) into a 500-ml round-bottomed flask.

537

Transfer the remainder, including the aqueous layer, into a separatory funnel: wash the residual solid with two 10 ml portions of ether and combine these washings with the liquid in the separatory funnel. Separate the ethereal portion and combine it with (*A*). Distil off the ether through an efficient fractionating column until the temperature of the vapour rises to about 50 °C. The residual crude nonanol contains a little of the corresponding formic ester. Remove the latter by refluxing for 3 hours with 25 ml of 15 per cent aqueous potassium hydroxide, and then isolate the purified nonanol by steam distillation (volume of distillate about 500 ml). Separate the upper layer of the secondary alcohol, dry it over anhydrous potassium carbonate or anhydrous calcium sulphate, and distil from a flask carrying a Claisen still-head under reduced pressure. Collect the pure nonan-5-ol at 97–98 °C/20 mmHg; the yield is 30 g (83%). The boiling point under atmospheric pressure is 195 °C.

Experiment 5.41 2-METHYLHEXAN-2-OL (*tertiary alcohol from a ketone*)

$$BuMgBr + Me \cdot CO \cdot Me \longrightarrow [Bu \cdot C(OMgBr)Me_2] \xrightarrow{H_3O^{\oplus}} Bu \cdot C(OH)Me_2$$

The apparatus and experimental details are similar to those given in the previous sections. Prepare a Grignard reagent from 24.5 g (1 mol) of magnesium turnings, 137 g (107 ml, 1 mol) of butyl bromide and 450 ml of sodium-dried ether. Add slowly with rapid stirring, and cooling with ice if necessary, a solution of 58 g (73.5 ml, 1 mol) of dry acetone in 75 ml of anhydrous ether. Allow the reaction mixture to stand overnight. Decompose the product by pouring it on to 500 g of crushed ice; dissolve the precipitated magnesium compounds by the addition of 10 per cent hydrochloric acid or of 15 per cent sulphuric acid. Transfer to a separatory funnel, remove the ether layer and extract the aqueous solution with three 50 ml portions of ether. Dry the combined ethereal solutions over anhydrous potassium carbonate or anhydrous calcium sulphate, filter, distil off the ether, and fractionally distil the residue. Collect the 2-methylhexan-2-ol at 137–141 °C. The yield is 105 g (90%). Assign the ^{13}C-n.m.r. absorptions which occur at δ 14.1, 23.4, 26.8, 29.3, 44.0 and 70.8.

Cognate preparations. *2-Methylpentan-2-ol* (*tertiary alcohol from a ketone*). Use propylmagnesium bromide and acetone. Collect the tertiary alcohol at 121–124 °C.

***trans*-1-Ethynyl-3,3,5-trimethylcyclohexan-1-ol** (*tertiary alcohol from a ketone*). Assemble in a fume cupboard a three-necked, 250-ml round-bottomed flask mounted on a magnetic stirrer unit and equipped with a gas inlet tube terminating in a glass frit for the passage of acetone-free acetylene (Expt 5.26), and a dropping funnel protected with a calcium chloride guard-tube; insert a calcium chloride tube in the third neck. Place in the flask 60 ml of pure tetrahydrofuran (Section 4.1.*19*, p. 406) and a magnetic follower. Saturate the tetrahydrofuran with acetylene by passing a rapid stream of the gas through the solvent. Then with the continued passage of acetylene, add dropwise from the funnel over 2 hours a previously prepared solution of ethylmagnesium bromide [from 16.4 g (0.15 mol) of ethyl bromide, 3.7 g (0.154 mol) of magnesium in 100 ml of tetrahydrofuran]. Cool the reaction mixture to 0 °C and add with stirring a solution of 7 g (0.05 mol) of 3,3,5-trimethylcyclohexanone in 20 ml of tetrahydrofuran; the slow passage of acetylene gas should be continued. Stir the mixture for 1 hour and then pour it into a mixture of 80 g of ammonium chloride and 200 g of crushed ice. Separate the organic layer and extract the aqueous layer with three 25 ml portions of tetrahydrofuran. Evaporate the tetrahydrofuran from the combined organic layers, dissolve the residue in 100 ml of ether, dry the solution over magnesium sulphate and evaporate the solvent on a rotary evaporator. Distil (Fig. 2.108) the residue and collect the alkynol at b.p. 78–79 °C/8 mmHg; this crystallises on standing and has m.p. 26–27 °C. The yield is 5.6 g (70%). The i.r. spectrum clearly shows characteristic group frequencies at v_{max} 3450 (OH), 3300 ($\equiv$C—H) and 2110 cm^{-1} (C$\equiv$C).

1,1,3-Triphenylprop-2-yn-1-ol (*tertiary alcohol from a ketone*). Prepare a solution of ethylmagnesium bromide in 50 ml of anhydrous ether from 27.3 g (19 ml, 0.25 mol) of ethyl bromide, 6.0 g (0.25 mol) of magnesium. Cool the solution and add dropwise a solution of 25.5 g (27 ml, 0.25 mol) of phenylacetylene (Expt 5.24) in 30 ml of anhydrous ether. Boil the reaction mixture gently under reflux for 2 hours and cool to room temperature. Start the stirrer, add slowly a solution of 45.5 g (0.25 mol) of benzophenone in 50 ml of anhydrous ether, and continue to stir at room temperature for 1.5 hours. Finally boil under reflux for 1 hour and cool in an ice bath. Liberate the product by adding slowly 55 g of ammonium chloride as a saturated aqueous solution, separate the ether layer and extract the aqueous phase with two 20 ml portions of ether. Dry the combined ether solutions over anhydrous sodium sulphate, and remove the ether on a rotary evaporator. Cool the residual oil in ice and triturate with light petroleum (b.p. 60–80 °C) until the triphenylpropynol crystallises (1), and recrystallise it from a mixture of benzene and light petroleum (b.p. 60–80 °C). The yield is 35 g (49%), m.p. 78–80 °C.

Note. (1) If the product fails to crystallise, purify it by distillation under reduced pressure, b.p. 190 °C/0.05 mmHg.

1-Vinylcyclobutanol (*tertiary alcohol from a ketone*).[63] A dry, three-necked, 250-ml flask under a nitrogen atmosphere is equipped with a mechanical stirrer, dry-ice condenser, and a small additional funnel. The apparatus is charged with magnesium (7.29 g, 0.30 mol) and dry tetrahydrofuran (25 ml). In the addition funnel is placed vinyl bromide (37.5 g, 0.350 mol) in dry tetrahydrofuran (75 ml) and about 2 ml of this solution is added to the flask with very rapid stirring to initiate the reaction. After initiation, the remaining vinyl bromide is added slowly to maintain steady reflux. Once all the magnesium has been consumed, the mixture is cooled to 35 °C and a solution of cyclobutanone (14.0 g, 0.2 mol) (Expt 7.13) in dry tetrahydrofuran (30 ml) is added slowly. The dry-ice condenser is replaced by a normal condenser and the solution is refluxed for 90 minutes. After the flask is cooled to 35 °C, saturated ammonium chloride solution (40 ml) is added dropwise with additional cooling and vigorous stirring to effect hydrolysis of the magnesium salts. These salts are collected by suction filtration and washed well with anhydrous ether. The combined filtrate is dried over magnesium sulphate and filtered and the solvent removed on the rotary evaporator. Distillation affords 12.89 g (66%) of 1-vinylcyclobutanol, b.p. 67–68 °C/45 mmHg; i.r. (thin film) 3360, 2290, 1246, 1150, and 920 cm^{-1}; p.m.r. (CCl$_4$, TMS) δ 1.4–2.2 (m, 6H, —(CH$_2$)$_3$—), 4.0 (broad s, 1H, OH) and 4.9–6.3 (ABX pattern, 3H, CH$_2$=CH—).

Experiment 5.42 2-METHYLPENTAN-2-OL (*tertiary alcohol from an ester*)

$$Pr \cdot CO_2Et \xrightarrow{\text{MeMgI}} [Pr \cdot CO \cdot Me] \xrightarrow{\text{MeMgI}}$$
$$Pr \cdot C(OMgI)Me_2 \longrightarrow Pr \cdot C(OH)Me_2$$

CAUTION: *Owing to the carcinogenic properties of methyl iodide and the toxicity of benzene all operations should be conducted with great care and in an efficient fume cupboard.*

Prepare an ether solution of methylmagnesium iodide from 49 g (2 mol) of dry magnesium turnings, 284 g (124.5 ml, 2 moles) of dry methyl iodide and 400 ml of sodium-dried ether; use the apparatus and procedure described in Expt 5.39, cognate preparation nonan-1-ol. Allow to cool, and slowly add a solution of 116 g (132 ml, 1 mol) of ethyl butanoate (1) in 100 ml of anhydrous ether into the vigorously stirred solution of the Grignard reagent. Reflux the mixture on a water bath for 1 hour to complete the reaction. Pour the ethereal solution into a mixture of 200 ml of approximately 2 M sulphuric acid and 750 g of crushed ice. Separate the upper ethereal layer and extract the aqueous solution with two 150 ml portions of ether. Wash the combined ethereal extracts with dilute sodium hydrogen carbonate solution, followed by a little water, then dry with anhydrous potassium carbonate or anhydrous calcium sulphate, distil off the ether on a water bath and distil the residue through a short fractionating column. Collect the 2-methylpentan-2-ol at 117–120 °C. A further small quantity of the tertiary alcohol may be obtained by redrying the low-boiling distillate, filtering and redistilling. The yield is 90 g (88%).

Note. (1) Ethyl butanoate may be prepared as described in Expt 5.142.

Cognate preparations. *2-Methylbutan-2-ol* (*tertiary alcohol from an ester*). From ethyl propanoate and methylmagnesium iodide. Collect the tertiary alcohol at 100–102 °C.

Triphenylmethanol (*tertiary alcohol from an ester*). Prepare a solution of phenylmagnesium bromide from 14 g (0.57 mol) of magnesium turnings, 90.5 g (60.5 ml, 0.57 mol) of dried redistilled bromobenzene and 250 ml of anhydrous ether. Treat this Grignard reagent with a solution of 37.5 g (36 ml, 0.25 mol) of dry ethyl benzoate (Expt 6.163) in 100 ml of sodium-dried benzene (**CAUTION**) following the procedure described above. Decompose the reaction product by pouring it slowly, with constant stirring, into a mixture of 750 g of crushed ice and 25 ml of concentrated sulphuric acid. Continue the stirring until all the solid dissolves; it may be necessary to add 25 g of solid ammonium chloride to facilitate the decomposition of the magnesium complex, and also a little more benzene to dissolve all the product. When all the solids have passed into solution, separate the benzene layer and wash it successively with 100 ml of water, 100 ml of 5 per cent sodium bicarbonate solution and 100 ml of water. Remove the benzene as completely as possible from a 1-litre round-bottomed flask using a rotary evaporator; steam distil the residue (Fig. 2.102) in order to separate unchanged bromobenzene and biphenyl (by-product). Filter the cold residue in the flask at the pump, wash it with water and dry. The resulting crude triphenylmethanol weighs 62 g. Recrystallise it from carbon tetrachloride (4 ml per gram of solid): the first crop of crystals, after drying in air to remove the solvent of crystallisation, weighs 56 g (86%) and melts at 162 °C. Treat the mother-liquid with 1 g of decolourising charcoal, concentrate to one-quarter of the original volume and cool in ice: a further 3 g of pure triphenylmethanol is obtained.

3-Ethylpentan-3-ol (*tertiary alcohol from diethyl carbonate*).

$$3EtMgBr + (EtO)_2CO \longrightarrow Et_3COMgBr \xrightarrow{H_3O^{\oplus}} Et_3COH$$

In a 1-litre three-necked flask, equipped as in Expt 5.39, prepare a solution of ethylmagnesium bromide from 36.5 g (1.50 mol) of magnesium turnings, 163 g (112 ml, 1.50 mol) of ethyl bromide and 600 ml of anhydrous ether, following the general procedure outlined in Expt 5.39. Run into the resulting ethereal Grignard reagent a solution of 52 g (53.5 ml, 0.44 mol) of pure diethyl carbonate (Section 4.2.*26*, p. 433) in 70 ml of anhydrous ether, with rapid stirring, over a period of about 1 hour. A vigorous reaction sets in and the ether refluxes continually. When the diethyl carbonate has been added, heat the flask on a water bath with stirring for another hour. Pour the reaction mixture, with frequent shaking, into a 2-litre round-bottomed flask containing 500 g of crushed ice and a solution of 100 g of ammonium chloride in 200 ml of water. Transfer to a separatory funnel, remove the ether layer and extract the aqueous solution with two 175 ml portions of ether. Dry the combined ethereal extracts with anhydrous potassium carbonate or with anhydrous calcium sulphate, and remove the ether on a water bath. Distil the alcohol through a short fractionating column and collect the fraction boiling at 139–142 °C as pure 3-ethylpentan-3-ol. A further small quantity may be obtained by drying the low boiling fraction with 2 g of anhydrous potassium carbonate or anhydrous calcium sulphate, filtering and redistilling. The total yield is 44 g (86%).

The following tertiary alcohols may be prepared similarly from the appropriate Grignard reagent and diethyl carbonate in yields of 75–80 per cent; 4-propylheptan-4-ol, b.p. 89–92 °C/20 mmHg; 5-butylnonan-5-ol, b.p. 129–131 °C/20 mmHg; 6-pentylundecan-6-ol, b.p. 160–163 °C/19 mmHg.

Experiment 5.43 1-ETHYNYLCYCLOHEXANOL

$$CH{\equiv}CH \xrightarrow[\text{NH}_3 \text{ liq.}]{\text{NaNH}_2} CH{\equiv}C^{\ominus}Na^{\oplus} + NH_3$$

Use the same technique as detailed for Expt 5.26 but with a 3-litre three-necked flask. Charge the flask with 1.5 litres of liquid ammonia. Prepare the sodamide using 0.7 g of iron(III) nitrate and 2 g of sodium, followed by 46 g (total 2.1 mol) of sodium, and convert it into a solution of sodium acetylide as before. Add, with stirring, a solution of 196 g (206 ml, 2 mol) of dry, redistilled cyclohexanone (1) in 256 ml of dry ether during 1 hour and continue the stirring for a further 2 hours. Decompose the sodium derivative of the product by the gradual addition of a slight excess (118 g) of powdered ammonium chloride. Allow to stand overnight, preferably with stirring, by which time all the ammonia will have evaporated. Extract the residue repeatedly with ether, i.e. until all the alkynol has been separated from the inorganic material (2). Wash the ethereal extract successively with water, dilute sulphuric acid and sodium hydrogen carbonate solution, dry with magnesium sulphate and distil. Collect the 1-ethynylcyclohexanol at 83 °C/20 mmHg (3); the yield is 210 g (85%).

Notes. (1) Dry the cyclohexanone over excess of anhydrous calcium chloride before distillation.
(2) A continuous ether extractor (Fig. 2.92) is recommended.
(3) The product has m.p. c. 25 °C, but the m.p. depends upon the purity of the cyclo-hexanone and the efficiency of the distillation. Pure 1-ethynylcyclohexanol has m.p. 32 °C.

5.4.3 THE HYDROBORATION–OXIDATION OF ALKENES

Borane, as a solution in tetrahydrofuran or generated *in situ* by the reaction of a metal hydride with boron trifluoride etherate, adds readily to alkenes to yield trialkylboranes. With a terminal alkene the reaction is highly (though not com-pletely) regioselective and gives a primary trialkylborane, since the mode of addition results from the electrophilic character of the boron atom.

In the case of internal alkenes, hydroboration may not yield the trialkylborane, and the reaction may cease at either the di- or mono-alkylborane stage. The con-trolling feature in these cases appears to be the steric hindrance to reaction generated by bulky alkyl groups in the alkene; furthermore the regioselectivity of these reactions is influenced by both steric and electronic effects. For example 2-methylbut-2-ene gives disiamylborane (Sia_2BH; 98% regioselectivity), and 2,3-dimethylbut-2-ene gives thexylborane (thexyl-BH_2). The preparation of these alkylboranes is given in Section 4.2.7, p. 419).

These two alkylboranes are important reagents for the hydroboration of other alkenes, since they frequently show in this further reaction a regioselectivity which is greater than that of borane itself.

The akylboranes may be converted into a vast range of other functionali-ties[64]; perhaps the most important is their conversion into alcohols by oxidation with alkaline peroxide.

The oxidation step is usually carried out *in situ* and is illustrated by the conver-sion of hex-1-ene into hexan-1-ol (Expt 5.44); trimethylamine *N*-oxide has been suggested as a safer alternative oxidising reagent.[65] The g.l.c. analysis of the alcohol product reveals the regioselectivity of the hydroboration step. The mechanism of the oxidation step involves a 1,2-nucleophilic shift of an alkyl group in the hydroperoxide intermediate as shown.

An important consequence of the mechanism of this oxidation is that the stereo-isomeric features, resulting from the *cis*-addition of borane formulated pre-viously (above), are retained in the final product, i.e. the overall reaction is highly stereospecific. Thus reaction of 1-methylcyclohexene with borane fol-lowed by oxidation gives *trans*-2-methylcyclohexan-1-ol.[66]

Reaction of borane (as a complex with dimethyl sulphide) with optically pure (+)-α-pinene (in 15% excess) results in the almost exclusive formation of di-isopinocampheylborane (Ipc$_2$,BH)[67] (Section 4.1.7, p. 421). The use of this optically pure borane in the hydroboration of the prochiral (Z)-but-2-ene, followed by oxidation, yields (R)-(−)-butan-2-ol (ee 98.1%) (Expt 5.45). Inter-estingly, monoisopinocampheylborane (IpcBH$_2$) with (Z)-but-2-ene gives (S)-(+)-butan-2-ol, and hence the importance of careful reagent preparation is emphasised.[68]

This method is clearly an attractive route to optically active alcohols; the use of other optically active alkylboranes has been reviewed.[69]

Experiment 5.44 HEXAN-1-OL

$$3NaBH_4 + 4BF_3 \cdot OEt_2 \longrightarrow 4BH_3 + 2NaBF_4 + 4Et_2O$$
$$12Bu \cdot CH{=}CH_2 + 4BH_3 \longrightarrow 4(Bu \cdot CH_2 \cdot CH_2)_3B$$
$$4(Bu \cdot CH_2 \cdot CH_2)_3B + 12H_2O_2 + 4NaOH \longrightarrow$$
$$12Bu \cdot CH_2 \cdot CH_2OH + 4NaB(OH)_4$$

CAUTION: *This experiment should be carried out in an efficient fume cup-board.*

Equip a 500-ml three-necked round-bottomed flask (1) with a sealed stirrer unit, a 100-ml pressure-equalising funnel fitted with an inlet adapter to allow flushing with dry nitrogen, and a two-necked adapter carrying a con-denser fitted with a calcium chloride guard-tube, and a thermometer reaching to the bottom of the flask. Arrange the apparatus so that it may be cooled occasionally in an ice-water bath supported on a laboratory jack.

Maintain a slow stream of dry nitrogen through the apparatus and place in the flask 90 ml of dry diglyme (Section 4.1.*18*, p. 406), and 3.41 g (0.090 mol, 20% excess over the theoretical requirement of 0.075 mol) of powdered sodium borohydride. Stir until most of the borohydride has dissolved then add a solution of 25.2 g (0.30 mol) of hex-1-ene in 50 ml of dry diglyme. Place 17.0 g (15.1 ml, 0.12 mol; 20% excess over the theoretical requirement of 0.10 mol) of purified boron trifluoride–etherate (48%, w/w, Section 4.2.8, p. 421) in the dropping funnel followed by 25 ml of dry diglyme. Adjust the flow of dry nitrogen so that a slow stream is maintained throughout the experiment. Add the boron trifluoride–etherate in portions to the rapidly stirred mixture during 30 minutes while maintaining the temperature at 20–25 °C by occasional cooling with the ice-water bath. Continue stirring at room temperature for 1 hour to ensure completion of the hydroboration reaction. Add 20 ml of water dropwise from the dropping funnel to the reaction mixture during about 25 minutes to decompose excess sodium borohydride; vigorous hydrogen evolution may cause foaming during this addition. When hydrogen evolution has stopped, place 40 ml (0.35 mol) of 30 per cent hydrogen peroxide in the dropping funnel. Add 40 ml (0.12 mol) of 3 M aqueous sodium hydroxide in one portion down the condenser to the reaction mixture and then add the hydrogen peroxide dropwise keeping the temperature at 30–50 °C by cooling in a cold-water bath; this addition should take about 25 minutes. Stir the reaction mixture at room temperature for a further 1 hour to ensure oxidation is complete and then pour it on to 250 ml of ice-water in a separatory funnel. Rinse the reaction vessel with 50 ml of water and add to the contents of the funnel. Extract the aqueous mixture with two 200 ml portions of ether (some insoluble inorganic material may separate in the aqueous layer at this stage) and then wash the combined ethereal extracts with eight 50 ml portions of water to remove diglyme (2). Dry the ether solution over magnesium sulphate, filter and remove the solvent by flash distillation or on the rotary evaporator. Transfer the residual colourless liquid to a 100-ml round-bottomed flask and fractionally distil through a well-lagged 14 cm column filled with glass helices, collecting the fraction, b.p. 154–157 °C. The product is hexan-1-ol containing approximately 6 per cent of hexan-2-ol (3); the yield is 24.7 g (81%). A complete separation of the isomers may be effected by using a more efficient fractionating column such as a spinning band column; hexan-1-ol and hexan-2-ol boil at 155–156 °C and 137–138 °C respectively at 760 mmHg. Assign the ^{13}C-n.m.r. absorptions which occur at δ 14.1, 22.9, 25.9, 32.0, 32.9, and 62.6.

Notes. (1) The apparatus should be thoroughly dried in an oven and assembled under a stream of dry nitrogen and allowed to cool.
(2) Diglyme, which has b.p. 162 °C at 760 mmHg, must be completely removed from the ether extract – otherwise it will contaminate the product. Its presence in the extract may be conveniently checked by g.l.c., using a 1.5-m column of 10 per cent Silicone oil on Chromosorb W held at 100 °C, with a nitrogen flow rate of 40 ml/minute, t_R 3.2 minutes.
(3) The g.l.c. analysis under the conditions specified in Note (2) gives hexan-2-ol and hexan-1-ol with t_R 1.4 minutes and t_R 2 minutes respectively.

Cognate preparation. *Octan-1-ol.* Use 1.70 g (0.045 mol, 20% excess over the theoretical requirement of 0.0375 mol) of sodium borohydride, 45 ml of dry diglyme and a solution of 16.8 g (0.15 mol) of oct-1-ene in 25 ml of dry diglyme in a 250-ml three-necked round-bottomed flask (1). In the pressure-equalising funnel place 8.5 g (7.55 ml, 0.06 mol, 20% excess over the theoretical requirement of 0.05 mol) of purified boron trifluoride–etherate. Proceed as for hexan-1-ol; use half quantities in the alkaline oxidation step. After work-up, distil and collect the fraction at 191–192 °C. A product, 15.8 g (81%), is obtained which consists of octan-1-ol contaminated with about 7 per cent of octan-2-ol (2). These may be separated, if required, by using a more efficient fractionating column. At 760 mmHg octan-1-ol has b.p. 194–195 °C and octan-2-ol has b.p. 179 °C.

Notes. (1) Alternatively the apparatus may be assembled with the funnel in the centre neck of the flask and the thermometer and condenser in the side-necks and stirred using a magnetic stirrer. However, the follower bar must be substantial as the reaction mixture becomes viscous towards the end of the experiment.
(2) The purity of the product may be determined by g.l.c. using a 1.5-m column of 10 per cent Silicone oil on Chromosorb W held at 128 °C with a nitrogen flow rate of 40 ml/minute. Octan-1-ol has t_R 3.3 minutes and octan-2-ol has t_R 2.4 minutes.

Experiment 5.45 (R)-(−)-BUTAN-2-OL[67]

To a stirred suspension of diisopinocampheylborane (50 mmol) (1) in tetra-hydrofuran (18 ml) is added 4.5 ml of (Z)-but-2-ene. The reaction mixture is stirred at 25 °C for 4.5 hours. The solid diisopinocampheylborane disappears and the formation of the trialkylborane is complete. The organoborane is treated with 4 ml of methanol, followed by 18.3 ml of 3 M sodium hydroxide and the careful addition of 20 ml of 30 per cent hydrogen peroxide, maintaining the temperature of the reaction below 40 °C. The reaction mixture is further stirred at 55 °C for 1 hour, cooled, and extracted with ether (3 × 50 ml). The extract is washed successively with water (2 × 25 ml) and brine (3 ml) and dried over magnesium sulphate. The organic layer is carefully fractionated to provide butan-2-ol, b.p. 96–98 °C, 2.9 g (73%), purity >95 per cent. The last traces of impurities are removed by preparative g.l.c. (2) to yield (R)-butan-2-ol, $[\alpha]_D^{23}$ −13.23° (neat), ee 98.1 per cent.

Notes. (1) See Section 4.2.7, p. 421; (+)-α-pinene $[\alpha]_D^{23}$ + 47.1° (neat), 92 per cent ee, distilled from a small excess of lithium aluminium hydride and stored under nitrogen, is used for the preparation of Ipc_2BH.
(2) For preparative g.l.c. a 1.8 m × 12.7 mm column packed with 10 per cent Carbowax 20M on Chromosorb W is used.

5.4.4 THE OXYMERCURATION–DEMERCURATION OF ALKENES

A mild and highly convenient procedure for the hydration of a carbon–carbon double bond involves the initial reaction of an alkene with mercury(II) acetate in aqueous tetrahydrofuran, the resulting mercurial intermediate is reduced *in situ* by alkaline sodium borohydride solution. The yields of alcohols which are

545

obtained with a wide variety of alkenes are usually excellent. The reaction is highly regioselective in that the orientation of addition to terminal and non-terminal alkenes proceeds via the more stable carbocation ion (Markownikoff). Thus in the preparation of hexan-2-ol from hex-1-ene none of the isomeric hexan-1-ol is detectable by g.l.c. analysis of the product. Furthermore the course of the reaction is tolerant to the presence of hydroxy, methoxy, acetoxy, and halogen substituents in the alkene.[70]

$$R\text{·CH·CH}_2\text{HgOAc} \xrightarrow[\text{NaBH}_4]{\ominus\text{OH}} R\text{·CH(OH)·Me}$$

Complicating side reactions may occasionally occur – as in the oxymercuration–demercuration of styrene to 1-phenylethanol for which experimental details are also given. In this case evidently some organomercurial compounds survive the reductive stage, and their subsequent decomposition during final distillation complicates the isolation of the pure product.

Experiment 5.46 HEXAN-2-OL

$$\text{Bu·CH=CH}_2 + \text{Hg(OAc)}_2 \longrightarrow$$
$$\text{Bu·CH(OAc)·CH}_2\text{HgOAc} \xrightarrow{\text{NaBH}_4} \text{Bu·CH(OH)·Me}$$

Place 31.9 g (0.1 mol) of mercury(II) acetate and 100 ml of water in a 1-litre three-necked flask fitted with an efficient mechanical stirrer, a dropping funnel and a thermometer. Stir until the acetate has dissolved and then run in rapidly 100 ml of tetrahydrofuran; an orange–yellow suspension forms almost immediately. After stirring for a further 15 minutes, add 8.4 g (12.5 ml, 0.1 mol) of hex-1-ene, whereupon the colour is rapidly discharged. Stir the mixture at room temperature for 1 hour to ensure completion of the oxy-mercuration step. Next add with vigorous stirring 100 ml of 3 M sodium hydroxide solution, followed by a solution of 1.9 g (0.05 mol) of sodium boro-hydride in 100 ml of 3 M sodium hydroxide. Control the rate of addition of both solutions so that the temperature of the reaction mixture remains at about 25 °C, cooling the flask in cold water from time to time if necessary. Reduction occurs readily with the separation of elemental mercury. Finally stir vigorously at ambient temperature for 3 hours and then allow the reac-tion mixture to remain overnight in a separating funnel supported over a large empty conical flask. Separate the mercury layer (19.25 g, 96%) and then the aqueous alkaline phase, retaining the organic layer. Saturate the aqueous phase with sodium chloride, remove the additional organic layer which separates and extract the aqueous phase with two 30 ml portions of ether. Combine both of the organic layers with the ether extracts and remove most of the organic solvent carefully under reduced pressure using a rotary evapor-ator; stop the evaporation when two phases begin to separate. Add 50 ml of ether and 20 ml of water, separate the ether layer and wash it with four 25 ml portions of water, and dry it over anhydrous calcium sulphate. Remove the ether by flash distillation and distil the residue, collecting the hexan-2-ol at 136–140 °C; the yield is 6.9 g (68%).

Check the purity by g.l.c. on a 10 per cent Silicone oil column at 100 °C, nitrogen flow rate 40 ml per minute. The retention time is 1.42 minutes (cf. hexan-1-ol, 1.96 minutes).

Cognate preparation. *1-Phenylethanol.* Use 10.4 g (11.5 ml, 0.1 mol) of styrene, and carry out the oxymercuration and reduction as described above. The yield of recovered mercury is 17.5 g (87%), and traces continue to separate during the work-up procedure. Distil the final crude product under reduced pressure and collect the 1-phenylethanol at 110–115 °C/25 mmHg. Towards the end of the distillation the decomposition of residual organomercurial compounds ensues, and co-distillation of mercury contaminates the product; collect the contaminated fraction separately. The first fraction, yield 6.2 g (51%), is 92 per cent pure by g.l.c. (retention time 5.33 minutes); the impurity is mainly styrene (t_R 2.16 minutes). The mercury-contaminated fraction (3.0 g, 25%) is 85 per cent pure by g.l.c.

5.4.5 THE HYDROXYLATION OF ALKENES

Hydroxylation of an alkene may be carried out using osmium tetroxide in an inert solvent (e.g. ether or dioxane), whereupon a cyclic osmate ester is formed. This undergoes hydrolytic cleavage under reducing conditions (e.g. aqueous sodium sulphite) to give the 1,2-diol resulting from a *cis*-hydroxylation process.[71]

Alternatively and more conveniently this *cis*-hydroxylation process can be effected using only catalytic amounts of osmium tetroxide together with hydrogen peroxide, which cleaves the first formed osmate ester to the diol and regenerates the osmium tetroxide. The reaction is carried out in t-butyl alcohol and is illustrated by the conversion of cyclohexene into *cis*-cyclohexane-1,2-diol (Expt 5.47).

Other variations in this hydroxylation procedure include the use of t-butyl hydroperoxide[72] in place of hydrogen peroxide, and the use of PTC conditions.[73]

Alkaline potassium permanganate also effects the *cis*-hydroxylation of an alkene. A useful procedure involves the hydroxylation reagent cetyltrimethylammonium permanganate in dichloromethane or in aqueous t-butyl alcohol.[74] The preparation of the reagent and the two alternative general procedures are given in the cognate preparation to Expt 5.47.

A 1,2-diol arising from a *trans*-hydroxylation process is formed from an alkene by way of an intermediate epoxide which is subjected to a ring-opening reaction and hydrolysis. The epoxides may be isolated when the alkene is reacted with perbenzoic acid or *m*-chloroperbenzoic acid (Section 4.2.56, p. 457) in a solvent such as chloroform or dichloromethane; the preparation of epoxides by this method and by other important procedures are discussed and illustrated

547

in Section 8.1.2, p. 1131. With performic acid (Section 4.2.*56*, p. 455) isolation of the intermediate epoxide is not possible since it is converted by the formic acid solvent into the hydroxyformate ester. This product, which is formed by nucleophilic ring-opening of the epoxide, is hydrolysed by treatment with dilute alkali to give the 1,2-diol. The formation of *trans*-cyclohexane-1,2-diol (Expt 5.48) is illustrative of this reaction.

Experiment 5.47 *CIS*-CYCLOHEXANE-1,2-DIOL

CAUTION: *This preparation involving the toxic osmium tetroxide must be carried out in a fume cupboard. For the precautions to be followed when using 30 per cent hydrogen peroxide see Section 4.2.*41, *p. 439.*

Prepare the *reagent* as follows. To a mixture of 100 ml of pure t-butyl alcohol (2-methylpropan-2-ol) and 25 ml of 30 per cent hydrogen peroxide, add anhydrous sodium sulphate or magnesium sulphate in small portions; two layers separate out. Remove the alcohol layer which contains most of the hydrogen peroxide, and dry it with magnesium sulphate, followed by anhydrous calcium sulphate. The resulting liquid is a stable solution of 6.3 per cent hydrogen peroxide in t-butyl alcohol.

Free cyclohexene from peroxides by treating it with a saturated solution of sodium metabisulphite, separate, dry and distil; collect the fraction having b.p. 81–83 °C. Mix 8.2 g (0.1 mol) of cyclohexene with 55 ml of the reagent (0.1 mol), add 3 ml of a 0.5 per cent solution of osmium tetroxide [**CAUTION**: (1)] in anhydrous t-butyl alcohol and cool the mixture to 0 °C. Allow to stand overnight, by which time the initial orange coloration will have disappeared. Remove the solvent and unreacted cyclohexene by distillation at atmospheric pressure and fractionate the residue under reduced pressure using an air condenser. Collect the fraction of b.p. 120–140 °C/15 mmHg; this solidifies almost immediately. Recrystallise from ethyl acetate. The yield of pure *cis*-cyclohexane-1,2-diol, m.p. 96 °C, is 5.0 g (45%).

Note. (1) *Osmium tetroxide is extremely irritating and toxic and constitutes a severe eye injury hazard.* It may be purchased in sealed ampoules, e.g. 100 mg; the solution in t-butyl alcohol must be prepared and dispensed in an efficient fume cupboard, with the added protection of gloves and goggles. This solution is reasonably stable (e.g. the decomposition after one month is about 20%), provided that no 2-methylprop-1-ene arising from the t-butyl alcohol is present as impurity. In the latter case formation of black colloidal osmium, which can catalyse the decomposition of hydrogen peroxide, is rapid.

Cognate preparations. *Hydroxylation using cetyltrimethylammonium permanganate (CTAP).*[74] The reagent is prepared as follows. To a stirred solution of potassium permanganate (3.168 g, 20 mmol) in water (100 ml) at 20 °C is added dropwise over 20 minutes a solution of cetyltrimethylammonium bromide (8.02 g, 22 mol) in water (100 ml). A fine violet precipitate forms immediately. Stirring is continued for 30 minutes, the product isolated by suction, washed thoroughly with water, and dried in a desiccator over phosphorus pentoxide *in vacuo* for 3 hours at room temperature to give a fluffy violet solid; yield 6.5 g (80%). The reagent is stored in a brown bottle in a refrigerator.

Method A. A solution of CTAP (2.02 g, 5 mmol) in dichloromethane (30 ml) is added dropwise to a stirred solution of the alkene (5 mmol) in dichloromethane (15 ml) at 20 °C. Stirring is continued for 1–5 hours and the mixture then concentrated to half its volume under reduced pressure. The residual solution is diluted with ether (50 ml) and filtered through a pad of Celite and magnesium sulphate. The filtrate is evaporated under reduced pressure and the remaining *vic*-diol purified by recrystallisation from ethyl acetate/light petroleum.

Method B. A solution of CTAP (2.02 g, 5 mmol) in t-butyl alcohol (20 ml) and water (5 ml) is added dropwise to a stirred solution of the alkene (5 mmol) in t-butyl alcohol (4 ml) at 20 °C, and stirring continued for 1–5 hours. Chloroform (50 ml) and 5 per cent aqueous sodium hydroxide solution (15 ml) are added, the mixture stirred for 30 minutes and the organic layer separated. The aqueous phase is extracted with chloroform (3 × 50 ml). The combined organic extracts are dried with magnesium sulphate, filtered, and evaporated to leave the *vic*-diol which is purified as in *Method A*.

The following alkenes have been hydroxylated by these methods (method, stirring time, yield %): dicyclopentadiene (*A*, 4 hours, 86%); oct-1-ene (*A*, 2 hours, 85%) (1); cyclohexene (*B*, 1 hour, 86%); cyclooctene (*B*, 1 hour, 73%).

Note. (1) In the case of the hydroxylation of oct-1-ene, the product was purified by column chromatography using ether:light petroleum 2:5 as eluate.

Experiment 5.48 TRANS-CYCLOHEXANE-1,2-DIOL

CAUTION: *All preparations and reactions with hydrogen peroxide and organic peroxyacids must be conducted behind a safety screen, because these reactions sometimes proceed with violence.*

In a 500-ml three-necked flask, equipped with a mechanical stirrer, a thermometer and a dropping funnel, place 300 ml (6 mol) of 88–90 per cent formic acid and add 70 ml (0.62 mol) of 30 per cent hydrogen peroxide (see Section

4.2.*41*, p. 439). Then introduce slowly 41 g (51 ml, 0.5 mol) of freshly distilled cyclohexene (Expt 5.12) over a period of 20–30 minutes; maintain the temperature of the reaction mixture between 40 and 45 °C by cooling with an ice bath and controlling the rate of addition. Keep the reaction mixture at 40 °C for 1 hour after all the cyclohexene has been added and then allow to stand overnight at room temperature. Remove most of the formic acid and water by distillation from a water bath under reduced pressure (rotary evaporator). Add an ice-cold solution of 40 g (1 mol) of sodium hydroxide in 75 ml of water in small portions to the residual mixture of the diol and its formate; take care that the temperature does not rise above 45 °C. Warm the alkaline solution to 45 °C and add an equal volume (*c*. 200 ml) of ethyl acetate. Extract thoroughly, separate the lower layer and extract at 45 °C six times with equal volumes of ethyl acetate. Combine the ethyl acetate extracts (total volume about 1 litre), distil off the solvent using a rotary evaporator until the residual volume is about 150 ml and solid commences to crystallise. Cool to 0 °C and separate the crude product (*c*. 4.5 g) by suction filtration. Concentrate the mother-liquor to 30–40 ml, when more solid crystallises (*c*. 8 g). Cool and filter the mixture as before. Distil the combined crude products under reduced pressure from an oil bath and using an air condenser (see Section 2.27); the pure *trans*-cyclohexane-1,2-diol passes over at 128–132 °C/15 mmHg (or at 120–124 °C/4 mmHg) and solidifies immediately, m.p. 102–103 °C. The yield is 40 g (69%). It may be recrystallised from acetone or from ethyl acetate.

5.4.6. SOME METHODS FOR THE PROTECTION OF THE HYDROXYL GROUP

The hydroxyl group of a primary, secondary, or tertiary acyclic or alicyclic alcohol may be protected by conversion into (*a*) an ether, (*b*) a silyl ether, or (*c*) an ester. The most important method for the protection of a 1,2- or 1,3-diol is conversion into (*d*) a cyclic acetal.

ETHERS
When the alcohol possesses no functionality which is susceptible to basic conditions, it may be readily converted into the corresponding *methyl ether* by treatment with a suitable methylating agent (e.g. methyl iodide or dimethyl sulphate). Typical conditions are described in Expt 5.73, which also include PTC procedures for those alcohols which are water insoluble. Polyols, as for example in the case of carbohydrate derivatives, may also be methylated by this PTC procedure. If the compound to be methylated is base-labile, methylation may be effected by treatment of a mixture of the compound and neutral silica gel in ether solution with diazomethane.[75]

Methyl ethers are stable to acidic and basic conditions, and oxidising or reducing reagents. *Deprotection* to regenerate the alcohol is difficult (see Section 9.6.10, p. 1254); a convenient mild procedure uses iodotrimethylsilane in chloroform solution at room temperature.[76a] The alkyl methyl ether under these conditions gives the alkyl silyl ether and methyl iodide; the former on treatment with methanol gives the deprotected alcohol.

$$\text{ROMe} + \text{Me}_3\text{SiI} \xrightarrow{-\text{MeI}} \text{ROSiMe}_3 \xrightarrow{\text{MeOH}} \text{ROH} + \text{MeOSiMe}_3$$

An alternative reagent is trichloromethylsilane/sodium iodide in dry aceto-nitrile.[76b]

Benzyl ethers are readily formed under PTC conditions described in Expt 5.73. These ethers are stable to acidic and basic media, and to oxidising reagents. *Deprotection* is effected by hydrogenolysis; the wide range of conditions appro-priate to the structural nature of the alcohol has been reviewed.[77]

Allyl ethers are also formed under PTC conditions (see Expt 5.73). These ethers are stable under moderately acidic and basic conditions. *Deprotection* is achieved by treatment with potassium t-butoxide in dimethyl sulphoxide, which results in rearrangement to the corresponding prop-1'-enyl ether, followed by reaction with a mercury(II) chloride/mercury(II) oxide reagent mixture.[78]

$$R \cdot O \cdot CH_2 \cdot CH{=}CH_2 \xrightarrow[\text{DMSO}]{\text{KOH}} R \cdot O \cdot CH{=}CH \cdot Me \xrightarrow[\text{HgO}]{\text{HgCl}_2} ROH$$

General procedure for deprotection of allyl ethers. A mixture of the allyl ether (0.05 mol) in dry dimethyl sulphoxide (50 ml) and potassium t-butoxide (0.025 mol) was heated at 100 °C, under dry nitrogen, for 15 minutes. After cool-ing, the mixture was diluted with water (50 ml) and extracted with dichloro-methane (3 × 50 ml). The organic layer was dried, evaporated and the residue distilled or recrystallised. The rearranged ether (0.0074 mol) and mercury(II) oxide (2.02 g) was dissolved in acetone:water (10:1, 67 ml). To this mixture was added dropwise over 8–10 minutes a solution of mercury(II) chloride (2.02 g) in acetone:water (10:1, 22.5 ml). The reaction mixture was stirred until the reaction was complete (t.l.c. monitoring). The solution was diluted with ether (20 ml), fil-tered through a filter-bed to remove mercury compounds, and the acetone evap-orated. Ether was added to the residue and washed with a semi-saturated aqueous solution of potassium iodide (3 × 10 ml), dried and evaporated. The regenerated alcohol was purified by distillation or recrystallisation as appro-priate.

An alternative deprotection method, which proceeds without prior iso-merisation, involves heating under reflux the allyl ether in aqueous methanol with palladium/activated charcoal in the presence of toluene-*p*-sulphonic acid.[79]

2-Methoxyethoxymethyl ethers, MEM-ethers, are formed by reaction of an alcohol (1 mol) with 2-methoxyethoxymethyl chloride (1.5 mol, MEM-chloride) in dichloromethane solution (10 ml/g of MEM-chloride) at room temperature in the presence of ethyldiisopropylamine (1.5 mol)[80]; this literature report also notes other preparative procedures.

$$ROH + ClCH_2 \cdot O \cdot CH_2 \cdot CH_2 \cdot OMe \longrightarrow RO \cdot CH_2 \cdot O \cdot CH_2 \cdot CH_2 \cdot OMe$$

These ethers are stable to reducing and oxidising agents, to organometallic reagents, and to mildly acidic conditions that deprotect tetrahydropyranyl and silyl ethers (see below). They are *deprotected* by the action of either anhydrous zinc bromide or titanium(IV) chloride in dichloromethane solution.[80]

Tetrahydropyranyl ethers, THP-ethers, are formed by reaction of an alcohol with 2,3-dihydropyran in the presence of an acid catalyst. The procedures below illustrate the use of concentrated hydrochloric acid or pyridinium toluene-*p*-sulphonate. Amberlyst H-15 resin has also been used as a reaction catalyst.[81]

Procedures for THP-ether formation. *Use of hydrochloric acid.* Concentrated hydrochloric acid (0.3 ml) was added to a mixture of 2,3-dihydropyran (0.75 mol) and the alcohol (0.5 mol). Reaction commenced immediately on shaking and was moderated by cooling in an ice-water bath. The mixture was shaken for a further 30 minutes, allowed to stand overnight, diluted with ether (150 ml), and the solution washed twice with aqueous sodium hydrogen carbonate solution. The ethereal solution was dried and evaporated, and the residue distilled under reduced pressure.

Use of pyridinium toluene-p-sulphonate (PPTS).[82] A solution of geraniol (154 mg, 1.0 mmol) and dihydropyran (126 mg, 1.0 mmol) (1) was stirred for 4 hours at room temperature. Then the solution was diluted with ether and washed once with half-saturated brine to remove the catalyst. The ethereal solution was dried, evaporated and distilled [b.p. 140 °C (bath temperature)/10 mmHg] to give an essentially quantitative yield of geraniol THP-ether (236 mg, 99%).

Note. (1) Pyridinium toluene-p-sulphonate was prepared as follows.[82] Toluene-p-sulphonic acid monohydrate (5.70 g, 30 mmol) was added to pyridine (12 ml, 150 mmol) with stirring at room temperature (slightly exothermic). After stirring for 20 minutes the excess of pyridine was removed with a rotary evaporator on a water bath at c. 60 °C to afford a quantitative yield of PPTS as slightly hygroscopic colourless crystals. Recrystallisation from acetone gave the pure salt (6.8 g, 90%), m.p. 120 °C.

Tetrahydropyranyl ethers are stable to strong bases, to Grignard reagents and alkyllithiums, and to acylating and alkylating reagents. *It has been reported that explosions have occurred during distillations of THP-ethers after a reaction involving borane–hydrogen peroxide–sodium hydroxide or epoxidation reagents.*[83] It is suggested that similar hazards might be experienced with similar protecting groups (e.g. 1,3-dioxolanes). Two frequently used methods of *deprotection* are as follows.

Procedures for deprotection of THP-ethers. *Use of methanolic hydrochloric acid.* The THP-ether (0.2 mol) was dissolved in methanol (200 ml) and concentrated hydrochloric acid (30 ml), and the mixture heated under reflux for 2 hours. After cooling, the solution was neutralised by the addition of an excess of sodium hydrogen carbonate, diluted with ether (200 ml), filtered and the ether evaporated. The residue was dissolved in ether, washed twice with water, dried and evaporated. The deprotected alcohol was distilled under reduced pressure.

Use of an ethanolic solution of PPTS.[82] A solution of geraniol THP-ether (119 mg, 0.5 mmol) and PPTS (12.6 mg, 0.05 mol) in ethanol (4 ml) was stirred at 55 °C (bath temperature) for 3 hours. The solvent was removed and the residue chromatographed on a column of silica gel to afford pure geraniol (77 mg, 100%).

SILYL ETHERS

There are many silylating reagents for the conversion of an alcohol into the corresponding silyl ether. The reader should consult the *Handbook of the Pierce Chemical Co.* for an authoritative survey. The procedure for the formation of *trimethylsilyl ethers* of carbohydrate derivatives with hexamethyldisilazane/chlorotrimethylsilyl chloride in pyridine and the subsequent examination by g.l.c. analysis is described in Section 2.31.

$$\text{ROH} + (\text{Me}_3\text{Si})_2\text{NH}/\text{Me}_3\text{SiCl} \xrightarrow{\text{C}_5\text{H}_5\text{N}} \text{R·O·SiMe}_3$$

Trimethylsilyl ethers are readily hydrolysed under acidic or basic conditions, e.g. with methanolic citric acid, or with potassium carbonate in dry methanol.

The *t-butyldimethylsilyl ether* (TBDMS-ether), formed from the alcohol with t-butyldimethylsilyl chloride in the presence of imidazole in dimethylformamide solution,[84a] is more stable to hydrolysis than the TMS-ether (see also Section 4.2.66, p. 461). Deprotection is readily effected by treatment with 2–3 equivalents of tetrabutylammonium fluoride in tetrahydrofuran at 25 °C,[84a] or tetrabutyl-ammonium chloride and potassium fluoride dihydrate in acetonitrile.[84b]

ESTERS

The alcoholic hydroxyl group may be protected by conversion into, for example, the *acetate, benzoate, p-nitrobenzoate, or toluene-p-sulphonate esters.* Suitable specific and general procedures are described in Expts 5.142, 6.46, 6.163 and in Section 9.6.4, p. 1241. Esters may be hydrolysed under basic or acidic conditions, suitable procedures are described in Section 9.6.17, p. 1266.

CYCLIC ACETALS

The reaction of a 1,2- or a 1,3-diol with an aldehyde or ketone under anhydrous conditions gives rise to a cyclic acetal. A discussion of this reaction is given in Section 5.10.3, p. 652, where some structural features for acetal formation and selectivity of removal are reviewed. Other instances of the value of this protective group are to be found in Expts 5.9 and 5.63 and in Section 5.8.8, p. 623. It should be pointed out that cyclic acetal formation is also an important procedure for the protection of the carbonyl group.

5.5 ALIPHATIC HALIDES

The halogen in a saturated alkyl halide may be located on a primary carbon atom (R·CH$_2$X), a secondary carbon atom (R$_2$CHX), or a tertiary carbon atom (R$_3$CX). The halogen-carrying carbon atom (the α-carbon) in a secondary or a tertiary halide may be a chiral site (*) as shown below in the case of (*R*)-2-chlorobutane (1) CH$_3$·CH$_2$·CHCl·CH$_3$. Compounds where the halogen is attached to an *sp*^{2}-hybridised carbon atom of an alkene are termed alkenyl (or vinyl) halides [R·CH=CHX, e.g. (*E*)-1-chlorobut-1-ene (2) CH$_3$·CH$_2$·CH=CHCl]. In those cases where the halogen is attached to an *sp*-hybridised carbon atom in an alkyne, the compounds are called alkynyl (or acetylenic) halides [R·C≡C·X, e.g. 1-bromobut-1-yne (3) CH$_3$·CH$_2$·C≡CBr]. Allylic halides (R·CH=CH·CH$_2$X) have an *sp*^{2}-hybridised carbon atom attached to a saturated halogen-carrying carbon [e.g. (*E*)-1-iodobut-2-ene (4) CH$_3$·CH=CH·CH$_2$I].

Me   Et

H-⟩*—Cl

Et   Cl   Et—≡—Br   I

                                                   Me

(1)   (2)   (3)   (4)

The synthesis of alkyl halides is exemplified by the following procedures.

1. The displacement of a hydroxyl group in an alcohol by halogen [(a) chlorides, Expts 5.49 to 5.52, (b) bromides, Expts 5.53 to 5.56, (c) iodides, Expts 5.57 to 5.60].
2. Displacement reactions involving (a) a halogen atom (Expts 5.61 and 5.62), (b) a methanesulphonyloxy group (Expt 5.63), and (c) an amino group.
3. The addition of hydrogen halides or halogens to alkenes (Expts 5.64 to 5.67).
4. The replacement of reactive allylic hydrogen atoms by bromine (Expts 5.68 and 5.69).

SUMMARY OF RETROSYNTHETIC STRATEGIES

Functional group interconversion (FGI) (methods 3(a) and (b))

R–CH₂–CH₂–X (TM) ⟹(3) R–CH=CH₂ ⟸(3) (R)(Me)CH–X or (R)(CH₂X)CH (TM) (TM)

Functional group removal (FGR) (method 4)

R–CH=CH–CH₂Br (TM) ⟹(4) R–CH=CH–Me

C—X Disconnection (methods 1(a)–(c) and 2(a)–(c))

R–CH₂CH₂–Y ⟸(2a) R–CH₂CH₂⊣X ⟹(1) R–CH₂CH₂–OH

Y = Cl, Br, I   X = Cl, Br, I
= F in (2a)
(TM)

(2c)   (2b)

R–CH₂CH₂–NH₂   R–CH₂CH₂–O·SO₂Me

SPECTROSCOPIC FEATURES

The i.r. spectrum of an alkyl halide shows the expected absorptions due to the vibrations of the carbon–hydrogen skeletal structure, together with strong absorption in the region below $700\,\text{cm}^{-1}$ due to the stretching of the carbon–halogen bond. Medium intensity absorptions for C—F bonds are found in the region 1150 to $1300\,\text{cm}^{-1}$ (p. 316). The structure of the alkyl group may often be established from the p.m.r. spectrum owing to the deshielding effect of the halogen (p. 324) which frequently leads to first order splitting patterns being obtained (p. 341). The rather weak intensity of the molecular ion in the m.s. of alkyl halides makes the recognition of the isotope ratios of chloro- and bromo-compounds more difficult than in the case of aryl halides (p. 382); the most intense fragment ions usually arise from loss of halogen or hydrogen halide from the molecular ion. The alkyl halides do not absorb in the accessible u.v.–visible region. Some of the preparations below provide illustrative descriptive structural interpretations based on these spectroscopic features.

5.5.1 PREPARATION OF ALKYL CHLORIDES FROM ALCOHOLS

The hydroxyl group in tertiary alcohols is most readily replaced, and this is effected by simply allowing the alcohol to react with concentrated hydrochloric acid at room temperature. The reaction is a nucleophilic displacement of the S_N1 type involving the formation of a relatively stable carbocation intermediate.

$$R_3COH \xrightarrow{+H^\oplus} R_3C\overset{\oplus}{O}H_2 \xrightarrow{-H_2O} R_3C^\oplus \xrightarrow{+Cl^\ominus} R_3CCl$$

The reaction is illustrated by the preparation of t-butyl chloride (Expt 5.49).

Secondary, and to a greater extent primary, alcohols require more vigorous conditions to effect the substitution reaction, which is usually achieved by heating the alcohol–acid mixture with anhydrous zinc chloride. Illustrative examples are given in Expt 5.50. In the case of alicyclic secondary alcohols anhydrous calcium chloride is recommended (e.g. the preparation of chlorocyclohexane, Expt 5.51). The unsaturated alcohol, allyl alcohol, gives a poor yield by the HCl–$ZnCl_2$ method, but an alternative procedure using copper(I) chloride as catalyst has proved to be more satisfactory (Expt 5.51, cognate preparation).

The hydrochloric acid–zinc chloride reaction may be an S_N2 type displacement, particularly in the case of primary alcohols.

$$ROH + HCl/ZnCl_2 \longrightarrow Cl_2\overset{\ominus}{Zn}-Cl \ {}^\curvearrowright R-\overset{\oplus}{O}H_2 \xrightarrow[-H_2O]{-ZnCl_2} RCl$$

An S_N1 mechanism is also possible, particularly in the case of secondary alcohols.

$$ROH + ZnCl_2 \longrightarrow R-\overset{\oplus}{\underset{H}{O}}-ZnCl_2 \longrightarrow \overset{\oplus}{R} + HOZnCl_2^\ominus \xrightarrow{-Zn(OH)Cl} RCl$$

The regioselectivity of this latter reaction pathway may be diminished owing to the tendency of carbocations to rearrange, particularly when branching of the carbon chain occurs in the β-position. Hence the method is preparatively useful only with secondary alcohols (e.g. butan-2-ol) where one unique secondary carbocation is involved (see also Section 5.5.2, p. 560).

Rearrangement may be largely (but not entirely) suppressed by preparing the alkyl chloride from a reaction of the alcohol with thionyl chloride, either (i) alone, or (ii) in the presence of catalytic or equimolar proportions of pyridine, or (iii) in the presence of dimethylformamide.[85a] In (i) a chlorosulphite is first formed which decomposes via two sequentially formed ion-pair species; the second ion pair collapses to yield the alkyl chloride.

$$R-\overset{\curvearrowleft}{O:} \ Cl-\underset{\underset{O}{\|}}{S}-Cl \xrightarrow{-HCl} R\cdot O\cdot SO\cdot Cl \longrightarrow$$

$$[R^\oplus]\left[\begin{matrix} O \\ Cl \end{matrix}\hspace{-0.5em}\underset{}{\overset{}{S}}{=}O\right]^\ominus \xrightarrow{-SO_2} [R^\oplus][Cl^\ominus] \longrightarrow RCl$$

In (ii) the pyridine liberates the chloride ion which then effects an S_N2 displacement on the chlorosulphite.

$$HCl + C_5H_5N \longrightarrow C_5H_5\overset{\oplus}{N}H + Cl^{\ominus}$$

$$Cl^{\ominus} \overset{\curvearrowright}{} R \overset{\frown}{-O} \overset{\curvearrowleft}{-S} \overset{\frown}{-Cl} \longrightarrow RCl + SO_2 + Cl^{\ominus}$$
$$\underset{\overset{\|}{O}}{}$$

In (iii) the interaction of thionyl chloride with dimethylformamide gives a chloroimine chlorosulphite (cf. p. 992), which then reacts with the alcohol to form an alkoxyimine; a subsequent S_N2 displacement with the chloride ion yields dimethylformamide and the alkyl chloride. This reaction has been studied in considerable detail[85a] and a g.l.c. procedure has been described to analyse the mixture of regiosomers.[85b] All these reaction procedures are described in Expt 5.52.

$$Me_2N \cdot CHO + SOCl_2 \longrightarrow [Me_2\overset{\oplus}{N}{=}CHCl] \, [O \cdot SO \cdot Cl]^{\ominus} \xrightarrow{ROH} Me_2\overset{\oplus}{N}{=}CHOR$$

$$Me_2\overset{\overset{\oplus}{\frown}}{N}{=}CH \overset{\curvearrowleft}{-O} {-}R \overset{\curvearrowright}{} {}^{\ominus}Cl \longrightarrow Me_2N \cdot CHO + RCl$$

Finally triphenylphosphine with carbon tetrachloride (or carbon tetra-bromide) results in the conversion of allylic alcohols into the corresponding chlorides (or bromides) with little or no rearrangement.[86]

Experiment 5.49   t-BUTYL CHLORIDE (2-Chloro-2-methylpropane)

$$Me_3COH + HCl \longrightarrow Me_3CCl + H_2O$$

In a 250-ml separatory funnel place 25 g (0.34 mol) of 2-methylpropan-2-ol (t-butyl alcohol, b.p. 82–83 °C, m.p. 25 °C) and 85 ml of concentrated hydro-chloric acid (1) and shake the mixture from time to time during 20 minutes. After each shaking, loosen the stopper to relieve any internal pressure. Allow the mixture to stand for a few minutes until the layers have separated sharply; draw off and discard the lower acid layer. Wash the halide with 20 ml of 5 per cent sodium hydrogen carbonate solution and then with 20 ml of water. Dry the preparation with 5 g of anhydrous calcium chloride or anhydrous calcium sulphate. Decant the dried liquid through a funnel supporting a fluted filter paper into a 100-ml distilling flask, add 2–3 chips of porous porcelain and distil. Collect the fraction boiling at 49–51 °C. The yield of t-butyl chloride is 28 g (90%). The p.m.r. spectrum (CCl₄, TMS) shows a signal at δ 1.67 (s, 9H, Me₃C—).

Note. (1) The addition of 10 g of anhydrous calcium chloride tends to concentrate the acid and assists the separation of the chloride; the yield is slightly improved.

Cognate preparation. *2-Chloro-2-methylbutane.* Use 22 g (27 ml, 0.25 mol) of 2-methylbutan-2-ol (t-pentyl alcohol) and 65 ml of concentrated hydro-chloric acid. Distil the chloride twice from a Claisen flask with fractionating side-arm or through a short column. Collect the 2-chloro-2-methylbutane at 83–85 °C; the yield is 18 g (68%). Record the p.m.r. spectrum (CCl₄, TMS) and assign the signals which appear at δ 1.01 (t, 3H), 1.51 (s, 6H) and 1.73 (q, 2H); note the overlap of the latter two signals.

Experiment 5.50 BUTYL CHLORIDE (*1-Chlorobutane*)

$$Me\cdot(CH_2)_2\cdot CH_2OH + HCl \xrightarrow{ZnCl_2} Me\cdot(CH_2)_2\cdot CH_2Cl + H_2O$$

Fit a 250-ml round-bottomed flask with a reflux condenser, the top of which is connected to a device for absorbing hydrogen chloride (Fig. 2.61). Place 68 g (0.5 mol) of anhydrous zinc chloride and 40 ml (47.5 g) of concentrated hydrochloric acid in the flask, add 18.5 g (23 ml, 0.25 mol) of butan-1-ol and reflux the mixture gently for 2 hours. Arrange the condenser for downward distillation, and distil the reaction product, collecting the material which boils below 115 °C. Separate the upper layer of the distillate, mix it with an equal volume of concentrated sulphuric acid (1) and transfer the mixture to a 250-ml flask fitted with a reflux condenser. Reflux gently for 15–30 minutes, and then distil the chloride from the acid; it will pass over at 76–79 °C. Wash the distillate successively with 25 ml of water, 10 ml of 5 per cent sodium hydroxide solution and 25 ml of water; dry over 1–2 g of anhydrous calcium chloride, filter and distil from a small distilling flask. Collect the butyl chloride at 75–78 °C. The yield is 15–16 g (65–69%). The p.m.r. spectrum (CCl₄, TMS) shows signals at δ 0.95 (t, 3H, Me), 1.10–2.00 (m, 4H, —CH₂·CH₂—) and 3.49 (t, 2H, —CH₂Cl). The base peak in the m.s. appears at m/z 56 (M − HCl).

Note. (1) The sulphuric acid treatment removes high-boiling impurities which are not easily separated by distillation.

Cognate preparations. The following alkyl chlorides may be prepared in similar yield by replacing the butan-1-ol in the above preparation by the appropriate quantity of the requisite alcohol: *1-chloropentane*, b.p. 104–107 °C, from 22 g of pentan-1-ol; *1-chloro-3-methylbutane*, b.p. 98–100 °C, from 22 g of 3-methylbutan-1-ol (isopentyl alcohol); *2-chlorobutane*, b.p. 67–69 °C, from 18.5 g of butan-2-ol.

Experiment 5.51 CHLOROCYCLOHEXANE (*Cyclohexyl chloride*)

In a 1-litre, round-bottomed two-necked flask, carrying a reflux condenser and mechanical stirrer, place 100 g (1 mol) of pure cyclohexanol, 250 ml of concentrated hydrochloric acid and 80 g of anhydrous calcium chloride: heat the mixture on a boiling water bath for 10 hours with stirring (1). Some hydrogen chloride is evolved, consequently the preparation should be conducted in the fume cupboard. Separate the upper layer from the cold reaction product, wash it successively with saturated salt solution, saturated sodium hydrogen carbonate solution, saturated salt solution, and leave the crude chlorocyclohexane over an excess of anhydrous calcium chloride for at least 24 hours. Distil from a 150-ml flask through a fractionating side-arm, and collect the pure product at 141.5–142.5 °C. The yield is 90 g (76%).

Note. (1) The refluxing period may be reduced to 6 hours and the yield improved slightly by mechanical stirring; a three-necked flask should be used.

An alternative method of conducting the preparation consists in treating 100 g of cyclohexanol with 250 ml of concentrated hydrochloric acid, refluxing slowly while a stream of hydrogen chloride gas is passed into the mechanically stirred mixture for 3 hours. (The apparatus required is similar to that described for a Clemmensen reduction in Expt 5.5.) Chlorocyclohexane, b.p. 141–143 °C, is isolated as above; the yield is 80 g.

Cognate preparations. *Chlorocyclopentane* (*cyclopentyl chloride*). Use 43 g (0.5 mol) of cyclopentanol (Expt 5.29), 125 ml of concentrated hydrochloric acid and 50 g of anhydrous calcium chloride. The yield of chlorocyclopentane, b.p. 113–115 °C, is 30 g (57%).

Allyl chloride. Place 87 g (100 ml, 1.5 mol) of allyl alcohol, 150 ml of concentrated hydrochloric acid and 2 g of freshly prepared copper(I) chloride (Section 4.2.22, p. 428) in a 750-ml round-bottomed flask equipped with a reflux condenser. Cool the flask in ice and add 50 ml of concentrated sulphuric acid dropwise through the condenser with frequent shaking of the flask. A little hydrogen chloride may be evolved towards the end of the reaction. Allow the turbid liquid to stand for 30 minutes in order to complete the separation of the allyl chloride. Remove the upper layer, wash it with twice its volume of water, and dry over anhydrous calcium chloride. Distil and collect the allyl chloride which passes over at 46–47 °C; the yield is about 100 g (87%). Interpret the following data from the m.s.; m/z 78 (RA 9.3%), 76 (RA 27.8%), 41 (RA 100%), 40 (RA 11.7%) and 39 (RA 72.5%).

Experiment 5.52 1-CHLOROHEXANE

$$Me \cdot (CH_2)_4 \cdot CH_2OH + SOCl_2 \longrightarrow Me \cdot (CH_2)_4 \cdot CH_2Cl + HCl + SO_2$$

Reaction in the absence of catalyst. Assemble in a fume cupboard a 500-ml three-necked flask equipped with a sealed stirrer unit, a double surface reflux condenser and a separatory funnel; fit the condenser and the funnel with calcium chloride guard-tubes. Place 179 g (109.5 ml, 1.5 mol) of redistilled thionyl chloride in the flask and 51 g (62.5 ml, 0.5 mol) of hexan-1-ol, b.p. 156–158 °C, in the separatory funnel. Add the alcohol with stirring during 2 hours; there is a slight evolution of heat, sulphur dioxide is evolved and the liquid darkens considerably. When all the alcohol has been added, reflux the mixture for 2 hours. Rearrange the apparatus for distillation, and distil slowly; the excess of thionyl chloride passes over below 80 °C, followed by a small fraction up to 120 °C; and finally the crude 1-chlorohexane at 132–134 °C. Wash the last-named successively with water, 10 per cent sodium carbonate solution, and twice with water. Dry with anhydrous calcium chloride and distil through a short fractionating column. Pure 1-chlorohexane passes over at 133–134 °C. The yield is 36 g (60%).

Cognate preparations. *1-Chloroheptane.* From 58 g (70.5 ml, 0.5 mol) of heptan-1-ol (b.p. 175–177 °C) and 179 g (109.5 ml, 1.5 mol) of redistilled thionyl chloride; refluxing period, 4 hours. The yield of 1-chloroheptane, b.p. 159–160 °C, is 52 g (77%).

1-Chlorododecane. From 46.5 g (0.25 mol) of dodecan-1-ol (lauryl alcohol), m.p. 24 °C, and 119 g (73 ml, 1 mol) of redistilled thionyl chloride; refluxing period, 6 hours. The crude chloride passes over at 252–257 °C, mainly at 255–257 °C. Upon redistillation under reduced pressure 35 g (68%) of 1-chlorododecane, b.p. 116.5 °C/5 mmHg, are obtained.

1,4-Dichlorobutane. Place 22.5 g (0.25 mol) of redistilled butane-1,4-diol and 3 ml of dry pyridine in the flask in an ice bath. Add 119 g (73 ml, 1 mol) of redistilled thionyl chloride dropwise to the vigorously stirred mixture at such a rate that the temperature remains at 5–10 °C. When the addition is complete, remove the ice bath, keep the mixture overnight and then reflux for 3 hours. Cool, add ice-water cautiously and extract with ether. Wash the ethereal extract successively with 10 per cent sodium hydrogen carbonate solution and water, and dry with magnesium sulphate. Remove the ether by flash distillation and distil the residue under reduced pressure. Collect the 1,4-dichlorobutane at 55.5–56.5 °C/14 mmHg; the yield is 18 g (58%). The b.p. under atmospheric pressure is 154–155 °C.

Isobutyl chloride (1-chloro-2-methylpropane). *Reaction in the presence of pyridine.* Place 37 g (46 ml, 0.5 mol) of 2-methylpropan-1-ol (b.p. 106–108 °C) and 40 g (41 ml, 0.5 mol) of pure pyridine (**CAUTION**) in the flask, and 119 g (73 ml, 1.0 mol) of redistilled thionyl chloride in the funnel. Introduce the thionyl chloride with stirring during 3–4 hours; a white solid separates, which partially dissolves as the reaction proceeds. Reflux for 45 minutes: the solid will dissolve completely. Allow to cool and remove the upper layer. Wash the latter cautiously with water, 5 per cent sodium hydroxide solution, and twice with water; dry with anhydrous calcium chloride. Distil the product through a short fractionating column and collect the isobutyl chloride at 68–69 °C. The yield is 26 g (56%). Record the p.m.r. spectrum (CCl_4, TMS) and assign the signals which appear at δ 1.03 (d, 6H), 1.91 (m, 1H) and 3.32 (d, 2H). The base peak in the m.s. appears at m/z 43 [M – (—$CH_2{}^{35,37}Cl$)] and corresponds to the isopropyl carbocation.

General procedure for reactions in the presence of dimethylformamide.[85] Thionyl chloride (1.1 mol) followed by the alcohol (1 mol) is added dropwise to dimethylformamide (40 ml) at 0–10 °C (1). The reaction is heated as specified below, cooled and excess water added. The upper layer is then washed, dried (K_2CO_3) and distilled. The following conditions are specified: neopentyl alcohol (150 °C, 2 hours), pentan-3-ol (100 °C, 0.5 hours), 3-methylbutan-2-ol (100 °C, 0.5 hours).

Note. (1) The dimethylformamide is dried over molecular sieves.

5.5.2 PREPARATION OF ALKYL BROMIDES FROM ALCOHOLS

The formation of alkyl bromides is more ready than that of the alkyl chlorides. Hence secondary as well as tertiary bromides can be obtained directly from the corresponding alcohols by heating with constant boiling hydrobromic acid (e.g. Expt 5.53), although in the case of primary alcohols the presence of sulphuric acid results, as a rule, in a more rapid reaction and in improved yields.

$$ROH + HBr \longrightarrow RBr + H_2O$$

The method is readily adapted for the preparation of dibromides from diols. Typical examples are provided in Expt 5.54. The cyclic ethers tetrahydrofuran and tetrahydropyran are readily cleaved by the hydrobromic acid–sulphuric acid medium, and this provides an alternative and convenient preparation of the corresponding α, ω-dihalides.

As with the corresponding formation of akyl chlorides however, the procedure is only preparatively satisfactory with primary alcohols which react by a predominantly S_N2 mechanism, or with secondary and tertiary alcohols (S_N1 mechanism) where carbocation rearrangements do not give rise to isomeric products. Some appreciation of the difficulty of attaining high regioselectivity, may be gained from the observation that both pentan-2-ol and pentan-3-ol give the same mixture of carbocations in the presence of hydrobromic acid, and yield 2-bromo- and 3-bromo-pentane in a 70:30 ratio as determined by a g.l.c. analysis procedure.[87]

2-Bromopentane (70)   3-Bromopentane (30)

Promotion of an S_N2 displacement mechanism, and hence greater regioselectivity, may be effected by the addition of liquid bromine to a warm suspension of purified red phosphorus in the appropriate alcohol. The reaction is of general application: with primary alcohols (isobutyl alcohol to hexadecan-1-ol) the yields are over 90 per cent of the theoretical, but with secondary alcohols the yields are in the range 50–80 per cent (Expt 5.55). This method is to be preferred to the direct use (rather than the *in situ* generation) of phosphorus tribromide which is the more hazardous reagent. The outline mechanism may be represented thus:

Other highly regioselective brominating reagents are chlorotrimethylsilane/lithium bromide and hexamethyldisilane/pyridinium perbromide.[88] An outline mechanism of the former reaction is formulated below and a preparative example is given in Expt 5.56.

Experiment 5.53 ISOPROPYL BROMIDE (*2-Bromopropane*)

$$\text{Me·CH(OH)·Me} + \text{HBr} \longrightarrow \text{Me·CHBr·Me} + \text{H}_2\text{O}$$

Mix 40 g (51 ml, 0.67 mol) of propan-2-ol (isopropyl alcohol) with 460 g (310 ml) of constant boiling-point hydrobromic acid in a 500-ml flask fitted with a double surface condenser, add a few boiling chips and distil slowly (1–2 drops per second) until about half of the liquid has passed over. Separate the lower alkyl bromide layer (70 g), and redistil the aqueous layer when a further 7 g of the crude bromide will be obtained (1). Shake the crude bromide in a separatory funnel successively with an equal volume of concentrated hydrochloric acid (2), water, 5 per cent sodium hydrogen carbonate solution and water, and dry with anhydrous calcium chloride. Distil from a 100-ml flask; the isopropyl bromide passes over at 59 °C. The yield is 66 g (81%). The p.m.r. spectrum (CCl₄, TMS) shows signals at δ 1.70 (d, 6H, Me₂CH—) and 4.21 (sept, 1H, —CH—).

Notes. (1) The residue in the flask may be mixed with the aqueous layer of the first distillate, 40 g of propan-2-ol added, and the slow distillation repeated. The yield of crude isopropyl bromide in the second distillation is only slightly less than that obtained in the original preparation. Subsequently most of the residual hydrobromic acid may be recovered by distillation as the constant boiling point acid (126 °C). (2) The hydrochloric acid washing removes any unchanged alcohol which may be present.

Cognate preparations. *Bromocyclohexane.* Use 50 g (0.5 mol) of cyclohexanol and 260 g (176 ml) of 48 per cent hydrobromic acid and distil *all* the mixture slowly (6 hours). Add a little water to the distillate, separate the lower layer of crude bromide, and purify as above. Collect the bromocyclohexane at 163–165 °C (60 g, 74%).

Bromocyclopentane. Use 43 g (0.5 mol) of cyclopentanol (Expt 5.29) and 260 g (176 ml) of 48 per cent hydrobromic acid. Collect the bromocyclopentane at 135–137 °C (55 g, 74%).

Experiment 5.54 BUTYL BROMIDE (*1-Bromobutane*)

$$\text{Me·(CH}_2)_2\text{·CH}_2\text{OH} + \text{HBr} \xrightarrow{\text{H}_2\text{SO}_4} \text{Me·(CH}_2)_2\text{·CH}_2\text{Br} + \text{H}_2\text{O}$$

To 250 g of 48 per cent hydrobromic acid contained in a 500-ml round-bottomed flask add 75 g (41 ml) of concentrated sulphuric acid in portions with shaking; some hydrogen bromide may be evolved. Add 88 g (110 ml, 1.2 mol) of butan-1-ol, followed by 60 g (32.5 ml) of concentrated sulphuric acid in several portions with shaking, and finally a few chips of porous porcelain. Attach a reflux condenser to the flask and reflux the mixture gently on a wire gauze for 2–3 hours; during this period the formation of butyl bromide is almost complete and a layer separates above the acid (1). If the preparation is carried out in the open laboratory, fit an absorption device (Fig. 2.61(a) or (b)) to the top of the condenser in order to absorb any hydrogen bromide and sulphur dioxide which may be evolved. Allow the contents of the flask to cool, remove the condenser and set it for downward distillation. Distil the mixture until no more oily drops of butyl bromide pass over (30–40 minutes). Transfer the distillate to a separatory funnel and remove the halide which forms the lower layer. Wash it successively with water, an equal volume of concentrated hydrochloric acid (2), water, 5 per cent sodium hydrogen carbonate or sodium carbonate solution, and water. Separate the water as completely as possible and dry with 2–3 g of anhydrous calcium chloride or magnesium sulphate; the desiccant should be left in contact with the bromide for at least 30 minutes and shaken occasionally. Filter the dried product through a small funnel supporting a fluted filter paper into a 200-ml flask, add a few chips of porous porcelain and distil either from an air bath (Fig. 2.46) or on a ceramic-centred wire gauze. Collect the portion boiling at 100–103 °C. The yield is 155 g (95%).

Notes. (1) A suitable hydrogen bromide medium can be prepared by dissolving 240 g of potassium bromide in 400 ml of warm water, cooling and adding 200 ml of concentrated sulphuric acid slowly and with constant stirring, so that the temperature does not rise above 40 °C. After further cooling to 15 °C, the mixture is filtered and the butan-1-ol is added to the filtrate. A further 120 ml of concentrated sulphuric acid is then added carefully and the mixture is heated under reflux for 3–4 hours.
(2) The crude bromide contains a little unchanged alcohol and is said to contain some dibutyl ether (b.p. 141 °C). The former is removed by washing with concentrated hydrochloric acid and this purification process is satisfactory for most purposes. Both the alcohol and the ether are removed by washing with 11–12 ml of concentrated sulphuric acid; the butyl bromide is not affected by this reagent.

Cognate preparations.   *s-Butyl bromide* (*2-bromobutane*).   The quantities required are as for butyl bromide but with butan-2-ol (b.p. 99–100 °C) replacing the butan-1-ol. Two to three washings with concentrated hydrochloric acid are necessary, i.e. until the volume of the acid layer remains unchanged on shaking with halide. The yield of s-butyl bromide, b.p. 90.5–92.5 °C, is 150 g (92%).

1-Bromopentane (pentyl bromide).   Use 210 g (142 ml) of 48 per cent hydrobromic acid, 60 g (33 ml) of concentrated sulphuric acid, followed by 88 g (108 ml, 1 mol) pentan-1-ol (b.p. 135–136 °C) and 10 g (5.5 ml) of concentrated sulphuric acid. Distil the product through a short fractionating column, and collect the 1-bromopentane at 127–130 °C (135 g, 89%).

1-Bromo-3-methylbutane.   Proceed as for *1-bromopentane*, but use 88 g (109 ml, 1 mol) of 3-methylbutan-1-ol, b.p. 129.5–131 °C. Distil the purified product through a fractionating column and collect the 1-bromo-3-methylbutane at 117–120 °C (125 g, 83%).

Allyl bromide.   Introduce into a 1-litre three-necked flask 250 g (169 ml) of 48 per cent hydrobromic acid and then 75 g (40.5 ml) of concentrated sulphuric acid in portions, with shaking; finally add 58 g (68 ml, 1 mol) of pure allyl alcohol. Fit the flask with a separatory funnel, a mechanical stirrer and an efficient condenser (preferably of the double surface type) set for downward distillation. Place 75 g (40.5 ml) of concentrated sulphuric acid in the separatory funnel, set the stirrer in motion and allow the acid to flow slowly into the warm solution. The allyl bromide will distil over (<30 minutes). Wash the distillate with 5 per cent sodium carbonate solution, followed by water, dry over anhydrous calcium chloride, and distil from a flask through a short fractionating column. The yield of allyl bromide, b.p. 69–72 °C, is 112 g (93%). There is a small high-boiling fraction containing 1,2-dibromopropane.

1,3-Dibromopropane (*trimethylene dibromide*). In a 1-litre round-bottomed flask place 500 g (338 ml) of 48 per cent hydrobromic acid and add 150 g (82 ml) of concentrated sulphuric acid in portions, with shaking. Then add 91 g of propane-1,3-diol (b.p. 210–215 °C), followed by 240 g (130.5 ml) of concentrated sulphuric acid slowly and with shaking. Attach a reflux condenser to the flask and reflux the mixture for 3–4 hours. Arrange for downward distillation and distil, using a wire gauze, until no more oily drops pass over (30–40 minutes). Purify the 1,3-dibromopropane as detailed for *butyl bromide* above. About 220 g (91%) of the pure dibromide, b.p. 162–165 °C, are obtained.

1,4-Dibromobutane (*from butane-1,4-diol*). In a 500-ml three-necked flask fitted with a stirrer, reflux condenser and dropping funnel, place 154 g (105 ml) of 48 per cent hydrobromic acid. Cool the flask in an ice bath. Add slowly, with stirring, 130 g (71 ml) of concentrated sulphuric acid. To the resulting ice-cold solution add 30 g (0.33 mol) of redistilled butane-1,4-diol dropwise. Leave the reaction mixture to stand for 24 hours; heat for 3 hours on a steam bath. The reaction mixture separates into two layers. Separate the lower layer, wash it successively with water, 10 per cent sodium carbonate solution and water, and then dry with magnesium sulphate. Distil and collect the 1,4-dibromobutane at 83–84 °C/12 mmHg. The yield is 55 g (76%).

1,4-Dibromobutane (*from tetrahydrofuran*). Place a mixture of 250 g (170 ml) of 48 per cent hydrobromic acid and 75 g (41 ml) of concentrated sulphuric acid in a 500-ml round-bottomed flask, add 18.1 g (20.5 ml, 0.25 mol) of redistilled tetrahydrofuran (Section 4.1.*19*, p. 406) (b.p. 65–66 °C), attach a reflux condenser and reflux gently for 3 hours. Separate the lower layer of dibromide and purify as in the previous preparation. The yield of 1,4-dibromobutane, b.p. 83–84 °C/12 mmHg is 40 g (74%).

1,5-Dibromopentane (*from pentane-1,5-diol*). Proceed as for 1,4-dibromobutane but use 35 g (0.33 mol) of redistilled commercial pentane-1,5-diol. The yield of 1,5-dibromopentane, b.p. 99 °C/13 mmHg, is 39 g (51%).

1,5-Dibromopentane (*from tetrahydropyran*). Proceed as for 1,4-dibromobutane (from tetrahydrofuran) but use 21.5 g (24.4 ml, 0.25 mol) of redistilled tetrahydropyran (b.p. 86.5–87.5 °C). The yield of 1,5-dibromopentane, b.p. 99 °C/13 mmHg, is 46 g (80%).

Experiment 5.55 ISOBUTYL BROMIDE (*1-Bromo-2-methylpropane*)

$$10 \quad \underset{Me}{\overset{Me}{\diagdown}} OH + 2P + 5Br_2 \longrightarrow 10 \quad \underset{Me}{\overset{Me}{\diagdown}} Br + 2H_3PO_4 + 2H_2O$$

Place 92.5 g (115 ml, 1.25 mol) of isobutyl alcohol (2-methylpropan-1-ol) and 8.55 g (0.275 mol) of purified red phosphorus (Section 4.2.*59*, p. 458) in a 500-ml three-necked flask fitted with a sealed mechanical stirrer, a reflux condenser and a dropping funnel containing 100 g (32 ml, 0.62 mol) of bromine (*for precautions in the use of bromine*, see Section 4.2.*9*, p. 422). Start the stirrer, heat the flask (e.g., in an oil bath) so that the contents reflux gently, and introduce the bromine at such a rate that it appears to react completely so that there is little bromine vapour above the surface of the reaction mixture, and the reaction is under control. When all the bromine has been added, reflux the mixture gently for 15–30 minutes more. Remove the stirrer, arrange the condenser for downward distillation and distil off most of the isobutyl bromide (1). Then add about 50 ml of water through the dropping funnel and continue the distillation to remove the remainder of the product. Separate the crude bromide and wash it successively with water, an approximately equal volume of concentrated hydrochloric acid, water, 10 per cent sodium carbonate solution, and finally water. Dry the product over anhydrous calcium chloride and distil, collecting the isobutyl bromide which passes over at 91–94 °C. The yield is 150 g (91%).

Note. (1) It is not advisable to distil the mixture almost to dryness since the formation of flammable alkene may then occur. This is avoided by conducting the distillation in two stages as described.

Cognate preparations.* *1-Bromohexane.* Use 152.5 g (186.5 ml, 1.49 mol) of hexan-1-ol, 9.3 g (0.3 mol) of purified red phosphorus and 120 g (38.5 ml, 0.95 mol) of bromine;† b.p. 154–156 °C.

1-Bromoheptane. Use 173 g (209 ml, 1.49 mol) of heptan-1-ol, 9.3 g (0.30 mol) of purified red phosphorus and 120 g (38.5 ml, 0.95 mol) of bromine; b.p. 180 °C.

1-Bromooctane. Use 81 g (98.5 ml, 0.623 mol) of octan-1-ol (b.p. 193–194 °C), 5.18 g (0.167 mol) of purified red phosphorus and 55 g (18 ml, 0.343 mol) of bromine; b.p. 198–201 °C.

1-Bromododecane. Use 116 g (0.623 mol) of dodecan-1-ol (lauryl alcohol), m.p. 24 °C, 5.18 g (0.167 mol) of purified red phosphorus and 55 g (18 ml, 0.343 mol) of bromine. Heat the alcohol–phosphorus mixture to about 250 °C with vigorous stirring and add the bromine slowly. Allow the mixture to cool after all the bromine has been introduced. Add ether, filter off the excess of phosphorus and wash the ethereal solution of the bromide with water and dry over anhydrous potassium carbonate. Remove the ether on a water bath, and distil the residue under reduced pressure, b.p. 149–151 °C/18 mmHg.

1-Bromotetradecane. Use 107 g (0.5 mol) of tetradecan-1-ol (m.p. 38 °C), 3.41 g (0.11 mol) of purified red phosphorus and 44 g (14.5 ml, 0.275 mol) of bromine and proceed as under *1-bromododecane*; b.p. 178.5–179.5 °C/20 mmHg, m.p. 5 °C.

1-Bromohexadecane. Use 121 g (0.5 mol) of hexadecan-1-ol (cetyl alcohol), m.p. 48 °C, 3.41 g (0.11 mol) of purified red phosphorus and 44 g (0.275 mol) of bromine, and proceed as for *1-bromododecane*; filter off the excess of phosphorus at 16–20 °C; b.p. 202–203 °C/21 mmHg; m.p. 14 °C.

1-Bromo-2-phenylethane. Use 152.5 g (148 ml, 1.25 mol) of 2-phenylethanol (Expt 5.37), b.p. 216.5–217 °C, 10.35 g (0.33 mol) of purified red phosphorus and 110 g (35.5 ml, 0.68 mol) of bromine. Isolate the 1-bromo-2-phenylethane as detailed for *1-bromododecane*; b.p. 98 °C/12 mmHg.

---

* Unless otherwise stated, the yields exceed 90 per cent of the theoretical.
† The slight excess of bromine over the theoretical equivalent to the alcohol in the preparation of high boiling point bromides ensures the absence of unchanged alcohol in the product; any excess of bromine may be removed by the addition of a little sodium metabisulphite.

1,4-Dibromobutane (*from butane-1,4-diol*). Use 45 g (0.5 mol) of redistilled butane-1,4-diol, 6.84 g (0.22 mol) of purified red phosphorus and 80 g (26 ml, 0.5 mol) of bromine. Heat the glycol–phosphorus mixture to 100–150 °C and add the bromine slowly; continue heating at 100–150 °C for 1 hour after all the bromine has been introduced. Allow to cool, dilute with water, add 100 ml of ether and remove the excess of red phosphorus by filtration. Separate the ethereal solution of the dibromide, wash it successively with 10 per cent sodium thiosulphate solution and water, then dry over the anhydrous potassium carbonate. Remove the ether on a water bath and distil the residue under diminished pressure. Collect the 1,4-dibromobutane at 83–84 °C/12 mmHg; the yield is 73 g (67%).

1,6-Dibromohexane. Proceed as for 1,4-dibromobutane but use 58 g (0.49 mol) of hexane-1,6-diol. The yield of 1,6-dibromohexane, b.p. 114–115 °C/12 mmHg, is 85 g (71%).

1,4-Dibromobutane (*from tetrahydrofuran*). Place 18.1 g (20.5 ml, 0.25 mol) of redistilled tetrahydrofuran (b.p. 65–66 °C), 3.41 g (0.11 mol) of purified red phosphorus and 4.5 g of water in the flask. Heat the mixture gently and add 40 g (13 ml, 0.25 mol) of bromine at such a rate that there is little bromine vapour above the surface of the reaction mixture. Heat at 100–150 °C for 45–60 minutes after all the bromine has been introduced. Work up as for the butane-1,4-diol preparation. The yield of 1,4-dibromobutane, b.p. 83–84 °C/12 mmHg, is 42 g (72%).

1,5-Dibromopentane (*from tetrahydropyran*). Proceed as in the previous preparation but replace the tetrahydrofuran by 21.5 g (24.4 ml, 0.25 mol) of redistilled tetrahydropyran (b.p. 86.5–87.5 °C). The yield of 1,5-dibromopentane, b.p. 99 °C/13 mmHg, is 43 g (75%).

Experiment 5.56   3-BROMO-1-PHENYLPROP-1-ENE (*Cinnamyl bromide*)[88]

$$\text{Ph·CH}=\text{CH·CH}_2\text{OH} + \text{Me}_3\text{SiCl/LiBr} \longrightarrow$$
$$\text{Ph·CH}=\text{CH·CH}_2\text{Br} + \text{Me}_3\text{SiOH}$$

Chlorotrimethylsilane (2.7 g, 25 mmol) (1) (**CAUTION**) is added to a solution of lithium bromide (1.74 g, 20 mmol) in dry acetonitrile (20 ml) (2) with good stirring under a nitrogen atmosphere. Cinnamyl alcohol (1.34 g, 10 mmol) is then added and the reaction mixture heated under reflux for 12 hours. The progress of the reaction is monitored by t.l.c. on silica gel plates with hexane as the eluant. On completion of the reaction (12 hours), the reaction mixture is taken up in ether (50 ml), washed successively with water (2 × 25 ml), sodium hydrogen carbonate solution (10%, 50 ml) and finally brine, and dried over anhydrous sodium sulphate. Evaporation of the ether affords the pure bromide in 93 per cent yield. The product may be recrystallised from ethanol and has m.p. 31–32 °C; **CAUTION** this compound is lachrymatory.

Notes. (1) Chlorotrimethylsilane is distilled over sodium hydroxide pellets.
(2) Acetonitrile is purified and stored over molecular sieves.

5.5.3 PREPARATION OF ALKYL IODIDES FROM ALCOHOLS

Alkyl iodides are the most easily formed of the alkyl halides and the slow distillation of the alcohol with constant boiling hydriodic acid is a general method of preparation (e.g. Expt 5.57). As with the corresponding chlorides and bromides (q.v.), the yields of the required alkyl iodides in this reaction may be diminished in the case of certain (tertiary and secondary) alcohols as a result of skeletal rearrangement.

$$\text{ROH} + \text{HI} \longrightarrow \text{RI} + \text{H}_2\text{O}$$

An alternative reagent, which is particularly effective for the conversion of diols into diiodo compounds, is a mixture of potassium iodide and 95 per cent orthophosphoric acid (Expt 5.58). The reagent also cleaves tetrahydrofuran and tetrahydropyran to yield the corresponding α, ω-diiodo compounds [cf. the hydrobromic acid–sulphuric acid reagent, Section 5.5.2, p. 559].

Regioselectivity may be greatly improved by the use of a varied range of phosphorus-based iodinating reagents. For example, the addition of iodine to a gently boiling suspension of purified red phosphorus in the alcohol gives excellent yields of alkyl iodides from primary and secondary alcohols (Expt 5.59). Diphosphorus tetraiodide converts alcohols to alkyl iodides at room temperature under argon with high regioselectivity.[89] The use of iodotrimethylsilane/sodium iodide in acetonitrile (cf. Expt 5.56) has also been reported.[90]

Reaction of an alcohol with the reagent o-phenylene phosphorochloridite followed by treatment of the alkyl o-phenylene phosphite so obtained with iodine in dichloromethane at room temperature results in a good yield of alkyl iodide. This method, exemplified by the preparation of 1-iodoheptane (Expt 5.60), is the preferred procedure when acid-sensitive functional groups are present.

Experiment 5.57 ISOPROPYL IODIDE (*2-Iodopropane*)

$$\text{Me·CH(OH)·Me} + \text{HI} \longrightarrow \text{Me·CHI·Me} + \text{H}_2\text{O}$$

Mix 30 g (38 ml, 0.5 mol) of propan-2-ol with 450 g (265 ml) of constant boiling point hydriodic acid (57%) (Section 4.2.32, p. 436) in a 500-ml distilling flask, attach a condenser for downward distillation, and distil slowly (1–2 drops per second) from an oil or air bath. When about half the liquid has passed over, stop the distillation. Separate the lower layer of crude iodide (70 g, 82%). Redistil the aqueous layer and thus recover a further 5 g of iodide from the first quarter of the distillate (1). Wash the combined iodides with an equal volume of concentrated hydrochloric acid, then, successively, with water, 5 per cent sodium carbonate solution and water. Dry with anhydrous calcium chloride and distil. The isopropyl iodide distils constantly at 89 °C.

Note. (1) A further quantity of isopropyl iodide, only slightly less than that obtained in the first distillation, may be prepared by combining the residues in the distilling flask, adding 30 g (38 ml) of propan-2-ol, and repeating the distillation. Finally, the residues should be distilled and the 57 per cent constant boiling point acid recovered.

Cognate preparations. *Isobutyl iodide* (*1-iodo-2-methylpropane*). Use 30 g (37.5 ml, 0.37 mol) of 2-methylpropan-1-ol and 273 g (161 ml) of 57 per cent hydriodic acid; 65 g (96%) of the crude iodide are obtained. If the crude iodide is dark in colour, add a little sodium metabisulphite. B.p. 119–120 °C.

s-Butyl iodide (*2-iodobutane*). Use 30 g (37.5 ml, 0.37 mol) of butan-2-ol and 273 g (161 ml) of 57 per cent hydriodic acid; 63 g of crude iodide are obtained, b.p. 117.5–119 °C.

Iodocyclopentane. Use 43 g (45.5 ml, 0.5 mol) of cyclopentanol and 340 g (200 ml) of 57 per cent hydriodic acid; 89 g (91%) of crude iodide are obtained, b.p. 58 °C/22 mmHg.

Allyl iodide. Use 29 g (34 ml, 0.5 mol) of allyl alcohol and 340 g (200 ml) of 57 per cent hydriodic acid; 74 g (88%) of crude iodide are obtained. Upon adding 29 g (34 ml) of allyl alcohol to the combined residue in the flask and the aqueous layer and distilling as before, a further 72 g of crude allyl iodide may be isolated, b.p. 99–101 °C (mainly 100 °C). The compound is very sensitive to light; the distillation should therefore be conducted in a darkened room and preferably in the presence of a little silver powder.

Experiment 5.58 1,4-DIIODOBUTANE

$$\text{HO}\diagdown\diagup\diagdown\diagup_{\text{OH}} + 2\text{KI} + 2\text{H}_3\text{PO}_4 \longrightarrow$$

$$\text{I}\diagdown\diagup\diagdown\diagup_{\text{I}} + 2\text{KH}_2\text{PO}_4 + 2\text{H}_2\text{O}$$

In a 500-ml three-necked flask, equipped with a thermometer, a sealed stirrer unit and a reflux condenser, place 32.5 g of phosphorus pentoxide and add 115.5 g (67.5 ml) of 85 per cent orthophosphoric acid (1). When the stirred mixture has cooled to room temperature, introduce 166 g (1 mol) of potassium iodide and 22.5 g (0.25 mol) of redistilled butane-1,4-diol (b.p. 228–230 °C or 133–135 °C/18 mmHg). Heat the mixture with stirring at 100–120 °C for 4 hours. Cool the stirred mixture to room temperature and add 75 ml of water and 125 ml of ether. Separate the ethereal layer, decolourise it by shaking with 25 ml of 10 per cent sodium thiosulphate solution, wash with 100 ml of cold saturated sodium chloride solution, and dry with magnesium sulphate. Remove the ether by flash distillation (Fig. 2.101) on a steam bath and distil the residue from a flask with fractionating side-arm under diminished pressure. Collect the 1,4-diiodobutane at 110 °C/6 mmHg, the yield is 65 g (84%).

Alternatively, add 18 g (20 ml, 0.25 mol) of redistilled tetrahydrofuran (b.p. 65–66 °C) to a mixture of 32.5 g of phosphorus pentoxide, 115.5 g (67.5 ml) of 85 per cent orthophosphoric acid and 166 g of potassium iodide, heat for 3–4 hours, cool and isolate the 1,4-diiodobutane as above. The yield of product, b.p. 110 °C/6 mmHg, is 70 g (90%).

Note. (1) The orthophosphoric acid must be adjusted to a concentration of 95 per cent H_3PO_4. Alternatively, the commercial 100 per cent orthophosphoric acid may be diluted with water to this concentration. The 95 per cent acid is claimed to be the most efficient for the preparation of iodides from alcohols and glycols, and for effecting cleavage of tetrahydrofuran and tetrahydropyran. Anhydrous orthophosphoric acid does not give such good results because of the limited solubility of hydrogen iodide in the reagent.

Cognate preparations. *1,5-Diiodopentane* (*from pentane-1,5-diol*). Proceed as for 1,4-diiodobutane but use 26 g (26.5 ml, 0.25 mol) of redistilled pentane-

1,5-diol (b.p. 238–239 °C) in place of the butane-1,4-diol. The yield of 1,5-di-iodopentane, b.p. 142–143 °C/16 mmHg, is 65 g (80%).

1,5-Diiodopentane (*from tetrahydropyran*). Use 21.5 g (24.4 ml, 0.25 mol) of redistilled tetrahydropyran (b.p. 86.5–87.5 °C) in place of the tetrahydrofuran, otherwise proceed as for 1,4-diiodobutane. The yield of 1,5-diiodopentane, b.p. 142–143 °C/16 mmHg, is 71 g (88%).

1,6-Diiodohexane. Proceed exactly as detailed for 1,4-diiodobutane but replace the butane-1,4-diol by 29.5 g (0.25 mol) hexane-1,6-diol, m.p. 41–42 °C. The yield of 1,6-diiodohexane, b.p. 150 °C/10 mmHg, m.p. 10 °C, is 70 g (83%).

Butyl iodide. Use 37 g (46 ml, 0.5 mol) of butan-1-ol together with the quantities of the other reactants used above; a 2-hour reaction time is sufficient. The yield of butyl iodide, b.p. 129–130 °C, is 64 g (70%).

Iodocyclohexane. Proceed as for butyl iodide using 50 g (0.5 mol) of redistilled cyclohexanol (b.p. 160–161 °C). Distil the iodocyclohexane under reduced pressure, b.p. 67–69 °C/9 mmHg; the yield is 90 g (86%).

Experiment 5.59   BUTYL IODIDE (*1-Iodobutane*)

$$10\ \text{Me}\diagdown\diagup\text{OH} + 2P + 5I_2 \longrightarrow 10\ \text{Me}\diagdown\diagup I + 2H_3PO_4 + 2H_2O$$

Place 12.5 g (0.4 mol) of purified red phosphorus (Section 4.2.59, p. 458) and 78 g (96 ml, 1.05 mol) of butan-1-ol in a 250-ml round-bottomed flask fitted with a Liebig-type reflux condenser. Heat the mixture to gentle refluxing with a Bunsen flame over a wire gauze. Remove or lower the flame and add 127 g (0.5 mol) of coarsely powdered iodine in approximately 2 g portions down the centre of the condenser. Complete the entire addition of the iodine fairly rapidly (i.e. in about 20–30 minutes), but allow the mildly exothermic reaction to subside after the addition of each portion. Finally continue heating under reflux for 30–60 minutes; little or no iodine should then be visible. Arrange the condenser for downward distillation and distil off most of the crude product. When the volume of liquid in the flask has been reduced to about 15–20 ml, add about 40 ml of water and continue the distillation until no more oily drops pass over into the receiver (1). Separate the crude alkyl iodide and wash it successively with approximately equal volumes of water, concentrated hydrochloric acid (2), water, 10 per cent sodium carbonate solution and water (3). Dry the product over anhydrous calcium chloride and distil, collecting the butyl iodide which passes over at 129–131 °C; the yield is 165 g (90%) (4).

Notes. (1) See Expt 5.55, Note (1).
(2) The washing with concentrated hydrochloric acid removes any unchanged alcohol which may be present.
(3) If the washed organic phase is darker than a pale brown-pink, add a few crystals of sodium thiosulphate to the final wash-water to remove traces of iodine.
(4) The purified iodide may be preserved in a bottle containing a short coil of clean copper wire.

Cognate preparations. The following may be prepared by the above procedure in similar yield using the same quantities of red phosphorus and of iodine:

Propyl iodide (*1-iodopropane*) – from 63 g (78 ml, 1.05 mol) of propan-1-ol. (B.p. 102–103 °C).

Isopropyl iodide (*2-iodopropane*) – from 63 g (80 ml, 1.05 mol) of propan-2-ol. A little hydrogen iodide is evolved. (B.p. 89–90 °C).

s-Butyl iodide (*2-iodobutane*) – from 78 g (97 ml, 1.05 mol) of butan-2-ol. A little hydrogen iodide is evolved. (B.p. 118–120 °C).

1-Iodopentane – from 92 g (113 ml, 1.05 mol) of pentan-1-ol. (B.p. 153–156 °C).

1-Iodohexane – from 107 g (130 ml, 1.05 mol) of hexan-1-ol. (B.p. 178–180 °C).

1-Iodoheptane – from 122 g (148 ml, 1.05 mol) of heptan-1-ol. (B.p. 198–201 °C, 62.5 °C/3.5 mmHg).

1-Iodooctane – from 137 g (166 ml, 1.05 mol) of octan-1-ol. (B.p. 219–222 °C, 86.5 °C/5 mmHg).

Iodocyclohexane – from 105 g (110 ml, 1.05 mol) of cyclohexanol. Carry out the reaction as above, dilute the cooled reaction product with ether and filter. Wash and dry the organic phase, remove the ether and distil under reduced pressure. (B.p. 81–83 °C/20 mmHg).

1-Iodo-2-phenylethane (*2-phenylethyl iodide*) – from 149 g (149 ml, 1.05 mol) of 2-phenylethanol. Proceed as for iodocyclohexane; a little hydrogen iodide is evolved towards the end of the reaction. (B.p. 114–116 °C/12 mmHg).

1,3-Diodopropane – from 40 g (38 ml, 0.525 mol) of propane-1,3-diol (tri-methylene glycol). Proceed as for iodocyclohexane; stop heating as soon as all the iodine has been added. (B.p. 88–89 °C/6 mmHg).

Experiment 5.60 1-IODOHEPTANE

o-Phenylene phosphorochloridite. Place 110 g (1.0 mol) of catechol (pyro-catechol) in a 500-ml three-necked flask supported on a water bath and moisten it with about 2 ml of water. Equip the flask with an efficient mechanical stirrer, a large double surface reflux condenser and a 250-ml dropping funnel. Connect the mouth of the condenser to an efficient gas absorption trap (Fig. 2.61(*d*)). Add 206 g (131 ml, 1.5 mol) of phosphorus trichloride from the dropping funnel during 15 minutes with vigorous stirring. There is a brisk evolution of hydrogen chloride and the contents of the flask soon solidify, stopping the stirrer. Leave the reaction mixture standing at room temperature for 1 hour, and then heat the water bath to boiling. The solid mass soon melts and stirring again becomes possible. Continue to heat on the boiling water bath with stirring for 2 hours. Arrange the flask for distillation under reduced pressure (water pump); insert a trap cooled in an acetone–Cardice mixture between the receiver and the pump. Distil the product and collect the *o*-phenylene phosphorochloridite as a fraction of b.p. 98 °C/25 mmHg (91 °C/18 mmHg). The yield is 165 g (95%); the product crystallises in the refrigerator, m.p. 30 °C. About 50 g of unreacted phosphorus trichloride is collected in the trap.

Heptyl *o*-phenylene phosphite. Place 87.0 g (0.5 mol) of *o*-phenylene phosphorochloridite, 39.5 g (0.05 mol) of dry redistilled pyridine and 500 ml of dry ether, in a 2-litre conical flask and cool to 0 °C. Add 58.1 g (0.5 mol) of heptan-1-ol dissolved in 400 ml of dry ether during about 5 minutes with occasional shaking. Stopper the flask and allow the reaction to proceed at room temperature overnight. Filter off the precipitated pyridinium chloride under suction, wash well with dry ether and remove the ether from the combined filtrate and washings on a rotary evaporator to obtain 127.0 g (100%) of heptyl *o*-phenylene phosphite as a colourless oil of sufficient purity for use in the next stage.

1-Iodoheptane. In a conical flask, stir magnetically at room temperature a solution of 50.9 g (0.20 mol) of heptyl *o*-phenylene phosphite in 500 ml of dry dichloromethane and add 50.8 g (0.20 mol) of iodine. After 1 hour (1) transfer the solution to a separatory funnel and wash the organic layer with 400 ml of 10 per cent aqueous sodium thiosulphate solution to remove any residual unreacted iodine. Wash the organic layer twice with 400 ml portions of 5 per cent aqueous sodium hydroxide solution to remove the *o*-phenylene phosphoroiodate by-product, followed by two 400 ml portions of 5 per cent aqueous sodium metabisulphite, and finally with 400 ml of a saturated solution of sodium chloride. Dry the organic layer with anhydrous calcium sulphate, filter and remove the solvent on a rotary evaporator. The yield of crude 1-iodoheptane is 41.8 g (92%). Fractionally redistil the product under reduced pressure using a nitrogen capillary leak when 39 g (86%) of the pure material having b.p. 96–104 °C/46–48 mmHg is obtained.

Note. (1) The reaction may be monitored by g.l.c. by the direct loading of a sample (0.1 μl) on to an S.E. 30-column, 1.5 m, held at 170 °C (nitrogen flow 40 ml/minute). The retention time of 1-iodoheptane is 52 seconds; that of heptyl *o*-phenylene phosphite under the same conditions is 4 minutes.

5.5.4 DISPLACEMENT REACTIONS INVOLVING A HALOGEN ATOM

Alkyl fluorides may be prepared in moderate yield by interaction of an alkyl bromide with anhydrous potassium fluoride in the presence of dry ethylene glycol as a solvent for the inorganic fluoride (e.g. Expt 5.61). A little alkene accompanies the alkyl fluoride produced and is readily removed by treatment with KBr/Br$_2$ solution.

$$\text{RBr} + \text{KF} \xrightarrow[\text{glycol}]{\text{ethylene}} \text{RF} + \text{KBr}$$

The most widely used example of halogen exchange is provided by the preparation of alkyl iodides from chlorides or bromides using sodium iodide in a solvent, such as acetone, in which sodium iodide is soluble but sodium chloride or bromide is relatively less so (e.g. Expt 5.62).

$$RCl + NaI \rightleftharpoons RI + NaCl\downarrow$$

These halogen displacement reactions may be carried out under PTC conditions, and have been reviewed.[91]

Experiment 5.61 1-FLUOROHEXANE

$$Me\cdot(CH_2)_4\cdot CH_2Br + KF \xrightarrow[\text{glycol}]{\text{ethylene}} Me\cdot(CH_2)_4\cdot CH_2F + KBr$$

CAUTION: *Alkyl fluorides are said to be highly toxic. Great care should be taken not to inhale the vapours; conduct the entire operation in an efficient fume cupboard.*

In a dry 500-ml three-necked flask, equipped with a sealed stirrer unit, a 100-ml dropping funnel and a short fractionating column (1), place a mixture of 116 g (2 mol) of anhydrous, finely powdered potassium fluoride (2) and 200 g of dry ethylene glycol (3). Connect the fractionating column (which carries a thermometer) to a downward double surface condenser fitted with a receiving flask with the aid of a side-arm adapter. Heat the flask in an oil bath at 160–170 °C and introduce 165 g (141 ml, 1 mol) of 1-bromohexane (Expt 5.55) dropwise, with stirring, during 5 hours. A liquid passes over intermittently at 60–90 °C. When the addition is complete, allow the bath temperature to fall to 110–120 °C; replace the dropping funnel by a tube of narrow bore dipping just below the surface of the liquid, attach the side-arm of the receiver adapter to a water pump, and draw a slow stream of air through the apparatus while maintaining the stirring. It is advisable to interpose a trap (e.g. a Drechsel bottle) cooled in ice between the water pump and receiver in order to recover any uncondensed liquid. Distil the combined distillates through an efficient fractionating column (4); after a small forerun (0.5 g) of hex-1-ene collect the crude 1-fluorohexane at 92–97 °C. Purify the crude product by cooling in ice and adding 1 ml portions of a solution containing 9.0 g of bromine and 6.0 g of potassium bromide in 50 ml of water until the organic layer acquires an orange colour: shake the mixture vigorously for a minute or so after each addition. The volume of KBr/Br₂ solution required is usually less than 5 ml. Separate the aqueous layer, wash the organic layer with saturated aqueous potassium bromide solution until colourless, and finally with water. Dry the liquid with magnesium sulphate and fractionate. Collect the fraction 92–94 °C: the yield is 44 g (42%). The colourless liquid keeps unchanged for long periods.

Notes. (1) Any fractionating column of moderate efficiency is satisfactory, e.g. a Dufton column (20 cm long containing a spiral 10 cm in length, 2 cm in diameter with 8 turns of the helix) or a Vigreux column (20–25 cm long).
(2) Grind finely pure laboratory grade, anhydrous potassium fluoride, and heat it in an electrically heated oven at 180–210 °C; store in a desiccator. Before use, dry the powdered salt at 180 °C for 3 hours and grind again in a warm (c. 50 °C) glass mortar.
(3) Redistil laboratory grade ethylene glycol under reduced pressure and collect the fraction of b.p. 85–90 °C/7 mmHg for use as a solvent for the potassium fluoride.
(4) A Widmer column (spiral 18 cm in length, 1.5 cm in diameter with 20 turns of the helix) is satisfactory.

Cognate preparation. *1-Fluoropentane.* Use 116 g (2 mol) of dry potassium fluoride in 200 g of dry ethylene glycol: heat in an oil bath at 140–150 °C and add 151 g (124 ml, 1 mol) of 1-bromopentane during 5 hours with stirring. The reaction product distils intermittently at 50–85 °C. The yield of 1-fluoropentane, b.p. 63.5–65 °C, is 25 g (28%).

Experiment 5.62   1-IODO-3-METHYLBUTANE

Dissolve 37.5 g (0.25 mol) of dry sodium iodide (1) in 250 ml of dry acetone in a 500-ml flask fitted with a reflux condenser protected by a calcium chloride guard-tube, and add 30.2 g (25 ml, 0.2 mol) of 1-bromo-3-methylbutane. A precipitate of sodium bromide soon begins to form; leave the reaction mixture at room temperature for 30 minutes, and then boil under reflux for 45 minutes to complete the reaction. Allow to cool and filter off the sodium bromide, washing the residue with a little acetone. Remove the acetone from the filtrate on a rotary evaporator, and shake the residual organic halide with 100 ml of water. Separate the lower dark-coloured layer and wash it twice more with 50 ml portions of water; incorporate sufficient crystals of sodium thiosulphate into the first portion of wash-water to decolourise the organic phase. Dry the product over anhydrous calcium sulphate, filter and distil, collecting the 1-iodo-3-methylbutane at 145–147 °C. The yield is 26 g (66%).

Note. (1) Dry the sodium iodide for 4 hours at 100 °C under reduced pressure (oil pump).

5.5.5   DISPLACEMENT REACTIONS INVOLVING A METHANESULPHONYLOXY GROUP

Alcohols, as their derived methanesulphonates, may be readily converted into alkyl halides by reaction with inorganic halides under PTC conditions. With chiral substrates the reaction is often highly stereospecific in that inversion of configuration (sometimes better than 90%) is obtained. The example in Expt 5.63, is the preparation of 2-bromomethyl-1,4-dioxaspiro[4,5]decane from glycerol. This reaction sequence is of interest in that two hydroxyl groups of glycerol are first protected by conversion into a cyclic acetal with cyclohexanone (cf. Section 5.4.6, p. 553); the remaining hydroxyl group is reacted with methanesulphonyl chloride before being displaced by a bromide ion.

Experiment 5.63   3-BROMO-1,4-DIOXASPIRO[4.5]DECANE

2-Hydroxymethyl-1,4-dioxaspiro[4.5]decane (*1,2-O-cyclohexylidenegly-cerol*). Concentrated sulphuric acid (15 ml) is added to an ice-cooled solution of glycerol (184 g, 2.0 mol) and cyclohexanone (196 g, 2.0 mol) in light petro-leum (b.p. 60–80 °C, 400 ml). The mixture is shaken for 18 hours when the petroleum layer is separated and dried (anhydrous potassium carbonate). Distillation of the residue after solvent evaporation yields a clear liquid (193 g, 56%), b.p. 108–109 °C/2.2 mmHg, t_R 18.8 minutes on a Carbowax 20M (10% on Chromosorb W) 1.5-m column held at 150 °C with a nitrogen carrier gas flow rate of 40 ml/minute. The p.m.r. spectrum (CDCl₃, TMS) shows signals at δ 1.55 (m, 10H, cyclohexylidene) and 3.30–4.40 (m, 6H, —CH₂·CH·CH₂OH).

2-Methanesulphonyloxymethyl-1,4-dioxaspiro[4.5]decane. The foregoing 1,2-*O*-cyclohexylideneglycerol (88 g, 0.5 mol) and dry pyridine (220 ml) (**CAUTION**) are placed in a 500-ml three-necked flask fitted with a stirrer, a pressure-equalising funnel and a calcium chloride guard-tube, and cooled to 0 °C. Methanesulphonyl chloride (68 g, 0.5 mol) (**CAUTION**) (1) is placed in the addition funnel and added dropwise with vigorous stirring over a period of 30 minutes. The reaction mixture is allowed to stand for 3 hours at 0 °C (2) before the addition of sufficient water to dissolve the precipitated pyridine hydrochloride. The reaction mixture is then poured into excess water and the oil which separates is washed repeatedly with fresh aliquots of water before being extracted into dichloromethane (300 ml). The organic layer is washed with dilute hydrochloric acid (3 × 50 ml) and then repeatedly with water until the washings are neutral. The organic layer is dried over magnesium sul-phate, the solvent evaporated and the residue distilled to give the product as a yellow/orange oil (109 g, 88%), b.p. 130–136 °C/2 × 10⁻² mmHg; the p.m.r. spectrum (CDCl₃, TMS) shows signals at δ 1.60 (m, 10H, cyclohexylidene), 3.08 (s, 3H, MeSO₂—) and 3.50–4.50 (m, 5H, —CH₂·CH·CH₂—).

2-Bromomethyl-1,4-dioxaspiro[4.5]decane. In a 50-ml two-necked flask fitted with a stirrer unit and a reflux condenser are placed the foregoing meth-anesulphonate (7.5 g, 0.03 mol), hexadecyltributylphosphonium bromide (0.75 g, 0.0015 mol), potassium bromide (10.7 g, 0.09 mol), toluene (10 ml) and water (3 ml). The reaction mixture is stirred vigorously and heated under reflux until t.l.c. analysis shows the reaction to be complete (2). Dichloro-methane (15 ml) is added, the organic layer separated and the residue extracted with a further portion of dichloromethane (15 ml). The combined organic extracts are dried with magnesium sulphate and the solvent evapor-ated. The residue is distilled under reduced pressure to give the product (4.9 g, 70%), b.p. 87–88 °C/1.5 mmHg, t_R 3.8 minutes using the g.l.c. system noted above at a temperature of 180 °C.

Notes. (1) *The handling of methanesulphonyl chloride, which is highly toxic and corros-ive, must be carried out in a fume cupboard and protective gloves must be worn. The reac-tion should also be conducted in a fume cupboard.*
(2) These reactions may be monitored by t.l.c. analysis on Silica gel G plates using toluene:methanol, 9:1, as solvent; the components may be located by iodine vapour (p. 204).

5.5.6 DISPLACEMENT REACTIONS INVOLVING THE AMINO GROUP

Until recently no satisfactory procedure for the direct conversion of a primary aliphatic amine into an alkyl halide was available. This contrasts sharply with the ready conversion of primary aromatic amines into aryl halides via the diazonium salt (Section 6.7, p. 922). However a useful method has now been discovered which is not only applicable to the synthesis of simple alkyl and aralkyl chlorides, bromides, iodides and fluorides, but could clearly be of value when it is required to replace an amino group in a naturally occurring amine by a halogen.[92]

The outline procedure involves the initial reaction of the 2,4,6-triphenylpyrylium halide with the primary amine to yield the corresponding 2,4,6-triphenylpyridinium halide (see Section 8.4.1, and also Section 5.15.3, p. 768); this reaction proceeds either at room temperature in a suitable solvent, or more efficiently under reflux in benzene with azeotropic removal of water. Pyrolysis of the pyridinium halide under controlled conditions then yields the alkyl (or aralkyl) halide in good yield. The mechanism of the reaction in this case is probably of the S_N2 type.

5.5.7 ADDITION OF HYDROGEN HALIDES OR HALOGENS TO ALKENES

Direct addition of a hydrogen halide to an alkene gives rise to an alkyl halide, the order of reactivity being $HI > HBr > HCl$. In the case of an unsymmetrical alkene, the regioselectivity of the reaction may be predicted from the mechanism of the reaction. Thus, the carbocation which is the most stabilised by charge dispersal will be the one which is formed preferentially. Classically the mode of addition is described as proceeding in the *Markownikoff* manner.

Addition of hydrogen halide (1 mol) to a diene (1 mol) is a method of greater preparative value. This reaction is illustrated by the addition of hydrogen bromide to isoprene (Expt 5.64); the overall 1,4-addition process (*thermodynamically controlled*), as opposed to the 1,2-addition (*kinetically controlled*), predominates under the conditions specified.

The addition of hydrogen bromide (but not the iodide or chloride) in the presence of an added peroxide catalyst proceeds by a radical mechanism, giving rise to a regioselectivity which is opposite to that of the ionic mechanism (*anti-Markownikoff*). An example is provided in the preparation of 11-bromo-undecanoic acid (Expt 5.65).

$$Ph-\overset{O\;O}{\underset{O-O}{C}}-Ph \longrightarrow 2\left[Ph-\overset{O}{\underset{O\cdot}{C}} \right] \xrightarrow{-2CO_2} 2P\dot{h} \xrightarrow{2HBr} 2PhH + 2\dot{B}r$$

$$R\overset{}{\underset{CH_2}{\diagup}} \cdot Br \longrightarrow R \overset{\cdot}{\diagup} \overset{Br}{\diagup} \xrightarrow{HBr} R \diagup \overset{Br}{\diagup} + \dot{B}r$$

Such 'anti-Markownikoff' regioselectivity may be achieved in the formation of iodides by proceeding via the corresponding organoborane, which is then decomposed with iodine in the presence of sodium methoxide[93] (Expt 5.66). More recently iodine monochloride has been found to be an economic alternative iodinating agent.[94]

Halogens add to alkenes to give vicinal dihalides.

$$R\cdot CH = CH_2 + X_2 \longrightarrow R\cdot CHX\cdot CH_2X$$

The addition of bromine usually proceeds the most smoothly, and is conveniently carried out in a solvent such as carbon tetrachloride. The examples of the addition of bromine to allyl bromide and the addition of bromine to undec-10-enoic acid are illustrative (Expts 5.67 and 5.23).

Experiment 5.64   1-BROMO-3-METHYLBUT-2-ENE

$$\underset{H_2C}{\overset{Me}{\diagup}}\overset{CH_2}{\diagdown} + HBr \longrightarrow \underset{Me}{\overset{Me}{\diagup}}\overset{}{\diagdown}_{Br}$$

CAUTION: *The entire preparation must be carried out in an efficient fume cupboard since the product is highly lachrymatory.*

Weigh a 100-ml three-necked flask fitted with stoppers. Remove the stoppers and attach to the flask a mechanical stirrer, a calcium chloride guard-tube to the outlet of which is connected a tube leading to the fume cupboard drain, and a gas inlet tube terminating in a glass frit and attached to a dry hydrogen bromide gas generator (Section 4.2.*37*, p. 437). Charge the flask with 34 g (0.5 mol) of redistilled 2-methylbuta-1,3-diene (*isoprene*) (1), cool in an ice–salt bath and pass dry hydrogen bromide gas slowly through the reaction mixture until an increase in weight of 40 g is obtained (2); this may require about 6 hours. Now fit the flask with a Vigreux column in the central joint and with a stopper and a suitable capillary air leak in the side joints. Fractionally distil the crude 1-bromo-3-methylbut-2-ene under reduced pressure (ensure that suitable potassium hydroxide traps are sited between the apparatus and the oil immersion rotatory pump) and collect the pure product of b.p. 56–57 °C/25 mmHg; the yield is 58 g (78%).

Notes. (1) Isoprene is purchased in sealed capsules. These are usually stored in a refrigerator and only removed just prior to opening and redistillation of the isoprene. A simple distillation unit may be employed using a double surface condenser, the rubber tubing water leads being immersed in a large container of ice to effect more efficient cooling. The distillation receiver should be cooled and the outlet protected with a calcium chloride guard-tube. Pure isoprene has a b.p. 33–34 °C.

(2) The reaction flask should be removed and replaced by a similar flask to ensure that the stirrer and gas inlet tube do not become unnecessarily contaminated with moisture. The reaction flask should be stoppered with the same stoppers used in the original weighing.

Experiment 5.65 11-BROMOUNDECANOIC ACID

$$H_2C \diagup (CH_2)_8 \diagdown C(O) \diagdown OH + HBr \xrightarrow{(Ph \cdot CO)_2O_2} Br \diagdown (CH_2)_8 \diagdown C(O) \diagdown OH$$

Equip a 500-ml three-necked round-bottomed flask with a sealed stirrer unit, a wide-bore gas inlet tube reaching to the bottom of the flask and a two-necked multiple adapter fitted with a thermometer and a condenser protected with a calcium chloride guard-tube; arrange the apparatus so that occasional cooling can be effected with an ice-water bath. In the flask place a solution of 27.6 g (0.15 mol) of undec-10-enoic acid in 220 ml of dry light petroleum (b.p. 40–60 °C) together with 1.5 g (0.006 mol) of benzoyl peroxide (air dried; **CAUTION**: see Section 4.2.6, p. 417). Pass a rapid stream of dry hydrogen bromide [from a cylinder or from 20 ml of tetralin and 17 ml of bromine (Section 4.2.37, p. 437)] through the stirred mixture until it is saturated (about 0.75 hour) while maintaining the temperature between 10 and 20 °C by occasional cooling. Should a white solid separate and tend to block the inlet tube towards the end of the reaction, either add a further small portion of dry light petroleum or maintain the temperature nearer to 20 °C when the material should dissolve.

Decant the pale straw-coloured solution into a conical flask and rinse the reaction vessel with 40 ml of light petroleum (b.p. 40–60 °C) and combine with the main solution. Cool the solution to −10 °C and collect the solid which separates by filtration under suction and wash with about 40 ml of similarly cooled light petroleum. A further quantity of the product may be obtained by concentrating the filtrate to about 40 ml and cooling. Dissolve the crude product in approximately 200 ml of boiling light petroleum (b.p. 40–60 °C), add decolourising charcoal, filter, concentrate to about 150 ml and cool. Collect the 11-bromoundecanoic acid which separates as microcrystalline platelike needles, m.p. 49–50 °C. A further quantity of slightly less pure acid may be obtained by concentrating the mother-liquors to about 30 ml and cooling. The yield is 27.9 g (70%).

Experiment 5.66 METHYL 11-IODOUNDECANOATE[93]

$$MeO \diagdown C(O) \diagdown (CH_2)_8 \diagdown CH_2 \xrightarrow{BH_3-Me_2S} MeO \diagdown C(O) \diagdown (CH_2)_8 \diagdown \Big]_3 B \xrightarrow[\ominus OMe]{I_2}$$

$$MeO \diagdown C(O) \diagdown (CH_2)_8 \diagdown I$$

A dry 500-ml flask equipped with a septum inlet, magnetic stirrer and gas connecting tube is flushed with dry nitrogen and maintained under static pressure until work-up. The flask is charged with tetrahydrofuran (100 ml) and methyl undec-10-enoate (33.7 ml, 150 mmol) and cooled to 0 °C in an ice bath. Conversion to the trialkylborane is achieved by the dropwise addition of neat borane–methyl sulphide (5.10 ml) over 40 minutes. The ice bath is removed and the reaction allowed to stir over 1 hour at room temperature. Then absolute methanol is added to destroy traces of residual hydride. Iodine (38.1 g, 150 mmol) is added all at once, followed by the dropwise addition of a solution of sodium methoxide in methanol (31.8 ml of a 4.72 M solution, 150 mmol) over a period of 10 minutes. The reaction mixture is allowed to stir for 24 hours. Gas–liquid chromatography analysis (10% Dow Corning on Chromosorb W, 1.8 m × 6.3 mm) of the reaction mixture using decane for an internal standard indicates an 80 per cent yield of methyl 11-iodoundecanoate. A saturated aqueous sodium thiosulphate solution is poured into the reaction mixture until excess iodine is decolourised. The reaction mixture is extracted with pentane (100 ml) and dried over magnesium sulphate. Distillation under vacuum gives methyl 11-iodoundecanoate; yield 33.1 g (68%), b.p. 139–141 °C/0.15 mmHg, n_D^{20} 1.4856.

Experiment 5.67 1,2,3-TRIBROMOPROPANE

Provide a 1-litre three-necked flask with a dropping funnel carrying a calcium chloride guard-tube, a mechanical stirrer and a thermometer reaching almost to the bottom of the flask, and cool the flask in a mixture of ice and salt. Place in the flask 182 g (132 ml, 1.5 mol) of allyl bromide (1) and 250 ml of dry carbon tetrachloride (**CAUTION**, p. 399), and introduce 255 g (80 ml, 1.6 mol) of dry bromine (**CAUTION**, p. 422) into the dropping funnel. Set the stirrer in motion and when the temperature has fallen to −5 °C, drop the bromine in slowly at such a rate that the temperature does not rise above 0 °C (about 90 minutes). Allow the orange-coloured solution (the colour is due to a slight excess of bromine) to warm to room temperature with constant stirring (about 30 minutes) and then remove the solvent under reduced pressure on a rotary evaporator. Distil the residue under reduced pressure; the residual carbon tetrachloride passes over first, followed by 1,2,3-tribromopropane at 92–93 °C/10 mmHg or 100–103 °C/18 mmHg as an almost colourless liquid. The yield is 400 g (95%).

Note. (1) The allyl bromide (Expt 5.54) should be dried over anhydrous calcium chloride and redistilled; the fraction, b.p. 69–72 °C, is collected for use in this preparation.

5.5.8 THE REPLACEMENT OF REACTIVE ALLYLIC HYDROGEN ATOMS BY BROMINE

The direct introduction of bromine into the allylic position of an alkene using *N*-bromosuccinimide is known as the *Wohl–Ziegler reaction*. Bromination is carried out in anhydrous reagents (to avoid hydrolysis of the bromoimide), usually boiling carbon tetrachloride or chloroform solution. The progress of the reaction can be followed by the fact that at first the dense *N*-bromosuccinimide is at the bottom of the flask and is gradually replaced by succinimide, which rises to the surface: the reaction is complete when all the crystals are floating at the surface (detected by stopping the boiling momentarily). This can be confirmed (when equimolar amounts are used) by transferring a drop of the solution to acidified potassium iodide–starch solution: iodine should not be liberated. After cooling, the insoluble succinimide is filtered off, washed with solvent and the product isolated, after removal of the solvent, by distillation or crystallisation.

The specific substitution into the allylic position is the result of a radical process which requires the generation of a low concentration of molecular bromine, probably by way of the action of traces of hydrogen bromide on the bromoimide.

$$Br_2 \rightleftharpoons 2\dot{Br}$$

$$R \cdot CH_2 \cdot CH{=}CH_2 + \dot{Br} \longrightarrow R \cdot \dot{CH} \cdot CH{=}CH_2 + HBr$$

$$R \cdot \dot{CH} \cdot CH{=}CH_2 + \dot{Br} \longrightarrow R \cdot CHBr \cdot CH{=}CH_2$$

Two simple applications may be mentioned. With cyclohexene, 3-bromo-cyclohexene is obtained in a satisfactory yield (Expt 5.68), the latter upon dehydrobromination with quinoline affords an 80–90 per cent yield of cyclohexa-1,3-diene (Expt 5.13). Methyl crotonate yields the valuable synthetic reagent methyl γ-bromocrotonate (Expt 5.69); this latter compound permits the introduction (in moderate yield) of a four-carbon atom chain at the site of the carbonyl group by the use of the Reformatsky reaction (compare Expt 5.170):

Experiment 5.68   3-BROMOCYCLOHEXENE

CAUTION: *Since many simple unsubstituted allylic compounds are powerful irritants by inhalation and by skin contact, this preparation should be carried out in an efficient fume cupboard and the product treated with appropriate caution.*

In a 500-ml round-bottomed flask place 39.4 g (0.20 mol) of N-bromosuccinimide (Section 4.2.10, p. 422), 49.2 g (0.60 mol) of redistilled cyclohexene, 150 ml of carbon tetrachloride and about 500 mg of benzoyl peroxide (**CAUTION**: Section 4.2.6, p. 417), previously dried by pressing between filter papers. Attach a double surface reflux condenser and allow the mixture to stand at room temperature when after a short induction period the reaction begins. The reaction mixture becomes warm and the heavy yellowish N-bromosuccinimide begins to be transformed into the light colourless succinimide which becomes suspended in the reaction mixture. When the reaction has moderated somewhat, transfer the flask to a steam bath and heat under reflux until all the N-bromosuccinimide has been converted to succinimide (about 1.5 hours) (1). Cool the flask, filter under suction and wash the residue with a little carbon tetrachloride. Distil the filtrate and washings on a boiling water bath from a flask fitted with a Claisen still-head to remove most of the carbon tetrachloride (b.p. 77 °C) and unreacted cyclohexene (b.p. 83 °C). Then fractionally distil the residue (2) under reduced pressure (water pump). Further quantities (about 50 g) of cyclohexene and carbon tetrachloride are obtained as a first fraction, followed by the main fraction (about 24 g) of b.p. 72–77 °C/32–35 mmHg, which is only about 50 per cent pure by g.l.c. analysis and which rapidly goes brown on standing. Subsequent redistillation under reduced pressure and using a short Vigreux column (12 cm) gives 14.4 g (45%) of colourless mobile 3-bromocyclohexene of b.p. 66–67 °C/20 mmHg, which is 99 per cent pure by g.l.c. analysis (10% Silicone oil on Chromosorb W, 1.5-m column held at 85 °C, nitrogen flow rate 40 ml/minute, t_R 10.4 minutes).

Notes. (1) The completion of the reaction may be confirmed by treating starch–iodide paper with a drop of the reaction solution. At the end of the reaction no coloration is observed.
(2) The flask and still-head should preferably have at least 24/29 sized joints to prevent losses due to foaming.

Experiment 5.69 METHYL γ-BROMOCROTONATE (*Methyl 4-bromobut-2-enoate*)

Methyl crotonate. Purify commercial crotonic acid by distilling 100 g from a flask attached to an air condenser, use an air bath (Fig. 2.46(*a*)). The pure acid passes over at 180–182 °C and crystallises out on cooling, m.p. 72–73 °C; the recovery is about 90 per cent. Place 75 g (2.34 mol) of absolute methanol, 5 g (2.7 ml) of concentrated sulphuric acid and 50 g (0.58 mol) of pure crotonic acid in a 500-ml round-bottomed flask, and heat under reflux for 12 hours. Add water, separate the precipitated ester and dissolve it in ether; wash with dilute sodium carbonate solution until effervescence ceases, dry with magnesium sulphate and remove the ether on a water bath. Distil and collect the methyl crotonate at 118–120 °C; the yield is 40 g (69%).

Methyl γ-bromocrotonate. Mix 36 g (0.2 mol) of N-bromosuccinimide, 20 g (0.2 mol) of methyl crotonate and 60 ml of dry, redistilled carbon tetrachloride in a 500-ml round-bottomed flask. Reflux on a water bath for 12 hours; by this time all the solid should have risen to the surface of the liquid. Filter off the succinimide at the pump and wash it with a little dry carbon tetrachloride. Remove the solvent on a water bath and distil the residue under reduced pressure through a short fractionating column. Collect the methyl γ-bromocrotonate at 77–78 °C/8 mmHg; the yield is 31 g (86%).

5.6 ALIPHATIC ETHERS

Saturated ethers may have symmetrical (R·O·R) or unsymmetrical (R^1·O·R^2) structures. Illustrative representations are given below for methyl propyl ether (1) $CH_3 \cdot CH_2 \cdot CH_2 \cdot O \cdot CH_3$, and for methyl isopropyl ether (2)

579

$(CH_3)_2CH \cdot O \cdot CH_3$. The most common cyclic ethers are tetrahydrofuran (3) and tetrahydropyran (4).

Benzylic ethers ($Ph \cdot CH_2 \cdot O \cdot R$), allylic ethers ($R \cdot CH = CH \cdot CH_2 \cdot O \cdot R$) and vinylic ethers [$R \cdot CH = CH(OR)$] together with the most commonly encountered tetrahydropyranyl ethers [THP-ethers, (5)] and β-methoxyethoxymethyl ethers [MEM-ethers, $RO \cdot CH_2 \cdot O \cdot (CH_2)_2 \cdot O \cdot CH_3$] play an important role in the protection of a hydroxyl group (p. 550). Macrocyclic ethers (the crown ethers) are important phase transfer catalysts [e.g. 18-Crown-6 (6)].

The synthesis of ethers is exemplified by two main general procedures.

1. The formation of ethers from alcohols under acidic conditions (Expts 5.70 and 5.71).
2. The interaction of an alcohol with a halogen compound under basic conditions (Expts 5.72 and 5.73).

SUMMARY OF RETROSYNTHETIC STRATEGIES

C—O Disconnection (methods 1 and 2)

$2 \cdot R \cdot CH_2OH$

$(R = R^1 = R^2)$

SPECTROSCOPIC FEATURES

The *i.r.* spectra of dialkyl ethers show the expected absorptions due to the vibrations of the carbon–hydrogen skeletal structure, together with the asymmetric stretching of the carbon–oxygen bond at $1150–1060 \text{ cm}^{-1}$ (cf. dibutyl ether, Fig. 3.24). Benzyl, allyl and vinylic ethers exhibit further appropriate absorption due to the presence of the aromatic ring or of the carbon–carbon double bond. The structure of the alkyl group may be assigned from the *p.m.r.* spectrum in those cases where a first-order, or near first-order, spectrum is possible (see p. 341 for splitting patterns of alkyl groups). The fragmentation pattern in the *m.s.* of dialkyl ethers is discussed on p. 376 and may provide further information on the structure of the alkyl groups. Saturated aliphatic ethers do not absorb in the accessible region of the *u.v.–visible* region of the spectrum.

5.6.1 THE FORMATION OF ETHERS FROM ALCOHOLS UNDER ACIDIC CONDITIONS

Diethyl ether (Et_2O) can be prepared by heating ethanol with sulphuric acid at about 140 °C, and adding more alcohol as the ether distils out of the reaction medium. A similar 'continuous etherification process' is used industrially. A more general procedure for the preparation of symmetrical ethers from primary alcohols (e.g. dibutyl ether, Expt 5.70) is to arrange for the water formed in the reaction to be removed azeotropically.

$$2R\cdot CH_2OH \xrightarrow{H_2SO_4} R\cdot CH_2\cdot O\cdot CH_2\cdot R + H_2O$$

Excessive heating of the reaction mixture must be avoided otherwise an alkene-forming elimination reaction is induced; this is particularly the case with secondary and especially tertiary alcohols.

Tetrahydropyranyl ethers are readily prepared from the alcohol and 2,3-tetrahydropyran in the presence of acid, and the reaction is widely used as a method of protection of hydroxyl groups. Preparative procedures and the methods of deprotection are given in Section 5.4.6, p. 551).

The synthesis of dichloromethyl methyl ether has been included because of its usefulness as a reagent for the preparation of aromatic aldehydes (Expt 6.115). It is readily obtained by the reaction of phosphorus pentachloride in admixture with phosphorus oxychloride with methyl formate (Expt 5.71).

Chloromethyl methyl ether and bis-chloromethyl ether ($ClCH_2\cdot O\cdot CH_2Cl$) in common with several alkylating agents possess carcinogenic properties in experimental animals and have been listed as assumed human carcinogens; similar hazardous properties associated with dichloromethyl methyl ether have not been reported but it would be prudent to handle this compound with due care.

Experiment 5.70 DIBUTYL ETHER

$$2Me\cdot CH_2\cdot CH_2\cdot CH_2OH \xrightarrow[-H_2O]{H_2SO_4} (Me\cdot CH_2\cdot CH_2\cdot CH_2)_2O$$

The success of this preparation depends on the use of the Dean and Stark apparatus [Fig. 2.31(*a*)] which permits the automatic separation of water produced in the reaction. The quantity of water which should be eliminated, assuming a quantitative conversion of the alcohol (0.67 mol) into the ether, is 6.0 g (0.33 mol). Equip a 250-ml two-necked round-bottomed flask with the Dean and Stark apparatus in the central socket; attach a reflux condenser in the upper joint. Fill the graduated side-arm with water and then drain off 6 ml (the quantity to be formed in the reaction). Place 49 g (61 ml, 0.67 mol) of butan-1-ol together with 16 g (9 ml) of concentrated sulphuric acid in the flask. Using a screw-capped cone adapter, insert a thermometer in the side socket of the flask with the bulb immersed in the liquid. Heat the flask gently so that the liquid refluxes from the condenser. Water and butan-1-ol will collect in the graduated side-arm and when this is full automatic separation of the two liquids will commence; the water will fall to the bottom of the tube and the lighter butan-1-ol will pass back into the flask. Continue the heating until the temperature in the flask rises to 134–135 °C (after about 30–40 minutes). At this stage 5–6 ml of additional water will have collected in the graduated side-arm and the reaction may be regarded as complete. Further heating will merely result in considerable darkening of the mixture in the flask and the forming of the highly flammable but-1-ene. Allow the reaction mixture to cool or cool the flask under running water from the tap. Pour the contents of the flask and water-separator tube into a separatory funnel containing 100 ml of water, shake well and remove the upper layer containing the crude ether mixed with a little unchanged butanol. Shake the crude ether with 25 ml of cold 50 per cent by weight sulphuric acid (from 20 ml of concentrated acid cautiously poured into 35 ml of water) (1) for 2–3 minutes, separate the upper layer and repeat the extraction with another 25 ml of the acid. Finally wash twice with 25 ml portions of water; dry with 2 g of anhydrous calcium chloride (2). Filter through a fluted filter paper into a 50 ml flask and distil. Collect the dibutyl ether at 139–142 °C. The yield is 15 g (34%). Record the i.r. spectrum and compare it with Fig. 3.24. Interpret the m.s. of dibutyl ether which shows principal fragmentation ions at m/z 87, 57 (base peak), 56, 41, 29 and 27.

Notes. (1) This separation utilises the fact that butan-1-ol is soluble in 50 per cent sulphuric acid by weight, whereas dibutyl ether is only slightly soluble.
(2) An alternative method for isolating dibutyl ether utilises the fact that butan-1-ol is soluble in saturated calcium chloride solution while dibutyl ether is slightly soluble. Cool the reaction mixture in ice and transfer to a separatory funnel. Wash cautiously with 100 ml of 2.5–3 M sodium hydroxide solution; the washings should be alkaline to litmus. Then wash with 30 ml of water, followed by 30 ml of saturated calcium chloride solution. Dry with 2–3 g anhydrous calcium chloride.

Cognate preparation. *Dihexyl ether.* Use 68 g (83 ml, 0.67 mol) of hexan-1-ol (b.p. 156–157 °C); heat until the temperature rises to 180 °C. Pour the reaction mixture into water, separate the upper layer, wash it twice with 5 per cent sodium hydroxide solution, then with water, and dry over anhydrous potassium carbonate. Distil and collect the fractions of b.p. (i) 160–221 °C (23 g) and (ii) 221–223 °C (23 g). Reflux fraction (i) with 5 g of sodium and distil from the excess of sodium when a further quantity of fairly pure dihexyl ether (13 g, fraction (iii)) is thus obtained. Combine fractions (ii) and (iii) and distil from a little sodium; collect the pure dihexyl ether (26 g, 42%) at 221.5–223 °C.

Experiment 5.71   DICHLOROMETHYL METHYL ETHER (*1,1-dichloro-dimethyl ether*)

$$H \cdot CO_2Me \xrightarrow{PCl_5/POCl_3} Cl_2CH \cdot OMe$$

In a fume cupboard equip a 500-ml three-necked flask with a dropping funnel, a mechanical stirrer, a thermometer and a reflux condenser, using a double-neck adapter (Fig. 2.15) to accommodate the last two items. Protect the condenser and dropping funnel with calcium chloride guard-tubes. Place 50 ml of phosphorus oxychloride and 156 g (0.75 mol) of phosphorus pentachloride in the flask and 48 g (49 ml, 0.8 mol) of methyl formate in the dropping funnel. Cool the mixture to 10 °C and add the methyl formate dropwise at such a rate that the reaction temperature does not rise above 20 °C (about

1 hour). When the addition is complete, remove the ice-bath, stir the mixture until all the phosphorus pentachloride has dissolved, keeping the temperature below 30 °C by occasional cooling. Remove the stirrer, condenser, thermometer and dropping funnel, stopper the two side-necks and insert a simple distillation head into the central neck. Introduce a pine splint to serve as an anti-bumping device (1) and attach a condenser leading to a receiver flask via a receiver adapter for vacuum distillation. Distil the reaction mixture under reduced pressure (water pump) on a water bath at about 60 °C and collect the distillate in a flask cooled to −25 to −30 °C in a cooling bath (acetone–Cardice or ice–salt). The distillate weighs 209 g and consists of a mixture of dichloromethyl methyl ether (b.p. 85 °C) and phosphorus oxychloride (b.p. 105 °C). Fractionally distil the mixture at atmospheric pressure through a 50-cm column filled with glass helices and surrounded by a heating jacket at 60 °C, using a reflux ratio of about 1:8 (Section 2.26). Collect the fraction which boils between 82 and 95 °C and refractionate to give 70 g (76%) pure dichloromethyl methyl ether, b.p. 85 °C. Protect the product from moisture.

Note. (1) A conventional capillary air leak, inserted into a side-neck, should only be used in this case if the air supply is pre-dried since the product is susceptible to decomposition by moisture.

5.6.2 THE INTERACTION OF AN ALCOHOL WITH A HALOGEN COMPOUND UNDER BASIC CONDITIONS

This involves the direct nucleophilic displacement of halogen in an alkyl halide by an alkoxide ion (the *Williamson synthesis*) (Expt 5.72), and the method is particularly useful for the preparation of mixed ethers. For an unsymmetrical ether [e.g. t-butyl ethyl ether (7)], the disconnection approach suggests two feasible routes.

$$(Me_3CX \equiv)Me_3C^{\oplus} \quad {}^{\ominus}OEt \xleftarrow{\text{(i)}} Me_3C \overset{\text{(i)}}{\underset{}{\cdots}} \overset{O}{\underset{}{\cdots}} \overset{\text{(ii)}}{\underset{}{\cdots}} Et \xrightarrow{\text{(ii)}} Me_3CO^{\ominus} \quad Et^{\oplus}(\equiv EtX)$$

$$(7)$$

The selection of reagents is governed by availability, cost, and, more importantly, the possible intrusion of side reactions. Thus in the above example, the action of the strongly basic ethoxide ion on t-butyl bromide would give rise to extensive alkene formation; on the other hand little or no elimination would occur by the alternative reaction route. In general therefore, secondary or tertiary alkyl groups can only be incorporated into ethers by the Williamson synthesis by way of the corresponding alkoxide ions in reaction with a primary halide.

The preparation of alkyl methyl ethers may be readily effected under PTC conditions from the alcohol, dimethyl sulphate and 50 per cent w/w aqueous sodium hydroxide, employing tetrabutylammonium hydrogen sulphate as catalyst.[95] The usefulness of this procedure has been extended, and optimum conditions have been described for the alkylation of a range of aliphatic alcohols using, for example, 1-chlorobutane or benzyl chloride.[96] The PTC preparative examples described in Expt 5.73 are for the methylation, allylation, but-2-enylation and benzylation of, for example, 2-hydroxymethyl-1,4-dioxaspiro[4.5]decane (Expt 5.63), and have been developed in the editors' laboratories. These methods have also been applied to the alkylation of protected monosaccharide derivatives (p. 652).

Experiment 5.72   ETHYL HEXYL ETHER

$$Bu \cdot CH_2 \cdot CH_2OH \xrightarrow{Na} Bu \cdot CH_2 \cdot CH_2O^{\ominus} \xrightarrow{EtI} Bu \cdot CH_2 \cdot CH_2 \cdot O \cdot Et$$

Place 204 g (250 ml, 2 mol) of dry hexan-1-ol in a 500-ml round-bottomed flask fitted with a Liebig-type reflux condenser and introduce 5.75 g (0.25 mol) of clean sodium (Section 4.2.*68*, p. 462) in small pieces and warm under reflux until all the sodium has reacted (*c.* 2 hours). Introduce 39 g (20 ml, 0.25 mol) of ethyl iodide down the condenser from a dropping funnel and reflux gently for 2 hours; sodium iodide gradually separates. Arrange the apparatus for downward distillation and collect the crude ether at 143–148 °C (27 g). When cold, refit the reflux condenser, add a further 5.75 g (0.25 mol) of clean sodium and warm until all has reacted: alternatively, allow the reaction to proceed overnight, by which time all the sodium will have reacted. Introduce a further 39 g (20 ml, 0.25 mol) of ethyl iodide and reflux for 2 hours; distil off the crude ether and collect the fraction passing over at 143–148 °C. Combine the two distillates. Remove most of the hexan-1-ol still present in the crude ether by heating under reflux for 2 hours with a large excess of sodium and then distil until no more liquid passes over. Distil the resulting liquid from a few grams of sodium using a short fractionating column, and collect the ethyl hexyl ether at 140–143 °C. The yield is 30 g (46%). If the sodium is appreciably attacked, indicating that all the alcohol has not been completely removed, repeat the distillation from a little fresh sodium. (See Section 2.3.2, p. 41), for instructions in the destruction of sodium residues.)

Experiment   5.73   2-METHOXYMETHYL-1,4-DIOXASPIRO[4.5]-DECANE

In a 50-ml three-necked round-bottomed flask fitted with a reflux condenser, a dropping funnel and a stirrer unit, place 4.0 g (0.025 mol) of 2-hydroxy-methyl-1,4-dioxaspiro[4.5]decane (Expt 5.63), 85 mg (1 mol %) of tetrabutyl-ammonium hydrogen sulphate, 10 ml of dichloromethane, 50 per cent w/w aqueous sodium hydroxide (from 2.6 g of sodium hydroxide and 2.6 g of water), and stir vigorously for 30 minutes. Cool the flask in an ice–water bath and from the funnel add dropwise over a period of 1 hour 3.8 g (0.03 mol) of dimethyl sulphate (**CAUTION** in handling, see p. 430). Stir the mixture vigorously at room temperature until t.l.c. analysis (1) reveals that the reaction is complete (usually about 3 hours). Add 1 ml of concentrated aqueous ammonia and stir for a further 30 minutes at room temperature to decompose excess dimethyl sulphate. Pour the reaction mixture into water, add 20 ml of dichloromethane, separate the organic layer and wash with water until the washings are neutral. Dry over magnesium sulphate and evaporate the solvent on a rotary evaporator. Distil the residue under reduced pressure to obtain the product, 3.4 g (78%), b.p. 76–77 °C/1.5 mmHg.

Note. (1) The t.l.c. analysis may be performed on Silica gel G plates using toluene: methanol, 9:1; the starting material has $R_F 0.26$, the product has $R_F 0.70$.

Cognate preparations. *Allylation, benzylation and but-2-enylation of 1,2-O-cyclohexylidene glycerol.* Use 4.0 g (0.025 mol) of starting material, 425 mg (5 mol %) of tetrabutylammonium hydrogen sulphate, 20 ml of *trans*-1,2-dichloroethylene, 0.03 mol of either allyl chloride, benzyl chloride or crotyl chloride (1-chlorobut-2-ene), and 6.0 g of sodium hydroxide dissolved in 6 g of water. Stir the mixture vigorously and heat under reflux until t.l.c. analysis shows the reaction to be complete (0.75–2 hours). Work-up the reaction mixture as above to isolate the products, which have the following characteristics: *O*-allyl ether 77 per cent yield, b.p. 103–105 °C/2.5 mmHg, $R_F 0.76$; *O*-benzyl ether 74 per cent yield, b.p. 156–160 °C/2 mmHg, $R_F 0.66$; *O*-but-2-enyl ether 84 per cent, b.p. 111–113 °C/2 mmHg, $R_F 0.86$.

5.7 ALIPHATIC ALDEHYDES

The carbonyl group in an aldehyde is by definition at the terminal position in the carbon chain $(R \cdot CH_2 \cdot CH_2 \cdot CHO)$. Branching in the carbon chain, or the presence of a functional substituent group, is located by systematic numbering; letters of the Greek alphabet may be used to specify in general terms substituents relative to the carbonyl carbon. Illustrative representations are shown below for:

2-methylpropanal (1) $CH_3 \cdot CH(CH_3) \cdot CHO$;
3-methylbutanal (2) $CH_3 \cdot CH(CH_3) \cdot CH_2 \cdot CHO$;
3-hydroxybutanal (3) $CH_3 \cdot CH(OH) \cdot CH_2 \cdot CHO$;

compound (3a) could be described in general terms as a β-hydroxyaldehyde, and the alternative representation (3b) emphasises the possibility of intra-molecular hydrogen bonding.

(1)   (2)   (3a)   (3b)

The presence of branching or of functional group substitution in the carbon chain leads to the possibility of a chiral site(*). Thus compound (3) is chiral but compounds (1) and (2) are achiral. α, β-Unsaturated aldehydes $(R \cdot CH{=}CH \cdot CHO)$ are considered in Section 5.18.2, p. 798.

The synthesis of aliphatic aldehydes is exemplified by the following typical procedures.

1. The controlled oxidation or dehydrogenation of primary alcohols (Expts 5.74 to 5.76).
2. The oxidative cleavage of 1,2-diols (Expts 5.77 and 5.78).
3. The ozonolysis of suitably substituted alkenes (Expt 5.79).
4. The reduction of nitriles, carboxylic acids and carboxylic acid chlorides (Expts 5.80 and 5.81).
5. Methods based upon alkyl halides (Expt 5.82).

585

6. The hydrolysis and decarboxylation of $\alpha\beta$-epoxy esters (glycidic esters) (Expt 5.83).
7. The oxidative hydrolysis of nitronate salts derived from primary nitroalkanes (Expt 5.84).
8. Methods for the alkylation of the carbon chain (Expt 5.85).

Methods for the protection of the carbonyl group are considered in Section 5.8.8, p. 623.

SUMMARY OF RETROSYNTHETIC STRATEGIES

Functional group interconversion (FGI) (methods 1, 4 and 7)

Reconnection (methods 2 and 3)

Disconnection (methods 5 and 8)

586

SPECTROSCOPIC FEATURES

The presence of the formyl group (—CHO) of an aldehyde is definitively recognised from the *i.r.* absorption spectrum and from the *p.m.r.* spectrum. In the former, the absorption frequency of the carbonyl group occurs in the region of $1745 \, \text{cm}^{-1}$, and that of the carbon–hydrogen bond as two weak absorptions near 2820 and $2720 \, \text{cm}^{-1}$ (Fig. 3.29, cyclohexanecarboxyaldehyde). These latter absorptions allow structural distinction to be made between aldehydes and other carbonyl-containing compounds. The aldehydic hydrogen may also be readily assigned in the *p.m.r.* spectrum as its absorption occurs at very low field (Fig. 3.72, crotonaldehyde). The skeletal structure of an aliphatic aldehyde may sometimes be deduced from a first-order analysis of the multiplicity of the signals corresponding to the alkyl groups (p. 341). The ^{13}C-*n.m.r.* spectrum of crotonaldehyde (Fig. 3.52) is illustrative of deshielding effects. The important fragmentations which are observed in the *m.s.* of aliphatic aldehydes are discussed on p. 377. Further descriptive accounts of structural assignments are to be found in some of the preparative examples below. Saturated aliphatic aldehydes show very weak and barely detectable absorption in the accessible region of the *u.v.–visible* spectrum near 270 nm; this absorption is attributed to the $n \rightarrow \pi^*$ transition.

5.7.1 THE CONTROLLED OXIDATION OR DEHYDROGENATION OF PRIMARY ALCOHOLS

Simple aldehydes may be obtained in reasonably good yield by oxidation of the corresponding primary alcohol with sodium dichromate in dilute sulphuric acid solution (e.g. butyraldehyde, Expt 5.74). To avoid further oxidation to the corresponding acid, the aldehyde is removed as rapidly as possible by distillation through a fractionating column. The main by-product is an ester which arises as the result of the oxidation of an intermediately formed hemiacetal.

$$R \cdot CH_2OH \xrightarrow{[O]} R \cdot CHO$$

$$R \cdot CH_2OH + R \cdot CHO \underset{}{\overset{H^{\oplus}}{\rightleftharpoons}} R \cdot CH_2O \cdot CH(OH) \cdot R \longrightarrow R \cdot CH_2O \cdot CO \cdot R$$

Satisfactory yields of simple aldehydes are also usually obtained when the vapour of the primary alcohol is dehydrogenated by passage over a heated catalyst of copper–chromium oxide deposited on pumice (Expt 5.75).

These rather vigorous conditions (high temperature and/or an aqueous strongly acidic environment) are however unsuitable for those primary alcohols which are insoluble in water, or owing to polyfunctionality, may be sensitive to acidic conditions. For these reasons several non-aqueous chromium(VI) oxidising reagents have been developed, including the Collins reagent[97] $[(C_5H_5N)_2CrO_3]$, pyridinium chlorochromate[98] $[PCC, C_5H_5NH\}Cl \cdot CrO_3]$ and pyridinium dichromate[99] $[PDC, (C_5H_5NH)_2^{\oplus} Cr_2O_7^{2\ominus}]$. The latter two are currently the reagents of choice, particularly for the oxidation of α, β-unsaturated primary and secondary alcohols to the corresponding unsaturated aldehydes and ketones; their relative merits and cautionary notes in handling are discussed in Section 4.2.*18*, p. 425. The use of pyridinium chlorochromate is illustrated in the preparation of heptanal from heptan-1-ol (Expt 5.76). The varied use of these reagents is further illustrated in Expt 5.88 (PCC supported on alumina)[100] and in Section 5.10, p. 653) (PDC in the presence of activated molecular sieves).[101] An alternative reagent for the oxidation of an alcohol to an aldehyde (or ketone) is periodinane, Section 6.3.5, p. 869. PCC oxidation of the alkylborane resulting from hydroboration of a terminal alkene with disiamylborane (Sia_2BH) provides a useful alternative route to an aldehyde.[102]

Since this hydroborating reagent is highly regioselective little of the alkyl methyl ketone ($<1\%$) is obtained. Furthermore with a non-conjugated diene having terminal and non-terminal carbon–carbon double bonds the terminal bonding system reacts preferentially giving finally an unsaturated aldehyde, e.g. limonene (4) to p-menth-1-en-9-al (5).

(4)

(5)

PTC methods for the oxidation of water-insoluble acid-sensitive alcohols have been described which use an aqueous sulphuric acid solution of potassium dichromate, dichloromethane and catalytic amounts of tetrabutylammonium hydrogen sulphate.[103]

Experiment 5.74 BUTYRALDEHYDE (*Butanal*)

$$Me \cdot (CH_2)_2 \cdot CH_2OH \longrightarrow Me \cdot (CH_2)_2 \cdot CHO$$

Equip a two-necked 500 ml flask with a dropping funnel so that the outlet extends to the bottom of the flask, and a Hempel column (15 cm × 1.5 cm) (Fig. 2.105(*c*)) filled with 6-mm glass or porcelain rings, to which is attached a simple distillation head. Fit to the distillation head a thermometer and a condenser set for downward distillation. Attach to the condenser outlet a receiver adapter and a flask immersed in a bath of crushed ice/water. Dissolve 56 g (0.188 mol) of sodium dichromate dihydrate in 300 ml of water and add cautiously, with stirring, 40 ml of concentrated sulphuric acid. Place 41 g (51 ml, 0.55 mol) of butan-1-ol together with a few small chips of porous porcelain in the flask, heat the butan-1-ol to boiling and run in the dichromate solution via the dropping funnel during about 20 minutes. The oxidation to butyraldehyde proceeds with the evolution of heat, but it is necessary to continue to heat the flask so that the mixture boils vigorously to maintain steady distillation. The temperature at the top of the column, however, should not exceed 80–85 °C. When all the oxidising agent has been added, continue heating the mixture for 15 minutes and collect all that passes over below 90 °C. Separate the water from the distillate and dry the residue (29 g) for 30–60 minutes with 3–4 g of anhydrous sodium sulphate. Meanwhile detach the fractionating column from the apparatus and dry the glass or porcelain rings by washing with acetone and blowing hot air through them. Fit the column into a 100-ml flask and arrange for distillation as before. Distil the dried distillate slowly (1–2 drops per second) through the column and collect as fairly pure butyraldehyde all that distils below 76 °C. The yield is 13 g (32%). Pure butyraldehyde boils at 74.5 °C. The i.r. spectrum reveals absorptions at 2700, 2800 (O=C—H) and 1705 cm^{-1} (C=O). Assign the p.m.r. absorptions which occur at δ 0.95 (t, 3H), 1.61 (m, 2H), 2.36 (d of t, 2H) and 9.68 (t, 1H). The m.s. shows principal fragment ions at m/z 72, 56, 44, 43, 41, 29.

Cognate preparation. *Propionaldehyde* (*propanal*). Use 34 g (42.5 ml, 0.567 mol) of propan-1-ol and a solution containing 56 g (0.188 mol) of sodium dichromate dihydrate, 300 ml of water and 40 ml of concentrated sulphuric acid. The experimental details are identical with those for butyraldehyde, except that the temperature at the top of the column is not allowed to rise above 70–75 °C, and during the subsequent heating for 15 minutes the liquid passing over below 80 °C is collected; the receiver must be cooled in ice. The yield of propionaldehyde, b.p. 47–50 °C, is 12 g (36%).

Experiment 5.75 HEXANAL

$$Me \cdot (CH_2)_4 \cdot CH_2OH \xrightarrow{-H_2} Me \cdot (CH_2)_4 \cdot CHO$$

Assemble the apparatus illustrated in Fig. 2.65(*c*) with the catalyst of copper–chromium oxide deposited on pumice packed into the Pyrex combustion tube in sections of about 25 cm long, each being separated by a small plug of glass wool. Place 100 g (122 ml, 0.98 mol) of hexan-1-ol in the dropping funnel. The gas outlet from the Drechsel bottle, E, should be led into a fume cupboard or to an outside window, since hydrogen is evolved in the reaction. Place 0.1 g of hydroquinone in the receiver to act as a 'stabiliser' for the aldehyde. Pass a gentle flow of nitrogen gas through the combusion tube and adjust the temperature of the furnace to 330 °C. After 2 hours, allow the alcohol to pass into the combusion tube at the rate of 1 drop every 3–4 seconds. The commencement of the dehydrogenation will be indicated by the production of white fumes at the point where the combustion tube enters the condenser; it will also be indicated by a gas flow (hydrogen) through the Drechsel bottles which continues after the nitrogen flow has been temporarily stopped. When all the hexanol has passed through the catalyst tube, remove the aqueous layer from the distillate, dry the organic layer with a little magnesium sulphate and distil from a flask carrying a fractionating side arm. Collect the fraction (30 g) (1) having b.p. 125–135 °C, and redistil to obtain 21 g (21%) of hexanal (2) having b.p. 127–129 °C.

Notes. (1) If the high boiling residue is transferred to a smaller flask and fractionally distilled, some hexanol passes over first, followed by hexyl hexanoate (Me·(CH$_2$)$_4$·CO$_2$(CH$_2$)$_5$·Me) (2 g) at 240–250 °C (mainly at 245 °C).
(2) About 0.1 per cent of hydroquinone should be added as a stabiliser since hexanal exhibits a marked tendency to polymerise. To obtain pure hexanal, treat the 21 g of the product with a solution of 42 g of sodium metabisulphite in 125 ml of water and shake; much bisulphite derivative will separate. Steam distil the suspension of the bisulphite complex until about 50 ml of distillate have been collected; this will remove any non-aldehyde impurities together with a little aldehyde. Cool the residual alde-

hyde–bisulphite solution to 40–50 °C, and add slowly a solution of 32 g of sodium hydrogen carbonate in 80 ml of water, and remove the free aldehyde by steam distillation. Separate the upper layer of hexanal, wash it with a little water, dry with magnesium sulphate and distil: the pure aldehyde passes over at 128–128.5 °C.

Cognate preparations. *Valeraldehyde* (*pentanal*). Use 100 g (123 ml, 1.14 mol) of pentan-1-ol, and fractionate the dried distillate. Collect the fraction of b.p. 98–110 °C (23 g); upon redistillation 20 g (20%) of valeraldehyde, b.p. 101–105 °C, are obtained. From the high boiling fractions 25 g of pentan-1-ol (b.p. 135–139 °C) may be recovered, together with 1.5 g of pentyl pentanoate (b.p. 205–210 °C).

Butyraldehyde (*butanal*). Use 100 g (123.5 ml, 1.35 mol) of butan-1-ol. The yield of butyraldehyde, b.p. 70–75 °C, is 38 g (39%), and of butyl butanoate (b.p. 165–170 °C) is 2 g; 40 g of butan-1-ol are recovered.

Propionaldehyde (*propanal*). Use 100 g (125 ml, 1.67 mol) of propan-1-ol and surround the receiver by a freezing mixture. The yield of propionaldehyde, b.p. 48–49.5 °C (mainly 49 °C), is 35 g (36%), and of propyl propanoate, b.p. 120–125 °C, is 1 g; 30 g of propan-1-ol are recovered.

Experiment 5.76 HEPTANAL[98]

$$Me\cdot(CH_2)_5\cdot CH_2OH \xrightarrow{PCC} Me\cdot(CH_2)_5\cdot CHO$$

In a 500-ml round-bottomed flask fitted with a reflux condenser is suspended pyridinium chlorochromate [PCC (Section 4.2.*18*, p. 425), 32.3 g, 150 mmol] in 200 ml of anhydrous dichloromethane. Heptan-1-ol (11.6 g, 100 mmol) in 20 ml of dichloromethane is added in one portion to the magnetically stirred solution. After 1.5 hours, 200 ml of dry ether is added and the supernatant solution decanted from the black gum. The insoluble residue is washed thoroughly with three 50-ml portions of anhydrous ether whereupon it becomes a black granular solid. The combined organic solutions are passed through a short pad of Florosil and the solvent removed by distillation. Distillation of the residual oil through a short Vigreux column gives 8.87 g (78%) of heptanal, b.p. 59–62 °C/30 mmHg (1).

Note. (1) These reaction conditions are used for the oxidation of decan-1-ol, hexane-1,6-diol and oct-2-yn-1-ol. In the case of alcohols where acid-sensitive groups are also present (e.g. tetrahydropyranyl ethers), sodium acetate is added to buffer the reaction mixture; alternatively pyridinium dichromate may be used.[99]

5.7.2 THE OXIDATIVE CLEAVAGE OF 1,2-DIOLS

Cleavage of 1,2-diols using either lead tetra-acetate or sodium metaperiodate is a general reaction which has important preparative applications.

$$R^1\cdot CH(OH)\cdot CH(OH)\cdot R^2 \xrightarrow[\text{or } NaIO_4/H^\oplus]{Pb(O\cdot CO\cdot Me)_4} R^1\cdot CHO + R^2\cdot CHO$$

Cleavage of open-chain symmetrical diols, which must yield a single aldehydic product, is clearly of most value. The preparation of butyl glyoxylate from dibutyl (+)-tartrate (Expt 5.77) using lead tetra-acetate in benzene solution is an interesting example. A second example is the cleavage of the 1,2-diol system in 1,2:5,6-di-*O*-isopropylidene-D-mannitol (6) to give (*R*)-2,3-isopropylidene-glyceraldehyde (7) (Expt 5.77, cognate preparation).[104] The reaction is of interest in that the readily available, naturally occurring, optically active D-mannitol provides an effective starting material to prepare the protected (*R*)-glyceraldehyde which is not easily obtained by other routes (e.g. by synthesis or by resolution). The enantiomeric, similarly protected, (*S*)-glyceraldehyde has been prepared from the isopropylidene derivative (8) of the readily available ascorbic acid (Vitamin C), or its reduced form, by a similar oxidative sequence.[105]

(6)   (7)   (8)

Periodic acid (or sodium metaperiodate) is often the reagent of choice when the substrate is soluble in an aqueous medium, as in the carbohydrate field; an example of its use is to be found in Expt 5.117. Sodium metaperiodate supported on silica gel (Section 4.2.55, p. 454) has been found to effectively oxidise water-insoluble α-diols,[106] and its value is illustrated in the preparation of hexanedial (Expt 5.78) from *trans*-cyclohexane-1,2-diol.

Experiment 5.77 BUTYL GLYOXYLATE

$$BuO_2C \cdot CH(OH) \cdot CH(OH) \cdot CO_2Bu \xrightarrow{Pb(OAc)_4}$$
$$2BuO_2C \cdot CHO + Pb(OAc)_2 + 2Me \cdot CO_2H$$

In view of the quantities of *benzene* (**CAUTION**) *required, the entire preparation must be carried out in the fume cupboard.* Place a mixture of 125 ml of pure benzene and 32.5 g (0.123 mol) of dibutyl (+)-tartrate (1) in a 500-ml three-necked flask, equipped with a Hershberg stirrer (Fig. 2.49) and a thermometer. Stir the mixture rapidly and add 58 g (0.13 mol) of lead tetra-acetate (Section 4.2.45, p. 441) in small portions over a period of 20 minutes while maintaining the temperature below 30 °C by occasional cooling with cold water. Continue the stirring for a further 60 minutes. Separate the salts by suction filtration and wash with two 25 ml portions of benzene. Remove the benzene and acetic acid from the filtrate by flash distillation and distil the residue under diminished pressure, preferably in a slow stream of nitrogen. Collect the butyl glyoxylate (2) at 66–69 °C/5 mmHg. The yield is 26 g (81%).

Notes. (1) The purified commercial dibutyl (+)-tartrate, m.p. 22 °C, may be used. It may be prepared by using the procedure described under isopropyl lactate (Expt 5.145). Place a mixture of 75 g of (+)-tartaric acid, 10 g of Zerolit 225/H⊕, 110 g (135 ml) of redistilled butan-1-ol and 150 ml of sodium-dried benzene in a 1-litre three-necked flask equipped with a sealed stirrer, a double surface condenser and an automatic water separator (cf. Fig. 2.31(*a*)). Reflux the mixture w th stirring for 10 hours: about 21 ml of water collects in the water separator. Filter ol the ion-exchange resin at the pump and wash it with two 30–40 ml portions of hot benzene. Wash the combined filtrate and washings with two 75 ml portions of saturated sodium hydrogen carbonate solution, followed by 100 ml of water, and dry over magnesium sulphate. Remove the benzene by distillation under reduced pressure (water pump) and finally distil the residue. Collect the dibutyl (+)-tartrate at 150 °C/1.5 mmHg. The yield is 90 g (69%).

(2) Store the butyl glyoxylate under nitrogen; it undergoes autoxidation in air. The product decomposes on boiling (159–161 °C) at atmospheric pressure.

Cognate preparation. (R)-2,3-O-Isopropylideneglyceraldehyde. To a solution of 1,2:5,6-di-O-isopropylidene-D-mannitol (50 g, 0.19 mol, Expt 5.114) in dry benzene (500 ml) **(CAUTION)** is slowly added lead tetra-acetate (90 g, 0.2 mol) with stirring. After 3 hours the mixture is filtered, and the filtrate concentrated below 25 °C to a syrup from which four 30-ml portions of carbon tetrachloride are evaporated at a temperature maintained below 25 °C. The syrup is distilled *in vacuo*. The first few millilitres are discarded and the main fraction is collected, 34 g (68%), b.p. 31 °C/5 mmHg (1).

Note. (1) The product is unstable on storage and changes are observed in the i.r. and p.m.r. spectra when it is kept at room temperature for several days.

Experiment 5.78 HEXANEDIAL[106]

trans-Cyclohexane-1,2-diol (232 mg, Expt 5.48) and sodium metaperiodate supported on silica gel (6.1 g, 10% loading. Section 4.2.55, p. 455) are stirred in dichloromethane (15 ml) (1) for 3 hours at 20 °C. The solids are then filtered off and washed with dichloromethane (2 × 10 ml). Evaporation of the solvent from the combined filtrate and washings gives a clear oil (185 mg, 80%) which is pure dial.

Note. (1) The dichloromethane is distilled from phosphorus pentoxide and stored over molecular sieves.

5.7.3 THE OZONOLYSIS OF SUITABLY SUBSTITUTED ALKENES

Oxidation of alkenes with ozone followed by cleavage of the resulting ozonides to carbonyl compounds is widely used for the determination of structure of unsaturated compounds. The ozonolysis technique is described in detail in Section 2.17.4, p. 103.

For preparative purposes the cleavage of the ozonide is best carried out by catalytic hydrogenation over palladium hydroxide-on-calcium carbonate, a catalyst system which does not hydrogenate the aldehydic products; the yield of the latter are usually fairly good. An alternative procedure for the decomposition of the ozonide is treatment with dimethylsulphide in methanol; the use of the less obnoxious thiourea is a good alternative.[107]

Two examples of preparative ozonolysis are given (Expt 5.79). In the first cyclohexene is subjected to ozonolysis in ethyl acetate to provide a convenient synthesis of the α,ω-dicarbonyl compound, adipaldehyde. In the second, the

substrate is oleic acid, which yields two aldehydic fragments, nonanal and azelaic hemialdehyde.

$$Me\cdot(CH_2)_7\cdot CH=CH\cdot(CH_2)_7\cdot CO_2H \longrightarrow$$
$$Me\cdot(CH_2)_7\cdot CHO + OCH\cdot(CH_2)_7\cdot CO_2H$$

The solvent used in the latter case is dry ethyl chloride, which must be replaced
by methanol before hydrogenation. Methanol when cooled to about $-20\,°C$ is
not attacked by ozone and can in many cases be used as the ozonisation solvent.
In such cases the main product of addition of ozone to the alkene is not the usual
ozonide, but is the methoxyhydroperoxide which, however, is also reductively
cleaved to the aldehyde on hydrogenation. This ozonisation process may be
mechanistically summarised as follows:

Ozonide

or $R^1R^2C-O\cdot O^\ominus \xrightarrow{MeOH} R^1R^2C\overset{O\cdot OH}{\underset{OMe}{<}}$ Methylhydroperoxide

Experiment 5.79 HEXANEDIAL (*Adipaldehyde*)

Dissolve 8.2 g (0.1 mol) of cyclohexene (Expt 5.12) in 200 ml of pure dry ethyl
acetate (Section 4.1.*24*, p. 409) contained in a 500-ml glass-stoppered wash
bottle, cool the solution to -20 to $-30\,°C$ or below (e.g. with solid carbon
dioxide–acetone) and attach the wash bottle through a calcium chloride drying tube to another containing acidified potassium iodide solution. Pass
ozonised oxygen (**CAUTION**: see Section 2.17.4, p. 104) until the reaction is
complete, i.e. until iodine is abundantly liberated. Then add 0.5 g of palladium hydroxide–calcium carbonate catalyst (Section 4.2.*54*, p. 453) (1) and
hydrogenate the cold solution of the ozonide in the usual manner (see Section
2.17.4, p. 106; cool the hydrogenation vessel in ice). Filter off the catalyst and
remove the solvent by distillation at normal pressure. Distil the residue under
reduced pressure and collect the adipaldehyde at 92–94 °C/12 mmHg. The
yield is 7 g (61%). This aldehyde oxidises readily and should be kept in a
sealed tube in an atmosphere of nitrogen or carbon dioxide. It may be converted into the dioxime by warming with aqueous hydroxylamine acetate
solution: after recrystallisation from water, the dioxime has m.p. 172 °C.

Cognate preparation. *Azelaic hemialdehyde semicarbazone.* Dissolve 7 g
(0.025 mol) of pure oleic acid in 30 ml of dry ethyl chloride (chloroform may
be used but is less satisfactory), and ozonise at about $-30\,°C$. Remove the

bulk of the solvent under reduced pressure, dissolve the residue in 50 ml of dry methanol and hydrogenate as for adipaldehyde in the presence of 0.5 g of palladium–calcium carbonate (2). Warm the resulting solution for 30 minutes with a slight excess of semicarbazide acetate and pour into water. Collect the precipitated semicarbazones and dry: the yield is 8.5 g. Separate the mixture of semicarbazones by either of the following methods: (a) Treat with dilute sodium hydrogen carbonate solution to extract the semicarbazone of azelaic hemialdehyde; upon acidifying the extract with dilute sulphuric acid, the semicarbazone of azelaic hemialdehyde is precipitated (4.4 g, 77%, m.p. 162 °C, after recrystallisation from methanol). The residue from the sodium hydrogen carbonate extraction consists of the semicarbazone of nonanal and melts at 101 °C after recrystallisation from methanol: yield 3.8 g (77%). (b) Extract the dry mixture of semicarbazones with ether: only the semicarbazone of nonanal dissolves easily.

Notes. (1) The Adams platinum oxide catalyst gives satisfactory results in the reduction of ozonides.
(2) An alternative work-up procedure after hydrogenation is to isolate the nonanal by steam distillation and to recover the azelaic hemialdehyde from the residual aqueous layer.

5.7.4 THE REDUCTION OF NITRILES, CARBOXYLIC ACIDS AND CARBOXYLIC ACID CHLORIDES

The reduction of a nitrile is achieved with anhydrous tin(II) chloride dissolved in ether or ethyl acetate saturated with dry hydrogen chloride (the *Stephen reaction*). The resulting aldimine hydrochloride (probably in the form of a complex with tin(IV) chloride) is then hydrolysed with warm water.

$$R{\cdot}C{\equiv}N + HCl \longrightarrow [R{\cdot}C{\equiv}\overset{\oplus}{N}H]\overset{\ominus}{Cl} \xrightarrow{[H]} [R{\cdot}CH{=}\overset{\oplus}{N}H_2]\overset{\ominus}{Cl} \xrightarrow{H_2O} R{\cdot}CHO$$

The Stephen reaction is used mainly for the synthesis of aromatic aldehydes but reduction of the higher aliphatic nitriles normally gives good yields. This method is illustrated by the synthesis of hexanal (Expt 5.80, *Method A*). Alternative and convenient reagents for the controlled reduction of the cyano group to the imino group are diisobutylaluminium hydride (DIBAL)[108] and lithium triethoxyaluminohydride;[109] the use of the latter reagent is described in Expt 5.80, *Method B*, where it is generated *in situ* from lithium aluminium hydride and ethyl acetate.

A most convenient procedure for the conversion of carboxylic acids to aldehydes results from their initial treatment with borane–dimethyl sulphide to give first, the triacyloxyborane (9), which is then reduced further to the intermediate (10); the trimer of (10) is the trialkyloxyboroxine (11). Oxidation of (11) with PCC then yields the aldehyde.[110] The method is illustrated for the preparation of octanal (Expt 5.81), but it has been applied to cyclohexanecarboxylic acid and to various aromatic carboxylic acids ($X{\cdot}C_6H_4{\cdot}CO_2H$, $X = p$-Cl, NO_2, OMe, m-CN).

The catalytic reduction of carboxylic acid chlorides by the Rosenmund procedure may be used for the preparation of aliphatic aldehydes but its application is mainly for the synthesis of aromatic aldehydes (e.g. Expt 6.120). Alternative procedures for the chemical reduction of acid chlorides include reduction with

$$3R \cdot CO_2H + BH_3 \cdot SMe_2 \xrightarrow[-Me_2S]{-3H_2} (R \cdot CO \cdot O)_3B \xrightarrow{2BH_3 \cdot SMe_2}$$

(9)

$$3(R \cdot CH_2OBO) \equiv$$

(10)

$$\xrightarrow{PCC} 3R \cdot CHO + 3B(OH)_3$$

(11)

lithium tri-t-butoxyaluminohydride,[111] or reduction with bis(triphenylphos-phine) copper(I) tetrahydroborate.[112] An example of the use of the latter reagent is given in cognate preparation in Expt 6.120, but the procedure may be applied to aliphatic acid chlorides giving superior yields of aliphatic aldehydes than does the former reagent.

Experiment 5.80 HEXANAL

$$C_5H_{11} \cdot CN \xrightarrow[SnCl_2]{HCl} C_5H_{11} \cdot CH{=}\overset{\oplus}{N}H_2\}Cl^{\ominus} \xrightarrow{H_2O} C_5H_{11} \cdot CHO$$

$$C_5H_{11} \cdot CN \xrightarrow{LiAl(OEt)_3H} C_5H_{11} \cdot CH{=}N{-}\overset{\ominus}{Al}(OEt)_3 \xrightarrow{H_3O^{\oplus}} C_5H_{11} \cdot CHO$$

Method A. Into a 500-ml three-necked flask, provided with a mechanical stirrer, a gas inlet tube and a reflux condenser, place 57 g of anhydrous tin(II) chloride (Section 4.2.73, p. 465) and 200 ml of anhydrous ether. Pass in dry hydrogen chloride gas (Section 4.2.38, p. 438) until the mixture is saturated and separates into two layers; the lower viscous layer consists of tin(II) chloride dissolved in ethereal hydrogen chloride. Set the stirrer in motion and add 19.4 g (0.2 mol) of hexanenitrile (Expt 5.160) through the separatory funnel. Separation of the crystalline aldimine hydrochloride commences after a few minutes; continue the stirring for 15 minutes. Filter off the crystalline solid, suspend it in about 50 ml of water and heat under reflux until it is completely hydrolysed. Allow to cool and extract with ether; dry the ethereal extract with anhydrous calcium sulphate or magnesium sulphate and remove the ether slowly by distillation through a short fractionating column. Finally, distil the residue and collect the hexanal at 127–129 °C. The yield is 19 g (95%).

Method B.[104] In a 1-litre flask equipped with a condenser, a thermometer and a dropping funnel is placed 0.3 mol of lithium aluminium hydride (**CAUTION**, see Section 4.2.49, p. 445) in dry ether (300 ml). A nitrogen atmosphere is maintained throughout the reaction. To this stirred solution 0.45 mol of ethyl acetate is added over a period of 75 minutes maintaining the temperature at 3–7 °C. The reaction is stirred for an additional 30 minutes. To this solution at −10 °C is added 29.1 g (0.3 mol) of hexanenitrile in 5 minutes. The reaction temperature rises to 12 °C in 10 minutes with the formation of a viscous solution. It is allowed to stir for another 50 minutes at 3 °C. The reaction mixture is decomposed by the addition of 300 ml of 2.5 M sulphuric acid. The ether layer is separated and the aqueous layer is extracted with ether. The combined ether extracts are washed with sodium hydrogen carbonate solution and water and dried over anhydrous sodium sulphate (1).

Distillation through a 30.4-cm Vigreaux column gives 16.6 g (55.4%) of hexanal, b.p. 51–55 °C/53–55 mmHg and 2.5 g (8.5%) of hexanenitrile, b.p. 73 °C/53–55 mmHg.

Note. (1) Gas-liquid chromatography analysis indicates a 68 per cent yield of aldehyde and the presence of the nitrile.

Experiment 5.81   OCTANAL[110]

$$C_7H_{15}{\cdot}CO_2H \xrightarrow{\text{BMS}} C_7H_{15}{\cdot}CH_2{\cdot}O{\cdot}B \underset{\substack{| \\ O}}{\overset{O}{\diagdown}} B{\cdot}OCH_2{\cdot}C_7H_{15} \xrightarrow{\text{PCC}} C_7H_{15}{\cdot}CHO$$

$$\underset{O{\cdot}CH_2{\cdot}C_7H_{15}}{\overset{B}{|}}$$

CAUTION: *This reaction should be conducted in an efficient fume cupboard as hydrogen is evolved.*

Trioctyloxyboroxine. In a dry 250-ml round-bottomed flask provided with a septum inlet, magnetic stirrer bar, and a reflux condenser attached to a connecting tube leading to a mercury bubbler, are placed octanoic acid (9.51 ml, 60 mmol) and diethyl ether (75 ml) under nitrogen. The mixture is stirred vigorously and borane–dimethyl sulphide (BMS, 6.12 ml, 60 mmol) is added dropwise from a syringe. Following the addition of the initial 2–3 ml of BMS, when gas evolution has ceased, the mixture is heated under gentle reflux to complete evolution of gas (hydrogen). The remainder of the BMS is added at such a rate as to maintain gentle reflux. Following completion of the addition, the mixture is heated under reflux for 1 hour. The solvent and dimethyl sulphide (**CAUTION**) are removed under vacuum (1) and dichloromethane (20 ml) is introduced to dissolve the product.

Octanal. To a well-stirred suspension of pyridinium chlorochromate [PCC (Section 4.2.18, p. 426), 14.3 g, 66 mmol] in dichloromethane (100 ml) contained in a 500-ml flask equipped as above, is added dropwise the above solution of trioctyloxyboroxine in dichloromethane. The stirred mixture is heated under reflux for 1 hour and then diluted with ether (150 ml). The supernatant liquid is filtered through Florosil (100 g) contained in a 300-ml sintered glass funnel; the solid residue is triturated with ether (3 × 50 ml) and passed through the same Florosil column. The colourless filtrate is concentrated and distilled under reduced pressure to give octanal: yield 5.9 g (77%), b.p. 64–65 °C/15 mmHg.

Note. (1) Very efficient cooling traps must be used in this distillation, and Section 4.2.7, p. 419 should be consulted for the treatment and disposal of the distillate and trap residues.

5.7.5 METHODS BASED UPON ALKYL HALIDES

The disconnection strategies summarised on p. 586 reveal that the nucleophilic organometallic reagents derived from alkyl halides (e.g. R·MgX, RLi) require reagents equivalent to the formyl cation ($H{\cdot}\overset{\oplus}{C}{=}O$), whereas the electrophilic halides themselves require reagents equivalent to the acyl anion ($H{\cdot}\overset{-\ominus}{C}{=}O$).

596

The reagent equivalents of the *acceptor* formyl synthon are a formate ester (e.g. H·CO·OEt), a trialkyl orthoformate (e.g. H·C(OR)₃), or the Vilsmeier reagent (see Section 6.10.1, p. 991). Although the addition of a Grignard reagent to a formate ester yields an aldehyde, further addition of the reagent to the more reactive aldehydic carbonyl–carbon results in the formation of a secondary alcohol (cf. synthesis of secondary alcohols, Section 5.4.2, p. 537). However, the reaction may be stopped at the aldehyde stage by an inverse addition procedure. More conveniently, the formation of a secondary alcohol may be prevented by using triethyl orthoformate in a reaction with a Grignard reagent, when the product is an acetal which is less reactive towards the organometallic compound. The acetal is a protected carbonyl group which may be revealed by hydrolysis with hot aqueous sulphuric acid. The reaction is illustrated by the preparation of hexanal (Expt 5.82).

$$R·MgX + HC(OEt)_3 \longrightarrow R·CH(OEt)_2 + Mg(OEt)X$$

The Grignard reaction has been improved by the use of phenyl diethyl orthoformate, which may be prepared by an ester exchange reaction between equimolar amounts of triethyl orthoformate and phenol.[113]

One of the most important reagent equivalents of the *donor* formyl synthon is 1,3-dithiane (see Chapter 1, p. 21). However the use of this reagent equivalent is not widely found for the synthesis of *simple* aldehydes since other methods may be economically more appropriate. An example of specific formylation by the use of 1,3-dithiane is illustrated in Expt 6.66. Disodium tetracarbonylferrate(II) is a synthetic equivalent of the carbon monoxide dianion ($\overset{2\ominus}{C}{=}O$), and as well as being used as a reagent for aldehyde synthesis it is also of value for the synthesis of ketones, esters, amides and carboxylic acids.[114] The toxicity and hazardous nature of the reagent however is such that only the most experienced synthetic chemists should attempt to use it.[115]

Experiment 5.82 HEXANAL

$$Bu·CH_2Br \xrightarrow{Mg} Bu·CH_2MgBr \xrightarrow{HC(OEt)_3} Bu·CH_2·CH(OEt)_2$$

$$Bu·CH_2·CH(OEt)_2 \longrightarrow Bu·CH_2·CHO + 2EtOH$$

In a suitably equipped 1-litre three-necked flask prepare a Grignard reagent from 15 g (0.675 mol) of dry magnesium turnings, 225 ml of sodium-dried ether, and 94.5 g (79 ml, 0.625 mol) of 1-bromopentane (Expt 5.54), using the detailed procedure described in Expt 5.40. When reagent formation is complete, cool the flask to about 5 °C and add 74 g (83 ml, 0.5 mol) of triethyl orthoformate (Expt 5.153) during about 10 minutes. Reflux the mixture for 6 hours; then arrange the condenser for distillation and remove the ether on a water bath. Allow the reaction mixture to cool. Add 375 ml of ice-cold 6 per cent hydrochloric acid with stirring; keep the contents of the flask cool by the occasional addition of a little crushed ice. When all the white solid has passed into solution, transfer to a separatory funnel and remove the upper layer of hexanal diacetal. Hydrolyse the acetal by distilling it with a solution of 50 g (27.5 ml) of concentrated sulphuric acid in 350 ml of water; collect the aldehyde, which distils over as an oil, in a solution of 50 g of sodium metabisulphite in 150 ml of water. Remove the oily layer (largely pentan-1-ol) insoluble in the bisulphite solution and discard it. Steam distil the bisulphite solution until 100 ml of the distillate have been collected: this will separate the remainder of the pentan-1-ol and other impurities. Cool the residual bisulphite solution to about 45 °C, cautiously add a suspension of 40 g of sodium hydrogen carbonate in 100 ml of water and separate the resulting free aldehyde by steam distillation. Remove the upper layer (crude aldehyde) of the distillate, wash it with three 25 ml portions of water and dry it with 10 g of calcium sulphate or magnesium sulphate. Distil through a short fractionating column, and collect the hexanal at 127–129 °C. The yield is 25 g (50%).

5.7.6. THE HYDROLYSIS AND DECARBOXYLATION OF α,β-EPOXY ESTERS (GLYCIDIC ESTERS

Glycidic esters are prepared by the reaction of an aldehyde or ketone with an α-haloester in the presence of base (sodium ethoxide, sodamide, finely divided sodium or potassium t-butoxide) (the *Darzens glycidic ester condensation*).

The use of ethyl chloroacetate as the starting ester results in the formation of aldehydes, as in the preparation of 2-phenylpropanal (Expt 5.83), whereas ketones are obtained using homologous ethyl α-chloroesters. A special procedure for the hydrolysis of a glycidic ester is used in Expt 5.83. This involves treatment of the ester with one equivalent of sodium ethoxide in absolute ethanol followed by the addition of one equivalent of water when precipitation of the sodium salt of the glycidic acid occurs; the precipitation of the sodium salts of glycidic acids is frequently facilitated by the addition of ether. With aqueous alkali the glycidic ester is hydrolysed to the corresponding epoxy acid which decarboxylates under acidic conditions to give a carbonyl compound.

A two-phase solid–liquid transfer catalytic system has been developed for aromatic aldehydes and for aliphatic aldehydes which are branched at the α-carbon.[116] The aldehyde and the α-chloroester are stirred in dimethylformamide in the presence of solid potassium carbonate and triethylbenzylammonium chloride. The isolated yield of glycidic ester compares favourably with that of the procedure above.

Experiment 5.83   2-PHENYLPROPANAL (*Hydratropaldehyde*)

Ethyl 3-phenyl-2,3-epoxybutanoate. In a 500-ml three-necked flask, fitted with a mechanical stirrer and a thermometer, place a mixture of 60 g (58.5 ml, 0.5 mol) of acetophenone, 61.5 g (53 ml, 0.5 mol) of ethyl chloroacetate (b.p. 142–143 °C) and 100 ml of sodium-dried benzene. Add, with stirring 23.6 g (0.6 mol) of finely powdered sodamide (Section 4.2.67, p. 462) over a period of 2 hours; maintain the temperature at 15–20 °C with the aid of external cooling. Ammonia is evolved. Stir for 2 hours at room temperature and pour the reddish mixture on to 350 g of crushed ice with hand stirring. Separate the organic layer and extract the aqueous layer with 100 ml of benzene. Wash the combined benzene solutions with three 150 ml portions of water, the last one containing 5 ml of acetic acid, and then dry with magnesium sulphate. After removal of the benzene on a rotary evaporator, distil the residue under reduced pressure through a short fractionating column. Collect the fraction of b.p. 111–114 °C/3 mmHg as the glycidic ester; the yield is 67 g (65%).

2-Phenylpropanal (*hydratropaldehyde*). Prepare a solution of sodium ethoxide in a 500-ml round-bottomed flask from 7.6 g (0.33 mol) of clean sodium and 150 ml of absolute ethanol (for experimental details see Expt 5.95). Add 66.5 g of the above glycidic ester slowly and with shaking. Cool the flask externally to 15 °C and add 6 ml of water slowly; much heat is evolved and the sodium salt soon separates. Keep the reaction mixture overnight. Collect the salt by suction filtration, wash it with 25 ml of ethanol followed by 25 ml of ether.

Add the salt to dilute hydrochloric acid (prepared from 28 ml of the concentrated acid and 150 ml of water) contained in a 500-ml flask fitted with a reflux condenser. Warm the mixture gently; carbon dioxide is evolved and an oil separates. Heat on a steam bath for 90 minutes, cool and extract the oil with 75 ml of benzene. Wash the extract with 100 ml of water, and distil the benzene solution under reduced pressure. Collect the 2-phenylpropanal at 90–93 °C/10 mmHg; the yield is 30 g (70%).

Cognate preparation. *Ethyl 2,1'-epoxycyclohexylacetate.* Add a mixture of 55 g (48 ml, 0.45 mol) of ethyl chloroacetate and 43 g (0.44 mol) of cyclohexanone dropwise to a suspension of finely divided sodium (11 g, 0.48 mol) in anhydrous xylene (165 ml) (Section 4.1.4, p. 399) with stirring and cooling in an ice–salt bath. Regulate the rate of addition so that the temperature of the reaction mixture does not exceed 8 °C. Pour the resulting dark-red clear solution into water, wash the organic layer repeatedly with water, dry with magnesium sulphate and distil. Collect the glycidic ester at 81–83 °C/ 0.04 mmHg or at 115–117 °C/10 mmHg. The yield is 37 g (46%).

5.7.7 THE OXIDATIVE HYDROLYSIS OF NITRONATE SALTS DERIVED FROM PRIMARY NITROALKANES

The hydrolysis of the conjugate base of a primary nitroalkane to an aldehyde under acidic conditions is known as the *Nef reaction* (cf. Section 5.8.7, p. 623).

$$R \cdot CH_2 \cdot N\underset{O}{\overset{O}{\lessgtr}} \xrightarrow{\ominus OH} \left[R \cdot \overset{\ominus}{CH} - N\underset{O}{\overset{O}{\lessgtr}} \longleftrightarrow R \cdot CH = N\underset{O}{\overset{O^\ominus}{\lessgtr}} \right] \xrightarrow{H_3O^\oplus} R \cdot CHO$$

However the acidic conditions often results in by-product formation which lowers the yield of aldehyde. Alternative procedures include the oxidation of the nitronate salt with potassium permanganate,[117] the oxidative cleavage of the carbon–nitrogen bond by ozonolysis,[118] or the reduction of the nitroalkane to the imine with titanium(III) chloride followed by hydrolysis to the aldehyde[119] (cf. Stephen reaction). The example given in Expt 5.84 for the preparation of octanal involves the conversion of 1-nitrooctane into its conjugate base with sodium hydride, followed by oxidation with potassium permanganate. Because of the susceptibility to further oxidation of the aldehydic product, the reaction is carried out under nitrogen, and a layer of pentane is placed over the aqueous phase into which the aldehyde is extracted as soon as it is formed.

The usefulness of the nitronate anion from a primary nitroalkane as an acyl anion equivalent (see Appendix A6.2) is further illustrated by the fact that it may be reacted with other electrophiles, prior to the conversion into the carbonyl group.[120] An example of its use in this manner is illustrated in Expt 5.178.

Experiment 5.84 OCTANAL[117]

$$C_6H_{13} \cdot CH_2 \cdot CH_2 \cdot NO_2 \longrightarrow C_6H_{13} \cdot CH_2 \cdot CH = NO_2^\ominus \longrightarrow C_6H_{13} \cdot CH_2 \cdot CHO$$

A 60 per cent oil suspension of sodium hydride (160 mg, 4 mmol) is washed free of oil with pentane and transferred to a nitrogen-filled flask with the aid of 20 ml of t-butyl alcohol. After stirring this mixture for 10 minutes, a solution of 1-nitrooctane (318 mg, 2 mmol) in 40 ml of t-butyl alcohol is added all at once; a heavy white precipitate formed immediately. The mixture is stirred for 20 minutes and then 600 ml of ice-cold pentane is added; this was followed at once by 50 g of ice and an ice-cold solution of 316 mg (2 mmol) of potassium permanganate and 248 mg (4 mmol) of boric acid in 80 ml of water. The mixture is stirred vigorously under nitrogen for 10 minutes and then 2 ml of aqueous 1 M sodium metabisulphite and 4 ml of 1 M sulphuric acid are added.

The pentane layer is separated and the aqueous layer is extracted twice with 60-ml portions of pentane. The combined pentane solutions are washed three times with 400-ml portions of ice-cold water and dried over magnesium sulphate under nitrogen. Removal of the pentane under vacuum gives 249 mg (97%) of crude aldehyde; 240 mg of this oil is dissolved in c. 2 ml of pentane and filtered through a 3 cm × 2.5 cm column of deactivated silica gel (1) and the column eluted with pentane. Removal of the pentane gives 215 mg (89%) of a very pale yellow oil. Spectra (n.m.r. and i.r.) show the oil to be pure octanal.

Note. (1) E. Merck Silica gel 60 (40–63 μm) stirred with acetone for 30 minutes, filtered, air-dried, and then oven-dried at 120 °C for 12 hours.

5.7.8 METHODS FOR THE ALKYLATION OF THE CARBON CHAIN

Although a retrosynthetic disconnection of an α-branched aldehyde to give an alkyl group and a mesomerically stabilised α-carbanion suggests a synthesis based on this strategy, in practice it is not feasible. This is because the α-carbanion, generated from the aldehyde by base, rapidly undergoes a self-aldol condensation (Section 5.18.2, p. 799) and this normally precludes any successful alkylation reaction.

$$R^1 \cdot CH(R^2) \cdot CHO \xleftarrow{\quad\xcancel{R^2X}\quad} R^1 \cdot \overset{\ominus}{CH}\!\!-\!\!C\!\!\overset{O}{\underset{H}{\diagdown}} \longrightarrow R^1CH_2 \cdot CH(OH) \cdot CH(R^1) \cdot CHO$$

α-Alkylation of the carbon chain is however possible if the carbonyl group is first converted into either the imine (12) (Section 5.16.5, p. 782), or the hydrazone (13).[121] Subsequently these nitrogen derivatives are converted into the corresponding lithio derivatives (which do not undergo self-condensation) by treatment with lithium diisopropylamide, and may then be alkylated by reaction with an alkyl halide.

$$R^1 \cdot CH_2 \cdot CHO \xrightarrow{R^2NH_2} R^1 \cdot CH_2 \cdot CH{=}NR^2 \xrightarrow{\text{base}}$$
$$(12)$$

$$[R^1 \cdot \overset{\ominus}{CH}\!\!-\!\!CH{=}NR^2] \xrightarrow[\text{(ii) H}_3\text{O}^\oplus]{\text{(i) R}^3X} R^1 \cdot CH(R^3) \cdot CHO$$

$$R^1 \cdot CH_2 \cdot CHO \xrightarrow{R^2_2N \cdot NH_2} R^1 \cdot CH_2CH{=}N \cdot NR^2_2 \xrightarrow{\text{base}}$$
$$(13)$$

$$[R^1 \cdot \overset{\ominus}{CH}\!\!-\!\!CH{=}N \cdot NR^2_2] \xrightarrow[\text{(ii) H}_3\text{O}^\oplus]{\text{(i) R}^3X} R^1 \cdot CH(R^3) \cdot CHO$$

The synthesis of *racemic* α-alkyl aldehydes by this method is feasible, although recourse to the methods previously described may be more convenient, as formulated below.

$$\overset{R^1}{\underset{R^2}{\diagup}}\!\!\!-Br \xrightarrow[\text{(ii) CH}_2\text{O}]{\text{(i) Mg}} \overset{R^1}{\underset{R^2}{\diagup}}\!\!\!-CH_2OH \xrightarrow{\text{[O]}} \overset{R^1}{\underset{R^2}{\diagup}}\!\!\!-CHO$$

However for the formation of optically active α-alkyl aldehydes, the imine or hydrazone routes have proved of considerable value.[122] In the preparative example[123] the aldehydes propanal and octanal are converted with (S)-(−)-2-amino-1-methoxy-3-phenylpropane (14) into the imines (15) and (16) respectively. Treatment with lithium diisopropylamide then yields the corresponding lithioenamines [only the (E)-isomers are formulated, since the (Z)-isomers would be less stable]. These intermediates have a topography which determines the subsequent direction of attack by the alkyl halide (see also Section 5.11.7, p. 688). In the formulation below, this stereoselection is from above the plane of the paper and leads to the (R)- and (S)-2-methyloctanals respectively.

Experiment 5.85   (R)- and (S)-2-METHYLOCTANAL[123]

$$R^1 \cdot CH_2 \cdot CHO \xrightarrow[\text{(ii) LDA}]{\text{(i) }(S)-H_2NCH(CH_2OMe) \cdot CH_2 \cdot Ph}$$

$$R^1 \cdot \overset{\ominus}{C}H \cdot CH{=}N \cdot CH(CH_2OMe) \cdot CH_2 \cdot Ph \xrightarrow[\text{(ii) H}_3O^{\oplus}]{\text{(i) R}^2X} R^1R^2CH \cdot CHO$$

R^1 = Me or C_6H_{13}
R^2 = C_6H_{13} or Me

(S)-Phenylalanine ethyl ester hydrochloride.[124] To a stirred, ice-cold suspension of (S)-phenylalanine (30 g, 0.182 mol) in absolute ethanol (800 ml), thionyl chloride (32.5 g, 0.273 mol) is added dropwise, and the reaction mixture is then heated under reflux for 3.5 hours. The pale yellow solution is allowed to stand at room temperature overnight, and then the ethanol is evaporated *in vacuo* to leave colourless crystals of the hydrochloride which are stirred with dry ether, filtered off and dried *in vacuo*.

(S)-2-Amino-3-phenylpropan-1-ol [(S)-phenylalaninol].[124] To a solution of sodium borohydride (3.5 g, 0.092 mol) in 50 per cent aqueous ethanol (50 ml) is added dropwise a solution of (S)-phenylalanine ethyl ester hydrochloride (5.0 g, 0.022 mol) in 50 per cent aqueous ethanol (50 ml). After the resulting mixture is refluxed for 4.5 hours, ethanol is evaporated *in vacuo*. The aqueous solution thus obtained is then extracted with ethyl acetate and the extract washed with saturated sodium chloride solution and dried over anhydrous sodium sulphate. Evaporation of the ethyl acetate under reduced pressure affords (S)-phenylalaninol (2.8 g, 84%) as a pale yellow solid m.p. 85–92 °C. Recrystallisation from ether gives colourless crystals of m.p. 91–93 °C, $[\alpha]_D^{25}$ − 25.6° (c. 1.037 in EtOH).

(S)-(−)-2-Amino-1-methoxy-3-phenylpropane. A solution of (S)-phenyl-alaninol (18.4 g, 0.122 mol) in 250 ml of anhydrous tetrahydrofuran is added dropwise to a stirred suspension of potassium hydride (5.23 g, 0.130 mol, pentane washed) in 100 ml of tetrahydrofuran at 25 °C under nitrogen. The resulting pale yellow gelatinous mixture is allowed to stand overnight and then a solution of methyl iodide (17.0 g, 0.119 mol, **CAUTION**) in 150 ml of tetrahydrofuran is added dropwise over 2 hours. Mixing is accomplished by external shaking since the gelatinous mixture would not stir with a magnetic stirrer bar. The reaction components are then allowed to mix for an additional 3 hours and then poured into 1 litre of cold saturated brine, extracted thrice with ether, dried over anhydrous sodium sulphate, and concentrated to give 24.9 g of crude product. Distillation gives 17.1 g, b.p. 55–59 °C/0.1 mmHg, of a clear oil which on standing becomes cloudy and rapidly produces a white precipitate which is found to be the carbonate. Since this experiment was performed it has been found that conversion of the freshly distilled methoxyamine to its hydrochloride salt is a more convenient way to store the compound. Thus the methoxyamine (17.0 g) immediately after distillation is dissolved in 700 ml of absolute ethanol and dry hydrogen chloride bubbled in for 10 minutes. The resulting solution is concentrated *in vacuo* to furnish 20.5 g of a colourless solid which is recrystallised from ethanol–ether (13:1), m.p. 151–152 °C, $[\alpha]_{578}^{25} + 19.7°$ (c. 2.5 in EtOH), δ (D$_2$O) 2.90 (d, 2H), 3.34 (s, 3H), 3.54 (s, 2H), 3.59 (m, 1H), and 7.37 (broad s, 5H). To release the free methoxyamine, it is dissolved in 5 per cent potassium carbonate solution and extracted with ether, dried over anhydrous sodium sulphate and concentrated. Bulb-to-bulb distillation at 52 °C/0.1 mmHg gives a clear oil, $[\alpha]_{578}^{23}$ −14.7° (c. 6 in C$_6$H$_6$); p.m.r. (CDCl$_3$) δ 1.75 (broad s, 2H, disappears on D$_2$O shake) 2.68 (m, 2H), 3.35 (broads, 6H) and 7.24 (s, 5H).

Formation of aldimines from propanal and octanal. To a stirred solution (0 °C) of 10 mmol of the methoxyamine dissolved in 30 ml of benzene **(CAUTION)** (previously washed with concentrated sulphuric acid and distilled) is added 10 mmol of the pure aldehyde. An immediate cloudiness usually results. The mixture is allowed to warm to room temperature and c. 15 g of anhydrous sodium sulphate added. After stirring an additional 30–40 minutes, it is filtered and the sodium sulphate washed thoroughly with dry ether. The solvent is removed first with aspirator pressure and then with a vacuum pump (0.5 mm) to generally furnish 9.5–10 mmol of the aldimine as a colourless oil. The aldimines are dissolved in tetrahydrofuran (0.4 M) and stored at −20 to −30 °C. Attempts to store the aldimines as neat liquids result in deterioration. The solutions of aldimines are conveniently transferred via a syringe to reaction vessels.

(R)- and (S)-2-Methyloctanal. (The editors suggest that the apparatus set-up shown in Fig. 2.77 could be used.) The flask is charged with 10 ml of anhydrous tetrahydrofuran under a nitrogen atmosphere. Freshly distilled diisopropylamine (1.54 ml, 11 mmol) is added via a syringe and the solution cooled to 0 °C. Butyllithium (4.6 ml of a 2.4 M solution in hexane) is added and the solution stirred at 0 °C for 40 minutes and then cooled to −23 °C (dry ice–carbon tetrachloride). The aldimine (10 mmol) in 25 ml of tetrahydrofuran is added via a syringe over 5 minutes. The resulting solution (generally yellow) is stirred at −23 °C for 30 minutes and then cooled to

−78 °C (dry ice–propanol). The alkyl halide (hexyl or methyl iodide) (10 mmol) dissolved in 5 ml of tetrahydrofuran is then added and the reaction mixture stirred for 2–7 hours at −78 °C (1). After warming to room temperature and addition of 50 ml of ether, the cloudy mixture is poured into 100 ml of water and the phases separated. The aqueous phase is extracted with ether and the combined organic phases are washed with brine, dried over anhydrous sodium sulphate and concentrated. The crude alkylated aldimines are hydrolysed by dissolving in 30 ml of pentane and shaking for 5 minutes in a separating funnel with an aqueous acetic acid–sodium acetate solution (prepared from 37.5 ml of acetic acid, 37.5 ml of water and 16.2 g of sodium acetate). The layers are separated and the aqueous layer extracted once with 30 ml of pentane. *Both layers are kept* (2). The combined pentane layers are washed successively with water, 10 per cent sodium hydrogen carbonate, water and then dried. Evaporation of the filtered pentane solution gives the crude aldehyde as a pale yellow oil. Bulb-to-bulb distillation at 90 °C/ 4 mmHg furnishes a clear colourless product which by g.l.c. analysis is shown to be 70–75 per cent pure 2-methyloctanal. The contaminant in the case of the reaction between the aldimine of propanal and hexyl iodide, is hexyl iodide, and in the case of the reaction between the aldimine of octanal and methyl iodide, is octanal. These impurities arise from an incomplete metallation reaction which leads in the first instance to generation of propanal and hexyl iodide in the work-up procedure, and in the second instance to the generation of octanal and methyl iodide. In the final bulb-to-bulb distillation the more volatile propanal or methyl iodide is removed from the respective 2-methyloctanal distillates, leaving the contaminants noted above (3).

Notes. (1) The reaction is monitored by withdrawing a 0.5 ml aliquot, quenching with water, and n.m.r. analysis of the extracted organic residue.
(2) The chiral methoxyamine is recovered from the aqueous solution by neutralisation with solid potassium hydroxide and extraction with ether. The ethereal extract is washed with brine, dried over potassium carbonate and concentrated to give the crude chiral amine in 80–88 per cent yield. Distillation affords the pure amine (70–75% recovery) with [α] values which indicate that no racemisation has occurred.
(3) The literature authors prepared mixtures of enantiomerically pure 2-methyloctanal ([α]$_D^{25}$-8.90 °C) (isolated by preparative g.l.c. from the sample prepared by the above procedure from the aldimine of propanal and hexyl iodide) containing w/w amounts of 12.7, 25.7 and 44.6 per cent of hexyl iodide. The [α]$_D$ values of these mixtures were plotted against the weight per cent to give a linear relationship. The g.l.c. composition of the reaction product and its extrapolated specific rotation thus allowed an *ee* per cent value to be calculated. In a similar way the *ee* per cent of the reaction of the aldimine of octanal and methyl iodide was calculated from the mixtures of octanal with the optically pure 2-methyloctanal of 55.4, 74.3 and 87.3 per cent and the plot weight per cent v. [α]$_D$ was again linear.

5.8   ALIPHATIC KETONES

Ketones may have symmetrical (R·CO·R) or unsymmetrical R^1·CO·R^2) structures. Branching in either of the alkyl groups or the presence of a functional group substituent is located by systematic numbering of the carbon chain; letters of the Greek alphabet may be used to locate positions relative to the carbonyl carbon atom. Illustrative representations are shown for:

3-methylhexan-2-one (1) $CH_3 \cdot CH_2 \cdot CH_2 \cdot CH(CH_3) \cdot CO \cdot CH_3$;
4-ethylhexan-2-one (2) $CH_3 \cdot CH_2 \cdot CH(C_2H_5) \cdot CH_2 \cdot CO \cdot CH_3$;
2,4-dimethylpentan-3-one (3) $(CH_3)_2CH \cdot CO \cdot CH(CH_3)_2$;
4-hydroxy-4-methylpentan-2-one (4) $(CH_3)_2C(OH) \cdot CH_2 \cdot CO \cdot CH_3$.

The presence of branching or functional group substitution leads to the possibility of a chiral site (*). Thus compounds (1) and (6) are chiral but compounds (2), (3) and (4) are achiral.

Cyclic ketones are those compounds in which the carbonyl carbon is part of the ring skeleton [e.g. cyclohexanone (5), and 3-methylcyclopentanone (6)]. α, β-Unsaturated ketones (e.g. $R^1 \cdot CH{=}CH \cdot CO \cdot R^2$) are considered in Section 5.18.2, p. 798.

The synthesis of aliphatic ketones and of alicyclic ketones (excluding ring synthesis) is exemplified by the following procedures.

1. The oxidation of secondary alcohols (Expts 5.86 to 5.90).
2. The hydration of alkynes (Expt 5.91).
3. The thermal decarboxylation of acids over a metal oxide catalyst (Expts 5.92 and 5.93).
4. The interaction of organometallic compounds with carboxylic acid chlorides and N,N-dimethylcarboxamides (Expt 5.94).
5. The hydrolysis and decarboxylation of β-keto esters and the hydrolysis of β-diketones (Expts 5.95 to 5.97).
6. The acid-catalysed rearrangement of 1,2-diols (Expt 5.98).
7. The oxidative hydrolysis of nitronate salts from secondary nitroalkanes (see Expt 5.84).

Methods for the protection of the carbonyl group in aldehydes and ketones are considered in Section 5.8.8.

605

SUMMARY OF RETROSYNTHETIC STRATEGIES

Functional group interconversion (FGI) (methods 1, 2 and 7)

$$\underset{\text{(TM)}}{\underset{R^1 \quad R^2}{\overset{\text{OH}}{\diagup}}} \overset{(1)}{\Longleftarrow} \underset{R^1 \quad R^2}{\overset{\text{O}}{\diagup}} \overset{(7)}{\Longrightarrow} \underset{R^1 \quad R^2}{\overset{NO_2^{\ominus}}{\diagup}}$$

$$\Big\Downarrow {(2)\ R^2 = Me}$$

$$R^1{-}C{\equiv}C{-}H$$

Disconnection (methods 3, 4, 5 and 8)

$$\underset{R^1 \overset{\oplus}{} \quad H_2\overset{\ominus}{C} \diagdown R^2}{\overset{\text{O}}{\diagup}} \overset{(i)}{\underset{(4)}{\Longleftarrow}} \underset{R^1 \quad \underset{H}{\overset{(i)}{\diagup}} R^2}{\overset{\text{O}}{\diagup}} \overset{(i)}{\underset{(3)}{\Longrightarrow}} \underset{R^1 \overset{\cdot}{} \quad H_2\overset{\cdot}{C} \diagdown R^2}{\overset{\text{O}}{\diagup}}$$

$$\big|\big|\big|$$

$$R^1{\cdot}COCl + Cd(CH_2{\cdot}R^2)_2$$
$$\text{or } (R^2{\cdot}CH_2)_2CuLi$$

$$R^1{\cdot}CO{\cdot}NMe_2 + R^2 CH_2 Li$$

(TM)

$$\big|\big| \text{(ii)} (5)$$

$$2(R{\cdot}CO_2H)$$
$$R = R^1 = R^2{\cdot}CH_2$$

$$\underset{R^1 \quad \overset{\ominus}{} \diagdown R^2}{\overset{\text{O}}{\diagup}}$$

$$\underset{R^1 \quad \underset{\overset{\text{O}}{\diagup}\ OH(Et)}{\overset{(iii)}{\diagup} R^2}}{\overset{\text{O}}{\diagup}} \qquad \underset{R^1 \quad \underset{\overset{\text{O}}{\diagup}\ Me}{\overset{(iii)}{\diagup} R^2}}{\overset{\text{O}}{\diagup}}$$

$$\big|\big| \text{(iii)} \qquad\qquad \big|\big| \text{(iii)}\ R^1 = Me$$

$$\underset{R^1 \quad \underset{CO_2Et}{\overset{\ominus}{\diagup}} \overset{\oplus}{R^2}}{\overset{\text{O}}{\diagup}} \qquad \underset{Me \quad \underset{CO{\cdot}Me}{\overset{\ominus}{\diagup}} \overset{\oplus}{R^2}}{\overset{\text{O}}{\diagup}}$$

$$\big|\big|\big| \qquad\qquad\qquad \big|\big|\big|$$

$$R^1{\cdot}CO{\cdot}CH_2{\cdot}CO_2Et + R^2X \qquad Me{\cdot}CO{\cdot}CH_2{\cdot}CO{\cdot}Me + R^2X$$

Rearrangement (method 6)

$$\underset{\underset{R^1 \quad R^2}{R^1} \quad R^2}{\overset{\text{O}}{\diagup}} \Longrightarrow \underset{\underset{R^1 \quad R^1}{R^2} \diagdown R^2}{\overset{\text{HO} \quad O{-}H}{\diagup}}$$

(TM)

606

SPECTROSCOPIC FEATURES

The presence of a carbonyl group in a compound is readily recognised by its strong characteristic *i.r.* absorption at *c.* 1700 cm^{-1} (Fig. 3.28, 4-methylpentan-2-one). Ketones may be definitively distinguished from other carbonyl containing compounds [RCHO, RCO_2H, $(RCO)_2O$, $R \cdot CO_2R$, $R \cdot CONH_2$] by the absence of absorption arising from groups associated with the carbonyl carbon (p. 296). The *p.m.r.* spectrum is of great value for the assignment of structure to the alkyl or aralkyl groups since the carbonyl group exerts a deshielding effect on the α-hydrogens, and to a lesser extent the β-hydrogens, resulting in differences in chemical shift values. Often the multiplicity of the signals of these groups of protons may be interpreted by first order analysis (p. 341). The fragmentation observed in the *m.s.* is discussed in detail on p. 377 (Fig. 3.84, 4-methylpentan-2-one). Further examples illustrating all these spectroscopic characteristics are descriptively analysed in some of the preparative examples below. The *u.v.–visible* absorption spectrum of saturated aliphatic and alicyclic ketones is of little structural value (CO, λ_{max} *c.* 240 nm, ε 50). However with chiral acyclic and cyclic ketones the optical rotatory dispersion curves are often of great value in the assignment of configurational and conformational features (p. 248).

5.8.1 THE OXIDATION OF SECONDARY ALCOHOLS

The oxidation of secondary alcohols with sodium dichromate in dilute sulphuric acid gives acceptable yields of ketones since these do not normally undergo extensive further oxidation under the reaction conditions (cf. Section 5.7.1, p. 587, the oxidation of primary alcohols to aldehydes).

$$3R_2CHOH + Cr_2O_7^{2\ominus} + 3SO_4^{2\ominus} \longrightarrow 3R_2CO + Cr_2(SO_4)_3 + 7H_2O$$

An excellent method for the conversion of ether-soluble secondary alcohols to the corresponding ketones is by chromic acid oxidation in a two-phase ether–water system. The reaction is carried out at 25–30 °C with the stoichiometric quantity of chromic acid calculated on the basis of the above equation, and is exemplified by the preparation of octan-2-one and cyclohexanone (Expt 5.86). The success of this procedure is evidently due to the rapid formation of the chromate ester of the alcohol, which is then extracted into the aqueous phase, followed by formation of the ketone which is then extracted back into the ether phase and is thus protected from undesirable side reactions.

A slightly modified procedure – oxidation with 100 per cent excess of chromic acid at 0 °C for a short period – is adopted for strained bicyclic alcohols (e.g. the oxidation of (−)-borneol to (−)-camphor, Expt 5.87) and gives excellent yields of the corresponding ketones. Cycle ketones which are susceptible to acid-catalysed epimerisation are moreover obtained by this procedure in a high degree of epimeric purity.

The use of pyridinium-based chromium(VI) oxidising agents (cf. Section 5.7.1, p. 587) is illustrated by the use of the supported reagent, pyridinium chlorochromate-alumina, for the conversion of the cyclic allylic alcohol, carveol, into the corresponding α, β-unsaturated ketone, carvone[125] (Expt 5.88).

The conversion of an alkene into the corresponding ketone may be effected by means of a convenient sequence which involves hydroboration followed by oxidation with chromic acid of the resulting organoborane (cf. Section 5.4.3,

p. 542); isomeric purity is dependent upon the regiospecificity of the hydroboration step. The sequence is illustrated by the conversion of 1-methyl-
cyclohexene into 2-methylcyclohexanone (Expt 5.89).

Secondary alcohols may be oxidised to the corresponding ketones by the use
of an aluminium alkoxide, frequently the t-butoxide, in the presence of a large
excess of acetone (the *Oppenauer oxidation*). The reaction involves an initial
alkoxy-exchange process followed by a hydride ion transfer from the so-formed
aluminium alkoxide of the secondary alcohol by a mechanism analogous to that
of the Meerwein–Ponndorf–Verley reduction (see Section 5.4.1, p. 520).

$$3R_2CHOH + [Me_3CO]_3Al \rightleftharpoons [R_2CHO]_3Al + 3Me_3COH$$
$$[R_2CHO]_3Al + 3Me_2CO \rightleftharpoons 3R_2CO + [Me_2CHO]_3Al$$

Acetone in conjunction with benzene as a solvent is widely employed. Alternatively cyclohexanone as the hydrogen acceptor, coupled with toluene or xylene
as solvent, permits the use of higher reaction temperatures and consequently the
reaction time is considerably reduced; the excess of cyclohexanone can be easily
separated from the reaction product by steam distillation. Usually at least
0.25 mol of aluminium alkoxide per mol of secondary alcohol is employed.
However, since an excess of alkoxide has no detrimental effect, the use of 1 to
3 mol of alkoxide is desirable, particularly as water, either present in the reagents
or formed during secondary reactions, will remove an equivalent quantity of the
reagent. It is recommended that 50 to 200 mol of acetone or 10 to 20 mol of
cyclohexanone be employed. Other oxidisable groups are usually unaffected in
the Oppenauer oxidation and the reaction has found wide application in the
steroid field.

The reaction is illustrated by the oxidation of cholesterol to cholest-4-en-3-
one (Expt 5.90); the migration of the double bond from the β,γ- to the α,β-
position is a commonly occurring side reaction associated with unsaturated steroids of this structural type.

The oxidation of primary or secondary alcohols to aldehydes or ketones
respectively with dimethyl sulphoxide activated by oxalyl chloride has wide
applicability (Swern oxidation).[243b] The initial reaction between the acid chloride and dimethyl sulphoxide in dichloromethane solvent is vigorous and exothermic at $-60\,°C$ and results in the formation of the complex (7); this complex
spontaneously decomposes, even at this low temperature, releasing carbon
dioxide and carbon monoxide to form the complex (8). The alcohol is added
within 5 minutes, followed after 15 minutes by triethylamine. After a further 5
minutes at low temperature the reaction mixture is allowed to warm to room
temperature and work-up follows standard procedures. The ratio of reactants is
dimethyl sulphoxide : oxalyl chloride : alcohol : triethylamine 2.2:1.1:1.0:5.

Experiment 5.86 OCTAN-2-ONE (*Hexyl methyl ketone*)

$$\text{Me·(CH}_2)_5\text{·CH(OH)·Me} \xrightarrow[\text{(C}_2\text{H}_5)_2\text{O}]{\text{Na}_2\text{Cr}_2\text{O}_7/\text{H}_2\text{O}/\text{H}_2\text{SO}_4} \text{Me·(CH}_2)_5\text{·CO·Me}$$

Equip a 500-ml three-necked flask with a mechanical stirrer, a thermometer and a two-way adapter carrying a dropping funnel and condenser; arrange for the occasional cooling of the flask in an ice-water bath. Place a solution of 32.5 g (0.25 mol) of octan-2-ol in 100 ml of ether in the flask and 125 ml (0.083 mol) of chromic acid solution (1) in the dropping funnel. Add the chromic acid solution dropwise during 15 minutes to the vigorously stirred ether solution of the ketone, keeping the temperature between 25 and 30 °C by cooling as necessary. Continue stirring at room temperature for a further 2 hours and then transfer the reaction mixture to a separating funnel. Separate the ether layer and extract the dark green aqueous layer with four 60 ml portions of ether. Combine the ether extracts and wash with 40 ml of saturated sodium hydrogen carbonate solution and then with 40 ml of saturated sodium chloride solution; finally dry the ether extract over anhydrous sodium sulphate. Filter, remove the ether on a rotary evaporator and distil the residue at atmospheric pressure. Collect the octan-2-one (2) at 170–172 °C. The yield is 26 g (81%). The purity may be checked by g.l.c. using a 1.5-m, 10 per cent Silicone oil column at 110 °C and with a nitrogen flow rate of 40 ml/minute. The ketone has a retention time of 3.1 minutes.

Notes. (1) The chromic acid solution may be prepared as follows. Dissolve 100 g (0.33 mol) of sodium dichromate dihydrate in 300 ml of water and slowly add 134 g (73 ml, 1.34 mol) of concentrated sulphuric acid (98%, d 1.84). Cool the solution and dilute to 500 ml with water in a graduated flask.
(2) The ketone has a cheese-like, rather pervading, odour.

Experiment 5.87 (−)-CAMPHOR

Assemble the apparatus described in Expt 5.86. Place a solution of 7.7 g (0.05 mol) of (−)-borneol in 25 ml of ether in the flask and cool to 0 °C. Precool 50 ml (0.033 mol) of chromic acid solution (1) to 0 °C before placing 25 ml in the dropping funnel. Add the cold chromic acid solution to the vigorously stirred borneol solution during 5 minutes, keeping the remainder cold in an ice-water bath. Add the remaining chromic acid over a further 5 minutes. Stir the mixture with cooling in the ice-water bath for a further 5 minutes and then transfer to a separating funnel. Carefully remove the ether layer (2) and extract the aqueous solution with two 25 ml portions of ether. Combine the ether extracts, and wash with 30 ml of 5 per cent sodium carbonate solution followed by four 25 ml portions of water. Dry the ether solution over anhydrous sodium sulphate, filter and remove the ether on the rotary evaporator. The yield of crude (−)-camphor, m.p. 159–164 °C (sealed tube), is 6.3 g (83%). It may be purified by sublimation at 80–90 °C/12 mmHg

609

(Section 2.21); after two sublimations 5.76 g (76%) of material is obtained, which melts at 173–175 °C (sealed tube).

Notes. (1) Prepared as described in Expt 5.86, Note (1).
(2) Both the ether and aqueous solutions are almost black and it is difficult to see the interface; it is best judged by noting the change in flow of the solutions through the tap.

Experiment 5.88 CARVONE (*p-Mentha-6,8-dien-2-one*)[125]

The pyridinium chlorochromate-on-alumina reagent (7.5 g, 6.1 mmol, Section 4.2.*18*, p. 426) is added to a flask containing a solution of carveol (0.60 g, 3.8 mmol) in hexane (10 ml). After stirring for 2 hours, the solution is filtered, and washed with three 10-ml portions of ether. The combined filtrates are evaporated and vacuum distilled to afford carvone; yield 0.54 g (93%), b.p. 104 °C/11 mmHg. Gas-liquid chromatography analysis is on a column of 10 per cent Carbowax 20M supported on 60–80 mesh Chromosorb W.

Experiment 5.89 2-METHYLCYCLOHEXANONE

CAUTION: *This experiment should be carried out in a fume cupboard.*

Equip a 250-ml three-necked round-bottomed flask with a thermometer, a reflux condenser protected by a calcium chloride guard-tube and (in the central neck) a 100-ml pressure-equalising dropping funnel fitted with an inlet adapter to allow flushing with a stream of dry nitrogen. The apparatus should be thoroughly dried in an oven and assembled under a stream of dry nitrogen. Adjust the nitrogen flow to a slight trickle, which should be maintained throughout the reaction, and in the flask place a magnetic follower bar, 30 ml of dry ether, 4.8 g (0.050 mol) of 1-methylcyclohexene (Expt 5.12) and 0.5 g (0.0225 mol, 20% excess) of lithium borohydride (1). In the dropping funnel place 5 ml of dry ether and 0.95 ml (0.075 mol, 20% excess) of purified boron trifluoride-etherate (Section 4.2.*8*, p. 421). Add the boron trifluoride-etherate solution to the stirred mixture dropwise during 15 minutes, keeping the temperature at 25–30 °C by cooling in a water bath. Stir for 2 hours at room temperature and then destroy excess hydride by adding carefully 5 ml of water. Place in the dropping funnel a solution of chromic acid, prepared from 11.0 g (0.0369 mol) of sodium dichromate dihydrate and 8.1 ml (0.1474 mol) of 98 per cent sulphuric acid made up to 45 ml with water. Add the chromic acid solution portionwise to the reaction mixture during 15 minutes while maintaining the temperature at 25–30 °C by cooling in an ice-water bath. Heat the dark mixture under reflux for 2 hours, cool and separate the upper ether layer. Extract the aqueous acid layer with two 15 ml portions of ether, combine the organic extracts, wash once with 5 ml of saturated sodium chloride solution and dry over magnesium sulphate. Filter, remove the ether from the filtrate by flash distillation and carefully distil the residual straw-coloured liquid. Collect the 2-methylcyclohexanone having b.p. 160–164 °C. The yield is about 3.5 g (63%). The purity may be checked by g.l.c. on a 1.5-m column of 10 per cent Silicone oil on Chromosorb W held at 128 °C, with a nitrogen flow rate of 40 ml/minute; t_R is about 1.7 minutes.

Note. (1) Reaction of lithium borohydride with water may be rapid and violent; do not expose to high humidity and avoid contact with eyes, skin and clothing (contact with cellulosic material may cause combustion). It should be handled with the same caution as is afforded to lithium aluminium hydride (Section 4.2.49, p. 445).

Experiment 5.90   CHOLEST-4-EN-3-ONE

Place a mixture of 20 g (0.052 mol) of pure cholesterol (m.p. 147–150 °C; dried to constant weight at 80–100 °C), 150 ml of dry acetone and 200 ml of sodium-dried benzene (**CAUTION**) in a dry, 1-litre, two-necked round-bottomed flask fitted with a reflux condenser and dropping funnel both pro-tected with calcium chloride tubes. Add a few carborundum chips and heat to boiling in an oil bath at 75–85 °C. Add from the dropping funnel, in one por-tion, a solution of 16 g (0.065 ml) of aluminium t-butoxide (Section 4.2.2, p. 414) in 100 ml of anhydrous benzene. The reaction mixture becomes cloudy and develops a yellow colour in about 10 to 15 minutes. Continue gentle boil-ing at a bath temperature of 75–85 °C for 8 hours. Treat the cold mixture with 40 ml of water and then with 100 ml of 10 per cent sulphuric acid. Shake vigorously and transfer to a 1-litre separating funnel. Dilute the mixture with 300 ml of water, shake for 5 minutes (filter, if necessary), then run off the yellow aqueous layer into a second separating funnel and extract the latter with 25 ml of benzene. Wash the combined benzene extracts thoroughly with water, dry with magnesium sulphate and remove the solvent on a rotary evaporator under reduced pressure. The yellow oily residue solidifies when it is cooled in an ice–salt bath and scratched with a glass rod; keep a small por-tion for seeding in the subsequent crystallisation. Dissolve the solid in a warm mixture of 14 ml of acetone and 20 ml of methanol, allow the solution to cool very slowly and seed, if necessary. When the bulk of the solid has crystallised, keep the mixture at 0 °C for 24 hours, filter with suction, wash with 20 ml of ice-cold methanol and dry in a vacuum desiccator. The yield of almost colourless cholest-4-en-3-one, m.p. 79–80 °C, is 17 g (85%).

5.8.2 THE HYDRATION OF ALKYNES

The direct hydration of a terminal alkyne, with dilute sulphuric acid in the presence of a mercury salt, yields initially an enol which rearranges to the more stable ketone. The regioselectivity of the reaction is consistent with that predicted on the basis of mechanistic theory.

$$R \cdot C \equiv CH \longrightarrow \underset{Hg^{2\oplus}}{R \cdot C = CH} \xrightarrow{\;H_2O^{\oplus}\;} \underset{Hg^{\oplus}}{R \cdot C = CH} \xrightarrow{-H^{\oplus}, +H^{\oplus}}$$

$$R \cdot C(OH) = CH_2 \longrightarrow R \cdot CO \cdot Me$$

The conversion of undec-10-ynoic aid to 10-oxoundecanoic acid (Expt 5.91) is the illustrative example.

With non-terminal alkynes the preparative applications are more limited since a mixture of ketones is usually obtained, the proportions being dependent upon the nature of R^1 and R^2.

$$R^1 \cdot C \equiv C \cdot R^2 \longrightarrow R^1 \cdot CO \cdot CH_2 \cdot R^2 + R^1 \cdot CH_2 \cdot CO \cdot R^2$$

Experiment 5.91 10-OXOUNDECANOIC ACID

$$HC \equiv C \cdot (CH_2)_8 \cdot CO_2H \xrightarrow[H_3O^{\oplus}]{Hg(O \cdot CO \cdot Me)_2} Me \cdot CO \cdot (CH_2)_8 \cdot CO_2H$$

Heat under reflux for 4 hours a solution of 3 g (0.0165 mol) of undec-10-ynoic acid (Expt 5.23) in 240 ml of glacial acetic acid containing 13 ml of concentrated sulphuric acid and 1.4 g of mercury(II) acetate. Dilute the dark coloured solution with 300 ml of water, filter and extract the solution with dichloromethane using a continuous extraction apparatus (Fig. 2.95) (1). Wash the dichloromethane extract carefully with distilled water until the washings are neutral, dry the extract and evaporate on a rotary evaporator. Recrystallise the solid residue from light petroleum (b.p. 60–80 °C) using a little decolourising charcoal. The yield of keto-acid, m.p. 56–57 °C, is 1.65 g (50%).

Note. (1) Batchwise extraction leads to the formation of stable emulsions which frequently take several days to break. When filling the continuous extraction apparatus, care should be taken to avoid forming an emulsion.

5.8.3 THE THERMAL DECARBOXYLATION OF ACIDS OVER A METAL OXIDE CATALYST

A long-established method of formation of symmetrical ketones (albeit in rather low yield) involves the pyrolysis of certain salts (usually calcium or barium) of carboxylic acids.

$$(R \cdot CO_2)_2 Ca \xrightarrow{heat} R \cdot CO \cdot R + CaCO_3$$

The method has been extended to include the synthesis of aldehydes or of unsymmetrical ketones by using appropriate mixtures of carboxylate salts.

$$(R^1 \cdot CO_2)_2 Ca + (R^2 \cdot CO_2)_2 Ca \xrightarrow{heat} 2R^1 \cdot CO \cdot R^2 + 2CaCO_3$$

The unsymmetrical carbonyl compound will clearly be accompanied by both possible symmetrical products, but the yield can be improved by the use of an excess of one of the carboxylate salts.

A more satisfactory process for the preparation of either symmetrical or unsymmetrical ketones involves the passage of the vapour of the acid or mixture of acids over heated manganese(II) oxide or thorium oxide deposited upon pumice. The yields of unsymmetrical ketones are satisfactory (c. 50%) and the technique is particularly well suited for laboratory preparations on a reasonably large scale. Several examples are included in Expts 5.92 and 5.93.

Experiment 5.92 DIETHYL KETONE (*Pentan-3-one*)

$$2\text{Me·CH}_2\text{·CO}_2\text{H} \xrightarrow{\text{catalyst}} \text{Me·CH}_2\text{·CO·CH}_2\text{·Me}$$

Preparation of manganese(II) carbonate–pumice catalyst. Dissolve 70 g (0.35 mol) of manganese(II) chloride tetrahydrate in 100 ml of water and add a solution of 38 g (0.35 mol) of anhydrous sodium carbonate in 120 ml of water with mechanical stirring. Filter the precipitated manganese(II) carbonate and wash well with distilled water. Transfer the solid to a large evaporating basin and add sufficient water to form a thick paste. Add sufficient pumice (4–8 mesh) with stirring so that most of the paste has been transferred to the pumice and heat cautiously (avoid local overheating) on an electric hotplate until the pumice lumps no longer cling together. If too much water is added to form the paste of manganese(II) carbonate initially, it will not adhere to the pumice satisfactorily.

Diethyl ketone. Pack the catalyst into the Pyrex glass combustion tube of the pyrolysis apparatus illustrated in Fig. 2.65(c) and assemble the remaining components. Displace the air in the apparatus with nitrogen, and while maintaining a continued gentle gas flow, heat the pumice for 8 hours at 360–400 °C in order to convert the manganese(II) carbonate into manganese(II) oxide. If necessary the catalyst may be allowed to cool in a stream of nitrogen if the preparation needs to be interrupted at this point. Place 740 g (746 ml, 10 mol) of redistilled propanoic acid, b.p. 139–141 °C, in the dropping funnel, and with the furnace at about 350 °C, add the acid to the catalyst dropwise (about 30 drops per minute); the stream of nitrogen is stopped when addition of acid is commenced. The apparatus requires little attention and the addition of acid occupies 48–72 hours. The distillate consists of two layers. Separate the lower aqueous layer, salt out the ketone with solid potassium carbonate and add it to the main ketonic layer. Treat the combined ketone fractions with small quantities of anhydrous potassium carbonate (1) until effervescence ceases (this both removes the excess of acid and dries the ketone), filter and distil through a short fractionating column. Collect the diethyl ketone at 101–103 °C. The yield is 252 g (29%). An improved yield may be obtained by recirculating the distillate over the catalyst, but in practice this is rarely worth while. It must be remembered that on each occasion that the catalyst is allowed to cool a slow stream of nitrogen must be passed through the apparatus to prevent the oxidation of the manganese(II) oxide catalyst. Assign the p.m.r. absorptions which occur at δ 1.00 (t, 3H) and 2.38 (q, 2H).

Note. (1) An alternative method of working up the distillate, which has its advantages when dealing with volatile ketones or when it is suspected that conversion into the ketone is incomplete, is to treat the combined fractions of ketones with sodium hydroxide pellets until the mixture is alkaline. Should solids separate, these may be dissolved by the addition of a little water. The ketone is then separated, dried over anhydrous potassium carbonate and fractionated.

Cognate preparations. *Heptan-4-one* (*dipropyl ketone*). Use 880 g (920 ml, 10 mol) of butanoic acid, b.p. 162–164 °C. The yield of ketone, b.p. 142–143 °C, is 285 g* (46%).

Pentan-2-one (*methyl propyl ketone*). Use 360 g (6 mol) of glacial acetic acid and 176 g (184 ml, 2 mol) of butanoic acid. The yield of methyl propyl ketone, b.p. 102–104 °C, is 75 g (43%); 75 g of acetone, b.p. 56–57 °C,† are also obtained. Assign the ^{13}C-n.m.r. absorptions which occur at δ 13.7, 17.5, 29.6, 45.6 and 208.3.

Hexan-3-one (*ethyl propyl ketone*). Use 296 g (298 ml, 4 mol) of propanoic acid and 352 g (368 ml, 4 mol) of butanoic acid. The yield is 214 g (53%) of ethyl propyl ketone, b.p. 122–124 °C; the by-products are 98 g of diethyl ketone, b.p. 100–102 °C, and 66 g of dipropyl ketone, b.p. 144–146 °C.

Undecan-6-one (*dipentyl ketone*). Use 400 g (428 ml, 3.45 mol) of hexanoic acid, b.p. 204–206 °C. The yield of ketone, b.p. 222–226 °C, is 225 g (76%).

Experiment 5.93 BENZYL METHYL KETONE

$$\text{Ph·CH}_2\text{·CO}_2\text{H} + \text{Me·CO}_2\text{H} \xrightarrow{\text{catalyst}} \text{Ph·CH}_2\text{CO·Me}$$

Preparation of thorium carbonate–pumice catalyst. Dissolve 294 g (0.5 mol) of thorium nitrate hexahydrate in the minimum of water (*c.* 450 ml) and add slowly a solution of 106 g (1 mol) of anhydrous sodium carbonate in 400 ml of water with stirring. Allow the thorium carbonate to settle, decant as much as possible of the mother-liquor and wash the sediment once by decantation with 500 ml of water. Make the resulting moist solid into a thick paste with distilled water and stir in pumice (4–8 mesh) until most of the suspension appears to be absorbed. Dry the impregnated pumice in quantities of 200 g by heating in a large evaporating dish upon an electric hotplate and stirring constantly with a glass rod. Stop the heating when the pumice particles no longer cling together. Sieve the resulting pumice, 250 g of a white powder (consisting largely of thorium carbonate but containing some oxide) are recovered and can be used for impregnating more pumice. The total weight of pumice catalyst thus prepared is about 1400 g; the exact weight will depend upon the grade of pumice used.

Benzyl methyl ketone. Fill the Pyrex combustion tube with catalyst and proceed as in Expt 5.92 but use a temperature of 400–450 °C for the conversion of thorium carbonate into the corresponding oxide; 6–12 hours are usually

---

* All the yields given refer to one circulation of the acid (or acids) over the catalyst, but can be improved by recirculating the product, from which the water layer has been removed, over the catalyst. With the higher ketones, the second circulation may result in carbonisation of the catalyst. thus rendering it inefficient.

† The symmetrical ketones, produced as by-products in the preparation of mixed ketones, are separated by distillation through an efficient fractionating column. If acetone is a by-product (as in the preparation of pentan-2-one, methyl propyl ketone), some is lost in the washing process.

required for complete conversion and a slow stream of nitrogen should be maintained through the combustion tube. Place a solution of 170 g (1.25 mol) of pure phenylacetic acid (m.p. 77 °C) in 225 g (3.75 mol) of glacial acetic acid in the funnel, and adjust its rate of flow into the catalyst tube to 1 drop every 2 or 3 seconds. Also pass a slow stream of nitrogen (1 bubble per second) through the apparatus in order to keep the gases in motion; the rate of flow may be estimated by passing the inert gas through a concentrated sulphuric acid wash-bottle or 'bubbler' before it enters the furnace. When all the acid mixture has passed through the catalyst tube, separate the lower aqueous layer of the product and treat the organic layer with 10–20 per cent sodium hydroxide solution until the washings are alkaline to litmus and then twice with water. Extract the aqueous layer twice with 50 ml portions of ether, wash the extracts successively with sodium hydroxide solution (until alkaline) and water, and add the resulting ether solution to the main product. Dry with magnesium sulphate, remove the ether on a rotary evaporator and distil the residue under reduced pressure preferably through a fractionating column. Collect the benzyl methyl ketone at 102–102.5 °C/20 mmHg; the yield is 85 g (51%). The residue in the flask is dibenzyl ketone; it may be purified by transferring to a smaller flask and redistilling (b.p. 200 °C/21 mmHg; m.p. 34–35 °C). The i.r. absorption for the carbonyl group appears at 1700 cm^{-1}. The p.m.r. (CDCl$_3$, TMS), shows signals at δ 2.02 (s, 3H, Me), 3.58 (s, 2H, CH$_2$) and 7.20 (s, 5H, Ph).

Cognate preparations. *Benzyl ethyl ketone.* Use 204 g (1.5 mol) of phenylacetic acid (m.p. 77 °C) and 333 g (335.5 ml, 4.5 mol) of propanoic acid (b.p. 139–141 °C), but omit the extraction with ether when working up the distillate. Distil the dried product from a 500-ml round-bottomed flask through an efficient fractionating column. Collect the diethyl ketone at 99.5–102.5 °C (160 g), and when the temperature rises to 130 °C (b.p. 103–130 °C: 7 g) transfer the residue to a 250-ml flask and distil fractionally under reduced pressure. The benzyl ethyl ketone passes over mainly at 118–123 °C/22 mmHg (105 g, 47%); the residue of high boiling point (34 g) consists largely of dibenzyl ketone. Pure benzyl ethyl ketone may be obtained by redistilling the fraction, b.p. 118–123 °C/22 mmHg, and collecting the fraction of b.p. 113–115 °C/17 mmHg.

Benzyl propyl ketone. Use 204 g (1.5 mol) of pure phenylacetic acid and 396 g (414 ml, 4.5 mol) of butanoic acid (b.p. 161–164 °C). Upon working up as for benzyl ethyl ketone, 180 g of dipropyl ketone, b.p. 140–145 °C (mainly 143–145 °C), 108 g (45%) of crude benzyl propyl ketone, b.p. 240–260 °C, and 49 g of crude dibenzyl ketone (residue in flask) are obtained. Redistil the fraction of b.p. 240–260 °C and collect the benzyl propyl ketone at 243–247 °C (the pure ketone boils at 244 °C).

4-Phenylbutan-2-one (*methyl 2-phenylethyl ketone*). Use 100 g (0.66 mol) of hydrocinnamic acid (3-phenylpropanoic acid) (m.p. 49–50 °C) (Expts 5.132 and 6.138) and 160 g (2.66 mol) of glacial acetic acid. The yield of methyl 2-phenylethyl ketone, b.p. 230–235 °C, is 70 g (71%) (the pure ketone boils at 234 °C).

1-Phenylpentan-3-one (*ethyl 2-phenylethyl ketone*). Use 100 g (0.66 mol) of pure hydrocinnamic acid and 200 g (201.5 ml, 2.7 mol) of pure propanoic acid. Fractionation of the distillate yields 70 g of diethyl ketone (b.p. 100–102 °C), 72 g (67%) of ethyl 2-phenylethyl ketone (b.p. 245–249 °C; the pure ketone boils at 248 °C) and 18 g of crude 1,5-diphenylpentan-3-one (high b.p. residue).

1-Phenylhexan-3-one (*propyl 2-phenylethyl ketone*). Use 100 g (0.66 mol) of pure hydrocinnamic acid and 235 g (245.5 ml, 2.66 mol) of pure butanoic acid. Upon working up as for benzyl ethyl ketone the following yields are obtained; 98 g of dipropyl ketone, b.p. 140–144 °C; 65 g (55%) of propyl 2-phenylethyl ketone, b.p. 139–143 °C/17 mmHg; and 22 g of crude 1,5-diphenylpentan-3-one (high b.p. residue). The required ketone, upon redistillation, boils almost completely at 138–139 °C/16 mmHg.

5.8.4   THE INTERACTION OF ORGANOMETALLIC COMPOUNDS WITH CARBOXYLIC ACID CHLORIDES AND *N,N*-DIMETHYLCARBOXAMIDES

Ketones are the intermediates in the synthesis of tertiary alcohols by the reaction of alkyl esters with organomagnesium or organolithium reagents (Section 5.4.2, p. 532). Since the ketonic carbonyl-carbon is more reactive than the ester carbonyl-carbon, the reaction cannot normally be stopped at the intermediate stage. This problem is solved by using the highly reactive acid chloride with the less reactive cadmium dialkyl, which reacts only sluggishly, or not at all, with the carbonyl group in a ketone. If a Grignard reagent must be used with an acid chloride, an inverse addition procedure is essential.

Cadmium dialkyls may be prepared by the addition of anhydrous cadmium chloride to the corresponding Grignard reagent in boiling ethereal solution. It is generally advantageous to replace the ether solvent by benzene before the addition of the acid halide; a higher reflux temperature is then possible, thus reducing the time required for the reaction. A procedure for the direct reaction of a highly reactive form of cadmium metal with alkyl and aryl halides may be more convenient.[126]

$$2R^1 \cdot MgBr + CdCl_2 \longrightarrow R_2^1 Cd + MgBr_2 + MgCl_2$$
$$R_2^1 Cd + 2ClCO \cdot CH_2 \cdot R^2 \longrightarrow 2R^1 \cdot CO \cdot CH_2 \cdot R^2 + CdCl_2$$

For most purposes the use of 1.0 mol of an alkyl or aryl bromide (for the preparation of the cadmium dialkyl or diaryl through the Grignard reagent) to 0.8 mol of the acid halide is recommended. This results in nearly equivalent molar ratios of the cadmium dialkyl and the acid halide, since the overall yield of the former is usually about 80 per cent. The entire preparation can be carried out in one flask without the isolation of intermediates. Experimental details are given for the preparation of 1-chlorohexan-2-one (Expt 5.94).

Because of the toxicity of cadmium compounds two alternative methods for the preparation of ketones from carboxylic acid derivatives are worthy of attention. The first involves the reaction of organocopper reagents [formed from copper(I) iodide and an alkyllithium] with a carboxylic acid chloride.[127a, b]

$$2R^1 Li + CuI \longrightarrow R_2^1 CuLi + LiI$$
$$R^2 \cdot COCl + R_2^1 CuLi \xrightarrow[-78\,°C]{Et_2O} R^2 \cdot CO \cdot R^1$$

An excess of lithium dialkylcuprate is employed and the presence of other functional groups (iodo, cyano, acyl and carbalkoxy) is tolerated. The illustrative example (cognate preparation, Expt 5.94) is the formation of tetradecane-5,10-dione from 6-oxodecanoyl chloride; in this preparation the description of the balloon technique for the maintenance of an inert atmosphere is instructive.

A second cognate preparation in Expt 5.94 describes a general procedure for the conversion of N,N-dimethylcarboxamides into ketones by reaction with primary alkyllithiums.[127c] As the reaction is not successful with secondary alkyllithiums, a branched chain ketone such as 4-methylheptan-3-one is prepared from ethyllithium and N,N,2-trimethylpentanamide, and not from 2-pentyllithium and N,N-dimethylpropanamide.

Experiment 5.94   1-CHLOROHEXAN-2-ONE

$$2BuBr \xrightarrow{Mg} 2BuMgBr \xrightarrow{CdCl_2} Bu_2Cd \xrightarrow{2CH_2Cl \cdot COCl} Bu \cdot CO \cdot CH_2Cl$$

CAUTION: *All operations in this preparation must be carried out in an efficient fume cupboard owing to the toxicity of benzene.*

Equip a 1-litre three-necked flask with a sealed Hershberg stirrer (preferably of tantalum wire) (see Fig. 2.49), a reflux condenser and a 250-ml dropping funnel (1). All apparatus must be thoroughly dry. Place 8.1 g (0.33 mol) of dry magnesium turnings in the flask, add 60 ml of anhydrous ether through the dropping funnel and charge the latter with a solution of 46 g (35.5 ml, 0.33 mol) of butyl bromide in 110 ml of dry ether. Start the stirrer and prepare the Grignard reagent in the usual manner (compare Expt 5.39). When the formation of the Grignard reagent is complete, cool the flask in an ice bath with stirring, remove the dropping funnel and, when cold, add 32.7 g (0.178 mol) of anhydrous cadmium chloride (Section 4.2.12, p. 423) in portions from a small conical flask during 5–10 minutes. Replace the dropping funnel, remove the ice bath, stir for 5 minutes and then heat the mixture under reflux with stirring for 45 minutes; at this point a test for the presence of Grignard reagent is made (2); continue stirring and refluxing until the test is negative. Replace the reflux condenser by a bend adapter connected to a condenser set for distillation, distil off the ether as stirring is continued; continue the distillation, with stirring, on a water bath until it becomes very slow and dark viscous residue remains. At this point add 120 ml of anhydrous benzene (**CAUTION**) from the dropping funnel, and continue the distillation until a further 35 ml of liquid has passed over. Then add 120 ml of dry benzene and replace the reflux condenser; reflux the mixture with vigorous stirring in order to break up the cake inside the flask. Remove the heating bath, cool the mixture to about 5 °C in an ice bath and add a solution of 38 g (25.5 ml, 0.33 mol) of chloroacetyl chloride (b.p. 105 °C) in 70 ml of anhydrous pure benzene from the dropping funnel during 2–3 minutes. After completion of the addition, stir the reaction mixture and hold the temperature at 15–20 °C for 3 hours and then at 20–25 °C for a further 1.5 hours. Add excess of crushed ice (c. 200 g) and dilute sulphuric acid. Separate the benzene and aqueous layers; extract the aqueous phase with two 30 ml portions of benzene. Wash the combined benzene layers successively with 70 ml of water, 70 ml of saturated sodium hydrogen carbonate solution, 70 ml of water and 35 ml of saturated sodium chloride solution. Filter the benzene solution through a little anhydrous sodium sulphate (this separates most of the suspended water), remove the benzene by flash distillation at atmospheric pressure and distil the residue under reduced pressure through a short fractionating column. Collect the 1-chlorohexane-2-one at 71–72 °C/15 mmHg; the yield is 24 g (54%).

Notes. (1) It is best to conduct the preparation in a nitrogen atmosphere; the apparatus shown in Fig 2.60 may be used.
(2) See Expt 5.8, Note (3).

Cognate preparations. *Tetradecane-5,10-dione* (*via lithium dialkylcuprates*).[127a] Into a dry 50-ml, two-necked round-bottomed flask, equipped with a rubber septum and a three-way stopcock bearing a wired-on balloon, is placed a magnetic stirrer bar and copper(I) iodide (571 mg, 3.00 mmol). The balloon is filled with nitrogen. The three-way stopcock is used to evacuate and then fill the flask with nitrogen. The purging procedure is repeated two more times, and the flask is gently flamed during the last evacuation. Anhydrous diethyl ether (8 ml) is added and the system is cooled to −40 °C (1). n-Butyllithium in pentane (4.5 ml of a 1.32 M solution, 6.0 mmol) is injected. After about 5 minutes at −40 °C, the temperature is lowered to −78 °C. A precooled ethereal solution (1 ml) of 6-oxodecanoyl chloride (213 mg, 1.04 mmol) (2) is injected. After 15 minutes at −78 °C, absolute methanol (352 mg, 11.0 mmol) is injected and the reaction mixture is allowed to reach room temperature. It is poured with stirring into an equal volume of saturated aqueous ammonium chloride; ether extraction followed by rotary evaporation gives 193 mg (83%) of tetradecane-5,10-dione, m.p. 59–62 °C, with consistent i.r. and n.m.r. spectroscopic properties. Recrystallisation from pentane gives white needles (162 mg, 70%), m.p. 65–66 °C.

Notes. (1) A temperature of 0 °C is used in the case of reactions involving methyllithium, i.e. for the preparation of alkyl methyl ketones.
(2) This acid chloride is prepared from oxalyl chloride and 6-oxodecanoic acid.[127d] The reaction procedure is applicable to a range of unsubstituted and substituted aliphatic and aromatic acid chlorides.

General procedure for the preparation of ketones from *N,N*-dimethylcarboxamides and alkyllithiums.[127c] A suspension of lithium ribbon (1.6 g-atom), cut into small pieces, in anhydrous ether (800 ml) is prepared in a flask fitted with a reflux condenser, nitrogen inlet tube, thermometer, a pressure-equalising addition funnel and a stirrer. The suspension is cooled to −10 °C, and, while the system is being swept with nitrogen, a solution of the required alkyl bromide (0.83 mol) in ether (200 ml) is added over a period of 2 hours. The mixture is stirred for 1 hour longer at −10 °C. Then the temperature is lowered to −20 °C and a solution of the required *N,N*-dimethylcarboxamide in anhydrous ether (200 ml) is added dropwise over a period of 1.5 hours. The temperature is then allowed to rise gradually to 25 °C over a period of 3 hours with continued stirring under a nitrogen atmosphere. At the end of this time, the solution is cooled to -10 °C and cold, saturated ammonium chloride solution (500 ml) is slowly added. The mixture is stirred for 30 minutes, the ether layer is separated, washed with 1 M hydrochloric acid, water and dried. Evaporation of the solvent under reduced pressure leaves a colourless liquid residue which is distilled through a 25-cm Vigreux column.

5.8.5 THE HYDROLYSIS AND DECARBOXYLATION OF β-KETO ESTERS AND THE HYDROLYSIS OF β-DIKETONES

An alternative representation of the disconnection strategy shown on p. 606 for these methods of synthesis of ketones is formulated below.

From this it could be envisaged that an electrophilic synthon from, for example, an alkyl halide could react at the anionic carbon site of the mesomeric species, generated from an alkyl methyl ketone with base. In practice such a procedure gives rise to many competing side reactions, e.g. self-condensation of the alkyl methyl ketone, uncontrolled di- or tri-alkylation, alkylation at the alternative α'-position, etc. α-Alkylation can be achieved by the use of an activating group such as the carbethoxy group ($-CO_2Et$). This group ensures regiospecific proton removal, usually with base under anhydrous conditions, and almost total conversion into the conjugate base. Following alkylation with an alkyl halide, the activating group is removed by the action of dilute alkali in the cold followed by acidification and boiling. An important extension is the further reaction of the alkylated β-keto ester with base, followed by another alkylation step to give after appropriate hydrolysis the ketone, $R^1 \cdot CO \cdot CHR^2R^3$.

Ketones of great structural variety may therefore be prepared by careful selection of the starting β-keto ester (cf. Section 5.14.3, p. 736) and the alkyl halides. The standard procedure is illustrated for the preparation of hexan-2-one from ethyl acetoacetate (Expt 5.95), and it may be suitably adapted to the preparation of most alkyl (straight-chain and branched-chain) methyl ketones.

A general convenient alkyl methyl ketone synthesis, which utilises the β-keto ester system as an intermediate, involves the acylation of a malonate ester by way of the ethoxymagnesium derivative. Hydrolysis and decarboxylation to the ketone is accomplished by heating in acid solution; the synthesis of cyclohexyl methyl ketone is the illustrative example (Expt 5.96).

$$R \cdot COCl + [CH(CO_2Et)_2]MgOEt \longrightarrow R \cdot CO \cdot CH(CO_2Et)_2 \xrightarrow{H_3O^{\oplus}} R \cdot CO \cdot Me$$

A related synthesis of alkyl methyl ketones involves the preparation and alkaline cleavage of a 3-alkylpentane-2,4-dione which can be readily achieved in

619

one step by refluxing a mixture of pentane-2,4-dione and the alkyl halide in alcoholic potassium carbonate. This method is illustrated by the preparation of 5-methylhex-5-en-one (Expt 5.97).

Experiment 5.95 HEXAN-2-ONE (*Butyl methyl ketone*)

$$
\underset{\underset{CO_2Et}{\overset{O}{\|}}}{Me}
\xrightarrow{\ominus OEt}
\underset{\underset{CO_2Et}{\overset{O}{\|}}}{Me}\!\!{}^{\ominus}
\xrightarrow{PrBr}
\underset{\underset{CO_2Et}{\overset{O}{\|}}}{Me}\!\!{\diagup}^{Pr}
\xrightarrow{\ominus OH}
$$

$$
\underset{\underset{CO_2^{\ominus}}{\overset{O}{\|}}}{Me}\!\!{\diagup}^{Pr}
\xrightarrow{H_3O^{\oplus}}
\underset{\overset{O}{\|}}{Me}\!\!{\diagup}^{Pr} + CO_2
$$

Fit a 2-litre three-necked flask with an efficient double surface condenser and a separatory funnel; close the central neck with a stopper. The apparatus must be perfectly dry. Place 34.5 g (1.5 mol) of clean sodium (Section 4.2.68, p. 462) cut into small pieces in the flask and clamp the flask by the wide central neck. Measure out 1 litre of super-dry ethanol (Section 4.1.9, p. 401), and place about 500 ml in the separatory funnel; insert calcium chloride guard-tubes at the top of the condenser and the separatory funnel respectively. Place a large bowl beneath the flask and have a large wet towel in readiness to control the vigour of the subsequent reaction. Run in about 200 ml of the absolute ethanol on to the sodium (1); a vigorous reaction takes place. If the ethanol refluxes violently in the condenser, cool the flask by wrapping it in the wet towel and also, if necessary, run a stream of cold water over it. As soon as the reaction moderates somewhat, introduce more alcohol to maintain rapid, but controllable, refluxing. In this manner most of the sodium reacts rapidly and the time required to produce the solution of sodium ethoxide is considerably reduced. Finally add the remainder of the ethanol and reflux the mixture on a water bath until the sodium has reacted completely. Remove the stopper in the central neck and introduce a sealed mechanical stirrer. Add 195 g (190 ml, 1.5 mol) of pure ethyl acetoacetate, stir the solution and heat to gentle boiling, then run in 205 g (151 ml, 1.66 mol) of propyl bromide over a period of about 60 minutes. Continue the refluxing and stirring until a sample of the solution is neutral to moist litmus paper (6–10 hours); the reaction is then complete.

Cool the mixture and decant the solution from the sodium bromide; wash the salt with two 20 ml portions of absolute ethanol and add the washings to the main solution. Distil off the ethanol, which contains a slight excess of propyl bromide, through a short fractionating column from a water bath. The residue (*A*) of crude ethyl propylacetoacetate may be used directly in the preparation of hexan-2-one. If the fairly pure ester is required, distil the crude product under diminished pressure and collect the fraction boiling at 109–113 °C/27 mmHg (183 g, 71%) (*B*).

To prepare hexan-2-one add the crude ester (*A*) or the redistilled ethyl propylacetoacetate (*B*) to 1500 ml of a 5 per cent solution of sodium hydroxide contained in a 4-litre flask equipped with a mechanical stirrer. Continue the stirring at room temperature for 4 hours; by this time the mono-substituted acetoacetic ester is completely hydrolysed and passes into solution. Transfer the mixture to a large separatory funnel, allow to stand and remove the small quantity of unsaponified material which separates as an upper oily layer. Place the aqueous solution of sodium propylacetoacetate in a 3-litre two-necked flask fitted with a small separatory funnel and a wide bent delivery tube connected to a condenser set for downward distillation. Add 150 ml of 50 per cent by weight sulphuric acid (d. 1.40) slowly through the separatory funnel with shaking; a vigorous evolution of carbon dioxide occurs. When the latter has subsided, heat the reaction mixture slowly to the boiling point and distil slowly until the total volume is reduced by about one-half; by this time all the hexan-2-one should have passed over. The distillate contains the ketone, ethanol and small quantities of acetic and valeric acids. Add small portions of solid sodium hydroxide to the distillate until it is alkaline and redistil the solution until 80–90 per cent has been collected; discard the residue.

Separate the ketone layer from the water, and redistil the latter until about one-third of the material has passed over. Remove the ketone after salting out any dissolved ketone with potassium carbonate (2). Wash the combined ketone fractions four times with one-third the volume of 35–40 per cent calcium chloride solution in order to remove the alcohol. Dry over 15 g of anhydrous calcium chloride; it is best to shake in a separatory funnel with 1–2 g of the anhydrous calcium chloride, remove the saturated solution of calcium chloride as formed, and then allow to stand over 10 g of calcium chloride in a dry flask. Filter and distil. Collect the hexan-2-one at 126–128 °C. The yield is 71 g (67%).

Notes. (1) The addition of the ethanol to the sodium, although attended by a very vigorous reaction which must be carefully controlled, is preferable to the reverse procedure of adding the sodium in small pieces to the ethanol. The latter method is longer and has the further disadvantage that it necessitates frequent handling and exposure to the air of small pieces of sodium.
(2) A more complete recovery of the ketone from the aqueous solution may be obtained by repeated distillation of the aqueous layer until no appreciable amount of ketone is found in the distillate. The procedure outlined is, however, quite satisfactory.

Cognate preparation. *Heptan-2-one.* Use 34.5 g (1.5 mol) of sodium, 1 litre of super-dry absolute ethanol, 195 g (1.5 mol) of redistilled ethyl acetoacetate and 225 g (177 ml, 1.63 mol) of dry butyl bromide (Expt 5.54). This yields 280 g of crude or 200 g (72%) of pure ethyl butylacetoacetate, b.p. 112–116 °C/16 mmHg. Upon hydrolysis 105 g (80%) of heptan-2-one, b.p. 149–151 °C, are isolated.

Experiment 5.96 CYCLOHEXYL METHYL KETONE

$$C_6H_{11}\cdot COCl + [CH(CO_2Et)_2]MgOEt \longrightarrow$$
$$C_6H_{11}\cdot CO\cdot CH(CO_2Et)_2 \xrightarrow[-CO_2,\ -2EtOH]{H_3O^{\oplus}} C_6H_{11}\cdot CO\cdot Me$$

Place 10.7 g (0.44 mol) of magnesium turnings in a 1-litre three-necked round-bottomed flask, equipped with a sealed stirrer unit, a dropping funnel and a double surface reflux condenser each protected with a calcium chloride guard-tube. Add in one portion a mixture of 10 ml of absolute ethanol and 1 ml of carbon tetrachloride. Allow the reaction, which commences almost

621

immediately, to proceed for about 5 minutes and then add carefully 150 ml of sodium-dried ether (Section 4.1.*15*, p. 404). Site the flask in a warm-water bath and allow the reaction mixture to reflux gently while a solution of 70 g (0.44 mol) of diethyl malonate in 50 ml of dry ether is added with stirring. On completion of the addition, heat the mixture under reflux for about 3 hours or until all the magnesium has reacted. Then add with vigorous stirring a solution of 58 g (0.4 mol) of cyclohexanecarbonyl chloride (Expt 5.138) in 50 ml of dry ether. Heat the reaction mixture under reflux for 2 hours and then cool and acidify with 50 ml of dilute sulphuric acid. Separate the ether layer and extract the residual aqueous solution with two 50 ml portions of ether. Wash the combined ether extracts with water and evaporate the solvent on a rotary evaporator. To the residue add a solution of 120 ml of glacial acetic acid, 15 ml of concentrated sulphuric acid and 80 ml of water and heat under reflux for 5 hours. Cool the reaction mixture, basify by the careful addition of 100 ml of 20 per cent sodium hydroxide solution and extract the solution with four 50 ml portions of ether. Dry the combined ether extracts over sodium sulphate and remove the ether on a rotary evaporator. Distil the crude product at atmospheric pressure through a short fractionating column. The yield of cyclohexyl methyl ketone of b.p. 178–180 °C is 35 g (70%).

Experiment 5.97 5-METHYLHEX-5-EN-2-ONE

Equip a 1-litre two-necked round-bottomed flask with a sealed stirrer unit and a reflux condenser protected with a guard-tube containing anhydrous calcium sulphate. Place in the flask 500 ml of anhydrous ethanol, 75 g (0.75 mol) of freshly distilled pentane-2,4-dione (b.p. 136–137 °C) (Expt 5.102), 63.4 g (0.70 mol) of 3-chloro-2-methylpropene (methallyl chloride) and 96.8 g (0.70 mol) of anhydrous potassium carbonate. Heat the stirred mixture under gentle reflux for 16 hours. Allow the mixture to cool a little and replace the condenser by a still-head and condenser arranged for downward distillation. Distil the stirred mixture until about 370 ml of ethanol and the ethyl acetate formed during the reaction has collected, then cool the residue and add sufficient ice-water to dissolve the suspended salts (about 550 ml is required). Transfer to a separatory funnel and extract with three 200 ml portions of ether. Wash the combined extracts with two 100 ml portions of saturated aqueous sodium chloride and then dry the ethereal solution over anhydrous sodium sulphate. Filter, and remove the ether by flash distillation. Fractionally distil the residue using a well-lagged fractionating column of about 12 cm length filled with glass helices. Collect the unsaturated ketone as a fraction of b.p. 148–153 °C; (mainly 148–150 °C) (1). The yield of 5-

methylhex-5-en-2-one is 33.1 g (39%); its purity may be checked by g.l.c. on a 10 per cent Silicone oil on Chromosorb W 1.5-m column, held at 82 °C, nitrogen flow rate 40 ml/minute, t_R 1 minute.

Note. (1) The forerun consists of residual ethanol and ethyl acetate together with some of the unsaturated ketone. A substantial high boiling residue remains.

5.8.6 THE ACID-CATALYSED REARRANGEMENT OF 1,2-DIOLS

The conversion of pinacol (Expt 5.35) to t-butyl methyl ketone (pinacolone, Expt 5.98) under acid conditions exemplifies a general reaction of 1,2-diols (the *pinacol–pinacolone rearrangement*). The mechanism, formulated below, involves loss of water from the protonated 1,2-diol accompanied by a 1,2-nucleophilic shift of a methyl group.

Experiment 5.98   t-BUTYL METHYL KETONE (*Pinacolone*)

$$\text{Me}_2\text{C(OH)}\cdot\text{C(OH)Me}_2 \xrightarrow{\text{H}^\oplus} \text{Me}\cdot\text{CO}\cdot\text{CMe}_3$$

In a 500-ml round-bottomed flask carrying a dropping funnel and a connection to a condenser set for distillation, place 50 g of pinacol hydrate (Expt 5.35) and 130 ml of 3 M sulphuric acid. Distil the mixture until the upper layer of the distillate no longer increases in volume (15–20 minutes). Separate the pinacolone layer from the water and return the latter to the reaction flask. Then add 12 ml of concentrated sulphuric acid to the water, followed by a second 50 g portion of pinacol hydrate. Repeat the distillation. Repeat the process twice more until 200 g of pinacol hydrate have been used.

Dry the combined pinacolone fractions over magnesium sulphate and distil. Collect the pinacolone at 103–107 °C. The yield is 62 g (70%).

5.8.7 THE OXIDATIVE HYDROLYSIS OF NITRONATE SALTS FROM SECONDARY NITROALKANES

Oxidation with potassium permanganate of the nitronate salts to yield carbonyl compounds has been discussed in Section 5.7.7, p. 599; the method is very suitable for the preparation of ketones from secondary nitroalkanes and the experimental details may be readily adapted from Expt 5.84.

5.8.8. SOME METHODS FOR THE PROTECTION OF THE CARBONYL GROUP

A range of methods has been developed for the protection of the carbonyl group in multifunctional aliphatic and alicyclic aldehydes and ketones. This has been necessary because in many multistage syntheses, modification of other functionalities (e.g. oxidation, reduction, hydrolysis, nucleophilic and electrophilic additions and displacements, etc.) requires a differing range of experimental conditions, and that protective group must be selected which is stable in the presence of the reaction medium. A further feature that should be noted is that

623

owing to differing reactivity of a carbonyl group in aliphatic (and alicyclic) aldehydes and ketones, and in their α, β-unsaturated and aromatic analogues, selective protection in, for example, a dicarbonyl compound may be possible and desirable.

The most widely used protective groups for aliphatic and alicyclic aldehydes and ketones are: (a) *cyclic acetals* and (b) *cyclic dithioacetals* and *hemithioacetals*. The use of *semicarbazones, oximes* and *2,4-dinitrophenylhydrazones* (for preparation see Section 9.6.13, p. 1257) is less common but has found application in certain instances where the more vigorous methods required for deprotection are not detrimental.

CYCLIC ACETALS

The reaction of the carbonyl group with ethane-1,2-diol or propane-1,3-diol in the presence of a suitable catalyst gives a *1,3-dioxolane* [(9); $R^1 = R^2 = H$] or a *1,3-dioxane* [(10); $R^1 = R^2 = H$] respectively (cf. Section 5.4.6, p. 553, and Section 5.10.3, p. 652). Aldehydes give 2-alkyl derivatives [(9) and (10), $R^1 = $ alkyl, $R^2 = H$] and ketones give 2,2-dialkyl derivatives [R^1 and $R^2 = $ alkyl].

The catalysts for derivative formation include toluene-*p*-sulphonic acid (with azeotropic removal of water,[128] or the use of a water scavenger such as triethyl orthoformate[129]), boron trifluoride-etherate,[130] or hydrogen chloride.[131] For carbonyl compounds possessing acid-labile functional groups, pyridinium toluene-*p*-sulphonate[132] (PPTS, p. 552), or passage of the mixture of the diol and carbonyl compound down a column of Amberlyst-15 may be more suitable.[133] The procedure given below[128a] has been widely used, and is illustrated by the protection of the carbonyl group in ethyl acetoacetate.

Procedure for 1,3-dioxolane formation with ethyl acetoacetate by azeotropic removal of water.[128a] Ethyl acetoacetate (30 g, 0.23 mol), ethane-1,2-diol (16 g, 0.248 mol), a crystal of toluene-*p*-sulphonic acid and benzene (50 ml) (**CAUTION**) were placed in a round-bottomed flask fitted with a Dean and Stark water separator (Fig. 2.31(a)) and a reflux condenser. The reaction mixture was heated until no more water collected. The product was fractionally distilled under reduced pressure to give the cyclic acetal (35 g, 87%), b.p. 99.5–101 °C/17–18 mmHg.

The 1,3-dioxolane and 1,3-dioxane systems are stable to an extremely wide range of reaction procedures provided that an acidic medium is avoided. *Deprotection* is effected under a wide variety of mild acid conditions such as, for example, aqueous tartaric acid,[134] perchloric acid,[135] or moist silica gel either alone or in the presence of oxalic or sulphuric acids.[136] The mild oxidative cleavage reaction with triphenylmethyl tetrafluoroborate in dichloromethane is

624

sometimes useful.[137] An illustrative procedure involving conversion of the 1,3-dioxane system into a dimethyl acetal (with methanolic hydrogen chloride) followed by hydrolysis with oxalic acid is described in Expt 5.9.

CYCLIC DITHIOACETALS AND HEMITHIOACETALS

The reaction of aldehydes or ketones with ethane-1,2-dithiol or propane-1,3-dithiol to form *1,3-dithiolanes* or *1,3-dithianes* is an important reaction, as these compounds under suitable conditions are acyl anion equivalents (see Section 5.9, p. 626). These cyclic dithioacetals have been less used as protective groups, though when required are formed in high yield in the presence of boron trifluoride-etherate.[138]

1,3-Oxathiolanes (11) are formed from aldehydes and ketones by reaction with 2-mercaptoethanol (HS·CH$_2$·CH$_2$OH) in the presence of, for example, zinc chloride–sodium acetate in dioxane solution at room temperature,[139] or boron trifluoride–etherate.[140] They are more stable to an acidic medium than the 1,3-dithianes, and therefore may be the protective group of choice in certain instances.

(11)

In the illustrative example below cycloheptanone is converted into the corresponding 1,3-oxathiolane, 2-oxa-5-thiaspiro[4.6]undecane.

General procedure for the preparation of oxathiolanes.[140] To a stirred refluxing solution of cycloheptanone (56.1 g, 0.5 mol) and 2-mercaptoethanol (29.1 g, 0.5 mol) in anhydrous ether (400 ml) was added dropwise over a 1-hour period boron trifluoride-etherate (71 g, 0.5 mol). After an additional hour of being heated under reflux, the solution was allowed to cool, washed with 0.1 M sodium hydrogen carbonate solution (2 × 100 ml) and once with saturated sodium chloride solution (100 ml), and dried over magnesium sulphate. After removal of the solvent under vacuum on a rotary evaporator, the residue was distilled under vacuum to yield a small forerun which was followed by 2-oxa-5-thiaspiro[4.6]undecane (79.2 g, 92%), b.p. 77–78 °C/1.2 mmHg, n$_D^{25}$ 1.5165.

Deprotection of 1,3-oxathiolanes may be effected in a buffered medium in the presence of either mercury(II) chloride or with Raney nickel,[141] or very conveniently with Chloramine-T.[142] In this latter deprotection reaction, 1,4-oxathiaspiro[4.4]nonane (12) affords cyclopentanone in 91 per cent yield when treated for 2 minutes with Chloramine-T in 85 per cent methanol–water at 25 °C.

(12)

5.9 DICARBONYL COMPOUNDS

The compounds described in this section are diketones; keto acids and keto esters, which may also be classified as dicarbonyl compounds, are considered in Section 5.14.3, p. 735. The relative location of the two carbonyl groups in the carbon chain may be designated numerically or by letters of the Greek alphabet. The alkyl groups may be the same or different; the formulae below also represent keto aldehydes and dialdehydes when one or both of the residues (R), are hydrogen.

1,2- (or α-)   1,3- (or β-)

1,4- (or γ-)   1,5- (or δ-)

These carbonyl compounds are important starting materials for the synthesis of alicyclic and heterocyclic ring systems (see Chapters 7, and 8 respectively). Their methods of preparation may be categorised by the use of a retrosynthetic analysis which includes both functional group interconversion and disconnection strategies. Table 5.2 is an illustrative summary of the synthons that are generated by disconnection of the various carbon–carbon bonds which connect the two carbonyl groups in the case of 1,2-, 1,3- and 1,4-diketones. Table 5.3 gives an *abbreviated* selection of reagent equivalents of these synthons. Thus a method of preparation of a 1,2-diketone, for example, could be the acylation of a 1,3-dithiane anion [(1) with (2)]; similarly that of a 1,4-diketone by the Michael reaction of a nitronate anion with an α, β-unsaturated ketone [(2) with (5)].

Table 5.2

Diketone	Disconnection (i)				Disconnection (ii)	
1,2-diketone	$\left[\begin{array}{cc} (1) & (2) \end{array} \right]$ or $\left[(1,3\text{-dithiane}) \right]$					
1,3-diketone	$\left[\begin{array}{cc} (1) & (4) \end{array} \right]$ or $\left[\begin{array}{cc} (2) & (3) \end{array} \right]$					
1,4-diketone	$\left[\begin{array}{cc} (1) & (6) \end{array} \right]$ or $\left[\begin{array}{cc} (2) & (5) \end{array} \right]$				$\left[\begin{array}{cc} (4) & (3) \end{array} \right]$	

626

Table 5.3

Synthon	Some reagent equivalents	Cross-references
$R \cdot \overset{\oplus}{C}=O$ (1)	$R \overset{O}{\underset{}{\overset{\|}{C}}} Cl$; $R \overset{O}{\underset{}{\overset{\|}{C}}} O \cdot CO \cdot R$; $R \overset{O}{\underset{}{\overset{\|}{C}}} OEt$	Appendix 6, Table A6.1; p. 519, p. 632
$R \cdot \overset{\ominus}{C}=O$ (2)	$R \!-\!\!\overset{S}{\underset{S}{<}}$; $R \cdot CH_2 \cdot NO_2$	Appendix 6, Table A6.2; p. 21, p. 739
$R \cdot CO \cdot \overset{\oplus}{C}H_2$ (3)	$R \cdot CO \cdot CH_2Br$; $R \cdot C(NO_2)=CH_2$	p. 21, p. 635
$R \cdot CO \cdot \overset{\ominus}{C}H_2$ (4)	$R \cdot CO \cdot \overset{\ominus}{C}H \cdot CO_2Et$; $R \cdot \underset{NR^2}{C}=CH_2$	p. 619, p. 632
$R \cdot CO \cdot CH_2 \cdot \overset{\oplus}{C}H_2$ (5)	$R \cdot CO \cdot CH=CH_2$	p. 635, p. 1094
$R \cdot CO \cdot CH_2 \cdot \overset{\ominus}{C}H_2$ (6)	$R \cdot C(OR^1)_2 \cdot CH_2 \cdot CH_2MgX$†	p. 478

† The carbonyl group requires protection prior to Grignard reagent formation.

Clearly there are a number of possible combinations, particularly when possible strategies for the preparation of unsymmetrical and substituted diketones [e.g. $R^1 \cdot CO \cdot CH_2 \cdot CO \cdot R^2$ and $R^1 \cdot CO \cdot CH(R^2) \cdot CO \cdot R^3$] are considered. A few of these are exemplified in the preparative methods which are discussed below.

5.9.1 1,2-DICARBONYL COMPOUNDS

The three preparative methods which are illustrated in this section are (a) the oxidation, or (b) the nitrosation, of the α-methylene group in a symmetrical ketone or an aryl alkyl ketone (Expts 5.99 and 5.100), and (c) the oxidation of alkynes under PTC conditions (Expt 5.101).

The oxidation of an aryl methyl ketone, with selenium dioxide [*Method (a)*] in a suitable solvent, to an aryl 1,2-ketoaldehyde is illustrated by the preparation

627

of phenylglyoxal from acetophenone (Expt 5.99). The mechanism probably involves the formation of a selenite ester of the enol form of the carbonyl compound.

With alkyl methyl ketones ($R \cdot CH_2 \cdot CO \cdot Me$) the reaction is complicated by the presence of two alternative sites of oxidation; in practice the methyl group appears to be oxidised in preference to the methylene group for reasons which have not been adequately clarified, but in any case the yields are usually poor. Unsubstituted, or symmetrically substituted cyclic ketones possessing of course an α-methylene group, are similarly converted into 1,2-diketones (e.g. the formation of cyclohexane-1,2-dione from cyclohexanone, Expt 5.99, cognate preparation); unsymmetrically substituted cyclic ketones would normally give rise to regioisomers.

   A further interesting case is provided by the synthesis of ninhydrin (Expt 5.99, cognate preparation) from indane-1,3-dione (Expt 7.9) in which the methylene group is activated by two adjacent carbonyl groups. Ninhydrin is the stable monohydrate of the triketone, indane-1,2,3-trione, and is a well-known colorimetric reagent for amino acids (Section 2.31).

   Alkyl methyl ketones undergo nitrosation at the reactive methylene group when treated with nitrous acid or an alkyl nitrite [*Method (b)*]. The presence of hydrogen on the α-carbon permits tautomeric rearrangement to the oxime of a 1,2-dicarbonyl compound. Acidic hydrolysis of the oxime, which is best carried out in the presence of a hydroxylamine acceptor such as laevulinic acid,[143] affords a further useful route to the 1,2-dicarbonyl system.

In the example included here (Expt 5.100) ethyl methyl ketone is nitrosated to butane-2,3-dione monoxime, which is then reacted with hydroxylamine to give the dioxime, dimethylglyoxime, the well-known reagent for nickel.

   A further general route to the 1,2-dicarbonyl system involves the oxidation of α-ketols (acyloins) (cf. the preparation of benzil from benzoin, Expt 6.143). The acyloins may be prepared from carboxylate esters by a radical coupling reaction involving the use of finely divided sodium metal in anhydrous ether, benzene, or toluene.[144]

A modification involves the addition to the reaction mixture of chlorotrimethylsilane which gives the bis-(trimethylsilyloxy)alkene which is hydrolysed to yield the acyloin.[145] A preparative example, which also illustrates an intramolecular acyloin reaction to form an alicyclic ring system, is given in Expt 7.10.

Finally, [*Method* (*c*)], treatment of an internal alkyne with potassium permanganate under PTC conditions leads to a good yield of a 1,2-diketone.[146] The reaction probably proceeds via hydroxylation of the triple bond, tautomeric rearrangement to the α-ketol, and oxidation to the diketone. The reaction is illustrated by the conversion of 1-phenylpent-1-yne to 1-phenylpentane-1,2-dione (Expt 5.101).

Terminal alkynes under these conditions undergo oxidative cleavage to the carboxylic acid, presumably because the intermediate keto aldehyde would yield the unstable α-keto acid.

Experiment 5.99 PHENYLGLYOXAL

$$Ph \cdot CO \cdot Me + SeO_2 \longrightarrow Ph \cdot CO \cdot CHO + Se + H_2O$$

CAUTION: *Selenium and its compounds are toxic (Section 2.3); carry out this and the following preparations in an efficient fume cupboard.*

Fit a 500-ml three-necked flask with a sealed stirrer, a reflux condenser and a thermometer. Place 300 ml of dioxane (1), 55.5 g (0.5 mol) of pure selenium dioxide (Section 4.2.*64*, p. 460) and 10 ml of water in the flask, heat the mixture to 50–55 °C and stir until the solid has dissolved. Remove the thermometer momentarily and add 60 g (0.5 mol) of acetophenone in one lot; replace the thermometer. Reflux the mixture, with stirring, for 4 hours; after about 2 hours the solution becomes clear and little further precipitation of selenium is observable. Decant the hot solution from the precipitated selenium through a fluted filter paper, and remove the dioxane and water by distillation through a short column. Distil the residual phenylglyoxal under reduced pressure and collect the fraction boiling at 95–97 °C/25 mmHg. The yield of pure phenylglyoxal (a yellow liquid) is 48 g (72%); this sets to a stiff gel on standing, probably as a result of polymerisation, but may be recovered without appreciable loss by distillation. The aldehyde is best preserved in the form of the hydrate, which is prepared by dissolving the yellow liquid in 3.5–4 volumes of hot water and allowing to crystallise. Phenylglyoxal hydrate (m.p. 91 °C) also crystallises from chloroform, ethanol or ether–light petroleum (b.p. 60–80 °C); upon distillation under diminished pressure, the free aldehyde is obtained.

Note. (1) Rectified spirit can also be used as solvent. The dioxane can, however, be recovered and used in a subsequent run (cf. Section 4.1.*20*, p. 407).

Cognate preparations. *Cyclohexane-1,2-dione.* Equip a 1-litre, three-necked flask with a reflux condenser, thermometer and dropping funnel. Place 250 g (2.55 mol) of pure cyclohexanone in the flask, heat to 70–80 °C and add a solution of 280 g (2.52 mol) of pure selenium dioxide in 1500 ml of rectified spirit from the dropping funnel over a period of 2 hours, maintaining the temperature at 70–80 °C. Reflux the reaction mixture for a further 2 hours. Distil off as much of the alcohol as possible and decant the liquid residue from the elemental selenium. Wash the latter several times with ether, and combine the ether extracts with the decanted liquid. Remove the ether by distillation and distil the residue under reduced pressure (c. 25 mmHg): about 200 g of an oil, consisting of cyclohexane-1,2-dione, cyclohexane and water, is obtained. Dissolve the oil in 1 litre of ether, and extract thrice with ice-cold 10 per cent potassium hydroxide solution; the total amount of potassium hydroxide solution should be equivalent to 1.5 times that necessary to react with the oil assumed to be the pure dione in the monoenol form (about 1.5 l). Shake the alkaline extract once with ether to remove cyclohexanone, acidify with ice-cold hydrochloric acid and then saturate with salt. Extract the hydrochloric acid solution with ether, dry the ethereal extract with magnesium sulphate, remove the ether by distillation at normal pressure and distil the residue under reduced pressure. Collect the cyclohexane-1,2-dione (a pale green liquid) at 96–97 °C/25 mmHg; the compound decomposes slightly on keeping. The yield is 55–56 g (19%).

It is important that the synthesis should be carried out as quickly as possible, particularly the washing with alkali at 0 °C, since the latter tends to convert the product into 1-hydroxycyclopentanecarboxylic acid.

Indane-1,2,3-trione hydrate (*ninhydrin*).

In a 500-ml three-necked flask, fitted with a reflux condenser and mechanical stirrer, place 11 g (0.1 mol) of pure selenium dioxide dissolved in 240 ml of dioxane and 5 ml of water. Heat the stirred solution to 60–70 °C, remove the source of heat, add 15 g (0.1 mol) of crude indane-1,3-dione (Expt 7.10) and reflux the resulting mixture for 6 hours. A solid separates during this period. Filter the mixture, transfer the filtrate to a distilling flask and distil off about 180 ml of dioxane; then add 100 ml of water, boil the solution to coagulate the red tarry precipitate and remove it by filtration. Concentrate the filtrate to about 50 ml and filter. Boil the filtrate with 0.2–0.3 g of decolourising carbon, filter again, concentrate to 20–25 ml and keep at room temperature. Collect the crystals of crude ninhydrin by suction filtration, and recrystallise from hot water with the addition of a little decolourising carbon, if necessary (1). The yield of colourless ninhydrin is 6 g (34%); the crystals turn red between 125 and 130 °C and melt at 242–243 °C.

Note. (1) Recrystallisation from water with the use of decolourising carbon should yield almost colourless crystals; selenium-containing contaminants may, however, give rise to discoloration. In this case a further recrystallisation with the addition of decolourising carbon and a little tin(II) chloride should be carried out.

Experiment 5.100 BUTANE-2,3-DIONE DIOXIME (*Dimethylglyoxime*)

In a 500-ml three-necked flask, supported on a water bath in a fume cupboard and provided with a dropping funnel, a reflux condenser and a thermometer, place 72 g (90 ml, 1 mol) of dried and redistilled ethyl methyl ketone (Section 4.1.22, p. 408). Introduce a follower bar and stir magnetically. Add 3 ml of concentrated hydrochloric acid and warm the liquid to 40 °C. Then add 103 g (115 ml, 1 mol) of butyl nitrite (b.p. 76–79 °C) or 117 g (134 ml, 1 mol) of isopentyl nitrite (b.p. 96–99 °C) (Section 4.2.1, p. 413) slowly, maintaining the temperature at 40–50 °C; the mixture must be stirred vigorously. Heat is generated in the reaction so that cooling may now be required. Continue the stirring, without cooling, for 30 minutes after all the nitrite has been added. The reaction mixture now consists of a solution of butane-2,3-dione monoxime in butan-1-ol or 3-methylbutan-1-ol. To remove any unused ketone, treat the mixture with a cold solution of 45 g of sodium hydroxide in 100 ml of water and stir for 20–30 minutes. Transfer the reaction mixture to a separatory funnel and extract the reddish-brown solution twice with 50 ml portions of ether: the alcohol may be recovered, if desired, by fractionation of the ethereal extracts. Keep the aqueous layer; it contains the sodium salt of butane-2,3-dione monoxime (1). Prepare a solution of 70 g (1 mol) of hydroxylamine hydrochloride or of 82 g (1 mol) of hydroxylamine sulphate in about three times its weight of water, and add sodium hydroxide solution until the solution is neutral to litmus. Place the aqueous solution of the sodium salt of the monoxime in a 1-litre round-bottomed flask and add the hydroxylamine solution with stirring. Heat the mixture on a water bath for about 45 minutes. Filter off the precipitated dioxime (2) while the solution is still hot, wash it with hot water and drain well. Recrystallise the crude product from about 10 times its weight of rectified spirit. The yield of pure dimethylglyoxime (a white, crystalline solid, m.p. 240 °C) is 55 g (47%).

Notes. (1) If it is desired to isolate the monoxime, cool a portion (say one-fifth part) of the aqueous solution of the sodium salt in an ice–salt bath and carefully neutralise by adding concentrated hydrochloric acid (about 20 ml) with vigorous stirring, keeping the temperature below 15 °C. During the addition, the resulting slurry becomes difficult to stir efficiently; filter off the solid product at this stage and continue to neutralise the filtrate to obtain the remainder of the crude monoxime. The pale brown solid thus obtained contains some sodium chloride; crystallisation from water gives almost colourless crystals, m.p. 73–74 °C.
(2) If the product is coloured, dissolve it in 2 M sodium hydroxide solution on a water bath. Filter the hot almost saturated solution, and to the hot filtrate add a concentrated solution of ammonium chloride in excess of the amount required to precipitate all the dimethylglyoxime, i.e. employ an amount greater than the equivalent of the sodium hydroxide used. Filter at once with suction, and wash with boiling water. Recrystallise the white product from rectified spirit.

Experiment 5.101 1-PHENYLPENTANE-1,2-DIONE[146]

$$\text{Ph-C}\equiv\text{C-Pr} \xrightarrow{\text{KMnO}_4} \left[\underset{\underset{\text{HO}}{}{\overset{\text{Ph}}{\diagup}}\!\!=\!\!\underset{\underset{\text{OH}}{}}{\overset{\text{Pr}}{\diagdown}} \rightleftharpoons \underset{\underset{\text{O}}{}}{\overset{\text{Ph}}{\diagup}}\!\!-\!\!\underset{\underset{\text{OH}}{}}{\overset{\text{Pr}}{\diagdown}} \right] \xrightarrow{\text{[O]}}$$

$$\text{Ph·CO·CO·Pr}$$

A 250-ml round-bottomed flask equipped with a reflux condenser is charged with dichloromethane (100 ml), acetic acid (5 ml) and 1-phenylpent-1-yne

(2.0 g, 0.014 mol). The solution is stirred magnetically and heated to reflux before finely powdered potassium permanganate (5.85 g, 0.037 mol) (1) and the phase transfer reagent, Adogen 464 (1.6 g) (2) are added. After being vigorously stirred for 4 hours the precipitated manganese dioxide is filtered off and washed with two 50-ml portions of dichloromethane. Residual manganese dioxide is reduced by addition to the combined filtrates of 20 per cent hydrochloric acid (40 ml) followed by small portions of sodium hydrogen sulphite until all brown colour has disappeared. The organic phase is separated, washed with water and dried over magnesium sulphate. Most of the solvent is removed on a rotary evaporator and the remaining oil distilled under vacuum to give a small forerun of starting material (0.14 g) followed by 1-phenylpentane-1,2-dione. The yield is 1.98 g (81%), b.p. 108–110 °C/5.5 mmHg; i.r. spectrum 2990, 2950, 2965, 1710, 1680 cm^{-1}; p.m.r. spectrum (CCl$_4$, TMS) δ 0.99 (t, 3H), 1.67 (m, 2H), 2.81 (t, 2H), and 7.57 (m, 5H).

Notes. (1) The finely powdered grade is available from Carus Chemical Company under the trade name Cairox M; this grade is also obtainable from other chemical manufacturers.
(2) Adogen 464 is a registered trademark of The Ashland Chemical Company and is obtainable commercially.

5.9.2 1,3-DICARBONYL COMPOUNDS

The disconnection strategies shown in Tables 5.2 and 5.3 suggest the acylation of a ketone, in either the α-carbanion (7) or enol (8) [or specific enol equivalent (9)] forms, as a route to a 1,3-diketone.

The acylating reagent may be an acid chloride or an acid anhydride. Symmetrical ketones (—CH$_2$·R^1 = R^2) yield only a single regioisomer. Thus acetone or cyclohexanone may be acylated with acetic anhydride in the presence of boron trifluoride–etherate to pentane-2,4-dione and 2-acetylcyclohexanone respectively (Expt 5.102). Both diketones are present in the reaction mixture as boron difluoride complexes [(10) and (11) respectively], from which they may be released by treatment with sodium acetate. Pentane-2,4-dione is appreciably water soluble and is isolated by means of its characteristic copper complex (12).

With unsymmetrical ketones the boron trifluoride catalysed reaction gives a mixture of regioisomers, although acylation appears to occur chiefly at the more highly substituted of the two alternative α-carbons.

632

Acylation of ketones with esters requires the presence of a strong base under anhydrous conditions. Ketones where only one unique mesomeric carbanion is formed (e.g. symmetrical ketones or alkyl aryl ketones) yield a single regio-isomer. The reaction is illustrated by the formation of benzoylacetone from acetophenone and ethyl acetate (Expt 5.103), and may be outlined mechanistically as follows:

Experiment 5.102 PENTANE-2,4-DIONE (*Acetylacetone*)

$$Me \cdot CO \cdot Me + (Me \cdot CO)_2O \xrightarrow{BF_3} Me \cdot CO \cdot CH_2 \cdot CO \cdot Me + Me \cdot CO_2H$$

Fit a 1-litre three-necked flask with a gas inlet tube and a gas outlet leading to a gas absorption device (Fig. 2.61) charged with aqueous alkali to trap excess boron trifluoride, and stopper the third neck. Place 58 g (73 ml, 1 mol) of pure, anhydrous acetone (Section 4.1.*21*, p. 407) and 255 g (236 ml, 2.5 mol) of acetic anhydride in the flask and cool in a freezing mixture of ice and salt. Connect the gas inlet tube through an empty wash-bottle to a cylinder of commercial boron trifluoride (**CAUTION**: see Section 4.2.8, p. 421), and bubble the gas through the reaction mixture at such a rate that 250 g is absorbed in about 5 hours (2 bubbles per second). Pour the reaction mixture into a solution of 400 g of crystallised sodium acetate in 800 ml of water contained in a 2.5-litre round-bottomed flask. Steam distil the mixture (Fig. 2.102), and collect the distillate in the following portions; 500 ml, 250 ml and 250 ml. In the meantime prepare a solution of 120 g of pure crystallised copper(II) acetate in 1500 ml of water at about 85 °C; if the solution is not clear, filter from any basic acetate. Precipitate the copper complex of acetylacetone by adding 700 ml of the hot copper(II) acetate solution to the first portion of the steam distillate, 350 ml to the second, 250 ml to the third and 200 ml to the fourth portion. Allow to stand for 3 hours, or better overnight, in the ice chest. Filter off the salt at the pump, wash once with water and suck as dry as possible. Transfer the copper complex to a separatory funnel, add 400 ml of 20 per cent by weight sulphuric acid and 400 ml of ether, and shake. Remove the ether layer. Extract the aqueous layer with two 150 ml portions of ether. Dry the combined extracts with 125 g of anhydrous sodium sulphate (or the equivalent quantity of magnesium sulphate), and distil off the ether. Distil the residue through a short fractionating column and collect the acetylacetone at 134–136 °C. The yield is 80 g (80%).

Cognate preparations. *2-Acetylcyclohexanone.* Place a mixture of 24.5 g (0.25 mol) of cyclohexanone (regenerated from the bisulphite compound) and 51 g (47.5 ml, 0.5 mol) of acetic anhydride in a 500-ml three-necked flask, fitted with an efficient sealed stirrer, a gas inlet tube reaching to within 1–2 cm of the surface of the liquid, and in the third neck a thermometer immersed in the liquid, combined with a gas outlet tube leading to a trap (1). Immerse the flask in a bath of Cardice–acetone, stir the mixture vigorously and pass in the boron trifluoride as fast as possible (10–20 minutes) until the mixture, kept at 0–10 °C, is saturated (copious evolution of white fumes when the outlet tube is disconnected from the trap). Replace the Cardice–acetone bath by an ice bath and pass the gas in at a slower rate to ensure maximum absorption. Stir for 3.5 hours while allowing the ice bath to attain room temperature slowly. Pour the reaction mixture into a solution of 136 g of hydrated sodium acetate in 250 ml of water, reflux for 60 minutes (or until the boron fluoride complexes are hydrolysed), cool in ice and extract with three 50 ml portions of light petroleum, b.p. 40–60 °C (2), wash the combined extracts free of acid with sodium hydrogen carbonate solution, dry over anhydrous calcium sulphate, remove the solvent by flash distillation and distil the residue under reduced pressure. Collect the 2-acetylcyclohexanone at 95–97 °C/10 mmHg. The yield is 27 g (77%).

Notes. (1) Alternatively the reaction may be effected by adding the ketone and the acetic anhydride to 100 g (0.75 mol) of a 1:1 acetic acid–boron trifluoride complex (Section 4.2.8, p. 421).

(2) Light petroleum is preferable to ether because it removes smaller amounts of acetic acid from the aqueous phase.

Experiment 5.103 BENZOYLACETONE

Sodium ethoxide. Prepare a suspension of 11.5 g (0.5 mol) of granulated sodium (Section 4.2.68, p. 462) in 75 ml of dry xylene, transfer it to a 1-litre three-necked flask, and decant the xylene. Wash the sodium by decantation with two 20 ml portions of dry ether and cover with 200 ml of dry ether. Set the flask on a water bath and fit it with a sealed stirrer unit, and with a reflux condenser and a dropping funnel, each protected by a calcium chloride guard-tube. Start the stirrer and run in 23 g (29 ml, 0.5 mol) of absolute ethanol from the dropping funnel during 1–2 hours with gentle refluxing, and continue to reflux the mixture with stirring until nearly all of the sodium has reacted (up to 6 hours; a little residual sodium does no harm). Stop the stirrer, set the condenser for downward distillation and distil off the ether as completely as possible. The residual sodium ethoxide should be white and finely divided. All moisture must be excluded during the preparation.

Benzoylacetone. Return the condenser (protected by the calcium chloride guard-tube) to the reflux position, surround the flask with ice and introduce 200 ml (2 mol) of pure, dry ethyl acetate (Section 4.1.24, p. 409). Start the stirrer and add 60 g (58 ml, 0.5 mol) of acetophenone from the dropping funnel; the reaction commences with the separation of the sodium salt of benzoylacetone. Continue stirring for 2 hours and then allow to stand in an ice box overnight. Filter the solid at the pump with the aid of the addition of a little dry ether. Dissolve the air-dried solid in cold water, and acidify the solution with acetic acid. Filter off the crude benzoylacetone, and dry in the air. Purify by distillation under reduced pressure; collect the benzoylacetone at 128–130 °C/10 mmHg. It solidifies on cooling to a colourless crystalline solid, m.p. 61 °C. The yield is 50 g (62%). Record the p.m.r. spectrum (CDCl₃) assign the absorptions and estimate the keto:enol ratio.

Section 5.9.3   1,4-DICARBONYL COMPOUNDS

The synthetic methods which are illustrated in this section are (a) the formation of symmetrical 1,4-diketones from 1,3- (or β-)keto esters (Expt 5.104), and (b) a Michael addition reaction involving nitroalkanes and α, β-unsaturated ketones (Expt 5.105). The synthesis of symmetrical 1,4-diketones from the sodio derivatives of β-keto esters, or their mono-alkyl derivatives, by treatment with iodine [*Method (a)*], may be formulated in the following general manner.

$$2R^1 \cdot CO \cdot CH(R^2) \cdot CO_2Et \xrightarrow{\text{NaOEt}} 2[R^1 \cdot CO \cdot \overset{\ominus}{C}(R^2) \cdot CO_2Et]Na^{\oplus}$$

$$R^1 \cdot CO \cdot \overset{R^2}{\underset{CO_2Et}{\overset{|}{\underset{|}{C}}}}{}^{\ominus} \quad I\!-\!I \xrightarrow{-I^{\ominus}} R^1 \cdot CO \cdot C(R^2)I \cdot CO_2Et$$

$$R^1 \cdot CO \cdot \overset{I}{\underset{\ominus}{\overset{|}{C}}}(R^2) \cdot CO_2Et \longrightarrow \overset{R^1 \cdot CO \cdot C(R^2) \cdot CO_2Et}{\underset{R^1 \cdot CO \cdot C(R^2) \cdot CO_2Et}{|}}$$
$$R^1 \cdot CO \cdot \overset{|}{C}(R^2) \cdot CO_2Et$$

The resulting diacylsuccinate ester is subjected to hydrolysis and decarboxylation by heating with aqueous potassium carbonate (cf. Section 5.14.3, p. 738). The reaction is illustrated by the preparation of hexane-2,5-dione (Expt 5.104).

The synthesis of unsymmetrical 1,4-diketones results from the conjugate addition of the nitronate anion (as the acyl anion equivalent) to an α, β-unsaturated ketone [*Method (b)*].[147a] The intermediate nitroketone is converted into the diketone by reduction with titanium(III) chloride at pH 1, or by ozonolysis of the nitronate anion[147b] (cf. Section 5.7.7, p. 599).

$$R^1 \cdot CH_2 \cdot NO_2 \rightleftharpoons \text{Base} [R^1 \cdot \overset{\ominus}{CH}\!-\!N(O)\!=\!O]$$
mesomeric anion

The illustrative example is the synthesis of heptane-2,5-dione (Expt 5.105).

A related reaction is the interaction of a silyl enol ether, derived from a ketone, with a nitro olefin in the presence of either titanium(IV) chloride or

635

tin(IV) chloride. The reaction is thought to proceed by a Michael addition followed by a Nef reaction on the intermediate nitronate ester.[148]

Experiment 5.104 HEXANE-2,5-DIONE (*Acetonylacetone*)

$$2Me \cdot CO \cdot CH_2 \cdot CO_2Et \xrightarrow{Na} 2[Me \cdot CO \cdot CH \cdot CO_2Et]^{\ominus}Na^{\oplus} \xrightarrow{I_2}$$

$$\begin{array}{l} Me \cdot CO \cdot CH \cdot CO_2Et \\ | \\ Me \cdot CO \cdot CH \cdot CO_2Et \end{array} \xrightarrow[\text{decarboxylation}]{\text{hydrolysis}} Me \cdot CO \cdot CH_2 \cdot CH_2 \cdot CO \cdot Me$$

Place 11.5 g (0.5 mol) of granular sodium (Section 4.2.68, p. 462) covered with 250 ml of dry ether in a 1-litre three-necked flask fitted with a sealed stirrer unit, a reflux condenser and a dropping funnel. Start the stirrer and add gradually a solution of 65 g (63.5 ml, 0.5 mol) of redistilled ethyl acetoacetate in 250 ml of dry ether, cooling the flask if the reaction becomes too vigorous. Continue stirring until all of the sodium has reacted, and then add steadily a solution of 63.5 g (0.25 mol) of powdered iodine in 350 ml of dry ether until the iodine colour persists. Filter off the sodium iodide, wash it with ether and evaporate the combined filtrate and washings (rotary evaporator). A somewhat sticky residue of diethyl 2,3-diacetosuccinate remains; crystallise a small portion from 50 per cent aqueous acetic acid to obtain a specimen, m.p. 88 °C. Boil the bulk of the crude product under reflux for 1 hour with 250 ml of 20 per cent aqueous potassium carbonate solution and treat the cooled clear yellow solution with 50 g of anhydrous potassium carbonate. Separate the organic phase, and extract the aqueous layer with four 25 ml portions of ether. Combine the ether extracts with the original organic layer and dry over anhydrous sodium sulphate. Remove the ether by flash distillation and distil the residue, collecting the hexanedione as a fraction of b.p. 185–192 °C. The yield is 7.5 g (26%).

Experiment 5.105 HEPTANE-2,5-DIONE[147a]

$$Et \cdot CH_2 \cdot NO_2 \xrightarrow{Pr_2^iNH} Et \cdot \overset{\ominus}{CH} \cdot NO_2$$

$$Et \cdot \overset{\ominus}{CH} \cdot NO_2 + CH_2 = CH \cdot CO \cdot Me \xrightarrow{+H^{\oplus}}$$

$$Et \cdot CH(NO_2) \cdot (CH_2)_2 \cdot CO \cdot Me \xrightarrow{TiCl_3} Et \cdot CO \cdot (CH_2)_2 \cdot CO \cdot Me$$

CAUTION: *All operations should be conducted in an efficient fume cupboard owing to the toxicity of aliphatic nitro compounds and of alkyl vinyl ketones.*
1-Nitropropane (17.8 g 0.2 mol) and diisopropylamine (10 ml) in 200 ml of chloroform are stirred at 60 °C under a nitrogen atmosphere. Methyl vinyl ketone (**CAUTION**) (7 g, 0.1 mol) is added dropwise to this solution. After 3 hours another portion of methyl vinyl ketone (7 g, 0.1 mol) is added and the solution stirred for 24 hours. The solution is washed sequentially with water, 10 per cent aqueous hydrochloric acid, 5 per cent sodium hydrogen carbon-

ate, and saturated brine; the organic solution is dried over anhydrous sodium sulphate, concentrated and distilled to yield the nitroketone. The yield is 17 g (55%), b.p. 120 °C/10 mmHg; p.m.r. spectrum (CCl$_4$, TMS) δ 0.97 (t, 3H), 2.7–2.5 (m, 6H), 2.13 (s, 3H) and 4.38 (m, 1H).

Reduction of nitroketones with titanium(III) chloride at pH 1. The foregoing nitroketone may be converted into the diketone by means of the following general procedure. The nitroketone in solution in tetrahydrofuran (0.2 M) is treated with 4 equivalents of titanium(III) chloride (1) (20% aqueous hydrochloric acid solution) and stirred under nitrogen at room temperature for 24 hours. The reaction mixture is then poured into ether and the phases separated. The aqueous phase is extracted several times with ether; the organic extracts are combined, washed with 5 per cent sodium hydrogen carbonate and with brine, and then dried over anhydrous sodium sulphate, concentrated and distilled. The yield of heptane-2,5-dione is 66 per cent; p.m.r. spectrum (CCl$_4$, TMS) δ 1.00 (t, 3H), 2.10 (s, 3H), 2.60 (s, 4H) and 3.41 (q, 2H).

Note. (1) A 20 per cent titanium(III) chloride solution in aqueous hydrochloric acid is available commercially. Solid titanium(III) chloride reagent is a flammable solid and should be handled only under an inert atmosphere, and if used in this reaction should be taken from a fresh bottle.

5.10 CARBOHYDRATES

The simplest carbohydrates are the monosaccharides which under specified conditions are structurally characterised as polyhydroxy aldehydes or polyhydroxy ketones; these are termed aldoses and ketoses respectively. Aldoses and ketoses are sub-classified, according to the number of carbons present in each molecule, into aldotetroses, aldopentoses, aldohexoses, etc., or ketotetroses, ketopentoses, etc.

The monosaccharides are the fundamental units for more complex carbohydrates. Thus disaccharides are compounds which yield two monosaccharide molecules upon dilute acid hydrolysis; trisaccharides given three monosaccharides upon hydrolysis; tetrasaccharides give four monosaccharide molecules, etc. The upper limit in this group (the oligosaccharides) is reached in the case of a polymer having ten monosaccharide units. Mono- and oligosaccharides are characterised by having an invariable molecular weight and high water solubility; they are usually sweet to taste and are frequently known as sugars. The non-sugar group of carbohydrates are polymers (polysaccharides) having more numerous monosaccharide units and the molecular weights of individual molecules in a given sample will not necessarily be of the same magnitude; acid hydrolysis yields the appropriate number of monosaccharide units.

The wide availability of various polysaccharides provides an important source of some of the monosaccharides. Such monosaccharides are now used in organic reactions as low-cost starting materials in the synthesis of a range of simpler optically pure compounds (e.g. Expt 5.77). These synthetic strategies have been made possible from earlier work on the development of numerous selective protection methods, on the application of new selective reagents for functional group modification within the monosaccharide molecule, and on the realisation of the role of conformation in the interpretation of a reaction course. The preparative examples in this section are illustrative of these developments.

Table 5.4   Aldoses

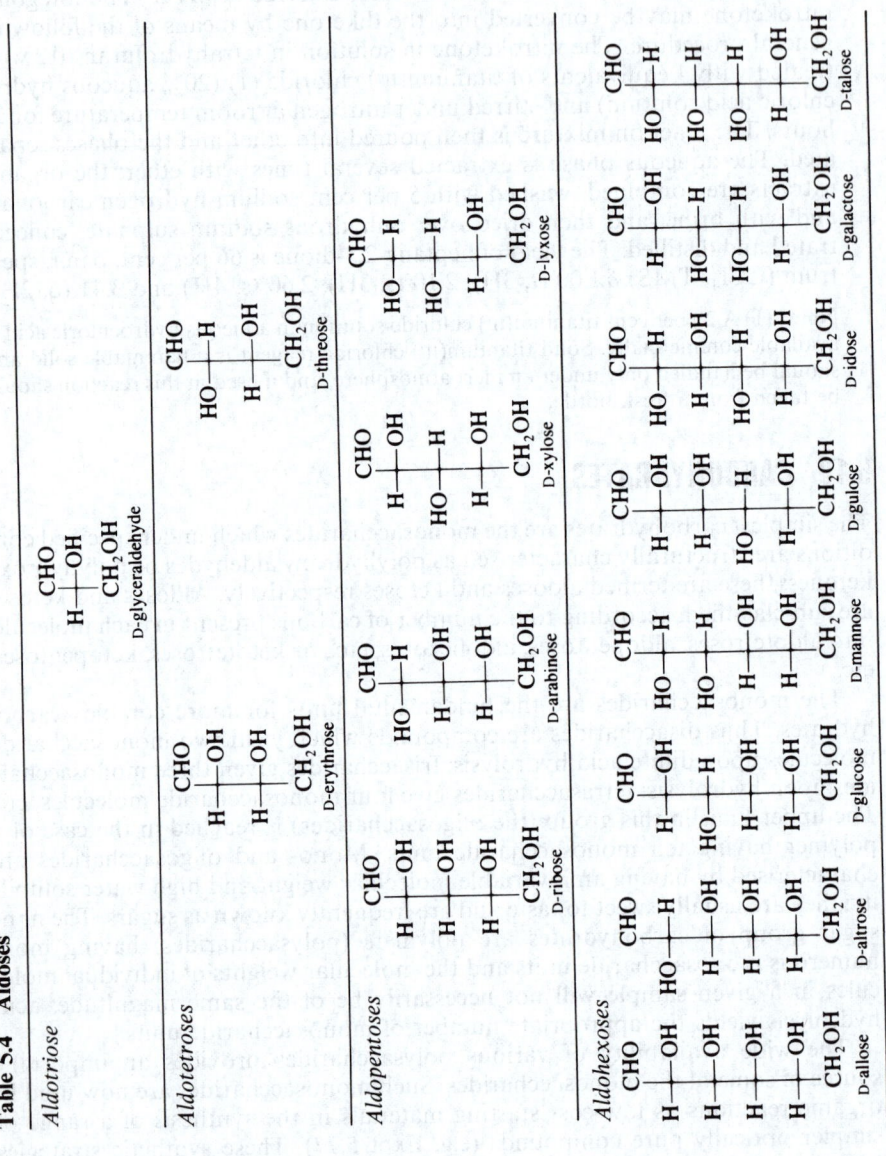

Aldotriose

Aldotetroses

Aldopentoses

Aldohexoses

Table 5.5   Ketoses

Ketotetrose

```
        CH₂OH
        |
        CO
   H ───┼─── OH
        CH₂OH
```
$$\text{CH}_2\text{OH} - \text{CO} - \overset{\text{H}}{\underset{}{|}} \text{OH} - \text{CH}_2\text{OH}$$

D-erythrulose
(*glycero*-tetrulose)

Ketopentoses

```
      CH₂OH              CH₂OH
      |                  |
      CO                 CO
  H ──┼── OH        HO ──┼── H
  H ──┼── OH         H ──┼── OH
      CH₂OH              CH₂OH
```

D-ribulose	D-xylulose
(*erythro*-pentulose)	(*threo*-pentulose)

Ketohexoses

```
   CH₂OH          CH₂OH          CH₂OH          CH₂OH
   |              |              |              |
   CO             CO             CO             CO
 H─┼─OH       HO─┼─H          H─┼─OH       HO─┼─H
 H─┼─OH        H─┼─OH        HO─┼─H        HO─┼─H
 H─┼─OH        H─┼─OH         H─┼─OH        H─┼─OH
   CH₂OH          CH₂OH          CH₂OH          CH₂OH
```

D-psicose	D-fructose	D-sorbose	D-tagatose

The structural relationships of the aldoses and ketoses are exemplified in Tables 5.4 and 5.5 respectively. The structures are formulated as Fischer projections to illustrate the convention D- (or L-) which is used to designate the configuration at the chiral site furthest from the carbonyl function. When the hydroxyl group at this site lies on the right side of the Fischer projection formula the molecule belongs to the D-series of monosaccharides and is related to D-glyceraldehyde. The L-series are mirror images of these structures. The tables also illustrate the epimeric relationships between the various structures. Thus erythrose and threose are C-2-epimers (i.e. of opposite configuration at C-2), as are the pairs ribose/arabinose; xylose/lyxose; allose/altrose; glucose/mannose; gulose/idose; and galactose/talose. Glucose and allose are C-3 epimers; glucose and galactose are C-4 epimers, etc.

Fischer projections are however, unsatisfactory when considering the physical properties and chemical reactivity of monosaccharides for which definitive spatial formulations are necessary. These are given below for D-glyceraldehyde, D-erythrose and D-threose, for which the (*R,S*)-configuration may be readily assigned at the appropriate chiral carbons.

639

(R)-2,3-dihydroxy-
propanal
(D-glyceraldehyde)

(2R,3R)-2,3,4-
trihydroxybutanal
(D-erythrose)

(2S,3R)-2,3,4-
trihydroxybutanal
(D-threose)

Similar representations may be drawn for the open-chain structures of the other compounds, and the (R,S)-convention applied.

A more detailed study on the structure of D-glucose based upon physical data and chemical reactivity has revealed that the open-chain formulation of the aldopentoses, aldohexoses, ketopentoses and ketohexoses is however an over-simplification. Thus, for example, in solution D-glucose exists as an equilibrium

(1)
α-D-glucopyranose

(2)
β-D-glucopyranose

(3a)

(3b)
aldehydo-D-glucose

(4)
α-D-glucofuranose

(5)
β-D-glucofuranose

640

mixture of five forms. The cyclic structures are those which arise from hemi-acetal formation involving the carbonyl function of C-1 of the open-chain form (3a/b) with either (i) the hydroxyl group on C-5 (in 3a) to give two pyranose ring structures (1), α, D-glucopyranose, and (2), β-D-glucopyranose, or (ii) the hydroxyl group on C-4 (in 3b) to give two furanose structures (4), α-D-gluco-furanose, and (5) β-D-glucofuranose. The open-chain structure has been drawn to emphasise the spatial relationships of the carbon-skeletal arrangement and the dispositions of the hydroxyl groups for hemi-acetal formation. The five-membered (furanose) ring structure is at right angles to the plane of the paper. The pyranose structure adopts the characteristic chair conformation of six-membered rings.

Structures (1) and (2) differ from one another only in the configuration at the anomeric carbon atom (C-1), the hydroxyl group being axial (α) in (1) and equa-torial (β) in (2). In the case of structures (3) and (4) the configuration at the anomeric carbon is that in which the hydroxyl group lies either below (α) [in (4)], or above (β) [in (5)] the general plane of the ring. In solution the equilib-rium ratio of the four cyclic and the acyclic structures is dependent upon the nature of the solvent; in aqueous solution at room temperature the percentage composition is (1), 36 per cent; (2), 64 per cent; (3), 0.0026 per cent; (4) + (5), < 1 per cent. The composition reflects the relative thermodynamic stabilities of the relevant molecules under these conditions. The change in optical rotation which is observed when crystalline α-D-glucopyranose is dissolved in a solvent, and which results from the attainment of this equilibrium is termed *mutarotation*.

X-ray crystallographic studies have shown that crystalline D-glucose as com-monly isolated exists in the α-D-glucopyranose form. Furthermore the stable chair conformation (1), in which the hydroxyl groups on C-2, C-3 and C-4, and the hydroxymethyl group on C-5 are equatorial, is preferred to the alternative chair conformation (6) in which these groups occupy axial positions.

There has been a considerable amount of discussion over recent years on the nomenclature to be used to describe the various conformations which these ring structures may adopt. Several authoritative accounts are available.[149]

Most, if not all, of the stable forms of crystalline aldose and ketose mono-saccharides exist in the pyranose structure. Each in solution, as with D-glucose, exists as an equilibrium mixture of open chain and of α- and β-anomers of the cyclic forms. The cyclic five- and six-membered structures formulated below are an illustrative selection of monosaccharides.

α-D-ribopyranose

β-D-ribofuranose

β-D-xylopyranose

α-D-xylofuranose

β-D-allopyranose

α-D-allofuranose

α-D-fructopyranose

α-D-fructofuranose

A selection of typical synthetic interconversions undergone by monosaccharides is contained in the following five sections (Sections 5.10.1 to 5.10.5).

5.10.1 CARBOHYDRATE INTERCONVERSIONS

α-D-glucose

$\xrightarrow[\text{catalyst}]{(\text{Me·CO})_2\text{O}}$

(7) β-anomer   or   (8) α-anomer

$\xrightarrow{\text{HBr}}$

(9)

$\xrightarrow{\text{MeOH/HgO}}$ (10)

$\xrightarrow{\text{MeO}^\ominus/\text{MeOH}}$

(11)

The hydroxyl groups at both the anomeric and non-anomeric carbon atoms may be readily acylated. For example, treatment of D-glucose with acetic anhydride in the presence of zinc chloride gives a reasonable yield of the penta-acetate, 1,2,3,4,6-penta-*O*-acetyl-α-D-glucopyranose (8) (Expt 5.106). On the other hand the β-anomer (7) predominates when D-glucose is treated with acetic anhydride in the presence of sodium acetate (Expt 5.107). Interestingly the β-anomer may be converted into the α-anomer by heating with an acetic anhydride/zinc chloride mixture. These reactions are reasonably interpreted on the basis that the β-anomer is the kinetically controlled product, being initially formed from the more rapid acetylation of the equatorial hydroxyl group at C-1 of β-D-glucopyranose. In the presence of the Lewis acid catalyst, zinc chloride, the β-penta-acetate (7) is thought to be readily converted into the mesomerically stabilised carbocation (12) by loss of the C-1-acetoxy group, which would then recombine to yield the thermodynamically more stable α-anomer (8).

Sodium acetate would not be expected to promote the formation of the species (12) and hence the β-anomer would accumulate if this catalyst were used.

D-Glucose may be benzoylated with benzoyl chloride in pyridine/chloroform solution (Expt 5.108). Here temperature control is essential to prevent the formation of an anomeric mixture.

The acetate (or benzoate) groups attached to the anomeric site may be readily replaced by a bromine atom to yield the synthetically useful acylglycosyl halides. Thus the formation of 2,3,4,6-tetra-*O*-acetyl-α-D-glucopyranosyl bromide (9) from D-glucose involves the formation of the mixed penta-acetates (7) and (8) followed by treatment of the reaction mixture with hydrogen bromide generated *in situ* from red phosphorus, bromine and water (Expt 5.109). Alternatively a solution of hydrogen bromide in glacial acetic acid is added to a solution of the acetylated or benzoylated monosaccharide (e.g. Expt 5.110, the preparation of 2,3,4,6-tetra-*O*-benzoyl-α-D-glucopyranosyl bromide). In either case the thermodynamically more stable anomer is formed. In the case of glucose (and of the other aldohexoses together with xylose and lyxose) this is the α-anomer, while in the case of arabinose and ribose the β-anomer predominates.

The halogen in the acylglycosyl halide is reactive and may be readily displaced, for example, by an alkoxy group on reaction with an alcohol under anhydrous conditions in the presence of a silver or mercury(II) salt. In this case the products are *glycosides* which are the mixed cyclic acetals related to the cyclic hemiacetal forms of the monosaccharides. In the case of the D-glucose derivative shown below (and of other 1,2-*cis* acylglycosyl halides) the replacement involves inversion of configuration at the anomeric site and the α-glucosyl halide (9) yields a β-glucoside (10) (e.g. the formation of methyl 2,3,4,6-tetra-*O*-acetyl-β-D-glucopyranoside, Expt 5.111).

The presence of the protecting acyl groups in the acylglycosyl halide ensures that the pyranose ring structure is retained. After glycoside formation, removal of these protecting groups (either acetyl or benzoyl) is readily effected by base (e.g. *Zemplen's method* which uses methanol containing a small amount of sodium methoxide) to give the alkali-stable glycoside (e.g. methyl β-D-gluco-pyranoside (11), (Expt 5.112).

5.10.2 CARBOHYDRATE INTERCONVERSIONS

α-D-galactose (13)   methyl α-D-galactopyranoside (14)

The above, indirect, procedure for glycoside synthesis is the *Koenigs–Knorr method*. Glycosidation may be effected from monosaccharides directly by treatment with an alcohol in the presence of a mineral acid catalyst. For example, when α-D-galactose (13) is heated in methanolic solution containing 2 per cent of hydrogen chloride the thermodymanically more stable methyl α-D-galacto-pyranoside (14) (Expt 5.113) is formed preferentially, and may be isolated from the reaction product by crystallisation as the monohydrate. The less abundant β-anomer may be recovered from the mother-liquors.

The commercially available methyl α-D-glucopyranoside may be obtained similarly. If however the glycosidation reaction of D-glucose is conducted at room temperature with a lower proportion of mineral acid, then the predominant products are the methyl α(and β)-D-glucofuranosides. It is considered that the furanosides are the products arising from kinetic control and the pyranosides from thermodynamic control. Clearly the composition of reaction mixtures arising from glycosidation of other monosaccharides will be dependent upon the nature of the monosaccharide and the proportions may be rationalised by the application of conformational analysis.[150]

Experiment 5.106   1,2,3,4,6-PENTA-*O*-ACETYL-α-D-GLUCOPYRANOSE
(α-D-*Glucopyranose penta-acetate*)

Method A. Into a 100-ml round-bottomed flask place 0.5 g of anhydrous zinc chloride (1) and 13.5 g (12.5 ml, 0.13 mol) of acetic anhydride; attach a Leibig reflux condenser and heat the mixture on a boiling water bath for 5–10 minutes with occasional shaking until the zinc chloride has largely dissolved.

Add slowly 2.5 g (0.014 mol) of powdered α-glucose, shaking the mixture gently during the addition to control the vigorous reaction which ensues. Finally heat the flask for 1 hour on a boiling water bath (2). Pour the contents of the flask into 125 ml of ice-water and stir vigorously to assist the hydrolysis of unreacted acetic anhydride. After about 30 minutes the oil which first separates will gradually solidify. Filter, wash well with cold water and recrystallise several times from industrial spirit until the m.p. is constant. The pure product melts at 110–111 °C; the yield is 3.5 g (63%).

Method B. To a mixture of 16.25 g (15 ml, 0.16 mol) of acetic anhydride and 25 ml of glacial acetic acid in a conical flask add 5 g (0.028 mol) of powdered glucose. Add dropwise (Pasteur pipette) and with shaking 1 ml of perchloric acid–acetic anhydride catalyst (3), at such a rate that the temperature of the mixture does not exceed 35 °C. Leave at room temperature for 30 minutes and then pour the liquid into a mixture of ice and water. Filter off the crystalline solid which separates on vigorous stirring and wash it thoroughly with cold water. Recrystallise from industrial spirit until the m.p. is constant; the pure product has m.p. 110–111 °C, $[\alpha]_D^{18}$ +101.6° (c 0.28 in $CHCl_3$). The yield is 8 g (72%).

Notes. (1) Zinc chloride is extremely deliquescent and it must therefore be introduced into the flask as rapidly as possible. Place a small stick of zinc chloride in a glass mortar, powder rapidly, and weigh out the required amount.
(2) Although the time of heating may be reduced to 30 minutes by heating in an air bath the product is somewhat discoloured and requires at least one recrystallisation involving the use of decolourising charcoal.
(3) The perchloric acid–acetic anhydride catalyst may be prepared by adding 1.0 g of 60 per cent perchloric acid to 2.3 g of acetic anhydride maintained at 0 °C.

Experiment 5.107   1,2,3,4,6-PENTA-*O*-ACETYL-β-D-GLUCOPYRA-NOSE (β-*D*-Glucopyranose penta-acetate)

Grind together in a porcelain mortar 4 g of anhydrous sodium acetate (Section 4.2.69, p. 464) and 5 g (0.028 mol) of dry α-D-glucose and place the powdered mixture in a 200-ml round-bottomed flask. Add 27 g (25 ml, 0.26 mol) of acetic anhydride, attach a double surface condenser and heat on a boiling water bath until a clear solution is obtained (1), shaking the mixture from time to time. Continue heating for a further 2 hours after a clear solution has been obtained and then pour the reaction mixture on to 250 ml of crushed ice. Allow to stand for 1 hour, stirring occasionally to break up the solid lumps which separate. Filter off the crystals, wash well with cold water and recrystallise from industrial spirit (or from methanol or ethanol) until the purified material has m.p. 131–132 °C, $[\alpha]_D^{18}$ +4.0° (c 4.5 in $CHCl_3$). The yield is 6.2 g (56%).

Conversion of β-into α-D-glucose penta-acetate. Add 0.5 g of anhydrous zinc chloride rapidly to 25 ml of acetic anhydride in a 100-ml round-bottomed flask, fitted with a Liebig condenser, and heat on a boiling water bath to dissolve the solid. Add 5 g of pure β-D-glucose penta-acetate, continue heating for 30 minutes, pour the mixture on to ice and purify the solid which separates as described above. The effectiveness of the conversion may be monitored by t.l.c. on silica gel plates using cyclohexane/acetone (7:3) and locating the two closely running spots by immersing the developed and dried plate in a tank of iodine vapour.

Note. (1) It is dangerous to scale up this experiment without modifying the preparative procedure. If 50 g of glucose is to be acetylated, a 2-litre round-bottomed flask should be fitted with two wide-bore Liebig condensers in series, and a large vessel filled with ice-water should be readily available to plunge the reaction flask into, should the vigorous reaction which ensues on heating need controlling. With a scale using 100 g of glucose a procedure involving the addition of α-D-glucose to a preheated sodium acetate–acetic anhydric mixture at such a rate as to keep the mixture under reflux but without the reaction getting out of control has been described.[151]

Experiment 5.108   1,2,3,4,6-PENTA-*O*-BENZOYL-α-D-GLUCOPYRANOSE (*α-D-Glucopyranose pentabenzoate*)

In a 1-litre flange flask fitted with a multiple socket head carrying a mechanical stirrer, a calcium chloride guard-tube, a 250-ml dropping funnel and a thermometer, place 126 ml of dry pyridine (**CAUTION**) and 105 ml of dry dichloromethane. Cool the flask well in an ice–salt bath and add from the dropping funnel, with stirring, a previously prepared and cooled solution of 127 g (105 ml, 0.9 mol) of benzoyl chloride in 105 ml of dry dichloromethane. Remove the dropping funnel and add 50 g (0.28 mol) of dry powdered α-D-glucose portionwise to the vigorously stirred benzoylating reagent at a rate which maintains the temperature of the reaction below 10 °C (1). Allow the pink-coloured solution to stand at 0 °C for 24 hours, dilute with 400 ml of dichloromethane and transfer the solution to a 2-litre separatory funnel. Wash the solution successively with several 300 ml portions of dilute aqueous sulphuric acid (2 M), water, saturated aqueous sodium hydrogen carbonate and water. Dry over anhydrous sodium sulphate and remove the dichloromethane on a rotary evaporator to give a yellow solid which is ground up with industrial spirit, filtered and washed well with spirit. Recrystallise the solid from acetone–water to give the pure product, m.p. 184–186 °C, $[\alpha]_D^{20} + 184.4°$ (*c* 1.75 in CHCl$_3$). The yield is 149 g (77%).

Note. (1) When the experiment is performed without effective cooling, i.e. if the temperature rises to 40–50 °C, a mixture of anomers is obtained which cannot be separated by simple recrystallisation. Frequently in large-scale preparations it is advisable to replace the ice–salt bath with one of acetone–Cardice to ensure good temperature control.

Experiment 5.109 2.3.4.6-TETRA-*O*-ACETYL-α-D-GLUCOPYRANOSYL BROMIDE (α-*Acetobromoglucose*)

Fit a 1-litre three-necked flask located in the fume cupboard with a mechanical stirrer unit using a Kyrides seal (Fig. 2.50), a dropping funnel and a thermometer to read the temperature of the reaction mixture. Immerse the flask in an ice–salt bath supported on a laboratory jack so that it may be easily removed if the reaction conditions so demand. Place 432 g (400 ml, 4.24 mol) of acetic anhydride in the flask, cool to 4 °C and add dropwise and with stirring 2.4 ml of 60 per cent perchloric acid. Remove the cooling bath and allow the reaction mixture to warm to room temperature; then add 100 g (0.56 mol) of dry powdered α-D-glucose in portions with stirring so that the temperature of the reaction mixture is maintained at between 30 and 40 °C. Cool to about 20 °C and add 31 g (1 mol) of red phosphorus followed by 181 g (58 ml, 2.26 mol) of bromine (**CAUTION**) dropwise at a rate that the temperature does not exceed 20 °C. Then add 36 ml of water over a period of about half an hour, the stirring and cooling being continued and the temperature being maintained below 20 °C. Allow the reaction mixture to stand for 2 hours at room temperature, transfer to a fume cupboard and dilute with 300 ml of dichloromethane, and filter through a large 60° glass funnel having a glass wool plug inserted not too tightly into the outlet (1). Finally rinse the reaction flask and funnel with small portions of dichloromethane, transfer the filtrate and washings to a 3-litre separatory funnel and wash it rapidly by shaking vigorously with two 800 ml portions of iced water (2). Run the lower dichloromethane layer from the second washing into 500 ml of a stirred saturated solution of aqueous sodium hydrogen carbonate to which has also been added some crushed ice. When the vigorous evolution of carbon dioxide has subsided transfer the mixture to a separatory funnel, run the dichloromethane layer into a large flask containing 10 g of powdered activated silica gel and filter after about 10 minutes (the bulk of the solution may be decanted from the silica gel and the remainder filtered under reduced pressure using a sintered glass funnel). Remove the solvent under reduced pressure using a rotary evaporator on a water bath maintained at 60 °C. Towards the conclusion of this operation the syrupy mass crystallises as a thick layer around the inside of the flask. At this stage remove the flask from the evaporator, break the crystalline cake away from the sides of the flask and remove the remaining solvent under reduced pressure without heating further. Transfer portions of the solid to a mortar and grind with a 2:1 mixture of light petroleum (b.p. 40–60 °C) and dry ether. Filter the combined slurry and wash the filter cake first with a light petroleum–ether solvent mixture and then with 50 ml of previously chilled (0 °C) dry ether. The crude product is obtained in a yield of 210 g (92%), and when recrystallised from ether–light petroleum (b.p. 40–60 °C) has m.p. 88–89 °C, $[\alpha]_D^{20}$ +197.5° (*c* 2 in $CHCl_3$). The glucosyl halide should be stored in a desiccator over sodium hydroxide pellets; whenever possible it should be used without delay.

Notes. (1) If care is used most of the solution may be decanted from the solid deposit so that the glass wool does not become blocked with material and hence slow down the filtration process. This filtration is best conducted in a fume cupboard.

(2) All the isolation operations must be conducted with the minimum of delay and under conditions which reduce the contact of the solutions of unstable glucosyl halide with moisture. Solutions to be used for washing the organic layer should have been previously prepared and contain sufficient ice to ensure that the temperature of the liquid is approximately 4 °C. To obtain good yields and to ensure that vessels do not become unduly 'sticky' as the result of residual carbohydrate deposits, the separatory funnels, receiver vessels and aqueous extracts before being discarded should be rinsed with dichloromethane at each stage and these washings combined with the main organic solution.

Cognate preparations. *2,3,4,6-Tetra-O-acetyl-α-D-galactopyranosyl bromide.* Use 100 g (0.56 mol) of dry D-galactose under precisely the same conditions; the product is obtained in a yield of 202 g (88%). When recrystallised from ether–light petroleum (b.p. 40–60 °C) it has m.p. 84–85 °C, $[\alpha]_D^{20} + 214°$ (c 1.2 in CHCl$_3$).

2,3,4-Tri-O-acetyl-β-L-arabinopyranosyl bromide. For this preparation use 10 g (0.067 mol) of L-(+)-arabinose, 40 ml (0.424 mol) of acetic anhydride, 0.24 ml of 60 per cent perchloric acid, 30 g (0.1 mol) of red phosphorus, 18.1 g (5.8 ml, 0.226 mol) of bromine and 3.6 ml of water. The yellow syrup which is obtained after the appropriate isolation procedure gives 21 g of crude crystalline product. Recrystallisation is effected by dissolving it in a mixture of benzene/ether (5:95) (**CAUTION**) warming and adding light petroleum (b.p. 40–60 °C) until a slight cloudiness is apparent, and then allowing the solution to cool. The pure product is obtained in a yield of 11 g (48%), m.p. 136–138 °C, $[\alpha]_D^{22} + 280°$ (c 3.13 in CHCl$_3$).

Experiment 5.110 2,3,4,6-TETRA-O-BENZOYL-α-D-GLUCOPYRA-NOSYL BROMIDE

In a 250-ml conical flask fitted with a ground glass stopper place 40 ml of 1,2-dichloroethane and 20 g (0.029 mol) of α-D-glucopyranose pentabenzoate (Expt 5.108) (1). When all the solid has dissolved add 40 ml (0.225 mol) of a solution of hydrogen bromide in glacial acetic acid (45% w/v HBr), stopper the flask and allow the reaction mixture to stand in the refrigerator overnight or at room temperature for about 2 hours. Pour the mixture into ice-water, rinse the flask with 1,2-dichloroethane, separate the organic layer and shake it with several portions of a saturated aqueous solution of sodium hydrogen carbonate until no further effervescence occurs. Wash the organic layer with water, dry over magnesium sulphate, filter and remove the 1,2-dichloroethane under reduced pressure on a rotary evaporator. Dissolve the crystalline solid which remains in dry ether, heating to 35 °C, and slowly add with further heating light petroleum (b.p. 40–60 °C) until a slight persistent cloudiness develops; then add a little more ether to give a clear solution, which is left to cool slowly to room temperature and finally refrigerated. Filter off the purified product and allow it to dry in the air; the yield is 16.5 g (88%), m.p. 129–130 °C, $[\alpha]_D^{20}$ +125° (c 2.0 in CHCl$_3$).

Note. (1) A mixture of anomeric glucose pentabenzoates such as might be obtained from a benzoylation reaction on glucose without careful temperature control gives equally good results.

Experiment 5.111 METHYL 2,3,4,6-TETRA-*O*-ACETYL-*β*-D-GLUCO-PYRANOSIDE

In a 500-ml two-necked flask fitted with a mechanical stirrer and calcium chloride guard-tube place 110 ml of dry methanol, 110 ml of pure chloroform (**CAUTION**), 22 g of anhydrous calcium sulphate, 7.2 g of yellow mercury(II) oxide and 0.55 g of mercury(II) bromide. Stir the suspension for 30 minutes and add 16.5 g (0.04 mol) of 2,3,4,6-tetra-*O*-acetyl-*α*-D-glucopyranosyl bromide (Expt 5.109) in one portion. The temperature of the mixture will rise to about 25–30 °C, the pH of the solution will fall from 7 to 2 and the yellow coloration of the mercury(II) oxide will disappear (1). Stir the suspension for a further 90 minutes, filter through a pad of Celite filter-aid and evaporate the filtrate on a rotary evaporator under reduced pressure. Dissolve the viscous oil which remains in 10 ml of chloroform, remove the inorganic salts which are precipitated by filtration and wash the residue well with further portions of chloroform. Evaporate the chloroform and triturate the resulting viscous oil with methanol until it solidifies. Recrystallise from methanol to give pure methyl 2,3,4,6-tetra-*O*-acetyl-*α*-D-glucopyranoside, m.p. 104–105 °C, $[\alpha]_D^{20}$ −18.2° (c 1 in CHCl$_3$). The yield is 13.7 g (95%).

Note. (1) The yellow coloration in the solution disappears within a few minutes of addition of the glucosyl halide and t.l.c. analysis (solvent system benzene–methanol, 98:2) (**CAUTION**) reveals virtual completion of the reaction.

Cognate preparation. *Methyl 2,3,4,6-tetra-O-acetyl-β-D-galactopyranoside.* Use 13.5 g (0.033 mol) of 2,3,4,6-tetra-*O*-acetyl-*α*-D-galactopyranosyl bromide (Expt 5.109), 19 g of anhydrous calcium sulphate, 5.6 g of yellow mercury(II) oxide, 0.5 g of mercury(II) bromide, 90 ml of dry chloroform and 90 ml of dry methanol under the reaction conditions and subsequent isolation procedure described above; 7.5 g (63%) of methyl 2,3,4,6-tetra-*O*-acetyl-*β*-D-galactopyranoside, m.p. 96–97 °C, $[\alpha]_D^{20}$ −28.0° (c 2.5 in CHCl$_3$), is obtained after several recrystallisations from ethanol.

Experiment 5.112   METHYL β-D-GLUCOPYRANOSIDE

In a 100-ml conical flask place 5.5 g (0.015 mol) of methyl 2,3,4,6-tetra-O-acetyl-β-D-glucopyranoside (Expt 5.111), 50 ml of dry methanol and 10 ml of a solution of sodium methoxide in methanol previously prepared by the cautious addition of 0.1 g of sodium to 20 ml of methanol. Stopper the flask and allow the solution to stand for 1 hour, then add sufficient ion exchange resin [Zeolite 225 (H⊕)] to render the solution neutral to moist universal indicator paper. Remove the resin by filtration, wash with methanol and evaporate the combined filtrate and washings under reduced pressure (rotary evaporator). Triturate the colourless syrup with absolute ethanol to cause it to solidify and recrystallise from absolute ethanol. The pure methyl β-D-glucopyranoside has m.p. 108–109 °C, $[\alpha]_D^{20}$ − 30.2° (c 2.8 in H_2O); the yield is 2.4 g (83%).

Cognate preparation. *Methyl β-D-galactopyranoside.* Use 3.6 g (0.01 mol) of methyl 2,3,4,6-tetra-O-acetyl-β-D-galactopyranoside. (Expt 5.111) and proceed as above. After recrystallisation from absolute ethanol, 1.4 g (73%) of the methyl galactopyranoside, m.p. 174–175 °C, $[\alpha]_D^{20}$ + 1.3° (c 1 in H_2O), is obtained.

Experiment 5.113   METHYL α-D-GALACTOPYRANOSIDE

In a 2-litre flask fitted with a reflux condenser place 100 g (0.56 mol) of dry α-D-galactose and 700 ml of an anhydrous methanolic solution of hydrogen chloride (about 0.6 M) (1). Heat the mixture under reflux for 14 hours, cool, add 150 ml of distilled water and treat the light brown solution with solid lead carbonate until all the acid has been neutralised (2). Filter off the inorganic salts and remove the solvent on a rotary evaporator under reduced pressure. Triturate the resulting brown syrup with absolute ethanol with cooling in ice to cause the product to crystallise and recrystallise it from the minimum quantity of absolute ethanol to obtain 62 g of crude material, m.p. 85–90 °C. Isolate the methyl α-D-galactopyranoside as the hydrate by dissolving the crude product in 30 ml of water and allowing the solution to stand for one day at room temperature and two days at 4 °C. Repeat the recrystallisations several times using proportionate amounts of water until pure hydrated product, m.p. 109–110 °C, $[\alpha]_D^{20}$ + 173.4° (c 1 in H_2O), is obtained; the yield is 38 g (35%) (3).

Notes. (1) Dry hydrogen chloride gas is passed into dry methanol (Section 4.1.8, p. 400) (contained in a flask protected by a calcium chloride guard-tube) until analysis of aliquot portions by titration with standard aqueous sodium hydroxide solution reveals the required concentration has been reached. It is usually more convenient to prepare initially a smaller volume of a more concentrated solution and dilute it to the appropriate concentration with dry methanol. The aliquot portions (say 5 ml) should be diluted with distilled water (20 ml) before titration.

(2) Universal indicator paper moistened with distilled water gives a satisfactory indication of neutralisation.

(3) Pure methyl β-D-galactopyranoside (m.p. 177–180 °C) may be isolated from the combined aqueous filtrates of these several crystallisations by removal of water and recrystallisation of the residue from absolute ethanol.

5.10.3 CARBOHYDRATE INTERCONVERSIONS

651

Reaction of an aldehyde or ketone with two molar proportions of a monohydric alcohol or with one molar proportion of a 1,2- or 1,3-diol in the presence of an acid catalyst yields an acetal (23) or cyclic acetal (24) or (25) respectively (see methods of protection, Section 5.4.6, p. 553, and Section 5.8.8, p. 624).

$$R^1_{R^2}{=}O \xrightarrow{2R^3OH} R^1_{R^2}\underset{OR^3}{\overset{OR^3}{<}} \qquad R^1_{R^2}{=}O \longrightarrow R^1_{R^2}\underset{O}{\overset{O}{<}}$$

(23)   (24)

$$R^1_{R^2}{=}O \longrightarrow R^1_{R^2}\underset{O}{\overset{O}{<}} \quad or \quad R^2\overset{R^1}{<}$$

(25)

The formation of acetals of the type (24) and (25) is an important reaction in the monosaccharide series. In general, ketones react with vicinal diol groups to give the five-membered cyclic products, and aldehydes react with 1,3-diols to give the six-membered cyclic products (see Section 5.10.4), although exceptions can be found with suitable, partially protected, monosaccharide derivatives. An example of the reaction of a monosaccharide with a ketone is the formation of 1,2:5,6-di-O-cyclohexylidene-α-D-glucofuranose (17) from D-glucose (15) and cyclohexanone in the presence of sulphuric acid (Expt 5.114). The formation of a furanose derivative in this case is promoted by the two favourable vicinal diol groups which are present in α-D-glucofuranose (4) (note the cis-orientation of the hydroxyl groups at C-1 and C-2), whereas there is only one vicinal cis-diol grouping in α-D-glucopyranose. Furthermore fusion of the five-membered acetal ring to the six-membered monosaccharide ring in (1) would be expected to introduce considerable ring strain. Reaction of D-glucose with acetone in the presence of zinc chloride and phosphoric acid gives the corresponding diisopropylidene derivative (16).

This reaction offers a convenient route to glucofuranose derivatives, particularly because the 5,6-acetal group may be selectively cleaved by hydrolysis to give the 1,2-mono acetal derivative (20) (Expt 5.115). The reaction conditions need to be carefully controlled to avoid extensive hydrolysis to the parent sugar. This is particularly important with the partial hydrolysis of 3-O-benzyl-1,2:5,6-di-O-cyclohexylidene-α-D-glucofuranose (cognate preparation in Expt 5.115), since the product (21) is then contaminated with 3-O-benzyl-D-glucose. Purification of the hydrolysis product, 3-O-benzyl-1,2-O-cyclohexylidene-α-D-glucofuranose (21), may be effected by column chromatography or by conversion of it into the more readily purified crystalline 5,6-dibenzoate derivative; the acyl groups may be then removed by the Zemplen method. The conversion of 1,2:5,6-di-O-cyclohexylidene-α-D-glucofuranose (17) into its 3-O-benzyl derivative (19) has been included (Expt 5.116) to illustrate the protection of hydroxyl groups in monosaccharides by conversion into ethers. The benzylation is effected by heating the diacetal with benzyl chloride in the presence of sodium hydroxide under PTC conditions. The formation of a benzyl ether is useful in synthetic interconversions with monosaccharides since it is easily and selectively cleaved by catalytic hydrogenolysis. Other means for the protection of hydroxyl groups include their conversion into methyl ethers (Expt 5.121), trimethylsilyl ethers (see Section 5.4.6, p. 552), triphenylmethyl ethers and allyl ethers (see Section 5.4.6, p. 551).

The preparation of 1,2-*O*-cyclohexylidene-α-D-xylofuranose (22) (Expt 5.117) from 1,2-*O*-cyclohexylidene-α-D-glucofuranose illustrates the use of sodium metaperiodate for the cleavage of carbon–carbon bonds in α-diols (see also Section 4.2.*55*, p. 454). In this case C-6 is lost as formaldehyde and C-5 is converted into an aldehyde group. This aldehydic product is isolated as a dimer, which is then reduced in methanol solution with sodium borohydride to the xylofuranose derivative (22).

The oxidation of the C-3-hydroxyl group in 1,2:5,6-di-*O*-isopropylidene-α-D-glucofuranose (16), with pyridinium chlorochromate, to a keto group (18) is a further example of the usefulness of recent oxidising agents for effecting functional group modification under mild conditions.[152] This keto-sugar derivative, and similar analogues, have become of importance for the development of synthetic routes to branched-chain monosaccharides (by reaction with organometallic reagents) and to rare monosaccharides by reduction with suitable reagents.

Experiment 5.114 **1,2:5,6-DI-*O*-CYCLOHEXYLIDENE-α-D-GLUCO-FURANOSE**

Fit a 3-litre flange flask with a multiple socket head carrying a mechanical stirrer capable of effecting vigorous agitation, a calcium chloride guard-tube, a 100-ml dropping funnel and a stoppered opening wide enough to allow for the addition of solid. Immerse the flask in a large plastic or metal container filled with an intimate mixture of ice and salt. Add 1000 g (1050 ml, 10 mol) of redistilled cyclohexanone to the flask and cool to 0 °C. Charge the separatory funnel with 62.5 ml of concentrated sulphuric acid and run the acid slowly into the vigorously stirred cyclohexanone; the final solution should be a light straw colour. Add slowly and portionwise with continued vigorous stirring 450 g (2.5 mol) of finely powdered dried α-D-glucose (1). Remove the cooling bath and allow the reaction mixture to reach ambient temperature with continual stirring; over a period of 8 hours the reaction mixture becomes progressively more viscous and finally sets into a solid off-white crystalline mass. Some caution should be exercised to prevent the stirrer motor from being overstrained. Allow the reaction mixture to stand at room temperature overnight, break up the crystalline mass, add 750 ml of heptane and a solution of 124 g of sodium carbonate in 375 ml of water, and heat on a boiling water bath with vigorous stirring. Decant as much of the upper heptane layer as possible from undissolved solid. Add a further 750 ml portion of heptane to the residue and heat under reflux until the remainder of the solid dissolves; decant the clear heptane layer and cool the combined heptane extracts in the refrigerator. Filter off the crystalline material, m.p. 121–124 °C, and recrystallise from heptane (2) using decolourising charcoal to clear the hot solution of traces of dark oily droplets. The purified 1,2:5,6-di-*O*-cyclohexylidene-α-D-glucofuranose has m.p. 131–132 °C, $[\alpha]_D^{20}$ −2.2° (c 1.8 in EtOH); the yield is 380 g (47%).

Notes. (1) If adequate cooling and stirring is not employed the final solution is dark yellow; furthermore the addition of glucose leads to an unacceptable local rise in temperature, and the final appearance of the reaction mixture is a dark red intractable oily mass. The powdered glucose should be dried in a vacuum desiccator over phosphorus pentoxide, not in an oven which apparently causes changes on the surface of the glucose particles which render them unreactive.
(2) Another solvent for recrystallisation is methylcyclohexane (0.17 g/ml).

Cognate preparations. *1,2:4,5-Di-O-cyclohexylidene-D-fructopyranose.* Add 200 g (1.11 mol) of finely powdered dry D-fructose with vigorous stirring to 419 g (440 ml, 4.49 mol) of ice-cooled cyclohexanone containing 30 ml of concentrated sulphuric acid; the reaction mixture becomes solid within 30 minutes. Leave the mixture overnight at room temperature, dissolve the product in 500 ml of chloroform and wash the solution with dilute aqueous sodium hydroxide, dilute hydrochloric acid and water and finally dry and evaporate. Solidify the residue by m.p. 145–156 °C, $[\alpha]_D^{20}$ −133.5° (c 1 in CHCl₃). The yield is 142 g (37%).

1,2:5,6-Di-O-isopropylidene-α-D-glucofuranose. A suspension of 150 g (0.83 mol) of dry D-glucose, 120 g (0.83 mol) of anhydrous zinc chloride and 7.5 g of phosphoric acid (88% v/v) in 1 litre of dry acetone is stirred at ambient temperature for 30 hours. Unchanged glucose is removed by filtration, and inorganic salts are precipitated by the addition of a solution of 85 g of sodium hydroxide in 85 ml of water. The resulting suspension is filtered, the residue washed with acetone and the acetone evaporated. The mass which remains is dissolved in 200 ml of water and extracted with five 100-ml portions of dichloromethane. The organic phase is dried and evaporated on a rotary evaporator. Recrystallisation from light petroleum (b.p. 80–100 °C) gives 70 g of product, m.p. 109–110 °C, $[\alpha]_D^{20}$ −18.5° (c 5 in H₂O).

1,2:5,6-Di-O-isopropylidene-D-mannitol. To a solution of 60 g of zinc chloride in 300 ml of acetone is added 10 g of finely powdered D-mannitol. The mixture is stirred vigorously at 20 °C until solution is complete (2–3 hours) and then allowed to stand for 16 hours. The reaction mixture is then poured into a solution of 70 g of potassium carbonate in 70 ml of water which is covered with 300 ml of ether. The mixture is stirred for half-an-hour when the organic layer is filtered from the agglomerated pellets of zinc carbonate. The latter is washed with 100 ml of 1:1 acetone–ether solution, and the combined filtrates evaporated to dryness on a rotary evaporator. The dry residue is successively extracted with five 250-ml portions of boiling light petroleum (b.p. 60–80 °C) and the combined filtrates slowly cooled to give the product, 7.9 g (55%), having m.p. 119 °C.

Experiment 5.115 1,2-*O*-CYCLOHEXYLIDENE-α-D-GLUCOFURANOSE

Heat a solution of 20 g (0.06 mol) of 1,2:di-*O*-cyclohexylidene-α-D-glucofura-nose (Expt 5.114) in 100 ml of aqueous acetic acid (75% v/v) in a round-bottomed flask immersed in a hot-water bath held at 70–80 °C for 90 minutes with intermittent shaking; then remove the solvent under reduced pressure on a rotary evaporator. To the residual syrup add 20 ml of hot water, then suffi-cient solid sodium hydrogen carbonate to neutralise the remaining acetic acid, and finally 90 ml of heptane. Heat the heterogeneous liquid mixture until two clear layers are obtained and then remove the upper heptane layer by careful decantation (1). Cool the aqueous layer to 0 °C, filter off the crys-tals of 1,2-*O*-cyclohexylidene-α-D-glucofuranose which separate and recrys-tallise from water to give the pure product, m.p. 149–150 °C, $[\alpha]_D^{20}$ +5.9° (*c* 1 in Me$_2$CO). The yield is 11.5 g (75%).

Note. (1) The solid which separates from the cooled heptane layer may be shown to be unchanged starting material by t.l.c. analysis on silica gel plates using methanol–benzene (4:96) (**CAUTION**) as the developing solvent.

Cognate preparation. *3-O-Benzyl-1,2-O-cyclohexylidene-α-D-glucofuranose.* Dissolve 100 g (0.23 mol) of 3-benzyl-1,2:5,6-di-*O*-cyclohexylidene-α-D-glucofuranose (Expt 5.116) in 400 ml of aqueous acetic acid (75% v/v) main-tained at 70–80 °C for 3 hours, remove the solvent under reduced pressure and dissolve the residual oil in 500 ml of dichloromethane. Wash this solution with aqueous sodium hydrogen carbonate and with water, dry over calcium sulphate and remove the dichloromethane by evaporation under reduced pressure with a rotary evaporator. Remove the last traces of solvent using an oil rotary immersion pump, transfer the warm fluid yellow syrup to the retort of a molecular still (Section 2.28) and distil using a vapour diffusion pump to give a pale yellow glass, b.p. 195–200 °C/2 × 10^{-3} mmHg, $[\alpha]_D^{20}$ −36.4° (*c* 4 in CHCl$_3$). The yield is 76 g (94%), and the product is pure enough for most purposes. However t.l.c. analysis on silica gel plates (solvent system; benzene–methanol 9:1) (**CAUTION**) reveals one major and two minor components. Purification may be effected by either of the two methods described below.

Chromatographic purification of 3-O-benzyl-1,2-O-cyclohexylidene-α-D-gluco-furanose. (The chromatographic column should be set up in a fume cupboard.) Prepare a silica gel column using benzene (**CAUTION**) as a solvent; use 10 g of adsorbent for each 1 g of monosaccharide derivative to be chromato-graphed. Dissolve the latter in the smallest volume of benzene and transfer the solution to the chromatographic column with a pipette. Elute the column with benzene and collect suitable-sized fractions; evaporate the solvent from each fraction and weigh the residues (Section 2.31) which consist of 3-*O*-benzyl-1,2:5,6-di-*O*-cyclohexylidene-α-D-glucofuranose. When all of this has

655

been eluted continue the development with methanol which elutes the required product. Evaporate the methanol and distil the residual syrup using a molecular still; about 50 per cent recovery of the purified product may be expected.

Purification of 3-O-benzyl-1,2-O-cyclohexylidene-α-D-glucofuranose by benzoylation. Dissolve 5 g of the crude product in 10 ml of pure dry pyridine and add 5 g of benzoyl chloride. Leave the reaction mixture overnight at room temperature, pour it on to ice and stir thoroughly. Extract the oil which separates into 50 ml of dichloromethane and wash the dichloromethane solution successively with 30 ml of ice-cold dilute aqueous hydrochloric acid (2 M), 30 ml of saturated aqueous sodium hydrogen carbonate, and 2 × 30 ml of water. Dry over sodium sulphate and evaporate the dichloromethane. The viscous oil crystallises on trituration with methanol and is recrystallised from methanol to give the pure derivative 5,6-di-O-benzoyl-3-O-benzyl-1,2-O-cyclohexylidene-α-D-glucofuranose, m.p. 104–106 °C, $[\alpha]_D^{20}$ −26.4° (c 1 in CHCl₃), in a yield of 3.6 g (57%). Remove the benzoyl groups by dissolving 3 g of the foregoing product in 20 ml of methanol and adding 20 ml of a solution of sodium methoxide in methanol (0.5%). After 2 hours neutralise the solution by adding ion exchange resin Zeolite 225 (H⊕), filter and evaporate. Distil the resulting colourless oil in a molecular still to obtain chromatographically pure 3-O-benzyl-1,2-O-cyclohexylidene-α-D-glucofuranose; the yield is 1.6 g (70% from the dibenzoate).

Experiment 5.116   3-O-BENZYL-1,2:5,6-DI-O-CYCLOHEXYLIDENE-α-D-GLUCOFURANOSE

In a 1-litre three-necked round-bottomed flask fitted with an efficient stirrer and a reflux condenser, place 170 g (0.5 mol) of 1,2:5,6-di-O-cyclohexylidene-α-D-glucofuranose, 8.1 g (0.025 mol) of tetrabutylammonium hydrogen sulphate, 400 ml of (E)-1,2-dichloroethylene, 76 g (0.6 mol) of benzyl chloride and a solution of 120 g (3 mol) of sodium hydroxide in 120 g of water. Stir the reaction mixture vigorously and heat under reflux. Thin-layer chromatography analysis (silica gel plates and ethyl acetate as developing solvent) shows the reaction to be complete within 2 hours. Cool the mixture and pour into 1 litre of water, separate the organic layer, wash, dry and evaporate. Distil the viscous yellow residue in a molecular still (Section 2.28) using a vapour diffusion pump. There is obtained a small forerun followed by the main fraction of 3-O-benzyl-1,2:5,6-di-O-cyclohexylidene-α-D-glucofuranose, which distils at a temperature of 210–220 °C/10 mmHg. The yield is 172 g (79%), $[\alpha]_D^{20} - 13.0°$ (c 5 in CHCl$_3$).

Experiment 5.117 1,2-O-CYCLOHEXYLIDENE-α-D-XYLOFURANOSE

Isolated as dimer

Di-(1,2-O-cyclohexylidene-α-D-xylo-pentodialdofuranose-5-hydrate)-5,5':3',5-dianhydride. Add a solution of 14.3 g (0.061 mol) of sodium metaperiodate in 220 ml of water dropwise to a well-stirred solution of 17.4 g (0.067 mol) of 1,2-O-cyclohexylidene-α-D-glucofuranose (Expt 5.115) in 50 ml of water (1). Stir for a further 30 minutes and remove the water at a temperature below 50 °C by evaporation under reduced pressure. Extract the solid residue with three 75 ml portions of dichloromethane and dry the combined extracts over magnesium sulphate. Filter and evaporate to give a residue which crystallises spontaneously. After recrystallisation from acetone the dimer has m.p. 182–183 °C, the yield is 7.9 g (51%).

1,2-O-Cyclohexylidene-α-D-xylofuranose. In a 250-ml two-necked round-bottomed flask fitted with a mechanical stirrer and dropping funnel, place 2 g (0.009 mol) of the dimer dissolved in 50 ml of aqueous ethanol (75% v/v). Add a solution of 1 g (0.053 mol) of sodium borohydride in 50 ml of ethanol from the dropping funnel with stirring. After a further 30 minutes remove the ethanol under reduced pressure (rotary evaporator) and extract the aqueous slurry which remains with three 25 ml portions of dichloromethane. Dry the combined extracts over magnesium sulphate and remove the solvent by evaporation under reduced pressure to give a viscous oil which crystallises on trituration with light petroleum (b.p. 40–60 °C). Recrystallise from benzene–light petroleum (b.p. 40–60 °C) to give the purified product m.p. 84–85 °C, $[\alpha]_D^{20} - 12°$ (c 1 in MeOH). The yield is 1.7 g (82%).

Note. (1) The progress of the oxidation may be followed by iodimetry. In this method unreacted periodate is reduced by arsenite solution in the presence of iodide at about pH 8. Excess arsenite is then determined by back titration with standard iodine solution.

$$IO_4^{\ominus} + AsO_2^{\ominus} \longrightarrow IO_3^{\ominus} + AsO_3^{\ominus}$$

For the oxidation described, remove 1 ml of solution, add 10 ml of saturated sodium hydrogen carbonate solution followed immediately by 10 ml of 0.05 M sodium arsenite solution and 1 ml of 20 per cent potassium iodide solution. Stand the solution in the dark for 15 minutes and titrate excess arsenite with 0.05 M iodine solution using a starch indicator.

A spectrophotometric method has been developed which utilises the stronger absorption of periodate ions at 222.5 nm compared to the iodate ion.[153]

5.10.4 CARBOHYDRATE INTERCONVERSIONS

This sequence serves to exemplify the formation and aspects of reactivity of toluene-p-sulphonate esters in monosaccharide systems, and further to illustrate the selective protection afforded to hydroxyl groups by the formation of cyclic acetals by reaction with carbonyl compounds. Thus reaction of methyl α-D-glucopyranoside (26) with benzaldehyde in the presence of zinc chloride gives the 4,6-acetal (27) (Expt 5.118), wherein two fused six-membered rings of the *trans*-decalin type are present. As a cognate preparation the reaction of benzaldehyde with methyl α-D-galactopyranoside results in a similar conversion to a 4,6-acetal, but in this case the product is the conformationally flexible system of the *cis*-decalin type, the most likely conformation being that shown below.

Since the 4,6-acetal grouping and the glycosidic grouping (27) are stable under basic conditions, though unstable in the presence of acid, the remaining two hydroxyl groups may be suitably protected by reactions that are base catalysed. Thus (27) may be converted into the 2,3-di-*O*-toluene-*p*-sulphonyl derivative (28) by reaction with toluene-*p*-sulphonyl chloride in the presence of pyridine (Expt 5.119); this reaction is analogous to the formation of sulphonate esters of alcohols or phenols.

$$R \cdot CH_2OH + ClSO_2 \cdot C_6H_4 \cdot Me \longrightarrow R \cdot CH_2 \cdot O \cdot SO_2 \cdot C_6H_4 \cdot Me$$

The value of the sulphonate esters in carbohydrate chemistry lies in their ability to undergo cleavage by reaction with nucleophiles in one of two possible ways, (i) cleavage of the S—O bond with regeneration of the hydroxyl group as a result of S_N2 attack at the sulphur atom, and (ii) cleavage of the C—O bond as a result of S_N2 attack at the carbon atom.

$$
\begin{array}{c}
\overset{(ii)\ (i)}{R \cdot CH_2 \overset{|}{+} O \overset{|}{+} SO_2 \cdot C_6H_4 \cdot Me}
\end{array}
\qquad
\begin{array}{l}
\xrightarrow[X^\ominus]{(i)} R \cdot CH_2O^\ominus + Me \cdot C_6H_4 \cdot SO_2X \\[2mm]
\xrightarrow[(ii)]{X^\ominus} R \cdot CH_2X + Me \cdot C_6H_4 \cdot SO_2 \cdot O^\ominus
\end{array}
$$

Both these processes are illustrated by the reaction of the 2,3-di-*O*-toluene-*p*-sulphonyl derivative with sodium methoxide in methanol (Expt 5.120). Here the sulphonyloxy group at position 2 undergoes cleavage (i) to yield the anion (31); this anionic site then participates in reaction at position 3 whereby cleavage of the sulphonyloxy group takes place by pathway (ii) with the formation of an oxirane ring system. This neighbouring group effect must proceed via the unfavourable boat conformation (32) in order that the stereoelectronic requirements for the reaction of the participating groups be met (i.e. *anti*-periplanar).

(28)   (31)

(32)   (29)

This reaction has led to an inversion of configuration at C-3 to yield the allose derivative (29). Ring opening of the oxirane system by treatment with alkali in aqueous media results in regeneration of a 2,3-vicinal diol but having the hydroxyl groups in the diaxial orientation, as a result of inversion of configuration at C-2 (Expt 5.120). Finally the isolation of methyl-α-D-altropyranoside (30) is described following the selective removal of the 4,6-acetal grouping by treatment with very dilute sulphuric acid, conditions which are insufficiently severe to result in hydrolysis of the glycosidic group.

5.10.5 CARBOHYDRATE INTERCONVERSIONS

The synthesis of the partially methylated glucoside, methyl 2,3-di-O-methyl-α-D-glucopyranoside (34) (Expt 5.121) also utilises the 4,6-O-benzylidene derivative (27), which is first converted into the 2,3-di-O-methyl derivative (33) by reaction with dimethyl sulphate in the presence of sodium hydroxide. Selective removal of the acetal grouping is achieved by mild acid hydrolysis; the methyl ether groups are stable under both acidic and basic conditions.

Experiment 5.118 METHYL 4,6-O-BENZYLIDENE-α-D-GLUCOPYRANOSIDE

Shake vigorously a mixture of 105 g (1.0 mol) of purified benzaldehyde (Expt 6.133), 38.8 g (0.2 mol) of methyl α-D-glucopyranoside (1) and 29.5 g of freshly fused and powdered anhydrous zinc chloride (0.22 mol) in a conical flask in a mechanical shaker for about 10 hours, until a clear solution is obtained. Allow the solution to stand at room temperature for a further period of 18 hours and then pour it into 700 ml of iced water (2). Stir the mixture vigorously, filter off the solid which separates and wash the compressed filter cake with light petroleum (b.p. 40–60 °C) to remove as much of the unreacted benzaldehyde as possible. Remove the solid from the Buchner funnel, stir or shake it vigorously with a solution of 12 g of sodium metabisulphite in 120 ml of water, filter and wash the filter cake with water. Crystallise the solid from hot water, or after drying in a vacuum desiccator, from a mixture of chloroform and ether. The pure product has m.p. 165 °C, $[\alpha]_D^{20} +112°$ (c 0.5 in $CHCl_3$), the yield is 29 g (52%).

Notes. (1) Methyl α-D-glucopyranoside is available commercially and it is not usually economic to prepare it in the laboratory; should, however, its preparation be necessary the method described in Expt 5.113 may be employed. The crude product (containing a mixture of anomeric glucosides) which is obtained after removal of methanol is stirred with cold (10 °C) methanol (0.6 ml/g), and the crude α-anomer removed by filtration. The pure compound, obtained after recrystallisation from ethanol (10 ml/g), has m.p. 167–169 °C, $[\alpha]_D^{20} + 157°$ (c 2 in H_2O). The yield is in the region of 30–40 per cent.

(2) An alternative isolation procedure is to extract the clear solution by shaking it with three successive 100 ml portions of light petroleum (b.p. 40–60 °C), which removes the unreacted benzaldehyde more effectively, and then to stir the viscous residue with ice-water until solidification occurs.

Cognate preparations. *Methyl 4,6-O-benzylidene-α-D-galactopyranoside* [recrystallised from ethanol/light petroleum (b.p. 60–80 °C)], m.p. 169–170 °C, $[\alpha]_D^{20} + 168.2$ (c 1.4 in $CHCl_3$), *methyl 4,6-O-benzylidene-β-D-gluco-pyranoside* (recrystallised from methanol), m.p. 194–196 °C, $[\alpha]_D^{20} - 74°$ (c 1 in EtOH), and *methyl 4,6-O-benzylidene-β-D-galactopyranoside* (recrystallised from methanol), m.p. 198–200 °C, $[\alpha]_D^{20} - 35.5°$ (c 2 in $CHCl_3$), may all be prepared similarly in yields of around 68 per cent.

Experiment 5.119   METHYL 4,6-O-BENZYLIDENE-2,3-DI-O-TOLUENE-p-SULPHONYL-α-D-GLUCOPYRANOSIDE

Place 40 ml of pure dry redistilled pyridine (**CAUTION**) in a conical flask and add 14.9 g (0.066 mol) of toluene-p-sulphonyl chloride (1) with cooling. Allow the yellow solution to stand for about half an hour at room termperature (2) before adding 8.5 g (0.03 mol) of methyl 4,6-O-benzylidene-α-D-gluco-pyranoside (Expt 5.118) with shaking and cooling. Leave the reaction mixture in the stoppered flask for five days at room temperature and then pour on to 75 g of crushed ice. Stir vigorously, extract the mixture of syrup and water with three 25 ml portions of dichloromethane and wash the combined extracts successively with cold dilute hydrochloric acid (2 M), water, saturated aqueous sodium hydrogen carbonate, and water, and then dry over magnesium sulphate. Remove the dichloromethane by evaporation under reduced pressure and triturate the gummy residue with ether until solid. Recrystallise from dichloromethane–ether to obtain the pure product, m.p. 152–154 °C, $[\alpha]_D^{20} + 11.8°$ (c 1 in $CHCl_3$). The yield is 13.5 g (75%).

Notes. (1) The toluene-p-sulphonyl chloride may be purified by following the procedure given in Section 4.2.78, p. 466.

(2) In all sulphonylations performed in these laboratories this procedure of allowing the mixture of toluene-*p*-sulphonyl chloride in pyridine to stand for about half an hour before the addition of the material to be sulphonylated has been found to be beneficial.

Experiment 5.120   METHYL α-D-ALTROPYRANOSIDE

Methyl 2,3-anhydro-4,6-*O*-benzylidene-α-D-allopyranoside. In a two-necked, 250-ml round-bottomed flask fitted with a mechanical stirrer unit and a pressure-equalising funnel protected with a calcium chloride guard-tube place a solution of 11.7 g (0.0195 mol) of methyl 4,6-*O*-benzylidene-2,3-di-*O*-toluene-*p*-sulphonyl-α-D-glucopyranoside (Expt 5.119) in 150 ml of dichloromethane. Cool the solution to 0 °C by means of an ice–salt bath and add a solution of sodium methoxide in methanol [prepared from 2.3 g (0.1 mol) of sodium and 40 ml of methanol] dropwise with stirring. When the addition is complete remove the stirrer and funnel, stopper the flask and leave it in a refrigerator for 48 hours and then at room temperature for a further 24 hours. Extract the dichloromethane solution with water until the aqueous washings are neutral, dry over magnesium sulphate and remove the solvent from dichloromethane to give the pure product, m.p. 195–199 °C, $[\alpha]_D^{20}$ +140° (*c* 2 in CHCl$_3$). The yield is 4.2 g (82%).

Methyl 4,6-*O*-benzylidene-α-D-altropyranoside. Triturate 4.0 g (0.015 mol) of the foregoing anhydro derivative in a mortar with a solution of 5 g of potassium hydroxide dissolved in 140 ml of water. Transfer the suspension to a round-bottomed flask and heat the mixture under reflux until all the solid has dissolved (about 28 hours). During this period solid material tends to creep up the inside of the flask surface; shake periodically to re-suspend material. Remove the trace of insoluble matter which remains and neutralise the cooled filtrate with carbon dioxide (use phenophthalein as an indicator). Extract the solution with five 25 ml portions of dichloromethane, wash the combined extracts with a little cold water, dry over anhydrous sodium sulphate and remove the solvent under reduced pressure (rotary evaporator). Crystallise the syrup by scratching a small portion on a watch glass with ether; stir the bulk syrup with ether and the seed crystals. Filter off and recrystallise the product from a small quantity of methanol to obtain 3.5 g (83%) of methyl 4,6-*O*-benzylidene-α-D-altropyranoside, m.p. 174 °C, $[\alpha]_D^{20}$ +115° (*c* 2 in CHCl$_3$).

Methyl α-D-altropyranoside. Hydrolyse the benzylidene protecting group by heating, at 60 °C for 1 hour, 3.5 g (0.025 mol) of the foregoing compound in a mixture of 140 ml of warm water and 7 ml of 0.05 M sulphuric acid. Concentrate the residual solution to 50 ml using a rotary evaporator under reduced pressure (benzaldehyde is removed during the process) and make just alkaline (phenolphthalein) with 0.1 M barium hydroxide solution (prepared by diluting a cold saturated solution with an equal volume of water). Remove the barium sulphate by filtration (Whatman No. 42 paper), wash the residue with water and concentrate the filtrate and washings to a syrup (rotary evaporator). Dissolve the residue in a little methanol, add ether until a slight turbidity is observed and set the solution on one side to crystallise. The resulting methyl α-D-altroside may be recrystallised from methanol/ether and has m.p. 107–108 °C, $[\alpha]_D^{20} + 126°$ (c 3 in H_2O); the yield is 2.1 g (88%).

Experiment 5.121 METHYL 2,3-DI-O-METHYL-α-D-GLUCOPYRANO-SIDE

Methyl 4,6-O-benzylidene-2,3-di-O-methyl-α-D-glucopyranoside. *Owing to the hazards in the use of dimethyl sulphate this experiment must be carried out in an efficient fume cupboard (see Section 4.2.24, p. 430 for precautions in the use of dimethyl sulphate).* Mount a 500-ml, three-necked round-bottomed flask in an electrically-heated water bath and fit an efficient stirrer unit in the central neck. Attach two 100-ml dropping funnels to the side-necks; one containing 38 g (29 ml, 0.30 mol) of purified dimethyl sulphate and the other containing 60 ml (0.6 mol) of 40 per cent aqueous sodium hydroxide solution. Place 14.1 g (0.05 mol) of methyl 4,6-O-benzylidene-α-D-glucopyranoside and 150 ml of acetone in the flask, run in 15 ml of sodium hydroxide solution, commence fairly rapid stirring and raise the temperature of the water bath to about 50 °C. Add dropwise and simultaneously the remainder of the sodium hydroxide solution and the dimethyl sulphate over a period of about 1.5 hours. Continue stirring at 50 °C for a further half an hour. Remove the dropping funnels, fit a condenser set for downward distillation into one of the side-necks and a nitrogen inlet tube into the other. Remove the acetone by distillation during 1 hour while maintaining a steady nitrogen flow. Pour the contents of the flask into 1500 ml of ice-cold water and collect the solid produced by filtration. Wash the product with cold water until the washings are neutral to litmus, dry overnight at 50 °C and recrystallise twice from light

663

petroleum (b.p. 60–80 °C), to give 12.5 g (81%) of product having m.p. 122–123 °C, $[\alpha]_D^{20}$ +94° (c 2 in CHCl$_3$), or $[\alpha]_D^{20}$ +97° (c 4 in Me$_2$CO).

Methyl 2,3-*O*-di-*O*-methyl-α-D-glucopyranoside. Dissolve 6.2 g (0.02 mol) of the foregoing derivative in 100 ml of acetone containing 0.3 per cent of concentrated sulphuric acid in a 250-ml round-bottomed flask fitted with a reflux condenser. Boil the solution under gentle reflux on a steam bath. Periodically remove the flask, cool to room temperature in a stream of cold water, remove an aliquot of suitable size and follow the progress of the hydrolysis by measuring the optical rotation, which changes from about 6.12 to about 5.74° for a 1-dm cell. When the reaction is complete (under 1 hour), add 100 ml of water and neutralise the solution with solid barium carbonate. Filter the reaction mixture and evaporate the filtrate on a rotary evaporator. If an odour of benzaldehyde remains, add 100 ml of water and repeat the evaporation. Distil the crude syrupy product under reduced pressure to give a colourless glass of b.p. 130–135 °C/0.1 mmHg which may be crystallised by trituration with dry benzene (**CAUTION**). Two recrystallisations from the same solvent give 3.2 g (72%) of methyl 2,3-di-*O*-methyl-α-D-glucopyranoside, m.p. 85 °C, $[\alpha]_D^{20}$ +146° (c 4 in Me$_2$CO).

5.11 ALIPHATIC CARBOXYLIC ACIDS

The carbon chain of carboxylic acids is numbered systematically starting with the carboxyl carbon. Illustrative representations are given for:

butanoic acid (1) CH$_3$·CH$_2$·CH$_2$·CO$_2$H;
4-methylpentanoic acid (γ-methylvaleric acid) (2)
    CH$_3$·CH(CH$_3$)·CH$_2$·CH$_2$·CO$_2$H;
2-methylpentanoic acid (α-methylvaleric acid) (3)
    CH$_3$·CH$_2$·CH$_2$·CH(CH$_3$)·CO$_2$H.

Branching in the carbon chain may lead to a chiral site (*); thus (3) is chiral but (1) and (2) are achiral.

The presence of two or more carboxyl groups in a carbon chain gives rise to dicarboxylic, tricarboxylic, etc., acids; many of these acids are designated by universally recognised and accepted trivial names (see Table 10.20). Unsaturated acids are considered in Section 5.18.3, p. 804; other functionally substituted acids (e.g. hydroxy acids, amino acids, etc.) are considered in Section 5.14.

The synthesis of aliphatic carboxylic acids is exemplified by the following typical procedures.

1. Oxidative methods (Expts 5.122 to 5.126).
2. The hydrolysis of alkyl cyanides (Expts 5.127 and 5.128).
3. The carboxylation of Grignard reagents (Expt 5.129).

4. The Arndt–Eistert method (Expt 5.130).
5. Electrolytic (anodic) coupling (Expt 5.131).
6. Methods utilising diethyl malonate (Expts 5.132 to 5.136).
7. The synthesis of optically active carboxylic acids (Expt 5.137).

Methods for the protection of the carboxyl group are considered in Section 5.11.8.

SUMMARY OF RETROSYNTHETIC STRATEGIES

Functional group interconversion (FGI) (methods 1 and 2)

R—CH₂CH₂—OH or R—CH₂—CHO ⟸(1) R—CH₂—CO₂H ⟹(1) R—C≡CH

(TM)

⇓ (2)

R—CH₂—CN

Disconnection (methods 1, 3, 5, 6, and 7)
Monocarboxylic acids

$R^1R^2CH^{\ominus}$ + $^{\oplus}CO_2H$ ⟸(3) (ii) { $R^1R^2C(i)$—CO_2H ⟹(1) R^1R^2C—CO·Me

(TM)

$\parallel$

$R^1R^2CHMgX + CO_2$ (ii)(6)   (7)(ii)

$\overset{\oplus}{R^1}; \overset{\oplus}{R^2}\ \overset{2\ominus}{CH}\cdot CO_2H$ $\overset{\oplus}{R^2}$ $R^1—CH_2—CO_2H^{\ominus}$

$\parallel$ $\parallel$

$R^1X; R^2X + CH_2(CO_2Et)_2$ $R^2X + R^1—CH_2—$ oxazoline (Me, Me, N)

1,2- and 1,3-Dicarboxylic acids and 1,2,3-tricarboxylic acids

HO_2C—CHR—CH₂—CO_2H ⟹(6) $HO_2C^{\ominus}$ + $^{\oplus}CHR$—CH₂—CO_2H ≡ $^{\ominus}CN$ + R—CH=CH—CO_2Et

(TM)

665

Alkyl substituted carboxylic acids may be prepared by these routes by using appropriately substituted reagents.

Long-chain dicarboxylic acids

Rearrangement (method 4)

SPECTROSCOPIC FEATURES

The characteristic profile of the i.r. spectral region ($2500–3300 \, cm^{-1}$) corresponding to a broad absorption from the intermolecular bonding of the carboxyl group is easily recognised in aliphatic carboxylic acids (hexanoic acid, Fig. 3.31). The frequency of absorption arising from the carbonyl group occurs in the region $1725–1700 \, cm^{-1}$ (p. 299). The p.m.r. spectrum of an aliphatic carboxylic acid will show a low field signal from the acidic proton, provided that the sample is dissolved in a solvent which does not undergo a deuterium exchange reaction. The structure of the alkyl group may be assessed from the spin–spin splitting patterns as illustrated in Fig. 3.57. The interpretation of the *m.s.* of mono-carboxylic acids is discussed on p. 379. The *u.v.–visible* spectra of saturated acids is not usually structurally informative. Further descriptive spectroscopic analyses are included in some of the following preparative examples.

5.11.1 OXIDATIVE METHODS

Saturated primary alcohols are readily oxidised to aldehydes, which in turn are further oxidised to monocarboxylic acids having the same number of carbon atoms.

$$R \cdot CH_2OH \longrightarrow R \cdot CHO \longrightarrow R \cdot CO_2H$$

The reaction is frequently effected using alkaline potassium permanganate solution (e.g. Expt 5.122). Aqueous sodium dichromate/sulphuric acid mixtures may be used, but yields are not always satisfactory because of the attendant production of appreciable amount of esters (cf. the preparation of aldehydes, Section 5.7.1, p. 587).

Secondary alcohols on oxidation give ketones which may be cleaved under vigorous oxidative conditions to a mixture of carboxylic acids.

Clearly such a method is of limited preparative value, but an important exception is the oxidation of cyclic secondary alcohols which on oxidation with nitric acid give good yields of dicarboxylic acids by way of the intermediate cyclic ketone, e.g. adipic acid from cyclohexanone, Expt 5.123.

The oxidation of a methyl ketone to a carboxylic acid can be effected by the use of the *haloform reaction*. This involves treatment of the methyl ketone with an alkaline hypohalite reagent. A trihalomethyl ketone is initially formed which then undergoes hydrolysis under the basic conditions used.

$$2\overset{\ominus}{HO} + Br_2 \rightleftharpoons \overset{\ominus}{OBr} + \overset{\ominus}{Br} + H_2O$$

$$R \cdot CO \cdot Me + \overset{\ominus}{OBr} \rightleftharpoons [R \cdot CO \cdot \overset{\ominus}{CH_2}] + HOBr$$
mesomeric anion

$$R \cdot CO \cdot \overset{\ominus}{CH_2} \quad Br \overset{\frown}{-} OH \longrightarrow R \cdot CO \cdot CH_2Br + \overset{\ominus}{OH}$$

$$R \cdot CO \cdot CH_2Br \longrightarrow R \cdot CO \cdot CBr_3 \xrightarrow{\ominus OH} R \cdot CO_2^{\ominus} + CHBr_3$$

In the examples given the preparative value depends upon the ready availability of the required methyl ketone; thus pinacolone (Expt 5.98), cyclopropyl methyl ketone (Expt 7.1) and mesityl oxide (Expt 5.213) are converted into 2,2-dimethylpropanoic acid, cyclopropanecarboxylic acid and 3,3-dimethylacrylic acid respectively (Expts 5.124 and 5.125).

A carboxylic acid group is formed when a carbon–carbon multiple bond is oxidatively cleaved. The main value of this is as a degradative process in structural elucidation, as for example in the ozonolysis of alkenes followed by oxidative decomposition of the ozonide (Section 2.17.4). Oxidation of a symmetrical alkene or alkyne gives rise to a single carboxylic acid product which could, however, probably be more conveniently synthesised by other routes. Unsymmetrical non-terminal alkenes or alkynes are rarely used as substrates since a mixture of acidic products would be produced. The method has, however, found application using terminal alkynes since the terminal carbon is lost as carbon dioxide when, for example, the oxidation is carried out with potassium permanganate. In the illustrative example the acetylenic carbinol, 1-ethynyl-3,3,5-trimethylcyclohexanol (Expt 5.41) gives a good yield of the corresponding hydroxy carboxylic acid on oxidation in aqueous acetone (Expt 5.126). Oxidation of appropriate alkynes (and alkenes) to carboxylic acids can also be carried out efficiently using potassium permanganate under PTC conditions.[154]

Experiment 5.122   2-METHYLPROPANOIC ACID (*Isobutyric acid*)

$$Me_2CH \cdot CH_2OH \xrightarrow[KMnO_4]{[O]} Me_2CH \cdot CO_2H + H_2O$$

Place a mixture of 52 g (0.7 mol) of 2-methylpropan-1-ol and a solution of 15 g of sodium carbonate in 150 ml of water in a 5-litre round-bottomed flask. Add a solution of 142 g (0.9 mol) of potassium permanganate in 2750 ml of water, with vigorous stirring, during 3–4 hours, cooling the mixture to 4–5 °C by immersion in a bath of ice-water. Then allow the reaction mixture to attain room temperature gradually. After 12 hours, filter off (or preferably, centrifuge) the precipitated manganese dioxide, concentrate the filtrate to about 150 ml under reduced pressure and then cool. Cover the solution with a layer of ether and acidify with dilute sulphuric acid. Separate the ether layer and extract the aqueous layer two or three times with 50 ml portions of ether. Dry the combined ethereal extracts over anhydrous sodium sulphate, remove the ether on a water bath and fractionate the residual liquid. Collect the 2-methylpropanoic acid at 153–155 °C. The yield is 45 g (76%); i.r. spectrum (thin film), 2200–3800 (br. OH) and 1700 cm^{-1} (br. C = O). Record the p.m.r. spectra and assign the signals; note the hydroxyl proton may not be observed.

Experiment 5.123   ADIPIC ACID (*Hexanedioic acid*)

$$\xrightarrow[HNO_3]{[O]} HO_2C \cdot (CH_2)_4 \cdot CO_2H$$

CAUTION: *This preparation must be carried out in an efficient fume cupboard and the apparatus should be sited behind a safety-screen for extra protection.*
    Into a 500-ml three-necked flask, fitted with a dropping funnel, a mechanical stirrer and an efficient reflux condenser, place 190 ml (270 g) of concentrated nitric acid, *d* 1.42. Since oxides of nitrogen are evolved in the subsequent oxidation, the reaction should be carried out in a fume cupboard, or the oxides of nitrogen are led by a tube from the top of the condenser to a water trap (Fig. 2.61). Heat the nitric acid to boiling, set the stirrer in motion, add a few drops of cyclohexanol and *make certain that these are acted upon by*

the acid before adding more; an explosion may result if cyclohexanol is allowed to accumulate in the acid. Once the reaction has started, add 50 g (0.5 mol) of cyclohexanol through the dropping funnel at such a rate that all is introduced in 2–3 hours. Keep the reaction mixture at the boiling point during the addition of the cyclohexanol and for a further period of about 15 minutes in order to complete the oxidation. Pour the warm reaction mixture into a beaker; upon cooling, the adipic acid crystallises. Filter on a large sintered glass funnel, and wash with 20 ml of cold water. Recrystallise the crude acid from 70 ml of concentrated nitric acid; filter and wash as above. The yield of recrystallised adipic acid, m.p. 152 °C, is 40 g (55%).

Experiment 5.124 2,2-DIMETHYLPROPANOIC ACID (*Trimethylacetic acid; pivalic acid*)

$$Me_3C \cdot CO \cdot Me \xrightarrow[\text{(ii) } H_3O^{\oplus}]{\text{(i) 3NaOBr}} Me_3C \cdot CO_2H + CHBr_3$$

In a 3-litre three-necked flask, fitted with a thermometer, a mechanical stirrer and dropping funnel, place a solution of 160 g (4 mol) of sodium hydroxide in 1400 ml of water. Cool to 0 °C in an ice–salt bath. Add 240 g (77 ml, 1.5 mol) of bromine with vigorous stirring at such a rate as to keep the temperature below 10 °C (15–20 minutes). Cool again to 0 °C, introduce 50 g (0.5 mol) of pinacolone (Expt 5.98) keeping the temperature below 10 °C. After the solution is decolourised (c. 1 hour), continue the stirring for 3 hours at room temperature. Replace the thermometer by a knee tube connected to a condenser for distillation and replace the stirrer by a steam inlet to allow the bromoform and carbon tetrabromide (if present) to be separated by steam distillation; heat the flask with a powerful Bunsen burner. Remove the burner, cool the reaction mixture to 50 °C and add 200 ml of concentrated sulphuric acid cautiously through the dropping funnel. Heat the flask again; the trimethylacetic acid passes over with about 200 ml of water. When all the trimethylacetic acid (the upper layer; 35–40 ml) has distilled, a liquid heavier than water (possibly brominated pinacolone) begins to pass over, Stop the distillation at this point, separate the trimethylacetic acid from the aqueous layer, and dry it by distillation with 25 ml of benzene (the latter carries over all the water) or with anhydrous calcium sulphate. Distil under reduced pressure and collect the trimethylacetic acid 75–80 °C/20 mmHg. The yield is 33 g (55%), m.p. 34–35 °C.

Cognate preparation. *Cyclopropanecarboxylic acid.* Use 42 g (47 mol, 0.5 mol) of cyclopropyl methyl ketone (Expt 7.1) and react with alkaline hypobromite solution exactly as in the above preparation except that the final period of stirring at room temperature need be only 1.5 hours. After removal of bromoform by steam distillation, cool and cautiously acidify the solution to Congo red with 250 ml of concentrated hydrochloric acid. Discharge the pale yellow colour by adding a little sodium metabisulphite solution. Saturate the solution with salt and extract with four 300 ml portions of ether; dry the combined extracts with anhydrous sodium sulphate, and distil off the ether on a water bath through a short fractionating column. Distil the residue under reduced pressure and collect the pure cyclopropanecarboxylic acid (a colourless liquid) at 92 °C/22 mmHg. The yield is 33 g (76%), ^{13}C-n.m.r. spectrum (CDCl$_3$, TMS) δ 9.2, 13.1 and 181.9.

Experiment 5.125   3,3-DIMETHYLACRYLIC ACID (*3-Methylbut-2-enoic acid*)

$$Me_2C{=}CH{\cdot}CO{\cdot}Me \xrightarrow[\text{(ii) } H_3O]{\text{(i) } 3NaOCl} Me_2C{=}CH{\cdot}CO_2H + CHCl_3$$

Fit a 1-litre three-necked flask with two double surface condensers and a sealed stirrer unit. Place 25 g (29 ml, 0.25 mol) of mesityl oxide (Expt 5.213), 50 ml of dioxane and a cold (10 °C) solution of sodium hypochlorite in 750 ml of water (1) in the flask, and stir the mixture. Heat is evolved in the reaction and after about 5 minutes chloroform commences to reflux. As soon as the reaction becomes very vigorous, stop the stirrer and cool the flask with water so that the chloroform refluxes gently; after 20–30 minutes, when the reaction has subsided, resume the stirring and continue it until the temperature of the mixture has fallen to that of the laboratory (2–3 hours). Decompose the slight excess of hypochlorite by the addition of sodium metabisulphite (about 1 g), i.e. until a test-portion no longer liberates iodine from potassium iodide solution.

Replace one of the reflux condensers by a dropping funnel and add 50 per cent sulphuric acid (about 50 ml) with stirring and cooling until the solution is acid to Congo red paper. Extract the cold solution with eight 50 ml portions of ether (2) and shake the mixture well during each extraction. Dry the combined ethereal extracts with anhydrous calcium sulphate, and remove the ether and chloroform slowly on a water bath. Distil the residue from a flask fitted with a Claisen still-head and a short fractionating column under diminished pressure and collect the acid at 100–106 °C/20 mmHg; this fraction solidifies on cooling and melts at 60–65 °C. The yield is 13 g (51%), Recrystallise from hot water (1 g of acid in 10 ml of water) (3), cool the solution in ice for 2–3 hours, filter and dry overnight in a vacuum desiccator. Alternatively, recrystallise from light petroleum, b.p. 60–80 °C. Pure 3,3-dimethylacrylic acid has m.p. 68 °C; p.m.r. spectrum (CCl₄, TMS) δ 1.95 (s, 3H), 2.19 (s, 3H), 5.69 (m, 1H) and 12.22 (s, 1H).

Notes. (1) This solution is prepared by diluting 300 ml of commercial sodium hypochlorite (containing 10–14% available chlorine) to 750 ml with water.
(2) A continuous ether extractor (Fig. 2.92) gives more satisfactory results.
(3) Do not boil the aqueous solution for a long time as the acid is markedly steam-volatile.

Experiment 5.126   1-HYDROXY-3,3,5-TRIMETHYLCYCLOHEXANE-CARBOXYLIC ACID

Equip a 100-ml round-bottomed two-necked flask with a magnetic follower bar, a dropping funnel and a thermometer positioned to dip into the reaction mixture. Site the flask in a plastic bowl on a magnetic stirrer unit. Place in the flask a solution of 1 g (0.0055 mol) of *trans*-1-ethynyl-3,3,5-trimethylcyclo-hexan-1-ol (Expt 5.41) in 10 ml of acetone and add a solution of 1.25 g

(0.0079 mol) potassium permanganate in 25 ml of water from the dropping funnel with stirring over 1 hour. Ensure that the temperature of the reaction mixture is maintained below 20 °C by surrounding the reaction flask with ice-water.

When the addition is complete, continue stirring at room temperature for 1 hour, and then heat the solution under reflux for a further hour. Cool the reaction mixture, filter, decolourise the filtrate by the passage of sulphur dioxide and acidify the solution with hydrochloric acid. The product obtained by filtration is dried in a vacuum desiccator before being recrystallised from light petroleum (b.p. 60–80 °C). The yield of pure product is 0.7 g (58%); m.p. 134–135 °C.

5.11.2 HYDROLYSIS OF ALKYL CYANIDES

Since alkyl cyanides are readily available from the interaction of alkyl halides with sodium or potassium cyanide in aqueous alcoholic solution (Expts 5.157 and 5.158), their hydrolysis to carboxylic acids is a valuable synthetic method. Aqueous alkaline or acidic conditions may be used. The reaction proceeds via the intermediate formation of an amide. Experimental conditions may be selected to interrupt the hydrolysis at the amide stage (Expt 6.167).

$$R \cdot C \equiv N$$

$$R \cdot CO_2H + NH_4 \xleftarrow{H_3O^{\oplus}} R \cdot CONH_2 \xrightarrow{{}^{\ominus}OH/H_2O} R \cdot CO_2^{\ominus} + NH_3$$

As well as the illustrative synthesis of a simple carboxylic acid (pentanoic acid, Expt 5.127), examples are given of the synthesis of some dicarboxylic acids, aryl substituted carboxylic acids, and the unsaturated acid, vinylacetic acid (Expt 5.128).

The nitrile group in the readily available α-hydroxynitriles (the cyanohydrins) may also be similarly hydrolytically converted into a carboxyl group to afford a convenient synthesis of α-hydroxy acids (Expt 5.168).

Experiment 5.127 PENTANOIC ACID (*Valeric acid*)

$$Bu \cdot CN + 2H_2O \xrightarrow{{}^{\ominus}OH} NH_3 \uparrow + \left[Bu \cdot CO_2^{\ominus} \xrightarrow{H_3O^{\oplus}} Bu \cdot CO_2H \right]$$

Place 100 g (125 ml, 1.2 mol) of valeronitrile (Expt 5.158) and a solution of 92 g of sodium hydroxide in 260 ml of water in a 1500-ml round-bottomed flask, attach a double surface condenser and boil under reflux until the nitrile layer disappears (5–10 hours). Add through the condenser 100 ml of water, then slowly, and with external cooling, 125 ml of 50 per cent (by volume) sulphuric acid. Separate the upper layer of valeric acid (it may be necessary to filter first from any solid present), and dry it with anhydrous calcium sulphate. Distil and collect the valeric acid at 183–185 °C (mainly 184 °C). The yield is 82 g (67%). A further 5 g of acid may be obtained by extracting the strongly acidified aqueous layer with ether, combining the ethereal extracts with the low and high boiling point fractions of the previous distillation, removing the ether on a water bath and distilling the residue.

Experiment 5.128 GLUTARIC ACID (*Pentanedioic acid*)

$$N{\equiv}C\diagdown\diagup\diagdown\diagup C{\equiv}N \xrightarrow[\text{H}_2\text{O}]{\text{H}_2\text{SO}_4} HO{-}\overset{O}{\underset{\|}{C}}\diagdown\diagup\diagdown\overset{O}{\underset{\|}{C}}{-}OH + (NH_4)_2SO_4$$

In a 2-litre round-bottomed flask, equipped with a double surface condenser, place 60 g (0.64 mol) of pentanedinitrile (Expt 5.157) and 900 g of 50 per cent sulphuric acid (by weight). Reflux the mixture for 10 hours and allow to cool. Saturate the solution with ammonium sulphate and extract with four 150 ml portions of ether; dry the ethereal extracts with anhydrous sodium sulphate. Distil off the ether on a water bath; the residual glutaric acid (69 g, 82%) crystallises on cooling and has m.p. 97–97.5 °C. Upon recrystallisation from chloroform, or benzene, the m.p. is 97.5–98 °C.

Cognate preparations. *Suberic acid* (*octanedioic acid*). Heat a mixture of octanedinitrile (Expt 5.157) with 15 times its weight of 50 per cent sulphuric acid by weight under reflux for 10 hours. The acid crystallises out on cooling. Filter off the suberic acid upon a sintered glass funnel, and recrystallise it from acetone: m.p. 141–142 °C. The yield is 90 per cent of the theoretical.

Pimelic acid (*heptanedioic acid*). Heat a mixture of 18 g (0.148 mol) of heptanedinitrile (Expt 5.157) and 250 g of 50 per cent sulphuric acid by weight in a 750-ml round-bottomed flask under reflux for 9 hours. Most of the pimelic acid separates from the cold reaction mixture. Filter off the crystalline acid upon a sintered glass funnel. Saturate the filtrate with ammonium sulphate and extract it with three 50 ml portions of ether. Dissolve the residue on the filter (which is slightly discoloured, but is fairly pure pimelic acid) in the combined ethereal extracts, dry with anhydrous sodium sulphate and remove the ether by distillation. Recrystallise the residual solid acid from benzene containing 5 per cent of ether. The yield of pure pimelic acid, m.p. 105–106 °C, is 22 g (93%).

Phenylacetic acid. Into a 500-ml round-bottomed flask, provided with a reflux condenser, place 100 ml of water, 100 ml of concentrated sulphuric acid and 100 ml of glacial acetic acid: add 100 g (98 ml, 0.85 mol) of benzyl cyanide. Heat under reflux for 45–60 minutes; hydrolysis is then complete. Pour the mixture into 2–3 volumes of water with stirring. Filter the crude acid at the pump. Melt the crude material under water, and wash it two or three times with small volumes of hot water; the acid solidifies on cooling. Test a small portion for the presence of phenylacetamide (m.p. 155 °C) by dissolving in sodium carbonate solution. If a clear solution results, phenylacetamide is absent; if the solution is not clear, shake the whole of the crude product with excess of sodium carbonate solution, filter and precipitate the phenylacetic acid from the clear filtrate by the addition of dilute sulphuric acid. Filter off the phenylacetic acid and recrystallise it from hot water or, better, light petroleum (b.p. 40–60 °C). The yield of pure acid, 77 °C, is 50 g (43%). Small quantities of acid may be recovered from the mother-liquors by extraction with ether, but this is rarely worth while. Alternatively the acid may be purified by distillation under reduced pressure, b.p. 140–150 °C/20 mmHg. The p.m.r. spectrum should be recorded and compared with that shown in Fig. 3.47. The ^{13}C-n.m.r. spectrum is shown in Fig. 3.55.

p-*Nitrophenylacetic acid*. Prepare a dilute solution of sulphuric acid by adding 150 ml of concentrated sulphuric acid cautiously to 140 ml of water. Place 50 g (0.31 mol) of *p*-nitrobenzyl cyanide (Expt 5.21) in a 500-ml round-bottomed flask, pour in about two-thirds of the sulphuric acid and shake well until all the solid is moistened with the acid. Wash down any nitrile adhering to the walls of the flask into the liquid with the remainder of the acid. Attach a reflux condenser to the flask and boil under reflux for 15 minutes. Dilute the rather dark reaction mixture with an equal volume of cold water and cool to 0 °C. Filter with suction, and wash several times with ice-water. Dissolve the solid in 800 ml of boiling water (add decolourising carbon, if necessary) and filter rapidly through a hot-water funnel supporting a fluted filter paper. If any solid remains on the filter, dissolve it in the minimum volume of boiling water and filter into the main filtrate. Collect the pale yellow needles of *p*-nitrophenylacetic acid which separate on cooling, and dry at 100 °C. The yield of acid, m.p. 151–152 °C, is 53 g (95%).

Vinylacetic acid (*but-3-enoic acid*). Place 134 g (161 ml, 2 mol) of allyl cyanide (Expt 5.157) and 200 ml of concentrated hydrochloric acid in a 1-litre round-bottomed flask attached to a reflux condenser. Warm the mixture cautiously with a small flame and shake from time to time. After 7–10 minutes, a vigorous reaction sets in and the mixture refluxes; remove the flame and cool the flask, if necessary, in cold water. Ammonium chloride crystallises out. When the reaction subsides, reflux the mixture for 15 minutes. Then add 200 ml of water, cool and separate the upper layer of acid. Extract the aqueous layer with three 100 ml portions of ether. Combine the acid and the ether extracts, and remove the ether under atmospheric pressure in a 250-ml flask fitted with a Claisen still-head and a short fractionating column: continue the heating on a water bath until the temperature of the vapour reaches 70 °C. Allow the apparatus to cool and distil under diminished pressure; collect the fraction (*a*) distilling up to 71 °C/14 mmHg and (*b*) at 72–74 °C/14 mmHg (chiefly at 72.5 °C/14 mmHg). A dark residue (about 10 ml) and some white solid (? crotonic acid) remains in the flask. Fraction (*b*) weighs 100 g (58%) and is analytically pure vinylacetic acid. Fraction (*a*) weighs about 50 g and separates into two layers: remove the water layer, dry the organic phase with anhydrous sodium sulphate and distil under reduced pressure; a further 15 g (8.7%) of reasonably pure acid, b.p. 69–70 °C/12 mmHg, is obtained.

5.11.3 CARBOXYLATION OF GRIGNARD REAGENTS

The addition of a Grignard reagent to carbon dioxide gives the salt of the corresponding carboxylic acid, which on acidification yields the free carboxylic acid.

$$\text{XMg-R} \quad \text{O=C=O} \longrightarrow \text{R·CO·OMgX} \xrightarrow{\text{H}_3\text{O}^{\oplus}} \text{R·CO}_2\text{H}$$

The reaction is best carried out by pouring the ethereal solution of the Grignard reagent directly on to an excess of coarsely powdered solid carbon dioxide. The alternative procedure of passing dry carbon dioxide gas into the Grignard reagent solution may give rise to the formation of ketonic by-products by further reaction of the Grignard reagent with the carboxylate salt. (cf. Expt 7.13).

$$\underset{R}{\overset{O}{\underset{}{\parallel}}}\!\!\!\!C\!\!-\!\!OMgX \ + \ R{-}MgX \ \longrightarrow \ R_2C(OMgX)_2 \ \longrightarrow \ R_2CO$$

The synthesis of 2-methylbutanoic acid (Expt 5.129) is illustrative of the method. Other organometallic reagents undergo a similar carboxylation reaction, and examples of the use of organolithium and organosodium reagents are included in the section on the synthesis of aromatic carboxylic acids (Section 6.13.3, p. 1069).

Experiment 5.129 2-METHYLBUTANOIC ACID

$$Me{\cdot}CH_2{\cdot}CHCl{\cdot}Me \ \xrightarrow[\text{ether}]{\text{Mg}} \ Me{\cdot}CH_2{\cdot}CH(Me)MgCl \ \xrightarrow{CO_2}$$

$$Me{\cdot}CH_2{\cdot}CH(Me){\cdot}CO_2MgCl \ \xrightarrow{H_3O^{\oplus}} \ Me{\cdot}CH_2{\cdot}CH(Me){\cdot}CO_2H$$

Fit a 1-litre three-necked flask with a mechanical stirrer, a double surface condenser and a separatory funnel and provide both the condenser and funnel with calcium chloride guard-tubes. Prepare an ethereal solution of but-2-ylmagnesium chloride from 12.5 g (0.51 mol) of dry magnesium turnings, 46 g (52.5 ml, 0.5 mol) of dry 2-chlorobutane (1) and 400 ml of anhydrous ether by the procedure described in Expt 5.39, cognate preparation. When the spontaneous reaction has subsided reflux the reaction mixture for a further 1 hour. Cool the flask in a mixture of ice and salt to − 12 °C and add a further 100 ml of anhydrous ether. Weigh out (rough balance) 125 g of Cardice (2) on a piece of stiff paper: wrap the Cardice in a stout cloth, and, by means of a pestle, break it into small lumps. Empty the Cardice into a dry 1500-ml beaker and at once pour in the Grignard reagent in a slow steady stream; any unreacted magnesium will adhere to the sides of the flask. A vigorous reaction occurs. Stir the mass well, and allow it to stand until all the Cardice has evaporated. Then add slowly a mixture of 300 g of crushed ice and 75 ml of concentrated hydrochloric acid. Stir until the gelatinous compound is decomposed and there is a clean separation into two layers. Pour the mixture into a separatory funnel; rinse the beaker with 50 ml of ether and transfer this to the funnel. Separate the upper layer and extract the aqueous layer with three 40 ml portions of ether. Cool the combined ether extracts by the addition of ice, and add cautiously 100 ml of 25 per cent sodium hydroxide solution; run off and keep the aqueous layer and repeat the extraction with a further 50 ml of alkali solution of the same strength. The organic acid is thus converted into the sodium salt and passes into the aqueous layer; test the extracts with phenolphthalein to make certain that all the acid has been removed. Distil the alkaline extract until its volume is reduced by about 10 per cent; this removes ether and other volatile impurities. Allow to cool, and cautiously acidify with concentrated hydrochloric acid; it is advisable to stir the mixture during the acidification process. Separate the upper layer of acid. Distil the water layer from a 1-litre flask until no more oily drops pass over; saturate the distillate with salt, remove the acid layer and combine it with the main product. Dry the combined acid fractions with anhydrous calcium sulphate, and distil. Collect the 2-methylbutanoic acid at 173–174 °C. The yield is 40 g (79%).

Notes. (1) 2-Chlorobutane is employed in preference to the bromide because it is cheaper and the yield of acid is slightly higher.

(2) Cardice should be handled with gloves or with a dry towel; if Cardice is held for a long time in the hand it may cause frost bite. The crushed Cardice should be used immediately otherwise it may absorb water which would react with some of the Grignard reagent.

5.11.4 THE ARNDT–EISTERT METHOD

The Arndt–Eistert reaction is a comparatively simple method for converting an acid into its next higher homologue, or to a derivative of the homologous acid, such as an amide or an ester. The reaction is applicable to aliphatic, aromatic, alicyclic and heterocyclic acids, and the yield is generally good.

The overall reaction is the insertion of a methylene group between the alkyl and carboxyl groups, and involves a rearrangement in the sequence of mechanistic steps. The reaction takes place in three stages:

1. Formation of the carboxylic acid chloride, e.g. with thionyl chloride or with phosphorus pentachloride.

$$R\text{—}\overset{O}{\underset{OH}{\overset{\|}{C}}} \xrightarrow{SOCl_2} R\text{—}\overset{O}{\underset{Cl}{\overset{\|}{C}}}$$

2. Formation of a diazoketone (4) by the gradual addition of the acid chloride to an excess of an ethereal solution of diazomethane. This reaction may be represented mechanistically as follows:

$$R\text{—}\overset{O}{\underset{Cl}{\overset{\|}{C}}} \overset{\ominus}{CH_2}\cdot\overset{\oplus}{N}\equiv N \xrightarrow{-Cl^{\ominus}} R\text{—}\overset{O}{\underset{\overset{\oplus}{CH_2}\cdot N\equiv N}{\overset{\|}{C}}} \xrightarrow[-Me\cdot N_2^{\oplus}]{+CH_2N_2} R\text{—}\overset{O}{\underset{\underset{\ominus}{CH}\text{—}\underset{\oplus}{N}\equiv N}{\overset{\|}{C}}}$$

$$\text{(4)}$$

$$Cl^{\ominus} + Me\cdot\overset{\oplus}{N_2} \longrightarrow MeCl + N_2\uparrow$$

If the reaction is carried out with the acid chloride in excess, e.g. by adding the diazomethane solution slowly to the acid chloride, some halomethyl ketone is produced:

$$R\cdot CO\cdot\overset{\ominus}{CH}\cdot\overset{\oplus}{N_2} + HCl \longrightarrow R\cdot CO\cdot CH_2Cl + N_2\uparrow$$

3. Rearrangement of the diazoketone, with loss of nitrogen, in the presence of suitable reagents and a catalyst (colloidal silver–silver oxide, or silver nitrate in the presence of ammonia solution). An acid is formed in the presence of water, an amide results when ammonia or an amine is used, and an ester is produced in the presence of an alcohol.

$$R^1\cdot CO\cdot CHN_2 \begin{array}{l} \xrightarrow{H_2O} R^1\cdot CH_2\cdot CO_2H + N_2 \\ \xrightarrow{NH_3} R^1\cdot CH_2\cdot CONH_2 + N_2 \\ \xrightarrow{R^2OH} R^1\cdot CH_2\cdot CO_2R^2 + N_2 \end{array}$$

This third operation, involving the conversion of the diazoketone into an acid or a simple derivative thereof, is known as the *Wolff rearrangement*. Loss of nitrogen from the diazoketone (4) is accompanied by a nucleophilic 1,2-shift of the alkyl group to yield a keten (5), which reacts with the solvent to give the carboxylic acid or the appropriate derivative.

$$O=\underset{\ominus}{\overset{R}{C}}-CH-\overset{\oplus}{N}\equiv N \longrightarrow O=C=CH\cdot R \xrightarrow{H_2O} \underset{HO}{\overset{O}{>}}C\cdot CH_2\cdot R$$

(4)   (5)

In order to prepare an acid, a dioxane solution of the diazoketone is added slowly to a suspension of silver oxide in a dilute solution of sodium thiosulphate. If the conversion to the acid yields unsatisfactory results, it is usually advisable to prepare the amide or ester, which are generally obtained in good yields; hydrolysis of the derivative gives the free acid.

The conversion of a diazoketone to an acid amide may be accomplished by treating a warm solution in dioxane with 10–28 per cent aqueous ammonia solution containing a small amount of silver nitrate solution, after which the mixture is heated to 60–70 °C for some time. Precautions should be taken (*by use of a safety glass shield*) when heating mixtures containing ammoniacal silver nitrate.

Esters of the homologous acids are prepared by adding silver oxide in portions rather than in one lot to a hot solution or suspension of the diazoketone in an anhydrous alcohol (methanol, ethanol or propan-1-ol): methanol is generally used and the silver oxide is reduced to metallic silver, which usually deposits as a mirror on the sides of the flask. The production of the ester may frequently be carried out in a homogeneous medium by treating a solution of the diazoketone in the alcohol with a solution of silver benzoate in triethylamine.

The reaction is illustrated here by the overall conversion of the dicarboxylic acid, sebacic acid, into dodecanedioic acid by a bis-homologation process (Expt 5.130).

A further illustrative use of a diazoketone, in this case as a source of a carbene, is given in the preparation of azulene (Expt 6.16).

Experiment 5.130 DODECANEDIOIC ACID

$$\underset{\overset{|}{CO_2H}}{\overset{CO_2H}{\overset{|}{(CH_2)_8}}} \xrightarrow{SOCl_2} \underset{\overset{|}{COCl}}{\overset{COCl}{\overset{|}{(CH_2)_8}}} \xrightarrow{CH_2N_2} \underset{\overset{|}{CO\cdot CHN_2}}{\overset{CO\cdot CHN_2}{\overset{|}{(CH_2)_8}}} \xrightarrow[H_2O]{Ag_2O} \underset{\overset{|}{CH_2\cdot CO_2H}}{\overset{CH_2\cdot CO_2H}{\overset{|}{(CH_2)_8}}}$$

Sebacoyl chloride. Convert 20 g (0.1 mol) of sebacic acid (Expt 5.131) into the corresponding acid chloride by heating it on a water bath in a flask fitted with a reflux condenser (protected with a calcium chloride tube) with 20 ml of thionyl chloride; the apparatus should be assembled in a fume cupboard. Purify the product by distillation under reduced pressure (use appropriate traps to protect the pump from the fumes of hydrogen chloride and sulphur dioxide). Collect the sebacoyl chloride as a fraction of b.p. 140–143 °C/ 2 mmHg; the yield is 18 g (77%).

1,8-Bis-diazoacetyloctane. Dissolve 7.4 g (0.033 mol) of the resulting sebacoyl chloride in anhydrous ether and add the solution slowly to an ethereal solution containing about 6.8 g of diazomethane, i.e. two portions of the ethereal solution prepared as described in Section 4.2.25, p. 432, *Method 3*, CAUTION (1). Allow the mixture to stand overnight and remove any excess reagent together with some of the ether by distillation from a warm-water bath. *To ensure that no undue hazard results from the possible presence of undecomposed excess diazomethane use a distillation assembly as described for the distillation of diazomethane–ethereal solutions (Section 4.2.25, p. 432, Method 1).* When the distillate is colourless, change the receivers, and complete the removal of solvent by distillation under reduced pressure (water pump). After recrystallisation from benzene the resulting 1,8-bis-diazoacetyloctane has m.p. 91 °C; the yield is 6.4 g (83%).

Dodecanedioic acid. Add, with stirring, a solution of 5 g (0.02 mol) of the bis-diazoketone in 100 ml of warm dioxane to a suspension of 6.0 g of freshly precipitated silver oxide (2) in 250 ml of water containing 8 g of sodium thiosulphate maintained at 75 °C. A brisk evolution of nitrogen occurs; after 1.5 hours at 75 °C, filter the liquid from the black silver residue. Acidify the almost colourless filtrate with nitric acid and extract the gelatinous precipitate with ether. Evaporate the dried ethereal extract: the residue of crude dodecanedioic acid weighs 3.3 g (72%), and has m.p. 116–117 °C. Recrystallisation from 20 per cent aqueous acetic acid raises the m.p. to 127–128 °C.

Alternatively, treat 3.9 g (0.0156 mol) of the bis-diazoketone in 50 ml of warm dioxane with 15 ml of 20 per cent aqueous ammonia and 3 ml of a 10 per cent aqueous silver nitrate solution under reflux in a 250- or 500-ml flask on a water bath. Nitrogen is evolved for a few minutes, followed by a violent reaction and the production of a dark brown opaque mixture. Continue heating for 30 minutes on the water bath and filter hot; the diamide of dodecanedioic acid is deposited on cooling. Filter the product and air dry; the yield is 3.1 g (87%), m.p. 182–184 °C, raised to 184–185 °C after recrystallisation from 20 per cent aqueous acetic acid. Hydrolyse the diamide by refluxing for 2–5 hours with a four molar excess of 3M potassium hydroxide solution. Acidify and recrystallise the precipitated acid from 20 per cent acetic acid. The yield of dodecanedioic acid, m.p. 127–128 °C, is almost quantitative.

Notes. (1) Precautions in the use of diazomethane are fully described in Section 4.2.25, p. 430, and should be carefully noted; the operations should be carried out in a fume cupboard.
(2) Prepare the silver oxide by adding dilute sodium hydroxide solution gradually to a stirred 10 per cent aqueous silver nitrate solution – until precipitation is just complete. Wash the product thoroughly with distilled water.

5.11.5 ELECTROLYTIC (ANODIC) COUPLING

The generation of radicals from carboxylate ions at the anode (Section 2.17.6, p. 115), and their coupling to form new carbon–carbon bonds is illustrated by the synthesis of hexacosane (Expt 5.11). The method has been usefully applied to the preparation of esters of long-chain carboxylic acids, from which of course the free acids may be prepared by hydrolysis.

Simple anodic coupling by electrolysis in anhydrous methanolic solution (containing a little sodium methoxide) of methyl hydrogen adipate (Expt 5.147) gives dimethyl sebacate; methyl hydrogen sebacate (Expt 5.147) in turn yields dimethyl octadecanedioate (Expt 5.131, cognate preparations).

$$2MeO_2C \cdot (CH_2)_4 \cdot CO_2^\ominus \xrightarrow{-2e} MeO_2C \cdot (CH_2)_8 \cdot CO_2Me + 2CO_2$$

$$2MeO_2C \cdot (CH_2)_8 \cdot CO_2^\ominus \xrightarrow{-2e} MeO_2C \cdot (CH_2)_{16} \cdot CO_2Me + 2CO_2$$

Electrolysis of a mixture of two carboxylic acids, $R^1 \cdot CO_2H$ and $R^2 \cdot CO_2H$, leads in addition to the products of normal coupling ($R^1 - R^1$ and $R^2 - R^2$) to the cross-coupled product ($R^1 - R^2$). Similarly if a mixture of a saturated carboxylic acid and a half-ester of an α, ω-dicarboxylic acid is electrolysed, there are three main products, viz. a hydrocarbon (6), a mono-ester (7) and a di-ester (8). Normally the three products are readily separable by distillation. Furthermore, by increasing the molar proportion of the monocarboxylic acid, the yield of (7) is improved at the expense of (8).

$$R \cdot CO_2H + HO_2C \cdot (CH_2)_n \cdot CO_2Me \longrightarrow R - R + R \cdot (CH_2)_n \cdot CO_2Me$$
$$ (6) (7)$$

$$+ MeO_2C \cdot (CH_2)_{2n} \cdot CO_2Me + CH_2{=}CH \cdot (CH_2)_{n-2} \cdot CO_2Me$$
$$ (8) (9)$$

The unsaturated ester (9) is also often present in small quantity and arises from the loss of a proton from the intermediate carbocation (11), which is produced when the radical species (10) (which is involved in the coupling reaction) undergoes further anodic oxidation.

$$MeO_2C \cdot (CH_2)_{n-2} \cdot CH_2 \cdot CH_2 \cdot C{\overset{O}{\underset{O}{\diagup\!\!\!\diagdown}}} \xrightarrow{-CO_2} MeO_2C \cdot (CH_2)_{n-2} \cdot CH_2 \cdot \dot{C}H_2 \xrightarrow{-e}$$
$$ (10)$$

$$MeO_2C \cdot (CH_2)_{n-2} \cdot CH_2 \cdot \overset{\oplus}{C}H_2 \xrightarrow{-H^\oplus} MeO_2C \cdot (CH_2)_{n-2} \cdot CH{=}CH_2$$
$$(11)$$

Two alternative syntheses of methyl myristate, and thence myristic acid, are described (Expt 5.131). In *Method A* hexanoic acid (2 mol) is coupled with methyl hydrogen sebacate (1 mol), the products being methyl myristate, decane and dimethyl octadecanedioate.

In *Method B* decanoic acid (2 mol) is coupled with methyl hydrogen adipate (1 mol) and the products are methyl myristate, octadecane and dimethyl sebacate.

Experiment 5.131 MYRISTIC ACID (*Tetradecanoic acid*)

Method A:

$$Me(CH_2)_4 \cdot CO_2^\ominus + {}^\ominus O_2C \cdot (CH_2)_8 \cdot CO_2Me \xrightarrow[-2CO_2]{-2e} Me \cdot (CH_2)_8 Me +$$
$$Me \cdot (CH_2)_{12} \cdot CO_2Me + MeO_2C \cdot (CH_2)_{16} \cdot CO_2Me$$

Method B:

$$Me(CH_2)_8 \cdot CO_2^\ominus + {}^\ominus O_2C \cdot (CH_2)_4 \cdot CO_2Me \xrightarrow[-2CO_2]{-2e} Me \cdot (CH_2)_{16} Me +$$
$$Me \cdot (CH_2)_{12} \cdot CO_2Me + MeO_2C \cdot (CH_2)_8 \cdot CO_2Me$$

Method A. Dissolve 23.2 g (0.184 mol) of redistilled hexanoic acid, b.p. 204.5–205.5 °C/760 mmHg, and 21.6 g (0.1 mol) of methyl hydrogen sebacate in

200 ml of absolute methanol to which 0.13 g of sodium has been added. Electrolyse at 2.0 amps (Section 2.17.6), while maintaining the temperature between 30 and 40 °C, until the pH is about 8.0 (c. 6 hours). Neutralise the contents of the electrolysis cell with a little acetic acid and distil off the methanol on a water bath. Dissolve the residue in 200 ml of ether, wash with three 50 ml portions of saturated sodium hydrogen carbonate solution, once with water, dry with magnesium sulphate, and distil through an efficient fractionating column (Sections 2.26 and 2.27). Collect the decane at 60 °C/10 mmHg (3.0 g), the methyl myristate at 158–160 °C/10 mmHg (12.5 g, 52%) and dimethyl octadecanedioate at 215–230 °C/7 mmHg (1.5 g).

Reflux a mixture of 7.3 g of methyl myristate with a solution of 4.8 g of sodium hydroxide in 200 ml of 90 per cent methanol for 2 hours, distil off the methanol on a water bath, dissolve the residue in 400 ml of hot water, add 15 ml of concentrated hydrochloric acid to the solution at 50 °C in order to precipitate the organic acid, and cool. Collect the acid by suction filtration, wash it with a little water and dry in a vacuum desiccator. The yield of myristic acid (tetradecanoic acid), m.p. 57–58 °C, is 5.9 g (87%).

Method B. Dissolve 55.2 g (0.32 mol) of pure decanoic acid, m.p. 31–32 °C, and 25.6 g (0.16 mol) of methyl hydrogen adipate in 200 ml of absolute methanol to which 0.25 g of sodium has been added. Electrolyse at 2.0 amps at 25–35 °C until the pH of the electrolyte is 8.2 (c. 9 hours). Neutralise the contents of the electrolytic cell with acetic acid, distil off the methanol on a water bath, dissolve the residue in about 200 ml of ether, wash with three 50 ml portions of saturated sodium hydrogen carbonate solution and remove the ether on a water bath. Treat the residue with a solution of 8.0 g of sodium hydroxide in 200 ml of 80 per cent methanol, reflux for 2 hours and distil off the methanol on a water bath. Add about 600 ml of water to the residue to dissolve the mixture of sodium salts: extract the hydrocarbon with four 50 ml portions of ether, and dry the combined ethereal extracts with magnesium sulphate. After removal of the ether, 23.1 g of almost pure octadecane, m.p. 23–24 °C, remains. Acidify the aqueous solution with concentrated hydrochloric acid (c. 25 ml), cool to 0 °C, filter off the mixture of acids, wash well with cold water and dry in a vacuum desiccator. The yield of the mixture of sebacic and myristic acids, m.p. 52–67 °C, is 26 g. Separate the mixture by extraction with six 50 ml portions of almost boiling light petroleum, b.p. 40–60 °C. The residue (5.2 g), m.p. 132 °C, is sebacic acid. Evaporation of the solvent gives 20 g (55%) of myristic acid, m.p. 52–53 °C; the m.p. is raised slightly upon recrystallisation from methanol.

Cognate preparations. *Sebacic acid* (*decanedioic acid*). Dissolve 40 g (0.25 mol) of methyl hydrogen adipate in 100 ml of absolute methanol to which 0.1 g of sodium has been added. Pass a current of about 2.0 amps until the pH of the solution is about 8 (c. 5 hours); test with narrow-range indicator paper. Transfer the contents of the electrolysis cell to a 500-ml round-bottomed flask, render neutral with a little acetic acid and distil off the methanol on a water bath. Dissolve the residue in 150 ml of ether, wash with three 50 ml portions of saturated sodium hydrogen carbonate solution, then with water, dry over magnesium sulphate and distil under reduced pressure. Collect the dimethyl sebacate at 155 °C/8 mmHg; it melts at 26 °C and the yield is 14.6–16.0 g (51–56%).

Reflux 14.6 g (0.064 mol) of the ester with a solution of 10 g of sodium hydroxide in 125 ml of 80 per cent methanol for 2 hours on a water bath. Add 200 ml of water to dissolve the solid which separates, extract with two 30 ml portions of ether and warm the aqueous solution on a water bath to remove dissolved ether. Acidify the ice-cold aqueous solution to litmus by the addition of concentrated hydrochloric acid. Collect the precipitated acid by suction filtration, wash it with a little cold water and dry at 100 °C. The yield of sebacic acid, m.p. 133 °C, is 11.5 g (89%).

Octadecanedioic acid.   Dissolve 31.5 g (0.145 mol) of methyl hydrogen sebacate in 140 ml of absolute methanol to which 0.4 g of sodium has been added. Electrolyse at 2.0 amps until the pH of the electrolyte is 7.8–8.0 (3.5–4 hours). Work up as described for sebacic acid. Upon distillation, an unsaturated ester passes over at 111–113 °C/20 mmHg (4.6 g), followed by dimethyl octadecanedioate at 212–219 °C/4 mmHg (mainly at 214–215 °C/4 mmHg), m.p. 56 °C (16.5 g, 66%).

Reflux 6.8 g of the dimethyl ester with a solution of 3.2 g of sodium hydroxide in 150 ml of 80 per cent methanol for 2 hours on a water bath. When cold, filter off the solid and wash it with a little cold methanol. Dissolve the solid in 350 ml of warm water, add concentrated hydrochloric acid to the solution at 60 °C until acidic to litmus, filter off the precipitated acid, wash with a little water and dry at 100 °C. The resulting octadecanedioic acid, m.p. 122 °C, weighs 5.3 g (84%). Recrystallisation from absolute methanol raises the m.p. to 124.5 °C.

5.11.6   METHODS UTILISING DIETHYL MALONATE

The retrosynthetic summary emphasises that diethyl malonate may be regarded as an acetic acid equivalent, and may be used when a disconnection gives rise to the synthons $^{\ominus}CH_2 \cdot CO_2H$ or $^{2\ominus}CH \cdot CO_2H$. The use of diethyl malonate in the preparation of carboxylic acids is illustrated in the sections below.

HYDROLYSIS OF ALKYLMALONIC ESTERS

When treated with one equivalent of sodium ethoxide, diethyl malonate is converted into the mono-sodio derivative, as a result of removal by base of one of the α-methylene protons to yield a mesomeric anion (12). This nucleophilic anion undergoes an S_N2 reaction with an alkyl halide to give a C-substituted malonic ester. A second, different, alkyl group can be similarly introduced on to the α-carbon atom, or alternatively two identical alkyl groups may be introduced in a one-step operation by using appropriate proportions of reactants.

Alkaline hydrolysis of an alkylmalonic ester followed by careful acidification at 0 °C gives the alkylmalonic acid (e.g. Expt 5.133). The alkylmalonic acids undergo smooth decarboxylation on heating under acidic conditions, thus providing a convenient synthesis of mono- or disubstituted acetic acids (Expt 5.132).

THE USE OF MICHAEL ADDITIONS OF MALONATE IONS

Typically the addition of the mesomeric anion (12) to the α, β-unsaturated ester, diethyl fumarate, proceeds in a 1,4-manner (the *Michael reaction* or *Michael addition*).

The resulting tetraethyl ester on hydrolysis and decarboxylation yields propane-1,2,3-tricarboxylic acid.[155] In this example the malonate anion is generated by using one molar proportion of sodium ethoxide; this is Michael's original method. However, these conditions sometimes lead to competing side reactions and the formation of abnormal reaction products. Better yields of the required product are often obtained with small amounts of sodium ethoxide (the so-called catalytic method) or in the presence of a secondary amine (e.g. diethyl-amine, see below).

Frequently the basic conditions used cause the initial Michael adduct to undergo intramolecular transformations, as for example in the synthesis of dimedone (Expt 7.11). This involves a Michael reaction between mesityl oxide and diethyl malonate followed by an internal Claisen ester condensation.

The α, β-unsaturated component for a Michael reaction may be formed *in situ* by an initial *Knoevenagel reaction*. An example is provided by the formation of tetraethyl propane-1,1,3,3-tetracarboxylate (14) from formaldehyde and diethyl malonate in the presence of diethylamine. Diethyl methylenemalonate (13) is first formed by the simple Knoevenagel reaction and this is followed by the Michael addition process.

Acid hydrolysis of (14) is accompanied by decarboxylation to give glutaric acid (Expt 5.134).

681

The synthesis of 2,2-dimethylsuccinic acid (Expt 5.135) provides a further variant of the synthetic utility of the Knoevenagel–Michael reaction sequence. Ketones (e.g. acetone) do not readily undergo Knoevenagel reactions with malonic esters, but will condense readily in the presence of secondary amines with the more reactive ethyl cyanoacetate to give an α, β-unsaturated cyanoester (e.g. 15). When treated with ethanolic potassium cyanide the cyanoester (15) undergoes addition of cyanide ion in the Michael manner to give a dicyanoester (16) which on hydrolysis and decarboxylation affords 2,2-dimethylsuccinic acid.

(15)   (16)

The synthesis of α-alkylglutaric acids (e.g. 2-propylglutaric acid, Expt 5.136), is conveniently achieved by allowing an alkylmalonic ester to react with the α, β-unsaturated nitrile, acrylonitrile, in the Michael manner and then subjecting the product to vigorous acidic hydrolysis.

Experiment 5.132   HEXANOIC ACID

Diethyl butylmalonate. Prepare a solution of sodium ethoxide from 34.5 g (1.5 mol) of clean sodium and 1 litre of super-dry ethanol (Section 4.1.9, p. 401) (1) in a 2-litre three-necked flask following the experimental conditions given for ethyl propylacetoacetate (Expt 5.95). When the sodium ethoxide solution, which is vigorously stirred, has cooled to about 50 °C, add 247.5 g (234.5 ml, 1.55 mol) of redistilled diethyl malonate slowly through the separatory funnel; to the resulting clear solution introduce gradually (60–90 minutes) 205.5 g (161.5 ml, 1.5 mol) of redistilled butyl bromide (Expt 5.54). Reaction occurs almost immediately and much heat is evolved; if the reaction becomes violent, cool the flask by directing a stream of cold water over it. Reflux the reaction mixture on a water bath until it is neutral to moist litmus (about 2 hours). Remove as much of the ethanol as possible by distillation under reduced pressure (rotary evaporator) on a water bath. Cool the contents of the flask to about 20 °C, add 600 ml of water and shake well. Separate the upper layer of crude ester, dry it with anhydrous sodium sulphate and distil under reduced pressure. A low boiling point fraction passes over first, followed by diethyl butylmalonate at 130–135 °C/20 mmHg. The yield is 285 g (88%). The distillation may also be conducted under normal pressure; the b.p. of the ester is 235–240 °C.

Hexanoic acid. Into a 2-litre three-necked flask, fitted with a separatory funnel, a mechanical stirrer and a reflux condenser, place a hot solution of 200 g of potassium hydroxide in 200 ml of water. Stir the solution and add slowly 200 g (0.925 mol) of diethyl butylmalonate. A vigorous reaction occurs and the solution refluxes. When all the ester has been added, boil the solution gently for 2–3 hours, i.e. until hydrolysis is complete: a test portion should dissolve completely in water. Dilute with 200 ml of water and distil off 200 ml of liquid in order to ensure the complete removal of the alcohol formed in the hydrolysis (2). To the cold residue in the flask add a cold solution of 320 g (174 ml) of concentrated sulphuric acid in 450 ml of water: add the acid slowly with stirring in order to prevent excessive foaming. The solution becomes hot. Reflux the mixture for 3–4 hours and allow to cool. Separate the upper layer of the organic acid and extract the aqueous portion with four 150 ml portions of ether (3). Combine the acid layer with the ether extracts, wash it with 25 ml of water and dry with anhydrous sodium sulphate. Distil off the ether (rotary evaporator), transfer the residue to a flask fitted with a short fractionating column (the latter should be well lagged and, preferably, electrically heated) and distil the product from an air bath. Collect the hexanoic acid at 200–206 °C. The yield is 80 g (75%). Record the i.r. spectrum and compare it with that shown in Fig. 3.31.

If desired, the distillation may be conducted under reduced pressure. The boiling points under various pressures are 99 °C/10 mmHg and 111 °C/20 mmHg; a 3 °C fraction should be collected.

Notes. (1) With commercial absolute ethanol, the yield is reduced to about 225 g.
(2) It is essential to remove the alcohol completely, otherwise some ethyl hexanoate, b.p. 168 °C, is formed which will contaminate the final product.
(3) Better results are obtained if a continuous extraction apparatus (e.g. Fig. 2.92) is employed.

Cognate preparations. *Diethyl propylmalonate.* Use 34.5 g (1.5 mol) of sodium and 345 g (440 ml) of super-dry ethanol, 240 g (227.5 ml, 1.5 mol) of diethyl malonate and 185 g (136.5 ml, 1.5 mol) of propyl bromide (Expt 5.54). The yield of diethyl propylmalonate, b.p. 218–225 °C, mainly 219.5–221.5 °C, is 220 g (72.5%).

Pentanoic acid (*valeric acid*). Convert the diethyl propylmalonate into valeric acid, b.p. 183–185 °C, following the procedure described for hexanoic acid. The yield is 75 per cent of theory.

3-Phenylpropanoic acid (*hydrocinnamic acid*). Use 11.5 g (0.5 mol) of sodium and 250 ml of dry ethanol, 80 g (75 ml, 0.49 mol) of diethyl malonate and 64 g (58 ml, 0.51 mol) of redistilled benzyl chloride (Expt 6.27). Follow the alkylation procedure described above and isolate the crude diethyl benzylmalonate. Hydrolyse the latter with a solution of 75 g of potassium hydroxide in 75 ml of water and isolate the resulting crude 3-phenylpropanoic acid as described previously, using 180 ml of 5 M sulphuric acid in the acidification stage. Purify the product by distillation under reduced pressure, collecting the fraction of b.p. 164–172 °C/25 mmHg which solidifies at room temperature. Recrystallise from light petroleum, b.p. 40–60 °C (or from water containing a little hydrochloric acid), to obtain 20 g (27%) of 3-phenylpropanoic acid of m.p. 47–48 °C.

Nonanoic acid (*pelargonic acid*). Equip a 1-litre three-necked flask with a reflux condenser, a sealed stirrer unit and a thermometer. Place 23 g (1 mol) of sodium, cut in small pieces, in the flask, and add 500 ml of anhydrous butan-1-ol (1) in two or three portions: follow the experimental details given in Expt 5.95 for the preparation of a solution of sodium ethoxide. When the sodium has reacted completely, allow the solution to cool to 70–80 °C and add 160 g (152 ml, 1 mol) of redistilled diethyl malonate rapidly and with stirring. Heat the solution to 80–90 °C, replace the thermometer with a dropping funnel and add 182.5 g (160 ml, 1.02 mol) of 1-bromoheptane (Expt 5.55) slowly at first until precipitation of sodium bromide commences, and subsequently at such a rate that the butanol refluxes gently. Reflux the mixture until it is neutral to moist litmus (about 1 hour).

Transfer the entire reaction mixture, including the precipitated sodium bromide and the small volume of water used to rinse the reaction flask, to a 3-litre flask. Add a solution of 140 g of potassium hydroxide in an equal quantity of water slowly and with shaking. Attach a reflux condenser to the flask, introduce a few fragments of porous porcelain and heat the mixture cautiously, with occasional shaking, until refluxing commences. Heat to gentle refluxing until hydrolysis is complete (about 5 hours, i.e. until a test portion is completely miscible with excess of water). Immediately equip the flask for steam distillation and steam distil the mixture until no more butanol passes over. Treat the residue cautiously with 270 ml of concentrated hydrochloric acid while shaking gently, and reflux the mixture for 1 hour; if sodium chloride separates as a solid cake, take care during the heating that the flask does not crack. When cold, transfer the mixture to a separatory funnel and remove the oil to a 750-ml round-bottomed flask. Heat it under an air-cooled reflux condenser in an oil bath at 180 °C until the evolution of carbon dioxide ceases (about 2 hours). Decant the oil into a Claisen flask with fractionating side-arm (the latter should be well lagged) and distil under reduced pressure. Collect the pelargonic acid at 140–142 °C/12 mmHg. The yield is 115 g (73%).

Note. (1) This is conveniently prepared by drying commercial butan-1-ol with anhydrous potassium carbonate or anhydrous calcium sulphate, distilling through a column and collecting the fraction, b.p. 117–118 °C.

Experiment 5.133   PROPYLMALONIC ACID

$$Me \cdot CH_2 \cdot CH_2 \cdot CH(CO_2Et)_2 \xrightarrow{\ominus OH} Me \cdot CH_2 \cdot CH_2 \cdot CH(CO_2H)_2$$

Dissolve 156 g (2.78 mol) of potassium hydroxide in 156 ml of water in a 1.5-litre round-bottomed flask and add 500 ml of rectified spirit to produce a homogeneous solution. Introduce 220 g (1.09 mol) of diethyl propylmalonate (Expt 5.132) slowly and with shaking. Attach a double surface reflux condenser and reflux the mixture for 3 hours; hydrolysis is then complete, i.e. a test portion dissolves completely in excess of water. Distil off as much ethanol as possible on a water bath, and dissolve the residue in a comparatively small volume of water. Cool the solution in a large beaker surrounded by ice; add dilute sulphuric acid slowly from a suitably supported dropping funnel, whilst stirring vigorously with a mechanical stirrer, until the solution is acid to Congo red paper. Extract the solution with three 150 ml portions of ether, dry the ethereal extract with anhydrous sodium sulphate and distil off the ether on a water bath. Spread the syrupy residue in thin layers upon large clock glasses (1); after 2–3 days, filter off the crystals at the pump, using light petroleum, b.p. 40–60 °C, to facilitate the transfer from the clock glasses to the sintered glass filter funnel. Spread the crystals on a porous tile to remove traces of oily impurities; the crude propylmalonic acid has m.p. 95–96 °C. Spread the filtrate and washings on large clock glasses as before and filter off the solid which crystallises after 1 day. Repeat the process until no further crystals are obtained. Recrystallise all the crystals from hot toluene. The yield of pure propylmalonic acid, m.p. 96 °C, is 110 g (69%).

Note. (1) An alternative procedure is to leave the syrupy residue in a vacuum desiccator over anhydrous calcium chloride and silica gel, and to filter off the successive crops of crystals as they separate. These are washed with light petroleum, b.p. 40–60 °C, spread on a porous tile and recrystallised.

Cognate preparations. *Butylmalonic acid.* This acid may be similarly prepared from diethyl butylmalonate (Expt 5.132) and melts at 102 °C after recrystallisation from benzene.

s-Butylmalonic acid. From diethyl s-butylmalonate (Expt 5.166); the acid melts at 76 °C after recrystallisation from benzene.

Experiment 5.134 GLUTARIC ACID (*Pentanedioic acid*)

$$\text{H·CHO} + 2\text{CH}_2(\text{CO}_2\text{Et})_2 \xrightarrow{\text{Et}_2\text{NH}} (\text{EtO}_2\text{C})_2\text{CH·CH}_2\text{·CH}(\text{CO}_2\text{Et})_2 \xrightarrow{\text{H}_3\text{O}^{\oplus}}$$

$$(\text{HO}_2\text{C})_2\text{CH·CH}_2\text{·CH}(\text{CO}_2\text{H})_2 \xrightarrow{-\text{CO}_2} \text{HO}_2\text{C·CH}_2\text{·CH}_2\text{·CH}_2\text{·CO}_2\text{H}$$

Tetraethyl propane-1,1,3,3-tetracarboxylate. Cool a mixture of 320 g (302 ml, 2 mol) of redistilled diethyl malonate and 80 g (1 mol) of 40 per cent formaldehyde solution ('formalin') contained in a 1-litre round-bottomed flask to 5 °C by immersion in ice, and add 5 g (7 ml) of diethylamine. Keep the mixture at room temperature for 15 hours and then heat under a reflux condenser on a boiling water bath for 6 hours. Separate the aqueous layer, dry the organic layer with anhydrous sodium sulphate and distil under reduced pressure. Collect the tetracarboxylate ester at 200–215 °C/20 mmHg. The yield is 250 g (75%).

Glutaric acid. Heat a mixture of 125 g (0.376 mol) of the preceding ester and 250 ml of 1:1-hydrochloric acid under reflux with stirring in a 1-litre two-necked flask equipped with a mechanical stirrer and reflux condenser. Continue the heating until the mixture becomes homogeneous (6–8 hours). Evaporate the contents of the flask to dryness on a steam bath (rotary evaporator) and distil the residual glutaric acid under reduced pressure. Collect the fraction boiling at 185–195 °C/10 mmHg; it crystallises on cooling. Moisten with a little water (to convert any glutaric anhydride present into the acid), heat gently and dry at 30 °C. Recrystallise from chloroform (or benzene) (**CAUTION**): the resulting practially pure glutaric acid, m.p. 96–97 °C, weighs 40 g (81%).

Experiment 5.135 2,2-DIMETHYLSUCCINIC ACID

$$Me_2CO + CH_2(CN)\cdot CO_2Et \xrightarrow{\text{base}} Me_2C{=}C(CN)\cdot CO_2Et \xrightarrow{\ominus CN}$$

$$Me_2C(CN)\cdot \overset{\ominus}{C}(CN)\cdot CO_2Et \xrightarrow{H_3O^{\oplus}} Me_2C(CO_2H)\cdot CH(CO_2H)_2 \xrightarrow{-CO_2}$$

$$Me_2C(CO_2H)\cdot CH_2\cdot CO_2H$$

CAUTION: *This preparation must be carried out in an efficient fume cupboard.*

Into a 500-ml round-bottomed flask, provided with a double surface condenser, place 50 g (63 ml, 0.86 mol) of pure, dry acetone, 50 g (47 ml, 0.44 mol) of ethyl cyanoacetate and 0.5 g of piperidine. Allow to stand for 60 hours and heat on a water bath for 2 hours. Treat the cold reaction mixture with 100 ml of ether, wash with dilute hydrochloric acid, then with water, and dry over anhydrous sodium sulphate. Distil under diminished pressure and collect the ethyl isopropylidenecyanoacetate (ethyl 2-cyano-3,3-dimethylacrylate) at 114–116 °C/14 mmHg (1). The yield is 39 g (58%).

Dissolve 20 g (0.13 mol) of the cyano ester in 100 ml of rectified spirit and add a solution of 19.2 g (0.295 mol) of pure potassium cyanide (**CAUTION**) in 40 ml of water. Allow to stand for 48 hours, then distil off the alcohol on a water bath. Add a large excess of concentrated hydrochloric acid and heat under reflux for 3 hours. (**CAUTION**: *hydrogen cyanide evolved.*) Dilute with water, saturate the solution with ammonium sulphate and extract with four 75 ml portions of ether. Dry the combined ethereal extracts with anhydrous sodium sulphate, and distil off the ether. Recrystallise the residual acid from excess concentrated hydrochloric acid, and dry in the air. The yield of pure 2,2-dimethylsuccinic acid, m.p. 141–142 °C, is 12 g (63%). The p.m.r. spectrum is recorded in trifluoracetic acid and reveals signals at δ 1.48 (s, 6H, Me$_2$) and 2.92 (s, 2H, CH$_2$); the hydroxyl proton is not observed.

Note. (1) Higher (including cycloaliphatic) ketones may be condensed with ethyl cyanoacetate under the following conditions. Mix 0.50 mol of ethyl cyanoacetate, 0.55–0.70 mol of the ketone, 0.02 mol of piperidine and 50 ml of dry benzene (**CAUTION**) and heat under reflux for 12–24 hours in an apparatus incorporating an automatic water separator (Fig. 2.31(*a*)). Piperidine may be replaced by a catalyst composed of 7.7 g (0.1 mol) of ammonium acetate and 24 g (0.4 mol) of glacial acetic acid. Wash the cold reaction mixture with three 25-ml portions of 10 per cent sodium chloride solution, and remove the benzene on a water bath under reduced pressure. Transfer the residue to a 1-litre bottle containing a solution of 65 g of sodium metabisulphite in 250 ml of water and shake mechanically for 2–6 hours. Dilute the turbid solution, which contains the sodium metabisulphite addition compound, with 400 ml of water, and extract the ethyl cyanoacetate with three 50 ml portions of benzene. Cool the bisulphite solution in ice, and add dropwise, with mechanical stirring, an ice-cold solution of 28 g of sodium hydroxide in 110 ml of water, Extract the regenerated unsaturated ester at once with four 25 ml portions of benzene, wash the extracts with 50 ml of 1 per cent hydrochloric acid and dry with anhydrous sodium sulphate. Filter and distil through a fractionating column under reduced pressure; the benzene may be conveniently removed by distilling at atmospheric pressure until the temperature rises to 90 °C. Diethyl ketone yields ethyl 2-cyano-3,3-diethylacrylate, b.p. 123–125 °C/12 mmHg or 96–97 °C/3 mmHg; dipropyl ketone gives ethyl 2-cyano-3,3-dipropylacrylate, b.p. 136–137 °C/11 mmHg or 116–117 °C/4 mmHg. The yield is 60–70 per cent.

The appropriate succinic acid can be prepared by condensation of the unsaturated cyano ester with alcoholic potassium cyanide and subsequent treatment with hydrochloric acid.

Experiment 5.136 2-PROPYGLUTARIC ACID

$$Pr \cdot CH(CO_2Et)_2 + CH_2{=}CH \cdot CN \longrightarrow \begin{array}{c} Pr \cdot C(CO_2Et)_2 \\ | \\ CH_2 \cdot CH_2 \cdot CN \end{array} \xrightarrow{H_3O^{\oplus}}$$

$$\begin{array}{c} Pr \cdot CH \cdot CO_2H \\ | \\ CH_2 \cdot CH_2 \cdot CO_2H \end{array}$$

Add 8.0 g (10.0 ml, 0.15 mol) of redistilled acrylonitrile (Expt 5.161, Note (1)) to a stirred solution of diethyl propylmalonate (30.2 g, 0.15 mol) (Expt 5.132) and of 30 per cent methanolic potassium hydroxide (4.0 g) in t-butyl alcohol (100 g). Keep the reaction mixture at 30–35 °C during the addition and stir for a further 3 hours. Neutralise the solution with 2 M-hydrochloric acid, dilute with water and extract with ether. Dry the ethereal extract with anhydrous sodium sulphate and distil off the ether: the residue [diethyl (2-cyanoethyl)-propylmalonate; 11 g] solidifies on cooling in ice, and melts at 31–32 °C after recrystallisation from ice-cold ethanol. Boil the cyanoethyl ester (10 g) under reflux with 40 ml of 48 per cent hydrobromic acid solution for 8 hours, and evaporate the solution almost to dryness under reduced pressure. Add sufficient water to dissolve the ammonium bromide, extract several times with ether, dry the ethereal extract and distil off the solvent. The residual oil (4.5 g, 66%) soon solidifies: upon recrystallisation from water, pure 2-propylglutaric acid, m.p. 70 °C, is obtained.

5.11.7 THE SYNTHESIS OF OPTICALLY ACTIVE CARBOXYLIC ACIDS

Before 1967 the preparation of 2-alkyl-and 2,2-dialkyl-acetic acids was achieved by the classical and important malonic ester route (Section 5.11.6 above). An alternative reagent equivalent of the synthon $R \cdot \overset{\ominus}{C}H \cdot CO_2H$ (or $\overset{2\ominus}{C}H \cdot CO_2H$) was proposed in the early 1970s.[156] Here a carboxylic acid in the form of a trialkyl orthoester, was first converted into a 2-alkyl-4,4-dimethyl-2-oxazoline with 2-amino-2,2-dimethylpropan-1-ol. Reaction of this oxazoline with butyllithium at -78 °C gave a lithio derivative which on treatment with an alkyl halide led, after hydrolysis, to a 2,2-dialkylacetic acid.

The most important development of this useful procedure has been the incorporation of an optically active amino alcohol, to provide a chiral adjuvant (or auxilliary), in the resulting oxazoline. The amino alcohol employed was (1S,2S)-

687

(+)-2-amino-1-phenylpropane-1,3-diol, a relatively inexpensive optically pure compound.[157] When reacted with a triethyl orthoalkanoate (or an iminoether hydrochloride of the corresponding carboxylic acid) an oxazoline was formed in good yield which was subsequently methylated to the 4-methoxymethyl compound.

In the subsequent reaction with lithium diisopropylamide the role of the methoxy group was seen to be crucial. Of the two lithio derivatives [(17) and (18)], which were probably in equilibrium with each other, the derivative (18) was considered to be the more stable owing to the unfavourable *cisoid* interaction in (17). In the subsequent addition of the alkyl halide to the lithio derivatives, co-ordination of the non-bonded electrons of the halogen on to the lithium cation of (19) was seen as the next mechanistic step. Realignment of the alkyl group then facilitated an S_N2 substitution reaction from the *underside* of the chelated complex (19). An *upperside* attack was considered to be unfavourable on steric and electronic grounds. Finally, hydrolysis gave rise to an optically active carboxylic acid in high optical purity, it having been established that no racemisation occurs during this hydrolytic step. In the example given in Expt 5.137, the orthoester used is triethyl orthopropanoate (R^1 = Me) and the alkyl halide is butyl iodide (R^2 = Bu); the product is (S)-(+)-2-methylhexanoic acid.

Experiment 5.137 (S)-(+)-2-METHYLHEXANOIC ACID[157]

$Me \cdot CH_2 \cdot C(OEt)_3 + (S, S)\text{-}Ph \cdot CH(OH) \cdot CH(NH_2) \cdot CH_2OH \longrightarrow$

Trans-(4S,5S)-2-ethyl-4-hydroxymethyl-5-phenyl-2-oxazoline. A mixture of 126.5 g (0.76 mol) of (1S,2S)-(+)-2-amino-1-phenylpropane-1,3-diol (1) and 160 g (0.91 mol) of triethyl orthopropanoate in 550 ml of 1,2-dichloroethane is heated under reflux for 7 hours. The solvent is removed leaving an oil (159 g) which crystallises on standing. This is treated with 70 ml of ether and cooled in a dry-ice–acetone bath and the partially purified product collected by filtration. The crystalline material is dissolved in ether (c. 600 ml), treated with charcoal, filtered, concentrated to about 300 ml and cooled to −78 °C. The recrystallised material is filtered and washed with a small amount (15–20 ml) of precooled (−78 °C) ether. The yield of product is 106.9 g (69%), m.p. 68–69 °C, $[\alpha]_{589}^{22}$ −135.1° (c 10.4 in $CHCl_3$).

(4S,5S)-2-Ethyl-4-methoxymethyl-5-phenyl-2-oxazoline (2). The foregoing compound (18.0 g, 88 mmol) in 150 ml of dry tetrahydrofuran is added dropwise at room temperature to a stirred heterogeneous suspension (under nitrogen), of sodium hydride [105.3 mmol; oil removed by washing with 50 ml of dry benzene (**CAUTION**)] at a rate to maintain a mild evolution of hydrogen. When addition is complete the mixture is heated at 50–60 °C for 1.5 hours and cooled to ambient temperature; a solution of methyl iodide (16.2 g, 114 mmol, **CAUTION**) in tetrahydrofuran (10 ml) is added dropwise. The reaction mixture is stirred for 2 hours and slowly poured into 300 ml of ice-water, and then extracted with two 200-ml portions of ether. The combined ether extracts are dried over anhydrous sodium sulphate and concentrated to give an oil which is distilled in vacuo to furnish the product in 87 per cent yield, b.p. 91–93 °C/0.25 mmHg.

(S)-(+)-2-Methylhexanoic acid. A solution of the foregoing compound (15.4 g, 70 mmol) in tetrahydrofuran (160 ml) under nitrogen is cooled to −78 °C in a dry-ice-acetone bath. To this is added a solution of 70 mmol of lithium diisopropylamide (from 9.8 ml of diisopropylamine and 33 ml of 2.2 M butyllithium in 75 ml of tetrahydrofuran) over a 20-minute period. The resulting solution is stirred for 40 minutes at −78 °C, and then the cooling bath is changed to methanol–liquid nitrogen and the reaction mixture allowed to cool to −98 °C for 30 minutes (3). A solution of butyl iodide (14.7 g, 80 mmol) in tetrahydrofuran (20 ml) is added over 15–20 minutes at −98 °C and the resulting almost colourless solution is stirred at this temperature for 2 hours, then slowly allowed to reach room temperature. The reaction mixture is poured into 300 ml of saturated brine and extracted with two 150 ml portions of ether, dried over magnesium sulphate and concentrated to give 17.7 g

(92%) of 2-(1-methylpentyl)-4-methoxymethyl-5-phenyl-2-oxazoline. A small portion is distilled (bulb to bulb) to give an analytically pure product $[\alpha]_{589}^{24}$ $-35.2°$ (c 10.1 in CHCl$_3$); i.r. (thin film) 1670 cm^{-1}; p.m.r. (CDCl$_3$, TMS) δ 7.33 (s, 5H), 5.33 (d, J = 7Hz, 1H), 4.33–3.93 (q, 1H), 3.80–3.33 (m, 2H), 3.43 (s, 3H), 2.87–2.33 (m, 1H), and 2.00–0.67 (m, 12H).

The crude oxazoline (17.2 g) is dissolved in 250 ml of 2 M sulphuric acid and heated to reflux for 3.5 hours, at which time the solution becomes homogeneous. When the heating of the solution is initiated a crystalline solid appears which slowly dissolves as the reflux progresses. After the heating is discontinued, the solution is cooled to room temperature and extracted with two 75-ml portions of ether. The combined ether extracts are washed with three 100-ml portions of 5 per cent aqueous potassium carbonate solution and the aqueous extracts neutralised to pH 1 with 12 M hydrochloric acid. The resulting turbid mixture is extracted with three 75-ml portions of ether and the extract dried over magnesium sulphate. Concentration of the ether solution leaves an oil which is distilled (bulb to bulb) to give 5.80 g (66%) of (S)-(+)-2-methylhexanoic acid. Gas-liquid chromatography analysis (10% UCW-98) indicates only a single compound; $[\alpha]_{589}^{24}$ $+14.5°$ (neat) (ee 78%, (4)].

Notes. (1) This amino alcohol may be recrystallised by dissolving 1 part in 1 part of methanol and adding 2 parts of ethyl acetate and cooling. The pure material has m.p. 112–113 °C, $[\alpha]_{589}^{22}$ 26.6° (c 10.0 in MeOH).
(2) The apparatus set-up for this, and the following reaction, should take into account the moisture sensitivity of the reagents and the reaction intermediates. The reagents should be added by syringe through a rubber septum, the reaction media should be held under an atmosphere of nitrogen via a suitable nitrogen bubbler device, and the solutions should be stirred magnetically (cf. Section 2.17.8, p. 120).
(3) Continued use of the cooling bath held at -78 °C leads to only a small diminution of optical purity in the final product.
(4) The specific rotation of optically pure acid is $[\alpha]_{D}^{25}$ $-18.7°$ (neat).

5.11.8 SOME METHODS FOR THE PROTECTION OF THE CARBOXYL GROUP

The carboxyl group of aliphatic and alicyclic carboxylic acids may be protected by conversion into a variety of organic esters; silylation affords a valuable alternative.

The wide range of standard procedures that are available for the formation of carboxylic esters of primary and secondary alcohols in the presence of suitable acid catalysts is discussed in detail in Section 5.12.3, p. 695. Also included is the mild method for *methyl ester* formation from the carboxylic acid and diazomethane, and a method appropriate for sterically hindered esters involving the acid, a secondary or tertiary alkyl halide, and the non-nucleophilic base DBU (Expt 5.151). An example of the formation of a *t-butyl ester* is noted in Expt 6.165.

For some purposes protection of the carboxyl group by conversion into *p-phenylphenacyl*, *p-bromophenacyl*, and *p-nitrobenzyl* esters is useful. General procedures for their formation are described in Section 9.6.15, p. 1261.

Methyl, ethyl (and to a lesser extent t-butyl) esters are perhaps the most commonly used, and are stable to a wide range of oxidising conditions providing that a moderate pH environment is applicable. Careful selection of hydride reducing agents is required to ensure chemoselectivity (see Section 4.2.49, p. 445) in the reduction of other functional groups (e.g. an aldehyde or ketone in the presence of a methyl or ethyl ester group). These esters react with lithium and magnesium organometallic reagents, but are stable to zinc and copper organometallics; this group is tolerant of Wittig-type reactants.

Deprotection is effected in the case of methyl or ethyl esters by hydrolysis, usually under basic conditions (Section 9.6.17, p. 1266). t-Butyl esters are deprotected by mild acidic conditions, for example, formic acid at room temperature,[159] toluene-*p*-sulphonic acid in benzene solution, or hydrogen bromide in acetic acid at 10 °C.[160]

The most appropriate silylating reagent for the formation of *silyl esters* is t-butyldimethylchlorosilane. A preparative example is given in Section 4.2.66, p. 461. Silyl esters are stable in non-aqueous media.

5.12 CARBOXYLIC ACID DERIVATIVES

1. Acyl halides (Expt 5.138).
2. Acid anhydrides (Expts 5.139 to 5.141).
3. Esters (Expts 5.142 to 5.153).
4. Acid amides (Expt 5.154 to 5.156).

Formal replacement of the hydroxyl moiety of the carboxyl group with a halogen (X), an acyloxy group (R·CO·O—), an alkoxy group (RO—), or an amino group (NH_2) leads to:

the acid (or acyl) halides [e.g. butanoyl chloride (1) $CH_3·CH_2·CH_2·COCl$];
the acid anhydrides [e.g. acetic anhydride (2) $(CH_3·CO)_2O$];
the esters [e.g. ethyl 2-methylpropanoate (3) $CH_3·CH(CH_3)·CO_2Et$];
the amides [e.g. pentanamide (4) $CH_3·CH_2·CH_2·CH_2·CONH_2$].

(1)   (2)   (3)   (4)

The realisable and synthetically useful interconversions between these carboxylic acid derivatives are summarised below.

A *retrosynthetic analysis* applied to each of these carboxylic acid derivatives suggests a fission of the carbon–heteroatom bond (C—O, C—X, C—N), i.e. *disconnection* processes. Such logic can be useful in the recognition of the reagents that could be used in the synthesis of these functional types, when these are present in more complex molecules, as illustrated below.

$$\text{Me·CHOH·CO}_2\text{H} + \text{Me}_2\text{CHOH}$$

$$+ \text{NH}_3$$

SPECTROSCOPIC FEATURES

All these derivatives show strong *i.r.* absorption arising from the stretching vibration of the carbonyl group. They may be distinguished from each other by inspection of the appropriate region for absorption arising from the carbon-bound halogen, acyloxy, alkyloxy, or amino (or imino) groups (p. 296). The *p.m.r.* spectrum as with the carboxylic acids (p. 666), allows an assessment of the skeletal structure of the alkyl groups to be made (p. 341). The interpretation of the *m.s.* of these derivatives is discussed on p. 380. Descriptive analyses of the spectra of specific compounds are given in the preparative sections below.

5.12.1 ACYL HALIDES

Acyl halides are invaluable acylating reagents and their preparation is therefore of great importance. The conversion of an aliphatic carboxylic acid into the corresponding acyl chloride is usually achieved by heating the acid with thionyl chloride.

$$\underset{R}{\overset{O}{\|}}\text{C—OH} + \text{SOCl}_2 \longrightarrow \underset{R}{\overset{O}{\|}}\text{C—Cl} + \text{HCl} + \text{SO}_2$$

This reagent is particularly convenient as the by-products of the reaction do not contaminate the product, and excess thionyl chloride is usually separable by fractional distillation. If the boiling point of the acyl chloride is too near to that of thionyl chloride the excess of the latter can be destroyed by the addition of pure formic acid.

$$\text{H·CO}_2\text{H} + \text{SOCl}_2 \longrightarrow \text{CO} + \text{SO}_2 + 2\text{HCl}$$

Phosphorus trichloride and phosphorus pentachloride are also suitable reagents for acid chloride formation, but their use is largely restricted to aromatic carboxylic acids. (Section 6.14.1, p. 1073).

Experiment 5.138 BUTYRYL CHLORIDE (*Butanoyl chloride*)

$$\text{Me·(CH}_2)_2\text{·CO}_2\text{H} + \text{SOCl}_2 \longrightarrow \text{Me·(CH}_2)_2\text{·COCl} + \text{SO}_2 + \text{HCl}$$

Fit a 100-ml two-necked flask with a dropping funnel and a reflux condenser connected at the top to a gas absorption trap (Fig. 2.61). Place 36 g (21.5 ml, 0.3 mol) of redistilled thionyl chloride in the flask and 22 g (23 ml, 0.25 mol) of butyric acid in the separatory funnel. Heat the flask gently on a water bath, and add the butyric acid during the course of 30–40 minutes (1). When all the acid has been introduced, heat on a water bath for 30 minutes. Rearrange the apparatus and distil: collect the crude acid chloride boiling between 70 and 110 °C. Finally, redistil from a flask provided with a short fractionating column and collect the butyryl chloride at 100–101 °C. The yield is 23 g (86%).

Note. (1) Wrap a piece of absorbent cotton wool around the stem of the reflux condenser above the joint of the reaction flask to prevent condensed moisture seeping into the flask.

Cognate preparations. *Hexanoyl chloride.* Place 58 g (62 ml, 0.5 mol) of hexanoic acid in the flask, heat on a water bath and add 72 g (43 ml, 0.6 mol) of redistilled thionyl chloride during 45 minutes; shake the flask from time to time to ensure mixing. Reflux for 30 minutes and isolate the hexanoyl chloride by distillation, b.p. 150–155 °C. The yield is 56 g (83%).

Valeryl chloride (*pentanoyl chloride*). Use 51 g (0.5 mol) of valeric acid and 72 g (0.6 mol) of redistilled thionyl chloride. Proceed as for hexanoyl chloride; the yield of valeryl chloride is 42 g (70%), b.p. 124–127 °C.

Isobutyryl chloride (*2-methylpropanoyl chloride*). Use 140 g (1.6 mol) of iso-butyric acid and 236 g (2 mol) of redistilled thionyl chloride. Proceed as for hexanoyl chloride; the yield is 121 g (71%), b.p. 90–93 °C, after distillation through a Vigreux column (36 cm).

Isovaleryl chloride (*3-methylbutanoyl chloride*). Use 34 g (0.4 mol) of iso-valeric acid and 47 g (0.5 mol) of thionyl chloride. Proceed as for hexanoyl chloride; the yield of isovaleryl chloride is 36 g (76%), b.p. 114–115 °C, after distillation through a Vigreux column.

Cyclohexanecarbonyl chloride. Use 91 g (0.7 mol) of cyclohexanecarboxylic acid and 166 g (1.4 mol) of thionyl chloride. Proceed as for hexanoyl chloride but heat under reflux for 2 hours. The yield of cyclohexanecarbonyl chloride is 100 g (78%), b.p. 76–78 °C/12 mmHg, after distillation through a Vigreux column.

5.12.2 ACID ANHYDRIDES

Symmetrical (but not unsymmetrical) anhydrides are useful acylating reagents as milder alternatives to acyl halides. The most convenient procedure for their preparation involves the interaction in benzene solution of the acyl halide with the corresponding carboxylic acid in the presence of pyridine, which removes the hydrogen chloride liberated as the insoluble pyridinium chloride.

$$
\underset{R}{\overset{O}{\|}}{-}Cl \;+\; HO{-}\underset{R}{\overset{O}{\|}} \;\xrightarrow{\;C_5H_5N\;}\; \underset{R}{\overset{O}{\|}}{-}O{-}\underset{R}{\overset{O}{\|}} \;+\; C_5H_5\overset{\oplus}{N}H\}\overset{\ominus}{Cl}
$$

Without pyridine, anhydride formation would be incomplete, as a result of an equilibrium reaction of the anhydride with hydrogen chloride which regenerates the acyl chloride and the carboxylic acid.

$$R\overset{O}{\underset{}{\parallel}}O\overset{O}{\underset{}{\parallel}}R \ + \ HCl \ \rightleftharpoons \ R\overset{O}{\underset{}{\parallel}}Cl \ + \ R\overset{O}{\underset{}{\parallel}}OH$$

Mixed anhydrides (i.e. $R^1 \cdot CO \cdot O \cdot CO \cdot R^2$) can also be readily prepared by this general route by choosing appropriate reactants.

The reaction of keten with a carboxylic acid gives a mixed anhydride as the first formed product.

$$R\overset{O}{\underset{}{\parallel}}OH \ + \ O{=}C{=}CH_2 \ \longrightarrow \ R\overset{O}{\underset{}{\parallel}}O\overset{O}{\underset{}{\parallel}}Me$$

Slow distillation at atmospheric pressure of the mixed anhydride with a second molar proportion of the carboxylic acid yields the symmetrical anhydride and acetic acid.

$$R\overset{O}{\underset{}{\parallel}}O\overset{O}{\underset{}{\parallel}}Me \ + \ HO\overset{O}{\underset{}{\parallel}}R \ \longrightarrow \ R\overset{O}{\underset{}{\parallel}}O\overset{O}{\underset{}{\parallel}}R \ + \ Me \cdot CO_2H$$

Cyclic anhydrides of dibasic acids such as succinic or glutaric acid are readily prepared by dehydrating the acid with an excess of acetic anhydride, e.g. as in the preparation of succinic anhydride, Expt 5.141; cf. 3-nitrophthalic anhydride, Expt 6.162.

Experiment 5.139 HEPTANOIC ANHYDRIDE

$$C_6H_{13} \cdot COCl \ + \ C_6H_{13} \cdot CO_2H \ \xrightarrow{C_5H_5N} \ C_6H_{13} \cdot CO \cdot O \cdot CO \cdot C_6H_{13}$$

CAUTION: *All operations should be conducted in an efficient fume cupboard owing to the toxicity of pyridine and benzene.*

In a 250-ml, round-bottomed three-necked flask, provided with a dropping funnel, stirrer and thermometer, place 15.8 g (16.1 ml, 0.2 mol) of dry pyridine (Section 4.1.*29*, p. 410) and 25 ml of dry benzene. Stir and add rapidly 14.8 g (15.5 ml, 0.1 mol) of heptanoyl chloride; the temperature rises slightly and a pyridinium complex separates. Introduce 13.0 g (14.1 ml, 0.1 mol) of heptanoic acid with stirring, over a period of 5 minutes; the temperature rises to 60–65 °C and pyridine hydrochloride is formed. Continue the stirring for 10 minutes and collect the hygroscopic pyridine hydrochloride as rapidly as possible on a chilled Buchner or sintered glass funnel, and wash it with two 25-ml portions of dry benzene. Remove the benzene from the filtrate under reduced pressure on a water bath, and distil the residue through a short fractionating column. Collect the heptanoic anhydride at 170–173 °C/ 15 mmHg; the yield is 20 g (83%).

Experiment 5.140 HEXANOIC ANHYDRIDE

$$2\ Me \cdot (CH_2)_4 \cdot CO_2H \ + \ CH_2{=}C{=}O \ \longrightarrow$$
$$[Me \cdot (CH_2)_4 \cdot CO]_2O \ + \ CH_3 \cdot CO_2H$$

Place 116 g (126 ml, 1 mol) of dry hexanoic acid in a 250-ml Drechsel bottle and cool in ice. Pass in 21–23 g of keten (Section 2.17.3, p. 100) (1). Carefully distil the reaction mixture through a highly efficient fractionating column (e.g. a well-lagged Widmer column; see Section 2.26) (2), using an oil bath for heating. A fraction of low boiling point, containing acetone, keten, acetic acid and a little acetic anhydride, is thus removed at atmospheric pressure. Raise the temperature of the bath to 220 °C over a period of 1 hour and maintain it at this temperature for 3 hours from the time distillation commences: this time is necessary to ensure that the conversion of the mixed anhydride to hexanoic anhydride and acetic acid is complete and that the acetic acid is completely removed. Discontinue the distillation, allow to cool somewhat and distil the residue in the flask under reduced pressure (3–10 mmHg). Discard the small fraction (20 g) of low boiling point and collect the hexanoic anhydride at 118–121 °C/6 mmHg (or 109–112 °C/3 mmHg). The yield is 90 g (84%). Record the i.r. spectrum and compare it with that of acetic anhydride (Fig. 3.34).

Notes. (1) Excess of keten over the calculated quantity does not increase the yield; it leads to more acetic anhydride being collected in the low boiling point fraction.
(2) The best results are obtained with a fractionating column surrounded by an electrically heated jacket but this is not essential for hexanoic anhydride. For the preparation of propanoic or butanoic anhydride, a highly efficient fractionating column must be used in order to obtain satisfactory results.

Experiment 5.141 SUCCINIC ANHYDRIDE

In a 500-ml round-bottomed flask, provided with a reflux condenser protected by a calcium chloride drying tube, place 59 g (0.5 mol) of succinic acid and 102 g (94.5 ml, 1 mol) of redistilled acetic anhydride. Reflux the mixture gently on a water bath with occasional shaking until a clear solution is obtained (c. 1 hour), and then for a further hour to ensure the completeness of the reaction. Remove the complete assembly from the water bath, allow it to cool (observe the formation of crystals) and finally cool in ice. Collect the succinic anhydride on a Buchner funnel or a sintered glass funnel, wash it with two 40 ml portions of anhydrous ether and dry in a vacuum desiccator. The yield is 45 g (90%), m.p. 119–120 °C.

5.12.3 ESTERS

DIRECT ESTERIFICATION PROCEDURES
The interaction between a carboxylic acid and an alcohol is a reversible process and proceeds very slowly:

Equilibrium is only attained after refluxing for several days. If, however, about 3 per cent (of the weight of the alcohol) of either concentrated sulphuric acid or of dry hydrogen chloride is added to the mixture, the same point of equilibrium can be reached after a few hours. When equimolecular quantities of the acid and alcohol are employed, only about two-thirds of the theoretically possible yield of ester is obtained. According to the law of mass action, the equilibrium may be displaced in favour of the ester by the use of an excess of one of the components. It is frequently convenient to use an excess of the acid, but if the acid is expensive a large excess of the alcohol is generally employed. This method of esterification, in general, gives good yields with primary alcohols and fairly good yields with secondary alcohols. The method is unsatisfactory for use with tertiary alcohols owing to competing alkene formation from an acid catalysed dehydration.

Esterification with alicyclic alcohols proceeds best when the alcohol is saturated with hydrogen chloride and treated with an excess of the carboxylic acid (the *Fischer–Speier method*); a very impure ester results if sulphuric acid is used as the catalyst.

Examples of the use of these methods for the preparation of simple esters are collected in Expt 5.142.

Esters of formic acid (Expt 5.143) are most simply prepared from the alcohol and an excess of formic acid, which, being a comparatively strong acid, does not require the use of added mineral acid to catalyse the esterification reaction. Sulphuric acid in any case should not be added since it causes the decomposition of formic acid to carbon monoxide.

The acid-catalysed esterification reaction usually proceeds via an acyl–oxygen fission process. This involves the cleavage of the bond between the original carbonyl-carbon atom and an oxygen of an hydroxyl group in the intermediate (6) arising from nucleophilic attack by an alcohol molecule on the protonated carboxylic acid group (5).

Several modifications of the simple direct esterification procedure described above have been developed. For example, it is sometimes convenient to prepare an ester by heating the organic acid, the alcohol and sulphuric acid in a solvent such as toluene. This method is illustrated by the preparation of diethyl adipate for which the procedure is particularly well suited (Expt 5.144).

The process of acid-catalysed esterification in the presence of benzene, or, better, of toluene, is greatly facilitated if the water produced in the reaction is removed by distillation in a Dean and Stark water separation unit. This allows the separation and removal of water from the azeotrope, the organic phase being returned continuously to the reaction flask. Examples are provided in Expt 5.145. These include examples of the synthesis of esters where either the alcohol or the carboxylic acid component is 'acid-sensitive'. The mineral acid catalyst is replaced by a cation exchange resin (e.g. Zerolit 225/H$^{\oplus}$), enabling good yields of the required esters to be obtained.

A recent procedure for the preparation of methyl esters involves refluxing the carboxylic acid with methanol and 2,2-dimethoxypropane in the presence of toluene-p-sulphonic acid as the catalyst (Expt 5.146). The water produced in the esterification process is effectively removed by acid-catalysed reaction with the ketal to give acetone and methanol.

$$R\cdot CO_2H + MeOH \overset{H^{\oplus}}{\rightleftharpoons} R\cdot CO_2Me + H_2O$$

$$Me\cdot C(OMe)_2\cdot Me + H_2O \longrightarrow Me\cdot CO\cdot Me + 2MeOH$$

Methyl esters are conveniently prepared on the small scale using diazomethane; a general procedure is given in Section 4.2.25, p. 433.

$$R\cdot CO_2H + CH_2N_2 \longrightarrow R\cdot CO_2Me + N_2$$

PREPARATION OF ACID ESTERS OF DICARBOXYLIC ACIDS

The acid-catalysed reaction of a dicarboxylic acid with an excess of alcohol yields the diester. However, the process may be adapted to prepare acid esters of dicarboxylic acids by using molar proportions of the diacid and alcohol (e.g. methyl hydrogen adipate, Expt 5.147). Alternatively the acid ester may be prepared by subjecting the diester to controlled partial hydrolysis with one molar proportion of potassium hydroxide.

The acid esters of 1,2-dicarboxylic acids are conveniently prepared by heating the corresponding cyclic anhydride with one molar proportion of the alcohol (see the preparation of alkyl hydrogen phthalates from phthalic anhydride and their use in the resolution of racemic alcohols, Section 5.19).

Acid esters are useful synthetic intermediates. For example, their use in the synthesis of long-chain dicarboxylic esters by electrolytic (anodic) synthesis has already been noted (Expt 5.131). Furthermore the reaction of the acid ester with thionyl chloride in the usual way will convert the carboxylic acid grouping to an acyl chloride group thus yielding the synthetically useful ester–acyl chloride; the products are usually purified by distillation under reduced pressure.

One of the uses of these ester–acyl chlorides is for the synthesis of ω-hydroxy esters which involves the selective reduction of the acyl chloride grouping with sodium borohydride[161]; the alkoxycarbonyl group is unaffected by this metal hydride reducing agent (cf. Section 5.4.1, p. 519).

697

THE USE OF ACYL CHLORIDES AND ACID ANHYDRIDES

Acyl chlorides react readily with primary and secondary alcohols to give esters in very good yields. With tertiary alcohols the presence of base (e.g. dimethylaniline) is essential to prevent acid-catalysed side reactions, such as dehydration or formation of the alkyl chloride.

Acylation may also be carried out with acid anhydrides in the presence of a suitable catalyst; either an acidic catalyst, such as sulphuric acid or zinc chloride, or a basic catalyst such as pyridine, may be used.

Examples of the use of these methods are given in Expts 5.148 and 5.149. The use of an acyl chloride or acid anhydride is the method of choice for the synthesis of phenyl esters (e.g. phenyl cinnamate; see also Section 9.6.6), which cannot be prepared by the direct esterification methods described above.

The synthesis of ethyl 2-bromopropanoate (Expt 5.150) illustrates the preparation of an acyl chloride and its ready bromination in the α-position in the presence of red phosphorus. The resulting bromoacyl chloride is converted into the α-bromoester on reaction with an alcohol.

THE USE OF ALKYL HALIDES AND CARBOXYLIC ACIDS UNDER BASIC CONDITIONS

The disconnection underlying this procedure is an alkyl-oxygen fission and not the acyl-oxygen fission as in the reactions discussed above. The non-nucleophilic base used in this reaction is 1,8-diazabicyclo[5.4.0]undec-7-ene (DBU) which converts the carboxylic acid into its carboxylate ion but does not interact in a competing substitution reaction with the alkyl halide. The ester-forming reaction may therefore be regarded as an S_N2 reaction between the carboxylate ion and the alkyl halide.[162]

The reaction [illustrated by the formation of 1-methylheptyl acetate (Expt 5.151)] proceeds in high yield and is particularly useful in those cases where a direct esterification of an acid with an alcohol is precluded on the grounds of undesirable side reactions (e.g. dehydration of the alcohol to an alkene).

THE ALCOHOLYSIS OF NITRILES

An ester is formed when a nitrile is heated with an alcohol in the presence of concentrated sulphuric acid, thus providing a two-step synthesis of an ester from an alkyl halide. Examples are to be found in Expt 5.152. The reaction proceeds by way of an intermediate imino-ester (7) which is not usually isolated, but may be if so required.[163]

$$R^1-C{\equiv}N + R^2OH \xrightarrow{H^{\oplus}} \left[\underset{R^1\quad OR^2}{\overset{\oplus NH_2}{\parallel}} \right] \xrightarrow{H_2O} \underset{R^1\quad OR^2}{\overset{O}{\parallel}} + NH_4^{\oplus}$$

$$(7)$$

ORTHO-ESTERS

These have the general formula $R^1 \cdot C(OR^2)_3$ and are stable derivatives of the unstable ortho acids $R \cdot C(OH)_3$. Important examples are the esters of orthoformic acid (the orthoformates), which may be readily prepared by the interaction of the appropriate sodium alkoxide with chloroform (Expt 5.153).

$$CHCl_3 + 3RO^{\ominus}Na^{\oplus} \longrightarrow CH(OR)_3 + 3NaCl$$

Experiment 5.142 BUTYL ACETATE

$$Me \cdot CO_2H + Me \cdot (CH_2)_2 \cdot CH_2OH \xrightarrow{H^{\oplus}} Me \cdot CO_2 \cdot (CH_2)_3 \cdot Me$$

Mix together 37 g (46 ml, 0.5 mol) of butan-1-ol and 60 g (60 ml, 1 mol) of glacial acetic acid in a 250- or 500-ml round-bottomed flask, and add cautiously 1 ml of concentrated sulphuric acid (use a small measuring cylinder or a calibrated dropper pipette). Attach a reflux condenser and reflux the mixture for 3–6 hours (1). Pour the mixture into about 250 ml of water in a separatory funnel, remove the upper layer of crude ester and wash it again with about 100 ml of water, followed by about 25 ml of saturated sodium hydrogen carbonate solution and 50 ml of water. Dry the crude ester with 5–6 g of anhydrous sodium sulphate. Filter through a small funnel containing a fluted filter paper and distil on a wire gauze or from an air bath. Collect the pure butyl acetate at 124–125 °C. The yield is 40 g (69%). The i.r. and mass spectra are shown in Figs. 3.33 and 3.85 respectively. The p.m.r spectrum (CCl₄, TMS) show absorptions at δ 0.93 (distorted t, 3H, Me·CH₂), 1.09–1.72 (m, 4H, ·CH₂·CH₂), 1.95 (s, 3H, Me·CO) and 3.99 (t, 2H, CH₂·O).

Note. (1) A slightly better yield of ester can be obtained by increasing the quantity of acetic acid to 90–120 g and refluxing for 12–18 hours.

Cognate preparations: sulphuric acid catalyst. *Ethyl butanoate.* Use a mixture of 88 g (92 ml, 1 mol) of butanoic acid, 23 g (29 ml, 0.5 mol) of ethanol and 9 g (5 ml) of concentrated sulphuric acid. Reflux for 14 hours. Pour into excess of water, wash several times with water, followed by saturated sodium hydrogen carbonate solution until all the acid is removed, and finally with water. Dry with anhydrous sodium sulphate, and distil. The ethyl butanoate passes over at 119.5–120.5 °C. Yield: 40 g (69%). An improved yield can be obtained by distilling the reaction mixture through an efficient fractionating column until the temperature rises to 125 °C, and purifying the crude ester as detailed above under methyl acetate.

Diethyl sebacate.   Reflux a mixture of 101 g (0.5 mol) of sebacic acid, 196 g (248 ml, 4.25 mol) of absolute ethanol and 20 ml of concentrated sulphuric acid for 12 hours. Distil off about half of the alcohol on a water bath, dilute the residue with 500–750 ml of water, remove the upper layer of crude ester and extract the aqueous layer with ether. Wash the combined ethereal extract and crude ester with water, then with saturated sodium hydrogen carbonate solution until effervescence ceases, and finally with water. Dry with magnesium sulphate or anhydrous sodium sulphate, remove the ether on a water bath and distil the residue under reduced pressure (b.p. 155–157 °C/6 mmHg). Yield: 110 g (85%).

Methyl crotonate.   Use 43 g (0.5 mol) redistilled crotonic acid (b.p. 180–182 °C, m.p. 72–73 °C), 75 g (95 ml, 2.33 mol) of absolute methanol, 3 ml of concentrated sulphuric acid and reflux for 12 hours. Isolate as for butyl acetate; the yield is 34 g (68%), b.p. 118–120 °C. Record the p.m.r. spectrum and by careful measurement of J values assign the absorptions.

Benzyl acetate.   Mix 31 g (29.5 ml, 0.287 mol) of benzyl alcohol (Expt 6.133) and 45 g (43 ml, 0.75 mol) of glacial acetic acid in a 500-ml round-bottomed flask; introduce 1 ml of concentrated sulphuric acid and a few fragments of 'porous pot'. Attach a reflux condenser to the flask and boil the mixture gently for 9 hours. Pour the reaction mixture into about 200 ml of water contained in a separatory funnel, add 10 ml of carbon tetrachloride (to eliminate emulsion formation owing to the slight difference in density of the ester and water, compare *methyl benzoate*, Expt 6.163) and shake. Separate the lower layer (solution of benzyl acetate in carbon tetrachloride) and discard the upper aqueous layer. Return the lower layer to the funnel, and wash it successively with water, concentrated sodium hydrogen carbonate solution (until effervescence ceases) and water. Dry over 5 g of magnesium sulphate, and distil from an air bath. Collect the benzyl acetate (a colourless liquid) at 213–215 °C. The yield is 16 g (37%).

Cognate preparations: hydrochloric acid catalyst.   *Cyclohexyl acetate.*   Pass dry hydrogen chloride (Section 4.2.38, p. 438) into 75 g (0.75 mol) of pure cyclohexanol until 1.5 g are absorbed, mix with 135 g (2.25 mol) of glacial acetic acid in a 500-ml round-bottomed flask, attach a reflux condenser and reflux for 14 hours. Pour into excess of water, wash the upper layer successively with water, saturated sodium hydrogen carbonate solution until effervescence ceases, and water. Dry with anhydrous calcium chloride. Distil through a well-lagged fractionating column (e.g. an all-glass Dufton column). A small fraction of low boiling point (containing cyclohexene) passes over first, followed by cyclohexyl acetate (57 g, 54%) at 168–170 °C. Upon redistillation, the boiling point is 170–172 °C, mainly 171–172 °C.

Cyclohexyl formate.   Use 103 g (84.5 ml, 2.24 mol) of formic acid (98/100%) and 75 g (0.75 mol) of cyclohexanol in which 1.5 g of dry hydrogen chloride gas are dissolved. Reflux for 14 hours. Work up as above and distil through a well-lagged column; 5.5 g of cyclohexene and 57 g (59%) of cyclohexyl formate, b.p. 156–158.5 °C (mainly 157–158.5 °C), are obtained. On redistillation the sample boils at 158–160 °C (mainly 159–160 °C).

s-Butyl acetate.   Pass dry hydrogen chloride gas into 37 g (46 ml, 0.5 mol) of butan-2-ol until 1.5 g is absorbed. Mix the solution with 60 g (1 mol) of glacial acetic acid, and reflux for 10 hours. Isolate the ester as for butyl acetate (b.p. 110–112 °C). Yield: 35 g (60%).

Ethyl p-aminobenzoate.   Saturate 80 ml (63.2 g, 1.37 mol) of absolute ethanol with dry hydrogen chloride, add 12 g (0.088 mol) of p-aminobenzoic acid and heat the mixture under reflux for 2 hours. Upon cooling, the reaction mixture sets to a solid mass of the hydrochloride of ethyl p-aminobenzoate. It is better, however, to pour the hot solution into excess of water (no hydrochloride separates) and add sodium carbonate to the clear solution until it is neutral to litmus. Filter off the precipitated ester at the pump and dry in the air. The yield of ethyl p-aminobenzoate, m.p. 91 °C, is 10 g (69%). Recrystallisation from rectified spirit does not affect the m.p.

Experiment 5.143   BUTYL FORMATE

$$H \cdot CO_2 H + Me \cdot (CH_2)_2 \cdot CH_2 OH \longrightarrow H \cdot CO_2 (CH_2)_3 \cdot Me + H_2 O$$

Into a 250- or 500-ml round-bottomed flask provided with a reflux condenser place 46 g (38 ml, 1 mol) of formic acid (98/100%) and 37 g (46 ml, 0.5 mol) of butan-1-ol. Reflux for 24 hours. Wash the cold mixture with small volumes of saturated sodium chloride solution, then with saturated sodium hydrogen carbonate solution in the presence of a little solid sodium hydrogen carbonate until effervescence ceases, and finally with saturated sodium chloride solution. Dry with anhydrous sodium sulphate, and distil through a short fractionating column. Collect the butyl formate at 106–107 °C. Yield: 38 g (74%).

Cognate preparations.   *Ethyl formate.*   Reflux a mixture of 61 g (50 ml, 1.33 mol) of formic acid (98/100%) and 31 g (39.5 ml, 0.67 mol) of absolute ethanol for 24 hours. Attach a fractionating column to the flask, distil and collect the liquid passing over below 62 °C. Wash the distillate with saturated sodium hydrogen carbonate solution and saturate with salt before removing the ester layer. Dry with anhydrous sodium sulphate, filter and distil. The ethyl formate passes over at 53–54 °C. The yield is 36 g (72%).

Propyl formate.   Use 46 g (38 ml, 1 mol) of formic acid (98/100%) and 30 g (37.5 ml, 0.5 mol) of propan-1-ol and reflux for 24 hours. Proceed as for ethyl formate, but collect the crude propyl formate up to 86 °C; b.p. 80.5–82 °C. Yield: 28 g (65%); p.m.r. spectrum (CCl₄, TMS) δ 0.99 (t, 3H), 1.67 (m, 2H), 4.08 (t, 2H), 7.94 (s, 1H).

Experiment 5.144   DIETHYL ADIPATE

$$HO_2 C \cdot (CH_2)_4 \cdot CO_2 H + 2EtOH \xrightarrow[\text{toluene}]{H^{\oplus}} EtO_2 C \cdot (CH_2)_4 \cdot CO_2 Et$$

Place 146 g (1 mol) of adipic acid, 360 ml (285 g, 6.2 mol) of absolute ethanol, 180 ml of toluene and 1.5 g of concentrated sulphuric acid in a 1-litre round-bottomed flask, attach a *short* fractionating column connected to a downward condenser and heat in an oil bath at 115 °C. When the acid has dissolved, an azeotropic mixture of alcohol, toluene and water commences to distil at 75 °C; the temperature of the oil bath may then be lowered to 100–

110 °C. Collect the distillate in a flask containing 150 g of anhydrous potassium carbonate. Continue the distillation until the temperature at the top of the column rises to 78 °C. Shake the distillate thoroughly with the potassium carbonate, filter through a Buchner funnel or fluted filter paper and return the filtrate to the flask. Heat the flask again until the temperature rises to 78–80 °C (1). Transfer the warm residue to a flask of suitable size and distil under reduced pressure. Alcohol and toluene pass over first, the temperature rises abruptly and the diethyl adipate distils at 138 °C/20 mmHg (2). The yield is 195 g (96%).

Notes. (1) The distillate contains ethanol, toluene and water, and may be dried with anhydrous potassium carbonate and used again for esterification after the addition of the necessary quantity of alcohol.
(2) The b.p. may rise several degrees towards the end of the distillation owing to superheating.

Experiment 5.145 ISOPROPYL LACTATE

$$Me \cdot CH(OH) \cdot CO_2H + Me_2CHOH \xrightarrow[\text{benzene}]{H^{\oplus}\text{(resin)}} Me \cdot CH(OH) \cdot CO_2CHMe_2$$

Place a mixture of 53 g (0.5 mol) of 'AnalaR' lactic acid (85–88% acid), 75 g (95.5 ml, 1.25 mol) of commercial anhydrous propan-2-ol (isopropyl alcohol), 300 ml of benzene (**CAUTION**) and 20 g of Zerolit 225 (acid form) (1) in a 1-litre flask, equipped with an automatic water separator (e.g., a large modified Dean and Stark apparatus with a stopcock at the lower end, see Fig. 2.31(a)) carrying an efficient reflux condenser at its upper end. Reflux the mixture using a magnetic stirrer/hotplate unit for 5 hours or until water no longer collects in appreciable amount in the water separator; run off the water from time to time. Filter off the resin at the pump and wash it with two 25 ml portions of benzene. Shake the combined filtrate and washings with about 5 g of precipitated calcium carbonate, filter, and wash with a little benzene. Distil the benzene solution under reduced pressure (water pump) through a short fractionating column; the isopropyl alcohol–benzene azeotrope (2) passes over first, followed by benzene. Collect the isopropyl lactate at 76 °C/24 mmHg; it is a colourless liquid and weighs 40 g (61%). The ester boils, with slight decomposition, at 157 °C/771 mmHg.

Notes. (1) This resin is available as the sodium form. It may be converted into the hydrogen form by treating it with about twice its volume of 1 M sulphuric acid and stirring frequently: the resin is thoroughly washed by decantation with distilled water until the washings have a pH of 6–7, filtered and dried in the air.
(2) The b.p. of the propanol–benzene azeotrope at atmospheric pressure is 71–72 °C.

Cognate preparations. *Butyl oleate.* Proceed as for isopropyl lactate using 28 g (0.1 mol) of redistilled oleic acid, 37.0 g (46 ml, 0.5 mol) of butan-1-ol (the excess of the latter acts as the water carrier) and 8.0 g of Zerolit 225/H$^{\oplus}$ in a 250-ml flask. Reflux the mixture with magnetic stirring for 4 hours, allow to cool, separate the resin by suction filtration and wash it with three 5 ml portions of butan-1-ol. Remove the butanol from the combined filtrate and washings by distillation under reduced pressure (water pump); the residue consists of crude ester. Distil the residue under diminished pressure (oil pump) and collect the butyl oleate at 232 °C/9 mmHg. The yield is 27 g (85%).

Furfuryl acetate. Reflux a mixture of 39.2 g (34.8 ml, 0.4 mol) of redistilled furfuryl alcohol, 48 g (0.67 mol) of glacial acetic acid, 150 ml of benzene (**CAUTION**) and 20 g of Zerolit 225/H$^\oplus$ in a 500-ml flask, using the apparatus described for isopropyl lactate. After 3 hours, when the rate of collection of water in the water separator is extremely slow, allow to cool, separate the resin by suction filtration and wash it with three 15 ml portions of benzene. Remove the benzene, etc., from the combined filtrate and washings under reduced pressure (water pump) and then collect the crude ester at 74–90 °C/10 mmHg; a small solid residue remains in the flask. Redistil the crude ester through a short fractionating column; pure furfuryl acetate passes over at 79–80 °C/17 mmHg. The yield is 14.5 g (26%).

Experiment 5.146 DIMETHYL ADIPATE

$$HO_2C\cdot(CH_2)_4\cdot CO_2H + 2MeOH \underset{}{\overset{H^\oplus}{\rightleftharpoons}} MeO_2C\cdot(CH_2)_4\cdot CO_2Me + 2H_2O$$

$$Me\cdot C(OMe)_2\cdot Me + H_2O \longrightarrow Me_2CO + 2MeOH$$

In a 500-ml single-necked flask containing a magnetic stirrer bar, place 58.5 g (0.4 mol) of adipic acid, 16 g (20 ml, 0.5 mol) of methanol, 83.2 g (0.8 mol) of 2,2-dimethoxypropane and 0.5 g of toluene-*p*-sulphonic acid. Fit a reflux condenser to the flask and stir the mixture magnetically for 4 hours in a water bath kept at 45 °C. Rearrange the condenser for distillation and distil off acetone (b.p. 56 °C) and methanol (b.p. 64 °C) on the water bath. Distil the residue under reduced pressure (water pump) and collect the dimethyl adipate, b.p. 130 °C/25 mmHg. The yield is 54.9 g (79%).

Experiment 5.147 METHYL HYDROGEN ADIPATE

$$HO_2C\cdot(CH_2)_4\cdot CO_2H + MeOH \overset{H^\oplus}{\longrightarrow} MeO_2C\cdot(CH_2)_4\cdot CO_2H + H_2O$$

Place 175 g (1.2 mol) of adipic acid, 50 ml (1.25 mol) of absolute methanol, 15 ml of concentrated hydrochloric acid and a few fragments of porous pot ('boiling chips') in a 500-ml round-bottomed flask provided with a reflux condenser (1). Heat cautiously at first until the mixture becomes homogeneous and then reflux for 8 hours. Transfer the mixture to a flask fitted with a fractionating column filled with glass helices and arrange to heat the column with a heating tape, the current to which is controlled by a Variac transformer. Careful fractionation under reduced pressure yields dimethyl adipate, b.p. 113–114 °C/6 mmHg (21 g), and methyl hydrogen adipate, b.p. 154–156 °C/6 mmHg (66 g, 34%). Unchanged adipic acid remains in the flask.

Cognate preparation. *Methyl hydrogen sebacate.* Place 115 g (0.56 mol) of sebacic acid, 20 ml (0.5 mol) of absolute methanol, 6 ml of concentrated hydrochloric acid and a few fragments of porous pot in a 500-ml round-bottomed flask fitted with a reflux condenser. Warm the mixture on a water bath until it becomes homogeneous and then reflux gently for 8 hours. Transfer the mixture to a flask fitted with a fractionating column as for methyl hydrogen adipate and fractionate under reduced pressure; due precautions must be taken so that the distillate does not solidify in the condenser or receiver. Collect the dimethyl sebacate at 153–154 °C/6 mmHg (20 g, m.p.

26 °C) and the methyl hydrogen sebacate at 185–186 °C/6 mmHg (46 g, 43%, m.p. 37 °C). The residue in the flask consists of unchanged sebacic acid.

Note. (1) The acid ester may also be prepared by either of two alternative procedures: (a) 1 mol of the diester is heated with 1 mol of the diacid for several hours;[164] (b) 1 mol of the diester is dissolved in 3 to 4 volumes of ethanol to which is added a solution of 1 mol of potassium hydroxide dissolved in the minimum amount of ethanol, and the solution allowed to stand at room temperature overnight. The ethanol is removed on a rotary evaporator, water is added to the residue and the solution extracted with ether to remove unreacted diester. The residual aqueous solution is then cautiously acidified at 0 °C, the acid ester extracted with ether and the ether extract washed, dried and evaporated. The residue is fractionally distilled under reduced pressure.

Experiment 5.148 t-BUTYL ACETATE

$$Me_3COH \xrightarrow[\text{or (Me·CO)}_2\text{O}]{\text{Me·COCl}} Me·CO_2CMe_3$$

Method A. Fit a 1-litre three-necked flask with a sealed stirrer, a reflux condenser and a dropping funnel. Place 57 g (73.5 ml, 0.77 mol) of dry 2-methylpropan-2-ol (t-butyl alcohol) (1), 101 g (106 ml, 0.84 mol) of pure dimethylaniline and 100 ml of anhydrous ether in the flask, set the stirrer in motion and heat the mixture to gentle refluxing on a water bath. Run in 62 g (56.5 ml, 0.79 mol) of redistilled acetyl chloride at such a rate that moderate refluxing continues after the source of heat is removed. When about two-thirds of the acetyl chloride has been introduced, the dimethylaniline hydrochloride commences to crystallise and the mixture refluxes very vigorously. Cool immediately in an ice bath, and, after refluxing ceases, add the remainder of the acetyl chloride; then heat the mixture on a water bath for 1 hour. Cool to room temperature, add about 100 ml of water and continue the stirring until all the precipitated solid has dissolved. Separate the ether layer and extract with 25-ml portions of cold 10 per cent sulphuric acid until the acid extract does not become cloudy when rendered alkaline with sodium hydroxide solution. Finally, wash with 15 ml of saturated sodium hydrogen carbonate solution and dry the ethereal solution with 5 g of anhydrous sodium sulphate overnight. Remove the ether by distillation through an efficient fractionating column and distil the residue through the same column. Collect the t-butyl acetate at 96–98 °C (mainly 97–98 °C). The yield is 55 g (62%).

Note. (1) The t-butyl alcohol should be dried over calcium oxide or anhydrous calcium sulphate and distilled.

Cognate Preparation. *t-Butyl propanoate.* Use 85.5 g (110.5 ml, 1.15 mol) of t-butyl alcohol, 151.5 g (159 ml, 1.26 mol) of pure dimethylaniline and 110 g (103 ml, 1.19 mol) of propanoyl chloride (compare Expt 5.138) and reflux for 3 hours (b.p. 117.5–118.5 °C). Yield: 92 g (62%).

Method B. Fit a 500-ml round-bottomed flask with a reflux condenser carrying a calcium chloride guard-tube. Place 100 ml (108 g, 1.06 mol) of redistilled acetic anhydride, 100 ml (1.07 mol) of dry t-butyl alcohol (see *Note* in *Method A*) and 0.3 g of anhydrous zinc chloride in the flask and shake. Heat the mixture gradually to the reflux temperature, maintain at gentle refluxing for 2 hours and then cool. Replace the reflux condenser by an efficient fractionating column and distil until the temperature reaches 110 °C. Wash the crude distillate, weighing 100–125 g, with two 25 ml portions of water, then with 25 ml portions of 10 per cent potassium carbonate solution until the ester layer is neutral to litmus, and finally dry with 10 g of anhydrous potassium carbonate. Filter off the desiccant, and distil through an efficient fractionating column (e.g. Widmer column) and collect the pure t-butyl acetate at 96–98 °C. The yield is 70 g (57%).

Experiment 5.149   ETHYL BUT-3-ENOATE (*Ethyl vinylacetate*)

$$H_2C \diagup\!\!\!\diagdown \overset{O}{\underset{}{\diagup}} OH \quad \xrightarrow{SOCl_2} \quad H_2C \diagup\!\!\!\diagdown \overset{O}{\underset{}{\diagup}} Cl \quad \xrightarrow{EtOH} \quad H_2C \diagup\!\!\!\diagdown \overset{O}{\underset{}{\diagup}} OEt$$

Prepare vinylacetyl chloride from 50 g (31 ml, 0.42 mol) of thionyl chloride and 30 g (0.35 mol) of vinylacetic acid (Expt 5.128) following the procedure described for butyryl chloride (Expt 5.138); 27 g (0.26 mol) of the acid chloride, b.p. 98–99 °C, are obtained. Place 12.6 g (16.0 ml, 0.27 mol) of absolute ethanol is a 250-ml two-necked flask provided with a reflux condenser and dropping funnel. Cool the flask in ice and introduce the vinylacetyl chloride into the dropping funnel; insert a calcium chloride guard-tube into the mouth of the funnel. Add the acid chloride dropwise (45 minutes) to the alcohol with frequent shaking. Remove the ice and allow to stand for 1 hour. Pour the reaction mixture into water, wash with a little sodium hydrogen carbonate solution, then with water, and dry with anhydrous calcium sulphate. Distil from a 50-ml flask through a short fractionating column, and collect the ethyl vinylacetate at 125–127 °C. The yield is 22 g (75%).

Cognate preparation.   *Phenyl cinnamate.*   Place 72 g (0.48 mol) of cinnamic acid (Expt 6.138) and 60 g (37 mol, 0.5 mol) of thionyl chloride in a 250-ml flask, fitted with a reflux condenser which is connected to a gas absorption trap. Heat the mixture on a water bath, cautiously at first, until hydrogen chloride ceases to be evolved (about 1 hour), allow to cool and add 47 g (0.5 mol) of pure phenol. Heat the mixture on a water bath until no further evolution of hydrogen chloride is observed (about 1 hour). Then place the apparatus on a ceramic-centred wire gauze and heat the flask until the contents are brought just to the reflux temperature in order to complete the reaction: do not heat unduly long as prolonged heating leads to loss of product due to decomposition and polymerisation. Allow the reaction mixture to cool and distil under diminished pressure; collect the fraction of b.p. 190–210 °C/ 15 mmHg. This solidifies to a pale yellow solid, m.p. 66–69 °C, weighing 98 g. Grind it to a powder in a glass mortar and wash the powder with 250 ml of cold 2 per cent sodium hydrogen carbonate solution. Recrystallise from recti-fied spirit (150 ml); 81 g (72%) of pure phenyl cinnamate (white crystals) of m.p. 75–76 °C are obtained.

Phenyl propanoate.   Slowly add 196 g (120 ml) of redistilled thionyl chloride to a mixture of 150 g (1.6 mol) of pure phenol and 132 g (133 ml, 1.7 mol) of propanoic acid (Fig. 2.55 with the addition of a gas absorption device), warm-ing to drive off all the sulphur dioxide and hydrogen chloride, and distilling; 190 g (79%) of phenyl propanoate, b.p. 202–212 °C (the pure substance boils at 211 °C), are obtained.

Experiment 5.150 ETHYL 2-BROMOPROPANOATE

$$Et \cdot CO_2H + SOCl_2 \longrightarrow Et \cdot COCl + HCl + SO_2$$

$$Et \cdot COCl \xrightarrow[P]{Br_2} Me \cdot CHBr \cdot COCl \xrightarrow{EtOH} Me \cdot CHBr \cdot CO_2Et$$

In a 1-litre two-necked round-bottomed flask, equipped with a dropping funnel and a double surface reflux condenser to which is attached a gas absorption trap (Fig. 2.61 (c)), place 220 g (135 ml, 1.86 mol) of redistilled thionyl chloride, and heat to boiling. Add 125 g (126 ml, 1.69 mol) of pure propanoic acid at such a rate that the mixture refluxes gently (c. 1 hour). Reflux the mixture for a further 30 minutes to expel the dissolved sulphur dioxide, allow to cool and add 0.5 g of purified red phosphorus. Introduce 310 g (100 ml, 1.93 mol) of dry bromine (**CAUTION**) during 5–7 hours to the gently boiling propanoyl chloride, and then reflux the mixture for 7 hours, by which time the evolution of hydrogen bromide almost ceases. Add the crude 2-bromopropanoyl chloride during 2 hours to 250 ml of absolute ethanol contained in a three-necked round-bottomed flask, equipped with a mechanical stirrer and a reflux condenser. Complete the reaction by heating on a water bath for 4 hours, when hydrogen chloride is slowly evolved. Filter the reaction liquid into 500 ml of distilled water, separate the oil and wash it successively with water, sodium hydrogen carbonate solution and water. Dry over calcium sulphate and distil at normal pressure to remove the low b.p. fraction (largely ethyl bromide: 75 g) and then under diminished pressure. Collect the ethyl 2-bromopropanoate as a colourless liquid at 69–70 °C/ 25 mmHg; the yield is 221 g (72%).

Experiment 5.151 1-METHYLHEPTYL ACETATE[162]

$$Me \cdot CO_2H + C_6H_{13} \cdot CHBr \cdot Me \xrightarrow{DBU} Me \cdot CO_2CH(Me) \cdot C_6H_{13}$$

A mixture of 2-bromooctane (3.86 g, 0.02 mol), acetic acid (1.8 g, 0.03 mol), DBU (4.56 g, 0.03 mol) and benzene (40 ml) (**CAUTION**), is refluxed for 10 hours. The reaction mixture is then washed with water, dried over magnesium sulphate and distilled to give 1-methylheptyl acetate, b.p. 87° C/ 20 mmHg, 3.1 g (91%); p.m.r. spectrum (CCl₄, TMS) δ 0.9–1.3 (m, 16H), 1.98 (s, 3H) and 4.8 (m, 1H).

Experiment 5.152 ETHYL VALERATE (*Ethyl pentanoate*)

$$Pr \cdot CH_2 \cdot CN + EtOH \xrightarrow[H_2O]{H_2SO_4} Pr \cdot CH_2 \cdot CO_2Et + NH_4 \cdot HSO_4$$

Place 200 g (250 ml) of rectified spirit in a 1-litre round-bottomed flask fitted with a reflux condenser. Cool in ice and run in, slowly and with frequent shaking, 200 g (109 ml) of concentrated sulphuric acid. Add 83 g (104 ml, 1 mol) of butyl cyanide (Expt 5.158) to the mixture and reflux the whole for 10 hours. Allow to cool, pour the reaction mixture into ice water, separate the upper layer of ester and alcohol, and dry over anhydrous calcium sulphate. Distil through a fractionating column and collect the ethyl valerate at 143–146 °C. A further amount of the pure ester may be obtained by redrying the fraction of low boiling point and redistilling. The yield is 100 g (85%).

Cognate preparation. *Ethyl phenylacetate.* Place 75 g (74 ml, 0.64 mol) of benzyl cyanide (Expt 5.157), 125 g (153 ml) of rectified spirit and 150 g (68 ml) of concentrated sulphuric acid in a round-bottomed flask, fitted with an efficient reflux condenser. Reflux the mixture, which soon separates into two layers, gently for 8 hours, cool and pour into 350 ml of water. Separate the upper layer. Dissolve it in about 75 ml of ether (1) in order to facilitate the separation of the layers in the subsequent washing process. Wash the ethereal solution carefully with concentrated sodium hydrogen carbonate solution until effervescence ceases and then with water. Dry over 10 g of anhydrous calcium sulphate for at least 30 minutes. Remove the solvent by flash distillation and distil the residue from an air bath. The ethyl phenylacetate passes over at 225–229 °C (mainly 228 °C) as a colourless liquid; the yield is 90 g (86%). Alternatively, the residue after removal of ether may be distilled under diminished pressure; collect the ester at 116–118 °C/20 mmHg.

Note. (1) Alternatively use 20 ml of carbon tetrachloride. The carbon tetrachloride solution then forms the lower layer in all washing operations.

Experiment 5.153 TRIETHYL ORTHOFORMATE (*Triethoxymethane*)

$$CHCl_3 + 3EtONa \longrightarrow CH(OEt)_3 + 3NaCl$$

CAUTION: *This experiment should be conducted in an efficient fume cupboard.*
     Fit a 1500-ml round-bottomed flask with a large double surface reflux condenser. Make sure that the apparatus is thoroughly dry. Place 750 ml of super-dry ethanol (Section 4.1.9, p. 401) and 123 g (82 ml, 1.03 mol) of dry chloroform (**CAUTION**) in the flask. Add 52 g (2.25 mol) of clean sodium, cut into small pieces, through the condenser in the course of 30 minutes; when the reaction becomes vigorous, cool the outside of the flask by running water from the condenser outlet. When all the sodium has reacted and the mixture has attained room temperature, filter off the sodium chloride through a sintered glass funnel. The filtration apparatus must be thoroughly dry, and a drying tube, filled with granular calcium chloride, should be placed between the filter flask and the pump. Wash the solid on the filter with 50 ml of absolute ethanol and allow the washings to run into the main filtrate. Distil the solution from a water bath through an efficient fractionating column in order to recover the excess of chloroform and most of the alcohol; collect the distillate (about 500 g) (*A*) in a filter flask protected by a drying tube. Decant the liquid remaining in the flask from a little salt which has separated, and distil it through an all-glass Dufton (or Widmer) column. A fraction (*B*) of low boiling point passes over first, followed by the triethyl orthoformate at 144–146 °C. The yield is 35 g (23%) but depends somewhat upon the efficiency of the fractionation.
     Carry out a second run with the recovered chloroform–alcohol mixture (*A*): add 100 g of dry chloroform and sufficient super-dry ethanol (200–250 ml) to give a total volume of 750 ml. Add 52 g of sodium as before. Remove the excess of chloroform and alcohol as before on a water bath through a fractionating column, add the intermediate fraction (*B*) from the first run, and fractionate again. The yield of product, b.p. 144–146 °C is 45 g (29%), p.m.r. spectrum (CCl₄, TMS) δ 1.19 (t, 9H), 3.52 (q, 6H) and 5.00 (δ, 1H).

5.12.4 ACID AMIDES

Primary aliphatic amides are formed on heating the ammonium salts of the corresponding carboxylic acids, or by heating an acid or its ammonium salt with urea.

$$R \cdot CO_2NH_4 \longrightarrow R \cdot CO \cdot NH_2 + H_2O$$

$$R \cdot CO_2H + NH_2 \cdot CO \cdot NH_2 \longrightarrow R \cdot CO \cdot NH_2 + CO_2 + NH_3$$

The preferred preparative routes to amides however start from the following carboxylic acid derivatives.

FROM ACYL HALIDES

The reaction of an acyl chloride with an excess of ammonia represents one of the best procedures for the preparation of primary amides (Expt 5.154).

$$\underset{R}{\overset{O}{\underset{\|}{\text{C}}}}_{Cl} + 2NH_3 \longrightarrow \underset{R}{\overset{O}{\underset{\|}{\text{C}}}}_{NH_2} + NH_4Cl$$

The acyl chloride (the crude material prepared by the thionyl chloride method is quite satisfactory) is added dropwise to well-stirred concentrated aqueous ammonia cooled in a freezing mixture. The amides of the higher carboxylic acids crystallise out on standing and need only to be filtered and recrystallised. Water-soluble amides are isolated by extraction with hot ethyl acetate following removal of water on a rotary evaporator.

A milder procedure[165] involves stirring a solution of the acyl chloride in acetone at room temperature with ammonium acetate. The filtered solution is evaporated to recover the required amide.

The use of primary or secondary amines in place of ammonia yields the corresponding secondary or tertiary amides in reaction with an acyl chloride. These compounds often serve as crystalline derivatives suitable for the characterisation of either the acyl chloride (and hence of the carboxylic acid itself) or the amine (Sections 9.6.16, p. 1265 and 9.6.21, p. 1273).

FROM ESTERS

Amides are very easily prepared by the interaction of carboxylic esters with concentrated aqueous ammonia (*ammonolysis*).

$$R^1 \cdot CO_2R^2 + NH_3 \longrightarrow R^1 \cdot CO \cdot NH_2 + R^2OH$$

The reaction usually proceeds readily in the cold, particularly when the methyl esters of the lower molecular weight carboxylic acids are involved. Sparingly soluble amides crystallise out from the reaction mixture upon standing, as in the case of succinamide (Expt 5.155).

FROM NITRILES

The interruption of the hydrolysis of a nitrile at the amide stage can often be achieved in a preparative manner, as for example in the preparation of phenylacetamide (Expt 5.156), where the nitrile is dissolved in concentrated hydrochloric acid at 40 °C and subsequently poured into water. The use of hot polyphosphoric acid has also been recommended.[166]

$$R \cdot CN + H_2O \xrightarrow{H^{\oplus}} R \cdot CO \cdot NH_2$$

Reaction conditions which are particularly applicable to aromatic nitriles involve the use of an aqueous solution of sodium hydroxide containing hydrogen peroxide, but alkyl cyanides do not always give good results; the method is illustrated by the preparation of toluamide (Expt 6.167).

Experiment 5.154   HEXANAMIDE

$$Me \cdot (CH_2)_4 \cdot COCl + 2NH_3 \longrightarrow Me \cdot (CH_2)_4 \cdot CO \cdot NH_2 + NH_4Cl$$

Place 125 ml of concentrated ammonia solution ($d\,0.88$) in a 600-ml beaker and surround the latter with crushed ice. Stir the ammonia solution mechanically, and introduce 56 g (0.42 mol) of hexanoyl chloride (Expt 5.138) slowly by means of a suitably supported separatory funnel. The rate of addition must be adjusted so that no white fumes are lost. The amide separates immediately. Allow to stand in the ice-water for 15 minutes after all the acid chloride has been introduced. Filter off the amide at the pump; use the filtrate to assist the transfer of any amide remaining in the beaker to the filter (1). Spread the amide on sheets of filter or drying paper to dry in the air. The crude hexanamide (30 g, 63%) has m.p. 98–99 °C and is sufficiently pure for conversion into the nitrile (Expt 5.160) (2). Recrystallise a small quantity of the amide by dissolving it in the minimum volume of hot water and allowing the solution to cool; dry on filter paper in the air. Pure hexanamide has m.p. 100 °C.

Notes. (1) The filtrate will deposit small amounts of hexanamide upon concentration to half its original volume.
(2) The process is of general application for higher (i.e. $> C_5$) fatty acids.

Cognate preparation. *Isobutyramide* (*2-methylpropanamide*). Add 106 g (1 mol) of isobutyryl chloride (Expt 5.138) to 400 ml of concentrated ammonia solution ($d\,0.88$) contained in a 2-litre two-necked flask fitted with a stirrer and dropping funnel, at such a rate that the temperature does not rise above 15 °C. Stir for 1 hour after the addition is complete and attach the flask to a rotary evaporator, connect to a water-jet pump and heat the flask on a boiling water bath until the crystalline deposit is quite dry. Boil the residue with 1200 ml of ethyl acetate, filter the hot solution and extract the residue with further quantities of hot ethyl acetate (2 × 750 ml). Combine the extracts, cool to 0 °C and collect the crystals of isobutyramide of m.p. 128–129 °C. The yield after recovery of further crops of crystalline material from the concentrated mother-liquors is 58 g (66%).

Experiment 5.155   SUCCINAMIDE

709

Add 5 g (4.8 ml, 0.034 mol) of dimethyl succinate to 25 ml of concentrated ammonia solution (*d* 0.88) in a 100-ml conical flask. Stopper the flask and shake the contents for a few minutes: allow to stand for 24 hours with occasional shaking. Filter off the crystals of succinamide, and wash with a little cold water. Recrystallise from a little hot water and dry in an oven. The yield is 3.5 g (88%). Pure succinamide melts at 254 °C with decomposition.

Experiment 5.156 PHENYLACETAMIDE

$$Ph \cdot CH_2 \cdot CN \xrightarrow[40°C]{H_2O, HCl} Ph \cdot CH_2 \cdot CO \cdot NH_2$$

In a 2-litre three-necked flask, provided with a thermometer, reflux condenser and *efficient* mechanical stirrer, place 100 g (98 ml, 0.85 mol) of benzyl cyanide (Expt 5.157) and 400 ml of concentrated hydrochloric acid. Immerse the flask in a water bath at 40 °C and stir the mixture vigorously: the benzyl cyanide passes into solution within 20–40 minutes and the temperature of the reaction mixture rises to about 50 °C. Continue the stirring for an additional 20–30 minutes after the mixture is homogeneous. Replace the warm water in the bath by tap water at 15 °C, replace the thermometer by a dropping funnel charged with 400 ml of cold distilled water and add the latter with stirring: crystals commence to separate after about 50–75 ml have been introduced. When all the water has been run in, cool the mixture externally with ice-water for 30 minutes (1), and collect the crude phenylacetamide by filtration at the pump. Remove traces of phenylacetic acid by stirring the wet solid for about 30 minutes with two 50 ml portions of cold water; dry the crystals at 50–80 °C. The yield of phenylacetamide, m.p. 154–155 °C, is 95 g (82%). Recrystallisation from rectified spirit raises the m.p. to 156 °C.

Note. (1) The suspension of phenylacetamide may be further hydrolysed to phenylacetic acid by refluxing with stirring until the solid dissolves. The mixture becomes turbid after 30 minutes and the product begins to separate as an oil: refluxing is continued for 6 hours, the mixture is cooled first with tap water and then by an ice-water bath for about 4 hours. The crude phenylacetic acid is filtered at the pump, washed with two 50 ml portions of cold water and dried in a desiccator. The resulting crude acid melts at 69–70 °C; it may be purified by recrystallisation from light petroleum (b.p. 40–60 °C) or, better, by vacuum distillation.

5.13 ALIPHATIC NITRILES (ALKYL CYANIDES)

The preparation of aliphatic nitriles is illustrated by the following four procedures.

1. Displacement with cyanide ion on an alkyl halide. (Expt 5.157 and 5.158).
2. Displacement with cyanide ion on an arylsulphonylhydrazone (Expt 5.159).
3. The dehydration of amides (Expt 5.160), and aldoximes (see Section 6.15.3, p. 1082).
4. Cyanoethylation (Expts 5.161 to 5.163), and the α-alkylation of nitriles.

SUMMARY OF RETROSYNTHETIC STRATEGIES

Functional group interconversion (FGI) (method 3)

$$R \underset{O}{\overset{NH_2}{\diagdown}} \underset{(3)}{\Longleftarrow} R \diagdown CN \underset{(3)}{\Longrightarrow} R \diagdown \underset{H}{\overset{NOH}{\diagup}}$$

$$\text{(TM)}$$

Disconnection (methods 1 and 4)

$$R \diagdown \{CN \underset{(1)}{\Longrightarrow} R \diagdown \overset{\oplus}{CH_2} \; \overset{\ominus}{CN} \equiv R{\cdot}CH_2{\cdot}CH_2Br + KCN$$

$$\text{(TM)}$$

$$Y \{ \diagdown CN \underset{(4)}{\Longrightarrow} \underset{Y^\ominus}{H_2C} \diagdown CN \equiv ROH \text{ etc.} + CH_2{=}CH{\cdot}CN$$

$$\text{(TM)}$$

$$Y = -OR, -NHR,$$
$$-NR_2$$
$$-CH(CO_2Et)_2$$

$$R^1 \underset{R^2}{\diagdown} CN \underset{(4)}{\Longrightarrow} R^1 \diagdown \overset{CN}{\underset{\ominus}{\diagdown}} {}^\oplus R^2 \equiv R^1CH_2{\cdot}CH_2{\cdot}CN + R^2X$$

SPECTROSCOPIC FEATURES

The sharp and intense characteristic *i.r.* absorption arising from the stretching of the carbon–nitrogen triple bond in the region of $2260 \, \text{cm}^{-1}$, is easily recognised and distinguished from the weaker absorption of the carbon–carbon triple bond, which also occurs in this region. The α-protons of alkyl cyanides are relatively more shielded due to magnetic anisotropic effects (*cf.* alkynes, p. 325), and therefore resonate in the *p.m.r.* spectra at higher field than might have been expected from the purely electron-withdrawing nature of the cyano group (*cf.* the alkyl halides, p. 324). The main features of the *m.s.* of alkyl cyanides have been outlined on p. 382. Examples of the interpretation of these spectral features are included in some of the preparations below.

5.13.1 DISPLACEMENT WITH CYANIDE ION ON AN ALKYL HALIDE

The classical procedure for the reaction involves heating the alkyl halide (usually the chloride or bromide) with sodium or potassium cyanide in methanolic or ethanolic solution. The method is clearly of value for the extension of the carbon chain by one carbon atom, since the cyano group may be converted into a carboxyl group by hydrolysis (Section 5.11.2, p. 671) or into an aminomethyl group ($-CH_2{\cdot}NH_2$) by reduction (Section 5.16.1, p. 771), or into a formyl group by controlled reduction to the imine followed by hydrolysis (Section 5.7.4, p. 594).

$$Br{-}R \overset{\frown}{} \overset{\ominus}{CN} \longrightarrow R{\cdot}CN + Br^\ominus$$

This method has been widely used for primary and secondary halides; tertiary halides undergo ready elimination under these conditions to yield the alkene and give little or no nitrile. Traces of the corresponding isonitriles ($R{\cdot}NC$) may be formed during these displacement reactions but may be removed by virtue of

711

their ready hydrolysis with aqueous mineral acid. In the case of the conversion of allyl bromide into allyl cyanide, copper(I) cyanide is used in place of an alkali metal cyanide. The procedures are illustrated with reference to the preparation of a range of mono- and dinitriles, the latter from the corresponding α,ω-dibromo compounds (Expt 5.157).

Aprotic solvents such as dimethyl sulphoxide or dimethyl formamide significantly improve the procedure in that the less reactive alkyl chlorides are more rapidly converted into the nitriles in good yield by sodium cyanide (Expt 5.158).

Three PTC methods are worthy of note. In the first the alkyl halide is treated with sodium cyanide in decane solution in the presence of catalytic amounts of hexadecyltributylphosphonium bromide.[167a] In the second tetraethylammonium cyanide in molar quantities is used in dichloromethane solution with the alkyl halide.[167b] In the third method, which is reported to be most satisfactory in the case of allylic and benzylic halides, a solid/liquid system of potassium cyanide and 18-crown-6 ether is employed.[168]

5.13.2 DISPLACEMENT WITH CYANIDE ION ON AN ARYLSULPHONYLHYDRAZONE

Reaction of cyanide ion with an alkyl halide is often the most convenient route to a nitrile, but in those cases where the corresponding aldehyde or ketone is more readily available than the alkyl halide, the following procedure is very convenient. The carbonyl compound is first converted into its 2,4,6-triisopropylbenzenesulphonyl hydrazone (1) (Expt 6.42 gives the method for the preparation of the reagent, TBSH), which without isolation is then reacted with potassium cyanide under gentle reflux.[169]

$$R^1R^2C{=}O + NH_2 \cdot NH \cdot SO_2Ar \longrightarrow R^1R^2C{=}N \cdot NH \cdot SO_2Ar$$
$$(1)$$

The method is applicable to a wide range of aliphatic and alicyclic (but not aromatic) carbonyl compounds, and is illustrated here by the conversion of cyclopentanone into cyclopentyl cyanide (Expt 5.159).

Experiment 5.157 PENTANEDINITRILE (*Glutaronitrile*)

$$BrCH_2 \cdot CH_2 \cdot CH_2Br + 2NaCN \longrightarrow NC \cdot CH_2 \cdot CH_2 \cdot CH_2 \cdot CN + 2NaBr$$

CAUTION: *Sodium and potassium cyanide are very poisonous and should be handled with great care; disposable plastic gloves should be worn. Alkyl cyanides are likewise very poisonous and all operations should be conducted in an efficient fume cupboard.*

Fit a 2-litre two-necked flask with a separatory funnel and a reflux condenser. Place 147 g (3 mol) of finely powdered sodium cyanide (1) and 150 ml of water in the flask and heat on a water bath until most of the solid passes into solution. Add a solution of 250 g (126 ml, 1.25 mol) 1,3-dibromopropane (Expt 5.54) in 500 ml of rectified spirit through the separatory funnel over a period of 30 minutes. Reflux the mixture on a water bath for 35 hours; then remove the solvent (rotary evaporator) using a boiling-water bath. The residue in the flask consists of sodium bromide, unreacted sodium cyanide and the dinitrile: the last-named alone is soluble in ethyl acetate. Extract the residue with 200 ml of ethyl acetate. Filter the solution through a sintered glass funnel and wash the solid with about 50 ml of ethyl acetate. Dry the filtrate, after removing the aqueous layer, with anhydrous calcium sulphate, distil off the ethyl acetate at atmospheric pressure (about 245 ml are recovered), allow to cool somewhat and distil the liquid under reduced pressure. Collect the pentanedinitrile at 139–140 °C/8 mmHg. The yield is 95 g (82%); ^{13}C-n.m.r. spectrum (CDCl$_3$, TMS) δ 16.3, 21.6 and 118.8; i.r. spectrum (thin film) 2200 cm^{-1} (CN).

Note. (1) Dry the powdered commercial material (c. 98% pure) in a vacuum desiccator over potassium hydroxide pellets. *Sodium cyanide is very poisonous and must be handled with great care.* Residual solutions containing alkali cyanides should be rendered innocuous by the addition of an excess of sodium hypochlorite before being washed down the main drain of the laboratory with a liberal supply of water; they should never be treated with acid.

Cognate preparations. *Heptanenitrile.* Use 30 g (0.61 mol) of sodium cyanide dissolved in 40 ml of water; 82 g (70 ml, 0.5 mol) of 1-bromohexane (Expt 5.55) in 150 ml of methanol. Remove the methanol through an efficient fractionating column, add 500 ml of water and separate the upper layer of crude nitrile. Purify the crude nitrile by shaking it twice with about half its volume of concentrated hydrochloric acid, and then successively with water, saturated sodium hydrogen carbonate solution and water. Dry with anhydrous calcium chloride and distil. Collect the heptanenitrile at b.p. 182–184 °C. The yield is 40 g (73%).

Heptanedinitrile. Use 29 g (0.446 mol) of potassium cyanide, 30 ml of water and 45 g (0.195 mol) of 1,5-dibromopentane (Expt 5.54) in 75 ml of rectified spirit. Reflux for 8 hours and work-up as for pentanedinitrile but using ether as the extracting solvent. Distil the crude heptanedinitrile under diminished pressure and collect the main fraction at b.p. 168–170 °C/15 mmHg. The yield is 18 g (75%).

Octanedinitrile. 1,6-Dibromohexane (Expt 5.55) may be converted into octanedinitrile, b.p. 178–180 °C/15 mmHg, as in the foregoing experiment.

Benzyl cyanide. Place 100 g (2 mol) of powdered sodium cyanide and 90 ml of water in a 1-litre round-bottomed flask provided with a reflux condenser. Warm on a water bath until the sodium cyanide dissolves. Add down the condenser a solution of 200 g (181.5 ml, 1.58 mol) of benzyl chloride (Expt 5.27) in 200 g of rectified spirit during 30–45 minutes. Heat the mixture in a water bath for 4 hours, cool and filter off the precipitated sodium chloride with suction; wash with a little alcohol. Distil off as much as possible of the alcohol using a rotary evaporator. Cool the residual liquid, filter if necessary and separate the layer of crude benzyl cyanide. (Sometimes it is advantageous to extract the nitrile with ether.) Dry over a little magnesium sulphate, and distil under diminished pressure. Collect the benzyl cyanide at 102–103 °C/10 mmHg. The yield is 160 g (86%).

This product is sufficiently pure for most purposes but it contains some benzyl isocyanide and usually develops an appreciable colour on standing.

The following procedure removes the isocyanide and gives a stable water-white compound. Shake the once-distilled benzyl cyanide vigorously for 5 minutes with an equal volume of warm (60 °C) 50 per cent sulphuric acid (prepared by adding 55 ml of concentrated sulphuric acid to 100 ml of water). Separate the benzyl cyanide, wash it with an equal volume of saturated sodium hydrogen carbonate solution and then with an equal volume of half-saturated sodium chloride solution. Dry with magnesium sulphate and distil under reduced pressure.

1-Naphthylacetonitrile. Place a mixture of 56 g (0.32 mol) of 1-(chloromethyl)naphthalene, 29 g (0.45 mol) of potassium cyanide, 125 ml of ethanol and 50 ml of water in a 500-ml round-bottomed flask fitted with a double surface reflux condenser, and reflux for 1 hour. Distil off the alcohol, transfer the residue to a separatory funnel, wash it with water, filter from a small amount of solid, transfer to a dish and dry under reduced pressure (vacuum desiccator charged with anhydrous calcium chloride). Distil under diminished pressure and collect the 1-naphthylacetonitrile at 155–160 °C/9 mmHg (1); the yield is 38 g (72%).

Note. (1) A little naphthalene may pass over first owing to impurities in the original chloromethylnaphthalene.

Allyl cyanide. Into a 2-litre three-necked flask, provided with a sealed stirrer and two long double surface condensers, place 293 g (210 ml, 2.42 mol) of freshly distilled allyl bromide, b.p. 70–71 °C (Expt 5.54) and 226 g (2.52 mol) of dry copper(I) cyanide (Section 4.2.23, p. 429). Warm the flask on a water bath so that the allyl bromide refluxes but do not stir at this stage. Immediately the vigorous reaction commences (after 15–30 minutes), remove the water bath and cool the flask in a bath of ice and water; the two double surface condensers will prevent any loss of product. When the reaction subsides, start the stirrer and heat the mixture on the water bath for 1 hour. Remove the condensers and arrange the apparatus for distillation: close one neck with a stopper. Heat the flask in an oil bath, and distil the allyl cyanide with stirring; it is advisable to reduce the pressure (water pump) towards the end of the distillation to assist the removal of the final portion of the allyl cyanide from the solid residue. Redistil and collect the pure allyl cyanide at 116–121 °C. The yield is 140 g (86%).

Experiment 5.158 VALERONITRILE (*Butyl cyanide*)

$$Me \cdot (CH_2)_2 \cdot CH_2Cl + NaCN \xrightarrow{DMSO} Me \cdot (CH_2)_2 \cdot CH_2 \cdot CN$$

CAUTION: *Note comments in Expt 5.157.*

Set up on a water bath a 500-ml three-necked flask fitted with a mechanical stirrer, a thermometer and a two-way adapter fitted with a dropping funnel and a reflux condenser protected by a calcium chloride guard-tube. Place 150 ml of dry dimethyl sulphoxide in the flask and add 30 g (0.61 mol) of dry powdered sodium cyanide. Heat the mixture with stirring to 90 °C and then remove the water bath. Add 46.3 g (0.5 mol) of butyl chloride (Expt 5.48) slowly from the dropping funnel so that the temperature of the exothermic reaction does not rise above 150 °C (about 10 minutes); continue to stir until the temperature falls to 50 °C (about 30 minutes). Pour the mixture into water and extract with three 300 ml portions of ether. Wash the combined extracts with two 100 ml portions of saturated sodium chloride solution, dry over magnesium sulphate and distil off the ether on a water bath. Distil the residue at atmospheric pressure and collect the butyl cyanide at 138–140 °C; the yield is 35.2 g (85%). The product has a nauseating odour, and should be handled in an efficient fume cupboard.

Cognate preparations. *Octanenitrile.* Use 250 ml of dry dimethyl sulphoxide, 89.6 g (0.5 mol) of 1-bromoheptane (Expt 5.55) and 30 g (0.61 mol) of dried sodium cyanide. The yield of octanenitrile is 50.8 g (81%), b.p. 199–203 °C.

Hexanedinitrile (*adiponitrile*). Use 150 ml of dry dimethyl sulphoxide, 21.2 g (0.17 mol) of 1,4-dichlorobutane (Expt 5.52) and 20 g (0.41 mol) of dried sodium cyanide. Maintain the reaction temperature at 90 °C for a further 15 minutes after the initial exothermic reaction has subsided. Add 150 ml of dichloromethane to the cooled reaction mixture and pour into an excess of saturated sodium chloride solution in a separatory funnel. Add just sufficient water to dissolve precipitated salts and separate the dichloromethane layer. Extract the acqueous layer once with dichloromethane, wash the combined extracts twice with salt solution, dry over magnesium sulphate and remove the solvent using a rotary evaporator. Fractionally distil the residue under reduced pressure and collect the adiponitrile as a fraction of b.p. 140–141 °C/1.5 mmHg. The yield is 14.6 g (81%). Some dimethyl sulphoxide (b.p. 40–42 °C/1.5 mmHg) may be obtained as a forerun.

Experiment 5.159 CYCLOPENTYL CYANIDE (*Cyclopentanecarbonitrile*)[169]

CAUTION: *Potassium cyanide is very poisonous and should only be handled with great care with the wearing of plastic gloves. See Note (1), Expt 5.157, for the disposal of alkali metal cyanides. All operations should be conducted in an efficient fume cupboard.*

Cyclopentanone (1.7 g, 20 mmol) and 2,4,6-triisopropylbenzenesulphonyl hydrazide (7.5 g, 25 mmol, Expt 6.42) are stirred together in methanol (30 ml) solution at room temperature for 1 hour. Potassium cyanide (3.9, 60 mmol) is then added and the reaction mixture is heated under reflux for 2 hours. Water (20 ml) is added and the products are extracted with dichloromethane (60 ml) and the extract is washed with a saturated solution of sodium hydrogen carbonate (100 ml) and water (50 ml). The organic layer is dried over magnesium sulphate and evaporated carefully under reduced pressure and the residue distilled to give cyclopentanecarbonitrile, b.p. 101 °C/100 mmHg, yield 1.28 g (67%); i.r. 2240 cm^{-1}, ^{13}C-n.m.r. (CDCl$_3$, TMS) δ 25.0, 27.9, 31.3, and 123.5.

5.13.3 DEHYDRATION OF AMIDES AND ALDOXIMES

Conversion of amides into nitriles may be effected by treating them with a variety of dehydrating reagents. Among those that have been employed are phosphorus pentoxide, phosphorus pentachloride,[170] ethyl polyphosphate,[171] thionyl chloride, and cyanuric chloride.[172] A method has been described which converts the acid chloride into the nitrile in a one-pot reaction; the reagent is sulphonamide with sulpholane as the solvent.[173] The reaction is applicable to aliphatic, alicyclic, aromatic and heteroaromatic carboxylic acid chlorides.

$$\underset{R}{\overset{O}{\parallel}}{\overset{}{C}}\text{--Cl} + H_2N\cdot SO_2\cdot NH_2 \longrightarrow \underset{R}{\overset{O}{\parallel}}{\overset{}{C}}\text{--NH}\cdot SO_2\cdot NH_2 \rightleftharpoons$$

$$\underset{R}{\overset{OH}{\parallel}}{\overset{}{C}}\text{=N}\cdot SO_2\cdot NH_2 \xrightarrow{H^\oplus} R\cdot CN + NH_2\cdot SO_3H$$

Some of these procedures are exemplified in Expt 5.160. Dehydration of aldoximes proceeds under milder conditions, acetic anhydride being frequently used. The method is particularly applicable to the synthesis of aromatic nitriles (see Expt 6.170).

Experiment 5.160 ISOBUTYRONITRILE (*Isopropyl cyanide*)

$$Me_2CH\cdot CONH_2 \xrightarrow[-H_2O]{P_2O_5} Me_2CH\cdot CN$$

Equip a 2-litre round-bottomed flask with a simple distillation head carrying a capillary air leak, and attach a double surface condenser fitted with a receiver adapter with vacuum connection and a 100-ml receiver flask. Remove the still-head, wrap some glazed paper around a glass tube and insert it into the flask neck until the lower end enters the bulb of the flask; upon removing the glass tube, the paper roll expands and thus lines the neck of the flask. Weigh out on pieces of glazed paper first 65 g (0.75 mol) of 2-methyl-propanamide (isobutyramide, Expt 5.154) and then, *as rapidly as possible* because of its extremely hygroscopic character, 114 g (0.8 mol) of phosphorus pentoxide (1). Immediately transfer, with the aid of a spatula, the phosphorus pentoxide down the glazed paper cylinder into the flask, then introduce the acid amide similarly, remove the paper, stopper the flask and mix the contents well by gentle shaking (2). Attach the flask to the distillation set-up and heat the flask cautiously with a luminous flame kept in constant motion and applied uniformly over the bottom of the flask. A reaction accompanied by much frothing takes place. After the reaction has subsided, apply the vacuum provided by a water-jet pump carefully and continue heating at such a rate that the isopropyl cyanide distils smoothly. Redistil the product from 10 g of phosphorus pentoxide in a flask fitted with a short Vigreux column and heated in an oil bath maintained at 145–155 °C; collect the pure product, b.p. 101–102 °C. The yield is 37 g (73%).

Notes. (1) Phosphorus pentoxide must be treated with great care since it produces painful burns if allowed to come in contact with the skin.
(2) Wet the papers thoroughly with water before throwing them away, as the residual phosphorus pentoxide may cause them to smoulder.

Cognate preparations. *Hexanenitrile* (*pentyl cyanide*). (*Use of thionyl chloride.*) Place 29 g (0.25 mol) of hexanamide (Expt 5.154) in a 100-ml flask fitted with a reflux condenser, add 45 g (27.5 ml, 0.38 mol) of redistilled thionyl chloride and connect the top of the condenser to a gas absorption device

[Fig. 2.61(a) or (b)]. Boil the mixture gently under reflux for 1 hour. Remove the reflux condenser and arrange it for downward distillation. Distil from an oil bath or an air bath; the excess thionyl chloride passes over below 90 °C. If the residue at this stage is dark and contains some solid matter, add a little anhydrous ether, filter, and remove the ether on a rotatory evaporator. Distil the crude hexanenitrile as a fraction of b.p. 161–163 °C. The yield is 21 g (86%). If the product is slightly turbid, shake it with a little anhydrous calcium sulphate, filter, and redistil.

General procedure for the one-pot conversion of acid chlorides into nitriles.[173] Heat a mixture of 0.01 mol of the acid chloride and 0.012 mol of sulphonamide in 10 ml of sulpholane for 3 hours at 120 °C. After cooling, pour the reaction mixture into 75 ml of 1 M sodium hydroxide solution and extract three times with ether–cyclohexane (1:1). Wash the organic layer three times with water to remove traces of sulpholane, dry over anhydrous sodium sulphate and evaporate to give the almost pure product. Distil under reduced pressure or recrystallise as appropriate.

5.13.4 CYANOETHYLATION PROCEDURES AND THE α-ALKYLATION OF NITRILES

Cyanoethylation is an important method for the introduction of a three-carbon residue in the form of a 2-cyanoethyl group ($-CH_2 \cdot CH_2 \cdot CN$), and conveniently leads to a range of polyfunctional compounds.[174] Disconnection in the manner shown in the retrosynthetic summary (p. 711) emphasises that this fragment arises from acrylonitrile, which undergoes a Michael 1,4-addition process with a range of nucleophilic addenda including carbanions, alcohols and amines. The first is illustrated by the reaction of acrylonitrile with diethylmalonate in the presence of sodium ethoxide to form diethyl (2-cyanoethyl)malonate (Expt 5.161). The mechanism of the reaction may be readily inferred from that given for the synthesis of α-alkylglutaric acids (Expt 5.136 and p. 682). Other carbanion sources include nitroalkanes, aldehydes and ketones, all of which must possess an α-hydrogen. Experimental conditions are given for the cyanoethylation of primary alcohols (Expt 5.162) and of secondary aliphatic amines (Expt 5.163); the reaction is equally applicable to phenols, thiols, ammonia, primary amines, etc.

Anion exchange resins of the quaternary ammonium hydroxide type [e.g. Zerolit FF or Amberlite (IRA-400)] are strong bases and are useful catalysts for cyanoethylation of alcohols. No additional basic catalyst is normally required in the case of the cyanoethylation of aliphatic amines.

As noted above in Section 5.13.1 the cyano group may be easily converted into other functionalities. The α-alkylation of a nitrile (1) by treatment with an appropriate base followed by reaction with an alkyl halide enables an α-branched chain aldehyde (2) or carboxylic acid (3), or a β-branched chain primary amine (4) to be readily synthesised. Furthermore an α-branched chain nitrile, or the corresponding acid or aldehyde may be further elaborated into an α-branched chain ketone (5) (see for example, Sections 5.8.4, p. 616, and 5.4.2, p. 531 with 5.8.1, p. 607).

In particular the synthesis of an α-alkylaldehyde or branched chain ketone by this overall strategy is of considerable merit since some of the problems associated with the α-alkylation of an enolate ion are largely avoided (e.g. competing aldol condensation in the case of an aldehyde, or polyalkylation and low regioselective alkylation in the case of a ketone).

Strong bases in an aprotic medium are used to generate a high concentration of the mesomeric anion of the nitrile, since this restricts the degree of dialkylation which otherwise might occur. The alkylation reagents may be primary or secondary alkyl bromides or iodides, or alkylsulphonates.[175a] The PTC procedure using concentrated sodium hydroxide solution with a phase transfer catalyst (e.g. benzyltriethylammonium chloride) is a particularly effective method for the α-alkylation of arylacetonitriles (Ar·CH$_2$·CN).[175b]

Experiment 5.161 DIETHYL (2-CYANOETHYL)MALONATE

$$(EtO_2C)_2CH_2 + CH_2{=}CH{\cdot}CN \longrightarrow (EtO_2C)_2CH{\cdot}CH_2{\cdot}CH_2{\cdot}CN$$

CAUTION: *Acrylonitrile vapour is highly toxic; it should therefore be handled with great caution and all operations with it should be conducted in a fume cupboard provided with an efficient extraction system.*

Assemble in a fume cupboard a 1-litre three-necked flask containing 480 g (455 ml, 3 mol) of diethyl malonate and fitted with a stirrer, a dropping funnel and a thermometer. Start the stirrer, add to the flask a solution of sodium ethoxide prepared from 3.5 g of sodium in 100 ml of absolute ethanol and then run in slowly 80 g (100 ml, 1.5 mol) of acrylonitrile (1) (CAUTION: *toxic and lachrymatory vapour*) at a rate such that the temperature does not exceed 35 °C. When the addition is complete, continue to stir for 1 hour more and then remove the stirrer. Equip the flask for distillation under reduced pressure through a short fractionating column (cf. Fig. 2.110) and remove the ethanol by distillation using a water pump. Continue the distillation of the residue using an oil pump at a recorded pressure of about 0.2 mmHg, and collect (i) recovered diethyl malonate at 52 °C (240 g), and (ii) somewhat crude diethyl (2-cyanoethyl)malonate at 130–140 °C (200 g). Redistil the latter to obtain 175 g (55%) of the purified product of b.p. 102–106 °C/0.2 mmHg (127–130 °C/3 mmHg); i.r. spectrum (thin film), 2250 (CN) and 1725 cm^{-1} (C=O); p.m.r. spectrum (CCl$_4$, TMS), δ 1.29 (t, 6H), 1.90–2.70 (m, 4H), 3.41 (t, 1H) and 4.19 (q, 4H).

Note. (1) Acrylonitrile forms an azeotropic mixture with water, b.p. 70.5 °C (12.5% water). The commercial product may contain the polymer; it should be redistilled before use and the fraction, b.p. 76.5–78 °C, collected separately as a colourless liquid.

Experiment 5.162 3-ETHOXYPROPIONITRILE

$$EtOH + CH_2{=}CH{\cdot}CN \xrightarrow[\text{resin}]{\overset{\ominus}{O}H} EtO{\cdot}CH_2{\cdot}CH_2{\cdot}CN$$

Regenerate the resin [Zerolit FF or Amberlite (IRA-400)] by washing it on a Buchner funnel with 5 per cent sodium hydroxide solution (5–6 times the volume of the resin); rinse the resin with distilled water until the washings are neutral and dry in the air. In a 500-ml three-necked flask equipped with a reflux condenser, stirrer and a dropping funnel, place 25 g of the regenerated resin and 46 g (58.5 ml, 1 mol) of ethanol. Immerse the flask in an ice bath to control the subsequent initial exothermic reaction and to hold the temperature below 15–20 °C throughout the experiment. Add 67 g (85 ml, 1.26 mol) of redistilled acrylonitrile [**CAUTION**: *see Expt 5.161*] slowly to the well-stirred mixture in the flask over a period of 1–2 hours; continue the stirring for a further 1.5 hours. Separate the resin by filtration. Distil the filtrate at atmospheric pressure to 100 °C in order to remove unreacted acrylonitrile and ethanol, and the residue under reduced pressure. Collect the 3-ethoxy-propionitrile at 77–78 °C/25 mmHg. The yield is about 110 g (90%); p.m.r spectrum (CCl$_4$, TMS), δ 1.20 (t, 3H), 2.50 (t, 2H), 3.52 (t, 2H) and 3.58 (q, 2H). The last two signals overlap, but may be clearly discerned.

Experiment 5.163 3-(DIETHYLAMINO)PROPIONITRILE

$$Et_2NH + CH_2{=}CH{\cdot}CN \longrightarrow Et_2N{\cdot}CH_2{\cdot}CH_2{\cdot}CN$$

Mix 42.5 g (60 ml, 0.58 mol) of fresly distilled diethylamine and 26.5 g (33 ml, 0.5 mol) of pure acrylonitrile (**CAUTION**: *see Expt 5.161*) in a 250-ml round-bottomed flask fitted with a reflux condenser. Heat at 50 °C in a water bath for 10 hours and then allow to stand at room temperature for 2 days. Distil off the excess of diethylamine on a water bath, and distil the residue under reduced pressure. Collect the 3-(diethylamino)propionitrile at 75–77 °C/11 mmHg; the yield is 54 g (86%).

5.14 SUBSTITUTED CARBOXYLIC ACIDS AND THEIR DERIVATIVES

The substituted carboxylic acids discussed in this section possess a halogen (X), a hydroxyl group (OH), a keto group (CO), or an amino group (NH$_2$) as the additional function. The location of the substituents may be designated systematically by numerals or by letters of the Greek alphabet. Illustrative representations are given for:

2-bromopentanoic acid (α-bromovaleric acid) (1) $CH_3{\cdot}CH_2{\cdot}CH_2{\cdot}CHBr{\cdot}CO_2H$;
3-hydroxypentanoic acid (2) $CH_3{\cdot}CH_2{\cdot}CH(OH){\cdot}CH_2{\cdot}CO_2H$;
4-oxodecanoic acid (3) $C_6H_{13}{\cdot}CO{\cdot}CH_2{\cdot}CH_2{\cdot}CO_2H$;
2-amino-3-hydroxypropanoic acid (serine) (4) $CH_2OH{\cdot}CH(NH_2){\cdot}CO_2H$.

The naturally occurring α-amino acids have universally recognised trivial names (see Table 10.36).

The presence of chiral sites (*) should be noted; thus (1), (2) and (4) are chiral but (3) is achiral.

SPECTROSCOPIC FEATURES

The characteristic absorptions in the *i.r.* spectra of lactones and amino acids are discussed on pp. 302 and 308 respectively; the spectra of (DL)-valine and L-tryptophan are given in Figs 3.36 and 3.37. Further descriptive spectral interpretations for a range of substituted carboxylic acids and their derivatives are given in appropriate preparative sections.

5.14.1 HALOGENO ACIDS

HALOGENATION OF A CARBOXYLIC ACID

Acetic acid can be chlorinated by gaseous chlorine in the presence of red phosphorus to yield successively mono-, di- and trichloroacetic acid ($CH_2Cl \cdot CO_2H$, $CHCl_2 \cdot CO_2H$ and $CCl_3 \cdot CO_2H$) respectively. This represents the most convenient commercial method for the formation of these acids. Chlorination of other aliphatic carboxylic acids is not usually of preparative value since mixtures of several monochlorinated products are obtained. Bromination, on the other hand, is highly selective and only the α-bromo acid is obtained when the reaction is carried out in the presence of a reagent which yields an acyl bromide, such as red phosphorus (the *Hell–Volhard–Zelinsky* reaction) or phosphorus trichloride or tribromide. The acyl bromide undergoes halogenation (in the α-position) much more readily than the parent acid.

$$6R \cdot CH_2 \cdot CO_2H + 3Br_2 + 2P \longrightarrow 6R \cdot CH_2 \cdot COBr + 2H_3PO_3$$

$$R \cdot CH_2 \cdot COBr \xrightarrow[-HBr]{Br_2} R \cdot CHBr \cdot COBr \xrightarrow{R \cdot CH_2 CO_2H}$$

$$R \cdot CHBr \cdot CO_2H + R \cdot CH_2 \cdot COBr$$

Examples of the formation of a range of α-bromo acids are given in Expt 5.164.

An alternative method, which has the advantage of being applicable to the formation of α-chloro, α-bromo and α-iodo acids, involves the initial formation of the acid chloride followed by treatment with N-chlorosuccinimide, N-bromo-succinimide, or iodine.[176]

Illustrative procedures are given in Expt 5.165 which also includes typical details for the transformation of the substituted acid chloride into the corresponding amide and carboxylic acid.

When it is required to prepare an α-bromo acid from a carboxylic acid which is not particularly readily available commercially, but which can be synthesised by the malonic acid route (Section 5.11.6, p. 680), advantage may be taken of the ease of bromination in the α-position of the intermediate alkylmalonic acid. The substituted bromomalonic acid undergoes ready decarboxylation on heating to yield the α-bromo acid (e.g. 2-bromo-3-methylpentanoic acid, Expt 5.166).

$$RBr + CH_2(CO_2Et)_2 \xrightarrow{\ominus OEt} R \cdot CH(CO_2Et)_2 \xrightarrow{\ominus OH} R \cdot CH(CO_2H)_2 \xrightarrow{Br_2}$$

$$R \cdot CBr(CO_2H)_2 \xrightarrow{heat} R \cdot CHBr \cdot CO_2H$$

HYDROGEN HALIDE ADDITION TO AN UNSATURATED CARBOXYLIC ACID

A halogeno acid will be formed when an unsaturated carboxylic acid undergoes addition of a halogen acid. In the case of α,β-unsaturated acids, the addition process yields the β-halogeno acid; when the double bond is more remote from the carboxyl group a mixture of regioisomers is likely to be obtained as a result of competing orientation effects and the process may have little preparative value. In the particular case of peroxide-catalysed addition of hydrogen bromide to a long-chain unsaturated acid having a terminal double bond, the reaction is essentially regiospecific and the product is an ω-bromoacid. This procedure is illustrated in Expt 5.65.

RING OPENING OF LACTONES

As noted in Section 5.14.2, p. 728, lactones may be readily prepared by the Baeyer–Villiger oxidative rearrangement of cyclic ketones, most usually of ring size of five carbon atoms or larger. Ring opening of these lactones with concentrated hydrobromic acid in the presence of concentrated sulphuric acid gives directly the α,ω-bromo acids (e.g. 6-bromohexanoic acid from 6-hexanolide, Expt 5.167, which further describes the esterification of the product). The reaction may proceed by (a) hydrolysis to the hydroxy acid followed by an S_N2 displacement of the hydroxyl group with bromine, or (b) by an alkyl-oxygen fission of the protonated lactone arising from direct attack of a bromide ion.

For methods for the conversion of lactones into ω-iodo acids see p. 733.

Experiment 5.164   2-BROMOHEXANOIC ACID

Place 100 g (107 ml, 0.86 mol) of freshly distilled, dry hexanoic acid (b.p. 202–205 °C) (1) and 150 g (48 ml, 0.94 mol) of dry bromine (Section 4.2.7, p. 422) (CAUTION) in a 500-ml flask equipped with a reflux condenser, the top of which is connected to a gas absorption device [compare Fig. 2.61(c)]. Momentarily remove the condenser and add cautiously 1.5 ml of phosphorus trichloride. Heat the mixture on a water bath to 65–70 °C, when reaction will commence and hydrogen bromide is smoothly evolved. Towards the end of the reaction allow the temperature of the bath to rise to 100 °C. The reaction is complete when all the bromine has reacted (about 4 hours). Distil the reaction mixture under reduced pressure using a water pump; much hydrogen bromide is evolved and a fraction of low boiling point passes over. When all the low boiling point fraction has distilled, connect the flask to an oil pump via a trap containing sodium hydroxide pellets and collect the 2-bromo-hexanoic acid at 116–125 °C/8 mmHg (or at 132–140 °C/15 mmHg). The yield is 145 g (86%). Upon redistillation the 2-bromohexanoic acid passes over almost entirely at 128–131 °C/10 mmHg.

Note. (1) The acid may be dried by adding toluene and distilling out the toluene–water azeotrope, before distilling the dried acid.

Cognate preparations. 2-Bromopropanoic acid. Proceed as detailed for 2-bromohexanoic acid using 64 g (64.5 ml, 0.86 mol) of freshly distilled, dry propanoic acid (b.p. 139–142 °C), 150 g (48 ml, 0.94 mol) of dry bromine and 1.5 mol of phosphorus trichloride. The reaction commences on warming to about 50 °C. Collect the 2-bromopropanoic acid at 95–97 °C/10 mmHg or at 100–102 °C/15 mmHg. The yield is 110 g (83%).

α-Bromovaleric acid (2-bromopentanoic acid). Use 88 g (0.86 mol) of valeric acid; allow reaction to proceed at 80 °C and finally heat for 2 hours at 100 °C. Collect the α-bromovaleric acid at 145 °C/15 mmHg; the yield is 125 g (80%).

α-**Bromoisovaleric acid** (*2-bromo-3-methylbutanoic acid*). Use 88 g of iso-valeric acid; proceed as for 2-bromohexanoic acid, allowing the reaction to proceed at 70–80 °C for about 10–20 hours (until bromine vapour is no longer evident), then add a further 2.5 ml of bromine and heat at 100 °C for about 2 hours. Distil and collect α-bromoisovaleric acid at 100–125 °C/ 15 mmHg. The yield is in the region of 80 per cent.

Experiment 5.165 2-CHLOROHEXANOYL CHLORIDE[176]

$$Bu \cdot CH_2 \cdot CO_2H \xrightarrow{SO_2Cl_2} Bu \cdot CH_2 \cdot COCl \xrightarrow{NCS} Bu \cdot CHCl \cdot COCl$$

Hexanoic acid (11.6 g, 0.1 mol) and thionyl chloride (28.8 ml, 0.4 mol) are placed in a 250-ml flask equipped with a magnetic stirrer bar and a condenser with a drying tube (**CAUTION**: *hydrogen chloride evolved*). The reaction mixture is stirred and heated in a 70 °C oil bath. After 0.5 hour, a portion of the reaction mixture is submitted to p.m.r measurement, which shows the disappearance of the triplet at δ 2.40 (—$CH_2 \cdot CO_2H$) and the emergence of a new triplet at δ 2.87 (—CH_2COCl). The flask is removed from the oil bath and cooled to room temperature. To the reaction mixture are added successively finely powdered N-chlorosuccinimide (26.7 g, 0.2 mol), thionyl chloride (20 ml) and concentrated hydrochloric acid (7 drops). The flask is then returned to the oil bath, the temperature of which is raised to 85 °C. The actual temperature of the reaction mixture is 70 °C. After 1.25 hours, the reaction is over, as indicated by the disappearance of the triplet at δ 2.87 and the emergence of two doublets at 4.77–4.55 (—CHCl·COCl). The solvent is removed under reduced pressure and the solid (succinimide) is collected and washed with tetrachloromethane. The filtrate is fractionally distilled to give 14.7 g (87%) of product, b.p. 174–176 °C/760 mmHg; n_D 1.4458; i.r. (neat) 1787 vs, 1721 cm^{-1} (shoulder); p.m.r. spectrum (neat, TMS) δ 4.51 (d of d, 1H, $J =$ 5.5 and 6 Hz, —CHCl·COCl); m/z 170, 168 (M), 135, 133 (M − Cl), 41 (base peak, CH=$\overset{\oplus}{C}$=O).

Conversion into 2-chlorohexanamide. The foregoing compound (0.8 g, 4.7 mol) is dissolved in anhydrous ether (10 ml) and held at 10 °C while bubbling dry ammonia into the stirred solution until there is a basic reaction to pH paper (10 minutes). Ether is evaporated, and the white precipitate (0.7 g, quantitative yield) is washed with ice-cold water, then sublimed at 30 °C/ 0.05 mmHg; m.p. 56–57 °C.

Cognate preparations. *2-Bromohexanoyl chloride and 2-bromohexanoic acid.* A mixture of hexanoic acid (11.6 g, 0.1 mol), tetrachloromethane (10 ml) and thionyl chloride (28.8 ml, 0.4 mol) is stirred at 65 °C for 0.5 hour. N-Bromo-succinimide (21.4 g, 0.12 mol), tetrachloromethane (50 ml) and 48 per cent hydrobromic acid (7 drops) are added to the mixture. The flask is heated at 70 °C for 10 minutes, then at 85 °C for 1.5 hours. After work-up and fractional distillation *in vacuo*, 2-bromohexanoyl chloride is obtained as a clear, slightly yellow oil (17.1 g, 80%), b.p. 44–47 °C/1.5 mmHg. Hydrolysis to the acid is achieved by treating its acetone solution (92 ml) with 115 ml of a saturated sodium hydrogen carbonate solution (c. 115 mmol) at 10 °C. After acidification with concentrated hydrochloric acid, extraction with chloroform, and drying over magnesium sulphate, the solvent is removed *in vacuo* to give a colourless liquid [9.36 g, 99.5% yield, 96% pure by g.l.c. (SE-30 column)]. Fractional distillation gives 7.76 g (83%) of *2-bromohexanoic acid*, b.p. 64–66 °C/0.075 mmHg, which is homogeneous by g.l.c.

2-Iodohexanoyl chloride. Hexanoic acid (11.6 g, 0.1 mol), resublimed iodine (15.23 g, 0.12 g-atom) and thionyl chloride (40 ml, 0.55 mol) are placed in a round-bottomed flask equipped as previously. The mixture is stirred and the heating bath adjusted to 130 °C. When the reflux is steady, the actual temperature of the reaction mixture is 85 °C. The reaction was complete after 1.5 hours, as indicated by p.m.r. measurement which was performed and analysed in the same way as with previous experiments. Thionyl chloride is evaporated under reduced pressure; excess iodine is filtered and washed with tetrachloromethane. The filtrate is shaken with concentrated sodium thiosulphate solution to remove the remaining iodine. The organic layer is separated and dried over magnesium sulphate. After solvent evaporation, the product is distilled to yield 21 g (80%) of 2-idohexanoyl chloride, b.p. 62–63 °C/0.5 mmHg; n_D^{21} 1.5179.

Experiment 5.166 2-BROMO-3-METHYLPENTANOIC ACID

$$\text{Et·CH(Me)·CH(CO}_2\text{Et)}_2 \xrightarrow{\ominus\text{OH}} \text{Et·CH(Me)·CH(CO}_2\text{H)}_2 \xrightarrow{\text{Br}_2}$$

$$\text{Et·CH(Me)·CBr(CO}_2\text{H)}_2 \xrightarrow{\Delta} \text{Et·CH(Me)·CHBr·CO}_2\text{H}$$

Hydrolyse 108 g (0.5 mol) of diethyl s-butylmalonate (1) with aqueous ethanolic potassium hydroxide solution following the procedure described for the hydrolysis of diethyl proplymalonate (Expt 5.133). Transfer the dried ether extract of the product to a 1-litre three-necked flask fitted with a sealed stirrer unit, an efficient reflux condenser and a dropping funnel charged with 80 g (26 ml, 0.5 mol) of bromine. Stir the solution, add about 2–3 ml of bromine, and when this has reacted completely drop in the remainder of the bromine at such a rate that the ether refluxes gently. When all the bromine has reacted, add 100 ml of water cautiously from the dropping funnel to decompose any acyl bromide. Separate the ether layer, dry it over anhydrous sodium sulphate and remove the ether (rotary evaporator). Decarboxylate the residue by heating it for 5 hours in an oil bath maintained at 130 °C and then distil the resulting bromo acid under reduced pressure, collecting the fraction of b.p. 125–140 °C/20 mmHg. The yield is 65 g (67%).

Note. (1) Diethyl s-butylmalonate, b.p. 110–120 °C/20 mmHg, is prepared by reacting 2-bromobutane with diethyl malonate. Use the general procedure described for the butyl isomer (Expt 5.133) but extend the period of heating under reflux to 48 hours to complete the alkylation.

Experiment 5.167 ETHYL 6-BROMOHEXANOATE

$$\text{O·CH}_2\text{·(CH}_2)_3\text{·CH}_2\text{·CO} \xrightarrow[\text{H}_2\text{SO}_4]{\text{HBr}} \text{Br(CH}_2)_5\text{·CO}_2\text{H} \xrightarrow[\text{H}^{\oplus}]{\text{EtOH}} \text{Br(CH}_2)_5\text{·CO}_2\text{Et}$$

Place a mixture of 140 ml of concentrated hydrobromic acid and 34 ml of concentrated sulphuric acid in a 500-ml, three-necked round-bottomed flask fitted with a sealed stirrer unit, a reflux condenser and a dropping funnel. Cool the flask contents in an ice–salt bath and add with stirring 28.5 g

(0.25 mol) of 6-hexanolide. Allow the reaction mixture to reach room temperature and then to stand for 2 hours; heat on a water bath for a further 4-hour period. Cool the mixture and pour on to 300 g of crushed ice and separate the organic layer; saturate the aqueous layer with ammonium sulphate and extract the solution with four 50 ml portions of ether. Wash the combined organic phases with three 25 ml portions of a saturated aqueous solution of ammonium sulphate, dry (anhydrous sodium sulphate) and evaporate the ether on a rotary evaporator. Boil the residual bromo acid under reflux for 8 hours with 85 ml of absolute ethanol containing 2 ml of concentrated sulphuric acid and then remove the ethanol on a rotary evaporator. Dissolve the residue in 100 ml of ether and wash the ethereal solution first with water and then with 5 per cent aqueous sodium carbonate to remove mineral acid. Dry the ethereal solution over anhydrous sodium sulphate, remove the ether (rotary evaporator) and distil the residue under reduced pressure. Collect the ethyl 6-bromohexanoate having b.p. 120–124 °C/14 mmHg; the yield is 28 g (50%).

5.14.2 HYDROXY ACIDS AND LACTONES

Lactones are formally products of the intramolecular cyclisation of hydroxy acids, e.g.

Lactonisation is an internal esterification process which in the case of γ- and δ-hydroxy acids takes place particularly readily, often simply on standing. Experimental conditions must therefore be carefully controlled if the free hydroxy acid is to be isolated from the reaction medium. Smaller or larger lactones cannot generally be prepared by direct intramolecular esterification; formation of linear polyesters can be a competing reaction and macrocyclic lactones have to be prepared by indirect routes. β-Hydroxy acids dehydrate to α,β-unsaturated acids rather than undergoing lactonisation. The three-membered lactone ring system is not known in the free state but α-lactones are believed to be the intermediates in the hydrolysis of α-halo acids, where a neighbouring group participation effect markedly increases the rate of hydrolysis. In the case of chiral α-halo acids the retention of configuration provides evidence for this mechanistic pathway.

β-Lactones and the macrocyclic lactones, like the γ- and δ-lactones, are of importance since these ring systems are found widespread in nature. The simpler substituted lactones are important building blocks in the synthesis of many natural products, e.g. in the fields of antibiotics, anticancer pharmaceuticals, pheromones and prostaglandins to name but a few.

725

α-HYDROXY ACIDS

As noted above, the hydrolysis of α-halogeno acids proceeds particularly readily, usually in boiling aqueous sodium carbonate solution. The preparation of the lower homologous α-hydroxy acids by this procedure, however, is not entirely satisfactory because the highly water-soluble and sometimes hygroscopic products are difficult to isolate efficiently. One method consists of treating an aqueous solution of the hydroxy acid with copper(II) acetate to precipitate the insoluble copper salt and decomposing an aqueous suspension of the latter with hydrogen sulphide.[177] The recovery of the α-hydroxy acid, however, is often poor.

A more generally satisfactory method is the hydrolysis of cyanohydrins. Here a mixture of the carbonyl compound and aqueous sodium cyanide is gradually acidified at room temperature resulting in the sequential formation of the cyanohydrin and its subsequent hydrolysis to the acid (the cyano group is of course a latent carboxyl group).

$$R^1R^2CO + HCN \longrightarrow R^1R^2C(OH)\cdot CN \xrightarrow{H_3O^\oplus} R^1R^2C(OH)\cdot CO_2H$$

An alternative procedure, suitable for the preparation of cyanohydrins which readily form bisulphite complexes, is illustrated by the preparation of mandelic acid described in Expt 5.168 (see also the preparation of acetone cyanohydrin). Here the procedure involves the addition of a saturated solution of sodium metabisulphite to a stirred solution of the carbonyl compound and aqueous sodium cyanide, and when applicable is usually to be preferred to the *in situ* generation of hydrogen cyanide or the use of the highly poisonous liquid hydrogen cyanide as a reagent.

The synthesis of optically active isomers of α-hydroxy acids is important since these acids are present in nature and they are also important intermediates in organic synthesis. The methods which have been used for their preparation (with varying success with respect to optical purity), include the asymmetric reduction of α-keto esters followed by hydrolysis (for acids of the type R·CHOH·CO_2H),[178] and the reaction of Grignard reagents with optically active α-keto esters,[179] or optically active keto oxazolines [for acids of the type R^1·C(R^2)(OH)·CO_2H].[180] An interesting preparative example given here is the novel use of an asymmetric halolactonisation reaction[181] (see also p. 729). In this reaction a suitable α,β-unsaturated acid is first converted into the acyl derivative (5) of (S)-proline (the chiral adjuvant, p. 15). Treatment of this derivative with N-bromosuccinimide in dimethylformamide gives a bromonium ion which lactonises to the bromolactone (6); reduction of (6) with tributyltin hydride yields the lactone (7) from which the optically active α-hydroxy acid (89–98% ee) is released by hydrolysis. The reaction sequence formulated below only gives the preferred conformations and the stereoselective routes that are suggested in the literature; the full account should be consulted for a detailed discussion of other, less likely, stereoisomeric possibilities. When tiglic acid [(E)-2-methylbut-2-enoic acid] is employed the product, (R)-2-hydroxy-2-methylbutanoic acid is formed in high optical purity (Expt 5.169).

β-HYDROXY ACIDS

A retrosynthetic disconnection of a simple β-hydroxy ester gives the carbocation (8) and the carbanion (9).

$$R \cdot CHO[+H^{\oplus}] + BrZnCH_2 \cdot CO_2R$$

As noted previously (p. 519) the reagent equivalent for the carbocation (8) is an aldehyde [or ketone in the case of a hydroxy ester target molecule of structure $R^1R^2 \cdot C(OH) \cdot CH_2 \cdot CO_2R$]. For the anionic synthon (9), the use of a base generated α-carbanion (e.g. from an ester, from diethyl malonate, or from malonic acid) would lead to spontaneous base-catalysed dehydration of the product to give the α,β-unsaturated ester or acid (Section 5.18.3, p. 804); subsequently further Michael addition reactions may also take place (Section 5.11.6, p. 681). Therefore, to arrest the reaction at the hydroxy ester stage, the carbanion source is the organozinc reagent as shown, which reacts similarly to a Grignard reagent, but is chemoselective in that it reacts with the carbonyl group of aldehydes and ketones but not esters. This is known as the *Reformatsky reaction*, (Expt 5.70).

$$BrCH_2 \cdot CO_2Et \xrightarrow[\text{(ii) } R^1R^2CO]{\text{(i) Zn}} R^1R^2C(OZnBr) \cdot CH_2 \cdot CO_2Et \xrightarrow{H_3O^{\oplus}}$$
$$R^1R^2C(OH) \cdot CH_2 \cdot CO_2Et$$

The success of the reaction is crucially dependent on the cleanliness of the zinc surface, prior to the formation of the organozinc reagent; 'sonication' has been suggested as a valuable aid to this reaction (cf. Expt 7.14). Under standard conditions, the hydroxy ester is the major product when the reaction is carried out in ether or benzene; the contaminants which frequently arise during distillation are the α,β- and β,γ-unsaturated esters. Dehydration can be deliberately effected with the aid of reagents such as fused potassium hydrogen sulphate or acetic anhydride. Catalytic hydrogenation of the mixture of unsaturated esters followed by hydrolysis yields a saturated carboxylic acid, and the whole procedure constitutes a useful two-carbon chain extension process.

Methyl 4-bromobut-2-enoate ($BrCH_2 \cdot CH=CH \cdot CO_2Me$, a vinyl analogue of ethyl bromoacetate, Expt 5.69) can also be used in the Reformatsky reaction, and its use permits an analogous four-carbon atom chain extension process.

γ-, δ-, AND ω-HYDROXY ACIDS AND THEIR RESPECTIVE LACTONES

One of the most important methods for lactone preparation, and hence of the corresponding hydroxy acids (or halogeno acids, Section 5.14.1 above) is the *Baeyer-Villiger rearrangment* of cyclic ketones by the action of peracids. A wide variety of peracids have been used in this reaction but currently the reagents of choice are pertrifluoroacetic acid, *m*-chloroperbenzoic acid, and permaleic acid. The mechanism is formulated below for the conversion of an acyclic ketone into an ester.

The migratory aptitudes of the groups is in the order t-alkyl > s-alkyl > aryl > n-alkyl > methyl. With cyclic ketones, leading to lactones, the more highly substituted group migrates. As with all intramolecular rearrangements if the migrating group is chiral, configuration is retained. The reaction is therefore regioselective and stereospecific.

A range of conditions has been selected in Expt 5.171 to illustrate this reaction. Thus, *m*-chloroperbenzoic is used with many cyclic ketones although the time of reaction increases with ring size,[182] pertrifluoroacetic acid in buffer solution is used in relatively small-scale oxidations and with less reactive ketones,[183] and permaleic acid appears suitable in large-scale oxidations where safety precautions preclude the use of 90 per cent hydrogen peroxide (permaleic acid is generated from maleic anhydride and the somewhat safer 30 per cent reagent).[184] The conditions used with the former two reagents have been widely employed for the conversion of variously substituted monocyclic and fused ring ketones into important lactone products.[185] Details are also given for the conversion of the lactones so prepared into ω-hydroxy acids by hydrolysis and into ω-iodo acids by treatment with hydriodic acid. An alternative convenient procedure for the formation of ω-iodo acids from lactones is reaction with trichloromethylsilane/sodium iodide in dry acetonitrile.[76b]

The conversion of β,γ- or γ,δ-unsaturated acids into iodolactones (*Halo-*

728

lactonisation reaction)[186] is an important reaction which has been widely used in natural product synthesis.[187] The simplest example is the conversion of pent-4-enoic acid into the iodolactone (11) by dissolution in aqueous sodium hydrogen carbonate followed by treatment with an aqueous solution of iodine in potassium iodide, the reaction proceeding via the intermediate iodonium cation (10).

The illustrative preparative example is the conversion of cyclohex-1-en-1-yl-acetic acid into the γ-iodolactone (12) (*cis*-1-iodo-7-oxabicyclo[4.3.0]nona-8-one,[188] Expt 5.172). In general the simpler iodolactones tend to be unstable; the preparative sequence therefore recommends conversion into the unsaturated lactone by a dehydrohalogenation reaction.

Optically active lactones have been synthesised by an adaptation of the oxazoline route to optically active carboxylic acids[189] (Section 5.11.7, p. 687).

Experiment 5.168 MANDELIC ACID

$$Ph\cdot CHO \xrightarrow{NaHSO_3} Ph\cdot CH(OH)\cdot SO_3^{\ominus}Na^{\oplus} \xrightarrow{NaCN} Ph\cdot CH(OH)\cdot CN \xrightarrow[conc.]{HCl}$$

$$Ph\cdot CH(OH)\cdot CO_2H + NH_4Cl$$

CAUTION: *This preparation must be carried out in an efficient fume cupboard.*

Prepare a saturated solution of sodium metabisulphite by stirring 250 g of finely powdered sodium metabisulphite with 335 ml of water for half an hour and then filtering to remove excess of the salt. In a 1-litre three-necked flask equipped with a mechanical stirrer and a dropping funnel, place a solution of 25 g (0.5 mol) of sodium cyanide (**CAUTION**) in 100 ml of water and 53 g (51 ml, 0.5 mol) of purified benzaldehyde (Expt 6.133). Add the sodium metabisulphite solution from the dropping funnel, slowly at first and then more rapidly (the addition occupies 10–15 minutes). During the initial stages of the addition, add 150 g of crushed ice to the reaction mixture in several portions through the third neck. Transfer the two-layer liquid mixture to a separatory funnel and remove the crude mandelonitrile (1). Place the crude product at once (2) in a large evaporating dish, add 75 ml of concentrated hydrochloric acid, cover with a clock glass and allow the hydrolysis to proceed at room temperature for 12 hours. Evaporate the solution to dryness on a steam bath, stirring from time to time to break up the deposit of ammonium chloride and mandelic acid which separates. Grind the residue of slightly discoloured mandelic acid and inorganic salts to a fine powder and wash it with two portions of 125 ml of cold benzene (**CAUTION**); this process will remove most of the colouring matter but a negligible quantity of mandelic acid. To separate the inorganic salts from the mandelic acid, extract the residue in a Soxhlet apparatus (Fig. 2.97) with about 200 ml of benzene. Allow the hot benzene extract to crystallise, collect the crystals on a Buchner funnel and dry in air. The yield of pure (±)-mandelic acid, m.p. 118 °C, is 35 g (46%).

Note. (1) For the safe disposal of the aqueous solution see Expt 5.158. A further small quantity of mandelonitrile may be obtained by extracting the aqueous solution with ether, evaporating the ether and adding the residue to the main portion of mandelonitrile. This extraction is hardly worth while except for large-scale preparations.

(2) It is important to mix the mandelonitrile with hydrochloric acid immediately it has been separated from the water. Standing results in rapid conversion to the acetal of benzaldehyde and mandelonitrile Ph·CH[OCH(CN)·Ph]$_2$; the yield of mandelic acid will, in consequence, be reduced.

Cognate preparation. *Acetone cyanohydrin.* Dissolve 110 g of sodium metabisulphite in 200 ml of cold water contained in a 1-litre round-bottomed flask. Add slowly 58 g of acetone while swirling the liquid mixture slowly, followed by a solution of 60 g of potassium cyanide in 200 ml of cold water (**CAUTION**). During this latter slow addition the cyanohydrin separates as the upper layer. When separation is complete the contents of the flask are transferred to a separatory funnel and the lower layer removed. The upper layer is transferred to a flask, sodium sulphate added to effect drying and the flask stoppered and kept in the dark. The dried cyanohydrin (60 g, 70%) is slightly discoloured; it may be distilled under reduced pressure when it distils at 80–82 °C/15 mmHg.

Experiment 5.169 **(R)-(−)-2-HYDROXY-2-METHYLBUTANOIC ACID**[181]

All reaction solvents were anhydrous.

Tigloyl chloride.[190] [(E)-2-*methylbut-2-enoyl chloride*.]. Tiglic acid (20 g) and phosphorus trichloride (15 g) are heated together at 70–80 °C for 2 hours. The upper, slightly yellow, layer is decanted from the lower syrupy layer and distilled at 64 °C/35 mmHg to give the acid chloride in 90 per cent yield.

(S)-(−)-N-Tigloylproline. (S)-Proline ([α]$_D^{20}$ −85.5° (c 4.00 in H$_2$O), 10.4 g, 0.090 mol) is dissolved in 2 M sodium hydroxide (53 ml, 0.106 mol) cooled in an ice bath, and the resulting alkaline solution is diluted with acetone (53 ml). An acetone solution (53 ml) of tigloyl chloride (16.0 g, 0.135 mol) and 2 M sodium hydroxide solution (80 ml, 0.160 mol) are simultaneously added over 70 minutes to the aqueous solution of (S)-proline with stirring in an ice bath. The pH of the mixture is kept at 10–11 during the addition of the acylating reagent. After the addition, stirring is continued for 2 hours at room temperature and the mixture is submitted to evaporation *in vacuo* to remove the acetone. The residual solution is washed with ether and acidified (pH 2) with concentrated hydrochloric acid. The acidic mixture is extracted with ethyl acetate after being saturated with sodium chloride, and the combined ethyl acetate extracts are washed with saturated sodium chloride solution. The organic layer is dried over anhydrous sodium sulphate and evaporated *in vacuo* to give the crude product as a colourless solid (19.5 g, quantitative yield). Recrystallisation from hexane–benzene (**CAUTION**), 2:3, gives pure (S)-N-tigloylproline as colourless pillars (15.3 g, 86%), m.p. 112.5–113.5 °C, $[\alpha]_D^{20}$ −72.7° (c 1.00 in MeOH); i.r. (Nujol) 1740 (acid) 1582 cm^{-1} (amide); p.m.r. (CDCl$_3$, TMS) δ 1.78 (s, 3H, $Me\cdot$CH=), 1.81 (s, 3H, =C(Me)CO), 1.50–2.50 (m, 4H, $CH_2CH_2CH_2$N), 3.35–3.75 (m, 2H, CH$_2$N), 4.53 (t, 1H, J = 7 Hz, NCHCO), 5.75 (br. s, 1H, CH=), 11.0 (s, 1H, CO$_2$H).

Bromolactonisation of (S)-N-Tigloylproline. A dimethylformamide solution (4 ml) of potassium t-butoxide (224 mg, 2.0 mmol) and a dimethylformamide solution (2 ml) of N-bromosuccinimide (712 mg, 4.0 mmol) are successively added to a stirred solution of (S)-N-tigloylproline (395 mg, 2.0 mmol) in dimethylformamide (2 ml) at −20 °C under nitrogen. The mixture is stirred at −20 °C for 2 hours, and then at room temperature for 48 hours. Extractive isolation with ethyl acetate followed by evaporation *in vacuo* gives the crude bromolactone as a thick yellow oil (562 mg, 95%). Thin-layer chromatography analysis (silica gel, solvent ether) of this oil shows three impurities (R_F 0.47, 0.28, and 0.22) in addition to the desired bromolactone (R_F 0.36). The pure bromolactone is obtained by recrystallisation from hexane–ether and has m.p. 111.5–112.5 °C, $[\alpha]_D^{20}$ −83.2° (c 0.754 in MeOH); i.r. (Nujol) 1760 (lactone), 1670 cm^{-1} (amide); p.m.r (CDCl$_3$, TMS) δ 1.70 (s, 3H, MeCCON), 1.93 (d, 3H, J = 7.2 Hz, MeCHBr), 1.40–2.70 (m, 4H, $CH_2CH_2CH_2$N), 3.37–3.92 (m, 2H, CH$_2$N), and 3.92–4.72 (m, 2H, NCHCO and MeCHBr).

Reductive debromination of bromolactone. A benzene solution (12 ml) of tri-butyltin hydride (2.33 g, 8.00 mmol) (**CAUTION**) is added dropwise over 1 minute to a stirred solution of pure bromolactone (1.70 g, 6.16 mmol) in benzene (10 ml) at 70 °C under nitrogen. The mixture is stirred at reflux for 15 hours and is evaporated *in vacuo* to afford a mixture of crystals and oil. Addition of hexane to the mixture, followed by cooling to −70 °C, precipitates the crude reduction product as colourless needles (1.29 g, quantitative yield), m.p. 99–101 °C. Recrystallisation from hexane–ether gives colourless needles (1.03 g, 85%), m.p. 105–106 °C, $[\alpha]_D^{20}$ −112° (c 0.760 in MeOH); i.r. (Nujol) 1740 (lactone), 1683 cm^{-1} (amide); p.m.r (CDCl$_3$, TMS) δ 1.00 (t, 3H, J = 7 Hz, MeCH$_2$), 1.57 (s, 3H, MeCCO), 1.92 (q, 2H, J = 7 Hz, MeCH_2), 1.80 (m, 4H, $CH_2CH_2CH_2$N), 3.40–3.90 (m, 2H, CH_2N), 4.10–4.50 (m, 1H, NHCO).

(R)-(−)-2-Hydroxy-2-methylbutanoic acid. A mixture of the foregoing pure lactone (985 mg, 5.00 mmol) and 36 per cent hydrochloric acid (10.3 ml) is refluxed for 8 hours. The acidic mixture is diluted with saturated sodium chloride solution (20 ml) and extracted with ethyl acetate. The combined organic layers are extracted with saturated sodium hydrogen carbonate solution after being washed with saturated brine. The sodium hydrogen carbonate extracts are combined, acidified (pH ≈ 2) with concentrated hydrochloric acid and extracted with ethyl acetate. The combined organic layers are washed with brine. After drying over anhydrous sodium sulphate, filtering, and evaporation *in vacuo*, (R)-(−)-2-hydroxy-2-methylbutanoic acid is obtained as colourless needles (535 mg, 91%), m.p. 72–74 °C, $[\alpha]_D^{25}$ −8.5° (c 3.01 in CHCl$_3$). Recrystallisation from hexane gives the optically pure acid, m.p. 78–79 °C, $[\alpha]_D^{25}$ −8.9° (c 2.97 in CHCl$_3$), p.m.r. (CDCl$_3$, TMS) 0.94 (t, 3H, J = 6 Hz, MeCH$_2$), 1.45 (s, 3H, MeCCO(, 1.42–2.02 (m, 2H, Me CH_2), 6.12–7.12 (broad s, CO$_2$H and OH).

Experiment 5.170   ETHYL 3-PHENYL-3-HYDROXYPROPANOATE

$$BrCH_2 \cdot CO_2Et + Zn \longrightarrow BrZnCH_2 \cdot CO_2Et \xrightarrow{Ph \cdot CHO}$$

$$Ph \cdot CH(OZnBr) \cdot CH_2 \cdot CO_2Et \xrightarrow{H_3O^\oplus} Ph \cdot CH(OH) \cdot CH_2 \cdot CO_2Et$$

It is essential that all the apparatus and the reagents be scrupulously dry for successful results (compare Grignard reaction). Equip a 500-ml three-necked flask with a 250-ml separatory funnel, a mechanical stirrer and a double surface condenser; insert calcium chloride guard-tubes in the funnel and condenser. Place 40 g (0.61 mol) of zinc dust (previously dried at 100 °C) (Section 4.2.80, p. 467) in the flask, and a solution of 83.5 g (55.5 ml, 0.5 mol) of ethyl bromoacetate (**CAUTION**: *lachrymatory*) (1) and 65 g (62 ml, 0.615 mol) of purified benzaldehyde (Expt 6.133) in 80 ml of sodium-dried benzene and 20 ml of sodium-dried ether in the separatory funnel. Add about 10 ml of the solution to the zinc and warm the flask gently until the reaction starts. When the reaction has commenced, but not before, stir the mixture and add the remainder of the solution at such a rate that moderate refluxing occurs (about 1 hour). Reflux the reaction mixture on a water bath for a further 30 minutes. Cool the flask in an ice bath, and add 200 ml of cold 10 per cent sulphuric acid with vigorous stirring. Transfer to a separatory funnel, remove the aqueous layer, wash the benzene layer twice with 50 ml portions of 5 per cent sulphuric acid, once with 25 ml of 10 per cent sodium carbonate solution and finally with two 25 ml portions of water. Extract the combined solutions with 100 ml of ether, and dry the combined benzene and ether solution with 5 g of anhydrous calcium sulphate. Filter from the desiccant, remove the solvent by distillation under atmospheric pressure and distil the residue under reduced pressure. Collect the ethyl 3-phenyl-3-hydroxypropanoate at 152–154 °C/ 12 mmHg. The yield is 60 g (62%).

Note. (1) Great care must be exercised in handling ethyl bromoacetate. Keep a 10 per cent aqueous ammonia solution available to react with any bromoester which may be spilled.

Cognate preparation. *Ethyl 1'-hydroxycyclohexylacetate.* Place 65 g (1 mol) of clean dry zinc dust and a few crystals of iodine in a 2.5-litre three-necked flask, equipped with an efficient reflux condenser with drying tube, a mechanical stirrer and a dropping funnel. Prepare a mixture of 400 ml of sodium-dried benzene and 350 ml of sodium-dried toluene with 167 g (111 ml, 1 mol) of ethyl bromoacetate and 98 g (103.5 ml, 1 mol) of pure dried and redistilled cyclohexanone. Transfer 150 ml of this mixture to the flask, start the stirrer and heat the flask in a boiling water bath. A vigorous reaction soon sets in. Add the remainder of the mixture through the dropping funnel at such a rate that gentle refluxing is maintained. Continue the stirring for an additional 2 hours: practically all the zinc dissolves. Cool the mixture, add sufficient 10 per cent sulphuric acid with stirring to dissolve all the zinc hydroxide. Separate the benzene-toluene layer, dry it with anhydrous sodium sulphate, remove the solvent using a rotary evaporator and distil the residue under reduced pressure. Collect ethyl 1'-hydroxycyclohexylacetate at 86–89 °C/2 mmHg. The yield is 125 g (67%).

Experiment 5.171   12-HYDROXYDODECANOIC ACID LACTONE
(*Dodecanolide*)[182]

Cyclododecanone (2.0 g, 11 mmol) is added to *m*-chloroperbenzoic acid (4.2 g, 21 mmol) in 25 ml of dry chloroform (**CAUTION**). The reaction mixture is heated at reflux for 48 hours after which time a starch–iodide test for peroxide is negative. After the mixture is cooled in ice-water, the precipitated *m*-chlorobenzoic acid is removed by filtration and the chloroform is evaporated to dryness. The residue is taken up in 60 ml of ether, washed three times with 20-ml portions of aqueous potassium carbonate, and once with brine. After drying and removal of solvent there remains crude dodecanolide (1.9 g, 9.6 mmol, 87%), which is used without further purification.

Conversion into 12-iodododecanoic acid. The above dodecanolide (1.9 g, 9.6 mmol) is added to a mixture of 67 per cent hydriodic acid (5 g, 24 mmol) and glacial acetic acid (3 g). The reaction mixture is held for 2 hours at 100 °C and after cooling is poured out into 50 ml of a cold 10 per cent aqueous solution of sodium thiosulphate. The reaction mixture is extracted three times with 25-ml portions of chloroform. The combined extracts are dried over magnesium sulphate, and the solvent is removed to leave crude 12-iodododecanoic acid, which after recrystallisation from ether–light petroleum (b.p. 40–60 °C) is obtained in 93 per cent yield (2.95 g), m.p. 61–62.5 °C.

Cognate preparations. *Dodecanolide* (*use of permaleic acid*).[184] (**CAUTION**: *see Section 4.2.41, p. 439 for the precautions that must be adopted when using 30 per cent hydrogen peroxide. Safety screens and extra personal protection measures should also be employed.*) Dichloromethane (1.6 litres) and acetic anhydride (1.25 litres) are stirred in a 5-litre flask fitted with a double surface condenser and an overhead stirrer, a dropping funnel, and cooled externally (ice-water) while 30 per cent hydrogen peroxide (1 litre) is added. After 1 hour maleic anhydride (1 kg) is added, the mixture is cooled and stirred for 1 hour, and then the cooling bath is removed, when the temperature rises during 1.5 hours and the mixture begins to reflux. External cooling is resumed when needed to moderate the reaction. When little more heat is evolved, cyclododecanone (250 g) is added; this does not greatly increase the rate of heating, and when spontaneous refluxing ceases, a heating mantle is used to maintain the mixture at its boiling point for 15 hours. The mixture is then cooled and the separated maleic acid is filtered off. The filtrate is washed in turn with water (3 × 600 ml), an aqueous solution containing 10 per cent each of potassium hydroxide and sodium sulphite (2 × 300 ml), and then water (600 ml); tests for peroxide are now negative. After being dried over anhydrous sodium sulphate the filtrate is evaporated to give the lactone (210.4 g, 77%); when potassium carbonate, rather than potassium hydroxide, is used for washing, the crude lactone contains a pungent contaminant, possibly peracetic acid.

·*Conversion into 12-hydroxydodecanoic acid.* The foregoing lactone is added to a solution of potassium hydroxide (150 g) in methanol (800 ml) and the mixture is heated under reflux for 1 hour. Most of the solvent is then removed on a rotary evaporator. Water (2 litres) is added and the solution is extracted with ether (2 × 400 ml). The aqueous layer is acidified (concentrated hydrochloric acid) and the precipitated acid is collected, dried *in vacuo*, and recrystallised from acetone–light petroleum (b.p. 60–80 °C) to afford the acid (185.8 g, 63% from cyclododecanone), m.p. 84 °C.

8-Hydroxynonanoic acid lactone.[183] (**CAUTION**: *see Section 4.2.41 for the precautions to be adopted in the use of 90% hydrogen peroxide. Safety screens and extra personal protection must be employed.*) To a mixture of 90 per cent hydrogen peroxide (1 ml, 40 mmol) and dichloromethane (5 ml) at 0 °C is added dropwise trifluoroacetic acid (7 ml, 50 mmol). The resulting homogeneous solution could be stored at − 20 °C for several weeks. To a solution of 2-methylcyclooctanone (1.12 g, 8 mmol) (1) in dichloromethane (8 ml) and buffered with anhydrous disodium hydrogen phosphate (7.1 g, 50 mmol), is added at 0 °C dropwise pertrifluoroacetic acid solution (7 ml, 22 mmol). The resulting mixture is stirred at room temperature for 5 hours, poured into water, washed with aqueous sodium hydrogen carbonate and dried over magnesium sulphate. After removal of the solvents at reduced pressure, the oily residue is flash chromatographed using 4 per cent ethyl acetate in light petroleum to yield 890 mg (72%) of the lactone as a transparent sweet-smelling oil (3% of starting material is recovered); i.r. (thin film) 2940, 1730, 1450 cm^{-1}; p.m.r. (CCl$_4$, TMS) δ 1.23 (d, 3H, $J = 7$ Hz), 1.1–2.0 (m, 10H), 2.15 (m, 2H), 5.00 (m, 1H); t.l.c. (silica gel, 5% ethyl acetate in pentane) R_F 0.36.

Note. (1) 2-Methylcyclooctanone may be prepared by the procedures noted in earlier sections. Thus cyclooctanone may be converted into 1-methylcyclooctanol by reaction with methylmagnesium bromide (cf. Expt 5.40); dehydration then gives 1-methylcyclooct-1-ene (cf. Expt 5.12); hydroboration gives *trans*-2-methylcyclooctanol (cf. Expt 5.44); finally, oxidation with PCC yields 2-methylcyclooctanone (cf. Expt 5.76, or alternatively Expt 5.86).

Experiment 5.172 *CIS*-1-IODO-7-OXABICYCLO[4.3.0]NONAN-8-ONE[188]

At room temperature, a solution of cyclohex-1-en-1-ylacetic acid (10 g) (1) and sodium hydrogen carbonate (19 g) in water (300 ml), is added to a solution of iodine (38 g) and potassium iodide (75 g) in water (600 ml). The oily precipitate is extracted after 4 hours with ether, and the ether solution washed with sodium hydrogen sulphite until the coloration of iodine has disappeared, and then with saturated sodium hydrogen carbonate solution. The dried ethereal extract is evaporated *in vacuo*; the oily residue gives on trituration with light petroleum 8.2 g (43%) of a white solid, m.p. 56 °C (2).

Dehydrohalogenation of γ-iodolactone. A solution of the foregoing lactone in pyridine (10 ml) (**CAUTION**) is kept at room temperature for 2 days. Ethanol (30 ml), zinc powder (5 g) and (during 1 hour) concentrated hydrochloric acid (10 ml) are added with stirring and cooling. The solution is filtered and extracted with benzene (**CAUTION**) and the organic layer washed successively with 10 per cent hydrochloric acid, 10 per cent sodium hydrogen carbonate, and finally water. Distillation of the benzene layer gives 1.4 g (54%) of the unsaturated lactone, b.p. 152–153 °C/30 mmHg, m.p. 28–30 °C; i.r. (thin film) 1750 (conjugated γ-lactone carbonyl), 1650 cm^{-1} (C=C).

Notes. (1) Cyclohex-1-en-1-ylacetic acid may be prepared from ethyl 1'-hydroxy-cyclohexylacetate (Expt 5.170) by dehydration to the unsaturated ester followed by acidic hydrolysis to the unsaturated acid. An alternative method from cyclohexanone has been described.[191]
(2) Iodolactones are unstable and should be used in further reactions without delay; recrystallisation may be carried out using an ethanol–light petroleum solvent mixture.

5.14.3 KETO ACIDS AND ESTERS

α-KETO ACIDS AND ESTERS
The simplest member of the series of aliphatic α-keto acids is pyruvic acid. It is conveniently prepared by the distillation of tartaric acid with a dehydrating agent such as postassium hydrogen sulphate (Expt 5.173). The reaction probably involves dehydration to the tautomeric oxaloacetic acid (13) intermediate, which then decarboxylates by virtue of its constitution as a β-keto acid.

(13)

A retrosynthetic disconnection of the ester of a higher homologue [e.g. (14)] leads to the carbanion and the carbocation synthons (15) and (16) respectively.

(14)   (15)   (16)

The reagent for the carbanion synthon (15) could be an organometallic reagent (RMgX or RLi), or an α-carbanion arising from a carboxylic ester. The reagent for the carbocation synthon (16) could be a derivative of oxalic acid [e.g. diethyl oxalate $(CO_2Et)_2$ or ethyl oxalyl chloride $(COCl\cdot CO_2Et)$]. Two methods which illustrate these strategies may thus be formulated. In the first, an initial mixed Claisen ester condensation (see p. 738) between a carboxylic ester and diethyl oxalate (cf. Expt 8.37, ethyl phenyl oxaloacetate) is followed by hydrolysis to the keto diacid (17) and subsequent decarboxylation of the β-keto acid function.

$$R \cdot CH_2 \cdot CO_2Et + {}^{\ominus}OEt \longrightarrow [R \cdot \overset{\ominus}{C}H \cdot CO_2Et] + EtOH$$

In the second method, ethyl oxalyl chloride is converted first into ethyl α-oxo-1H-imidazole-1-acetate, which is then treated with a Grignard reagent.[192] The precedent for this reaction is that acylimidazolides are highly reactive towards Grignard reagents to yield ketones without further significant reaction with the carbonyl function to give tertiary alcohols.[193]

The yield of α-keto ester is best when an aromatic Grignard reagent is used, and the reaction is illustrated by the formation of ethyl benzoylformate (Expt 5.174).

A general route to aryl-substituted pyruvic acids (e.g. phenylpyruvic acid, Expt 5.175) is the acid hydrolysis of 2-acetamido-3-arylacrylic acids (19), which are themselves formed by hydrolysis of the corresponding azlactones (18) (cf. Expt 8.21) with water.

β-KETO ACIDS AND ESTERS

The β-keto acids are thermally unstable and readily decompose in the presence of hot acid or alkali into ketones (see Section 5.8.5, p. 617). On the other hand the β-keto esters are stable and are extremely important synthetic reagents.

The typical β-keto ester is ethyl acetoacetate (ethyl 3-oxo-butanoate). A retrosynthetic disconnection on this compound points to one of the most common methods for its synthesis, namely the *Claisen ester condensation*.

736

The condensation proceeds under the influence of strong base catalysts of which sodium ethoxide is the most common example. This is usually formed from the ethanol present in ordinary samples of ester by the action of the sodium used in the condensation. The first step in the mechanism is the removal of the α-hydrogen in ethyl acetate by the base catalyst to produce the mesomerically stabilised α-carbanion (20). The nucleophilic carbanion so formed then attacks the carbonyl carbon of a second ester molecule to produce the anion (21) which is converted into the β-keto ester (22) by loss of an ethoxide ion. Finally (22) reacts with the ethoxide ion to produce the mesomerically stabilised β-keto ester anion (23).

The equilibrium in the last step (iii) is far to the right because of the greater basic strength of the ethoxide ion compared to the anion (20), and this largely assists the forward reactions in (i) and (ii). The reaction product is therefore the sodio-derivative of the β-keto ester from which the free ester is obtained upon careful acidification.

With ethyl esters of higher straight-chain carboxylic acids (e.g. ethyl propanoate, ethyl butanoate, etc.) the products are α-alkylated β-keto esters, and the condensation may also be affected by the action of sodium alkoxides.

$$2R \cdot CH_2 \cdot CO_2Et \xrightarrow{-EtOH} R \cdot CH_2 \cdot CO \cdot CHR \cdot CO_2Et$$

For esters with only one α-hydrogen, such as ethyl 2-methyl propanoate (ethyl isobutyrate), a more powerful base (e.g. sodium triphenylmethide, $Ph_3C^{\ominus}Na^{\oplus}$) is required to affect the condensation reaction [the *forced* Claisen ester condensation, e.g. the synthesis of ethyl 2,2,4-trimethyl-3-oxopentanoate (ethyl iso-butyrylisobutyrate), Expt 5.176]. In this case the reaction sequence is completed in the step corresponding to (ii) above since the β-keto ester (24) has no α-hydrogen for step (iii), and the powerful base is required to force the equilibrium (i) to the right.

(24)

When a mixture of two dissimilar esters ($R^1 \cdot CH_2 \cdot CO_2Et$ and $R^2 \cdot CH_2 \cdot CO_2Et$) is treated with sodium as the ethoxide, a 'mixed' β-keto ester ($R^1 \cdot CH_2 \cdot CO \cdot CH(R^2) \cdot CO_2Et$ or $R^2 \cdot CH_2 \cdot CH(R^1) \cdot CO_2Et$) may be formed. However the disadvantage which is common to all mixed reactions, i.e. the formation of both symmetrical and crossed products, reduces its preparative value. Exceptions are provided in those cases where one of the esters contains no α-hydrogens. An important example is provided by the use of esters of oxalic acid in such mixed Claisen ester condensations (e.g. Section 5.14.3, p. 735, and Expt 8.37).

Retrosynthetic disconnection of a β-keto ester of the type $R \cdot CO \cdot CH_2 \cdot CO_2Et$, gives the carbocation $R \cdot \overset{\oplus}{C}O$ and the carbanion $\overset{\ominus}{C}H_2 \cdot CO_2Et$. The reagent equivalent of the former synthon could be the acid chloride $R \cdot COCl$ in place of the ester in the Claisen condensation above. The reagent equivalent of the carbanion synthon could be diethyl malonate (see Section 5.11.6, p. 680). The overall reaction then becomes an acylation of malonic ester followed by suitable removal of the carbethoxy group.

One preparative procedure for achieving this reaction involves the acylation of the magnesium enolate of diethyl malonate with an acid chloride in benzene solution (cf. Expt 5.96), and the resulting acylmalonic ester is then heated to 200 °C with an arylsulphonic acid to effect the decarbethoxylation step. An illustrative example is the preparation of ethyl 3-oxopentanoate (ethyl propionylacetate, Expt 5.177).

In an alternative method (cognate preparation in Expt 5.177), the dilithio complex of ethyl hydrogen malonate in tetrahydrofuran at low temperature is reacted with the acid chloride[194]; the decarboxylation process takes place spontaneously when the reaction mixture is quenched with dilute acid.

γ-KETO ACIDS AND ESTERS

The retrosynthetic disconnection for the γ-keto acid shown below generates an acyl anion synthon and a three-carbon carbocation.

Convenient reagent equivalents could be a nitroalkane (for the acyl anion synthon, p. 627), and acrolein or an acrylic ester (for the carbocation). The reaction would then be a Michael addition process (Section 5.11.6, p. 681), followed by functional group modification (e.g. C—NO$_2$ to C=O, p. 599, and CHO to CO$_2$H, p. 667).

This method is illustrated (Expt 5.178) by the reaction of nitroethane with acrolein in the presence of basic alumina[195]; this catalyst apparently avoids the induction of undesirable side reactions. The Michael adduct is treated with hydrogen peroxide which effects the functional group modifications.

The alternative charge distribution of the synthons generated in the above disconnection leads to the acyl carbocation (R·$\overset{\oplus}{C}$O) and the carbanion $\overset{\ominus}{C}$H$_2$·CH$_2$·CO$_2$Et. From the latter the three-carbon unit of diethyl succinate may be recognised, and the *Stobbé condensation* is a suitable illustration of this process. Thus the reaction of aldehydes or ketones with succinic esters in the presence of potassium t-butoxide (or with sodium hydride) gives alkylidenesuccinic acid esters (25). The reaction is somewhat slower in the case of the less basic sodium ethoxide.

(25)

When the alkylidenesuccinic ester is derived from an aldehyde [e.g. (26)], functional group modification provides a route to γ-keto acids and the overall process relates to the disconnection strategy outlined above. The required modifications are hydrolysis of the diester to the diacid and photocatalysed bromination to yield the dibromo acid (27), which on treatment with alkali is converted to the γ-keto acid. This sequence is illustrated by the preparation of 4-oxodecanoic acid (Expt 5.179).

Experiment 5.173 PYRUVIC ACID

$$HO_2C(CHOH)_2CO_2H \xrightarrow[\Delta]{KHSO_4} Me \cdot CO \cdot CH_2 \cdot CO_2H + CO_2 + H_2O$$

Grind together in a glass mortar 200 g (1.33 mol) of powdered tartaric acid and 300 g (2.2 mol) of freshly *fused* potassium hydrogen sulphate to form an intimate mixture. Place the mixture in a 1.5-litre round-bottomed flask, and fit the latter with a still-head and a long air condenser. Heat the flask in an oil bath maintained at 210–220 °C until liquid no longer distils over. If foaming is considerable and there is danger of the mixture frothing over, heat the upper part of the flask with a free flame. Fractionate the distillate under reduced pressure and collect the pyruvic acid at 75–80 °C/25 mmHg. The yield is 60 g (51%).

Experiment 5.174 ETHYL BENZOYLFORMATE[192]

Ethyl α-oxo-1H-imidazole-1-acetate. Ethyl oxalyl chloride (20.0 ml, 0.179 mol) (**CAUTION**) (Aldrich) (1) in tetrahydrofuran (80 ml) (2) is added to a stirred solution of imidazole (24.3 g, 0.35 mol) in tetrahydrofuran (500 ml) at 0 °C under nitrogen over a 1-hour period. After an additional 1 hour of being stirred at 0 °C, the mixture is filtered and the precipitate washed with anhydrous tetrahydrofuran (100 ml). The tetrahydrofuran is removed from the filtrate and washings under vacuum, and the residue is distilled to furnish 27.2 g (90%) of product, b.p. 100 °C/1 mmHg; i.r. (neat) 1730 (ester) and 1770 (amide) cm^{-1}; p.m.r. (CDCl$_3$, TMS) δ 1.45 (t, 3H), 4.45 (q, 2H), 7.05 (s, 1H), 7.60 (s, 1H), and 8.40 (s, 1H); m/z 168 (M). This material undergoes slow discoloration at room temperature and is stored at 5 °C.

Ethyl benzoylformate. The foregoing imidazolide (3.03 g, 24 mmol) is dissolved in tetrahydrofuran (75 ml) under nitrogen and cooled to $-50\,°C$ in a dry-ice–acetone bath. Phenylmagnesium bromide (1.0 equiv. in 50 ml of tetrahydrofuran) is added by a dropping funnel over 1 hour with stirring. The solution is allowed to come to room temperature over 3 hours and poured into ice-water (200 ml). The solution is extracted with ether (a few drops of acetic acid are added to break up emulsions), and the ether extract is washed with brine and dried over magnesium sulphate. Removal of solvents under reduced pressure and distillation affords ethyl benzoylformate (72%), b.p. $88\,°C/0.5\,mmHg$.

Notes. (1) Ethyl oxalyl chloride should be handled with care in a fume cupboard since it is lachrymatory.

(2) The tetrahydrofuran should be distilled from sodium benzophenone ketyl immediately prior to use.

Experiment 5.175 PHENYLPYRUVIC ACID

$$Ph\cdot CH{=}\underset{\underset{\displaystyle CO_2H}{|}}{C}\cdot NH\cdot CO\cdot Me \xrightarrow{+2H_2O} Ph\cdot CH_2\cdot CO\cdot CO_2H + Me\cdot CO_2NH_4$$

Place 10.3 g (0.05 mol) of α-acetamidocinnamic acid (Expt 8.21) and 200 ml of 1 M hydrochloric acid in a 500-ml round-bottomed flask and boil the mixture steadily under reflux for 3 hours. Remove a small quantity of green oil by rapidly filtering the hot reaction mixture through a small plug of cotton wool loosely inserted into the stem of a preheated glass filter funnel, cool the filtrate to room temperature and leave it at $0\,°C$ for 48 hours. Collect the crystalline product by filtration, wash it with a small quantity of ice-cold water and dry it in a vacuum desiccator over anhydrous calcium chloride and potassium hydroxide pellets. The yield of phenylpyruvic acid, which is sufficiently pure for most purposes, is 4.4 g, m.p. $157\,°C$ (decomp.). A further 1.7 g of product of comparable purity (total yield 74%) separates from the aqueous acidic filtrate when this is set aside at $0\,°C$ for about one week.

Experiment 5.176 ETHYL 2,2,4-TRIMETHYL-3-OXOPENTANOATE
(*Ethyl isobutyrylbutyrate*)

$$2Me_2CH\cdot CO_2Et \xrightarrow{Ph_3C^{\ominus}Na^{\oplus}} Me_2CH\cdot CO\cdot C(Me)_2\cdot CO_2Et$$

Triphenylmethyl sodium. Prepare a 1.5 per cent sodium amalgam from 15 g (0.65 mol) of sodium and 985 g of mercury (Section 4.2.70, p. 464). Place a mixture of 1000 g of the amalgam and 74 g (0.265 mol) of triphenylchloromethane (Section 4.2.79, p. 466) in a 2-litre Pyrex glass-stoppered bottle and add 1500 ml of sodium-dried ether. Grease the glass stopper with a little Silicone grease, insert it firmly, clamp the bottle in a mechanical shaker and shake. **CAUTION**: *the reaction is strongly exothermic; cool the bottle with wet rags and stop the shaking from time to time, if necessary.* A characteristic red colour appears after about 10 minutes' shaking. After shaking for 4 to 6 hours, *cool the bottle to room temperature*, remove it from the shaker, wire the stopper down and allow the mixture to stand undisturbed; sodium chloride and particles of mercury settle to the bottom.

Separate the ether solution of triphenylmethyl sodium as follows. Remove the glass stopper and replace it immediately by a tightly fitting two-holed bung carrying a short glass tube that protrudes about 1 cm into the bottle, and a long glass tube bent into an inverted U-shape. Connect the bottle through a drying train to a cylinder of nitrogen. Lead the other arm of the U-tube into a 2-litre, two-necked round-bottomed flask (which has been previously filled with nitrogen) via a suitable screw-capped adapter and fit a dropping funnel into the other neck. Open the stopcock of the dropping funnel slightly and force the ether solution of triphenylmethyl sodium slowly and steadily into the nitrogen-filled flask by means of a small pressure of nitrogen from the cylinder. By carefully adjusting the depth of the siphon tube in the bottle, all but 50–75 ml of the clear ether solution may be removed.

If pure triphenylchloromethane and freshly prepared sodium amalgam are used, the yield of triphenylmethyl sodium should be almost quantitative and the concentration is usually 0.15 mol per litre (1). The reagent should be used as soon as possible after its preparation.

Ethyl 2,2,4-trimethyl-3-oxopentanoate. Add 24 g (28 ml, 0.21 mol) of ethyl 2-methylpropanoate b.p. 110–111 °C, to the solution of c. 0.21 mol of triphenylmethyl sodium in approximately 1400 ml of ether contained in the 2-litre two-necked flask. Stopper the flask, shake well to effect complete mixing and keep at room temperature for 60 hours. Acidify the reaction mixture by adding, with shaking, 15 ml of glacial acetic acid, and then extract with 100 ml of water. Wash the ethereal solution with 50 ml portions of 10 per cent sodium carbonate solution until free from excess acid, dry over anhydrous sodium sulphate; remove the ether under reduced pressure with a rotary evaporator. Distil the residue under reduced pressure through a short fractionating column. Collect the ethyl 2,2,4-trimethyl-3-oxopentanoate at 95–96 °C/18 mmHg; the yield is 14.5 g (74%). The b.p. at atmospheric pressure is 201–202 °C.

Note. (1) The solution may be analysed approximately as follows. Remove 25 ml of ether solution, run it into 25 ml of water contained in a small separatory funnel and shake. Run off the aqueous layer into a 250-ml conical flask and extract the ether layer with two 25 ml portions of water. Titrate the combined aqueous extracts with 0.95 M-sulphuric acid, using methyl red as indicator.

Ketonic hydrolysis to di-isopropyl ketone. Mix 14 g (0.75 mol) of the ester with 30 ml of glacial acetic acid, 10 ml of water and 10 ml of concentrated sulphuric acid, and boil under reflux until evolution of carbon dioxide ceases. Dilute the cooled solution with 180 ml of water, add 100 ml of ether and render alkaline to phenolphthalein with 20 per cent sodium hydroxide solution. Separate the ether layer, extract the aqueous layer with two 50 ml portions of ether, dry the combined ether layer and extracts with anhydrous sodium sulphate, distil off the ether and fractionate the residue. The yield of di-isopropyl ketone (2,4-dimethylpentan-3-one), b.p. 123–124 °C, is 6.5 g (76%).

Experiment 5.177 ETHYL 3-OXOPENTANOATE (*Ethyl propionylacetate*)

$$\text{Et·COCl} + \underset{\underset{\text{MgOEt}}{|}}{\text{CH(CO}_2\text{Et)}_2} \longrightarrow \text{Et·CO·CH(CO}_2\text{Et)}_2 \xrightarrow[\Delta]{\text{Ar·SO}_3\text{H}} \text{Et·CO·CH}_2\text{·CO}_2\text{Et}$$

Prepare the ethoxymagnesium diethyl malonate derivative from 13 g (0.53 mol) of magnesium and 80 g (0.5 mol) of diethyl malonate following the procedure described in Expt 5.96. Then add with vigorous stirring a solution of 49 g (46 ml, 0.53 mol) of propanoyl chloride in 50 ml of anhydrous ether. Reflux the reaction mixture for 30 minutes and then cool and acidify with 60 ml of dilute sulphuric acid. Separate the ether layer, extract the residual aqueous solution with two 50 ml portions of ether. Wash the combined organic phases with water, dry over anhydrous sodium sulphate and remove the ether on a rotary evaporator. Add to the residue 8 g of naphthalene-2-sulphonic acid monohydrate and heat the mixture slowly to 200 °C in an oil bath. A vigorous evolution of gas sets in at about 120 °C; when gas evolution has subsided, cool the reaction mixture and dissolve it in about 150 ml of ether. Wash the ethereal extract with four 25 ml portions of 10 per cent sodium carbonate solution and then with water; back extract the combined aqueous solutions with three 25 ml portions of ether. Dry the combined ether extracts with anhydrous sodium sulphate, remove the ether on a rotary evaporator and distil the residue under reduced pressure. Collect the ethyl 3-oxopentanoate as a fraction of b.p. 100–105 °C/22 mmHg; the yield is 34 g (47%).

Cognate preparation. *General procedure for the acylation of the dianion of ethyl hydrogen malonate* (1).[194] To 250 ml of tetrahydrofuran (distilled from sodium/benzophenone under nitrogen prior to use) under nitrogen with stirring is added 13.47 g (0.1 mol) of monoethyl malonate (2) and several milligrams of 2,2'-bipyridyl as an indicator. After cooling to − 70 °C, butyllithium (hexane) is added slowly while allowing the temperature to rise to c. − 5 °C near the end of the addition (∼ 130 ml of 1.6 M solution, 0.2 mol). After the pink indicator persists at − 5 °C the heterogeneous solution is recooled to − 65 °C and the acid chloride (0.057 mol) is added dropwise over 5 minutes. After the appropriate reaction time (3), the reaction solution is poured into 400 ml of ether and 200 ml of 1 M hydrochloric acid. After mixing and separating the aqueous phase, the organic phase is washed with 2 × 100 ml of saturated sodium hydrogen carbonate solution and 100 ml of water, dried over anhydrous sodium sulphate and concentrated to yield the β-keto ester.

Notes. (1) See Section 2.17.8, p. 120, for a suitable apparatus set-up for this moisture-sensitive reaction.

(2) Ethyl hydrogen malonate may be prepared from diethyl malonate by controlled hydrolysis with potassium hydroxide (see also Expt 5.147, Note (1)).

(3) The following acid chlorides have been used and the reaction times are given in parentheses: Pr·COCl (5 minutes); Me₂CH·CH₂COCl (5 minutes); Bu·COCl (5 minutes); Ph·CH₂COCl (5 minutes); and Ph·COCl (30 minutes).

Experiment 5.178 4-OXOPENTANOIC ACID[195]

743

CAUTION: *All procedures should be conducted behind a safety screen; particular care should be taken in the handling of acrolein; see Section 4.2.41, p. 439, for the measures to be taken in the handling of 30 per cent hydrogen peroxide.*

4-Nitropentanal. A 100-ml two-necked flask equipped with a mechanical stirrer is charged with nitroethane (3.75 g, 0.05 mol) and cooled with an ice-water bath. Acrolein (2.8 g, 0.05 mol) is added and the mixture stirred for 5 minutes. Chromatographic alumina (activity I on the Brockmann scale; 10 g) is added and stirring is continued for 6 hours. The reaction is monitored by t.l.c. (ethyl acetate–hexane, 2:8 as eluent). At the end of the reaction, the alumina is washed with ether (4 × 50 ml) and the filtered extract evaporated under reduced pressure. The product is purified by distillation to give 4-nitropentanal (50%), b.p. 71 °C/0.4 mmHg; i.r. 1720 (C=O) and 1545 cm^{-1} (NO$_2$).

4-Oxopentanoic acid. Aqueous hydrogen peroxide (30 ml, 30%) is added to a cooled (0 °C) and stirred solution of the foregoing 4-nitroalkanal (1.96 g, 0.015 mol) in methanol (70 ml). To the resultant solution, potassium carbonate (12 g) is added, and stirring is continued for 15 hours at room temperature. The solution is then acidified with 2 M hydrochloric acid and extracted with dichloromethane (3 × 40 ml). The organic layer is washed with water (40 ml), and dried with magnesium sulphate. The solvent is removed under vacuum to leave the crude product which can be used without further purification. The pure 4-oxopentanoic acid (60%) is obtained by distillation and has b.p. 120 °C/10 mmHg.

Experiment 5.179   4-OXODECANOIC ACID

$$C_6H_{13}\cdot CHO + \underset{\underset{CH_2\cdot CO_2Et}{|}}{CH_2\cdot CO_2Et} \xrightarrow{\overset{\oplus\ominus}{KOBu^t}} \underset{\underset{CO_2Et}{|}}{C_6H_{13}\cdot CH=C\cdot CH_2\cdot CO_2H} \xrightarrow[\text{(ii) } H_3O^\oplus]{\text{(i) NaOH}}$$

$$\underset{\underset{CO_2H}{|}}{C_6H_{13}\cdot CH=C\cdot CH_2\cdot CO_2H} \xrightarrow{Br_2} \underset{\underset{CO_2H}{|}}{C_6H_{13}\cdot CHBr\cdot CBr\cdot CH_2\cdot CO_2H} \xrightarrow{NaOH}$$

$$\xrightarrow[\text{(ii) } H_3O^\oplus]{\text{(i) NaOH}} C_6H_{13}\cdot CO\cdot CH_2\cdot CH_2\cdot CO_2H$$

3-Carboxydec-3-enoic acid. (*Heptylidenesuccinic acid.*) *Potassium t-butoxide.* Prepare a solution of potassium t-butoxide in t-butyl alcohol using the following procedure. Equip a two-necked, round-bottomed 1-litre flask with a reflux condenser and a pressure-equalising dropping funnel. Attach to the top of the condenser a nitrogen inlet system with nitrogen escape valve (cf. Fig. 2.60) and place the flask in a magnetic stirrer-heating mantle unit. Flush the flask with a stream of dry nitrogen, charge the flask with 375 ml of dry t-butanol (Section 4.1.12, p. 402) and add 19.6 g (0.5 mol) of potassium [CAUTION: (1)]. Heat the mixture under reflux until the potassium completely dissolves (c. 4 hours). To the cooled solution add a mixture of 122 g

(0.7 mol) of diethyl succinate and 63 g (0.5 mol) of heptanal over half an hour; the reaction is exothermic. Finally heat the mixture under reflux for 1 hour. Rearrange the condenser for distillation and remove the t-butyl alcohol by vacuum distillation using a water pump. Acidify the residue with dilute hydrochloric acid and extract with three 200 ml portions of ether. Extract the acidic compounds from the ethereal solution by shaking with 50 ml portions of saturated sodium hydrogen carbonate solution until no more carbon dioxide is evolved. Acidify the combined aqueous solutions by the careful addition of concentrated hydrochloric acid. Extract the 3-ethoxycarbonyldec-3-enoic acid which separates as an oil with ether, dry the ethereal extract over magnesium sulphate and evaporate the ether solution on a rotary evaporator. Saponify the crude half ester by heating it under reflux for 1 hour with 400 ml of 10 per cent sodium hydroxide solution. Cool the solution, acidify with concentrated hydrochloric acid and filter the precipitated acid with suction. Dissolve the crude acid in the minimum volume of ether (about 200 ml are required) and add the solution to an equal volume of light petroleum (b.p. 60–80 °C). Filter the precipitate with suction and recrystallise from benzene. The yield of 3-carboxydec-3-enoic acid, m.p. 128–130 °C, 37.3 g (31.6%). The infrared spectrum shows absorptions at 3500–2300 cm^{-1} (O—H stretch of carboxylic acid), 1700 cm^{-1} (C=O) and 1640 cm^{-1} (C=C).

3-Carboxy-3,4-dibromodecanoic acid. *This reaction should be carried out in a fume cupboard.* Place 5.3 g (0.025 mol) of 3-carboxydec-3-enoic acid, 6.0 g (0.038 mol) of bromine and 60 ml of carbon tetrachloride in a 100-ml round-bottomed flask equipped with a magnetic stirrer and reflux condenser. Irradiate the stirred mixture with a 100-watt lamp for 6 hours; the dibromide forms and precipitates out during this period. Filter the product with suction and wash thoroughly with hexane. The yield of 3-carboxy-3,4-dibromodecanoic acid is 7.9 g (85%). The acid can be recrystallised from toluene, m.p. 142–143 °C. The infrared spectrum shows absorptions at 3400–2400 cm^{-1} (O—H stretch of CO$_2$H) and 1730 cm^{-1} (C=O).

4-Oxodecanoic acid. Dissolve 4.0 g of 3-carboxy-3,4-dibromodecanoic acid in 60 ml of 2M sodium hydroxide solution and heat the solution at 80–90 °C for 2 hours. Cool to room temperature and acidify with dilute sulphuric acid; carbon dioxide is evolved and a white precipitate is formed. Filter the precipitated keto acid from the cold solution and recrystallise from light petroleum (b.p. 40–60 °C). 4-Oxodecanoic acid, m.p. 68–69 °C, is obtained; the yield is 1.6 g (80%). The infrared spectrum shows absorptions at 3400–2400 cm^{-1} (OH stretch of CO$_2$H) and 1700 cm^{-1} (C=O).

If the dibromo acid is treated with alkali under milder conditions, for example with 1 M sodium hydroxide solution at 20–25 °C for 0.5 hour, the intermediate γ-hexylaconic acid (m.p. 123–125 °C) can be isolated after acidification. The infrared spectrum shows absorptions at 3110 cm^{-1} (C—H stretch, alkene), 1715 and 1745 cm^{-1} (C=O stretch of carboxylic acid and lactone) and 1630 cm (C=C stretch).

Note. (1) *Great care must be taken in the handling of potassium.* See Section 4.2.62, p. 460, for the precautions that must be adopted.

5.14.4 AMINO ACIDS AND PEPTIDES

The α-aminocarboxylic acids are of particular importance as the result of their involvement in the primary structure of protein molecules. Although customarily represented by the general formula $NH_2·CHR·CO_2H$ amino acids are actually zwitterionic compounds, $\overset{\oplus}{N}H_3·CHR·CO_2^{\ominus}$, and in the solid state have obvious polar properties. In solution, the following pH-dependent equilibria are established; the pH at which the molecule has no net charge is the *isoelectric point*, characteristic for a particular amino acid.

$$\overset{\oplus}{H_3}N·CHR·CO_2H \underset{+H^{\oplus}}{\overset{-H^{\oplus}}{\rightleftharpoons}} \overset{\oplus}{N}H_3·CHR·CO_2^{\ominus} \underset{+H^{\oplus}}{\overset{-H^{\oplus}}{\rightleftharpoons}} NH_2·CHR·CO_2^{\ominus}$$

Amino group reactions (e.g. acylation) proceed best under basic conditions, whereas carboxyl group reactions (e.g. esterification) are effected under acidic conditions with the amino function protected by protonation.

The main general routes for the synthesis of the α-amino acids are illustrated in the following sections (pp. 746–750). Some of the natural amino acids can be isolated from the hydrolsates of suitable proteins; examples are provided on p. 750. They are optically active having the L [(S)] configuration. The synthetic products, formed by routes not involving asymmetric syntheses, are of of course racemic, from which the optically active forms may be obtained by the process of resolution. Some resolution procedures are discussed in Section 5.19.

THE AMINATION OF α-HALOGENOCARBOXYLIC ACIDS

This method is generally suitable for those α-amino acids $[H_2N·CH(R)·CO_2H]$ where R is an unsubstituted alkyl group. The examples included in Expt 5.180 are alanine ($R = CH_3$), glycine ($R = H$), valine ($R = (CH_3)_2CH$), norvaline ($R = CH_3·CH_2·CH_2$), norleucine ($R = CH_3·CH_2·CH_2·CH_2$) and isoleucine ($R = CH_3·CH_2·CH(CH_3)$).

Chloroacetic acid is readily converted into glycine by treatment with concentrated aqueous ammonia solution, but in general an α-bromocarboxylic acid is preferred. This can usually be prepared in good yield by a Hell–Volhard–Zelinsky bromination (Section 5.14.1, p. 720) of the corresponding carboxylic acid; if the carboxylic acid is not readily available it can usually be obtained by the synthesis and bromination of the appropriate alkylmalonic acid (see Expt 5.166).

$$R·CH_2·CO_2H \qquad CH_2(CO_2Et)_2 \xrightarrow[\text{(ii) RBr}]{\text{(i) }^{\ominus}OEt} R·CH(CO_2Et)_2$$

R·CH₂·CO₂H → (Br₂, P) → R·CHBr·CO₂H

CH₂(CO₂Et)₂ → R·CH(CO₂Et)₂ → (i) ⁻OH, (ii) H₃O⁺ → R·CH(CO₂H)₂ → (Br₂) → R·CBr(CO₂H)₂

R·CHBr·CO₂H → (NH₃) → R·CH(NH₂)·CO₂H

R·CBr(CO₂H)₂ → (−CO₂) → R·CHBr·CO₂H

A large excess of concentrated ammonia has to be used to accomplish the amination step, otherwise the amino acid which is formed reacts further with the bromo acid to give substantial amounts of the secondary and tertiary amine derivatives. The use of an excess of ammonium carbonate in aqueous ammonia is also recommended. The reaction is usually carried out at room temperature, but temperatures of about 50 °C and advantageous when the substitution process is retarded as the result, for example, of steric effects. The higher homologous amino acids are not particularly water-soluble, and they may be isolated in the crude state by concentrating and filtering the reaction product. Care must be taken during the isolation of the readily water-soluble lower members of the series to free them from accompanying inorganic salts; in such cases advantage is taken of the solubility of ammonium bromide in methanol to achieve the required purification.

THE FORMATION AND HYDROLYSIS OF α-AMINO NITRILES (*the Strecker synthesis*)

The original Strecker procedure is the reaction of an aldehyde with ammonia and then with hydrogen cyanide to form the α-amino nitrile. This intermediate may also be obtained by reacting the aldehyde cyanohydrin with ammonia, but a more convenient method is to treat the aldehyde in one step with ammonium chloride and sodium cyanide. The α-amino acid is obtained when the amino nitrile is hydrolysed under either acidic or basic conditions; the former are usually preferred. The preparation of α-phenylglycine (R = Ph) from benzaldehyde is typical of the general procedure (cognate preparation in Ept 5.181).

$$R \cdot CH(NH_2)OH \xrightarrow{HCN} R \cdot CH(NH_2) \cdot CN \xrightarrow{H_3O^{\oplus}} R \cdot CH(NH_2) \cdot CO_2H$$

$$R \cdot CHO \xrightarrow{HCN} R \cdot CH(OH) \cdot CN$$

When applied to formaldehyde, however, the reaction is somewhat anomalous in that methyleneaminoacetonitrile (CH₂=N·CH₂·CN), the condensation product derived from the aldehyde and the amino nitrile, is formed (Expt 5.181). The free amino nitrile is obtained by careful basification of its sulphate salt, which is formed when methyleneaminoacetonitrile is treated with concentrated sulphuric acid in ethanol. Details of the hydrolysis of the amino nitrile (as the sulphate) under basic conditions are given. Barium hydroxide is used, the excess of which is finally removed by precipitation as the sulphate to facilitate the isolation of the glycine formed.

THE REDUCTIVE HYDROLYSIS OF ARYLIDENEOXAZOLONES (*azlactones – the Erlenmeyer synthesis*)

This method is mainly restricted to the synthesis of amino acids with aromatic side-chains since the required unsaturated azlactones [e.g. (30)] are most readily prepared using aromatic aldehydes. Typically, benzaldehyde condenses under the influence of base with the reactive methylene group in the azlactone (29) which is formed by the dehydration of benzoylglycine (28) when the latter is heated with acetic anhydride in the presence of sodium acetate (cf. Expt 8.21). The azlactone ring is readily cleaved hydrolytically and compounds of the type (30) yield substituted acylaminoacrylic acids [e.g. (31)] on boiling with water. Reduction and further hydrolysis yields the amino acid [e.g. phenylalanine,

(32)], but this result can be achieved in one step (Expt 5.182) by heating the azlactone (30) in acetic anhydride solution with red phosphorus and hydriodic acid.

THE FORMATION AND HYDROLYSIS OF C-SUBSTITUTED ACYL-AMINOMALONIC ESTERS

Acylaminomalonic esters and related reagents are widely used for the synthesis of α-amino acids. The method differs from those syntheses already discussed in that the amino group is incorporated into the system from the outset. A popular reagent is diethyl acetamidomalonate (35). The acetamido group can readily be introduced into the reactive methylene position in diethyl malonate by first converting the latter into the hydroxy-imino derivative (33) by reaction with nitrous acid or an alkyl nitrite (cf. Section 4.2.1, p. 413). This derivative is then reduced catalytically to diethyl aminomalonate (34) which is acetylated using acetic anhydride.

$$CH_2(CO_2Et)_2 \xrightarrow[\text{RONO}]{\text{HNO}_2 \text{ or}} O=N-CH(CO_2Et)_2 \rightleftharpoons HON=C(CO_2Et)_2 \xrightarrow[\text{catalyst}]{H_2}$$
$$(33)$$

$$H_2N \cdot CH(CO_2Et)_2 \xrightarrow{(MeCO)_2O} Me \cdot CO \cdot NH \cdot CH(CO_2Et)_2$$
$$(34) \qquad\qquad\qquad (35)$$

In common with other malonate derivatives, the α-hydrogen atom in the acetamidomalonate is reactive; on treatment with a base the reagent forms a mesomerically stabilised carbanion (36) from which a variety of substituted acetamidomalonic esters can be made. For example, C-alkylation ensues when the anion is allowed to react with an alkyl halide; the resulting product (37) is then subjected to the hydrolytic and decarboxylative sequence shown to yield a simple α-amino acid.

$$Me \cdot CO \cdot NH \cdot CH(CO_2Et)_2 + B^\ominus \rightleftharpoons BH + Me \cdot CO \cdot NH \cdot \overset{\ominus}{C}(CO_2Et)_2$$

(36)

$$X \overset{\frown}{-} R \overset{\frown}{\underset{NH \cdot CO \cdot Me}{\overset{\ominus}{C}(CO_2Et)_2}} \longrightarrow R \cdot \underset{NH \cdot CO \cdot Me}{C(CO_2Et)_2} \xrightarrow[\text{(ii) }H_3O^\oplus]{\text{(i) }^\ominus OH}$$

(37)

$$R \cdot \underset{NH \cdot CO \cdot Me}{C(CO_2H)_2} \xrightarrow[-CO_2]{\text{heat}} R \cdot \underset{NH \cdot CO \cdot Me}{CH \cdot CO_2H} \xrightarrow[\text{(ii) }H_3O^\oplus]{\text{(i) }^\ominus OH} R \cdot \underset{NH_2}{CH \cdot CO_2H}$$

The two examples illustrative of the section, however, involve alternative procedures for introducing the required substituent into the α-position of the acetamidomalonate reagent. In the first (Expt 5.183) the hydroxymethyl group is introduced by a simple base-catalysed condensation with formaldehyde; subsequent hydrolysis and decarboxylation yields serine (R = CH$_2$OH). In this case, acidic conditions are preferred for the final hydrolytic stage, and the use of a weakly basic ion exchange resin to obtain the halide-free amino acid from a solution of its hydrochloride is described.

The second example is an interesting synthesis of the heterocyclic amino acid tryptophan (R = 3-indolymethyl) which involves the initial base catalysed 1,4-addition (the Michael reaction, Section 5.11.6, p. 681) of diethyl acetamidomalonate to the α,β-unsaturated aldehyde, acrylaldehyde, yielding the aldehydic derivative (38). The derived phenylhydrazone (39) is then cyclised under acidic conditions (see Expt 8.26) to form the indolylacetamidomalonate derivative (40) which is then converted into the corresponding α-amino acid (i.e. tryptophan) in the usual way (Expt 5.184).

$$(EtO_2C)_2\overset{\ominus}{\underset{Me \cdot CO \cdot NH}{C}} \quad H_2C \overset{\frown}{\underset{O}{\diagdown}} H \xrightarrow[-H^\oplus, +H^\oplus]{\text{Base;}} (EtO_2C)_2\underset{Me \cdot CO \cdot NH}{C \cdot CH_2 \cdot CH_2 \cdot CHO}$$

(38)

(39)   →   (40)

Although it does not involve the use of an acylaminomalonic ester a related convenient synthesis of the heterocyclic amino acid, proline, is included here (Expt 5.185).

The synthesis starts with the preparation of diethyl (2-cyanoethyl)malonate (Expt 5.161) by the Michael addition of diethyl malonate to acrylonitrile. Hydrogenation over Raney nickel converts the cyanoethyl compound to the corresponding primary amine (41) which is converted into proline (isolated initially as the hydrochloride) by the reaction sequence shown. Liberation of the free amino acid from its salt is achieved in this case by treatment with triethylamine (cf. serine, Expt 5.183).

ISOLATION FROM PROTEIN HYDROLYSATES

The hydrolysis of proteins yields a complex mixture of amino acids having closely related physical properties. Although a quantitative separation of the individual amino acids can be achieved by chromatographic methods, only in a few cases is the preparative isolation of an individual member worth while. Provided that racemisation is avoided during hydrolysis, however, amino acids are obtained in this way in the optically active form. The normal synthetic products are of course racemic, but they may be resolved by appropriate procedures (e.g. the enzymic method described in Expt 5.221).

The amino acid composition of keratin, the protein of hair and wool, includes a greater-than-average proportion of the sulphur-containing amino acid, cystine. Since this is the least soluble of the protein amino acids it can readily be isolated after carefully neutralising an acid hydrolysate of hair (Expt 5.187). Protein hydrolysis is usually effected by boiling for about 10–20 hours with 20 per cent hydrochloric acid. The hydrolysis of hair for the isolation of cystine is, however, best achieved using a mixture of hydrochloric and formic acids.

Many relatively specific reagents, particularly a variety of metal complexes, have been developed as an aid to the isolation of individual amino acids. An example is provided by the isolation of L-proline from a gelatine hydrolysate using the chromium complex ammonium rhodanilate {ammonium dianiline-tetrathiocyanatochromate (III), $[Cr(CNS)_4(C_6H_5NH_2)_2]^{\ominus}\overset{\oplus}{N}H_4$. Proline is isolated as the rhodanilate salt, which is purified and then treated with pyridine to form the less soluble pyridine rhodanilate, thus liberating the proline into solution.

PEPTIDE SYNTHESIS

In a polypeptide chain the α-amino acid residues are linked 'head to tail' by means of amide (peptide) bonds.

$$NH_2 \cdot CHR^1 \cdot CO \cdot NH \cdot CHR^2 \cdot CO \cdot NH \cdot CH_2R^3 \cdot CO \cdot NH \sim\sim CO \cdot NH \cdot CHR^X \cdot CO_2H$$

The formation of a *dipeptide*, for example, thus involves the condensation of the carboxyl group of one amino acid with the amino group of another. To achieve this apparently simple synthetic objective efficiently however, a number of reaction steps, requiring the use of appropriate functional group protection procedures (see p. 13), must be carried out. An illustrative synthesis is that of L-prolylglycine described in Expt 5.188, where the sequence is formulated in full. Note that proline is actually a secondary rather than a primary α-amino acid, the side chain R being linked to the nitrogen atom to form a pyrrolidine ring. The principal phases in the synthetic sequence are as follows.

Functional group protection. The NH— group in proline is protected by acylation in the usual Schotten–Baumann manner with benzyl chloroformate to yield the benzyloxycarbonyl derivative (42). Correspondingly the —CO$_2$H group in glycine is protected by esterification in ethanol to form the ethyl ester, obtained as the hydrochloride (43) under Fischer–Speier conditions.

$$Ph \cdot CH_2 \cdot O \cdot CO \cdot N—CH \cdot CO_2H \qquad \overset{\ominus}{Cl} \; \overset{\oplus}{NH_3} \cdot CH_2 \cdot CO_2Et$$
$$(42) \hspace{9em} (43)$$

Peptide bond formation. The process requires that the derivative (42), a carboxylic acid, should be caused to acylate the free base liberated from the hydrochloride (43). Activation of the carboxyl group is effected by conversion into a type of acid anhydride; the mixed carbonic anhydride (44) is used here, and is prepared by reaction of the acid (42) with ethyl chloroformate.

$$Ph \cdot CH_2 \cdot O \cdot CO \cdot N—CH \cdot \overset{O}{\overset{\|}{C}}—O—\overset{O}{\overset{\|}{C}}—OEt$$
$$(44)$$
$$H_2N \cdot CH_2 \cdot CO_2Et$$

The free glycine ethyl ester nucleophilically attacks the mixed anhydride specifically at the amino acid carbonyl, rather than the carbethoxycarbonyl group, liberating carbon dioxide and ethanol and forming the acylated dipeptide ester in good yield.

Deprotection. Subsequent treatment of the fully protected dipeptide illustrates, in an elementary way, an important aspect of functional group protection, namely the *selective* removal of specific protecting groups. Here the ester group is smoothly removed by hydrolysis with aqueous alkali at room temperature, conditions which do not affect the benzyloxycarbonyl group. The latter can readily be cleaved by catalytic hydrogenolysis without affecting the ethyl ester group. Performance of these two procedures sequentially affords the required dipeptide. The intermediate mono-protected derivatives can of course be used in similar synthetic sequences for the construction of longer peptide chains.

Experiment 5.180 DL-ALANINE (*2-Aminopropanoic acid*)

$$Me \cdot CHBr \cdot CO_2H \xrightarrow{NH_3} Me \cdot CH(NH_2) \cdot CO_2H$$

Place 2 litres (1760 g, 36 mol) of concentrated ammonia solution (*d* 0.88, 35% w/w) (1) in a large (e.g. Winchester) bottle, cool thoroughly in an ice bath and pour in slowly 77 g (0.5 mol) of 2-bromopropanoic acid (Expt 5.164). Close the bottle with a rubber bung held in place with wire, and leave at room temperature for 4 days. Concentrate the solution to about 250 ml by distillation under reduced pressure using a rotary evaporator; apply the vacuum with caution in the initial stages when most of the excess of ammonia is being removed. Filter, concentrate further to 150 ml, cool in ice and add 750 ml of

751

methanol with swirling. Leave the resulting suspension overnight in a refrigerator, and then filter off the crude alanine with suction and wash it with 200 ml of methanol. Dissolve the product in 150 ml of water, reprecipitate the alanine by adding 750 ml of methanol and filter and wash as before. The yield of almost pure DL-alanine, m.p. 295–296 °C (decomp.), is 30 g (67%).

Note. (1) Alternatively use 225 g of 'ammonium carbonate', 175 ml of water and 250 ml (4.5 mol) of concentrated ammonia solution. 'Ammonium carbonate' is a mixture of roughly equimolar amounts of ammonium hydrogen carbonate and ammonium carbamate ($NH_4HCO_3 \cdot NH_2CO_2NH_4$).

Cognate preparations. *Glycine* (*aminoacetic acid*). Use 47 g (0.5 mol) of chloroacetic acid (**CAUTION:** *the compound causes blistering if it is allowed to come into contact with the skin*) and 2 litres of concentrated ammonia solution. Allow the reaction to proceed for 2 days, concentrate to 60 ml and precipitate the crude glycine by adding 360 ml of methanol. This material contains ammonium chloride as the chief impurity; remove most of this by stirring the crystals with 150 ml of methanol and refiltering. Finally purify the glycine by dissolving it in 50 ml of hot water and adding 250 ml of methanol; the yield is 25 g (67%), m.p. *c.* 252–254 °C (decomp.).

DL-Valine (*2-amino-3-methylbutanoic acid*). Use 60 g (0.33 mol) of 2-bromo-3-methylbutanoic (Expt 5.164) and 400 ml (7.25 mol) of concentrated ammonia solution. Allow reaction to proceed at room temperature for 7 days. Concentrate the solution to 50 ml and filter the resulting thin paste. Dissolve the solid in 150 ml of hot water, decolourise with 1 g of charcoal, filter hot and dilute the filtrate with 150 ml of ethanol. Cool at 0 °C overnight, filter off the purified DL-valine and wash with 10 ml of cold ethanol. The yield is 12.5 g (32%); m.p. 280–282 °C (decomp.). A further 2 g may be isolated by concentrating the mother-liquor to about 25 ml and adding an equal volume of ethanol.

DL-Norvaline (*2-aminopentanoic acid*). Prepare as for valine, using 2-bromo-pentanoic acid (Expt 5.164); m.p. *c.* 300 °C (decomp.).

DL-Norleucine (*2-aminohexanoic acid*). Use 65 g (0.33 mol) of 2-bromo-hexanoic acid (Expt 5.164) and 400 ml of concentrated ammonia. Ensure that the bung is securely wired to the reaction bottle and allow the latter to stand in a warm place (50–55 °C) for 30 hours. Filter the amino acid at the pump and keep the filtrate (A) separately. Wash the amino acid well with methanol to remove the ammonium bromide present. Concentrate the filtrate (A) almost to dryness and add 150 ml of methanol. A second crop of amino acid contaminated with ammonium bromide is thus obtained; wash it with methanol and recrystallise from hot water, thus affording a further 6 g of pure DL-norleucine. The total yield is 28 g (65%); the decomposition point is about 325 °C.

DL-Isoleucine (*2-amino-3-methylpentanoic acid*). Allow 65 g (0.33 mol) of 2-bromo-3-methylpentanoic acid (Expt 5.166) to react with 400 ml of concentrated ammonia solution as for valine. Concentrate the resulting solution to about 130 ml, filter off a first crop of crude product and wash with 20 ml of ethanol. Further concentrate the aqueous filtrate to about 60 ml to obtain a second crop of crude product, and wash it with 10 ml of water followed by

10 ml of ethanol. Dissolve the combined product (28 g) in 400 ml of hot water, decolourise with charcoal and add 200 ml of rectified spirit. Cool well in ice, and filter off the pure DL-isoleucine; yield 16.5 g (38%), m.p. 278–280 °C (decomp.). A further 5 g may be recovered by concentrating the recrystallisation mother-liquor to 40 ml and diluting with an equal volume of ethanol.

Experiment 5.181   AMINOACETONITRILE AND GLYCINE (*Amino-acetic acid*)

$$2\text{H·CHO} + \overset{\ominus}{\text{C}}\text{N} + \overset{\oplus}{\text{NH}}_4 \xrightarrow{-\text{H}_2\text{O}} \text{CH}_2\text{=N·CH}_2\text{·CN} \xrightarrow[-\text{CH}_2(\text{OEt})_2]{2\text{EtOH}; \text{H}^{\oplus}}$$

$$\overset{\oplus}{\text{NH}}_3\text{·CH}_2\text{·CN} \xrightarrow{\ominus\text{OMe}} \text{NH}_2\text{·CH}_2\text{·CN}$$

$$(\overset{\oplus}{\text{NH}}_3\text{·CH}_2\text{·CN})_2\text{SO}_4^{2\ominus} \xrightarrow[\text{Ba(OH)}_2]{\text{H}_2\text{O}} (\text{NH}_2\text{·CH}_2\text{·CO}_2)_2\text{Ba} \xrightarrow{\text{H}_2\text{SO}_4} 2\text{NH}_2\text{·CH}_2\text{·CO}_2\text{H}$$

N-**Methyleneaminoacetonitrile.** Place 160 g (3 mol) of ammonium chloride in a 2-litre flange or three-necked flask surrounded by a large bath containing a cooling mixture of ice and salt (*fume cupboard*). Add 450 ml (6 mol) of filtered 40 per cent w/v formaldehyde solution and stir the mixture with an efficient mechanical stirrer. When the temperature has reached 0 °C, begin the drop-wise addition of a solution of 150 g (3 mol) of sodium cyanide (98% pure) (**CAUTION**) in 250 ml of water; the addition should take about 5 hours and the temperature throughout should be kept between 0 and 5 °C. When half of the cyanide solution has been added and all of the ammonium chloride has dissolved begin the simultaneous gradual addition of 100 ml of glacial acetic acid, and adjust the rate so that addition is complete by the end of the remaining 2.5 hours. Stir the mixture for a further period of 1 hour, and then filter off the product and wash it with a little cold water. Transfer the filter-cake to a beaker, stir thoroughly with 500 ml of water to remove soluble salts, filter, wash with a little more water and dry in a vacuum desiccator. The yield of *N*-methyleneaminoacetonitrile (1) is 120 g (59%), m.p. 127–128 °C.

Aminoacetonitrile sulphate. Cautiously add 85 ml of concentrated sulphuric acid to 400 ml of rectified spirit in a 1-litre conical flask and adjust the temperature to 50 °C. Add rapidly 102 g (1.5 mol) of dried powdered *N*-methyleneaminoacetonitrile, shake vigorously until the solid has dissolved and continue shaking while the product crystallises. Cool in an ice bath for 4 hours, filter and wash the product with a little cold rectified spirit. Dissolve the crude product in a minimum of water, filter off traces of insoluble matter and run the solution with stirring into 400 ml of cold rectified spirit. Cool well, filter off the purified aminonitrile salt and wash it with a little cold rectified spirit. The yield is 105 g (67%), m.p. 164 °C (decomp.).

Hydrolysis to glycine. Boil a suspension of 79 g (0.25 mol) of barium hydroxide octahydrate in 175 ml of water in a 500-ml round-bottomed flask and add in portions 21 g (0.1 mol) of aminoacetonitrile sulphate. Fit a reflux condenser and continue boiling until no more ammonia is evolved (about 3 hours). Transfer the suspension to a beaker, add 50 per cent v/v aqueous sulphuric acid (about 20 ml) until precipitation of the barium is complete and the solution is slightly acidic, and digest on a steam bath. Filter the suspension through a medium-speed filter paper (e.g. Whatman No. 30) using gentle suction, and adjust the filtrate to neutrality by carefully adding saturated barium hydroxide solution. Digest further and decant the supernatant solution through a similar filter paper (this is best done under gravity), suspend the precipitate in a little hot water and add this to the filter. Concentrate the filtrate under reduced pressure using a rotary evaporator until a thick suspension is obtained, and complete the precipitation of the glycine by adding 100 ml of methanol. Cool well and filter off the crude glycine (about 12 g). Dissolve the product in 25 ml of hot water, add gradually 125 ml of methanol, cool, filter and dry the crystals in an oven at 50 °C. The yield of glycine, m.p. *c.* 250 °C (decomp.), is 11.2 g (75%).

Aminoacetonitrile. Stir a suspension of 88 g (0.4 mol) of the nitrile sulphate in 100 ml of dry methanol in a 500-ml three-necked flask cooled in crushed ice. Add a few crystals of phenolphthalein as an indicator, pass a slow stream of nitrogen through the flask and run in during 1 hour a solution of sodium methoxide in methanol prepared from 17 g (0.75 mol) of sodium and 350 ml of dry methanol; the suspension should at no time be allowed to become permanently alkaline to phenolphthalein. Filter, and remove the methanol and distil the residue under reduced pressure under nitrogen. The yield of aminoacetonitrile, b.p. 73.5 °C/15 mmHg, is 35 g (83% based on sodium). Store the product at 0 °C under nitrogen.

Note. (1) The product in fact has a trimeric structure.

Cognate preparation. *DL-2-Aminophenylacetic acid* (α-phenylglycine). Dissolve 10 g (0.2 mol) of sodium cyanide in 40 ml of water, add 11 g (0.2 mol) of ammonium chloride and shake until dissolved. Add a solution of 21 g (0.2 mol) of redistilled benzaldehyde in 40 ml of methanol and shake vigorously. The mixture soon becomes warm; allow it to stand at ambient temperature for 2 hours, shaking occasionally. Then add 100 ml of water and shake well, and extract out the oily aminonitrile which separates using two portions (60 ml and 40 ml) of toluene. Combine the toluene layers, wash twice with water and extract with two 60 ml portions of 5 M hydrochloric acid. Boil the acid extract under reflux for 22 hours (**CARE**: *hydrogen cyanide is evolved*). Cool and filter through a small plug of cotton wool to remove a little tarry matter. Basify the solution by adding about 40 ml of concentrated ammonia solution (*d* 0.88) with stirring and cooling, collect the resulting precipitate by filtration and wash it with 100 ml of cold water and then with 15 ml of warm ethanol. The crude material is almost colourless; the yield is 11.5 g after drying in an oven at 50 °C. To purify the phenylglycine, dissolve it in 80 ml of 1 M sodium hydroxide, add 50 ml of ethanol and clarify the solution by adding a little decolourising charcoal, warming and filtering. Heat the filtrate almost to boiling and neutralise by slowly adding with stirring 16 ml of 5 M hydrochloric acid. Filter off the purified DL-phenylglycine, wash it with 10 ml of ethanol followed by 20 ml of water and dry at 50 °C; the yield of colourless glistening plates is 9 g (30%). The compound has no m.p.; when placed in a rapidly heated melting point apparatus when the temperature reaches 275 °C, it sublimes between 300 and 310 °C, depending upon the rate of heating.

Experiment 5.182 DL-**PHENYLALANINE** (*2-Amino-3-phenylpropanoic acid*)

In a 1-litre three-necked flask, fitted with a reflux condenser, a sealed stirrer unit and dropping funnel, place 25 g (0.1 mol) of 4-benzylidene-2-phenyloxazol-5-one (Expt 8.21) 20 g (0.65 mol) of purified red phosphorus (Section 4.2.59, p. 458) and 135 g (125 ml, 1.32 mol) of acetic anhydride. Add with stirring over a period of 1 hour 125 ml of hydriodic acid (*d* 1.56; 50%). Reflux the mixture for 3 hours, cool and filter with suction; wash the unreacted phosphorus on the filter with two 5 ml portions of glacial acetic acid. Evaporate the filtrate and washings on a water bath (rotary evaporator) and collect the distillate (which may be used for another reduction) in a flask cooled in ice. Add 100 ml of water to the dry residue and repeat the evaporation to dryness. Shake the residue in the flask with 150 ml of water and 150 ml of ether until solution is complete; separate the aqueous layer and extract it with three 75 ml portions of ether. Discard the ether extracts. Introduce 2–3 g of decolourising carbon and a trace of sodium sulphite into the aqueous phase, heat on a water bath until the dissolved ether has been removed, filter, heat the filtrate to boiling and neutralise to Congo red with ammonia solution (*d* 0.88; about 25 ml are required). When cold, filter the colourless DL-phenylalanine at the pump and wash with two 30-ml portions of cold water and finally with a little cold ethanol. The yield is 11 g (67%), m.p. 284–288 °C (decomp., rapid heating).

Experiment 5.183 DL-**SERINE** (*2-Amino-3-hydroxypropanoic acid*)

Prepare a suspension of 43.5 g (0.2 mol) of diethyl acetamidomalonate in 25 ml of water and add in one portion 17 g of neutral (1) 37–41 per cent w/v aqueous formaldehyde (0.21–0.23 mol). Add 0.5 ml of 1 M sodium hydroxide solution as the catalyst, shake vigorously and leave at room temperature for 2 hours. (Most of the solid goes into solution within 30 minutes; the mixture may require gentle warming on a steam bath to complete the dissolution of the solid.) Then add a solution of 18 g (0.45 mol) of sodium hydroxide in 350 ml of water and leave at room temperature overnight. Acidify the solution by adding 40 g (38 ml, 0.67 mol) of glacial acetic acid and heat almost to boiling, when brisk decarboxylation sets in. Continue to heat on a boiling water bath under reflux for 1 hour to complete the decarboxylation and then evaporate the solution to a syrup under reduced pressure on a rotary evaporator. Dissolve the syrup in 120 ml of concentrated hydrochloric acid, boil under reflux (*fume cupboard*) for 1 hour and evaporate to dryness under reduced pressure. Extract the dry residue with 200 ml of boiling absolute ethanol, filter off the sodium chloride and extract the latter with a further 100-ml portion of hot ethanol. Evaporate the ethanol extracts on the rotary evaporator and boil the residue under reflux with 100 ml of concentrated hydrochloric acid for 1 hour (*fume cupboard*). Evaporate again to a syrup, and dissolve the latter in water and re-evaporate twice more to remove most of the excess hydrochloric acid. Finally dissolve the residual gum in about 200 ml of distilled water and pass the solution through a 50 × 2.75 cm column of a weakly basic anion exchange resin (e.g. Amberite IR 45, $^{\ominus}$OH form, about 300 ml of moist granules). Continue to elute the column with distilled water (about 1.5 litres) until the eluate gives no purple coloration when a portion is tested by boiling with a few mg of ninhydrin (2). Combine all the ninhydrin-positive eluates, which should be free from chloride ion, and evaporate to dryness under reduced pressure (rotary evaporator). Dissolve the straw-coloured residue in 150 ml of hot water, boil with a little decolourising charcoal, filter and add to the filtrate 750 ml of hot ethanol. Cool in ice, filter off the purified DL-serine which crystallises, wash it with a little cold ethanol and dry in an oven at 50 °C. The yield is 12 g (57%), m.p. c. 235 °C (decomp.). The product thus obtained is homogeneous (R_F 0.57) on t.l.c. (silica gel, 15 cm run; butan-1-ol–formic acid–water, 6:3:1) (3).

Notes. (1) If necessary, the formaldehyde solution should be neutralised (narrow range pH paper) by the careful dropwise addition of 1 M sodium hydroxide solution. (2) After use, regenerate the resin in the following way. Firstly exhaust the column by passing 0.25 M hydrochloric acid through it until the pH of the eluate is about 2. Wash the column with 2 or 3 bed-volumes if distilled water, and then regenerate the resin by passing through it 0.25 M sodium hydroxide solution until the eluate is strongly alkaline. Finally wash the column thoroughly with much distilled water until the pH of the eluate is within the range 5.5–6.5 (narrow range indicator paper). (3) Using this solvent system, serine is not completely separable from glycine (R_F 0.58) which is a possible contaminant. The latter may be distinguished, however, by the characteristic brownish-pink spot which it gives on spraying with ninhydrin; that of serine is purple.

Experiment 5.184   DL-TRYPTOPHAN [2-Amino-3-(3-indolyl)propanoic acid]

$$Me \cdot CO \cdot NH \cdot CH(CO_2 Et)_2 + CH_2{=}CH \cdot CHO \longrightarrow$$

Phenylhydrazone of 4-acetamido-4,4-diethoxycarbonylbutanal. CAUTION: *Carry out all operations in an efficient fume cupboard.* Place 43.5 g (0.2 mol) of diethyl acetamidomalonate and 70 ml of benzene (CAUTION) in a 250-ml three-necked flask fitted with a stirrer and dropping funnel and surrounded by a bath of water at room temperature. Stir mechanically, add about 0.5 ml of a concentrated solution of sodium ethoxide in ethanol and then add slowly from the dropping funnel a solution of 12 g (14 ml, 0.215 mol) of acrylaldehyde [acrolein – CAUTION: *highly toxic and irritant vapour* (1)] in 14 ml of benzene; adjust the rate of addition so that the temperature of the reaction mixture does not exceed 35 °C. When the addition is complete, stir for 2 hours more and filter off any traces of insoluble material. Add 5 ml of glacial acetic acid and 24 g (22 ml, 0.22 mol) of redistilled phenylhydrazine [CAUTION: (2)], warm to 50 °C and leave the resulting orange solution at room temperature for 2 days. Collect the crystalline phenylhydrazone by filtration and wash it thoroughly by trituration with two 40 ml portions of benzene. The yield of off-white crystals, m.p. 141 °C, is 50 g (69%). If the yield is low, warm the filtrate to 50 °C and set it aside for a further 2 days, when a further crop of the product may be obtained.

Diethyl (3-indolylmethyl)acetamidomalonate. Add 47 g (0.13 mol) of the phenylhydrazone to 300 ml of water containing 14 ml of concentrated sulphuric acid in a 500-ml two-necked flask fitted with a sealed stirrer unit and a reflux condenser. Boil the mixture under reflux with vigorous stirring for 4.5 hours; the suspended solid liquefies and then solidifies during this time. Cool, filter off the resulting product (in the form of hard nodules) and wash it thoroughly by grinding it with water and re-filtering. Recrystallise the product from 1:1 aqueous ethanol to obtain the purified malonate derivative; yield 32 g (71%). The product melts at 143 °C, re-solidifies and then melts at 159 °C.

DL-**Tryptophan.** Boil 31 g (0.09 mol) of the above product under reflux for 4 hours with a solution of 18 g (0.45 mol) of sodium hydroxide in 180 ml of water. Add a little decolourising charcoal, filter and cool the filtrate in an ice–salt bath. Acidify the filtrate by adding about 55 ml of concentrated hydrochloric acid slowly and with shaking, keeping the temperature below 20 °C. Cool the resulting suspension at 0 °C for 4 hours and then collect the crude (indolylmethyl)malonic acid, a pale buff solid, by filtration. Boil the crude product under reflux with 130 ml of water for 3 hours; decarboxylation ensues and some N-acetyltryptophan separates. Add a solution of 16 g (0.4 mol) of sodium hydroxide in 30 ml of water, continue to boil under reflux for 20 hours and then add about 1 g of decolourising charcoal and filter. Cool, acidify the filtrate by adding 24 g (23 ml, 0.4 mol) of glacial acetic acid and cool the mixture at 0 °C for 5 hours. Collect the crude tryptophan by filtration, dissolve it in a solution of 5 g of sodium hydroxide in 200 ml of water and warm to 70 °C. Dilute the solution with 100 ml of ethanol at 70 °C and decant it from a little gummy precipitate which separates. Acidify the hot solution with 7.5 ml of glacial acetic acid and allow to cool slowly. When crystallisation is complete, filter off the purified DL-tryptophan and wash it successively with ice-cold water (2 × 40 ml), ethanol (2 × 40 ml) and ether (2 × 40 ml). The yield of colourless plates is 15 g (82%), m.p. 283–284 °C (decomp.). The i.r. spectrum of the L-enantiomer is shown in Fig. 3.37.

Notes. (1) Acrolein is usually supplied in sealed amber-coloured ampoules stabilised by the addition of a little hydroquinone. The ampoule should be cooled thoroughly before being opened with great care. It has been recorded that opened samples of acrolein stored in screw-capped bottles may explode violently, presumably as the result of rapid exothermic polymerisation.

(2) Phenylhydrazine is highly poisonous and produces unpleasant burns in contact with the skin. Always wear disposable gloves. If any liquid does come in contact with the skin wash off immediately with 2 per cent acetic acid, then with soap and water.

Experiment 5.185 DL-PROLINE (DL-*Pyrrolidine-2-carboxylic acid*)

$$(EtO_2C)_2CH_2 + CH_2{=}CH{\cdot}CN \xrightarrow{EtO^{\ominus}} (EtO_2C)_2CH{\cdot}CH_2{\cdot}CH_2{\cdot}CN \xrightarrow{Ni/H_2}$$

3-Ethoxycarbonyl-2-piperidone. Dissolve 160 g (0.75 mol) of diethyl (2-cyanoethyl)malonate (Expt 5.161) in 600 ml of ethanol and hydrogenate in the presence of about 4–5 g of Raney nickel catalyst (Section 4.2.50, p. 450) at 80 °C and 75 atmospheres pressure of hydrogen in an autoclave (Section 2.17.1); uptake of hydrogen is complete in about 2 hours. Remove the catalyst by filtration and the solvent by distillation under reduced pressure (rotary evaporator) and pour the residue with stirring into 500 ml of light petroleum (b.p. 60–80 °C). Filter off the precipitated piperidone and allow it to dry in the air; yield 115 g (90%), m.p. 74 °C. A specimen crystallised from ethanol/light petroleum has m.p. 80 °C.

3-Chloro-3-ethoxycarbonyl-2-piperidone. Assemble in a fume cupboard a 500-ml three-necked flask fitted with a sealed stirrer unit, a dropping funnel and a reflux condenser protected by a calcium chloride guard-tube. Charge the flask with a solution of 111 g (0.65 mol) of 3-ethoxycarbonyl-2-piperidone in 175 ml of dry chloroform (**CAUTION**), and the dropping funnel with 90 g (54 ml, 0.67 mol) of redistilled sulphuryl chloride dissolved in 125 ml of dry chloroform; close the neck of the funnel with a calcium chloride guard-tube. Start the stirrer, and slowly run in the solution of sulphuryl chloride so that the reaction mixture refluxes gently. When the addition is complete, warm on a steam bath until hydrogen chloride evolution ceases. Remove the solvent under reduced pressure using a rotary evaporator, and cool the residue, scratching the sides of the flask to induce crystallisation. Dissolve the solid in 70 ml of hot ethyl acetate, add hot light petroleum (b.p. 80–100 °C) until the solution is slightly turbid, and cool while scratching the side of the vessel vigorously until the product crystallises (it helps to add a few crystals of the crude solid as a seed). Add 30 ml more of light petroleum, cool further and filter. The chloropiperidone is obtained as a somewhat sticky white solid, m.p. 64–68 °C; yield is 100 g (82%).

DL-Proline hydrochloride. Boil 103 g (0.5 mol) of the above chloropiperidone under reflux with 200 ml of concentrated hydrochloric acid (*fume cupboard*) for 5 hours during which time decarboxylation ensues. Boil the solution with about 2 g of decolourising charcoal (acid-washed grade), filter and evaporate under reduced pressure using a rotary evaporator; dissolve the residue in water and re-evaporate to assist the removal of most of the excess hydrochloric acid. Dissolve the resulting pale golden syrup in 80 ml of water, add a solution of 60 g of sodium hydroxide in 120 ml of water and leave at room temperature for 2 days. Acidify the solution to Congo red paper with concentrated hydrochloric acid (about 130 ml are needed) and evaporate to dryness under reduced pressure (rotary evaporator). Dry the resulting solid completely by leaving it overnight in a vacuum desiccator over phosphorus pentoxide and potassium hydroxide pellets, and then extract it with 200 ml of boiling absolute ethanol. Filter off the sodium chloride and re-extract it with 150 ml more of boiling ethanol. Evaporate the ethanol from the combined filtrates and boil the residue under reflux for 1 hour with 1.25 litres of 2 M hydrochloric acid to hydrolyse any proline ester formed in the extraction process. Evaporate to dryness, redissolve the residue in water and re-evaporate, and dry the resulting crude DL-proline hydrochloride in a vacuum desiccator over phosphorus pentoxide and potassium hydroxide; the yield is 70 g. Recrystallise the product from 175 ml of hot propan-2-ol, cool in ice, filter and wash the crystals with 20 ml of ice-cold propan-2-ol. The yield of the purified product, m.p. 148–150 °C, is 50 g (66%).

Conversion to DL-proline. Suspend 15 g (0.1 mol) of dry proline hydrochloride in 70 ml of dry chloroform, stir vigorously and run in dropwise 15 g (0.15 mol) of dried, redistilled triethylamine. Continue stirring for 1 hour, filter off the product and wash it with a little cold chloroform. The yield of DL-proline is 10 g (87%); a sample crystallised from absolute ethanol has m.p. 206–207 °C (with preliminary sintering).

Experiment 5.186 L-PROLINE (L-*Pyrrolidine-2-carboxylic acid*)

$$PhNH_2 \xrightarrow[Me \cdot CO_2H]{Cr^{3\oplus}, \ CNS^{\ominus}} PhNH_3^{\oplus}[Cr(CNS)_4(PhNH_2)_2]^{\ominus} \xrightarrow{NH_3}$$

$$NH_4^{\oplus}[Cr(CNS)_4(PhNH_2)_2]^{\ominus} \xrightarrow[hydrolysate]{gelatine}$$

Ammonium rhodanilate

$$\text{L-Proline rhodanilate} \xrightarrow{pyridine} \text{L-Proline + Pyridine rhodanilate}$$

Ammonium rhodanilate. Heat a mixture of 250 g (0.5 mol Cr$^{3\oplus}$) of hydrated chromium(III) potassium sulphate (chrome alum), 290 g (3.0 mol) of potassium thiocyanate and 250 ml of water in a 5-litre flask on a steam bath for 4 hours. Cool, add 235 g (230 ml, 2.53 mol) of redistilled aniline and heat in a water bath at 60 °C with stirring for 3 hours. Cool again, add while still stirring 3 litres of water containing 300 ml of glacial acetic acid and leave at 0 °C overnight. Filter off the purple precipitate with suction, wash it with water and suck as dry as possible. Extract the filter cake with 750 ml of methanol, filter and run into the filtrate, with stirring, 3 litres of water. Cool at 0 °C for 1 hour and filter off the crystalline purple mass of aniline rhodanilate. Dissolve

759

the damp product in 400 ml of methanol, add 200 ml of concentrated ammonia solution (d 0.88), cool to 0 °C and then run in slowly, with stirring, 2 litres of water. Collect the precipitate by filtration, and again treat it with 250 ml of methanol, 125 ml of ammonia solution and 1250 ml of water in the above manner. Dry the resulting ammonium rhodanilate sesquihydrate in an oven at 50 °C; the yield is 130 g (50.5%).

L-**Proline.** Place 150 g of good quality sheet gelatin (cut into conveniently sized pieces) in a 1-litre flask and add 450 ml of concentrated hydrochloric acid. Boil gently under reflux for 8 hours (*fume cupboard*), or boil under reflux for about 3 hours and leave on a steam bath overnight; complete hydrolysis is indicated by a negative biuret reaction (1). Concentrate the hydrolysate to a syrup by distillation under reduced pressure (rotary evaporator), and remove excess hydrochloric acid by dissolving the syrup in water and evaporating twice more. Dissolve the residual syrup finally in 500 ml of water, boil briefly with 3 g of decolourising charcoal, filter, cool and dilute with water to 1200 ml. Add this solution slowly with stirring to a filtered solution of 125 g of ammonium rhodanilate in 750 ml of methanol and keep at 0 °C for 2 hours to allow the complete separation of proline rhodanilate. Filter off the latter, wash it with water and suck as dry as possible. Dissolve the damp crude product in 400 ml of methanol, filter and add 800 ml of 0.5 M hydrochloric acid slowly with stirring. Cool at 0 °C for 2 hours, filter off the purified proline rhodanilate and wash it with 250 ml of cold water. Dry the product in an oven at 50 °C; the yield is about 100 g, m.p. c. 130 °C, with preliminary softening and blackening.

Suspend the purified salt in 850 ml of water in a stoppered bottle, add 25 ml of pure pyridine and shake the mixture for 4–5 hours. Remove the insoluble pyridine rhodanilate by filtration and wash it with 100 ml of cold water (2). Combine the pale pink filtrate and washings, and add glacial acetic acid dropwise until the formation of a small pink precipitate is complete. Filter, evaporate the almost colourless filtrate to dryness (rotary evaporator) and suspend the residue in absolute ethanol and re-evaporate twice. Dry the resulting faintly pink crude proline in a vacuum desiccator over silica gel; the yield is about 18 g. Recrystallise from the minimum volume of absolute ethanol to obtain 11 g (7.3% based on gelatin) of L-proline, m.p. 218–219 °C (decomp.), $[\alpha]_D^{18}$ −85.6° (c 3.0 in H_2O). Check the purity of the product by t.l.c. on silica gel using the solvent system butan-1-ol–acetic acid–water, 4:1:1; R_F 0.26 (yellow spot with ninhydrin).

Notes. (1) Remove about 0.25 ml of the hydrolysate, cool it and basify it with 5 M sodium hydroxide solution. To a portion add a few drops of very dilute copper(II) sulphate solution, and note the absence of any colour change. As a control, prepare a specimen of biuret ($NH_2 \cdot CO \cdot NH \cdot CO \cdot NH_2$) by heating about 10 mg of urea just above its melting point for about 2 minutes. Add a little basified hydrolysate warm to dissolve, cool and add a trace of copper(II) sulphate. A deep pink colour superimposed upon the pale brown colour of the hydrolysate should be observed. If the hydrolysate gives a similar colour originally, it contains peptide material and hydrolysis should be continued until the biuret test is negative.

(2) About 100 g of pyridine rhodanilate is obtained. To convert this into the ammonium salt for re-use, suspend it in 175 ml of methanol and add 90 ml of concentrated ammonia solution. Stir at 0 °C for 30 minutes and then dilute gradually with 900 ml of water. Filter off the precipitated ammonium salt, wash it with water and repeat the

methanol–ammonia treatment once more. The final yield of dried ammonium rhodanilate is about 80 g.

Experiment 5.187 L-CYSTINE (*3,3'-Dithiodialanine*)

$$\text{Keratin (hair)} \xrightarrow[\text{H}_3\text{O}^\oplus]{\text{hydrolysis}} [\text{S}\cdot\text{CH}_2\cdot\text{CH(NH}_2)\cdot\text{CO}_2\text{H}]_2$$

Place 500 g of washed, dried human hair (1) in a 3-litre round-bottomed flask and add 1 litre of a 1:1 mixture of concentrated hydrochloric acid and formic acid. Fit a reflux condenser and boil the mixture gently under reflux (fume cupboard) until the biuret test [Expt 5.186, Note (1)] is negative (about 20 hours). Treat the hydrolysate with about 12 g of acid-washed decolourising charcoal, filter hot and concentrate the filtrate to a thick syrup under reduced pressure using a rotary evaporator. Dissolve the residue in 250 ml of water and add with shaking warm 50 per cent aqueous sodium acetate until the solution is no longer acid to Congo red. Leave the mixture at room temperature for 3 days, and then filter off the precipitate of crude cystine and wash it with 50 ml of warm water. Dissolve the product in 750 ml of 1M hydrochloric acid, treat the hot solution with 5 g of acid-washed decolourising charcoal and filter. If the filtrate is more than faintly yellow, repeat the treatment with a further 5 g of charcoal. Neutralise the filtrate to Congo red and with sodium acetate solution as previously and leave at room temperature for 5–6 hours (2). Filter off the colourless plates of purified L-cystine, wash with two 50 ml portions of hot water, then with ethanol and finally with ether. The yield is about 25 g (5%); m.p. 260–262 °C (decomp.); $[\alpha]_D^{20}$ $-216°$ (c 0.69 in 1 M HCl).

Notes. (1) Raw material can usually be obtained from barbers' shops. The washing procedure must remove effectively grease and natural oils. A warm aqueous solution of a good quality detergent is satisfactory; it is important that the washing medium should not be alkaline as this may markedly reduce the yield of cystine. Raw sheep's wool, adequately cleaned, may also be used as the starting material if available, but processed material (e.g. old woollen garments, etc.) may give poor yields.
(2) The crude product contains some of the aromatic amino acid tyrosine which is present in the original hydrolysate. A portion of this is removed by the charcoal treatment and by the hot-water washing, but the final recrystallised cystine may be contaminated with tyrosine if the suggested 5–6 hour period for crystallisation is greatly exceeded.

Experiment 5.188 L-PROLYLGLYCINE

$$\text{H}_2\text{N}\diagup\diagdown\text{CO}_2\text{H} \xrightarrow[\text{HCl}]{\text{EtOH}} \text{H}_3\overset{\oplus}{\text{N}}\diagup\diagdown\text{CO}_2\text{Et}\;\;\overset{\ominus}{\text{Cl}} \xrightarrow{\text{Et}_3\text{N}} \text{H}_2\text{N}\diagup\diagdown\text{CO}_2\text{Et} \qquad (1)$$

$$\xrightarrow[\text{(ii) H}^\oplus]{\text{(i) Ph}\cdot\text{CH}_2\text{O}\cdot\text{COCl}/\overset{\ominus}{\text{OH}}}$$

$$\xrightarrow[\text{Et}_3\text{N}]{\text{Cl}\cdot\text{CO}_2\text{Et}}$$

(2)

CAUTION: *Benzyl chloroformate and ethyl chloroformate are highly lachrymatory and should be handled with great care in a fume cupboard; disposable plastic gloves should be worn.*

Glycine ethyl ester hydrochloride. Esterify 15 g (0.2 mol) of glycine in 100 ml of absolute ethanol with the aid of about 10 g of dry hydrogen chloride in the manner described for the preparation of glycylglycine ethyl ester hydrochloride (Expt 8.38). Recrystallise the product from the minimum of absolute ethanol. The yield of glycine ethyl ester hydrochloride, m.p. 144 °C, is 18.5 g (66%).

Benzyloxycarbonyl-L-proline. Dissolve 4.6 g (0.04 mol) of L-proline (Expt 5.186) in 10 ml (0.04 mol) of 4 M sodium hydroxide solution, cool to 0 °C in an ice bath, and add 7.0 g (0.041 mol) of benzyl chloroformate and 10 ml (0.04 mol) of 4 M sodium hydroxide solution alternately and portionwise, with shaking and cooling, during 30 minutes. Allow the mixture to warm to room temperature with intermittent shaking during 1 hour, extract with 2 × 15-ml portions of ether, and acidify the aqueous phase (Congo red) with dilute (1:1) hydrochloric acid. Extract the liberated oil with 4 × 10-ml portions of ether, dry over anhydrous sodium sulphate, and remove the ether to obtain 8.8 g (88%) of benzyloxycarbonyl-L-proline which crystallises slowly. A sample recrystallised from ether–light petroleum (b.p. 40–60 °C) has m.p. 75 °C.

Benzyloxycarbonyl-L-prolyglycine. Dissolve 6.2 g (25 mmol) of benzyloxycarbonyl-L-proline in 15 ml of dry chloroform containing 2.5 g (3.4 ml, 25 mmol) of dried redistilled triethylamine and cool the mixture in ice while adding 2.7 g (2.4 ml, 25 mmol) of ethyl chloroformate steadily and with shaking. Set the mixture aside at 0 °C for 15 minutes and then stir into the semi-solid mass a suspension of 3.5 g (25 mmol) of glycine ester hydrochloride in 20 ml of dry chloroform containing a further 2.5 g of dry triethylamine. Carbon dioxide is evolved and the solids go into solution. Set the mixture aside at room temperature for 30 minutes, warm at 50 °C for 10 minutes, and then wash it successively with 15 ml of water, 10 ml of 1 M hydrochloric acid, 2 × 10-ml portions of 0.5 M sodium hydrogen carbonate solution and 10 ml of water. Dry the solution over anhydrous sodium sulphate and remove the chloroform under reduced pressure to obtain 8.3 g (99%) of benzyloxycarbonyl-L-prolylglycine ethyl ester as an oil. Dissolve the entire product in a mixture of 26 ml of 1 M sodium hydroxide solution and 10 ml of acetone and set it aside at room temperature for 1 hour. Concentrate the solution somewhat under reduced pressure to remove most of the acetone and then acidify (Congo red) with dilute (1:1) hydrochloric acid. Extract the precipitated oil with ethyl acetate, dry over anhydrous sodium sulphate and remove the solvent under reduced pressure. The resulting oil crystallises slowly on cooling and scratching under a small quantity of ethyl acetate. To recrystallise dissolve the product in the minimum of hot ethyl acetate, cool somewhat and add an equal volume of ether, and then seed and set aside overnight. The yield of benzyloxycarbonyl-L-prolyglycine, m.p. 122–123 °C, is 5.3 g (69%). A sample crystallised from water has m.p. 125 °C.

L-**Prolyglycine.** Dissolve 4.6 g (15 mmol) of the foregoing product in 75 ml of methanol and add 75 ml of water and 2 ml of glacial acetic acid. Place 0.4 g of palladium black (Section 4.2.54, p. 453) into a Drechsel bottle and add the above solution, taking care to rinse all traces of catalyst down below the surface of the liquid. Connect the Drechsel bottle inlet tube to a hydrogen cylinder and vent the exit tube to the external atmosphere or to a fume cupboard via a second Drechsel bottle charged with lime-water. Suspend the bottom of the inlet tube just above the surface of the liquid in the reaction bottle and pass in hydrogen to displace the air. Lower the Drechsel head to close the system and allow hydrogen to bubble through the reaction mixture at a rate sufficient to agitate the catalyst. Continue the passage of hydrogen for about 2 hours, shaking periodically. The onset of the reaction, accompanied by the liberation of carbon dioxide, is indicated by typical carbonate precipitation in the lime-water bubbler; continue to pass hydrogen until carbon dioxide evolution ceases (confirm using a fresh charge of lime-water). Then filter off the catalyst (**CARE** – pyrophoric when dry) and evaporate the solution to dryness under reduced pressure. Recrystallise the residue by dissolving it in the minimum of hot water, adding hot methanol and cooling. L-Prolylglycine is obtained as the monohydrate; yield 2.4 g (86%), m.p. 226–227 °C (decomp.), $[\alpha]_D^{17}$ −21.9° (c 3.9 in H_2O). Check the homogeneity of the product by t.l.c. on silica gel using the system butan-1-ol–ethanol–acetic acid–water, 9:3:2:4; R_F 0.14.

5.15 NITROALKANES

Primary ($R \cdot CH_2 \cdot NO_2$), and also secondary ($R_2CH \cdot NO_2$), nitroalkanes exist in tautomeric equilibrium with the corresponding nitronic acids, the two species being interconvertible via the mesomerically stabilised α-carbanion. These nitroalkanes are therefore soluble in alkali.

Tertiary nitroalkanes ($R_3C \cdot NO_2$) are neutral. The most important unsaturated analogues are the vinyl nitro compounds (e.g. $R \cdot CH = CH \cdot NO_2$). All these nitro compounds have in recent years become of great interest as versatile intermediates in organic synthesis.[196] Some of these aspects are noted in Chapter 1, p. 21, in Section 5.7.7, p. 599, and in Section 5.9.3, p. 635, and are further exemplified in the following discussion.

Laboratory routes to the nitroalkanes (and nitroalkenes) include the following procedures.

1. The displacement of a halogen by a nitrite ion in an alkyl halide (Expt 5.189).
2. The oxidation of oximes and of primary amines (Expts 5.190 and 5.191).
3. C-Alkylation of nitroalkanes and other homologation procedures (Expt 5.192).

SUMMARY OF RETROSYNTHETIC STRATEGIES

Functional group interconversion (FGI) (method 2)

Disconnection (methods 1 and 3)

$R^1 \cdot CH_2 \cdot NO_2 + R^2X$ $\qquad$ $R^1R^2CHX + NaNO_2$

$R \cdot CH_2 \cdot NO_2 +$

SPECTROSCOPIC FEATURES

The characteristic *i.r.* frequency for the nitro group is observed as twin absorptions which occur in the region of 1560 and 1388 cm^{-1} and are attributed to the asymmetric and symmetric stretching of the nitrogen–oxygen bonds respectively. The nitro group, being strongly electron withdrawing, causes the α-protons in the alkyl group to be deshielded, and hence the signals occur in the δ 4.4 region of the *p.m.r* spectrum (see Appendix 3). The structure of the alkyl group may be inferred from the spin–spin coupling patterns (cf. p. 341). Further illustrative and descriptive spectroscopic interpretations are given in some of the preparations below.

5.15.1 THE DISPLACEMENT OF A HALOGEN BY A NITRITE ION IN AN ALKYL HALIDE

Classically, primary nitroalkanes may be prepared by heating the alkyl bromide (or iodide) but not the chloride, with silver nitrite, frequently in anhydrous ether. The method is not satisfactory with secondary or tertiary halides, when the major products are the alkyl nitrites.

$$O=\ddot{N}\rightarrow O \quad R\!-\!Br \longrightarrow R\cdot NO_2 + Br^{\ominus}$$

This method is clearly expensive, and nowadays the cheaper sodium nitrite is employed with the alkyl halide in dimethyl sulphoxide or dimethylformamide as solvent.[197] Although the yields are a little lower than in the silver nitrite method, a further feature is that secondary halides may be converted into secondary nitroalkanes, although even this modification fails with tertiary halides. The reaction is illustrated by the preparation of 2-nitrooctane (Expt 5.189).

This method offers therefore a means of preparation of primary and secondary nitroalkanes which then provides a facile route to primary amines (by reduction, Section 5.16.2, p. 774), aldehydes and ketones (the Nef reaction, Sections 5.7.7, p. 599, and 5.8.7, p. 623), and to the rapidly developing strategy of carbon–carbon bond construction by way of the mesomerically stabilised anions (Section 5.15.3, p. 768).

Experiment 5.189   2-NITROOCTANE[197]

$$C_6H_{13}\cdot CHI\cdot Me + NaNO_2 \xrightarrow{\text{DMSO}} C_6H_{13}\cdot CH(NO_2)\cdot Me + NaI$$

CAUTION: *Nitroalkanes should always be distilled in apparatus sited behind a safety screen.*

2-Iodooctane (71.2 g, 0.30 ml) is poured into a stirred solution of 225 ml of dimethyl sulphoxide and 36 g of sodium nitrite (0.52 mol) contained in a 500-ml flask immersed in a water bath held at room temperature. Stirring is continued for 4 hours. The reaction mixture is poured into 600 ml of ice-water layered over with light petroleum (b.p. 35–37 °C). After separation the aqueous phase is further extracted with four 100-ml portions of light petroleum. The combined organic extracts are washed with four 100–ml portions of water and dried over magnesium sulphate. The drying agent is removed by filtration and the light petroleum solution is distilled through a small fractionating column at atmospheric pressure, after which the residue is distilled under reduced pressure. The following fractions are obtained: (i) 14.0 g (30%) of 2-octyl nitrite, b.p. 32 °C/2 mmHg, n_D^{21} 1.4089; (ii) 3.9 g, b.p. 53–56 °C/1 mmHg, n_D^{21} 1.4111–1.4382; and (iii) 27 g (58%), b.p. 61 °C/1 mmHg of 2-nitrooctane, n_D^{21} 1.4281.

Cognate preparation. *Nitrocyclopentane.* Use cyclopentyl bromide (22.0 g, 0.15 mol), sodium nitrite (18 g, 0.26 mol) and 100 ml of dimethyl sulphoxide; allow the reaction mixture to stir at 15 °C for 3 hours. The product (9.9 g, 58%), b.p. 62 °C/8 mmHg, n_D^{21} 1.4538, was isolated as above.

5.15.2 THE OXIDATION OF OXIMES AND AMINES

The oxidation of oximes offers an attractively simple route to nitroalkanes from carbonyl compounds. The most effective reagent is pertrifluoroacetic acid in acetonitrile in the presence of sodium hydrogen carbonate as a buffer. Yields are improved by the addition of small quantities of urea to remove oxides of nitrogen. The reaction is illustrated by the conversion of dipropyl ketoxime into 4-nitroheptane (Expt 5.190).

$$R_2C=N\diagdown_{OH} \xrightarrow{[O]} R_2C=N\diagup^{O}_{\diagdown OH} \xrightarrow{-H^{\oplus},\,+H^{\oplus}} R_2CH\cdot N\diagdown^{O}_{\diagdown O}$$

Since many aliphatic primary amines are usually prepared by methods not involving the reduction of a nitroalkane (cf. aromatic amines), their oxidation may provide a useful route to this class of compound. For amines which lead to tertiary nitroalkanes, potassium permanganate appears to be the most satisfactory reagent[198] (the preparation of 2-nitro-2-methylpropane, Expt 5.191). In the case of amines which lead to primary or secondary nitroalkanes the oxidant of choice is m-chloroperbenzoic acid, and Expt 5.191 includes a general procedure for this reaction.[199]

Experiment 5.190   4-NITROHEPTANE

$$(CF_3\cdot CO)_2O + H_2O_2 \longrightarrow CF_3\cdot CO_3H + CF_3\cdot CO_2H$$

$$Pr_2C=O \xrightarrow{NH_2OH} Pr_2C=NOH \xrightarrow{[O]} Pr_2CH\cdot NO_2$$

CAUTION: *The reactions involving hydrogen peroxide and pertrifluoroacetic acid should be carried out in a fume cupboard behind a safety screen. Adequate precautions should be observed in handling the hydrogen peroxide solution (1).*

Dipropyl ketoxime (*heptan-4-one oxime*). Heat a mixture of 20 g (0.18 mol) of heptan-4-one, 17 g (0.25 mol) of hydroxylamine hydrochloride, 19.6 g (20 ml, 0.25 mol) of pyridine and 150 ml of ethanol under reflux for 1 hour in a 500-ml round-bottomed flask. Rearrange the condenser for downward distillation and remove the ethanol on the water bath. Allow the residue in the flask to cool and add 150 ml of water. Extract the oxime with three 50 ml portions of ether. Wash the combined extracts with water and dry over magnesium sulphate. Remove the ether on the rotary evaporator and distil the residue at atmospheric pressure. Collect the fraction having b.p. 192–195 °C; the yield is 18.2 g (80%). The oxime shows infrared absorptions at 3300 cm^{-1} (O—H stretch) and 1655 cm^{-1} (C=N stretch).

4-Nitroheptane. Prepare a solution of pertrifluoroacetic acid in acetonitrile as follows. Place 50 ml of acetonitrile in a two-necked, 250-ml round-bottomed flask fitted with a dropping funnel and a reflux condenser. Insert a plastic-covered magnetic stirrer follower bar and cool the flask in an ice bath sited on the stirrer unit. To the cooled and stirred solution add 5.8 ml (0.2 mol) of 85 per cent hydrogen peroxide (1) and then 39.0 ml (58.1 g, 0.24 mol) of trifluoroacetic anhydride. Stir the solution for 5 minutes and then allow to warm to room temperature. In a three-necked, 500-ml round-bottomed flask fitted with a sealed stirrer unit, dropping funnel and reflux condenser place 200 ml of acetonitrile, 47 g (0.56 mol) of sodium hydrogen carbonate, 2 g of urea and 12.9 g (0.1 mol) of dipropyl ketoxime. Heat the stirred suspension under reflux on the water bath and add dropwise over 90 minutes the prepared solution of pertrifluoroacetic acid. When the addition is complete heat the mixture under reflux for 1 hour. Pour the cooled reaction mixture into 600 ml of cold water and extract with four 100 ml portions of dichloromethane (note: the organic layer is the upper layer in the first extraction, but subsequently it is the lower layer). Wash the combined extracts with three 100 ml portions of saturated sodium hydrogen carbonate solution and dry over magnesium sulphate. Remove the solvent on the rotary evaporator and distil the residue under reduced pressure through a short fractionating column packed with glass helices. *Site the apparatus behind a safety screen.* 4-Nitroheptane distils at 66–69 °C/2 mmHg; the yield is 8.4 g (58%). A small amount of dipropyl ketone is obtained as a forerun. The infrared spectrum of the nitroalkane shows absorption at 1555 and 1385 cm^{-1} attributable to the antisymmetric and symmetric stretching of the nitro group. The purity of the product may be investigated by g.l.c. using a 10 per cent Silicone oil on Chromosorb W column held at 100 °C; the nitroalkane has a slightly shorter retention time than the oxime; the ketone is rapidly eluted from the column.

Note. (1) Considerable care should be exercised when carrying out reactions with 85 per cent w/v hydrogen peroxide (see also Section 4.2.41, p. 439). Rubber gloves and a face mask should be worn and reactions carried out behind a safety screen. Plenty of water should be at hand to wash away any spillages; fire may result if the peroxide is spilled on to combustible material. Any spillage on the skin similarly should be washed with plenty of water. Care should be taken to avoid the formation of potentially explosive emulsions with organic materials. Spirit thermometers and not mercury thermometers should be used; grease on taps and joints should be kept to the absolute minimum; safety pipettes must be used for pipetting. Disposal of high test perioxide solution may be effected by diluting with a large excess of water. The peroxide used in these experiments was supplied gratis by Interox Chemicals Ltd.

Experiment 5.191 2-NITRO-2-METHYLPROPANE[198]

$$\text{Me}_3\text{C}\cdot\text{NH}_2 \xrightarrow{\text{KMnO}_4} \text{Me}_3\text{C}\cdot\text{NO}_2$$

CAUTION: *All operations should be conducted behind a safety screen.*

In the course of 15 minutes t-butylamine (1,1-dimethylethylamine, 100 g) is added to a stirred solution of potassium permanganate (650 g) in 3 litres of water; the temperature rises to 45 °C. The stirring is continued for 8 hours without external heating, and the solution is then held at 55 ± 5 °C for a further 8 hours. The product is isolated by steam distillation and washed with dilute hydrochloric acid, then with water and dried. Distillation gives 117 g (83%) of the tertiary nitro compound, b.p. 127–128 °C, n_D^{28} 1.3980, m.p. 25–26 °C.

Cognate preparation. *General procedure for the oxidation of primary and secondary amines using* **m-*chloroperbenzoic acid.***[199] m-Chloroperbenzoic acid (4.1 g, 0.020 mol, 85% pure) is dissolved in 30 ml of 1,2-dichoroethane in a three-necked flask equipped with a condenser and a pressure-equalising dropping funnel. The amine (0.0050 mol) in 3–5 ml of solvent is added dropwise to the refluxing peracid solution. Refluxing temperature 83 °C, time 3 hours. After the addition the reaction mixture is cooled, filtered and washed with three 50-ml portions of 1 M sodium hydroxide solution and dried over magnesium sulphate. Removal of solvent gives a product which is distilled or recrystallised as appropriate. The following amines have been oxidised by this procedure: cyclohexylamine, 2-aminobutane, hexylamine, propylamine and 2-phenyl-1-aminoethane.

5.15.3 C-ALKYLATION OF NITROALKANES AND OTHER HOMOLOGATION PROCEDURES

Primary nitroalkanes when treated with two equivalents of butyllithium, in a mixed tetrahydrofuran–hexamethylphosphoramide solvent system held at $-78\,°C$, may be doubly deprotonated to give the dilithio derivative (2) via the monolithio derivative (1).

$$R \cdot CH_2 NO_2 \xrightarrow{BuLi} \left[\begin{array}{c} R \\ \diagdown \\ H \end{array} C\!=\!N \begin{array}{c} OLi \\ \diagup \\ O \end{array} \right] \xrightarrow{BuLi} \left[\begin{array}{c} R \\ \diagdown \\ Li \end{array} C\!=\!N \begin{array}{c} OLi \\ \diagup \\ O \end{array} \right]$$

$$(1) \qquad\qquad\qquad (2)$$

The dilithio derivative (2), in contrast to (1), may be C-alkylated[200] or C-acylated[201] to give homologous products.

An alternative and convenient α-alkylation reaction uses a 1-alkyl-2,4,6-triphenylpyridinium tetrafluoroborate with the monosodio derivative of a primary or secondary nitroalkane.[202]

$$(3) \qquad\qquad\qquad (4)$$

The reaction is thought to proceed by a radical chain mechanism ($S_{RN}1$), probably first involving a charge transfer complex between cationic species (3) and the anionic species (4). The illustrative example is the synthesis of 2-phenylnitroethane from nitromethane and 1-benzyl-2,4,6-triphenylpyridinium tetrafluoroborate (Expt 5.192); this latter compound is prepared from benzylamine and 2,4,6-triphenylpyrylium tetrafluoroborate (Expt 8.31).

Other major homologation processes are the nitroaldol reaction, arising from the reaction between a nitronate anion and a carbonyl compound, and a Michael addition reaction.

$$R^1 \cdot CH_2 \cdot NO_2 + R^2 \cdot CHO \longrightarrow R^1 \cdot \underset{\underset{NO_2}{|}}{CH} \cdot CH(OH) \cdot R^2 \xrightarrow{-H_2O} R^1 \cdot \underset{\underset{NO_2}{|}}{C}\!=\!CH \cdot R^2$$

$$R^1 \cdot CH_2 \cdot NO_2 + CH_2\!=\!CH \cdot CO_2 Me \longrightarrow R^1 \cdot \underset{\underset{NO_2}{|}}{CH} \cdot CH_2 \cdot CH_2 \cdot CO_2 Me$$

Both reaction processes are base catalysed, and the use of alumina (Brockmann activity I) at room temperature is a particularly mild method leading to good yields of products.[203]

The nitroaldol may be dehydrated to a nitroalkene; in the case of nitroaldols arising from aromatic aldehydes this dehydration reaction occurs spontaneously as is illustrated in the preparation of ω-nitrostyrene (Expt 6.136). The nitroalkenes are important dienophiles in the Diels–Alder reaction (Section 7.6, p. 1117). An example of the Michael addition reaction is illustrated in the synthesis of 1,4-dicarbonyl compounds (Section 5.9.3, p. 635) and γ-keto carboxylic acids (Section 5.14.3, p. 738); these examples exemplify the value of nitroalkanes as acyl anion equivalents.

Experiment 5.192   2-PHENYLNITROETHANE[202]

$$Ph \cdot CH_2 \cdot CH_2 \cdot NO_2 \; +$$

Sodium hydride (0.72 g, 0.03 mol) is dissolved in ethanol (30 ml). Nitromethane (1.83 g, 0.030 mol) is added with stirring, followed by 1-benzyl-2,4,6-triphenylpyridinium tetrafluoroborate (Expt 8.31) (4.85 g, 0.01 mol). The solution is refluxed with stirring for 1 hour: the reaction is followed by t.l.c. (silica gel; chloroform). When no further pyridinium compound remains, the solution is cooled; the resulting crystalline by-product 2,4,6-triphenylpyridine is filtered off. The filtrate is added to water (50 ml), extracted with ether (3 × 25 ml) and the extract dried over magnesium sulphate. Dry hydrogen chloride is passed into the ethereal extract to remove residual 2,4,6-triphenylpyridine, and the solution is evaporated under reduced pressure (25 °C/ 15 mmHg). The crude product is distilled in vacuo (15 mmHg) to yield 2-phenylnitroethane, b.p. 120–130 °C/15 mmHg, yield 78 per cent, pure by g.l.c. (Carbowax 220M, 200 °C); p.m.r. (CDCl$_3$, TMS) δ 3.45 (t, 2H), 4.75 (t, 2H), and 7.50 (s, 5H).

5.16 ALIPHATIC AMINES

Amines are classified as primary (RNH_2), secondary (R_2NH), or tertiary (R_3N). The alkyl groups in secondary and tertiary amines can of course be the same or different. These latter classes of amines are usually prepared by alkylation at a nitrogen atom, as is also the case with quaternary ammonium salts ($R_4\overset{\oplus}{N}\}\overset{\ominus}{X}$). Quaternary ammonium salts having four different alkyl groups are chiral; tertiary amines with three different groups, although chiral, undergo rapid pyrimidal inversion and hence autoracemisation, and enantiomeric forms are not isolable. Chirality in amines may arise from the presence of a chiral carbon atom in an alkyl chain (e.g. (+)-Ph·CH(NH$_2$)·Me).

Imines (e.g. R·CH$_2$·CH=N·R or R·CH$_2$·C(R)=N·R) are derived from the reaction of primary amines with aldehydes or ketones respectively, and may exist in tautomeric equilibrium with the corresponding enamine form, but in which the imine form predominates.

769

$$R \cdot CH_2 \cdot CHO + H_2N \cdot R \xrightarrow{-H_2O} R \cdot CH_2 \cdot CH = N \cdot R \rightleftharpoons R \cdot CH = CH - NH \cdot R$$

$$R \cdot CH_2 \cdot CO \cdot R + H_2N \cdot R \xrightarrow{-H_2O} R \cdot CH_2 \cdot C(R) = N \cdot R \rightleftharpoons R \cdot CH = C(R) - NH \cdot R$$

imine   enamine

The reaction of carbonyl compounds with secondary amines gives the stable enamine form.

$$R \cdot CH_2 \cdot CHO + HNR_2 \longrightarrow [R \cdot CH_2 \cdot CH(OH) \cdot NR_2] \xrightarrow{-H_2O} R \cdot CH = CH - NR_2$$

$$R \cdot CH_2 \cdot CO \cdot R + HNR_2 \longrightarrow [R \cdot CH_2 \cdot CR(OH) \cdot NR_2] \xrightarrow{-H_2O} R \cdot CH = C(R) - NR_2$$

The most important general methods for the preparation of aliphatic amines are as follows.

1. The reduction of alkyl azides, alkyl cyanides and amides (Expts 5.193 to 5.195).
2. The reduction of nitro compounds and oximes (Expt 5.196).
3. The reductive alkylation of ammonia or amines (Expt 5.197).
4. The alkylation of ammonia and its derivatives (Expts 5.198 and 5.199).
5. Imine and enamine formation (Expt 5.200 and 5.201).
6. Molecular rearrangements of the Hofmann type (see Expt 6.53).

Methods for the protection of the amino and imino groups are considered in Section 5.16.7, p. 784.

SUMMARY OF RETROSYNTHETIC STRATEGIES

Functional group interconversion (FGI) (methods 1 and 2)

Primary amines

$$R{\diagup}{\diagdown}CN \underset{(1)}{\Longleftarrow} R{\diagup}{\diagdown}{\diagup}NH_2 \underset{(1)}{\Longrightarrow} R{\diagup}{\diagdown}\overset{O}{\underset{}{\diagup}}NH_2$$

(TM)

$$\underset{R^1 \;\;\;\; R^2}{\overset{NOH}{\diagup\!\!\diagdown}} \underset{(2)}{\Longleftarrow} \underset{R^1 \;\;\;\; R^2}{\overset{NH_2}{\diagup\!\!\diagdown}} \underset{\substack{(1)\text{ and}\\(2)}}{\Longrightarrow} \underset{R^1 \;\;\;\; R^2}{\overset{NO_2}{\diagup\!\!\diagdown}} \;\; \text{or} \;\; \underset{R^1 \;\;\;\; R^2}{\overset{N_3}{\diagup\!\!\diagdown}}$$

R^2 = H or alkyl

(TM)

Secondary and tertiary amines

$$R^1{\diagup}{\diagdown}{\diagup}NHR^2 \underset{(1)}{\Longrightarrow} R^1{\diagup}{\diagdown}\overset{O}{\underset{}{\diagup}}NHR^2$$

(TM)

$$R^1{\diagup}{\diagdown}{\diagup}NR_2^2 \underset{(1)}{\Longrightarrow} R^1{\diagup}{\diagdown}\overset{O}{\underset{}{\diagup}}NR^2$$

(TM)

C–N Disconnection (method 4)

Primary amines

Secondary amines

FGI/C—N Disconnection (method 3)

R^2, R^3 = H or alkyl

Rearrangement (method 6)

SPECTROSCOPIC FEATURES

The characteristic i.r. absorption of the stretching vibration of the amino and imino groups of primary and secondary amines is clearly visible in the region of $3300 \, cm^{-1}$; the bending vibration and the out-of-plane deformation of the amino group are usually easily recognised, as the example of the spectrum of butylamine (Fig. 3.26) illustrates. The nitrogen-bound protons are frequently observed as broad signals in the p.m.r. spectrum of primary and secondary amines and their ready exchange with, for example, deuterium oxide is discussed on p. 348. The structure of the alkyl groups may be often assigned from the spin-spin splitting patterns (p. 341). In the m.s. of aliphatic amines the molecular ion is usually very weak or negligible (p. 368). The fragmentation reactions are noted in detail on p. 381 where the spectrum of diethylamine (Fig. 3.87) is given. The u.v.–visible absorption is not structurally informative. Further illustrative spectroscopic interpretations are given in appropriate preparative sections below.

5.16.1 THE REDUCTION OF ALKYL AZIDES, ALKYL CYANIDES AND AMIDES

As these compounds all possess a nitrogen-containing functional feature, it is worth pointing out that they arise from alkyl halides (or methanesulphonates) in the case of azides and cyanides, or from carboxylic acid chlorides (in the case of amides).

A one-pot PTC reaction procedure for the overall conversion of an alkyl halide into a primary amine via an azide is particularly illustrative.[204] Thus the reduction of the azide is effected by the addition of sodium borohydride to a reaction mixture arising from the PTC displacement reaction of an alkyl halide with sodium azide (the preparation of 1-octylamine, Expt 5.193). The reaction appears to be applicable to primary and secondary alkyl halides, alkyl methanesulphonates and benzylic halides.

Alkyl cyanides in general are smoothly reduced to primary amines (giving a product having one carbon atom more than the alkyl halide starting material) by the action of sodium and ethanol (e.g. the preparation of pentylamine, Expt 5.194). The use of a metal–acid reducing medium is unsatisfactory since extensive hydrolysis of the cyano group occurs. Catalytic hydrogenation over a Raney nickel catalyst (Section 2.17.1, p. 87), in the presence of excess ammonia to suppress secondary amine formation, is illustrated by the preparation of 2-phenylethylamine (cognate preparation in Expt 5.194). The secondary amine may be formed by the following sequence.

$$R \cdot CN \longrightarrow [R \cdot CH = NH] \longrightarrow R \cdot CH_2 \cdot NH_2$$

$$R \cdot CH = NH \xrightarrow{R \cdot CH_2 NH_2} \underset{NH \cdot CH_2 \cdot R}{R \cdot CH \cdot NH_2} \underset{-NH_3}{\rightleftharpoons} \underset{N \cdot CH_2 \cdot R}{R \cdot CH} \longrightarrow (R \cdot CH_2)_2 NH$$

Lithium aluminium hydride has also been used for the reduction of nitriles to amines; a recommended procedure involves the slow addition of the nitrile to at least one molar proportion of the reducing agent in a cooled ethereal solution.[205] The reduction of nitriles is also effected by the use of sodium trifluoroacetoxyborohydride (from sodium borohydride and trifluoroacetic acid) in tetrahydrofuran solution.[206]

Amides may be reduced to the corresponding primary, secondary and tertiary amines with lithium aluminium hydride with a varying degree of success. The reagent sodium acetoxyborohydride (1) has been used to reduce a wide range of aliphatic and aromatic primary and secondary amides to the corresponding primary and secondary amines; in the case of tertiary amides $(R \cdot CO \cdot NR_2)$ sodium trifluoroacetoxyborohydride is the reagent of choice.[207]

$$Me \cdot CO_2H + NaBH_4 \longrightarrow NaBH_3(O \cdot CO \cdot Me)$$
$$(1)$$

$$R^1 \cdot CO \cdot NH \cdot R^2 + (1) \longrightarrow R^1 \cdot CH_2 \cdot NH \cdot R^2$$

The reduction of benzamide to benzylamine is the illustrative example (Expt 5.195). Sodium borohydride with methanesulphonic acid is also a convenient and effective reductant for amides,[208] and a general procedure is included in Expt 5.195.

Experiment 5.193 OCTYLAMINE[204]

$$C_6H_{13} \diagdown Br \xrightarrow[PTC]{NaN_3} C_6H_{13} \diagdown N_3 \xrightarrow[PTC]{NaBH_4} C_6H_{13} \diagdown NH_2$$

A mixture of 1-bromooctane (19.3 g, 0.1 mol), hexadecyltributylphosphonium bromide (5.1 g, 0.01 mol), sodium azide (16.2 g, 0.25 mol) and water (50 ml) is stirred at 80 °C for 8 hours. The aqueous phase is carefully removed, the organic phase is diluted with toluene (15.5 ml), and then a solution of sodium borohydride (11.7 g, 0.3 mol) in water (30 ml) is added in 30 minutes. The mixture is stirred at 80 °C for 16 hours. The layers are separated and the organic phase is extracted with 10 per cent hydrochloric acid. The catalyst is recovered from the organic phase by removal of the solvent and is reused as isolated. Basification of the acid phase and extraction with ether affords octylamine, 11.4 g (88%), b.p. 172–174 °C.

Experiment 5.194 PENTYLAMINE

$$\text{Pr·CH}_2\text{·CN} \xrightarrow{\text{4[H]}} \text{Pr·CH}_2\text{·CH}_2\text{·NH}_2$$

Equip a three-necked 1-litre flask with a dropping funnel, an efficient mechanical stirrer and a reflux condenser. Place 55 g (2.4 mol) of clean sodium and 200 ml of sodium-dried toluene in the flask, heat the mixture until the toluene commences to boil and then stir the molten sodium vigorously thus producing an emulsion. Run in through the dropping funnel a mixture of 33 g (41.5 ml) of butyl cyanide (Expt 5.158) and 60 g (76 ml) of absolute ethanol during 1 hour. During the addition and the subsequent introduction of ethanol and of water, the stirring should be vigorous and the temperature adjusted so that the refluxing is continuous; the heat of reaction will, in general, be sufficient to maintain the refluxing. After the butyl cyanide solution has been added, introduce gradually a further 60 g (76 ml) of absolute ethanol. In order to destroy any residual sodium, treat the reaction mixture slowly with 40 g (50 ml) of rectified spirit and then with 20 g of water. Steam distil the contents of the flask (compare Fig. 2.102) (about 2 hours) and add 40 ml of concentrated hydrochloric acid to the distillate. Separate the toluene layer; evaporate the aqueous layer, which contains alcohol and amine hydrochloride, to dryness under reduced pressure (rotary evaporator). Treat the resulting amine hydrochloride with a solution of 40 g of sodium hydroxide in 200 ml of water. Separate the amine layer, dry it by shaking with sodium hydroxide pellets (prolonged contact is required for complete drying) and distil. Collect the fraction boiling at 102–105 °C as pure pentylamine. Dry the fraction of low boiling point again over sodium hydroxide and redistil; this gives an additional quantity of amine. The total yield is 30 g (86%).

Cognate preparation. *2-Phenylethylamine.* Saturate commercial absolute methanol with ammonia (derived from a cylinder) at 0 °C; the resulting solution is c. 10 M. Dissolve 58 g (0.5 mol) of benzyl cyanide (Expt 5.157) (1) in 300 ml of the cold methanolic ammonia, and place the solution in a high-pressure hydrogenation bomb (Section 2.17.1, p. 95), add 10 ml of settled Raney nickel catalyst (Section 4.2.50, p. 450), securely fasten the cap and introduce hydrogen until the pressure is 500–1000 lb. Set the mechanical stirring device in motion, and heat at 100–125 °C until absorption of hydrogen ceases (about 2 hours). Allow the bomb to cool, open it and remove the contents. Rinse the bomb with two 100 ml portions of anhydrous methanol and pour the combined liquids through a fluted filter paper to remove the catalyst; do not permit the catalyst to become dry since it is likely to ignite. Remove the solvent and ammonia by distillation (*fume cupboard*), and fractionate the residue through a short column. Collect the 2-phenylethylamine at 92–93 °C/18 mmHg. The yield is 54 g (90%).

Note. (1) Minute amounts of halide have a powerful poisoning effect upon the catalyst; it is advisable to distil the benzyl cyanide from Raney nickel.

Experiment 5.195 BENZYLAMINE[207]

$$Ph \cdot CONH_2 \xrightarrow{NaBH_3(O \cdot CO \cdot Me)} Ph \cdot CH_2 \cdot NH_2$$

CAUTION: *Hydrogen gas is evolved in these reactions.*

To a stirred suspension of sodium borohydride (1.89 g, 50 mmol) and benzamide (1.21, 10 mmol) in dioxane (20 ml) is added acetic acid (3.0 g, 50 mmol) in dioxane (10 ml) over a period of 10 minutes at 10 °C; the reaction mixture is stirred at reflux for 2 hours. The reaction mixture is concentrated to dryness *in vacuo*, excess reagent is decomposed with water and the solution extracted with chloroform. The extract is washed with water and dried over anhydrous sodium sulphate. The chloroform layer is treated with dry hydrogen chloride, evaporated *in vacuo* and the residue recrystallised from methanol–ether to give benzylamine hydrochloride (1.09 g, 76.2%).

Cognate preparation. *General procedure for the reduction of amides with sodium borohydride/methanesulphonic acid in dimethyl sulphoxide.*[208] To a vented 100-ml two-necked flask equipped with a magnetic stirrer, are added 16 mmol of amide, 1.61 g (40 mmol) of sodium borohydride, and 20 ml of dimethyl sulphoxide. Methanesulphonic acid (**CAUTION**) (freshly distilled prior to use, b.p. 112–114 °C/0.5 mmHg) (3.4 ml, 55 mmol) and 20 ml of dimethyl sulphoxide are mixed and added dropwise to the reaction mixture by means of an addition funnel over a 30-minute period. During this addition the reaction mixture is constantly stirred with the magnetic stirrer. Gas is evolved during the addition of acid and a gelatinous substance is formed, which dissolves when the acid comes to exceed the sodium borohydride on a molar basis. At the end of this time, if the reduction product is to be a primary or secondary amine, the reaction mixture is quenched by the addition of 20 ml of 10 per cent sodium hydroxide. If the product is to be a tertiary amine, the reaction mixture is kept at 70 °C for 2 hours to liberate the product from what appears to be a complex with borane. After this additional reaction period tertiary amines are isolated in the same way as primary and secondary amines. After the reaction is quenched, the product is extracted from the reaction mixture with three 10-ml portions of dichloromethane. The product solution is then washed with three 10-ml portions of 0.1 M sodium hydroxide to remove most of the residual dimethyl sulphoxide, and the product is extracted into three 10-ml portions of 10 per cent hydrochloric acid. Neutralisation with 10 per cent sodium hydroxide, followed by extraction with three 10-ml portions of dichloromethane, drying over calcium sulphate and fractional distillation affords the amine. In this way hexanamide gives hexylamine, b.p. 38–40 °C/25 mmHg, 74 per cent yield; acetanilide gives N-ethylaniline, b.p. 90–91 °C/15 mmHg, 77 per cent yield; N,N-dimethylbenzamide gives N,N-dimethylbenzylamine, b.p. 73–77 °C/22 mmHg, 59 per cent yield.

5.16.2 THE REDUCTION OF NITRO COMPOUNDS AND OXIMES

These methods are particularly useful for branched chain primary amines.

Although the reduction of nitro compounds is more widely applied in the aromatic series using procedures discussed in Section 6.5.1, p. 890, these are equally applicable to the reduction of primary or secondary nitroalkanes when these are readily available. α,β-Unsaturated nitro compounds (arising for example from the Claisen–Schmidt reaction, Section 6.12.2, p. 1032) may be reduced to the saturated amine very effectively with sodium dihydrobis(2-methoxyethoxy)-aluminate ('Red-Al').

The ready conversion of aldehydes and ketones into oximes and their subsequent reduction is more generally applicable to the synthesis of straight and branched chain primary aliphatic amines. Reduction with sodium and alcohol is convenient and effective and two examples of its use are given in Expt 5.196.

Other methods of reduction include the use of metal–acid systems, catalytic hydrogenation over Raney nickel or palladium-on-charcoal, and modified metal hydride reducing agents such as 'Red-Al'.[209]

Experiment 5.196   HEPTYLAMINE

$$\text{Me·(CH}_2)_5\text{·CH=NOH} \xrightarrow{4[\text{H}]} \text{Me·(CH}_2)_5\text{·CH}_2\text{·NH}_2$$

Heptaldoxime. Fit a 1-litre three-necked flask with an efficient mechanical stirrer, a double surface condenser and a thermometer. Place 115 g (141 ml, 1.25 mol) of heptanal (1) and a solution of 87 g of hydroxylamine hydro-chloride in 150 ml of water in the flask, and stir the mixture vigorously (2). Introduce, from a separatory funnel down the reflux condenser, a solution of 67 g (0.63 mol) of anhydrous sodium carbonate in 250 ml of water at such a rate that the temperature of the reaction mixture does not rise above 45 °C. Continue the stirring for 1 hour at room temperature. Separate the upper layer and wash the oil with two 25 ml portions of water; dry with magnesium sulphate. Distil from a flask fitted with a short fractionating column. A small fraction of low boiling point (containing heptanenitrile and heptaldoxime) passes over first, and as soon as the temperature is constant the heptaldoxime is collected (e.g. at 103–107 °C/6 mmHg); the temperature of the oil bath is maintained at about 30 °C above the boiling point of the liquid. The yield is about 110 g, and the liquid slowly solidifies on cooling and melts at 44–46 °C; it is sufficiently pure for conversion into heptylamine. If required pure, the heptaldoxime may be recrystallised from 60 per cent ethanol (25 g of solid to 70 ml of solvent) and then melts at 53–55 °C (the m.p. depends somewhat upon the rate of heating).

Heptylamine. In a 3-litre round-bottomed flask, equipped with two large Liebig condensers (34/35 joints) joined in series, place a solution of 64.5 g (0.5 mol) of heptaldoxime in 1 litre of super-dry ethanol (Section 4.1.9, p. 401) and heat on a water bath. Immediately the alcohol boils, remove the flask from the water bath and introduce 125 g (5.4 mol) of sodium, cut in small pieces, as rapidly as possible through the condenser consistent with keeping the vigorous reaction under control. The last 30 g of sodium melts in the hot mixture and may be added very rapidly without appreciable loss of alcohol or

775

of amine. As soon as the sodium has completely dissolved (some warming may be necessary), cool the contents of the flask and dilute with 1250 ml of water. At once equip the flask with a condenser set for downward distillation and arrange for the distillate to be collected in a solution of 75 ml of concentrated hydrochloric acid in 75 ml of water contained in a 3-litre flask. Continue the distillation as long as amine passes over. Towards the end of the reaction considerable frothing sets in; then add a further 750 ml of water to the distillation flask. The total distillate is 2–2.2 litres and contains alcohol, water and some unreacted oxime as well as the amine hydrochloride. Evaporate the solution under reduced pressure using a rotary evaporator; the amine hydrochloride will crystallise out in the flask. Cool the flask, attach a reflux condenser and introduce 250 ml of 40 per cent potassium hydroxide solution. Rotate the flask to wash down the hydrochloride from the sides of the flask, cool the mixture to room temperature and transfer it to a separatory funnel. Run off the lower alkaline layer and add solid potassium hydroxide to the amine in the funnel. Again remove the lower aqueous layer, add more solid potassium hydroxide and repeat the process until no further separation of an aqueous layer occurs. Finally, transfer the amine to a small flask and leave it in contact with potassium hydroxide pellets for 24 hours. Decant the amine into a flask and distil through a well-lagged fractionating column. Collect the heptylamine at 153–157 °C. The yield is 40 g (70%).

Notes. (1) The heptanal should be dried and redistilled: b.p. 150–156 °C or 54–59 °C/16 mmHg.
(2) The solution may be rendered homogeneous by the addition of ethanol but the yield appears to be slightly diminished and more high boiling point material is produced.

Cognate preparation. *Hexyl methyl ketoxime.* From hexyl methyl ketone (*octan-2-one*) (Expt 5.86) in 90 per cent yield; b.p. 106–108 °C/12 mmHg.

1-Methylheptylamine. Reflux a solution of 50 g (0.35 mol) of the oxime in 200 ml of super-dry ethanol on a water bath while adding 75 g (3.25 mol) of sodium; introduce more alcohol (about 300 ml) to maintain a vigorous reaction. When all the sodium has passed into solution, cool, dilute with 250 ml of water and distil gently until the b.p. reaches 96 °C; add a further 200 ml of water and repeat the distillation to ensure the complete removal of the alcohol. The amine remains as a layer on the strongly alkaline solution: extract it with ether, dry the ethereal solution with sodium hydroxide or anhydrous calcium sulphate, remove the ether on a water bath, and distil the residue under diminished pressure. Collect the 1-methylheptylamine at 58–59 °C/13 mmHg; the b.p. under atmospheric pressure is 163–164 °C. The yield is 31 g (69%).

5.16.3 REDUCTIVE ALKYLATION OF AMMONIA OR AMINES

The process of reductive alkylation involves the treatment of ammonia or primary or secondary amines with an aldehyde or ketone under reducing conditions.
   The conversion in the case of *ammonia* or *primary amines* and a carbonyl compound probably involves the following stages.

$$R^1 \overset{O}{\underset{}{\overset{\|}{C}}} R^2 \xrightarrow{NH_3} \left[R^1 \overset{OH}{\underset{NH_2}{\overset{|}{\underset{|}{C}}}} R^2 \right] \xrightarrow{-H_2O} \left[R^1 \overset{NH}{\underset{}{\overset{\|}{C}}} R^2 \right] \xrightarrow{[H]} R^1 \overset{NH_2}{\underset{}{C}} R^2$$

$$R^1 \overset{O}{\underset{}{\overset{\|}{C}}} R^2 \xrightarrow{R^3 \cdot NH_2} \left[R^1 \overset{OH}{\underset{NH \cdot R^3}{\overset{|}{\underset{|}{C}}}} R^2 \right] \xrightarrow{-H_2O} \left[R^1 \overset{N \cdot R^3}{\underset{}{\overset{\|}{C}}} R^2 \right] \xrightarrow{[H]} R^1 \overset{NH \cdot R^3}{\underset{}{C}} R^2$$

A retrosynthetic analysis for a target molecule of either a primary or secondary amine is therefore a two-stage process; first an FGI to reveal an imine, and then a disconnection to ammonia (or primary amine) and the carbonyl compound.

When *secondary amines* are used the intermediate aminol cannot give rise to an imine and the tertiary amine product is formed by a hydrogenolytic pathway.

$$R^1 \overset{O}{\underset{}{\overset{\|}{C}}} R^2 \xrightarrow{R_2^3 NH} \left[R^1 \overset{OH}{\underset{NR_2^3}{\overset{|}{\underset{|}{C}}}} R^2 \right] \xrightarrow{[H]} R^1 \overset{NR_2^3}{\underset{}{C}} R^2$$

The reduction is usually effected catalytically in ethanol solution using hydrogen under pressure in the presence of Raney nickel. As in the reduction of nitriles (Section 5.16.1, p. 771), which also involves the intermediate imine, ammonia or the amines should be present in considerable excess to minimise the occurrence of undesirable side reactions leading to the formation of secondary and tertiary amines. These arise from the further reaction of the carbonyl compound with the initially formed amine product. Selected experimental conditions for these reductive alkylation procedures have been well reviewed.[210] Sodium borohydride has also been used as an *in situ* reducing agent and is particularly effective with mixtures of primary amines and aliphatic aldehydes and ketones.[211]

$$R^1 \overset{O}{\underset{}{\overset{\|}{C}}} R^2 \xrightarrow[-H_2O]{R^3 \cdot NH_2} \left[R^1 \overset{N \cdot R^3}{\underset{}{\overset{\|}{C}}} R^2 \right] \xrightarrow{NaBH_4} R^1 \overset{NH \cdot R^3}{\underset{}{C}} R^2$$

A major variation is the use of formic acid or one of its derivatives as the reductant (the *Leuckart reaction*). In the synthesis of 1-phenylethylamine (Expt 5.197), ammonium formate is heated with acetophenone while the water formed in the reaction is carefully removed by fractional distillation to give the required amine as its N-formyl derivative, (1-phenylethyl)formamide. This is then hydrolysed with acid to yield the primary amine. The procedure has been satisfactorily applied to many aliphatic–aromatic, alicyclic and aliphatic–heterocyclic ketones, some aromatic ketones and aldehydes, and to some aliphatic aldehydes and ketones boiling at about 100 °C or higher.

$$R^1 \overset{O}{\underset{}{\overset{\|}{C}}} R^2 \xrightleftharpoons{HCO_2NH_4} \left[R^1 \overset{OH}{\underset{NH_2}{\overset{|}{\underset{|}{C}}}} R^2 \right] \xrightarrow{-H_2O} \left[R^1 \overset{NH}{\underset{}{\overset{\|}{C}}} R^2 \right] \xrightarrow[-CO_2]{H \cdot CO_2H}$$

$$R^1 \overset{NH_2}{\underset{}{C}} R^2 \xrightarrow{HCO_2NH_4} R^1 \overset{NH \cdot CHO}{\underset{}{C}} R^2 + H_2O + NH_3$$

Experiment 5.197 1-PHENYLETHYLAMINE (α-*Methylbenzylamine*)

Place 126 g (2.0 mol) of ammonium formate, 72 g (0.6 mol) of acetophenone and a few chips of porous porcelain in a 250-ml flask fitted with a Claisen still-head carrying a short fractionating column; insert a thermometer extending nearly to the bottom of the flask, and attach a short condenser set for downward distillation to the side-arm. Heat the flask with a heating mantle or in an air bath; the mixture first melts to two layers and distillation occurs. The mixture becomes homogeneous at 150–155 °C and reaction takes place with slight frothing. Continue the heating, more slowly if necessary, until the temperature rises to 185 °C (about 2 hours); acetophenone, water and ammonium carbonate distil. Stop the heating at 185 °C, separate the upper layer of acetophenone from the distillate and return it without drying to the flask. Heat the mixture for 3 hours at 180–185 °C and then allow to cool; the acetophenone may be recovered from the distillate by extraction with 20 ml portions of toluene (1). Transfer the reaction mixture to a 250-ml separatory funnel and shake it with two 75 ml portions of water to remove formamide and ammonium formate. Transfer the crude (1-phenylethyl)formamide into the original reaction flask; extract the aqueous layer with two 20 ml portions of toluene, transfer the toluene extracts to the flask, add 75 ml of concentrated hydrochloric acid and a few chips of porous porcelain. Heat the mixture cautiously until about 40 ml of toluene are collected, and boil gently under reflux for a further 40 minutes; hydrolysis proceeds rapidly to 1-phenylethylamine hydrochloride except for a small layer of unchanged acetophenone. Allow the reaction mixture to cool, remove the acetophenone by extraction with four 20 ml portions of toluene (1). Transfer the aqueous acid solution to a 500-ml round-bottomed flask equipped for steam distillation, cautiously add a solution of 6.5 g of sodium hydroxide in 125 ml of water, and steam distil; heat the distillation flask so that the volume remains nearly constant. Most of the amine is contained in the first 500 ml of distillate; stop the operation when the distillate is only faintly alkaline. Extract the distillate with five 25 ml portions of toluene, dry the extract with sodium hydroxide pellets and fractionally distil (2). Toluene distils over at 111 °C, followed by the phenylethylamine. Collect the latter as a fraction of b.p. 180–190 °C (the bulk of the product distils at 184–186 °C (3); the yield is 43 g (59%).

Notes. (1) The acetophenone may be recovered by washing the toluene solution with dilute alkali, drying with anhydrous calcium sulphate and distilling; the fraction, b.p. 198–205 °C, is collected.

(2) Ground glass apparatus must be used as the amine attacks rubber (and cork). Since the product absorbs carbon dioxide from the air, attach the receiving flask to the condenser with a take-off adapter carrying a soda-lime guard-tube.

(3) The b.p. under diminished pressure is 80–81 °C/18 mmHg. To obtain a very pure sample of the amine, dissolve 1 part (by weight) of the above product in a solution of 1.04 parts of crystallised oxalic acid in 8 parts of hot water, add a little decolourising

carbon and filter. The filtered solution deposits crystals of the oxalate salt; about 5 g of this salt remains in each 100 ml of mother-liquor, but most can be recovered by evaporation and further crystallisation. The amine may be liberated from its oxalate salt with sodium or potassium hydroxide, steam distillation and purification as described above. The salt provides a convenient method of obtaining a known weight of the amine in water, since it can be weighed out and decomposed with alkali hydroxide.

5.16.4 THE ALKYLATION OF AMMONIA AND ITS DERIVATIVES

A C—N disconnection of a primary amine gives rise to the carbocation and amide anion synthons. It might be predicted therefore that treatment of an alkyl halide with ammonia (reagents equivalent to the above synthons) under pressure would constitute a suitable synthesis of a primary amine. In practice, however, the yield is poor since a mixture of all three classes of amines, together with some of the quaternary ammonium salt, is obtained, owing to more ready further alkylation of the sequentially formed products.

$$RX + NH_3 \rightleftharpoons R\overset{\oplus}{N}H_3\}\overset{\ominus}{X} \overset{NH_3}{\rightleftharpoons} RNH_2 + NH_4X$$

$$RX + RNH_2 \rightleftharpoons R_2\overset{\oplus}{N}H_2\}\overset{\ominus}{X} \overset{NH_3}{\rightleftharpoons} R_2NH + NH_4X$$

$$RX + R_2NH \rightleftharpoons R_3\overset{\oplus}{N}H\}\overset{\ominus}{X} \overset{NH_3}{\rightleftharpoons} R_3N + NH_4X$$

$$RX + R_3N \rightleftharpoons R_4\overset{\oplus}{N}\}\overset{\ominus}{X}$$

A similar ammonolysis of alcohols in the presence of certain metallic oxide catalysts is, however, extensively used on the large scale for the manufacture of all classes of amines.

An alternative reagent equivalent for the amide anion synthon is the potassium salt of phthalimide which can only react with one molecular proportion of alkyl halide. The resulting N-alkylphthalimide is then cleaved to the primary amine (the *Gabriel synthesis*). The preliminary preparation of potassium phthalimide (from a solution of phthalimide in absolute ethanol and potassium hydroxide in 75% ethanol) may be avoided in some cases by boiling phthalimide with the halide in the presence of anhydrous potassium carbonate. The cleavage of the N-substituted phthalimide is best effected by reaction with hydrazine hydrate and then heating the reaction mixture with hydrochloric acid. The insoluble phthalylhydrazide is filtered off, leaving the amine hydrochloride in solution from which the amine may be liberated and isolated in the appropriate manner.

The Gabriel synthesis is illustrated by the preparation of benzylamine and 2-phenylethylamine (Expt 5.198).

The preparation of pure *symmetrical* secondary amines (e.g. dibutylamine, Expt 5.199) is conveniently achieved by the hydrolysis of dialkylcyanamides with dilute sulphuric acid. The appropriate dialkyl cyanamide is prepared by treating sodium cyanamide (itself obtained from calcium cyanamide and aqueous sodium hydroxide solution) with an alkyl halide. In this case the reagent $[:\ddot{N}—C≡N]^{2\ominus}$ may be regarded as a masked NH group.

$$\text{CaN·CN} \xrightarrow{\text{NaOH}} \text{Na}_2\text{N·CN} \xrightarrow{\text{2RBr}} \text{R}_2\text{N·CN}$$

$$\text{R}_2\text{N·CN} \xrightarrow[-\text{NH}_3]{\text{H}_3\text{O}^\oplus} \text{R}_2\text{N·CO}_2\text{H} \xrightarrow{-\text{CO}_2} \text{R}_2\text{NH}$$

Experiment 5.198 BENZYLAMINE

N-Benzylphthalimide. Grind together 76 g (0.55 mol) of finely powdered, anhydrous potassium carbonate and 147 g (1 mol) of phthalimide (Expt 6.155) in a glass mortar, transfer the mixture to a round-bottomed flask and treat it with 151 g (1.2 mol) of redistilled benzyl chloride. Heat in an oil bath at 190 °C under a reflux condenser for 3 hours. While the mixture is still hot, remove the excess of benzyl chloride by steam distillation. The benzylphthalimide commences to crystallise near the end of the steam distillation. At this point, cool the mixture rapidly with vigorous swirling so that the solid is obtained in a fine state of division. Filter the solid with suction on a Buchner funnel, wash well with water and drain as completely as possible; then wash once with 200 ml of 60 per cent ethanol and drain again. The yield of crude product, m.p. 100–110 °C, is 180 g (76%). Recrystallise from glacial acetic acid to obtain pure benzylphthalimide, m.p. 116 °C: the recovery is about 80 per cent.

Benzylamine. Warm an alcoholic suspension of 118.5 g (0.5 mol) of finely powdered benzylphthalimide with 25 g (0.5 mol) of 100 per cent hydrazine hydrate (**CAUTION**: *corrosive liquid, see Section 4.2.31, p. 436*): a white, gelatinous precipitate is produced rapidly. Decompose the latter (when its formation appears complete) by heating with excess of hydrochloric acid on a steam bath. Collect the phthalylhydrazide which separates by suction filtration, and wash it with a little water. Concentrate the filtrate by distillation on a rotary evaporator to remove alcohol, cool, filter from the small amount of precipitated phthalylhydrazide, render alkaline with excess of sodium hydroxide solution and extract the liberated benzylamine with ether. Dry the ethereal solution with potassium hydroxide pellets, remove the solvent (rotary evaporator) and finally distil the residue. Collect the benzylamine at 185–187 °C: the yield is 50 g (94%).

Cognate preparation. *2-Phenylethylamine.* Prepare 2-phenylethylphthalimide as above by substituting 2-phenylethyl bromide (Expt 5.55) for benzyl chloride: recrystallise the crude product from glacial acetic acid; m.p. 131–132 °C. Convert it into 2-phenylethylamine by treatment with hydrazine hydrate and hydrochloric acid as described for benzylamine. The yield of 2-phenylethylamine, b.p. 200–205 °C, is about 95 per cent.

Experiment 5.199 DIBUTYLAMINE

$$CaNCN + 2NaOH \longrightarrow Na_2NCN + Ca(OH)_2$$

$$Na_2NCN + 2BuBr \longrightarrow Bu_2NCN + 2NaBr$$

$$Bu_2NCN + 2H_2O \longrightarrow Bu_2NH + CO_2 + \overset{\oplus}{N}H_4$$

Dibutyl cyanamide. Equip a 2-litre three-necked flask with a reflux condenser and a sealed stirrer unit. Place 220 ml of water and 50 g of finely crushed ice in the flask and add slowly, with vigorous stirring, 70 g (0.46 mol) of commercial calcium cyanamide (1). As soon as the solid is thoroughly suspended, fit a separatory funnel into the third neck of the flask and introduce through it a cold solution of 34 g (0.85 mol) of sodium hydroxide in 70 ml of water; replace the funnel by a thermometer. Continue the vigorous stirring for 1 hour to complete the decomposition of the calcium cyanamide; if the temperature rises above 25 °C, add a little more ice. Add to the resulting solution of sodium cyanamide a solution of 134 g (195 ml, 1 mol) of butyl bromide (Expt 5.54) in 220 ml of rectified spirit. Heat the mixture, with stirring, on a water bath until it refluxes gently; continue the refluxing and stirring for 2.5 hours. Replace the reflux condenser by one set for downward distillation and distil the mixture until 165–170 ml of liquid are collected: stir during distillation. Cool the residue in the flask and filter it, with suction, through a Buchner or sintered glass funnel, and wash the residue with rectified spirit. Extract the filtrate, which separates into two layers, first with 90 ml and then with 45 ml of benzene. Dry the combined benzene extracts with anhydrous calcium sulphate, and remove the benzene on a rotary evaporator. Distil under reduced pressure and collect the dibutul cyanamide at 147–161 °C/35 mm. The yield is 33 g (47%).

Dibutylamine. Into a 1-litre round-bottomed flask furnished with a reflux condenser place a solution of 34 g (18.5 ml) of concentrated sulphuric acid in 100 ml of water: add 33 g (0.2 mol) of dibutyl cyanamide and a few fragments of porous porcelain. Reflux gently for 6 hours. Cool the resulting homogeneous solution and pour in a cold solution of 52 g of sodium hydroxide in 95 ml of water down the side of the flask so that most of it settles at the bottom without mixing with the solution in the flask. Connect the flask with a condenser for downward distillation and shake it to mix the two layers; the free amine separates. Heat the flask, when the amine with some water distils: continue the distillation until no amine separates from a test portion of the distillate. Estimate the weight of water in the distillate and add about half this amount of potassium hydroxide in the form of pellets so that it dissolves slowly. Cool the solution in ice while the alkali hydroxide is dissolving; some ammonia gas is evolved. When the potassium hydroxide has dissolved, separate the amine, and dry it for 24 hours over sodium hydroxide pellets. Filter and distil from a flask fitted with a Claisen still-head. Collect the dibutyl-amine at 157–160 °C. The yield is 21 g (75%).

Note. (1) Also known as 'nitrolim'. The fresh product contains approximately 55 per cent of calcium cyanamide, 20 per cent of lime, 12 per cent of graphite and small amounts of other impurities. It should be protected from moisture when stored in order to prevent slow polymerisation to dicyanodiamide.

5.16.5   IMINE AND ENAMINE FORMATION[212]

Imines are formed by the reaction of a primary amine with aldehydes or ketones with the simultaneous removal of water, for example by azeotropic distillation,[213] by the addition of anhydrous sodium sulphate,[214] by the addition of molecular sieves,[215] or by the use of titanium(IV) chloride.[216] When one, or both, of the reactants is aromatic, the imine is quite stable and usually known as a Schiff base (see Section 6.5.5, p. 902). In the case of wholly aliphatic reactants the imines tend to decompose or polymerise; in these cases their further reaction is carried out without delay.

A typical procedure involving water removal with anhydrous potassium carbonate[217] is described in Expt 5.200 for the preparation of N-(1,1-methylethyl)-2-methylpropylimine. The imines are important synthetic intermediates and some of these aspects have been outlined in Section 5.16.3 above, Section 6.5.5, p. 902, and Section 5.7.8, p. 600.

Enamines are the stable products of a similar reaction between secondary amines (such as pyrrolidine or morpholine) and aldehydes and ketones.[218] These vinylamines are reactive reagents of value in synthesis; they function as specific enol equivalents of carbonyl compounds, readily undergoing alkylation and acylation processes (e.g. Section 5.9.2, p. 632).

As with imines, enamine formation may be achieved using azeotropic distillation,[219] or with the aid of molecular sieves. The procedure described in Expt 5.201 for the preparation of the enamine from morpholine and diisobutyl ketone utilises titanium(IV) chloride as a catalyst and water scavenger.[220]

The reaction of an amine with, principally, formaldehyde, followed by further reaction with compounds having active hydrogen atoms, give rise to the *Mannich bases* which are discussed in Sections 5.18.2, p. 801, and 6.12.7, p. 1050.

Experiment 5.200   *N*-(1,1-DIMETHYLETHYL)-2-METHYLPROPYL-IMINE[217]

2-Methylpropanal (72 g, 1.0 mol) is added dropwise and with stirring during 2 hours to 1,1-dimethylethylamine (t-butylamine) (73 g 1.0 mol). During the addition the temperature rises from 25 to 40 °C and an aqueous layer separates near the end of the addition. The organic layer is treated with anhydrous potassium carbonate (15 g), stirred at 25 °C for 17 hours, and then decanted on to barium oxide (12 g). After the mixture has been stirred for 10 hours, it is filtered and the organic filtrate is distilled to separate the imine as a colourless liquid, b.p. 56 °C/75 mmHg, 88.3 g (70%); i.r. (CCl$_4$) 1675 cm^{-1} (C=N); p.m.r. (CCl$_4$, TMS) 1.05 [d, 6H $J = 7$ Hz, (Me)$_2$CH−], 1.10 [s, 9H, −C(Me)$_3$], 2.0–2.6 (m, 1H, CH), and 7.49 (d, 1H, $J = 4.5$ Hz, CH=N); m/z 127, 112, 72, 57, 56, 55, and 41.

Experiment 5.201   4-(2,6-DIMETHYL-3-HEPTEN-4-YL)MORPHO-LINE[220]

To a vigorously stirred, cold (0 °C) solution of morpholine (680 g, 7.8 mol) in 1500 ml of hexane is added a solution of titanium(IV) chloride (146 ml, 1.33 mol) in 300 ml of hexane. After the addition is complete diisobutyl ketone (142.2 g, 1.0 mol) is added in one portion. The cooling bath is removed and the reaction allowed to proceed under reflux for 2 hours. After cooling, the mixture is filtered through a sintered glass filter and the solvent removed under reduced pressure, giving a yellow oil which is distilled under reduced pressure. The yield of enamine is 171.4 g (87.7%), b.p. 106–108 °C/10 mmHg (1).

Note. (1) This method has been successfully applied to a very large range of straight and branched chain aliphatic ketones.

5.16.6 MOLECULAR REARRANGEMENTS OF THE HOFMANN TYPE

By treatment of an amide with sodium hypobromite or sodium hypochlorite solution (or with halogen admixed with aqueous alkali), a primary amine having one less carbon atom is produced. Good yields are obtained when the reaction is applied to most aliphatic and aromatic amides. Examples are provided by the preparation of anthranilic acid and 3-aminopyridine (Expt 6.53).

The conversion of an amide in this way is termed the *Hofmann reaction* or the *Hofmann rearrangement*. The mechanism of the reaction involves an intramolecular 1,2-carbon-to-nitrogen nucleophilic shift of the alkyl (or aryl) group in the bromo amide anion (2) to form an isocyanate, which is hydrolysed to the primary amine by the aqueous base present.

$$R\cdot CONH_2 \xrightarrow[\ominus OH]{Br_2} \left(R \overset{\ominus}{\underset{\underset{O}{\|}}{-C-\overset{\cdot\cdot}{N}}}-Br \right) \xrightarrow{-Br^\ominus} R-N=C=O \xrightarrow[H_2O]{\ominus OH} R\cdot NH_2 + CO_3^{2\ominus}$$
$$(2)$$

An important consequence of the mechanism of the rearrangement step is that if the migrating group R is attached to the amide carbonyl via a chiral carbon atom the configuration is retained in the product, thus generating a fully optically active amine from optically active starting material (e.g. 2-phenylpropanamide).

If the reaction is carried out by treating the amide with bromine and sodium methoxide in methanol, the alkyl isocyanate intermediate is trapped as the *N*-alkylurethane (3) which can subsequently be hydrolysed to the primary amine.

$$R\cdot N=C=O \xrightarrow{MeOH} R\cdot NH\cdot CO_2 Me \xrightarrow{\overset{\ominus}{OH/H_2O}} R\cdot NH_2$$
$$(3)$$

With long chain (C > 8) alkanamides this modification gives better yields than those obtained by the standard procedure.[221]

A closely related reaction of general applicability is the *Curtius rearrangement*[222] of acyl azides. The rearranging species in the Schmidt reaction (see p. 898) is in fact also a protonated acyl azide; these azides are readily prepared by the action of nitrous acid on acyl hydrazides which are themselves formed from esters and hydrazine (Section 9.6.17, p. 1269). On heating in aprotic solvents the acyl azides decompose to yield the corresponding isocyanates in good yield.

5.16.7 SOME METHODS FOR THE PROTECTION OF THE AMINO AND IMINO GROUPS

Only a restricted selection of the very great number of methods for the protection of the amino and imino groups is possible here. These have often been developed for the protection of the amino group in an amino acid for the purpose of peptide synthesis. Those considered below are: (*a*) *N*-acyl derivatives; (*b*) carbamates; and (*c*) phthalimides.

N-ACYL DERIVATIVES
The commonly encountered compounds in this class are the formyl, acetyl and benzoyl derivatives. The formation of the N-*formyl* compound is readily accomplished by heating the amino compound with formic acid in the presence of acetic anhydride.[223] Alternatively, formic acid in the presence of dicyclohexylcarbodiimide (DCC) in pyridine solution is a suitable method for the protection of amino acid esters.[224]

General procedure for the formation of *N*-formyl derivatives of α-amino acids.[223]
Acetic anhydride (83 ml) is added dropwise to a mixture of approximately 0.10 mol of the amino acid in 250 ml of 85 per cent formic acid at a rate to maintain the temperature of the mixture between 50 and 60 °C. After the addition is complete, the mixture is stirred at room temperature for 1 hour at the end of which time 80 ml of icewater is introduced, and the mixture is concentrated

under reduced pressure. The crystalline residue is recrystallised from water or aqueous ethanol.

The *N*-formyl group is stable providing hot aqueous acid or alkaline conditions are avoided. *Deprotection* is affected by prolonged hot alkaline hydrolysis, or in the case of protected amino acids, by oxidative[225] or reductive procedures.[226]

N-*Acetyl* and N-*benzoyl* derivatives may be readily prepared from the corresponding acid chloride or, in the case of acetyl derivatives, acetic anhydride. General procedures are detailed in Sections 9.6.21, p. 1273, and 9.6.23, p. 1279, and the use of *N*-acetyl derivatives in amino acid synthesis is to be found, for example, in Expts 5.183 and 5.184.

Deprotection may be effected by vigorous acidic or alkaline hydrolysis (p. 918). Alternative methods of deprotection are sodium hydride in refluxing dimethoxyethane,[227] or by heating with 85 per cent hydrazine hydrate (*hydrazinolysis*).[228]

CARBAMATES

The two commonest examples are the *benzyl carbamates* (CBZ derivatives) (4) and the *t-butyl carbamates* (BOC derivatives) (5) where the amino group is protected by the benzyloxycarbonyl or the t-butyloxycarbonyl groups respectively.

$$R^1R^2N \overset{\displaystyle O}{\underset{\displaystyle (4)}{\|}} O\!\!-\!\!Ph \qquad\qquad R^1R^2N \overset{\displaystyle O}{\underset{\displaystyle (5)}{\|}} O\!\!-\!\!Bu^t$$

An illustrative example of the introduction of the benzyloxycarbonyl group, and its removal by hydrogenolysis is given in the synthesis of L-prolylglycine (Expt 5.188). An alternative reagent for the removal of this protecting group is iodotrimethylsilane in acetonitrile at room temperature.[229]

The conversion of the amino group in amino acids into a t-butyloxycarbonyl group is a process which is widely used in automated peptide synthesis carried out on solid (polymeric) supports (*solid phase synthesis; the Merrifield technique).*[230]

t-Butyloxycarbonyl derivatives may be prepared by reacting the amino acid with t-butyl azidoformate while maintaining the pH at about 9.5 by the controlled addition of aqueous sodium hydroxide solution using an autotitrator (pH stat), and then liberating the product by careful acidification of the reaction mixture to pH 3 with citric acid.[231]

$$NH_2 \cdot CHR \cdot CO_2H \xrightarrow[\ominus OH]{Bu^tO \cdot CON_3}$$

$$Bu^tO \cdot CO \cdot NH \cdot CHR \cdot CO_2^{\ominus} \xrightarrow{H^{\oplus}} Bu^tO \cdot CO \cdot NH \cdot CHR \cdot CO_2H$$

Although it has been on the market for some time, the reagent is no longer commercially available however. One convenient route for its preparation, starting from phenyl chloroformate, proceeds as formulated below; concise details have been published.[232] The reagent is toxic and is best prepared as required and used in the crude state, since it may explode on distillation or storage.

$$ClCO \cdot OPh \xrightarrow{Bu^tO^{\ominus}} Bu^tO \cdot CO \cdot OPh \xrightarrow{N_2H_4} Bu^tO \cdot CONH \cdot NH_2 \xrightarrow{HNO_2} Bu^tO \cdot CON_3$$

785

Alternatively, acylation with di-t-butyl pyrocarbonate[233] has been recommended as a general procedure for the preparation of t-butyloxycarbonyl amino acids.[234]

$$Bu^tO \cdot CO \cdot O \cdot CO \cdot OBu^t + NH_2 \cdot CHR \cdot CO_2H \xrightarrow[-Bu^tOH]{-CO_2} Bu^tO \cdot CO \cdot NH \cdot CHR \cdot CO_2H$$

Procedure for BOC-amino acids. A solution of the amino acid (10 mol) in a mixture of dioxane (20 ml), water (10 ml) and 1 M sodium hydroxide (10 ml) is stirred and cooled in an ice-water bath. Di-t-butyl pyrocarbonate (1) (2.4 g, 11 mmol) is added and stirring is continued at room temperature for 30 minutes. The solution is concentrated *in vacuo* to about 10–15 ml, cooled in an ice-water bath, covered with a layer of ethyl acetate (30 ml) and acidified with dilute aqueous potassium hydrogen sulphate solution to pH 2–3 (Congo red). The aqueous phase is extracted with ethyl acetate (2 × 15 ml). The ethyl acetate extracts are pooled, washed with water (2 × 30 ml), dried over anhydrous sodium sulphate and evaporated *in vacuo*. The residue is recrystallised with a suitable solvent (e.g. ethyl acetate–hexane).

Note. (1) Commercially available, designated as di-t-butyl-dicarbonate. It melts at 20–22 °C and should be stored in a refrigerator.

Deprotection. The BOC group is easily removed under quite mildly acidic conditions, a feature which underlines its value in selective deprotection in peptide synthesis. Typically, treatment at room temperature for 30–60 minutes with a 1 M solution of hydrogen chloride in acetic acid, or with neat trifluoroacetic acid, is used. It is of interest that after t.l.c. of BOC-amino acids, brief exposure of the plates to hydrogen chloride fumes enables the ninhydrin reaction to be used to detect the presence of the liberated free amino acids.

PHTHALIMIDES
These may be prepared from a primary amine by reaction with phthalic anhydride. The general procedure described in Section 9.6.21, p. 1276 uses glacial acetic acid; reaction conditions using chloroform as the solvent medium have been reported.[235] Deprotection using hydrazine is described in Expt 5.198.

5.17 ALIPHATIC SULPHUR COMPOUNDS

The simpler aliphatic sulphur derivatives considered in this section are the sulphur analogues of oxygen compounds, e.g. the thiols (RSH), and the symmetrical or unsymmetrical thioethers (R·S·R or R¹·S·R²). Compounds derived from these two groups, where the sulphur atom can attain a higher oxidation state, are the sulphinic and sulphonic acids (R·SO·OH and R·SO₂·OH), the sulphoxides (R·SO·R), and the sulphones (R·SO₂·R). Other compounds of interest which are considered in the following discussion are:

the dialkyldisulphides (R·S·S·R: sulphur analogues of the peroxides);
sulphonium salts [e.g. trimethylsulphonium halides (1)];
thioacetals [e.g. 2-ethyl-1,3-dithiane (2)];
O,S-dialkyldithiocarbonates (xanthate esters, e.g. *O*-ethyl *S*-ethyl dithiocarbonate (3)].

The chirality exhibited by the pyramidal sulphur compounds (e.g. sulphonium salts, sulphoxides and sulphinic esters) should be noted, but cannot be considered within the scope of this book.

SPECTROSCOPIC FEATURES

The characteristic absorption frequencies of this varied group of sulphur compounds may be inferred by reference to the data given in Appendices 2, 3 and 4.

5.17.1 THIOLS AND THIOACETALS

The thiols are the sulphur analogues of the alcohols and were formerly called the *mercaptans*. Although they may be prepared by the interaction of an alkyl halide and sodium hydrosulphide in ethanolic solution, a better method involves the interaction of an alkyl bromide and thiourea to form an S-alkylisothiouronium salt, followed by hydrolysis of the latter with sodium hydroxide solution.

$$RBr + NaSH \longrightarrow RSH + NaBr$$

$$2RBr + 2S{=}C(NH_2)_2 \longrightarrow$$

$$2RS{\cdot}C({=}NH_2){\cdot}NH_2\}Br \xrightarrow{NaOH} 2RSH + H_2N{\cdot}C({=}NH){\cdot}NH{\cdot}CN$$

This preparative method is quite general as indicated by the range of examples included in Expt 5.202.

In many cases it is not necessary to prepare the alkyl bromide; the S-alkylisothiouronium salt may be prepared directly from the alcohol by heating with thiourea and concentrated aqueous hydrobromic acid.[236]

The lower members of the thiol series have remarkably disageeable odours, but the offensive odour diminishes with increasing carbon content; for example, dodecane-1-thiol is not noticeably unpleasant.

The preparation and isolation of S-benzylisothiouronium chloride by the interaction of benzyl chloride and thiourea is described in Expt 5.203. On recrystallisation the compound separates in either, or sometimes as both, of two dimorphic forms, m.p. 150 and 175 °C respectively. The former may be converted into the higher m.p. form by dissolving it in ethanol and seeding with crystals of the form, m.p. 175 °C: the low m.p. form when warmed to 175 °C gives, after solidification, a m.p. of 175 °C. The compound is of particular interest as a reagent for the characterisation of carboxylic acids or sulphonic acids with which it forms the S-benzylisothiouronium carboxylate (or sulphonate) salts (Sections 9.6.15, p. 1264, and 9.6.26, p. 1285). Both dimorphic forms give identical derivatives.

Thioacetals, arising from the reaction of an aldehyde or ketone with either two moles of a thiol or one mole of a 1,2- or 1,3-dithiol, are useful both as a means of protection of a carbonyl group (Section 5.8.8, p. 625) and as important intermediates in organic synthesis.

$$R^1 \cdot CHO \xrightarrow{R^2SH} R^1 \cdot CH(SR^2)_2 \qquad R^1 \cdot CHO \; + \; \begin{matrix} HS \\ \\ HS \end{matrix} \longrightarrow R^1 \begin{matrix} S \\ \\ S \end{matrix}$$

In particular, 1,3-dithiane prepared from dimethoxymethane (methylal) and propane-1,3-dithiol in the presence of boron trifluoride-etherate,[237] and 2-alkyl-1,3-dithianes prepared similarly from aldehydes,[238a] are important acyl anion equivalents. These and other uses are discussed in Sections 5.7.5, p. 596, and 6.6.1, p. 909. A wide-ranging review of the reversal of polarity of the carbonyl group through the formation of these sulphur-containing reagents has emphasised their value in organic synthesis.[238b]

Experiment 5.202 HEXANE-1-THIOL

$$2Me \cdot (CH_2)_4 \cdot CH_2Br + 2H_2N \cdot CS \cdot NH_2 \longrightarrow$$

$$2[Me \cdot (CH_2)_4 \cdot CH_2 \cdot S \cdot C(NH_2){=}\overset{\oplus}{N}H_2]\overset{\ominus}{Br} \xrightarrow{NaOH}$$

$$2Me \cdot (CH_2)_4 \cdot CH_2 \cdot SH + H_2N \cdot C({=}NH) \cdot NH \cdot CN$$

CAUTION: *This preparation must be carried out in an efficient fume cupboard; disposable plastic gloves should be worn when handling thiourea.*

Into a 500-ml two-necked flask, equipped with a sealed stirrer unit and a reflux condenser, place 62.5 g (53.5 ml, 0.38 mol) of 1-bromohexane (Expt 5.55) and a solution of 35 g (0.5 mol) of thiourea in 25 ml of water. Connect a tube from the top of the condenser leading to an inverted funnel just immersed in potassium permanganate solution in order to prevent the escape of unpleasant odours. Stir the mixture vigorously and heat under reflux for 2 hours; the mixture becomes homogeneous after about 30 minutes and the additional heating ensures the completeness of the reaction. Add a solution of 30 g of sodium hydroxide in 300 ml of water and reflux, with stirring, for a further 2 hours; during this period the thiol separates since it is largely insoluble in the alkaline medium. Allow to cool and separate the upper layer of almost pure hexane-1-thiol (35 g). Acidify the aqueous layer with a cold solution of 75 ml of concentrated sulphuric acid in 50 ml of water, and extract it with 75 ml of ether. Combine the ethereal extract with the crude thiol, dry with anhydrous sodium sulphate and remove the ether on a water bath. Distil the residue using an air bath (Fig. 2.46(a) and (b)) and collect the hexane-1-thiol at 150–152 °C. The yield is 37.5 g (84%).

Cognate preparation. *Butane-1-thiol.* Use 51 g (40 ml, 0.372 mol) of butyl bromide (Expt 5.54), 38 g (0.5 mol) of thiourea and 25 ml of water. Reflux, with stirring, for 3 hours; the mixture becomes homogeneous after 1 hour. Allow to cool and separate the upper layer of the thiol (*A*). Acidify the aqueous layer with a cold solution of 7 ml of concentrated sulphuric acid in 50 ml of water, cool and saturate with salt; remove the upper layer of butane-1-thiol (*B*) and combine it with (*A*). Extract the aqueous liquid with 75 ml of ether, dry the ethereal extract with sodium sulphate or calcium sulphate and distil off the ether from a water bath through a fractionating column. Combine the residue with (*A*) and (*B*), and distil. Collect the butane-1-thiol at 97–99 °C. The yield is 24 g (72%).

General remarks on the preparation of thiols. The above method is of quite general application. If the bromoalkane is inexpensive, the extraction with ether may be omitted, thus rendering the preparation of thiols a comparatively easy and not unduly unpleasant operation. The following thiols may be prepared in yields of the same order as hexane-1-thiol and butane-1-thiol; ethanethiol, b.p. 35–36 °C; propane-1-thiol, b.p. 66–67 °C; propane-2-thiol, b.p. 51–52 °C; 2-methylpropane-1-thiol, b.p. 87–88 °C; pentane-1-thiol, b.p. 124–125 °C; heptane-1-thiol, b.p. 175–176 °C; octane-1-thiol, b.p. 198–200 °C or 98–100 °C/22 mmHg; nonane-1-thiol, b.p. 220–222 °C or 98–100 °C/ 15 mmHg; decane-1-thiol, b.p. 96–97 °C/5 mmHg or 114 °C/13 mmHg; undecane-1-thiol, b.p. 103–104 °C/3 mmHg; dodecane-1-thiol, b.p. 111– 112 °C/3 mmHg or 153–155 °C/24 mmHg; tetradecane-1-thiol, b.p. 176– 180 °C/22 mmHg; phenylmethanethiol, b.p. 195 °C.

Experiment 5.203 S-BENZYLISOTHIOURONIUM CHLORIDE

$$Ph \cdot CH_2Cl + (NH_2)_2C{=}S \longrightarrow Cl^\ominus \overset{\oplus}{N}H_2{=}C(NH_2) \cdot S \cdot CH_2 \cdot Ph$$

Dissolve 76 g (1 mol) of thiourea in 200 ml of warm water in a 1-litre round-bottomed flask. Dilute the solution with 135 ml of rectified spirit and add 126.5 g (1 mol) of benzyl chloride. Heat the mixture under reflux on a water bath until the benzyl chloride dissolves (about 15 minutes) and for a further 30 minutes taking care that the mixture is well shaken from time to time. Cool the mixture in ice: there is a tendency to supersaturation so that it is advisable to stir (or shake) the cold solution vigorously, when the substance crystallises suddenly. Filter off the solid at the pump. Evaporate the filtrate to about half bulk in order to recover a further small quantity of product. Dry the compound upon filter paper in the air. The yield of S-benzylisothiouronium chloride, m.p. 174 °C, is 200 g (99%). Recrystallise the salt from 400 ml of 0.2 M hydrochloric acid; filter off the solid which separates on cooling. The yield of recrystallised salt, m.p. 175 °C, is 185 g (91%); some of the dimorphic form, m.p. 150 °C, may also separate.

5.17.2 DIALKYL SULPHIDES (THIOETHERS) AND TRIALKYLSULPHONIUM SALTS

Retrosynthetic C—S disconnection reveals two possible routes to dialkyl sulphides.

(a) $R{+}S{+}R \Longrightarrow 2R^\oplus + S^{2\ominus}$ (b) $R^1{-}S{+}R^2 \Longrightarrow R^1{-}S^\ominus + {}^\oplus R^2$

Formation of a symmetrical sulphide (a) (e.g. dipropyl sulphide, Expt 5.204), is conveniently effected by boiling an alkyl halide (the source of carbocations) with sodium sulphide in ethanolic solution. Mixed sulphides (b) are prepared by alkylation of a thiolate salt (a *mercaptide*) with an alkyl halide (cf. Williamson's ether synthesis, Section 5.6.2, p. 583). In the case of an alkyl aryl sulphide (R·S·Ar) where the aromatic ring contains activating nitro groups (see Section 6.5.3, p. 900), the aryl halide is used with the alkyl thiolate salt. The alternative alkylation of a substituted thiophenol is described in Section 8.3.4, p. 1160. The former procedure is illustrated by the preparation of isobutyl 2,4-dinitrophenyl sulphide (Expt 5.205) from 1-chloro-2,4-dinitrobenzene and 2-methylpropane-1-thiol.

A variant of this procedure is provided by the preparation of S-benzyl-L-cysteine (Expt 5.206). The required thiolate salt is prepared by the reductive cleavage with sodium in liquid ammonia of the disulphide linkage in the amino acid, (S)-cystine, and is alkylated *in situ* with benzyl chloride. The preparation of this S-benzyl derivative constitutes a method of protection of the thiol grouping in cysteine.

The use of a cyanoethylation procedure (Section 5.13.3, p. 717) or related Michael type addition reactions for the synthesis of functionalised sulphides should be noted as an important preparative route.

This general procedure has been used for the preparation of sulphides before their conversion into five-membered heterocyclic sulphur systems.[239]

Dialkyl sulphides are converted into trialkylsulphonium salts by treatment with an alkyl halide (the bromide or iodide is usually the reactant of choice). An important example of this group is trimethylsulphonium iodide, which is used as a methylene transfer reagent by virtue of its being converted in the presence of base into a sulphur ylide, which is a nucleophilic carbene equivalent.

$$Me_2S + MeI \xrightarrow{} Me_3\overset{\oplus}{S}\overset{\ominus}{I} \xrightarrow{\text{base}} Me_2\overset{\oplus}{S}{-}\overset{\ominus}{C}H_2 \equiv Me_2S + [:CH_2]$$

Examples of its use are to be found in Expt 8.4.

Dialkyl disulphides, of which (S)-cystine above is an example, may be prepared from thiols by mild oxidation, usually with iodine in the presence of alkali. A convenient synthesis of unsymmetrical analogues results when a symmetrical disulphide is heated with thiol to establish an equilibrium mixture. By careful choice of the reactants, fractional distillation then removes the more volatile thiol leaving the mixed disulphide as a residue.[240]

$$R^1 {\cdot} S {\cdot} S {\cdot} R^1 + R^2 SH \rightleftharpoons R^1 {\cdot} S {\cdot} S {\cdot} R^2 + R^1 SH$$

Experiment 5.204 DIPROPYL SULPHIDE

$$2\ Me{\cdot}CH_2{\cdot}CH_2Br + Na_2S \longrightarrow (Me{\cdot}CH_2{\cdot}CH_2)_2S + 2NaBr$$

CAUTION: *This preparation must be carried out in an efficient fume cupboard.*

Place 56 g (0.5 mol) of finely powdered, fused sodium sulphide and 100 ml of rectified spirit in a 500-ml round-bottomed flask equipped with a reflux condenser. To the boiling mixture add 46 g (34 ml, 0.374 mol) of propyl bromide slowly and reflux for 6 hours. Distil off the ethanol on a water bath, and add a large excess of water to the distillate. Separate the upper layer of crude sulphide, wash it with three 40-ml portions of 5 per cent sodium hydroxide solution, then with water until the washings are neutral, and dry over anhydrous calcium chloride or anhydrous calcium sulphate. Distil, and collect the dipropyl sulphide at 141–143 °C. The yield is 20 g (91%). If the sulphide is required perfectly pure, it should be redistilled from a little sodium.

Cognate preparation. *Dibenzyl sulphide.* Heat a solution of 63 g (0.5 mol) of benzyl chloride in 160 ml of rectified spirit on a steam bath and stir while adding a solution of 29 g (0.25 mol) of fused sodium sulphide in about 50–60 ml of water. Continue stirring and heating for 3 days, remove the ethanol on a rotary evaporator and pour the residue on to 350 g of crushed ice. Separate the oil and triturate with a little 70 per cent ethanol to crystallise the product. Recrystallise from the same solvent; the yield of dibenzyl sulphide is 26 g (83%), m.p. 49 °C.

Experiment 5.205   ISOBUTYL 2,4-DINITROPHENYL SULPHIDE

$$2,4\text{-}(NO_2)_2C_6H_3Cl + HS\cdot CH_2\cdot CHMe_2 \longrightarrow$$
$$2,4\text{-}(NO_2)_2C_6H_3\cdot S\cdot CH_2\cdot CHMe_2$$

Dissolve 1 g (0.005 mol) of 1-chloro-2,4-dinitrobenzene in 5 ml of rectified spirit with warming and add a solution of 0.5 ml (0.005 mol) of 2-methylpropane-1-thiol in 5 ml of rectified spirit containing 2 ml of 10 per cent aqueous sodium hydroxide. Heat under reflux for 10 minutes and decant the hot solution from any insoluble material into a clean conical flask. Allow the solution to cool, filter and recrystallise the sulphide twice from methanol. The product is obtained as yellow flakes, m.p. 75–76 °C; the yield is 440 mg (35%).

Experiment 5.206   *S*-BENZYL-L-CYSTEINE [L-*2-Amino-3-(benzylthio)-propanoic acid*]

$$[NH_2\cdot CH(CO_2H)\cdot CH_2\cdot S]_2 \xrightarrow{Na/NH_3} 2NH_2\cdot CH(CO_2Na)\cdot CH_2SNa \xrightarrow[\text{(ii) HCl}]{\text{(i) Ph}\cdot CH_2Cl}$$
$$2NH_2\cdot CH(CO_2H)\cdot S\cdot CH_2\cdot Ph$$

Collect about 750 ml of liquid ammonia (Section 2.17.7, p. 116) in a 1-litre three-necked flask, surrounded by a lagging bath of cork chips and fitted with a sealed stirrer unit, a soda-lime guard-tube and a stopper. Weigh out 24 g (0.1 mol) of L-cystine (Expt 5.187), and about 10 g (0.48 mol) of sodium cut into small pieces under dry light petroleum. Start the stirrer, add about 2 g of the sodium followed by L-cystine in small portions until the blue colour has disappeared. Repeat this addition sequence until all of the cystine has been added and a permanent blue colour remains. Discharge the blue colour by gradually adding powdered ammonium chloride, and then add dropwise 25.3 g (23 ml, 0.2 mol) of benzyl chloride. Remove the stirrer and lagging bath and allow the ammonia to evaporate overnight. Dissolve the residue in 100 ml of cold water and add concentrated hydrochloric acid until the resulting mass is acid to Congo red. Heat the mixture gradually to boiling to dissolve the precipitated product and allow to cool. Filter off the long needles of *S*-benzyl-L-cysteine which separate, wash with a little cold water and allow to dry in the air. The yield is 38 g (90%), m.p. 214 °C (decomp.).

5.17.3 SULPHOXIDES AND SULPHONES

Dialkyl sulphides may be oxidised to the sulphoxides (4) (alkylsulphinylalkanes), and thence to the sulphones (5) (alkylsulphonylalkanes).

$$R\cdot S\cdot R \xrightarrow{[O]} R\cdot SO\cdot R \xrightarrow{[O]} R\cdot SO_2\cdot R$$
$$\qquad\qquad (4) \qquad\qquad\quad (5)$$

Sulphoxides are obtained by using a variety of mild oxidising agents; hydrogen peroxide has been widely used, but aqueous sodium metaperiodate and sodium perborate[241] have the advantage of being easily handled reagents. The use of the former reagent is described in Expt 5.207.

More vigorous oxidising agents yield sulphones; an excess of potassium permanganate in aqueous acetic acid is often employed. The preparation of isobutyl 2,4-dinitropenyl sulphone (Expt 5.208) may be regarded as typical. Frequently the vigour of this oxidant (and indeed others such as organic peracids) results in other functional groups which may be present in the sulphide (e.g. carbon–carbon multiple bonds) being oxidatively degraded. A reagent which is chemoselective for the sulphide to sulphone conversion is potassium hydrogen persulphate.[242]

The simplest sulphoxide, dimethyl sulphoxide, is an important aprotic solvent (Section 4.1.*33*, p. 412). Its use as a reagent in carbon–carbon forming reactions and as a reagent for the oxidation of alcohols to carbonyl compounds (p. 608) (the *Pfitzner–Moffatt* and *Swern oxidations*) has been extensively reviewed.[243] An illustrative example of carbon–carbon bond formation using dimethyl sulphoxide is noted in Expt 7.3.

Experiment 5.207 DIBENZYL SULPHOXIDE

$$(Ph \cdot CH_2)_2 S \xrightarrow{NaIO_4} (Ph \cdot CH_2)_2 SO$$

Stir 2.35 g (0.011 mol) of sodium metaperiodate (Section 4.2.*55*, p. 454) in 45 ml of a 1:1 mixture of water and methanol held at 0 °C. Add portionwise 2.14 g (0.01 mol) of dibenzyl sulphide (Expt 5.204) and continue to stir the mixture at 0 °C for several hours, preferably overnight. Extract the reaction mixture (which contains precipitated sodium iodate) with three 20 ml portions of chloroform. Dry the combined chloroform extracts over magnesium sulphate and remove the solvent on a rotary evaporator. Recrystallise the product from ethanol. The yield of dibenzyl sulphoxide is 2.2 g (96%), m.p. 135 °C.

Experiment 5.208 ISOBUTYL 2,4-DINITROPHENYL SULPHONE

$$2,4\text{-}(NO_2)_2 C_6 H_3 \cdot S \cdot CH_2 \cdot CHMe_2 \xrightarrow{KMnO_4} 2,4\text{-}(NO_2)_2 C_6 H_3 \cdot SO_2 \cdot CH_2 \cdot CHMe_2$$

Dissolve 340 mg of isobutyl 2,4-dinitrophenyl sulphide (Expt 5.205) in 10 ml of acetic acid, warm to about 50 °C and add dropwise 8 ml of 3 per cent aqueous potassium permanganate solution. Maintain the solution at about 50 °C for 20 minutes, and then pass through the solution a stream of sulphur dioxide to decompose excess potassium permanganate. Add crushed ice to the yellow solution to precipitate the crude sulphone. Filter the precipitate, dry and recrystallise from rectified spirit to give the yellow crystalline product, m.p. 105–106 °C; the yield is 190 mg (50%).

5.17.4 *O,S*-DIALKYL DITHIOCARBONATES (XANTHATE ESTERS)

Alkali metal salts of the *O*-alkyl dithiocarbonates (e.g. $RO \cdot CS \cdot S^{\ominus} K^{\oplus}$, the xanthates) are prepared by the reaction of carbon disulphide with an alcohol and an alkali metal hydroxide.

$$KOH + CS_2 + R'OH \longrightarrow R'O \cdot CS \cdot \overset{\ominus \oplus}{SK} + H_2O$$

Alkylation of this xanthate salt with an alkyl halide in absolute ethanol leads to an *O,S*-dialkyl dithiocarbonate.

$$R'O \cdot CS \cdot \overset{\ominus}{S} \frown R^2 \!\!-\!\! I \longrightarrow R'O \cdot CS \cdot SR^2 + \overset{\ominus}{I}$$

The xanthate esters are of interest in that if the *O*-alkyl group contains at least one α-hydrogen atom, pyrolysis produces an olefin, a thiol and carbon oxysulphide (the Chugaev reaction, Section 5.2.1, p. 489).

Isolation of the intermediate xanthate salts in the preparation of the xanthate esters is not essential. The formation of the latter may be achieved, in one step, by converting the alcohol into the corresponding alkoxide by reaction with a potassium derivative of a tertiary alcohol (e.g. potassium t-pentoxide), followed by reaction successively with carbon disulphide and the alkyl halide. Both preparative procedures are illustrated in the following section.

Experiment 5.209 *O*-ETHYL S-ETHYL DITHIOCARBONATE

$$EtO^{\ominus}K^{\oplus} + CS_2 \longrightarrow EtO \cdot C(=S) \cdot S^{\ominus}K^{\oplus} \xrightarrow{EtI} EtO \cdot C(=S) \cdot S \cdot Et$$

CAUTION: *These preparations must be conducted in an efficient fume cupboard.*

Potassium *O*-ethyl dithiocarbonate. Into a 500-ml round-bottomed flask, fitted with a reflux condenser, place 42 g (0.75 mol) of potassium hydroxide pellets and 120 g (152 ml) of absolute ethanol. Heat under reflux for 1 hour. Allow to cool and decant the liquid from the residual solid into another dry 500-ml flask; add 57 g (45 ml, 0.75 mol) of carbon disulphide (1) slowly and with constant shaking. Filter the resulting almost solid mass, after cooling in ice, on a sintered glass funnel at the pump, and wash it with three 25 ml portions of ether. Dry the potassium *O*-ethyl dithiocarbonate in a vacuum desiccator over silica gel. The yield is 74 g. If desired, it may be recrystallised from absolute ethanol but this is usually unnecessary.

***O*-Ethyl *S*-ethyl dithiocarbonate.** Place 32 g (0.2 mol) of potassium *O*-ethyl dithiocarbonate and 50 ml of absolute ethanol in a 500-ml round-bottomed flask provided with a double surface condenser. Add 32 g (16.5 ml, 0.205 mol) of ethyl iodide. No reaction appears to take place in the cold. Heat on a water bath for 3 hours: a reaction sets in within 15 minutes and the yellow reaction mixture becomes white owing to the separation of potassium iodide. Add about 150 ml of water, separate the lower layer and wash it with water. Dry it with anhydrous calcium chloride or anhydrous calcium sulphate and distil collecting *O*-ethyl *S*-ethyl dithiocarbonate at 196–198 °C. The yield is 23 g (77%).

Note. (1) Carbon disulphide is toxic and has a dangerously low flash point (Section 2.3.2, p. 40).

Cognate preparation. O-*Ethyl S-butyl dithiocarbonate*. Use 32 g (0.2 mol) of potassium *O*-ethyl dithiocarbonate, 37 g (23 ml, 0.2 mol) of butyl iodide (Expt 5.58) and 50 ml of absolute ethanol. Reflux on a water bath for 3 hours. Pour into 150 ml of water, saturate with salt (in order to facilitate the separation of the upper layer), remove the upper xanthate layer, wash it once with 25 ml of saturated salt solution and dry with anhydrous calcium chloride or anhydrous calcium sulphate. Distil under reduced pressure and collect the pale yellow *O*-ethyl *S*-butyl dithiocarbonate at 90–91 °C/4 mmHg. The yield is 34 g (95%).

O-[1,2,2-Trimethylpropyl] S-methyl dithiocarbonate. In a 2-litre three-necked flask equipped with a stirrer, reflux condenser and dropping funnel, prepare potassium t-pentoxide by dissolving 48.5 g (60 ml, 0.55 mol) of 2-methylbutan-2-ol (t-pentyl alcohol) in 750 ml of dry toluene and adding in portions 21.5 g (0.55 mol) of potassium metal (1) and refluxing gently until reaction is complete. Then add 51 g (63 ml, 0.5 mol) of 3,3-dimethylbutan-2-ol to the hot solution slowly and with stirring. Cool and add slowly 57 g (0.75 mol) of carbon disulphide. When the reaction has subsided, cool the resulting orange xanthate suspension to room temperature, add 78 g (34 ml, 0.55 mol) of methyl iodide and heat on a water bath for 4–5 hours. Filter, remove toulene and the residual alcohols on a rotary evaporator (*fume cupboard*) and distil the product under reduced pressure collecting the xanthate ester at 85–87 °C/6 mmHg (2). The yield is 63 g (70%).

Note. (1) Great care must be taken in the handling of potassium (see Section 4.2.62, p. 460) and of methyl iodide (Section 2.3).
(2) Use a 500-ml flask fitted with a short fractionating column; the distillation is accompanied by considerable frothing.

5.18 UNSATURATED COMPOUNDS

This section reviews some of the futher important methods of synthesis of (i) unsaturated alcohols, (ii) unsaturated carbonyl compounds, and (iii) unsaturated carboxylic acids. It also provides an opportunity to consider those methods which were not conveniently included elsewhere, but which are of great importance and interest.

The importance of these three classes of unsaturated compounds is that they contain many naturally occurring examples having specific biological activity. For this reason their general methods of synthesis are of interest. Furthermore, their particular functional features are to be found embedded in the structures of more complex acyclic, alicyclic and polycyclic molecules. Thus the incorporation of the simpler structures as 'building blocks' in multistage syntheses of, for example, antibiotics has been one of the synthetic challenges of recent decades.

5.18.1 UNSATURATED ALCOHOLS

The two important groups of unsaturated alcohols are the allylic alcohols [e,g, (1)], and the homoallylic alcohols [e.g. (2)]. The (*E*)-isomers are formulated for (1) and (2) but the (*Z*)-isomers are of course also known; the hydroxyl-carrying carbon may also be chiral [(*R*)- and (*S*)-isomers].

(1)   (2)

A synthetic survey embracing all these structural and isomeric features cannot be attempted here, and the discussion is restricted to the allylic alcohols.

Retrosynthetic strategies, limited to the more obvious but widely used FGI and disconnection transforms in the case of an allylic alcohol target molecule, are summarised below.

The synthetic *method (a)* is the regioselective reduction of an α,β-unsaturated aldehyde or ketone (Section 5.18.2, p. 798), which is most conveniently effected by the Meerwein–Ponndorf–Verley procedure (Section 5.4.1, p. 520). The further disconnection shown of the α,β-carbonyl compound is a *retro*-aldol condensation (Section 5.18.2, p. 799); however it should be emphasised that other routes to the unsaturated carbonyl compound, such as the Horner–Emmons reaction (Section 5.18.2, p. 799), may also be feasible.

Asymmetric reduction of α,β-unsaturated carbonyl compounds using chiral complexes (Section 5.4.1, p. 521) could feasibly lead to optically active allylic alcohols. Other reducing agents which have some merit of regioselectivity, but not stereoselectivity, are sodium cyanoborohydride,[244] and sodium borohydride in the presence of lanthanide salts.[245]

The synthetic *method (b)* combines the formation of a primary or secondary alkynol [from formaldehyde or an aldehyde respectively and an organometallic acetylenic reagent (Section 5.4.2, p. 532)] with the semihydrogenation of the triple bond to a double bond. As noted in Section 5.2.2, p. 493, appropriate selection of catalyst is necessary in the hydrogenation step to ensure the formation of either the (*E*)- or the (*Z*)-isomer. The specific formation of the allylic alcohol, $CH_2{=}CH{\cdot}CH(OH){\cdot}R$, is from a vinylmagnesium halide (Expt 5.41) and an aldehyde.

The synthetic *method (c)*, is a base-catalysed ring opening of an oxirane.[246] Since the oxirane may be formed by the epoxidation of an olefin (Section 8.1.3, p. 1132), or by a methylene insertion reaction into a carbonyl group (Section 8.1.2, p. 1131), this method is of some versatility.

This reaction is used as one step [(7) to (6)] of the multistage synthesis of the monocyclic sesquiterpenic hydrocarbon, dehydro-α-curumene (3) [6-methyl-2-(4-methylphenyl)hepta-1,5-diene].[247] The overall strategy is revealed from a retrosynthetic analysis for this compound, and is of interest since it emphasises the logic of the disconnection approach and draws together a number of important reactions which are relevant to this discussion of unsaturated compounds.

Thus the initial disconnection of (3) reveals the ylide, isopropylidenetriphenyl-phosphorane, and the γ,δ-unsaturated aldehyde (4) (see the Wittig reaction, Sections 5.2.3, p. 495). A *retro*-Claisen–Cope rearrangement (a [3,3] sigmatropic shift, see also the Claisen rearrangement, Section 6.9.2, p. 978) of the aldehyde (4) discloses the allyl vinyl ether (5). Disconnection of this ether reveals the allylic alcohol and an alkyl vinyl ether. The forward synthetic step [(6) to (5)] is a transetherification reaction with ethyl vinyl ether. Reconnection of (6) gives the epoxide (7). Various retrosynthetic strategies for epoxides are considered in Section 8.1, but the most suitable in the case of (7) is disconnection to *p*-methyl-acetophenone and a carbene insertion reagent. The methodology of the synthesis of dehydro-α-curumene is formulated in Expt 5.210.

Experiment 5.210   2-(4-METHYLPHENYL)-PROP-2-EN-1-OL AND 6-METHYL-2-(4-METHYLPHENYL)HEPTA-1,5-DIENE[247]

1,2-Epoxy-2-(4-methylphenyl)propane. To a solution of dimethyloxosulphonium methylide [prepared from sodium hydride (1.5 g, 50% dispersion), finely powdered trimethyloxosulphonium iodide (6.6 g) and dimethyl sulphoxide (30 ml) under a nitrogen atmosphere] is added with stirring, a solution of *p*-methylacetophenone (3.4 g) in dimethyl sulphoxide (10 ml). The reaction mixture is left overnight at room temperature and then heated to 50 °C for 1 hour. After cooling and adding water (60 ml), the mixture is extracted with ether, the combined ethereal extracts washed with water and then dried over anhydrous sodium sulphate. Evaporation of the solvent followed by chromatography over neutral alumina using light petroleum (b.p. 40–60 °C) as eluant, affords the epoxide (3.3 g, 90%).

2-(4-methylphenyl)prop-2-en-1-ol. To a solution of diisopropylamine (3 g) in ether is added with stirring butyllithium (15 ml, 2.1 M in ether) under a nitrogen atmosphere. After 30 minutes, the solution of the foregoing epoxide (3 g) in ether (40 ml) is added slowly with stirring and the mixture left overnight at room temperature. The mixture is refluxed for 4 hours and the disappearance of epoxide is monitored by t.l.c. The solution, after cooling, is partitioned between ether and water, the organic phase washed successively with water and brine and then dried over anhydrous sodium sulphate. Solvent evaporation followed by chromatography over neutral alumina affords 2-(4-methylphenyl)prop-2-en-1-ol (2.6 g, 85%), pure by t.l.c. analysis; b.p. 105–107 °C/ 5–7 mmHg; i.r. (film) 3380, 1050 (OH), 3070, 1650, 890 ($>$C$=$CH$_2$), and 820 cm^{-1} (*para*-substituted benzene).

Conversion into 6-methyl-2-(4-methylphenyl)hepta-1,5-diene. *2-(4-methylphenyl)prop-2-enyl vinyl ether.* A mixture of the foregoing allylic alcohol (2.5 g), ethyl vinyl ether (75 ml) and freshly crystallised mercury(II) acetate (600 mg) is refluxed continuously for 12 hours on a water bath. The reaction mixture is chilled in ice and mixed with 10 per cent aqueous sodium carbonate solution (25 g) and stirred well for 30 minutes at 0 °C. The organic layer is separated, washed with water and dried over anhydrous potassium carbonate. Evaporation of the solvent and chromatography of the residue on neutral alumina (50 g) furnished, on elution with light petroleum (b.p. 40–60 °C), the allyl vinyl ether which is further purified by distillation under reduced pressure, b.p. 90–92 °C/5–7 mmHg; yield 2.0 g (80%); 1630, 1610 (C=C), 1200 (vinyl ether), 900 (>C=H$_2$) and 820 cm^{-1} (*para*-substituted benzene).

Claisen-Cope rearrangement of the allyl vinyl ether. The ether (1.1 g) is heated for 30 minutes at 180–185 °C under nitrogen in a fully immersed half-filled sealed Pyrex glass tube (**CAUTION**). After cooling, the product is distilled under reduced pressure to furnish 4-(4-methylphenyl)pent-4-enal (850 mg, 87%), b.p. 100–102 °C/5–7 mmHg; i.r. (film) 2700, 1710 (CHO), 3070, 1650, 890 (>C=CH$_2$), and 825 cm^{-1} (*para*-substituted benzene).

Wittig reaction. To the phosphorane [prepared from isopropyltriphenylphosphonium iodide (2.6 g) in dimethyl sulphoxide (6 ml) and sodium hydride (300 mg, 50% dispersion) under a nitrogen atmosphere], is added the foregoing aldehyde (700 mg) in tetrahydrofuran (1 ml). The contents are stirred for 2 hours and then left overnight at room temperature and warmed to 50 °C for 2 hours. The cooled reaction mixture is poured on to crushed ice and extracted with light petroleum (b.p. 40–60 °C). The organic extract is dried over anhydrous sodium sulphate, the solvent removed and the residue chromatographed over neutral alumina (20 g) when elution with light petroleum (b.p. 40–60 °C) gives 6-methyl-2-(4-methylphenyl)hepta-1,5-diene (450 mg, 66%). This is further purified by distillation under reduced pressure, b.p. 110–113 °C/5–7 mmHg; p.m.r. (CCl$_4$, TMS) δ 1.52 (s, 6H, C=C(Me)$_2$), 2.34 (s, 3H, Ar.Me), 5.05 (t, 1H, —CH=C—), 5.24 (broad s, 2H, >C=CH$_2$), and 7.23 (4H, aromatic protons).

5.18.2 UNSATURATED CARBONYL COMPOUNDS

By far the most important compounds of this type are the α,β-unsaturated aldehydes and ketones, of which the simplest general representation is (8) and (9) respectively. Branching may be present at either or both of the α- or β- positions as for example in 2-ethylhex-2-enal (10) and 4-methylpent-3-en-2-one (11). Compounds where the carbon–carbon double bond is terminal are acraldehyde [(8), R=H], and the alkyl (or aryl) vinyl ketones [(9), R′ = H, R = alkyl, aryl etc.].

The preparation of α,β-unsaturated aldehydes and ketones is exemplified by the following four important methods: (*a*) the oxidation of the corresponding unsaturated primary or secondary alcohol; (*b*) the Horner–Emmons or Wadsworth–Emmons modification of the Wittig reaction[248]; (*c*) the aldol condensation followed by dehydration; and (d) the specific formation of alkyl (or aryl) vinyl ketones from Mannich bases.

THE OXIDATION OF α,β-UNSATURATED PRIMARY OR SECONDARY ALCOHOLS.

This method is of value when the alcohol is readily available from natural sources, or when it can be prepared, for example, by the reaction of an alkenyl-organometallic reagent with an aldehyde. An example of the former is the oxidation of the terpenoid alcohol carveol to carvone (Expt 5.88) using pyridinium chlorochromate-on-alumina reagent.

THE HORNER–EMMONS OR WADSWORTH–EMMONS MODIFICATION OF THE WITTIG REACTION

This reaction is discussed in Section 5.2.3, p. 496, where the broad mechanism of the reaction and its preparative value is considered. The example included here is the preparation of 2-oxoundec-3-ene from octanal and diethyl 2-oxopropane-phosphonate,[249] (Expt 5.211). The interest of the procedure is that the reaction is carried out under heterogeneous liquid–liquid conditions using a high concentration of potassium carbonate, but in the absence of an organic solvent and a phase transfer catalyst. This may be contrasted with the conditions which employ an aprotic solvent and a base such as sodium hydride (cf. Expt 5.17).

THE ALDOL REACTION FOLLOWED BY DEHYDRATION

A retrosynthetic analysis of an α,β-unsaturated aldehyde or ketone involves an initial functional group interconversion into a β-hydroxycarbonyl compound, followed by a disconnection into the carbocation (12) and the carbanion (13) synthons. The reagent equivalents of these two synthons are the corresponding carbonyl compounds.

$R^2, R^3 = H$ or alkyl

The forward synthetic reaction is a base-catalysed condensation reaction between two carbonyl compounds, the *aldol condensation* leading to β-hydroxy-aldehydes or β-hydroxyketones followed by dehydration. This sequence is one of the most important carbon–carbon bond forming reactions, and *aldol-type* condensation reactions are considered in a number of other sections of the text, for example, the Doebner reaction (Section 5.18.3, p. 805), the Knoevenagel reaction (Section 5.11.6, p. 681), the Perkin reaction (Section 6.12.3, p. 1036) and the Robinson annelation reaction (Section 7.2).

The mechanism is illustrated in the simple case of the self-condensation of an aldehyde in the presence of base. Here the nucleophilic, mesomerically stabilised α-carbanion (the enolate ion) formed by the action of base, attacks the electrophilic carbonyl carbon of a second aldehyde molecule to form, after proton exchange with the solvent, the β-hydroxyaldehyde.

All the steps of this reaction are reversible but the position of the equilibrium is significantly in favour of the aldol, which generally may be obtained when the reaction is carried out at room temperature or below, followed by extraction and careful distillation under reduced pressure. When the required product is the unsaturated aldehyde the reaction is carried out at a higher temperature, and dehydration of the aldol occurs readily (e.g. 2-ethylhex-2-enal, Expt 5.212). In the case of aldehydes with only one α-hydrogen atom, aldol formation occurs but the resulting β-hydroxyaldehyde cannot undergo the dehydration step.

In a similar way a ketone may undergo a self-condensation reaction to give a ketol, which may be dehydrated in a separate reaction to an α,β-unsaturated ketone.

In this case however the position of equilibrium is not in favour of the condensation product (which rapidly dissociates into the ketone in the presence of base), and a technique has to be employed which continuously removes the ketol product, as it is formed, from the presence of base. A satisfactory procedure for converting acetone into diacetone alcohol is described in Expt 5.213. The dehydration step to give mesityl oxide is subsequently effected by the presence of a trace of iodine.

'Mixed' aldol reactions may be broadly classified as the reaction between two different aldehydes or ketones, or the reaction of an aldehyde with a ketone. Apart from the concomitant self-condensation, not less than two possible 'crossed' products can be envisaged. Such reactions are therefore only preparatively useful either if appropriate structural conditions are present, or if certain experimental conditions are used to effect a *directed* aldol condensation.

In the former instance, the major structural condition is that one of the carbonyl components (either the aldehyde or the ketone) has no α-hydrogen and thus is unable to form an enolate ion, but is nevertheless sufficiently electrophilic at its carbonyl carbon that it reacts with the carbanion of the second carbonyl component. Examples are provided in the formation of 4-phenylbut-3-en-2-one from benzaldehyde (no α-hydrogens) and acetone in the presence of base, and the formation of benzylideneacetophenone from benzaldehyde and acetophenone (Expt 6.135).

Recent decades have seen the enormous development of a great range of procedures for the achievement of directed self- or cross-coupling reactions. The interested reader should consult the relevant review articles which cover the field comprehensively.[250] One example is used here to illustrate, in part, the complexity and interest of these reactions. This is the reaction between cyclo-hexanone and benzaldehyde where a conventional aldol-type reaction is unlikely to be successful owing to the low concentration of carbanions arising from the ketone. However, if the cyclohexanone is first converted into the corresponding silyl enol ether (14), and then treated with the aldehyde in the presence of titanium chloride, the reaction gives a chelated titanium complex of 2-(1'-hydroxybenzyl)cyclohexanone (15); hydrolysis with water gives the aldol[251] (16) (Expt 5.214).

The stereoisomeric features of the intermediates and of the product of this reaction should be noted. Thus the silyl enol ether (14) can only have the (E)-configuration. The product (16) has two chiral sites (*) and the aldol is therefore a mixture of *erythro/threo* ketols; in this case the ratio is 23:69.

It should be noted that several procedures are available for the preparation of 1-(trimethylsilyloxy)cyclohexene. In an original report[252] it was prepared by heating a dimethylformamide solution of cyclohexanone under reflux with chlorotrimethylsilane in the presence of triethylamine. This procedure is generally satisfactory for aldehydes and symmetrical ketones, or ketones which can enolise in one direction only. The method described in Expt 5.214 employs lithium sulphide as an additive to increase the silylating power of the chlorotrimethylsilane, so that the reaction proceeds at room temperature; acetonitrile is used as the solvent.[253]

THE MANNICH REACTION

The mechanism of the reaction is considered in Section 6.12.7, p. 1050, where acetophenone, which can only enolise in one direction, reacts with the formaldehyde equivalent (17) formed from formaldehyde and dimethylamine hydrochloride, to give the Mannich base (18), which in this case is isolated as the hydrochloride. The free Mannich bases are obtained by treatment with base and solvent extraction or crystallisation (e.g. gramine, Expt 6.147).

Conversion of a Mannich base hydrochloride into an α,β-unsaturated carbonyl compound, is illustrated by the formation of phenyl vinyl ketone, which is obtained directly by steam distillation (Expt 6.147). Alternatively the Mannich base may be treated with methyl iodide to form the quaternary salt, which then gives the α,β-unsaturated carbonyl compound by a base-catalysed elimination reaction.

Experiment 5.211   2-OXOUNDEC-3-ENE[249]

$$Me \cdot (CH_2)_5 \cdot CH_2 \cdot CHO + (EtO)_2 P(O) \cdot CH_2 \cdot CO \cdot Me \longrightarrow$$
$$Me \cdot (CH_2)_5 \cdot CH_2 \cdot CH = CH \cdot CO \cdot Me$$

Potassium carbonate (8.3 g, 60 mmol) in water (10 ml) is added to a vigorously stirred mixture of octanal (3.84 g, 30 mmol) and diethyl 2-oxopropanephosphonate (7.0 g, 36 mmol) at room temperature. The mixture is stirred for 18 hours at room temperature, water (15 ml) is then added, and the organic phase is extracted with hexane (3 × 25 ml). The extract is dried with magnesium sulphate (1 g), evaporated, and the residue distilled to give 2-oxoundec-3-ene; 4,4 g (87%), b.p. 80–83 °C/0.7 mmHg.

Experiment 5.212   2-ETHYLHEX-2-ENAL

$$2Et \cdot CH_2 \cdot CHO \xrightarrow{\ominus OH} Et \cdot CH_2 \cdot CH = C(Et) \cdot CHO$$

Place 100 ml of 1 M sodium hydroxide solution in a 500-ml three-necked flask fitted with a sealed stirrer unit and an efficient reflux condenser. Heat the solution to 80 °C and attach to the flask a dropping funnel containing 216 g (264 ml, 3.0 mol) of redistilled butyraldehyde (compare Expt 5.74). With vigorous stirring, add the butyraldehyde as rapidly as the efficiency of the reflux condenser will allow and then boil the reaction mixture under reflux for 1 hour. Cool, separate the organic layer and distil it without further treatment under reduced pressure through a fractionating column (e.g. of the Vigreux type). Collect the pure 2-ethylhex-2-enal as a fraction of b.p. 66–67 °C/25 mmHg; the yield is 160 g (85%).

Cognate preparation.   2-Methylpent-2-enal. Add 174 g (215 ml, 3.0 mol) of propionaldehyde with vigorous stirring to 100 ml of 1 M sodium hydroxide solution, but without initial heating, during 15 minutes. Cool rapidly in an ice bath, isolate the organic product with the aid of a little ether and fractionally distil. The yield is 70 per cent; b.p. 38–39 °C/25 mmHg, 136–137 °C/760 mmHg.

Experiment 5.213   MESITYL OXIDE (4-Methylpent-3-en-2-one)

$$Me_2CO + Me \cdot CO \cdot Me \underset{}{\overset{Ba(OH)_2}{\rightleftharpoons}} Me_2C(OH) \cdot CH_2 \cdot CO \cdot Me$$

$$Me_2C(OH) \cdot CH_2 \cdot CO \cdot Me \xrightarrow[heat]{I_2} Me_2C = CH \cdot CO \cdot Me + H_2O$$

4-Methyl-4-hydroxypentan-2-one (diacetone alcohol). Fit a 1-litre round-bottomed flask with a large Soxhlet extractor (Fig. 2.96) and attach an efficient double surface condenser to the latter. Place 595 g (750 ml, 10.25 mol) of commercial acetone, preferably dried over anhydrous potassium carbonate, and a few fragments of porous porcelain in the flask. Select as large a Soxhlet thimble as the extractor will accommodate and three-quarters fill it with barium hydroxide (1). Fill the remaining space in the thimble with glass wool. Insert the charged thimble into the extractor. Heat the flask on a water bath or steam bath so that the acetone refluxes back into the extractor rather rapidly. Continue the heating until the acetone no longer refluxes when the flask is almost completely immersed in the boiling water bath (72–120 hours). The refluxing may be interrupted at any time for as long as desired without influencing the preparation. Equip the flask with a fractionating column attached to an efficient double surface condenser set for downward distillation. Immerse the flask in an oil bath and raise the temperature gradually to 125 °C; maintain this temperature as long as acetone distils over. The recovery of acetone is complete when the temperature at the top of the column is about 70 °C. Distil the residue (2) under diminished pressure (3); a little acetone passes over first, followed by the diacetone alcohol at 71–74 °C/23 mmHg (or 62–64 °C/13 mmHg). The yield is 450 g (75%).

Mesityl oxide. Fit a 750-ml round-bottomed flask with a fractionating column attached to a condenser set for downward distillation. Place 400 g (3.44 mol) of diacetone alcohol (the crude product is quite satisfactory), 0.1 g of iodine and a few fragments of porous porcelain in the flask. Distil slowly with a small free flame (best in an air bath) and collect the following fractions: (a) 56–80 °C (acetone and a little mesityl oxide); (b) 80–126 °C (two layers, water and mesityl oxide); and (c) 126–131 °C, which is almost pure mesityl oxide. Separate the water from fraction (b), dry with anhydrous potassium carbonate or anhydrous sodium sulphate and fractionate from a small flask. A further quantity of mesityl oxide is thus obtained. The total yield is about 320 g (95%).

Notes. (1) If crystallised barium hydroxide [Ba(OH)$_2$,8H$_2$O] is employed, this becomes dehydrated after one run; the anhydrous compound is just as satisfactory and may be used repeatedly.
(2) The residual liquid contains about 95 per cent of diacetone alcohol and is satisfactory for the preparation of mesityl oxide.
(3) Diacetone alcohol partially decomposes when distilled under normal pressure.

Experiment 5.214 *ERYTHRO-* AND *THREO*-2-(1'-HYDROXYBENZYL) CYCLOHEXANONES[251]

1-Trimethylsilyloxycyclohexene.[253] To a well-stirred suspension of lithium sulphide (1.5 g, 30 mmol) (1) in dry acetonitrile (25 ml) in a 100-ml round-bottomed flask, fitted with a water condenser and a nitrogen inlet, is added chlorotrimethylsilane (6.3 ml, 50 mmol). To this mixture are added cyclohexanone (1.96 g, 20 mmol) and triethylamine (3 ml, 20 mmol) in succession, and the solution is allowed to stir at room temperature (25 °C). The progress of the reaction is monitored by removing aliquots periodically and analysing them after work-up by t.l.c. (silica gel plates with hexane as an eluant). Soon after the completion of the reaction (c. 16 hours), the mixture is taken up in ether (50 ml) and washed thoroughly with ice-cold aqueous 5 per cent hydrochloric acid solution (4 × 50 ml) to remove all basic and water-soluble materials. The ethereal extract is washed with ice-cold aqueous 5 per cent sodium hydrogen carbonate solution (50 ml), water (50 ml), and brine (25 ml). It is dried over anhydrous sodium sulphate and subjected to distillation under reduced pressure to obtain the crude enol silyl ether. The product is further purified by distillation (b.p. 70–71 °C/12 mmHg) to obtain spectrally (p.m.r. and i.r.) pure 1-trimethylsilyloxycyclohexene (95%).

Erythro- **and** ***Threo-*2-(1′-Hydroxybenzyl)cyclohexanones.** A dichloromethane (10 ml) solution of 1-trimethylsilyloxycyclohexene (0.426 g, 2.5 mmol) is added dropwise into a mixture of benzaldehyde (0.292 g, 2.75 mmol) and titanium(IV) chloride (0.55 g, 2.75 mmol) (2) in dry dichloromethane (20 ml) under an argon atmosphere at −78 °C, and the reaction mixture is stirred for 1 hour. After hydrolysis (with water) at that temperature, the resulting organic layer is extracted with ether, and the extract is washed with water and dried over anhydrous sodium sulphate. The extract is evaporated under reduced pressure, and the residue is purified by column chromatography (silica gel). Elution with dichoromethane affords 115 mg (23%) of *erythro*-2-(1′-hydroxybenzyl)cyclohexanone, m.p. 103 °C (recrystallised from propan-2-ol, m.p. 103.5–104.5 °C); i.r. 3530 (OH), 1700 (C=O) cm^{-1}; p.m.r. (CDCl$_3$, TMS) 1.1–2.7 (broad, 9H, aliphatic CH), 3.05 (s, 1H, OH, exchangeable with D$_2$O), 5.40 (d, 1H, $J = 2.5$ Hz, O—CH), and 7.27 (s, 5H, aryl CH).

From the last fraction, 346 mg (69%) of *theo*-2-(1′-hydroxybenzyl)cyclohexanone are obtained, m.p. 74 °C (recrystallised from hexane–ether, m.p. 75 °C); i.r. 3495 (OH), 1695 (C=O) cm^{-1}; p.m.r. (CCl$_4$, TMS) 1.1–2.9 (broad, 9H, aliphatic CH), 3.77 (s, 1H, OH, exchangeable with D$_2$O), 4.83 (d, 1H, $J = 9.0$ Hz, O—CH), 7.29 (s, 5H, aryl CH).

Notes. (1) Lithium sulphide is obtained from Alfa Ventron. Acetonitrile is purified and stored over molecular sieves.
(2) Titanium(IV) chloride is distilled under an argon atmosphere before use **(CAUTION)**.

5.18.3 UNSATURATED ACIDS AND ESTERS

As for the unsaturated carbonyl compounds, the most significant groups are the α,β-unsaturated acids or esters. Some of their methods of formation are closely analogous, for example, (*a*) the Horner–Emmons or Wadsworth–Emmons reaction, and (*b*) the aldol-type reactions. Rather more specific methods are considered below for the formation of α,β-acetylenic acids, and for the conversion of the unsaturated dicarboxylic acid, maleic acid, into its anhydride or into fumaric acid.

THE HORNER–EMMONS OR WADSWORTH–EMMONS REACTION
This reaction is considered in Sections 5.2.3, p. 496, and 5.18.2, p. 799, and the specific examples that are included in this section are: (i) the formation of ethyl cyclohexylideneacetate from cyclohexanone and triethyl phosphonoacetate in the solvent 1,2-dimethoxyethane and using sodium hydride as the base for the formation of the phosphoryl-stabilised anion;[254] and (ii) the formation of ethyl (*E*)-but-2-enoate from acetaldehyde and triethyl phosphonoacetate under PTC conditions[255] (Expt 5.215). In the latter example the (*E*)-configuration is to be expected from the general features of the reaction that are summarised in Section 5.2.3, p. 496.

THE ALDOL-TYPE REACTIONS

This reaction type leading to α,β-unsaturated acids and esters is exemplified in the Perkin reaction (Section 6.12.3, p. 1036) and the Knoevenagel reaction (Section 5.11.6, p. 681). The *Doebner reaction*, which is illustrated in this section, is the condensation of an aldehyde with malonic acid in pyridine solution, often in the presence of a trace of piperidine. The reaction mechanism involves the addition of a malonate anion to the aldehydic carbonyl carbon atom followed by the elimination of water accompanied by decarboxylation.

Examples given in Expt 5.216 include the preparation of non-2-enoic acid starting from hexanal, but-2-enoic acid (crotonic acid) from acetaldehyde, and also hexa-2,4-dienoic acid (sorbic acid) starting from the conjugated aldehyde crotonaldehyde.

MISCELLANEOUS METHODS

A useful route to α,β-acetylenic acids, illustrated by the preparation of but-2-ynoic acid (Expt 5.217), involves the base-induced decomposition of the dibromopyrazolone which is obtained by brominating pyrazol-5-one, which is itself prepared by the reaction of a β-keto ester with hydrazine.

A likely mechanism for the last stage may be formulated as follows:

If the bromination of the pyrazol-5-one is interrupted when only one of the methylenic hydrogens has been substituted, and the monobromopyrazol-5-one is similarly treated with alkali, the α,β-ethylenic acid is obtained.

The simplest unsaturated dicarboxylic acids are maleic acid and fumaric acid, both of which are cheap, commercially available, materials. They are geometric isomers; maleic acid is the (Z) isomer (19), and fumaric acid is the (E) isomer (20). Maleic acid forms an internal anhydride, maleic anhydride (21), which is widely used to form adducts with conjugated dienes (the *Diels–Alder* reaction, Section 7.6). The formation of the anhydride from maleic acid and the conversion of maleic acid into fumaric acid are described in Expt 5.218. The hydrogenation of maleic acid to succinic acid is of value as a means of evaluating the activity of a catalyst for use in hydrogenations at atmospheric pressure; the experimental procedure is given in Section 2.17.1, p. 87.

(19)   (20)   (21)

Experiment 5.215 ETHYL CYCLOHEXYLIDENEACETATE[254]

$$\text{\large O} + (EtO)_2P(O){\cdot}CH_2{\cdot}CO_2Et \longrightarrow \text{\large =CH}{\cdot}CO_2Et$$

CAUTION: *Hydrogen gas is evolved in the first part of this experiment, hence the apparatus should be sited in an efficient fume-cupboard.*

Triethyl phosphonoacetate (11.2 g, 0.05 mol) is added dropwise at 20 °C to a slurry of 50 per cent sodium hydride (2.4 g, 0.05 mol) in 100 ml of dry 1,2-dimethoxyethane. After the addition, the reaction mixture is stirred for 1 hour at room temperature until gas evolution has ceased. Cyclohexanone (4.9 g, 0.05 mol) is added dropwise at such a rate that the temperature is maintained below 30 °C. After the addition, the solution is stirred for 15 minutes at room temperature during which time a viscous semi-solid appears. The mixture is taken up in a large excess of water, and the aqueous solution extracted with ether. The ether layer, after being dried over magnesium sulphate and evaporated, gives a liquid residue, b.p. 88–90 °C/10 mmHg, 5.8 g (70%), n_D^{25} 1.4704. The i.r. spectrum shows a strong absorption band at 1660 cm^{-1}.

Cognate preparation. *Ethyl (E)-but-2-enoate*[255] (*PTC procedure*). A solution of triethyl phosphonoacetate (35 mmol) and acetaldehyde (35 mmol) in dichloromethane (5 ml) is added dropwise to a stirred two-phase system consisting of dichloromethane (35 ml), aqueous sodium hydroxide (20 ml, 50%) and tetrabutylammonium iodide (0.7 g) (1). The strongly exothermic reaction is complete in 15 minutes. The organic layer is separated, washed with water (5 ml), and dried with magnesium sulphate. Evaporation of the solvent and distillation of the residue affords the product, b.p. 51–52 °C/25 mmHg, in 54 per cent yield.

Note. To avoid undesirable reactions of the aldol-type, both substrates should be added simultaneously to the reaction mixture.

Experiment 5.216 3-HEXYLACRYLIC ACID (*Non-2-enoic acid*)

$$Me{\cdot}(CH_2)_5{\cdot}CHO + CH_2(CO_2H)_2 \xrightarrow{\text{pyridine}}$$
$$Me{\cdot}(CH_2)_5{\cdot}CH{=}CH{\cdot}CO_2H + CO_2 + H_2O$$

Dissolve 57 g (0.55 mol) of malonic acid in 92.5 ml of dry pyridine contained in a 500-ml round-bottomed flask, cool the solution in ice and add 57 g (70 ml, 0.5 mol) of freshly distilled heptanal with stirring or vigorous shaking. After a part of the aldehyde has been added, the mixture yields a semi-solid slurry of crystals. Insert a calcium chloride tube into the mouth of the flask and allow the mixture to stand at room temperature for 60 hours with periodic shaking. Finally, warm the mixture on a water bath until the evolution of carbon dioxide ceases (about 8 hours) and then pour into an equal volume of water. Separate the oily layer and shake it with 150 ml of 25 per cent hydrochloric acid to remove pyridine. Dissolve the product in benzene, wash with water, dry with anhydrous sodium sulphate and distil under reduced pressure. Collect the nonenoic acid at 130–132 °C/2 mmHg. The yield is 62 g (79%).

Cognate preparations. *Crotonic acid* [(E)]-*But-2-enoic acid*]. Mix together in a 250-ml flask carrying a reflux condenser and a calcium chloride drying tube 25 g (32 ml, 0.57 mol) of freshly distilled acetaldehyde with a solution of 59.5 g (0.57 mol) of dry, powdered malonic acid in 67 g (68.5 ml, 0.85 mol) of dry pyridine to which 0.5 ml of piperidine has been added. Leave in an ice chest or refrigerator for 24 hours. Warm the mixture on a steam bath until the evolution of carbon dioxide ceases. Cool in ice, add 60 ml of 1:1 sulphuric acid (by volume) and leave in the ice bath for 3–4 hours. Collect the crude crotonic acid (c. 27 g) which has separated by suction filtration. Extract the mother-liquor with three 25 ml portions of ether, dry the ethereal extract, and evaporate the ether; the residual crude acid weighs 6 g. Recrystallise from light petroleum, b.p. 60–80 °C; the yield of crude crotonic acid, m.p. 72 °C, is 20 g (41%).

Sorbic acid (*Hexa-2,4-dienoic acid*). Place 40 g (46.5 ml, 0.57 mol) of crotonaldehyde (b.p. 101–103 °C), 60 g (0.575 mol) of malonic acid and 60 g (61 ml, 0.76 mol) of dry pyridine (b.p. 113–115 °C) in a 500-ml round-bottomed flask, attach a reflux condenser and heat on a water bath for 3 hours. At the end of this period the vigorous evolution of carbon dioxide will have ceased. Cool the mixture in ice and cautiously acidify it by the addition of a solution of 21.3 ml of concentrated sulphuric acid in 50 ml of water with shaking. Most of the sorbic acid separates out immediately; a more complete separation is obtained by cooling the solution in ice for 3–4 hours. Filter the acid at the pump and wash it with a little ice-cold water. Recrystallise from about 125 ml of boiling water; the maximum recovery of purified acid is achieved by leaving the solution in an ice chest or a refrigerator overnight and then filtering. The yield of sorbic acid, m.p. 134 °C, is 20 g (31%).

Experiment 5.217 BUT-2-YNOIC ACID

3-Methylpyrazol-5-one. Place 65 g (0.5 mol) of ethyl acetoacetate in a conical flask and stir magnetically during the slow dropwise addition of a solution of 25 g (0.5 mol) of hydrazine hydrate (98–100%) in 40 ml of absolute ethanol. The temperature rises during this addition which should be regulated so that a temperature of about 60 °C is maintained; a crystalline deposit separates.

After further stirring for 1 hour at room temperature, cool the reaction mixture in an ice bath to complete the crystallisation, and filter. Wash the product with ice-cold ethanol; it is then pure enough for use in the next stage. The yield is 43 g (90%), m.p. 222 °C (phase change at 195 °C; microscope m.p. apparatus).

4,4-Dibromo-3-methylpyrazol-5-one. Dissolve 20.0 g (0.2 mol) of 3-methyl-pyrazol-5-one in 80 ml of glacial acetic acid and stir magnetically during the slow dropwise addition of a solution of 32 g (0.2 mol) of bromine in 20 ml of glacial acetic acid (1). On completion of this addition, add 50 ml of water and continue the dropwise addition of a further 32 g (0.2 mol) of bromine dissolved in 20 ml of glacial acetic acid. On completion of this second addition of bromine solution allow the mixture to stand at room temperature overnight. Add water to precipitate the dibromopyrazolone, filter and wash the solid product under suction with distilled water until the washings are neutral. The air-dried product, sufficiently pure for use in the next statge, has m.p. 130–132 °C, the yield is 41 g (79%).

But-2-ynoic acid. Prepare a solution of 20 g of sodium hydroxide in 500 ml of water and stir magnetically in an ice bath until the temperature reaches 0–5 °C. Add portionwise over 10 minutes 34 g (0.132 mol) of 4,4-dibromo-3-methylpyrazol-5-one. The bromoketone dissolves to give an orange–red solution which evolves nitrogen gas; the temperature of the solution during the addition shows only a slight tendency to rise. Stir the reaction mixture for 1 hour at 0–5 °C and then at room temperature for 1 hour. Cool the solution again and acidify it with concentrated hydrochloric acid. Continuously extract the acidified solution with ether overnight (Section 2.22), dry the ethereal extract with magnesium sulphate and remove the solvent on a rotary evaporator. Place the flask containing the orange oil in a vacuum desiccator and allow to stand until it solidifies. Extract the orange crystalline deposit with successive portions of boiling light petroleum (b.p. 60–80 °C) and concentrate the combined extracts to about 50 ml. Filter the slightly off-white product, m.p. 74–75 °C; recrystallise by dissolving in the minimum volume of light petroleum (b.p. 80–100 °C), adding an equal volume of light petroleum (b.p. 40–60 °C) and allowing to cool. The pure but-2-ynoic acid has m.p. 75–76 °C, the yield is 5.9 g (54%). The i.r. spectrum shows absorption at 2950 (broad, —OH), 2240 (sharp, disubstituted —C≡C—), 1690 cm^{-1} (broad, —C=O in carboxylic acid).

Note. (1) Removal of a portion of the reaction mixture when 1 mol of bromine has been added and addition to it of water results in the precipitation of the monobromo compound, m.p. 180–182 °C.

Experiment 5.218 FUMARIC ACID AND MALEIC ANHYDRIDE

Conversion of maleic acid into fumaric acid. Dissolve 10 g of maleic acid in

10 ml of warm water, add 20 ml of concentrated hydrochloric acid and boil gently under reflux for 30 minutes. Crystals of fumaric acid soon crystallise out from the hot solution. Allow to cool, filter off the fumaric acid and recrystallise it from hot 1 M hydrochloric acid. The m.p. in a sealed capillary tube is 286–287 °C.

Conversion of maleic acid into maleic anhydride. CAUTION: *All operations must be conducted in an efficient fume cupboard, owing to the highly toxic nature of the solvent.* Mix 100 g of maleic acid with 1,1,2,2-tetrachloroethane (100 ml) in a distillation flask fitted with a Claisen still-head, a thermometer and a condenser set for downward distillation. Heat the mixture on an air bath; when the temperature reaches 150 °C, 75 ml of 1,1,2,2-tetrachloroethane and between 15 and 15.5 ml of water are present in the receiver. Continue the distillation using an air condenser and change the receiver flask when the temperature reaches 190 °C. Collect the maleic anhydride at 195–197 °C. Recrystallise the crude anhydride from chloroform. The yield of pure maleic anhydride, m.p. 54 °C, is 70 g (83%).

5.19 RESOLUTION OF RACEMATES

The development of the various strategies for the synthesis of optically active compounds in high enantiomeric excess has made great advances over the last decade or so (p. 15). Nevertheless, resolution is still necessary to prepare optically pure chiral auxiliaries, to purify products of low enantiomeric excess and as a valid strategy for chiral synthesis in its own right. The most important general procedure for such resolutions, originally proposed and explored by Pasteur, involves the conversion of the mixture of enantiomers into a pair of diastereoisomeric derivatives by reaction with a pure, optically active, reagent, e.g.

$$2(\pm)\text{-X} + 2(+)\text{-Y} \longrightarrow (+)\text{-X}(+)\text{-Y} + (-)\text{-X}(+)\text{-Y}$$
$$\text{(1)} \qquad\qquad\qquad\qquad \text{(2)} \qquad\qquad \text{(3)}$$

It should be recalled that whereas the enantiomers in the mixture (or racemate) (1) have *identical* physical properties (except for their action on the plane of polarised light), the diastereoisomers (2) and (3) have physical properties (e.g. solubility, boiling points, chromatographic behaviour, etc.) which are frequently significantly different. Resolution of the mixture (or racemate) can then be achieved provided that one of the diastereoisomers may be obtained in a pure state, and that regeneration from it of the pure enantiomorphous form is not accompanied by any degree of racemisation.

Pasteur's original chemical method of resolution, which is still widely used at the present time, involves the formation of diastereoisomeric salts from racemic acids or bases by neutralisation with available optically pure bases or acids respectively. The required optically pure reactants are often available from natural sources and include tartaric, malic and mandelic acids, and alkaloids such as brucine, strychnine, morphine and quinine. Ideally, by appropriate choice of the resolving reagent, the diastereoisomeric salts are crystalline and have solubilities sufficiently different to permit the separation and ready purification of the less soluble salt by fractional crystallisation from a suitable solvent. The regeneration of the optically pure enantiomorph, and incidentally the recovery of the resolving reagent, normally presents no problems.

809

The procedure is illustrated in Expt 5.219 by the resolution of $(\pm)$-α-methyl-benzylamine (1-phenylethylamine) with the aid of tartaric acid.

The Pasteur method can also be applied to the resolution of neutral racemates, if these can be first converted into an acidic or basic derivative from which eventually a mixture of crystalline diastereoisomeric salts may be prepared by appropriate neutralisation. Thus, a racemic alcohol (e.g. $(\pm)$-octan-2-ol, Expt 5.220) may be converted into the corresponding racemic hydrogen phthalate ester by heating with phthalic anhydride, and the ester is then resolved by the Pasteur procedure using an optically active base. The resulting optically active hydrogen phthalate ester is then carefully hydrolysed with aqueous sodium hydroxide to regenerate one of the optically active forms of the alcohol.

A racemic aldehyde or ketone may similarly be resolved by conversion into a diastereoisomeric mixture of hydrazones (or semicarbazones) using, for example, the optically active 1-phenylethylhydrazine.[256]

The most important development in the last decade for the resolution of mixtures of enantiomers is in the application of g.l.c. and h.p.l.c. techniques. These techniques have been used in the analytical mode to determine the ratio of enantiomers, for example in the case of products arising from an asymmetric synthesis. Indeed this method of analysis has several advantages over the conventional procedure of optical rotation measurement and comparison with the maximum known rotation of one of the pure enantiomers (p. 34). Thus the analysis is rapid and is feasible on quantities of sample which would be inadequate for measurement of optical rotation. These chromatographic methods may also be used in the preparative mode for the realistic isolation of the pure enantiomers. Furthermore, by a close study of the chromatographic behaviour (i.e. elution times) of a series of enantiomeric pairs of compounds having like functionality and known configurations, some judgement is possible for the assignment of configuration to a similar member of the series but of unknown configuration.

Both techniques can be applied in two ways. In the first method the enantiomeric mixture (or racemate) is converted into a diastereoisomeric mixture with a suitable optically pure reagent, and this mixture chromatographed on a column having an achiral stationary phase. Separation then depends on the differential molecular interactions of the diastereoisomers with the stationary phase. In the second method, the stationary phase on the support material (usually chemically bonded) contains a chiral, optically pure residue. In this case the mixture of enantiomers which is loaded directly on to the column is separated by virtue of differential diastereoisomeric molecular interactions between each enantiomer and the optically pure stationary phase.

Research in this area of resolution methods was first in g.l.c. However, the need to derivatise many of the compounds of interest to make them sufficiently volatile for analysis tended to increase the complexity of sample preparation. High performance liquid chromatography does not have this drawback (p. 232) and research activity in this field is now extensive. One illustrative example of the elegance of this technique is noted here[257] but the reader's attention is directed to an excellent summary to 1983 of both these chromatographic methods.[258]

A commercially packed h.p.l.c. column (25 cm × 4.6 mm) of γ-aminopropyl silanised silica [e.g. 5 μm Spherisorb (Regis Chemical Co.), or 7 μm Zorbax (Dupont Co.), or 10 μm Lithosorb (Merck), or 5 μm irregular (J. T. Baker Chemical Co.)] was sequentially treated, at a pumping rate of 2 ml/min, with the following solutions: 2 ml of triethylamine in 40 ml of dry tetrahydrofuran, 2 g of (R)-N-(3,5-dinitrobenzoyl)phenylglycine in 40 ml of dry tetrahydrofuran, 20 ml of dry tetrahydrofuran, and finally 10 per cent propan-2-ol in hexane, until the base line stabilised. The chiral amino acid derivative (which is available from Aldrich Chemical Co.) becomes ionically bonded to the amino residues on the stationary phase.

This chiral column has been shown successfully to resolve enantiomeric mixtures (or racemates) of aromatic alcohols including 1,1'-bi-2-naphthol and its analogues (p. 836), aromatic hydroxy (or alkoxy) carboxylic esters and amides, amino acid derivatives, sulphoxides, cyclic imides and amides, lactones, etc. Even this list should not be regarded as limiting. The potential of this method, coupled with the simplicity of operation will undoubtedly be extensively developed and explored in the coming years.

Other chiral immobilised stationary phases are available commercially, for example covalently bonded bovine serum albumin or chemically bonded L-hydroxyproline–$Cu^{2\oplus}$ complexes to wide-bore Nucleosil (both from Macherey-Nagel).

Several useful methods of resolution, which are particularly applicable to the α-amino acids, involve the use of enzymes. Their success depends upon the fact that enzyme-catalysed reactions are stereospecific, only one of the enantiomorphous forms (actually the form having the 'natural', L, configuration) taking part in the reaction. For example, the conversion of an acyl-L-amino acid (best results are usually obtained with the benzoyl derivative) into the corresponding anilide by reaction with aniline in a buffered medium* is catalysed by the enzyme papain which occurs in papaya latex. Therefore, if the racemic acyl-amino acid is used, only the L-form is converted into the neutral anilide which can be readily separated from the unreacted acidic D-form.

This method represents a resolution by asymmetric enzymic synthesis (e.g. L-alanine, $H_2N·CH(CH_3)·CO_2H$, Expt 5.221). Related procedures involve other types of enzymic reactions (e.g. hydrolysis, oxidation, etc.). Asymmetric enzymic hydrolysis, for example, proceeds according to the following reaction sequence.

Experiment 5.219   RESOLUTION OF (±)-α-METHYLBENZYLAMINE
[(±)-1-Phenylethylamine]

(−)-α-**Methylbenzylamine.**   Add 450 ml of methanol to 31.5 g (0.21 mol) of (+)-tartaric acid in a 1-litre conical flask and heat the mixture almost to boiling on a water bath. Then add cautiously with swirling 24.2 g (0.20 mol) of (±)-α-methylbenzylamine (Expt 5.197); too rapid an addition may cause the mixture to boil over. Allow the mixture to cool to room temperature and then to stand for 24 hours to allow slow separation of the (−)-amine-(+)-hydrogen tartrate as prismatic crystals (1). Filter off the product (17.9 g); concentrate the filtrate to 225 ml under reduced pressure on a rotary evaporator and allow it to stand at room temperature for 24 hours to obtain a second crop. The total yield of the (−)-amine-(+)-hydrogen tartrate is about 21 g (77%).

Shake the total product with 90 ml of water in a 250-ml separating funnel and basify the mixture by cautiously adding 50 per cent aqueous sodium hydroxide. Extract out the liberated amine with three 40 ml portions of ether, dry the extract over anhydrous sodium sulphate, filter and concentrate to about 25 ml using a rotary evaporator. Remove the remainder of the ether in

---

* The optimum pH of the buffer varies within the approximate range 4.5–5.5 depending upon the substrate used.

a distillation apparatus and fractionally distil the residue at atmospheric pressure (2), collecting the (−)-α-methylbenzylamine as a fraction of b.p. 184–186 °C; the yield is about 5 g (53%). Measure the optical rotation of the neat liquid and calculate the specific rotation (Section 2.36). Pure (−)-α-methylbenzylamine has d_4^{22} 0.950, $[\alpha]_D^{22}$ −40.3° (neat); $[\alpha]_D^{20}$ −31.5° (c 3.2 in EtOH).

(+)-α-**Methylbenzylamine.** This enantiomer may be recovered using the following procedure. Evaporate the methanolic filtrate from the isolation of the (−)-amine salt to dryness using a rotary evaporator. Convert the residual salt to the free amine by treatment with sodium hydroxide solution followed by ether extraction as described above. Do not distil the recovered amine but remove the last traces of ether completely by warming under reduced pressure. Weigh the resulting product (x g), measure its optical rotation and calculate the specific rotation. The ratio of this value to that of pure (+)-amine is the optical purity of the sample; the weight of (+)-amine in the sample in excess of that present in the racemic modification is given by:

$$\text{excess (+)-amine} = \frac{x \times \text{observed } [\alpha]_D \text{ for neat liquid}}{40.3} \text{ grams}$$

For *each* gram of excess (+)-amine present add firstly 10.0 ml of rectified spirit, bring to the boil and then add, for each gram, a hot solution of 0.44 g of 98 per cent sulphuric acid (1.03 times the theoretical amount) in 21.5 ml of rectified spirit. Allow the solution to cool slowly to room temperature, filter off the crystalline (+)-amine sulphate and wash it with cold rectified spirit. The yield is about 1 g of sulphate per gram of (+)-amine (71%). Liberate the free (+)-amine from the sulphate as described for the (−)-amine from the tartrate, but use 4 ml of water and 0.5 ml of 50 per cent aqueous sodium hydroxide for each gram of sulphate. The yield of (+)-amine, b.p. 184–186 °C, is 60 per cent of theory; its optical purity is 95 per cent.

Notes. (1) If fine needles separate, the mixture should be warmed until they redissolve, and the solution allowed to cool. The solution should be seeded with the prismatic crystals if these are available.
(2) The free amine rapidly absorbs carbon dioxide. It is therefore essential to protect the distillation apparatus from the atmosphere with a guard-tube filled with soda-lime. As the product tends to foam excessively during distillation, the apparatus used should be larger than is customary for the volume of liquid to be distilled.

Experiment 5.220 RESOLUTION OF (±)-OCTAN-2-OL*

Heat a mixture of 65 g (0.5 mol) of dry octan-2-ol (b.p. 178–180 °C), 74 g (0.5 mol) of pure phthalic anhydride (1) and 40 g of dry pyridine (**CAUTION**) on a water bath for 1 hour, and allow to cool. Dissolve the resulting viscous mass in an equal volume of acetone. Add slowly, preferably with stirring, 55 ml of concentrated hydrochloric acid diluted with an approximately equal volume of crushed ice: if an oil separates before all the hydrochloric acid has been added, introduce more acetone to render the mixture homogeneous. Add ice-water until the oil is completely precipitated; this usually sets to a hard mass within 1–2 hours. If the resulting mass is semi-solid or pasty (2),

---

* The following experimental details were kindly supplied by the late Dr J. Kenyon F.R.S.

transfer it to a large flask and pass steam through it until the octan-2-one is removed, i.e. until the steam distillate is clear; pour the contents of the flask while still warm into a beaker. The ($\pm$)-2-octyl hydrogen phthalate solidifies on cooling. Filter the octyl hydrogen phthalate at the pump, wash it with water, grind it thoroughly in a mortar with water, filter again and dry in the air. The crude material is quite satisfactory for the subsequent resolution (3).

Introduce 197 g (0.5 mol) of anhydrous brucine (**CAUTION**: *poisonous*) or 215 g of the air-dried dihydrate (4) into a warm solution of 139 g of ($\pm$)-2-octyl hydrogen phthalate in 300 ml of acetone and warm the mixture under reflux on a water bath until the solution is clear. Upon cooling, the brucine salt [($+$)-A, ($-$)-B] separates as a crystalline solid. Filter this off on a sintered glass funnel, press it well to remove mother-liquor and wash it in the funnel with 123 ml of acetone. Set the combined filtrate and washings (W) aside. Cover the crystals with acetone and add, slowly and with stirring, a slight excess (to Congo red) of dilute hydrochloric acid (1:1 by volume; about 60 ml); if the solution becomes turbid before the introduction of the acid is complete, add more acetone to produce a clear liquid. Add ice-water until the precipitation of the active 2-octyl hydrogen phthalate [crude ($+$)-A] is complete; filter (5), wash with cold water and dry in the air. The yield is about half that of the ($\pm$)-ester originally taken (6).

Concentrate the combined filtrate and washings (W) to about half the original volume, and pour it into slightly more than the calculated amount of dilute hydrochloric acid (use a mixture of 30 ml of concentrated hydrochloric acid and 30 ml of ice-water); then add about 300 ml of water. Collect the active 2-octyl hydrogen phthalate (crude($-$)-A) as above (5). The weight of the air-dried ester is about half that of the ($\pm$)-ester originally used (7).

Crystallise the two lots of crude active 2-octyl hydrogen phthalates separately twice from 90 per cent acetic acid; use 2 g of acetic acid to each gram of solid. The recrystallised esters, if optically pure (8), will melt sharply at 75 °C; if the melting points are below 75 °C, further recrystallisation is necessary. The yields of optically pure products, m.p. 75 °C, are 48 g and 49 g respectively.

To obtain optically pure ($+$)- and ($-$)-octan-2-ol, steam distil the respective esters with 30 per cent sodium hydroxide solution; use the proportions 1 mol of ester to 2 mols of sodium hydroxide. Separate the alcohols from the steam distillate, dry over anhydrous potassium carbonate and distil under diminished pressure. Both samples boil at 86 °C/20 mmHg (9) and have the following rotations:

$$[\alpha]_D^{17°} +9.9°, \ [\alpha]_{5461}^{17°} +11.8°; \ [\alpha]_D^{17°} -9.9°, \ [\alpha]_{5461}^{17°} -11.8°.$$

The yields from the 2-octyl hydrogen phthalates are almost quantitative.

Notes. (1) If the presence of phthalic acid is suspected, it may be readily removed by mixing with cold chloroform; phthalic anhydride dissolves readily, but the acid is insoluble.

(2) This is due to octan-2-one in the original octan-2-ol; it is most easily separated by steam distillation as described.

(3) The inactive 2-octyl hydrogen phthalate may be recrystallised from light petroleum, b.p. 60–80 °C, or from glacial acetic acid, and then melts at 55 °C. If the octan-2-ol is pure, the yield of pure material is almost quantitative.

(4) Commercial brucine is usually the tetrahydrate $C_{23}H_{26}O_4N_2,4H_2O$; upon air drying, this loses two molecules of water of crystallisation and passes into the dihydrate.
(5) The filtrates from the decomposition of the brucine salts with dilute hydrochloride acid should be carefully preserved. The brucine is recovered by the addition of an excess of dilute ammonia solution (1:4); if the solution becomes turbid before all the ammonia solution is added, introduce a little ethanol until the solution becomes clear. After several hours in an open beaker, filter off the brucine, wash it well with cold water and dry it in the air.
(6) The rotation in absolute ethanol is about $[\alpha]_D +44°$, $[\alpha]_{5461} +47°$.
(7) The rotation in absolute ethanol is about $[\alpha]_D -44°$, $[\alpha]_{5461} -47°$.
(8) The optically pure esters have rotations in ethanol of $[\alpha]_D -48.4°$, $[\alpha]_{5461} -58.5°$, and $[\alpha]_D +48.4°$, $[\alpha]_{5461} +58.5°$ respectively. A preliminary check of the optical purity is, however, more simply made by a m.p. determination; the rotation is determined, if desired, when the m.p. is 75 °C.
(9) The boiling point under atmospheric pressure is 179 °C.

Experiment 5.221 RESOLUTION OF DL-ALANINE

Benzoyl DL-alanine. Dissolve 100 g (1.1 mol) of DL-alanine (Expt 5.180) in 400 ml of water containing 44.5 g (1.1 mol) of sodium hydroxide and cool the solution in an ice bath. Add 175 g (1.2 mol) of benzoyl chloride and a solution of 49 g (1.2 mol) of sodium hydroxide in 200 ml of water to the stirred, cooled, amino acid solution, alternately and in portions during 2 hours; continue to stir for a further 2-hour period. Boil the reaction mixture with 10 g of decolourising charcoal, filter, cool the clear yellow filtrate to 0 °C and acidify carefully to Congo red with concentrated hydrochloric acid. Triturate a portion of the oil which separates with water to induce crystallisation and then seed the bulk of the acidified solution with crystals and leave in an ice bath to complete the crystallisation process. Filter off the product, wash the filter cake with 500 ml of ice-cold water and recrystallise from about 3.5 litres of boiling water. The yield of benzoyl-DL-alanine, m.p. 162–164 °C, is 194.5 g (90%).

Benzoyl-L-alanine anilide. Use freshly boiled, cooled, distilled water throughout this stage. Prepare an 0.1 M citrate buffer solution by dissolving 48 g of anhydrous citric acid and 16.5 g of sodium hydroxide in 2.5 litres of water. Stir together 50 g of technical powdered papain (1) and 4 g of potassium cyanide in 500 ml of the buffer solution, adjust the pH to 5 (narrow range pH paper) with glacial acetic acid, and stir for a further 75 minutes. Filter this enzyme extract through a Celite filter bed. Dissolve 193 g (1 mol) of benzoyl-DL-alanine by warming it in 300 ml of the citrate buffer to which has been added 120 ml of 2.5 molar sodium hydroxide, 360 ml of 3 M sodium acetate solution and 93 g (91 ml, 1 mol) of redistilled aniline; adjust the pH of this solution to 5 with sodium hydroxide solution. Cool the solution to 45 °C, add the filtered enzyme extract and transfer the mixture to several conical flasks of suitable size such that each is filled to the neck and tightly stoppered with a rubber bung to exclude air. Leave the flasks in an incubator held at 37 °C, shaking them occasionally during the early stages.

Product begins to separate within 5 minutes and the contents of the flasks becomes almost immobile within 2 hours. After 24 hours filter the mixture and return the filtrate to the incubator for a further 24 hours and remove the additional crop of solid which separates. Wash the combined solids with

250 ml of water and recrystallise from 1 litre of 50 per cent aqueous ethanol with the aid of decolourising charcoal. The yield of benzoyl-L-alanine anilide is 122 g (91%), m.p. 175–176 °C, $[\alpha]_D^{20}$ −7.8° (c 5 in Me·CO_2H).

L-Alanine. Heat a mixture of 50 g (0.187 mol) of benzoyl-L-alanine anilide and 250 ml of 6 M hydrochloric acid under reflux for 5 hours. Leave the mixture at room temperature overnight, remove the precipitated benzoic acid by filtration and evaporate the filtrate to dryness under reduced pressure (rotary evaporator). Dissolve the brown oily residue in 100 ml of water and boil it with decolourising charcoal. Filter and pass the filtrate through 450 g of a weakly basic anion exchange resin, e.g. Amberlite IR4B (which has been washed free from soluble colour with dilute hydrochloric acid and regenerated with dilute aqueous ammonia) in the form of a column 60 cm long. Collect the effluent (in all about 2 litres) until it gives no colour when boiled with ninhydrin (Expt 5.99). Evaporate the effluent to dryness under reduced pressure (rotary evaporator) and boil the yellow solid residue with 20 ml of water and a little decolourising charcoal. Add ethanol to the hot filtered solution until crystallisation begins and cool in ice to complete the separation of the L-alanine. The yield is 13.9 g (76%), $[\alpha]_D^{20}$ +12.0° (c 4 in 1 M HCl).

Note. (1) An inexpensive crude commercial product (dried papaya latex) was used; the activity was not determined. Purified highly active enzyme preparations may, however, be obtained[259] (e.g. Koch–Light, Sigma, etc.).

REFERENCES

1. H. C. Brown and C. A. Brown (1963). *J. Am. Chem. Soc.*, **85**, 1004.
2. C. A. Brown and V. K. Ahuja (1973). *J. Org. Chem.*, **38**, 2226.
3. S.-K. Chung (1979). *J. Org. Chem.*, **44**, 1015.
4. C. E. Waring, E. E. Kern and W. A. Blann (1941). *J. Am. Chem. Soc.*, **63**, 1767.
5. K. Alder and G. Stein (1931). *Liebigs Ann. Chem.*, **485**, 241.
6. H. C. Brown and S. Krishnamurthy (1983). *J. Org. Chem.*, **48**, 3089 and 3091.
7. R. Vanderesse, J.-J. Brunet and P. Caubere (1981). *J. Org. Chem.*, **46**, 1270.
8. S. Krishnamurthy and H. C. Brown (1976). *J. Org. Chem.*, **41**, 3064.
9. F. Rolla (1981). *J. Org. Chem.*, **46**, 3909.
10. S. Krishnamurthy (1980). *J. Org. Chem.*, **45**, 2551.
11. S. Yamamura, M. Toda and Y. Hirata (1973). *Organic Syntheses*, **53**, 86.
12. R. O. Hutchins, C. Milewski and B. Marganott (1973). *J. Am. Chem. Soc.*, **96**, 3662; R. O. Hutchins, M. Kacher and L. Rua (1975). *J. Org. Chem.*, **40**, 923.
13. R. O. Hutchins and N. R. Natale (1978). *J. Org. Chem.*, **43**, 2299.
14. (a) G. W. Kabala and J. D. Baker (1975). *J. Org. Chem.*, **40**, 1834.
    (b) G. W. Kabala and J. H. Chandler (1979). *Synth. Commun.*, **9**, 275.
15. J. C. Stowell and B. T. King (1984). *Synthesis*, 278.
16. (a) J. C. Stowell (1976). *J. Org. Chem.*, **41**, 560.
    (b) J. C. Stowell, D. R. Kieth and B. T. King (1984). *Organic Syntheses*, **62**, 140.
17. F. C. Whitmore and E. Rohrmann (1941). *J. Am. Chem. Soc.*, **43**, 2033; R. V. Hoffman, R. D. Bishop, P. M. Fitch and R. Hardenstein (1980). *J. Org. Chem.*, **45**, 917.
18. P. Wolkoff (1982). *J. Org. Chem.*, **47**, 1944.
19. Y. Kimura and S. L. Regen (1983). *J. Org. Chem.*, **48**, 195; J. Barry, G. Bram, G. Decodts, A. Loupy, P. Pigeon and J. Sansoulet (1984). *J. Org. Chem.*, **49**, 1138.
20. A. C. Cope, T. T. Foster and P. H. Towle (1949). *J. Am. Chem. Soc.*, **71**, 3929.
21. J. L. Roberts, P. S. Borromeo and C. D. Poulter (1977). *Tetrahedron Lett.*, 1299.

22. T. A. Bryson, G. H. Bonitz, C. J. Reichel and R. E. Dardis (1980). *J. Org. Chem.*, **45**, 524.

23. W. E. Gore, G. T. Pearce and R. M. Silverstein (1975). *J. Org. Chem.*, **40**, 1705.

24. A. W. Johnson (1946). *J. Chem. Soc.*, 1014.

25. E. N. Marvell and T. Li (1973). *Synthesis*, 468.

26. J.-J. Brunet, P. Gallois and P. Caubere (1980). *J. Org. Chem.*, **45**, 1937 and 1946.

27. N. Suzuki, Y. Kaneko, T. Tsukanaka, T. Nomoto, Y. Ayagachi and Y. Izawa (1985). *Tetrahedron*, **41**, 2387.

28. E. Winterfeldt (1975). *Synthesis*, 617.

29. H. Dissenlnkötter and K. Eiter (1976). *Tetrahedron*, **32**, 1591.

30. E. F. Silversmith (1986). *J. Chem. Ed.*, **63**, 645.

32. W. S. Wadsworth, Jr (1977). 'Synthetic applications of phosphoryl-stabilised anions', in *Organic Reactions* (25). Ed. W. G. Dauben. New York: Wiley, p. 73.

32. J. Fuhrhop and G. Penzlin (1984). *Organic Synthesis*. Weinheim: Verlag Chemie.

33. L. Horner, H. Hoffmann and H. G. Wippel (1958). *Chem. Ber.*, **91**, 61.

34. A. D. Buss and S. Warren (1981). *J. Chem. Soc., Chem. Commun.*, 100.

35. A. D. Buss and S. Warren (1985). *J. Chem. Soc., Perkin Trans I*, 2307.

36. D. J. Peterson (1968). *J. Org. Chem.*, **33**, 780.

37. D. J. Ager (1984). *Synthesis*, 384; E. Colvin (1981). *Silicon in Organic Synthesis*, London: Butterworth.

38. T. H. Chan and E. Chang (1974). *J. Org. Chem.*, **39**, 3264.

39. A. Gorgues and A. Le Coq (1976). *Tetrahedron Lett.*, 4723.

40. E. V. Dehmlow and M. L. Lissel (1980). *Liebigs Ann. Chem.*, 1.

41. E. V. Dehmlow (1981). *Tetrahedron*, **37**, 1653.

42. T. Makaiyama, H. Nambu and T. Kumamoto (1964). *J. Org. Chem.*, **29**, 2243.

43. A. C. Cope, D. S. Smith and R. J. Cotter. *Organic Syntheses*, Coll. Vol., IV, p. 377.

44. J. Tsuji, H. Kezuka, Y. Toshida, H. Takayanagi and K. Yamamoto (1983). *Tetrahedron*, **39**, 3279.

45. E. Negishi and S. Baba (1975). *J. Am. Chem. Soc.*, **97**, 7386.

46. K. Eiter, F. Lieb, H. Disselnkötter and H. Oediger (1978). *Liebigs Ann. Chem.*, **658**, 65.

47. D. J. Raber and W. C. Guida (1976). *J. Org. Chem.*, **41**, 690; J. P. Masse and E. R. Porayre (1976). *J. Chem. Soc., Chem. Commun.*, 438. D. J. Raber and W. C. Guida (1981). *Tetrahedron Lett.*, 5105.

48. W. von E. Doering and R. W. Young (1950). *J. Am. Chem. Soc.*, **72**, 631.

49. E. R. Grandbois, S. I. Howard and J. D. Morrison (1983). 'Reduction with chiral modifications of lithium aluminium hydride', in *Asymmetric Synthesis*. Ed. J. D. Morrison. Orlando, Florida: Academic Press, Vol. 2A, p. 71.

50. R. Noyori, I. Tomino, M. Yamada and M. Nishizawa (1984). *J. Am. Chem. Soc.*, **106**, 6717; R. Noyori, I. Tomino, Y. Tanimoto and M. Nishizawa (1984). Ibid., **106**, 6709.

51. R. Noyori (1981). *Pure and Appl. Chem.*, **53**, 2315.

52. S. R. Landor, B. Miller and A. R. Tatchell (1966). *J. Chem. Soc.*, C, 2250; (1967). Ibid., 197.

53. S. Krishnamurthy and H. C. Brown (1976). *J. Am. Chem. Soc.*, **98**, 3383.

54. S. Kim, Y. C. Moon and K. H. Ahn (1982). *J. Org. Chem.*, **47**, 3311.

55. E. J. Corey and A. L. Danhuser (1976). *J. Org. Chem.*, **41**, 260.

56. W. A. Lazier, J. W. Hill and W. J. Amend. *Organic Syntheses*, Coll. Vol., II, p. 325.

57. K. Soai and H. Oyamada (1984). *Synthesis*, 605.

58. J.-P. Mazaleyrat and D. J. Cram (1981). *J. Am. Chem. Soc.*, **103**, 4585.

59. T. Mukaiyama, K. Soai, T. Sato, H. Shimizu and K. Suzukai (1979). *J. Am. Chem. Soc.*, **101**, 1455; T. Mukaiyama (1981). *Tetrahedron*, **37**, 4111.

60. G. Solladie (1983). 'Addition of chiral nucleophiles to aldehydes and ketones', in *Asymmetric Synthesis*. Ed. J. D. Morrison. Orlando, Florida: Academic Press, Vol. 2A, p. 157.

61. S. V. Frye and E. L. Eliel (1985). *J. Org. Chem.*, **50**, 3402.

62. P. A. Wade, D. T. Price, J. P. McCauley and P. J. Carroll (1985). *J. Org. Chem.*, **50**, 2804.
63. R. P. Thummel and W. Nutakul (1977). *J. Org. Chem.*, **42**, 300.
64. H. C. Brown (1975). *Organic Synthesis via Organoboranes*. New York: Wiley.
65. G. W. Kabalka and H. C. Heathcock (1975). *J. Org. Chem.*, **40**, 1776.
66. H. C. Brown and G. Zweifel (1961). *J. Am. Chem. Soc.*, **83**, 2544.
67. H. C. Brown, M. C. Desai and P. K. Jadhav (1982). *J. Org. Chem.*, **47**, 5065.
68. H. C. Brown and B. Singram (1984). *J. Org. Chem.*, **49**, 945.
69. H. C. Brown and P. K. Jadhav (1983), 'Asymmetric hydroboration', in *Asymmetric Synthesis*. Ed. J. D. Morrison. Orlando, Florida: Academic Press, Vol. 2A, p. 1.
70. H. C. Brown and G. J. Lynch (1981). *J. Org. Chem.*, **46**, 531 and 930.
71. M. Schroder (1980). *Chem. Rev.*, **80**, 187.
72. K. Akashi, R. E. Palmero and K. B. Sharpless (1978). *J. Org. Chem.*, **43**, 2064.
73. C. Venturello, E. Alneri and M. Ricci (1983). *J. Org. Chem.*, **48**, 3832; R. D. Bach, M. W. Klein, R. A. Ryntz and J. W. Holubka (1979). *J. Org. Chem.*, **44**, 2569.
74. V. Bhushan, R. Rathore and S. Chandrasekaran (1984). *Synthesis*, 431.
75. K. Ohno, H. Nishiyama and H. Nagase (1979). *Tetrahedron Lett.*, 4405.
76. (a) M. E. Jung and M. A. Lyster (1977). *J. Org. Chem.*, **42**, 3761.
    (b) G. A. Olah, A. Husain, B. P. Singh and A. K. Mehrotra (1983). *J. Org. Chem.*, **48**, 3667.
77. W. H. Hartung and R. Simonoff (1953). 'Hydrogenolysis of benzyl groups attached to oxygen, nitrogen or sulphur', in *Organic Reactions* (VII). Ed. R. Adams. New York: Wiley, p. 263.
78. J. Gigg and R. Gigg (1966). *J. Chem. Soc.*, C, 82; R. Gigg and C. D. Warren (1968). Ibid., 1903.
79. R. Boss and R. Scheffold (1976). *Angew. Chem., Int. Ed. Engl.*, **15**, 558.
80. E. J. Corey, J.-L. Gras and P. Urlich (1976). *Tetrahedron Lett.*, 809.
81. A. Bongini, G. Cardillo, M. Orena and S. Sandri (1979). *Synthesis*, 618.
82. M. Miyashita, A. Yoshikoshi and P. A. Grieco (1977). *J. Org. Chem.*, **42**, 3772.
83. A. I. Meyers, S. Schwartzman, G. L. Olson and H.-C. Cheung (1976). *Tetrahedron Lett.*, 2417.
84. (a) E. J. Corey and A. Venkateswarlu (1972). *J. Am. Chem. Soc.*, **94**, 6190.
    (b) L. A. Carpino and A. C. Sau (1979). *J. Chem. Soc., Chem. Commun.*, 514.
85. (a) H. R. Hudson and G. R. de Spinoza (1976). *J. Chem. Soc., Perkin Trans I*, 104.
    (b) H. R. Hudson, B. A. Chaudri and W. S. Murphy (1967). *J. Chromatogr.*, **29**, 218; D. R. Hepburn and H. R. Hudson (1975). Ibid., **103**, 166.
86. J. Hooz and S. S. H. Gilani (1968). *Can. J. Chem.*, **46**, 86; E. H. Axelrod, G. M. Milne and E. E. van Tamelen (1970). *J. Am. Chem. Soc.*, **92**, 2139; E. I. Snyder (1972). *J. Org. Chem.*, **37**, 1466.
87. B. A. Howell and R. E. Kohrman (1984). *J. Chem. Ed.*, **61**, 932.
88. G. A. Olah, B. G. B. Gupta, R. Malhotra and S. C. Narang (1980). *J. Org. Chem.*, **45**, 1638.
89. M. Lauwers, B. Regnier, M. van Eenoo, J. N. Denis and A. Kriel (1979). *Tetrahedron Lett.*, 1801.
90. G. A. Olah, S. C. Narang, B. G. B. Gupta and R. Malhotra (1979). *J. Org. Chem.*, **44**, 1247.
91. C. M. Starks and C. Liotta (1978). *Phase Transfer Catalysis*. New York: Academic Press.
92. A. R. Katritzky and C. M. Mason (1984). *Angew. Chem., Int. Ed. Engl.*, **23**, 420, and the references cited therein.
93. N. R. de Lue and H. C. Brown (1976). *Synthesis*, 114.
94. G. W. Kabalka and E. E. Gooch (III) (1980). *J. Org. Chem.*, **45**, 3578.
95. A. Merz (1973). *Angew. Chem. Int. Ed. Engl.*, **12**, 846.
96. H. H. Freedman and R. A. Dubois (1975). *Tetrahedron Lett.*, 3251.
97. J. C. Collins, W. W. Hes and F. J. Franck (1968). *Tetrahedron Lett.*, 3363.

98. E. J. Corey and J. W. Suggs (1975). *Tetrahedron Lett.*, 2647.
99. E. J. Corey and G. Schmidt (1979). *Tetrahedron Lett.*, 399.
100. Y.-S. Cheng and W.-L. Liu and S. Chen (1980). *Synthesis*, 223.
101. S. Czernecki, C. Georgoulis, C. L. Stevens and K. Vijayakumaran (1985). *Tetrahedron Lett.*, 1699.
102. H. C. Brown, S. U. Kulkarni and C. G. Rao (1980). *Synthesis*, 151.
103. D. Pletcher and S. J. D. Tait (1979). *J. Chem. Soc., Perkin Trans II*, 788; D. Landini, F. Montanari and F. Rolla (1979). *Synthesis*, 134.
104. D. Horton, J. B. Hughes and J. K. Thomson (1968). *J. Org. Chem.*, **33**, 728.
105. M. E. Jung and T. J. Shaw (1980). *J. Am. Chem. Soc.*, **102**, 6304; C. Hubschwerlen (1986). *Synthesis*, 962.
106. D. N. Gupta, P. Hodge and J. E. Davies (1981). *J. Chem. Soc., Perkin Trans I*, 2970.
107. J. J. Pappas, W. P. Keaveney, E. Gancher and M. Berger (1966). *Tetrahedron Lett.*, 4273; D. Gupta, R. Soman and S. Dev (1982). *Tetrahedron*, **38**, 3013.
108. J. A. Marshall, N. H. Andersen and J. W. Schlicher (1970). *J. Org. Chem.*, **35**, 858.
109. H. C. Brown and C. P. Garg (1964). *J. Am. Chem. Soc.*, **86**, 1085.
110. H. C. Brown, C. G. Rao and S. U. Kulkarni (1979). *Synthesis*, 704.
111. H. C. Brown and B. C. S. Rao (1958). *J. Am. Chem. Soc.*, **80**, 5377; J. E. Siggins, A. Larsen, J. H. Ackermann and C. D. Carabateas (1973). *Organic Syntheses*, **53**, 52.
112. G. W. J. Fleet, C. J. Fuller and P. J. C. Harding (1978). *Tetrahedron Lett.*, 1437; T. N. Sorrell and R. J. Spillane (1978). Ibid., 2473.
113. H. Stetter and E. Reske (1970). *Chem. Ber.*, **103**, 643.
114. M. P. Cooke, Jr (1970). *J. Am. Chem. Soc.*, **92**, 6080.
115. R. G. Finke and T. N. Sorrell (1979). *Organic Syntheses*, **59**, 102.
116. S. Gladiali and F. Soccolini (1982). *Synth. Commun.*, **12**, 355.
117. N. Kornblum, A. S. Erickson, W. J. Kelly and B. Henggeler (1982). *J. Org. Chem.*, **47**, 4534.
118. J. E. McMurry, J. Melton and H. Padgett (1974). *J. Org. Chem.*, **39**, 259.
119. J. E. McMurry and J. Melton (1973). *J. Org. Chem.*, **38**, 4367.
120. D. Seebach, E. W. Colvin, F. Lehr and T. Weller (1979). *Chimia*, **33**, 1.
121. J. K. Whitesell and M. A. Whitesell (1983). *Synthesis*, 517.
122. D. E. Bergbreiter and M. Newcombe (1983). 'Alkylation of imine and enamine salts', in *Asymmetric Synthesis*, Ed. J. D. Morrison. Orlando, Florida: Academic Press. Vol. 2A, p. 243; D. Enders (1984). 'Alkylation of chiral hydrazones', in *Asymmetric Synthesis*. Ed. J. D. Morrison. Orlando, Florida: Academic Press, Vol. 3, p. 275.
123. A. I. Meyers, G. S. Poindexter and Z. Brich (1978). *J. Org. Chem.*, **43**, 892.
124. H. Seki, K. Koga, H. Matsuo, S. Ohki, I. Matsuo and S. Yamada (1965). *Chem. Pharm. Bull.*, **13**, 995; S. Yamada, K. Koga and H. Matsuo (1963). *Chem. Pharm. Bull.*, **11**, 1140.
125. See reference 100.
126. E. R. Burkhardt and R. D. Rieke (1985). *J. Org. Chem.*, **50**, 416.
127. (a) G. H. Posner, C. E. Whitten and P. E. McFarland (1972). *J. Am. Chem. Soc.*, **94**, 5106.
   (b) G. H. Posner and C. E. Whitten (1970). *Tetrahedron Lett.*, 4647.
   (c) P. T. Izzo and S. R. Safir (1959). *J. Org. Chem.*, **24**, 701.
   (d) D. G. M. Diaper (1955). *Can. J. Chem.*, **33**, 1720.
128. (a) E. J. Salmi (1938). *Chem. Ber.*, **71B**, 1803.
   (b) R. A. Daignault and E. L. Eliel (1973). *Organic Syntheses*, Coll. Vol. 5, p. 303.
129. F. F. Caserio and J. D. Roberts (1958). *J. Am. Chem. Soc.*, **80**, 5837.
130. L. L. Fieser and Stevenson (1954). *J. Am. Chem. Soc.*, **76**, 1728.
131. E. G. Howard and R. V. Lindsberg (1960). *J. Am. Chem. Soc.*, **82**, 158.
132. R. Sterzycki (1979). *Synthesis*, 724.
133. A. E. Dann, J. B. Davis and M. J. Nagler (1975), *J. Chem. Soc., Perkin Trans. I*, 158.
134. B. Willham, U. Steiner and H. Schinz (1958). *Helv. Chim. Acta*, **41**, 1359.
135. P. A. Greico, T. Oguri, S. Gilman and G. R. DeTitta (1978). *J. Am. Chem. Soc.*, **100**,

1616; J. A. Zderic and D. C. Limon (1959). Ibid., **81**, 4570; H. M. Walkorsky, R. H. Davis and D. R. Howton (1951). Ibid., **73**, 2590.

136. F. Huet, A. Lechevallier, M. Pellet and J. M. Conia (1978). *Synthesis*, 63.

137. D. H. R. Barton, P. D. Magnus, G. Smith and D. Zarr (1971). *J. Chem. Soc., Chem. Commun.*, 861.

138. R. P. Hatch, J. Shringapure and S. M. Weinreb (1978). *J. Org. Chem.*, **43**, 4172; L. F. Fieser (1954), *J. Am. Chem. Soc.*, **76**, 1945.

139. J. Romo, G. Rosenkranz and C. Djerassi (1951). *J. Am. Chem. Soc.*, **73**, 4961.

140. G. E. Wilson, Jr, M. G. Huang and W. W. Schloman, Jr (1968). *J. Org. Chem.*, **33**, 2133.

141. C. Djerassi, M. Shumna and T. Y. Khan (1958). *J. Am. Chem. Soc.*, **80**, 4723.

142. W. F. J. Huurdeman, H. Wynberg and D. W. Emerson (1971). *Tetrahedron Lett.*, 3449; D. W. Emerson (1971). Ibid., 3445.

143. C. H. Depuy and B. W. Ponder (1959). *J. Am. Chem. Soc.*, **81**, 4629.

144. S. M. McElvain (1948). 'The acyloins', in *Organic Reactions* (IV). Ed. R. Adams. New York: Wiley, p. 256.

145. J. J. Bloomfield, D. C. Owsley and J. M. Nelke (1976). 'The acyloin condensation', in *Organic Reactions* (23). Ed. W. G. Dauben. New York: Wiley, p. 259.

146. D. G. Lee and V. S. Chang (1978). *Synthesis*, 462.

147. (a) See reference 119.
     (b) J. E. McMurry and J. Melton (1977). *Organic Syntheses*, **56**, 36.

148. M. Miyashita, T. Yanami and A. Yoshikoshi (1981). *Organic Syntheses*, **60**, 117.

149. L. Hough and A. C. Richardson (1969). 'The monosaccharides', in Rodd's *Chemistry of Carbon Compounds*, 2nd edn. Ed. S. Coffey. Amsterdam: Elsevier, p. 67.

150. E. L. Eliel, N. L. Allinger, S. A. Angyal and G. A. Morrison (1966). *Conformational Analysis*. New York: Interscience.

151. M. L. Wolfrom and A. Thompson (1963). *Methods in Carbohydrate Chemistry*, Vol. II. Eds. R. L. Whistler and M. L. Wolfrom. New York: Academic Press, p. 21.

152. See reference 101.

153. G. O. Aspinal and R. J. Ferrier (1957). *Chem. Ind. (London)*, 1216.

154. See reference 91, p. 298.

155. H. T. Clarke and T. F. Murray (1932). *Organic Syntheses*, Coll. Vol. I, 2nd edn, p. 523.

156. A. I. Meyers, D. L. Temple, R. L. Nolen and E. D. Mihelich (1974). *J. Org. Chem.*, **39**, 2778.

157. A. I. Meyers, G. Knauss, K. Kamata and M. E. Ford (1976). *J. Am. Chem. Soc.*, **98**, 567.

158. P. A. Levene and R. E. Marker (1932). *J. Biol. Chem.*, **98**, 1.

159. S. Chandrasekaran, A. F. Kluge and J. A. Edwards (1977). *J. Org. Chem.*, **42**, 3972.

160. G. W. Anderson and F. M. Callahan (1960). *J. Am. Chem. Soc.*, **82**, 3359.

161. J. Dale (1965). *J. Chem. Soc.*, 72.

162. N. Ono, T. Yamada, T. Saito, K. Tanaka and A. Kaji (1978). *Bull. Chem. Soc (Japan)*, **51**, 2401.

163. A. W. Dox. *Organic Syntheses*, Coll. Vol., I, p. 5.

164. S. Swann, R. Oehler and B. J. Buswell. *Organic Syntheses*, Coll. Vol. II, p. 276.

165. P. A. Finnan and G. A. Fothergill (1962). *J. Chem. Soc.*, 2824.

166. H. R. Snyder and C. T. Elston (1954). *J. Am. Chem. Soc.*, **76**, 3039.

167. (a) C. M. Starks (1971). *J. Am. Chem. Soc.*, **93**, 195; D. Savoia, E. Tayliarini, C. Trombini and A. Umani-Ronchi (1980). *J. Org. Chem.*, **45**, 3227.
     (b) G. Simchen and H. Kobler (1975). *Synthesis*, 606.

168. J. W. Zubriek, D. I. Dunbar and H. D. Durst (1975). *Tetrahedron Lett.*, 71.

169. J. Jiricny, D. M. Orere and C. B. Reese (1980). *J. Chem. Soc., Perkin Trans. I*, 1487.

170. B. B. Corson, R. W. Scott and C. E. Vose. *Organic Syntheses*, Coll. Vol. II, p. 379.

171. T. Imanoto, T. Takaoka and M. Yokoyama (1983). *Synthesis*, 142.

172. G. A. Olah, S. C. Narang, A. P. Fung and B. G. B. Gupta (1980). *Synthesis*, 657.

173. A. Hulkenberg and J. J. Troost (1982). *Tetrahedron Lett.*, 1505.
174. H. A. Brason (1949). 'Cyanoethylation', in *Organic Reactions* (V). Ed. R. Adams. New York: Wiley, p. 79.
175. (a) S. Arseniyadis, K. S. Kyler and D. S. Watt (1984). 'Addition and substitution reactions of nitrile-stabilised carbanions', in *Organic Reactions* (31). Ed. W. D. Dauben. New York: Wiley, p. 1.
     (b) See Reference 91, p. 170.
176. D. N. Harpp, L. Q. Bao, C. J. Black, J. G. Gleason and R. A. Smith (1975). *J. Org. Chem.*, **40**, 3420.
177. C. S. Marvel, D. W. MacCorquodale, F. E. Kendall and W. A. Lazier (1924). *J. Am. Chem. Soc.*, **46**, 2838.
178. H. B. Kagan and J. C. Fiaud (1978). *Topics in Stereochemistry*, **10**, 175; J. W. Ap Simon and R. P. Sequin (1979). *Tetrahedron*, **35**, 2797.
179. J. D. Morrison and H. S. Mosher (1971). *Asymmetric Organic Reactions*. American Chemical Society publication.
180. A. I. Meyers and J. Slade (1980). *J. Chem. Soc.*, **45**, 2785.
181. S.-S. Jew, T. Terashimao and K. Kogo (1979). *Tetrahedron*, **35**, 2337 and 2345.
182. H. Kruizinga and R. M. Kellogg (1981). *J. Am. Chem. Soc.*, **103**, 5183.
183. W. C. Still and I. Galynker (1981). *Tetrahedron*, 3981; W. D. Emmons and G. B. Lucas (1955). *J. Am. Chem. Soc.*, **77**, 2287.
184. I. Bidd, D. J. Kelly, P. M. Ottley, O. I. Paynter, D. J. Simmonds and M. C. Whiting (1983). *J. Chem. Soc., Perkin Trans. 1*, 1369.
185. G. R. Krow (1981). *Tetrahedron*, **37**, 2697.
186. M. J. Bougault (1904). *Compte rendu.*, **139**, 864; M. J. Bougault (1908). *Ann. Chim, Phys.*, **14**, 145; M. J. Bougault (1911). Ibid., **22**, 125.
187. M. D. Dowle and D. I. Davies (1979). *Chem. Soc. Rev.*, **8**, 171.
188. J. Klein (1959). *J. Am. Chem. Soc.*, **81**, 3611.
189. A. I. Meyers, Y. Yamamoto, E. D. Mihelich and R. A. Bell (1980). *J. Org. Chem.*, **45**, 2792.
190. G. Barger, W. F. Martin and W. Mitchell (1973). *J. Chem. Soc.*, 1820.
191. A. C. Cope, A. A. D'Addieco, D. E. Whyte and S. A. Glickman. *Organic Syntheses*, Coll. Vol. IV, p. 234.
192. J. S. Nimitz and H. S. Mosher (1981). *J. Org. Chem.*, **46**, 211.
193. H. A. Staab and E. Jost (1962). *Liebigs Ann. Chem.*, **655**, 90.
194. W. Wierenga and H. I. Skulnick (1979). *J. Org. Chem.*, **44**, 310.
195. R. Ballini and M. Petrini (1986). *Synthesis*, 1024.
196. See reference 120.
197. N. Kornblum and J. W. Powers (1957). *J. Org. Chem.*, **22**, 455.
198. N. Kornblum (1962). 'Synthesis of aliphatic and alicyclic nitro compounds', in *Organic Reactions* (12). Ed. A. C. Cope. New York: Wiley, p. 133.
199. K. E. Gilbert and W. T. Borden (1979). *J. Org. Chem.*, **44**, 659.
200. D. Seebach, R. Henning, F. Lehr and J. Gonnermann (1977). *Tetrahedron Lett.*, 1161.
201. D. Seebach and F. Lehr (1976). *Angew. Chem., Int. Ed. Engl.*, **88**, 505.
202. A. R. Katritzky, G. De Ville and R. C. Patel (1981). *Tetrahedron*, **37**, Suppl. 1, 25.
203. G. Rosini, R. Ballini and P. Sorrenti (1983). *Synthesis*, 1014; G. Rosini, E. Maretta, R. Ballini and M. Petrini (1986). Ibid., 239; R. Ballini and P. Petrini (1986). Ibid., 1024.
204. F. Rolla (1982). *J. Org. Chem.*, **47**, 4327.
205. L. H. Amunsden and L. S. Nelson (1951). *J. Am. Chem. Soc.*, **73**, 242.
206. N. Umino, T. Iwakuma and N. Itoh (1976). *Tetrahedron Lett.*, 2875.
207. N. Umino, T. Iwakuma and N. Itoh (1976). *Tetrahedron Lett.*, 763.
208. S. R. Wann, P. T. Thorsen and M. M. Kreevoy (1981). *J. Org. Chem.*, **46**, 2579.
209. M. Cĕrný, J. Mălek, M. Căpka and V. Chvalovský (1969). *Collect. Czech. Chem. Commun.*, **34**, 1033.
210. W. S. Emerson (1948). 'The preparation of amines by reductive alkylation', in *Organic Reactions* (IV). Ed. R. Adams. New York: Wiley, p. 174.

211. K. A. Schellenberg (1963). *J. Org. Chem.*, **28**, 3259.
212. See reference 121.
213. K. Jewers and J. McKenna (1958). *J. Chem. Soc.*, 2209.
214. D. Savoia, C. Trombini and A. Umani-Ronchi (1978). *J. Org. Chem.*, **43**, 2907.
215. D. P. Roelofsen and H. van Bekkun (1972). *Recl. Trav. Chim. Pays-Bas.*, **91**, 605.
216. H. Weingarten, J. P. Chupp and W. A. White (1967). *J. Org. Chem.*, **32**, 3246.
217. H. O. House, W. C. Liang and P. D. Weeks (1974). *J. Org. Chem.*, **39**, 3104.
218. G. Stork, R. Terrell and J. Szmuszkovicz (1954). *J. Am. Chem. Soc.*, **76**, 2029; G. Stork and H. Landesman (1956). *J. Am. Chem. Soc.*, **78**, 5128; G. Stork, A. Brizzolara, H. Landesman, J. Szmuszkovicz and R. Terrell (1963). *J. Am. Chem. Soc.*, **85**, 207.
219. F. Johnson, N. H. Starkovsky, A. C. Paton and A. D. Carlson (1966). *J. Am. Chem. Soc.*, **88**, 149.
220. R. Carlson, A. Nilsson and M. Stromqvist (1983). *Acta Chem. Scand., Ser. B*, **37**, 7.
221. P. Radlick and L. R. Brown (1974). *Synthesis*, 290.
222. P. A. S. Smith (1946). 'The Curtius reaction', in *Organic Reactions* (III). Ed. R. Adams. New York: Wiley, p. 337.
223. J. C. Sheenan and D.-D. A. Young (1958). *J. Am. Chem. Soc.*, **80**, 1154.
224. M. Waki and J. Meienhofer (1977). *J. Org. Chem.*, **42**, 2019.
225. G. Losse and W. Zonnchen (1960). *Liebigs Ann. Chem.*, **636**, 140.
226. G. Losse and D. Nadolski (1964). *J. Pract. Chem.*, **24**, 118.
227. J. C. Power, R. Seidner and J. G. Parsons (1965). *Tetrahedron Lett.*, 1713.
228. R. Geiger and W. Seidel (1968). *Chem. Ber.*, **101**, 3386.
229. R. S. Lott, V. S. Chauchan and Stanmer (1979). *J. Chem. Soc., Chem. Commun.*, 495.
230. J. M. Stewart and J. D. Young (1969). *Solid Phase Peptide Synthesis*. San Francisco: Freeman.
231. M. Bodanszky and A. Bodanszky (1984). *The Practice of Peptide Synthesis*. Berlin: Springer-Verlag.
232. See reference 231, pp. 241–2.
233. D. S. Tarbell, Y. Yamamoto and B. M. Pope (1972). *Proc. Nat. Acad. Sci. USA*, **69**, 730.
234. See reference 231, p. 20.
235. T. Sasuki, K. Minamoto and W. Itoh (1978). *J. Org. Chem.*, **43**, 2320.
236. R. L. Franck and P. V. Smith (1946). *J. Am. Chem. Soc.*, **68**, 2103.
237. E. J. Corey and D. Seebach (1970). *Organic Syntheses*, **50**, 72.
238. (a) E. J. Corey and D. Seebach (1975). *J. Org. Chem.*, **40**, 231.
     (b) B.-T. Grobel and D. Seebach (1977). *Synthesis*, 357.
239. P. A. Rossy, W. Hoffman and N. Muller (1980). *J. Org. Chem.*, **45**, 617.
240. D. T. McAllan, T. V. Cullum, R. A. Dean and F. A. Fidler (1951). *J. Am. Chem. Soc.*, **73**, 3627.
241. A. McKillop and J. A. Tarbin (1983). *Tetrahedron Lett.*, 1505.
242. B. M. Trost and D. P. Curran (1981). *Tetrahedron Lett.*, 1287.
243. (a) D. Martin and H. G. Hauthal (1975). *Dimethyl Sulphoxide*. Wokingham, Berkshire: Reinhold.
     (b) A. J. Mancuso and D. Swern (1981), *Synthesis*, 165.
244. R. O. Hutchins and D. Kandasamy (1975). *J. Org. Chem.*, **40**, 2530.
245. J.-L. Luche, L. Rodriguez-Hahn and P. Crabbé (1978). *J. Chem. Soc., Chem. Commun.*, 601.
246. J. K. Crandall and M. Apparu (1983). 'Base-promoted isomerisation of epoxides', in *Organic Reactions* (29). Ed. W. G. Dauben. New York: Wiley, p. 345.
247. O. P. Vig, S. S. Bari, S. D. Sharma and S. S. Rana (1977). *Indian J. Chem., Sect. B*, **15**, 1076.
248. See reference 31.
249. J. Villieras and M. Rambauld (1983). *Synthesis*, 300.
250. T. Mukaiyama (1982). 'The directed aldol reaction', in *Organic Reactions* (29). Ed. W. G. Dauben. New York: Wiley, p. 203.

251. T. Mukaiyama, K. Banno and K. Narasaka (1974). *J. Am. Chem. Soc.*, **96**, 7503.
252. H. D. House, L. J. Czuba, M. Gall and H. D. Olmstead (1969). *J. Org. Chem.*, **34**, 2324; see also Y. Ito, S. Fujii, M. Nakatsuka, F, Kawamoto and T. Saegusa (1981). *Organic Syntheses*, **60**, 113.
253. G. A. Olah, B. G. B. Gupta, S. C. Narang and R. Malhotra (1979). *J. Org. Chem.*, **44**, 4272.
254. W. S. Wadsworth, Jr, and W. D. Emmons (1961). *J. Am. Chem. Soc.*, **83**, 1733.
255. C. Piechucki (1974). *Synthesis*, 869.
256. F. Nerdel and H. Henkel (1952). *Chem. Ber.*, **85**, 1138.
257. W. H. Pirkle, J. M. Finn, J. L. Schreiner and B. C. Hamper (1981). *J. Am. Chem. Soc.*, **103**, 3964.
258. (a) V. Schurig (1983). 'Gas chromatographic methods', in *Asymmetric Synthesis*. Ed. J. D. Morrison. New York: Academic Press, Vol. 1, p. 59.
(b) W. H. Prikle and J. Finn (1983). 'Separation of enantiomers by liquid chromatographic methods', reference (a), p. 97.
259. J. R. Mohrig and S. M. Shapiro (1976). *J. Chem. Ed.*, **53**, 586.

CHAPTER 6
AROMATIC COMPOUNDS

6.1 AROMATIC HYDROCARBONS

Benzene (1) is the simplest aromatic hydrocarbon upon which our knowledge of aromatic chemistry is based. This hydrocarbon, the alkylbenzenes (2), the arylmethanes [e.g. diphenylmethane (3)], the biphenyls [e.g. biphenyl (4)] and the condensed polycyclic systems [e.g. naphthalene (5) and anthracene (6)] all exhibit chemical reactivity and spectroscopic features which are markedly different from their aliphatic and alicyclic hydrocarbon counterparts. Indeed the term aromatic character was introduced to specify the chemistry of this group of hydrocarbons and their substituted functional derivatives, and it was soon used to summarise the properties of certain groups of heterocyclic compounds having five- and six-membered ring systems and the associated condensed polycyclic analogues (Chapter 8).

The crucial structural feature which underlies the aromatic character of benzenoid compounds is of course the cyclic delocalised system of six π-electrons. Other carbocyclic systems similarly possessing this 'aromatic sextet' of electrons include, for example, the ion $C_5H_5^{\ominus}$ formed from cyclopentadiene under basic conditions. The cyclopentadienide anion is centrosymmetrical and strongly resonance stabilised, and is usually represented as in (7). The analogous cycloheptatrienylium (tropylium) cation (8), with an aromatic sextet delocalised over a symmetrical seven-membered ring, is also demonstrably aromatic in character. The stable, condensed, bicyclic hydrocarbon azulene ($C_{10}H_8$) possesses marked aromatic character; it is usually represented by the covalent structure (9). The fact that the molecule has a finite dipole moment, however, suggests that the ionic form (10) [a combination of (7) and (8)] must contribute to the overall hybrid structure.

$C_6H_6 \equiv PhH$

(1)

$Ph \cdot R$

(2)

$(Ph)_2CH_2$

(3)

(4)

(5)

(6)

(7)   (8)

(9)   (10)

SPECTROSCOPIC FEATURES

Aromatic hydrocarbons give rise to spectra which differ markedly from those of aliphatic and alicyclic hydrocarbons owing to the mesomeric stabilisation of the aromatic ring system. In the *i.r.* absorption spectrum of an aromatic hydrocarbon there are four wavenumber regions which should be inspected to provide definitive evidence for the presence of an aromatic nucleus, and for the possible state of substitution: (*a*) at *c*. 3050 cm^{-1} (C_{AR}—H stretching); (*b*) at 1600, 1590, 1500 and 1450 cm^{-1} (ring-breathing vibrations); (*c*) at 650–820 cm^{-1} (C_{AR}—H out-of-plane vibrations); and (*d*) at 2000–1600 cm^{-1} (summation bands) (p. 278). The *p.m.r.* spectrum (e.g. Fig. 3.42 of toluene) also provides definitive evidence of the presence of an aromatic nucleus, since the aromatic protons are deshielded by the induced ring current sustained by the π-electrons and they appear as signals at *c*. δ 7.00 (p. 325). The ^{13}C-*n.m.r.* is discussed on p. 329 (see also Fig. 3.43). The *u.v.–visible* spectrum is more useful in the aromatic field than the aliphatic, since the energy associated with the π-electron transitions is markedly affected by the presence of a fused ring system or of substituent interaction in a mesomeric or inductive manner (p. 390). The profile of the *m.s.* of an aromatic compound also provides invaluable confirmation of the presence of an aromatic nucleus in that the relative abundance of the molecular ion and closely associated fragment ions is usually high compared to the low relative abundance found in the aliphatic series (p. 368).

The isomeric xylenes provide a useful illustration of these spectroscopic features. Thus the i.r. spectrum of *o*-xylene (Fig. 3.16) shows absorption bands at *c*. 3050 cm^{-1} and *c*. 2950 cm^{-1} for the stretching vibrations of the aromatic and aliphatic carbon-bound hydrogens respectively. The ring-breathing vibrations are found at 1600, 1590, 1500 cm^{-1} (the 1450 cm^{-1} band is obscured by the methyl deformation absorption). The *meta* and *para* isomers show almost identical absorptions in these regions. However, the absorption in the 700–800 cm^{-1} region clearly defines the state of substitution; for the *ortho* isomer the band at *c*. 740 cm^{-1} indicates four adjacent hydrogens; for the *meta* isomer the bands are at *c*. 690 and 770 cm^{-1} for three adjacent hydrogens and at *c*. 870 cm^{-1} for one lone hydrogen; for the *para* isomer the band at *c*. 790 cm^{-1} indicates two adjacent hydrogens. The three isomers each give a characteristic pattern in the summation band region when the spectra are appropriately recorded. The p.m.r. spectra of the three isomers do not clearly define the substitution pattern, the aromatic protons appearing as a four-proton singlet in the spectra of the *ortho* and *para* isomers, and as a multiplet in the case of the *meta* isomer. The m.s. of these isomers are very similar (*ortho* isomer, Fig. 3.76) and may be interpreted as *m/z* 106 (M), 105 (M − H), 91 (M − Me, base peak), 65 (91 − C_2H_2), 77 (91 − CH_2), 51 (77 − C_2H_2) and 39 (65 − C_2H_2).

The synthesis of aromatic hydrocarbons (arenes) is for convenience organised under the following five headings.

1. Alkylbenzenes (Expts 6.1 to 6.5).
2. Di- and triarylmethanes (Expts 6.6 and 6.7).
3. Biphenyl systems (Expts 6.8 to 6.10, see also Expt 6.79).
4. Condensed polycyclic systems (Expts 6.11 to 6.14).
5. Some non-benzenoid systems (Expts 6.15 and 6.16).

6.1.1   ALKYLBENZENES

The synthesis of alkylbenzenes may be classified under: (*a*) the Wurtz–Fittig reaction; (*b*) the reduction of aryl aldehydes and ketones; and (*c*) and Friedel–Crafts alkylation reaction.

SUMMARY OF RETROSYNTHETIC STRATEGIES

The Wurtz–Fittig reaction (cf. the Wurtz reaction, Section 5.1.4, p. 478). The interaction of an aryl halide, an alkyl halide and sodium gives a reasonable yield of an alkylbenzene.

$$ArBr + RBr + 2Na \longrightarrow Ar\cdot R + 2NaBr$$

The by-products of the reaction, e.g. R—R and Ar—Ar, can usually readily be separated by distillation.

A likely mechanism for this reaction involves the formation of an arylsodium and its subsequent reaction with the alkyl halide.

$$ArBr + 2Na \longrightarrow Ar^{\ominus}Na^{\oplus} + NaBr$$

$$Ar^{\ominus} \frown R \frown Br \longrightarrow Ar\cdot R + {:}Br^{\ominus}$$

The reaction of the arylsodium with the more reactive alkyl bromide occurs in preference to reaction with the less reactive aryl bromide, which would lead to the biaryl Ar—Ar. Furthermore, the formation of the aryl anion, which is likely to be stabilised by electron delocalisation, probably occurs more readily than the formation of the alkyl anion by reaction of the sodium with the alkyl bromide. For these reasons acceptable yields of the alkylbenzene are usually obtained.

Better yields of alkylbenzenes result if the arylsodium is first prepared and then subjected to a suitable alkylation reaction. In the preparative example of butylbenzene (Expt 6.1), benzylsodium is conveniently obtained by first forming phenylsodium by reaction between sodium and chlorobenzene in a toluene medium, and then heating the toluene suspension of the phenylsodium at 105 °C for about 35 minutes when a transmetalation process occurs (formulated at the beginning of Expt 6.1).

The Grignard reagent derived from benzyl chloride undergoes ready alkylation with an alkyl toluene-p-sulphonate, a reaction which provides a further useful synthesis of an alkylbenzene (e.g. pentylbenzene, Expt 6.2).

$$Ph \cdot CH_2 Cl \xrightarrow{\text{Mg}} Ph \cdot CH_2 MgCl \xrightarrow{2[p\text{-}Me \cdot C_6H_4 \cdot SO_3R]} Ph \cdot CH_2 \cdot R + RCl$$

The reduction of aldehydes and ketones. Aromatic hydrocarbons are the main products when aromatic aldehydes or ketones are reduced with amalgamated zinc and concentrated hydrochloric acid (the Clemmensen reduction, e.g. hexylbenzene, Expt 6.3).

$$Ar \cdot CO \cdot R \longrightarrow Ar \cdot CH_2 \cdot R$$

Some features of the Clemmensen reduction are discussed in Section 5.1.3, p. 476. Purely aromatic ketones generally do not give satisfactory results: pinacols and resinous products often predominate. The reduction of ketonic compounds of high molecular weight and very slight solubility is facilitated by the addition of a solvent, such as ethanol, acetic acid or dioxane, which is miscible with aqueous hydrochloric acid. With some carbonyl compounds, notably keto acids, poor yields are obtained even in the presence of ethanol, etc., and the difficulty has been ascribed to the formation of insoluble polymolecular reduction products, which coat the surface of the zinc. The addition of a hydrocarbon solvent, such as toluene, is beneficial because it keeps most of the material out of contact with the zinc and the reduction occurs in the aqueous layer at such a high dilution that polymolecular reactions are largely inhibited (see Expt 6.123).

Aryl alkyl ketones are readily prepared by the Friedel–Crafts acylation process (see Section 6.11.1, p. 1006) and their Clemmensen reduction constitutes a more efficient procedure for the preparation of monoalkylbenzenes than the alternative direct Friedel–Crafts alkylation reaction (see below). Alternatively aldehydes and ketones may be reduced to the corresponding hydrocarbon by the Wolff–Kishner method which involves heating the corresponding hydrazone or semicarbazone with potassium hydroxide or with sodium ethoxide solution.

$$Ar \cdot CO \cdot R \xrightarrow{N_2H_4} Ar \cdot (R)C = N \cdot NH_2 \xrightarrow{KOH} Ar \cdot CH_2 \cdot R$$

The *Huang–Minlon modification* of the reaction has the following advantages: (i) the actual isolation of the hydrazone is unnecessary; (ii) the reaction time is considerably reduced; (iii) the reaction can be carried out at atmospheric pressure and on a large scale and; (iv) the yields are usually excellent. The hydrazone is first formed *in situ* by refluxing a solution of the carbonyl compound in a moderate amount of diethylene glycol or triethylene glycol with the commercial 85 or 90 per cent hydrazine hydrate and about 3 equivalents of potassium hydroxide for 1 hour; the water and excess of hydrazine are removed by distillation until a favourable temperature for the decomposition of the hydrazone is attained (170–190 °C) and the solution is refluxed for 3–5 hours longer.

An alternative milder procedure is the reduction of the corresponding toluene-*p*-sulphonylhydrazones with catecholborane, followed by decomposition of the intermediate with sodium acetate in the presence of dimethyl sulphoxide, or with tetrabutylammonium acetate.[1] These methods, which do not have the disadvantages of the Clemmensen reduction, are illustrated by the preparation of ethylbenzene from acetophenone (Expt 6.4, *Methods A and B*). Outline mechanisms for these reactions are given below.

$$R_2C=N-NH_2 \rightleftharpoons \xrightarrow{-H^\oplus} [R_2C=N-\ddot{N}H]^\ominus \xrightarrow{-H^\oplus, +H^\oplus} R_2CH-N=\ddot{N}^\ominus$$

$$R_2\overset{\curvearrowright}{C}H-N=\ddot{N}^\ominus \xrightarrow{-N_2} R_2\ddot{C}H^\ominus \xrightarrow{+H^\oplus} R_2CH_2$$

$$R_2C=N-NHTs \xrightarrow{\text{(O,O-BH)}} \underset{\text{H}}{\overset{}{R_2C}}-\underset{\text{B}}{\overset{}{N}}-NHTs \xrightarrow{AcO^\ominus} R_2CH-N\!-\!NH\!-\!Ts$$
$$Ts = -O_2S\cdot C_6H_4\cdot Me \qquad\qquad AcO-B$$

$$\xrightarrow{-Ts} R_2\overset{\curvearrowright}{C}H-N=\overset{\curvearrowright}{N}-H \xrightarrow{-N_2} R_2\ddot{C}H^\ominus \xrightarrow{+H_2} R_2CH_2$$

The Friedel–Crafts alkylation reaction. An alkyl halide reacts with an aromatic hydrocarbon in the presence of aluminium chloride to yield in the first instance a hydrocarbon, thus:

$$ArH + RX \xrightarrow{AlCl_3} Ar\cdot R + HX$$

The reaction does not, however, stop at the stage of mono-substitution, since the alkylbenzene (Ar·R) initially produced undergoes alkylation more easily than the original hydrocarbon ArH, owing to the electron-releasing effect of the alkyl group. Mixtures of substances therefore often result and extensive purification may be required in order to isolate the monosubstituted compound. Some mono-alkylbenzenes may be prepared by using an excess of the hydrocarbon, which also acts as a diluent in moderating the violence of the reaction and prevents the undue formation of poly-alkylbenzenes.

The mechanism of the reaction is generally considered to proceed by way of carbocations which attack the aromatic nucleus:

$$R-X:\curvearrowright AlCl_3 \rightleftharpoons R^\oplus[XAlCl_3]^\ominus \rightleftharpoons R^\oplus + [XAlCl_3]^\ominus$$

mesomeric
cation

$$+ HX$$
$$+ AlCl_3$$

In many cases the alkyl group may undergo extensive skeletal rearrangement under the reaction conditions employed. For example, when benzene is alkylated with propyl chloride in the presence of aluminium chloride over a wide temperature range, a mixture of propylbenzene and isopropylbenzene, in approximately 1:2 ratio, is obtained. On the other hand isobutyl bromide or chloride with benzene in the presence of aluminium chloride gives exclusively t-butylbenzene. This alkylbenzene is, however, more conveniently prepared by using t-butyl chloride as the alkylating reagent (Expt 6.5).

The rearrangement processes mentioned above involve a 1,2-nucleophilic shift of a hydride ion to form a more stable secondary or tertiary carbocation.

Other catalysts which may be used in the Friedel–Crafts alkylation reaction include iron(III) chloride, antimony pentachloride, zirconium tetrachloride, boron trifluoride, zinc chloride and hydrogen fluoride but these are generally not so effective in small-scale preparations. The alkylating agents include alkyl halides, alcohols and alkenes.

Experiment 6.1   BUTYLBENZENE (2-Phenylbutane)

$$PhCl \xrightarrow{2Na} PhNa \xrightarrow{Ph\cdot Me} Ph\cdot CH_2Na \xrightarrow{PrBr} C_6H_4\cdot CH_2\cdot Pr$$

Equip a 500-ml three-necked flask as detailed for p-*toluic acid* (Expt 6.158, *Method A*) and pass a slow stream of nitrogen through the apparatus. Charge the flask with 150 ml of sodium-dried, sulphur-free toluene (Section 4.1.3, p. 398) and 13.8 g (0.6 mol) of sodium wire. Place 34 g (31 ml, 0.3 mol) of chlorobenzene in the dropping funnel and add it dropwise through the condenser during 1 hour, with vigorous stirring, while maintaining the temperature inside the flask at 30–35 °C. The start of the reaction is indicated by the appearance of black specks on the sodium surface. (If the reaction is slow to start, it may be instantly initiated by a few drops of butanol.) Complete the formation of phenylsodium by stirring for 2–3 hours at 30 °C. Attach a calcium chloride tube to the top of the reflux condenser and reflux the mixture for 40 minutes. The reflux temperature, initially 107 °C, gradually falls to 103 °C as benzene is formed by the exchange reaction. Remove the heating bath and add 27.6 g (20.5 ml, 0.224 mol) of redistilled propyl bromide during 20–25 minutes at 103–105 °C; the reaction is strongly exothermic. Allow the reaction mixture to cool to room temperature: maintain the stirring and the slow stream of nitrogen. Add water *slowly* to destroy the excess of sodium. Separate the toluene layer, dry it (magnesium sulphate) and distil it through a short, jacketed column filled with glass helices (19 cm packed length, 14 mm diameter; compare Fig. 2.104). After removal of the toluene (up to 111 °C) and a small intermediate fraction (111–179 °C), pure butylbenzene passes over at 179.5–181 °C/752 mmHg (23 g, 77%). A brown residue (4 g) remains in the flask.

The i.r. spectrum shows absorptions at c. 3050 and 2950 cm⁻¹ for the aromatic and alkyl carbon–hydrogen stretching vibrations respectively, at

1600, 1590 and 1500 cm^{-1} for the ring breathing vibrations (the 1450 cm^{-1} absorption is obscured by the alkyl carbon–hydrogen deformation vibrations), at c. 700 and 750 cm^{-1} characteristic of monosubstitution, and well-defined summation bands at 1600–2000 cm^{-1}. The p.m.r. spectrum (CCl$_4$, TMS), shows signals at δ 0.92 (t, 3H, Me), 1.10–1.80 (m, 4H, —CH$_2$CH$_2$—), 2.58 (t, 2H, ArCH$_2$) and 7.09 (s, 5H, C$_{AR}$—H). The m.s. shows significant fragment ions at m/z 134 (M), 105 (M – C$_2$H$_5$), 91 (M – C$_3$H$_7$, base peak), and 65 (91 – C$_2$H$_2$).

Experiment 6.2 PENTYLBENZENE (*1-Phenylpentane*)

$$Ph\cdot CH_2Cl \xrightarrow{Mg} Ph\cdot CH_2MgCl \xrightarrow{2[p\text{-}Me\cdot C_6H_4\cdot SO_3Bu]} Ph\cdot CH_2\cdot Bu + BuCl$$

In a 1500-ml three-necked flask equipped with a dropping funnel, a sealed stirrer unit and a double surface condenser to which is attached a guard-tube filled with a mixture of calcium chloride and soda-lime, prepare an ethereal solution of benzylmagnesium chloride from 24.3 g (1 mol) of clean, dry magnesium turnings (under 100 ml of anhydrous ether) and a solution of 126.5 g (115 ml, 1 mol) of freshly distilled benzyl chloride in 500 ml of anhydrous ether. Use a crystal of iodine as a catalyst and proceed as described under the general procedure noted in Expt 5.39. Finally, cool the flask by immersion in a bath of ice-water. Place a solution of 456 g (2 mol) of butyl toluene-*p*-sulphonate (Expt 6.46) in about twice the volume of anhydrous ether in the dropping funnel, and add it slowly to the vigorously stirred benzylmagnesium chloride solution, at such a rate that the ether just boils; a white solid soon forms. The addition is complete after about 2 hours. Pour the reaction product slowly into a mechanically stirred mixture of 1 kg of finely crushed ice, 1 litre of water and 125 ml of concentrated hydrochloric acid contained in a 4- or 5-litre beaker; the precipitated magnesium toluene-*p*-sulphonate will ultimately pass into the solution. Separate the ether layer, extract the aqueous layer with 250 ml of ether and wash the combined ether solutions with about 100 ml of water. Dry the ether solution with about 10 g of anhydrous potassium carbonate. Distil off the ether on a rotary evaporator, add to the mixture 5–7 g of sodium cut into small pieces and heat under reflux for about 2 hours in order to remove any benzyl alcohol which may have formed by atmospheric oxidation of benzylmagnesium chloride. Decant the solution and distil it from an air bath through a well-lagged and efficient fractionating column; collect the fraction, b.p. 190–210 °C. Redistil and collect the pentylbenzene at 198–203 °C. The yield is 90 g (61%).

Record the i.r. spectrum and the p.m.r. spectrum and assign the absorptions using the spectra quoted above for butulbenzene (Expt 6.1) as a guide. Interpret the m.s. which shows principal fragment ions at m/z 148, 105, 91, 77 and 65.

Experiment 6.3 HEXYLBENZENE (*1-Phenylhexane*)

Clemmensen reduction. Prepare 200 g of amalgamated zinc in a 2-litre three-necked flask as detailed in Section 4.2.*80*, p. 467. Fit the flask with a reflux condenser, a sealed stirrer and a gas entry tube reaching to within 1 cm of the bottom; connect the last-named through an intermediate empty wash bottle to a Kipp's apparatus supplying hydrogen chloride gas (Section 4.2.38, p. 438). Place a mixture of 500 ml of concentrated hydrochloric acid and 100 ml of water in the flask and introduce 100 g (0.57 mol) of 1-phenylhexan-3-one (Expt 5.93). Stir the mixture and pass a slow stream of hydrogen chloride gas while warming the flask on a suitable wire gauze by means of a small flame. If the reaction becomes unduly vigorous, stop the supply of hydrogen chloride until it subsides somewhat. Most of the zinc dissolves after 6 hours, by which time the reaction is almost complete; allow to stand overnight. Arrange the apparatus for steam distillation (Fig. 2.102) and pass steam into the flask, heated by means of a small flame, until the distillate is clear. Separate the upper hydrocarbon layer, wash it with 5 per cent sodium hydroxide solution, then with water and dry over magnesium sulphate. Distil from a 100 ml flask and at 218–230 °C collect the crude hexylbenzene, which contains some unsaturated compounds. These can be removed by repeated shaking with 5 per cent of the volume of concentrated sulphuric acid until the latter is colourless or, at most, very pale yellow. The hydrocarbon is then washed with 5 per cent sodium carbonate solution, then with water and dried over magnesium sulphate. It is then distilled twice from sodium when pure hexylbenzene, b.p. 220–225 °C is obtained. The yield is 40 g (43%).

Experiment 6.4 ETHYLBENZENE

$$\text{Ph·CO·Me} \xrightarrow{\text{NH}_2\text{·NHX}} \text{Ph·C(=N·NHX)·Me} \xrightarrow[\text{(B); [H]}]{\text{(A); KOH}} \text{Ph·CH}_2\text{·Me}$$

Method (A); X = H   *Method (B); X = p-Me·C_6H_4·SO$_2$*

Method A. **Huang–Minlon modification of the Wolff–Kishner reduction.** Place 36.0 g (0.3 mol) of redistilled acetophenone, b.p. 201 °C, 300 ml of diethylene glycol, 30 ml of 90 per cent hydrazine hydrate (**CAUTION**) and 40 g of potassium hydroxide pellets in a 500-ml two-necked round-bottomed flask fitted with a reflux condenser; insert a thermometer supported in a screw-capped adapter in the side-neck so that the bulb dips into the reaction mixture. Warm the mixture on a boiling water bath until most of the potassium hydroxide has dissolved and then heat under reflux for 1 hour either by means of a free flame or by using a heating mantle. Remove the reflux condenser and fit a still-head and condenser for downward distillation. Distil until the temperature of the liquid rises to 175 °C (1). Separate the upper hydrocarbon layer from the distillate and extract the aqueous layer twice with 20 ml portions of ether. Dry the combined upper layer and ethereal extracts with magnesium sulphate, remove the ether on a water bath and distil the residue. Collect ethylbenzene at 135–136 °C; the yield is 20 g (62.5%).

The i.r. spectrum shows absorptions at c. 3050 and 2950 cm^{-1} for the stretching of the aromatic and alkyl carbon–hydrogen bonds respectively. Monosubstitution is confirmed from the absorption at c. 690 and 745 cm^{-1} arising from five adjacent hydrogens. The p.m.r. spectrum (CDCl$_3$, TMS) shows signals at δ 1.19 (q, 3H, Me), 2.53 (t, 2H, CH$_2$) and 7.07 (s, 5H, C$_{AR}$—H). The ^{13}C-n.m.r. spectrum (CDCl$_3$, TMS) shows signals at δ 15.8, 29.1, 125.8, 127.9, 128.4 and 144.1.

Note. (1) The reduction takes place at a comparatively low temperature and is fairly rapid for acetophenone. With higher ketones, the upper layer of the distillate should be returned to the contents of the flask and the heating under reflux continued for 3–5 hours. The reaction mixture and the aqueous distillate are then combined, extracted with ether and the ether extract treated as described above.

Method B. The toluene-*p*-sulphonylhydrazone of acetophenone (0.721 g, 2.5 mmol) (1), m.p. 140–141.5 °C, is placed in a flame-dried, nitrogen-filled flask containing 5 ml of chloroform. Catecholborane (0.52 ml, 5.0 mmol) (Section 4.2.7, p. 420) is added and the reduction allowed to proceed for 2 hours at room temperature (2). Methanol (1 ml) is added to destroy the excess of hydride followed by the addition of tetrabutylammonium acetate (0.7 g, 2.5 mmol). The reaction mixture is stirred for 4 hours when g.l.c. analysis indicates a 94 per cent yield of ethylbenzene. The product is isolated by distillation, yield 0.21 g (79%), b.p. 132–136 °C.

Notes. (1) The general procedure for the preparation of toluene-*p*-sulphonylhydrazones is given in Expt 5.6.
(2) The reaction may be monitored by removing aliquot portions with a syringe, mixing with deuterochloroform in an n.m.r. tube, and recording the spectra.

Experiment 6.5 t-BUTYLBENZENE (*2-Methyl-2-phenylpropane*)

$$Me_3CCl + PhH \text{ (excess)} \xrightarrow{AlCl_3} Ph \cdot CMe_3$$

CAUTION: *This preparation should be conducted in an efficient fume cupboard.*

Into a 1-litre three-necked flask, equipped as in Expt 6.121, place 50 g (0.33 mol) of anhydrous aluminium chloride (1) and 200 ml (2.25 mol) of dry benzene (**CAUTION**); cool in a bath of crushed ice. Stir the mixture and add 50 g (59 ml, 0.54 mol) of t-butyl chloride (Expt 5.49) from the dropping funnel during 4–5 hours; the first addition should be 3–4 ml in order to prevent the benzene from freezing. Maintain the mixture at a temperature of 0–5 °C by the addition of salt to the ice, if necessary. When all the t-butyl chloride has been run in, continue the stirring for 1 hour longer. Remove the separatory funnel and add 200 g of finely crushed ice in small portions with stirring; finally add 100 ml of cold water to complete the decomposition of the intermediate addition compound. Arrange the flask for steam distillation (Fig. 2.102) and steam distil the resulting reaction mixture. Transfer the steam distillate to a separatory funnel, remove the upper hydrocarbon layer, extract the water layer with two 50 ml portions of ether and combine the extracts with the upper layer. Dry with magnesium sulphate, distil off the ether on a water bath and fractionally distil the residue twice, using a well-lagged column (Fig. 2.104). Collect the t-butylbenzene at 165–170 °C. The yield is 45 g (62%). Pure t-butylbenzene boils at 168.5 °C.

Note the characteristic absorptions for the aromatic system at *c*. 3050 cm^{-1}, at 1600, 1590 and 1500 cm^{-1} and at *c*. 700 and 765 cm^{-1} for a monosubstituted nucleus. The t-butyl group shows characteristic carbon–hydrogen stretching absorptions at *c*. 2950 cm^{-1}.

Note. (1) In an alternative procedure 25 g of anhydrous iron(III) chloride replace the aluminium chloride, the mixture is cooled to 10 °C and the 50 g of t-butyl chloride are added. The mixture is slowly warmed to 25 °C and maintained at this temperature until no more hydrogen chloride is evolved. The reaction mixture is then washed with dilute hydrochloric acid and with water, dried and fractionally distilled. The yield of t-butyl benzene, b.p. 167–170 °C, is 60 g.

6.1.2 DI- AND TRIARYLMETHANES

Alkylation with benzyl chloride by the Friedel–Crafts procedure yields diphenylmethane (Expt 6.6). The reactive nature of the halide obviates the need for heating the reaction mixture, and a smaller proportion of aluminium chloride is employed compared to the standard conditions as illustrated by Expt 6.5.

A Friedel–Crafts reaction between chloroform and an excess of benzene affords a convenient route to triphenylmethane (cognate preparation in Expt 6.6).

$$3PhH + CHCl_3 \longrightarrow Ph_3CH$$

The synthesis of the triptycene system illustrates a synthetic use of the important reactive intermediate benzyne (13) which is generated by the thermal decomposition of benzenediazonium-2-carboxylate (see Section 6.5.3, p. 900). The latter is produced when anthranilic acid (11) is treated with an alkyl nitrite in an aprotic solvent, and probably exists mainly as the zwitterionic form (12).

(11)   (12)   (13)

The benzyne functions as a dienophile towards reactive diene systems. The reactivity of the 9,10-positions in anthracene is well known (Diels–Alder reaction, Section 7.6), and addition of benzyne to 9-bromoanthracene yields the interesting cage-ring alkyl halide: 9-bromotriptycene (9-bromo-9,10-*o*-benzenoanthracene). The reaction is incomplete and some unreacted 9-bromoanthracene remains in the crude reaction products, but may be removed by virtue of its ready conversion into a maleic anhydride adduct in a further Diels–Alder-type reaction.

The environment of the halogen in the ring structure renders it almost unreactive in displacement reactions by nucleophiles.

Experiment 6.6 DIPHENYLMETHANE

$$Ph \cdot CH_2Cl + PhH \text{ (excess)} \xrightarrow{AlCl_3} Ph \cdot CH_2 \cdot Ph$$

CAUTION: *This preparation should be conducted in an efficient fume cupboard.*

Fit a 500-ml three-necked round-bottomed flask with a sealed mechanical stirrer, attach a gas absorption device to one of the side-necks and stopper the third neck. Place 38 g (35 ml, 0.3 mol) of redistilled benzyl chloride and 150 ml of dry benzene (**CAUTION**) (Section 4.1.2, p. 398) in the flask. Weigh out 2 g (0.015 mol) of anhydrous aluminium chloride (Section 4.2.4, p. 416) into a dry capped specimen tube with the minimum exposure to the atmosphere. Cool the flask in a bath of crushed ice and add about one-fifth of the aluminium chloride. Stir the mixture; a vigorous reaction will set in within a few minutes and hydrogen chloride is evolved. When the reaction has subsided, add a further portion of the aluminium chloride and repeat the process until all has been introduced. The mixture should be kept well shaken and immersed in the ice bath during the addition. After 15 minutes cautiously add 100 g of crushed ice, followed by 100 ml of water in order to decompose the aluminium complex. Shake the mixture well, transfer to a separatory funnel and run off the lower aqueous layer. Wash the upper layer successively with dilute hydrochloric acid and water and dry it with anhydrous calcium chloride. Remove the benzene with the aid of the apparatus shown in Fig. 2.101. Distil the remaining liquid through an air condenser either with a free flame or from an air bath. Collect the diphenylmethane at 250–275 °C (the pure substance boils at 262 °C) (1). The distillate should solidify on cooling in ice and scratching with a glass rod, or by seeding with a crystal of the pure material. If it does not crystallise, redistil from a small flask and collect the fraction, b.p. 255–267 °C; this generally crystallises on cooling and has m.p. 24–25 °C. The yield is 25 g (50%).

Note. (1) Alternatively the distillation may be conducted under diminished pressure; the fraction, b.p. 125–130 °C/10 mmHg, is collected.

Cognate preparation. *Triphenylmethane.* The apparatus is similar to that described above, but incorporating a reflux condenser to the outlet of which is fitted the gas absorption device. Place a mixture of 200 g (230 ml, 2.57 mol) of dry benzene (**CAUTION**) and 40 g (26 ml, 0.33 mol) of dry chloroform (**CAUTION**) (Section 4.1.6, p. 399) in the flask, and add 35 g (0.26 mol) of anhydrous aluminium chloride in portions of about 6 g and at intervals of 5 minutes with constant stirring. The reaction sets in upon the addition of the aluminium chloride and the liquid boils with the evolution of hydrogen chloride. Complete the reaction by refluxing for 30 minutes on a water bath. When cold, pour the contents of the flask very cautiously on to 250 g of crushed ice and 10 ml of concentrated hydrochloric acid. Separate the upper benzene layer, dry it with anhydrous calcium chloride or with magnesium sulphate and remove the benzene by flash distillation (Fig. 2.101). Attach a Claisen still-head connected to a short air condenser and distil the remaining oil under reduced pressure; collect the fraction, b.p. 190–215 °C/10 mmHg. This is crude triphenylmethane which solidifies on cooling. Recrystallise it from about four times its weight of ethanol; triphenylmethane separates in needles and melts at 92 °C. The yield is 30 g (37%).

Experiment 6.7 9-BROMOTRIPTYCENE

Place 2.3 g (0.02 mol) of isopentyl nitrite (isoamyl nitrite), 2.6 g (0.01 mol) of 9-bromoanthracene and 25 ml of 1,2-dimethoxyethane ('glyme', Section

4.1.*18*, p. 406) in a 250-ml round-bottomed flask fitted with a reflux condenser. Heat the mixture to gentle reflux on an electric mantle and add down the condenser, dropwise during 30 minutes, a solution of 3.4 g (0.025 mol) of anthranilic acid in 15 ml of glyme. Remove the mantle, cool the mixture to about 40 °C and add 2.3 g more of isopentyl nitrite dissolved in 5 ml of glyme. Resume the gentle refluxing and add during 15 minutes another solution of 3.4 g of anthranilic acid in 10 ml of glyme. Heat for an additional 15 minutes, add 10 ml of 95 per cent ethanol and pour the reaction mixture into a solution of 3.0 g of sodium hydroxide in 100 ml of water. Cool the resulting brown suspension thoroughly in ice-water and filter under suction. Wash the residue with a chilled methanol/water mixture (4:1 v/v) and transfer to a 100-ml round-bottomed flask which is then evacuated (rotary evaporator) on a steam bath until the weight is constant.

To the flask containing the crude product, add 2.0 g (0.02 mol) of maleic anhydride and 25 ml of triethyleneglycol dimethyl ether ('triglyme', b.p. 222 °C), fit a reflux condenser and boil under reflux for 10 minutes over a Bunsen flame. Cool the solution to about 100 °C, add 10 ml of 95 per cent ethanol and pour into a solution of 3.0 g of sodium hydroxide in 75 ml of water. Stir for a few minutes, then cool in ice-water and filter under suction. Wash the residue with chilled methanol/water (4:1 v/v) and recrystallise from methylcyclohexane or a chloroform/methanol mixture to give colourless plates of 9-bromotriptycene, m.p. 251–256 °C; yield is 1.5 g (45%). Further recrystallisation gives the pure product, m.p. 258–262 °C.

6.1.3 BIPHENYL SYSTEMS

A reaction which is reminiscent of the Wurtz coupling procedure, and which is particularly valuable in the syntheiss of biphenyl and its symmetrically substituted derivatives, is that of *Ullmann*. It involves heating an aryl halide with copper powder, or better, with activated copper bronze.

$$2ArX + 2Cu \longrightarrow Ar - Ar + 2CuX$$

Aryl iodides and bromides are more reactive than the corresponding chlorides but the latter may be used when activating substituents (e.g. the nitro group) are present, as for example in the synthesis of 2,2′-dinitrobiphenyl (Expt 6.8). The coupling reaction can be effectively carried out in the absence of a solvent, but the use of dimethylformamide as a solvent and diluent often results in an increase of yield, particularly in the case of reactive halides, when the vigour of the exothermic reaction is moderated.

The symmetrically substituted biphenyl system is also formed when aryl radicals generated from an appropriate precursor are allowed to undergo a self-coupling process. An interesting reaction involving the use of the diazonium salt derived from anthranilic acid affords an excellent method of preparing biphenyl-2,2′-dicarboxylic acid (diphenic acid, Expt 6.9). The diazotised anthranilic acid is reduced with the aid of a cupro-ammonia reagent obtained by dissolving copper(II) sulphate in aqueous ammonia and treating the solution with hydroxylamine hydrochloride.

$$2Ar \cdot \overset{\oplus}{N} \equiv N \xrightarrow{+2e} 2Ar \cdot + 2N_2$$

$$2Ar \cdot \longrightarrow Ar - Ar$$

Unsymmetrically substituted biphenyls may be prepared from diazonium salts by means of the Gomberg–Bachmann–Hey reaction.[2] Some preparative examples of the classical method and of a PTC procedure are described in Expt 6.79.[3]

Reactive aromatic systems, in particular phenols, readily undergo oxidation by a single electron transfer process, leading to a mesomerically stabilised radical species, which then dimerises. The coupling of phenols in this way is frequently encountered in biogenetic pathways leading to naturally occurring molecules. A reaction illustrative of the process is the oxidation of 2-naphthol with iron(III) chloride (Expt 6.12) to give 2,2'-dihydroxy-1,1'-binaphthyl (1,1'-bi-2-naphthol).

This compound, in common with other suitable substituted biphenyls, possesses a chiral axis (p. 6) and is isolated from the reaction as a racemate. Although several resolution procedures have been reported, the superior method to date[4] is that in which the binaphthol is first converted by treatment with phosphorus oxychloride into the binaphthyl phosphoric acid (14). Resolution is then effected by formation of diastereoisomeric salts with (+)-cinchonine, appropriate fractional crystallisation and recovery of the (S)-(+)-binaphthyl phosphoric acid. Suitable hydrolysis gives (S)-(−)-1,1'-bi-2-naphthol (15).

(14)   (15)

Earlier attempts to effect an asymmetric coupling to mimic biological coupling reactions by using chiral copper(II)–amine complexes gave only low enantiomeric excesses.[5] The use of (S)-(+)-amphetamine sulphate [Ph·CH$_2$·CH(NH$_3^{\oplus}$)·Me}HSO$_4^{\ominus}$] with copper(II) nitrate was more successful in that the (S)-(−)-enantiomer (15) was formed in 85 per cent chemical yield and 95 per cent optical purity.[6] (S)-(−)-10,10'-Dihydroxy-9,9'-biphenanthryl has been similarly prepared from the coupling of 9-phenanthrol, using (R)-(−)-1,2-diphenylethylamine and copper(II) nitrate.[7] Enantioselective syntheses of binaphthyls using chiral oxazolines in the coupling reaction have also been reported.[8] These chiral biaryls may be complexed with metal hydrides and used in the chiral reduction of, for example, the carbonyl group (p. 521).

Experiment 6.8   2,2'-DINITROBIPHENYL

Place 50 g (0.32 mol) of o-chloronitrobenzene and 75 g of clean dry sand in a 250-ml three-necked flask equipped with a mechanical stirrer and an air reflux condenser. Heat the mixture in an oil or fusible metal bath to 215–225 °C and add, during 40 minutes, 50 g (0.78 mol) of copper bronze or, better, of activated copper bronze (Section 4.2.*19*, p. 426) (1). Maintain the temperature at 215–225 °C for a further 90 minutes and stir continuously. Pour the hot mixture into a Pyrex beaker containing 125 g of sand and stir until small lumps are formed; if the reaction mixture is allowed to cool in the flask, it will set to a hard mass, which can only be removed by breaking the flask. Break up the small lumps by powdering in a mortar, and boil them for 10 minutes with two 400 ml portions of ethanol; filter after each extraction. Cool the filtered extracts in ice, and collect the crude product on a Buchner funnel. Concentrate the filtrate to about half the original volume and thus obtain a second crop of crystals. The total yield of crude solid should be about 24 g; if it is less than this, a third extraction of the reaction product should be made. Dissolve the crude solid in about 400 ml of hot ethanol, add a little decolourising charcoal, boil for a few minutes, filter and cool in ice. Recrystallise again from hot ethanol. The yield of pure 2,2'-dinitrobiphenyl, m.p. 123–124 °C, is 20–22 g (54%).

The experimental conditions for conducting the above reaction in dimethylformamide as solvent are as follows. In a 250-ml three-necked flask, equipped with a reflux condenser and a tantalum wire Hershberg-type stirrer, place 20 g of o-chloronitrobenzene and 100 ml of dimethylformamide (dried over anhydrous calcium sulphate). Heat the solution to reflux and add 20 g of activated copper bronze in one portion. Heat under reflux for 4 hours, add another 20 g portion of copper powder and continue refluxing for a second 4-hour period. Allow to cool, pour the reaction mixture into 2 litres of water and filter with suction. Extract the solids with three 200 ml portions of boiling ethanol: alternatively, use 300 ml of ethanol in a Soxhlet apparatus. Isolate the 2,2'-dinitrobiphenyl from the alcoholic extracts as described above: the yield of product, m.p. 124–125 °C, is 11.5 g (75%).

Note. (1) If the temperature is allowed to rise above 240 °C, reduction of the nitro groups will occur and carbazole will be formed.

Experiment 6.9   BIPHENYL-2,2'-DICARBOXYLIC ACID (*Diphenic acid*)

The reducing agent (a solution containing cupro-ammonia ions) is first prepared. Dissolve 63 g (0.25 mol) of crystallised copper(II) sulphate in 250 ml of water in a 1-litre beaker, add 100 ml of concentrated ammonium hydroxide solution (d 0.88) and cool the solution to 10 °C. Dissolve 17.8 g (0.256 mol) of hydroxylammonium chloride or 21 g (0.256 mol) of hydroxylammonium sulphate in 60 ml of water, cool to 10 °C and add 42.5 ml of 6 M sodium hydroxide solution; if the resulting solution of hydroxylamine is not clear, filter it at the pump. Without delay add the hydroxylamine solution, with stirring, to the ammoniacal copper(II) sulphate solution. Reduction occurs at once, a gas is evolved and the solution assumes a pale blue colour. Protect the reducing agent from the air if it is not used immediately.

Grind 25 g (0.18 mol) of anthranilic acid (Expt 6.53) with 46 ml of concentrated hydrochloric acid and 75 ml of water in a glass mortar, and transfer the suspension to a 500-ml round-bottomed flask which is provided with a mechanical stirrer. Cool the contents of the flask in an ice bath to 0–5 °C, and add a solution of 13.0 g (0.19 mol) of sodium nitrite in 175 ml of water from a dropping funnel during about 20 minutes. Keep the diazonium solution below 5 °C and, if it is not clear, filter it by suction through a chilled Buchner funnel immediately before use.

Surround the reducing solution in the 1-litre beaker (which is equipped with a mechanical stirrer) with a bath of crushed ice so that the temperature of the solution is about 10 °C. Attach, by means of a short length of rubber tubing, to the stem of a dropping funnel a glass tube which dips well below the surface of the solution and is bent upwards at the end and constricted so that the opening is about 2 mm (this arrangement ensures that the diazonium solution reacts with the ammoniacal solution in the beaker and prevents the latter rising in the stem of the funnel). Place about 45 ml of the cold diazonium solution in the funnel and add it at the rate of about 10 ml per minute while the mixture is stirred. Add the remainder of the diazonium solution at the same rate; continue the stirring for 5 minutes after the addition is complete. Heat the solution rapidly to boiling and carefully acidify with 125 ml of concentrated hydrochloric acid; the diphenic acid precipitates as pale brown crystals. Allow to stand overnight and filter with suction; wash the crude diphenic acid with about 25 ml of cold water. Suspend the crude acid in 100 ml of water and add 20 g of solid sodium hydrogen carbonate. Filter the resulting solution by gravity, and then boil with about 0.5 g of decolourising carbon; filter and acidify the filtrate while still hot with excess of dilute hydrochloric acid (1:1). Collect the precipitated diphenic acid on a Buchner funnel, wash it with 20 ml of cold water and dry at 100 °C. The yield of diphenic acid is 18 g (82%); it melts at 227–228 °C and usually possesses a light cream colour.

Experiment 6.10 1,1′-BI-2-NAPHTHOL

In a 1-litre three-necked flask, provided with a dropping funnel, a sealed stirrer and a reflux condenser, place 14.4 g (0.1 mol) of 2-naphthol and 600 ml of water, and heat to the boiling point. To the boiling liquid containing liquid 2-naphthol in suspension, add slowly through the dropping funnel and with vigorous stirring a solution of 28 g (0.1 mol) of crystallised iron(III) chloride in 60 ml of water. The oily drops of 2-naphthol will disappear and the bis-2-naphthol separates out in flakes. Boil for 5–10 minutes, filter the hot suspension at the pump through a previously warmed Buchner funnel, wash with boiling water and dry in the air upon filter paper. The crude product weighs 9 g. Recrystallise from toluene (about 150 ml); almost colourless crystals (7.5 g, 52%), m.p. 218 °C, are obtained.

6.1.4 CONDENSED POLYCYCLIC SYSTEMS

Dehydrogenation (the conversion of alicyclic or hydroaromatic compounds into their aromatic counterparts by removal of hydrogen – and also, in some cases, of other atoms or groups) has found wide application in the determination of structure of natural products containing complex hydroaromatic systems. Dehydrogenation is employed also for the synthesis of polycyclic hydrocarbons and their derivatives from readily accessible synthetic hydroaromatic compounds. The general process is illustrated by the conversion of tetralin into naphthalene.

The principal dehydrogenation agents are (i) sulphur, (ii) selenium and (iii) catalytic metals. With *sulphur*, the general method is to heat the compound at 200–260 °C with the theoretical amount of sulphur required to convert it into an aromatic system. In the case of *selenium*, the substance is heated with a large excess of selenium at 280–350 °C for 36–48 hours. Better yields (and less side reactions) are usually obtained than with sulphur, but, owing to the higher temperature, rearrangements are more likely. Oxygen-containing groups are particularly prone to elimination. *Palladium* and *platinum catalysts* are generally employed with a charcoal carrier. The dehydrogenation can be conducted in the vapour phase by distilling the compound through a tube containing the catalyst heated to 300–350 °C, but the liquid phase method is generally more convenient. Charcoal is employed containing 10–30 per cent of the metal. It has been established that the best results are obtained by conducting the process in an actively boiling medium (e.g. mesitylene, b.p. 165 °C; p-cymene, b.p. 177 °C; naphthalene, b.p. 218 °C; and 1-methylnaphthalane, b.p. 242 °C) and to provide for the removal of the hydrogen as it is formed (e.g. by sweeping the system with a stream of carbon dioxide).

Examples of dehydrogenation processes are to be found in the synthetic sequences described in Expts 6.11 and 6.12. In the former instance, 1-methylnaphthalene is synthesised from α-tetralone by first introducing a 1-methyl substituent by reaction with methylmagnesium iodide. The resulting tertiary alcohol undergoes dehydration and dehydrogenation on heating with a palladium-on-charcoal catalyst. The preparation of the ketonic starting material from benzene is described in Expt 6.123 (see also Section 6.11.1, p. 1006); the conversion of α-tetralone into 1-methylnaphthalene completes the reaction sequence known as the *Haworth procedure* for the synthesis of polycyclic aromatic hydrocarbons.

Intramolecular cyclisation of a 4-arylbutanoic acid system is also an important step in a convenient synthesis of the polycyclic system, chrysene, which is formulated and described in Expt 6.12. Here, methyl cinnamate is first subjected to reductive dimerisation to give methyl *meso*-β,γ-diphenyladipate, which is accompanied by some of the (±)-form. The *meso* isomer (16) is the most easily isolable and cyclisation occurs smoothly in sulphuric acid to yield the diketone 2,11-dioxo-1,2,9,10,11,18-hexahydrochrysene, which is obtained as the *trans* form (17) as shown in the following formulation. Clemmensen reduction of this ketone followed by dehydrogenation (in this case using selenium) completes the synthesis of chrysene.

The synthesis of a tetraphenyl derivative (rubrene, Expt 6.13) of the linearly fused tetracyclic aromatic hydrocarbon naphthacene involves an interesting intermolecular cyclisation process between two molecules of 1-chloro-1,3,3-triphenylpropa-1,2-diene. This substituted allene is formed *in situ* from 1,1,3-triphenylprop-2-yn-1-ol (Expt 5.41) when the latter is allowed to react with thionyl chloride and the resulting chlorosulphite ester heated with a little quinoline; cyclisation occurs spontaneously under these reaction conditions to give rubrene which has an intense red colour.

The synthesis of 9-phenylphenanthrene (Expt 6.14) illustrates the formation of the phenanthrene system by the cyclisation of a 1,2-diphenylethylene (stilbene). The process involves an allowed photochemical cyclisation which gives initially a dihydrophenanthrene. This is readily dehydrogenated *in situ* by molecular oxygen in the presence of iodine.

The required stilbene is usually readily prepared by dehydration of the appropriate alcohol obtained by a Grignard reaction, e.g.

The method has been applied to the synthesis of a range of substituted phenanthrenes and has the merit that it involves fewer steps than, for example, the Haworth synthesis. Good results are not always obtained, however, when some electron-withdrawing substituents (e.g. NO_2, Me·CO) are present.

Experiment 6.11 1-METHYLNAPHTHALENE

841

1-Hydroxy-1-methyl-1,2,3,4-tetrahydronaphthalene. Prepare an ethereal solution of methylmagnesium iodide (cf. Expt 5.39) from 1.8 g (0.075 mol) of magnesium, 10.6 g (4.7 ml, 0.075 mol) of methyl iodide (**CAUTION**) and 30 ml of ether in a 100-ml two-necked flask fitted with a dropping funnel and a reflux condenser protected by a calcium chloride guard-tube. Slowly add a solution of 7.3 g (0.05 mol) of α-tetralone (Expt 6.123) in 10 ml of ether, swirling the contents of the flask from time to time, and finally heat under reflux on a steam bath for 1 hour to complete the reaction. Cool in ice, and decompose the reaction mixture with a cold saturated aqueous solution of ammonium chloride. Separate the ether layer, and extract the aqueous phase with 10 ml of ether. Wash the combined ethereal solutions with aqueous ammonium chloride solution, dry over anhydrous sodium sulphate and evaporate off the ether. A solid residue of 1-hydroxy-1-methyl-1,2,3,4-tetrahydronaphthalene of sufficient purity for use in the next stage is obtained; the yield is 7.5 g (92%), m.p. 77–79 °C. Recrystallise a sample from light petroleum (b.p. 60–80 °C); the purified material has m.p. 86–87 °C.

Fig. 6.1

1-Methylnaphthalene. This preparation illustrates the general procedure for catalytic dehydrogenation. The apparatus used is shown in Fig. 6.1. Heat a mixture of 3.2 g (0.02 mol) of the above hydroaromatic compound with 0.3 g of palladised charcoal (Section 4.2.*54*, p. 452) at 250–270 °C in a slow current of dry carbon dioxide in a Silicone oil or fusible metal bath for 3 hours (1). Cool, dissolve the residue in ether and filter off the catalyst. Wash the extract with dilute aqueous sodium hydroxide and dry it over anhydrous sodium sulphate. Remove the ether and distil the residual oil under reduced pressure; use a small-scale distillation apparatus (cf. Fig. 2.111). Collect the 1-methylnaphthalene, b.p. 121–123 °C/20 mmHg. The yield is 2.5 g (89%).

Note. (1) If it is desired to follow the progress of the dehydrogenation, attach the side-tube through a U-tube packed with self-indicating soda-lime to a nitrometer filled with potassium hydroxide solution: almost the theoretical quantity of hydrogen will be collected.

Experiment 6.12   CHRYSENE

***meso*-Dimethyl 3,4-diphenyladipate.**   Prepare aluminium amalgam from 75 g of the foil (Section 4.2.*3*, p. 415) and cover with 1 litre of moist ether (1) in a 2-litre flask fitted with a large double surface relux condenser. Immediately add a solution of 48.5 g (0.3 mol) of recrystallised methyl cinnamate (Expt 6.163) in 100 ml of ether and shake the flask to mix the contents thoroughly. Evolution of hydrogen ceases and after an induction period of up to 1 hour the ether boils; moderate the reaction if necessary by cooling the flask. Towards the end of the reaction, which should be complete in 6 hours, warm the flask to maintain gentle reflux of the ether. Cool, filter off the aluminium hydroxide sludge (do not discard this residue) and dry the ethereal solution over anhydrous sodium sulphate. Remove ether on a rotary evaporator and cool the residual oil thoroughly in an ice bath (or overnight in a refrigerator). Filter the resulting crystalline mass and wash the crystals with a little cold ether; about 4 g of the crude *meso*-adipate, m.p. 170 °C, are obtained (2). Much of the product adheres to the aluminium hydroxide residue, however. Cover this residue with 1 litre of water and add 1 litre of concentrated hydrochloric acid slowly with stirring to dissolve most of the aluminium hydroxide. Extract with three 150 ml portions of chloroform, dry the extract over anhydrous sodium sulphate and remove the chloroform on a rotary evaporator; a further 7 g of the crude *meso*-adipate is obtained. Recrystallise the combined crops of crude material from rectified spirit (about 250 ml). Pure *meso*-dimethyl 3,4-diphenyladipate, m.p. 175 °C, is obtained; the yield is about 9 g (18%).

***trans*-2,11-Dioxo-1,2,9,10,11,18-hexahydrochrysene.**   Add 8.2 g (0.025 mol) of *meso*-dimethyl 3,4-diphenyladipate to a stirred mixture of 100 ml of concentrated sulphuric acid previously added to 33 ml of water and heat on a water bath for 3 hours. Cool the mixture and pour it carefully with stirring into 1 litre of water. Filter off the resulting precipitate and digest it on the steam bath with hot sodium carbonate solution. Filter again, wash the product with water and dry in an oven. The dioxohexahydrochrysene is obtained as a pale buff powder, m.p. 293–300 °C, yield 5.5 g (85%). Crystallise a small specimen from butan-1-ol; colourless plates, m.p. 303 °C (sealed tube), are obtained.

trans-**1,2,9,10,11,18-Hexahydrochrysene.**   Prepare amalgamated zinc from 35 g of zinc wool (Section 4.2.*80*, p. 467) and cover it with 25 ml of concentrated hydrochloric acid. Add 5.2 g (0.02 mol) of the dioxohexahydrochrysene and boil the mixture under reflux for 8 hours. Add 20 ml more of concentrated hydrochloric acid and reflux for a further 8 hours. Cool the mixture, when the oily product solidifies. Extract the product, together with any solid material which has formed in the reflux condenser, with two 50 ml portions of hot benzene. Cool, separate the benzene layer and dry over anhydrous sodium sulphate. Remove the benzene with a rotary evaporator, transfer the residue to a small-scale distillation unit (Fig. 2.111) and distil under reduced pressure. Collect the hexahydrochrysene which distils at 230 °C/14 mmHg as a colourless oil which readily crystallises, m.p. 113 °C, yield 3.5 g (75%).

Chrysene.   Since selenium and selenium compounds are *toxic* this dehydrogenation and the associated work-up procedure *must be carried out in an efficient fume cupboard.* Mix 3.5 g (0.015 mol) of hexahydrochrysene with 16 g (0.2 mol) of selenium in a boiling tube and heat in a fusible metal bath at 300 °C for 20 hours (*fume cupboard*). From time to time, melt the crystalline sublimate which gradually forms so that it runs back into the reaction mixture. Remove the cooled product and grind it in a mortar to a fine powder. [**CAUTION**: (3).] Extract by boiling under reflux for 30 minutes with 200 ml of benzene, filter and reflux the filtered extract over a little clean sodium wire (or thin narrow slices of sodium metal); this treatment removes traces of selenium. Evaporate the benzene solution using a rotary evaporator and crystallise the residue from toluene (about 20 ml per 1 g) (4). Colourless plates with a bluish fluorescence, m.p. 254 °C, are obtained. The yield of chrysene is about 2 g (59%).

Notes.   (1) The ether should be saturated by shaking it with a little water in a separatory funnel.
(2) If the filtrate is distilled under reduced pressure, 30 g of methyl 3-phenylpropionate, b.p. 111–130 °C/15 mmHg, are obtained. When the distillation residue is dissolved in the minimum of hot ether and cooled, a further 1 g of the *meso*-adipate is obtained. The ether filtrate from this contains racemic dimethyl 3,4-diphenyladipate; the latter may be recovered by evaporating the ether and crystallising the residue from the minimum of methanol; m.p. 70–71 °C.
(3) Do not inhale any of the finely divided material; it is advisable to wear a face mask or alternatively to grind the material in an enclosed (i.e. glove) box.
(4) The crude product may also be purified by sublimation under reduced pressure (*c.* 0.1 mmHg) from an oil bath maintained at 200 °C.

Experiment 6.13   5,6,11,12-TETRAPHENYLNAPHTHACENE (*Rubrene*)

$$2\text{Ph}\cdot\text{C}\equiv\text{C}\cdot\text{C(Ph)}_2\text{OH} \xrightarrow{\text{SOCl}_2} 2\text{Ph}\cdot\text{CCl}=\text{C}=\text{C(Ph)}_2 \xrightarrow{-2\text{HCl}}$$

Add 28.5 g (0.1 mol) of 1,1,3-triphenylprop-2-yn-1-ol (Expt 5.41) in small portions with shaking to 24 g (14.5 ml, 0.2 mol) of redistilled thionyl chloride cooled to $-10\,°C$ in an ice–salt bath. Set the resulting solution aside at room temperature for 1 hour and then remove excess chloride by warming under reduced pressure (water pump); the residue solidifies (1). Add 0.5 ml of redistilled quinoline and heat the mixture in an oil bath at $120\,°C$ under reduced pressure (water pump) for 2 hours. Wash the dark red residue with ether and then with boiling acetone and dissolve it in benzene. Filter off any insoluble impurity, dilute with light petroleum (b.p. $100–120\,°C$) and remove most of the benzene carefully by evaporation under reduced pressure. Cool and filter off the bright red crystals of rubrene; the yield is 7 g (26%), m.p. $320\,°C$. Store the product in a specimen tube protected from the light by a covering of black paper.

Note. (1) If the product does not solidify it must be purified by distillation under reduced pressure using an efficient oil-immersion pump; b.p. $190\,°C/0.05\,mmHg$.

Experiment 6.14 9-PHENYLPHENANTHRENE

Triphenylethylene. Equip a 2-litre, three-necked round-bottomed flask with a sealed stirrer unit, a 250-ml dropping funnel and a double surface condenser. All the apparatus should be rigorously dried and the dropping funnel and reflux condenser protected by calcium chloride guard-tubes. Place in the flask 12.2 g (0.50 mol) of magnesium turnings and about 250 ml of ether previously dried over anhydrous calcium chloride. In the funnel place a solution of 57.5 ml (63.3 g, 0.50 mol) of freshly distilled benzyl chloride in about 100 ml of dry ether. Without stirring the contents of the flask run in about 20 ml of the benzyl chloride solution and then add a large crystal of iodine and allow the mixture to stand undisturbed. Reaction commences after about 5 minutes and the magnesium begins to dissolve. Add the benzyl chloride solution dropwise at such a rate as to maintain a steady reflux of the ether solvent. When the reaction is well established, start the stirrer motor and stir the reaction mixture at a moderate rate. When all the benzyl chloride solution has been added (about half an hour), heat the flask in an electric heating mantle and with continued stirring maintain a vigorous rate of reflux for a further half an hour. Turn off the heating mantle but continue stirring fairly rapidly, and add from the dropping funnel a solution of 91.1 g (0.50 mol) of benzophenone in about 250 ml of dry ether. The rate of addition should be sufficient to maintain rapid reflux. On completion of the addition (about 15 minutes) heat under reflux with stirring for a further half an hour before allowing the reaction mixture to stand at room temperature for at least 2 hours or preferably overnight. Cool the reaction mixture in an ice-water bath, add 400 g of crushed ice to the flask contents, followed by 250 ml of cold 2.5 M sulphuric acid. Transfer the reaction mixture to a 3-litre separating funnel, run off the aqueous layer and extract with two 150 ml portions of ether. Combine the ether layer and extracts, dry over calcium sulphate and evaporate the ether on a rotary evaporator. To the syrupy residue add 100 ml of 2.5 M sulphuric acid and boil vigorously under reflux for 2 hours. Cool the mixture to room temperature, add 50 ml of ether and shake vigorously. Transfer the resulting emulsion to a 1-litre separatory funnel, wash the flask thoroughly with a mixture of 50 ml of water and 50 ml of ether and add this to the separatory funnel. After a few minutes the emulsion separates into two layers and the aqueous fraction can be run off and extracted with two 50 ml portions of ether. Combine the ether layer and extracts (1), and evaporate the ether using a rotary evaporator. Distil the residue under reduced pressure collecting the main fraction (93 g, 73%) at 196–200 °C/2.5 mmHg. Dissolve the greenish-yellow syrupy distillate in 450 ml of hot 95 per cent aqueous ethanol and allow to cool slowly. Trituration is usually required to induce crystallisation. Complete the crystallisation process by cooling in ice-water and filter under suction to obtain 85 g (66%) of colourless crystals having m.p. 68–69 °C. Further recrystallisations from ethanol or acetic acid gives pure triphenylethylene, m.p. 72–73 °C.

9-Phenylphenanthrene. *The photochemical reactor should be completely screened with metal foil to avoid hazards from stray radiation* (see Section 2.17.5, p. 113). Dissolve 2.56 g (0.01 mol) of triphenylethylene and 0.127 g (0.0005 mol) of iodine in 1 litre of redistilled cyclohexane contained in a photochemical reactor vessel of 1-litre capacity and fitted with a 100-W medium-pressure mercury arc lamp in a water-cooled quartz immersion well (Section 2.17.5, p. 110, Fig. 2.67(*f*)). Place the apparatus on a magnetic stirrer unit, insert a magnetic follower and with both the side-necks of the vessel open to the atmosphere irradiate the stirred solution until all the triphenyl-ethylene has reacted (2); this should take approximately 21–22 hours. There is no observable change in the colour of the cyclohexane solution during the irradiation. Remove the reaction mixture and evaporate to dryness under reduced pressure on a rotary evaporator to obtain the crude phenylphenan-threne as a pale fawn solid. Purify by column chromatography as follows. Dissolve the solid in 60 ml of warm cyclohexane, allow to cool and pour on to a short (*c.* 2 cm × 7 cm) column of neutral alumina (*c.* 16–17 g, Activity I, prepared as a slurry in cyclohexane). Collect the eluate, and when most of the solution has passed down the column rinse the flask which contained the solid with two 10 ml portions of cyclohexane and add to the column. When this has also been adsorbed, elute with a further 100–120 ml of cyclohexane or until no more phenylphenanthrene is obtained in the eluate. It is best to monitor the elution of the product by collecting and evaporating 40 ml fractions; a band of yellow material should remain near the top of the column and elution of this must be avoided. Crystallise the solid so obtained from 60 ml of ethanol and collect the crystals which separate by suction filtration; a second crop may be obtained by concentrating the filtrate to *c.* 20 ml and cooling in ice. The yield of 9-phenylphenanthrene (white needles), m.p. 104.5–105.5 °C, is 2.16 g (85%) (3).

Notes. (1) Do not wash the ethereal fraction with water or with aqueous sodium hydrogen carbonate solution, since the presence of a trace of sulphuric acid during the final distillation is necessary to complete the dehydration.

(2) The irradiation time for complete reaction will depend on the power of the mercury arc used; in particular the light output of a lamp will vary with its age and use of a higher wattage lamp will, of course, shorten the reaction time required. The extent of the reaction may be most conveniently followed by g.l.c. of 0.5 μl samples of the reaction solution on a 1.5-m column of Methyl Silicone Gum S.E.30 on Chromosorb W held at 240 °C and with a nitrogen flow rate of 40 ml/minute. Triphenylethylene and 9-phenylphenanthrene have t_R 2.3 minutes and 5.2 minutes respectively, cyclohexane has t_R 0.5 minutes. Alternatively the reaction may be followed by evaporating 5–10 ml portions of the solution and recording the infrared spectrum of the residue as a mull in Nujol. Bands at 887(m), 750(m) and 738(s) cm^{-1} characteristic of 9-phenylphenanthrene gradually appear and increase in intensity relative to bands due to the stilbene at 872(w) and 757(s) cm^{-1} which gradually disappear. Other regions of the spectrum of the individual compounds are very similar and it is advisable to use the scale expansion adjustment of the instrument to enable the bands in the low wavenumber region to be readily identified.

(3) For the preparation of larger amounts of the phenanthrene the reaction should be repeated; photolysis of more concentrated solutions is to be avoided in order to minimise the possibility of photodimerisation to give a cyclobutane derivative.

6.1.5 SOME NON-BENZENOID SYSTEMS

Two examples, namely ferrocene and azulene, are selected to illustrate this group of aromatic compounds.

The former is prepared (Expt 6.15) by the reaction of cyclopentadiene with iron(II) chloride in the presence of diethylamine. Iron(II) chloride is prepared *in situ* from iron(III) chloride in tetrahydrofuran by reduction with iron metal. This preparation is one of the simplest for ferrocene, although not of general application for substituted ferrocenes. The acetylation of ferrocene in a Friedel–Crafts manner to yield the monoacetyl derivative is described in Expt 6.122.

The aromatic character of ferrocene originates in the anionic cyclopentadienide system (18), which has six π-electrons delocalised over a symmetrical cyclic five-carbon system. Coordination of two such ring systems with a metal ion (iron in the case of ferrocene) in a sandwich-like structure (20) gives rise to a remarkably stable compound.

(18)   (19)   (20)

(21)

It is possible that the cyclopentadienide system also contributes to the structure of the undoubtedly aromatic azulene molecule (21) which may be regarded as a condensed bicyclic system comprised of the anion (18) in combination with the cycloheptatrienylium ion (19) – the latter also has an aromatic sextet of six π-electrons, disposed in this case over a seven-membered ring. Alternatively azulene may be viewed as a conjugated cyclodecapentaene with an additional transannular C—C single bond; such a structure is predicted to be aromatic by the Huckel $(4n + 2)$ rule, having ten π-electrons in a cyclic system.

The preparation of azulene (Expt 6.16) provides an interesting series of rearrangements.[9] The key step in the synthesis is the intramolecular insertion reaction, carried out under conditions of high dilution, of the ketocarbene (22) into the 1,2-position of the benzene ring. The unstable norcaradiene (23) which is formed then ring-opens to the bicyclic trienone (24), which isomerises to the more stable cross-conjugated trienone (25) during isolation. Dehydration of the trienone with a mixture of phosphorus pentoxide and methanesulphonic acid yields azulene.

The ketocarbene is formed from 3-phenylpropanoyl chloride (dihydrocinnamic acid chloride) and diazomethane; ring substituted dihydrocinnamic acids provide a route to substituted azulenes. Alternatively the keto group in (25) may be treated with a Grignard reagent for the introduction of alkyl substituents into the five-membered ring.

Experiment 6.15 FERROCENE*

$$2FeCl_3 + Fe \longrightarrow 3FeCl_2$$

$$2C_5H_6 + FeCl_2 \xrightarrow[-2HCl]{Et_2NH} Fe^{2\oplus}, (\overset{\ominus}{C_5H_5})_2$$

Equip a 250-ml, three-necked round-bottomed flask with a stirrer, a reflux condenser with a nitrogen inlet (cf. Fig. 2.60), and a stopper in the third neck. Flush the flask with nitrogen and charge it with dry tetrahydrofuran (100 ml) (Section 4.1.*19*, p. 406). Add with stirring anhydrous iron(III) chloride (27.1 g, 0.166 mol) followed by iron powder (4.7 g, 0.84 g-atom) which should preferably be 300 mesh. Heat under reflux with stirring and under nitrogen for 4.5 hours. The reaction mixture should then consist of a grey powder and a brown supernatant liquid. Remove the tetrahydrofuran under reduced pressure by setting the condenser for downward distillation, and attaching a vacuum take-off receiver adapter and receiver flask; stir until the residue is

---

* The editors are indebted to Dr R. Thomas, Thames Polytechnic, for these experimental details; this experiment, and the preparation of acetylferrocene (Expt 6.122), are very suitable as undergraduate exercises.

almost dry. Cool the flask in an ice bath, re-assemble the apparatus as before but with a pressure-equalising funnel in the third neck. Add dropwise with stirring and cooling under nitrogen, first cyclopentadiene (42 ml, 0.5 mol, Expt 7.24) and then diethylamine (100 ml). Stir the mixture at room temperature for 6–8 hours or overnight. Remove the excess amine under reduced pressure. Extract the residue repeatedly with light petroleum (b.p. 60–80 °C) under reflux, filter the hot extracts and evaporate the solvent to leave crude ferrocene. Recrystallise the crude product from cyclohexane to give pure ferrocene (34 g, 73%), m.p. 173–174 °C. The p.m.r. spectrum (CDCl₃, TMS) exhibits a single absorption at δ 4.09.

Experiment 6.16 AZULENE[9]

(25)

3-Phenylpropanoyl chloride. A mixture of 15 g (0.1 mol) of 3-phenylpropanoic acid and 18.0 ml (0.25 mol) of thionyl chloride, protected from the atmosphere by a drying tube, is heated under reflux for 2 hours. Excess thionyl chloride is removed under reduced pressure and the residue is vacuum distilled giving 16.3 g (97%) of 3-phenylpropanoyl chloride as a colourless liquid, b.p. 43–48 °C/0.04 mmHg; i.r. (neat) 1812 cm⁻¹ (C=O); δ (CCl₄, TMS) 7.04 (s, 5H), 2.93 (m, 4H).

1-Diazo-4-phenylbutan-2-one. CAUTION: *All operations involving diazomethane are hazardous and must be conducted in an efficient fume cupboard and using adequate screens, Section 4.2.25, p. 430.* A solution of diazomethane (c. 100 mmol) in 300 ml of anhydrous ether is prepared in the standard way from 30 g N-methyl-N-nitrosotoluene-p-sulphonamide (Section 4.2.25, p. 431). To this solution, cooled in an ice–salt bath under a drying tube, is added an additional 35 ml of anhydrous ether containing 4.94 ml (33 mmol) of 3-phenylpropanoyl chloride dropwise during 1 hour with magnetic stirring. At the end of the addition the reaction mixture is stirred 1 hour more in the cold bath and then 1 hour more while being warmed to room temperature. Concentration of the reaction mixture under reduced pressure gives 5.8 g (100%) of the diazoketone as a yellow oil, i.r. (neat) 2083 and 1631 cm⁻¹; δ (CCl₄, TMS) 7.07 (s, 5H), 5.07 (s, 1H), 3.00–2.29 (m, 4H).

3,4-Dihydro-1(2H)-azulenone (25). A suspension of 0.8 g of anhydrous copper(I) chloride, in 3 litres of bromobenzene freshly distilled from calcium hydride, is heated to 80 °C under nitrogen in an oven-dried, 5-litre, three-necked round-bottomed flask fitted with a mechanical stirrer, a constant addition funnel and a reflux condenser. To this vigorously stirred suspension is added 1 litre of bromobenzene containing 20.4 g (0.117 mol) of the foregoing diazoketone dropwise over 7 hours. Heating and stirring are

849

continued for an additional 1 hour; the pale yellow reaction mixture is cooled and filtered through 30 g of alumina to remove copper salts and to isomerise the unconjugated trienone (24) to the more stable isomer (25). The alumina is washed with 500 ml of ethyl acetate, and the combined filtrates are concentrated at 40 °C under reduced pressure, first at 20 mm to remove the more volatile solvent, and then at 0.5 mm to remove the bromobenzene; there is left finally a brown oil. Chromatography on 200 g of silica gel using 15 per cent ethyl acetate-light petroleum as the eluant gives 8.9 g (52%) of (25) as a light brown oil; i.r. (neat) 1690 (C=O) and 1620 cm^{-1} (C=C); u.v. (EtOH) 232 (ε 15 200) and 265 nm (ε 4800); δ (CDCl$_3$, TMS) 6.68 (d, 1H, $J = 10$ Hz, C$_8$—H), 6.48 (d of d, 1H, $J = 10$ and 6 Hz, C$_7$—H), 6.09 (d of d, 1H, $J = 6$ and 9 Hz, C$_6$—H), 5.39 (d of t, $J = 9$ and 6 Hz, C$_5$—H), 2.83 (d, 2H, $J = 6$ Hz, C$_4$ (CH$_2$)), and 2.58 (sym. m, 4H, —CH$_2$·CH$_2$—).

Azulene. A mixture of 5 g (35 mmol) of phosphorus pentoxide (weighed under nitrogen) and 34 ml of redistilled methanesulphonic acid (**CAUTION**) is heated under nitrogen to 60 °C with mechanical stirring. When the mixture is homogeneous, 0.48 g (3.3 mmol) of (25) is added neat, dropwise during 7 minutes. The solution rapidly darkens to orange–brown. Progress of the reaction is followed by removing one-drop aliquot portions and quenching them into water–hexane. The u.v. spectrum of the hexane layer shows a gradual decrease in the absorption at 226 nm due to (25) and the corresponding growth of the characteristic three-line pattern due to azulene around 274 nm. When the 226 nm absorption is no longer discernible (about 6 hours), the reaction mixture is poured into 500 ml of ice-cold water and extracted twice with pentane. The dark-blue pentane layers are combined and washed with water until neutral, washed with saturated sodium chloride solution, dried over magnesium sulphate and concentrated under reduced pressure. The oily blue residue is purified by preparative layer chromatography on silica gel (1) using pentane as the eluant to give 130–210 mg (30–50%) of azulene as beautiful dark-blue crystals, m.p. 98–98.5 °C.

Note. (1) Silica gel PF 254-366 (Brinkmann) is used for the preparative layer chromatography. The editors suggest that the use of flash chromatography (Section 2.31) could be explored in this case.

6.2 AROMATIC NITRO COMPOUNDS

The importance of aromatic nitro compounds arises in particular from the ready conversion of the nitro group into other functional groups, principally by routes involving initial reduction to the amino group. The following procedures, of which the first is by far the most important, are available for the synthesis of aromatic nitro compounds.

1. Direct nitration (Expts 6.17 to 6.21).
2. The oxidation of amines (Expt 6.22).
3. The replacement of a diazo group by a nitro group (see Expt 6.78).

SUMMARY OF RETROSYNTHETIC STRATEGIES

Functional group interconversion (FGI) (method 2), e.g.

NO$_2$ NH$_2$

(2)

(TM)

C—N Disconnection (methods 1 and 3), e.g.

$\overset{\oplus}{NO_2}$ (1)   NO$_2$ **(TM)**   (3) $\overset{\ominus}{NO_2}$

HNO$_3$/H$_2$SO$_4$ + PhH   Ph·$\overset{\oplus}{N_2}$ + NaNO$_2$/Cu

SPECTROSCOPIC FEATURES

The presence of a nitro group in a compound is readily recognised from its strong characteristic *i.r.* absorption frequencies (Fig. 3.38). The aromatic substitution pattern is less easily assigned by inspection of the longer wavelength region (p. 310). However, this may be frequently assessed by examination of the *p.m.r.* spectrum (Fig. 3.64, *m*-dinitrobenzene). Further examples are given in descriptive style in the preparative examples. The fragmentation pattern observed in the *m.s.* of a nitro compound is discussed on p. 855 and further illustrated in some of the preparations described below. The nitro group exerts a bathochromic shift on the aromatic absorption frequencies in the *u.v.–visible* spectrum (p. 392).

6.2.1 DIRECT NITRATION

Aromatic hydrocarbons may be nitrated, i.e. the hydrogen atoms replaced by nitro (NO$_2$) groups, with concentrated nitric acid in the presence of concentrated sulphuric acid ('mixed acid reagent').

$$ArH + HNO_3/H_2SO_4 \longrightarrow Ar\cdot NO_2$$

The function of the sulphuric acid is to convert the nitric acid into the highly reactive, electrophilic, nitronium ion, $\overset{\oplus}{NO_2}$, which is the effective nitrating agent.

$$HONO_2 + H_2SO_4 \longrightarrow \overset{\oplus}{NO_2} + \overset{\oplus}{H_3O} + 2HSO_4^{\ominus}$$

The mechanism of aromatic nitration, which is illustrated below in the case of benzene, is a two-step process involving electrophilic attack of the nitronium ion on the benzene molecule to form the intermediate mesomeric ion (1), followed by removal of a proton by the hydrogen sulphate ion, which is the most basic species in the reaction mixture.

(1)

Nitration of aromatic hydrocarbons is usually carried out with the above mixed acid reagent at comparatively low temperatures (e.g. about 50 °C, as used in the preparation of nitrobenzene and 1-nitronaphthalene, Expt 6.17). Unnecessarily high temperatures should be avoided since polynitration is then more likely and oxidative breakdown of the aromatic ring system may occur.

The nitration of substituted aromatic compounds (e.g. Ar·Z) presents a range of practical and theoretical problems which need to be briefly surveyed. These include: (a) the severity of the reaction conditions; (b) the sensitivity of the group Z to the oxidising conditions of the nitrating agents; (c) the regioselectivity of the reaction (i.e. the orientating influence of the group Z to either the *meta* or to the *ortho/para* positions); (d) the sequence of reactions (i.e. either to nitrate Ar·Z, or to introduce the substituent Z into the nitro compound Ar·NO$_2$); (e) the use of a disubstituted compound Z·C$_6$H$_4$·Y, wherein the orientating influence of Y predominates, and thus directs the nitro group to the desired position, but is then removed in a subsequent reaction step.

Thus the nitration of aromatic compounds containing an electron-withdrawing group (e.g. —NO$_2$, —SO$_3$H, —CHO, —CO·R, —CO$_2$H, —CO$_2$R) does not occur readily under the above conditions, in which case forcing conditions which require the use of fuming nitric acid and concentrated sulphuric acid need to be employed. Nitrobenzene, for example, is converted by a mixture of fuming nitric acid and concentrated sulphuric acid into about 90 per cent of *m*-dinitrobenzene (Expt 6.18) and small amounts of the *o*- and *p*-isomers; the latter are eliminated in the process of recrystallisation. *p*-Nitrotoluene is similarly converted largely into 2,4-dinitrotoluene (Expt 6.18). A further example is the nitration of benzaldehyde to give *m*-nitrobenzaldehyde (Expt 6.19) in reasonable yield in spite of the fact that the reaction conditions are strongly oxidising. The nitro group is thus seen to be *meta*-directing in common with the other deactivating groups specified above.

The deactivating effect of the nitro group is largely the result of its mesomeric interaction (-*M* effect) with the π-electron system of the benzene ring which is supplemented by the inductive (-*I*) effect (2).

(2)

The overall electron withdrawal from the ring system results in the rate of attack of the nitronium ion being substantially retarded compared to benzene. Moreover, the representations of the canonical forms of nitrobenzene formulated above show that the *ortho* and *para* positions are subject to the greatest reduction in electron density. In addition the reaction intermediate resulting from attack at a position *para* to the nitro group would be represented by the following hybrid species (similar formulations being possible as a result of *ortho* attack).

(a)   (b)   (c)
(3)

It follows that since (3(b)) is energetically unfavourable the mesomeric stabilisation of this intermediate is less than that of the corresponding intermediate resulting from attack in the *meta* position (4).

(a)   (b)   (c)
(4)

The nitration of aromatic compounds containing electron-releasing groups (e.g. —R, —OH, —NH$_2$, —NHCO·Me) needs to be conducted under conditions which are milder than those specified for benzene, since in these cases the aromatic nucleus is activated towards attack by the electrophilic species, and furthermore the ring system is more likely to be oxidatively cleaved. An example is provided by the nitration of acetanilide (Expt 6.68) to yield a mixture of *o*- and *p*-nitroacetanilides (see also Section 6.6.2, p. 916); the overall nitration of phenol to a mixture of *o*- and *p*-nitrophenols (Expt 6.102) is anomalous in that the reagent is dilute nitric acid alone; the mechanism is outlined in Section 6.9.2, p. 975. These activating groups are therefore seen to be *ortho* and *para* directing. By contrast to the effect of electron-withdrawing groups, activating and *ortho/para* directing groups will stabilise by electron release the reaction intermediate formed as a result of attack at any of the three possible positions, but particularly so when attack is at the *ortho* and *para* positions.

An interesting case is provided by the nitration of aryl halides where the effect of the halogen is to deactivate the aromatic nucleus (by the -*I* effect) but to direct the incoming nitronium ion to the *ortho* and *para* positions as a result of the mesomeric interaction of the halogen lone electron pair with the charge developed in the corresponding intermediates [e.g. (5), in the formation of *p*-bromonitrobenzene from bromobenzene, Expt 6.20].

(5)

Experimental details for the nitration of benzyl cyanide are also included (Expt 6.21). The product is largely *p*-nitrobenzyl cyanide (some of the *ortho* isomer is also formed): the cyanomethyl substituent (as a substituted alkyl group) is thus an *ortho/para* directing and weakly activating group.

The electronic effect of groups in aromatic electrophilic substitution processes is fully discussed in all standard organic chemistry textbooks.

Experiment 6.17 NITROBENZENE

$$PhH + HNO_3 \xrightarrow{\text{H}_2\text{SO}_4} Ph \cdot NO_2$$

CAUTION: *This preparation should be conducted in an efficient fume cupboard.*

Place 50 g (35 ml, *c*. 0.5 mol) of concentrated nitric acid in a 500-ml round-bottomed flask, and add, in portions with shaking, 74 g (40 ml) of concentrated sulphuric acid. Keep the mixture cool during the addition by immersing the flask in cold water. Place a thermometer (110 °C range) in the acid mixture. Introduce 26 g (30 ml, 0.33 mol) of benzene (**CAUTION**) in portions of 2–3 ml; shake the flask well, to ensure thorough mixing, after each addition of the benzene. Do not allow the temperature of the mixture to rise above 55 °C; immerse the flask, if necessary, in cold water or in ice-water. When all the benzene has been added, fit a reflux condenser to the flask and heat it in a water bath maintained at 60 °C (but not appreciably higher) for 40–45 minutes; remove the flask from time to time from the bath and shake it vigorously to ensure good mixing of the immiscible layers. Pour the contents of the flask into about 500 ml of cold water in a beaker, stir the mixture well in order to wash out as much acid as possible from the nitrobenzene and allow to stand. When the nitrobenzene has settled to the bottom, pour off the acid liquor as completely as possible, and transfer the residual liquid to a separatory funnel. Run off the lower layer of nitrobenzene and reject the upper aqueous layer; return the nitrobenzene to the separatory funnel and shake it vigorously with about 50 ml of water. Separate the nitrobenzene as completely as possible and run it into a small conical flask containing about 5 g of anhydrous calcium chloride. If the nitrobenzene does not become clear on shaking because of the presence of emulsified water, warm the mixture, with shaking, for a short period on a water bath; the cloudiness will soon disappear. Filter the cold product through a small fluted filter paper into a small (50- or 100-ml) distilling flask and attach a still-head and air condenser. Heat the flask on a ceramic-centred wire gauze or preferably in an air bath, and collect the fraction which boils at 206–211 °C (1). Do not distil quite to dryness nor allow the temperature to rise above 214 °C, for there may be a residue of *m*-dinitrobenzene and higher nitro compounds and an explosion may result. The yield of nitrobenzene is 35 g (85%). Pure nitrobenzene is a clear, pale yellow liquid, b.p. 210 °C.

The i.r. spectrum of nitrobenzene is reproduced on p. 313. The p.m.r. spectrum (CCl$_4$, TMS) shows signals at δ 7.3–7.8 (m, 3H, C$_{3,4,5}$—H) and a lower field signal at δ 8.0–8.3 (m, 2H, C$_{2,6}$—H). The m.s. reveals significant fragment ions at m/z 123 (M), 93 (M − NO), 77 (M − NO$_2$; base peak), and 65 (93 − CO).

Note. (1) Nitrobenzene (and many other liquid organic compounds containing nitrogen) is appreciably toxic and its vapour should not be allowed to escape into the atmosphere of the laboratory. Site the distillation apparatus in a fume cupboard, use the receiver assembly illustrated in Fig. 2.98, and attach to the outlet of the receiver adapter a piece of rubber tubing leading to the extraction system. The liquid is also a skin poison; if it is accidentally spilled on the skin, it should be removed by washing with a little methylated spirit, followed by soap and warm water.

Cognate preparation. *1-Nitronaphthalene.* Use 40 ml of concentrated nitric acid and 40 ml of concentrated sulphuric acid with 50 g (0.39 mol) of *finely powdered* naphthalene as detailed for nitrobenzene. The subsequent heating at 60 °C should be for 30–40 minutes or until the smell of naphthalene has disappeared. Pour the reaction mixture into 500 ml of water; decant the washings from the product and then boil the solid with 200 ml of water for 20 minutes. Decant the water and subject the oil to steam distillation (Fig. 2.102) to remove unreacted naphthalene. Pour the warm residue into a large volume of water with vigorous stirring. Filter and recrystallise from dilute alcohol to give pure 1-nitronaphthalene (60 g, 89%), m.p. 61 °C.

Experiment 6.18 *m*-DINITROBENZENE

$$\text{Ph·NO}_2 \xrightarrow[\text{conc. H}_2\text{SO}_4]{\text{fuming HNO}_3} m\text{-C}_6\text{H}_4(\text{NO}_2)_2$$

Place 37.5 g (21 ml) of concentrated sulphuric acid and 22.5 g (15 ml) of fuming nitric acid, d 1.5, in a 250- or 500-ml round-bottomed flask; add a few fragments of unglazed porcelain. Attach a reflux condenser and place the apparatus in a fume cupboard. Add slowly, in portions of about 3 ml, 15 g (12.5 ml, 0.122 mol) of nitrobenzene; after each addition, shake the flask to ensure thorough mixing. Heat the mixture, with frequent shaking, on a boiling water bath for 30 minutes. Allow the mixture to cool somewhat and pour it cautiously with vigorous stirring into about 500 ml of cold water; the dinitrobenzene soon solidifies. Filter with suction, wash thoroughly with cold water and allow to drain as completely as possible.

Transfer the crude dinitrobenzene to a 250-ml flask fitted with a reflux condenser, add 80–100 ml of industrial (or rectified) spirit and heat on a water bath until all the crystalline solid dissolves. If the resulting solution is not quite clear, filter it through a fluted filter paper on a large funnel which has previously been warmed or through a warm Buchner funnel. Colourless crystals of *m*-dinitrobenzene (15 g, 73%) are deposited on cooling. If the m.p. is below 89–90 °C, recrystallisation is necessary.

The i.r. absorption bands of the nitro group (KBr disc) should be noted at c. 1525 and 1340 cm^{-1} (see Fig. 3.38). The p.m.r. is noted on p. 349 (cf. *ortho* and *para* isomers, both being noted on p. 942). The ^{13}C-n.m.r. spectrum shows signals at δ 118.6, 128.8, 130.8 and 148.1. The m.s. shows significant fragment ions at m/z 168 (M), 122 (M − NO$_2$), 76 (M − 2NO$_2$), 92 (M − [NO$_2$ + NO]), 64 (92 − CO), and 30 (NO$^\oplus$, base peak).

Cognate preparation. *2,4-Dinitrotoluene.* Use 18 g (12 ml, *c.* 0.36 mol) of fuming nitric acid, *d* 1.5, and 30 g (16.5 ml) of concentrated sulphuric acid in a two-necked flask (reflux condenser, stopper side-arm). Add, in small portions, 14 g (0.1 mol) of *p*-nitrotoluene through the side-arm keeping the temperature below 50 °C. Heat the reaction mixture and isolate the product as described above. Recrystallise from methanol. The yield of pure 2,4-dinitrotoluene, m.p. 71 °C, is 12.5 g (69%). The stretching vibrations of the nitro groups are clearly visible in the i.r. spectrum (KBr disc) at *c.* 1520 and 1340 cm^{-1}. The p.m.r. spectrum (CDCl$_3$, TMS) may be interpreted by first-order analysis: δ 2.75 (s, 3H, Me), 7.68 (d, 1H, C$_6$—H), 8.38 (d of d, 1H, C$_5$—H), and 8.74 (d, 1H, C$_3$—H).

Experiment 6.19 *m*-NITROBENZALDEHYDE

$$\text{Ph·CHO} \xrightarrow[\text{conc. H}_2\text{SO}_4]{\text{fuming HNO}_3} m\text{-O}_2\text{N·C}_6\text{H}_4\text{·CHO}$$

Place 250 ml of concentrated sulphuric acid and 21.5 ml of fuming nitric acid, *d* 1.5, in a 500-ml two-necked flask fitted with a mechanical stirrer (unsealed) and a dropping funnel. Stir and cool to 0 °C in a bath of ice and salt. Add 62.5 g (60 ml, 0.59 mol) of benzaldehyde (see Expt 6.133) dropwise from the dropping funnel; do not allow the temperature to rise above 5 °C. Then warm the mixture gradually to 40 °C, cool to room temperature and pour in a thin stream with vigorous stirring on to finely crushed ice. Filter through a sintered glass funnel, wash with a little water, press out the oil with a wide glass stopper and dry the solid in the air upon absorbent paper. The resulting crude *m*-nitrobenzaldehyde weighs 55 g and melts at 48–50 °C. Melt the crude solid under excess of 10 per cent sodium carbonate solution, stir, cool, filter and dry in the air; the product has m.p. 51–52 °C. Dissolve the solid in 120 ml of hot toluene (water bath; *fume cupboard*), decant from any solid present and add light petroleum, b.p. 40–60 °C, until a slight turbidity results and cool. Collect the pure *m*-nitrobenzaldehyde and dry in the air; the yield is 45 g (50%), m.p. 58 °C.

Note the i.r. absorptions at *c.* 3100 cm^{-1} (C$_{AR}$—H), *c.* 2800 cm^{-1} (H—CO), and at *c.* 1700 cm^{-1} (C=O), together with well-observed aromatic ring vibrations (at 1600, 1590, 1500 and 1450 cm^{-1}) and the absorptions arising from the vibrations of the nitro group. The p.m.r. spectrum (CDCl$_3$, TMS) shows δ 7.85 (distorted t, 1H, C$_4$—H), 8.2–8.9 (broad m, 3H, C$_{2,5,6}$—H), and 10.19 (s, 1H, H—CO).

Experiment 6.20 *p*-BROMONITROBENZENE

$$\text{PhBr} + \text{HNO}_3 \xrightarrow{\text{H}_2\text{SO}_4} p\text{-Br·C}_6\text{H}_4\text{·NO}_2$$

Prepare a mixture of 28.5 g (20 ml) of concentrated nitric acid and 37 g (20 ml) of concentrated sulphuric acid in a 250-ml round-bottomed flask (see Expt 6.17) and cool it to the laboratory temperature. Attach a reflux condenser to the flask. Add 16 g (10.5 ml, 0.1 mol) of bromobenzene (Expt 6.23) in portions of 2–3 ml during about 15 minutes; shake the flask vigorously during the whole process and do not allow the temperature to rise above 50–60 °C by cooling in running water, if necessary. When the temperature no longer tends to rise owing to the heat of reaction, heat the flask on a boiling water bath for

30 minutes. Allow to cool to room temperature and pour the reaction mixture with stirring into 200 ml of cold water. Filter the bromonitrobenzene at the pump, wash well with cold water and finally drain as far as possible. Recrystallise from 100 to 125 ml of industrial spirit (flask, reflux condenser and water bath; see Expt 6.18). When cold, filter the almost pure *p*-bromonitrobenzene, m.p. 125 °C. The yield is 14 g (70%). The mother-liquor contains the *o*-bromonitrobenzene, contaminated with some of the *p*-isomeride.

The *para*-substitution pattern is clearly revealed by the p.m.r. spectrum (CDCl$_3$, TMS), which shows the characteristic pair of two-proton doublets, δ 7.26 (d, 2H, *ortho*-H's to Br), and 8.03 (d, 2H, *ortho*-H's to NO$_2$).

Experiment 6.21   *p*-NITROBENZYL CYANIDE

$$\text{Ph·CH}_2\text{·CN} + \text{HNO}_3 \xrightarrow{\text{H}_2\text{SO}_4} p\text{-O}_2\text{N·C}_6\text{H}_4\text{·CH}_2\text{·CN}$$

Place a mixture of 275 ml of concentrated nitric acid with an equal volume of concentrated sulphuric acid in a 2-litre three-necked flask, fitted with a thermometer, a mechanical stirrer and a dropping funnel and assembled in the fume cupboard. Cool the mixture to 10 °C in an ice bath, and run in 100 g (98 ml, 0.85 mol) of benzyl cyanide (Expt 5.157) at such a rate (about 1 hour) that the temperature remains at about 10 °C and does not rise above 20 °C. Remove the ice bath, stir the mixture for 1 hour and pour it on to 1200 g of crushed ice. A pasty mass slowly separates; more than half of this is *p*-nitrobenzyl cyanide, the other components being the *ortho* isomeride and a variable amount of an oil. Filter the mass on a sintered glass funnel, press well to remove as much oil as possible and then dissolve in 500 ml of boiling rectified spirit. The *p*-nitrobenzyl cyanide crystallises on cooling. Filter this off at the pump and recrystallise from 80 per cent ethanol. The yield of *p*-nitrobenzyl cyanide, m.p. 115–116 °C, is 75 g (54%). Another recrystallisation raises the m.p. to 116–117 °C.

The i.r. spectrum (KBr disc) shows absorptions at *c.* 3050 (C$_{AR}$—H str.), *c.* 2950 (C$_{AL}$—H str.), 2250 (C≡N str.), 1500 and 1380 cm^{-1} (NO$_2$ str.). The *para* substitution is confirmed by the p.m.r. spectrum (CDCl$_3$, TMS), δ 3.89 (s, 2H, CH$_2$), 7.48 (d, 2H, C$_{2,6}$—H), and 8.18 (d, 2H, C$_{3,5}$—H).

6.2.2 THE OXIDATION OF AMINES

Various reagents are available for the oxidation of an aromatic amine to the corresponding nitro compound. For example, peroxymonosulphuric acid (Caro's acid) and other peroxyacids have been quite widely used in the past, although the yields of nitro compounds are rather variable owing to the concomitant formation of azoxycompounds. Pertrifluoroacetic acid is the reagent of choice since it generally gives improved yields of purer products; its use is illustrated by the conversion of *p*-toluidine into *p*-nitrotoluene (Expt 6.22). Further examples are the formation of *o*-chloronitrobenzene and 2,6-dichloronitrobenzene from the corresponding primary amines. The method is not, however, generally suitable for amines having ring systems which are highly activated as the result, for example, of the presence of an alkoxy substituent.

$$\text{ArNH}_2 \xrightarrow{\text{CF}_3\text{·CO}_3\text{H}} \text{ArNO}_2$$

6.2.3 THE REPLACEMENT OF A DIAZO GROUP BY A NITRO GROUP

This replacement is achieved by the decomposition of the aryldiazonium fluoroborate with aqueous sodium nitrite in the presence of copper powder and is described in Expt 6.78. This procedure gives better yields and thus replaces the former method of reacting an acidic aryldiazonium salt solution with nitrous acid in the presence of copper(I) oxide.

Experiment 6.22 *p*-NITROTOLUENE

$$(CF_3 \cdot CO)_2O + H_2O_2 \longrightarrow CF_3 \cdot CO_3H + CF_3 \cdot CO_2H$$

$$p\text{-Me} \cdot C_6H_4 \cdot NH_2 \xrightarrow{CF_3 \cdot CO_3H} p\text{-Me} \cdot C_6H_4 \cdot NO_2$$

CAUTION: *This reaction should be carried out in a fume cupboard behind a safety screen. Adequate precautions should be observed in handling the hydrogen peroxide solution (Expt 5.190, Note (1)).*

Prepare a solution of pertrifluoroacetic acid in dichloromethane as follows. Place 50 ml of dichloromethane in a two-necked, 250-ml round-bottomed flask fitted with a reflux condenser and a dropping funnel. Insert a plastic-covered magnetic stirrer follower bar, and cool the flask in an ice bath sited on the stirrer unit. To the cooled and stirred solution add 3.5 ml (0.12 mol) of 85 per cent hydrogen peroxide and then 22.7 ml (33.9 g, 0.14 mol) of trifluoroacetic anhydride. Stir for 5 minutes after the addition is completed and then allow the solution to warm to room temperature. To this stirred solution add over 15 minutes a solution of 3.2 g (0.03 mol) of *p*-toluidine in 10 ml of dichloromethane. The reaction is exothermic but no cooling is necessary in this case. When the addition is complete heat the solution under reflux on the water bath for 1 hour. Cool, wash the dichloromethane solution with two 100 ml portions of water, dry (MgSO$_4$) and remove the solvent on a rotary evaporator to obtain *p*-nitrotoluene, 2.9 g (71%), having m.p. *c.* 49 °C. Recrystallisation from a small volume of ethanol gives a purer product, m.p. 51–52 °C.

In the i.r. spectrum (KBr disc) note absorptions at *c.* 1555 and 1355 cm^{-1} (NO$_2$ str.). The *para* substitution is confirmed by the p.m.r. spectrum (CDCl$_3$, TMS) which shows δ 2.45 (s, 3H, Me), 7.30 (d, 2H, C$_{2,6}$—H), and 8.10 (d, 2H, C$_{3,5}$—H). The m.s. reveals principal fragment ions at *m/z* 137 (M), 107 (M − NO), 91 (M − NO$_2$, base peak), 79 (107 − CO), and 77 (79 − H$_2$).

6.3 AROMATIC HALOGEN COMPOUNDS

The compounds described and discussed below are those in which the halogen is directly attached to the aromatic nucleus (e.g. C$_6$H$_5$Cl, or PhCl, chlorobenzene) and those in which the halogen is substituted into an alkyl side chain (e.g. C$_6$H$_5$·CH$_2$Cl, or Ph·CH$_2$Cl, benzyl chloride).

The following procedures are available for the preparation of aromatic halogen compounds.

1. Direct halogenation by (*a*) substitution in the nucleus, and (*b*) substitution in the side chain (Expts 6.23 to 6.28).

2. Chloromethylation (Expt 6.29).
3. The replacement of a diazo group by a halogen (see Expts 6.70 to 6.72).
4. The replacement of a hydroxyl group by a halogen (Expt 6.30).
5. Methods leading to polyvalent iodine compounds (Expts 6.31 to 6.36).

SUMMARY OF RETROSYNTHETIC STRATEGIES

C—X Disconnection into an aryl anionic site (methods 1(a) and 2), e.g.

$X^{\oplus} \equiv PhH + X_2$

(1a)

(TM)

$\overset{\ominus}{C}H_2Cl \equiv PhH + CH_2O/HCl$

(2)

(TM)

C—X Disconnection into an aryl cationic site (methods 3 and 4), e.g.

$X^{\ominus} \equiv Ph\cdot N_2^{\oplus} + Cu^IX$

(3)

(TM)

$X^{\ominus} \equiv Ph\cdot O\cdot \overset{\oplus}{P}Ph_3\}X^{\ominus}$

(4)

(TM)

C—X Disconnection in the side-chain (method 1(b)), e.g.

$\dot{X} \equiv Ph\cdot Me + X_2(h\nu)$

(1b)

(TM)

SPECTROSCOPIC FEATURES

The *i.r.* spectrum of an aromatic halogen compound shows the expected absorptions due to the aromatic nucleus and strong absorption in the longer wavelength region due to the stretching of the C—X bond (p. 316). The *p.m.r.* spectrum and the m.s. are more useful for definitive structural characterisation. The former may frequently enable the substitution pattern to be assigned; examples are to be found in the preparative accounts below. By virtue of the natural isotope ratios of bromine and chlorine (p. 366) the *m.s.* often enables the various possible combinations in mono- and poly-halogenated compounds to be assessed (see for example bromobenzene and *p*-dibromobenzene, Expt 6.23; *p*-bromochlorobenzene, Expt 6.30; 2,4,6-tribromoaniline Expt 6.59). The *u.v.–visible* spectrum is not sufficiently informative to be of value in structural assignments.

6.3.1   DIRECT HALOGENATION

NUCLEAR SUBSTITUTION

Benzene does not react appreciably with chlorine and bromine in the cold, but in the presence of catalysts, such as aluminium amalgam, pyridine or iron, reaction takes place readily, affording in the first instance the mono-halogenated derivative as the main product. Di-substituted products (largely the *para* isomer) are obtained if the proportion of the halogen is increased. A typical procedure is given in the preparation of bromobenzene (Expt 6.23). The function of the catalyst is to increase the electrophilic activity of the halogen and the mechanism of the bromination of benzene with pyridine as the catalyst can be represented by the following scheme (cf. the nitration of benzene, Section 6.2.1, p. 851).

Since iodine is the least reactive of the halogens, iodination does not occur unless the reaction is carried out in the presence of an oxidising agent. When using fuming nitric acid, the nature of the electrophile which functions as the $I^{\oplus}$ donor is thought to be $[O{=}N(I)OH]^{\oplus}$.[10] More recently a procedure has been reported using copper(II) chloride as the source of oxidant in the presence of aluminium chloride.[11] Both methods give good yields of iodobenzene (Expt 6.24), but the latter method is more suitable for the iodination of alkylbenzenes (e.g. *p*-xylene, which gives a good yield of 2-iodo-1,4-dimethylbenzene). Benzene derivatives with electron-withdrawing groups are not iodinated under these conditions.

Condensed aromatic hydrocarbons are more reactive towards electrophilic reagents, and naphthalene, for example, may be brominated quite readily in solution in carbon tetrachloride without the need for a catalyst; electrophilic attack takes place at the more reactive α-position to yield 1-bromonaphthalene (Expt 6.25).

The effects of substituents in electrophilic aromatic substitution processes, which are outlined in Section 6.2.1, p. 852, are further illustrated here with reference to the bromination of nitrobenzene, where a substantially higher reaction temperature is required (the nitro group is deactivating) and the product is predominantly *m*-bromonitrobenzene (Expt 6.26). Nitration of an aryl halide gives a mixture of *o*- and *p*-halonitrobenzenes (Expt 6.20).

SIDE-CHAIN HALOGENATION

In the absence of catalysts, treatment of toluene with chlorine (or bromine) at the boiling point, preferably with exposure to sunlight or other bright light source, results in halogenation in the side chain. The introduction of the first chlorine atom, for example, proceeds at a much faster rate than the introduction of the second chlorine atom so that in practice the major portion of the toluene is converted into benzyl chloride (1) before appreciable chlorination of benzyl chloride occurs to give benzylidene chloride (2) and benzotrichloride (3).

$$Ph{\cdot}Me + Cl_2 \xrightarrow[hv]{heat} Ph{\cdot}CH_2Cl \longrightarrow Ph{\cdot}CHCl_2 \longrightarrow Ph{\cdot}CCl_3$$
$$\quad\qquad\qquad\qquad\qquad (1) \qquad\qquad\qquad (2) \qquad\qquad (3)$$

The reaction proceeds by the radical mechanism shown; the first step is facilitated by the mesomeric stabilisation of the benzylic radical.

$$Cl_2 \rightleftharpoons 2Cl{\cdot}$$
$$Ph{\cdot}Me + Cl{\cdot} \xrightarrow{-HCl} Ph{\cdot}\dot{C}H_2 \xrightarrow{Cl_2} Ph{\cdot}CH_2Cl + Cl{\cdot} \text{ etc.}$$

Rapid side-chain chlorination of toluene proceeds in the dark with sulphuryl chloride in the presence of benzoyl peroxide (0.001–0.005 mol per mol of SO_2Cl_2) as catalyst (Expt 6.27). With an excess of sulphuryl chloride, benzylidene chloride is formed, but in this case chlorination does not proceed beyond this stage.

Side-chain halogenation is illustrated also by the bromination of *p*-nitrotoluene (Expt 6.28). The radical mechanism with molecular bromine is similar to that of chlorination above; *N*-bromosuccinimide has also found use as a side-chain brominating reagent and its application is described in Expts 6.119 and 6.152.

Experiment 6.23   BROMOBENZENE

$$PhH + Br_2 \xrightarrow{pyridine} C_6H_5Br$$

This preparation should be conducted in an efficient fume cupboard.

Place 50 g (57 ml, 0.64 mol) of dry benzene (**CAUTION**) and 0.5 ml of dry pyridine (1) (dried over potassium hydroxide pellets) in a 500-ml round-bottomed flask. Attach a reflux condenser to the flask and fit a device for absorbing the hydrogen bromide gas subsequently evolved (Fig. 2.61(*a*)). Partially immerse the flask in a bath of cold water, supported upon a tripod and gauze. Carefully pour 125 g (40 ml) of bromine (for precautions to be taken with bromine, see Section 4.2.9, p. 422) through the condenser and immediately insert the absorption device into the upper end of the condenser. A vigorous reaction soon occurs and hydrogen bromide is evolved which is absorbed by the water in the beaker; when the reaction slackens, warm the bath to 25–30 °C for 1 hour. Finally raise the temperature of the bath to 65–

70 °C for a further 45 minutes or until all the bromine has disappeared (no red vapours visible) and the evolution of hydrogen bromide has almost ceased. Transfer the dark-coloured reaction product to a separatory funnel and shake successively with water, with sufficient 5–10 per cent sodium hydroxide solution to ensure that the washings are alkaline to litmus, and finally with water. Dry with magnesium sulphate or anhydrous calcium chloride. Filter through a fluted filter paper into a small distilling flask and distil slowly. Collect the crude bromobenzene at 150–170 °C; pour the residue while still hot into a small porcelain basin. Redistil the liquid of b.p. 150–170 °C (2) and collect the bromobenzene at 154–157 °C; the yield is about 60 g (60%).

In the i.r. spectrum (thin film) note absorptions at c. 3050 cm^{-1} (C_{AR}—H, stretching), and at 1600, 1590, 1500 and 1450 cm^{-1} (ring-breathing vibrations). The p.m.r. spectrum (CCl$_4$, TMS) shows two complex signals at δ 7.13 (m, 3H, $C_{3,4,5}$—H) and at 7.41 (m, 2H, $C_{2,6}$—H). The presence of bromine is clearly revealed in the m.s. which shows principal fragment ions at 158 (M, ^{81}Br) and 156 (M, ^{79}Br) of nearly equal intensity, and at 77 (M − Br, base peak) and 51 (77 − C_2H_2).

Isolate the *pure p-dibromobenzene* from the residue in the basin by recrystallisation from hot ethanol with the addition of 1–2 g of decolourising charcoal; use about 4 ml of ethanol (industrial spirit) for each gram of material. Filter the hot solution through a fluted filter paper, cool in ice and filter the crystals at the pump. The yield of *p*-dibromobenzene, m.p. 89 °C, is about 12 g.

The p.m.r. spectrum (CDCl$_3$, TMS) shows a sharp signal at δ 7.29 (s, 4H, C_{AR}—H) characteristic of an A$_4$ aromatic substitution pattern. The m.s. shows the typical molecular ion cluster due to the presence of two bromine atoms (p. 366), namely, m/z 238 (RA 48.8%), 236 (RA 100%) and 234 (RA 51.4%); the other ions are observable at m/z 157 ($C_6H_4{}^{81}$Br), 155 ($C_6H_4{}^{79}$Br), 75 (157 − H^{81}Br, and 155 − H^{79}Br), 76 (157 − ^{81}Br and 155 − ^{79}Br) and 50 (76 − C_2H_2).

Notes. (1) Other halogen carriers may be used, e.g. 1–2 g of iron filings, or 1 g of aluminium amalgam. The bromine must then be added slowly from a dropping funnel to the benzene warmed on a water bath; a suitable two- or three-necked flask should be used. After all the bromine has been introduced, the mixture is heated on a water bath until no red vapours are visible above the liquid. The subsequent procedure is as above.

(2) The best results are obtained by distillation from a small flask through a short fractionating column: a Hempel column filled with glass rings (Fig. 2.105(c)) and lagged with several thicknesses of linen cloth is quite satisfactory.

Experiment 6.24 IODOBENZENE

$$2PhH + I_2 \xrightarrow[HNO_3]{|O|} 2PhI + H_2O$$

This preparation should be conducted in an efficient fume cupboard.

Equip a 500-ml three-necked flask with a reflux condenser, a sealed mechanical stirrer and separatory funnel, and support it on a water bath. Attach an absorption device (Fig. 2.61) to the top of the condenser. Place 134 g (152 ml, 1.72 mol) of benzene (**CAUTION**) and 127 g (0.5 mol) of iodine in the flask, and heat the water bath to about 50 °C; add 92 ml of fuming nitric acid, d 1.50, slowly from the separatory funnel during 30 minutes (1). Oxides of nitrogen are evolved in quantity. The temperature rises slowly without the application of heat until the mixture boils gently. When all the nitric acid has been introduced, reflux the mixture gently for 15 minutes. If iodine is still present, add more nitric acid to the warm solution until the purple colour (due to iodine) changes to brownish-red.

Separate the lower oily layer, mix with it an equal volume of 10 per cent sodium hydroxide solution and steam distil from a 1-litre flask until no more oil passes over. A yellow solid, consisting of nitro compounds, may collect towards the end of the distillation; remove this by mechanical stirring of the oil for about 3 hours with 7 ml of concentrated hydrochloric acid, 100 ml of water and 70 g of iron filings in a 1-litre three-necked flask connected with a reflux condenser. Allow the mixture to cool and filter. Render the filtrate distinctly acid to Congo red with hydrochloric acid and again steam distil. Separate the oil, dry it with anhydrous calcium chloride or magnesium sulphate, distil through a suitably lagged fractionating column and collect the fraction of b.p. 180–190 °C. Upon redistillation, pure iodobenzene, b.p. 184–186 °C, is obtained. The yield is 180 g (87%).

Note. (1) In an alternative procedure,[11] iodine (3.8 g, 15 mmol) is dissolved in benzene (50 ml), to this solution is added a mixture of anhydrous aluminium chloride (4.0 g, 30 mmol) and copper(II) chloride (4.0 g, 30 mmol) previously dried at 120 °C/3 mmHg for more than 3 hours. The mixture is stirred at 40 °C for 2 hours, and then poured into water. The organic layer is washed with aqueous sodium hydrogen sulphite, dried and fractionally distilled. The yield of iodobenzene is 79 per cent.

Experiment 6.25 1-BROMONAPHTHALENE

$$C_{10}H_8 + Br_2 \xrightarrow{CCl_4} 1\text{-}C_{10}H_7Br + HBr$$

Use a 500-ml three-necked flask equipped as in Expt 6.26 but mounted on a water bath. Place 128 g (1 mol) of naphthalene and 45 ml of dry carbon tetrachloride in the flask, and 177 g (55 ml, 1.11 mol) of bromine (**CAUTION**) in the separatory funnel. Heat the mixture to gentle boiling and run in the bromine at such a rate that little, if any, of it is carried over with the hydrogen bromide into the trap; this requires about 3 hours. Warm gently, with stirring, for a further 2 hours or until the evolution of hydrogen bromide ceases. Replace the reflux condenser by a condenser set for downward distillation, stir and distil off the carbon tetrachloride as completely as possible. Mix the residue with 8 g of sodium hydroxide pellets and stir at 90–100 °C for 3 hours; this treatment will remove impurities which gradually evolve hydrogen bromide. Distil the product under diminished pressure and collect the following fractions: (i) up to 131 °C/12 mmHg (or 144 °C/20 mmHg); (ii) 132–135 °C/12 mmHg (or 145–148 °C/20 mmHg); and (iii) above 135 °C/12 mmHg (or 148 °C/20 mmHg). Fraction (ii) is almost pure 1-bromonaphthalene. Fraction (i) contains unchanged naphthalene, while (iii) contains dibromonaphthalene. Cool fraction (i) in ice when most of the naphthalene will crystallise out; filter this off on a sintered glass funnel, combine the filtrate with fraction (iii), redistil and collect the 1-bromonaphthalene fraction separately. The total yield of colourless product is 150 g (72.5%).

Experiment 6.26 *m*-BROMONITROBENZENE

$$Ph \cdot NO_2 + Br_2 \xrightarrow{\text{Fe}} m\text{-}Br \cdot C_6H_4 \cdot NO_2 + HBr$$

Equip a 1-litre three-necked flask with a separatory funnel, a sealed mechanical stirrer (1) and a double surface reflux condenser carrying an outlet tube connected to a gas trap (Fig. 2.62). Support the flask in an oil bath. Place 90 g (75 ml, 0.73 mol) of dry, freshly distilled nitrobenzene in the flask. Weigh out 10 g of pure iron powder (95%) 'reduced by hydrogen'. Heat the oil bath to 135–145 °C and introduce 3 g of the iron powder by temporarily removing the separatory funnel. Into the latter place 62.5 g (20 ml, 1.17 mol) of dry bromine (**CAUTION**) (Section 4.2.9, p. 422) and run it into the flask at such a rate that bromine vapours do not rise appreciably in the condenser (*c.* 20 minutes). Continue stirring and heating for 1 hour before adding a further 3 g of iron powder and 20 ml of dry bromine in a similar manner. Stir for a further hour, add another 3 g of iron powder and 20 ml of bromine. When there is no more bromine vapour in the condenser, make a final addition of 1 g of iron powder and heat for 1 hour longer.

Pour the resulting dark reddish-brown liquid into 500 ml of water to which 17 ml of saturated sodium metabisulphite solution has been added (the latter to remove the excess of bromine). Steam distil the resulting mixture (Fig. 2.102); collect the first portion of the distillate, which contains a little unchanged nitrobenzene, separately. Collect about 4 litres of distillate. Filter the yellow crystalline solid at the pump, and press well to remove the adhering liquid. The resulting crude *m*-bromonitrobenzene, m.p. 51–52 °C, weighs 110 g (74%). If required pure, distil under reduced pressure (Section 2.27) and collect the fraction of b.p. 117–118 °C/9 mmHg; it then melts at 56 °C and the recovery is about 85 per cent.

Note. (1) Mechanical stirring, although not essential and replaceable by occasional shaking by hand, is advantageous.

Experiment 6.27 BENZYL CHLORIDE

$$Ph \cdot Me + SO_2Cl_2 \xrightarrow{\text{peroxide}} Ph \cdot CH_2Cl + SO_2 + HCl$$

In an efficient fume cupboard attach a double surface condenser to a 500-ml round-bottomed flask containing 92 g (106 ml, 1 mol) of toluene, 68 g (41 ml, 0.5 mol) of redistilled sulphuryl chloride and 1 g of benzoyl peroxide (Section 4.2.6, p. 417). Reflux gently and remove the heat source if necessary to moderate the vigorous reaction which takes place. The reaction is complete in 30 minutes. Distil the reaction mixture using a fractionating column, first under atmospheric pressure until the temperature reaches 135–140 °C, and continue the distillation under diminished pressure collecting the benzyl chloride at 64–69 °C/12 mmHg (1). The latter on redistillation boils at 63–65 °C/12 mmHg. The yield of benzyl chloride is 50 g (79%).

Note. (1) The benzyl chloride may also be isolated by distillation under atmospheric pressure. The material boiling between 165 and 185 °C is collected and redistilled; the final product is collected at 178–182 °C (pure benzyl chloride has b.p. 179 °C). The resulting benzyl chloride is, however, of lower purity unless an efficient fractionating column is used.

Experiment 6.28 *p*-NITROBENZYL BROMIDE

$$p\text{-}O_2N\cdot C_6H_4\cdot Me \xrightarrow{\text{Br}_2} p\text{-}O_2N\cdot C_6H_4\cdot CH_2Br$$

Place 150 g (1.1 mol) of *p*-nitrotoluene, m.p. 51–52 °C, in a 500-ml three-necked flask, fitted with a reflux condenser, a sealed mechanical stirrer and a separatory funnel with stem reaching nearly to the bottom of the flask. Attach a gas absorption trap (Fig. 2.62) to the top of the condenser; the whole essembly should be sited in a fume cupboard. Heat the flask in an oil bath at 145–150 °C and add 184 g (59 ml, 1.15 mol) of bromine during 2 hours (1). Continue the stirring for an additional 10 minutes after all the bromine has been added. Pour the contents of the flask while still liquid (**CAUTION**) (2) into a 2.5-litre round-bottomed flask containing 2 litres of hot light petroleum, b.p. 80–100 °C, and 8 g of decolourising carbon. Attach a reflux condenser to the flask, heat it with the aid of a heating mantle until the material dissolves, boil for 10 minutes and filter rapidly through a pre-heated Buchner funnel. Cool the filtrate to 20 °C, filter the crystals with suction, press well and wash with two 25 ml portions of cold light petroleum. The crude *p*-nitrobenzyl bromide, m.p. 95–97 °C (150 g), is sufficiently pure for many purposes. Purify by dissolving in 1500–1700 ml of light petroleum, b.p. 80–100 °C, boil with 8 g of decolourising carbon, and filter through a preheated Buchner or sintered glass funnel. Cool the filtrate in ice, filter at the pump, drain well and wash with two 15 ml portions of cold light petroleum. The yield of pure *p*-nitrobenzyl bromide (pale yellow crystals, m.p. 98–99 °C) is 135 g (57%).

Structural characterisation is confirmed by the p.m.r. spectrum (CDCl₃, TMS) which shows signals at δ 4.51 (s, 2H, CH₂), 7.57 (d, 2H, *ortho*-H's to CH₂Br) and 8.18 (d, 2H, *ortho*-H's to NO₂). In the i.r. spectrum the absorption due to the nitro group (*c.* 1340 and 1550 cm⁻¹) is clearly visible.

Notes. (1) Improved yields may be obtained by exposing the flask to the light of two 300-watt tungsten lamps during the bromination.
(2) Care must be taken in manipulating the lachrymatory solutions of *p*-nitrobenzyl bromide. If the substance should come into contact with the skin, bathe the affected part with alcohol.

6.3.2 CHLOROMETHYLATION

This is the replacement of a hydrogen atom in an aromatic compound by a chloromethyl (CH₂Cl) group in a single operation. The original classical reaction consists essentially of the interaction of formaldehyde and hydrogen chloride in the presence of a catalyst such as zinc chloride or aluminium chloride with an aromatic system (*Blanc chloromethylation reaction*). The reaction is similar in some respects to that of Friedel and Crafts (see Sections 6.1.1, p. 828, and 6.11.1, p. 1006) and involves the hydroxymethyl cation as the electrophilic species. This reacts with the aromatic ring to give the benzylic alcohol which is then converted into the chloromethyl derivative by hydrogen chloride.

$$H_2C{=}O + HCl \rightleftharpoons Cl^{\ominus} + [H_2C{=}\overset{\oplus}{O}H \longleftrightarrow H_2\overset{\oplus}{C}{-}OH]$$

$$ArH + \overset{\oplus}{C}H_2OH \longrightarrow Ar\cdot CH_2OH + H^{\oplus}$$

$$Ar\cdot CH_2OH + HCl \longrightarrow Ar\cdot CH_2Cl + H_2O$$

However, it has been found that a by-product arising from the interaction of formaldehyde and hydrogen chloride is *bis(chloromethyl)ether* (*BCME*), *which is a potent carcinogen*. For this reason the classical chloromethylation route to side-chain halogenated products should not be regarded as a desirable synthetic procedure, and in general should only be used if the required compound cannot be readily prepared by other methods. When used, effective precautions should be taken during the reaction and in the disposal of the reaction residues.[12, 13]

Recently two alternative procedures have been described. In one case[14] methoxyacetyl chloride (MeO·CH$_2$·COCl) in the presence of anhydrous aluminium chloride is thought to provide the source of the cation (4), which substitutes in the aromatic nucleus to give the benzyl methyl ether (5), which is subsequently converted into the chloromethylated product (6).

$$\text{MeO·CH}_2\text{·COCl} \xrightarrow{\text{AlCl}_3} \text{MeÖ-CH}_2\text{--}\overset{\oplus}{\text{C}}\text{=O\}} \overset{\ominus}{\text{AlCl}_4} \xrightarrow{-\text{CO}} \text{MeÖ=CH}_2\} \overset{\ominus}{\text{AlCl}_4}$$

$$\text{(4)}$$

$$\text{ArH} + \text{MeÖ=CH}_2 \longrightarrow \text{Ar·CH}_2\text{·O·Me} \longrightarrow \text{Ar·CH}_2\text{Cl}$$
$$\text{(5)} \qquad\qquad\qquad \text{(6)}$$

A wide range of substituted aromatic systems have been chloromethylated by this procedure, the only disadvantage being the expense of the methoxyacetyl chloride.

In the second alternative method a two-step procedure has been described.[15] The first step involves amidomethylation of the aromatic nucleus by reaction with paraformaldehyde and acetamide in the presence of concentrated sulphuric acid. The isolated intermediate is then treated with phosphorus oxychloride in dimethylformamide and xylene and converted into the chloromethylated product [e.g. see 2-(chloromethyl)-4-nitrotoluene, Expt 6.29, for formulation]. The mechanisms of these reactions are currently only speculative.

Experiment 6.29   2-(CHLOROMETHYL)-4-NITROTOLUENE[15]

N-(2-Methyl-5-nitrobenzyl)acetamide. To a solution of 6.85 (0.05 mol) of *p*-nitrotoluene and 1.5 g (0.05 mol) of paraformaldehyde in 60 ml of concentrated sulphuric acid is added 8.85 g (0.15 mol) of acetamide in portions. The solution is heated at 55 °C for 8 hours and poured over ice and water. The resulting solid is collected and heated in 10–12 parts of butyl acetate at 90 °C and filtered. Cooling the filtrate gives 7.8 g (75%) of the acetamide as a colourless solid, m.p. 141–143 °C; i.r. 3275, 1640 and 1660 (sh) cm^{-1}; m/z 208 (M); p.m.r. (CDCl$_3$, TMS) δ 2.02 (s, 3H, CO·Me), 2.37 (s, 3H, Ar·Me), 4.39 (d, 2H, CH$_2$), 6.32 (broad s, 1H, NH), 7.18 (d, 1H, $J = 8$ Hz, C$_3$—H), and 7.86 (d, 2H, $J = 10$ Hz, C$_4$—H and C$_6$—H). Material insoluble in butyl acetate is recrystallised from chloroform–methanol to give bis(2-methyl-5-nitrophenyl)methane, m.p. 153 °C.

2-(Chloromethyl)-4-nitrotoluene. A solution of 6.24 g (0.03 mol) of the foregoing acetamide, 9.67 g (0.063 mol) of phosphorus oxychloride and 2.19 g (0.03 mol) of dimethylformamide in 50 ml of xylene is refluxed for 1 hour. The cooled solution is washed with water and evaporated to give 4.73 g (85%) of product [**CAUTION** (1)] which after recrystallation from hexane has m.p. 61–62 °C.

Note. (1) This compound is lachrymatory and may be a skin irritant.

6.3.3 THE REPLACEMENT OF A DIAZO GROUP BY A HALOGEN

The most generally applicable route to nuclear substituted aromatic halogen compounds involves decomposition of a diazonium salt under suitable conditions. These reactions are discussed in Section 6.7.1, p. 922.

6.3.4 THE REPLACEMENT OF A HYDROXYL GROUP BY A HALOGEN

Replacement of the hydroxyl group in a phenol by halogen cannot be accomplished by reaction with the hydrogen halides as in the case of alcohols, and reaction with phosphorus halides gives only low yields of halogenobenzenes (except in the case of nitrophenols), the main product being a phosphite or phosphate ester.

$$ArOH + PCl_3 \longrightarrow (ArO)_3P + 3HCl$$

However, if the phenol is first treated with the complex formed from triphenylphosphine and a halogen in acetonitrile solution, an aryloxytriphenylphosphonium halide is formed which on thermal decomposition yields the aryl halide in good yield (e.g. the preparation of *p*-bromochlorobenzene, Expt 6.30).

$$Ph_3P + X_2 \longrightarrow Ph_3PX_2 \xrightarrow[-HX]{ArOH} Ph_3\overset{\oplus}{P}(OAr)\}X^{\ominus} \xrightarrow[-Ph_3PO]{} ArX$$

Experiment 6.30 *p*-BROMOCHLOROBENZENE

$$p\text{-}Cl\cdot C_6H_4OH + Ph_3\overset{\oplus}{P}Br\}\overset{\ominus}{Br} \longrightarrow p\text{-}Cl\cdot C_6H_4Br + HBr + Ph_3PO$$

CAUTION: *This preparation should be carried out in the fume cupboard.*

Equip a 250-ml three-necked round-bottomed flask with a reflux condenser and a dropping funnel (both protected by calcium chloride guard-tubes), and a sealed stirrer unit. In the flask place 29 g (0.11 mol) of finely powdered triphenylphosphine and 100 ml of dry acetonitrile (Section 4.1.27, p. 410). Cool the flask in an ice-water bath and add with stirring over a period of 20 minutes 17.3 g (5.5 ml, 0.108 mol) of bromine. After the addition is complete remove the cooling bath, set the condenser for downward distillation and replace the dropping funnel with a stopper. Remove the acetonitrile under reduced pressure using a water pump; warm the flask to 40 °C on a water bath and use an oil pump to remove last traces of solvent. Add 10 g (0.078 mol) of redistilled, powdered p-chlorophenol to the solid residue, replace the condenser (together with its guard-tube) in the reflux position and site the flask on a sand bath. Raise the temperature of the sand bath to between 250 and 280 °C and slowly stir the contents of the flask which soon melt; hydrogen bromide is evolved over a period of 3 hours. When evolution of gas ceases, cool the residue, add water and steam distil. Collect the solid p-bromochlorobenzene by filtration of the distillate; it has m.p. 63–65 °C. Recrystallise from ethanol to obtain the pure product, m.p. 65–66 °C; the yield is 12.5 g (83%).

The p.m.r. spectrum (CCl$_4$, TMS) shows a pair of doublets characteristic of para substitution at δ 7.13 (d, 2H, ortho-H's to Cl) and at 7.37 (d, 2H, ortho-H's to Br). The m.s. provides an excellent illustration of the isotope combinations of chlorine and bromine (p. 366); principal fragment ions are at m/z 194 (M, ^{81}Br^{37}Cl, RA 23.6%), 192 (M, ^{81}Br^{35}Cl and ^{79}Br^{37}Cl, RA 100%), 190 (M, ^{79}Br^{35}Cl, RA 76.1%), 113 (194 – ^{81}Br, 192 – ^{79}Br), 111 (192 – ^{81}Br, 190 – ^{79}Br) and 75 (113 – H^{37}Cl, 111 – H^{35}Cl).

6.3.5 METHODS LEADING TO POLYVALENT IODINE COMPOUNDS

Aryl iodides are exceptional among the aromatic halogen compounds in that they form a series of derivatives in which the iodine exhibits a covalency greater than one. Some typical interconversions for which experimental conditions are given (Expts 6.31 to 6.36) are summarised below.

Two polyvalent iodine compounds arising from o-iodobenzoic acid have been found to be useful synthetic reagents. Thus oxidation of the iodo acid (7) with potassium persulphate, followed by the addition of benzene and treatment with potassium iodide gives the iodonium iodide (8) which is converted into diphenyliodonium-2-carboxylate [DPIC (9)] with aqueous alkali (Expt 6.35).[16]

(8) (9)

(10)

DPIC has been used as a benzyne precursor[17] into which it decomposes at 160–220 °C. At lower temperatures (80–100 °C) and in the presence of copper(II) acetate, it reacts with amino and phenolic nucleophiles to give *ortho*-substituted benzoic acids.[16]

The periodinane (10) may also be prepared from *o*-iodobenzoic acid by oxidation with potassium bromate and then treatment with acetic anhydride[18a] (see Expt 6.36 for detailed formulation). It should be noted that the organic derivatives of pentacoordinate iodine(v) are termed periodinanes.[18b] This compound (the systematic name is 1,1,1-triacetoxy-2,1-benzoxiodol-3(3*H*)-one) has found use as an oxidant of primary alcohols to aldehydes and alicyclic secondary alcohols to ketones; it is claimed to have advantages over the chromium-based oxidation reagents.

Experiment 6.31   (DICHLOROIODO)BENZENE (*Iodobenzene dichloride*)

$$PhI + Cl_2 \longrightarrow PhICl_2$$

Equip a 500-ml three-necked flask with a mechanical stirrer, an adjustable inlet tube at least 10 mm in diameter for the introduction of chlorine (see Section 4.2.*15*, p. 424) and an outlet tube carrying a calcium chloride guard-tube. Charge the flask with 75 ml of chloroform (dried with anhydrous calcium chloride: see Section 4.1.*6*, p. 399) (**CAUTION**) and 51 g (0.25 mol) of iodobenzene (Expt 6.24); adjust the inlet tube so that it terminates about 5 mm above the surface of the liquid. Set up the apparatus in the fume cupboard and protect it from the light. Cool the flask in an ice–salt mixture and pass in dry chlorine as rapidly as the solution will absorb it until an excess is present (1.5–2 hours). Filter the yellow, crystalline iodobenzene dichloride at the pump, wash it sparingly with chloroform and dry it in the air upon filter paper. The yield is 65 g (93.5%). The substance decomposes slowly upon standing; it may be kept unchanged for a short period in a well-fitting, ground glass stoppered bottle.

Experiment   6.32   IODOSYLBENZENE (*Iodosobenzene*)   **AND IODOBENZENE DIACETATE** [(*Diacetoxyiodo)benzene*]

$$PhICl_2 \xrightarrow{2NaOH} PhIO \xrightarrow{2Me \cdot CO_2H} PhI(O \cdot CO \cdot Me)_2$$

Cool a large glass mortar in ice and then place in it 50 g of anhydrous sodium carbonate, 55 g (0.2 mol) of iodobenzene dichloride (Expt 6.31) and 100 g of finely crushed ice. Grind the mixture thoroughly until all the ice has melted and a thick paste results. Add 140 ml of 5 M sodium hydroxide in 20 ml portions and triturate vigorously after each addition; finally add 120 ml of water and allow to stand overnight. Filter with suction, press well with a large flat glass stopper on the filter, transfer to a beaker and stir with 300 ml of water (1). Filter again at the pump, transfer again to a beaker containing 300 ml of water, filter and wash with about 200 ml of water on the filter. Dry in the air upon filter papers, stir with a little chloroform (to dissolve a little iodobenzene which is present), filter with suction and dry on filter paper in the air. The yield is 27 g (61%).

Iodosobenzene explodes violently at about 220 °C, so that determinations of the melting point should not be attempted. It may, however, be converted into *iodobenzene diacetate* in the following manner. Dissolve 2 g of iodoso-benzene in 6 ml of glacial acetic acid; boiling is usually necessary. Cool. The resulting diacetate is readily soluble in acetic acid but is insoluble in ether. Add about 50 ml of ether in order to precipitate the iodobenzene diacetate. Filter and wash with ether. The yield is 2 g, m.p. 157 °C. It may be recrystallised from benzene, and will keep indefinitely (unlike the iodoben-zene dichloride).

Note. (1) The filtrate contains some diphenyliodonium salts; these may be recovered as the sparingly soluble diphenyliodonium iodide (about 8 g) (Expt 6.34) by the addition of potassium iodide.

Experiment 6.33 IODYLBENZENE (*Iodoxybenzene*)

$$2PhIO \xrightarrow{heat} PhIO_2 + PhI$$

Fit up a 1-litre round-bottomed flask for steam distillation (Fig. 2.102) and place in it 22 g (0.1 mol) of iodosylbenzene (Expt 6.32) made into a thin paste with water (1). Steam distil until almost all the iodobenzene has been removed (about 9 g); cool the residue in the flask at once, filter the white solid with suction and dry in the air. Wash it with a little chloroform, filter with suction and dry in the air upon filter paper. The yield is 10.5 g (89%). It may be recrystallised from 800–900 ml of water. Iodylbenzene melts with explosive decomposition at 237 °C.

Note. (1) Iodosylbenzene when heated directly may decompose with explosive violence, particularly when dry.

Experiment 6.34 DIPHENYLIODONIUM IODIDE

$$PhIO + PhIO_2 \longrightarrow [Ph_2\overset{\oplus}{I}]\overset{\ominus}{IO_3} \xrightarrow{KI} [Ph_2\overset{\oplus}{I}]I^{\ominus}$$

Grind together 12 g (0.05 mol) of iodylbenzene (Expt 6.33), and 11 g (0.05 mol) of iodosylbenzene (Expt 6.32) with 25 ml of water, add 100 ml of 1 M sodium hydroxide solution and stir for 24 hours in a 1-litre vessel. Dilute with 500 ml of cold water, stir thoroughly, allow to settle and decant the supernatant solution of diphenyliodonium iodate, through a fluted filter paper. Extract the solid residue with two 250 ml portions of water, and decant the extract through a fluted filter paper: a small tarry residue remains. To the combined filtrates add an aqueous solution containing 10 g of potassium iodide. Allow the bulky white precipitate of diphenyliodonium iodide to stand for 1.5 hours with occasional shaking, and then filter it with suction. Dry on a porous tile. The yield is 15 g (74%). The product melts at 173–175 °C with vigorous decomposition.

Experiment 6.35 DIPHENYLIODONIUM-2-CARBOXYLATE[16]

Potassium persulphate (42.0 g, 0.156 mol) is added at 10 °C over 40 minutes to a solution of o-iodobenzoic acid (20.0 g, 0.0807 mol) in 80 ml of concentrated sulphuric acid, and the mixture is kept at this temperature for an additional 20 minutes. Benzene (75 ml) **(CAUTION)** is added, and the mixture is stirred for 3 hours at 25 °C and then poured on to ice. A small amount of 2-carboxydiphenyliodonium bisulphate is collected. The addition of 40 g of potassium iodide in a minimum of water precipitates the iodonium iodide. The salts are combined and stirred with 200 ml of 5 M sodium hydroxide, and the product (DPIC) is collected and washed with water. Recrystallisation from c. 350 ml of water gives 20.1 g (77%) of DPIC as coarse prisms (not hydrated), m.p. 223–225 °C.

Experiment 6.36 1,1,1-TRIACETOXY-2,1-BENZOXIODOL-3(H)-ONE
(A periodinane)[18a]

2-Iodoxybenzoic acid. Potassium bromate (76.0 g, 0.45 mol) is added over a half-an-hour period to a vigorously stirred mixture of o-iodobenzoic acid (85.2 g, 0.34 mol) and 730 ml of 0.73 M sulphuric acid. During the addition the reaction mixture is kept below 55 °C. The mixture is warmed to 65 °C and stirred for 3.6 hours. Cooling to 0 °C, filtering, and washing with 1000 ml of water and two 50-ml portions of ethanol give the product (89.1 g, 93%).

The periodinane. A stirred slurry of the foregoing compound (25.0 g, 0.089 mol) in acetic anhydride (84.0 g, 0.83 mol) and acetic acid (70 ml) is heated to 100 °C. After 40 minutes the solid will dissolve. The solvent is removed under vacuum at room temperature until a thick slurry remains.

Filtering the slurry in an inert atmosphere and washing with 180 ml of ether gives the product (35.1 g), 93 per cent in overall yield from *o*-iodobenzoic acid. This compound has an indefinite shelf-life when stored in a sealed container.

6.4 AROMATIC SULPHONIC ACIDS AND THEIR DERIVATIVES

The following structures indicate the range of aromatic sulphonic acids (1), sulphinic acids (2), sulphonyl chlorides (3), sulphonamides (4) and sulphonate esters (5) to be discussed below.

The reaction processes described in this section are as follows.

1. The preparation of arylsulphonic acids by direct sulphonation (Expts 6.37 to 6.40).
2. The preparation of arylsulphonyl chlorides (Expt 6.41).
3. The preparation of arylsulphonamides from arylsulphonyl chlorides (Expts 6.42 to 6.45).
4. The preparation of arylsulphonate esters from arylsulphonyl chlorides (Expt 6.46).
5. Reduction products from arylsulphonyl chlorides (Expt 6.47).

SUMMMARY OF RETROSYNTHETIC STRATEGIES

C—S Disconnection at the aryl nucleus (methods 1 and 2), e.g.

S—N and S—O Disconnections within the functional group (methods 3 and 4), e.g.

Functional group interconversion (FGI) (method 5), e.g.

SH   SO$_2$H   SO$_2$Cl

(TM)   (TM)

SPECTROSCOPIC FEATURES

The strong characteristic *i.r.* absorption of the SO$_2$ (or SO) group exhibited by all these compounds is clearly apparent in the spectrum of toluene-*p*-sulphonamide (Fig. 3.40). In addition, the absorption arising from the presence of the OH, Cl, NH$_2$ or OR groups is usually easily assigned. The confirmation of aromatic substitution patterns by inspection of the *p.m.r.* spectra is described in the preparative examples below, wherein the fragmentation patterns observable in the *m.s.* are also discussed.

6.4.1 DIRECT SULPHONATION

Aromatic hydrocarbons may be monosulphonated by heating with a slight excess of concentrated sulphuric acid; for benzene, oleum (7–8% SO$_3$) gives somewhat better results. The reaction is usually complete when all the hydrocarbon has dissolved.

$$ArH + H_2SO_4 \rightleftharpoons Ar \cdot SO_3H + H_2O$$

The mechanism of aromatic sulphonation is broadly analogous to that previously described for aromatic nitration and halogenation and may be represented in the following way, the neutral sulphur trioxide molecule functioning as the electrophilic species. Sulphonation differs from nitration and halogenation, however, in that the overall reaction is reversible.

$$2H_2SO_4 \rightleftharpoons SO_3 + H_3O^\oplus + HSO_4^\ominus$$

mesomeric

Because of their high solubility in water the sulphonic acids are not usually isolated from aqueous solution in the free state, but are converted into and isolated as their sodium salts. The simplest procedure is to partially neutralise the reaction mixture (say, with sodium hydrogen carbonate) and then to pour it into water and add an excess of sodium chloride, when the following equilibrium is established.

$$Ar \cdot SO_3H + NaCl \rightleftharpoons Ar \cdot SO_3Na + HCl$$

The high sodium ion concentration results in crystallisation of the sodium salt. This process of salting out with common salt may be used for recrystallisation, but, for example, sodium benzenesulphonate (and salts of other acids of comparable molecular weight) is so very soluble in water that the solution

873

must be almost saturated with sodium chloride, and consequently the product is likely to be contaminated with it. In such a case a pure product may be obtained by crystallisation from, or Soxhlet extraction with, absolute ethanol; the inorganic salts are almost insoluble, but the sulphonate salt is slightly soluble and may be recovered by evaporation of the ethanol. Very small amounts of sulphones are formed as by-products, but since these are insoluble in water, they separate when the reaction mixture is poured into water.

$$Ar \cdot SO_3H + ArH \longrightarrow Ar \cdot SO_2 \cdot Ar + H_2O$$

The sulphonation of toluene (Expt 6.37) with concentrated sulphuric acid at 100–120 °C results in the formation of toluene-p-sulphonic acid as the chief product, accompanied by small amounts of the *ortho* and *meta* isomers; these are easily removed by crystallisation of the sodium salt of the *para* isomer in the presence of sodium chloride. Sulphonation of naphthalene at about 160 °C yields largely the 2-sulphonic acid (the product of thermodynamic control) (Expt 6.38); at lower temperatures (0–60 °C) the 1-sulphonic acid (the product of kinetic control) is produced almost exclusively. In both cases the product is isolated as its sodium salt. In anthraquinone the carbonyl groups deactivate the aromatic nucleus towards electrophilic attack and vigorous conditions of sulphonation are required, i.e. oleum at about 160 °C. The product is largely sodium anthraquinone-2-sulphonate (Expt 6.39).

The free sulphonic acids [e.g. toluene-p-sulphonic acid, for preparation see, Expt 6.37, *Note* (1)], as opposed to their sodium salts, may sometimes be obtained directly if the sulphonation reaction is carried out with continuous removal of the water formed in the reaction, conveniently by using a Dean and Stark water separator. p-Xylene-2-sulphonic acid (Expt 6.40) is an example of a sulphonic acid whose solubility in water is such that it crystallises directly from the reaction medium and hence it may readily be isolated.

Examples of the sulphonation of functionally substituted aromatic compounds are given in Expt 6.37, cognate preparation of sodium p-bromobenzenesulphonate; Expt 6.63, sulphanilic acid; and Expt 6.104, as an intermediate in the preparation of 2-nitroresorcinol.

Experiment 6.37   SODIUM TOLUENE-p-SULPHONATE AND TOLUENE-p-SULPHONIC ACID

$$Ph \cdot Me + H_2SO_4 \longrightarrow p\text{-}Me \cdot C_6H_4 \cdot SO_3H \xrightarrow{NaCl} p\text{-}Me \cdot C_6H_4 \cdot SO_3^{\ominus}Na^{\oplus}$$

Into a 500-ml three-necked flask, provided with a sealed mechanical stirrer and a reflux condenser, place 60 g (69 ml, 0.65 mol) of pure toluene (Section 4.1.3, p. 398) and 60 g (33 ml) of concentrated sulphuric acid. Heat the mixture, with stirring, in an oil bath maintained at 110–120 °C. When the toluene layer has disappeared (c. 1 hour), allow the reaction mixture to cool to room temperature. Pour it with stirring into 250 ml of cold water; filter from any solid substance which may separate. Partly neutralise the acid solution by adding cautiously and in small portions 30 g of sodium hydrogen carbonate. Heat the solution to boiling and saturate it with sodium chloride (about 100 g of salt are required), filter hot through a hot water funnel, or through a Buchner funnel previously warmed to about 100 °C. Transfer the hot filtrate to a beaker and cool the solution, with stirring, in ice. Filter the crystals at the pump (rinse any residual crystals out of the beaker with a little of the filtered mother-liquor), press well with a large glass stopper and wash with 30 ml of saturated salt solution. To recrystallise the crude sodium toluene-*p*-sulphonate, dissolve it in 200–250 ml of water, heat to boiling, saturate with salt, allow to cool somewhat, stir with 2–3 g of decolourising charcoal (if the solution is coloured) and filter the hot solution with suction through a previously warmed Buchner funnel. Transfer the warm filtrate to a beaker and cool in ice: collect the sulphonate with suction on a Buchner funnel, wash it with 20 ml of saturated sodium chloride solution, press well and finally wash with a little alcohol. Dry the hydrated crystals in air upon filter papers, powder in a mortar and then dry in an oven or in an air oven at 100–110 °C. The yield of anhydrous sodium toluene-*p*-sulphonate is 50 g (40%). It still contains traces of sodium chloride and other salts; these can be removed by recrystallisation from rectified spirit (1 g of solid to about 40 ml of alcohol) or by extraction with boiling alcohol in a Soxhlet apparatus (Figs 2.96 and 2.97).

Note. (1) Toluene-*p*-sulphonic acid may be prepared by gently boiling a mixture of 87 g (100 ml, 0.95 mol) of pure toluene and 37 g (20 ml) of concentrated sulphuric acid (92% H_2SO_4 by weight) in a 250-ml bolt-head flask fitted with a Dean and Stark tube (Fig. 2.31(*a*)), which should be filled such that a further 9 ml of water may be collected (4–5 hours). After this period of heating add 6.3 ml of water to the cold contents of the flask, and when crystallisation is complete, filter on a sintered-glass funnel. Dissolve the solid (47 g) in 22 ml of water, saturate with hydrogen chloride gas and allow the solution to stand for several hours. Filter the product of toluene-*p*-sulphonic acid and dry in a desiccator (KOH and $CaCl_2$ as absorbents). The yield is 35 g (22%), m.p. 105–106 °C (sealed tube). Note the i.r. absorption (KBr disc) at *c.* 3350 (OH), 3050 (C_{AR}—H), 2900 cm (C_{AL}—H), multiple bands around 1250–1160 and 1080–1000 cm^{-1} (SO_2 asym. and sym. stretching respectively), and at *c.* 685 cm^{-1} (S—O stretching). The SO_2 stretching vibration is observed at lower frequencies in sulphonic acids compared to sulphonamides due, it is thought, to hydrogen bonding.

Cognate preparation. *Sodium p-bromobenzenesulphonate.* Equip a 500-ml three-necked flask with a separatory funnel, an unsealed stirrer guide fitted with a mechanical stirrer and a thermometer. Place 75 g (40 ml) of fuming sulphuric acid, *d* 1.88 (7–8% SO_3), in the flask and 40 g (27 ml, 0.25 mol) of bromobenzene (Expt 6.23) in the separatory funnel. Add the bromobenzene in small portions so that the temperature does not rise above 100 °C. If any bromobenzene remains unattacked, warm the mixture on a water bath until all of it has passed into solution. Allow to cool, and pour the reaction mixture in a thin stream with stirring into 140 ml of cold water. If a precipitate separates (dibromodiphenylsulphone, a by-product), filter the warm solution at the pump. Add 55 g of sodium chloride to the filtrate and heat (with stirring) until the salt dissolves. Cool the solution rapidly with stirring, filter the separated crude sulphonate at the pump and press the crystals as dry as possible. Upon drying in the air, the yield is 47 g. To purify the crude sodium *p*-bromobenzenesulphonate, powder the crystals in a mortar, transfer to a beaker, add 75 ml of a filtered, saturated solution of sodium chloride, stir, heat on a water bath for 30 minutes, allow to cool, filter and press the crystals as dry as possible; finally wash with a little alcohol. Dry in the air by spreading upon filter papers. The yield of purified sodium *p*-bromobenzene-sulphonate is 45 g (68%). The product, although pure enough for most practical purposes, contains traces of sodium chloride and other salts: these can be removed either by recrystallisation from hot rectified spirit (1 g of salt requires *c*. 25 ml of alcohol) or, more economically, by extraction with alcohol in a Soxhlet apparatus (Figs 2.96 and 2.97).

Experiment 6.38 SODIUM NAPHTHALENE-2-SULPHONATE

$$C_{10}H_8 + H_2SO_4 \longrightarrow 2\text{-}C_{10}H_7 \cdot SO_3H \xrightarrow[-HCl]{NaCl} 2\text{-}C_{10}H_7 \cdot SO_3^{\ominus}Na^{\oplus}$$

Equip a 500-ml three-necked flask with a separatory funnel, a thermometer with its bulb about 2 cm from the bottom and a mechanical stirrer; the bearing for the stirrer consists of a glass tube lubricated with a little glycerine. Place 100 g (0.78 mol) of naphthalene in the flask and heat it either in an air bath or by means of a free flame. When the naphthalene melts, start the stirrer and adjust the heating so that the temperature is $160 \pm 5\,°C$. Run in 166 g (90 ml) of concentrated sulphuric acid from the funnel during 5–6 minutes: take care to maintain the temperature at $160\,°C$ and remove the flame if necessary. Stir for 5 minutes and pour the solution into 750 ml of cold water. If the sulphonation has been properly conducted, there will be no precipitate of naphthalene but about 4 g of insoluble di-2-naphthyl sulphone may separate. Boil with 3–4 g of decolourising carbon and filter with suction through a Buchner funnel. Partly neutralise the clear solution by carefully adding 40 g of sodium hydrogen carbonate in small portions. Heat the solution to the boiling point, saturate with sodium chloride (about 70 g are required) and then set aside to crystallise. Filter the crude sodium naphthalene-2-sulphonate at the pump and recrystallise from hot 10 per cent sodium chloride solution; dry by heating on a water bath or in an oven. The yield is 140 g (78%).

Experiment 6.39 SODIUM ANTHRAQUINONE-2-SULPHONATE

Place 50 g of fuming sulphuric acid (40–50% SO_3, Section 4.2.75, p. 465) in a 250- or 500-ml two-necked flask equipped with a thermometer with the bulb within 2 cm of the bottom and add 50 g (0.24 mol) of dry, finely powdered anthraquinone (Expt 6.128). Fit an air condenser to the flask and heat the mixture slowly in an oil bath, with occasional shaking, so that at the end of 1 hour the temperature has reached $160\,°C$. Allow to cool and pour the warm mixture carefully into a 2-litre beaker containing 500 g of crushed ice. Boil for about 15 minutes and filter off the unchanged anthraquinone at the pump. Neutralise the filtrate by carefully adding concentrated (50% w/w) aqueous sodium hydroxide and allow to cool, when the greater part of the sodium anthraquinone-2-sulphonate separates as silvery glistening plates ('silver salt'). Filter these with suction and dry upon filter paper or upon a porous plate. The yield is 40–45 g (54–60%).

Experiment 6.40 **2,5-DIMETHYLBENZENESULPHONIC ACID**
(p-*Xylene-2-sulphonic acid*)

$$p\text{-Me·C}_6\text{H}_4\text{·Me} \longrightarrow 2,5\text{-(Me)}_2\text{C}_6\text{H}_3\text{·SO}_3\text{H}$$

Place 5.2 g (6.0 ml, 0.05 mol) of p-xylene in a 25- or 50-ml round-bottomed flask and add, with gentle swirling agitation, 10 ml of concentrated sulphuric acid. Heat the mixture on a water bath for 10–15 minutes; remove the flask from the bath and mix the contents with a circular motion every two minutes. The reaction is complete when the xylene layer on the surface of the acid has disappeared. Cool to room temperature, add 5.0 ml of water cautiously with gentle swirling. Pour the warm reaction mixture into a 100-ml beaker and cool in ice. Filter off the crystalline solid with suction on a sintered glass funnel and press the crystals down with a glass stopper. Recrystallise the crude product from 5 ml of water and dry on filter paper. The yield of pure 2,5-dimethylbenzenesulphonic acid, m.p. 82 °C, is 8.2 g (90%).

After drying the crystals over phosphorus pentoxide, record the i.r. spectrum (KBr disc) and assign the principal absorptions. Record the p.m.r. spectrum in deuterium oxide using DSS as reference; note the solvent signal (HOD) at δ 5.45 and signals at δ 2.31 (s, 3H, C_5—Me), 2.69 (s, 3H, C_2—Me), 7.20 (sharp m, 2H, $C_{3,4}$—H), and 7.82 (broad s, 1H, C_6—H).

6.4.2 THE PREPARATION OF ARYLSULPHONYL CHLORIDES

Arylsulphonic acids, either free or in the form of their sodium salts, are converted into the acid chloride by reaction with phosphorus pentachloride (or phosphorus oxychloride).

$$3\text{Ar·SO}_3\text{Na} + \text{PCl}_5 \xrightarrow{170-180\,°C} 3\text{Ar·SO}_2\text{Cl} + \text{NaPO}_3 + 2\text{NaCl}$$

The arylsulphonyl chlorides may be obtained more conveniently from the aromatic hydrocarbon by reaction with an excess of chlorosulphonic acid (*chlorosulphonation*)

$$\text{ArH} + 2\text{Cl·SO}_3\text{H} \xrightarrow{20-25\,°C} \text{Ar·SO}_2\text{Cl} + \text{H}_2\text{SO}_4 + \text{HCl}$$

The latter general procedure is illustrated by the chlorosulphonation of toluene which yields a mixture of toluene-o- and p-sulphonyl chlorides which may be separated by cooling to −10 to −20 °C when most of the *para* isomer, which is a solid, m.p. 69 °C, separates out (Expt 6.41).

Experiment 6.41 **TOLUENE-o AND p-SULPHONYL CHLORIDES**

$$\text{Ph·Me} \xrightarrow{2\text{Cl·SO}_3\text{H}} o\text{- and } p\text{-Me·C}_6\text{H}_4\text{·SO}_2\text{Cl} + \text{H}_2\text{SO}_4 + \text{HCl}$$

In a 750-ml three-necked flask equipped as in Fig. 6.2 place 400 g (228 ml, 3.44 mol) of chlorosulphonic acid (**CAUTION**: see Section 4.2.*16*, p. 424) and cool to 0 °C in a freezing mixture of ice and salt. Introduce 100 g (115 ml, 1.09 mol) of pure dry toluene from the dropping funnel dropwise at such a rate that the temperature of the well-stirred mixture does not rise above 5 °C. When all the toluene has been added (about 3 hours), stir the reaction mixture for 4 hours, and then allow to stand overnight in a refrigerator. Pour

877

To gas
adsorption trap

Fig. 6.2

the liquid on to 1 kg of crushed ice, separate the aqueous solution from the oily layer (mixture of toluene-*o*- and -*p*-sulphonyl chlorides) and wash the latter several times by decantation with cold water. To separate the *ortho* and *para* isomers, cool the oil at −10 to −20 °C (e.g. with ice and calcium chloride) for several hours; the almost pure toluene-*p*-sulphonyl chloride will crystallise out. Filter at the pump upon a sintered glass funnel. The toluene-*p*-sulphonyl chloride (30 g, 14%) may be purified by recrystallisation from light petroleum (b.p. 40–60 °C) and then melts at 69 °C. The filtrate consists largely of toluene-*o*-sulphonyl chloride: it may be obtained pure by dissolving it in carbon tetrachloride, removing the solvent and fractionating under reduced pressure; it is an oil, b.p. 126 °C/10 mmHg. The yield is about 120 g (58%).

The i.r. spectra of these two isomers may be recorded as a KBr disc (*para*) and as a thin film (*ortho*); both spectra exhibit strong absorption at *c.* 1380 and 1170 cm^{-1} arising from the stretching of the sulphur–oxygen double bonds. The p.m.r. spectrum (CDCl$_3$, TMS) of the *para* isomer confirms the aromatic substitution pattern, having signals at δ 2.49 (s, 3H, Me), 7.41 (d, 2H, *ortho*-H's to Me) and 7.90 (d, 2H, *ortho*-H's to SO$_2$Cl). The *ortho* isomer has δ 2.79 (s, 3H, Me), 7.21–7.73 (m, 3H, C$_{4,5,6}$—H), and 8.0 (m, 1H, C$_3$—H). The m.s. of the *ortho* isomer has principal fragment ions at *m/z* 192 (M, ^{37}Cl), 190 (M, ^{35}Cl), 155 (M − ^{37}Cl or ^{35}Cl), 91 (M − SO$_2$Cl, base peak), and 65 (91 − C$_2$H$_2$).

Cognate preparation. *2,4,6-Triisopropylbenzenesulphonyl chloride.*[19] To a mixture of triisopropylbenzene (100 ml) in chloroform (500 ml), magnetically stirred at 0 °C, is added chlorosulphonic acid (120 ml) within 15 minutes under cooling and exclusion of moisture (**CAUTION**). The stirring is continued for 45 minutes at room temperature and the mixture poured on to crushed ice. The product is extracted with chloroform (500 ml) (**CAUTION**), dried over anhydrous sodium sulphate and freed from solvent *in vacuo* at 30 °C. The white residue is dissolved in warm pentane, and the solution is filtered and concentrated to about 200 ml. Triisopropylbenzenesulphonyl chloride crystallises slowly on standing and is washed with pre-chilled ($-30\,°C$) pentane; after drying *in vacuo* over phosphorus pentoxide the yield is 75.1 g, m.p. 96–97 °C. Further concentration of the mother-liquor gives a second crop, 31.4 g, m.p. 94–96 °C. Total yield 106.5 g (86%). The product is recrystallised from pentane and dried over phosphorus pentoxide *in vacuo*.

6.4.3 THE PREPARATION OF ARYLSULPHONAMIDES

Arylsulphonyl chlorides may be readily converted into the corresponding arylsulphonamides (e.g. by treatment with solid ammonium carbonate or with concentrated ammonia solution) or into a substituted arylsulphonamide (by reaction with the appropriate amino compound). Arylsulphonylhydrazides (for use in the preparation of aliphatic nitriles, Section 5.13.2, p. 712) may be prepared by a similar reaction using anhydrous hydrazine.[20]

$$Ar\cdot SO_2Cl + NH_3 \longrightarrow Ar\cdot SO_2NH_2 + HCl$$

$$Ar\cdot SO_2Cl + RNH_2 \longrightarrow Ar\cdot SO_2NHR + HCl$$

$$Ar\cdot SO_2Cl + N_2H_4 \longrightarrow Ar\cdot SO_2NH\cdot NH_2 + HCl$$

These sulphonamides are highly crystalline, and in particular the derivatives from benzenesulphonyl chloride and toluene-*p*-sulphonyl chloride are used in the separation and identification of aromatic amines (Sections 9.6.21, p. 1273 and 9.7.1, p. 1287).

Toluene-*o*-sulphonamide is an intermediate in the synthesis of saccharin (Expt 6.42). Upon oxidising toluene-*o*-sulphonamide with potassium permanganate in alkaline solution, the sodium salt of *o*-sulphonamidobenzoic acid is formed, which on acidification with concentrated hydrochloric acid passes spontaneously into the cyclic imide of *o*-sulphobenzoic acid or saccharin. Saccharin itself is sparingly soluble in cold water, but the imino hydrogen is acidic and the compound forms a water-soluble sodium salt. The latter is about 500 times as sweet as cane sugar.

When toluene-*p*-sulphonamide is dissolved in excess calcium hypochlorite solution and then acidified with acetic acid, the *N,N*-dichloro derivative [(6) dichloramine-T] separates rapidly. When this is heated with sodium hydroxide solution the sodium salt of the *N*-monochloro derivative [(7) chloramine-T)] is formed and crystallises out on cooling at a suitable concentration (Expt 6.43).

$$Ar\cdot SO_2NH_2 + {}^{\ominus}OCl \longrightarrow Ar\cdot SO_2\overset{\ominus}{\ddot{N}}Cl + H_2O$$

$$Ar\cdot SO_2\overset{\ominus}{N}Cl + {}^{\ominus}OCl + 2H^{\oplus} \longrightarrow Ar\cdot SO_2NCl_2 + H_2O$$

$$(6)$$

$$Ar\cdot SO_2NCl_2 + 2NaOH \longrightarrow Ar\cdot SO_2\overset{\ominus}{N}Cl\}\overset{\oplus}{Na} + NaOCl + H_2O$$

$$(7)$$

Both chloramine-T and dichloramine-T slowly liberate hypochlorous acid in contact with water and are therefore employed as antiseptics; the former is employed in the form of a dilute (e.g. 0.2%) aqueous solution, and the latter (which is insoluble in water) as a solution in an organic solvent, such as a chloroalkane.

The chlorosulphonation of acetanilide with excess chlorosulphonic acid affords mainly *p*-acetamidobenzenesulphonyl chloride (Expt 6.44). This is the essential intermediate in the synthesis of a range of sulphanilamide drugs. The simplest example, *p*-aminobenzenesulphonamide (sulphanilamide), is obtained by converting *p*-acetamidobenzenesulphonyl chloride into the amide with aqueous ammonia, and then selectively removing the protecting acetyl grouping with boiling aqueous hydrochloric acid. 2-(*p*-Aminobenzenesulphonamido)-pyridine (sulphapyridine, M and B 693, Expt 6.45) is prepared by reacting *p*-acetamidobenzenesulphonyl chloride with 2-aminopyridine and then removing the acetyl group as before. The synthesis of the required 2-aminopyridine is of interest in that it represents an example of a nucleophilic aromatic substitution by the amide ion at the 2-position in a pyridine ring system, under the influence of the powerful electron-withdrawing influence of the heteroatom.

Experiment 6.42 TOLUENE-*o*-SULPHONAMIDE AND SACCHARIN

Place 20 g (0.105 mol) of toluene-*o*-sulphonyl chloride (Expt 6.41) in a large evaporating dish mounted on a water bath. Add powdered ammonium carbonate cautiously with stirring until the mass is quite hard and solid and the unpleasant odour of the sulphonyl chloride has disappeared. Allow to cool, and extract with cold water to remove the excess of ammonium carbonate. Recrystallise the crude toluene-*o*-sulphonamide first from hot water (add a little decolourising carbon if it is dark in colour) and then from ethanol. The yield of pure product, m.p. 154 °C, is 16 g (89%).

Record the i.r. spectrum and compare it with that of the *para* isomer shown on p. 317. The p.m.r. spectrum (DMSO-d_6 at 60 °C, TMS) shows signals at δ 2.65 (s, 3H, Me), 7.21 (broad s, 2H, NH$_2$), 7.25–7.50 (m, 3H, C$_{4,5,6}$—H) and 7.91 (m, 1H, C$_3$—H); this spectrum should be contrasted with that of the *para* isomer detailed below. The m.s. shows principal fragment ions at m/z 171 (M), 155 (M − NH$_2$), 91 (M − SO$_2$NH$_2$ or 155 − SO$_2$), and 90 [M − (SO$_2$ + NH$_3$), base peak].

Oxidation of toluene-*o*-sulphonamide to saccharin.   In a 600-ml beaker, mounted on an electric hot plate and provided with a mechanical stirrer, place 12 g (0.07 mol) of toluene-*o*-sulphonamide, 200 ml of water and 3 g of pure sodium hydroxide. Stir the mixture and warm to 34–40 °C until nearly all has passed into solution (about 30 minutes). Introduce 19 g (0.12 mol) of finely powdered potassium permanganate in small portions at intervals of 10–15 minutes into the well-stirred liquid. At first the permanganate is rapidly reduced, but towards the end of the reaction complete reduction of the permanganate is not attained. The addition occupies 4 hours. Continue the stirring for a further 2–3 hours, and then allow the mixture to stand overnight. Filter off the precipitated manganese dioxide at the pump and decolourise the filtrate by the addition of a little sodium metabisulphite solution. Exactly neutralise the solution with dilute hydrochloric acid (use methyl orange or methyl red as external indicator). Filter off any *o*-sulphonamidobenzoic acid (and/or toluene-*o*-sulphonamide) which separates at this point. Treat the filtrate with concentrated hydrochloric acid until the precipitation of the saccharin is complete. Cool, filter at the pump and wash with a little cold water. Recrystallise from hot water. The yield of pure saccharin, m.p. 228 °C, is 7.5 g (58%).

Conversion of saccharin into pseudosaccharin chloride.   Mix intimately in a glass mortar 35 g (0.19 mol) of saccharin and 70 g (0.336 mol) of phosphorus pentachloride (**CAUTION**), transfer to a 250-ml round-bottomed flask connected by a ground glass joint to a reflux condenser; attach the latter through a calcium chloride guard-tube to a gas absorption trap. Heat the mixture in an oil bath at 175–180 °C for 90 minutes; at the end of this period the vigorous evolution of hydrogen chloride will have subsided. Replace the reflux condenser by a fractionating column, distil off the phosphorus oxychloride and pour the warm residue upon finely crushed ice. Extract the crude solid pseudosaccharin chloride with chloroform, dry the chloroform solution with anhydrous magnesium sulphate and distil off the solvent. Recrystallise the residue from chloroform or from dry benzene. The yield of pure pseudosaccharin chloride, m.p. 143–145 °C (decomp.), is 26 g (67.5%). It is best kept in a sealed glass tube or in a glass-stoppered bottle.

Cognate preparation.   *2,4,6-Triisopropylbenzenesulphonyl hydrazide.*[20]   A magnetically stirred solution of 2,4,6-triisopropylsulphonyl chloride (30.3 g, 0.1 mol) (Expt 6.41), in tetrahydrofuran (60 ml), is cooled to − 10 °C in an ice–salt freezing mixture and anhydrous hydrazine hydrate (10.2 g, 0.22 mol, 100%) (**CAUTION**) is added dropwise over a period of 15 minutes. The temperature rises to 0 °C, and after stirring at 0 °C for 3 hours, water (*c.* 2 ml) is added dropwise to dissolve the precipitated solids. The product is then transferred to a separatory funnel and the lower aqueous layer discarded. The organic layer is washed with ice-cold brine (3 × 20 ml), dried (anhydrous sodium sulphate) for 3 hours, filtered through hyflosupercel and then concentrated under reduced pressure below room temperature. Light petroleum (100 ml, b.p. 30–40 °C) is added to the crystalline mass so obtained. The material is collected by filtration, washed several times with light petroleum (b.p. 30–40 °C) and any remaining solvent removed by evaporation under reduced pressure at room temperature. The product is triturated with ice-cold water (3 × 100 ml), and finally dried *in vacuo* over phosphorus pentoxide at room temperature for 24 hours to give 2,4,6-triisopropylbenzenesulphonyl hydrazide (TBSH) as a colourless solid, m.p. 118–120 °C (decomp.); yield 28.9 g (96%) (1), R_F 0.7 (chloroform–methanol, 9:1).

Note. (1) TBSH should be stored in a dark bottle preferably at 4 °C. Under these conditions decomposition occurs to only a small extent (5% in 9 months). The purity of TBSH may be assessed by t.l.c. analysis (CHCl$_3$–MeOH, 9:1 v/v). Impure TBSH may be purified as follows: TBSH (60 g) is dissolved in ether (600 ml) and the solution washed first with aqueous sodium hydroxide solution (3 × 50 ml, 0.75 M), and then with water (25 ml). Light petroleum (1 litre, b.p. 30–40 °C) is added to the dried (magnesium sulphate) organic layer. The precipitated TBSH (50 g) is collected by filtration and dried *in vacuo* over phosphorus pentoxide for 24 hours at room temperature.

Experiment 6.43 TOLUENE-*p*-SULPHONAMIDE, DICHLORAMINE-T AND CHLORAMINE-T

$$p\text{-Me·C}_6\text{H}_4\text{·SO}_2\text{Cl} \xrightarrow{\text{NH}_3} p\text{-Me·C}_6\text{H}_4\text{·SO}_2\text{·NH}_2 \xrightarrow[\text{Me·CO}_2\text{H}]{\ominus\text{OCl}}$$

$$p\text{-Me·C}_6\text{H}_4\text{·SO}_2\text{·NCl}_2 \xrightarrow[\text{H}_2\text{O}]{\text{NaOH}} p\text{-Me·C}_6\text{H}_4\text{·SO}_2\text{·}\overset{\ominus}{\text{NCl}}\}\text{Na}^{\oplus}$$

Toluene-*p*-sulphonamide. Grind together 10 g (0.0525 mol) of toluene-*p*-sulphonyl chloride (Expt 6.41) and 20 g of ammonium carbonate in a mortar until a fine uniform powder is obtained. Heat the mixture in an evaporating dish on a water bath for 1–2 hours and stir the mixture frequently with a glass rod. Allow to cool and extract with a little cold water to remove the excess of ammonium salts. Recrystallise the crude toluene-*p*-sulphonamide from boiling water (200–250 ml), and dry the colourless crystals at 100 °C. The yield of pure product, m.p. 138 °C, is 7.9 g (88%).

The i.r. spectrum is given on p. 317. The *para*-substitution pattern is clearly shown by the p.m.r. spectrum (TFA, TMS) having signals at δ 2.46 (s, 3H, Me), *c*. 4.05 (broad low signal, 2H, NH$_2$), 7.37 (d, 2H, *ortho*-H's to Me), and 7.84 (d, 2H, *ortho*-H's to SO$_2$NH$_2$). The m.s. is similar to that of the *ortho* isomer noted above with some variation of relative abundance of individual fragment ions.

Alternatively, grind 10 g of toluene-*p*-sulphonyl chloride to a fine powder and add it to 30 ml of concentrated ammonia solution (*d* 0.88). Heat the mixture to boiling (*fume cupboard*)) and cool. Filter and recrystallise the toluene-*p*-sulphonamide from boiling water (add 1 g of decolourising carbon, if necessary). The yield of pure product, m.p. 138 °C, is almost theoretical.

Dichloramine-T(*N,N*-dichlorotoluene-*p*-sulphonamide). Prepare about 200 ml of a saturated solution of calcium hypochlorite by grinding a fresh sample of bleaching powder with water and filtering with slight suction. Dissolve 5 g (0.029 mol) of toluene-*p*-sulphonamide in as small a volume of the calcium hypochlorite solution as possible (about 150 ml) and filter the solution if necessary. Cool in ice, and add about 50 ml of a mixture of equal volumes of glacial acetic acid and water *slowly* and with stirring until precipitation is complete. The dichloramine-T separates out first as a fine emulsion, which rapidly forms colourless crystals. Filter the latter at the pump, wash with a little cold water, drain and dry immediately either between pads of filter paper or upon a porous tile. The yield is 5.3 g (76%), m.p. 81 °C. Upon recrystallisation from light petroleum (b.p. 60–80 °C) or from chloroform–light petroleum, pure dichloramine-T, m.p. 83 °C, is obtained with negligible loss.

Chloramine-T (sodium *N*-chlorotoluene-*p*-sulphonamide). For this preparation use dichloramine-T which has been prepared as above and thoroughly drained but not necessarily dried. Heat 45 ml of 10 per cent aqueous sodium hydroxide solution in a beaker to a temperature of about 80 °C, add 3.5 g (0.015 mol) of dichloramine-T in small quantities, stirring the mixture gently after each addition until a clear solution is obtained. When the addition is complete, filter the hot solution if turbid, and then allow it to cool spontaneously. Filter the crystals with suction, wash with a little saturated sodium chloride solution and dry upon filter paper or in a desiccator over anhydrous calcium chloride. The resulting chloramine-T weighs 3 g (75%) and is almost pure. It may be recrystallised, if desired, from twice its weight of hot water.

Chloramine-T is a salt and has no definite m.p.: upon heating it loses water of crystallisation and decomposes violently at 175–180 °C.

Experiment 6.44   *p*-AMINOBENZENESULPHONAMIDE
(*Sulphanilamide*)

$$p\text{-Me·CO·NH·C}_6\text{H}_5 \xrightarrow{\text{Cl·SO}_3\text{H}} p\text{-Me·CO·NH·C}_6\text{H}_4\text{·SO}_2\text{Cl} \xrightarrow{\text{NH}_3}$$

$$p\text{-Me·CO·NH·C}_6\text{H}_4\text{·SO}_2\text{NH}_2 \xrightarrow{\text{H}_3\text{O}^{\oplus}} p\text{-NH}_2\text{·C}_6\text{H}_4\text{·SO}_2\text{NH}_2$$

***p*-Acetamidobenzenesulphonyl chloride.** Equip a 500-ml two-necked flask with a dropping funnel and a reflux condenser: attach the top of the latter to a device for the absorption of hydrogen chloride (e.g. Fig. 2.61). Place 20 g (0.148 mol) of dry acetanilide in the flask and 50 ml (90 g, 0.77 mol) of a good grade of chlorosulphonic acid (**CAUTION**: see Section 4.2.*16*, p. 424) in the dropping funnel and insert a calcium chloride guard-tube into the latter. Add the chlorosulphonic acid in small portions and shake the flask from time to time to ensure thorough mixing (1). When the addition has been made, heat the reaction mixture on a water bath for 1 hour in order to complete the reaction. Allow to cool and pour the oily mixture in a thin stream with stirring into 300 g of crushed ice (or ice-water) contained in a 1-litre beaker. Carry out this operation carefully in the fume cupboard since the excess of chlorosulphonic acid reacts vigorously with the water. Rinse the flask with a little ice-water and add the rinsings to the contents of the beaker. Break up any lumps of solid material and stir the mixture for several minutes in order to obtain an even suspension of the granular white solid. Filter off the *p*-acetamidobenzenesulphonyl chloride at the pump and wash it with a little cold water; press and drain well. *Use the crude product (2) immediately in the next stage.*

***p*-Acetamidobenzenesulphonamide.** Transfer the crude *p*-acetamidobenzenesulphonyl chloride to the rinsed reaction flask, and add a mixture of 70 ml of concentrated ammonia solution (*d* 0.88) and 70 ml of water. Mix the contents of the flask thoroughly, and heat the mixture with occasional swirling (*fume cupboard*) to just below the boiling point for about 15 minutes. The sulphonyl chloride will be converted into a pasty suspension of the corresponding sulphonamide. Cool the suspension in ice, and then add dilute sulphuric acid until the mixture is just acid to Congo red paper. Collect the product on a Buchner funnel, wash with a little cold water and drain as completely as possible. It is desirable, but not essential, to dry the crude *p*-acetamidobenzenesulphonamide at 100 °C: the yield is about 18 g. The material is sufficiently pure (3) for the next stage.

p-Aminobenzenesulphonamide. Transfer the crude *p*-acetamidobenzene-sulphonamide to a 500-ml flask, add 10 ml of concentrated hydrochloric acid and 30 ml of water. Boil the mixture gently under reflux for 30–45 minutes. The solution, when cooled to room temperature, should deposit no solid amide; if a solid separates, heat for a further short period. Treat the cooled solution with 2 g of decolourising carbon, heat the mixture to boiling and filter with suction through a hardened filter paper. Place the filtrate (a solution of sulphanilamide hydrochloride) in a litre beaker and cautiously add 16 g of solid sodium hydrogen carbonate in portions with stirring. After the evolution of gas has subsided, test the suspension with litmus paper and if it is still acid, add more sodium hydrogen carbonate until neutral. Cool in ice, filter off the sulphanilamide with suction and dry. The yield is 15 g (59% overall yield), m.p. 161–163 °C. A pure product, m.p. 163–164 °C, may be obtained by recrystallisation from water or from alcohol.

Notes. (1) The reaction may be more easily controlled and the chlorosulphonic acid added all at once if the acetanilide is employed in the form of a hard cake. The latter is prepared by melting the acetanilide in the flask over a free flame and causing the compound to solidify over the lower part of the flask by swirling the liquid. If the reaction becomes too vigorous under these conditions, cool the flask momentarily by immersion in an ice bath.

(2) The crude sulphonyl chloride, even if dry, cannot be kept without considerable decomposition. It may be purified by triturating with dried acetone, filtering any undissolved *p*-acetamidosulphonic acid and evaporating the filtrate. The pure chloride has m.p. 149 °C.

(3) A small portion may be recrystallised from water, with the addition of a little decolourising carbon if necessary. The pure compound has m.p. 218 °C.

Experiment 6.45 2-(*p*-AMINOBENZENESULPHON-AMIDO)PYRIDINE (*Sulphapyridine*)

p-Me·CO·NH·C$_6$H$_4$·SO$_2$Cl +

$\xrightarrow{C_5H_5N}$

p-Me·CO·NH·C$_6$H$_4$·SO$_2$·NH

$\xrightarrow{HCl\ conc.}$

p-H$_2$N·C$_6$H$_4$·SO$_2$·NH

CAUTION: *This preparation should be conducted in an efficient fume cupboard and great care exercised in the handling of pyridine.*

2-Aminopyridine. In a 1-litre four-necked flange flask, equipped with a sealed mechanical stirrer, reflux condenser, thermometer and inlet tube for nitrogen, place 300 ml of dry toluene (1) and 75 g (1.92 mol) of fine granular sodamide (2); bubble a steady stream of nitrogen through the toluene. Stir the mixture vigorously and heat the flask in an oil bath until the internal temperature is 110 °C (the bath temperature required is approximately 130 °C). Add 100 g (1.26 mol) of pure dry pyridine (**CAUTION**: Section 4.1.*29*, p. 410) dropwise through the condenser fitted with a pressure-equalising funnel over a period of 4 hours: maintain the very efficient stirring and the stream of nitrogen. After 1 hour the reaction mixture becomes black in colour, and after 3 hours becomes viscous, and bubbling and slight frothing occurs, due to liberation of hydrogen. When all the pyridine has been introduced, continue the heating for a further 5 hours while maintaining the internal temperature at 110 °C. Towards the end of the reaction, stirring may become difficult owing to the separation of a solid or viscous cake. Allow the reaction mixture to cool (without the stream of nitrogen and without stirring); then introduce 175 ml of water very slowly through the condenser over a period of 2 hours while continuing the passage of the stream of nitrogen. During the addition the temperature rises to about 50 °C; resume the stirring as soon as possible. Transfer the contents of the flask to a separatory funnel, separate the lower aqueous solution and extract it with two 150 ml portions of toluene. Dry the combined main toluene layer and toluene extracts over anhydrous potassium carbonate for 2 hours; filter and remove the toluene by distillation. Distil the syrupy residue from an oil bath under diminished pressure through an air condenser: adjust the bath temperature to 120–130 °C. Collect the 2-aminopyridine at 95 °C/10 mmHg; this solidifies on cooling to a colourless solid, m.p. 55 °C (3). The yield is about 80 g (67%).

Sulphapyridine. Dissolve 4.7 g (0.05 mol) of 2-aminopyridine in a mixture of 40 ml of anhydrous acetone and 6 ml of dry pyridine (**CAUTION**) in a 250-ml flask, and add 11.7 g (0.05 mol) of pure *p*-acetamidobenzenesulphonyl chloride (4). The reaction mixture is set aside overnight and 5.5 g of the almost pure 2-(*p*-acetamidobenzenesulphonamido)pyridine (≡ acetylsulpha-pyridine) is filtered off; by diluting the filtrate with water a further crop (4 g) is obtained. The total product is recrystallised from acetone to give white needles of pure product, m.p. 224 °C. The yield is 8 g (55%).

The acetylsulphapyridine (7.3 g) is hydrolysed by heating it under reflux with 75 ml of ethanol containing 15 ml of concentrated hydrochloric acid for 20 min. The cooled solution is diluted with water and made just alkaline with concentrated ammonia solution (*d* 0.880). The sulphapyridine is isolated by filtration and recrystallised from ethanol; yield 4.9 g (75%), m.p. 190–191 °C.

Notes. (1) Technically pure toluene can be conveniently dried by distilling 350 ml from a litre flask and rejecting the first 50 ml.

(2) It is important to use recently prepared pure sodamide, which must be of fine granular form. Old material of irregular lumpy form, even if ground gives poor results, and should not be employed. The sodamide may be prepared as detailed in Section 4.2.*67*, p. 462. A satisfactory grade is marketed by May and Baker Ltd.

(3) The residue in the flask is said to contain 4-amino- and 2,6-diaminopyridine, 4,4'-bipyridyl and di(2-pyridyl) amine in varying amounts.

(4) The p-acetamidobenzenesulphonyl chloride (Expt 6.44) must be pure: under no circumstances should it contain more than 1–2 per cent of the corresponding sulphonic acid. This may be ensured by lixiviating the sulphonyl chloride with pure anhydrous acetone and filtering the solution from the acid.

6.4.4 THE PREPARATION OF ARYLSULPHONATE ESTERS

These are prepared by the interaction of the arylsulphonyl chloride and an alcohol or phenol in the presence of sodium hydroxide solution or of pyridine (Expt 6.46).

$$ROH + Ar \cdot SO_2Cl \xrightarrow{\text{NaOH}} RO \cdot SO_2Ar + HCl$$

These sulphonate esters cannot be made by direct esterification of the sulphonic acids since further reaction of the ester with the alcohol gives rise to an ether by means of an alkyl-oxygen fission process.

$$R\overset{\cdots}{O}H \quad R\underset{\curvearrowright}{-}O \cdot SO_2Ar \xrightarrow{-H^{\oplus}, +H^{\oplus}} R \cdot O \cdot R + HO \cdot SO_2Ar$$

For this reason the arylsulphonate esters find use as alkylating agents (cf. dimethyl sulphate and dialkyl sulphates, Section 4.2.24, p. 430).

Experiment 6.46 BUTYL TOLUENE-p-SULPHONATE

$$p\text{-Me} \cdot C_6H_4 \cdot SO_2Cl + BuOH \xrightarrow[\text{or pyridine}]{\text{NaOH}} p\text{-Me} \cdot C_6H_4 \cdot SO_2 \cdot OBu$$

Equip a 1-litre three-necked flask with a separatory funnel, a mechanical stirrer and a thermometer, the bulb of which reaches within 2 cm from the bottom. Place 74 g (89 ml, 1 mol) of butan-1-ol and 105 g (0.55 mol) of toluene-p-sulphonyl chloride (Expt 6.41) in the flask and 160 ml of 20 per cent sodium hydroxide solution in the separatory funnel; immerse the flask in a bath of cold water. Run in the sodium hydroxide solution, with stirring, at such a rate that the temperature does not rise above 15 °C (3–4 hours). Now add another portion of 105 g (0.55 mol) of toluene-p-sulphonyl chloride, and introduce 160 ml of 20 per cent sodium hydroxide solution slowly, keeping the temperature below 15 °C. Continue the stirring for 4 hours longer. Separate the oily layer and treat it with enough light petroleum (b.p. 60–80 °C) to cause it to float on water; then wash it well with 25 ml of 10 per cent sodium hydroxide solution, and dry by allowing it to stand over 10 g of anhydrous potassium carbonate. Filter and distil off the solvent using a 250-ml flask (rotary evaporator). Distil the residual ester under reduced pressure (1) (oil pump) and collect the butyl toluene-p-sulphonate at 132–133 °C/3 mmHg. The yield is 130 g (57%).

Record the i.r. spectrum (thin film) and note the strong absorptions at c. 1325 and 1165 cm^{-1} arising from the SO_2-group. The p.m.r. spectrum (CCl$_4$, TMS) shows signals at δ 0.88 (distorted t, 3H, —CH$_2$·Me), 1.08–2.70 (m, 4H, —CH$_2$·CH$_2$—), 2.42 (s, 3H, C$_{AR}$—Me), 3.94 (t, 2H, —OCH$_2$—), 7.27 (d, 2H, ortho-H's to Me), and 7.71 (d, 2H, ortho-H's to SO$_2$).

Note. (1) It is best to distil under greatly reduced pressure; slight decomposition occurs even at 10 mm pressure (b.p. 170–171 °C/10 mmHg).

Cognate preparation. *Dodecyl toluene-p-sulphonate* (*pyridine method*) (*fume cupboard*). In a 500-ml three-necked flask, equipped with a stirrer and thermometer, place 46.5 g (0.25 mol) of dodecan-1-ol (lauryl alcohol), m.p. 22–23 °C, and 79 g (81 ml, 1 mol) of dry pyridine (**CAUTION**). Surround the flask by a bath sufficiently cold to lower the temperature of the mixture to 10 °C. Add 52.5 g of toluene-p-sulphonyl chloride in portions during 20 minutes, or at such a rate that the temperature does not rise above 20 °C. Stir the mixture for 3 hours at a temperature below 20 °C, then dilute with 150 ml of concentrated hydrochloric acid in 500 ml of ice-water. Collect the ester on a chilled Buchner funnel and suck as dry as possible. Transfer the solid to a 400-ml beaker, add 150 of methanol and warm the mixture on a steam bath until the ester melts. Cool in a freezing mixture while stirring vigorously; the ester separates in a finely divided state. Collect it on a chilled funnel and allow to dry in the air, preferably below 20 °C. The yield of ester, m.p. 24–25 °C, is 78 g (92%). Recrystallise by dissolving in 100 ml of light petroleum, b.p. 40–60 °C, drying the solution over magnesium sulphate to remove traces of water, and cool to 0 °C. Collect the pure dodecyl toluene-p-sulphonate, m.p. 29–30 °C, in a chilled funnel (Section 2.20).

The pyridine procedure may be applied to the preparation of other esters; they are isolated by ether extraction. The yields are generally better than by the sodium hydroxide method.

Decyl toluene-p-sulphonate (*pyridine–chloroform method*).[21] Decan-1-ol (1.58 g, 10 mmol) is dissolved in chloroform (10 ml) and cooled in an ice bath (0 °C). Pyridine (1.62 ml, 20 mmol; **CAUTION**) is then added, followed by the addition of toluene-p-sulphonyl chloride (2.85 g, 15 mmol) in small portions with constant stirring. The reaction is completed in 2.5 hours (t.l.c. monitoring). Ether (30 ml) and water (7 ml) are added, and the organic layer is washed successively with hydrochloric acid (2 M), 5 per cent sodium hydrogen carbonate and water, and dried (magnesium sulphate). The solvent is removed under reduced pressure and the crude tosylate is chromatographed on silica gel (eluant, 2% ether–light petroleum) to yield the product as an oil, 3.06 g (98%); p.m.r. spectrum (CDCl$_3$, TMS) δ 0.87 (t, 3H, Me), 1.22 (broad s, 14H, alkane), 1.59 (m, 2H, —CH$_2$CH$_2$O), 2.44 (s, 3H, Me), 4.02 (t, 2H, CH$_2$O), 7.56 (AA'BB', 4H, ArH); ^{13}C-n.m.r. spectrum (CDCl$_3$) δ 144.46, 133.12, 129.65, 127.64 (Ar carbons) 70.53 (CH$_2$O), 31.70, 29.29, 29.24, 29.10, 28.75, 28.64, 25.17, 22.49, 21.38 and 13.94 (alkane).

6.4.5 REDUCTION PRODUCTS FROM ARYLSULPHONYL CHLORIDES

Reduction of an arylsulphonyl chloride with zinc dust and water affords the zinc salt of the sulphinic acid, converted by sodium carbonate to the sodium salt, in which form it is conveniently isolated (e.g. sodium toluene-p-sulphinate, Expt 6.47) and thence by hydrochloric acid into the somewhat unstable sulphinic acid.

$$\text{Ar·SO}_2\text{Cl} \xrightarrow[\substack{\text{(ii) Na}_2\text{CO}_3 \\ \text{(iii) HCl}}]{\text{(i) Zn/H}_2\text{O}} \text{Ar·SO}_2\text{H}$$

Excessive drying of the free sulphinic acid must be avoided since it leads to partial conversion into the sulphonic acid and the thiolsulphonic ester.

$$3Ar \cdot SO_2H \xrightarrow{-H_2O} Ar \cdot SO_3H + Ar \cdot SO_2 \cdot SAr$$

An alternative synthesis of benzenesulphinic acid by way of the diazonium salt is given in Expt 6.75.

More vigorous reduction of arylsulphonyl chlorides (or of sulphinic acids), for example with zinc and dilute sulphuric acid, gives rise to thiophenols, which are conveniently isolated by steam distillation.

$$Ar \cdot SO_2Cl \xrightarrow{Zn, HCl} ArSH$$

The thiophenols have an extremely unpleasant and repulsive odour and their preparation in the undergraduate laboratory is *not* recommended. In those cases where preparative details are required, reference to reliable procedures is advised.[22]

Experiment 6.47 SODIUM TOLUENE-*p*-SULPHINATE

$$p\text{-Me} \cdot C_6H_4 \cdot SO_2Cl \xrightarrow[\text{(ii) Na}_2\text{CO}_3]{\text{(i) Zn/H}_2\text{O}} p\text{-Me} \cdot C_6H_4 \cdot SO_2Na \xrightarrow{HCl} p\text{-Me} \cdot C_6H_4 \cdot SO_2H$$

In a 2-litre bolt-necked flask place 300 ml of water and insert a mechanical stirrer. Warm on a boiling water bath until the temperature reaches 70 °C and add 40 g (0.61 mol) of zinc powder (90–100% pure). Stir the mixture and add 50 g (0.26 mol) of finely powdered toluene-*p*-sulphonyl chloride in portions during about 10 minutes; the temperature rises to about 80 °C. Stir for a further 10 minutes and then heat the mixture further until the temperature is 90 °C. Add 25 ml of 12 M sodium hydroxide solution, followed by finely powdered sodium carbonate in 5 g portions until the mixture is strongly alkaline. Considerable frothing occurs. Filter at the pump and extract the residue by heating on a steam bath with 75 ml of water until excessive frothing occurs; continue stirring for a further 10 minutes. Filter with suction and add the filtrate to the main solution, concentrate to a volume of about 100 ml with a rotary evaporator and cool in ice-water. Filter at the pump and dry the crystals upon filter or drying paper until efflorescence just commences, then place in a tightly stoppered bottle. The yield of sodium toluene-*p*-sulphinate dihydrate is 35 g (63%).

The free sulphinic acid may be precipitated from a solution of the sodium salt in cold water by cautious acidification with hydrochloric acid. The toluene-*p*-sulphinic acid is filtered and sucked as dry as possible at the pump and dried rapidly between sheets of filter paper, m.p. 85 °C.

6.5 AROMATIC AMINES

Aromatic amines may be divided into three classes:

(*a*) *Primary amines*: where the amino group is directly attached to the aromatic ring, e.g. aniline Ph·NH₂. Amines with the amino group in the side chain, e.g. benzylamine Ph·CH₂NH₂, possess properties similar to those of simple aliphatic amines (Section 5.16).

(b) *Secondary amines*: (i) purely aromatic amines, e.g. diphenylamine Ph_2NH, and (ii) aromatic–aliphatic amines, e.g. *N*-methylaniline, $Ph \cdot NH \cdot Me$.

(c) *Tertiary amines*: (i) purely aromatic amines, e.g. triphenylamine Ph_3N, and (ii) aromatic–aliphatic amines, e.g. *N,N*-dimethylaniline, $Ph \cdot NMe_2$.

Primary aromatic amines are prepared by the following procedures.
1. The reduction of nitro compounds (Expts 6.48 to 6.52).
2. Molecular rearrangements of the Hofmann type (Expts 6.53 and 6.54).
3. The replacement of a halogen by an amino group (see Expt 6.93).
4. The replacement of a phenolic hydroxyl group by an amino group (Expt 6.55).
   Secondary and tertiary amines are prepared by:
5. (a) Alkylation and (b) reductive alkylation procedures (Expts 6.56 to 6.58).

Methods for the protection of the amino and imino groups are discussed in Section 6.5.6.

SUMMARY OF RETROSYNTHETIC STRATEGIES

Functional group interconversion (FGI) (method 1)

(TM)

C—N Disconnection (methods 3 and 5(a)), e.g.

(TM)

Benzyne
from PhX $+ NaNH_2$

Y = electron-
withdrawing
group
(TM)

(TM)

FGI followed by C—N disconnection (method 5(b)), e.g.

(TM)

Rearrangement (method 2), e.g.

(TM)

SPECTROSCOPIC FEATURES

The characteristic *i.r.* absorption of the stretching vibrations of the NH_2 and NH groups of primary and secondary amines are clearly visible in the $3300\,cm^{-1}$ region (p. 290), as in the spectrum of *N*-methylaniline (Fig. 3.27). The substitution pattern may often be assigned from inspection of the longer wavelength region (p. 280). The provisional assignment may be supported by consideration of the *p.m.r.* spectrum; descriptive illustrations are given in the preparative examples below. The frequently observed broad signals observed in the p.m.r. spectra of the nitrogen-bound protons, and also their ready exchange with, for example, deuterium oxide have been discussed on pp. 350 and 348. Some typical fragmentation patterns in the m.s. are descriptively analysed in the appropriate preparative sections. The *u.v.–visible* absorption of the aromatic system is slightly affected by the presence of an amino or imino group (bathochromic shift), but the additional presence of electron-withdrawing groups (e.g. NO_2) in the *ortho* or *para* positions can cause a marked shift into the longer wavelength (visible) region of the spectrum.

6.5.1 THE REDUCTION OF NITRO COMPOUNDS

Primary arylamines are generally prepared by the reduction of nitro compounds. When only small quantities are to be reduced and cost is a secondary consideration, tin and hydrochloric acid may be employed.

$$Ar \cdot NO_2 + 6[H] \longrightarrow Ar \cdot NH_2 + H_2O$$

Theoretically 1.5 mol of tin are needed for the reduction of the nitro group, the metal being oxidised to the tin(IV) state.

$$2Ar \cdot NO_2 + 3Sn + 12H^{\oplus} \longrightarrow 2Ar \cdot NH_2 + 3Sn^{4\oplus} + 4H_2O$$

When reduction is complete, a complex amine chlorostannate may separate from which the amine is liberated by basification, using enough alkali to dissolve the tin hydroxides formed (e.g. the preparation of aniline, Expt 6.48).

$$[Ar \cdot NH_3]_2^{\ominus}[SnCl_6]^{2\ominus} + 8^{\ominus}OH \longrightarrow Ar \cdot NH_2 + SnO_3^{2\ominus} + 6Cl^{\ominus}$$

Alternatively decomposition of the complex is achieved by precipitation of the tin as its sulphide by passing hydrogen sulphide into the mixture at the end of the reaction (e.g. *m*-phenylenediamine, cognate preparation in Expt 6.48).

Other metal–acid reducing systems may be used; reduction with iron and hydrochloric acid is employed on the technical scale for the manufacture of aniline.

$$Ph{\cdot}NO_2 + 2Fe + 6H^\oplus \longrightarrow Ph{\cdot}NH_2 + 2Fe^{3\oplus} + 2H_2O$$

$$Ph{\cdot}NO_2 + 3Fe + 6H^\oplus \longrightarrow Ph{\cdot}NH_2 + 3Fe^{2\oplus} + 2H_2O$$

In practice, however, the amount of hydrochloric acid used is less than 5 per cent of the amounts indicated by either of the above equations. Various explanations have been advanced to account for this; one is that the following reaction is catalysed by acid or by hydroxonium ions.

$$Ph{\cdot}NO_2 + 2Fe + 4H_2O \xrightarrow{\ H^\oplus\ } Ph{\cdot}NH_2 + 2Fe(OH)_3$$

The use of iron and a little hydrochloric acid (or alternatively of iron and acetic acid) is to be preferred when the use of the more strongly acidic medium leads to the formation of undesirable by-products; nuclear chlorination, for example, often results when tin or zinc is used in association with concentrated hydrochloric acid. A suitable experimental procedure for the reduction of 2,4-dinitrotoluene to 2,4-diaminotoluene has been described.[23]

Compounds which are sensitive to acidic conditions may sometimes be successfully reduced to amines under alkaline conditions; in these cases iron(II) sulphate is often used as the reducing agent. Reduction with metal under alkaline (or alternatively essentially neutral) conditions leads to the formation of intermediate products of the reduction of the nitro group (see Section 6.8.1, p. 953). Vigorous reduction under alkaline conditions using an excess of the reducing metal (e.g. zinc) may, however, lead to the complete reduction of the nitro group with the formation of the corresponding amine; this *o*-phenylenediamine (Expt 6.49) is conveniently obtained by reducing *o*-nitroaniline in this way.

Aromatic and heterocyclic nitro compounds are readily reduced in good yield to the corresponding amines (e.g. *o*-aminophenol, Expt 6.50) by sodium borohydride in aqueous methanol solution in the presence of a palladium-on-carbon catalyst. In this reduction there is no evidence for the formation of intermediates of the azoxybenzene or azobenzene type, although if the reaction is carried out in a polar aprotic solvent, such as dimethyl sulphoxide, azoxy compounds may sometimes be isolated as the initial products.

Ammonium or alkali metal sulphides or polysulphides exhibit a useful selective reducing action in that they smoothly reduce one nitro group in a polynitro compound to yield the corresponding nitroamine (e.g. *m*-nitroaniline, Expt 6.51).

The two methods described for the synthesis of ethyl *p*-aminobenzoate (Expt 6.52) from *p*-nitrobenzoic acid offer interesting comparison. Prior esterification of the carboxylic acid group (*Method 2*) yields a nitroester which cannot be reduced by a metal–acid reducing system owing to the possibility of hydrolytic cleavage of the ester grouping; conversion of the nitro into the amino group in this case is therefore by catalytic hydrogenation with Adams' platinum dioxide catalyst. In *Method 1*, metal–acid reduction of the nitro group yields the amino acid, p-aminobenzoic acid, a reaction in which the final mixture requires careful basification to effect precipitation of the product, which is then esterified to give ethyl p-aminobenzoate. This compound (Benzocaine) is used as a local anaesthetic.

Experiment 6.48 ANILINE

$$Ph \cdot NO_2 + 6[H] \xrightarrow{Sn/HCl} Ph \cdot NH_2 + 2H_2O$$

Into a 500-ml round-bottomed flask equipped with a reflux condenser, place 25 g (21 ml, 0.25 mol) of nitrobenzene and 45 g (0.38 mol) of granulated tin. Measure out 100 ml of concentrated hydrochloric acid. Pour about 15 ml of this acid down the condenser and shake the contents of the flask steadily. The mixture becomes warm and before long the reaction should be quite vigorous; if it boils very vigorously, moderate the reduction somewhat by temporarily immersing the flask in cold water. When the initial reaction slackens of its own accord, pour another 15 ml of hydrochloric acid down the condenser, shake the flask steadily to ensure thorough mixing and cool again if the reduction becomes too violent. Do not cool more than is necessary to keep the reaction under control; keep the mixture well shaken. Proceed in this way until all the 100 ml of acid has been added. Finally heat the mixture on a boiling water bath for 30–60 minutes, i.e. until the odour of nitrobenzene is no longer perceptible and a few drops of the reaction mixture when diluted with water yield a perfectly clear solution. During the course of the reduction, particularly during the cooling, aniline chlorostannate may separate as a white or yellow crystalline complex.

Cool the reaction mixture to room temperature and add gradually a solution of 75 g of sodium hydroxide in 125 ml of water; if the mixture boils during the addition of the alkali, cool again. The hydroxide of tin which is first precipitated should all dissolve and the solution should be strongly alkaline: the aniline separates as an oil. Equip the flask for steam distillation as in Fig. 2.102, and pass steam into the warm mixture until, after the distillate has ceased to pass over as a turbid liquid, a *further* 120 ml of clear liquid are collected. Since aniline is appreciably soluble (c. 3%) in water, it must be 'salted out' by saturating the distillate with salt. Use about 20 g of commercial salt for each 100 ml of liquid. Transfer the distillate, saturated with salt, to a separatory funnel, add about 40 ml of ether and shake to ensure intimate mixing of the solution and the ether; relieve the pressure within the funnel by momentarily lifting the stopper. (*All flames in the vicinity must be extinguished during the extraction.*) Allow the two layers to separate; run off the lower aqueous layer into a beaker, and pour the remaining ethereal layer through the mouth of the funnel into a 200-ml flask. Return the aqueous solution to the funnel and extract with a further 40 ml of ether. Proceed as before, and pour the ethereal extract into the flask. Dry the combined ethereal solutions with a few grams of anhydrous potassium carbonate (1): shake the well-stoppered flask for several minutes.

Filter the ethereal solution through a fluted filter paper and remove the ether by flash distillation, using a 50-ml round-bottomed flask to which has been added a few boiling chips. Since ether is extremely volatile and also highly flammable, the flask must be heated by means of an electrically heated water bath. When all the ethereal solution has been introduced into the flask, and no more ether distils on a boiling water bath, run out the water from the condenser, and distil the aniline either by direct heating over a wire gauze or, preferably, using an air bath (Fig. 2.46). A small quantity of ether may pass over during the early part of the distillation; it is therefore advisable to interpose a uralite board between the receiver and the flame. Collect the fraction b.p. 180–184 °C, in a weighed conical flask. The yield of aniline is 18 g (97%).

Record the i.r. spectrum and note the absorptions at c. 3400 cm^{-1} (NH$_2$, sym. and asym. stretching vibrations), 1625 cm^{-1} (NH$_2$ bending vibration), 1600, 1590, 1500, 1450 cm^{-1} (ring breathing vibrations), and 750 and 690 cm^{-1} (monosubstitution). The p.m.r. spectrum (CCl$_4$ or CDCl$_3$, TMS) shows a singlet (c. δ 3.32) for the NH$_2$-group which disappears when the sample is shaken with deuterium oxide; the aromatic region of the spectrum reveals overlapping multiplets. The m.s. shows principal fragment ions at m/z 93 (M, base peak), 92 (M − H), 66 (M − HCN), 65 (66 − H or 92 − HCN), and 39 (65 − C$_2$H$_2$).

Pure aniline has a b.p. of 184 °C. When freshly distilled it is a colourless liquid, but becomes discoloured on standing, particularly when exposed to light, owing to atmospheric oxidation. The colour may usually be removed by distillation from a little zinc dust.

Note. (1) Calcium chloride cannot be used to dry the ethereal solution because it combines with aniline (and other amines) to form molecular compounds. The best drying agent is sodium or potassium hydroxide (pellet form).

Cognate preparation. m-*Phenylenediamine.* In a 2-litre round-bottomed flask, provided with a reflux condenser, place 25 g (0.15 mol) of m-dinitrobenzene (Expt 6.18) and 100 g (0.84 mol) of granulated tin; add 200 ml of concentrated hydrochloric acid in 15 ml portions. When all the acid has been introduced, complete the reduction by heating on a water bath for 1 hour. Dilute with 750 ml of water, heat nearly to boiling and pass hydrogen sulphide into the liquid until all the tin is precipitated as the sulphide. Filter a small quantity from time to time and test for completeness of precipitation with hydrogen sulphide. Allow the precipitate to settle overnight, decant the clear liquid and filter the residue through a large fluted filter paper (1). Add sodium hydroxide solution to the filtrate until the latter is strongly alkaline, and extract several times with ether. Dry over anhydrous potassium carbonate or sodium hydroxide pellets, remove the ether and then distil the residue: use an air condenser after all the ether has passed over. Collect the portion boiling between 280 and 284 °C: this solidifies on standing to crystalline m-phenylenediamine, m.p. 63 °C. The yield is 13 g (74%).

Note. (1) The dihydrochloride may be obtained by evaporating the filtrate on a water bath until crystals appear, and then cooling in ice. The crystals are filtered at the pump, washed with a little concentrated hydrochloric acid and dried in a vacuum desiccator over sodium hydroxide.

Experiment 6.49 *o*-PHENYLENEDIAMINE

$$o\text{-NH}_2\cdot\text{C}_6\text{H}_4\cdot\text{NO}_2 \xrightarrow[\text{Zn/}\overset{\ominus}{\text{OH}}]{\text{[H]}} o\text{-NH}_2\cdot\text{C}_6\text{H}_4\cdot\text{NH}_2$$

Equip a 750-ml three-necked flask with a reflux condenser and sealed mechanical stirrer, and place in it 46 g (0.33 mol) of o-nitroaniline, 27 ml of 20 per cent sodium hydroxide solution and 170 ml of rectified spirit. Stir the mixture vigorously and heat it on a water bath to gentle boiling. Remove the source of heat from beneath the bath, and introduce 5 g portions of zinc powder at such a rate that the solution is kept boiling (1); add 90 g (1.4 mol) of zinc powder (2) in all. Reflux the mixture, with stirring, for 1 hour; the colour of the solution changes from deep red to nearly colourless. Filter the hot mixture at the pump; return the zinc residue to the flask and extract it with two 100 ml portions of hot rectified spirit. Combine the extracts with the filtrate, add 2 g of sodium dithionite ($Na_2S_2O_4$) and concentrate the solution under reduced pressure (water pump) on a steam bath to a volume of 80–100 ml; use a rotary evaporator (Fig. 2.112). Cool the solution in a freezing mixture of ice and salt, collect the pale yellow crystals on a Buchner funnel, wash once with 10–15 ml of ice-water and dry in a vacuum desiccator. The yield of crude o-phenylenediamine, m.p. 98–100 °C, is 33 g. This is sufficiently pure for most practical purposes. If a pure material is required (3), dissolve the crude product in 100–115 ml of hot water containing 1 g of sodium dithionite and add a few grams of decolourising carbon, filter and cool in an ice–salt mixture. Collect the colourless crystals of pure o-phenylenediamine on a Buchner funnel, wash with 10 ml of ice-water and dry in a vacuum desiccator; the yield is 28.5 g (79%), m.p. 100–101 °C. It darkens rapidly upon exposure to light.

The i.r. spectrum (KBr disc) shows absorptions at c. 3400 cm^{-1} (NH_2 stretching), 1625 cm^{-1} (NH_2 bend), 1600, 1500, 1450 cm^{-1} (ring breathing) and 750 cm^{-1} (four adjacent hydrogens). The p.m.r. spectrum (CDCl$_3$, TMS) shows signals at δ 3.28 [s, 4H, (NH_2)$_2$] and 6.67 (s, 4H, C_{AR}—H); interestingly the chemical shift values of the aromatic hydrogens are all similarly affected by the presence of the amino groups, and this region is not structurally informative. The m.s. shows principal fragment ions at m/z 108 (M, base peak), 107 (M − H), 81 (M − HCN), 80 (107 − HCN) and 54 (81 − HCN).

Notes. (1) Sometimes the reaction stops suddenly; it is then necessary to add a further 10 ml of 20 per cent sodium hydroxide solution and warm to the boiling point: this causes the reaction to continue. Occasionally, the reduction becomes very vigorous: a wet towel and a bath of ice-water should be kept close at hand.

(2) This weight of zinc powder assumes 100 per cent purity: an equivalent amount of less pure material may be used (see Section 4.2.80, p. 467).

(3) The crude o-phenylenediamine may be converted into the dihydrochloride and the salt purified in the following manner. Dissolve it in 60 ml of concentrated hydrochloric acid and 40 ml of water containing 2 g of tin(II) chloride, and treat the hot solution with 2–3 g of decolourising carbon. Filter, add 100 ml of concentrated hydrochloric acid to the hot colourless filtrate and cool in a freezing mixture of ice and salt. Collect the colourless crystals of the dihydrochloride on a Buchner or sintered glass funnel, wash with a small volume of concentrated hydrochloric acid and dry in a vacuum desiccator over sodium hydroxide. The yield is 51 g.

Experiment 6.50 o-AMINOPHENOL

$$o\text{-}HO\cdot C_6H_4\cdot NO_2 \xrightarrow{\text{NaBH}_4/\text{Pd-C}} o\text{-}HO\cdot C_6H_4\cdot NH_2$$

CAUTION: *This preparation should be conducted in an efficient fume cupboard since hydrogen is evolved during the reaction.*

Place a suspension of 0.5 g of 10 per cent palladium-on-charcoal (cf. Section 4.2.54, p. 452) in 50 ml of water in a 500-ml conical flask and add 3.9 g (0.10 mol) of sodium borohydride in 75 ml of water (1); introduce a magnetic follower bar. Pass a slow stream of nitrogen through the stirred mixture and add a solution of 7.0 g (0.05 mol) of *o*-nitrophenol in 250 ml of 2 M sodium hydroxide, dropwise from a separatory funnel, during 5 minutes. Stir at room temperature until the yellow colour disappears (about 10 minutes) and then filter. Acidify the filtrate with 2 M hydrochloric acid to destroy excess borohydride and then neutralise the solution with dilute sodium hydroxide. Extract the product with four 50 ml portions of ether and evaporate the dried ($MgSO_4$) extract on the rotary evaporator. *o*-Aminophenol is obtained as an off-white solid, m.p. 167–169 °C; the yield is 4.3 g (75%). The product may be recrystallised from water to give the purified phenol, m.p. 169–171 °C. The infrared spectrum of *o*-aminophenol as a Nujol mull shows strong absorptions at 3380 and 3300 cm^{-1} due to the antisymmetric and symmetric N—H stretching vibrations. The p.m.r. spectrum (TFA) is complex and shows signals corresponding to the aromatic protons as overlapping multiplets at δ 7.53–6.92. This should be contrasted with the p.m.r. spectrum of the *para* isomer, p. 956.

Note. (1) Addition of the reagents in this order prevents the possible ignition of hydrogen which can take place on addition of dry palladium–charcoal to solutions of sodium borohydride.

Experiment 6.51 *m*-NITROANILINE

$$m\text{-}O_2N \cdot C_6H_4 \cdot NO_2 \xrightarrow{\text{[H]}} m\text{-}O_2N \cdot C_6H_4 \cdot NH_2$$

Dissolve 18 g (0.075 mol) of crystallised sodium sulphide, $Na_2S \cdot 9H_2O$ (1), in 50 ml of water; add 6.0 g (0.0714 mol) of finely powdered sodium hydrogen carbonate in small portions with constant stirring. When the carbonate has dissolved completely, add 50 ml of methanol and cool below 20 °C. Filter off the precipitated sodium carbonate at the pump, using a small Buchner funnel; if necessary add 2–3 g of filter-aid to the reaction mixture before filtering, and prepare a bed of filter-aid in the Buchner funnel since this will aid the retention of the finely divided solid. Wash the precipitate with three 8 ml portions of methanol. Retain the filtrate and washings: these contain about 3.9 g of NaSH in solution and must be used forthwith for the reduction.

Dissolve 6.7 g (0.04 mol) of *m*-dinitrobenzene in 50 ml of hot methanol in a 250-ml round-bottomed flask and add, with shaking, the previously prepared methanolic solution of sodium hydrogen sulphide. Attach a reflux condenser and boil the mixture for 20 minutes; ignore any further sodium carbonate which may precipitate. Allow the reaction mixture to cool and fit the condenser for distillation. Distil off most of the methanol (100–120 ml) from a water bath. Pour the liquid residue with stirring into about 200 ml of cold water. Collect the yellow crystals of *m*-nitroaniline by suction, wash with water and recrystallise from 75 per cent aqueous methanol. The yield of bright yellow crystals, m.p. 114 °C, is 3.7 g (69%).

Note in the i.r. spectrum the absorptions arising from the vibrational modes of the NO_2 and NH_2 groups. The p.m.r. spectrum ($CDCl_3$ at 60 °C, TMS) shows a very complex pattern of overlapping multiplets at δ 6.8–7.7 arising from the four dissimilar aromatic protons. The spectrum should be contrasted with those given by the *ortho* and *para* isomers (the latter is noted on p. 920). The m.s. of the three isomers show the same fragment ions but having different relative abundance: the *ortho* isomer has m/z 138 (M, RA < 10%), 92 (M − NO_2, RA 80%), 65 (92 − HCN, RA 100%): the *meta* isomer has ions at m/z 138 (RA 40%), 108 (RA 100%), 92 (RA 50%), and 65 (RA 70%); the *para* isomer has ions at m/z 138 (RA 100%), 108 (RA 30%), 92 (RA 42%), and 65 (RA 30%).

Note. (1) Crystallised sodium sulphide is very deliquescent and only a sample which has been kept in a tightly stoppered bottle should be used.

Cognate preparation. *Reduction of 2,4-dinitrophenol.* It is an interesting student exercise to carry out the reduction of 2,4-dinitrophenol under the conditions described above for *m*-dinitrobenzene. The spectroscopic features of the isolated and purified product, together with the melting point, in comparison with the literature values for the possible isomers, should enable a deduction to be made on the regioselectivity of the reaction.

Experiment 6.52 ETHYL *p*-AMINOBENZOATE (*Benzocaine*)

Method 1

$$p\text{-}O_2N\cdot C_6H_4\cdot CO_2H \xrightarrow{\text{Sn/HCl}} p\text{-}H_2N\cdot C_6H_4\cdot CO_2H \xrightarrow[\text{HCl}]{\text{EtOH}} p\text{-}H_2N\cdot C_6H_4\cdot CO_2Et$$

Method 2

$$p\text{-}O_2N\cdot C_6H_4\cdot CO_2H \xrightarrow[\text{HCl}]{\text{EtOH}} p\text{-}O_2N\cdot C_6H_4\cdot CO_2Et \xrightarrow{\text{H}_2/\text{Pt}} p\text{-}H_2N\cdot C_6H_4\cdot CO_2Et$$

Method 1. p-Aminobenzoic acid. Place 15 g (0.09 mol) of *p*-nitrobenzoic acid (Expt 6.149) in a 1-litre round-bottomed flask fitted with a reflux condenser. Introduce 35 g (0.295 mol) of powdered tin and 75 ml of concentrated hydrochloric acid. Heat the mixture gently until the reaction commences, and remove the flame. Shake the flask frequently and take care that the insoluble acid adhering to the sides of the flask is transferred to the reaction mixture: occasional gentle warming may be necessary. After about 20 minutes, most of the tin will have reacted and a clear solution remains. Allow to cool somewhat and decant the liquid into a 1-litre beaker; wash the residual tin by decantation with 15 ml of water, and add the washings to the contents of the beaker. Add concentrated ammonia solution (*d* 0.88) until the solution is *just* alkaline to litmus and digest the suspension of precipitated hydrated tin oxide on a steam bath for 20 minutes. Add 10 g of filter-aid ('Celite'), stir well, filter at the pump and wash with hot water. Transfer the filter cake to a beaker, heat on a water bath with 200 ml of water to ensure extraction of the product and refilter. Concentrate the combined filtrate and washings until the volume has been reduced to 175–200 ml: filter off any solid which separates. Acidify the liquid to litmus with glacial acetic acid and evaporate on a water bath until crystals commence to separate; cool in ice, filter the crystals at the pump and dry in the steam oven. The yield of *p*-aminobenzoic acid, m.p. 192 °C, is 9.5 g (77%).

Ethyl p-aminobenzoate (*esterification of* p-*aminobenzoic acid*). Place 80 ml of absolute ethanol in a 250-ml two-necked flask equipped with a double surface reflux condenser and a gas inlet tube. Pass dry hydrogen chloride (Section 4.2.*38*, p. 438) through the alcohol until saturated – the increase in weight is about 20 g – remove the gas inlet tube, introduce 12 g (0.088 mol) of p-aminobenzoic acid and heat the mixture under reflux for 2 hours. Upon cooling, the reaction mixture sets to a solid mass of the hydrochloride of ethyl p-aminobenzoate. It is better, however, to pour the hot solution into *c*. 300 ml of water (no hydrochloride separates) and add solid sodium carbonate carefully to the clear solution until it is neutral to litmus. Filter off the precipitated ester at the pump and dry in the air. The yield of ethyl p-aminobenzoate, m.p. 91 °C, is 10 g (69%). Recrystallisation from rectified (or methylated) spirit does not affect the m.p.

Method 2. **Ethyl p-nitrobenzoate.** Place 21 g (0.125 mol) of p-nitrobenzoic acid (Expt 6.149), 11.5 g (0.25 mol) of absolute ethanol, 3.8 g of concentrated sulphuric acid and 30 ml of sodium-dried benzene in a 250-ml round-bottomed flask, fit a reflux condenser and heat the mixture under reflux for 16 hours. Add 50 ml of ether to the cold reaction mixture, wash the extract successively with sodium hydrogen carbonate solution and water, dry with magnesium sulphate or calcium chloride and distil off the solvent on a water bath. Remove the last traces of benzene by heating in a bath at 100–110 °C. The residual ethyl p-nitrobenzoate (21 g, 86%) solidifies completely on cooling and melts at 56 °C.

Ethyl p-aminobenzoate (*catalytic reduction of ethyl* p-*nitrobenzoate*). The general experimental details may be adapted from those described in Section 2.17.1, p. 87. Place a solution of 9.75 g (0.05 mol) of ethyl p-nitrobenzoate in 100 ml of rectified spirit together with 0.1 g of Adams' platinum dioxide catalyst in the hydrogenation flask, and shake in hydrogen in the usual manner. The theoretical volume of hydrogen (*c*. 3600 ml at 24 °C and 760 mmHg) is absorbed in 2.5 hours. Filter off the platinum with suction and rinse the reaction vessel with rectified spirit. Evaporate the alcohol from the combined filtrate and washings on a water bath; the residue solidifies on cooling and weighs 8.2 g. Dissolve the crude ethyl p-aminobenzoate in rectified spirit, add a little decolourising charcoal, boil and filter; heat the filtrate to the boiling point, add hot water to incipient crystallisation and allow to cool. The resulting pure benzocaine has m.p. 90 °C; the yield is 7 g (85%).

The p.m.r. spectrum (CDCl$_3$, TMS) shows signals at δ 1.31 (t, 3H, Me), 4.24 (q, 2H, —CH$_2$Me), 4.28 (s, 2H, NH$_2$), 6.51 (d, 2H, *ortho*-H's to NH$_2$), and 7.81 (d, 2H, *ortho*-H's to CO$_2$Et). The m.s. reveals principal fragment ions at *m/z* 165 (M), 137 (M – C$_2$H$_4$), 120 (M – OEt), 92 (M – CO$_2$Et), and 65 (92 – HCN).

6.5.2 MOLECULAR REARRANGEMENTS OF THE HOFMANN TYPE

The general procedure of treating carboxylic acid amides with alkaline hypohalite solution is illustrated and discussed in the aliphatic series (Section 5.16.6, p. 783) and is equally applicable to the synthesis of primary aromatic amines (e.g. anthranilic acid and 3-aminopyridine, Expt 6.53). It is of particular use in those cases where it is required to introduce an amino group into a position on the aromatic ring which, because of the orienting effects of the substituents, cannot be effectively nitrated directly in order to use the direct nitration–reduction sequence.

For the preparation of anthranilic acid the starting material is phthalimide, the cyclic imide ring of which is opened by alkaline hydrolysis in the first step of the reaction to give the sodium salt of phthalimidic acid (the half amide of phthalic acid). The intermediate undergoes the Hofmann reaction in the manner outlined on p. 783 yielding o-aminobenzoic acid (anthranilic acid).

The conversion of a carboxylic acid into an amine by treatment with hydrazoic acid in concentrated sulphuric acid is known as the *Schmidt reaction* or *rearrangement*, which often gives higher yields than the related Hofmann rearrangement procedure

$$R \cdot CO_2H + HN_3 \xrightarrow{H_2SO_4} RNH_2 + CO_2 + N_2$$

The use of the toxic and hazardous hydrazoic acid is avoided by generating it *in situ* by adding sodium azide gradually to the carboxylic acid in the presence of concentrated sulphuric acid and chloroform (eg.. 3,5-dinitroaniline, Expt 6.54). The reaction involves the hydrolysis of an intermediate isocyanate (RNCO), which is formed by a mechanistic pathway analogous to that involved in the Hofmann reaction.

It should be noted that amines may be formed by hydrolysis of amides arising from the intramolecular Beckmann rearrangement of ketoximes (p. 1047); this rearrangement is a further example of the migration of a nucleophilic carbon species from a carbon to an electron-deficient nitrogen.

Experiment 6.53 ANTHRANILIC ACID

Prepare a solution of 30 g of sodium hydroxide in 120 ml of water in a 350-ml conical flask and cool to 0 °C or below in a bath of ice and salt. Add 26.2 g

(8.4 ml, 0.16 mol) of bromine in one portion and shake (or stir) until all the bromine has reacted. The temperature will rise somewhat; cool again to 0 °C or below. Meanwhile, prepare a solution of 22 g of sodium hydroxide in 80 ml of water. Add 24 g (0.163 mol) of finely powdered phthalimide (Expt 6.155) in one portion to the cold sodium hypobromite solution in the form of a smooth paste with water, rapidly with stirring. Remove the flask from the cooling bath and shake vigorously until a clear yellow solution is obtained (c. 5 minutes). Add the prepared sodium hydroxide solution rapidly and in one portion, heat the solution to 80 °C for about 2 minutes and filter if necessary. Cool in ice and add concentrated hydrochloric acid slowly and with stirring until the solution is *just* neutral (about 60 ml are required). (It is re-commended that a little of the alkaline solution be set aside in case too much acid is added.) Precipitate the anthranilic acid completely by the gradual addition of glacial acetic acid (20–25 ml are required): it is advisable to transfer the mixture to a 1-litre beaker as some foaming occurs. Filter off the acid at the pump and wash with a little cold water. Recrystallise from hot water with the addition of a little decolourising carbon; collect the acid on a Buchner funnel and dry at 100 °C. The yield of pure anthranilic acid, m.p. 145 °C, is 14 g (62%).

Cognate preparation. *3-Aminopyridine.* Prepare a cold sodium hypobromite solution from 32 g (10 ml, 0.2 mol) of bromine and 25 g (0.62 mol) of sodium hydroxide in 250 ml of water. Add in one portion 20 g (0.163 mol) of finely powdered nicotinamide (Expt 6.169) and stir vigorously for 15 minutes. Warm the solution in a water bath at 75 °C for 45 minutes. Isolate the crude product by continuous ether extraction (Section 2.22) of the cooled reaction mixture after saturation with sodium chloride. Dry the extract over potassium hydroxide pellets and remove the ether. Crystallise the dark residue from a 4:1 mixture of benzene–light petroleum (b.p. 60–80 °C) with the aid of decolourising charcoal. The yield of almost colourless product, m.p. 63 °C, is 9.3 g (61%).

Experiment 6.54 3,5-DINITROANILINE

$$3,5\text{-}(NO_2)_2C_6H_3 \cdot CO_2H + HN_3 \xrightarrow{H_2SO_4} 3,5\text{-}(NO_2)_2C_6H_3 \cdot NH_2$$

Place a solution of 50 g (0.25 mol) of 3,5-dinitrobenzoic acid (Expt 6.160) in 90 ml of oleum (10% SO_3) and 20 ml of concentrated sulphuric acid in a 1-litre four-necked flange flask equipped with a reflux condenser, mechanical stirrer, a dropping funnel and thermometer (fume cupboard). Add 100 ml of chloroform and raise the temperature to 45 °C. Stir rapidly and add 17.5 g (0.27 mol) of sodium azide in small portions while maintaining the tempera-ture at 35–45 °C. The reaction is accompanied by foaming, which usually commences after about 3 g of sodium azide has been introduced. After all the sodium azide has been added raise the temperature so that the chloroform refluxes vigorously and maintain this temperature for 3 hours. Then cool the reaction mixture, pour it cautiously on to 500 g of crushed ice and dilute with 3 litres of water. After 1 hour, separate the yellow solid by filtration at the pump, wash well with water and dry at 100 °C. The yield of 3,5-dinitroaniline, m.p. 162–163 °C, is 39 g (90%). The m.p. is unaffected by recrystallisation from dilute ethanol.

The i.r. spectrum (KBr disc) clearly shows absorptions at c. 3400 and 1625 cm^{-1} (stretching and bending modes of the NH$_2$), 1525 and 1340 cm^{-1} (NO$_2$ group), 1600, 1590, 1450 cm^{-1} (ring breathing); the presence of the NO$_2$-group makes the assignment of the substitution pattern unreliable. The p.m.r. spectrum (Me$_2$CO-d_6, TMS) has signals at δ 5.96 (broad s, 1H, NHD), 7.84 (d, 2H, *ortho*-H's to NH$_2$), and 8.05 (t, 1H, *ortho*-H's to both NO$_2$ groups).

6.5.3 THE REPLACEMENT OF A HALOGEN BY AN AMINO GROUP

Unlike the ready replacement of a halogen in an alkyl halide with an amino group (Section 5.16.4, p. 779), the corresponding reaction with an aryl halide only proceeds under vigorous conditions, and generally the reaction is not preparatively useful. Thus aniline may be formed from chlorobenzene only by treatment with sodamide in liquid ammonia. A detailed study of the mechanism of the reaction led to the recognition for the first time of the reaction intermediate, *benzyne*. Other more convenient methods for the formation of benzyne have subsequently been developed, and the investigation of the trapping reactions of this intermediate has led to the introduction of a range of new synthetic reactions. An illustrative example is to be found in Expt 6.7.

The presence of electron-withdrawing groups (notably the nitro group) sited in the *ortho* or *para* positions of an aryl halide facilitates the halogen replacement. Here the mechanism proceeds by an addition–elimination sequence and it is discussed and illustrated in the formation of 2,4-dinitroaniline and 2,4-dinitrophenylhydrazine (Section 6.8.2, p. 959), and in the use of 1-fluoro-2,4-dinitrobenzene in the derivatisation of an amino group in an amino acid (Section 9.6.23, p. 1279).

6.5.4   THE REPLACEMENT OF A PHENOLIC HYDROXYL GROUP BY AN AMINO GROUP

The direct replacement of the hydroxyl group in simple phenols by an amino or substituted amino group requires drastic conditions and the method is not suitable for laboratory preparations. With the polyhydric phenols, and more particularly with the naphthols, such replacements occur more readily. Thus 2-naphthol is converted into 2-naphthylamine by heating with ammoniacal ammonium sulphite solution at 150 °C in an autoclave. The reaction (the *Bucherer reaction*) depends upon the addition of the hydrogen sulphite ion to the keto form of the naphthol and the subsequent reaction with ammonia.

The reaction is reversible; thus 2-naphthylamine can be converted into 2-

naphthol by heating with aqueous sodium hydrogen sulphite solution and then adding alkali and boiling until all the ammonia is expelled.

2-Naphthylamine is no longer manufactured and its laboratory preparation should never be attempted because of its potent carcinogenic properties. For many preparative purposes (e.g. see 2-bromonaphthalene, cognate preparation in Expt 6.72, and 2-naphthoic acid, Expt 6.154), 2-naphthylamine-1-sulphonic acid may be used. This is obtained commercially by cautious treatment of 2-naphthol with sulphuric acid – the sulphonic acid group entering the 1-position – followed by a Bucherer reaction.

The Bucherer reaction is illustrated and its reversibility demonstrated in Expt 6.55, wherein the amino group of 2-amino-5-naphthol-7-sulphonic acid is replaced by the p-tolylamino group via the corresponding naphthol intermediate.

Experiment 6.55   2-p-TOLYLAMINO-5-NAPHTHOL-7-SULPHONIC ACID

Reflux a mixture of 10.8 g (0.1 mol) of pure p-toluidine, 10.8 g (0.057 mol) of 2-amino-5-naphthol-7-sulphonic acid ('J' acid), 8.4 g (0.08 mol) of sodium metabisulphite and 25 ml of water for 30 hours in a 250-ml three-necked flask, equipped with a reflux condenser and mechanical stirrer. Add sodium carbonate until the mixture is alkaline and remove the excess of p-toluidine by steam distillation. Keep the residual solution in a refrigerator until crystallisation is complete, filter with suction on a Buchner funnel and wash with 10 ml of saturated sodium chloride solution. Dissolve the product in c. 35 ml of hot water to which sufficient hydrochloric acid is added to render the mixture acid to Congo red. Keep in a refrigerator until crystallisation is complete, filter with suction, wash with a little ice-cold hydrochloric acid, followed by a small volume of ice-cold water. Dry the residual 2-p-tolylamino-5-naphthol-7-sulphonic acid at 100 °C; the yield is 9.5 g (60%).

6.5.5 ALKYLATION AND REDUCTIVE ALKYLATION PROCEDURES FOR THE PREPARATION OF SECONDARY AND TERTIARY AMINES

Simple N-alkyl- and N,N-dialkylanilines are readily prepared commercially by the alkylation of aniline with the appropriate alcohol. For example, N-methylaniline is prepared by heating a mixture of aniline hydrochloride (55 parts) and methanol (16 parts) at 120 °C in an autoclave. For N,N-dimethylaniline, aniline and methanol are mixed in the proportion 80:78, 8 parts of concentrated sulphuric acid are added and the mixture heated in an autoclave at 230–235 °C at a pressure of 25–30 atmospheres. N-Ethyl- and N,N-diethylaniline are prepared similarly. In the laboratory, alkylation of the amino group, to yield secondary and tertiary amines, is effected by reaction with the appropriate alkyl halide; Expt 6.56 includes the preparation of N-benzylaniline and general procedures for the synthesis of a range of N-alkyl- and N,N-dialkylanilines.

N-Alkylanilines may be purified by converting them into the N-nitroso derivative with nitrous acid followed by reduction of the separated nitroso compound with tin and hydrochloric acid, thus regenerating the N-alkylaniline. For example:

$$Ph·NH·Me \xrightarrow[HCl]{NaNO_2} Ph·N(NO)·Me \xrightarrow{Sn/HCl} Ph·NH·Me$$

N,N-Dialkylanilines may be purified by refluxing with an excess of acetic anhydride: any unchanged aniline and N-alkylaniline is converted into the relatively non-volatile acetyl derivative.

$$Ph·NH_2 + (Me·CO)_2O \longrightarrow Ph·NH·CO·Me + Me·CO_2H$$

$$Ph·NH·R + (Me·CO)_2O \longrightarrow Ph·N(CO·Me)·R + Me·CO_2H$$

Upon fractionation the acetic acid and unreacted acetic anhydride pass over first, followed by the pure N,N-dialkylaniline.

A convenient method for preparing in good yield a pure N,N-dialkylaniline or substituted aniline (Expt 6.57) directly from the corresponding amine consists on heating the latter with the appropriate trialkyl phosphate.

$$3Ar·NH_2 + 2(RO)_3PO \longrightarrow 3Ar·NR_2 + 2H_3PO_4$$

Secondary amines can be prepared from the primary amine and carbonyl compounds by way of the reduction of the derived Schiff bases, with or without the isolation of these intermediates. This procedure represents one aspect of the general method of reductive alkylation discussed in Section 5.16.3, p. 776. With aromatic primary amines and aromatic aldehydes the Schiff bases are usually readily isolable in the crystalline state and can then be subsequently subjected to a suitable reduction procedure, often by hydrogenation over a Raney nickel catalyst at moderate temperatures and pressures. A convenient procedure, which is illustrated in Expt 6.58, uses sodium borohydride in methanol, a reagent which owing to its selective reducing properties (Section 5.4.1, p. 519) does not affect other reducible functional groups (particularly the nitro group) which may be present in the Schiff base; contrast the use of sodium borohydride in the presence of palladium-on-carbon, p. 894.

$$Ar^1·NH_2 + OHC·Ar^2 \longrightarrow Ar^1·N{=}CH·Ar^2 \xrightarrow{NaBH_4} Ar^1·NH·CH_2·Ar^2$$

Experiment 6.56 N-BENZYLANILINE

$$Ph·CH_2Cl + Ph·NH_2 \xrightarrow{Na_2CO_3} Ph·CH_2·NH·Ph$$

Equip a 500-ml three-necked flask with a separatory funnel, a mechanical stirrer and a reflux condenser; mount the assembly on a water bath. Place 35 g of pure sodium hydrogen carbonate, 35 ml of water and 124 g (121 ml, 1.33 mol) of aniline in the flask, and 42 g (38 ml, 0.33 mol) of freshly distilled benzyl chloride (b.p. 177–179 °C) in the separatory funnel protected by a calcium chloride guard-tube. Heat the flask and contents to 90–95 °C, stir vigorously and run in the benzyl chloride slowly (about 1 hour). Continue the heating and stirring for a further 3 hours. Allow to cool. Filter with suction, separate the organic layer from the filtrate and wash it with 25 ml of saturated salt solution. Dry with magnesium sulphate and filter again with suction. Distil from a flask with fractionating side-arm (compare Fig. 2.108) under reduced pressure: aniline (about 80 g) distils at 81 °C/12 mmHg and the temperature rises rapidly. Collect the benzylaniline at 170–190 °C/12 mmHg (most of it distils at 178–180 °C/12 mmHg); this solidifies on cooling, melts at 34–36 °C, and is sufficiently pure for most purposes. The yield is 52 g (85%). If required perfectly pure, it may be recrystallised from about 35 ml of light petroleum, b.p. 60–80 °C; cool the solution in a freezing mixture to induce crystallisation, filter at the pump, wash with a little cold light petroleum, press and dry. The recrystallised *N*-benzylaniline has m.p. 36 °C.

Notes on the preparation of secondary alkylarylamines. The preparation of *N-propyl-*, *N-isopropyl-* and *N-butylanilines* can be conveniently carried out by heating the alkyl bromide with an excess (2.5–4 mol) of aniline for 6–12 hours. The tendency for the alkyl halide to yield the corresponding tertiary amine is thus repressed and the product consists almost entirely of the secondary amine and the excess of primary amine combined with the hydrogen bromide liberated in the reaction. The separation of the primary and secondary amines is easily accomplished by the addition of an excess of 50 per cent zinc chloride solution: aniline and its homologues form sparingly soluble additive compounds of the type B_2ZnCl_2 whereas the alkylanilines do not react with zinc chloride in the presence of water. The excess of primary amine can be readily recovered by decomposing tthe chlorozincate with sodium hydroxide solution followed by steam distillation or solvent extraction. The yield of secondary amine is about 70 per cent of the theoretical.

The experimental details for *N-propylaniline* are as follows. Reflux a mixture of 230 g (2.5 mol) of aniline and 123 g (1 mol) of propyl bromide for 8–10 hours. Allow to cool, render the mixture alkaline and add a solution of 150 g (1.1 mol) of zinc chloride in 150 g of water. Cool the mixture and stir: after 12 hours, filter at the pump and drain well. Extract the thick paste several times with boiling light petroleum, b.p. 60–80 °C (it is best to use a Soxhlet apparatus), wash the combined extracts successively with water and dilute ammonia solution, and then dry over anhydrous potassium carbonate or magnesium sulphate. Remove the solvent on a water bath, and distil the residue through a well-lagged fractionating column. Collect the *N*-propylaniline at 218–220 °C; the yield is 80 g (59%). Treat the pasty solid chlorozincate with an excess of sodium hydroxide solution and steam distil: 130 g of pure aniline are recovered.

N-Isopropylaniline, b.p. 206–208 °C, and *N*-butylaniline, b.p. 235–237 °C, may be similarly prepared.

Notes on the preparation of tertiary alkylarylamines. *Pure dialkylanilines* may be prepared by refluxing the monoalkylaniline (1 mol) with an alkyl bromide (2 mol) for 20–30 hours; the solid product is treated with excess of sodium hydroxide solution, the organic layer separated, dried and distilled. The excess of alkyl bromide passes over first, followed by the dialkylaniline. *N,N-*

dipropylaniline, b.p. 242–243 °C, and *N,N-dibutylaniline*, b.p. 269–270 °C, are thus readily prepared.

If the tertiary amines are suspected of being contaminated with primary and/or secondary amines, they may be purified by treatment with acetic anhydride: the following procedure is illustrative. Into a 250-ml round-bottomed flask, fitted with a reflux condenser, place 50 g (52.5 ml, 0.414 mol) of a good commercial sample of dimethylaniline and 25 g (23 ml, 0.245 mol) of acetic anhydride. Heat under reflux for 3 hours and allow to cool. Transfer to a 100-ml distillation flask and distil using an air bath. Some acetic acid and the excess acetic anhydride passes over first, followed by pure dimethylaniline (a colourless liquid) at 193–194 °C. There is a small dark residue in the flask. The yield depends upon the purity of the commercial sample but is not usually less than 40 g.

Purification of N-methylaniline. The laboratory preparation of N-methyl- and N-ethylanilines is hardly worth while since commercial grades of good quality (97–99% pure) are available. The following procedure, however, illustrates a useful and instructive method of purifying crude samples of secondary alkylarylamines via the derived N-nitroso compound [**CAUTION**: (1)].

N-*Nitroso*-N-*methylaniline* (*methylphenylnitrosamine*). Place 53.5 g (0.5 mol) of commercial N-methylaniline, 72.5 ml of concentrated hydro-chloric acid and 200 g of crushed ice in a 500-ml beaker equipped with a mechanical stirrer. Support a separatory funnel with a long bent stem containing a solution of 36 g (0.52 mol) of sodium nitrite in 125 ml of water over the beaker. Stir the solution and run in the sodium nitrite solution during 10 minutes; do not allow the temperature to rise about 10 °C and add more ice if necessary. Continue the stirring for a further hour. Separate the oily layer, wash it once with 50 ml of water and dry it with anhydrous calcium sulphate. Distil under reduced pressure. Collect the N-nitroso-N-methylaniline (a pale yellow liquid) at 120 °C/13 mmHg. The yield is about 65 g (96%).

Reduction of N-*nitroso*-N-*methylaniline*. Into a 1-litre round-bottomed flask, fitted with a reflux condenser, place 39 g (0.29 mol) of N-nitroso-N-methylaniline and 75 g of granulated tin. Add 150 ml of concentrated hydrochloric acid in portions of 25 ml (compare Expt 6.48); do not add the second portion until the vigorous action produced by the previous portion has subsided, etc. Heat the reaction mixture on a water bath for 45 minutes, and allow to cool. Add cautiously a solution of 135 g of sodium hydroxide in 175 ml of water, and steam distil (see Section 2.25); collect about 500 ml of distillate. Saturate the distillate with salt, separate the organic layer, extract the aqueous layer with 50 ml of ether and combine the extract with the organic layer. Dry with anhydrous potassium carbonate, remove the ether on a water bath and distil the residual liquid collecting the pure methylaniline at 193–194 °C as a colourless liquid. The yield is 23 g (74%).

Note. (1) The potentially carcinogenic nature of N-nitroso compounds is again emphasised (see Section 2.3.4, p. 49).

Experiment 6.57 *N,N*-DIMETHYLANILINE

$$3Ph \cdot NH_2 + 2(MeO)_3PO \longrightarrow 3Ph \cdot NMe_2 + 2H_3PO_4$$

Place 28 g (27.5 ml, 0.3 mol) of pure aniline and 28 g (23 ml, 0.2 mol) of purified trimethyl phosphate in a 500-ml round-bottomed flask equipped with a reflux condenser. Heat gently at first and remove the flame when the vigorous and exothermic reaction commences. When the latter subsides, two layers are present; heat under gentle reflux for two hours. Cool the mixture to about 50 °C, add a solution of 25 g of sodium hydroxide in 100 ml of water, reflux the mixture for 1 hour, then pour into a 600-ml beaker and allow to cool to room temperature. Pour off the oily layer of amine from the solid sodium phosphate, add water to the latter and extract the aqueous solution with ether. Dry the combined oil and ether extract with magnesium sulphate, distil off the ether, treat the residue with an equal volume of acetic anhydride and allow to stand overnight. (The acetic anhydride treatment will remove any monoalkylaniline present.) Then add hydrochloric acid (20 ml of the concentrated acid and 30 ml of water), shake until the base dissolves, extract the solution with two 30 ml portions of ether and add 25 per cent sodium hydroxide solution to the water layer to liberate the base. Collect the oil by extracting the mixture with ether, dry the ethereal solution with magnesium sulphate and remove the ether on a water bath. Distil the residue, using an air condenser, and collect the dimethylaniline at 192–193 °C. The yield is 28 g (76.5%).

The m.s. shows principal fragment ions at m/z 121 (M), 120 (M − H, base peak), 105 (120 − Me or 121 − CH_4), 104 (120 − CH_4), 77 (M − NMe_2), and 51 (77 − C_2H_2).

Cognate preparation. N,N-*Diethylaniline*. Use 28 g of pure aniline and 36 g (34 ml, 0.2 mol) of purified triethyl phosphate, and proceed exactly as described for dimethylaniline. The reaction is not so vigorous initially. Separation into two layers occurs after 30 to 90 minutes. The yield of diethylaniline, b.p. 215–216 °C, is 41–45 g (91–100%).

Experiment 6.58 *N*-(*m*-NITROBENZYL)ANILINE

$$m\text{-}O_2N \cdot C_6H_4 \cdot CHO + H_2N \cdot Ph \longrightarrow$$

$$m\text{-}O_2N \cdot C_6H_4 \cdot CH{=}N \cdot Ph \xrightarrow{\text{NaBH}_4} m\text{-}O_2N \cdot C_6H_4 \cdot CH_2 \cdot NH \cdot Ph$$

N-(*m*-**Nitrobenzylidene**)**aniline**. In a 100-ml round-bottomed flask fitted with a reflux condenser, place 7.5 g (0.05 mol) of *m*-nitrobenzaldehyde, 4.6 g (0.05 mol) of aniline and 20 ml of rectified spirit (1). Heat the solution under reflux, using a water bath, for 20 minutes, add water until a slight cloudiness persists and set the solution on one side to cool. The oil which separates may be induced to crystallise by rubbing with a glass rod. Collect the solid deposit by filtration and wash well with cold aqueous ethanol; 10 g (88%) of air-dried crude Schiff base is obtained which is sufficiently pure for conversion into the amine. Recrystallise a small portion from aqueous methanol to give light-fawn crystals having m.p. 65–66 °C.

N-(*m*-**Nitrobenzyl**)**aniline**. Fit a two-necked round-bottomed flask with a reflux condenser, place a stopper in the side-neck and insert a magnetic follower. Mount the flask on a water bath sited on a magnetic-stirrer unit. In the flask place 10 g (0.044 mol) of the above Schiff base and add 100 ml of methanol. Warm the solution to about 40 °C and with stirring add portionwise, over a period of 30 minutes, 1.7 g (0.044 mol) of sodium borohydride; a steady evolution of hydrogen occurs. Now heat the solution under reflux for a further 15 minutes, then add 100 ml of water and cool. Collect the solid amine which, after air-drying, has m.p. 80–81 °C; upon recrystallisation from aqueous methanol 9 g (90%) of pure N-(m-nitrobenzyl)aniline, m.p. 84–85 °C, is obtained.

Note. (1) The Schiff base may be prepared by heating the components in the absence of solvent at 100 °C for 15 minutes, cooling, and then stirring the product with methanol to induce crystallisation.

Cognate preparations. N-(p-*Methoxybenzyl*)aniline. Prepare the Schiff base, N-(p-methoxybenzylidene)aniline from 6.8 g (0.05 mol) of anisaldehyde and 4.6 g (0.05 mol) of aniline in 20 ml of rectified spirit under the conditions described above. The yield of crude product is 8.5 g (81%); recrystallisation of a small portion gives white plates, m.p. 57–58 °C. Reduce 7 g (0.034 mol) of crude product with 1.1 g (0.034 mol) of sodium borohydride. The yield of pure N-(p-methoxybenzyl)aniline, m.p. 46–47 °C, is 6.1 g (85%).

N-*Benzyl*-m-*nitroaniline*. Prepare the Schiff base from 5.3 g (0.05 mol) of benzaldehyde and 6.9 g (0.05 mol) of m-nitroaniline in 30 ml of rectified spirit. Yield of crude product is 10 g (88%); the pure compound has m.p. 71–72 °C. Reduce 7.8 g (0.034 mol) of crude Schiff base with 1.1 g (0.034 mol) of sodium borohydride. Pure N-benzyl-m-nitroaniline is obtained as orange-yellow crystals, m.p. 106–107 °C; the yield is 7.0 g (90%).

6.5.6 SOME METHODS FOR THE PROTECTION OF THE AMINO AND IMINO GROUPS

The most commonly used protecting group for the aromatic amino and imino groups is the N-acetyl group. Furthermore this protecting group restricts the activating influence of the amino group towards electrophilic ring substitution reactions. The general methods of formation are detailed in Section 9.6.21, p. 1273 (see also Section 6.6.2, p. 916). Other methods of protection which are generally suitable are described under aliphatic amines (Section 5.16.7, p. 784).

6.6 SUBSTITUTION PRODUCTS OF AROMATIC AMINES

1. Nuclear substitution products (Expts 6.59 to 6.66).
2. Acylated amines and their substitution reactions (Expts 6.67 and 6.68).

The important *spectroscopic features* (i.r., p.m.r., m.s., and u.v.–visible) which are observed with the varied range of substitution products of aromatic amines are descriptively discussed under appropriate preparative sections.

6.6.1 NUCLEAR SUBSTITUTION PRODUCTS

The nuclear substitution reactions that are considered in this section are (a) halogenation, (b) nitrosation, (c) sulphonation and (d) *ortho* formylation. All these reactions involve the introduction of the substituent into an arylamine, and the regioselectivity is governed by the *ortho/para* orientating effect of the amino (or alkylamino) group. As appropriate, experimental conditions are described which control the extent of substitution. The examples provide interesting comparison with the substitution reactions of acylated amines (Section 6.6.2, p. 916).

HALOGENATION
The free amino group strongly activates the aromatic ring towards electrophilic attack and aromatic substitution of amines often results in polysubstitution. For example, the bromination of aniline yields largely 2,4,6-tribromoaniline (Expt 6.59).

Monosubstitution of the free amine may be achieved by using a less reactive electrophile. Thus aniline and *o*-toluidine may be mono-iodinated (Expt 6.60) by treatment with iodine (in the presence of sodium hydrogen carbonate or calcium carbonate to remove the liberated hydrogen iodide), the substituent entering the position *para* to the amino group. Direct iodination can also be effected by using the more powerfully electrophilic reagent, iodine monochloride $(\overset{\delta+}{I} \rightarrow \overset{\delta-}{Cl})$; with *p*-aminobenzoic acid both of the positions *ortho* to the amino group are substituted to give 4-amino-3,5-diiodobenzoic acid (cognate preparation in Expt 6.60).

Several reactive chloro compounds have been used to attempt to effect the controlled monochlorination of aromatic amines. One such reagent is *N*-chlorosuccinimide, with which the chlorination of aniline, for example, can be largely restricted to monosubstitution, although a mixture of isomers (*ortho–para*, 1.9:1) is obtained.[24] One approach to the achievement of specific *ortho* chlorination is illustrated by the synthesis of *o*-chlorobenzanilide (Expt 6.61), readily hydrolysable to *o*-chloroaniline. The anilide is formed, by a type of S_Ni' mechanism indicated below, when *N*-phenylhydroxylamine is benzoylated and the product is treated with thionyl chloride.[25] The reaction has been successfully applied to several substituted *N*-phenylhydroxylamines, prepared by the controlled reduction of the corresponding substituted nitro compounds (cf. Expt 6.87).

NITROSATION
The nitrosonium ion ($\overset{\oplus}{N}O$), generated *in situ* from sodium nitrite in the presence of hydrochloric acid at 0–5 °C, is also a weak electrophile and with tertiary amines, e.g. *N,N*-dimethylaniline, ring substitution occurs leading to the *p*-nitroso derivative (Expt 6.62).

907

$$C_6H_5 \cdot NR_2 \xrightarrow{\text{NaNO}_2/\text{HCl}} p\text{-ON} \cdot C_6H_4 \cdot NR_2$$

Secondary aromatic amines under these conditions form initially the *N*-nitroso derivative (see notes on the purification of secondary amines in Expt 6.56), which when treated with hydrogen chloride in anhydrous ethanol–ether solution rearranges to the nuclear substituted nitrosoamine (eg.. *p*-nitroso-*N*-methylaniline, cognate preparation in Expt 6.62). This rearrangement proceeds via the intermediate formation of the electrophilic nitrosyl chloride:

SULPHONATION

When conditions for the electrophilic substitution are strongly acidic, extensive protonation of the nitrogen lone electron pair occurs and its activating influence is considerably diminished. For example, if aniline is treated with an excess of concentrated sulphuric acid, and the resulting mixture (which contains aniline sulphate) is heated at 180 °C until a test portion when mixed with sodium hydroxide solution no longer liberates aniline, *p*-aminobenzenesulphonic acid (sulphanilic acid) is formed; this separates as the dihydrate upon pouring the cooled mixture into water. The mechanism of this reaction is uncertain; a possible pathway is the rearrangement of the intermediate phenylsulphamic acid (1). The product is more appropriately represented by the zwitterionic structure (2) (Expt 6.63).

2-Aminobenzenesulphonic acid (orthanilic acid, Expt 6.64) is readily prepared by the reduction of 2-nitrobenzenesulphonic acid. The latter may be prepared by the hydrolysis of the corresponding sulphonyl chloride which is obtained from di-*o*-nitrophenyl disulphide. The preparation of this disulphide involves the use of the reactive aryl halide, 2-chloronitrobenzene (cf. Expts 6.93

and 6.100) in a disulphide-forming nucleophilic displacement using sodium disulphide.

A straightforward route to the *meta* isomer (metanilic acid, Expt 6.65) is provided by the sulphonation of nitrobenzene followed by reduction of the nitro group.

ORTHO FORMYLATION

o-Aminobenzaldehyde and its ring-substituted analogues are important starting materials in the formation of various heterocyclic ring systems. The parent compound has been prepared by the reduction of *o*-nitrobenzaldehyde or by the oxidation of *o*-toluidine. Such procedures are obviously accompanied by undesirable side reactions resulting from the susceptibility of the carbonyl and amino groups respectively to the reaction conditions. The presence of other nuclear substituents could clearly provide further problems in chemoselectivity. A procedure involving the specific *ortho* formylation of a *para*-substituted amine, using 1,3-dithiane as the formyl equivalent, is worthy of note[26] (cf. *ortho* formylation of phenols, p. 993). The reaction sequence is formulated and illustrated by the conversion of *p*-toluidine into 2-acetamido-5-methylbenzaldehyde (Expt 6.66).

Experiment 6.59 2,4,6-TRIBROMOANILINE

$$Ph \cdot NH_2 + 3Br_2 \longrightarrow H_2N \cdot C_6H_2Br_3$$

Dissolve 10 g (0.11 mol) of aniline in 40 g of glacial acetic acid and stir well with a mechanical stirrer while running in slowly a solution of 52.8 g (17 ml, 0.33 mol) of bromine (**CAUTION**) in 34 ml of glacial acetic acid. The beaker should be cooled in ice during the addition as the reaction is exothermic. The final product (a pasty mass) should be coloured yellow by the addition of a little more bromine if necessary. Pour into excess of water, filter at the pump, wash well with water, press thoroughly and dry. The yield of tribromoaniline, m.p. 119–120 °C, is quantitative. Recrystallise a small portion from industrial (or rectified) spirit; m.p. 120 °C.

The m.s. provides a good example of the fragmentation pattern of a compound having three bromine atoms; m/z 333 (M, RA 29%), 331 (M, RA 99%), 329 (M, RA 100%), 327 (M, RA 32%), 170 and 168 (M—Br_2, two each of RA 22.5%). The p.m.r. spectrum ($CDCl_3$, TMS) has δ 4.59 (broad s, 2H, NH_2), and 7.50 (s, 2H, $C_{3,5}$—H).

Conversion to 2,4,6-tribromoacetanilide. Dissolve 1 g of 2,4,6-tribromoaniline in 20 ml of acetic anhydride and add 2 drops of concentrated sulphuric acid. After 10 minutes, pour the reaction mixture into excess of warm water. Filter off the tribromoacetanilide, wash and dry; the m.p. is 231 °C. Recrystallise from alcohol; the m.p. is raised to 232 °C.

Experiment 6.60 *p*-IODOANILINE

$$Ph \cdot NH_2 + I_2 + NaHCO_3 \longrightarrow p\text{-}I \cdot C_6H_4 \cdot NH_2 + NaI + CO_2 + H_2O$$

Into a 1-litre beaker, provided with a mechanical stirrer, place 37 g (36 ml, 0.4 mol) of aniline, 50 g (0.6 mol) of sodium hydrogen carbonate and 350 ml of water; cool to 12–15 °C by the addition of a little crushed ice. Stir the mixture, and introduce 85 g (0.33 mol) of powdered, resublimed iodine in portions of 5–6 g at intervals of 2–3 minutes so that all the iodine is added during 30 minutes. Continue stirring for 20–30 minutes, by which time the colour of the free iodine in the solution has practically disappeared and the reaction is complete. Filter the crude p-iodoaniline with suction on a Buchner funnel, drain as completely as possible and dry it in the air. Save the filtrate for the recovery of the iodine (1). Place the crude product in a 750-ml round-bottomed flask fitted with a reflux double surface condenser, add 325 ml of light petroleum, b.p. 60–80 °C, and heat in a water bath maintained at 75–80 °C. Shake the flask frequently and after about 15 minutes, slowly decant the clear hot solution into a beaker set in a freezing mixture of ice and salt, and stir constantly. The p-iodoaniline crystallises almost immediately in almost colourless needles; filter and dry the crystals in the air. Return the filtrate to the flask for use in a second extraction as before (2). The yield of p-iodoaniline, m.p. 62–63 °C, is 60 g (82%).

Notes. (1) The *iodine may be recovered* from the aqueous filtrate, containing sodium iodide, in the following manner. Add 33 ml of concentrated sulphuric acid and a solution of 65 g of sodium dichromate in 65 ml of water. Allow the iodine to settle, wash it three times by decantation, filter and allow to dry on a clock glass. The weight of crude iodine is about 50 g.

(2) Two extractions usually suffice, but if much organic material remains, a third extraction should be made. If the p-iodoaniline from the second and third extractions is coloured, it should be refluxed for a short period in light petroleum solution with a little decolourising carbon and filtered through a hot water funnel (**CAUTION:** flammable).

Cognate preparations.   2-Amino-5-iodotoluene.   Triturate 20 g (0.14 mol) of dry o-toluidine hydrochloride and 35.5 g (0.14 mol) of powdered iodine in a mortar and then grind in 17.5 g of precipitated calcium carbonate. Transfer the mixture to a conical flask, and add 100 ml of distilled water with vigorous shaking of the flask. Allow the mixture to stand for 45 minutes with occasional agitation, then heat gradually to 60–70 °C for 5 minutes, and cool. Transfer the contents of the flask to a separatory funnel, extract the amine with three 80 ml portions of ether, dry the extract with anhydrous calcium chloride or magnesium sulphate and remove the excess of solvent. The crude 2-amino-5-iodotoluene separates in dark crystals. Recrystallise from 50 per cent alcohol; nearly white crystals, m.p. 87 °C, are obtained, 26 g (80%).

4-Amino-3,5-diiodobenzoic acid.   In a 2-litre beaker, provided with a mechanical stirrer, dissolve 10 g (0.073 mol) of pure p-aminobenzoic acid, m.p. 192 °C (Expt 6.52), in 450 ml of warm (75 °C) 12.5 per cent hydrochloric acid. Add a solution of 48 g (0.295 mol) of iodine monochloride (Section 4.2.44, p. 440) in 40 ml of 25 per cent hydrochloric acid and stir the mixture for one minute: during this time a yellow precipitate commences to appear. Dilute the reaction mixture with 1 litre of water whereupon a copious precipitate is deposited. Raise the temperature of the well-stirred mixture gradually and maintain it at 90 °C for 15 minutes. Allow to cool to room temperature, filter, wash thoroughly with water and dry in the air; the yield of crude acid is 24 g. Purify the product by dissolving it in dilute sodium hydroxide solution and precipitate with dilute hydrochloric acid: the yield of air-dried 4-amino-3,5-diiodobenzoic acid, m.p. >350 °C, is 23 g (81%).

Experiment 6.61 *o*-**CHLOROBENZANILIDE**[25]

HO·N·CO·Ph   HN·CO·Ph

$\xrightarrow{\text{SOCl}_2}$

A solution of thionyl chloride (1.42 g, 12 mmol) in dry benzene (**CAUTION**) or ether (10 ml) is added dropwise to a stirred solution of *N*-phenyl-benzohydroxamic acid (2.13 g, 10 mmol, Expt 6.87) in dry benzene (**CAUTION**) or ether (100 ml) at 0–5 °C during 20 minutes. The reaction mixture is stirred at the same temperature for 30 minutes and at room temperature for 1 hour. Water (50 ml) is added. The solvent layer is washed with a 5 per cent solution of sodium hydroxide (10 ml) and then with water until free of alkali. After drying over anhydrous sodium sulphate and concentrating, *o*-chlorobenzanilide (1.90 g, 84%) is isolated, and is recrystallised from aqueous 30 per cent ethanol. The m.p. is 104 °C (1).

Note. (1) The benzanilide may be hydrolysed to *o*-chloroaniline using the conditions noted on p. 1050.

Experiment 6.62 *p*-**NITROSO-*N*,*N*-DIMETHYLANILINE**

$$C_6H_5 \cdot N(CH_3)_2 \xrightarrow{\text{HNO}_2} p\text{-ON} \cdot C_6H_4 \cdot N(CH_3)_2$$

Dissolve 30 g (31.5 ml, 0.25 mol) of *N*,*N*-dimethylaniline in 105 ml of concentrated hydrochloric acid contained in a 600-ml beaker, and add finely crushed ice until the temperature falls below 5 °C. Stir the contents of the beaker mechanically (or, less satisfactorily, with a thermometer) and slowly add (*c*. 10 minutes) a solution of 18 g (0.26 mol) of sodium nitrite in 30 ml of water from a separatory funnel, the stem of which dips beneath the surface of the liquid. Maintain the temperature below 8 °C by the addition of ice, if necessary. When all the nitrite solution has been added, allow the mixture to stand for 1 hour, filter the yellow crystalline *p*-nitrosodimethylaniline hydrochloride at the pump, wash it with 40 ml of dilute hydrochloric acid (1:1), drain well and finally wash with a little alcohol. The yield is good and depends upon the purity of the original dimethylaniline. If the pure hydrochloride is required, it may be recrystallised from hot water in the presence of a little dilute hydrochloric acid; yellow needles, m.p. 177 °C. Recrystallisation is, however, unnecessary if the free base is to be prepared.

Transfer 30 g of the hydrochloride to a 500-ml separatory funnel, add 100 ml of water and shake until a thin paste of uniform consistency is obtained; add cold 10 per cent aqueous sodium hydroxide solution with shaking until the whole mass has become bright green (the colour of the free base) and the mixture has an alkaline reaction. Extract the free base by shaking with two 60 ml portions of benzene (1). Dry the combined benzene extracts with a little anhydrous potassium carbonate, and filter into a distilling flask fitted with a water condenser. Distil off about half of the benzene. Upon cooling the residual solution, the *p*-nitrosodimethylaniline crystallises in deep green leaflets. Filter these off and dry them in the air. The yield of *p*-nitrosodimethylaniline, m.p. 85 °C, from the hydrochloride is almost quantitative.

Note. (1) The base is only slightly soluble in ether, thus rendering its use uneconomical. It may be extracted with chloroform and precipitated from the dried chloroform solution with carbon tetrachloride.

Cognate preparation

p-*Nitroso*-N-*methylaniline*.   Dissolve 5 g of N-nitroso-N-methylaniline [**CAUTION**: *see Expt 6.56, Note (1)*] in 10 ml of anhydrous ether, and add 20 g of a saturated solution of hydrogen chloride in absolute ethanol. Allow to stand. After some time a mass of crystalline needles of the hydrochloride of p-nitroso-N-methylaniline separates. Filter with suction on a sintered glass funnel and wash with a mixture of alcohol and ether. Dissolve the solid in water and add a slight excess of sodium carbonate solution or dilute ammonia solution. Filter off the blue-green free base, and recrystallise it from benzene. The yield of p-nitroso-N-methylaniline, m.p. 118 °C, is 4.5 g (90%).

Experiment 6.63   SULPHANILIC ACID

$$Ph\cdot NH_2 + H_2SO_4 \longrightarrow p\text{-}H_3\overset{\oplus}{N}\cdot C_6H_4\cdot SO_3^{\ominus}$$

Place 20.4 g (20 ml, 0.22 mol) of aniline in a 250-ml round-bottomed flask and cautiously add 74 g (40 ml) of concentrated sulphuric acid in small portions; swirl the mixture gently during the addition and keep it cool by occasionally immersing the flask in cold water. Support the flask in an oil bath, and heat the mixture at 180–190 °C (*fume cupboard*) for about 5 hours (1). The sulphonation is complete when a test portion (2 drops) is completely dissolved by 3–4 ml of c. 2 M sodium hydroxide solution without leaving the solution cloudy. Allow the product to cool to about 50 °C and pour it carefully with stirring into 400 g of cold water or of crushed ice. Allow to stand for 10 minutes, and collect the precipitated sulphanilic acid on a Buchner funnel, wash it well with water and drain. Dissolve the crude sulphanilic acid in the minimum volume of boiling water (450–500 ml); if the resulting solution is coloured, add about 4 g of decolourising carbon and boil for 10–15 minutes. Filter through a hot-water funnel or through a preheated Buchner funnel. Upon cooling, the sulphanilic acid dihydrate separates in colourless crystals. When the filtrate is quite cold, filter the crystals with suction, wash with about 10 ml of cold water and press thoroughly with a wide glass stopper. Dry between sheets of filter paper or in a desiccator containing anhydrous calcium chloride; in the latter case, the water of crystallisation (and hence the crystalline form) is lost. The yield of sulphanilic acid is 20–22 g (52–58%). The substance does not melt sharply and no attempt should be made to determine the melting point; the crystals are efflorescent.

Note. (1) If 40 ml of 10 per cent oleum is cautiously added to the aniline sulphate mixture, sulphonation proceeds much more rapidly and the time of heating is reduced from 5 hours to 1 hour.

Experiment 6.64 ORTHANILIC ACID (*2-Aminobenzenesulphonic acid*)

$$o\text{-}O_2N \cdot C_6H_4 \cdot SO_3H \xrightarrow[\text{Me} \cdot CO_2H]{\text{Fe}} o\text{-}H_2N \cdot C_6H_4 \cdot SO_3H$$

Di-*o*-nitrophenyl disulphide. Place 120 g (0.5 mol) of crystallised sodium sulphide (1) and 500 ml of rectified spirit in a 1-litre round-bottomed flask provided with a reflux condenser. Heat the flask on a water bath until the sulphide dissolves. Then add 16 g (0.5 mol) of finely powdered sulphur and continue the heating until all the sulphur dissolves forming a brownish-red solution of sodium disulphide (2). Prepare a solution of 105 g (0.66 mol) of *o*-chloronitrobenzene in 175 ml of rectified spirit in a 2-litre round-bottomed flask equipped with a reflux condenser; by means of a pressure-equalising dropping funnel, add the sodium disulphide solution down the condenser slowly and at such a rate that the reaction is under control. Heat the mixture on a water bath, gently at first until the violent reaction subsides, and then with the water boiling vigorously for 2 hours. Allow to cool. Filter with suction on a Buchner funnel. Transfer the mixture of organic disulphide and sodium chloride to a 400-ml beaker and stir thoroughly with 175 ml of water to remove the salt. Filter at the pump, drain well and wash the crystalline residue on the filter with 35 ml of alcohol to remove any unreacted *o*-chloronitrobenzene. The residual di-*o*-nitrophenyl disulphide melts at 193–195 °C and weighs 70 g (68%).

Notes. (1) Crystallised sodium sulphide $Na_2S \cdot 9H_2O$ is very deliquescent, and only a sample which has been kept in a tightly stoppered bottle should be used; crystals as dry as possible should be selected. Alternatively, an equivalent amount of analysed fused sodium sulphide may be employed; this dissolves somewhat more slowly in alcohol.

(2) If some sodium disulphide separates at the bottom of the flask, this should be dissolved in a little more rectified spirit and added to the chloronitrobenzene solution.

***o*-Nitrobenzenesulphonyl chloride.** Equip a 1-litre three-necked flask with an inlet tube for introducing chlorine well beneath the surface of the liquid, an efficient mechanical stirrer and a reflux condenser. Set up the assembly (cf. Fig. 2.59) in the fume cupboard and absorb the excess of chlorine in sodium hydroxide solution. Place 60 g (0.195 mol) of di-*o*-nitrophenyl disulphide, 300 ml of concentrated hydrochloric acid and 60 ml of concentrated nitric acid in the flask, pass a stream of chlorine from a cylinder into the mixture at the rate of 2 bubbles per second and warm the solution to 70 °C on a water bath. After about 30 minutes, the disulphide melts and the solution assumes an orange-red colour; after the melting stage has been reached, the passage of the chlorine and the heating are continued for 1 hour. Immediately separate

913

the sulphonyl chloride from the supernatant liquid by decantation, wash with two 90 ml portions of water at about 70 °C and allow to solidify. Drain the water from the solid mass as completely as possible. Dissolve the sulphonyl chloride in 45 ml of glacial acetic acid at 50–60 °C, and rapidly filter the solution at the pump. Cool the filtrate in cold water and stir it vigorously so that the sulphonyl chloride separates in small crystals. Triturate the mixture well with 300 ml of cold water and decant through a Buchner funnel; repeat the process twice. Finally add 300 ml of cold water and 3 ml of concentrated ammonia solution to the mixture, stir well and filter immediately, through a Buchner funnel, wash with 60 ml of water, drain well and dry in the air. The yield of moderately pure o-nitrobenzenesulphonyl chloride, m.p. 64–65 °C, is 72 g (84%). The undried material may be used in the preparation of orthanilic acid.

Orthanilic acid. Fit a 1-litre three-necked flask with a sealed mechanical stirrer and a reflux condenser. Place 60 g of o-nitrobenzenesulphonyl chloride, 30 g of anhydrous sodium carbonate and 180 ml of water in the flask. Heat the mixture to boiling, with stirring; the hydrolysis of the sulphonyl chloride to the sulphonic acid is complete within 40 minutes after the compound has melted. Filter the orange-red solution and acidify (to litmus) with acetic acid (about 7.5 ml are required). Transfer the solution to the original flask (which has been thoroughly rinsed with water) equipped as before. Heat the solution to boiling, and add 105 g of finely divided iron filings (about 20 mesh) with vigorous stirring at the rate of about 7.5 g every 15 minutes. The mixture soon becomes deep brown and exhibits a tendency to froth. Complete the reaction by stirring for a further 4 hours, i.e. until a test portion when filtered yields an almost colourless filtrate; if the filtrate is orange or red, the heating and stirring must be continued. When the reduction is complete, add 2 g of decolourising carbon, filter the hot reaction mixture at the pump and wash the residue with three 15 ml portions of hot water: combine the washings with the main solution. Cool the filtrate to about 15 °C, and add 28.5 ml of concentrated hydrochloric acid slowly, and cool to 12–15 °C. Filter the acid with suction on a Buchner funnel, wash with a little cold water, followed by a little ethanol and dry upon filter paper in the air. The yield is 27 g (57%); the orthanilic acid has a purity of 97–99 per cent. If required perfectly pure it may be recrystallised from hot water; it decomposes at about 325 °C.

Experiment 6.65 METANILIC ACID (*3-Aminobenzenesulphonic acid*)

$$\text{Ph·NO}_2 \xrightarrow{\text{oleum}} m\text{-O}_2\text{N·C}_6\text{H}_4\text{·SO}_3\text{H} \xrightarrow{\text{Fe/H}_3\text{O}^{\oplus}} m\text{-H}_3\overset{\oplus}{\text{N}}\text{·C}_6\text{H}_4\text{·SO}_3^{\ominus}$$

CAUTION: *This preparation should be carried out in a fume cupboard.*

In a 500-ml bolt-head flask, provided with a mechanical stirrer, place 70 ml of oleum (20% SO₃) and heat it in an oil bath to 70 °C. By means of a dropping funnel, supported so that the stem is just above the surface of the acid, introduce 41 g (34 ml, 0.33 mol) of nitrobenzene slowly and at such a rate that the temperature of the well-stirred mixture does not rise above 100–105 °C. When all the nitrobenzene has been introduced, continue the heating at 110–115 °C for 30 minutes. Remove a test portion and add it to excess of water. If the odour of nitrobenzene is still apparent, add a further 10 ml of fuming sulphuric acid, and heat at 110–115 °C for 15 minutes: the reaction mixture should then be free from nitrobenzene. Allow the mixture to cool and pour it carefully with vigorous stirring on to 200 g of finely crushed ice contained in a beaker. All the nitrobenzenesulphonic acid passes into solution; if a little sulphone is present, remove this by filtration. Stir the solution mechanically and add 70 g of sodium chloride in small portions: the sodium salt of *m*-nitrobenzenesulphonic acid separates as a pasty mass. Continue the stirring for about 30 minutes, allow to stand overnight, filter and press the cake well. The latter will retain sufficient acid to render unnecessary the addition of acid in the subsequent reduction with iron.

Place 84 g (1.5 mol) of iron filings and 340 ml of water in a 1.5- or 2-litre bolt-head flask equipped with a mechancial stirrer. Heat the mixture to boiling, stir mechanically and add the sodium *m*-nitrobenzenesulphonate in small portions during 1 hour. After each addition the mixture foams extensively: a wet cloth should be applied to the neck of the flask if the mixture tends to froth over the sides. Replace from time to time the water which has evaporated so that the volume is approximately constant. When all the sodium salt has been introduced, boil the mixture for 20 minutes. Place a small drop of the suspension upon filter paper and observe the colour of the 'spot': it should be a pale brown but not deep brown or deep yellow. If it is not appreciably coloured, add anhydrous sodium carbonate cautiously, stirring the mixture, until red litmus paper is turned blue and a test drop upon filter paper is not blackened by sodium sulphide solution. Filter at the pump and wash well with hot water. Concentrate the filtrate to about 200 ml, acidify with concentrated hydrochloric acid to Congo red and allow to cool. Filter off the metanilic acid and dry upon filter paper. A further small quantity may be obtained by concentrating the mother-liquor. The yield is 55 g (95%).

Experiment 6.66 2-ACETAMIDO-5-METHYLBENZALDEHYDE[26]

2-Amino-5-methylbenzaldehyde trimethylene mercaptal. To a rapidly stirred solution of 5.4 g (0.050 mol) of *p*-toluidine and 6.0 g (0.050 mol) of 1,3-dithiane in 600 ml of dry dichloromethane under nitrogen at −78 °C is added dropwise 5.4 g (0.050 mol) of t-butyl hypochorite (Section 4.2.*11*, p. 422) in 25 ml of dry dichloromethane in diffuse light. The reaction mixture is stirred for 6 hours at −78 °C. A solution of 5.4 g (0.100 mol) of sodium methoxide in

50 ml of absolute methanol at 25 °C is added while keeping the temperature below −70 °C. The reaction mixture is stirred for 30 minutes at −78 °C, then the cooling bath is removed, and the solution is warmed to room temperature. The solvent is removed *in vacuo*, 300 ml of dry toluene is added and the suspension heated under reflux for 12 hours. After cooling, the solution is concentrated *in vacuo* and the residue dissolved in 300 ml of dichloromethane, washed with two 100-ml portions of water, 100 ml of saturated sodium chloride solution and dried over magnesium sulphate. The solution is filtered and concentrated *in vacuo* leaving 12.1 g of a brown oil. After low-boiling materials are removed by distillation the residue is chromatographed on 450 g of Fisher basic alumina (Activity I) using 0.5 per cent methanol in ether. This yields 3.8 g of a mixture which is distilled under reduced pressure to give 2.8 g (25%) of product, b.p. 188–191 °C/0.10 mmHg; p.m.r. (CDCl$_3$, TMS) δ 7.14 (d, 1H, $J = 2$ Hz), 6.92 (d of d, 1H $J = 2$ and 8 Hz), 6.57 (d of d, 1 H, $J = 8$ Hz), 5.27 (s, 1H), 3.93 (broad s, 2H), 3.2–2.7 (m, 4H), 2.22 (s, 3H), and 2.3–1.7 (m, 2H).

2-Acetamido-5-methylbenzaldehyde trimethylene mercaptal. Into a 25-ml round-bottomed flask are placed 300 mg of the foregoing amino compound, 5 ml of acetic anhydride and 3 ml of pyridine. The solution is heated on a steam bath for 5 minutes, allowed to cool and poured into 50 ml of water, whereupon the product precipitates. Sodium carbonate is added until the solution is basic. The product is collected by filtration, washed with water and dried under vacuum (320 mg, 90%). Further recrystallisation affords fine, white crystals, m.p. 159–160 °C.

2-Acetamido-5-methylbenzaldehyde. Into a 5-ml round-bottomed flask equipped with a mechanical stirrer and a nitrogen atmosphere are placed 0.7–0.8 mmol of red mercury(II) oxide, 0.7–0.8 mmol of boron trifluoride–etherate (Section 4.2.8, p. 421), and 2 ml of 15 per cent aqueous tetrahydrofuran. To this stirred suspension under nitrogen is added 100 mg of the foregoing mercaptal dissolved in the minimum of tetrahydrofuran. The reaction mixture is stirred for 1 hour at 25 °C, poured into 10 ml of ether and 2 ml of 10 per cent sodium carbonate solution is added. After mixing, the organic phase is separated, and the aqueous layer is extracted with two 5-ml portions of ether. The combined organic phases are washed with two 5-ml portions of water, and 5 ml of saturated sodium chloride solution and dried over anhydrous magnesium sulphate. The reaction mixture is then filtered and concentrated under reduced pressure to give 70 mg, m.p. 72–77 °C, of 2-acetamido-5-methylbenzaldehyde. Recrystallisation from cyclohexane–tetrahydrofuran followed by sublimation gives 61 mg (92% yield) of product, m.p. 77–78.5 °C.

6.6.2 ACYLATED AMINES AND THEIR SUBSTITUTION REACTIONS

Acylation of an aromatic primary or secondary amine may be readily achieved by using an acid chloride in the presence of base; however, acetylation is more usually effected with acetic anhydride rather than the more obnoxious acetyl chloride.

In general, benzoylation of aromatic amines finds less application than acetylation in preparative work, but the process is often employed for the identification and characterisation of aromatic amines (Section 9.6.21, p. 1273). In the *Schotten–Baumann method* of benzoylation, the amine, or its salt, is dissolved or suspended in a slight excess of 8–15 per cent sodium hydroxide solution, a small excess (about 10–15% more than the theoretical quantity) of benzoyl chloride is then added and the mixture vigorously shaken in a stoppered vessel (or else the mixture is stirred mechanically). Benzoylation proceeds smoothly and the sparingly soluble benzoyl derivative separates as a solid. The use of the aqueous medium is possible because the sodium hydroxide only slowly hydrolyses the excess of benzoyl chloride to yield sodium benzoate and sodium chloride which remain in solution.

$$Ar \cdot NH_2 + Ph \cdot COCl + NaOH \longrightarrow Ar \cdot NH \cdot CO \cdot Ph + NaCl + H_2O$$

$$Ar_2NH + Ph \cdot COCl + NaOH \longrightarrow Ar_2N \cdot CO \cdot Ph + NaCl + H_2O$$

The benzoyl compounds frequently occlude traces of unchanged benzoyl chloride, which thus escapes hydrolysis by the alkali; it is therefore advisable, wherever possible, to recrystallise the benzoyl derivatives from methanol, ethanol or rectified spirit, since these solvents will esterify the unchanged acid chloride and so remove the latter from the recrystallised material. Sometimes the benzoyl compound does not crystallise well; this difficulty may be frequently overcome by the use of *p*-nitrobenzoyl chloride or 3,5-dinitrobenzoyl chloride (Expt 6.161), which usually give highly crystalline derivatives of high melting point. Benzoyl compounds are readily hydrolysed by heating with about 70 per cent sulphuric acid (alkaline hydrolysis is very slow for anilides).

$$Ar \cdot NH \cdot CO \cdot Ph + H_2SO_4 + H_2O \longrightarrow Ar \cdot \overset{\oplus}{NH_3}\}HSO_4^{\ominus} + Ph \cdot CO_2H$$

Primary amines react readily upon warming with acetic anhydride to yield, in the first instance, the monoacetyl derivative.

$$Ar \cdot NH_2 + (Me \cdot CO)_2O \longrightarrow Ar \cdot NH \cdot CO \cdot Me + Me \cdot CO_2H$$

If heating is prolonged and excess of acetic anhydride is employed, variable amounts of the diacetyl derivative are formed.

$$Ar \cdot NH \cdot CO \cdot Me + (Me \cdot CO)_2O \longrightarrow Ar \cdot N(CO \cdot Me)_2 + Me \cdot CO_2H$$

In general, however, the diacetyl derivatives are unstable in the presence of water, undergoing hydrolysis to the monoacetyl compound, so that when they (or a mixture of mono- and di-acetyl derivatives) are crystallised from an aqueous solvent, e.g. dilute ethanol, only the monoacetyl derivative is obtained. Highly substituted amines (e.g. Expt 6.59) react extremely slowly with acetic anhydride, but in the presence of a few drops of concentrated sulphuric acid as catalyst, acetylation occurs rapidly.

The disadvantages attending the use of acetic anhydride alone are absent when the acetylation is conducted in aqueous solution (see Section 9.6.21, p. 1273).

Conversion of the amino group into the acetamido group by acetylation modifies the interaction of the nitrogen lone pair with the π-electron system of the aromatic ring so that the ring is less powerfully activated towards electrophilic attack.

917

Protection of the amino group by acetylation, as in acetanilide, therefore usually permits monosubstitution reactions with appropriate electrophilic reagents to proceed smoothly. Thus with bromine, *p*-bromoacetanilide is the main product; the small quantity of the *ortho* isomer simultaneously formed can be easily eliminated by recrystallisation (Expt 6.67); hydrolysis of *p*-bromoacetanilide gives *p*-bromoaniline. Nitration leads similarly to *p*-nitroacetanilide which can be hydrolysed to *p*-nitroaniline (Expt 6.68).

Experiment 6.67 *p*-BROMOACETANILIDE AND *p*-BROMOANILINE

$$Ph\cdot NH\cdot CO\cdot Me \xrightarrow{Br_2} p\text{-}Br\cdot C_6H_4\cdot NH\cdot CO\cdot Me \xrightarrow{H_3O^\oplus} p\text{-}Br\cdot C_6H_4\cdot NH_2$$

Bromination of acetanilide.   Dissolve 13.5 g (0.1 mol) of finely powdered acetanilide in 45 ml of glacial acetic acid in a 350-ml conical flask. In another small flask dissolve 17 g (5.3 ml, 0.106 mol) of bromine in 25 ml of glacial acetic acid, and transfer the solution to a burette or a separatory funnel supported over the flask. (For precautions attending the use of bromine, see Section 4.2.9, p. 422. The preparation should be conducted in a fume cupboard.) Add the bromine solution slowly and with constant shaking to ensure thorough mixing: stand the flask in cold water. When all the bromine has been added, the solution will have an orange colour due to the slight excess of bromine; a part of the reaction product may crystallise out. Allow the final reaction mixture to stand at room temperature for 30 minutes with occasional shaking. Pour the reaction product into 400 ml of water; rinse the flask with about 100 ml of water. Stir the mixture well and if it is appreciably coloured, add just sufficient sodium metabisulphite solution to remove the orange colour. Filter the crystalline precipitate with suction on a Buchner funnel, wash thoroughly with cold water and press as dry as possible with a wide glass stopper. Recrystallise from dilute methanol or ethanol (industrial spirit). The yield of *p*-bromoacetanilide, colourless crystals m.p. 167 °C, is 18 g (84%).

The *para*-substitution pattern is confirmed by the p.m.r. spectrum (DMSO-d_6, TMS) which shows signals at δ 2.10 (s, 3H, Me), 7.45 and 7.61 (d of d, 4H, the two pairs of aromatic protons), and 10.04 (broad s, 1H, NH). The m.s. has principal fragment ions at m/z 215 (M, ^{81}Br), 213 (M, ^{79}Br), 173 (215 − CH$_2$CO), 171 (213 − CH$_2$CO), 92 (173 − ^{81}Br and 171 − ^{79}Br) and 43 (CH$_2$CO).

Hydrolysis of *p*-bromoacetanilide.   Dissolve 18 g (0.084 mol) of *p*-bromoacetanilide in 35 ml of boiling ethanol contained in a 500-ml round-bottomed flask equipped with a reflux condenser. With the aid of a pressure-equalising dropping funnel add 22 ml of concentrated hydrochloric acid down the condenser in small portions to the boiling solution. Reflux for 30–

40 minutes or until a test portion remains clear when diluted with water. Dilute with 150 ml of water, and fit the flask with a condenser set for downward distillation. Distil the mixture from an air bath (Fig. 2.46) and collect about 100 ml of distillate; the latter consists of ethyl acetate, ethanol and water. Pour the residual solution of p-bromoaniline hydrochloride into 100 ml of ice-water, and add, with vigorous stirring, 5 per cent sodium hydroxide solution until just alkaline. The p-bromoaniline separates as an oil, which soon crystallises. Filter the crystals at the pump, wash with cold water and dry in the air upon pads of filter paper. The yield is 14 g (97%), m.p. 66 °C. Recrystallisation from dilute alcohol, which results in appreciable loss, is usually unnecessary.

The assignment of signals in the p.m.r. spectrum (CDCl$_3$, TMS) is made possible from the greater difference, compared to the amide above, in the chemical shifts of the aromatic protons thus: δ 3.53 (broad s, 2H, NH$_2$), 6.57 (d, 2H, ortho-H's to NH$_2$) and 7.21 (d, 2H, ortho-H's to Br).

Experiment 6.68 p-NITROACETANILIDE and p-NITROANILINE

$$\text{Ph·NH·CO·Me} \xrightarrow[\text{H}_2\text{SO}_4]{\text{HNO}_3} p\text{-O}_2\text{N·C}_6\text{H}_4\text{·NH·CO·Me} \xrightarrow{\text{H}_3\text{O}^\oplus} p\text{-O}_2\text{N·C}_6\text{H}_4\text{·NH}_2$$

Nitration of acetanilide. Add 25 g (0.185 mol) of finely powdered, dry acetanilide to 25 ml of glacial acetic acid contained in a 500-ml beaker; introduce into the well-stirred mixture 92 g (50 ml) of concentrated sulphuric acid. The mixture becomes warm and a clear solution results. Surround the beaker with a freezing mixture of ice and salt, and stir the solution mechanically. Support a separatory funnel, containing a cold mixture of 15.5 g (11 ml) of concentrated nitric acid and 12.5 g (7 ml) of concentrated sulphuric acid, over the beaker. When the temperature of the solution falls to 0–2 °C, run in the acid mixture gradually while the temperature is maintained below 10 °C. After all the mixed acid has been added, remove the beaker from the freezing mixture, and allow it to stand at room temperature for 1 hour. Pour the reaction mixture on to 250 g of crushed ice (or into 500 ml of cold water), whereby the crude nitroacetanilide is at once precipitated. Allow to stand for 15 minutes, filter with suction on a Buchner funnel, wash it thoroughly with cold water until free from acids (test the wash water) and drain well (1). Recrystallise the pale yellow product from ethanol or industrial spirit, filter at the pump, wash with a little cold alcohol and dry in the air upon filter paper (2). (The yellow o-nitroacetanilide remains in the filtrate.) The yield of p-nitroacetanilide, a colourless crystalline solid of m.p. 214 °C, is 20 g (60%).

Investigate by thin-layer chromatography the effectiveness of the recrystallisation process in the following way. Load a 20 × 5 cm thin-layer plate (Silica Gel G) with approximately 3 mm diameter spots of concentrated solutions (in acetone) of the crude and the recrystallised product. Concentrate a portion of the ethanolic mother-liquor and similarly apply to the plate. Develop the chromatogram with a toluene–ethyl acetate mixture (4:1) and dry the plate. Mark the positions of the visible spots and leave the plate in a tank of iodine vapour to reveal the rest of the components. The recrystallised p-nitroacetanilide (R_F 0.07) should be free from the pale yellow o-isomer (R_F 0.36); these compounds are revealed by the iodine treatment.

The mother-liquor contains two readily visible yellow components, which are p-nitroaniline (R_F 0.24) and o-nitroaniline (R_F 0.45), as well as both p- and o-nitroacetanilide.

The *para*-substitution pattern is confirmed by the p.m.r. spectrum (CDCl$_3$, TMS) which shows signals at δ 2.48 (s, 3H, Me), 7.82 (d, 2H, *ortho*-H's to NH), 8.35 (d, 2H, *ortho*-H's to NO$_2$), and 9.01 (s, 1H, NH). The m.s. shows principal fragment ions at m/z 180 (M), 138 (M $-$ CH$_2$CO, base peak), 92 (138 $-$ NO$_2$), and 65 (92 $-$ HCN).

Hydrolysis of p-nitroacetanilide. Boil a mixture of 15 g (0.083) of p-nitroacetanilide and 75 ml of 70 per cent w/w sulphuric acid (3) under a reflux condenser for 20–30 minutes or until a test sample remains clear upon dilution with 2–3 times its volume of water. The p-nitroaniline is now present in the liquid as the sulphate. Pour the clear hot solution into 500 ml of cold water and precipitate the p-nitroaniline by adding excess of 10 per cent sodium hydroxide solution or of concentrated ammonia solution. When cold (cool the mixture in ice-water, if necessary), filter the yellow crystalline precipitate at the pump, wash it well with water and drain thoroughly. Recrystallise it from a mixture of equal volumes of rectified (or industrial) spirit and water or from hot water. Filter, wash and dry. The yield of p-nitroaniline, m.p. 148 °C, is 11 g (96%).

The p.m.r spectrum (DMSO-d_6, TMS) shows signals at δ 6.67 (broad s, 2H, NH$_2$), 6.79 (d, 2H, *ortho*-H's to NH$_2$), and 8.09 (d, 2H, *ortho*-H's to NO$_2$). The m.s. of the three isomers are noted on p. 896.

Notes. (1) Washing is accomplished most effectively by transferring the crude solid to a beaker, stirring well with wash water and refiltering.
(2) The recrystallised material and the crude product should be examined by t.l.c. analysis.
(3) The 70 per cent sulphuric acid is prepared by adding 60 ml of concentrated sulphuric acid cautiously and in a thin stream with stirring to 45 ml of water.

6.7 FORMATION OF DIAZONIUM SALTS AND THEIR USES

Primary aromatic amines on reaction with nitrous acid in the presence of hydrochloric acid (or other mineral acid) at about 0 °C yield diazonium salts as discrete intermediates. The diazonium salts similarly derived from aliphatic primary amines decompose readily even at this temperature to yield the corresponding alcohol (and other products) with the evolution of nitrogen.

$$\text{Ar·NH}_2 + \text{NaNO}_2 + 2\text{HCl} \longrightarrow \text{Ar·N} \overset{\oplus}{\equiv} \text{N} \overset{\ominus}{\}} \text{Cl} + \text{NaCl} + \text{H}_2\text{O}$$

The acidified nitrite solution provides a source of the nitrosonium ion (1) which electrophilically replaces the hydrogen in the primary amino group to form the N-nitroso derivative (2). This has a tautomeric structure, the hydroxydiazo form (3) yielding the diazonium ion (4) under acidic conditions.

$$\text{Ar·}\overset{..}{N}\text{H}_2 \quad \overset{\oplus}{N}=\text{O} \quad \xrightarrow{-H^{\oplus}} \quad \text{Ar·N—N}=\text{O} \rightleftharpoons \text{Ar·N}=\text{N—OH}$$

$$\begin{matrix} & & \text{|} & \\ & & \text{H} & \\ (1) & & (2) & (3) \end{matrix}$$

920

$$Ar \cdot \overset{\curvearrowright}{N} = N \overset{\frown}{-OH} \overset{\curvearrowleft}{\underset{\oplus}{H}} \longrightarrow Ar \cdot \overset{\oplus}{N} \equiv N + H_2O$$

(4)

The experimental conditions necessary for the preparation of a solution of a diazonium salt, *diazotisation of a primary amine*, are as follows. The amine is dissolved in a suitable volume of water containing 2.5–3 equivalents of hydrochloric acid (or of sulphuric acid) by the application of heat if necessary, and the solution is cooled in ice when the amine hydrochloride (or sulphate) usually crystallises. The temperature is maintained at 0–5 °C and an aqueous solution of sodium nitrite is added portionwise until, after allowing 3–4 minutes for reaction, the solution gives an *immediate* positive test for excess of nitrous acid with an external indicator – moist potassium iodide—starch paper.* The precipitated amine hydrochloride (or sulphate), if any, dissolves during the diazotisation to give a clear solution of the highly soluble diazonium salt. The excess of acid (0.5–1 equivalents) maintains the proper condition of acidity required to stabilise the diazonium salt and hence to minimise secondary reactions, e.g. the interaction of some of the diazonium salt with unchanged amine to form a diazoamino compound, a reaction which occurs readily in neutral solution (see Expt 6.86). The reaction mixture must be kept very cold during the process (which is exothermic in character), otherwise the diazonium salt may be partially hydrolysed to the corresponding phenol (see below).

Some amines, such as the nitroanilines, react rather slowly at low temperatures, but since the diazonium compounds formed are somewhat more stable the diazotisation may be conducted at room temperature, when the reaction proceeds more rapidly. If the amine salt is only sparingly soluble in water, it should be suspended in the acid in a fine state of division (this is generally attained by cooling a hot solution and stirring vigorously), and it passes into solution as the soluble diazonium salt is formed.

A solution of sodium nitrite in concentrated sulphuric acid, which provides a nitrosonium hydrogen sulphate reagent, is a very effective diazotising medium which is particularly valuable for even more weakly basic amines, such as 2,4-dinitroaniline or the corresponding trinitro compound, picramide.

$$HNO_2 + H_2SO_4 \rightleftharpoons [H_2NO_2]^{\oplus}[HSO_4]^{\ominus} \xrightarrow{H_2SO_4}$$

$$[\overset{\oplus}{NO}][HSO_4^{\ominus}] + H_3O^{\oplus} + HSO_4^{\ominus}$$

This reagent is used here for the diazotisation of 4-amino-3,5-diiodobenzoic acid in the preparation of 3,4,5-triiodobenzoic acid (cognate preparation in Expt 6.70). It is also used for the bis-diazotisation of *m*-phenylenediamine (the

---

* In actual practice it is found that some time before the theoretical quantity of sodium nitrite has been added, the solution will give a blue coloration (presumably, in part, by atmospheric oxidation) within a few seconds of being placed upon the test paper. It must, however, be remembered that towards the end of the diazotisation the reaction with nitrous acid is somewhat slow, and it is imperative to wait a few minutes before making the test, and furthermore only an immediate blue coloration has any significance. It is advisable to dilute the drop of the test solution with a few drops of water on a watch glass before making the test. It is recommended that about 10 per cent excess of sodium nitrite of good quality (>96% NaNO₂: e.g. sodium nitrite recryst.) be employed; this will serve as an additional check. If a slight excess of sodium nitrite is accidentally added, it may be decomposed by the addition of a little urea or sulphamic acid; alternatively a small amount of the primary amine, dissolved in the acid used, may be added.

preparation of 1,3-diiodobenzene in Expt 6.70) which must be carried out in strongly acidic conditions. Unless the amino groups are extensively protonated in this way, partial diazotisation and self-coupling occurs (see below) with the formation of the azo dye Bismarck Brown.

To prepare the *solid benzenediazonium chloride or sulphate*, the reaction is conducted in the absence of water as far as possible. The source of nitrous acid is one of its organic esters (e.g. pentyl nitrite) and a solution of hydrogen chloride gas in absolute ethanol; upon the addition of ether only the diazonium salt is precipitated as a crystalline solid.

$$Ph\cdot\overset{\oplus}{N}H_3 + Pn\cdot O\cdot NO \xrightarrow{HCl} Ph\cdot\overset{\oplus}{N}\equiv N\}\overset{\ominus}{Cl} + PnCl + 2H_2O$$

Solid diazonium salts are very sensitive to shock when perfectly dry, and detonate violently upon gentle heating: they are, therefore, of little value for preparative work. Happily, most of the useful reactions of diazonium compounds can be carried out with the readily-accessible aqueous solutions, so that the solid (explosive) diazonium salts are rarely required.

6.7.1 REACTIONS INVOLVING REPLACEMENT OF THE DIAZO GROUP

Diazonium salts undergo a large number of reactions in which the diazo group is lost as molecular nitrogen and is replaced by a variety of other groups (e.g. OH, I, Br, Cl, F, CN, NO_2, SO_2H, Ar and H) which become attached to the aromatic ring. This section describes the experimental conditions necessary to effect such conversions.

When a solution of a diazonium salt is heated, nitrogen is evolved and the diazo group is replaced by a hydroxyl group in an S_N1 type of displacement reaction.

$$Ar\overset{\oplus}{\frown N}\equiv N \xrightarrow{-N_2} [Ar^{\oplus}] \xrightarrow[-H^{\oplus}]{+H_2O} ArOH$$

The diazonium sulphate is used in preference to the diazonium chloride, since the presence of chloride ions gives rise to small quantities of the aryl chloride as a by-product. The solution must be acidic in order to avoid the coupling reaction between unreacted diazonium salt and the phenol (see Section 6.7.2, p. 946). For the preparation of phenols and cresols, the aqueous solution of the diazonium compound is warmed to about 50 °C; at higher temperatures the reaction may become unduly vigorous and lead to appreciable quantities of tarry compounds. For certain substituted amines, a higher temperature (e.g. boiling 40–60% sulphuric acid) is necessary to decompose the diazonium salt completely (e.g. *m*-nitrophenol, Expt 6.69). These strongly acidic conditions at elevated temperatures are clearly unsuitable where there are present in the diazonium salt acid-sensitive groups. In such cases the decomposition of the diazonium salt has been found to proceed smoothly at 0 °C in the presence of a copper(II) nitrate–copper(I) oxide reagent.[27] As an illustration of this method, the conversion of *p*-toluidine into *p*-cresol is given in the cognate preparation of Expt 6.69. The method has been applied successfully to those cases where the diazonium salt undergoes undesirable side reactions, rather than conversion to the phenolic product. An example is provided by the reactions of the 6-nitro-2-methylbenzenediazonium ion:

When an aqueous solution of an aryldiazonium salt is treated with an equivalent of potassium iodide and warmed on a water bath, the aryl iodide is formed in good yield (e.g. iodobenzene, Expt 6.70).

$$\overset{\oplus}{Ar \cdot N} \equiv N\}\overset{\ominus}{Cl} + KI \longrightarrow ArI + N_2 + KCl$$

This simple procedure cannot be applied in the preparation of the corresponding chloro and bromo compounds. Sandmeyer (1884) found that the replacement of the diazonium group by halogen can be successfully accomplished in the presence of the appropriate copper(I) salt, thus providing an excellent method for the preparation of nuclear-substituted aromatic compounds from the corresponding amines. The reaction has been extended to groups other than halogens, for example the cyano (—CN) and the thiocyanate (—SCN) group. A general procedure for carrying out the *Sandmeyer reaction* is as follows. The amine is diazotised in the presence of hydrochloric acid with sodium nitrite at 0–5 °C, and a solution of an equimolecular quantity of copper(I) chloride in hydrochloric acid is added: a deep brown, sparingly soluble complex of copper(I) chloride and the diazonium salt is formed, and when the temperature of the reaction mixture is raised, decomposition ensues accompanied by the evolution of nitrogen, the disappearance of the solid and the separation of an oily layer of the aryl chloride.

$$Ar \cdot NH_2 \xrightarrow[HCl]{NaNO_2} \overset{\oplus}{ArN} \equiv N\}\overset{\ominus}{Cl} \xrightarrow[HCl]{CuCl} [Complex] \xrightarrow{heat} ArCl + N_2$$

The aryl chloride is formed when the diazonium–copper(I) chloride complex decomposes by a radical mechanism summarised below. Copper catalyses this decomposition because it can undergo interconversion between the +1 and +2 oxidation states as a result of electron transfer.

Details of the preparation of p-chlorotoluene are given in Expt 6.71, which also includes o-chlorotoluene, chlorobenzene, m-chloronitrobenzene and o-chlorobenzoic acid as examples of cognate preparations.

In the preparation of bromo compounds by the Sandmeyer reaction (e.g. p-bromotoluene and the cognate preparations in Expt 6.72), the amine is generally diazotised in sulphuric acid solution (or in hydrobromic acid solution), and the resulting aryldiazonium sulphate (or bromide) is treated with a solution of copper(I) bromide in an excess of hydrobromic acid; the addition complex is then decomposed by gentle heating and the bromo compound isolated by steam distillation. For the preparation of 2-bromonaphthalene the use of 2-naphthylamine (which is a potent carcinogen) as a starting material is avoided by using 2-naphthylamine-1-sulphonic acid (see Section 6.5.4, p. 900). Diazotisation and reaction with copper(I) bromide yields 2-bromonaphthalene-1-sulphonic acid; heating with aqueous sulphuric acid eliminates the sulphonic acid group (see also Expt 6.104) to give 2-bromonaphthalene.

One of the drawbacks of the Sandmeyer reaction is the number of competing side reactions leading to the formation of biaryls, azo compounds and phenols. A recent procedure to maximise the yield of aryl halide involves treatment of the arylamine with t-butyl nitrite and anhydrous copper(II) halide (the chloride or the bromide) in acetonitrile at $65\,°C$.[28] The method is illustrated by the preparation of p-chloronitrobenzene (Expt 6.73); the overall reaction may be represented as:

$$2Ar\cdot NH_2 + 2R\cdot O\cdot NO + CuX_2 \longrightarrow 2ArX + 2ROH + CuO + H_2O + N_2$$

Gattermann (1890) found that finely divided (i.e. freshly precipitated) copper or copper bronze acts catalytically in the decomposition of solutions of diazonium salts.

$$Ar\cdot \overset{\oplus}{N}{\equiv}N\}\overset{\ominus}{X} \xrightarrow{\text{Cu}} ArX + N_2$$

The yields in the *Gattermann reaction*, however (e.g. o-bromotoluene, Expt 6.74), are usually not as high as those obtained by the Sandmeyer method. Copper powder is also employed in the preparation of sulphinic acids (e.g. benzenesulphinic acid, Expt 6.75) which are obtained when a solution of a diazonium sulphate is saturated with sulphur dioxide and decomposed by the addition of copper powder.

$$Ar\cdot \overset{\oplus}{N}{\equiv}N\}\overset{\ominus}{HSO_4} + SO_2 + Cu \longrightarrow Ar\cdot SO_2H + N_2 + CuSO_4$$

The Sandmeyer reaction may also be applied to the preparation of aryl

924

nitriles. The solution of the diazonium salt, which should preferably be carefully neutralised with sodium or calcium carbonate to avoid excessive evolution of hydrogen cyanide in the subsequent stage, is added to a solution of copper(I) cyanide in excess sodium or potassium cyanide solution (e.g. p-tolunitrile, Expt 6.76); sometimes improved yields are obtained by substituting nickel cyanide for copper(I) cyanide. Hydrolysis of the aryl nitrile with sodium hydroxide solution, followed by acidification, yields the corresponding acid (see Section 6.13.2, p. 1062). The Sandmeyer reaction thus affords an important indirect method of introducing the carboxylic acid group into an aromatic ring.

$$Ar{\cdot}NH_2 \xrightarrow{NaNO_2/HX} Ar{\cdot}\overset{\oplus}{N}_2\}\overset{\ominus}{X} \xrightarrow{CuCN/KCN} Ar{\cdot}CN \xrightarrow[\text{(ii) } H_3O^{\oplus}]{\text{(i) } {}^{\ominus}OH} Ar{\cdot}CO_2H$$

The controlled thermal decomposition of dry aryldiazonium fluoroborates to yield an aryl fluoride, boron trifluoride and nitrogen is known as the *Balz–Schiemann reaction* (Expt 6.77).

$$Ar{\cdot}\overset{\oplus}{N}{\equiv}N\}\overset{\ominus}{B}F_4 \xrightarrow{heat} ArF + N_2 + BF_3$$

In general the required diazonium fluoroborate is obtained as a precipitate when a concentrated solution of sodium fluoroborate is added to a solution of a diazonium salt. In an alternative procedure (e.g. the preparation of p-fluoroanisole, cognate preparation in Expt 6.77), the amine is diazotised in solution in aqueous fluoroboric acid. The diazonium fluoroborates are less sensitive to shock and heat than most diazonium salts and may be prepared and handled in the dry state with relative safety. Most diazonium fluoroborates have definite decomposition temperatures and the rates of decomposition, with few exceptions, are easily controlled. Diazonium fluoroborates containing the nitro group, however, usually decompose suddenly and with violence on heating; in such cases the fluoroborate should be mixed with 3–4 times its weight of dry sand and heated cautiously until decomposition commences.

The diazonium hexafluorophosphates, prepared similarly from the appropriate diazonium chloride solution and hexafluorophosphoric acid, may in general be used instead of the fluoroborates with advantage. The thermal decomposition of diazonium hexafluorophosphates to aryl fluorides generally proceeds smoothly and in better yield.[29]

A further interesting application of the diazo reaction is in the preparation of the otherwise difficultly accessible o- and p-dinitrobenzenes (Expt 6.78). The requisite nitroaniline is converted into the diazonium fluoroborate which is then decomposed in aqueous suspension in the presence of sodium nitrite with the aid of copper powder.

It is frequently observed that in the replacement reactions discussed above significant amounts of biphenyl derivatives are present in the reaction product. Compounds of this type may be prepared deliberately by adding the aqueous diazonium salt solution to a liquid aromatic compound and then basifying the vigorously stirred two-phase system by adding sodium hydroxide (or sodium acetate) solution.

$$Ar^1{\cdot}\overset{\oplus}{N}_2\}\overset{\ominus}{X} + Ar^2H + {}^{\ominus}OH \longrightarrow Ar^1{-}Ar^2 + N_2 + X^{\ominus} + H_2O$$

The reaction (the *Gomberg reaction* or *Gomberg–Bachmann–Hey* reaction) probably involves the intermediate formation of the diazohydroxide, which

925

because it has a largely covalent structure passes substantially into the non-polar organic phase. In this phase it decomposes into free aryl radicals which displace hydrogen from the added aromatic reactant.

$$Ar^1-\overset{\oplus}{N}\equiv N\overset{\ominus}{:}OH \rightleftharpoons Ar^1-N=N-OH \rightleftharpoons Ar^1-N=N-O^{\ominus}$$

$$Ar^1-N=N-O^{\ominus}\overset{\oplus}{N}\equiv\overset{\oplus}{N}-Ar^1 \longrightarrow Ar^1-N=N-O-N=N-Ar^1 \longrightarrow$$

$$Ar^1N_2O\cdot + Ar^1\cdot + N_2$$

$$Ar^1\cdot + Ar^2H \xrightarrow{Ar^1N_2O\cdot} Ar^1-Ar^2 + Ar^1N_2OH$$

In the example (Expt 6.79) the reaction of the diazonium salt from o-chloroaniline with benzene to yield 2-chlorobiphenyl is illustrative. It should be noted, however, that when the liquid aromatic compound in which substitution is to occur is of the type ArZ, the directive influences which are used to explain electrophilic substitution processes are not operative. Thus irrespective of the nature of the substituent Z, ortho–para substitution predominates; this result supports the assumption that the substitution process is radical in type. Although the classical reaction occurs in a two-phase system, the use of the more stable diazonium fluoroborates together with the phase-transfer catalyst 18-crown-6 can sometimes be more convenient. The literature method for the preparation of 4-chlorobiphenyl in this way is given as a cognate preparation in Expt 6.79.[3]

The process of *deamination* involves the replacement of the diazonium group by hydrogen, thus effecting the overall removal of the primary amino group. In a simple procedure illustrated by the preparation of 1,3,5-tribromobenzene from 2,4,6-tribromoaniline (Expt 6.80), the amine is converted into the diazonium sulphate in ethanol solution. Heating the solution brings about the reductive removal of the diazo group, the ethanol being oxidised to acetaldehyde.

$$Ar\cdot\overset{\oplus}{N}\equiv N\}HSO_4^{\ominus} + EtOH \longrightarrow ArH + Me\cdot CHO + H_2SO_4 + N_2$$

The value of the deamination process in synthesis is illustrated by a classical example – the synthesis of m-bromotoluene, described and formulated in Expt 6.81. The key to the sequence is the bromination of p-acetotoluidide which occurs at the position *ortho* to the more strongly electron-releasing acetamido group. This is *meta* to the methyl group, a position which is virtually unattacked in the direct bromination of toluene. The acetamido group can then be readily removed by sequential hydrolysis, diazotisation and reduction. Other reducing agents may be used in place of ethanol, e.g. an alkaline sodium stannite solution is quite effective; this reagent is prepared by adding sodium hydroxide to an aqueous solution of tin(II) chloride until the initial precipitate just redissolves.

An effective reagent is aqueous hypophosphorus acid, and an example of its use has been described in detail for the synthesis of 3,3′-dimethylbiphenyl from 4,4′-diamino-3,3′-dimethylbiphenyl (o-tolidine).[30] As the latter compound is carcinogenic its preparation is ill-advised. More recently the deamination of diazonium salts with t-butyl nitrite in dimethylformamide has been shown to have wide applicability.[31] The reaction proceeds smoothly at 65 °C with the evolution of nitrogen gas, the volume of which may be measured and used to monitor the reaction. The mechanism of the reaction is thought to involve the following radical sequence in which the solvent (DMF = solH), is the hydrogen donor.

$$Ar \cdot NH_2 + RO \cdot NO \longrightarrow Ar \cdot N=N-OR + H_2O$$

$$Ar \cdot N=\overset{\frown}{N}-OR \longrightarrow [\cdot OR] + Ar \cdot N=N \cdot \longrightarrow [N_2 \uparrow] + Ar\cdot$$

$$Ar \cdot + solH \longrightarrow ArH + sol\cdot$$

The general procedure is quoted in Expt 6.80 (cognate preparations) with reference to the preparation of m-nitrotoluene and 2,5-dichloroaniline.

CAUTION: Diazonium compounds have been used for the preparation of:

(a) Thiophenols – by treatment with a solution of sodium hydrogen sulphide, for example:

Anthranilic acid   Thiosalicylic acid

(b) O-Alkyl-S-aryl dithiocarbonates – by reaction with aqueous potassium O-ethyl dithiocarbonate (see Expt 5.209), and thence to thiophenols by treatment with potassium hydroxide, for example:

$$C_6H_5\overset{\oplus}{N}\equiv N\}\overset{\ominus}{Cl} + C_2H_5O \cdot CS \cdot \overset{\ominus\oplus}{SK} \longrightarrow C_6H_5S \cdot CS \cdot OC_2H_5 \longrightarrow C_6H_5SH$$

(c) Disulphides – by interaction with a solution of sodium disulphide.

It cannot be too strongly emphasised that in all these reactions violently explosive diazo sulphides and related compounds may be formed, and another less hazardous method for the preparation of the desired compound should be used, if possible. *The following reactions are known to lead to dangerous explosions:*

(i) diazotised o-nitroaniline, m-chloroaniline, 5-chloro-2-aminotoluene or 2-naphthylamine and sodium disulphide;
(ii) diazotised m-nitroaniline and potassium O-ethyl dithiocarbonate; and
(iii) diazotised aniline, p-bromoaniline, toluidines and naphthylamines and sodium hydrogen sulphide.

Experiment 6.69 m-NITROPHENOL

$$m\text{-}O_2N \cdot C_6H_4 \cdot NH_2 \xrightarrow[H_2SO_4]{NaNO_2} m\text{-}O_2N \cdot C_6H_4 \cdot \overset{\oplus}{N}\equiv N\}\overset{\ominus}{HSO_4} \xrightarrow[H_2SO_4]{H_2O}$$

$$m\text{-}O_2N \cdot C_6H_4OH$$

Add 101 g (55 ml) of concentrated sulphuric acid cautiously to 75 ml of water contained in a 1-litre beaker and introduce 35 g (0.25 mol) of finely powdered m-nitroaniline (Expt 6.51). Add 100–150 g of finely crushed ice and stir until the m-nitroaniline has been converted into the sulphate and a homogeneous paste results. Cool to 0–5 °C by immersion of the beaker in a freezing mixture, stir mechanically and add a cold solution of 18 g (0.26 mol) of sodium nitrite in 40 ml of water over a period of 10 minutes until a permanent colour is immediately given to potassium iodide–starch paper: do not allow the temperature to rise above 5–7 °C during the diazotisation. Continue the stirring for 5–10 minutes and allow to stand for 5 minutes; some mnitrobenzenediazonium sulphate may separate. Decant the supernatant liquid from the solid as far as possible.

While the diazotisation is in progress, cautiously add 165 ml of concentrated sulphuric acid to 150 ml of water in a 1-litre round-bottomed flask. Heat the mixture just to boiling. Add the supernatant liquid (diazonium solution) from a separatory funnel supported over the flask at such a rate that the mixture boils very vigorously (about 30 minutes). Then add the residual damp solid (or suspension) in small portions; avoid excessive frothing. When all the diazonium salt has been introduced, boil for a further 5 minutes and pour the mixture into a 1-litre beaker set in ice-water, and stir vigorously to obtain a homogeneous crystal magma. When cold, filter at the pump, drain well and wash with four 20 ml portions of ice-water. Recrystallise by dissolving the crude product in hot dilute hydrochloric acid (1:1 by volume), decant from any residual dark oil, filter and cool to 0 °C, when light yellow crystals separate (1). Spread these upon a large sheet of filter paper, and dry in the air in a warm room. The mother-liquor deposits a further crop (about 2 g) upon standing for 24 hours. The yield of *m*-nitrophenol, m.p. 96 °C, is 23 g (66%). Record the i.r. and p.m.r. spectra and compare them with those of the *ortho* and *para* isomers.

Note. (1) When working with larger quantities of material, it is more convenient (and a better yield is obtained) to purify the air-dried product by distillation under diminished pressure using a short air condenser of wide bore and a few fragments of porous porcelain (or alternatively a pine wood splinter) to prevent bumping. Collect the pure *m*-nitrophenol at 160–165 °C/12 mmHg; allow the flask to cool before admitting air otherwise the residue may decompose with explosive violence. The recovery is over 90 per cent of the pure *m*-nitrophenol.

Cognate preparation. p-*Cresol*. *Use of copper ions in the decomposition.*[27] *p*-Toluidine (5.34 g, 50 mmol) is dissolved in 50 ml of hot 35 per cent sulphuric acid and then allowed to cool to below 15 °C. Ice (50 g) is added and the amine bisulphate precipitated. A solution of 4.4 g (64 mmol) of sodium nitrite in 50 ml of ice-water is added dropwise under the surface of the ice-cooled solution with stirring at such a rate as to maintain the temperature at 0–5 °C. After the solution has been stirred for an additional 5 minutes, a few crystals of urea are added to decompose any excess sodium nitrite. To the cold (0 °C) solution of the *p*-toluenediazonium bisulphate is added a solution of 187 g (775 mol) of copper(II) nitrate trihydrate in 1.75 litres of water. With vigorous stirring 6.63 g (45 mmol) of copper(I) oxide is added to the solution. The liquid becomes dark blue and rapidly changes to green. About 1 minute after the addition of the copper(I) oxide the nitrogen evolution ceases and a negative test with alkaline 2-naphthol indicates that the reaction is complete. The mixture is extracted with ether and the combined ethereal extracts are extracted with aqueous alkali. Acidification of the alkaline extract and re-extraction with ether, followed by drying of the ether extract and evaporation gives *p*-cresol in 88 per cent yield.

Experiment 6.70 IODOBENZENE

$$Ph \cdot NH_2 \xrightarrow[HCl]{NaNO_2} Ph \cdot \overset{\oplus}{N} \equiv N \} Cl^{\ominus} \xrightarrow{KI} PhI + N_2 + KCl$$

Dissolve 20 g (19.6 ml, 0.125 mol) of aniline in a mixture of 55 ml of concentrated hydrochloric acid (1) and 55 ml of water contained in a 500-ml round-bottomed flask. Place a thermometer in the solution and immerse the flask in a bath of crushed ice (2); cool until the temperature of the stirred solution falls below 5 °C. Dissolve 16 g (0.23 mol) of sodium nitrite in 75 ml of water and chill the solution by immersion in the ice bath; add the sodium nitrite solution (3) in small volumes (2–3 ml at a time) to the cold aniline hydrochloride solution, and keep the latter well shaken. Heat is evolved by the reaction. The temperature should not be allowed to rise above 10 °C (add a few grams of ice to the reaction mixture if necessary) otherwise appreciable decomposition of the diazonium compound and of nitrous acid will occur. Add the last 5 per cent of the sodium nitrite solution more slowly (say, about 1 ml at a time) and, after shaking for 3–4 minutes, test a drop of the solution diluted with 3–4 drops of water with potassium iodide–starch paper (4); if no *immediate* blue colour is obtained at the point of contact with the paper, add a further 1 ml of the nitrite solution, and test again after 3–4 minutes. Continue until a slight excess of nitrous acid is present.

To the solution of benzenediazonium chloride add a solution of 36 g (0.216 mol) of potassium iodide in 40 ml of water slowly and with shaking. Nitrogen is evolved. Allow the mixture to stand for a few hours. Fit the flask with an air condenser and heat it cautiously in a boiling water bath until evolution of gas ceases. Allow to cool. Decant as much as possible of the upper aqueous layer and render the residual aqueous and organic layers alkaline by the cautious addition of 10 per cent sodium hydroxide solution, i.e. until a drop of the well-shaken mixture withdrawn on a glass rod imparts a blue colour to red litmus paper. The alkali converts any phenol present into sodium phenoxide, which, unlike phenol itself, is not volatile in steam. Steam distil until no more oily drops pass over (Fig. 2.102). Transfer the distillate to a separatory funnel and run off the lower layer of iodobenzene into a small conical flask. The crude iodobenzene should have a pale yellow colour; if it is dark in colour, return it to the separatory funnel and shake it with a little sodium metabisulphite solution until a pale yellow colour is obtained, then remove the heavy layer as before. Dry with about 1 g of anhydrous calcium chloride or magnesium sulphate: filter through a fluted filter paper into a small distilling flask equipped with a short air condenser. Distil using an air bath (Fig. 2.46) and collect the fraction, b.p. 185–190 °C (5). The yield of iodobenzene (an almost colourless liquid) is 33 g (75%); the compound gradually develops a yellow colour upon exposure to light.

Notes. (1) In computing the volume of acid required in the diazotisation process, it is helpful to remember that 100 ml of concentrated hydrochloric acid, d 1.18, contain 42.4 g of HCl, and 100 ml of concentrated sulphuric acid, d 1.84, contain 176 g H_2SO_4.
(2) For preparations on a larger scale, the lowering of temperature may be conveniently achieved by the addition of a quantity of crushed ice equal in weight to that of the hydrochloric acid and water. The mixture should be stirred mechanically.
(3) It is advisable to add the sodium nitrite solution, particularly in preparations on a larger scale, with the aid of a dropping funnel with the tip of the stem extending well below the surface of the liquid: this will prevent loss of nitrous acid by surface decomposition into oxides of nitrogen.
(4) It is advisable to test the potassium iodide–starch paper with acidified sodium nitrite solution: the commercial test paper is, particularly if it has been kept for a considerable period, sometimes almost useless. The solution must contain an excess of acid at all times, i.e. it must give a blue colour on Congo red paper.

(5) The iodobenzene is conveniently distilled under reduced pressure and the fraction, b.p. 77–80 °C/20 mmHg or 63–64 °C/8 mmHg, collected. The product has a higher degree of purity than that obtained directly from benzene (Expt 6.24).

Cognate preparations. **p-*Iodotoluene*.** Use 27 g (0.25 mol) of p-toluidine, 63 ml of concentrated hydrochloric acid and 63 ml of water: warm, if necessary, until all the amine dissolves. Cool the solution with vigorous stirring to 0–5 °C by immersion in a freezing mixture of ice and salt and the addition of a little crushed ice. Diazotise by the introduction, with stirring, of a solution of 18.5 g (0.27 mol) of sodium nitrite in 40 ml of water; maintain the temperature of the solution at 0–5 °C if possible, but do not allow it to rise above 10 °C. Add a solution of 44 g (0.265 mol) of potassium iodide in an equal weight of water gradually and with stirring. Allow to stand for 1 hour at the laboratory temperature and then heat cautiously on a water bath until evolution of nitrogen ceases. Allow to cool: a dark-coloured oil settles to the bottom and soon solidifies. Pour off as much of the aqueous layer as possible, add 1–2 g of sodium metabisulphite to remove the dark colour (gentle warming may be necessary) and then render the mixture alkaline with 10 per cent sodium hydroxide solution in order to retain any cresol which may be formed. Steam distil the mixture; if the p-iodotoluene solidifies in the condenser, turn off the condenser water for a few moments until the solid melts and runs down into the receiver. Filter off the solid in the receiver and recrystallise it from ethanol. The yield of p-iodotoluene (colourless plates), m.p. 35 °C, b.p. 211–212 °C, is 50 g (92%).

p-*Iodonitrobenzene*. Stir a mixture of 50 g (0.36 mol) of p-nitroaniline (Expt 6.68), 75 g (41 ml) of concentrated sulphuric acid and 300 ml of water for 1 hour. Cool the mixture to 0–5 °C, and diazotise with a solution of 25 g (0.36 mol) of sodium nitrite in 75 ml of water. Filter the cold solution, and add the filtrate with stirring to a solution of 100 g (0.6 mol) of potassium iodide in 300 ml of water. Collect the precipitated solid by suction filtration and recrystallise it from ethanol. The yield of p-iodonitrobenzene, m.p. 171 °C, is 73 g (82%).

p-*Iodophenol*. Dissolve 54.5 g (0.5 mol) of p-aminophenol (Expt 6.87) in a mixture of 60 g (32.5 ml) of concentrated sulphuric acid, 250 ml of water and 250 g of crushed ice in a large beaker or bolt-head flask. Cool the solution in a freezing mixture, stir mechanically and add during 1 hour a solution of 34.5 g (0.5 mol) of sodium nitrite in 75 ml of water. Stir for a further 20 minutes, and then add 18.5 g (10 ml) of concentrated sulphuric acid. Pour the cold diazonium solution into an ice-cold solution of 100 g (0.6 mol) of potassium iodide in 100 ml of water contained in a beaker provided with a mechanical stirrer. After 5 minutes, add 1 g of copper bronze (which has been washed with ether), with continued stirring, and warm the solution slowly on a water bath. Maintain the temperature at 75–80 °C until the evolution of nitrogen ceases; the iodophenol separates as a dark heavy oil. Cool to room temperature, extract the reaction mixture with three 80 ml portions of dichloromethane, wash the combined extracts with dilute sodium meta-bisulphite solution or sodium thiosulphate solution and dry with magnesium sulphate. Remove the solvent on a water bath (rotary evaporator) and distil the residue under diminished pressure. Collect the p-iodophenol at 138–140 °C/5 mmHg; this solidifies on cooling. Recrystallise from about 1 litre of light petroleum (b.p. 80–100 °C). The yield of colourless product, m.p. 94 °C, is 78 g (70%).

o-Iodobenzoic acid. Dissolve 14 g (0.1 mol) of anthranilic acid (Expt 6.53) in 100 ml of water containing 14 ml of concentrated sulphuric acid, cool to 5 °C and diazotise by the gradual addition of a cold solution of 7 g (0.101 mol) of sodium nitrite in 25 ml of water to an end-point with starch–iodide paper. Introduce into the clear solution, with stirring, a solution of 26 g (0.156 mol) potassium iodide in 50 ml of 1 M sulphuric acid, heat the mixture to boiling for 10 minutes and then cool. Collect the *o*-iodobenzoic acid by suction filtration, and recrystallise from hot water. The yield is almost quantitative; m.p. 162 °C.

3,4,5-Triiodobenzoic acid. Dissolve 6.8 g (0.0175 mol) of 4-amino-3,5-diiodobenzoic acid (Expt 6.60) in 30 ml of cold concentrated sulphuric acid, add a large excess (3.0 g, 0.0435 mol) of powdered sodium nitrite and allow the mixture to stand at 0 °C for 2 hours. Treat the cold diazonium solution with a solution of 17.0 g (0.12 mol) of potassium iodide in 40 ml of water; a dark red precipitate separates. Warm the mixture on a water bath until evolution of nitrogen ceases, and remove any residual iodine with a little sodium metabisulphite. Filter the yellow precipitate of crude 3,4,5-triiodobenzoic acid, and recrystallise from dilute ethanol. The yield of pure acid, m.p. 289–290 °C, is 6.8 g (78%).

1,3-Diiodobenzene. Add 45 g (0.65 mol) of sodium nitrite carefully and with stirring to 470 ml of concentrated sulphuric acid at 70 °C contained in a 2-litre beaker; ensure that the temperature does not exceed 75 °C during the addition. Cool the resulting solution in an ice bath with stirring and add dropwise a solution of 32.4 g (0.3 mol) of *m*-phenylenediamine (Expt 6.48) in 215 ml of glacial acetic acid; do not allow the temperature to rise above 25 °C. When diazotisation is complete, replace the ice in the cooling bath by an ice–salt mixture, and when the temperature of the tetrazonium salt solution falls below 0 °C (1), run in steadily 500 ml of distilled water with vigorous stirring, keeping the temperature of the solution below 10 °C. Then run the cold diluted solution with rapid stirring into a solution of 100 g (0.6 mol) of potassium iodide in 200 ml of water. The temperature of the reaction mixture rises to about 30 °C; allow it to remain with vigorous stirring for 45 minutes. Finally heat the mixture at 75 °C for 2 hours and leave it at room temperature overnight. Isolate the product by steam distillation; collect about 2 litres of distillate, make it just alkaline with 50 per cent aqueous sodium hydroxide and extract with ether. Remove the ether from the washed, dried extract and crystallise the residual solid from a mixture of ethanol and ether with the aid of decolourising charcoal. A further crystallisation gives almost white crystals of pure 1,3-diiodobenzene, yield 60 g (61%), m.p. 36–37 °C.

Note. (1) Excessive cooling may cause the tetrazonium salt to crystallise out and make stirring difficult.

Experiment 6.71   *p*-CHLOROTOLUENE

$$p\text{-Me·C}_6\text{H}_4\text{·NH}_2 \xrightarrow[\text{HCl}]{\text{NaNO}_2} p\text{-Me·C}_6\text{H}_4\text{·}\overset{\oplus}{\text{N}}\equiv\text{N}\}\overset{\ominus}{\text{Cl}} \xrightarrow[\text{HCl}]{\text{CuCl}} p\text{-Me·C}_6\text{H}_4\text{Cl}$$

In a 1.5- or 2-litre round-bottomed flask, prepare copper(I) chloride from

105 g (0.42 mol) of crystallised copper(II) sulphate as detailed in Section 4.2.22, p. 428. Either wash the precipitate once by decantation or filter it at the pump and wash it with water containing a little sulphurous acid; dissolve it in 170 ml of concentrated hydrochloric acid. Stopper the flask loosely (to prevent oxidation) and cool it in an ice–salt mixture whilst the diazotisation is being carried out.

Dissolve 36 g (0.33 mol) of *p*-toluidine in 85 ml of concentrated hydrochloric acid and 85 ml of water contained in a 750-ml conical flask or beaker. Cool the mixture to 0 °C in an ice–salt bath with vigorous stirring or shaking and the addition of a little crushed ice. The salt, *p*-toluidine hydrochloride, will separate as a finely divided crystalline precipitate. Add during 10–15 minutes a solution of 24 g (0.35 mol) of sodium nitrite in 50 ml of water (1); shake or stir the solution well during the diazotisation, and keep the mixture at a temperature of 0–5 °C by the addition of a little crushed ice from time to time. The hydrochloride will dissolve as the very soluble diazonium salt is formed; when all the nitrite solution has been introduced, the solution should contain a trace of free nitrous acid. Test with potassium iodide–starch paper (see footnote, p. 921).

Pour the cold diazonium chloride solution slowly and with shaking into the cold copper(I) chloride solution (2). The mixture becomes very thick, owing to the separation of an addition product between the diazonium salt and the copper(I) chloride ($Me \cdot C_6H_4 \cdot N_2^{\oplus} Cl^{\ominus} \cdot CuCl$). Allow the mixture to warm up to room temperature without external heating, and shake occasionally (3). When the temperature reaches about 15 °C, the solid addition complex commences to break down with the liberation of nitrogen and the formation of an oily layer of *p*-chlorotoluene. Warm the mixture on a water bath to about 60 °C to complete the decomposition of the double salt; shake occasionally. When the evolution of nitrogen ceases, steam distil the mixture (compare Fig. 2.102) until no more oily drops are present in the distillate. Transfer the distillate to a separatory funnel, and remove the layer of *p*-chlorotoluene. Wash it successively with 30 ml of 10 per cent sodium hydroxide solution (to remove any *p*-cresol which may be present), water, an equal volume of concentrated sulphuric acid (to remove a trace of azo compound that usually colours the crude product and cannot be removed by distillation) and water (to remove the acid). Dry with 3–4 g of anhydrous calcium chloride or magnesium sulphate, decant or filter through a small fluted filter paper and distil from an air bath (Fig. 2.46) using an air condenser. Collect the *p*-chlorotoluene at 158–162 °C (a colourless liquid; m.p. 6–7 °C); the yield is 33 g (78%).

The p.m.r. spectrum (CDCl$_3$, TMS) shows signals at δ 2.29 (s, 3H, Me), and a four-proton multiplet, that has the appearance of an AB spectrum, in which the difference in the chemical shift (Δv) approaches the value of the coupling constant (see Fig. 3.58); the values for these pairs of protons have been calculated as 6.98 and 7.12. The p.m.r. spectrum of the *ortho* isomer (cognate preparation below) shows signals at δ 2.37 (s, 3H, Me; note the more significant deshielding influence of the *ortho* chlorine in this case), and 6.9–7.4 (m, 4H, C$_{AR}$—H). The m.s. of the two isomers are similar with variations in RA of particular fragment ions: for the *para* isomer the values of m/z are 128 (M, ^{37}Cl), 126 (M, ^{35}Cl), 127 and 125 (M − H), 91 (M − ^{37}Cl or ^{35}Cl, base peak).

Notes. (1) The sodium nitrite solution is conveniently added from a dropping funnel; it is recommended, particularly for preparations on a larger scale, that the tip of the stem of the funnel dip well below the surface of the liquid.

(2) The diazonium salt solution decomposes on standing and hence must be mixed with the copper(I) chloride solution without delay. Mechanical stirring is an advantage.

(3) For preparations on a larger scale, mechanical stirring is essential and should be continued for 2–3 hours after the solution has attained room temperature.

Cognate preparations. *o-Chlorotoluene.* Proceed as for *p*-chlorotoluene, but use 36 g of *o*-toluidine. Collect the *o*-chlorotoluene at 155–158 °C; the yield is 33 g.

m-Chloronitrobenzene. This preparation is very similar to that of *p*-chlorotoluene, but certain modifications must be introduced. The quantities required are: 46 g (0.35 mol) of *m*-nitroaniline (Expt 6.51), 85 ml of concentrated hydrochloric acid, 85 ml of water, a solution of 24 g (0.35 mol) of sodium nitrite in 50 ml of water (if the resulting diazonium salt solution is not clear, it must be filtered), and copper(I) chloride from 105 g (0.42 mol) of crystallised copper(II) sulphate (Section 4.2.22, p. 428), dissolved in 170 ml of concentrated hydrochloric acid. Run the diazonium salt solution into the solution of copper(I) chloride while the temperature is kept at 25–30 °C (water bath); at lower temperatures the decomposition of the unstable addition compound proceeds too slowly and would cause too violent an evolution of nitrogen upon warming, and at a higher temperature the formation of tarry by-products increases. Warm the mixture under a reflux condenser on a water bath until the evolution of nitrogen ceases. Steam distil (1); if the *m*-chloronitrobenzene solidifies in the condenser, turn off the condenser water for a few moments until the solid melts and runs down into the receiver. Allow the steam distillate to cool, decant the water and shake the solid with 200 ml of 1 per cent sodium hydroxide solution at 50 °C (to remove *m*-nitrophenol, if present). Allow the mixture to cool, filter with suction, wash with a little cold water and dry in the air. Determine the m.p. If this is not satisfactory, i.e. if it is appreciably below 44–45 °C, purify the product either by recrystallisation from a small volume of ethanol or preferably by distillation under diminished pressure and collect the fraction of b.p. 124–125 °C/18 mmHg or 116–117 °C/12 mmHg; the distillate solidifies to a pale yellow solid, m.p. 44–45 °C. The yield is 35 g (67%), depending upon the purity of the original *m*-nitroaniline.

Note. (1) The steam distillation may be omitted, if desired, by utilising the following method of purification. Allow the reaction mixture to cool, decant the aqueous layer and dissolve the residue in about 150 ml of toluene. Wash the toluene solution with water, 1 per cent sodium hydroxide solution and finally with water; dry with magnesium sulphate, distil off the toluene on a water bath (rotary evaporator) and distil the residue under diminished pressure.

o-Chlorobenzoic acid. Dissolve 14 g (0.1 mol) of anthranilic acid (Expt 6.53) in a solution of 20 ml of concentrated hydrochloric acid and 100 ml of water. Cool to about 5 °C, and diazotise by the gradual addition of a cold solution of 7 g (0.1 mol) of sodium nitrite in 25 ml of water to an end-point with starch–potassium iodide paper (see footnote, p. 921). In the meantime prepare a solution of copper(I) chloride as follows. Dissolve 26 g (0.104 mol) of crystallised copper(II) sulphate and 12 g of sodium chloride in 50 ml of water in a 750-ml round-bottomed flask. Heat the solution to boiling, then add 80 ml of concentrated hydrochloric acid and 14 g of copper turnings, and continue the heating under reflux until the solution is practically colourless. (Alternatively, prepare the copper(I) chloride by the method given in Section 4.2.22, p. 428.) Cool in ice, and then add the cold diazonium solution slowly and with shaking. The reaction proceeds rapidly and with frothing: allow the mixture to stand for 2–3 hours with frequent shaking. Filter the precipitated *o*-chlorobenzoic acid and wash it with a little cold water. Recrystallise the crude acid from hot water containing a little alcohol to which a little decolourising carbon has been added. The yield of pure *o*-chlorobenzoic acid, m.p. 138–139 °C, is 14 g (87%).

Experiment 6.72   *p*-BROMOTOLUENE

$$p\text{-Me·C}_6\text{H}_4\text{·NH}_2 \xrightarrow[\text{H}_2\text{SO}_4]{\text{NaNO}_2} p\text{-Me·C}_6\text{H}_4\text{·}\overset{\oplus}{\text{N}}\equiv\text{N}\}\overset{\ominus}{\text{HSO}}_4 \xrightarrow{\text{CuBr}} p\text{-Me·C}_6\text{H}_4\text{Br}$$

Prepare a *solution of copper(I) bromide* in a 2.5-litre two-necked flask by heating under reflux 31.5 g (0.124 mol) of copper(II) sulphate pentahydrate, 10 g (0.158 mol) of copper turnings, 77 g (0.55 mol) of crystallised sodium bromide, 15 g (8.2 ml) of concentrated sulphuric acid as described in Section 4.2.21, p. 428.

In a 1-litre flask mix 53.5 g (0.5 mol) of *p*-toluidine and 400 ml of water, and then add cautiously 98 g (53.5 ml) of concentrated sulphuric acid; warm until the *p*-toluidine dissolves. Cool the flask in a bath of ice and salt to 0–5 °C; add about 100 g of crushed ice to the contents of the flask in order to accelerate the cooling. Add slowly and with frequent shaking a solution of 35 g (0.5 mol) of sodium nitrite in 60 ml of water until a slight excess of sodium nitrite is present (see footnote, p. 921); keep the temperature of the mixture below 10 °C.

Equip the 2.5-litre two-necked flask containing the copper(I) bromide solution for steam distillation (compare Fig. 2.102) and insert into the side-neck a tube (7–8 mm in diameter) leading almost to the bottom of the flask via a screw-capped adapter; attach a short-stemmed separatory funnel to this tube by means of a short length of rubber tubing and support the funnel in a ring clamped to a retort stand. Heat the copper(I) bromide solution to boiling, add the toluene-*p*-diazonium sulphate solution from the separatory funnel while steam is passed rapidly through the mixture. In order to reduce the amount of decomposition of the diazonium salt solution, transfer only about one-fourth to the separatory funnel (the remainder being kept in the freezing mixture) and run this into the copper(I) bromide solution: when the funnel is nearly empty, transfer a further portion of the cold diazonium solution to it without interrupting the addition. Add all the diazonium solution in this way during 20–30 minutes. Continue the steam distillation until no more organic matter distils. Render the distillate alkaline with 20 per cent sodium hydroxide solution (to remove any *p*-cresol present), shake well and separate the crude *p*-bromotoluene. In order to obtain a colourless product, wash the crude substance with 40–50 ml of warm (30 °C) concentrated sulphuric acid, then with water, sodium hydroxide solution, and finally with water. If the *p*-bromotoluene solidifies, warm the wash liquids to 30 °C before use; unless this is done, considerable loss may occur. Dry over magnesium sulphate or anhydrous calcium chloride, warm, filter and distil through an air-cooled condenser. Coolect the *p*-bromotoluene at 182–184 °C. The yield is 60 g (70%); m.p. 25–26 °C.

The *para*-substitution pattern is clearly confirmed by the p.m.r. spectrum ($CDCl_3$, TMS) which shows signals at δ 2.25 (s, 3H, Me), 6.92 (d, 2H, *ortho*-H's to Me), and 7.30 (d, 2H, *ortho*-H's to Br). The *ortho* and *meta* isomers show multiplets in the aromatic proton region. All the isomers show absorption in the 650–830 cm^{-1} region of their i.r. spectra (thin film); i.e. at *c.* 810 cm^{-1} for the *para* isomer (2 adj. H's), at *c.* 750 cm^{-1} for the *ortho* isomer (5 adj. H's), and at *c.* 770 cm^{-1} (3 adj. H's) and at 830 cm^{-1} (1H) for the *meta* isomer. The m.s. of the isomers are noted in Expt 6.81.

Cognate preparations. o-*Bromotoluene*. Use 53.5 g of *o*-toluidine and other components as above. The yield of *o*-bromotoluene, b.p. 178–181 °C, is of the same order.

o-*Bromochlorobenzene*. Place a mixture of 64 g (0.5 mol) of *o*-chloroaniline and 175 ml of constant boiling point hydrobromic acid (*d* 1.48; 100 ml contains 71 g of HBr) in a 1-litre flask set in an ice–salt bath, and cool it to 0–5 °C by the addition of a little ice. Add, with shaking or stirring, a solution of 35 g (0.5 mol) of sodium nitrite in 70 ml of water until a slight excess of nitrous acid is present (starch–potassium iodide paper test; see footnote, p. 921); maintain the temperature below 10 °C by the addition of ice if necessary.

Prepare copper(I) bromide from 75 g (0.33 mol) of crystallised copper(II) sulphate as detailed in Section 4.2.21, p. 428, and dissolve it in 40 ml of 48 per cent hydrobromic acid; heat the solution to boiling and add *o*-chlorobenzene-diazonium bromide solution as detailed above. When all the latter has been introduced, continue to pass steam through the mixture until no more organic material distils. Follow the procedure, including purification, given for *p*-bromotoluene. Collect the *o*-bromochlorobenzene (a colourless liquid) at 200–202 °C. The yield is 85 g (89%).

This procedure may also be employed for m-*bromochlorobenzene*, b.p. 191–194 °C, from *m*-chloroaniline; m-*dibromobenzene*, b.p. 215–217 °C, from *m*-bromoaniline; and o-*bromoanisole*, b.p. 114–116 °C/29 mmHg, from *o*-anisidine (the sulphuric acid washing is omitted in the last example).

2-*Bromonaphthalene*. Dissolve 112 g (0.5 mol) of 2-naphthylamine-1-sulphonic acid (1), with stirring, in 850 ml of 0.6 M sodium hydroxide solution: add, with stirring, an aqueous solution of 35 g (0.5 mol) of sodium nitrite, and filter the resulting solution. Place 250 ml of concentrated hydrochloric acid and 100 g of crushed ice in a 2-litre beaker, and equip the latter with a mechanical stirrer. Introduce the filtered solution of sodium nitrite and sodium 2-naphthylamine-1-sulphonate (2) slowly with stirring, and maintain the temperature at 0–5 °C by adding crushed ice. Collect the reddish-brown precipitate which forms on a large Buchner funnel and wash it with about 500 ml of ice-water. While the diazotisation is in progress, suspend 160 g (1.1 mol) of copper(I) bromide (from 300 g of crystallised copper(II) sulphate; Section 4.2.21, p. 428) in 75 ml of 48 per cent hydrobromic acid and 200 ml of water. Add the damp cake of the diazonium compound portionwise and with vigorous stirring to the copper(I) bromide suspension contained in a 2-litre beaker. After the vigorous evolution of nitrogen has subsided heat the mixture to 95–100 °C on a steam bath and then filter the hot mixture through a large Buchner funnel. Pour the filtrate back into the beaker and add 112 g of potassium chloride with stirring. Allow the resulting paste to cool to room temperature, filter with suction and wash with 250 ml of 20 per cent aqueous potassium chloride. Dry the reddish-brown precipitate of 2-bromo-naphthalene-1-sulphonic acid in the air overnight, and transfer it to a 2-litre round-bottomed flask. Add dilute sulphuric acid (prepared from 200 ml of the concentrated acid and 200 g of crushed ice), attach a reflux condenser and reflux the mixture gently, using an electric heating mantle, for 12–16 hours. Cool to room temperature; pour on to about 500 g of crushed ice. Transfer the mixture with the aid of 500 ml of toluene to a large separatory funnel, shake well, remove the toluene layer and wash the latter with water until the washings are neutral to litmus. Dry the toluene solution with magnesium sulphate, remove the toluene using a rotary evaporator and distil the residue under reduced pressure. Collect the bromonaphthalene at 100–101 °C/2 mmHg or at 140 °C/20 mmHg; this solidifies to a pale yellow solid, m.p. 56–57 °C. The yield is 67 g (65%).

The pale yellow colour cannot be removed by redistillation or recrystallisation; the coloured product probably contains some amino compound rendering it unsuitable for conversion into a Grignard reagent. A pure white product may be obtained by the following procedure. Dissolve 50 g of the coloured compound in 200 ml of hexane and pass the solution through a column of activated alumina (80–200 mesh; dimensions about 9 cm × 3 cm); wash the column with 750 ml of hexane. Remove the hexane by distillation: 49 g of pure 2-bromonaphthalene, m.p. 58 °C, remains. This is sufficiently pure for use in Grignard reactions.

Notes. (1) Tobias acid, available from Fluka AG Chemische Fabrik; see Section 6.5.4, p. 900.

(2) If the solid sodium salt is available, 123 g may be dissolved in 850 ml of distilled water and a solution containing 35 g of sodium nitrite added.

Experiment 6.73 *p*-CHLORONITROBENZENE[28]

$$2p\text{-}O_2N\cdot C_6H_4\cdot NH_2 + 2Bu^t\cdot O\cdot NO + CuCl_2 \longrightarrow$$

$$2p\cdot O_2N\cdot C_6H_4Cl + Bu^tOH + CuO + H_2O + 2N_2$$

Anhydrous copper(II) chloride (1.62, 12 mmol) (1), t-butyl nitrite (1.55 g, 15 mmol) (2), and anhydrous acetonitrile (40 ml) (3) are added to a three-necked round-bottomed flask equipped with a reflux condenser, an addition funnel and a gas outlet tube. The resulting, rapidly stirred (magnetic stirrer) mixture is warmed to 65 °C. *p*-Nitroaniline (1.38 g, 10 mmol) in 2 ml of acetonitrile is slowly added over a period of 5 minutes to the reaction solution. During this addition the reaction solution turns completely black from an initial green colour as nitrogen is evolved (4). After gas evolution is complete the reaction is allowed to reach room temperature; the reaction solution is then poured into 200 ml of 20 per cent aqueous hydrochloric acid and extracted with 200 ml of ether. The organic layer is washed once with 200 ml of 20 per cent aqueous hydrochloric acid. The resulting ether solution is dried over magnesium sulphate and the ether removed under reduced pressure. Recrystallisation from ethanol gives 1.43 g (92%) of *p*-chloronitrobenzene, m.p. 82 °C. The p.m.r. (CDCl$_3$, TMS) should be recorded and interpreted.

Notes. (1) Copper(II) chloride is dried at 110 °C prior to use.
(2) t-Butyl nitrite is prepared from t-butyl alcohol by the method given for butyl nitrite (Section 4.2.*1*, p. 414).
(3) Reagent grade acetonitrile is distilled from calcium hydride prior to its use.
(4) Total gas evolution (at 65 °C complete generally within 10 minutes) is measured on the closed system by water displacement from a calibrated gas burette; the yield of gaseous products in this reaction is 220 ± 20 ml.

Experiment 6.74 *o*-BROMOTOLUENE

$$o\text{-Me·C}_6\text{H}_4\text{·NH}_2 \xrightarrow[\text{HBr}]{\text{NaNO}_2} o\text{-Me·C}_6\text{H}_4\text{·}\overset{\oplus}{\text{N}}\equiv\text{N}\}\overset{\ominus}{\text{Br}} \xrightarrow{\text{Cu}} o\text{-Me·C}_6\text{H}_4\text{Br}$$

In a 1- or 1.5-litre round-bottomed flask prepare a solution of 53.5 g (0.5 mol) of *o*-toluidine in 170 ml of 40 per cent w/w hydrobromic acid; cool to 5 °C by immersion in a bath of ice and salt. Diazotise by the gradual addition of a solution of 36.5 g (0.53 mol) of sodium nitrite in 50 ml of water; stopper the flask after each addition and shake until all red fumes are absorbed. Keep the temperature between 5 and 10 °C. When the diazotisation is complete, add 2 g of copper powder or copper bronze (Section 4.2.*19*, p. 426), attach a reflux condenser to the flask and heat very cautiously on a water bath. *Immediately* evolution of gas occurs, cool the flask in crushed ice; unless the flask is rapidly removed from the water bath, the reaction may become so violent that the contents may be shot out of the flask. When the vigorous evolution of nitrogen moderates, heat the flask on a water bath for 30 minutes. Then dilute with 400 ml of water, and steam distil the mixture until about 750 ml of distillate are collected. Render the distillate alkaline with 10 per cent sodium hydroxide solution (about 50 ml) and separate the lower red layer of crude *o*-bromotoluene. Wash it with two 20 ml portions of concentrated sulphuric acid (which removes most of the colour) and then twice with water. Dry with magnesium sulphate or anhydrous calcium chloride, and distil from a flask fitted with a lagged fractionating column. Collect the *o*-bromotoluene at 178–181 °C. The yield is 40 g (47%). The spectral features are noted in Expts 6.72 and 6.81.

Experiment 6.75 BENZENESULPHINIC ACID

$$\text{Ph·NH}_2 \xrightarrow[\text{H}_2\text{SO}_4]{\text{NaNO}_2} \text{Ph·}\overset{\oplus}{\text{N}}\equiv\text{N}\}\text{HSO}_4^{\ominus} \xrightarrow[\text{Cu}]{\text{SO}_2} \text{Ph·SO}_2\text{H}$$

Dissolve 9.3 g (9.1 ml, 0.1 mol) of aniline in a mixture of 19.6 g (10.7 ml) of concentrated sulphuric acid and 100 ml of water, and cool to about 5 °C. Diazotise by the addition of a solution of 7.0 g (0.1 mol) of sodium nitrite in 15 ml of water to an end-point with potassium iodide–starch paper; maintain the temperature below 10 °C. Add an ice-cold mixture of 40 g (22 ml) of concentrated sulphuric acid and 30 ml of water, cool in ice and pass sulphur dioxide into the solution until there is no further increase in weight (about 25 g). The solution should not develop any appreciable colour during this operation and should remain quite clear. When the solution is saturated with

937

sulphur dioxide, transfer it to a beaker provided with a mechanical stirrer, and add copper powder (Section 4.2.*19*, p. 426) or copper bronze (previously washed with ether) gradually until no more nitrogen is evolved (about 50 g of copper powder are required). Filter at the pump and wash the precipitate with several small amounts of dilute ammonia solution to remove any sulphinic acid which may have separated: add the washings to the filtrate. The combined filtrate and washings should be acid to Congo red paper. Treat it with concentrated iron(III) chloride solution as long as any precipitate forms. Filter the precipitate of iron(III) benzenesulphinate, and wash it with a little water. Decompose the iron(III) salt with a slight excess of 5 per cent sodium hydroxide solution, and filter the precipitated iron(III) hydroxide. Acidify the filtrate and extract the sulphinic acid with ether. Upon evaporation of the solvent, pure benzenesulphinic acid, m.p. 84 °C, is obtained as a colourless crystalline solid. The yield is 10 g (70%). It oxidises in the air.

Experiment 6.76 *p*-TOLUNITRILE (p-*Tolyl Cyanide*)

$$p\text{-Me·C}_6\text{H}_4\text{·NH}_2 \xrightarrow[\text{HCl}]{\text{NaNO}_2} p\text{-Me·C}_6\text{H}_4\text{·}\overset{\oplus}{\text{N}}\equiv\text{N}\}\overset{\ominus}{\text{Cl}} \xrightarrow[\text{KCN}]{\text{CuCN}} p\text{-Me·C}_6\text{H}_4\text{·CN}$$

CAUTION: *This and related cognate preparations must be carried out in an efficient fume cupboard.*

Prepare copper(I) cyanide from 100 g (0.4 mol) of hydrated copper(II) sulphate following the procedure described in Section 4.2.*23*, p. 429, transfer the product to a 1-litre round-bottomed flask and dissolve it in a solution of 52 g of potassium cyanide in 125 ml of water (**CAUTION**).

Diazotise 36 g (0.33 mol) of *p*-toluidine, following the method given under p-*chlorotoluene* (Section 6.71). While keeping the solution cold, carefully add about 20 g of powdered anhydrous sodium carbonate with constant stirring until the solution is neutral to litmus. Warm the copper(I) cyanide solution on a water bath to about 60 °C, and add the cold neutralised diazonium salt solution in small quantities at a time, shaking vigorously (1) after each addition and taking care to maintain the temperature of the mixture at 60–70 °C. Attach a reflux condenser to the flask and heat on a boiling water bath for 15–20 minutes in order to complete the reaction. Equip the flask for steam distillation (Fig. 2.102), and pass steam into the mixture until no more yellow oil passes over; if the oil solidifies in the condenser tube, turn off the condenser water, and, after the material melts and flows through, slowly turn on the water again. Cool the distillate in ice-water, and when the crude *p*-tolunitrile has solidified, filter it at the pump and press well to remove liquid impurities. Dry upon filter paper or in a desiccator. Mix the dried product with 2–3 g of decolourising carbon, transfer to a small distilling flask and distil using an air condenser. Collect the pure *p*-tolunitrile at 215–219 °C (2); this solidifies on cooling and melts at 29 °C. The yield is 26 g (67%).

The *para*-substitution pattern is confirmed by the i.r. and p.m.r. spectra. Thus in the i.r. spectrum (melt) absorption may be observed at c. 2200 (CN) and 820 cm^{-1} (2 adj. H's) together with the characteristic absorption from the methyl group and the aromatic ring. The p.m.r. spectrum (CDCl$_3$, TMS) shows signals at δ 2.41 (s, 3H, Me), 7.23 (d, 2H, *ortho*-H's to Me), and 7.47 (d,

2H, *ortho*-H's to CN). These spectra should be compared to those obtained with *o*-tolunitrile (cognate preparation below).

Notes. (1) Mechanical stirring is preferable.
(2) The crude substance may also be distilled under diminished pressure and the *p*-tolunitrile collected at 104–106 °C/20 mmHg.

Cognate preparation. o-*Tolunitrile*. The preparation is similar to that described for p-*tolunitrile* except that *p*-toluidine is replaced by *o*-toluidine. The *o*-tolunitrile is isolated by steam distillation; the oil, which may be dissolved in a little toluene, is distilled. The *o*-tolunitrile passes over as an almost colourless liquid at 94–96 °C/20 mmHg. Note the i.r. absorption (thin film) at *c*. 2200 (CN) and 760 cm^{-1} (4 adj. H's) together with the expected aromatic ring and methyl group absorptions. The p.m.r. spectrum (CDCl$_3$, TMS) has δ 2.51 (s, 3H, Me), and 7.02–7.66 (m, 4H, C$_{AR}$—H).

Experiment 6.77 FLUOROBENZENE

$$Ph \cdot NH_2 \longrightarrow Ph \cdot \overset{\oplus}{N} \equiv N \} \overset{\ominus}{Cl} \xrightarrow{NaBF_4} Ph \cdot \overset{\oplus}{N} \equiv N \} \overset{\ominus}{BF_4} \xrightarrow[-BF_3, -N_2]{\Delta} PhF$$

CAUTION: *This preparation should be carried out in an efficient fume cupboard and behind a suitable safety screen.*

Dissolve 46.5 g (45.5 ml, 0.5 mol) of aniline in a mixture of 126 ml of concentrated hydrochloric acid and 126 ml of water contained in a 1-litre beaker. Cool to 0–5 °C in a bath of ice and salt, and add a solution of 36.5 g (0.53 mol) of sodium nitrite in 75 ml of water in small portions; stir vigorously with a thermometer and maintain the temperature below 10 °C, but preferably at about 5 °C by the addition of a little crushed ice if necessary. The diazotisation is complete when a drop of the solution diluted with 3–4 drops of water gives an immediate blue coloration with potassium iodide–starch paper; the test should be performed 3–4 minutes after the last addition of the nitrite solution. Prepare a solution of 76 g (0.69 mol) of sodium fluoroborate (1) in 150 ml of water, cool and add the chilled solution slowly to the diazonium salt solution; the latter must be kept well stirred and the temperature controlled so that it is below 10 °C. Allow to stand for 10 minutes with frequent stirring. Filter the precipitated benzenediazonium fluoroborate with suction on a Buchner funnel, drain well and wash the yellow solid with about 30 ml of ice-water, 15 ml of methanol and 30–40 ml of ether; suck the solid as free as possible from liquid after each washing (2). Spread the salt upon absorbent filter paper and allow to dry overnight, if possible in a current of air. The yield of benzenediazonium fluoroborate is 60–65 g; the pure salt melts with decomposition at 119–120 °C.

Assemble the apparatus shown in Fig. 6.3; this is self-explanatory. The distilling flask has a capacity of 250 ml and the beaker contains 150 ml of 10 per cent sodium hydroxide solution. Place half of the yield of the dry benzenediazonium fluoroborate in the distilling flask. Heat the solid gently with a small luminous flame at one point near its surface until decomposition begins; withdraw the flame and allow the reaction to continue as long as it will (3). Continue the cautious heating from time to time as may be necessary to keep the reaction going. When the decomposition appears to be complete, heat the flask more strongly to drive off any remaining fluorobenzene. Allow

939

Fig. 6.3

to cool, add the other half of the benzenediazonium fluoroborate through a glazed paper funnel and decompose it as before; finally heat the flask strongly until no more fumes of boron trifluoride are evolved in order to drive off the last traces of fluorobenzene. Most of the fluorobenzene collects in the first, cooled, receiver. Wash the combined distillates three times with an equal volume of 10 per cent sodium hydroxide solution (4) or until the washings are almost colourless; this will remove any phenol present. Remove the last sodium hydroxide washing as completely as possbile, and then shake with an equal volume of almost saturated salt solution. Dry over anhydrous calcium chloride or magnesium sulphate, and distil the fluorobenzene (a colourless liquid) at 84–85 °C. The yield is 24 g (50%).

Notes. (1) The use of sodium fluoroborate solution supersedes the less convenient *fluoroboric acid* and permits the preparation to be carried out in ordinary glass vessels. If it is desired to employ *fluoroboric acid* HBF_4, it can be prepared by adding 100 g of AR boric acid in small proportions to 325 g of AR hydrofluoric acid (40% HF) cooled in ice; the hydrofluoric acid is contained in a polypropylene beaker, in a beaker coated with wax or in a lead vessel. One-third of the above solution should be employed in the preparation. *Handle with great care.*

   Note on precautions to be adopted when using hydrofluoric acid. Attention is directed to the fact that hydrofluoric acid in contact with the skin produces extremely painful burns. In case of accident, the burned surface, which becomes white, is held under running water until the natural colour returns. A paste made from magnesium oxide and glycerine should be applied immediately; this is said to be helpful in preventing the burn becoming serious. It is essential to wear acid-resisting rubber gloves and protective goggles.

(2) Careful washing with methanol and ether is necessary to remove from the crude product any moisture which tends to make the material unstable and liable to spontaneous decomposition.

(3) If the reaction becomes too vigorous, it may be necessary to cool the flask by covering it with a damp cloth. Normally the decomposition proceeds smoothly under the intermittent heating. If the salt is damp, the reaction may proceed more vigorously and unless the flask is cooled, it may pass beyond control.

(4) The density of fluorobenzene is about 1.025 at room temperature; it is important to use the correct strength of sodium hydroxide solution in order to obtain a clear separation of the layers.

Cognate preparations.   p-*Fluorotoluene.* Diazotise 53.5 g (0.5 mol) of *p*-toluidine in a mixture of 126 ml of concentrated hydrochloric acid and 126 ml of water contained in a 1-litre beaker following the procedure given in Expt 6.71. Add a chilled solution of 76 g (0.69 mol) of sodium fluoroborate in 150 ml of water slowly and with good stirring to the cold diazonium salt solution. Continue stirring for about 15 minutes. Filter the toluene-*p*-diazonium fluoroborate on a Buchner or sintered glass funnel, wash with about 30 ml of ice-water, 15 ml of methanol and 30–40 ml of ether. Dry overnight upon absorbent paper in a vacuum desiccator or, if possible, in a current of air. The yield of toluene-*p*-diazonium fluoroborate is 78 g (76%); it melts with decomposition at 114 °C. Decompose the salt in two equal lots, and work up as for *fluorobenzene*. The yield of pure *p*-fluorotoluene (a colourless liquid), b.p. 116–117 °C, is 27 g (50%).

p-*Fluoroanisole*. To 105 ml (2 mol) of *c*. 42 per cent fluoroboric acid (**CAUTION**: *corrosive chemical*) diluted with an equal volume of water, contained in a 600 ml beaker, add 31 g (0.25 mol) of *p*-anisidine. Place the beaker in an ice bath and stir the solution mechanically. Add a solution of 17.5 g (0.25 mol) of sodium nitrite in 35 ml of water slowly and maintain the temperature at about 10 °C. Stir the solution vigorously towards the end of the reaction, cool the mixture to 0 °C and filter with suction on a sintered glass funnel. Wash the precipitate successively with 30–40 ml of cold 5 per cent fluoroboric acid, 40 ml of ice-cold methanol and several times with ether. Dry overnight by spreading the salt thinly on absorbent paper supported upon a screen or wire netting allowing circulation underneath. The yield of *p*-methoxybenzenediazonium fluoroborate is 54 g (98%). Decompose the dry salt as detailed for *fluorobenzene*. Return the small amount of product in the receiver to the distilling flask and steam distil. Extract the steam distillate with two 50 ml portions of ether, wash the ethereal solution with 50 ml of 10 per cent sodium hydroxide solution, followed by water and dry over magnesium sulphate. Remove the ether on a steam bath and distil the residue. Collect the *p*-fluoroanisole at 156–157 °C. The yield is 16 g (51%).

Experiment 6.78 *o*-DINITROBENZENE

$$o\text{-}O_2N\cdot C_6H_4\cdot NH_2 \xrightarrow[\text{HCl}]{\text{NaNO}_2} o\text{-}O_2N\cdot C_6H_4\cdot \overset{\oplus}{N}{\equiv}N\}\overset{\ominus}{Cl} \xrightarrow{\text{NaBF}_4}$$
$$o\text{-}O_2N\cdot C_6H_4\cdot \overset{\oplus}{N}{\equiv}N\}\overset{\ominus}{BF_4} \xrightarrow[\text{Cu}]{\text{NaNO}_2} o\text{-}O_2N\cdot C_6H_4\cdot NO_2$$

CAUTION: *This preparation should be carried out behind a safety screen.*

Dissolve 34 g (0.25 mol) of *o*-nitroaniline in a warm mixture of 63 ml of concentrated hydrochloric acid and 63 ml of water contained in a 600-ml beaker. Place the beaker in an ice–salt bath, and cool to 0–5 °C while stirring mechanically; the *o*-nitroaniline hydrochloride will separate in a finely divided crystalline form. Add a cold solution of 18 g (0.26 mol) of sodium nitrite in 40 ml of water slowly and with stirring to an end-point with potassium iodide–starch paper; do not allow the temperature to rise above 5–7 °C. Introduce, while stirring vigorously, a solution of 40 g (0.36 mol) of sodium fluoroborate in 80 ml of water. Stir for a further 10 minutes, and filter the solid diazonium fluoroborate with suction on a Buchner funnel. Wash it immediately once with 25 ml of cold 5 per cent sodium fluoroborate solution, then twice with 15 ml portions of rectified (or industrial) spirit and several times with ether; in each washing stir the fluoroborate well before applying suction. The *o*-nitrobenzenediazonium fluoroborate weighs about 50 g (86%); the pure substance melts with decomposition at 135 °C.

Dissolve 200 g (2.9 mol) of sodium nitrite in 400 ml of water in a 2-litre

941

beaker provided with an efficient mechanical stirrer, and add 40 g of copper powder (either the precipitated powder or copper bronze which has been washed with a little ether). Suspend the fluoroborate in about 200 ml of water and add it slowly to the well-stirred mixture. Add 4–5 ml of ether from time to time to break the froth. The reaction is complete when all the diazonium compound has been added. Transfer the mixture to a large flask and steam distil until no more solid passes over (about 5 litres of distillate). Filter off the crystalline solid in the steam distillate and dry upon filter paper in the air; this *o*-dinitrobenzene (very pale yellow crystals) has m.p. 116 °C (i.e. is practically pure) and weighs 29 g (69%). It may be recrystallised from ethanol; the recrystallised solid melts at 116.5 °C.

The p.m.r. spectrum (DMSO-d_6, TMS) has δ 8.08 and 8.26 (A_2B_2 multiplet, values calculated from a mathematical analysis); this should be compared with the *para* isomer (Me$_2$CO), cognate preparation below, which has δ 8.50 (s, 4H, A$_4$ singlet), and with the *meta* isomer (Fig. 3.64).

Cognate preparation. **p-*Dinitrobenzene*.** Use 34 g (0.25 mol) of *p*-nitroaniline (Expt 6.68) and proceed exactly as above to the point where all the suspension of *p*-nitrobenzenediazonium fluoroborate has been added. Filter the reaction mixture with suction, wash the residue well with water, twice with 25 ml of 5 per cent sodium hydroxide solution and finally with water. Dry the solid at 100–110 °C, powder it and extract it with four 150 ml portions of boiling toluene. Remove the toluene with a rotary evaporator and recrystallise the residue from about 120 ml of boiling glacial acetic acid. The yield of *p*-dinitrobenzene (reddish-yellow crystals), m.p. 173 °C, is 30 g (71.5%). Further recrystallisation from ethanol affords pale yellow crystals of the same m.p. The p.m.r. spectrum is noted above.

Experiment 6.79 2-CHLOROBIPHENYL

CAUTION: *This preparation should be conducted in an efficient fume cupboard.*

Diazotise 32 g (0.25 mol) of *o*-chloroaniline in the presence of 40 ml of concentrated hydrochloric acid and 22.5 ml of water in the usual manner (compare Expt 6.70) with a concentrated solution of 18.5 g sodium nitrite in water. Transfer the cold, filtered diazonium solution to a 1.5-litre bolt-head flask surrounded by ice-water, introduce 500 ml of cold benzene (**CAUTION**), stir vigorously and add a solution of 80 g of sodium acetate trihydrate in 200 ml of water dropwise, maintaining the temperature at 5–10 °C. Continue the stirring for 48 hours: after the first 3 hours, allow the reaction to proceed at room temperature. Separate the benzene layer, wash it with water and remove the benzene by distillation at atmospheric pressure; distil the residue under reduced pressure and collect the 2-chlorobiphenyl at 150–155 °C/10 mmHg. The yield is 18 g (76%). Recrystallise from aqueous ethanol; m.p. 34 °C.

Cognate preparations. *4-Bromobiphenyl.* Diazotise 43 g (0.25 mol) of *p*-

bromoaniline (Expt 6.67) in the presence of 40 ml of concentrated hydro-chloric acid and 22.5 ml of water with a concentrated solution of 18.5 g of sodium nitrite in water. Mix the filtered diazonium solution with 500 ml of cold benzene (**CAUTION**), stir vigorously and add a solution of 30 g of sodium hydroxide in 150 ml of water dropwise (during 30–45 minutes) while maintaining the temperature at 5–10 °C. Complete the reaction as for 2-chlorobiphenyl. The yield of 4-bromobiphenyl, b.p. 170–175 °C/8 mmHg, m.p. 90 °C (from ethanol), is 25 g (86%).

4-Chlorobiphenyl (*PTC procedure*).[3] Potassium acetate (1.20 g, 0.0122 mol) is added in one portion to a stirred colourless mixture of 4-chlorobenzene-diazonium fluoroborate (1.37 g, 0.061 mol) (1) and 18-crown-6 (0.08 g, 0.003 mol) in benzene (60 ml, **CAUTION**) at 20 °C in a vessel protected from the light and purged with nitrogen. The mixture turns yellow immediately, and then red within a few minutes. Stirring is continued for 90 minutes, followed by filtration to remove the solid. The resulting solution is washed with brine, water, and then dried over anhydrous sodium sulphate and the solvent removed *in vacuo*. The resulting red oil is chromatographed on a short column of alumina using hexane as eluant from which is obtained 4-chlorobiphenyl (0.920 g, 80%), m.p. 76–77 °C.

Note. (1) The 4-chlorobenzenediazonium fluoroborate may be prepared using the conditions described for *p*-methoxybenzenediazonium fluoroborate, Expt 6.77, *p*-fluoroanisole. This methoxy derivative could also be used in the above preparation to give 4-methoxybiphenyl, m.p. 89 °C.

Experiment 6.80 1,3,5-TRIBROMOBENZENE

$$2,4,6\text{-Br}_3\text{C}_6\text{H}_2 \cdot \text{NH}_2 \longrightarrow 2,4,6\text{-Br}_3\text{C}_6\text{H}_2 \cdot \overset{\oplus}{\text{N}}{\equiv}\text{N}\}\text{HSO}_4^{\ominus} \xrightarrow[-\text{N}_2]{\text{EtOH}} 1,3,5\text{-C}_6\text{H}_3\text{Br}_3$$

Dissolve 10 g (0.03 mol) of 2,4,6-tribromoaniline (Expt 6.59) by heating on a water bath with 60 ml of rectified spirit and 15 ml of benzene in a 200-ml two-necked flask fitted with a reflux condenser, the second neck being closed with a stopper. Add, from a burette or small graduated pipette, 5.3 g (3.5 ml) of concentrated sulphuric acid to the hot solution via the side-neck and gently swirl the liquid, replace the stopper and heat on a water bath until the clear solution boils. Remove the flask from the water bath, and add 3.5 g (0.05 mol) of powdered sodium nitrite in two approximately equal portions via the side-neck; after each addition, replace the stopper and shake the flask vigorously; when the reaction subsides, add the second portion of the sodium nitrite. Heat the flask on a boiling water bath as long as gas is evolved; shake well from time to time. Allow the solution to cool for 10 minutes, and then immerse the flask in an ice bath. A mixture of tribromobenzene and sodium sulphate crystallises out. Filter with suction on a Buchner funnel, wash with a small quantity of ethanol and then repeatedly with water to remove all the sodium sulphate. Dissolve the crude tribromobenzene (7.5 g) in a boiling mixture of 120 ml of glacial acetic acid and 30 ml of water (1), boil the solution with 2.5 g of decolourising carbon and filter through a hot water funnel or a preheated Buchner funnel: allow the solution to cool. Collect the crystals on a Buchner funnel and wash with a small quantity of chilled rectified spirit to remove the acetic acid. Dry in the air upon filter paper. The yield of 1,3,5-tribromobenzene (colourless crystals), m.p. 122 °C, is 6.5 g (68%).

Note. (1) Rectified spirit may also be employed for crystallisation.

Cognate preparation. *Reductive deamination in dimethylformamide.*[31] The following procedure has been applied, *inter alia*, to the deamination of 4-methyl-2-nitroaniline to *m*-nitrotoluene, and 2,5-dichloroaniline to *p*-dichlorobenzene. To a rapidly stirred solution of t-butyl nitrite (15 mmol) (1) in anhydrous dimethylformamide (10 ml, redistilled from calcium hydride) heated at 65 °C in a three-necked round-bottomed flask equipped with a reflux condenser, addition funnel and gas outlet tube is added the aromatic amine (10 mmol) dissolved in the minimum amount of dimethylformamide. Gas evolution is immediate, continues throughout the addition steadily, and generally is complete 10 minutes after the addition of the amine. Total gas evolution is measured on a closed system by water displacement from a calibrated burette and is usually 220 ± 20 ml per 10 mmol of amine. After complete gas evolution, the reaction solution, which turns deep red from an initial yellow, is cooled and poured into 200 ml of 20 per cent aqueous hydrochloric acid. The aqueous solution is extracted with 200 ml of ether and the ether extract washed with 200 ml of 20 per cent aqueous hydrochloric acid. The resulting ether solution is dried over magnesium sulphate and the ether removed under reduced pressure. Solid products are suitably recrystallised and liquid products distilled through a 12.5 cm Vigreux column.

Note. (1) See Expt 6.73, *Note* (2).

Experiment 6.81 *m*-BROMOTOLUENE

p-Acetotoluidide and 4-acetamido-3-bromotoluene. Prepare a solution of *p*-acetotoluidide in glacial acetic acid by boiling 107 g (1 mol) of *p*-toluidine with 400 ml of glacial acetic acid in a 1-litre, round-bottomed three-necked flask, provided with a reflux condenser, stirrer and thermometer, for 2 hours. Cool the solution when some *p*-acetotoluidide may separate as small crystals as the temperature falls (1). When the temperature has fallen to about 45 °C, add 162.5 g (52.5 ml, 1.01 mol) of bromine from a separatory funnel at such a rate that the temperature of the well-stirred mixture is maintained at 50–55 °C. A precipitate may separate during the addition which requires 30–40 minutes, but this dissolves later. Continue the stirring for a further 30 minutes after all the bromine has been added. Then pour the reaction mixture in a thin stream into a well-stirred mixture of 1 kg of crushed ice and 1 kg of water to which 14 g of solid sodium metabisulphite has been added. If the colour of the bromine persists, add a little more sodium metabisulphite. Filter the crystalline 4-acetamido-3-bromotoluene with suction on a Buchner funnel, wash thoroughly with water and press well. Dry in the air until the weight does not exceed 250 g (2); further purification is unnecessary before proceeding to the next stage.

4-Amino-3-bromotoluene hydrochloride. Transfer the partially dried 4-acetamido-3-bromotoluene to a 1.5-litre round-bottomed flask, add 250 ml of rectified spirit and reflux on a water bath until the solid dissolves completely. Introduce through the condenser 250 ml of concentrated hydrochloric acid to the boiling solution and continue the refluxing for a further 3 hours. During this time crystals of 4-amino-3-bromotoluene hydrochloride separate. Pour the hot mixture into a 1-litre beaker and cool thoroughly. Filter the crystals of the hydrochloride at the pump through a Buchner funnel and wash rapidly with two 50 ml portions of chilled rectified spirit. The yield of the hydrochloride is 150 g (67.5%).

4-Amino-3-bromotoluene. Suspend the hydrochloride in 400 ml of water in a 1-litre beaker equipped with a mechanical stirrer. Add a solution of 70 g of sodium hydroxide in 350 ml of water. The free base separates as a dark heavy oil. After cooling to 15–20 °C, transfer the mixture to a separatory funnel and run off the crude 4-amino-3-bromotoluene. This weighs 125 g (67%) and can be used directly in the next step (3).

m-Bromotoluene. To a cold mixture of 400 ml of rectified spirit and 100 ml of concentrated sulphuric acid contained in a 2.5-litre three-necked flask, provided with an efficient mechanical stirrer, add 125 g (0.67 mol) of crude 4-amino-3-bromotoluene. Stir the solution and cool to 5 °C; then add slowly a solution of 74 g (1.07 mol) of pure sodium nitrite in 135 ml of water from a separatory funnel taking care that the temperature does not rise above 10 °C. Continue the stirring for 20 minutes after all the nitrite solution has been added in order to complete the diazotisation (test with potassium iodide–starch paper for the presence of free nitrous acid). Add 17.5 g (0.28 mol) of copper bronze (which has been washed with ether) or copper powder (Section 4.2.*19*, p. 426) to the diazotised solution, and replace the stirrer by a long double surface condenser. Have an ice bath at hand to cool the flask if the reaction becomes too vigorous. Warm the flask *cautiously* on a water bath until a vigorous evolution of gas commences, then immerse at once in an ice bath to prevent loss through the condenser by too rapid evolution of nitrogen and acetaldehyde. When the reaction has subsided, again warm the flask gently, and finally heat on a boiling water bath for 10 minutes. At the end of the reaction, the colour of the solution changes from reddish-brown to yellow. Add 1 litre of water and steam distil the mixture as long as oily drops pass over. Separate the heavy yellow oil, wash it with two 100 ml portions of 10 per cent sodium hydroxide solution, once with 50 ml of water, twice with 75 ml portions of ice-cold concentrated sulphuric acid, once with 50 ml of water, and finally with 50 ml of 5 per cent sodium carbonate solution. Dry with 2–3 g of magnesium sulphate or anhydrous calcium chloride, and filter through a little glass wool into a distilling flask. Distil, using an air condenser, and collect the m-bromotoluene (a colourless liquid) at 180–183 °C. The yield is 65 g (38% overall).

The essential i.r. and p.m.r. features for m-bromotoluene are noted in Expt 6.72. The m.s. of all the isomers shows principal fragment ions at m/z 172 (M, ^{81}Br), 170 (M, ^{79}Br), 91 (M − $^{81, 79}$Br), and 65 (91 − C_2H_2).

Notes. (1) If the mixture is cooled in ice, most of the p-acetotoluidide separates out in a crystalline form. It may be recrystallised from ethanol.
(2) Unless the material is at least partly dried before hydrolysis, the yield of hydrochloride is reduced because of its solubility. If pure 4-acetamido-3-bromotoluene is required, the crude material may be recrystallised from 50 per cent ethanol with the addition of a little decolourising carbon; it separates as colourless needles, m.p. 116–117 °C (180 g, 79%).
(3) If pure 4-amino-3-bromotoluene is required, the crude base may be purified either by steam distillation or, more satisfactorily, by distillation under reduced pressure. The oil is dried with 5 g of sodium hydroxide pellets, and fractionally distilled under reduced pressure: a little p-toluidine may be present in the low boiling point fraction, and the pure substance is collected at 92–94 °C/13 mmHg or at 120–122 °C/30 mmHg. The purified amine solidifies on cooling and melts at 17–18 °C.

6.7.2 COUPLING REACTIONS

Azo compounds are prepared by the interaction of a diazonium salt with a phenol in the presence of sodium hydroxide or with an amine in the presence of sodium acetate. The coupling reaction is an electrophilic substitution involving the diazonium ion which reacts at the position of greatest electron availability, i.e. the position *ortho* or *para* to the electron releasing phenoxy or amino groups.

2-Naphthol couples in the more reactive 1-position as in the synthesis of 1-phenylazo-2-naphthol (Expt 6.82). 1-Naphthol couples almost exclusively in position 4; when the diazo component is the p-nitrobenzenediazonium ion the product is Magneson II (Expt 6.83), which is employed as a test reagent for magnesium.

These simpler azo compounds are not of great practical value as dyestuffs owing to their slight solubility in water. The introduction of a sulphonic acid group into the molecule has no effect upon the colour, but renders the dye water-soluble – a fact of great commercial value. The simplest way of achieving this is to employ an amine, e.g. sulphanilic acid, in which the —SO_3H group is already present.

Sulphanilic acid, which has a dipolar or zwitterion structure (Section 6.6.1, p. 908), is sparingly soluble in water. It is best diazotised by bringing it into solution as the sodium salt by adding the calculated quantity of sodium carbonate, introducing the requisite quantity of sodium nitrite and pouring the solution on to a mixture of hydrochloric acid and ice; nitrous acid and the dipolar sulphanilic acid are liberated together and immediately react, and after a short time the internal diazonium salt separates from solution. Coupling with 2-naphthol in sodium hydroxide solution yields the useful dyestuff Orange II (Expt 6.84). When, *N,N*-dimethylaniline is used as the coupling component the product is methyl orange (Expt 6.85). This latter substance is more useful as an indicator than as a dye, for it changes colour at a certain concentration of hydrogen ions (pH 3.1–4.4). Treatment of a solution of methyl orange with a strong acid gives rise to a red form – which is essentially an internal salt stabilised by electron delocalisation.

It is interesting to note that azo compounds may be conveniently reduced either by a solution of tin(II) chloride in hydrochloric acid or by sodium dithionite. Thus 1-phenylazo-2-naphthol yields both aniline and 1-amino-2-naphthol, and methyl orange gives *p*-amino-*N,N*-dimethylaniline and sulphanilic acid.

947

$$NaO_3S-\langle\rangle-N{=}N-\langle\rangle-NMe_2 \longrightarrow$$

$$HO_3S-\langle\rangle-NH_2 + H_2N-\langle\rangle-NMe_2$$

Attention has previously been drawn (Section 6.7.1, p. 922) to the fact that unless an excess of hydrochloric (or mineral) acid is used in the diazotisation process, coupling occurs between the diazonium salt and the amino group in the amine to give diazoamino compounds. Thus benzenediazonium chloride and aniline yield diazoaminobenzene. This substance may be conveniently prepared by dissolving two equivalents of aniline in three equivalents of hydrochloric acid, and adding one equivalent of sodium nitrite in aqueous solution followed by two equivalents of sodium acetate (Expt 6.86).

$$\langle\rangle-\overset{\oplus}{N}{\equiv}N\ \ H_2\overset{..}{N}-\langle\rangle \xrightarrow{-H^\oplus} \langle\rangle-N{=}N-NH-\langle\rangle$$

If diazoaminobenzene is dissolved in aniline, to which a small quantity of aniline hydrochloride has been added, and the mixture kept at about 40 °C for a short time, it is converted into *p*-aminoazobenzene (Expt 6.86). The mechanism of this *diazoamino–aminoazo rearrangement* is dependent on (*a*) the heterolytic cleavage of a protonated diazoaminobenzene molecule to yield the benzenediazonium ion and aniline, and (*b*) a recoupling reaction, under weakly acidic conditions, of the diazonium ion at the *para* position of aniline.

$$\langle\rangle-N{=}N-NH-\langle\rangle + H^\oplus \rightleftharpoons$$

$$\langle\rangle-\overset{..}{N}{=}N-\overset{\overset{H}{|}}{\underset{\underset{H}{|}}{\overset{\oplus}{N}}}-\langle\rangle \rightleftharpoons \langle\rangle-\overset{\oplus}{N}{\equiv}N \langle\rangle\overset{..}{N}H_2 \xrightarrow{-H^\oplus}$$

$$\langle\rangle-N{=}N-\langle\rangle-NH_2$$

Experiment 6.82 1-PHENYLAZO-2-NAPHTHOL

$$\langle\rangle-\overset{\oplus}{N_2}\}\overset{\ominus}{Cl} + \overset{OH}{\langle\rangle} \xrightarrow{NaOH} \langle\rangle-N{=}N-\overset{OH}{\langle\rangle}$$

Dissolve 5.0 g (4.9 ml, 0.054 mol) of aniline in 16 ml of concentrated hydrochloric acid and 16 ml of water contained in a small beaker or conical flask. Diazotise by the addition of a solution of 4.0 g (0.058 mol) of sodium nitrite in 20 ml of water; follow the method given in Expt 6.70. Prepare a solution of 7.8 g (0.054 mol) of 2-naphthol in 45 ml of 10 per cent sodium hydroxide solution in a 250-ml beaker; cool the solution to 5 °C by immersion in an ice bath, assisted by the direct addition of about 25 g of crushed ice. Stir the naphthol solution vigorously and add the cold diazonium salt solution *very slowly*: a red colour develops and red crystals of 1-phenylazo-2-naphthol soon separate. When all the diazonium salt solution has been added, allow the mixture to stand in an ice bath for 30 minutes with occasional stirring. Filter the solution through a Buchner funnel with *gentle suction*, wash well with water and drain thoroughly by pressing the crystals with the back of a large glass stopper. Recrystallise one-fourth of the product from glacial acetic acid (30–35 ml): retain the remainder for reduction by stannous chloride. Filter the recrystallised product with suction, wash with a little ethanol (or industrial spirit) to eliminate acetic acid and dry upon filter paper. The yield of deep red crystals is about 3 g. Pure 1-phenylazo-2-naphthol has m.p. 131 °C; if the m.p. is low, recrystallise the dry product from ethanol.

Reduction with tin(II) chloride. *1-Amino-2-naphthol hydrochloride.* Into a 350- or 500-ml round-bottomed flask, provided with a reflux condenser and containing 100 ml of industrial spirit, place the crude 1-phenylazo-2-naphthol reserved above and boil gently until most of the azo compound has dissolved. Meanwhile dissolve 20 g of a good grade of tin(II) chloride in 60 ml of concentrated hydrochloric acid (warming is necessary to produce a clear solution) (1), add this to the contents of the flask and boil under reflux for a further 30 minutes. All the azo compound dissolves rapidly and is reduced by the tin(II) chloride; the solution acquires a very pale brown colour. Decant the solution to a beaker and cool in ice: the 1-amino-2-naphthol hydrochloride separates as fine greyish-white crystals. Filter with suction, and wash with dilute hydrochloric acid (1:4). Recrystallise from the minimum volume of hot water which contains a few drops of tin(II) chloride solution in an equal weight of hydrochloric acid (this reduces atmospheric oxidation), cool the clear solution in an ice bath and collect the recrystallised product as before. Dry the colourless crystals in a desiccator. The yield is 3–4 g. The compound will remain colourless, or nearly so, if protected from light during storage.

Note. (1) Sodium dithionite, $Na_2S_2O_4$, may also be used for the reduction; see under *methyl orange*, Expt 6.85.

Experiment 6.83   4-(4'-NITROBENZENEAZO)-1-NAPHTHOL (*Magneson II*)

Dissolve 5.0 g (0.036 mol) of *p*-nitroaniline (Expt 6.68) in a warm mixture of 13 ml of concentrated hydrochloric acid and 13 ml of water contained in a 250-ml beaker. Place the beaker in an ice–salt bath and cool to 0–5 °C whilst stirring vigorously; *p*-nitroaniline hydrochloride will separate in a finely divided crystalline form. Add a cold solution of 3.7 g (0.054 mol) of sodium nitrite in 8 ml of water slowly and with stirring to an end-point with potassium iodide–starch paper: do not allow the temperature of the solution to rise above 8 °C. Dissolve 5.2 g (0.035 mol) of 1-naphthol in a solution of 7 g of sodium hydroxide in 25 ml of water, cool in ice and add the diazotised solution slowly and with stirring. Then add concentrated hydrochloric acid slowly and with vigorous stirring to the cold mixture until it is strongly acid to Congo red paper. The colour will change from violet to dark red-brown. Filter with *gentle suction*, wash with water until free from acid and dry upon filter-paper in the air. The yield is 8 g (74%).

2,4-Dihydroxy-4'-nitroazobenzene ('*Magneson I*') may be similarly prepared by substituting resorcinol for 1-naphthol; it may be recrystallised from methanol and melts at 199–200 °C.

Experiment 6.84 ORANGE II (*β-Naphthol Orange*)

Diazotise 10.5 g (0.05 mol) of sulphanilic acid dihydrate as described under *methyl orange* (Expt 6.85), and keep the suspension of the diazonium compound in ice-water until required. Dissolve 7.2 g (0.05 mol) of a good grade of 2-naphthol in 40 ml of cold 10 per cent sodium hydroxide solution in a 600-ml beaker, cool to 5 °C and pour in, with stirring, the well-mixed suspension of diazotised sulphanilic acid. Coupling takes place readily and the dyestuff separates as a crystalline paste. Stir well and, after 10 minutes, heat the mixture until all the solid has dissolved. Add 20 g of sodium chloride (to decrease the solubility of the product further) and warm until this dissolves. Allow the solution to cool spontaneously in the air for 1 hour, and then cool in ice until crystallisation is complete. Collect the product on a Buchner funnel and apply gentle suction; wash with a little saturated salt solution, and dry at 80 °C. The product weighs about 22 g, and contains about 20 per cent of sodium chloride; further purification is unnecessary for dyeing purposes. To obtain pure, crystalline Orange II, dissolve the crude substance in the minimum volume of boiling water, allow to cool to about 80 °C, add about twice the volume of rectified (or industrial) spirit and allow crystallisation to proceed spontaneously. When cold, filter at the pump, wash the pure dyestuff (it is a dihydrate) with a little ethanol and dry in the air. The yield is 14 g (80%) (1).

Note. (1) For the reduction of Orange II to 1-amino-2-naphthol and its conversion to 1,2-naphthoquinone, see Expt 6.131.

Experiment 6.85 METHYL ORANGE

$$NaO_3S{-}\!\!\left\langle\!\!\bigcirc\!\!\right\rangle\!\!{-}NH_2 \xrightarrow{NaNO_2;\ HCl} {}^{\ominus}O_3S{-}\!\!\left\langle\!\!\bigcirc\!\!\right\rangle\!\!{-}N{\equiv}N \xrightarrow[NaOH]{PhNMe_2}$$

$$NaO_3S{-}\!\!\left\langle\!\!\bigcirc\!\!\right\rangle\!\!{-}N{=}N{-}\!\!\left\langle\!\!\bigcirc\!\!\right\rangle\!\!{-}NMe_2$$

In a 250-ml conical flask place 10.5 g (0.05 mol) of sulphanilic acid dihydrate, 2.65 g (0.025 mol) of anhydrous sodium carbonate and 100 ml of water, and warm until a clear solution is obtained. Cool the solution under the tap to about 15 °C, and add a solution of 3.7 g (0.059 mol) of sodium nitrite in 10 ml of water. Pour the resulting solution slowly and with stirring into a 600-ml beaker containing 10.5 ml of concentrated hydrochloric acid and 60 g of crushed ice (1). Test for the presence of free nitrous acid with potassium iodide–starch paper after 15 minutes. Fine crystals of the diazobenzene sulphonate will soon separate; do not filter these off as they will dissolve during the next stage of the preparation. Dissolve 6.05 g (6.3 ml, 0.05 mol) of dimethylaniline in 3.0 ml of glacial acetic acid, and add it with vigorous stirring to the suspension of diazotised sulphanilic acid. Allow the mixture to stand for 10 minutes; the red or acid form of methyl orange will gradually separate. Then add slowly and with stirring 35 ml of 20 per cent sodium hydroxide solution: the mixture will assume a uniform orange colour due to the separation of the sodium salt of methyl orange in fine particles. Direct filtration of the latter is slow, hence, while stirring the mixture with a thermometer, heat it almost to the boiling point. Most of the methyl orange will dissolve. Add about 10 g of sodium chloride (to assist the subsequent separation of the methyl orange) and warm at 80–90 °C until the salt has dissolved. Allow the mixture to cool undisturbed for 15 minutes and then cool in ice-water; this gives a fairly easily filterable product. Filter off the methyl orange at the pump, but apply only gentle suction so as to avoid clogging the pores of the filter paper; rinse the beaker with a little saturated salt solution and drain well. Recrystallise from hot water (about 150 ml are required); filter the hot solution, if necessary, through a hot water funnel or through a preheated Buchner funnel. Reddish-orange crystals of methyl orange separate as the solution cools. Filter these at the pump, drain well, wash with a little ethanol, and finally with a small volume of ether. The yield is 13 g (80%). Methyl orange, being a salt, has no well-defined m.p.

Note. (1) An alternative procedure is to cool the solution containing the sodium sulphanilate and sodium nitrite in a bath of crushed ice to about 5 °C and then add 10.5 ml of concentrated hydrochloric acid diluted with an equal volume of water slowly and with stirring; the temperature must not be allowed to rise above 10 °C and an excess of nitrous acid should be present (the solution is tested after standing for 5 minutes). The subsequent stages in the preparation – addition of dimethylaniline solution, etc. – are as above.

Reduction of methyl orange to *p*-aminodimethylaniline. *Method 1.* Dissolve 2.0 g of methyl orange in the minimum volume of hot water and to the hot solution add a solution of 8 g of tin(II) chloride in 20 ml of concentrated hydrochloric acid until decolourisation takes place; gentle boiling may be necessary. Cool the resulting solution in ice; a crystalline precipitate

951

consisting of sulphanilic acid and some *p*-aminodimethylaniline hydrochloride separates out. In order to separate the free base, add 10 per cent sodium hydroxide solution until the precipitate of tin hydroxide redissolves. Extract the cold solution with three or four 20 ml portions of ether, dry the extract with anhydrous potassium carbonate and remove the ether by distillation. The residual base soon crystallises, particularly if it is stirred with a glass rod; it melts at 41 °C.

Method 2. Suspend 2.0 g of methyl orange in 4 ml of water, and add a small quantity of sodium dithionite ($Na_2S_2O_4$). Heat the mixture and add more sodium dithionite until the colour is discharged. The sulphanilic acid remains in the solution as sodium sulphanilate and the *p*-aminodimethylaniline may be extracted with ether as in *Method 1.*

Experiment 6.86   DIAZOAMINOBENZENE AND *p*-AMINOAZOBENZENE

$$\overset{\oplus}{Ph\cdot N}\equiv\overset{\ominus}{N\}Cl} \xrightarrow{Ph\cdot NH_2} Ph\cdot N{=}N\cdot NH\cdot Ph \longrightarrow Ph\cdot N{=}N\cdot C_6H_4\cdot NH_2\text{-}p$$

Diazoaminobenzene. In a 250-ml flask place 75 ml of water, 24 g (20 ml) of concentrated hydrochloric acid and 14 g (13.7 ml, 0.15 mol) of aniline. Shake vigorously (1) and then add 50 g of crushed ice. Run in a solution of 5.2 g (0.075 mol) of sodium nitrite in 12 ml of water, with constant shaking, during a period of 5–10 minutes. Allow to stand with frequent shaking (1) for 15 minutes, and add a solution of 21.0 g of crystallised sodium acetate in 40 ml of water during 5 minutes. A yellow precipitate of diazoaminobenzene begins to form immediately; allow to stand with frequent shaking for 45 minutes and do not allow the temperature to rise above 20 °C (add ice, if necessary). Filter the yellow diazoaminobenzene on a Buchner funnel, wash it with 250 ml of cold water, drain as completely as possible and spread it on a sheet of filter paper to dry. The yield of crude diazoaminobenzene, m.p. 91 °C, is 13 g (87%) (2). Recrystallise a small portion from light petroleum, b.p. 60–80 °C: the pure compound, m.p. 97 °C, is obtained.

Notes. (1) For preparations on a larger scale, mechanical stirring is recommended. (2) The crude compound may be employed in the preparation of *p*-aminoazobenzene.

Conversion into **p**-*aminoazobenzene.* Dissolve 5 g (0.025 mol) of finely powdered diazoaminobenzene in 12–15 g of aniline in a small flask and add 2.5 g of finely powdered aniline hydrochloride (1). Warm the mixture, with frequent shaking, on a water bath at 40–45 °C for 1 hour. Allow the reaction mixture to stand for 30 minutes. Then add 15 ml of glacial acetic acid diluted with an equal volume of water: stir or shake the mixture in order to remove the excess of aniline in the form of its soluble acetate. Allow the mixture to stand, with frequent shaking, for 15 minutes: filter the aminoazobenzene at the pump, wash with a little water and dry upon filter paper. Recrystallise the crude *p*-aminoazobenzene (3.5 g, 70%; m.p. 120 °C) from 15–20 ml of carbon tetrachloride to obtain the pure compound, m.p. 125 °C. Alternatively, the compound may be recrystallised from dilute ethanol, to which a few drops of concentrated ammonia solution have been added.

   To prepare the hydrochloride dissolve about 1 g of the compound (which need not be perfectly dry) in about 8 ml of ethanol. Add this solution to boiling dilute hydrochloric acid (10 ml of the concentrated acid and 80 ml of water). Boil for 5 minutes, filter the hot solution if necessary and allow to cool. p-Aminoazobenzene hydrochloride separates in steel-blue crystals. Filter, wash with a little dilute hydrochloric acid, and dry.

To recover the free base, dissolve the hydrochloride in the minimum volume of boiling ethanol, add concentrated ammonia solution dropwise until a clear solution results and the blue colour has become light brown. Add water carefully until a cloudiness appears, warm on a water bath until the cloudiness just disappears and allow to cool. Yellow crystals of p-aminoazobenzene separate on cooling.

Note. (1) The aniline hydrochloride may be prepared by treating 2 g of aniline with an excess (about 3 ml) of concentrated hydrochloric acid in a small beaker, cooling, filtering at the pump, washing with a *small* volume of ether and drying between filter paper.

6.8 MISCELLANEOUS AROMATIC NITROGEN COMPOUNDS

1. Intermediate products in the reduction of nitro compounds (Expts 6.87 to 6.91).
2. Arylhydrazines (Expts 6.92 to 6.95).
3. Arylureas and related compounds (Expts 6.96 to 6.98).

The important *spectroscopic features* (*i.r.*, *p.m.r.*, *m.s.*, and *u.v.–visible*) which are observed in this varied range of nitrogen compounds are discussed descriptively under appropriate preparative sections.

6.8.1 INTERMEDIATE PRODUCTS IN THE REDUCTION OF NITRO COMPOUNDS

The reduction of an aromatic nitro compound with a powerful reducing agent (tin or tin(II) chloride and hydrochloric acid; iron and dilute hydrochloric acid; hydrogen and a platinum catalyst) leads to a good yield of primary amine, e.g. aniline from nitrobenzene. By the use of milder reducing agents and by the control of the hydrogen ion concentration of the solution, a number of intermediate products may be isolated, some of which are products of direct reduction and others are formed through secondary reactions. The various stages of the reduction of nitrobenzene have been established by investigating the process electrolytically under conditions of varying pH, current density and electrode construction and composition. The sequence is as follows.

$$\text{Ph·NO}_2 \xrightarrow{2H} \text{Ph·NO} \xrightarrow{2H} \text{Ph·NHOH} \xrightarrow{2H} \text{Ph·NH}_2$$

Nitrobenzene   Nitrosobenzene   *N*-Phenylhydroxylamine   Aniline

The initial product, nitrosobenzene, is so easily reduced to *N*-phenylhydroxylamine that it has not been isolated from the reduction medium, but its presence has been established by reaction in solution with hydroxylamine to yield a benzenediazonium salt, which couples readily with 1-naphthylamine to form the dyestuff 2-phenylazo-1-naphthylamine.

$$\text{Ph·NO} + \text{H}_2\text{NOH} + \text{HX} \longrightarrow \text{Ph·}\overset{\oplus}{\text{N}}\!\!\equiv\!\!\text{N}\}\overset{\ominus}{\text{X}} + 2\text{H}_2\text{O}$$

953

Under the catalytic influence of alkali, nitrosobenzene and N-phenylhydroxyl-amine react to yield azoxybenzene.

$$Ph \cdot NO + \overset{H}{\underset{HO}{\diagdown}} N \cdot Ph \xrightarrow[-H_2O]{\ominus OH} Ph \cdot N {=} N \cdot Ph$$
$$\downarrow$$
$$O$$

Further reduction in alkaline solution (e.g. with zinc powder) leads to azobenzene and hydrazobenzene.

$$Ph \cdot N {=} N \cdot Ph \xrightarrow{2H} Ph \cdot N {=} N \cdot Ph \xrightarrow{2H} Ph \cdot NH \cdot NH \cdot Ph$$
$$\downarrow$$
$$O$$

Electrolytic reduction of hydrazobenzene gives aniline.

$$Ph \cdot NH \cdot NH \cdot Ph \xrightarrow{2H} 2Ph \cdot NH_2$$

The various intermediate compounds may be prepared in the laboratory, and convenient methods are described below.

N-Phenylhydroxylamine (Expt 6.87) is formed when nitrobenzene is treated with a 'neutral' reducing agent, e.g. zinc powder and aqueous ammonium chloride solution. The compound rearranges, in the presence of acids, with the formation of p-aminophenol (Expt 6.87).

N-Phenylhydroxylamine may be converted into an N-benzoyl derivative by treating it in aqueous solution with benzoyl chloride while simultaneously removing, with sodium hydrogen carbonate, the liberated hydrogen chloride.[32] On reaction with thionyl chloride this benzoyl derivative is converted into o-chlorobenzanilide (Expt 6.61) by way of an intermediate chlorosulphite ester, which then undergoes a specific *ortho* rearrangement.[25]

$$Ph \cdot NHOH + Ph \cdot COCl \xrightarrow{-HCl} Ph \cdot N(OH) \cdot CO \cdot Ph \xrightarrow[-HCl, -SO_2]{SOCl_2}$$
$$o\text{-}Cl \cdot C_6H_4 \cdot NH \cdot CO \cdot Ph$$

Nitrosobenzene (Expt 6.88) may be obtained by the oxidation of N-phenylhydroxylamine with acid dichromate solution at $0\,°C$. The solid product is colourless and is probably a dimer; it dissociates to a green monomer upon melting or in solution.

Azoxybenzene is readily prepared by reduction of nitrobenzene in an alkaline medium with a variety of mild reducing agents. Reducing sugars have been used successfully for the reduction of substituted nitro compounds to the correspond-ing azoxyarenes[33] and the use of D-glucose for the reduction of nitrobenzene is illustrated in Expt 6.89.

Reduction of nitrobenzene in methanolic or ethanolic sodium hydroxide solution with zinc powder leads to azobenzene or hydrazobenzene according to the proportion of zinc powder employed (Expts. 6.90 and 6.91). Hydrazoben-zene may be oxidised to azobenzene by sodium hypobromite solution at $0\,°C$.

In the presence of acids, hydrazobenzene rearranges to give a mixture containing about 70 per cent of benzidine (4,4′-diaminobiphenyl) and about 30 per cent of 2,4′-diaminobiphenyl (diphenyline), the *benzidine rearrangement*. Benzidine is carcinogenic and its preparation and storage is under strict control.

Experiment 6.87   N-PHENYLHYDROXYLAMINE

$$Ph \cdot NO_2 \xrightarrow[\text{Zn/NH}_4\text{Cl}]{4[\text{H}]:} Ph \cdot NHOH + H_2O$$

In a 2-litre beaker, equipped with a thermometer and mechanical stirrer, place 25 g of ammonium chloride, 800 ml of water and 50 g (41.6 ml, 0.41 mol) of redistilled nitrobenzene. Stir the mixture vigorously, and add 59 g (0.83 mol) of zinc powder of 90 per cent purity (Section 4.2.80, p. 467) during about 15 minutes; the rate of addition should be such that the temperature rapidly rises to 60–65 °C and remains in this range until all the zinc has been added. Continue the stirring for a further 15 minutes, by which time the reduction is complete as is shown by the fact that the temperature commences to fall. Filter the warm reaction mixture at the pump to remove the zinc oxide, and wash it with 100 ml of hot water. Place the filtrate in a conical flask, saturate it with common salt (about 300 g) and cool in an ice bath for at least 1 hour to ensure maximum crystallisation of the desired product. Filter the pale yellow crystals of phenylhydroxylamine with suction and drain well. The yield of crude, dry product is about 38 g; this contains a little salt and corresponds to about 29 g (66%) of pure phenylhydroxylamine as determined by its separation from inorganic materials by dissolution in ether. The substance deteriorates upon storage and is therefore used immediately for a secondary preparation (e.g. nitrosobenzene, Expt 6.88). If required perfectly pure, it may be recrystallised from benzene–light petroleum (b.p. 40–60 °C) or from benzene alone; the resulting pure compound is somewhat more stable and has a melting point of 81 °C.

N-Benzoylation of phenylhydroxylamine.[32]   Phenylhydroxylamine (10 g, 0.105 mol) is dissolved in warm water (400 ml) and the solution cooled to room temperature and filtered if necessary. Benzoyl chloride (15 g, 0.106 mol, **CAUTION**) is added in small portions to the solution which is vigorously shaken; the acidity of the solution is monitored with litmus paper and neutralised by the addition of appropriate quantities of sodium hydrogen carbonate solution (1). After several minutes an off-white solid separates. When the smell of benzoyl chloride has disappeared and the aqueous solution remains alkaline, the benzoylation reaction is complete; a small portion of the solution gives a negative reaction on warming with Fehling's solution. The solid is filtered and washed with water and finally well triturated with sodium hydrogen carbonate solution to remove any occluded acid (2). The crude product is boiled with water in which the monobenzoate is soluble but the dibenzoate remains as a dark brown oil. After decantation and cooling the monobenzoate separates as white needles, m.p. 121–122 °C.

Notes. (1) An alternative procedure[34] is to add the benzoyl chloride in ether to a stirred suspension of phenylhydroxylamine in ether at 0 °C. When the reaction is complete the ether solution is washed with dilute alkali and water, dried and evaporated.

(2) Further purification may be effected by dissolving the crude product in aqueous ammonia (28%), in which the dibenzoate is insoluble, filtering, and acidifying with sulphuric acid,[32] or better hydrochloric acid.[35]

Conversion of phenylhydroxylamine into p-aminophenol.   Add 4.4 g of recrystallised phenylhydroxylamine to a mixture of 20 ml of concentrated

955

sulphuric acid and 60 g of ice contained in a 1-litre beaker cooled in a freezing mixture. Dilute the solution with 400 ml of water, and boil until a sample, tested with dichromate solution, gives the smell of quinone and not of nitrosobenzene or nitrobenzene (c. 10–15 minutes). Neutralise the cold reaction mixture with sodium hydrogen carbonate, saturate with salt, extract twice with ether and dry the ethereal extract with magnesium sulphate or anhydrous sodium sulphate. Distil off the ether; p-aminophenol, m.p. 186 °C, remains. The yield is 4.3 g (98%).

The i.r. spectrum shows similarities to that of the *ortho* isomer with respect to the absorptions due to the NH and OH groups (p. 894). The p.m.r. spectrum (TFA, TMS) confirms the substitution pattern and shows signals at δ 7.12 (d, 2H, *ortho*-H's to NH_2), 7.45 (d, 2H, *ortho*-H's to OH), and 8.70 (OH and NH_2).

Experiment 6.88 NITROSOBENZENE

$$Ph \cdot NO_2 \xrightarrow{[H]} Ph \cdot NHOH \xrightarrow{[O]} Ph \cdot NO + H_2O$$

In a 2-litre beaker, equipped with a thermometer and mechanical stirrer, place 30 g (0.56 mol) of ammonium chloride, 1 litre of water and 61.5 g (51 ml, 0.5 mol) of pure nitrobenzene. Stir the mixture vigorously, and add 75 g (1.03 mol, 90% purity; see Section 4.2.80, p. 467) of zinc powder during about 15 minutes; the rate of addition should be such that the temperature rises rapidly to 60–65 °C and remains in this range until all the zinc has been added. Continue the stirring for a further 15 minutes, by which time the reduction is complete as shown by the fact that the temperature commences to fall. Filter the warm reaction mixture at the pump to remove the zinc oxide, and wash it with 600–700 ml of boiling water. Transfer the filtrate and washings to a 4-litre round-bottomed flask or beaker and cool *immediately* to 0–1 °C by the addition of sufficient crushed ice and leave at least 250 g unmelted. Without delay, add with stirring a cold solution of concentrated sulphuric acid (150 ml of the concentrated acid added to sufficient ice to reduce its temperature to −5 °C). Then add an ice-cold solution of 34 g (0.114 mol) of crystallised sodium dichromate in 125 ml of water as rapidly as possible to the stirred solution. After 2–3 minutes, filter the straw-coloured precipitate of nitrosobenzene on a Buchner funnel and wash it with 200 ml of water. Steam distil the nitrosobenzene as rapidly as possible; the nitrosobenzene tends to decompose at the elevated temperature. Cool the receiver in ice because the compound has a high vapour pressure at room temperature. The nitrosobenzene condenses to a green liquid, which solidifies to a white solid; care should be taken that the solid does not clog the condenser by turning off the water supply from time to time. Stop the distillation when yellow oily material appears in the condenser. Filter; grind the nitrosobenzene in a glass mortar with a little water. Filter at the pump, wash it with water until the washings are no longer brown and drain as completely as possible. Dry the solid between layers of filter paper. The yield of nitrosobenzene, m.p. 66–67 °C, is 30 g (56%). A pure product, m.p. 68 °C, may be obtained by recrystallisation from a small volume of ethanol with good cooling: the compound should be dried over anhydrous calcium chloride at atmospheric pressure. The substance may be kept 1–2 days at room temperature and for longer periods at 0 °C.

The i.r. spectrum offers an interesting contrast with that of nitrobenzene in that only one stretching absorption of the NO group is observed (c. 1425 cm^{-1}). The m.s. reveals significant fragment ions at m/z 107 (M), 77 (M − NO, base peak), and 51 (77 − C$_2$H$_2$).

Experiment 6.89 AZOXYBENZENE

$$Ph \cdot NO_2 + 6[H] \longrightarrow Ph \cdot N = N \cdot Ph + 3H_2O$$
$$\downarrow$$
$$O$$

Equip a 500-ml three-necked flask with an efficient stirrer (e.g. a Hershberg stirrer, Fig. 2.49) and a reflux condenser; stopper the third neck. Place a solution of 30 g of sodium hydroxide in 100 ml of water, and also 20.5 g (17.1 ml, 0.167 mol) of pure nitrobenzene in the flask, immerse it in a water bath maintained at 55–60 °C, and add 21 g (0.117 mol) of anhydrous glucose in small portions, with continuous stirring, during 1 hour. Then heat on a boiling water bath for 2 hours. Pour the hot mixture into a 1-litre round-bottomed flask and steam distil (Fig. 2.102) to remove aniline and nitrobenzene. When the distillate is clear (i.e. after about 1 litre has been collected), pour the residue into a beaker cooled in an ice bath. The azoxybenzene soon solidifies. Filter with suction, grind the lumps of azoxybenzene in a mortar, wash with water and dry upon filter paper or upon a porous plate. The yield of material, m.p. 35–35.5 °C, is 13 g (79%). Recrystallise from 7 ml of rectified spirit or of methanol; the m.p. is raised to 36 °C.

The u.v.–visible absorption spectra (MeOH) of (a) hydrazobenzene (Expt 6.91), (b) azobenzene (Expt 6.90) and (c) azoxybenzene offer interesting comparison. This shows the bathochromic shifts which result from meso-meric interaction between the two phenyl groups via the nitrogen linkages: (a) λ_{max} 229 nm (ε 13 000), 235 nm (ε 12 500), 244 nm (ε 11 600), 315 nm (ε 11 700), 435 nm (ε 446); (b) λ_{max} 228 nm (ε 13 400), 314 nm (ε 18 200), 433 nm (ε 727); (c) λ_{max} 232 nm (ε 8680), 239 nm (ε 7530), 260 nm (ε 7530), 322 nm (ε 14 700).

Experiment 6.90 AZOBENZENE

$$2Ph \cdot NO_2 \xrightarrow[8NaOH]{4Zn} Ph \cdot N = N \cdot Ph + 4Na_2[ZnO_2] + 4H_2O$$

Support a 1-litre three-necked flask, equipped with a sealed stirrer unit and a reflux condenser, on a water bath, and place a solution of 65 g of sodium hydroxide in 150 ml of water, 50 g (41.5 ml, 0.41 mol) of pure nitrobenzene and 500 ml of methanol in the flask. Add 59 g (0.9 mol) of zinc powder (90% purity; see Section 4.2.80, p. 467) to the mixture, start the stirrer and reflux for 10 hours (1). Filter the mixture while hot, and wash the precipitate of sodium zincate with a little methanol. The strongly alkaline filtrate is not always clear: render it neutral to litmus by the cautious addition of concentrated hydrochloric acid, and filter again. Distil off the methanol from the filtrate, cool the residue in ice and filter off the solid azobenzene. The crude azobenzene contains occluded zinc salts. To remove these, add the crude product to 100 ml of 2 per cent hydrochloric acid, warm to about 70 °C in order to melt the azobenzene and stir mechanically for 5 minutes; continue the stirring while the mixture is immersed in ice-water in order to solidify the azobenzene. Filter, wash well with water, drain thoroughly and recrystallise from a mixture of 145 ml of rectified spirit and 12 ml of water; collect the azobenzene and dry in the air. The yield of pure azobenzene (reddish-orange crystals), m.p. 67–68 °C, is 31 g (86%) (2). The u.v.–visible spectrum is quoted in Expt 6.89.

Notes. (1) At the end of this time, the reddish mixture should be free from the odour of nitrobenzene; if it is not, reflux for 2–3 hours longer.
(2) Frequently the recrystallized azobenzene has m.p. 61 °C, which is unaffected by recrystallisation from ethanol. Upon distillation from a 50-ml distilling flask fitted with a short air condenser, the m.p. is raised to 67.5 °C and the recovery is about 90 per cent: one recrystallisation from diluted ethanol (as above) then gives perfectly pure azobenzene of m.p. 68.5 °C.

Experiment 6.91 HYDRAZOBENZENE (sym.-*Diphenylhydrazine*)

$$2Ph \cdot NO_2 \xrightarrow[\text{10NaOH}]{\text{5Zn}} Ph \cdot NH \cdot NH \cdot Ph + 5Na_2[ZnO_2] + 4H_2O$$

Support a 1500-ml three-necked flask, equipped with a sealed stirrer unit and a double surface reflux condenser, on a water bath, and place a solution of 84 g of sodium hydroxide in 185 ml of water, 50 g (41.5 ml, 0.406 mol) of nitrobenzene and 500 ml of methanol in the flask. Add 78 g (1.07 mol, 90%) of zinc powder, start the stirrer and reflux for 10 hours. The solution gradually assumes the reddish colour of azobenzene and then, on further reduction, turns to a pale yellow (due to hydrazobenzene). If the colour is not almost completely discharged at the end of the refluxing period, add a further 11 g (0.15 mol; 90%) of zinc powder, and reflux for 2–3 hours longer. Filter the hot solution through a preheated Buchner funnel and wash the sodium zincate upon the filter with a little hot methanol. Pour the filtrate into a large flask (1), stopper it loosely and cool it in a freezing mixture of ice and salt to accelerate crystallisation. After 1 hour filter off the almost colourless crystals of hydrazobenzene at the pump as rapidly as possible (it is helpful to displace the air above the solution undergoing filtration in the funnel with a stream of nitrogen), wash with 50 per cent methanol to which a little sulphurous acid has been added until the filtrate is no longer alkaline. Dry in a vacuum desiccator. The resulting almost colourless hydrazobenzene (15 g; 40%; m.p. 125 °C) is sufficiently pure for most preparative purposes. If it is required pure (m.p. 126 °C with production of a yellow colour), it may be recrystallised from hot methanol containing a little ammonium sulphide or sulphurous acid (these assist in preventing atmospheric oxidation).

Owing to the great tendency of hydrazobenzene to undergo oxidation, all operations involving filtration should be carried out as rapidly as possible and air should not be drawn through it unnecessarily. The substance should be dried in a vacuum desiccator: it can only be preserved in a colourless condition if it is kept in an atmosphere of carbon dioxide or nitrogen or in sealed vessels. The u.v.–visible spectrum is quoted in Expt 6.89.

Notes. (1) If the methanol is distilled off before thorough cooling in a freezing mixture, the yield of hydrazobenzene is appreciably increased, but the product is considerably more coloured due to admixture with a trace of azobenzene. About 12 g of impure hydrazobenzene may be recovered by distilling off the methanol from the filtrate after the colourless hydrazobenzene has been collected.

6.8.2 ARYLHYDRAZINES

Arylhydrazines may be prepared by reducing diazonium salts with excess warm sodium sulphite solution, followed by acidification with hydrochloric acid. The hydrochloride usually crystallises out on cooling and treatment of the latter with excess sodium hydroxide solution liberates the free base. The preparation of *p*-nitrophenylhydrazine by this method is illustrated in Expt 6.92.

The mechanism of this reduction probably involves the initial addition of a sulphite ion to the diazonium group to give an azosulphonate which undergoes further conjugate (1,4-) addition of the nucleophilic sulphite ion. The resulting intermediate is protonolytically cleaved on heating under acidic conditions:

This method cannot be applied to polynitro amines, since these are so weakly basic that they can be diazotised only under special conditions in strongly acidic media (Section 6.7.1, p. 922). In such cases use may be made of the susceptibility to nucleophilic displacement of halogen when activated by *ortho* and *para* nitro groups. Thus the valuable reagent 2,4-dinitrophenylhydrazine (Expt 6.93) is readily prepared by reacting 1-chloro-2,4-dinitrobenzene with hydrazine. Reaction with ammonia similarly gives 2,4-dinitroaniline (cognate preparation in Expt 6.93).

A further example is provided by the reaction of 1-chloro-2,4,6-trinitrobenzene (picryl chloride) with *N,N*-diphenylhydrazine to give *N,N*-diphenylpicrylhydrazine (Expt 6.94). This compound is of interest in that oxidation with lead dioxide yields the highly stable *N,N*-diphenylpicrylhydrazyl radical, which is obtained as an intensely coloured, paramagnetic solid. Stabilisation of the radical is promoted by the strongly electron-attracting trinitrophenyl residue resulting in extensive delocalisation of the odd electron. The *N,N*-diphenylhydrazine required for this preparation is obtained by reduction of *N*-nitroso-*N,N*-diphenylamine with zinc and acetic acid (Expt 6.95).

Experiment 6.92   *p*-NITROPHENYLHYDRAZINE

$$p\text{-}O_2N\cdot C_6H_4\cdot NH_2 \xrightarrow[\text{HCl}]{\text{NaNO}_2} p\text{-}O_2N\cdot C_6H_4\cdot \overset{\oplus}{N}{\equiv}N\}Cl^{\ominus} \xrightarrow{+4e, +4H^{\oplus}}$$

$$p\text{-}O_2N\cdot C_6H_4\cdot NH\cdot \overset{\oplus}{N}H_3\}Cl^{\ominus} \xrightarrow{\ominus OH} p\text{-}O_2N\cdot C_6H_4\cdot NH\cdot NH_2$$

Dissolve 10 g (0.075 mol) of *p*-nitroaniline (Expt 6.68) in a mixture of 21 ml of concentrated hydrochloric acid and an equal volume of water, and cool rapidly to 0 °C in order to obtain the hydrochloride of the base in a fine state of division. Diazotise in the usual way (see Expt 6.70) by the gradual addition of a solution of 5.2 g (0.075 mol) of sodium nitrite in 12 ml of water. Continue the stirring for a few minutes, filter the solution rapidly and add it from a separatory funnel to an ice-cold solution of 41 g (0.156 mol) of sodium sulphite (96% Na$_2$SO$_3$·7H$_2$O) in 100 ml of water containing 4 g of sodium hydroxide; stir the mixture during the addition which requires about 5 minutes. (If the diazonium solution is added too rapidly, an orange-red precipitate of sodium *p*-nitrobenzenediazosulphonate is produced, and is apt to form a resin.) Allow the solution to stand for 5 minutes, acidify with 70 ml of concentrated hydrochloric acid and heat on a water bath at 25 °C for 3 minutes, when yellow needles commence to separate. Allow to stand overnight, filter off the crystals, heat them with 20 ml of concentrated hydrochloric acid on a water bath for 7 minutes and allow to cool. Filter off the precipitate, consisting of *p*-nitrophenylhydrazine hydrochloride and sodium salts, dissolve it in water and treat the solution with a concentrated solution of sodium acetate: the free base will separate out in an almost pure state (7–8 g, 63–72%). The *p*-nitrophenylhydrazine may be recrystallised from ethanol and is obtained as light brown crystals, m.p. 158 °C (decomp.).

Experiment 6.93   2,4-DINITROPHENYLHYDRAZINE

$$2,4\text{-}(O_2N)_2C_6H_3Cl + NH_2NH_2 \longrightarrow 2,4\text{-}(O_2N)_2C_6H_3\cdot NH\cdot NH_2$$

Suspend 35 g (0.27 mol) of finely powdered hydrazine sulphate in 125 ml of hot water contained in a 400-ml beaker, and add, with stirring, 118 g (0.87 mol) of sodium acetate hydrate or 85 g of potassium acetate. Boil the mixture for 5 minutes, cool to about 70 °C, add 80 ml of rectified spirit, filter at the pump and wash with 80 ml of hot rectified spirit. Keep the filtered hydrazine solution for the next stage in the preparation.

Equip a 1-litre three-necked flask with a reflux condenser and a sealed mechanical stirrer. Dissolve 50.5 g (0.25 mol) of commercial 1-chloro-2,4-dinitrobenzene (1) in 250 ml of rectified spirit in the flask, add the hydrazine solution and reflux the mixture with stirring for an hour. Most of the reaction product separates during the first 10 minutes. Cool, filter with suction and wash with 50 ml of warm (60 °C) rectified spirit to remove unchanged chlorodinitrobenzene, and then with 50 ml of hot water. The resulting 2,4-dinitrophenylhydrazine (30 g, 60%) melts at 191–192 °C (decomp.), and is pure enough for most purposes. Distil off half the alcohol from the filtrate and thus obtain a less pure second crop (about 12 g): recrystallise this from butan-1-ol (30 ml per gram). If pure 2,4-dinitrophenylhydrazine is required, recrystallise the total yield from butan-1-ol or from dioxane (10 ml per gram): this melts at 200 °C (decomp.).

The following alternative method of preparation is recommended. Dissolve 50 g of purified chlorodinitrobenzene (1) in 100 ml of triethylene glycol (gentle warming may be necessary; alternatively, 125 ml of warm diethylene glycol may be used) in a 600-ml beaker and cool, with mechanical stirring, in an ice bath to 15–18 °C. Place 15 ml of commercial 60–65 per cent hydrazine solution in a small separatory funnel supported over the beaker. Add the hydrazine solution to the stirred solution in the beaker at such a rate that the temperature is maintained between 15 and 20 °C (20–30 minutes). When the exothermic reaction is over, digest the paste on a boiling water bath with 50 ml of methanol for 15–20 minutes. Cool the reaction mixture, filter with suction and wash with a little methanol. Dry at 100 °C. The yield of 2,4-dinitrophenylhydrazine, m.p. 192–193 °C (decomp.), is 46 g (93%). The product is pure enough for most purposes: the pure compound may be obtained by recrystallisation from butan-1-ol or from dioxane as described above.

Cognate preparation. *2,4-Dinitroaniline.* Place a mixture of 18 g of ammonium acetate and 50 g (0.246 mol) of commercial 1-chloro-2,4-dinitrobenzene (1) in a 250-ml two-necked flask, and fit it with a reflux condenser and inlet tube (at least 2 cm diameter in order to prevent clogging) which terminates just above the surface of the reaction mixture. Half immerse the flask in an oil bath. Pass ammonia gas (from a cylinder) through a bubble counter, which contains a solution of 3 g of potassium hydroxide in 2.5 ml of water, into the mixture. Heat the oil bath to 170 °C, and pass the ammonia gas at the rate of 3–4 bubbles per second for 6 hours. Allow the reaction mixture to cool, break up the solid cautiously with a glass rod, add 100 ml of water, heat to boiling and filter while hot. Dissolve the residue in 500 ml of boiling rectified (or industrial) spirit, and add water (c. 150 ml) until the solution becomes turbid; heat until the turbidity disappears and allow the clear solution to cool overnight. Filter the crystals at the pump and dry in an oven. The yield is 35 g (78%), m.p. 176–177 °C. To obtain a perfectly pure product, recrystallise again from ethanol and water; use 20 ml of ethanol per gram of solid: 31.5 g of pure 2,4-dinitroaniline, m.p. 180 °C, are thus obtained. The p.m.r. spectrum (Me_2CO-d_6; saturated solution, TMS) provides an excellent example of a first-order analysis from which coupling constants may be measured and interrelated; δ 3.08 (s, 2H, NH_2), 7.22 (d, 1H, C_6—H), 8.20 (d of d, 1H, C_3—H) and 8.93 (d, 1H, C_5—H).

Note. (1) It is advisable to recrystallise the commercial chlorodinitrobenzene from ethanol; m.p. 51–52 °C.

Experiment 6.94   N,N-DIPHENYLPICRYLHYDRAZINE AND N,N-DIPHENYLPICRYLHYDRAZYL

$$2Ph_2N\cdot NH_2 + Cl\cdot C_6H_2(NO_2)_3 \xrightarrow{-Ph_2N\cdot NH_3Cl} Ph_2N\cdot NH\cdot C_6H_2(NO_2)_3\text{-}2,4,6$$

Dissolve 7.4 g (0.04 mol) of N,N-diphenylhydrazine (Expt 6.95) in 10 ml of dry dichloromethane and add a solution of 5 g (0.02 mol) of picryl chloride [1-chloro-2,4,6-trinitrobenzene, (1)] in 40 ml of dry dichloromethane. Shake the mixture, which becomes dark and warm, and set it aside for 1 hour. Cool the suspension in an ice bath, filter off the amine salt and wash it with a little dichloromethane. Concentrate the filtrate and washings to 30 ml (rotary evaporator) and treat the hot concentrate with 60 ml of boiling ethanol. Cool, collect the almost pure diphenylpicrylhydrazine which crystallises and wash it with a little cold ethanol. The yield is 6 g (76%), m.p. 172–173 °C (decomp.). If desired, the product may be recrystallised from ethyl acetate.

Conversion to N,N-diphenylpicrylhydrazyl. Dissolve 3.95 g (0.01 mol) of the above hydrazine in 60 ml of dry dichloromethane, add 50 g of lead dioxide and 4 g of anhydrous sodium sulphate and shake the mixture mechanically for 1 hour. Filter, and concentrate the deep-violet filtrate on a rotary evaporator. Dilute the residual solution with two volumes of ether, and allow the product to crystallise. Filter off the large black-violet crystals of diphenylpicrylhydrazyl and wash them with ether; the yield is 3.5 g (89%), m.p. 137–138 °C. The product may be recrystallised from a mixture of dichloromethane and ether.

Note. (1) To prepare picryl chloride, dissolve 5 g (0.022 mol) of picric acid in 50 ml of phosphoryl chloride and add dropwise with shaking 3.8 g (4.0 ml, 0.025 mol) of N,N-diethylaniline. Allow the resulting brown solution to remain at room temperature for 15 minutes and then pour it into 500 ml of iced water. Stir vigorously until the excess of phosphoryl chloride has hydrolysed and the product has solidified; do not allow the temperature to rise above 35–40 °C during hydrolysis, cooling the mixture when necessary in an ice–salt bath. Filter off the almost colourless solid, wash it with cold water and dry it in a desiccator over calcium chloride. The yield of almost pure picryl chloride, m.p. 82 °C, is 5.0 g (92%).

Experiment 6.95   N,N-DIPHENYLHYDRAZINE

$$Ph_2NH \xrightarrow[HCl]{NaNO_2} Ph_2N\cdot NO \xrightarrow[Zn:\ Me\cdot CO_2H]{4[H]} Ph_2N\cdot NH_2$$

CAUTION: *The intermediate in this reaction, as an N-nitrosoamine, is potentially carcinogenic; see Section 2.3.4, p. 49).*

Diphenylnitrosamine. Dissolve 17 g (0.1 mol) of pure diphenylamine in 140 ml of warm ethanol; also 8 g (0.116 mol) of sodium nitrite in 12 ml of water. Cool each solution in ice until the temperature falls to 5 °C. Add 12 ml of concentrated hydrochloric acid slowly and with stirring to the diphenylamine solution, and immediately (otherwise diphenylamine hydrochloride may crystallise out) pour the sodium nitrite solution rapidly into the well-stirred mixture. The temperature soon rises to 20–25 °C and the diphenylnitrosamine crystallises out. Cool the mixture in ice-water for 15–20 minutes, filter with suction on a Buchner funnel, wash with water to remove sodium chloride and press well with a wide glass stopper. Recrystallise from rectified spirit. The yield of pure diphenylnitrosamine (pale yellow crystals), m.p. 68 °C, is 17 g (86%).

N,N-Diphenylhydrazine. Dissolve 15.8 g (0.08 mol) of diphenylnitrosamine in 72 ml of ethanol in a 250-ml three-necked flask equipped with a stirrer, reflux condenser and dropping funnel. Add 36 g (0.55 mol) of zinc powder and stir vigorously. From the dropping funnel slowly add about 22 ml of glacial acetic acid; cool the flask in a bath of cold water from time to time to moderate the reaction, which is complete when the addition of acetic acid no longer causes an increase in temperature, and when a sample of the supernatant liquid no longer gives a deep blue colour when concentrated hydrochloric acid is added. Filter the warm reaction mixture, wash the residue on the filter with warm ethanol and concentrate the filtrate and washings to 40 ml on a rotary evaporator. Cool in an ice bath, and add slowly with shaking 36 ml of concentrated hydrochloric acid. Collect the blue needles of the phenyl-hydrazine hydrochloride by suction filtration on a sintered glass funnel, wash them with a little cold 0.5 M hydrochloric acid and suck dry. Suspend the crystals in a little water, cool in an ice bath and add slowly, with shaking and cooling, 36 ml of 25 per cent aqueous sodium hydroxide solution. Extract the liberated base with three 15 ml portions of toluene, dry the combined extracts over anhydrous potassium carbonate and remove the toluene on a rotary evaporator. Distil the residue under reduced pressure, and collect the diphenylhydrazine as a pale yellow oil of b.p. 136–137 °C/1 mmHg; the yield is 7.2 g (50%) (1). The product solidifies on cooling at 0 °C, and after crystallisation at low temperature from light petroleum (b.p. 40–60 °C) has m.p. 35 °C.

Note. (1) The product should not be allowed to come into contact with the skin.

6.8.3 ARYLUREAS AND RELATED COMPOUNDS

N-Arylureas are obtained when salts of primary aromatic amines react with solutions of alkali metal cyanates. The process involves the rearrangement of an amine cyanate, and is analogous to Wöhlers' classical synthesis of urea from ammonium cyanate.

$$\text{Ar·}\overset{\oplus}{\text{N}}\text{H}_3\}\overset{\ominus}{\text{C}}\text{NO} \longrightarrow \text{Ar·NH·CONH}_2$$

The reaction is most conveniently carried out by warming the amine in aqueous solution with the equivalent quantity of sodium cyanate and an excess of acetic acid (Expt 6.96, *Method 1*).

In an alternative synthesis (Expt 6.96, *Method 2*), which is also convenient for the synthesis of alkylureas, the amine hydrochloride is heated in aqueous solution with urea. This reaction also probably involves the amine cynate since in aqueous solution urea serves as a source of ammonium cyanate.

$$H_2N \cdot CO \cdot NH_2 \rightleftharpoons NH_4 \cdot CNO$$

In the case of the synthesis of phenylurea, some *sym.*-diphenylurea (carbanilide) is also formed and the quantity increases with continued refluxing.

$$Ar \cdot NH \cdot CO \cdot NH_2 + Ar\overset{\oplus}{N}H_3\}\overset{\ominus}{Cl} \longrightarrow Ar \cdot NH \cdot CO \cdot NH \cdot Ar + \overset{\oplus}{N}H_4\overset{\ominus}{Cl}$$

The diarylurea is very sparingly water-soluble and is therefore easily separated from the monoarylurea which is readily soluble. Diarylurea formation is less extensive when ring-substituted anilines are used.

sym.-Diarylthioureas (3) (e.g. *sym.*-diphenylthiourea or thiocarbanilide, Expt 6.97) are prepared by heating a mixture of a primary aromatic amine and carbon disulphide in absolute ethanol. Intermediates in the reaction sequence are the aryldithiocarbamic acid (1) and the aryl isothiocyanate (2), thus:

$$Ar \cdot NH_2 + S{=}C{=}S \longrightarrow Ar \cdot NH \cdot \overset{\overset{\displaystyle S}{\|}}{C} \cdot SH \xrightarrow{-H_2S} Ar \cdot N{=}C{=}S$$

$$\text{(1)} \qquad\qquad\qquad \text{(2)}$$

$$Ar{-}\overset{\frown}{N}{=}C{=}S \xrightarrow[+H^{\oplus}]{-H^{\oplus}} Ar \cdot NH \cdot CS \cdot NH \cdot Ar$$

$$\underset{H_2\ddot{N}Ar}{} \qquad\qquad \text{(3)}$$

Upon heating the diarylthiourea with concentrated hydrochloric acid, it is partly converted into the aryl isothiocyanate (e.g. phenyl isothiocyanate or phenyl mustard oil, Expt 6.98, *Method 1*). A little hydrogen sulphide is evolved in a side reaction forming diphenylcarbodi-imide (4) which undergoes nucleophilic addition of aniline to yield triphenylguanidine (5), isolated from the reaction mixture as the hydrochloride.

$$Ph \cdot N{=}C{=}N \cdot Ph \xleftarrow{-H_2S} Ph \cdot NH \cdot CS \cdot NH \cdot Ph \xrightarrow{-Ph \cdot NH_2} Ph \cdot N{=}C{=}S$$

$$\text{(4)}$$

$$Ph \cdot N{=}\overset{\frown}{C}{=}\overset{}{N} \cdot Ph \longrightarrow Ph \cdot N{=}C(NH \cdot Ph)_2 \xrightarrow{HCl} [Ph \cdot NH{=}C(NH \cdot Ph)_2]^{\oplus}Cl^{\ominus}$$

$$\underset{Ph \cdot \ddot{N}H_2}{} \qquad\qquad \text{(5)}$$

Phenyl isothiocyanate may be prepared in quantity (Expt 6.98, *Method 2*) by allowing aniline to react with carbon disulphide to form phenyldithiocarbamic acid (cf. 1), which is isolated as the ammonium salt. Treatment of the latter with lead nitrate removes the elements of hydrogen sulphide to produce phenyl isothiocyanate. As indicated in the preparation of *p*-bromophenyl isothiocyanate which is given as a further example, a slightly modified procedure which requires the use of rectified spirit as a reaction solvent is necessary in order to obtain good yields of isothiocyanates from substituted anilines.

Experiment 6.96   PHENYLUREA

Method 1

$$Ph \cdot NH_2 + NaCNO + Me \cdot CO_2H \longrightarrow Ph \cdot NH \cdot CO \cdot NH_2 + Me \cdot CO_2Na$$

Method 2

$$\overset{\oplus}{Ph \cdot NH_3} \overset{\ominus}{Cl} + NH_2 \cdot CO \cdot NH_2 \longrightarrow Ph \cdot NH \cdot CO \cdot NH_2 + NH_4Cl$$

Method 1. Dissolve 9.3 g (9.1 ml, 0.1 mol) of aniline in 10 ml of glacial acetic acid diluted to 100 ml contained in a 250-ml beaker or conical flask, and add with stirring or shaking a solution of 6.5 g (0.1 mol) of sodium cyanate in 50 ml of warm water. Allow to stand for 30 minutes, then cool in ice, and allow to stand for a further 30 minutes. Filter at the pump, wash with water and dry at 100 °C. The resulting phenylurea is generally colourless and has a m.p. of 148 °C (i.e. is pure): the yield is 11 g (81%). If the colour or the m.p. of the product is not quite satisfactory, recrystallise it from boiling water (10 ml per gram) with the aid of decolourising charcoal.

Method 2. Dissolve 65 g (0.5 mol) of aniline hydrochloride and 120 g (2 mol) of urea in 200 ml of water contained in a 1-litre round-bottomed flask; filter the solution, if necessary. Add 4 ml of concentrated hydrochloric acid and 4 ml of glacial acetic acid. Fit a reflux condenser to the flask, introduce a few fragments of broken porcelain and boil the mixture for 30 minutes. Fine white crystals (largely *sym.*-diphenylurea) appear after about 15 minutes and gradually increase in amount as the refluxing is continued. Cool the flask in ice and filter with suction. Separate the mixture of phenylurea and diphenylurea (*c.* 42 g) by boiling with 500 ml of water and filter at the pump through a preheated Buchner funnel into a warm flask; cool the filtrate, collect the phenylurea, drain well and dry in the steam oven. The phenylurea melts at 146–147 °C and weighs 30 g (44%); recrystallisation from hot water raises the m.p. to 148 °C. The crude diphenylurea (residue from first recrystallisation after drying at 100 °C) has m.p. 241 °C and weighs 10 g (19%); recrystallisation from glacial acetic acid or ethyl acetate with the addition of a little decolourising carbon gives a colourless product, m.p. 242 °C.

Cognate preparations. p-*Tolylurea* (*Method 1*). Dissolve 10.7 g (0.1 mol) of *p*-toluidine in a warm mixture of 10 ml of glacial acetic acid and 50 ml of water, and then dilute with 150 ml of hot water. Introduce, with stirring or shaking, a solution of 6.5 g (0.1 mol) of sodium cyanate in 50 ml of hot water. The *p*-tolylurea precipitates almost immediately. Allow to stand several hours, filter at the pump, wash with water and dry. The yield of *p*-tolylurea, m.p. 180–180.5 °C, is 14 g (85%). Recrystallise from aqueous ethanol; the resulting *p*-tolylurea melts sharply at 181 °C.

p-*Bromophenylurea*. Proceed as for *p-tolylurea* (*Method 1*), but use 17.2 g (0.1 mol) of *p*-bromoaniline dissolved in a mixture of 50 ml of glacial acetic acid and 100 ml of water at 35 °C; add gradually a solution of 6.5 g (0.1 mol) of sodium cyanate in 50 ml of water at 35 °C. The yield of crude *p*-bromophenylurea is 19 g (88%); m.p. 227 °C. Recrystallise from 90 per cent aqueous ethanol; m.p. 228 °C. The m.p. depends somewhat upon the rate of heating.

p-*Methoxyphenylurea*. Proceed as for *phenylurea, Method 2*, but use 79 g (0.5 mol) of *p*-anisidine hydrochloride in place of 65 g of aniline hydrochloride; reflux the mixture for 1 hour. Cool the reaction mixture slowly to 0 °C, filter and recrystallise from boiling water. The yield of *p*-methoxyphenylurea, m.p. 168 °C, is 60 g (72%).

p-Ethoxyphenylurea. Proceed as for *phenylurea, Method 2*, but use 87 g (0.5 mol) of *p*-phenetidine hydrochloride; reflux the mixture for 45–90 minutes. The product commences to separate after 20–30 minutes and increases rapidly until the entire contents of the flask suddenly set to a solid mass: withdraw the source of heat *immediately* at this point. Cool to room temperature, add 150 ml of water, stir, filter with suction and wash with cold water. Suspend the solid in 2 litres of boiling water, add 1 g of decolourising carbon, boil for 5 minutes and filter through a hot water funnel; cool the colourless filtrate slowly to 0 °C, collect the solid which separates and dry at 100 °C. The yield of *p*-ethoxyphenylurea, m.p. 174 °C, is 60 g (67%).

Experiment 6.97 THIOCARBANILIDE (sym.-*Diphenylthiourea*)

$$2Ph \cdot NH_2 + CS_2 \xrightarrow{EtOH} S = C(NH \cdot Ph)_2 + H_2S$$

In a 1-litre round-bottomed flask provided with an efficient double surface condenser, place 40 g (39 ml, 0.43 mol) of aniline, 50 g (40 ml, 0.66 mol) of carbon disulphide (**CAUTION**: flammable) (1) and 50 g (63.5 ml) of absolute ethanol (2). Set up the apparatus in the fume cupboard or attach an absorption device to the top of the condenser (see Fig. 2.61) to absorb the hydrogen sulphide which is evolved. Heat upon an electrically heated water bath or upon a steam bath for 8 hours or until the contents of the flask solidify. When the reaction is complete, arrange the condenser for downward distillation and remove the excess of carbon disulphide and alcohol (**CAUTION**: *flammable; there must be no flame near the receiver*). Shake the residue in the flask with excess of dilute hydrochloric acid (1:10) to remove any aniline present, filter at the pump, wash with water and drain well. Dry in the steam oven. The yield of crude product, which is quite satisfactory for the preparation of phenyl isothiocyanate (Expt 6.98), is 40–45 g (81–91%). Recrystallise the crude thiocarbanilide by dissolving it, under reflux, in boiling rectified spirit (filter through a hot water funnel if the solution is not clear), and add hot water until the solution just becomes cloudy and allow to cool. Pure *sym*.-diphenylthiourea separates in colourless needles, m.p. 154 °C.

Notes. (1) No flames may be present in the vicinity: read Section 2.3.2, p. 39.
(2) The addition of powdered potassium hydroxide (about 20 per cent of the weight of the carbon disulphide) reduces the refluxing period necessary to complete the reaction.

Experiment 6.98 PHENYL ISOTHIOCYANATE

Method 1

$$S = C(NH \cdot Ph)_2 \xrightarrow{HCl} Ph \cdot N = C = S + Ph \cdot NH_2$$

Method 2

$$Ph \cdot NH_2 + CS_2 \longrightarrow Ph \cdot NH \cdot CS \cdot SNH_4 \xrightarrow{Pb(NO_3)_2}$$
$$Ph \cdot N = C = S + NH_4NO_3 + HNO_3 + PbS$$

Method 1. Place 25 g (0.11 mol) of crude thiocarbanilide (Expt 6.97) and 100 ml of concentrated hydrochloric acid in a 250-ml flask; fit a reflux condenser and reflux gently in the fume cupboard for 30 minutes. Distil the mixture until the oily phenyl isothiocyanate has all passed over; the volume remaining in the flask will be 25–30 ml. Crystals of triphenylguanidine hydrochloride may appear in the distilling flask during the latter part of the distillation. Dilute the distillate with an equal volume of water, and extract the isocyanate with ether; wash the extract with a little sodium carbonate solution, and dry over anhydrous calcium chloride or magnesium sulphate. Remove the ether on a rotary evaporator and distil the residual oil, collecting the phenyl isothiocyanate at 217–220 °C (1). The yield is 10 g (67.5%).

To isolate the *triphenylguanidine* formed as a by-product dilute the residue in the flask with 50 ml of water, add 2–3 g of decolourising carbon, warm and filter. Cool the solution in ice, and filter off the hydrochloride at the pump. Dissolve it in the minimum volume of hot water, render the solution alkaline with sodium hydroxide and allow to cool. Filter off the free base (triphenyl-guanidine), and recrystallise it from ethanol; it separates in colourless crystals, m.p. 144 °C. The yield is 3 g.

Note. (1) It may also be distilled under diminished pressure, b.p. 95 °C/12 mmHg.

Method 2. Equip a 500-ml three-necked flask with a powerful mechanical stirrer and a separatory funnel; leave the third neck open or loosely stoppered. Introduce, while the flask is cooled in a freezing mixture of ice and salt, 90 ml of concentrated ammonia solution (d 0.88) and 54 g (43 ml, 0.71 mol) of pure carbon disulphide (1). Stir the mixture and run in 56 g (55 ml, 0.60 mol) of aniline from the separatory funnel during about 20 minutes; stir for a further 30 minutes, and allow to stand for another 30 minutes. A heavy precipitate of ammonium phenyldithiocarbamate separates. Transfer the salt to a 5-litre round-bottomed flask by four extractions with 200 ml portions of water. Add to the resulting solution, with constant stirring, a solution of 200 g (0.605 mol) of lead nitrate in 400 ml of water; lead sulphide precipitates. Steam distil the mixture into a receiver containing 10 ml of c. 0.5 M sulphuric acid as long as organic material passes over (2–3 litres of distillate). Separate the oil, dry it over anhydrous calcium chloride or magnesium sulphate and distil under diminished pressure. Collect the phenyl isothiocyanate at 120–121 °C/35 mmHg or at 95 °C (12 mmHg). The yield is 62 g (76%).

Note. (1) **CAUTION**, see Expt 6.97.

Cognate preparation. **p-*Bromophenyl isothiocyanate*** (*Method 2*). Add 41 ml of concentrated ammonia solution (d 0.88) slowly with stirring to a solution of 45 g (0.26 mol) of p-bromoaniline (Expt 6.67), 30 g (24 ml, 0.396 mol) of carbon disulphide and 40 ml of rectified spirit at 10–15 °C. Considerable heat is evolved; cool the flask in a freezing mixture from time to time so that the temperature does not rise above 30 °C. The original milky suspension becomes clear and the intermediate dithiocarbamate soon crystallises out. Allow to stand overnight, filter the crystals, wash with a little ether, dissolve in 1500 ml of water and stir mechanically while a solution of 87 g (0.262 mol) of lead nitrate in 175 ml of water is slowly added. Continue the stirring for 20 minutes, and isolate the p-bromophenyl isothiocyanate by steam distillation into a receiver containing 5 ml of c. 0.5 M sulphuric acid; if the substance solidifies in the condenser, stop the cooling water until the solid has melted and run into the receiver. Filter the cold solid product, wash with a little water and dry in the air upon filter paper. The yield is 15 g (50%), m.p. 61 °C.

6.9 PHENOLS AND PHENYL ETHERS

1. Methods for the introduction of a hydroxyl group into an aromatic ring by: (a) replacement of a sulphonic acid group (Expt 6.99); (b) replacement of a halogen (Expts 6.100 and 6.101); and (c) replacement of a diazo group (see Expt 6.69).
2. Substitution reactions of phenols: (a) nitrosation and nitration (Expts 6.102 to 6.104); (b) halogenation (Expts 6.105 and 6.106); and (c) acylation and alkylation (Expts 6.107 and 6.108).
3. Formation of phenyl ethers (Expts 6.109 to 6.111).

Methods for the protection of the phenolic hydroxyl group are considered in Section 6.9.4, p. 988.

SUMMARY OF RETROSYNTHETIC STRATEGIES

C—O Disconnections (methods 1(a)–(c) and 3), e.g.

$$\text{(TM)} \quad \Longrightarrow_{(1a-c)} \quad \overset{\oplus}{\quad} \overset{\ominus}{OH} \equiv Ph\cdot SO_3H, \; PhX, \; or \; Ph\cdot\overset{\oplus}{N}_2 + \overset{\ominus}{OH}/H_2O$$

$$\text{(TM)} \quad \Longrightarrow_{(3)} \quad \overset{\oplus}{\quad} R \equiv PhOH + RX$$

C—N, C—X, and C—C Disconnections (methods 2(a)–(c)), e.g.

$$\text{(TM)} \quad \overset{OH}{\quad}\!\!-NO_2(NO) \quad \Longrightarrow_{(2a)} \quad \overset{OH}{\quad} + \overset{\oplus}{N}O_2(HNO_2/HNO_3)$$

$$\text{(TM)} \quad \overset{OH}{\quad}\!\!-X \quad \Longrightarrow_{(2b)} \quad \overset{OH}{\quad} + \overset{\oplus}{X}(X_2)$$

SPECTROSCOPIC FEATURES

Phenols, wherein the hydroxyl group is directly attached to the aromatic nucleus, exhibit in their *i.r.* absorption spectra strong absorption (3400–3600 cm^{-1}) arising from the O—H stretching vibrations. A full discussion of this and other characteristic absorptions due to the C—O stretching and the O—H bending vibrations, and the assignment of the substitution pattern from the absorption in the longer wavelength region is given on p. 286. The important feature of hydrogen bonding (intra- and inter-molecular) must be noted, *m*-cresol being an illustrative example (Fig. 3.22). The absorption frequency of the low field signal of the oxygen-bound proton in the *p.m.r.* spectrum of a phenol depends on the extent of hydrogen bonding, but it may be distinguished from other low-field signals since the absorption disappears after appropriate treatment of the sample with deuterium oxide (p. 348). Aromatic substitution patterns are confirmed by inspection of the appropriate region of the p.m.r. spectrum. Further descriptive accounts of structural assignments are to be found in the preparative examples below which also include an analysis of the important fragmentation patterns observed in the *m.s.* (see also p. 375). The presence of a hydroxyl group in an aromatic nucleus results in a bathochromic shift of the absorption of the aromatic system in the *u.v.–visible*; this shift becomes enhanced when a nitro group is also present p. 392).

Concerning *phenyl ethers*, the *i.r.* spectrum of anisole (Fig. 3.25), its *p.m.r.* spectrum (Fig. 3.46) and ^{13}C-*n.m.r.* spectrum (Fig. 3.54) are discussed on pp. 325, and 337 respectively. Further illustrative analyses are included in some of the preparative examples below which also include a discussion of relevant *m.s.*

6.9.1 METHODS FOR THE INTRODUCTION OF A HYDROXYL GROUP INTO AN AROMATIC RING

REPLACEMENT OF A SULPHONIC ACID GROUP

A fairly general procedure, which has also been used on the industrial scale, involves heating the alkali metal sulphonate with either sodium or potassium hydroxide in the presence of a small amount of water to aid the fusion process. The reaction mechanism may be formulated as a bimolecular nucleophilic addition–elimination sequence.

mesomeric anion

The reaction is illustrated by the synthesis of 2-naphthol (Expt 6.99).

Occasionally in the synthesis of phenols by this route oxidation products are formed. A particular example is provided by the alkali fusion of sodium anthraquinone-2-sulphonate during which a second hydroxyl group is introduced into the 1-position, forming the dyestuff alizarin (1) (cognate preparation in Expt 6.99). In the procedure described the oxidation step is promoted by the deliberate introduction of potassium chlorate as an oxidant.

REPLACEMENT OF A HALOGEN

Direct nucleophilic displacement of the halogen in an aryl halide is difficult and hydrolysis to phenols requires high temperatures and pressures; the method is therefore only suitable on the large scale. The presence of a nitro group in the *ortho* or *para* position, however, makes the halogen more labile since electron withdrawal by the nitro group in these positions stabilises the intermediate anion by electron delocalisation. *p*-Chloronitrobenzene, for example, is hydrolysed to *p*-nitrophenol when heated with 15 per cent sodium hydroxide solution at about 150 °C.

mesomeric anion

When two activating nitro groups are present hydrolysis takes place readily with dilute aqueous alkali solution (e.g. 2,4-dinitrophenol, Expt 6.100).

Aryl halides of many different types, including simple unsubstituted halides, may be conveniently converted into phenols by an indirect route involving the preparation of an arylboronic acid and its subsequent oxidation with hydrogen peroxide. The arylboronic acid (3) is normally prepared by reaction of the corresponding arylmagnesium halide with a borate ester (typically tributyl borate) at between -60 and $-80\,°C$, to yield the dialkyl boronate ester (2) which is then hydrolysed to the arylboronic acid (3). The latter may be isolated, purified and then oxidised with hydrogen peroxide as described in the preparation of m-cresol (Expt 6.101). Alternatively the crude reaction mixture from the preparation of (3) may be treated directly with hydrogen peroxide.[36]

$$\text{Ar·MgX} \xrightarrow[-70\,°C]{\text{B(OBu)}_3} \underset{(2)}{\text{Ar·B(OBu)}_2} \xrightarrow{\text{H}_3\text{O}^\oplus} \underset{(3)}{\text{Ar·B(OH)}_2} \xrightarrow{\text{H}_2\text{O}_2} \text{ArOH}$$

A convenient purification procedure for an arylboronic acid is to convert it into the trimeric anhydride (4) by removal of water as a benzene azeotrope (see Expt 6.101, Note (3)).

$$3\text{Ar·B(OH)}_2 \longrightarrow \quad (4) \quad + 3\text{H}_2\text{O}$$

REPLACEMENT OF A DIAZO GROUP

The preparation of phenols by the hydrolysis of diazonium salts with hot aqueous acid, and by a recent milder procedure suitable for diazonium salts having additional acid-sensitive groups, is discussed in Section 6.7.1, p. 922, and illustrated in Expt 6.69. Although these methods enable an aromatic hydrocarbon system to be converted in good yield into a phenol via the corresponding nitro and amino derivatives, the shorter route involving the alkaline fusion of the sulphonic acid discussed above may often be preferred.

Experiment 6.99 2-NAPHTHOL

$$2\text{-C}_{10}\text{H}_7\text{·SO}_3\text{Na} \xrightarrow[\text{(ii) H}^\oplus]{\text{(i) KOH}} 2\text{-C}_{10}\text{H}_7\text{OH}$$

CAUTION: *This preparation should be carried out in a fume cupboard with the window protecting the face.*

Support a 250-ml nickel, copper (better silver-plated copper) or iron crucible or beaker in a large circular hole in a sheet of heat resistant board resting on a tripod. Prepare a case of nickel or copper to surround a 360 °C thermometer for about two-thirds of its length; this may be done either by cutting a suitable length of nickel or copper tubing already closed at one end, or by hammering down the end of the open tube and folding over the flat part in a vice. Fit a large cork around the top of the tube; this will serve for handling the tube containing the thermometer when it is subsequently used for stirring the molten alkali. Since some splattering of the latter cannot generally be avoided, goggles, gloves and a well-fitting laboratory coat must be worn.

Place 120 g (2.15 mol) of potassium hydroxide sticks or pellets (1) together with 5 ml of water in the crucible, and heat with a Bunsen burner until it melts. When the temperature reaches about 250 °C, remove the flame, and quickly add with stirring 50 g (0.22 mol) of finely powdered sodium naphthalene-2-sulphonate (Expt 6.38). Replace the flame, stir the stiff pasty mass and continue the heating so that the temperature rises to 300 °C in 5–10 minutes. Stir the mixture continuously; there is some frothing at first and at about 300 °C the mass suddenly becomes a clear, mobile, brown oil of the potassium salt of 2-naphthol floating on a pasty mass of alkali. Raise the temperature during 5 minutes to 310 °C, remove the flame, push down the material from the side of the crucible and reheat to 310 °C for about 2 minutes, and then allow the melt to cool. Do not permit the melt to solidify completely. When it becomes pasty, ladle it out in small portions (with a nickel spatula, 'spoon' end) into a 1-litre beaker half-filled with crushed ice. Extract the residual material in the crucible with water and add it to the contents of the beaker. Precipitate the 2-naphthol by adding concentrated hydrochloric acid slowly and with stirring (*fume cupboard*: SO_2); if the 2-naphthol separates in a finely divided form, warm until the particles coagulate. Cool in ice, filter at the pump and transfer the precipitate to a beaker containing cold water. Add just sufficient 5 per cent sodium hydroxide solution to dissolve the solid and also 1 g of sodium dithionite ($Na_2S_2O_4$) to prevent oxidation, and filter from traces of insoluble matter. Precipitate the 2-naphthol with acetic acid, warm to produce a more readily filterable form of the precipitate, cool in ice and filter the product. Dry in the air upon filter paper. The yield is 25 g (80%), m.p. 122 °C. If the m.p. is unsatisfactory, recrystallise from water, dilute ethanol or carbon tetrachloride.

Note. (1) Sodium hydroxide may replace potassium hydroxide in this preparation; 150 g together with 15 ml of water are required. The sulphonate is stirred in when the temperature reaches 280 °C and the reaction is complete at 310–320 °C.

Cognate preparation. *Alizarin.* Dissolve successively in 75 ml of water 6 g (0.049 mol) of potassium chlorate, 20 g (0.065 mol) of sodium anthraquinone-2-sulphonate (Expt 6.39) and 75 g of sodium hydroxide. Transfer the mixture to a 500-ml autoclave (compare Section 2.17.2, p. 97) and heat for 20 hours at 170 °C. After cooling, scrape out the violet-coloured mass and extract it three or four times with 100 ml portions of boiling water. Acidify the filtered extract with hydrochloric acid. When cold, filter the orange precipitate of alizarin at the pump, wash it thoroughly with cold water and dry at 100 °C. The yield of alizarin is 14 g (90%). It may be purified by recrystallisation from glacial acetic acid or by sublimation. The pure compound has m.p. 289 °C.

Experiment 6.100 2,4-DINITROPHENOL

$$2,4\text{-}(NO_2)_2C_6H_3Cl \longrightarrow 2,4\text{-}(NO_2)_2C_6H_3OH$$

In a 1-litre round-bottomed flask equipped with a reflux condenser place a solution of 62.5 g (0.6 mol) of anhydrous sodium carbonate in 500 ml of water and add 50 g (0.25 mol) of commercial 1-chloro-2,4-dinitrobenzene. Reflux the mixture for 24 hours or until the oil passes into solution. Acidify the yellow solution with hydrochloric acid and, when cold, filter the crystalline dinitrophenol which has separated. Dry the product upon filter paper in the air. The yield is 42 g (91%). If the m.p. differs appreciably from 114 °C, recrystallise from ethanol or from water.

The p.m.r. spectrum (DMSO-d_6, TMS) provides an excellent example for the assignment of absorptions to specific hydrogens from the correlation of coupling constants; the signals appear at δ 7.29 (d, 1H, C_6—H), 8.30 (d of d, 1H, C_5—H), 8.63 (d, 1H, C_3—H), and 12.62 (s, 1H, OH).

Experiment 6.101 *m*-CRESOL

$$m\text{-}Me\cdot C_6H_4\cdot MgX \xrightarrow[-75\,°C]{B(OBu)_3} m\text{-}Me\cdot C_6H_4\cdot B(OBu)_2 \xrightarrow{H_2O}$$
$$m\text{-}Me\cdot C_6H_4\cdot B(OH)_2 \xrightarrow{H_2O_2} m\text{-}Me\cdot C_6H_4OH$$

m-Tolyboronic acid. Equip a thoroughly dried 1-litre flange flask with a low temperature reading thermometer, and an efficient sealed stirrer unit, and attach a dropping funnel and a reflux condenser both protected with calcium chloride guard-tubes. Mount the flask in a Cardice–acetone cooling bath and add to the flask 40.5 g (0.17) of pure tributyl borate (1) dissolved in 150 ml of ether (previously dried over sodium wire). Then, with fairly rapid stirring, add slowly from the dropping funnel a solution of *m*-tolylmagnesium bromide [0.175 mol; (2)] in 450 ml of dry ether. It is essential that the rate of addition throughout the reaction should be such that the temperature of the reaction mixture is kept between −70 and −75 °C. As the ethereal solution of the Grignard reagent is added, a white precipitate is formed immediately and slowly dissolves; when all of the reagent has been added (about 3 hours), continue to stir the mixture at −75 °C for a further 2 hours. With continued stirring, allow the reddish-orange solution to warm slowly to 0 °C, remove the dropping funnel, condenser and stirrer, stopper the flask and place it in a refrigerator to attain a temperature of about 5 °C over a period of 12 hours. Slowly add the cold reaction mixture with efficient stirring to 120 ml of chilled (0–5 °C) 10 per cent sulphuric acid. Transfer the resultant mixture to a separating funnel, separate the layers and thoroughly extract the aqueous phase with four 100 ml portions of ether. Concentrate the combined ether solutions on a rotary evaporator using a water-bath temperature of 40–45 °C. To the residual *m*-tolylboronic acid in butan-1-ol add 50 ml of water, followed by 10 per cent aqueous potassium hydroxide solution until the solution is alkaline to litmus paper. Remove the butanol as an azeotrope with water by concentrating the mixture on a rotary evaporator (water-bath temperature 40–45 °C). Add further portions of water (25 ml) and repeat the evaporation until no smell of butanol remains. During these evaporation stages, solid material often separates but this should not be filtered off. Render the residual solution acid to litmus by adding dilute aqueous sulphuric acid and, if necessary, make up the volume to 60 ml by the addition of distilled water. Without separating the precipitated solid, heat the mixture to boiling when the precipitate dissolves and a heavy dark brown oil appears. Decant the hot supernatant solution through a fluted filter paper and extract the oil with four 40 ml portions of boiling water. Combine the hot filtered extracts and cool. Filter off the crystalline *m*-tolylboronic acid and dry at room temperature by spreading upon filter papers. The yield is 5.9 g (25%) (3).

m-Cresol. In a 250-ml, two-necked round-bottomed flask fitted with an efficient stirrer and a dropping funnel, place 5.4 g (0.040 mol) of *m*-tolylboronic acid and 100 ml of ether. While stirring the slurry add 30 ml of 10 per cent hydrogen peroxide solution from the dropping funnel over a period of about 5 minutes. The two-phase system in the flask becomes warm as the reaction proceeds. When all the hydrogen peroxide has been added, continue stirring until the contents of the flask have cooled to room temperature (20–30 minutes) and then transfer to a separating funnel. Run off and discard the aqueous layer and wash the ether layer thoroughly with three 30 ml portions of 10 per cent iron(II) ammonium sulphate solution to remove remaining traces of hydrogen peroxide. Extract the product from the ether layer by shaking with three 30 ml portions of 10 per cent aqueous sodium hydroxide solution. Acidify the combined alkaline extracts with concentrated hydrochloric acid and extract the product with three 70 ml portions of ether. Dry the combined ether extracts with calcium sulphate, filter and evaporate on a rotary evaporator. Distil the crude product and collect the *m*-cresol at 198–202 °C. The yield is 2.5 g (58%).

The i.r. spectrum of *m*-cresol is given on p. 288. The substitution pattern is confirmed from the absorptions in the 600–800 cm^{-1} [the *ortho* isomer shows one band at *c*. 750 cm^{-1} (5 adj. H's); the *para* isomer shows absorption at *c*. 815 cm^{-1} (2 adj. H's)]. In the p.m.r. spectra (CDCl$_3$, TMS) of the three isomers the substitution pattern is only clearly defined in the case of the *para* isomer which shows δ 2.20 (s, 3H, Me), 6.45 (broad s, 1H, OH), 6.69 (d, 2H, *ortho*-H's to OH), and 6.92 (d, 2H, *ortho*-H's to Me). The *meta* and *ortho* isomers both show complex multiplets in the δ 6.3–7.20 region. The m.s. of all three isomers are very similar having significant fragment ions at *m/z* 108 (M, base peak), 107 (M − H), 79 (107 − CO), 77 (79 − H$_2$), and 51 (77 − C$_2$H$_2$).

Notes. (1) The tributyl borate is available commercially (Aldrich) and packaged in Sure/Seal^{TM} bottles under nitrogen.
(2) The ethereal solution of *m*-tolylmagnesium bromide may be prepared by using 20 g (0.175 mol) of *m*-bromotoluene, 4.2 g (0.175 mol) of magnesium turnings and 450 ml of dry ether (cf. Expt 5.39). When nearly all the magnesium has reacted the solution should be quickly decanted into a dry dropping funnel and addition to the cooled tributyl borate solution begun immediately.
(3) It is not possible to obtain an analytically pure sample of arylboronic acids since, on drying, partial conversion to the trimeric anhydride occurs. A quantitative conversion to the anhydride is achieved in the case of *m*-tolylboronic acid by heating with forty times its weight of benzene in a flask fitted with a Dean and Stark water separator (Fig. 2.31(*a*)). When no further water droplets separate the benzene solution

is concentrated to one-quarter volume and cooled. *m*-Tolylboronic anhydride crystallises out and has m.p. 161–162 °C. The anhydride may be reconverted to the acid by dissolving it in the minimum quantity of hot water and allowing the solution to cool, whereupon the acid crystallises.

6.9.2 SUBSTITUTION REACTIONS OF PHENOLS

The phenolic group is activating and *ortho–para* directing. The electrophilic substitution reactions in the nucleus in: (*a*) nitrosation and nitration; (*b*) halogenation and; (*c*) acylation and alkylation, are therefore particularly facile, and various experimental procedures need to be adopted to control the extent of substitution (cf. substitution reactions of aromatic amines and their acylated derivatives, Sections 6.6.1 and 6.6.2, pp. 906 and 916 respectively).

NITROSATION AND NITRATION

Phenol may be converted into a mixture of *o*- and *p*-nitrophenols (Expt 6.102) by reaction with dilute nitric acid; the yield of *p*-nitrophenol is increased if a mixture of sodium nitrate and dilute sulphuric acid is employed. Upon steam distillation of the mixture of nitrophenols, the *ortho* isomer passes over in a substantially pure form; the *para* isomer remains in the distillation flask, and can be readily isolated by extraction with hot 2 per cent hydrochloric acid. The mechanism of the substitution probably involves an electrophilic attack (cf. Section 6.2.1, p. 851) by a nitrosonium ion at a position either *ortho* or *para* to the activating hydroxyl group, to yield a mixture of *o*- and *p*-nitrosophenols, which are then oxidised by the nitric acid to the corresponding nitrophenols. The reaction depends upon the presence in the nitric acid of traces of nitrous acid which serve as the source of the nitrosonium ion.

If the phenol is allowed to react with nitrous acid (generated in an acidified solution of sodium nitrite), the nitrosophenol may be obtained in good yield. An example is provided by the nitrosation of 2-naphthol which yields 1-nitroso-2-naphthol (Expt 6.103).

By suitably introducing sulphonic acid 'blocking groups', which may subsequently be removed by heating under aqueous acidic conditions, control over the orientation of other substituents introduced into the aromatic nucleus of a phenol may be achieved. The procedure is illustrated by the synthesis of 2-nitroresorcinol (Expt 6.104). In this reaction disulphonation of the dihydric phenol can readily be achieved, the sulphonic acid groups taking up the 4- and 6-positions. When the disulphonic acid is then nitrated the nitro group enters the remaining active site, i.e. the 2-position, removal of the sulphonic acid residues then yields the required 2-nitroresorcinol. In this sequence it is essential that the experimental conditions for the nitration step are as mild as possible (i.e. below 20 °C), since more vigorous conditions may result in electrophilic replacement of a sulphonic acid group by a nitro group. This is apparent in the

975

classical synthesis of 2,4,6-trinitrophenol (picric acid). Thus phenol is first sulphonated to yield a mixture of o- and p-phenolsulphonic acids; nitration of this product with a hot mixture of concentrated nitric acid and concentrated sulphuric acid results in the introduction of a nitro group into all the *ortho* and *para* activated positions with displacement of the sulphonic acid group. The direct nitration of phenol to the trinitro derivative in good yield is not possible since much of the starting material is oxidatively destroyed.

HALOGENATION

When treated with bromine water an aqueous solution of phenol gives an immediate precipitate of 2,4,6-tribromophenol (Section 9.6.6, p. 1251), owing to the powerfully activating influence of the negatively charged oxygen in the phenoxide ion.

The monobromination of phenol can, however, be achieved by using solutions of bromine in non-polar solvents such as carbon disulphide and carbon tetrachloride at low temperature (0–5 °C). The product is almost exclusively the *para* isomer (Expt 6.105).

o-Bromophenol is conveniently prepared by first sulphonating phenol with excess of concentrated sulphuric acid to yield phenol-2,4-disulphonic acid, neutralising with sodium hydroxide, heating the solution of the sodium salt with 1 mol of bromine and then removing the sulphonic acid groups by treatment with aqueous sulphuric acid at 200 °C. The sequence is described and formulated in Expt 6.106.

ACYLATION AND ALKYLATION

The various methods for introducing a formyl group (—CO·H) into a phenolic nucleus are discussed in Section 6.10.1, p. 990. The formation of phenolic ketones (e.g. HO·C$_6$H$_4$·CO·R) by the standard Friedel–Crafts acylation procedure (i.e. the reaction of a phenol with an acid chloride in the presence of aluminium chloride) does not always give acceptable yields except in the case of polyhydroxyphenols (p. 1006). The preferred method is to convert the phenol into the phenyl ester and to subject this to rearrangement (the *Fries reaction*) in the presence of aluminium chloride.

Experimental procedures are given in Expt 6.107 for *o*- and *p*-hydroxy-propiophenones (R = Et). The *ortho–para* ratio in the product is influenced by the nature of the alkyl residue, the temperature, the solvent and the amount of aluminium chloride used: generally low temperatures favour the formation of *p*-hydroxyketones. It is usually possible to separate the two hydroxyketones by fractional distillation under reduced pressure through an efficient fractionating column or by steam distillation; the *ortho* isomers, being chelated, are more steam volatile. It may be mentioned that Clemmensen reduction (cf. Sections 5.1.3, p. 476 and 6.1.1, p. 826) of the hydroxyketones affords an excellent route to alkyl phenols.

$$HO \cdot C_6H_4 \cdot CO \cdot R \xrightarrow[\text{HCl}]{\text{Zn/Hg}} HO \cdot C_6H_4 \cdot CH_2 \cdot R$$

2,5-Dihydroxyacetophenone (6) (cognate preparation in Expt 6.107), which cannot be prepared by a Friedel–Crafts acetylation of hydroquinone, is obtained in good yield when hydroquinone diacetate (5) is heated in the presence of 3.3 mol of aluminium chloride.

(5)   (6)

The details of the mechanism of the Fries rearrangement are uncertain but the reaction probably involves the formation and migration of the acylium ion (7).

[*Continued overleaf*]

(7)

When phenol is treated with allyl bromide in the presence of potassium carbonate and acetone, the product is almost entirely allyl phenyl ether (Expt 6.110). This undergoes ready thermal rearrangement to give 2-allylphenol (Expt 6.108), which is an example of the *Claisen rearrangement*. The mechanism of this intramolecular rearrangement involves a cyclic transition state (8) as formulated below, and is designated as a [3,3] sigmatropic shift; the numbers give the location of the atoms which are joined by the new σ-bond relative to the bond undergoing fission.

A protropic shift may be induced to effect double bond migration in the side chain by heating the 2-allylphenol with methanolic potassium hydroxide when 2-(prop-1-enyl) phenol is formed (Expt 6.108).

(8)

Experiment 6.102   o- AND p-NITROPHENOLS

$$\text{PhOH} \xrightarrow{\text{dil. HNO}_3} \textit{o-} \text{ and } \textit{p-}\text{O}_2\text{N·C}_6\text{H}_4\text{OH}$$

Cautiously add 250 g (136 ml, c. 2.5 mol) of concentrated sulphuric acid in a thin stream and with stirring to 400 ml of water contained in a 1-litre three-necked flask, and then dissolve 150 g (1.75 mol) of sodium nitrate in the diluted acid. Cool in a bath of ice or iced water. Melt 94 g (1 mol) of phenol (**CAUTION**)[37] with 20 ml of water, and add this dropwise from a separatory funnel to the vigorously stirred mixture in the flask; maintain the temperature at about 20 °C. Continue the stirring for a further 2 hours after all the phenol has been added. Pour off the mother-liquor from the resinous mixture of nitro compounds. Melt the residue with 500 ml of water, shake and allow the contents of the flask to settle. Pour off the wash liquor and repeat the washing at least two or three times to ensure the complete removal of any residual acid. Steam distil the mixture (Fig. 2.102) until no more *o*-nitrophenol passes over; if the latter tends to solidify in the condenser, turn off the cooling water temporarily. Collect the distillate in cold water, filter at the pump and drain thoroughly. Dry upon filter paper in the air. The yield of o-nitrophenol, m.p. 46 °C (1), is 50 g (36%).

Allow the residue in the flask to cool during 2 hours and then cool in ice for 15–30 minutes. Filter off the crude p-nitrophenol and boil it with 1 litre of 2 per cent hydrochloric acid (2) together with about 5 g of decolourising charcoal for at least 10 minutes. Filter through a hot water funnel (or through a preheated Buchner funnel): allow the filtrate to crystallise overnight. Filter off the almost colourless needles and dry them upon filter paper. The yield of p-nitrophenol, m.p. 112 °C, is 35 g (25%). Further small quantities may be obtained by concentrating the mother-liquor and also by repeating the extraction of the residue with 2 per cent hydrochloric acid.

The p.m.r. spectra of the two isomers confirm their substitution patterns. In the case of p-nitrophenol (Me_2CO, TMS) signals are observed at δ 7.06 (d, 2H, ortho-H's to OH), 8.15 (d, 2H, ortho-H's to NO_2) and 9.32 (broad s, 1H, OH); with o-nitrophenol the four dissimilar aromatic protons give rise to complex coupling leading to signals at δ 6.8–7.2 (m, 2H, $C_{5,6}$—H), 7.51 (m, 1H, C_4—H), 8.00 (d of d, 1H, C_3—H), and 10.46 (s, 1H, OH).

Notes. (1) If the m.p. is not quite satisfactory, dissolve the o-nitrophenol in hot ethanol (or industrial spirit) under reflux, add hot water drop by drop until a cloudiness just appears and allow to cool spontaneously. Filter off the bright yellow crystals and dry between filter paper.

(2) It is not advisable to treat the crude p-nitrophenol with sodium hydroxide solution in order to convert it into the sodium derivative: alkali causes extensive resinification.

Experiment 6.103 1-NITROSO-2-NAPHTHOL

Dissolve 100 g (0.7 mol) of 2-naphthol (Expt 6.99) in a warm solution of 28 g (0.7 mol) of sodium hydroxide in 1200 ml of water contained in a 2.5-litre round-bottomed flask fitted with a mechanical stirrer. Cool the solution to 0 °C in a bath of ice and salt, and add 50 g (0.725 mol) of powdered sodium nitrite. Start the stirrer and add, by means of a separatory funnel supported above the flask, 220 g (166.5 ml) of 5.6 M sulphuric acid at such a rate that the whole is added during 90 minutes and the temperature is kept at 0 °C: add crushed ice (about 200 g in all) from time to time in order to maintain the temperature at 0 °C. The solution should react acid to Congo red paper after all the sulphuric acid has been introduced. Stir the mixture for an additional hour; keep the temperature at 0 °C. Filter off the 1-nitroso-2-naphthol at the pump and wash it thoroughly with water. Dry the pale yellow product upon filter paper in the air for four days; the colour changes to dark brown and the 1-nitroso-2-naphthol, m.p. 97 °C, weighs 130 g. It contains about 10 per cent of its weight of moisture, but is otherwise almost pure. The moisture may be removed by leaving the air-dried compound in a desiccator for 24 hours; the yield is 115 g (96%), m.p. 106 °C.

If 1-nitroso-2-naphthol is required in the crystalline condition, recrystallise it from light petroleum (b.p. 60–80 °C, 7.5 ml per gram); the recovery is almost quantitative, m.p. 106 °C.

Experiment 6.104 2-NITRORESORCINOL

Carefully add 25 ml of concentrated sulphuric acid (98%) to 5.5 g (0.05 mol) of resorcinol contained in a 150-ml beaker while stirring the mixture continuously with a glass rod; then warm the mixture to 60–65 °C on a water bath and allow to stand for 15 minutes. Cool the slurry of the 4,6-disulphonic acid which is obtained to 0–10 °C and add carefully from a well-supported dropping funnel a cooled mixture of 4 ml of concentrated nitric acid (72%) and 5.6 ml of concentrated sulphuric acid. It is essential that the temperature of the reaction mixture should not be allowed to exceed 20 °C. When the addition is complete, allow the mixture to stand for a further 15 minutes, and then cautiously add 15 g of crushed ice. External cooling may be also necessary to keep the temperature below 20 °C.

Transfer the resulting yellow-brown solution to a 250-ml round-bottomed flask and steam distil (Fig. 2.102); collect about 250 ml of distillate, cool and filter off the precipitated yellow-orange 2-nitroresorcinol. A further quantity of crude material may be obtained by extracting the filtered steam distillate with ether. The yield of crude material is 2.2 g (28%) of m.p. 76 °C; the m.p. may be raised to 85 °C by recrystallisation from aqueous ethanol. The p.m.r. spectrum (polysol-*d*) should be recorded and the signals assigned to the aromatic protons.

Experiment 6.105 *p*-BROMOPHENOL

$$C_6H_5OH \xrightarrow{Br_2, \ CS_2} (o) + p\text{-}Br \cdot C_6H_4OH$$

CAUTION: *This preparation should be conducted in an efficient fume cupboard.*

Equip a 500-ml three-necked flask with a reflux condenser, a mechanical stirrer and a separatory funnel. Attach to the top of the condenser a calcium chloride guard-tube leading by means of a glass tube to a funnel just immersed in a beaker holding about 150 ml of water for absorption of hydrogen bromide (compare Fig. 2.61(*a*)) (1). Place 94 g (1 mol) of phenol dissolved in 100 ml of dry carbon disulphide (CAUTION: *flammable*) in the flask, set the stirrer in motion and cool the flask in a mixture of ice and salt. When the temperature falls below 5 °C, add slowly (during about 2 hours) from the separatory funnel a solution of 160 g (51 ml, 1 mol) of bromine (CAUTION) in 50 ml of carbon disulphide. Then arrange the flask for distillation under reduced pressure by inserting a Claisen still-head which incorporates a short fractionating side-arm (compare Fig. 2.108) into the central socket of the three-necked flask; stopper the remaining sockets. Connect a condenser set for downward distillation to the still-head and attach the device for absorbing the hydrogen bromide evolved to the side-arm of the receiver adapter. Distil off the carbon disulphide at atmospheric pressure on a water bath held at 60 °C. (**CAUTION**: *very low flash point, see Section 2.3.2*, p. 40.) Remove the absorption device, insert a capillary leak and a thermometer into the Claisen still-head sockets and continue distillation under reduced pressure (oil bath). Collect two fractions: (*a*) b.p. below 145 °C/25–30 mmHg which is an inseparable mixture of *o*- and *p*-bromophenols (24–33 g), and; (*b*) b.p. 145–150 °C/25–30 mmHg, which is fairly pure *p*-bromophenol. The residue in the flask contains some higher boiling 2,4-dibromophenol. The *p*-bromophenol solidifies on cooling to a solid white mass, which usually contains traces of an oil; this may be removed by spreading on a porous tile or by centrifuging. The dry crystals have m.p. 63 °C; the yield is 140–145 g (81–84%).

The p.m.r. spectrum (CDCl$_3$, TMS) is characteristic of *para* substitution showing signals at δ 5.20 (s, 1H, OH), 6.69 (d, 2H, *ortho*-H's to OH), and 7.30 (d, 2H, *ortho*-H's to Br); with *o*-bromophenol (Expt 6.106) the p.m.r. spectrum (CDCl$_3$, TMS) shows a complex multiplet at δ 6.74–7.43 for the four dissimilar aromatic protons. The m.s. of *p*-bromophenol shows significant fragment ions at m/z 174, 172 (M, ^{81}Br, ^{79}Br), 93 (M − Br), 65 (93 − CO), and 39 (65 − C$_2$H$_2$).

Note. (1) A considerable quantity of constant boiling point hydrobromic acid may be obtained by distilling these solutions.

Experiment 6.106   *o*-BROMOPHENOL

In a 1-litre flask, equipped as in the preceding experiment, but omitting the gas absorption device, place a mixture of 31 g (0.33 mol) of phenol (**CAUTION**) and 116 g (63 ml) of concentrated sulphuric acid, and heat in a boiling water bath for 3 hours with mechanical stirring. Cool to room temperature or below by immersing the flask in ice-water, and then add slowly a solution of 95 g of sodium hydroxide in 235 ml of water: a solid salt may separate, but this will dissolve at a later stage. Replace the separatory funnel by a thermometer, which dips well into the liquid, and support a small dropping funnel over the top of the condenser. Cool the alkaline solution to room temperature, and add 53 g (17 ml) of bromine (**CAUTION**) from the dropping funnel down the condenser during 20–30 minutes while stirring constantly; permit the temperature to rise to 40–50 °C. Continue the stirring for 30 minutes after the bromine has been introduced: the reaction mixture should still be alkaline and contain only a small amount of suspended matter. The solution must now be evaporated. Arrange the flask assembly so that a rapid stream of air can be passed through the stirred reaction mixture, i.e. replace the thermometer by a wide air leak, connect the condenser for downward distillation using a knee bend and fit a receiver adapter with the take-off arm connected to the water pump. Heat the flask in an oil bath at 150–155 °C while maintaining a brisk current of air until a thick pasty mass remains (30–40 minutes). Allow to cool and then add 270 ml of concentrated sulphuric acid (*fume cupboard*: much hydrogen bromide is evolved). Heat the flask in an oil bath at 195–205 °C and pass a current of steam into the mixture (compare Fig. 2.102); this results in the hydrolysis of the sulphonate groups and the bromophenol distils over as a heavy, colourless (or pale yellow) oil. When the distillate is clear, extract it with ether. Dry the ethereal extract with a little magnesium sulphate, remove the ether on a water bath (Fig. 2.101) and distil the residue as rapidly as possible since the bromophenol is somewhat unstable and decomposes appreciably at the high temperature. Collect the fraction, b.p. 195–200 °C (a colourless liquid with a characteristic odour), which is practically pure *o*-bromophenol. The yield is 25 g (43%). The compound is somewhat unstable and decomposes on standing, becoming brown or red in colour. The p.m.r. spectrum is noted above in comparison with that of *p*-bromophenol.

Experiment 6.107   *o*- AND *p*-HYDROXYPROPIOPHENONES

$$\text{Ph·O·CO·Et} \xrightarrow{\text{AlCl}_3} \text{\textit{o}- and \textit{p}-HO·C}_6\text{H}_4\text{·CO·Et}$$

CAUTION: *This preparation should be carried out in an efficient fume cupboard.*

Equip a 1-litre three-necked flask with a dropping funnel, a sturdy mechanical stirrer and an efficient double surface reflux condenser, and place 187 g (1.4 mol) of anhydrous aluminium chloride and 200 ml of carbon disulphide in it (**CAUTION**: *see Section 2.3.2, p. 40*); attach a gas absorption trap (Fig. 2.61) to the top of the condenser. Stir the suspension and add 188 g (179 ml, 1.25 mol) of phenyl propanoate (Expt 5.149) slowly and at such a rate that the solvent boils vigorously (about 90 minutes). Much hydrogen chloride is evolved and is absorbed by the trap. When all the phenyl propanoate has been introduced, gently reflux the reaction mixture on a water bath until the evolution of hydrogen chloride ceases (about 2 hours). Turn the reflux condenser downwards (compare Fig. 2.98), and distil off the solvent from the water bath (**CAUTION**: *carbon disulphide*). Then replace the latter by an oil bath maintained at 140–150 °C and heat, with stirring, for 3 hours. During this period more hydrogen chloride is evolved, the mixture thickens and finally becomes a brown resinous mass; continue the stirring as long as possible. Allow the reaction mixture to cool and decompose the aluminium chloride complex by *slowly* adding first 150 ml of dilute hydrochloric acid (1:1) and then 250 ml of water; much heat is evolved and a dark oil collects on the surface. Allow to stand overnight, when most of the *p*-hydroxypropiophenone in the upper layer solidifies. Filter this off at the pump, and recrystallise it from 200 ml of methanol; 74 g (39%) of *p*-hydroxypropiophenone (a pale yellow solid), m.p. 147 °C, are obtained.

Remove the methanol from the mother-liquors using a rotary evaporator and combine the residue with that obtained by extracting the original filtrate with ether and similarly evaporating. Dissolve the combined residues in 250 ml of 10 per cent sodium hydroxide solution, and extract with two 50 ml portions of ether to remove non-phenolic products. Acidify the alkaline solution with hydrochloric acid, separate the oily layer, dry it over magnesium sulphate and distil under diminished pressure, preferably from a flask with fractionating side arm (Fig. 2.108). Collect the o-hydroxypropiophenone (65 g, 35%) at 110–115 °C/6 mmHg and a further quantity (20 g, 11%) of crude p-hydroxypropiophenone at 140–150 °C/11 mmHg.

Record the i.r. spectrum of each isomer and note the absorptions due to the OH and CO stretching vibrations (in particular the effect of hydrogen bonding in the case of the *ortho* isomer). Each spectrum shows absorptions in the 700–800 cm^{-1} region which confirms the respective state of substitution. The p.m.r. spectra are easily differentiated; the *ortho* isomer (CDCl$_3$, TMS) has signals at δ 1.19 (t, 3H, Me), 2.93 (q, 2H, CH$_2$), 6.60–7.75 (m, 5H, C$_{AR}$—H), and 12.13 (s, 1H, OH); the *para* isomer (DMSO-d_6, TMS) has signals at δ 1.10 (t, 3H, Me), 2.92 (q, 2H, CH$_2$), 6.90 (d, 2H, *ortho*-H's to COEt) and 7.88 (d, 2H, *ortho*-H's to OH). In the latter case the hydroxyl proton is not observed.

Cognate preparation. *2,5-Dihydroxyacetophenone.* Finely powder a mixture of 40 g (0.2 mol) of dry hydroquinone diacetate (1) and 87 g (0.65 mol) of anhydrous aluminium chloride in a glass mortar and introduce it into a 500-ml round-bottomed flask, fitted with an air condenser protected by a calcium chloride tube and connected to a gas absorption trap (Fig. 2.61). Immerse the flask in an oil bath and heat slowly so that the temperature reaches 110–120 °C at the end of about 30 minutes: the evolution of hydrogen chloride then begins. Raise the temperature slowly to 160–165 °C and maintain this temperature for 3 hours. Remove the flask from the oil bath and allow to cool. Add 280 g of crushed ice followed by 20 ml of concentrated hydrochloric acid in order to decompose the excess of aluminium chloride. Filter the resulting solid with suction and wash it with two 80 ml portions of cold water. Recrystallise the crude product from 200 ml of 95 per cent ethanol. The yield of pure 2,5-dihydroxyacetophenone, m.p. 202–203 °C, is 23 g (58%). Record the p.m.r. spectrum (DMSO-d_6, TMS) and by careful inspection of the magnitude of the coupling constants confirm the following assignment of signals at δ 2.59 (s, 3H, Me), 6.80 (d, 1H, C$_3$—H, measure $J_{3,4}$ and note that $J_{3,6}$ is negligible), 7.06 (d of d, 1H, measure $J_{3,4}$ and $J_{4,6}$), 7.24 (d, 1H, measure $J_{4,6}$), 9.14 (s, 1H, C$_5$—OH), and 11.40 (s, 1H, C$_2$—OH, hydrogen-bonded). Interpret the m.s. which shows significant ions at m/z 152, 137, 123, 109 and 43.

Note. (1) Hydroquinone diacetate may be prepared as follows. Add 1 drop of concentrated sulphuric acid to a mixture of 55 g (0.5 mol) of hydroquinone and 102 g (95 ml, 1 mol) of acetic anhydride in a 500-ml conical flask. Stir the mixture gently by hand; it warms up rapidly and the hydroquinone dissolves. After 5 minutes, pour the clear solution on to 400 ml of crushed ice, filter with suction and wash with 500 ml of water. Recrystallise the solid from 50 per cent aqueous ethanol (c. 400 ml are required). The yield of pure hydroquinone diacetate, m.p. 122 °C, is 89 g (91%).

Experiment 6.108   2-ALLYLPHENOL AND 2-(PROP-1-ENYL)PHENOL

2-Allylphenol. Boil 50 g of allyl phenyl ether (Expt 6.110) gently in a round-bottomed flask fitted with an air reflux condenser. Determine the refractive index of the mixture at intervals; the rearrangement is complete and the boiling is stopped when the refractive index (n_D^{25}) has risen to 1.55 (about 6 hours are required). Dissolve the product in 100 ml of 5 M sodium hydroxide solution and extract with two 30 ml portions of light petroleum (b.p. 40–60 °C) which removes the small amount of 2-methyldihydrobenzofuran formed as by-product and which is neutral. Carefully acidify the alkaline solution with 5 M hydrochloric acid with cooling and extract the mixture with one 50 ml portion and two 25 ml portions of ether. Dry the extract over anhydrous sodium sulphate and remove the ether with a rotary evaporator. Distil the residue under reduced pressure to give 2-allylphenol, b.p. 103–106 °C/19 mmHg or b.p. 96 °C/13 mmHg; n_D^{25} 1.5440. The yield is 35 g (70%).

Rearrangement to 2-(prop-1-enyl)phenol. Prepare a saturated solution (about 50% w/v) of potassium hydroxide in 60 ml of methanol. Place this solution together with 20 g of 2-allylphenol in a round-bottomed flask fitted with a still-head and condenser set for downward distillation. Arrange a thermometer so that the bulb dips into the mixture and distil the latter slowly until the temperature reaches 110 °C. Remove the still-head, attach an air condenser to the flask and boil the reaction mixture gently under reflux for 6 hours. Cool the mixture, cautiously acidify with concentrated hydrochloric acid and extract the product with three 30-ml portions of ether. Dry the extract, remove the ether and distil the residue under reduced pressure. Collect the 2-(prop-1-enyl)phenol as a fraction, b.p. 110–115 °C/15 mmHg; it crystallises on cooling in an ice bath, yield 15 g (75%). Recrystallisation from dry light petroleum (b.p. 60–80 °C) gives shining needles, m.p. 37 °C. The i.r. and p.m.r. spectra of allyl phenyl ether, 2-allylphenol and 2-(prop-1-enyl)phenol should be recorded and critically compared.

6.9.3   FORMATION OF PHENYL ETHERS

Examples of the preparation of alkyl benzyl ethers by the Williamson synthesis are included in Section 5.6.2, p. 583. An example of an alkyl phenyl ether is provided by the synthesis of phenacetin (Expt 6.109) where *p*-aminophenol is first converted into its *N*-acetyl derivative by reaction with slightly more than one equivalent of acetic anhydride. Treatment of the product with ethanolic sodium ethoxide solution followed by ethyl iodide then yields the ethyl ether of *N*-acetyl-*p*-phenetidine (phenacetin). This compound is biologically active and has been widely employed for example as an antipyretic and analgesic: however, owing to undesirable side reactions, its use is now restricted.

  The initial preparation of the sodium derivative of the phenol by treatment with sodium ethoxide may be avoided in a number of instances by heating directly the phenol, the alkyl halide and anhydrous potassium carbonate in acetone solution. Examples are provided by the preparation of allyl phenyl ether and butyl 2-nitrophenyl ether (Expt 6.110). A further cognate preparation, that of 2,4-dichlorophenoxyacetic acid, is of interest since the product is an important plant growth hormone and selective weed-killer. The conversion of phenols into phenoxyacetic acids by this route is of value in that these crystalline derivatives are useful for the characterisation of the phenolic compounds (Section 9.6.6, p. 1248).

Conversion of phenols into their methyl or ethyl ethers by reaction with the corresponding alkyl sulphates in the presence of aqueous sodium hydroxide affords a method which avoids the use of the more expensive alkyl halides (e.g. the synthesis of methyl 2-naphthyl ether and veratraldehyde, Expt 6.111). Also included in Expt 6.111 is a general procedure for the alkylation of phenols under PTC conditions.[38, 39] The method is suitable for 2,6-dialkylphenols, naphthols and various functionally substituted phenols. The alkylating agents include dimethyl sulphate, diethyl sulphate, methyl iodide, allyl bromide, epichlorohydrin, butyl bromide and benzyl chloride.

Experiment 6.109 PHENACETIN

$$p\text{-HO·C}_6\text{H}_4\text{·NH}_2 \xrightarrow{\text{(Me·CO)}_2\text{O}} p\text{-HO·C}_6\text{H}_4\text{·NH·CO·Me} \xrightarrow[\text{EtI}]{^\ominus\text{OEt}}$$

$$p\text{-EtO·C}_6\text{H}_4\text{·NH·CO·Me}$$

Suspend 11 g (0.1 mol) of *p*-aminophenol in 30 ml of water contained in a 250-ml beaker or conical flask and add 12 ml (0.127 mol) of acetic anhydride. Stir (or shake) the mixture vigorously and warm on a water bath. The solid dissolves. After 10 minutes, cool, filter the solid acetyl derivative at the pump and wash with a little cold water. Recrystallise from hot water (about 75 ml) and dry upon filter paper in the air. The yield of *p*-hydroxyacetanilide, m.p. 169 °C (1), is 14 g (93%).

Place 1.55 g (0.0675 mol) of clean sodium in a 250-ml round-bottomed flask equipped with a reflux condenser. Add 40 ml of absolute alcohol (or rectified spirit). If all the sodium has not disappeared after the vigorous reaction has subsided, warm the flask on a water bath until solution is complete. Cool the mixture and add 10 g (0.066 mol) of *p*-hydroxyacetanilide. Introduce 15 g (8 ml, 0.1 mol) of ethyl iodide slowly through the condenser and reflux the mixture for 45–60 minutes. Pour 100 ml of water through the condenser at such a rate that the crystalline product does not separate; if crystals do separate, reflux the mixture until they dissolve. Then cool the flask in an ice bath: collect the crude phenacetin with suction and wash with a little cold water. Dissolve the crude product in 80 ml of rectified spirit; if the solution is coloured, add 2 g of decolourising carbon, boil and filter. Treat the clear solution with 125 ml of hot water and allow to cool. Collect the pure phenacetin at the pump and dry in the air. The yield is 9.5 g (80%), m.p. 137 °C.

The p.m.r. spectrum (DMSO-d_6, TMS) shows signals at δ 1.30 (t, 3H, Me), 2.0 (s, 3H, COMe), 3.92 (q, 2H, CH$_2$), 6.80 (d, 2H, *ortho*-H's to OEt), 7.42 (d, 2H, *ortho*-H's to NH) and 9.68 (s broad, 1H, NH).

Note. (1) If the m.p. is unsatisfactory, dissolve the product in dilute alkali in the cold

985

and then reprecipitate it by the addition of acid to the neutralisation point. This procedure will eliminate traces of the diacetate of p-aminophenol which may be present; the acetyl group attached to nitrogen is not affected by cold dilute alkali, but that attached to oxygen is readily hydrolysed by the reagent.

Experiment 6.110 ALLYL PHENYL ETHER

$$PhOH + Br\diagup\diagdown\diagup CH_2 \xrightarrow{K_2CO_3} Ph\diagdown O\diagup\diagdown\diagup CH_2$$

Place 47 g (0.5 mol) of phenol, 60.5 g (0.5 mol) of allyl bromide (Expt 5.54), 69.1 g (0.5 mol) of anhydrous potassium carbonate and 100 ml of acetone in a 250-ml, two-necked round-bottomed flask fitted with a reflux condenser and sealed stirrer unit, and boil on a steam bath for 8 hours with stirring. Pour the reaction mixture into 500 ml of water, separate the organic layer and extract the aqueous layer with three 20 ml portions of ether. Wash the combined organic layer and ether extracts with 2 M sodium hydroxide solution, and dry over anhydrous potassium carbonate. Remove the ether with a rotary evaporator and distil the residue under reduced pressure. Collect the allyl phenyl ether, b.p. 85 °C/19 mmHg; the yield is 57 g (85%).

Cognate preparations. *Butyl 2-nitrophenyl ether* (o-Butoxynitrobenzene). Place a mixture of 28 g (0.2 mol) of o-nitrophenol (Expt 6.102), 28 g (0.2 mol) of anhydrous potassium carbonate, 30 g (23.5 ml, 0.22 mol) of butyl bromide and 200 ml of dry acetone in a 1-litre round-bottomed flask fitted with an efficient reflux condenser, and reflux on a steam bath for 48 hours. Distil off the acetone, add 200 ml of water and extract the product with two 100 ml portions of benzene (**CAUTION**). Wash the combined benzene extracts with three 90 ml portions of 10 per cent sodium hydroxide solution, remove the benzene by distillation at atmospheric pressure and distil the residue under reduced pressure. Collect the o-butoxynitrobenzene at 171–172 °C/19 mmHg (or at 127–129 °C/2 mmHg); the yield is 30 g (77%).

2,4-Dichlorophenoxyacetic acid.

$$2,4\text{-}Cl_2C_6H_3OH + Cl\cdot CH_2\cdot CO_2H \xrightarrow{\ominus OH} 2,4\text{-}Cl_2C_6H_3\cdot O\cdot CH_2\cdot CO_2H$$

Place 8.1 g (0.05 mol) of 2,4-dichlorophenol and 4.7 g (0.05 mol) of chloro-acetic acid (**CAUTION**: *causes skin blisters*) in a 400-ml beaker; add slowly, with stirring, a solution of 4.5 (0.112 mol) of sodium hydroxide in 25 ml of water. Considerable heat is developed during the reaction. Heat the reaction mixture on a wire gauze until most of the liquid has evaporated: treat the residue with 150 ml of water, cool and filter if necessary. Acidify the clear solution with dilute hydrochloric acid (litmus). Extract the dense oil which separates with two 25 ml portions of ether, wash the combined extracts with 10–15 ml of water, dry with 1 g of magnesium sulphate, filter through a fluted filter paper and distil off the ether on a rotary evaporator. Recrystallise the residue of crude, 2,4-dichlorophenoxyacetic acid from about 35 ml of benzene (fume cupboard). The yield of pure acid, m.p. 138 °C, is 6.0 g (54%). The p.m.r. spectrum (DMSO-d_6, TMS) provides a good example for assignment of signals on the basis of measurement of coupling constants; the signals appear at δ 4.75 (s, 2H, OCH$_2$), 7.00 (d, 1H, C$_6$—H, measure $J_{5,6}$ and note that $J_{3,6}$ is negligible), 7.26 (d of d, 1H, C$_5$—H, measure $J_{5,6}$ and $J_{3,5}$), 7.41 (d, 1H,

C_3—H, only $J_{3,5}$ observed) and 11.90 (broad s, 1H, CO_2H). Note some overlapping of the signals arising from the protons at C_3 and C_5.

Experiment 6.111 METHYL 2-NAPHTHYL ETHER (*Nerolin*)

$$2\text{-}C_{10}H_7OH + NaOH + Me_2SO_4 \longrightarrow 2\cdot C_{10}H_7\cdot O\cdot Me + Na(Me)SO_4 + H_2O$$

CAUTION: *This preparation should be carried out in an efficient fume cupboard.*

Equip a 500-ml three-necked flask with a separatory funnel, a sealed mechanical stirrer and a reflux condenser. Place 36 g (0.25 mol) of 2-naphthol in the flask, add a solution of 10.5 g of sodium hydroxide in 150 ml of water and stir the mixture; cool the warm mixture to about 10 °C by immersing the flask in an ice bath. Place 31.5 g (23.5 ml, 0.25 mol) of dimethyl sulphate (**CAUTION**: *see Section 4.2.24, p. 430*) in the separatory funnel and add dropwise, during 1 hour, while stirring the mixture vigorously. Warm for 1 hour at 70–80 °C with stirring in order to complete the methylation. Allow to cool and filter off the methyl 2-naphthyl ether at the pump, wash with 10 per cent sodium hydroxide solution, then liberally with water and drain thoroughly. Recrystallise from industrial spirit. The yield is 33 g (84%), m.p. 72 °C.

Cognate preparations. *Veratraldehyde* (3,4-dimethoxybenzaldehyde). Place 152 g (1 mol) of a good sample of commercial vanillin, m.p. 81–82 °C, in a 1-litre three-necked flask equipped with a reflux condenser, a mechanical stirrer and two separatory funnels (one of which is supported over the top of the reflux condenser). Melt the vanillin by warming on a water bath and stir vigorously. Charge one funnel with a solution of 82 g of pure potassium hydroxide in 120 ml of water and the other funnel with 160 g (120 ml, 1.04 mol) of purified dimethyl sulphate (**CAUTION**). Run in the potassium hydroxide solution at the rate of two drops a second, and 20 seconds after this has started add the dimethyl sulphate at the same rate. Stop the external heating after a few minutes; the mixture continues to reflux gently from the heat of the reaction. The reaction mixture should be pale reddish-brown since this colour indicates that it is alkaline; should the colour change to green, an acid reaction is indicated and this condition should be corrected by slightly increasing the rate of addition of the alkali. When half to three-quarters of the reagents have been added, the reaction mixture becomes turbid and separates into two layers. As soon as all the reagents have been run in (about 20 minutes), pour the yellow reaction mixture into a large porcelain basin and allow to cool without disturbance, preferably overnight. Filter the hard crystalline mass of veratraldehyde, grind it in a glass mortar with 300 ml of ice-cold water, filter at the pump and dry in a vacuum desiccator. The yield of veratraldehyde, m.p. 43–44 °C, is 160 g (96%). This product is sufficiently pure for most purposes; it can be purified without appreciable loss by distillation under reduced pressure, b.p. 158 °C/8 mmHg; m.p. 46 °C. The aldehyde is easily oxidised in the air and should therefore be kept in a tightly stoppered bottle.

Butyl phenyl ether. Weigh out 11.5 g (0.5 mol) of clean sodium into a dry, 1-litre round-bottomed flask provided with a double surface condenser, and

987

add 250 ml of absolute ethanol. If the reaction becomes so vigorous that the alcohol cannot be held back by the condenser, direct a stream of cold water or place a wet towel on the outside of the flask until it is again under control: do not cool the alcohol unduly otherwise the last traces of sodium will take a considerable time to dissolve. Add a solution of 47 g (0.5 mol) of pure phenol (**CAUTION**) in 50 ml of absolute ethanol and shake. Into a small separatory funnel supported in the top of the condenser, place 133 g (82.5 ml, 0.72 mol) of butyl iodide (Expt 5.59) or an equivalent quantity of butyl bromide (Expt 5.54) and add it, with shaking, during 15 minutes. Boil the solution gently for 3 hours, arrange the apparatus for downward distillation and distil off as much as possible of the alcohol on a water bath; this process is facilitated by wrapping the exposed part of the flask in a cloth. Add water to the residue in the flask, separate the organic layer and wash it twice with 25 ml portions of 10 per cent sodium hydroxide solution, then successively with water, dilute sulphuric acid and water: dry with magnesium sulphate. Distil and collect the butyl phenyl ether at 207–208 °C. The yield is 60 g (80%).

General procedure for the preparation of phenolic ethers under PTC conditions.[38] A mixture of dichloromethane (50 ml), water (50 ml), the phenol (10 mmol), sodium hydroxide (15 mmol), the alkylating agent (20–30 mmol) and benzyltributylammonium chloride, bromide or iodide (0.1–1.0 mmol) is agitated with a vibromix (Section 2.14), at room temperature for 2–12 hours. The organic layer is then separated and the aqueous layer extracted twice with dichloromethane (20 ml). The combined extracts are evaporated, the residue mixed with water and the mixture extracted with ether or pentane. The organic extract is washed twice with 2 M ammonia solution to remove dimethyl sulphate if necessary (methanolic ammonia in the case of diethyl sulphate), then 2 M sodium hydroxide solution (ethanolic sodium hydroxide in the case of hindered phenols) and finally with saturated sodium chloride solution. After drying with sodium sulphate, the solvent is evaporated and the residual phenolic ether purified by distillation or crystallisation.

When methyl iodide is used as the alkylating agent it is found that the reaction proceeds most satisfactorily when a stoichiometric amount of quaternary ammonium salt is used. Use of less than one equivalent of catalyst results in slow and incomplete reaction due to the inhibition of the catalytic effect by iodide.[39] Reactions with benzyl chloride are carried out similarly.

6.9.4 SOME METHODS FOR THE PROTECTION OF THE PHENOLIC HYDROXYL GROUP

The important methods for the protection of phenols are very similar to those used for the alcoholic hydroxyl group (Section 5.4.6, p. 550), namely (a) ether formation, and (b) ester formation.

ETHERS
Section 6.9.3 provides experimental details for the formation of *alkyl, allyl* and *benzyl ethers* under a range of conditions, including PTC procedures.

Deprotection of aryl methyl ethers may be effected under strongly acidic conditions (e.g. hydriodic acid, Section 9.6.11, p. 1253), but the milder methods employing either iodotrimethylsilane in chloroform solution at 20–50 °C for several hours,[40] or boron tribromide at room temperature may be preferable.[41]

General procedure for deprotection of mono- and polymethyl-aryl ethers with boron tribromide.[41] To a 10-ml flask fitted with a septum and magnetic stirrer bar are added reactant (3.6 mmol) and 5 ml of dichloromethane. An inert atmosphere is established and maintained. This mixture is cooled in a dry ice/propan-2-ol bath and boron tribromide [0.13 ml, 1.32 mmol (for monomethyl ethers), or 0.38 ml, 4 mmol (for dimethyl ethers)] is added through the septum by use of a syringe. The cold bath is removed and the mixture stirred for 30 minutes, poured into ice water, stirred for 30 minutes, saturated with salt and extracted with dichloromethane. The extract is dried and concentrated. The purity of the product is established by h.p.l.c. analysis on a Waters Associates 6000A model using both refractive index and u.v. absorbance detectors with a Waters 3.9 mm i.d. × 30 cm μ-Bondapack C_{18} reverse phase column.

Deprotection of allyl aryl ethers is accomplished by protonolysis with palladium on activated charcoal in methanol solution in the presence of toluene-*p*-sulphonic acid,[42] or by reduction with sodium bis(2-methoxyethoxy)aluminium hydride in toluene solution[43] (Aldrich). This latter reagent also cleaves aryl benzyl ethers.

Catechol and other *o*-dihydroxy aromatic systems may be protected by the formation of the *methylenedioxy* compound, most conveniently achieved by the following PTC procedure.[44] A mixture of water (20 ml), dibromomethane (0.15 mol) and Adogen 464 (1 mmol), is vigorously stirred and heated to reflux. The air in the system is displaced by nitrogen. A solution of the appropriate *o*-dihydroxybenzene (0.1 mol) and sodium hydroxide (0.25 mol) in water (50 ml) is added at such a rate that addition is complete after 2 hours. After addition is complete, the reaction mixture is stirred and refluxed for a further hour. The product is isolated by standard work-up procedures.

ESTERS

The general procedures for the formation of *acetates*, *benzoates* and *toluene-p-sulphonates* are described in Section 9.6.6, p. 1248. Toluene-*p*-sulphonates are stable to the presence of lithium aluminium hydride, to the acidic conditions used in aromatic nitrations and to the high temperatures that might be necessary in an Ullmann reaction.

The most convenient procedure for *deprotection* of these esters is aqueous alkaline hydrolysis; a general procedure for this operation, including possible variations in technique which might be required in certain instances, is discussed in Section 9.6.17, p. 1266.

6.10 AROMATIC ALDEHYDES

Aromatic aldehydes may be prepared by the following general procedures.

1. Aromatic formylation reactions (Expts 6.112 to 6.116).
2. Reactions involving modification of aromatic ring substituents (Expts 6.117 to 6.120).

Methods for the protection of the carbonyl group are considered in Section 6.12.8, p. 1056.

SUMMARY OF RETROSYNTHETIC STRATEGIES

Disconnection (method 1), e.g.

R—C₆H₄—CHO $\Longrightarrow$ R—C₆H₄ $\overset{\ominus}{}$ CHO $\overset{\oplus}{}$ $\equiv$ R·C₆H₅ + CO/HCl
(1)

R = H or alkyl
(TM)

Z—C₆H₄—CHO $\Longrightarrow$ Z—C₆H₄ $\overset{\ominus}{}$ CHO $\overset{\oplus}{}$ $\equiv$ Z·C₆H₅ + formaldehyde equivalents
(1)

Z = OH, OR
(TM)

Functional group removal/interconversion (FGR/FGI) (method 2), e.g.

Z—C₆H₄—COCl or Z—C₆H₄—OH $\overset{FGI}{\underset{(2)}{\longleftarrow}}$ Z—C₆H₄—CHO $\overset{FGR}{\underset{(2)}{\longrightarrow}}$ Z—C₆H₄—Me

Z = X, NO₂

(TM)

SPECTROSCOPIC FEATURES

The electron-withdrawing mesomeric interaction of the carbonyl group with the aromatic system results in the *i.r.* absorption frequency of its stretching vibration being lower than that exhibited in aliphatic aldehydes (p. 299). The absorption arising from the stretching vibration of the H—CO bond is also clearly observed (Fig. 3.30; anisaldehyde). This hydrogen may also be readily assigned in the *p.m.r.* spectrum of aromatic aldehydes and distinguished from other low field signals arising from acidic or phenolic protons, since the signal is unaffected by treatment of the sample with deuterium oxide. Aromatic substitution patterns are frequently assignable by inspection of both i.r. and p.m.r. spectra. The fragmentation patterns observed in the *m.s.* of aromatic aldehydes is illustrated and discussed in the case of benzaldehyde on p. 378. Further descriptive accounts of structural assignments are to be found in the preparative examples below.

6.10.1 AROMATIC FORMYLATION REACTIONS

Aromatic aldehydes may be obtained by passing a mixture of carbon monoxide and hydrogen chloride into the aromatic hydrocarbon in the presence of a mixture of copper(I) chloride and aluminium chloride which acts as a catalyst (*Gatterman–Koch reaction*). It is probable that the electrophilic species is the formyl cation [H—$\overset{\oplus}{C}$=O] formed from the mixture of gases in the presence of the Lewis acid. The preparation of *p*-tolualdehyde by this method has been described.[45]

$$ArH + [H—\overset{\oplus}{C}=O] \xrightarrow[\text{CuCl}]{\text{AlCl}_3} Ar \cdot CHO + H^{\oplus}$$

The Gatterman–Koch formylation is unsuitable for the preparation of aldehydes from phenols and phenolic ethers owing to the formation of complexes with the Lewis acid.

A convenient method for the introduction of a formyl group, specifically in the *ortho* position of a phenol, involves reaction with tin(IV) chloride in the presence of tributylamine followed by treatment with paraformaldehyde.[46] It is applicable to phenols with other electron-donating or -withdrawing substituents, and is illustrated by the conversion of *p*-cresol into 5-methyl-2-hydroxybenzaldehyde (Expt 6.112). The reaction is thought to proceed via two successive six-membered coordinated tin complexes.

$$X \cdot C_6H_4 \cdot OH + SnCl_4 \xrightarrow[-HCl]{Bu_3N} X \cdot C_6H_4 \cdot O \cdot SnCl_3$$

Other methods which are variously suitable for the formylation of aromatic hydrocarbons, phenols, phenolic ethers and heterocyclic systems, employ a range of alternative *formaldehyde equivalents*. These latter include the species $H—\overset{\oplus}{C}=NH$ (from hydrogen cyanide/aluminium chloride, *Gatterman aldehyde synthesis*), $Me_2\overset{\oplus}{N}=CHCl$ (from dimethylformamide/phosphorus oxychloride, *Vilsmeier reaction*), $Cl\overset{\oplus}{C}HOMe$ (from dichloromethyl methyl ether/titanium(IV) chloride), :CCl$_2$ (from chloroform/sodium hydroxide, *Reimer–Tiemann* reaction), and 1,3-dithiane.

The Gatterman aldehyde synthesis employs hydrogen cyanide and unlike the Gatterman–Koch procedure, this method is successful in the case of phenols and phenolic ethers, although only the more reactive hydrocarbons are formylated in good yield. The mechanism probably involves the species $[H—\overset{\oplus}{C}=NH]$ as the effective electrophile.

$$ArH + [H—\overset{\oplus}{C}=NH] \xrightarrow[\text{ZnH}_2]{\text{AlCl}_3} [Ar \cdot CH=NH] \xrightarrow{H_2O} Ar \cdot CHO$$

991

One alternative which avoids the use of the hazardous hydrogen cyanide is passing dry hydrogen chloride either into a mixture of zinc cyanide, aluminium chloride, the hydrocarbon or phenolic ether and a solvent (such as tetrachloroethane or benzene), or into a mixture of zinc cyanide, the phenol and anhydrous ether or benzene. The zinc cyanide is converted by the hydrogen chloride into hydrogen cyanide (which reacts *in situ*), and zinc chloride which is known to be an effective catalyst in this reaction. A second alternative, which is applicable to hydrocarbons only, is to use acetone cyanohydrin as an *in situ* source of hydrogen cyanide.[47] The preparation of 2,4,6-trimethylbenzaldehyde (mesitaldehyde) from mesitylene and the related cognate preparations (Expt 6.113) provides a varied range of examples.

Certain reactive aromatic hydrocarbons are formylated by dimethylformamide in the presence of phosphorus oxychloride (the Vilsmeier reaction, e.g. 9-formylanthracene, Expt 6.114). This method can also be used with advantage for the formylation of π-excessive heteroaromatic systems (e.g. 2-formylthiophene, cognate preparation in Expt 6.114).

A generally applicable method of formylation involves the reaction of an aromatic hydrocarbon and dichloromethyl methyl ether under Friedel–Crafts conditions (cf. Section 6.11.1, p. 1006). The intermediate chloroacetal (1) thus formed is readily hydrolysed to the corresponding aldehyde (e.g. p-t-butylbenzaldehyde, Expt 6.115).

$$ArH + Cl_2CHOMe \xrightarrow[-HCl]{TiCl_4} Ar \cdot CH(OMe)Cl \xrightarrow{H_2O} Ar \cdot CHO + MeOH$$
$$(1)$$

The procedure is of value for the formylation of polycyclic aromatic and heteroaromatic systems, phenols and phenolic ethers.

Phenols are smoothly converted into phenolic aldehydes by reaction with chloroform in the presence of base (the Reimer–Tiemann reaction). This overall formylation reaction is of interest in that it involves the generation from chloroform and alkali of the reactive intermediate, dichlorocarbene (2). This effects electrophilic substitution in the reactive phenolate ions giving the benzylidene dichloride (3) which is hydrolysed by the alkaline medium to the corresponding hydroxyaldehyde. The phenolic aldehyde is isolated from the reaction medium after acidification.

$$HO^\ominus \frown H-CCl_3 \longrightarrow H_2O + :CCl_3^\ominus \longrightarrow Cl^\ominus + :CCl_2$$

(2)

(3)

In the case of phenol the main product is *o*-hydroxybenzaldehyde (salicylalde-hyde, Expt 6.116), but some of the *para* isomer is also formed. The isomers are separated by steam distillation in which only the *ortho* isomer is steam volatile owing to intramolecular hydrogen bonding. In the cognate preparation (2-hydroxy-1-naphthaldehyde) the preferential reaction at the 1-position should be noted.

Finally, it should be noted that a specific *ortho* formylation process which uses 1,3-dithiane as a formaldehyde equivalent has been described.[48] The reaction process appears to be of wide applicability and is similar to the *ortho* formylation of primary aromatic amines which has been described by the same authors (see p. 909).

Experiment 6.112 5-METHYL-2-HYDROXYBENZALDEHYDE[46]

To a four-necked round-bottomed flask (2.5 litres) equipped with a reflux condenser, mechanical stirrer, thermometer and nitrogen supply, are added anhydrous toluene (200 ml), *p*-cresol (108 g, 1.0 mol), tin(IV) chloride (26 g, 0.1 mol) (1) and tributylamine (74 g, 0.4 mol). The mixture is stirred for 20 minutes at room temperature, then paraformaldehyde (66 g) is added. The resulting yellowish solution is heated at $100 \pm 0.5\,°C$ for 8 hours. After cooling, the reaction mixture is poured into water (5 litres), acidified to pH 2 with 2 M hydrochloric acid and extracted with ether. The ether extract is washed with saturated sodium chloride solution, dried over anhydrous sodium sulphate and concentrated to leave the crude product, which is subjected to steam distillation. The final product is obtained in 86 per cent yield, m.p. 55–56 °C.

Note. (1) The commercially available anhydrous reagent should be used.

Experiment 6.113   2,4,6-TRIMETHYLBENZALDEHYDE (*Mesitaldehyde*)

$$1,3,5\text{-}C_6H_3Me_3 \xrightarrow[\text{HCl, AlCl}_3]{\text{Zn(CN)}_2} 2,4,6\text{-}Me_3C_6H_2\cdot CH\!\!\stackrel{\oplus}{=}\!\!NH_2\}Cl^{\ominus} \xrightarrow{H_3O^{\oplus}}$$

$$2,4,6\text{-}Me_3C_6H_2\cdot CHO$$

Method A (*use of zinc cyanide*). **CAUTION**: *The entire preparation should be conducted in an efficient fume cupboard. Tetrachloroethane is a potent poison and hydrogen cyanide is liberated during the reaction.*

Equip a 500-ml multi-necked flange flask with a reflux condenser, an efficient stirrer, a gas-inlet tube and a thermometer; the thermometer should be so sited that during the subsequent reaction the bulb is well immersed in the liquid, but the gas inlet tube need extend only just below the surface. Place 51 g (59 ml, 0.425 mol) of redistilled mesitylene (b.p. 163–166 °C), 73.5 g (0.625 mol) of zinc cyanide (Section 4.2.81, p. 467) and 200 ml of 1,1,2,2-tetrachloroethane in the flask, and stir the mixture while a rapid stream of dry hydrogen chloride (Section 4.2.38, p. 438) is passed through it until the zinc cyanide is decomposed (about 3 hours). Immerse the flask in a bath of crushed ice, remove the inlet tube and replace it by means of the arrangement depicted in Fig. 2.57, having previously charged the conical flask with 197 g of finely ground, anhydrous aluminium chloride. Stir the mixture *very vigorously* and add the aluminium chloride over a period of 10 minutes. Remove the ice bath, and resume the passage of hydrogen chloride gas for 3.5 hours; the heat of reaction will raise the temperature to about 70 °C at the end of an hour. Maintain the temperature at 67–72 °C for the remainder of the reaction period. Cool, and pour the reaction mixture, with hand stirring, into a 2-litre beaker about half-full of crushed ice to which 50 ml of concentrated hydrochloric acid has been added. Allow to stand overnight, transfer to a 1.5-litre round-bottomed flask, fit a condenser and reflux for 3 hours. Allow to cool, separate the organic layer and extract the aqueous layer once with 25 ml of tetrachloroethane. Wash the combined tetrachloroethane solutions with 75 ml of 10 per cent sodium carbonate solution, and steam distil (Fig. 2.102). Set the first 400–450 ml of distillate aside for the recovery of the solvent (1), and collect the second portion (about 4.5 litres) as long as oily drops pass over. Extract the distillate with 250 ml of toluene, dry the extract with a little magnesium sulphate and remove the solvent on an oil bath or a rotary evaporator. Distil the residue from a 150-ml flask with a fractionating side-arm (compare Fig. 2.108), and collect the mesitaldehyde at 118–121 °C/16 mmHg; the yield is 50 g (79%) (2).

The p.m.r. spectrum (CDCl$_3$, TMS) shows signals at δ 2.30 (s, 3H, C$_4$—Me), 2.59 (s, 6H, C$_2$—Me and C$_6$—Me), 6.85 (s, 2H, C$_3$—H, and C$_5$—H), and 10.50 (s, 1H, CHO). The m.s. gives significant fragment ions at m/z 148 (M), 147 (M – H, base peak), 119 (147 – CO), 105 (119 – CH$_2$), 91 (105 – CH$_2$), and 77 (91 – CH$_2$).

Notes. (1) The first portion of the steam distillate consists almost entirely of tetrachloroethane and water. The solvent is recovered by separating the organic layer, drying with anhydrous calcium chloride or magnesium sulphate and distilling.
(2) The following procedure is more convenient and less time-consuming, but the yield is lower (about 40 g). Mix the powdered aluminium chloride and zinc cyanide by shaking, add the mesitylene and immerse the flask in an oil bath at 100 °C. Stir the mixture and pass in a fairly rapid stream of dry hydrogen chloride for 4 hours; continue

the heating and stirring for a further 2 hours, but discontinue the passage of the gas. Decompose the reaction mixture, and complete the preparation as above.

Method B (use of acetone cyanohydrin).[47] A mixture of mesitylene (6 g, 0.05 mol) and acetone cyanohydrin (4.2, 0.05 mol) (Expt 5.168) in 1,2-dichloroethane (20 ml) is cooled to 0 °C. Aluminium chloride (13.3 g, 0.1 mol) is slowly added to the stirred mixture which is then allowed to warm to room temperature. After refluxing for 24 hours the mixture is hydrolysed and extracted with 1,2-dichloroethane. The organic layer is washed, neutralised and dried over magnesium sulphate. The solvent is evaporated and the crude product distilled to yield 3.5 g (48%) of 2,4,6-trimethylbenzaldehyde, b.p. 71 °C/9 mmHg.

Cognate preparations. *Method A 2,4-Dihydroxybenzaldehyde (β-resorcylaldehyde).* Equip a 500-ml three-necked flask with a reflux condenser, an efficient sealed stirrer and a wide inlet tube (to prevent clogging by the precipitate) extending nearly to the bottom of the vessel. Attach the inlet tube to an empty (safety) wash bottle and to this a generator producing hydrogen chloride (Section 4.2.38, p. 438); connect the top of the condenser by means of a tube to a wash bottle containing concentrated sulphuric acid, then to an empty bottle, and finally to the surface of sodium hydroxide solution (Fig. 2.61(a)). Place 20 g (0.18 mol) of resorcinol, 175 ml of sodium-dried ether and 40 g (0.34 mol) of powdered anhydrous zinc cyanide in the flask, start the stirrer and pass in a rapid stream of hydrogen chloride. The zinc cyanide gradually disappears with the formation of a cloudy solution; further passage of hydrogen chloride results in the separation of the imine hydrochloride condensation product as a thick oil which solidifies after 10–30 minutes. When the ether is saturated with hydrogen chloride (after about 1.5 hours), pass the gas more slowly and continue the stirring for a further half an hour to ensure the completeness of the reaction. Decant the ether from the solid material, add 100 ml of water to the latter, heat to the boiling point, filter the hot solution through a hot water funnel and allow the filtrate to cool. Filter the resorcylaldehyde (12 g) which separates as soon as the mixture is cold; allow the filtrate to stand for 15 hours when a further 11.5 g of the aldehyde is obtained (total yield 94%). The β-resorcylaldehyde, after drying, has m.p. 135–136 °C and is very faintly coloured. The colour may be removed by recrystallisation from hot water with the addition of a little decolourising carbon.

2-Hydroxy-1-naphthaldehyde. Proceed as for β-resorcylaldehyde except that 20 g (0.138 mol) of 2-naphthol replaces the resorcinol. Recrystallise the crude product (20 g, 83%) from water with the addition of a little decolourising carbon; the pure aldehyde has m.p. 80–81 °C.

Anisaldehyde (p-methoxybenzaldehyde). Use the apparatus described for β-resorcylaldehyde. Place 30 g (30 ml, 0.28 mol) of anisole, 75 ml of sodium-dried benzene (**CAUTION**) and 52 g (0.44 mol) of powdered zinc cyanide in the flask. Cool the mixture in a bath of cold water, start the stirrer and pass in a rapid stream of hydrogen chloride for 1 hour. Remove the gas inlet tube, and without stopping the stirrer, add 45 g of finely powdered anhydrous aluminium chloride slowly. Replace the gas inlet and pass in a slow steam of hydrogen chloride while heating the mixture at 40–45 °C for 3–4 hours. Allow to cool somewhat and pour the reaction mixture with stirring into excess of dilute hydrochloric acid; the imine hydrochloride separates as a heavy precipitate. Reflux the mixture for half an hour in order to decompose the imine hydrochloride and steam distil. Separate the organic layer in the distillate, dry with a little anhydrous magnesium sulphate and distil off the benzene. Continue distillation with an air bath and collect the anisaldehyde as a fraction which has a b.p. 246–248 °C; the yield is 35 g (92%). If required the product may be redistilled under reduced pressure, b.p. 134–135 °C/12 mmHg. The i.r. spectrum is given on p. 301; the p.m.r. spectrum should be recorded and interpreted.

Experiment 6.114   9-FORMYLANTHRACENE (9-Anthraldehyde)

$$+ \ Me_2N\cdot CHO + POCl_3 \ \xrightarrow{-HCl}$$

$$Me_2N\cdot CH\cdot O\cdot POCl_2$$

$$\xrightarrow[-4H^{\oplus}(CH_3\cdot CO_2^{\ominus})]{3H_2O}$$

CHO

$$+ \ Me_2NH + H_2PO_4^{\ominus} + 3Cl^{\ominus}$$

Equip a 500-ml three-necked flask with a sealed stirrer unit, a reflux condenser and a dropping funnel. Assemble the apparatus on a water bath in a fume cupboard. Place in the flask a mixture of 17.8 g (0.1 mol) of anthracene (1), 19 g (20 ml, 0.26 mol) of dimethylformamide and 20 ml of o-dichlorobenzene (2), and charge the dropping funnel with 27 g (16 ml, 0.175 mol) of phosphorus oxychloride; close the condenser and dropping funnel with calcium chloride guard-tubes. Start the stirrer, run in the phosphorus oxychloride steadily and then heat on a boiling water bath for 2 hours. Cool the reaction flask in an ice–salt bath and neutralise the contents to Congo red by running in aqueous sodium acetate solution (about 100 g of the trihydrate in 175 ml of water are required). Dilute with more water to about 2 litres and allow the mixture to stand at 0 °C for 2 hours. Filter off the yellow crystalline product and recrystallise it from aqueous acetic acid; the yield of 9-formylanthracene is 12 g (58%), m.p. 104 °C.

Notes. (1) Good quality material should be used; commercial fluorescent grade of m.p. c. 215 °C is suitable.
(2) The use of o-dichlorobenzene as a solvent is recommended. If the reaction is carried out in excess dimethylformamide alone, the product is contaminated with unreacted anthracene. It is then best to extract the crude material with cold methanol, remove the anthracene by filtration and recover the product by dilution with water.

Cognate preparation. *2-Formylthiophene* (thiophene-2-aldehyde). Use 21 g (19.3 ml, 0.25 mol) of thiophene, 23 g (24 ml, 0.315 mol) of dimethylformamide and 80 ml of 1,2-dichloroethane as solvent. Cool to 0 °C, add 48 g (29 ml, 0.313 mol) of phosphorus oxychloride slowly with stirring, and then heat, carefully at first, and then under reflux for 2 hours. Cool, pour on to crushed ice, neutralise with sodium acetate (c. 200 g of the hydrate), separate the organic phase and extract the aqueous phase with ether. Wash the combined organic phases with aqueous sodium hydrogen carbonate, dry over magnesium sulphate and remove the solvent on a rotary evaporator. Distil the residue under reduced pressure and collect the 2-formylthiophene as a fraction of b.p. 85–86 °C/16 mmHg; yield 20 g (71%). Record the p.m.r. spectrum (CCl$_4$, TMS) and assign the signals which appear at δ 7.12 (t, 1H), 7.67 (d, 2H), and 9.86 (s, 1H), bearing in mind that the protons on C$_3$ and C$_5$ appear equivalent (cf. furan-2-aldehyde, cognate preparation in Expt 6.133).

Experiment 6.115 *p*-t-BUTYLBENZALDEHYDE

$$\text{Ph·CMe}_3 \xrightarrow[\text{TiCl}_4]{\text{Cl}_2\text{CH·O·Me}} p\text{-Me}_3\text{C·C}_6\text{H}_4\text{·CHO}$$

Equip a 250-ml three-necked flask with a thermometer, reflux condenser, dropping funnel (protected with a calcium chloride guard-tube) and magnetic stirrer, and attach a gas absorption trap to the top of the condenser; assemble the apparatus in the fume cupboard. Place 15.1 g (0.12 mol) of t-butylbenzene (Expt 6.5) and 60 ml of dry dichloromethane (Section 4.1.5, p. 399) in the flask and cool to 0–5 °C in an ice–salt bath. To the stirred solution add 38 g (22 ml, 0.2 mol) of titanium (IV) chloride rapidly from the dropping funnel (2 to 3 minutes); the mixture becomes orange. Then add 11.5 g (0.1 mol) of dichloromethyl methyl ether (Expt 5.71) (**CAUTION**) during 20 minutes to the stirred and cooled solution. Hydrogen chloride is evolved after the first few drops of the ether are added. Stir the mixture for 5 minutes after completion of the addition, remove the cooling bath, allow the mixture to warm to room temperature (about half an hour) and then heat at 35 °C for 15 minutes. Pour the mixture into a separating funnel containing 100 g of ice and shake thoroughly. Separate the lower organic layer and extract the aqueous layer with three 25 ml portions of dichloromethane. Wash the combined dichloromethane extracts with three 25 ml portions of water, add a crystal of hydroquinone to prevent oxidation of the aldehyde and dry over magnesium sulphate. Filter the solution, remove the solvent by flash distillation and distil the residue under reduced pressure through a short fractionating column. The fraction which distils at 52 °C/4 mmHg is t-butylbenzene; collect the *p*-t-butylbenzaldehyde as a fraction of b.p. 98 °C/4 mmHg; the yield is 10.8 g (67%).

Experiment 6.116 SALICYLALDEHYDE

$$\text{PhOH} \xrightarrow[\ominus\text{OH}]{\text{CHCl}_3} o\text{- and } p\text{-HO·C}_6\text{H}_4\text{·CHO}$$

Equip a 1-litre three-necked flask with an efficient double surface reflux condenser, a mechanical stirrer and a thermometer, the bulb of which is within 2 cm of the bottom of the flask. Place a warm solution of 80 g of sodium hydroxide in 80 ml of water in the flask, add a solution of 25 g (0.266 mol) of phenol (**CAUTION**) in 25 ml of water and stir. Adjust the temperature inside the flask to 60–65 °C (by warming on a water bath or by cooling, as may be found necessary); do not allow the crystalline sodium phenoxide to separate out. Introduce 60 g (40.5 ml, 0.5 mol) of chloroform (**CAUTION**) in three portions at intervals of 15 minutes down the condenser.

Maintain the temperature of the well-stirred mixture at 65–70 °C during the addition by immersing the flask in hot or cold water as may be required. Finally heat on a boiling water bath for 1 hour to complete the reaction. Remove the excess of chloroform from the alkaline solution by steam distillation (Fig. 2.102). Allow to cool, acidify the orange-coloured liquid cautiously with dilute sulphuric acid and again steam distil the almost colourless liquid until no more oily drops are collected. Set aside the residue in the flask for the isolation of p-hydroxybenzaldehyde. Extract the distillate at once with ether, remove most of the ether from the extract by distillation on a water bath using a rotary evaporator. Transfer the residue, which contains phenol as well as salicylaldehyde, to a small glass-stoppered flask, add about twice the volume of saturated sodium metabisulphite solution, and shake vigorously (preferably mechanically) for at least half an hour, and allow to stand for 1 hour. Filter the paste of bisulphite compound at the pump, wash it with a little alcohol, and finally with a little ether (to remove the phenol). Decompose the bisulphite compound by warming in a round-bottomed flask on a water bath with dilute sulphuric acid, allow to cool, extract the salicylaldehyde with ether and dry the extract with anhydrous magnesium sulphate. Remove the ether by flash distillation and distil the residue collecting the salicylaldehyde (a colourless liquid) at 195–197 °C. The yield is 12 g (37%).

To isolate the p-*hydroxybenzaldehyde*, filter the residue from the steam distillation while hot through a fluted filter paper in order to remove resinous matter, and extract the cold filtrate with ether. Distil off the ether, and recrystallise the yellow solid from hot water to which some aqueous sulphurous acid is added. The yield of p-hydroxybenzaldehyde (colourless crystals), m.p. 116 °C, is 2–3 g (6–9%).

Record the i.r. spectrum of both isomers and assign the OH and CHO bands. The substitution patterns may be confirmed by comparison of their p.m.r. spectra (Me_2CO-d_6, TMS); for the *para* isomer the signals appear at δ 7.04 (d, 2H, *ortho*-H's to OH), 7.86 (d, 2H, *ortho*-H's to CHO), 8.73 (s, 1H, OH), and 8.73 (s, 1H, CHO); for the *ortho* isomer, wherein hydrogen bonding exists, the two signals at δ 9.80 and 10.98 may be assigned to either the OH or CHO groups, and the aromatic protons appear as two groups of complex multiplets at δ 6.7–7.1 and 7.3–7.7. The m.s., which shows the following principal fragment ions for either isomer, should be interpreted; m/z 122, 121, 104, 93, 65, and 39.

Cognate preparation.   *2-Hydroxy-1-naphthaldehyde.*   Equip a 1-litre three-necked flask with a separatory funnel, a sealed mechanical stirrer and a double surface reflux condenser. Place 50 g of 2-naphthol and 150 ml of rectified spirit in the flask, start the stirrer and rapidly add a solution of 100 g of sodium hydroxide in 210 ml of water. Heat the resulting solution to 70–80 °C on a water bath, and place 62 g (42 ml) of pure chloroform (**CAUTION**) in the separatory funnel. Introduce the chloroform dropwise until reaction commences (indicated by the formation of a deep blue colour), remove the water bath and continue the addition of the chloroform at such a rate that the mixture refluxes gently (about 1.5 hours). The sodium salt of the phenolic aldehyde separates near the end of the addition. Continue the stirring for a further 1 hour. Set the condenser for downward distillation (but retaining the stirrer) and distil off the excess chloroform and alcohol. Treat the residue, with stirring, dropwise with concentrated hydrochloric acid until the contents of the flask are acid to Congo red paper (about 88 ml are required); a dark oil, accompanied by a considerable amount of sodium chloride, separates. Add sufficient water to dissolve the salt, extract the oil with ether, wash the ethereal solution with water, dry with anhydrous magnesium sulphate and remove the solvent. Distil the residue under reduced pressure and collect the slightly coloured aldehyde at 177–180 °C/20 mmHg; it solidifies on cooling. Recrystallise the solid from about 40 ml of ethanol. The yield of 2-hydroxy-1-naphthaldehyde, m.p. 80 °C, is 28 g (47%).

6.10.2 REACTIONS INVOLVING MODIFICATIONS OF AROMATIC RING SUBSTITUENTS

Several procedures for the synthesis of aromatic aldehydes are available which involve the selective oxidation of a methyl group attached to an aromatic ring. A useful general reagent is a solution of chromium trioxide in acetic anhydride and acetic acid. The aldehyde is converted into the *gem*-diacetate as it is formed and is thus protected from further oxidation. The aldehyde is liberated from the diacetate by hydrolysis under acid conditions; the yields, however, are frequently only moderate (e.g. *p*-nitrobenzaldehyde, Expt 6.117).

$$Ar \cdot Me \xrightarrow[\text{(Me} \cdot \text{CO)}_2\text{O}]{\text{CrO}_3} Ar \cdot CH(O \cdot CO \cdot Me)_2 \xrightarrow{H_3O^{\oplus}} Ar \cdot CHO$$

Aldehydes may also be obtained by the hydrolysis of *gem*-dihalogen compounds obtained by the side-chain halogenation of a methylarene.

$$Ar \cdot Me \xrightarrow{2Br_2} Ar \cdot CHBr_2 \xrightarrow{H_2O} Ar \cdot CHO$$

Side-chain bromination occurs under the influence of light (cf. Expt 6.28) and the extent of bromination is controlled by ensuring that the bromine (used in the theoretical amount) is added no faster than the rate at which it is consumed. The halogen in the benzylidene halide is reactive and hydrolysis occurs readily under mild conditions. In the example cited (*p*-bromobenzaldehyde, Expt 6.118) the use of a boiling aqueous suspension of calcium carbonate gives good results.

Aromatic and heteroaromatic aldehydes can alternatively be prepared from the corresponding methyl compound by subjecting the chloromethyl or bromomethyl derivative to the *Sommelet reaction*. This procedure involves an initial reaction between the halomethyl compound and hexamethylenetetramine (hexamine), and hydrolysing the resulting quaternary hexamine salt (4) with hot aqueous acetic acid.

$$Ar \cdot Me \xrightarrow{X_2} Ar \cdot CH_2X \xrightarrow{(CH_2)_6N_4} Ar \cdot CH_2\overset{\oplus}{N}(CH_2)_6N_3\}\overset{\ominus}{X} \xrightarrow{H_2O} Ar \cdot CHO$$
$$(4)$$

The mechanism of the reaction is not certain but hydrolysis of the salt may yield the primary amine (5), formaldehyde and ammonia. A hydride ion transfer then probably occurs between the benzylamine and the protonated aldimine (6), derived from formaldehyde and ammonia. Hydrolysis of the resulting aromatic aldimine (7) then yields the required aldehyde.

$$Ar \cdot CH_2 \overset{\oplus}{N}(CH_2)_6 N_3 \} \overset{\ominus}{X} \xrightarrow{H_2O} Ar \cdot CH_2 NH_2 + 4H \cdot CHO + 3NH_3$$

(5)

$$Ar \cdot CH-H \quad CH_2 = \overset{\oplus}{N}H_2 \longrightarrow Ar \cdot CH = \overset{\oplus}{N}H_2 + Me \cdot NH_2$$
$$\underset{:NH_2}{|}$$

(5)   (6)   (7)

$$Ar \cdot CH = \overset{\oplus}{N}H_2 \xrightarrow{H_2O} Ar \cdot CHO + NH_3$$

A typical procedure is that described in Expt 6.119 for the synthesis of 1-naphthaldehyde. The synthesis of p-nitrobenzaldehyde provides an example in which the intermediate crystalline hexamine salt is isolated prior to hydrolysis. 2-Naphthaldehyde is prepared from the bromomethyl compound, the preparation of which illustrates the use of N-bromosuccinimide for effecting benzylic bromination of 2-methylnaphthalene.

Acid chlorides can be selectively hydrogenated in the presence of a catalyst (palladium deposited on a carrier, which is usually barium sulphate but is occasionally charcoal). The reaction, which involves the hydrogenolysis of the carbon–halogen bond, is known as the *Rosenmund reduction* and has been widely used for the synthesis of aromatic and heterocyclic aldehydes.

$$Ar \cdot COCl + H_2 \xrightarrow{catalyst} Ar \cdot CHO + HCl$$

The procedure is to pass purified hydrogen through a hot solution of the pure acid chloride in toluene or xylene in the presence of a catalyst; the exit gases are bubbled through water to absorb the hydrogen chloride, and the solution is titrated with standard alkali from time to time so that the reduction may be stopped when the theoretical quantity of hydrogen chloride has been evolved. Further reduction of the aldehyde, leading to the corresponding alcohol and thence to the methylarene, can usually be prevented by using the appropriate catalyst poison or regulator, which inactivates the catalyst towards reduction of the aldehyde but not the acid chloride. The regulator usually contains sulphur, e.g. quinoline-sulphur or thiourea; its use is not always necessary, however, and it has been stated that the decisive factors are to keep the reaction mixture at the lowest temperature at which hydrogen chloride is liberated and to arrest the reaction as soon as 1 mol of hydrogen chloride is evolved. The reduction is illustrated by the synthesis of 2-naphthaldehyde (Expt 6.120).

An alternative procedure[49] for effecting the conversion of acid chlorides into aldehydes is chemical reduction with bis(triphenylphosphine)copper(I) tetrahydroborate (see also Section 5.7.4, p. 594). The procedure is illustrated by the synthesis of 3,4-dimethoxybenzaldehyde which is isolated as the 2,4-dinitrophenylhydrazine derivative[50] (cognate preparation in Expt 6.120).

Experiment 6.117   p-NITROBENZALDEHYDE

$$p\text{-}O_2N \cdot C_6H_4 \cdot Me \xrightarrow[\text{(Me} \cdot \text{CO)}_2\text{O}]{CrO_3} p\text{-}O_2N \cdot C_6H_4 \cdot CH(O \cdot CO \cdot Me)_2 \xrightarrow{H_3O^{\oplus}}$$
$$p\text{-}O_2N \cdot C_6H_4 \cdot CHO$$

Equip a 1-litre three-necked flask with a mechanical stirrer, a dropping funnel and a thermometer, and immerse the flask in a bath of ice and salt. Place

200 g (185 ml, 2 mol) of acetic anhydride and 25 g (0.18 mol) of *p*-nitrotoluene in the flask, and add slowly, with stirring, 40 ml of concentrated sulphuric acid. When the temperature has fallen to 0 °C introduce slowly, with stirring, a solution of 50 g (0.5 mol) of chromium trioxide in 225 ml of acetic anhydride (1), at such a rate that the temperature does not exceed 10 °C; continue stirring for 2 hours after all the chromium trioxide solution has been added. Pour the contents of the flask into a 3-litre beaker one-third filled with crushed ice and almost fill the beaker with cold water. Filter the solid at the pump and wash it with cold water until the washings are colourless. Suspend the product in 150 ml of cold 2 per cent sodium carbonate solution and stir mechanically for about 10–15 minutes; filter, wash with cold water, and finally with 10 ml of ethanol. Dry in a vacuum desiccator; the yield of crude *p*-nitrobenzylidene diacetate is 30 g (65%) (2).

Reflux the crude *p*-nitrobenzylidene diacetate with a mixture of 70 ml of ethanol, 70 ml of water and 7 ml of concentrated sulphuric acid for 30 minutes, filter through a fluted filter paper and cool the filtrate in ice. Collect the crystals by suction filtration, wash with cold water and dry in a vacuum desiccator. The yield of *p*-nitrobenzaldehyde, m.p. 106 °C, is 15 g (55% overall). The spectroscopic characteristics are discussed with those of the *ortho* isomer, cognate preparation below.

Notes. (1) The solution is prepared by adding the chromium trioxide portionwise *to the well-cooled acetic anhydride. Addition of the anhydride to the oxide in bulk may lead to explosive decomposition.*
(2) The pure diacetate may be isolated by dissolving in 100 ml of hot ethanol, filtering from any insoluble impurities and allowing to cool: 28 g (61%), m.p. 125–126 °C, are obtained.

Cognate preparations. *o-Nitrobenzaldehyde.* Use 25 g (0.18 mol) of *o*-nitrotoluene and proceed as for *p*-nitrobenzaldehyde, but allow a period of 3 hours stirring at 5–10 °C after the addition of the chromium trioxide solution. In the work-up, omit the final ethanol washing; to remove unchanged *o*-nitrotoluene boil the crude product under reflux for 30 minutes with 120 ml of light petroleum (b.p. 60–80 °C). The yield of *o*-nitrobenzylidene diacetate of m.p. 82–84 °C is 16 g (36%).

Suspend 16 g of the diacetate in a mixture of 85 ml of concentrated hydrochloric acid, 140 ml of water and 25 ml of ethanol and boil under reflux for 45 minutes. Cool the mixture to 0 °C, filter the solid with suction and wash with water. Purify the crude aldehyde by rapid steam distillation; collect about 1 litre of distillate during 15 minutes, cool, filter and dry in a vacuum desiccator over calcium chloride. The yield of pure *o*-nitrobenzaldehyde, m.p. 44–45 °C, is 7.5 g (28% overall). The crude solid may also be purified after drying either by distillation under reduced pressure (the distillate of rather wide b.p. range, e.g. 120–144 °C/3–6 mmHg, is quite pure) or by dissolution in toluene (2–2.5 ml per gram) and precipitation with light petroleum, b.p. 40–60 °C (7 ml per ml of solution). The substitution pattern of the *ortho* and *para* isomers is confirmed by their p.m.r. spectra ($CDCl_3$, TMS) which should be compared with that of the *meta* isomer (p. 856). For the *para* isomer signals appear at δ 8.09 (d, 2H, *ortho*-H's to CHO), 8.39 (d, 2H, *ortho*-H's to NO_2), and 10.19 (s, 1H, CHO); for the *ortho* isomer signals appear at δ 7.63–8.2 (m, 4H, C_{AR}—H), and 10.31 (s, 1H, CHO).

Experiment 6.118 *p*-BROMOBENZALDEHYDE

$$p\text{-Br·C}_6\text{H}_4\text{·Me} \xrightarrow{\text{Br}_2} p\text{-Br·C}_6\text{H}_4\text{·CHBr}_2 \xrightarrow[\text{CaCO}_3]{\text{H}_2\text{O}} p\text{-Br·C}_6\text{H}_4\text{·CHO}$$

CAUTION: *This preparation should be carried out in the fume cupboard.*

Equip a 1-litre multi-necked flange flask with a reflux condenser, a mechanical stirrer, a dropping funnel and a thermometer which reaches nearly to the bottom of the flask; connect the upper end of the condenser to an absorption trap (Fig. 2.61). Place 100 g (65 ml, 0.58 mol) of *p*-bromotoluene (Expt 6.72) in the flask and immerse the latter in an oil bath (colourless oil in a large beaker). Heat the bath until the temperature of the stirred *p*-bromotoluene reaches 105 °C. Illuminate the liquid with an unfrosted 150-watt tungsten lamp, and add 200 g (1.25 mol) of bromine (**CAUTION**) slowly from the dropping funnel: do not allow a large excess of bromine to accumulate in the reaction mixture. Add about one half of the bromine during 1 hour while the temperature is kept at 105–110 °C, and add the remainder during 2 hours while the temperature is slowly raised to 135 °C. Raise the temperature slowly to 150 °C when all the bromine has been introduced. Transfer the crude *p*-bromobenzylidene dibromide (1) to a 2-litre flask, mix it intimately with 200 g of precipitated calcium carbonate and then add about 300 ml of water. Attach a reflux condenser to the flask, heat the mixture first on a water bath and then on a wire gauze over a free flame with continuous shaking until the liquid boils (2); reflux the mixture for 15 hours to complete the hydrolysis. Steam distil the reaction mixture rapidly (3); collect the first 1 litre of distillate separately, filter off the product and dry in a vacuum desiccator; 60 g (56%) of pure *p*-bromobenzaldehyde, m.p. 56–57 °C, are thus obtained. Collect a further 2 litres of distillate (4); this yields about 15 g of a less pure product, m.p. 52–56 °C. Purify this by trituration with saturated sodium bisulphite solution (2 ml per gram) and, after about 3 hours, filter off the pasty mixture at the pump, wash it with alcohol and then with ether. Transfer the bisulphite compound to a flask fitted for steam distillation (Fig. 2.102), add excess of sodium carbonate solution and isolate the aldehyde by steam distillation; 13 g (12%) of *p*-bromobenzaldehyde, m.p. 56–57 °C, are thus collected. The p.m.r. spectrum should be recorded and interpreted.

Notes. (1) This compound is lachrymatory and also produces a burning sensation on the skin; the latter is relieved by washing the affected parts with alcohol.
(2) This gradual heating reduces the risk of breaking the flask.
(3) The best results are obtained by conducting the steam distillation in a large three-necked flask provided with a sealed stirrer unit in the central aperture; the aldehyde distils slowly unless the mixture is well stirred.
(4) If the solution in the flask is acidified with hydrochloric acid, about 8 g of crude *p*-bromobenzoic acid may be isolated.

Experiment 6.119 1-NAPHTHALDEHYDE

$$1\text{-C}_{10}\text{H}_7\text{·CH}_2\text{Cl} \xrightarrow[\text{Me·CO}_2\text{H}]{\text{(CH}_2)_6\text{N}_4} 1\text{-C}_{10}\text{H}_7\text{·CH}_2\text{·C}_6\text{H}_{12}\overset{\oplus}{\text{N}}_4\}\text{Cl}^{\ominus} \longrightarrow 1\text{-C}_{10}\text{H}_7\text{·CHO}$$

In a 500-ml flask, fitted with a reflux condenser, place 53 g (0.3 mol) of 1-(chloromethyl)naphthalene, 84 g (0.6 mol) of hexamethylenetetramine and

250 ml of 50 per cent aqueous acetic acid. [**CAUTION**: 1-(*Chloro-methyl*)*naphthalene and, to a lesser degree,* 1-*naphthaldehyde have lachry-matory and vesicant properties; adequate precautions should therefore be taken to avoid contact with these substances.*] Heat the mixture under reflux for 2 hours; it becomes homogeneous after about 15 minutes and then an oil commences to separate. Add 100 ml of concentrated hydrochloric acid and reflux for a further 15 minutes; this will hydrolyse any Schiff's bases which may be formed from amine and aldehyde present and will also convert any amines into the ether-insoluble hydrochlorides. Cool, and extract the mixture with 150 ml of ether. Wash the ether layer with three 50 ml portions of water, then cautiously with 50 ml of 10 per cent sodium carbonate solution, followed by 50 ml of water. Dry the ethereal solution with anhydrous magnesium sulphate, remove the ether by distillation on a steam bath and distil the residue under reduced pressure. Collect the 1-naphthaldehyde at 160–162 °C/18 mmHg; the yield is 38 g (81%).

Cognate preparations. p-*Nitrobenzaldehyde*. This preparation is an example of the Sommelet reaction in which the hexaminium salt is isolated. Dissolve 11 g (0.13 mol) of hexamethylenetetramine in 70 ml of chloroform (**CAUTION**) and add 11.4 g (0.067 mol) of *p*-nitrobenzyl chloride or 14.4 g of *p*-nitrobenzyl bromide (Expt 6.28). Heat the mixture under reflux on a steam bath for 4 hours; a precipitate gradually separates. Replace the reflux condenser by a condenser set for distillation and distil off about 35 ml of solvent. Add 35 ml of acetone, cool in ice, collect the precipitate by suction filtration and dry it in the air. Heat the hexaminium salt thus obtained under reflux for 1 hour with 100 ml of 50 per cent acetic acid; then add 100 ml of water and 25 ml of concentrated hydrochloric acid and continue the refluxing for 5–10 minutes. Cool the solution in ice, collect the crystals of *p*-nitrobenzaldehyde and dry them in a vacuum desiccator. The yield is 6.4 g (63%), m.p. 106 °C. The p.m.r. spectrum is noted in Expt 6.117.

2-Naphthaldehyde.

$$2\text{-}C_{10}H_7\cdot CH_3 \xrightarrow{\text{NBS}} 2\text{-}C_{10}H_7\cdot CH_2Br \xrightarrow[\text{(ii) } H_3O^\oplus]{\text{(i) }(CH_2)_6N_4} 2\text{-}C_{10}H_7\cdot CHO$$

Dissolve 71 g of 2-methylnaphthalene in 450 g (283 ml) of carbon tetra-chloride and place the solution in a 1-litre three-necked flask equipped with a mechanical stirrer and reflux condenser. Introduce 89 g of *N*-bromosuccinimide (Section 4.2.*10*, p. 422) through the third neck, close the latter with a stopper and reflux the mixture with stirring for 16 hours. Filter off the succinimide and remove the solvent under reduced pressure on a water bath. Dissolve the residual brown oil (largely 2-(bromomethyl)naphthalene) in 300 ml of chloroform, and add it to a rapidly stirred solution of 84 g of hexamethylenetetramine in 150 ml of chloroform contained in a 2-litre three-necked flask, fitted with a reflux condenser, mechanical stirrer and dropping funnel: maintain the rate of addition so that the mixture refluxes vigorously. A white solid separates almost immediately. Heat the mixture to reflux for 30 minutes, cool and filter. Wash the crystalline hexaminium bromide with two 100 ml portions of light petroleum, b.p. 40–60 °C, and dry; the yield of solid, m.p. 175–176 °C, is 147 g. Reflux the hexaminium salt for 2 hours with 750 ml of 50 per cent acetic acid, add 150 ml of concentrated hydrochloric acid, continue the refluxing for 5 minutes more and cool. Extract the aldehyde from the solution with ether, evaporate the ether and recrystallise the residue from hot hexane. The yield of 2-naphthaldehyde, m.p. 59–60 °C, is 50 g (64% overall).

Experiment 6.120   2-NAPHTHALDEHYDE

$$2\text{-}C_{10}H_7\cdot COCl \xrightarrow[\text{Pd-BaSO}_4]{H_2} 2\text{-}C_{10}H_7\cdot CHO$$

Fit a 250-ml three-necked flask with a reflux condenser, a high-speed sealed stirrer (1) and a gas inlet tube extending to a point just above the bottom of the stirrer. Place 28.5 g (0.15 mol) of 2-naphthoyl chloride (Expt 6.161), 100 ml of sodium-dried xylene, 3 g of palladium–barium sulphate catalyst and 0.3 ml of the stock poison solution (Section 4.2.54, p. 452) in the flask. Connect the top of the condenser by a rubber tube to a 6-mm glass tube extending to the bottom of a 250-ml conical flask containing 200 ml of distilled water and a few drops of phenolphthalein indicator; arrange a burette charged with c. 1 M sodium hydroxide solution (prepared from the pure solid) for delivery into the flask. The apparatus must be sited in the fume cupboard.

Displace the air in the reaction flask with hydrogen from a cylinder of the gas, heat the flask in an oil bath at 140–150 °C, and stir the mixture vigorously. Continue to pass hydrogen at such a rate that 1–2 bubbles per second emerge in the conical flask. Follow the course of the reaction by the rate of hydrogen chloride evolution. The first 25 ml of alkali should be neutralised in 12–15 minutes, and the reaction should be complete in about 2 hours. About 92 per cent of the theoretical amount of hydrogen chloride ($\equiv 142.5$ ml of 1 M NaOH solution) is recovered; the end of the reaction is indicated by a rather abrupt cessation of hydrogen chloride evolution. Cool the flask add 1 g of decolourising carbon with stirring and filter the solution with suction through a hardened filter paper and keep the spent catalyst for recovery. Remove the xylene by flash distillation using a 50–75 ml flask with fractionating side-arm and then distil under reduced pressure with the aid of an oil bath: a small fraction, consisting largely of naphthalene, passes over first, followed by 2-naphthaldehyde at 147–149 °C/11 mmHg (temperature of bath, 170–180 °C). This (19 g, 81%) solidifies on cooling to a white solid, m.p. 59–60 °C.

Note. (1) Rapid stirring is desirable in order to obtain the maximum reaction rate; absorption of hydrogen occurs chiefly at the rapidly agitated surface.

Cognate preparation.   3,4-Dimethoxybenzaldehyde 2.4-dinitrophenylhydrazone (use of bis(triphenylphosphine)copper(I) tetrahydroborate)[50] 3,4-Dimethoxybenzoyl chloride (1.14 g, 5.7 mmol) in acetone (100 ml) is treated with triphenylphosphine (3.04 g, 11.6 mmol). To this solution at room temperature, bis(triphenylphosphine)copper(I) tetrahydroborate (3.47 g, 5.8 mmol) (Section 4.2.49, p. 449) is added and the reaction mixture stirred for 45 minutes. The white precipitate of triphenylphosphine copper chloride (4.9 g, 5.5 mmol) is removed by filtration and the filtrate evaporated to dryness. The residue is extracted with ether (the ether-insoluble residue is shown to be triphenylphosphine borane). The ether is removed, the residue redissolved in chloroform (60 ml) and the resulting solution stirred over copper(I) chloride (1 g) to remove the remaining triphenylphosphine. The reaction mixture is filtered, the chloroform evaporated and the residue extracted with methanol (1). The methanol extract is treated with a solution of 2,4-dinitrophenylhydrazine sulphate to give 1.66 g (86%) of 3,4-dimethoxy-benzaldehyde 2,4-dinitrophenylhydazone, m.p. 269–271 °C.

Note. (1) Triphenylphosphine may be removed from a mixture of triphenylphosphine and organic compounds by stirring a chloroform solution over copper(I) chloride, the reaction mixture is filtered, the chloroform removed and the organic compound dissolved in ether or methanol. A 1:1 Ph₃P:CuCl complex is formed which is soluble in chloroform but insoluble in ether or methanol.

6.11 AROMATIC KETONES AND QUINONES

1. Friedel–Crafts-type acylation processes (Expts 6.121 to 6.125).
2. Synthesis of aromatic ketones from carboxylic acid derivatives (Expts 6.126 and 6.127).
3. Synthesis of quinones (Expts 6.128 to 6.132).

Methods for the protection of the carbonyl group are considered in Section 6.12.8, p. 1056.

SUMMARY OF RETROSYNTHETIC STRATEGIES FOR AROMATIC KETONES

Disconnection (methods 1 and 2), e.g.

Z = H, R, X, OR
(TM)

$RCO \equiv Z \cdot C_6H_5$ + acyl cation equivalents

Z = H, R
(TM)

$RCO \equiv (Z \cdot C_6H_4)_2Cd + R \cdot COCl$

$R^{\ominus} \equiv Z \cdot C_6H_4 \cdot COCl + R_2CuLi$

SPECTROSCOPIC FEATURES
The spectroscopic characteristics exhibited by acetophenone (Ph·CO·Me) are fully discussed on p. 268 (*i.r.* spectrum, Fig. 3.9), p. 329 (*p.m.r.* spectrum, Fig. 3.48), and p. 378 (*m.s.*, Fig. 3.83(*b*)) and may be regarded as being typical of

aromatic ketones. Further substitution in the aromatic nucleus, or the presence of other alkyl or aralkyl groups, or of other aromatic residues attached to the carbonyl carbon gives rise to predictable changes in these spectral features. Descriptive assignments are given in some of the preparative examples below.

6.11.1 FRIEDEL–CRAFTS-TYPE ACYLATION PROCESSES

The reaction of a carboxylic acid chloride or anhydride with an aromatic hydrocarbon in the presence of anhydrous aluminium chloride generally gives a good yield of the aromatic ketone:

$$ArH + R \cdot COCl \xrightarrow{AlCl_3} Ar \cdot CO \cdot R + HCl$$

$$Ar^1 H + Ar^2 \cdot COCl \xrightarrow{AlCl_3} Ar^1 \cdot CO \cdot Ar^2 + HCl$$

$$ArH + (R \cdot CO)_2 O \xrightarrow{AlCl_3} Ar \cdot CO \cdot R + R \cdot CO_2 H$$

It should be noted that *Friedel–Crafts acylation* differs from Friedel–Crafts alkylation (see Section 6.1.1, p. 826) in one important respect in that the alkylation process requires relatively small (catalytic) quantities of aluminium chloride. With acylations, however, at least one molar equivalent of aluminium chloride is necessary for each carbonyl group present in the acylating agent. This is because aluminium chloride is capable of forming rather stable complexes with the carbonyl group (see formulation below). This complex formation therefore requires an equivalent quantity of metal halide, and hence a slight excess over this amount is employed in order to ensure that the free reagent may be present to act as the catalyst: thus 1.2 and 2.2 molar equivalents of aluminium chloride are generally employed for acid chlorides and acid anhydrides respectively. Excess of benzene or of toluene may be used as a solvent (when either of these substances constitutes one of the reactants), otherwise carbon disulphide or nitrobenzene is usually employed. Friedel–Crafts acylation is free of two features which complicate the alkylation reaction, namely, (i) polysubstitution and (ii) rearrangements. There is usually no difficulty in arresting the acylation with the introduction of a single acyl group into the aromatic nucleus as the acyl group deactivates the nucleus to further electrophilic attack. In the case of benzene homologues which may show a tendency to isomerise or disproportionate under the influence of aluminium chloride (see Section 6.1.1, p. 826) preliminary mixing of the acyl and aluminium halides is recommended.

The mechanism of the Friedel–Crafts acylation reaction, formulated below for reactions using acid chlorides, probably involves the acylium ion (2) as the reactive electrophilic species, although an electrophilic complex (1) between the acid chloride and aluminium chloride may also be involved.

The use of aliphatic carboxylic acid anhydrides in place of the corresponding acid chlorides offers many advantages; these include:

(a) the greater ease of obtaining the anhydrides in a state of purity, and their availability as commercial products (acetic, propanoic, butanoic and succinic anhydrides);

(b) the handling of disagreeable acid chlorides is avoided;

(c) the absence of any appreciable quantities of by-products and of resinous substances;

(d) the reaction is smooth and the yield is generally good.

The preparations of butyrophenone, propiophenone, benzophenone and methyl 2-naphthyl ketone, given in Expt 6.121, provide examples of acylation reactions carried out under standard Friedel–Crafts conditions using the appropriate acid chloride as the acylating reagent. In the case of the acetylation of naphthalene, two isomeric acetyl derivatives are possible and the composition of the product is dependent upon the reaction conditions (cf. the sulphonation of naphthalene, Section 6.4.1, p. 873). With nitrobenzene as the solvent the product is largely the thermodynamically more stable 2-acetonaphthalene, whereas in carbon tetrachloride the 1-isomer is formed as the kinetically controlled product. Also included in the cognate preparations are several acylation reactions of phloroglucinol which, unlike simple phenols, proceed with acceptable yields, together with a convenient alternative preparation of benzophenone from benzene and excess carbon tetrachloride; this latter reaction involves the intermediate formation of dichlorodiphenylmethane which is hydrolysed to the ketone.

The use of acetic anhydride as the acylating reagent is described in the preparation of p-methylacetophenone (Expt 6.122), and in a range of cognate preparations in which the substrates are benzene derivatives possessing electron-releasing groups (Br, Cl, OMe, Ph). In these cases the product is largely the para isomer. The presence of deactivating groups (CHO, CN, NO_2) renders the aromatic ring inactive towards acylation under Friedel–Crafts conditions, which therefore permits the use of nitrobenzene as a reaction solvent in suitable cases. Also included here is the acetylation of thiophene and of ferrocene. Electrophilic substitution in these cases proceeds with great ease and the catalyst is a small amount of phosphoric acid. With thiophene the most reactive site is the 2-position, and 2-acetylthiophene is the predominant isomer formed; ferrocene gives a unique mono-acetyl derivative.

Reaction of succinic anhydride with benzene in the presence of anhydrous aluminium chloride (slightly over two equivalents; see above) yields 3-benzoylpropanoic acid. This may be reduced by the Clemmensen method in the presence of a solvent (toluene) immiscible with the hydrochloric acid to 4-phenylbutanoic acid. Cyclisation to α-tetralone (Expt 6.123) is then effected smoothly by treatment with hot polyphosphoric acid. This reaction sequence represents the first stages in the Haworth procedure for the synthesis of polycyclic aromatic hydrocarbons (see Section 6.1.4, p. 839).

Aroylation of an aromatic system by reaction with phthalic anhydride under Friedel–Crafts conditions yields the o-aroylbenzoic acid. These readily available compounds have characteristic melting points which make them useful as derivatives in the characterisation of aromatic hydrocarbons and of aryl halides (Section 9.6.3, p. 1238).

With benzene the product is o-benzoylbenzoic acid (Expt 6.124), and with toluene, o-(p-toluoyl)benzoic acid (Expt 6.132). The aroylbenzoic acids are important starting materials for the synthesis of quinones (Section 6.11.3, p. 1022), into which they are cyclised using polyphosphoric acid (Expt 6.132).

Phenolic ketones may be prepared by the *Hoesch* acylation reaction, which may be regarded as an extension of the Gattermann aldehyde synthesis (Section 6.10.1, p. 990). The procedure involves reaction of a nitrile with a phenol (or phenolic ether) in the presence of zinc chloride and hydrogen chloride; best results are usually obtained with polyhydric phenols or their ethers, as for example in the preparation of phloroacetophenone (Expt 6.125). The formation of phenolic ketones by means of the Fries rearrangement of phenolic esters with aluminium chloride is discussed on p. 976.

Experiment 6.121 BUTYROPHENONE (*Phenyl propyl ketone–use of the acyl chloride*)

CAUTION: *All operations should be conducted in an efficient fume cupboard.*

Equip a 500-ml three-necked flask with a double surface reflux condenser, a sealed stirrer unit and a dropping funnel protected by a calcium chloride guard-tube. Connect the top of the condenser to a trap for absorbing the hydrogen chloride evolved (Fig. 2.61). Place 60 g (0.45 mol) of anhydrous, finely powdered aluminium chloride (see Section 4.2.4, p. 416) and 88 g (100 ml) of dry benzene (**CAUTION**) in the flask. Start the stirrer, and add 39.5 g (38 ml, 0.37 mol) of butanoyl chloride (Expt 5.138) slowly during about half-an-hour. In the early stages it may be necessary to warm the flask gently to initiate the reaction (hydrogen chloride is evolved) which should then proceed without further heating while the remainder of the acid chloride is added. Finally, heat the flask in a water bath at about 50 °C for 1 hour in order to complete the reaction. Cool and pour the reaction mixture into about 250 ml of water and a little crushed ice contained in a 750-ml flask; decomposition occurs with the evolution of heat and a dark oil (largely a solution of butyrophenone in benzene) separates on the surface. Stopper the flask and shake to complete the decomposition; if any solid remains undissolved, add a little concentrated hydrochloric acid to dissolve it. Pour the mixture into a separatory funnel, run off and discard the lower layer, wash the benzene with water, then with dilute sodium hydroxide solution (to remove the hydrogen chloride), again with water, and finally dry over magnesium sulphate or calcium chloride. Remove excess benzene by flash distillation from a boiling water bath; use the apparatus shown in Fig. 2.101 (100-ml distilling flask). When most of the benzene has been removed, replace the dropping funnel by a 360 °C thermometer and the water condenser by a short air condenser. Continue the distillation by careful heating over a gauze or, better, in an air bath – (**CAUTION**): there may be some benzene in the residual oil – and collect the butyrophenone at 227–230 °C or at 110 °C/10 mmHg. The yield is 25 g (51%).

The i.r. spectrum should be recorded and the absorptions assigned using the spectrum of acetophenone (p. 268) as a guide. The p.m.r. spectrum (CDCl$_3$, TMS) shows signals at δ 0.95 (t, 3H, Me), 1.92 (m, 2H, CH$_2$Me), 2.82 (t, 2H, CO·CH$_2$), 7.12–7.55 (m, 3H, C$_{3,4,5}$—H), and 7.80–8.1 (m, 2H, C$_{2,6}$—H). The multiplet at δ 1.92 provides a good example of the (N + 1) (M + 1) rule (p. 341) for the multiplicity of protons when coupled to N and M non-equivalent adjacent protons.

Cognate preparations. *Propiophenone* (*ethyl phenyl ketone*). Proceed as above but use 34.5 g (32 ml, 0.37 mol) of propanoyl chloride. The yield of propiophenone, b.p. 214–217 °C, is 30 g (60%). An improved yield is obtained by the following process. Add a mixture of 75 g (70.5 ml, 0.81 mol) of propanoyl chloride and 90 g (103 ml, 1.15 mol) of sodium-dried benzene (**CAUTION**) to a vigorously stirred suspension of 75 g (0.56 mol) of finely powdered anhydrous aluminium chloride in 100 ml of dry carbon disulphide (**CAUTION**: *see Sections 2.3.2, p. 40 and 4.2.32, p. 411*). Then introduce more of the aluminium chloride (about 35 g) until no further evolution of hydrogen chloride occurs. The yield of propiophenone, b.p. 123 °C/25 mmHg, is about 90 g (90%).

Benzophenone. A. Into a 500-ml round bottomed flask place 120 ml (105 g, 1.35 mol) of dry benzene (**CAUTION**) and 35 g (29 ml, 0.25 mol) of redistilled benzoyl chloride. Weigh out 37 g (0.275 mol) of finely powdered, anhydrous aluminium chloride into a dry stoppered conical flask, and add the solid, with frequent shaking, during 10 minutes to the contents of the flask. Fit a reflux condenser with a gas absorption trap attachment to the flask, and heat on a water bath for 3 hours or until hydrogen chloride is no longer evolved. Pour the contents of the flask while still warm into a mixture of 200 g of crushed ice and 100 ml of concentrated hydrochloric acid. Separate the upper benzene layer (filter first, if necessary), wash it with 50 ml of 5 per cent aqueous sodium hydroxide solution, then with water, and dry with magnesium sulphate. Remove the benzene after filtration by flash distillation and distil the residue under diminished pressure through a short fractionating side-arm (Fig. 2.108). Collect the benzophenone at 187–190 °C/15 mmHg; it solidifies to a white solid on cooling, m.p. 47–48 °C. The yield is 30 g (66%).

B.

$$2\text{PhH} + \text{CCl}_4 \xrightarrow{-2\text{HCl}} \text{Ph·CCl}_2\text{·Ph} \xrightarrow{\text{H}_2\text{O}} \text{Ph·CO·Ph}$$

Use a four-necked 1-litre flange flask fitted with a double surface condenser, a sealed stirrer unit, a thermometer and a dropping funnel protected by a calcium chloride guard-tube. Attach a gas absorption trap to the reflux condenser outlet. Place 91 g (0.68 mol) of powdered anhydrous aluminium chloride and 200 ml (319 g, 2.07 mol) of dry carbon tetrachloride (**CAUTION**, p. 399) in the flask, surround the latter with an ice bath and, when the temperature has fallen to 10–15 °C, introduce 10 ml (9 g) of sodium-dried benzene (**CAUTION**). The reaction commences immediately (hydrogen chloride is evolved and the temperature rises); add salt to the ice bath to get more efficient cooling. When the temperature commences to fall after the reaction has once started, add a mixture of 100 ml (97 g) of dry benzene (total 1.37 mol) and 110 ml (1.14 mol) of dry carbon tetrachloride at such a rate that the temperature is maintained between 5 and 10 °C (1). The addition usually requires 1–2 hours; continue the stirring for a further 3 hours while maintaining the temperature at 10 °C, and then allow to stand overnight.

Immerse the flask in ice, start the stirrer and add about 500 ml of water through the separatory funnel; the excess carbon tetrachloride usually refluxes during the addition. Distil off as much as possible of the carbon tetrachloride on a water bath, and then distil the mixture with steam (Fig. 2.102) during 30 minutes to remove the residual carbon tetrachloride (2) and to hydrolyse the dichlorodiphenylmethane to benzophenone. Separate the benzophenone layer and extract the aqueous layer with 40 ml of benzene. Dry the combined benzene extract and benzophenone with magnesium sulphate. Remove the benzene by flash distillation and isolate the pure benzophenone, m.p. 47–48 °C, as described in A above. The yield is 105 g (85%).

Notes. (1) Below 5 °C, the reaction is too slow; above 10 °C, appreciable amounts of tarry matter are formed.
(2) About 200 ml of carbon tetrachloride are recovered; this contains some benzene, but may be used after drying and redistillation, in another run.

Methyl 2-naphthyl ketone (*2-acetylnaphthalene*). Equip a 1-litre three-necked flask with a sealed mechanical stirrer unit and a pressure-equalising dropping funnel fitted with a calcium chloride guard-tube; stopper the third neck. Place 64 g (0.5 mol) of resublimed naphthalene and 350 g (291 ml) of pure nitrobenzene in the flask and stir until dissolved. To the homogeneous solution add 43.5 g (38.5 ml, 0.55 mol) of redistilled acetyl chloride from the dropping funnel. Cool to −5 °C in a freezing mixture of ice and salt and introduce, while stirring vigorously, 73.5 g (0.55 mol) of finely powdered anhydrous aluminium chloride in small portions during 90 minutes; do not allow the temperature to rise above 0 °C. The aluminium chloride dissolves and a deep green solution results. Remove the stirrer and stopper the central neck; into the side-necks of the flask fit respectively a drawn-out capillary tube and a tube leading through a filter flask trap to a water filter pump. Reduce the pressure to 15–20 mmHg; hydrogen chloride is copiously evolved and a vigorous ebullition occurs in the mixture. When no more gas is evolved, add an excess of crushed ice and separate the nitrobenzene layer. Wash the latter successively wtih two 100 ml portions of dilute hydrochloric acid and 100 ml of 5 per cent sodium carbonate solution. Use either of the following methods for isolating the pure 2-acetylnaphthalene from the accompanying 1-isomer (about 10%) (1).

1. Steam distil from a 1.5-litre flask until the odour of nitrobenzene is no longer perceptible in the distillate (6–12 hours). Extract the cold residue with three 100 ml portions of ether, dry the combined extracts with magnesium sulphate and distil off the ether. The residue solidifies and consists of almost pure methyl 2-naphthyl ketone, m.p. 52 °C; the yield is 30 g (35%). Upon recrystallisation from glacial acetic acid, the m.p. is raised to 54 °C.

2. Distil the dried (magnesium sulphate) nitrobenzene solution under reduced pressure. Nitrobenzene passes over at 95–100 °C/16 mmHg and the

temperature rises rapidly to 170 °C/15 mmHg; collect the fraction of b.p. 170–180 °C/15 mmHg. Transfer while still liquid to a porcelain basin; it solidifies on cooling. Spread it on a porous tile to absorb the small proportion of liquid methyl 1-naphthyl ketone which is present: the resulting yield of crude methyl 2-naphthyl ketone, m.p. 40–42 °C, is 50 g (59%). Two recrystallisations from glacial acetic acid (or from glacial acetic acid–water) give the almost pure 2-isomer, m.p. 53 °C.

Note. (1) Acetylation in carbon tetrachloride solution gives the 1-isomer as the major component. Add 70 g (0.52 mol) of powdered dry aluminium chloride to a vigorously stirred mixture of 41.9 g (38 ml, 0.53 mol) of acetyl chloride and 100 ml of carbon tetrachloride. The mixture becomes warm; cool to 20 °C and then run in slowly a solution of 32 g (0.25 mol) of naphthalene in 100 ml of carbon tetrachloride. Complete the reaction by warming to 30 °C for 30 minutes. Decompose the reaction mixture and work up the product, distilling the final product under reduced pressure, b.p. 165 °C/15 mmHg. The yield is 38.5 g (90%).

2,4,6-Trihydroxyisobutyrophenone (*Phloroisobutyrophenone*). Stir a mixture of 15 g (0.12 mol) of anhydrous phloroglucinol (1), 48 g (0.56 mol) anhydrous powdered aluminium chloride and 60 ml carbon disulphide (**CAUTION**) for 30 minutes in a three-necked flask fitted with a mechanical stirrer, a dropping funnel and a double surface condenser protected by a calcium chloride guard-tube. Add 45 ml of nitrobenzene, stir the reaction mixture for a further 30 minutes and then surround the flask with a water bath maintained at 60 °C. Add a solution of 13 g (0.12 mol) of 2-methylpropanoyl chloride (Expt 5.138) in 5 ml of nitrobenzene from the dropping funnel over a period of 30 minutes keeping the temperature of the water bath at 60 °C (2). After a further 30 minutes of stirring at this temperature pour the viscous residue on to crushed ice, add 425 g (1.5 mol) of Rochelle salt (sodium potassium tartrate) and neutralise the solution with approximately 40 per cent aqueous sodium hydroxide solution (3). Remove the nitrobenzene and carbon disulphide within a period of 20 minutes by vigorous steam distillation without allowing the volume of residual liquor to increase above about 700 ml (4). Filter off the crystals which separate from the residual solution on cooling and standing (usually overnight) and recrystallise twice from water to give 9.3 g (40%) of pure phloroisobutyrophenone hydrate, m.p. 78–80 °C; the anhydrous product, m.p. 138–140 °C, may be obtained by allowing the hydrate to stand in a vacuum desiccator over phosphorus pentoxide for a few days.

Notes. (1) Phloroglucinol contains two molecules of water of crystallisation; these are removed by heating for 12 hours at 120 °C.
(2) When carried out on a larger scale the volume of hydrogen chloride gas which is evolved justifies the attachment of a gas adsorption trap (Fig. 2.61) to the outlet of the calcium chloride guard-tube. On the scale suggested in this experiment, the apparatus should be sited within an efficient fume cupboard and the vapours led via a tube to a drain to prevent corrosion of the stirrer motor.
(3) Phloroisobutyrophenone is unstable when heated in the presence of acid; the addition of Rochelle salt prevents the precipitation of aluminium hydroxide when the free acid is neutralised with sodium hydroxide. If any precipitate does appear more Rochelle salt should be added; the exact amount of aqueous sodium hydroxide will vary with each experimental sequence; the point of neutrality is ascertained with universal indicator paper.
(4) If a bench supply of steam is not available, a large steam-can heated with three

large Bunsen burners, or with the large flame of an air–gas blow lamp may be employed. Two efficient double surface water condensers connected in series will be required to cope with the volume of steam and vapour produced and *great care* must be exercised to ensure that the outlet to the receiver vessel is suitably trapped to prevent the escape of the highly flammable carbon disulphide vapour.

2,4,6-Trihydroxyvalerophenone. Use 15 g (0.12 mol) of phloroglucinol, 48 g (0.56 mol) of anhydrous powdered aluminium chloride and 14.4 g (0.12 mol) of valeryl chloride (Expt 5.138) with the same volumes of nitrobenzene and carbon disulphide and under the conditions described for phloroisobutyro-phenone. The yield of hydrated product, m.p. 88–90 °C (m.p. 152–154 °C, anhydrous), is 11.5 g (42%).

2′,4′,6′-Trihydroxy-2-methylbutanophenone. Use 15 g (0.12 mol) of phloro-glucinol, 48 g (0.56 mol) of anhydrous powdered aluminium chloride and 14.4 g (0.12 mol) of 2-methylbutanoyl chloride (Expt 5.138) under the conditions specified above. The yield of hydrated product, m.p. 61–63 °C, is 14.3 g (52%). The anhydrous product is hygroscopic.

Experiment 6.122 p-METHYLACETOPHENONE (*Acetylation with acetic anhydride*)

$$\text{Ph·Me} + (\text{Me·CO})_2\text{O} \xrightarrow{\text{AlCl}_3} p\text{-Me·C}_6\text{H}_4\text{·CO·Me} + \text{Me·CO}_2\text{H}$$

Equip a 500-ml three-necked flask with a double surface condenser, a sealed stirrer unit and a dropping funnel protected with a calcium chloride guard-tube. Connect the top of the condenser to a trap for absorbing the hydrogen chloride evolved (Fig. 2.61). Place 75 g (0.56 mol) of anhydrous, finely powdered aluminium chloride (see Section 4.2.4, p. 416) and 120 g (140 ml, 1.30 mol) of pure dry toluene (Section 4.1.3, p. 398) in the flask and cool the latter in a bath of ice-water. Add 26 g (24 ml, 0.25 mol) of redistilled acetic anhydride during half an hour while the contents of the flask are thoroughly stirred; much heat is evolved in the reaction. Heat on a boiling water bath for about 30 minutes (or until the evolution of hydrogen chloride almost ceases) to complete the reaction, cool and pour the contents of the flask into a mixture of 150 g of crushed ice and 150 ml of concentrated hydrochloric acid contained in a beaker or flask. Stir or shake until all the aluminium salts are dissolved. Transfer the mixture to a separatory funnel, add 25–30 ml of ether, shake and separate the upper (largely toluene) layer. Extract the aqueous layer with 25 ml of ether and add this to the toluene solution. Wash the combined toluene and ether extracts with 50 ml of 10 per cent sodium hydroxide solution (or until the washings remain alkaline), then with water, separate the organic layer and dry it with magnesium sulphate or anhydrous calcium chloride. Remove the ether and toluene by distillation under atmospheric pressure and isolate the *p*-methylacetophenone by distillation at reduced pressure through a short fractionating column. The product (29 g, 86%) has b.p. 93–94 °C/7 mmHg. The i.r. and p.m.r. spectra of *p*-methyl-acetophenone and the compounds of the following cognate preparations should be recorded and interpreted using the spectra of acetophenone as a guide (pp. 268 and 329).

Cognate preparations. **p-*Bromoacetophenone.*** In a 1-litre three-necked flask, equipped as above, place 78.5 g (52.5 ml, 0.5 mol) of dry bromobenzene (Expt 6.23), 200 ml of dry carbon disulphide (**CAUTION**: *see Section 2.3.2, p. 40*) and 150 g of finely powdered anhydrous aluminium chloride. Stir the mixture and heat on a water bath until gentle refluxing commences; add 51 g (47.5 ml, 0.5 mol) of redistilled acetic anhydride slowly through the dropping funnel (30–60 minutes). Maintain gentle refluxing during the addition of the acetic anhydride and for 1 hour afterwards. Distil off most of the carbon disulphide on a water bath, allow the reaction mixture to cool somewhat and while still warm pour it slowly with stirring into a mixture of 500 g of crushed ice and 300 ml of concentrated hydrochloric acid. Decompose any residue in the flask and add it to the main product. Extract with 150 and 100 ml portions of ether, wash the combined extracts twice with water, once with 10 per cent sodium hydroxide solution and twice with water. Dry the extract with magnesium sulphate or anhydrous calcium chloride, remove the ether and distil the residue under reduced pressure. The *p*-bromoacetophenone boils at 130 °C/15 mmHg or at 117 °C/7 mmHg and a 3 °C fraction should be collected; it crystallises to a white solid, m.p. 50 °C. The yield is 75 g (75%).

The b.p. under atmospheric pressure has been given as 255.5 °C/736 mmHg.

p-*Chloroacetophenone*. Use 56 g (51 ml, 0.5 mol) of chlorobenzene and proceed as for *p*-bromoacetophenone. The yield of product, b.p. 124–126 °C/24 mmHg, m.p. 20–21 °C, is 60 g (78%). The b.p. under atmospheric pressure is 237 °C.

p-*Methoxyacetophenone*. Use 54 g (54.5 ml, 0.5 mol) of anisole and proceed as for *p*-bromoacetophenone. The yield of *p*-methoxyacetophenone, b.p. 139 °C/15 mmHg, is 70 g (93%). The b.p. under atmospheric pressure is 265 °C.

p-*Phenylacetophenone*. In a 1-litre three-necked flask provided with a dropping funnel, a mechanical stirrer and a reflux condenser, place 77 g (0.5 mol) of biphenyl, 150 g (1.125 mol) of finely powdered anhydrous aluminium chloride and 350 ml of anhydrous carbon disulphide (**CAUTION**). Charge the dropping funnel with 51 g (47.5 ml, 0.5 mol) of pure acetic anhydride and close the mouth of the funnel with a calcium chloride guard-tube. Heat the mixture on a water bath until gentle refluxing commences, and add the acetic anhydride during 1 hour; the addition product makes its appearance as a curdy mass when about three-quarters of the anhydride has been added. Reflux the reaction mixture gently for a further hour. Allow to cool and pour the reaction product slowly and with stirring on to crushed ice to which hydrochloric acid has been added. Filter the precipitated *p*-phenylacetophenone on a Buchner funnel, wash repeatedly with water until free from acid, dry, and distil under reduced pressure. There is usually a small fraction of low boiling point; the main product passes over at 196–210 °C/18 mmHg and solidifies on cooling. The yield of crude *p*-phenylacetophenone, m.p. 118 °C, is 85 g (86%). Upon recrystallisation from rectified spirit, the m.p. is raised to 120–121 °C; the recovery is about 80 per cent.

2-Acetylthiophene

Place 84 g (79 ml, 1 mol) of thiophene and 51 g (47.5 ml, 0.5 mol) of acetic anhydride in a 500-ml three-necked flask, fitted with a thermometer, mechanical stirrer and reflux condenser. Heat the stirred solution to 70–75 °C, remove the source of heat and add 5 g (4 ml) of 85–89 per cent orthophosphoric acid. An exothermic reaction occurs after 2–3 minutes and the temperature may rise to 90 °C; immerse the flask in a bath of cold water to control the reaction. When the boiling subsides (c. 5 minutes), reflux the mixture for 2 hours. Add 125 ml of water, stir for 5 minutes, transfer the cold reaction mixture to a separatory funnel, remove the water layer, wash with two 50 ml portions of 5 per cent sodium carbonate solution and dry over magnesium sulphate. Distil the orange-red liquid through a short fractionating column (Fig. 2.108) at atmospheric pressure and thus recover 38 g of unchanged thiophene at 83–84 °C. Distil the residue under reduced pressure and collect the 2-acetylthiophene at 89–90 °C/10 mmHg; this solidifies on cooling in ice, m.p. 10 °C. The yield is 44 g (70%).

Acetylferrocene[51] [*acetylcyclopentadienyl*(*cyclopentadienyl*)*iron*(II)]. Using a Pasteur pipette add 85 per cent phosphoric acid (1 ml) to a mixture of ferrocene (Expt 6.15) (1.5 g) and acetic anhydride (5 ml) contained in a small round-bottomed flask with constant shaking. Protect the mixture by the attachment of a calcium chloride drying tube, and heat it on a water bath for 10 minutes. Pour the mixture on to ice (20 g) contained in a 500 ml beaker. Add water to melt the remaining ice, and then neutralise the mixture by adding solid sodium hydrogen carbonate until carbon dioxide is no longer evolved. Cool the mixture in an ice bath for 15 minutes and filter on a No. 2 or No. 3 sintered glass crucible. Wash the collected solid with water to remove any white inorganic salts. Dry the orange-brown product by drawing air through it for at least 15 minutes. The product consists mainly of the acetyl derivative and unreacted ferrocene.

Purification by column chromatography. Push a small wad of glass wool to the bottom of a dry 50 ml burette with the aid of a glass rod; pour in enough acid-washed sand to give a 5 mm layer. Prepare a solvent mixture of ethyl acetate (20 ml) and light petroleum (180 ml, b.p. 60–80 °C). Make a slurry of column grade silica gel with a little of the prepared solvent and transfer to the burette. Repeat this operation until the silica gel level reaches the 20 ml mark, and then drain the solvent to *just* coincide with this level. Mix the above crude acetyl derivative (0.4 g) with a little silica gel and slurry with a small portion of solvent mixture. Transfer the slurry to the burette using a *small* amount of additional solvent, and again lower the solvent level to the silica gel layer. Add another 5 mm-layer of sand. Fill the burette with solvent taking care not to disturb the upper sand layer. Elute the products at the rate of 1–2 drops per second when the difference in colour of ferrocene and its acetyl derivative will enable them to be collected as separate fractions. Evaporate each collected fraction to dryness and confirm their identity by melting point determinations, and by recording their i.r. spectra in chloroform solution, and their p.m.r. spectra. The p.m.r. spectrum of ferrocene exhibits a single absorption peak, δ 4.09; that of the acetyl derivative (CDCl$_3$, TMS) has absorptions at δ 2.39 (s, 3H, Me), 4.19 (s, 5H, unsubstituted ring H's), 4.48 (m, 2H, C$_{3,4}$—H), and 4.77 (m, 2H, C$_{1,5}$—H).

Experiment 6.123 α-TETRALONE

3-Benzoylpropanoic acid. Place 175 g (2.25 mol) of sodium-dried benzene (**CAUTION**) and 34 g (0.34 mol) of succinic anhydride (Expt 5.141) in a 1-litre three-necked flask equipped with a sealed stirrer unit and two efficient reflux condensers, the tops of which are connected through a Y-junction to a single efficient gas absorption device (Fig. 2.61). Stir the mixture and add 100 g (0.75 mol) of powdered, anhydrous aluminium chloride all at once. The reaction usually starts immediately – hydrogen chloride is evolved and the mixture becomes hot; if there is no apparent reaction, warm gently. Heat in an oil bath to gentle refluxing, with continued stirring, for half an hour. Allow to cool, immerse the flask in a bath of cold water and slowly add 150 ml of water from a separatory funnel inserted into the top of one of the condensers. Introduce 50 ml of concentrated hydrochloric acid and separate the benzene by steam distillation (Fig. 2.102). Transfer the hot mixture to a 600-ml beaker; the 3-benzoylpropanoic acid separates as a colourless oil, which soon solidifies. Cool in ice, filter off the acid at the pump and wash with 100 ml of cold dilute hydrochloric acid (1:3 by volume) and then with 100 ml of cold water. Dissolve the crude acid in a solution of 40 g of anhydrous sodium carbonate in 250 ml of water by boiling for 10–15 minutes; filter the solution with suction to remove the small amount of aluminium hydroxide and wash with two 25 ml portions of hot water. Treat the *hot* filtrate with 2 g of decolourising carbon, stir for 5 minutes and filter at the pump through a preheated Buchner funnel. Transfer the hot filtrate to a 1-litre beaker, cool to about 50 °C and cautiously acidify with 65–70 ml of concentrated hydrochloric acid. Cool to 0 °C in a freezing mixture of ice and salt, filter, wash thoroughly with cold water, dry for 12 hours upon filter papers, and then to constant weight at 45–50 °C. The yield of practically pure 3-benzoylpropanoic acid, m.p. 115 °C, is 57 g.

4-Phenylbutanoic acid. Prepare amalgamated zinc from 120 g of zinc wool contained in a 1-litre round-bottomed flask (Section 4.2.*80*, p. 467), decant the liquid as completely as possible and add in the following order 75 ml of water, 180 ml of concentrated hydrochloric acid, 100 ml of pure toluene and 50 g (0.28 mol) of 3-benzoylpropanoic acid. Fit the flask with a reflux condenser connected to a gas absorption device (Fig. 2.61), and boil the reaction mixture vigorously for 30 hours; add three or four 50 ml portions of concentrated hydrochloric acid at approximately six-hour intervals during the refluxing period in order to maintain the concentration of the acid. Allow to cool to room temperature and separate the two layers. Dilute the aqueous portion with about 200 ml of water and extract with three 75 ml portions of ether. Combine the toluene layer with the ether extracts, wash with water and dry over anhydrous magnesium or calcium sulphate. Remove the solvents by distillation under diminished pressure using a rotary evaporator and distil the residue under reduced pressure (Fig. 2.108). Collect the 4-phenylbutanoic acid at 178–181 °C/19 mmHg; this solidifies on cooling to a colourless solid (40 g, 89%) and melts at 47–48 °C.

α-**Tetralone.** Heat 120 g of polyphosphoric acid (Section 4.2.58, p. 458) to 90 °C in a 1-litre beaker on a steam bath. Liquefy 33 g (0.20 mol) of 4-phenylbutanoic acid by heating to 70 °C and add this in one portion to the polyphosphoric acid with manual stirring. Remove the beaker from the steam bath and continue stirring for 3 minutes; the temperature should remain at about 90 °C. Then add 100 g more of polyphosphoric acid and warm on a steam bath with vigorous stirring for 4 minutes. Cool to 60 °C, add 300 g of crushed ice and stir until the polyphosphoric acid is completely hydrolysed and a yellow oil has separated. Extract the mixture with three 150 ml portions of ether and wash the combined extracts with water, with 5 per cent aqueous sodium hydroxide solution and then with water until the washings are neutral. Dry the ethereal solution over magnesium sulphate and remove the ether on a rotary evaporator. Distil the residue under reduced pressure through a short fractionating column and collect the α-tetralone at 105–107 °C/2 mmHg or 135–137 °C/15 mmHg. The yield is 23 g (79%).

Experiment 6.124 o-BENZOYLBENZOIC ACID

Equip a 750-ml three-necked flask with a sealed mechanical stirrer and a reflux condenser connected with a gas absorption trap (Fig. 2.61); insert a stopper in the third neck. Place 25 g (0.17 mol) of pure phthalic anhydride [see Expt 5.220, Note (1)] and 100 ml (1.16 mol) of sodium-dried benzene (**CAUTION**) in the flask; start the stirrer and add 50 g (0.375 mol) of powdered anhydrous aluminium chloride from a stoppered test tube in four portions or, alternatively, use the device shown in Fig. 2.57. If the reaction does not commence after the addition of the first 12 g of aluminium chloride, warm for a few seconds on a water bath. When all the aluminium chloride has been added and the evolution of hydrogen chloride slackens, warm on a water bath and ultimately reflux the mixture until the evolution of gas practically ceases. Cool the flask, add crushed ice slowly until the dark mass is completely decomposed and then run in concentrated hydrochloric acid (35–40 ml) until the solution clears. Steam distil (Fig. 2.102) to remove the excess of benzene; the residue in the flask, when cooled in ice, largely solidifies and consists of crude *o*-benzoylbenzoic acid. Filter off the solid product and wash it well with 74 ml of cold water; dissolve the solid in 150 ml of warm 10 per cent sodium carbonate solution. Treat the solution of the sodium salt with 2 g of decolourising charcoal, boil for 2 minutes and filter through a preheated Buchner funnel. Place the filtrate in a 1-litre beaker, cool in ice and cautiously acidify with concentrated hydrochloric acid while stirring well (*c.* 20 ml are required). The acid separates as an oil but it soon crystallises on stirring and cooling. Filter when ice cold, and wash with a little water. Dry in the air upon filter paper; the product, which is somewhat efflorescent, consists largely of the monohydrate, m.p. 94 °C.

To prepare pure anhydrous *o*-benzoylbenzoic acid, dissolve the air-dried (or the moist) product in about 175 ml of toluene contained in a 500-ml round-bottomed flask fitted with a reflux condenser and heat on a water bath. Transfer the toluene solution to a separatory funnel, run off any water present and dry with magnesium sulphate. Concentrate the toluene solution to about half its volume and add light petroleum (b.p. 60–80 °C) to the hot solution until a slight turbidity is produced. Allow to cool spontaneously to room temperature, then cool in ice to about 5 °C, collect the crystals and dry. The yield of pure, anhydrous *o*-benzoylbenzoic acid, m.p. 128 °C, is 32 g (84%).

Experiment 6.125 2,4,6-TRIHYDROXYACETOPHENONE
(*Phloroacetophenone*)

Place 25.2 g (0.2 mol) of dry phloroglucinol (1), 16.4 g (20.9 ml, 0.4 mol) of anhydrous acetonitrile (2), 100 ml of sodium-dried ether and 5 g of finely powdered, fused zinc chloride in a 500-ml Buchner flask fitted with a wide gas inlet tube. Protect the side-arm of the flask with a calcium chloride guard-tube. Cool the flask in an ice–salt mixture in the fume cupboard and pass a rapid stream of dry hydrogen chloride (Section 4.2.*38*, p. 438) through the solution for 2 hours with occasional shaking. Allow the flask to stand in an ice chest for 24 hours, and again pass dry hydrogen chloride into the pale orange mixture for a further 2 hours. Stopper the flask and leave it in an ice chest (or refrigerator) for 3 days. A bulky orange-yellow precipitate of the ketimine hydrochloride is formed. Decant the ether and wash the solid with two 25 ml portions of anhydrous ether. Transfer the solid with the aid of about 1 litre of hot water to a 2-litre round-bottomed flask provided with a reflux condenser. Boil the yellow solution vigorously for 2 hours, allow to cool somewhat, add 4–5 g of decolourising carbon, boil the solution for 5 minutes longer and filter the hot solution with suction through a preheated Buchner funnel. Extract the decolourising carbon with two 100 ml portions of boiling water and add the filtrate to the main product. Allow to stand overnight, and filter the pale yellow or colourless needles of phloroacetophenone at the pump, dry at 120 °C to remove the molecule of water of crystallisation and preserve in a tightly stoppered bottle. The yield is 29 g (85%), m.p. 217–219 °C. This product is pure enough for many purposes, but may be obtained absolutely pure by recrystallisation from hot water (35 ml per gram) and drying at 120 °C; m.p. 218–219 °C.

Notes. (1) See acetylation of phloroglycinol, Expt 6.121, cognate preparation, Note (1).
(2) The acetonitrile may be dried over anhydrous calcium sulphate or by distilling from phosphorus pentoxide.

6.11.2   SYNTHESIS OF AROMATIC KETONES FROM CARBOXYLIC ACID DERIVATES

Grignard reagents derived from aryl bromides are readily prepared and may be converted into organocadmium compounds by treatment with cadmium chloride (cf. Section 5.8.4, p. 616). Reaction of an organocadmium with a carboxylic acid chloride constitutes a convenient synthesis of aryl alkyl ketones.

$$2ArBr + 2Mg \longrightarrow 2ArMgBr \xrightarrow{CdCl_2} Ar_2Cd + 2MgClBr$$

$$Ar_2Cd + 2R \cdot COCl \longrightarrow 2Ar \cdot CO \cdot R + CdCl_2$$

The reaction is illustrated by the formation of propiophenone from diphenyl-cadmium and propanoyl chloride (Expt 6.126). Better yields are obtained by carrying out the synthesis in this manner rather than attempting the alternative combination of diethylcadmium with benzoyl chloride. Alternatively an alkyl-lithium 'cuprate' could be used with an aromatic acid chloride (p. 616).

Several examples of the synthesis of aryl alkyl ketones by the thermal decarboxylation of mixtures of carboxylic acids over heated metal salts are included under the preparation of aliphatic ketones (Expt 5.93). In this section the preparation of dibenzyl ketone (Expt 6.127) by the pyrolysis of the barium salt of phenylacetic acid, which proceeds in good yield, is included as a further example of this general type of synthesis.

Experiment 6.126   PROPIOPHENONE (*Ethyl phenyl ketone*)

$$Ph_2Cd + 2Et \cdot COCl \longrightarrow 2Ph \cdot CO \cdot Et + CdCl_2$$

CAUTION: *All operations should be conducted in an efficient fume cupboard.*
Prepare a solution of diphenylcadmium in 110 ml of dry benzene (**CAUTION**) using 4.9 g (0.2 mol) of magnesium, 31.4 g (0.2 mol) of bromo-benzene and 19.5 g (0.106 mol) of anhydrous cadmium chloride following the experimental procedure given for the preparation of dibutylcadmium (Expt 5.94). Cool the solution to 10 °C, and add during 3 minutes a solution of 14.8 g (0.16 mol) of propanoyl chloride (b.p. 78–79 °C) in 30 ml of dry benzene (**CAUTION**); use external cooling with an ice bath to prevent the tempera-ture rising above 40 °C. Stir the mixture for 2 hours at 25–35 °C. Add crushed ice (c. 200 g) and sufficient dilute (1 M) sulphuric acid to give a clear aqueous layer. Separate the benzene from the aqueous layer and extract the latter with two 20 ml portions of benzene. Wash the combined extracts successively with 50 ml portions of water, 5 per cent sodium carbonate solution, water and saturated sodium chloride solution. Dry over anhydrous sodium sulphate, remove the benzene by flash distillation and distil the residue under reduced

pressure. The yield of propiophenone, b.p. 100–102 °C/16 mmHg, is 17.5 g (82%).

Experiment 6.127 DIBENZYL KETONE

$$(Ph \cdot CH_2 \cdot CO_2)_2 Ba \xrightarrow{heat} (Ph \cdot CH_2)_2 CO + BaCO_3$$

Place 40 g (0.127 mol) of barium hydroxide octahydrate with 60 ml of water in a 250-ml round-bottomed flask and add 34 g (0.25 mol) of phenylacetic acid (Expt 5.128) slowly with swirling; warm the mixture until a clear solution is obtained. Evaporate the solution on a water bath under reduced pressure using a rotary evaporator, to yield a pasty mass of moist barium phenylace-tate. Fit the flask with a Claisen still-head carrying a gas inlet tube for nitrogen extending well into the flask and a 360 °C thermometer, and attach an air condenser with a receiver flask connected by means of an adapter with side arm. Lag the Claisen head with suitable insulating tape. Pass a slow stream of nitrogen into the flask and heat the latter gently in an air bath. When the residual water has been expelled, change the receiver. Now heat more strongly; dibenzyl ketone passes over at 320–325 °C as a pale yellow oil (24 g) which solidifies on standing. Redistil under reduced pressure and collect pure dibenzyl ketone at 210 °C/35 mmHg as a colourless oil (21 g, 80%); this completely crystallises on standing and has m.p. 33–34 °C.

6.11.3 QUINONES

The simplest quinones are *o*- and *p*-benzoquinone [(3) and (4) respectively]. This quinonoid structural feature is widespread in naturally occurring com-pounds isolated from moulds, fungi, lichens, plants and insects,[52] which include not only substituted benzoquinones but also substituted polycyclic quinones [i.e. the substituted analogues of, for example, 1,2-naphthoquinone (5), 9,10-anthraquinone (6), and 9,10-phenanthraquinone (7)].

(3) (4) (5)

(6) (7)

These quinonoid systems play a vital role in biosynthetic routes, are found as structural units in antibiotics and pigments, and are found in compounds having antihaemorrhagic activity (e.g. the Vitamin K group). In the laboratory substituted quinones are used as oxidising agents, and as starting materials for the synthesis of polycyclic compounds by virtue of their dienophilic reactivity in the Diels–Alder reaction (Section 7.6).

An excellent survey of the various methods of synthesis is available[53]; this section includes some of the illustrative methods of synthesis of the simpler analogues. These methods may be considered under the following headings: (a) the oxidation of hydrocarbons; (b) the oxidation of phenols and the oxidative coupling of phenols; (c) the oxidation of dihydric phenols and aminophenols; and (d) the cyclisation of aroylbenzoic acids.

OXIDATION OF HYDROCARBONS
Quinones of the more reactive, polycyclic, aromatic systems can usually be obtained by direct oxidation, which is best carried out with chromium(VI) compounds under acidic conditions. In this way 1,4-naphthoquinone, 9,10-anthraquinone and 9,10-phenanthraquinone are prepared from naphthalene, anthracene and phenanthrene respectively (Expt 6.128). Also included in this section is the reduction of anthraquinone with tin and acid to give anthrone, probably by the sequence of steps formulated below.

OXIDATION OF PHENOLS AND THE OXIDATIVE COUPLING OF PHENOLS
The classical and the most useful laboratory method for the preparation of quinones is the oxidation of monohydric phenols with the radical reagent, potassium nitrosodisulphonate [$(KO_3S)_2NO$] (Fremy's salt) (the *Teuber reaction*).[54] Details for the conversion of 3,4-dimethylphenol into 3,4-dimethyl-1,2-benzoquinone may be regarded as typical[55]; the probable mechanistic pathway is formulated below.

The *ortho* coupling of phenoxy radicals is noted in the formation of 1,1'-bi-2-naphthol from 2-naphthol by oxidation with iron(III) chloride (Expt 6.10). In the case of hindered 2,6-dialkylphenols, a very wide range of oxidants have been found to be variously successful in effecting *para* coupling; further oxidation then occurs to give an extended quinone.

Oxidation of hindered phenols bearing additionally a *para* methyl group leads to substituted stilbenequinones. These may be reduced in the presence of zinc and acid to the corresponding dihydroxystilbenes.

The reagents which have been used here to illustrate these interesting reactions are periodic acid for the oxidation of 2,6-di-t-butylphenol,[56] and a Celite-supported silver carbonate reagent for the oxidation of 2,6-dimethyl phenol, and 2,4,6-trimethylphenol[57] (Expt 6.129).

OXIDATION OF DIHYDRIC PHENOLS AND AMINOPHENOLS

p-Benzoquinone is conveniently prepared in the laboratory by the oxidation of hydroquinone (Expt 6.130) with either chromium trioxide in acetic acid or with sodium chlorate in the presence of vanadium pentoxide as a catalyst.

p-Benzoquinone and hydroquinone form a well-defined molecular complex known as quinhydrone. This complex, in the form of dark green crystals having a glistening metallic lustre, is conveniently prepared (Expt 6.130) by the partial oxidation of hydroquinone with a solution of iron alum.

The behaviour of p-benzoquinone on reaction with acetic anhydride in the presence of sulphuric acid is of interest. The eventual product is 1,2,4-triacetoxybenzene (the *Thiele acetylation*), which is formed by the following reaction sequence, initiated by a 1,4-addition of acetic anhydride across an α,β-unsaturated carbonyl system.

The formation of o-quinones by the above oxidative methods is less reliable since the *ortho* quinonoid system is more susceptible to attack by electrophilic and nucleophilic species; mild conditions are therefore essential. The use of the silver carbonate/Celite reagent noted above for phenol coupling reactions is particularly suitable; the conditions are those described in Expt 6.129, and they have been applied to the oxidation of catechol, 4-methylcatechol, 4-t-butylcatechol, and 3,5-di-t-butylcatechol to yield the corresponding o-quinones in almost quantitative yield.

A fairly general procedure for the synthesis of o- and p-quinones consists in coupling a phenol with a diazonium salt and reducing the resulting azo compound to an aminophenol with sodium dithionite. Mild oxidation with, for example, iron(III) chloride results in the formation of the corresponding quinone (e.g. the preparation of 1,2-naphthoquinone described and formulated in Expt 6.131).

CYCLISATION OF AROYLBENZOIC ACIDS

The aroylation of an aromatic system by reaction with phthalic anhydride under Friedel–Crafts conditions is described in Section 6.11.1, p. 1006. The cyclisation of the derived o-aroylbenzoic acid with polyphosphoric acid is a convenient route to substituted anthraquinones. The reaction is illustrated by the formation of 2-methylanthraquinone from o-(p-toluoyl)benzoic acid (Expt 6.132).

Experiment 6.128   1,4-NAPHTHOQUINONE

Place a solution of 120 g (1.2 mol) of pure chromium trioxide in 150 ml of 80 per cent aqueous acetic acid in a 2-litre three-necked flask, fitted with a thermometer, mechanical stirrer and 1-litre dropping funnel. Surround the flask by a mixture of ice and salt and, when the temperature has fallen to 0 °C, add a solution of 64 g (0.5 mol) of pure naphthalene in 600 ml of glacial acetic acid, with constant stirring, over a period of 2–3 hours while maintaining the internal temperature at 10–15 °C. Continue the stirring overnight, during which time the reaction mixture and bath attain room temperature. Allow the dark green solution to stand for 3 days and stir occasionally. Pour the reaction mixture into 5–6 litres of water, collect the crude naphthoquinone by suction filtration, wash with 200 ml of water and dry in a desiccator. Recrystallise from 500 ml of petroleum ether (b.p. 80–100 °C). The yield of pure 1,4-naphthoquinone, m.p. 124–125 °C, is 17 g (22%).

Cognate preparations. *9,10-Anthraquinone.* Place 5.0 g of powdered anthracene and 50 ml of glacial acetic acid in a 250-ml, two-necked round-bottomed flask with a reflux condenser and a dropping funnel. Mix the flask contents thoroughly by a swirling action and heat the mixture to reflux when most of the anthracene dissolves. Dissolve 10.0 g of chromium trioxide in 7–8 ml of water, add 25 ml of glacial acetic acid and pour the well-stirred mixture into the dropping funnel. Remove the heat source from the flask and add slowly the oxidising reagent at such a rate that the mixture continues to reflux (7–10 minutes); then reflux for a further 10 minutes when all the anthracene will have reacted completely. Cool the solution and pour into 250 ml of cold water. Stir the mixture vigorously, filter off the precipitated anthraquinone under gentle suction, wash it thoroughly on the filter with hot water, then with 50 ml of hot 1 M sodium hydroxide solution and finally with much cold water; drain well. Dry the anthraquinone by pressing it between several sheets of filter paper and leave it overnight in a desiccator over calcium chloride. The yield is 5.5 g (94%).

Purify the anthraquinone by either of the following methods:

1. Recrystallise the crude product from boiling glacial acetic acid with the aid of decolourising charcoal, wash the resulting crystals on the Buchner funnel with a little cold rectified spirit and dry in the air.
2. Sublime the dry solid using the procedure described in Section 2.21. The purified anthraquinone is obtained as yellow crystals having m.p. 273 °C.

Reduction of anthraquinone to anthrone. Place 52 g (0.25 mol) of anthraquinone, 50 g (0.42 mol) of granulated tin and 375 ml of glacial acetic acid in a 1-litre round-bottomed flask fitted with a reflux condenser. Heat the contents of the flask to boiling and slowly run in 125 ml of concentrated hydrochloric acid from a dropping funnel down the condenser over a period of 2 hours. By this time all the anthraquinone should have passed into solution; if not, add more tin and hydrochloric acid. Filter the liquid with suction through a sintered glass funnel, and add 50 ml of water. Cool the solution to about 10 °C when the anthrone will crystallise out. Filter the crystals at the pump on a Buchner funnel and wash with water. Dry upon filter-paper or upon a porous tile: the yield of crude anthrone, m.p. about 153 °C, is 40 g (82%). Recrystallise from a 3:1 mixture of benzene and light petroleum, b.p. 60–80 °C (10–12 ml per gram); this gives 30 g (61%) of pure anthrone, m.p. 155 °C.

Phenanthraquinone. Add 20 ml of concentrated sulphuric acid cautiously and with stirring to 40 ml of water contained in a 250-ml beaker. Heat to 90–95 °C on a water bath (it may be necessary to place the beaker in the boiling water bath), add 2.0 g (0.011 mol) of purified phenanthrene, and then 12.0 g of potassium dichromate in 0.5 g quantities until a vigorous reaction sets in: the latter usually occurs by the time about half of the oxidising agent has been added. Remove the beaker from the water bath – the temperature of the mixture will be 110–115 °C – and continue adding the potassium dichromate in small portions to maintain the reaction. Do not allow the temperature to fall below 85 °C as the reaction will cease: if necessary, heat on a water bath. When the addition is completed, heat on a boiling water bath for a further 30 minutes.

Cool the beaker in a bath of cold water and add 150 ml of cold water. Filter off the crude phenanthraquinone with suction and wash it with water until free from chromium salts. Suspend the solid in 20 ml of rectified spirit and add, with stirring, 20 ml of saturated sodium metabisulphite solution. Break up the lumps of the addition product with a glass rod and allow to stand, with frequent stirring, for 10 minutes. Add 150 ml of water to dissolve the addition product and filter with suction. Reject the precipitate which consists of the impurities present in the phenanthrene. Add saturated sodium carbonate solution to the filtrate until the bisulphite addition product is completely decomposed: allow the precipitate to settle for 1 minute, then add a few drops of sodium carbonate solution and note whether any further precipitation occurs. Stir the precipitate for 2–3 minutes, filter with suction, wash with three 20 ml portions of water and drain well. Dry the product between filter papers and then in a desiccator over calcium chloride. The yield of phenanthraquinone, m.p. 206 °C, is 1.4 g (60%). The product may be recrystallised from glacial acetic acid (about 20 ml), but the m.p. is unaffected.

Experiment 6.129   3,3′,5,5′-TETRA-t-BUTYLDIPHENOQUINONE[56]

2,6-Di-t-butylphenol (2.06 g, 10 mmol) is dissolved in dimethylformamide (10 ml), an aqueous solution of 4 M periodic acid (3.5 g, 4 ml) added, and the solution stirred at 85–95 °C (hot water bath) for 4–5 minutes. The coloured product partially crystallises out by this time. After 5–10 minutes at room temperature, the reaction mixture is diluted with 50 per cent aqueous methanol (8–10 ml), and the product isolated by filtration and washed with cold methanol. The resultant is recrystallised from aqueous acetone to give the product (94%) as brown-red needles, m.p. 245–247 °C.

Cognate preparation. *General procedure using Celite-supported silver carbonate.*[57] *Preparation of reagent.* The Celite is purified by washing it successively with methanol containing 10 per cent concentrated hydrochloric acid and then with distilled water until neutral; it is finally dried at 120 °C. Purified Celite (30 g) is added to a mechanically stirred solution of silver nitrate (34 g, 200 mmol) in distilled water (200 ml). A solution of sodium carbonate decahydrate (30 g, 105 mmol) (or 21 g, 210 mmol of potassium hydrogen carbonate) in distilled water (300 ml) is then added slowly to the resulting homogeneous suspension. When the addition is complete, stirring is continued for a further 10 minutes. The yellow-green precipitate which is formed is then filtered off and finally dried in a rotary evaporator over a period of several hours. The silver carbonate/Celite reagent contains about 1 mmol of silver carbonate in 0.57 g.

Oxidation procedure. The reagent is freed from residual water by azeotropic distillation with benzene (**CAUTION**). The compound to be oxidised is then added and refluxed in benzene (*c.* 200 ml for 0.5–2.0 g of compound). At the end of the reaction (determined by t.l.c. monitoring), the solid phase is filtered off and the solvent evaporated. The product is usually highly pure and recrystallisation is unnecessary. With *2,6-dimethylphenol* the molar ratio of phenol to silver carbonate is 1:4.4, and the reaction time is half an hour; in this case 3,3′,5,5′-tetramethyldiphenoquinone is obtained in 98 per cent yield and has m.p. 217–218 °C. With *2,4,6-trimethylphenol*, using the same molar ratio of phenol to oxidant, and a reaction time of 2 hours, 3,3′,5,5′-tetramethylstilbenequinone is obtained in 93 per cent yield and having m.p. 227–228 °C.

Reduction of a stilbenequinone. A solution of 3,3′5,5′-tetramethylstilbene-quinone (0.05 g) in acetic acid (100 ml) is shaken with zinc dust (4 g) for about 1 hour. The colourless mixture is filtered and the filtrate neutralised with sodium hydrogen carbonate. Extraction with ether and evaporation of solvent gives 3,3′,5,5′-tetramethyl-4,4′-dihydroxystilbene as yellow crystals, m.p. 237–240 °C; p.m.r. (Me$_2$CO-d_6, TMS) δ 2.24 (s, 12H), 3.00 (s, 2H), 6.87 (s, 2H), and 7.12 (s, 4H).

Experiment 6.130 *p*-BENZOQUINONE

Method 1. Cool a solution of 33 g (0.33 mol) of hydroquinone in 150 ml of 60 per cent acetic acid contained in a 600-ml beaker to below 5 °C in an ice bath. Dissolve 42 g (0.42 mol) of chromium trioxide in 70 ml of water, and add 30 ml of glacial acetic acid. By means of a separatory funnel with bent stem and supported over the beaker, add the chromium trioxide solution to the mechanically stirred hydroquinone solution at such a rate that the temperature does not rise above 10 °C; the addition takes about 2 hours. Filter the mixture at once and wash the quinone several times with 10 ml portions of ice-cold water. Spread the material upon filter paper until dry, but no longer or the quinone will be lost through sublimation. The yield of quinone (a bright yellow crystalline solid), m.p. 115 °C, is 21 g (66%); it darkens when exposed to light.

Impure quinone may be purified by placing it in a distilling flask attached to a condenser and passing a rapid current of steam into the flask: the quinone sublimes and collects in the receiver. It is separated from the water by filtration and dried; the m.p. is 116 °C. The vapour has a penetrating odour and attacks the eyes.

Method 2. In a 1-litre round-bottomed flask, provided with a mechanical stirrer, place 0.5 g of vanadium pentoxide (catalyst), 500 ml of 2 per cent sulphuric acid, 55 g (0.5 mol) of hydroquinone and 30 g of sodium chlorate. Stir the mixture vigorously for about 4 hours. Greenish-black quinhydrone is first formed and this is converted into yellow quinone; the temperature of the mixture rises to about 40 °C (do not allow it to exceed this temperature). Cool the flask in running water, filter the mixture at the pump and wash it with 50 ml of cold water. Dry the quinone upon filter paper in the air (see *Method 1*) or in a desiccator over anhydrous calcium chloride. The yield is 45 g (83%), m.p. 111–112 °C. The crude quinone may be purified by steam distillation as in *Method 1*, or by recrystallisation from boiling light petroleum, b.p. 100–120 °C (12 ml per gram): the resulting pure, bright yellow quinone has m.p. 115 °C and the recovery is about 95 per cent.

Conversion of hydroquinone into quinhydrone. Dissolve 100 g of iron alum [iron(III) ammonium sulphate] in 300 ml of water at 65 °C. Pour the solution, with stirring, into a solution of 25 g (0.228 mol) of hydroquinone in 100 ml of water contained in a 600-ml beaker. The quinhydrone is precipitated in fine needles. Cool the mixture in ice, filter with suction and wash three or four times with cold water. Dry in the air between filter paper. The yield of quinhydrone, m.p. 172 °C, is 15 g (60%).

Conversion of *p*-benzoquinone into 1,2,4-triacetoxybenzene (*Thiele acetylation*). Add 11 g (0.1 mol) of *p*-benzoquinone in small portions to a mechanically stirred mixture of 33 g (0.32 mol) of acetic anhydride and 0.25 ml of concentrated sulphuric acid. The temperature of the mixture rises to 40–50 °C and is kept within this range by regulating the rate of addition of the quinone. When the addition is complete allow the solution to cool to about 25 °C and pour into 150 ml of cold water. Collect the precipitated triacetate and recrystallise it from about 50 ml of rectified spirit; the yield is 22 g (86%), m.p. 97 °C.

Experiment 6.131   1,2-NAPHTHOQUINONE

Place 20 g (0.057 mol) of Orange II (Expt 6.84) in a 600-ml beaker and dissolve it in 250 ml of water at 40–50 °C. Add, with stirring, 24–25 g (0.114 mol) of hydrated sodium dithionite; this discharges the colour and yields a pink or cream-coloured, finely divided precipitate of 1-amino-2-naphthol (compare Expt 6.82). Heat the mixture nearly to boiling until it commences to froth considerably, then cool to 25 °C in ice, filter on a Buchner funnel and wash with a little cold water. Transfer the precipitate to a beaker containing a solution of 0.25 g of tin(II) chloride in 5 ml of concentrated hydrochloric acid diluted with 100 ml of water; upon stirring the aminonaphthol dissolves and a small amount of insoluble matter remains. The function of the tin(II) chloride is as an antioxidant, preventing the readily oxidisable aminonaphthol hydrochloride from undergoing appreciable change. Stir the solution for 5 minutes with 2 g of decolourising carbon, and filter at the pump. If crystalline material should separate at any stage, dissolve it by warming and by the addition of a little water if necessary. Transfer the clear solution to a beaker, add 25 ml of concentrated hydrochloric acid and warm until the solid dissolves. Cool to 0 °C, filter the almost colourless crystals of the aminonaphthol hydrochloride with suction and wash with 25 ml of dilute hydrochloric acid (1:4 by volume). *From this point all operations must be carried out rapidly.* In the meantime, prepare the oxidising solution by dissolving 30 g (0.11 mol) of crystallised iron(III) chloride in a mixture of 10 ml of concentrated hydrochloric acid and 25 ml of water by heating, cool to room temperature by adding *c*. 30 g of crushed ice and filter the solution at the pump. Wash the crystalline 1-amino-2-naphthol hydrochloride into a 600-ml beaker with water, add 150 ml of water and a few drops of concentrated hydrochloric acid and dissolve the precipitated solid by stirring and warming to about 35 °C. If necessary, filter rapidly by suction from a trace of residue, transfer to a 500-ml round-bottomed flask, add the iron(III) chloride solution all at once while shaking the flask vigorously. The quinone separates rapidly as a voluminous micro-crystalline yellow precipitate. Filter on a Buchner funnel and wash it thoroughly with water at 30 °C to remove all traces of acid. Dry the product upon filter paper in an atmosphere free from acid fumes. The yield of 1,2-naphthoquinone, which melts with decomposition at 145–147 °C, is 7 g (78%).

Experiment 6.132   2-METHYLANTHRAQUINONE

***o*-(*p*-Toluoyl)-benzoic acid.**   Using the conditions described in Expt 6.124, react 100 g (115.5 ml, 1.09 mol) of toluene, 25 g (0.17 mol) of phthalic anhydride and 50 g (0.375 mol) of anhydrous aluminium chloride to obtain the corresponding *o*-aroylbenzoic acid. The air-dried product consists largely of the monohydrate; this becomes anhydrous upon drying at 100 °C and melts at 138–139 °C. The yield of anhydrous product is 39 g (95%). It may be recrystallised from toluene.

2-Methylanthraquinone.   Place 25 ml of polyphosphoric acid (Section 4.2.58, p. 458) and 2.0 g of the foregoing aroylbenzoic acid in a 150-ml conical flask fitted with an air condenser to prevent water vapour from the boiling water bath entering the flask during the subsequent heating period of 2 hours. Cool the reaction product in ice and add 40 ml of water with stirring. Filter with suction and wash with water. Boil the residue with 10 ml of concentrated ammonia solution for 5 minutes (to remove unchanged acid) and filter at the pump. Recrystallise from ethanol in the presence of a little decolourising charcoal; filter off the crystals, wash with a little rectified spirit and dry at 100–120 °C. The yield of pure 2-methylanthraquinone, m.p. 175 °C, is 1.7 g (92%).

6.12 SOME REACTIONS OF AROMATIC CARBONYL COMPOUNDS

1. The Cannizzaro reaction (Expts 6.133 and 6.134).
2. The Claisen–Schmidt and related reactions (Expts 6.135 to 6.137).
3. The Perkin (Expt 6.138) and Doebner (Expt 6.139) reactions.
4. The synthesis of diphenylpolyenes (Expts 6.140 and 6.141).
5. The benzoin condensation (Expts 6.142 to 6.144).
6. Oxime formation (Expt 6.145).
7. Some reactions of alkyl aryl ketones (Expts 6.146 to 6.148).
8. Methods for the protection of the aromatic carbonyl group.

6.12.1 THE CANNIZZARO REACTION

Aromatic aldehydes (and other aldehydes in which α-hydrogen atoms are absent, e.g. formaldehyde and trimethylacetaldehyde) under the influence of strong aqueous or alcoholic alkali undergo simultaneous oxidation and reduction yielding the alcohol and corresponding carboxylate salt. Thus:

$$2Ar\cdot CHO \xrightarrow{KOH} Ar\cdot CH_2OH + Ar\cdot CO_2^{\ominus}K^{\oplus}$$

This dismutation or disproportionation reaction is known as the *Cannizzaro reaction*. The mechanism of the reaction involves the production of the anion (1) which may transfer a hydride ion to a carbonyl carbon atom in another aldehyde molecule. The reaction sequence is completed by a proton transfer to yield the carboxylate anion and the alcohol.

$$Ar\cdot CH_2OH + Ar\cdot CO_2^{\ominus}$$

The reaction is illustrated here by the conversion of benzaldehyde into benzoic acid and benzyl alcohol (Expt 6.133) and by a similar converson of furfural into furfuryl alcohol and furoic acid (cognate preparation in Expt 6.133).

A preparatively more useful form of this reaction is the crossed Cannizzaro reaction which ensues when a mixture of an aromatic aldehyde and formaldehyde is allowed to react under the influence of strong base (e.g. the preparation of *p*-methylbenzyl alcohol, Expt 6.134). A substantial proportion of the aromatic aldehyde is reduced to the corresponding alcohol while the formaldehyde is oxidised to formate. This is a reflection of the fact that nucleophilic attack of the hydroxide ion takes place preferentially at the more electrophilic carbonyl carbon atom in formaldehyde.

Experiment 6.133 BENZYL ALCOHOL AND BENZOIC ACID

$$\text{Ph·CHO} + \text{Ph·CHO} \xrightarrow{\text{KOH}} \text{Ph·CH}_2\text{OH} + \text{Ph·CO}_2^{\ominus}\text{K}^{\oplus}$$

Dissolve 29 g of potassium hydroxide in 27 ml of water contained in a beaker or conical flask, and cool the solution to about 20 °C in ice-water. Pour the solution into a 250-ml reagent bottle, and add 32 g (30 ml, 0.3 mol) of pure benzaldehyde (1); *cork* the bottle securely and shake the mixture vigorously until it has been converted into a thick emulsion. Allow the mixture to stand overnight or for 24 hours in the stoppered bottle. Add *just sufficient water* (about 105 ml) to dissolve the potassium benzoate. Pour the liquid into a separatory funnel, rinse out the bottle with about 30 ml of ether and add this ether to the solution in the funnel. Shake the solution in order to thoroughly extract the benzyl alcohol with the ether, separate the lower aqueous solution and carry out two further extractions each with about 25 ml of ether. *Save the aqueous solution.* Combine the ether extracts and distil the ether from a water bath (rotary evaporator) until the volume is about 25 ml. Cool and shake the ether solution twice with 5 ml portions of saturated sodium metabisulphite solution in order to remove any benzaldehyde which may be present. Separate the ethereal solution, wash it with 10 ml of 10 per cent sodium carbonate solution (to ensure complete removal of the bisulphite), then with 10 ml of water, and dry with anhydrous magnesium sulphate or anhydrous potassium carbonate. Remove the ether (Fig. 2.98; 50-ml distilling flask) on a water bath, and distil the residual liquid from an air bath; replace the water condenser by an air condenser or empty the water completely from the condenser jacket. Collect the benzyl alcohol at 204–207 °C (the pure compound boils at 205.5 °C). The yield is 13 g (86.5%).

The p.m.r. and ^{13}C-n.m.r. spectra of benzyl alcohol are given in Figs 3.65 and 3.53 respectively. The i.r. spectrum (thin film) shows absorptions at c. 3300 cm^{-1} (broad, OH stretching), c. 3050 and 2950 cm^{-1} (C_{AR}—H and C_{AL}—H, stretching), 1600, 1590 and 1500 cm^{-1} (ring breathing vibrations), c. 690 and 735 cm^{-1} (C_{AR}—H, out-of-plane vibrations), and 2000–1600 cm^{-1} (summation bands).

Pour the aqueous solution remaining from the ether extraction with stirring into a mixture of 80 ml of concentrated hydrochloric acid, 80 ml of water and about 100 g of crushed ice. Filter the precipitated benzoic acid at the pump, wash it with a little cold water, drain and recrystallise from boiling water. The yield of benzoic acid (colourless crystals), m.p. 121 °C, is 13.5 g (79%) (2).

The i.r. spectrum (KBr disc) should be recorded and interpreted; the p.m.r. spectrum (CCl$_4$, TMS) should also be recorded and the two groups of aromatic protons assigned, bearing in mind the electron-withdrawing nature of the carboxyl group.

Notes. (1) The benzaldehyde should be free from benzoic acid; it may be purified by washing with aqueous 10 per cent sodium carbonate solution, drying over anhydrous magnesium sulphate, adding a few crystals of hydroquinone or catechol, and distilling under reduced pressure in a stream of nitrogen. Benzaldehyde has a b.p. 79 °C/25 mmHg or 62 °C/10 mmHg.
(2) The reaction may alternatively be carried out by boiling the benzaldehyde vigorously under reflux for 2 hours with a solution of 20 g of potassium hydroxide in 90 ml of water, and then working up the cooled reaction mixture as described above. Reaction is normally complete under these conditions; the extent of the reaction may be determined by subjecting the crude ether extract, before the latter is washed with bisulphite solution, to g.l.c. analysis on a 1.5 m column of Chromosorb W with 10 per cent of Carbowax as the stationary phase, at 156 °C with a nitrogen flow rate of 40 ml per minute. The retention times of benzaldehyde and benzyl alcohol are approximately 2.7 and 9.7 minutes respectively.

Cognate preparation. *Furfuryl alcohol and 2-furoic acid.*

Place 200 g (172.5 ml, 2.08 mol) of redistilled furfural (1) in a 1-litre beaker provided with a mechanical stirrer and surrounded by an ice bath. Start the stirrer and, when the temperature has fallen to 5–8 °C, add a solution of 50 g (1.2 mol) of sodium hydroxide in 100 ml of water from a separatory funnel at such a rate that the temperature of the reaction mixture does not rise above 20 °C (20–25 minutes); continue the stirring for a further 1 hour. Much sodium furoate separates during the reaction. Allow to cool to room temperature, and add just enough water to dissolve the precipitate (about 65 ml). Extract the solution at least five times with 60 ml portions of ether in order to remove the furfuryl alcohol: the best results are obtained by the use of the continuous extraction apparatus (charged with 350 ml of ether) depicted in Fig. 2.92. Keep the aqueous layer. Dry the ethereal extract with a little magnesium sulphate, and remove the ether on a rotary evaporator. Distil the residue under reduced pressure (Fig. 2.108) and collect the furfuryl alcohol (a very pale yellow liquid) at 75–77 °C/15 mmHg; the yield is 65 g (64%). Because of the tendency to undergo polymerisation, add about 1 per cent of its weight of urea as stabiliser if the furfuryl alcohol is to be stored.

Treat the aqueous solution, containing the sodium furoate, with 40 per cent sulphuric acid until it is acid to Congo red paper, and cool. Filter off the furoic acid, contaminated with a little sodium hydrogen sulphate, at the pump. Dissolve it in 240 ml of boiling water, add 12 g of decolourising carbon, boil the solution for about 45 minutes, filter hot and cool the filtrate with stirring to 16–20 °C; below 16 °C, sodium hydrogen sulphate also separates. Filter off the furoic acid with suction, and dry. The yield is 65 g (55%), m.p. 123–124 °C. It may be further purified either by recrystallisation from carbon tetrachloride to which a little decolourising carbon is added or by distillation under reduced pressure, b.p. 142–144 °C/20 mmHg; the resulting pure acid softens at 125 °C and is completely melted at 132 °C.

The i.r. spectra of 2-furaldehyde, furfuryl alcohol (thin films) and 2-furoic acid (KBr disc) should be recorded and the absorption bands of the functional groups assigned. The p.m.r. spectra are excellent examples of proton coupling in the 2-substituted furan nucleus. In the case of the alcohol and acid the $J_{3,5}$ coupling constant is negligible and $J_{4,5} < J_{3,4}$. Bearing these facts in mind the signals should be assigned from the following data; 2-furaldehyde (CCl_4, TMS) δ 6.52 (d of d, 1H), 7.11 (d, 1H), 7.61 (d, 1H) and 9.63 (s, 1H); 2-furoic acid (Me_2CO, TMS) δ 6.71 (d of d, 1H), 7.32 (d, 1H), 7.89 (d, 1H) and 10.51 (s, 1H). The lesser electron-withdrawing nature of the hydroxymethyl group in furfuryl alcohol results in the C_3 and the C_4 protons having very similar chemical shift values; in the spectrum (CCl_4, TMS) the signals thus appear at δ 3.32 (s, 1H, disappears on treatment with D_2O), 4.50 (s, 2H), 6.29 (m, 2H), and 7.38 (d, exhibiting small coupling, 1H).

Note. (1) Furfural (2-furaldehyde) is best purified by distillation under reduced pressure: b.p. 54–55 °C/17 mmHg.

Experiment 6.134 *p*-METHYLBENZYL ALCOHOL (*Crossed Cannizzaro reaction*)

$$p\text{-Me}\cdot C_6H_4\cdot CHO \xrightarrow[\ominus OH]{H\cdot CHO} p\text{-Me}\cdot C_6H_4\cdot CH_2OH + H\cdot CO_2^{\ominus}\cdot K^{\oplus}$$

Equip a 1-litre three-necked flask with a reflux condenser, a sealed mechanical stirrer and a thermometer; the bulb of the thermometer should reach almost to the bottom of the flask. Place 170 g of commercial potassium hydroxide pellets (about 85% KOH, *c.* 2.6 mol) and 250 ml of methanol in the flask and set the stirrer in motion. Most of the alkali dissolves in a few minutes and the temperature rises considerably. Immerse the flask in a large cold-water bath and, when the temperature has fallen to 60–65 °C, add down the condenser a mixture of 120 g (118 ml, 1 mol) of *p*-tolualdehyde and 100 ml (*c.* 1.3 mol) of formalin at such a rate (during about 15 minutes) that the internal temperature remains at 60–70 °C: maintain the internal temperature at 60–70 °C for a further 3 hours. Replace the reflux condenser by a condenser set for downward distillation, and distil off the methanol, while stirring, until the temperature reaches about 100 °C. Add 300 ml of water to the warm residue, cool the mixture and separate the resulting two layers at once; if the upper layer is allowed to stand, it will solidify. Extract the aqueous layer with four 50 ml portions of toluene. Wash the combined oil and toluene extracts with five 25 ml portions of water, extract the combined washings with 25 ml of toluene and add the toluene layer to the washed extract. Dry the toluene solution by shaking with a few grams of magnesium sulphate, distil off the toluene and finally distil under reduced pressure (Fig. 2.108) and collect the *p*-methylbenzyl alcohol at 116–118 °C/20 mmHg (1). The product solidifies in the receiver to a mass (110 g) of oily crystals, m.p. 54–55 °C. Recrystallise from an equal weight of technical heptane (b.p. 90–100 °C); 88 g (72%) of pure *p*-methylbenzyl alcohol, m.p. 61 °C, are obtained. The p.m.r. spectrum ($CDCl_3$, TMS) should be recorded and interpreted.

Note. (1) The b.p. at atmospheric pressure is 217 °C.

6.12.2 CLAISEN–SCHMIDT AND RELATED REACTIONS

A retrosynthetic analysis of α,β-unsaturated ketones leading to various methods of synthesis is outlined in Section 5.18.2, p. 798. These methods are equally applicable to aromatic aldehydes. Aromatic aldehydes condense with aliphatic or mixed alkyl aryl ketones in the presence of aqueous alkali to form α,β-unsaturated ketones (the *Claisen–Schmidt reaction*).

The first step is a condensation of the aldol type (see Section 5.18.2, p. 799) involving the nucleophilic addition of the carbanion derived from the methyl ketone to the carbonyl-carbon of the aromatic aldehyde. Dehydration of the hydroxyketone to form the conjugated unsaturated carbonyl compound occurs spontaneously.

Experiment 6.135 describes the preparation of a range of α,β-unsaturated ketones, including benzylideneacetone, furfurylideneacetone and benzylidene-acetophenone. The conversion of this latter compound into β-phenylpropio-phenone is readily achieved by hydrogenation at atmospheric pressure over an active platinum catalyst.

The formation of ω-nitrostyrenes (illustrated in Expt 6.136) by reaction of nitroalkanes with aromatic aldehydes in the presence of aqueous alkali may be classified with reactions of the Claisen–Schmidt type.

A further example of the above reaction type is provided by the condensation between an aromatic aldehyde and an ester (the *Claisen reaction*, e.g. the synthesis of ethyl cinnamate, Expt 6.137), which requires a more powerfully basic catalyst (e.g. sodium ethoxide) to effect conversion of the ester into the corresponding anion.

$$EtO^\ominus \quad H-CH_2 \cdot CO_2Et \rightleftharpoons EtOH + {}^\ominus CH_2 \cdot CO_2Et$$

Experiment 6.135 *A* **BENZYLIDENEACETONE** (*4-Phenylbut-3-en-2-one*)
B **DIBENZYLIDENEACETONE** (*1,5-Diphenylpenta-1,4-dien-3-one*)

A

$$Ph \cdot CHO + Me \cdot CO \cdot Me \xrightarrow{-H_2O} Ph \cdot CH=CH \cdot CO \cdot Me$$

B

$$2Ph \cdot CHO + Me \cdot CO \cdot Me \xrightarrow{-2H_2O} Ph \cdot CH=CH \cdot CO \cdot CH=CH \cdot Ph$$

A. Place 42.5 g (40.5 ml, 0.4 mol) of pure benzaldehyde (Expt 6.133) and 63.5 g (80 ml, 1.1 mol) of pure acetone in a 250-ml flask equipped with a mechanical stirrer. Immerse the reaction vessel in a bath of cold water and add slowly (during about 30 minutes) from a dropping funnel 10 ml of 10 per cent sodium hydroxide solution: adjust the rate of addition so that the temperature remains between 25 and 30 °C. Stir the mixture at room temperature for a further 2 hours; alternatively, securely stopper the flask and shake mechanically for the same period. Render the mixture just acid to litmus paper by the addition of dilute hydrochloric acid. Transfer to a separatory funnel. Remove the upper organic layer, extract the lower aqueous layer with 20 ml of toluene and add the extract to the yellow upper layer. Wash the latter with 20 ml of water, and dry with a little magnesium sulphate, and transfer to a distillation assembly incorporating a Claisen still-head and a short fractionating side-arm (compare Fig. 2.108). Remove the toluene by distillation at atmospheric pressure and distil the residue under diminished pressure. The benzylideneacetone distils at 133–143 °C/16 mmHg (or at 120–130 °C/7 mmHg or at 150–160 °C/25 mmHg) and solidifies to a crystalline mass on standing, m.p. 38–39 °C; the yield is 45 g (77%). This is pure enough for most practical purposes, but may be further purified by redistillation (b.p. 137–142 °C/16 mmHg) or by recrystallisation from light petroleum (b.p. 40–60 °C): the pure benzylideneacetone melts at 42 °C. The residue in the distilling flask contains some dibenzylideneacetone.

The i.r. spectrum of benzylideneacetone should be recorded and the absorption bands assigned appropriately. The p.m.r. spectrum (CDCl$_3$, TMS) shows signals at δ 2.32 (s, 3H, Me), 6.67 (d, 1H, =CH·CO), 7.44 (d, 1H, Ph·CH=) and 7.21–7.62 (m, 5H, C$_{AR}$—H); careful inspection of the coupling constants observable with the signal at δ 6.67 enables the signal at δ 7.44 to be differentiated from those of the aromatic protons.

B. In a 500-ml round-bottomed flask place a cold solution of 25 g of sodium hydroxide in 250 ml of water and 200 ml of ethanol (1); equip the flask with a mechanical stirrer and surround it with a bath of water. Maintain the temperature of the solution at 20–25 °C, stir vigorously and add one-half of a previously prepared mixture of 26.5 g (25.5 ml, 0.25 mol) of pure benzaldehyde and 7.3 g (9.3 ml, 0.125 mol) of acetone. A flocculent precipitate forms in 2–3 minutes. After 15 minutes add the remainder of the benzaldehyde–acetone mixture. Continue the stirring for a further 30 minutes. Filter at the pump and wash with cold water to eliminate the alkali as completely as possible. Dry the solid at room temperature upon filter paper to constant weight; 27 g (93%) of crude dibenzylideneacetone, m.p. 105–107 °C, are obtained. Recrystallise from hot ethyl acetate (2.5 ml per gram) or from hot rectified spirit. The recovery of pure dibenzylideneacetone, m.p. 122 °C, is about 80 per cent.

The p.m.r. spectrum ($CDCl_3$, TMS) should be recorded and the signals assigned, using that of benzylideneacetone as a guide.

Note. (1) Sufficient ethanol is employed to dissolve the benzaldehyde and to retain the initially-formed benzylideneacetone in solution until it has time to react with the second molecule of benzaldehyde.

Cognate preparations. *Furfurylideneacetone*

In a 1-litre bolt-head flask, equipped with a mechanical stirrer, mix 75 g (65 ml, 0.78 mol) of redistilled furfural (1) and 600 ml of water. Add 100 g (126 ml, 1.73 mol) of acetone. Stir the mixture, cool to 10 °C and add a solution of 5 g of sodium hydroxide in 10 ml of water; some heat is generated. Continue the stirring, without cooling, for 4 hours. Then add 10 per cent sulphuric acid (about 70 ml) until the mixture is acid to litmus, whereupon the milkiness disappears and the liquid separates out into layers. Separate the lower organic layer, dry it with a little anhydrous magnesium sulphate and distil under reduced pressure from a flask with fractionating side-arm (compare Fig. 2.108). Collect the furfurylideneacetone at 114–118 °C/10 mmHg; it solidifies on cooling (m.p. 38–39 °C) and weighs 65 g (62%). The residue of high boiling point material in the flask contains much difurfurylideneacetone.

Note. (1) Furfural is best purified by distillation under reduced pressure: b.p. 54–55 °C/17 mmHg.

Benzylideneacetophenone (*Chalcone* or *1,3-diphenylprop-2-en-1-one*)

$$Ph \cdot CHO + Me \cdot CO \cdot Ph \xrightarrow{NaOH} Ph \cdot CH{=}CH \cdot CO \cdot Ph$$

Place a solution of 22 g of sodium hydroxide in 200 ml of water and 100 g (122.5 ml) of rectified spirit in a 500-ml bolt-head flask provided with a mechanical stirrer. Immerse the flask in a bath of crushed ice, pour in 52 g (0.43 mol) of freshly distilled acetophenone, start the stirrer and then add 46 g (44 ml, 0.43 mol) of pure benzaldehyde. Keep the temperature of the mixture at about 25 °C (the limits are 15–30 °C) and stir vigorously until the mixture is so thick that stirring is no longer effective (2–3 hours). Remove the stirrer and leave the reaction mixture in an ice chest or refrigerator overnight. Filter the product with suction on a Buchner funnel or a sintered glass funnel, wash with cold water until the washings are neutral to litmus, and then with 20 ml of ice-cold rectified spirit. The crude chalcone, after drying in the air, weighs 88 g and melts at 50–54 °C. Recrystallise from rectified spirit warmed to 50 °C (about 5 ml per gram). The yield of pure benzylideneacetophenone (a pale yellow solid), m.p. 56–57 °C, is 77 g (85%). This substance should be handled with great care since it acts as a skin irritant.

Hydrogenation to β-phenylpropiophenone.

$$Ph \cdot CH{=}CH \cdot CO \cdot Ph \xrightarrow{H_2/Pt} Ph \cdot CH_2 \cdot CH_2 \cdot CO \cdot Ph$$

Place a solution of 10.4 g (0.05 mol) of benzylideneacetophenone, m.p. 57 °C, in 75 ml of pure ethyl acetate (Section 4.1.*24*, p. 409) in the reaction bottle of the atmospheric pressure hydrogenation apparatus (Section 2.17.1, p. 89) and add 0.2 g of Adams' platinum oxide catalyst (Section 4.2.*61*, p. 459). Displace the air with hydrogen, and shake the mixture with hydrogen until 0.05 mol is absorbed (i.e. *c.* 1100 ml at s.t.p.). Filter off the platinum, and remove the ethyl acetate by distillation. Recrystallise the residual *β*-phenylpropiophenone from about 12 ml of ethanol. The yield of pure product, m.p. 73 °C, is 9 g (86%).

Experiment 6.136   *ω*-NITROSTYRENE

$$\text{Ph·CHO} + \text{Me·NO}_2 \xrightarrow{-\text{H}_2\text{O}} \text{Ph·CH}=\text{CH·NO}_2$$

Equip a 1500-ml three-necked flask with a thermometer, mechanical stirrer and a dropping funnel. Place 61 g (54 ml, 1 mol) of nitromethane (1), 106 g (101 ml, 1 mol) of purified benzaldehyde (Expt 6.133) and 200 ml of methanol in the flask and cool it with a mixture of ice and salt to about −10 °C. Dissolve 42 g of sodium hydroxide in 40–50 ml of water, cool and dilute to 100 ml with ice and water; place this cold solution in the dropping funnel. Add the sodium hydroxide solution, with vigorous stirring, to the nitromethane mixture at such a rate that the temperature is held at 10–15 °C. Introduce the first few ml cautiously since, after a short induction period, the temperature may rise to 30 °C or higher; check the rise in temperature, if necessary, by adding a little crushed ice to the reaction mixture. A bulky white precipitate forms; if the mixture becomes so thick that stirring is difficult, add about 10 ml of methanol. After standing for about 15 minutes, add 700 ml of ice-water containing crushed ice; the temperature should be below 5 °C. Run the resulting cold solution immediately from a dropping funnel and with stirring into 500 ml of 4 M hydrochloric acid contained in a 3-litre flask; adjust the rate of addition so that the stream just fails to break into drops. A pale yellow crystalline precipitate separates almost as soon as the alkaline solution mixes with the acid. The solid settles to the bottom of the vessel when the stirrer is stopped. Decant most of the cloudy liquid layer, filter the residue by suction and wash it with water until free from chlorides. Transfer the solid to a beaker immersed in hot water; two layers form and on cooling again, the lower layer of nitrostyrene solidifies; pour off the upper water layer. Dissolve the crude nitrostyrene in 85 ml of hot ethanol. (**CAUTION**: *nitrostyrene vapours are irritating to the nose and eyes, and the skin of the face is sensitive to the solid.*) Filter through a hotwater funnel and cool until crystallisation is complete. The yield of pure *ω*-nitrostyrene, m.p. 57–58 °C, is 125 g (85%).

Note. (1) The commercial material may be redistilled and the fraction having b.p. 100–102 °C collected.

Cognate preparations.   *3,4-Methylenedioxy-ω-nitrostyrene.* In a 250-ml round-bottomed flask mix 30 g (0.20 mol) of 3,4-methylenedioxybenzaldehyde (*piperonal*), 13.4 g (0.22 mol) of nitromethane, 7.8 g (0.1 mol) of ammonium acetate and 50 ml of glacial acid. Attach a reflux condenser, and boil the mixture under gentle reflux for 1 hour. Pour the reaction mixture with stirring into a large excess of ice-water (about 1 litre). When all the ice has melted, filter off the crude product under suction and recrystallise from a mixture of absolute ethanol and acetone (about 2:1 v/v). Almost pure yellow crystals of the nitrostyrene, m.p. 161 °C, are obtained. The yield is 23.3 g (60%). Further recrystallisation from the same solvent yields the pure compound, m.p. 162 °C.

2,4-Dimethoxy-ω-nitrostyrene. Follow the above procedure, but use 33.2 g (0.20 mol) of 2,4-dimethoxybenzaldehyde as the starting material. The yield of recrystallised product (yellow crystals, m.p. 103 °C) is 28.5 g (68%). Further recrystallisation gives pure product of m.p. 105 °C.

Experiment 6.137 ETHYL CINNAMATE

$$Ph \cdot CHO + Me \cdot CO_2Et \xrightarrow[\text{EtOH}]{\text{Na}} Ph \cdot CH{=}CH \cdot CO_2Et + H_2O$$

Prepare powdered (or 'molecular') sodium from 14.5 g (0.63 mol) of clean sodium and 150–200 ml of sodium-dried xylene contained in a 1-litre three-necked flask (Section 4.2.6, p. 417) fitted with a mechanical stirrer and a reflux condenser. When cold, pour off the xylene as completely as possible, and then add 220 g (240 ml, 2.5 mol) ethyl acetate (Section 4.1.24, p. 409) containing 2 ml of absolute ethanol (1). Cool the flask rapidly to 0 °C and add 53 g (51 ml, 0.5 mol) of pure benzaldehyde (Expt 6.133) slowly (during 90 minutes) from a dropping funnel while the mixture is stirred. Keep the temperature between 0 and 5 °C; do not allow it to rise above 10 °C otherwise a poor yield will be obtained. The reaction commences as soon as the benzaldehyde is added, as is indicated by the production of a reddish substance on the particles of sodium. Continue the stirring until practically all the sodium has reacted (about 1 hour after all the benzaldehyde has been introduced). Then add 45 ml of glacial acetic acid followed by an equal volume of water (**CAUTION**: *some sodium may be present*). Separate the layer of ester, extract the aqueous layer with 25 ml of ethyl acetate, wash the combined organic layers with 150 ml of 1:1 hydrochloric acid and dry with magnesium sulphate or sodium sulphate. Distil off the ethyl acetate on a water bath. Distil the residue under diminished pressure (Fig. 2.108). Collect the ethyl cinnamate (a colourless liquid) at 126–131 °C/6 mmHg; the yield is 65 g (74%) (2).

The p.m.r. spectrum (CDCl₃) should be recorded and the signals which are apparent at δ 1.30 (t, 3H), 4.80 (q, 2H), 6.33 (d, 1H), 7.73 (d, 1H), and 7.2–7.6 (m, 5H) assigned.

Notes. (1) A little ethanol (*c.* 1%) is required to start the reaction; the yield is consistently lower in its absence.
(2) Ethyl cinnamate may also be prepared by the esterification of cinnamic acid (cf. *methyl cinnamate*, Expt 6.163). The pure compound boils at 127 °C/6 mmHg.

6.12.3 THE PERKIN AND DOEBNER REACTIONS

A retrosynthetic analysis of α,β-unsaturated acids leading to various methods of synthesis is outlined in Section 5.18.3, p. 804. The following methods are particularly applicable to aromatic aldehydes.

The condensation of an aromatic aldehyde with an acid anhydride in the presence of the sodium or potassium salt of the acid corresponding to the anhydride to yield an α,β-unsaturated acid is known as the *Perkin reaction*.

$$\text{Ar·CHO} + (\text{Me·CO})_2\text{O} \xrightarrow{\text{Me·CO}_2^{\ominus}\text{Na}^{\oplus}} \text{Ar·CH=CH·CO}_2\text{H} + \text{Me·CO}_2\text{H}$$

The mechanism of the reaction, which is of the aldol type, involves the carbonyl group of the aldehyde and an active methylene group of the anhydride; the function of the basic catalyst (acetate anion, $\text{Me·CO}_2^{\ominus}$, or triethylamine, Et_3N) is to form an anion by removal of a proton from the anhydride:

$$\text{B: } + \text{Me·CO·O·CO·Me} \rightleftharpoons \overset{\oplus}{\text{B}}\text{H} + \overset{\ominus}{\text{C}}\text{H}_2\text{·CO·O·CO·Me}$$

The standard procedure is illustrated by the preparation of cinnamic acid and furylacrylic acid (Expt 6.138). The cinnamic acid obtained is the more stable (E)-isomer. It may be readily reduced to the saturated acid (3-phenylpropanoic acid) and two procedures are described. Catalytic hydrogenation is a convenient method, but the conjugated double bond may also be reduced with, for example, sodium amalgam in the presence of alkali.

A modified Perkin procedure is illustrated in the synthesis of α-phenylcinnamic acid in which benzaldehyde is condensed with phenylacetic acid in the presence of acetic anhydride and triethylamine. Presumably equilibria are set up which result in the formation of either a mixed anhydride, phenylacetic acetic anhydride or the symmetrical phenylacetic anhydride.

$$\text{Ph·CH}_2\text{·CO}_2\text{H} + (\text{Me·CO})_2\text{O} \rightleftharpoons \text{Ph·CH}_2\text{·CO·O·CO·Me} + \text{Me·CO}_2\text{H}$$

$$2\text{Ph·CH}_2\text{·CO}_2\text{H} + (\text{Me·CO})_2\text{O} \rightleftharpoons (\text{Ph·CH}_2\text{·CO})_2\text{O} + 2\text{Me·CO}_2\text{H}$$

Coumarin is formed from acetic anhydride and salicylaldehyde in the presence of triethylamine as the base catalyst. It is the lactone of the (Z)-form of o-hydroxycinnamic acid; some of the (E)-isomer in the form of its acetyl derivative (o-acetoxycinnamic acid) is also obtained (Expt 6.138).

Arylacrylic acids may alternatively be conveniently prepared by the *Doebner modification* of the *Knoevenagel reaction*. The Knoevenagel reaction embraces a number of base-catalysed condensations between a carbonyl compound and a component having an active methylene group (see also Section 5.11.6, p. 681). Examples of the Doebner modification, which usually involves the reaction of an aldehyde with malonic acid in the presence of pyridine or possibly a little piperidine, are given in Expt 6.139. The reaction mechanism is of the aldol type and involves the formation of a hydroxymalonic acid and then an α,β-unsaturated malonic acid, which undergoes decarboxylation at the temperature of refluxing pyridine.

$$Ar \cdot CHO + CH_2(CO_2H)_2 \xrightarrow{base} [Ar \cdot CH(OH) \cdot CH(CO_2H)_2] \xrightarrow{-H_2O}$$

$$Ar \cdot CH{=}C(CO_2H)_2 \xrightarrow[-CO_2]{heat} Ar \cdot CH{=}CH \cdot CO_2H$$

Experiment 6.138   CINNAMIC ACID

$$Ph \cdot CHO + (Me \cdot CO)_2O \xrightarrow[\text{(ii) } H_3O^{\oplus}]{\text{(i) base}} Ph \cdot CH{=}CH \cdot CO_2H + Me \cdot CO_2H$$

Place 21 g (20 ml, 0.2 mol) of pure benzaldehyde (1), 30 g (28 ml, 0.29 mol) of acetic anhydride and 12 g (0.122 mol) of freshly fused and finely powdered potassium acetate (2) in a dry, 250-ml round-bottomed flask fitted with an air condenser carrying a calcium chloride guard-tube. Mix well and heat the reaction mixture in an oil bath at 160 °C for 1 hour and at 170–180 °C for 3 hours. Pour the mixture while still hot (80–100 °C) into about 100 ml of water contained in a 1-litre round-bottomed flask which has previously been fitted for steam distillation (Fig. 2.102); rinse the reaction flask with a little hot water. Now add with vigorous shaking a saturated aqueous solution of sodium carbonate (3) until a drop of the liquid withdrawn on the end of a glass rod turns red litmus a distinct blue. Steam distil the solution until all the unchanged benzaldehyde is removed and the distillate is clear. Cool the residual solution and filter at the pump from resinous by-products. Acidify the filtrate by adding concentrated hydrochloric acid slowly and with vigorous stirring until the evolution of carbon dioxide ceases. When cold, filter the cinnamic acid at the pump, wash with cold water and drain well. Recrystallise either from hot water or from a mixture of 3 volumes of water and 1 volume of rectified spirit. The yield of dry cinnamic acid (colourless crystals), m.p. 133 °C, is 18 g (62%).

The p.m.r. spectrum (CDCl$_3$, TMS) shows signals at δ 6.41 (d, 1H, $=$CH$\cdot$CO$_2$), 7.73 (d, 1H, Ph$\cdot$CH), 7.17–7.69 (m, 5H, C$_{AR}$—H) and 11.90 (s, 1H, CO$_2$H). The m.s. shows principal fragment ions at m/z 148 (M), 147 (M − H), 131 (M − OH), 130 (M − H$_2$O), 103 (M − CO$_2$H), 102 (130 − CO), 77 (103 − C$_2$H$_2$) and 51 (77 − C$_2$H$_2$).

Notes. (1) The benzaldehyde must be free from benzoic acid; it may be purified as detailed in Expt 6.133.
(2) Fused potassium acetate should be freshly prepared following the procedure described for sodium acetate (Section 4.2.69, p. 464). It may, however, be replaced by an equivalent quantity of freshly fused sodium acetate, but the reaction is slower and a further 3–4 hours heating is necessary.
(3) Sodium hydroxide solution cannot be used at this stage since it may produce benzoic acid by the Cannizzaro reaction (Expt 6.133) from any unchanged benzaldehyde. If, however, the reaction mixture is diluted with 3–4 volumes of water, steam distilled to remove the unreacted benzaldehyde, the residue may then be rendered alkaline with sodium hydroxide solution. A few grams of decolourising carbon are added, the mixture boiled for several minutes, and filtered through a fluted filter paper. Upon acidifying carefully with concentrated hydrochloric acid, cinnamic acid is precipitated. This is collected, washed and purified as above.

Reduction of cinnamic acid to 3-phenylpropanoic acid (*hydrocinnamic acid*).
Method A. Carry out the hydrogenation of 14.8 g (0.1 mol) of pure cinnamic acid dissolved in 100 ml of ethanol using 0.1 g of Adams' catalyst (Section 4.2.61, p. 459) according to the procedure detailed in Section 2.17.1, p. 89

until hydrogen uptake ceases. Record the volume of hydrogen required, filter off the platinum and evaporate the filtrate on a rotary evaporator. The resulting oil solidifies on cooling to a colourless solid, m.p. 47–48 °C; the yield is 14.3 g (95%). Upon recrystallisation from light petroleum (b.p. 60–80 °C) pure hydrocinnamic acid, m.p. 48–49 °C, is obtained.

Method B.   Dissolve 20 g (0.135 mol) of cinnamic acid in 145 ml of approximately 1 M sodium hydroxide solution contained in a 500-ml two-necked flask equipped with a mechanical stirrer and situated within a fume cupboard. Add 350 g (0.38 mol) of 2.5 per cent sodium amalgam (Section 4.2.70, p. 464) gradually during 1 hour through the open side-neck while the mixture is well stirred. When hydrogen is no longer evolved, separate the mercury and wash it with water: add the washings to the solution and acidify the whole with dilute hydrochloric acid (1:1). Hydrocinnamic acid is precipitated, at first in the form of an oil, which solidifies on cooling and rubbing with a glass rod. Filter at the pump and recrystallise as in *Method A*. The yield of hydrocinnamic acid, m.p. 46–48 °C, is 17 g (85%). The p.m.r. spectrum should be recorded and the signals assigned.

Cognate preparations.   *Furylacrylic acid*

Place 48 g (41.5 ml, 0.5 mol) of freshly distilled furfural (see Note (1) to cognate preparation in Expt 6.133), 77 g (71 ml, 0.75 mol) of pure acetic anhydride and 49 g (0.5 mol) of dry, powdered, freshly fused potassium acetate in a 500-ml two- or three-necked flask, provided with a mechanical stirrer and a long air condenser. Heat the flask, with stirring, in an oil bath at 150 °C (bath temperature) for 4 hours: when the temperature approaches 145–150 °C, a vigorous exothermic reaction sets in and must be controlled by the application of cold wet towels (or cloths) to the flask in order to avoid too vigorous boiling. Allow to cool slightly, transfer the reaction mixture to a 1-litre round-bottomed flask and add 600 ml of water: use part of this to rinse out the reaction flask. Boil the mixture with 6 g of decolourising charcoal for 10 minutes, and filter hot through a preheated Buchner funnel into a preheated filter flask. Transfer the hot filtrate to a beaker, add dilute hydrochloric acid (1:1) until it is acid to Congo red paper and cool to about 10 °C with stirring. Allow to stand for at least 1 hour, filter at the pump and wash with a little ice-water. The yield of crude furylacrylic acid (a light tan solid), m.p. 138–139 °C, is 41 g (59%). A perfectly pure acid (white solid), m.p. 140 °C, is obtained by recrystallisation from benzene or light petroleum, b.p. 80–100 °C, with the addition of a little decolourising carbon; the loss is about 20 per cent.

α-Phenylcinnamic acid

$$Ph \cdot CH_2 \cdot CO_2H + (Me \cdot CO)_2O \rightleftharpoons (Ph \cdot CH_2 \cdot CO)_2O + Me \cdot CO_2H$$

$$Ph \cdot CHO + (Ph \cdot CH_2 \cdot CO)_2O \xrightarrow{Et_3N} Ph \cdot CH = C(Ph) \cdot CO_2H$$

Place 42.5 g (40.5 ml, 0.4 mol) of purified benzaldehyde (Expt 6.133), 54.5 g (0.4 mol) of phenylacetic acid, 80 ml (0.83 mol) of redistilled acetic anhydride and 40 ml of anhydrous triethylamine in a 500-ml round-bottomed flask fitted with a reflux condenser and drying tube. Boil the mixture gently for 5 hours. Steam distil the mixture directly from the reaction flask until the distillate passing over is no longer cloudy, and collect a further 50 ml of distillate; discard the distillate. Cool the residue in the flask and decant the solution from the solid; make up the volume of the solution to 500 ml with water (A). Dissolve the solid in 500 ml of hot 95 per cent ethanol, add the solution (A) followed by 2 g of decolourising carbon; heat the mixture to boiling, filter and acidify the filtrate immediately to Congo red with 1:1-hydrochloric acid. Cool. Collect the separated crystals by suction filtration and recrystallise from 60 per cent ethanol. The yield of α-phenylcinnamic acid (1), m.p. 172–173 °C, is 55 g (61%).

Note. (1) The product is the isomer with the two phenyl groups *cis* to each other, since decarboxylation with quinoline–copper–chromium oxide at 210–220 °C yields (Z)-stilbene.

Coumarin

In a 250-ml round-bottomed flask, provided with a small reflux condenser and a calcium chloride drying tube at the top, place 2.1 g (0.17 mol) of salicylaldehyde, 2.0 ml of anhydrous triethylamine and 5.0 ml (0.052 mol) of acetic anhydride, and reflux the mixture gently for 12 hours. Steam distil the mixture from the reaction flask and discard the distillate. Render the residue in the flask basic to litmus with solid sodium hydrogen carbonate, cool, filter the precipitated crude coumarin at the pump and wash it with a little cold water. Acidify the filtrate to Congo red with 1:1-hydrochloric acid, collect the precipitated o-acetoxycinnamic acid and recrystallise it from 70 per cent propan-2-ol; the yield is 0.40 g (11%), m.p. 153–154 °C.

Boil the crude coumarin with 200 ml of water to which 0.2 g of decolourising carbon is added, filter the hot solution and concentrate it to a volume of 80 ml. Cool, collect the coumarin which separates and recrystallise it from 40 per cent aqueous methanol. The yield of coumarin, m.p. 68–69 °C, is 1.0 g (40%). The p.m.r. spectrum ($CDCl_3$, TMS) shows signals at δ 6.33 (d, 1H, $=CH \cdot CO$), 7.02–7.52 (m, 4H, C_{AR}—H) and 7.65 (d, 1H, Ar·CH$=$).

Experiment 6.139   3,4-METHYLENEDIOXYCINNAMIC ACID (β-*Piperonylacrylic acid*)

Dissolve 50 g (0.33 mol) of piperonal and 75 g (0.72 mol) of malonic acid (1) in a mixture of 150 ml of pyridine (**CAUTION**) (2) and 2.5 ml of piperidine contained in a 500-ml round-bottomed flask, and heat under reflux for 1 hour on a water bath. A rapid evolution of carbon dioxide takes place. Complete the reaction by boiling the solution for 5 minutes. Cool, pour into excess of water containing enough hydrochloric acid to combine with the pyridine, filter off the piperonylacrylic acid, wash with a little water, and dry. The yield (64 g) is almost quantitative and the acid is practically pure. It may be recrystallised from glacial acetic acid; m.p. 238 °C.

Record the p.m.r. spectrum (DMSO-d_6, TMS) and assign the signals which appear at δ 6.10 (s, 2H), 6.41 (d, 1H), 6.96 (d, 1H), 7.19 (d of d, 1H), 7.36 (d, 1H) and 7.58 (d, 1H); note that the acidic proton does not appear in the spectrum, and that the $J_{2,5}$ value is zero.

Cognate preparations. p-*Methylcinnamic acid.* From *p*-tolualdehyde; heat for 6 hours. Recrystallise from glacial acetic acid; m.p. 198 °C. Yield: 87 per cent.

m-*Nitrocinnamic acid.* From *m*-nitrobenzaldehyde. Recrystallise from alcohol; m.p. 197 °C. Yield: 80 per cent.

p-*Methoxycinnamic acid.* From anisaldehyde. Recrystallise from alcohol; m.p. 172 °C. Yield: 80 per cent.

Furylacrylic acid. Place 48 g (41.5 ml, 0.5 mol) of freshly distilled furfural, 52 g (0.5 mol) of dry malonic acid (1) and 24 ml (0.31 mol) of dry pyridine (**CAUTION**) (2) in a 500-ml round-bottomed flask, fitted with a reflux condenser. Heat the flask on a boiling water bath for 2 hours, cool the reaction mixture and dilute with 50 ml of water. Dissolve the acid by the addition of concentrated ammonia solution, filter the solution and wash the filter paper with a little water. Add dilute hydrochloric acid (1:1), with stirring, to the combined filtrate and washings until acid to Congo red paper, and cool in an ice bath for at least 1 hour. Filter the furylacrylic acid and wash it with a little ice-water; it weights 63 g (91%) after drying and melts at 139–140 °C. A purer acid may be obtained by recrystallisation from light petroleum (b.p. 80–100 °C), with the addition of a little decolourising carbon; the loss is about 20 per cent.

Notes. (1) Commercial malonic acid is dried at 90–100 °C for 2 hours.
(2) The pyridine is dried by allowing it to stand, with frequent shaking, over potassium hydroxide pellets and then filtering.

6.12.4 THE SYNTHESIS OF DIPHENYLPOLYENES

1,8-Diphenylocta-1,3,5,7-tetraene can be prepared (Expt 6.140) by condensing two mols of cinnamaldehyde with succinic acid in the presence of acetic anhydride and lead oxide, in a reaction which bears similarities with those of the Perkin type discussed above.

1,4-Diphenylbuta-1,3-diene (Section Expt 6.141) is prepared by a variant of the general method in which cinnamaldehyde is similarly condensed with phenylacetic acid.

The bathochromic shift which results from the increasing length of the conjugated system is evident from the fact that the substituted butadiene is colourless (λ_{max} 344 nm), while the substituted octatetraene is bright yellow (λ_{max} 402 nm); see also p. 388.

Experiment 6.140 1,8-DIPHENYLOCTA-1,3,5,7-TETRAENE

Heat a mixture of 13.2 g (11.9 ml, 0.1 mol) of cinnamaldehyde, 5.9 g (0.05 mol) of succinic acid, 11.2 g of lead oxide and 14.3 ml of acetic anhydride to 140 °C for 10 minutes with frequent shaking in a flask fitted with a reflux condenser. Boil the resulting clear solution under reflux for 2 hours; some of the tetraene crystallises at this stage. Cool the solution to 40 °C and filter the solid rapidly using a large Buchner funnel. Boil the filtrate under reflux with a further 8 ml portion of acetic anhydride for 2 hours to obtain a second crop of product on cooling and filtering. Combine the crystalline material and wash with first a little acetic anhydride and then with a little glacial acetic acid to remove brown resins. Wash the tetraene with alcohol and finally water to obtain 1.9 g (15%) of fairly pure product. Recrystallisation from chloroform gives a specimen as yellow plates, m.p. 232 °C.

Experiment 6.141 1,4-DIPHENYLBUTA-1,3-DIENE

In a 100-ml round-bottomed flask fitted with a reflux condenser place 10 g (0.07 mol) of phenylacetic acid, 10 g (9 ml, 0.075 mol) of redistilled cinnamaldehyde, 10 ml of acetic anhydride and 8.5 g of lead oxide. Heat the mixture slowly to boiling with intermittent shaking so that a clear solution is obtained, and then boil under reflux for 5 hours. Pour the hot solution into a beaker and set aside for 12 hours. Filter the semi-solid product under suction and wash the filter-cake with two 5 ml portions of ethanol, stirring the solid thoroughly with the wash liquid before applying suction. Transfer the solid to a small beaker and triturate with 8 ml of ethanol, refilter, suck dry and repeat the trituration procedure with a further 8-ml portion of ethanol. Dissolve the filtered solid in 20 ml of hot toluene, treat with 1 g of charcoal and filter the hot solution (*fume cupboard*). Add 35 ml of hot ethanol to the filtrate, boil and cool the solution in ice with shaking. Filter the purified diene, wash with 5 ml of cold ethanol and dry at 50 °C; the final product has m.p. 153 °C. The yield is 3.5 g (23%).

6.12.5 THE BENZOIN CONDENSATION

Aromatic aldehydes when treated with an alkali metal cyanide, usually in aqueous solution, undergo condensation to the α-hydroxyketone or benzoin. The examples in Expt 6.142 are benzoin and furoin.

$$2Ar \cdot CHO \xrightarrow{\text{NaCN aq.}} Ar \cdot CH(OH) \cdot CO \cdot Ar$$

By use of 1 mol each of two different aldehydes, an unsymmetrical or mixed benzoin is obtained (for example, the formation of 4-methoxybenzoin).

$$Ar^1 \cdot CHO + Ar^2 \cdot CHO \xrightarrow[\text{KCN aq.}]{\text{NaCN or}} Ar^1 \cdot CH(OH) \cdot CO \cdot Ar^2$$

The reaction depends upon the catalytic influence of the cyanide ion and the mechanism may be represented in the following way.

Oxidation of the α-hydroxyketone with concentrated nitric acid, or by catalytic amounts of copper(II) salts in acetic acid solution which are regenerated continuously by ammonium nitrate, yields the diketone (e.g. benzil and furil, Expt 6.143).

$$Ar \cdot CH(OH) \cdot CO \cdot Ar \xrightarrow{[O]} Ar \cdot CO \cdot CO \cdot Ar$$

α-Diketones (Ar·CO·CO·Ar) upon refluxing with aqueous–alcoholic potassium hydroxide undergo the *benzilic acid rearrangement* and are converted into the salt of a benzilic acid (Expt 6.144).

$$Ar \cdot CO \cdot CO \cdot Ar + KOH \longrightarrow (Ar)_2 C(OH) CO_2^{\ominus} K^{\oplus}$$

The mechanism involves nucleophilic attack of the hydroxide ion at a carbonyl carbon atom to yield the oxyanion (1) which undergoes a 1,2-nucleophilic shift of an aryl group as shown. Proton transfer completes the reaction sequence.

$$\underset{O}{\overset{O}{\underset{Ar}{\bigwedge}}}\overset{\ominus OH}{\underset{Ar}{}} \rightleftharpoons \underset{\ominus O}{\overset{O^{\ominus}}{\underset{Ar}{\bigwedge}}}\overset{OH}{\underset{Ar}{}} \rightleftharpoons \underset{O^{\ominus}}{\overset{Ar\ \ O}{\underset{Ar}{\bigwedge}}}\overset{}{\underset{OH}{}} \rightleftharpoons \underset{HO}{\overset{Ar\ \ O}{\underset{Ar}{\bigwedge}}}\overset{}{\underset{O^{\ominus}}{}}$$

(1)

Direct conversion of a benzoin into the corresponding benzilic acid may be accomplished conveniently and in good yield by reaction with alkaline bromate solution at 85–90 °C (see Expt 6.144, *Method 2*).

Experiment 6.142 BENZOIN

$$2\text{Ph·CHO} \xrightarrow[\text{EtOH/H}_2\text{O}]{\text{NaCN}} \text{Ph·CH(OH)·CO·Ph}$$

In a 500-ml round-bottomed flask place 65 ml of rectified spirit, 50 g (47.5 ml, 0.47 mol) of pure benzaldehyde (1) and a solution of 5 g of sodium cyanide (96–98%) [**CAUTION**: *see Expt 5.157, Note (1)*] in 50 ml of water. Attach a reflux condenser (preferably of the double surface type) and boil the mixture gently for half an hour (2). Cool the contents of the flask (preferably in an ice bath). Filter the crude benzoin, wash it with cold water, drain well (3) and dry. The yield of crude benzoin, which is white or pale yellow in colour, is 45 g (90%).

Recrystallise 5.0 g from about 40 ml of hot rectified (or industrial) spirit; upon cooling, 4.5 g of pure benzoin (a white, crystalline solid, m.p. 137 °C) separates. Reserve the remainder of the preparation for benzil and benzilic acid (Expts 6.143 and 6.144 respectively).

Notes. (1) For the purification of commercial benzaldehyde, see Expt 6.133.
(2) The reaction sometimes takes place with considerable violence and material may be lost through the condenser unless a large flask (e.g. at least of the size given) is employed.
(3) The filtrate contains sodium cyanide, and should be washed down the sink with a liberal quantity of water; see also Expt 5.157, Note (1).

Cognate preparations. *Furoin*

$$2\ \underset{O}{\overset{}{\bigcirc}}\overset{}{\text{CHO}} \xrightarrow{\ominus\text{CN}} \underset{O}{\overset{}{\bigcirc}}\overset{}{\text{CH(OH)·CO}}\underset{O}{\overset{}{\bigcirc}}$$

In a 1-litre three-necked flask, equipped with a mechanical stirrer, a reflux condenser and a separatory funnel, place 400 ml of water, 200 g (172.5 ml, 2.08 mol) of freshly distilled furfural (see Expt 6.133) and 150 ml of rectified spirit. Heat the reaction mixture to boiling, remove the source of heat and, when the liquid has just ceased to boil, add with stirring a solution of 10 g of potassium cyanide (**CAUTION**) in 30 ml of water from the separatory funnel as rapidly as the vigour of the reaction permits. When the ebullition subsides (exothermic reaction), heat to boiling for a further 5 minutes. Acidify the reaction mixture with glacial acetic acid (use litmus paper) and allow to cool overnight, preferably in an ice chest or a refrigerator. Filter off the dark crystals at the pump, wash with cold water and then with cold methanol to remove as much of the tar (colouring matter) as possible. Recrystallise from methanol with the addition of about 10 g of decolourising carbon. The yield of furoin, m.p. 135–136 °C, is 75 g (37.5%). If the m.p. is slightly low, another recrystallisation from toluene–ethanol will give satisfactory results.

4-Methoxybenzoin. Dissolve 25 g of potassium cyanide (**CAUTION**) in 175 ml of water in a 1500-ml round-bottomed flask, and add 136 g (121.5 ml, 1 mol) of redistilled *p*-methoxybenzaldehyde (anisaldehyde), 108 g (103 ml, 1.02 mol) of redistilled benzaldehyde and 350 ml of 95 per cent ethanol. Reflux the mixture (which becomes homogeneous at the boiling temperature) for 90 minutes. Remove all the unreacted aldehydes and the ethanol by steam distillation. Decant the water and set the residue aside to crystallise. Press the product as free as possible from oily material on a suction funnel and wash it with a little ethanol. Recrystallise the crude product (*c*. 125 g) by dissolving it in hot ethanol and allowing to crystallise slowly. The *p*-methoxybenzoin separates out first in large clumps of long needles, while the little benzoin present crystallises in small compact balls of needles. With a little experience it is possible to filter off a good yield of the former before the appearance of the benzoin. The yield of 4-methoxybenzoin is about 55 g (23%). Recrystallise again until the m.p. is 105–106 °C.

Experiment 6.143 BENZIL

$$Ph \cdot CH(OH) \cdot CO \cdot Ph \xrightarrow{[O]} Ph \cdot CO \cdot CO \cdot Ph$$

Method 1. Place 20 g (0.094 mol) of crude benzoin (Expt 6.142) and 100 ml of concentrated nitric acid in a 250-ml round-bottomed flask. Heat on a boiling water bath (in the fume cupboard) with occasional shaking until the evolution of oxides of nitrogen has ceased (about 1.5 hours). Pour the reaction mixture into 300–400 ml of cold water contained in a beaker, stir well until the oil crystallises completely as a yellow solid. Filter the crude benzil at the pump, and wash it thoroughly with water to remove the nitric acid. Recrystallise from ethanol or rectified spirit (about 2.5 ml per gram). The yield of pure benzil, m.p. 94–96 °C, is 19 g.

Method 2. Place 0.2 g of copper(II) acetate, 10 g (0.125 mol) of ammonium nitrate, 21.2 g (0.1 mol) of benzoin and 70 ml of an 80 per cent v/v aqueous acetic acid solution in a 250-ml flask fitted with a reflux condenser. Heat the mixture with occasional shaking (1). When solution occurs, a vigorous evolution of nitrogen is observed. Reflux for 90 minutes, cool the solution, seed the solution with a crystal of benzil (2) and allow to stand for 1 hour. Filter at the pump and keep the mother-liquor (3): wash well with water and dry (preferably in an oven at 60 °C). The yield of benzil, m.p. 94–95 °C, is 19 g (90%); the m.p. is unaffected by recrystallisation from alcohol or from carbon tetrachloride (2 ml per gram). Dilution of the mother-liquor with the aqueous washings gives a further 1.0 g of benzil.

Notes. (1) For large-scale preparations use a three-necked flask equipped with two reflux condensers and a sealed mechanical stirrer.
(2) Stirring or vigorous shaking also induces crystallisation.
(3) The mother-liquor should not be concentrated as an explosion may result.

Cognate preparation. *Furil.* Proceed exactly as for *Method 2*, using 19.2 g (0.1 mol) of furoin (Expt 6.142) but use 250 ml of the aqueous acetic acid. The yield of furil, yellow needles of m.p. 165–166 °C, after recrystallisation from methanol is 17 g (89%).

Experiment 6.144 BENZILIC ACID

$$Ph \cdot CO \cdot CO \cdot Ph \xrightarrow[\text{Method 1}]{\ominus OH} Ph \cdot \underset{\underset{\text{CO}_2\text{H}}{|}}{C}(OH) \cdot Ph \xleftarrow[\text{Method 2}]{[O]; \ominus OH} Ph \cdot CH(OH) \cdot CO \cdot Ph$$

Method 1. In a 500-ml round-bottomed flask, place a solution of 35 g of potassium hydroxide in 70 ml of water, then add 90 ml of rectified spirit and 35 g (0.167 mol) of recrystallised benzil (Expt 6.143). A deep bluish-black solution is produced. Fit a reflux condenser to the flask and heat the mixture on a boiling water bath for 10–15 minutes. Pour the contents of the flask into a porcelain dish and allow to cool, preferably overnight. The potassium salt of benzilic acid crystallises out. Filter off the crystals at the pump and wash with a little ice-cold alcohol. Dissolve the potassium salt in about 350 ml of water, and add 1 ml of concentrated hydrochloric acid from a burette slowly and with stirring. The precipitate thus produced is coloured red-brown and is somewhat sticky. Filter this off; the filtrate should be nearly colourless. Continue the addition of hydrochloric acid with stirring until the solution is acid to Congo red paper. Filter off the benzilic acid with suction, wash it thoroughly with cold water until free from chlorides and allow to dry. The yield of crude benzilic acid, which is usually light pink or yellow in colour, is 30 g (79%). Purify the product either by recrystallisation from hot benzene (about 6 ml per gram) or from hot water with the use of a little decolourising carbon. The coloured and sticky material obtained by the first precipitation may be recrystallised from hot water with the addition of a little decolourising carbon, and a further 1–2 g obtained. Pure benzilic acid has m.p. 150 °C.

Method 2. Prepare a solution of 50 g of sodium hydroxide and 11.5 g of sodium bromate (or 12.5 g of potassium bromate) in 90 ml of water in an evaporating dish. Add 42 g (0.2 mol) of benzoin (1) in portions to this solution while stirring (preferably with a mechanical stirrer) and heating on a water bath at 85–90 °C (2). Add small quantities of water from time to time to prevent the mixture becoming too thick; about 80 ml of water are required. Continue the heating and stirring until a test portion is completely or almost completely soluble in water; this usually requires 3–4 hours. Dilute the mixture with 400 ml of water and allow to stand, preferably overnight. Filter off the solid or oil impurity (benzhydrol). Set aside 5 ml of the filtrate (3) and to the bulk add dilute sulphuric acid (4) slowly and with stirring to a point just short of the liberation of bromine; about 130 ml are required. If the end-point is overstepped, add the 5 ml of the filtrate which was set aside and then sufficient sulphuric acid to the end-point. Filter off the product at the pump, wash it well with water and dry. The benzilic acid weighs 39 g (85%) and has a m.p. of 149–150 °C, i.e. is practically pure. If desired, it may be recrystallised from benzene, or from water.

Notes. (1) The crude benzoin (Expt 6.142) gives satisfactory results.
(2) The reaction mixture should not be heated to boiling since this leads to the formation of much benzhydrol. The temperature attained by heating on a boiling water bath is 85–90 °C.

(3) This precaution is generally unnecessary if the addition of sulphuric acid is made carefully.

(4) Prepared by adding 1 volume of concentrated sulphuric acid to 3 volumes of water.

6.12.6 OXIME FORMATION

Benzaldehyde reacts with hydroxylamine in the presence of sodium hydroxide to yield an oxime of low m.p. (α- or *syn*-benzaldoxime) which is stable to alkali, but is rapidly rearranged by acids to give an isomeric oxime of higher m.p. (β- or *anti*-benzaldoxime) (Expt 6.145).

α-Benzaldoxime
m.p. 35 °C

β-Benzaldoxime
m.p. 130 °C

Two isomeric oximes of benzil, i.e. the α- and β-forms (1) and (2), are obtained in a similar manner (cognate preparation in Expt 6.145).

α-Benzilmonoxime
(1)

β-Benzilmonoxime
(2)

Details of the preparation of α-benzoinoxime ('cupron') and salicylaldoxime are also included in Expt 6.145; these are employed as analytical reagents for molybdenum and for copper and nickel respectively.

Oximes of ketones undergo rearrangement (the *Beckmann rearrangement*) to amides under the influence of a variety of acidic reagents (e.g. sulphuric acid, hydrogen fluoride, acetic anhydride, phosphorus pentachloride, thionyl chloride, etc.). The process is illustrated by the conversion of benzophenone oxime to benzanilide in the presence of phosphorus pentachloride.

$$Ph \cdot CO \cdot NH \cdot Ph$$

In the case of unsymmetrically substituted oximes (e.g. $Ar^1 \cdot C(=NOH) \cdot Ar^2$) the structure of the oxime (*syn* or *anti*) may be deduced by hydrolysis of the anilide and characterisation of the isolated carboxylic acid and amine. Thus if the products are $Ar^1 \cdot CO_2H$ and Ar^2NH_2, the hydroxyl group in the oxime must of necessity be *anti* to the group Ar^2 owing to the *trans* intramolecular rearrangement step specified in the formulation above. Illustrative details for the hydrolysis of anilides and the isolation of the acidic and basic constituents are given in Expt 6.145.

Experiment 6.145 α- AND β-BENZALDOXIMES

$$Ph \cdot CHO + NH_2OH \longrightarrow Ph \cdot CH = N \cdot OH + H_2O$$

In a 250-ml conical flask mix a solution of 14 g of sodium hydroxide in 40 ml of water and 21 g (20 ml, 0.2 mol) of pure benzaldehyde (Expt 6.133). Add 15 g (0.22 mol) of hydroxylamine hydrochloride in small portions, and shake the mixture continually (mechanical stirring may be employed with advantage). Some heat is developed and the benzaldehyde eventually disappears. Upon cooling, a crystalline mass of the sodium derivative of the oxime separates out. Add sufficient water to form a clear solution, and pass carbon dioxide into the solution until saturated. A colourless emulsion of the α- or *syn*-aldoxime separates. Extract the oxime with ether, dry the extract over magnesium or anhydrous sodium sulphate and remove the ether on a water bath using a rotary evaporator. Distil the residue under diminished pressure (Fig. 2.108). Collect the pure *syn*-benzaldoxime (α-benzaldoxime) at 122–124 °C/12 mmHg; this gradually solidifies on cooling in ice and melts at 35 °C. The yield is 12 g (49%).

To prepare the β-benzaldoxime, dissolve 10 g of α-benzaldoxime in 50 ml of pure anhydrous ether and pass dry hydrogen chloride (Section 4.2.38, p. 438) through a *wide* delivery tube into the solution with constant shaking. Colourless crystals of the hydrochloride of the β-aldoxime separate. Filter these at the pump through a sintered glass funnel, wash with dry ether, transfer to a separatory funnel and cover with a layer of ether. Add a concentrated solution of sodium carbonate gradually and with constant shaking until efferevescence ceases. Separate the ethereal layer, which contains the β-oxime, dry over magnesium sulphate or sodium sulphate and remove the ether using a rotary evaporator. The residue crystallises; remove the small amount of oily matter by pressing on a porous tile. Recrystallise by dissolving the product in the minimum volume of ether and then adding light petroleum (b.p. 60–80 °C). The yield of β-benzaldoxime (*anti*-benzaldoxime), m.p. 130 °C, is 7–8 g (70–80%).

The p.m.r. spectra of these aldoximes should be recorded and compared, particularly with respect to the chemical shift values of the —CH=N proton.

Cognate preparations. *α-Benzilmonoxime.* Grind 42 g (0.2 mol) of pure benzil (Expt 6.143) to a thin paste with a little ethanol, and add a concentrated aqueous solution of 17.5 g (0.25 mol) of hydroxylamine hydrochloride. Cool to −5 °C in an ice–salt bath, and add 30 g of sodium hydroxide as a 20 per cent aqueous solution dropwise with rapid mechanical stirring: do not allow the temperature to rise above 0 °C. After 90 minutes dilute the mixture with water and filter off the small quantity of unchanged benzil on a sintered glass funnel. *Just* acidify the filtrate with glacial acetic acid, allow to stand for 30 minutes, filter off the crude pinkish α-monoxime and recrystallise it from aqueous ethanol (60 vol. % alcohol); the resulting oxime weighs 37 g (82%) and melts at 137 °C. To obtain the pure α-benzilmonoxime, recrystallise twice from benzene; the final yield is 28 g (62%) of the pure product, m.p. 140 °C. Animal charcoal must not be used in the recrystallisation (see below).

β-Benzilmonoxime. The α-oxime is converted into the β-form by treatment with a solution of hydrogen chloride in benzene (**CAUTION**) (or ether) at room temperature. From benzene, solvated crystals which melt on rapid heating at about 65 °C are obtained. Removal of benzene of crystallisation in an oven at 50 °C and recrystallisation from carbon disulphide (**CAUTION**) yields pure β-benzilmonoxime, m.p. 112 °C. The product gives no colour change with aqueous–alcoholic copper acetate solution; if it is contaminated with the α-form a greenish colour is produced. (Conversion of the α-form into the β-form may also be effected by boiling in benzene solution in the presence of animal charcoal, which presumably contains adsorbed acidic catalysts.)

α-Benzoinoxime. In a 250-ml round-bottomed flask, fitted with a reflux condenser, place a mixture of 10 g (0.047 mol) of benzoin (Expt 6.142) and 20 g (25 ml) of rectified spirit together with an aqueous solution of 8.0 g (0.087 mol) of hydroxylamine hydrochloride which has previously been neutralised with 4.4 g (0.091 mol) of sodium hydroxide. Reflux for 60 minutes. Add water to precipitate the benzoinoxime, and cool in an ice bath. Filter the solid with suction at the pump, wash it with water and recrystallise from dilute ethanol. Alternatively, the dry solid may be recrystallised from ether. The yield of pure α-benzoinoxime, m.p. 151 °C, is 5 g (47%).

Salicylaldoxime. Dissolve 20.0 g (0.164 mol) of salicylaldehyde (Expt 6.116) in 30 ml of rectified spirit, add a solution of 15 g (0.216 mol) of hydroxylamine hydrochloride in 10 ml of water and render the mixture just alkaline with 10 per cent sodium carbonate solution while cooling in ice. Allow to stand overnight. Acidify with acetic acid, distil off the alcohol under reduced pressure on a rotary evaporator, dilute with twice the volume of water and extract with two 50 ml portions of ether. Dry the ethereal extract with sodium sulphate or magnesium sulphate, distil off most of the ether and allow the residue to crystallise. Recrystallise from chloroform–light petroleum (b.p. 40–60 °C). The yield of salicylaldoxime, m.p. 57 °C, is 12 g (40.5%).

Benzophenone oxime. Place a mixture of 25 g (0.137 mol) of benzophenone (Expt 6.121), 15 g (0.216 mol) of hydroxylamine hydrochloride, 50 ml of rectified spirit and 10 ml of water in a 500-ml round-bottomed flask. Add 28 g (0.7 mol) of sodium hydroxide (pellet form) in portions with shaking; if the reaction becomes too vigorous, cool the flask with running tap water. When all the sodium hydroxide has been added, attach a reflux condenser to the flask, heat to boiling and reflux for 5 minutes. Cool, and pour the contents of the flask into a solution of 75 ml of concentrated hydrochloric acid in 500 ml of water contained in a 1-litre beaker. Filter the precipitate at the pump, wash thoroughly with cold water and dry in an oven at 40 °C or in a vacuum desiccator. The yield of benzophenone oxime, m.p. 142 °C, is 26.5 g (98%). It may be recrystallised from methanol (4 ml per gram) but the m.p. is unaffected. The oxime is gradually decomposed by oxygen and traces of moisture into benzophenone and nitric acid; it should be preserved in a vacuum desiccator filled with pure dry carbon dioxide or nitrogen.

Beckmann rearrangement of benzophenone oxime to benzanilide. Dissolve 2 g of benzophenone oxime in 20 ml of anhydrous ether in a small conical flask and add 3 g of powdered phosphorus pentachloride (or 3 ml of pure thionyl chloride). Distil off the solvent and other volatile products on a water bath on a rotary evaporator, add 25 ml of water, boil for several minutes and break up any lumps which may be formed. Decant the supernatant liquid, and recrystallise, in the same vessel, from boiling ethanol. The product is benzanilide, 1.6 g (80%), m.p. 163 °C; confirm this by a mixed m.p. determination with an authentic specimen.

Hydrolysis of benzanilide. Place 5 g of benzanilide and 50 ml of 70 per cent w/w sulphuric acid (1) in a small flask fitted with a reflux condenser, and boil gently for 30 minutes. Some of the benzoic acid will vaporise in the steam and solidify in the condenser. Pour 60 ml of hot water down the condenser: this will dislodge and partially dissolve the benzoic acid. Cool the flask in ice-water; filter off the benzoic acid (aniline sulphate does not separate at this dilution), wash well with water, drain, dry upon filter paper and identify by m.p. (121 °C). Render the filtrate alkaline by cautiously adding 10 per cent sodium hydroxide solution, cool and isolate the aniline by ether extraction. Remove the ether and distil the aniline, b.p. 184 °C.

Note. (1) For preparation see Expt 6.68, Note (3).

6.12.7 SOME REACTIONS OF ALKYL ARYL KETONES

Aryl methyl ketones can only give a single enol (or enolate ion) and subsequent reactions are therefore entirely regioselective. These include halogenation (cf. Section 5.11.1, p. 667), and the Mannich reaction (cf. Section 5.18.2, p. 801).

Bromination of the methyl group can be restricted to monosubstitution when the reaction is carried out in acidic media. The mechanism involves protonation of the carbonyl-oxygen, followed by proton loss to give the enol. After monobromination, protonation of the bromoketone is less favourable owing to the presence of the electron-withdrawing halogen atom. Further enolisation does not occur therefore and halogenation ceases (contrast the behaviour of methyl ketones on bromination under alkaline conditions, p. 667); the product is an aryl bromomethyl ketone or phenacyl bromide.

The preparation of *p*-bromophenacyl bromide, which is a useful reagent for the characterisation of carboxylic acids (Section 9.6.15, p. 1261), is described in Expt 6.146.

Condensation of the active methyl group in acetophenone with formaldehyde and dimethylamine (in the form of its hydrochloride) is an example of the *Mannich reaction* (e.g. the synthesis of dimethylaminopropiophenone, Expt 6.147). The probable mechanism of the reaction involves the intermediate formation of the hydroxymethyldimethylamine which eliminates water to form the reactive species (1). This condenses with the α-carbon atom of acetophenone reacting in its enol form.

$$Me_2\overset{\oplus}{N}H_2\}\overset{\ominus}{Cl} + H{\cdot}CHO \rightleftharpoons Me_2NH + H{\cdot}CH{=}\overset{\oplus}{O}H + \overset{\ominus}{Cl}$$

$$Me_2\ddot{N}H \frown CH_2{=}\overset{\oplus}{O}H \rightleftharpoons Me_2\overset{\oplus}{N}H{\cdot}CH_2OH \xrightarrow{-H_2O} Me_2\overset{\oplus}{N}{=}CH_2$$

$$(1)$$

Other compounds containing active hydrogen atoms similarly undergo the Mannich reaction, and a further interesting example is provided by the synthesis of dimethylaminomethylindole (gramine), in which indole is the reactive component (cognate preparation in Expt 6.147).

The quaternary salts produced by the Mannich reaction undergo a number of conversions useful in synthesis. For example, although stable at room temperature they eliminate an amine hydrochloride on heating to yield an α,β-unsaturated ketone (e.g. the conversion of dimethylaminopropiophenone to phenyl vinyl ketone, Expt 6.147).

This ready decomposition makes Mannich bases convenient *in situ* sources of α,β-unsaturated carbonyl compounds (see also Section 7.2, p. 1094).

A further example of the usefulness of Mannich bases is illustrated by the reaction of dimethylaminopropiophenone hydrochloride on heating with aqueous potassium cyanide, which results in the ready replacement of the dimethylamino group by the nitrile group forming β-benzoylpropionitrile. This replacement occurs even more readily when the dimethylamino compound is of the benzylic type, as in gramine. When the latter is boiled for a long time with aqueous potassium cyanide, the plant growth hormone 3-indolylacetic acid is formed by way of hydrolysis of the intermediate nitrile.

When an alkyl aryl ketone is heated with yellow ammonium polysulphide solution at elevated temperature, an aryl-substituted aliphatic acid amide is formed; the product actually isolated is the amide of the ω-arylcarboxylic acid together with a smaller amount of the corresponding ammonium salt of the carboxylic acid.

This conversion of a carbonyl compound by ammonium polysulphide solution into an amide with the same number of carbon atoms is known as the *Willgerodt reaction*. The procedure has been improved by the addition of about 40 per cent of dioxane or of pyridine to increase the mutual solubility of the ketone and aqueous ammonium polysulphide; the requisite temperature is lowered to about 160 °C and the yield is generally better.

A further improvement is embodied in the *Kindler variation* of the Willgerodt reaction which is illustrated by several examples in Expt 6.148. This consists of heating the ketone with approximately equal amounts of sulphur and a dry amine (e.g. morpholine) instead of aqueous ammonium polysulphide. The principal product is a thioamide, and subsequent hydrolysis with acid or alkali affords the carboxylic acid, usually in good yield.

An evaluation of this complex and variable reaction concludes that it cannot be described by a single mechanism.[58]

Experiment 6.146 α,p-DIBROMOACETOPHENONE (p-*Bromophenacyl bromide*)

$$p\text{-BrC}_6\text{H}_4\cdot\text{CO}\cdot\text{Me} \xrightarrow[\text{Me}\cdot\text{CO}_2\text{H}]{\text{Br}_2} p\text{-Br}\cdot\text{C}_6\text{H}_4\cdot\text{CO}\cdot\text{CH}_2\text{Br}$$

Place a solution of 50 g (0.25 mol) of p-bromoacetophenone (Expt 6.122) in 100 ml of glacial acetic acid in a 500-ml flask. Add very slowly (about 30 minutes) from a dropping funnel 40 g (12.5 ml, 0.25 mol) of bromine (**CAUTION**); shake the mixture vigorously during the addition and keep the temperature below 20 °C. p-Bromophenacyl bromide commences to separate as needles after about half of the bromine has been introduced. When the addition is complete, cool the mixture in ice-water, filter the crude product at the pump and wash it with 50 per cent alcohol until colourless (about 100 ml are required). Recrystallise from rectified (or industrial) spirit (c. 400 ml). The yield of pure p-bromophenacyl bromide (colourless needles, m.p. 109 °C) is 50 g (72%).

Cognate preparation. α-Bromo-p-phenylacetophenone (p-*phenylphenacyl bromide*).   Suspend 36 g (0.183 mol) of p-phenylacetophenone in 200 ml of glacial acetic acid in a 500-ml flask, warm gently on a water bath until a clear solution results, then cool as far as possible without the formation of crystals. To this solution add 29.5 g (9.5 ml, 0.184 mol) of bromine; do not allow the temperature to rise above 45 °C during the addition. The brominated product separates from the solution when about three-quarters of the bromine has been added. After 2 hours, cool the flask in a bath of ice and salt, filter the product, wash with a little cold glacial acetic acid, followed by small volumes of water until all the acid has been removed. The yield of crude material, m.p. 124.5–125.5 °C, is 42 g (83%). Recrystallise from hot rectified spirit (600–700 ml) and add a little decolourising carbon to remove the colour: pure, colourless p-phenylphenacyl bromide, m.p. 125.5 °C, is obtained.

Experiment 6.147   DIMETHYLAMINOPROPIOPHENONE HYDRO-CHLORIDE (Mannich reaction)

$$\text{Ph·CO·Me} + \text{H·CHO} + \text{Me}_2\overset{\oplus}{\text{NH}_2}\}\text{Cl}^{\ominus} \longrightarrow \text{Ph·CO·(CH}_2)_2\text{·}\overset{\oplus}{\text{NHMe}_2}\}\text{Cl}^{\ominus}$$

Place 26.5 g (0.326 mol) of dry dimethylamine hydrochloride, 10 g (0.33 mol) of powdered paraformaldehyde and 30 g (29.3 ml, 0.25 mol) of acetophenone in a 250-ml round-bottomed flask attached to a reflux condenser. Introduce 40 ml of 95 per cent ethanol to which 0.5 ml of concentrated hydrochloric acid has been added, and reflux the mixture on a water bath for 2 hours; the reaction mixture should ultimately be almost clear and homogeneous. Filter the yellowish solution (if necessary) through a hot-water funnel: transfer the filtrate to a 500-ml wide-mouthed conical flask and, while still warm, add 200 ml of acetone. Allow to cool to room temperature and leave in a refrigerator overnight. Filter the crystals at the pump, wash with 10 ml of acetone and dry for 6 hours at 40–50 °C: the yield of crude product, m.p. 152–155 °C, is 38 g (71%). Recrystallise the crude product by dissolving in 45 ml of hot rectified spirit and slowly adding 225 ml of acetone to the solution; collect the solid which separates by suction filtration and dry at 70 °C. The purified material melts at 155–156 °C and the recovery is about 90 per cent.

Conversion into 3-benzoylpropionitrile

$$\text{Ph·CO·(CH}_2)_2\text{·}\overset{\oplus}{\text{NHMe}_2}\}\text{Cl}^{\ominus} \xrightarrow{\text{KCN}} \text{Ph·CO·(CH}_2)_2\text{·CN} + \text{Me}_2\text{NH} + \text{KCl}$$

To a mixture of 21.4 g (0.1 mol) of dimethylaminopropiophenone hydrochloride and 13.0 g (0.2 mol) of potassium cyanide in a 500-ml flask, add 260 ml of boiling water; heat the heterogeneous mixture under reflux for 30 minutes. Part of the dimethylamine, which is eliminated in the reaction, distils: collect this in dilute hydrochloric acid. Cool the reaction mixture in ice; the oil solidifies and crystals form in the aqueous layer. Collect the solid [crude 3-benzoylpropionitrile, 10.5 g (66%)] by suction filtration and recrystallise it from benzene–light petroleum (b.p. 40–60 °C); the product separates as almost colourless blades, m.p. 76 °C.

Conversion to phenyl vinyl ketone.

Place an intimate mixture of 21.4 g (0.1 mol) of β-dimethylaminopropiophenone hydrochloride and 0.2 g of hydroquinone in a 100-ml round-bottomed flask. Attach a Claisen still-head fitted with a stout capillary air leak and condenser arranged for distillation under reduced pressure; place a few crystals of hydroquinone in the receiving flask. Pyrolyse the amine hydrochloride by heating the flask in an electric mantle at 2 mmHg (oil immersion pump), and collect the crude ketone which distils between 70 and 90 °C. On redistillation 7 g (51%) of pure phenyl vinyl ketone, b.p. 72–73 °C/3 mmHg (115 °C/18 mmHg), are obtained. Characterise the product by reaction with phenylhydrazine in the following way. Dissolve 0.5 g of phenylhydrazine hydrochloride and 0.5 g of sodium acetate trihydrate in the minimum of water, and add 0.5 g of the ketone followed by a little ethanol to give a homogeneous solution. Heat the mixture on a steam bath for 5–10 minutes, collect the product which separates on cooling and crystallise it from ethanol. Yellow needles of 1,3-diphenyl-2-pyrazoline, m.p. 154 °C, are obtained.

Cognate preparation. *Dimethylaminomethylindole* (*gramine*). Cool 42.5 ml (0.236 mol) of aqueous dimethylamine solution (*c.* 25% w/v) contained in a 100-ml flask in an ice bath, add 30 g of cold acetic acid, followed by 17.2 g (0.21 mol) of cold, 37 per cent aqueous formaldehyde solution. Pour the solution on to 23.4 g (0.2 mol) of indole; use 10 ml of water to rinse out the flask. Allow the mixture to warm up to room temperature, with occasional shaking as the indole dissolves. Keep the solution at 30–40 °C overnight and then pour it, with vigorous stirring, into a solution of 40 g of potassium hydroxide in 300 ml of water; crystals separate. Cool in an ice bath for 2 hours, collect the crystalline solid by suction filtration, wash with three 50 ml portions of cold water and dry to constant weight at 50 °C. The yield of gramine is 34 g (97.5%); this is quite suitable for conversion into 3-indolylacetic acid. The pure compound may be obtained by recrystallisation from acetone–hexane; m.p. 133–134 °C.

Conversion into 3-indolylacetic acid. In a 1-litre flask, fitted with a reflux condenser, place a solution of 35.2 g (0.72 mol) of sodium cyanide (**CAUTION**) in 70 ml of water, then add 25 g (0.144 mol) of gramine and 280 ml of 95 per cent ethanol. Boil the mixture under reflux for 80 hours. Dilute the cooled reaction mixture with 350 ml of water, shake with a little activated charcoal (e.g., Norit), filter and concentrate to about 350 ml under reduced pressure (water pump) in order to remove most of the alcohol. Cool to about 5 °C, filter off the solid and wash it with a little cold water; keep the filtrate (*A*). Recrystallise the solid from ethanol–ether to obtain 5.0 g (20%) of 3-indolylacetamide, m.p. 150–151 °C.

Concentrate the filtrate (*A*) to about 300 ml, cool to 5–10 °C and add concentrated hydrochloric acid dropwise and with vigorous stirring (*fume cupboard: hydrogen cyanide is evolved*) to a pH of 1–2 (about 50 ml); a crude, slightly pink 3-indolylacetic acid is precipitated. The yield of crude acid, m.p. 159–161 °C, is 20 g. Recrystallise from 1,2-dichloroethane containing a small amount of ethanol; 17.5 g (70%) of pure 3-indolylacetic acid, m.p. 167–168 °C, are obtained.

Experiment 6.148 2-NAPHTHYLACETIC ACID (*Willgerodt reaction*)

$$2\text{-}C_{10}H_7\cdot CO\cdot Me \xrightarrow[\text{morpholine}]{S} 2\text{-}C_{10}H_7\cdot CH_2\cdot CS\cdot N\!\!\diagup\diagdown\!\!O \xrightarrow{H_3O^{\oplus}}$$

$$2\text{-}C_{10}H_7\cdot CH_2\cdot CO_2H + H_2S + HN\!\!\diagup\diagdown\!\!O$$

In a conical or round-bottomed flask, fitted with a reflux condenser by means of a ground glass joint, place a mixture of 128 g (0.75 mol) of methyl 2-naphthyl ketone (Expt 6.121), 35 g (1.1 mol) of sulphur and 97 g (97 ml,

1.1 mol) of morpholine (b.p. 126–128 °C). Reflux in the *fume cupboard* gently at first until the evolution of hydrogen sulphide subsides and then more vigorously for a total period of 14 hours. Pour the hot reaction mixture, which has separated into two layers, into 400 ml of warm ethanol and leave to crystallise. The 2-naphthylthioacetomorpholide separates as pale buff crystals. Filter at the pump and wash with a little cold ethanol; the yield of crude thiomorpholide, m.p. 103–108 °C, is 178 g (87%).

Mix 130 g (0.48 mol) of the crude thiomorpholide with 270 ml of glacial acetic acid, 40 ml of concentrated sulphuric acid and 60 ml of water; raise the temperature of the mixture carefully to the boiling point and reflux for 5 hours. Decant the solution from a little tarry matter into 2 litres of water and keep overnight. Collect the solid by suction filtration and wash it well with cold water. Digest the solid with a solution of 50 g of sodium hydroxide in 1 litre of water, filter and acidify the filtrate with hydrochloric acid; filter off the crude 2-naphthylacetic acid, wash with water and dry. The yield of the crude acid, m.p. 137–140 °C, is 75 g (84%). Recrystallisation from benzene raises the m.p. to 142–143 °C; the loss is about 10 per cent.

Cognate preparations. *3-Phenylpropanoic acid* (*hydrocinnamic acid*). Reflux a mixture of 53.5 g (0.4 mol) of propiophenone (Expt 6.126), 20.5 g (0.64 mol) of sulphur and 46 g (46 ml, 0.53 mol) of morpholine for 6 hours. Pour the reaction product into 400 ml of 10 per cent ethanolic sodium hydroxide solution and reflux for 7 hours. Distil off the ethanol, dilute with water, acidify with hydrochloric acid (to Congo red paper) and extract three times with ether. Wash the ether extracts with water, dry, remove the ether and distil. Collect the hydrocinnamic acid at 125–129 °C/6 mmHg; it solidifies completely on cooling, m.p. 46–47 °C. The yield is 39 g (65%).

p-*Methoxyphenylacetic acid*. Reflux a mixture of 42 g (0.28 mol) of p-methoxyacetophenone (Expt 6.122), 13.5 g (0.42 mol) of sulphur and 36.5 g (36 ml, 0.42 mol) of morpholine for 5 hours. Pour the reaction mixture slowly into water, allowing the first addition to crystallise before the bulk of the mixture is added. Filter off the crude yellow solid, grind it up thoroughly with water, filter again and dry in the air. The yield of crude thioacetomorpholide, m.p. 65–67 °C, is 56 g (80%). Recrystallisation from dilute methanol raises the m.p. to 71–72 °C.

Add 50 g of the crude thioacetomorpholide to 400 ml of 10 per cent ethanolic sodium hydroxide solution and reflux the mixture for 10 hours. Distil off most of the ethanol, add 100 ml of water to the residue and strongly acidify the alkaline solution with hydrochloric acid. Cool, extract thrice with ether, dry the combined ether extracts, evaporate the solvent and recrystallise the residue from water or dilute ethanol. The yield of p-methoxyphenylacetic acid, m.p. 85–86 °C, is 26 g (63% overall). A further quantity of acid may be obtained by extracting the mother-liquors with ether.

p-*Bromophenylacetic acid*. Reflux a mixture of 50 g (0.25 mol) of p-bromoacetophenone (Expt 6.122), 12.8 g (0.4 mol) of sulphur and 30 ml (0.35 mol) of morpholine for 8 hours. Saponify the crude reaction product with 250 ml of 10 per cent ethanolic sodium hydroxide solution and work up as described for the p-methoxy acid. The yield of crude p-bromophenylacetic acid, m.p. 107–109 °C, is 25 g (46%). Recrystallisation from water gives the pure acid, m.p. 113–114 °C.

6.12.8 SOME METHODS FOR THE PROTECTION OF THE AROMATIC CARBONYL GROUP

The methods which are available for the protection of the aromatic carbonyl group are similar to those for the aliphatic and alicyclic analogues (Section 5.8.8, p. 623). It should be noted however, that when a cyclic acetal is used as the protecting group, a Birch reduction (Section 7.5) on the protected compound usually results in hydrogenolysis of the protecting acetal.

6.13 AROMATIC CARBOXYLIC ACIDS

This section is principally concerned with the synthesis of carboxylic acids in which the carboxyl group is directly attached to the aromatic nucleus. Several examples of the preparation of acids in which the carboxyl group is located in the side chain, i.e. aryl-substituted aliphatic acids, are included in Section 5.12.

1. Oxidative methods (Expts 6.149 to 6.152).
2. The hydrolysis of nitriles (Expts 6.153 to 6.155).
3. Carboxylation of the aromatic ring system (Expts 6.156 to 6.158).
4. Nuclear nitration of aromatic carboxylic esters and acids (Expts 6.159 and 6.160).

Methods for the protection of the aromatic carboxyl group are considered in Section 6.13.5.

SUMMARY OF RETROSYNTHETIC STRATEGIES

Functional group removal/interconversion (FGR/FGI) (methods 1 and 2)

$Z = Hal, NO_2$

(TM)

Disconnection (methods 3 and 4)

$\equiv Z \cdot Ar \cdot MgX(Li)$ when $Z = R, Hal$
$\equiv Z \cdot ArH$ when $Z = OH$
$+ \overset{\oplus}{HOCO}$ (CO_2)

(TM)

$+ \overset{\oplus}{NO_2}$ (HNO_3/H_2SO_4)

(TM)

SPECTROSCOPIC FEATURES

The characteristic profile of the *i.r.* spectral region $(2500-3300\,cm^{-1})$, corresponding to absorption arising from the intermolecular hydrogen bonding of the carboxyl group, is easily observed (p. 299). The frequency of absorption arising from the carbonyl group is a little lower than in aliphatic carboxylic acids owing to aromatic mesomeric interaction. The substitution pattern in the aromatic nucleus may be inferred from the longer wavelength region of the i.r. spectrum and may be confirmed by inspection of the *p.m.r.* spectrum. The latter will also show a low field signal from the acidic proton if the sample is dissolved in a solvent with which it does not undergo a deuterium exchange reaction (e.g. phenylacetic acid, Fig. 3.47). (The ^{13}C-*n.m.r.* spectrum is given in Fig. 3.55.) The *m.s.* of *ortho*-substituted acids (and their derivatives) is discussed on p. 383, and analyses of fragmentation patterns of suitable compounds are included in some of the preparative examples below.

6.13.1 OXIDATIVE METHODS

The oxidation of an alkyl group attached to an aromatic system is a frequently used method for the preparation of the corresponding carboxylic acid.

$$Ar \cdot R \xrightarrow{\ [O]\ } Ar \cdot CO_2H$$

The conversion can be accomplished most readily in the laboratory by using either a solution of sodium dichromate in concentrated sulphuric acid or aqueous potassium permanganate. The method is not applicable to those cases where activating groups are attached to the aromatic system, since these render the ring susceptible to oxidative cleavage. The use of acid dichromate is illustrated in Expt 6.149 by the oxidation of *p*-nitrotoluene and of *p*-xylene. The examples illustrating the use of potassium permanganate in Expt 6.150 are the oxidation of two isomeric chlorotoluenes and the three isomeric picolines (methylpyridines).

Diphenic acid (Expt 6.151) is obtained when phenanthrene is oxidised with 30 per cent hydrogen peroxide in glacial acetic acid solution at 85 °C. No phenanthraquinone is formed under these conditions (compare the oxidation of phenanthrene with acid dichromate, Expt 6.128); the reaction is essentially an oxidation by peracetic acid of the reactive 9,10-positions in phenanthrene.

Experiment 6.152 describes a synthesis of quinaldinic acid from quinaldine (2-methylquinoline). The method depends upon the reactive nature of the 2-methyl group in quinaldine which can readily be brominated using bromine in acetic acid. Hydrolysis of the resulting tribromo derivative by boiling with dilute sulphuric acid occurs smoothly to give the corresponding carboxylic acid in good yield.

The use of the haloform reaction for the conversion of methyl ketones into carboxylic acids (Sections 5.11.1, p. 667) is also applicable in the aromatic field. The conditions may be readily adapted from those described for the conversion of aliphatic methyl ketones into carboxylic acids (Expt 5.124).

Experiment 6.149 *p*-NITROBENZOIC ACID

$$p\text{-}O_2N \cdot C_6H_4 \cdot Me \xrightarrow[H_2SO_4]{Na_2Cr_2O_7} p\text{-}O_2N \cdot C_6H_4 \cdot CO_2H$$

Place 46 g (0.33 mol) of *p*-nitrotoluene, 136 g of sodium dichromate dihydrate and 300 ml of water in a 1-litre, two-necked round-bottomed flask equipped with an unsealed mechanical stirrer. By means of a dropping funnel, add 340 g (185 ml) of concentrated sulphuric acid during about 30 minutes to the well-stirred mixture. The heat of dilution of the acid causes the *p*-nitrotoluene to melt and oxidation takes place; if the reaction shows signs of becoming vigorous, the rate of addition must be reduced. When all the sulphuric acid has been introduced and the temperature of the mixture commences to fall, attach a reflux condenser to the flask, and heat to gentle boiling for half an hour. Cool and pour the reaction mixture into 400–500 ml of water. Filter the crude *p*-nitrobenzoic acid at the pump and wash it with about 200 ml of water. Transfer the solid to a 1-litre beaker, add about 200 ml of 5 per cent sulphuric acid (11 g or 6 ml of concentrated sulphuric acid added to 200 ml of water) and digest on a water bath, with agitation, in order to remove the chromium salts as completely as possbile; allow to cool and filter again. Transfer the acid to a beaker, break up any lumps of material and treat it with 5 per cent sodium hydroxide solution until the liquid remains alkaline. The *p*-nitrobenzoic acid passes into solution, any unchanged *p*-nitrotoluene remains undissolved and chromium salts are converted into chromic hydroxide and/or sodium chromite. Add about 5 g of decolourising carbon, warm to about 50 °C with stirring for 5 minutes and filter with suction. Run the alkaline solution of sodium *p*-nitrobenzoate into about 450 ml of well-stirred 15 per cent sulphuric acid (74 g or 40 ml of concentrated sulphuric acid in 400 ml of water). Do not add the acid to the alkaline solution, for in this way the acid is liable to be contaminated by the sodium salt. Filter the purified acid at the pump, wash it thoroughly with cold water and dry it in the oven. The yield of *p*-nitrobenzoic acid, m.p. 237 °C, is 48 g (86%): this is sufficiently pure for most purposes. Upon recrystallisation from glacial acetic acid, the m.p. is raised to 239 °C.

The i.r. spectrum should be recorded and the absorptions of the functional groups assigned. The p.m.r. spectrum (DMSO-d_6, TMS) is a good example of a case where the Δv value for the two pairs of aromatic protons approaches that of their coupling constant (p. 343); δ 8.21 (d, 2H, *ortho*-H's, to CO$_2$H), and 8.30 (d, 2H, *ortho*-H's to NO$_2$). This spectrum should be compared with that of the *meta* isomer, Expt 6.159.

Cognate preparation. *Terephthalic acid.* Use 18 g (0.169 mol) of *p*-xylene in place of the *p*-nitrotoluene and proceed as above to the stage of isolation of the crude product. Wash this with 40 ml of water followed by 20 ml of ether and purify, isolate and dry the acid as detailed above for *p*-nitrobenzoic acid; the yield of colourless terephthalic acid is 12 g (44%); it sublimes without melting at 300 °C and is almost insoluble in water and ethanol.

Experiment 6.150 *o*-CHLOROBENZOIC ACID

$$o\text{-Cl·C}_6\text{H}_4\text{·Me} \xrightarrow{\text{KMnO}_4} o\text{-Cl·C}_6\text{H}_4\text{·CO}_2\text{H}$$

Place 1250 ml of water, 75 g of pure potassium permanganate and 50 g (0.4 mol) of *o*-chlorotoluene (Expt 6.71) in a 2.5-litre three-necked flask equipped with a sealed mechanical stirrer and reflux condenser. Stir the mixture and reflux gently until practically all the permanganate colour has disappeared (about 2 hours). At this point add 37.5 g more of potassium permanganate and reflux the mixture again until the permanganate colour disappears (about 2 hours); the colour of the solution can easily be seen by removing the flame and stopping the refluxing. Finally, add a second 37.5 g (0.95 mol total) of potassium permanganate and continue refluxing until the permanganate colour has disappeared (about 2–4 hours) (1). Steam distil the mixture (Fig. 2.102) to remove unreacted o-chlorotoluene (about 12 g). Filter the hot contents of the flask from the manganese dioxide with suction (2) and wash with two 125-ml portions of hot water. Concentrate the filtrate to about 800 ml using a rotary evaporator (3), and precipitate the o-chlorobenzoic acid by cautiously adding 75 ml of concentrated hydrochloric acid with continual stirring. When cold, filter with suction, wash the acid with cold water and dry at 100 °C. The yield of o-chlorobenzoic acid, m.p. 138–139 °C, is 42 g (68%). Upon recrystallisation from hot water or from toluene (c. 4 ml per gram), the m.p. is raised to 139–140 °C.

The p.m.r. spectra (DMSO-d_6, TMS) of the three isomeric chlorobenzoic acids show the following signals for the aromatic protons: *ortho* isomer δ 6.8–8.00 (m, 4H); *meta* isomer δ 7.27–7.72 (m, 2H) and 7.80–8.2 (m, 2H); *para* isomer δ 7.56 (d, 2H, *ortho*-H's to Cl) and 8.0 (d, 2H, *ortho*-H's to CO$_2$H).

Notes. (1) A somewhat lower yield is obtained if all the potassium permanganate (150 g) is added all at once and, furthermore, the reaction may become violent. Addition in three portions results in a more controllable reaction.
(2) The addition of a Whatman filter tablet or of a little diatomaceous earth (Super Cel, etc.) assists in the filtration of the finely divided manganese dioxide.
(3) If the acid is precipitated before the solution is concentrated, the yield is considerably reduced (c. 25 g). If the concentrated solution is not clear, it may be clarified by the addition of 1 g of decolourising charcoal.

Cognate preparations. p-*Chlorobenzoic acid.* Proceed exactly as for o-chlorobenzoic acid. Use 1250 ml of water, 50 g (0.4 mol) of p-chlorotoluene (Expt 6.71) and 75 g, 37.5 g and 37.5 g (0.95 mol total) of potassium permanganate. When the oxidation is complete, steam distil the mixture to recover any unreacted p-chlorotoluene (3–4 g). Filter the reaction mixture from hydrated manganese dioxide and wash the precipitate with two 100 ml portions of water. Precipitate the p-chlorobenzoic acid in the filtrate (1) by the addition of 75 ml of concentrated hydrochloric acid. Filter the cold solution with suction, wash with cold water and dry in an oven at 100 °C. The yield of p-chlorobenzoic acid, m.p. 234–235 °C, is 55 g (89%). Recrystallisation from hot water raises the m.p. to 238–239 °C.

Note. (1) If the filtrate has a faint permanganate colour, add a few drops of sodium metabisulphite solution until the solution is colourless. In this case (compare o-chlorobenzoic acid) concentration of the solution before precipitation only increases the yield by about 1 g and may cause occlusion of inorganic salts.

Picolinic acid (*pyridine-2-carboxylic acid*). Equip a 3-litre three-necked flask with a thermometer, sealed stirrer unit and a reflux condenser (Liebig pattern with a wide inner tube). Place a solution of 100 g (106 ml, 1.08 mol) of 2-methylpyridine in 1 litre of water in the flask and heat to 70 °C on a water bath. Add 450 g (2.84 mol) of potassium permanganate in 10 equal portions through the condenser over a period of 3–4 hours; maintain the temperature at 70 °C for the first five additions and at 85–90 °C for the last five. Make each successive addition of potassium permanganate only after the preceding amount is decolourised and wash it down with 20–25 ml of water. After the last charge of potassium permanganate is decolourised, raise the temperature to 95 °C, filter the hot reaction mixture with suction and wash the manganese dioxide cake on the filter with four 200 ml portions of hot water: allow each portion to soak into the cake without application of vacuum and finally suck dry before adding fresh wash water. Evaporate down the combined filtrate and washings to a volume of about 300 ml: allow to cool and adjust to a pH of 3.2 (the isoelectric point) using narrow-range indicator paper (about 125 ml of concentrated hydrochloric acid are required). Picolinic acid is very soluble in water (90 g in 100 ml of water at 9 °C) and therefore does not separate at this stage. Subject this aqueous acidic solution to continuous extraction with 1 litre of dichloromethane for 6 hours (Fig. 2.95). Dry the dichloromethane extract with anhydrous sodium sulphate, filter and remove the dichloromethane under reduced pressure on a rotary evaporator, removing the residual traces of dichloromethane as completely as possible. Transfer the solid to a vacuum desiccator and leave to dry. Place the powdered solid in the thimble of a large Soxhlet extraction apparatus (Fig. 2.97) with 1250 ml of benzene (**CAUTION**) in the flask assembled in an *efficient fume cupboard* (1). Continue the extraction until crystals of picolinic acid separate from the boiling solution. Stop the extraction, cool the benzene to 15 °C and filter to remove the first crop of picolinic acid. Return the benzene filtrate to the Soxhlet extraction flask and exhaustively extract the remainder of the product from the residue in the Soxhlet thimble. The combined yield of picolinic acid of m.p. 137–138 °C is 81 g (61%). The m.p. is unaffected by recrystallisation from ethanol.

Note. (1) The use of toluene as the extraction solution could be usefully explored in this case.

Nicotinic acid (*pyridine-3-carboxylic acid*). Dissolve 100 g (104.5 ml, 1.08 mol) of 3-methylpyridine (99% purity) in 1 litre of water and oxidise it with 450 g (2.84 mol) of potassium permanganate: follow the experimental details given under *picolinic acid*. Wash the manganese dioxide cake with four 500 ml portions of water; evaporate the combined filtrate and washings to about 1250 ml. Adjust the pH to 3.4 (the isoelectric point) with the aid of narrow-range indicator paper; 120–130 ml of concentrated hydrochloric acid are required. Allow to cool overnight, collect the voluminous precipitate of nicotinic acid by suction filtration, wash with three 50 ml portions of cold water and dry at 90–100 °C. Concentrate the filtrate to about 650 ml and cool *slowly* to 5 °C and so obtain a second crop of nicotinic acid: the purpose of the slow cooling is to reduce the contamination by potassium chloride. The first crop of acid weights 90 g and has a purity of about 90 per cent (1); the second crop weighs 10 g and the purity is about 80 per cent. Recrystallise from hot water (2) and dry at 100 °C; the yield of pure nicotinic acid, m.p. 235 °C, from 90 g of the crude acid is 67 g (51%). A further quantity may be obtained by concentrating the mother-liquor.

Notes. (1) The impurity is potassium chloride. The approximate acid content is determined by heating a weighed sample of the acid in a crucible gently at first and finally at a red heat until no trace of black residue remains, and weighing the white residual potassium chloride.

(2) The solubility of pure nicotinic acid in 1000 ml of water at 0, 40, 80 and 100 °C is 1.0, 2.6, 8.2 and 12.7 g respectively.

Isonicotinic acid (*pyridine-4-carboxylic acid*). Use 100 g (104.5 ml, 1.08 mol) of 4-methylpyridine (98% purity) and oxidise it with 450 g (2.84 mol) of potassium permanganate: follow the experimental details given for *picolinic acid*. Evaporate the combined filtrate and washings to about 1500 ml and add concentrated hydrochloric acid until the pH is 3.6; isonicotinic acid precipitates. Heat to 90–95 °C (not all the acid dissolves) and allow the mixture to crystallise slowly. Collect the crude isonicotinic acid by suction filtration, wash well with water and dry at 100 °C. Concentrate the mother-liquor to about half the original volume and so obtain a second crop of acid. The first crop of acid weighs 85 g (64%) (99% pure) and the second crop weighs 7 g (80% pure). Recrystallise from hot water: the resulting isonicotinic acid is pure and has a m.p. of 311 °C (sealed tube).

The solubility of isonicotinic acid in 1000 ml of water at 0, 40, 80 and 100 °C is 3, 9, 24 and 34 g respectively. The solubility is appreciably less in the presence of potassium chloride.

Experiment 6.151 DIPHENIC ACID (*Biphenyl-2,2'-dicarboxylic acid*)

Equip a 3-litre three-necked flask with a sealed mechanical stirrer, a reflux condenser and a thermometer. Dissolve 89 g (0.5 mol) of pure phenanthrene (1) in 1 litre of glacial acetic acid in the flask and warm to 85 °C on a water bath. Introduce 345 ml of 30 per cent hydrogen peroxide solution (4 mol) during 40 minutes [**CAUTION**: *see Section 4.2.41, p. 439*]; the temperature falls to about 80 °C and some phenanthrene may precipitate. After the addition is complete, heat the mixture with stirring on a water bath for a further 3–4 hours. Reduce the volume of the solution to about half by distillation under reduced pressure with a rotary evaporator and allow to cool. Filter off the considerable amount of diphenic acid which crystallises out on cooling. Keep the filtrate and evaporate it almost to dryness under reduced pressure: extract the residue with 375 ml of 10 per cent sodium carbonate solution by warming on a water bath, boil the extract with a little decolourising carbon, filter and add dilute hydrochloric acid until the pH is 4.5 (use narrow-range indicator paper). Stir the solution with a further small amount of active charcoal and filter off the tarry material; cool the clear solution to 0 °C and acidify with dilute hydrochloric acid. Collect the precipitate by suction filtration, wash with water and dry at 110 °C. The total yield of crude diphenic acid, m.p. 228 °C, is 83 g (69%). Recrystallisation from glacial acetic acid raises the m.p. to 230 °C.

Note. (1) Technical phenanthrene may be purified as follows. Dissolve 500 g of technical 90 per cent phenanthrene in 3 litres of ethanol in a 4-litre flask on a steam bath and decant the hot solution from any insoluble material: collect the solid which crystallises upon cooling the solution. Dissolve 250 g of the crystallised product in

550 ml of hot glacial acetic acid in a 1-litre three-necked flask provided with an efficient reflux condenser and a dropping funnel. To the boiling solution add gradually 18 ml of an aqueous solution containing 15 g of chromium trioxide; then add slowly 7.5 ml of concentrated sulphuric acid from the dropping funnel. Reflux the solution for 15 minutes, and then pour it with vigorous stirring into 1225 ml of water in a 3-litre round-bottomed flask. Filter when cold, wash with water and dry in the air. Distil the product under reduced pressure (oil pump) using a short air condenser and collect the phenanthrene at 148–149 °C/1 mmHg. Recrystallise the solidified distillate from ethanol: 200–225 g of nearly white phenanthrene, m.p. 99 °C, are obtained.

Experiment 6.152 QUINALDINIC ACID (*Quinoline-2-carboxylic acid*)

Fit a 500-ml, two-necked round-bottomed flask with a reflux condenser and dropping funnel. Place a mixture of 50 g (0.61 mol) of anhydrous, powdered sodium acetate (Section 4.2.69, p. 464), 100 g of glacial acetic acid and 14 g (0.1 mol) of pure 2-methylquinoline in the flask, and a solution of 48 g (15.5 ml, 0.3 mol) of bromine (**CAUTION**) in 100 g of glacial acetic acid in the dropping funnel. Heat the flask to 70 °C in a water bath, and add the bromine solution during 10–15 minutes while keeping the mixture thoroughly shaken. Remove the flask from the water bath, bring the solution to boiling for a few minutes (until the separation of sodium bromide causes violent bumping) and then heat for 30 minutes on a boiling water bath and allow to cool. Pour the reaction mixture into 300 ml of ice-water, collect the precipitate by suction filtration and wash thoroughly with water. The yield of crude product, after drying at 100 °C, is 36 g (95%). Recrystallise from ethanol or glacial acetic acid: the pure ω-tribromoquinaldine has m.p. 128 °C.

Hydrolyse the ω-tribromoquinaldine by boiling it under reflux with excess of dilute (1:10) sulphuric acid until a test portion, on neutralisation, yields no unchanged halogen compound. The quinaldinic acid is best isolated, via the copper salt, in the following manner. Cool, nearly neutralise the solution and add excess of copper(II) sulphate solution. Collect the pale green copper quinaldinate by suction filtration and wash it well with cold water. Suspend the copper salt in hot water and subject it to prolonged treatment with hydrogen sulphide gas. Filter off the copper sulphide and evaporate the clear filtrate to dryness on a water bath. Recrystallise the residual quinaldinic acid from glacial acetic acid; it then melts at 157 °C. The yield is almost quantitative.

6.13.2 THE HYDROLYSIS OF NITRILES

The hydrolysis of nitriles under either acidic or basic conditions, which has already been discussed in Section 5.11.2, p. 671, for alkyl and aralkyl nitriles, is equally applicable to the synthesis of aromatic carboxylic acids (Expt 6.153). The aromatic nitriles are readily obtained by the Sandmeyer reaction (see Section 6.7.1, p. 923).

In the preparation of 2-naphthoic acid (Expt 6.154) the preferred starting material is 2-naphthylamine-1-sulphonic acid (see Section 6.5.4, p. 900). After replacement of the amino by the cyano group using the Sandmeyer procedure, subsequent treatment with aqueous acid removes the sulphonic acid group (see Section 6.7.1, p. 924 and Section 6.9.2, p. 975) and hydrolyses the cyano group in one step.

The formation of a carboxyl group by hydrolysis of the corresponding nitrile constitutes the last step in an interesting multi-stage synthesis of homophthalic acid (Expt 6.155). The starting material is phthalic anhydride which is first converted into the cyclic imide, phthalimide (1), by treatment with aqueous ammonia, or more conveniently with urea. Reduction of phthalimide with a zinc–copper couple in the presence of alkali is the most convenient laboratory preparation of phthalide (3), although the latter can also be obtained by the direct reduction of phthalic anhydride.[59] The immediate reduction product is the salt of o-hydroxymethylbenzoic acid (2); on acidification this gives the free acid which cyclises to the γ-lactone, phthalide (3). Nucleophilic attack of a cyanide ion on the methylene group in phthalide results in alkyl–oxygen fission of the lactone ring to yield o-cyanomethylbenzoic acid, which on hydrolysis yields homophthalic acid.

Experiment 6.153 o-TOLUIC ACID

$$o\text{-Me·C}_6\text{H}_4\text{·CN} \xrightarrow[\text{or } B\ ^\ominus\text{OH}]{A\ \text{H}_3\text{O}^\oplus} o\text{-Me·C}_6\text{H}_4\text{·CO}_2\text{H}$$

Method A (acidic hydrolysis). Prepare o-tolunitrile, b.p. 94–96 °C/20 mmHg, from o-toluidine following the method given in Expt 6.76 under p-toluidine. Also prepare 600 g of 75 per cent sulphuric acid by adding 450 g (245 ml) of concentrated sulphuric acid cautiously, with stirring and cooling, to 150 ml of water. Place the latter in a 1-litre three-necked flask, equipped with a dropping funnel, a mechanical stirrer and reflux condenser. Heat the solution in an oil bath to about 150 °C, stir and add 220 g (1.88 mol) of o-tolunitrile during 2 hours. Continue the stirring for a further 2 hours while the temperature is maintained at 150–160 °C; finally raise the temperature to 190 °C and stir for another hour. Some crystalline solid will appear in the condenser at this stage. Allow the reaction mixture to cool, pour into ice-cold water and filter off the precipitated acid. Dissolve the crude acid in an excess of 10 per cent sodium hydroxide solution, filter off any insoluble material (probably o-toluamide, m.p. 141 °C) through a sintered glass funnel while still hot and acidify the filtrate with dilute sulphuric acid. Collect the o-toluic acid on a Buchner funnel, dry in the air and recrystallise from benzene [or better toluene (CAUTION) – fume cupboard]. The yield of pure o-toluic acid, m.p. 102–103 °C, is 200 g (78%).

Method B (basic hydrolysis). Boil a mixture of 5 g of o-tolunitrile, 80 ml of 10 per cent aqueous sodium hydroxide solution and 15 ml of alcohol under a reflux condenser; the alcohol is added to increase the rate of hydrolysis. The solution becomes clear after heating for about 1 hour, but continue the boiling for a total period of 1.5 hours to ensure complete hydrolysis. Detach the condenser and boil the solution for a few minutes in the open flask to remove dissolved ammonia and some of the alcohol (**CAUTION**). Cool, and add concentrated hydrochloric acid until precipitation of the *o*-toluic acid is complete. When cold, filter off the *o*-toluic acid with suction and wash with a little cold water. Recrystallise from benzene (or toluene). The yield of *o*-toluic acid, m.p. 102–103 °C, is 5 g (86%).

The p.m.r. spectrum (CDCl$_3$, TMS) shows signals at δ 2.65 (s, 3H, Me), 7.08–7.50 (m, 3H, C$_{4,5,6}$—H), 8.03 (m, 2H, *ortho*-H's to CO$_2$H), and 11.69 (s, 1H, OH). This should be contrasted with the p.m.r. spectrum (TFA, TMS) of *p*-toluic acid (cognate preparation below) which shows signals at δ 2.47 (s, 3H, Me), 7.30 (d, 2H, *ortho*-H's to Me), and 8.02 (d, 2H, *ortho*-H's, to CO$_2$H); the hydroxyl proton is obscured by the TFA solvent. The m.s. of the isomers are similar, but with some differences in the RA of particular ions arising from the *ortho* effect (p. 383); the *ortho* isomer gives principal fragment ions at m/z 136 (M), 119 (M − OH), 118 (M − H$_2$O, base peak), 91 (M − CO$_2$H), 90 (118 − CO), and 65 (91 − C$_2$H$_2$).

Cognate preparations.   p-*Toluic acid* may be similarly prepared by either *Method A* or *Method B* using *p*-tolunitrile (Expt 6.76). The spectral features are noted in the main preparation above.

1-Naphthoic acid (by *Method A*). In a 750-ml or 1-litre flask equipped with a reflux condenser, place 50 g (0.327 mol) of 1-naphthonitrile (Expt 6.168), 100 ml of glacial acetic acid, 100 ml of water and 100 ml of concentrated sulphuric acid. Heat in an oil bath at 115–120 °C for 1.5 hours: do not allow the temperature to rise above 120 °C as the 1-naphthoic acid formed tends to lose carbon dioxide at higher temperatures and the yield will be reduced. Dilute the cold reaction mixture, which contains much crystalline solid, with an equal volume of water and filter at the pump; if the product consists of large lumps, transfer it first to a glass mortar and thoroughly grind it to a fine paste. Wash with water until free from mineral acid. Dissolve the crude acid in dilute aqueous sodium carbonate solution, heat for a short time to separate the resinous impurities and filter the hot solution. Acidify the clear filtrate with a slight excess of dilute sulphuric acid and collect the voluminous precipitate of almost pure 1-naphthoic acid, wash until free from inorganic salts and dry at 100 °C. Recrystallise from toluene or from light petroleum (b.p. 80–100 °C). The yield of pure 1-naphthoic acid, m.p. 160–161 °C, is 50 g (89%).

Experiment 6.154   2-NAPHTHOIC ACID

Diazotise 122 g (0.5 mol) of 2-naphthylamine-1-sulphonic acid (1) as detailed under *2-bromonaphthalene* in Expt 6.72. Prepare copper(I) cyanide from

62.5 g (0.25 mol) of copper(II) sulphate pentahydrate (Section 4.2.23, p. 429) and dissolve it in a solution of 32.5 g (0.5 mol) of potassium cyanide (**CAUTION**) in 250-ml of water contained in a 1-litre three-necked flask. Cool the potassium cuprocyanide solution in ice, stir mechanically and add the damp cake of the diazonium compound in small portions whilst maintaining the temperature at 5–8 °C. Nitrogen is soon evolved and a red precipitate forms gradually. Continue the stirring for about 10 hours in the cold, heat slowly to the boiling point, add 125 g of potassium chloride, stir and allow to stand. Collect the orange crystals which separate by suction filtration; recrystallise first from water and then from ethanol; dry at 100 °C. The product is almost pure potassium 2-cyanonaphthalene-1-sulphonate. Transfer the product to a 1-litre round-bottomed flask, add a solution prepared from 200 ml of concentrated sulphuric acid and 200 g of crushed ice, and heat the mixture under reflux for 12 hours. Collect the 2-naphthoic acid formed (some of which sublimes from the reaction mixture) by suction filtration on a sintered glass funnel, wash well with water and dry at 100 °C; recrystallise from rectified spirit. The yield of 2-naphthoic acid, m.p. 184–185 °C, is 65 g (75%).

Note. (1) Tobias acid, available from Fluka AG Chemische Fabrik.

Experiment 6.155 HOMOPHTHALIC ACID (o-*Carboxyphenylacetic acid*)

Phthalimide. *Method 1.* Place 100 g (0.675 mol) of phthalic anhydride and 105 ml of concentrated ammonia solution (*d* 0.88) in a 1-litre round-bottomed flask fitted with a wide air condenser (≮ 10 mm in diameter). Heat on a sand bath, gradually at first until the mixture is in a state of quiet fusion and forms a homogeneous melt (the temperature reaches 300 °C in about 1.5–2 hours; all the water is evaporated during the first hour). Shake the flask occasionally during the heating and push down any material which sublimes into the condenser with a glass rod. Pour the contents of the flask while still hot into a porcelain basin, allow to cool and grind to a fine powder in a mortar. The phthalimide (95 g, 96%) is practically pure and melts at 233–234 °C. It may be recrystallised from ethanol, but the solubility is only slight (about 5%).

Method 2. Intimately mix 99 g (0.67 mol) of pure phthalic anhydride and 20 g

1065

(0.33 mol) of urea, and place the mixture in a 1-litre, long-necked, round-bottomed flask. Heat the flask in an oil bath at 130–135 °C. When the contents have melted, effervescence commences and gradually increases in vigour: after 10–20 minutes, the mixture suddenly froths up to about three times the original volume (this is accompanied by a rise in temperature to 150–160 °C) and becomes almost solid. Remove the heat from beneath the bath and allow to cool. Add about 80 ml of water to disintegrate the solid in the flask, filter at the pump, wash with a little water and then dry at 100 °C. The yield of phthalimide, m.p. 233 °C (i.e. it is practically pure), is 86 g (87%). If desired, the phthalimide may be recrystallised from 1200 ml of industrial spirit; the first crop consists of 34 g of m.p. 234 °C, but further quantities may be recovered from the mother-liquor.

Phthalide. In a 1-litre three-necked flask stir 90 g (1.37 mol) of a high quality zinc powder to a thick paste with a solution of 0.5 g of crystallised copper(II) sulphate in 20 ml of water (this serves to activate the zinc), and then add 165 ml of 20 per cent sodium hydroxide solution. Cool the flask in an ice bath to 5 °C, stir the contents mechanically and add 73.5 g (0.5 mol) of phthalimide in small portions at such a rate that the temperature does not rise above 8 °C (about 30 minutes are required for the addition). Continue the stirring for half an hour, dilute with 200 ml of water, warm on a water bath until the evolution of ammonia ceases (about 3 hours) and concentrate to a volume of about 200 ml by distillation under reduced pressure (rotary evaporator). Filter, cool in ice and render the filtrate acid to Congo red paper with concentrated hydrochloric acid (about 75 ml are required). Much of the phthalide separates as an oil, but, in order to complete the lactonisation of the hydroxymethylbenzoic acid, boil for an hour: transfer while hot to a beaker. The oil solidifies on cooling to a hard red-brown cake. Leave overnight in an ice chest or refrigerator, and then filter at the pump. The crude phthalide contains much sodium chloride. Recrystallise it in 10 g portions from 750 ml of water: use the mother-liquor from the first crop for the recrystallisation of the subsequent portion. Filter each portion while hot, cool in ice below 5 °C, filter and wash with small quantities of ice-cold water. Dry in the air upon filter paper. The yield of phthalide (transparent plates), m.p. 72–73 °C, is 47 g (70%).

o-**Cyanomethylbenzoic acid.** *This preparation must be conducted in an efficient fume cupboard.* Into a 1-litre three-necked flask, provided with a mechanical stirrer and a thermometer, place 40 g (0.33 mol) of phthalide and 40 g (0.615 mol) of powdered potassium cyanide (**CAUTION**). Heat the stirred mixture to 180–190 °C (internal temperature) in an oil bath for 4–5 hours. Allow to cool, add 400 ml of distilled water and stir the mixture until all the solids are dissolved (about 1 hour). Filter off any unreacted phthalide. Add dilute hydrochloric acid (1:1) to the dark aqueous solution (**CAUTION**: *hydrogen cyanide is evolved*) until it becomes turbid (about 20 ml are required), and continue the addition until the solution is slightly acid: filter off any dark impurities which may separate. Neutralise the solution carefully with sodium hydrogen carbonate, add a few grams of decolourising carbon, stir the mixture for several minutes and filter. Acidify the nearly colourless filtrate with about 20 ml of concentrated hydrochloric acid, cool in ice and filter at the pump. The resulting *o*-cyanomethylbenzoic acid (36 g) melts at

114–115 °C and is satisfactory for most purposes. It may be crystallised from benzene or glacial acetic acid, but with considerable loss. See Section 4.2.*17* for disposal of aqueous solutions containing alkaline cyanides.

Homophthalic acid. Place a mixture of 25 g (0.155 mol) of *o*-cyanomethyl benzoic acid and 25 g of 50 per cent sulphuric acid in a 100-ml flask, heat the mixture on a boiling water bath for 10–12 hours and then pour it into twice its volume of ice and water. Filter the precipitate at the pump and dry in the air. The yield of crude homophthalic acid is 21 g. Recrystallise by dissolving it in 500 ml of boiling water, add decolourising carbon, filter the hot solution through a hot water funnel and cool the filtrate in an ice bath: collect the acid and dry at 100 °C. The yield of practically colourless acid, m.p. 181 °C, is 17 g (61%). The melting point depends upon the rate of heating; immersion of the capillary in a bath at 170 °C gives a m.p. of 182–183 °C.

6.13.3 CARBOXYLATION OF THE AROMATIC RING SYSTEM

In general a phenol will undergo direct carboxylation of the nucleus when the dry sodium salt is heated under pressure with carbon dioxide (the *Kolbé– Schmidt reaction*). Addition of the weakly electrophilic carbon dioxide is promoted by electron release from the oxyanionic site. With phenol itself the ultimate product is salicylic acid (*o*-hydroxybenzoic acid); predominantly *ortho* attack may be attributable to stabilisation of the transition state through chelation.

It is of interest to record that *p*-hydroxybenzoic acid may be prepared by the vigorous thermal rearrangement of potassium salicylate (Expt 6.157).

The carboxylation reaction is particularly facile with di- and tri-hydric phenols. Thus 2,4-dihydroxybenzoic acid (Expt 6.156) is readily obtained by passing carbon dioxide through a boiling aqueous solution of the potassium or sodium salt of resorcinol.

The standard carboxylation reaction of a Grignard reagent (Section 5.11.3, p. 673) is also applicable in the aromatic series. The similar carboxylation of organosodium or organolithium compounds is described in the preparation of *p*-toluic acid (Expt 6.158). The organosodium compound is prepared by the direct reaction of sodium metal with *p*-chlorotoluene; the organolithium compound is similarly obtained from *p*-bromotoluene and lithium metal. The preparation of *m*-chlorobenzoic acid (cognate preparation in Expt 6.158) illustrates an alternative preparation of the required organolithium compound by means of a transmetalation process between butyllithium and *m*-bromochlorotoluene.

Experiment 6.156   2,4-DIHYDROXYBENZOIC ACID (*β-Resorcylic acid*)

$$1,3\text{-}(HO)_2C_6H_4 \xrightarrow[KHCO_3]{CO_2} 2,4\text{-}(HO)_2C_6H_3\cdot CO_2H$$

Place a solution containing 40 g (0.364 mol) of resorcinol, 200 g of potassium hydrogen carbonate and 400 ml of water in a litre flask fitted with a reflux condenser and gas inlet tube. Heat gently on a steam bath for 4 hours; then reflux vigorously over a flame for 30 minutes while passing a rapid stream of carbon dioxide through the solution. Acidify the solution while still hot by adding 180 ml of concentrated hydrochloric acid from a separatory funnel with a long tube delivering acid to the bottom of the flask. Allow to cool to room temperature, chill in an ice bath and collect the crude β-resorcylic acid by filtration with suction. Recrystallise by boiling the crude acid with 180–200 ml of water in the presence of a little decolourising carbon, filter through a hot water funnel and cool in an ice–salt mixture with stirring. Collect and dry the pure β-resorcylic acid; the yield is 36 g (64%), m.p. 216–217 °C. The p.m.r. spectrum should be recorded and the aromatic region expanded to allow a correlation of coupling constants; assign the signals bearing in mind that $J_{3,6} = 0$.

Experiment 6.157   *p*-HYDROXYBENZOIC ACID

Place 100 g (0.725 mol) of salicylic acid and 150 ml of water in a 20-cm porcelain dish and slowly stir in 60 g of potassium carbonate. Evaporate the solution on a steam bath to a thick, pasty solid; break this up into small pieces and dry at 105–110 °C for 2 hours. Finely grind the solid, dry for a further 2 hours at 105–110 °C and grind again to a fine powder. Transfer the powder (a mixture of potassium salicylate and potassium carbonate) to a 500-ml round-bottomed flask fitted with an air condenser set for downward distillation and immersed in an oil bath. Heat the oil bath to 240 °C and maintain this temperature for 90 minutes, stirring the solid occasionally with a glass rod; phenol formed in the reaction distils out of the mixture. When the reaction is complete (1), transfer the contents of the flask while still hot to a 2-litre flask containing 1 litre of hot water; rinse the reaction flask with several portions of the hot solution. Acidify with concentrated hydrochloric acid (*c*. 75 ml are required), heat nearly to boiling, add 5 g of decolourising carbon, filter, cool and collect the brown solid by suction filtration. Concentrate the filtrate to about 300 ml cool and collect a second crop of the acid. Dissolve the crude acid in 300 ml of hot water, boil for a few minutes with 5 g of decolourising carbon and filter. Cool the filtrate under the tap, filter the solid with suction, wash with 15 ml of cold water and dry. The yield of *p*-hydroxybenzoic acid, m.p. 211–212 °C, is 40 g (40%). The p.m.r. spectrum should be recorded and interpreted.

Note. (1) This may be determined roughly by treating a small test portion with 3–4 ml of hot water and acidifying with concentrated hydrochloric acid; the absence of a precipitate in the warm solution indicates the essential completeness of the reaction. Salicylic acid is sparingly soluble and p-hydroxybenzoic acid is relatively soluble under these conditions.

Experiment 6.158 p-TOLUIC ACID

$$p\text{-Me·}C_6H_4X \xrightarrow{\text{Na or Li}} p\text{-Me·}C_6H_4\text{Na(Li)} \xrightarrow[\text{(ii) } H_3O^\oplus]{\text{(i) } CO_2} p\text{-Me·}C_6H_4\text{·}CO_2H$$

Method A (use of organosodium reagent). Equip a 250-ml, three-necked round-bottomed flask with a reflux condenser carrying a pressure-equalising dropping funnel to which is attached a calcium chloride guard-tube, a sealed stirrer unit and a thermometer combined with a gas inlet tube to allow the air in the apparatus assembly to be displaced by nitrogen (cf. p. 85). Introduce 50 ml of dry light petroleum (b.p. 40–60 °C) and 4.6 g (0.2 mol) of sodium wire and pass a slow stream of nitrogen through the apparatus. Add 12.6 g (0.1 mol) of redistilled p-chlorotoluene (Expt 6.71) from the dropping funnel, while stirring vigorously, during 90 minutes: maintain the temperature at 25 °C and continue stirring for a further 2 hours. Pour the reaction mixture on to 200 g of crushed Cardice in the form of a slurry with 200 ml of dry ether contained in a large beaker. After 30–45 minutes, while some of the solid carbon dioxide still remains, add water cautiously to destroy the excess of sodium and to dissolve the sodium salt of the acid. Separate the aqueous layer, extract it once with 50 ml of ether and warm the aqueous solution on a boiling water bath to remove the dissolved solvent. Filter if necessary, and acidify the aqueous solution with dilute hydrochloric acid. Collect the precipitated acid by suction filtration, wash it with a little water and dry at 100 °C. The yield of p-toluic acid, m.p. 175–176 °C, is 9.8 g (72%).

Method B (use of organolithium reagent). Fit a 250-ml three-necked flask with a reflux condenser, protected with a calcium chloride guard-tube, a sealed mechanical stirrer and a dropping funnel combined with a T-connection to provide for the inlet of nitrogen.

Place 35 ml of anhydrous ether in the flask, displace the air by nitrogen and continue passing the nitrogen in a slow stream throughout the duration of the experiment. Introduce 1.90 g (0.274 mol) of lithium in the form of fine shavings (1) into the ether and start the stirrer. Place a solution of 21.5 g (0.125 mol) of p-bromotoluene (Expt 6.72) in 35 ml of ether in the dropping funnel. Run in about 1 ml of the solution into the stirred mixture. The ether in the flask soon becomes turbid; if the ether does not reflux within 10 minutes, immerse the flask in a beaker of warm water and remove it immediately refluxing commences. Add the remainder of the p-bromotoluene solution dropwise or at such a rate that the solvent refluxes continuously (60–90 minutes). Stir the mixture while refluxing gently (warm water bath) for a further 45–60 minutes; at the end of this period most of the lithium will have disappeared. Cool the reaction mixture in ice-water, dilute it with 50–60 ml of anhydrous ether and cool (with stirring) to about − 50 °C with the aid of an acetone–Cardice bath. Pour the contents of the flask slowly and with stirring (use a long glass rod) on to 200 g of crushed Cardice in the form of a slurry with 200 ml of dry ether contained in a large beaker. Rinse the flask with a little of the solid carbon dioxide–ether slurry and add the rinsings to the contents of the beaker. Allow the Cardice to evaporate (3–4 hours or preferably overnight). Add about 200 ml of water to the contents of the beaker; rinse the reaction flask with 10 ml of 10 per cent sodium hydroxide solution and pour the rinsings into the beaker. A white solid appears which dissolves upon stirring. (If most of the ether has evaporated on standing, add a further 50 ml.) Separate the two layers, extract the aqueous solution with 50 ml of ether (to remove traces of neutral products) and combine the extract with the ether layer. Shake the combined ethereal solutions with 10 per cent sodium hydroxide solution and add the alkaline extract to the aqueous layer. Warm the combined aqueous layers to 60–70 °C (water bath) to drive off the dissolved ether, then cool to about 5 °C and strongly acidify with hydrochloric acid. Collect the precipitated p-toluic acid by suction filtration and wash it with a little cold water. The yield of the crude acid, m.p. 174–176 °C, is 11.9 g (70%); recrystallisation from dilute alcohol gives pure p-toluic acid, m.p. 176–177 °C.

Evaporate the dried ethereal extract; the residue, m.p. 85–90 °C, weighs 3.3 g. Recrystallise it from alcohol: pure di-p-tolyl ketone, m.p. 95 °C, is obtained.

Note. (1) A convenient method of preparing the lithium shavings is as follows. Place a piece of lithium weighing about 3 g and slightly moist with paraffin oil on a dry surface (slate or tiles) and pound it with a clean hammer or 500-g weight into a thin sheet about 0.5 mm thick. Cut the sheet into thin strips about 2–3 mm wide and transfer it to a beaker containing anhydrous ether. Weigh out the quantity of lithium required under dry ether or paraffin oil. Dry each strip with filter paper, cut it by means of a pair of scissors into small pieces about 1 mm wide and allow the small pieces to fall directly into the anhydrous ether in the reaction flask. The lithium thus retains its bright lustre.

The lithium may also be pressed into wire of about 0.5 mm diameter; a rather sturdy press is necessary. The wire may be collected directly in sodium-dried ether.

Cognate preparation. m-*Chlorobenzoic acid*. *Prepare a solution of butyllithium in anhydrous ether as follows.*

Place 100 ml of sodium-dried ether in a 500-ml three-necked flask equipped as in *Method A*. Displace the air and maintain a slow stream of nitrogen throughout the experiment. Introduce 4.3 g (0.62 mol) of fine lithium shavings into the reaction flask. Place a solution of 34.5 g (26.5 ml, 0.25 mol) of butyl bromide (Expt 5.54) in 50 ml of anhydrous ether in the dropping funnel, start the stirrer and run in 1–2 ml of the solution into the reaction flask cooled to about −10 °C (Cardice–acetone bath). The reaction has commenced when bright spots appear on the lithium and the reaction mixture becomes slightly cloudy. Add the remainder of the butyl bromide solution during about 30 minutes while the internal temperature is maintained at about −10 °C. Then allow the reaction mixture to warm up to 0–10 °C during 1 hour (with stirring) in order to complete the formation of butyllithium (1).

Cool the solution of butyllithium to −35 °C in a Cardice–acetone bath and add, while stirring vigorously, a solution of 48 g of m-bromochlorobenzene (Expt 6.72) in 75 ml of anhydrous ether. Stir for 15 minutes and pour the mixture with stirring on to a large excess of solid carbon dioxide in the form of a Cardice–ether slurry contained in a large beaker. Isolate the acid as detailed above for p-*toluic acid* and recrystallise it from hot water. The yield of *m*-chlorobenzoic acid, m.p. 150–151 °C, is 27 g.

Note. (1) If a clear solution of butyllithium is required for any purpose, it may be decanted through a glass wool plug as detailed under *2-phenylpyridine* [Expt 8.32, Note (2)].

6.13.4 NUCLEAR NITRATION OF AROMATIC CARBOXYLIC ACIDS AND ESTERS

The carboxyl and alkoxycarbonyl groups exert an electron-withdrawing influence when attached to the aromatic ring system, and are thus deactivating and *meta* directing in electrophilic substitution reactions. Electrophilic substitution is illustrated by the nitration of methyl benzoate (Expt 6.159) and of benzoic acid and phenylacetic acid (Expt 6.160). In the former instance nitration is effected with 'mixed acid' at a temperature below 15 °C, conditions which effect monosubstitution. The resulting methyl *m*-nitrobenzoate may be hydrolysed to the corresponding nitro acid with aqueous alkali. Nitration of benzoic acid with a hot mixture of concentrated sulphuric acid and fuming nitric acid results in substitution in both *meta* positions to yield 3,5-dinitrobenzoic acid. Dinitration of phenylacetic acid is achieved using fuming nitric acid alone, the directive influence of the —CH$_2$·CO$_2$H group is to the *ortho* and *para* positions.

Experiment 6.159 *m*-NITROBENZOIC ACID

$$\text{Ph·CO}_2\text{Me} \xrightarrow[\text{H}_2\text{SO}_4]{\text{HNO}_3} m\text{-O}_2\text{N·C}_6\text{H}_4\text{·CO}_2\text{Me} \xrightarrow{\ominus\text{OH}} m\text{-O}_2\text{N·C}_6\text{H}_4\text{·CO}_2\text{H}$$

Methyl *m*-nitrobenzoate. In a 1-litre, round-bottomed three-necked flask, fitted with a mechanical stirrer and a thermometer, place 102 g (94 ml, 0.74 mol) of pure methyl benzoate (Expt 6.163). Prepare a mixture of 62.5 ml of concentrated sulphuric acid and 62.5 ml of concentrated nitric acid in a dropping funnel, cool the flask in an ice bath to 0–10 °C and then run in the nitrating mixture, with stirring, while maintaining the temperature of the reaction mixture between 5 and 15 °C; the addition requires about 1 hour. Continue the stirring for 15 minutes longer, and pour the mixture upon 700 g of crushed ice. Filter off the crude methyl *m*-nitrobenzoate at the pump and wash it with cold water. Transfer the solid to a 500-ml bolt-head flask and stir it with 100 ml of ice-cold methanol in order to remove a small amount of the *ortho* isomer and other impurities. Filter the cooled mixture with suction, wash it with 50 ml of ice-cold methanol and dry in the air. The practically colourless methyl *m*-nitrobenzoate weighs 115 g (84%) and melts at 75–76 °C; it is sufficiently pure for conversion into *m*-nitrobenzoic acid. The pure ester, m.p. 78 °C, may be obtained by recrystallisation from an equal weight of methanol.

Hydrolysis of methyl *m*-nitrobenzoate to *m*-nitrobenzoic acid. Place 90.5 g (0.5 mol) of methyl *m*-nitrobenzoate and a solution of 40 g of sodium hydroxide in 160 ml of water in a 1-litre round-bottomed flask equipped with a reflux condenser. Heat the mixture to boiling during 5–10 minutes or until the ester has disappeared. Dilute the reaction mixture with an equal volume of water. When cold pour the diluted reaction product, with vigorous stirring, into 125 ml of concentrated hydrochloric acid. Allow to cool to room temperature, filter the crude acid at the pump and wash it with a little water. Upon drying at 100 °C, the crude *m*-nitrobenzoic acid, which has a pale brownish colour, weighs 80 g (96%) and melts at 140 °C. Recrystallisation from 1 per cent hydrochloric acid affords the pure acid, m.p. 141 °C, as a pale cream solid; the loss of material is about 5 per cent.

The p.m.r. spectra of *m*-nitrobenzoic acid (CDCl$_3$ + TFA) and its ester (CDCl$_3$) should be recorded and the signals assigned appropriately.

Experiment 6.160   3,5-DINITROBENZOIC ACID

$$Ph \cdot CO_2H \xrightarrow[\text{H}_2\text{SO}_4]{\text{HNO}_3 \text{ (fuming)}} 3,5\text{-}(NO_2)_2C_6H_3 \cdot CO_2H$$

CAUTION: *This preparation must be carried out in the fume cupboard since nitrous fumes are evolved.*

Dissolve 50 g (0.41 mol) of pure benzoic acid in 230 ml of concentrated sulphuric acid in a litre flask equipped with a reflux condenser. Add 73 ml of fuming nitric acid (*d* 1.5) a few ml at a time (**CAUTION**). Shake the flask well and cool in ice-water during the addition; much heat is evolved and a clear yellow solution results. Add a few fragments of porous porcelain and heat the mixture gradually on a water bath to 100 °C during 45 minutes. At 70–80 °C the reaction may (and usually does) become vigorous; moderate, when necessary, by cooling the flask in cold water. Maintain the mixture at 100 °C for 15 minutes with occasional shaking, and then transfer it to an oil bath at 100 °C; raise the temperature to 130 °C over 30 minutes and keep it at 130–140 °C for 1 hour. Allow the flask to cool: crystals commence to separate at about 90 °C. When cold, pour the reaction mixture into 3–4 litres of ice-water, filter the separated crystals, wash with water and dry. The yield of 3,5-dinitrobenzoic acid, m.p. 204 °C, is 50 g (57%): this acid is pure enough for most purposes. Upon recrystallisation from 50 per cent alcohol (4.5 ml per gram), the m.p. is raised to 207 °C.

Cognate preparation. *2,4-Dinitrophenylacetic acid.* Place 25 g (0.184 mol) of phenylacetic acid (Expt 5.128) in a 500-ml round-bottomed flask, cool the latter in running water and add from a suitably supported dropping funnel 250 ml of fuming nitric acid, rather slowly at first and then more rapidly. The addition occupies about 15 minutes. Attach a reflux condenser to the flask and heat the mixture under reflux for 1 hour, and then carefully pour the cooled solution into 500 ml of cold water. When cold, filter the crude product at the pump and wash it with a little cold water: the resulting acid, after drying at 100 °C, is almost pure (m.p. 181 °C) and weighs 31 g. Recrystallise it from 300 ml of 20 per cent ethanol. Collect the first main crop (25 g), and allow the mother-liquor to stand overnight when a further 2 g of pure acid is obtained; dry at 100 °C. The yield of pure 2,4-dinitrophenylacetic acid, m.p. 183 °C, is 27 g (64%).

6.13.5 SOME METHODS FOR THE PROTECTION OF THE CARBOXYL GROUP

The esterification procedures suitable for the protection of aliphatic carboxylic acids (Section 5.11.8, p. 690) are equally applicable in the aromatic series. These methods for protection, and the methods for *deprotection* are exemplified under aliphatic acids and further illustrated in Sections 6.14.3, p. 1076, 9.6.15, p. 1261 and 9.6.17, p. 1266.

6.14 AROMATIC CARBOXYLIC ACID DERIVATIVES

1. Acid halides (Expt 6.161).
2. Acid anhydrides (Expt 6.162).
3. Esters (Expts 6.163 to 6.166).
4. Acid amides (Expt 6.167).

The structural interrelationships of these aromatic carboxylic acid derivatives and the retrosynthetic analyses that indicate the broad methods of synthesis are similar to the aliphatic analogues. The methods of preparation described in these sections are those which appear to work best in the aromatic field.

SPECTROSCOPIC FEATURES

All these derivatives show strong *i.r.* absorption arising from the stretching vibration of the carbonyl group. They may be distinguished from each other by inspection of the appropriate region for absorption arising from the carbonyl-bound halogen, acyloxy, alkyloxy, or amino groups (p. 296). The *p.m.r.* spectra provide predictable evidence for the pattern of aromatic substitution, for the structure of the alkyl group and for the presence of nitrogen-bound hydrogens. Suitable *ortho*-substituted compounds give expected fragmentation patterns in the *m.s.* (p. 383). Further descriptive analyses of the spectra of specific compounds are given in the preparative examples below.

6.14.1 ACID HALIDES

General methods for the preparation of acid halides from aliphatic carboxylic acids are described in Section 5.12.1, p. 692. Phosphorus pentachloride is the preferred chlorinating agent for aromatic acids which contain electron-withdrawing substituents, and which do not react readily with thionyl chloride. The preparation of both *p*-nitrobenzoyl chloride and 3,5-dinitrobenzoyl chloride is described in Expt 6.161. These particular acid chlorides are valuable reagents for the characterisation of aliphatic alcohols and simple phenols, with which they form crystalline esters (see Section 9.6.4, p. 1241 and Section 9.6.6, p. 1248).

Experiment 6.161   *p*-NITROBENZOYL CHLORIDE

$$p\text{-}O_2N\cdot C_6H_4\cdot CO_2H + PCl_5 \longrightarrow p\text{-}O_2N\cdot C_6H_4\cdot COCl + POCl_3 + HCl$$

Mix 100 g (0.6 mol) of pure *p*-nitrobenzoic acid (Expt 6.149) and 126 g (0.6 mol) of pure phosphorus pentachloride in a 500-ml round-bottomed flask. Fit the flask with a calcium chloride guard-tube and connect the latter to a gas absorption device (e.g. Fig. 2.61). Heat the flask on a water bath, with occasional shaking, until the reaction commences and then for a further 30 minutes or until the vigorous evolution of hydrogen chloride has almost ceased: a pale yellow homogeneous liquid is formed. Attach a Claisen still-head connected with a water-cooled condenser, and remove the phosphorus oxychloride (b.p. 107 °C) at ordinary pressure either by heating in an oil bath gradually to 200–220 °C or by heating in an air bath until the boiling point is about 150 °C. Allow to cool, replace the water condenser by a *short* air-cooled condenser and distil the residual liquid under reduced pressure (water pump) (1). A small quantity of phosphorus oxychloride passes over first and the temperature rises rapidly to about 150 °C/20 mmHg; change the receiver and collect the *p*-nitrobenzoyl chloride at 155 °C/20 mmHg. Pour the product while still fluid into a small wide-mouthed bottle and allow it to solidify: this prevents any moisture in the air from decomposing more than the surface layer of acid chloride. The yield of *p*-nitrobenzoyl chloride (a yellow crystalline solid, m.p. 71 °C) is 105 g (95%) and is pure enough for most purposes. A perfectly pure product, m.p. 73 °C, is obtained by recrystallising from carbon tetrachloride.

Note. (1) Either an oil bath (maintained at 210–215 °C for a pressure of 20 mmHg) or an air bath must be used. If the flask is heated with a free flame, superheating will occur leading to decomposition (sometimes violent) of the *p*-nitrobenzoyl chloride.

Cognate preparations. *3,5-Dinitrobenzoyl chloride.* Place a mixture of 30 g (0.141 mol) of 3,5-dinitrobenzoic acid (Expt 6.160) and 33 g (0.158 mol) of phosphorus pentachloride in a round-bottomed flask: fit a reflux condenser, and heat the mixture in an oil bath at 120–130 °C for 75 minutes. Allow to cool. Remove the phosphorus oxychloride by distillation under reduced pressure (25 °C/20 mmHg); raise the temperature of the bath to 110 °C. The residual 3,5-dinitrobenzoyl chloride solidifies on cooling to a brown mass; the yield is quantitative. Recrystallise from carbon tetrachloride: the yield is 25 g (77%), m.p. 67–68 °C, and this is satisfactory for most purposes. Further recrystallisation from a large volume of light petroleum, b.p. 40–60 °C, gives a perfectly pure product, m.p. 69.5 °C.

3,5-Dinitrobenzoyl chloride reacts readily with water and it should be kept in sealed tubes or under light petroleum. When required for the preparation of derivatives it is usually best prepared in small quantities from 3,5-dinitrobenzoic acid immediately before use.

2-Naphthoyl chloride. This compound may be prepared from 57.4 g (0.33 mol) of 2-naphthoic acid and 69 g (0.33 mol) of phosphorus penta-chloride following the procedure described above for *p*-nitrobenzoyl chloride. After removing the phosphorus oxychloride by distillation, the product is collected as a fraction of b.p. 160–162 °C/11 mmHg. This solidifies on cooling to a colourless solid, m.p. 51–52 °C; The yield is 60 g (95%).

CAUTION: *The preparation of* o-*nitrobenzoyl chloride,* o-*nitrophenacetyl chloride and all* o-*nitroacid chlorides should not be attempted by the above methods: a violent explosion may occur upon distilling the product.*

6.14.2 ACID ANHYDRIDES

o-Dicarboxylic acids (e.g. phthalic acid) readily form intramolecular anhydrides on heating. In Expt 6.162 the nitration of phthalic anhydride to yield a mixture of the isomeric 3- and 4-nitrophthalic acids is described, which can be separated by fractional crystallisation from water. The 3-nitrophthalic acid is efficiently converted into the corresponding anhydride by warming with acetic anhydride and allowing the product to crystallise.

Phthalic anhydrides readily form hydrogen phthalate esters on reaction with alcohols; the derivatives from 3-nitrophthalic anhydride are usually nicely crystalline compounds and are hence suitable for purposes of characterisation. Hydrogen phthalate esters are also useful in appropriate instances for the resolution of racemic alcohols (Section 5.19).

Experiment 6.162 3-NITROPHTHALIC ANHYDRIDE

CAUTIONARY NOTE: *The nitration of phthalic anhydride has been reported to proceed with explosive violence.[60a] The reaction may be carried out with safety on the scale adopted here provided that strict attention to detail is paid to the rate of addition of the acid and to the control of the temperature. A modified procedure for use in large-scale nitration has been published.[60b]*

3-Nitrophthalic acid. Equip a 500-ml, three-necked round-bottomed flask, supported on a water bath, with a dropping funnel, a thermometer and a stirrer supported in the central (open) neck connected by means of a flexible drive to a stirrer motor. The latter should not be sited near the open neck since the nitrous fumes evolved in the subsequent reaction may otherwise cause damage. Place 100 g (0.675 mol) of technical phthalic anhydride and 100 ml of concentrated sulphuric acid in the flask and heat it until the temperature of the mixture rises to 80 °C. Remove the water bath, and add a mixture of 42 ml of fuming nitric acid (*d* 1.5) and 30 ml of concentrated sulphuric acid slowly from the dropping funnel at such a rate as to maintain the temperature of the stirred mixture at 100–110 °C (about 1 hour). Then add 180 ml of concentrated nitric acid (*d* 1.42) as rapidly as possible without causing the temperature to rise above 110 °C. Heat the mixture on the water bath, with stirring, for 2 hours. Allow the reaction mixture to stand overnight and then pour it into 300 ml of water contained in a 2-litre beaker. Cool and filter the mixture of 3- and 4-nitrophthalic acids through a sintered glass funnel. Return the wet cake of acids to the rinsed-out beaker and stir it thoroughly with 40 ml of water, which dissolves a large amount of the 4-nitrophthalic acid (1). Filter again at the pump and dissolve the solid in 40–60 ml of boiling water; filter the hot solution and stir until crystallisation commences and then leave overnight until crystallisation is complete. Filter again with suction and dry upon filter paper. The yield of crude 3-nitrophthalic acid, m.p. 208–210 °C (sealed tube), is 44 g. Recrystallisation from about 100 ml of boiling water (2) gives about 36 g (25%) of the pure acid, m.p. 216–218 °C (sealed tube).

Notes. (1) The mother-liquors from the washings and recrystallisations are saved for the recovery of 4-nitrophthalic acid. The combined mother-liquors are concentrated to small bulk and the organic acids extracted into ether. Upon esterification of the residue after evaporation of the ether by the Fischer–Speier method (Section 5.12.3, p. 695), the 3-nitro acid forms the acid ester and may be removed by shaking the product with sodium carbonate solution, while the 4-nitrophthalic acid yields the neutral diester. Hydrolysis of the neutral ester gives the pure 4-nitrophthalic acid, m.p. 165 °C. (2) The acid may also be recrystallised from glacial acetic acid.

3-Nitrophthalic anhydride. In a 100-ml round-bottomed flask fitted with a reflux condenser, place 21 g (0.1 mol) of 3-nitrophthalic acid and 20 g (18.5 ml, 0.2 mol) of redistilled acetic anhydride. Heat the mixture to gentle boiling until a clear solution is obtained, and then for about 10 minutes longer. Pour the hot mixture (fume cupboard) into a large porcelain dish and allow to cool. Grind the crystalline mass thoroughly in a mortar and filter at the pump through a sintered glass funnel. Return the crystals to the mortar, grind them with 15 ml of sodium-dried ether and filter. Again return the crystals to the mortar and wash once more with 15 ml of dry, alcohol-free ether. Dry in air for a short time, and then to constant weight at 100 °C. The yield of 3-nitrophthalic anhydride, m.p. 163–164 °C, is 17 g (88%). If the m.p. is unsatisfactory, recrystallise the anhydride from benzene or from benzene–light petroleum (b.p. 40–60 °C).

6.14.3 ESTERS

Aromatic esters may be prepared by direct esterification methods similar to those already described for aliphatic esters (Section 5.12.3, p. 695). A large range of examples of simple alkyl esters of aromatic carboxylic acids is included in Expt 6.163. Corresponding esterification of a simple aliphatic acid (e.g. acetic acid) with benzyl alcohol is illustrated in Expt 5.142.

Methyl esters may be prepared by reaction of the aromatic carboxylic acid with diazomethane (cf. Section 4.2.25, p. 433) or, more conveniently, by reaction with a boron trifluoride–methanol reagent. The latter procedure is illustrated by the preparation of methyl m-chlorobenzoate and dimethyl terephthalate (Expt 6.164). t-Butyl esters may be prepared by conversion of the acid into an N-acylimidazole by reaction with N,N'-carbonyldiimidazole, followed by reaction with t-butyl alcohol in the presence of DBU[62] (Expt 6.165).

Esterification of aromatic carboxylic acids with phenols, however, cannot be accomplished by a direct esterification procedure and resort must be made to the greater reactivity exhibited by the acid chlorides. Reaction is usually carried out in dilute aqueous alkali (Schotten–Baumann conditions, Section 6.6.2, pp. 916 and 1248). The preparation of 2-naphthyl benzoate is a typical example of this procedure (Expt 6.166).

The preparation of a carboxylic ester by a variant of the Grignard carboxylation route has been described.[61] The arylmagnesium bromide is first prepared and added to an excess of diethyl carbonate, conditions which minimise the possibility of further reaction of the Grignard reagent with the ester initially produced to form the tertiary alcohol.

$$Ar\cdot MgBr + (EtO)_2CO \longrightarrow Ar-\underset{\underset{OEt}{|}}{\overset{\overset{OEt}{|}}{C}}-O\cdot MgX \xrightarrow{-Mg(OEt)X} Ar\cdot CO_2Et$$

Experiment 6.163 METHYL BENZOATE

$$Ph\cdot CO_2H + MeOH \xrightarrow{H^{\oplus}} Ph\cdot CO_2Me + H_2O$$

In a 500-ml round-bottomed flask place a mixture of 30 g (0.246 mol) of benzoic acid, 80 g (101 ml, 2.5 mol) of absolute methanol and 5 g (2.7 ml) of concentrated sulphuric acid. Add a few small chips of porous porcelain, attach a reflux condenser and boil the mixture gently for 4 hours (1). Distil off the excess of alcohol on a water bath (rotary evaporator) and allow to cool. Pour the residue into about 250 ml of water contained in a separatory funnel and rinse the flask with a few ml of water which are also poured into the separatory funnel. If, owing to the comparatively slight difference between the density of the ester and of water, difficulty is experienced in obtaining a sharp separation of the lower ester layer and water, add 10–15 ml of carbon tetrachloride (2) and shake the mixture in the funnel vigorously; upon standing, the heavy solution of methyl benzoate in the carbon tetrachloride separates sharply and rapidly at the bottom of the separatory funnel. Run off the lower layer carefully, reject the upper aqueous layer, return the methyl benzoate to the funnel and shake it with a strong solution of sodium hydrogen carbonate until all free acid is removed and no further evolution of carbon dioxide occurs. Wash once with water, and dry by pouring into a small dry conical flask containing about 5 g of magnesium sulphate. Stopper the flask, shake for about 5 minutes and allow to stand for at least half an hour with occasional shaking. Filter the methyl benzoate solution through a small fluted filter paper directly into a round-bottomed flask fitted with a still-head carrying a 360 °C thermometer and an air condenser. Add a few boiling chips and distil from an air bath; raise the temperature *slowly* at first until all carbon tetrachloride has passed over and then heat more strongly. Collect the methyl benzoate (a colourless liquid) at 198–200 °C. The yield is 31 g (92%).

Notes. (1) Slightly improved results may be obtained by increasing the time of heating.
(2) Alternatively, the ester may be extracted with two 50 ml portions of ether. The ethereal solution is washed with concentrated sodium hydrogen carbonate solution (handle the separatory funnel cautiously as carbon dioxide is evolved) until effervescence ceases, then with water, and dried over magnesium sulphate. The ether is removed by flash distillation and the residual ester distilled.

Cognate preparations. [The p.m.r. spectra (CDCl$_3$ or CCl$_4$, TMS) of the ethyl, propyl and butyl esters should be recorded and interpreted, since the signals corresponding to the protons of these alkyl groups, afford good examples of the (N + 1)(M + 1) rule for multiplicity (p. 341).]

Ethyl benzoate (sulphuric acid as a catalyst). Use 30 g (0.246 mol) of benzoic acid, 115 g (145 ml, 2.5 mol) of absolute ethanol and 5 g (2.7 ml) of concentrated sulphuric acid. Reflux the mixture for 4 hours and work up as for *methyl benzoate*. The yield of ethyl benzoate, b.p. 212–214 °C, is 32 g (86%).

Propyl benzoate. Into a 500-ml round-bottomed flask place 30 g (0.246 mol) of benzoic acid, 30 g (37.5 ml, 0.5 mol) of propan-1-ol, 50 ml of sodium-dried toluene and 10 g (5.4 ml) of concentrated sulphuric acid. Reflux the mixture for 10 hours. Pour the reaction product into about 250 ml of water, and extract with ether. Wash the ethereal extract with saturated sodium hydrogen carbonate solution and then with water: dry over magnesium sulphate. Distil off the ether and some of the toluene through a fractionating column, and distil the residue from a Claisen flask. Collect the propyl benzoate at 229–230 °C. The yield is 37 g (91%).

Butyl benzoate. Use 30 g (0.246 mol) of benzoic acid, 37 g (46 ml, 0.5 mol) of butan-1-ol, 50 ml of sodium-dried toluene and 10 g (5.4 ml) of concentrated sulphuric acid, and reflux the mixture for 12 hours. Work up the product as for *propyl benzoate*; after the ether and toluene have been removed with the aid of a rotary evaporator, distil the residue under reduced pressure. The yield of butyl benzoate, b.p. 119–120 °C/11 mmHg, is 35 g (80%).

Methyl salicylate. Use 28 g (0.2 mol) of salicylic acid, 64 g (81 ml, 2 mol) of dry methanol and 8 ml of concentrated sulphuric acid. Reflux the mixture for at least 5 hours and work up as for *methyl benzoate*. Collect the pure methyl salicylate (a colourless oil of delightful fragrance, 'oil of wintergreen') at 221–224 °C; the yield is 25 g (81%). The ester may also be distilled under reduced pressure; the b.p. is 115 °C/20 mmHg and a 2 °C fraction should be collected.

Ethyl salicylate. This colourless ester, b.p. 231–234 °C, is similarly obtained in 75 per cent yield from salicylic acid, ethanol and sulphuric acid as catalyst. It is best to distil the ester under reduced pressure; the boiling points under various pressures are given in Table 2.10.

Methyl cinnamate. Use 59 g (0.4 mol) of cinnamic acid (Expt 6.138), 128 g (162 ml, 4 mol) of absolute methanol and 6 ml of concentrated sulphuric acid; reflux the mixture for 5 hours. Remove excess methanol, pour the residue into about 500 ml of water and add 300 ml of ether. Separate, wash and dry the ether solution in the usual way. Remove the ether on a rotary evaporator; the residue crystallises on cooling, yielding 58 g (90%) of methyl cinnamate, m.p. 33–34 °C. To obtain a pure specimen, m.p. 36 °C, dissolve a sample in the minimum of methanol maintained at 30 °C in a water bath, and add water slowly from a dropping pipette with stirring until the oily ester just begins to separate. Seed the solution and transfer rapidly to an ice bath with a glass rod, scratching the sides of the vessel vigorously with a glass rod. Filter the resulting colourless needles rapidly.

Ethyl nicotinate (*ethyl pyridine-3-carboxylate*). Reflux a mixture of 37 g (0.3 mol) pure nicotinic acid (Expt 6.150), 92 g (115 ml, 2 mol) of absolute ethanol and 90 g (50 ml) of concentrated sulphuric acid on a steam bath. Cool the solution and pour it slowly and with stirring on to 200 g of crushed ice. Add sufficient ammonia solution to render the resulting solution strongly alkaline: generally, some ester separates as an oil but most of it remains dissolved in the alkaline solution. Extract the mixture with five 25 ml portions of ether, dry the combined ethereal extracts over magnesium sulphate, remove the ether by flash distillation and distil the residue under reduced pressure. The ethyl nicotinate distils at 117–118 °C/16 mmHg; the yield is 32 g (71%). The boiling point under atmospheric pressure is 222–224 °C.

Experiment 6.164   METHYL *m*-CHLOROBENZOATE

$$m\text{-Cl·C}_6\text{H}_4\text{·CO}_2\text{H} \xrightarrow{\text{BF}_3/\text{Me·CO}_2\text{H}} m\text{-Cl·C}_6\text{H}_4\text{·CO}_2\text{Me}$$

Place 9.4 g (0.06 mol) of *m*-chlorobenzoic acid and 66 ml (0.12 mol) of boron trifluoride–methanol complex (14% w/v of BF_3; Section 4.2.8, p. 421) in a 250-ml round-bottomed flask. Heat the mixture under reflux on an oil bath for 2 hours, cool and pour into about 250 ml of saturated sodium hydrogen carbonate solution. Extract the organic product with three 50 ml portions of ether, dry the ethereal extract over magnesium sulphate and evaporate on a rotary evaporator. Distil the residue under reduced pressure and collect the methyl *m*-chlorobenzoate as a colourless liquid of b.p. 63 °C/3 mmHg; the yield is 9.3 g (91%).

Cognate preparation. *Dimethyl terephthalate.* Use 9.97 g (0.06 mol) of terephthalic acid, 132 ml of boron trifluoride–methanol complex (14% w/v) and 100 ml of dry methanol (1) and heat the mixture under reflux for 6 hours. Cool the reaction mixture and pour into excess (500 ml) of saturated sodium hydrogen carbonate solution. Filter the precipitated dicarboxylic ester under suction and recrystallise from methanol. The yield is 9.8 g (84%); m.p. 139–140 °C. The p.m.r. spectrum (CDCl$_3$, TMS) of this ester should be recorded and compared with those of the *ortho* and *meta* isomers. It should be noted that dimethyl terephthalate gives an A_4 pattern for the aromatic protons, the *ortho* isomer gives an A_2B_2 splitting pattern, and the *meta* isomer shows signals at δ 7.50 (t, 1H, C$_5$—H), 8.19 (d of d, 2H, C$_{4,6}$—H) and 8.63 (t, 1H, C$_2$—H) for the aromatic region of the spectrum.

Note. (1) Additional methanol is used in this case because of the low solubility of the di-acid and di-ester. Dry the methanol by distillation from magnesium (Section 4.1.8).

Experiment 6.165   t-BUTYL *o*-CHLOROBENZOATE[62]

N,N'-Carbonyldiimidazole (1.65 g, 10 mmol) was added to a solution of *o*-chlorobenzoic acid (1.57 g, 10 mmol) in dimethylformamide (10 ml) held under nitrogen, and the mixture was stirred for 1 hour at 40 °C. t-Butyl alcohol (1.48 g, 20 mmol) and DBU (1,8-diazabicyclo[5.4.0]undec-7-ene, 1.52 g, 10 mmol) were added and the mixture stood for 24 hours at 40 °C. Ether (100 ml) was then added, and the solution was washed with 10 per cent hydrochloric acid (20 ml), water (20 ml), aqueous potassium hydrogen carbonate (20 ml) and dried over anhydrous sodium sulphate. The solvent was removed and the oily residue distilled *in vacuo* to yield the ester (1.80 g, 85%), b.p. 135–140 °C/2 mmHg.

Experiment 6.166 2-NAPHTHYL BENZOATE

$$2\text{-}C_{10}H_7OH + Ph\cdot COCl \xrightarrow{\text{NaOH}} 2\text{-}C_{10}H_7O\cdot CO\cdot Ph$$

Dissolve 7.2 g (0.05 mol) of 2-naphthol in 40 ml of 5 per cent sodium hydroxide solution in the cold; add a little more water if necessary. If the solution is highly coloured, add 1.5 g of decolourising carbon and filter the cold solution through a hardened filter paper. Pour the solution into a 100-ml conical flask and run in 7.0 g (5.8 ml, 0.05 mol) of benzoyl chloride. Stopper the flask and shake vigorously until the odour of benzoyl chloride has disappeared (10–15 minutes). Filter off the solid product on a Buchner funnel and wash it with a little cold water. Recrystallise it from about 60 ml of rectified spirit. Filter off the crystals which separate and dry them upon filter paper in the air. The yield of pure 2-naphthyl benzoate, m.p. 110 °C, is 11 g (89%).

6.14.4 ACID AMIDES

The preparation of toluamide (Expt 6.167) illustrates a useful procedure for the conversion of aromatic nitriles into acid amides with the aid of alkaline hydrogen peroxides (see discussion, Section 5.12.4, p. 708).

A further example of the preparation of amides by the ammonolysis of esters (cf. succinamide, Expt 5.155) is provided by the preparation of nicotinamide described in Expt 6.169 as a stage in the synthesis of 3-cyanopyridine.

Experiment 6.167 *o*-TOLUAMIDE

$$2\ o\text{-}Me\cdot C_6H_4\cdot CN + 2H_2O_2 \xrightarrow{\ominus OH} 2\ o\text{-}Me\cdot C_6H_4\cdot CO\cdot NH_2 + O_2$$

Place 29 g (0.25 mol) of *o*-tolunitrile (Expt 6.76), 130 ml of rectified spirit (1) and 10 ml of 25 per cent sodium hydroxide solution in a 1-litre bolt-necked flask. Set up the flask inside an aluminium bowl placed on a magnetic stirrer/hotplate unit which is supported on an adjustable laboratory jack. Insert into the flask a magnetic stirrer follower and a thermometer, start the stirrer motor and run in steadily 100 ml of 30 per cent hydrogen peroxide (**CAUTION**: *see Section 4.2.41, p. 439 for precautions in the use of this reagent*) (2). Oxygen is soon evolved and the mixture becomes warm; when the temperature approaches 40 °C add an ice-water slurry to the bowl in order that the temperature may be controlled within the 40–50 °C range. If the temperature tends to fall below this range it may be necessary to remove the cooling bath for a while. If the temperature is permitted to rise above 50 °C not only may the evolution of oxygen become so rapid as to cause the mixture to foam out of the flask, but also there is a danger of a violent explosion due to the ignition of an oxygen–ethanol vapour mixture (3). The exothermic reaction is complete after about 1 hour and then the temperature of the reaction mixture is kept at 50 °C for a further 3 hours by external heating. While still warm, add 5 per cent sulphuric acid until exactly neutral to litmus, and remove the ethanol and concentrate the residue to about

200 ml under reduced pressure using a rotary evaporator. Cool the residue to 20 °C, filter off the crystals at the pump and grind them to a paste with 30 ml of cold water in a mortar. Filter again, wash the product in the filter with a further 30 ml of water and dry in the air upon filter paper. The yield of *o*-toluamide (white crystals), m.p. 141 °C, is 30 g (90%). It may be recrystallised from hot water (1 g per ml), but the m.p. is unchanged.

Notes. (1) This volume of rectified spirit is required to produce a homogeneous solution.

(2) Difficultly hydrolysable nitriles, such as *o*-tolunitrile, require 30 per cent hydrogen peroxide. For most nitriles, however, both aromatic and aliphatic, an equivalent amount of 6–12 per cent hydrogen peroxide gives more satisfactory results; the above procedure must, however, be modified, according to the solubility of the nitriles and amides.

(3) To minimise this explosion danger, the reaction should be carried out in a well-ventilated fume cupboard in the absence of free flames; brisk stirring should be maintained throughout the exothermal stage.

6.15 AROMATIC NITRILES

1. The Sandmeyer procedure (see Expt 6.76).
2. The displacement of halogen by cyanide in an aryl halide (Expt 6.168).
3. The dehydration of amides and aldoximes (Expts 6.169 and 6.170).

SUMMARY OF RETROSYNTHETIC STRATEGIES

Functional group interconversion (method 3), e.g.

Disconnection (methods 1 and 2), e.g.

The *i.r.* spectrum of benzonitrile is reproduced on p. 315. An illustrative spectral analysis for *o*- and *p*-tolunitrile is to be found under Expt 6.76.

6.15.1 THE SANDMEYER PROCEDURE

This valuable method for the preparation of aryl nitriles via the diazonium salt is discussed in Section 6.7.1, p. 922, and offers one of the most convenient routes for obtaining this class of compound. Experimental procedures are described in Expt 6.76.

6.15.2 THE DISPLACEMENT OF HALOGEN BY CYANIDE IN AN ARYL HALIDE

The ready replacement of the halogen in an alkyl or an aralkyl halide illustrated in Expt 5.157 by reaction with sodium or potassium cyanide is inapplicable in the case of aryl halides wherein the halogen is relatively inert. However, aryl bromides can be converted into nitriles in good yield by heating them for several hours at about 200 °C with copper(I) cyanide in the presence of pyridine (e.g. 1-naphthonitrile, Expt 6.168). This displacement may be achieved more readily by using dimethylformamide as the solvent, when reaction is usually completed in a few hours at reflux temperature.[63]

Experiment 6.168 1-NAPHTHONITRILE

$$1\text{-}C_{10}H_7Br \xrightarrow[\text{pyridine}]{\text{CuCN}} 1\text{-}C_{10}H_7 \cdot CN$$

Place 80 g (54 ml, 0.386 mol) of redistilled 1-bromonaphthalene (Expt 6.125), 43 g (0.48 mol) of dry powdered copper(I) cyanide (Section 4.2.23, p. 429) and 36 g (37 ml, 0.457 mol) of dry pure pyridine (**CAUTION**) (1) (Section 4.1.29, p. 410) in a 250-ml round-bottomed flask fitted with a reflux condenser carrying a calcium chloride guard-tube, and heat the mixture in a metal bath at 215–225 °C for 15 hours (2). Pour the resulting dark brown solution while still hot (c. 100 °C) into a litre flask containing 180 ml of concentrated ammonia solution (d 0.88) and 180 ml of water. Add 170 ml of toluene (**CAUTION**), stopper the flask and shake until all the lumps have disintegrated. When cold, add 100 ml of ether and filter through a sintered glass funnel (3). Add a further 50 ml of ether, transfer to a separatory funnel, separate the ether–toluene layer and wash it successively with: (i) four 125 ml portions of dilute ammonia solution (or until the organic layer is colourless); (ii) two 125 ml portions of dilute hydrochloric acid (1:1) (any precipitate which separates should be filtered off); (iii) two 125 ml portions of water; and (iv) two 125 ml portions of saturated sodium chloride solution. Finally, dry with magnesium sulphate, remove the ether and toluene by distillation on a rotary evaporator and distil the residue under reduced pressure (water pump) through a short fractionating column (compare Fig. 2.108). Collect the 1-naphthonitrile at 166–169 °C/18 mmHg as a colourless liquid. The yield is 50 g (84%).

Notes. (1) Much heat is liberated when pyridine is added to the mixture.
(2) The metal bath may be replaced by a bath of Silicone oil.
(3) The cuprammonium solution attacks filter paper.

6.15.3 THE DEHYDRATION OF AMIDES AND ALDOXIMES

Two examples of the dehydration of aromatic carboxamides using phosphorus pentoxide (cf. Section 5.13.3, p. 715) are given in Expt 6.169; these are the preparation of benzonitrile and 3-cyanopyridine.

The one-pot conversion of an aromatic carboxylic acid into the corresponding nitrile may be effected by reaction with ammonia in the presence of ethyl polyphosphate[64]; the initial stage is the formation of the amide and this is followed by dehydration to the nitrile. The reaction has been used successfully with a range of aromatic carboxylic acids.

The indirect conversion of an aromatic aldehyde into the corresponding nitrile by dehydration of an oxime is illustrated by the synthesis of veratronitrile (Expt 6.170). The dehydrating agent is acetic anhydride which probably effects an initial acetylation of the oximino group followed by the elimination of acetic acid.

$$Ar \cdot CHO \longrightarrow Ar \cdot CH{=}NOH \longrightarrow Ar \cdot CH{=}N \cdot O \cdot CO \cdot Me$$

$$\longrightarrow Ar \cdot C{\equiv}N$$

Conditions for effecting this conversion in one step have been described.[65] The method involves heating a mixture of the aldehyde, hydroxylamine hydrochloride, sodium formate and formic acid; and the reaction is considered to proceed through the intermediate formation of an oxime formate.

Experiment 6.169 BENZONITRILE

$$Ph \cdot CO \cdot NH_2 \xrightarrow{P_2O_5} Ph \cdot CN$$

Place 45 g (0.37 mol) of benzamide (prepared from benzoyl chloride, cf. Expt 5.154) and 80 g (0.56 mol) of phosphorus pentoxide in a 250-ml round-bottomed flask (for exact experimental details on the handling and weighing out of phosphoric oxide, see *isobutyronitrile*, Expt 5.160). Mix well. Fit the flask with a Claisen distillation head (Fig. 2.13(*c*)) and distil under reduced pressure using a water pump with an air leak in the system so that a pressure of about 100 mmHg is attained. Heat the flask with a free flame until no more liquid distils: the nitrile will pass over at 126–130 °C/100 mmHg. Wash the distillate with a little sodium carbonate solution, then with water, and dry over anhydrous calcium chloride or magnesium sulphate. Distil under normal pressure from a 50-ml flask: the benzonitrile passes over as a colourless liquid at 188–189 °C. The yield is 28 g (74%). The i.r. spectrum of benzonitrile is shown in Fig. 3.39.

Cognate preparation. *3-Cyanopyridine*

Nicotinamide. Place 50 g (0.33 mol) of pure ethyl nicotinate (Expt 6.163) in a 350-ml flask and add 75 ml of cold concentrated aqueous ammonia saturated at 0 °C. Keep the flask loosely stoppered for 18 hours, after which time the lower layer generally dissolves on shaking. Again saturate the solution with ammonia and allow it to stand for a further 4 hours. Repeat the saturation with ammonia; crystals of the amide commence to appear in the solution. Evaporate to dryness in a dish on the steam bath and dry at 120 °C. The yield of nicotinamide, m.p. 130 °C, is usually quantitative.

3-Cyanopyridine. Mix 24 g (0.2 mol) of powdered nicotinamide with 30 g of phosphoric oxide in a 150-ml round-bottomed flask by shaking. Immerse the flask in a Silicone oil or fusible metal bath and arrange for distillation under a pressure of about 30 mmHg. Raise the temperature of the bath rapidly to 300 °C, then remove the bath and continue the heating with a free flame as long as a distillate is obtained. The nitrile crystallises on cooling to a snow-white solid. Redistil the solid at atmospheric pressure; practically all of it passes over at 201 °C and crystallises completely on cooling. The yield of 3-cyanopyridine, m.p. 49 °C, is 18 g (86%).

General procedure for the direct conversion of a carboxylic acid into a nitrile.[64] In a 50-ml three-necked flask are placed the carboxylic acid (0.01 mol), ethyl polyphosphate (6 g, PPE) and purified chloroform (5 ml). The mixture is cooled in an ice bath and the flask is connected to a balloon containing ammonia gas (≈ 3 litres). Air in the flask is replaced with ammonia and the mixture is mechanically stirred at 0–5 °C for 30 minutes and then at room temperature for one and a half hours whereupon the mixture turns very viscous (1). The balloon is removed and PPE (10 g) is added. The stirring is continued at 80 °C until the reaction is complete (usually within several hours); the dehydration is monitored by t.l.c. analysis (1). The mixture is stirred with aqueous 25 per cent sodium carbonate solution (150 ml), and then extracted with benzene (3 × 40 ml; **CAUTION**). The combined organic extracts are dried with sodium sulphate and evaporated. The residual oil is passed through a short column packed with silica gel (≈ 20 g) and the product eluted with benzene. The eluate is evaporated and the residue purified by short path distillation under reduced pressure (Kugelrohr apparatus).

Note. (1) Formation of the amide and the nitrile may be monitored by t.l.c. analysis.

Experiment 6.170 VERATRONITRILE

Dissolve 83 g (0.5 mol) of veratraldehyde (Expt 6.111) in 200 ml of warm rectified spirit in a 1-litre round-bottomed flask, and add a warm solution of 42 g (0.6 mol) of hydroxylamine hydrochloride in 50 ml of water. Mix thoroughly and run in a solution of 30 g of sodium hydroxide in 40 ml of water. Allow the mixture to stand for 2.5 hours, add 250 g of crushed ice and saturate the solution with carbon dioxide. The aldoxime separates as an oil: allow the mixture to stand for 12–24 hours in an ice chest or refrigerator when the oil will solidify. Filter off the crystalline aldoxime at the pump, wash well with cold water and dry in the air upon filter paper. The yield of *veratraldoxime* is 88 g (96%).

Into a 250-ml round-bottomed flask, fitted with a reflux air condenser, place 88 g of veratraldoxime and 100 g (92.5 ml) of redistilled acetic anhydride. Heat cautiously. Immediately the vigorous reaction commences, remove the flame. When the reaction subsides, boil the solution gently for 20 minutes and then pour it carefully with stirring into 300 ml of cold water.

Continue the stirring and cool in ice. Filter off the almost colourless crystals of veratronitrile and dry in the air. The resulting nitrile (60 g, 73.5% overall) is quite pure and melts at 67 °C.

REFERENCES

1. G. Kabalka and J. H. Chandler (1979). *Synth. Commun.*, **9**, 275.
2. W. E. Bachmann and R. A. Hofmann (1944). 'The preparation of unsymmetrical biaryls by the diazo reaction and the nitrosoacetylamine reaction', in *Organic Reactions* (2). Ed. R. Adams. New York: Wiley, p. 224.
3. S. H. Korzeniowski, L. Blum and G. W. Gokel (1977). *Tetrahedron Lett.*, 1871.
4. J. Jacques and C. Fouguay (1971). *Tetrahedron Lett*, 4617.
5. B. Feringa and H. Wynberg (1978). *Bioorg. Chem.*, **7**, 397.
6. J. Brassec and A. C. A. Jansen (1983). *Tetrahedron Lett.*, 3261.
7. K. Yamanoto, H. Fushima and M. Nakazaki (1984). *J. Chem. Soc., Chem. Commun.*, 1490.
8. A. I. Meyers and K. A. Lutomski (1982). *J. Am. Chem. Soc.*, **104**, 879.
9. L. T. Scott, M. A. Minton and M. A. Kirma (1980). *J. Am. Chem. Soc.*, **102**, 6311.
10. A. R. Butler and A. P. Sanderson (1971). *J. Chem. Soc. B*, 2264.
11. T. Sugita, M. Idei, Y. Ishibashi and Y. Takegami (1982). *Chem. Lett.*, 1481.
12. (a) H. Stephen, W. F. Short and G. Gladding (1920). *J. Chem. Soc.*, 510.
    (b) S. R. Buc. *Organic Syntheses*, Coll. Vol. IV, p. 101.
13. Anon (1972). In *Organic Reactions* (19). Ed. W. G. Dauben. New York: Wiley, p. 422.
14. A. McKillop, F. A. Madjdabadi and D. A. Long (1983). *Tetrahedron Lett.*, 1933.
15. D. R. Maudling, K. D. Lotts and S. A. Robinson (1983). *J. Org. Chem.*, **48**, 2938.
16. R. A. Sherrer and H. R. Beatty (1980). *J. Org. Chem.*, **45**, 2127.
17. L. F. Fieser and M. J. Haddadin. *Organic Syntheses*, Coll. Vol. V, p. 1037.
18. (a) D. B. Dess and J. C. Martin (1983). *J. Org. Chem.*, **48**, 4155.
    (b) R. L. Amey and J. C. Martin (1979). *J. Am. Chem. Soc.*, **101**, 5294.
19. R. Lohrmann and H. G. Khorana (1966). *J. Am. Chem. Soc.*, **88**, 829.
20. N. J. Cusak, C. B. Reese, A. C. Risius and B. Roozpeikar (1976). *Tetrahedron*, **32**, 2157.
21. G. W. Kabalka, M. Varma, R. S. Varma, P. C. Srivastava and F. F. Knapp, Jr (1986). *J. Org. Chem.*, **51**, 2386.
22. R. Adams and C. S. Marvel. *Organic Syntheses*, Coll. Vol. I, p. 504.
23. S. A. Mahood and P. V. L. Schaffner. *Organic Syntheses*, Coll. Vol. II, p. 160.
24. R. S. Neale, R. G. Schepers and M. R. Walsh (1964). *J. Org. Chem.*, **29**, 3390.
25. N. R. Ayyanger, U. R. Kalkote and P. V. Nikrand (1982). *Tetrahedron Lett.*, 1099.
26. P. G. Gassman and H. R. Drewes (1978). *J. Am. Chem. Soc.*, **100**, 7600.
27. T. Cohen, A. G. Dietz, Jr, and J. R. Miser (1977). *J. Org. Chem.*, **42**, 2053.
28. M. P. Doyle, B. Siegfried and J. F. Dellavia, Jr (1977). *J. Org. Chem.*, **42**, 2426.
29. K. G. Rutherford, W. Redmond and J. Rigamondi (1961). *J. Org. Chem.*, **26**, 5149.
30. N. Kornblum. *Organic Syntheses*, Coll. Vol. III, p. 295.
31. M. P. Doyle, J. F. Dellaria, Jr, B. Siegfried and S. W. Bishop (1977). *J. Org. Chem.*, **42**, 3494.
32. E. Bamberger (1919). *Chem. Ber.*, **52**, 1116.
33. B. T. Newbold and R. P. Leblanc (1965). *J. Chem. Soc.*, 1547.
34. S. Oae and T. Sakurai (1975). *Bull. Chem. Soc. (Japan)*, **48**, 3759.
35. D. E. Ryan and G. D. Lutwick (1953). *Can. J. Chem.*, **31**, 9.
36. M. F. Hawthorne (1957). *J. Org. Chem.*, **22**, 1001.
37. J. B. Hamlin (1984). *Chem. Ber.*, **117**, 402.
38. A. McKillop, J.-C. Fiaud and R. P. Hug (1974). *Tetrahedron*, **30**, 1379.
39. J. Dockx (1973). *Synthesis*, 441.
40. M. E. Jung and M. A. Lyster (1977). *J. Org. Chem.*, **42**, 3761.

41. (a) E. H. Vickery, L. F. Pahler and E. J. Eisenbraun (1979). *J. Org. Chem.*, **44**, 4444.
    (b) J. F. W. McOmie, M. L. Watts and D. E. West (1968). *Tetrahedron*, **24**, 2289.
42. R. Ross and R. Scheffold (1976). *Angew. Chem., Int. Ed. Engl.*, **15**, 558.
43. T. Kametani, S-P. Huang, M. Ihara and K. Fukumoto (1976). *J. Org. Chem.*, **41**, 2545.
44. A. Bashall and J. F. Collins (1975). *Tetrahedron Lett.*, 3489.
45. G. H. Coleman and D. Craig. *Organic Syntheses*, Coll. Vol. II, p. 583.
46. G. Casiraghi, G. Casnati, G. Puglia, G. Sartori and G. Terenghi (1980). *J. Chem. Soc., Perkin Trans I*, 1862.
47. A. Rahm, R. Guilhemat and M. Pereyre (1982). *Synth. Commun.*, **12**, 485.
48. P. G. Gassman and D. R. Amick (1978). *J. Am. Chem. Soc.*, **100**, 7611.
49. Chapter 4, p. 469, see reference 28.
50. G. W. J. Fleet, C. J. Fuller and P. J. C. Harding (1978). *Tetrahedron Lett.*, 1437.
51. The editors are grateful to Dr R. Thomas, School of Chemistry, Thames Polytechnic, for the experimental details of the acetylation of ferrocene.
52. R. H. Thomson (1971). *Naturally Occurring Quinones*, 2nd edn. London: Academic Press.
53. R. H. Thomson (1974). 'Synthesis', in *The Chemistry of Quinonoid Compounds*. Ed. S. Patai, Pts 1 and 2. London: Wiley.
54. H. Zimmer, D. C. Lankin and S. W. Horgan (1971). *Chem. Rev.*, **71**, 229.
55. H.-J. Teuber (1972). *Organic Syntheses*, **52**, 88.
56. A. J. Fatiadi (1973). *Synthesis*, 357.
57. V. Balogh, M. Fetizon and M. Golfier (1971). *J. Org. Chem.*, **36**, 1339.
58. W. Walter and K.-D. Bode (1966). *Angew. Chem., Int. Ed. Engl.*, **5**, 457.
59. J. H. Brewster, A. M. Fusco, L. E. Carosino and B. G. Corman (1963). *J. Org. Chem.*, **28**, 498.
60. (a) J. H. P. Tyman and A. A. Durrani (1972). *Chem. Ind. (London)*, 664.
    (b) R. K. Bentley (1972). *Chem. Ind. (London)*, 767.
61. F. C. Whitmore and D. J. Loder. *Organic Syntheses*, Coll. Vol. II, p. 282.
62. S. Ohta, A. Shimabayashi, M. Aono and M. Okamoto (1982). *Synthesis*, 833.
63. L. Friedman and H. Shechter (1963). *J. Org. Chem.*, **26**, 2522.
64. T. Imamoto, T. Takaoka, M. Yokoyama (1983). *Synthesis*, 142.
65. T. van Es (1961). *J. Chem. Soc.*, 1564.

CHAPTER 7
SELECTED ALICYCLIC COMPOUNDS

In this chapter some illustrative and interesting examples of ring-forming reactions leading to a selection of compounds having a range of ring sizes are discussed. The reactions are grouped under the following general procedures.

1. Intramolecular nucleophilic displacement reactions.
2. Intramolecular addition reactions of carbanions to a carbonyl group.
3. Insertion reactions.
4. Ring expansion and ring contraction reactions.
5. Reduction of aromatic compounds.
6. Cycloaddition reactions.

Procedures 1 and 2, in which the key reaction step is the formation of a carbon–carbon bond by an intramolecular process, provide important syntheses for ring sizes from three carbon atoms upwards. For a discussion of the kinetic and thermodynamic factors of these ring-forming reactions, and of the Baldwin Rules,[1] the reader is referred to standard and specialised texts. These aspects are briefly referred to here in the context of particular syntheses. The insertion of a carbon species (procedure 3) into a carbon–carbon multiple bond uniquely provides a route to three-membered rings. Ring contraction and ring expansion reactions (procedure 4) are of value for the conversion of a readily available ring compound into a smaller or a larger ring that could be difficult to form by the other general methods. The reduction of aromatic compounds (procedure 5) is an important route to six-membered ring compounds. The examples of cyclo-addition reactions (procedure 6), which illustrate the formation of four- and six-membered rings, involve the probable simultaneous formation of two carbon–carbon bonds under the influence of either thermal or photochemical conditions.

7.1 INTRAMOLECULAR NUCLEOPHILIC DISPLACEMENT REACTIONS

Three target molecules, namely, cyclopropyl methyl ketone (1), cyclobutane-carboxylic acid (2) and 3-benzylcyclobutanone (3), are used to illustrate appropriate retrosynthetic disconnection strategies that may be applied to devise suitable synthetic procedures for these compounds.

In the case of cyclopropyl methyl ketone, disconnection of either the 1,2- or 1,3-carbon–carbon bond of the cyclopropane ring results in the preferred charge distribution shown in (4), namely, the carbanion site is adjacent to the meso-merically stabilising carbonyl group, and the carbocation site may be viewed as a halide-carrying carbon. The reagent equivalent may therefore be 5-chloro-pentan-2-one.

A retrosynthetic analysis for this latter compound (as shown below) shows that it could arise from ethylene oxide [reagent equivalent to (5), see Section 5.4.2, p. 533], and ethyl acetoacetate [reagent equivalent to (6), see Section 5.8.5, p. 619].

The synthetic route (Expt 7.1) uses 2-acetylbutyrolactone, which is available commercially, as a convenient source of the required chloropentanone.

The three-membered ring system, although the most highly strained of all the ring sizes, is readily formed in the cyclisation step. This arises from the fact that carbon–carbon bond formation in this case is an irreversible intramolecular S_N2 displacement of halogen reaction, which is facilitated by the thermodynamically favourable *anti*-periplanar conformation of the anion.

A retrosynthetic disconnection for cyclobutanecarboxylic acid (2) gives rise to a synthon having the charge distribution shown in (7).

(2)   (7)   (8)

The carbocation site may be seen to arise from a carbon-carrying halogen; the reagent equivalent of the carbanion site is a substituted malonic ester (see Section 5.11.6, p. 680). The reagent equivalent of the synthon could therefore reasonably be 3-bromopropylmalonic ester (8). Disconnection of this latter compound would yield, as reagent equivalents, 1,4-dibromobutane and malonic ester. Overall, therefore, the synthetic route may be seen to be the reaction of the dianion of malonic ester with an α, ω-dihalide, followed by hydrolysis and decarboxylation (Expt 7.2). This general strategy of ring synthesis is an important method which is applicable to ring sizes from three- to seven-membered rings, and is only limited by the availability of the appropriate dihalogen compounds.

The synthesis of 3-benzylcyclobutanone (3) is an illustration of an overall intramolecular alkylation of an acyl anion equivalent (Section 5.9). The α,ω-dihalide is 2-benzyl-1,3-dibromopropane, and the acyl anion equivalent is methyl methylthiomethyl sulphoxide[2]; the product is 1-methylsulphinyl-1-methylthio-3-benzylcyclobutane which is obtained as a mixture of cis/trans isomers [(9) and (10)] (Expt 7.3). Aqueous acid hydrolysis in ethereal solution unmasks the carbonyl group. The possible mechanism of the reaction is via a Stevens-type rearrangement of the intermediate sulphur ylide, which may proceed in a pericyclic, radical or ion pair fashion.

(9)   (10)

Experiment 7.1 CYCLOPROPYL METHYL KETONE

1089

5-Chloropentan-2-one. Assemble a distillation unit consisting of a 500-ml round-bottomed flask, a still-head, a large double surface condenser and a 250-ml receiving flask; cool the latter in an ice bath. Place into the flask 64 g (0.5 mol) of 2-acetylbutyrolactone, 90 ml of water and a few chips of broken porcelain, and then add 75 ml of concentrated hydrochloric acid and shake to mix. Heat the flask gently on a wire gauze until carbon dioxide is evolved briskly, and continue to heat cautiously until effervescence moderates and the contents of the flask become very dark (about 5–10 minutes). Then heat the flask strongly, preferably using two burners, so that rapid distillation ensues. When about 125 ml of distillate has been collected, add 75 ml of water steadily to the flask from a dropping funnel fitted to the still-head without interrupting the distillation. Collect a total of 200 ml of distillate; the entire distillation should be completed within about 75 minutes. Separate the pale yellow organic phase from the distillate, extract the aqueous phase with three 30 ml portions of ether and dry the combined organic layers over anhydrous calcium chloride. Remove the ether by distillation through a lagged 10 × 1.5 cm fractionating column packed with glass helices and distil the residue (1) under reduced pressure (water-pump). Collect the 5-chloropentan-2-one at 70–73 °C/20 mmHg, the yield is 45 g (75%).

Cyclopropyl methyl ketone. Place a solution of 20 g (0.5 mol) of sodium hydroxide in 20 ml of water in a 250-ml three-necked flask fitted with a dropping funnel, a reflux condenser and an efficient sealed stirrer unit. With vigorous stirring add 42 g (0.35 mol) of 5-chloropentan-2-one during 30 minutes; heat the flask gently during the addition so that the reaction mixture refluxes steadily. Continue heating under reflux for a further period of 1 hour, and then add slowly 45 ml of water and continue to reflux for 1 hour more, maintaining vigorous stirring throughout. Rearrange the condenser for distillation and distil out the reaction product until an organic layer no longer remains in the flask. Saturate the distillate with potassium carbonate, separate the organic phase and extract the aqueous phase with three 15 ml portions of ether. Dry the combined organic extracts over anhydrous calcium chloride and then over anhydrous calcium sulphate. Remove the ether by flash distillation through a lagged 10 × 1.5 cm fractionating column (glass helices), and continue distilling collecting the cyclopropyl methyl ketone as a fraction of b.p. 111–112 °C; the yield is 24 g (82%).

Note. (1) The crude residue may be used directly in the next stage.

Experiment 7.2 CYCLOBUTANE-1,1-DICARBOXYLIC ACID AND CYCLOBUTANECARBOXYLIC ACID

Equip a 3-litre three-necked flask with a thermometer, a sealed mechanical stirrer and a double surface reflux condenser. It is important that all the apparatus be thoroughly dry. Place 212 g (1.05 mol) of 1.3-dibromopropane (Expt 5.54) and 160 g (1 mol) of diethyl malonate (dried over anhydrous calcium sulphate) in the flask. Start the stirrer and add a solution of 46 g (2 mol) of sodium in 800 ml of super-dry ethanol (Section 4.1.9, p. 401) (1) down the condenser from a dropping funnel at such a rate that the temperature of the reaction mixture is maintained at 60–65 °C (50–60 minutes). When the addition is complete, allow the mixture to stand until the temperature falls to 50–55 °C, and then heat on a water bath until a few drops of the liquid when added to water are no longer alkaline to phenolphthalein (about 2 hours). Add sufficient water to dissolve the precipitate of sodium bromide, and remove the ethanol by distillation from a water bath. Steam distil the residue until all the diethyl cyclobutane-1,1-dicarboxylate and unchanged diethyl malonate are removed; collect about 4 litres of distillate during 9–10 hours. Extract the entire steam distillate with three 350 ml portions of ether; remove the ether from the combined extracts on a water bath (rotary evaporator). Reflux the residual liquid with a solution of 112 g of potassium hydroxide in 200 ml of ethanol for 2 hours. Distil off most of the ethanol and then evaporate the residue to dryness on a water bath (rotary evaporator). Dissolve the solid residue in 100 ml of hot water, and add concentrated hydrochloric acid (c. 80 ml) cautiously until the solution is just acid to litmus. Boil for a few minutes to remove carbon dioxide, render slightly alkaline with ammonia solution and add a slight excess of aqueous barium chloride to the boiling solution. Filter the hot solution to remove the barium malonate, cool the filtrate and render it strongly acid with concentrated hydrochloric acid (90–100 ml of acid: use Congo red paper). Extract the solution with four 250 ml portions of ether. Dry the combined extracts with anhydrous calcium chloride and remove the ether on a rotary evaporator. Spread the solid on a porous tile to remove oily impurities. The beautifully crystalline product (55 g, 38%) consists of pure cyclobutane-1,1-dicarboxylic acid, m.p. 158 °C. It may be recrystallised from hot ethyl acetate, but the m.p. is unchanged.

Place 30 g (0.28 mol) of cyclobutane-1,1-dicarboxylic acid in a 100-ml flask fitted with a still-head carrying a thermometer and leading to a cooled receiver flask via a short air condenser. Heat the flask in a Silicone oil bath at 160–170 °C until all effervescence ceases. Then raise the temperature of the bath to 210 °C; the cyclobutanecarboxylic acid passes over at 191–197 °C. Redistil the acid; the pure acid distils at 195–196 °C. The yield of cyclobutanecarboxylic acid (a colourless liquid) is 18 g (86%).

Note. (1) See Expt 5.95 for experimental details pertaining to the preparation of an ethanolic solution of sodium ethoxide.

Experiment 7.3 3-BENZYLCYCLOBUTANONE[2]

1-Methylsulphinyl-1-methylthio-3-benzylcyclobutane. To a solution of methyl methylthiomethyl sulphoxide (1.01 g, 8.15 mmol) in dry tetrahydrofuran (15 ml), is added a 1.44 mol/litre hexane solution (6 ml) of butyllithium (8.64 mmol) at − 10 °C, and the mixture is stirred for 2 hours. After the addition of 2-benzyl-1,3-dibromopropane (1.03 g, 3.53 mmol) (1) at $-10\,°C$, the mixture is stirred at $-10\,°C$ for 100 minutes and then at room temperature for 3 hours. Dichloromethane (100 ml) is added and the organic layer washed with water (30 ml). The aqueous layer is extracted with two 50 ml portions of dichloromethane. The organic layers are combined, washed with brine (20 ml) and dried (sodium sulphate). The solvents are evaporated and the residue subjected to column chromatography on silica gel using dichloromethane–ethyl acetate (9:1) as eluant to give the product as a pale yellow oil, 686 mg, 77 per cent; i.r. 1055, 1035 cm^{-1}. The product consists of two diastereoisomers in the ratio 2:1; the methyl signals of the major isomer appeared at δ 2.12 and 2.49, and of the minor isomer at δ 2.08 and 2.42.

3-Benzylcyclobutanone. The foregoing compound (2.70 g, 10.6 mmol) is dissolved in ether (140 ml), and 4.5 M sulphuric acid (2.8 ml) is added. The resulting mixture is stirred at room temperature for two days, and then heated under reflux for two days. After the addition of sodium hydrogen carbonate and magnesium sulphate, the mixture is stirred for a while and then the insoluble solid filtered off. The filtrate is concentrated and the residue subjected to column chromatography on silica gel using dichloromethane and hexane as eluants to give 3-benzylcyclobutanone as a colourless oil, 1.13 g, 66 per cent; i.r. (neat) 1785 cm^{-1}; p.m.r. (CDCl$_3$, TMS), δ 2.43–3.49 (m, 7H), and 7.21 (s, 5H).

Note. (1) 2-Benzyl-1,3-dibromopropane may be prepared from diethyl benzylmalonate (Expt 5.132), by lithium aluminium hydride reduction to give 2-benzylpropane-1,3-diol (for conditions compare Expt 5.38), and subsequent conversion into the dibromo derivative using the conditions described in Expt 5.54 as a guide.

7.2 INTRAMOLECULAR ADDITION REACTIONS OF CARBANIONS TO A CARBONYL GROUP

The intramolecular carbon–carbon bond-forming reactions considered in this section are based on the aldol condensation (see Section 5.18.2, p. 799), the Claisen–Schmidt reaction (see Section 6.12.2, p. 1032), the Claisen ester condensation (see Section 5.14.3, p. 736), and the Claisen reaction (see Section 6.12.2, p. 1032). Since these carbonyl addition reactions are reversible, the methods of synthesis are most successful for the formation of the thermodynamically stable five- and six-membered ring systems. The preparation of the starting materials for some of these cyclisation reactions further illustrates the utility of the Michael reaction (see Section, 5.11.6, p. 681).

The *first group* of target molecules that are considered are the cyclic α,β-unsaturated ketones, 3-methylcyclopent-2-enone (11), 3-methylcyclohex-2-enone (12), (+)-7a(S)-7,7a,-dihydro-7a-methyl-1,5(6H)-indanedione (13), and tetraphenylcyclopentadienone (14).

(11)   (12)   (13)   (14)

In the case of (11), retrosynthetic functional group interconversion into the aldol followed by disconnection of the α, β-bond gives the dipolar synthon (15), of which the reagent equivalent is the 1,4-dicarbonyl compound, hexane-2,5-dione (i.e. a *retro*-aldol condensation). The action of base on this diketone effects the forward aldol reaction followed by spontaneous dehydration (see Expt 7.4 for formulation).

(15)

A similar retrosynthetic analysis for (12), gives the 1,5-diketone, heptane-2,6-dione. Disconnection on this diketone, similar to those described in Section 5.9 for 1,2- to 1,4-diketones, would lead to a number of possible methods of synthesis. In the synthesis described in Expt 7.5, 3,5-diethoxycarbonylheptane-2,6-dione (ethyl methylenebisacetoacetate) (16) is formed from formaldehyde (as paraformaldehyde) and ethyl acetoacetate in the presence of piperidine (Knoevenagel/Michael reactions) (cf. the synthesis of pyridines, Expt 8.29). Cyclisation of (16) is effected on heating in the presence of base to give 4,6-diethoxycarbonyl-3-methylcyclohex-2-enone (17). Both ethoxycarbonyl groups are removed when the diester is heated for some time with aqueous acid, forming 3-methylcyclohex-2-enone (12). The further reaction of this compound with a Grignard reagent, which proceeds by 1,4-conjugate addition to the enone system, gives a 3-alkyl-3-methylcyclohexanone (18).

(16)

(17)   (12)   (18)

The chiral target molecule (13) is of particular interest. Setting aside asymmetric aspects of the reaction, a retrosynthetic analysis, similar to the cases above, is as follows.

(13a)   (19)   (20)

The forward synthetic sequence would therefore involve the Michael reaction of 2-methylcyclopentane-1,3-dione with methyl vinyl ketone to give (20), followed by cyclisation to the hydroxyketone (19), and then dehydration to the target molecule (13a). The overall process of addition and cyclisation is known as the *Robinson annelation reaction*.[3] In this preparative example (Expt 7.6) the methyl vinyl ketone is used directly under conditions which minimise its polymerisation;[4a] it should be noted, however, that many literature examples of the annelation reaction use Mannich bases or the corresponding methiodides as an *in situ* source of the α, β-unsaturated carbonyl component (see Section 5.18.2, p. 801).

When the cyclisation step (20) to (19) is carried out in the presence of (S)-(−)-proline as the chiral inducing reagent, the bicyclic hydroxydiketone (22) is obtained in 93.4 per cent optical purity.[4b] It is suggested that proline adds to one of the carbonyl groups of the cyclopentanedione, and the formation of two hydrogen bonds as shown, imposes topographical rigidity on the conformation of the tricyclic transition state (21). The proline residue is thus *trans*-related to the angular methyl group, and carbon–carbon bond formation would then occur from the side opposite to the methyl group to give the *cis*-fused hydroxy diketone. Dehydration by azeotropic distillation in benzene then gives the target molecule (13) (Expt 7.6).

(21)   (22)   (13)

The disconnection of both carbon–carbon double bonds in tetraphenylcyclopentadienone (14) ('tetracyclone'), in the manner of the examples above, leads to recognition of benzil and dibenzyl ketone as the reagents for its synthesis.

(14)

This double Claisen–Schmidt reaction takes place under the influence of ethanolic potassium hydroxide (Expt 7.7) and presumably proceeds in the stepwise manner (cf. formulation in Section 6.12.2, p. 1032). The four aryl groups in tetracyclopentadienone effectively stabilise the cyclopentadienone system, which otherwise has only a transient existence and readily undergoes dimerisation by way of a diene–dienophile interaction (Diels–Alder reaction, Section 7.6). The use of tetracyclone as a dienophile for the preparation of 3,4,5,6-tetraphenyldihydrophthalic anhydride is noted on p. 1121.

The *second group* of cyclic compounds which illustrate the intramolecular carbanion addition to a carbonyl group, include 2-ethoxycarbonylcyclopentanone (23) and indane-1,3-dione (24).

(23)   (24)

Disconnection of the cyclic β-keto ester (23) gives the dipolar synthon (25), the reagent equivalent of which is diethyl adipate.

(23)   (25)

The synthesis may thus be seen to be an intramolecular Claisen ester condensation, which is known as the *Dieckmann reaction*. The procedure is an important method for the synthesis of five- and six-membered ring systems, and the cyclic β-keto ester product may be converted into the corresponding cyclic ketone by hydrolysis followed by decarboxylation (ketonic hydrolysis, see Section 5.8.5, p. 619). The base catalyst used in Expt 7.8 is sodium ethoxide, but sodium hydride as a 50 per cent dispersion in oil is a recommended alternative.

A similar disconnection for indane-1,3-dione (24) gives the dipolar synthon (26), from which it may be inferred that the reagent equivalent could be *o*-ethoxycarbonylacetophenone.

(24)   (26)

The synthesis of this starting material may prove troublesome, but if an activating group (CO_2Et) is formally added to the methyl group, further disconnection leads to the recognition of diethyl phthalate and ethyl acetate as the reagent equivalents of the synthons (27) and (28) respectively.

The forward synthetic reaction (formulated in Expt 7.9) is thus an initial mixed Claisen ester condensation, followed by a Dieckmann cyclisation, hydrolysis and decarboxylation. Indane-1,3-dione is used for the synthesis of the trione, ninhydrin (Expt 5.99).

The *radical reductive cyclisation* of diesters to acyloins (see also Section 5.9.1, p. 628) is an important method of synthesis for ring sizes from four-membered upwards. The example selected here is 2-hydroxy-3-methylcyclopent-2-enone ('corylone') (29) (Expt 7.10), which is an important perfumery and flavouring material.[5a] In the first step (i), methyl acrylate is converted into its dimer with tris(cyclohexyl)phosphine in pyridine solution.[5b] Step (ii) is the protection of the double bond by conversion into the dimethylamino adduct. The acyloin reaction is step (iii), and the product is trapped as its bis(trimethylsilyl)ether. Finally, in step (iv), the protecting dimethylamino and trimethylsilyl groups are removed by passage down a column of silica gel.

The *third group* of cyclic compounds formed by intramolecular carbanion addition to the carbonyl group are the target molecules 5,5-dimethylcyclohexane-1,3-dione (dimedone, 30) and 3-methylcyclopentane-1,2,4-trione (31).

(30)　　　　(31)

Disconnection of either of the carbonyl-carbon bonds to the common methylene group in dimedone gives the δ-keto ester (32) as the reagent equivalent. A further disconnection of this ester reveals the synthons (33) and (34), the reagent equivalents of which are 4-methylpent-3-en-2-one (mesityl oxide) and malonic ester. The overall synthesis (see Expt 7.11 for formulation) is thus a Michael reaction between these latter reagents followed by a Claisen condensation in the cyclisation step, and then hydrolysis and decarboxylation to remove the activating ethoxycarbonyl group originally sited in the malonic ester molecule.

(30)　　　　　　　　(32)　　　　(33)　　(34)

Suitable retrosynthetic disconnections of 3-methylcyclopentane-1,2,4-trione may not be immediately obvious owing to the presence of the three carbonyl groups. However, disconnection in the manner shown gives the reagent equivalents diethyl oxalate and butan-2-one.

In practice the synthesis requires two molar equivalents of diester to one molar equivalent of ketone in the presence of sodium ethoxide.

(35)

The initial step is a Claisen condensation involving both the α-methyl and α'-methylene groups, followed by a Dieckmann-type cyclisation to form ethyl 4-methyl-2,3,5-trioxocyclopentylglyoxalate (35). When heated with hot aqueous

1097

phosphoric acid the ester grouping is hydrolysed and this is followed by the overall loss of a molecule of oxalic acid, to yield finally the trione product (Expt 7.12).

Cyclobutanone is an important synthetic starting material (e.g. see Expt 5.41); recently a simple synthesis from readily available materials has been reported.[6] The synthesis (Expt 7.13) involves the formation of 1,3-bis[bromo-magnesio]propane and its further reaction with carbon dioxide to form the complex (36), which precipitates from solution thus simplifying the purification procedure. Although the overall yield is low (13%), this is compensated for by the cheapness of the reagents and the simplicity of the procedure.

Experiment 7.4 3-METHYLCYCLOPENT-2-ENONE

Dissolve 2.5 g of sodium hydroxide in 250 ml of water in a 500-ml two-necked flask fitted with a reflux condenser and a dropping funnel. Bring the solution to the boil, add rapidly from the dropping funnel 28.5 g (0.25 mol) of hexane-2,5-dione (Expt 5.104) and continue to boil steadily under reflux for exactly 15 minutes (1). Cool the resulting dark-brown solution rapidly in an ice–salt bath, saturate with sodium chloride and extract with one 100 ml and two 50 ml portions of ether. Wash the ether extract with three 5 ml portions of water, dry over anhydrous sodium sulphate and remove the ether on a rotary evaporator. Distil the residual dark oil under reduced pressure and collect the colourless 3-methylcyclopent-2-enone as a fraction of b.p. 74–76 °C/16 mmHg, n_D^{20} 1.4818; yield 9.5 g (40%). The product thus obtained is pure enough for most purposes; when perfectly pure the refractive index is 1.4893. The product may darken on storage.

Note. (1) The reaction conditions are critical; excessive boiling or the use of more concentrated alkali increases the formation of tarry by-products.

Experiment 7.5 3-METHYLCYCLOHEX-2-ENONE

4,6-Diethoxycarbonyl-3-methylcyclohex-2-enone. Place 130 g (1.0 mol) of ethyl acetoacetate, 15.0 g (0.5 mol) of powdered paraformaldehyde and 5 g (5.8 ml) of piperidine in a 500-ml round-bottomed flask. Allow the reaction to proceed at room temperature; after a short period (about 5 minutes) the contents of the flask rapidly heat up and the solid paraformaldehyde begins to dissolve. Moderate the reaction as required by cooling in iced water. When the vigorous reaction is over and the reaction mixture is homogeneous (about 20 minutes) heat the mixture on a water bath for 1 hour. The flask now contains the crude product together with a little water formed in the reaction. The crude product may be used directly for conversion into 3-methylcyclo-hex-2-enone by acid hydrolysis as described below. If required, it may be purified and dried in the following way. Dissolve the crude material in 200 ml of dichloromethane in a separatory funnel and wash successively with two 100 ml portions of 2.5 M hydrochloric acid, then two 100 ml portions of saturated aqueous sodium hydrogen carbonate and two 200 ml portions of water. Dry the organic phase over anhydrous calcium sulphate and remove the solvent on a rotary evaporator. The yield of crude 4,6-diethoxycarbonyl-3-methylcyclohex-2-enone is 106.5 g (84%). This compound cannot be purified by vacuum distillation since extensive decomposition takes place. It is, however, pure enough for most purposes.

3-Methylcyclohex-2-enone. Dissolve the crude product from the previous preparation in a mixture of 300 ml of glacial acetic acid, 30 ml of concentrated sulphuric acid and 200 ml of water and boil the solution under reflux for 6 hours. Add a solution of 254 g of sodium hydroxide in 700 ml of water carefully and with cooling to the cooled reaction solution. Extract with three 150 ml portions of ether, dry the ethereal extract over anhydrous sodium sulphate and remove the ether on a rotary evaporator. Distil the residue under reduced pressure through a short fractionating column and collect the 3-methylcyclohex-2-enone as a fraction, b.p. 95 °C/25 mmHg. The yield is 24 g (44%).

Conversion into 3-ethyl-3-methylcyclohexanone. Prepare a solution of ethyl-magnesium iodide in 120 ml of ether from 18.7 g (0.77 mol) of dry magnesium turnings and 120 g (0.77 mol) of ethyl iodide in a 500-ml, three-necked round-bottomed flask equipped with a sealed stirrer unit, a pressure-equalising funnel and a reflux condenser. When all the magnesium has dissolved, clamp the flask in a cooling bath of Cardice–acetone and replace the condenser with a low-temperature-reading thermometer. When the temperature has fallen to −10 °C, add 1.5 g (0.15 mol) of thoroughly dried copper(I) chloride (1). Then while maintaining the internal temperature of the reaction mixture between −10 and −5 °C (2), and with rapid stirring add dropwise a solution of 55 g (0.50 mol) of 3-methylcyclohex-2-enone in 150 ml of ether. After all the ketone has been added, continue stirring for a further 2 hours at −5 °C. Decompose the reaction complex by adding 150 ml of a cold saturated ammonium chloride solution, with continued stirring and cooling, followed by dilute hydrochloric acid to give a clear solution. Separate the ethereal layer and thoroughly extract the aqueous phase with six portions of ether. Dry the combined ethereal extracts with anhydrous calcium sulphate, remove the ether on a rotary evaporator and distil the residue under reduced pressure. Collect the 3-ethyl-3-methylcyclohexanone as a fraction, b.p. 97–98 °C/

22 mmHg, n_D^{21} 1.4594; the yield is 48 g (68%). The product is sufficiently pure for most purposes; the semicarbazone has m.p. 181–182 °C.

Notes. (1) Temperature control is important at this stage, otherwise the yield of product is considerably reduced.
(2) Careful temperature control is important, or the yield of product is reduced and considerable quantities of the by-product 1-methyl-3-ethyl-1,3-cyclohexadiene are produced.

Experiment 7.6 (+)-(7aS)-7,7a-DIHYDRO-7a-METHYL-1,5(6H)-INDANEDIONE[4]

2-methyl-2-(3-oxobutyl)cyclopentane-1,3-dione. To a suspension of 2-methyl-cyclopentane-1,3-dione (65 g) in demineralised water (136 ml) is added at once methyl vinyl ketone (96 ml), and the mixture stirred under nitrogen at 20 °C for 5 days. It is then extracted with benzene (**CAUTION**) and treated with sodium sulphate, charcoal and magnesium sulphate. After filtration, the solids are extracted with boiling benzene (100 ml). Evaporation *in vacuo* of the combined benzene extract gives the crude product (100.9 g). Fractional distillation gives the pure triketone (92.5 g, 87.6%) as a pale yellow oil; b.p. 100–109 °C/0.08–0.1 mmHg); i.r. 1770 (C=O) and 1725 cm^{-1} (C=O); p.m.r. (CDCl$_3$, TMS) δ 1.12 (s, 3H, 2-Me), 2.22 (s, 3H, Me.CO), 2.82 (m, 4H, —CO·CH$_2$·CH$_2$·CO—).

(+)-(3aS, 7aS)-3a, 4, 7, 7a-Tetrahydro-3a-hydroxy-7a-methyl-1, 5(6H)-indane-dione. The foregoing triketone (1.82 g, 10 mmol) and (S)-(−)-proline (34.5 mg, 0.3 mmol) are stirred in anhydrous N,N-dimethylformamide (10 ml; distilled from calcium hydride) under argon for 20 hours. The brown-coloured reaction mixture is filtered, and the filtrate evaporated under high vacuum at 22 °C (bath temperature) to give 2.4 g of an oil. This is dissolved in 10 ml of ethyl acetate and filtered through 8.0 g of silica gel. The absorbent is eluted with 150 ml of ethyl acetate, and the solvent evaporated *in vacuo* to give 2.0 g of an oil, which crystallises on seeding. The crystalline mass is broken up and placed under high vacuum at 55 °C (bath temperature) for 1 hour to remove traces of dimethylformamide to give 1.82 g (100%) of crude product as a tan-coloured solid, $[\alpha]_D^{25}$ +56.1° (c 1.0 in CHCl$_3$). Recrystallisation from ether gives the analytically pure compound, m.p. 119–119.5 °C, $[\alpha]_D^{25}$ +60.4° (c 1.06 in CHCl$_3$); i.r. 3600 and 3350–3508 (OH), 1742 (5-ring C=O), and 1722 cm^{-1} (6-ring C=O); p.m.r. (CDCl$_3$, TMS) δ 1.26 (s, 3H, 7a-Me), 2.63 (s, 2H, —CO·CH$_2$·COH), and 2.92 (s, 1H, OH).

Dehydration to (+)-(7aS)-7,7a-dihydro-7a-methyl-1,5(6H)-indanedione. The above dextrorotatory ketol (1.79 g) obtained as above is refluxed in 0.01 M toluene-*p*-sulphonic acid–benzene (15 ml) with stirring under nitrogen for 15 minutes. Water is removed from the azeotrope by a Dean and Stark water separator filled with Linde Type 4A molecular sieves. After cooling to room temperature, it is stirred with 1 M aqueous sodium hydrogen carbonate solution (0.3 ml) for 5 minutes, dried with magnesium sulphate, and filtered, and the solids are rinsed with chloroform. The filtrate is evaporated *in vacuo* to give 1.6 g (99.4%) of crude product as an oil, which crystallises rapidly on seeding with an authentic sample, $[\alpha]_D^{25}$ +322° (c 0.94 in C_6H_6), u.v. 233 nm (ε 10 200). The quotient of 87.7 per cent optical yield and 92.4 per cent chemical purity by u.v. represents 94.9 per cent optical yield.

A portion of 1.56 g of the crude crystalline product is broken up in a small amount of ether on a coarse sintered funnel. Removal of this ether by suction gives 1.11 g of a colourless crystalline product (70.2% yield based on the triketone), m.p. 64–65.5 °C, $[\alpha]_D^{25}$ +356° (c 0.99 in C_6H_6), u.v. 233 nm (ε 11 540). This represents a chemically pure sample of 97 per cent purity. Recrystallisation from ether gives optically pure (+)-(7aS)-7,7a-dihydro-7a-methyl-1,5(6H)-indandione, m.p. 66–66.5 °C, $[\alpha]_D^{25}$ +367° (c 1.0 in C_6H_6) (1).

Note. (1) Physical constants for this compound, obtained by a chemical resolution process,[7] are: m.p. 66–66.5 °C, $[\alpha]_D^{25}$ +362° (c 1.0 in C_6H_6), i.r. ($CHCl_3$) 1746 (5-membered ketone), 1665 cm^{-1} (α,β-unsaturated ketone).

Experiment 7.7   TETRAPHENYLCYCLOPENTADIENONE
(*Tetracyclone*)

Dissolve 4.2 g (0.02 mol) of benzil (Expt 6.143) and 4.2 g (0.02 mol) of dibenzyl ketone (Expt 6.127) in 30 ml of hot absolute ethanol in a 100-ml round-bottomed flask fitted with a reflux condenser. Heat the solution to near its boiling point on a steam bath and then add in portions a solution of 0.6 g (0.011 mol) of potassium hydroxide in 6 ml of absolute ethanol. Some foaming may occur. Heat the reaction mixture under reflux for 15 minutes (1) and cool to below 5 °C in an ice bath. Collect the dark crystalline product by filtration with suction, wash with three 5 ml portions of rectified spirit and dry in an oven at 50 °C. The tetraphenylcyclopentadienone has m.p. 217–220 °C and is sufficiently pure for most purposes; the yield is 7.0 g (91%). Recrystallise a portion from toluene–ethanol (1:1) to obtain the pure compound as deep purple crystals, m.p. 219–200 °C; λ_{max} 340 and 510 mm, ε 1.26 × 10^3 and 0.33 × 10^3 respectively.

Note. (1) In a modified procedure[8] using triethylene glycol as solvent and benzyltrimethylammonium hydroxide ('Triton B') as the base catalyst, the reaction may be completed in a very short time.

Experiment 7.8 2-ETHOXYCARBONYLCYCLOPENTANONE

Prepare 25 g (1.09 mol) of granulated sodium in a 1500-ml round-bottomed flask (Section 4.2.68, p. 462, *Method 2*). Cover the sodium with 625 ml of sodium-dried benzene (**CAUTION**); fit the flask with an efficient reflux condenser protected from the air by means of a calcium chloride guard-tube. Add 151.5 g (0.75 mol) of diethyl adipate (Expt 5.144) in one lot, followed by 1.5 ml of absolute ethanol. Warm the flask on a water bath until, after a few minutes, a vigorous reaction sets in and a cake of the sodio compound commences to separate. Keep the flask well shaken by hand during the whole of the initial reaction. After the spontaneous reaction has subsided, reflux the mixture on a water bath overnight, and then cool in ice. Decompose the product with ice and dilute hydrochloric acid (1:1); add the acid until Congo red paper is turned blue. Separate the benzene layer, and extract the aqueous layer with 100 ml of benzene. Wash the combined extracts with 100 ml of 5 per cent sodium carbonate solution and 150 ml of water: dry over magnesium sulphate. Remove the benzene on a rotary evaporator (*fume cupboard*), and fractionate the residue under reduced pressure. Collect the 2-ethoxycarbonyl-cyclopentanone at 108–111 °C/15 mmHg (95 g, 81%). Upon redistillation, the product boils at 102 °C/11 mmHg.

Experiment 7.9 INDANE-1,3-DIONE

Place 125 g (106.5 ml, 0.563 mol) of diethyl phthalate and 25 g (1.09 mol) of sodium wire (Section 4.2.68, p. 462) in a 500-ml round-bottomed flask fitted with a reflux condenser and dropping funnel, each protected with a calcium chloride tube. Heat the flask on a steam bath and add a mixture of 122.5 g (136 ml, 1.39 mol) of dry ethyl acetate and 2.5 ml of absolute ethanol over a period of 90 minutes. Continue the heating for 6 hours, cool and add 50 ml of ether. Filter the sodium salt on a sintered glass funnel and wash by stirring with ethyl acetate; filter. Dissolve the sodium salt (96 g) in 1400 ml of hot water in a 3-litre beaker, cool the solution to 70 °C, stir vigorously and add 100 ml of sulphuric acid (3 parts of concentrated acid to 1 part of water). Cool the mixture to 15 °C in an ice bath, collect the indane-1,3-dione by suction filtration, wash with a little water and dry at 100 °C; the yield is 58 g (71%). Recrystallisation from a dioxane–benzene mixture by the addition of light petroleum (b.p. 80–100 °C) gives the pure compound, m.p. 130 °C.

Experiment 7.10   2-HYDROXY-3-METHYLCYCLOPENT-2-ENONE[5]

Dimethyl 2-methylenepentanedioate. Methyl acrylate (30.0 g, 349 mmol) (distilled immediately before use) and dry pyridine (30 ml, **CAUTION**) containing tris(cyclohexyl)phosphine–carbon disulphide complex (2.0 g, 6 mmol) (1) are refluxed under nitrogen for 16 hours. The deep red solution is cooled and the pyridine removed under reduced pressure. The residue is taken up in ether (400 ml) and the solution washed with aqueous 1 M hydrochloric acid (3 × 40 ml). The combined aqueous layers are extracted with ether (2 × 50 ml) and the combined organic layers washed with 1 M hydrochloric acid (30 ml), saturated brine (40 ml) and saturated aqueous sodium hydrogen carbonate (2 × 30 ml), dried over sodium sulphate and evaporated. Distillation of the oil gives dimethyl 2-methylenepentanedioate (23.8 g, 79%) as a liquid, b.p. 66–68 °C/1 mmHg; i.r. (thin film) 1738, 1715, 1635 cm^{-1}.

Dimethyl 2-(dimethylaminomethyl)pentanedioate. To a stirred solution of dimethylamine (28%) in anhydrous methanol (100 ml) held at 0 °C (ice bath), under nitrogen, is added, via a syringe, dimethyl 2-methylenepentanedioate (10.3 g, 60 mmol) in one portion. The solution is stirred for 4 hours, during which period the temperature must never exceed 5 °C. Removal of excess amine and methanol under reduced pressure gives a pale yellow oil, which is distilled under reduced pressure to give dimethyl 2-(dimethylaminomethyl)-pentanedioate (12.8 g, 99%) as an oil b.p. 81–83 °C/0.4 mmHg; i.r. (thin film) 1735 cm^{-1}; p.m.r (CDCl$_3$, TMS) δ 1.8–3.0 (m, 7H), 2.22 (s, 6H), 3.67 (s, 3H) and 3.69 (s, 3H).

3-(Dimethylaminomethyl)-1,2-bis(trimethylsilyloxy)-cyclopentene. In a 500-ml three-necked flask, equipped with a 'football' stirrer, a reflux condenser and a 100-ml pressure-equalising dropping funnel (maintained under oxygen-free nitrogen), are added toluene (200 ml) (2) and sodium (4.0 g, 174 mmol). The toluene is brought to reflux on an oil bath (bath temperature 135 °C) and the mixture stirred until a fine suspension of sodium is produced. Dimethyl 2-(dimethylaminomethyl)pentanedioate (10.6 g, 48 mmol) and chlorotrimethyl-silane (20.0 g, 184 mmol, **CAUTION**) (distilled from calcium hydride, under nitrogen, immediately prior to use) in dry toluene (80 ml) are added dropwise

1103

over 2 hours. A dark purple precipitate appears within a few minutes of the ester being added. After heating and stirring for a further 12 hours the contents of the flask are cooled and filtered, under nitrogen, through a 76.2 mm number three sinter. The precipitate is washed with anhydrous light petroleum (3 × 40 ml, b.p. 40–60 °C). The solvent is removed under reduced pressure to give a pale yellow oil that is purified by short-path distillation (bath temperature 85 °C at 0.1 mmHg) to give the bis-silyl ether (11.2 g, 78%); i.r. (CCl$_4$) 1700 cm^{-1}; p.m.r. (CDCl$_3$, TMS) δ 0.14 (s, 18H), 2.17 (s, 6H), 1.62–2.80 (m, 7H).

2-Hydroxy-3-methylcyclopent-2-enone. The bis-silyl ether (1.0 g, 3 mmol) is dissolved in anhydrous ether (50 ml) under nitrogen. The solution is added to a column of silica gel [150 g, 100–200 mesh (BDH)] [made in 2:1, light petroleum (b.p. 40–60 °C):ether] over 15 minutes. The material is eluted with light petroleum–ether (2:1; 200 ml). Elution is halted for 6 hours; further elution after that time (1:1, light petroleum–ether) gives the ketone (0.26 g, 76%); i.r. (CHCl$_3$) 3500, 1710 and 1605 cm^{-1}; p.m.r. (CDCl$_3$, TMS) δ 2.00 (s, 3H), 2.41 (s, 4H) and 6.79 (broad s, 1H, removed by D$_2$O); m/z 112 (M, base peak), 83 (22%), 69 (44%), 55 (26%), 43 (27%) and 41 (39%).

Notes. (1) Pyridine was stored over potassium hydroxide and distilled immediately before use. The tris(cyclohexyl)phosphine–carbon disulphide complex is prepared by the method of K. Issleib and A. Brack.[9] This involves the addition of carbon disulphide to an ethereal solution of tricyclohexylphosphine, the precipitate is washed with light petroleum (b.p. 50–60 °C), and recrystallised under a nitrogen atmosphere from either methanol, ethanol or dioxane; the complex has m.p. 118 °C.
(2) Toluene is distilled and the first 10 per cent and last 15 per cent are discarded. The distillate is refluxed over sodium for 24 hours, under nitrogen, before being distilled (under nitrogen) for use.

Experiment 7.11 5,5-DIMETHYLCYCLOHEXANE-1,3-DIONE
(*Dimedone*)

Equip a dry 1-litre three-necked flask with a dropping funnel, a sealed stirrer unit and an efficient double surface condenser. Place 11.5 g (0.5 mol) of sodium in the flask, cool in an ice bath and add 200 ml of absolute ethanol in one portion. When the initial vigorous reaction has subsided, remove the ice bath and allow the reaction to proceed until all the sodium has reacted: warming on a water bath is sometimes necessary to dissolve the last traces of sodium. Place a calcium chloride guard-tube at the top of the condenser. Introduce 85 g (0.53 mol) of diethyl malonate, and then add through the dropping funnel 50 g (0.51 mol) of freshly distilled mesityl oxide (Expt 5.213) slowly. Reflux the mixture with stirring for 2 hours, then add a solution of 62.5 g (1.11 mol) of potassium hydroxide in 300 ml of water, and reflux again on a water bath with stirring for 6 hours. Acidify the reaction mixture (to litmus) while still hot with dilute hydrochloric acid (1:2 by volume): about 275 ml are required. Fit the flask with a condenser for distillation, and distil off as much alcohol as possible by heating with stirring on a water bath. Allow the residue in the flask to cool somewhat, add 8 g of decolourising carbon slowly, boil for 10 minutes and filter; repeat the treatment with decolourising carbon. Neutralise the residue to litmus by the addition of dilute hydrochloric acid (about 75 ml) and boil again with 8 g of decolourising charcoal. Filter and render the hot yellow filtrate distinctly acid to methyl orange with dilute hydrochloric acid (25–50 ml), boil for a few minutes and allow to cool whereupon the dimedone crystallises out. Filter at the pump, wash with ice-cold water and dry in the air. The yield of dimedone, m.p. 147–148 °C, is 60 g (84%). Recrystallisation from acetone (about 8 ml per gram) raises the m.p. to 148–149 °C, but this is generally unnecessary.

Experiment 7.12 3-METHYLCYCLOPENTANE-1,2,4-TRIONE

Ethyl 4-methyl-2,3,5-trioxocyclopentylglyoxalate. Equip a 2-litre, three-necked round-bottomed flask with an efficient sealed stirrer unit and a reflux condenser protected with a calcium chloride tube, and insert a stopper in the third neck. Place 315 ml of absolute ethanol in the flask and add portionwise 23 g (1 mol) of sodium metal; if necessary moderate the initial vigour of the reaction by external cooling (see Expt 5.95). When all the sodium metal has reacted cool the flask in an ice–salt cooling bath, replace the stopper with a dropping funnel protected with a calcium chloride guard-tube and add dropwise over a period of about half an hour a mixture of 36 g (0.5 mol) of dry redistilled butan-2-one and 160 g (1.1 mol) of dry redistilled diethyl oxalate. Allow the red solution to attain room temperature and, with continued stirring, heat the reaction mixture under reflux for 30 minutes. Cool the reaction mixture again in an ice–salt bath and slowly add from the dropping funnel with efficient stirring 55 ml of dilute sulphuric acid (prepared by adding concentrated sulphuric acid to an equal volume of water). Remove the sodium sulphate which crystallises out by filtration and wash it with ethanol; concentrate the combined filtrate and washings on a rotary evaporator to a volume of about 100 ml. Cool the residual acqueous solution in an ice–salt cooling bath for several hours, filter the glyoxalate ester and wash with a little ice-cold water. The yield of crude product is 39.5 g (35%), which may be used for the next stage. A small portion of the product may be recrystallised from ethyl acetate (decolourising charcoal), when it has m.p. 162–164 °C.

1105

3-Methylcyclopentane-1,2,4-trione. In a 500-ml round-bottomed flask fitted with a reflux condenser place 300 ml of dilute phosphoric acid (prepared by diluting phosphoric acid with an equal volume of water) and add 30 g (0.133 mol) of the foregoing glyoxalate ester. Heat the solution under reflux for 30 minutes; cool the solution to about $-10\,°C$ and allow to stand at this temperature for 2 hours. Remove the oxalic acid which crystallises out by filtration and wash it with small portions of iced water. Extract the filtrate with eight 120 ml portions of ether, combine the extracts and rinse with a little water, dry and evaporate to dryness. Dissolve the residue in about 10 ml of water, cool and filter off crystals of the monohydrate which have a wide melting range of about 72–79 °C according to the rate of heating. Anhydrous 3-methylcyclopentane-1,2,4-trione may be obtained as colourless plates, m.p. 118–119 °C, by sublimation under reduced pressure (≈ 0.1 mmHg); the yield is 9 g (54%).

Experiment 7.13 CYCLOBUTANONE[6]

A 4-litre, three-necked round-bottomed flask, equipped with a nitrogen inlet, a reflux condenser, a mechanical stirrer and a pressure-equalising dropping funnel, is dried by heating with a flame under evacuation. Under nitrogen, it is charged with magnesium (264 g, 11 mol) and anhydrous ether (1.5 litres). Under rapid stirring, a solution of 1,3-dibromopropane (212 g, 1.05 mol) in ether (1.5 litres) is added during 15 hours. Afterwards, the reaction mixture is transferred by nitrogen pressure through a glass tube plugged with glass wool into another dry 4-litre, three-necked round-bottomed flask equipped with a gas inlet, a reflux condenser and a mechanical stirrier. At room temperature, a gaseous mixture (2:1) of nitrogen and carbon dioxide (30.2 litres, carbon dioxide from dry ice, dried by leading the gas through a tube filled with a mixture of calcium chloride and phosphorus pentoxide) is introduced above the surface of the rapidly stirred solution during 3 hours. Then the supernatant solution is decanted; to the white, solid residue, water (150 ml) is added dropwise, followed by 4 M hydrochloric acid (250 ml). The mixture is extracted with pentane (400 ml) by continuous extraction (14 hours); g.l.c. analysis (Carbowax 20M, isothermal, 90 °C) of the pentane solution and comparison with authentic samples reveals the presence of cyclobutanone (41%) and cyclobutanol (2%); p.m.r. spectroscopy shows that glutaric acid (17%) is also present. The pentane solution is subjected to distillation on a spinning band column (0.65 m) at normal pressure; after removal of the pentane, cyclobutanone is obtained by vacuum distillation. The yield is 9.7 g (13%, based on dibromopropane), b.p. 44 °C/140 mbar, purity >98 per cent by g.l.c.; p.m.r. ($CDCl_3$, TMS, 90 MHz) δ 1.93 (m, 2H) and 3.02 (m, 4H).

7.3 INSERTION REACTIONS

A most convenient method for the formation of cyclopropane ring systems is the insertion of a carbon residue into a carbon–carbon double bond. The reagents may be carbenes or carbenoids generated from diazomethane and related diazo compounds (cf. also Sections 5.11.4, p. 675 and 6.1.5, p. 848), diiodo- or dibromo-methane and zinc–copper couple, or chloroform and alkali (cf. also Section 6.10.1, p. 991), or the sulphur ylide generated from trimethyl-sulphoxonium iodide and sodium hydride (cf. also Section 8.1.2, p. 1131).

The illustrative examples of these reactions are the synthesis of $(-)$-2,7,7-trimethyltricyclo[4.1.1.0^{2,4}]octane (37), 7,7-dichlorobicyclo[4.1.0]heptane (38), and ethyl *trans*-2-methylcyclopropane-1-carboxylate (39).

(37)   (38)   (39)

Diiodomethane with zinc–copper couple is a particularly good reagent for the insertion of a methylene group into an olefinic bond. The reaction is known as the *Simmons–Smith procedure*,[10] and is likely to proceed as follows:

(40)

Since the free carbene (:CH_2) is not thought to be present, the reagent (40) is termed a carbenoid. The activity of the zinc surface is of crucial importance to the smoothness and success of the reaction. Failure to ensure appropriate activation procedures may cause the reaction to fail, or the onset of the exothermic reaction may be delayed and then proceed with excessive vigour. The recent use of ultrasonic activation of the zinc surface from the outset of the reaction results in a smoother, less unpredictable reaction rate, and leads to satisfactory yields.[11a] Furthermore, sonication enables the cheaper dibromomethane to be employed.[11b]

The reaction is illustrated by the cyclopropanation of α-pinene (Expt 7.14); a general procedure which is applicable to a wide range of olefins is also noted.

Dichlorocarbene is the reactive intermediate formed by the reaction of alkali on chloroform, and typically it adds to olefins to give 1,1-dichlorocyclopropanes. The PTC procedure for the generation of dichlorocarbene is particularly useful and is illustrated by its reaction with cyclohexene to form (38) (Expt 7.15). The mechanism is formulated below and probably involves the reaction of the quaternary ammonium hydroxide with chloroform at the phase boundary, and dissolution into the organic phase of the quaternary ammonium derivative of the trichloromethyl anion (41). This species breaks down to form dichlorocarbene and the quaternary ammonium chloride. The latter returns to the aqueous phase to maintain the cycle of events, while the dichlorocarbene reacts rapidly with the cyclohexene in the organic phase.

$$\overset{\oplus}{R_4}\overset{\ominus}{N}Cl \xrightarrow[\text{phase}]{\ominus OH; \text{ aqueous}} \overset{\oplus}{R_4}\overset{\ominus}{N}OH + \overset{\ominus}{Cl}$$

$$\overset{\oplus}{R_4}\overset{\ominus}{N}OH + CHCl_3 \xrightarrow[\text{boundary}]{\text{phase}} \overset{\oplus}{R_4}\overset{\ominus}{N}CCl_3 + H_2O$$

$$\underset{(41)}{\overset{\oplus}{R_4}\overset{\ominus}{N}CCl_3} \xrightarrow[\text{phase}]{\text{organic}} :CCl_2 + \overset{\oplus}{R_4}\overset{\ominus}{N}Cl$$

$$[\overset{\oplus}{R_4}N \equiv Et_3\overset{\oplus}{N}\cdot CH_2\cdot Ph]$$

The use of dimethylsulphoxonium methylide, as a specific methylene insertion reagent for α, β-unsaturated ketones and esters, is illustrated by its reaction with ethyl crotonate in dimethylformamide solution to form (39) (Expt 7.16).[12] The sulphur ylide initially attacks the β-carbon of the conjugated system (Michael acceptor site), and this is followed by cyclisation and loss of dimethyl sulphoxide.

Experiment 7.14 (−)-2,7,7-TRIMETHYLTRICYCLO[4.1.1.0²,⁴]-OCTANE[11a]

A 1-litre, four-necked round-bottomed flask, equipped with a thermometer, two condensers, mechanical stirrer and nitrogen-inlet tube, is charged with 1,2-dimethoxyethane (200 ml) (1) and mossy zinc (35.6 g, 0.544 mol), and irradiated with ultrasound for 2 hours (2). (−)-α-Pinene (21.0 g, 0.154 mol) is added rapidly, and the mixture heated to reflux. Diiodomethane (44.8 ml, 0.556 mol) (3) is added with stirring. The mixture is heated to reflux (the reaction temperature rises from 85 to 96 °C, therefore an efficient condenser is required) and continuously irradiated with ultrasound, until g.l.c. indicates that the reaction has stopped at 90 per cent completion (4 hours); [g.l.c. conditions: column (2 m × 2 mm) of 3% OV-1 on 80/100 Supercelcopport, isothermal 100 °C, helium 30 ml/minute; retention time, starting material 1.7 minutes, product 3.2 minutes]. The mixture is cooled to 25 °C, pentane (200 ml), saturated ammonium chloride (200 ml), and solid ammonium chloride are added slowly until all solids dissolve. The aqueous phase is extracted with two 150 ml portions of pentane, and the combined organic layers are washed with two 150 ml portions of saturated aqueous sodium chloride. The solution is dried over magnesium sulphate, filtered and evaporated at atmospheric pressure. The residual yellow oil (30 ml) is fractionally distilled at 78–80 °C/36 mmHg to obtain the product as an oil, 15.5 g (67%; 97% pure by g.l.c.); $[\alpha]_D^{25}$ −70.0° (c 1.67 in EtOH); p.m.r. (CDCl$_3$, TMS) δ 0.20 (m, 1H), 0.72 (m, 2H), 0.93 (d, 1H), 1.04 (s, 3H), 1.08 (s, 3H), 1.26 (s, 3H), 1.65 (m, 2H), 1.84 (d of d, 1H), and 2.02 (m, 2H); ^{13}C-n.m.r. (CDCl$_3$, TMS) δ 16.0 d, 19.1 t, 20.3 s, 21.1 q, 25.5 q, 27.0 q, 27.1 t, 27.5 t, 41.0 s, 41.6 d, and 45.5 d.

Notes. (1) 1,2-Dimethoxyethane (Aldrich) is used without purification or drying.
(2) A laboratory ultrasonic cleaner (150 W, 50–55 kHz) manufactured by Branson Co. is employed.
(3) Before use diiodomethane is fractionally distilled at 5 mmHg, fractionally crystallised by partial freezing and stored over mercury.

Cognate preparation. *General procedure for cyclopropanations using dibromomethane.*[11b] A 500-ml, three-necked round-bottomed flask is fitted with a Graham (coil) condenser and drying tube over an Allihn condenser, a pressure-equalising dropping funnel, and equipped with mechanical stirring. Into the flask are added zinc dust (52 g, 0.80 mol), copper(I) chloride (8 g, 0.08 mol), anhydrous ether (75 ml) and dibromomethane (70 g, 0.40 mol). To the addition funnel is added the alkene (0.20 mol) in anhydrous ether (50 ml). The apparatus is positioned in a 125 W Branson ultrasonic bath which is filled to about 3 cm from the top with water preheated to 45–50 °C. Sonication and stirring are started, and the position of the bath beneath the reaction vessel is varied so as to achieve maximum cavitation. Then the alkene is added dropwise to the reaction mixture over a 5–10 minute period. After approximately a 1 hour induction period, the reaction of the dibromomethane and zinc–copper couple usually starts. This is evidenced by the change in the colour of the reaction mixture from grey to a purple grey and by the onset of rapid refluxing. The stirring and sonication are continued for an additional 3 hours.

As an alternative procedure, which results in a shorter induction period but otherwise has no major effect on the reaction, addition of the alkene is postponed until after the reaction of the zinc–copper couple and dibromomethane starts (c. 30 minutes). Sonication is continued for an additional 60 minutes and stirring without sonication for a further 60 minutes.

After the reaction is complete, the ultrasonic bath is removed, and the reaction mixture cooled in an ice bath, diluted with pentane (200 ml), and, while being stirred, treated by dropwise addition of saturated ammonium chloride solution (150 ml). The organic layer is separated and the aqueous layer washed with pentane (1 × 50 ml). The combined organic layer is washed with 10 per cent sodium hydroxide solution (1 × 100 ml) and dried over magnesium sulphate. The solvents are removed by distillation on a steam bath through a Vigreux column, and the remaining oil fractionally distilled.

This procedure has been used for the cyclopropanation of cyclohexene, cyclooctene, α-pinene, β-pinene, hex-1-ene, oct-1-ene, (E)-but-2-en-1-ol, and 2,3-dihydropyran.

Experiment 7.15 7,7-DICHLOROBICYCLO[4.1.0]HEPTANE

Place 8.2 g (10.1 ml, 0.1 mol) of freshly distilled cyclohexene, 36 g (24 ml, 0.3 mol) of chloroform (1) and 0.4 g (0.0017 mol) of triethylbenzylammonium chloride in a two-necked, round-bottomed 100-ml flask fitted with a sealed stirrer unit and a reflux condenser. Stir the solution vigorously and add to it a solution of 16 g (0.4 mol) of sodium hydroxide in 16 ml of water in portions down the condenser during 5 minutes. Within 10 minutes an emulsion is formed, and the temperature of the mixture increases slowly during 25 minutes to a maximum of about 50–55 °C; thereafter the temperature decreases while the colour changes from white to pale brown. After stirring for 2.5 hours add 40 ml of ice-cold water to the reaction mixture, transfer to a separating funnel and collect the lower chloroform layer. Extract the aqueous alkaline solution with 30 ml of ether and combine the ether extract with the chloroform solution and wash with 25 ml of 2 M hydrochloric acid followed by two 25 ml portions of water. Dry the organic solution over magnesium sulphate, filter and remove the solvents on a rotary evaporator (water bath at 35–40 °C). Transfer the residual deep golden-coloured liquid to a 25-ml flask to which is attached a short (7 cm) fractionating column, and distil under reduced pressure. Collect the 7,7-dichlorobicyclo[4.1.0]heptane at 80–82 °C/16 mmHg; the yield is 10.2 g (62%). The purity may be checked by g.l.c. using a 5-ft column of Silicone oil on Chromosorb W held at 110 °C and with a nitrogen flow rate of 40 ml/minute; the retention time is 6 minutes.

Note. (1) Ethanol-free chloroform should be used (**CAUTION**). Ethanol present in chloroform (as stabiliser) can be removed by shaking chloroform several times with an equal volume of water, followed by drying over anhydrous calcium chloride and distilling. Alternatively stand the chloroform over a few grams of the molecular sieve 4A.

Experiment 7.16 ETHYL *TRANS*-2-METHYLCYCLOPROPANE-1-CARBOXYLATE[12]

CAUTION: *This experiment should be conducted in a well-vented fume cupboard owing to the volume of hydrogen gas which is evolved.*

To a suspension of sodium hydride (1.2 g, 0.05 mol) in dimethylformamide (100 ml) is added in one portion solid trimethylsulphoxonium iodide (11.05 g, 0.052 mol) (1). An exothermic reaction takes place with copious evolution of hydrogen. After all the hydrogen has been evolved (5 minutes), the mixture is stirred for another 15 minutes, and ethyl crotonate (5.7 g, 0.05 mol) in dimethylformamide (15 ml) is added to the methylide in one portion. An exothermic reaction takes place and the mixture turns slightly yellow. Stirring is continued for another hour. The mixture is poured into hydrochloric acid–ice-water (100 ml, 3%), extracted with ether (3 × 50 ml), the ether extract washed with water (4 × 50 ml) and the aqueous phase counter-extracted with ether (2 × 25 ml). The combined ether layer is dried (magnesium sulphate) and distilled giving ethyl *trans*-2-methylcyclopropane-1-carboxylate (4.0 g, 63%), b.p. 74–76 °C 70 mmHg; i.r. 1740 (C=O), 1020 cm^{-1} (cyclopropane); g.l.c. analysis shows a single peak (t_R 17 minutes) on a 10 per cent dinonyl phthalate column held at 70 °C.

Note. (1) *N,N*-Dimethylformamide is azeotropically distilled from toluene (Section 4.1.*26*, p. 409).

Trimethylsulphoxonium iodide is prepared by refluxing methyl iodide with dimethyl sulphoxide according to the method of Kuhn and Trichmann,[13] and washed with acetone to remove traces of iodine, dried in an oven at 110 °C for 2 hours, and kept in a vacuum desiccator before use; it may also be obtained commercially (Aldrich). All reactions are carried out under dry, oxygen-free nitrogen; calcium chloride guard-tubes are used to ensure anhydrous conditions. A magnetic stirrer is used to maintain brisk stirring.

7.4 RING EXPANSION AND RING CONTRACTION REACTIONS

Cyclohexanone and pulegone provide useful examples to illustrate ring expansion and ring contraction procedures.

Ring expansion of cyclohexanone to cycloeheptanone may be effected by reaction with diazomethane (Expt 7.17). The ring-expanded ketone is obtained in about 60 per cent yield and is accompanied by some epoxide and some cyclooctanone which results from further ring expansion of the cycloheptanone. Mechanistically the reaction may be represented in the following manner.

The cycloheptanone is readily separated by taking advantage of the fact that it alone forms a solid bisulphite compound.

The *Favorskii rearrangement* results from the action of base on an α-halo ketone. When applied to cyclic ketones, ring contraction results, as shown in the classic case of cyclohexanone formulated below.

It should be noted that in this case either of the carbonyl-carbon bonds in the symmetrical intermediate cyclopropanone system could be cleaved. With unsymmetrically substituted cyclic ketones (or indeed open chain ketones), the direction of cleavage is that which would lead to the more stable carbanion.

The example given in Expt 7.18 is the formation of a mixture of *cis* and *trans* ethyl pulegenates (ethyl 5-methyl-2-isopropylidenecyclopentanecarboxylate) from pulegone by reaction with bromine to give the dibromide, followed by treatment with sodium ethoxide.[14]

In this case the cleavage of the cyclopropanone system is directed by the concerted loss of the second bromine. The *cis/trans* ratio appears to be determined by a subsequent epimerisation in which the thermodynamically more stable *trans* isomer predominates. Hydrolysis of the mixed isomers with aqueous alkali leads to the exclusive formation of *trans*-pulgenic acid, resulting from the rapid base catalysed equilibration of the *cis/trans* esters and the retarded rate of hydrolysis of the sterically hindered *cis* ester.

Experiment 7.17   CYCLOHEPTANONE

CAUTION: *This preparation must be carried out in an efficient fume cupboard (see Section 4.2.25, p. 430).*

In a 1-litre three-necked flask equipped with a thermometer, a mechanical stirrer and a dropping funnel, place 49 g (0.5 mol) of redistilled cyclohexanone, 125 g (0.585 mol) of *N*-methyl-*N*-nitrosotoluene-*p*-sulphonamide, 150 ml of 95 per cent ethanol and 10 ml of water. The nitrosamide is largely undissolved. Adjust the height of the stirrer so that only the upper part of the solution is stirred and the precipitate moves slightly; place the thermometer so that the bulb is in the liquid. Cool the mixture to about 0 °C in an ice–salt bath. While stirring gently, add a solution of 15 g of potassium hydroxide in 50 ml of 50 per cent aqueous ethanol dropwise very slowly from the dropping funnel: after 0.5–1 ml of the solution has been added, a vigorous evolution of nitrogen commences and the temperature rises (1). Adjust the rate of addition so that the temperature is maintained at 10–20 °C; the duration of the addition of alkali is about 2 hours and the nitroso compound ultimately disappears. Stir the orange-yellow solution for a further 30 minutes, and then add 2 M hydrochloric acid until the solution is acidic to litmus paper (c. 50 ml).

Introduce a solution of 100 g of sodium metabisulphite in 200 ml of water and continue the stirring, preferably for 10 hours with exclusion of air. A thick precipitate separates after a few minutes. Collect the bisulphite compound by suction filtration, wash it with ether until colourless and then decompose it in a flask with a lukewarm solution of 125 g of sodium carbonate in 150 ml of water. Separate the ketone layer, extract the aqueous layer with four 30 ml portions of ether, dry the combined organic layers over magnesium sulphate, remove the ether at atmospheric pressure and distil the residual oil under reduced pressure through a short fractionating side-arm. Collect the cycloheptanone at 64–65 °C/12 mmHg; the yield is 17 g (31%).

Note. (1) If the reaction does not start at this stage remove the flask from the cooling bath and allow the mixture to warm to 10 °C; do not add any further alkali until the reaction has started.

Experiment 7.18 ETHYL *CIS*- AND *TRANS*-PULEGENATES AND *TRANS*-PULEGENIC ACID[14]

Pulegone dibromide. To a stirred and cooled solution of (+)-pulegone (21.6 g, 0.14 mol) in glacial acetic acid (30 ml) is added, dropwise, bromine (20 g, 0.125 mol). After the addition is complete (c. 0.5 hour) the solution is stirred for another 30 minutes and then poured on to crushed ice. The resulting oily mixture is extracted with eight 20 ml portions of light petroleum (b.p. 35–37 °C). The combined organic extracts (160 ml) are washed with dilute sodium hydrogen carbonate solution and dried over magnesium sulphate. No further attempt is made to purify the unstable dibromide.

Ethyl *cis*- and *trans*-pulegenates. The foregoing dried light petroleum solution of pulegone dibromide is added dropwise to a heated and stirred solution of sodium ethoxide [from sodium (9.5 g)] in carefully dried ethanol (200 ml). As the addition proceeds the light petroleum is distilled from the reaction mixture. After the addition is complete and the hydrocarbon has been distilled, the mixture is kept at reflux for 2 hours. The mixture is cooled and poured rapidly into 10 per cent hydrochloric acid (300 ml). The heavy oil which separates is taken up in ether and the aqueous phase is extracted thoroughly with ether. The combined ether extracts are washed successively with water, sodium hydrogen carbonate solution, and water and finally dried. Distillation gives after a forerun of pulegone, b.p. 54–56 °C/0.60–0.65 mmHg, the mixture of isomeric ethyl pulegenates, b.p. 56–62 °C/0.60–0.65 mmHg (13.8 g,

64%). Gas–liquid chromatography analysis on a Carbowax 20M column indicates the presence of 26 per cent of the *cis* isomer and 74 per cent of the *trans* isomer.

trans-**Pulegenic acid.** To a solution of the ethyl esters (10.0 g, 51 mol) in absolute ethanol (50 ml) is added potassium hydroxide (5.71 g, 102 mmol) in water (20 ml). The resulting solution is kept at reflux for 3 hours. After dilution with water the solution is extracted with ether to remove neutral products. The alkaline solution is acidified with dilute hydrochloric acid and the mixture extracted with ether. *trans*-Pulegenic acid may be isolated from this extract by drying, evaporation and distillation; the product may be isolated as a colourless liquid, b.p. 124–128 °C/3 mmHg, or b.p. 95–97 °C/0.45 mmHg. A portion of the ethereal extract is treated with an ethereal solution of diazomethane (Section 4.2.25, p. 430) and converted into the corresponding methyl ester; g.l.c. analysis indicates the exclusive presence of methyl *trans*-pulegenate.

7.5 REDUCTION OF AROMATIC COMPOUNDS

A convenient route to the preparation of substituted cylohexanes is the reduction of appropriately substituted benzenes. Total reduction is effected by heterogeneous catalytic hydrogenation at temperatures in the region of 100–200 °C and usually under pressure. Under these conditions functional side-chain substituents may be variously affected (as a result of reduction or hydrogenolysis), and specialised texts should be consulted for coverage of this vast field.[15]

Reduction of substituted benzenes with sodium (or lithium) in liquid ammonia in the presence of a *proton source* (such as methanol, ethanol, etc.) leads to a substituted, non-conjugated cyclohexadiene as a result of 1,4-addition of hydrogen (*the Birch reduction*).[16] With benzene the product is cyclohexa-1,4-diene as a result of the following mechanistic pathway.

With substituted benzenes the regioselectivity and the rate of the reduction are crucially dependent on the electron-donating or electron-withdrawing characteristics of the substituent. Thus with anisole, the rate of reaction is decreased and the product is 1-methoxycyclohexa-1,4-diene (2,5-dihydroanisole) (42); with benzoic acid the rate of reaction is increased and the product is cyclohexa-2,5-dienecarboxylic acid (1,4-dihydrobenzoic acid) (43).

(42)   (43)

The reaction rate differences are readily rationalised from a consideration of fundamental reaction mechanism principles (e.g. electron-donating groups increase electron density in the aromatic ring and thus hinder acceptance of an electron); the regioselectivity of the reduction is explained from a consideration of the relative stabilities of the possible alternative mesomeric intermediates, these being (44) in the case of anisole and (45) in the case of benzoic acid.

(44)   (45)

The Birch reduction of 3,4,5-trimethoxybenzoic acid [17a, b] (cognate preparation in Expt 7.19) is of interest in that the product is 1,4-dihydro-3,5-dimethoxy-benzoic acid (46) thus revealing that in this reduction the 4-methoxy group is removed by hydrogenolysis. The product possesses two methyl vinyl ether residues which are labile under aqueous acidic conditions to yield the saturated diketone, 3,5-diketocyclohexanecarboxylic acid (47).

(46)   (47)

There is no standard, universal, procedure for the Birch reduction. Experiment 7.19 illustrates some of the variants which have been reported in the literature. The original Birch procedure is to add small pieces of sodium metal to a solution of the aromatic compound in a mixture of liquid ammonia and the proton source (ethanol).[18] After completion of the reaction, which is usually indicated by the disappearance of the blue colour, it is quenched by the addition of ammonium chloride and the ammonia allowed to evaporate before the cautious addition of water, and isolation of the product by ether extraction. In a modified procedure a co-solvent (ether, tetrahydrofuran, etc.) is initially added to the solution of aromatic compound/liquid ammonia prior to the addition of metal; lithium metal is often used in place of sodium.[19a, b] In general these latter procedures are used for substrates which are more difficult to reduce. Redistilled liquid ammonia is found to be beneficial since the common contaminant iron, in collodial form or in the form of its salts, has a deleterious effect on the reaction.[20] A representative selection of procedures is given in Expt 7.19 for the reduction of o-xylene, anisole, benzoic acid, and 3,4,5-trimethoxybenzoic acid.

Experiment 7.19 1,2-DIMETHYLCYCLOHEXA-1,4-DIENE[21]

CAUTION: *This experiment must be conducted in an efficient fume cupboard and the precautions and techniques for handling liquid ammonia described in Section 2.17.7, p. 116, must be noted.*

To liquid ammonia (400 ml) (1) held at $-70\,°C$ is added with brisk stirring o-xylene (53 g, 0.5 mol) followed by small pieces of sodium metal (23 g, 1 g-atom) (2). Stirring is continued until all the sodium has dissolved and a blue coloured solution is obtained. Dry methanol (32 ml, 1 mol) is *cautiously* added dropwise at $-70\,°C$, when a vigorous reaction ensues. When the solution is colourless the ammonia is allowed to evaporate, and water cautiously added. The organic layer is separated (3), dried and distilled to give 1,2-dimethylcyclohexa-1,4-diene (49 g, 92.5%), b.p. 141.5–143 °C (4).

Note. (1) In general the apparatus used in these metal–liquid ammonia reductions is essentially that shown in Fig. 2.69 (*d*), but without the dropping funnel tube extension. The dropping funnel is replaced with the soda-lime guard-tube, which may then be removed at appropriate intervals for the periodic addition of metal fragments. After the flask has been charged with liquid ammonia, the inlet tube is replaced with a suitable thermometer, the bulb of which should be immersed in the reaction solution. A dry ice–acetone cooling bath should be used.
(2) Details of a large-scale preparation of 1,2-cyclohexa-1,4-diene have been published.[22]
(3) The quenched reaction solution may alternatively be extracted with ether to facilitate the isolation of the reduction product.
(4) Gas–liquid chromatography analysis of the product may be effected using didecyl phthalate as the stationary phase.[19a]

Cognate preparations. *Cyclohexa-2,5-dienecarboxylic acid* (*1,4-dihydrobenzoic acid*).[17a] A solution of benzoic acid (10 g, 0.082 mol) in ethanol (100 ml) and liquid ammonia (600 ml) is stirred and sodium (6.2 g, 0.27 g-atom) added in small pieces, followed by ammonium chloride (14.6 g, 0.27 mol). The ammonia is evaporated and the residual material dissolved in ice-water (500 ml). After acidification with 10 per cent hydrochloric acid the solution is extracted with four 100 ml portions of ether, the ether washed once with saturated sodium chloride solution, dried over magnesium sulphate and concentrated *in vacuo*. The remaining pale yellow oil is distilled at 96–98 °C (0.01 mmHg) to give 9.0 g (89%) of 1,4-dihydrobenzoic acid; this product shows no u.v. absorption above 220 nm.

1,4-Dihydro-3,5-dimethoxybenzoic acid.[17a] To solution of 3,4,5-trimethoxybenzoic acid (31.8 g, 0.15 mol) in ethanol (225 ml) and liquid ammonia (1.5 litres) is added sodium (18.0 g, 0.78 mol) in small pieces, followed by ammonium chloride (75 g, 1.45 mol). The ammonia is evaporated and the residue taken up in 2 litres of ice-water. Alternate additions of small portions of 10 per cent hydrochloric acid and immediate extractions with dichloromethane are carried out until the solution becomes acidic to Congo red. The combined extracts are washed several times with water, dried over magnesium sulphate, filtered and concentrated *in vacuo* at room temperature. Trituration of the residue with a small amount of ether and filtration gives 17.7 g of product. Concentration of the ether filtrate yields an additional 6.3 g of white crystalline product. The crude material, m.p. 100–105 °C (87%), could be recrystallised from ether–hexane. A sample prepared for elemental analysis shows m.p. 105 °C (decomp) or 118 °C (decomp) depending on the rate of heating.

Conversion to 3,5-diketocyclohexanecarboxlic acid.[17a]   A suspension of 1,4-dihydro-3,5-dimethoxybenzoic acid (80 g, 0.43 mol) in 2 per cent hydro-chloric acid (800 ml) is heated on a steam bath for 15 minutes. The clear solu-tion is then concentrated to dryness *in vacuo*, the residue is triturated with a small amount of ether and filtered, giving the diketone (66 g, 97%) as a white crystalline compound, m.p. 182 °C.

General procedure for reduction with lithium.[19b]   The apparatus used for all these reductions in liquid ammonia is a three-necked flask equipped with a stirrer, dropping funnel and soda-lime guard-tube. The flask is placed in a box stuffed with insulating wool or cotton. At the front of the box is a sealed double window with a cork spacer ring and rubber gasket against which the flask is placed. This arrangement prevents condensation of moisture on the flask or window, and with illumination from the top permits observation of the reaction mixture at all times. The flask is filled from one-third to no more than one-half of its capacity, to aid in control of foaming which occurs in some instances towards the end of alcohol addition.

To a solution of the compound to be reduced in dry ether or 1,2-dimethoxyethane is added liquid ammonia with stirring, and to the homo-geneous solution is added lithium wire in small pieces over a period of 1 to 10 minutes, depending on the quantity of metal. After stirring for 10 minutes, absolute ethanol is added dropwise over a period of about 20 minutes. In some cases there is a tendency for foaming to occur near the end of this addi-tion. This is easily controlled by stopping the stirrer momentarily. When the blue colour has disappeared the ammonia is evaporated, ether and water are added, separated and the aqueous layer re-extracted. After washing the com-bined extract with saturated salt solution and drying over potassium carbon-ate, the ether is removed and the product isolated by crystallisation or distillation as appropriate.

1-Methoxycyclohexa-1,4-diene (*2,5-dihydroanisole*).   A solution of anisole (15 g) in ether (50 ml) and liquid ammonia is treated by the general procedure outlined above with lithium (4.5 g, 4.6 equivalents per mol) and finally with absolute ethanol (35 g) (30 minute addition). After ether extractions, the con-centrated aqueous layer (75 ml) is acidified and treated with bromine, giving tribromophenol (0.90 g), m.p. 93–94.5 °C, which indicates only 2 per cent demethylation. From the ether extracts is obtained the product (12.8 g, 84%), b.p. 148–149 °C, n_D^{25} 1.4782. The u.v. absorption spectrum (95% ethanol), λ 268–269 nm (ε 800) indicates the presence of about 20 per cent of 2,3-di-hydroanisole in the 2,5-dihydroanisole [2,3-dihydroanisole is reported[23] as having λ_{max} 268 nm (ε 4270); see also *Fieser–Woodward rules for conjugated diene absorption*, p. 390].

7.6   CYCLOADDITION REACTIONS

Cycloaddition reactions represent a very versatile route to alicyclic compounds. The most important for six-membered rings is the *Diels–Alder reaction*, and its great utility lies in the fact that it is both regioselective and stereospecific. The reaction involves compounds containing a double or triple bond, usually activ-ated by conjugation with additional multiply-bonded systems (carbonyl, cyano, nitro, phenyl, etc.), which add to the 1,4-positions of a conjugated diene system (e.g. buta-1,3-diene) with the formation of a six-membered ring. The ethylenic or acetylenic compound is known as the *dienophile* and the second reactant as the *diene*; the product is the *adduct*. The product in the case of an ethylenic dienophile is a substituted cyclohexene, and in that of an ethynyl dienophile it is a substituted cyclohexa-1,4-diene. The active unsaturated portion of the dienophile, or that of the diene, or those in both, may be involved in rings; the adduct is then polycyclic.

(e.g. Y=CO·R, C≡N, NO$_2$, etc.)

This formulation emphasises that the Diels–Alder reaction may be represented in the retrosynthetic sense as a two-group disconnection (p. 22).

(TM)

Thus the two key features to be identified in the target molecule are a cyclohexene ring and an appropriately placed electron-withdrawing substituent. It should be noted that both structural features may be constituent parts of a larger, more complex molecule.

The Diels–Alder reaction is a concerted reaction in which four π-electrons from the diene and two π-electrons from the dienophile participate in the transition state. The Woodward–Hoffmann Rules provide a theoretical framework for these reactions.[24] They suggest that those reactions are thermally allowed which have $4n + 2$ pericyclic electrons, i.e. 6, 10, 14, etc. The Diels–Alder reaction is an example where n = 1, i.e. $(4 + 2)$ π-electrons.

The Woodward–Hoffmann rules also allow the prediction of the stereochemistry of pericyclic reactions. The Diels–Alder reaction is an example of $(\pi 4_s + \pi 2_s)$ cycloaddition. The subscript s, meaning *suprafacial*, indicates that both elements of the addition take place on the same side of the π-system. Addition to opposite sides is termed *antarafacial*. The Woodward–Hoffmann rules apply only to concerted reactions and are derived from the symmetry properties of the orbitals involved in the transition state. These rules may be summarised as shown in Table 7.1.

Table 7.1 Woodward–Hoffmann rules for cycloaddition reactions

p + q	Thermal diene, dienophile	Photochemical diene, dienophile
4n	supra, antara	supra, supra
	antara, supra	antara, antara
4n + 2	supra, supra	supra, antara
	antara, antara	antara, supra

The number of electrons in the dienophile and diene are represented by the letters p and q respectively.

For a more detailed treatment of the theory of pericyclic reactions the reader is referred to modern texts on reaction mechanism.[25]

The preparative procedure for the Diels–Alder reaction is extremely simple and the cycloaddition usually merely requires warming the reactants together either alone, or in the presence of a suitable solvent.

The reaction has been shown to involve a stereospecific *syn*-addition with respect to the dienophile. For example, the reaction of 2,3-dimethylbuta-1,3-diene with maleic anhydride gives *cis*-1,2,3,6-tetrahydro-4,5-dimethylphthalic anhydride (Expt 7.20). An example of the use of a quinone as the dienophile is provided by the synthesis of *cis*-1,4,4a,9a-tetrahydro-2,3-dimethyl-9,10-anthraquinone which upon dehydrogenation (most simply by the action of oxygen upon its solution in alcoholic potassium hydroxide) yields 2,3-dimethylanthraquinone (Expt 7.21). Tetraphenylcyclopentadienone (tetracyclone) readily undergoes the addition of dienophiles, such as maleic anhydride, to give an adduct, which then extrudes a molecule of carbon monoxide on heating, as in the preparation of 3,4,5,6-tetraphenyldihydrophthalic anhydride (Expt 7.22).

With less reactive dienophiles more extensive heating may be required in which case the reaction is best carried out in a suitable pressure vessel (e.g. the preparation of 4-nitro-5-phenylcyclohexene, Expt 7.23).

Four-membered rings can be synthesised by [2 + 2] cycloadditions. However, thermal [2 + 2] cycloadditions occur only with difficulty; they are not concerted but involve diradicals. Photochemical [2 + 2] reactions are common and although some of these may occur by a stepwise mechanism many are concerted. An example of a [2 + 2] reaction is the photodimerisation of cyclopent-2-enone. This compound, as the neat liquid, or in a variety of solvents, on irradiation with light of wavelength greater than 300 nm (the $n \to \pi^*$ excited state is involved) is converted to a mixture of the 'head-to-head' (48) and 'head-to-tail' (49) dimers, both having the *cis,anti,cis* stereochemistry as shown. It is believed that the reaction proceeds by attack of an $n \to \pi^*$ triplet excited species on a ground state molecule of the unsaturated ketone (Section 2.17.5, p. 106). In the reaction described (Expt 7.24) the cyclopent-2-enone is irradiated in methanol and the 'head-to-tail' dimer further reacts with the solvent to form the di-acetal which conveniently crystallises from the reaction medium as the irradiation proceeds; the other dimer (the minor product under these conditions) remains in solution. The di-acetal is converted to the diketone by treatment with the two-phase dilute hydrochloric acid-dichloromethane system.

(48)   (49)

The cyclopent-2-enone required for the photodimerisation is prepared by the hydrolysis and oxidation of 3-chlorocyclopentene, which is obtained by the low temperature addition of hydrogen chloride to cyclopentadiene. The latter is obtained by heating dicyclopentadiene. This depolymerisation is an example of a reverse (or *retro*) Diels–Alder cycloaddition reaction; the diene readily reforms the dicyclopentadiene on standing at room temperature.

1119

The (2 + 2) cycloaddition reactions of ketenes with alkenes are synthetically useful routes to cyclobutanones. Ketenes are particularly useful due to the low steric hindrance at the carbonyl carbon. An example is the reaction of dichloroketene with cyclopentadiene which, after reductive dechlorination of the product, gives bicyclo[3.2.0]hept-2-en-6-one (Expt 7.25).

A further example of a photodimerisation reaction is provided by the formation, in poor yield, of '1,4-naphthoquinone photodimer' (50) on irradiation of 1,4-naphthoquinone in benzene solution (Expt 7.26). The dimerisation may be effected by sunlight or by means of a mercury arc lamp.

(50)

Experiment 7.20   CIS-1,2,3,6-TETRAHYDRO-4,5-DIMETHYL-PHTHALIC ANHYDRIDE

Add 4.1 g (0.05 mol) of freshly distilled 2,3-dimethylbuta-1,3-diene (Expt 5.12) to 4.9 g (0.05 mol) of finely powdered maleic anhydride (Expt 5.218) contained in a small conical flask. Reaction occurs in a few minutes (indicated by evolution of heat). Allow to stand until the mixture attains room temperature. Remove the excess of maleic anhydride by extraction with cold water until the aqueous extract no longer gives an acid reaction to Congo red paper. Dry the residual white crystals upon filter paper in the air, and then recrystallise from light petroleum (b.p. 40–60 °C). The yield of the tetrahydrophthalic anhydride, m.p. 78–79 °C, is almost quantitative.

Experiment 7.21   CIS-1,4,4a,9a-TETRAHYDRO-2,3-DIMETHYL-9,10-ANTHRAQUINONE

In a small round-bottomed flask, fitted with a reflux condenser, place a solution of 8.2 g (0.1 mol) of freshly distilled 2,3-dimethylbuta-1,3-diene (Expt 5.12) and 7.9 g (0.05 mol) of 1,4-naphthoquinone (Expt 6.128) in 30 ml of ethanol, and reflux for 5 hours. Keep the resulting solution in a refrigerator for 12 hours: break up the crystalline mass, filter and wash with 5 ml of ethanol.

The yield of crude adduct, m.p. 147–149 °C, is 11.5 g (91%); recrystallisation from methanol raises the m.p. to 150 °C.

Conversion into 2,3-dimethylanthraquinone. Dissolve 10 g (0.0417 mol) of the adduct in 150 ml of 5 per cent potassium hydroxide solution (prepared by dissolving 7.5 g of potassium hydroxide pellets in 142.5 g of 95% ethanol) in a 250-ml three-necked flask equipped with a reflux condenser, gas inlet tube and a take-off adapter leading to a water pump. Bubble a current of air through the solution by means of gentle suction for 24 hours; the initial green colour changes to yellow and much heat is generated. Filter the yellow solid at the pump, wash successively with 50 ml of water, 25 ml of ethanol and 10 ml of ether and dry in the air. The yield of 2,3-dimethylanthraquinone, m.p. 209–210 °C, is 7.5 g (76%).

Experiment 7.22 3,4,5,6-TETRAPHENYLDIHYDROPHTHALIC ANHYDRIDE

Place a mixture of 7.0 g (0.018 mol) of tetraphenylcyclopentadienone (Expt 7.7) 1.9 g (0.019 mol) of maleic anhydride (Expt 5.218) and 5 ml of bromobenzene in a 100-ml round-bottomed flask fitted with a reflux condenser. Reflux gently in a fume cupboard (carbon monoxide is evolved) for 1.5 hours; during this period the dark brown colour of the cyclic ketone disappears. Cool to room temperature, add about 15 ml of light petroleum, b.p. 60–80 °C, break up the solid and filter. Wash with light petroleum and dry. The yield of crude product, m.p. 237–240 °C, is 7.5 g (90%). Dissolve the solid in hot benzene (120 ml), filter and add light petroleum, b.p. 60–80 °C (120 ml), to the hot filtrate and cool. Collect the solid which separates, wash with light petroleum and dry. The yield of pure tetraphenyldihydrophthalic anhydride, m.p. 235–240 °C, is 6.2 g (75%). The m.p. depends on the rate of heating.

Experiment 7.23 4-NITRO-5-PHENYLCYCLOHEXENE

This reaction is carried out using the steel pressure vessel and techniques described in Section 2.17.2, p. 97. To the dry pressure vessel add 7.46 g (0.05 mol) of ω-nitrostyrene (Expt 6.136), about 0.1 g of hydroquinone (a polymerisation inhibitor) and 15 ml of dry toluene. Fit a rubber bung carrying a calcium chloride guard-tube and cool the vessel to −78 °C in an acetone–Cardice bath. During the cooling process set up the acetone–Cardice condenser assembly (see p. 98) and charge the receiving flask with about 100 g of fresh potassium hydroxide pellets. Condense into the flask about 45 ml (27 g, 0.5 ml) of buta-1-3-diene from a cylinder, swirl well with the potassium hydroxide pellets and transfer the dried diene directly to the pressure vessel following the procedure described in Section 2.17.2, p. 98. Assemble the vessel, evacuate with an oil pump until a constant pressure is obtained (a few minutes only) and close the exit valve. Allow the apparatus to reach room temperature, wrap with heating tape and heat to 120–125 °C behind a safety screen for 48 hours. After allowing the vessel to cool to room temperature, cool further to about −15 °C and open the valve to the atmosphere. Remove the vessel cap and wash the contents of the vessel into a 1-litre round-bottomed flask with the aid of a little acetone. Remove the solvent on a rotary evaporator and crystallise the semi-solid residue from industrial spirit. The total yield (three crops) of crystallised material, m.p. 102–104 °C, is about 9 g (88%) and a pure colourless sample, m.p. 103 °C, may be obtained by further recrystallisation from industrial spirit.

Experiment 7.24 PHOTODIMERS OF CYCLOPENT-2-ENONE

Cyclopentadiene. Place 200 ml of liquid paraffin in a 500-ml, three-necked round-bottomed flask fitted with a large (30-cm) Vigreux column, a dropping funnel and a thermometer dipping into the paraffin. Attach a distillation head carrying a thermometer and a double surface condenser arranged for distillation. Heat the liquid paraffin to 200–240 °C (electric heating mantle) and add dicyclopentadiene portionwise, from the dropping funnel, and collect the cyclopentadiene, b.p. 40–42 °C, which distils over in a cooled receiver, protected from moisture. The dicyclopentadiene must be added slowly to ensure complete breakdown of the dimer; the temperature at the top of the still-head rises above 42 °C when addition is too rapid. Continue adding dicyclopentadiene (c. 300 ml) until 230 g of cyclopentadiene is obtained. The diene dimerises readily at room temperature, hence it should be used immediately or stored in the ice compartment of a refrigerator overnight.

3-Chlorocyclopentene. Place 230 g (*c.* 285 ml) of cyclopentadiene in a 500-ml measuring cylinder which is fitted with a rubber bung through which passes two glass tubes. One of the tubes (for introducing hydrogen chloride) should reach to near the bottom of the cylinder, but the other (outlet) tube should terminate just below the bung with its other end attached to a calcium chloride guard-tube. Cool the cylinder and contents in a Cardice–acetone bath maintained at −15 to −20 °C (note the volume of the liquid), and pass in a rapid stream of hydrogen chloride gas until the volume of the reaction mixture increases by 42 ml. This takes about 4 hours when a Kipp's apparatus is used to generate the hydrogen chloride gas. Transfer the reaction mixture to a distillation flask and distil under reduced pressure, collecting in a receiver cooled in Cardice–acetone the 3-chlorocyclopentene as a colourless liquid of b.p. 30 °C/20 mmHg. The yield is 222 g (62%). The compound is unstable at room temperature, polymerising slowly to a black tar. It should be used immediately for the next stage although it may be kept overnight in a Dewar flask packed with Cardice.

Cyclopent-2-enone. Place a solution of 180 g (0.6 mol) of sodium dichromate dihydrate in 630 ml of water in a 3-litre, three-necked round-bottomed flask equipped with an efficient stirrer, a dropping funnel and a thermometer. Cool the solution to 0 °C in an ice–salt bath and slowly add, in portions over about 1 hour, 157 g (1.53 mol) of 3-chlorocyclopentene to the rapidly stirred solution while keeping the temperature at 0–10 °C. Stir the reaction mixture at 0 °C for 0.5 hour then add 300 ml of 50 per cent sulphuric acid dropwise while maintaining the temperature at 0–10 °C by cooling. Saturate the dark brown mixture with sodium chloride and extract with five 250 ml portions of ether. Wash the combined extracts with two 150 ml portions of saturated sodium chloride, dry over magnesium sulphate and remove the ether by flash distillation. Distil the residual yellow oil under nitrogen under reduced pressure and collect the cyclopent-2-enone as a colourless liquid of b.p. 42–44 °C/ 13 mmHg. The yield is 58.8 g (47%). The compound has a retention time of 4.1 minutes on a 1.5-m column of Carbowax 20M on Chromosorb W maintained at 80 °C and with a nitrogen flow rate of 40 ml/minute.

Photodimerisation of cyclopent-2-enone. cis,trans,cis-*3,3,8,8-Tetramethoxytricyclo[5.3.0.0²,⁶]decane* and cis,trans,cis-*tricyclo[5.3.0.0²,⁶]decane-3,10-dione*. Place a solution of 49.9 g (0.61 mol) of cyclopent-2-enone in 800 ml of methanol (distilled from solid potassium hydroxide) in a photochemical reactor vessel of 1-litre capacity equipped with a 100-W medium pressure mercury arc lamp surrounded by a Pyrex cooling jacket (Section 2.17.5, p. 111), flush the magnetically stirred solution with nitrogen for 10 minutes and then irradiate under nitrogen overnight. Filter the white crystalline solid which separates and continue irradiating until no further crystalline material separates from the methanol solution. About 21.5 g of material is obtained after approximately 40 hours' irradiation. Crystallisation from methanol gives cis,trans,cis-3,3,8,8-tetramethoxytricyclo[5.3.0.0²,⁶]decane as white plates, m.p. 173–174 °C. The yield is 19.3 g (30%).

Remove the methanol from the filtrate remaining from the photolysis using a rotary evaporator and remove any cyclopent-2-enone which remains (*c.* 8.6 g) by distillation under reduced pressure. Several recrystallisations of the residual gum from carbon tetrachloride-hexane followed by a final crystallisation from hexane gives *cis,trans,cis*-tricyclo[5.3.0.0^{2,6}]decane-3,10-dione, m.p. 66–67 °C. The yield is about 2.24 g (5.5%).

Cis, trans,cis--tricyclo[5.3.0.0^{2,6}]decane-3,8-dione. Dissolve 15 g (0.058 mol) of *cis,trans,cis*-3,3,8,8-tetramethoxytricyclo[5.3.0.0^{2,6}]decane in 225 ml of dichloromethane in a 500-ml two-necked flask equipped with a sealed stirrer unit and reflux condenser. Add 30 ml of 2 M hydrochloric acid and heat the stirred mixture under reflux for 1 hour. Cool, separate the organic layer and extract the acid solution with two 40 ml portions of dichloromethane. Combine the dichloromethane solutions and wash with 50 ml of 10 per cent aqueous sodium hydrogen carbonate and then 50 ml of water. Dry over magnesium sulphate, filter and remove the solvent on a rotary evaporator. Crystallise the residual solid from carbon tetrachloride; *cis,trans,cis*-tricyclo-[5.3.0.0^{2,6}]decane-3,8-dione is obtained as white plates, m.p. 125–126.5 °C. The yield is about 9.1 g (95%).

Experiment 7.25 BICYCLO[3.2.0]HEPT-2-EN-6-ONE[26]

All solvents and reagents should be dried before use

7,7-Dichlorobicyclo[3.2.0]hept-2-en-6-one. The reaction is conducted in a three-necked round-bottomed flask fitted with a mechanical stirrer, a condenser and a pressure-equalising dropping funnel. A solution of triethylamine (96 g, 0.95 mol) in pentane (500 ml) is added over a period of 3.5 hours to a refluxing solution of dichloroacetyl chloride (129 g, 0.875 mol) and freshly distilled cyclopentadiene (216 ml, 2.62 mol) (Expt 7.24) in pentane (1 litre). The reaction is exothermic and precipitation of triethylamine hydrochloride is instantaneous. The mixture is refluxed for an additional 2 hours and then allowed to stand overnight at room temperature. The mixture is treated with cold water and the organic phase extracted successively with 5 per cent aqueous hydrochloric acid and with several portions of 5 per cent aqueous sodium hydrogen carbonate and then dried over anydrous sodium sulphate. After removal of the solvent *in vacuo*, distillation gives 150.6 g of a fraction boiling between 50 and 65 °C/2 mmHg which still contains an appreciable amount of dicyclopentadiene. Redistillation gives 119.2 g (77%) of 7,7-di-chlorobicyclo[3.2.0]hept-2-en-6-one (b.p. 67–68 °C/2 mmHg, n_D^{25} 1.5136) which appears homogeneous on t.l.c. and on three different g.l.c. columns (SE-30, QF-1 and 1500-Carbowax); i.r. 3050(w), 2955(m), 2855(m), 1807(s), 1608(w), 752(s) and 728(s) cm^{-1}.

Bicyclo[3.2.0]hept-2-en-6-one. A solution of freshly distilled tributyltin hydride (31 g, 0.108 mol) (1) in cyclohexane (30 ml) is placed in a three-necked flask equipped with a reflux condenser, a nitrogen inlet tube and a dropping funnel. The flask is flushed with nitrogen. The solution is refluxed while a solution of 7,7-dichlorobicyclo[3.2.0]hept-2-en-6-one (5.83 g, 0.054 mol) in cyclohexane containing α,α'-azobisisobutyronitrile (AIBN) (0.1 g, 0.6 mmol) is gradually added. After completion of the addition the mixture is distilled to give 4.38 g (75%) of bicyclo[3.2.0]hept-2-en-6-one (b.p. 62 °C/20 mmHg), n_D^{25}

1.4800; i.r. 3060(w). 2950(m), 2920(m), 2860(w), 1786 (s), 1608(w), 746(m) and 702(m) cm^{-1}. The semicarbazone recrystallises from a mixture of methanol–water, m.p. 218–219 °C.

Note. (1) Tributyltin hydride is prepared according to a slight modification of the literature procedure. A solution of lithium aluminium hydride (4.75 g, 0.136 mol) in anhydrous ether (400 ml) is placed in a three-necked round-bottomed flask equipped with a mechanical stirrer, a condenser, a nitrogen inlet tube and a pressure-equalising dropping funnel containing tributyltin chloride (100 g, 0.308 mol). The system is flushed with nitrogen and then tributyltin chloride is added dropwise at a rate which maintains a gentle reflux of ether. The resulting mixture is stirred for 7 hours. Careful addition of water to the cooled mixture destroys the excess lithium aluminium hydride. The ether solution is decanted, washed with three 100 ml portions of water and dried over calcium sulphate. Evaporation of the solvent and distillation under high vacuum gives 68 g (78%) of tributyltin hydride (b.p. 68–69 °C/0.3 mmHg, n_D^{25} 1.4688) which is kept under an argon atmosphere.

Experiment 7.26 1,4-NAPHTHOQUINONE PHOTODIMER

Method A. Place a solution of 5 g of pure, 1,4-naphthoquinone (1) in 115 ml of dry thiophene-free benzene (**CAUTION**) in the appropriate sized photo-chemical reactor vessel (Section 2.17.5, p. 111; note also the precautions to be observed for photochemical reactions) fitted with a 100-W medium pressure mercury arc lamp surrounded by Pyrex cooling jackets. Stir the solution by means of a magnetic follower bar and pass dry nitrogen through the solution for 10 minutes before switching on the lamp; adjust the nitrogen flow to a slow trickle and continue irradiation for about 6 hours, by which time a quantity of the solid dimer will have separated from the solution. Switch off the lamp, remove the lamp insert and collect the solid by suction filtration and wash with a little benzene. Gently remove and collect any of the solid which may have separated on the surface of the lamp insert and thoroughly clean the latter with a little acetone on a paper tissue before replacing the ben-zene filtrate in the reactor and continuing the irradiation process. Repeat these operations as appropriate and collect the dimer over a total irradiation period of about 28 hours; about 0.84 g of solid, m.p. 243–246 °C (decomp.) (2), is obtained. Recrystallise from 150–160 ml of glacial acetic acid with the aid of a little decolourising carbon and dry the pale straw-coloured crystals in a desiccator over solid sodium hydroxide pellets; the yield of pure naphtho-quinone dimer, m.p. 245–249 °C (decomp.), is 0.62 g (12%).

Method B. Place a solution of 2.5 g of pure 1,4-naphthoquinone (1) in 50 ml of dry thiophene-free benzene in a 50-ml round-bottomed flask and pass a slow stream of dry nitrogen through the solution for 5 minutes. Insert a lightly greased stopper in the flask taking care that the solution does not penetrate the joint. Place the flask in a sunny position – if situated in the open air wrap some aluminium foil around the stopper to prevent the possibility of rainwater seeping through the joint into the reaction mixture. Solid slowly separates from the solution over a period of days; collect the product by suction filtration, wash with a little benzene and dry. Return the filtrate to the flask, flush with dry nitrogen and place in the sunlight until a further portion of solid is ready for collection. The amount of material obtained in a given period is dependent on light conditions; typically, a total of 0.33 g (13%) of crude naphthoquinone photodimer, m.p. 241–246 °C (decomp.), has been obtained during a 31-day irradiation period (April–May).

Notes. (1) Technical grade naphthoquinone may be purified as described in Expt 6.128.
(2) The melting point is best observed on a microscope–hot stage apparatus.

REFERENCES

1. J. E. Baldwin (1976). *J. Chem. Soc., Chem. Commun.*, 734.
2. K. Ogura, M. Yamashita, M. Suzuki, S. Furukawa and G. Tsuchiashi (1984). *Bull. Chem. Soc. (Japan)*, **57**, 1637.
3. R. E. Gawley (1976). *Synthesis*, 777; M. E. Jung (1976). *Tetrahedron*, **32**, 1.
4. (a) Z. G. Hajos and D. R. Parrish (1974). *J. Org. Chem.*, **39**, 1612.
    (b) Z. G. Hajos and D. R. Parrish (1974). *J. Org. Chem.*, **39**, 1615.
5. (a) R. C. Cookson and S. A. Smith (1979). *J. Chem Soc. Perkin Trans I*, 2447.
    (b) N. Nagato, M. Ogawa and T. Naito (1974). *Chem. Abs.*, **80**, 59 497.
6. J. W. F. L. Seetz, R. Tol, O. S. Akkerman and F. Bickelhaupt (1983). *Synthesis*, 721.
7. Z. G. Hajos, D. R. Parrish and E. P. Oliveto (1968). *Tetrahedron*, **24**, 2039.
8. L. F. Fieser (1964). *Organic Experiments*. Boston: Heath and Co, p. 301.
9. K. Issleib and A. Brack (1954). *Z. anorg. Chem.*, **277**, 258.
10. H. E. Simmons, T. L. Cairns, S. A. Viaduchick and C. M. Hoiness (1973). *Organic Reactions* (20). Ed. W. G. Dauben. New York: Wiley, p. 1.
11. (a) O. Repič and S. Vogt (1982). *Tetrahedron Lett.*, 2729.
     (b) E. C. Friedrich, J. M. Domek and R. Y. Pong (1985). *J. Org. Chem.*, **50**, 4640.
12. S. R. Landor and N. Punja (1967). *J. Chem. Soc., C*, 2495.
13. R. Kuhn and H. Trischmann (1958). *Liebigs Ann. Chem.*, **611**, 117.
14. J. Wolinsky and D. Chan (1965). *J. Org. Chem.*, **30**, 41; S. A. Achmad and G. W. K. Cavill (1963). *Aust. J. Chem.*, **16**, 858.
15. P. N. Rylander (1985). *Hydrogenation Methods*. London: Academic Press.
16. A. J. Birch and G. Subba Rao (1972). 'Reduction by metal–ammonia solutions and related reagents', in *Advances in Organic Chemistry*. Ed. E. C. Taylor, New York: Wiley-Interscience, Vol. 8, p. 1. See also A. A. Akhrem, I. G. Reshetova and Y. A. Titov (1972). *Birch Reduction of Aromatic Compounds*. New York: IFI/Plenum.
17. (a) M. E. Kuehne and B. F. Lambert (1959). *J. Am. Chem. Soc.*, **81**, 4278.
     (b) A. J. Birch, P. Hextall and S. Sternhell (1954). *Aust. J. Chem.*, **7**, 256.
18. A. J. Birch (1946). *J. Chem. Soc.*, 593.
19. (a) A. P. Krapcho and A. A. Bothner-By (1959). *J. Am. Chem. Soc.*, **81**, 3658.
     (b) A. L. Wilds and N. A. Nelson (1953). *J. Am. Chem. Soc.*, **75**, 5360.
20. H. L. Dryden, Jr, G. M. Webber, R. R. Burtner and J. A. Cella (1961). *J. Org. Chem.*, **26**, 3237.
21. W. Hückel and U. Wörffel (1955). *Chem. Ber.*, **88**, 338.
22. L. A. Paquette and J. H. Bassett. *Organic Syntheses*, Coll. Vol. V, p. 467.
23. A. J. Birch (1950). *J. Chem Soc.*, 1551.
24. R. B. Woodward and R. Hoffmann (1970). *The Conservation of Orbital Symmetry*. New York: Academic Press.
25. See references 3(a), 3(c) and 3(d) in Chapter 1 of this volume.
26. L. Ghosez, R. Montaigne, A. Roussel, H. Vanlierde and P. Mollet (1971). *Tetrahedron*, **27**, 615.

CHAPTER 8
SELECTED HETEROCYCLIC COMPOUNDS

The methods of preparation which are considered in this chapter involve, principally, ring-forming reactions leading to a restricted and subjective selection of various typical and interesting heterocyclic systems. These cyclic systems may be classified as (a) saturated or partly unsaturated heterocyclic, or (b) heteroaromatic, according to their physical, chemical and spectroscopic properties and to the reactivity of the functional groups attached to the ring atoms. In this way such properties and reactivity are broadly related to those of alicyclic and aromatic compounds.

In this chapter the material is subdivided first on the basis of ring size (8.1, three-membered; 8.2, four-membered; etc.) and thence according to the ring-forming reaction type. Where appropriate, routes to selected target molecules are analysed in a retrosynthetic sense. The reader's attention is drawn to a book on heterocyclic chemistry[1] which includes a more detailed theoretical treatment of the kinetic and thermodynamic factors of these ring-forming reactions; this book also cites many examples of heterocyclic compounds which play an important role in biological systems.

The spectroscopic features of the heterocyclic ring systems may be inferred from the data noted in the correlation tables (Appendices 2, 3 and 4). Since the range of one- and multi-heteroatom systems is so diverse, a generalised discussion of these features would be too extensive for the scope of this book. Illustrative and descriptive accounts are included in appropriate preparative sections.

8.1 THREE-MEMBERED HETEROCYCLES

The parent saturated three-membered ring systems containing nitrogen, or oxygen, or sulphur as the heteroatom are known as aziridine (1), oxirane (2) and thiirane (3).

NH   O   S

(1)   (2)   (3)

Four illustrative methods for the synthesis of these systems or their substituted analogues are as follows.

1. Intramolecular cyclisation of a 1,2-bifunctional compound.
2. Insertion of a methylene group into a carbon–heteroatom double bond.
3. Insertion of a heteroatom into a carbon–carbon double bond.
4. Heteroatom exchange.

8.1.1 INTRAMOLECULAR CYCLISATION OF A 1,2-BIFUNCTIONAL COMPOUND

A generalised retrosynthetic analysis involving disconnection of the carbon–heteroatom bond may be depicted as follows.

The reagent equivalent therefore requires an anionic site at the heteroatom, and a good leaving group at the vicinal position.

An illustrative example of oxirane formation by the action of alkali on a β-halohydrin is to be found in the reaction sequence involved in the Darzens glycidic ester synthesis (Section 5.7.6, p. 598). Three target molecules, namely 2-phenylaziridine (4), methyl (S)-thiiranecarboxylate (5) and cyclooctene sulphide (6), are selected here to exemplify this intramolecular cyclisation reaction type.

In the case of 2-phenylaziridine (4) the leaving group may originate from the hydroxyl function of 2-amino-1-phenylethanol by treatment with sulphuric acid to convert it into a sulphate ester which is then cyclised by treatment with base[2] (Expt 8.1).

This method is generally useful for the formation of various substituted aziridines by suitable selection from a range of readily available alkanolamines.

The formation of methyl (S)-thiiranecarboxylate (5) is of interest,[3] since in this case the leaving group is a diazo residue originating from an amino group. The starting material is therefore the chiral amino acid ester, methyl (R)-cysteinate, and the stereoelectronic requirements of the reaction lead to an optically pure product of predictable stereoisomeric form (Expt 8.2).

A further example of this general type of reaction is provided by the synthesis of cyclooctene sulphide (6).[4] Here an alkene is converted into an adduct with succinimide-N-sulphenyl chloride (reaction formulated in full in Expt 8.3), which is reduced with lithium aluminium hydride at $-78\,^{\circ}$C to form the product.

1128

$$(CH_2)_n \overset{O}{\underset{O}{\parallel}} NSCl \xrightarrow[\text{(ii) LiAlH}_4]{\text{(i)}} (CH_2)_n \overset{Cl}{\underset{S^{\ominus}}{\diagup}} \longrightarrow (CH_2)_n \triangleright S$$

Experiment 8.1 2-PHENYLAZIRIDINE[2]

$$Ph\cdot CH(OH)\cdot CH_2NH_2 \xrightarrow{H_2SO_4} Ph\cdot CH(O\cdot SO_3^{\ominus})\cdot CH_2\overset{\oplus}{N}H_3 \xrightarrow{\text{base}} \underset{Ph}{\triangle}\!\!NH$$

CAUTION: *All three-membered nitrogen heterocycles must be handled with great care as they may be highly toxic, and frequently are foul smelling. All operations therefore should be conducted in an efficient fume cupboard and disposable plastic gloves must be worn.*

An aqueous solution of 2-amino-1-phenylethanol (0.5 mol) is neutralised to a methyl red end-point with 50 per cent aqueous sulphuric acid, followed by the addition of a further equal volume of aqueous acid. Water is removed by heating the solution at 10–15 mmHg, finally at 130 °C. The product crystallises and is coarsely ground and heated at 120–130 °C under reduced pressure to constant weight.

The conversion of the sulphate ester (0.5 mol) into the aziridine is carried out by dissolving it in 1 litre of 2 M sodium hydroxide held at 0 °C, and then slowly heating the resulting solution in an oil bath. At approximately 90 °C, the aziridine product separates from the solution as an upper layer. The reaction mixture is then steam distilled until a fresh portion of the distillate gives a neutral reaction. The product is separated from the aqueous distillate and dried over potassium hydroxide pellets. Fractional distillation using a 24-inch Podbielniak tantalum spiral column affords a 90 per cent yield of 2-phenylaziridine, b.p. 94–94 °C/10 mmHg. Gas–liquid chromatography analysis through a 10-ft column of 20 per cent Carbowax 20M reveals a single peak.

Experiment 8.2 METHYL (*S*)-THIIRANECARBOXYLATE[3]

$$\underset{NH_2}{\overset{HS}{\diagdown}} \overset{H}{\underset{}{\diagup}} CO_2Me \xrightarrow{HNO_2} \underset{CO_2Me}{\overset{S}{\triangle}\!\!\cdot\!H}$$

Sodium nitrite (2.07 g, 30 mmol) is added in one portion to a solution of methyl (*R*)-cysteinate (1.35 g, 10 mmol) (1) in 1 M hydrochloric acid (50 ml) at 0 °C. The solution develops a deep red coloration which partly fades within a few minutes. After 5 minutes the mixture is extracted three times with chloroform and the organic layer dried over magnesium sulphate and evaporated. Purification of the product by chromatography on aluminium (Brockmann activity III, light petroleum (b.p. 40–60 °C) as eluant) affords methyl (*S*)-thiiranecarboxylate (0.555 g, 47%) as a chromatographically homogeneous liquid; $[\alpha]_D$ −25° (*c* 2.8 in CHCl$_3$); p.m.r. (CDCl$_3$, TMS) δ 2.68 (d of d, 1H, $J = 1.0$ and 5.9 Hz, SCHH*C*H), 2.90 (d of d, 1H, $J = 1.0$ and 4.8 Hz, SCH*H*CH), 3.45 (d of d, 1H, $J = 5.9$ and 4.8 Hz, SCH*CH$_2$*), and 3.80 (s, 3H, Me); *m/z* 118 (M) and 86 (C$_3$H$_2$OS$^{\oplus}$, base peak).

1129

Note. (1) This amino ester is prepared as the corresponding hydrochloride by heating (R)-cysteine hydrochloride in methanol saturated with hydrogen chloride, followed by evaporation (cf. Expt 8.7). The hydrochloride is partitioned between dichloromethane and sodium hydrogen carbonate solution; evaporation of the organic layer affords the pure amino ester which is used immediately.

Experiment 8.3 CYCLOOCTENE SULPHIDE[4]

CAUTION: *The use of sulphur monochloride and of chlorine gas in these reactions makes it imperative that they are carried out in an efficient fume cupboard.*

N,N′-Dithiobis(succinimide). A hot solution of succinimide (100 g) in absolute ethanol (1 litre) is poured into an ice-cold solution of potassium hydroxide (58 g) in absolute ethanol (300 ml). The mixture is stirred and cooled to 0 °C. Potassium succinimide crystallises from the mixture as colourless crystals, and is collected by suction filtration. After washing with light petroleum (b.p. 60–80 °C) the crystals are dried (110.3 g, 79.1%). A solution of sulphur monochloride (54 g) in dichloromethane (50 ml) is added over 5 minutes to a vigorously stirred suspension of the potassium succinimide (110.3 g) in dichloromethane (400 ml) at 0 °C. After the initial exothermic reaction, the mixture is stirred for a further 5 minutes, filtered and the filtrate evaporated *in vacuo* to leave a solid residue, which is washed three times with light petroleum (b.p. 60–80 °C) and once with water. Crystallisation from methanol gives colourless plates of N,N′-dithiobis(succinimide) (83 g, 79.3%), m.p. 192–193 °C; m/z 260, 162, 134, 102, 99 and 55.

Succinimide-N-sulphenyl chloride. Chlorine gas (**CAUTION**) is passed for 8 hours through a stirred solution of the foregoing compound (53.7 g) in chloroform (250 ml) warmed to 50 °C. Nitrogen gas is passed through the mixture to remove excess chlorine, followed by evaporation *in vacuo* to give succinimide-N-sulphenyl chloride (65.1 g, 93.3%) as yellow crystals, m.p. 65–67 °C.

Cyclooctene sulphide. The foregoing imidosulphenyl chloride (6.6 g, 0.04 mol) is dissolved in dichloromethane (25 ml) to form a yellow solution which is added dropwise at room temperature to cyclooctene (4.5 g, 0.041 mol) in dichloromethane (50 ml). The mixture is stirred until no yellow colour persists.

Removal of the solvent under reduced pressure gives the adduct, *trans-N*-(2-chlorocyclooctylthio)succinimide (9.4 g, 85.3%), m.p. 83–85 °C (from ether); *m/z* 277, 275, 260, 239, 211, 176, 141, 109 and 99.

The foregoing adduct (9.4 g, 0.034 mol) is dry tetrahydrofuran (25 ml) is added dropwise to a stirred suspension of lithium aluminium hydride (1.1 g) in tetrahydrofuran (10 ml) at − 78 °C under nitrogen. After stirring for 10 minutes, the mixture is allowed to warm to room temperature and then quenched with water. The episulphide is extracted with ether and, after drying over magnesium sulphate, the solvent is removed and the residue distilled under reduced pressure to afford cyclooctene sulphide (4.7 g, 97.35%), b.p. 50 °C/2 mmHg.

8.1.2 INSERTION OF A METHYLENE GROUP INTO A CARBON–HETEROATOM DOUBLE BOND

This method is illustrated by the synthesis of the target molecules 1,2-diphenyl-aziridine (7) and phenyloxirane (8).

A retrosynthetic disconnection in the case of each of these molecules gives a methylene group synthon, and benzylideneaniline and benzaldehyde from (7) and (8) respectively.

$$\text{(7)} \Longrightarrow Ph\cdot CH{=}N\cdot Ph + [:CH_2]$$

$$\text{(8)} \Longrightarrow Ph\cdot CH{=}O + [:CH_2]$$

The most convenient reagent equivalent is a methylene transfer reagent which arises from the action of base on either a trimethyloxosulphonium salt (see Section 7.3) or a trimethylsulphonium salt. In this latter case dimethyl-sulphonium methylide ($Me_2\overset{\oplus}{S}\cdot\overset{\ominus}{C}H_2$) may be generated by the action of sodium hydride or butyllithium.[5] However, a PTC method which is applicable to the synthesis of both target molecules (Expts 8.4 and 8.5) offers a convenient alternative. Here benzylideneaniline[6] or benzaldehyde[7] in dichloromethane solution containing the phase transfer catalyst is stirred with a suspension of tri-methylsulphonium iodide in aqueous sodium hydroxide. The mechanism of the reaction is illustrated with reference to the aziridine synthesis.

$$I^{\ominus}\}Me_2\overset{\oplus}{S}\cdot Me \xrightarrow{\text{base}} Me_2\overset{\oplus}{S}\cdot\overset{\ominus}{C}H_2$$

Experiment 8.4 1,2-DIPHENYLAZIRIDINE[6]

$$Ph \cdot CH = N \cdot Ph + Me_3\overset{\oplus}{S}\}\overset{\ominus}{I} \xrightarrow[PTC]{\ominus OH}$$

Ph—N—Ph (aziridine)

CAUTION: *See cautionary note in Expt 8.1.*

Benzylideneaniline (18.1 g, 0.1 mol) and tetrabutylammonium hydrogen sulphate (0.5 g, 1.35 mol) are dissolved in dichloromethane (100 ml) and a layer of 50 per cent aqueous sodium hydroxide introduced under this solution. Trimethylsulphonium iodide (20.4 g, 0.1 mol) is then added and the whole warmed at 50 °C with vigorous stirring for 2 hours, whereupon the originally undissolved sulphonium salt disappears. The mixture is poured on to ice, the organic phase separated, washed with water and dried. The solvent is evaporated and the residue distilled under reduced pressure to afford 1,2-diphenylaziridine (94%), b.p. 120 °C/0.05 mmHg.

Experiment 8.5 PHENYLOXIRANE (*1,2-Epoxyethylbenzene*)

$$\underset{Ph \quad H}{\overset{O}{\|}} + Me_3\overset{\oplus}{S}\}\overset{\ominus}{I} \xrightarrow[PTC]{\ominus OH} \underset{Ph}{\overset{O}{\triangle}}$$

Dissolve 5.3 g (0.05 mol) of benzaldehyde (previously shaken with sodium hydrogen carbonate solution) and 0.25 g (0.67 mmol) of tetrabutylammonium iodide in 50 ml of dichloromethane. Place this solution in a 250-ml, three-necked round-bottomed flask equipped with an efficient sealed stirrer unit, a reflux condenser and a thermometer sited in a screw-capped adapter, and supported in an oil bath mounted on an electric hot plate. Introduce 50 ml of a 50 per cent (w/v) aqueous solution of sodium hydroxide, and then 10.2 g (0.05 mol) of finely powdered trimethylsulphonium iodide. Adjust the electric hot plate so that the oil bath is maintained at a constant temperature of 55 °C for 60 hours and during this period stir the reaction mixture rapidly (1). Pour the reaction mixture on to ice, separate the organic phase and extract the aqueous solution with one 20 ml portion of dichloromethane. Wash the combined organic phases successively with four 20 ml portions of water, two 10 ml portions of a saturated solution of sodium metabisulphite and finally two 20 ml portions of water. Dry the organic phase over anhydrous calcium sulphate, remove the dichloromethane on a rotary evaporator and distil the residue. Collect the phenyloxirane as a fraction having b.p. 191–192 °C; the yield is 4.7 g (78%).

Note. (1) The reaction may be monitored by observing the diminution of carbonyl absorption in the infrared spectrum of successive samples withdrawn from the organic phase, after allowing the dichloromethane to evaporate from a portion placed on a sodium chloride plate. Alternatively samples of the reaction medium may be analysed by g.l.c. using a 10 per cent Carbowax column on Chromosorb W held at 150 °C with a nitrogen carrier gas flow rate of 40 ml/minute; benzaldehyde has t_R 2 minutes and phenyloxirane has t_R 2.75 minutes.

8.1.3 INSERTION OF A HETEROATOM INTO A CARBON–CARBON DOUBLE BOND

The most important example of this method is the epoxidation of an olefinic bond. The reaction is illustrated by the synthesis of the target molecules *trans-*

1,2-dibutyloxirane (9), phenyloxirane (8) and (2S,3S)-epoxygeraniol (10) formed from *trans*-dec-5-ene,[8] styrene, and geraniol [(E)-3,7-dimethylocta-2,6-dien-1-ol][10b] respectively (Expt 8.6).

(8)   (9)   (10)

The epoxidation reaction is achieved most conveniently by employing *m*-chloroperbenzoic acid (or perbenzoic acid) in a solvent such as chloroform. The use of other peracids, such as peracetic acid or pertrifluoroacetic acid, give lower yields of the oxirane since the oxide may be readily cleaved to form the mono-ester of the diol (e.g. Section 5.4.5, p. 547).

The regioselectivity of this reaction in the case of certain dienes is of interest, since it has been found that the most highly substituted double bond is preferentially attacked, for example as in the mono-epoxidation of 1,2-dimethylcyclo-hexa-1,4-diene[9] (formed by the Birch reduction of *o*-xylene, Section 7.5, p. 1114).

It should be noted that although oxiranes (8) and (9) are chiral, this epoxidation methodology gives the racemate since no chiral inducing agent is present.

However, the asymmetric synthesis of optically active oxiranes has been the subject of much research activity in recent years. This arises from the fact that the synthesis of stereoisomerically pure structural units, required for the total synthesis of a range of biologically active compounds, may be effected by using the many ring-opening reactions of chiral oxiranes (e.g. reduction, acidic hydrolysis, base-catalysed reactions, reaction with organometallic compounds, etc.). In particular the asymmetric epoxidation of allylic alcohols with t-butyl hydroperoxide in the presence of titanium(IV) isopropoxide–diethyl tartrate catalyst is a significant and important step forward in this methodology (the *Sharpless epoxidation reaction*). A preparative example, for which precise experimental details have been published,[10a] is the conversion of (E)-hex-2-en-1-ol into (2S,3S)-3-propyloxiranemethanol in 96.8 per cent enantiomeric purity using diethyl (2R,3R)-(+)-tartrate [(R,R)-DET]; the enantiomeric oxirane is obtained by the use of diethyl (−)-tartrate.

This reaction has now been applied to a very great number of substituted allylic alcohols, and the mechanistic and stereoisomeric features of the reaction are becoming clearer.[11] In broad outline, it would appear that the initial step is an alkoxy-exchange reaction between two alkoxy residues in the titanium complex and the two hydroxyl groups in the tartrate ester, thus:

$$Ti(OR^1)_4 + DET \rightleftharpoons Ti(DET)(OR^1)_2 + 2R^1OH$$

(11)

Subsequently the remaining isopropoxide residues in the resulting complex (11) undergo exchange with the hydroxyl group of the allylic alcohol and the hydroxyl group of the peroxide to give the complex (12); further coordination activates the peroxide and the topography of this complex (13) determines the favourable enantioselective transfer of oxygen to the carbon–carbon double bond to give complex (14). The epoxidised allylic alcohol and the t-butyl alcohol are then released by further exchange reactions with the allylic alcohol and peroxide.

The methodology used for the asymmetric epoxidation of geraniol to give (10), (2S,3S)-epoxygeraniol (Expt 8.6, cognate preparation), illustrates a general procedure which employs 1.5 equivalents of t-butyl hydroperoxide with catalytic amounts ($<10\%$) of titanium(IV) isopropoxide and of diethyl tartrate.[10b] The success of this procedure depends on the presence of 3A or 4A molecular sieves since moisture, from no matter what source, contributes to a lowering of enantioselectivity and a slowing of reaction rate.

Experiment 8.6 *TRANS*-1,2-DIBUTYLOXIRANE (trans-*Dec-5-ene* oxide)[8]

CAUTION: *All operations, particularly distillations, should be conducted behind a safety screen.*

Into a stirred solution of m-chloroperbenzoic acid (15.0 g, 87 mmol) in chloroform (150 ml) held at 0 °C is added *trans*-dec-5-ene (10 g, 70 mmol) dissolved in chloroform (50 ml) over a 30-minute period. The mixture is stirred overnight at room temperature, filtered, washed with 10 per cent sodium hydrogen carbonate solution (1) and dried over anhydrous sodium sulphate. The solvent is removed by rotary evaporation and the epoxide distilled under reduced pressure, affording 9.6 g (88%) of product having b.p. 71–72 °C/

12 mmHg; p.m.r. (CDCl$_3$, TMS) δ 0.97 (t, 6H, 2 Me), 1.45 (m, 12H, 6 CH$_2$), 2.63 (m, 2H, 2 CH—O).

Note. (1) It is advisable to test for the complete disappearance of peracid (see below) before proceeding.

Cognate preparations. *Phenyloxirane* (*styrene oxide*). *Use of perbenzoic acid.* Use 42 g (0.29 mol) of perbenzoic acid (**CAUTION**, *see Section 4.2.56, p. 455*), in 350 ml of dichloromethane and 30 g (0.29) of styrene in 100 ml of dichloromethane. Stir the reaction mixture at 0 °C for 24 hours. Only a slight excess of peracid remains; confirm this by mixing an aliquot portion with excess of acidified potassium iodide and titrating with standard sodium thiosulphate solution. Work-up as described above and distil the dried dichloromethane solution through an efficient fractionating column. Styrene oxide passes over at 189–192 °C (or at 101 °C/40 mmHg); the yield is 25 g (72%).

Catalytic asymmetric epoxidation of (E)-3,7-dimethylocta-2,6-dien-1-ol (*geraniol*).[10b] *Preparation of anhydrous t-butyl hydroperoxide (TBHP) in dichloromethane.* **CAUTION** (1). Solutions of TBHP in dichloromethane are prepared as described previously for toluene solutions with minor modifications.[10a] Two litres of aqueous 70 per cent TBHP (2) and 2 litres of dichloromethane are shaken in a separatory funnel. The lower, organic phase is transferred to a 5-litre flask fitted with a heavier-than-water solvent Dean-Stark trap (Ace Glass Co.) with condenser. *Although the literature authors report that they have never experienced a problem with this procedure, they advise that all heating should be done behind an adequate blast shield in a well-ventilated fume hood.* After addition of a few boiling chips, the mixture is brought to a gentle reflux by using a heating mantle set on a low voltage. Periodically the collected water is removed from the trap. After 10 hours about 50 ml have been removed, and no more water is observed in the azeotrope. The TBHP solution (c. 2.5 litres) is divided into two batches, and each is finally dried in a refrigerator for several hours (usually overnight) over 200–300 g of activated 3A sieve pellets (3) either in a flask covered with cellophane or in a polyethylene bottle. The solutions (about 50% v/v TBHP, 5–6 M) are then transferred to high density polyethylene bottles and stored over activated 3A molecular sieve pellets at 0–5 °C. When properly capped, polyethylene bottles develop negative pressure upon cooling in the refrigerator and compress. Such solutions have been stored for months without loss of effectiveness and only slight loss of titre (5–10%, possibly due to constant use, and thus warming). Assay of the solution is effected by iodometric titration as follows. An 0.1 M aqueous sodium thiosulphate solution is prepared (12.4 g of sodium thiosulphate pentahydrate with enough water to make 500 ml suffices for 15 to 20 titrations), and 50 ml of this solution is placed in a 50-ml graduated burette. A 250-ml Erlenmeyer flask is charged with 25 ml of propan-2-ol and 1 ml of glacial acetic acid. To this is added 10 ml of a freshly prepared, cooled solution of 20 g of sodium iodide in 100 ml of warm propan-2-ol. After addition of 0.25 ml of anhydrous TBHP/dichloromethane solution (4), the mixture is heated to reflux (with stirring on a hot plate) and refluxed for 30–45 *seconds.* Failure to reflux the solution will result in a low titre. After dilution with 100 ml of distilled water, the warm solution is titrated rapidly with the 0.1 M sodium thiosulphate (25–30 ml required) to the disappearance of the yellow iodine colour. Starch indicator may be used towards the end of the titration to enhance the end-point. The concentration is calculated according to the equation [(molarity of titrant) × (ml of titrant)/(ml of TBHP solution) × 2], and should be in the range 5–6 M. The active oxygen content of a 5.0 M (45 wt %) TBHP/dichloromethane solution is about 7 wt per cent). Solutions of lower molarity are obtained either by dilution just prior to titration or by addition of less than 70 per cent TBHP at the start of the procedure. In any case, a flask size should be selected which ensures that the liquid remains above the top of the heating mantle throughout the azeotropic process, with the addition of more dichloromethane if necessary.

(2S,3S)-Epoxygeraniol. A mixture of powdered, commercially activated 4A molecular sieves (5) (1.8 g, Aldrich, 15–20 wt % based on substrate) and 100 ml of dichloromethane (6) is cooled to −10 °C (7). (2R,3R)-(+)-Diethyl tartrate (8) (1.00 g, 4.8 mmol), titanium(IV) isopropoxide (9) (0.91 g, 3.2 mmol), and t-butyl hydroperoxide (4) (15.6 ml, 97 mmol, 6.2 M in dichloromethane) are added sequentially. After 10 minutes, the mixture is cooled to −20 °C and freshly distilled geraniol (10.0 g, 65 mmol, in 10 ml of dichloromethane) is added dropwise, *with vigorous overhead stirring*, over a 15 minute period.

After 45 minutes of stirring at −20 °C to −15 °C, the reaction is warmed to 0 °C (5 minutes) and quenched with water [20 ml, c. 20 times the weight of $Ti(OPr^i)_4$ used in the reaction]. Upon warming to room temperature (10 minutes), phase separation is apparent (aqueous suspension above a clear to slightly cloudy organic phase). Without separation, hydrolysis of tartrates is effected by adding 4.5 ml of a 30 per cent aqueous solution of sodium hydroxide saturated with sodium chloride. After 10 minutes of vigorous stirring, sudden, dramatic phase separation occurs. The lower (organic) phase is removed and combined with two extractions of the aqueous phase (dichloromethane, 2 × 10 ml) (10). The combined organic phases are dried over magnesium sulphate and filtered through analytical grade Celite to give a clear, colourless solution, which turns bluish (TiO_2) on standing (11). Concentration, followed by Kugelrohr distillation (140 °C, 1.0/mmHg) gives (2S,3S)-epoxygeraniol as a colourless oil [10.95 g, 99%, purity c. 95% by n.m.r. (12), $[\alpha]_D^{25} -5.3°$ (c 3.0 in $CHCl_3$)].

Notes. (1) The literature authors note that this procedure has been carried out many times without incident. However, they warn that solutions of oxidants and oxidisable substances are potentially hazardous and possibly subject to violent decomposition by adventitious catalysts. They state that the following rules should be applied when handling solutions of TBHP. The first rule is never to add a strong acid (not even a drop) to high strength TBHP solutions. The second rule is never to add transition-metal salts known to be good autoxidation catalysts to high strength TBHP solutions (Mn, Fe and Co are particularly bad). Alkyl hydroperoxides are sensitive to metal-catalysed radical-chain decomposition. Among other things, this produces oxygen gas. The third rule is never to work with pure TBHP and avoid using high strength solutions of it whenever possible. They do not recommend storing TBHP solutions in glass bottles due to the slight danger of gas evolution. Instead they recommend high density polyethylene bottles, even though there may be some solvent migration through the walls of the bottle.
(2) Aqueous 70 per cent t-butyl hydroperoxide (TBHP) is obtained from Aldrich Chemical Co.

(3) Activation of crushed or pelleted 3A molecular sieves involves heating in a vacuum oven at 160 °C and 0.05 mm pressure for at least 3 hours.

(4) Cold stock solutions of TBHP in dichloromethane should be warmed to room temperature prior to opening (warm water baths are convenient), in order to minimise exposure to moisture. Somewhat more than the required amount of solution should then be dispensed into a small flask or graduated cylinder containing activated 3A or 4A sieve pellets and stoppered. After a few minutes, the desired volume of solution is transferred to the reaction flask, either by syringe, addition funnel, or direct addition. Syringe needles should never be inserted into any stock solution of TBHP which is to be stored.

(5) Preactivated, powdered 4A sieves are available from Aldrich Chemical Co.

(6) Dichloromethane (EM Reagent) is not distilled but is stored over activated 3A molecular sieves. 4A sieves should not be used since pressurisation of bottles of dichloromethane containing 4A sieves has been observed.

(7) The flask used for this reaction is three-necked and round-bottomed, fitted with an overhead mechanical stirrer, thermometer, and either a septum or an addition funnel. The internal temperature is carefully monitored. All equipment is either flame dried under vacuum or cooled under vacuum after storage in an oven at 125 °C. The reaction is carried out under an inert atmosphere (nitrogen or argon) in order to exclude atmospheric moisture. All additions are made either by addition funnel or syringe. Cooling is effected by using a water–ethylene glycol (70:80) dry ice bath.

(8) Diethyl tartrate is used as obtained from Aldrich Chemical Co.

(9) Titanium(IV) isopropoxide is distilled under vacuum and stored under an inert atmosphere. Neither the tartrate ester nor the titanium isopropoxide should be stored over sieves. Reagents handled by syringe are measured by weight rather than by volume.

(10) During any separations, if the phases do not immediately separate, c. 5 per cent v/v methanol should be added. After very brief shaking, clean phase separation generally occurs, leaving an almost clear organic phase below a milky aqueous phase.

(11) If the mixture is allowed to stand for a longer period after the addition of magnesium sulphate, no titanium will be found in solution after filtration.

(12) The c. 5 per cent impurity seen in the n.m.r. is related to an impurity in the geraniol, possibly a double bond isomer.

8.1.4 HETEROATOM EXCHANGE

Since the above methods provide a range of procedures for the formation of oxiranes, the exchange reaction of oxygen to sulphur is a convenient method for thiirane synthesis. A simple procedure involves stirring the oxirane in toluene solution with a potassium thiocyanate-on-silica gel reagent.[8] A drawback of this reaction is that the rate is slow in the case of disubstituted oxiranes. The reaction is highly stereospecific in that, for example, trans-oxiranes yield trans-thiiranes. A probable mechanism is given below and involves a double inversion process which explains the high stereospecificity.

8.2 FOUR-MEMBERED HETEROCYCLES

An important group of four-membered heterocyclic compounds are the derivatives of the β-lactam [azetidin-2-one (15)] system. Many of the compounds are biologically active, for example, the monocyclic nocardicins [e.g. (16)] and the bicyclic penicillin and cephalosporin antibiotics [e.g. (17) and (18) respectively], and the β-lactamase inhibitors of the clavam group (19).

(15)   (16)   (17)

(18)   (19)

The numerous methods which are available for the synthesis of substituted β-lactams involve a variety of ring-forming strategies. Two categories only are selected and exemplified below to illustrate some of the interesting chemistry involved: Cyclisation reactions; and Cycloaddition reactions.

8.2.1 CYCLISATION REACTIONS

The target molecule ethyl α-[p-(benzyloxy)phenyl]-2-oxo-1-azetidineacetate (20) (Expt 8.7) is an intermediate in the synthesis of one of the compounds of nocardicin group.[12] It is an interesting example with which to illustrate the application of retrosynthetic analysis in this field.

(20)   (21)

(22)

The retrosynthetic analysis given above involves an initial disconnection at the N—C_4 bond, although it should be pointed out that disconnections at some of the other bonds have been variously explored in the literature. The synthesis of the reagent equivalent (22) of the synthon (21) may be seen to require reaction of a doubly protected *p*-hydroxyphenylglycine derivative with 3-bromopropanoyl chloride in the presence of base. The protection sequence (formulated in Expt 8.7) for the amino acid is: (i) esterification of the carboxyl group; (ii) conversion of the amino group into a benzylidene derivative; (iii) benzylation of the phenolic hydroxyl group; and (iv) deprotection of the amino group by hydrolysis. The protected amino acid is finally isolated as the toluene-*p*-sulphonate salt.

The cyclisation of (22) to the β-lactam system, which is a thermodynamically unfavourable reaction, is promoted by using sodium hydride as the base in a dichloromethane–dimethylformamide solvent system at high dilution.

Experiment 8.7   ETHYL   α-[*p*-(BENZYLOXY)PHENYL]-2-OXO-1-AZETIDINEACETATE[12]

2-(*p*-Hydroxyphenyl)glycine ethyl ester.   To a suspension of 2-(*p*-hydroxyphenyl)glycine (10.0 g, 59.8 mmol) in absolute ethanol (40 ml) is added concentrated sulphuric acid (8 ml). The resulting yellow solution is heated at reflux for 2 hours, cooled to 0 °C and neutralised with concentrated aqueous ammonia solution. The precipitated product is then collected by filtration, washed with cold water, and recrystallised from aqueous ethanol to give 4.9 g (42%) of pure amino acid ester, m.p. 160–162 °C; i.r. (KBr disc) 3250, 2800–2100, 1700, 1575 cm^{-1}; p.m.r. (Me$_2$SO-d_6/CDCl$_3$, TMS) δ 1.14 (t, $J = 7$ Hz, 3H), 4.05 (q, $J = 7$ Hz, 2H), 6.68 (d, $J = 9$ Hz, 2H) and 7.12 (d, $J = 9$ Hz, 2H).

2-[(*p*-Benzyloxy)phenyl]glycine ethyl ester hydrotosylate.   A mixture of 2-(*p*-hydroxyphenyl)glycine ethyl ester (2.7 g, 14.6 mmol), benzaldehyde (1.6 g, 15.1 mmol), and a catalytic amount of toluene-*p*-sulphonic acid monohydrate is added to dimethylformamide (25 ml) containing 4 A molecular sieves (5 g). The mixture is stirred at 40 °C under nitrogen for 12 hours, decanted to remove the sieves and treated with anhydrous potassium carbonate (2.8 g, 20.3 mmol) followed after 15 minutes by dry benzyl chloride (1.9 g, 15 mmol). After being heated at 60 °C for 16 hours, the reaction mixture is poured on to 10 per cent hydrochloric acid (50 ml), stirred for 15 minutes and then extracted with ether (2 × 50 ml). Following neutralisation with solid sodium carbonate, the aqueous layer is re-extracted with fresh ether (3 × 75 ml), and the combined ethereal portions are washed with water (5 × 100 ml), dried over anhydrous magnesium sulphate and concentrated to a yellow oil (1.3 g). The oily residue is then dissolved in ethyl acetate (10 ml) and treated with toluene-*p*-sulphonic acid monohydrate (0.9 g, 4.8 mmol) in ethyl acetate (20 ml). The resulting precipitate is collected by filtration and dried under vacuum to give 2.5 g (37%) of the amino acid ester hydrotosylate, m.p. 174–175 °C; i.r. (KBr disc) 3400, 1740, 1600, 1500, 1380 cm^{-1}; p.m.r. (Me$_2$SO-d_6, TMS) 1.16 (t, J = 7 Hz, 3H), 2.29 (s, 3H), 4.17 (q, J = 7 Hz, 2H), 5.11 (s, 3H), 7.02–7.55 (m, 13H) and 8.75 (broad s, 3H). The product may be recrystallised from ethanol–ether–heptane.

2-[*p*-(Benzyloxy)phenyl]-*N*-(3-bromopropanoyl)glycine ethyl ester. To a stirred solution of the foregoing hydrotosylate (2.5 g, 5.5 mmol) and *N,N*-dimethylaniline (2.4 g, 19.9 mmol) in dry dichloromethane (65 ml) at –20 °C (ice–methanol bath) is added over 20 minutes 3-bromopropanoyl chloride (1.2 g, 6.3 mmol) in dry dichloromethane (5 ml). After completion of the addition, the reaction mixture is allowed to come to room temperature and stirred for 1.5 hours. The clear solution is diluted with dichloromethane–ether (2:1) and washed successively with 1 M hydrochloric acid, 5 per cent sodium hydrogen carbonate solution and saturated sodium chloride solution. The organic layer is dried over anhydrous sodium sulphate and concentrated to give 2.5 g of a light blue solid. Following column chromatography over silica gel with ether elution, the product is recrystallised from aqueous ethanol to provide 2.0 g (87%) of white crystals, m.p. 98–99 °C; i.r. (CHCl$_3$) 3330, 1730, 1660, 1500, 1220 cm^{-1}; p.m.r. (CDCl$_3$, TMS) δ 1.18 (t, J = 7 Hz, 3H), 2.75 (t, J = 6 Hz, 2H), 3.55 (t, J = 6 Hz, 2H), 4.16 (q (split), J = 7 Hz, 2H), 5.00 (s, 2H), 5.46 (d, J = 7 Hz, 2H), 6.58 (d, J = 7 Hz, 1H), 6.88 (d, J = 9 Hz, 2H), 7.26 (d, J = 9 Hz, 2H) and 7.32 (s, 5H). The product may be recrystallised from methyl acetate–heptane–ether.

Ethyl α-[(*p*-benzyloxy)phenyl]-2-oxo-1-azetidineacetate. To a suspension of 0.375 g (7.8 mmol) of a 50 per cent mineral oil dispersion of sodium hydride (prewashed with pentane) and dry dimethylformamide (15 ml) in dichloromethane (55 ml) is added over 3.5 hours a solution containing the foregoing substituted glycine ethyl ester (2.98 g, 7.1 mmol) and dry dimethylformamide (15 ml) in dry dichloromethane (55 ml). After completion of the addition, the reaction mixture is stirred for 3 hours at room temperature and then quenched with a saturated ammonium chloride solution (10 ml). The resulting mixture is diluted with ether and dichloromethane. The organic phase is separated, washed with water and with saturated aqueous sodium chloride and then dried over anhydrous magnesium sulphate. Removal of the solvent under reduced pressure gives an orange oil which is crystallised from ether–hexane to give 1.9 g (80%) of white solid product, m.p. 52–54 °C (1).

Note. (1) This substituted azetidinone was also synthesised by the same workers using a different strategy when the product had m.p. 57–59 °C; i.r. (CHCl$_3$) 1740 (split), 1610, 1510 cm^{-1}; p.m.r. (CDCl$_3$, TMS, 270 MHz) δ 1.18 (t, J = 6.6 Hz, 3H), 2.76 (m, 1H), 2.92 (m, 1H), 2.99 (m, 1H), 3.54 (m, 1H), 4.14 (q (split), J = 6.6 Hz, 2H), 4.99 (s, 2H), 5.46 (s, 1H), 6.91 (d, J = 8.8 Hz, 2H), 7.14 (d, J = 8.8 Hz, 2H) and 7.33 (m, 5H).

8.2.2 CYCLOADDITION REACTIONS.

To illustrate the two alternative $(2 + 2)$ cycloaddition processes, three target molecules are selected, namely, 4-acetoxy-3-ethylazetidin-2-one (23), 1-azaspiro[3.5]nonan-2-one (24) and 3-chloro-1-cyclohexyl-4-cyclohexylimino-3-phenylazetidin-2-one (25).

(23)   (24)   (25)

For the synthesis of compounds (23) and (24), bond formation is between N to C_4 and C_2 to C_3; for compound (25) bond formation is between N to C_2 and C_3 to C_4. These reactions may proceed in two steps via a dipolar intermediate (formulated below) rather than in the pericyclic sense.

Compound (23) (Expt 8.8) is formed by the reaction of but-1-enyl acetate with chlorosulphonyl isocyanate (Section 4.2.*17*, p. 425).[13] The reaction intermediate is the *N*-chlorosulphonyl derivative which is subjected to reduction with alkaline sodium sulphide to yield the product which is isolated as a *cis*: *trans* (1:1) mixture. But-1-enyl acetate is readily synthesised by acetylation of the enol form of butanal.

Compound (24) (Expt 8.8, cognate preparation) is similarly prepared from chlorosulphonyl isocyanate and methylenecyclohexane followed by reduction with sodium sulphite.[14] Chlorosulphonyl isocyanate has been widely used in cycloaddition reactions of this type to yield β-lactams having a range of functional groups.[15] These functional groups enable further structural modifications to be carried out leading to many mono- and bicyclic compounds containing the β-lactam structural feature. A straightforward example is the conversion of 4-acetoxyazetidin-2-one (26) into clavam (28) (Expt 8.9). Compound (26), prepared from vinyl acetate and chlorosulphonyl isocyanate followed by reductive removal of the chlorosulphonyl group,[16] or obtained commercially, is submitted to a Lewis-acid-catalysed displacement of the acetoxy group by reaction with 2-bromoethanol to give 4-(2-bromoethoxy)azetidin-2-one (27). Cyclisation to the bicyclic system is then effected by using the non-nucleophilic base 1,8-diazabicyclo[5.4.0]undec-7-ene (DBU).

(26)   (27)   (28)

An illustration of a cycloaddition reaction resulting in the formation of the N—C_2 and C_3—C_4 bonds is the reaction of 2-chlorophenylketene (29) with dicyclohexylcarbodiimide, to give 3-chloro-1-cyclohexyl-4-cyclohexylimino-3-phenylazetidin-2-one (25) (Expt 8.10).[17] The ketene is generated *in situ* from 2-chloro-2-phenylacetyl chloride by the action of triethylamine (cf. Expt 7.25).

(29)   (25)

Experiment 8.8 4-ACETOXY-3-ETHYLAZETIDIN-2-ONE[13]

But-1-enyl acetate. A mixture of butanal (100 g), acetic anhydride (330 g) and sodium acetate (14 g) is heated at 80 °C for 12 hours. After cooling the resulting mixture is diluted with pentane (100 ml), washed with water, saturated aqueous sodium hydrogen carbonate, and water, and dried over anhydrous sodium sulphate. Evaporation of the solvent gives a colourless oil, which is purified by distillation to afford but-1-enyl acetate [(E):(Z) c. 3:2] (56 g, 38%), b.p. 45–55 °C/20 mmHg, i.r. $(CHCl_3)$ 1750 cm^{-1}.

4-Acetoxy-3-ethylazetidin-2-one. To a stirred solution of the foregoing enol acetate (10 g) in dry dichloromethane (10 ml) is added dropwise at 0 °C chlorosulphonyl isocyanate (7 ml) (Aldrich) (1) (**CAUTION**). After stirring for a further 2 hours at 0 °C, the mixture is poured into an aqueous solution (300 ml) of sodium hydrogen carbonate (20 g) and sodium sulphide (10 g) at 0 °C with stirring. Stirring is again continued at 0 °C for 0.5 hour and the mixture extracted with dichloromethane. Evaporation of the solvent gives a yellow oil, which is chromatographed on silica gel using dichloromethane as eluant to afford 4-acetoxy-3-ethylazetidin-2-one (*trans:cis* c. 1:1) (5.4 g, 48%) as a yellow oil; i.r. $(CHCl_3)$ 3400 (NH) and 1780 cm^{-1} (C=O); p.m.r. $(CDCl_3$, TMS) δ 1.06 (t, $J = 7$ Hz, 6H, —CH_2·Me), 1.73 (q, $J = 7$ Hz, —CH_2·Me), 2.10 (s, 6H, O·CO·Me), 3.13 (broad t, $J = 7$ Hz, 2H, C_3—H), 5.53 (broad s, 1H, C_4—H), 5.85 (d, $J = 4$ Hz, 1H, C_4—H) and 6.93 (broad s, 2H, NH).

Note. (1) See Section 4.2.*17*, p. 425; **CAUTION**: *it must be emphasised that this reagent is corrosive and lachrymatory and appropriate protective measures are essential.*

Cognate preparation. *1-Azaspiro[3.5]nonan-2-one.* *1-Chlorosulphonyl-1-*

azaspiro[3.5]nonan-2-one.[14] Chlorosulphonyl isocyanate (3.5 g, **CAUTION**: *see Note (1) above*) is added dropwise to methylenecyclohexane (2.4 g) in ether (10 ml) at 10 °C. The reaction mixture becomes semi-solid with fine needles. The product is filtered and recrystallised from ether (yield 5.1 g, 96%), m.p. 88–90 °C; i.r. shows the lactam band at 5.53 μm (1810 cm^{-1}).

General procedure for the reduction of **N**-*chlorosulphonyl* β-*lactams with sodium sulphite.* A solution of *N*-chlorosulphonyl β-lactam dissolved in ether is added slowly to a stirred mixture of about two parts of 25 per cent aqueous sodium sulphite and one part of ether. The aqueous phase is kept slightly basic by addition of 10 per cent potassium hydroxide solution as the reduction proceeds. The reaction course could easily be followed by t.l.c. in which the product has a considerably smaller R_F value than the starting material. At the end of the reaction (usually less than 15 minutes) the ether layer is separated and dried and evaporated. The products are of greater than 95 per cent purity as determined by p.m.r. spectroscopy. The reaction may be carried out at 25 °C or at 0 °C.

1-Azaspiro[3.5]nonan-2-one is prepared by the above general procedure in 98 per cent yield, and is isolated as a colourless oil, b.p. 123 °C/4.2 mmHg; i.r. (CHCl$_3$) 5.72 μm (1750 cm^{-1}) p.m.r. (CDCl$_3$, TMS), δ 1.3–2.0 (m, 10H), 2.61 (d, J = 1.5 Hz, 2H, collapses to a singlet on the addition of D$_2$O) and 7.3–8.0 (broad s, 1H).

Experiment 8.9 4-OXA-1-AZABICYCLO[3.2.0]HEPTAN-7-ONE (*Clavam*)[16]

4-(2-Bromoethoxy)azetidin-2-one.   4-Acetoxyazetidin-2-one (10 g) and 2-bromoethanol (10 g) are dissolved in dry benzene (150 ml, **CAUTION**) and finely powdered zinc acetate dihydrate (9 g) is added. The mixture is stirred and refluxed with azeotropic removal of water for 24 hours. The mixture is cooled, diluted with ethyl acetate (200 ml), and washed twice with saturated sodium hydrogen carbonate solution and three times with water. The solution is dried and the solvent removed to yield a yellow oil (4.54 g) which is chromatographed (see below) to give 4-(2-bromoethoxy)azetidin-2-one as a pale yellow gum (2.6 g); i.r. 3390, 3230 and 1780 cm^{-1}; p.m.r. (CDCl$_3$, TMS) δ 3.10 (m, 2H), 3.60 (m, 2H), 3.95 (m, 2H), 5.27 (d of d, 1H, J = 3 and 2 Hz) and 7.60 (broad s, 1H); m/z 196 (M + 1, RA 0.1%), 194 (M + 1, 0.1), 167 (3), 165 (3), 144 (5), 142 (32), 140 (27), 109 (80), 107 (82) and 43 (100).

4-Oxa-1-azabicyclo[3.2.0]heptan-7-one. DBU (1.8 g) is added dropwise to a stirred solution of the foregoing substituted β-lactam (1.4 g) in ether (15 ml). The mixture is stirred for 18 hours and then concentrated to 5 ml. The concentrated solution is chromatographed (Silica gel 60, ethyl acetate–light petroleum, b.p. 60–80 °C mixture) to give clavam as a colourless oil (270 mg); i.r. 1785 cm^{-1}; p.m.r. (CDCl$_3$, TMS) δ 2.84 (d, J = 17 Hz, 1H), 2.9–3.6 (m, 2H), 3.7–4.4 (m, 3H) and 5.25 (d, J = 2 Hz, 1H).

Experiment 8.10 3-CHLORO-1-CYCLOHEXYL-4-CYCLOHEXYLIMINO-3-PHENYLAZETIDIN-2-ONE[17]

A solution of 2-chloro-2-phenylacetyl chloride (13.3 g, 0.070 mol) (1), in hexane (30 ml) (2) is added dropwise to a refluxing solution of dicyclohexyl-carbodiimide (14.5 g, 0.070 mol) **(CAUTION)** and triethylamine (14.2 g, 0.141 mol) in dry hexane (200 ml) (3). After the addition is complete, the mixture is allowed to continue to reflux for 2 hours. The amine salt is removed by filtration and the hexane evaporated to yield 16 g (65%) of the product which is recrystallised from methanol and has a m.p. 86–88 °C; i.r. 1822 (C=O) and 1700 cm^{-1} (C=N); p.m.r. (CCl$_4$, TMS) δ 1.5 (m, 20H), 3.4 (m, 2H) and 7.42 (m, 5H).

Notes. (1) This compound is available from Aldrich; alternatively it may be prepared from mandelic acid and phosphorus pentachloride according to the method of Walden.[18]
(2) Dicyclohexyldicarbodiimide is toxic and may cause sensitisation and subsequent allergic reactions. The vapour particularly irritates the eyes severely.
(3) Hexane is dried by refluxing and distilling from calcium hydride.

8.3 FIVE-MEMBERED HETEROCYCLES

The commonest five-membered heteroaromatic compounds containing one heteroatom are pyrrole (30), furan (31) and thiophene (32). Their substituted derivatives and the substituted derivatives of the corresponding benzo ana-logues [indole (33), benzofuran (34) and benzothiophene (35)] are widespread naturally-occurring compounds.

N H	O	S
(30)	(31)	(32)
N H	O	S
(33)	(34)	(35)

Examples of five-membered heterocyclic systems having more than one heteroatom, the derivatives of which have been selected to illustrate some important methods of ring formation, are pyrazole (36), pyrazolone (37), hydantoin (38), oxazolone (39), thiazole (40), benzimidazole (41) and benzotriazole (42).

(36)   (37)   (38)

(39)   (40)

(41)   (42)

A useful initial guide to an overall strategy of ring formation is to formulate the ring atoms as various skeletal fragments. For example, in the case of the five-membered heterocyclic system (43), ring construction may be seen to result from the union of the skeletal fragments shown in (44) to (46). With the benzo analogue (47) a limited selection of possibilities is given by (48) and (49).

(43)   (44)   (45)   (46)

(47)   (48)   (49)

This time-honoured view of ring construction preceded the retrosynthetic approach; it is still of value since it provides an indication of which bonds could be selected for disconnection. The more rigorous application of the principles of retrosynthetic analysis leads of course to the formulation of synthons and their reagent equivalents.

A selection of cyclisation reactions is detailed below to exemplify these principles. They are conveniently discussed in relation to the synthesis of compounds belonging to the following groups of heterocyclic systems.

1. Pyrroles, furans, thiophenes and pyrazolones from dicarbonyl compounds.
2. Hydantoins from urea, and thiazoles from thiourea or thioamides.
3. Oxazolones (and azlactones) from α-amino acids.
4. Indoles, benzofurans and benzothiophenes.
5. Benzimidazoles and benzotriazole from o-phenylene diamines.

1145

8.3.1 PYRROLES, FURANS, THIOPHENES AND PYRAZOLONES FROM DICARBONYL COMPOUNDS

FROM 1,4-DICARBONYL COMPOUNDS

The first group of illustrative target molecules comprises 2,5-dimethylpyrrole (50), 2,5-dimethylfuran (51) and 2,5-dimethylthiophene (52).

A retrosynthetic analysis of (50) and (52) involving disconnection at both carbon–heteroatom bonds reveals hexane-2,5-dione as the four-carbon fragment needed for ring assembly; ring construction is thus of type (45). The most convenient reagents for the appropriate heteroatom synthons are ammonium carbonate and phosphorus pentasulphide (Expts 8.11 and 8.13).

In the case of 2,5-dimethylfuran (51) the dicarbonyl compound is revealed by the disconnection of one carbon–oxygen bond; ring construction is of type (44). The forward synthetic reaction is a cyclodehydration reaction effected by the action of acetic anhydride in the presence of zinc chloride (Expt 8.12).

These methods [which are known as the *Paal–Knorr* (pyrrole and furan) or *Paal* (thiophene) syntheses] are applicable to other 1,4-dicarbonyl compounds, the limitation being the accessibility of the dicarbonyl starting material.

FROM 1,3-DICARBONYL COMPOUNDS

The second group of target molecules selected to illustrate the use of dicarbonyl compounds consists of 3,5-dimethylpyrazole (53), 3-methyl-1-phenylpyrazol-5-one (54) and 3,5-diethoxycarbonyl-2,4-dimethylpyrrole (55).

Disconnection of both carbon–nitrogen bonds in (53) in the manner depicted above, reveals hydrazine and a synthon which is equivalent to pentane-2,4-

dione. In a similar way disconnection of both carbon–nitrogen bonds in (54) reveals phenylhydrazine and ethyl acetoacetate (see also p. 805).

(53)

(54)

Synthesis of (53) and (54) proceeds with ease on mixing the relevant reagents (Expts 8.14 and 8.15). N-Methylation of (54) with dimethyl sulphate in the presence of alkali gives 2,3-dimethyl-1-phenylpyrazol-5-one (antipyrin), which was one of the earliest synthetic antipyretics.

An example of the *Knorr pyrrole synthesis* is provided by the formation of 3,5-diethoxycarbonyl-2,4-dimethylpyrrole (55). Overall ring construction in this case may be related to (46) above. A retrosynthetic analysis involving disconnection of the N—C_2 bond, appropriate prototropic shifts, and finally a *retro*-aldol reaction to effect disconnection of the C_3—C_4 bond, reveals ethyl acetoacetate and ethyl α-aminoacetoacetate (ethyl 2-amino-3-oxo-butanoate) (56) as reagents. An FGI transform on this latter compound generates the corresponding nitroso (oximino) compound which may also be derived from ethyl acetoacetate.

(55)

(56)

(56)

In practice the synthesis (Expt 8.16) is a one-pot reaction in which one-half of the ethyl acetoacetate is converted to the oximino compound by treatment with nitrous acid in acetic acid, which is then reduced by the addition of zinc dust. The resulting amino compound spontaneously reacts with the remaining ethyl acetoacetate to yield the pyrrole derivative (55).

The Knorr synthesis is generally suitable for the synthesis of a range of substituted pyrroles utilising α-aminoketones (or keto esters as illustrated above) with carbonyl compounds containing an active methylene group.

Experiment 8.11   2,5-DIMETHYLPYRROLE

In a 250-ml flask, fitted with an air condenser of wide bore, place 50 g (51.5 ml, 0.44 mol) of hexane-2,5-dione (Expt 5.104) and 100 g of ammonium carbonate (lump form). Heat the mixture in an oil bath at 100 °C until effervescence stops (60–90 minutes); some ammonium carbonate (or carbamate) sublimes into the condenser and this must be pushed back into the reaction mixture by means of a stout glass rod. Replace the air condenser by a Liebig condenser with wide bore inner tube and reflux the mixture gently (bath temperature, 115 °C) for a further 30 minutes; dissolve the solid which has sublimed into the condenser in about 5 ml of hot water and return the solution to the reaction mixture. Cool and separate the upper yellow layer of crude dimethylpyrrole, extract the lower layer with 10 ml of dichloromethane and combine it with the crude dimethylpyrrole; carry out the foregoing operations in apparatus which has been flushed with nitrogen. Dry over anhydrous sodium sulphate in a tightly stoppered flask filled with nitrogen. Transfer to a flask fitted with a fractionating column. Displace the air from the apparatus by nitrogen and distil under reduced pressure, preferably in a stream of nitrogen. Collect the 2,5-dimethylpyrrole at 78–80 °C/25 mmHg. The yield is 36 g (86%). Store the product in an inert atmosphere in a sealed, dark glass container. Record the p.m.r. spectrum (CDCl₃) and assign the signals that occur at δ 2.20 (s, 6H), 5.71 (d, 2H) and 7.49 (broad s, 1H).

Experiment 8.12   2,5-DIMETHYLFURAN

Dissolve 0.5 g of anhydrous zinc chloride in 16.8 g (15.5 ml, 0.165 mol) of acetic anhydride contained in a 50-ml round-bottomed flask, and to this solution add 17.1 g (17.6 ml, 0.15 mol) of redistilled hexane-2,5-dione (Expt 5.104). Attach a reflux condenser fitted with a calcium chloride drying tube and warm the mixture carefully. When a vigorous reaction commences remove the source of heat until the reaction subsides and then boil the mixture under gentle reflux for 2 hours. Transfer the cooled dark brown mixture to a 250-ml round-bottomed flask and add 6 M aqueous sodium hydroxide until alkaline (c. 50 ml), keeping the temperature below 50 °C by cooling in an ice-water bath. Steam distil the mixture until no more oily drops appear in the distillate (about 100 ml). Transfer the distillate to a separating funnel, collect the upper layer of product and dry over anhydrous calcium chloride. Filter the straw-coloured crude product into a 25-ml pear-shaped flask and distil. Collect the colourless 2,5-dimethylfuran (1) as a fraction, b.p. 93–95 °C. The yield is 8.9 g (62%). Record the p.m.r. spectrum and compare it with that of 2,5-dimethyl-pyrrole.

Note. (1) The purity of the product may be checked by g.l.c. on a 1.5-m Silicone oil column held at 105 °C. At this temperature and with a nitrogen flow rate of 40 ml/minute, 2,5-dimethylfuran has a retention time of 48 seconds.

Experiment 8.13   2,5-DIMETHYLTHIOPHENE

CAUTION: *This preparation should be carried out in an efficient fume cupboard.*

Place 55 g (0.25 mol) of phosphorus pentasulphide in a 500-ml three-necked flask equipped with a long Leibig-type condenser, a dropping funnel and a stopper. Add from the dropping funnel 23 g (24 ml) of hexane-2,5-dione (Expt 5.104) and heat the reaction mixture with a small luminous flame until the exothermic reaction commences. Remove the flame and add dropwise 91 g (94 ml, total 1 mol) of hexane-2,5-dione over a period of about 45 minutes so that the mixture refluxes gently. A bath of cold water should be to hand should the reaction show signs of becoming too vigorous. When the addition is complete, and the reaction has subsided, heat the mixture under reflux for a further 1 hour, cool to room temperature and pour into 200 ml of ice-water (*fume cupboard*). Extract the aqueous mixture with four 40 ml portions of ether and wash the combined ethereal extracts with aqueous sodium carbonate solution and then with water. Dry the ether solution over magnesium sulphate, remove the ether by flash distillation and distil the residue. 2,5-Dimethylthiophene is obtained after redistillation as a very unpleasant-smelling colourless liquid, b.p. 135–136 °C; the yield is 52 g (46%).

Experiment 8.14   3,5-DIMETHYLPYRAZOLE

Dissolve 65 g (0.5 mol) of hydrazine sulphate in 400 ml of 2.5 M sodium hydroxide solution contained in a 1-litre three-necked flask, equipped with a thermometer, mechanical stirrer and dropping funnel. Immerse the flask in an ice bath and when the temperature reaches 15 °C (some sodium sulphate may separate at this point), add 50 g (51.5 ml, 0.5 mol) of pentane-2,4-dione (Expt 5.102) dropwise, with stirring, while maintaining the temperature at 15 °C. When the addition is complete (after about 30 minutes), stir for 1 hour at 15 °C; the dimethylpyrazole separates during this period. Add 200 ml of water, stir to dissolve inorganic salts, transfer the contents of the flask to a separatory funnel and shake with 100 ml of ether. Separate the layers and extract the aqueous layer with four 40 ml portions of ether. Wash the combined ethereal extracts with saturated sodium chloride solution, dry over anhydrous potassium carbonate and remove the ether on a rotary evaporator. The yield of pale yellow solid, m.p. 107–108 °C, is 38 g. Recrystallise from about 250 ml of light petroleum, b.p. 80–100 °C; the yield of 3,5-dimethylpyrazole, of unchanged m.p., is 36 g (75%). Record and interpret the p.m.r. spectrum.

Experiment 8.15 3-METHYL-1-PHENYLPYRAZOL-5-ONE
(*Antipyrine*)

Mix together 50 g (49 ml, 0.384 mol) of redistilled ethyl acetoacetate and 40 g (36.5 ml, 0.37 mol) of phenylhydrazine (**CAUTION** in handling) in a large evaporating dish. Heat the mixture on a boiling water bath in the fume cupboard for about 2 hours and stir from time to time with a glass rod (1). Allow the heavy reddish syrup to cool somewhat (2), add about 100 ml of ether and stir the mixture vigorously. The syrup, which is insoluble in ether, will solidify within 15 minutes. Filter the solid at the pump and wash it thoroughly with ether to remove coloured impurities. Recrystallise it from hot water or from a mixture of equal volumes of ethanol and water. The yield of methylphenylpyrazolone (colourless crystals, m.p. 127 °C) is 52 g (80%). Record and interpret the p.m.r. spectrum.

Formation of 2,3-dimethyl-1-phenylpyrazol-5-one by *N*-methylation. In a 500-ml three-necked flask, equipped with a dropping funnel, a sealed stirrer unit and a double surface condenser and set up in the fume cupboard, place a solution of 10 g of sodium hydroxide in a small volume of water and also a solution of 43.5 g (0.25 mol) of 3-methyl-1-phenylpyrazol-5-one in 20 ml of methanol. Warm the mixture on a water bath and add 36 g (27 ml, 0.285 mol) of dimethyl sulphate (**CAUTION**: *toxic, see discussion in Section 4.2.24, p. 430*). Reflux the mixture for 1 hour and allow to cool, with continuous stirring. Distil off the methanol. Add hot water to the residue, filter from impurities, extract the antipyrin with benzene (*fume cupboard*) and evaporate the solvent. Recrystallise the crude product from benzene (**CAUTION**) or benzene–light petroleum, or from hot water with the addition of a little decolourising carbon. The yield of antipyrin (white crystalline solid, m.p. 113 °C) is 35 g (74%).

Notes. (1) The heating should be continued until a test portion solidifies completely when it is rubbed with a little ether; when freshly redistilled phenylhydrazine is used, a period of 1 hour's heating may be adequate. If the product is reluctant to crystallise inoculation with seeds of previously prepared material is advantageous.

(2) If the product is already solid at this stage it should be ground thoroughly with the 100 ml portion of ether and filtered.

Experiment 8.16   3,5-DIETHOXYCARBONYL-2,4-DIMETHYL-PYRROLE

In a 1500-ml three-necked flask, fitted with a dropping funnel and a sealed mechanical stirrer, place 195 g (190 ml, 1.5 mol) of ethyl acetoacetate and 450 ml of glacial acetic acid. Cool the solution in an ice–salt mixture to 5 °C; add a cold solution of 52 g (0.75 mol) of sodium nitrite in 75 ml of water dropwise and with vigorous stirring at such a rate that the temperature remains between 5 and 7 °C (about 30 minutes), stir for a further 30 minutes, and keep at room temperature for 4 hours. Replace the dropping funnel by a wide-bore condenser: close the third neck with a stopper. Stir the solution vigorously and add 100 g (1.5 mol) of zinc powder (Section 4.2.*80*, p. 467) via the third neck in portions of about 10 g; introduce the first 3 or 4 portions quickly so that the liquid boils. Keep a bath of ice-water and also wet towels at hand to control the reaction should it become violent or foam badly. When all the zinc has been added (about 45 minutes), reflux the mixture for 1 hour; if stirring becomes difficult, add some acetic acid. While still hot, decant the contents of the flask into 5 litres of water in a large beaker with vigorous stirring. Wash the zinc residue with two 25 ml portions of hot glacial acetic acid and decant the washings into the water also. Keep overnight, collect the crude product by suction filtration, wash with two 250 ml portions of water and dry in the air to constant weight. The yield of crude product is 114 g (64%), m.p. 127–130 °C. Recrystallisation from hot 95 per cent ethanol gives pure 3,5-diethoxycarbonyl-2,4-dimethylpyrrole as pale yellow crystals, m.p. 136–137 °C; the recovery is about 80 per cent. Record the p.m.r. spectrum (CDCl$_3$, TMS) and assign the signals which occur at δ 1.38 (t, 6H), 2.52 (s, 3H), 2.57 (s, 3H), 4.29 (q, 2H) and 4.32 (q, 2H).

8.3.2 HYDANTOINS FROM UREA, AND THIAZOLES FROM THIOUREA OR THIOAMIDES

The illustrative target molecules are 5,5-diphenylhydantoin (57), 2-amino-4-methylthiazole (58), 2-aminothiazole (59) and 2,4-dimethylthiazole (60).

(57)   (58)   (59)   (60)

Urea may be recognised as a structural unit in 5,5-diphenylhydantoin (57) and it constitutes one of the reagents in the synthesis. Although it is not obvious, the appropriate reagent equivalent corresponding to the accompanying synthon is in fact benzil.

(57)

The base-catalysed reaction sequence (formulated in Expt 8.17) is thought to proceed via an intermediate heterocyclic pinacol which on acidification yields the required hydantoin as a result of a pinacolic rearrangement. The procedure is applicable in general to diaryl-1,2-diketones.

The thiourea fragment may be readily recognised in (58); appropriate disconnection thus reveals a chloroacetone as the other reagent; the synthesis is readily effected by heating the two reagents together (Expt 8.18). In a similar manner disconnection of 2-aminothiazole (59) reveals thiourea and chloroacetaldehyde; in this case the carbonyl reagent is generated *in situ* from α,β-dichloroethyl ethyl ether (formulated in Expt 8.19).

(58)

(59)

Chloroacetone is also a recognisable reagent equivalent following disconnection of 2,4-dimethylthiazole (60); the reagent equivalent of the other synthon is thioacetamide, which is formed *in situ* from acetamide and phosphorus pentasulphide (Expt 8.20).

(60)

Experiment 8.17 5,5-DIPHENYLHYDANTOIN

Place 5.3 g (0.025 mol) of benzil (Expt 6.143), 3.0 g (0.05 mol) of urea, 15 ml of 30 per cent aqueous sodium hydroxide solution and 75 ml of ethanol in a 100-ml round-bottomed flask. Attach a reflux condenser and boil under reflux using an electric heating mantle for at least 2 hours. Cool to room temperature, pour the reaction product into 125 ml of water and mix thoroughly. Allow to stand for 15 minutes and then filter under suction to remove an insoluble by-product. Render the filtrate strongly acidic with concentrated hydrochloric acid, cool in ice-water and immediately filter off the precipitated product under suction. Recrystallise at least once from industrial spirit to obtain about 2.8 g (44%) of pure 5,5-diphenylhydantoin, m.p. 297–298 °C.

Experiment 8.18 2-AMINO-4-METHYLTHIAZOLE

Suspend 76 g (1 mol) of thiourea (**CAUTION**: *see Section 2.3.4, p. 50*) in 200 ml of water in a 500-ml three-necked flask sited in a fume cupboard and equipped with a sealed stirrer unit, a reflux condenser and a dropping funnel. Stir and add 92.5 g (80 ml, 1 mol) of chloroacetone (1) over a period of 30 minutes. The thiourea dissolves as the reaction proceeds and the temperature rises. Reflux the yellow solution for 2 hours. To the cold solution immersed in an ice bath add, with stirring, 200 g of solid sodium hydroxide. Transfer to a separatory funnel, add a little ice-water, separate the upper oil layer and extract the aqueous layer with three 100 ml portions of ether. Dry the combined oil and ether extracts with anhydrous sodium sulphate, remove the ether using a rotary evaporator and distil the residual oil under diminished pressure. Collect the 2-amino-4-methylthiazole at 130–133 °C/18 mmHg; it solidifies on cooling in ice to a solid, m.p. 44–45 °C. The yield is 84 g (74%).

Note. (1) See Expt 8.20, Note (1).

Experiment 8.19 2-AMINOTHIAZOLE

Place a solution of 76 g (1 mol) of thiourea (**CAUTION**: *see Section 2.3.4, p. 50*) in 200 ml of warm water in a 500-ml three-necked flask equipped with a dropping funnel, sealed mechanical stirrer and reflux condenser. Add 143 g (122 ml, 1 mol) of α, β-dichloroethyl ethyl ether, and heat the mixture under gentle reflux with stirring for 2 hours. As the reaction proceeds, the two layers gradually merge. To the cold solution add sufficient solid sodium hydroxide to liberate the 2-aminothiazole from its salt. Add ether to dissolve the product, dry the ethereal extract with anhydrous sodium sulphate and evaporate the ether. Recrystallise the crude 2-aminothiazole from ethanol; the resulting yellow crystalline solid has m.p. 90 °C. The yield is 80 g (80%).

Experiment 8.20 2,4-DIMETHYLTHIAZOLE

$$\text{Me·CONH}_2 \xrightarrow{\text{P}_2\text{S}_5} \text{Me·CSNH}_2$$

Equip a 1-litre, two-necked round-bottomed flask with a reflux condenser and a dropping funnel. Prepare a mixture of 150 g (2.54 mol) of finely powdered acetamide and 100 g of powdered phosphorus pentasulphide quickly, transfer it rapidly into the flask and immediately add 100 ml of dry benzene (**CAUTION**). Set up the apparatus in a fume cupboard. Prepare a mixture of 232 g (200 ml, 2.5 mol) of chloroacetone (1) (**CAUTION**: *the compound is lachrymatory*) and 75 ml of dry benzene; place it in the dropping funnel and insert a calcium chloride drying tube in the mouth. Add about 10 ml of the chloroacetone–benzene mixture to the contents of the flask and warm gently on a water bath: remove the water bath immediately the exothermic reaction commences. Introduce the remainder of the chloroacetone in *c.* 10 ml portions at such intervals that the reaction is under control. When all the chloroacetone has been added, reflux the mixture on a water bath for 30 minutes. Then add 400 ml of water to the reaction mixture with shaking; after 20 minutes, transfer the contents of the flask to a separatory funnel, run off the lower layer into a beaker and discard the reddish upper layer containing the benzene. Make the lower layer alkaline by the addition of 20 per cent sodium hydroxide solution: test the highly coloured aqueous solution (and not the dark dimethylthiazole floating on top of the liquid) with universal indicator paper. Separate the black upper layer of crude dimethylthiazole with 50 ml of ether, and extract the aqueous layer with five 60 ml portions of ether. Dry the combined ethereal extracts over magnesium sulphate, and filter through glass wool. Remove the ether by flash distillation through a short fractionating column; insert a calcium chloride drying tube into the dropping funnel since the thiazole is hygroscopic and fractionate the residue. Collect the fraction boiling at 140–150 °C and redistil. The yield of 2,4-dimethylthiazole, b.p. 143–145 °C, is 115 g (40%). Record and interpret the p.m.r. spectrum (CDCl₃).

Note. (1) Distil and store commercial chloroacetone, b.p. 118–120 °C over calcium carbonate. It is prepared *inter alia* by the chlorination of acetone in the cold.

8.3.3 OXAZOLONES (AND AZLACTONES) FROM α-AMINO ACIDS

The retrosynthetic analysis outlined below for 4-benzylidene-2-methyloxazol-5-one (61) follows familiar general principles.

The synthetic steps are thus the conversion of N-acetylglycine with acetic anhydride into 2-methyloxazol-5-one (62), followed by reaction of the active methylene group with benzaldehyde to afford the corresponding benzylidene derivative (61) (Expt 8.24; see also Section 5.14.3, p. 736). In a similar manner N-benzoylglycine gives 4-benzylidene-2-phenyloxazol-5-one. These azlactones have been used as starting materials for the preparation of phenylpyruvic acid and phenylalanine (Expts 5.175 and 5.182 respectively).

Experiment 8.21   4-BENZYLIDENE-2-METHYLOXAZOL-5-ONE

Acetylglycine. Place 37.5 g (0.5 mol) of glycine (Expt 5.180) and 150 ml of water in a 500-ml conical flask. Introduce a mechanical stirrer and stir vigorously until the solid has almost completely dissolved. Add 102 g (95 ml, 1 mol) of acetic anhydride in one portion and stir vigorously for 15–20 minutes; the solution becomes hot and some acetylglycine may crystallise. Cool in a refrigerator, preferably overnight; collect the precipitate on a Buchner funnel, wash with ice-cold water and dry at 100 °C. The product weighs 40 g and melts at 207–208 °C. Evaporate the combined filtrate and washings to dryness under reduced pressure on a water bath at 50–60 °C, and recrystallise the residue from 40 ml of boiling water: collect the solid which separates, wash and dry it as before. The second fraction of acetylglycine weighs 15 g (total yield 55 g, 94%) and melts at 207–208 °C.

4-Benzylidene-2-methyloxazol-5-one. Warm a mixture of 29 g (0.25 mol) of acetylglycine, 39.5 g (37.5 ml, 0.37 mol) of redistilled benzaldehyde [Expt 6.133, Note (1)], 15 g (0.183 mol) of anhydrous sodium acetate and 63.5 g (59 ml, 0.62 mol) of acetic anhydride in a 500-ml flask (equipped with a reflux condenser) on a water bath with occasional stirring until solution is complete (10–20 minutes). Boil the resulting solution for 1 hour, cool and leave in a refrigerator overnight. Stir the solid mass of yellow crystals with 60 ml of cold water, transfer to a Buchner funnel and wash well with cold water. (If the odour of benzaldehyde is still apparent, wash with a little ether.) Recrystallise from carbon tetrachloride or from ethyl acetate–light petroleum. The yield of the oxazolone, m.p. 150 °C, is 35 g (76%).

1155

Hydrolysis to α-acetamidocinnamic acid. Boil a mixture of 23.5 g (0.125 mol) of 4-benzylidene-2-methyloxazol-5-one (the crude product is satisfactory), 90 ml of water and 225 ml of acetone in a 500-ml round-bottomed flask under reflux for 4 hours. Remove most of the acetone with a rotary evaporator, dilute the residual solution with 200 ml of water, heat to boiling for 5–10 minutes and filter through a hot-water funnel. Dissolve any crystals which separate from the filtrate by heating, add 5 g of decolourising carbon, boil for 5 minutes, filter with gentle suction through a warm Buchner funnel, and wash the residue with four 25 ml portions of boiling water. Place the combined filtrate and washings in a refrigerator overnight. Collect the colourless crystals by suction filtration, wash with about 100 ml of cold water and dry at 100 °C. The yield of 2-acetamidocinnamic acid, m.p. 191–192 °C, is 22 g (86%).

Cognate preparation. *Benzoylglycine* (*hippuric acid*). Dissolve 25 g (0.33 mol) of glycine (Expt 5.180) in 250 ml of 10 per cent sodium hydroxide solution contained in a conical flask. Add 54 g (45 ml, 0.385 mol) of benzoyl chloride in five portions to the solution. Stopper the vessel and shake vigorously after each addition until all the chloride has reacted. Transfer the solution to a beaker and rinse the conical flask with a little water. Place a few grams of crushed ice in the solution and add concentrated hydrochloric acid slowly and with stirring until the mixture is acid to Congo red paper. Collect the resulting crystalline precipitate of benzoylglycine, which is contaminated with a little benzoic acid, upon a Buchner funnel, wash with cold water and drain well. Place the solid in a beaker with 100 ml of carbon tetrachloride, cover the beaker with a watch glass and boil gently for 10 minutes (*fume cupboard*); this extracts any benzoic acid which may be present. Allow the mixture to cool slightly, filter under gentle suction and wash the product on the filter with 10–20 ml of carbon tetrachloride. Recrystallise the dried product from boiling water (about 500 ml) with the addition of a little decolourising charcoal if necessary, filter through a hot-water funnel and allow to crystallise. Collect the benzoylglycine in a Buchner funnel and dry it in an oven. The yield is 45 g (76%), m.p. 187 °C.

4-Benzylidene-2-phenyloxazol-5-one. Place a mixture of 27 g (26 ml, 0.25 mol) of redistilled benzaldehyde, 45 g (0.25 mol) of benzoylglycine, 77 g (71.5 ml, 0.75 mol) of acetic anhydride and 20.5 g (0.25 mol) of anhydrous sodium acetate in a 500-ml conical flask and heat on an electric hotplate with constant shaking. As soon as the mixture has liquefied completely, transfer the flask to a water bath and heat for 2 hours. Then add 100 ml of ethanol slowly to the contents of the flask and allow the mixture to stand overnight. Filter the crystalline product with suction, wash with two 25 ml portions of ice-cold alcohol and then wash with two 25 ml portions of boiling water: dry at 100 °C. The yield of almost pure oxazolone, m.p. 165–166 °C, is 40 g (64%). Recrystallisation from benzene raises the m.p. to 167–168 °C.

8.3.4 INDOLES, BENZOFURANS AND BENZOTHIOPHENES

The four target molecules are indole [(63), $R^1 = R^2 = H$], 5-chloro-2-phenyl-indole [(63), $R^1 = Ph$, $R^2 = Cl$], benzofuran (64) and 5-chloro-3-methylbenzo-thiophene (65). The retrosynthetic analyses shown below individually illustrate a possible disconnection from which the method of synthesis is devised. The ring-forming strategy is either of type (48) or (49).

(63)

(64)

(65)

In the synthesis of indole [(63), R^1 = H, R^2 = H], the cyclisation step is effected when N-formyl-o-toluidine [N-(2-methylphenyl)-formamide] is treated with strong base (e.g. potassium alkoxide or sodamide) at high temperatures.[19] The reaction, the *Madelung indole synthesis*, may be regarded mechanistically as an intramolecular Claisen condensation. In the preparative example (Expt 8.22) a simplified procedure is described in which the somewhat hazardous preparation of a potassium alkoxide from potassium metal and the alcohol is avoided by using a mixture of sodium methoxide and anhydrous potassium acetate. The yields are however poor; there is extensive loss of carbon monoxide and about half the starting material is recovered as the parent amine. Furthermore, the reaction is limited in that the presence of halogen or alkoxy ring substituents causes the reaction to fail, and the use of amides having N-aroyl or branched N-alkanoyl groups [e.g. (63), R^1 = Ar or Bu^t], leading to the corresponding 2-aryl- or 2-alkylindoles, gives poor yields. This drawback has been overcome by the use of 2 or 3 molar equivalents of butyllithium in tetrahydrofuran at −20 to +25 °C, followed by standing overnight at room temperature.[20] The reaction sequence is formulated in Expt 8.23 which describes the preparation of 5-chloro-2-phenylindole.

The analysis depicted for (64) reveals that the synthesis of benzofuran involves the conversion of salicylaldehyde into the corresponding aryloxyacetic acid by reaction with sodium chloroacetate in the presence of alkali, followed by heating with a mixture of acetic anhydride, acetic acid and sodium acetate (Expt 8.24). The ensuing cyclisation may be regarded as an internal Perkin reaction (Section 6.12.3, p. 1036) accompanied by a decarboxylative dehydration step.

The analysis for (65) illustrates the value of disconnecting the bond which joins the aromatic ring to a side-chain or functional substituent. The synthesis therefore involves reaction of thiophenol with α-chloroacetone under basic conditions and cyclisation of the resulting substituted ketone by heating with phosphorus pentoxide (Expt 8.25).

An important general method for the preparation of substituted indoles (low yields with indole itself), known as the *Fischer indole synthesis*, consists of heating the phenylhydrazone of an aldehyde, ketone, or keto acid in the presence of

1157

an acid catalyst. Zinc chloride, hydrochloric acid or glacial acetic acid may be used, but polyphosphoric acid is often preferred. Thus acetophenone phenyl-hydrazone [(66), $R^1 = H$, $R^2 = Ph$] gives 2-phenylindole [(66), $R^1 = H$, $R^2 = Ph$] (Expt 8.26). The reaction sequence involves an intramolecular condensation with the elimination of ammonia and the mechanism of the reaction is outlined below (see also the synthesis of tryptophan, Expt 5.184).

(66)

(67)

A further interesting application is the preparation of 1,2,3,4-tetrahydrocarbazole (69) (Expt 8.26, cognate preparation) which is formed when phenylhydrazine is added to a boiling solution of cyclohexanone in acetic acid; the phenylhydrazone (68), intermediately produced, undergoes ring closure directly.

(68)   (69)

Experiment 8.22 INDOLE

$$o\text{-Me·C}_6\text{H}_4\text{·NH}_2 + \text{H·CO}_2\text{H} \longrightarrow o\text{-Me·C}_6\text{H}_4\text{·NH·CHO} + \text{H}_2\text{O}$$

$$+ \text{H}_2\text{O} + \text{CO} \quad [+ o\text{-toluidine}]$$

N-Formyl-o-toluidine [N-(2-methylphenyl)formamide]. Mix together 43 g (43 ml, 0.4 mol) of o-toluidine and 21 g (17.5 ml, 0.4 mol $\text{H·CO}_2\text{H}$) of 90 per cent w/w formic acid in a 100-ml round-bottomed flask fitted with a reflux condenser, and heat the mixture on a boiling water bath for 3 hours. Replace the reflux condenser by a Claisen still-head and an air condenser arranged for distillation under reduced pressure, and distil the product using a water pump, collecting the formyl-o-toluidine as a fraction of b.p. 173–175 °C/25 mmHg, which solidifies on cooling, m.p. 57–59 °C; the yield is 43 g (80%).

A pure specimen, m.p. 61 °C, may be obtained by crystallisation from a mixture of benzene and light petroleum (b.p. 40–60 °C).

Indole. Prepare a solution of sodium methoxide in 125 ml of anhydrous methanol using 5.75 g (0.25 mol) of sodium in a 250-ml flask fitted with a reflux condenser protected by a calcium chloride guard-tube, and add 34 g (0.25 mol) of N-formyl-o-toluidine. Then add rapidly 50 g (0.51 mol) of coarsely ground, freshly fused potassium acetate (1) and heat under reflux with shaking until all has dissolved. Remove the methanol under reduced pressure (rotary evaporator), transfer the flask to a fume cupboard and fit a still-head and condenser set for downward distillation. Surround the flask with a bath of molten Wood's metal (Section 2.13) and raise the temperature steadily to about 300–350 °C. The subsequent reaction is accompanied by the distillation of o-toluidine and the evolution of carbon monoxide; continue to heat until no further distillation occurs (about 30 minutes) and finally remove traces of o-toluidine by carefully applying partial vacuum. Remove the heating bath, allow the flask to cool and decompose the residue by adding 100 ml of water and steam distilling. Colourless plates of indole separate from the cooled distillate; make the latter slightly acidic with hydrochloric acid, collect the crystals by suction filtration and wash them with a little cold water. The yield of indole, m.p. 48–49 °C, is 5 g (17%). A purer specimen, m.p. 52 °C, may be obtained by crystallisation from light petroleum, b.p. 40–60 °C.

Note. (1) Heat the potassium acetate in a porcelain dish until a tranquil melt is obtained, and allow to cool in a desiccator (cf. anhydrous sodium acetate, Section 4.2.69, p. 464).

Experiment 8.23 5-CHLORO-2-PHENYLINDOLE[20]

N-(4-chloro-2-methylphenyl)benzamide. A mixture of 4-chloro-2-methyl-aniline (14.2 g, 0.10 mol), benzoyl chloride (14.1 g, 0.10 mol) and triethylamine (10.1 g, 0.10 mol) in anhydrous toluene (150 ml) is stirred and refluxed for 3 hours and then allowed to stand for c. 16 hours at room temperature. The resulting solids (mixture of product and salts) are filtered off and stirred at room temperature for c. 1.5 hours with water (150 ml). The solid product is filtered off and recrystallised from dichloromethane–toluene (yield 87%, m.p. 170–171 °C).

5-Chloro-2-phenylindole. A stirred solution of the foregoing amide (12.5 g, 0.05 mol) in dry tetrahydrofuran (100 ml) (from storage over 3 A molecular sieves) is maintained under a nitrogen atmosphere at an internal temperature of − 20 °C and treated dropwise with butyllithium in hexane (0.10 mol using either a 1.4 or a 1.6 M solution). The stirred solution is kept at ambient temperature for 15 hours, cooled in an ice bath and treated dropwise with 2 M hydrochloric acid (60 ml). The organic layer is separated and the aqueous layer washed with benzene. The combined organic layers are dried with anhydrous magnesium sulphate, filtered, and concentrated in vacuo. The residue is recrystallised from ether–benzene: the yield of 5-chloro-2-phenylindole is 94 per cent, m.p. 195–196 °C.

Experiment 8.24 BENZOFURAN

o-**Formylphenoxyacetic acid.** To a mixture of 35 ml (40 g, 0.33 mol) of salicyl-aldehyde, 31.5 g (0.33 mol) of chloroacetic acid and 250 ml of water contained in a 500-ml, two-necked round-bottomed flask fitted with a stirrer unit, add slowly with stirring a solution of 26.7 g (0.66 mol) of sodium hydroxide in 700 ml of water. Heat the mixture to boiling with stirring and reflux for 3 hours. The solution acquires a red-brown colour. Cool and acidify the solution with 60 ml of concentrated hydrochloric acid and steam distil to remove unreacted salicylaldehyde; 12 ml (14 g) are thus recovered. Cool the residual liquor which first deposits some dark red oil which then solidifies; on stand-ing, almost colourless crystals appear in the supernatant solution. Decant the supernatant solution and crystals and filter off the crystals, and air dry; the yield of almost pure product, m.p. 132–133 °C, is 21 g. The solidified red oil may be extracted with small quantities of hot water, the extracts treated with decolourising charcoal and cooled, to yield a further 6 g of product; total yield 27 g (71% calculated on salicylaldehyde consumed in reaction).

Benzofuran. Heat under reflux for 8 hours a mixture of 20 g (0.11 mol) of *o*-formylphenoxyacetic acid, 40 g of anhydrous sodium acetate, 100 ml of acetic anhydride and 100 ml glacial acetic acid. Pour the light brown solution into 600 ml of iced water, and allow to stand for a few hours with occasional stir-ring to aid the hydrolysis of acetic anhydride. Extract the solution with three 150 ml portions of ether and wash the combined ether extracts with 5 per cent aqueous sodium hydroxide until the aqueous layer is basic; the final basic washing phase acquires a yellow colour. Wash the ether layer with water until the washings are neutral, dry the ethereal solution over anhydrous calcium chloride and remove the ether on a rotary evaporator. Distil the residue and collect the benzofuran as a fraction of b.p. 170–172 °C. The yield of colourless product is 9.5 g (91%).

Experiment 8.25 5-CHLORO-3-METHYLBENZOTHIOPHENE

CAUTION: *This preparation should be carried out in an efficient fume cupboard.*

(*p*-Chlorophenylthio)acetone. To a solution of 5.6 g (0.14 mol) of sodium hydroxide in 200 ml of water in a 500-ml conical flask equipped with a stirrer, add 20.2 g (0.14 mol) of *p*-chlorothiophenol followed by 13.0 g (0.14 mol) of chloroacetone (1). Stir the mixture at room temperature for about 45 minutes and then extract with ether. Wash the combined ether extracts with water, dry over magnesium sulphate and evaporate the ether on the rotary evaporator. Distil the residue under reduced pressure and collect (*p*-chlorophenylthio)-acetone as a fraction of b.p. 180–183 °C/16 mmHg; the yield is 23 g (82%).

5-Chloro-3-methylbenzothiophene. Heat 10 g (0.05 mol) of *p*-chlorophenylthio)acetone and 100 g of polyphosphoric acid (Section 4.2.*58*, p. 458) in a 250-ml flask on an oil bath maintained at 120–140 °C. When the contents of the flask initially reach the reaction temperature swirl vigorously to ensure thorough mixing of the reactants. After 2 hours allow the reaction mixture to cool and add 100 ml of water. Extract the organic product with ether, wash the combined extracts with water and dry over magnesium sulphate. Distil the residue under reduced pressure and collect 5-chloro-3-methylbenzothiophene as a fraction of b.p. 98–100 °C/0.1 mmHg. The yield is 4.7 g (52%).

Note. (1) See Expt 8.20, Note (1); the compound is lachrymatory.

Experiment 8.26 2-PHENYLINDOLE

Prepare *acetophenone phenylhydrazone* by warming a mixture of 20 g (0.167 mol) of acetophenone and 18 g (0.167 mol) of phenylhydrazine (**CAUTION**, *in handling*) with 60 ml of ethanol and a few drops of glacial acetic acid. Filter the cold reaction mixture, wash the solid with dilute hydrochloric acid followed by about 12 ml of cold rectified spirit. Recrystallise a small portion from ethanol and thus obtain a sample of pure acetophenone phenylhydrazone as a white solid, m.p. 106 °C.

Place 28 g of the crude phenylhydrazone in a 250-ml beaker containing 180 g of polyphosphoric acid (Section 4.2.*58*, p. 458). Heat on a boiling water bath, stir with a thermometer and maintain at 100–120 °C for 10 minutes (the reaction is exothermic). Add 450 ml of cold water and stir well to complete solution of the polyphosphoric acid. Filter at the pump and wash well with water. Heat the crude solid under reflux with 300 ml of rectified spirit, add a little decolourising charcoal and filter through a preheated Buchner funnel; wash the residue with 40 ml of hot rectified spirit. Cool the combined filtrates to room temperature, filter off the 2-phenylindole and wash it three times with 10 ml portions of cold alcohol. Dry in a vacuum desiccator over anhydrous calcium chloride. The yield of pure 2-phenylindole, m.p. 188–189 °C, is 20 g (79%).

Cognate preparation. *1,2,3,4-Tetrahydrocarbazole.* In a 500-ml three-necked flask fitted with a dropping funnel, a sealed stirrer unit and reflux condenser, place a mixture of 49 g (0.5 mol) of cyclohexanone and 180 g of glacial acetic acid. Heat under reflux with stirring and add 54 g (49 ml, 0.5 mol) of redistilled phenylhydrazine (**CAUTION**) during 1 hour; continue the stirring for a further hour. Pour the reaction mixture into a 1-litre beaker and stir vigorously while it solidifies. Cool to 5 °C and filter at the pump through a Buchner funnel; cool the filtrate in ice and refilter through the same Buchner funnel. Wash the solid on the filter with 50 ml of water, suck almost dry and then wash with 50 ml of 75 per cent ethanol. Spread the crude solid upon absorbent paper and dry in the air overnight. Recrystallise the slightly damp solid from 350 ml of methanol: add a little decolourising carbon and filter through a hot-water funnel. The yield of 1,2,3,4-tetrahydrocarbazole, m.p. 116–117 °C, is 65 g (76%). A further 5 g of product may be obtained by concentrating the mother-liquor to one-quarter of the original volume.

8.3.5 BENZIMIDAZOLES AND BENZOTRIAZOLE FROM *O*-PHENYLENE DIAMINE

Disconnection of the two carbon–nitrogen bonds in benzimidazole (41) reveals the *ortho*-diamine and formic acid.

(41)

The synthesis is effected by simply heating the reagents together. 2-Methylbenz-imidazole and 2-benzylbenzimidazole are similarly formed from *o*-phenylene diamine and acetic acid or phenylacetic acid respectively (Expt 8.27).

Disconnection of benzotriazole (42) is almost trivial; in fact conversion of the diamine into the mono-diazonium derivative is followed by spontaneous cyclisation (Expt 8.28).

(42)

Experiment 8.27 BENZIMIDAZOLE

Place 27 g (0.25 mol) of *o*-phenylenediamine (Expt 6.49) in a 250-ml round-bottomed flask and add 17.5 g (16 ml, 0.34 mol) of 90 per cent formic acid. Heat the mixture on a water bath at 100 °C for 2 hours. Cool, add 10 per cent sodium hydroxide solution slowly, with constant rotation of the flask, until the mixture is just alkaline to litmus. Filter off the crude benzimidazole at the pump, wash with ice-cold water, drain well and wash again with 25 ml of cold water. Dissolve the crude product in 400 ml of boiling water, add 2 g of decolourising carbon and digest for 15 minutes. Filter rapidly at the pump through a preheated Buchner funnel and flask. Cool the filtrate to about 10 °C, filter off the benzimidazole, wash with 25 ml of cold water and dry at 100 °C. The yield of pure benzimidazole, m.p. 171–172 °C, is 25 g (85%).

Cognate preparations. *2-Methylbenzimidazole*. Heat together a mixture of 5.43 g (0.03 mol) of *o*-phenylenediamine dihydrochloride (Expt 6.49, Note (3)), 20 ml of water and 5.4 g (0.09 mol) of acetic acid under reflux for 45 minutes. Make the cooled reaction mixture distinctly basic by the gradual addition of concentrated ammonia solution, collect the precipitated product and recrystallise it from 10 per cent aqueous ethanol. The yield is 2.2 g (56%), m.p. 176 °C.

2-Benzylbenzimidazole.   Use 5.43 g (0.03 mol) of *o*-phenylenediamine dihydrochloride, 20 ml of water, 12.3 g (0.09 mol) of phenylacetic acid (Expt 5.128) and proceed as for 2-methylbenzimidazole. Recrystallise the crude product from 40 per cent aqueous ethanol. The yield is 3.4 g (55%), m.p. 191 °C.

Experiment 8.28 BENZOTRIAZOLE

Dissolve 10.8 g (0.1 mol) of *o*-phenylenediamine (1) in a mixture of 12 g (11.5 ml, 0.2 mol) of glacial acetic acid and 30 ml of water contained in a 250-ml beaker; slight warming may be necessary. Cool the clear solution to 15 °C, stir magnetically and then add a solution of 7.5 g (0.11 mol) of sodium nitrite in 15 ml of water in one portion. The reaction mixture becomes warm and within 2–3 minutes reaches a temperature of about 85 °C and then begins to cool while the colour changes from deep red to pale brown. Continue stirring for 15 minutes, by which time the temperature will have dropped to 35–40 °C, and then thoroughly chill in an ice-water bath for 30 minutes. Collect by vacuum filtration the pale brown solid which separates and wash with three 30 ml portions of ice-cold water. Dissolve the solid in about 130 ml of boiling water, add decolourising charcoal, filter and allow the filtrate to cool to about 50 °C before adding a few crystals of the crude benzotriazole which have been retained for seeding. Allow the mixture to attain room temperature slowly (to avoid separation of the material as an oil) and then thoroughly chill in ice and collect the benzotriazole which separates as pale straw-coloured needles, m.p. 99–100 °C. A second crop may be obtained by concentrating the filtrate. The yield is about 8 g (67%). The benzotriazole crystallises much more readily from benzene (c. 55 ml) but the material is still slightly coloured. A pure white product can be obtained by sublimation at 90–95 °C at 0.2 mmHg.

Note. (1) *o*-Phenylenediamine is generally contaminated by highly coloured impurities. A somewhat more pure benzotriazole may be obtained by using purified *o*-phenylenediamine (see Expt 6.49).

8.4 SIX-MEMBERED HETEROCYCLES

The six-membered nitrogen heteroaromatic compound is pyridine (70); the benzo analogues are quinoline (71) and isoquinoline (72).

(70)   (71)   (72)

The selected examples of six-membered heterocyclic rings containing two nitrogen atoms are based on the pyrimidine [(73), 1,3-diazabenzene], pyrazine [(74), 1,4-diazabenzene] and quinoxaline [(75), benzopyrazine] systems.

(73)   (74)   (75)

The pyrylium salts (76) are the six-membered oxygen heteroaromatic compounds. The 2- and 4-pyrones [(77) and (78) respectively] and the corresponding benzo analogues [coumarins (79) and chromones (80)] are represented by examples which further illustrate the usefulness of retrosynthetic analysis in the derivation of preparative methods.

(76)   (77)   (78)   (79)   (80)

The 2- or 4-hydroxypyridines [(81) and (82)] are tautomeric and in solution exist preferentially in the corresponding pyridone forms. Their synthesis from the corresponding pyrones by a heteroatom exchange reaction is possible.

(81)   (82)

Barbituric acid (83), which is formally 2,4,6-trihydroxyprimidine, is extensively tautomeric (p. 1175). The barbiturate drugs are derivatives of the triketo form. Piperazine-2,5-dione (84) is a diamide based upon the pyrazine ring system.

(83) (84)

As in the five-membered heterocyclic section the discussion of the relevant ring-forming reaction is related, as appropriate, to an overall strategy of skeletal assembly, and to a more rigorous retrosynthetic analysis. The reactions are conveniently discussed in relation to the synthesis of compounds belonging to the following groups of heterocycles.

1. Pyridines and pyrylium salts.
2. Pyrones and pyridones.
3. Barbituric acid and barbiturates.
4. Quinolines, isoquinolines and quinoxalines.
5. Coumarins and chromones.

8.4.1 PYRIDINES AND PYRYLIUM SALTS

The two target molecules, which illustrate cyclisation reactions involving substituted 1,5-dicarbonyl compounds in the presence of appropriate reagents, are diethyl 2,6-dimethylpyridine-3,5-dicarboxylate (85) and 2,4,6-triphenylpyrylium fluoroborate (86). The latter is of specific interest since it provides a simple example to illustrate a general procedure for the conversion of pyrylium salts into pyridinium derivatives; these latter compounds are important reagents for a variety of functional group interconversions (see Sections 5.5.6, p. 574, and 5.15.3, p. 768).

(85) (86)

In the case of (85) an initial reduction transform to the corresponding 1,4-dihydropyridine (87) facilitates recognition of the synthons arising from subsequent disconnections. This reduction step is legitimate since the reverse oxidation step in the synthesis is particularly facile owing to mesomeric stabilisation of the aromatic ring system. Disconnection of both carbon–nitrogen bonds in the 1,4-dihydropyridine, in a manner similar to that formulated for pyrrole derivatives, reveals the bis-keto ester (88) and ammonia. Further rational disconnection of the bis-keto ester then finally discloses two molecules of ethyl acetoacetate and one of formaldehyde as the other starting materials.

(85) (87) (88)

The preparation of (85) (Expt 8.29) is an example of the *Hantzsch pyridine synthesis*. This is a widely used general procedure since considerable structural variation in the aldehydic compound (aliphatic or aromatic) and in the 1,3-dicarbonyl component (β-keto ester or β-diketone) is possible, leading to the synthesis of a great range of pyridine derivatives. The precise mechanistic sequence of ring formation may depend on the reaction conditions employed. Thus if, as implied in the retrosynthetic analysis above, ethyl acetoacetate and the aldehyde are first allowed to react in the presence of a base catalyst (as in Expt 8.29), a bis-keto ester [e.g. (88)] is formed by successive Knoevenagel and Michael reactions (Section 5.11.6, p. 681). Cyclisation of this 1,5-dione with ammonia then gives the dihydropyridine derivative. Under different reaction conditions condensation between an aminocrotonic ester and an alkylidene acetoacetate may be involved.

Oxidation of the dihydropyridine derivative is most conveniently carried out with an aqueous nitric–sulphuric acid mixture. Removal of the ethoxycarbonyl groups may be achieved by a stepwise hydrolysis and decarboxylation sequence, but the one-step reaction described in Expt 8.29 using soda-lime is convenient.

The retrosynthetic analysis of 2,4,6-triphenylpyrylium tetrafluoroborate (86), involving an initial reduction followed by a disconnection of one carbon–oxygen bond (cf. disconnection of 2,5-dimethylfuran, Section 8.3.1, p. 1146), reveals the substituted 1,5-dicarbonyl compound (89). Further rational disconnection then reveals acetophenone and 1,3-diphenylprop-2-en-1-one (chalcone); clearly the latter may originate from acetophenone and benzaldehyde (cf. Section 6.12.2, p. 1032).

(86)   (89)

A convenient synthetic procedure[21] (Expt 8.30) involves stirring and warming acetophenone and benzaldehyde with boron trifluoride–etherate; hydrogen is evolved during the final oxidation stage which leads to the pyrylium tetra-fluoroborate derivative. An alternative method using chalcone, acetophenone and fluoroboric acid has been published.[22] 2,4,6-Triphenylpyrylium perchlorate is formed when chalcone and acetophenone are heated in the presence of per-chloric acid[23]; the perchlorates tend to decompose with explosive violence if not handled with care.

Either the pyrylium tetrafluoroborate or the perchlorate salt may be con-verted into the corresponding halide by initial conversion into the pseudobase [(90), 1,3,5-triphenylpent-2-en-1,5-dione] with dilute alkali, followed by recycli-sation in the presence of the hydrogen halide.[24]

(90)

All these various pyrylium salts are converted by ammonia, or alkyl, arylkyl or aryl primary amines into the corresponding pyridinium salts.[25] The probable reaction sequence is given in the illustrative example 1-benzyl-2,4,6-triphenyl-pyridinium tetrafluoroborate[26] (Expt 8.31).

The preparations of 2-alkyl- and 2-arylpyridines from pyridine itself are of interest in that their syntheses illustrate the activating influence of the ring nitro-gen on the 2- and 4-ring carbons and on the 2- and 4-alkyl substituents. A pre-parative example to exemplify the former is the synthesis of 2-phenylpyridine by the reaction of pyridine with phenyllithium (Expt 8.32). Addition of the organo-metallic derivative to the azomethine linkage in pyridine gives an intermediate derived from a 1,2-dihydropyridine system. This undergoes thermal elimination of lithium hydride to form 2-phenylpyridine. The formation of 2-alkylpyridines may be effected similarly using the appropriate alkyllithium.

Activation of alkyl substituents is illustrated by the synthesis of the isomeric pentylpyridines (Expt 8.33) and ethyl 2-pyridylacetate (Expt 8.32, cognate pre-paration). Thus removal of a proton from the 2-methyl group of 2-methyl-pyridine (2-picoline) with the aid of base (e.g. sodamide) gives a mesomerically

1167

stabilised anion (91) which may then be alkylated with an alkyl halide. Alternatively, the anion may be reacted with carbon dioxide to give 2-pyridylacetic acid, which may be subsequently esterified by treatment with ethanolic hydrogen chloride.

4-Methylpyridine, and even the less reactive 3-methyl derivative, may also be alkylated in a similar manner.

Experiment 8.29 DIETHYL 2,6-DIMETHYLPYRIDINE-3,5-DICARBOXYLATE AND 2,6-DIMETHYLPYRIDINE

Diethyl 1,4-dihydro-2,6-dimethylpyridine-3,5-dicarboxylate. Cool 52 g (51 ml, 0.4 mol) of ethyl acetoacetate to 0 °C and add 15 ml (0.2 mol) of 40 per cent aqueous formaldehyde solution, followed by a few drops of diethylamine as a catalyst. Keep the mixture at 0 °C for 6 hours and then at room temperature for 40 hours. Separate the lower organic layer, extract the aqueous phase with ether and dry the combined organic fractions over anhydrous calcium chloride. Remove the ether under reduced pressure (rotary evaporator) and transfer the residue together with an equal volume of ethanol to a stout reagent bottle cooled in an ice bath. Pass a steady stream of ammonia gas (from a cylinder) into the solution held at 0 °C for 1 hour, close the bottle with a bung securely attached with wire and set the bottle and contents aside at room temperature for 40 hours. Filter the resulting yellow solution to remove a small quantity of almost colourless material and heat the filtrate on a boiling water bath in an evaporating dish until most of the ethanol has been removed, and then cool and crystallise the residue from about 400 ml of rectified spirit. The yield of the pale yellow crystalline dihydropyridine derivative is 36 g (71%), m.p. 181–183 °C.

Diethyl 2,6-dimethylpyridine-3,5-dicarboxylate. Place 35.5 (0.14 mol) of the above dihydropyridine derivative in a 1-litre round-bottomed flask and add carefully a cold mixture of 50 ml of water, 9 ml of concentrated nitric acid (d 1.42) and 7.5 ml of concentrated sulphuric acid. Swirl the mixture and heat it cautiously on a boiling water bath until a vigorous reaction, accompanied by much foaming, sets in. When the reaction has moderated continue to heat cautiously for 15 minutes until oxidation is complete and a deep red solution is obtained. Cool the solution, add 100 ml of water and 100 g of crushed ice, and make it distinctly alkaline with concentrated aqueous ammonia solution (d 0.88). Filter off the solid product, wash it with a little cold water and recrystallise it from aqueous ethanol. The yield of colourless crystals of the pyridine derivative, m.p. 71–72 °C, is 22.5 g (64%).

2,6-Dimethylpyridine.   Place an intimate mixture of 10 g (0.04 mol) of the above pyridine di-ester and 60 g of soda-lime (10–14 mesh) in a 100-ml round-bottomed flask fitted with a still-head and condenser arranged for distillation. Heat the flask gradually in an oil bath to about 250 °C, and maintain this temperature until no further material distils below 105 °C (about 2 hours may be required). Remove the oil bath, clean the outside of the flask and continue to heat more strongly with a Bunsen burner held in the hand, keeping the flame moving over the surface of the flask. Collect the product which now distils, and continue to heat strongly until the flask reaches dull red heat and no further distillate is obtained. Treat the distillate with potassium hydroxide pellets so that the pyridine separates, and isolate the latter by extraction with ether. Dry the ether extract over fresh potassium hydroxide pellets, and remove the ether and distil the residue at atmospheric pressure . Collect the dimethylpyridine as a fraction of b.p. 142–145 °C; the yield is 2.8 g (65%).

Experiment 8.30   2,4,6-TRIPHENYLPYRYLIUM TETRAFLUORO-BORATE[21]

CAUTION: *As hydrogen is evolved in this reaction the preparation should be conducted in a well-vented fume cupboard.*

To a mixture of acetophenone (60 g, 0.5 mol) and benzaldehyde (26.5 g, 0.25 mol) is added, with stirring, boron trifluoride–etherate (84 g, 0.65 mol); the mixture becomes coloured and the temperature rises, often to about 70 °C. On maintaining the temperature at 100 °C for 2 hours ether distils from the reaction mixture. The product of the reaction is a viscous brown oil which must be dissolved in acetone (300 ml) and the tetrafluoroborate salt precipitated by the addition of ether (3 litres). 2,4,6-Triphenylpyrylium tetrafluoroborate (yield 40%) is obtained as a yellow crystalline solid by filtration and recrystallisation from acetone; m.p. 253–255 °C.

Experiment 8.31 1-BENZYL-2,4,6-TRIPHENYLPYRIDINIUM TETRAFLUOROBORATE[26]

2,4,6-Triphenylpyrylium tetrafluoroborate (12 g, 0.03 mol) and benzylamine (2.8 g, 0.03 mol) are stirred in absolute ethanol (50 ml) with triethylamine (0.3 g, 0.003 mol) for 2 hours. The crude product is filtered off, washed with ether (25 ml) and recrystallised as white needles from absolute ethanol (15 ml). The product is obtained in 81 per cent yield; m.p. 193–195 °C.

Experiment 8.32 2-PHENYLPYRIDINE

The first stage is the *preparation of a solution of phenyllithium in dry ether*. Equip a 1-litre three-necked flask as shown in Fig. 2.60. Flush the apparatus with dry, oxygen-free nitrogen gas. Place 7.35 g (1.06 mol) of lithium shavings (1) in the flask, and introduce a solution of 78.5 g (52.5 ml, 0.5 mol) of dry, redistilled bromobenzene in 250 ml of anhydrous ether into the dropping funnel. Start the stirrer. Run in about 2 ml of the solution; when the reaction starts, as indicated by an initial cloudiness, add the remainder at such a rate that the solvent refluxes gently (about 45 minutes). Finally, add 50 ml of anhydrous ether through the dropping funnel. Continue the stirring until all or most of the lithium disappears (1–1.5 hours) (2).

Now introduce slowly, and with stirring, 79 g (1 mol) of pure anhydrous pyridine (Section 4.1.*29*, p. 410) (**CAUTION**) dissolved in 200 ml of anhydrous toluene: remove the ether by distillation, replace the dropping funnel by a thermometer and stir the residual suspension at 110 °C (internal temperature) for 8 hours. Then cool to about 40 °C, and add cautiously 75 ml of water through the condenser. Filter the liquids if necessary, separate the upper toluene layer, dry it by shaking for an hour with 20 g of potassium hydroxide pellets and distil slowly through a short fractionating column. When the temperature reaches 150 °C at ordinary pressure (thus indicating the removal of most of the toluene, etc.), distil the residue under reduced pressure and collect the liquid passing over at 138–142 °C/12 mmHg. Upon redistillation 38 g (49%) of pure 2-phenylpyridine, b.p. 140 °C/12 mmHg, is obtained.

Notes. (1) See *p-toluic acid*, Expt 6.158, Note (1).
(2) The yield of phenyllithium generally exceeds 95 per cent. One interesting and instructive method of determination is to allow the phenyllithium to react with excess benzophenone and to weigh the triphenylmethanol formed which is assumed to be

formed quantitatively. A better method is to hydrolyse a 2 ml aliquot portion of the filtered solution with distilled water and to titrate the hydrolysate with standard acid, using phenolphthalein as indicator. To obtain the filtered solution, the dropping funnel is replaced by a short L-shaped tube loosely plugged with glass wool, and the solution is decanted through this tube into a graduated flask that has been swept out with nitrogen.

Cognate preparation. *Ethyl 2-pyridylacetate.* Prepare a solution of phenyllithium in anhydrous ether as detailed above for 2-phenylpyridine, using 7.35 g (1.06 mol) of lithium. Introduce 46.5 g (49 ml, 0.5 mol) of dry, redistilled 2-methylpyridine, with continued stirring dropwise during about 10 minutes. Stir the dark red-brown solution of 2-pyridylmethyllithium for a further 30 minutes, and then pour it slowly (1) and with shaking on to about 400 g of solid carbon dioxide contained in a 1.5-litre round-bottomed flask. Break up the lumpy residue of lithium salts before adding 375 ml of absolute ethanol. Cool the solution in ice, saturate it with dry hydrogen chloride and then insert a calcium chloride drying tube into the neck of the flask. Allow the mixture to stand overnight, remove the ethanol under reduced pressure (rotary evaporator) on a water bath. Dissolve the syrupy residue in 375 ml of dichloromethane and transfer to a three-necked flask fitted with a mechanical stirrer and a reflux condenser. Prepare a paste from 112.5 g (0.815 mol) of potassium carbonate and 70 ml of water, and add it slowly through the third neck to the dichloromethane solution with constant stirring. Stir the almost boiling solution vigorously for 1 hour. Decant the dichloromethane solution from the inorganic salts, remove the solvent by distillation from a water bath (rotary evaporator) and distil the residue under diminished pressure from a flask with a short fractionating column. 2-methylpyridine (*c.* 20 g) passes over first, followed by ethyl 2-pyridylacetate as a pale yellow liquid at 135–137 °C/ 28 mmHg, or 110–112 °C/6 mmHg. The yield is 30 g (36%).

Note. (1) It is advisable to filter the 2-pyridylmethyllithium solution rapidly through a thin layer of glass wool (to remove any unreacted lithium) before it is added to the solid carbon dioxide.

Experiment 8.33 2-PENTYLPYRIDINE

Into a 500-ml three-necked flask fitted with a dropping funnel, a sealed stirrer unit and reflux condenser protected by a drying tube, place a fine suspension of 40 g (1 mol) of good quality sodamide (Section 4.2.67, p. 462) in about 150 ml of anhydrous xylene. Introduce 37.5 g (40 ml, 0.4 mol) of 2-methylpyridine through the dropping funnel and rinse the latter with a few ml of dry xylene. Set the stirrer in motion and add 44.5 g (50.5 ml, 0.48 mol) of butyl chloride (Expt 5.50) over a period of 1 hour: reflux the mixture with stirring for 2–3 hours. When cold, destroy the excess of sodamide by the cautious addition of 100 ml of water. Transfer the contents of the flask to a separatory funnel and discard the lower aqueous layer. Extract the xylene solution with four 50 ml portions of 1:1 hydrochloric acid. Steam distil the acid extracts to remove traces of xylene, cool the aqueous solution and render strongly alkaline by the addition of solid sodium hydroxide: a brown oil appears. Steam distil again and collect about 700 ml of distillate. Separate the upper layer in the steam distillate, extract the aqueous layer with ether and dry the combined upper layer and ether extract with anhydrous potassium carbonate. After removing the ether, distil through a Fenske-type column (15 cm diameter and packed with glass helices for a length of 12–15 cm) at a pressure of 50 mmHg (*manostat*, see Section 2.30) and collect the 2-pentylpyridine (42 g, 70%) at 122.5–124.5 °C/50 mmHg. Upon redistillation, the product boils almost entirely at 105 °C/17 mmHg.

Cognate preparations. *4-Pentylpyridine.* Charge a 1-litre three-necked flask (equipped with a sealed stirrer unit, a dropping funnel and a short air condenser) with 600 ml of liquid ammonia (Section 2.17.7, p. 116). Stir vigorously, add 0.5 g of powdered iron(III) nitrate followed, after 1 minute, by 11.9 g (0.52 mol) of clean sodium in small pieces through the short air condenser over a period of half an hour; continue the stirring until the initial blue colour is replaced by a colourless or pale grey suspension of sodamide (see p. 462). Introduce 42.0 g (44.0 ml, 0.45 mol) of pure 4-methylpyridine through the air condenser; a green colour develops immediately. Stir for 15–20 minutes and add 46.3 g (52.6 ml, 0.5 mol) of butyl chloride (or an equivalent amount of butyl bromide) from the dropping funnel at such a rate that the reaction does not become unduly vigorous (*c.* 10 minutes): upon completion of the addition the green colour will have been discharged. Stir for a further 10–15 minutes, pour the reaction mixture into a 2-litre beaker: allow the liquid ammonia to evaporate overnight. Rinse the reaction flask with 100 ml of water and add the rinsings to the residue in the beaker; two layers form. Separate them and keep the upper layer of 4-pentylpyridine: extract the lower layer with a little xylene and wash the xylene extract with 25 ml of 1:1 hydrochloric acid. Dissolve the 4-pentylpyridine in 1:1 hydrochloric acid, combine it with the acid washings of the xylene extract and steam distil to remove traces of xylene; cool, add solid sodium hydroxide until strongly alkaline and steam distil again. Isolate the 4-pentylpyridine as described above for the 2-pentyl compound. The yield of the pure base, b.p. 95 °C/6 mmHg, is 46 g (69%).

3,Pentylpyridine. Proceed exactly as described for 4-pentylpyridine using 11.9 g (0.52 mol) of sodium, 42 g (0.45 mol) of 3-methylpyridine and 46.3 g (0.5 mol) of butyl chloride. The yield of pure 3-pentylpyridine, b.p. 100.5 °C/9 mmHg, is 46 g (69%).

8.4.2 PYRONES AND PYRIDONES

In this section the synthesis of coumalic acid (92), representing the $\alpha(2)$-pyrone system, is described. α-Pyrones readily undergo a simple lactone–amide transformation, and the conversion of coumalic acid into the tautomeric 6-hydroxynicotinic acid (93) provides a further illustration of the conversion of a six-membered oxygen heterocycle into the corresponding nitrogen system.

(92)   (93)

Application of the principles of disconnection in heterocyclic rings to coumalic acid (92) reveals two molecules of formylacetic acid.

In the synthesis (Expt 8.34), formylacetic acid is formed *in situ* by the action of a concentrated sulphuric acid–fuming sulphuric acid mixture on malic acid ($HO_2C \cdot CHOH \cdot CH_2 \cdot CO_2H$). This α-hydroxydicarboxylic acid undergoes decarbonylation and dehydration under these conditions before the acid-catalysed self-condensation of formylacetic acid. The cyclisation step is generally applicable to β-keto esters.

Conversion of the derived methyl coumalate into methyl 6-hydroxy-nicotinate (Expt 8.35) is effected by the action of concentrated aqueous ammonia; subsequent hydrolysis with aqueous alkali yields 6-hydroxynicotinic acid.

Experiment 8.34 COUMALIC ACID

Add 180 ml of sulphuric acid to 202 g (1.5 mol) of finely powdered malic acid contained in a 2-litre round-bottomed flask. At intervals of 45 minutes, add three 50 ml portions of fuming sulphuric acid (25% SO_3); a slight exothermic reaction sets in with the steady evolution of gas. Swirl the mixture frequently to obviate excessive foaming, and when the evolution of gas has slackened, heat the reaction mixture on a water bath for 2 hours. Cool the mixture and pour on to 800 g of ice with stirring. Set the mixture aside in a refrigerator for

24 hours and then filter the crude coumalic acid and wash with small portions of iced water. Recrystallise the crude product from methanol to give coumalic acid as light yellow crystals, m.p. 206–208 °C. The yield is 65 g (62%). Record the p.m.r. spectrum (Me$_2$SO-d_6) and assign the signals which appear at δ 6.40 (d of d, 1H), 7.81 (d of d, 1H), 8.49 (d of d, 1H) and 10.32 (broad, 1H).

Experiment 8.35 6-HYDROXYNICOTINIC ACID

Methyl coumalate. Place 100 ml of concentrated sulphuric acid in a 250-ml round-bottomed flask and add with swirling 35 g (0.224 mol) of finely powdered coumalic acid (Expt 8.34). The solution acquires a deep red-brown colour and the temperature tends to rise but should be kept below 30 °C by cooling in an ice-water bath. Add 50 ml of methanol with continued swirling in small portions so that the temperature of the reaction mixture does not rise above 35 °C. Heat the reaction mixture on a boiling water bath for 1.5 hours. Cool the mixture and pour slowly into 500 ml of an ice-water slurry with stirring. Add about 150 g of sodium carbonate as a slurry in water until the mixture is just neutral. If too much carbonate has been added the supernatant liquid will acquire a red colour; in these circumstances add a few drops of sulphuric acid. The final solution containing a suspension of methyl coumalate and inorganic salts should be an orange colour. Filter the solution, and slurry the residue with four 70 ml portions of cold water containing a drop of concentrated sulphuric acid. Filter off the methyl coumalate; the yield after air drying is 18 g (47%). Although the product may be used for the next stage, a sample may be purified by vacuum sublimation; pure methyl coumalate has m.p. 74 °C.

6-Hydroxynicotinic acid. Place a mixture of 20 ml of ammonia solution (d 0.880) and 30 ml of water in a 250-ml beaker sited in an ice-water cooling bath on a magnetic stirrer unit. Insert a magnetic follower and add portionwise 18 g (0.105 mol) of methyl coumalate over a period of about 5 minutes, ensuring that the temperature is kept below 20 °C. Stir the resulting dark red solution for a further 45 minutes at 20 °C and then add it to boiling aqueous sodium hydroxide prepared from 40 g of sodium hydroxide and 250 ml of water. Boil the reaction mixture for 5 minutes, cool to 10 °C in an ice-water cooling bath and add concentrated hydrochloric acid with stirring to precipitate the product; the temperature of the solution during neutralisation should not rise above 30 °C. Allow the acidified solution to stand in the ice-water bath for 1 hour and then collect the bright yellow crystalline solid by filtration. Wash the crystals with water and then air dry. The yield of 6-hydroxynicotinic acid, m.p. 299–300 °C (decomp.), is 8.5 g (58%).

8.4.3 BARBITURIC ACID AND BARBITURATES

The target molecules selected to illustrate some further cyclisation reactions are barbituric acid (83) and 5-ethyl-5-phenylbarbituric acid (94), the latter being an example of the well-known barbiturate drugs. Piperazine-2,5-dione [diketo-piperazine (84)] has also been included here because of the similar disconnection involved in retrosynthetic analysis.

(83)   (94)   (84)

Barbituric acid (83) may be regarded as 2,4,6-trihydroxypyrimidine, but in the crystalline state it exists as the triketo-form (95). In aqueous solution the compound is remarkably acidic as the result of ionisation of the mono-enolic form (96) with the formation of a resonance stabilised anion (97).

(95)   (96)   (97)

Urea may be recognised as a structural feature in the triketo form; disconnection then reveals diethyl malonate as the other reagent. The synthesis of barbituric acid is therefore effected (Expt 8.36) by condensation of diethyl malonate with urea in the presence of sodium ethoxide. Barbituric acid undergoes nitration in the 5-position on treatment with fuming nitric acid, and reduction of the nitro derivative (98) yields 5-aminobarbituric acid (99) (uramil).

$+ (NH_2)_2CO$

(98)   (99)

A similar retrosynthetic analysis for 5-ethyl-5-phenylbarbituric acid (94) discloses urea and diethyl ethylphenylmalonate.

$+ (NH_2)_2CO$

(94)

The synthesis of the disubstituted malonic ester cannot be achieved by success-ive ethylation and phenylation procedures [cf. Section 5.11.6, p. 680] since for the latter reaction the aryl halide is not sufficiently reactive. The procedure de-scribed below (formulated in Expt 8.37) provides a suitable indirect alternative route. This involves a mixed Claisen ester condensation (cf. Section 5.14.3, p. 738) between ethyl phenylacetate and diethyl oxalate, followed by decarbony-lation of the α-keto diester on heating to 175 °C.

Piperazine-2,5-dione (84) is a cyclic diamide based upon the alternative 1,4-diaza heterocyclic system and appropriate disconnection of the N—CO bond reveals two molecules of glycine.

(84)

The synthesis of (84) is effected by heating glycine in solvent such as ethylene glycol (Expt 8.38). Brief treatment with hot concentrated aqueous hydrochloric acid cleaves one of the amide linkages with the formation of the dipeptide, glycylglycine, which is isolated as the hydrochloride monohydrate and may be converted into the ester hydrochloride under Fischer–Speier conditions (Section 5.12.3, p. 696).

Experiment 8.36 BARBITURIC ACID

In a 2-litre round-bottomed flask, fitted with a double surface reflux con-denser, place 11.5 g (0.5 mol) of clean sodium. Add 250 ml of absolute ethanol in one portion: if the reaction is unduly vigorous, immerse the flask moment-arily in ice. When all the sodium has reacted, add 80 g (76 ml, 0.5 mol) of diethyl malonate, followed by a solution of 30 g (0.5 mol) of dry urea in 250 ml of hot (c. 70 °C) absolute ethanol. Shake the mixture well, fit a calcium chlor-ide guard-tube to the top of the condenser and reflux the mixture for 7 hours in an oil bath heated to 110 °C. A white solid separates. Treat the reaction mixture with 450 ml of hot (50 °C) water and then with concentrated hydro-chloric acid, with stirring, until the solution is acid (about 45 ml). Filter the resulting almost clear solution and leave it in the refrigerator overnight. Filter the solid at the pump, wash it with 25 ml of cold water, drain well and then dry at 100 °C for 4 hours. The yield of barbituric acid is 50 g (78%). It melts with decomposition at 245 °C.

Conversion to aminobarbituric acid (*uramil*). *Nitrobarbituric acid.* Place 72 ml of fuming nitric acid, *d* 1.52, in a 1-litre flask equipped with a mechanical stirrer and surrounded by an ice bath. Add 50 g (0.39 mol) of barbituric acid with stirring, over a period of 2 hours; keep the temperature below 40 °C during the addition. Stir for a further 1 hour, and continue the stirring while 215 ml of water is added and the solution is cooled to 10 °C. Filter with suction through a sintered glass funnel, wash with cold water and dry on a clock glass at 60–80 °C. Dissolve the crude nitrobarbituric acid in 450 ml of boiling water, filter and allow to stand overnight. Collect the crystals by suction filtration, wash well with cold water, and dry at 90–95 °C for 2–3 hours. The product is the trihydrate, m.p. 181–183 °C (decomp., rapid heating), and weighs 70 g. Drying at 110–115 °C for 2–3 hours gives 47 g (70%) of anhydrous nitrobarbituric acid, m.p. 176 °C (decomp.).

Uramil. In a 3-litre flask place 38 g (0.22 mol) of anhydrous nitrobarbituric acid and 300 ml of concentrated hydrochloric acid; heat the mixture on a boiling water bath. Add 125 g of granulated tin and 200 ml of concentrated hydrochloric acid over a period of about 30 minutes: continue the heating until the yellow colour, due to the nitro compound, in the liquid is no longer visible. Introduce 1500 ml more of concentrated hydrochloric acid and heat until all the white solid dissolves; add a little decolourising charcoal, and filter the hot mixture through a sintered glass funnel. Keep the filtrate at 0 °C overnight, collect the uramil by filtration with suction, wash well with dilute hydrochloric acid and finally with water. Concentrate the filtrate under reduced pressure (rotary evaporator) to about 500 ml and cool overnight. Collect the second crop of uramil, wash it as before and combine it with the first product. Dry in a vacuum desiccator over concentrated sulphuric acid. The resulting uramil (23 g, 73%) is a fine white powder, it does not melt below 400 °C, and becomes pink to red on standing, particularly if ammonia is present in the air.

Experiment 8.37 5-ETHYL-5-PHENYLBARBITURIC ACID (*Phenobarbitone*)

Diethyl phenylmalonate. In a thoroughly dried 1-litre flange flask, equipped with a dropping funnel, reflux condenser (both protected with calcium chloride guard-tubes) and a sealed stirrer unit, place 11.5 g (0.5 mol) of clean sodium pieces (see Section 4.2.*68*, p. 462); add 250 ml of super-dry ethanol

(Section 4.1.9, p. 400) and allow the vigorous reaction to proceed, cooling only if the reaction appears to be beyond control. When all the sodium has reacted, cool the solution to 60 °C, and add 73 g (67 ml, 0.5 mol) of pure, freshly distilled diethyl oxalate from the dropping funnel in a rapid stream with vigorous stirring. Wash this down with 5 ml of absolute ethanol and add immediately 87.5 g (85 ml, 0.535 mol) of pure ethyl phenylacetate (Expt 5.152). Discontinue stirring and raise the stirrer clear of the reaction mixture. Within 4–7 minutes after the ethyl phenylacetate has been added, crystallisation commences. Allow the nearly solid paste of the sodio derivative to cool to room temperature, remove the flange lid and stir manually with 400 ml of dry ether. Collect the solid by suction filtration and wash it repeatedly with dry ether. Transfer the solid to a beaker and liberate the ethyl phenyloxaloacetate with ice-cold dilute sulphuric acid (14–15 ml of concentrated sulphuric acid in 250 ml of water). Separate the almost colourless oil and extract the aqueous layer with three 50 ml portions of ether; dry the combined oil and ethereal extracts with magnesium sulphate, remove the ether with a rotary evaporator. Equip the flask containing the residue for distillation under reduced pressure and heat the flask under a pressure of about 15 mm of mercury (water pump) in an oil bath. Raise the temperature of the bath gradually to 175 °C and maintain this temperature until the evolution of carbon monoxide is complete (*fume cupboard*); if the pressure rises unduly during the heating (owing to a rather rapid evolution of gas), discontinue the heating momentarily. When the reaction is complete (5–6 hours), return the oil which has passed over to the flask, and distil under reduced pressure. Collect the diethyl phenylmalonate at 159–161 °C/10 mmHg (or at 165–166 °C/15 mmHg). The yield is 95 g (80.5%).

Diethyl ethylphenylmalonate. In a dry 500-ml round-bottomed flask, fitted with a reflux condenser and guard-tube, prepare a solution of sodium ethoxide from 7.23 g (0.315 mol) of clean sodium and 150 ml of super-dry ethanol in the usual manner; add 1.5 ml of pure ethyl acetate (dried over anhydrous calcium sulphate) to the solution at 60 °C and maintain this temperature for 30 minutes. Meanwhile equip a 1-litre three-necked flask with a dropping funnel, a sealed mechanical stirrer and a double surface reflux condenser: *the apparatus must be perfectly dry* and guard-tubes should be inserted in the funnel and condenser respectively. Place a mixture of 74 g (0.315 mol) of diethyl phenylmalonate and 60 g (0.385 mol) of ethyl iodide in the flask. Heat the apparatus in a bath at 80 °C and add the sodium ethoxide solution, with stirring, at such a rate that a drop of the reaction mixture when mixed with a drop of phenolphthalein indicator is never more than faintly pink (1). The addition occupies 2–2.5 hours; continue the stirring for a further 1 hour at 80 °C. Allow the flask to cool, and remove the ethanol under reduced pressure using a rotary evaporator. Add 100 ml of water to the residue in the flask and extract the ester with three 100 ml portions of toluene. Dry the combined extracts with anhydrous sodium sulphate, distil off the toluene at atmospheric pressure and the residue under diminished pressure. Collect the diethyl ethylphenylmalonate at 159–160 °C/8 mmHg. The yield is 72 g (87%).

5-ethyl-5-phenylbarbituric acid. In a 250-ml round-bottomed flask, fitted with an efficient reflux condenser and guard-tube, prepare a solution of sodium methoxide from 4.6 g (0.2 mol) of clean sodium and 50 ml of anhydrous methanol (Section 4.1.8, p. 400). Add 15 g (0.25 mol) of urea (previously dried at 60 °C for 4 hours), and then 26.4 g (0.1 mol) of diethyl ethylphenylmalonate dropwise down the condenser. Reflux the mixture for 6 hours and then remove the excess methanol under reduced pressure (rotary evaporator) from a water bath held below 60 °C. Transfer the residue to a small beaker cooled in a freezing mixture and add 100 ml of ice-water with mechanical stirring: the temperature of the reaction mixture must be kept below 5 °C since barbiturates are decomposed by concentrated alkali into the salt of the corresponding malonic acid, sodium carbonate and ammonia. Filter and extract the filtrate with two 50 ml portions of toluene in order to remove esters; acidify the aqueous solution cautiously to Congo red, allow to stand for a few hours and filter off the crude phenobarbitone at the pump. The yield after drying at 90–100 °C is 13 g (56%). Recrystallisation from hot water yields reasonably pure ethylphenylbarbituric acid, m.p. 171 °C. A somewhat higher m.p. (175–176 °C) is obtained if rectified spirit is employed for recrystallisation, but the recovery is considerably less.

Note. (1) The rate of addition should be controlled so that the reaction mixture does not become strongly basic.

Experiment 8.38 PIPERAZINE-2,5-DIONE (*Diketopiperazine*)

Place 100 g (1.33 mol) of glycine and 500 ml of ethylene glycol in a 1-litre three-necked flask fitted with an air-cooled reflux condenser and a sealed mechanical stirrer; fit a thermometer so that the bulb is in the liquid. Heat the mixture in the fume cupboard to 175 °C and maintain the temperature at this level with continuous stirring for 1 hour. Cool the dark brown reaction product to room temperature and leave overnight in the refrigerator. Centrifuge the resulting suspension (1) and decant the mother-liquor. Transfer the solid to a Buchner funnel with the aid of cold methanol and wash on the filter with gentle suction with more methanol, using about 200 ml in all. Crystallise the product from 300 ml of boiling water but do not attempt to filter the hot solution at this stage; collect the light brown crystals which separate on cooling in ice. Dissolve the crude material in 350 ml of hot water, add 4 g of decolourising carbon and boil for 3 minutes. Filter the hot suspension through a preheated Buchner funnel (2) and cool the filtrate in ice. Collect the colourless crystals of pure piperazine-2,5-dione, wash with a little ice-cold water and dry in the oven at 50 °C; the yield is 34 g (45%). The m.p. is rather indefinite and depends on the rate of heating; when introduced into a rapidly heated (10 °C/minute) melting point apparatus at 280 °C, piperazine-2,5-dione has m.p. 310–312 °C (decomp.).

Boil a few mg each of the product and of ninhydrin (Expt 5.99) in 0.5 ml of water; the absence of a blue-purple coloration shows that the product is free from contaminating peptide material.

Notes. (1) If a centrifuge of suitable capacity is not available, the product may be isolated by decanting the suspension through a pad of glass wool in an ordinary filter funnel and allowing the crystals to drain as much as possible. In this case the methanol washing procedure is best omitted, and the product together with the glass wool subjected directly to the preliminary crystallisation from hot water described in the text.

(2) If the filtrate from this treatment with decolourising carbon is not colourless it should be boiled with more of the carbon and re-filtered.

Conversion into glycylglycine ethyl ester hydrochloride. Heat 150 ml of concentrated hydrochloric acid almost to boiling, add in one portion 28.5 g (0.25 mol) of piperazine-2,5-dione and boil for exactly 90 seconds. Cool the resulting solution immediately with swirling in an ice–salt bath, and scratch the sides of the vessel to induce crystallisation. Leave at 0 °C for 1 hour, filter off the resulting glycylglycine hydrochloride monohydrate with suction using an acid-resistant paper (e.g. Whatman No. 50–54) and wash it with three 15 ml portions of cold ethanol. Suspend the entire product in 600 ml of absolute ethanol in a 3-litre three-necked flask fitted with a sealed stirrer unit, a gas inlet tube and a reflux condenser leading to a gas absorption trap (Fig. 2.61(c). Surround the flask with an ice bath, start the stirrer and pass into the suspension a steady stream of dry hydrogen chloride (Section 4.2.*38*, p. 438) until about 60 g have been absorbed. Replace the cooling bath with a heating mantle and boil the mixture under reflux for 10 minutes. Cool the flask in an ice bath while still stirring and keep at 0 °C for 1 hour. Collect the crystals by filtration and wash them with 12 ml of cold ethanol. Recrystallise the product from about 290 ml of absolute ethanol, filter, wash the crystals with a little cold ethanol and dry in the air. The yield of glycylglycine ethyl ester hydrochloride, m.p. 183 °C, is 30 g (61%).

8.4.4 QUINOLINES, ISOQUINOLINES AND QUINOXALINES

The examples selected to illustrate some of the important cyclisation reactions leading to substituted quinolines and isoquinolines are quinoline [(100), X = H], 8-nitroquinoline [(100), X = NO_2], 8-hydroxyquinoline [(100), X = OH], 2-methylquinoline (101), 2,4,6-trimethylquinoline (102), 2-phenylquinoline-4-carboxylic acid (103) and 6-methylisoquinoline (104).

X = H, NO_2, OH
(100)

(101)

(102)

(103)

(104)

The quinoline derivatives (100) to (103) exemplify two important overall strategies by which the heterocyclic ring may be constructed, indicated by (105) and (106). The isoquinoline derivative (104) has been selected to illustrate the overall strategy shown in (107). It should be pointed out however that a range of other strategies have been employed for the synthesis of quinolines and isoquinolines and the reader's attention is directed to specialist texts.[27]

(105)   (106)   (107)

With quinoline [(100), X = H] the first retrosynthetic transform is reduction to 1,2-dihydroquinoline (cf. pyridines above) to facilitate recognition of the derived synthons obtained in the subsequent disconnection.

(100)   (108)

The *Skraup synthesis*, the above analysis reveals, is an important method for quinoline and many of its derivatives (e.g. 8-nitro- and 8-hydroxyquinoline). The reaction involves heating a mixture of an aromatic amine, glycerol, concentrated sulphuric acid and a suitable oxidising reagent. The original Skraup procedure employed nitrobenzene as the oxidant, in which case the reaction is frequently very vigorous and requires moderation by the addition of iron(II) sulphate. Other oxidising agents which may be used include, for example, iodine and arsenic pentoxide. The overall sequence for [(100), X = H] involves first the conversion of glycerol into acrolein by the action of sulphuric acid, then the conjugate 1,4-addition of aniline to acrolein to yield 3-phenylaminopropanal [(108), X = H], followed by an acid-catalysed cyclisation and dehydration to form the 1,2-dihydroquinoline, which is finally oxidised to quinoline.

The preparation of quinoline itself (Expt 8.39) proceeds smoothly without becoming excessively violent when iodine is employed as the oxidising agent. In the preparation of 8-nitroquinoline (from *o*-nitroaniline, Expt 8.39, cognate preparation), the oxidant is arsenic pentoxide. In the case of 8-hydroxyquinoline (oxine; Expt 8.39, cognate preparation), a pre-mix of *o*-aminophenol, glycerol and sulphuric acid held at 80 °C is added portionwise to a mixture of *o*-nitrophenol and iron(II) sulphate maintained at 100–120 °C; this procedure serves to control the vigour of the reaction. Oxine is a useful reagent in qualitative and quantitative analysis since it forms water-insoluble complexes with such metals as aluminium, magnesium, bismuth, zinc, etc.

A final point to note with the Skraup reaction is that although *ortho*- and *para*-substituted amines lead to 8-substituted or 6-substituted quinolines respectively, *meta*-substituted amines give a mixture of regioisomers, namely the 5- and 7-substituted quinolines.

In a reaction which is mechanistically related to the Skraup reaction an α,β-unsaturated carbonyl compound, generated by way of an acid-catalysed aldol condensation, reacts with a primary aromatic amine in the presence of acid to yield a quinoline derivative (*Doebner–Miller reaction*). For example, when aniline is heated with paraldehyde (which depolymerises to acetaldehyde during the reaction) in the presence of hydrochloric acid the final product is 2-methyl-quinoline (101) (quinaldine, Expt 8.40). Retrosynthetic analysis for the 1,2-dihydroquinoline reveals crotonaldedhyde as the unsaturated carbonyl component which is in turn formed from acetaldehyde (see Section 5.18.2, p. 799).

In this case the final dehydrogenation of 1,2-dihydroquinaldine to quinaldine is effected by anils formed by the condensation of aniline with either acetaldehyde or crotonaldehyde during the course of the reaction. This yields secondary amines as by-products; these together with excess aniline are separated from the quinaldine by acetylation of the reaction mixture. The acetylated primary and secondary amines thus formed are less steam volatile than quinaldine which forms the basis of the isolation of the latter.

The starting materials for the synthesis of 2-phenylquinoline-4-carboxylic acid (Expt 8.41) may be discovered by a similar retrosynthetic analysis for the 1,2-dihydro derivative.

This is an example of the Doebner synthesis of quinoline-4-carboxylic acids (cinchoninic acids), which consists of the condensation of a primary aromatic amine with pyruvic acid and an aldehyde. The mechanism is probably similar to the synthesis of quinolines discussed above and involves the intermediate formation of a dihydroquinoline derivative, which is subsequently dehydrogenated by the anils derived from the aromatic amine and the aldehyde.

A retrosynthetic analysis for 2,4,6-trimethylquinoline (102) reveals *p*-toluidine and pentane-2,4-dione. This is an example of the *Combes reaction.*

(102)

The reaction proceeds by way of the imine (109) which is converted into the enamine (110) by prototropic rearrangement before being cyclised with concentrated sulphuric acid.

(109)   (110)

Formation and isolation of the enamine are achieved by heating the amine and the diketone in xylene under reflux using a Dean and Stark apparatus until the theoretical amount of water has separated, or alternatively at 100 °C in the presence of anhydrous calcium sulphate. The latter is the most convenient and is the procedure used in Expt 8.42.

Retrosynthetic analysis for 6-methylisoquinoline (104) reveals *p*-tolu-aldehyde and α-aminoacetaldehyde.

(104)

The synthesis of isoquinolines by this overall strategy represents the *Pomeranz–Fritsch reaction*. In fact the dimethyl (or diethyl) acetal of the aminoaldehyde is used to effect the formation of the aldimine (111) before the cyclisation step. In the classical procedure rather vigorous conditions for cyclisation are required. The milder, one-pot reaction sequence which is described below, and illustrated in Expt 8.43, has considerable advantages.[28] The aldimine (111) is first treated with ethyl chloroformate which activates the imine, and then with trimethyl phosphite to form the carbamate phosphonate (112) by a type of Arbuzov reaction; subsequent treatment with titanium(IV) chloride in dichloromethane effects cyclisation with loss of the carbamate and phosphonate groups and the formation of the isoquinoline ring. Although several solvents are used during the course of the reaction sequence, the original publication indicates that chloroform may be used throughout without loss of yield.

(111)

Quinoxalines (113) transform into *o*-phenylenediamine and 1,2-dicarbonyl compounds.

For quinoaxaline itself (Expt 8.44) glyoxal is the dicarbonyl component. The interest in this synthesis is that 2,3-dihydroxy-1,4-dioxane is employed as a stable reagent equivalent for anhydrous glyoxal. It is prepared by the azeotropic removal of water from a benzene solution of aqueous glyoxal and ethylene glycol.[29]

Also included in Expt 8.44, is the preparation of 2,3-diphenylquinoxaline (114) and 3-benzyl-2-oxo-1,2-dihydroquinoxaline (115) from *o*-phenylenedi-amine and benzil or phenylpyruvic acid respectively.

Experiment 8.39 QUINOLINE

CAUTION: *This preparation should be carried out in a fume cupboard.*

Equip a 250-ml three-necked flask with a double surface condenser, a sealed stirrer and a screw-capped adapter carrying a thermometer positioned so that subsequently the temperature of the reaction mixture may be noted. Place 10.0 g (9.8 ml, 0.107 mol) of pure aniline, 15.0 g (0.163 mol) of glycerol (1) and 0.5 g of iodine in the flask. Stir the reaction mixture and add down the condenser from a dropping funnel 30 g (16.4 ml, 0.306 mol) of concentrated sulphuric acid. Reaction soon commences, the temperature rises to 100–105 °C. Heat the flask gradually, with stirring, in an air bath or oil bath to 140 °C; the reaction proceeds with the evolution of sulphur dioxide and a little iodine vapour and the liquid refluxes. Continue heating at 170 °C for 1 hour, allow to cool and then add cautiously with stirring sufficient 5 M sodium hydroxide solution (about 85 ml) to render the mixture alkaline. Rearrange the apparatus for steam distillation (Fig. 2.102) and steam distil until no more oily drops pass over. The distillate contains quinoline and a little aniline. Extract the distillate with three 25 ml portions of ether, combine the ethereal extracts and remove the ether on a rotary evaporator.

To remove the aniline present in the residual crude quinoline, advantage is taken of the fact that bis-quinolinium tetrachlorozincate(II) $[(C_9H_8N)_2^{\oplus} (ZnCl_4)^{2\ominus}]$ is almost insoluble in water and crystallises out, while under the same experimental conditions, bis-anilinium tetrachlorozincate(II) $[(C_6H_5 \cdot NH_3)_2^{\oplus} (ZnCl_4)^{2\ominus}]$ remains in solution (2). Dissolve the crude quinoline in 100 ml of dilute hydrochloric acid (1:4 by volume), warm the solution to 60 °C and add, with stirring, a solution of 13 g (0.095 mol) of zinc chloride in a 22 ml portion of the diluted hydrochloric acid. Cool the well-stirred mixture thoroughly in ice-water and, when crystallisation is complete, filter the bis-quinolinium tetrachlorozincate(II) with suction, wash with two 10 ml portions of dilute hydrochloric acid and drain well. Transfer the solid to a 250-ml beaker, add a little water and then 10 per cent sodium hydroxide solution until the initial precipitate of zinc hydroxide dissolves completely. Extract the quinoline with three 25 ml portions of ether in a separatory funnel and dry the combined ethereal extracts with anhydrous calcium sulphate. Remove the ether by flash distillation using a 10-ml flask and finally distil the residue from an air bath using an air condenser. Collect the quinoline at 236–238 °C as a colourless liquid; the yield is 6.9 g (50%). If distilled under reduced pressure quinoline has b.p. 118–120 °C/20 mmHg.

Notes. (1) Laboratory grade glycerol may be used since anhydrous glycerol is not necessary when iodine is used as the oxidising agent. Anhydrous glycerol may be prepared by heating commercial glycerol in a porcelain evaporating dish carefully over a wire gauze (preferably in a fume cupboard), stirring it steadily with a thermometer until the temperature rises to 180 °C, allowing it to cool to about 100 °C, pouring it into a Pyrex beaker and transferring the beaker to a large desiccator containing concentrated sulphuric acid. It must be remembered that glycerol is a very hygroscopic substance.

(2) *An alternative method of removing the aniline* is to add 8 ml of concentrated sulphuric acid carefully to the steam distillate, cool the solution to 0–5 °C and add a concentrated solution of sodium nitrite until a drop of the reaction mixture colours potassium iodide–starch paper a deep blue instantly. As the diazotisation approaches completion, the reaction becomes slow; it will therefore be necessary to test for excess of nitrous acid after an interval of 5 minutes, stirring all the while. About 3 g of sodium nitrite are usually required. The diazotised solution is then heated on a boiling water bath for an hour (or until active evolution of nitrogen ceases), treated with a solution of 15 g of sodium hydroxide in 50 ml of water, the mixture steam-distilled and the quinoline isolated from the distillate by extraction with ether as above.

Cognate preparations. *8-Nitroquinoline.* Place a mixture of 69 g (0.5 mol) of *o*-nitroaniline, 86 g (0.375 mol) of arsenic pentoxide (**CAUTION**) and 184 g (2 mol) of *anhydrous* glycerol in a 500-ml three-necked flask, fitted with a sealed stirrer unit, a thermometer and a reflux condenser. Set the stirrer in motion, heat to 100 °C (oil bath) and add 220 g (120 ml, 2.24 mol) of concentrated sulphuric acid gradually down the condenser at such a rate that the temperature does not rise above 120 °C (about 20 minutes). Insert a calcium chloride guard-tube into the top of the condenser, gradually raise the temperature to 130–135 °C and maintain this temperature for 7–8 hours. Watch the reaction during the first hour of heating: should the reaction become very vigorous, lower the oil bath momentarily. Allow the contents of the flask to cool and pour into 1500 ml of water contained in a 2-litre beaker. Add 15 g of decolourising carbon, stir mechanically, heat at 90 °C for 1 hour and filter. Neutralise the cold filtrate slowly with aqueous ammonia solution (1 vol concentrated ammonia, *d* 0.88 + 1 vol water), filter off the crude nitro compound at the pump and wash with a little water. Recrystallise from hot water or from methanol. The yield of 8-nitroquinoline, m.p. 92 °C, is 45 g (52%).

8-Hydroxyquinoline ('oxine'). Place 170 ml (3.18 mol) of concentrated sulphuric acid in a 1-litre three-necked flask provided with a stirrer, and add 112.5 g (1.03 mol) of *o*-aminophenol, followed by 287 g (3.12 mol) of anhydrous glycerol: maintain the temperature below 80 °C by cooling, if necessary. Keep the mixture in a fluid state by placing the flask on a steam bath.

In a 3-litre three-necked flask, fitted with a thermometer, stirrer and reflux condenser, place 72.5 g (0.52 mol) of *o*-nitrophenol and 10 g of crystallised iron(II) sulphate, and heat to 100–120 °C. Add the liquid amine–glycerol–sulphuric acid pre-mix in about ten portions over 2 hours: allow the reaction to proceed at 135–150 °C before adding the subsequent portions. Reflux the mixture for a further 4 hours, during which time the temperature drops to about 130 °C. Neutralise the cooled reaction mixture with sodium hydroxide solution (250 g in 250 ml of water) with rapid stirring and addition of ice so that the temperature does not rise above 40 °C. The pH of the resulting solution is about 7, and the 8-hydroxyquinoline together with tarry by-products precipitates. Filter the precipitate at the pump, dry at 50–60 °C and then distil under reduced pressure from a flask through a fractionating side-arm. A little water passes over first and this is followed by 8-hydroxyquinolone at 100–110 °C/5 mmHg. It crystallises on cooling to a white solid, m.p. 74–75 °C. The yield of 'oxine' is 140 g (94%).

Experiment 8.40 2-METHYLQUINOLINE (*Quinaldine*)

Place 62 g (61 ml, 0.67 mol) of aniline in a 1-litre round-bottomed flask fitted with a reflux condenser the top of which is connected to a gas absorption trap (Fig. 2.61(*a*)). Cool the flask in an ice bath, add 120 ml of concentrated hydrochloric acid slowly, followed by 90 g (2.04 mol) of paraldehyde: swirl the contents of the flask to ensure thorough mixing. Remove the flask from the ice bath and shake it frequently at room temperature during 1–2 hours. Heat cautiously to the boiling point: keep an ice-water bath at hand in case the reaction should become unduly vigorous and require moderating. Reflux the mixture for 3 hours and allow to cool. Render alkaline with about 100 ml of 12 M sodium hydroxide solution and steam distil the mixture: collect about 2.4 litres of distillate. Separate the upper oily layer, extract the aqueous phase with a little chloroform (or with ether) and combine the extract with the crude oil. Dry the combined oil and extract with magnesium sulphate, remove the solvent and heat the residue under reflux for 20 minutes with 20 ml of acetic anhydride. After cooling, render alkaline with sodium carbonate solution and steam distil; collect about 2.4 litres of distillate. Extract the latter with two 50 ml portions of toluene. Distil off the toluene from the combined toluene extracts and distil the residue with the aid of an air bath. Collect the pure quinaldine at 245–248 °C: the yield is 40 g (42%). Alternatively, distil the quinaldine under reduced pressure; b.p. 116–118 °C/12 mmHg. Keep the colourless liquid in a well-stoppered bottle since it darkens on exposure to air.

Experiment 8.41   2-PHENYLQUINOLINE-4-CARBOXYLIC ACID (*Atophan*)

In a 1-litre round-bottomed flask, equipped with a reflux condenser, place 25 g (24 ml, 0.236 mol) of purified benzaldehyde (Expt 6.133, Note (1)), 22 g (0.25 mol) of freshly distilled pyruvic acid (Expt 5.173) and 200 ml of absolute ethanol. Heat the mixture to the boiling point on a water bath and add slowly, with frequent shaking, a solution of 23 g (22.5 ml, 0.248 mol) of pure aniline in 100 ml of absolute ethanol. The addition usually occupies about 1 hour. Reflux the mixture on a water bath for 3 hours, and allow to stand over-night. Filter off the crude atophan (1) at the pump and wash the crystals with a little ether. Recrystallise from ethanol (about 20 ml per gram). The yield of pure 2-phenylquinoline-4-carboxylic acid, m.p. 210 °C, is 30 g (51%).

Note. (1) If the atophan does not crystallise – this is rarely the case unless pyruvic acid which has been standing for some time is employed – pour the reaction mixture into a solution of 25 g of potassium hydroxide in 1 litre of water, and extract the resulting solution two or three times with ether. Treat the aqueous layer with 70 ml of glacial acetic acid with vigorous stirring. Allow to stand for several hours and collect the crude atophan by filtration with suction.

Experiment 8.42   2,4,6-TRIMETHYLQUINOLINE

4-(p-Tolylamino)pent-3-en-2-one. Add 10 g of granular anhydrous calcium sulphate to a mixture of 5.32 g (0.05 mol) of p-toluidine and 5.1 g (0.05 mol) of pentane-2,4-dione (Expt 5.102) contained in a 100-ml round-bottomed flask. Attach an air condenser fitted with a calcium chloride guard-tube to the flask and heat the mixture on a steam bath for 1 hour (1), with occasional shaking. Cool, add 40 ml of ether to the reaction mixture and filter. Wash the calcium sulphate in the filter funnel with 40 ml of ether and evaporate the combined ether filtrates. Crystallise the solid so obtained from about 25 ml of hexane. The enamine is obtained as deep straw-coloured plates, m.p. 67–69 °C. The yield is 7.2 g (76%).

2,4,6-Trimethylquinoline. Add 6 g (0.032 mol) of the above enamine, in portions, to 25 ml of concentrated sulphuric acid (d 1.84) contained in a 250-ml conical flask. Swirl the mixture occasionally to ensure thorough mixing. The first portions of the enamine dissolve rather slowly but solution occurs more rapidly with the later portions as the temperature of the mixture increases to 60–70 °C. Heat the reaction mixture on the steam bath for 30 minutes, and then cool the brown solution to room temperature and add it slowly to 250 ml of ice-water in a 1-litre beaker. Add solid sodium carbonate to the solution until it is alkaline. During this addition the quinoline salt tends to separate and the whole mixture may solidify; if this happens the mass should be broken up and stirred with a stout glass rod or spatula while the sodium carbonate is added. The quinoline eventually separates as an oil from the alkaline solution. Cool the mixture in an ice-water bath until the quinoline solidifies; avoid over-cooling as this will result in the separation of a large amount of solid hydrated sodium sulphate. Collect the quinoline by filtration and wash with a little cold water. Dissolve the product in about 20 ml of hot ethanol, add decolourising carbon, filter and add warm water (c. 20 ml) to the solution until it becomes slightly cloudy. Add a few drops of ethanol to remove the cloudiness and allow the solution to cool slowly to room temperature and finally chill in the refrigerator. 2,4,6-Trimethylquinoline dihydrate separates as long glistening white needles, m.p. 63–65 °C. The yield is 5.2 g (79%). Record the p.m.r. spectrum ($D_2O + H_2SO_4$) and assign the signals δ 2.55 (s, 3H), 2.70 (s, 3H), 2.90 (s, 3H), 7.64 and 7.70 (m, 4H).

Note. (1) The progress of the reaction may be followed by removing samples of the reaction mixture and noting the disappearance of either the pentanedione or p-toluidine on g.l.c. Using a 1.5-m column of 10 per cent Silicone oil on Chromosorb W at 100 °C, with a nitrogen flow rate of 40 ml/minute, the retention times are 1.3 minutes and 6.5 minutes respectively. On the same column, at 170 °C, the enamine has a retention time of 10.5 minutes.

Experiment 8.43 6-METHYLISOQUINOLINE[28]

A benzene solution (50 ml, **CAUTION**) of equimolar amounts of p-tolualdehyde (8.4 g, 0.07 mol) and aminoacetaldehyde dimethyl acetal (7.35 g, 0.07 mol) is refluxed overnight into a Dean and Stark trap. The solution is evaporated in vacuo and then twice evaporated again with added benzene and the viscous oil dissolved in dry tetrahydrofuran. This solution is cooled to −10 °C and 1 equivalent of ethyl chloroformate (7.1 ml, 0.07 mol) is added with rapid stirring and the mixture stirred for a further 5 minutes. The cooling bath is removed and 1.2 equivalents of trimethyl phosphite (10.5 ml, 0.09 mol) is added with stirring. The solution is stirred at room temperature for 15 hours and then evaporated to an oil. The oil is then re-evaporated twice with added toluene to remove traces of trimethyl phosphite. The oil is dissolved in dry dichloromethane, 6 equivalents of titanium(IV) chloride (50 ml, 0.45 mol) is added and the solution refluxed for 36 hours under a drying tube. The cooled solution is shaken with 1 equivalent of aqueous sodium hydroxide solution to neutrality, whereupon titanium(IV) oxide precipitates as a white solid. The filtered organic layer is extracted with 3 M hydrochloric acid, and the extract is washed with dichloromethane, basified strongly with aqueous alkali, and extracted with dichloromethane. This organic extract is dried over anhydrous sodium sulphate and evaporated to afford 6-methylisoquinoline (yield 71%, m.p. 88 °C).

Experiment 8.44 QUINOXALINE[29]

2,3-Dihydroxy-1,4-dioxane. A solution of 40 per cent aqueous glyoxal (528 g, 3.62 mol) and ethane-1,2-diol (230 g, 3.70 mol) in benzene (2 litres) **(CAUTION)** is heated under reflux with efficient stirring and with azeotropic removal of water via a Dean and Stark trap. After 10 hours, removal of water is complete, and the reaction mixture is allowed to cool when it separates into two layers. The supernatant layer is decanted away, the residual amber syrup is thoroughly triturated with cold acetone (500 ml), and then refrigerated. The solid material is collected by filtration, washed with cold acetone (500 ml) and dried *in vacuo* over phosphorus pentoxide for 72 hours to give a lumpy solid, which is converted to a fine white powder by spinning in a rotary evaporator: yield 197 g (47%), m.p. 100–104 °C; i.r. (KBr disc) 3350 cm^{-1}; p.m.r. (Me$_2$SO-d_6, TMS) δ 3.20–4.05 (m, 4H), 4.32 (d, $J = 5$ Hz, 2H) and 6.45 (d, $J = 5$ Hz, 2H).

Quinoxaline. To a solution of 2,3-dihydroxy-1,4-dioxane (15.6 g, 130 mmol) in ethanol (250 ml) is added *o*-phenylenediamine (10.8 g, 100 mmol). The brown solution is stirred for 30 minutes at room temperature, at which time t.l.c. analysis (silica gel GF plates, dichloromethane–ethyl acetate 3:1 as solvent, ammonium molybdate char development) shows complete conversion to the product. The reaction mixture is evaporated, and the residue distilled to give quinoxaline, 12.4 g (95%), b.p. 220–223 °C/760 torr; the product solidifies on cooling, m.p. 29–31 °C; p.m.r. (CDCl$_3$, TMS) δ 7.70–8.05 (m, 2H), 8.10–8.35 (m, 2H) and 8.95 (s, 2H).

Cognate preparations. *2,3-Diphenylquinoxaline.* To a warm solution of 2.1 g (0.01 mol) of benzil in 8 ml of rectified spirit add a solution of 1.1 g (0.01 mol) of *o*-phenylenediamine in 8 ml of rectified spirit. Warm in a water bath for 30 minutes, add water until a slight cloudiness persists and allow to cool. Filter and recrystallise from aqueous ethanol to give 1.43 g (51%) of 2,3-diphenyl-quinoxaline, m.p. 125–126 °C.

3-Benzyl-2-oxo-1,2-dihydroquinoxaline. Dissolve 2.7 g (0.025 mol) of *o*-phenylenediamine in 40 ml of hot rectified spirit, boil the solution with 0.5 g of decolourising carbon, filter and cool to room temperature. Add a solution of 4.1 g (0.025 mol) of phenylpyruvic acid (Expt 5.175) in 20 ml of rectified spirit, swirl to mix and set aside for 2 hours. Filter, wash the crystalline product with a little cold rectified spirit and recrystallise from about 100 ml of rectified spirit to obtain the oxodihydroquinoxaline as long silky colourless needles, m.p. 203 °C. The yield is 3.2 g (54%).

8.4.5 COUMARINS (BENZO-2-PYRONES) AND CHROMONES (BENZO-4-PYRONES)

Three target molecules, namely coumarin (116), 4-methylcoumarin (117) and flavone [2-phenylbenzo-4-pyrone, (118)], are used to illustrate typical cyclisation reactions.

(116)   (117)   (118)

Overall strategies for cyclisation are of the type illustrated (119), (120) and (121).

(119)   (120)   (121)

A retrosynthetic analysis for coumarin reveals salicylaldehyde and acetic anhydride (under basic conditions) as suitable reagent equivalents of the derived synthons.

The reaction involves a spontaneous cyclisation of o-hydroxycinnamic acid obtained by a Perkin reaction using triethylamine as the base catalyst (Expt 6.138).

A more general synthesis of coumarins involves the interaction of a phenol with a β-ketoester in the presence of an acid condensing agent (the *Pechmann reaction*). In the case of 4-methylcoumarin (117) appropriate disconnection reveals ethyl acetoacetate and phenol.

(117)

Concentrated sulphuric acid is generally used as the condensing agent for simple monohydric phenols and β-ketoesters, although phenol itself reacts better in the presence of aluminium chloride (Expt 8.45). The mechanism of the reaction is thought to involve the initial formation of a β-hydroxy ester, which then cyclises and dehydrates to yield the coumarin.

Polyhydric phenols, particularly when the two hydroxyl groups are *meta* oriented, react with great ease. If sulphuric acid is used as the condensing agent, careful temperature control is needed to ensure a good yield (e.g. 4-methyl-7-hydroxycoumarin, cognate preparation in Expt 8.45). In these cases the use of polyphosphoric acid is recommended and this alternative process is illustrated. Good yields are also obtained from polyhydric phenols by condensation in the presence of trifluoroacetic acid.[30]

A retrosynthetic analysis for flavone (118) reveals *o*-benzoyloxyacetophenone (122), readily formed by the benzoylation of *o*-hydroxyacetophenone.

(118) (122)

In practice the cyclisation of (122) takes place in two stages. Initially a base-catalysed rearrangement converts (122) into *o*-hydroxydibenzoylmethane (123) which may be isolated, and then cyclised in the presence of acid to flavone (118) (Expt 8.46).

(122) (123)

(118)

Experiment 8.45 4-METHYLCOUMARIN

Place 94 g (1 mol) of phenol and 134 g (130.5 ml, 1.03 mol) of ethyl acetoacetate in 150 ml of redistilled nitrobenzene in a 3-litre three-necked flask, fitted with a dropping funnel, a sealed stirrer unit and an air condenser, the open end of which is connected to a gas absorption trap (Fig. 2.61). Heat the mixture to 100 °C in an oil bath, stir and add a solution of 266 g (2 mol) of anhydrous aluminium chloride in 1 litre of nitrobenzene (1) from the dropping funnel over a period of 45 minutes. Replace the dropping funnel by a thermometer, raise the temperature of the solution to 130 °C and maintain this temperature, with stirring, for 3 hours, by which time evolution of hydrogen chloride will have almost ceased. Cool the reaction mixture to room temperature and add 250 ml of 1:1 hydrochloric acid with stirring in order to decompose the excess of aluminium chloride. Equip the flask for steam distillation, warm it and pass steam into the reaction mixture: this will remove any unchanged keto ester and some of the nitrobenzene: collect about 100 ml of distillate. Transfer the residue in the flask while hot to a large separatory funnel; separate and discard the aqueous layer. Filter the organic layer (with the addition of a filter-aid, if necessary) through a Buchner or slit-sieve funnel to remove tarry matter and distil under reduced pressure. The nitrobenzene passes over first, followed by crude 4-methylcoumarin at 180–195 °C/15 mmHg (75 g) as a red-yellow oil which solidifies on cooling. Dissolve the crude product in ether, shake the ether solution with small volumes of 5 per cent sodium hydroxide solution until the aqueous layer is colourless. Dry, evaporate the ether and recrystallise the residue from a 4:1 mixture of light petroleum (b.p. 60–80 °C) and benzene. The resulting 4-methylcoumarin (62 g, 39%) is almost colourless and melts at 83–84 °C.

Note. (1) Add the aluminium chloride in 25 g portions to the 1 litre of dry nitrobenzene contained in a 2.5-litre round-bottomed flask; stir after each addition. The temperature may rise to about 80 °C during the addition: cool the flask occasionally under running water. When all the aluminium chloride has been added, cool the solution to room temperature: a little solid may settle to the bottom.

Cognate preparation. *4-Methyl-7-hydroxycoumarin.* Place 1 litre of concentrated sulphuric acid in a 3-litre three-necked flask fitted with a thermometer, mechanical stirrer and a dropping funnel. Immerse the flask in an ice bath. When the temperature falls below 10 °C, add a solution of 100 g (0.91 mol) of resorcinol in 134 g (130.5 ml, 1.03 mol) of redistilled ethyl acetoacetate dropwise and with stirring. Maintain the temperature below 10 °C by means of an ice–salt bath during the addition (c. 2 hours). Keep the reaction mixture at room temperature for about 18 hours, then pour it with vigorous stirring into a mixture of 2 kg of crushed ice and 3 litres of water. Collect the precipitate by suction filtration and wash it with three 25 ml portions of cold water. Dissolve the solid in 1500 ml of 5 per cent sodium hydroxide solution, filter and add dilute 2 M sulphuric acid (about 550 ml) with vigorous stirring until the solution is acid to litmus. Collect the crude 4-methyl-7-hydroxycoumarin by filtration at the pump, wash it with four 25 ml portions of cold water and dry at 100 °C: the yield is 155 g (97%). Recrystallise from 95 per cent ethanol: the pure compound separates in colourless needles, m.p. 185 °C.

Polyphosphoric acid procedure. Add 160 g of polyphosphoric acid (Section 4.2.58, p. 458) to a solution of 11 g (0.1 mol) of resorcinol in 13 g (0.1 mol) of ethyl acetoacetate. Stir the mixture and heat at 75–80 °C for 20 minutes, and then pour into ice-water. Collect the pale yellow solid by suction filtration, wash with a little cold water and dry at 60 °C. The yield of crude 4-methyl-7-hydroxycoumarin, m.p. 178–181 °C, is 17 g (97%). Recrystallisation from dilute ethanol yields the pure, colourless compound, m.p. 185 °C.

Experiment 8.46 FLAVONE (*2-Phenylbenzo-4-pyrone*)

o-**Benzoyloxyacetophenone.** Place 34 g (0.25 mol) of *o*-hydroxyacetophenone in a flask and add 49 g (40 ml, 0.35 mol) of benzoyl chloride and 50 ml of dry, redistilled pyridine. Shake to mix the contents, which become warm. After 20 minutes, pour the reaction mixture with stirring into 1200 ml of 1 M hydrochloric acid containing 500 g of crushed ice. Filter off the product with suction and wash it with 50 ml of ice-cold methanol and then with 50 ml of water. Recrystallise from methanol (60 ml), cooling thoroughly in ice before collecting the purified product by filtration. The yield is 48 g (80%), m.p. 87–88 °C.

o-**Hydroxydibenzoylmethane.** Dissolve 48 g (0.2 mol) of benzoyloxyacetophenone in 180 ml of dry pyridine in a 1-litre bolt-necked flask and heat the solution to 50 °C. Add with mechanical stirring 17 g (0.3 mol) of potassium hydroxide which has been powdered rapidly in a mortar preheated in an oven at 100 °C. Continue to stir for 15 minutes; if the separation of the yellow potassium salt of the product makes mechanical stirring impossible, stir by hand. Cool the reaction mixture to room temperature and acidify it by adding with stirring 250 ml of 10 per cent aqueous acetic acid. Collect the pale yellow precipitate by suction filtration and dry it in an oven at 50 °C. The yield of *o*-hydroxydibenzoylmethane, m.p. 117–120 °C, which is sufficiently pure for use in the next stage, is 38.5 g (80%). The pure product has m.p. 121 °C after crystallisation from methanol.

Flavone. Dissolve 36 g (0.15 mol) of *o*-hydroxydibenzoylmethane in 200 ml of glacial acetic acid in a 500-ml flask and add with shaking 8 ml of concentrated sulphuric acid. Attach a reflux condenser and heat the mixture on a boiling water bath with intermittent shaking for 1 hour. Pour the reaction mixture with stirring on to about 1 kg of crushed ice, and allow the ice to melt. Filter off the flavone which has separated, wash it with water until the washings are no longer acidic (about 2 litres are required) and dry in an oven at 50 °C. The yield of material having m.p. 95–97 °C is 31.5 (95%). Recrystallisation of a portion from a large volume of light petroleum (b.p. 60–80 °C) gives pure flavone, m.p. 98 °C, as tufts of white needles.

REFERENCES

1. T. L. Gilchrist (1985). *Heterocyclic Chemistry*. Harlow: Longman.
2. S. J. Brois (1962). *J. Org. Chem.*, **27**, 3532.
3. C. D. Maycock and R. J. Stodley (1979). *J. Chem. Soc., Perkin Trans I*, 1852.
4. M. V. Bombala and S. V. Ley (1979). *J. Chem. Soc., Perkin Trans I*, 3013.
5. E. J. Corey and M. Chaykovsky (1965). *J. Am. Chem. Soc.*, **87**, 1353.
6. R. S. Tewari, A. K. Awasthi and A. Awasthi (1983). *Synthesis*, 331.
7. C. H. Merz and G. Markl (1973). *Angew. Chem. Int. Ed. Engl.*, **12**, 845.
8. M. O. Brimeyer, A. Mehrota, S. Quici, A. Nigam and S. L. Regen (1980). *J. Org. Chem.*, **45**, 4254.
9. L. A. Pacquette and J. H. Barrett. *Organic Syntheses*, Coll. Vol V, p. 467.
10. (a) J. G. Hill, K. B. Sharpless, C. M. Exon and R. Regenye (1984). *Organic Syntheses*, **63**, 66.
    (b) R. H. Hanson and K. B. Sharpless (1986). *J. Org. Chem.*, **51**, 1922.
11. B. E. Rossiter (1985). 'Synthetic aspects and application of asymmetric epoxidation', in *Asymmetric Synthesis*. Ed. J. Morrison. Orlando: Academic Press, p. 194; M. G. Finn and K. B. Sharpless 'On the mechanism of asymmetric epoxidation with titanium-tartrate catalysts'. Ibid., p. 247.

12. H. H. Wasserman, D. J. Hlasta, A. W. Tremper and J. S. Wu (1981). *J. Org. Chem.*, **46**, 2999.
13. T. Kametani, T. Honda, A. Nakayama, Y. Sasakai, T. Mochizuki and K. Fukumoto (1981). *J. Chem. Soc., Perkin Trans I*, 2228.
14. T. Durst and M. J. O'Sullivan (1970). *J. Org. Chem.*, **35**, 2043.
15. K. Clauss, D. Grimm and G. Prossel (1974). *Liebigs Ann. Chem.*, 539; W. A. Szabo (1977). *Aldrichimica Acta*, **10**, 23; S. Mickel (1985). *Aldrichimica Acta*, **18**, 95.
16. P. H. Bentley and E. Hunt (1980). *J. Chem. Soc., Perkin Trans I*, 2222.
17. W. T. Brady, E. D. Dorsey and F. H. Parry, III (1969). *J. Org. Chem.*, **34**, 2846.
18. P. Walden (1895). *Chem. Ber.*, **28**, 1287.
19. C. F. Allen and J. VanAllan. *Organic Syntheses*, Coll. Vol. III, p. 597.
20. W. J. Houlihan, V. A. Parrino and Y. Uike (1981). *J. Org. Chem.*, **46**, 4511.
21. R. Lombard and J.-P. Stéphan (1958). *Bull. Soc. Chim. Fr.*, 1458.
22. K. Dimroth, C. Reichardt and K. Vogel. *Organic Syntheses*, Coll. Vol. V, p. 1135.
23. A. T. Balaban (1963). *C. R. Acad. Sci., Paris*, **256**, 4239.
24. T. C. Chadwick (1973). *Anal. Chem.*, **45**, 985; see also reference 21.
25. A. R. Katritzky, K. Horvath and B. Plau (1979). *Synthesis*, 437; see also A. R. Katritzky and C. M. Mason (1984). *Angew. Chem., Int. Ed. Engl.*, **23**, 420.
26. A. R. Katritzky, G. de Ville and R. C. Patel (1981). *Tetrahedron*, **37**, Suppl. 1, 25.
27. *Comprehensive Organic Chemistry* (1979). Vol. 4, *Heterocyclic Chemistry*. Ed. P. G. Sammes, Oxford: Pergamon.
28. J. B. Hendrickson and C. Rodriguez (1983). *J. Org. Chem.*, **48**, 3344.
29. M. C. Venuti (1982). *Synthesis*, 61.
30. L. L. Woods and J. Sapp (1962). *J. Org. Chem.*, **27**, 3707.

CHAPTER 9
INVESTIGATION AND CHARACTERISATION OF ORGANIC COMPOUNDS

The organic chemist is frequently faced with the problem of characterising and ultimately elucidating the structure of unknown organic compounds. The worker in the field of natural products, for example, has the prospect of isolating such compounds from their sources in a pure state and then of determining their structures. On the other hand the synthetic organic chemist encounters new, or unexpected, compounds in the course of investigations into the applicability of new reagents or techniques or as by-products of established reactions. Not infrequently a literature report of a long-forgotten investigation comes to light, and the structure of the compounds is submitted to new scrutiny based upon the current understanding of reaction mechanisms.

In each of these and similar related instances, the elucidation of the structure requires the identification of the molecular framework, the nature of the functional groups which are present and their location within the skeletal structure, and finally the establishment of any stereochemical relationships which might exist. This challenging exercise is in a sense complementary to the question posed in Chapter 1, where the key objective was 'how may this compound be efficiently synthesised, given that its structure is X?' The key objective in this chapter is 'what is the structure of this substance (solid, liquid, oil, syrup or gum)?' Just as the former objective is, for the organic chemist, rooted in developing experience and understanding of methodology, strategy, and not infrequently impressive flights of intuition, so in the elucidation of structure the experienced organic chemist can often take 'short cuts' and does not usually need to proceed by a set pattern of procedures. The experience of the chemist (based both on intuitive deduction and on hard-won 'know-how') of the interrelated aspects of the typical behaviour of organic compounds and their spectroscopic properties is often overlooked when the success of the final structural achievement is reported.

Broadly speaking such a worker is frequently selecting from the topics and areas surveyed in this chapter the most appropriate aspects ('tools') to apply to the problem in hand. The undergraduate student needs therefore to consider carefully the relative ease with which some preliminary information of structure may be achieved at the bench, and its interrelationship with that obtained from various spectroscopic data, so that in due time a comprehensive appreciation of all aspects of the physical and chemical properties of particular compounds may be gained.

It is undeniable that for the student the problem of the elucidation of structure has been revolutionised by the progressive adoption, over the last thirty years or so, of the wide range of spectroscopic techniques which are now available; these are discussed in Chapter 3 and have been applied extensively in the preparative sections to confirm the structure of the expected products. Nevertheless, a valuable background to this area is provided by the performance of appropriate investigative exercises, involving the study of selected, relatively simple 'unknown' compounds to accomplish their classification and characterisation, and most systematic courses in chemistry include practical exercises of this type. Although structural elucidation projects in the interpretation of spectra may be conducted as a tutorial exercise, most benefit will be gained when the principal spectroscopic features are considered in conjunction with the practical results obtained at the bench.

With this in mind, it is suggested that the preliminary investigations outlined in the following Sections 9.1 to 9.3 should be carefully carried out, particularly the study of solubility. Accurate classification of a compound as acidic, basic, or neutral greatly assists the subsequent search to identify the characteristic functional group or groups present. This involves a careful consideration of the results of selected classifying chemical tests in conjunction with the recognition of adsorptions of diagnostic value in the appropriate spectra. Such cross-correlation of information may obviate a diversion into a blind 'cul-de-sac' owing to inexperience in spectroscopic interpretation, resulting in an invalid conclusion.

Having established the functionality, the spectroscopic information may reveal information of skeletal features, and the compound may then be converted into an appropriate solid derivative, which if known, may confirm the identity of the compound, or provide further compounds on which reactivity and spectroscopic studies could be relevant.

Derivative preparation involves the application of a concisely described general procedure to a specific compound (possibly at this stage of uncertain identity), and its success may well depend upon the use of intuition and initiative in the preparative and purification stages. As such it must be regarded as making an important contribution to the training of the organic chemist.

9.1 PHYSICAL CONSTANTS

The most widely used physical constants in the characterisation of organic compounds are melting points and boiling points. The technique of the *determination of melting point* is discussed in detail in Section 2.33. In general, a sharp melting point (say, within 0.5 °C) is a characteristic property of a pure organic compound. The purity should not, however, be assumed but must be established by observation of any changes in the melting point (or in the melting range) when the compound is subjected to purification by recrystallisation (the technique of recrystallisation is discussed fully in Section 2.20). If the melting point is unaffected by at least one recrystallisation, then the purity of the substance may be regarded as established. In some cases purification may be effected by sublimation at atmospheric pressure or under reduced pressure (see Section 2.21). Confirmation of purity may be obtained by t.l.c. analysis (see Section 2.31).

The various methods for the experimental *determination of boiling point* of a liquid by distillation are discussed in Section 2.24 for sample sizes ranging down to about 0.5 ml. If the liquid is shown by distillation to have a wide boiling range (10–30 °C), it will be necessary to subject it to fractional distillation, or to one of the preparative chromatographic procedures, in order to obtain a reasonably pure sample of the compound; suitable procedures are to be found in Section 2.26 and 2.31. If at all possible it is desirable to submit a sample of the liquid to g.l.c. analysis to establish homogeneity. The boiling point of very small samples may be determined by Siwoloboff's method (Section 2.34).

The *determination of refractive index* is useful in assisting the characterisation of a pure liquid, particularly in the case of compounds which do not possess functional groups suitable for conversion into solid derivatives. The refractive index is conveniently determined with an Abbé refractometer (Section 2.37). An additional characteristic is provided by the density of the liquid, which is determined with the aid of a pycnometer.

9.2 THE STUDY OF SOLUBILITY BEHAVIOUR

GENERAL DISCUSSION

When a mixture of a specified amount of a given solute and a specified amount of a given solvent forms a homogeneous liquid, the former is said to be soluble in the latter. The arbitrary standard employed in this book is 0.10 g of solid or 0.20 ml of liquid to 3.00 ml of solvent. The study of the solubility behaviour of an unknown substance in various liquids, viz. water, ether, 5 per cent sodium hydroxide solution, 5 per cent sodium hydrogen carbonate solution, 5 per cent hydrochloric acid and cold concentrated sulphuric acid, may provide useful preliminary information about the nature of the compound.

The substance should be tested for solubility in the various solvents in the order cited above since, for example, when solubility in dilute acid or base is being considered, it is important to note whether the unknown is more soluble in aqueous acid or base than it is in water: this increased solubility is the positive test for a basic or acidic functional group. Acidic compounds are detected by their solubility in 5 per cent sodium hydroxide solution. Strong and weak acids are differentiated by the solubility of the former, but not the latter, in the weakly basic 5 per cent sodium hydrogen carbonate solution. Nitrogenous bases are frequently detected by their solubility in 5 per cent hydrochloric acid. Many compounds that are neutral even in concentrated aqueous acidic solutions behave as bases in strongly acidic solvents, such as concentrated sulphuric acid; these include compounds that are neutral in water and contain oxygen in any form. The presence of acidic or basic functional groups in water-soluble compounds is detected by testing their aqueous solutions with litmus or other indicator paper.

The subsequent systematic search for functional groups is based primarily on the knowledge as to whether the compound is *neutral, acidic* or *basic* in character as determined by the application of these preliminary solubility tests.

Solubility in water. Since water is a polar compound, it is a poor solvent for hydrocarbons of all types. Salts are usually extremely polar, and are generally water-soluble. Other compounds fall between these two extremes; these include alcohols, esters, aldehydes, ketones, acids, ethers, amides, nitriles and amines. Acids and amines are generally more soluble than neutral compounds.

For homologous series of mono-functional alcohols, esters, aldehydes, ketones, acids, ethers, amides, nitriles and amines, the upper limit of water solubility is found at about the member containing four carbon atoms. The solubility in water is due largely to the polar group, and as the homologous series is ascended the hydrocarbon (non-polar) part of the molecule increases while the polar function remains substantially unchanged; this accounts for the decrease in solubility in polar solvents such as water. This behaviour is an illustration of a general rule that increased structural similarity between the solute and the solvent results in increased solubility in that solvent.

It must be emphasised that the particular region (that of the member containing four carbon atoms) of water solubility for many homologous series is determined by the arbitrary proportions of solute and solvent defined in the previous general discussion. The limit would be elsewhere for a different ratio of solute to solvent.

Solubility in ether. Non-polar and slightly polar compounds will, in general, dissolve in ether because they are largely unassociated. Ionic compounds, such as salts, are not soluble in ether. The solubility of a polar compound in ether will depend upon the influence of the polar group or groups relative to that of the non-polar part of the molecule. Usually, compounds that have but one polar group per molecule will dissolve in ether unless they are highly associated or of extreme polarity (e.g. the sulphonic acids).

Many organic compounds that are insoluble in water dissolve in ether. If a compound is soluble in both ether and water, it probably (i) is non-ionic, (ii) contains five or less carbon atoms, (iii) has a functional group that is polar and capable of forming hydrogen bonds and (iv) does not contain more than one strongly polar group. If a compound dissolves in water but not in ether, it may (i) be ionic (a salt) or (ii) contain two or more polar groups but not more than four carbon atoms per polar group. There are, of course, exceptions to these statements.

Solubility in dilute hydrochloric acid. Most compounds that are soluble in dilute hydrochloric acid contain a basic nitrogen atom (incorporating an unshared electron pair) in the molecule. Thus most aliphatic amines (primary, secondary and tertiary) form salts (polar, water-soluble compounds) with hydrochloric acid. Aryl groups reduce the basicity of the nitrogen atom. Primary aromatic amines (e.g. aniline), although more weakly basic than primary aliphatic amines, are soluble, but in secondary and tertiary purely aromatic amines (e.g. diphenylamine, carbazole and triphenylamine) the basic character of the nitrogen atom has been diminished to such an extent that they do not form salts with dilute hydrochloric acid and consequently do not dissolve. Alkylarylamines (containing not more than one aryl group) and alicyclic amines, however, do dissolve. A few types of oxygen-containing compounds (such as the pyrones and the anthocyanidin pigments of certain flowers), which form oxonium salts, dissolve in dilute hydrochloric acid. Amides which are insoluble in water are generally unaffected by 5 per cent hydrochloric acid but may dissolve in higher concentrations (10–20%) of acid: this emphasises the importance of employing the correct strength of acid in the solubility tests. Many disubstituted amides which are of sufficiently high molecular weight to be water-insoluble dissolve in 5 per cent hydrochloric acid.

It may be noted that some aromatic amines react with 5 per cent hydrochloric acid to form insoluble hydrochlorides: the latter sometimes dissolve upon warming slightly and diluting with water. The appearance of the solid will usually show whether the arylamine has undergone a change: the solid should be separated and its melting point compared with that of the original compound. A test with ethanolic silver nitrate solution would indicate the formation of a hydrochloride.

Solubility in dilute sodium hydroxide solution and in dilute sodium hydrogen carbonate solution. Carboxylic acids, sulphonic acids, sulphinic acids, phenols, thiophenols, thiols, amides, arylsulphonamides, arylsulphonyl derivatives of primary amines, oximes, primary and secondary nitro compounds and some enols (e.g. of 1,3-diketones or β-keto esters) dissolve in dilute sodium hydroxide solution, i.e. they contain an acidic group of sufficient strength to react with the alkali. Carboxylic acids, sulphinic acids and sulphonic acids are soluble in dilute solutions of sodium hydrogen carbonate with the evolution of carbon dioxide; some phenols substituted with electron-withdrawing groups (for example, picric acid, 2,4,6-tribromophenol and 2,4-dinitrophenol) are strongly acidic and also dissolve in sodium hydrogen carbonate solution with the evolution of carbon dioxide. Primary and secondary nitro compounds, imides, arylsulphonamides and oximes are insoluble in sodium hydrogen carbonate solution. Some of the sodium salts of highly substituted phenols are insoluble in sodium hydroxide solution but may dissolve upon dilution and warming with water.

Certain substituents (e.g. the amino group) may markedly affect the solubility and other properties of a sulphonic acid or a carboxylic acid. Thus such sulphonic acids as the aminobenzenesulphonic acids and the pyridine- and quinoline-sulphonic acids exist in the form of inner salts or dipolar ions that result from the interaction of the basic amino group and the acidic sulphonic acid group. Sulphanilic acid, for example, is more accurately represented by formula (1) than by formula (2):

$$p\text{-}H_3\overset{\oplus}{N}\cdot C_6H_4\cdot SO_3^{\ominus} \qquad\qquad p\text{-}H_2N\cdot C_6H_4\cdot SO_3H$$
$$\qquad(1)\qquad\qquad\qquad\qquad\qquad(2)$$

These aminosulphonic acids posses the high melting points usually associated with salts, but are sparingly soluble or insoluble in water. They all dissolve readily in dilute alkali but not in dilute acid, i.e. they appear to exhibit the reactions of the sulphonic acid group but not of the amino group. The aliphatic aminocarboxylic acids (3), because of the presence of the strongly basic amino group, exist as dipolar ions (4):

$$H_2N\cdot CHR\cdot CO_2H \qquad\qquad H_3\overset{\oplus}{N}\cdot CHR\cdot CO_2^{\ominus}$$
$$\qquad(3)\qquad\qquad\qquad\qquad\qquad(4)$$

they are soluble in water but not in ether, and dissolve in both dilute acid and dilute alkali but react slowly or not at all with dilute sodium hydrogen carbonate solution. The carboxyl derivatives of the arylamines (e.g. p-aminobenzoic acid) are also amphoteric, but the diminution of the basic character of the amino group because of its attachment to the aryl group prevents the formation of inner salts to any degree; they react normally with sodium hydrogen carbonate.

Solubility in concentrated sulphuric acid. Solubility in cold concentrated sulphuric acid is used to characterise further those compounds which by virtue of the results of the previous solubility tests are considered to be neutral.

The most important group of compounds to exhibit solubility in this reagent are those containing oxygen. The initial solubility of these compounds is due to the basic character of one or more of the oxygen atoms that are present in the molecules, and results from oxonium ion formation; dilution of the sulphuric acid solution often results in recovery of the compound in an unchanged form. More fundamental changes resulting from subsequent transformations of the oxonium ion may, however, occur.

Unsaturated hydrocarbons dissolve through formation of soluble alkyl hydrogen sulphates, e.g.

$$R^1 \cdot CH{=}CH \cdot R^2 \xrightarrow{H^\oplus} [R^1 \cdot \overset{\oplus}{C}H \cdot CH_2 \cdot R^2] \xrightarrow{HSO_4^\ominus} R^1 \cdot CH(O \cdot SO_3H) \cdot CH_2R^2$$

Polyalkylated aromatic hydrocarbons and alkyl phenyl ethers undergo sulphonation, e.g.

$$ArH + 2H_2SO_4 \longrightarrow Ar \cdot SO_3H + HSO_4{}^\ominus + H_3O^\oplus$$

Compounds which dissolve in concentrated sulphuric acid may be further classified as those which are soluble in syrupy phosphoric acid and those which are insoluble in this solvent: in general dissolution takes place without the production of appreciable heat or colour. Those compounds soluble in phosphoric acid include alcohols, esters, aldehydes, methyl ketones and cyclic ketones provided that they contain less than nine carbon atoms. The solubility limit is somewhat lower than this for ethers; thus dipropyl ether dissolves in 85 per cent phosphoric acid but dibutyl ether and anisole do not. Ethyl benzoate and diethyl malonate are insoluble.

SUMMARY OF SOLUBILITY BEHAVIOUR

It is convenient to summarise the solubility characteristics of the commoner classes of organic compounds into seven groups as specified in Table 9.1. The compounds are grouped according to: (a) their solubility towards the reagents specified above, and (b) the elements, other than carbon and hydrogen, that they contain (see Section 9.3).

1. *Compounds soluble in both water and ether.* This includes the lower members of the various homologous series (4–5 atoms in a normal chain) that contain oxygen and/or nitrogen in their structures: they are soluble in water because of their low carbon content. If the compound is soluble in both water and in ether, it would also be soluble in other solvents so that further solubility tests are generally unnecessary: the aqueous solution should be tested with indicator paper. The test with sodium hydrogen carbonate solution should also be performed.

2. *Compounds soluble in water but insoluble in ether.* These compounds with the exception of *Salts*, are usually also soluble in dilute alkali and acid. The behaviour of salts to alkaline or acidic solvents may be informative. Thus, with a salt of a water-soluble base, the characteristic odour of an amine is usually apparent when it is treated with dilute alkali: likewise, the salt of a water-soluble, weak acid is decomposed by dilute hydrochloric acid or by concentrated sulphuric acid. The water-soluble salt of a water-insoluble acid or base will give a precipitate of either the free acid or the base when treated with dilute acid or dilute alkali. The salts of sulphonic acids and of quaternary bases are unaffected by dilute sodium hydroxide or hydrochloric acid.

3. *Compounds insoluble in water, but soluble in dilute sodium hydroxide.* It should be recalled that some of the compounds belonging to this group are

Table 9.1   Classification of organic compounds according to solubility behaviour

1. Soluble in both ether and water	2. Soluble in water but insoluble in ether	3. Soluble in 5% sodium hydroxide solution	4. Soluble in 5% hydrochloric acid	5. Not containing N or S. Soluble only in concentrated sulphuric acid	6. Not containing N or S. Insoluble in concentrated sulphuric acid	7. Neutral compounds containing N or S
The lower members of the homologous series of: Alcohols; Aldehydes; Ketones; Acids; Esters; Phenols; Anhydrides; Amines; Nitriles; Polyhydroxyphenols.	Polybasic acids and hydroxy acids. Glycols, polyhydric alcohols, polyhydroxy aldehydes and ketones (sugars). Some amides, amino acids, di- and polyamino compounds, amino alcohols. Sulphinic acids. Salts.	Acids. Phenols. Imides. Some primary and secondary nitro compounds; oximes. Thiols and thiophenols. Sulphonic acids, sulphinic acids, aminosulphonic acids and sulphonamides. Some diketones and β-keto esters.	Primary amines. Secondary aliphatic and aryl-alkyl amines. Aliphatic and some aryl-alkyl tertiary amines. Hydrazines.	Unsaturated hydrocarbons. Some polyalkylated aromatic hydrocarbons. Alcohols. Aldehydes. Ketones. Esters. Anhydrides. Ethers and acetals. Lactones. Acyl halides.	Saturated aliphatic hydrocarbons. Cycloalkanes. Aromatic hydrocarbons. Halogen derivatives of the above hydrocarbons. Diaryl ethers.	Nitro compounds (tertiary). Amides and derivatives of aldehydes and ketones. Nitriles. Negatively substituted amines. Nitroso, azo, hydrazo and other intermediate reduction products of nitro compounds. Sulphones, sulphonamides of secondary amines, sulphides, sulphates and other sulphur compounds.

sufficiently strong acids to release carbon dioxide from sodium hydrogen carbonate (e.g. carboxylic acids, sulphonic and sulphinic acids, and certain substituted phenols, see above).

4. *Compounds insoluble in water, but soluble in dilute hydrochloric acid.* It should be remembered that the hydrochlorides of some bases are sparingly soluble in cold water and one should therefore not be misled by an apparent insolubility of a compound (containing nitrogen) in dilute hydrochloric acid. The suspension in dilute hydrochloric acid should always be filtered and the filtrate made alkaline. A precipitate will indicate that the compound is indeed a base and should be included in this group.

5. *Water-insoluble hydrocarbons and oxygen compounds that do not contain N or S and are soluble in cold concentrated sulphuric acid.* Any changes – colour, excessive charring, evolution of gases or heat, polymerisation and precipitation of an insoluble compound – attending the dissolution of the substance should be carefully noted.

   Alcohols, esters (but not ethyl benzoate, diethyl malonate or diethyl oxalate), aldehydes, methyl ketones and cyclic ketones containing less than nine carbon atoms as well as ethers containing less than seven carbon atoms are also soluble in 85 per cent phosphoric acid.

6. *Compounds, not containing N or S, insoluble in concentrated sulphuric acid.* This test provides for a differentiation *inter alia* between alkanes and cycloalkanes and also simple aromatic hydrocarbons which are insoluble, and unsaturated hydrocarbons which are soluble in the reagent.

7. *Compounds that contain N or S which are not in groups 1–4;* many of the compounds in this group are soluble in concentrated sulphuric acid.

   It will be observed that halogen compounds are not listed separately, but appear in each of the seven categories in accordance with their solubility behaviour.

PROCEDURE FOR SOLUBILITY TESTS

All solubility determinations are carried out at the laboratory temperature in small test tubes (e.g. 100×12 mm) but of sufficient size to permit vigorous shaking of the solvent and the solute.

Amount of material required. It is convenient to employ an arbitrary ratio of 0.10 g of solid or 0.20 ml of liquid for 3.0 ml of solvent. Weigh out 0.10 g of the *finely powdered* solid to the nearest 0.01 g: after some experience, subsequent tests with the *same* compound may be estimated by eye. Measure out 0.20 ml of the liquid either with a calibrated dropper or a small graduated pipette. Use either a calibrated dropper of a graduated pipette to deliver 3.0 ml of solvent.

Much time will be saved if each of the solvents (water, ether, 5% sodium hydroxide, 5% sodium hydrogen carbonate and 5% hydrochloric acid) be contained in a 30- or 60-ml reagent bottle fitted with a ground-in calibrated dropper.

Solubility in water. Treat a 0.10 g portion of the solid with successive 1.0 ml portions of water, shaking vigorously after each addition, until 3.0 ml have been added. If the compound does not dissolve completely in 3.0 ml of water, it may be regarded as insoluble in water. When dealing with a liquid, add 0.20 ml of the compound to 3.0 ml of water and shake. In either case, test the contents of the small test-tube with Universal indicator paper: it is best to remove a little of the solution or supernatant liquid with a dropper. It is usually convenient at this stage to test for the presence of water-soluble enolic compounds by observing any coloration resulting from the addition of neutral aqueous iron(III) chloride solution.

Solubility in ether. Use 0.10 g of solid or 0.20 ml of a liquid in a dry test tube and proceed exactly as in testing the solubility in water, but do not employ more than 3.0 ml of solvent.

Solubility in 5 per cent sodium hydroxide solution. Note whether there is any rise in temperature. If the compound appears insoluble, remove some of the super-natant liquid by means of a dropper to a semimicro test tube (75 × 10 mm), add 5 per cent hydrochoric acid dropwise until acid and note whether any precipit-ate (or turbidity) is formed. The production of the latter will confirm the pres-ence of an acidic compound.

Solubility in 5 per cent sodium hydrogen carbonate solution. If the compound is soluble in 5 per cent sodium hydroxide solution, test its solubility in a 5 per cent solution of sodium hydrogen carbonate. Observe whether it dissolves and par-ticularly whether carbon dioxide is evolved either immediately (carboxylic acids, sulphonic acids, negatively substituted phenols) or after a short time (some amino acids).

Solubility in 5 per cent hydrochloric acid. Add the acid to 0.10 g of the solid or 0.20 ml of the liquid in quantities of 1.0 ml until 3.0 ml have been introduced. Some organic bases form hydrochlorides that are soluble in water but are pre-cipitated by an excess of acid: if solution occurs at any time, the unknown is classified as a basic compound. If the compound appears insoluble, remove some of the supernatant liquid by means of a dropper to a semimicro test tube (75 × 10 mm), and add 5 per cent sodium hydroxide solution until basic and ob-serve whether any precipitate is produced: the formation of a precipitate will confirm that the compound is a base.

Solubility in concentrated sulphuric acid. Place 3.0 ml of pure concentrated sul-phuric acid in a dry test tube and add 0.10 g of a solid or 0.20 ml of a liquid. If the compound does not dissolve immediately, agitate for some time but do not heat. Observe any change in colour, charring, evolution of gaseous products, poly-merisation accompanied by precipitation, etc.

Solubility in syrupy phosphoric acid. This test should only be applied if the com-pound is soluble in concentrated sulphuric acid. Place 3.0 ml of 85 per cent orthophosphoric acid in a dry test tube and add 0.10 g of a solid or 0.20 ml of a liquid. If the compound does not dissolve immediately, agitate for some time but do not boil.

9.3 DETECTION OF ELEMENTS PRESENT

The most commonly occurring elements in organic compounds are carbon, hydrogen, oxygen, nitrogen, sulphur and the halogen elements; less common ele-ments are phosphorus, arsenic, antimony, mercury or other metals which may be present as salts of organic acids. There is no direct method for the detection of oxygen.

There are three straightforward tests which will provide useful information as to the general properties of the compound, and the presence of the more important of the elements noted above other than carbon, hydrogen and oxygen; these tests are: the ignition test; heating with soda-lime; and the sodium fusion test (*Lassaigne's test*).

IGNITION TEST
Procedure. Place about 0.1 g of the compound on a metal spatula. Heat it gently at first and finally to dull redness. Observe:
(a) whether the substance melts, is explosive or is flammable and note the nature of the flame;
(b) whether gases or vapour are evolved, and their odour (**CAUTION**);
(c) whether the residue fuses.

Evidence of the organic nature of the compound, i.e. the presence of *carbon and hydrogen* (1), and the absence of metals, is indicated by burning with a more or less smoky flame, and a black residue consisting largely of carbon, which 'burns off' on prolonged heating, leaving no residue. In general, aromatic compounds characteristically burn with a very smoky flame.

If an appreciable amount of residue remains, note its colour. Add a few drops of water and test the solution (or suspension) with Universal indicator paper. Then add a little dilute hydrochloric acid and observe whether effervescence occurs and the residue dissolves. Apply a flame test with a platinum wire, which may determine the metal present.

Note. (1) If it is desired to test *directly* for carbon and hydrogen in a compound, mix 0.1 g of the substance with 1–2 g of freshly ignited, fine copper (II) oxide powder in a dry test tube, and fit the latter with a bung carrying a tube bent at an angle so that escaping gases can be bubbled below the surface of lime water contained in a second test tube. Clamp the test tube near the top and heat the mixture gradually. If carbon is present, carbon dioxide will be evolved which will produce a turbidity in the lime water. If hydrogen is present, small drops of water will collect in the cooler part of the tube.

HEATING WITH SODA-LIME
This is often a useful preliminary test. Mix thoroughly about 0.2 g of the substance with about 1 g of powdered soda-lime. Place the mixture in a Pyrex test tube; close the tube by a bung and delivery tube. Incline the test tube so that any liquid formed in the reaction cannot run back on the hot part of the tube. Heat the test tube gently at first and then more strongly. Collect any condensate produced in a test tube containing 2–3 ml of water. Nitrogenous compounds will usually evolve ammonia or vapours alkaline to indicator paper and possessing characteristic odours; hydroxybenzoic acids yield phenols; formates and acetates yield hydrogen; simple carboxylic acids yield hydrocarbons (methane from acetic acid, benzene from benzoic or phthalic acid, etc.); amine salts and aromatic amino carboxylic acids yield aromatic amines, etc.

SODIUM FUSION TEST
In order to detect *nitrogen, sulphur and halogen* in organic compounds, it is necessary to convert them into ionisable inorganic substances so that ionic tests of inorganic analysis may be applied. This may be accomplished by several methods, but the best procedure is to fuse the organic compound with metallic sodium (Lassaigne's test). In this way sodium cyanide, sodium sulphide and

sodium halides, which are readily identified, are formed respectively if the above elements are present. It is essential to use an excess of sodium, otherwise if sulphur and nitrogen are both present sodium thiocyanate, NaCNS, may be produced; this may give a red coloration with iron(III) ions but will not of course respond to tests for cyanide or sulphide ions. With excess of sodium the thiocyanate, if formed, will be decomposed thus:

$$NaCNS + 2Na \longrightarrow NaCN + Na_2S$$

Cyanide ion, and hence nitrogen in the sample, may be detected by the Prussian Blue test. The filtered alkaline solution, resulting from the action of water upon the sodium fusion, is treated with iron(II) sulphate and thus forms sodium hexacyanoferrate(II). Upon boiling the alkaline iron(II) salt solution, some iron(III) ions are inevitably produced by the action of air; upon addition of dilute sulphuric acid, thus dissolving the iron(II) and (III) hydroxides, the hexacyanoferrate(II) reacts with the iron(III) salt producing iron(III) hexacyanoferrate(II), Prussian blue:

$$FeSO_4 + 6NaCN \longrightarrow Na_4[Fe(CN)_6] + Na_2SO_4$$
$$3Na_4[Fe(CN)_6] + 2Fe_2(SO_4)_3 \longrightarrow Fe_4[Fe(CN)_6]_3 + 6Na_2SO_4$$

Hydrochloric acid should not be used for acidifying the alkaline solution since the yellow colour, due to the iron(III) chloride formed, causes the Prussian blue to appear greenish. For the same reason, iron(III) chloride should not be added – as is frequently recommended: a sufficient concentration of iron(III) ions is produced by atmospheric oxidation of the hot alkaline solution. An alternative, sensitive, test for cyanide ion is the formation of a deep purple coloration when the fusion solution is treated with an alkaline p-nitrobenzaldehyde–o-dinitrobenzene reagent.

Sulphur, as sulphide ion, may be detected by precipitation as black lead sulphide with lead acetate solution and acetic acid or by the purple colour produced on addition of disodium pentacyanonitrosoferrate(III). Halogens are detected as the characteristic silver halides by the addition of dilute nitric acid and silver nitrate solution. Cyanide and sulphide ions both interfere with this test for halide by forming silver cyanide and silver sulphide precipitates. If nitrogen or sulphur has been detected, therefore, the interfering ions must be removed by boiling the acidified fusion solution as detailed later, before the silver nitrate solution is added to detect the halogen.

Procedure for Lassaigne's test. Support a Pyrex test tube (150 × 12 mm) vertically in a clamp lined with sheet cork. Place a cube (c. 4 mm = 0.04 g) of freshly cut sodium in the tube (*for precautions in the use of sodium see Section 4.2.68, p. 462*). Have in readiness about 0.05 g of the compound (if a solid) on a microspatula or tip of a knife blade, or about 3 drops of the compound (if a liquid) in a dropping pipette. Heat the tube steadily until sodium vapour, which is *dark grey* in colour, rises 2–3 cm in the test tube and drop the sample, preferably portionwise, directly on to the molten sodium (**CAUTION**: *there may be a slight explosion*). Do not allow the sample to fall on to the side of the tube above the vapour layer. Heat the tube to redness for about 2 minutes, and then allow it to cool. Add about 3 ml of methanol to decompose any unreacted sodium and then halffill the tube with distilled water. Boil gently for a few minutes to remove the methanol and extract inorganic salts from the fusion residue; *gently* crush the residue with a glass rod to assist the extraction process. Filter and use the clear, colourless filtrate for the various tests detailed below. If the filtrate is dark coloured, decomposition was probably incomplete: repeat the entire fusion procedure. Keep the test tube for further sodium fusions; it will become discoloured and should be cleaned from time to time with a little scouring powder.

Test for nitrogen. Either of the following methods may be used.

1. Pour 2–3 ml of the filtered fusion solution into a test tube containing 0.1–0.2 g of powdered iron(II) sulphate crystals. Heat the mixture gently with shaking until it boils, then, without cooling, add just sufficient dilute sulphuric acid to dissolve the iron hydroxides and give the solution an acid reaction. {The addition of 1 ml of 5% potassium fluoride solution is beneficial (possibly owing to the formation of potassium hexafluoroferrate(III), $K_3[FeF_6]$) and usually leads to a purer Prussian blue.} A Prussian blue precipitate or coloration indicates that nitrogen is present. If no blue precipitate appears at once, allow to stand for 15 minutes, filter through a small filter and wash the paper with water to remove all traces of coloured solution: any Prussian blue present will then become perceptible in the cone of the filter paper. If in doubt, repeat the sodium fusion, preferably using a mixture of the compound with pure sucrose or naphthalene. In the absence of nitrogen, the solution should have a pale yellow colour due to iron salts.

    If sulphur is present, a *black* precipitate of iron(II) sulphide is obtained when the iron(II) sulphate crystals dissolve. Boil the mixture for about 30 seconds, and acidify with dilute sulphuric acid; the iron(II) sulphide dissolves and a precipitate of Prussian blue remains if nitrogen is present.

2. Mix together 5 drops of a 1.5 per cent solution of *p*-nitrobenzaldehyde in 2-methoxyethanol, 5 drops of a 1.7 per cent solution of *o*-dinitrobenzene in 2-methoxyethanol and 2 drops of a 2.0 per cent aqueous solution of sodium hydroxide. To this mixture add 1 drop of fusion solution. A deep purple coloration is positive for cyanide ions, a yellow or tan colour is negative. Neither halide ions nor sulphide ions interfere.

Test for sulphur. This element may be tested for by either of the following two methods.

1. Acidify 2 ml of the fusion solution with dilute acetic acid, and add a few drops of lead acetate solution. A black precipitate of lead sulphide indicates the presence of sulphur.

2. To 2 ml of the fusion add 2–3 drops of a freshly prepared dilute solution (c 0.1%) of di-sodium pentacyanonitrosyl ferrate $Na_2[Fe(CN)_5NO]$. (The latter may be prepared by adding a minute crystal of the solid to about 2 ml of water.) A purple coloration indicates sulphur; the coloration slowly fades on standing.

Test for halogens. *Nitrogen and sulphur absent.* 1. Acidify a portion of the fusion solution with dilute nitric acid and add an excess of silver nitrate solution. A precipitate indicates the presence of a halogen. Decant the mother-liquor and treat the precipitate with dilute aqueous ammonia solution. If the precipitate is white and readily soluble in the ammonia solution, chlorine is present; if it is pale yellow and difficultly soluble, bromine is present; if it is yellow and insoluble, then iodine is indicated. Iodine and bromine may be confirmed by tests (2) or (3); these tests may also be used if it is suspected (e.g. from be-haviour in the silver nitrate test) that more than one halogen is present.

2. Acidify 1–2 ml of the fusion solution with a moderate excess of glacial acetic acid and add 1 ml of dichloromethane. Then introduce 20 per cent sodium nitrite solution drop by drop with constant shaking. A purple or violet colour in the organic layer indicates the presence of *iodine*. The reaction is:

$$2NaI + 2NaNO_2 + 4Me \cdot CO_2H \longrightarrow I_2 + 2NO + 4Me \cdot CO_2Na + 2H_2O$$

This solution may also be employed in the test for *bromine*. If iodine has been found, add further additional quantities of sodium nitrite solution, warm and by means of a dropper pipette remove and replace the organic phase with fresh portions of dichloromethane; repeat until the organic phase is colour-less. Boil the acid solution until no more nitrous fumes are evolved and cool. Add a small amount of lead dioxide, place a strip of fluorescein paper (1) across the mouth of the tube and warm. If bromine is present, it will colour the test paper rose-pink (eosin is formed). If *iodine* has been found to be *absent* use 1 ml of the fusion solution, acidify strongly with glacial acetic acid, add lead dioxide and proceed as above.

In this test for bromine, lead dioxide in acetic acid solution gives lead tetra-acetate which oxidises hydrogen bromide (and also hydrogen iodide), but has practically no effect under the above experimental conditions upon hydrogen chloride:

$$2NaBr + PbO_2 + 4Me \cdot CO_2H \longrightarrow$$
$$Br_2 + (Me \cdot CO_2)_2Pb + 2Me \cdot CO_2Na + 2H_2O$$

To test for *chlorine* in the presence of iodine and/or bromine, acidify 1–2 ml of the fusion solution with glacial acid, add a slight excess of lead dioxide (say, 0.5 g) and boil gently until all the iodine and bromine is liberated. Dilute, filter off excess lead dioxide and test for chloride ions with dilute nitric acid and silver nitrate solution.

3. Acidify 1–2 ml of the fusion solution with dilute sulphuric acid, cool and add 1 ml of dichloromethane. Prepare the equivalent of 'chlorine water' by acidifying 10 per cent sodium hypochlorite solution with one-fifth of its volume of dilute hydrochloric acid. Add this solution dropwise with vigorous shaking to the mixture. If *iodine is present* the organic phase first becomes purple in colour. As the addition of chlorine water is continued, the purple colour disappears (owing to oxidation of iodine to iodate) and, if *bromine is present*, is replaced by a brown or reddish colour. If bromine is absent, the organic layer will be colourless. It is, of course, evident that if the dichloro-methane layer remains uncoloured and the results of test (1) were positive, the halogen present is *chlorine*.

Nitrogen and/or sulphur present. To remove cyanide and sulphide ions, make 2–3 ml of the fusion solution just acidic with dilute nitric acid, and evaporate to half of the original volume in order to expel hydrogen cyanide and/or hydrogen sulphide which may be present (**CAUTION**). Dilute with an equal volume of water and proceed as in test (1), (2) and (3) above.

Alternatively, add 1–2 drops of 5 per cent nickel(II) nitrate solution to 2–3 ml of the fusion solution, filter off the nickel(II) cyanide and/or nickel sulphide, acidify the filtrate with 2 M nitric acid and test for halides as above.

The presence of halogens may be further confirmed by the *Beilstein test*. This test serves to detect the presence of halogen in many organic compounds. In consists in heating the substance in contact with pure copper oxide in the Bunsen flame: the corresponding copper halide is formed, which, being volatile, imparts an intense green or bluish-green colour to the mantle of the flame.

Push one end of a 20-cm length of stout copper wire into a cork (this will serve as a holder); coil the other end by making two or three turns about a thin glass rod. Heat the coil in the outer mantle of a Bunsen flame until it ceases to impart any colour to the flame. Allow the wire to cool somewhat and, while still warm, dip the coil into a small portion of the substance to be tested and heat again in the non-luminous flame. If the compound contains a halogen element, a green or bluish-green flame will be observed (usually after the initial smoky flame has disappeared). Before using the wire for another compound, heat it until the material from the previous test has been destroyed and the flame is not coloured.

It has been stated that many halogen-free compounds, e.g. certain derivatives of pyridine and quinoline, purines, acid amides and cyano compounds, when ignited on copper oxide impart a green colour to the flame, presumably owing to the formation of volatile copper cyanide. The test is therefore not always trustworthy. The test is not given by fluorides since copper fluoride is not volatile.

The detection of the following elements, which occur in organic less frequently compounds, is included here for the sake of completeness.

Test for fluorine. Use either of the following tests.

1. Strongly acidify about 2 ml of the fusion filtrate with glacial acetic acid, and boil until the volume is reduced to about one-half. Cool. Place one drop of the solution upon zirconium-alizarin red S test paper (2). A yellow colour on the red paper indicates the presence of fluoride. Large amounts of sulphates and phosphates may interfere with this test.

2. If nitrogen and/or sulphur is present, acidify 3–4 ml of the fusion solution with dilute nitric acid and evaporate to half the original volume in order to expel any HCN and/or H_2S which may be present. If nitrogen and sulphur are absent, proceed directly with 2 ml of the sodium fusion filtrate. Render the solution just neutral to litmus by the addition of dilute (5 M) aqueous ammonia solution, then add 5 drops of 5 M acetic acid and 20 mg of lanthanum chloranilate (2,5-dichloro-3,6-dihydroxy-*p*-benzoquinone, lanthanum salt) and shake intermittently for 10–15 minutes. Filter. A pink-violet coloration of the filtrate is a positive test for fluorine.

Test for phosphorus. The presence of phosphorus may be indicated by a smell of phosphine during the sodium fusion and the immediate production of a jet-black colour when a piece of filter paper moistened with silver nitrate solution is placed over the mouth of the ignition tube after the sample has been dropped on the hot sodium. Treat 1.0 ml of the fusion solution with 3 ml of concentrated nitric acid and boil for 1 minute. Cool and add an equal volume of ammonium molybdate reagent (3). Warm the mixture to 40–50 °C, and allow to stand. If phosphorus is present, a yellow crystalline precipitate of ammonium 12-molybdophosphate, $(NH_4)_3[PMo_{12}O_{40}]$, will separate.

It is usually preferable to oxidise the compound directly as follows. Intimately mix 0.02–0.05 g of the compound with 3 g of sodium peroxide and 2 g of anhydrous sodium carbonate in a nickel crucible. Heat the crucible and its contents with a small flame, gently at first, afterwards more strongly until the contents are fused, and continue heating for a further 10 minutes. Allow to stand, extract the contents of the crucible with water and filter. Add excess of concentrated nitric acid to the filtrate and test with ammonium molybdate reagents as above. A yellow precipitate indicates the presence of phosphorus. It must be borne in mind that the above treatment will convert any arsenic present into arsenate.

Notes. (1) Fluorescein test paper is prepared by dipping filter papers into a dilute solution of fluorescein in ethanol; it dries rapidly and is then ready for use. The test paper has a lemon yellow colour.

(2) Prepare the zirconium–alizarin red S paper as follows. Soak dry filter paper in a 5 per cent solution of zirconium nitrate in 5 per cent hydrochloric acid and, after draining, place it in a 2 per cent aqueous solution of sodium alizarin sulphonate (BDH 'Alizarin Red S'). The paper is coloured red-violet by the zirconium lake. Wash the paper until the wash water is nearly colourless and then dry in the air.

(3) Ammonium molybdate reagent may be prepared by dissolving 45 g of pure ammonium molybdate in a mixture of 40 ml of aqueous ammonia (d 0.88) and 60 ml of water and then adding 120 g of ammonium nitrate and diluting the solution to 1 litre with water.

9.4 SPECTROSCOPIC INFORMATION

The techniques for the determination of the spectroscopic features of organic compounds, and of their relationship to the presence of skeletal and functional group structures are considered in Chapter 3. Chapters 5 and 6 provide spectroscopic summaries of classes of compounds, and more detailed spectroscopic information on specific known compounds as examples in which the interpretation of spectra may be further explored.

When considering the use of spectra for the classification of an unknown compound with the aim of elucidating its structure, some careful judgement on the relative merits of each spectroscopic tool is required.

In general, the main value of *i.r.* spectroscopic features is for the detection of functional groups, although the information on skeletal structure is clearly of importance. To use a very simple illustration, the technique would clearly distinguish between functional isomers (e.g. alcohols and ethers), or aromatic and aliphatic compounds, but would be inadequate, for example, to distinguish between adjacent straight-chain members of a homologous series of four carbons upwards of a particular functional type.

With *n.m.r.* spectroscopy (^{13}C and ^{1}H), the main value lies in the provision of information on skeletal structure (e.g. aliphatic straight or branched chain, aromatic, heterocyclic, etc.); confirmatory evidence on the nature of the functional group is deduced from a consideration of the chemical shift values, and from the definitive information in the case of groups having replaceable hydrogens, or groups with highly deshielded protons.

The *m.s.* data, particularly now that recent techniques give a 'best-fit' elemental composition of molecular ions and fragment ions which facilitates speculation of fission processes, provides in association with the above techniques, further information of structural value. Polyhalogenated compounds are particularly easily handled by this technique.

The principal merit of *u.v.–visible* spectroscopy is in the assessment of the degree of conjugation (not readily deduced from the previous methods) in polyenes, polyenynes, and polyenones, and in the recognition of the presence of aromatic structures.

Chapter 3 includes a recommended sequence for the systematic examination of *i.r.* spectra (p. 271), *p.m.r.* spectra (p. 359) and *m.s.* (p. 373).

9.5 FUNCTIONAL GROUP IDENTIFICATION

The solubility behaviour of an unknown compound will serve to classify it into one of the three main divisions, namely, *acidic, basic* or *neutral*. This information, supplemented by elemental analysis if deemed necessary, and as noted above cross-correlated with spectroscopic inferences, forms the basis for the subsequent systematic search to identify definitively the functional group or groups present. It cannot be too clearly emphasised that inexperience in spectroscopic interpretation can lead to erroneous conclusions of structure. The value of chemical tests is that they reduce the chance of this happening, furthermore they are frequently easily and quickly performed and provide experience in accurate and reliable observation and reporting.

When an organic compound contains more than one functional group, the classification is generally based upon the one which is most easily detected and manipulated. Thus benzoic acid, *p*-chlorobenzoic acid, *p*-methoxybenzoic acid (anisic acid) and *p*-nitrobenzoic acid will be classified as *acids* both by the solubility tests and the class reactions, and the identifiction of, say, the nitrogen-containing acid *may* be completed by the preparation of derivatives of the carboxyl group without the absolute necessity of applying the class reactions that would discover the nitro group; however, if possible, it is always desirable to establish the nature of the subsidiary functional group or groups (including the presence of unsaturation), the presence of which may become apparent when the preliminary identification of the compound has been completed.

9.5.1 ACIDIC COMPOUNDS

The principle classes of acidic compounds are listed in Table 9.1 (Column 3). The distinction between true acids and the weakly acidic pseudo acids (e.g. phenols, enols, nitroalkanes) should be made by observing the nature of the reaction with sodium hydrogen carbonate. To ensure that evolution of carbon dioxide does not go unnoticed in those cases where reaction appears sluggish, add a solution of the compound in methanol carefully to a saturated solution of sodium hydrogen carbonate solution, when a vigorous effervescence at the interface will be observed.

CARBOXYLIC ACIDS

The presence of a carboxylic acid group is indicated by strong infrared absorption in the region of $1720 \, \text{cm}^{-1}$ (C=O str.) and broad absorption between $3400 \, \text{cm}^{-1}$ and $2500 \, \text{cm}^{-1}$ (OH str.); in the nuclear magnetic resonance spectrum the acidic hydrogen (replaceable by D_2O) will appear at very low field (δ 10-13).

1211

Confirmatory test for carboxylic acids. *Ester formation.* Warm a small amount of the acid with 2 parts of absolute ethanol and 1 part of concentrated sulphuric acid for 2 minutes. Cool, and pour cautiously into aqueous sodium carbonate solution contained in an evaporating dish, and smell immediately. An acid usually yields a sweet, fruity smell of an ester. (Acids of high molecular weight often give almost odourless esters, however.)

For derivative preparations for carboxylic acids see Section 9.6.15, p. 1261.

CARBOXYLIC ACID HALIDES
With these compounds the presence of the halogen will have been detected in the tests for elements. Most acid halides undergo ready hydrolysis with water to give an acidic solution and the halide ion produced may be detected and confirmed with silver nitrate solution. The characteristic carbonyl adsorption at about $1800 \, \text{cm}^{-1}$ in the infrared spectrum will be apparent. Acid chlorides may be converted into esters as a confirmatory test: to 1 ml of absolute ethanol in a dry test tube add 1 ml of the acid chloride dropwise (use a dropper pipette; keep the mixture cool and note whether any hydrogen chloride gas is evolved). Pour into 2 ml of saturated salt solution and observe the formation of an upper layer of ester; note the odour of the ester. Acid chlorides are normally characterised by direct conversion into carboxylic acid derivatives (e.g. substituted amides) or into the carboxylic acid if the latter is a solid (see Section 9.6.16, p. 1265).

CARBOXYLIC ACID ANHYDRIDES
The simpler examples are readily hydrolysed in aqueous solution, and therefore react with sodium hydrogen carbonate and also give the ester test; they may be confirmed by applying the hydroxamic ester test (Section 9.5.3, p. 1222). Carbonyl adsorption is apparent in the infrared spectrum at about $1820 \, \text{cm}^{-1}$ and at about $1760 \, \text{cm}^{-1}$. It should be noted that aromatic anhydrides and higher aliphatic anhydrides are not readily hydrolysed with water and are therefore effectively neutral (Section 9.5.3, p. 1218). The final characterisation of the acid anhydride is achieved by conversion into a crystalline carboxylic acid derivative as for acid halides.

SULPHONIC AND SULPHINIC ACIDS
These compounds are suggested if sulphur is present. If nitrogen is also present the compound may be an aminosulphonic acid. The infrared spectrum will show absorption at $3400–3200 \, \text{cm}^{-1}$ (OH str.) and 1150 and $1050 \, \text{cm}^{-1}$ (S=O str. in a sulphonic acid) or at $1090 \, \text{cm}^{-1}$ (S=O str. in a sulphinic acid). For derivative preparations for sulphonic acids see Section 9.6.26, p. 1284. The presence of an aromatic sulphinic acid may be further confirmed by dissolving in cold concentrated sulphuric acid and adding one drop of phenetole or anisole when a blue colour is produced (Smiles's test), due to formation of a *para*-substituted aromatic sulphoxide. The reaction is:

$$\text{Ar·SO}_2\text{H} + \text{PhOR} \longrightarrow \text{Ar·SO·C}_6\text{H}_4\text{OR} + \text{H}_2\text{O}$$

Aromatic sulphinic acids are oxidised by potassium permanganate to sulphonic acids and are reduced by zinc and hydrochloric acid to thiophenols.

PHENOLS
These do not usually liberate carbon dioxide from 5 per cent sodium hydrogen

carbonate solution; most are crystalline solids although notable exceptions are
m-cresol and *o*-bromophenol. The monohydric phenols generally have characteristic odours. The solubility in water increases with the number of hydroxyl
groups in the molecule. The infrared spectrum shows broad strong absorption at
3400–$3200\,\text{cm}^{-1}$ (OH str.).

Confirmatory tests for phenols. *Iron(III) chloride solution.* Dissolve about $0.05\,\text{g}$
of the compound in $5\,\text{ml}$ of water; if the compound is sparingly soluble, prepare
a hot saturated aqueous solution, filter and use $1\,\text{ml}$ of the cold filtrate. Place the
solution in a test tube and add 1 drop of neutral 1 per cent iron(III) chloride solution and observe the colour; add another drop after 2–3 seconds. If a transient or
permanent coloration (usually purple, blue or green) other than yellow or
orange is observed, the substance is probably a phenol (or an enol). If no coloration is obtained, repeat the test as above but substitute absolute ethanol or
methanol for water as solvent.

Prepare the *neutral ferric chloride* solution (i.e. free from hydrochloric acid)
by adding dilute sodium hydroxide solution to the bench reagent until a slight
precipitate of iron(III) hydroxide is formed. Filter off the precipitate and use the
clear filtrate for the test.

Bromine water. Many phenols (with the exception of those with strong reducing
properties) yield crystalline bromo compounds on the addition of bromine
water. Dissolve or suspend $0.25\,\text{g}$ of the compound in $10\,\text{ml}$ of dilute hydrochloric acid or of water, and add bromine water dropwise until decolourisation
is slow: a white precipitate of the bromophenol may form.

Phthalein test. Many phenols yield phthaleins, which give characteristic colorations in alkaline solution, when fused with phthalic anhydride and a little concentrated sulphuric acid. Place in a dry test tube $0.5\,\text{g}$ of the compound and an
equal bulk of pure phthalic anhydride, mix well together and add 1 drop of concentrated sulphuric acid. Stand the tube for 3–4 minutes in a small beaker of
Silicone oil (or paraffin oil) previously heated to $160\,^\circ\text{C}$. Remove from the bath,
allow to cool, add $4\,\text{ml}$ of 5 per cent sodium hydroxide solution and stir until the
fused mass has dissolved. Dilute with an equal volume of water, filter and
examine the colour of the filtrate against a white background: if the solution exhibits a fluorescence, observe the colour against a black background.

It must be borne in mind that there are many nitrogen-containing phenols
and acids; of these the nitro and amino derivatives are the most common. The
influence of the nitro and other groups in the *ortho-* and *para*-positions upon the
acidity of a phenol has already been noted: such groups tend to produce a
marked deepening in the colour of alkaline solutions of the phenol. Amino substituents in water-insoluble phenols and acids cause these compounds to be
soluble in both dilute acid and dilute alkali, i.e. to be amphoteric. Frequently it is
helpful to destroy the basic character of the nitrogen by conversion of the amino
group into a neutral amide group by acetylation or benzoylation in aqueous
alkaline solution: the resulting compound is not amphoteric.

For derivative preparations for phenols see Section 9.6.6, p. 1248.

ENOLS
These compounds (e.g. *β*-keto esters and 1,3-diketones) usually respond to the
following tests.

Iron(III) chloride. Add a few drops of neutral iron(III) chloride solution to a solution of 0.1 g of the compound in water or in methanol. Most enols give a red coloration.

Copper derivative. Shake 0.2 g of the substance vigorously with a little cold, saturated, aqueous copper(II) acetate solution. Many enols give a solid, green or blue, copper derivative, which can be crystallised from ethanol and often has a definite m.p. (e.g. from ethyl acetoacetate, m.p. 192 °C; from diethyl acetonedicarboxylate, m.p. 142 °C.

For derivative preparations for enols see Section 9.6.7.

THIOLS

These are generally liquids with an unmistakable penetrating, disagreeable and characteristic odour, which persists even at extremely low concentrations in air. Alkanethiols are partly soluble in concentrated solutions of sodium hydroxide but their salts are hydrolysed to the free thiols on dilution with water. Thiophenols are soluble in sodium hydroxide solution but like the alkanethiols do not evolve carbon dioxide from sodium hydrogen carbonate solution. Treatment of a dry thiophenol with sodium results in the evolution of hydrogen (cf. *Alcohols*, Section 9.5.3, p. 1223).

For derivative preparations for thiols see Section 9.6.25, p. 1283.

NITROGEN-CONTAINING COMPOUNDS

Those which are soluble in sodium hydroxide include *aliphatic primary and secondary nitro compounds, oximes, imides and primary sulphonamides.* Aliphatic primary and secondary nitro compounds dissolve in sodium hydroxide solution to give, in general, a yellow solution; on acidification with hydrochloric acid the nitro compound is regenerated. The nitro compounds show pronounced absorption due to the nitro group in the infrared at about 1530 and 1370 cm^{-1} (NO_2, asymmetric and symmetric str.). The presence of the nitro group is confirmed by reduction to the corresponding hydroxylamine (see *nitro compounds*, Section 9.5.3, p. 1227), which can be detected by its action upon Tollen's reagent. To distinguish between *primary, secondary and tertiary aliphatic nitro compounds* (the latter are neutral) the following test should be performed. Dissolve a few drops of the nitro compound in concentrated sodium hydroxide solution, and add excess of sodium nitrite solution. Upon cautiously acidifying with dilute sulphuric acid, added a drop at a time, the following effects may be observed.

(a) Primary nitro compound: intense red colour, disappearing upon acidification. The coloration is that of the alkali salt of the nitrolic acid.

$$R \cdot CH{=}N \underset{O=N}{\overset{O}{\longrightarrow}} \quad R \cdot CH \cdot NO_2 \rightleftharpoons R \cdot C \cdot NO_2$$
$$\underset{N=O}{|} \quad \underset{NOH}{||}$$

(b) Secondary nitro compound: dark blue or dark green colour due to nitronitroso derivatives. The coloured compound is soluble in chloroform.

$$R^1 R^2 C{=}NO_2^{\ominus} + \overset{\oplus}{N}{=}O \longrightarrow R^1 R^2 C(NO)NO_2$$

(c) Tertiary compound: no coloration.

For characterisation of aliphatic nitro compounds by reduction see details under *aromatic nitro compounds*, Section 9.5.3, p. 1227.

For further characterisation of oximes, imides and primary sulphonamides see p. 1226 under the miscellaneous neutral and acidic *compounds containing nitrogen*.

9.5.2 BASIC COMPOUNDS

Organic compounds that dissolve in dilute hydrochloric acid usually contain nitrogen: the rarely encountered pyrones and anthocyanidin pigments are exceptions. The most important basic nitrogen compounds are the primary, secondary and tertiary amines.* The only hydrazines commonly encountered in the group are the monoaryl hydrazines, which are recognised by their ability to reduce Fehling's solution with the evolution of hydrogen. They are also conveniently detected by their condensation with benzaldehyde or some other suitable carbonyl compound as a reagent. The lower aliphatic amines and diamines are soluble in water and possess characteristic ammoniacal odours which distinguish them from water-insoluble amines. The reactions to be described below apply to both water-soluble and water-insoluble amines.

PRIMARY, SECONDARY AND TERTIARY AMINES

Primary amines may be readily distinguished from secondary and tertiary analogues by the presence of two absorption bands in the infrared spectrum between 3320 and 3500 cm^{-1} (symmetric and antisymmetric NH str.). Secondary amines exhibit a single absorption band at about 3350 cm^{-1} (NH str.). In both cases deformation modes for the NH bond appear at about 1600 cm^{-1}. There is no satisfactory absorption to allow a definitive characterisation in the case of tertiary amines. In the nuclear magnetic resonance spectrum of primary and secondary amines, the nitrogen-bound hydrogens are recognisable by their replaceability on the addition of deuterium oxide.

Chemical classification of the amine function.   The classification of primary, secondary or tertiary amines should be carried out by means of the reaction with nitrous acid.

Nitrous acid test.   Dissolve 0.2 g of the substance in 5 ml of 2 M hydrochloric acid: cool in ice and add 2 ml of ice-cold 10 per cent aqueous sodium nitrite solution slowly by means of a dropper and with stirring until, after standing for 3–4 minutes, an immediate positive test for nitrous acid is obtained with starch–iodide paper (see Section 6.7). If a clear colution is obtained with a continuous evolution of nitrogen gas the substance is a *primary aliphatic or aralkyl amine*. If there is apparently no evolution of nitrogen from the clear solution, add one-half of the solution to a cold solution of 0.4 g of 2-naphthol in 4 ml of 5 per cent sodium hydroxide solution. The formation of a coloured (e.g. orange-red) azo-dye indicates the presence of a *primary aromatic amine*; in which case warm the other half of the diazotised solution and note the evolution of nitrogen and the strong phenolic aroma which is produced. If a colourless solution is obtained which gives an immediate and sustained positive test with starch–iodide paper when only a little sodium nitrite solution has been added, the compound is a *tertiary aliphatic amine*.

---

* Many amines are regarded as being potentially carcinogenic: those which are on the restricted list have been specified in the Tables of Physical Constants.

The formation of *N*-nitrosamines* which usually separate as orange-yellow oils or low melting solids indicates the presence of a *secondary amine*. Confirm the formation of the nitrosamine by the *Liebermann nitroso reaction*. This consists in warming the nitrosamine with phenol and concentrated sulphuric acid. The sulphuric acid liberates nitrous acid from the nitrosamine, and the nitrous acid reacts with the phenol to form *p*-nitrosophenol, which then combines with another molecule of phenol to give red indophenol. In alkaline solution the red indophenol yields a blue indophenol anion.

$$R_2N \cdot NO \xrightarrow{H_2SO_4} R_2NH + HONO \xrightarrow{PhOH} HO-\underset{}{\bigcirc}-NO \xrightarrow{PhOH}$$

$$HO-\bigcirc-N=\bigcirc=O \xrightarrow{\ominus OH} \ominus O-\bigcirc-N=\bigcirc=O$$

red   blue

Extract the oil obtained in the nitrous acid test with about 5 ml of ether and wash the extract successively with water, dilute sodium hydroxide and water, and evaporate off the ether. Apply Liebermann's nitroso reaction to the residual oil or solid. Place 1 drop of 0.01–0.02 g of the nitroso compound in a dry test tube, add 0.05 g of phenol and warm together for 20 seconds; cool, and add 1 ml of concentrated sulphuric acid. An intense green (or greenish-blue) coloration will be developed, which changes to pale red upon pouring into 30–50 ml of cold water; the colour becomes deep blue or green upon adding excess of sodium hydroxide solution.

If the unknown base is a *tertiary aromatic amine* the treatment with nitrous acid will yield a dark orange-red solution or an orange crystalline precipitate resulting from the formation of the hydrochloride of the *C*-nitrosamine. Basification of the solution or of the isolated orange precipitate with either sodium hydroxide or carbonate solution yields the bright green nitrosamine base. In favourable cases this may be isolated by extraction with ether, recrystallised and used for characterisation purposes (see p. 1278).

Confirmatory tests for primary amines. *Carbylamine test.* To 1 ml of 0.5 M-alcoholic potassium hydroxide solution (or to a solution prepared by dissolving a fragment of potassium hydroxide half the size of a pea in 1 ml of ethanol) add 0.05–0.1 g of the amine and 3 drops of chloroform, and heat to boiling. A carbylamine (isocyanide) is formed and will be readily identified by its extremely nauseating odour (**CAUTION**):

$$RNH_2 + CHCl_3 + 3KOH \longrightarrow RNC + 3KCl + 3H_2O$$

When the reaction is over, add concentrated hydrochloric acid to decompose the isocyanide and pour it away after the odour is no longer discernible. The test is extremely delicate and will often detect traces of primary amines in secondary and tertiary amines; it must therefore be used with due regard to this and other factors.

5-Nitrosalicylaldehyde reagent test. This test is based upon the fact that 5-nitrosalicylaldehyde and nickel ions when added to a primary amine produce an immediate precipitate of the nickel derivative of the Schiff's base.

---

* Potentially carcinogenic, see Section 2.3.4, p. 49.

To 5 ml of water add 1–2 drops of the amine; if the amine does not dissolve, add a drop or two of concentrated hydrochloric acid. Add 0.5–1 ml of this amine solution to 2–3 ml of the reagent; an almost immediate precipitate indicates the presence of a primary amine. A slight turbidity indicates the presence of a primary amine as an impurity. (Primary aromatic amines generally require 2–3 minutes for the test. Urea and other amides, as well as amino acids, do not react.)

The 5-*nitrosalicylaldehyde reagent* is prepared as follows. Add 0.5 g of 5-nitrosalicylaldehyde (m.p. 124–125 °C) to 15 ml of pure triethanolamine and 25 ml of water; shake until dissolved. Then introduce 0.5 g of crystallised nickel(II) chloride dissolved in a few ml of water, and dilute to 100 ml with water. If the triethanolamine contains some ethanolamine (thus causing a precipitate), it may be necessary to add a further 0.5 g of the aldehyde and to filter off the resulting precipitate. The reagent is stable for long periods.

Rimini's test (for primary aliphatic amines). To a suspension or solution of 1 drop of the compound or to an equivalent quantity of its solution in 3 ml of water, add 1 ml of pure acetone and 1 drop of a freshly prepared 1 per cent aqueous solution of disodium pentacyanonitrosoferrate(III) (nitroprusside). A violet-red colour will develop within 1 minute.

Confirmatory tests for secondary aliphatic amines. *Simon's test.* To a solution or suspension of 1 drop of the compound or to an equivalent quantity of its solution in 3 ml of water, add 2 drops of freshly prepared acetaldehyde solution, followed by 1 drop of a 1 per cent aqueous solution of disodium pentacyanonitrosoferrate(III). A blue coloration is produced within 5 minutes, after which the colour gradually changes through greenish-blue to yellow.

Carbon disulphide reagent test. This test is based upon the formation from a secondary amine and carbon disulphide of a dialkyldithiocarbamate; the latter readily forms a nickel derivative with a solution of a nickel salt:

$$R_2NH + CS_2 \xrightarrow{NH_3} R_2N \cdot C \underset{\underset{SNH_4}{\ominus \oplus}}{\overset{S}{\diagup}} \xrightarrow{NiCl_2} \left(R_2N \cdot C \underset{S}{\overset{S}{\diagup}} \right)_2 Ni$$

To 5 ml of water add 1–2 drops of the secondary amine; if it does not dissolve, add a drop or two of concentrated hydrochloric acid. Place 1 ml of the reagent in a test tube, add 0.5–1 ml of concentrated ammonia solution, followed by 0.5–1 ml of the above amine solution. A precipitate indicates a secondary amine. A slight turbidity points to the presence of a secondary amine as an impurity. The test is very sensitive; it is not given by primary amines.

The *carbon disulphide reagent* is prepared by adding to a solution of 0.5 g crystallised nickel(II) chloride in 100 ml of water enough carbon disulphide so that after shaking a globule of carbon disulphide is left at the bottom of the bottle. The reagent is stable for long periods in a well-stoppered bottle. If all the carbon disulphide evaporates, more must be added (**CAUTION**).

It should be noted that aliphatic and aromatic primary and secondary amines may be distinguished from tertiary amines by their reaction with acetyl chloride, benzoyl chloride and benzenesulphonyl chloride. In the latter case a primary amine yields an alkali-soluble derivative which distinguishes it from a secondary amine when the derivative is neutral and insoluble in acid and alkali.

The separation of primary, secondary and tertiary amines using toluene-*p*-sulphonyl chloride (Hinsberg's method) is described in Section 9.7.

For the preparation of derivatives of primary, secondary and tertiary amines see Sections 9.6.21, and 9.6.22.

9.5.3 NEUTRAL COMPOUNDS

COMPOUNDS CONTAINING CARBON, HYDROGEN AND POSSIBLY OXYGEN

It is convenient to consider the indifferent or neutral oxygen derivatives of the hydrocarbons – (*a*) *aldehydes* and *ketones*, (*b*) *esters* and *anhydrides*, (*c*) *alcohols* and *ethers* – together. All of these, with the exception of the water-soluble members of low molecular weight, are soluble only in concentrated sulphuric acid, as are *alkenes* and readily sulphonated *arenes*. The above classes of compounds must be tested for in the order in which they are listed, otherwise erroneous conclusions may be drawn from the reactions for the functional group about to be described.

ALDEHYDES AND KETONES

These compounds contain the carbonyl group, hence a general test for carbonyl compounds will immediately identify both classes of compounds. The preferred reagent is 2,4-dinitrophenylhydrazine, which gives sparingly soluble dinitrophenylhydrazones with carbonyl compounds (including many quinones). Add 2 drops or 0.05–0.1 g of the substance to be tested to 3 ml of 2,4-dinitrophenylhydrazine reagent, and shake. If no precipitate forms immediately allow to stand for 5–10 minutes. A crystalline precipitate indicates the presence of a carbonyl compound. Occasionally the precipitate is oily at first, but this becomes crystalline upon standing.

2,4-Dinitrophenylhydrazine reagent may be prepared by either of the following methods.

1. Suspend 2.0 g. of 2,4-dinitrophenylhydrazine in 100 ml of methanol; add cautiously and slowly 4.0 ml of concentrated sulphuric acid. The mixture becomes warm and the solid usually dissolves completely. Filter, if necessary.
2. Dissolve 0.25 g of 2,4-dinitrophenylhydrazine in a mixture of 42 ml of concentrated hydrochloric acid and 50 ml of water by warming on a water bath: dilute the cold solution to 250 ml with distilled water. This reagent is more suitable for water-soluble aldehydes and ketones since alcohol is absent.

The above reagent is very dilute and is intended for qualitative reactions. It is hardly suitable for the preparation of crystalline derivatives except in very small quantities (compare Section 9.6.13, p. 1257).

The *acetals* $R^1 \cdot CH(OR^2)_2$ are so readily hydrolysed by acids that they may give a positive result in the above test:

$$R^1 \cdot CH(OR^2)_2 + H_2O \xrightarrow{H^\oplus} R^1 \cdot CHO + 2R^2OH$$

(For a more detailed discussion on *Acetals*, see Section 9.6.12, p. 1257).

If the unknown compound gives a positive test with 2,4-dinitrophenylhydrazine it then becomes necessary to decide whether it is an aldehyde or a ketone. The infrared spectrum of the compound should be very informative; both aldehydes and ketones show strong absorption at 1740–1700 cm^{-1} (C=O str.), but only aldehydes exhibit two absorption bands at about 2720 and 2820 cm^{-1}

(C—H str.). In the nuclear magnetic resonance spectrum of an aldehyde a low-field signal for the aldehydic hydrogen (δ 9–10) is characteristic.

Chemical differentiation between aldehydes and ketones. *Schiff's reagent.* Aldehydes produces a pink colour, while ketones are without effect. Use 2 drops (or 0.05 g) of the compound and 2 ml of Schiff's reagent and shake the mixture in the cold. Some aromatic aldehydes (e.g. vanillin) give a negative result.

The *reagent* may be prepared by either of the following methods.

1. Dissolve 0.2 g of pure *p*-rosaniline hydrochloride in 20 ml of a cold, freshly prepared, saturated aqueous solution of sulphur dioxide; allow the solution to stand for a few hours until it becomes colourless or pale yellow. Dilute the solution of 200 ml and keep it in a tightly stoppered bottle. If the bottle is not adequately stoppered, the reagent will gradually lose sulphur dioxide and the colour will return. The solution keeps well if not unnecessarily exposed to light and air.
2. Add 2 g of sodium metabisulphite to a solution of 0.2 g of *p*-rosaniline hydrochloride and 2 ml of concentrated hydrochloric acid in 200 ml of water.

By way of caution it should be noted that free alkali or the alkali salts of weak acids will redden the reagent like an aldehyde. It is also, of course, reddened by heat or when exposed in small quantities to the air for some time. Mineral acids greatly reduce the sensitivity of the test.

Ammoniacal silver nitrate solution (Tollen's solution). Aldehydes alone reduce Tollen's reagent and produce a silver mirror on the inside of the test tube. Add 2–3 drops (or 0.05 g) of the compound to 2–3 ml of Tollen's solution contained in a clean test tube (the latter is preferably cleaned with hot nitric acid). If no reaction appears to take place in the cold, warm in a beaker of hot water. (**CAUTION**: *After the test, pour the contents of the test tube into the sink and wash the test tube with dilute nitric acid. Any silver fulminate present, which is highly explosive when dry, will thus be destroyed.*)

Tollen's reagent is prepared as follows: Dissolve 3 g of silver nitrate in 30 ml of water (solution *A*) and 3 g of sodium hydroxide in 30 ml of water (solution *B*). When the reagent is required, mix equal volumes (say, 1 ml) of solutions *A* and *B* in a clean test tube, and add dilute ammonia solution drop by drop until the silver oxide is just dissolved. Great care must be taken in the preparation and use of this reagent, which must not be heated. Only a small volume should be prepared just before use, any residue washed down the sink with a large quantity of water, and the test tubes rinsed with dilute nitric acid.

Fehling's solution. Aldehydes alone reduce Fehling's solution to yellow or red copper(I) oxide. Use 2 drops (or 0.05 g) of the compound and 2–3 ml of Fehling's solution: heat on a boiling water bath for 3–4 minutes. This test is positive for aliphatic aldehydes, but is often indecisive for aromatic aldehydes.

Preparation of *Fehling's solution. Solution No.* 1. Dissolve 34.64 g of copper(II) sulphate crystals in water containing a few drops of dilute sulphuric acid, and dilute the solution to 500 ml.

Solution No. 2. Dissolve 60 g of pure sodium hydroxide and 173 g of pure Rochelle salt (sodium potassium tartrate) in water, filter if necessary through a sintered glass funnel and make up the filtrate and washings to 500 ml.

Keep the two solutions separately in tightly stoppered bottles and mix exactly equal volumes immediately before use.

Further classification tests for aldehydes and ketones *Sodium metabisulphite test.* Aldehydes and simple ketones react with a saturated solution of sodium meta-bisulphite to yield crystalline bisulphite-addition compounds:

$$R^1R^2CO + NaHSO_3 \rightleftharpoons R^1R^2C(OH)SO_3Na$$

A condition of equilibrium is reached (70–90 per cent of bisulphite compound with equivalent quantities of the reagents in 1 hour), but by using a large excess of bisulphite almost complete conversion into the addition compound results. Since the reaction is reversible, the carbonyl compound can be recovered by adding to an aqueous solution of the bisulphite compound sufficient sodium carbonate solution or hydrochloric acid to react with the free sodium meta-bisulphite present in the equilibrium mixture. Bisulphite compounds may therefore be employed for the purification of carbonyl compounds or for their separation from other organic substances.

The most satisfactory reagent is a saturated solution of sodium meta-bisulphite containing ethanol; it must be prepared as required since it oxdises and decomposes on keeping. Frequently, a saturated aqueous solution is used without the addition of ethanol.

Prepare 10 ml of saturated sodium metabisulphite solution and add 4 ml of the carbonyl compound; shake thoroughly and observe the rise in temperature. Filter the crystalline precipitate at the pump, wash it with a little alcohol, followed by ether and allow it to dry. The *sodium metabisulphite reagent* is prepared by treating a saturated aqueous solution of sodium metabisulphite with 70 per cent of its volume of industrial spirit, and then adding just sufficient water to produce a clear solution.

Iodoform test. Methyl ketones and acetaldehyde, i.e. compounds containing the Me·CO— grouping, give a positive iodoform reaction. Alcohols having the structure Me·CH(OH)·R, which undergo oxidation to the corresponding methyl ketone, also slowly give a positive test. Dissolve 0.1 g or 4–5 drops of the compound in 2 ml of water; if it is insoluble in water, add sufficient dioxane to produce a homogeneous solution. Add 2 ml of 5 per cent sodium hydroxide solution and then introduce a potassium iodide–iodine reagent dropwise with shaking until a definite dark colour of iodine persists. Allow to stand for 2–3 minutes; if no iodoform separates at room temperature, warm the test tube in a beaker of water at 60 °C. Add more drops of the iodine reagent if the faint iodine colour disappears: continue the addition of the reagent until the dark colour is not discharged after 2 minutes heating at 60 °C. Remove the excess of iodine by the addition of a few drops of dilute sodium hydroxide solution with shaking, dilute with an equal volume of water and allow to stand for 10–15 minutes. The test is positive if a yellow precipitate of iodoform is deposited. Filter off the yellow precipitate, dry upon pads of filter paper and determine the m.p.: iodoform melts at 120 °C.

The *potassium iodide–iodine reagent* is prepared by dissolving 20 g of potassium iodide and 10 g of iodine in 100 ml of water.

For the preparation of derivatives of aldehydes and ketones see Section 9.6.13, p. 1257.

QUINONES
The following tests will be found useful for the detection of simple quinones.

Quinones are coloured (generally yellow) crystalline solids; they are usually insoluble in water, soluble in ether and sublime on heating. Frequently the vapour has a penetrating odour and attacks the eyes. The carbonyl groups of quinones often do not react in a normal way with carbonyl group reagents, because of their oxidising properties: thus quinones are reduced by sodium metabisulphite. Crystalline products are usually formed with one molecule of phenylhydrazine or with one molecule of 2,4-dinitrophenylhydrazine, but these are not always of normal structure. Thus p-benzoquinone reacts with 2,4-dinitrophenylhydrazine hydrochloride in hot alcoholic solution to give 2′,4′-dinitrophenyl-4-azophenol, m.p. 185–186 °C.

$$(NO_2)_2C_6H_3 \cdot NH \cdot NH_2 + O{=}\!\!\big\langle\ \big\rangle\!\!{=}O \xrightarrow{-H_2O} (NO_2)_2C_6H_3 \cdot N{=}N \cdot C_6H_4OH$$

Tests for quinones *Hydriodic acid.* Compounds of the p-benzoquinone type liberate iodine from hydriodic acid. Dissolve 0.1 g. of the quinone in a little rectified spirit. Add 10 ml of 10 per cent aqueous potassium iodide solution to a mixture of 5 ml of ethanol and 5 ml of concentrated hydrochloric acid, and then introduce the quinone solution. Iodine is liberated immediately. This test is also given by other oxidising agents.

Reduction with zinc powder and acid. Simple p-quinones are reduced to hydroquinones in the following manner. Dissolve or suspend 0.5 g of the quinone in dilute hydrochloric acid (1 : 5) and add a little zinc powder. When the solution is colourless, filter, neutralise with sodium hydrogen carbonate, extract the dihydric phenol with ether, remove the solvent and identify (Section 9.6.6, p. 1248).

Reduction with zinc powder and sodium hydroxide. Compounds of the anthraquinone type are reduced to oxanthrols. Treat 0.1 g of the quinone with dilute sodium hydroxide and zinc powder. Upon boiling the mixture a red colour is produced: this disappears when the solution is shaken owing to aerial oxidation to the original quinone.

Distillation with zinc powder. Quinones derived from polycyclic hydrocarbons may be reduced to the parent hydrocarbon as follows. Grind 0.5 g of the compound with 3–4 g of zinc powder, pour the mixture into a Pyrex test tube and cover it with an equal volume of zinc powder. Clamp the tube horizontally at the open end. Heat the zinc powder first, then the mixture of zinc powder and the compound to a dull red heat: the hydrocarbon sublimes into the cooler part of the tube. Remove the sublimate; determine the m.p. and characterise it by the preparation of the picrate (Section 9.6.3, p. 1238).

Reaction with semicarbazide hydrochloride. Many simple quinones yield crystalline mono-semicarbazones by the following procedure. Dissolve 0.2 g of semicarbazide hydrochloride in a little water, add 0.2 g of the quinone and warm. The mono-semicarbazone is immediately formed as a yellow precipitate. Filter and recrystallise from hot water; any bis-semicarbazone will remain undissolved.

Reaction with o-phenylenediamine. o-Quinones (and also aromatic α-diketones, e.g. benzil) react with o-phenylenediamine to yield quinoxalines. Dissolve the substance in ethanol or glacial acetic acid, add an equivalent amount of o-phenylenediamine in ethanolic solution and warm for 15 minutes on a water bath. Cool, dilute with water, filter and recrystallise from dilute ethanol.

For the preparation of derivatives of quinones see Section 9.6.14, p. 1261.

ESTERS AND ACID ANHYDRIDES

When a compound fails to respond to the 2,4-dinitrophenylhydrazine test for aldehydes and ketones, yet exhibits carbonyl absorption in the infrared region, it may be either an ester, an acid anhydride or, possibly, a lactone. The infrared absorption for esters is in the region of 1750–1730 cm^{-1} (C=O str.) and in the region 1000–1300 cm^{-1} (C—O str.). Anhydrides exhibit two absorption bands at about 1820 and 1750 cm^{-1} (C=O str.) together with absorption in the range of 1100–1200 (C—O str.). Lactones absorb at about 1750 cm^{-1} (C=O str., the frequency depends upon the ring size), and in the range 1000–1300 cm^{-1} (C—O str.).

The presence of any of these functional types may be established chemically by applying the *hydroxamic acid test*. These compounds react with hydroxylamine in the presence of sodium hydroxide to form the sodium salt of the corresponding hydroxamic acid. On acidification and addition of iron(III) chloride solution the magenta coloured iron(III) complex of the hydroxamic acid is formed.

$$R^1 \cdot CO_2 R^2 + H_2NOH \xrightarrow{\ominus OH} R^1 \cdot CO \cdot NHOH + R^2OH$$

$$R^1 \cdot CO \cdot O \cdot CO \cdot R^1 + H_2NOH \xrightarrow{\ominus OH} R^1 \cdot CO \cdot NHOH + R^1 \cdot CO_2H$$

$+ H_2NOH \xrightarrow{\ominus OH} R \cdot CH(OH) \cdot CH_2 \cdot CH_2 \cdot CO \cdot NHOH$

The hydroxamic acid test is also given by acid chlorides and some primary aliphatic amides which are readily converted by hydroxylamine hydrochloride into hydroxamic acids. Some esters, mainly of carbonic, carbamic, sulphuric and other inorganic acids, give only a yellow colour.

It is always advisable to ensure that the sample does not give a colour with iron(III) chloride before carrying out the hydroxamic acid test.

Hydroxamic acid test. A. Dissolve a drop or a few small crystals of the compound in 1 ml of rectified spirit (95% ethanol) and add 1 ml of M hydrochloric acid. Note the colour produced when 1 drop of 5 per cent iron(III) chloride solution is added to the solution. If a pronounced violet, blue, red or orange colour is produced, the hydroxamic acid test described below is not applicable and should not be used.

B. Mix 1 drop or several small crystals (c. 0.05 g) of the compound with 1 ml of 0.5 M hydroxylamine hydrochloride in 95 per cent ethanol and add 0.2 ml of 6 M aqueous sodium hydroxide. Heat the mixture to boiling and, after the solution has cooled slightly, add 2 ml of M hydrochloric acid. If the solution is cloudy, add 2 ml of 95 per cent ethanol. Observe the colour produced when 1 drop of 5 per cent iron(III) chloride solution is added. If the resulting colour does not persist, continue to add the reagent dropwise until the observed colour pervades the entire solution. Usually only 1 drop of the iron(III) chloride solution is necessary. Compare the colour with that produced in test *A*. A positive test will be a distinct burgundy or magenta colour as compared with the yellow colour observed when the original compound is tested with iron(III) chloride solution in the presence of acid. It is often advisable to conduct in parallel the test with, say, ethyl acetate, to ensure that the conditions for this test are correct.

Differentiation between esters and anhydrides. The following simple test relies on the fact that hydrolysis of acid anhydrides is more rapid than that of esters under basic conditions. Add 1 ml of the compound to 2 ml of water to which has been added 1 drop of 1 M sodium hydroxide solution and a trace of phenolphthalein indicator. Warm the solution gently on a water bath; with anhydrides the pink colour is discharged within about 1 minute and the further dropwise addition of alkali enables the rate of hydrolysis to be monitored. With most esters hydrolysis is very slow under these conditions. For the preparation of derivatives of esters and anhydrides see Sections 9.6.17, p. 1266 and 9.6.16, p. 1265 respectively.

ALCOHOLS AND ETHERS

If the unknown, neutral, oxygen-containing compound does not give the class reactions for aldehydes, ketones, esters and anhydrides, it is probably either an alcohol or an ether. Alcohols are readily identified by the intense characteristic hydroxyl adsorption which occurs as a broad band in the infrared spectrum at $3600-3300\,cm^{-1}$ (O—H str.). In the nuclear magnetic resonance spectrum, the adsorption by the proton in the hydroxyl group gives rise to a broad peak the chemical shift of which is rather variable; the peak disappears on deuteration.

Chemically, alcohols and ethers may be simply distinguished by the use of two reagents – metallic sodium and acetyl chloride.

Reaction with sodium. Metallic sodium reacts with alcohols with the evolution of hydrogen.

$$2ROH + 2Na \longrightarrow 2RO^{\ominus}Na^{\oplus} + H_2$$

The most common interfering substance, especially with alcohols of low molecular weight, is water; this may result in an inaccurate interpretation of the test if applied alone. Most of the water may usually be removed by shaking with a little anhydrous calcium sulphate. Dry ethers (and also the saturated aliphatic and the simple aromatic hydrocarbons) do not react with sodium. Treat 1.0 ml of the dried compound with a small thin slice of freshly cut sodium (handle with the tongs or with a penknife) in a small dry test tube. Observe whether hydrogen is evolved and the sodium reacts.

The acetyl chloride test. Acetyl chloride reacts vigorously with primary and secondary alcohols with the evolution of hydrogen chloride. Ethers are unaffected by acetyl chloride. In a small dry test tube place 0.5 ml of the dried compound with 0.3–0.4 ml of redistilled acetyl chloride and note whether reaction occurs. Add 3 ml of water and neutralise the aqueous layer with solid sodium hydrogen carbonate and note whether the smell of the products is different from that of the original alcohol. The product in the case of a tertiary alcohol is mainly the alkyl chloride.

Differentiation between primary, secondary and tertiary alcohols. The three classes of alcohols differ in their behaviour on oxidation with hot acidic dichromate solution. Primary alcohols yield aldehydes and secondary alcohols yield ketones; these products are readily distinguishable. Tertiary alcohols yield no carbonyl product under the conditions of the reaction; experimental details are given in Section 9.6.4, p. 1241.

For the preparation of derivatives of alcohols and ethers see Section 9.6.4, p. 1241 and Sections 9.6.10, p. 1254, and 9.6.11, p. 1255 respectively.

POLYHYDRIC ALCOHOLS AND CARBOHYDRATES

If the neutral substance containing carbon, hydrogen and possibly oxygen is insoluble in ether but freely soluble in water, a polyhydric alcohol or a simple mono- or di-saccharide (or related compound) is indicated. Treatment with concentrated sulphuric acid usually produces excessive charring.

Polyhydric alcohols are colourless viscous liquids, or crystalline solids. Upon heating with a little potassium hydrogen sulphate, they may yield aldehydes (e.g. ethylene glycol yields acetaldehyde; glycerol gives the irritating odour of acrolein which can additionally be detected with Schiff's reagent). Two confirmatory tests for polyhydric alcohols are as follows.

Boric acid test. Add a few drops of phenolphthalein to a 1 per cent solution of borax; a pink coloration is produced. The addition of the polyhydric alcohol causes the pink colour to disappear, but it reappears on warming and vanishes again on cooling. This reaction is due to the combination of two *cis*-hydroxyl groups of the compound with boric acid to form reversibly a much stronger monobasic acid.

$$
\begin{array}{c}
-\overset{|}{\underset{|}{C}}-OH \\
-\overset{|}{\underset{|}{C}}-OH
\end{array}
\begin{array}{c}
HO \\
\quad\;\;B-OH \\
HO
\end{array}
\quad
\begin{array}{c}
HO-\overset{|}{C}- \\
HO-\overset{|}{\underset{|}{C}}-
\end{array}
\xrightarrow{\;-3H_2O\;}
\left[
\begin{array}{c}
-\overset{|}{C}-O \\
\qquad\quad B \\
-\overset{|}{C}-O \quad O-\overset{|}{\underset{|}{C}}-
\end{array}
\right]
\;H^{\oplus}
$$

Periodic acid test. (For 1,2-glycols and α-hydroxyaldehydes and ketones, Section 4.2.55, p. 454). Add 1 drop (0.05 ml) of concentrated nitric acid to 2.0 ml of a 0.5 per cent aqueous solution of paraperiodic acid (H_5IO_6) contained in a small test tube and shake well. Then introduce 1 drop or a small crystal of the compound. Shake the mixture for 15 seconds and add 1–2 drops of 5 per cent aqueous silver nitrate. The immediate production of a *white* precipitate (silver iodate) constitutes a positive test and indicates that the organic compound has been oxidised by the periodic acid. The test is based upon the fact that silver iodate is sparingly soluble in dilute nitric acid whereas silver periodate is very soluble: if too much nitric acid is present the silver iodate will not precipitate.

An alternative procedure for the above test is as follows. Mix 2–3 ml of 2 per cent aqueous paraperiodic acid solution with 1 drop of dilute sulphuric acid (*c.* 1.25 M) and add 20–30 mg of the compound. Shake the mixture for 5 minutes, and then pass sulphur dioxide through the solution until it acquires a pale yellow colour (to remove the excess of periodic acid and also iodic acid formed in the reaction). Add 1–2 ml of Schiff's reagent: the production of a violet colour constitutes a positive test.

For the preparation of derivatives of polyhydric alcohols see under *Alcohols and polyhydric alcohols*, Section 9.6.4, p. 1241.

Mono- and di-saccharides are colourless solids or syrupy liquids, which are freely soluble in water, practically insoluble in ether and other organic solvents, and neutral in reaction. Polysaccharides possess similar properties, but are generally insoluble in water because of their high molecular weights. Both poly- and di-saccharides are converted into monosaccharides upon hydrolysis. The following are confirmatory tests for carbohydrates.

Molisch's test. This is a general test for carbohydrates. Place 5 mg of the substance in a test tube containing 0.5 ml of water and mix it with 2 drops of a 10 per cent solution of 2-naphthol in ethanol or in chloroform. Allow 1 ml of concentrated sulphuric acid to flow down the side of the inclined tube (it is best to use a dropper pipette) so that the acid forms a layer beneath the aqueous solution without mixing with it. If a carbohydrate is present, a red ring appears at the common surface of the liquids; the colour quickly changes on standing or shaking, a dark purple solution being formed. Shake and allow the mixture to stand for 2 minutes, then dilute with 5 ml of water. In the presence of a carbohydrate, a dull-violet precipitate will appear immediately.

Barfoed's reagent. This reagent may be used as a general test for monosaccharides. Heat a test tube containing 1 ml of the reagent and 1 ml of a dilute solution of the carbohydrate in a beaker of boiling water. If red copper(I) oxide is formed within 2 minutes, a monosaccharide is present. Disaccharides on prolonged heating (about 10 minutes) may also cause reduction, owing to partial hydrolysis to monosaccharides.

The *reagent* is prepared by dissolving 13.3 g of crystallised neutral copper(II) acetate in 200 ml of 1 per cent acetic acid solution. The reagent does not keep well.

Fehling's solution (test for reducing sugars). Place 5 ml of Fehling's solution (prepared by mixing equal volumes of Fehling's solution No. 1 and solution No. 2) in a test tube and heat to gentle boiling. Add a solution of 0.1 g of the carbohydrate in 2 ml of water and continue to boil gently for a minute or two, and observe the result. A yellow or red precipitate of copper(I) oxide indicates the presence of a reducing sugar. An alternative method of carrying out the test is to add the hot Fehling's solution dropwise to the boiling solution of the carbohydrate; in the presence of a reducing sugar the blue colour will disappear and a yellow precipitate, changing to red, is thrown down.

Benedict's solution. This is a modification of Fehling's solution and consists of a single test solution which does not deteriorate appreciably on standing. To 5 ml of Benedict's solution add 0.4 ml of a 2 per cent solution of the carbohydrate, boil for 2 minutes and allow to cool spontaneously. If no reducing sugar is present, the solution remains clear; in the presence of a reducing sugar, the solution will contain copper(I) oxide. The test may also be carried out according to the experimental details given under *Fehling's solution.*

The *reagent* is prepared as follows. Dissolve 86.5 g of crystallised sodium citrate ($2Na_3C_6H_5O_7,11H_2O$) and 50 g of anhydrous sodium carbonate in about 350 ml of water. Filter, if necessary. Add a solution of 8.65 g of crystallised copper(II) sulphate in 50 ml of water with constant stirring. Dilute to 500 ml. The resulting solution should be perfectly clear; if it is not, pour it through a fluted filter paper.

Non-reducing sugars may be hydrolysed by boiling with dilute hydrochloric acid; if the solution is then neutralised with aqueous sodium hydroxide the reduction of Fehling's solution or Benedict's solution occurs readily.

1225

For the preparation of derivatives of mono- and di-saccharides see Section 9.6.5, p. 1245.

HYDROCARBONS

Hydrocarbons may be differentiated by their solubility in sulphuric acid since *unsaturated hydrocarbons* are soluble in concentrated sulphuric acid as are those *arenes* which are readily sulphonated, whereas the saturated alkanes and lesser reactive arenes are insoluble in this reagent. The presence of an alkene, alkyne or arene is usually readily apparent from an inspection of the infrared and nuclear magnetic resonance spectra, the characteristic features of which are fully discussed in Sections 5.2., 5.3 and 6.1 respectively and in Chapter 3.

The two tests employed for the *detection of unsaturation* are decolourisation of a dilute solution of bromine in dichloromethane, and reaction with dilute aqueous potassium permanganate. It is essential to apply both tests since some symmetrically substituted alkenes (e.g. stilbene, $C_6H_5 \cdot CH{=}CH \cdot C_6H_5$) react only slowly under the conditions of the bromine test. With dilute potassium permanganate solution the double bond is readily attacked, probably through the intermediate formation of a *cis*-diol.

$$Ph \cdot CH{=}CH \cdot Ph \xrightarrow[\text{H}_2\text{O}]{[O]} Ph \cdot CH(OH) \cdot CH(OH) \cdot Ph \longrightarrow 2Ph \cdot CO_2H$$

Bromine test. Dissolve 0.2 g or 0.2 ml of the compound in 2 ml of dichloromethane, and add a 2 per cent solution of bromine in dichloromethane dropwise until the bromine colour persists for one minute. Blow across the mouth of the tube to detect any hydrogen bromide which may be evolved.

Potassium permanganate test. Dissolve 0.2 g or 0.2 ml of the substance in 2 ml of water or in 2 ml of acetone (which gives a negative test with the reagent), and add 2 per cent potassium permanganate solution dropwise. The test is negative if no more than 3 drops of the reagent are decolourised.

Differention between alkanes and aromatic hydrocarbons. The most satisfactory reagent is fuming sulphuric acid. Place 2 ml of 20 per cent fuming sulphuric acid in a dry test tube, add 0.5 ml of the hydrocarbon and shake vigorously. Only the aromatic hydrocarbon dissolves completely, heat is evolved, but excessive charring should be absent. Warm the solution gently, cool and pour cautiously on to crushed ice; the aromatic hydrocarbon which undergoes sulphonation gives a homogeneous aqueous solution.

For the preparation of derivatives and for further characterisation of alkanes, alkenes, alkynes and arenes see Sections 9.6.1, p. 1235, 9.6.2, p. 1235, and 9.6.3, p. 1238 respectively.

COMPOUNDS CONTAINING NITROGEN

The *neutral* nitrogen compounds include: tertiary aliphatic nitro compounds and aromatic nitro compounds; amides (simple and substituted); nitrogen derivatives of aldehydes and ketones (hydrazones, semicarbazones, etc.); nitriles; nitroso, azo, hydrazo and other intermediate reduction products of aromatic nitro compounds. The imides, primary and secondary nitro compounds, oximes and primary sulphonamides are *weakly acidic* nitrogen compounds. All the above nitrogen compounds and also the secondary sulphonamides, with few exceptions, respond to the same classification reactions (reduction and hydrolysis) and hence will be considered together (experimental procedures below).

α-Amino acids are appreciably water-soluble giving solutions which are essentially neutral towards indicator paper. Water-insoluble α-amino acids are readily soluble in both dilute alkali and dilute acid as the result of their amphoteric character.

In most instances, confirmation of the presence of any of the above functional groups may be obtained from a consideration of the spectroscopic properties of the compound.

Nitro compounds and their reduction products. Tertiary aliphatic nitro compounds and aromatic nitro compounds are reduced by zinc and ammonium chloride solution to the corresponding hydroxylamines, which may be detected by their reducing action upon an ammoniacal solution of silver nitrate or Tollen's reagent:

$$R \cdot NO_2 + 4[H] \xrightarrow[NH_4Cl \text{ aq.}]{Zn} R \cdot NHOH + H_2O$$

It must be remembered, however, that nitroso, azoxy and azo compounds (which are usually more highly coloured than nitro compounds) may be reduced by zinc powder to the corresponding hydroxylamine, hydrazo and hydrazine compounds respectively, all of which reduce Tollen's reagent in the cold.

Nitro compounds are reduced in acid solution (for example, by tin and hydrochloric acid) to the corresponding primary amines.

$$R \cdot NO_2 + 6[H] \longrightarrow RNH_2 + 2H_2O$$

Nitrosamines are similarly reduced to secondary amines:

$$R_2N \cdot NO + 6[H] \longrightarrow R_2NH + NH_3 + H_2O$$

The *N*-nitrosamines (and some *C*-nitroso compounds that yield nitrous acid when treated with concentrated sulphuric acid) may be detected by Liebermann's reaction (p. 1216).

Azo compounds may be identified by examination of the amine(s) formed on reduction in acid solution.

$$R^1 \cdot N = N \cdot R^2 + 4[H] \longrightarrow R^1NH_2 + H_2NR^2$$

They are always coloured but give colourless products upon reduction. *Hydrazo and azoxy compounds* are reduced in acid solution to the parent amine.

Amides. *Simple (primary) amides* when warmed with dilute sodium hydroxide solution give ammonia readily, together with the salt of the corresponding acid:

$$R \cdot CONH_2 + H_2O \longrightarrow R \cdot CO_2H + NH_3$$

Complete hydrolysis may be effected by boiling either with 10 per cent sodium hydroxide solution or with 10 per cent sulphuric acid for 1–3 hours. It is preferable to employ the non-volatile sulphuric acid for acid hydrolysis; this acid should also be used for acidification of the solution resulting from alkaline hydrolysis so that any volatile organic acid (formic acid, acetic acid, etc.) may be distilled off.

Substitute amides undergo hydrolysis with greater difficulty. The choice of an acid or an alkaline medium will depend upon the solubility of the compound in the medium, and the effect of the reagent upon the products of hydrolysis. Substituted amides of comparatively low molecular weight (e.g. acetanilide) may be hydrolysed by boiling either with 10 per cent sodium hydroxide solution or with 10 per cent sulphuric acid for 2–3 hours. Other substituted amides are so insoluble in water that little reaction occurs when they are refluxed with dilute acid or dilute alkali for several hours. These include such substances as benzanilide or the benzotoluidides. For these substances satisfactory results may be obtained with 70 per cent sulphuric acid*: this hydrolysis medium is a much better solvent for the substituted amide than is water or more dilute acid; it also permits a higher reaction temperature.

$$R^1 \cdot CONHR^2 + H_2O \xrightarrow{H_2SO_4} R^1 \cdot CO_2H + R^2 \overset{\oplus}{N}H_3\}HSO_4^{\ominus}$$

Nitriles. These are best hydrolysed by boiling either with 30–40 per cent sodium hydroxide solution or with 50–70 per cent sulphuric acid during several hours, but the reaction takes place less readily than for primary amides. Indeed the latter are intermediate products in the hydrolysis:

$$R \cdot CN + H_2O \xrightarrow{\ominus OH} R \cdot CONH_2 \xrightarrow{\ominus OH} R \cdot CO_2^{\ominus} + NH_3$$

Nitriles and simple amides differ in physical properties: the former are liquids or low-melting solids, while the latter are generally solids. If the amide is a solid and insoluble in water, it may be readily prepared from the nitrile by dissolving in concentrated sulphuric acid and pouring the solution into water:

$$R \cdot CN \xrightarrow{H_2SO_4} [R \cdot \overset{\oplus}{C} = NH]HSO_4^{\ominus} \xrightarrow{H_2O} R \cdot CONH_2$$

Oximes, hydrazones and semicarbazones. The hydrolysis products of these compounds, i.e. aldehydes and ketones, may be sensitive to alkali (this is particularly so for aldehydes): it is best, therefore, to conduct the hydrolysis with strong mineral acid. After hydrolysis the aldehyde or ketone may be isolated by distillation with steam, extraction with ether or, if a solid, by filtration, and then identified. The acid solution may be examined for hydroxylamine or hydrazine or semicarbazide; substituted hydrazines of the aromatic series are precipitated as oils or solids upon the addition of alkali.

$$R^1R^2C = NOH + H_2O \xrightarrow{HCl} R^1R^2C = O + NH_2OH$$

Imides. Imides are generally water-soluble, consequently they are much more readily hydrolysed in an alkaline medium, e.g. by refluxing with 10 per cent sodium hydroxide solution:

$$(R \cdot CO)_2NH \xrightarrow{\ominus OH} 2R \cdot CO_2^{\ominus} + NH_3$$

Sulphonamides. Sulphonamides are very resistant to the normal reagents for hydrolysis. Heating with 80 per cent sulphuric acid at 160–170 °C results in rapid hydrolysis.

$$Ar \cdot SO_2NR^1R^2 + H_2O \xrightarrow{H_2SO_4} Ar \cdot SO_3H + R^1R^2\overset{\oplus}{N}H_2\}HSO_4^{\ominus}$$

---

* Prepared by adding 40 ml of concentrated sulphuric acid cautiously and with stirring and cooling to 30 ml of water.

The following methods represent typical procedures for the reduction or hydrolysis of these nitrogen-containing compounds.

(a) Reduction of a nitro compound to a hydroxylamine. Dissolve 0.5 g of the compound in 10 ml of 50 per cent ethanol, add 0.5 g of solid ammonium chloride and about 0.5 g of zinc powder. Heat to boiling and allow the ensuing chemical reaction to proceed for 5 minutes. Filter from the excess of zinc powder and test the filtrate with Tollen's reagent (see p. 1219). An immediate black or grey precipitate or a silver mirror indicates the presence of a hydroxylamine formed by the reduction of the nitro compound. Alternatively, warm the filtrate with Fehling's solution: a hydroxylamine will precipitate red copper(I) oxide. (A blank test should be performed with the original compound.)

(b) Reduction of a nitro compound to a primary amine. In a 50-ml round-bottomed flask fitted with a reflux condenser, place 1 g of the nitro compound and 2 g of granulated tin. Measure out 10 ml of concentrated hydrochloric acid and add it in three equal portions to the mixture; shake thoroughly after each addition. When the vigorous reaction subsides, heat under reflux on a water bath until the nitro compound has completely reacted (20–30 minutes). Shake the reaction mixture from time to time; if the nitro compound appears to be very insoluble, add 5 ml of ethanol. Cool the reaction mixture, and add 20–40 per cent sodium hydroxide solution until the precipitate of tin hydroxide dissolves. Extract the resulting amine from the cooled solution with ether, and remove the ether by distillation. Examine the residue with regard to its solubility in 5 per cent hydrochloric acid and its reaction with acetyl chloride or benzenesulphonyl chloride.

(c) Reduction of a nitrosamine to a secondary amine. Proceed as for a nitro compound. Determine the solubility of the residue after evaporation of the ether and also its behaviour towards benzenesulphonyl (or toluene-p-sulphonyl) chloride.

(d) Hydrolysis of simple (primary) amides in alkaline solution. Boil 0.5 g of the compound with 5 ml of 10 per cent sodium hydroxide solution and observe whether ammonia is evolved.

(e) Hydrolysis of a substituted amide. 1. *With 10 per cent sulphuric acid.* Reflux 1 g of the compound (e.g. acetanilide) with 20 ml of 10 per cent sulphuric acid for 1–2 hours. Distil the reaction mixture and collect 10 ml of distillate: this will contain any volatile organic acids which may be present. Cool the residue, render it alkaline with 20 per cent sodium hydroxide solution, cool and extract with ether. Distil off the ether and examine the ether-soluble residue for an amine.
2. *With 70 per cent sulphuric acid.* Reflux 1 g of the substance (e.g. benzanilide) with 10–15 ml of 70 per cent sulphuric acid (4:3 by volume) for 30 minutes. Allow to cool and wash down any acid which has sublimed into the condenser with hot water. Filter off the acid, wash it with water and examine for solubility, etc. Render the filtrate alkaline with 10–20 per cent sodium hydroxide solution, cool and extract with ether. Examine the residue, after evaporation of the ether, for an amine.

(f) Hydrolysis of a niltrile to an acid. Reflux 1 g of the nitrile with 5 ml of 30–40 per cent sodium hydroxide solution until ammonia ceases to be evolved (2–3 hours). Dilute with 5 ml of water and add, with cooling, 7 ml of 50 per cent sulphuric acid. Isolate the acid by ether extraction, and examine its solubility and other properties.

(g) Hydrolysis of a nitrile to an amide. Warm a solution of 1 g of the nitrile in 4 ml of concentrated sulphuric acid to 80–90 °C, and allow the solution to stand for 5 minutes. Cool and pour the solution cautiously into 40 ml of cold water. Filter off the precipitate; stir it with 20 ml of cold 5 per cent sodium hydroxide solution and filter again. Recrystallise the amide from dilute ethanol, and determine its m.p. Examine the solubility behaviour and also the action of warm sodium hydroxide solution upon the amide.

(h) Hydrolysis of a sulphonamide. Mix 2 g of the sulphonamide with 3.5 ml of 80 per cent sulphuric acid* in a test tube and place a thermometer in the mixture. Heat the test tube, with frequent stirring by means of the thermometer, at 155–165 °C until the solid passes into solution (2–5 minutes). Allow the acid solution to cool and pour it into 25–30 ml of water. Render the resulting solution alkaline with 20 per cent sodium hydroxide solution in order to liberate the free amine. Two methods may be used for isolating the base. If the amine is volatile in steam, distil the alkaline solution and collect about 20 ml of distillate: extract the amine with ether, dry the ethereal solution with anhydrous potassium carbonate and distil off the solvent. If the amine is not appreciably steam-volatile, extract it from the alkaline solution with ether. The sulphonic acid (as sodium salt) in the residual solution may be identified by conversion into a suitable derivative (Section 9.6.26, p. 1284).

For further methods for the characterisation of these nitrogen-containing compounds see Sections 9.6.18, p. 1270, 9.6.20, p. 1271, and 9.6.24, p. 1281 and 9.6.27, p. 1286.

α-Amino acids. These are in general insoluble (or very sparingly soluble) in organic solvents such as ether or toluene, sparingly soluble in ethanol, usually soluble in water and neutral in reaction. They have no true melting points, but decompose on heating at temperatures between 120 and 300 °C; the apparent melting points vary considerably according to the conditions of heating and are therefore of no great value for purposes of identification. The following tests may be used to confirm the presence of an α-amino acid.

The ninhydrin test. Heat a solution of the compound with a few drops of a 0.25 per cent aqueous solution of ninhydrin (Expt 5.99). α-Amino acids give a blue-violet coloration. This highly sensitive test is also given by some β-amino acids and by some peptides and proteins, particularly on warming. The colour test is of great value in the characterisation of the α-amino acids separated by t.l.c., and the R_F values are useful aids to identification.

Copper complex formation. Add a few drops of aqueous copper(II) sulphate solution to an aqueous solution of the amino acid. A deep blue coloration is obtained. The deep blue copper derivative may be isolated by boiling a solution of the amino acid with precipitated copper(II) hydroxide or with copper(II) carbonate, filtering and concentrating the solution. These blue complexes are co-ordination compounds of the structure:

---

* Prepared by cautiously adding 3 vol. of concentrated sulphuric acid to 1 vol. of water.

$$\text{R} \diagdown \overset{\text{H}_2}{\underset{\text{O}}{\text{N}}} \diagup \overset{\text{O}}{\underset{\text{Cu}}{\text{O}}} \diagup \overset{\text{O}}{\underset{\text{N}}{\text{R}}} \overset{}{\underset{\text{H}_2}{}}$$

Nitrous acid test. The conditions of the test are similar to those described for the classification of primary amines (Section 9.5.2, p. 1215), but using acetic acid in place of hydrochloric acid. An α-amino acid yields nitrogen and an α-hydroxy acid.

For the preparation of derivatives of α-amino acids, see Section 9.6.23, p. 1279.

COMPOUNDS CONTAINING SULPHUR

The *neutral* sulphur compounds include sulphides or thioethers, disulphides, sulphoxides and sulphones, sulphate and sulphonate esters, and isothiocyanates. *Acidic* sulphur compounds, i.e. sulphonic and sulphinic acids, thiols and thiophenols, and the primary sulphonamides have already been discussed. The sulphates of amines are converted by aqueous sodium hydroxide into the free bases; the sulphate anion can be detected in the resulting aqueous solution as barium sulphate in the usual manner.

Those compounds which contain the highly polar S═O group exhibit characteristic strong adsorption bands in the fingerprint region of the i.r. spectra. Two such bands originate from the symmetrical and asymmetrical stretching vibrations of the S═O groups (see Sections 5.17 and 6.4).

Sulphides (thioethers). The organic sulphides are usually liquids with penetrating and disagreeable odours. In contrast to the oxygen analogues (ethers), they are readily oxidised; thus sulphoxides are produced with hydrogen peroxide, and sulphones with nitric acid or with potassium permanganate in glacial acetic acid solution (see Section 5.17.3, p. 791).

$$\overset{\text{R}^1}{\underset{\text{R}^2}{\diagup}}\text{S} \xrightarrow{\text{H}_2\text{O}_2} \overset{\text{R}^1}{\underset{\text{R}^2}{\diagup}}\text{S}{=}\text{O} \xrightarrow{\text{KMnO}_4} \overset{\text{R}^1}{\underset{\text{R}^2}{\diagup}}\text{S}\overset{\diagup\text{O}}{\underset{\diagdown\text{O}}{}}$$

Thioethers usually yield sulphonium salts when warmed with ethyl iodide and allowed to cool. The physical properties (b.p. density and refractive index) are useful for identification purposes.

Disulphides. Disulphides are liquids or low m.p. solids and have unpleasant odours, particularly if liquid. They are reduced by zinc and dilute acids to the thiols.

$$\text{R S·S R} + 2[\text{H}] \longrightarrow 2\text{RSH}$$

Sulphoxides. These are usually solids of low m.p. They may be oxidised in glacial acetic acid solution by potassium permanganate to the corresponding sulphones, and reduced to the sulphides by boiling with tin or zinc and hydrochloric acid.

Sulphones. Sulphones are usually crystalline solids, and are extremely stable to most oxidising, reducing and hydrolytic reagents.

Esters of sulphuric acid. These compounds are generally water-insoluble liquids and are saponified by boiling with water or dilute alkali to the corresponding alcohols and sulphuric acid:

$$R_2SO_4 + 2H_2O \longrightarrow 2ROH + H_2SO_4$$

The simple dialkyl sulphates, in particular dimethyl sulphate, are markedly toxic (see Section 4.2.*24*, p. 430). They may be characterised as the alkyl 2-naphthyl ethers (cf. alkyl halides, Section 9.6.8, p. 1251; and also Expt 6.111).

The *alkyl esters of sulphonic acids* exhibit properties similar to those of the alkyl sulphates, and are hydrolysed, by boiling with aqueous alkalis, to the corresponding alcohols and sulphonic acids. Thus with ethyl toluene-*p*-sulphonate:

$$p\text{-Me·C}_6\text{H}_4\text{·SO}_2\text{OEt} + H_2O \longrightarrow p\text{-Me·C}_6\text{H}_4\text{·SO}_3\text{H} + EtOH$$

The salts of *monoalkyl sulphates* are frequently encountered as commercial detergents: these are usually sodium salts, the alkyl components contain 12 or more carbon atoms, and they give colloidal solutions. They are hydrolysed by boiling with dilute sodium hydroxide solution:

$$RO·SO_2·ONa + NaOH \longrightarrow ROH + NaSO_4$$

Isothiocyanates. These compounds, also known as *mustard oils* are oils or low melting point solids, and usually possess irritating odours. Upon boiling with acids, for example with concentrated hydrochloric acid, they are hydrolysed to the primary amines and hydrogen sulphide is evolved:

$$R·NCS + HCl + 2H_2O \longrightarrow R\overset{\oplus}{N}H_3\}\overset{\ominus}{Cl} + CO_2 + H_2S$$

They react with amines to form substituted thioureas:

$$R^1·NCS + R^2NH_2 \longrightarrow R^1NH·CS·NHR^2$$

This reaction is also employed for the characterisation of amines (see Section 9.6.21, p. 1273).

COMPOUNDS CONTAINING HALOGEN

The spectroscopic features (i.r., p.m.r., and m.s.) of aliphatic and aromatic halogen-containing compounds are summarised in Section 5.5 and 6.3 respectively.

Reactivities of halogen compounds. A halogen substituent in an organic compound is an inert functional group in the sense that it has relatively little effect on the solubility characteristics of the parent compound or upon the properties of other functional groups which may be present. For the purposes of classification and characterisation therefore the presence of the halogen is to a certain extent immaterial. Halogen may of course be present in the ionic form (i.e. in a salt), and then may readily be detected in the usual way with aqueous silver nitrate. Highly reactive halogen compounds, typically acid halides, will also respond to this test. With other, neutral, halogen compounds, it is useful to examine their behaviour towards ethanolic silver nitrate, and to categorise their reactivity in the following way:

Reaction with ethanolic silver nitrate. Treat 2 ml of a 2 per cent solution of silver nitrate in ethanol with 1 or 2 drops (or 0.05 g) of the compound. If no appreciable precipitate appears at the laboratory temperature, heat in a boiling water bath for a few minutes. Some organic acids give insoluble silver salts, hence it is advisable to add 1 drop of *dilute* (5%) nitric acid at the conclusion of the test: most silver salts of organic acids are soluble in aqueous nitric acid. (**CAUTION**: *On no account should concentrated nitric acid be used; a dangerous explosion may result.*)

Organic compounds containing halogen react with ethanolic silver nitrate in the following order of *decreasing* reactivity:

1. Water-soluble compounds containing ionisable halogen or compounds such as acyl halides of low molecular weight which readily yield ionisable compounds with water will react immediately, even with aqueous silver nitrate.
2. Acyl and sulphonyl halides, α-halogeno-ethers and alkyl iodides react rapidly.
3. Alkyl chlorides and bromides, aromatic compounds containing halogen in the side chain, or nuclear-halogenated aromatic compounds with nitro groups in the *ortho* and/or *para* position, do not usually react readily at room temperature but react fairly rapidly on heating. The order of reactivity of alkyl halides is tertiary > secondary > primary, and indeed some tertiary halides may react in the cold.
4. Aromatic compounds in which the halogen is attached directly to the aromatic nucleus and polyhalogenated compounds with three or more halogens on the same carbon atom do not react even on heating.

Further classification of halogen compounds. Two further reactions aid the classification of halogen compounds.

Reaction with ethanolic potassium hydroxide. Boil 0.5 ml of the compound with 4 ml of 0.5 M ethanolic potassium hydroxide under reflux for 15 minutes. Most alkyl halides and benzyl halides give a crystalline precipitate of the potassium halide. Dilute with 5 ml of water, acidify with dilute nitric acid and test with silver nitrate solution.

The 0.5 M ethanolic potassium hydroxide solution is prepared by dissolving 16 g of potassium hydroxide pellets in 500 ml of ethanol in a bottle closed with a cork. After standing for 24 hours, the clear solution is decanted and filtered from the residue of potassium carbonate. It is said that a solution in methanol has better keeping qualities than that in ethanol.

Halogen exchange reaction. This is based upon the fact that sodium chloride and sodium bromide are sparingly soluble in pure acetone:

$$\text{RCl(Br)} + \text{NaI} \longrightarrow \text{RI} + \text{NaCl(Br)}\downarrow$$

The test consists in treating a solution of sodium iodide in pure acetone with the organic compound. The reaction is probably of the S_N2 type involving a bimolecular attack of the iodide ion upon the carbon atom carrying the chlorine or bromine; the order of reactivities of halides is: primary > secondary > tertiary and Br > Cl.

Primary bromides give a precipitate of sodium bromide within 3 minutes at 25 °C; chlorides react only when heated at 50 °C for up to 6 minutes. Secondary and tertiary bromides must be heated at 50 °C for up to 6 minutes, but tertiary chlorides do not react within this time.

1,2-Dichloro and dibromo compounds give a precipitate with the reagent and also liberate free iodine:

$$R^1 \cdot CHBr \cdot CHBr \cdot R^2 + 2NaI \longrightarrow R^1 \cdot CHI \cdot CHI \cdot R^2 + 2NaBr$$

$$\Updownarrow$$

$$R^1 CH = CH \cdot R^2 + I_2$$

Polybromo compounds (bromoform, s-tetrabromoethane) react at 50 °C, but simple polychloro compounds (chloroform, carbon tetrachloride and trichloro-acetic acid) do not.

Sulphonyl chlorides give an immediate precipitate and also liberate iodine:

$$Ar \cdot SO_2 Cl + NaI \xrightarrow{-NaCl} Ar \cdot SO_2 I \xrightarrow{NaI} Ar \cdot SO_2 Na + I_2$$

Acid chlorides and bromides, allyl halides, and α-halo-ketones, -esters, -amides and -nitriles react at 25 °C within 3 minutes. Vinyl and aryl halides are inert.

Prepare the *reagent* by dissolving 7.5 g of sodium iodide in 50 ml of AnalaR acetone. The colourless solution gradually acquires a yellow colour. Keep it in a dark bottle. When a red-brown colour develops, it should be discarded.

For the preparation of derivatives of alkyl and aryl halides see Sections 9.6.8, p. 1251 and 9.6.9, p. 1253 respectively.

9.6 PREPARATION OF DERIVATIVES

Before the development and widespread application of spectroscopic methods for the elucidation of structure, confirmation of the class type of an unknown organic compound was completed by the preparation of two or more crystalline functional derivatives. If the compounds had been previously reported in the literature, agreement between the published physical constants of the derivatives with those prepared by the worker was accepted as proof of identity. In many cases, and particularly in natural product chemistry, functional group recogni-tion led to oxidative, reductive, or hydrolytic breakdown into smaller carbon-containing fragments. These were, if necessary, separated, characterised and identified by derivative preparation. The reassembly of the 'jig-saw' of fragments inferred by the identity of the fission products, then led to postulated structures.

Derivative preparation provides the developing organic chemist with an im-portant area of study for the following reasons.

Firstly, the combination of spectroscopic information and reactivity tests enables the classification of the compound to be made with greater certainty, leading to derivative selection to be made with greater confidence. The require-ments of a satisfactory derivative include the following:

1. The derivative should be easily and quickly prepared in good yield by an unambiguous reaction, and be easily purified. In practice, this generally means that the derivative must be a solid, because of the greater ease of manipulation of small quantities of solids, and the fact that melting points are more accurately and more easily determined than boiling points. The melting point should preferably be above 50 °C, but below 250 °C; compounds which melt below 50 °C are frequently difficult to recrystallise.
2. The derivative should be prepared preferably by a general reaction, which under the same experimental conditions would yield a definite derivative with

the other likely compounds. Rearrangements and side reactions should be avoided.

3. The properties (physical and chemical) of the derivatives should be markedly different from those of the original compound.

4. It should be borne in mind that when a compound has several functional groups, that functional group should be chosen for the preparation of the derivative which gives the least ambiguous reaction.

Secondly, the description of the *general* procedures given below, as distinct from the specific experimental procedures of the preparations described in earlier chapters, provides an excellent opportunity for the student to explore on the small scale the optimum reaction conditions, the chromatographic monitoring of the reaction, the methods of isolation and purification procedures (solvent extraction, recrystallisation, etc.) for the successful completion of the preparation. The small-scale nature of the experiments is of particular importance in providing experience of those techniques of reaction work-up in which mechanical loss is frequently the reason for failure. Such experience is vital to the synthetic chemist since many of the new chemo-, regio- and stereo-specific reagents are expensive and used in small-scale reactions.

Thirdly, some of the derivatives provide protection of the functional group, and their preparation, chemical reactivity, and deprotection procedures may be of value in the design of synthetic strategies.

The methods of preparation of some of the more important derivatives of a number of classes of organic compounds are described in the following sections. These sections are cross-referenced with tables incorporating the melting points and boiling points of the compounds themselves, and also the melting points of selected derivatives. For convenience, the references to the various derivative preparations and tables are collected in Table 9.2.

9.6.1 SATURATED ALIPHATIC HYDROCARBONS

Because of the chemical inertness of the saturated aliphatic hydrocarbons and of the closely related cycloalkanes, no satisfactory crystalline derivatives can be prepared. A pure sample may be characterised by consideration of such physical properties such as the boiling point, the refractive index (and/or the density), and these physical constants are listed in Table 10.1. If required, confirmation of structure should be sought from a more detailed study of appropriate spectra, particularly ^{13}C-n.m.r., and mass spectra.

9.6.2 UNSATURATED ALIPHATIC HYDROCARBONS

DERIVATIVES OF ALKENES

The alkenes are distinguished from the alkanes by their solubility in concentrated sulphuric acid and their characteristic reactions with dilute potassium permanganate solution and with bromine. Characterisation may be based upon the determination of their physical and/or spectral properties. Characterisation by way of solid adducts with nitrosyl chloride has been quite widely used in the terpene field; the preparation of adducts with 2,4-dinitrobenzenesulphenyl chloride is described below (see also Section 8.1.1, p. 1128).

1235

Table 9.2 Index of derivative preparations

Class of compound	Derivative preparations (Section no.)	Physical constants (Table no.)
Saturated aliphatic hydrocarbons	9.6.1	10.1
Unsaturated aliphatic hydrocarbons	9.6.2	10.2
Aromatic hydrocarbons	9.6.3	10.3
Aliphatic alcohols	9.6.4	10.4
Aromatic alcohols	9.6.4	10.5
Phenols	9.6.6	10.6
Enols	9.6.7	10.7
Polyhydric alcohols	9.6.4	10.8
Carbohydrates (sugars)	9.6.5	10.9
Aliphatic halogen compounds	9.6.8	10.10
Aromatic halogen compounds	9.6.9	10.11
Aliphatic ethers	9.6.10	10.12
Aromatic ethers	9.6.11	10.13
Acetals	9.6.12	10.14
Aliphatic aldehydes	9.6.13	10.15
Aromatic aldehydes	9.6.13	10.16
Aliphatic ketones	9.6.13	10.17
Aromatic ketones	9.6.13	10.18
Quinones	9.6.14	10.19
Aliphatic carboxylic acids	9.6.15	10.20
Aromatic carboxylic acids	9.6.15	10.21
Acid chlorides (aliphatic)	9.6.16	10.22
Acid anhydrides (aliphatic)	9.6.16	10.23
Acid chlorides and acid anhydrides of aromatic acids	9.6.16	10.24
Aliphatic esters	9.6.17	10.25
Aromatic esters	9.6.17	10.26
Primary aliphatic amides	9.6.18	10.27
Primary aromatic amides	9.6.18	10.28
Substituted aromatic amides	9.6.19	10.29
Aliphatic nitriles	9.6.20	10.30
Aromatic nitriles	9.6.20	10.31
Primary and secondary aliphatic amines	9.6.21	10.32
Primary aromatic amines	9.6.21	10.33
Secondary aromatic amines	9.6.21	10.34
Tertiary amines	9.6.22	10.35
Amino acids	9.6.23	10.36
Aromatic nitro compounds	9.6.24	10.37
Aliphatic nitro compounds	9.6.24	10.38
Thiols	9.6.25	10.39
Sulphonic acids	9.6.26	10.40
Aromatic sulphonamides	9.6.27	10.41
Imides	—	10.42
Nitroso, azo, azoxy and hydrazo compounds	—	10.43
Miscellaneous sulphur compounds	—	10.44
Miscellaneous phosphorus compounds	—	10.45
Esters or inorganic acids	—	10.46

ADDUCTS WITH 2,4-DINITROBENZENESULPHENYL CHLORIDE

2,4-Dinitrobenzenesulphenyl chloride reacts in polar solvents (acetone, 1,2-dichloroethane, acetic acid and dimethylformamide) with alkenes to yield crystalline adducts, the β-chloroalkyl-2,4-dinitrophenyl sulphides, e.g.:

$$R \cdot CH{=}CH_2 + 2,4\text{-}(NO_2)_2C_6H_3SCl \longrightarrow R \cdot CHCl \cdot CH_2 \cdot S \cdot C_6H_3(NO_2)_2\text{-}2,4$$

Addition of the reagent is stereospecific (*trans* addition) and one can thus differentiate between *cis* and *trans* isomers: thus *cis*-butene and *trans*-butene give products of m.p. 129 and 77 °C respectively.

Heat a solution of 0.2 g of the reagent and 0.2–0.3 g of the alkene in glacial acetic acid on the steam bath for 15 minutes or until the potassium iodide test shows that the reaction is complete. Cool the mixture in ice. If a solid separates, filter it off; if not, pour the reaction mixture on to 5–10 g of crushed ice. Recrystallise the resulting solid or oil from ethanol. *Test:* Add a drop of the reaction solution to a drop of potassium iodide solution on a spot plate; the presence of unreacted reagent is revealed by the liberation of iodine:

$$2RSCl + 2I^{\ominus} \longrightarrow RSSR + I_2 + 2Cl^{\ominus}$$

DERIVATIVES OF ALKYNES

ADDITION PRODUCTS WITH 2,4-DINITROBENZENESULPHENYL CHLORIDE

The reagent reacts with symmetrical alkynes as follows:

$$R \cdot C{\equiv}C \cdot R + ArSCl \longrightarrow R \cdot C(Cl){=}C(SAr) \cdot R; \quad Ar = 2,4\text{-}(NO_2)_2C_6H_3$$

Dissolve 1.60 g of the reagent in 15 ml of 1,2-dichloroethane at 0 °C and add 3.0 ml of the ice-cold alkyne. Keep at 0 °C for 2 hours, remove the solvent by aspiration and keep the clear yellow oil in a refrigerator until crystallisation occurs. Dissolve the crystals in 25 ml of absolute ethanol, decolourise with charcoal and filter. Concentrate the filtrate, collect the crystals which separate and recrystallise from ethanol.

MERCURIDES OF MONOSUBSTITUTED ALKYNES

Monosubstituted alkynes form mercurides which are suitable for identification purposes:

$$2R \cdot C{\equiv}CH + K_2[HgI_4] + 2KOH \longrightarrow (R \cdot C{\equiv}C)_2Hg + 4KI + 2H_2O$$

The procedure consists in adding a dilute solution of the alkyne in ethanol to an excess of an alkaline mercury(II) iodine reagent: a white or greyish-white precipitate forms immediately, which is filtered off, washed with dilute ethanol and recrystallised. The yield of mercuride is 85–95 per cent.

The mercury(II) iodide reagent is *prepared* by dissolving 6.6 g of mercury(II) chloride (*POISONOUS*) in a solution of 16.3 g of potassium iodide in 16.3 ml of water and adding 12.5 ml of 10 per cent sodium hydroxide solution.

Into a cooled dilute solution of 2 equivalents of alkaline mercury(II) iodide reagent, drop slowly, with mechanical stirring, a solution of 1 equivalent of the monosubstituted alkyne in 20 volumes of 95 per cent ethanol. A white crystalline precipitate separates at once. Stir for 2–3 minutes, filter rapidly with suction and wash with 50 per cent ethanol. Recrystallise from ethanol or benzene.

Data for a selection of alkenes and alkynes are collected in Table 10.2.

9.6.3   AROMATIC HYDROCARBONS

For characterisation, aromatic hydrocarbons can be sulphonated, chlorosulphonated, carboxybenzoylated and nitrated. Polynuclear aromatic hydrocarbons, and many of their derivatives, yield crystalline adducts with picric acid, styphnic acid, 1,3,5-trinitrobenzene and 2,4,7-trinitrofluorenone.

SULPHONAMIDES

Aromatic hydrocarbons react with chlorosulphonic acid to yield the corresponding sulphonyl chlorides (the process is known as *chlorosulphonation*). These do not usually crystallise well and are therefore converted into the sulphonamides by treatment with concentrated ammonia solution or with solid ammonium carbonate. (See also Section 6.4.2, p. 877.)

$$ArH + 2HOSO_2Cl \longrightarrow Ar \cdot SO_2Cl + H_2SO_4 + HCl$$
$$Ar \cdot SO_2Cl + (NH_4)_2CO_3 \longrightarrow Ar \cdot SO_2NH_2 + NH_4Cl + CO_2 + H_2O$$

Dissolve 1.0 g of the compound in 5 ml of dry $(CaCl_2)$ chloroform in a dry test tube, cool it in a beaker of ice and add 3–5 ml of chlorosulphonic acid (**CAUTION**) dropwise. When the evolution of hydrogen chloride has subsided, remove the test tube from the ice bath and allow to stand at room temperature for 20–30 minutes; then pour on to crushed ice (30 g). Separate the chloroform layer, wash it with water, dry $(CaCl_2)$, and evaporate the solvent.

Boil the arenesulphonyl chloride (0.5 g) with 5 ml of aqueous ammonia (*d* 0.88) for 10 minutes (*fume cupboard*). Cool the reaction mixture and dilute it with 10 ml of water. Filter off the sulphonamide, wash it with water and recrystallise from dilute ethanol.

Alternatively, heat a mixture of 0.5 g of the arenesulphonyl chloride with 2.8 g of dry powdered ammonium carbonate at 100 °C during 30 minutes. Wash the residue with several portions (10 ml) of cold water, filter and recrystallise from dilute ethanol.

If the presence of a sulphone is suspected (cf. *Aromatic halogen compounds*, Section 9.6.9, p. 1256), treat the product with 6 M sodium hydroxide solution (only the sulphonamide dissolves), filter and reprecipitate the sulphonamide with 6 M hydrochloric acid.

o-AROYLBENZOIC ACIDS

Aromatic hydrocarbons react with phthalic anhydride in the presence of anhydrous aluminium chloride producing aroylbenzoic acids in good yields. The process is termed *carboxybenzoylation*. See formulation on p. 1008.

Place a mixture of 1.0 g of the hydrocarbon, 10 ml of dry dichloromethane or 1,2-dichloroethane, 2.5 g of powdered anhydrous aluminium chloride and 1.2 g of pure phthalic anhydride in a 25–50 ml round-bottomed flask fitted with a reflux condenser (127 mm jacket). Heat on a water bath for 30 minutes (or until no more hydrogen chloride fumes are evolved). Cool in ice and add 10 ml of concentrated hydrochloric acid cautiously and with constant shaking. When the reaction has subsided, add 20 ml of water and shake vigorously. (All the solid material should pass into solution.) Transfer the two-phase system to a separatory funnel, add 25 ml of ether and shake. Discard the lower aqueous phase. Wash the ethereal layer with 25 ml of 2.5 M hydrochloric acid to ensure removal of any aluminium salts present. Shake the ethereal solution cautiously with 25 ml of M sodium carbonate solution, and run the aqueous phase slowly into 30 ml of M hydrochloric acid. Collect the aroylbenzoic acid by suction filtration, wash it with 25–50 ml of water and recrystallise it from dilute ethanol or from acetic acid. The derivatives prepared from benzene and toluene crystallise with water of crystallisation; the latter is removed by drying at 100 °C.

NITRO DERIVATIVES
No general experimental details for the preparation of nitro derivatives can be given, as the ease of nitration and the product formed frequently depend upon the exact experimental conditions. Moreover some organic compounds react violently so that nitrations should *always be conducted on a small scale*. Typical procedures for benzene hydrocarbons are illustrated by the following concise notes for the nitration of benzene and toluene to yield the solid dinitro compounds. (Full experimental details are given in Expts 6.17 and 6.18).

Benzene. Add 0.5 ml of benzene slowly and with shaking and cooling to a mixture of 4 ml each of concentrated sulphuric and nitric acids. Heat the mixture carefully until it just boils, cool and pour into excess of cold water. Filter off the precipitate, wash it free from acid and recrystallise it from dilute alcohol. *m*-Dinitrobenzene, m.p. 90 °C, is formed.

Toluene. Proceed as for *benzene* but use 0.5 ml of toluene and a mixture of 3 ml of concentrated sulphuric acid and 2 ml of fuming nitric acid. Gently warm the mixture over a free flame for 1–2 minutes, cool and pour into 20 ml of ice-water. Recrystallise the product from dilute alcohol. 2,4-Dinitrotoluene, m.p. 71 °C, is obtained.

Conditions for the mononitration of *naphthalene*, typical of the more reactive polynuclear aromatic hydrocarbons, are to be found in Expt 6.17. Naphthalene can also be nitrated by a mixture of nitric acid and glacial acetic acid, a reagent also suitable for some polyalkylated benzenes.

OXIDATION OF A SIDE CHAIN BY ALKALINE PERMANGANATE
Aromatic hydrocarbons containing side chains may be oxidised to the corresponding acids: the results are generally satisfactory for compounds with one side chain (e.g. toluene or ethylbenzene → benzoic acid) or with two side chains (e.g. *o*-xylene → phthalic acid).

Suspend in a round-bottomed flask 1 g of the substance in 75–80 ml of boiling water to which about 0.5 g of sodium carbonate crystals have been added, and introduce slowly 4 g of finely powdered potassium permanganate. Heat under reflux until the purple colour of the permanganate has disappeared (1–4 hours). Allow the mixture to cool and carefully acidify with dilute sulphuric acid. Heat the mixture under reflux for a further 30 minutes and then cool. Remove any excess of manganese dioxide by the addition of a little sodium metabisulphite. Filter the precipitated acid and recrystallise it from a suitable solvent (e.g. toluene, ethanol, dilute ethanol or water). If the acid does not separate from the solution, extract it with ether, toluene, or dichloromethane.

PICRATES

Many polynuclear aromatic hydrocarbons form 1:1 molecular compounds (π-complexes) with picric acid, for example, naphthalene picrate $C_{10}H_8 \cdot C_6H_2(NO_2)_3OH$. Some picrates, e.g. anthracene picrate, are so unstable as to be decomposed by many, particularly hydroxylic, solvents; they therefore cannot be easily recrystallised but may be washed with a little ether and dried on a porous title. Their preparation may often be accomplished in such non-hydroxylic solvents as chloroform, benzene or ether. The picrates of hydrocarbons can be readily separated into their constituents by warming with dilute ammonia solution and filtering (if the hydrocarbon is a solid) through a moist filter paper. The filtrate contains the picric acid as the ammonium salt, and the hydrocarbon is left on the filter paper.

Picrates are usually prepared by adding a hot solution of the compound in ethanol to a cold saturated ethanolic solution of picric acid, warming and allowing to cool; the derivative separates in a crystalline condition. It is filtered off, washed with a little ether and pressed on a porous tile. If the picrate is stable, it is recrystallised from ethanol, ethyl acetate, benzene or ether. Do not mistake the recrystallised reagent (m.p. 122 °C) for a picrate.

The following are typical experimental details for the preparation of naphthalene picrate. Dissolve 0.1 g of naphthalene in the minimum of hot ethanol and add to 1 ml of a saturated solution of picric acid in ethanol. Warm the mixture and then cool to allow the product to crystallise.

OTHER ADDITION COMPOUNDS

Suitably reactive aromatic hydrocarbons also form crystalline 1:1 π-complexes ('styphnates') with styphnic acid (2,4,6-trinitroresorcinol). These derivatives do not crystallise quite so well as the corresponding picrates, but are frequently of great value. Benzene and its simple homologues do not give stable derivatives.

Dissolve equimolecular amounts of the hydrocarbon and styphnic acid in the minimum volume of hot acetic acid and allow to cool. Filter off the crystalline derivative which separates, wash it with a little acetic acid and dry in the air. Determine the m.p. Recrystallise from acetic acid and again determine the m.p.

Other π-complexing reagents are *1,3,5-trinitrobenzene* and *2,4,7-trinitro-9-fluorenone* (1). The crystalline adducts are usually formed in ethanol, glacial acetic acid, or toluene solution and recrystallised from similar solvents. They are useful both for the characterisation and also for the isolation of appropriately reactive aromatic hydrocarbons.

Data for a number of typical aromatic hydrocarbons are collected in Table 10.3.

(1)

9.6.4 ALCOHOLS AND POLYHYDRIC ALCOHOLS

OXIDATION WITH 'CHROMIC ACID'
A primary alcohol is oxidised by 'chromic acid' to the corresponding aldehyde while a secondary alcohol yields a ketone: tertiary alcohols are generally unaffected or are decomposed into non-ketonic products. Oxidation therefore provides a method for distinguishing between primary, secondary and tertiary alcohols and characterisation of the carbonyl compound provides a means of identifying the alcohol:

$$R \cdot CH_2OH \xrightarrow{[O]} R \cdot CHO \qquad R^1 \cdot CH(OH) \cdot R^2 \xrightarrow{[O]} R^1 \cdot CO \cdot R^2$$

To an ice-cold mixture of 1.0 ml of concentrated sulphuric acid and 5 ml of saturated aqueous potassium dichromate solution, add 2 ml of the alcohol or its concentrated aqueous solution. If the alcohol is not miscible with the reagent, shake the reaction mixture vigorously. After 5 minutes, dilute with an equal volume of water, distil and collect the first few ml of the aqueous distillate in a test tube cooled in ice. (Aldehydes and ketones are volatile in steam.) Test a portion of the distillate for a carbonyl compound with 2,4-dinitrophenylhydrazine reagent (p. 1218). If a solid derivative is obtained, indicating that the compound was a primary or secondary alcohol, test a further portion with Schiff's reagent (p. 1219) to distinguish between the two possibilities. The derivative may be recrystallised; the m.p. may give a preliminary indication of the identity of the alcohol.

3,5-DINITROBENZOATES
3.5-Dinitrobenzoyl chloride reacts with alcohols to form solid esters which possess sharp melting points and are therefore admirably suited for purposes of characterisation:

$$3,5\text{-}(NO_2)_2C_6H_3 \cdot COCl + ROH \longrightarrow 3,5\text{-}(NO_2)_2C_6H_3 \cdot CO_2R + HCl$$

The acid chloride is available commercially, but it is preferable to prepare it from the acid as and when required since 3,5-dinitrobenzoyl chloride tends to undergo hydrolysis if kept for long periods, particularly if the stock bottle is frequently opened. The substance may, however, be stored under dry light petroleum.

Prepare the reagent in a fume cupboard in the following way. Mix 1.0 g of 3,5-dinitrobenzoic acid with 1.5 g of phosphorus pentachloride in a small, dry test tube. Warm the mixture gently over a small smoky flame to start the reaction; when the reaction has subsided (but not before), boil for 1–2 minutes or until the solid matter has dissolved. Pour the mixture while still liquid on a dry watch glass (**CAUTION**: *the fumes are irritating to the eyes*). When the product has solidified, remove the liquid by-product (phosphorus oxychloride) by transferring the pasty mixture to a pad of several thicknesses of filter paper or to a small piece of porous tile. Spread the material until the liquid has been absorbed and the residual solid is dry and transfer the 3,5-dinitrobenzoyl chloride to a test tube. Add 0.5–1 ml of the alcohol, cork the tube loosely and heat on a boiling water bath for 10 minutes: secondary and tertiary alcohols require longer heating (up to 30 minutes). Cool the mixture, add 10 ml of 5 per cent (or saturated) sodium hydrogen carbonate solution, break up the resulting solid ester with a

1241

stirring rod (alternatively, stir until crystalline) and filter at the pump; wash with a little sodium hydrogen carbonate solution, followed by water, and then suck as dry as possible. Recrystallise the crude ester from light petroleum (of suitable b.p. range) or from aqueous ethanol. Collect the crystals, dry and determine the melting point.

The above procedure may also be carried out in the presence of 1 ml of dry pyridine; with some alcohols improved yields may be obtained by this modification.

p-NITROBENZOATES
Alcohols react readily with *p*-nitrobenzoyl chloride to yield *p*-nitrobenzoates:

$$p\text{-}NO_2\text{·}C_6H_4\text{·}COCl + ROH \longrightarrow p\text{-}NO_2\text{·}C_6H_4\text{·}CO_2R + HCl$$

The melting points of these esters are usually much lower than those of the corresponding 3,5-dinitrobenzoates: their preparation, therefore, offers no advantages over the latter except for alcohols of high molecular weight and for polyhydroxy compounds. The reagent is, however, cheaper than 3,5-dinitrobenzoyl chloride; it hydrolyses in the air so that it should either be stored under light petroleum or be prepared from the acid, when required, by the phosphorus pentachloride method.

The experimental technique is similar to that given under 3,5-*dinitrobenzoates* above.

BENZOATES
Alcohols react with benzoyl chloride in the presence of pyridine or of sodium hydroxide solution to produce esters of benzoic acid:

$$Ph\text{·}COCl + ROH \longrightarrow Ph\text{·}CO_2R + HCl$$

These derivatives are generally liquids and hence are of little value for characterisation; the polyhydric alcohols, on the other hand, afford solid benzoates. Thus the benzoates of ethylene glycol, trimethylene glycol and glycerol melt at 73, 58 and 76 °C respectively

Mix together 0.5–0.8 ml of the polyhydroxy compound, 5 ml of pyridine and 2.5 ml of redistilled benzoyl chloride in a 50-ml flask, and heat under reflux for 30–60 minutes. Add 25 ml of 5 per cent sodium hydrogen carbonate solution to the cold reaction mixture and cool in ice until the precipitate solidifies. Filter and wash with a little water. Recrystallise from dilute ethanol.

PHENYL- AND 1-NAPHTHYL-URETHANS (PHENYL- AND 1-NAPHTHYL-CARBAMATES)
Both phenyl isocyanate and 1-naphthyl isocyanate (**CAUTION**: *see p. 51*) react with alcohols to yield phenylurethans and 1-naphthylurethans respectively:

$$Ph\text{·}N{=}C{=}O + ROH \longrightarrow Ph\text{·}NH\text{·}CO_2R$$
$$C_{10}H_7{}^1\text{·}N{=}C{=}O + ROH \longrightarrow C_{10}H_7{}^1\text{·}NH\text{·}CO_2R$$

If the alcohol is not anhydrous, reaction also occurs between the water and the reagent to produce diphenylurea (m.p. 238 °C) and di-1-naphthylurea (m.p. 297 °C) respectively, for example:

$$2Ph\text{·}N{=}C{=}O + H_2O \longrightarrow Ph\text{·}NH\text{·}CO\text{·}NH\text{·}Ph + CO_2$$

The ureas are less soluble than the corresponding urethans, but their separation is not always easy. For this reason the urethans are generally prepared from alcohols which are insoluble in water and can therefore be easily obtained in the anhydrous condition.

1-Naphthyl isocyanate is usually preferred to phenyl isocyanate for the following reasons: (*a*) it is much less lachrymatory; (*b*) it is not so readily decomposed by cold water and thus possesses better keeping qualities; and (*c*) the melting points of the 1-naphthylurethans are generally higher than those of the corresponding phenylurethans. Furthermore, with primary alcohols, which react readily in the cold, only small amounts of the urea are produced and these may be removed by taking advantage of the extreme insolubility of di-1-naphthylurea in hot ligroin. (See also Section 9.6.8, p. 1251.)

Place 1 g of the anhydrous alcohol in a dry test tube and add 0.5 ml of 1-naphthyl isocyanate* (if the molecular weight is known, use a 10% excess of the reagent); insert a loose plug of cotton wool in the mouth of the tube. If no solid separates after shaking and standing for 5 minutes, warm on a water bath for 5–10 minutes, and then cool in ice. If no solid is now obtained, 'scratch' the sides of the tube with a glass rod to induce crystallisation. Extract the solid with 5–10 ml of boiling light petroleum (b.p. 100–120 °C); this rapidly dissolves the 1-naphthylurethan but not the di-1-naphthylurea. Remove the urea (if any) by filtration and allow the hot solution to cool. If the urethan does not crystallise out, evaporate the solution to half its original volume, and allow to cool. Collect the crystals on a filter, dry and determine the melting point. If the latter is not sharp, recrystallise from light petroleum (b.p. 100–120 °C), ethanol, chloroform or dichloromethane.

HYDROGEN 3-NITROPHTHALATES
3-Nitrophthalic anhydride, a yellow crystalline powder of m.p. 163–164 °C, reacts with alcohols to yield monoesters of 3-nitrophthalic acid. Although two isomeric esters are theoretically possible, the main product is the 2-ester; traces of the isomeric 1-ester are eliminated during purification.

The reagent must be carefully protected from moisture as it is comparatively easily hydrated to the acid, m.p. 216–218 °C (sealed capillary tube). Dilute aqueous solutions of an alcohol should be treated with solid potassium carbonate and the alcohol layer used for the test.

Phthalic anhydride reacts similarly (see Section 5.19, p. 809), but the acid phthalates are somewhat more difficult to isolate and the melting points are considerably lower.

For alcohols of b.p. below 150 °C, mix 0.5 g of 3-nitrophthalic anhydride (Expt 6.162) and 0.5 ml (0.4 g) of the alcohol in a test tube fitted with a short condenser, and heat under reflux for 10 minutes after the mixture liquefies. For alco-

---

* The procedure for phenyl isocyanate is similar, but great care must be taken to protect both the reagent and the reaction mixture from moisture.

hols boiling above 150 °C, use the same quantities of reactants, add 5 ml of dry toluene, heat under reflux until all the anhydride has dissolved and then for 20 minutes more: remove the toluene under reduced pressure (suction with water pump). The reaction product usually solidifies upon cooling, particularly upon rubbing with a glass rod and standing. If it does not crystallise, extract it with dilute sodium hydrogen carbonate solution, wash the extract with ether and acidify. Recrystallise from hot water, or from 30 to 40 per cent ethanol or from toluene. It may be noted that the m.p. of 3-nitrophthalic acid is 218 °C.

3,4,5-TRIIODOBENZOATES
The derivatives enumerated above are unsatisfactory for hydroxy ethers, e.g. the mono-ethers of ethyleneglycol ('cellosolves') and the mono-ethers of diethyleneglycol ('carbitols') (see Table 10.4). Crystalline derivatives of hydroxy ethers are readily obtained with 3,4,5-triiodobenzoyl chloride, for example:

$$3,4,5\text{-}I_3C_6H_2COCl + HOCH_2 \cdot CH_2OR \xrightarrow{-HCl} 3,4,5\text{-}I_3C_6H_2 \cdot CO_2CH_2 \cdot CH_2OR$$

Place 0.5 g of 3,4,5-triiodobenzoyl chloride in a small test tube, add 0.25 ml of the hydroxy ether and heat the mixture gently over a micro burner until the evolution of hydrogen chloride ceases (3–5 minutes). Pour the molten mass into 10 ml of 20 per cent ethanol to which crushed ice has been added. Some derivatives solidify instantly; those which separate as oils change to solids in a few minutes without further manipulation. Recrystallise from rectified spirit (use 50% ethanol for esters of methyl and butyl carbitol).

3,4,5-Triiodobenzoyl chloride is prepared by refluxing 5 g of 3,4,5-triiodobenzoic acid (Expt 6.70) with 10 ml of thionyl chloride for 2 hours. The excess thionyl chloride is removed by distillation and the residue recrystallised from carbon tetrachloride–light petroleum. The acid chloride has m.p. 138 °C; the yield is 3.8 g. It should be kept in a well-stoppered bottle.

PSEUDOSACCHARIN ETHERS
Pseudosaccharin chloride (Expt 6.42) reacts with alcohols to give ethers (*O*-alkyl derivatives of saccharin):

Heat a little pseudosaccharin chloride with excess of the anhydrous alcohol in a test tube until hydrogen chloride is no longer evolved. Recrystallise from ethanol or other organic solvent.

With the lower primary alcohols, heating at 100 °C for 10 minutes suffices: for higher alcohols, a temperature of 125 °C is preferable. Secondary alcohols require longer heating at 125 °C. A large excess of alcohol should be used when identifying the lower alcohols and the excess removed by evaporation; for the higher alcohols, it is better to employ an excess of pseudosaccharin chloride and the product washed free from the reagent with dilute aqueous alkali.

The melting points of derivatives of selected aliphatic alcohols and polyhydric alcohols are collected in Tables 10.4 and 10.8 respectively.

The melting points of some derivatives of aromatic alcohols are collected in Table 10.5.

9.6.5 CARBOHYDRATES

The melting points (more accurately termed the decomposition points) of sugars and some of their derivatives, e.g. osazones, are not so definite as those of other classes of organic compounds: they vary with the rate of heating and the differences between individual members are not always large. There are, however, a number of reactions and derivatives which will assist in the characterisation of the simple sugars.

OSAZONE FORMATION

The carbohydrates containing a potential aldehyde or keto group in their cyclic form react with one molecular proportion of phenylhydrazine in the cold to form the corresponding phenylhydrazones (compare *Aldehydes and ketones*, Section 9.6.13, p. 1257); these are usually soluble in water and consequently are of little value for purposes of separation and identification. If, however, the carbohydrate is heated in the presence of excess (3–4 moles) of phenylhydrazine, the $>$CHOH in an aldose or the $-$CH$_2$OH in a ketose adjacent to the phenylhydrazone group is effectively oxidised by one molecule of phenylhydrazine to the corresponding carbonyl group, which then reacts with a further molecule of phenylhydrazine to give a bisphenylhydrazone or *osazone*; aniline and ammonia are by-products of the reaction.

Phenylhydrazones

Osazones

Glucose and fructose (and also mannose) form the same osazone. The osazones are usually yellow, well-defined crystalline compounds and are sparingly soluble in cold water. The characteristic crystalline forms of the osazones of the commonly occurring sugars, when examined under the microscope, may be employed for their identification (Fig. 9.1); the melting or decomposition

Glucosazone

Galactosazone

Arabinosazone

Fig. 9.1

Xylosazone

1246

points are less satisfactory since these depend to a marked degree on the rate of heating.

Certain carbohydrates (sugars) may be identified by the length of time required to form osazones upon treatment with phenylhydrazine under standard experimental conditions. Monosaccharides give precipitates at 100 °C within 20 minutes. The disaccharides maltose and lactose give no osazone at 100 °C even after 2 hours, but osazones are obtained on cooling after 10–15 minutes heating. With sucrose an osazone commences to separate after about 30 minutes, due to gradual hydrolysis into glucose and fructose, but no osazone is produced on cooling after heating for 10–15 minutes.

Place 0.20 g of the carbohydrate, 0.40 g of pure *white* phenylhydrazine hydrochloride, e.g. of AnalaR quality; (**CAUTION**: *handle with great care*), 0.60 g of crystallised sodium acetate and 4.00 ml of water in a dry test tube. (Weigh the quantities within an accuracy of 0.01 g.) Stopper the tube *loosely* with a cork, and stand or clamp it upright in a beaker containing boiling water. Note the time of immersion and the time when the osazone first separates. Shake the tube occasionally (without removing it from the boiling water) in order to prevent supersaturation. The precipitate separates quite suddenly: duplicate experiments should agree within 0.5 minute. Note whether the precipitate is white (mannose), yellow or orange yellow, and whether it is crystalline or 'oily'.

The approximate times of osazone formation in minutes are given in Table 10.9. The product from mannose is the simple hydrazone and is practically white. Arabinose osazone separates first as an oil, while that from galactose is highly crystalline. Lactose and maltose give no precipitate from hot solution.

p-NITROPHENYLHYDRAZONES
This reagent has been used in the characterisation of a number of monosaccharides.

Heat 0.25 g of the compound with 3 ml of ethanol, add 0.25 g of *p*-nitrophenylhydrazine and heat the suspension until the reaction appears complete. The *p*-nitrophenylhydrazone soon separates. Filter, preferably after standing overnight, wash with a little cold ethanol and then recrystallise from ethanol.

ACETATES
Complete acetylation of all the hydroxyl groups is desirable in order to avoid mixtures. In some cases, the completely acetylated sugars may be obtained in the α- or β-forms depending upon the catalyst, e.g. zinc chloride or sodium acetate, that is employed in the acetylation. The experimental details for acetylation may be easily adapted from those already given for α- and β-glucose penta-acetates (Expts 5.106 and 5.107 respectively).

BENZOATES
Benzoyl chloride has a limited application as a reagent in the sugar series. Details for the benzoylation of D-glucose are given in Expt 5.108.

TRIMETHYLSILYLATION
The conversion of monosaccharides and the smaller oligosaccharides (di, tri, and tetra-) into their trimethylsilyl derivatives, which are sufficiently volatile to be analysed by g.l.c., has greatly simplified the problem of sugar identification. The method described in Section 2.31 uses hexamethyldisilazane and chlorotri-

methylsilane in pyridine solution, and may conveniently be applied to a sample size down to 1 mg; the trimethylsilylation reaction is complete within about 20 minutes.

$$> CHOH + [Me_3Si]_2NH/Me_3SiCl \xrightarrow{pyridine} > CHO \cdot SiMe_3$$

The melting points of some derivatives of carbohydrates are included in Table 10.9.

9.6.6 PHENOLS

ACETATES
The acetates of monohydric phenols are usually liquids, but those of di- and tri-hydric phenols and also of many substituted phenols are frequently crystalline solids.

Acetates may be prepared by adding acetic anhydride to somewhat dilute solutions of compounds containing hydroxyl (or amino) groups in aqueous caustic alkalis. The amount of alkali used should suffice to leave the liquid slightly basic at the end of the operation, so much ice should be added that a little remains unmelted, and the acetic anhydride should be added quickly.

Dissolve 0.01 mol (or 1 g if the molecular weight is unknown) of the compound in 5 ml of 3 M sodium hydroxide solution, add 10–20 g of crushed ice followed by 1.5 g (1.5 ml) of acetic anhydride. Shake the mixture vigorously for 30–60 seconds. The acetate separates in a practically pure condition either at once or after acidification by the addition of a mineral acid. Collect the acetyl derivative, and recrystallise it from hot water or from dilute ethanol.

BENZOATES
The benzoates of a few phenols (e.g. o-cresol) are liquids. Many phenols do, however, yield crystalline benzoyl derivatives: these are useful for purposes of characterisation.

The Schotten–Baumann method of benzoylation with benzoyl chloride in the presence of aqueous sodium hydroxide may be used. Full details are given under *Primary and secondary amines*, Section 9.6.21, p. 1273. Alternatively, dissolve 1.0 g of the phenol in 3 ml of dry pyridine and add 0.5 g of benzoyl chloride. After the initial reaction has subsided, warm the mixture over a small flame for a minute or two and pour, with vigorous stirring, into 10–15 ml of water. Allow the precipitate to settle, decant the supernatant liquid, stir the residue thoroughly with 5–10 ml of M sodium carbonate solution, filter and recrystallise from ethanol or from light petroleum.

TOLUENE-p-SULPHONATES
Toluene-p-sulphonyl chloride reacts readily with phenols to yield toluene-p-sulphonates:

$$p\text{-}Me \cdot C_6H_4 \cdot SO_2Cl + ArOH \longrightarrow p\text{-}Me \cdot C_6H_4 \cdot SO_2OAr + HCl$$

Mix 1.0 g of the phenol with 2.5 ml of pyridine, add 2 g of toluene-p-sulphonyl chloride and heat on a water bath for 15 minutes. Pour into 25 ml of cold water and stir until the oil solidifies. Filter, wash with cold dilute hydrochloric acid (to remove pyridine), with cold dilute sodium hydroxide solution (to remove any phenol present), and then with cold water. Recrystallise from methanol or ethanol.

p-NITROBENZOATES AND 3,5-DINITROBENZOATES
Both p-nitrobenzoyl chloride and 3,5-dinitrobenzoyl chloride react with phenols, best in pyridine solution, to yield crystalline p-nitrobenzoates and 3,5-dinitrobenzoates respectively:

$$p\text{-}NO_2 \cdot C_6H_4 \cdot COCl + ArOH \longrightarrow p\text{-}NO_2 \cdot C_6H_4 \cdot CO_2Ar + HCl$$
$$3,5\text{-}(NO_2)_2C_6H_3 \cdot COCl + ArOH \longrightarrow 3,5\text{-}(NO_2)_2C_6H_3 \cdot CO_2Ar + HCl$$

For properties of these reagents and their preparation from the corresponding acids, see under *Alcohols and polyhydric alcohols*, Section 9.6.4, p. 1241.

Dissolve 0.5 g of the phenol in 4–5 ml of dry pyridine, add 1.3 g of 3,5-dinitrobenzoyl chloride and reflux for 25–30 minutes. Pour the cold reaction mixture into 40 ml of *c.* 2 M-hydrochloric acid. Decant the supernatant aqueous liquid from the precipitated solid or oil and stir it vigorously with about 10 ml of M sodium carbonate solution. Filter off the solid derivative and wash it with water. Recrystallise from ethanol, dilute ethanol, toluene–acetone or toluene–light petroleum (b.p. 60–80 °C).

ARYLOXYACETIC ACIDS
Phenols, in the presence of alkali, react with chloroacetic acid to give aryloxy-acetic acids, which are generally crystalline compounds with sharp melting points. (See also Expt 6.110.)

$$Ar \cdot O^\ominus + Cl \cdot CH_2 \cdot CO_2^\ominus \xrightarrow{-Cl^\ominus} Ar \cdot O \cdot CH_2 \cdot CO_2^\ominus \xrightarrow{H^\oplus} Ar \cdot O \cdot CH_2 \cdot CO_2H$$

To a mixture of 1.0 g of the compound and 3.5 ml of 33 per cent sodium hydroxide solution in a test tube, add 2.5 ml of 50 per cent chloroacetic acid solution. If necessary, add a little water to dissolve the sodium salt of the phenol. Stopper the test tube loosely and heat on a gently boiling water bath for an hour. After cooling, dilute with 10 ml of water, acidify to Congo red with dilute hydrochloric acid and extract with 30 ml of ether. Wash the ethereal extract with 10 ml of water, and extract the aryloxyacetic acid by shaking with 25 ml of 5 per cent sodium carbonate solution. Acidify the sodium carbonate extract (to Congo red) with dilute hydrochloric acid, collect the aryloxyacetic acid which separates and recrystallise it from water or from aqueous ethanol.

DIPHENYLURETHANS
Phenols react with diphenylcarbamoyl chloride to yield diphenylurethans (or aryl N,N-diphenylcarbamates):

$$Ph_2N \cdot COCl + HOAr \xrightarrow{C_5H_5N} Ph_2N \cdot CO_2Ar + HCl$$

The reagent is unsuitable for a number of phenolic acids.

Dissolve 0.5 g of the phenol in 2.5 ml of pyridine, and add one equivalent of diphenylcarbamoyl chloride (or 0.4–0.5 g if the molecular weight is uncertain). Reflux the mixture for 30–60 minutes on a boiling-water bath, and then pour into about 25 ml of water. Filter the derivative, wash with a little sodium hydrogen carbonate solution and recrystallise from ethanol, toluene, light petroleum (b.p. 60–80 °C) or dichloromethane.

1-NAPHTHYLURETHANS

1-Naphthyl isocyanate reacts smoothly with monohydric, but not with poly-hydric, phenols to give 1-naphthylurethans (or *N*-1-naphthylcarbamates):

$$1\text{-}C_{10}H_7N{=}C{=}O + ArOH \longrightarrow 1\text{-}C_{10}H_7NH{\cdot}CO_2Ar$$

(compare *Alcohols and polyhydric alcohols*, Section 9.6.4, p. 1242). Some phenols, e.g. nitrophenols and halogeno-phenols, react with difficulty with the reagent alone; the addition of a few drops of pyridine or 1 drop of an ethereal solution of trimethylamine or triethylamine generally results in the rapid formation of the urethan.

Place 0.25 g of the phenol together with an equal weight of 1-naphthyl iso-cyanate in a *dry* test tube closed with a stopper carrying a calcium chloride guard-tube. If a spontaneous reaction does not occur, boil the mixture gently for 2–3 minutes, and cool; if the reaction mixture does not solidify, rub the walls of the tube vigorously with a glass rod. If no crystalline solid is obtained, add 2 drops of dry pyridine or 1 drop of an ethereal solution of triethylamine, and warm on a water bath for 5 minutes. Extract contents of the tube with boiling light petroleum (b.p. 80–100 °C or 100–120 °C) to separate any insoluble di-1-naphthyl urea. Recrystallise the crystals which separate on cooling from the same solvent.

The following alternative method may be used. Dissolve 0.01 mol of the phe-nol and 0.01 mol of 1-naphthyl isocyanate in 20 ml of light petroleum (b.p. 60–80 °C), add 2 drops of triethylamine (or, less satisfactorily, 2 drops of pyridine), reflux for 5 minutes and allow to crystallise. Filter off the crystalline solid through a sintered glass funnel.

2,4-DINITROPHENYL ETHERS

1-Chloro-2,4-dinitrobenzene reacts with the sodium salts of phenols to yield crystalline 2,4-dinitrophenyl ethers:

$$2,4\text{-}(NO_2)_2C_6H_3Cl + ArONa \longrightarrow (NO_2)_2C_6H_3{\cdot}O{\cdot}Ar + NaCl$$

Dissolve 1 g (or 0.01 mol) of the phenol in a solution of 0.40 g of sodium hydroxide in 5 ml of water. Add the resulting solution to 2.0 g of 1-chloro-2,4-dinitrobenzene dissolved in 30 ml of 95 per cent ethanol; add more ethanol, if necessary, to effect solution. Heat the solution under reflux on a water bath until the colour (usually red) is discharged and a copious precipitate of sodium chlor-ide appears (30–60 minutes). Dilute the reaction mixture with an equal volume of water, filter off the precipitated 2,4-dinitrophenyl ether, wash with water and recrystallise from ethanol.

Note. The chlorodinitrobenzene must be handled cautiously (use disposable gloves). If any touches the skin, wash it with industrial spirit and then copiously with water.

PSEUDOSACCHARIN ETHERS

When pseudosaccharin chloride is heated with an excess of a phenol, *O*-aryl derivatives of saccharin are produced (compare Section 9.6.4, p. 1244).

Heat 0.5 g of pseudosaccharin chloride with an excess of the phenol to 125–140 °C for 15–20 minutes; hydrogen chloride is evolved. Wash the product with dilute sodium hydroxide solution and then with water. Recrystallise the derivat-ive from ethanol.

BROMO DERIVATIVES

The presence of the hydroxyl group in phenols facilitates the substitution of the nuclear hydrogen atoms by halogen; the number and position of the substituent atoms varies with the nature of the phenol. This method is an indirect means of identification, as the formation of a substitution derivative is not a characteristic reaction of the phenol group but of the benzene nucleus. Phenol reacts with bromine to give 2,4,6-tribromophenol:

$$PhOH + 3Br_2 \longrightarrow 2,4,6\text{-}Br_3C_6H_3OH + 3HBr$$

Bromo derivatives are often difficult to prepare, particularly in the case of polyhydroxy phenols which oxidise easily.

Dissolve 1.0 g of the compound in 10–15 ml of glacial acetic acid, cautiously add a solution of 3–4 ml of liquid bromine in 10–15 ml of glacial acetic acid until the colour of bromine persists and allow the mixture to stand for 15–20 minutes. Pour into 50–100 ml of water, filter off the bromo compound at the pump and wash with a little cold water. Recrystallise from dilute ethanol. Alternatively dissolve 1.0 g of the phenol in water, ethanol or acetone and add slowly, with constant shaking, just sufficient of a bromine solution (prepared by adding 5 g of bromine to a solution of 7.5 g of potassium bromide in 50 ml of water) to impart a yellow colour to the mixture. Allow to stand for 5 minutes. Add about 50 ml of water, and shake vigorously to break up any lumps. Filter and wash the bromo derivative with a dilute solution of sodium metabisulphite. Recrystallise from ethanol or from dilute ethanol.

The melting points of the derivatives of a number of selected phenols are collected in Table 10.6.

9.6.7 ENOLS

β-Keto esters and some 1,3-diketones may be characterised by conversion into semicarbazones (see *Aldehydes and ketones*, Section 9.6.13, p. 1258).

Heating with an equivalent amount of phenylhydrazine often yields characteristic derivatives. Thus β-keto esters afford 1-phenylpyrazalones (cf. Expt 8.15), while 1,3-diketones yield 1-phenylpyrazoles (cf. Expt 8.14).

Heat a mixture of 0.5 g. of the β-keto ester and an equivalent amount of phenylhydrazine in an oil bath at 100–110 °C for 2 hours. Water and alcohol vapours are evolved. Cool and recrystallise the product from ethanol.

For 1,3-diketones, excellent results are obtained by refluxing the reactants in ethanolic solution for 2–3 hours; the product separates on cooling.

The physical properties as well as the melting points of the derivatives of a number of enols (β-keto esters and 1,3-diketones) are given in Table 10.7.

9.6.8 ALIPHATIC HALOGEN COMPOUNDS

ANILIDES AND 1-NAPHTHALIDES

The Grignard reagents prepared from alkyl halides react with phenyl isocyanate or with 1-naphthyl isocyanate to yield addition products that are converted by hydrolysis into anilides and naphthalides respectively:

$$Ar\cdot NCO \xrightarrow{\text{RMgX}} Ar\cdot N{=}C(OMgX)R \xrightarrow{\text{H}_2\text{O}} Ar\cdot N{=}C(OH)R \rightleftharpoons Ar\cdot NH\cdot CO\cdot R$$

Phenyl isocyanate is a colourless liquid, b.p. 164 °C or 55 °C/13 mmHg; its vapour is lachrymatory. The liquid reacts readily with water, yielding diphenyl urea, m.p. 238 °C, and hence must be protected from atmospheric moisture:

$$Ph \cdot NCO + H_2O \longrightarrow Ph \cdot NHCONH \cdot Ph + CO_2$$

1-Naphthyl isocyanate, b.p. 269–270 °C or 153 °C/18 mmHg, is not quite so irritant and is somewhat more stable towards water (di-1-naphthyl urea has m.p. 297 °C). It is therefore to be preferred as a reagent; furthermore the 1-naphthalides are less soluble than the corresponding anilides.

In a small dry flask, fitted with a short reflux condenser and a calcium chloride guard-tube, place 0.4 g of dry magnesium turnings, a minute crystal of iodine and a solution of 1 ml (or 0.01 mol) of the alkyl halide in 10–15 ml of anhydrous ether. If the reaction does not start immediately (as indicated by the disappearance of the iodine colour), warm for a short period in a beaker of warm water; allow the reaction to proceed spontaneously, moderating it if necessary by immersing the flask in cold water. When the reaction has ceased, decant the nearly clear liquid from any solid material into another flask, and fit the reflux condenser into it. Add, portionwise, through the condenser a solution of 0.5 ml of phenyl or 1-naphthyl isocyanate in 15 ml of anhydrous ether, shaking the flask after each addition. Allow the mixture to stand for 10 minutes and then add 30 ml of M hydrochloric acid dropwise and with vigorous shaking and cooling in ice. (Alternatively, pour the reaction mixture cautiously into 20 ml of ice water containing 1 ml of concentrated hydrochloric acid, and shake the mixture well.) Transfer to a separatory funnel, shake well, then discard the lower aqueous layer. Dry the ethereal solution with a little magnesium sulphate and distil off the ether. Recrystallise the residue: methanol, ethanol, light petroleum, ether or hot water are suitable recrystallisation solvents.

ALKYL MERCURY(II) HALIDES
Grignard reagents, prepared from alkyl halides, react with a mercury(II) halide that contains the *same halogen* as the reagent to form alkyl mercury(II) halides:

$$RMgX + HgX_2 \longrightarrow RHgX + MgX_2$$

The reaction is applicable to primary and secondary halides only; tertiary halides do not react.

Filter the Grignard solution, prepared as in *Anilides and 1-naphthalides*, rapidly through a little glass wool into a test tube containing 4–5 g of mercury(II) chloride, bromide or iodide (**CAUTION**), depending upon the halogen in the original alkyl halide. Shake the reaction mixture vigorously for a few minutes and then evaporate the ether. Boil the residue with 20 ml of rectified spirit, filter the solution, dilute it with 10 ml of distilled water, reheat to dissolve any precipitated solid and allow to cool. Recrystallise the alkyl mercury(II) halide from dilute ethanol.

S-ALKYLISOTHIOURONIUM PICRATES
Alkyl bromides or iodides react with thiourea in ethanolic solution to produce S-alkylisothiouronium salts, which yield picrates of sharp melting point:

$$RX + S=C{<}^{NH_2}_{NH_2} \longrightarrow \left[R\cdot S\cdot C{<}^{NH_2}_{NH_2}\right]^{\oplus} X^{\ominus} \xrightarrow{HOC_6H_2(NO_2)_3}$$

$$\left[R\cdot S\cdot C{<}^{NH_2}_{NH_2}\right]^{\oplus} \overset{\ominus}{O}C_6H_2(NO_2)_3$$

Alkyl chlorides react slowly and the yield of the derivative is poor. Tertiary halides give anomalous results.

Place a mixture of 0.5 g of finely powdered thiourea (**CAUTION**), 0.5 g of the alkyl halide and 5 ml of ethanol in a test tube or small flask equipped with a reflux condenser. Reflux the mixture for a period depending upon the nature of the halide: primary alkyl bromides and iodides, 10–20 minutes (according to the molecular weight); secondary alkyl bromides or iodides, 2–3 hours; alkyl chlorides, 3–5 hours*; polymethylene dibromides or di-iodides, 20–50 minutes. Then add 0.5 g of picric acid, boil until a clear solution is obtained and cool. If no precipitate is obtained, add a few drops of water. Recrystallise the resulting S-alkyl isothiouronium picrate from ethanol.

PICRATES OF ALKYL 2-NAPHTHYL ETHERS

Alkyl halides react with the sodium or potassium derivative of 2-naphthol in alcoholic solution to yield the corresponding alkyl 2-naphthyl ethers (which are usually low m.p. solids) and the latter are converted by picric acid into the crystalline picrates:

$$RX + 2\text{-}C_{10}H_7ONa \longrightarrow 2\text{-}C_{10}H_7OR + NaX$$

Mix together 1.0 g of pure 2-naphthol and the theoretical quantity of 50 per cent potassium hydroxide solution, add 0.5 g of the halide, followed by sufficient rectified spirit to produce a clear solution. For alkyl chlorides, the addition of a little potassium iodide is recommended. Heat the mixture under reflux for 15 minutes, and dissolve any potassium halide by the addition of a few drops of water. The 2-naphthyl ether usually crystallises out on cooling; if it does not, dilute the solution with 10 per cent sodium hydroxide solution until precipitation occurs. Dissolve the 2-naphthyl ether in the minimum volume of hot ethanol and add the calculated quantity of picric acid dissolved in hot ethanol. The picrate separates out on cooling. Recrystallise it from rectified spirit.

The 2-naphthyl ethers from methylene halides have m.p. 133 °C, from ethylene halides 217 °C and trimethylene halides 148 °C.

Table 10.10, deals with a number of aliphatic halogen compounds together with their crystalline derivatives.

9.6.9 AROMATIC HALOGEN COMPOUNDS

NITRATION PRODUCTS

Although no general method of nitration can be given, the following procedure is widely applicable.

---

* Alkyl chlorides often react more rapidly (50–60 minutes) upon adding 0.5 g of potassium iodide to the original reaction mixture, followed by sufcient water or ethanol to produce a clear solution at the boiling point. After refluxing, 0.5 g of picric acid is added, etc.

Add 1 g of the compound to 4 ml of concentrated sulphuric acid and *cautiously* introduce, drop by drop, 4 ml of fuming nitric acid. Warm the mixture on a water bath for 10 minutes, then pour it on to 25 g of crushed ice (or 25 ml of ice-water). Collect the precipitate by filtration at the pump, and recrystallise it from dilute ethanol.

Twenty per cent oleum may be substituted for the concentrated sulphuric acid for compounds which are difficult to nitrate.

REACTION WITH CHLOROSULPHONIC ACID – SULPHONAMIDES
Many aryl halides, either alone or in chloroform solution, when treated with excess of chlorosulphonic acid afford the corresponding sulphonyl chlorides in good yield (use the experimental details given in Section 9.6.3, p. 1238); the latter may be readily converted into the aryl sulphonamides by reaction with concentrated ammonia solution or with solid ammonium carbonate.

The following give abnormal results when treated with chlorosulphonic acid alone, preferably at 50 °C for 30–60 minutes: fluorobenzene (4,4'-difluoro-diphenylsulphone, m.p. 98 °C); iodobenzene (4,4'-diiododiphenylsulphone, m.p. 202 °C); *o*-dichlorobenzene (3,4,3',4'-tetrachlorodiphenylsulphone, m.p. 176 °C); and *o*-dibromobenzene (3,4,3',4'-tetrabromodiphenylsulphone, m.p. 176–177 °C). The resulting sulphones may be crystallised from glacial acetic acid, toluene or ethanol, and are satisfactory for identification of the original aryl halide. In some cases sulphones accompany the sulphonyl chloride; they are readily separated from the final sulphonamide by their insolubility in cold 6 M sodium hydroxide solution; the sulphonamides dissolve readily and are reprecipitated by 6 M hydrochloric acid.

OXIDATION OF SIDE CHAINS
The oxidation of halogenated toluenes and similar compounds and of compounds with side chains of the type $-CH_2Cl$ and $-CH_2OH$ proceeds comparatively smoothly with alkaline permanganate solution (for experimental details, see under *Aromatic hydrocarbons*, Section 9.6.3, p. 1239 and also Expt 6.149). The resulting acid may be identified by a m.p. determination and by the preparation of suitable derivatives (see Section 9.6.15, p. 1261).

PICRATES
Some halogen derivatives of polynuclear aromatic hydrocarbons form picrates (for experimental details, see under *Aromatic hydrocarbons*, Section 9.6.3, p. 1240), for example, 1-chloronaphthalene (m.p. 137 °C), 1-bromonaphthalene (m.p. 134 °C) and 2-bromonaphthalene (m.p. 86 °C).

The properties of a number of aromatic halogen compounds together with the melting points of their derivatives are collected in Table 10.11.

9.6.10   ALIPHATIC ETHERS

The low reactivity of aliphatic ethers renders the problem of the preparation of suitable crystalline derivatives a somewhat difficult one. There are, however, two reactions based upon the cleavage of the ethers which are useful for characterisation.

REACTION WITH 3,5-DINITROBENZOYL CHLORIDE

Ethers undergo cleavage with 3,5-dinitrobenzoyl chloride in the presence of zinc chloride:

$$ROR + (NO_2)_2C_6H_3COCl \xrightarrow{ZnCl_2} (NO_2)_2C_6H_3CO_2R + RCl$$

The resulting alkyl 3,5-dinitrobenzoate may be employed for the characterisation of the ether. The method is only applicable to symmetrical ethers; a mixed aliphatic ether R^1OR^2 would yield a mixture of solid esters.

Add 1 ml of the ether to 0.1–0.15 g of finely powdered anhydrous zinc chloride and 0.5 g of pure 3,5-dinitrobenzoyl chloride (Section 9.6.4, p. 1241) contained in a test tube; attach a small water condenser and reflux gently for 1 hour. Treat the reaction product with 10 ml of 0.75 M sodium carbonate solution, heat and stir the mixture for 1 minute upon a boiling water bath, allow to cool and filter at the pump. Wash the precipitate with 5 ml of 0.75 M sodium carbonate solution and twice with 5 ml of ether. Dry on a porous tile or upon a pad of filter paper. Transfer the crude ester to a test tube and boil it with 10 ml of dichloromethane; filter the hot solution, if necessary. If the ester does not separate on cooling, evaporate to dryness on a water bath, and recrystallise the residue from 2–3 ml of either aqueous ethanol or light petroleum. Determine the melting point of the resulting 3,5-dinitrobenzoate (see Table 10.4).

CLEAVAGE OF ETHERS WITH HYDRIODIC ACID

Aliphatic ethers suffer fission when boiled with constant boiling point hydriodic acid. (See also Section 5.4.6, p. 550.)

$$R^1OR^2 + 2HI \longrightarrow R^1I + R^2I + H_2O$$

If the ether is a simple one ($R^1{=}R^2$), the identification of the resulting alkyl iodide presents no difficulties. If, however, it is a mixed alphatic ether, the separation of the two alkyl iodides by fractional distillation is generally difficult unless R^1 and R^2 differ considerably in molecular weight and sufficient material is available.

Reflux 1 ml of the ether with 5 ml of freshly distilled, constant boiling point hydriodic acid (Section 4.2.32, p. 436), b.p. 126–128 °C, for 2–3 hours. Add 10 ml of water, distil and collect about 7 ml of liquid. Decolourise the distillate by the addition of a little sodium metabisulphite, and separate the two layers by means of a dropper pipette. Determine the b.p. of the resulting iodide by the Siwoloboff method (Section 2.34) and prepare a crystalline derivative (Section 9.6.8, p. 1251).

The physical properties of a number of aliphatic ethers, including cyclic ethers, are collected in Table 10.12.

9.6.11 AROMATIC ETHERS

Commonly encountered purely aromatic ethers (e.g. diphenyl ether) are limited in number; most aromatic ethers are of the mixed aliphatic–aromatic type. The following procedures may be used for their characterisation.

CLEAVAGE WITH A HYDRIODIC ACID

Aromatic ethers undergo fisson when heated with constant boiling point hydriodic acid:

$$\text{ArOR} + \text{HI} \longrightarrow \text{ArOH} + \text{RI}$$

The cleavage products are a phenol and an alkyl iodide, which will serve to characterise the ether (see also Section 6.9.4, p. 988).

Experimental details can easily be adapted from those given under *Aliphatic ethers*, Section 9.6.10, p. 1254.

To isolate the phenol, treat the residue in the flask with aqueous sodium carbonate until alkaline and extract the mixture with ether. Wash the ethereal extract with saturated aqueous sodium carbonate, and then with 2 M sodium hydroxide solution. Acidify the sodium hydroxide solution (to Congo red paper) and extract the liberated phenol with ether. Characterise the phenol as in Section 9.6.6, p. 1248.

PICRATES
Ethers of many polynuclear aromatic systems are conveniently characterised as their picrates prepared by the method described for the corresponding derivates in *Aromatic hydrocarbons*, Section 9.6.3, p. 1238.

DERIVATIVES BY NUCLEAR SUBSTITUTION
Nitration. These may generally be prepared as detailed under *Aromatic hydrocarbons*, Section 9.6.3, p. 1238; the following experimental procedure for anisole may be regarded as typical. Add 0.5 g of anisole to a mixture of equal volumes of concentrated nitric acid and concentrated sulphuric acid keeping the temperature below 25 °C by cooling in an ice bath. Finally warm to 40 °C until dilution of a small portion with water gives a solid product. Pour the whole of the reaction mixture into water; collect the resulting 2,4-dinitroanisole and recrystallise from ethanol.

Bromination. These may be prepared as described under *Phenols*, Section 9.6.6, p. 1251, using glacial acetic acid as solvent. In some cases dichloromethane is a satisfactory solvent; the dichloromethane is separated by distillation and the residue is recrystallised from dilute ethanol.

Formation of sulphonamides. These may be prepared as described for *Aromatic hydrocarbons*, Section 9.6.3, p. 1238.

OXIDATION OF SIDE CHAINS
General conditions for the oxidation of an alkyl side chain attached to an aromatic ring are given under *Aromatic hydrocarbons*, Section 9.6.3, p. 1239. The following procedure for the oxidation of *p*-cresyl methyl ether to anisic acid is illustrative.

Prepare a solution of 6 g of potassium permanganate in a mixture of 20 ml of 5 per cent sodium hydroxide solution and 150 ml of water, add 2.0 g of *p*-cresyl methyl ether and heat under reflux for 2–3 hours. If any permanganate remains at the end of this period, destroy it by the addition of a few drops of ethanol. Remove the precipitated manganese dioxide by filtration at the pump, evaporate the filtrate to a volume of 25–30 ml and acidify it (to Congo red) with dilute sulphuric acid. Anisic acid, m.p. 183–184 °C, crystallises out on cooling.

Table 10.13 contains data referring to a number of selected aromatic ethers.

9.6.12 ACETALS

Acetals may be characterised by reference to the alcohol and aldehyde (or ketone if a ketal) which they yield readily when hydrolysed in dilute acid solution (e.g. with 3–5% acid):

$$R^1 \cdot CH(OR^2)_2 + H_2O \xrightarrow{H^{\oplus}} R^1 \cdot CHO + 2R^2OH$$

The rate of hydrolysis depends upon the solubility of the acetal in the hydrolysis medium. Acetals of low molecular weight are completely hydrolysed by refluxing for 5–10 minutes; those of higher molecular weight, and therefore of low solubility, may require 30–60 minutes, but the rate of hydrolysis may be increased by the addition of dioxane which increases the solubility of the acetal.

The experimental procedure to be followed depends upon the products of hydrolysis. If the alcohol and aldehyde are both soluble in water, the reaction product is divided into two parts. One portion is used for the characterisation of the aldehyde by the preparation of a suitable derivative (e.g. the 2,4-dinitrophenylhydrazone, semicarbazone or dimethone, see *Aldehydes and ketones*, Section 9.6.13, below). The other portion is employed for the preparation of a 3,5-dinitrobenzoate, etc. (see *Alcohols and polyhydric alcohols*, Section 9.6.4, p. 1241): it is advisable first to concentrate the alcohol by distillation or to attempt to salt out the alcohol by the addition of solid potassium carbonate. If one of the hydrolysis products is insoluble in the reaction mixture, it is separated and characterised. If both the aldehyde and the alcohol are insoluble, they are removed from the aqueous layer; separation is generally most simply effected with sodium metabisulphite solution (compare Expt 5.82), but fractional distillation may sometimes be employed.

The formulae and physical properties of a number of common acetals are collected in Table 10.14.

9.6.13 ALDEHYDES AND KETONES

2,4-DINITROPHENYLHYDRAZONES

Small quantities may be prepared with the class reagent described on p. 1218. The following procedure is generally more satisfactory.

Suspend 0.25 g of 2,4-dinitrophenylhydrazine in 5 ml of methanol and add 0.4–0.5 ml of concentrated sulphuric acid cautiously. Filter the warm solution and add a solution of 0.1–0.2 g of the carbonyl compound in a small volume of methanol or of ether. If no solid separates within 10 minutes, dilute the solution carefully with M sulphuric acid. Collect the solid by suction filtration and wash it with a little aqueous methanol. Recrystallise the derivative from ethanol, dilute ethanol, ethyl acetate, acetic acid, dioxane, nitromethane, nitrobenzene or xylene.

Alternatively, to the clear solution obtained by warming 0.5 g of 2,4-dinitrophenylhydrazine, 1 ml of concentrated hydrochloric acid and 8–10 ml of ethanol, add 0.25 g of the carbonyl compound and heat just to boiling. Allow to cool to room temperature, filter off the 2,4-dinitrophenylhydrazone and recrystallise it from ethanol or glacial acetic acid.

The following reagent, a 0.25 M solution of *2,4-dinitrophenylhydrazine*, may be used for the preparation of derivatives of keto compounds. Dissolve 25 g of 2,4-

dinitrophenylhydrazine in 300 ml of 85 per cent phosphoric acid in a 600-ml beaker on a steam bath, dilute the solution with 200 ml of 95 per cent ethanol, allow to stand and filter through a sintered glass funnel. It must be emphasised that *this reagent is not suitable for the routine detection of carbonyl compounds* since it also gives a precipitate in the cold with certain amines, esters and other compounds: if, however, a dilute solution of the ketonic compound in ethanol is treated with a few drops of the reagent and the mixture diluted with water and heated, the precipitate produced with non-ketonic compounds generally dissolves.

For the preparation of 2,4-dinitrophenylhydrazones, dissolve the carbonyl compound (say, 0.5 g) in 5 ml of ethanol and add the calculated volume of the reagent. If a precipitate does not form immediately, dilute with a little water. Collect the derivative and recrystallise it as above.

p-NITROPHENYLHYDRAZONES
Reflux a mixture of 0.5 g of *p*-nitrophenylhydrazine, 0.5 g of the aldehyde (or ketone), 10–15 ml of ethanol and 2 drops of glacial acetic acid for 10 minutes. Add more ethanol if the boiling solution is not homogeneous. Cool the clear solution, filter off the *p*-nitrophenylhydrazone and recrystallise it from ethanol or acetic acid.

Alternatively, dissolve approximately equivalent amounts of the aldehyde (or ketone) and the solid reagent in the minimum volume of cold glacial acetic acid, and reflux for 15 minutes. The *p*-nitrophenylhydrazone separates on cooling or upon careful dilution with water.

PHENYLHYDRAZONES
Dissolve 0.5 g of *colourless* phenylhydrazine hydrochloride (**CAUTION**) and 0.8 g of sodium acetate in 5 ml of water, and add a solution of 0.2–0.4 g of the aldehyde (or ketone) in a little ethanol (free from aldehydes and ketones). Shake the mixture until a clear solution is obtained and add a little more ethanol, if necessary. Warm on a water bath for 10–15 minutes and cool. Filter off the crystalline derivative, and recrystallise it from dilute ethanol or water; sometimes benzene or light petroleum (b.p. 60–80 °C) may be used.

SEMICARBAZONES
Dissolve 1 g of semicarbazide hydrochloride and 1.5 g of crystallised sodium acetate in 8–10 ml of water, add 0.5 g of the aldehyde or ketone and shake. If the mixture is turbid, add alcohol (acetone-free) or water until a clear solution is obtained; shake the mixture for a few minutes and allow to stand. Usually the semicarbazone crystallises from the cold solution on standing, the time varying from a few minutes to several hours. The reaction may be accelerated, if necessary, by warming the mixture on a water bath for a few minutes and then cooling in ice-water. Filter off the crystals, wash with a little cold water and recrystallise from water or from methanol or ethanol either alone or diluted with water.

Note. When semicarbazide is heated in the absence of a carbonyl compound for long periods, condensation to *biurea*, $NH_2CONH\cdot NHCONH_2$, m.p. 247–250 °C (decomp.), may result; occasionally this substance may be produced in the normal preparation of a semicarbazone that forms slowly. Biurea is sparingly soluble in alcohol and soluble in hot water, whereas semicarbazones with melting points in the same range are insoluble in water: this enables it to be readily distinguished from a semicarbazone.

OXIMES

The method given for semicarbazones (above) may be employed: use 1 g of hydroxylamine hydrochloride, 2 g of crystallised sodium acetate and 0.5 g of the aldehyde or ketone. It is usually advisable to warm on a water bath for 10 minutes.

For water-insoluble aldehydes or ketones, the following alternative procedure may be used. Reflux a mixture of 0.5 g of the aldehyde or ketone, 0.5 g of hydroxylamine hydrochloride, 5 ml of ethanol and 0.5 ml of pyridine on a water bath for 15–60 minutes. Remove the ethanol either by distillation (water bath) or by evaporation of the hot solution in a stream of air (water pump). Add 5 ml of water to the cooled residue, cool in an ice bath and stir until the oxime crystallises. Filter off the solid, wash it with a little water and dry. Recrystallise from ethanol (95% or more dilute), benzene or benzene–light petroleum (b.p. 60–80 °C).

Note. All aldehydes, and also those ketones which have two different groups attached to the carbonyl grouping, are capable of yielding two stereoisomeric oximes, hydrazones or semicarbazones. As a general rule, however, one of the stereoisomerides is formed in much greater amount than the other, and no doubt therefore arises as to the purity of the ketonic compound under investigation; occasionally a mixture of stereoisomerides is obtained, which may be difficult to separate by recrystallisation. The formation, therefore, of one of the above derivatives having indefinite melting point and obvious heterogeneity does not necessarily imply the presence of an impure ketonic substance.

DIMEDONE DERIVATIVES (aldehydes only)

Dimedone (5,5-dimethylcyclohexane-1,3-dione) in saturated aqueous solution or in 10 per cent alcohol solution gives crystalline derivatives (2) with aldehydes, but not with ketones. The reaction is:

(2)

(3)

The condensation products (conveniently called alkylidene dimethones) are almost insoluble in water, but can be crystallised from *dilute* ethanol. Dimedone is therefore a good reagent for the detection and characterisation of aldehydes.

The alkylidene dimethone (2) upon boiling with glacial acetic acid, acetic anhydride, hydrochloric acid and other reagents frequently loses water and passes into the anhydride, or dimethone anhydride (3) (a substituted octahydro-xanthene) which often serves as another derivative. The derivatives (2) are soluble in dilute alkali and the resulting solutions give colorations with ferric chloride solution; on the other hand, the anhydrides (3) are insoluble in dilute alkali and hence can easily be distinguished from the alkylidene dimethones (2).

Add 0.1 g of the aldehyde in 5 ml of 50 per cent ethanol to 2 ml of a 10 per cent or saturated ethanolic solution of dimedone. If a precipitate does not form immediately, warm for 5 minutes; if the solution is still clear at the end of this period, add hot water until the mixture is just cloudy and cool to about 5 °C. Collect the crystalline derivative and recrystallise it from methanol–water or ethanol–water.

To prepare the anhydride, boil a solution of 0.1 g of the dimethone derivative (2) in 5 ml of 80 per cent ethanol to which 1 drop of concentrated hydrochloric acid has been added, for 5 minutes, then add hot water until the mixture is just turbid, cool and collect the anhydride by filtration. Recrystallise it from dilute methanol.

BENZYLIDENE DERIVATIVES

Compounds containing the ketomethylene group ($-CH_2 \cdot CO-$) react with benzaldehyde to yield benzylidene derivatives:

$$R^1 \cdot CO \cdot CH_2 \cdot R^2 + Ph \cdot CHO \longrightarrow R \cdot CO \cdot C(=CH \cdot Ph)R^2 + H_2O$$

$$R^1 \cdot CH_2 \cdot CO \cdot CH_2 \cdot R^2 + 2Ph \cdot CHO \longrightarrow$$
$$R^1 \cdot C(=CH \cdot Ph) \cdot CO \cdot C(=CH \cdot Ph)R^2 + 2H_2O$$

Dissolve 1 g of the ketomethylene compound and 1.1 g or 2.2 g of pure benzaldehyde (according as to whether the compound may be regarded as $R \cdot CO \cdot CH_2 \cdot R$ or as $R \cdot CH_2 \cdot CO \cdot CH_2 \cdot R$) in about 10 ml of rectified (or industrial) spirit, add 0.5 ml of 5 M sodium hydroxide solution, shake and allow the mixture to stand for about an hour at room temperature. The benzylidene derivative usually crystallises out or will do so upon 'scratching' the walls of the vessel with a glass rod. Filter off the solid, wash it with a little cold ethanol and recrystallise it from absolute ethanol. (See also Expt 6.135.)

AZINES

Aldehydes react with hydrazine to yield azines: the reaction cannot usually be arrested at the hydrazone stage. This reaction may be illustrated by the preparation of benzylideneazine from benzaldehyde:

$$2Ph \cdot CHO + N_2H_4 \cdot H_2SO_4 \xrightarrow[-H_2SO_4]{aq. \ NH_3} Ph \cdot CH=N \cdot N=CH \cdot Ph + 2H_2O$$

Stir a mixture of 2.4 g of powdered hydrazine sulphate, 18 ml of water and 2.4 ml of concentrated aqueous ammonia (d 0.88), and add 4.6 g (4.4 ml) of benzaldehyde (free from benzoic acid) dropwise, with stirring, over a period of 30–60 minutes. Stir the mixture for a further hour, collect the solid by suction filtration and wash it with water. Recrystallise from 8 ml of rectified spirit. The yield of benzylideneazine (yellow needles), m.p. 92–93 °C, is 3.6 g.

The physical constants of the various derivatives of aliphatic and aromatic aldehydes and ketones are given in Tables 10.15, 10.16, 10.17 and 10.18.

9.6.14 QUINONES

REDUCTION TO THE HYDROQUINONE
Dissolve, or suspend, 0.5 g of the quinone in 5 ml of ether or benzene and shake vigorously with a solution of 1.0 g of sodium dithionite ($Na_2S_2O_4$) in 10 ml of M sodium hydroxide until the colour of the quinone has disappeared. Separate the alkaline solution of the hydroquinone, cool it in ice and acidify with concentrated hydrochloric acid. Collect the product (extract with ether, if necessary) and recrystallise it from ethanol or water.

REDUCTIVE ACETYLATION
Suspend 0.5 g of the quinone in 2.5 ml of pure acetic anhydride, and add 0.5 g of zinc powder and 0.1 g of powdered, anhydrous sodium acetate. Warm the mixture gently until the colour of the quinone has largely disappeared and then boil for 1 minute. Add 2 ml of glacial acetic acid and boil again to dissolve the product and part of the precipitated zinc acetate. Decant the hot solution from the zinc acetate and zinc, and wash the residue with 3–4 ml of hot glacial acetic acid. Combine the solutions, heat to boiling, carefully add sufficient water to hydrolyse the acetic anhydride and to produce a turbidity. Cool the mixture in ice, filter off the diacetate of the hydroquinone and recrystallise it from dilute ethanol or from light petroleum.

THIELE ACETYLATION
Quinones, when treated with acetic anhydride in the presence of perchloric acid or of concentrated sulphuric acid (strong acid catalyst), undergo simultaneous reductive acetylation and substitution to yield triacetoxy derivatives, e.g. benzoquinone gives 1,2,4-triacetoxybenzene (cf. Expt 6.130).

Add 0.1 ml of concentrated sulphuric acid or of 72 per cent perchloric acid cautiously to a cold solution of 0.01 mol (or 1.0 g) of the quinone in 3–5 ml of acetic anhydride. Do not permit the temperature to rise above 50 °C. Allow to stand for 15–30 minutes and pour into 15 ml of water. Collect the precipitated solid and recrystallise it from ethanol.

SEMICARBAZONES
The preparation of these derivatives is described on p. 1258.

QUINOXALINES FROM *o*-QUINONES
The preparation of these derivatives is described in Expt 8.44.

The melting points of the derivatives of a selection of quinones are collected in Table 10.19.

9.6.15 CARBOXYLIC ACIDS

AMIDES, ANILIDES AND *p*-TOLUIDIDES
The dry acid is first converted by excess of thionyl chloride into the acid chloride:

$$R\cdot CO_2H + SOCl_2 \longrightarrow R\cdot COCl + SO_2 + HCl$$

The by-products are both gaseous and the excess of thionyl chloride (b.p. 78 °C) may be readily removed by distillation. Interaction of the acid chloride with ammonia solution, aniline or *p*-toluidine yields the amide, anilide or *p*-toluidide respectively:

$$R \cdot COCl + 2NH_3 \longrightarrow R \cdot CONH_2 + NH_4Cl$$
$$R^1 \cdot COCl + 2R^2NH_2 \longrightarrow R^1 \cdot CONHR^2 + R^2NH_2,HCl$$

Place 0.5–1.0 g of the dry acid (finely powdered if it is a solid) into a 25-ml flask fitted with a reflux condenser, add 2.5–5.0 ml of redistilled thionyl chloride and reflux gently for 30 minutes; it is advisable to place a plug of cotton wool* in the top of the condenser to exclude moisture. Rearrange the condenser and distil off the excess of thionyl chloride† (b.p. 78 °C). The residue in the flask consists of the acid chloride and can be converted into any of the derivatives given below.

Amides. Treat the acid chloride cautiously with about 20 parts of concentrated ammonia solution (*d* 0.88) and warm for a few moments. If no solid separates on cooling, evaporate to dryness on a water bath. Recrystallise the crude amide from water or dilute ethanol.

Alternatively stir the acid chloride with an equivalent weight of ammonium acetate in 10 ml of acetone at room temperature for one hour, filter the mixture and evaporate the acetone, and crystallise the residual amide from water or from dilute ethanol.

Anilides. Dilute the acid chloride with 5 ml of pure ether (or benzene), and add a solution of 2 g of pure aniline in 15–20 ml of the same solvent until the odour of the acid chloride has disappeared; excess of aniline is not harmful. Shake with excess of dilute hydrochloric acid to remove aniline and its salts, wash the ethereal (or benzene) layer with 3–5 ml of water and evaporate the solvent (**CAUTION**). Recrystallise the anilide from water, dilute ethanol or toluene–light petroleum (b.p. 60–80 °C).

p-*Bromoanilides* are similarly prepared with *p*-bromoaniline.

***p*-Toluidides.** Proceed as under *Anilides*, but substitute *p*-toluidine for aniline.

Anilides and p-*toluidides* may also be *prepared directly from the acids*‡ by heating them with aniline or *p*-toluidine respectively:

$$R^1 \cdot CO_2H + R^2NH_2 \longrightarrow R^1 \cdot CONHR^2 + H_2O$$

Place 1.0 g of the monobasic acid and 2 g of aniline or *p*-toluidine in a dry test tube, attach a short air condenser and heat the mixture in an oil bath at 140–160 °C for 2 hours: do not reflux too vigorously an acid that boils below this temperature range and only allow steam to escape from the top of the condenser. For a sodium salt, use the proportions of 1 g of salt to 1.5 g of the base. If the acid is dibasic, employ double the quantity of amine and a reaction temper-

---

* This is more convenient than the conventional calcium chloride guard-tube and possesses the advantages of cheapness and hence can easily be renewed for each experiment.
† If the boiling point of the acid chloride is too near that of thionyl chloride to render separation by distillation practicable, the excess of the reagent can be destroyed by the addition of pure formic acid:

$$HCO_2H + SOCl_2 \longrightarrow CO + SO_2 + 2HCl$$

‡ Alternatively, the alkali metal salts of the acids may be heated with the hydrochloride of the appropriate base.

ature of 180–200 °C: incidentally, this procedure is recommended for dibasic acids since the latter frequently give anhydrides with thionyl chloride. Powder the cold reaction mixture, triturate it with 20–30 ml of 10 per cent hydrochloric acid* and recrystallise from dilute ethanol.

p-BROMOPHENACYL ESTERS

p-Bromophenacyl bromide reacts with the alkali metal salts of acids to form crystalline *p*-bromophenacyl esters:

$$R \cdot CO_2Na + BrCH_2 \cdot CO \cdot C_6H_4Br \, (p) \xrightarrow{-NaBr} R \cdot CO_2CH_2 \cdot CO \cdot C_6H_4Br(p)$$

Dissolve or suspend 0.5 g of the acid in 5 ml of water in a small conical flask, add a drop or two of phenolphthalein indicator, and then 4–5 per cent sodium hydroxide solution until the acid is just neutralised. Add a few drops of very dilute hydrochloric acid so that the final solution is *faintly* acid (litmus).† Introduce 0.5 g of *p*-bromophenacyl bromide (m.p. 109 °C) dissolved in 5 ml of rectified (or industrial) spirit, and heat the mixture under reflux for 1 hour: if the mixture is not homogeneous at the boiling point or a solid separates out, add just sufficient ethanol to produce homogeneity. (Di- and tri-basic acids require proportionately larger amounts of the reagent and longer refluxing periods.) Allow the solution to cool, filter the separated crystals at the pump, wash with a little alcohol and then with water. Recrystallise from dilute ethanol: dissolve the solid in hot ethanol, add hot water until a turbidity just results, clear the latter with a few drops of ethanol and allow to cool. Acetone may sometimes be employed for recrystallisation.

p-NITROBENZYL ESTERS

p-Nitrobenzyl bromide (m.p. 100 °C) reacts with the alkali metal salts of acids to give *p*-nitrobenzyl esters:

$$R \cdot CO_2Na + BrCH_2 \cdot C_6H_4 \cdot NO_2(p) \xrightarrow{-NaBr} R \cdot CO_2CH_2 \cdot C_6H_4 \cdot NO_2(p)$$

It is important that the solution of the sodium salt be *faintly* acid in order that the formation of coloured by-products in the subsequent reaction may be prevented. If the molecular weight of the monobasic acid is known, it is desirable to employ a slight excess of the sodium salt, since excess of the latter is more easily removed than the unchanged reagent.

Use the procedure given above for *p*-bromophenacyl esters. If the ester does not crystallise out on cooling, reheat the reaction mixture, and add small portions of hot water to the point of incipient cloudiness and allow to cool.

p-PHENYLPHENACYL ESTERS

p-Phenylphenacyl bromide reacts with soluble salts of organic acids to yield crystalline *p*-phenylphenacyl esters.

---

* When the derivative is appreciably soluble in ether, the following alternative procedure may be employed. Dissolve the cold reaction mixture in about 50 ml of ether, wash it with 20–30 ml of 10 per cent hydrochloric acid (to remove the excess of base), followed by 20 ml of 10 per cent sodium hydroxide solution, separate the ether layer and evaporate the solvent (**CAUTION**). Recrystallise the residue from dilute ethanol.

† If the sodium salt of the acid is available, dissolve 0.5 g in 5 ml of water, add a solution of 0.5 g of the reagent in 5 ml of ethanol and proceed as detailed in the text after just acidifying (litmus) with dilute hydrochloric acid.

$$R \cdot CO_2Na + BrCH_2 \cdot CO \cdot C_6H_4 \cdot Ph(p) \xrightarrow{-NaBr} R \cdot CO_2CH_2 \cdot CO \cdot C_6H_4 \cdot Ph(p)$$

The procedure is similar to that given under p-*Bromophenacyl esters* and p-*Nitrobenzyl esters* above. Add a weighed amount of acid (0.005 mol) to 5 ml of water in a small conical flask and neutralise it with 0.5 M sodium carbonate or M sodium hydroxide. The final solution should be faintly acid to litmus (add more of the organic acid or a few drops of dilute hydrochloric acid); unless this precaution is taken, coloured by-products are formed which are very difficult to remove. (If the alkali metal salt is available, dissolve 0.005 mol in 5 ml of water, and render the solution just acid to litmus by the addition of dilute hydrochloric acid.) Introduce 10 ml of ethanol, and if the salt of the organic acid is not thrown out of solution, add 0.005 mol of *p*-phenylphenacyl bromide. Dibasic and tribasic acids will require 0.01 and 0.015 mol respectively. Reflux the mixture for periods of up to 1, 2 or 3 hours according to the basicity of the acid. If the salt of the organic acid is precipitated by the ethanol, add more water until the salt dissolves. Some of the esters are sparingly soluble in the reaction mixture and crystallise from the boiling solution; in most cases, however, crystal formation does not occur until the mixture is cooled. In some instances it may be necessary to concentrate the solution before crystallisation occurs. Recrystallise the crude *p*-phenylphenacyl ester from ethanol, dilute ethanol, acetone or toluene.

Certain dibasic acids, of which the sodium or potassium salts are sparingly soluble in dilute ethanol, cause difficulty; these should be neutralised with ethylamine solution.

S-BENZYLISOTHIOURONIUM SALTS

S-Benzylisothiouronium chloride reacts with the alkali metal salts of organic acids to produce crystalline *S*-benzylisothiouronium salts:

$$\left[Ph \cdot CH_2 \cdot S \cdot C \underset{NH_2}{\overset{NH_2}{\diagdown}} \right]^{\oplus} Cl^{\ominus} + R \cdot CO_2^{\ominus}Na^{\oplus} \xrightarrow{-NaCl} \left[Ph \cdot CH_2 \cdot S \cdot C \underset{NH_2}{\overset{NH_2}{\diagdown}} \right]^{\oplus} R \cdot CO_2^{\ominus}$$

It is important not to allow the reaction mixture to become appreciably alkaline, since the free base then decomposes rapidly yielding phenylmethanethiol, which has an unpleasant odour.

Dissolve (or suspend) 0.25 g of the acid in 5 ml of warm water, add a drop or two of phenolphthalein indicator and neutralise carefully with c. M sodium hydroxide solution. Then add 2–3 drops of c. 0.1 M hydrochloric acid to ensure that the solution is almost neutral (*pale* pink colour). (Under alkaline conditions the reagent tends to decompose to produce the evil-smelling phenylmethanethiol.) If the sodium salt is available, dissolve 0.25 g in 5 ml of water and add 2 drops of c. 0.1 M hydrochloric acid. Introduce a solution of 1 g of *S*-benzylisothiouronium chloride in 5 ml of water, and cool in ice until precipitation is complete. Recrystallise the crude derivative from dilute ethanol or from hot water.

With some acids (e.g. succinic acid and sulphanilic acid) more satisfactory results are obtained by reversing the order of mixing, i.e. by adding the solution of the sodium salt of the acid to the reagent. In view of the proximity of the melting points of the derivatives of many acids, the mixed m.p. test (Section 2.33) should be applied.

ANHYDRIDES

1,2-Dicarboxylic acids are readily converted into cyclic anhydrides when heated alone or in acetic anhydride. Heat 0.5 g of the acid in 2–3 ml of refluxing acetic anhydride for 30 minutes. Remove most of the excess reagent by distillation and crystallise the residual cyclic anhydride from chloroform or toluene. The melting points of aliphatic and aromatic anhydrides are to be found in Tables 10.23 and 10.24 respectively. For conversion of the anhydrides into anilic acids see Section 9.6.16, below.

FUSION WITH SODA-LIME

An additional useful test for aromatic carboxylic acids is to distil the acid or its sodium salt with soda-lime. Heat 0.5 g of the acid or its sodium salt with 0.5 g of soda-lime in an ignition tube to make certain that there is no danger of explosion. Then grind together 0.5 g of the acid with 3 g of soda-lime, place the mixture in a Pyrex test tube and cover it with an equal bulk of soda-lime. Fit a wide delivery tube dipping into an empty test tube. Clamp the tube near the mouth. Heat the soda-lime first and then the mixture gradually to a dull-red heat. Examine the product: this may consist of aromatic hydrocarbons or derivatives, e.g. phenol from salicylic acid, anisole from anisic acid, toluene from toluic acid, etc.

The melting points of the derivatives of aliphatic and aromatic carboxylic acids are collected in Tables 10.20 and 10.21.

9.6.16   CARBOXYLIC ACID CHLORIDES AND ANHYDRIDES

HYDROLYSIS TO THE ACIDS

A general procedure is to hydrolyse the acid chloride (or anhydride) by warming with dilute alkali and acidifying the resulting solution with dilute hydrochloric acid to Congo red. If the acid is sparingly soluble, filter it off and characterise it in the usual way. If no precipitate of carboxylic acid is obtained, adjust the pH of the solution to neutrality to phenolphthalein and evaporate to dryness. Use the mixture of the sodium salt of the acid and sodium chloride thus obtained for the preparation of a suitable derivative (e.g. the p-bromophenacyl ester).

CONVERSION INTO ANILIDES

Acid chlorides are converted directly to the corresponding anilides by reaction with aniline, as described in Section 9.6.15, p. 1261. For anhydrides, heat a mixture of 1 g of the anhydride and 1 g of aniline in a boiling water bath for 5 minutes, add 5 ml of water, boil and cool. Crystallise the resulting product from water or from aqueous ethanol.

ANILIC ACIDS FROM CYCLIC ANHYDRIDES

Dissolve 0.5 g of the anhydride in 15 ml of toluene by heating on a water bath, and add a solution of 0.5 ml of aniline in 3 ml of toluene. If the anilic acid does not separate after a short time, cool the solution, wash it with a little dilute hydrochloric acid to remove the excess of aniline and evaporate the solvent; the anilic acid will then usually crystallise. Recrystallise from aqueous ethanol. When heated above their melting points, anilic acids dehydrate to form cyclic imides, e.g.

$$O{=}\!\!\underset{O}{\diagdown}\!\!{=}O \xrightarrow{\text{PhNH}_2} HO_2C\diagup\diagdown CO\cdot NH\cdot Ph \xrightarrow{-H_2O} O{=}\!\!\underset{\underset{Ph}{N}}{\diagdown}\!\!{=}O$$

The melting points of the derivatives of some dicarboxylic acids are:

Succinic; anilic acid 148 °C, imide 156 °C.
Phthalic; anilic acid 169 °C, imide 205 °C.
3-Nitrophthalic;   —   , imide 136 °C.

The physical properties of a number of carboxylic acid chlorides and anhydrides are given in Tables 10.22, 10.23 and 10.24.

9.6.17 ESTERS

HYDROLYSIS
Esters are usually characterised by hydrolysis followed by identification of the alcoholic and acidic components, although it is possible to prepare derivatives of both these components directly from the ester.

In the routine examination of esters it is often a good plan to carry out two hydrolyses, [(A) and (B) below] one for the isolation and characterisation of the parent acid, and the other for the isolation and identification of the parent alcohol.

A. Drop 1 g of sodium into 10 ml of methanol in a small flask provided with a small water condenser; heat the mixture until all the sodium has dissolved. Cool, and add 1 g of the ester and 0.5 ml of water. Frequently the sodium salt of the acid will be deposited either at once or after boiling for a few minutes. If this occurs, filter off the solid at once, wash it with a little methanol and convert it into the p-bromophenacyl ester, p-nitrobenzyl ester or S-benzyl isothiouronium salt (for experimental details, see Section 9.6.15, p. 1261). If no solid separates, continue the boiling for 30–60 minutes, boil off the alcohol, allow to cool, render the product just neutral to phenolphthalein with dilute sulphuric or hydrochloric acid, convert the sodium salt present in solution into a crystalline derivative (Section 9.6.15, p. 1261) and determine its melting point.

B. Boil 2 g of the ester with 30 ml of 10 per cent sodium or potassium hydroxide solution under reflux for at least 1 hour. If the alcohol formed is water (or alkali) soluble, the completion of the hydrolysis will be indicated by the disappearance of the ester layer. Distil off the liquid through the same condenser and collect the first 3–5 ml of distillate. If a distinct layer separates on standing (or upon saturation of half the distillate with potassium carbonate), remove this layer with a capillary dropper, dry it with a little anhydrous potassium carbonate or anhydrous calcium sulphate and determine the b.p. by the Siwoloboff method (Section 2.34). Whether an insoluble alcohol separates out or not, prepare a crystalline derivative (e.g. the 3,5-dinitrobenzoate, Section 9.6.4, p. 1241) and determine its m.p.

The residue in the flask will contain the sodium (or potassium) salt of the acid together with excess of alkali. Just acidify with dilute sulphuric acid and observe whether a crystalline acid separates; if it does, filter, recrystallise and identify (Section 9.6.15, p. 1261). If no crystalline solid is obtained, the solution may be just neutralised to phenolphthalein and the solution of the akali salt used for the preparation of a crystalline derivative. This will confirm, if necessary, the results of hydrolysis by method *A*. If the time factor is important, either method *A* or the product of the caustic alkali hydrolysis may be used for the identification of the acid.

The following notes may be useful:

1. Some esters, e.g. methyl formate, dimethyl oxalate, dimethyl succinate, dimethyl and diethyl tartrate, are appreciably soluble in water. These are usually easily hydrolysed by alkali.

2. If the original ester is a fat or oil and produces an odour of acrolein when heated, it may be a *glyceride*. Esters of ethylene glycol and of glycerol with simple fatty acids are viscous and of high b.p. They are hydrolysed (method A) and the alcohol distilled off. The residue is diluted (a soap may be formed) and acidified with hydrochloric acid (Congo red paper). The acid is filtered or extracted with ether. If no acid can be isolated by these methods, it must be simple and volatile, and should be separated by distillation. The residual aqueous solution of glycol or glycerol is neutralised, evaporated to a syrup on a water bath and extracted with ethanol or with ethyl acetate; the solvent is evaporated and the glycol or glycerol in the residue is identified as usual.

3. *β-Keto esters* (e.g. ethyl acetoacetate) are soluble in solutions of caustic alkalis but not in sodium carbonate solution. They give colours with freshly prepared ferric chloride solution; a little ethanol should be added to bring the ester into solution. Sodium ethoxide solution reacts to yield sodio compounds, which usually crystallise out in the cold. They are hydrolysed by boiling sulphuric acid to the corresponding ketones, which can be identified as usual (Section 9.6.13, p. 1257).

A slight modification in the procedure for isolating the products of hydrolysis is necessary for phenolic (or phenyl) esters since the alkaline solution will contain both the alkali phenate and the alkali salt of the organic acid: upon acidification, both the phenol and the acid will be liberated. Two methods may be used for separating the phenol and the acid.

(i) Acidify the cold alkaline reaction mixture with dilute sulphuric acid (use litmus or Congo red paper) and extract both the acid and the phenol with ether. Remove the acid by washing the ethereal extract with saturated sodium hydrogen carbonate solution until effervescence ceases; retain the aqueous washings. Upon evaporating the ether, the phenol remains; it may be identified by its action upon ferric chloride solution, by the formation of a crystalline derivative with bromine water, or by any of the methods given in Section 9.6.6, p. 1248. Acidify the aqueous washings with dilute sulphuric acid while stirring steadily, and investigate the organic acid (Section 9.6.15, p. 1261).

(ii) Add dilute sulphuric acid, with stirring, to the cold alkaline solution until the solution is acid to litmus or Congo red paper, and the acid, if a solid, commences to separate as a faint permanent precipitate. Now add dilute sodium carbonate solution until the solution is alkaline (litmus paper) and any precipitate has completely redissolved. Extract the clear solution twice with ether; evaporate or distil the ether from the ethereal solution on a water bath, and identify the residual phenol as under (i). Remove the dissolved ether from the aqueous solution by boiling, acidify with dilute sulphuric acid and identify the organic acid present (see Section 9.6.15, p. 1261).

DIRECT IDENTIFICATION OF THE ALCOHOLIC COMPONENT OF AN ESTER

The alcohol components of many simple esters may be identified as the crystalline 3,5-dinitrobenzoates (compare Section 9.6.15, p. 1261) by heating them with 3,5-dinitrobenzoic acid in the presence of a little concentrated sulphuric acid:

$$R^1 \cdot CO_2R^2 + 3,5\text{-}(NO_2)_2C_6H_3 \cdot CO_2H \rightleftharpoons$$
$$R^1 \cdot CO_2H + 3,5\text{-}(NO_2)_2C_6H_3 \cdot CO_2R^2$$

The reaction does not appear to be applicable if either of the groups R^1 or R^2 reacts readily with concentrated sulphuric acid; esters of molecular weight in excess of about 250 react with difficulty.

Dissolve 2 drops of concentrated sulphuric acid in 2 ml of the ester and add 1.5 g of 3,5-dinitrobenzoic acid. If the b.p. is above 150 °C heat the mixture, with frequent shaking at first, in an oil bath at about 150 °C. If the 3,5-dinitrobenzoic acid dissolves within 15 minutes, heat the mixture for 30 minutes, otherwise 60 minutes heating is required. Allow the reaction mixture to cool, dissolve it in 25 ml of ether and extract thoroughly with 5 per cent sodium carbonate solution (c. 25 ml). Wash the ethereal solution with water, and remove the ether. Dissolve the residue (which is usually an oil) in 5 ml of hot ethanol, add hot water cautiously until the 3,5-dinitrobenzoate commences to separate, cool and stir. Recrystallise the derivative from dilute ethanol: the yield is 0.1–0.2 g.

DIRECT IDENTIFICATION OF THE ACIDIC COMPONENT OF AN ESTER

The following procedures may be regarded as alternative to that described above involving hydrolysis of the ester.

Anilides or _p_-toluidides of acids from esters. Esters are converted into the corresponding anilides or _p_-toluidides by treatment with anilino- or with _p_-toluidinomagnesium bromide, which are readily obtained from any simple Grignard reagent and aniline or _p_-toluidine:

$$ArNH_2 + RMgX \longrightarrow ArNHMgX + RH$$
$$2ArNHMgX + R^1 \cdot CO_2R^2 \longrightarrow R^1 \cdot C(OMgX)(NHAr)_2 + Mg(OR^2)X \xrightarrow{2H_2O}$$
$$R^1 \cdot CONHAr + ArNH_2 + Mg(OH)X + R^2OH$$

This procedure is speedy, economical and employs materials which are readily available. It is not satisfactory for esters of dibasic acids.

Add 4.0 g (4.0 ml) of pure aniline dropwise to a cold solution of ethylmagnesium bromide prepared from 1.0 g of magnesium, 5.0 g (3.5 ml) of ethyl bromide and 30 ml of pure, sodium-dried ether. When the vigorous evolution of ethane has ceased, introduce 0.02 mol of the ester in 10 ml of anhydrous ether, and warm the mixture on a water bath for 10 minutes; cool. Add dilute hydrochloric acid to dissolve the magnesium compounds and excess of aniline. Separate the ethereal layer, dry it with magnesium sulphate and evaporate the ether. Recrystallise the residual anilide, which is obtained in almost quantitative yield, from dilute ethanol or other suitable solvent.

Alternatively, add a solution of 4.5 g of _p_-toluidine in dry ether to the Grignard reagent prepared from 1.0 g of magnesium as detailed above. Then introduce 1.0 g (or 0.02 mol) of the ester and proceed as described for anilides.

N-Benzylamides of acids from esters. Esters are converted into the N-benzyl-amides of the corresponding acids by heating with benzylamine in the presence of a little ammonium chloride as catalyst:

$$R^1 \cdot CO_2R^2 + Ph \cdot CH_2NH_2 \longrightarrow R^1 \cdot CONHCH_2 \cdot Ph + R^2OH$$

The reaction (which is essentially the direct aminolysis of esters with benzyl-amine) proceeds readily when R^2 is methyl or ethyl. Esters of higher alcohols should preferably be subjected to a preliminary methanolysis by treatment with sodium methoxide in methanol:

$$R^1 \cdot CO_2R^2 + MeOH \xrightarrow{MeCO_2Na} R^1 \cdot CO_2Me + R^2OH$$

N-Benzylamides are recommended when the corresponding acid is liquid and/or water-soluble so that it cannot itself serve as a derivative. The benzyl-amides derived from the simple fatty acids or their esters are not altogether satis-factory since they are often low melting; those derived from most hydroxy acids and from polybasic acids or their esters are formed in good yield and are easily purified. The esters of aromatic acids yield satisfactory derivatives but the method must compete with the equally simple process of hydrolysis and precipi-tation of the free acid, an obvious derivative when the acid is a solid. The pro-cedure fails with esters of keto acids, sulphonic acids and inorganic acids and some halogenated aliphatic esters.

Reflux a mixture of 1 g of the ester, 3 ml of benzylamine and 0.1 g of powdered ammonium chloride for 1 hour in a Pyrex test tube fitted with a short condenser. Wash the cold reaction mixture with water to remove the excess of benzylamine. If the product does not crystallise, stir it with a little water containing a drop or two of dilute hydrochloric acid. If crystallisation does not result, some un-changed ester may be present: boil with water for a few minutes in an evapor-ating dish to volatilise the ester. Collect the sold N-benzylamide on a filter, wash it with a little petroleum, b.p. 100–120 °C, and recrystallise it from dilute ethanol, ethyl acetate or acetone.

If the ester does not yield a benzylamide by this procedure, convert it into the methyl ester by refluxing 1 g for 30 minutes with 5 ml of absolute methanol in which about 0.1 g of sodium has been dissolved. Remove the methanol by distil-lation and treat the residual ester as above.

The melting points of the N-benzylamides are collected in Tables 10.20 and 10.21.

Acid hydrazides from esters. Methyl and ethyl esters react with hydrazine to give acid hydrazides:

$$R \cdot CO_2Me + H_2NNH_2 \longrightarrow R \cdot CONHNH_2 + MeOH$$

The hydrazides are often crystalline and then serve as useful derivatives. Esters of higher alcohols should be converted first to the methyl esters by boiling with sodium methoxide in methanol (see under N-benzylamides).

Place 1.0 ml of hydrazine hydrate (**CAUTION**: *corrosive chemical*) in a test tube fitted with a short reflux condenser. Add 1.0 g of the methyl or ethyl ester dropwise (or portionwise) and heat the mixture gently under reflux for 15 minutes. Then add just enough absolute ethanol through the condenser to pro-duce a clear solution, reflux for a further 2–3 hours, distil off the ethanol and cool. Filter off the crystals of the acid hydrazide, and recrystallise from ethanol, dilute ethanol or from water.

The melting points of the hydrazides of some aliphatic and aromatic acids are collected in Tables 10.20 and 10.21.

Tables 10.25 and 10.26 list the boiling points, densities and refractive indices of a number of selected esters.

9.6.18 PRIMARY AMIDES

HYDROLYSIS

Amides may be hydrolysed by boiling with 10 per cent sodium hydroxide solution to the corresponding acid (as the sodium salt). The alkaline solution should be acidified with dilute hydrochloric acid; the liberated acid if sparingly water-soluble is isolated by filtration, otherwise any water-soluble acidic component may be isolated by extraction with ether or by distillation from the acidic aqueous solution. The procedure is illustrated by the following experimental details for benzamide. Place 1.5 g of benzamide and 25 ml of 10 per cent sodium hydroxide solution in a 100-ml conical or round-bottomed flask equipped with a reflux condenser. Boil the mixture gently for 30 minutes; ammonia is freely evolved. Detach the condenser and continue the boiling in the open flask for 3–4 minutes to expel the residual ammonia. Cool the solution in ice, and add concentrated hydrochloric acid until the mixture is strongly acid; benzoic acid separates immediately. Leave the mixture in ice until cold, filter at the pump, wash with a little cold water and drain well. Recrystallise the benzoic acid from hot water.

Hydrolysis may also be effected (but usually rather less readily) with 20 per cent sulphuric acid.

XANTHYLAMIDES

Xanthhydrol reacts with primary amides with the formation of crystalline xanthylamides or 9-aclyaminoxanthens.

Dissolve 0.25 g of xanthhydrol* in 3.5 ml of glacial acetic acid; if an oil separates (as is sometimes the case with commercial material), allow to settle for a

---

* Best results are obtained with reagent freshly prepared by the reduction of xanthone with sodium amalgam. Prepare an amalgam from 0.9 g of clean sodium and 75 g (5.5 ml) of mercury as described in Section 4.2.70, p. 464, and then warm it in a stoppered Pyrex reagent bottle to 50 °C. Add a cold suspension of 2.5 g of commercial xanthone in 20 ml of rectified spirit, stopper the flask and shake vigorously releasing the pressure from time to time. The temperature rises rapidly to 60–70 °C, the xanthone passes into solution and a transient blue colour develops. After 5 minutes the solution is clear and colourless; after a further 10 minutes shaking, separate the mercury and wash it with 3 ml of ethanol. Filter the warm solution into 200 ml of cold distilled water with shaking. Filter the xanthhydrol under suction, wash with water and dry. The yield of xanthhydrol, m.p. 122–123 °C, is 2.4 g. The product may be recrystallised from ethanol; it is comparatively unstable but may be kept in an alcoholic solution.

short time and decant the supernatant solution. Add 0.25 g of the amide, shake and allow to stand. If a crystalline derivative does not separate in about 10 minutes, warm on a water bath for a period not exceeding 30 minutes, and allow to cool. Filter off the solid xanthylamide (9-acylaminoxanthen) and recrystallise it from dioxane-water or from acetic acid–water, dry at 80 °C for 15 minutes and determine the m.p.

Some amides do not dissolve in glacial acetic acid; in such cases a mixture of 2 ml of glacial acetic acid and 3 ml of water may be used as a solvent for the reaction. Urea may be characterised as the di-xanthyl derivative (m.p. 274 °C) prepared in acetic acid. Di- and tri-chloroacetamide, oxamide, and salicylamide do not give satisfactory results.

The melting points of aliphatic and aromatic primary carboxylic acid amides and those of the corresponding xanthylamides are collected in Tables 10.27 and 10.28.

9.6.19 SUBSTITUTED AMIDES

HYDROLYSIS
Substituted aromatic carboxylic acid amides of the type Ar·CONHR and Ar·CONR$_2$ are only slowly attacked by aqueous alkali and are characterised by hydrolysis under acidic conditions; 70 per cent sulphuric acid (prepared by carefully adding 4 parts of acid to 3 parts of water) is the preferred reagent. Use the general procedure which has been outlined on p. 1229; characterise the acidic and basic components.

FUSION WITH SODA-LIME
The basic component is liberated upon fusion with sode-lime and at the same time the aroyl group yields a hydrocarbon. Use the experimental details described for carboxylic acids, Section 9.6.15, p. 1261; characterise the base by the preparation of a suitable derivative.

The melting points of some typical substituted aromatic amides are collected in Table 10.29. Other examples will be found in the appropriate columns of Tables 10.20, 10.33 and 10.34.

9.6.20 NITRILES

HYDROLYSIS TO THE CARBOXYLIC ACID
Hydrolysis with alkali. When nitriles are treated with 20–40 per cent sodium or potassium hydroxide solution, there is no reaction in the cold; upon prolonged boiling hydrolysis proceeds comparatively slowly (compare primary amides which are rapidly hydrolysed) to the sodium salt of the acid and ammonia. The reaction is complete when ammonia is no longer evolved:

$$R\cdot CN + H_2O + NaOH \longrightarrow R\cdot CO_2Na + NH_3$$

The excess of alkali is then neutralised to phenolphthalein or to Congo red with dilute hydrochloric acid and the solution is evaporated to dryness on the water bath. The acid may then be characterised as the S-benzylisothiouronium salt or as the p-bromophenacyl ester (Section 9.6.15, p. 1263). In many instances the derivative may be prepared directly from the neutralised solution.

1271

Hydrolysis with acid. Most nitriles are hydrolysed by boiling with 5–8 times the weight of 50–75 per cent sulphuric acid under reflux for 2–3 hours:

$$2R \cdot CN + H_2SO_4 + 4H_2O \longrightarrow 2R \cdot CO_2H + (NH_4)_2SO_4$$

If the acid is a simple aliphatic monobasic acid it can usually be distilled directly from the reaction mixture. If this procedure is not possible, the reaction mixture is poured into excess of crushed ice, and the acid is isolated by ether extraction or by other suitable means. The acid is then characterised (Section 9.6.15, p. 1261). The addition of hydrochloric acid (as sodium chloride; say, 5% of the weight of sulphuric acid) increases the rate of the reaction.

A mixture of 50 per cent sulphuric acid and glacial acetic acid may be used with advantage in the case of difficultly-hydrolysable aromatic nitriles. The reaction product is poured into water, and the organic acid is separated from any unchanged nitrile or from amide by means of sodium carbonate solution.

Those nitriles which yield water-insoluble amides (e.g. the higher alkyl cyanides), *conversion to the amide* often leads to a satisfactory derivative. The hydration if effected by warming a solution of the nitrile in concentrated sulphuric acid for a few minutes, cooling and pouring into water. For experimental details see p. 1229.

REDUCTION TO A PRIMARY AMINE AND CONVERSION INTO A SUBSTITUTED PHENYLTHIOUREA.

Reduction of a nitrile with sodium and ethanol yields the primary amine, which may be identified by direct conversion into a substitiuted phenylthiourea.

$$R \cdot CN + 2H_2 \xrightarrow[\text{ROH}]{\text{Na}} R \cdot CH_2NH_2 \xrightarrow{\text{PhNCS}} R \cdot CH_2NH \cdot CS \cdot NH \cdot Ph$$

Dissolve 1.0 g of the nitrile in 20 ml of absolute ethanol in a dry 200-ml round-bottomed flask fitted with a reflux condenser. Add through the top of the condenser 1.5 g of clean sodium (previously cut into small pieces) at such a rate that the reaction, although vigorous, remains under control. When all the sodium has reacted (10–15 minutes), cool the reaction mixture to about 20 °C, and add 10 ml of concentrated hydrochloric acid dropwise through the condenser while swirling the contents of the flask vigorously: the final solution should be acid to litmus. Connect the flask to a still-head and condenser, and distil off about 20 ml of liquid (dilute ethanol). Cool the flask and fit a small dropping funnel into the top of the still-head. Place 15 ml of 40 per cent sodium hydroxide solution in the dropping funnel, attach an adapter to the end of the condenser and so arrange it that the end dips into about 3 ml of water contained in a 50-ml conical flask. Add the sodium hydroxide solution dropwise and with shaking: a vigorous reaction ensues. When all the alkali has been added, separate the amine by distillation until the contents of the flask are nearly dry.

Add 0.5 ml of phenyl isothiocyanate to the distillate and shake the mixture vigorously for 3–4 minutes. If no derivative separates, crystallisation may be induced by cooling the flask in ice and 'scratching' the walls with a glass rod. Filter off the crude product, wash it with a little 50 per cent ethanol and recrystallise from hot dilute ethanol.

α-IMINOALKYLMERCAPTOACETIC ACID HYDROCHLORIDES

Mercaptoacetic acid (thioglycollic acid) reacts with nitriles in the presence of hydrogen chloride to give α-iminoalkylmercaptoacetic acid hydrochlorides:

$$R{\cdot}CN + HSCH_2{\cdot}CO_2H + HCl \longrightarrow R{\cdot}C\underset{SCH_2{\cdot}CO_2H}{\overset{\overset{\oplus}{N}H_2\}\overset{\ominus}{Cl}}{<}}$$

These salts have sharp and reproducible decomposition temperatures but no true melting points. They act as dibasic acids when titrated with standard alkali, thymol blue being used as indicator.

Dissolve 1.0 g of the nitrile and 2.0 g of mercaptoacetic acid in 25 ml of sodium-dried ether in a dry test tube or small flask. Cool the solution in ice and saturate it with dry hydrogen chloride (5–10 minutes). Stopper the test tube or flask and keep it at 0 °C until crystallisation is complete (15–60 minutes). Collect the crystals by suction filtration, wash with anhydrous ether and dry in a vacuum desiccator over potassium hydroxide pellets (to remove hydrogen chloride) and paraffin wax shavings (to remove ether).

ACYL PHLOROGLUCINOLS

Crystalline derivatives of aliphatic nitriles may be prepared by an application of the Hoesch reaction. Equimolecular proportions of phloroglucinol and the nitrile react in dry ethereal solution in the presence of anhydrous zinc chloride and hydrogen chloride to give an imine hydrochloride, which is converted into a solid alkyl trihydroxyphenyl ketone by hydrolysis. The alkyl 2,4,6-trihydroxyphenyl ketones are usually highly crystalline solids of sharp melting point and are purified by recrystallisation from hot water. Many contain water of crystallisation which can be removed by drying *in vacuo* at about 100 °C; the melting points of both the hydrated and anhydrous compound should be determined.

A detailed preparative procedure is described in Expt 6.125, where the reaction is formulated. The following concise general instructions are suitable for small-scale working.

Add 0.4 g of powdered, anhydrous zinc chloride to a solution of 1.1 g of anhydrous phloroglucinol in 25 ml of sodium-dried ether, and introduce the nitrile (0.01 mol) dissolved in 5 ml of dry ether. Pass a steady stream of dry hydrogen chloride through the solution for 25–30 minutes; it becomes turbid after 2–3 minutes but the turbidity subsequently disappears. Decant the supernatant liquid, dissolve the residual oil or crystals in 25 ml of water and shake the aqueous solution with two 20 ml portions of ether. Concentrate the aqueous layer to about 10–12 ml. The hydroxy ketone separates upon cooling; recrystallise it from hot water and dry in the air. The hydrate is thus produced.

Physical data for aliphatic and aromatic nitriles are collected in Tables 10.30 and 10.31 respectively.

9.6.21 PRIMARY AND SECONDARY AMINES

ACETYL DERIVATIVES

Primary and secondary amines are best acetylated with acetic anhydride:

$$RNH_2 + (Me{\cdot}CO)_2O \longrightarrow Me{\cdot}CO{\cdot}NHR + Me{\cdot}CO_2H$$
$$R^1R^2NH + (Me{\cdot}CO)_2O \longrightarrow Me{\cdot}CO{\cdot}NR^1R^2 + Me{\cdot}CO_2H$$

Acetyl chloride is not so satisfactory since an equivalent quantity of the amine hydrochloride is simultaneously produced:

$$2RNH_2 + Me \cdot COCl \longrightarrow Me \cdot CO \cdot NHR + R\overset{\oplus}{N}H_3\}\overset{\ominus}{Cl}$$

Reflux gently in a test tube under a short air condenser 1 g of the base with 2.5 moles [or 3.0 g (3.0 ml) if the molecular weight is unknown] of acetic anhydride for 10–15 minutes. Cool the reaction mixture and pour it into 20 ml of cold water. Boil to decompose the excess of acetic anhydride. When cold, filter the residual insoluble acetyl derivative and wash it with a little cold water. Recrystallise from water or from dilute ethanol.

Certain *ortho*-substituted derivatives of aromatic amines are difficult to acetylate under the above conditions owing to steric hindrance. The process is facilitated by the addition of a few drops of concentrated sulphuric acid (compare Expt 6.59), which acts as a catalyst, and by the use of a large excess of acetic anhydride.

Excellent results may be obtained by conducting the acetylation in aqueous solution. Dissolve 0.5 g of the amine in 2 M hydrochloric acid, and add a little crushed ice. Introduce a solution of 5 g of hydrated sodium acetate in 25 ml of water, followed by 5 ml of acetic anhydride. Shake the mixture in the cold until the smell of acetic anhydride disappears. Collect the solid acetyl derivative, and recrystallise it from water or dilute ethanol.

BENZOYL DERIVATIVES

Both primary and secondary amines form benzoyl derivatives under the conditions of the Schotten–Baumann reaction (see Section 6.6.2, p. 916, for a discussion).

Suspend 1 g (or 1 ml) of the substance in 20 ml of 5 per cent sodium hydroxide solution in a well-corked boiling tube or small conical flask, and add 2 ml of benzoyl chloride, *c*. 0.5 ml at a time, with constant shaking, and cooling in water (if necessary). Shake vigorously for 5–10 minutes until the odour of the benzoyl chloride has disappeared. Make sure that the mixture has an alkaline reaction. Filter off the solid benzoyl derivative, wash it with a little cold water and recrystallise it from ethanol or dilute ethanol.

If the benzoyl derivative is soluble in alkali, precipitate it together with the benzoic acid derived from the reagent by the addition of hydrochloric acid: filter and extract the product with cold ether or light petroleum (b.p. 40–60 °C) to remove the benzoic acid.

The following alternative procedure is sometimes useful.

To a solution of 0.5 g of the amine in 4 ml of dry pyridine and 10 ml of dry benzene (**CAUTION**), add dropwise 0.5 ml of benzoyl chloride. Heat the mixture under reflux on a water bath at 60–70 °C for 20–30 minutes and then pour into 80–100 ml of water. Separate the benzene layer and extract the aqueous layer with 10 ml of benzene. Wash the combined benzene solutions with 5 ml of 5 per cent sodium carbonate solution, followed by 5 ml of water, and dry with magnesium sulphate. Filter off the desiccant through a small fluted filter paper and concentrate the benzene solution to a small volume (3–4 ml). Stir 15–20 ml of hexane into the residue: the crystalline product separates. Filter and wash with a little hexane. Recrystallise from a mixture of cyclohexane with hexane or with ethyl acetate; alternatively use ethanol or dilute ethanol for recrystallisation.

BENZENESULPHONYL AND TOLUENE-*p*-SULPHONYL DERIVATIVES

Treat 1 g (1 ml) of the amine with 4 molar equivalents of 10 per cent sodium or potassium hydroxide solution (say, 20 ml), and add 1.5 moles (or 3 g if the molecular weight is unknown) of benzenesulphonyl or toluene-*p*-sulphonyl chloride in small portions with constant shaking. To remove the excess of acid chloride, either shake vigorously or warm gently. Acidify with dilute hydrochloric acid and filter off the sulphonamide. Recrystallise it from ethanol or dilute ethanol.

If the presence of a disulphonyl derivative from a primary amine is suspected (e.g. formation of a precipitate in alkaline solution even after dilution), reflux the precipitate, obtained after acidifying, with a solution of 1 g of sodium in 20 ml of rectified spirit for 15 minutes. Evaporate the ethanol, dilute with water and filter if necessary; acidify with dilute hydrochloric acid. Collect the sulphonyl derivative and recrystallise it from ethanol or dilute ethanol.

It is generally more convenient to employ the solid toluene-*p*-sulphonyl chloride (m.p. 69 °C) rather than the liquid benzenesulphonyl chloride. Moreover, the benzenesulphonamides of certain secondary amines are oils or low melting point solids that may be difficult to crystallise: the toluene-*p*-sulphonamides usually have higher melting points and are more satisfactory as derivatives.*

Feebly basic amines, e.g. the nitroanilines, generally react so slowly with benzenesulphonyl chloride that most of the acid chloride is hydrolysed by the aqueous alkali before a reasonable yield of the sulphonamide is produced; indeed, *o*-nitroanaline gives little or no sulphonamide under the above conditions. Excellent results are obtained by carrying out the reaction in *pyridine solution*:

$$o\text{-}NO_2 \cdot C_6H_4 \cdot NH_2 + Ph \cdot SO_2Cl \xrightarrow[-HCl]{\text{pyridine}} o\text{-}NO_2 \cdot C_6H_4 \cdot NH \cdot SO_2 \cdot Ph$$

Reflux a mixture of 1 g (1 ml) of the amine, 2–3 g of benzenesulphonyl chloride and 6 ml of pyridine (**CAUTION**) for 30 minutes. Pour the reaction mixture into 10 ml of cold water and stir until the product crystallises. Filter off the solid and recrystallise it from ethanol or dilute ethanol.

Most amines react so rapidly in pyridine solution that the reaction is usually complete after refluxing for 10–15 minutes.

FORMYL DERIVATIVES

Formic acid condenses with primary and secondary amines to yield formyl derivatives:

$$Ar \cdot NHR + H \cdot CO_2H \longrightarrow Ar \cdot N(R) \cdot CHO + H_2O$$

Reflux 0.5 g of the amine with 5 ml of 90 per cent formic acid (**CAUTION**: *in handling*) for 10 minutes, and dilute the hot solution with 10 ml of cold water. Cool in ice and, in some cases, saturate with salt if the derivative does not separate immediately. Filter, wash with cold water and recrystallise from water, ethanol or light petroleum (b.p. 60–80 °C).

---

* For the separation of mixtures of primary, secondary and tertiary amines using benzenesulphonyl or toluene-*p*-sulphonyl chloride (Hinsberg's method), see p. 1288.

DERIVATIVES WITH 3-NITROPHTHALIC ANHYDRIDE

3-Nitrophthalic anhydride reacts with primary and secondary amines to yield nitrophthalamic acids; it does not react with tertiary amines. The phthalamic acid derived from a primary amine undergoes dehydration when heated to 145 °C to give a neutral N-substituted 3-nitrophthalimide. The phthalamic acid from a secondary amine is stable to heat and is, of course, soluble in alkali. The reagent therefore provides a method for distinguishing between, and separating a mixture of, primary and secondary amines.

Heat 0.5 g (or 0.5 ml) of the amine with 0.5 g of pure 3-nitrophthalic anhydride (Expt 6.162) in an oil bath at 145–150 °C for 10–20 minutes, pour the reaction mixture into a small mortar or Pyrex dish and allow it to solidify. Recrystallise from ethanol, aqueous ethanol or ethanol–acetone.

N-SUBSTITUTED PHTHALIMIDES (FROM PRIMARY AMINES)

Phthalic anhydride reacts with primary amines similarly to yield N-substituted phthalimides.

Dissolve 0.5 g of the primary amine and 0.5 g of phthalic anhydride in 5 ml of glacial acetic acid and reflux for 20–30 minutes. (If the amine salt is used, add 1 g of sodium acetate.) The N-substituted phthalimide separates out on cooling. Recrystallise it from ethanol or from glacial acetic acid.

2,4-DINITROPHENYL DERIVATIVES

The halogen atom in 1-chloro-2,4,-dinitrobenzene is reactive and coloured crystalline compounds (usually yellow or red) are formed with primary and with secondary amines:

$$2,4\text{-}(NO_2)_2C_6H_3Cl + R\cdot NH_2 \longrightarrow 2,4\text{-}(NO_2)_2C_6H_3\cdot NH\cdot R + HCl$$

Dissolve 1.0 g (or 1.0 ml) of the amine and 1.0 g of 1-chloro-2,4-dinitrobenzene in 5–10 ml of ethanol, add a slight excess of anhydrous potassium carbonate or of powdered fused sodium acetate, reflux the mixture on a water bath for 20–30 minutes and then pour into water. Wash the precipitated solid with dilute sodium carbonate solution, followed by dilute hydrochloric acid. Recrystallise from ethanol, dilute ethanol or glacial acetic acid.

Note. Chlorodinitrobenzene must be handled with care (use disposable gloves). It if touches the skin, wash it off with industrial spirit and then copiously with water.

PHENYLTHIOUREAS

Primary and secondary amines react with phenyl isothiocyanate to yield phenylthioureas:

$$Ph \cdot N{=}C{=}S + RNH_2 \longrightarrow Ph \cdot NH \cdot CS \cdot NHR$$
$$Ph \cdot N{=}C{=}S + R_2NH \longrightarrow Ph \cdot NH \cdot CS \cdot NR_2$$

Phenyl isothiocyanate is not sensitive to water; the reaction may be carried out with an aqueous solution of an amine.

Dissolve equivalent quantities of the reagent and of the amine in a small amount of rectified spirit. If no reaction appears to take place in the cold, reflux the mixture for 5–15 minutes. Upon cooling (and 'scratching' with a glass rod, if necessary) the crystalline thiourea separates. Recrystallise it from rectified spirit or from 60–80 per cent ethanol.

Alternatively, mix equal amounts (say, 0.2 g of each) of the amine and phenyl isothiocyanate in a test tube and shake for 2 minutes. If no reaction occurs, heat the mixture gently for 2 minutes and then cool in ice until the mass solidifies. Powder the solid, wash it with a little light petroleum (b.p. 100–120 °C) and recrystallise from rectified spirit.

1-Naphthyl isothiocyanate yields crystalline 1-naphthylthioureas and is similarly applied.

PICRATES

Picric acid combines with amines to yield salts (picrates), which usually possess characteristic melting points. Most picrates have the composition 1 mol amine : 1 mol picric acid. The picrates of the amines, particularly of the more basic ones, are generally more stable than the molecular π-complexes formed between picric acid and the hydrocarbons (compare Section 9.6.3, p. 1238).

If the amine is soluble in water, mix it with a slight excess (about 25%) of a saturated solution of picric acid in water (the solubility in cold water is about 1%). If the amine is insoluble in water, dissolve it by the addition of 2–3 drops of dilute hydrochloric acid (1 : 1) for each 2–3 ml of water, then add a slight excess of the reagent. If a heavy precipitate does not form immediately after the addition of the picric acid solution, allow the mixture to stand for some time and then shake vigorously. Filter off the precipitated picrate and recrystallise it from boiling water, ethanol or dilute ethanol, boiling 10 per cent acetic acid, chloroform, or, best, toluene.

The following alternative procedure may sometimes be employed. Dissolve 0.5 g of the amine in 5 ml of rectified spirit and add 5 ml of a cold saturated solution of picric acid in ethanol. Warm on a *water bath* for 5 minutes and allow to cool. Collect the precipitated picrate and recrystallise it as above.

BENZYLIDENE DERIVATIVES

Primary aromatic amines generally condense directly with benzaldehyde to form benzylidene derivatives (*Schiff's bases* or *anils*).

$$R \cdot NH_2 + Ph \cdot CHO \longrightarrow R \cdot N{=}CH \cdot Ph + H_2O$$

These are often crystalline and therefore useful for the characterisation of primary amines. Diamines may, of course, yield di-benzylidene derivatives.

Heat the amine with 1 or 2 moles of redistilled benzaldehyde (according to whether the base is a monamine or diamine) to 100 °C for 10 minutes; if the molecular weight is unknown, use 1 g of the base and 1 or 2 g of benzaldehyde. Sometimes a solvent, such as methanol (5 ml) or acetic acid, may be used. Recrystallise from ethanol, dilute ethanol or toluene.

The melting points of the derivatives of a number of aliphatic and aromatic primary and secondary amines are collected in Tables 10.32, 10.33 and 10.34.

9.6.22 TERTIARY AMINES

PICRATES
Experimental details are given under *Primary and secondary amines*, Section 9.6.21, above.

METHIODIDES
Methyl iodide (**CAUTION**) reacts with tertiary amines to form the crystalline quaternary ammonium iodide (methiodide):

$$R^1R^2R^3N + MeI \longrightarrow R^1R^2R^3NMe\}^\oplus I^\ominus$$

Some of these derivatives are hygroscopic.

Allow a mixture of 0.5 g of the tertiary amine and 0.5 ml of colourless methyl iodide* to stand for 5 minutes. If reaction has not occurred, warm under reflux for 5 minutes on a water bath and then cool in ice-water. The mixture will generally set solid: if it does not, wash it with a little dry ether and 'scratch' the sides of the tube with a glass rod. Recrystallise the solid product from absolute ethanol or methanol, ethyl acetate, glacial acetic acid or ethanol–ether.

Alternatively, dissolve 0.5 g of the tertiary amine and 0.5 ml of methyl iodide in 5 ml of dry ether or toluene, and allow the mixture to stand for several hours. The methiodide precipitates, usually in a fairly pure state. Filter, wash with a little of the solvent and recrystallise as above.

The ethiodide is prepared similarly, using ethyl iodide.

METHOTOLUENE-*p*-SULPHONATES
Methyl toluene-*p*-sulphonate combines with many tertiary amines to yield crystalline derivatives:

$$R^1R^2R^3N + p\text{-}Me\cdot C_6H_4\cdot SO_3Me \longrightarrow R^1R^2R^3NMe\}^\oplus p\text{-}Me\cdot C_6H_4\cdot SO_3\}^\ominus$$

Dissolve 2–3 g of methyl toluene-*p*-sulphonate in 10 ml of dry toluene, add 1 g of the amine and boil the mixture for 20–30 minutes. Cool, and filter the precipitated quaternary salt. Recrystallise by dissolving the solid in the minimum volume of boling ethanol and then adding ethyl acetate until crystallisation commences. Filter the cold mixture, dry rapidly on a porous plate and determine the m.p. immediately.

The benzyltrialkylammonium salts ($R_3N\cdot CH_2\cdot Ph\}^\oplus Cl^\ominus$) are prepared similarly; 3 g of redistilled benzyl chloride replaces the methyl toluene-*p*-sulphonate.

REACTION WITH NITROUS ACID
N,*N*-Dialkylanilines yield green solid *p*-nitroso derivatives on treatment with

---

* Keep a coil of copper wire (prepared by winding copper wire round a glass tube) or a little silver powder in the bottle, which should be of brown or amber glass; the methyl iodide will remain colourless indefinitely. Ethyl iodide may sometimes give more satisfactory results.

nitrous acid (see p. 1215). Illustrative preparative details for *p*-nitrosodimethyl-aniline are as follows. Dissolve 1.0 g of dimethylaniline in 10 ml of dilute hydro-chloric acid (1 : 1), cool to 0–5 °C and slowly add, with stirring, a solution of 0.70 g of sodium nitrite in 4 ml of water. After 20–30 minutes, filter off the pre-cipitated yellow hydrochloride,* and wash it with a little dilute hydrochloric acid. Dissolve the precipitate in the minimum volume of water, add a solution of sodium carbonate or sodium hydroxide to decompose the hydrochloride (i.e. until alkaline) and extract the free base with ether. Evaporate the ether, and recrystallise the residual green crystals of *p*-nitrosodimethylaniline from light petroleum (b.p. 60–80 °C) or from toluene. The pure compound has m.p. 85 °C.

The melting points of the derivatives of a number of aliphatic and aromatic tertiary amines are collected in Table 10.35.

9.6.23 α-AMINO ACIDS

2,4-DINITROPHENYL DERIVATIVES
The reaction between 1-fluoro-2,4-dinitrobenzene and amino acids leads to 2,4-dinitrophenyl derivatives: these are often crystalline and possess relatively sharp melting points.

$$2,4\text{-}(NO_2)_2C_6H_3F + NH_2 \cdot CHR \cdot CO_2H \xrightarrow[\text{(ii) HCl}]{\text{(i) NaHCO}_3}$$
$$2,4\text{-}(NO_2)_2 \cdot C_6H_3 \cdot NH \cdot CHR \cdot CO_2H$$

To a solution or suspension of 0.25 g of the amino acid in 5 ml of water and 0.5 g of sodium hydrogen carbonate, add a solution of 0.4 g of 1-fluoro-2,4-dinitrobenzene in 3 ml of ethanol. Shake the reaction mixture vigorously and allow to stand for 1 hour with intermittent vigorous shaking. Add 3 ml of satur-ated sodium chloride solution and extract with ether (2 × 5 ml) to remove unchanged reagent. Pour the aqueous layer into 12 ml of cold 5 per cent hydro-chloric acid with vigorous agitation: this mixture should be distinctly acid to Congo red indicator paper. If the product separates as an oil, try to induce crys-tallisation by 'scratching' or stirring. Collect the derivative by suction filtration and recrystallise it from 50 per cent ethanol.

BENZOYL DERIVATIVES
Dissolve 0.5 g of the amino acid in 10 ml of 10 per cent sodium hydrogen carbon-ate solution and add 1 g of benzoyl chloride. Shake the mixture vigorously in a stoppered test tube; remove the stopper from time to time since carbon dioxide is evolved. When the odour of benzoyl chloride has disappeared, acidify with dilute hydrochloric acid to Congo red and filter. Extract the solid with a little cold ether to remove any benzoic acid which may be present. Recrystallise the benzoyl derivative which remains from hot water or from dilute ethanol.

3,5-DINITROBENZOYL DERIVATIVES
The following experimental details are for glycine (aminoacetic acid) and may be easily adapted for any other amino acid. Dissolve 0.75 g of glycine in 20 ml of

---

* The hydrochloride may not separate with other dialkylanilines. Add a slight excess of sodium car-bonate or sodium hydroxide to the solution, extract the free base with ether, etc.

M sodium hydroxide solution and add 2.32 g of finely powdered 3,5-dinitro-benzoyl chloride. Shake the mixture vigorously in a stoppered test tube; the acid chloride soon dissolves. Continue the shaking for 2 minutes, filter (if necessary) and acidify with dilute hydrochloric acid to Congo red. Recrystallise the derivative immediately from water or 50 per cent ethanol.

Excess of the reagent should be avoided, if possible. If excess of dinitro-benzoyl chloride is used, this appears as the acid in the precipitate obtained upon acidification: the acid can be removed by shaking in the cold with a mixture of 5 volumes of light petroleum (b.p. 40–60 °C) and 2 volumes of ethanol. The glycine derivative is insoluble in this medium. For some amino acids (leucine, valine and phenylalanine) acetic acid should be used for acidification.

TOLUENE-*p*-SULPHONYL DERIVATIVES

Amino acids react with toluene-*p*-sulphonyl chloride (compare Section 9.6.21, p. 1275) under the following experimental conditions to yield, in many cases, crystalline toluene-*p*-sulphonylamino acids.

Dissolve 0.01 g equivalent of the amino acid in 20 ml of M sodium hydroxide solution and add a solution of 2 g of toluene-*p*-sulphonyl chloride in 25 ml of ether; shake the mixture mechanically or stir vigorously for 3–4 hours. Separate the ether layer: acidify the aqueous layer to Congo red with dilute hydrochloric acid. The derivative usually crystallises out rapidly or will do so on standing in ice. Filter off the crystals and recrystallise from 4–5 ml of 60 per cent ethanol.

With phenylalanine and tyrosine, the sodium salt of the derivative is sparingly soluble in water and separates during the initial reaction. Acidify the suspension to Congo red: the salts pass into solution and the mixture separates into two layers. The derivative is in the ethereal layer and crystallises from it within a few minutes. It is filtered off and recrystallised.

2.4-DICHLOROPHENOXYACETYL DERIVATIVES

Amino acids react with 2,4-dichlorophenoxyacetyl chloride to give crystalline derivatives:

$$2,4\text{-}Cl_2C_6H_3 \cdot O \cdot CH_2 \cdot COCl + H_2N \cdot CHR \cdot CO_2H \xrightarrow{-HCl}$$
$$2,4\text{-}Cl_2C_6H_3 \cdot O \cdot CH_2 \cdot CO \cdot NH \cdot CHR \cdot CO_2H$$

Dissolve 0.01 mol of the amino acid in 30 ml of M sodium hydroxide solution and cool to 5 °C in a bath of ice. Add, with rapid stirring, 0.01 mol of 2,4-dichlorophenoxyacetyl chloride dissolved in 5 ml of dry toluene at such a rate (5–10 minutes) that the temperature of the mixture does not rise above 15 °C, if the reaction mixture gels after the addition of the acid chloride, add water to thin it. Remove the ice bath and stir for 2–3 hours. Extract the resulting mixture with ether, and acidify the aqueous solution to Congo red with dilute hydrochloric acid. Collect the precipitate by filtration and recrystallise it from dilute ethanol.

Commercial 2,4-dichlorophenoxyacetic acid may be recrystallised from toluene; m.p. 139–140 °C. Reflux 10 g of the acid with 15 ml of thionyl chloride on a steam bath for 1 hour, distil off the excess of thionyl chloride at atmospheric pressure and the residue under reduced pressure: 2,4-dichlorophenoxyacetyl chloride (8 g) passes over at 155–157 °C/22–23 mmHg. It occasionally crystallises (m.p. 44.5–45.5 °C), but usually tends to remain as a supercooled liquid.

PHTHALOYL DERIVATIVES

Many amino acids condense with phthalic anhydride at 180–185 °C to yield crystalline phthaloyl derivatives (*N*-substituted phthalimides).

$$\text{phthalic anhydride (CO—O—CO)} + H_2N\cdot CHR\cdot CO_2H \xrightarrow{-H_2O} \text{phthalimide (CO—N—CO)}\cdot CHR\cdot CO_2H$$

Place 0.5 g of the amino acid and 1.0 g of phthalic anhydride in a Pyrex test tube and immerse the lower part of the tube in an oil bath, which has previously been heated to 180–185 °C. Stir the mixture occasionally during the first 10 minutes and push down the phthalic anhydride which sublimes on the walls into the reaction mixture with a glass rod. Leave the mixture undisturbed for 5 minutes. After 15 minutes, remove the test tube from the bath: when the liquid mass solidifies, invert the test tube and scrape out the excess of phthalic anhydride on the walls. Recrystallise the residue from 10 per cent ethanol or from water.

1-NAPHTHYLUREIDO ACIDS (or 1-naphthylhydantoic acids)

Amino acids react in alkaline solution with 1-naphthyl isocyanate to yield the sodium salts of the corresponding 1-naphthylureido acids, which remain in solution: upon addition of a mineral acid, the ureido acid is precipitated.

$$1\text{-}C_{10}H_7\cdot NCO + H_2N\cdot CHR\cdot CO_2^{\ominus} \longrightarrow 1\text{-}C_{10}H_7\cdot NH\cdot CO\cdot NH\cdot CHR\cdot CO_2^{\ominus}$$

Dissolve 0.5 g of the amino acid in slightly more than the equivalent quantity of M sodium hydroxide solution in a stoppered flask. Add a quantity of 1-naphthyl isocyanate just equivalent to the alkali (if the molecular weight of the compound is not known, use 1 g of the reagent and the corresponding quantity of alkali), stopper the flask and shake vigorously until the odour of the reagent has disappeared. Filter off any insoluble di-1-naphthylurea (resulting from the action of water on any excess reagent), and acidify the filtrate to Congo red with dilute hydrochloric acid. Filter the naphthylureido acid at the pump, wash it with a little cold water and recrystallise from hot water or dilute ethanol.

The phenylureido acid is prepared similarly, using phenyl isocyanate. The latter is more sensitive to water than 1-naphthyl isocyanate and therefore does not keep so well.

The melting points of the derivatives of a number of amino acids are collected in Table 10.36. Most α-amino acids decompose on heating so that the melting points would be more accurately described as decomposition points: the latter vary somewhat with the rate of heating and the figures given are those obtained upon rapid heating.

9.6.24 NITRO COMPOUNDS

REDUCTION TO THE PRIMARY AMINE

Add 10 ml of concentrated hydrochloric acid in small portions to a mixture of 1.0 g of the nitro compound and 3 g of granulated tin contained in a small (say, 50-ml) flask fitted with a reflux condenser. Shake the flask well to ensure thorough mixing during the addition of the acid. After 10 minutes warm under reflux at 100 °C with vigorous shaking until the nitro compound has dissolved and its odour is no longer apparent. (If the nitro compound dissolves slowly, add a few ml of ethanol.) Cool the reaction mixture thoroughly and cautiously make it alkaline with 20–40 per cent sodium hydroxide solution. Isolate the liberated amine by steam distillation or by ether extraction. Characterise the amine by the preparation of a suitable crystalline derivative (see *Primary and secondary amines*, Section 9.6.21, p. 1273).

Some aromatic dinitro compounds, e.g. *m*-dinitrobenzene, may be characterised by partial reduction to the nitroamine. Experimental details using sodium polysulphide as the reducing agent are to be found in Expt 6.51.

OXIDATION OF SIDE CHAINS

Aromatic nitro compounds that contain a side chain (e.g. nitro derivatives of alkylbenzenes) may be oxidised to the corresponding acids either by alkaline potassium permanganate (*Aromatic hydrocarbons*, Section 9.6.3, p. 1238) or, preferably, with a sodium dichromate–sulphuric acid mixture in which medium the nitro compound is more soluble.

Mix 1.0 g of the nitro compound with 4 g of sodium dichromate and 10 ml of water in a 50-ml flask, then attach a reflux condenser to the flask. Add slowly and with shaking 7 ml of concentrated sulphuric acid. The reaction usually starts at once; if it does not, heat the flask gently to initiate the reaction. When the heat of reaction subsides, boil the mixture, cautiously at first, under reflux for 20–30 minutes. Allow to cool, dilute with 30 ml of water and filter off the precipitated acid. Purify the crude acid by extraction with sodium carbonate solution, precipitation with dilute mineral acid and recrystallisation from hot water, toluene, etc.

NITRATION TO A POLY-NITRO COMPOUND

Aromatic mononitro compounds may sometimes be characterised by conversion into the corresponding dinitro or trinitro derivatives. It may be noted that many poly-nitro compounds form characteristic addition compounds with naphthalene.

The nitration of an aromatic compound, especially if its composition is unknown, must be conducted with great care, preferably behind a safety screen, since many aromatic compounds react violently.

A. Add about 0.5 g of the compound to 2.0 ml of concentrated sulphuric acid. Introduce 2.0 ml of concentrated nitric acid drop by drop, with shaking after each addition. Attach a small reflux condenser to the flask and heat in a beaker of water at 50 °C for 5 minutes. Pour the reaction mixture on to 15 g of crushed ice and collect the precipitated solid by suction filtration. Recrystallise from dilute ethanol.

B. Proceed as in *A*, but use 2.0 ml of fuming nitric acid instead of the concentrated nitric acid, and warm the mixture on a boiling water bath for 5–10 minutes.

The physical constants of a number of selected aromatic and alphatic nitro compounds are collected in Tables 10.37 and 10.38 respectively. It will be noted that a few nitro aromatic esters have been included in the tables.

9.6.25 THIOLS

2,4-DINITROPHENYL SULPHIDES AND SULPHONES

Thiols react with 1-chloro-2,4-dinitrobenzene in alkaline solution to yield crystalline thioethers (2,4-dinitrophenyl sulphides) (1). These sulphides (1) can be readily oxidised in glacial acetic acid solution by potassium permanganate to the corresponding sulphones (2); the latter exhibit a wide range of melting points and are therefore particularly valuable for the characterisation of thiols. (See also Expt 5.208.)

$$2,4\text{-}(NO_2)_2C_6H_3Cl + RSNa \xrightarrow[]{-NaCl} 2,4\text{-}(NO_2)_2 \cdot C_6H_3 \cdot SR \xrightarrow[KMnO_4]{[O]}$$

$$(1)$$

$$2,4\text{-}(NO_2)_2C_6H_3 \cdot SO_2R$$

$$(2)$$

Preparation of 2,4-dinitrophenyl sulphides. Dissolve about 0.5 g (or 0.005 mol) of the thiol in 10–15 ml of rectified spirit (or in the minimum volume necessary for solution; warming is permissible) and add 2 ml of 10 per cent sodium hydroxide solution. Mix the resulting solution with a solution of 1 g of 1-chloro-2,4-dinitrobenzene (**CAUTION**: *see Section 9.6.21, p. 1276*) in 5 ml of rectified spirit. Reaction may occur immediately with precipitation of the thioether. In any case reflux the mixture for 10 minutes on a water bath in order to ensure the completeness of the reaction. Filter the hot solution rapidly; allow the solution to cool when the sulphide will crystallise out. Recrystallise from ethanol.

Preparation of the sulphones.   Dissolve the 2,4-dinitrophenyl sulphide (0.005 mol) in the minimum volume of warm glacial acetic acid and add 3 per cent potassium permanganate solution with shaking as fast as decolourisation occurs. Use a 50 per cent excess of potassium permanganate: if the sulphide tends to precipitate, add more acetic acid. Just decolourise the solution with sulphur dioxide (or with sodium metabisulphite or ethanol) and add 2–3 volumes of crushed ice. Filter off the sulphone, dry and recrystallise from ethanol.

3,5-DINITROTHIOBENZOATES

Thiols react with 3,5-dinitrobenzoyl chloride in the presence of pyridine as a catalyst to yield 3,5-dinitrothiobenzoates (compare *Alcohols and polyhydric alcohols*, Section 9.6.4, p. 1241).

$$3,5\text{-}(NO_2)_2C_6H_3 \cdot COCl + RSH \xrightarrow[-HCl]{pyridine} 3,5\text{-}(NO_2)_2C_6H_3 \cdot COSR$$

Mix 0.2 g of 3,5-dinitrobenzoyl chloride, 6 drops of the thiol and 1–3 drops of pyridine in a test tube, and heat the mixture in a beaker of boiling water until fumes of hydrogen chloride cease to appear (10–15 minutes). Add a few drops of water, followed by a drop or two of pyridine to eliminate the excess of the reagent. The product solidifies upon stirring with a glass rod. Add water, filter and recrystallise from dilute ethanol or dilute acetic acid.

HYDROGEN 3-NITROTHIOPHTHALATES

Thiols react with 3-nitrophthalic anhydride to yield hydrogen 3-nitrothiophthalates (compare Section 9.6.4, p. 1243).

Mix 3-nitrophthalic anhydride (0.005 mol or 1.0 g) and the thiol (0.0075 mol, or 1.0 g if the molecular weight it not known) in a test tube and heat gently over a free flame for about 30 seconds. Allow the mixture to cool, and add 0.5 ml of 2 M sodium hydroxide solution dropwise and with cooling in an ice bath. Then add about 0.3 ml of 2 M hydrochloric acid and shake the reaction mixture vigorously. Collect the solid which separates by suction filtration and dry it upon a porous tile. Recrystallise from dilute acetic acid or from aqueous acetone.

The melting points are determined using a bath preheated to about 100 °C. The compounds decompose slightly at the m.p. The melting points of the derivatives of the more commonly occurring thiols are collected in Table 10.39.

9.6.26 SULPHONIC ACIDS

SULPHONAMIDES
Mix together 1.0 g of the dry acid of 1.2 g of the anhydrous salt with 2.5 g of phosphorus pentachloride and heat under a reflux condenser in an oil bath at 150 °C for 30 minutes. Cool the mixture, add 20 ml of dry benzene (**CAUTION**), warm in a steam bath (*fume cupboard*) and stir the solid mass well to extract the sulphonyl chloride: filter.* Add the benzene solution slowly and with stirring to 10 ml of concentrated ammonia solution. If the sulphonamide precipitates, separate it by filtration; if no solid is obtained, evaporate the benzene on a steam bath. Wash the sulphonamide with a little cold water, and recrystallise from water, aqueous ethanol or ethanol to constant m.p.

The procedure is not usually applicable to aminosulphonic acids owing to the interaction between the amino group and the phosphorus pentachloride. If, however, the chlorosulphonic acid is prepared by diazotisation and treatment with a solution of cuprous chloride in hydrochloric acid, the crystalline chlorosulphonamide and chlorosulphonanilide may be obtained in the usual way. With some compounds, the amino group may be protected by acetylation. Sulphonic acids derived from a phenol or naphthol cannot be converted into the sulphonyl chlorides by the phosphorus pentachloride method.

The *sulphonanilides* may be prepared by either of the following methods:

1. Reflux the solution of the sulphonyl chloride in benzene obtained as above, with 2.5 g of aniline for 1 hour. Concentrate the benzene solution to half its volume and cool in ice. Collect the solid which separates on a filter, wash with hot water and recrystallise from ethanol or dilute ethanol.
2. Treat the crude sulphonyl chloride (isolated by evaporating the solvent after extraction with benzene (or ether or chloroform) as above) with 1 g of p-toluidine and 30 ml of c. 2 M sodium hydroxide solution. Shake for 10–15 minutes.

---

* The sulphonyl chloride may be isolated by evaporating the benzene solution and recrystallising the residue from chloroform–light petroleum (b.p. 40–60 °C) or toluene–light petroleum (b.p. 40–60 °C).

Extract the alkaline solution with ether to remove excess of *p*-toluidine, acidify, filter and recrystallise the residue as in (1).

$$R \cdot SO_2ONa + PCl_5 \longrightarrow R \cdot SO_2Cl + POCl_3 + NaCl$$
$$R \cdot SO_2Cl + 2NH_3 \longrightarrow R \cdot SO_2NH_2 + NH_4Cl$$
$$R \cdot SO_2Cl + 2Ph \cdot NH_2 \longrightarrow R \cdot SO_2NH \cdot Ph + Ph \cdot NH_3^{\oplus}Cl^{\ominus}$$

S-BENZYLISOTHIOURONIUM SALTS

(See *Carboxylic acids*, Section 9.6.15, p. 1264.) If the substance is the free sulphonic acid, dissolve 0.5 g of it in 5–10 ml of water, add a drop or two of phenolphthalein indicator and neutralise with *c*. M sodium hydroxide solution. Then add 2–3 drops of 0.1 M hydrochloric acid to ensure that the solution is almost neutral (*pale* pink colour); under alkaline conditions the reagent tends to decompose to produce the evil-smelling phenylmethanethiol.

To a solution of 0.5 g of the salt in 5 ml of water and 2–3 drops of 0.1 M hydrochloric acid (or to a solution of the acid treated as above), add a slight excess of a cold, 15 per cent aqueous solution of *S*-benzylisothiouronium chloride (if the molecular weight of the compound is not known, use a solution of 1 g of the reagent in 5 ml of water), and cool in ice. Filter off the crystalline derivative and recrystallise it from 5 per cent ethanol.

SULPHONACETAMIDES

Sulphonacetamides are derivatives of sulphonamides (Section 9.6.27, p. 1286), but since the latter are readily prepared from the sulphonic acids or their salts, sulphonacetamides may be employed for the characterisation of sulphonic acids; for this reason they are included in this Section.

Sulphonamides upon heating with acetyl chloride are converted into the *N*-acetyl derivatives or sulphonacetamides:

$$Ar \cdot SO_2NHR + Me \cdot COCl \xrightarrow{-HCl} Ar \cdot SO_2N(R) \cdot CO \cdot Me \quad \text{(R = H or alkyl)}$$

The sulphonacetamides (R=H) are freely soluble in sodium hydrogen carbonate solution thus rendering purification facile. Sulphonacetamides are moderately strong acids, and can generally be titrated in aqueous or aqueous–alcoholic solution with phenolphthalein as indicator. The acidic properties of sulphonacetamides may be used to effect a *separation of a sulphonamide from an N-alkylsulphonamide*. Acetylation of such a mixture gives a sulphonacetamide and an *N*-alkylsulphonacetamide, of which only the former is soluble in sodium hydrogen carbonate solution. Both sulphonacetamides and *N*-alkylsulphonacetamides are readily hydrolysed by boiling with excess of 5 per cent potassium hydroxide solution for about 1 hour, followed by acidification with dilute hydrochloric acid, giving the corresponding sulphonamides and *N*-alkylsulphonamides respectively.

Reflux 1 g of the sulphonamide with 2.5 ml of acetyl chloride for 30 minutes; if solution is not complete within 5 minutes, add up to 2.5 ml of glacial acetic acid. Remove the excess of acetyl chloride by distillation on a water bath, and pour the cold reaction mixture into water. Collect the product, wash with water and dissolve it in warm sodium hydrogen carbonate solution. Acidify the filtered solution with glacial acetic acid; filter off the precipitated sulphonacetamide and recrystallise it from aqueous ethanol.

p-TOLUIDINE SALTS OF SULPHONIC ACIDS

These are prepared by the interaction of the sulphonic acid or its sodium salt with *p*-toluidine hydrochloride in aqueous solution:

$$R \cdot SO_3^{\ominus} Na^{\oplus} + p\text{-Me} \cdot C_6H_4 \cdot NH_2 \xrightarrow{+HCl} (p\text{-Me} \cdot C_6H_4 NH_3)^{\oplus} (R \cdot SO_3)^{\ominus} + NaCl$$

Dissolve 1.0 g of the alkali metal salt of the sulphonic acid in the minimum volume of hot water and add 0.5 g of *p*-toluidine and 2 ml of concentrated hydrochloric acid. If a solid separates or the *p*-toluidine does not dissolve completely, add more hot water and a few drops of concentrated hydrochloric acid until a clear solution is obtained at the boiling point. Cool the solution; if crystallisation does not occur immediately, 'scratch' the walls of the test tube to induce crystallisation. Collect the product by suction filtration, and recrystallise it from hot water containing a drop of concentrated hydrochloric acid, or from dilute ethanol.

The melting points of the derivatives of a number of selected sulphonic acids are collected in Table 10.40; the melting points of the corresponding sulphonyl chlorides are included for purposes of reference. The acids do not possess sharp melting points; the sulphonic acids are therefore arranged in groups of related compounds. In a subsidiary Table (10.40A), a number of sulphonic acids are arranged in the order of increasing melting points of the *S*-benzylisothiouronium salts.

9.6.27 SULPHONAMIDES

HYDROLYSIS

Sulphonamides are most readily identified by hydrolysis with 80 per cent sulphuric acid (for experimental details see p. 1230).

$$R^1 \cdot SO_2NHR^2 + H_3O^{\oplus} \longrightarrow R^1 \cdot SO_3H + R^2 \cdot \overset{\oplus}{N}H_3$$

$$Ar \cdot SO_2NHR + H_3O^{\oplus} \longrightarrow Ar \cdot SO_3H + R \cdot \overset{\oplus}{N}H_3$$

$$Ar \cdot SO_2NR^1R^2 + H_3O^{\oplus} \longrightarrow Ar \cdot SO_3H + R^1R^2\overset{\oplus}{N}H_2$$

The amine is liberated by the addition of alkali and then characterised by a suitable derivative; the sulphonic acid may then be recovered as the sodium salt and converted into a suitable crystalline derivative, e.g. the *S*-benzylisothiouronium salt.

XANTHYLSULPHONAMIDES

Primary sulphonamides (R·SO$_2$NH$_2$) may be most simply characterised by reaction with xanthhydrol to yield the corresponding *N*-xanthylsulphonamides (cf. *Primary amides*, Section 9.6.18, p. 1270). Dissolve 0.25 g of xanthhydrol and 0.25 g of the primary sulphonamide in 10 ml of glacial acetic acid. Shake for 2–3 minutes at the laboratory temperature and allow to stand for 60–90 minutes. Filter off the derivative, recrystallise it from dioxane–water (3:1) and dry at room temperature.

SULPHONACETAMIDES

Sulphonamides may also be characterised as sulphonacetamides: for experimental details see Section 9.6.26, above).

The melting points of sulphonamides, both aliphatic and aromatic, together with the melting points of some *N*-xanthylsulphonamides and sulphonacetamides are collected in Tables 10.41 and 10.40.

9.7 SEPARATION OF MIXTURES OF ORGANIC COMPOUNDS

9.7.1 INTRODUCTION

The organic chemist is frequently faced with the task of separating mixtures of compounds. Since few organic reactions give a single product, this commonly occurs in the work-up stage of a reaction when it is desired to isolate a single pure compound from a mixture which may contain unreacted starting materials and reagents, inorganic material produced from reagents, and expected and unexpected by-products (Section 2.18). The chemist may also need to separate mixtures of naturally occurring substances obtained from plant or animal sources. Earlier chapters have illustrated many of the techniques used by the organic chemist to separate mixtures as part of a reaction work-up. The particular method used will depend on the chemical and physical properties of the substances involved. A simple, standard procedure is as follows. Pour the reaction mixture into water; extract the organic material with a low-boiling solvent such as ether or dichloromethane by shaking in a separatory funnel with several successive portions of the solvent; combine the organic solutions and dry with a suitable drying agent (Section 2.23); remove the solvent by distillation or by using a rotary evaporator; purify the residual organic product by distillation (Section 2.24) or recrystallisation (Section 2.20). The reaction work-up procedure may be modified to take advantage of particular physical or chemical properties of one or more components of the mixture, such as volatility, solubility characteristics and acidic or basic properties, and some of these methods are described in this section. The mixture may be examined by one or more chromatographic methods (Section 2.31), commonly thin-layer chromatography (t.l.c.), gas–liquid chromatography (g.l.c.) or high performance liquid chromatography to determine the number of components. The mixture may also be investigated by spectroscopic techniques (Section 9.4) and chemical methods (Section 9.5) to give an indication of the nature of the components. In the light of all this information the chemist can decide on the appropriate method of separation. The greater the similarity in chemical and physical properties, the more challenging does the task of separation become. Examples are included in earlier chapters of the separation of structural isomers (for example: toluene-*p*- and *o*-sulphonyl chlorides, Expt 6.41; *o*- and *p*-hydroxypropiophenones, Expt 6.107; *o*- and *p*-hydroxybenzaldehydes, Expt 6.116; 3- and 4-nitrophthalic acids, Expt 6.162), geometric isomers (for example, (*E*)- and (*Z*)-azobenzenes, Section 2.31.2, p. 223) and optical isomers (for example: (±)-α-methylbenzylamine, Expt 5.219; (±)-octan-2-ol, Expt 5.220; and (±)-alanine, Expt 5.221).

In order to gain an understanding of the principles involved and of the variety of methods which may be employed, it can be useful practice for the organic chemistry student to carry out exercises on the investigation and separation of given mixtures. It is impossible, however, to give a set of procedures which will be applicable, without modification, to the great variety of combinations which may be encountered. Advantage should be taken of any facts which have emerged in the preliminary examination and to adapt, if necessary, the general schemes given below to the mixture under examination. The preliminary examination is therefore of fundamental importance.

Broadly speaking, the separation of the components of mixtures may be divided into three main groups.

SEPARATIONS BASED UPON DIFFERENCES IN THE CHEMICAL PROPERTIES OF THE COMPONENTS

A mixture of toluene and aniline may be separated by extraction with dilute hydrochloric acid: the aniline passes into the aqueous layer in the form of the salt, aniline hydrochloride, and may be recovered by neutralisation. Similarly, a mixture of phenol and toluene may be separated by treatment with dilute sodium hydroxide. The above examples are, of course, simple applications of the fact that the various components fall into different solubility groups (compare Section 9.2). Another example is the separation of a mixture of dibutyl ether and chlorobenzene: concentrated sulphuric acid dissolves only the dibutyl ether and it may be recovered from solution by dilution with water. With some classes of compounds, e.g. unsaturated compounds, concentrated sulphuric acid leads to polymerisation, sulphonation, etc., so that the original component cannot be recovered unchanged: this solvent, therefore, possesses limited application.

Phenols may be separated from acids (for example, o-cresol from benzoic acid) by a dilute solution of sodium hydrogen carbonate: the weakly acidic phenols (and also enols) are not converted into salts by this reagent and may be removed by ether extraction or by other means; the acids pass into solution as the sodium salts and may be recovered after acidification. For experimental details see Sections 2.22 and 9.6.17, p. 1266.

Mixtures of primary, secondary and tertiary amines can usually be separated by Hinsberg's method. This is based on the fact that reaction with benzenesulphonyl (or toluene-p-sulphonyl) chloride converts primary amines into alkali-soluble sulphonamides, secondary amines into alkali-insoluble sulphonamides and leaves tertiary amines unaffected.

$$\text{Ph·SO}_2\text{Cl} + \text{H}_2\text{N·R} \xrightarrow{\text{NaOH}} [\text{Ph·SO}_2\overset{\ominus}{\text{N}}\text{·R}]\text{Na}^{\oplus} \text{ (water-soluble)}$$

$$\xrightarrow{\text{HCl}} \text{Ph·SO}_2\text{NH·R} \text{ (water-insoluble)}$$

$$\text{Ph·SO}_2\text{Cl} + \text{NHR}_2 \xrightarrow{\text{NaOH}} \text{Ph·SO}_2\text{NR}_2 \text{ (insoluble in alkali)}$$

The following experimental details illustrate how the Hinsberg separation of amines may be carried out in practice.

Treat 2.0 g of the mixture of amines with 40 ml of 10 per cent of sodium hydroxide solution and add 4 g (3 ml) of benzenesulphonyl chloride (or 4 g of toluene-p-sulphonyl chloride) in small portions. Warm on a water bath to complete the reaction. Acidify the alkaline solution with dilute hydrochloric acid when the sulphonamides of the primary and secondary amines are precipitated. Filter off the solid and wash it with a little cold water; the tertiary amine will be present in the filtrate. To convert any disulphonamide that may have been formed from the primary amine into the sulphonamide, boil the solid under reflux with 2.0 g of sodium dissolved in 40 ml of absolute ethanol for 30 minutes. Dilute with a little water and distil off the alcohol: filter off the precipitate of the sulphonamide of the secondary amine. Acidify the filtrate with dilute hydrochloric acid to precipitate the derivative of the primary amine. Recrystallise the respective derivatives from ethanol or from dilute ethanol.

Aldehydes, e.g. benzaldehyde, may be separated from liquid hydrocarbons and other neutral, water-insoluble liquid compounds by shaking with a solution of sodium metabisulphite: the aldehyde forms a solid bisulphite compound, which may be filtered off and decomposed with dilute acid or with sodium bicarbonate solution in order to recover the aldehyde.

A valuable method of separating ketones from other neutral and water-insoluble compounds utilises the *Girard-T reagent.* (Section 4.2.*30*, p. 434). It reacts with the carbonyl compound to yield another quaternary ammonium salt:

$$R^1R^2CO + [Me_3\overset{\oplus}{N}\cdot CH_2\cdot CO\cdot NH\cdot NH_2]Cl^{\ominus} \longrightarrow$$

$$[Me_3\overset{\oplus}{N}\cdot CH_2\cdot CO\cdot NH\cdot N=CR^1R^2]Cl^{\ominus} + H_2O$$

The latter is a polar compound and is therefore soluble in water. Extraction with ether removes the water-insoluble compounds, leaving the salt in the aqueous layer. The ketone is easily regenerated by hydrolysis with dilute hydrochloric acid (p. 434).

SEPARATIONS BASED UPON DIFFERENCES IN THE VOLATILITIES OF THE COMPONENTS IN AQUEOUS SOLUTION

This procedure is generally employed for the water-soluble compounds, and may also be applied to mixtures in which one of the components is slightly soluble in water. The water-soluble compounds include the lower members of the homologous series of alcohols, aldehydes, ketones, acids, esters, amines and nitriles; compounds containing two or more hydroxyl or amino groups, hydroxy-, amino-, di- and poly-basic acids, sulphonic acids and salts. The compounds with one functional group are usually volatile in steam and distil with the water: compounds with two or more functional groups (amino, hydroxyl or carboxyl) are not generally steam-volatile. The stable salts of steam-volatile bases and acids may be decomposed by a stronger non-volatile mineral acid (sulphuric acid or phosphoric acid) or base (sodium or potassium hydroxide), and the organic base or acid separated by steam distillation from the aqueous solution or suspension. The salts of weaker bases and acids (e.g. the amine or ammonium salts of carboxylic acids and the alkali metal salts of some phenols) are sufficiently hydrolysed by boiling water to permit the basic or acidic compound to distil with the water. It should be noted that sulphonic acids and their salts are not volatile in steam. The only disadvantage of this procedure is that certain compounds may decompose or polymerise or otherwise undergo change under the influence of aqueous alkali or acid at 100 °C, and this fact must be borne in mind when the method is employed. Thus sugars are decomposed by alkali; these may sometimes be isolated by evaporation of the solvent at pH 7, preferably under diminished pressure.

The essential basis of the scheme for the separation of water-soluble compounds is, therefore, distillation of (*a*) an aqueous solution of the mixture, (*b*) an alkaline (with sodium hydroxide) solution of the mixture and (*c*) an acidic (with sulphuric or phosphoric acid) solution of the mixture. The residue will contain the non-volatile components, which must be separated from inorganic salts and from each other by any suitable process.

The following are examples of the above procedure. A mixture of diethyl-amine and butan-1-ol may be separated by adding sufficient dilute sulphuric acid to neutralise the base: steam distillation will remove the alcohol. The amine can be recovered by adding sodium hydroxide to the residue and repeating the distillation. A mixture of diethyl ketone and acetic acid may be treated with sufficient dilute sodium hydroxide solution to transform the acid into sodium acetate and distilling the aqueous mixture. The ketone will pass over in the steam and the non-volatile, stable salt will remain in the flask. Acidification with dilute sulphuric acid liberates acetic acid, which can be isolated by steam distillation or by extraction.

SEPARATIONS BASED UPON DIFFERENCES IN THE PHYSICAL PROPERTIES OF THE COMPONENTS

When the separation procedures described in detail above are unsatisfactory for the separation of a mixture of organic compounds, purely physical methods may be employed. Thus a mixture of volatile liquids may be fractionally distilled (compare Section 2.26); or a mixture of non-volatile solids may frequently be separated by making use of the differences in solubilities in inert solvents. The progress of such separations may be monitored by application of the various chromatographic techniques detailed in Section 2.31, or indeed these techniques may be employed on the preparative scale for effecting the separation itself (e.g. flash chromatography, p. 217). The techniques of counter current distribution, fractional crystallisation or fractional sublimation (Section 2.21) may also be employed where appropriate.

9.7.2   PRELIMINARY EXAMINATION OF A MIXTURE

LIQUID MIXTURES
Physical properties. Examine the mixture with regard to odour, viscosity and colour.

Solubility in water. Transfer 1.0 ml of the mixture by means of a calibrated dropper or a small pipette into a small graduated test tube: add 1 ml of water and shake. Observe whether there is complete or partial solution and if there is any sign of chemical reaction. If solution is not complete, add more water (in 1 ml portions) and note (a) if the mixture dissolves completely, and (b) if a portion is insoluble and, if so, whether it is heavier or lighter than the aqueous layer. If an emulsion is formed, it may be assumed that at least one component of the mixture is insoluble in water and at least one component is soluble. Test the aqueous layer obtained with litmus and with phenolphthalein: if there is an acid reaction, test also with 5 per cent sodium hydrogen carbonate solution.

Presence or absence of water. Determine the absence or presence of water in the mixture by one or more of the following tests:

1. Investigate its miscibility with ether or with toluene.
2. Observe its action upon anhydrous copper(II) sulphate.
3. Distil a small portion and note the b.p. and properties of the distillate.
4. Distil 3 ml of the mixture with 3 ml of dry toluene from a dry 10-ml distilling flask. Collect 2 ml of the distillate and dilute it with 5 ml of dry toluene; the

formation of two layers or of distinct drops suspended in the toluene indicates the presence of water.

Behaviour upon distillation. If the original mixture is not an aqueous solution, place 5 ml (or 10 ml) of it in a 10-ml (or 25-ml) flask set for downward distillation, immerse the latter in a small beaker of cold water and gradually heat to boiling. Observe the b.p. of any liquid which passes over and set it aside for subsequent examination. Determine the water-solubility of any residue. If it dissolves in water, examine it by Table 9.4; if insoluble in water, apply Table 9.3 (p. 1294 and p. 1296 respectively).

Test for elements. If the mixture is an aqueous solution, evaporate a small portion (c. 1 ml) to dryness upon platinum foil or in a small crucible. Use a portion of the residue to test for elements (Section 9.3) and another portion for the *Ignition test*, below.

If no water is present in the mixture, use it directly in the tests for elements.

Ignition test. Place 0.1–0.2 g of the mixture on a porcelain crucible cover: heat gently at first over a small flame and finally ignite strongly. Observe:

1. The inflammability and nature of the flame (e.g. smoky or otherwise).
2. Whether a residue is left after ignition; moisten with hydrochloric acid and test with a platinum wire.

Spectroscopic investigation. See Section 9.4.

Miscellaneous tests. Treat 1 ml of the mixture with 5 per cent sodium hydroxide solution until strongly alkaline: note whether an oil or solid separates, whether any ammonia is evolved and any colour changes which occur. Heat to boiling and cool: compare odour with that of the original mixture (a change in odour may indicate the presence of esters). Add dilute hydrochloric acid and observe the result.

Treat 1 ml of the mixture with dilute hydrochloric acid until strongly acid. Note any evolution of gas or the separation of a solid. Add dilute sodium hydroxide solution and observe the effect.

Miscellaneous class reactions. (Section 9.5.) Determine the effect of the following class reagents upon small portions of the original mixture: (i) bromine in dichloromethane solution; (ii) potassium permanganate solution; (iii) alcoholic silver nitrate solution; (iv) 2,4-dinitrophenylhydrazine reagent; (v) Schiff's reagent; (vi) acetyl chloride; (vii) benzoyl chloride (in the presence of aqueous sodium hydroxide); (viii) sodium (only if water is absent); (ix) iron(III) chloride solution; and (x) bromine water.

SOLID MIXTURES
Physical properties. Observe the colour, odour and crystalline form. Examine with a lens or a microscope, if available.

Solubility in water. Determine the solubility of 1.0 g of the sample in water. If in doubt as to whether a portion of the mixture dissolves, remove the supernatant liquid with a dropper and evaporate to dryness on a water bath. Determine the reaction of the aqueous solution or suspension to litmus and to phenolphthalein.

Test for elements, Section 9.3.

Ignition test. Place 0.1 g–0.2 g of the mixture upon a porcelain crucible cover or upon a piece of platinum foil; heat gently at first and finally ignite strongly. Note:

1. Whether the mixture melts and if decomposition occurs.
2. The flammability and the nature of the flame (e.g. if smoky or otherwise).
3. Whether a residue is obtained after ignition (moisten with hydrochloric acid and test with a platinum wire).

Spectroscopic examination. See Section 9.4.

Miscellaneous tests. Test the solubility behaviour of 0.5–1 g of the mixture to 5 per cent sodium hydroxide solution, 5 per cent sodium hydrogen carbonate solution and to 5 per cent hydrochloric acid (for details, see under *Liquid mixtures*).

Miscellaneous class reactions. (Section 9.5.) Determine the effect of the following class reagents upon small portions of the original mixture (for some of the tests an aqueous solution or suspension may be used): (i) bromine in dichloromethane solution; (ii) potassium permanganate solution; (iii) alcoholic silver nitrate solution; (iv) 2,4-dinitrophenylhydrazine reagent; (v) Schiff's reagent; (vi) acetyl chloride; (vii) benzoyl chloride (in the presence of aqueous sodium hydroxide); (viii) iron(III) chloride solution; and (ix) bromine water.

A careful consideration of the results of the above tests will provide much useful information and will indicate which of the following general procedures should be applied and the modifications which are necessary. Thus if nitrogen is absent, it is doubtful whether the separation for bases would be necessary.

9.7.3 SEPARATION OF WATER-INSOLUBLE MIXTURES

If the water-insoluble mixture is a liquid, evaporate a small sample (say, 4 ml) in an evaporating dish on a water bath in order to determine the amount of volatile components, if any. If the solvent distils at the temperature of the boiling water bath, it is advisable to distil off this solvent on a water bath and to replace it by ether.

Place 15–20 ml of the liquid mixture in a 50-ml distilling flask arranged for distillation, and heat the flask on a boiling water bath until no more liquid passes over: redistil the distillate and if it is a single substance, identify it in the usual manner. Dissolve the residue (*R*) in ether and employ the same proportions as given for a solid mixture.

STEP 1. EXTRACTION AND SEPARATION OF THE ACIDIC COMPONENTS

Shake 5–10 g of the solid mixture (or of the residue *R* obtained after the removal of the solvent on a water bath) with 50 ml of pure ether, or other appropriate water-immiscible solvent (e.g. petroleum, dichloromethane, etc.). If there is a residue, separate it by filtration, preferably through a sintered glass funnel, wash it with a little ether and examine it appropriately. Shake the resulting ethereal solution in a small separatory funnel with 15 ml portions of 5 per cent aqueous sodium hydroxide solution until *all* the acidic components have been removed. Three portions of alkali are usually sufficient. Set aside the residual ethereal solution (E_1) for *Step 2*. Combine the sodium hydroxide extracts and wash the resulting mixture with 15–20 ml of ether; discard these washings. Render the alkaline extract acid to litmus with dilute sulphuric acid and then add excess of solid sodium hydrogen carbonate. Alternatively treat the sodium hydroxide extract with Cardice until no more is absorbed.

Separate any *phenolic* or *enolic compounds* which may be present by extracting the sodium hydrogen carbonate solution with two 20 ml portions of ether; remove the ether from the extract and examine any residue for phenols (or enols).

Strongly acidify the residual sodium hydrogen carbonate solution to Congo red with dilute sulphuric acid. If a solid acid forms, filter. Extract the filtrate or the acidified solution with two 20 ml portions of ether: keep the aqueous solution (*A*). Distil off the ether, and add the residual acid (if a solid) to the solid separated by filtration. Identify the acid.

Now distil the filtrate (*A*) and collect the distillate as long as it is acid to litmus. Should any solid separate out in the distilling flask during the distillation, add more water to dissolve it. Set aside the residue (*B*) in the flask. Identify the volatile acid in the distillate. A simple method is to just neutralise it with sodium hydroxide solution, evaporate to dryness and convert the residual sodium salt into the *S*-benzylisothiouronium salt (Section 9.6.15, p. 1264).

The residue (*B*) in the distilling flask may still contain a water-soluble, non-volatile acid. Cool the acid solution, neutralise it with dilute sodium hydroxide solution to Congo red and evaporate to dryness on a water bath under reduced pressure (water pump). Heat a little of the residual salt (*C*) upon the tip of a nickel spatula in a Bunsen flame and observe whether any charring takes place. If charring occurs, thus indicating the presence of organic matter, extract the solid residue with 25 ml portions of hot absolute ethanol. Evaporate the ethanolic extract and identify the material which remains. The residue (*C*) contains the sodium salt of a water-soluble, non-volatile acid, which may be characterised as the *S*-benzylisothiouronium salt.

STEP 2. EXTRACTION OF THE BASIC COMPONENTS
Extract the ethereal solution (E_1) with 15 ml portions of 5 per cent hydrochloric acid until *all* the basic components have been removed: two or three portions of acid are usually sufficient. Preserve the residual ethereal solution (E_2) for the separation of the neutral components. Wash the combined acid extracts with 15–20 ml of ether: discard the ether extract as in *Step 1*. Make the acid extract alkaline with 10–20 per cent sodium hydroxide solution: if any basic component separates, extract it with ether, evaporate the ether and characterise the residue. If a water-soluble base is also present, it may be recognised by its characteristic ammoniacal odour; it may be isolated from the solution remaining after the separation of the insoluble base by ether extraction by distilling the aqueous solution as long as the distillate is alkaline to litmus. Identify the base with the aid of phenyl isothiocyanate (compare Section 9.6.21, p. 1277) or by other means.

STEP 3. THE NEUTRAL COMPONENTS
The ethereal solution (E_2) remaining after the acid extracton of *Step 2* should contain only the neutral compounds of solubility groups 5, 6 and 7 (see Table 9.1). Dry it with a little magnesium sulphate, and distil off the ether. If a residue is obtained, neutral compounds are present in the mixture. Test a portion of this

Table 9.3 General scheme for separation of a water-insoluble mixture

Treat the mixture with pure ether* and filter, if necessary.

Residue. Examine for polysaccharides, etc., according to nature of original mixture.	Filtrate or ether solution. Extract with 5 per cent NaOH solution and separate the ethereal layer.		Ether solution (E_1). Extract with 5 per cent HCl. Separate ether layer.		Ether solution (E_2). This will contain any neutral compounds present. Dry with magnesium sulphate, and distil off the ether. A residue indicates the presence of a neutral component. Determine the solubility of a portion in conc. H_2SO_4. Apply any other suitable tests.
	Sodium hydroxide extract. This will contain the acids and phenols (or enols) present. Acidify (litmus) with dilute H_2SO_4, add excess of solid $NaHCO_3$. Extract with ether.		**Hydrochloric acid extract.** This will contain any basic components present. Render alkaline with 10–20 per cent NaOH and extract with ether.		
	Ether solution. Contains *phenolic* (or *enolic*) *compounds.*	**Sodium hydrogen carbonate solution.** Acidify with dilute H_2SO_4. (i) Filter or extract *acid* with ether. (ii) Distil aqueous solution (*A*) from (i) as long as distillate is acid to recover *water-soluble volatile acids.* (iii) Neutralise aqueous solution (*B*) from (ii), evaporate to recover *water-soluble, non-volatile compounds.* The residue may contain the sodium salt of a water-soluble, non-volatile acid.	**Ether solution.** Contains *water-insoluble amines.*	**Aqueous solution.** Will possess ammoniacal odour of water-soluble amines present. Distil as long as distillate is alkaline to recover *volatile water-soluble amines.*	

* Or other appropriate solvent.

with respect to its solubility in concentrated sulphuric acid; if it dissolves in the acid, pour the solution slowly and cautiously into ice-water and note whether any compound is recovered. Examine the main residue for homogeneity and if it is a mixture devise procedures, based for example upon differences in volatility, solubility in inert solvents, reaction with hydrolytic and other reagents, to separate the components.

The above procedure for water-insoluble mixtures is shown, in outline, in tabular form in Table 9.3. If the mixture is a liquid, the volatile solvent is assumed to have been removed.

9.7.4 SEPARATION OF WATER-SOLUBLE MIXTURES

A water-soluble mixture may be in the form of a mixture of water-soluble solids or in the form of a liquid. The liquid mixtures are frequently aqueous solutions. The preliminary examination of a liquid mixture (p. 1290) will indicate whether a volatile solvent (i.e. removable on a boiling water bath) is present. If a volatile solvent is present, distil 20 g of the mixture from a water bath until no more liquid passes over: set aside the volatile solvent for identification. Dissolve the residue (R) in water as detailed below for a mixture of solids.

STEP 1. DISTILLATION FROM ACID SOLUTION AND THE SEPARATION OF THE VOLATILE ACIDIC AND NEUTRAL COMPOUNDS

Dissolve 6–10 g of a solid mixture in 50–75 ml of water: for a liquid mixture, use 6–10 g of the residue (R) (non-volatile from a boiling water bath) and dilute with 50–75 ml of water: for an aqueous solution use sufficient of it to contain 6–10 g of the dissolved components and dilute, if necessary, to 50–75 ml. Acidify the solution with sufficient 20 per cent sulphuric acid to decompose the salts of all the acidic components and to ensure the presence of a slight excess of acid: many organic acids give an acid reaction with Congo red, hence it is necessary to add the sulphuric acid somewhat beyond the point at which the mixture is acid to this indicator. If an insoluble acidic component separates, filter it off and identify it. Distil the acid solution so long as the distillate appears turbid, or is acid to litmus, or until 100–150 ml are collected: in the last case, add more water to the contents of the distilling flask if the volume has been reduced below one-fourth of the original volume. Keep the residue (R_1) in the distilling flask for *Step 2*.

The distillate may contain volatile neutral compounds as well as volatile acids and phenols. Add a slight excess of 10–20 per cent sodium hydroxide solution to this distillate and distil until it is judged that all volatile organic compounds have passed over into the distillate. If necessary, the determination of the refractive index of the distillate or the application of specific chemical tests (e.g. for carbonyl compounds use the 2,4-dinitrophenylhydrazine reagent) should be used to confirm completion of distillation. Keep this distillate (S_1) for *Step 4*.

Cool the alkaline solution resulting from the distillation of the volatile neutral compounds, make it acid to litmus with dilute sulphuric acid and add an excess of *solid* sodium hydrogen carbonate. Extract this hydrogen carbonate solution with two 20 ml portions of ether; remove the ether from the combined ether extracts and identify the residual phenol (or enol). Then acidify the hydrogen carbonate solution cautiously with dilute sulphuric acid; if an acidic compound separates, remove it by two extractions with 20 ml portions of ether; if the

Table 9.4 General scheme for separation of a water-soluble mixture

Acidify the aqueous solution (50–75 ml) prepared from (a) 6–10 g of the solid mixture, (b) 6–10 g of the liquid residue (R) after distillation from a boiling water bath, or (c) sufficient of original aqueous solution to contain 6–10 g of solute, with 20 per cent H_2SO_4 and distil.

Distillate. This will contain the steam-volatile acidic and neutral components present. Render alkaline with 10–20 per cent NaOH and distil.

Aqueous acid solution (R_1). Render alkaline with 10–20 per cent NaOH and distil.

Under **Distillate**:

Distillate (S_1). This will contain the *volatile neutral components* present. Concentrate by distillation and saturate with solid K_2CO_3; the neutral component may separate.

Aqueous alkaline solution. This will contain any acids or phenols present. Cool, acidify (litmus) with dilute H_2SO_4, and add excess of solid $NaHCO_3$. Extract with ether.

   Ether solution. Contains *phenolic compounds*.

   Sodium hydrogen carbonate solution. Acidify with dilute H_2SO_4. Extract with ether.

      Ether solution. Contains *volatile water-insoluble acids*.

      Aqueous solution. Distil. The distillate may contain *volatile water-soluble acids*.

Under **Aqueous acid solution**:

Distillate. Extract with ether.

   Ether solution. Contains *volatile water-insoluble amines.**

   Aqueous solution. Concentrate by distillation contains *volatile water-soluble amines.**

Aqueous alkaline solution (S_2). Neutralise with dilute H_2SO_4 (Congo red). Evaporate to dryness and extract with absolute ethanol. The alcoholic extract contains the *water-soluble, non-volatile components*.

---

* The separation of a mixture of amines by means of benzenesulphonyl chloride or toluene-*p*-sulphonyl chloride (Hinsberg's method) is described in Section 9.7.1, p. 1288.

acidified solution remains clear, distil and collect any water-soluble, volatile acid in the distillate. Characterise the acid as described in Section 9.6.15, p. 1261.

STEP 2. DISTILLATION FROM ALKALINE SOLUTION

Treat the solution (R_1) remaining in the distilling flask after the volatile acidic and neutral compounds have been removed with 10–20 per cent sodium hydroxide solution until distinctly alkaline. If a solid separates, filter it off and identify it. Distil the alkaline solution until no more volatile bases pass over (distillate no longer turbid, or not basic to litmus: water-soluble bases also possess characteristic odours): add more water to the contents of the flask if the solution becomes too concentrated during this distillation. [Set aside the solution in the distilling flask (S_2) for *Step 3*.] If the volatile basic compounds are insoluble in water, remove them by extraction with two 20 ml portions of ether, and identify the bases (compare Sections 9.6.21, p. 1273 and 9.6.22, p. 1278) after evaporation of the ether. The water-soluble amines may be identified with phenyl isothiocyanate: it is best to concentrate the bases by redistilling and collecting the first half of the distillate separately.

STEP 3. THE NON-STEAM-VOLATILE COMPOUNDS

The alkaline solution (S_2) remaining in the distilling flask from *Step 2* may contain water-soluble, non-volatile acidic, basic or neutral compounds. Add dilute sulphuric acid until the solution is just acid to Congo red, evaporate to dryness and extract the residual solid with boiling absolute ethanol: extraction is complete when the undissolved salt exhibits no sign of charring when heated on a metal spatula in the Bunsen flame. Evaporate the alcoholic solution to dryness and identify the residue.

STEP 4. THE STEAM-VOLATILE NEUTRAL COMPOUNDS

The solution (S_1) containing water-soluble neutral compounds obtained in *Step 1* is usually very dilute. It is advisable to concentrate it by distillation until about one-third to one half of the original volume is collected as distillate; the process may be repeated if necessary. It is frequently possible to salt out the neutral components from the concentrated distillate by saturating it with solid potassium carbonate. If a layer of neutral compound makes its appearance, remove it. Treat this upper layer (which usually contains much water) with solid anhydrous potassium carbonate: if another aqueous layer forms, separate the upper organic layer and add more anhydrous potassium carbonate to it. Identify the neutral compound.

Note. Some neutral compounds (e.g. methanol) cannot be salted out with potassium carbonate: distillation of the saturated aqueous potassium carbonate solution frequently yields the organic compound in a comparatively pure state, or at least in sufficiently concentrated a form to enable certain derivatives to be prepared.

The above procedure for the separation of a water-soluble mixture is summarised in Table 9.4.

CHAPTER 10
PHYSICAL CONSTANTS OF ORGANIC COMPOUNDS

TABLES OF DERIVATIVES

The tables in this chapter contain lists of the more common members of the various classes of organic compounds arranged, as far as possible, in the order of increasing boiling points or melting points, together with the melting points of selected derivatives. Some of the tables are subdivided: thus in Table 10.6, **Phenols**, the additional sub-headings are *Halogeno-phenols* and *Nitro-phenols*. This subdivision renders the tables less cumbersome and facilitates their use.

In describing compounds in the literature, the range of a certain number of degrees is usually recorded for the boiling point or the melting point; the highest point of the boiling point or the melting point range is listed in the tables and the value is given to the nearest whole degree. For those compounds the author has purified or prepared, the actual observed boiling point or melting point is recorded. Densities are given for a temperature of 20 °C referred to water at 4 °C unless otherwise indicated. Refractive indices are recorded for the sodium D line at 20 °C. It must be remembered that the value obtained for a melting point depends upon the observer and upon the method which was used in the determination: this often accounts for the slightly different values found in the literature for the same compound.

Reference should be made to Section 2.3.4, p. 44, for those classes of compounds which are potentially carcinogenic; those marked **C** should be handled with great caution; some are subject to legal control in Great Britain or the USA or both.

INDEX TO TABLES

Table 10.1 Saturated aliphatic hydrocarbons

Hydrocarbon	B.P. °C (/mmHg)	$d_4^{20°}$	$n_D^{20°}$
Pentane	36	0.627	1.358
Hexane	68.5	0.659	1.374
Heptane	98	0.683	1.388
Octane	125	0.703	1.397
Nonane	150.5	0.717	1.405
Decane	173	0.730	1.412
Undecane	196 (87/20)	0.740	1.417
Dodecane	216 (94/14)	0.750	1.422
Tridecane	92.5/4.5	0.756	1.425
Tetradecane	252 (123/12)	0.762	1.429
Pentadecane	270 (120/4.5)	0.769	1.432
Hexadecane	143.5/9 (m.p. 18)	0.774	1.435
Octadecane	308 (m.p. 28)	—	—
2-Methylbutane (isopentane)	28	0.620	1.354
2-Methylpentane	60	0.653	1.372
2,2,4-Trimethylpentane	99	0.688	1.389
2,7-Dimethyloctane	160	0.725	1.409
Cyclopentane	49	0.745	1.406
Cyclohexane	81	0.779	1.426
Methylcyclohexane	101	0.769	1.423
Cycloheptane	118	0.811	1.445
Ethylcyclohexane	130	0.784	1.432
Isopropylcyclohexane	154.5	0.802	1.441
Propylcyclohexane	155	0.790	1.436
Butylcyclohexane	177	0.800	1.440
Isopentylcyclohexane	193	0.802	1.442
Pentylcyclohexane	200	0.804	1.444
Bicyclohexyl	237 (m.p. 3)	0.889	1.480
trans-Decahydronaphthalene (trans-Decalin)	185	0.870	1.470
cis-Decahydronaphthalene (cis-Decalin)	194	0.895	1.481
1,2,3,4-Tetrahydronaphthalene (Tetralin)	207	0.971	1.540
trans-p-Menthane (1)	161	0.792	1.439
cis-p-Menthane (1)	169	0.816	1.451

(1) 1-Isopropyl-4-methylcyclohexane.

Table 10.2   Unsaturated aliphatic hydrocarbons

Hydrocarbon	B.P. °C (/mmHg)	$d_4^{20°}$	$n_D^{20°}$	Adducts with 2,4-dinitrophenyl sulphenyl chloride	Mercurides
Pent-1-ene	30	0.641	1.371	—	—
2-Methylbut-1-ene	31	0.650	1.378	—	—
cis-Pent-2-ene	35	0.659	1.382	—	—
trans-Pent-2-ene	37	0.649	1.379	—	—
Trimethylethylene	38	0.662	1.388	—	—
Hex-1-ene	64	0.674	1.388	62	—
Hept-1-ene	93	0.697	1.400	—	—
Oct-1-ene	121	0.716	1.409	—	—
Non-1-ene	146	0.731	1.413	—	—
Dec-1-ene	169	0.742	1.422	—	—
Undec-1-ene	193	0.779	1.444	—	—
Dodec-1-ene	213	0.760	1.430	—	—
Tetradec-1-ene	125/15	0.773	1.437	—	—
Hexadec-1-ene	153/14 (m.p. 15)	0.782	1.441	—	—
Octadec-1-ene	180/18 (m.p. 18)	0.789	1.445	—	—
Isoprene (2-methylbuta-1,3-diene)	34	0.681	1.419	—	—
Penta-1,3-diene (piperylene)	42	0.680	1.431	—	—
Hexa-1,5-diene (biallyl)	59	0.690	1.402	—	—
2,3-Dimethylbuta-1,3-diene	69	0.726	1.439	—	—
Cyclopentene	45	0.772	1.420	—	—
Cyclohexene	83	0.810	1.445	117	—
Cyclopenta-1,3-diene	42	0.803	1.443	—	—
Dicyclopentadiene	170 (m.p. 32)	—	—	—	—
Cyclohexa-1,3-diene	81	0.841	1.474	—	—
α-Pinene	156	0.860	1.456	—	—
Camphene	160 (m.p. 51)	—	—	—	—
Dipentene	178	0.840	1.473	195	—
Pent-1-yne	39	0.695	1.385	—	118
Pent-2-yne	56	0.712	1.404	—	—
Hex-1-yne	71	0.715	1.399	—	99
Hept-1-yne	100	0.734	1.409	—	61
Oct-1-yne	126	0.746	1.416	—	—
Non-1-yne	151	0.758	1.423	—	—
Phenylacetylene	142	0.925	1.552	—	125

Table 10.3 Aromatic hydrocarbons

Hydrocarbon	B.P. °C	M.P. °C	$d_4^{20°}$	$n_D^{20°}$	Picrate °C
Benzene **C**	80	6	0.879	1.501	—
Toluene	111	—	0.867	1.497	—
Ethylbenzene	135	—	0.868	1.496	—
p-Xylene	138	13	0.861	1.496	—
m-Xylene	139	—	0.864	1.497	—
Phenylacetylene	142	—	0.925	1.552	—
o-Xylene	144	—	0.880	1.505	—
Styrene (1)	146	—	0.909	1.546	—
Isopropylbenzene (2)	153	—	0.862	1.491	—
Propylbenzene	159	—	0.864	1.493	—
Mesitylene (3)	164	—	0.865	1.499	—
t-Butylbenzene	169	—	0.867	1.493	—
Pseudocumene (4)	169	—	—	1.504	—
s-Butylbenzene	172	—	0.861	1.490	—
p-Cymene (5)	177	—	0.857	1.490	—
Indane	177	—	0.965	1.538	—
m-Diethylbenzene	182	—	0.865	1.496	—
Indene	182	—	0.992	1.576	—
Butylbenzene	182	—	0.861	1.490	—
Isodurene (6)	197	—	—	—	—
Prehnitene (7)	204	—	0.901	1.523	—
Pentylbenzene	204	—	0.859	1.488	—
Tetralin (8)	207	—	0.971	1.540	—
1,3,5-Triethylbenzene	218	—	0.863	1.497	—
Cyclohexylbenzene	238	7	0.950	1.533	—
1-Methylnaphthalene	241	—	1.019	1.618	141
Diphenylmethane	262	25	—	—	—
2-Methylnaphthalene	241	34	—	—	115
Bibenzyl	284	52	—	—	—
Pentamethylbenzene	231	54	—	—	131
Biphenyl†	255	70	—	—	—
Durene (9)	193	79	—	—	—
Naphthalene	218	80	—	—	150
Triphenylmethane	358	92	—	—	—
Acenaphthene	278	95	—	—	162
Retene (10)	390	99	—	—	123
Phenanthrene	340	100	—	—	143
Fluorene (11)	294	114	—	—	84
trans-Stilbene (12)	306	124	—	—	94
cis-Stilbene	148/17 mm	5	—	—	—
Pyrene	—	149	—	—	227
1,1'-Binaphthyl	—	160	—	—	145
2,2'-Binaphthyl	—	188	—	—	184
Anthracene	340	216	—	—	138
Chrysene (13)	448	254	—	—	273

(1) Phenylethylene.
(2) Cumene.
(3) 1,3,5-Trimethylbenzene.
(4) 1,2,4-Trimethylbenzene.
(5) 4-Isopropyl-1-methylbenzene.
(6) 1,2,3,5-Tetramethylbenzene.
(7) 1,2,3,4-Tetramethylbenzene.

(8) 1,2,3,4-Tetrahydronaphthalene.
(9) 1,2,4,5-Tetramethylbenzene.
(10) 7-Isopropyl-1-methylphenanthrene.
(11) Diphenylenemethane.
(12) trans-1,2-Diphenylethylene.
(13) 1,2-Benzphenanthrene.

Aroyl-benzoic acid °C	Compound with 2,4,7-trinitro-9-fluorenone	Compound with 1,3,5-trinitro-benzene	Styphnate	Other derivatives* °C
28	—	—	—	1,3-Dinitro, 90
38	—	—	—	2,4-Dinitro, 71
28	—	—	—	2,4,6-Trinitro, 37
32	—	—	—	2,3,5-Trinitro, 139
26	—	—	—	2,4,6-Trinitro, 182
				Mercuride, 125
67	—	—	—	4,5-Dinitro, 71
—	—	—	—	2,4-Dinitrobenzene sulphenyl chloride
				adduct, 143
34	—	—	—	2,4,6-Trinitro, 109
26	—	—	—	—
12	—	—	—	2,4,6-Trinitro, 235
—	—	—	—	2,4-Dinitro, 62
49	—	—	—	3,5,6-Trinitro, 185
—	—	—	—	$CrO_3 \rightarrow C_6H_5COCH_3$
24	—	—	—	2,6-Dinitro, 54
—	—	—	—	—
14	—	—	—	2,4,6-Trinitro, 62
—	—	102	—	—
—	—	—	—	—
13	—	—	—	4,6-Dinitro, 157
—	—	—	—	5,6-Dinitro, 176; Dibromo, 208
—	—	—	—	Dibromo, 64
54	—	—	—	5,7-Dinitro, 95
—	—	—	—	2,4,6-Trinitro, 108
—	—	—	—	4-Nitro, 59
68	—	154	135	4-Nitro, 71
—	—	—	—	2,2',4,4'-Tetranitro, 172
90	127	123	130	1-Nitro, 81
—	—	102	—	4,4'-Dinitro, 180
—	—	121	—	—
20	132	—	—	4,4'-Dibromo, 164
64	—	—	—	3,6-Dinitro, 207
73	153	156	168	1-Nitro, 61
—	—	—	—	4,4',4''-Trinitro, 212; Triphenylcarbinol, 162
00	175	168	154	5-Nitro, 101
—	—	139	141	—
—	196	164	142	Phenanthraquinone, 202
28	179	105	134	2-Nitro, 156; 2,7-Dibromo, 165
—	—	120	142	—
—	—	—	—	—
—	242	245	191	—
—	—	—	—	—
—	194	164	180	Anthraquinone, 286
14	248	186	—	—

For melting points of sulphonamides derived from some of these aromatic hydrocarbons see Table 10.41.

Biphenyl should not be nitrated since 4-nitrobiphenyl is a proven powerful carcinogen.

Table 10.4 Aliphatic alcohols

Alcohol	B.P. °C	M.P. °C	3,5-Dinitro-benzoate °C
Methanol (methyl alcohol)	65	—	109
Ethanol (ethyl alcohol)	78	—	94
Propan-2-ol (isopropyl alcohol)	82	—	122
2-Methylpropan-2-ol (t-butyl alcohol)	83	25	142
Allyl alcohol (prop-2-en-1-ol)	97	—	50
Propan-1-ol (propyl alcohol)	97	—	75
Butan-2-ol (s-butyl alcohol)	99	—	76
2-Methylbutan-2-ol (t-pentyl alcohol)	102	—	118
2-Methylbut-3-yn-2-ol	105	—	—
2-Methylpropan-1-ol (isobutyl alcohol)	108	—	88
3-Methylbutan-2-ol	113	—	76
2,2-Dimethylpropan-1-ol- (neopentyl alcohol)	113	52	—
Prop-2-yn-1-ol (propargyl alcohol)	114	—	—
Pentan-3-ol	116	—	100
Butan-1-ol (butyl alcohol)	118	—	64
Pentan-2-ol	119	—	62
2-Methylpentan-2-ol	121	—	72
3-Methylpentan-3-ol	128	—	—
2-Methylbutan-1-ol	129	—	70
3-Methylbutan-1-ol (isopentyl alcohol)	132	—	62
Hexan-3-ol	135	—	77
Pentan-1-ol (pentyl alcohol)	138	—	46
Cyclopentanol	141	—	115
2-Ethylbutan-1-ol	149	—	52
Heptan-4-ol	156	—	64
Hexan-1-ol	156	—	61
Heptan-2-ol	160	—	49
Cyclohexanol	161	25	113
2-Methylcyclohexanol	165	—	115
4-Hydroxypentan-2-one (diacetone alcohol)	166	—	55
Furfuryl alcohol	170	—	81
4-Methylcyclohexanol	174	—	134
3-Methylcyclohexanol	175	—	98
Heptan-1-ol	176	—	48
Tetrahydrofurfuryl	177	—	84
Octan-2-ol	179	—	32
Octan-1-ol	194	—	62
(−)-Linalool	199	—	—
Nonan-1-ol	214	—	52
(−)-Isoborneol	216	—	—
(±)-α-Terpineol	219	35	79
Geraniol	230	—	63
Decan-1-ol	231	6	57
Undecan-1-ol	243	16	55
Dodecan-1-ol	259	24	60

ɒ-Nitro-benzoate °C	Phenyl-urethan °C	1-Naphthyl-urethan °C	Hydrogen 3-nitro-phthalate °C	O-Alkyl saccharin °C	Other derivatives °C
96	47	124	153	182	—
57	52	79	157	219	
10	86	106	153	137	—
16	136	101	—	—	—
29	70	109	124	—	—
35	57	80	145	125	—
26	64	98	131	66	—
85	42	72	—	—	—
—	102	—	—	—	d_4^{20} 0.807, n_D^{20} 1.421
68	86	104	179	100	—
—	69	109	—	—	—
—	144	100	—	—	—
—	63	—	—	—	d_4^{20} 0.948, n_D^{20} 1.432
17	49	95	121	—	—
36	61	72	147	96	—
17	—	76	103	—	—
—	239	—	—	—	—
—	50	—	—	—	—
—	—	82	158	—	—
21	57	68	166	64	—
—	—	—	—	—	—
11	46	68	136	62	—
62	132	118	—	—	—
—	—	61	147	—	—
35	—	80	—	—	—
5	42	59	124	60	—
—	—	54	—	—	—
50	82	129	160	—	—
65	103	155	—	—	—
48	—	—	—	—	2,4-Dinitrophenylhydrazone, 203
76	45	129	—	—	—
—	125	160	—	—	—
—	94	122	—	—	—
10	65	62	127	55	—
47	61	90	—	—	Diphenylcarbamate, 81
28	114	64	—	—	—
12	74	66	128	46	—
70	66	53	—	—	—
10	69	65	125	49	—
29	138	—	130	—	—
97	113	152	—	—	—
35	—	48	117	—	Diphenylcarbamate, 82
30	60	71	123	48	—
29	62	73	123	59	—
45	74	80	124	54	—

Table 10.4 Aliphatic alcohols (*continued*)

Alcohol	B.P. °C	M.P. °C	3,5-Dinitrobenzoate °C
Tetradecan-1-ol	160/10	39	67
(−)-Menthol	216	43	153
Hexadecan-1-ol	190/15	50	66
But-2-yne-1,4-diol	144/15	55	191
Octadecan-1-ol	—	59	66
(−)-Borneol	212	205	154
2-Chloroethanol (ethylene chlorohydrin)	129	—	92
3-Chloropropan-1-ol (trimethylene chlorohydrin)	161d	—	77
1,3-Dichloropropan-2-ol (glycerol αγ-dichlorohydrin)	176	—	129
2,3-Dichloropropan-1-ol (glycerol βγ-dichlorohydrin)	183	—	—
3-Chloropropane-1,2-diol (glycerol α-monochlorohydrin)	213	—	—
1-Chloropropan-2-ol	127	—	83
2-Chloropropan-1-ol	132	—	76
Trichloroethanol	151 (m.p. 19)	—	—
2-Bromoethanol (ethylene bromohydrin)	149	—	86
1,3-Dibromopropan-2-ol (glycerol αγ-dibromohydrin)	219d	—	—
Propane-1,2-diol (propylene glycol)	187	—	—
Ethane-1,2-diol (ethylene glycol)	198	—	169
Propane-1,3-diol (trimethylene glycol)	215	—	178
Butane-1,4-diol (tetramethylene glycol)	230	19	—
Pentane-1,5-diol (pentamethylene glycol)	239	—	—
Di-2-hydroxyethyl ether (diethylene glycol)	244	—	149
Hexane-1,6-diol (hexamethylene glycol)	250	42	—
Glycerol	290d	—	—
Ethyleneglycol monomethyl ether	124	—	—
Ethyleneglycol monoethyl ether	135	—	75
Ethyleneglycol monoisopropyl ether	142	—	—
Ethyleneglycol monopropyl ether	151	—	—
Ethyleneglycol monobutyl ether	168	—	—
Ethyleneglycol monophenyl ether	245	—	—
Diethyleneglycol monomethyl ether	194	—	—
Diethyleneglycol monoethyl ether	202	—	—
Diethyleneglycol monopropyl ether	—	—	—
Diethyleneglycol monobutyl ether	232	—	—
2-Aminoethanol (monoethanolamine)	171	—	—
2,2′-Dihydroxydiethylamine (diethanolamine)	270	28	—
2,2′,2″-Trihydroxytriethylamine (triethanolamine)	360	—	—

p-Nitro benzoate °C	Phenyl-urethan °C	1-Naphthyl-urethan °C	Hydrogen 3-nitro-phthalate °C	O-Alkyl saccharin °C	Other derivatives °C
51	74	82	123	62	—
62	112	126	—	—	Benzoate, 54
52	73	82	120	70	—
—	132	—	—	—	—
64	80	89	119	75	—
137	138	127	—	—	—
—	51	101	98	—	d_4^{20} 1.202, n_D^{20} 1.442
—	—	76	—	—	d_4^{20} 1.131, n_D^{20} 1.447
—	73	115	—	—	d_4^{20} 1.353, n_D^{20} 1.480
38	73	93	—	—	—
108	—	—	—	—	—
—	—	—	—	—	—
—	—	—	—	—	—
—	87	120	—	—	—
—	86	—	172	—	d_4^{20} 1.763, n_D^{20} 1.492
78	81	—	—	—	d_4^{20} 2.120, n_D^{20} 1.550
127	153	—	—	—	—
141	157	176	—	—	Dibenzoate, 73
119	137	164	—	—	Dibenzoate, 59
175	183	198	—	—	Dibenzoate, 82
105	176	147	—	—	—
—	—	122	—	—	d_4^{20} 1.116, n_D^{20} 1.448
—	—	—	—	—	—
188	180	192	—	—	Tribenzoate, 72
50	—	113	129	—	d_4^{20} 0.966, n_D^{20} 1.402; 3,4,5-triiodobenzoate, 152
—	—	67	118	—	d_4^{20} 0.930, n_D^{20} 1.408; 3,4,5-triiodobenzoate, 128
—	—	—	—	—	d_4^{20} 0.903, n_D^{20} 1.410 3,4,5-triiodobenzoate, 80
120	—	—	—	—	d_4^{20} 0.911, n_D^{20} 1.413
113	—	—	—	—	d_4^{20} 0.902, n_D^{20} 1.420; 3,4,5-triiodobenzoate, 85
—	—	—	—	—	d_4^{20} 1.104, n_D^{20} 1,534; 3,4,5-triiodobenzoate, 145; toluene-p-sulphonate, 80
—	—	—	89	—	d_4^{20} 1.036, n_D^{20} 1.424; 3,4,5-triiodobenzoate, 82
—	—	—	—	—	d_4^{20} 1.024, n_D^{20} 1.430; 3,4,5-triiodobenzoate, 76
—	—	—	—	—	d_4^{20} 0.958, n_D^{20} 1,434 3,4,5-triiodobenzoate, 54
—	—	—	—	—	d_4^{20} 1.022, n_D^{20} 1.454; Picrate, 160
—	—	—	—	—	d_4^{20} 1.097, n_D^{20} 1.478; Picrate, 110
—	—	—	—	—	d_4^{20} 1.124, n_D^{20} 1.485; Hydrochloride, 177

Table 10.5 Aromatic alcohols

Alcohol	B.P. (/mmHg) °C	M.P. °C	3,5-Dinitrobenzoate °C
1-Phenylethanol	203	20	94
Benzyl alcohol	205	—	113
α-Hydroxy-m-xylene	217	—	111
1-Phenylpropan-1-ol	219	—	—
2-Phenylethanol	220	—	108
1-Phenylbutan-1-ol	118/18	16	—
1-Phenylpentan-1-ol	137/21	—	—
3-Phenylpropan-1-ol	237	—	92
o-Methoxybenzyl alcohol	249	—	—
m-Methoxybenzyl alcohol	252	—	—
p-Methoxybenzyl (anisyl alcohol)	259	25	—
Cinnamyl alcohol (3-phenylprop-2-en-1-ol)	257	33	121
α-Hydroxy-o-xylene	219	39	—
Piperonyl (3,4-dimethylenedioxybenzyl) alcohol	—	58	—
α-Hydroxy-p-xylene	217	60	118
Diphenylmethanol	298	69	142
m-Hydroxybenzyl alcohol	—	73	—
o-Hydroxybenzyl alcohol	—	87	—
p-Hydroxybenzyl alcohol	—	125	—
Benzoin	—	137	—
Triphenylmethanol	380	162	—

Halogeno-alcohols

Alcohol	B.P. (/mmHg) °C	M.P. °C	3,5-Dinitrobenzoate °C
m-Chlorobenzyl alcohol	234	—	—
m-Bromobenzyl alcohol	254	—	—
m-Iodobenzyl alcohol	165/16	—	—
p-Iodobenzyl alcohol	—	72	—
o-Chlorobenzyl alcohol	230	74	—
p-Chlorobenzyl alcohol	235	75	—
p-Bromobenzyl alcohol	—	77	—
o-Bromobenzyl alcohol	—	80	—
o-Iodobenzyl alcohol	—	90	—

Nitro- and amino-alcohols

Alcohol	B.P. (/mmHg) °C	M.P. °C	3,5-Dinitrobenzoate °C
m-Nitrobenzyl alcohol	—	27	—
p-Aminobenzyl alcohol	—	65	—
o-Nitrobenzyl alcohol	270	74	—
o-Aminobenzyl alcohol	—	82	—
p-Nitrobenzyl alcohol	185/12	93	—
m-Aminobenzyl alcohol	—	97	—

-Nitro-enzoate C	Phenyl-urethan °C	1-Naphthyl-urethan °C	Hydrogen 3-nitro-phthalate °C	Other derivatives °C
43	92	106	—	—
86	76	134	176	—
89	—	116	—	—
60	—	102	—	—
63	80	119	123	—
58	—	99	—	—
—	75	—	—	—
46	48	—	117	—
—	—	136	—	Benzoyl, 59
—	—	—	—	—
94	93	—	—	Benzoyl, 38; anisic acid, 184
78	91	114	—	—
01	79	—	—	o-Toluic acid, 104
—	102	—	—	Benzoyl, 66
—	79	—	—	—
31	140	136	—	Acetyl, 42
—	—	—	—	Acetyl, 84; diacetyl, 75
—	—	—	—	Benzoyl, 51
—	—	—	—	—
23	165	140	—	Acetyl, 83; benzoyl, 125; semicarbazone, 206d; 2,4-dinitrophenylhydrazone, 234
—	—	—	—	Acetyl, 88; triphenylmethane, 92
—	—	—	—	—
—	—	—	—	—
—	—	—	—	—
94	—	—	—	—
—	—	—	—	—
—	—	—	—	Acetyl, 23
—	—	—	—	o-Bromobenzoic acid, 150 (KMnO₄)
—	—	—	—	—
—	—	—	—	Benzoyl, 72
—	—	—	—	Diacetyl, 188
—	—	—	—	Benzoyl, 102
—	—	—	—	N-Acetyl, 114; picrate, 110
—	—	—	—	Benzoyl, 95; acetyl, 78
—	—	—	—	N-Acetyl, 107; dibenzoyl, 114

Table 10.6   Phenols

Phenol	B.P. °C	M.P. °C	Bromo compound °C	Acetate °C	Benzoate °C	p-Nitro-benzoate °C
Salicylaldehyde (1)	197	—	—	39	—	128
m-Cresol	202	12	84†	Liq.	55	90
o-Ethylphenol	207	—	—	—	39	56
m-Ethylphenol	217	—	—	Liq.	52	68
Methyl salicylate	223	—	—	49	92	128
Ethyl salicylate	231	—	—	—	80	108
Carvacrol (2)	238	—	46	Liq.	—	51
Propyl salicylate	239	—	—	—	—	—
Isopropyl salicylate	241	—	—	—	—	—
m-Methoxyphenol	244	—	104†	Liq.	—	—
Eugenol (3)	254	—	118‡	30	70	81
Butyl salicylate	260	—	—	—	—	—
Isoeugenol (4)	266	—	—	80	106	109
p-Butylphenol	248	22	—	—	127	68
o-Methoxyphenol (5)	205	28	116†	Liq.	58	93
2,4-Dimethylphenol (6)	211	28	179†	Liq.	38	105
o-Cresol	191	30	56*	Liq.	Liq.	94
p-Cresol	202	36	49*	Liq.	71	98
Phenol	182	42	95†	Liq.	69	126
Phenyl salicylate (7)	—	43	—	98	81	111
p-Ethylphenol	219	47	—	Liq.	60	81
2,6-Dimethylphenol (8)	203	49	79	—	—	—
Thymol (9)	233	51	55	Liq.	33	70
o-Cyclohexylphenol	—	55	—	—	—	—
p-Methoxyphenol (10)	243	56	—	32	87	—
o-Hydroxybiphenol	275	58	—	63	76	—
Orcinol hydrate (11)	289	58	104†	25	88	214
3,4-Dimethylphenol (12)	228	62	171†	—	58	—
3,5-Dimethylphenol (13)	219	68	166†	Liq.	24	109
2,4,6-Trimethylphenol (14)	220	69	158*	—	62	—
4-Hexylresorcinol	335	69	—	—	—	—
2,4,5-Trimethylphenol (15)	232	71	35	34	63	—
2,3-Dimethylphenol (16)	218	75	—	—	—	—
2,5-Dimethylphenol (17)	211	75	178†	Liq.	61	87
Vanillin (18)	—	81	160	102	78	—
o-Hydroxybenzyl alcohol (19)	—	87	—	—	51	—
1-Naphthol	279	94	105*	49	56	143
2-Naphthyl salicylate	—	95	—	136	—	—
p-t-Pentylphenol	266	96	—	Liq.	61	—
2,3,5-Trimethylphenol	233	96	—	—	50	—
p-t-Butylphenol	237	99	—	Liq.	82	—
Catechol (1,2-dihydroxybenzene)	240	105	192‡	65	84	169
2,5-Dihydroxybiphenyl (20)	—	103	—	—	—	—
Orcinol (21)	289	108	104†	25	88	214
m-Hydroxybenzaldehyde	240	108	—	Liq.	38	—
1,2-Dihydroxynaphthalene (22)	—	108	—	106	—	—
2,2'-Dihydroxybiphenol	—	109	—	95	101	—
Resorcinol (1,3-dihydroxybenzene)	280	110	112*	Liq.	117	182
Ethyl-p-hydroxybenzoate	—	116	—	—	94	—
p-Hydroxybenzaldehyde	—	116	—	Liq.	90	—
2-Naphthol	285	123	84	72	107	169
1,3-Dihydroxynaphthalene (23)	—	124	—	56	—	—
3,3'-Dihydroxybiphenyl	—	124	—	83	92	—
Methyl p-hydroxybenzoate	—	131	—	85	135	—
p-Cyclohexylphenol	—	132	—	—	118	137
Pyrogallol (1,2,3-trihydroxybenzene)	309	133	158*	173	90	230
1,6-Dihydroxynaphthalene	—	138	—	73	104	—
Hydroxyhydroquinone (24)	—	140	—	97	120	—

3,5-Dinitro-benzoate °C	Aryloxy-acetic acid °C	N,N-diphenyl urethan °C	1-Naphthyl urethan °C	Toluene-p-sulphonate °C	2,4-Dinitro-phenyl ether °C	O-Aryl saccharin °C
—	132	—	—	64	—	—
165	102	101	128	56	74	146
108	141	—	—	—	—	—
—	75	—	—	—	—	—
—	—	—	—	—	—	—
77	151	—	116	—	—	—
—	—	—	—	—	—	—
—	114	—	129	—	—	—
131	80	—	122	85	115	—
—	—	—	150	—	130	—
—	94	—	—	—	—	—
—	81	—	—	—	—	—
142	119	118	118	85	97	—
164	142	—	135	—	—	—
138	152	73	142	55	90	163
189	136	94	146	70	94	172
146	99	105	133	96	69	182
—	—	144	—	—	—	—
132	97	—	128	—	—	—
159	140	—	—	—	—	—
103	148	—	160	71	67	—
—	111	—	—	—	—	—
—	—	—	—	65	—	—
190	217	154	160	—	—	—
181	163	—	142	—	—	—
195	86	—	—	—	—	—
—	142	—	—	—	—	—
—	—	—	—	—	—	—
—	132	—	—	—	—	—
—	187	—	—	—	—	—
137	118	—	173	—	—	—
—	189	—	—	115	131	—
—	120	—	—	—	—	—
217	192	—	152	88	128	—
—	—	—	—	—	—	—
—	—	—	—	54	—	—
—	—	—	—	—	—	—
—	86	—	110	—	—	—
152	—	—	175	—	—	—
—	—	—	—	—	—	—
190	217	154	160	—	—	—
—	148	—	—	—	—	—
—	—	—	—	—	—	—
—	—	—	—	190	—	—
201	195	130	206	81	194	—
—	—	—	—	—	—	—
—	198	—	—	—	—	—
210	154	141	157	125	95	—
—	—	—	—	—	—	—
—	—	—	—	—	—	—
168	—	—	—	—	—	—
—	—	—	—	—	—	—
205	198	212	—	—	—	—
—	—	—	—	—	—	—
—	—	—	—	—	—	—

Table 10.6 Phenols (*continued*)

Phenol	B.P. °C	M.P. °C	Bromo compound °C	Acetate °C	Benzoate °C	p-Nitro-benzoate °C
1,8-Dihydroxynaphthalene	—	142	—	155	175	—
Salicyclic acid (25)	—	159	—	135	132	205
p-Hydroxybiphenyl	306	165	—	88	151	—
Hydroquinone (1,4-dihydroxybenzene)	286	170	186*	124	199	250
1,4-Dihydroxynaphthalene (26)	—	176	—	128	169	—
2,7-Dihydroxynaphthalene	—	190	—	136	139	—
m-Hydroxybenzoic acid	—	200	—	131	—	—
p-Hydroxybenzoic acid	—	214	—	187	—	—
Phloroglucinol (1,3,5-trihydroxybenzene)	—	218	151†	104	174	283
1,5-Dihydroxynaphthalene	—	265	—	160	235	—
4,4'-Dihydroxybiphenyl	—	274	—	161	241	—

Halogeno-phenols

Phenol	B.P. °C	M.P. °C	Bromo compound °C	Acetate °C	Benzoate °C	p-Nitro-benzoate °C
o-Chlorophenol	176	9	—	Liq.	Liq.	115
o-Bromophenol	195	5	95†	Liq.	—	—
m-Chlorophenol	214	33	—	Liq.	71	99
m-Bromophenol	236	33	—	Liq.	86	—
2,4-Dibromophenol	239	40	—	36	98	184
m-Iodophenol	—	40	—	38	—	133
o-Iodophenol	—	43	—	—	34	—
p-Chlorophenol	217	43	—	Liq.	89	168
2,4-Dichlorophenol	210	45	68	—	96	—
p-Bromophenol	235	64	95†	21	102	180
2,4,5-Trichlorophenol	249	68	—	—	93	—
2,4,6-Trichlorophenol	246	69	—	—	75	106
2,4-Di-iodophenol	—	72	—	71	98	—
p-Iodophenol	—	94	—	32	119	—
2,4,6-Tribromophenol	—	95	120‡	87	81	153
Chlorohydroquinone	—	106	—	72	—	—
Bromohydroquinone	—	111	—	72	—	—
2,4,6-Tri-iodophenol	—	159	—	156	—	—
Pentachlorophenol	310d	191	—	150	164	—

Nitro-phenols

Phenol	B.P. °C	M.P. °C	Bromo compound °C	Acetate °C	Benzoate °C	p-Nitro-benzoate °C
o-Nitrophenol△	216	45	117*	41	59	141
m-Nitrophenol	—	97	91*	56	95	174
2,4-Dinitrophenol	—	113	118	72	132	139
p-Nitrophenol△	—	114	142*	83	142	159
Picric acid	—	122	—	76	—	143
Styphnic acid (27)	—	179	—	—	—	—

* Dibromo.   † Tribromo.   ‡ Tetrabromo.

△ O-Aryl saccharin derivatives of o-nitrophenol and p-nitrophenol have m.p.s. 236 °C and 192 °C respectively.

(1) o-Hydroxybenzaldehyde.
(2) 2-Methyl-5-isopropylphenol.
(3) 4-Allyl-2-methoxy phenol.
(4) 2-Methoxy-4-propenylphenol (*cis* and *trans*).
(5) Guaiacol.
(6) m-4-Xylenol.
(7) Salol.

(8) m-2-Xylenol.
(9) 3-Hydroxy-4-isopropyltoluene.
(10) Hydroquinone monomethyl ether.
(11) 3,5-Dihydroxytoluene.
(12) o-4-Xylenol.
(13) m-5-Xylenol.
(14) Mesitol.

,5-Dinitro-enzoate °C	Aryloxy-acetic acid °C	NN-diphenyl urethan °C	1-Naphthyl urethan °C	Toluene-p-sulphonate °C	2,4-Dinitro-phenyl ether °C	O-Aryl saccharin °C
—	—	—	220	—	—	—
—	191	—	—	—	—	—
—	—	—	—	177	118	—
17	250	230	247	159	—	—
—	149	176	—	150	—	—
—	206	—	—	—	—	—
—	278	—	—	—	—	—
62	—	—	—	—	—	—
—	—	—	—	—	—	—
—	274	—	—	—	—	—
43	145	—	120	74	99	—
—	143	—	129	78	89	—
56	110	—	158	—	75	—
—	108	—	108	53	—	—
—	153	—	—	120	135	—
83	115	—	—	61	—	—
—	135	—	—	—	95	—
86	156	97	166	71	126	—
—	140	—	—	125	119	—
91	159	99	169	94	141	—
—	157	—	—	—	—	—
36	182	143	188	—	136	—
—	—	—	—	—	—	—
—	156	127	—	99	156	—
74	200	153	153	113	135	—
—	—	—	—	—	—	—
81	—	—	—	—	—	—
—	196	—	—	145	—	—
55	158	114	113	83	142	—
59	156	—	167	113	138	—
—	—	—	—	121	248	—
86	187	112	151	97	120	—
—	—	—	—	—	—	—
—	—	—	—	—	—	—

15) Pseudo-cuminol.
16) o-3-Xylenol.
17) p-2-Xylenol.
18) 4-Hydroxy-3-methoxybenzaldehyde.
19) Saligenin.
20) Phenylhydroquinone.
21) 3,5-Dihydroxytoluene.

(22) 1,2-Naphthohydroquinone.
(23) Naphthoresorcinol.
(24) 1,2,4-Trihydroxybenzene.
(25) o-Hydroxybenzoic acid.
(26) 1,4-Naphthohydroquinone.
(27) 2,4,6-Trinitroresorcinol.

1313

Table 10.7   Enols

Compound	B.P. °C	M.P. °C	d_4^{20}	n_D^{20}	Semicar- bazone °C	Pyrazo- lone °C
Pentane-2,4-dione (acetylacetone)	139	—	0.977	1.452	—	100*
Methyl acetoacetate	170	—	1.077	1.419	152	127
Methyl methylacetoacetate	177	—	1.030	1.418	138	120
Ethyl acetoacetate	180	—	1.028	1.419	129d	127
Ethyl methylacetoacetate	181	—	1.006	1.419	86	120
Methyl ethylacetoacetate	189	—	0.989	—	98	108
Hexane-2,5-dione (acetonylacetone)	194	—	0.974	1.428	220	92
Ethyl ethylacetoacetate	198	—	0.972	1.422	154d	108
Diethyl acetonedicarboxylate	250d	—	1.113	—	95	85
Ethyl benzoylacetate	262d	—	1.117	—	125	63
Diethyl oxalacetate	131/34	—	1.131	1.454	162	—
Benzoylacetone	261	61	—	—	—	63
Dibenzoylmethane	—	78	—	—	—	137

* 1-(p-Nitrophenyl)-3,5-dimethylpyrazole: with aqueous solution of p-nitrophenylhydrazine hydrochloride. Phenylhydrazine yields a liquid pyrazole, b.p. 273 °C.

Table 10.8   Polyhydric alcohols

Alcohol	B.P. °C	Benzoate °C	p-Nitro- benzoate °C	3,5-Dinitro- benzoate °C	Phenyl- urethan °C	1-Naphthyl- urethan °C	Other derivatives °C
Butane-2,3-diol	182	76	—	—	201	—	—
Propane-1,2-diol	189	—	127	—	153	—	—
Ethylene glycol	198	73	141	169	157	176	—
Butane-1,3-diol	208	—	—	—	123	—	—
Propane-1,3-diol	215	59	119	178	137	164	—
Butane-1,4-diol	230	82	175	—	183	199	—
Pentane-1,5-diol	239	—	105	—	176	147	—
Di-2-hydroxyethyl ether (1)	244	—	—	149	—	122	—
Hexane-1,6-diol	250*	—	—	—	—	—	—
Di-β-hydroxyethoxy ethane (2)	285	—	—	—	108	—	—
2-Butyne-1,4-diol	145/15**	—	—	190	131	—	—
Glycerol	290d	76	188	—	180	192	—
Sorbitol	m.p. 110	129	—	—	—	—	Acetate, 99
Mannitol	m.p. 166	149	—	—	303	—	Acetate, 121
Pentaerythritol	m.p. 253	99	—	—	—	—	Acetate, 84

(1) Diethylene glycol.   (2) Triethylene glycol.

* m.p. 42 °C.
** m.p. 55 °C.

Table 10.9 Carbohydrates (sugars)

Carbohydrate	M.P.* °C	$[\alpha]_D^{20}$ in water °C	Osazone M.P. °C	Time of formation (minutes)	Other derivatives °C
†D-Glucose (hydrated)		90 ⎫	+52	205	4	Penta-acetate, α- 112, β- 132; pentabenzoate, 179
D-Glucose (anhydrous)	146	⎬				
D-Ribose		95	−21.5	166	—	—
D-Fructose		104	−92	205	2	Penta-acetate, α- 70, β- 109
L-Rhamnose (hydrated)		105 ⎫	+9	190	9	Penta-acetate, 99
L-Rhamnose (anhydrous)	125	⎬				
L-Lyxose		106	+13.5	163	—	—
D-Galactose (hydrated)		120 ⎫	+82	201	15–19	Penta-acetate, α- 95, β- 142; D-galactaric acid, 213
D-Galactose (anhydrous)	170	⎬				
D-Mannose		132	+14.5	205	0.5	Penta-acetate, α- 74, β- 115
D-Xylose		145	+19	164	7	Penta-acetate, α- 59, β- 126
L-Arabinose		160	+105	166	9	Penta-acetate, α- 94, β- 86
L-Sorbose		161	−43	162	4	—
Maltose (hydrated)		100 ⎫	+130	206	—	Octa-acetate, α- 125, β- 160
Maltose (anhydrous)	165	⎬				
Sucrose		185	+66.5	205	30	Octa-acetate, 69
Gentiobiose		190	+9.5	162	—	Octa-acetate, α- 189, β- 193
Lactose (hydrated)		203 ⎫	+52.5	200	—	Octa-acetate, α- 152, β- 90; D-galactaric acid, 213
Lactose (anhydrous)	233	⎬				
Cellobiose		225	+35	198	—	Octa-acetate, α- 230, β- 192

* The melting points of carbohydrates (sugars) are not usually sharp and they are perhaps best expressed as decomposition points.
† The small capital letter prefix refers to configuration, related to D-glyceraldehyde, and not to the direction of optical rotation.

Table 10.10   Aliphatic halogen compounds

Halide	B.P. °C	M.P. °C	$d_4^{20°}$	$n_D^{20°}$
Chlorides				
Ethyl	12	—	—	—
Isopropyl	35	—	0.863	1.378
Allyl	45	—	0.940	1.416
Propyl	46	—	0.889	1.388
t-Butyl	51	—	0.846	1.386
s-Butyl	68	—	0.874	1.397
Isobutyl	69	—	0.881	1.398
Butyl	77	—	0.886	1.402
1-Chloro-2,2-dimethylpropane	85	—	0.879	—
2-Chloro-2-methylbutane	85	—	0.865	1.405
3-Chloropentane	96	—	0.872	1.408
2-Chloropentane	97	—	0.873	1.408
1-Chloro-3-methylbutane	99	—	0.872	1.409
1-Chloropentane	106	—	0.882	1.412
Chlorocyclopentane	114	—	1.005	1.451
1-Chlorohexane	134	—	0.878	1.420
Chlorocyclohexane	142	—	0.989	1.462
1-Chloroheptane	159	—	0.877	1.426
1-Chloro-octane	182	—	0.875	1.431
1-Chlorononane	202	—	0.870	1.434
1-Chlorodecane	223	—	0.868	1.437
1-Chloroundecane	241	—	0.868	1.440
1-Chlorododecane	244	—	0.867	1.443
Benzyl chloride	179	—	1.100	1.539
1-Chloro-1-phenylethane	195	—	1.063	1.530
1-Chloro-2-phenylethane	198	—	1.073	—
Benzilidene chloride	207	—	—	—
Benzotrichloride	218	—	—	—
Bromides				
Ethyl	38	—	1.460	1.425
Isopropyl	59	—	1.425	1.314
Allyl	70	—	1.432	1.470
Propyl	71	—	1.435	1.355
s-Butyl	91	—	1.256	1.437
Isobutyl	91	—	1.253	1.435
Butyl	101	—	1.274	1.440
1-Bromo-2,2-dimethylpropane	109	—	1.225	—
2-Bromopentane	117	—	1.212	1.442
3-Bromopentane	118	—	1.211	1.443
1-Bromo-3-methylbutane	119	—	1.213	1.442
1-Bromopentane	129	—	1.219	1.445
Bromocyclopentane	137	—	1.387	1.489
1-Bromohexane	154	—	1.175	1.448
Bromocyclohexane	164	—	1.336	1.495
1-Bromoheptane	178	—	1.140	1.451
1-Bromo-octane	200	—	1.112	1.453
1-Bromononane	220	—	1.090	1.454
1-Bromodecane	103/6	—	1.066	1.455
1-Bromoundecane	114/5	—	1.054	1.457
1-Bromododecane	130/6	—	1.038	1.458

Anilide °C	1-Naphthalide °C	Alkyl mercury(II) halide °C	S-Alkylthiouronium picrate °C	Picrate of alkyl 2-naphthyl ether °C
04	126	193	188	102
04	—	—	196	95
14	—	—	154	99
92	121	140	177	81
28	147	—	—	—
08	129	39	166	85
10	126	—	—	84
63	112	128	177	67
26	—	—	—	—
92	138	—	—	—
—	—	—	—	—
—	—	—	—	—
10	111	86	173	94
96	112	110	154	67
—	—	—	—	—
69	106	125	—	—
46	188	—	—	—
57	95	119	—	—
57	91	151	—	—
—	—	—	—	—
—	—	—	—	—
—	—	—	—	—
—	—	—	—	—
17	166	—	188	123
33	—	—	—	—
97	—	—	—	84
66	—	—	—	—
—	—	—	—	—
04	126	194	188	102
04	—	94	196	95
14	—	—	154	99
92	121	138	177	81
08	129	39	166	85
10	126	56	167	84
63	112	129	177	67
26	—	—	—	—
93	—	—	—	—
24	—	—	—	—
10	111	80	173	94
96	112	122	154	67
—	—	—	—	—
69	106	119	157	—
46	188	153	—	—
57	95	115	142	—
57	91	109	134	—
—	—	109	—	—
—	—	—	—	—
—	—	—	—	—
—	—	108	—	—

Table 10.10   Aliphatic halogen compounds (*continued*)

Halide	B.P. °C	M.P. °C	$d_4^{20°}$	$n_D^{20°}$
Bromides (*continued*)				
1-Bromotetradecane	179/20	5	1.017	1.460
1-Bromohexadecane	201/19	14	1.001	1.462
Benzyl bromide	198	—	1.438	—
1-Bromo-1-phenylethane	203	—	1.361	1.561
1-Bromo-2-phenylethane	218	—	1.359	1.556
Iodides				
Methyl **C**	42	—	2.282	1.532
Ethyl	73	—	1.940	1.514
Isopropyl	89	—	1.703	1.499
Allyl	100	—	1.777	1.578
Propyl	102	—	1.743	1.505
s-Butyl	118	—	1.592	1.499
Isobutyl	119	—	1.602	1.496
2-Iodo-2-methylbutane	128	—	1.479	—
Butyl	129	—	1.616	1.499
3-Iodopentane	142	—	1.511	1.497
2-Iodopentane	142	—	1.510	1.496
1-Iodo-3-methylbutane	147	—	1.503	1.493
1-Iodopentane	155	—	1.512	1.496
1-Iodohexane	180	—	1.437	1.493
1-Iodoheptane	201	—	1.373	1.490
1-Iodo-octane	221	—	1.330	1.489
Iodocyclopentane	58/22	—	1.709	1.547
Iodocyclohexane	82/20	—	1.624	1.547
Benzyl iodide	93/10	24	—	—
1-Iodo-2-phenylethane	116/12	—	1.632	1.602
Chloro compounds				
Dichloromethane	42	—	1.336	1.425
trans-1,2-Dichloroethylene	48	—	1.257	1.444
Ethylidene chloride (1,1-dichloroethane)	57	—	1.176	1.416
cis-1,2-Dichloroethylene	60	—	1.282	1.446
Chloroform **C**	61	—	1.489	1.446
2,2-Dichloropropane	70	—	1.092	1.412
1,1,1-Trichloroethane	75	—	1.349	1.438
Carbon tetrachloride **C**	77	—	1.594	1.461
1,2-Dichloroethane (ethylene dichloride)	84	—	1.256	1.445
1,1,2-Trichloroethylene	87	—	1.465	1.478
1,2-Dichloropropane	96	—	1.155	1.439
1-Bromo-1-chloroethane	107	—	1.689	1.491
1,1,2-Trichloroethane	114	—	1.443	1.471
1,3-Dichloropropane	120	—	1.183	1.449
1,1,2,2-Tetrachloroethylene	121	—	1.623	1.506
1-Bromo-3-chloropropane	143	—	1.593	1.471
1,1,2,2-Tetrachloroethane	147	—	1.597	1.495
1,4-Dichlorobutane	153	—	1.139	1.455
1,2,3-Trichloropropane	157	—	1.394	1.486
Pentachloroethane	162	—	1.680	1.503
1,5-Dichloropentane	178d	—	1.100	1.457
1,6-Dichlorohexane	204d	—	1.069	1.457

Anilide °C	1-Naphthalide °C	Alkyl mercury(II) halide °C	S-Alkylthiouronium picrate °C	Picrate of alkyl naphthyl ether °C
—	—	—	—	—
—	—	—	137	—
117	166	119	188	123
133	—	—	—	—
97	—	169	—	84
114	160	145	224	117
104	126	182	188	102
104	—	—	196	95
114	121	112	154	99
92	121	113	177	81
108	129	—	166	85
110	126	72	167	84
92	138	—	—	—
63	112	117	177	67
—	—	—	—	—
—	—	—	—	—
110	111	122	173	94
96	112	110	154	—
69	106	110	157	—
57	95	103	—	—
—	—	—	—	—
—	—	—	—	—
—	—	—	—	—
116	—	—	—	123
—	—	—	—	84

Table 10.10 Aliphatic halogen compounds (*continued*)

Halide	B.P. °C	M.P. °C	$d_4^{20°}$	$n_D^{20°}$
Chloro compounds (*continued*)				
Hexachloroethane	—	187 (sub.)	—	—
Bromo compounds				
Dibromomethane	97	—	2.496	1.541
Ethylidene bromide	113	—	2.055	1.513
1,2-Dibromomethane **C** (1)	131	—	2.183	1.539
1,2-Dibromopropane (2)	141	—	1.932	1.520
Bromoform	150	—	2.887	1.598
1,2-Dibromo-2-methylpropane (3)	150	—	1.783	1.512
2,3-Dibromobutane	157	—	1.792	1.515
1,3-Dibromopropane (4)	165	—	1.982	1.523
1,2-Dibromobutane	166	—	1.820	—
1,4-Dibromobutane	198	—	1.826	1.519
1,2,3-Tribromopropane	220	—	2.402	1.582
1,5-Dibromopentane	221	—	1.702	1.513
1,6-Dibromohexane	240	—	1.603	1.506
1,1,2,2-Tetrabromoethane	124/19	—	2.967	1.638
Carbon tetrabromide	—	92	—	—
Iodo compounds				
Di-iodomethane	80/25	—	3.324	1.741
1.3-Di-iodopropane	90/9	—	2.576	1.642
1,4-Di-iodobutane	110/10	—	2.358	1.621
1,5-Di-iodopentane	132/10	—	2.182	1.602
1,6-Di-iodohexane	140/10	—	2.040	1.586
1,2-Di-iodoethane (ethylene di-iodide)	—	81	—	—
Iodoform	—	119	—	—

The melting points of the *S*-alkylisothiouronium picrates of (1), (2), (3) and (4) are 260, 232, 223 and 229 °C respectively.

Table 10.11   Aromatic halogen compounds

Compound °C	B.P. °C	M.P. °C	$d_4^{20°}$	$n_D^{20°}$
m-Difluorobenzene	83	—	1.153	1.439
Fluorobenzene	85	—	1.024	1.466
p-Difluorobenzene	89	—	1.166	1.441
o-Difluorobenzene	93	—	1.157	1.443
o-Fluorotoluene	114	—	0.998	1.470
m-Fluorotoluene	116	—	0.990	—
p-Fluorotoluene	116	—	0.998	1.469
Chlorobenzene	132	—	1.107	1.525
Bromobenzene	156	—	1.494	1.560
o-Chlorotoluene	159	—	1.082	1.527
m-Chlorotoluene	162	—	1.072	1.522
p-Chlorotoluene	162	7	1.070	1.521
m-Dichlorobenzene	173	—	1.288	1.546
Benzyl chloride	179	—	1.100	1.539
o-Dichlorobenzene	180	—	1.305	1.551
o-Bromotoluene	181	—	1.425	—
m-Bromotoluene	183	—	1.410	—
p-Bromotoluene	185	26	1.390	—
Iodobenzene	188	—	1.831	1.620
o-Bromochlorobenzene	195	—	1.646	1.580
Benzyl bromide	198	—	1.438	—
2,4-Dichlorotoluene	199	—	1.249	1.549
2,6-Dichlorotoluene	199	—	1.269	1.551
m-Iodotoluene	204	—	1.698	—
Benzyl iodide	93/10	24	—	—
Benzylidene chloride	205	—	1.250	1.550
o-Iodotoluene	207	—	1.698	—
3,4-Dichlorotoluene	207	—	1.251	1.549
Benzylidene bromide	156/23	—	1.460	1.541
1,2,4-Trichlorobenzene	213	17	1.468	1.554
2-Fluoronaphthalene	212	61	—	—
1-Fluoronaphthalene	214	—	1.134	1.594
m-Dibromobenzene	219	—	1.952	1.606
Benzotrichloride	220	—	1.173	—
o-Dibromobenzene	224	7	1.956	1.609
Bromomesitylene	225	−1	—	—
2-Bromocymene	234	—	1.267	—
2,5-Dibromotoluene	236	—	1.811	—
3,4-Dibromotoluene	240	—	1.811	1.600
m-Bromoiodobenzene	252	—	—	—
Chloronaphthalene	256	61	—	—
o-Bromoiodobenzene	257	—	2.262	1.665
1-Chloronaphthalene	259	—	1.192	1.633
1-Bromonaphthalene	281	—	1.484	1.658
m-Di-iodobenzene	285	40	—	—
o-Di-iodobenzene	287	27	—	—
2-Bromobiphenyl	297	—	1.233	—
3-Bromobiphenyl	300	—	—	1.641
1-Iodonaphthalene	302	—	1.729	—
2-Iodobiphenyl	158/6	—	1.609	1.662
2-Chlorobiphenyl	273	32	—	—

Nitration product		Sulphonamide ($-SO_2NH_2$, 1)		Other derivatives °C
Position	M.P. °C	Position	M.P. °C	
—	—	—	—	—
—	—	4, F	125	Sulphone, 98
—	—	—	—	—
—	—	3, CH$_3$; 4, F	105	o-Fluorobenzoic acid, 127
—	—	2, CH$_3$; 4, F	173	m-Fluorobenzoic acid, 124
—	—	2, CH$_3$; 5, F	141	p-Fluorobenzoic acid, 186
2,4	52	4, Cl	143	—
2,4	75	4, Br	162	—
3,5	64	3, CH$_3$; 4, Cl	126	o-Chlorobenzoic acid, 141
4,6	91	2, CH$_3$; 4, Cl	185	m-Chlorobenzoic acid, 158
2	38	2, CH$_3$; 5, Cl	143	p-Chlorobenzoic acid, 242
4,6	103	2,4, diCl	180	—
—	—	—	—	S-Benzylisothiouronium picrate, 188
4,5	110	3,4, diCl	135	Sulphone, 176
3,5	82	3, CH$_3$; 4, Br	146	o-Bromobenzoic acid, 150
4,6	103	2, CH$_3$; 4, Br	168	m-Bromobenzoic acid, 155
2	47	2, CH$_3$; 5, Br	165	p-Bromobenzoic acid, 251
4	174	—	—	Sulphone, 202
—	—	—	—	—
3,5	104	2,4, diCl; 5, Me	176	2,4-Dichlorobenzoic acid, 164
3	53	2,4, diCl; 3, Me	204	2,6-Dichlorobenzoic acid, 139
—	—	—	—	m-Iodobenzoic acid, 186
—	—	—	—	—
—	—	—	—	Benzaldehyde phenylhydrazone, 156
6	103	—	—	o-Iodobenzoic acid, 162
6	64	3,4, diCl; 6, Me	190	3,4-Dichlorobenzoic acid, 208
—	—	—	—	—
5	56	—	—	—
—	—	—	—	Picrate, 101
—	—	—	—	Picrate, 113
4	62	2,4, diBr	189	—
—	—	—	—	Benzoic acid, 121
4,5	114	3,4, diBr	176	Sulphone, 177
—	—	—	—	—
—	—	—	—	2,5-Dibromobenzoic acid, 157
—	—	—	—	3,4-Dibromobenzoic acid, 235
—	—	—	—	—
1,8	175	7, Cl	232	Picrate, 81
—	—	—	—	—
4,5	180	4, Cl	186	Picrate, 137
4	85	4, Br	193	Picrate, 134
—	—	—	—	—
—	—	—	—	o-Bromobenzoic acid, 150 (CrO$_3$)
—	—	—	—	—
—	—	—	—	Picrate, 127
—	—	—	—	—
—	—	—	—	o-Chlorobenzoic acid, 141

1323

Table 10.11 Aromatic halogen compounds (*continued*)

Compound	B.P. °C	M.P. °C	d_4^{20}	n_4^{20}
p-Iodotoluene	211	35	—	—
1,2-Dichloronaphthalene	296	35	—	—
m-Di-iodobenzene	285	40	—	—
1,2,4-Tribromobenzene	275	44	—	—
p-Dichlorobenzene	174	53	—	—
1,2,3-Trichlorobenzene	218	53	—	—
2-Iodonaphthalene	309	54	—	—
2-Bromonaphthalene	282	59	—	—
2-Fluoronaphthalene	—	60	—	—
2,2'-Dichlorobiphenyl	—	60	—	—
2-Chloronaphthalene	256	61	—	—
1,3,5-Trichlorobenzene	208	63	—	—
p-Bromochlorobenzene	195	67	—	—
1,2-Dibromonaphthalene	—	68	—	—
4-Fluorobiphenyl	253	74	—	—
4-Chlorobiphenyl	291	77	—	—
2,2'-Dibromobiphenyl	—	81	—	—
1,2,3-Tribromobenzene	—	88	—	—
4-Bromobiphenyl	310	89	—	—
p-Dibromobenzene	219	89	—	—
p-Bromoiodobenzene	251	92	—	—
4,4'-Difluorobiphenyl	255	95	—	—
4-Iodobiphenyl	—	114	—	—
1,3,5-Tribromobenzene	271	120	—	—
p-Di-iodobenzene	285	129	—	—
1,2,4,5-Tetrachlorobenzene	240	140	—	—
4,4'-Dichlorobiphenyl	—	149	—	—
4,4'-Dibromobiphenyl	—	164	—	—
1,2,4,5-Tetrabromobenzene	—	181	—	—
1,2,3,4-Tetrachloronaphthalene	—	183	—	—
Hexachlorobenzene	—	229	—	—

| Nitration product | | Sulphonamide ($-SO_2NH_2$, 1) | | Other derivatives °C |
Position	M.P. °C	Position	M.P. °C	
—	—	—	—	p-Iodobenzoic acid, 269
—	—	—	—	—
—	—	—	—	—
2	54	2,5 diCl	180	—
4	56	2,3,4, triCl	230	—
—	—	—	—	Picrate, 95
—	—	7, Br	208	Picrate, 86
—	—	—	—	Picrate, 101
1,8	175	7, Cl	232	Picrate, 81
2	68	2,4,6, triCl	212	—
2	72	—	—	—
—	—	—	—	—
—	—	—	—	—
—	—	—	—	p-Chlorobenzoic acid, 242
—	—	—	—	—
—	—	—	—	p-Bromobenzoic acid, 251 (CrO$_3$)
2,5	84	2,5, diBr	195	—
—	—	—	—	—
—	—	—	—	—
—	—	2,4,6, triBr	222d	—
2,5	171	—	—	—
3	99	—	—	—
—	—	—	—	p-Chlorobenzoic acid, 242 (CrO$_3$)
—	—	—	—	p-Bromobenzoic acid, 251 (CrO$_3$)
3	168	—	—	—
—	—	—	—	—

Table 10.12 Aliphatic ethers

Ether	B.P. °C	$d_4^{20°}$	$n_D^{20°}$
Diethyl	34	0.714	1.353
Ethyl vinyl	36	0.759	1.377
Allyl ethyl	67	0.765	1.388
Di-isopropyl	68	0.726	1.368
Butyl methyl	70	0.774	1.374
Dipropyl	90	0.749	1.381
Butyl ethyl	92	0.749	1.382
Diallyl	94	0.803	—
Pentyl methyl	99	0.761	1.387
Cyclopentyl methyl	105	0.862	1.420
Pentyl ethyl	118	0.762	1.393
Di-s-butyl	121	0.764	1.396
Cyclopentyl ethyl	122	0.853	1.423
Di-isobutyl	123	0.756	—
Hexyl methyl	126	0.772	1.397
Cyclohexyl methyl	134	0.875	1.435
Dibutyl	141	0.770	1.399
Ethyl hexyl	142	0.772	1.401
Cyclohexyl ethyl	149	0.864	1.435
Di-isopentyl	171	0.778	1.409
Cineole	176	0.923	1.458
Dipentyl	188	0.785	1.412
Dihexyl	229	0.793	1.420
Diheptyl	259	0.801	1.427
Dioctyl	288	0.806	1.433
Didecyl	185/5 mmHg	0.815	1.441
Chloromethyl methyl C	59	1.070	1.397
1-Chloroethyl methyl	73	0.991	1.400
Chloromethyl ethyl	83d	1.026	1.404
2-Chloroethyl methyl	91	1.035	1.411
sym-Dichlorodimethyl	105	1.310	1.436
1,1'-Dichlorodiethyl	116	1.111	1.423
Epichlorohydrin (1) C	117	1.181	1.438
1,2-Dichlorodiethyl	140	1.177	1.444
2,2'-Dichlorodiethyl	118	1.210	1.457
Di-2-chloropropyl	188	1.109	1.447
Di-3-chloropropyl	215	1.139	1.452
Ethyleneglycol dimethyl (2)	85	0.866	1.379
Ethyleneglycol diethyl (2)	123	0.848	—
Diethyleneglycol diethyl (2)	187	0.906	1.411
Tetraethyleneglycol dimethyl (2)	266	1.009	1.432
Tetrahydrofuran	65	0.889	1.407
2-Methyltetrahydrofuran	79	0.855	1.407
Dihydropyran	86	0.923	1.440
Tetrahydropyran	88	0.881	1.421
Dioxane	102	1.034	1.422

(1) 1-Chloro-2,3-epoxypropane.
(2) For alternative names see Section 4.1.18, p. 406.

Table 10.13 Aromatic ethers

Ether	B.P. °C (/mmHg)	M.P. °C	$d_4^{20°}$
Furan	32	—	0.937
2-Methylfuran	64	—	0.913
Anisole	154	—	0.996
Phenetole	170	—	0.965
Benzyl methyl ether	171	—	0.965
Methyl o-tolyl ether (1)	171	—	0.985
Methyl p-tolyl ether	175	—	0.970
Methyl m-tolyl ether	177	—	0.972
Ethyl o-tolyl ether	184	—	0.953
Benzyl ethyl ether	186	—	0.948
Phenyl n-propyl ether	188	—	0.949
Ethyl p-tolyl ether	190	—	0.949
Ethyl m-tolyl ether	191	—	0.949
Guaiacol (2)	205	28	1.129
Veratrole (3)	206	22	—
Butyl phenyl ether	208	—	0.934
Thymol methyl ether	212	—	—
Resorcinol dimethyl ether	217	—	1.050
Safrole (4)	232	11	1.100
Anethole (5)	235	22	0.989
Resorcinol diethyl ether	235	12	—
Eugenol methyl ether (6)	244	—	1.050
Isosafrole (7)	248	7	1.122
Diphenyl ether	259	28	—
Isoeugenol methyl ether (8)	264	—	1.053
Methyl 1-naphthyl ether	271	—	1.092
Ethyl 1-naphthyl ether	280	5	1.060
Dibenzyl ether	296	—	1.034
2-Methoxybiphenol	274	29	—
Ethyl 2-naphthyl ether	282	37	—
Pyrogallol triethyl ether	—	39	—
Catechol diethyl ether	217	43	—
Pyrogallol trimethyl ether	241	47	—
Hydroquinone dimethyl ether	212	56	—
Methyl 2-naphthyl ether	274	72	—
Hydroquinone diethyl ether	—	72	—
Benzyl 1-naphthyl ether	—	77	—
4-Methoxybiphenyl	—	90	—
Benzyl 2-naphthyl ether	—	99	—
Halogeno-ethers			
m-Chloroanisole	194	—	—
o-Chloroanisole	195	—	1.191
p-Chloroanisole	198	—	—
m-Chlorophenetole	205	—	1.171
o-Chlorophenetole	208	—	1.134
o-Bromoanisole	210	—	—
m-Bromoanisole	211	—	—
p-Chlorophenetole	212	21	1.121
p-Bromoanisole	215	11	—
o-Bromophenetole	218	—	—
p-Bromophenetole	233	4	—
o-Iodoanisole	242	—	—

$n_D^{20°}$	Sulphonamide °C	Picrate °C	Other derivatives °C
1.442	—	—	—
1.434	—	—	—
1.518	111	—	Dinitro, 87; 2,4-dibromo, 61
1.507	150	—	p-Nitro, 58
1.501	—	—	—
1.505	137	—	o-Methoxybenzoic acid, 101
1.512	182	—	Anisic acid, 184
1.513	130	—	m-Methoxybenzoic acid, 110
1.505	149	—	o-Ethoxybenzoic acid, 25
1.496	—	—	—
1.510	—	—	—
1.505	138	—	p-Ethoxybenzoic acid, 196
1.506	111	—	m-Ethoxybenzoic acid, 137
1.544	—	—	Tribromo, 116
—	136	—	Dibromo, 93; nitro, 95
1.497	—	—	—
—	—	—	Trinitro, 92
—	167	—	Dibromo, 140; trinitro, 124
1.538	—	—	Piperonylic acid, 228; pentabromo, 169
1.558	—	—	Anisic acid, 184; tribromo, 108
—	184	—	—
1.532	—	—	Tribromo, 78; veratric acid, 179
1.578	—	—	Tribromo, 109; piperonylic acid, 228
—	159	—	Dibromo, 55; dinitro, 144
1.569	—	—	—
1.696	157	129	Dibromo, 55
1.597	165	119	4-Bromo, 48
—	—	—	Nitro, 95
—	163	100	1-Bromo, 66
—	—	—	—
—	162	—	Trinitro, 122
—	124	—	—
—	148	—	Nitro, 72; dibromo, 142
—	151	117	Bromo, 63
—	155	—	Nitro, 49
—	—	—	—
—	—	—	—
—	—	—	—
1.545	131	—	Nitro, 95
—	151	—	Nitro, 98
—	—	—	—
1.530	133	—	Nitro, 82
—	140	—	Nitro, 106
—	—	—	—
1.522	134	—	Nitro, 61
—	148	—	Nitro, 88
—	135	—	Nitro, 98
—	145	—	Nitro, 47
—	—	—	—

Table 10.13 Aromatic ethers (*continued*)

Ether	B.P. °C (/mmHg)	M.P. °C	d_4^{20}
Halogeno-ethers (*continued*)			
m-Iodoanisole	244	—	—
o-Iodophenetole	246	—	—
m-Iodophenetole	134/15	—	—
p-Iodophenetole	252	29	—
p-Bromodiphenyl ether	168/15	—	—
2,4,6-Trichlorophenetole	246	44	—
p-Iodoanisole	240	52	—
2,4,6-Trichloroanisole	—	62	—
2,4,6-Tribromophenetole	—	73	—
2,4,6-Tribromoanisole	—	88	—
Nitro-ethers			
o-Nitrophenetole	267	—	—
o-Nitroanisole	272	10	1.254
m-Nitrophenetole	284	34	—
m-Nitroanisole	258	39	—
p-Nitroanisole	259	54	—
p-Nitrophenetole	283	60	—
2,4,6-Trinitroanisole	—	68	—
2,4,6-Trinitrophenetole	—	78	—
2,4-Dinitrophenetole	—	87	—
2,4-Dinitroanisole	—	94	—

(1) *o*-Methoxytoluene.
(2) *o*-Methoxyphenol.
(3) Catechol dimethyl ether.
(4) 1-Allyl-3,4-methylenedioxybenzene.

Table 10.14 Acetals

Name	Formula	B.P. °C	d_4^{20}	n_D^{20}
Methylal	$CH_2(OCH_3)_2$	43	0.859	1.353
Dimethylacetal	$CH_3CH(OCH_3)_2$	64	0.852	1.366
Ethylal	$CH_2(OC_2H_5)_2$	87	0.831	1.373
Acetal	$CH_3CH(OC_2H_5)_2$	103	0.826	1.381
1,3-Dioxane	$H_2C{\overset{O}{\underset{O}{\diamondsuit}}}(CH_2)_3$	105	1.034	1.420
Isopropylal	$CH_2(OC_3H_7^i)_2$	122	0.818	1.384
Ethylpropylal	$CH_3CH_2CH(OC_2H_5)_2$	124	0.833	1.390
Acrolein acetal	$CH_2{=}CHCH(OC_2H_5)_2$	125	0.850	—
Propylal	$CH_2(OC_3H_7)_2$	137	0.834	1.393

n_D^{20}	Sulphonamide °C	Picrate °C	Other derivatives °C
—	—	—	—
—	—	—	Nitro, 96
—	—	—	Nitro, 96
—	131	—	—
—	—	—	Dinitro, 100
—	—	—	—
—	—	—	Dinitro, 95
—	—	—	Nitro, 79
—	—	—	—
—	—	—	o-Phenetidine, 228
1.562	—	—	o-Anisidine, 225
—	—	—	m-Phenetidine, 248
—	—	—	m-Anisidine, 251
—	—	—	p-Anisidine, 246
—	—	—	p-Phenetidine, 254
—	—	—	—
—	—	—	—
—	—	—	—

(5) p-(1-propenyl)anisole.
(6) 1-Allyl-3,4-dimethoxybenzene.
(7) 1,2-Methylenedioxy-4-(1-propenyl)benzene.
(8) 3,4-Dimethoxy-1-(1-propenyl)benzene.

Name	Formula	B.P. °C	d_4^{20}	n_D^{20}
Ethylbutylal	$CH_3CH_2CH_2CH(OC_2H_5)_2$	143	0.921	1.402
Propylacetal	$CH_3CH(OC_3H_7)_2$	147	0.830	1.397
Isobutylal	$CH_2(OC_4H_9{}^i)_2$	164	0.824	1.400
Isobutylacetal	$CH_3CH(OC_4H_9{}^i)_2$	176	0.821	1.403
Butylal	$CH_2(OC_4H_9)_2$	181	0.835	1.406
Butylacetal	$CH_3CH(OC_4H_9)_2$	187	0.833	1.409
Pentylal	$CH_2(OC_5H_{11})_2$	219	0.838	1.416
Pentylacetal	$CH_3CH(OC_5H_{11})_2$	222	0.839	1.418
Hexylal	$CH_2(OC_6H_{13})_2$	255	0.841	1.423
Benzaldehyde diethyl acetal	$C_6H_5CH(OC_2H_5)_2$	222	0.983	1.480

Table 10.15   Aliphatic aldehydes

Aldehyde	B.P. °C	M.P. °C	Alkylidene dimethone °C	Dimethone anhydride °C
Formaldehyde	−21	—	189	171
Acetaldehyde	20	—	141	174
Propionaldehyde (propanal)	49	—	155	143
Glyoxal	50	—	186	224
Acrolein (propenal)	52	—	192	163
Isobutyraldehyde (2-methylpropanal)	64	—	154	144
2-Methylpropenal	73	—	—	—
Butyraldehyde (butanal)	75	—	142	141
Trimethylacetaldehyde	75	—	—	—
Isovaleraldehyde (3-methylbutanal)	92	—	155	173
Chloral (trichloroacetaldehyde)	98	—	—	—
Crotonaldehyde (but-2-enal)	102	—	184	167
Valeraldehyde (pentanal)	104	—	105	113
Diethylacetaldehyde	117	—	102	—
4-Methylpentanal	121	—	—	133
Paraldehyde	124	—	—	—
Hexanal	131	—	109	—
Tetrahydrofurfural	145	—	—	—
Heptanal	155	—	103	112
Furfural	161	—	162	164
Hexahydrobenzaldehyde (1)	162	—	—	—
2-Ethylhexanal	163	—	—	—
Octanal	170	—	90	101
Bromal (tribromoacetaldehyde)	174	—	—	—
Nonanal	190	—	86	—
(+)-Citronellal	207	—	79	173
Decanal	208	—	92	—
Citral	229d	—	—	—
β-Hydroxybutyraldehyde	83/20	—	147	126
Chloral hydrate	—	53	56	—
Lauraldehyde (2)	238	45	—	—
Myristaldehyde (3)	155/10	23	—	—
Palmitaldehyde (4)	201/29	34	—	—
Stearaldehyde (5)	—	38	—	—
(±)-Glyceraldehyde	—	142	—	—

(1) Cyclohexanecarboxaldehyde.
(2) Dodecanal.
(3) Tetradecanal.
(4) Hexadecanal.
(5) Octadecanal.

2,4-Dinitro-phenylhydrazone °C	Semicarbazone °C	p-Nitro-phenylhydrazone °C	Other derivatives °C
166	169d	182	Methylene di-2-naphthol, 190
168	163	129	Oxime, 47
155	154(89)	124	Oxime, 40
328	270	311	Oxime, 178
165	171	151	—
187	126	131	—
206	198	—	—
123	106	87	—
209	190	119	Oxime, 41
123	132	110	Oxime, 48
131	—	—	$d_4^{20°}$ 1.512, $n_D^{20°}$ 1.457
190	199	185	Phenylhydrazone, 56; oxime, 119
107	—	—	Oxime, 52
130	99	—	—
99	127	—	—
—	—	—	$d_4^{20°}$ 0.994, $n_D^{20°}$ 1.420
107	106	—	Oxime, 51
204	166	—	$d_4^{20°}$ 1.107, $n_D^{20°}$ 1.436
108	109	73	Oxime, 57
230 (213)	203	154	Phenylhydrazone, 98
—	173	—	Oxime, 91
120	254d	—	—
106	101	80	Oxime, 60
—	—	—	—
100	100	—	Oxime, 64
78	84	—	$d_4^{20°}$ 0.855, $n_D^{20°}$ 1.449
104	102	—	Oxime, 69
110	164	—	$d_4^{20°}$ 0.887, $n_D^{20°}$ 1.488
—	110	—	—
131	—	—	Oxime, 56
106	106	—	Oxime, 78
—	107	95	Oxime, 83
108	109	97	Oxime, 88
—	109	101	Oxime, 89
170	160d	—	Oxime, 118

Table 10.16 Aromatic aldehydes

Aldehyde	B.P. °C	M.P. °C	Dimethone °C	Dimethone anhydride °C
Benzaldehyde	179	—	195	200
Phenylacetaldehyde	194	34	165	126
Salicylaldehyde (1)	197	—	—	208
m-Tolualdehyde	199	—	172	206
o-Tolualdehyde	200	—	167	215
p-Tolualdehyde	204	—	—	—
Phenoxyacetaldehyde	215d	38	—	—
3-Phenylpropanal	224	—	—	—
m-Methoxybenzaldehyde	230	—	—	—
Cuminaldehyde (2)	235	—	171	173
Anisaldehyde (3)	248	2	145	243
Cinnamaldehyde	252	—	213	175
1-Naphthaldehyde	292	34	—	—
Piperonal (4)	263	37	178	220
o-Methoxybenzaldehyde	236	38	—	—
2,3-Dimethoxybenzaldehyde	—	54	—	—
Veratraldehyde (5)	285	58	—	—
2-Naphthaldehyde	—	61	—	—
2,4-Dimethoxybenzaldehyde	—	69	—	—
Vanillin (6)	—	81	197	228
m-Hydroxybenzaldehyde	240	108	—	—
p-Hydroxybenzaldehyde	—	116	189	246
Terephthaldialdehyde	246	116	—	—
β-Resorcylaldehyde (7)	—	136	—	—
Protocatechuicaldehyde (8)	—	153	—	145

Halogeno-aldehydes

o-Chlorobenzaldehyde	213	11	205	225
m-Chlorobenzaldehyde	214	18	—	—
o-Bromobenzaldehyde	230	22	—	—
m-Bromobenzaldehyde	234	—	—	—
o-Iodobenzaldehyde	—	37	—	—
3,4-Dichlorobenzaldehyde	248	44	—	—
p-Chlorobenzaldehyde	214	47	—	—
m-Iodobenzaldehyde	—	57	—	—
p-Bromobenzaldehyde	—	67	—	—
2,6-Dichlorobenzaldehyde	—	71	—	—
2,4-Dichlorobenzaldehyde	—	72	—	—
p-Iodobenzaldehyde	—	78	—	—

Nitro- and amino-aldehydes

o-Aminobenzaldehyde	—	40	—	—
o-Nitrobenzaldehyde	—	44	—	
m-Nitrobenzaldehyde	—	58	—	—
p-Aminobenzaldehyde	—	72	—	—
p-Dimethylaminobenzaldehyde	—	74	—	—
p-Nitrobenzaldehyde	—	106	—	—
m-Aminobenzaldehyde	—	Amorphous	—	—

(1) o-Hydroxybenzaldehyde.
(2) p-Isopropylbenzaldehyde.
(3) p-Methoxybenzaldehyde.
(4) 3,4-Methylenedioxybenzaldehyde.

2,4-Dinitrophenyl hydrazone °C	Semicarbazone °C	Oxime °C	Phenylhydrazone °C	p-Nitrophenyl hydrazone °C
237	224	35d	158	192
121	156	99	63	151
252	231	63	143	228
194	223	60	91	157
194	212	49	106	222
233	234	80	112	201
—	145	95	86	—
149	127	94	—	123
—	—	40	—	171
241	211	52	129	190
254	209	132(65)	121	161
255d	215	139	168	195
—	221	98	80	234
265	234	110	106	200
253	215	92	—	205
—	231	99	138	—
264	177	95	121	—
270	245	156	206	230
—	—	106	—	—
269	239	117	105	228
259	198	90	130	222
280	224	72	178	266
—	—	200	278d	281
286	260	192	160	285
275	230	157	176	—
209	229 (146)	76d	86	249
248	229	71d	134	216
—	214	102	—	240
—	205	72d	141	220
—	206	108	79	—
—	—	119	—	277
265	232	107 (140)	127	220
—	226	62	155	212
—	228	111	113	208
—	—	150	—	—
—	—	—	—	—
—	224	—	121	201
—	247	135	221	220
265	256	103	156	263
292	246	122	121	247
—	173	124	156	—
325	222	185	148	182
320	221	133	159	249
—	280d	195	162	226

(5) 3,4-Dimethoxybenzaldehyde.
(6) 4-Hydroxy-3-methoxybenzaldehyde.
(7) 2,4-Dihydroxybenzaldehyde.
(8) 3,4-Dihydroxybenzaldehyde.

Table 10.17 Aliphatic ketones

Ketone	B.P. °C	M.P. °C	2,4-Dinitro-phenylhydrazone °C
Acetone	56	—	128
Ethyl methyl ketone (butan-2-one)	80	—	115
Methyl vinyl ketone (but-3-en-2-one)	80	—	—
Biacetyl (butane-2,3-dione)	88	—	315 (Di)
Isopropyl methyl ketone (3-methylbutan-2-one)	94	—	120
Methyl propyl ketone (pentan-2-one)	102	—	144
Diethyl ketone (pentan-3-one)	102	—	156
Pinacolone (t-butyl methyl ketone)	106	—	125
Isobutyl methyl ketone (4-methylpentan-2-one)	117	—	95
Di-isopropyl ketone (2,4-dimethylpentan-3-one)	124	—	88
Ethyl propyl ketone (hexan-3-one)	124	—	130
Butyl methyl ketone (hexan-2-one)	128	—	107
4-Methylpent-3-en-2-one	130	—	203
Cyclopentanone	131	—	146
Pentane-2,4-dione	139	—	209
2-Methylcyclopentanone	139	—	—
Dipropyl ketone (heptan-4-one)	144	—	75
Acetyl methyl carbinol (3-hydroxybutan-2-one)	145	—	318
Acetol (hydroxyacetone)	146	—	129
Heptan-2-one	151	—	89
Cyclohexanone	156	—	162
2-Methylcyclohexanone	165	—	137
4-Hydroxy-4-methylpentan-2-one	166	—	203
Di-isobutyl ketone (2,6-dimethylheptan-4-one)	168	—	92
Methyl acetoacetate	170	—	—
3-Methylcyclohexanone	170	—	155
4-Methylcyclohexanone	171	—	134
Hexyl methyl ketone (octan-2-one)	173	—	58
Cycloheptanone	180	—	148
Cyclohexyl methyl ketone	180	—	140
Ethyl acetoacetate (ethyl 3-oxobutanoate)	181	—	93
Dibutyl ketone (nonan-5-one)	188	—	—
(+)-Fenchone	193	—	140
Hexane-2,5-dione	194	—	257 (Di)
Methyl levulinate (methyl 4-oxopentanoate)	196	—	142
Phorone (1)	199	28	118
β-Thujone	202	—	114
Ethyl levulinate	206	—	102
(−)-Menthone	209	—	146
Isophorone (2)	215	—	130
Pulegone	224	—	147
Undecan-6-one	226	14	—
(+)-Carvone	230	—	191
Tridecan-7-one	255	33	—
α-Ionone	130/13	—	151
β-Ionone	139/18	—	128
2,2′-Furoin	—	135	217
2,2′-Furil	—	165	215
(+)-Camphor	209	179	177
Chloroacetone	119	—	125
1,1-Dichloroacetone	120	—	—
1,3-Dichloroacetone	173	45	133

(1) 2,6-Dimethylhepta-2,5-dien-4-one.   (2) 3,3,5-Trimethylcyclohex-2-enone.

Semi-carbazone °C	Benzylidene derivative °C	Phenyl-hydrazone °C	p-Nitro-phenyl-hydrazone °C	Other derivatives °C
190	112	42	149	Oxime, 59
146	—	—	129	—
141	—	—	—	—
279 (Di)	53	243 (Di)	230	Dioxime, 234
114	—	—	109	—
112	—	—	117	Oxime, 58
139	31	—	144	Oxime, 69
158	41	—	—	Oxime, 78
132	—	—	79	Oxime, 58
160	—	—	—	Oxime, 34
112	—	—	—	—
125	—	—	88	Oxime, 49
164	—	142	134	Oxime, 49
210	190	55	154	Oxime, 57
—	—	—	—	Oxime, 149
184	—	—	—	—
133	—	—	—	—
185	—	—	—	—
196	—	103	—	Oxime, 71
127	—	207	73	—
167	118	81	147	Oxime, 91
197	—	—	132	Oxime, 43
—	—	—	209	Oxime, 58
122	—	—	—	—
152	—	—	—	d_4^{20} 1.077, n_D^{20} 1.420
191	122	94	119	—
203	99	110	128	Oxime, 39
123	—	—	93	—
162	108	—	—	—
177	—	—	154	Oxime, 60
133	—	—	—	d_4^{20} 1.025, n_D^{20} 1.420
90	—	—	—	—
184	—	—	—	Oxime, 167
220	—	120 (Di)	—	Dioxime, 137
143	—	96	—	d_4^{20} 1.050, n_D^{20} 1.423
221	—	—	—	Oxime, 48
174	—	—	—	Oxime, 55
148	—	104	—	d_4^{20} 1.011, n_D^{20} 1.423
189	—	53	—	Oxime, 59
199	77	68	—	Oxime, 79
174	—	—	—	Oxime, 119
—	—	—	—	d_4^{20} 0.825, n_D^{20} 1.429
163	—	110	175	Oxime, 73
—	—	—	97	—
143 (108)	—	—	113	Oxime, 90
149	—	—	173	—
—	—	81	—	Oxime, 161
—	—	184	199	Dioxime, 100
238	98	233	217	Oxime, 119
150	—	—	—	—
163	—	—	—	—
120	—	—	—	—

Table 10.18   Aromatic ketones

Ketone	B.P. °C	M.P. °C	2,4-Dinitro-phenylhydrazone °C
Acetophenone	202	20	250 (237)
2-Hydroxyacetophenone	215	—	—
2-Methylacetophenone (1)	216	—	159
Benzyl methyl ketone	216	27	156
Propiophenone (2)	218	19	191
3-Methylacetophenone (3)	220	—	207
Isobutyrophenone (4)	222	—	163
4-Methylacetophenone (5)	224	28	258
Benzyl ethyl ketone	226	—	—
Butyrophenone (6)	230	12	190
m-Methoxyacetophenone	240	—	—
Valerophenone (7)	242	—	166
o-Methoxyacetophenone	245	—	—
α-Tetralone	129/12	—	257
β-Tetralone	138/16	18	—
1-Acetylnaphthalene	302	34	—
Phenyl o-tolyl ketone (8)	310	—	190
Phenyl m-tolyl ketone	314	—	221
Dibenzyl ketone	331	35	100
p-Methoxyacetophenone	258	39	220
α-Hydrindone	242	42	258
Benzalacetone	262	42	227
Benzophenone	306	49	238
2-Acetylnaphthalene	301	56	262
Phenyl styryl ketone (9)	347	58	245
Phenyl p-tolyl ketone	326	60	200
Deoxybenzoin (10)	320	60	206
p-Methoxybenzophenone	355	62	180
Fluorenone (11)	341	83	284
α-Hydroxyacetophenone (12)	—	86	—
Di-p-tolyl ketone	335	95	229
Benzil	347d	95	189
m-Hydroxyacetophenone	—	96	—
p-Hydroxyacetophenone	—	109	261
Dibenzylideneacetone	—	112	180
p-Benzoquinone	—	116	186
Acenaphthenone	—	121	—
1,4-Naphthoquinone	—	125	—
Benzoin	344	137	245
1,2-Naphthoquinone	—	146	—
Resacetophenone (13)	—	147	—
9,10-Phenanthraquinone	—	207	313
Phloroacetophenone (14)	—	219	—
Anthraquinone	—	285	—

Halogeno-ketones

m-Chloroacetophenone	228	—	—
o-Chloroacetophenone	229	—	206
p-Chloroacetophenone	236	20	231
o-Bromoacetophenone	112/10	—	—
m-Bromoacetophenone	131/16	8	—
α-Bromoacetophenone (15)	—	51	—
p-Bromoacetophenone	256	51	230
p-Chlorobenzophenone	323	78	185

Semi-carbazone	Oxime °C	Phenyl-hydrazone °C	p-Nitro-phenyl-hydrazone °C	Other derivatives °C
199	59	105	185	Benzylidene, 58
210	117	110	—	—
203	61	—	—	—
198	69	87	145	—
174	53	147	—	—
198	55	—	—	—
181	94	73	—	—
205	88	96	198	—
136	—	—	—	—
188	50	—	—	—
196	—	—	—	—
166	52	162	—	—
183	83	114	—	—
217	89 (103)	84	231	Benzylidene, 105
215	88	109	—	—
229	139	149	—	Picrate, 116; benzylidene, 126
—	—	—	—	—
—	101	—	—	—
146	125	129	—	Benzylidene, 162
198	87	142	—	—
233	146	128	235	Benzylidene, 113
186	116	157	166	Benzylidene, 112
165	144	137	155	—
236	145	177	—	Picrate, 85
168	115	119	—	Picrate, 97
122	154	109	—	—
148	98	116	163	Benzylidene, 102
—	138	132	199	—
—	195	152	269	—
146	70	112	—	Benzoyl, 118; acetyl, 49
—	163	100	—	—
244 (Di)	237	235 (Di)	290	Quinoxaline, 126
195	—	—	—	—
199	145	151	—	—
189	143	153	173	Picrate, 114
243 (Di)	140	—	—	Picrate, 79
—	175	90	—	Picrate, 113
247	198	—	278	—
206	151 (99)	159	—	Benzoyl, 125; acetyl, 83
184	162	138	235	—
218	199	159	—	Dibenzoyl, 81; diacetyl, 38
—	158	165	245	—
—	—	—	—	Tribenzoyl, 118; triacetyl, 103
—	224	183	—	—
232	88	—	176	—
160 (179)	113	—	215	—
201	95	114	239	—
—	177	—	—	—
238	—	—	—	—
146	89	—	—	—
208	129	126	—	—
—	156 (95)	106	—	—

Table 10.18   Aromatic ketones (*continued*)

Ketone	B.P. °C	M.P. °C	2,4-Dinitro-phenylhydrazone °C
Halogeno-ketones (*continued*)			
p-Bromobenzophenone	350	82	230
p-Iodoacetophenone	—	85	—
α,p-Dibromoacetophenone (16)	—	109	—
Nitro- and amino-ketones			
o-Nitroacetophenone	159/16	—	—
o-Aminoacetophenone	251	20	—
m-Nitroacetophenone	—	81	228
p-Nitroacetophenone	—	81	—
m-Aminoacetophenone	—	99	—
p-Aminoacetophenone	294	106	—

(1) Methyl o-tolyl ketone.
(2) Ethyl phenyl ketone.
(3) Methyl m-tolyl ketone.
(4) Isopropyl phenyl ketone.
(5) Methyl p-tolyl ketone.
(6) Phenyl propyl ketone.
(7) Butyl phenyl ketone.
(8) 2-Methylbenzophenone.

Semi-carbazone	Oxime °C	Phenyl-hydrazone °C	p-Nitro-phenyl-hydrazone °C	Other derivatives °C
—	169	126	—	
—	115	—	—	
290	109	108	—	
257	132	135	—	
—	—	132	—	
196	148	—	—	
250	—	—	—	

(9) Chalcone.
10) Benzyl phenyl ketone.
11) Diphenylene ketone.
12) Phenacyl alcohol.
13) 2,4-Dihydroxyacetophenone.
14) 2,4,6-Trihydroxyacetophenone.
15) Phenacyl bromide.
16) p-Bromophenacyl bromide.

Table 10.19   Quinones

Quinone	M.P. °C	Semicarbazone °C	Oxime °C
Thymoquinone	45	204	162
2-Methyl-1,4-benzoquinone	69	179	135
2-Methyl-1,4-naphthoquinone	106	247	167
Duroquinone (1)	112	—	—
p-Benzoquinone	116	243d	240d
1,4-Naphthoquinone	125	247	198
1,2-Naphthoquinone	116–120	184	162
1-Methylanthraquinone	172	—	—
2-Methylanthraquinone	177	—	—
3-Methyl-1,2-benzoquinone	195	—	140
Camphorquinone	199	236	170
Quinizarin (2)	201	—	—
9,10-Phenanthraquinone	206	220d	162
Acenaphthenequinone	261	192	222 Di
Anthraquinone	286	—	224
Chloranil (3)	290*	—	—
Alizarin	290	—	—

* Sealed tube.   (1) 2,3,5,6-Tetramethyl-1,4-benzoquinone.

Table 10.20   Aliphatic carboxylic acids

Acid	B.P. °C	M.P. °C	Anilide °C	p-Tolu-idide °C	Amide °C
Formic	101	8	50	53	3
Acetic	118	16	114	153	82
Acrylic (propenoic)	140	13	105	141	85
Propionic (propanoic)	141	—	106	126	79
Propiolic (propynoic)	144d	18	87	—	62
Isobutyric (2-methylpropanoic)	154	—	105	109	129
Butyric (butanoic)	163	—	96	75	115
Pivalic (2,2-dimethylpropanoic)	164	34	133	120	154
Pyruvic (2-oxopropanoic)	165d	13	104	130	125
Crotonic (cis) (but-2-enoic)	165	15	102	—	102
Isovaleric (3-methylbutanoic)	176	—	110	109	136
2-Methylbutanoic	177	—	112	93	112
Valeric (pentanoic)	186	—	63	74	106
2-Ethylbutanoic	193	—	127	116	112
4-Methylpentanoic	199	—	112	63	121
Methoxyacetic	203	—	58	—	96
Hexanoic	205	—	—	74	100
Ethoxyacetic	207	—	95	—	82
Heptanoic	223	—	71	80	96
2-Ethylhexanoic	228	—	—	—	103
Cyclohexanecarboxylic	233	31	144	—	186
Octanoic	239	16	57	70	107
Levulinic (4-oxopentanoic)	246	33	102	109	108

Hydroquinone °C	Diacetate of hydroquinone °C	Thiele acetylation product °C	Other derivatives °C
143	74	—	—
124	52	114	—
—	—	113	—
239	207	—	—
171	123	97	Picrate, 179
176	128	135	—
103	105	135	—
—	217	—	—
—	—	—	—
—	—	—	Quinoxaline, 78
—	—	207	—
148	202	—	Quinoxaline, 220
—	—	—	Quinoxaline, 241
180	260	—	—
232	251	—	—
—	182	—	—

(2) 1,4-Dihydroxyanthraquinone.   (3) 2,3,5,6-Tetrachloro-1,4-benzoquinone.

p-Bromophenacyl ester °C	p-Nitrobenzyl ester °C	p-Phenylphenacyl ester °C	S-Benzylthiouronium salt °C	p-Bromoanilide °C	Hydrazide* °C	N-Benzylamide* °C
140	31	74	151	119	54	60
86	78	111	136	166	77	61
—	—	—	—	—	—	—
63	31	102	152	148	40	44
—	—	—	—	—	—	—
77	—	89	149	151	104	87
63	35	82	149	111	44	38
76	—	—	—	—	—	—
—	—	—	—	—	—	—
68	—	78	159	129	68	54
55	—	—	—	—	—	—
75	—	63	156	106	—	42
—	—	77	—	—	—	—
77	—	70	—	—	—	—
—	—	—	—	—	—	—
72	—	70	159	105	—	53
104	—	—	—	—	—	—
72	—	62	—	95	—	—
—	—	54	—	—	—	—
—	—	—	—	—	—	—
67	—	67	157	102	—	—
84	61	—	—	—	—	—

Table 10.20 Aliphatic carboxylic acids (continued)

Acid	B.P. °C	M.P. °C	Anilide °C	p-Tolu-idide °C	Amide °C
Nonanoic	254	12	57	84	99
Decanoic	269	31	70	78	108
Undec-10-enoic	275	25	67	68	87
Undecanoic	164/15	29	71	80	103
(±)-Lactic (2-hydroxypropanoic)	122/15	18	59	107	79
Dodecanoic	225/100	43	78	87	99
Myristic (tetradecanoic)	250/100	58	84	93	103
Palmitic (hexadecanoic)	268/100	63	91	98	106
Oleic (cis-octadec-9-enoic)	233/10	16	41	43	76
Cyanoacetic	—	66	198	—	120
Stearic (octadecanoic)	291/100	70	94	102	109
Crotonic (trans) (but-2-enoic)	189	72	118	132	160
Glycollic (hydroxyacetic)	—	79	97	143	120
Citraconic (cis-methylbutenedioic)	—	93	175	—	186
Glutaric (pentanedioic)	—	98	224	218	175
Citric (hydrated)	—	100	199	189	215
(−)-Malic (±, m.p. 133 °C)	—	101	197	207	157
Oxalic (dihydrate)	—	101	246	268	419d
Pimelic (heptanedioic)	—	105	156	206	—
Azelaic (nonanedioic)	—	106	187	202	172
Sebacic (decanedioic)	—	133	202	201	209
Sorbic (hexa-2,4-dienoic)	—	134	153	—	—
Furoic	—	134	124	108	142
Maleic	—	135	187	142	181
Malonic	—	135d	225	253	170
meso-Tartaric	—	140	—	—	190
2-Furylacrylic	—	141	—	—	169
Suberic (octanedioic)	—	142	187	219	217
Adipic (hexanedioic)	—	152	239	241	220
Itaconic (methylenesuccinic)	—	165	190	—	192
(+)-Tartaric	—	170	264	—	196
Succinic	—	185	230	255	260
(+)-Camphoric	—	187	226	—	193
Aconitic (trans-propene-1,2,3-tricarboxylic)	—	191	—	—	250
Mesaconic (trans-methylbutenedioic)	—	204	186	212	176
(±)-Tartaric	—	206	—	—	226
D-Galactaric	—	214	—	—	—
Fumaric	—	286	314	—	266
Thioacetic	93	—	76	131	115

Halogeno-acids

Acid	B.P. °C	M.P. °C	Anilide °C	p-Tolu-idide °C	Amide °C
2-Chloropropanoic	186	—	92	124	80
Dichloroacetic	194	10	119	153	97
2-Bromopropanoic	206	25	99	125	123
Bromoacetic	208	50	130	91	91
Trichloroacetic	196	58	95	113	141
Chloroacetic	189	63	137	162	120
Iodoacetic	—	84	144	—	95
Fluoroacetic	167	35	—	—	108
Difluoroacetic	134	—	—	—	52
Trifluoroacetic	72	—	—	—	—

p-Bromo-phenacyl ester °C	p-Nitro-benzyl ester °C	p-Phenyl-phenacyl ester °C	S-Benzyl-thiouron-ium salt °C	p-Bromo-anilide °C	Hydrazide* °C	N-Benzyl-amide* °C
69	—	71	—	100	—	—
67	—	—	—	102	—	—
—	—	—	149	—	—	—
68	—	79	—	—	—	—
113	—	145	153	—	—	—
76	—	86	141	—	105	83
81	—	90	139	—	—	90
86	43	94	141	—	111	95
45	—	61	—	—	—	—
—	—	—	—	—	—	124
90	—	97	143	—	—	97
95	67	—	172	—	—	114
138	107	—	146	—	—	104
—	70	109	—	—	177	—
137	69	152	161	—	176	170
148	102	146	—	—	—	170
179	124	106	124	—	178	157
242	204	165d	198	—	243	223
137	—	146d	—	—	182	154
131	44	141	—	—	—	—
147	73	140	155	—	—	167
129	—	—	—	—	—	—
139	134	86	211	—	80	—
168	89	168	163	—	—	150
—	86	175	147	—	154	142
—	93	—	—	—	—	205
—	—	—	—	—	—	—
144	85	151	—	—	—	—
155	106	148	163	—	171	189
117	90	—	—	—	—	—
216	163	204	—	—	—	199
211	88	208	154	—	168	206
—	67	—	—	—	—	—
186	—	—	—	—	—	—
—	—	—	—	—	—	—
—	147	—	—	—	—	210
—	—	149	178	—	215	201
—	151	—	195	—	—	205
—	—	—	—	—	—	—
99	—	—	178	—	—	—
—	—	—	—	—	—	—
—	—	—	—	—	—	—
—	80	—	148	—	—	—
105	—	116	160	—	—	—
—	—	—	—	—	—	—
—	—	—	—	—	—	—
—	—	—	—	—	—	—

* See Section 9.6.17 for details of the preparation of hydrazides and N-benzylamides.

Table 10.21 Aromatic carboxylic acids

Acid	M.P. °C	Anilide °C	*p*-Toluidide °C	Amide °C
o-Ethoxybenzoic	25	—	—	132
3-Phenylpropanoic (1)	48	98	135	105
Phenylacetic	76	118	136	157
Phenoxyacetic	99	101	—	101
o-Methoxybenzoic	101	131	—	129
o-Toluic	105	125	144	143
m-Methoxybenzoic	110	—	—	—
m-Toluic	111	126	118	95
(±)-Mandelic (2)	120	152	172	134
Benzylmalonic	120d	217	—	225
Benzoic	121	162	158	129
o-Benzoylbenzoic	128	195	—	165
Cinnamic	133	153	168	147
1-Naphthylacetic	133	156	—	181
Acetylsalicylic	135	136	—	138
Phenylpropiolic	136	126	142	109
m-Ethoxybenzoic	137	—	—	139
2-Naphthylacetic	142	—	—	200
Diphenylacetic	148	180	173	168
Benzilic	150	175	190	155
2-Hydroxy-5-methylbenzoic (3)	153	—	—	178
Salicylic	158	135	156	139
1-Naphthoic	162	163	—	202
2-Hydroxy-3-methylbenzoic (4)	169	—	—	112
2-Hydroxy-4-methylbenzoic (5)	177	—	—	—
p-Toluic	178	146	160	159
p-Methoxybenzoic (6)	184	169	186	162
2-Naphthoic	185	170	191	192
p-Ethoxybenzoic	198	170	—	202
3,4-Dihydroxybenzoic (7)	199d	167	—	212
3-Hydroxybenzoic	201	157	163	167
Phthalic (benzene-1,2-dicarboxylic)	c. 208d	251	—	219
4-Hydroxy-3-methoxy benzoic (8)	210	—	—	—
4-Hydroxybenzoic	213	197	204	162
2,4-Dihydroxybenzoic (9)	213	127	—	221
3-Hydroxy-2-naphthoic	223	244	222	218
1-Hydroxy-2-naphthoic	226	—	—	218
Diphenic	229	230	—	212
3,4-Methylenedioxybenzoic (10)	229	—	—	169
Gallic (3,4,5-trihydroxybenzoic)	c. 240d	207	—	245
Isophthalic (benzene-1,3-dicarboxylic)	347	—	—	280
Terephthalic (benzene-1,4-dicarboxylic)	sub. > 300	337	—	—
Benzene-1,3,5-tricarboxylic (11)	380	—	—	365

Halogeno-carboxylic acids

Acid	M.P. °C	Anilide °C	*p*-Toluidide °C	Amide °C
m-Chlorophenoxyacetic	100	—	—	—
m-Fluorobenzoic	124	—	—	130
o-Fluorobenzoic	127	—	—	116
o-Chlorobenzoic	141	118	131	141
o-Chlorophenoxyacetic	146	121	—	150
o-Bromobenzoic	150	141	—	155
m-Bromobenzoic	155	146	—	155

p-Bromophenacyl ester °C	p-Nitrobenzyl ester °C	p-Phenylphenacyl ester °C	S-Benzylthiouronium salt °C	N-Benzylamide °C	Other derivatives °C
—	—	—	—	—	—
04	36	95	—	85	—
89	65	88	165	122	—
48	—	—	—	—	—
—	113	131	—	—	—
57	91	95	146	—	Hydrazide, 124
—	—	—	—	—	—
08	87	136	140	75	Hydrazide, 97
—	123	—	166	—	—
—	120	—	—	—	—
19	89	167	167	106	Hydrazide, 112
—	100	—	—	—	—
46	117	182	183	225	—
—	—	—	—	—	—
—	90	—	144	—	—
—	83	—	—	—	—
—	—	—	—	—	—
—	—	—	—	—	—
—	—	111	—	—	—
52	100	122	—	—	Acetyl, 98
—	147	—	185	—	Acetyl, 153
40	98	148	148	136	Benzoyl, 132; p-nitrobenzoyl, 205
—	99	—	204	—	Acetyl, 113
—	175	—	165	—	Acetyl, 139
53	104	165	190	133	Hydrazide, 117
52	132	160	185	132	—
—	110	—	—	—	—
—	188	—	—	—	—
76	108	—	—	142	Acetyl, 131
53	155	167	158	179	—
—	141	—	—	—	Acetyl, 146; benzoyl, 178
91	192	240	145	—	Acetyl, 187
—	189	—	—	—	—
—	—	—	—	—	—
—	186	—	—	—	—
—	—	—	—	—	—
—	—	198d	—	—	Triacetyl, 172; tribenzoyl, 192
79	203	280	216	—	Hydrazide, 220
25	264	—	204	266	—
97	—	—	—	—	Tri-Me-ester, 144; tri Et-ester, 135
—	—	—	—	—	Hydrazide, 139
—	—	—	—	—	Hydrazide, 73
07	106	123	—	—	Hydrazide, 110
—	—	—	—	—	—
02	110	98	171	—	—
26	105	155	168	—	—

Table 10.21   Aromatic carboxylic acids (*continued*)

Acid	M.P. °C	Anilide °C	p-Toluidide °C	Amide °C
Halogeno-carboxylic acids (*continued*)				
p-Chlorophenoxyacetic	157	125	—	133
m-Chlorobenzoic	158	124	—	134
o-Iodobenzoic	162	141	—	184
2,4-Dichlorobenzoic	164	—	—	194
p-Fluorobenzoic	185	—	—	154
m-Iodobenzoic	187	—	—	186
3,4-Dichlorobenzoic	209	—	—	169
p-Chlorobenzoic	243	194	—	179
p-Bromobenzoic	252	197	—	189
p-Iodobenzoic	270	210	—	218
Nitro- and amino-carboxylic acids				
m-Nitrophenylacetic	120	—	—	110
m-Nitrobenzoic	141	154	162	142
o-Nitrophenylacetic	141	—	—	161
Anthranilic (o-aminobenzoic)	146	131	151	109
o-Nitrobenzoic	147	155	—	175
p-Nitrophenylacetic	152	212	210	198
4-Nitrophthalic	165	—	—	200
m-Aminobenzoic	174	140	—	111
2,4-Dinitrobenzoic	183	—	—	204
N-Acetylanthranilic	185	167	—	171
Hippuric (12)	187	208	—	183
p-Aminobenzoic	188	—	—	114
m-Nitrocinnamic	205	—	—	196
3,5-Dinitrobenzoic	207	234	—	183
3-Nitrophthalic	219	234	223	201
2,4,6-Trinitrobenzoic	228	—	—	264
Nicotinic	235	85	150	128
p-Nitrobenzoic	239	211	203	201
o-Nitrocinnamic	240	—	—	185
β-Phenylalanine (13)	273	—	—	140
p-Nitrocinnamic	287	—	—	217

(1) Hydrocinnamic acid.
(2) 2-Hydroxy-2-phenylacetic acid.
(3) 6-Hydroxy-*m*-toluic acid.
(4) 2-Hydroxy-*m*-toluic acid.
(5) 2-Hydroxy-*p*-toluic acid.
(6) Anisic acid.
(7) Protocatechuic acid.

p-Bromo-phenacyl ester °C	p-Nitro-benzyl ester °C	p-Phenyl-phenacyl ester °C	S-Benzyl-thio-uronium salt °C	N-Benzyl-amide °C	Other derivatives °C
136	—	—	—	—	—
117	107	154	155	—	Hydrazide, 158
110	111	143	—	110	—
—	—	—	—	—	Dinitro, 211
—	—	—	—	—	Hydrazide, 162
128	121	—	—	—	—
—	—	—	—	—	—
126	130	160	—	—	Hydrazide, 163
134	141	160	—	—	Hydrazide, 164
146	141	171	—	—	—
—	—	—	—	—	—
132	142	153	163	101	—
—	—	—	—	—	—
—	205	—	149	—	N-Benzoyl, 81; N-toluene-p-sulphonyl, 217
107	112	140	159	—	—
207	—	—	—	—	—
—	—	120	—	—	—
—	201	—	—	—	N-Acetyl, 248
158	142	—	—	—	—
—	—	—	—	—	—
151	136	163	—	—	Hydrazide, 162
—	—	—	—	90	N-Acetyl, 250; N-benzoyl, 278
178	174	—	—	—	—
159	157	154	—	—	—
—	190	149	—	—	—
—	—	—	—	—	—
—	—	—	—	—	—
136	169	182	182	142	—
142	132	146	—	—	—
—	222	—	—	—	N-Benzoyl, 188
191	187	192	—	—	—

(8) Vanillic acid.
(9) β-Resorcylic acid.
(10) Piperonylic acid.
(11) Trimesic acid.
(12) Benzoylaminoacetic acid.
(13) 2-Amino-3-phenylpropanoic acid.

Table 10.22   Acid chlorides (aliphatic)

Acyl chloride	B.P. °C (/mmHg)	M.P. °C	$d_4^{20°}$	$n_D^{20°}$
Acetyl	52	—	1.104	1.390
Propionyl (propanoyl)	80	—	1.056	1.404
Isobutyrl (2-methylpropanoyl)	92	—	1.017	1.408
Butyryl (butanoyl)	102	—	1.028	1.412
Chloroacetyl	105	—	1.420	1.454
Dichloroacetyl	108	—	—	—
Methoxyacetyl	113	—	1.187	1.419
3-Methylbutanoyl	115	—	0.987	1.416
Trichloroacetyl	118	—	1.620	1.470
Crotonoyl	126	—	—	—
Valeryl (pentanoyl)	127	—	1.000	1.420
4-Methylpentanoyl	144	—	0.973	—
Hexanoyl	152	—	0.975	1.426
Heptanoyl	175	—	0.962	1.432
Octanoyl	195	—	0.949	1.432
Nonanoyl	215	—	0.942	1.433
Decanoyl	232	—	—	—
Oxalyl	64	—	1.479	1.432
Succinyl	192	17	1.375	1.468
Glutaryl	218	—	1.324	1.473
Adipoyl	125/11	—	—	—
Pimeloyl	137/15	—	—	—
Suberoyl	150/12	—	1.171	1.468
Azelaoyl	165/13	—	—	—
Sebacoyl	182/16	—	1.212	1.468

Table 10.23   Acid anhydrides (aliphatic)

Anhydride	B.P. °C (/mmHg)	M.P. °C	$d_4^{20°}$	$n_D^{20°}$
Acetic	140	—	1.081	1.390
Propionic	168	—	1.022	1.404
Isobutyric	182	—	0.956	—
Butyric	198	—	0.968	1.413
Citraconic	213	7	—	—
Isovaleric (3-methylbutanoic)	215	—	0.933	1.404
Valeric	218	—	0.925	—
4-Methylpentanoic	139/19	—	—	—
Hexanoic	245	—	0.920	1.430
Crotonic	248	—	1.040	1.474
Heptanoic	258	17	0.917	1.433
Octanoic	285	—	0.910	1.434
Maleic	198	56	—	—
Glutaric	150/10	56	—	—
Itaconic	139/30	68	—	—
Succinic	261	120	—	—
(+)-Camphoric	270	221	—	—
Trifluoroacetic	39	—	1.490	1.269
Dichloroacetic	101/16	—	—	—
Trichloroacetic	223	—	—	—
Chloroacetic	109/11	46	—	—

Table 10.24 Acid chlorides and acid anhydrides of aromatic acids

Acid chloride	B.P. °C (/mmHg)	M.P. °C
Benzoyl	197	—
Phenylacetyl	210	—
o-Toluoyl	212	—
m-Toluoyl	219	—
p-Toluoyl	227	—
m-Methoxybenzoyl	244	—
o-Methoxybenzoyl	254	—
Phthaloyl	281	16
Anisoyl	145/14	24
1-Naphthoyl	163/10	24
Cinnamoyl	131/11	36
2-Naphthoyl	305	53
Diphenylcarbamoyl	—	86
p-Chlorobenzoyl	222	16
m-Chlorobenzoyl	225	—
o-Chlorobenzoyl	238	—
m-Bromobenzoyl	243	—
o-Bromobenzoyl	245	11
p-Bromobenzoyl	245	42
o-Nitrobenzoyl	148/9	20
m-Nitrobenzoyl	278	35
2,4-Dinitrobenzoyl	—	46
3,5-Dinitrobenzoyl	196/11	74
p-Nitrobenzoyl	—	75
3-Nitrophthaloyl	—	77

Anhydride	B.P. °C	M.P. °C
o-Toluic	—	39
Benzoic	360	42
m-Toluic	—	71
Phenylacetic	—	72
p-Toluic	—	95
Anisic	—	99
Phthalic	284	132
2-Naphthoic	—	135
Cinnamic	—	136
1-Naphthoic	—	146
Naphthalene-1,2-dicarboxylic	—	169
Diphenic	—	217
(+)-Camphoric	270	222
Naphthalene-2,3-dicarboxylic	—	246d
Naphthalene-1,8-dicarboxylic	—	274
o-Chlorobenzoic	—	79
m-Chlorobenzoic	—	95
p-Chlorobenzoic	—	194
Tetrachlorophthalic	—	255
Tetrabromophthalic	—	280
Tetra-iodophthalic	—	325
3,5-Dinitrobenzoic	—	109
4-Nitrophthalic	—	119
o-Nitrobenzoic	—	135
2,4-Dinitrobenzoic	—	160
m-Nitrobenzoic	—	163
3-Nitrophthalic	—	164
p-Nitrobenzoic	—	190

Table 10.25 Aliphatic esters

It is considered that the table will be of greatest use if the esters are subdivided under the various acids rather than arranged in order of increasing b.p. or m.p. irrespective of the nature of the carboxylic acid. The latter procedure leads to an unwieldy, heterogeneous table which has relatively little pedagogic or, indeed, practical value.

Ester	B.P. °C (/mmHg)	$d_4^{20°}$	$n_D^{20°}$
Methyl formate	32	0.974	1.344
Ethyl formate	53	0.923	1.360
Isopropyl formate	71	0.873	1.368
Propyl formate	81	0.904	1.377
t-Butyl formate	83	—	—
Allyl formate	84	0.946	—
s-Butyl formate	97	0.884	1.384
Isobutyl formate	98	0.876	1.386
Butyl formate	106	0.892	1.389
Isopentyl formate	124	0.882	1.398
Pentyl formate	131	0.885	1.400
Cyclopentyl formate	138	1.000	1.432
Hexyl formate	154	0.879	1.407
Cyclohexyl formate	161	0.994	1.443
Ethylene glycol diformate	177	1.229	—
Methyl acetate	56	0.939	1.362
Ethyl acetate	77	0.901	1.372
Isopropyl acetate	88	0.872	1.377
t-Butyl acetate	97	0.867	1.386
Propyl acetate	101	0.887	1.384
Allyl acetate	104	0.928	1.404
s-Butyl acetate	112	0.872	1.389
Isobutyl acetate	116	0.871	1.390
Butyl acetate	124	0.881	1.394
t-Pentyl acetate	124	0.873	1.392
Isopentyl acetate	141	0.872	1.400
Pentyl acetate	148	0.875	1.402
Cyclopentyl acetate	153	0.975	1.432
Hexyl acetate	169	0.872	1.409
Cyclohexyl acetate	172	0.970	1.442
Heptyl acetate	192	0.865	1.414
Tetrahydrofurfuryl acetate	195	1.061	1.438
Octyl acetate	210	—	—
Methyl 'cellosolve' acetate	144	1.088	—
'Cellosolve' acetate	156	0.976	—
Ethylene glycol diacetate	190	1.104	1.415
Propylene glycol diacetate	191	1.059	1.417
Trimethylene glycol diacetate	210	1.069	—
'Carbitol' acetate	217	1.013	—
Butyl 'carbitol' acetate	246	0.983	—
α-Monoacetin (glycerol 1-acetate)	158/15	1.206	1.416
Diacetin (mixture of αγ and αβ)	143/12	1.180	—
Triacetin (glyceryl triacetate)	153/22	1.161	1.430
Methyl propionate (propanoate)	79	0.915	1.377
Ethyl propionate	98	0.892	1.384
Isopropyl propionate	111	—	—
Propyl propionate	122	0.882	1.393
Allyl propionate	123	0.914	1.410

Table 10.25 Aliphatic esters (*continued*)

Ester	B.P. °C (/mmHg)	$d_4^{20°}$	$n_D^{20°}$
Butyl propionate	145	0.875	1.401
Isopentyl propionate	160	0.859	1.412
Pentyl propionate	169	0.881	—
Hexyl propionate	190	0.870	1.419
Methyl butyrate (butanoate)	102	0.898	1.387
Ethyl butyrate	120	0.879	1.392
Isopropyl butyrate	128	—	—
Propyl butyrate	142	0.872	1.400
Allyl butyrate	142	0.902	1.416
Butyl butyrate	165	0.869	1.406
Isopentyl butyrate	179	0.864	1.411
Pentyl butyrate	185	0.866	1.412
Hexyl butyrate	208	0.866	1.420
Methyl isobutyrate (2-methylpropanoate)	91	0.888	1.383
Ethyl isobutyrate	110	0.869	1.387
Isopropyl isobutyrate	121	—	—
Propyl isobutyrate	134	0.864	1.396
Butyl isobutyrate	156	0.862	1.402
Methyl valerate (pentanoate)	127	0.890	1.397
Ethyl valerate	144	0.874	1.400
Isopropyl valerate	154	0.858	1.401
Propyl valerate	164	0.870	1.407
Butyl valerate	184	0.868	1.412
Methyl isovalerate (3-methylbutanoate)	116	0.881	1.393
Ethyl isovalerate	133	0.865	1.396
Propyl isovalerate	156	0.862	1.403
Isobutyl isovalerate	171	0.853	1.406
Butyl isovalerate	176	0.861	1.409
Methyl hexanoate	149	0.885	1.405
Ethyl hexanoate	168	0.871	1.407
Propyl hexanoate	187	0.867	1.417
Butyl hexanoate	208	0.865	1.421
Pentyl hexanoate	226	0.863	1.426
Methyl cyclohexanecarboxylate	183	0.990	1.451
Ethyl cyclohexanecarboxylate	196	0.962	1.448
Methyl heptanoate	171	0.882	1.412
Ethyl heptanoate	186	0.870	1.413
Propyl heptanoate	208	0.866	1.421
Butyl heptanoate	226	0.864	1.426
Methyl octanoate	192	0.878	1.417
Ethyl octanoate	206	0.869	1.418
Methyl nonanoate	214	—	—
Ethyl nonanoate	227	0.866	1.422
Methyl decanoate	228	0.873	1.426
Ethyl decanoate	242	0.865	1.426
Propyl decanoate	115/5	0.862	1.428
Butyl decanoate	123/4	0.861	1.430

Table 10.25 Aliphatic esters (*continued*)

Ester	B.P. °C (/mmHg)	$d_4^{20°}$	$n_D^{20°}$
Methyl dodecanoate	262	0.870	1.432
Ethyl dodecanoate	273	0.862	1.431
Propyl dodecanoate	140/4	0.862	1.434
Butyl dodecanoate	154/5	0.860	1.436
Methyl stearate	m.p. 39	—	—
Ethyl stearate	m.p. 33	—	—
Methyl chloroformate	73	1.223	1.387
Ethyl chloroformate	94	1.136	1.397
Propyl chloroformate	115	1.090	1.404
Isobutyl chloroformate	129	1.040	1.406
Butyl chloroformate	138	1.079	1.412
Pentyl chloroformate	61/15	—	1.417
Hexyl chloroformate	63/10	—	—
Methyl chloroacetate	129	1.234	1.422
Ethyl chloroacetate	142	1.150	1.422
Methyl dichloroacetate	143	1.377	1.443
Ethyl dichloroacetate	156	1.283	1.438
Methyl trichloroacetate	152	1.488	1.457
Ethyl trichloroacetate	164	1.380	1.450
Methyl bromoacetate	144d	—	—
Ethyl bromoacetate	169	1.506	1.451
Methyl iodoacetate	170	—	—
Ethyl iodoacetate	180	1.818*	1.508*
Methyl methoxyacetate	130	1.051	1.396
Ethyl methoxyacetate	132	1.007	—
Methyl ethoxyacetate	148	1.006	—
Ethyl ethoxyacetate	158	0.970	1.403
Methyl acrylate	80	0.960	1.398
Ethyl acrylate	101	0.909	1.406
Methyl crotonate	119	0.946	1.425
Ethyl crotonate	137	0.918	1.425
Propyl crotonate	157	0.908	1.428
Butyl crotonate	55/4	0.899	1.432
Isopentyl crotonate	60/4	0.891	1.434
Pentyl crotonate	72/5	0.894	1.436
Methyl lactate	145	1.089	1.414
Ethyl lactate	154	1.030	1.415
Methyl glycollate	151	1.166	—
Ethyl glycollate	160	1.082	—
Methyl pyruvate	138	—	—
Ethyl pyruvate	155	1.055	1.406
Methyl levulinate	196	1.049	1.423
Ethyl levulinate	206	1.011	1.423

Table 10.25 Aliphatic esters (*continued*)

Ester	B.P. °C (/mmHg)	$d_4^{20°}$	$n_D^{20°}$
Methyl furoate	181	1.180	1.486
Ethyl furoate	197 (m.p. 34)	1.117*	1.480*
Trimethyl orthoformate	105	0.968	1.379
Triethyl orthoformate	143	0.893	1.390
Tripropyl orthoformate	91/17	0.879	1.407
Tributyl orthoformate	127/16	0.871	1.416
Dimethyl carbonate	90	1.071	1.369
Diethyl carbonate	126	0.976	1.384
Dipropyl carbonate	165	0.943	1.400
Diisobutyl carbonate	188	0.914	1.407
Dibutyl carbonate	205	0.925	1.412
Dimethyl oxalate	m.p. 54	—	—
Diethyl oxalate	183	1.079	1.410
Diisopropyl oxalate	191	0.995	1.413
Dipropyl oxalate	212	1.019	1.416
Dibutyl oxalate	241	0.987	1.423
Diisopentyl oxalate	127/7	0.961	1.427
Dipentyl oxalate	139/9	0.966	1.429
Dimethyl malonate	179	1.119	1.420
Diethyl malonate	197	1.055	1.414
Diallyl succinate	104	1.051	1.452
Dimethyl succinate	195	1.120	1.420
Diethyl succinate	218	1.042	1.420
Diisopropyl succinate	82/3	0.985	1.418
Dipropyl succinate	102/3	1.006	1.425
Diisobutyl succinate	116/4	0.968	1.427
Dibutyl succinate	120/3	0.977	1.430
Diisopentyl succinate	130/4	0.958	1.434
Dipentyl succinate	129/2	0.960	1.434
Dimethyl glutarate	109/21	1.087	1.424
Diethyl glutarate	118/15	1.023	1.424
Dimethyl adipate	121/17	1.063	1.428
Diethyl adipate	134/17	1.009	1.428
Diisopropyl adipate	120/6	0.966	1.425
Dipropyl adipate	146/9	0.981	1.431
Dibutyl adipate	159/17	0.945	1.435
Diisopentyl adipate	184/13	0.945	1.437
Dipentyl adipate	186/10	0.948	1.439
Dimethyl pimelate	128/16	1.038	1.431
Diethyl pimelate	149/18	0.993	1.430
Dimethyl suberate	120/6	1.024	1.434
Diethyl suberate	131/5	0.981	1.432
Dipropyl suberate	165/8	0.962	1.435
Dibutyl suberate	176/4	0.948	1.439

* Values at 21 °C with supercooled liquid.

Table 10.25 Aliphatic esters (continued)

Ester	B.P. °C (/mmHg)	M.P. °C	$d_4^{20°}$	$n_D^{20°}$
Dimethyl azelate	156/20	—	1.007	1.436
Diethyl azelate	291	—	0.973	1.435
Dimethyl sebacate	293	27	—	—
Diethyl sebacate	307	—	0.964	1.437
Dipropyl sebacate	179/5	—	0.950	1.439
Dimethyl maleate	201	—	1.150	1.442
Diethyl maleate	220	—	1.066	1.440
Dipropyl maleate	126/12	—	1.025	1.443
Dibutyl maleate	147/12	—	0.994	1.445
Dimethyl fumarate	193	102	—	—
Diethyl fumarate	214	—	1.052	1.441
Dipropyl fumarate	110/5	—	1.013	1.444
Dibutyl fumarate	139/5	—	0.987	1.447
Dimethyl itaconate	208	38	—	—
Diethyl itaconate	229	—	1.047	1.439
Dimethyl mesaconate	205	—	1.120	1.454
Diethyl mesaconate	225	—	1.043	1.448
Dimethyl citraconate	210	—	1.112	1.448
Diethyl citraconate	228	—	1.041	1.444
Dimethyl (+)-tartrate	280	61	—	—
Diethyl (+)-tartrate	280	18	1.203	1.447
Dipropyl (+)-tartrate	297	—	1.139	—
Dibutyl (+)-tartrate	200/18	22	—	—
Dimethyl (±)-tartrate	282	90	—	—
Diethyl (±)-tartrate	280	18	1.203	1.447
Dipropyl (±)-tartrate	286	25	—	—
Dibutyl (±)-tartrate	320	—	1.086	—
Dimethyl malate	242	—	1.233	1.442
Diethyl malate	253	—	1.129	1.436
Dimethyl galacturate	—	167	—	—
Diethyl galacturate	—	164	—	—
Trimethyl citrate	—	76	—	—
Triethyl citrate	294	—	1.137	1.466

Table 10.26 Aromatic esters

It is considered that the table will be of greatest use if the esters are in the main subdivided under the various acids rather than be arranged in order of increasing b.p. or m.p. irrespective of the nature of the carboxylic acid. The latter procedure leads to an unwieldy, heterogeneous table which has relatively little pedagogic or, indeed, practical value.

Ester	B.P. °C (/mmHg)	M.P. °C	$d_4^{20°}$	$n_D^{20°}$
Methyl benzoate	199	—	1.089	1.517
Ethyl benzoate	212	—	1.047	1.505
Isopropyl benzoate	218	—	1.015	1.491
Propyl benzoate	230	—	1.023	1.500
Allyl benzoate	230	—	1.052	—
Isobutyl benzoate	242	—	0.997	—
Butyl benzoate	248	—	1.005	1.497
Isopentyl benzoate	262	—	0.986	1.495
Pentyl benzoate	137/15	—	—	—
Ethylene glycol dibenzoate	—	73	—	—
Methyl phenylacetate	215	—	1.068	1.507
Ethyl phenylacetate	228	—	1.033	1.497
Propyl phenylacetate	241	—	1.010	1.493
Butyl phenylacetate	256	—	0.994	1.489
Methyl o-toluate	213	—	1.068	—
Ethyl o-toluate	227	—	1.034	1.508
Methyl m-toluate	215	—	1.061	—
Ethyl m-toluate	227	—	1.028	1.506
Methyl p-toluate	217	34	—	—
Ethyl p-toluate	228	—	1.025	1.507
Methyl salicylate	223	—	1.184	1.537
Ethyl salicylate	234	—	1.125	1.522
Propyl salicylate	240	—	1.098	1.516
Butyl salicylate	260	—	1.073	1.512
Methyl m-hydroxybenzoate	—	70	—	—
Ethyl m-hydroxybenzoate	295	73	—	—
Ethyl p-hydroxybenzoate	297	116	—	—
Methyl p-hydroxybenzoate	—	131	—	—
Methyl o-methoxybenzoate	248	—	1.156	1.534
Ethyl o-methoxybenzoate	261	—	1.104	1.525
Methyl m-methoxybenzoate	237	—	1.131	1.522
Ethyl m-methoxybenzoate	251	—	1.100	1.515
Methyl anisate	255	49	—	—
Ethyl anisate	269	7	1.103	1.524
Methyl o-chlorobenzoate	234	—	—	1.536
Ethyl o-chlorobenzoate	243	—	1.190	1.522
Methyl m-chlorobenzoate	231	20	—	1.492
Ethyl m-chlorobenzoate	242	—	1.182	1.520
Ethyl p-chlorobenzoate	238	—	1.181	1.524
Methyl p-chlorobenzoate	—	44	—	—
Methyl o-bromobenzoate	246	—	—	—
Ethyl o-bromobenzoate	255	—	—	—
Ethyl m-bromobenzoate	259	—	—	—
Methyl m-bromobenzoate	—	32	—	—
Ethyl p-bromobenzoate	263	—	—	—
Methyl p-bromobenzoate	—	81	—	—

Table 10.26 Aromatic esters (*continued*)

Ester	B.P. °C (/mmHg)	M.P. °C	$d_4^{20°}$	$n_D^{20°}$
Ethyl o-iodobenzoate	275	—	—	—
Methyl o-iodobenzoate	278	—	—	—
Ethyl m-iodobenzoate	150/15	—	—	—
Methyl m-iodobenzoate	277	54	—	—
Ethyl p-iodobenzoate	153/14	—	—	—
Methyl p-iodobenzoate	—	114	—	—
Ethyl o-nitrobenzoate	—	30	—	—
Methyl o-nitrobenzoate	275	—	1.286	—
Ethyl m-nitrobenzoate	297	47	—	—
Methyl m-nitrobenzoate	279	79	—	—
Ethyl p-nitrobenzoate	—	57	—	—
Methyl p-nitrobenzoate	—	96	—	—
Ethyl 3,5-dinitrobenzoate	—	94	—	—
Methyl 3,5-dinitrobenzoate	—	108	—	—
Ethyl 2,4-dinitrobenzoate	—	41	—	—
Methyl 2,4-dinitrobenzoate	—	70	—	—
Ethyl anthranilate	267	13	1.117	1.565
Methyl anthranilate	300	24	—	—
Ethyl m-aminobenzoate	294	—	—	—
Methyl m-aminobenzoate	—	38	—	—
Ethyl p-aminobenzoate	—	92	—	—
Methyl p-aminobenzoate	—	112	—	—
Ethyl cinnamate	273	—	1.049	1.560
Propyl cinnamate	284	—	1.028	1.551
Butyl cinnamate	162/12	—	1.013	1.544
Methyl cinnamate	261	36	—	—
Methyl hydrocinnamate	232	—	1.043	1.503
Ethyl hydrocinnamate	248	—	1.016	1.495
Propyl hydrocinnamate	262	—	0.998	1.491
Butyl hydrocinnamate	123/11	—	0.984	1.489
Ethyl o-nitrocinnamate	—	44	—	—
Methyl o-nitrocinnamate	—	73	—	—
Ethyl m-nitrocinnamate	—	79	—	—
Methyl m-nitrocinnamate	—	124	—	—
Ethyl p-nitrocinnamate	—	142	—	—
Methyl p-nitrocinnamate	—	161	—	—
Methyl o-aminocinnamate	—	65	—	—
Ethyl o-aminocinnamate	—	78	—	—
Ethyl m-aminocinnamate	—	64	—	—
Methyl m-aminocinnamate	—	84	—	—
Ethyl p-aminocinnamate	—	69	—	—
Methyl p-aminocinnamate	—	129	—	—
Methyl phenoxyacetate	245	—	1.147	—
Ethyl phenoxyacetate	251	—	1.101	—
Ethyl (±)-mandelate	255	37	—	—
Methyl (±)-mandelate	—	58	—	—
Methyl o-benzoylbenzoate	352	52	—	—
Ethyl o-benzoylbenzoate	—	58	—	—

Table 10.26 Aromatic esters (*continued*)

Ester	B.P. °C (/mmHg)	M.P. °C	$d_4^{20°}$	$n_D^{20°}$
Ethyl diphenylacetate	—	58	—	—
Methyl diphenylacetate	—	60	—	—
Dimethyl phthalate	282	—	1.191	1.516
Diethyl phthalate	298	—	1.118	1.502
Dipropyl phthalate	130/1	—	—	—
Diisopropyl phthalate	154/10	—	—	—
Dibutyl phthalate	205/20	—	—	—
Diethyl isophthalate	285	11	1.121	1.507
Dimethyl isophthalate	—	68	—	—
Diethyl terephthalate	302	44	—	—
Dimethyl terephthalate	—	142	—	—
Diethyl 3-nitrophthalate	—	45	—	—
Dimethyl 3-nitrophthalate	—	69	—	—
Diethyl 4-nitrophthalate	—	34	—	—
Dimethyl 4-nitrophthalate	—	66	—	—
Methyl 1-naphthoate	116/1	—	1.163	1.612
Ethyl 1-naphthoate	309	—	1.121	1.594
Ethyl 2-naphthoate	304	32	—	—
Methyl 2-naphthoate	290	77	—	—
Diethyl diphenate	—	42	—	—
Dimethyl diphenate	—	74	—	—
Furfuryl acetate	176	—	1.118	—
Phenyl acetate	196	—	1.078	1.503
Phenyl propanoate	211	20	1.050	—
Phenyl butanoate	228	—	1.023	—
Diphenyl oxalate	190/15	—	—	—
Phenyl salicylate (salol)	—	43	—	—
Diphenyl succinate	330	121	—	—
Phenyl benzoate	299	68	—	—
Phenyl cinnamate	—	73	—	—
Diphenyl carbonate	306	78	—	—
o-Cresyl acetate	208	—	1.045	—
p-Cresyl acetate	212	—	1.050	1.500
m-Cresyl acetate	212	12	1.043	1.498
Guaiacol acetate	240	—	1.133	1.512
Thymyl acetate	243	—	—	—
Carvacryl acetate	245	—	0.994	—
Resorcinol diacetate	278	—	—	—
Eugenol acetate	282	30	—	—
1-Naphthyl acetate	—	49	—	—
Catechol diacetate	—	63	—	—
2-Naphthyl acetate	—	70	—	—
Benzoin acetate	—	83	—	—
Phloroglucinol triacetate	—	104	—	—
Hydroquinone diacetate	—	124	—	—
Pyrogallol triacetate	—	165	—	—

Table 10.26 Aromatic esters (*continued*)

Ester	B.P. °C (/mmHg)	M.P. °C	$d_4^{20°}$	$n_D^{20°}$
o-Cresyl benzoate	307	—	—	—
Thymyl benzoate	—	33	—	—
m-Cresyl benzoate	—	54	—	—
1-Naphthyl benzoate	—	56	—	—
p-Cresyl benzoate	316	72	—	—
Catechol dibenzoate	—	84	—	—
Pyrogallol tribenzoate	—	90	—	—
2-Naphthyl benzoate	—	107	—	—
Resorcinol dibenzoate	—	117	—	—
Phloroglucinol tribenzoate	—	185	—	—
Hydroquinone dibenzoate	—	199	—	—
Di-o-cresyl carbonate	—	60	—	—
Diphenyl carbonate	306	78	—	—
Diguaiacol carbonate	—	87	—	—
Di-m-cresyl carbonate	—	111	—	—
Di-p-cresyl carbonate	—	115	—	—
Benzyl formate	203	—	1.082	—
Benzyl acetate	214	—	1.057	1.523
Benzyl salicylate	186/10	—	1.180	1.581
Benzyl benzoate	323	21	—	—
Dibenzyl succinate	—	45	—	—
1-Phenylethyl acetate	222	—	—	—
2-Phenylethyl acetate	224	—	1.059	1.512

Table 10.27 Primary aliphatic amides

Amide	M.P. °C	Xanthylamide °C
Formamide	2	184
	(b.p. 193d)	
Propionamide	79	214
Acetamide	82	245
Acrylamide	86	—
Heptanamide	96	154
Dichloroacetamide	98	—
Lauramide	99	—
Hexanamide	101	160
Myristamide	103	—
Palmitamide	106	142
Valeramide	106	167
Octanamide	107	148
Decanamide	108	—
Stearamide	109	141
Butyramide	115	187
Chloroacetamide	120	209
Cyanoacetamide	120	223
Isobutyramide	129	211
Isovaleramide	136	183
Trichloroacetamide	141	—
Furoamide	142	210
Trimethylacetamide	154	—
Cyclohexanecarboxamide	185	—
N-Allylurea	80	—
N-Methylurea	102	230
Urea	132	274
N,N-Dimethylurea	182	250
N-Acetylurea	218	—
Thiourea	182	—
Ethyl carbamate (urethan)	49	169
Methyl carbamate	54	193
Butyl carbamate	54	—
Isobutyl carbamate	55	—
Pentyl carbamate	57	—
Propyl carbamate	61	—
Isopentyl carbamate	67	—
Isopropyl carbamate	92	—
Malonamide	170	270
Azelamide	172	—
Glutaramide	175	—
Maleamide	180	—
Sebacamide	209	—
Suberamide	217	—
Adipamide	220	—
(±)-Tartaramide	226	—
Succinamide	260d	275
Oxamide	419d	—
Succinimide	126	246

Table 10.28 Primary aromatic amides

Amide	M.P. °C	Xanthylamide °C
2-Phenylpropanamide	92	158
m-Toluamide	95	—
3-Phenylpropanamide	105	189
Hydrobenzamide	110	—
Benzamide	129	224
o-Methoxybenzamide	129	—
(±)-Mandelamide	133	—
m-Chlorobenzamide	134	—
Salicylamide	139	—
o-Chlorobenzamide	141	—
N-*m*-Tolylurea	142	—
m-Nitrobenzamide	142	—
o-Toluamide	143	200
N-Phenylurea	147	225
Cinnamamide	148	—
N-Benzylurea	149	—
o-Bromobenzamide	155	—
m-Bromobenzamide	155	—
Phenylacetamide	157	196
p-Toluamide	159	225
p-Hydroxybenzamide	162	—
Anisamide	162	—
Diphenylacetamide	167	—
m-Hydroxybenzamide	167	—
Piperonylamide	169	—
p-Phenetylurea (*N*-*p*-ethoxyphenylurea)	173	—
o-Nitrobenzamide	175	—
p-Chlorobenzamide	179	—
N-*p*-Tolylurea	183	—
3,5-Dinitrobenzamide	183	—
o-Iodobenzamide	184	—
m-Iodobenzamide	186	—
N,*N*-Diphenylurea	189	180
p-Bromobenzamide	189	—
N-*o*-Tolylurea	191	228
2-Naphthamide	192	—
p-Nitrobenzamide	201	232
1-Naphthamide	202	—
p-Ethoxybenzamide	202	—
p-Iodobenzamide	218	—
Phthalamide	219d	—
Phthalimide	235	177

Table 10.29 Substituted aromatic amides

Amide	M.P. °C	Amide	M.P. °C
Formanilide	50	N-Propylacetanilide	50
Nonananilide	57	N-Ethylacetanilide	54
Octananilide	57	m-Chloroacetanilide	79
Lactanilide	59	m-Methoxyacetanilide	80
Valeranilide	63	o-Chloroacetanilide	88
Decananilide	70	m-Bromoacetanilide	88
Heptananilide	71	o-Methoxyacetanilide	88
Lauranilide	78	m-Aminoacetanilide	88
Myristanilide	84	o-Nitroacetanilide	94
Acetoacetanilide	85	o-Bromoacetanilide	99
Palmitanilide	91	N-Methylacetanilide	103
Stearanilide	94	o-Iodoacetanilide	110
Hexananilide	95	Acetanilide	114
Butyranilide	96	N-Ethyl-p-nitroacetanilide	118
Isobutyranilide	105	Phenylacetanilide	118
Acrylanilide	105	m-Iodoacetanilide	119
Propionanilide	106	p-Methoxyacetanilide	130
Isovaleranilide	110	o-Aminoacetanilide	132
Acetanilide	114	2,4-Dimethylacetanilide	133
Furoanilide	124	2,5-Dimethylacetanilide	142
o-Toluanilide	125	4-Methyl-3-nitroacetanilide	148
m-Toluanilide	126	m-Hydroxyacetanilide	149
o-Methoxybenzanilide	131	N-Methyl-p-nitroacetanilide	153
Salicylanilide	135	m-Nitroacetanilide	155
p-Toluanilide	146	p-Aminoacetanilide	163
Cinnamanilide	153	p-Bromoacetanilide	167
m-Nitrobenzanilide	154	p-Hydroxyacetanilide	168
o-Nitrobenzanilide	155	p-Chloroacetanilide	179
Benzanilide	162	p-Iodoacetanilide	184
1-Naphthanilide	163	2-Methyl-4-nitroacetanilide	196
Anisanilide	169	o-Hydroxyacetanilide	209
2-Naphthanilide	171	p-Nitroacetanilide	216
p-Nitrobenzanilide	211	p-Hydroxy-N-methylacetanilide	240
Pimelic dianilide	156	Acetyl-m-toluidine	66
Suberic dianilide	187	Acetyl-o-phenetidine	79
Maleic dianilide	187	Acetyl-m-anisidine	80
Azelaic dianilide	187	Acetyl-o-anisidine	88
Sebacic dianilide	202	Acetyl-m-phenetidine	96
Glutaric dianilide	224	Acetyl-o-toluidine	112
Malonic dianilide	225	Acetyl-p-anisidine	130
Succinic dianilide	230	Acetyl-p-phenetidine (phenacetin)	137
Adipic dianilide	239	Acetyl-p-toluidine	154
Oxanilide	246	Acetyl-1-naphthylamine	160

Table 10.29 Substituted aromatic amides (*continued*)

Amide	M.P. °C	Amide	M.P. °C
NN'-Diacetyl-*o*-phenylenediamine	186	Benzoylpiperidine	48
NN'-Diacetyl-*m*-phenylenediamine	191	*N*-Phenylsuccinimide	156
NN'-Diacetyl-*p*-phenylenediamine	304	*N*-Phenylphthalimide	205
		Phthalimide	235
Benzoyl-*o*-anisidine	60	Triphenylguanidine	145
Benzoyl-*m*-anisidine	—	Diphenylguanidine	147
Benzoyl-*m*-phenetidine	103	Saccharin	220
Benzoyl-*o*-phenetidine	104		
Benzoyl-*m*-toluidine	125	*N*-Phenylurethane	53
Benzoyl-*o*-toluidine	144	Ethyl oxanilate	67
Benzoyl-*p*-anisidine	154		
Benzoyl-*p*-toluidine	158	*N,N'*-Di-*m*-tolylurea	218
Benzoyl-1-naphthylamine	161	*N,N'*-Diphenylurea (carbanilide)	238
Benzoyl-*p*-phenetidine	173	*N,N'*-Di-*o*-tolylurea	250
		N,N'-Di-*p*-tolylurea	268
		N,N'-Di-1-naphthylurea	297
NN'-Dibenzoyl-*m*-phenylenediamine	240		
NN'-Dibenzoyl-*o*-phenylenediamine	301	Ethylbutylbarbituric acid	125
NN'-Dibenzoyl-*p*-phenylenediamine	> 300	Ethylhexylbarbituric acid	127
		Ethylisopentylbarbituric acid	154
Acetyl-*N*-methyl-*o*-toluidine	56	Ethylphenylbarbituric acid	172
Acetyl-*N*-methyl-*m*-toluidine	66	Diallylbarbituric acid	172
Acetyl-*N*-methyl-*p*-toluidine	83	Diethylbarbituric acid	198
Acetyl-*N*-methyl-1-naphthylamine	94	Ethylisopropylbarbituric acid	201
		Barbituric acid	245
N-Formyldiphenylamine	74		
N-Acetyldiphenylamine	101	Butyl oxamate	88
N-Benzoyldiphenylamine	180	Ethyl oxamate	131

Table 10.30 Aliphatic nitriles (cyanides)

Cyanide	Nitrile	B.P. °C (/mm Hg)
Vinyl	Acrylo-	78
Methyl	Aceto-	82
Ethyl	Propiono-	97
Isopropyl	Isobutyro-	108
Propyl	Butyro-	118
Allyl	Vinylaceto-	118
Chloromethyl	Chloroaceto-	127
Isobutyl	Isovalero-	131
Butyl	Valero-	141
Isopentyl	4-Methylpentane-	154
Pentyl	Hexane-	162
Hexyl	Heptane-	183
Heptyl	Octane-	199
Octyl	Nonane-	224
Nonyl	Decane-	244
Decyl	Undecane-	254
Undecyl	Dodecane-	275
Methylene	Malono-	220 (1)
Ethylene	Succino-	276d (2)
Trimethylene	Glutaro-	286 (3)
Tetramethylene	Adipo-	295
Pentamethylene	Pimelo-	169/15
Hexamethylene	Subero-	185/15
Acetaldehyde cyanohydrin	α-Hydroxypropiono-	183
Ethylene cyanohydrin	β-Hydroxypropiono-	221
Trimethylene cyanohydrin	γ-Hydroxybutyro-	240
Trimethylene chlorocyanide	γ-Chlorobutyro-	197
Methyl cyanoacetate		200
Ethyl cyanoacetate		207
2-Furyl	Furo-	147
Phenyl	Benzo-	191
Benzyl	Phenylaceto-	109/15
o-Tolyl	o-Tolu-	205
m-Tolyl	m-Tolu-	212

* Decomposition temperature. Sample placed in bath at 105–110 °C.
† 15 °C.

$d_4^{20°}$	$n_D^{20°}$	Acyl phloroglucinol °C	α-Iminoalkyl-mercaptoacetic acid hydrochloride* °C
0.806	1.391	—	—
0.784	1.344	218	115
0.783	1.366	176	128
—	—	—	137
0.791	1.384	181	136
0.838	1.406	—	—
1.193	—	—	—
0.788	—	—	—
0.799	1.397	149 (hydrate 88)	138
0.803	1.406	122 (hydrate 104)	128
0.805	1.407	121 (hydrate 96)	136
0.810	1.414	—	—
0.817†	1.422†	—	—
0.822†	—	—	—
0.829†	1.432†	—	—
—	—	—	—
0.827†	—	—	—
—	—	—	—
—	—	—	—
0.988	1.429	—	—
0.962	1.439	—	—
0.945	1.441	—	—
0.933	1.445	—	—
0.988	—	—	—
—	—	—	—
—	—	—	—
1.079	—	—	—
1.101	—	—	—
1.063	1.418	—	—
1.082	1.480	—	—
1.006	1.528	—	—
1.016	1.523	—	146
0.996	1.530	—	—
1.032	1.525	—	—

(1) m.p. 31 °C.　　(2) m.p. 54 °C.　　(3) m.p. 9 °C.

Table 10.31 Aromatic nitriles

Nitrile	B.P. °C	M.P. °C
(±)-Mandelonitrile	170d	22
Benzonitrile	191	—
o-Tolunitrile	205	—
m-Tolunitrile	212	—
β-Phenylpropiononitrile	232	—
Phenylacetonitrile	234	—
γ-Phenylpropiononitrile	261	—
Cinnamonitrile	255	20
p-Tolunitrile	218	29
m-Bromobenzonitrile	225	38
1-Naphthonitrile	299	36
m-Chlorobenzonitrile	—	41
o-Chlorobenzonitrile	232	43
o-Bromobenzonitrile	252	53
o-Iodobenzonitrile	—	55
2-Naphthonitrile	306	66
p-Chlorobenzonitrile	223	96
o-Nitrobenzonitrile	—	111
p-Bromobenzonitrile	236	113
p-Nitrophenylacetonitrile	—	116
m-Nitrobenzonitrile	—	118
Phthalonitrile	—	141
p-Nitrobenzonitrile	—	149

Table 10.32 Primary and secondary aliphatic amines

Amine	B.P. °C	$d_4^{20°}$	$n_D^{20°}$	Benzene-sulphonamide °C
Methylamine	−7	—	—	30
Ethylamine	17	—	—	58
Isopropylamine	35	0.689	1.374	26
t-Butylamine	46	—	—	—
Propylamine	49	0.717	1.388	36
Allylamine	55	0.762	1.420	39
s-Butylamine	63	0.725	1.393	70
Isobutylamine	68	0.735	1.397	53
Butylamine	77	0.741	1.401	—
Isopentylamine	97	0.749	1.408	—
Pentylamine	105	0.754	1.411	—
Hexylamine	129	0.766	1.418	96
Cyclohexylamine	134	0.867	1.459	—
Heptylamine	155	0.775	1.425	—
Ethanolamine	171	1.022	1.454	—
Octylamine	177	0.782	1.429	—
(−)-Menthylamine	212	0.854	—	—
Benzylamine	185	0.982	1.544	88
1-Phenylethylamine	187	—	—	—
2-Phenylethylamine	198	0.854	—	69
1,2-Diaminoethane	117 (1)	0.898	1.457	168
1,2-Diaminopropane	120	0.874	—	—
1,3-Diaminopropane	136	0.889	1.460	—
1,4-Diaminobutane	159 (2)	—	—	—
1,5-Diaminopentane	180	—	—	119
1,6-Diaminohexane	205 (3)	—	—	154
Dimethylamine	7	—	—	47
Diethylamine	56	0.707	1.386	42
Di-isopropylamine	84	0.717	1.392	94
Dipropylamine	110	0.738	1.405	51
Diallylamine	111	—	—	—
Di-s-butylamine	135	0.753	1.411	—
Di-isobutylamine	137	0.746	1.409	—
Dibutylamine	159	0.760	1.418	—
Di-isopentylamine	186	0.771	1.423	—
Dipentylamine	205	0.777	1.427	—
Dicyclohexylamine	255d (4)	—	—	—
Diethanolamine (5)	270d (6)	1.097	1.478	130
Pyrrolidine	89	0.854	1.424	—
Piperidine	106	0.861	1.453	94
2-Methylpiperidine	118	—	—	—
3-Methylpiperidine	126	—	—	—
4-Methylpiperidine	128	—	—	—
Morpholine	130	1.000	1.455	119
Pyrrole	131	0.969	1.509	—
Piperazine	140 (7)	—	—	292 (di)
1,2,3,4-Tetrahydroisoquinoline	232	—	—	154
1,2,3,4-Tetrahydroquinoline	250 (m.p. 20)	—	—	67

Toluene-*p*-sulphon-amide °C	Phenyl-thiourea °C	1-Naphthyl-thiourea °C	Picrate °C	*N*-Substi-tuted phthal-imide °C	Benzamide °C	Acetamide °C
75	113	192	215	134	80	—
63	106	121	165	78	71	—
51	101	143	150	86	100	—
—	—	—	198	—	134	—
52	63	103	135	66	84	—
64	98	—	140	70	—	—
55	101	—	140	—	76	—
78	82	137	151	93	57	—
—	65	109	151	34	42	—
65	102	97	138	—	—	—
—	69	103	139	—	—	—
—	77	79	127	—	40	—
—	—	142	—	158	149	104
—	75	—	121	—	—	—
—	—	—	160	127	—	—
—	—	—	112	—	—	—
—	135	—	—	—	157	145
116	147	172	196	115	105	60
—	—	—	189	—	120	57
66	135	—	174	130	116	114
160	102	—	233	—	244	172
—	—	—	137	—	193	139
—	—	—	250	—	148	101
224	168	—	255d	—	177	137
—	148	—	—	—	135	—
—	—	—	220	—	158	—
79	135	168	158	—	41	—
60	34	108	155	—	42	—
—	—	—	140	—	—	—
—	—	161	75	—	—	—
—	—	—	—	—	—	—
—	—	—	—	—	—	—
—	86	123	59	—	—	—
—	—	118	—	—	—	—
119	—	—	173	153	—	103
99	—	—	110	—	—	—
123	—	—	112	—	—	—
96	—	—	152	—	48	—
55	—	—	135	—	45	—
—	—	—	138	—	—	—
147	136	—	148	—	75	—
—	—	—	69d	—	—	—
173 (mono)	—	—	280	—	196 (di)	144 (di)
—	—	—	195	—	129	46
—	—	—	—	—	76	—

Table 10.32 Primary and secondary aliphatic amines (*continued*)

Amine	B.P. °C	M.P. °C
Ester-amides (*derivatives of aminoformic acid*, NH_2COOH)		
Methyl carbamate	177	54
Ethyl carbamate (urethan)	184	50
Propyl carbamate	195	61
Butyl carbamate	204d	54
Pentyl carbamate	—	57
Isopentyl carbamate	—	67
N-Methylurethan (Ethyl *N*-methyl carbamate)	170	—
N-Ethylurethan (Ethyl *N*-ethyl carbamate)	170	—
N-Propylurethan	192	—
N-Butylurethan	202	—
N-s-Butylurethan	194	—
N-Phenylurethan (Ethyl *N*-phenyl carbamate)	237	53
Ethyl oxanilate	—	67

Note. Esters of carbamic acid upon boiling with aniline yield carbanilide (m.p. 238 °C), ammonia and the corresponding alcohol.

Table 10.33 Primary aromatic amines*

Amine °C	B.P. °C	M.P. °C	Acetamide °C	Benzamide °C	Benzene-sulphon-amide °C
Aniline	183	—	114	163	112
Benzylamine	185	—	60	105	88
1-Phenylethylamine	187	—	57	120	—
2-Phenylethylamine	198	—	51	116	69
o-Toluidine	200	—	112	144	124
m-Toluidine	203	—	66	125	95
p-Xylidine (1)	214	15	142	140	—
p-Ethylaniline	215	—	94	151	—
m-2-Xylidine (2)	215	11	177	168	—
o-Ethylaniline	216	—	112	147	—
m-4-Xylidine (3)	216	—	130	192	130
m-5-Xylidine (4)	220	10	144	136	—
o-Anisidine (5)	225	5	88	60	89
3,4-Dimethylaniline	226	49	99	—	118
N,N-Methylphenylhydrazine	227	—	92	153	132
o-Phenetidine (6)	229	—	79	104	102
Mesidine (7)	232	—	216	206	—
Phenylhydrazine	242	23	128	168	—
m-Phenetidine	248	—	96	103	—
m-Anisidine	251	—	80	—	—
o-Aminoacetophenone	251d	20	77	98	—
p-Phenetidine	254	—	135	173	143

* All aromatic amines should be treated as potentially carcinogenic (see Section 2.3.4, p. 49).

$d_4^{20°}$	$n_D^{20°}$		Derivatives °C
—	—		—
—	—		N-p-Nitrobenzoyl, 152; Benzylidene, 179
—	—		—
—	—		—
—	—		—
—	—		—
0.981	1.422		—
—	—		—
—	—		—
—	—		—
—	—		N-Acetyl, 59; N-Benzoyl, 161; N-Nitroso, 62
—	—		N-Acetyl, 65

(1) m.p. 8 °C.
(2) m.p. 28 °C.
(3) m.p. 42 °C.
(4) m.p. 20 °C.
(5) Di-2-hydroxyethylamine.
(6) m.p. 28 °C.
(7) m.p. 104 °C; Hydrate, 6H$_2$O, m.p. 44 °C.

Toluene-p-sulphon-amide °C	Benzylidene derivative °C	Picrate °C	3-Nitro-phthalimide °C	2,4-Dinitro-phenyl derivative °C	Formyl derivative °C	Phenyl thiourea °C
103	54	—	138	156	47	154
116	—	199	143	—	—	156
—	—	—	—	—	—	135
—	—	167	—	—	—	135
110	—	213	150	126	59	136
114	—	200	130	161	—	104
232	—	—	—	150	—	—
212	—	180	—	—	176	148
—	—	—	—	—	—	—
181	—	209	—	156	114	152
—	—	209	—	—	77	153
127	—	200	185	151	83	136
154	—	—	—	—	—	—
164	—	—	164	164	62	137
167	—	193	—	—	—	193
—	—	—	—	—	145	172
157	—	158	—	—	52	138
68	—	169	158	138	57	—
148	—	—	—	—	—	—
107	76	69	173	118	76	148

Table 10.33   Primary aromatic amines (*continued*)

Amine °C	B.P. °C	M.P. °C	Acetamide °C	Benzamide °C	Benzene-sulphon-amide °C
Methyl anthranilate	255	24	101	100	—
p-Aminodiethylaniline	261	—	104	172	—
Ethyl anthranilate	266d	13	61	98	93
Ethyl m-aminobenzoate	294	—	—	114	—
p-Aminodimethylaniline	262	41	132	228	—
p-Toluidine	200	45	154	158	120
o-Aminobiphenyl (8)	299	50	121	102	—
1-Naphthylamine C	300	50	160	161	169
p-Aminobiphenyl (9) C	302	51	171	230	—
p-Anisidine	246	57	130	154	96
2-Aminopyridine	204	58	71	87	—
2,5-Diaminotoluene (10)	273	64	220	307	—
3-Aminopyridine	252	64	133	119	—
p-Tolylhydrazine	244d	65	130	146	—
m-Phenylenediamine	283	64	191	240	194
o-Nitroaniline	—	71	94	98	104
4-Amino-2-nitrotoluene	—	78	145	172	160
m-Aminoacetanilide	—	88	191	—	—
3,4-Diaminotoluene (11)	265	90	210	264	179
Ethyl p-aminobenzoate (12)	—	92	110	148	—
6-Nitro-2-aminotoluene	—	92	158	168	—
3-Nitro-2-aminotoluene	—	97	158	—	—
m-Aminoacetophenone	—	99	129	—	—
2,4-Diaminotoluene (13)	292	99	224	224	191
o-Phenylenediamine	257	102	186	301	186
p-Aminoacetophenone	294	106	167	205	128
2-Amino-4-nitrotoluene	—	107	151	186	172
2-Naphthylamine C	294	113	134	162	102
m-Nitroaniline	—	114	155	157	136
4-Amino-3-nitrotoluene	—	117	96	148	102
5-Nitro-1-naphthylamine	—	119	220	—	183
1-Nitro-2-naphthylamine	—	126	123	168	156
p-Aminoazobenzene	—	126	145	211	—
Benzidine C	—	126	317	352	235
o-Tolidine (14) C	—	129	314	265	—
2-Amino-5-nitrotoluene	—	129	202	174	159
o-Aminoacetanilide	—	132	—	—	—
2,6-Dinitroaniline	—	138	197	—	—
p-Phenylenediamine	267	141	304	>300	247
2-Nitro-1-naphthylamine	—	144	199	175	—
Anthranilic acid	—	146	185	181	214
p-Nitroaniline	—	148	216	199	139
p-Nitrophenylhydrazine	—	157d	205	193	—
4-Aminopyridine	—	158	150	202	—
p-Aminoacetanilide	—	163	304	—	—
Sulphanilamide (15)	—	166	219	284	211
m-Aminobenzoic acid	—	174	250	—	—
2,4-Dinitroaniline	—	180	121	220	—
p-Aminobenzoic acid	—	187	251	278	212
Picramide (16)	—	190	230	196	211
4-Nitro-1-naphthylamine	—	195	190	224	173
2,4-Dinitrophenylhydrazine	—	198d	198	207	—
2-Aminoanthraquinone	—	302	257	228	271

Toluene-p-sulphonamide °C	Benzylidene derivative °C	Picrate °C	3-Nitro-phthalimide °C	2,4-Dinitro-phenyl derivative °C	Formyl derivative °C	Phenyl thiourea °C
—	—	106	—	—	58	—
112	—	—	—	—	—	—
112	—	—	—	—	57	—
—	98	188	—	168	108	—
118	—	181	156	137	53	141
—	—	—	—	—	75	—
157	73	163	223	190	139	165
255	—	—	—	—	172	—
114	62	—	197	141	81	144
216	—	221	—	—	—	—
—	—	—	—	—	—	—
—	—	—	—	—	—	—
172	105	184	—	172	155	—
110	—	73	171	—	122	—
163	—	—	—	—	—	—
241	—	—	—	—	—	—
—	—	131	—	—	—	—
—	—	—	—	—	—	—
130	—	—	—	—	—	—
192	175	—	—	184	177	—
202	106	208	—	—	170†	—
203	—	—	—	—	—	—
—	116	—	—	—	179	—
133	—	195	212	179	129	129
139	73	143	219	—	134	160
146	78	—	—	—	—	—
—	—	—	—	—	199	—
160	—	—	—	—	—	—
—	130	—	—	—	162	—
243	238	—	—	—	—	—
—	152	185	—	—	254	—
174	—	—	—	—	—	—
—	—	—	—	—	—	—
266	140	—	—	177	206	—
—	—	—	—	—	—	—
217	127	—	—	—	168	—
191	115	100	255	—	194	—
—	—	120	—	—	—	—
—	—	216	—	—	—	—
—	—	—	—	—	—	—
—	—	—	—	—	—	—
—	119	—	—	—	225	—
219	—	—	—	—	—	—
—	193	—	—	—	268	—
—	—	—	—	—	—	—
185	—	—	—	—	—	—
—	—	—	—	—	—	—
304	—	—	—	—	—	—

Table 10.33 Primary aromatic amines (*continued*)

Amine °C	B.P. °C	M.P. °C	Acetamide °C	Benzamide °C	Benzene-sulphon-amide °C
C-Halogeno-amines					
o-Chloroaniline	209	—	88	99	130
2-Amino-3-chlorotoluene	215	—	120	—	—
4-Amino-3-chlorotoluene	223	7	113	137	110
m-Chloroaniline	230	—	79	122	121
2-Amino-4-chlorotoluene	237	22	140	—	—
4-Amino-3-bromotoluene	240	26	117	149	—
2-Amino-5-chlorotoluene	241	29	140	142	125
2-Amino-6-chlorotoluene	245	—	159	173	—
m-Bromoaniline	251	18	88	120	—
m-Iodoaniline	—	25	119	151	—
2-Amino-4-bromotoluene	255	32	165	—	—
o-Bromoaniline	229	32	99	116	—
2,5-Dichloroaniline	255	50	132	120	—
2-Amino-5-bromotoluene	240	59	157	115	—
o-Iodoaniline	—	60	110	139	—
2,4-Dichloroaniline	245	63	146	117	128
p-Iodoaniline	—	63	184	222	—
p-Bromoaniline	—	66	167	204	134
p-Chloroaniline	232	71	179	193	122
2,4,6-Trichloroaniline	263	78	206	174	—
2,4-Dibromoaniline	—	79	146	134	—
2,6-Dibromoaniline	—	84	210	—	—
2,4-Diaminochlorobenzene	—	88	243	178	—
2-Chloro-4-nitroaniline	—	108	139	161	—
4-Chloro-2-nitroaniline	—	116	104	—	—
2,4,6-Tribromoaniline	—	120	—	232	198
Amino-phenols‡					
2,4-Diaminophenol (17)	—	79d	222 (di)	231	—
m-Aminophenol	—	123	101 (di)	153 (di)	—
Picramic acid (18)	—	169	201 (*N*)	229 (*N*)	—
5-Amino-2-hydroxytoluene	—	173	103 (di)	194	—
o-Aminophenol	—	174	124 (di)	184 (di)	141
p-Aminophenol	—	186d	150 (di)	234 (di)	125
8-Amino-2-naphthol	—	207	165	208	—
1-Amino-2-naphthol	—	dec.	206	235	—

† This compound is benzimidazole.
‡ See also Table 10.34 and Table 10.35 for secondary and tertiary amines having a nuclear hydroxyl substituent.

(1) 2,5-Dimethylaniline.
(2) 2,6-Dimethylaniline.
(3) 2,4-Dimethylaniline.
(4) 3,5-Dimethylaniline.

(5) *o*-Methoxyaniline.
(6) *o*-Ethoxyaniline.
(7) 2,4,6-Trimethylaniline.

Toluene-p-sulphonamide °C	Benzylidene derivative °C	Picrate °C	3-Nitrophthalimide °C	2,4-Dinitrophenyl derivative °C	Formyl derivative °C	Phenyl thiourea °C
105	34	134	136	150	77	156
—	—	—	—	—	—	—
—	—	—	—	—	—	—
138	—	177	172	184	58	124
—	—	—	—	—	—	—
—	—	—	—	—	—	—
—	—	—	—	—	—	—
—	—	180	187	—	—	143
128	—	—	—	—	—	—
—	—	—	—	—	—	—
90	—	129	—	161	—	146
—	—	—	—	—	—	166
—	—	—	—	—	—	—
—	—	112	—	—	—	—
—	—	106	—	116	—	—
—	86	—	—	—	109	153
101 (141)	67	180	202	158	—	148
96 (121)	62	178	199	167	102	152
—	—	83	—	—	180	—
—	—	124	—	—	146	171
—	—	124	—	—	—	—
215	—	—	—	—	—	—
164	—	—	—	—	—	—
110	—	—	—	—	—	—
—	95	—	—	—	222	—
—	—	—	—	—	—	—
—	—	—	—	—	—	156
191 (N)	—	—	—	—	—	—
110	—	—	—	—	—	—
139	89	—	—	199	129	146
252	182	—	—	190	140	150
—	—	—	—	—	—	—

(8) o-Xenylamine.
(9) p-Xenylamine.
(10) 2,5-Tolylenediamine.
(11) 3,4-Tolylenediamine.
(12) Benzocaine.
(13) 2,4-Tolylenediamine.

(14) 4,4′-Diamino-3,3′-dimethylbiphenyl.
(15) p-Aminobenzenesulphonamide.
(16) 2,4,6-Trinitroaniline.
(17) Hydrochloride = *Amidol*.
(18) 2-Amino-4,6-dinitrophenol.

Table 10.34 Secondary aromatic amines

Amine	B.P. °C	M.P. °C	Acetamide °C	Benzamide °C
N-Methylbenzylamine	181	—	—	—
N-Methylaniline	194	—	103	63
N-Ethylbenzylamine	199	—	—	—
N-Ethylaniline	205	—	55	60
N-Methyl-m-toluidine	206	—	66	—
N-Methyl-o-toluidine	208	—	56	66
N-Methyl-p-toluidine	210	—	83	53
N-Ethyl-o-toluidine	214	—	—	72
N-Ethyl-p-toluidine	217	—	—	39
N-Ethyl-m-toluidine	221	—	—	72
N-Propylaniline	222	—	47	—
N-Butylaniline	240	—	—	56
N-Methyl-1-naphthylamine	294	—	94	—
Dibenzylamine	300d	—	—	112
N-Methyl-2-naphthylamine	317	—	51	84
N-Ethyl-2-naphthylamine	315	—	49	—
N-Ethyl-1-naphthylamine	325	—	68	—
o-Nitro-N-methylaniline	—	37	70	—
N-Benzylaniline	306	38	58	107
Diphenylamine	302	54	103	180
m-Nitro-N-ethylaniline	—	60	89	—
N-Phenyl-1-naphthylamine	—	62	115	152
m-Nitro-N-methylaniline	—	68	95	155
Di-p-tolylamine	330	79	88	125
p-Hydroxy-N-methylaniline	—	86	43 (mono)	174 (mono)
o-Hydroxy-N-methylaniline	—	96	64 (di)	160 (mono)
p-Nitro-N-ethylaniline	—	96	119	—
N-Phenyl-2-naphthylamine	—	108	93	136
p-Nitro-N-methylaniline	—	152	152	111
Indole	254	52	—	68
Carbazole	355	246	69	98

Benzene-sulphonamide °C	Toluene-*p*-sulphonamide °C	Picrate °C	Formyl derivative °C	Other derivatives °C
—	95	—	—	—
79	95	145	—	Phthalamic acid, 194
—	50	118	—	Urea (with PhNCO), 81
—	88	138	—	Phthalamic acid, 204
—	—	—	—	—
—	120	90	—	—
—	60	131	—	*N*-Nitroso*, 52 C
—	75	—	—	—
—	71	—	—	—
54	—	—	—	Phthalamic acid, 225
—	56	—	—	Phthalamic acid, 204
—	164	—	—	—
68	—	—	52	—
—	78	145	—	*N*-Nitroso, 88 C
—	—	—	—	—
—	—	—	—	—
—	—	—	—	*N*-Nitroso, 36 C
119	140	—	48	*N*-Nitroso, 58 C
123	142	182	74	*N*-Nitroso, 67 C
—	—	—	—	—
—	—	—	—	—
83	—	—	—	*N*-Nitroso, 76 C
—	—	—	—	*N*-Nitroso, 101 C
—	135 (mono)	—	—	*N*-Nitroso, 136 C
—	—	—	—	*N*-Nitroso, 130 C
—	—	—	—	*N*-Nitroso, 120 C
—	—	—	—	—
120	—	—	—	*N*-Nitroso, 104 C
—	—	187	52	*N*-Nitroso, 171 C
—	137	185	—	—

* The particularly powerful carcinogenic properties of these compounds should be noted.

Table 10.35 Tertiary amines

Amine	B.P. °C (/mmHg)	M.P. °C	Methiodide °C
Trimethylamine	3	—	230
Triethylamine	89	—	—
Triallylamine	155	—	—
Tripropylamine	156	—	208
Tributylamine	212	—	186
Tri-isopentylamine	245	—	—
Tripentylamine	257	—	—
N,N-Dimethylbenzylamine	184	—	179
N,N-Dimethyl-o-toluidine	185	—	210
N,N-Dimethylaniline	193	—	228
N-Methyl-N-ethylaniline	201	—	125
N,N-Diethyl-o-toluidine	210	—	224
N,N-Dimethyl-p-toluidine	211	—	220
N,N-Dimethyl-m-toluidine	212	—	177
N,N-Diethylaniline	218	—	102
N,N-Diethyl-p-toluidine	229	—	184
N,N-Diethyl-m-toluidine	231	—	—
N,N-Di-n-propylaniline	245	—	156
N,N-Di-n-butylaniline	271	—	—
N,N-Dimethyl-1-naphthylamine	273	—	—
N-Benzyl-N-methylaniline	306	—	164
N-Benzyl-N-ethylaniline	186/22	—	161
N,N-Dimethyl-2-naphthylamine	305	47	—
p-Bromo-N,N-dimethylaniline	264	55	—
Dibenzylaniline	300	70	135
p-Hydroxy-N,N-dimethylaniline	—	76	201
p-Nitroso-N,N-dimethylaniline	—	87	—
Tribenzylamine	380	92	184
Triphenylamine	365	127	—
p-Nitro-N,N-dimethylaniline	—	163	—
Pyridine	115	—	118
α-Picoline (1)	129	—	227
2,6-Lutidine (2)	142	—	238
γ-Picoline	143	—	152
β-Picoline	144	—	92
2,4-Lutidine	159	—	113
2,5-Lutidine	160	—	—
2,3-Lutidine	164	—	—
2,4,6-Trimethylpyridine (3)	172	—	—
5-Ethyl-2-methylpyridine	178	—	—
3-Ethyl-4-methylpyridine	196	—	—
Ethyl nicotinate	223	—	—
Nicotine	246	—	—
Methyl nicotinate	204	38	—
2,2'-Bipyridyl	273	70	—
Quinoline	238	—	72* (133)†
Isoquinoline	242	24	159
Quinaldine (4)	247	—	195
8-Methylquinoline	248	—	—
6-Methylquinoline	258	—	219
Lepidine (5)	262	—	174
2,4-Dimethylquinoline	264	—	264

Picrate °C	Methotoluene-p-sulphonate °C	Other derivatives °C
216	—	—
173	—	$d_4^{20°}$ 0.728; $n_D^{20°}$ 1.401
117	—	Ethiodide, 238; $d_4^{20°}$ 0.756; $n_D^{20°}$ 1.417
106	—	Benzyl chloride, 185; $d_4^{20°}$ 0.778; $n_D^{20°}$ 1.430
125	—	$d_4^{20°}$ 0.785; $n_D^{20°}$ 1.433
—	80d	$d_4^{20°}$ 0.791; $n_D^{20°}$ 1.437
93	—	—
122	—	—
164	161	Ethiodide, 136
134	—	Ethiodide, 102; p-nitroso, 66
180	—	—
130	85	Benzyl chloride, 171
131	—	—
142	—	p-Nitroso, 84; benzyl chloride, 104
110	—	—
97	—	—
—	—	—
125	180	—
145	—	—
127	—	—
121	—	—
206	—	—
—	—	—
132	—	p-Nitroso, 91
—	—	o-Acetyl, 79
140	—	—
190	—	Ethiodide, 190
—	—	—
—	—	—
167	139	Ethiodide, 90
169	150	Ethiodide, 123; picolinic acid, 136
163	—	Dipicolinic acid, 226
167	—	Isonicotinic acid, 308
150	—	Nicotinic acid, 228
183	—	—
169	—	—
188	—	—
156	—	—
166	—	—
150	—	—
—	—	—
218	—	Nicotinic acid, 228
—	—	—
158	—	—
203	126	Ethiodide, 158
223	163	Ethiodide, 148
195	161	Ethiodide, 234
200	—	—
229	154	Benzyl chloride, 239
211	—	—
194	—	Ethiodide, 214

Table 10.35 Tertiary amines (*continued*)

Amine	B.P. °C	M.P. °C	Methiodide °C
6-Methoxyquinoline	284	26	236
7-Methylquinoline	252	39	—
8-Methoxyquinoline	283	50	160
2,6-Dimethylquinoline	267	60	237
8-Hydroxyquinoline	267	76	143
8-Nitroquinoline	—	92	—
6-Nitroquinoline	—	154	245
6-Hydroxyquinoline	—	193	—
3-Chloropyridine	149	—	—
3-Bromopyridine	170	—	165
2-Chloropyridine	170	—	—
2-Bromopyridine	194	—	—
3,5-Dibromopyridine	222	112	274
2,6-Dibromopyridine	249	119	—
6-Bromoquinoline	278	19	278
2-Chloroquinoline	267	38	—
6-Chloroquinoline	262	41	248
2-Bromoquinoline	—	49	210
Acridine	—	111	224
Hexamethylenetetramine	—	280 Sub.	190

* Monohydrate. † Anhydrous.

Picrate °C	Methotoluene-*p*-sulphonate °C	Other derivatives °C
—	—	—
237	—	—
143	—	—
191	175	Ethiodide, 227
204	—	—
—	—	—
236	—	—
135	—	—
—	156	—
—	120	—
—	127	—
—	219	—
—	—	—
217	—	—
122	—	—
—	143	Ethiodide, 169
—	—	—
208	—	Trinitrobenzene, 115
179	205	—

(1) 2-Methylpyridine.
(2) 2,6-Dimethylpyridine.
(3) γ-Collidine.
(4) 2-Methylquinoline.
(5) 4-Methylquinoline.

Table 10.36 Amino acids

Amino acid	M.P.* °C	Benzoate °C	3,5-Dinitro-benzoate °C	Phenylureido acid °C
N-Phenylglycine	126	63	—	195
Anthranilic acid	145	182	278	181
m-Aminobenzoic acid	174	248	270	270
(±)-3-Amino-2-methylpropanoic	177	—	—	—
p-Aminobenzoic acid	186	278	290	300
3-Aminopropanoic acid	196	165	202	174
(+)- or (−)-Glutamic acid	198	138	217	—
p-Aminophenylacetic acid	200	206	—	—
(±)-Proline	203	—	217	170
Sarcosine	210	103	153	—
(+)- or (−)-Proline	222	—	—	170
(+)- or (−)-Lysine	224	150	169	184
(+)- or (−)-Asparagine	227	189	196	164
(±)-Glutamic acid	227	156	—	—
(+)-Serine	228	—	—	—
Glycine	232	187	179	163
(±)-Threonine	235	148	—	—
(±)-Arginine	238	230	—	—
(±)-Serine	246	171	183	169
(+)- or (−)-Threonine	253	148	—	—
(+)- or (−)-Cystine	260	181	180	160
(+)- or (−)-Aspartic acid	272	185	—	162
(±)-Methionine	272	151	—	—
(±)-Phenylalanine	274	188	93	182
(±)-Tryptophan	275	188	240	—
(+)- or (−)-Histidine	277	249	189	—
2-Amino-2-methylpropanoic	Sub. 280	202d	—	—
(±)-Aspartic acid	280	165	—	—
(+)- or (−)-Methionine	283	150	95	—
(+)- or (−)-Isoleucine	284	117	—	120
(+)- or (−)-Tryptophan	289	104	233	166
(±)-Isoleucine	292	118	—	—
(±)-α-Alanine	295	166	177	174
(+)- or (−)-α-Alanine	297	151	—	190
(±)-Valine	298	132	—	164
(±)-Norvaline	303	—	182	—
(±)-2-Aminobutanoic	307	147	—	170
(+)- or (−)-Valine	315	127	181	147
(±)-Tyrosine	318	197	254	—
(+)- or (−)-Phenylalanine	320	146	93	181
(±)-Norleucine	327	—	—	—
(±)-Leucine	332	141	—	165
(+)- or (−)-Leucine	337	107	187	115
(+)- or (−)-Tyrosine	344	166	—	104
(±)-Asparagine	>300	—	—	—
(±)-Histidine	—	—	—	—
(±)-Lysine	—	249	—	196

* These melting points are probably better described as decomposition points and their values will depend somewhat upon the rate of heating. Many of the naturally-occurring amino acids are (−)-rotatory.

Toluene-p-sulphonate °C	2,4-Dichloro-phenoxyacetate °C	1-Naphthyl-ureido acid °C	Phthalyl derivative °C	2,4-Dinitro-phenyl derivative °C
217	—	—	—	—
—	—	—	—	—
223	—	—	—	—
—	—	236	—	146
117	—	236	159	—
—	—	—	—	—
—	145	—	—	181
102	—	—	—	—
133	106	—	—	138
—	87	199	—	171
175	—	199	—	181
213	192	—	—	149
—	—	—	—	174
150	235	191	192	204
—	139	—	103	178
—	—	—	—	—
213	195	191	—	201
—	—	—	—	145
205	216	—	—	109
140	202	115	193	187
105	145	—	—	117
135	180	—	175	186
176	148	—	—	—
204	—	—	296	233
—	—	198	—	—
—	217	—	221	196
—	134	186	—	—
132	—	178	121	113
176	—	158	—	221
141	143	—	—	175
139	213	198	161	—
139	199	202	—	—
110	159	204	102	184
—	—	—	—	—
—	—	194	96	143
149	—	—	115	132
—	—	—	268	—
165	155	—	—	189
124	—	—	112	—
—	138	—	141	—
124	150	163	116	94
119	—	205	—	180
—	—	—	—	—
—	129	—	—	—
—	176	—	171	—

Table 10.37 Aromatic nitro compounds

Nitro compound	B.P. °C	M.P. °C	Nitro compound	B.P. °C	M.P. °C
Nitrobenzene (1)	221	6	m-Nitrobenzyl chloride	—	46
o-Nitrotoluene (2)	222	—	o-Nitrobenzyl bromide	—	47
2-Nitro-m-xylene	226	—	o-Nitrobenzyl chloride	—	49
m-Nitrotoluene (3)	229	16	2,4-Dinitrocholorobenzene	315	51
2-Nitro-p-xylene	237	—	o-Nitroiodobenzene	—	54
3-Nitro-o-xylene	240	15	m-Nitrobromobenzene	256	56
4-Nitro-m-xylene	244	—	2,5-Dichloronitrobenzene	267	56
2-Nitro-p-cymene (4)	264	—	m-Nitrobenzyl bromide	—	59
o-Nitroanisole	265	10	p-Nitrobenzyl chloride	—	71
m-Nitrophenetole	267	2	o-Nitrobenzyl iodide	—	75
m-Nitrobenzyl alcohol	—	27	2,4-Dinitrobromobenzene	—	75
4-Nitro-o-xylene	254	30	Picryl chloride	—	83
m-Nitrophenetole	284	34	p-Nitrochlorobenzene	242	83
2-Nitrobiphenyl	320	37	2-Nitro-p-dibromobenzene	—	84
m-Nitroanisole	258	39	m-Nitrobenzyl iodide	—	86
Nitromesitylene	255	44	2,4-Dinitroiodobenzene	—	88
p-Nitrotoluene	238	54	p-Nitrobenzyl bromide	—	100
p-Nitroanisole	259	54	p-Nitrobromobenzene	256	127
ω-Nitrostyrene	260d	58	p-Nitrobenzyl iodide	—	127
p-Nitrophenetole	283	60	p-Nitroiodobenzene	—	174
1-Nitronaphthalene	304	61			
m-Nitrobenzyl cyanide	—	62			
2,4,6-Trinitroanisole	—	68	Methyl-o-nitrobenzoate	275	—
2,4-Dinitrotoluene	—	71	Ethyl-o-nitrobenzoate	—	30
o-Nitrobenzyl alcohol	270	74	Diethyl 4-nitrophthalate	—	34
5-Nitro-m-xylene	273	74	Ethyl o-nitrocinnamate	—	44
2,4,6-Trinitrophenetole	—	79	Diethyl 3-nitrophthalate	—	46
2-Nitronaphthalene	—	79	Ethyl m-nitrobenzoate	297	47
2,4,6-Trinitrotoluene	—	82	Ethyl p-nitrobenzoate	—	57
o-Nitrobenzyl cyanide	—	84	Dimethyl 4-nitrophthalate	—	66
2,4-Dinitrophenetole	—	87	Dimethyl 3-nitrophthalate	—	69
m-Dinitrobenzene	—	90	Methyl o-nitrocinnamate	—	73
p-Nitrobenzyl alcohol	—	93	Methyl m-nitrobenzoate	—	78
2,4-Dinitroanisole	—	95	Ethyl m-nitrocinnamate	—	79
p-Nitrobenzyl cyanide	—	117	Ethyl 3,5-dinitrobenzoate	—	94
o-Dinitrobenzene	—	118	Methyl p-nitrobenzoate	—	96
1,3,5-Trinitrobenzene	—	122	Ethyl 3,5-dinitrosalicylate	—	99
1,8-Dinitronaphthalene	—	173	Ethyl 5-nitrosalicylate	—	102
p-Dinitrobenzene	—	173	Methyl 3,5-dinitrobenzoate	—	112
1,5-Dinitronaphthalene	—	217	Ethyl 3-nitrosalicylate	—	118
			Methyl 5-nitrosalicylate	—	119
o-Nitrochlorobenzene	245	33	Methyl m-nitrocinnamate	—	124
m-Nitroiodobenzene	—	38	Methyl 3,5-dinitrosalicylate	—	127
o-Nitrobromobenzene	261	42	Methyl 3-nitrosalicylate	—	132
3,4-Dichloronitrobenzene	255	43	Ethyl p-nitrocinnamate	—	142
m-Nitrochlorobenzene	236	46	Methyl p-nitrocinnamate	—	161

(1) $d_4^{20°}$ 1.204; $n_D^{20°}$ 1.553.   (3) $d_4^{20°}$ 1.157; $n_D^{20°}$ 1.547.
(2) $d_4^{20°}$ 1.168; $n_D^{20°}$ 1.546.   (4) $d_4^{20°}$ 1.074; $n_D^{20°}$ 1.531.

Table 10.38 Aliphatic nitro compounds

Nitro compound	B.P. °C (/mmHg)	$d_4^{20°}$	$n_D^{20°}$
Nitromethane	101	1.137	1.381
Nitroethane	114	1.050	1.392
2-Nitropropane	120	0.988	1.394
1-Nitropropane	131	1.001	1.401
1-Nitrobutane	152	0.971	1.410
1-Nitropentane	66/16	0.953	1.418
1-Nitrohexane	82/15	0.940	1.423
Phenylnitromethane	227	1.160	1.532

Table 10.39 Thiols

Thiol	B.P. °C (/mmHg)	M.P. °C	2,4-Dinitro-phenyl-thioether °C	2,4-Dinitro-phenyl-sulphone °C	3,5-Dinitro-thio-benzoate °C	Hydrogen 3-nitro-thio-phthalate
Methanethiol	6	—	128	190	—	—
Ethanethiol	36	—	115	160	62	149
Propane-2-thiol	58	—	95	141	84	145
Propanethiol	67	—	81	128	52	137
2-Methylpropanethiol	88	—	76	106	64	136
Prop-2-ene-1-thiol	90	—	72	105	—	—
Butanethiol	97	—	66	92	49	144
3-Methylbutanethiol	117	—	59	95	43	145
Pentanethiol	126	—	80	83	40	132
Hexanethiol	151	—	74	97	—	—
Cyclohexanethiol	159	—	148	172	—	—
Heptanethiol	176	—	82	101	53	132
Octanethiol	199	—	78	98	—	—
Nonanethiol	220	—	86	92	—	—
Decanethiol	114/13	—	85	93	—	—
Dodecanethiol	154/24	—	89	101	—	—
Hexadecanethiol	—	51	91	105	—	—
1,2-Ethane dithiol	146	—	248	—	—	—
1,3-Propane dithiol	173	—	194	—	—	—
1,4-Butane dithiol	196	—	—	—	—	—
1,5-Pentane dithiol	217	—	170	—	—	—
1,6-Hexane dithiol	237	—	218	—	—	—
Thiophenol	169	—	121	161	149	130
Phenylmethanethiol	194	—	130	183	120	137
o-Thiocresol	194	15	101	155	—	—
m-Thiocresol	195	—	91	145	—	—
p-Thiocresol	195	44	103	190	—	—
1-Phenylethanethiol	199	—	90	133	—	—
2-Phenylethanethiol	105/23	—	—	—	—	—
1-Thionaphthol	161/20	—	176	—	—	—
2-Thionaphthol	162/20	81	145	—	—	—
4-Mercaptobiphenyl	—	111	146	170	—	—
2-Furylmethanethiol	84/65	—	130	—	—	—
2-Thienylmethanethiol	166	—	119	143	—	—

Table 10.40 Sulphonic acids

Note. Aromatic sulphonic acids are usually hygroscopic solids and do not generally have sharp melting points: they are frequently supplied in the form of their sodium (or other metal) salts. It is therefore not possible to classify them in order of increasing melting points. In this table

Acid	Sulphonamide ArSO$_2$NH$_2$ °C	S-Benzylisothio-uronium salt °C
Benzenesulphonic	153	150
Toluene-*o*-sulphonic	156	170
Toluene-*m*-sulphonic	108	—
Toluene-*p*-sulphonic	137	182
o-Chlorobenzenesulphonic	188	—
m-Chlorobenzenesulphonic	148	—
p-Chlorobenzenesulpohonic	144	175
o-Bromobenzenesulphonic	186	—
m-Bromobenzenesulphonic	154	—
p-Bromobenzenesulphonic	166	170
o-Nitrobenzenesulphonic	193	—
m-Nitrobenzenesulphonic	168	146
p-Nitrobenzenesulphonic	179	—
Sulphanilic	164	187
Orthanilic	153	132
Metanilic	142	148
o-Sulphobenzoic (salt)	—	206
m-Sulphobenzoic	170	163
p-Sulphobenzoic	236	213
Phenol-*p*-sulphonic	177	169
Thymolsulphonic	—	213
o-Xylene-4-sulphonic	144	208
m-Xylene-4-sulphonic	138	146
p-Xylenesulphonic	148	184
Naphthalene-1-sulphonic	150	137
Naphthalene-2-sulphonic	217	191
Anthraquinone-1-sulphonic	—	191
Anthraquinone-2-sulphonic	261	211
1-Naphthylamine-4-sulphonic	206	195
1-Naphthylamine-5-sulphonic	260	180
1-Naphthylamine-6-sulphonic	219	191
1-Naphthylamine-7-sulphonic	181	—
1-Naphthylamine-8-sulphonic	—	300
2-Naphthylamine-1-sulphonic	—	139
2-Naphthylamine-6-sulphonic	—	184
1-Naphthol-2-sulphonic	—	170
1-Naphthol-4-sulphonic	—	104
1-Naphthol-5-sulphonic	—	—
2-Naphthol-1-sulphonic	—	136
2-Naphthol-6-sulphonic	238	217
2-Naphthol-8-sulphonic	—	218
Benzene-*o*-disulphonic	254	206
Benzene-*m*-disulphonic	229	214
Benzene-*p*-disulphonic	288	—
Naphthalene-1,4-disulphonic	273	—
Naphthalene-1,5-disulphonic	310	257
Naphthalene-1,6-disulphonic	298	235
Naphthalene-2,6-disulphonic	305	256
Naphthalene-2,7-disulphonic	243	211
2-Naphthylamine-4,8-disulphonic	—	210

related compounds are grouped together. For convenience a subsidiary table (Table 10.40A) is given in which sulphonic acids are listed in the order of increasing melting points of the *S*-benzylisothiouronium salts; these derivatives are easily prepared either from the free acid or from the salt.

Sulphonanilide ArSO₂NHPh °C	p-Toluidine salt °C	Sulphonyl Chloride, ArSO₂Cl °C	Sulphonacetamide ArSO₂NHCOCH₃ °C
110	205	—	125
136	204	68	—
96	—	12	—
103	198	71	137
—	—	28	—
—	—	—	—
104	209	53	—
—	—	51	—
119	216	75	203
115	—	69	—
126	222	64	189
136	—	80	192
200	—	—	—
—	—	—	—
—	200	79	—
—	—	20	—
—	—	57	—
—	202	—	—
—	—	52	—
110	—	34	—
—	—	25	—
112	181	68	185
132	221	79	146
216	—	217	—
193	—	197	—
—	—	—	—
—	—	—	—
—	—	—	—
140	—	—	—
—	—	—	—
—	—	—	—
200	196	—	—
201	—	—	—
—	162	124	—
—	248	—	—
195	232	—	—
241	—	143	—
144	—	63	—
—	—	131	—
179	—	—	—
249	332	183	—
—	315	129	—
—	360	225	—
—	300	159	—
—	—	—	—

Table 10.40 Sulphonic acids (*continued*)

Acid	Sulphonamide ArSO₂NH₂ °C	S-Benzylisothio-uronium salt °C
2-Naphthylamine-5,8-disulphonic	—	276
2-Naphthylamine-6,7-disulphonic	—	—
1-Naphthylamine-3,6-disulphonic	—	—
1-Naphthylamine-3,8-disulphonic	—	—
1-Naphthol-3,6-disulphonic	—	217
1-Naphthol-4,8-disulphonic	—	205
2-Naphthol-3,6-disulphonic	—	233
2-Naphthol-6,8-disulphonic	—	228
(+)-Camphorsulphonic	132	210

Aliphatic sulphonic acids

Sulphonic acid	B.P. °C	Sulphonyl chloride, B.P. °C	Sulphonamide, M.P. °C	S-Benzyl-isothio-uronium salt M.P. °C	Sulphon-anilide M.P. °C
Methane	167/10	163	90	—	99
Ethane	—	177	59	115	58
Propane-2-	—	79/18	60	—	84
Propane-1-	—	78/13	52	—	—
Butane-1-	—	75/10	45	—	—

Table 10.40A Sulphonic acids (*continued*)

(*Arranged in the order of increasing melting points of the S-benzylisothiouronium salts.*)

Sulphonic acid	M.P. °C	Sulphonic acid	M.P. °C
1-Naphthol-4-	104	p-Bromobenzene-	170
Ethane-	115	1-Naphthol-2-	170
Orthanilic acid	132	o-Toluene-	170
2-Naphthol-1-	136	p-Chlorobenzene-	175
Naphthalene-1-	137	1-Naphthylamine-5-	180
2-Naphthylamine-1-	139	p-Toluene-	182
m-Nitrobenzene-	146	2-Naphthylamine-6-	184
m-Xylene-4-	146	p-Xylene-	184
Metanilic acid	148	Sulphanilic acid	187
Benzene-	150	Naphthalene-2-	191
m-Sulphobenzoic acid	163	Anthraquinone-1-	191
Phenol-p-	169	1-Naphthylamine-6-	191

Sulphonanilide ArSO$_2$NHPh °C	p-Toluidine salt °C	Sulphonyl Chloride ArSO$_2$Cl °C	Sulphonacetamide ArSO$_2$NHCOCH$_3$ °C
—	—	—	—
—	—	—	—
—	—	—	—
—	—	—	—
—	—	—	—
—	—	—	—
202	—	—	—
195	—	162	—
—	—	88	—

Sulphonic acid	M.P. °C	Sulphonic acid	M.P. °C
1-Naphthylamine-4-	195	2-Naphthol-6-	217
1-Naphthol-4,8-di-	205	1-Naphthol-3,6-di-	217
Benzene-o-di-	206	2-Naphthol-8-	218
o-Sulphobenzoic acid	206	2-Naphthol-6,8-di-	228
o-Xylene-4-	208	2-Naphthol-3,6-di-	233
(+)-Camphor-	210	Naphthalene-1,6-di-	235
2-Naphthylamine-4,8-di-	210	Naphthalene-2,6-di-	256
Naphthalene-2,7-di-	211	Naphthalene-1,5-di-	257
Anthraquinone-2-	211	2-Naphthylamine-5,8-di-	276
p-Sulphobenzoic acid	213	1-Naphthylamine-8-	300
Thymol-	213		
Benzene-m-di-	214		

Table 10.41 Sulphonamides, R·SO₂NH₂

Sulphonamide	M.P. °C	N-Xanthyl-sulphon-amide	Sulphonamide	M.P. °C	N-Xanthyl-sulphon-amide
Butane-1-	45	—	2-Naphthalene-	217	—
Propane-1-	52	—	o-Sulphobenzimide		
Ethane-	59	—	(saccharin)	226d	198
Propane-2-	60	—	1,3-Benzenedi-	229	170
m-Ethylbenzene-	86	—	2,7-Naphthalenedi-	242	—
Methane-	90	—	1,2-Benzenedi-	254	—
2,6-Dimethylbenzene-	96	—	2-Anthraquinone-	261	—
o-Ethylbenzene-	100	—	1,4-Naphthalenedi-	273	—
Phenylmethane-	105	—	p-Sulphonamidobenzoic		
Toluene-m-	108	—	acid	280	—
p-Ethylbenzene-	109	196	1,4-Benzenedi-	288	—
p-Methoxybenzene-	111	—	1,6-Naphthalenedi-	298	—
3,5-Dimethylbenzene-	135	—	1,5-Naphthalenedi-	310	—
(+)-Camphor-8-	137	—	1,3,5-Benzenetri-	312	—
2,4-Dimethylbenzene-	137	188	1,8-Anthraquinonedi-	340	—
Toluene-p-	137	197			
2,4,6-Trimethylbenzene-	142	203	*Halogeno-sulphonamides*		
m-Aminobenzene-	142	—			
3,4-Dimethylbenzene	144	—	p-Fluorobenzene-	125	—
2,5-Dimethylbenzene-	147	176	3,4-Dichlorobenzene-	135	—
p-Ethoxybenzene-	150	—	p-Chlorobenzene-	144	—
Naphthalene-1-	150	—	m-Chlorobenzene-	148	—
Benzene-	153	200	m-Bromobenzene-	154	—
o-Aminobenzene-	153	—	p-Bromobenzene-	166	—
Toluene-o-	156	183	3,4-Dibromobenzene-	175	—
m-Nitrophenylmethane-	159	—	2,4-Dichlorobenzene-	180	—
p-Aminobenzene-			2,5-Dichlorobenzene-	181	—
(sulphanilamide)	165	208	o-Bromobenzene	186	—
2,3-Dimethylbenzene-	167	—	p-Bromophenylmethane-	188	—
m-Nitrobenzene	168	—	o-Chlorobenzene-	188	—
p-Nitrobenzene-	179	—	2,4-Dibromobenzene-	190	—
2,4,5-Trimethylbenzene-	181	—	2,5-Dibromobenzene-	194	—
o-Nitrobenzene-	193	—	p-Iodophenylmethane-	206	—
p-Nitrophenylmethane-	204	—	2,4,6-Trichlorobenzene-	212d	—
1-Nitro-2-naphthalene-	214	—	2,3,4-Trichlorobenzene-	230d	—

Table 10.42 Imides

Compound	M.P. °C
N-2-Bromoethylphthalimide	82
N-Phenylmaleimide	91
Maleimide	93
Allyl-(1-methylbutyl) barbituric acid	100
Succinimide	125
Ethylhexylbarbituric acid (Ortal)	126
Ethylbutylbarbituric acid (Neonal)	128
N-2-Hydroxyethylphthalimide	128
Ethyl-(1-methylbutyl)barbituric acid (Pentobarbital)	130
Allylisopropylbarbituric acid (Alurate)	137
Ethylisopentylbarbituric acid (Amytal)	155
N-Phenylsuccinimide	156
Alloxan (4H₂O)	170d
Ethylphenylbarbituric acid (Phenobarbital)	172
Diallylbarbituric acid (Dial)	173
Diethylbarbituric acid (Veronal)	190
Ethylisopropylbarbituric acid (Ipral)	201
N-Phenylphthalimide	205
3-Nitrophthalimide	216
o-Sulphobenzimide (saccharin)	226d
Phthalimide	233
Barbituric acid	245d
Naphthalimide	300

Table 10.43 Nitroso, azo, azoxy and hydrazo compounds*

Compound	M.P. °C
Nitroso compounds	
Methylphenylnitrosoamine	B.P. 120 °C/13 mmHg —
Ethylphenylnitrosoamine	B.P. 134 °C/16 mmHg —
p-Nitrosotoluene	48
m-Nitrosotoluene	53
N-Nitrosodiphenylamine	66
Nitrosobenzene	68
o-Nitrosotoluene	72
p-Nitroso-N-ethylaniline	78
p-Nitroso-NN-diethylaniline	84
p-Nitroso-NN-dimethylaniline	87
1-Nitrosonaphthalene	98
1-Nitroso-2-naphthol	109
p-Nitroso-N-methylaniline	118
p-Nitrosophenol	125d
p-Nitrosodiphenylamine	144
2-Nitroso-1-naphthol	152d
4-Nitroso-1-naphthol	198
Azo compounds	
2,2′-Dimethylazobenzene	55
3,3′-Dimethylazobenzene	55
Azobenzene	68
4-Anilinoazobenzene	82
3,3′-Diethoxyazobenzene	91

Table 10.43 Nitroso, azo, azoxy and hydrazo compounds*
(*continued*)

Compound	M.P. °C
Azo compounds (*continued*)	
3,3'-Dichloroazobenzene	101
p-Dimethylaminoazobenzene **C**	117
p-Aminoazobenzene	126
4-Hydroxy-3-methylazobenzene	128
2,2'-Diethoxyazobenzene	131
1-Phenylazo-2-naphthol	134
2,2'-Dichloroazobenzene	137
2-Phenylazo-1-naphthol	138
4,4'-Dimethylazobenzene	144
o-Azobiphenyl	145
p-Hydroxyazobenzene	152
4,4-Diethoxyazobenzene	160
4,4'-Dichloroazobenzene	188
1,1'-Azonaphthalene	190
4-Phenylazo-1-naphthol	206d
2,2'-Azonaphthalene	208
p-Azobiphenyl	250
Azoxy compounds	
Azoxybenzene	36
3,3'-Dimethylazoxybenzene	39
3,3'-Diethoxyazoxybenzene	50
3,3'-Dimethoxyazoxybenzene	52
2,2'-Dichloroazoxybenzene	56
2,2'-Dimethylazoxybenzene	60
4,4'-Dimethylazoxybenzene	70
2,2'-Dimethoxyazoxybenzene	81
3,3'-Dichloroazoxybenzene	97
2,2'-Diethoxyazoxybenzene	102
4,4'-Dimethoxyazoxybenzene	119
1,1'-Azoxynaphthalene	127
4,4'-Diethoxyazoxybenzene	138
4,4'-Dichloroazoxybenzene	158
o-Azoxybiphenyl	158
2,2'-Azoxynaphthalene	168
p-Azoxybiphenyl	212
Hydrazo compounds	
3,3'-Dimethylhydrazobenzene	38
4,4'-Diethoxyhydrazobenzene	86
2,2'-Diethoxyhydrazobenzene	89
2,2'-Dimethoxyhydrazobenzene	102
3,3'-Diethoxyhydrazobenzene	119
Hydrazobenzene	127
4,4'-Dimethylhydrazobenzene	134
2,2'-Hydrazonaphthalene	141
2,2'-Hydrazodiphenol	148
1,1'-Hydrazonaphthalene	153
2,2'-Dimethylhydrazobenzene	165
4,4'-Hydrazodibiphenyl	169
2,2'-Hydrazodibiphenyl	182

* All these compounds should be regarded as potential carcinogens; those marked **C** are subject to legal control.

Table 10.44 Miscellaneous sulphur compounds

Compound	B.P. °C (/mmHg)	M.P. °C	$d_4^{20°}$	$n_D^{20°}$
Dimethyl sulphide	38	—	0.849	1.436
Ethyl methyl sulphide	66	—	0.846	1.440
Diethyl sulphide	92	—	0.837	1.442
Di-isopropyl sulphide	119	—	0.817	1.440
Di-allyl sulphide	140	—	—	—
Dipropyl sulphide	142	—	0.839	1.449
Di-isobutyl sulphide	169	—	0.826	1.447
Di-s-butyl sulphide	165	—	0.835	1.451
Dibutyl sulphide	187	—	0.840	1.453
Di-isopentyl sulphide	86/5	—	0.834	1.453
Dipentyl sulphide	85/4	—	0.841	1.456
Dihexyl sulphide	114/4	—	0.841	1.459
Diheptyl sulphide	142/4	—	0.842	1.461
Dioctyl sulphide	162/4	—	0.845	1.469
Diphenyl sulphide	145/8	—	1.114	1.633
Dibenzyl sulphide	—	50	—	—
Di-p-tolyl sulphide	—	57	—	—
Dimethyl disulphide	109	—	1.065	1.526
Diethyl disulphide	153	—	0.992	1.507
Di-isopropyl disulphide	176	—	0.944	1.492
Dipropyl disulphide	194	—	0.960	1.498
Di-allyl disulphide	100/48	—	—	—
Di-isobutyl disulphide	215	—	0.928	1.487
Di-t-butyl disulphide	65/5	—	0.923	1.490
Dibutyl disulphide	231	—	0.938	1.493
Dipentyl disulphide	119/7	—	0.922	1.489
Di-isopentyl disulphide	115/9	—	0.919	1.496
Di-p-tolyl disulphide	—	48	—	—
Diphenyl disulphide	—	60	—	—
Dibenzyl disulphide	—	73	—	—
Dimethyl sulphoxide	189	18.5	1.101	1.477
Diethyl sulphoxide	88.9/5	14(4–6)	—	—
Dipropyl sulphoxide	82/15	22–3	0.965	1.466
Di-isopropyl sulphoxide	—	68.5	—	—
Dibutyl sulphoxide	—	32	0.832	1.467
Diphenyl sulphoxide	—	70	—	—
Di-p-tolyl sulphoxide	—	95	—	—
Dibenzyl sulphoxide	—	134	—	—
Dipropyl sulphone	—	29	—	—
Dibutyl sulphone	—	44	—	—
Diethyl sulphone	248	74	—	—
Trional	—	76	—	—
Dimethyl sulphone	238	109	—	—
Sulphonal	—	126	—	—
Diphenyl sulphone	—	128	—	—
Dibenzyl sulphone	—	150	—	—
Di-p-tolyl sulphone	—	159	—	—
Methyl thiocyanate	131	—	1.082	—
Ethyl thiocyanate	147	—	1.024	1.465
Isopropyl thiocyanate	151	—	—	—
Propyl thiocyanate	165	—	0.981	1.463
Butyl thiocyanate	184	—	0.961	1.464

Table 10.44 Miscellaneous sulphur compounds (*continued*)

Compound	B.P. °C (/mmHg)	M.P. °C	$d_4^{20°}$	$n_D^{20°}$
Benzyl thiocyanate	—	38	—	—
Allyl isothiocyanate	152	—	1.010	1.524
Phenyl isothiocyanate	221	—	1.134	1.651
Thiophene	84	—	1.062	1.525
Methyl benzenesulphonate	150/15	—	1.273	
Ethyl benzenesulphonate	156/16	—	1.219	
Propyl benzenesulphonate	162–3/15	—	1.180	
Methyl toluene-*p*-sulphonate	—	28	—	
Ethyl toluene-*p*-sulphonate	173/15	33	—	
Propyl toluene-*p*-sulphonate	165/10	—	—	
Butyl toluene-*p*-sulphonate	175/10	—	—	
Phenyl toluene-*p*-sulphonate	—	96	—	
N-Allylthiourea	—	78	—	
N,*N*′-Di-*m*-tolylthiourea	—	112	—	
N-Phenylthiourea	—	154	—	
N,*N*′-Diphenylthiourea (thiocarbanilide)	—	154	—	
N,*N*′-Di-*o*-tolylthiourea	—	166	—	
N,*N*′-Di-*p*-tolylthiourea	—	178	—	
Thiourea **C**	—	180	—	
Thiosemicarbazide	—	182	—	

Table 10.45 Miscellaneous phosphorus compounds

Phosphate	B.P. °C (/mmHg)	M.P. °C	$d_4^{20°}$	$n_D^{20°}$
Trimethyl	62/5 (197)	—	1.214	1.396
Triethyl	76/5 (216)	—	1.070	1.405
Tri-isopropyl	84/5	—	0.987	1.406
Tripropyl	108/5	—	1.012	1.416
Tri-isobutyl	117/5	—	0.968	1.419
Tributyl	139/6	—	0.977	1.425
Tri-isopentyl	143/3	—	—	—
Tripentyl	167/5	—	0.961	1.432
Tri-o-cresyl	264/20	—	—	—
Tri-m-cresyl	274/17	26	—	—
Triphenyl	245/11	50	—	—
Tribenzyl	—	65	—	—
Tri-p-cresyl	—	78	—	—
Tri-2-naphthyl	—	111	—	—

Phosphite	B.P. °C (/mmHg)	M.P. °C	$d_4^{20°}$	$n_D^{20°}$
Trimethyl	112	—	1.052	1.410
Triethyl	157	—	0.969	1.414
Tri-isopropyl	60/9	—	0.918	1.412
Tripropyl	207	—	0.952	1.427
Tri-isobutyl	235	—	0.917	1.425
Tributyl	120/10	—	0.923	1.432
Tripentyl	123/6	—	0.901	1.433
Triphenyl	228/12	24	—	—
Tri-o-cresyl	238/11	—	—	—
Tri-m-cresyl	235/7	—	—	—
Tri-p-cresyl	238/7	—	—	—
Dimethyl hydrogen	72/25	—	1.200	1.404
Diethyl hydrogen	66/6	—	1.079	1.408
Di-isopropyl hydrogen	90/25	—	0.996	1.407
Dipropyl hydrogen	92/11	—	1.019	1.416
Di-isobutyl hydrogen	106/12	—	0.976	1.420
Dibutyl hydrogen	119/7	—	0.995	1.423

Table 10.46 Esters of inorganic acids

Ester	B.P. °C (/mmHg)	$d_4^{20°}$	$n_D^{20°}$
Nitrites			
Methyl nitrite	−12	—	—
Ethyl nitrite	17	0.907 (10 °C)	1.331 (10 °C)
Propyl nitrite	48	0.886	1.360
Isopropyl nitrite	45	0.856	—
Butyl nitrite	76	0.882	1.377
Isobutyl nitrite	67	0.871	1.373
s-Butyl nitrite	68	0.872	1.371
t-Butyl nitrite	63	0.867	1.369
Pentyl nitrite	104	0.882	1.389
Isopentyl nitrite	99	0.871	1.387
t-Pentyl nitrite	93	0.896	1.387
Nitrates			
Methyl nitrate	65	1.208	1.375
Ethyl nitrate	88	1.108	1.385
Propyl nitrate	111	1.054	1.397
Isopropyl nitrate	102	1.035	1.391
Butyl nitrate	136	1.023	1.407
Isobutyl nitrate	124	1.015	1.403
s-Butyl nitrate	124	1.026	1.402
Pentyl nitrate	157	0.996	—
Isopentyl nitrate	148	0.998	1.413
Sulphites			
Dimethyl sulphite	126	1.213	1.409
Diethyl sulphite	157	1.083	1.414
Dipropyl sulphite	191	1.028	1.424
Di-isopropyl sulphite	170	1.006	1.415
Dibutyl sulphite	91/5	0.996	1.431
Di-isobutyl sulphite	210	0.986	1.427
Dipentyl sulphite	111/5	0.978	1.436
Di-isopentyl sulphite	98/4	0.973	1.436
Sulphates			
Dimethyl sulphate C	188	1.328	1.387
Diethyl sulphate	208	1.177	1.400
Dipropyl sulphate	94/5	1.110	1.414
Dibutyl sulphate	116/6	1.062	1.421
Di-isobutyl sulphate	133/19	1.045	1.415
Dipentyl sulphate	117/4	1.029	1.429

See Table 10.45 for alkyl phosphates and alkyl phosphites.

APPENDICES

APPENDIX 1
THE LITERATURE OF ORGANIC CHEMISTRY

A1.1 INTRODUCTION

The annual volume of published chemical papers has increased roughly five-fold over the past forty years. This volume of information imposes two sorts of tasks on the organic chemist. Firstly, to keep abreast of developments in aspects of the subject which are of particular interest by sampling the flow of published information as it appears, and secondly, to retrieve information of interest from the accumulation of published material. The information which is to be retrieved may be the physical properties of a specific compound or a reliable method for its preparation, a comprehensive list of preparative methods, a survey of recent work on a particular group of compounds, etc. Although some chemists, particularly those in large pharmaceutical companies, may have access to the services of information scientists, many still need to maintain current awareness and to carry out literature searches themselves, and all organic chemists need to be familiar with the nature of the chemical literature. Computerised methods for storage and retrieval of information are becoming increasingly common and accessible to the individual chemist and this trend is likely to continue. Given access to a good chemical library with a range of on-line facilities the experienced chemist should be able to obtain the information required without too much difficulty. The principal sources of such information are described in this appendix. The coverage is necessarily selective and readers who require a more detailed treatment are recommended to consult one of the specialised monographs on the subject.[1-5]

A.1.2 PRIMARY SOURCES – JOURNALS

Scientific journals are the principal method for communicating scientific information. The majority of papers abstracted by *Chemical Abstracts* is in English. These original papers are complemented by a range of other less-used primary literature such as patents, dissertations and theses, and reports of various kinds. Although a great number of journals are published which contain information of interest to the chemist, only a small proportion of them regularly contain material which is directly and immediately useful to the organic chemist. The most important journals dealing with organic chemistry are listed below. The standard abbreviation listed in the *Chemical Abstracts Service Source Index* (*CASSI*) is given in parenthesis.

Acta Chemica Scandinavica. Series B. Organic Chemistry and Biochemistry (*Acta Chem. Scand. Ser. B*).

Angewandte Chemie. International Edition in English (*Angew. Chem., Int. Ed. in Engl.*).

Australian Journal of Chemistry (*Aust. J. Chem.*).

Bulletin of the Academy of Sciences of the USSR, Division of Chemical Sciences (English Translation). (Izvestia Akademii Nauk SSSR, Ser. Khim.) (*Bull. Acad. Sci. USSR. Div. Chim. Sci. (Engl. Transl.)*).

Bulletin of the Chemical Society of Japan (*Bull. Chem. Soc. Japan*).

Bulletin de la Société Chimique de France (*Bull. Soc. Chim. Fr.*).

Canadian Journal of Chemistry (*Can. J. Chem.*).

Chemical Communications (*J. Chem. Soc., Chem. Commun.*).

Chemische Berichte (*Chem. Ber.*).

Helvetica Chimica Acta (*Helv. Chim. Acta*)

Journal of the American Chemical Society (*J. Am. Chem. Soc.*).

Journal of Chemical Research, Synopses and microfiche or miniprint (*J. Chem. Res. (S) and (M)*).

Journal of the Chemical Society, Perkin Transactions 1, Organic and Bio-organic Chemistry, and Perkin Transactions 2, Physical Organic Chemistry (*J. Chem. Soc., Perkin Trans 1 and 2*).

Journal of General Chemistry of the USSR (English Translation) (Zhurnal Obschei Khimii) (*J. Gen. Chem. USSR. (Engl. Transl.)*).

Journal of Heterocyclic Chemistry (*J. Heterocycl. Chem.*).

Journal of Organic Chemistry (*J. Org. Chem.*).

Journal of Organic Chemistry of the USSR (English Translation) (Zhurnal Organischeskoi Khimii) (*J. Org. Chem. USSR (Engl. Transl.)*).

Journal of Organometallic Chemistry (*J. Organomet. Chem.*).

Organic Preparations and Procedures International (*Org. Prep. Proced. Int.*).

Synthesis (*Synthesis*).

Synthetic Communications (*Synth. Commun.*).

Tetrahedron (*Tetrahedron*).

Tetrahedron Letters (*Tetrahedron Lett.*).

A.1.3 SECONDARY SOURCES – ABSTRACTS

Abstracting journals are publications giving contemporaneous, concise summaries of the various original journals and other contributions to knowledge. Each abstract usually supplies the title of the original paper, names of authors, original reference (i.e. name of journal, year of publication, volume or series number, page) and generally a brief summary of the paper. The value of the abstract in the first instance will depend on how detailed the summary is. It must be emphasised, however, that the reader should never be satisfied with the account to be found in the abstract. If at all possible the original work should be consulted. The only abstracting journal in English covering all fields of chemistry and chemical technology is *Chemical Abstracts* published by the American Chemical Society. All organic chemists should be familiar with the use of *Chemical Abstracts*. Other abstracting journals such as *Biological Abstracts* may contain information of interest to the organic chemist.

Chemical Abstracts was first published in 1907 and has continued ever since.

About 14 000 journals are monitored but only about 300 are abstracted completely. *Chemical Abstracts Service Source Index (CASSI)* provides a comprehensive list of journals currently abstracted by *Chemical Abstracts*. Two volumes of *Chemical Abstracts* are published each year, in 1988 volumes 108 and 109, and every five years a Collective Index is published covering the ten volumes published during the period. The publication of the Eleventh Collective Index, covering Volumes 96–105 (1982–86), was completed in 1988. The documents abstracted by *Chemical Abstracts* include journal articles, patents, reviews, technical reports, monographs, conference proceedings, symposia, dissertations and books. The abstracts are divided into eighty sections. The reader should consult *Subject Coverage and Arrangement of Abstracts by Sections in Chemical Abstracts – 1982* published by Chemical Abstracts Services for a description of the subject coverage of each of the eighty sections. *Chemical Abstracts* appears weekly, each week containing alternately Sections 1–34 (Biochemistry and Organic Chemistry sections) and Sections 35–80 (Macromolecular Chemistry; Applied Chemistry and Chemical Engineering; and Physical, Inorganic and Analytical Chemistry sections). Each weekly issue of *Chemical Abstracts* contains two parts, the abstracts and the issue index (keyword, patent and author index). An Illustrative Key is included at the start of each new volume (which covers a six-month period) to assist the reader in the most effective use of the abstracts, and readers are recommended to consult this key for more detailed information than can be included in the present summary. Abstracts are numbered consecutively throughout each volume. Since 1976 (Volume 84), each abstract is immediately preceded by the volume number in lightface type, and followed by a computer-generated check-letter by which each reference is computer validated. Check letters have been in use since 1966 (Volume 66) and should not be confused with the column fractional designations (a–g) previously used to indicate the position of the abstract on the page.

The indexes of *Chemical Abstracts* are a key to the world's chemical and chemical engineering literature. There are six different indexes and care and practice are required to make effective use of them. In the *Chemical Substance Index* specified substances are identified and ordered by name; the *General Subject Index* lists classes of substances and general subjects; the *Formula Index* lists specific substances according to their molecular formula; in the *Index of Ring Systems* cyclic skeletons and stereoparents are identified and ordered by ring analysis; the *Patent Index*, which replaces the *Numerical Patent Index and Patent Concordance*, contains information on patent documents; and the *Author Index* links the names of authors to the corresponding abstract. Explanatory and illustrative matter relating to indexing is published separately as the *Index Guide*. Each Collective Index has its own *Index Guide*. The initial *Index Guide* for the Twelfth Collective Index (1987–91) was published in 1987 and will be revised and reissued three times during the five-year period. The *Index Guides* provide complete instructions on the use of *Chemical Abstracts* and also a comprehensive account of the policies for selecting the names of chemical substances in *Chemical Abstracts*. Appendix II of the *Index Guide* describes the organisation and use of the indexes and the relationship between them. In addition, Chemical Abstracts Service produces a workbook which provides further instruction on the use of *Chemical Abstracts*. The *Ring Systems Handbook* (1984) is a tool for searching *Chemical Abstracts* for specific ring systems, it provides a file of diagrams of ring systems (59 510 in the 1984 edition) together with indexes of molecular formulae and ring names. The *Ring Systems Handbook* replaced the *Parent Compounds Handbook* which in turn replaced the *Ring Index*.

An additional method for the cross-referencing of chemical substances is the Chemical Abstracts Service (CAS) registry system. It is a computer-based system that assigns to each substance a unique identifying number called a registry number. The numbers have no chemical significance, but are assigned in sequential order as substances are entered into the system for the first time. Operation of the CAS registry system began in 1965. By August 1980 5 million compounds had been registered and new substances are being recorded at the rate of about 300 000 per year. The *Registry Handbook* includes in registry number order all substances that have been recorded since the system began, together with the molecular formula of each and one complete *Chemical Abstracts* index name. Registry numbers thus provide a means of structure identification without the need to resort to complex and sometimes ambiguous chemical nomenclature. A *Registry Number Update* is published annually. It provides a means of determining whether a registry number is still valid, and should always be consulted before using the *Registry Handbook*.

A useful abstracts journal specifically devoted to organic synthesis is the *Journal of Synthetic Methods*. This developed as an expansion and adaptation of Theilheimer's *Methods of Organic Chemistry* (see Section A1.4.3 below). About 6000 reactions reported in the literature are included annually. Of these, about 3000 are presented as abstracts and the remainder, mainly modifications and improvements, are given as supplementary references.

A1.4 TERTIARY SOURCES – REVIEWS, MONOGRAPHS, REFERENCE WORKS

There is a wide range of publications which gather together information from the published primary literature and organise and present it in a way which may be of value to the organic chemist. These publications range from comprehensive reference works such as Beilstein to the more recent series of specialist reports, such as those published by the Royal Society of Chemistry, which review the recent literature on one topic of specialist interest.

A1.4.1 BEILSTEIN

Beilstein's Handbook of Organic Chemistry (*Beilstein's Handbuch der Organischen Chemie*) is the largest compilation of information on organic compounds. Earlier volumes were published in German, but since 1984 it has been published in English; an indispensible guide to its use is available as *Notes for Users of Beilstein's Handbook of Organic Chemistry*, of which the following account is an extract.[6]

INTRODUCTION

The Beilstein Handbook of Organic Chemistry (*Beilstein's Handbook der Organischen Chemie*) is a reference work containing information and data relating to the structure, preparation and properties of organic compounds reported in the primary literature. The information is condensed into short entries, each of which records the important facts and numerical data relating to a particular compound, complete with literature citations, to achieve a very high density of information in the Handbook.

The Handbook is published in several series, each of which deals with the organic compounds reported in the literature during a given period; for each compound there is only one entry within a series. Publication of the Fourth edition (*Vierte Auflage*) started in 1918; it consists of the Basic Series (*Hauptwerk*), which covers all the compounds reported up to 1910, and Supplementary Series (*Ergänzungswerke*) which cover succeeding periods as indicated in Table A1.1.

Table A1.1 The series of the Beilstein Handbook

Series	Abbreviation	Literature covered	Colour of label on spine	Language
Basic Series	H	up to 1910	Green	German
Supplementary Series I	E I	1910–1919	Dark Red	German
Supplementary Series II	E II	1920–1929	White	German
Supplementary Series III	E III	1930–1949	Blue	German
Supplementary Series III/IV*	E III/IV	1930–1959	Blue/Black	German
Supplementary Series IV	E IV	1950–1959	Black	German
Supplementary Series V	E V	1960–1979	Red	English

* Volumes 17 to 27 of Series E III and E IV are combined in a joint issue.

Within each series of the Handbook the entries are ordered using the Beilstein System of Compound Classification; this is a set of rules which enables each organic compound to be assigned a position within the Handbook solely on the basis of its molecular structure; conversely, knowledge of the rules enables any compound from the *c.* 1 million compounds already described in the Handbook to be found rapidly, purely on the basis of its structure. A further advantage of the system is that many structurally related compounds are brought together within the same volume of the Handbook.

The Handbook is divided into 27 volumes (volume = *Band*), many of which run to several subvolumes; each compound is assigned to its particular volume, firstly on the basis of its skeletal structure, and secondly according to its functional groups. Within the individual volumes the ordering is based on structural features such as the type and number of functional groups, the degree of unsaturation, the number of carbon atoms and other similar criteria, until each compound has been assigned its position within the series. To use the Handbook effectively, however, it is not necessary to know *all* the rules of the Beilstein System, but merely to be able to identify in which of the 27 volumes the compound is described; this is outlined in the following sections. A more detailed description with worked examples is to be found in the booklet *How to Use Beilstein* (also available in German and Japanese) which, together with other informative material about the Beilstein Handbook, may be obtained free of charge by writing to

Beilstein-Institut
    für Literatur der Organischen Chemie
Varrentrappstrasse 40–42
D-6000 Frankfurt/M. 90

Springer-Verlag
Abt. 4005
Heidelberger Platz 3
D-1000 Berlin 33

For those users of the Beilstein Handbook who are unfamiliar with the German language, a pocket-size *Beilstein Dictionary* (German/English) has been compiled by the Beilstein editorial staff and is also available free of charge. The contents of this dictionary are also to be found on the green pages in either of the subvolumes E III/IV 22/7 or E IV 6/4.

LOCATING COMPOUNDS IN THE BEILSTEIN HANDBOOK

The best way of finding the entry for a particular compound in the Beilstein Handbook is to identify in which volume the compound is listed, and then to consult the appropriate volume index. In order to identify the volume of interest a few basic principles of the Beilstein System need to be understood and these are described below.

For the purpose of classification in the Beilstein System, all organic compounds are divided into three basic types, viz. acyclic compounds (Volumes 1–4), isocyclic (i.e. carbocyclic) compounds (Volumes 5–16), and heterocyclic compounds (Volumes 17–27). Heterocyclic compounds are further subdivided according to the nature and number of the ring heteroatoms. All heterocycles

Table A1.2 Registry compounds in the Beilstein Handbook – volume numbers

Functional groups		Acyclic compounds	Isocyclic compounds	Heterocyclic compounds Heteroatom: type and number (n)					
				Oxygen only		Nitrogen only			All other types of heterocycle
				n=1	n⩾2	n=1	n=2	n⩾3	
Compounds without functional groups			5			20			
OH	Hydroxy-compounds	1	6	17			23		
=O	Oxo-compounds		7			21	24		
	Hydroxy–Oxo-compounds		8						
	Carboxylic acids	2	9						
Carboxylic acids + hydroxy- and oxo-functions		3	10						
SO₂H, —SO₃H, —SeO₂H, —SeO₃H, —TeO₂H			11	19			25	26	27
NH₂	Monoamines		12						
	Polyamines and hydroxy-amines		13	18		22			
	Amines + other functional groups		14						
NH(OH), —N(OH)₂, —NHNH₂		4	15						
N=NH, N≡N⁺, —NH—N=O, NH—NO₂,									
Polynitrogen functional groups*			16						
Compounds containing C—M bonds; M=P, As, Sb, Bi; Si, Ge, Sn, Pb; B, Al and other metals									

* Functional groups containing three or more nitrogen atoms, substituted by —H, —OH and —O only.

containing chalcogen (oxygen, sulphur, etc.) atoms as the only type of ring heteroatom are covered in Volumes 17–19; those with one chalcogen ring atom (hetero: 1 O) in Volumes 17 and 18, and those with two or more (hetero: 2 O, 3 O, ... etc.) in Volume 19. Heterocycles containing nitrogen as the only type of ring heteroatom are covered in Volumes 20–26; all other heterocyclic compounds, including those with both chalcogen and nitrogen ring heteroatoms, are dealt with in Volume 27 (further details are given in Table A1.2).

The position of the entry for any compound is determined by that fragment of its structure which is classified latest in the Beilstein System. Thus, for example, a compound containing a heterocyclic ring, a carbocyclic ring and an aliphatic chain is classified as a heterocycle, irrespective of the other structural elements, since all heterocycles are ordered after acyclic and isocyclic compounds in the Beilstein System. This feature of the systematic classification of compounds is called the Principle of Latest Systematic Entry.

Further classification of compounds within the divisions described above is based upon the type and number of the functional groups which they contain. At this level it is necessary to distinguish between

1. registry compounds, and
2. derivatives of registry compounds.

Registry compounds are defined as hydrocarbons or heterocycles which bear either no functional groups (i.e. parent compounds), or one or more of the functional groups listed in Table A1.2, bonded to carbon atoms; no carbon atom, however, may bear more than one functional group.* Three further restrictions apply to heterocyclic rings: the ring must contain at least one carbon atom; ring heteroatoms may neither be substituted nor bear functional groups; the ring must not contain chalcogen atoms other than oxygen as ring heteroatoms. Compounds which do not conform to these criteria are classified as derivatives (see below).

The volume of the Handbook containing the entry for any particular registry compound may be identified directly from Table A1.2 on the basis of its skeletal framework and functional groups. For compounds containing two or more different groups, the Principle of Latest Systematic Entry is again applied, and each compound is classified under the functional group which appears lowest in the list in Table A1.2. For example, 4-hydroxy-benzenesulphonic acid is classified as an isocyclic sulphonic acid (Volume 11), and not as an isocyclic hydroxy-compound (Volume 6). In some cases, e.g. isocyclic amines, the second type of functional group is important to determine in which volume the compound will be dealt with; thus benzene-1,4-diamine (without a second type of functional group) is in Volume 13, whereas 4-amino-benzoic acid is in Volume 14.

---

* In carboxylic acids, the functional group consists only of the —OH and =O groups.

1407

Further examples of registry compounds are given below:

| | I | II | III | IV |

Registry compound (formula no.)	I	II	III	IV
Structural type	Iso-cyclic	Acyclic	Isocyclic	Heterocyclic Hetero: 2 N
Functional groups (in the order of latest systematic entry)	None	(i) Carboxylic acid (ii) Oxo-function	(i) Amine (ii) Sulphonic acid (iii) Hydroxy-group	Oxo-function
Volume (determined from Table A1.2)	5	3	14	24

All other compounds which cannot be classified as registry compounds are derivatives of registry compounds; they are not necessarily derived in the chemical sense but they are structurally related.

Derivatives of registry compounds are defined as compounds in which one or more of the following types of structural modification of the registry compounds occur:

(a) Modification of a functional group or the carbon atom bearing a functional group, e.g. acetyl chloride is classified as a derivative of the registry compound acetic acid.
(b) Substitution of hydrogen atoms, which are not bonded to carbon atoms bearing functional groups, by the following atoms and groups (which are classed as substituents and not as functional groups)
F   Cl   Br   I   NO NO_2 N_3 N_5 (=pentazolyl)
e.g. 4-chloro-phenol is classified as a derivative of phenol.
(c) Replacement of one or more oxygen atoms in the registry compound by other bivalent chalcogen atoms, e.g. thiophene is the sulphur analogue of furan and is thus classified as a 'derivative'.

Derivatives are entered in the Handbook following the entry for the registry compound; the order is again determined by the Beilstein System. (If the registry compound was not reported in the literature covered by a particular series, its entry is omitted, but this does not affect the ordering of its derivatives in the Handbook.)

A.1.4.2   COMPILATIONS OF DATA

Collections of spectroscopic data are listed in the references to Chapter 3, p. 393. On-line databases are described in references 4 and 7. There is, in addition, a range of useful printed collections of data which are of great value to the organic chemist. Some of these are listed below.

Dictionary of Organic Compounds, 5th edn. Ed. J. Buckingham. London: Chapman and Hall, 1982 (seven volumes). Supplements published annually, 1982–7.

Dictionary of Organometallic Compounds. Ed. J. E. MacIntyre. London: Chapman and Hall, 1984. Supplements 1985, 1986 and 1987.

Kirk–Othmer Encyclopedia of Chemical Technology, 3rd edn. Ed. A. Standen. New York: Interscience, 1978–84.

Merck Index: an encyclopedia of chemicals, drugs and biologicals (1983), 10th edn. Ed. M. Windholz. Rahway, New Jersey: Merck and Co.

Handbook of Chemistry and Physics (1987–8), 68th edn. Ed. R. C. Weast. Boca Raton, Florida: CRC Press. A ready-reference book of chemical and physical data; revised annually.

Handbook of Tables for Organic Compound Identification (1967), 3rd edn. Ed. Z. Rappoport. Cleveland, Ohio: CRC Press.

Handbook of Data on Organic Compounds. Eds R. C. Weast and M. J. Astle. Boca Raton, Florida: CRC Press. Data include physical and spectral properties and references. Chemical Abstracts Service registry number and Beilstein reference.

Lange's Handbook of Chemistry (1985), 13th edn. Ed. J. A. Dean. New York: McGraw-Hill. Includes data on 7600 organic compounds; 13th edn. includes ^{13}C n.m.r. data for the first time.

Atlas of Spectral Data and Physical Constants for Organic Compounds (1975), 2nd edn. Eds J. G. Grassell and W. M. Ritchey. Cleveland, Ohio: CRC Press. Six volumes containing the following data on 21 000 compounds: Wiswesser Line Notation, i.r., u.v., ^{1}H and ^{13}C n.m.r. and m.s. data.

A1.4.3 SYNTHETIC METHODS AND TECHNIQUES

There are a number of valuable publications which provide accounts, of a more or less comprehensive nature, of aspects of the methodology and techniques of synthetic organic chemistry. A number of these are listed below.

Methoden der Organischen Chemie (Houben-Weyl). Stuttgart: G. Thieme, (in German). Comprehensive and critical coverage of experimental procedure and appropriate theoretical background.

Rodd's Chemistry of Carbon Compounds, 2nd edn. Ed. S. Coffey and (in part) M. F. Ansell; supplements to 2nd edn. edited by M. F. Ansell. Amsterdam: Elsevier. Each supplementary chapter stands on its own as a review of recent advances in the particular field surveyed.

Organic Syntheses, many editions. New York: Wiley. An annual publication of satisfactory methods for the preparation of organic compounds. The emphasis is on model procedures. 'Collective Volumes' are published after every ten annual volumes.

Theilheimer's Synthetic Methods of Organic Chemistry. Ed. A. F. Finch. Basel: Karger. Annual publication containing an abstract and classification of new synthetic methods and improvements on known methods. A guide entitled *Getting the Best out of Theilheimer's Synthetic Methods of Organic Chemistry* is available from the publishers.

Reagents for Organic Synthesis, L. F. Fieser and M. Fieser. New York: Wiley-Interscience. Volumes 1–12, 1967–86. These twelve volumes provide a wealth of information drawn from the current literature, on the nature and use of hundreds of reagents used in organic synthesis. References are given to the original literature, and each succeeding volume provides references to the mention of each of the reagents in earlier volumes. This series is an essential reference manual for the practical organic chemist.

Organic Reactions. Began publication in 1942, Volume 34 (1986). Each volume contains detailed surveys of a small number of named reactions, with a range of typical experimental procedures.

Comprehensive Organic Chemistry (1982), in six volumes. Eds D. H. R. Barton and W. D. Ollis. Oxford: Pergamon.

Comprehensive Heterocyclic Chemistry (1984), in eight volumes. Eds A. R. Katritzky and C. W. Rees. Oxford: Pergamon.

Chemistry of Heterocyclic Compounds. Eds A. Weissberger and E. C. Taylor. Very detailed accounts of chemistry of particular heterocyclic systems, totalling over 100 volumes.

C. A. Buehler and D. E. Pearson (1970). *Survey of Organic Syntheses.* New York: Wiley.

Techniques of Chemistry. Ed. A. Weissberger. London: Interscience, 1971 onwards. (Successor to *Techniques of Organic Chemistry.*) Comprehensive treatment covering theoretical background, description of techniques and tools, their modifications, merits and limitations, and their handling.

A1.4.4 REVIEWS

With the increasing volume of primary literature, reviews and surveys of particular areas of activity are essential for the organic chemist. Some of the important review publications are Chemical Society Reviews, Chemical Reviews, Accounts of Chemical Research, Angewandte Chemie (International Edition in English), Russian Chemical Reviews, Survey of Progress in Chemistry. In addition there are a number of series of Advances in ... and Progress in ... various aspects of organic chemistry. The Specialist Periodical Reports, published by the Royal Society of Chemistry, London, provide systematic and comprehensive reviews of progress in major areas of research. Current titles include: General and Synthetic Methods; Carbohydrate Chemistry; Amino Acids and Peptides; Heterocyclic Chemistry; Organometallic Chemistry; Organophosphorus Chemistry.

A1.5 COMPUTERISED STORAGE AND RETRIEVAL OF INFORMATION

This is a rapidly developing subject. There is an increasing number of on-line databases readily available to the organic chemist. It is likely that all organic chemists will require skills to conduct searches of such databases in future. Any advanced course in organic chemistry should therefore provide an introduction to on-line searching techniques in order to provide the basis for development later in the student's career. A detailed account of this field is beyond the scope of this appendix and would in any case date rapidly. The reader is referred to recent monographs for further information.[4,7]

A1.6 CURRENT AWARENESS

A general level of scientific awareness can be maintained from perusal of such periodicals as *Nature* and *Science*. In chemistry, such journals as the *Journal of the American Chemical Society* and *Chemical Communications* will keep the organic chemist in touch. At the specific level of particular aspects of organic chemistry it is necessary to scan a limited range of journals such as those listed in Section A1.2 above. The chemist is helped in this task by current awareness publications such as *Current Contents* published by ISI (Institute for Scientific Information) and *Current Titles* published by Chemical Abstracts Service. Each of these provides information on the contents of research journals in a particular field of interest. Neither gives direct access to the research journals but they provide an indicator of material of possible interest. Chemical Abstracts also publishes a range of current awareness bulletins known as *CA Selects*. Titles include Asymmetric Synthesis and Induction, Electrochemical Organic Synthesis, Natural Product Synthesis, etc. In the field of organic synthesis, *Methods of Organic Synthesis* published by the Royal Society of Chemistry provides coverage of about 200 items each month.

REFERENCES

1. J. Ash, P. Chubb, S. Ward, S. Welford and P. Willett (1985). *Communication, Storage and Retrieval of Chemical Information*. Chichester: Ellis Horwood.
2. M. G. Mellon (1982). *Chemical Publications – Their Nature and Use*, 5th edn. New York: McGraw-Hill.
3. R. T. Bottle (1979). *The Use of the Chemical Literature*, 3rd edn. London: Butterworths.
4. Y. Wolman (1983). *Chemical Information – A Practical Guide to Utilization*. Chichester: Wiley.
5. K. Subramanyam (1981). *Scientific and Technical Information Resources*. New York: Marcel Dekker.
6. The *Notes for Users* are a shortened version of those appearing in Beilstein and are reproduced by permission of Springer-Verlag, Berlin.
7. *Modern Approaches to Chemical Reaction Searching* (1986). Ed. P. Willett. Aldershot: Gower Publishing Co.

APPENDIX 2
INFRARED CORRELATION TABLES

INDEX TO TABLES

Table A2.1 Alkanes, cycloalkanes and alkyl groups

Group/vibration	cm^{-1}	μm	Comments*
C—H stretching			
CH$_3$—	2972–2953	3.36–3.39	(s) asym.
$>$CH$_2$	2936–2916	3.41–3.43	(s) asym.
CH$_3$—	2882–2862	3.47–3.49	(s) sym.
$>$CH$_2$	2863–2843	3.49–3.52	(s) sym.
$\geqq$CH	2900–2880	3.45–3.47	(w)
cyclopropane	~3060–3040	3.27–3.29	(w) cf. alkene C—H str.
CH$_3$—N	2825–2765	3.54–3.62	(m) precise pattern depends on whether amine is aliphatic or aromatic and if NCH$_3$ or N(CH$_3$)$_2$
CH$_3$—O	2830–2810	3.53–3.56	(v)
CH$_3$—CO—	3000–2900	3.33–3.40	(w) much reduced intensity
C—H deformation			
CH$_3$—	1470–1430	6.80–7.00	(m) asym.
$>$CH$_2$	1485–1445	6.73–6.92	(m) scissoring, normally overlaps above
CH$_3$—$\overset{\vert}{\underset{\vert}{C}}$—	1380–1370	7.25–7.30	(m) sym. very useful
$\overset{CH_3}{\underset{CH_3}{>}C<}$	1385–1380 1370–1365	7.22–7.25 7.30–7.33	(m)⎫approximately of (m)⎬equal intensity
(CH$_3$)$_3$C—	1395–1385, 1365	7.17–7.22 7.32	(m)⎫intensity ratio (s) ⎬approx. 1:2
$\geqq$CH	~1340	7.46	(m) no practical value
CH$_3$—N	1440–1410	6.95–7.09	
CH$_3$—O	1460–1440	6.85–6.95	
CH$_3$—CO—	1364–1354	7.33–7.39	(s) very much enhanced intensity
—CH$_2$—CO—	1420–1410	7.04–7.09	(s) compared to CH$_3$ and CH$_2$ attached to saturated C
Skeletal			
(CH$_3$)$_3$—C	1255–1245 1250–1200	7.97–8.03 8.00–8.33	(s) (s)
$\overset{CH_3}{\underset{CH_3}{>}C}$	1175–1165 1170–1140	8.51–8.59 8.55–8.77	(s) (s)
—(CH$_2$)$_n$—	720–725	13.88–13.80	(m) for n $\geqslant$ 4, doublet in solid state; frequency increases the shorter the chain
cyclopropane	1020–1000	9.80–10.0	(m)

* Abbreviations used for the intensity of absorptions are: s, strong; m, medium; w, weak; v, variable.

Table A2.2   Alkenes

Group/vibration	cm^{-1}	μm	Comments
=C—H stretching and deformation			
R·CH=CH$_2$	3040–3010	3.29–3.32	(m) CH str.
	3095–3075	3.23–3.25	(m) CH$_2$ str.
	995–985	10.05–10.15	(s) CH out-of-plane def.
	915–905	10.93–11.05	(s) CH$_2$ out-of-plane def. (wag)— often overtone at $\sim$1830 cm^{-1}
	1420–1410	7.04–7.09	(s) CH$_2$ in-plane def.
	1300–1290	7.69–7.75	(s–w) CH in-plane def.
R^{1}R^{2}C=CH$_2$	3095–3075	3.23–3.25	(m) CH str.
	895–885	11.17–11.30	(s) out-of-plane def.; often an overtone at 1780 cm^{-1}
	1420–1410	7.04–7.09	(s) CH$_2$ in-plane def.
CH=CH (*trans*)	3040–3010	3.29–3.32	(m) CH str.
	970–960	10.31–10.42	(s) CH out-of-plane def.
	1310–1295	7.64–7.72	(s–w) CH in-plane def.
CH=CH (*cis*)	3040–3010	3.29–3.32	(m) CH str.
	728–675	13.74–14.92	(s) often near 690 cm^{-1}
R^{1}R^{2}C=CHR^3	3040–3010	3.29–3.32	(m) CH str.
	840–790	11.90–12.66	(s) CH out-of-plane def.
C=C stretching			
non conjugated	1680–1625	5.95–6.15	(v) more substituted appear at higher frequency with lower intensity; frequency lowered by attached polar groups such as Br, O
Ar conjugated	$\sim$1625	$\sim$6.15	enhanced intensity
C=O or C=C conjugated	$\sim$1600	$\sim$6.25	enhanced intensity

Table A2.3 Aromatic compounds

Group/vibration	cm^{-1}	μm	Comments
$\stackrel{\vdots}{=}$C—H stretching	~3030	~3.03	(v) sharp
C—H out-of-plane summation bands	2000–1660	5.00–6.02	(w) weak bands, pattern depends on substitution pattern, see Fig. 3.18
C$=$C skeletal vibrations	~1600	~6.25	(v) ⎫ bands are characteristic of
	~1500	~6.67	(v) ⎬ the aromatic ring
	~1580	~6.33	(m) ⎫ intensities variable;
	~1450	~6.90	(m) ⎭ 1580 cm^{-1} band only just perceptible – intensity increased when polar group conjugated with ring; 1450 cm^{-1} band masked if alkyl groups present
$\stackrel{\vdots}{=}$C—H in-plane deformation	1225–950	8.16–10.53	series of weak bands, positions characteristic of substitution pattern, rarely used diagnostically
—C—H out-of-plane deformation			
five adjacent hydrogen atoms	770–730 710–690	12.99–13.7 14.09–14.5	(s) ⎫ (s) ⎪
four adjacent hydrogen atoms	770–735	12.99–13.6	(s) ⎪ very useful, also for
three adjacent hydrogen atoms	810–750	12.35–13.34	(s) ⎬ polycylics; position of bands disturbed however
two adjacent hydrogen atoms	860–800	11.63–12.5	(s) ⎪ when polar groups, e.g. —NO$_2$ attached to ring
one hydrogen atom	900–860	11.11–11.63	(w) ⎭

Table A2.4   Alkynes and allenes

Group/vibration	cm^{-1}	μm	Comments
ALKYNES			
≡C—H stretching			
	3320–3310	3.01–3.02	(s) sharp; NH and OH broad in this region
C≡C stretching			
monosubstituted	2140–2100	4.67–4.76	(m) 2130–2120 cm^{-1} for alkyl substituted
disubstituted	2260–2190	4.43–4.57	(w) 2240–2230 cm^{-1} for alkyl substituted; may be v. weak or absent
≡C—H deformation			
	680–610	14.70–16.39	(m) near 630 cm^{-1} for alkyl substituted; broad overtone near 1250 cm^{-1}
ALLENES			
C=C=C stretching			
monosubstituted	1980–1945	5.05–5.14	(m) asym. str.
disubstituted	1955–1930	5.12–5.18	(w) asym. str.
=CH$_2$ deformation			
	875–840	11.43–11.91	(s) overtone near 1700 cm^{-1}

Table A2.5   Alcohols and phenols

Group/vibration	cm^{-1}	μm	Comments
O—H stretching			
Free OH	3650–3590	2.74–2.79	(v) sharp; only in dilute solution; frequency decreases for primary > sec > tert > phenol
Bonded OH			
intermolecular dimeric	3550–3450	2.82–2.90	(v) sharp ⎫ intensity changes and
			frequency
intermolecular polymeric	3400–3200	2.94–3.13	(s) broad ⎭ increases on dilution
intramolecular	3570–3450	2.80–2.90	(v) sharp ⎫ not affected
chelate compounds	3200–2500	3.13–400	(w) v. broad ⎭ on dilution
C—O stretching			
and **O—H deformation** (in-plane)			
primary alcohols	~1050	~9.52	(s) both types of absorptions
	1350–1260	7.41–7.93	(v) are sensitive to change
secondary alcohols	~1100	~9.09	(s) of state; these values
	1350–1260	7.41–7.94	(v) are for H bonded state
tertiary alcohols	~1150	~8.69	(s)
	1410–1310	7.09–7.64	(v)
phenols	~1200	~8.33	(s)
	1410–1310	7.09–7.64	(v)

Table A2.6 Ethers and cyclic ethers

Group/vibration	cm⁻¹	μm	Comments
C—O stretching			
dialkyl	1150–1060	8.7–9.43	(s) asym.
aralkyl	1270–1230	7.87–8.13	(s) aryl—O
	1075–1020	9.3–9.8	(s) alkyl—O, also strong band at 1176 cm⁻¹
diaryl	1250–1150	8.0–8.7	(s)
vinyl	1225–1200	8.16–8.33	(s)
cyclic:			
6-membered ring	1100	9.09	(s) for tetrahydropyran
5-membered ring	1100–1075	9.09–9.3	(s)
4-membered ring	980–970	10.2–10.31	(s)
3-membered ring	~1250	~8.0	(s) ring breathing
	~890	~11.23	*trans* } tentative
	~830	~12.05	(m) *cis* } tentative

Table A2.7 Amines

Group/vibration	cm⁻¹	μm	Comments
N—H stretching			
aliphatic, primary	3398–3381	2.94–2.96	(w) values for
	3344–3324	2.99–3.01	(w) dilute solution; in
aromatic, primary	3509–3460	2.85–2.89	(m) associated state all the
	3416–3382	2.93–2.96	(m) bands intensify and
dialkyl > NH	3360–3310	3.07–3.02	(w) move to lower
aralkyl > NH	~3450	~2.9	(m) frequencies
imines	3350–3320	2.99–3.01	(m)
N—H deformation			
primary amines	1650–1590	6.06–6.29	(m–s)
secondary amines	1650–1550	6.06–6.45	(m) for aryl, weak or absent in alkyl
C—N stretching			
aromatic amines			
primary	1340–1250	7.46–8.00	(s)
secondary	1350–1280	7.41–7.81	(s)
tertiary	1360–1310	7.35–7.64	(s)
aliphatic amines	1220–1020	8.20–9.8	(m–w)

Table A2.8 Compounds containing the carbonyl group

Group/vibration	cm^{-1}	μm	Comments*
1. KETONES			
C=O stretching			
saturated, acyclic	1725–1705	5.80–5.87	(s)
α,β-unsaturated, acyclic	1690–1675	5.92–5.97	(s)
α,β-α',β'-unsaturated, acyclic	1670–1660	5.99–6.02	(s)
aryl	1715–1695	5.92–5.95	(s) ⎫ modified by nature and
diaryl	1670–1660	5.99–6.02	(s) ⎭ position of substituents
6-membered ring and higher	1725–1705	5.80–5.87	(s)
5-membered ring	1750–1740	5.71–5.75	(s)
4-membered ring	~1775	~5.63	(s)
α-halogeno	1745–1725	5.73–5.08	(s) two bands
α-diketones	1730–1710	5.78–5.85	(s)
β-diketones	1640–1540	6.10–6.49	(s) enolic, H-bonded
o-hydroxy and o-amino aryl ketones	1655–1635	6.04–6.12	(s) modified by nature and position of substituents
1,4-quinones	1690–1660	5.92–6.02	(s)
2. ALDEHYDES			
C=O stretching			
saturated, acyclic	1740–1730	5.75–5.78	(s)
α,β-unsaturated, acyclic	1705–1680	5.87–5.95	(s)
α,β-$\gamma\delta$-unsaturated, acyclic	1680–1660	5.95–6.02	(s)
aryl	1715–1695	5.83–5.90	(s) modified by nature and position of substituents
C—H stretching	2900–2700	3.45–3.70	(w) 2 bands near 2820 and 2720 cm^{-1}
C—H deformation	975–780	10.26–12.82	(m)
3. ACIDS			
O—H stretching			
free	3560–3500	2.81–2.86	(m) very dilute solution
bonded	3300–2500	3.03–4.00	(w) very broad
C=O stretching			
saturated, acyclic	1725–1700	5.80–5.89	(s)
α,β-unsaturated	1705–1690	5.86–5.92	(s)
aryl conjugated	1700–1680	5.89–5.95	(s)
C—O stretching and O—H deformation	1440–1395	6.94–7.17	(w) coupled vibrations
	1320–1211	7.57–8.26	(s)
O—H deformation (out-of-plane)	950–900	10.53–11.11	(v) acid dimer
—C(=O)O$^{\ominus}$ stretching	1610–1550	6.21–6.45	(s) salts, asym.
	1420–1300	7.04–7.69	(s) and sym. vibrations

* All ketone and aldehyde carbonyl band positions refer to dilute solutions except where indicated.

Table A2.8 Compounds containing the carbonyl group (*continued*)

Group/vibration	cm⁻¹	μm	Comments

Group/vibration	cm^{-1}	μm	Comments
4. ESTERS AND LACTONES			
C=O stretching			
saturated, acyclic	1750–1735	5.71–5.76	(s)
α,β-unsaturated and aryl	1730–1717	5.78–5.82	(s)
α-keto	1755–1740	5.70–5.74	(s)
β-keto	~1650	~6.06	(s) enolic
o-amino and o-hydroxyaryl	1690–1670	5.91–5.98	(s) H-bonding
δ-lactones	1750–1735	5.71–5.76	(s)
γ-lactones	1780–1760	5.62–5.68	(s)
β-lactones	~1820	~5.49	(s)
vinyl esters			
(—CO·O·CH=CH—)	1775–1755	5.63–5.70	(s)
C—O stretching			
formates	1200–1180	8.33–8.48	(s)
acetates	1250–1230	8.00–8.13	(s)
propanoates and higher	1200–1150	8.33–8.70	(s)
α,β-unsaturated	1300–1200	7.69–8.33	(s)
	1180–1130	8.47–8.85	(s)
aryl conjugated	1310–1250	7.63–8.00	(s)
	1150–1100	8.69–9.09	(s)
phenolic	~1205	~8.30	(s)
5. ACID ANHYDRIDES			
C=O stretching			
acyclic, saturated	1850–1800	5.40–5.56	(s)
	1790–1740	5.58–5.75	(s) frequency lowered by c. 20 cm⁻¹ when conjugated
cyclic, 5-membered ring	1870–1820	5.35–5.49	(s)
	1800–1750	5.56–5.71	(s)
C—O stretching			
acyclic	1170–1050	8.55–9.52	(s)
cyclic	1300–1200	7.69–8.33	(s)
6. ACYL HALIDES			
C=O stretching			
	1815–1770	5.51–5.65	(s) conjugated compounds absorb at lower end of range
7. AMIDES			
N—H stretching			
primary	~3520	~2.84	(m) free NH
	~3410	~2.93	(m)
primary	~3350	~2.98	(m) bonded NH
	~3180	~3.14	(m)
secondary	3480–3440	2.87–2.91	(m) free NH, *trans-*
	3435–3395	2.91–2.95	(m) free NH, *cis-*
	3320–3270	3.11–3.06	(m) bonded NH, *trans-*
	3180–3140	3.15–3.18	(m) bonded NH, *cis-*
	3100–3070	3.23–3.26	(w) bonded NH, *cis-* and *trans-*
C=O stretching (Amide I)			
primary	~1650	~6.06	(s) solid phase
	~1690	~5.92	(s) dilute solution

1419

Table A2.8 Compounds containing the carbonyl group (continued)

Group/vibration	cm⁻¹	μm	Comments
secondary	1680–1630	5.95–6.14	(s) solid phase
	1700–1680	5.88–5.95	(s) dilute solution
tertiary	1670–1630	5.98–6.14	(s) solid phase and dilute solution
cyclic amides			
(a) δ-lactams	∼1680	∼5.95	(s) dilute solution
(b) γ-lactams	∼1700	∼5.88	(s) dilute solution, shifted to higher frequency when fused to another ring
(c) β-lactams	1760–1730	5.68–5.78	(s)
imides (CO—NH—CO)			
acyclic	1740–1720	5.74–5.81	(s) bands not always resolved
	1720–1700	5.81–5.88	(s)
cyclic	1790–1735	5.58–5.76	(s) lower frequency band more intense
	1745–1680	5.73–5.95	(s)
ureas (NH—CO—NH)			
acyclic, monoalkyl	∼1605	∼6.23	(s)
acyclic, dialkyl	∼1640	∼6.10	(s)
N—H deformation (Amide II)			
primary	1650–1620	6.06–6.17	(s) solid phase
	1620–1590	6.17–6.29	(s) solution
secondary (non cyclic)	1570–1515	6.37–6.60	(s) solid phase
	1550–1510	6.45–6.62	(s) solution

Table A2.9 Amino acids

Group/vibration	cm⁻¹	μm	Comments
$\overset{\oplus}{N}H_3$ **stretching**			
	3130–3030	3.19–3.30	(m) asym.
	3030–2500	3.30–4.00	(m) forms continuous series of overlapping bands with above, also combination and overtone bands
$\overset{\oplus}{N}H_3$ **deformation**			
	1660–1610	6.02–6.21	(w) often appears as shoulder on $CO_2^{\ominus}$ band
	1550–1485	6.45–6.73	(m) sym.
$CO_2^{\ominus}$ **stretching**			
	1600–1560	6.25–6.41	(s) asym.
	∼1410	∼7.09	sym.
C=O stretching			
α-amino acids	1754–1720	5.70–5.81	(s) in hydrochlorides – normal carboxyl C=O
α-amino acids; β,γ and lower amino acids	1730–1695	5.78–5.90	(s)
others	∼2130	∼4.69	(w) found in all α-amino acids, displaced in others; a combination band

Table A2.10 Nitro compounds, nitroso compounds and nitrites

Group/vibration	cm^{-1}	μm	Comments
1. NITRO COMPOUNDS			
NO$_2$ stretching			
aliphatic, C—NO$_2$	1560–1534	6.41–6.52	(s) asym.
	1388–1344	7.20–7.44	(s) sym.
aromatic, C—NO$_2$	1555–1487	6.43–6.72	(s) asym.
	1357–1318	7.37–7.59	(s) sym
C—N stretching			
	857–830	11.43–12.05	(m–s) alkyl and aryl
2. NITRITES			
N=O stretching			
	1681–1653	5.95–6.05	(s) *trans* form
	1625–1613	6.15–6.20	(s) *cis* form
3. NITROSO COMPOUNDS			
N=O stretching			
C-nitroso	1600–1500	6.25–6.66	(s) in monomeric state
trans	1290–1190	7.75–8.40	dimer
cis	1425–1370	7.02–7.30	dimer
N-nitroso	1460–1430	6.85–6.99	(s) in solution

Table A2.11 Unsaturated nitrogen compounds

Group/vibration	cm^{-1}	μm	Comments
C≡N stretching			
nitriles			
alkyl	2260–2240	4.42–4.46	(s) ⎫
aryl	2240–2220	4.46–4.50	(s) ⎬ large intensity
α,β-unsaturated alkyl	2235–2215	4.47–4.51	(s) ⎭ variations
isonitriles	2180–2120	4.59–4.72	(s)
isocyanates	2275–2240	4.40–4.46	(s)
C=N stretching			
imines			
alkyl	1690–1590	5.92–6.29	(v)
α,β-unsaturated alkyl	1660–1590	6.02–6.29	(v)
oximes	1690–1620	5.92–6.17	(v) also broad O—H stretch at 3300–3150 cm^{-1}
N=N stretching			
azo compounds	1630–1575	6.13–6.35	(v) frequency lowered by conjugation

Table A2.12 Organo-sulphur compounds

Group/vibration	cm^{-1}	μm	Comments
S—H stretching			
	2590–2550	3.86–3.92	(w) smell!
C=S stretching			
thioketones, dithioesters	1270–1190	7.88–8.4	unlike C=O, is not strong
S=O stretching			
sulphoxides	1070–1035	9.35–9.66	(s)
sulphones	1350–1300	7.41–7.69	(s)⎱little affected by
	1160–1120	8.62–8.93	(s)⎰conjugation
sulphonamides	1358–1336	7.37–7.49	(s)⎱primary and secondary
	1169–1152	8.56–8.68	(s)⎰ also show N—H str.
sulphonyl chlorides	1410–1360	7.09–7.36	(s)
	1195–1168	8.37–8.56	(s)
sulphonic acids	1350–1340	7.41–7.46	(s)⎱also broad H-bonded
	1165–1150	8.59–8.70	(s)⎰O—H str.
sulphonates	1380–1347	7.25–7.43	(s)
	1193–1170	8.38–8.55	(s)
sulphates (organic)	1415–1380	6.92–7.25	(s)
	1200–1185	8.33–8.44	(s)

Table A2.13 Halogen compounds

Group/vibration	cm^{-1}	μm	Comments
C—F stretching			
monofluoroalkanes	1100–1000	9.09–10.0	(s)
polyfluoroalkanes	1400–1000	7.15–10.0	(s) series of bands
C—Cl stretching			
monochloroalkanes	760–540	13.15–18.52	(s) or more bands in solution
equatorial	780–740	12.82–17.24	(s)⎱cyclohexanes and
axial	730–580	13.70–17.24	(s)⎰steroids
C—Br stretching			
monobromoalkanes	600–500	16.66–20.0	(s) 2 or more bands in solution
equatorial	750–690	13.33–14.5	(s)⎱cyclohexanes and
axial	690–550	14.5–18.18	(s)⎰steroids
C—I stretching			
	600–465	16.67–21.5	(s) limited value

APPENDIX 3
NUCLEAR MAGNETIC RESONANCE CORRELATION TABLES

INDEX TO TABLES

Table A3.1   Chemical shifts of CH₃, CH₂, CH groups (δ values, TMS)*

X	CH₃—X	R·CH₂—X	R¹R²CH—X
(a) Carbon substituents			
alkyl	0.90	1.25	1.50
—C=C—	} 1.70	1.95	2.6
—C=C—C=C—			
C=C—C=C	1.95	2.2	
—C≡C—R	1.8		
—C≡C—Ph	2.9		
—C=N·R	2.0		
—C≡N	2.0	2.48	
—CO₂R	2.0	2.10	
—CO₂H	2.07	2.34	2.57
—CO·NR₂	2.02	2.05	
—CO·R	2.10	2.40	2.48
—CHO	2.17	2.2	2.4
—C₆H₅	2.34	2.6	2.87
—CO·C₆H₅	2.62	—	3.58
(b) Nitrogen substituents			
—NH₂, —NR₂	2.15	2.50	2.87
—NH·CO·R	2.9	3.3	3.5
—NR₃⊕	3.33	3.40	3.5
—NO₂	4.33	4.40	4.60
—N≡C	2.9	3.3	4.9
(c) Oxygen substituents			
—OR	3.30	3.36	3.80
—OH	3.38	3.56	3.85
—O·SO₂R	3.58	—	
—O·CO·R	3.65	4.15	5.01
—OC₆H₅	3.73	3.90	4.0
—O·CO·C₆H₅	3.90	4.23	5.12
(d) Halogen substituents			
I	2.16	3.15	4.2
Br	2.65	3.34	4.1
Cl	3.02	3.44	4.02
F	4.26	4.35	—
(e) Sulphur substituents			
—SH	3.2	3.4	
—SR	2.10	2.40	3.1
—SO₂R	2.6	3.1	

(f) Alicyclic rings		
Ring size	—CH₂—	—CH—
3	0.2	0.4
4	2.0	—
5	1.5	—
6	1.4	1.70
7	1.2	—

* It should be pointed out that although these δ values are typical, as with other spectroscopic data, some variation is possible in individual cases; a range of say ±0.05 Hz is feasible.

Table A3.2* Shielding constants for aliphatic methylene groups, X·CH₂·Y

Substituent	Shielding constant	Substituent	Shielding constant	Substituent	Shielding constant	Substituent	Shielding constant
—CH_3	0.47	—CO·R	1.70	—I	1.82	—OC_6H_5	3.23
—C=C—	1.32	—CO·C_6H_5	1.84	—NR_2	1.57	—O·CO·R	3.13
—C≡C—	1.44	—CO_2R	1.55	—NHCO·R	2.27	—O·SO_2R	3.13
—C_6H_5	1.85	—$CONH_2$	1.59	—N_3	1.97	—SR	1.64
—CF_3	1.14	—Br	2.33	—OH	2.56		
—C≡N	1.70	—Cl	2.53	—OR	2.36		

To calculate the δ value (TMS) for disubstituted methylene groups, add the sum of the Shielding Constants to 0.23 (which is the δ value for methane).

* Data reproduced from R. M. Silverstein, C. G. Bassler and T. C. Merrill (1974), *Spectroscopic Identification of Organic Compounds*, 3rd edn. Wiley International Edition, p. 220.

Table A3.3 Chemical shifts of protons attached to unsaturated systems (δ, TMS)

Group	δ	Group	δ
HC≡C—R	1.80*		
H—C≡C—C≡C—R	2.80*		
H—C≡C·C_6H_5	2.13*		6.6
H_2C=CR_2	4.65		
	5.3		
	a 5.55 b 5.15 c 6.7		7.8
	5.6		7.85
—C=C—C=C	6.2		
=C—C=O	5.8	RO·C	8.03
	6.0	R·C	9.6
	6.2		9.8
	6.8	Ar—C	9.9

* Signals shifted to lower field by a trace of pyridine, and removed on deuteration.

Table A3.4 Chemical shifts (TMS) of protons attached to aromatic and heteroaromatic rings

Benzene 7.27

Pyridazine: 7.55, 9.24

Indene: 6.84, 6.50

Naphthalene: 7.73, 7.42, 7.36

Furan: 6.28, 7.36

Anthracene: 8.30, 7.91, 7.39

Pyridine: 6.99, 7.36, 8.50

Thiophene: 7.06, 7.17

Quinoline: 7.68, 8.00, 7.44, 7.25, 7.61, 8.80, 8.05, 8.05

Pyrazine: 8.63

Pyrrole: 6.15, 6.36, 7.31 (N–H)

Pyrimidine: 9.26, 7.37, 8.77

Pyrazole: 6.33, 7.62, 12.52 (N–H)

Indole: 6.40, 6.70, 7.40 (N–H)

Imidazole: 7.13, 7.13, 7.70

Table A3.5 Effect of substituents in benzenoid compounds

Substituent	ortho	meta	para
H	0	0	0
CH$_3$	−0.2	−0.1	−0.2
C$_6$H$_5$	0.2	0	−0.1
C=C	0.2	0.2	0.2
F	−0.3	0	−0.2
Cl	0	0	0
Br	0.2	−0.1	0
I	0.4	−0.3	0
OH	−0.5	−0.1	−0.4
OR	−0.4	−0.1	−0.4
SR	0.1	−0.1	−0.2
O·CO·R	0.2	−0.1	−0.2
NH$_2$	−0.8	−0.2	−0.6
NMe$_2$	−0.6	−0.1	−0.6
NH·CO·R	0.4	−0.2	−0.3
NO$_2$	1.0	0.2	0.3
C≡N	0.3	0.1	0.3
CO$_2$H/CO$_2$R	0.8	0.1	0.2
CHO	0.6	0.2	0.3
CONH$_2$	0.5	0.2	0.2
CO·R	0.6	0.3	0.3

The calculated δ-value (TMS) for the substituted benzene is obtained by adding the appropriate substituent parameters to 7.27, the value for benzene.

Table A3.6 Chemical shift of protons attached to atoms other than carbon (all signals are sensitive to solvent, concentration and temperature and are removed by deuteration)

Compound type	δ (TMS)	Comments
Oxygen: ROH	0.5–4.0	Often appears as broad singlet; trace of acid gives sharp singlet
enols	11–16	
ArOH	4.5–9	Position depends on extent of hydrogen bonding
$R \cdot CO_2H$	10–13	
$R \cdot SO_3H$	10–12	
H_2O	4–5	
$R \cdot C = NOH$	9–12	
Nitrogen: RNH_2, RNHR and cyclic amines	0.5–3	Usually broad absorption
$ArNH_2$; ArNHR	3.5–6	
$R \cdot CONH_2$	5–8.5	Usually very broad; may not be observable
Sulphur: RSH	1–1.5	
ArSH	2.5–4	

Table A3.7   Characteristic proton–proton coupling constants (Hz)

Saturated compounds

Acyclic			

CH_2 10–18

$\searrow CH{-}OH$
$\searrow CH{-}NH$ 4–8
$\searrow CH{-}SH$

$\searrow CH{-}CH \diagup$ 6–8

$\searrow CH{-}\underset{|}{\overset{|}{C}}{-}CH \diagup$ 0–1

$\searrow CH{-}CHO$ 1–3

Alicyclic	gem	cis	trans
cyclopropanes epoxides aziridines	4–6	4–9	3–6
cyclobutanes	10–17	6–11	5–9
cyclopentanes	10–17	7–11	2–8
cyclohexanes	gem 10–17; diaxial 8–11; axial–equatorial 2–4; diequatorial 2–4		

Unsaturated

$J_{a,b}$ 0–3; $J_{a,c}$ 12–18; $J_{b,c}$ 6–12; $J_{c,d}$ 5–10; $J_{a,d}$, $J_{b,d}$ 0–2.

$-N{=}C\diagup H \diagdown H$ 8–16

$\searrow C{=}C{-}C{=}C\diagup$ 10–13
$\quad\quad\underset{H}{|}\ \underset{H}{|}$

$\searrow C{=}CH{-}CHO$ 5–8

$\searrow CH{-}C{=}C{-}CH\diagup$ 0–2

$-HC{=}C{=}CH-$ 6–7

$\searrow CH{-}C{\equiv}CH$ 2–3

$\searrow CH{-}C{\equiv}C{-}CH\diagup$ 2–3

	ring size	J
	3	0.5–2
	4	2.5–4
	5	5–7
	6	9–11
	7	9–13
	8	10–13

1428

Table A3.7 Characteristic proton–proton coupling constants (Hz) (*continued*)

Aromatic and heteroaromatic compounds

Benzene derivatives	*ortho* 5–9	*meta* 2–3			*para* 0–1
		$J_{2,3}$	$J_{3,4}$	$J_{2,4}$	$J_{2,5}$
	X = O	1.8	3.5	0.8	1.6
	X = S	5.2	3.6	1.3	2.7
	X = NH	2.7	3.3	1.4	1.9

$J_{2,3}$ 5.5; $J_{3,4}$ 7.5; $J_{2,4}$ 1.9; $J_{2,5}$ 0.9; $J_{3,5}$ 0.9; $J_{2,6}$ 0.4

$J_{2,5}$ 1.5; $J_{2,4}$ –; $J_{4,5}$ 5.0; $J_{4,6}$ 2.5

$J_{2,5}$ 1–2; $J_{2,4}$ 2; $J_{4,5}$ 3–4

$J_{3,4}$ 4.9; $J_{3,5}$ 2.0; $J_{3,6}$ 3.0; $J_{4,5}$ 8.4

$J_{2,3}$ 1.8; $J_{2,6}$ 0.5; $J_{2,5}$ 1.8

$J_{3,4}$ 1.9

1429

Table A3.8 Characteristic coupling constants of protons with other nuclei (Hz)

Proton–fluorine coupling constants

44–81

$-CH-CF-$ 3–25

$-CH=CF-$ cis 2–20; trans 20–50

ortho 6–10; meta 5–8; para ~2

Proton–phosphorus coupling constants

P—H coupling	range, Hz	examples			
P^{III}	180–200	$(CH_3)_2P-H$	192		
P^{IV}	450–550	$(CH_3)_3P^{\oplus}-H$	505		
P^{V}	450–1050		468;		1030

P—C—H coupling			
P^{III}	1–15	$(CH_3)_3P$	2.7
P^{IV}	12–18	$(C_2H_5)_4P^{\oplus}$	12.6
P^{V}	5–20		19

P—C—C—H coupling			
P^{III}	10–16	$(C_2H_5)_3P$	13.7
P^{IV}	15–20	$(C_2H_5)_3\overset{\oplus}{P}H$	20.0
P^{V}	14–25	$(C_2H_5)_3P=O$	18

Table A3.9 Fluorine–fluorine coupling constants

System	range, Hz
	155–225
$-\overset{\mid}{C}F-\overset{\mid}{C}F-$	16–18
	28–87
$\underset{\mid}{C}F=\underset{\mid}{C}F$	cis 20–58 trans 95–120
	ortho ~20 meta 2–4 para 11–15

Table A3.10 Chemical shifts of residual protons in deuterated solvents

Compound	Formula	Residual absorption (δ from TMS)
Acetic acid—d_4	$CD_3 \cdot CO_2D$	2.06, 12.0
Acetone—d_6	$CD_3 \cdot CO \cdot CD_3$	2.07
Benzene—d_6	C_6D_6	7.24
Chloroform—d_1	$CDCl_3$	7.25
Cyclohexane—d_{12}	C_6D_{12}	1.42
Deuterium oxide	D_2O	8.5*
Dimethyl sulphoxide—d_6	$CD_3 \cdot SO \cdot CD_3$	2.50
Methanol—d_4	CD_3OD	3.34, 4.1*
Pyridine—d_5	C_5D_5N	7.0–7.8, 8.57
Trifluoroacetic acid—d_1	$CF_3 \cdot CO_2D$	11.34

* Positions of these absorptions vary according to temperature and solvent.

Table A3.11 Spin–spin systems

Spin systems	Description	Example
AX	$J_{AX} \ll \delta_A - \delta_X$	$C_6H_5 \cdot CHCl \cdot CHCl_2$
AB	$J_{AB} \simeq \delta_A - \delta_B$	
AMX	$J_{AM} \ll \delta_A - \delta_M$ $J_{XM} \ll \delta_M - \delta_X$	 (Fig. 3.61)
ABX	$J_{AB} \simeq \delta_A - \delta_B$ $J_{AX} \ll \delta_A - \delta_X$ $J_{BX} \ll \delta_B - \delta_X$	 (Fig. 3.62)
ABC	$(\delta_A - \delta_B), (\delta_B - \delta_C), (\delta_A - \delta_C)$ all of the same order as J_{AB}, J_{BC}, J_{AC}	
AA'BB'	Nuclei A,A' have same chemical shift but couple differently with B,B' and vice versa	$CH_2Cl \cdot CH_2Br$

Table A3.12 Effect on ^{13}C chemical shifts caused by replacing a methyl group by a polar substituent (δ values, TMS)

Substituent	C-1	C-2	C-3
OR	+45	−3	−1
OH	+40	+1	−1
O·CO·R	+43	−2	−1
NH₂	+20	+2	−1
Cl	+23	+2	−1
F	+61	−1	−2
CO·X	+15	−5	0
CO₂R	+10	−1	−1
CO₂H	+12	−3	−1
CN	−2	−1	−1

R = alkyl; X = Cl or NR₂

Data reproduced from G. Levy and G. L. Nelson (1972).
Carbon-13 Nuclear Magnetic Resonance for Organic Chemists, New York: Wiley-Interscience, p. 47.

Table A3.13 ^{13}C Chemical shifts (TMS) in some monosubstituted alkenes ($\overset{2}{C}H_2 = \overset{1}{C}H - X$)

X	C-1	C-2
H	123.3	123.3
CH$_3$	133.1	115.0
CH$_2$Br	133.2	117.7
C$_2$H$_5$	140.2	113.3
C$_6$H$_5$	136.7	113.2
CO$_2$R	129.7	130.4
CO$_2$H	128.0	131.9
CHO	136.4	136.1, 136.0
CO·CH$_3$	137.5	128.6
I	85.3	130.4
Br	115.5	122.0
Cl	126.0	117.3
N·COR	130.0	94.3
O·CO·CH$_3$	141.6	96.3
OCH$_3$	153.2	84.1

Table A3.14 ^{13}C Substituent effects for substituted benzenes

Substituent	Position			
	C-1	*ortho*	*meta*	*para*
Br	−5.5	+3.4	+1.7	−1.6
CF$_3$	−9.0	−2.2	+0.3	+3.2
CH$_3$	+8.9	+0.7	−0.1	−2.9
CN	−15.4	+3.6	+0.6	+3.9
CO·CF$_3$	−5.6	+1.8	+0.7	+6.7
CO·CH$_3$	+9.1	+0.1	0.0	+4.2
CO·Cl	+4.6	+2.4	0.0	+6.2
CHO	+8.6	+1.3	+0.6	+5.5
CO$_2$H	+2.1	+1.5	0.0	+5.1
Cl	+6.2	+0.4	+1.3	−1.9
F	+34.8	−12.9	+1.4	−4.5
H	0.0	—	—	—
NH$_2$	+18.0	−13.3	+0.9	−9.8
NO$_2$	+20.0	−4.8	+0.9	+5.8
OCH$_3$	+31.4	−14.4	+1.0	−7.7
OH	+26.9	−12.7	+1.4	−7.3
C$_6$H$_5$	+13.1	−1.1	+0.4	−1.2

Data reproduced from G. C. Levy and G. L. Nelson (1972), *Carbon-13 Nuclear Magnetic Resonance for Organic Chemists*, New York: Wiley-Interscience, p. 63.

Table A3.15 ^{13}C **Chemical shift (TMS) for some heteroaromatic compounds**

furan: 109.6, 142.6 (O)

2-methylfuran: 110.3, 140.8, 105.5, 151.8, CH₃ 12.7

pyrrole: 108.2, 118.5 (N–H)

2-methylpyrrole: 108.1, 116.7, 105.4, 127.2, CH₃ 12.4 (N–H)

imidazole: 122.3, 136.2 (N, N–H)

pyrazole: 105.2, 134.3 (N, N–H)

1-methylpyrazole: 105.7, 139.2, 128.7 (N–N), CH₃ 38.1

thiophene: 127.5, 125.4 (S)

2-methylthiophene: 127.0, 125.5, 123.2, 139.2, S, CH₃ 14.8

pyridine: 136.4, 124.5, 150.6 (N)

pyrazine: N, 145.6 (N)

pyrimidine: 157.5, 122.1, 159.5 (N, N)

pyridazine: 127.6, 152.8 (N–N)

indole: 128.8, 121.3, 102.6, 122.3, 125.2, 120.3, 111.8, 136.1 (N–H)

APPENDIX 4
MASS SPECTROMETRY CORRELATION TABLES

INDEX TO TABLES

Table A4.1 Atomic weights of isotopes of some common elements

Isotope	Atomic mass	Isotope	Atomic mass
^{1}H	1.007 825	^{28}Si	27.976 927
^{2}H	2.014 102	^{29}Si	28.976 491
^{12}C	12.000 000	^{30}Si	29.973 761
^{13}C	13.003 354	^{31}P	30.993 763
^{14}N	14.003 074	^{32}S	31.972 094
^{15}N	15.000 108	^{33}S	32.971 461
^{16}O	15.994 915	^{34}S	33.967 865
^{17}O	16.999 133	^{36}S	35.967 09
^{18}O	17.999 160	^{35}Cl	34.968 855
^{19}F	18.998 405	^{37}Cl	36.965 896
		^{79}Br	78.918 348
		^{81}Br	80.916 344
		^{127}I	126.904 352

Table A4.2 Natural isotopic abundances of some common elements

Isotope	Natural abundance	
	Per cent of total isotopes present	Per cent relative to most abundant isotope
^{1}H	99.985	100
^{2}H	0.015	0.016
^{12}C	98.89	100
^{13}C	1.11	1.08
^{14}N	99.63	100
^{15}N	0.37	0.36
^{16}O	99.79	100
^{17}O	0.037	0.04
^{18}O	0.204	0.20
^{19}F	100	100
^{28}Si	92.21	100
^{29}Si	4.70	5.09
^{30}Si	3.09	3.35
^{31}P	100	100
^{32}S	95.0	100
^{33}S	0.76	0.80
^{34}S	4.22	4.44
^{35}Cl	75.33	100
^{37}Cl	24.47	32.40
^{79}Br	50.54	100
^{81}Br	49.46	97.85
^{127}I	100	100

Table A4.3 Some common losses from molecular ions*

Ion	Groups commonly associated with the mass lost	*Possible* inference
M − 1	H	—
M − 2	H$_2$	—
M − 14	—	Homologue?
M − 15	CH$_3$	—
M − 16	O	Ar—NO$_2$· $\geqslant$ N—O, sulphoxide
M − 16	NH$_2$	ArSO$_2$NH$_2$, —CONH$_2$
M − 17	OH	—
M − 17	NH$_3$	—
M − 18	H$_2$O	Alcohol, aldehyde, ketone, etc.
M − 19	F	} Fluorides
M − 20	HF	
M − 26	C$_2$H$_2$	Aromatic hydrocarbon
M − 27	HCN	{ Aromatic nitriles / Nitrogen heterocycles
M − 28	CO	Quinones
M − 28	C$_2$H$_4$	{ Aromatic ethyl ethers / Ethyl esters, propyl ketones
M − 29	CHO	—
M − 29	C$_2$H$_5$	Ethyl ketones, Ar—C$_3$H$_7$
M − 30	C$_2$H$_6$	—
M − 30	CH$_2$O	Aromatic methyl ether
M − 30	NO	Ar—NO$_2$
M − 31	OCH$_3$	Methyl ester
M − 32	CH$_3$OH	Methyl ester
M − 32	S	—
M − 33	H$_2$O + CH$_3$	—
M − 33	HS	} Thiols
M − 34	H$_2$S	
M − 41	C$_3$H$_5$	Propyl ester
M − 42	CH$_2$CO	{ Methyl ketone / Aromatic acetate, Ar·NHCO·CH$_3$
M − 42	C$_3$H$_6$	{ Butyl ketone, isobutyl ketone, / Aromatic propyl ether, Ar—C$_4$H$_9$
M − 43	C$_3$H$_7$	Propyl ketone, Ar—C$_4$H$_9$
M − 43	CH$_3$·CO	Methyl ketone
M − 44	CO$_2$	{ Ester (skel rearr.) / Anhydride
M − 44	C$_3$H$_8$	—
M − 45	CO$_2$H	Carboxylic acid
M − 45	OC$_2$H$_5$	Ethyl ester
M − 46	C$_2$H$_5$OH	Ethyl ester
M − 46	NO$_2$	Ar—NO$_2$
M − 48	SO	Aromatic sulphoxide
M − 55	C$_4$H$_7$	Butyl ester
M − 56	C$_4$H$_8$	{ Ar—C$_5$H$_{11}$, ArO—C$_4$H$_9$ / Ar—^{i}C$_5$H$_{11}$, ArO—^{i}C$_4$H$_9$ Pentyl ketone
M − 57	C$_4$H$_9$	Butyl ketone
M − 57	C$_2$H$_5$·CO	Ethyl ketone
M − 58	C$_4$H$_{10}$	—
M − 60	CH$_3$·CO$_2$H	Acetate

* Data reproduced from D. H. Williams and I. Howe (1972), *Principles of Organic Mass Spectrometry*, London: McGraw-Hill, p. 103.

Table A4.4 Masses of some possible compositions of common fragment ions*

m/e	Groups commonly associated with the mass	Possible inference
15	$CH_3^{\oplus}$	—
18	$H_2O^{\ominus}$	—
26	$C_2H_2^{\ominus}$	—
27	$C_2H_3^{\oplus}$	—
28	$CO^{\ominus}$, $C_2H_4^{\ominus}$, $N_2^{\ominus}$	—
29	$CHO^{\oplus}$, $C_2H_5^{\oplus}$	—
30	$CH_2{=}\overset{\oplus}{N}H_2$	Primary amine?
31	$CH_2{=}\overset{\oplus}{O}H$	Primary alcohol?
36/38(3:1)	$HCl^{\ominus}$	—
39	$C_3H_3^{\oplus}$	—
40†	$Argon^{\ominus}$, $C_3H_4^{\ominus}$	—
41	$C_3H_5^{\oplus}$	—
42	$C_2H_2O^{\ominus}$, $C_3H_6^{\ominus}$	—
43	$CH_3{\cdot}CO^{\oplus}$	$CH_3{\cdot}COX$
43	$C_3H_7^{\oplus}$	C_3H_7X
44	$C_2H_6N^{\oplus}$	Some aliphatic amines
44	$O{=}C{=}\overset{\oplus}{N}H_2$	Primary amides
44	$CO_2^{\ominus}$, $C_3H_8^{\ominus}$	—
44	$CH_2{=}CH(OH)^{\ominus}$	Some aldehydes
45	$CH_2{=}\overset{\oplus}{O}CH_3$ $CH_3{\cdot}CH{=}\overset{\oplus}{O}H$	} Some ethers and alcohols
47	$CH_2{=}\overset{\oplus}{S}H$	Aliphatic thiol
49/51(3:1)	$CH_2Cl^{\oplus}$	—
50	$C_4H_2^{\ominus}$	Aromatic compound
51	$C_4H_3^{\oplus}$	C_6H_5X
55	$C_4H_7^{\oplus}$	—
56	$C_4H_8^{\ominus}$	—
57	$C_4H_9^{\oplus}$	C_4H_9X
57	$C_2H_5{\cdot}CO^{\oplus}$	{ Ethyl ketone { Propionate ester
58	$CH_2{=}C(OH)CH_3^{\ominus}$	{ Some methyl ketones { Some dialkyl ketones
58	$C_3H_8N^{\oplus}$	Some aliphatic amines
59	$CO_2CH_3^{\oplus}$	Methyl ester

* Data reproduced from D. H. Williams and I. Howe (1972). *Principles of Organic Mass Spectrometry*. London: McGraw-Hill, p. 105.
† Appears as a doublet in the presence of argon from air; useful as a reference point in counting the mass spectrum.

Table A4.4 Masses of some possible compositions of common fragment ions (*continued*)

m/e	Groups commonly associated with the mass	*Possible* inference
59	$CH_2=C(OH)NH_2^{+\bullet}$	Some primary amides
59	$C_2H_5CH=\overset{\oplus}{O}H$	$C_2H_5CH(OH)-X$
59	$CH_2=\overset{\oplus}{O}-C_2H_5$ and isomers	Some ethers
60	$CH_2=C(OH)\overset{\oplus}{O}H^{\bullet}$	Some carboxylic acids
61	$CH_3CO(OH_2)^{\oplus}$	$CH_3CO_2C_nH_{2n+1}(n>1)$
61	$CH_2\cdot CH_2SH^{\oplus}$	Aliphatic thiol
66	$H_2S_2^{+\bullet}$	Dialkyl disulphide
69	$CF_3^{\oplus}$	—
68	$CH_2\cdot CH_2\cdot CH_2\cdot CN^{\oplus}$	—
69	$C_5H_9^{\oplus}$	—
70	$C_5H_{10}^{+\bullet}$	—
71	$C_5H_{11}^{\oplus}$	$C_5H_{11}X$
71	$C_3H_7\cdot CO^{\oplus}$	$\begin{cases}\text{Propyl ketone}\\\text{Butyrate ester}\end{cases}$
72	$CH=C(OH)C_2H_5^{+\bullet}$	Some ethyl alkyl ketones
72	$C_3H_7\cdot CH=\overset{\oplus}{N}H_2$ and isomers	Some amines
73	$C_4H_9O^{\oplus}$	—
73	$CO_2C_2H_5^{\oplus}$	Ethyl ester
73	$(CH_3)_3Si^{\oplus}$	$(CH_3)_3SiX$
74	$CH_2=C(OH)OCH_3^{+\bullet}$	Some methyl esters
75	$(CH_3)_2Si=\overset{\oplus}{O}H$	$(CH_3)_3SiOX$
75	$C_2H_5\cdot CO(OH_2)^{\oplus}$	$C_2H_5\cdot CO_2C_nH_{2n+1}(n>1)$
76	$C_6H_4^{+\bullet}$	$\begin{cases}C_6H_5X\\XC_6H_4Y\end{cases}$
77	$C_6H_5^{\oplus}$	C_6H_5X
78	$C_6H_6^{+\bullet}$	C_6H_5X
79	$C_6H_7^{\oplus}$	C_6H_5X
79/81 (1:1)	$Br^{\oplus}$	—
80/82 (1:1)	$HBr^{+\bullet}$	—
80	$C_5H_6N^{\oplus}$	
81	$C_5H_5O^{\oplus}$	
83/85/87 (9:6:1)	$HCCl_2^{\oplus}$	$CHCl_3$
85	$C_6H_{13}^{\oplus}$	$C_6H_{13}X$
85	$C_4H_9\cdot CO^{\oplus}$	$C_4H_9\cdot COX$

Table A4.4 Masses of some possible compositions of common fragment ions (*continued*)

m/e	Groups commonly associated with the mass	*Possible* inference
85		
85		
86	$CH_2=C(OH)C_3H_7^{\oplus \cdot}$	Some propyl alkyl ketones
86	$C_4H_9 \cdot CH = \overset{\oplus}{N}H_2$ and isomers	Some amines
87	$CH_2=CH-\overset{\overset{\displaystyle \overset{\oplus}{O}H}{\|}}{C}-OCH_3$	$XCH_2 \cdot CH_2 \cdot CO_2CH_3$
91	$C_7H_7^{\oplus}$	$C_6H_5 \cdot CH_2X$
92	$C_7H_8^{\oplus \cdot}$	$C_6H_5 \cdot CH_2$-alkyl
92	$C_6H_6N^{\oplus}$	
91/93 (3:1)		Alkyl chloride ($\geqslant$ hexyl)
93/95 (1:1)	$CH_2Br^{\oplus}$	
94	$C_6H_6O^{\oplus \cdot}$	C_6H_5O-alkyl (alkyl $\neq CH_3$)
94		
95		
95	$C_6H_7O^{\oplus}$	
97	$C_5H_5S^{\oplus}$	
99		
99		

Table A4.4 Masses of some possible compositions of common fragment ions (*continued*)

m/e	Groups commonly associated with the mass	*Possible* inference
105	$C_6H_5 \cdot CO^{\oplus}$	$C_6H_5 \cdot COX$
105	$C_8H_9^{\oplus}$	$CH_3-C_6H_4 \cdot CH_2X$
106	$C_7H_8N^{\oplus}$	
107	$C_7H_7O^{\oplus}$	
107/109 (1:1)	$C_2H_4Br^{\oplus}$	—
111		
121	$C_8H_9O^{\oplus}$	
122	$C_6H_5 \cdot CO_2H^{\oplus}$	} Alkyl benzoates
123	$C_6H_5 \cdot CO_2H_2^{\oplus}$	
127	$I^{\oplus}$	—
128	$HI^{\cdot\oplus}$	—
135/137 (1:1)		Alkyl bromide ($\geqslant$ hexyl)
130	$C_9H_8N^{\oplus}$	
141	$CH_2I^{\oplus}$	—
147	$(CH_3)_2Si\!=\!\overset{\oplus}{O}-Si(CH_3)_3$	—
149		Dialkyl phthalate
160	$C_{10}H_{10}NO^{\oplus}$	
190	$C_{11}H_{12}NO_2^{\oplus}$	

APPENDIX 5
USEFUL SOLVENT CHARACTERISTICS

Solvent[a]	Boiling point (°C)	Melting point	Molecular weight
	(760 mmHg)	(°C)	(g)
Ether (15)	35	−116	74
Pentane (1)	36	−130	72
Dichloromethane (5)	40	−95	85
Carbon disulphide (32)	46	−111	76
Acetone (21)	56	−95	58
Chloroform (6) **C**	61	−64	119
Methanol (8)	65	−98	32
Tetrahydrofuran (9)	66	−109	72
Hexane (1)	69	−95	86
Trifluoroacetic acid	72	−15	114
Carbon tetrachloride (7)	77	−23	154
Ethyl acetate (24)	77	−84	88
Ethanol (9)	78	−114	46
Cyclohexane (1)	81	6.5	84
Benzene (2) **C**	80	5.5	78
Methyl ethyl ketone (22)	80	−87	72
Acetonitrile (27)	82	−44	41
Propan-2-ol (11)	82	−88	60
t-Butanol (12)	82	26	74
Ethylene glycol dimethyl ether (18)	83	−58	90
Triethylamine	90	−115	101
Propan-1-ol (10)	97	−126	60
Water	100	0	18
Methylcyclohexane (1)	101	−127	98
Formic acid	101	8	46
Nitromethane	101	−29	61
1,4-Dioxane (20)	101	12	88
Toluene (3)	111	−95	92
Pyridine (29)	115	−42	79
Butan-1-ol (12)	118	−89	74
Acetic acid	118	17	60
Ethylene glycol monomethyl ether (13)	125	−85	76
Morpholine	129	−3	87
Chlorobenzene	132	−46	113
Acetic anhydride	140	−73	102
Xylenes (mixed) (4)	138–142	13[d]	106
Dibutyl ether (17)	142	−95	130
sym-Tetrachloroethane	146	−44	168
Anisole	154	−38	108

Density at 20 °C	Dielectric constant	Solubility in water (g/100 g)^b	Azeotrope with water b.p. (°C)	Azeotrope with water %H$_2$O	Flash point (°C)	RL^c (p.p.m.)	RL^c (mg m^{-3})
0.71	4.3	6.0	34	1	−45	400	1200
0.63	1.8	Insol.	35	1	−49	600	1800
1.33	8.9	1.30	39	2	None	100 (CL)	350
1.26	2.6	0.29 (20 °C)	44	2	−30	10 (CL)	30
0.79	20.7	∞	None	—	−18	1000 (CL)	2400
1.49	4.8	0.82 (20 °C)	56	3	None	10	50
0.79	32.7	∞	None	—	11	200	260
0.89	7.6	∞	64	5	−18	200	590
0.66	1.9	Insol.	62	6	−23	100	360
1.49	39.5	∞	105	21	None	—	—
1.59	2.2	0.08	66	4	None	10	65
0.90	6.0	8.1	71	8	−4	400	1400
0.79	24.6	∞	78	4	12	1000	1900
0.78	2.0	0.01	70	8	−20	300	1050
0.88	2.3	0.18	69	9	−11	10	30
0.80	18.5	24.0 (20 °C)	73	11	−6	200	590
0.78	37.5	∞	77	16	6	40	70
0.79	19.9	∞	80	12	12	400	980
0.78 (30 °C)	12.5	∞	80	12	10	100	300
0.86	7.2	∞	77	10	1	—	—
0.73	2.4	∞	75	10	−7	10	40
0.80	20.3	∞	88	28	5	200	500
1.00	80.2	—	—	—	None	—	—
0.77	2.0	0.01	80	24.1	−6	400	1600
1.22	58.5	∞	107	26	—	5	9
1.14	35.9	11.1	84	24	35	100	250
1.03	2.2	∞	88	18	12	50	180
0.87	2.4	0.05	85	20	4	100	375
0.98	12.4	∞	94	42	20	5	15
0.81	17.5	7.45	93	43	29	50	150
1.05	6.2	∞	None	—	40	10	25
0.96	16.9	∞	100	85	42	5 (CL)	16
1.00	7.4	∞	None	—	35	20	70
1.11	5.6	0.05 (30 °C)	90	28	24	75	350
1.08	20.7	Reacts	—	—	54	5	20
0.86	2^e	0.02	93	33	17	100	435
0.77	3.1	0.03 (20 °C)	93	33	38	—	—
1.59	8.2	0.29 (20 °C)	94	34	None	5 (TLV)	35
0.99	4.3	1.04	96	41	—	—	—

Solvent[a]	Boiling point (°C)	Melting point	Molecular weight
	(760 mmHg)	(°C)	(g)
Dimethylformamide (26)	153	−60	73
Diethylene glycol dimethyl ether (18)	160 (dec.)	—	134
Mesitylene	165	−45	120
Dimethyl sulphoxide (33)	189	18	78
Diethylene glycol monomethyl ether (14)	194	−76	120
Ethylene glycol (ethane-1,2-diol)	197	−16 to −13[f]	62
N-Methyl-2-pyrrolidone (28)	202	−24	99
Nitrobenzene (31)	211	6	123
Formamide (25)	210 (dec.)	3	45
Hexamethylphosphoric triamide (35) **C**	233	7	179
Quinoline (30)	237	−15	129
Diethylene glycol	245	−7	106
Diphenyl ether	258	27	170
Triethylene glycol	288	−4	150
Sulpholane (34)	287 (dec.)	28	120
Glycerol	290	18	92
Triethanolamine	335	22	149
Dibutyl phthalate	340	−35	278

Density at 20 °C	Dielectric constant	Solubility in water (g/100 g)^b	Azetrope with water b.p. (°C)	%H$_2$O	Flash point (°C)	RL^c (p.p.m.)	(mg m^{-3})
0.95	36.7	∞	None	—	58	10	30
0.94	—	∞	100	78	63	—	—
0.87	2.3	0.03 (20 °C)	97	—	—	—	—
1.10	46.7	25.3	None	—	95	—	—
1.02	—	∞	None	—	93	—	—
1.11	37.7	∞	None	—	112	—	60
1.03	32.0	∞	—	—	96	—	—
1.20	34.8	0.19 (20 °C)	99	88	88	1	5
1.13	111	∞	—	—	154	20	30
1.03	30	∞	—	—	—	—^{g}	—
1.09	9.0	0.6 (20 °C)	—	97	—	—	—
1.11	31.7	∞	None	—	143	—	—
1.07	3.7 (>27 °C)	0.39	100	96	205	—	—
1.12	23.7	∞	None	—	166	—	—
1.26 (30 °C)	43	∞ (30 °C)	None	—	177	—	—
1.26	42.5	∞	None	—	160	—	—
1.12 (25 °C)	29.4	∞	—	—	179	—	—
1.05	6.4	Insol.	None	—	171	—	5

Notes. (*a*) The numbers in parentheses enable the solvent to be located in Section 4.1, where alternative names and methods of purification are to be found.

(*b*) Values for 25 °C unless otherwise indicated. Values <0.01 per cent described as insoluble.

(*c*) Recommended Limit, Control Limit and Threshold Limit Values, Section 2.3.4, p. 45.

(*d*) Value for *p*-xylene (isomer of highest m.p.).

(*e*) Approximate value.

(*f*) Value in doubt because of strong tendency to supercool and form a glass.

(*g*) No TLV value set.

HAZARD SYMBOLS

Legislation on packaging and labelling of dangerous substances defines hazardous chemicals under the following categories:

 Toxic
T

 Explosive
E

 Corrosive
C

 Oxidising
O

 Highly Flammable/ Flammable
F

 Irritant
Xi

 Harmful
Xn

Toxic These products can cause death or serious illness when small amounts enter the body by ingestion, inhalation of vapour, fumes or dust, or by absorption through the skin; hygiene considerations should be rigorously observed.

Corrosive These products may destroy living tissue; eyes are particularly susceptible. Emergency showers should be available. If swallowed plenty of water should be given after immediate mouth rinsing.

Explosive These products may explode by the action of heat, sources of ignition, shock or friction. The compounds are often packaged wet to reduce the risk of explosion; they will become dangerous if allowed to dry. Some compounds form sensitive explosive salts on contact with metals.

Oxidising These compounds may cause fire and will always assist combustion. They produce heat on contact with organic matter and reducing agents.

Flammable These compounds have a low flash point, and those which react with water or damp air to give rise to flammable gases (e.g. hydrogen) from metal hydrides. Ignition sources include Bunsen burners, hot metal surfaces, electric sparks, etc. Fire fighting equipment should be readily available and frequently checked.

Harmful *Irritant* chemicals cause inflammation of the skin, mucous membranes, or discomfort of the respiratory system. *All* laboratory chemicals should be regarded as harmful; some are specifically harmful by skin contact, inhalation or swallowing.

APPENDIX 6
COMMON SYNTHONS AND THEIR REAGENT EQUIVALENTS

Table A6.1 Cationic synthons (acceptor species)

Synthon	Some reagent equivalents (masked or latent functionality, LF)	Page ref.
$R^{\oplus}$ or $R \cdot \overset{\oplus}{C}H_2$	RX or $R \cdot CH_2X(X = \overset{\oplus}{O}H_2$, Halogen, OTs, OMs); $(RO)_2SO_2$; $R^{\oplus}\overset{\ominus}{AlCl_4}$, or a protonated alkene	554 826
$R \cdot CH = CH \cdot \overset{\oplus}{C}H_2$ $\longleftrightarrow R \cdot \overset{\oplus}{C}H \cdot CH = CH_2$	$R \cdot CH = CH \cdot CH_2X$ (X = Halogen, etc.)	969
$R \cdot CH_2 \cdot C \equiv \overset{\oplus}{C}$	$R \cdot CH_2 \cdot C \equiv CBr$	508
	PhX (X = Halogen); $Ph \cdot SO_3H$; $Ph\overset{\oplus}{N_2}\}\overset{\ominus}{X}$	968
$Ph \cdot \overset{\oplus}{C}H_2$	$Ph \cdot CH_2X$ (X = $\overset{\oplus}{O}H_2$, Halogen, OTs, OMs, etc.)	554
	H_2CO	518
	$R \cdot CHO$	518
	R_2CO	518
	$R \cdot CH = CH \cdot CO \cdot R$	635 1093

Table A6.1 Cationic synthons (acceptor species) (*continued*)

Synthon	Some reagent equivalents (masked or latent functionality, LF)	Page ref.
OH–CH$_2^{\oplus}$	(epoxide: ethylene oxide)	518
R–CH(OH)–CH$_2^{\oplus}$	(epoxide with R)	518
H–CHO$^{\oplus}$	H·CO$_2$Et; HC(OEt)$_3$; CO/HCl/AlCl$_3$; HC=NH}$^{\oplus}$Cl$^{\ominus}$; Me$_2$N=CHOPOCl$_2$; :CCl$_2$/(LF)	586 990
H–CO–CH$_2^{\oplus}$	(RO)$_2$CH·CH$_2$Br	21
H–CO–CH$_2$–CH$_2^{\oplus}$	CH$_2$=CH·CHO	749
R–CO$^{\oplus}$	R·CO$_2$Et; R·C(OEt)$_3$; R·COCl/AlCl$_3$ or (R·CO)$_2$O/AlCl$_3$	518 632 1006
R–CO–CH$_2^{\oplus}$	R·CO·CH$_2$Br; R·C(NO$_2$)=CH$_2$(LF)	21 768
R–CO–CH$_2$–CH$_2^{\oplus}$	R·CO·CH=CH$_2$	1094
(R)HO–CO$^{\oplus}$	CO$_2$; (RO)$_2$CO; RO·COCl	665
(R)HO–CO–CH$_2^{\oplus}$	RO$_2$C·CH$_2$Br; or HO$_2$C·CH$_2$Cl	720
(R)HO–CO–CH$_2$–CH$_2^{\oplus}$	RO$_2$C·CH=CH$_2$	665

Table A6.2 Anionic synthons (donor species)

Synthon	Some reagent equivalents (masked or latent functionality, LF)	Page ref.
$\overset{\ominus}{R}$ or $R{\cdot}\overset{\ominus}{C}H_2$	$RMgX$ or RLi	
$R{\cdot}CH{=}CH{\cdot}\overset{\ominus}{C}H_2$	$R{\cdot}CH{=}CH{\cdot}CH_2MgX$ (or Li)	
$R{\cdot}CH_2{\cdot}CH{=}\overset{\ominus}{C}H$	$R{\cdot}CH_2{\cdot}CH{=}CHLi$ (or MgX)	532
$R{\cdot}CH_2{\cdot}C{\equiv}\overset{\ominus}{C}$	$R{\cdot}CH_2{\cdot}C{\equiv}CNa$ (or Li or MgX)	
(cyclohexadienyl anion)$^{\ominus}$	PhLi (or MgX)	826
$Ph{\cdot}\overset{\ominus}{C}H_2$	$Ph{\cdot}CH_2Li$ (or MgX)	1056
$R{\cdot}CH_2{\cdot}\overset{\ominus}{O}$	$R{\cdot}CH_2{\cdot}ONa$ (or K)	583
$R{\cdot}CH_2{\cdot}\overset{\ominus}{S}$	$R{\cdot}CH_2{\cdot}SNa$	789
$R{\cdot}CH_2{\cdot}\overset{\ominus}{N}H(R)$	$R{\cdot}CH_2NH_2$ or $R_2\overset{\ominus}{N}\}\overset{\oplus}{L}i$	771
$\overset{\ominus}{N}H_2$	$NaNH_2$ or NH_3, or potassium phthalimide (LF)	779
$\overset{\ominus}{C}H_2NH_2$	$\overset{\ominus}{C}N$(LF); or $\overset{\ominus}{C}H_2{\cdot}NO_2$(LF)	770
$\overset{\ominus}{C}H_2{\cdot}NO_2$	$Me{\cdot}NO_2$	635
$R_2\overset{\ominus}{C}{\cdot}NO_2$	$R_2CH{\cdot}NO_2$	768
$\underset{H}{\overset{O}{\|}}{\overset{\ominus}{}}$ (acyl anion, formyl)	$Fe(CO)_4^{2\ominus}$, $H{-}\overset{\ominus}{\langle S,S \text{ dithiane}\rangle}$, $\overset{\ominus}{:}CN$ (LF)	586
$\underset{R}{\overset{O}{\|}}{\overset{\ominus}{}}$ (acyl anion)	$R{-}\overset{\ominus}{\langle S,S \text{ dithiane}\rangle}$ (LF), $R{\cdot}\overset{\ominus}{C}H_2{\cdot}NO_2$ (LF)	597 739
$\underset{R}{\overset{O}{\|}}{\overset{}{C}H_2}$	$R{\cdot}CO{\cdot}Me$ (kinetic control) or $R{\cdot}CO{\cdot}CH_2{\cdot}CO_2Et$, or $R{\cdot}C(NR_2){=}CH_2$	606
$\underset{\ominus}{R}\overset{O}{\underset{}{\|}}{Me}$	$R{\cdot}CH_2{\cdot}CO{\cdot}Me$ (thermodynamic control)	619 632
$\underset{R}{\overset{O}{\|}}{\overset{\ominus}{C}H_2}$	$R{\cdot}C(OR')_2{\cdot}CH_2{\cdot}CH_2MgX$ (carbonyl protection required)	478

Table A6.2 Anionic synthons (donor species) (*continued*)

Synthon	Some reagent equivalents (masked or latent functionality, LF)	Page ref.
(R)HO–CO⁻ (O)	KCN (LF)	671
(Et)HO–CO–CH₂⁻	$CH_2(CO_2Et)_2$ (LF), $Me \cdot CO_2Et$, or $BrZnCH_2CO_2Et$	665 727
EtO–CO–⁻–CO–OEt	$CH_2(CO_2Et)_2$, or $CH_2(CN)_2$ (LF)	665 680
EtO–CO–⁻–R	$R \cdot CH(CO_2Et)_2$ (LF)	
Me–CO–⁻–CO–OEt	$Me \cdot CO \cdot CH_2 \cdot CO_2Et$	606
Me–CO–⁻–CO–Me	$Me \cdot CO \cdot CH_2 \cdot CO \cdot Me$	606
EtO–CO–⁻–CH₂–CO–OEt	$EtO_2C \cdot CH_2 \cdot CH_2 \cdot CO_2Et$	739

APPENDIX 7
SELECTION OF MANUFACTURERS AND SUPPLIERS OF LABORATORY APPARATUS, EQUIPMENT AND CHEMICALS*

Ace Glass, 1430 Northwest Blvd, PO Box 688, Vineland, New Jersey, 08360, USA.

A.I. Scientific, London Road, Pampisford, Cambridge, CB2 4EF, UK.

Aldrich Chemical Co. Ltd, The Old Brickyard, New Road, Gillingham, Dorset, SP8 4JL, UK.

Applied Photophysics Ltd, 18/21 Corsham Street, London, N1 6DR, UK.

Baird & Tatlock Ltd, PO Box 1, Romford, Essex, RM1 1HA, UK.

Baskerville & Lindsay Ltd, 324c Barlow Moor Road, Chorlton-cum-Hardy, Manchester, M21 2AX, UK.

BDH Chemicals Ltd, Broom Road, Poole, BH12 4NN, UK.

Beckmann-RIIC Ltd, Progress Road, Sands Industrial Estate, High Wycombe, Buckinghamshire, HP12 4JL, UK.

Bellingham & Stanley Ltd, Longfield Road, North Farm Industrial Estate, Tunbridge Wells, Kent, TN2 3EY, UK.

J. Bibby Science Products Ltd, Tilling Drive, Stone, Staffordshire, ST15 0SA, UK.

BOC Ltd, The Priestley Centre, 10 Priestley Road, The Surrey Research Park, Guildford, Surrey, GU2 5XY, UK.

Cambridge Instruments Ltd, Viking Way, Bar Hill, Cambridge, CB3 8EL, UK.

Camlab Ltd, Nuffield Road, Cambridge, CB4 1TH, UK.

Ciba–Corning Diagnostics Ltd, Halstead, Essex, CO9 2DX, UK.

W. Coles & Co. Ltd, PO Box 42, Plastic Works, 47/49 Tanner Road, London, SE1, UK.

C. W. Cook & Sons Ltd, 190 Camden Street, Birmingham, B1 3DS, UK.

Corning Glass Works, Science Products, MP-21-5-8, Corning, New York 14831, USA.

Decon Laboratories Ltd, Conway Street, Hove, East Sussex, BN3 3LY, UK.

Detectawl (Gastec) Ltd, Unit 61, Garamonde Drive, Wymbush, Milton Keynes, Buckinghamshire, MK8 8DE, UK.

Draeger Ltd, The Willows, Mark Road, Hemel Hempstead, Hertfordshire, HP2 7BW, UK.

Eastman Kodak Co., 343 State Street, Rochester, New York, 14650, USA.

Edwards High Vacuum International, Manor Royal, Crawley, West Sussex, RH10 2LW, UK.

Electrothermal Engineering Ltd, 419 Sutton Road, Southend-on-Sea, Essex, SS2 5PH, UK.

Fisher Scientific Co., 711 Forbes Avenue, Pittsburgh, PA, USA.

Fluka AG, CH-9470 Buchs, Switzerland

Gallenkamp (Fisons Scientific Equipment Division), Belton Road West, Loughborough, Leicestershire, LE11 0TR, UK.

---

* Mostly UK addresses are given in this list. The *International Directory* (*Buyers' Guide*) is an invaluable source for addresses of companies in other countries, and is available from International Scientific Communications, Inc. 808 Kings Highway, PO Box 827, Fairfield, Connecticut, 06430-0827, USA.

Glas-Col Apparatus Co., 711, Hulman Street, PO Box 2128, Terre Haute, Indiana, 47802, USA.

Gow-Mac Instrument Co., PO Box G.13, Gillingham, Kent, ME7 4HA, UK.

Griffin and George Ltd, Bishop Meadow Road, Loughborough, Leicestershire, LE11 0RG, UK.

Hanovia Lamps Ltd, 145 Farnham Road, Slough, Berkshire, SL1 4XB, UK.

Interox Chemicals Ltd, PO Box 7, Warrington, Cheshire, WA4 6HB, UK.

Isopad Ltd, Isopad House, Shenley Road, Borehamwood, Hertfordshire, WD6 1TE, UK.

Jasco International Co. Ltd, through Ciba Corning Diagnostics Ltd, Halstead, Essex, CO9 2DX, UK.

Jencons (Scientific) Ltd, Cherrycourt Way Industrial Estate, Stanbridge Road, Leighton Buzzard, Bedfordshire, LU7 8UA, UK.

Jeol & Co. Ltd, Jeol House, Grove Park, Colindale, London, NW9 0JN, UK.

JJ's (Chromatography) Ltd, Hardwick Industrial Estate, King's Lynn, Norfolk, PE30 4JG, UK.

Jones Chromatography Ltd, New Road, Hengoed, Mid Glamorgan, CF8 8AU, UK.

Just Plastics Ltd, Cromwell House, Staffa Industrial Estate, Staffa Road, Leyton, London, E10 7PY, UK.

Koch-Light Ltd, Hollands Road, Haverhill, Suffolk, CB9 8PU, UK.

Lab-Marc, Icknield House, 1 Icknield Green, Letchworth, Hertfordshire, SG6 4DL, UK.

Macherey-Nagel AG, through Field Analytical, PO Box 113, Weybridge, Surrey, KT13 9UZ, UK.

Manostat Corporation, 519 Eight Avenue, New York, NY 10018, USA.

Mattson Instruments Ltd, Linford Forum, Rockingham Drive, Linford Wood, Milton Keynes. MK14 6LY, UK.

May & Baker Ltd, see Rhône-Poulenc

E. Merk through BDH.

Miller Howe Ltd, Watlington Industrial Park, Watlington, Oxfordshire, OX9 5LU, UK.

Millipore (UK) Ltd, see Waters Chromatography Division.

Optical Activity Ltd, Industrial Estate, Bury Road, Ramsey, Huntingdon, Cambridgeshire, PE17 1NA, UK.

Parr Instrument Co., through Scientific & Medical Products, Shirley Institute, Didsbury, Manchester, M20 8RX, UK.

PergaBase Inc., 12 Vandy Street, London, EC2A 2DE, UK.

Perkin-Elmer Ltd, Post Office Lane, Beaconsfield, Buckinghamshire, HP9 1QA, UK.

Phase Separations Ltd, Deeside Industrial Estate, Queensferry, Clwyd, CH5 2LR, UK.

Phillips Analytical, Pye Unicam Ltd, York Street, Cambridge, CB1 2PX, UK.

Pierce Chemical Co, through Life Science Laboratories Ltd, Sedgewick Road, Luton, LU4 9DT, UK.

P. B. Radley & Co. Ltd, 53 London Road, Sawbridgeworth, Hertfordshire, UK.

C. Reichert Optische Werke AG, through Cambridge Instruments Ltd, Viking Way, Bar Hill, Cambridge, CB3 8EL, UK.

Rhône-Poulenc, May and Baker Laboratory Products Ltd, Liverpool Road, Barton Moss, Eccles, Manchester, M30 7RT, UK.

Romil Chemicals Ltd, 63 Ashby Road Central, Shepshed, Loughborough, Leicestershire, LE12 9BS, UK.

Scientific Furnishings Ltd, Terminus Road, Chichester, Sussex, PO19 2UJ, UK.

Scientific and Medical Products Ltd, Shirley Institute, Didsbury, Manchester, M20 8RX, UK.

Shandon Southern Products Ltd, Chadwick Road, Astmoor, Runcorn, Cheshire, WA7 1PR, UK.

Sigma Chemical Co. Ltd, Fancy Road, Poole, Dorset, BH17 7NH, UK.

The Southern New England Ultraviolet Co., PO Box 4134, Hamden, Connecticut, 06514, USA.

TC Research, 9 West Parade, Norwich, Norfolk, NR2 3DN, UK.

Ultra-Violet Products Ltd, Science Park, Milton Road, Cambridge, CB4 4FH. UK.

Varian Associates Ltd, 28 Manor Road, Walton-on-Thames, Surrey, KT12 2QF, UK.

Vickers Laboratories, Greenholme Industrial Estate, Burley in Wharfedale, West Yorkshire, LS29 7DB, UK.

Vinten Instruments Ltd, Jessamy Road, Weybridge, Surrey, KT13 8LE, UK.

Waters Chromatography Division, Millipore (UK) Ltd, 11/15 Peterborough Road, Harrow, Middlesex, HA1 2YH, UK.

Whatman Ltd, Springfield Mill, Maidstone, Kent, ME14 2LE, UK.

J. Young (Scientific Glassware) Ltd, 11 Colville Road, Acton, London, W3 8BS, UK.

ATOMIC WEIGHTS*

Aluminium	Al	26.9815	Manganese	Mn	54.938
Antimony	Sb	121.75	Mercury	Hg	200.59
Arsenic	As	74.9216	Molybdenum	Mo	95.94
Barium	Ba	137.34	Nickel	Ni	58.71
Beryllium	Be	9.0122	Nitrogen	N	14.0067
Bismuth	Bi	208.9806	Oxygen	O	15.9994
Boron	B	10.81	Palladium	Pd	106.4
Bromine	Br	79.904	Phosphorus	P	30.9738
Cadmium	Cd	112.40	Platinum	Pt	195.09
Calcium	Ca	40.08	Potassium	K	39.102
Carbon	C	12.011	Selenium	Se	78.96
Cerium	Ce	140.12	Silicon	Si	28.086
Chlorine	Cl	35.453	Silver	Ag	107.868
Chromium	Cr	51.996	Sodium	Na	22.9898
Cobalt	Co	58.9332	Strontium	Sr	87.62
Copper	Cu	63.546	Sulphur	S	32.06
Fluorine	F	18.9984	Tellurium	Te	127.60
Germanium	Ge	72.59	Thorium	Th	232.0381
Gold	Au	196.9665	Tin	Sn	118.69
Hydrogen	H	1.008	Titanium	Ti	47.90
Iodine	I	126.9045	Tungsten	W	183.85
Iron	Fe	55.847	Uranium	U	238.029
Lead	Pb	207.20	Vanadium	V	50.9414
Lithium	Li	6.941	Zinc	Zn	65.37
Magnesium	Mg	24.305	Zirconium	Zr	91.22

* These atomic weights are those adopted by the International Union of Pure and Applied Chemistry and are based on a relative atomic mass of $C^{12} = 12.000$.

INDEX

Page numbers in **bold-faced** type refer to details of preparative procedures.
Page numers in *italics* refer to mechanistic discussions.

REFERENCE DATA

GENERAL

ABBREVIATIONS

NOMENCLATURE

R	alkyl	n straight-chain
Me or CH$_3$—	methyl	s secondary
Et or C$_2$H$_5$	ethyl	t tertiary
Pr or C$_3$H$_7$—	n-propyl	i iso
Pr^i or (CH$_3$)$_2$CH—	isopropyl (1-methylethyl)	
Bu or C$_4$H$_9$—	n-butyl	
Bu^i or (CH$_3$)$_2$CH·CH$_2$—	isobutyl	
Bu^t or (CH$_3$)$_3$C—	t-butyl (1,1-dimethylethyl)	
Bu^s or CH$_3$·CH$_2$·CH(CH$_3$)—	s-butyl (1-methylpropyl)	
Sia or (CH$_3$)$_2$CH·CH(CH$_3$)—	s-isoamyl (1,2-dimethylpropyl)	
Thex or (CH$_3$)$_2$CH·C(CH$_3$)$_2$—	t-hexyl (1,1,2-trimethylpropyl)	
Ph or C$_6$H$_5$	phenyl	

Ac or Me·CO—	acetyl
Bz or Ph·CO—	benzoyl
Bn or Ph·CH$_2$—	benzyl
Boc or Me$_3$C·O·CO—	t-butoxycarbonyl
Cbz or Ph·CH$_2$·O·CO—	benzyloxycarbonyl
MEM or Me·O·CH$_2$·CH$_2$·O·CH$_2$—	2-methoxyethoxymethyl
Ms or Me·SO$_2$—	methanesulphonyl (mesyl)
Ts or p-Me·C$_6$H$_4$·SO$_2$—	toluene-p-sulphonyl (tosyl)

THP or   tetrahydropyranyl

REAGENTS AND SOLVENTS

AIBN	α,α′-azobisisobutyronitrile
9-BBN	9-borabicyclo[3.3.1]nonane
BDSC	t-butyldimethylchlorosilane
BTMSA	N,O-bis(trimethylsilyl)acetamide
DBU	1,8-diazabicyclo[5.4.0]undec-7-ene
DCC	dicyclohexylcarbodiimide
DDQ	2,3-dichloro-5,6-dicyano-1,4-benzoquinone
DEG	diethylene glycol
DET	diethyl tartrate
DIBAL	diisobutylaluminium hydride
DME	1,2-dimethoxyethane (glyme)

DMF	dimethylformamide
DMSO	dimethyl sulphoxide
DNFB	2,4-dinitrofluorobenzene
DSS	sodium 3-(trimethylsilyl)-1-propanesulphonate
EAA	ethyl acetoacetate
HMDS	hexamethyldisilazane [bis(trimethylsilyl)amine]
HMPT	N,N,N',N',N'',N''-hexamethylphosphotriamide
LAH	lithium aluminium hydride
LDA	lithium isopropylamide
MCPBA	m-chloroperbenzoic acid
MEK	ethyl methyl ketone
MVK	methyl vinyl ketone
NBS	N-bromosuccinimide
NCS	N-chlorosuccinamide
PCC	pyridinium chlorochromate
PDC	pyridinium dichromate
PEG	polyethylene glycol
PPA	polyphosphoric acid
PPTSA	pyridinium toluene-p-sulphonate
Py	pyridine
SBH	sodium borohydride
TEG	triethylene glycol
TFA	trifluoroacetic acid
THF	tetrahydrofuran
TMEDA	N,N,N',N'-tetramethylethylenediamine
TMCS	chlorotrimethylsilane
TMS	tetramethylsilane